Advanced Materials

Ajit Behera

Advanced Materials

An Introduction to Modern Materials Science

Springer

Ajit Behera
Department of Metallurgical and Materials Engineering
National Institute of Technology Rourkel
Rourkela, India

ISBN 978-3-030-80358-2 ISBN 978-3-030-80359-9 (eBook)
https://doi.org/10.1007/978-3-030-80359-9

This Springer imprint is published by the registered company Springer Nature Switzerland AG
The registered company address is: Gewerbestrasse 11, 6330 Cham, Switzerland

About the Book

Materials always bring a new horizon with their development to influence significantly on society's expectation. Advancement of materials has become primary objective of the researchers as the property of materials is largely dependent upon types of materials. In order to incorporate the high-end techniques of modern decade, the discovery and application of advanced materials has become important extensively. Hence, in the past few years there are many improvements and applications of advanced materials in various industries of engineering, aviation, defence, biomedical, etc. Therefore, this book named "Advanced Materials" tries to enlighten the various applications of advanced materials in the current era. Each of the 20 chapters of this book is focused on a particular cutting-edge material. This book will open someone's eye towards the current materials status that will help to develop more number of industries in near future.

Contents

About the Author

Ajit Behera is currently working in Metallurgical and Materials Department at the National Institute of Technology, Rourkela. He was born in 1987 in Odisha. He completed his Ph.D. from IIT-Kharagpur in 2016. He got National "Yuva Rattan Award" in 2020 for his contribution towards his significant academic carrier as well as social work. In addition, he got "young faculty award" in 2017 and "C.V. Raman Award" in 2019. He has published more than 100 publications including Books, Book Chapters and Journals. Currently, he is involved with many reputed leading scientific organizations throughout the world (as visiting scientist/material develop committee member). More than 10 Ph.D. students (National/International) are working on different projects with him. He has significant contribution in developing device-based application to empower the society for mankind.

Web Links:

LinkedIn: https://www.linkedin.com/in/ajitbehera1/

Academia: https://nitrkl.academia.edu/AjitBehera

Google Scholar: https://scholar.google.com/citations?user=R-G7pSoAAAAJ&hl=en

ORCID: http://orcid.org/0000-0001-5357-7733

Facebook: https://www.facebook.com/ajit.behera8

Shape-Memory Materials

1

Abstract

In the era of materials technology, a fundamental understanding of microstructure, material properties, and processing techniques is required for the suitable application that blooming the advanced multifunctional materials: shape-memory materials (SMM). The ability to design different specialized subgroups of multifunctional SMM that have stimulating and deflection ability is recognized as active materials. This chapter summarizes the most advanced and active shape-memory materials. Emphasis is placed on the new functionalities of SMM that allow the researcher to investigate with functionalities customized for a remarkable application. SMM fabrication, phase transformation phenomenon, and different influencing parameters have been discussed in this chapter. Various categories of SMM with their emerging applications have been presented here.

Keywords

Shape memory material (SMM) · Shape memory alloy (SMA) · Shape memory polymer (SMP) · Shape memory composite (SMC) · Shape memory hybrid (SMH) · Shape memory effect (SME) · Superelasticity (SE) or pseudoelasticity (PE) · Martensite transformation

1.1 Introduction

The introduction of shape memory technology (SMT) to the materials community was found in the 1990s. The shape-memory material (SMM) has the exceptional property of being able to regain its original shape when subjected to a particular stimulus. After severe deformation and quasi-plastically distortion, SMMs can regain their original shape in the interaction of specific stimulus (such as thermal energy, magnetic energy, electrical energy, intensity of light, P^H in solution, stress, etc.) [1, 2]. Depending on the material, SMMs are classified into shape memory alloys (SMA), shape memory polymers (SMP), shape memory ceramics (SMC), and shape memory hybrids (SMH) [3, 4] as shown in Fig. 1.1. The technological orientation toward "smart" systems endowed with adaptive and/or intelligent functions and characteristics requires greater use of these materials in sensors, actuators, and microcontrollers. In sensors, the mechanical signal is turned to a nonmechanical output (e.g., voltage), while an actuator converts a nonmechanical input to a mechanical output (deflection by shape change) [5]. The active structure generally expresses a mechanical response as stimulated by nonmechanical fields, such as thermal, electrical, magnetic, and optical. The mechanical response of the active structure is generally one or more multiplicative magnitude than the conventional action of the material, such as thermal expansion and so on. [6]. The shape memory effect (SME) and pseudoelasticity (in alloys) or viscoelasticity (in polymers) are important properties that play a vital role under certain conditions in different SMMs. The historical background of the gradual developments of SMM devices is given in Table 1.1.

1.2 Shape Memory Alloy

Shape memory alloys (SMA) are a type of intelligent material that has already taken the forefront position of research in recent decades. They have the potential to restore default configuration when stimulated above a threshold-phase transformation temperature that explores a variety of industrial applications. Austenite and martensite are two different stable phases that have an identical chemical composition with three different crystallographic structures (twin-martensite, detwin-martensite, and austenite). The austenitic parent phase is the high-temperature phase, and the martensite product phase is the low-temperature phase. The change in shape implies a phase change in the solid-state without diffusion which implies a molecular orientation between martensite

A. Behera, *Advanced Materials*, https://doi.org/10.1007/978-3-030-80359-9_1

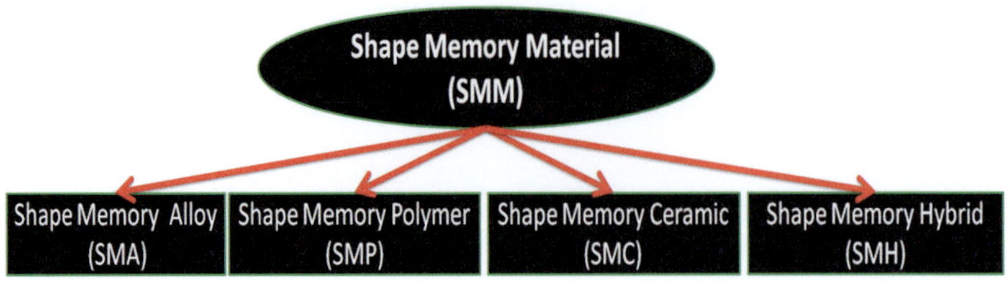

Fig. 1.1 The classification of shape-memory material

Table 1.1 Gradual development of various SMMs given in the form of a specific device [7–15]

Year of development	Devices	Year of development	Devices
1963	Nitinol discovery	1971	Orthodontic braces
1976	Scoliosis Harrington	1977	Simon vena cava filter
1981	Orthopedic staple	1983	NiTi stent, proshetic joint
1990	Thin-film SMA and microdevices, SMA coil springs electrical connector	1991	Spatula (variable curvature), SMA disc drive slider lifter
1992	SMA damping body	1993	Laparoscopic hernia repair mesh
1994	SMA actuated rod for endoscopic instruments	1995	Laparoscopic clamp, reactor, thin-film microgripper
1996	RF ablation device, hernia repair reactor	1998	Atrial septal occluder
1999	Thin film SMA microvalve, laparoscopic suturing clip	2000	Abdominal wall lift, vascular ligation clip, multipoint injector, self-tightening biodegradable SMP, gastric loop snare
2001	Microwrapper, and microstent (thin film), drug-eluting stent	2002	Ischemic stroke preventing SMP microactuator, laser-assisted SMP for preventing attacks
2003	SMP stent with drug delivery system, SMP spinal fixation	2004	SMP microfluidic reservoir, anastomosis ring in laparoscopic
2005	Light assisted SMP, SMA thin heart valve, SMP nanoparticle, single crystal SMA devices	2006	SMP neurovascular stent, two-way SMP fiber, SMP microtag, triple shape SMP, magnetic SMP device
2007	Reactable SMP stent, laser-activated SMP vascular stent	2008	SMA actuators jet nozzle, superelastic polycrystalline SMA microwires
2009	Bone clamp SMA, light-induced SMP display screen, shape memory riblet, SMA active hatch vent	2010	SMA actuated aerostructure, SMA vehicle energy harvesting device, SMP ophthalmological implant
2011	SMA Li-ion battery electrode, SMA heat engines and energy harvesting systems, SMA flexible actuator	2012	SMA active spars for blade twist, SMP intraocular lens, SMA thermostat for subsea equipment

*Many developed form of devices of SMM has been observed in the year 2000–2020 and patented similar to the above-discussed devices

and austenite. The phase change in SMA has two main characteristics: "shape memory effect" and "superelasticity effect." Within the large family of SMA (AgCd, MnCu, CuZn, InTl, FePt, TiNb, NiMnGa, NiMnSn, CoNiAl, CoNiGa, etc.), NiTi-based alloys (with 48–52% Ni wt.%) are associated with better shape memory effect (SME), pseudoelastic effects (PE), good workability, corrosion resistance, and fatigue property [16]. The application of NiTi-based SMA in biomedical industries demonstrates its excellent biocompatibility, good magnetic resonance imaging, and its compatibility with computed tomography. Iron-based and copper-based SMAs, such as FeNiGa, FeMnSi, CuAlNi, and CuZnSn, are inexpensive and commercially available due to their instability and poor thermomechanical performance; whereas, NiTi-based SMA is much more suitable in most applications. Of all the SMAs that have been discovered so far, NiTi (nitinol) has proven to be the most flexible and beneficial in engineering applications with the best combination of properties [17]. Nitinol is a combination of words: (nickel (Ni), titanium (Ti), and US Naval Ordinance Laboratory (NOL)) and is discovered at the beginning of 1960s by William Buehler. Nitinol is distinguished from other SMAs due to its behaviors such as greater ductility, greater recoverable movement, excellent corrosion resistance, stable transformation temperatures, high biocompatibility, and low manufacturing costs.

1.2.1 Historical Background of SMAs

The discovery of the term martensite in steels in the 1890s by Adolf Martens was an important step toward the eventual discovery of shape memory alloys. Martensitic transformation was perhaps the most studied metallurgical phenomenon in the early nineteenth century. Diffusionless martensitic transformation, as observed in the Fe-C system, has been

Table 1.2 Properties of different polycrystal and single crystal NiTi, NiTiAl, and AlCuZn SMA [26–28]

Properties	NiTi (polycrystal)	NiTi (single crystal)	Ni–Cu–Al	Al–Cu–Zn
Melting point (°C)	1310	1250	1050	1020
Density (kg/m^3)	6450	6450	7150	7900
Grain size (m × 10E^{-6})	30–140	Depends on fabrication size	30–100	50–150
Electrical resistivity (Ω × m × 10E^{-6})	0.01	0.01	0.1–0.14	0.07–0.12
Thermal conductivity, RT (W/m*K)	18	18	75	120
Thermal expansion coeff. (10E^{-6}/K)	11	11	17	17
Specific heat (J/kg × K)	837.36	837.36	440	390
Transformation enthalpy (J/kg)	–	–	9000	7000
E-modulus (GPa)	83	95	80–100	70–100
UTS, Mart (MPa)	895	800–1000	1000	800–900
Elongation at fracture, Mart (%)	25–50	30–50	8–10	15
Fatigue strength N = 10E^{+6} (MPa)	350	300	350	270
Transformation temperature range (°C)	−100 to 110		−150 to +200	−200 to +110
Max one-way memory (%)	7		6	4
Normal two-way memory (%)	3.2		1	8
Normal working stress (MPa)	100–130		70	40
Normal number of thermal cycles	100,000		+5000	+10,000
Max. Overheating temperature (°C)	400		300	150
Damping capacity (SDC %)	High [20]	Comparatively lower [≈10]	20	85
Corrosion resistance	Excellent	Good	Good	Fair
Biological compatibility	High	High	Bad	Bad

established as an irreversible process. The SME and the PE properties had been discovered as early as 1932 by a Swedish physicist Arne Olander in an alloy of gold-cadmium (AuCd) [18]. In 1932, Chang and Read detected the reversibility of the AuCd SMA not only by metallographic observations but also by observing changes in resistivity. Greninger and Mooradian in 1938 observed the shape memory effect in Cu-Zn and Cu-Sn alloys. The concept of thermoelastic martensite transformation, which explained the reversible transformation of martensite, was introduced in 1949 by Kurdjumov and Khandros [19] based on experimental observations of the thermally reversible martensite structure in CuZn and CuAl alloys. All the associate transformation in shape memory effect is also diffusionless transformation. Similar effects on other alloys such as In–Tl and Cu–Al–Ni were also observed in the 1950s. In 1953, the appearance of thermoelastic martensitic transformation was demonstrated in other alloys such as In–Tl and Cu–Zn [20]. In 1958, SME was presented at the universal exhibition in Brussels, showing a cyclic lifting application [21] and then remained silent on the reversible transformation of martensitic until 1962 [22]. William J. Buehler and Frederic Wang in 1962–1963 observed the shape memory effect on a nitinol alloy at the U.S. Naval Ordnance Laboratory by accident, during the research on corrosion behavior of $Ni_{50}Ti_{50}$ [23]. The advance in engineering applications came with the discovery of NiTi by Buehler and his colleagues when studying materials useful for thermal shielding. In 1965, studies showed that the addition of a third alloying element such as Co or Fe to the existing NiTi system caused a dramatic drop in the transformation temperatures of SMA [24].

The new alloys inspired the first commercial application of SMA, known as Cryofit, where the SMA material was used for pipe fittings in F-14 fighter aircraft [25]. Cryofit's transformation temperature was low so as to prevent actuation before assembly, and the pipe fittings were transported in liquid nitrogen. Various mechanical applications such as air conditioning vents, electronic cable connectors, valves, and a variety of other products are also gradually being developed. Table 1.2 gives the competitive properties of few shape memory alloys.

1.2.2 Fabrication Processes of SMAs

It is now recognized that shape-memory materials derive their outstanding inherent anisotropic and nonlinear properties from small-scale phase rearrangements, and the strain produced by the pseudoelasticity effect depends on crystal orientations. Specially oriented single crystals of certain SMMs can produce considerable deformation (≈10%) due to phase transformations [29]. In applications, SMMs are generally polycrystalline in nature and are generally processed by casting, followed by hot work (drawing for rods and wires and laminate for sheets), and appropriate heat treatments. The polycrystalline SMAs thus produced generally have a strong texture, and several researchers have recently emphasized that the crystallographic texture is very important in determining the overall properties of the SMAs. Several fabrication processes have been discussed subsequently.

1.2.2.1 Vacuum Melting

Conventional vacuum casting methods can be used to fabricate various SMAs. But it is very difficult to avoid contamination due to atmospheric C, O_2, etc., which can alter the mechanical properties [30]. The contamination problem can be avoided by melting the SMA constituents in an inert chamber. Two methods facilitate the inert atmospheric melt which include:

1. Vacuum induction melting (VIM).
2. Vacuum arc remelting (VAR).

In VIM (Fig. 1.2a), melting will be carried out by electromagnetic induction. Here the electric eddy currents are induced in the graphite crucible and metallic charges. The electrodynamics force controls the melt stirring. With the use of graphite crucible to melt NiTi in VIM, the carbon contamination will increase resulting in the production of TiC in the melt. The presence of TiC can drop the transformation temperature. VIM has the advantage to homogenize the chemical composition of the melt [31]. But VAR provides high purity and there is no carbon contamination problem. VAR is of two types with respect to the heating system: (1) use of nonconsumable electrode with Cu mold and W electrode to produce a button shape melt product and (2) use of consumable electrode (Fig. 1.2b), comprising the Ar arc heating source that heats the electrode and drops the molten metal in the cylindrical mold. The limitation of the second method is the vacuum leak problem that contaminates the melt by O_2 and C. The dual fusion technique using primary VIM followed by VAR is often used to obtain additional refinement [32]. Better homogenization can avoid the microsegregation in the melt but cannot escape out from O_2 contamination in VAR. The main limitation of the VIM and VAR is the more reactivity between the molten material and the segregation [33]. The main reason for segregation is due to the difference in density in the reaction melt.

1.2.2.2 Powder Metallurgy

Powder metallurgy is a cost-saving process to develop various components with an almost near net shape of SMA. In these methods, the required weight percentage of Ni and Ti was mixed mechanically followed by compaction to obtain a green structure. The green compact is sintered near the melting temperature. Different starting metallurgy methods for mechanically alloyed powders are hot isostatic pressing (HIP), microwave sintering (MS), spark plasma sintering (SPS), conventional sintering (CS), self-propagating high-temperature synthesis (SHS), metal or powder injection molding (MIM), space holder technique (SHT), hybriding, and so on. The basic steps to follow in powder metallurgy methods include; (a) powder mixing, (b) mold compacting, and (c) sintering in a furnace (Fig. 1.3). In this process, if elementary powders are used, the Kirkendall porosity can occur in the component because the diffusivity of the Ti in Ni is greater than the Ni in Ti [35]. HIP is the most widely adopted powder sintering technique in which the pressure is applied uniformly in all the direction to produce a dense structure. In HIP, the mixture of powder is subjected to high temperature and pressure in an inert chamber. HIP gives better control in density, homogeneity, and pore size with respect to the geometry and thermodynamic stability. Metal injection molding (MIM) is a combination of polymer injection molding and powder metallurgy which has chosen to prepare the near net shape. MIM is used to produce a large structure of high-density NiTi in the form of a small and complex mesh [36]. The MIM process takes place in four stages, namely, raw material, injection molding, unwinding, and finally sintering. In the first step, the powder mixture takes place with the binder in a suitable ratio. The second stage consists of the fusion and the injection of metallic

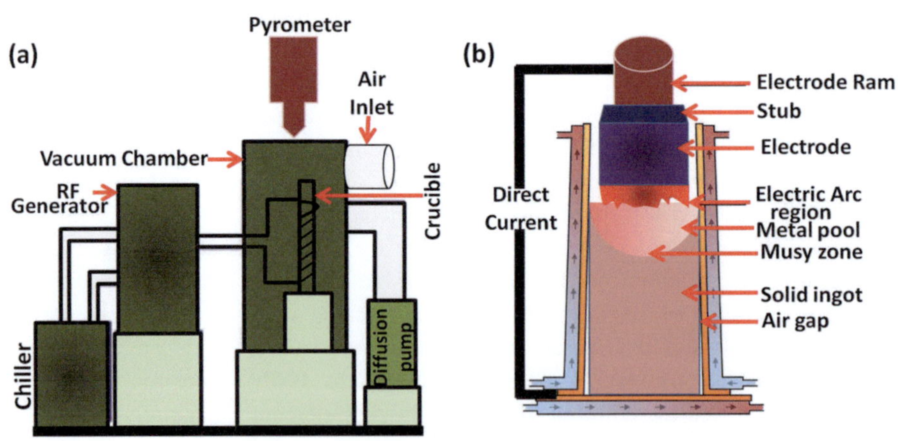

Fig. 1.2 The casting set up of (**a**) VIM and (**b**) VAR processes [34]

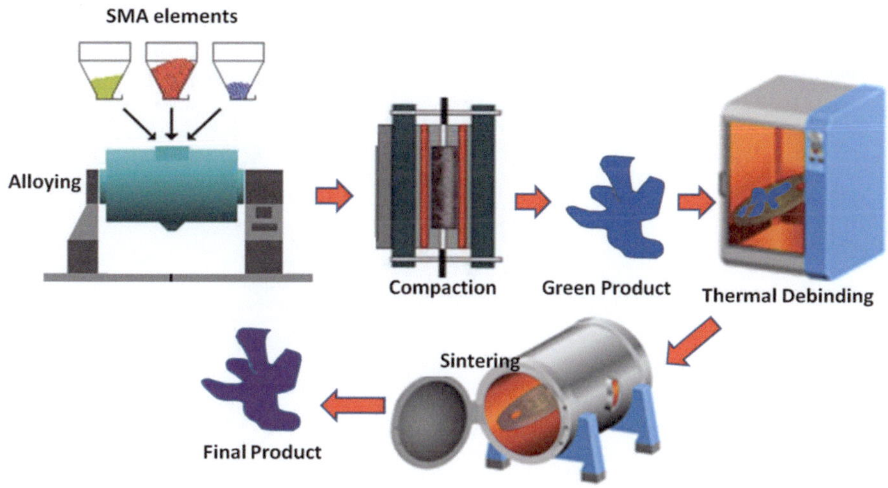

Fig. 1.3 Graphical representation of powder metallurgy process of SMA

powder through a mold. Mold design is important to transfer the temperature and pressure throughout the structure during compaction with the ability to adequately withdraw the injected sample. Compaction followed by the thermal debinding in a vacuum-assisted high-temperature furnace to vaporize the binder. Gradual increase in temperature in sintering furnace at 1200 °C results in a dense structure. In SPS, the pre-alloyed powders are filled and compressed in a die followed by electric current heating. Finally, fusion occurs at low temperature with a lesser time than that of the HIP process. The porosity obtained by this method is less than the HIP process. Microwave sintering has aroused the interest of researchers because of its energy-saving properties, its short treatment time, its improved mechanical properties, and its faster and more ecological heating rate. Porous NiTi can be made using microwave sintering process even at low temperature (850 °C) with less sintering time (15 min). The porosity percentage can be improved by increasing the holding time in microwave sintering [37].

1.2.2.3 Additive Manufacturing

Most powder metallurgy method lacks homogeneous porosity control (e.g., the amount of porosity, pore size, and pore arrangement), chemistry (contaminant percentage, homogeneity, intermetallic), and flexible dimension. The formation of secondary phases by using elementary powder is inevitable since its formation is much more thermodynamically favorable. Undesirable secondary phases and contamination not only affect mechanical properties but also make materials brittle and also change phase transformation temperatures, which are important for any practical applications. Another important limitation of these processes (with the exception of MIM) is that they cannot produce complex geometries. Therefore, most NiTi SMA devices are manufactured in the form of simple geometries, such as wire, sheet, tube, and bar. Additive manufacturing (AM) is a very attractive process to overcome these problems mentioned. Additive manufacturing is also called rapid prototyping (RP) or rapid manufacturing (RM) or 3D printing, which produces near-net parts with the addition of powder layers in the interval [38]. In these processes, a thin layer of NiTi powder is spread individually followed by the binding treatment by some source that is operated by the computer-aided designing process. The product does not require any post-treatment such as machining, milling, and so on. It is an energy-efficient and ecofriendly process. The common methods used to produce NiTi components via additive metallurgy include selective laser sintering (SLS), selective laser melting (SLM), laser-engineered net shaping (LENS), and electron beam melting (EBM). Figure 1.4 shows a schematic diagram of the SLM process which is a laser-assisted process. SLM process can fabricate the SMA objects. SMA exhibits pseudoelastic behavior directly after manufacture without any post-training procedure or heat-treatment process. LENS is used to fabricate the SMA results in a higher microhardness compared to conventional alloy. The major limitation of the LENS process is higher cost and poor surface finish [39]. In the EBM process, the source is high-voltage electron beams that use to binding the powder in the thin spread layer. EBM process needs a high-vacuum chamber to operate the electron beam from the electron gun. Electron gun can produce very small spot sizes with high scanning speed [40]. This method is mainly adopted for SMA that does not need precise or definitive control of the transformation temperature. In addition, rapid manufacturing processes also use to produce the porous SMA with great control over the required volume fraction of porosity with acceptable mechanical properties, such as resistance to fatigue and corrosion.

Fig. 1.4 Schematic illustration to represent laser-assisted additive manufacturing process

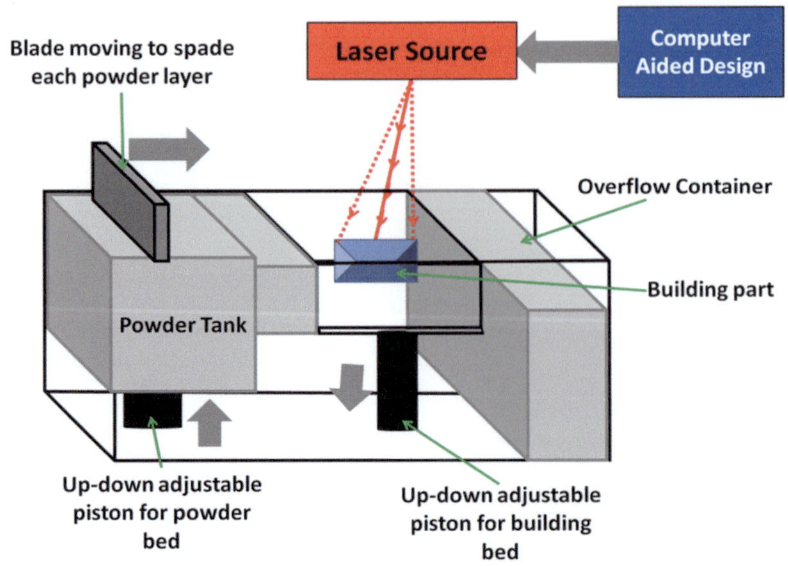

Fig. 1.5 Schematic diagram of the plasma spray process

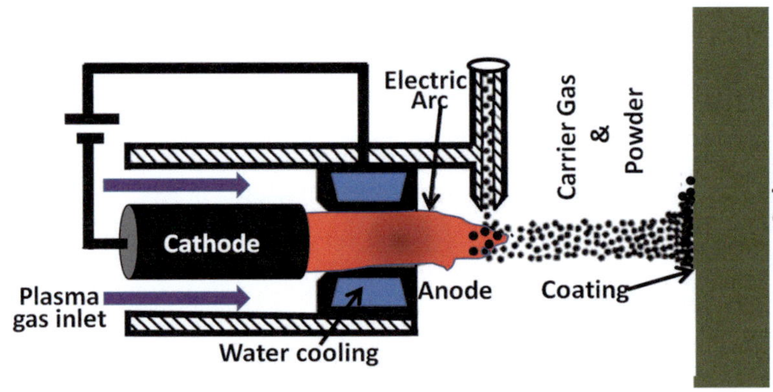

1.2.2.4 Thermal Spray

Thermal spraying technique can be used to produce SMA sheets or tubes, thin-walled structure, and varieties of 3D shapes [41]. With an optimized parameter set up, it can produce a near-net shape. Parameter optimization is the most important thing in thermal spraying. There are two methods suitable under thermal spraying for SMA: (1) low-pressure wire arc spraying and (2) vacuum plasma spraying technique. Here, the final product is dense and of greater strength. The low-pressure thermal spraying process will minimize the delivery period and the manufacturing cost of NiTi semifinished plates and tubes. Figure 1.5 shows the plasma spraying process to fabricate the NiTi thin sheet. Higher powder loss is observed in this process.

1.2.2.5 Plasma Melting

Figure 1.6a shows the plasma melting method, in which a low-viscosity electron beam discharges from hollow plasma cathode to melt the materials. Less loss of materials and uniform distribution of the composition occurs in this process when compared to thermal spray process.

1.2.2.6 Magnetron Sputtering Deposition

SMA thin film can be fabricated by magnetron sputter deposition technique. This technique can produce a dense film and have better control over composition (Fig. 1.6b). Chemical composition depends on the target composition, process parameter, and plasma ionization and is responsible for the phase transition temperature of the thin-film SMA. Magnetron sputtering can produce a deposited layer from nanometer to few micrometers. Homogenization of composition can be obtained by optimizing the substrate rotation, substrate bias, and substrate temperature during the deposition. In NiTi SMA, two types of target arrangement can be used: (1) single mosaic target and (2) more than one pure target with suitable power supply (RF/DC) for co-deposition or batch-wise deposition. The batch-wise deposition needs a post-annealing process to homogenize the composition. In magnetron sputtering, generally, the conducting target is connected with the DC power supply, whereas nonconducting materials are connected by the RF power supply. In the case of Ni and Ti co-deposition, the connection of Ni with the DC and Ti with RF is preferable to get equal sputter yield for

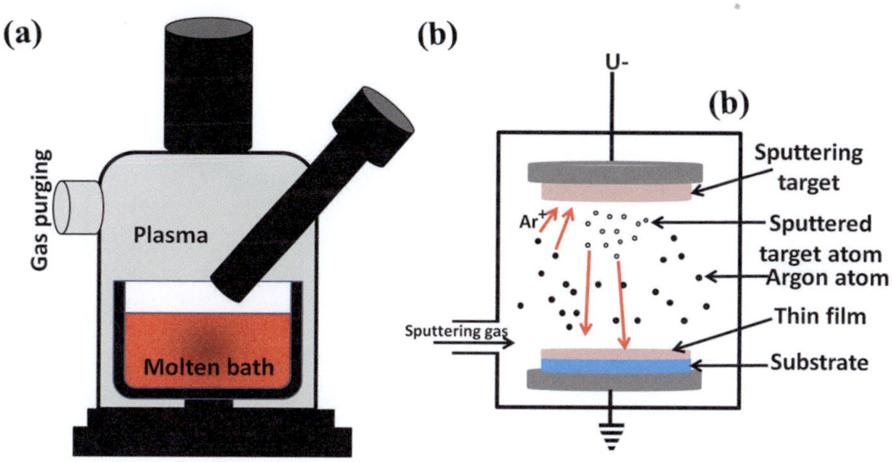

Fig. 1.6 (**a**) Plasma melting method and (**b**) sputtering deposition process [42]

Table 1.3 The comparison of different manufacturing methods [46–53]

Manufacturing technique	Classification	Properties: quality of product (surface finish and dimensional accuracy)	Defect	Porosity (Max. %)	Strength of as-get specimen	Requirement of posttreatment	Product cost
Melting and casting	VAR	Average	Average (C and O_2, segregation)	35	Good	Machining	Low
	VIM	Average	Average (C and O_2, segregation)	40	Good	Machining	Low
Powder metallurgy	Conventional sintering	Low	High (mixing and sintering defects)	40	Good	Surface treatment	High
	SHS	High	Average (microstructure control)	65	Good	Surface treatment	High
	HIP	High	Average (decrease internal microporosity)	50	Good	Surface treatment	High
	MIM	High	Average (larger part difficult to fill in furnace)	41	Good	Surface treatment	High
	SPS	High	High (complex part can not be fabricated)	12	Good	Surface treatment	High
Additive manufacturing	SLM	Very high	Low/minimum	15	Very good	Not required	Very high
	SLS	Very high	Low/minimum	79	Very good	De-burring	Very high
	LENS	Low	Low/minimum	16	Good	De-burring	Very high
	EBM	High	Low/minimum	70	Very good	De-burring	Very high
Thermal spray		Average	High	20–40	Good	Not required	High
Thin-film fabrication process	PVD	Very high	Low	2–10	Good	Not required	Low
	CVD	High	Low	2–15	Good	Not required	Low

homogenized composition [43]. The martensite transformation of SMA is also influenced by sample thickness. Reducing the thickness of the sample reduces the transformation temperature of NiTi.

1.2.2.7 Post Fabrication Process

Post fabrication, that is, heat treatment is necessary for SMAs, when there is a major question about the homogenization of the fabricated bulk composition. In the case of superelastic NiTi, heat treatment around 500 °C is required, whereas 350–450 °C is required for shape memory effect [44]. In some cases, solution treatment is followed by aging to alter the martensite transformation temperature [45].

Table 1.3 shows the effect of various parameters of various techniques used to fabricate the SMA. A little alternation in the production process is responsible for SMA response

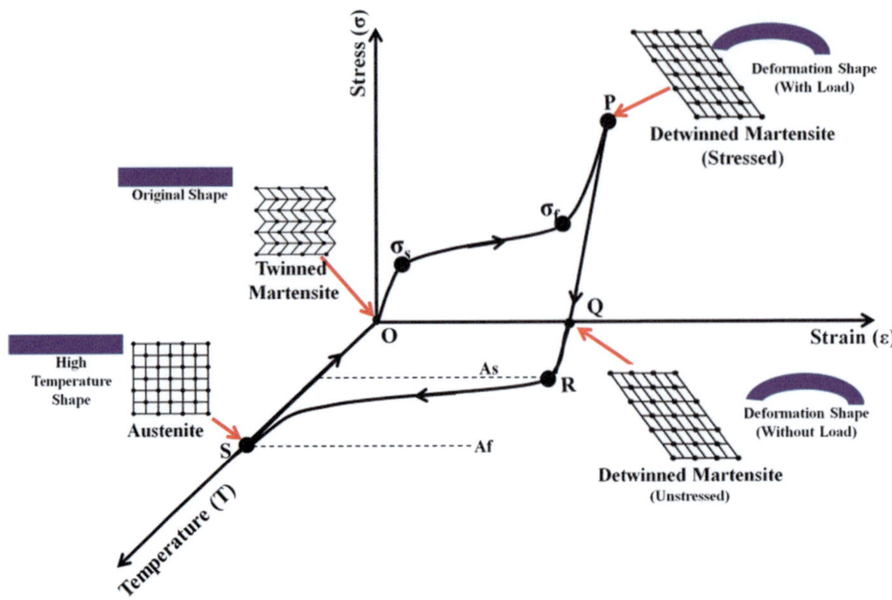

Fig. 1.7 Stress–strain–temperature plot for SME in NiTi shape-memory material

and is directly related to the SMA's physical, mechanical, and biocompatible properties.

1.2.3 Shape Memory Effect

The shape memory effect (SME) is a peculiar property of SMM and is utilized in commercial applications in a broad range of industries. The SME is based on the phase transformation between the martensite phase and the austenite phase that occurs without diffusion. This phase transformation occurs following stress or change in temperature. By regaining their shape, alloys can produce displacement or force, or a combination of the two, depending on the temperature. The relationship between stress–strain–temperature can be understood by Fig. 1.7. The low-temperature phase is martensite with a twin atomic configuration as shown in point O in Fig. 1.7. When stress is applied to the structure, the twin-structured martensite transforms to a detwin structured martensite at point P. It is a shear dominant transformation. Here, the twinned martensite is subjected to applied stress which exceeds the start stress level (σ_s), and the reorientation process begins, causing the growth of certain favorably oriented martensite variants that develop at the expense of other less favorable variants. The stress level for reorientation of the variants is far lower than the permanent plastic yield stress of martensite. The detwinning process is completed at a stress level, σ_f, that is characterized by the end of the plateau in the stress–strain diagram. By releasing the stress on that structure at point P, the material retains the detwined structured martensite at point Q. To recover the SMA shape, now the role of stimuli is to be presented. For thermal-induced SMA, the temperature increases above the threshold temperature (above austenite finish temperature, A_f) following a path from Q $\rightarrow$ R to R $\rightarrow$ S. At high temperature at S, the structure attains fully austenite phase (cubic structure). Then, by subsequent cooling up to M_f temperature, the cubic structure of austenite transfers to detwined martensite following the path from S to O in Fig. 1.7. The undeformed structure of the SMA is restored. All the transformation associated here is diffusionless transformation known as military transformation (transformation is in a coordinated manner similar to military movement in marching). The transformation does not depend on time [54]. Transformation occurs due to nucleation, which occurs when a small crystal seed begins to crystallize in a polymorphous zone by overcoming a large energy barrier. The strain recovered due to the phase transformation of the detwinned martensite into austenite is called the transformation strain [55]. Unconstrained cooling of the austenite below forward transformation temperatures (M_s and M_f) results in the formation of self-accommodated twinned martensite variants without associated shape change, and the entire SME cycle can be repeated. The phenomenon described earlier is called SME.

1.2.4 Superelasticity or Pseudoelasticity

Another important peculiar characteristic of SMA is superelasticity or pseudoelasticity (PE) transformation. It is an elastic (reversible) response to the applied stress. Unlike the SME, this property allows SMA to withstand large amounts of stress without undergoing permanent deformation. The transformation occurs without temperature change. For this behavior, the SMA temperature is

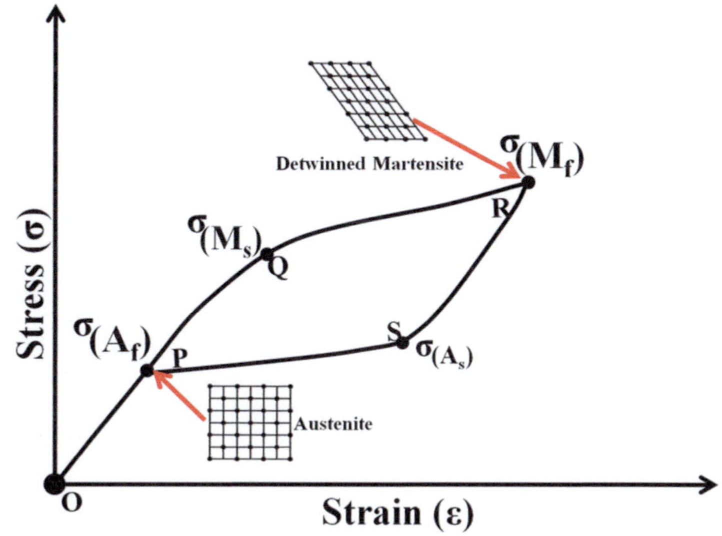

Fig. 1.8 Schematic of a pseudoelastic stress–strain diagram

maintained above the transition temperature, which is above the austenite finish (A_f) temperature. The stress–strain curve for superelasticity is given in Fig. 1.8. The stress increases until the austenite becomes martensite up to point R. The result of this loading is a fully detweened martensite created from austenite. When the load is reduced, the martensite transformations return to austenite at point P. The SMA returns to its original form because the temperature is always higher than the transition temperature. This is a consequence of a stress-induced transformation from austenite to martensite and vice versa. The resulting stress–strain curve (P-Q-R-S) is called pseudoelastic or superelastic curve. This stress-induced transformation is also known as the rubber-like effect to emphasize the similarities with the rubber that experienced the nonlinear elastic behavior. In SMA exhibiting the rubber-like effect, the stress required to detwin martensite is very small compared to $\sigma_{(Ms)}$ [56]. Hysteresis exists there, and the mechanism responsible for hysteresis is the movement of clear interfaces between the two phases. For a given material, the size and other qualitative characteristics of "flag-type" hysteresis loops generally change with the speed of loading and the temperature at which the work is performed [57].

There is no adequate capture of the asymmetry observed between the tension and the compression in pseudoelastic behavior. At a given test temperature, it is found that (a) the load need to nuclei the martensitic phase from austenitic phase is considerably higher in compression than in tension; (b) the transformation deformation measured in compression is less than the stress; and (c) the hysteresis curve generated by compression is wider (along the tension axis) than the hysteresis circuit generated by tension. These main differences between the stress and compression response of an SMA for SE are given in Fig. 1.9.

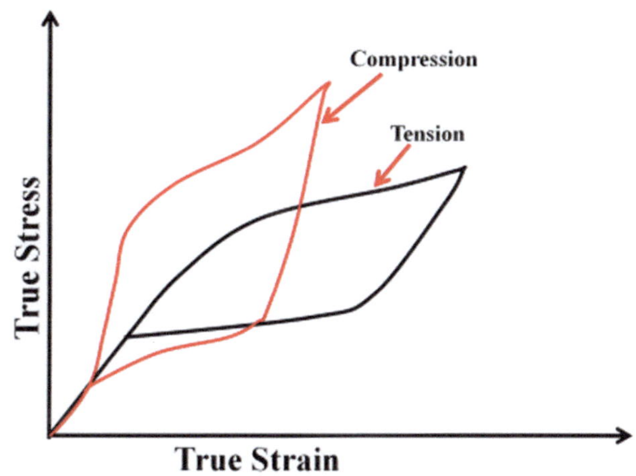

Fig. 1.9 Tension and the compression response of an SMA for SE

1.2.5 Phase Transformation Phenomenon in SMA

The martensite transformation explains the recovery of the shape of SMA. An SMA has two main phases: (1) parent phase (austenite) and (2) product phase (martensite). The high-temperature phase is austenite (cubic crystal structure) and low-temperature phase is martensite (orthorhombic, monoclinic, or tetragonal crystal structure). The transformation from one structure to another does not occur by diffusion of atoms but lattice distortion by shear in SMA. Such a transformation is known as a martensite transformation. Each martensite crystal generated has a preferential direction of orientation called a variant. The set of martensite variants presents two forms that are twin martensite (M_t), which is formed by a combination of self-adaptive martensite variants, and unfolded or detwinned or reoriented martensite in which a specific variant is dominant. The *forward phase transformation* from parent phase (austenite) to product phase

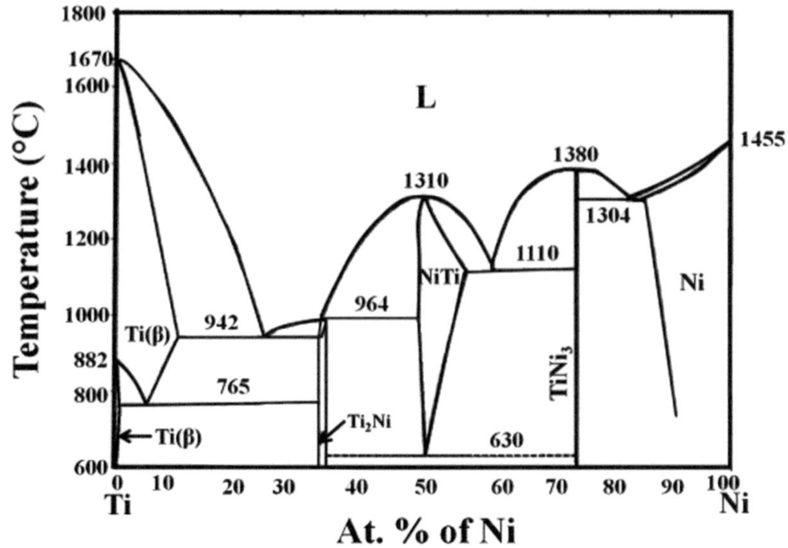

Fig. 1.10 NiTi shape memory alloy binary phase diagram

(martensite) and vice versa forms the basis for the unique behavior of SMA results in the formation of several martensite variants (up to 24 for NiTi) [58]. All the associated phase changes directly depend on the composition of the alloy. For example, in Fig. 1.10, around 50/50 at. % of Ni/Ti gives the best SME. The phase diagram of an alloy is a schematic representation of equilibrium conditions between co-existence phases. Phase diagrams are made up of phase boundaries or equilibrium lines that distinguish various phases during fabrication with respect to the control variable, such as composition, stress, temperature, and so on. The transformation from austenite to martensite occurs in two stages: The lattice deformation stage produces the new structure via small layered dislocations, while the lattice invariant shear/accommodation stage involves accommodating the new structure by altering the surrounding austenite either by twinning (unable to accommodate volume changes) or slip (permanent and common). In the case of NiTi SMA, the parent cubic (B2) phase may transform to any of three different crystal structures, that is, the monoclinic (B19') phase, the rhombohedral (R) phase, or the orthorhombic (B19) phase, depending on various conditions. The phase transformation can follow different paths for various SMA: B2 → B19', B2 → R → B19', B2 → B19 → B19' and B2 → R → B19 → B19' [59, 60]. The conditions determining the abovementioned paths are: (a) variation of Ni percentage, (b) addition of other elements in NiTi, (c) aging followed by solution treatment, and (d) thermomechanical treatments.

1.2.6 Martensite Reorientation

The reorientation of martensite is a transformation between variants of martensite, that is, no phase change is associated. This transformation is always related to stress and the corresponding recoverable strain is associated with the reformation of variants. If a fully twinned SMA is subjected to a gradual increase in uniaxial loading, the transformation of the variant in each of the groups of twin produces a preferred variant growth. This process minimizes the elastic energy, as the change in internal stress partially neutralizes the externally applied stress. As the load increases, similar conversions will take place among martensite groups until the entire structure converts to only one variant in a single crystal or in polycrystal. Mostly, each crystal is preferred by a certain variant or variant group. This single preferential orientation is called detwinning and the product martensite is called detwinned martensite (D_M), shown in Fig. 1.11a. If the maximum stress level is not high enough, the material will be only partially detwinned, and there will still be several different variants/variant groups in each crystal as shown in Fig. 1.11b. If the local stress field is changed, for example, by unloading the specimen, the martensite variants may convert, according to the local stress, to minimize the total energy in the local area [61]. The positions of the martensite interfaces change under the influence of stress, creating a balance of variants whose shears best accommodate the direction of the applied strain. Upon heating, the material reverts to the beta phase and recovers its original shape or parent phase of a crystal (Fig. 1.11c). The interfaces between variants move to "grow" the most favorably oriented variants and shrink the least favorably oriented. Some variants become dominant in the configuration. This process creates a macroscopic strain that is recoverable as the crystal structure reverts to austenite during reverse transformation.

1.2.7 Crystallography of Phases

NiTi and NiTi-based SMAs show three categories of transformation path as shown in Fig. 1.12: (1) Parent phase (B2 phase, cubic structure) → martensite phase (B19′

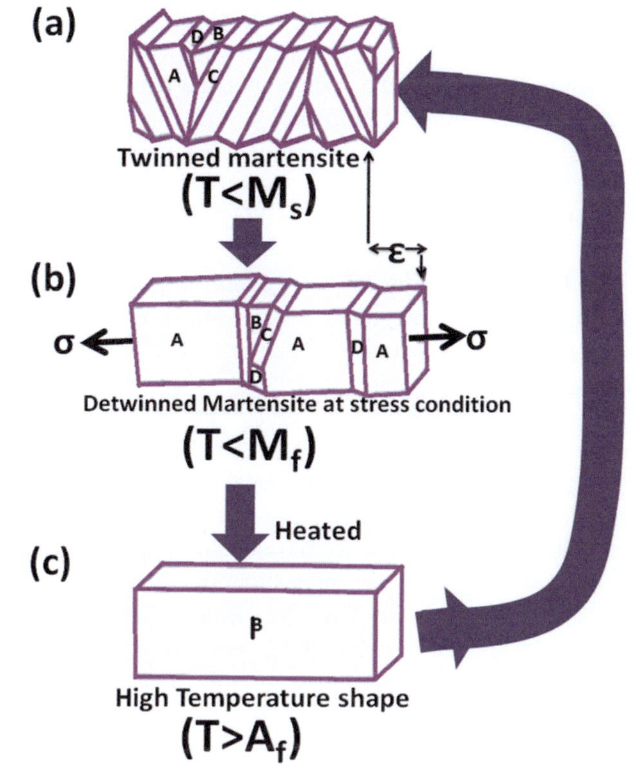

Fig. 1.11 (**a**) Self-accommodating twin martensite, (**b**) Variant A becomes dominant with the application of load; (**c**) Heating revert the material to the beta phase and recovers its parent phase [62]

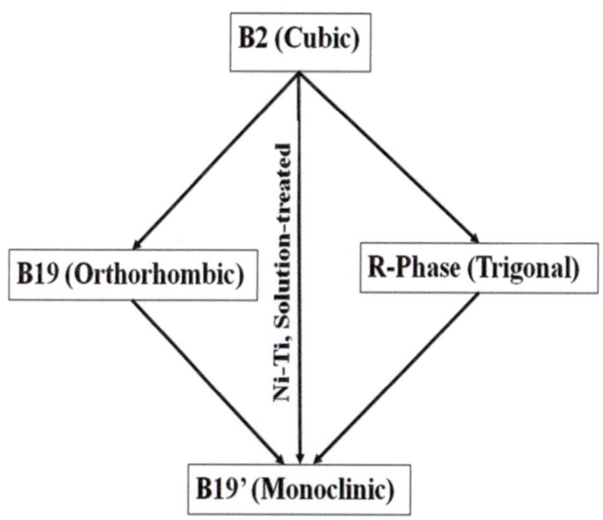

Fig. 1.12 Three-phase transformation paths in NiTi-based alloys

phase, M phase, monoclinic structure), (2) B2 phase → B19 phase (O phase, orthorhombic structure) → B19′ phase, and (3) B2 phase → R-phase → B19′ phase. The Austenite phase is the high energy associated phase. In SMA, austenite is a BCC structure that has a B2 crystal structure and high symmetry with a lattice parameter of 0.3015 nm at normal temperature. There is an order–disorder transition from B2 to BCC at 1090 °C [63]. B2 phase is the high symmetry phase. The martensite B19′ structure has lower symmetry than B2. The most common low-temperature (low energy) phase martensite of nitinol is a monoclinic structure with a = 0.2889 nm, b = 0.4120 nm, c = 0.4622 nm, and β = 96.8°. Wang et al. reported martensite formation by the shear in <111> direction, but the structure is nondefined since the amount of displacement of atoms is a function of specimen temperature [64]. B19 is an intermediate stage in between B2 and B19′. The structure of the martensite in the first stage of $Ti_{50}Ni_{50-x}Cu_x$(x = 10 − 30) alloys are orthorhombic (B19) in structure. The structure is simple and is essentially the same as those of martensite in $Au_{47.5}Cd$ and in CuAlNi alloy and so on. Another intermediate phase is R-phase that appears under certain conditions and comes prior to the transformation to B19′ phase [65]. Aging by solution treatment and addition of the third element is responsible for the formation of the R-phase between B2 and B19' phases. The R-phase can be formed by deformation along <111> direction of the B2 structure. The R phase can be stabilized with the incorporation of ternary alloys such as Fe, Al, or Co in addition to equiatomic NiTi. The R-phase transition provides perfect stability against cyclic deformation because of its small change in dimension with no slip deformation, while the martensite transition temperatures increased and temperature hysteresis decreased during cycling because of the formation of internal stress which is due to the introduction of dislocations. The measured lattice constants of the R phase are a_h = 7.3580 Å and c_h = 5.2855 Å, and they may deviate by 1.4% and 2.4%, respectively, from the calculated value [66].

In B2→B19′ transformation, the lattice invariant shear (i.e., twinning modes) is an important parameter for the phenomenological theoretical calculations. Experiments clearly showed that the phenomenological crystallographic theory applies to the B2→B19′ transformation in NiTi alloys with the lattice invariant shear of <011> Type II twinning. In B2→B19→B19′ transformation, B2 → B19 transformation appears in $Ti_{50}Ni_{50-x}Cu_x$ ($x \geq 5$) and $Ti_{50}Ni_{44}Cu_5A_{l1}$ alloys and is soon followed by B19 → B19′ transformation. Thus, it is difficult to observe the B2→B19 transformation separately. The twinning modes in B19 martensite were studied by electron microscopy. There are two twinning modes; {111} o Type I and {011}o compound ones, and both are associated with small twinning shears in B19 [67]. In B2 →R →B19′ transformation, R-phase transformation is characterized by a sharp increase in electrical resistivity with extremely small temperature hysteresis (1–2 K). The structure of R-phase is trigonal, and thus, it is convenient to describe it with hexagonal indices. A typical morphology of R-phase transformation is not associated with a lattice invariant shear. R-phase transformation appears prior to the martensite transformation upon cooling. B2→R transformation follows immediately the beginning of B2 + B19′ and possibly is initiated by it. Generally, we can write the transformation path as

B2→B2 + B19′→R + B19′→B19′, with its second stage observed in a narrow (60 − 55 °C) temperature interval [68]. Thermal cycling under a stress-free condition is favorable for the formation of R-phase in NiTi. Other conditions are the introduction of dislocations and the internal stress by the thermal cycling that provides the site for nucleation of R-phase.

1.2.8 Thermodynamics of Phase Transformation

Transformation temperatures (M_s, M_f, A_s, A_f) can be determined by measuring some physical properties as a function of temperature, since many physical properties often drastically change upon starting or finishing martensitic transformation. The free energy curves of both austenite and martensite phases as a function of temperature may be represented as schematically shown in Fig. 1.13. The transformation begins to take place below the critical temperature T_0, at which the free enthalpies of both phases are the same. With decreasing temperature, the transformation continues until the temperature M_f [69]. The $M_s - M_f$ temperature interval is an important parameter in defining the memory behavior. T_0 is the thermodynamic equilibrium temperature between the two phases, G^m is the Gibbs free energy of martensite, and G^p is the Gibbs free energy of austenite. So, the driving force for the nucleation of martensite is:

$$\Delta G^{p-m}|_{Ms} = G^m - G^p \qquad (1.1)$$

and,

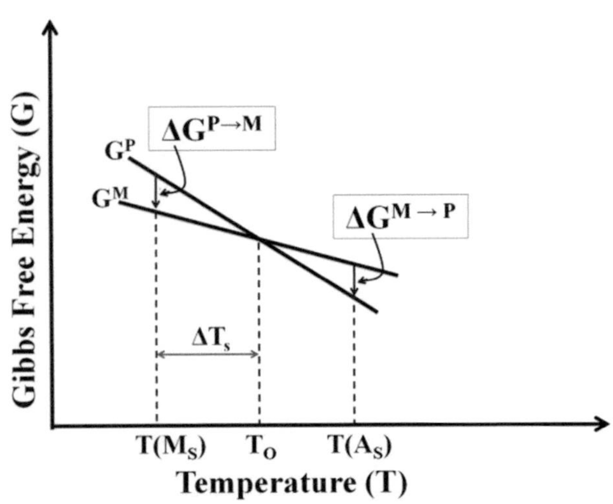

Fig. 1.13 Schematic representation of free energy relationship in forward transformation in SMA. ΔT_s is the supercooling driving energy for the forward transformation

$$T_0 \approx ½\,(M_s + A_s) \qquad (1.2)$$

A Gibbs free energy of a system upon martensite transformation may be written as follows.

$$\Delta G = \Delta G_c + \Delta G_s + \Delta G_e = \Delta G_c + \Delta G_{nc}, \qquad (1.3)$$

where ΔG_c is a chemical energy term originating in the structural change from parent to martensite, ΔG_s is a surface energy term between parent and martensite, ΔG_e is an elastic (plus plastic in the nonthermoelastic case) energy term around the martensite, and ΔG_{nc} is a nonchemical energy term.

In most martensite transformations, ΔG_{nc} is equally as large as ΔG_c, which is an essential point when we discussed martensite transformations. Because of this, supercooling of ΔT_s is necessary for the nucleation of martensite, and superheating is necessary for the reverse transformation [70]. But, M_s is not the same with M_f, since the elastic energy around the martensite resists the growth of the martensite unless a further driving force is given (i.e., cooling).

When we discuss martensitic transformations, we classify them into two categories, thermoelastic and nonthermoelastic. In the case of the Au–Cd alloy, the transformation hysteresis is as small as 15 K, while in Fe–Ni alloy, it is as large as approximately 400 K. In the first case, the driving force for the transformation is very small (as evidenced by the small temperature hysteresis), the interface between austenite and martensite is very mobile upon cooling and heating, and the transformation is crystallographically reversible in the sense that the martensite reverts to the parent phase in the original orientation. This type of transformation is called thermoelastic. In the second case, the driving force is very large, the interface between austenite and martensite is immobile once the martensite grows to some critical size, the reverse transformation occurs by the renucleation of the parent phase, and thus the transformation is not reversible. It is known that shape memory effect and superelasticity are generally characteristic of thermoelastic transformations [71].

1.2.9 Training and Stability of SMA

Whether the SMA is used for its SME or SE behavior, a preferred shape suitable to the application must be obtained. Training is a process to maintain the required suitable shape of the SMM in a certain way. For the classical one-way shape memory effect, there is a sequence of processes that give the desired parent phase shape as shown in Fig. 1.14. Shape setting refers to a procedure used to form the desired shape that follows cold working. This is done in the following way: The sample is formed to the desired shape and is held in that

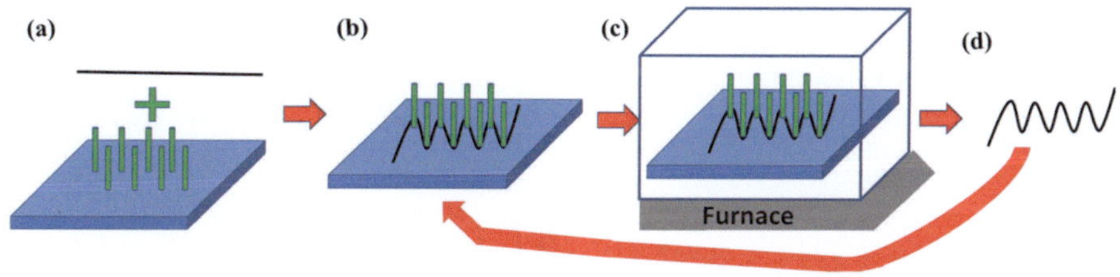

Fig. 1.14 Training process showing (**a**) as-casted SMA bar and ceramic board fixed with pin for a required shape, (**b**) fixing the bar in the board, (**c**) keeping in a vacuum furnace with a constant temperature, and (**d**) repetition of the process as per the arrow mark carried out to get the required final shape

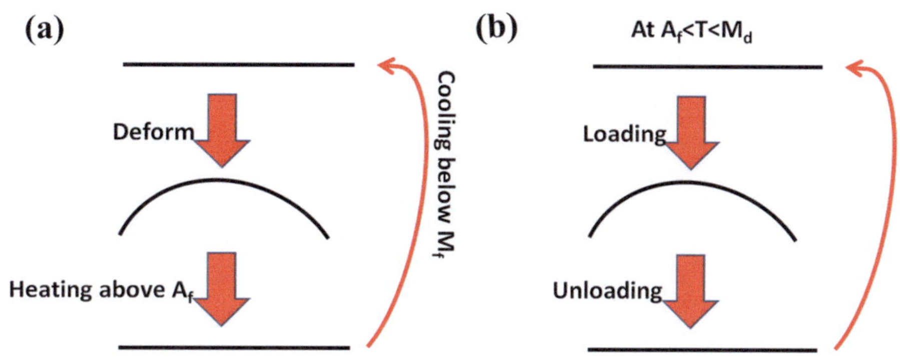

Fig. 1.15 Schematic representation of (**a**) shape memory training and (**b**) stress-induced martensite transformation training

position tightly (Fig. 1.14b). Afterward, the component is heated to a certain temperature that is enough to obtain a uniform temperature on its entire mass. The heat treatment time depends on the sample's dimension. It is quite essential that the sample remains still while being heated; otherwise, the required shape will not be set (Fig. 1.14c). Once at room temperature the alloy has transformed into the martensite state, the twinning has occurred with maintaining the macroscopic shape. At the first completion of the cycle, as shown in Fig. 1.14 procedure, a partial strain recovery will occur that is some permanent (irrecoverable or plastic) strain generated upon heating [72]. To cease the plastic deformation in the SMA, a repetitive cycle, as in Fig. 1.14 (d) to (b), have to be followed. By each consecutive cycle, there is a gradual decrease in permanent strain. The number of training cycles varies from material to material with respect to the chemical composition and variants. Hence, the training can be defined as a process of loading–unloading thermomechanical cycle at a predefined temperature until the hysteretic response of the material stabilizes and the inelastic strain saturates.

In the case of two-way SMA, the training process followed an additional setup of keeping another temperature medium for maintaining the shape adjacent to the parent shape at lower temperature and above M_f. The repetitive cycle is carried out to train the material. The training can be done in a variety of different ways, and all are focused on optimizing the TWSME behavior. This behavior is known as two-way SME. An SMA with the TWSME has the ability to remember and eventually recover both its hot (austenite) and its cold (martensite) phase shapes simply upon heating and cooling, respectively.

Again, the training procedures involve repetitive phase transformations between austenite and a preferred configuration of martensite. Furthermore, certain martensite variants are favored and grown upon cooling at the expense of others, eventually leading to a specific cold shape. The preferentially oriented martensite is introduced either upon systematically loading–unloading cycles or by an overdeformation. The training procedures are presented subsequently (Fig. 1.15a). Another useful training process is stress-induced martensitic transformation training, that is, pseudoelasticity training procedure (Fig. 1.15b) [73]. The material is maintained above the A_f temperature but below the M_d temperature where stress-induced martensite is feasible. Upon loading, stress-induced martensite is introduced and the microstructure is consisted of detwinned martensite. After that, the load is removed, and the material gets back into its austenite phase. This procedure continues.

During training, the SMM has experienced a nonlinear behavior with significant hysteresis, as shown in Fig. 1.16. Stabilizing the shape by reducing the hysteresis is an important parameter when designing an actuator. When the movement of the actuator needs to be controlled, for example, the movements of the actuator produce linear displacement; hysteresis and nonlinearity are to be maintained. Consequently, many SMA actuators are "on/off" controlled with only two movement positions. This is easily achieved with continuous heating to maintain the complete austenite phase or continuous cooling to obtain the complete martensite phase. The amount of hysteresis is influenced by the composition of the alloy. Hysteresis can be beneficial in some applications, such as temperature-controlled thermostats. When the

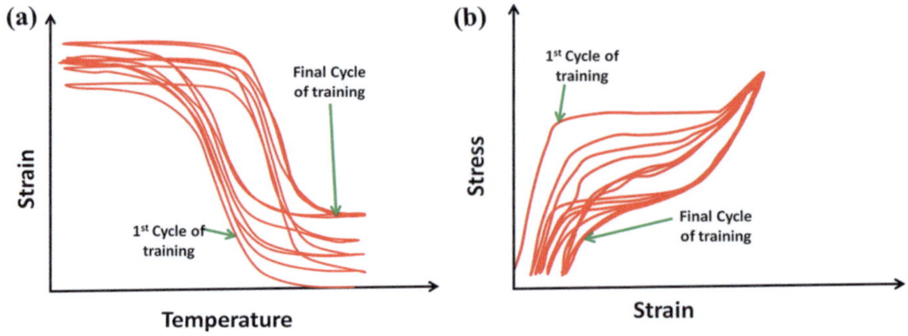

Fig. 1.16 Thermal cyclic loading of a shape memory alloy (**a**) SME under constant load and (**b**) SE under constant temperature

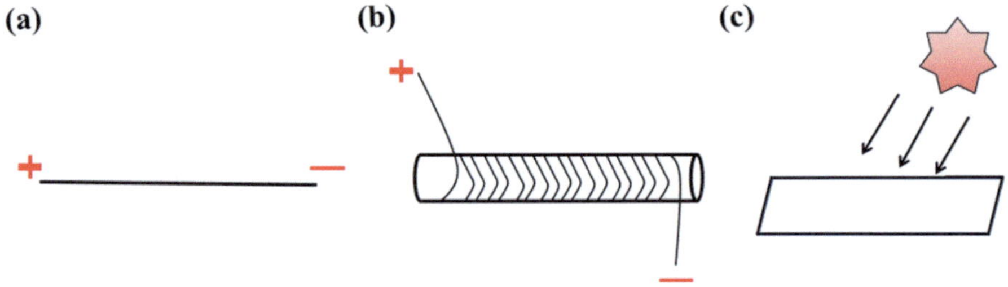

Fig. 1.17 Various heating methods for temperature-induced SMA (**a**) by passing a current through the structure, (**b**) by external winding of the coil, and (**c**) by thermal radiation

temperature rises sufficiently, the SMA stops heating or begins to cool. The hysteresis prevents the immediate "on/off" toggle of the heater/cooler and creates the appropriate thermostat function [74]. Figure 1.16 represents the reduction of hysteresis loop with respect to the number of cycles in both SME and SE behavior.

1.2.10 Heating Methods of Temperature-Induced SMA

The heating methods of SMA can be facilitated in three different ways: One is by passing an electrical current through them, Fig. 1.17a. This way is only suitable when an SMA wire or spring with a small diameter is used; otherwise, the electrical resistance is too small to generate the required heating. The simplicity of this method is an advantage, whereas a big disadvantage is the requirement of electrical insulation. The second one is by passing an electrical current through a high-resistance wire or tape coiled around the SMA element (Fig. 1.17b). This method is suitable for SMA bars or tubes. Here the electrical wire also needs electric insulation, but the insulation material should have good thermal conductivity to let the heat go to the SMA structure. The third one is by the operation of SMA structure by exposing to thermal radiation as shown in Fig. 1.17c. This is the simplest way of heating in space, exposed to sunlight that can raise the temperature 150 °C or more. No additional heating system is required for this third way. The main disadvantage of this method is that it is inexible, and it could be very difficult to retract the structure.

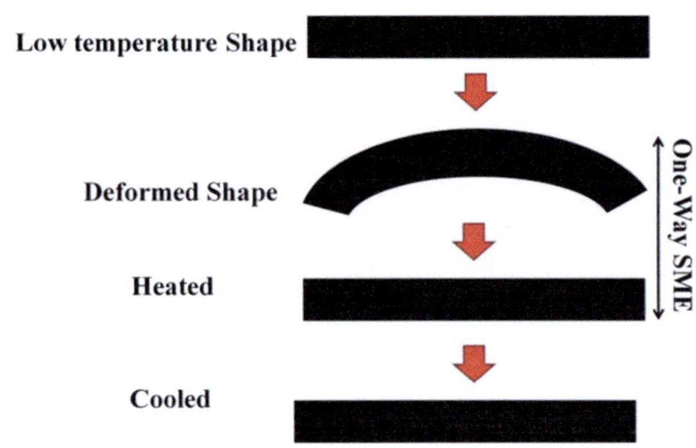

Fig. 1.18 One-way shape memory bar showing the recovery after deformation

1.2.11 Types of Shape Memory Alloys

Based on the training procedure, the shape memory alloys can be categorized into two types: (1) one-way SMA and (2) two-way SMA.

1.2.11.1 One-Way Shape Memory Alloy

Recovery of a particular shape above the transformation temperature is shown in Fig. 1.18. As an actuator, the SMA structure can actuate in one direction. For example, an SMA wire that compresses does not stretch without the external stimuli after cooling down. A return mechanism from the external is required to gain the undeformed shape. This is one limitation of the SMA actuators. A return (bias) mechanism, shown in Fig. 1.19a, is used to revert back the cold shape during the stretching–compression cycle. Here, gravity

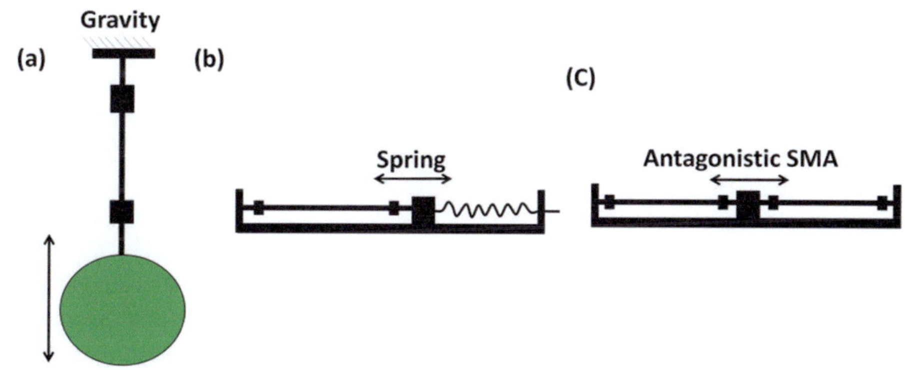

Fig. 1.19 Bias mechanisms in SMA actuators: (**a**) gravity is act as a bias force, (**b**) bias force by steel coil spring, and (**c**) SMA bias force

acts as a bias force. Again, some possible conditions for actuator are shown in Fig. 1.19(a–c). In Fig. 1.19b, the bias mechanism is applied by a conventional steel coil spring. Figure 1.19c shows a bias force, that is, SMA elements operating in both directions of movement. One SMA has a property of compression at the parent phase whereas the second one has the product phase. Alternatively, heating and cooling are done to actuate the system continuously. This is referred to as an antagonistic SMA [75]. This setup results in output force in both directions. Here, the heating and cooling processes must be arranged separately. The two limitations of this setup are: (1) due to the force of the bias setup, opposes the force of the SMA, the net output force decreases, and (2) if each of the bias element is very close to each other in a system, then the heat transfer between elements can result in undesired forces.

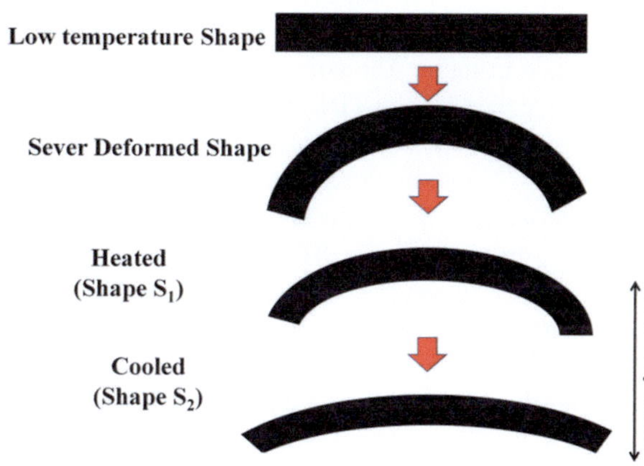

Fig. 1.20 Sequence of actuation of two-way shape memory bar showing two shapes (i.e. higher temperature shape and lower temperature shape) after severe deformation

1.2.11.2 Two-Way Shape Memory

This type of SMA is trained in such a way to remember two shapes. It has ability to regain the specific shape upon heating and then return to an alternate shape when cooling below the transformation temperature. The two-way shape memory effect could provide force in both heating and cooling phases as shown in Fig. 1.20. Here, the memory is between the parent and the product phases with two intermediate memory temperatures. Two-way shape memory alloy does not require any bias force mechanism [76].

However, there are limitations that reduce the usability of the two-way effect, such as smaller strains (2%), extremely low cooling transformation forces, and unknown long-term fatigue and stability. Even slight overheating removes the SME in two-way devices. Setting shapes in two-way SMAs is a more complex procedure than the one used with one-way SMA. Hence, one-way shape memory devices with return (bias) mechanisms are preferred over two-way memory device mechanisms.

Again, we can classify the SMA based on dominating element, that is, (1) Copper-based and Nickel-based. Cu-based (Cu–Al–Zn, Cu–Ni–Al, etc.) are used for narrow hysteresis and adaptability and (2) Ni-based (Ni–Ti–X, X = ternary element) are used more than 90% of new SMA applications. Ni-alloys have higher fabrication cost, higher corrosion resistance, and higher biocompatibility than Cu-based alloy. Ni-based alloys experience more resistance to electric heating compared to Cu-based alloy, which is mostly used in actuator applications. The SME in Fe-based SMA is traditionally known to be relatively much weaker, and Fe-based SMAs were most likely to be used only as a fastener/clamp for one-time actuation largely due to the extremely low cost. Generally, Ni elements are almost found in all Fe-based alloy. Based on stimuli response, SMAs are of different types, such as temperature–response SMA, magnetic response SMA, electric response SMA, photochemical SMA, and so on. Also, we can distinguish the SMA as polycrystalline and single crystal. Polycrystalline SMAs show huge superelasticity (13%) and high tensile strength (over 1 GPa) when compared to single-crystal SMA.

1.2.12 Different Parameters of NiTi SMA

1.2.12.1 Effect of Thermomechanical Treatment

Moderate cold working, followed by annealing below recrystallization temperatures, of NiTi alloys leads to a reduction in transformation hysteresis. NiTi when heavily cold-worked eliminates the martensitic plateau on the stress–strain curve and exhibits poor shape memory. This is because high densities of random dislocations are introduced, which obstruct the mobility of twin boundaries. However, the yield strength is considerably improved during cold working. Annealing restores shape memory but decreases yield strength. NiTi alloys are sensitive to temperature and time in a heat treatment process, owing to compositional changes that take place in them. The composition can vary between TiNi, $Ti_{11}Ni_{14}$, Ti_2Ni_3, and $TiNi_3$ by varying either temperature or time. When Ni-rich alloys are aged at lower temperatures (<773 K) or slowly cooled after heat treatment, the alloy decomposes to a more Ti-rich matrix with a Ni-rich phase finely dispersed in it. This causes an elevation in transformation temperatures. Hence, the aging of Ni-rich alloys is done at 773–973 K to preserve the lower transformation temperatures. Additionally, aging at higher temperatures suppresses the R-phase transformation and superelasticity in Ni-rich binary alloys. The instability occurs on the Ni-rich side because of the solubility range of NiTi extending to higher Ni contents at temperatures above 773 K. However, for Ti-rich compositions, there is very little variation involving heat treatment conditions [77]. Thermal cycling of solution-treated NiTi alloys (without aging) results in a gradual shift of transformation temperatures after each cycle. However, the transformation temperatures remain constant for aged Ni-rich alloys and those annealed at temperatures lower than the recrystallization temperature after cold working. The final steps in the manufacturing process of a NiTi alloy are cold working and heat treatment. Cold working is done for dimensional control. The introduced dislocations by cold working strengthen the alloy but reduce ductility and impede martensite interfaces. Thus, heat treatment must be followed to rearrange these dislocations, increase ductility, and restore the SME behavior. For pseudoelasticity, the dislocations should not be eliminated so that the parent phase maintains high yield strength. Heat treatment leads to phase transformations, where precipitates are formed that contribute to the alloy's strengthening.

1.2.12.2 Effects of Aging

The precipitation characteristics and the corresponding shape memory properties are highly dependent on aging temperature, aging time, and cooling rate. The aging time and temperature control the composition, quality, and quantity of precipitates in the NiTi matrix. For instance, with low temperature or short time aging, metastable Ni_4Ti_3 precipitates are formed, while the stable Ni_3Ti phase is introduced into matrix after high temperature or long duration aging. Ni-rich alloys are even more sensitive to aging. When the Ni content is lower than 50.5 at. %, the nucleation rate is slow and the precipitation process is affected by the grain boundaries because the precipitate nucleation rate is higher at grain boundaries rather than in the grain interiors. Therefore, the formation of precipitation is inhomogeneous. However, for high-Ni content alloys, nucleation rates at grain boundaries and grain interiors are similar, and the precipitation happens homogenously throughout the microstructure. In Ni-rich alloys, Ni-rich metastable to stable secondary phases such as Ni_4Ti_3, Ni_3Ti_2, and Ni_3Ti precipitates can be formed, where they improve the strength of material by acting as obstacles for dislocation motion [78]. The strength and transformation temperature are also highly dependent on the precipitation properties, that is, size, volume fraction, and the distance between particles. Another effective mechanism is the created local stress fields around the precipitates. Transformation temperature is increased with correctly oriented stress fields that help to nucleate the martensite. However, whenever the stress fields are around the fine precipitates with small interparticle distance, they generate obstacles and resist the martensite transformation. Attention should be paid during the heat treatment, since high aging conditions lead to growth and eventually to dissolution of the precipitates, leading to a weaker parent phase.

1.2.12.3 Effect of Grain Size

Grain size and texture are important parameters and can be adjusted properly for optimum performance. Reduction in grain size appears to improve superelasticity and other mechanical properties. Grain size and defect structures change with the type of heat treatment. For example, increasing heat treatment time can decrease grain size and defects.

1.2.12.4 Effect of Deviation from Equiatomic Stoichiometry

The stoichiometric range of NiTi is very narrow at low temperatures. Hence, at low temperatures, the alloy tends to form precipitates of a secondary intermetallic phase when there is a stoichiometric deviation from equiatomic NiTi. The microstructure will thus contain small amounts of secondary phases distributed in the matrix. Martensite start temperature depends strongly on the composition, especially for Ni-rich alloys. On the other hand, Ti-rich alloys exhibit less sensitivity, mainly because of the formation of Ti-rich precipitates, leaving the matrix composition essentially the same. Increasing Ni content strongly depresses the transformation temperature. NiTi alloys exhibit superelasticity when the Ni content is above 50.6 at% [79]. The effects of thermal cycling and annealing are substantial when the Ni content is high. Furthermore, excess Ni increases the yield strength of the austenite.

1.2.12.5 Effect of Additive Elements

NiTi is fairly sensitive to alloy element additives. The addition of Fe, Al, Cr, Co, and V tend to substitute Ni and depresses the martensite start temperature. The depression of martensite start temperature is strongest with the addition of Cr and weakest with V and Co. Their additions have practical importance in creating cryogenic SMAs, stiffening a superelastic alloy, or increasing the separation of the R-phase from the martensite phase. The addition of Nb and Cu alters the hysteresis and martensitic strength. Hysteresis is increased with the addition of Nb, while with Cu it is reduced. Cu can substitute Ni for up to 30 atomic % without any changes in the shape memory effect. Additionally, the addition of Cu causes two-step transformations, cubic (B2) → orthorhombic (B19) → monoclinic (B19′). The effect of substitution of Cu makes the transformation temperature less sensitive to the Ni content, in addition to reducing the hysteresis to about 15 °C [80]. Most benefits of adding Cu are shown only up to 10%, and extra additions show only marginal improvements.

1.2.12.6 Effect of Precipitation

Heat treatments have a significant influence on the final microstructure of the NiTi alloys and therefore affect their mechanical and functional properties. As far as Ni-rich alloys are concerned, there may exist various precipitation sequences depending on the heat treatment conditions and on the alloy's composition. In particular, high aging temperatures and time lead to the formation of Ni_3Ti precipitates, intermediate aging conditions lead to Ni_3Ti_2 precipitates, and low aging conditions lead to Ni_4Ti_3 precipitates [81]. In the early stages of aging, it possesses coherency with the matrix. Most of the biomedical applications use Ni-rich alloys, since the transformation temperatures are strongly dependent on Ni content. Precipitation of Ni-rich particles in Ni-rich alloys leads to a decrease of Ni at the matrix since Ni atoms participate in the formation of these precipitates, resulting eventually in an increase in transformation temperatures. The size and the dispersion of the precipitates have significant effects on mechanical properties, shape memory effect, and pseudoelastic behavior. These parameters depend on the heat treatment conditions such as temperature and time. As the aging time increases coarser dispersion, larger precipitates, and coherency loss are observed. Fine dispersion and small sizes strengthen the alloy since dislocation's motion is more effectively blocked than with coarser dispersion and larger particles. Moreover, coherent precipitates introduce a stress field that strengthens the alloy and affects the mobility of the martensitic interfaces [82]. Precipitation of the new phase takes place by nucleation and growth. The total kinetic of the precipitation follows the C-kinetics in which the nucleation and diffusion rates are optimum at intermediate temperatures. The precipitation may take place either homogeneously or heterogeneously depending on the supersaturation. High supersaturation brings about homogeneous nucleation which is not assisted by lattice defects, and the precipitates are dispersed at the grain interior [83]. Whenever the driving force is not high enough to enable homogeneous nucleation, lattice defects such as grain boundaries contribute to the precipitation. Regarding the morphology of the new phase, both the interfacial and the strain energy are considered and the shape of the particles is one that minimizes the sum of these energies. Whether the NiTi alloy is used for its SME or SE behavior, its mechanical properties combined with these unique properties must be optimized to meet the application demands.

1.2.13 Potential Applications

The demand for SMA in engineering and technical applications has been increasing in numerous commercial fields, such as in consumer products and industrial applications, structures and composites, automotive, aerospace, mini-actuators and micro-electro-mechanical-systems (MEMS), robotics, biomedical, and even in fashion. Shape memory alloys market was valued at USD 9.46 billion in 2016 and is predicted to reach USD 18.97 billion by 2022, at a CAGR of 12.3% between 2017 and 2022 [84]. The SMA value chain includes raw material suppliers that manufacture metals, such as Ni, Cu, Ti, Al, and other materials. These metals are used by SMA manufacturers to develop various types of shape memory alloys that are further used in the biomedical, aerospace & defense, automotive, consumer electronics and home appliances, and other end-use industries. Some of the key players operating in the shape memory alloys global market are SAES (Italy), Nippon Steel & Sumitomo Metals (Japan), ATI Specialty Alloys & Components (US), Furukawa Electric Company (Japan), Johnson Matthey (UK), Fort Wayne Metals (US), Nippon Seisen (Japan), Saite Metal Materials (China), DYNALLOY, Inc. (US), and Seabird Metal Material (China) [85]. Competition among these players is high, and they are investing in strategic developments, such as expansions, agreements, and acquisitions. The history of the application of NiTi SMAs can be traced back to the late 1960s when SMA couplings ("*Cryofit*® *couplings*") were successfully used to join the high-pressure hydraulic system piping on Grumman F-14 fighter airplanes. The application of Nitinol in the medical field was first reported in the early 1970s [86]. Some of the potential current applications are given subsequently.

1.2.13.1 In Space and Aero-Industries

Shape memory alloy has a great contribution to developing many important parts in space industries and aero-vehicle industries. In 1997, SMA was used in Mars Pathfinder Sojourner Rover to assist with Martian dust measurement on the rover's upper surface [87]. Here the SMA-connected

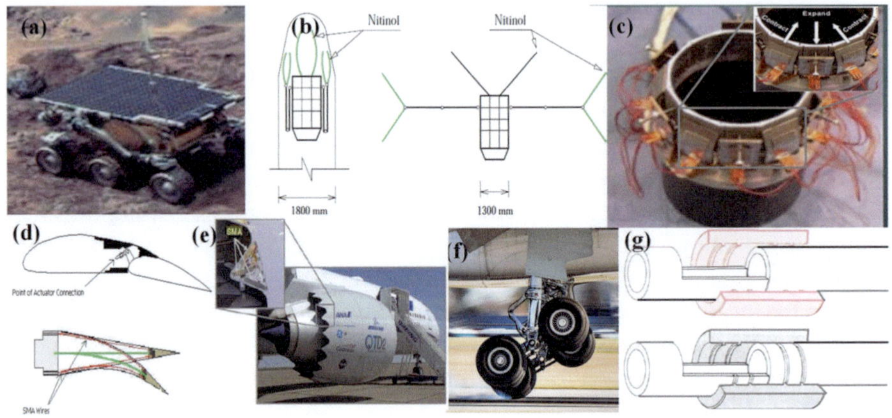

Fig. 1.21 (**a**) Mars Pathfinder Sojourner Rover, (**b**) Satellite antenna opening, (**c**) Nozzle exit (**d**) Plane wings, (**e**) Vibration reducer, (**f**) Landing gear, and (**g**) Hydraulic tube coupling

solar cell operates to drag the glass plate on top of the rover for various investigations as shown in Fig. 1.21a. NiTi SMA is used to open the satellite antenna at the window as shown in Fig. 1.21b. The spontaneous unfolding of the antenna is the beauty of SMA [88]. Figure 1.21c shows the nozzle. The larger diameter of the nozzle at takeoff can decrease the jet velocity by minimizing the noise. Manipulating the diameter in a cruise also minimizes the fuel consumption. By using SMA in Boeing tested VAFN capable of 20% area change [89]. SMA actuators were used to position 12 panels at the nozzle exit. A small resistive heater is bonded at the surface of each actuator. Figure 1.21d shows the SMA wire attached in the actuator to operate the wing angle to increase or decrease the speed and for takeoff. In Fig. 1.21e, the chevron is used to reduce the noise [90]. Their principle of operation is that as high-temperature air (from the engine core) mixes with low-temperature air (blowing through the engine fan), the SMA edges serve to get a smooth mixing condition resulting in minimization of turbine noise. These can replace hydraulic and electromechanical actuators. SMA is applied to actuate the landing gear in the aeroplane as shown in Fig. 1.21f. Hydraulic-tube couplings for the Grumman F-14 jet fighter are fabricated from a nitinol alloy that has a martensite transformation temperature in the cryogenic region below −120 °C [91]. The sleeve-like coupling is machined at normal temperature to have an inner diameter 4% less than the outer diameter of the tubes to be joined. The sleeve is then cooled below the transformation temperature and is mechanically expanded to have an inner diameter 4% greater than the tubes' outer diameter. Still held at a temperature below the martensite transformation temperature, the sleeve is placed around the ends of the tubes. When the coupling is warmed, it shrinks to form a tight seal. Ribs machined into the sleeve enhance seal by biting into tubes, as shown in Fig. 1.21g. Again, in the aeroplane, SMAs are used in wing, winglet, vortex generator, flap edge, inlet, nozzle, rotor, landing gear, electromechanical control, hydraulic lines, and so on. [12, 92].

1.2.13.2 In Automobile Industries

Currently, the progress in the utilization of SMA in automobile industries is higher. The utilized parts are radiator, fan clutch, engine control (sensor and actuator), start-up clutch, tumble flaps, fuel injector, fuel system, piston ring, booster/charger, valves, battery, transmission control, break, absorber, tire, headlight/lamps, wiper, sunroof/sunshade, door and locking mechanism, side mirror, boot, engine hood, petrol cap, bumpers, crash structures, air dams, grill/louver, spoiler, structural parts/panels, interior dashboard, rear view mirror, seats, airbags, structural part, and impact structure. The new 2014 Chevrolet uses a lightweight heat-activated SMA wire in place of a heavier motorized part to open a vent that allows the trunk lid to close more easily [93]. Like this, various applications of SMA are shown in Fig. 1.22a. SMA spoke-attached wheels can give a high resistance capacity to catastrophic failure during continuous jerk, as shown in Fig. 1.22b [94]. In the new Corvette (Fig. 1.22c), an SMA wire opens the hatch vent whenever the deck lid is opened, using heat from an electrical current in a similar manner to the trunk lights. When activated, the wire contracts and moves a lever arm to open the vent, allowing the trunk lid to close [95]. Once the trunk lid is closed, the electrical current switches off, allowing the wire to cool and return to its normal shape, which closes the vent to maintain cabin temperature. In BMW, the control of the airflow into the engine compartment uses an SMA-activated louver system. Reducing the cooling airflow into the engine compartment reduces aerodynamic drag. The result is improved aerodynamics and drag reduction and rapid warm-up during cold starts (Fig. 1.22d). The actuator provided 1 in. of powered stroke in both directions. To overcome wind resistance at high speed, the actuators needed to provide more than 5 pounds of force. Designed for automotive 12-volt power bus, each modular actuator used more than 48″ of 0.015″ diameter Flexinol HT by Dynalloy [96]. Figure 1.22e shows the air dams, which are important to reducing aerodynamic drag at highway speeds and are frequently damaged by low-speed impacts with parking

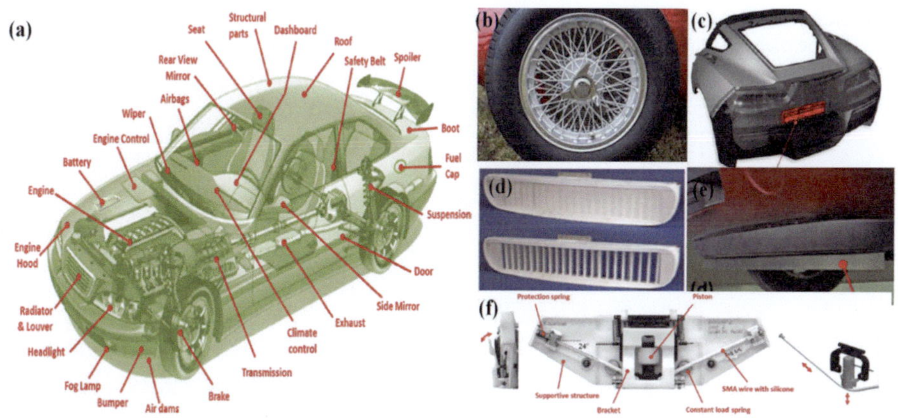

Fig. 1.22 (**a**) SMA employed in different parts, (**b**) SMA spoked wheel, (**c**) Hatch vent, (**d**) Louver system, (**e**) air dams, and (**f**) automotive actuator

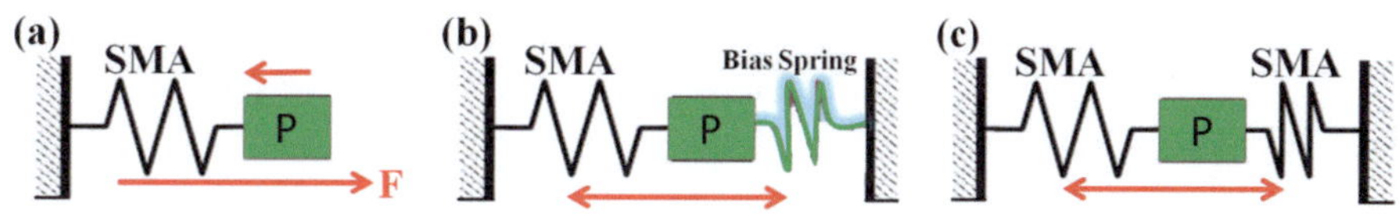

Fig. 1.23 Basic types of SMA actuators. (**a**) External load is necessary to operate it again, (**b**) biased actuator with a conventional steel coil bias, and (**c**) use of two SMA actuator

bumpers, ramps, and snow and ice. The "active" air dam, activated by shape memory alloy, can monitor vehicle speed, the use of four-wheel drive, and the presence of snow to intuitively lower or raise the dam to optimize aero drag. The use of SMA-assisted electrically actuated antiglare rear-view (EAGLE) mirror has been shown in Fig. 1.22f, which is an excellent actuator for automotive application [97]. Due to high recovery strain as well as the high actuation stress, SMA has a higher growing application in vehicle manufacturing.

Future applications are envisioned to include engines in cars and airplanes and electrical generators utilizing the mechanical energy resulting from the shape transformations. Nitinol with its shape memory property is also envisioned for use as car frames. Other possible automotive applications using SMA springs include engine cooling, carburetor and engine lubrication controls, and the control of a radiator blind ("to reduce the flow of air through the radiator at start-up when the engine is cold and hence to reduce fuel usage and exhaust emissions").

1.2.13.3 In Electrical and Electronics

There is a wide demand for SMA in electrical and electronic industries. From nano-scaled component to micro-scale component, SMA is applied to resist the damping and catastrophic failure along with its application. For example, the use of SMA-based actuator. In this actuator, both one-way SMA and two-way SMA may be used. Although two-way SMA can perform in two directions due to its two-way shape memory mechanism, the transformation strain associated with it is normally only half of that in one-way SMA. An alternative solution is to put two one-way SMAs. Biased actuators one against another to generate mechanical two-way performance, that is, heating SMA in one actuator to get forward motion and heating SMA in another actuator to reverse. The advantage of the mechanical two-way actuator is higher motion and higher force than that in material two-way actuator, while the advantage of the material two-way actuator is simpler, compacter, and much less elements involved. Figure 1.23 shows three basic types of SMA actuators using one-way SMA as the mechanism associated with Fig. 1.19. Figure 1.23a shows a one-way actuator [98]. The SMA element is elongated initially, at low temperature, and is then heated to move element P in the direction of the arrow. Figure 1.23b shows a biased actuator, which is capable of moving the element P back and forth [99]. The SMA element is deformed at low temperature, before being connected to the spring. When it is heated, the recovery force that is generated pulls the spring, thus storing energy in it. When the SMA element is cooled, the energy stored in the spring is released and the SMA element deforms back, thus completing the cycle. Figure 1.23c shows a two-way actuator that includes two SMA elements [100]. Two opposing SMA elements are used instead of the SMA element and bias spring of the biased actuator. Any motion can be obtained by appropriately cooling or heating the two SMA elements. Two-way SMA-based actuator is similar to, Fig. 1.23a, one-way actuator in shape, while its behavior is more similar to, Fig. 1.23b, biased actuator. Most of the robotic hand wires generally use SMA wires for smooth operation to replicate the motions of a human hand as given in Fig. 1.24. Various MEMS devices use SMA thin films for actuation due to their higher work per unit volume. In MEMS and NEMS, SMA gives outstanding performance among all currently available micro-actuator

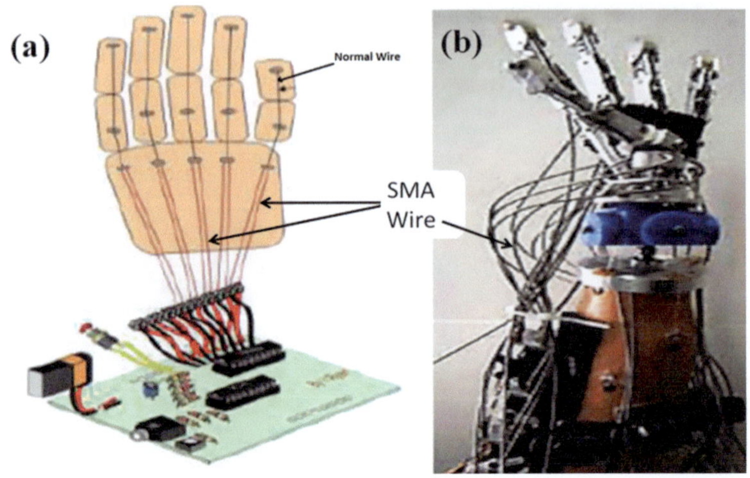

Fig. 1.24 (**a**) The electric-driven and (**b**) the electric- and mechanical-driven SMA wire used in robotic hand

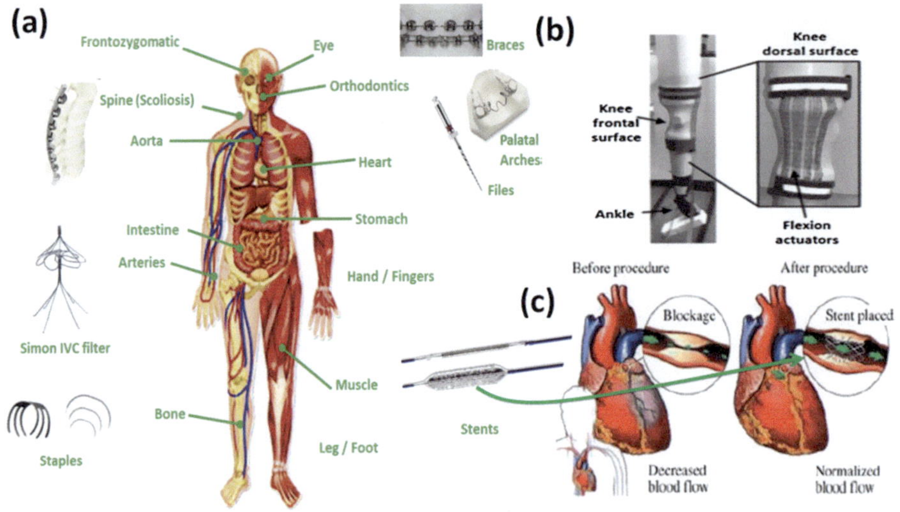

Fig. 1.25 (**a**) SMA applications in various parts of the human body, (**b**) flexion support at the knee, and (**c**) use of stent at the blood flowing passage

materials. SMA-based thin films can be fabricated not only by magnetron sputtering but also by laser ablation, ion beam deposition, arc plasma ion plating, flash evaporation, and so on. The magnetron-sputtered SMA thin film is the most widely accepted technique in terms of its uniformity, high quality, high deposition rate, and good compatibility [101].

1.2.13.4 In Biomedical Industries

SMAs have exhibited excellent behavior for biomedical applications, such as high corrosion resistance, biocompatibility, nonmagnetic, the unique physical properties that replicate those of human tissues and bone, and can be manufactured to respond and change at the temperature of the human body. SMAs are mainly used in most of the medical equipment and devices including orthopedics, neurology, cardiology, and interventional radiology; and other medical applications include endodontics, stents, medical tweezers, sutures, anchors for attaching tendon to bone, implants, aneurism treatments, eyeglass frames and guidewires, braces/brackets, palatal arches, files, head, spine, bone, muscles, hands, fingers, vascular, aorta, arteries, vena cava filter, ventricular septal defect, vessels, and biomedical instruments (in catheters, scopes, suture, cardiology, hepatology, otorhinolaryngology, gastroenterology, urology, plastic, reconstructive and aesthetic surgery, and ophthalmology) [102]. Potential applications in body parts are given in Fig. 1.25a. The first cardiovascular device developed with shape memory was the Simon filter. The Simon filter (Fig. 1.25a) represents a new generation of devices that are used for blood vessel interruption to prevent pulmonary embolism. Persons who cannot take anticoagulant medicines are the major users of the Simon filter. The purpose of this device is to filter clots that travel inside the bloodstream. The Simon filter traps these clots that in time are dissolved by the bloodstream. The insertion of the filter inside the human body is done by exploiting the shape memory effect. From its original shape in the martensitic state, the filter is deformed and placed on a catheter tip. Saline solution flowing through

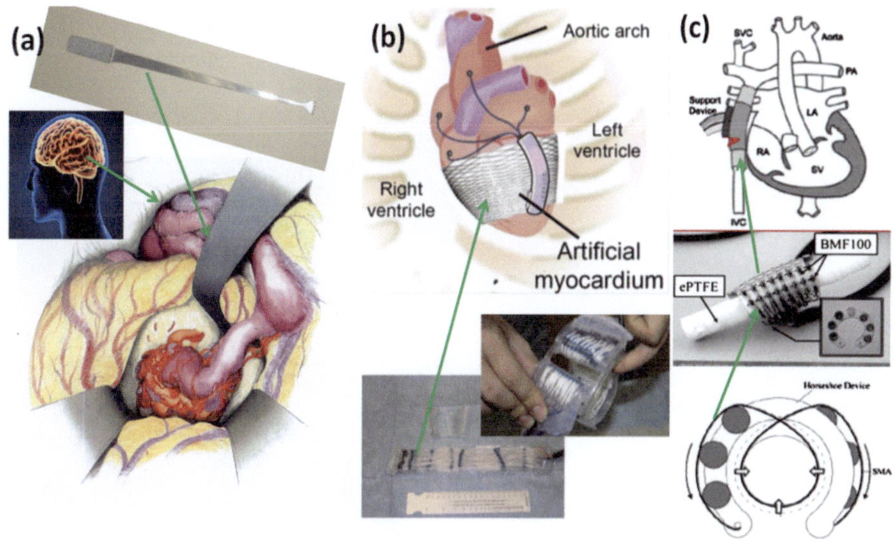

Fig. 1.26 SMA use in (**a**) brain spatula, (**b**) atrial septal occlusion device, and (**c**) artificial circulation support

the catheter is used to keep a low temperature, while the filter is placed inside the body. When the catheter releases the filter, the flow of the saline solution is stopped. As a result, the bloodstream promotes the heating of the filter that returns to its former shape [103]. Figure 1.25b shows the SMA flexion actuator use to support knee implants. This structure can resist stress concentration during activities. Figure 1.25c shows the most used structure stent. The stent helps the easy flow of blood. In our day-to-day life, due to cholesterol deposition in blood passage develops blood pressure, which is a cause of many diseases. The stent is used to stretch the path to clear the blood passage. Figure 1.25c shows the stent application in the blocked vein in the heart. The deformed squeezed shape of the stent is placed at the blocked place, and then by getting the body temperature, it attains the trained shape (expanded shape) to provide the blood flow passage.

A brain spatula or a brain retractor is used as an instrument in a surgical operation of a brain. This instrument is used to hold the opened state of the brain during the operation of the cerebral tumor which is located in the inner part of the brain [104]. An image of the brain spatula used in the operation is shown in Fig. 1.26a. The brain spatula is used in the bent form, fitting it to the shape and depth of each patient's brain. After the operation, the brain spatula is struck with a mallet to recover the original flat plane followed by the treatment in a sterilizer by heating and is used again thereafter. The main material used in the existing brain spatula is copper. Since the irrecoverable unevenness appears on the surface of the brain spatula after use, it is disposed of after using several times. If the SMA is used for the brain spatula, the brain spatula used in the bent form regains its original flat shape automatically based on the SME through the treatment of sterilization by heating in the autoclave. Therefore, since the SMA–brain spatula not only saves the time to strike with a mallet to regain the original flat shape but also recovers its original shape automatically, the appearance of the uneven plastic deformation is inhibited and the spatula can be used several times [105]. Figure 1.26b shows the atrial septal occlusion device is employed to seal the atrial hole. The atrial hole is located between the two upper heart chambers upon the surface that splits the upper part of the heart into the right and left atria. The anomaly that occurs when this hole opens can reduce life. The atrial septal occlusion device is an alternative to this surgery. This device is composed of SMA wires and a waterproof film of polyurethane [106]. Figure 1.26c has shown artificial circulation support for a weak region of the heart. The SMA wire-assisted fiber around the weak part is fixed for the easy rhythm of blood flow [107]. The intra-aortic balloon pump is used to unblock blood vessels during angioplasty. The device has an SMA tube whose diameter is reduced compared to polymer materials due to its pseudoelastic effect. Moreover, it also allows greater flexibility and torsion resistance when compared to the same tube made of stainless steel. Laparoscopy is another procedure where SMA has been employed. In laparoscopy surgical tools, the actions of grippers, scissors, tongs, and other mechanisms are performed by SMA [20]. These devices allow smooth movements tending to mimic the continuous movement of muscles. Moreover, these devices facilitate access to intricate regions [108].

SMA has a large number of orthopedic applications. The spinal vertebra spacer is shown in Fig. 1.27a. The insertion of this spacer between two vertebrae assures the local reinforcement of the spinal vertebrae, preventing any traumatic motion during the healing process [109]. The use of a shape memory spacer permits the application of a constant load regardless of the position of the patient who preserves some degree of motion. This device is used in the treatment of scoliosis. Here, the spacer shows the pseudoelastic phenomenon to recover the shape to support the backbone activity. Several

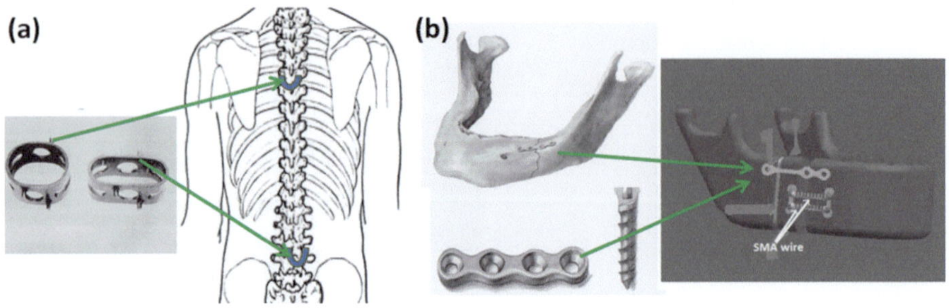

Fig. 1.27 SMA used in (**a**) backbone spacer and (**b**) Lower jaw bone

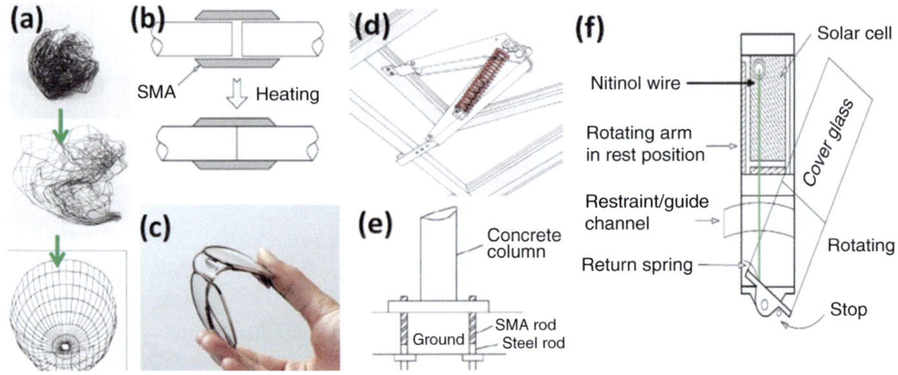

Fig. 1.28 SMAs used in (**a**) Military antenna, (**b**) coupling agent, (**c**) eyeglass frame, (**d**) window opener, (**e**) vibration stabilizer, and (**f**) solar panel opener

types of shape memory orthopedic staples are used to accelerate the healing process of bone fractures, exploiting the shape memory effect. The shape memory staple, in its opened shape, is placed at the site where one desires to rebuild the fractured bone. Through heating, this staple tends to close, compressing this separate part of bones. It should be pointed out that an external device performs this heating and not the temperature of the body. The force generated by this process accelerates healing, reducing the time of recovery. Figure 1.27b presents an application of these staples during the healing process of a patient's lower jaw fracture. Shape memory plates are primarily used in situations where a normal metal cannot be applied to the injured area, that is, facial areas, nose, jaw, and eye socket [110]. They are placed on the fracture and fixed with screws, maintaining the original alignment of the bone and allowing cellular regeneration. Because of the shape memory effect, when heated, these plates tend to recover their former shape, exerting a constant force that tends to join parts separated by fractures, helping with the healing process [111]. Orthopedic treatment also exploits the properties of SMA in the physiotherapy of semi-standstill muscles. SMA gloves are composed of shape memory wires on regions of the fingers. These wires reproduce the activity of hand muscles, promoting the original hand motion. The two-way SME is exploited in this situation. When the glove is heated, the length of the wires is shortened. On the other hand, when the glove is cooled, the wires return to their former shape, opening the hand. As a result, semi-standstill muscles are exercised [112].

1.2.13.5 Other Industries

SMAs find a variety of applications in civil structures such as bridges and buildings. One such application is intelligent reinforced concrete, which incorporates SMA wires embedded within the concrete. Another application is the active tuning of structural natural frequency using SMA wires to dampen vibrations. The most consumer commercial application was a shape-memory coupling for piping, for example, oil-line pipes for industrial applications and water pipes [113, 114]. A NiTi SMA antenna (Fig. 1.28a) is used in military and spy applications. The antenna can be a squeeze and can be kept in a packet. At the time of use, just expose to small heat or sunlight, it immediately opens to a larger hemispherical structure, which can provide mobile connectivity in any place [115]. A memory alloy coupling is expanded so that it fits over the tubing, Fig. 1.28b. When the coupling is reheated, it shrinks back to its original diameter, squeezing the tubing for a tight fit. Similar devices include SMA rivets, SMA clamps, SMA seals, and so on. [116]. Figure 1.28c shows the eye glass frames made from NiTi SMA are marketed under the trademarks Flexon and TITANflex. These frames are usually made out of SMA that has their transition temperature set below the expected room temperature [117]. This allows the frames to undergo large deformation under stress yet regain their intended shape once the metal is unloaded again. Fig. 1.28d shows the simple window opener that would be suitable for a green house is actuated by a spring fabricated with a Cu-rich SMA. At a temperature below about 18 °C, the spring is fully contracted and the window remains closed. When the temperature rises above 18 °C, the shape-memory spring overcomes the force of a bias spring and begins to open the window. At 25 °C, the actuating spring is fully extended [118]. Figure 1.28e shows the NiTi rod attachment below the concrete column to reduce the damping behavior [119]. The automatic solar panel opening and closing are operated by the NiTi wires as shown in Fig. 1.28f [120]. Currently, SMA-based actuated toys and micro-vehicles are in demand in the market. Mostly, SMAs are used as thin wires, and in some cases, steel strip serves as the elastic spring to stretch the SMA wire at the time of cooling when power is off. It is quite complicated but virtually more efficient in terms of much longer distance in one stoke and much faster than that of two one-way wheels are used together with an SMA coil spring [121]. A rolling car is based on the operational principle of moving the gravity center. SMA springs are alternately joule heated to shift the center of the mass accordingly to move the equipment.

1.2.14 Advantages of Shape Memory Alloy

Various advantages in the application of shape memory alloys are given subsequently [122, 123]:

- Noiseless and clean operation: When compared to many actuator materials, SMA actuators worked with no friction and no vibration and results in a silent operation. These factors are more useful in prosthetic applications due to nongeneration of dust particles.
- Compact, lightweight with high power/mass ratio and energy: As competitive with a large selection of actuation technologies, SMA actuators proved the highest power-to-mass ratio at less than 0.45 kg masses. The energy/work density of SMA is also very high, between 5000 and 25,000 KJ/m^3 while human muscles in general exhibit between 40 and 70 kJ/m^3.
- Higher strength: SMA alloys exhibit higher mechanical strength along with its phase transformation.
- Compactness, allowing for a reduction in overall actuator size: When using SMA as actuators, there is no need for complex and bulky transmission systems. SMA mechanism minimizes the complexity of the hand's driving operation.
- High corrosion resistance and high biocompatibility: Most of the SMAs perform higher resistance to corrosion and are also biocompatible in nature. For example, NiTi and NiTi-based alloys.
- Ease of actuation and low voltage requirement: Various techniques operate using thermal activation, but SMA or intermetallic of SMA can be easily driven by electrical current via Joule heating. This small voltage-assisted work is more susceptible for safe human-oriented applications.
- Long actuation life: Operation within the confined limit of strain and stress, SMA actuators can be expected to last hundreds of thousands of operation cycles.
- Combo-capability: Both actuating and sensing functions can be combined by measuring changes in electrical resistance associated with the phase transformation.

Despite the above-discussed advantages, SMA faces some limitations: the need for a cooling mechanism. High heat dissipation requires an additional heat extraction system. Relatively higher manufacture cost when compared to other materials such as steel and Al. The final product properties depend on the process parameter, process environmental condition, and dimension of the product and hence sensitive to materials properties. Additional treatment is necessary to reduce the internal stress developed [124]. Nonlinearity of actuation force is observed in SMA [125]. The level of cycles per minute is based on the rate of cooling and rate of heating. Various techniques are proposed to increase heat extraction speeds, such as water immersion, heat sinking, and forced air. However, even if these methods improve the bandwidth, they also cause an increase in power consumption as more heat is required to actuate the wire within the cooling medium [126]. Hysteresis, nonlinearities, parameter uncertainties, and unmodeled dynamics present difficulties in controlling the SMA accurately.

1.3 Shape Memory Polymer

Shape memory polymers (SMP) are one of the categories of smart materials. SMPs are stimuli-responsive polymers that can respond according to the environmental changes and are self-accommodated as per the environmental stimuli. Similar to that of SMAs, SMPs also involve the mechanisms of shape recovery in interaction with appropriate stimuli. In SMP, the stimuli may be temperature, electric field, magnetic field, light, pH, specific ions, or enzymes [127, 128]. The shape recovery in SMP is a result of alternation in polymer morphology and polymer functionalization in it. The polymer functionalization mostly depends on polymer processing. The shape change in SMP to temporary shape is by shrinkage or bending when they are exposed to particular stimuli and achieve the original shape as removal of that stimulus. There are two responsible segments found in SMP for shape recovery: one is the elastic segment and the second is the transition segment. The elastic segments control the high elasticity in the structure during the SME cycle. The transition segments change its stiffness significantly with a change in exact stimulus. The SME is not an intrinsic property in SMP, that is, the polymer does not display SME by them. Shape memory effect results from a combination of polymer morphology

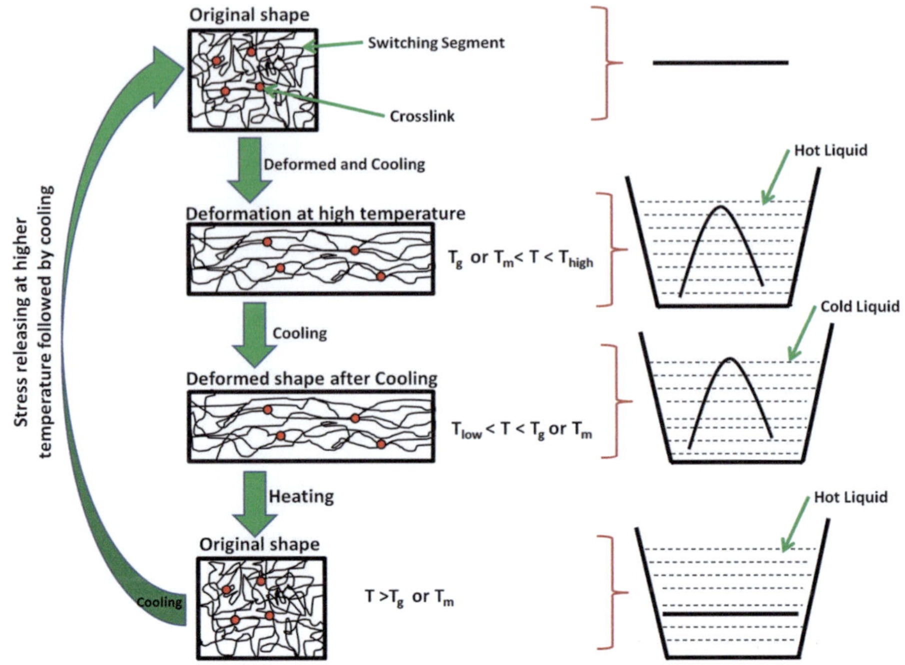

Fig. 1.29 The molecular mechanism of thermal-sensitive SME

and specific processing and can be understood as a polymer functionalization [129].

1.3.1 Thermo-Stimulated SMP

Thermo-responsive SMPs can recover their shape with the help of heat stimuli. Thermo-sensitive SMP has a wide demand in most industries, among all the types of SMP. This SMP consist of a physical cross-linking structure, crystalline/amorphous hard phase, or chemical cross-linking structure, and a low-temperature transition of crystalline, amorphous, or liquid-crystal phase as a switching phase. Figure 1.29 shows the mechanism involved with the thermosensitive SMP work. Initially, the permanent shape (straight strip) has no stress as shown in the figure. T_0 deforms the material, the temperature increases above the glass transition temperature (T_g) or melting temperature (T_m). At this higher temperature, the transition segment becomes soft and facilitates to deform or bend as given in Fig. 1.29. During bending at higher temperatures, the resistance is largely from the elastic segment. Then, the bending is followed by cooling below its switch transition temperature and the stress retained in the bending shape. During cooling, the cross-linking structure stored a large internal stress [130] in the formed temporary shape. The elastic spring energy is stored in the elastic segment, which is the driving energy for shape recovery in SMP. Now, to recover the original shape of the strip, heat energy is employed. When the temperature of the structure increases above the softening temperature or T_g or T_m, the crosslinking structure released the stored energy and the elastic segment recovers the original shape or permanent shape and completes an SME cycle [131]. A stable polymer network and a reversible switching transition of the polymer are the two prerequites for the SME. Polymer network consists of molecular switches and net points. The net points determine the permanent shape of the polymer network and can be of a chemical (covalent bonds) or a physical (intermolecular interactions) nature. The stable network of SMPs determines the original shape, which can be formed by molecule entanglement, crystalline phase, chemical cross-linking, or interpenetrated network. The lock in the network represents the reversible switching transition responsible for fixing the temporary shape, which can be crystallization/melting transition, vitrification/glass transition, liquid crystal anisotropic/isotropic transition, reversible molecule cross-linking, and supramolecular association/disassociation [132, 133].

Typical reversible molecule cross-linking reactions such as switching transitions include photodimerization, Dielse Alder reaction, and oxidation/redox reaction of the mercapto group. Typical switching transition per supramolecular association/disassociation includes hydrogen bonding, self-assembly metal–ligand coordination, and self-assembly of b-CD [134]. Carefully examining the mechanism behind the SME reveals that opposite to that in SMA, SMPs normally are hard at low temperatures and become soft at high temperatures. Therefore, SMP alone is normally not applicable in cyclic actuation. If the sample has been previously deformed by the application of external stress, it snaps back into its initial shape once the external stress is released. Various SMPs with different structures have been developed are polyurethane (DiAPLEX, SMP Technologies Inc., originally from Mitsubishi Heavy Industries), polystyrene-based SMP (Veriflex, Verilyte, Veritex, Cornerstone Research Group, Inc.), aliphatic polyurethane (Tecoflex, Lubrizol Advanced Materials), epoxy-based SMP (TEMBO, Composite Technology Development, Inc.), and UV-curable polyurethane (NOA-63, Norland Products Inc.) [135].

1.3.2 Electric-Stimulated SMP

In electric-responsive SMP, electricity is used as a stimulating agent to activate the SME of polymers. This polymer has demand in applications, where direct thermal activation is impractical. Typical polymers are nonconductive in nature. To induce the conductivity in polymer, several processes are adopted to blend conductive particles or fibers, such as CNT, short carbon fibers, carbon black, or various metallic powders. Chemical modification is necessary between the polymer and conducting reinforcement elements. For example, multiwalled CNT is reinforced in a polymer followed by a chemical surface modification in a mixed solvent of nitric acid and sulfuric acid results in an increase in the interface bond strength. Another technique is used for surface modification of superparamagnetic nanoparticles reinforced polymer by oligo (e-caprolactone) dimethacrylate/butyl acrylate composite with between 2% and 12% magnetite nanoparticles [136]. In electro-responsive polymer, the SME depends on the percentage of filler/reinforcement, degree of surface modification, and type of filler [137]. Here, the electric current is used to generate heat. Regain of the original shape is achieved when an electric current is passed through the SMP. By the application of electric current in SMP, the Joule heating may raise the internal temperature. When the temperature reached just above the switch transition temperature, the shape recovery happened.

1.3.3 Light-Stimulated SMP

The shape memory polymers which are to be activated by light should have to be equipped with photosensitive functional groups or fillers, for example, cinnamic acid or azobenzenes that act as molecular switches. Light is a particularly fascinating stimulus because it can be precisely modulated in terms of wavelength, polarization direction, and intensity, allowing non-contact control. The incorporation of such photosensitive groups or molecules into a tailored polymer surrounding in combination with the functionalization process is a well-established strategy for transferring effects from the molecular level into effects that are macroscopically visible. The shape memory polymers are stretched and illuminated by a light of wavelength greater than a fixed wavelength and the photosensitive groups form cross-links. Light-activated SMPs use processes of photo-crosslinking and photo-cleaving to change T_g. Photo-crosslinking is achieved by using one wavelength of light, while a second wavelength of light reversibly cleaves the photo-crosslinked bonds. The effect achieved is that the material may be reversibly switched between an elastomer and a rigid polymer. Light does not change the temperature but only the cross-linking density within the material. For example, it has been reported that polymers containing cinnamic groups can be fixed into predetermined shapes by UV light illumination (>260 nm) and then recover their original shape when exposed to UV light of a different wavelength (<260 nm). Examples of photoresponsive switches include cinnamic acid and cinnamylidene acetic acid [138].

Light can also increase SMP temperature above the switch transition temperature to trigger shape recovery. The heat absorbance of SMPs can be enhanced by incorporating fillers such as CNT. These materials may be used for an intravascular laser-activated therapeutic device that can mechanically retrieve thrombus and restore blood flow. In a temporary straight rod form, the microactuator is delivered through a catheter distal to the thrombotic vascular occlusion by minimally invasive surgery. Then, the microactuator recovers its primary corkscrew form by laser heating. As a result, the thrombus can be captured by the deployed microactuator and removed. There are also other kinds of light-active SMEs which are induced by photoisomerization such as azobenzenes, photoinduced ionic dissociation such as triphenylmethane leuco derivatives, or photoreactive molecules such as cinnamates. Some photochemically reactive groups can form photo-induced reversible covalent cross-linking in polymers. Photodimerization moieties such as cinnamic acid, cinamylidene acetic acid, coumarin, anthracene, and their derivatives have been incorporated in polymer macromolecules to act as the photoactive shape memory switches of the polymers. Another type of photoactive SMP is based on the metal–ligand coordinative interactions of metallo supramolecular networks [139].

1.3.4 Magnetically Stimulated SMP

Noncontact triggering of shape changes in polymers has been realized by incorporating magnetic nanoparticles in shape memory polymers and inductive heating of these compounds in alternating magnetic fields. By using ferromagnetic fillers, magnetic-sensitive SMP composites have also been prepared. Magnetic field-enabled remote actuation of SMP composites can be used by incorporating nano iron(II,III) oxide nanoparticles into thermoplastics and thermosetting SMP.

To improve the particle dispersion of the matrix, the iron(III) oxide nanoparticles could be coated with silica [140]. The magnetic-sensitive shape-memory materials have good potential biomedical applications.

1.3.5 Humid-Stimulated SMP

The actuation of the polymer can be achieved by immersion in water. The glass transition temperature of T_g-type SMP drops if the SMPs are exposed to a high humidity environment or solution after a certain period because water molecules have plasticizing effect on polymeric materials and can then increase the flexibility of macromolecule chains. The abovementioned two functions lead to the shape recovery of a deformed SMP. Another method to obtain moisture/water active SME is by using a hydrophilic or water-soluble ingredient in SMP. A water-sensitive biodegradable stent made of chitosan films cross-linked with an epoxy compound is prepared following this strategy. The raw materials (chitosan and PEG-400) used in the fabrication of the developed stent are relatively hydrophilic [141]. The degradable stent is proposed to be used as an alternative to metallic stents and for local drug delivery. An elastic polymer network combined with cellulose nanowhisker should be enough to construct a water-active SMP.

1.3.6 Shape Memory Effect of SMP

One-way SME: A better understanding of the shape memory effect of polymer can be seen in Fig. 1.30. The deformation is initiated at point O, at a temperature greater than the softening temperature (T_g or T_m) and deformation followed a path O-P-Q. By cooling that deformed structure, a very small percentage of recovery happened, that is, at point R. Here a stimulus is required to recover the rest percentage. The shape recovery process started at a temperature above T_g. Here the polymer enters the soft rubber phase. When cooled below T_g, the material autonomously returns to the original shape following the path R-O. This property of the polymer that repeatedly recovers its permanent shape is called shape memory effect. Two important quantities that are used to describe shape-memory effects are the strain recovery rate (R_r) and strain fixity rate (R_f) [142]. The strain recovery rate describes the ability of the material to memorize its permanent shape, while the strain fixity rate describes the ability of switching segments to fix the mechanical deformation.

Two-way SMEs: Most of the SMEs reported for polymers are one-way SMEs. Researchers have made many efforts to prepare SMP with two-way shape memory, where heating and cooling change the shapes of SMP reversibly. For example, cross-linked poly(cyclooctene) gives a two-way SME. Here the mechanism is the cooling-induced crystallization of cross-linked poly(cyclooctene) films under tensile stress gives stretching and subsequent heating to melt the network yields contracting and the shape recovered. Ahir et al. and Qin and Mather fabricated the liquid-crystalline polymers with two-way SMEs. Reversible nematic isotropic transition causes contraction at heating and stretching on the cooling of the polymer [141]. The stretching or contraction of the polymer is based on the extent to which the nanocomposite is strained. If the material is slightly pulled, it stretches when exposed to stimuli (UV ray). Conversely, if the material is pulled to a strain higher than 10%, it contracts under identical exposure to stimuli. This process is completely reversible and persists after many cycles [143].

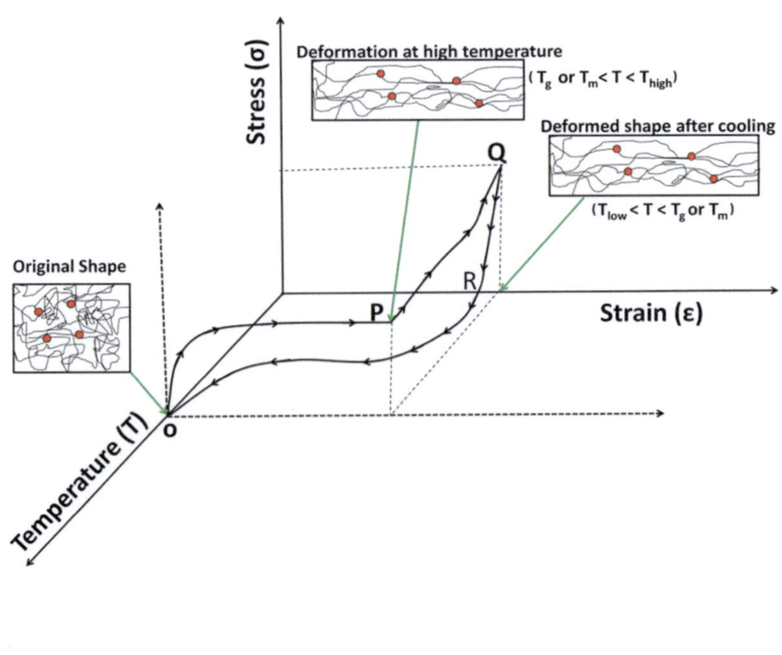

Fig. 1.30 Stress–strain–temperature diagram of thermo-induced one-way SMP

Triple-shape memory effect: A recent technological advance permits the introduction of triple shape-memory materials. Much as a traditional double shape-memory polymer will change from a temporary shape back to a permanent shape at a particular temperature, triple shape-memory polymers will switch from one temporary shape to another at the first transition temperature and then back to the permanent shape at another, higher activation temperature. This is usually achieved by blending two double shape-memory polymers with different glass transition temperatures or when heating a trained shape-memory polymer first above the T_g and then above the T_m of the switching segment [144].

Multiple shapes memory effect: More shapes can be remembered by the polymer using a suitable blending mechanism such as triple shape memory polymer. If an SMP blend has several crystalline phases and amorphous phases with well-separated thermal transition temperatures, several switch temperatures may be obtained in the polymer. A "multiple shape" memory effect maybe observed on the SMP blend if the blend is deformed by a multiple step and recovers properly. The blended linear high-density polyethylene, ethylene-1-octene copolymers, and short-chain branched polyethylenes are used to obtain multiple crystallization and melting behavior of the material [145].

1.3.7 Basics of Reinforcement in SMP

According to the reinforced phase, SMP may be classified into three types: particulate reinforced composites, short fiber-reinforced composites, and fiber-reinforced composites. The advantages of particulate-reinforced composites are large recovery stress and high recovery strain; however, their mechanical properties are poor. Fiber-reinforced composites are widely used as a structural material due to their remarkable mechanical properties, but the main direction of deformation cannot be along the direction of reinforced fiber due to the small effective strain of the reinforced fiber. Microfiber, fabrics, and mats made of carbon fibers, glass fibers, and Kevlar fibers have high elastic modulus and high strength and can remarkably increase the mechanical strength of SMPs. These composites, in the fiber (axial) direction, can bear a much higher mechanical load; while in the transverse direction, the SME can be mostly maintained. Chopped and continuous, especially continuous, microfibers and fabrics are superior to micro- and nano-sized particles to improve the mechanical strength of SMP. Due to the high elastic modulus of fibers or fabrics, good SME cannot be observed in the fiber direction of the composite. SME is usually determined by bending the composite in the fiber transverse direction but not by tensile testing in the fiber direction. Due to the exceptional mechanical strength with high elastic modulus and high aspect ratios, CNT and CNF are also effective in improving the mechanical strength and shape recovery stress of SMP. Another advantage of CNT/CNF as fillers of SMP is the potential electrical and infrared-light-active SME of the prepared SMP composites. Efforts to improve the CNT/CNF dispersion in polymer matrix include the use of micro-scale twins crew extruder, surfactants, oxidation, or chemical functionalization of CNT/CNF surface and in situ polymerization [146].

Generally, the mechanical strength of a NiTi SMA is 700–1100 MPa (annealed) or 1300–2000 MPa (not annealed), that of a Cu-based SMA is about 800 MPa, while that of an SMP is in the range of 5 MPa to 100 MPa, depending on constituents of the SMP. The recovery stress of an SMA can reach as high as 800 MPa. The recovery stress of an SMP is usually from a few tenths of MPa to a few tens of MPa [147]. Organic-exfoliated nanoclay is a good material in some circumstances to prepare functional composites with advanced properties. Organic-exfoliated nanoclay is also a good candidate for improving shape recovery stress and has good mechanical properties of SMPs. Silicon carbide as a filler material can improve the elastic modulus of SMP. However, it deteriorates the SME of SMP even at a very low loading. Polymer composites filled with conducting carbon black have broad applications in the field of electrical and electronic industry as polymer conductors, semiconductors, and the media for heat transferring. Poly(D,L-lactide) is a good shape memory biomaterial having good biodegradability and biocompatibility [148].

1.3.8 Fabrication and Shaping Techniques of SMP

Chemical crosslinking of a high-molecular-weight thermoplastic polymer: It allows the base polymer to be shaped by conventional thermoplastic processing methods, such as extrusion and injection molding, followed by a crosslinking process. In this process, the high-molecular-weight polymers have at least one thermal transition that can be used as the switching mechanism for shape memory. Chemical crosslinking in polymer introduces a network structure that defines the permanent shape or memory. Two common crosslinking methods used to prepare SMP are (1) organic peroxides and (2) high-energy radiation. However, the crosslinking efficiency can vary significantly among different polymers. Typically, saturated polymers are less susceptible to free radical crosslinking compared to unsaturated polymers, resulting in incomplete crosslinking and low gel fractions. This can hurt the shape memory performance, leading to both low recovery and permanent deformation. One possible solution is to physically blend the polymer with an unsaturated species that functions as a "sensitizer," before crosslinking. For example, Zhu et al. blended poly (e-caprolactone) (PCL, a saturated polymer) with small fractions of polymethylvinylsiloxane (PMVS) before subjecting the system to γ-ray irradiation [149]. The resulting materials showed a monotonic increase in gel fraction with increasing amounts of PMVS at identical radiation dosages. The main disadvantage of this approach is its rather limited control over shape memory properties. For a given SMP prepared under this approach, the transition temperature (either T_m or T_g) is largely inherited from the starting polymer and can only be adjusted in a small range. Moreover, such adjustment would inevitably change other properties. Possible side reactions (e.g., main chain degradation) during the crosslinking process must also be considered.

One-step polymerization of monomers/prepolymers and crosslinking agents: It is possible to prepare an SMP from the "bottom-up" by reacting properly selected monomers, reactive prepolymers, and crosslinking agents in which polymerization and chemical crosslinking occur in a single step. The system usually undergoes a sol-gel transition similar to many thermosetting resins. For example, copolymer networks are prepared from the free radical polymerization of acrylate/methacrylate monomers. SMPs prepared under this approach exhibit excellent shape memory properties (e.g., high fixing and recovery ratios, good cycle lifetime/stability) owing to both the thermoset (permanent network) nature of such SMP and their well-defined structures down to the molecular level. Control of shape memory properties is reasonably easy, usually involving adjusting variables such as monomer composition, prepolymer M_w, and crosslinker concentration and functionality. The biggest disadvantage of this strategy may be constraints for processing. Because of the thermosetting nature, many SMPs exhibit relatively narrow processing windows (before the gel time) and require processing techniques usually relegated to those of coatings and adhesives [150].

One-step synthesis of phase-segregated block copolymers: In this method, the goal is to yield a thermoplastic polymer (rather than a thermoset) that can be processed using more conventional plastics processing techniques. To obtain shape memory, SMPs are usually phase-segregated block copolymers, with two blocks exhibiting different transition temperatures. The most extensively studied SMPs prepared by this strategy are shape memory polyurethanes, owing to the simplicity and versatility of urethane chemistry compared to other polymerization techniques as well as the good availability of raw materials (polyols and isocyanates). In shape memory polyurethanes, soft and hard segments form a multiblock structure. The hard segment functions as physical crosslinks and defines the permanent shape ("memory"), whereas the soft segment provides the switching mechanism for shape memory [151].

Direct blending of different polymers: Physically blending different thermoplastic polymers (each with no or limited shape memory) to prepare blends with shape memory properties represents another major design strategy. The fundamental concept is to obtain "memory" from one polymer and a switching mechanism from the other. Following this concept, two different approaches have been demonstrated. As mentioned earlier, all polymers contain at least one thermal transition. What many conventional thermoplastics lack is an effective network structure when the transition temperature is exceeded? Therefore, the first approach is to blend in a second thermoplastic polymer with a higher transition temperature (than the host/matrix polymer) that functions as physical crosslinks for temperatures between the two transitions. For example, an SMP blend consists of poly (vinyl acetate) (PVAc) and poly(−lactic acid) (PLA). PVAc is an amorphous polymer with a Tg of 40 °C, whereas PLA is a semi-crystalline polymer with a high melting temperature of 165 °C. PLA crystals physically crosslink the PVA, and a well-defined rubbery plateau is obtained when the blend is heated between the T_g of PVA and the T_m of PLA. Although conceptually simple, implementing the blending strategy in reality can be complicated and challenging. Blends from different polymers can vary significantly in morphology, thermal, and phase behavior, depending on a large number of variables including miscibility, molecular weights, and specific interactions, among others. Factors such as thermal history, interfacial strength, and morphology all have a pronounced impact on shape memory properties. Even more, it is often difficult to control one material property without affecting others [152].

1.3.9 Application of SMPs

SMPs can be used widely in many areas such as biomedical devices, aerospace, textiles, energy, bionics engineering, electronic engineering, civil engineering, and household products. Some of the potential applications are discussed subsequently.

1.3.9.1 Medical Applications

Most medical applications of SMP are yet to be developed, but devices with SMP are now beginning to hit the market. SMPs are now being used in various devices including punctal plugs, glaucoma shunts, aneurysm occlusion device, assembly/disassembly tool, Bio-MEMS, bone defect filler, orthodontic, orthopedic cast, orthopedic cardio valve repair, cell manipulating and capturing, clot removal device, controlled drug release, endoscopic surgery suture, selective desalination material, self-healing, skincare product, hair treatment, kidney dialysis needles, measuring tool for complex cavities, micro tweezers in medicine, artificial muscle,

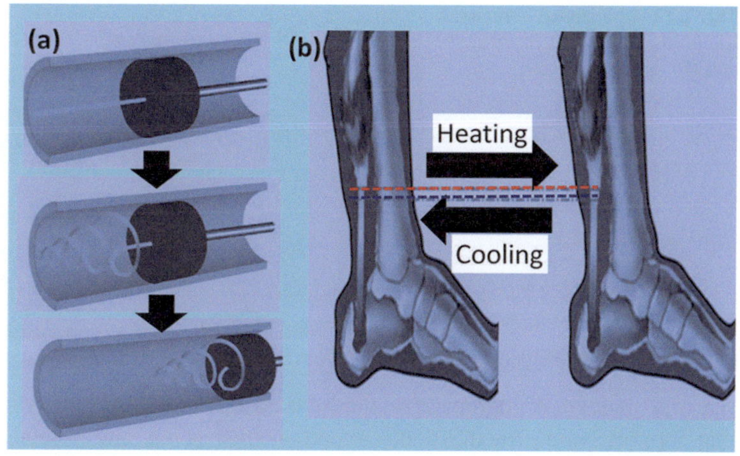

Fig. 1.31 (**a**) Depiction of removal of a clot in a blood vessel using the laser-activated shape memory polymer actuator and (**b**) SMP artificial tendon [154]

surgery inside living cell, wound dressing, self-adjusting orthodontic wires, selectively pliable tools for small-scale surgical procedures, and so on. Shape-memory technologies have shown great promise for cardiovascular stents, since they allow a small stent to be inserted along a vein or artery and then expanded to open. After activating the shape memory by temperature increase or mechanical stress, it would assume its permanent shape. Certain classes of shape-memory polymers possess an additional property: biodegradability. This offers the option to develop temporary implants. In the case of biodegradable polymers, after the implant has fulfilled its intended use, for example, healing/tissue regeneration has occurred, the material degrades into substances that can be eliminated by the body. Thus, full functionality would be restored without the necessity for a second surgery to remove the implant. Examples of this development are vascular stents and surgical sutures [153]. When used in surgical sutures, the shape-memory property of SMP enables wound closure with self-adjusting optimal tension, which avoids tissue damage due to over tightened sutures and does support healing and regeneration. SMP laser-activated device is used for mechanical removal of blood clots (Fig. 1.31a). The device is inserted by minimally invasive surgery into the blood vessel and, upon laser activation, the shape-memory material coils into its permanent shape, enabling the mechanical removal of the thrombus (blood clot). In its temporary straight rod form, the microactuator is delivered through a catheter distal to the blood clot. The microactuator is then transformed into its permanent corkscrew form by laser heating. The deployed microactuator is retracted to capture the thrombus. Figure 1.31b shows the SMP as an artificial tendon helping to make the movement in a problematic leg. Another example of a medical challenge to be addressed is obesity, which is one of the major health problems in developed countries. In most cases, overeating is the key problem, which can be circumvented by methods for curbing appetite. One solution may be biodegradable intragastric implants that inflate after an approximate predetermined time and provide the patient with a feeling of satiety after only a small amount of food has been eaten.

The usefulness of SMPs in wound closure has been recently investigated and a design of smart surgical sutures has been examined, where a temporary shape is obtained by elongating the fiber under controlled stress and his suture can be applied loosely in its temporary shape under elongated stress. When the temperature is raised above Tg, the suture will shrink and then tighten the knot in which case it will apply an optimal force. Shape memory polymer foam called "cold hibernated elastic memory" (CHEM) based on SMPs received much attention in recent years. This CHEM material utilizes an SMP foam structure or sandwich structures made of SMP foam cores and polymeric composite skins. This SMP foam was proposed for space-bound structural applications. This SMP foam exhibits micron-sized cells that are uniformly and evenly distributed within the cellular structure and, therefore, is suitable for ultra-lightweight porous membrane or thin-film space applications. A study on the influence of long-term storage in cold hibernation on the recovery strain–stress of polyurethane SMP foam revealed its excellent stability. As one of these applications, a spring-lock truss-element concept for large boom structures was proposed. It involved a unique hybrid design of SMP foams and normal polymer composites. Porous polymer scaffolds are heavily used in tissue-engineering applications, where the polymer matrix can serve as a scaffold for tissue growth and provide temporary structural support for the cells during regeneration of the target tissues [155]. While traditional porous polymers have provided utility as tissue-engineering scaffolds, they lack the shape-changing properties of SMP. There is a growing interest in SMP for tissue reconstruction due to their tunable biocompatibility, shape recovery, Tg, and mechanical properties. Additionally, SMP can be implanted in the body via minimally invasive surgery that requires miniaturized devices to pass through small incisions. A porous SMP can be procedures and its shape-changing capacity to fill the defect. The PCL-based

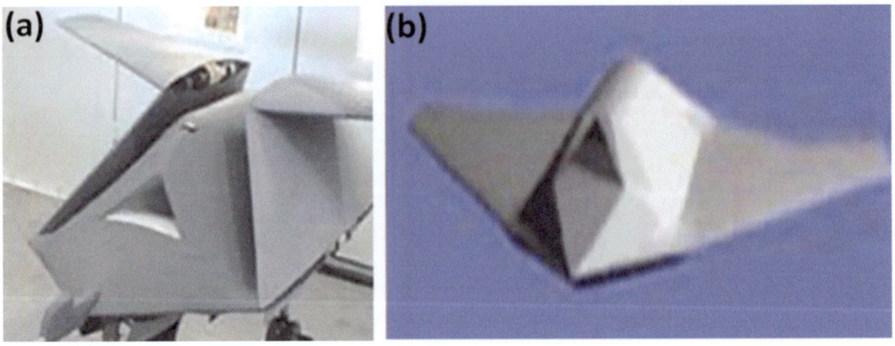

Fig. 1.32 Unmanned Z-Shaped morphing wing: (**a**) Closed state and (**b**) open state [160]

scaffold maintained comparable modulus to other systems that support bone growth, highly interconnected pores, and allowed osteoblast adhesion and proliferation [156].

1.3.9.2 In Aerospace

SMPs can be used in selective parts of the aerospace vehicle structure to reduce the weight as well as to increase operational efficiency. For the traditional aerospace deployable devices, the change of structural configuration in-orbit is accomplished through the use of a mechanical hinge, stored energy devices, or motor-driven tools. There are some intrinsic drawbacks for the traditional deployment devices, such as complex assembling process, massive mechanisms, large volumes, and undesired effects during deployment. In contrast, the deployment devices fabricated using SMP and their composites may overcome certain inherent disadvantages [157]. Morphing structures used in flight vehicles (morphing of aircraft wings and helicopter) as shown in Fig. 1.32 are envisioned to be multifunctional so that they can perform more missions during a single flight, such as an efficient cruising and a high maneuverability mode [164]. When the airplane moves toward other portions of the flight envelope, its performance and efficiency may deteriorate rapidly. To solve this problem, researchers have proposed to radically change the shape of the aircraft during flight. By applying this kind of technology, both the efficiency and the flight envelope can be improved. This is because different shapes correspond to different trade-offs between beneficial characteristics, such as speed, low energy consumption, and maneuverability. For instance, the defense advanced research projects agency (DARPA) is also developing morphing technology to demonstrate such radical shape changes [158]. As illustrated in the figure, Lockheed Martin is addressing technologies to achieve a z-shaped morphing change under the DARPA's program fund during the development of morphing aircraft, finding a proper skin under certain criteria is crucial. Generally, a wing skin is necessary, especially for the wing of a morphing aircraft. Researchers focus their works on investigating proper types of materials that are currently available to be used as a skin material for a morphing wing. In this case, the SMPs show more advantages for this application. It becomes flexible when heated to a certain degree and then returns to a solid state when the stimulus is terminated. Since SMPs hold the ability to change its elastic modulus, they could potentially be used in the mentioned concept designs [159].

1.3.9.3 In Textile Industries

Significant progress has been made in manufacturing textile fibers from shape memory polymers through common spinning methods. It is easy to make a polymer fiber as long as the polymer has an adequate molecular weight and sufficient viscosity. Shape-memory films and foams have a number of applications in laminated smart fabrics. The functions of SMP films applied to textiles include waterproofing, seam sewing, crease recoverability, and crease fixing. SMP can be used for breathable textiles [161]. A higher comfort level can be achieved using shape memory polymer-coated garments (Fig. 1.33a). By coating with shape memory polymers like shape memory polyurethanes, we can control the permeability to a large extent. It can have high water vapor permeability at a higher temperature and a lower water vapor permeability at a lower temperature, thus making it useful to be used in both the conditions. A shape memory fabric shirt with long sleeves could be programmed so that the sleeves shorten as room temperature becomes hotter (Fig. 1.33b). The fabric can be rolled up, pleated, creased, and returned to its former shape by applying heat. Heat can be supplied by blowing air through hairdryer. A suit was developed to help the sailors on the ocean and sea. It adapts to the temperature variations and maintains a person's body temperature constant. The membrane gives optimal breathability in any given atmospheric condition [162].

1.3.9.4 Automobile

SMPs have been used in automobile engineering, and many interesting products have been developed. Some interesting applications of SMP include seat assemblies, reconfigurable storage bins, energy-absorbing assemblies, tunable vehicle structures, hood assemblies, releasable fastener systems, airflow control devices, adaptive lens assemblies, and morphable automotive body molding. An automotive seat

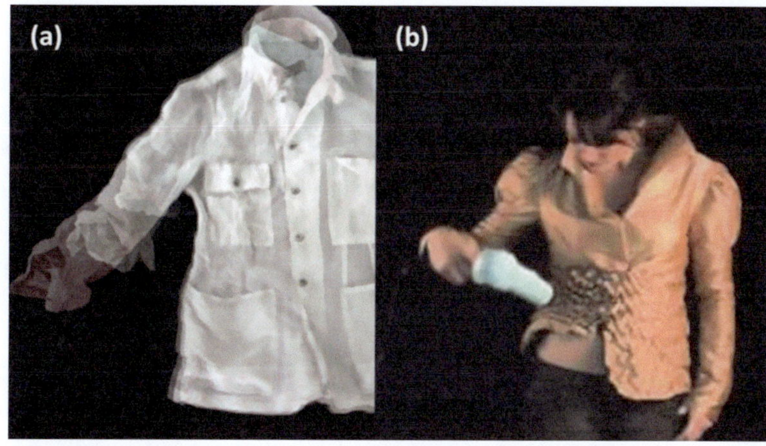

Fig. 1.33 Shape memory jacket: (**a**) Folded at hand and (**b**) folded at the belly side

belt uses SMP fibers that absorb kinetic energy and increases safety. The reasons for using SMPs are due to their excellent advantages such as shape-memory behavior, easy manufacturing, high deformed strain, and low cost. That is why they have attracted a lot of attention in automobile engineering and have even been used to replace the traditional structural materials, actuators, or sensors. As a typical example, SMPs are proposed to be used for the reversible attachments in this embodiment; one of the two surfaces to be engaged contains smart hooks, at least one portion of which is made from SMP materials [163]. By actuating the hook and/or the loop, the on-demand remote engagement and disengagement of joints/attachments can be realized. With a "memorized" hook shape, the release is effective, and the pull-off force can be dramatically reduced by heating above the Tg. It can be used for a reversible lockdown system in the lockdown regions between the vehicle body and closure. SMP can also be used in an airflow control system to solve a long-time problem for automobiles. As we know, airflow over, under, around, and/or through a vehicle can affect many aspects of vehicle performance, including vehicle drag, vehicle lift and downforce, and cooling/heating exchange. Reduction in vehicle drag reduces the consumption of fuel. A vehicle airflow control system, which comprises an activation device made of SMP material, actively responds to the external activation signal and alters the deflection angle accordingly. Thus, the airflow is under control based on the environmental changes. Self-healing composite system, a healable composite system for use as primary load-bearing aircraft components, has been developed by Cornerstone Research Group [164]. The composite system consists of a piezoelectric structural health monitoring system and thermal activation systems based on SMP. Upon damaging, the monitoring system will sense the location and magnitude of damage, send the corresponding signals to the controlling system, resistively heat the SMP at the location of damage, and finally, the induced shape recovery of SMPs will heal the damage [165].

1.3.9.5 Electric, Electronics, and Robotics

There are an almost infinite amount of potential applications for shape memory polymers in the electrical and electronic sector. Currently, the most demand is found in wearable electronic devices. One fascinating field in which SMPs are impacting quite significantly nowadays is photonics. Due to the shape-changing capability, SMPs enable the production of functional and responsive photonic gratings. In fact, by using modern soft lithography techniques such as replica molding, it is possible to imprint periodic nanostructures, with sizes of the order of magnitude of visible light, onto the surface of shape memory polymeric blocks. As a result of the refractive index periodicity, these systems diffract light. By taking advantage of the polymer shape memory effect, it is possible to reprogram the lattice parameter of the structure and consequently tune its diffractive behavior. Another application example of SMP in photonics is in shape-changing random lasers [166]. By doping SMP with highly scattering particles such as titania ones, it is possible to tune the light transport properties of the composite. Additionally, the optical gain may be introduced by adding a molecular dye to the material. By configuring both the amount of scatters and the organic dye, a light amplification regime may be observed when the composites are optically pumped. Shape memory polymers have also been used in conjunction with nanocellulose to fabricate composites exhibiting both chiroptical properties and thermo-activated shape memory effect. One of the first conceived industrial applications was in robotics, where shape memory foams were used to provide initial soft pretension in gripping. These shape memory foams could be subsequently hardened by cooling making a shape adaptive grip. SMP is used in packaging, damping materials, deodorant fabrics, electro-active shape memory hinge, heat shrinkable package for electronics, hot-shrinkage micro-tag, light-modulated and display device, micro-valve in microdevice, and temperature sensor [167].

1.3.9.6 Other Industrial Applications

There are an almost infinite amount of potential applications for shape memory polymers. The ability to return to a shape after different stimuli are applied allows a dynamic element to product design that could revolutionize certain product types. There are already a series of applications in use in real-world settings, such as morph suture anchor, phase change fabrics, physiological monitoring, pressure garments, recordable and erasable memories, damping fabric, self-peeling dry adhesive, shape-changing nanofiber, surface wetting, fashion design, rotor blades, microfluidic device, toys, wrinkle-free cotton fabric, foam that expands with warmth to seal window frames, sportswear (helmets, judo, and karate suits), and thermochromic additives (Polyurethane SMP) [168, 169].

1.3.10 Advantages and Disadvantages of SMP

In comparison to SMAs, the major advantages of SMPs are [170, 171]:

Low density: As an example, the typical bulk density of PU SMP is 1.25 g/cm^3, while that of NiTi SMA is 6.4 g/cm^3.

Low cost of materials: In addition, SMP products can be easily produced with high quality into almost any specified shapes (including thin/ultrathin films/wires, foam with different porosity, etc.) at different scales.

Less cost of fabrication: SMPs are fabricated by various traditional and advanced polymer processing technologies, such as injection molding, extrusion, dip coating, spin coating, water float casting, and so on with very little cost.

Higher recoverable strain: Recoverable strain is normally an order higher than that in SMA. A piece of PU SMP can fully recover from a few hundred percent of prestrain after uniaxial tension at well above T_g. For PU SMP foam, more than 95% precompression is fully recoverable.

Easy to tailoring the thermomechanical properties: It is easy to tailor the thermomechanical properties of SMPs using, for instance, blending with different types of fillers or varying the compositions.

Easy variation of shape: Shape recovery temperature range can be easily altered within a wide range and even gradient. For instance, after immersing into water for a different period of time, a gradient T_g is introduced into the PU SMP wire. As such, after programming into a shape, the wire recovers its original shape in a segment by segment manner upon heating.

Good biodegradability: Many SMPs have excellent chemical stability, biocompatibility, and even biodegradability. The degradation rate can be adjusted if required. SMPs can be used as a drug carrier as an additional function to effectively prevent infection.

Flexibility in stimuli process: Stimuli for SMP include heat by means of direct heating, Joule heating, induction heating, infrared/radiation heating, laser heating; moisture or solvent or change in p^H value, light; and so on. It is possible for an SMP to be activated by more than one type of stimulus.

Flexible in control media: It is possible for truly wireless/contactless operation inside a human body [294, 320], offered by remote contactless actuation by means of, for instance, applying an alternating magnetic field for induction heating.

Broad training process: The multiple SME can be achieved using synthesis or training. Multistimuli or functionally gradient SMPs have more advantages and flexibility for the multi-SME following a required recovery sequence.

Damping resistance: Damping ratio in particular within the transition range is higher.

Potential of recycling: The potential for recycle and reuse at low cost is higher.

1.4 Shape Memory Ceramic

Shape memory effect is observed not only in special alloys or polymers but also in ceramics. A number of brittle ceramics exhibit martensitic transformations and thus can be included in the SMM family. For example, zirconia has a well-studied martensitic transformation between tetragonal and monoclinic phases with associated shear strains of up to 15%. However, the mismatch stresses in polycrystalline zirconia prevent shape memory behavior. At strains of only about ~2%, cracking is observed after only a few transformation cycles. This is unlike NiTi SMA, which can access large strains up to ~8% and, at lower strain levels, can be reversibly transformed up to millions of cycles [172]. The prospect of a new class of SMM based on fine-scale ceramics is potentially technologically interesting because ceramics occupy a different region in property space than any other SMM as shown in Fig. 1.34. The very high strength of ceramics permits access to shape memory and superelasticity at very high stress levels relative to shape memory metals. Another property axis on which ceramics are generally differentiated from metals is temperature; ceramics are generally more refractory than metals, and as such, shape memory ceramics (SMC) should be thought of as a class of potential high-temperature shape-memory materials [173]. A final interesting prospect for the use of shape memory ceramics is as a structural material subject to a one-time mechanical loading event, in which repeatable transformations are not required. In such cases, the ceramic can be loaded all the way to failure, and unlike a conventional brittle ceramic, the superelastic ceramic can

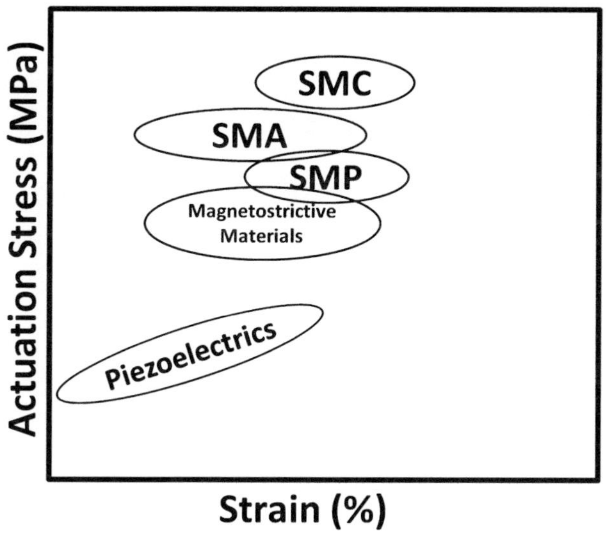

Fig. 1.34 Position of SMC actuation strength

exhibit substantial apparent ductility (or malleability) before fracture, by virtue of the phase transformation [174].

For application, these polycrystalline ceramic specimens, micron-scale pillars were fabricated using focused ion beam milling. The resulting pillars were sized to be smaller than or near the average grain size of the ceramics and so were either single crystalline or oligocrystalline (having more surface area than grain boundary area) in every case [191]. We applied loads to pillars both in a mode of axial compression and in bending, using a Hysitron nano-mechanical test platform with a blunt conospherical tip. The shape memory effect of SMC can be calculated by Eq. 1.4 [172]:

$$\mathrm{SME} = \frac{L_H - L_C}{L_O - L_C} \tag{1.4}$$

where, L_O = The length of the specimen, L_C = The length after compression, and L_H = The length after reheating the specimen to above austenite temperature then cooling down to room temperature.

Unlike SMA and SMP, there is a fundamental cause of cracking in ceramics with martensitic transformations is transformation mismatch stresses, our strategy is to reduce these stresses through two main approaches. First, we reduce the size scale of the specimen itself (a smaller ceramic has a higher surface area to volume ratio, and the free surfaces can contribute to stress relaxation). Second, we strive to reduce the number of crystals within the volume of the specimen—an "oligocrystalline" or even single crystalline material contains fewer grain junctions where the individual grain transformation strains will compete. Both of these concepts have been explored, which extend their application in the case of ceramics [175].

1.4.1 Various Shape-Memory Ceramics

1.4.1.1 Zirconia-Based SMC

Zirconia-based SMC is a unique family of smart materials with many potential applications in sensing, actuation, and mechanical energy damping. Zirconia (ZrO_2) has a well-known martensitic phase transformation between the tetragonal and monoclinic phases, which is accompanied by a large shape strain (≈15%) and volume change (≈5%) [176]. In pure ZrO_2, this transformation occurs at approximately 1000 °C on cooling, causing sintered polycrystals to undergo significant cracking, precluding their use as a structural ceramic. However, dopants such as ceria (CeO_2), yttria (Y_2O_3), or magnesia (MgO) have been added to suppress the martensitic transformation, allowing uncracked tetragonal or cubic phase to be retained at room temperature [177]. This strategy underpins the highly successful tetragonal zirconia polycrystal (TZP), partially stabilized zirconia (PSZ), and fully stabilized zirconia (FSZ) materials that have been used as structural ceramics, solid oxide fuel cell electrolytes, and biomaterials, among other applications [178]. Another emerging use case for these materials is as SMCs, which are attractive functional materials for their significantly higher transformation temperatures, work output, and environmental resistance compared to their metallic counterparts, SMA. The reversible thermoelastic martensitic transformation from a high-temperature tetragonal phase (referred to as austenite in the context of shape-memory materials) to a low-temperature monoclinic phase (referred to as martensite) [179]. When these ceramics are in the high-temperature phase, the transformation can be triggered through shearing of the austenite phase with applied stress, giving rise to a large shear strain as the lattice transforms to martensite. This crystallographic transformation strain can be fully recovered by releasing the stress to induce reverse transformation, leading to the "superelasticity." The shape-memory and superelastic behavior with large strains are of the order of 2% in CeO_2-doped TZP (Ce-TZP); however, these materials experienced a fracture after four or fewer cycles [180], which precludes their use in bulk polycrystalline form. Recently, it has been shown that by reducing the grain constraint in micron-scale single- or oligo-crystal specimens, ZrO_2-based SMCs can be subjected to hundreds of superelastic cycles without failure. Nonetheless, even in these single-crystal micro-specimens, clear dislocation build-up, and eventually catastrophic cracking are observed during superelastic cycling [181].

An early observation of superelasticity in zirconia ceramics was made by Reyes-Morel et al. who reported that ceria-stabilized tetragonal zirconia polycrystalline (Ce-TZP) ceramics exhibited superelasticity with a stress-induced transformation strain of ~0.5e1.0%. However, this strain is far below the full transformation strain of ~8%, and the

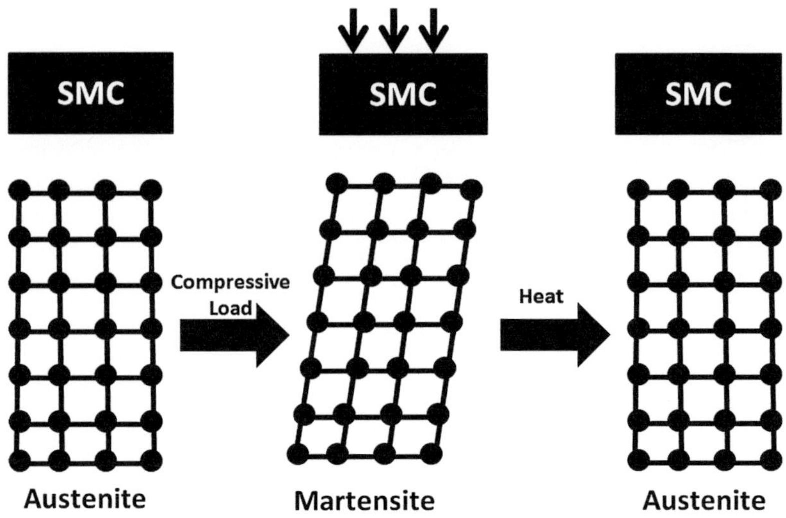

Fig. 1.35 Micro-compression stress–strain curve for superelastic ceramic (zirconia) pillar in which initial elastic loading of austenite is followed by forward transformation during the formation of martensite. This is followed by elastic unloading and reverse transformation with a reversion to austenite

superelastic behavior could only be repeated for five to six loading cycles before the samples failed due to intergranular cracking [182]. Currently, to obtain the oligocrystalline SMC in the form of isolated spherical particles, it can be mass produced by spray drying. Specifically, we explore a series of SMC particles of average grain size of ~1.7 mm. By using spray drying, micro-scale shape memory ceramic particles of ceria-doped zirconia comprising one grain or a few grains have been produced [183].

CeO_2-ZrO_2 superelastic specimen loaded in compression is shown in Fig. 1.35. Loading begins in the tetragonal phase (which here, in the context of SMM, is referred to as austenite) and after an initial linear elastic response, a critical stress is reached that induces the phase changes into the monoclinic phase (which is referred to as martensite). This is seen as a slope reduction in the curve, which evolves as the transformation progresses to large strains of $\approx 7\%$ [184]. Upon unloading, there is a linear elastic response of the martensite phase until another critical stress—that for reversion to austenite—is reached and a second lower plateau is seen. Complete unloading leads to a full recovery of the strain. The total energy dissipated during the cycle is quantified by the area within the curve. The stress–strain curve shown in Fig. 1.35 is characteristic of the many hundreds that we have measured on a variety of fine-scale austenitic zirconia ceramic pillars. We find these samples exhibit all the attributes of a good superelastic material. This includes large strains up to 7% in some pillars as well as the ability to cycle reversibly through the transformation many times. A series of typical stress–strain curves during cycling are shown in Fig. 1.35 and show that superelastic properties are present over many cycles. There is a gradual evolution of the stress–strain hysteresis loop with cycling that is expected as a superelastic material becomes "trained" to a particular kinematic transformation pathway [172]. In addition to the results in Fig. 1.35, a variety of other specimens were cyclically loaded without failing to dozens of cycles; up to 50 cycles have been applied to a single pillar. More stress–strain curves exhibiting a range of different achievable stress and strain combinations for different pillars can be found. Some pillars did experience cracking but only after many more cycles than previously reported in polycrystalline specimens [185].

1.4.1.2 Lanthanum-Niobium oxide SMC

Rare-earth orthoniobate, LaNbO, undergoes a tetragonal-to-monoclinic phase transformation at temperature around 520 °C. Both high- and low-temperature phases possess the scheelite structure, and the low-temperature monoclinic phase is heavily preferred in two orientations [186]. The domains are separated by highly mobile boundaries, exhibiting a rubber-like behavior in the monoclinic $LaNbO_4$ single crystals. This undergoes a martensitic transformation and is situated in the low-temperature martensitic condition. The stress–strain behavior of polycrystalline ceramic LaNbO is significantly different from that of LaNbO single crystal, in which a distinct yielding point shows the beginning of the off-elastic portion of the curve. Usually, the low-temperature monoclinic phase of $LaNbO_4$ is heavily domained in single crystals or in some grains of polycrystalline ceramics. For a single crystal, while the specimen is deformed, the domain walls move in response to the applied stresses. During unloading, the walls move back to where they were, and the specimen assumes its original shape. However, in the case of polycrystalline ceramics, when stress is applied new domains form, in addition to the existing domain boundaries moving, and the newly formed domains do not disappear and the specimen remains deformed after unloading [187]. On the other hand, the mobility of the domain boundary in grains of polycrystal ceramics is limited due to the interaction with grain boundaries and lower than that in single crystals.

1.4.1.3 Advantage and Disadvantage of SMCs

Shape memory ceramics are a unique family of shape-memory materials with many potential applications in sensing, actuation, and mechanical energy damping. Compared to metallic shape memory alloys, SMCs offer many advantages such as high operating temperature, high strength, and chemical inertness. In shape memory ceramics such as zirconia, the cracking problems are associated with the reversible martensitic transformation from the tetragonal phase to the monoclinic phase, which can be induced by either stress or temperature [188]. Due to pseudoelasticity and shape memory effect, SMC can resist catastrophic failure and protect the structure at higher temperatures. Tensile load on the structure is the only problem during various applications.

1.5 Shape Memory Hybrids

Shape memory hybrid (SMH) can provide an alternative solution that is based only on some simple concepts and utilizes only ordinary materials, which have well-understood properties but do not have shape memory as an individual. Overall, similar to hydrophobic SMP, the design of an SMH should also meet two structural requirements. One is the hard segment and another is the switching segment [189]. The hard segment represents a stable network constructed by net points that remain stable during the actuation of SME. Hard segment defines the permanent shape and is responsible for driving the recovery of the deformation from the temporary shape to the original shape (or permanent shape). The switching segment serves as a stimuli-sensitive switch to fix or recover the temporary shapes. Both segments contribute to the network architecture with the formed temporary shape. Typically, chain segments or domains with a transition temperature (T_{trans}) can serve as switching segments to endow a polymeric network with the ability of SME. T_{trans} can either be a glass transition, T_g, for an amorphous polymer or a melting temperature, T_m, for a crystalline. Similar to triple- or multi-SMP, SMH can also fix two or more temporary shapes under different temperatures via distinct T_{trans} [190].

1.5.1 Basic Mechanism Behind SHM

The mechanism of SMH is identical to that of SMP. However, in SMH, both the matrix and the additives are not limited to polymers only but can be selected from any materials as given in Fig. 1.36. To work out a right SMH that meets the required function(s) in a particular application, both matrix and inclusion can be chosen from metal, organic, and inorganic. Ideally, in an SMH, there is not any chemical interaction between the matrix and the inclusion, so the

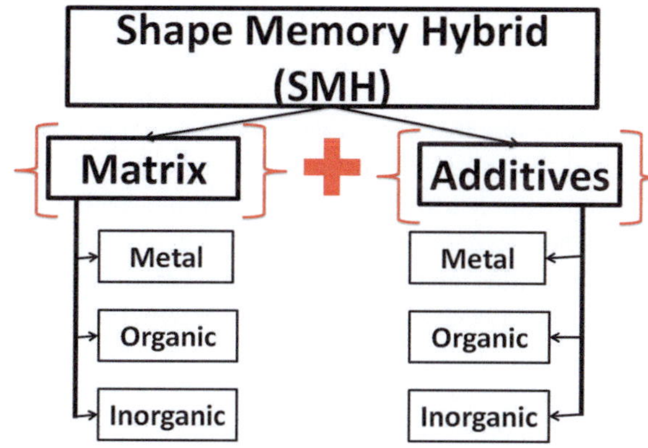

Fig. 1.36 Constituting addition in the synthesis of SHM

properties of individual materials are largely maintained. Therefore, we can well predict the properties of SMH from the very beginning [191]. For instance, the transition temperature of SMH is the softening temperature of the original inclusion material. For example, we can select silicone and wax, both are biocompatible, as the elastic matrix and the transition inclusion, respectively. While silicone normally keeps its high elasticity characteristic within a wide range of temperatures, wax melts upon heating to its melting temperature and becomes very soft. As such, we can easily compress the sample. The silicone matrix is elastically deformed so that an elastic energy is stored in it. When the sample is cooled back to room temperature, wax becomes solid again, which can effectively prevent the release of the elastic energy in the silicone matrix [192]. Consequently, after the constraint is fully removed, the sample largely maintains its deformed shape until it is heated again to above the melting temperature of the wax.

Furthermore, one can predict the major properties and behaviors of the SMH even in the early design stage by means of materials selection and simple estimation. As such, anyone can conveniently and quickly design the SMH and fabricate it shortly according to the requirement. On the other hand, SMHs are convenient to be tailored to meet the exact requirement. For instance, high recovery strain and high recovery stress are not easy to be satisfied simultaneously in SMP by means of polymer modification. However, SMH stands to configure a right processing technique with suitable matrix material and additives. The SME of an SMH is highly dependent on the switching segments introduced in the network as was already been described. The type of switching segment plays a dominant role in determining which types of applied stimuli can activate the SME. Due to the reversible nature, the crosslinks formed by different molecular switches in the SMH are switchable (on-off) when exposed to corresponding stimuli such as light, heat, ion species, and p^H. Generally, a typical dual-SMH contains only one switching segment while integrating two or more types of switching segments into one ensemble to fix two or more temporary shapes via applying non-interfering stimuli to exhibit triple or multiple SME [193]. For example, a hydrogel with a single network constructed via chemical cross-links and physical cross-links could exhibit a dual SME, while triple SME will be realized when two noninterfering physical cross-links are employed in the network. Similarly, introducing more non-interfering switching segments can endow the hydrogel with multi-SME. Also, an SMH that was fabricated with two types of molecular switches—Schiff base bonds and chitosan–Ag + (CS–Ag+) coordination interactions showed a triple-SME [194].

In the SMH case, the fabrication process is more complex because functional fillers such as nano-gold, nano-platinum, graphene oxide, and Fe_3O_4 particles, acting as energy conversion medium, are required to incorporate into the SMHs [195].

1.5.2 Responses in SMH

In SMH material, the controlling stimuli may be one or more than one. Thermo-responsive SMH actuated with the variation in operating temperature. For example, silicone and two types of wax can be prepared as a thermo-responsive SMH. A piece of pre-bent silicone/wax (two types of waxes with melting temperatures of 45 °C and 65 °C, respectively) sample can actuate upon exposure in two different temperatures [196].

Similar to cupric sulfate pentahydrate, sodium acetate trihydrate is dissolvable in the water as well. In addition, unlike cupric sulfate pentahydrate, which does not melt when heated to release the water molecules, sodium acetate trihydrate melts at around 70 °C (i.e., sodium acetate dissolves within its own released water). It is an ideal additive material to demonstrate the thermo/water responsive feature. While shape recovery can be finished within a couple of seconds upon heating to above the melting temperature of sodium acetate trihydrate, it takes much longer for water-induced shape recovery because sodium acetate trihydrate micro powders are embedded inside silicone so that the penetration of water into the sample to dissolve sodium acetate trihydrate is slower [197].

One example of water-responsive SMH is hydrogels. Hydrogels are three-dimensional soft networks formed by physical or chemical crosslinking of hydrophilic polymers, which are able to swell absorbing and retaining a substantial amount of water. This stimuli-responsiveness allows hydrogels to be processed to vary spatially and/or temporally their properties as a response against external changes. Different external triggers have been studied to induce reversible hydrogel solution or swollen-collapsed transitions, such as p^H, temperature, radiation, redox reactions, or chemical triggers. As a consequence of these switchable properties, these materials can behave as actuators or sensors, and therefore, they have been intensively investigated in the last decades in a great variety of fields, among which it is worth highlighting, biomedicine, agriculture, and wastewater treatment. However, hydrogels generally show poor mechanical properties, and consequently, mechanical damages and cracks limit their correct function over a long period of time, which is especially problematic for biomedical applications [198]. A supramolecular shape memory hydrogel refers to shape memory polymer, in which temporary shapes are stabilized by reversible crosslinks such as supramolecular interactions and dynamic covalent bonds. Because of the reversible and dynamic nature of supramolecular chemistry, supramolecular shape memory hydrogels can exhibit excellent cycled shape memory behavior at room temperature. Hydrogels have been intensively studied due to their relevant properties, such as their similarity to body tissues, low surface friction, ability to encapsulate and release (molecules, ions, cells), appropriate morphology for cell proliferation, and stimuli-responsive properties.

1.6 Summary

This chapter explained all the types of shape-memory materials along with their different peculiar properties. With the innovative ideas for applications of SMA and the number of products in the market using SMA continually is growing. Advances in the field of polymer for use in many different fields of study seem very promising because SMPs can respond to temperature, light, pH, and moisture, and there are many very interesting possibilities is likely the range of applications will continue to grow. The study of SMC has only just begun in the past several years. Further investigations on the improvement in the induced strain magnitude, the stability of the strain characteristics with respect to temperature change, mechanical strength, and durability after repeated driving are required to produce practical and reliable materials. This category of ceramic actuators will be a vital new element in the next generation of "micromechatronic" or electromechanical actuator devices. SMHs have gained much attention in recent decades. Although the scientific research in the SMH area is developing day by day, still there are a number of challenges and new avenues to be explored. In this chapter, all the SMMs are described among the scientific community to explore all the future potential applications.

References

1. Huang, W.M., Ding, Z., Wang, C.C., Wei, J., Zhao, Y., Purnawali, H.: Shape memory materials. Mater. Today. **13**(7–8), 54–61 (2010). https://doi.org/10.1016/S1369-7021(10)70128-0
2. Sun, L., Huang, W.M., Ding, Z., Zhao, Y., Wang, C.C., Purnawali, H., Tang, C.: Stimulus-responsive shape memory materials: a review. Mater. Des. **33**, 577–640 (2012). https://doi.org/10.1016/j.matdes.2011.04.065
3. Othmane Benafan, G.S., Bigelow, A., Young, W.: Shape memory materials database tool-a compendium of functional data for shape memory materials. Adv. Eng. Mater. **22**(7), 1901370 (2020). https://doi.org/10.1002/adem.201901370
4. Vasina, M., Solc, F., Hoder, K.: Shape memory alloys - unconventional actuators. IEEE International Conference on Industrial Technology. (2003) https://doi.org/10.1109/ICIT.2003.1290266
5. Copaci, D., Blanco, D., Moreno, L.E.: Flexible shape-memory alloy-based actuator: mechanical design optimization according to application. Actuators. **8**(3), 63 (2019). https://doi.org/10.3390/act8030063
6. Vollach, S., Shilo, D.: The mechanical response of shape memory alloys under a rapid heating pulse. Exp. Mech. **50**, 803–811 (2010). https://doi.org/10.1007/s11340-009-9320-z
7. Kauffman, G.B., Mayo, I.: The story of nitinol: the serendipitous discovery of the memory metal and its applications. Chem. Educator. **2**, 1–21 (1997). https://doi.org/10.1007/s00897970111a
8. https://en.wikipedia.org/wiki/Shape-memory_alloy. 08 Jan 2019
9. Zainal, M.A., Sahlan, S., Ali, M.S.M.: Micromachined shape-memory-alloy microactuators and their application in biomedical devices. Micromachines. **6**, 879–901 (2015). https://doi.org/10.3390/mi6070879
10. Simon, M., Kaplow, R., Salzman, E., Freiman, D.: A vena cava filter using thermal shape memory alloy. Experimental aspects. Radiology. **125**(1), 87–94 (1977)
11. Stoeckel, D., Pelton, A., Duerig, T.: Self-expanding nitinol stents: material and design considerations. Eur. Radiol. **14**(2), 292–301 (2004). https://doi.org/10.1007/s00330-003-2022-5
12. Jani, J.M., Leary, M., Subic, A., Gibson, M.A.: A review of shape memory alloy research, applications and opportunities. Mater. Des.. (1980–2015). **56**, 1078–1113 (2014). https://doi.org/10.1016/j.matdes.2013.11.084
13. Ma, N., Lu, Y., He, J., Dai, H.: Application of shape memory materials in protective clothing: a review. J. Text. Inst. **110**, 6 (2019). https://doi.org/10.1080/00405000.2018.1532783
14. https://en.wikipedia.org/wiki/Shape-memory_polymer. 08 Jan 2019
15. Song, G., Ma, N., Li, H.-N.: Applications of shape memory alloys in civil structures. Eng. Struct. **28**(9), 1266–1274 (2006). https://doi.org/10.1016/j.engstruct.2005.12.010
16. Swain, B.K., Bajpai, S., Behera, A.: Microstructural evolution of NITINOL and their species formed by atmospheric plasma spraying. Surf. Topogr.: Metrol. Prop. **7**, 015006 (2019). https://doi.org/10.1088/2051-672X/aaf30e
17. Thompson, S.A.: An overview of nickel-titanium alloys used in dentistry. Int. Endod. J. **33**(4), 297–310 (2000). https://doi.org/10.1046/j.1365-2591.2000.00339.x
18. https://ethw.org/Arne_Olander_discovers_the_shape-memory_effect_in_an_alloy_of_gold_and_cadmium#:~:text=Technology%20History%20Wiki-Arne%20Olander%20discovers%20the%20shape%2Dmemory%20effect%20in%20an%20alloy,in%20medical%20and%20other%20applications. 08 Jan 2019
19. Kumar, P., Lagoudas, D.: Introduction to shape memory alloys. In: Shape Memory Alloys. Springer, Boston, MA (2008). https://doi.org/10.1007/978-0-387-47685-8_1
20. Machado, L.G., Savi, M.A.: Medical applications of shape memory alloys. Braz. J. Med. Biol. Res. **36**(6), 683–691 (2003). https://doi.org/10.1590/S0100-879X2003000600001
21. Bil, C., Massey, K., Abdullah, E.J.: Wing morphing control with shape memory alloy actuators. J. Intell. Mater. Syst. Struct. **24**(7), 879–898 (2013). https://doi.org/10.1177/1045389X12471866
22. Dunne, D.: 4 - Shape memory in ferrous alloys, Phase transformations in steels. In: Diffusionless transformations, high strength steels, modelling and advanced analytical techniques Woodhead Publishing Series in Metals and Surface Engineering, vol. 2, pp. 83–125. Woodhead Publishing, Sawston (2012). https://doi.org/10.1533/9780857096111.1.83
23. Khanlari, K., Ramezani, M., Kelly, P.: 60NiTi: a review of recent research findings, potential for structural and mechanical applications, and areas of continued investigations. Trans. Indian Inst. Metals. **71**, 781–799 (2018). https://doi.org/10.1007/s12666-017-1224-5
24. Soares, R.L., de Castro, W.B.: Effects of composition on transformation temperatures and microstructure of Ni-Ti-Hf shape memory alloys. REM – Int. Eng. J. **72**(2), 227–235 (2019). https://doi.org/10.1590/0370-44672018720072
25. Auricchio, F., Sacco, E.: A one-dimensional model for superelastic shape-memory alloys with different elastic properties between austenite and martensite. Int. J. Non-Linear Mech. **32**(6), 1101–1114 (1997). https://doi.org/10.1016/S0020-7462(96)00130-8
26. Patoor, E., Lagoudas, D.C., Pavlin, B., Entchev, L., Brinson, C., Gao, X.: Shape memory alloys, Part I: General properties and modeling of single crystals. Mech. Mater. **38**(5–6), 391–429 (2006). https://doi.org/10.1016/j.mechmat.2005.05.027
27. Eggeler, G., Hornbogen, E., Yawny, A., Heckmann, A., Wagner, M.: Structural and functional fatigue of NiTi shape memory alloys. Mater. Sci. Eng. A. **378**(1–2), 24–33 (2004). https://doi.org/10.1016/j.msea.2003.10.327
28. Reginald DesRoches, M.ASCE, McCormick, J., Delemont, M.: Cyclic properties of superelastic shape memory alloy wires and bars. J. Struct. Eng. **130**(1) (2004). https://doi.org/10.1061/(ASCE)0733-9445
29. Follador, M., Cianchetti, M., Arienti, A., Laschi, C.: A general method for the design and fabrication of shape memory alloy active spring actuators. Smart Mater. Struct. **21**(11). https://doi.org/10.1088/0964-1726/21/11/115029
30. Otubo, J., Rigo, O.D., Neto, M., Mei, P.R.: The effects of vacuum induction melting and electron beam melting techniques on the purity of NiTi shape memory alloys. Mater. Sci. Eng. A. **438–440** (25), 679–682 (2006). https://doi.org/10.1016/j.msea.2006.02.171
31. Frenzel, J., Zhang, Z., Neuking, K., Eggeler, G.: High quality vacuum induction melting of small quantities of NiTi shape memory alloys in graphite crucibles. J. Alloys Comp. **385**(1–2), 214–223 (2004). https://doi.org/10.1016/j.jallcom.2004.05.002
32. Schöller, E., Krone, L., Bram, M., et al.: Metal injection molding of shape memory alloys using prealloyed NiTi powders. J. Mater. Sci. **40**, 4231–4238 (2005). https://doi.org/10.1007/s10853-005-2819-5
33. Köhl, M., Habijan, T., Bram, M., Buchkremer, H.P., Stöver, D., Köller, M.: Powder metallurgical near-net-shape fabrication of porous NiTi shape memory alloys for use as long-term implants by the combination of the metal injection molding process with the space-holder technique. Adv. Eng. Mater. **11**(12), 959–968 (2009). https://doi.org/10.1002/adem.200900168
34. Gilberto, H.T., da Silva, A., Otubo, J.: Designing NiTiAg shape memory alloys by vacuum arc Remelting: first practical insights on melting and casting. Shap. Mem. Superelasticity. **4**, 402–410 (2018). https://doi.org/10.1007/s40830-018-0190-z
35. Xenos-Despina, S., Karamichailidou: The Unique Properties, Manufacturing Processes and Applications of Near Equatomic

Ni-Ti Alloys. University of Thessaly Department of Mechanical Engineering Laboratory of Materials. (2016). http://www.mie.uth.gr/ekp_yliko/TheUniqueProperties,ManufacturingProcessesandApplicationsofNearEquatomicNi-TiAlloys.pdf
36. Hu, G., Lixiang, Z., Yunliang, F., Yanhong, L.: Fabrication of high porous NiTi shape memory alloy by metal injection molding. J. Mater. Process. Technol. **206**(1–3), 395–399 (2008). https://doi.org/10.1016/j.jmatprotec.2007.12.044
37. Khanlari, K., Ramezani, M., Kelly, P., Cao, P., Neitzerta, T.: Synthesis of as-sintered 60NiTi parts with a high open porosity level. Mater. Res. **21**(5), e20180088 (2018). https://doi.org/10.1590/1980-5373-MR-2018-0088
38. Panel, S., Shivaa, I.A., Palania, S.K., Mishrab, C.P., Kukreja, P.L. M.: Investigations on the influence of composition in the development of Ni–Ti shape memory alloy using laser based additive manufacturing. Opt. Laser Technol. **69**, 44–51 (2015). https://doi.org/10.1016/j.optlastec.2014.12.014
39. Elahinia, M.H., Hashemi, M., Tabesh, M., Bhaduri, S.B.: Manufacturing and processing of NiTi implants: a review. Prog. Mater. Sci. **57**(5), 911–946 (2012). https://doi.org/10.1016/j.pmatsci.2011.11.001
40. Walker, J., Andani, M.T., Haberland, C., Elahinia, M.: Additive manufacturing of nitinol shape memory alloys to overcome challenges in conventional nitinol fabrication. In: Proceedings of the ASME 2014 International Mechanical Engineering Congress and Exposition Advanced Manufacturing, vol. 2A, pp. 14–20. ASME, Montreal, Quebec, Canada (2014. V02AT02A037). https://doi.org/10.1115/IMECE2014-40432
41. Swain, B., Mallick, P., Bhuyan, S., Mishra, S.C., Mohapatara, S.S., Behera, A.: Mechanical properties of NiTi plasma spray coating. J. Therm. Spray Technol. (2020). https://doi.org/10.1007/s11666-020-01017-6.
42. Swain, B., Mallick, P., Gupta, R., Mohapatra, S.S., Yasin, G., Nguyen, N.A., Behera, A.: Mechanical and tribological properties evaluation of plasma-sprayed shape memory alloy coating. J. Alloys Comp. **863**, 158599 (2021). https://doi.org/10.1016/j.jallcom.2021.158599
43. Behera, A., Aich, S.: Characterization and properties of magnetron sputtered nanoscale NiTi thin film and the effect of annealing temperature. Surf. Interface Anal. **47**, 805–814 (2015). https://doi.org/10.1002/sia.5777
44. Hosodaa, H., Kinoshitaa, Y., Fukuia, Y., Inamuraa, T., Wakashimaa, K., Kimb, H.Y., Miyazaki, S.: Effects of short time heat treatment on superelastic properties of a Ti–Nb–Al biomedical shape memory alloy. Mater. Sci. Eng. A. **438–440**(25), 870–874 (2006). https://doi.org/10.1016/j.msea.2006.02.151
45. Yeung, K.W.K., Cheung, K.M.C., Lu, W.W., Chung, C.Y.: Optimization of thermal treatment parameters to alter austenitic phase transition temperature of NiTi alloy for medical implant. Mater. Sci. Eng. A. **383(2)**(15), 213–218 (2004). https://doi.org/10.1016/j.msea.2004.05.063
46. Foroozmehr, A., Kermanpur, A., Ashrafizadeh, F., Kabiri, Y.: Investigating microstructural evolution during homogenization of the equiatomic NiTi shape memory alloy produced by vacuum arc remelting volume 528. Mater. Sci. Eng. A. **27**(15), 7952–7955 (2011). https://doi.org/10.1016/j.msea.2011.07.024
47. S. Badakhshan Raz & S. K. Sadrnezhaad, Effects of VIM frequency on chemical composition, homogeneity and microstructure of NiTi shape memory alloy, Materials Science and Technology, Volume 20, 2004 - Issue 5, doi: https://doi.org/10.1179/026708304225016680
48. Yuan, B., Chung, C.Y., Zhu, M.: Microstructure and martensitic transformation behavior of porous NiTi shape memory alloy prepared by hot isostatic pressing processing. Mater. Sci. Eng. A. **382**(1–2), 181–187 (2004). https://doi.org/10.1016/j.msea.2004.04.068
49. Vajpai, S.K., Dube, R.K., Chatterjee, P., Sangal, S., Novel Powder, A.: Metallurgy processing approach to prepare fine-grained cu-Al-Ni shape-memory alloy strips from elemental powders. Metall. Mater. Trans. A. **43a** (2012). https://doi.org/10.1007/S11661-012-1081-0
50. Sharma, M., Vajpai, S.K., Dube, R.K.: Processing and characterization of Cu-Al-Ni shape memory alloy strips prepared from elemental powders via a novel powder metallurgy route. Metall. Mater. Trans. A. **41a** (2010). https://doi.org/10.1007/s11661-010-0351-y.
51. Haberland, C., Elahinia, M., Walker, J.M., Meier, H., Frenzel, J.: On the development of high quality NiTi shape memory and pseudoelastic parts by additive manufacturing. Smart Mater. Struct. **23**(10). https://doi.org/10.1088/0964-1726/23/10/104002
52. Ajit Behera, R., Suman, S., Aich, S., Mohapatra, S.: Sputter-deposited Ni/Ti double-bilayer thin film and the effect of intermetallics during annealing. Surf. Interf. Anal. **49**(7), 620–629 (2016). https://doi.org/10.1002/sia.6201
53. Patel, S.K., Swain, B., Roshan, R., Sahu, N.K., Behera, A.: A brief review of shape memory effects and fabrication processes of NiTi based alloys. Mater. Today Proc. **33**(8), 5552–5556 (2020). https://doi.org/10.1016/j.matpr.2020.03.539
54. Ren, X., Otsuka, K.: Origin of rubber-like behaviour in metal alloys. Nature. **389**, 579–582 (1997). https://doi.org/10.1038/39277
55. Nayan, N., Buravalla, V., Ramamurty, U.: Effect of mechanical cycling on the stress–strain response of a martensitic nitinol shape memory alloy. Mater. Sci. Eng. A. **525**(1–2), 60–67 (2009). https://doi.org/10.1016/j.msea.2009.07.038
56. Thamburaja, P., Anand, L.: Superelastic behavior in tension-torsion of an initially-textured Ti-Ni shape-memory alloy. Int. J. Plast. **18**(11), 1607–1617 (2002). https://doi.org/10.1016/S0749-6419(02)00031-1
57. Behera, A., Aich, S., Behera, A., Sahu, A.: Processing and characterization of magnetron sputtered Ni/Ti thin film and their annealing behaviour to induce shape memory effect. Mater. Today Proc. **2**, 1183–1192 (2015). https://doi.org/10.1016/j.matpr.2015.07.030
58. Simon, T., Kröger, A., Somsen, C., Dlouhy, A., Eggeler, G.: On the multiplication of dislocations during martensitic transformations in NiTi shape memory alloys. Acta Mater. **58**(5), 1850–1860 (2010). https://doi.org/10.1016/j.actamat.2009.11.028
59. Wang, X., Xu, B., Yue, Z.: Phase transformation behavior of pseudoelastic NiTi shape memory alloys under large strain. J. Alloys Compd. **463(1–2)**(8), 417–422 (2008). https://doi.org/10.1016/j.jallcom.2007.09.029
60. Nomsa, M., Sisa, M., Sisa, P., Patricia, P., Patricia, P., Tebogo, P., Tebogo, M.: NiTi intermetallic surface coatings by laser metal deposition for improving wear properties of Ti-6Al-4V substrates. Adv. Mater. Sci. Eng. **10**, 1–8 (2014). https://doi.org/10.1155/2014/363917
61. Laplanche, G., Birk, T., Schneider, S., Frenzel, J., Eggeler, G.: Effect of temperature and texture on the reorientation of martensite variants in NiTi shape memory alloys. Acta Mater. **127**(1), 143–152 (2017). https://doi.org/10.1016/j.actamat.2017.01.023
62. Laplanche, G., Pfetzing-Micklich, J., Eggeler, G.: Sudden stress-induced transformation events during nanoindentation of NiTi shape memory alloys. Acta Mater. **78**, 144–160 (2014). https://doi.org/10.1016/j.actamat.2014.05.061
63. Pushin, V., Kuranova, N., Marchenkova, E., Pushin, A.: Design and development of Ti–Ni, Ni–Mn–Ga and Cu–Al–Ni-based alloys with high and low temperature shape memory effects. Materials. **12**(16), 2616 (2019). https://doi.org/10.3390/ma12162616

64. Otsuka, K., Ren, X.: Physical metallurgy of Ti–Ni-based shape memory alloys. Prog. Mater. Sci. **50**, 511–678 (2005). https://doi.org/10.1016/j.pmatsci.2004.10.001
65. Delaey, L.: Diffusionless transformations. Mater. Sci. Technol. (2013). https://doi.org/10.1002/9783527603978.mst0392
66. Chluba, C., Ge, W., Dankwort, T., Bechtold, C., Lima de Miranda, R., Kienle, L., Wuttig, M., Quandt, E.: Effect of crystallographic compatibility and grain size on the functional fatigue of sputtered TiNiCuCo thin films. Philos. Trans. A Math. Phys. Eng. Sci. **374** (2074), 20150311 (2016). https://doi.org/10.1098/rsta.2015.0311
67. Liu, Y., Xie, Z.L.: Twinning and detwinning of ⟨0 1 1⟩ type II twin in shape memory alloy. Acta Mater. **51**(18), 5529–5543 (2003). https://doi.org/10.1016/S1359-6454(03)00417-8
68. Liang, Y., Jiang, S., Zhang, Y., Yu, J.: Microstructure, mechanical property, and phase transformation of quaternary NiTiFeNb and NiTiFeTa shape memory alloys. Metals. **7**(8), 309 (2017). https://doi.org/10.3390/met7080309
69. Müller, I., Seelecke, S.: Thermodynamic aspects of shape memory alloys. Math. Comp. Model. **34**(12–13), 1307–1355 (2001). https://doi.org/10.1016/S0895-7177(01)00134-0
70. Paulo da Silva, E.: The influence of loading on the heat of transformation in shape memory alloys. J. Braz. Soc. Mech. Sci. Eng. **27**, 3. https://doi.org/10.1590/S1678-58782005000300012
71. Omori, T., Kainuma, R.: Martensitic transformation and superelasticity in Fe–Mn–Al-based shape memory alloys. Shap. Mem. Superelasticity. **3**, 322–334 (2017). https://doi.org/10.1007/s40830-017-0129-9
72. Wang, J., Zhang, W., Zhu, J., Xu, Y., Gu, X., Moumni, Z.: Finite element simulation of thermomechanical training on functional stability of shape memory alloy wave spring actuator. J. Intell. Mater. Syst. Struct. **30**(8), 1239–1251 (2019). https://doi.org/10.1177/1045389X19831356
73. Pieczyska, E.A., Staszczak, M., Dunić, V., et al.: Development of stress-induced martensitic transformation in TiNi shape memory alloy. J. Mater. Eng. Perform. **23**, 2505–2514 (2014). https://doi.org/10.1007/s11665-014-0959-y
74. Luo, H., Liao, Y., Abel, E., Wang, Z., Liu, X.: Hysteresis behaviour and modeling of SMA actuators. In: Shape Memory Alloys. IntechOpen, London (2010). https://doi.org/10.5772/9982
75. Doroudchi, A., Zakerzadeh, M.R., Baghani, M.: Developing a fast response SMA-actuated rotary actuator: modeling and experimental validation. Meccanica. **53**, 305–317 (2018). https://doi.org/10.1007/s11012-017-0726-x
76. Lexcellent, C., Leclercq, S., Gabry, B., Bourbon, G.: The two way shape memory effect of shape memory alloys: an experimental study and a phenomenological model. Int. J. Plast. **16**(10–11), 1155–1168 (2000). https://doi.org/10.1016/S0749-6419(00)00005-X
77. Behera, A.: Ni-Ti-based smart micro- and nanoalloys: an introduction. In: Nickel-Titanium Smart Hybrid Materials: from Micro- to Nano-Structured Alloys for Emerging Applications. Elsevier Publisher, Netherland (2021) (in press)
78. Shu-yong, J., Zhang, Y.-Q., Zhao, Y.-N., Hu, S.-W.L.L., Zhao, C.-Z.: Influence of Ni_4Ti_3 precipitates on phase transformation of NiTi shape memory alloy. Trans. Nonferrous Metals Soc. China. **25**(12), 4063–4071 (2015). https://doi.org/10.1016/S1003-6326(15)64056-0
79. Adharapurapu, R.R., Vecchio, K.S.: Superelasticity in a new BioImplant material: Ni-rich 55NiTi alloy. Exp. Mech. **47**, 365–371 (2007). https://doi.org/10.1007/s11340-006-9004-x
80. Yang, X., Ma, L., Shang, J.: Martensitic transformation of Ti50(Ni50−xCux) and Ni50(Ti50−xZrx) shape-memory alloys. Sci. Rep. **9**, 3221 (2019). https://doi.org/10.1038/s41598-019-40100-z
81. Behera, A., Aich, S., Thevasanthi, T.: Magnetron sputtering for development of nanostructured materials. In: Design, Fabrication and Characterization of Multifunctional Nanomaterials., ISBN: 9780128205587 (In Press). Elsevier Publisher, Netherland (2021)
82. Michutta, J., Carroll, M.C., Yawny, A., Somsen, C., Neuking, K., Eggeler, G.: Martensitic phase transformation in Ni-rich NiTi single crystals with one family of Ni_4Ti_3 precipitates. Mater. Sci. Eng. A. **378**(1–2), 152–156 (2004). https://doi.org/10.1016/j.msea.2003.11.061
83. Evirgen, A., Karaman, I., Noebe, R.D., Santamart, R., Pons, J.: Effect of precipitation on the microstructure and the shape memory response of the Ni50.3Ti29.7Zr20 high temperature shape memory alloy. Scr. Mater. **69**(5), 354–357 (2013). https://doi.org/10.1016/j.scriptamat.2013.05.006
84. https://www.industryarc.com/Report/16235/shape-memory-alloy-market.html. Accessed 15 Jan 2019
85. https://www.marketsandmarkets.com/PressReleases/shape-memory-alloy.asp. Accessed 15 Jan 2019
86. https://www.grandviewresearch.com/industry-analysis/shape-memory-alloys-market. Accessed 15 Jan 2019
87. https://www.nnin.org/sites/default/files/files/NNIN-1030_0.pdf
88. Eggleston, M.S., Messer, K., Zhang, L., Yablonovitch, E., Wu, M.C.: Optical antenna enhanced spontaneous emission. Proc. Natl. Acad. Sci. U. S. A. **112**(6), 1704–1709 (2015). https://doi.org/10.1073/pnas.1423294112
89. Calkins, F.T., Mabe, J.H.: Shape memory alloy based morphing aerostructures. J. Mech. Des. **132**(11), 111012 (2010. ASME). https://doi.org/10.1115/1.4001119
90. Costanza, G., Tata, M.E.: Shape memory alloys for aerospace, recent developments, and new applications: a short review. Materials. **13**, 1856 (2020). https://doi.org/10.3390/ma13081856
91. Wanhill, R.J.H., Ashok, B.: Shape memory alloys (SMAs) for aerospace applications. In: Prasad, N., Wanhill, R. (eds.) Aerospace Materials and Material Technologies Indian Institute of Metals Series. Springer, Singapore (2017). https://doi.org/10.1007/978-981-10-2134-3_21
92. Swain, B., Priyadarshini, M., Mohapatra, S.S., Gupta, R.K., Behera, A.: Parametric optimization of atmospheric plasma spray coating using fuzzy TOPSIS hybrid technique. J. Alloys Comp. **867**, 159074 (2021). https://doi.org/10.1016/j.jallcom.2021.159074
93. https://media.gm.com/media/us/en/gm/home.detail.html/content/Pages/news/us/en/2013/Feb/0212-corvette.html. Accessed 18 Jan 2019
94. Nematollahi, M., Baghbaderani, K.S., Amerinatanzi, A., Zamanian, H., Elahinia, M.: Application of NiTi in assistive and rehabilitation devices: a review. Bioengineering (Basel). **6**(2), 37., Published online 2019 Apr 29 (2019). https://doi.org/10.3390/bioengineering6020037
95. https://www.greencarcongress.com/2013/02/sma-20130212.html. Accessed 18 Jan 2019
96. https://nitinolactuators.com/page/6?cat=-1. Accessed 18 Jan 2019
97. https://www.saesgetters.com/sites/default/files/Shape%20Memory%20Actuators%20for%20Automotive%20Applications_0.pdf. Accessed 18 Jan 2019
98. https://www.furukawa.co.jp/en/product/electronic/ele_parts/nt_alloy.html#:~:text=Shape%20memory%20alloys%20are%20alloys,like%20they%20have%20a%20brain. Accessed 18 Jan 2019
99. Manik, R., Behera, A.: Thermal encoding using shape memory alloy. Mater. Today Proc. **33**, 219471266 (2020) https://www.sciencedirect.com/science/article/pii/S2214785320331515
100. https://researchbank.rmit.edu.au/eserv/rmit:161736/Mohd_Jani.pdf. Accessed 18 Jan 2019
101. Behera, A., Aich, S., Ghosh, S.: Simulation of magnetron sputtered Ni/Ti thin film and the effect of annealing. Emerg. Mater. Res. **6**(2), 1–6 (2017). https://doi.org/10.1680/jemmr.16.00093

102. BMorgan, N.: Medical shape memory alloy applications—the market and its products. Mater. Sci. Eng. A. **378**(1–2), 16–23 (2004). https://doi.org/10.1016/j.msea.2003.10.326
103. Petrini, L., Migliavacca, F.: Biomedical applications of shape memory alloys. J. Metall. **501483**, 15 (2011). https://doi.org/10.1155/2011/501483
104. Tobushi, H., Kitamura, K., Yoshimi, Y., Date, K.: Bending deformation and fatigue properties of precision-casted TiNi shape-memory alloy brain spatula. In: Shape Memory Alloys. IntechOpen, London (2010). https://doi.org/10.5772/intechopen.83982
105. Kazuhiro, K., Hisaaki, T., Yukiharu, Y., Kousuke, D., Kouji, M.: Fatigue properties of cast TiNi shape-memory alloy brain spatula. J. Solid Mech. Mater. Eng. **4**(6), 796–805 (2010). https://doi.org/10.1299/jmmp.4.796
106. Melly, S.K., Liu, L., Liu, Y., Leng, J.: Active composites based on shape memory polymers: overview, fabrication methods, applications, and future prospects. J. Mater. Sci. **55**, 10975–11051 (2020). https://doi.org/10.1007/s10853-020-04761-w
107. Pittaccio, S., Viscuso, S.: Shape Memory Actuators for Medical Rehabilitation and Neuroscience. IntechOpen, London (2012). https://doi.org/10.5772/50201
108. Auricchio, F., Boatti, E., Conti, M.: SMA biomedical applications. In: Shape Memory Alloy Engineering. Elsevier. ISBN: 978-0-08-099920-3. https://doi.org/10.1016/B978-0-08-099920-3.00011-5
109. Assell, R., Ainsworth, S., Cragg, A., Dickhudt, E. Kits for enabling axial access and procedures in the spine. Patent no. US20050165406A1
110. Gil, F.J., Planell, J.A.: Shape memory alloys for medical applications. Proc. Inst. Mech. Eng. H. **212**, 6 (1998). https://doi.org/10.1243/0954411981534231
111. Tarniţă, D., Tarniţă, D.N., Bîzdoacă, N., Mîndrilă, I., Vasilescu, M.: Properties and medical applications of shape memory alloys. Romanian J. Morphol. Embryol. **50**(1), 15–21 (2009)
112. https://www.yumpu.com/en/document/view/48437742/finite-strain-shape-memory-alloys-modeling-scuola-di-dottorato-in-. Accessed 19 Jan 2019
113. https://www.kelloggsresearchlabs.com/?gclid=Cj0KCQjw4f35BRDBARIsAPePBHyQ9bLf9N9zri9d0YmLlwIcLoiQ_5yeGZv4ukWyvLT7MOjGZJ6QSnkaAjC7EALw_wcB. Accessed 19 Jan 2019
114. http://www.123seminarsonly.com/ME/Shape-Memory-Alloy.html. Accessed 19 Jan 2019
115. https://www.assemblymag.com/articles/94935-shape-memory-alloys-new-ways-of-using-heat-for-a-technology-advantage
116. Izadi, M., Motavalli, M., Ghafoori, E.: Iron-based shape memory alloy (Fe-SMA) for fatigue strengthening of cracked steel bridge connections. Constr. Build. Mater. **227**, 116800 (2019). https://doi.org/10.1016/j.conbuildmat.2019.116800
117. https://www.youtube.com/watch?v=XPrg8EZlD1E,
118. Liu, S.H., Huang, T.S., Yen, J.Y.: Tracking control of shape-memory-alloy actuators based on self-sensing feedback and inverse hysteresis compensation. Sensors (Basel). **10**(1), 112–127 (2010). https://doi.org/10.3390/s100100112
119. Varela, S., Saiidi, M.: Resilient deconstructible columns for accelerated bridge construction in seismically active areas. J. Intell. Mater. Syst. Struct. **28**(13), 1751–1774 (2017). https://doi.org/10.1177/1045389X16679285
120. Sharma, N., Raj, T., Jangra, K.K.: Applications of nickel-titanium alloy. J. Eng. Technol. **5**, 1–7 (2015). https://doi.org/10.4103/0976-8580.149472
121. Ishii, T.: 5 - Design of shape memory alloy (SMA) coil springs for actuator applications. In: Yamauchi, K., Ohkata, I., Tsuchiya, K., Miyazaki, S. (eds.) Woodhead Publishing Series in Metals and Surface Engineering, Shape Memory and Superelastic Alloys, pp. 63–76. Woodhead Publishing, Cambridge (2011). https://doi.org/10.1533/9780857092625.1.63
122. David Johnson, A., Martynov, V., Gupta, V.: Applications of shape memory alloys: advantages, disadvantages, and limitations. In: Micromachining and Microfabrication Process Technology VII, vol. 4557, pp. 341–351. Proceedings of the SPIE, San Francisco (2001). https://doi.org/10.1117/12.442964
123. Behera, A., Parida, P., Kumar, A.: NiTi thin film shape memory alloy and their industrial application. In: Functional Materials and Advanced Manufacturing. CRC Press, Boca Raton, FL (2020). https://doi.org/10.1201/9780429298042-9. https://www.taylorfrancis.com/books/e/9780429298042/chapters/10.1201/9780429298042-9
124. Scirè Mammanoa, G., Dragoni, E.: Functional fatigue of shape memory wires under constant-stress and constant-strain loading conditions. Proc. Eng. **10**, 3692–3707 (2011). https://doi.org/10.1016/j.proeng.2011.04.607
125. Moallem, M., Lu, J.: Application of shape memory alloy actuators for flexure control: theory and experiments. IEEE/ASME Trans. Mechatron. **10**(5), 495–501 (Oct. 2005). https://doi.org/10.1109/TMECH.2005.856220
126. Mohd Jani, J., Leary, M., Subic, A.: Designing shape memory alloy linear actuators: a review. J. Intell. Mater. Syst. Struct. **28** (13), 1699–1718 (2017). https://doi.org/10.1177/1045389X16679296
127. Patel, S.K., Swain, B., Behera, A.: Advanced processing of superalloys for aerospace industries. In: Functional and Smart Materials. Taylor & Francis Publisher, Boca Raton, FL (2020). https://doi.org/10.1201/9780429298035. eISBN: 9780429298035. https://books.google.co.in/books?id=NFT7DwAAQBAJ&lr=&source=gbs_navlinks_s
128. Zhang, H., Wang, H., Zhong, W., Qiangguo, D.: A novel type of shape memory polymer blend and the shape memory mechanism. Polymer. **50**(6), 1596–1601., ISSN 0032-3861 (2009). https://doi.org/10.1016/j.polymer.2009.01.011
129. Andreas Lendlein, Steffen Kelch, Shape-Memory Polymers, John Wiley & Sons, Inc., Hoboken, NJ 41, 12, 2002, 2034-2057, doi: https://doi.org/10.1002/1521-3773(20020617)41:12<2034::AID-ANIE2034>3.0.CO;2-M
130. Sun, L., Huang, W.M.: Thermo/moisture responsive shape-memory polymer for possible surgery/operation inside living cells in future. Mater. Des. (1980–2015). **31**(5), 2684–2689 (2010). https://doi.org/10.1016/j.matdes.2009.11.036
131. Huang, W.M., Yang, B., Zhao, Y., Ding, Z.: J. Mater. Chem. **20**, 3367 (2010). https://doi.org/10.1039/B922943D
132. H. Lv, J. Leng, Y. Liu, S. Du, Shape-Memory Polymer in Response to Solution, John Wiley & Sons, Inc. Hoboken, NJ, 10, 6 2008, Pages 592-595, doi: https://doi.org/10.1002/adem.200800002
133. Lu, H., Liu, Y., Leng, J., Du, S.: Qualitative separation of the physical swelling effect on the recovery behavior of shape memory polymer. Eur. Polym. J. **46**(9), 1908–1914 (2010). https://doi.org/10.1016/j.eurpolymj.2010.06.013
134. Meng, H., Mohamadian, H., Stubblefield, M., Jerro, D., Ibekwe, S., Pang, S.-S., Li, G.: Various shape memory effects of stimuli-responsive shape memory polymers. Smart Mater. Struct. **22**, 093001 (2013). https://doi.org/10.1088/0964-1726/22/9/093001
135. Biswas, A., Singh, A.P., Rana, D., Aswalc, V.K., Maiti, P.: Biodegradable toughened nanohybrid shape memory polymer for smart biomedical applications. Nanoscale. **10**, 9917–9934 (2018). https://doi.org/10.1039/C8NR01438H
136. Garcia Rosales, C.A., Garcia Duarte, M.F., Kim, H., Chavez, L., Hodges, D., Mandal, P., Lin, Y., Tseng, T.-L.(.B.).: 3D printing of shape memory polymer (SMP)/carbon black (CB) nanocomposites with electro-responsive toughness enhancement. Mater. Res. Exp. **5**, 6. https://doi.org/10.1088/2053-1591/aacd53

137. Lu, H., Gou, J.: Study on 3-D high conductive graphene buckypaper for electrical actuation of shape memory polymer. Nanosci. Nanotechnol. Lett. **4**(12), 1155–1159 (2012). https://doi.org/10.1166/nnl.2012.1455
138. Snyder, E., Tong, T.: Towards Novel light-activated shape memory polymer: thermomechanical properties of photo-responsive Polymers. MRS Proc. **872**, J18.6 (2005). https://doi.org/10.1557/PROC-872-J18.6
139. Liang, F., Fang, T., Liu, X., Ni, Y., Lu, C., Xu, Z.: Precise stimulation of near-infrared light responsive shape-memory polymer composites using upconversion particles with photothermal capability. Comp. Sci. Technol. **152**, 190–197 (2017). https://doi.org/10.1016/j.compscitech.2017.09.021
140. Narendra Kumar, U., Kratz, K., Behl, M., Lendlein, A.: Shape-memory properties of magnetically active triple-shape nanocomposites based on a grafted polymer network with two crystallizable switching segments. Express. Polym. Lett. **6**(1), 26–40 (2012). https://doi.org/10.3144/expresspolymlett.2012.4
141. Huang, W.M., Yang, B.: Water-driven programmable polyurethane shape memory polymer: demonstration and mechanism. Appl. Phys. Lett. **86**, 114105 (2005). https://doi.org/10.1063/1.1880448
142. Ohki, T., Ni, Q.-Q., Ohsako, N., Iwamoto, M.: Mechanical and shape memory behavior of composites with shape memory polymer. Compos. A: Appl. Sci. Manuf. **35**(9), 1065–1073 (2004). https://doi.org/10.1016/j.compositesa.2004.03.001
143. Chen, S., Hu, J., Zhuo, H., Zhu, Y.: Two-way shape memory effect in polymer laminates. Mater. Lett. **62**(25), 4088–4090 (2008). https://doi.org/10.1016/j.matlet.2008.05.073
144. Wu a, Y., Hu, J., Zhang, C., Han a, J., Wang, Y., Kumar, B.: A facile approach to fabricate a UV/heat dual-responsive triple shape memory polymer. J. Mater. Chem. A. **3**, 97–100 (2015). https://doi.org/10.1039/C4TA04881D
145. Xie, T.: Tunable polymer multi-shape memory effect. Nature. **464**, 267–270 (2010). https://doi.org/10.1038/nature08863
146. Zheng, Y., Shen, J., Guo, S.: Optical, electrical, and magnetic properties of shape-memory polymers, polymer blends, and composites. In: Parameswaranpillai, J., Siengchin, S., George, J., Jose, S. (eds.) Shape Memory Polymers, Blends and Composites. Advanced Structured Materials, vol. 115. Springer, Singapore (2020). https://doi.org/10.1007/978-981-13-8574-2_11
147. Meng, Q., Hu, J.: A review of shape memory polymer composites and blends. Compos. A: Appl. Sci. Manuf. **40**(11), 1661–1672 (2009). https://doi.org/10.1016/j.compositesa.2009.08.011
148. Zheng, X., Zhou, S., Li, X., Weng, J.: Shape memory properties of poly(d,l-lactide)/hydroxyapatite composites. Biomaterials. **27**(24), 4288–4295 (2006). https://doi.org/10.1016/j.biomaterials.2006.03.043
149. Xiao, X., Kong, D., Qiu, X., Zhang, W., Zhang, F., Liu, L., Liu, Y., Zhang, S., Hu, Y., Leng, J.: Shape-memory polymers with adjustable high glass transition temperatures. Macromolecules. **48** (11), 3582–3589 (2015). https://doi.org/10.1021/acs.macromol.5b00654
150. Hearon, K., Gall, K., Ware, T., Maitland, D.J., Bearinger, J.P., Wilson, T.S.: Post-polymerization crosslinked polyurethane shape memory polymers. J. Appl. Polym. Sci. **121**(1), 144–153 (2011). https://doi.org/10.1002/app.33428
151. Behl, M., Ridder, U., Feng, Y., Kelch, S., Lendlein, A.: Shape-memory capability of binary multiblock copolymer blends with hard and switching domains provided by different components. Soft. Matter. **5**, 676–684 (2009). https://doi.org/10.1039/B810583A
152. Qi, X., Wang, Y.: Novel techniques for the preparation of shape-memory Polymers, polymer blends and composites at micro and nanoscales. In: Parameswaranpillai, J., Siengchin, S., George, J., Jose, S. (eds.) Shape Memory Polymers, Blends and Composites. Advanced Structured Materials, vol. 115. Springer, Singapore (2020). https://doi.org/10.1007/978-981-13-8574-2_3
153. Wong, Y., Kong, J., Widjaja, L.K., et al.: Biomedical applications of shape-memory polymers: how practically useful are they? Sci. China Chem. **57**, 476–489 (2014). https://doi.org/10.1007/s11426-013-5061-z
154. Behl, M., Lendlein, A.: Shape-memory polymers. Mater. Today. **10**(4), 20–28 (2007). https://doi.org/10.1016/S1369-7021(07)70047-0
155. Hasan, S.M., Nash, L.D., Maitland, D.J.: Porous shape memory polymers: design and applications. Inc. J. Polym. Sci. Part B: Polym. Phys. **54**, 1300–1318 (2016). https://doi.org/10.1002/polb.23982
156. Pate, S.K., Behera, B., Swain, B., Roshan, R., Sahoo, D., Behera, A.: A review on Ni-Ti alloys for biomedical applications and their biocompatibility. Materials Today: Proceedings. **33** (2020). https://doi.org/10.1016/j.matpr.2020.03.538
157. Santo, L., Quadrini, F., Accettura, A., Villadei, W.: Shape memory composites for self-deployable structures in aerospace applications. Proc. Eng. **88**, 42–47 (2014). https://doi.org/10.1016/j.proeng.2014.11.124
158. Vasista, S., Mierheim, O., Kintscher, M.: Morphing structures, applications of. In: Altenbach, H., Öchsner, A. (eds.) Encyclopedia of Continuum Mechanics. Springer, Berlin, Heidelberg (2020). https://doi.org/10.1007/978-3-662-55771-6_247
159. Lewis, C.L., Dell, E.M.: A review of shape memory polymers bearing reversible binding groups. Shaping Shape-Memory. **54** (14), 1340–1364 (2016). https://doi.org/10.1002/polb.23994
160. Zheng, M., Kien, V.K., Richard, L.J.Y.: Aircraft morphing wing concepts with radical geometry change. IES J. Part A: Civil Struct. Eng. **3**, 3188–3195 (2010). https://doi.org/10.1080/19373261003607972
161. Thakur, S.: Shape Memory Polymers for Smart Textile Applications. IntechOpen, London (2017). https://doi.org/10.5772/intechopen.69742
162. Anis, A., Faiz, S., Luqman, M., Poulose, A.M., Gulrez, S.K.H., Shaikh, H.: Developments in shape memory polymeric materials. Polym. Plast. Technol. Eng. **52**, 15 (2013). https://doi.org/10.1080/03602559.2013.824466
163. Xin, X., Liu, L., Liu, Y., et al.: Mechanical models, structures, and applications of shape-memory polymers and their composites. Acta Mech. Solida Sin. **32**, 535–565 (2019). https://doi.org/10.1007/s10338-019-00103-9
164. Mauldin, T.C., Kessler, M.R.: Self-healing polymers and composites. Int. Mater. Rev. **55**(6), 317–346 (2010). https://doi.org/10.1179/095066010X12646898728408
165. Sun, L., Wang, T.X., Chen, H.M., et al.: A brief review of the shape memory phenomena in polymers and their typical sensor applications. Polymers (Basel). **11**(6), 1049. Published 2019 Jun 15 (2019). https://doi.org/10.3390/polym11061049
166. Moirangthem, M., Engels, T.A.P., Murphy, J., Bastiaansen, C.W.M., Schenning, A.P.H.J.: Photonic shape memory polymer with stable multiple colors. ACS Appl. Mater. Interf. **9**(37), 32161–32167 (2017). https://doi.org/10.1021/acsami.7b10198
167. Pilate, F., Toncheva, A., Dubois, P., Raquez, J.-M.: Shape-memory polymers for multiple applications in the materials world. Eur. Polym. J. **80**, 268–294 (2016). https://doi.org/10.1016/j.eurpolymj.2016.05.004
168. Li, J., Duan, Q., Zhang, E., Wang, J.: Applications of shape memory Polymers in kinetic buildings. Adv. Mater. Sci. Eng. **13**, 7453698 (2018). https://doi.org/10.1155/2018/7453698
169. Chakraborty, J.N., Dhaka, P.K., Sethi, A.V., Arif, M.: Technology and application of shape memory polymers in textiles. Res. J. Text. Appar. **21**(2), 86–100 (2017). https://doi.org/10.1108/RJTA-12-2016-0034

170. Rousseau, I.A.: Challenges of shape memory polymers: a review of the progress toward overcoming SMP's limitations. Polym. Eng. Sci. **48**(11), 2075–2089 (2008). https://doi.org/10.1002/pen.21213
171. Liu, T., Zhou, T., Yao, Y., Zhang, F., Liu, L., Liu, Y., Leng, J.: Stimulus methods of multi-functional shape memory polymer nanocomposites: a review. Comp. Part A: Appl. Sci. Manufact. **100**, 20–30 (2017). https://doi.org/10.1016/j.compositesa.2017.04.022
172. Lai, A., Du, Z., Gan, C.L., Schuh, C.A.: Shape memory and superelastic ceramics at small scales. Science. **341**(6153), 1505–1508 (2013). https://doi.org/10.1126/science.1239745
173. Uchino, K.: Antiferroelectric shape memory ceramics. Actuators. **5**(2), 11 (2016). https://doi.org/10.3390/act5020011
174. Schurch, K., Ashbee, K.: A near perfect shape-memory ceramic material. Nature. **266**, 706–707 (1977). https://doi.org/10.1038/266706a0
175. Zeng, X., Zehui, D., Schuh, C.A., Gan, C.L.: MRS Communications. **7**(4), 747–754 (2017). https://doi.org/10.1557/mrc.2017.99
176. Du, Z., Zhou, X., Ye, P., Zeng, X., Gan, C.L.: Shape-memory actuation in aligned zirconia nanofibers for artificial muscle applications at elevated temperatures. ACS Appl. Nano Mater. **3**(3), 2156–2166 (2020). https://doi.org/10.1021/acsanm.9b02073
177. Zehui, D., Zeng, X.M., Liu, Q., Schuh, C.A., Gan, C.L.: Superelasticity in micro-scale shape memory ceramic particles. Acta Mater. **123**, 255–263 (2017). https://doi.org/10.1016/j.actamat.2016.10.047
178. Zeng, X.: Development and Characterization of Shape Memory Ceramics at Micro/Nanoscale. Master's thesis, Nanyang Technological University, Singapore (2017) https://dr.ntu.edu.sg/handle/10356/69895
179. Crystal, I.R., Lai, A., Schuh, C.A.: Cyclic martensitic transformations and damage evolution in shape memory zirconia: single crystals vs polycrystals. J. Am. Ceramic Soc. **103**(8), 4678–4690 (2020). https://doi.org/10.1111/jace.17117
180. Lai, A.: Shape Memory Ceramics in Small Volumes. Department of Materials Science and Engineering, Massachusetts Institute of Technology, Massachusetts (2016) http://dspace.mit.edu/handle/1721.1/7582
181. Pang, E.L., McCandler, C.A., Schuh, C.A.: Reduced cracking in polycrystalline ZrO_2-CeO_2 shape-memory ceramics by meeting the cofactor conditions. Acta Mater. **177**, 230–239 (2019). https://doi.org/10.1016/j.actamat.2019.07.028
182. Wilkes, K.E., Liaw, P.K., Wilkes, K.E.: The fatigue behavior of shape-memory alloys. JOM. **52**, 45–51 (2000). https://doi.org/10.1007/s11837-000-0083-3
183. Zhao, X., Lai, A., Schuh, C.A.: Shape memory zirconia foams through ice templating. Scripta Mater. **135**, 50–53 (2017). https://doi.org/10.1016/j.scriptamat.2017.03.032
184. Zeng, X.M., Du, Z., Schuh, C.A., Tamura, N., Gan, C.L.: Microstructure, crystallization and shape memory behavior of titania and yttria co-doped zirconia. J. Eur. Ceram. Soc. **36**(5), 1277–1283 (2016). https://doi.org/10.1016/j.jeurceramsoc.2015.11.042
185. Zeng, X.M., Du, Z., Tamura, N., Liu, Q., Schuh, C.A., Gan, C.L.: In-situ studies on martensitic transformation and high-temperature shape memory in small volume zirconia. Acta Mater. **134**, 257–266 (2017). https://doi.org/10.1016/j.actamat.2017.06.006
186. Li Jian, C., Wayman, M.: Domain boundary and domain switching in a ceramic rare-earth orthoniobate $LaNbO_4$. J. Am. Ceramic. Soc. **79**(6), 1642–1648 (1996). https://doi.org/10.1111/j.1151-2916.1996.tb08776.x
187. Kondrat'eva, O.N., Nikiforova, G.E., Tyurin, A.V., Khoroshilov, A.V., Gurevich, V.M., Gavrichev, K.S.: Thermodynamic properties of, and fergusonite-to-scheelite phase transition in, gadolinium orthoniobate GdNbO4 ceramics. J. Alloys Compd. **779**, 660–666 (2019). https://doi.org/10.1016/j.jallcom.2018.11.272
188. Stefanowicz, Ł., Adamski, M., Wiśniewski, R., Lipiński, J.: Application of hypergraphs to SMCs selection. In: Camarinha-Matos, L.M., Barrento, N.S., Mendonça, R. (eds.) Technological Innovation for Collective Awareness Systems. DoCEIS 2014 IFIP Advances in Information and Communication Technology, vol. 423. Springer, Berlin, Heidelberg (2014). https://doi.org/10.1007/978-3-642-54734-8_28
189. Lu, H., Huang, W.M., Ding, Z., Wang, C.C., Cui, H.P., Tang, C., Wei, J., Zhao, Y., Song, C.L.: 10 - Rubber-like polymeric shape memory hybrids with repeatable heat-assisted, self-healing, and joule heating functions. In: Li, G., Meng, H. (eds.) Woodhead Publishing Series in Composites Science and Engineering, Recent Advances in Smart Self-Healing Polymers and Composites, pp. 263–292. Woodhead Publishing, Sawston (2015). https://doi.org/10.1016/B978-1-78242-280-8.00010-8
190. Wang, T.X., Renata, C., Chen, H.M., Huang, W.M.: Elastic shape memory hybrids programmable at around body-temperature for comfort fitting. Polymers. **9**(12), 674 (2017). https://doi.org/10.3390/polym9120674
191. Fan, K., Huang, W.M., Wang, C.C., Ding, Z., Zhao, Y., Purnawali, H., Liew, K.C., Zheng, L.X.: Water-responsive shape memory hybrid: design concept and demonstration. Exp. Polym. Lett. **5**(5), 409–416 (2011). https://doi.org/10.3144/expresspolymlett.2011.40
192. Yang, D.: Shape memory alloy and smart hybrid composites — advanced materials for the 21st century. Mater. Des. **21**(6), 503–505 (2000). https://doi.org/10.1016/S0261-3069(00)00008-X
193. Kode, V.R.C., Cavusoglu, M.C.: Design and characterization of a Novel hybrid actuator using shape memory alloy and DC micromotor for minimally invasive surgery applications. IEEE/ASME Trans. Mechatron. **12**(4), 455–464 (2007). https://doi.org/10.1109/TMECH.2007.901940
194. Ran, Y., Yang, X., Zhang, Y., Zhao, X., Wu, X., Zhao, T., Zhao, Y., Huang, W.: Three-dimensional printing of shape memory composites with epoxy-acrylate hybrid photopolymer. ACS Appl. Mater. Interf. **9**(2), 1820–1829 (2016). https://doi.org/10.1021/acsami.6b13531
195. Wei, Z.G., Sandstrom, R., Miyazaki, S.: Shape memory materials and hybrid composites for smart systems: part II shape-memory hybrid composites. J. Mater. Sci. **33**, 3763–3783 (1998). https://doi.org/10.1023/A:1004674630156
196. Turner, T.L., Lach, C.L., Cano, R.J.: Fabrication and characterization of SMA hybrid composites. In: Proceedings Volume 4333, Smart Structures and Materials 2001: Active Materials: Behavior and Mechanics. SPIE, Newport Beach, CA (2001). https://doi.org/10.1117/12.432774
197. Faiella, G., Antonucci, V., Daghia, F., Fascia, S., Giordano, M.: Fabrication and thermo-mechanical characterization of a shape memory alloy hybrid composite. J. Intell. Mater. Syst. Struct. **22**(3), 245–252 (2011). https://doi.org/10.1177/1045389X10396954
198. Wang, C.C., Huang, W.M., Ding, Z., Zhao, Y., Purnawali, H.: Cooling-/water-responsive shape memory hybrids. Compos. Sci. Technol. **72**(10), 1178–1182 (2012). https://doi.org/10.1016/j.compscitech.2012.03.027

2 Piezoelectric Materials

Abstract

This chapter describes the historical background, basic phenomenon associated with the piezoelectric materials. Both direct and inverse piezoelectric effect has been explained with their mechanisms. Various piezoelectric operational constants and parameters have been discussed. Various materials such as ceramic, polymer, metal, and composites in the bulk form as well as in the thin film have been explained with their basic flow diagram. The basic difference between the electrostrictive material which is a material nearly similar to piezoelectric materials has been explained. Various piezoelectric material properties and potential applications in aero-industries, marine industries, automobiles, biomedical, energy harvest, household, and other applications have been described. Various advantages and limitations of piezoelectric materials have been given briefly.

Keywords

Direct piezoelectric effect · Inverse piezoelectric effect · Dielectric · Ceramics piezoelectric · Materials · Polymer piezoelectric materials · Composite piezoelectric materials · Single crystal piezoelectric

2.1 Introduction

Piezoelectric materials are one of the advanced materials that are widespread around us, although they are unknown to the public at large. The word "piezo" stems from the ancient Greek word piezein, which means "to press" or "to squeeze". "Piezoelectricity" thus literally means "pressure-induced electricity". Mobile phones, automotive electronics, medical technology, and industrial systems are only a few areas where piezoelectric components are indispensable. Echoes to capture the image of an unborn baby in a womb make use of piezoelectricity [1]. Even in a parking sensor at the back of our car, the piezoelectric material is present. Piezoelectric materials constitute various types of ceramics, polymers, crystals, and composites that can generate a voltage when being subjected to external pressure or, conversely, expand upon the application of a voltage. A piezoelectric material's response to mechanical forces/pressures resulting in the generation of electric charges/voltages is referred to as the direct piezoelectric effect [2, 3]. In contrast, the application of electric charges/fields causing the induction of mechanical stresses or strains is termed the inverse piezoelectric effect (Fig. 2.1). Asia-Pacific leads the global market in terms of demand for piezoelectric materials, which is likely to be sustained by burgeoning auto and consumer electronics industries. The use of piezoelectric materials in the automotive sector has grown by leaps and bounds and they are finding application in fuel injectors, tire pressure sensors, engine knock sensors, backup sensors, dynamic pressure sensors, and several other crucial safety features common in modern cars [4, 5]. In order for a material to show piezoelectric behavior, the material should be dielectric, that is, non-conducting, and the structure should lack a center of symmetry. Figure 2.2 illustrated the ionic arrangement of quartz (silicon dioxide, SiO_2) that shows a simplified view of the crystal lattice, that is, the arrangement of the positively charged silicon ions and the negatively charged oxygen ions. Upon application of a compressive force, the centers of positive and negative charge shift with respect to each other, and an electric dipole results within the material. Figure 2.3 shows the energy generation and storage by the piezoelectric material.

2.2 History of Piezoelectric

In 1880, while working with Rochelle salt and Quartz, Curie brothers discovered that some electrical charge is developed with the application of pressure [6]. Later, in 1881, from fundamental mathematical models and fundamental

A. Behera, *Advanced Materials*, https://doi.org/10.1007/978-3-030-80359-9_2

Fig. 2.1 Input/output of piezoelectric materials

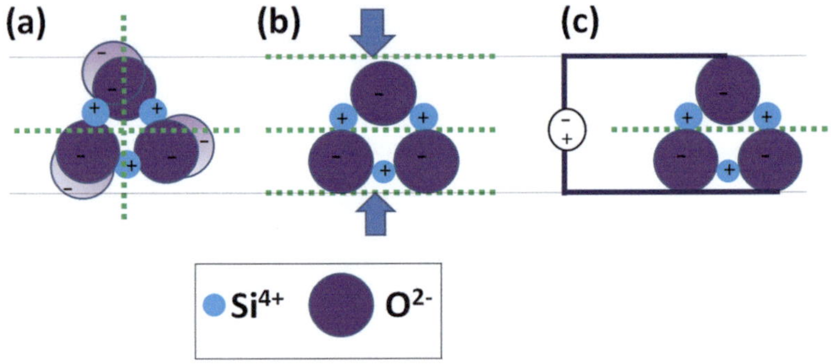

Fig. 2.2 (**a**) Arrangement of silicon and oxygen ion in quartz, (**b**) Direct piezoelectric effect, and (**c**) Inverse piezoelectric effect

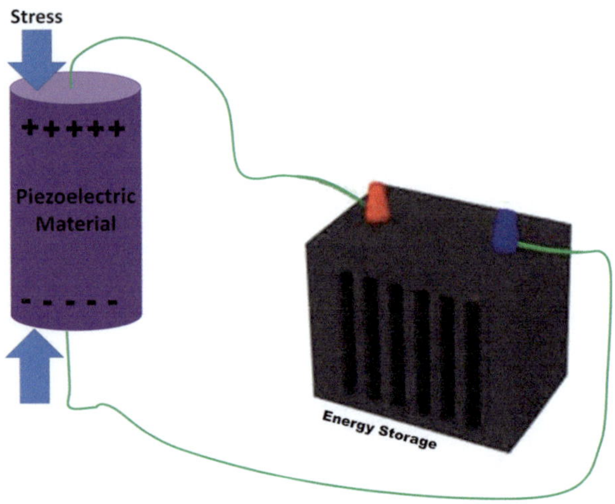

Fig. 2.3 Energy storage by piezoelectric material by stress

thermodynamic principles, Gabriel Lippman concluded that while applying an electric field to these materials, an internal mechanical strain is generated [7]. Curies experimented on these findings. This emerged an interest in the European community and later 1910, it got a place in the textbooks on Crystal Physics [8]. Some countries during World War II discovered ferroelectric materials with higher values of piezoelectric constants. Quartz has always been the favorite piezoelectric material but dedicated researchers developed lead zirconate titanate and barium titanate, which have specific application-based properties. The first experimental demonstration of a connection between macroscopic piezoelectric phenomena and crystallographic structure was published in 1880 by Pierre and Jacques Curie [9]. Their experiment consisted of a conclusive measurement of surface charges appearing on specially prepared crystals (such as tourmaline, quartz, topaz, cane sugar, and Rochelle salt), which were subjected to mechanical stress. These results were a credit to the Curies' imagination and perseverance. In the scientific circles of the day, this effect was considered quite a "discovery," and was quickly dubbed as "piezoelectricity" to distinguish it from other areas of scientific phenomenological experience such as "contact electricity" (friction generated static electricity) and "pyroelectricity" (electricity generated from crystals by heating) [10]. The Curie brothers asserted, however, that there was a one-to-one correspondence between the electrical effects of temperature change and mechanical stress in a given crystal, and that they had used this correspondence not only to pick the crystals for the experiment but also to determine the cuts of those crystals. To them, their demonstration was a confirmation of predictions that followed naturally from their understanding of the microscopic crystallographic origins of pyroelectricity (i.e., from certain crystal asymmetries). The Curie brothers did not, however, predict that crystals exhibiting the direct piezoelectric effect (electricity from applied stress) would also exhibit the converse piezoelectric effect (stress in response to the applied electric field). The discovery of the direct piezoelectric effect is, therefore, credited to the Curie brothers. They did not, however, discover the inverse piezoelectric effect. Rather, it was mathematically predicted from fundamental laws of thermodynamics by Lippmann in 1881 [11]. The Curie brothers are recognized for experimental and immediately confirmed the existence of the inverse effect, and continued to obtain quantitative proof of the complete reversibility of electro-elasto-mechanical deformations in piezoelectric crystals. At this point, after only two years of interactive work within the European scientific community, the core of piezoelectric applications science was established. The identification of piezoelectric crystals has been done on the basis of asymmetric crystal structure, the reversible exchange of electrical and mechanical energy, and the usefulness of thermodynamics in quantifying complex relationships among mechanical, thermal, and electrical variables. In the following 25 years (leading up to 1910), much more work was done to make this core grow into a versatile and complete framework which defined completely the 20 natural crystal classes in which piezoelectric effects occur. The first serious applications work on piezoelectric devices took place during World War I. In 1917, Paul Langevin and French coworkers began to perfect an ultrasonic submarine detector [12]. The transducer they built was made of a mosaic of thin quartz crystals that was glued between two steel plates in a way that the composite system had a resonance frequency of 50 kHz. The device was used to transmit a high-frequency chirp signal into the water and to measure the depth by timing the return echo. Their invention, however, was not perfected until the end of the war. Following their successful use in sonar transducers, and between the two World Wars, piezoelectric crystals were employed in many applications. In the current era, piezoelectric applications are well recognized in engineering structures.

2.3 Piezoelectric Effect

Both the effects of piezoelectric materials depend on electric polarization as shown in Fig. 2.2, described subsequently:

2.3.1 Direct Piezoelectric Effect

For a *linear, isotropic* dielectric medium, the polarization P is given in Eq. (2.1) [13]:

$$\vec{P} = \varepsilon_0 \chi_e \vec{E} \tag{2.1}$$

where, ε_0 is permittivity of free space ($8.854 \times 10r^{-12}$) [F/m], χ_e is electric susceptibility of the material, and $\vec{E}$ is electric field [V/m] = ration between the voltage or difference in electric potential between the plates and the distance between the conducting plates.

For a linear, isotropic dielectric medium (Eq. (2.1)), this may be condensed to the following relation, describing the electric behavior of an unstressed dielectric material (Eq. (2.2):

The electric displacement,

$$\vec{D} = \varepsilon \vec{E} \tag{2.2}$$

where, ε is (absolute) permittivity of the material [F/m] = (permittivity of free space x dielectric constant).

The mechanical, elastic behavior of a dielectric material, placed in a zero electric field, is given by the following relation:

The strain,

$$\vec{S} = s\vec{T} \tag{2.3}$$

where, $\vec{T}$ is stress, and s is the reciprocal of the modulus of elasticity (or Young's modulus) [N/m^2] of the material.

For piezoelectric materials, the electric and mechanical constitutive Eqs. ((2.2) and ((2.3) are coupled. Application of mechanical stress causes a deformation Eq. ((2.3), but induces electric polarization as well, through the direct piezoelectric effect:

$$\vec{P}_{\text{piezo}} = d\vec{T} \tag{2.4}$$

where, $\vec{P}_{\text{piezo}}$ is mechanically induced polarization (C/m^2) and d is piezoelectric charge constant (C/N = m/V).

If the electric field is kept zero ($\vec{E} = 0$, short-circuited condition), the electric displacement is equal to piezoelectric polarization:

$$\vec{D}_{E=0} = \vec{P}_{\text{piezo}} = d\vec{T} \tag{2.5}$$

In terms of the capacitor, the free charge (or short circuit charge) at the electrodes equals the polarization charge that is induced by the mechanical stress.

For a nonzero electric field, the total electric displacement is given by:

$$\vec{D} = \varepsilon^T \vec{E} + d\vec{T} \tag{2.6}$$

Here the superscript T for the permittivity denotes the mechanical boundary condition:

ε^T= permittivity under constant stress $\vec{T}$.

2.3.2 Inverse Piezoelectric Effect

Inversely, the application of an electric field to a piezoelectric material causes an electric displacement Eq. ((2.2), but the associated mechanical stress within the material induces a deformation as well, that is linearly related to the applied electric field. For zero external mechanical stress ($\vec{T} = 0$, unloaded condition), the inverse piezoelectric effect is described by:

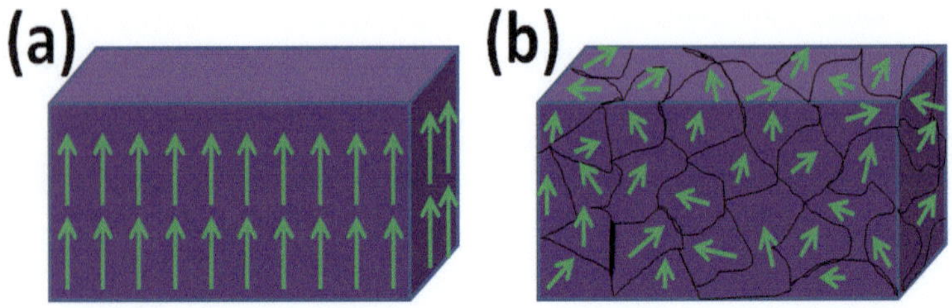

Fig. 2.4 Orientation of poles in (**a**) monocrystalline showing one-directional dipole and (**b**) polycrystalline showing multidirectional dipole

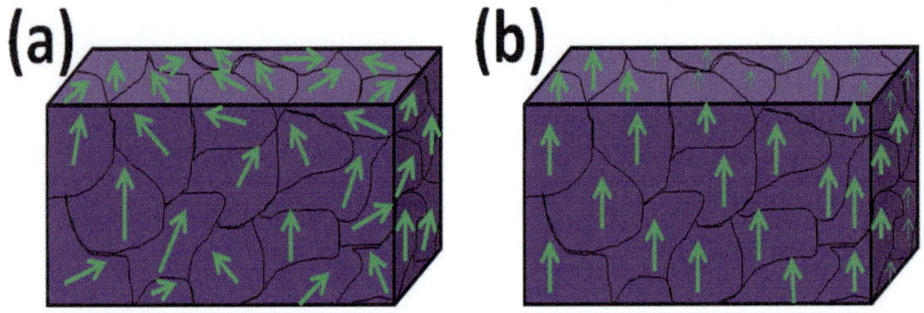

Fig. 2.5 In polycrystalline material, (**a**) the random electric dipole moment before application of stress or electric field and (**b**) uniform dipole moment after application of stress or electric field

$$\vec{S}_{T=0} = d^T\vec{E} \tag{2.7}$$

For nonzero external mechanical stress, the total strain is given by:

$$\vec{S} = s^E\vec{T} + d^T\vec{E} \tag{2.8}$$

Here the superscript E for the compliance denotes the electric boundary condition:

s^E= compliance for constant electric field $\vec{E}$.

In Eq. ((2.7), the same material parameter d as in Eq. ((2.8) appears for describing the electromechanical interaction.

2.4 Mechanism and Working of Piezoelectric Effect

The piezoelectric effect converts electrical field to mechanical strain. In piezoelectric crystals, the unit cells are not symmetrical but are electrical balanced. Piezoelectricity depends on the orientation of dipole density, crystal symmetry, and the applied mechanical stress. When the crystal is squeezed or stretched and if the structure is deformed, pushing some of the atoms closer together or further apart, upsetting the balance of positive and negative, and causing net electrical charges to appear [14]. This effect carries through the whole structure so net positive and negative charges appear on the opposite, outer faces of the crystal as shown in Fig. 2.2. The reverse piezoelectric effect occurs in the opposite way. Put a voltage across a piezoelectric crystal and the atoms are subjected to electrical pressure inside it. They have to move to rebalance themselves and that's what causes piezoelectric crystals to deform (slightly change shape) when the voltage across them is applied.

In monocrystals, the polar axes of all of the dipoles are aligned in one direction. They demonstrate symmetry even if the crystal is cut into pieces. However, in polycrystalline, there are different regions within the material that have a different polar axis and there is no net polarization within the crystal. This difference has been shown in Fig. 2.4. The nature of the piezoelectric effect is closely related to the occurrence of *electric dipole moments* in solids. The *dipole density or polarization* (dimensionality [Cm/m^3]) may easily be calculated for crystals by summing up the dipole moments per volume of the crystallographic unit cell [15]. The piezoelectric effect is the change of polarization when applying mechanical stress. This might either be caused by a reconfiguration of the dipole-inducing surrounding or by reorientation of molecular dipole moments under the influence of the external stress (Fig. 2.5).

Input mechanical energy may have various origins, such as shocks or vibrations, with various frequency spectrums. This mechanical energy is transmitted to the piezoelectric material through an important element of the device, so-called "mechanical structure," which may act as a band-pass filter (in steady state operation), but also as intermediate mechanical energy storage tank (in impulse mode of operation) [16]. Variations of the strain/stress of the piezoelectric material enable conversion from mechanical to electrical energy as shown in Fig. 2.6. It must be noted that resulting AC piezoelectric voltage and current are neither suitable for energy storage devices nor small electronic or electrical devices. These devices actually require DC voltage. So, another element must be included in the energy conversion chain that is the electrical interface, whose basic function is AC to DC conversion of electrical energy. In addition, this

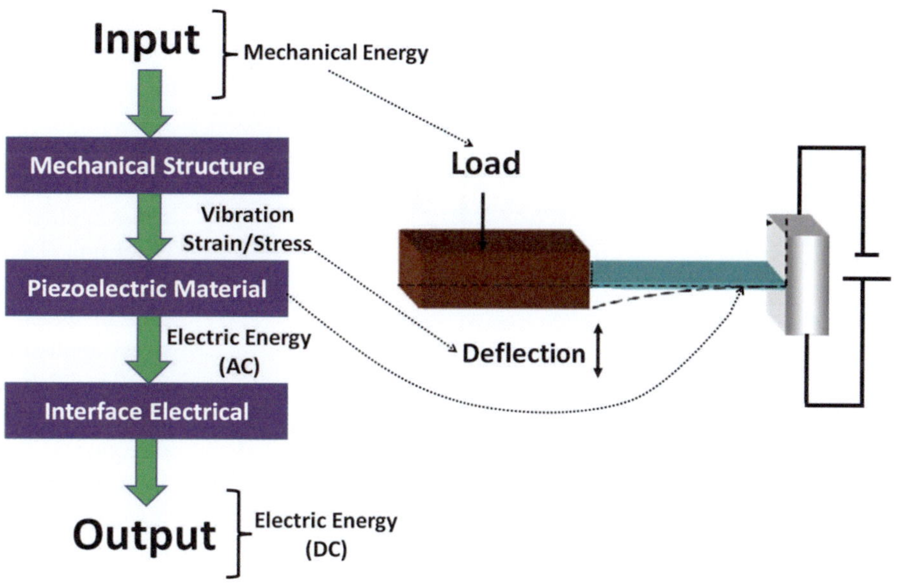

Fig. 2.6 Energy conversion flow from input to output in a direct piezoelectric material

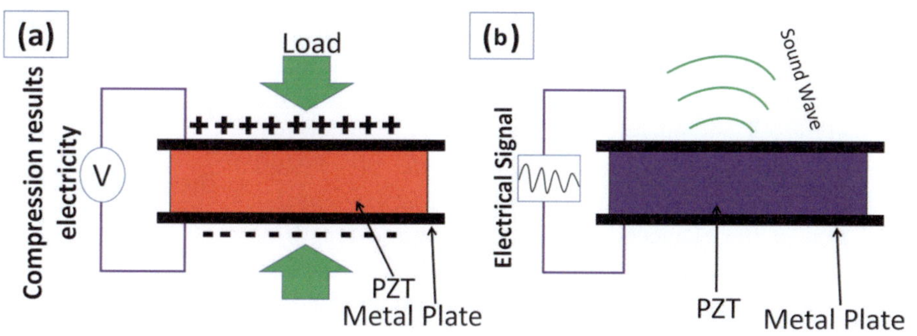

Fig. 2.7 Schematics of application of (**a**) direct and (**b**) inverse piezoelectric effect

electrical interface may ensure the function of output voltage regulation. But the most profitable effect which may be induced by this last interface is overall optimization and improvement of the electromechanical energy conversion [17].

Indirect piezoelectric effect (Fig. 2.7a), the piezoelectric crystal is placed between two metal plates. At this point, the material is in perfect balance and does not conduct an electric current. Mechanical pressure is then applied to the material by the metal plates, which forces the electric charges within the crystal out of balance. Excess negative and positive charges appear on opposite sides of the crystal face. The metal plate collects these charges, which can be used to produce a voltage and send an electrical current through a circuit [18]. In inverse piezoelectric effect, the piezoelectric crystal is placed between two metal plates. The crystal's structure is in perfect balance. Electrical energy is then applied to the crystal, which shrinks and expands the crystal dimension [19]. As the crystal dimension expands and contracts, it converts the received electrical energy and releases mechanical energy in the form of a sound wave as in Fig. 2.7b. The inverse piezoelectric effect is used in a variety of applications. Take a speaker for example, which applies a voltage to a piezoelectric ceramic, causing the material to vibrate the air as sound waves.

2.5 Various Piezoelectric Constants

The constitutive equations for linear piezoelectric materials under low stress (X) are written according to the following formulas, these equations define how the piezoelectric material's stress, strain, charge density displacement, and electric field interact [20]:

$$S = sX + dE \tag{2.9}$$

$$D = \varepsilon X + dX \tag{2.10}$$

where, S is the strain, s the elastic compliance, E the electric field, D the dielectric displacement, ε permittivity. The sX and εX of the equations apply to all the materials, which strain is in relation to elastic compliance and stress. The

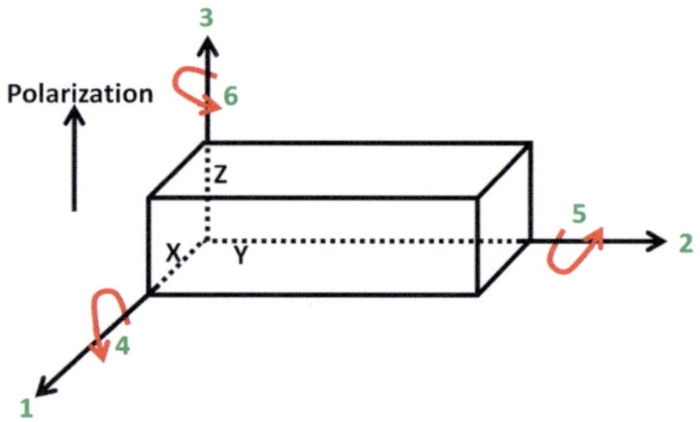

Fig. 2.8 Direction of forces affecting a piezoelectric element. *The direction of positive polarization usually is made to coincide with the Z-axis*

second part of the equations containing piezoelectric charge constant (d) are only properties of piezoelectric material. These equations are important when it comes to making piezoelectric material.

Because a piezoelectric ceramic is anisotropic, physical constants relate to both the direction of the applied mechanical or electric force and the directions perpendicular to the applied force. Consequently, each constant generally has two subscripts that indicate the directions of the two related quantities, such as stress (force on the ceramic element/surface area of the element) and strain (change in length of element/original length of element) for elasticity. The direction of positive polarization usually is made to coincide with the Z-axis of a rectangular system of X, Y, and Z axes (Fig. 2.8). Direction X, Y, or Z is represented by the subscript 1, 2, or 3, respectively, and shear about one of these axes is represented by the subscript 4, 5, or 6, respectively [25, 26]. Definitions of the most frequently used constants, and equations for determining and interrelating these constants, are summarized here. The *piezoelectric charge constant, d,* the *piezoelectric voltage constant, g,* and the *permittivity, e,* are temperature-dependent factors.

2.6 Piezoelectric Charge Constant

The *piezoelectric charge constant, d,* is defined as the electric polarization generated in material per unit of mechanical stress (T) applied to a piezoelectric material or, alternatively, is the mechanical strain (S) experienced by a piezoelectric material per unit of an electric field applied. As per the below description, the first subscript to d indicates the direction of polarization generated in the material when the electric field, $E = 0$ or, alternatively, is the direction of the applied field strength. The second subscript is the direction of the applied stress or the induced strain, respectively. Because the strain induced in a piezoelectric material by an applied electric field is the product of the value for the electric field and the value for d, d is an important indicator of a material's suitability for strain-dependent (actuator) applications. The nonzero piezoelectric constants are [21]:

$\mathbf{d_{33}}$: Induced polarization in direction 3 (parallel to the direction in which ceramic element is polarized) per unit stress applied in direction 3 **or** induced strain in direction 3 per unit electric field applied in direction 3

$\mathbf{d_{31}}$: Induced polarization in direction 3 (parallel to the direction in which ceramic element is polarized) per unit stress applied in direction 1 (perpendicular to the direction in which ceramic element is polarized) **or** induced strain in direction 1 per unit electric field applied in direction 3

$\mathbf{d_{15}}$: Induced polarization in direction 1 (perpendicular to the direction in which ceramic element is polarized) per unit shear stress applied about direction 2 (direction 2 perpendicular to the direction in which ceramic element is polarized) **or** induced shear strain about direction 2 per unit electric field applied in direction 1

2.7 Piezoelectric Voltage Constant

The piezoelectric voltage constant (g) is the electric field achieved by a piezoelectric ceramic per unit of mechanical stress applied or alternatively, is the mechanical strain experienced by a piezoelectric material per unit of electric displacement applied. For the below description, the first subscript to (g) shows the direction of the electric field generated in the material or the direction of the applied electric displacement. The second subscript is the direction of the applied stress or the induced strain, respectively [22]. Because the strength of the induced electric field produced by a piezoelectric material in response to applied physical stress is the product of the value for the applied stress and the value for g. The g is important for assessing a material's suitability for sensing (sensor) applications.

$\mathbf{g_{33}}$: Induced electric field in direction 3 (parallel to the direction in which ceramic element is polarized) per unit stress applied in direction 3 **or** induced strain in direction 3 per unit electric displacement applied in direction 3

$\mathbf{g_{31}}$: Induced electric field in direction 3 (parallel to the direction in which ceramic element is polarized) per unit stress applied in direction 1 (perpendicular to the direction in which ceramic element is polarized) **or** induced strain in direction 1 per unit electric displacement applied in direction 3

$\mathbf{g_{15}}$: Induced electric field in direction 1 (perpendicular to the direction in which ceramic element is polarized) per unit shear stress applied about direction 2 (direction 2 perpendicular to the direction in which ceramic element is polarized) **or** induced shear strain about direction 2 per unit electric displacement applied in direction 1

2.8 Permittivity Constant

The *permittivity* or *dielectric constant*, ε, for a piezoelectric ceramic material is the dielectric displacement per unit electric field. ε^T is the permittivity at constant stress, ε^S is the permittivity at constant strain [23]. The first subscript to ε indicates the direction of the dielectric displacement and the second is the direction of the electric field. The relative dielectric constant, K, is the ratio of, the amount of charge that an element constructed from the ceramic material can store, relative to the absolute dielectric constant, the charge that can be stored by the same electrodes when separated by a vacuum, at the equal voltage the value will be 8.85×10^{-12} farad/meter).

$\boldsymbol{\varepsilon^T}_{11}$: Permittivity for dielectric displacement and electric field in direction 1 (perpendicular to the direction in which ceramic element is polarized), under constant stress

$\boldsymbol{\varepsilon^S}_{33}$: Permittivity for dielectric displacement and electric field in direction 3 (parallel to the direction in which ceramic element is polarized), under constant strain

2.9 Elastic and Compliance

Elastic compliance, *s*, is the strain produced in a piezoelectric material per unit of stress applied and, for the 11 and 33 directions, is the reciprocal of the modulus of elasticity (Young's modulus, *Y*). s^D is the compliance under a constant electric displacement; s^E is the compliance under a constant electric field. The first subscript indicates the direction of strain, the second is the direction of stress [24].

$\mathbf{s^E}_{11}$: elastic compliance for stress in direction 1 (perpendicular to the direction in which ceramic element is polarized) and accompanying strain in direction 1, under constant electric field (short circuit)

$\mathbf{s^D}_{33}$: elastic compliance for stress in direction 3 (parallel to the direction in which ceramic element is polarized) and accompanying strain in direction 3, under constant electric displacement (open circuit)

2.9.1 Electromechanical Coupling Factor

The *electromechanical coupling factor*, *k*, is an indicator of the effectiveness with which a piezoelectric material converts electrical energy into mechanical energy, or converts mechanical energy into electrical energy [25]. The first subscript to *k* denotes the direction along which the electrodes are applied; the second denotes the direction along which the mechanical energy is applied, or developed. *k* values quoted in ceramic suppliers' specifications typically are theoretical maximum values. At low input frequencies, a typical piezoelectric ceramic can convert 30–75% of the energy delivered to it in one form into the other form, depending on the formulation of the ceramic and the directions of the forces involved. A high *k* usually is desirable for efficient energy conversion, but *k* does not account for dielectric losses or mechanical losses, nor recovery of unconverted energy. The accurate measure of efficiency is the ratio of converted, useable energy delivered by the piezoelectric element to the total energy taken up by the element. By this measure, piezoelectric technology in well-designed systems can exhibit efficiencies that exceed 90% [26]. The dimensions of a ceramic element can dictate unique expressions of *k*. For a thin disc of piezoelectric ceramic the planar coupling factor, k_p, expresses radial coupling- the coupling between an electric field parallel to the direction in which the ceramic element is polarized (direction 3) and mechanical effects that produce radial vibrations, relative to the direction of polarization (direction 1 and direction 2). For a disc or plate of material whose surface dimensions are large relative to its thickness, the thickness coupling factor, k_t, a unique expression of k_{33}, expresses the coupling between an electric field in direction 3 and mechanical vibrations in the same direction. The resonance frequency for the thickness dimension of an element of this shape is much higher than the resonance frequency for the transverse dimensions. At the same time, strongly attenuated transverse vibrations at this higher resonance frequency, a result of the transverse contraction/expansion that accompanies the expansion/contraction in thickness, make k_t lower than k_{33}, the corresponding factor for longitudinal vibrations of a thin rod of the same material, for which a much lower longitudinal resonance frequency more closely matches the transverse resonance frequency.

$\mathbf{k_{33}}$: factor for the electric field in direction 3 (parallel to the direction in which ceramic element is polarized) and longitudinal vibrations in direction 3 (ceramic rod, length > 10× diameter)

$\mathbf{k_t}$: factor for the electric field in direction 3 and vibrations in direction 3 (thin disc, surface dimensions large relative to thickness; $k_t < k_{33}$)

$\mathbf{k_{31}}$: factor for the electric field in direction 3 (parallel to the direction in which ceramic element is polarized) and longitudinal vibrations in direction 1 (perpendicular to the direction in which ceramic element is polarized) (ceramic rod)

$\mathbf{k_p}$: factor for the electric field in direction 3 (parallel to the direction in which ceramic element is polarized) and radial vibrations in direction 1 and direction 2 (both perpendicular to the direction in which ceramic element is polarized) (thin disc)

2.9.2 Young's Modulus

Young's modulus, *Y*, is an indicator of the stiffness (elasticity) of a ceramic material. *Y* is determined from the value for the stress applied to the material divided by the value for the resulting strain in the same direction [27].

2.9.3 Dielectric Dissipation Factor

The δ, tan δ, for a ceramic material is the tangent of the dielectric loss angle. Tan δ is determined by the ratio of effective conductance to effective susceptance in a parallel circuit, measured by using an impedance bridge. Values for tan δ typically are determined at 1 kHz [28].

2.9.4 Piezoelectric Frequency Constant

When an unrestrained piezoelectric ceramic element is exposed to a high frequency alternating electric field, an impedance minimum, the planar or radial resonance frequency, coincides with the series resonance frequency (f_s) [29].

2.10 Materials Used for Piezoelectricity

There are a variety of piezoelectric materials that can conduct an electric current, both man-made and natural. The most well-known, and the first piezoelectric material used in electronic devices is the quartz crystal. Other naturally occurring piezoelectric materials include Rochelle salt, topaz, tourmaline, berlinite, cane sugar, and even bone [30, 31]. Some metallic natural piezoelectric materials are shown in Fig. 2.9. Manmade crystals include Gallium orthophosphate, a colorless crystal with similar properties to quartz, but double the piezoelectric effect, Langasite, a piezoelectric crystal similar to quartz, and so on. Various materials (ceramic, polymer, metal, etc.) used in piezoelectric activities are discussed below:

2.10.1 Ceramics Piezoelectric Materials

To prepare a piezoelectric ceramic, fine lead-zirconate titanate (PZT) powders of the component metal oxides are mixed in specific proportions, then heated to form a uniform powder. The piezo powder is mixed with an organic binder and is formed into structural elements having the desired shape (discs, rods, plates, etc.). Some examples of piezoelectric materials are PZT, barium titanate, and lithium niobate. These man-made materials have a more pronounced effect than quartz and other natural piezoelectric materials. For example, PZT is more effective than quartz. Ceramics with

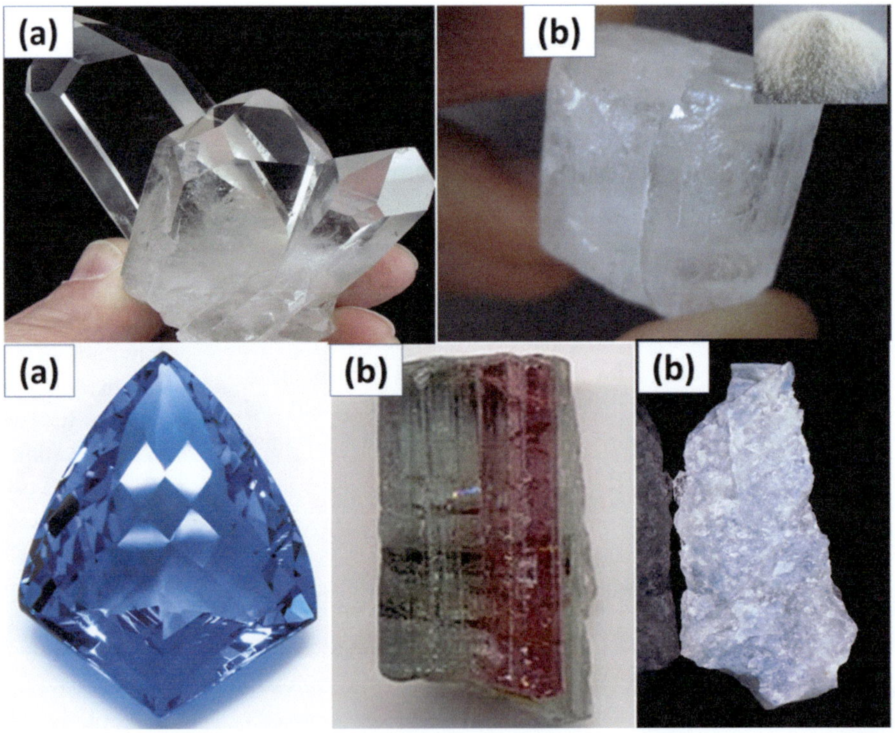

Fig. 2.9 Natural piezoelectric materials: (**a**) Quartz crystal, (**b**) Rochelle salt, (**c**) Topaz, (**d**) Tourmaline, and (**e**) Berlinite.

randomly oriented grains must be ferroelectric to exhibit piezoelectricity. The macroscopic piezoelectricity is possible in textured polycrystalline non-ferroelectric piezoelectric materials, such as AlN and ZnO. The family of ceramics with perovskite, tungsten-bronze, and related structures exhibits piezoelectricity are: Barium titanate ($BaTiO_3$), Lead zirconate titanate ($Pb[Zr_xTi_{1-x}]O_3$ with $0 \leq x \leq 1$), Potassium niobate ($KNbO_3$), Sodium tungstate (Na_2WO_3), $Ba_2NaNb_5O_5$, $Pb_2KNb_5O_{15}$, Zinc oxide (ZnO), Lead titanate ($PbTiO_3$), Lithium niobate ($LiNbO_3$), and Lithium tantalite ($LiTaO_3$) [41–43]. Barium titanate was the first piezoelectric ceramic discovered. PZT is the most common piezoelectric ceramic in use today. Zinc oxide structure is Wurtzite, while single crystals of ZnO are piezoelectric and pyroelectric, polycrystalline (ceramic) ZnO with randomly oriented grains exhibits neither piezoelectric nor pyroelectric effect. Not being ferroelectric, polycrystalline ZnO cannot be poled like barium titanate or PZT. Ceramics and polycrystalline thin films of ZnO may exhibit macroscopic piezoelectricity and pyroelectricity only if they are textured (grains are preferentially oriented), such that the piezoelectric and pyroelectric responses of all individual grains do not cancel. This is readily accomplished in polycrystalline thin films [32].

A piezoelectric ceramic is a mass of perovskite crystals. Each crystal is composed of a small, tetravalent metal ion placed inside a lattice of larger divalent metal ions and O_2, as shown in Fig. 2.2. To prepare a piezoelectric ceramic, fine powders of the component metaloxides are mixed in specific proportions. This mixture is then heated to form a uniform powder. The powder is then mixed with an organic binder and is formed into specific shapes, for example, discs, rods, plates, and so on. These elements are then heated for a specific time and under a predetermined temperature. As a result of this process, the powder particles sinter and the material form a dense crystalline structure. The elements are then cooled and, if needed, trimmed into specific shapes. Finally, electrodes are applied to the appropriate surfaces of the structure. Above a critical temperature, known as the Curie temperature, each perovskite crystal in the heated ceramic element exhibits a simple cubic symmetry with no dipole moment. However, at temperatures below the Curie temperature, each crystal has tetragonal symmetry and, associated with a dipole moment. Adjoining dipoles form regions of local alignment called domains. This alignment gives a net dipole moment to the domain, and thus a net polarization [33]. As demonstrated in Fig. 2.2, the direction of polarization among neighboring domains is random. Subsequently, the ceramic element has no overall polarization. The domains in a ceramic element are aligned by exposing the element to a strong DC electric field, usually at a temperature slightly below the Curie temperature. This is referred to as the poling process. After the poling treatment, domains most nearly aligned with the electric field expand at the expense of domains that are not aligned with the field, and the element expands in the direction of the field. When the electric field is removed most of the dipoles are locked into a configuration of near alignment. The element now has a permanent polarization, the remnant polarization, and is permanently elongated. The increase in the length of the element, however, is very small, usually within the micrometer range. The mechanical compression or tension on the element changes the dipole moment associated with that element. This creates a voltage. Compression along the direction of polarization, or tension perpendicular to the direction of polarization, generates a voltage of the same polarity as the poling voltage. Tension along the direction of polarization, or compression perpendicular to that direction, generates a voltage with polarity opposite to that of the poling voltage. When operating in this mode, the device is being used as a sensor. That is, the ceramic element converts the mechanical energy of compression or tension into electrical energy [34].

2.10.2 Polymer Piezoelectric Materials

Some polymer expands or contracts in an electrical field or a property of a certain *polymer* that generates an electrical charge when pressure is applied. PVDF is a chain of CH_2CF_2 and is a semicrystalline *polymer* [35]. It is produced in large thin clear sheets, which are then stretched and poled to give it the piezoelectric properties. The stretch direction is the direction along the sheet in which most of the carbon chains run and is visible to the naked eye when the sheet is held up to the light. The poled direction is either to the top or bottom of the sheet. The hydrogen atoms, which have a net positive charge, and the fluorine atoms, which have a net negative charge end up on opposite sides of the sheet. This creates a pole direction and is either directed to the top or bottom of the sheet. The PVDF currently used is 32 μm thick and comes in long rolls 16 inches wide from measurement specialties Inc. The dielectric constant is around 10. The piezo-response of polymers is not as high as the response for ceramics; however, polymers hold properties that ceramics do not. Over the last few decades, nontoxic, piezoelectric polymers have been studied and applied due to their flexibility and smaller acoustical impedance. Other properties that make these materials significant include their biocompatibility, biodegradability, low cost, and low power consumption compared to other piezo-materials (ceramics, etc.) [36]. Piezoelectric polymers and nontoxic polymer composites can be used for their different physical properties.

Piezoelectric polymers can be classified by bulk polymers, voided charged polymers, and polymer composites. A piezoresponse observed by bulk polymers is mostly due to its molecular structure. Bulk polymers are solid polymer films and they have the piezoelectric effect due to their structural orientation. There are two types of bulk polymers: amorphous and semicrystalline. Examples of semicrystalline polymers are Polyvinylidene Fluoride (PVDF) and its copolymers, Polyamides, and Paralyne-C. PVDF is the high-energy density material among piezoelectric polymers. It contains about 50% crystals that are embedded in an amorphous matrix. Besides their high-energy density, they are easy to deform under mechanical shock, which makes them resilient and suitable for curved surfaces. PVDF exhibits piezoelectricity several times greater than quartz. The piezo-response observed from PVDF is about 20–30 pC/N. That is an order of 5–50 times less than that of piezoelectric ceramic lead zirconate titanate (PZT). The thermal stability of the piezoelectric effect of polymers in the PVDF family (i.e., vinylidene fluoride co-poly trifluoroethylene) goes up to 125 °C [37]. Some applications of PVDF are pressure sensors, hydrophones, and shock wave sensors. Due to their flexibility, piezoelectric composites have been proposed as energy harvesters and nanogenerators [38]. Non-crystalline polymers, such as Polyimide and Polyvinylidene Chloride (PVDC), fall under amorphous bulk polymers. In amorphous polymers, the polarization is not in a state of thermal equilibrium, but rather a quasi-stable state due to the freezing of molecular dipoles. The result is a piezoelectric-like effect. In amorphous material, piezoelectricity is the result of the orientation polarization of molecular dipoles. Voided charged polymers exhibit the piezoelectric effect due to the charge induced by poling of a porous polymeric film. This structure was first invented by Gerhard Sessler in the early 1960s and is sometimes called cellular polymers. Voided charged polymers are polymer materials that contain internal gas voids. When the surface of the polymer, surrounding the voids are charged which can add extra voltage to the generated voltage. Such structures can have a high piezoelectric coefficient $d33$ which can reach up to 20,000 pC N − 1 in some cases [39]. However, the piezoelectric coefficient in such materials relies on factors that are distinct from the regular piezoelectric materials. Factors related to voids can have crucial effects on the value of a piezoelectric constant. Also, density and shape of the voids can affect the distribution of the final dipoles, which is not desirable in the harvesting system. Pressure of gas inside the voids can also affect the amount of ionization occurring during the poling process. The piezoelectric effect can also be observed in polymer composites by integrating piezoelectric ceramic particles into a polymer film. Composite polymers are a group of polymer materials with embedded inorganic piezoelectric material. One of the importance of combining piezoelectric ceramics with polymers is to achieve the advantages of both materials, which includes the higher coupling factor and dielectric constant from ceramics and the mechanical flexibility from polymers. These types of composite polymers are preferable for acoustic devices because of the polymer's low acoustic impedance and fewer spurious modes.

2.10.3 Composite Piezoelectric Materials

Polymer piezoelectric materials such as PVDF are very suitable for sensor applications. However, because of its small piezoelectric constants and very small elastic stiffness, PVDF cannot be used by itself in fabricating actuators or high-power transducers. Piezocomposites composed of a piezoelectric ceramic and a polymer are promising materials because of their excellent tailorable properties. The geometry for two-phase composites can be classified according to the connectivity of each phase (1-, 2-, or 3-dimensionally) into 10 structures. In particular, a 1–3 piezocomposite (PZT-rod/polymer-matrix composite) is considered most useful [40]. The advantages of this composite are high coupling factors, low acoustic impedance, good matching human tissue, mechanical flexibility, and broad bandwidth in combination with a low mechanical quality factor. Piezoelectric composite materials are especially useful for underwater sonar and medical diagnostic ultrasonic transducer applications. Piezoelectric passive dampers comprise a piezoelectric ceramic particle, polymer, and carbon black, which suppress the noise vibration more effectively than traditional rubbers. A composite with a magnetostrictive ceramic and a piezoelectric ceramic produces an intriguing product effect-the magnetoelectric effect in which an electric field is produced in the material in response to an applied magnetic field [41].

2.10.4 Single Crystal Piezoelectric

Single crystals of natural or man-made materials exhibit the desirable piezoelectric properties that might be offered by a polycrystalline ceramic element if all of its domains were perfectly aligned. An expanding variety of single crystals is being developed for acoustical, optical, wireless communication, and other applications. Materials used to fabricate single-crystal piezoelectric elements include lead magnesium niobate/lead titanate (PMN-PT), lead zirconateniobate/lead titanate (PZN-PT), lithium niobate ($LiNbO_3$), lithium niobate with dopants, lithium tetraborate ($Li_2B_4O_7$), and quartz. Barium titanate ($BaTiO_3$) is a potential non-lead source of piezoelectric crystals for low temperature and room temperature applications [42–44]. Single-crystal PMN-PT and PZN-PT elements exhibit 10 times the strain of comparable polycrystalline lead-zirconate-titanate elements. Applications for single-crystal materials include actuators and diagnostic and therapeutic medical devices. As the name of this group of material suggests, piezoelectric single crystals are counterparts of piezoelectric ceramics. Ferroelectric single crystals such as lead-nickel-niobate [$Pb(Ni_{1/3}Nb_{2/3})O_3$ or PNN], and lead-titanate [PbTiO3] are the two most popular and widely used materials based on their performance in piezoelectric energy harvesting. In these materials, the arrangement of the positive and negative ions is highly ordered, causing an alignment of the dipoles across the entire material, which makes them have a higher piezoelectric strain constant than ceramics. For small size devices, ferroelectric single crystals are favorable, because they have higher Young's moduli than the ceramics which results in the material having a lower resonance frequency [45].

2.10.5 Thin Films Piezoelectric Materials

Thin films of perovskite materials are fabricated by sputtering, pulsed laser ablation, MOCVD, and sol-gel deposition. The ferroelectric properties of these perovskite thin films are different from bulk ceramic properties probably due to the difference in microstructure and the presence of the growth stress in the thin films [46].

2.10.6 Piezoelectric Material Properties

Piezoelectric materials' chemical composition majorly affects the properties of the materials. For example, in comparison to metallic oxide-based piezoelectric material Barium Titanate ($BaTiO_3$), PZT materials exhibit greater sensitivity and have a higher operating temperature. On the basis of chemical composition, electric behavior, dielectric behavior, elasticity behavior, electrochemical behavior, and so on, of the piezoelectric materials has been discussed below.

2.10.7 Electric Behavior

All piezoelectric materials are nonconductive for the piezoelectric effect to occur and work. For a dielectric medium in between the two parallel conducting plates, the relation between the electric field and the distribution of electric charge is somewhat more complicated. This is due to the presence of so-called bound electric charges within the dielectric material, that is, charges that cannot move freely [47]. The application of an electric field results in a redistribution of these bound charges, thus locally generating electric dipoles within the material. On a macroscopic level, the accumulation of dipoles is referred to as polarization (Fig. 2.2). Due to the polarization, some of the free charges will be bound to the dipole of the dielectric medium, leaving less net charge on the conducting plates (Q_{net}):

$$Q_{net} = Q_{free} - Q_{pol} \tag{2.11}$$

where, Q_{free} is the free charge on the conducting plate, related to the electric displacement,

$$\overrightarrow{(D)} = A\left|\vec{D}\right| \tag{2.12}$$

Again

$$Q_{net} = A\varepsilon_0\left|\vec{E}\right| \tag{2.13}$$

Q_{pol} = Charge on the conducting plates that compensates

the polarization of the dielectric medium $= A\left|\vec{P}\right|$ (2.14)

where, $\vec{P}$ = Polarization [c/m^2]

For a linear, isotropic dielectric medium, the polarization is given by:

$$\vec{P} = \varepsilon_0\chi_e\vec{E} \tag{2.15}$$

where, χ_e= Electric susceptibility of the material

By writing Eq. ((2.6) to

$$Q_{free} = Q_{net} + Q_{pol} \tag{2.16}$$

The following relation is found for the electric displacement Eq. ((2.2):

$$\vec{D} = \varepsilon_0\vec{E} + \vec{P} \tag{2.17}$$

For the linear, isotropic dielectric medium Eq. ((2.15), this may be condensed to the following relation, describing the electric behavior of an unstressed dielectric material:

$$\vec{D} = \varepsilon\vec{E} \tag{2.18}$$

where, ε = (absolute) permittivity of the material [F/m] $=\varepsilon_0\varepsilon_r$, ε_r is the dielectric constant (or relative permittivity) of the material,

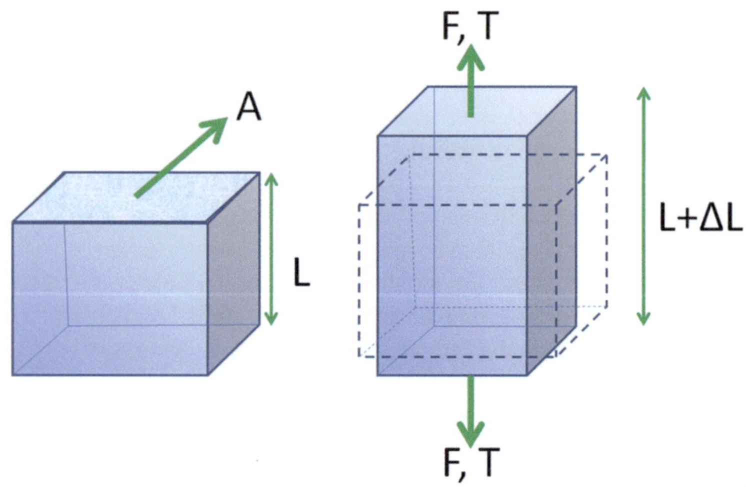

Fig. 2.10 Elastic behavior of piezoelectric materials [69]

$$1 + \chi_e \quad (2.19)$$

For dielectric materials, the latter material parameter is often given in the datasheet.

2.10.8 Dielectric Behavior

A dielectric material is one that is electrically insulating (nonmetallic) and exhibits or may be made to exhibit an electric dipole structure; that is, there is a separation of positive and negative electrically charged entities on a molecular or atomic level. The dielectric materials ordinarily exhibit at least one of the polarization types discussed in the previous section depending on the material and also the manner of the external field application. There are two types of dielectrics. The first type is polar dielectrics, which are dielectrics that have permanent electric dipole moments. Here the orientation of polar molecules is random in the absence of an external field. When an external electric field E is present, a torque is set up and causes the molecules to align with E. However, the alignment is not complete due to random thermal motion. The second type of dielectrics is nonpolar dielectrics, which are dielectrics that do not possess permanent electric dipole moment. Electric dipole moments can be induced by placing the materials in an externally applied electric field [48]. Dielectric materials are electrically insulating, yet susceptible to polarization in the presence of an electric field. This polarization phenomenon accounts for the ability of the dielectrics to increase the charge storing capability of capacitors, the efficiency of which is expressed in terms of a dielectric constant.

2.10.9 Elasticity Behavior

The mechanical, elastic behavior of a dielectric material, placed in a zero electric field, is given by the following relation [49]:

$$\vec{S} = s\vec{T} \quad (2.20)$$

where $\vec{S} =$ strain, that is, the relative deformation= $\Delta l/l$, $\Delta l =$ the absolute deformation [m] and $l =$ the original length of the component [m], $\vec{T} =$ stress [N/m^2], that is, the force per unit area $=F/A$, $F =$ the force applied to the component [N] and A is the cross section of the component [m^2] and s is the compliance of the material (strain per unit stress) [m^2/N]. The compliance s is the reciprocal of the modulus of elasticity (or Young's modulus) c[N/m^2] of the material i.e. C^{-1}.

The elastic behavior is also affected by Poisson's ratio of the materials as illustrated in Fig. 2.10. Upon applying an axial load in a certain direction, the material gets strained along that axis, but also deforms in a perpendicular direction [50]. The ratio of the perpendicular strained (e.g., concentration) to the axial strain (e.g., elongation) is given by Poisson's ratio, a material property that is usually denoted by the Greek symbol v.

2.10.10 Electromechanical Behavior

Before describing the electromechanical interaction in piezoelectric materials, let's first consider electrostatics in free space, taking the example of two conducting plates (Fig. 2.11). When assuming a homogeneous electric displacement field perpendicular to the plates, in free space in between those plates the electric field is related to the distribution of electric charge as follows [51]:

$$\vec{D} = \varepsilon_0 \vec{E} \quad (2.21)$$

where $\vec{D}$ is electric displacement (the electric charge per unit area) [C/m^2], ($|\vec{D}| = Q/A$), and Q is the charge on the conducting plates [C], A is the cross section of the conducting

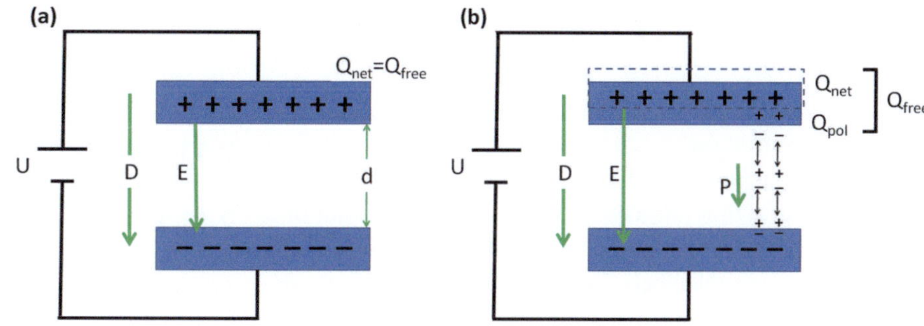

Fig. 2.11 Electric behavior of parallel plate capacitor (**a**) in free space and (**b**) with a dielectric medium

plate [m^2], $\vec{E}$ is electric field [V/m], ($|E^{\rightarrow}| = U/d$), U is the voltage, or difference in electric potential, between the plate [V] and d is the distance between the conducting plate [m], ε_0 is permittivity of free space $= 8.854 \times 10r^{-12}$[F/m].

2.10.11 Coupling Coefficient

Electromechanical coupling coefficient is a numerical measure of the conversion efficiency between electrical and acoustic energy in piezoelectric materials. Qualitatively the electromechanical coupling coefficient (k) can be determined as [52]:

$$k^{-2} = \frac{1}{\text{energy converted per input energy}} \tag{2.22}$$

2.10.12 Material Damping

Piezoelectric materials (piezo-materials) convert mechanical energy into electrical energy via the mechanism of piezoelectricity, and this electrical energy may be dissipated through a connected load resistance. The energy dissipation through this mechanism is termed piezo-damping [53].

2.10.13 Mechanical Loss

Although the total power dissipation density and total power dissipation remain the same for all piezoelectric constitutive forms, the contributing terms vary significantly. There is therefore no means of extracting a unique mechanical, electrical, and piezoelectric contribution with any individual physical meaning from this phenomenological model. Hence, the phase of the material constants does not reflect the energy loss from one specific physical loss mechanism, but rather a complicated combination of a number of these physical loss mechanisms [54].

2.10.14 Sound Velocity

The natural resonant frequency of a material is based on its physical dimensions and the speed of sound in the material (c). The frequency constant (N) is defined as the controlling dimension (t) multiplied by the frequency (F_r) [55]:

$$N = F_r \times t \tag{2.23}$$

The time required to complete a full cycle (i.e., one wavelength) is the period (T), measured in seconds. The relation between frequency and period in a continuous wave is given in Eq. ((2.24).

$$f = \frac{1}{T} \tag{2.24}$$

The velocity of ultrasound (c) in a perfectly elastic material at a given temperature and pressure is constant. The relation between c, f, λ, and T is given by Eqs. ((2.25) and ((2.26):

$$\lambda = \frac{c}{f} \tag{2.25}$$

$$\lambda = cT \tag{2.26}$$

where λ is the wavelength, c is material sound velocity, f is frequency, and t is period of time

The nondestructive ultrasonic testing process is based on the introduction of high-frequency sound waves into an object to obtain information. These high-frequency sound waves do not alter or damage the object in any way. The "time of flight" (the amount of time required for the sound to travel through the object) and the "amplitude of the received signal" are the two primary measurements critical in ultrasonic testing [75]. Thickness can be determined based on the velocity and the round trip time of flight through the object, as follows:

$$t = cT\frac{f}{2} \tag{2.27}$$

where t is material thickness, c is material sound velocity, and T_f is the time of flight.

Measurements of the relative change in signal amplitude can be used in sizing flaws or measuring the attenuation of a material. The relative change in signal amplitude is commonly measured in decibels. Decibel values are the logarithmic value of the ratio of two signal amplitudes. This can be calculated using the following equation. Some useful relationships are also displayed in the table below:

$$\mathrm{dB} = 20\log_{10}\left(\frac{A1}{A2}\right) \tag{2.28}$$

where dB is decibels, $A1$ is amplitude of signal 1, and $A2$ is amplitude of signal 2.

2.10.15 Acoustic Impedance

For the analysis of acoustic waves, propagating within a piezoelectric material and interacting with different media, the acoustic impedance Z of a material is a useful parameter. The acoustic impedance depends on the modulus of elasticity c and the density ½ of the material as follows [56]:

$$Z = \sqrt{\rho c} \tag{2.29}$$

In analogy to electric impedance, the transfer of acoustic energy from one medium to another is maximal when the two media have the same acoustic impedance. The larger the impedance difference, the larger the part of the acoustic wave that is reflected, and the smaller the part that is transmitted across the interface

2.10.16 Two-port Description

In the present section, we will merely describe the (quasi)-static behavior. Dynamic behavior of piezoelectric components is described in Sect. 2.4. The analysis of a piezoelectric component will furthermore be limited to a single direction, that is, the electric behavior (voltage and current) is considered for a single direction; this direction is determined by the way in which the electrodes are applied to the component; the mechanical behavior (force and deformation) is also considered in a single direction; this direction may be the same as the electric direction that is considered, as for the operation of poled ceramics in d_{33}-mode, but does not necessarily have to be so, as for the operation of poled ceramics in d_{31}- or d_{15}-mode [57].

2.11 Piezoelectric Material Parameter

2.11.1 Temperature

Piezoactuators can be designed to operate at very high temperatures as well as at extremely low temperatures (cryo-temperature). The extreme upper limit of operation is the Curie temperature of the piezo-material. At this temperature, the piezo-material loses its piezoelectric effect. Curie temperatures of piezo actuator materials fall in the range of 140–350 °C [58]. However, piezoelectric properties are temperature dependent. In precision positioning applications, temperatures approaching this can cause serious detrimental effects to the accuracy and performance of the piezo-stage. The crystals in piezoelectric material remain in their piezoelectric configuration at the lower temperature, for example, standard commercially available stack actuators can operate down to −40 °C [59]. The biggest issue in cold environments is not the piezo itself, but induced stress from thermally contracting mechanisms. For extremely cold environments, special design considerations are required for the actuator to survive the cooling process. Carefully chosen electrodes and extremely homogeneous ceramic must be used to prevent cracking due to mismatch of the thermal expansion coefficients. Piezo-ceramics operate at low temperatures. At low temperatures, the ceramic stiffens, which causes a decrease in the amount of strain generated per volt. This is offset by increased electrical stability in the crystal structure, allowing fully bipolar operation. Other advantages of low-temperature operation include lower hysteresis, better linearity, lower capacitance, and smaller dielectric loss. For the highest accuracy, it is recommended that the operation be at or near 20 °C because that is the temperature at which the nanopositioning stages are built and calibrated [60].

2.11.2 Accuracy/Linearity

Piezo actuators exhibit hysteresis and nonlinearity when operated in open-loop mode. When operating in closed-loop mode, the non-repeatabilities due to piezo actuator hysteresis are eliminated. However, the piezo-stage may still exhibit nonlinearity and hysteresis that affect the overall positioning accuracy of the device. The magnitude of these nonlinearities is a function of the quality of the closed-loop feedback sensor and electronics used in the design, as well as the quality of the mechanical stage design. With our high-resolution capacitance sensors, advanced electronics, and optimized flexure designs, linearity errors below 0.02% are achievable

[61]. Accuracy and linearity are measured with precise laser interferometers at a distance of ~15 mm above the moving carriage of the piezo nanopositioner. The terms accuracy and linearity are sometimes used synonymously when describing the positioning capability of piezoelectric nanopositioners. However, they can have subtle differences in meaning. *Accuracy* is defined as the measured peak-to-peak error (in units of micrometers, nanometers, etc.) from the nominal commanded position that results from a positioning stage as it is commanded to move bidirectionally throughout travel. *Linearity* is defined as the maximum deviation from a best-fit line of the position input and position output data. Linearity is reported as a percentage of the measurement range or travel of the positioning stage. The deviation of a best-fit line to the measurement data is used to calculate the linearity error. In conclusion, the term accuracy is used to quantify both sensitivity effects (slope of measured vs. actual position) as well as nonlinearities in positioning and is reported as a peak-to-peak value. The term linearity is used to quantify the effects of nonlinearities in positioning only and is reported as a maximum error or deviation of the residuals from the best fit line through the measured versus actual position data. The positioning accuracy can be approximated from the linearity specification by doubling the linearity specification. For example, a 0.02% linearity for a 100 μm stage is a 20 nm maximum deviation. The approximated accuracy error is 2 × 20 nm or 40 nm peak-to-peak [62].

2.11.3 Resolution

Resolution is defined as the smallest detectable mechanical displacement of a piezo nanopositioning stage. Many piezo-stage manufacturers will state that the resolution of a piezoactuator is theoretically unlimited because even the smallest change in an electric field will cause some mechanical expansion or contraction of the piezo stack. Although theoretically true, this fact is largely impractical because all piezo actuators and stages are used with electronics and sensors that produce some amount of noise. The noise in these devices generally rises with increasing measurement sensor bandwidth. As a result, the resolution (or noise) of a piezo nanopositioner is a function of the sensor bandwidth of the feedback device. Because the noise is primarily Gaussian, taking six times the 1 sigma value gives an approximation of the peak-to-peak noise. Unless specified, the measurement point is centered and at a height of approximately 15 mm above the output carriage. In noise critical applications, measuring at a lower servo bandwidth will result in lower noise (jitter). Values are specified for open-loop and closed-loop resolution. Open-loop resolution is governed only by the noise in the power electronics whereas closed-loop resolution contains feedback sensor and electronics noise as well as power amplifier noise [63].

2.11.4 Stiffness

The stiffness of a piezo actuator or nanopositioner is specified in the direction of travel of the output carriage. The stiffness is a function of the piezo stack, stage flexure, and amplification mechanisms used in the design. Higher stiffness piezo stages allow for higher dynamics in positioning such as faster move and settle times and better dynamic tracking performance. Most longer-travel (>50 μm) piezo flexure stages use lever amplification to achieve longer travels in more compact package size. Lever amplification designs cause the stiffness in the direction of travel (inversely proportional to the square of the lever amplification ratio) to be reduced when compared to a directly coupled design. Also, most lever amplification designs cause the stiffness of the actuator to change depending on location in travel due to the nonlinear nature of the amplification gain. For this reason, along with manufacturing and device tolerances, the stiffness of piezo nanopositioning stages is specified at a nominal value of ±20% [64].

2.11.5 Resonant Frequency

The resonant frequency of a nanopositioning stage can be estimated as follows [65]:

$$f_n = \left(\frac{1}{2\pi}\right) \cdot \left(\frac{k}{m_{\text{eff}}}\right) \tag{2.30}$$

where f_n is the resonant frequency (Hz), k is the stiffness of the piezo nanopositioner (N/m) and m_{eff} is the effective mass of the stage (kg). In a very general sense, it is typically the first (lowest) resonant frequency of the positioning system that limits the achievable servo bandwidth. The design of the flexure, supporting mechanics, and piezo actuator stiffness govern the location of this resonant frequency. By adding an applied mass to the piezo stage, the resonant frequency will decrease by the following relationship:

$$f'_n = \left(\frac{1}{2\pi}\right) \cdot \sqrt{\left(\frac{k}{m_{\text{eff}} + m_{\text{load}}}\right)} \tag{2.31}$$

where m_{load} is the mass of the applied load. In lever amplification designs, the stiffness can change throughout travel, as mentioned above. As a result, the resonant frequency will change by the square root of the change in stiffness. For example, if the stiffness changes by 7%, the resonant frequency will shift by approximately 3.4% throughout travel. Equations (2.30 and (2.31 will provide a first-order approximation of resonant frequency in piezo nanopositioning systems [66]. Complex interactions of the dynamics due to damping, nonlinear stiffnesses, and mass/inertia effects cause these calculations to provide only an approximation of the resonant frequency.

2.11.6 Mechanical Amplification

The use of multiple layers may be regarded as a way to modify the behavior at the electric port of a piezoelectric component, that is, to modify the electric impedance. Similarly, it may also be attractive to modify the mechanical impedance of a piezoelectric component. Piezoelectric ceramic components, for example, are characterized by high stiffness, that is, by high forces and small deformations. For actuation purposes, it is often desired to increase the stroke of a piezoelectric component. To realize this, some kind of lever mechanism may be designed, like conceptually illustrated in Fig. 2.12. The lever mechanism is characterized by the mechanical amplification factor or transmission ratio a_m. The behaviour at the mechanical port changes as follows [67]:

$$\Delta l_a = a_m \Delta l \tag{2.32}$$

$$F_a = \frac{1}{a_m} F \tag{2.33}$$

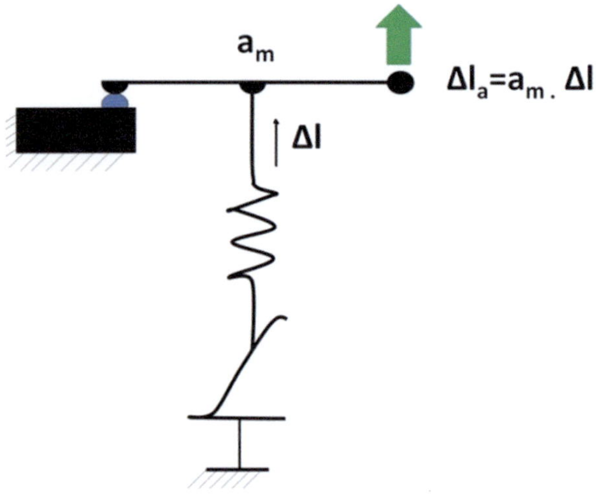

Fig. 2.12 Mechanical amplification

Thus, whereas the output stroke increases with a_m, the output force decreases with the same factor. For the piezoelectric component, the amplification causes an increase in the effective piezoelectric (charge) constant, for the inverse effect:

$$\Delta l_a = a_m dU \tag{2.34}$$

As well as for the direct effect:

$$\Delta q = d a_m F_a \tag{2.35}$$

Finally, the effective stiffness decreases (or the compliance increases) for the mechanically amplified piezoelectric component, quadratically with a_m:

$$k_{m.a}{}^{U} = \frac{1}{a_m{}^2} k_m{}^{U} \tag{2.36}$$

The latter effect is similar to the multilayer capacitance increasing quadratically with the number of layers. Mechanical amplification can thus be regarded as the mechanical counterpart of the electric multilayer technique. Also similar to the case of the multilayer, the coupling coefficient of the levered piezoelectric component remains unchanged.

2.11.7 Quality Factor

The mechanical quality factor (Q_m) characterizes the sharpness of the electromechanical resonance spectrum. Its value can be determined by dividing the $(\Delta f)_{3\text{dB}}$ width of the piezoelectric resonance by the resonance frequency (f_s) [68]:

$$Q_m = \frac{f_s}{(\Delta f)_{3\text{dB}}} \tag{2.37}$$

The mechanical Q (also referred to as Q_M) is the ratio of the reactance to the resistance in the series equivalent circuit representing the piezoelectric resonator. The Q_M is also related to the sharpness of the resonance frequency.

$$Q_M = \frac{1}{2\pi F_r Z_m C_o}\left(\frac{F_a{}^2}{F_a{}^2 F_r{}^2}\right) \tag{2.38}$$

where F_r is resonance frequency (Hz), F_a is the antiresonance frequency (Hz), Z_m is impedance at F_r (ohm) and C_o is static capacitance (Farad).

2.11.8 Bandwidth

The frequency span between the series and the parallel resonance frequency is referred to as the bandwidth of a piezoelectric transducer [69]. In this frequency band, which is proportional to the coupling coefficient *k*2, energy is transmitted in the most efficient way, that is, with minimum losses.

$$B = f_p - f_s \tag{2.39}$$

Knowing the rate at which a piezo is capable of changing length is essential in many high-speed applications. The bandwidth of a piezo controller and stack can be estimated if the following is known:

1. The maximum amount of current the controllers can produce. This is 0.5 A for the BPC series piezo controllers.
2. The higher the capacitance, the slower the system.
3. The desired signal amplitude (V), which determines the length that the piezo extends.
4. The absolute maximum bandwidth of the driver, which is independent of the load being driven.

To drive the output capacitor, current is needed to charge it and to discharge it. The change in charge, *dV/dt*, is called the slew rate. The larger the capacitance, the more current needed:

$$\text{Slew rate} = \frac{dV}{dt} = \frac{I_{\max}}{C} \tag{2.40}$$

For example, if a 100 μm stack with a capacitance of 20 μF is being driven by a BPC series piezo controller with a maximum current of 0.5 A, the slew rate is given by

$$\text{Slew rate} = \frac{0.5\,A}{20\mu\text{F}} = 25\ \text{V/ms} \tag{2.41}$$

Hence, for an instantaneous voltage change from 0 V to 75 V, it would take 3 min. for the output voltage to reach 75 V. (For these calculations, it is assumed that the absolute maximum bandwidth of the driver is much higher than the bandwidths calculated, and thus, driver bandwidth is not a limiting factor.) Also, these calculations only apply to open-loop systems. In closed-loop mode, the slow response of the feedback loop puts another limit on the bandwidth.

2.11.9 Frequency Constant

When an alternating electric field is applied to a piezoelectric component, elastic waves are created. The frequency at which mechanical resonance occurs is determined by the geometry of the component in relation to the wave speed (or velocity of sound) v [m/s] of the material in a particular operation mode [70]:

$$v = \frac{1}{\sqrt{S^E \rho}} \tag{2.42}$$

where ρ is the density of the material [kg/m^3]

In data sheets of piezoelectric materials, rather than the velocity of sound, the so-called frequency constant N [Hz.m] is given to enable calculation of the fundamental resonance frequency for various operation modes, for components of a given geometry. With the exception of the so-called planar (or radial) vibration mode, the frequency constant can be shown to equal half the velocity of sound in the piezoelectric material:

$$\text{N}^*{}_i = \frac{v^*{}_i}{2} \tag{2.43}$$

where the subscript i denotes the particular vibration mode considered and the superscript (*) indicates the electric boundary condition.

2.11.10 Humidity

One of the most important factors for ensuring long life is to protect the piezo actuator against humidity. For this reason, most industries use specially sealed coatings on the actuators that protect the actuator from moisture. Operation at 60% or lower relative humidity (RH) environments is preferred as it helps further prolong the life of the actuator [71].

2.11.11 Load Ratings

Piezo actuators are ceramic materials and are brittle. As with most ceramics, PZTs have a higher compressive strength than tensile strength. The actuators used in the stage designs are preloaded so as to always maintain a compressive load state during standard operational limits. For some stages, the load rating may be different depending on the direction of the applied load [72].

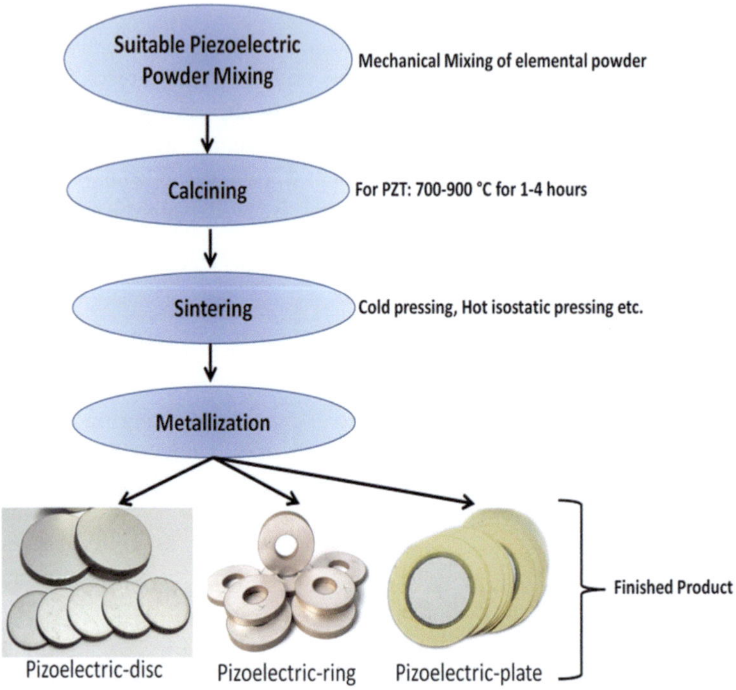

Fig. 2.13 Flow diagram of bulk ceramic fabrication

2.11.12 Vacuum

Low-voltage (<200 V) piezo actuators are particularly well-suited for vacuum operation. Piezo actuators do not require lubrication that typically requires great care when selecting for ultra-high vacuum applications. Vacuum pressures from 10 to 10^{-2} Torr need to be avoided because the insulation resistance of air dramatically decreases in this range (known as the corona area), thus allowing easier dielectric breakdown [73].

2.12 Manufacturing of Piezoelectric Components

Piezoelectric components are used in wide variety of applications. According to the application, piezoelectric structural materials are manufactured as (1) bulk ceramics: disks, rings, plates; (2) benders: unimorphs and bimorphs-actuators and sensors; multilayer actuators and multilayer benders; thin films for piezo-MEMS. Commercially, the focus will be on PZT based ceramics.

2.12.1 Bulk Ceramics: Disks, Rings, Plates

Piezoceramic materials are available in a large variety of shapes or forms. Mostly, piezoelectric components are processed in two steps: first step, the powders are mixed and calcining to form the perovskite and in the second step, the reacted powder is formed into the desired shape and hereafter sintered to obtain fully dense ceramics (Fig. 2.13). The process starts with weighing the appropriate oxides like PbO, TiO_2, ZrO_2, and the other constituent oxides-the dopants with the required ratio [74]. In recent years bead mills and pearl mills are used instead, to obtain more efficiently smaller particle sizes. As milling media, balls of steatite, YSZ (yttria-stabilized zirconia), alumina, or even steel balls are used [75]. For PZT based materials the calcination is done at temperatures of 700–900 °C for 1–4 h [76]. At too low temperatures the reaction to form the perovskite will not occur. Sintering takes place at a high temperature which is somewhat below the melting temperature of the ceramic material. Generally, the sintering of PZT ceramics is done at temperatures of 1100–1250 °C for 1–4 h. In most cases closed crucibles of refractory materials like alumina or zirconia are used to prevent too much lead loss during the sintering process. Lead tends to evaporate from the material at temperatures above 800 °C [77]. If binders are used during the processing of the powders these must be removed first during heating. After the sintering process, the ceramic component—which now has a density close to the theoretical density-is ready for further processing. In most cases, the outer surface of the sintered PZT ceramic component is removed in a mechanical way by grinding or lapping. The outside of the sintered part is generally deficient in lead because of evaporation during sintering, which is inevitable. After mechanical processing of the ceramic component into its final shape by for example grinding, lapping, sawing, dicing, and polishing, the ceramics are ready to be metalized.

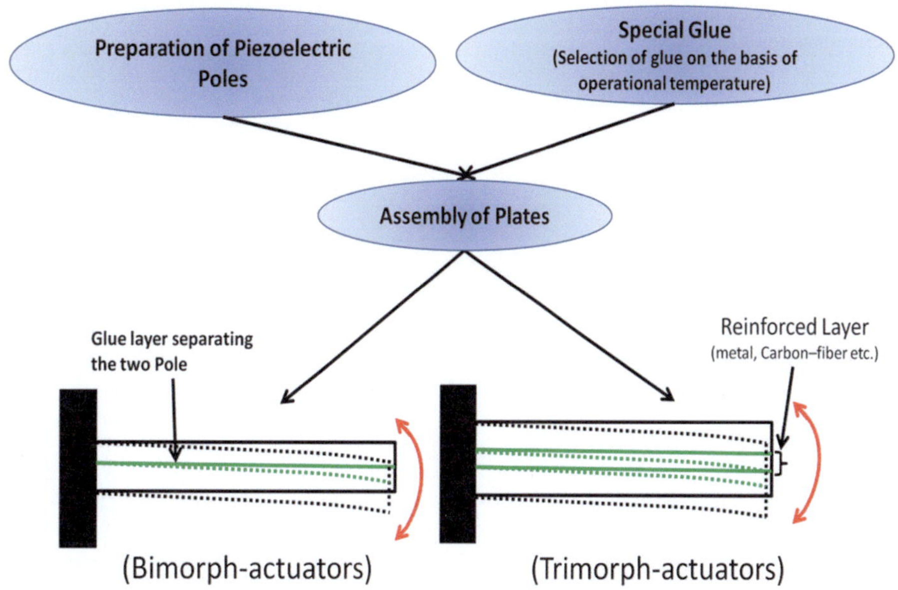

Fig. 2.14 Benders piezo-actuator preparation process

Metalization is the process during which electrodes are applied to the material. After the metalization, the piezoelectric component can finally be poled by the application of a high electric field at elevated temperatures to make the material piezoelectrical active.

2.12.2 Benders: Unimorphs and Bimorphs–Actuators and Sensors

In many applications, a larger displacement is required than is possible by simple PZT based ceramics. For most commercial ceramic PZT-based materials that are used in the d_{33} mode, the strain is limited to about 0.1% at a driving voltage of 2 kV/mm [78]. So the high voltage requirement for obtaining relatively high displacement and the limited strain limit the number of applications. In cases where more displacement is required and no high forces are required, the use of benders or bimorphs is a good solution. The working field of piezoelectric actuators is presented where typical forces and displacements or strains are presented for multilayer actuators and bimorphs. A piezoelectric bimorph consists basically of two thin polarized piezoelectric plates which are bonded together. If one of the thin piezoplates is being actuated by applying a voltage over the plate, the actuated plate will expand in the thickness direction (d_{33} mode). However, it will shrink over the two other perpendicular and in-plane directions, which will force the total bimorph assembly to bend. For this reason, the bimorph actuators are also called benders. The most simple form of a bimorph consists only of these two metalized polarized piezoelectric plates. The plates can be placed together in such a way that they are poled in the same direction, which is called a parallel bimorph. The other way is the series bimorph in which the polarization of the two plates is in opposite directions. In some cases, a third plate is placed in the middle of the two piezoelectric plates (Fig. 2.14). This makes up the reinforced bimorph, or sometimes known as the trimorph. The third plate can be a metal plate but also carbon- or glass fiber composite plates are used for reinforcement. The bonding is in most cases done by special glues. Note that the maximum application temperature of the bimorph is related to the properties of the glue which is used in the bimorph assembly. The glue will have a glass transition temperature or will soften at elevated temperatures which will limit the application at too high temperatures because creep will occur. Typically the maximum temperatures at which bimorphs can be used is about 100 °C [79]. By using specially developed glues this limit can be increased up to 150 °C. In a similar way, a unimorph or amonomorph can be made out of one piezoelectric plate which is glued onto the metal of the carbon fiber plate. The unimorph has one big disadvantage: since the thermal expansion coefficient or the piezoelectric material will not be the same as the metal or carbon fiber plate, the temperature stability of this assembly is not very good. The deflection of the unimorph is not only dependent on the applied voltage but also the temperature. In the case of the bimorph, there will be no temperature-dependent properties. Note that bimorphs are also used as sensor elements. The production of the bimorph actuator starts with the making of a thin piezoelectric plate. This plate is typically made out of a block of ceramics by wire slicing or sawing [80]. This plate is metalized and then poled. Hereafter, the glue is applied by a screen printing process and the plates can be assembled together. After curing the glue at elevated temperatures the assembly is poled. Now an insulation layer consisting of

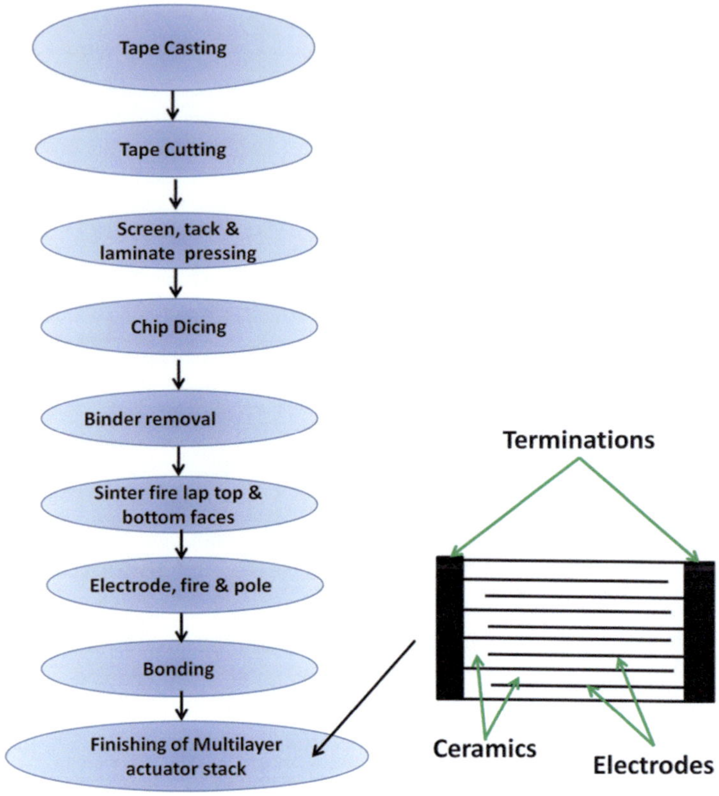

Fig. 2.15 Fabrication procedure of multilayer piezo-actuator

lacquer is applied to protect the electrode. Consequently, the individual bimorphs are made by dicing, and insulation lacquer is applied at the sides.

2.12.3 Multilayer Actuators

Multilayer actuators (MLA) can provide relatively high forces but low displacements or strains. The multilayer actuator has been invented to come to a solution to create an actuator that can be operated at moderate voltages. For a monolithic ceramics actuator, a driving voltage of 2 kV/mm is required. As a consequence, to obtain a displacement of 30 μm an actuator with a thickness of 30 mm is required which would be working with an extremely high operating voltage of 60 kV [81]. The solution for this issue is to make use of internal electrodes. In almost all actuators that are currently being produced the so-called soft piezoelectric materials are used. The making of the multilayer actuator starts with ceramic powder which is prepared in a similar way as described in Fig. 2.13. To this powder, a liquid like water or an organic solvent is added together with a dispersant to make a slurry. The milling of slurries is normally carried out in large ball mills, bead mills, or pearl mills, to obtain smaller particle sizes in an efficient way. When the target particle size distribution has been reached, the slurries are separated from the milling beads and an appropriate amount of binder solution is added. After the suspension, often also called slip, is homogeneously mixed and degassed, foils of it are cast in the tape casting process. The thickness of the casted foils in piezoelectric actuator applications is typically between 5 and 200 μm [82]. Note that the same technology is used to make multilayer ceramic capacitors where the minimum foil thickness reaches 0.5 μm. After the foils have been dried, a pattern of rectangular metal electrodes is printed on the sheets via screen printing. The metal which is used for the inner electrodes can be Pt, Pd, an Ag–Pd alloy, or even Cu. The printed sheets are stacked layer-by-layer with precise alignment to produce alternating layers of electrodes in a ceramic body as shown in Fig. 2.15. The stack then is uniaxially pressed to the layers to form a compact plate. Now the part can be diced into single parts. To remove organic materials the plates are subjected to binder burnout at moderate temperatures. This process must be carefully controlled because if organic residues are leftover this can result in the reduction of the metal ions (e.g., lead) in the powders. After that, the plates are sintered at high temperatures to obtain dense bodies. In some production processes, the dicing is done after sintering of the stack. After sintering, the parts are lapped and metalized. Poling of the multilayer actuator is again performed at elevated temperatures and voltages of 2–3 kV/mm [83]. In some cases, the poling is done under a mechanical preload. Out of the multilayer actuator components, a higher *stack* can be made by placing these

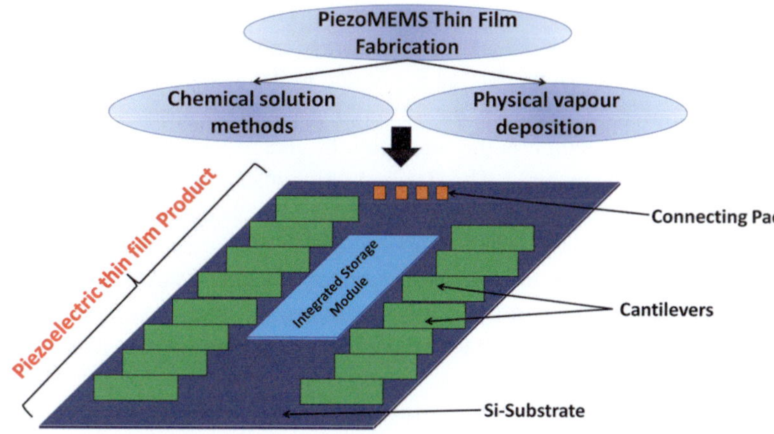

Fig. 2.16 Fabrication techniques of Piezo-MEMS thin film

parts on top of each other and making the *stack* a common electrode along the side. The process for making a multilayer actuator is summarized in Fig. 2.15.

2.12.4 Thin Films for Piezo-MEMS

Recent development and application of piezo-MEMS have been a major driver for research in the field of integration of PZT thin films with silicon technology. Both the sensing and actuating capabilities of piezo-MEMS are used in many applications, such as inkjet print heads, energy harvesting, micro-pumps, bio-sensors, and piezoelectric Micromachined Ultrasound Transducer (pMUT) [84]. Using piezo-MEMS, miniaturization of the size of, for instance, the inkjet printing nozzles increases the printing resolution and decreases the ink consumption. Using deep reactive ion etching, arrays of membranes with a small pitch can be fabricated. This is advantageous in case of limited space, for example in pMUT that are situated in the tip of an endoscopic probe. The small device volume and low leakage currents offer low-power consumption, which is an added value for wireless applications. PZT films for piezo-MEMS are deposited by chemical solution methods or physical vapor deposition. These techniques allow the fabrication with high-throughput and low cost and have removed a major economical obstacle for the application of these complex oxide-based devices. For example, PZT thin chemical films are prepared from precursor solutions, which are spin coated on Pt/Ti/SiO deposition 2/Si wafers, followed by a pyrolysis step [85]. This process is repeated until the required PZT film thickness is obtained. Finally, thermal annealing steps are carried out to obtain the piezoelectric phase in the PZT thin film. Physical deposition technique, for example, pulsed laser deposition (PLD) is used. PLD is a physical vapor deposition method that delivers dense and homogenous films. PLD has been established in recent years as a very versatile technique for the controlled synthesis of functional materials, including piezoelectric PZT films. PLD utilizes the transient particle flows of a laser-induced plasma for the growth of thin films. The complete multilayer, buffer layers, electrodes, and PZT, can be deposited in a single process run. An additional advantage single process of PLD is the option to deposit atomic monolayers to improve the device functionality. For example, the use of $SrRuO_3$ and $LaNiO_3$ between the platinum electrode and functional PZT increases the long-term stability drastically [86]. The development of large area pulsed laser deposition has allowed the deposition of PZT thin films on a wafer scale. Typical piezo-MEMS devices consist of a silicon membrane covered by a piezoelectric film and electrodes in parallel plate geometry. The size of the elements will influence the deflection amplitude, resonance frequencies, and Q-factors. In Fig. 2.16, piezo-MEMS fabrication technique on products have been shown.

2.13 Difference Between Piezoelectric and Electrostrictive Materials

Piezoelectric materials generate strain when an electric field is applied in a prescribed direction. This effect is also called the inverse piezoelectric effect. The most common form of piezoceramics is based on Lead-Zirconate-Titanate (PZT) compounds. In piezoceramics, the unit cell has a certain degree of symmetry, which gives it a permanent dipole [87]. A poling operation is done on a bulk material to orient all the dipoles in a preferred direction to produce a net polarization. Once polarized, an applied electric field in the poling direction produces a temporary expansion in the poling direction, thus producing induced strain.

Similar to the piezoelectric effect, all dielectric materials become strained upon application of an electric field. This phenomenon is known as electrostriction and the material is called electrostrictive material. Electrostrictive materials also exhibit strain on the application of the electric field. The phenomenon of electrostriction, however, is fundamentally different from the inverse piezoelectric effect. The unit cell in an electrostrictive material is centrosymmetric and hence, the strain exhibited by such material is not due to the change in structure but is inherent in the material itself. The basic mechanism of actuation is the separation of charged ions in the unit cell of the material. The electrostrictive stack used for example Lead-Magnesium-Niobate (PMN) compound [88]. Electrostriction is fundamentally similar to magnetostriction and exhibits many common characteristics. Typically, the strain exhibited by an electrostrictive material has a quadratic relationship with the applied field and shows the property of frequency doubling. Ideally, an electrostrictive stack can operate under both positive and negative electric fields. The electrical impedance of an electrostrictive stack, like a piezoelectric stack, is capacitive in nature and has a value of 0.37 mF. Because the PMN stack has a lower capacitance than the PZT stacks, its energy consumption at a given voltage and frequency is lower. Application of preload does not have a marked effect on the strain output of piezoelectric materials. Preload is generally applied to these stacks to ensure that the stack always remains in contact with the actuator piston during high-frequency operation. Application of preload also offsets the effect of tensile stress in the stack. Extensional strains are highly detrimental to the integrity of the stacks. Special care is required while mounting the stack in the actuator body as the slightest misalignment can produce a bending moment on the stack [89]. Especially in stacks with a large aspect ratio (ratio of length to cross-sectional width), this bending moment can result in extensional stresses during high-frequency operation. To mitigate the effect of extensional strains produced by bending moments, ball ends are used as the contact points on either end of the stack.

Thus, the piezoelectric effect describes linear, mutual electromechanical coupling, electrostriction merely describes a mechanical response to an electric field, where the strain varies quadratically with the applied field. The electrostrictive effect is in general rather small. However, in materials with a high dielectric constant, especially in ferroelectrics, electrostriction can be of sufficient magnitude to be used as an actuator. By applying an electric field (nonzero bias), it is possible to induce piezoelectric behavior in electrostrictives, enabling use as sensors [90].

2.14 Applications of Piezoelectric Devices

Piezo technology is used in advanced technology markets, such as in medical industries, mechanical and automotive industries, or semiconductor industries. In 2019, piezoelectric ceramics holds the largest market size for piezoelectric devices and is expected to continue to hold a significant market share in line with market demand. According to the research report, the market for piezoelectric devices is expected to increase from $ 28.9 billion in 2020 to $ 34.7 billion by 2025 at a CAGR of 3.7% [91]. The growing demand for piezoelectric devices for energy harvesting and aerospace and defense applications are the factors that encourage the growth of this market. Piezoelectric ceramics are widely used and are commercially acceptable piezoelectric material. In addition, the production cost of piezoelectric ceramics is low and these materials can be easily adapted to meet the requirement of specific application such as medical industry, engineering, automotive, and semiconductor industry. The piezoelectric converter or piezoelectric sensor senses power, force, temperature, acceleration, and so on, by converting the associated energy to electrical energy. The electrical energy can be measured by means of a detector which in turn is calibrated to know the input value. So, the magnitude of force, acceleration, and so on, can be determined. The voltage (electrical energy) is a function of the input value. Hence the input parameters such as pressure and force can be known by measuring voltage in a piezoelectric transducer [92]. But a piezoelectric actuator is the opposite of a piezoelectric sensor. Here when electrical energy is provided by means of voltage the material undergoes mechanical deformation. It may be stretched or bent [93]. Based on these mechanisms, different types of applications in different industries are discussed subsequently:

2.14.1 Aero Industries

Piezoelectric materials are used in microprocessors for satellites as given in Fig. 2.17a, where they are used to determine the position and stabilize the satellite. Of course, great emphasis is placed on the reliability and functionality of products used in space. Micro-thrusters can use a "cold gas micro-thruster." In this process, the propellants create a very small and controlled force (<500 μN) by emitting a gas, usually nitrogen, stored in a high pressure tank. This requires precise regulation of the propulsive pressure in the circuit and fast and accurate "distribution". Piezoactuators built into the valves provide both functions. The movement of the piezo controls ensures fast and accurate flow control during operation. From the low pressure side, piezoelectric components integrated into the micro-actuators are selected for low energy requirements, reduced size and weight, high stroke, and low power. For this application, a multilayer bending activator is preferred, which provides fast and accurate movement in a small package [94]. Figure 2.17b shows the interferometry used in space. Interferometry is the process by which one can analyze the waves by studying the pattern of interference. Interference is detected by using an interferometer. The interferometer can be used in the field of optical

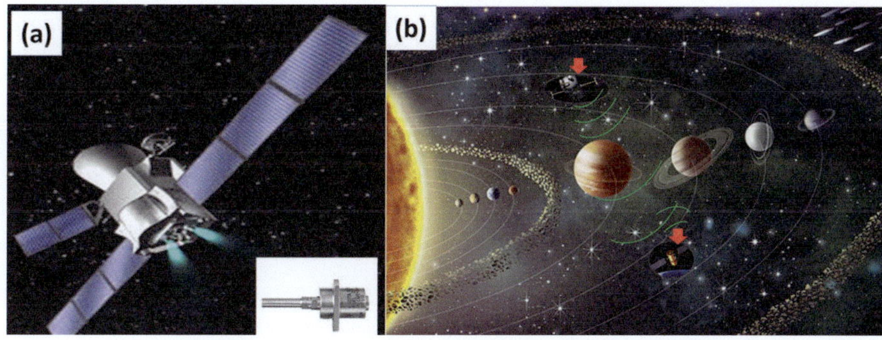

Fig. 2.17 (**a**) Micro-thrusters of satellites (**b**) Showing the interferometry used in space

Fig. 2.18 Piezoelectric materials used in (**a**) Structural health monitoring, and (**b**) Active vibration damping

metrology, quantum mechanics, seismology, plasma research, and so on. The piezoelectric actuators are used to have a perfect alignment of mirrors inside the interferometer. It operates at a low voltage having no electromagnetic interference (EMI). Most importantly it has a higher sub-nanometer resolution. For interferometer applications, the actuators such as NAC20XX controller actuator could be preferred. With an unloaded frequency above 500 kHz, they allow precise positioning at high frequencies. For dimensions from $2 \times 2 \times 2\ \text{mm}^3$ to $15 \times 15 \times 2\ \text{mm}^3$, the controls adapt to the available size [95].

Piezos are used to monitor the condition of a structure, during which the integrity of the mechanical structures is checked. This is very important in cases where security is an important issue, for example, an aircraft that requires scheduled service and assembly of selected parts (Fig. 2.18). The sensors used to monitor the condition of the structure are mounted or integrated into structures that are constantly monitored. Structural health monitoring can be active or passive. The "active" system produces a high-frequency sound from the transducer. When high-frequency sound hits a material with different acoustic property the sound may be absorbed or reflected. There are several methods for "active" monitoring of structural health, the most common being pitch-catch analysis or pitch-echo listening [96]. If there is an error in the structure, the transmitted signal is reduced and the reflected signal will increase. The "passive" system only responds to any sound produced by cracks or by any changes in the structure. It is generally applied for composite structures.

Structural health monitoring applications are usually manufactured with single-layer piezoelectric material that provides low power, high sensitivity, and a wide range of frequencies. Piezoelectric actuators are also used to actively reduce vibrations. Vibration damping may be designed to reduce fuel consumption, noise and waste. The decrement in noise levels in helicopter construction has recently aroused great interest. The technique can be applied to airplanes, engines, machine tools, wind turbines, laboratory tables, and so on. [97]. There are two ways to damp unwanted vibrations in constructions using pressure components: active and passive vibration damping. With the active method, the piezoelectric actuators create opposite motions in a control loop that effectively reduces vibrations. This is the preferred method because it is the most effective one. Active vibration damping in the aerospace industry usually focuses on high amplitude, low-frequency vibrations that require strong large stroke controls. Stacked multilayer piezoelectric actuators are beneficial for their high voltage and are often integrated into reinforcement structures. For example, the amplified NAC2643 controller provides 550 μm of displacement with high stiffness, that is, high resonance (discharged 1100 Hz)

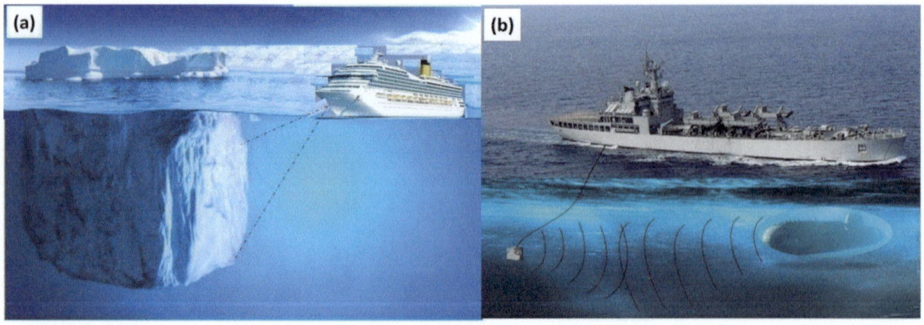

Fig. 2.19 (**a**) Underwater imaging and (**b**) Underwater communication

[98]. Passive vibration damping is a method in which vibrations are converted into electricity by a piezoelectric effect and then stored or dissolved in heat by resistors.

2.14.2 Marine Industries

The most popular equipment in the marine industry is sonar equipment for underwater imaging, object detection, and communication. Sonar equipment and depth sounders rely heavily on piezoelectric sensors to transmit and receive "pings" in 50–200 kHz range [99]. Piezoelectric transducers have high power density which allows a large amount of acoustic power to be transmitted from a small packet. For example, a transducer with a diameter of only 4″(100 mm) can withstand more than 500 watts of power. Piezos uses underwater imaging system modulators as shown in Fig. 2.19a [100]. Their objective is to detect substances underwater and knowing the texture at the bottom. They are also used to know the signs of lakes, oceans, and so on. Underwater imaging uses downward sonar and other imaging techniques. They may provide accurate 2D or 3D images having detailed information about the study area. All sonar-based techniques are used to create a sound wave or pulse that is based on sonar converters. In underwater imaging, the echoes of repeated sound waves appear on a detailed 2D or 3D map, which is considered the most complex. The range of piezoelectric format molds ranges from simple single-format molds to advanced multistage crossover arrays. Inside underwater imaging devices, piezoelectric components made of soft piezoelectric materials are used [101]. Figure 2.19b shows sonar for underwater communications. Piezoelectric materials are used for sonar underwater communications, where you can use sonar sound waves. Acoustic wave is the most convenient underwater communication for sending and receiving data underwater. Underwater sonar can also be used to assess deep oceans for tsunami reporting. All sonar-based systems use a formatter to generate sound waves or pulses. Special signal configuration types are used to achieve high signal reliability. This allows long-distance data transmission with acceptable energy consumption in a high-pressure environment [102]. Underwater communication device builders are usually equipped with pressure fittings of various shapes made of soft piezoelectric materials. However, electrically solid materials must also be considered in heavy-duty applications. A multilayer piezoelectric element used as a transmitter can significantly reduce the drive voltage level.

Piezo is used in fish finders which use sonar. The fish finder may sound an alarm if there is an object in the search area below (fish, shipwreck, vegetation, sudden changes in water temperature, etc.). Acoustic technology works particularly well for fish detectors and other underwater applications, as the sound travels well underwater without much attenuation [103]. The fishfinder is a sonar, which consists of a transducer that generates a sound wave or pulses of sound. The transducer also receives the waves reflected from the target. The sound waves of the fish tracker are sent back mainly because the fish are in a bladder full of air. The echo of this return appears on the screen. The same system, equipped with a horizontally oriented and rotating beam of sound, makes it possible to determine the herds of fish and their distance from the boat using the reflected sound. Fish growers usually consist of piezo components of different shapes and sizes depending on the application [104]. Simple sonar materials could be a solid type of piezoceramic. They provide high power density and high sensitivity. However, for high-resolution smart sonar, with a diaphragm composition, it would be better to use soft piezoelectric materials. The soft material has low mechanical Q and provides higher resolution. Multilayer pie could be used effectively as a transmitter of a wave. This reduces the need for high voltage. Compared to conventional single-layer piezoceramic elements, the driving voltage could be reduced by increasing the number of layers in the case of multilayer elements [105].

2.14.3 Automobiles

Engine manufacturers are constantly faced with the challenges of controlling engine parameters. In unfavorable conditions, petrol engines tend to knock which is called detonation. Detonation is very hazardous for the engine. Hence the manufacturer takes a conservative operational

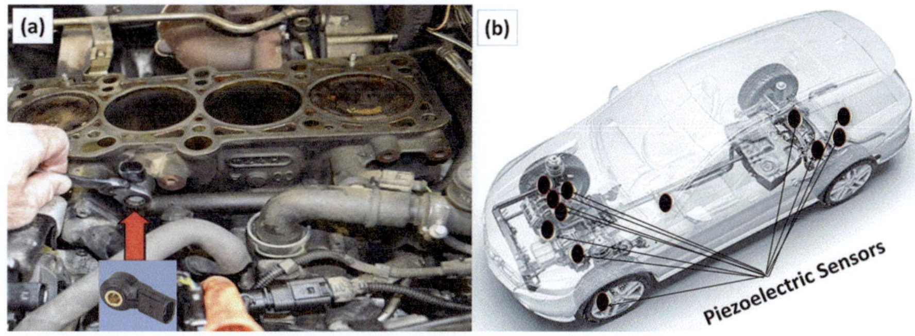

Fig. 2.20 (**a**) Engine knock sensors and (**b**) various piezoelectric sensors in automobile

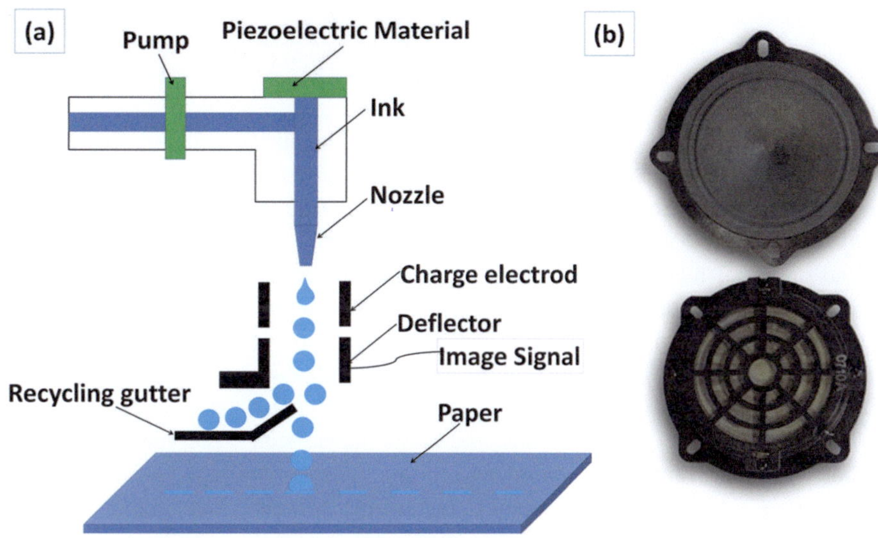

Fig. 2.21 (**a**) Piezoelectric material used in a printer and (**b**) piezoelectric speakers

margin to design the engine at the expense of cost. But nowadays as better control techniques are developed so the engine parameters can be adjusted to maximize power and efficiency. The control system can detect detonation well before so that necessary changes can be done beforehand as shown in Fig. 2.20a. This gives management systems time to make the necessary adjustments. Piezoelectric devices have a high-frequency response. It doesn't require blowers, diaphragms, or any type of mechanical connection in combination with a voltage sensor or displacement sensor [106]. Hence piezoelectric pressure sensors are used to measure dynamic pressure changes and provide more accurate results than conventional sensors.

Over the last decade, the rules on diesel engine emissions have been given increasing attention. In addition, it is demanded to have more power and torque in the engine. To meet the desired performance requirements, engine producers have used precise timers and measured fuel injections in the combustion process. As incredible as it sounds, the fuel injection nozzle can change fuel flow at pressures above 26,000 psi (1800 bar) multiple times, quickly and sequentially, in a simple power supply path. By using piezoelectric actuators that control the small valves inside the fuel injection nozzles such precise control of the high-pressure fluid is possible [107]. Different piezoelectric sensors in different general positions are shown in Fig. 2.20b.

2.14.4 Electrical and Electronics

Printers may use piezoelectric sensors. Generally, printers are of two types: dot matrix printer and inkjet printer. The dot matrix piezoelectric printer uses needle-like pins that strike on the paper to impart the impression of ink on the paper. The dot matrix piezoelectric printer is the only printer that can generate duplicate and triplicate carbon copy printouts. But it is somehow obsolete nowadays. The inkjet printer incorporates piezoelectric actuators as the printer head. It uses a diaphragm type head which changes the geometry of an ink well so that it will produce various shapes of ink droplets on the paper. This is the most versatile type of printer in the market. Figure 2.21a shows the piezoelectric material used in a continuous ink launch system. Figure 2.21b shows piezoelectric speakers, appears in almost every application that needs to produce sound from electronic gadget efficiently. The speakers are very efficient and use very little

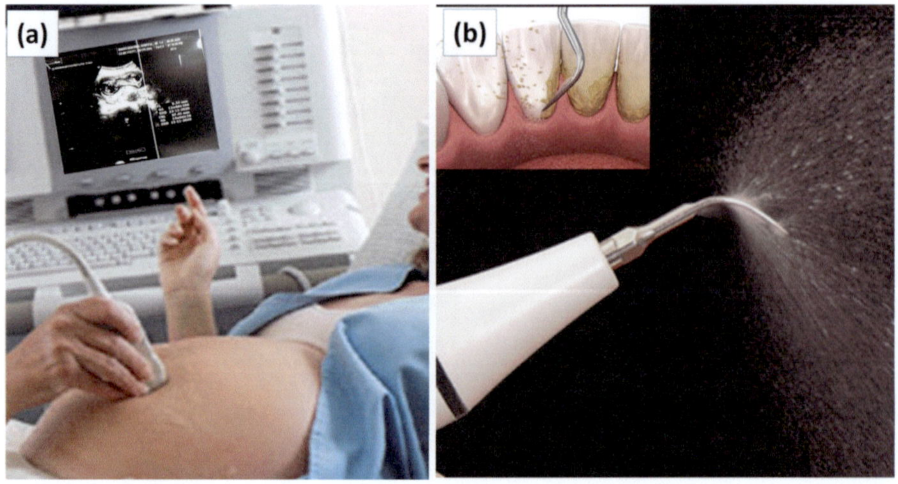

Fig. 2.22 (a) Ultrasound scanners for medical imaging of pregnancy monitoring and (b) ultrasonic dental scalers removal of plaque and calculus

power. These are used in mobiles, earbuds, toys, musical greeting cards, and many more. They can produce more sound in a smaller frequency range. Buzzers are used in many electronic devices, humidifiers, electronic toothbrush, and so on. Cool mist humidifier uses a piezoelectric transducer. It uses ultrasonic sound energy to form a pool of water. The high-frequency ultrasonic vibration splits the pool of water into very small droplets resulting in the atomization of droplets. The minute particles mix with airstream and enter into desired space. Linear piezoelectric units have been introduced to vibrate the hairs of certain electronic toothbrushes [108]. In the case of fast-acting tubular reactions, some procedures require a quick and accurate mechanical start-up, which is not easy to achieve with electromagnetic tubes. Here in this case piezoelectric actuators can fulfill the demand. The piezoelectric actuators provide fast response and lower power consumption, which is a required characteristic. Again, in piezoelectric motors, the expansion and contraction of the piezoelectric actuator can be precisely controlled as long as the supply voltage is controlled [109]. Some engine designs use this to use piezoelectric components to accurately increase the rotor or linear element. With some engine piezoelectric designs, nanometer accuracy can be achieved. Piezo engines operate in a wide range of frequencies but generally, work best in a low-frequency range. In addition to their inherent accuracy, piezoelectric motors can be used in environments with strong magnetic fields or cryogenic temperatures in environments where conventional motors are unlikely to operate. These unique problems are found in NMRI machines, particle accelerators, and other similar environments [110]

2.14.5 Biomedical

Piezoelectric components for medical technology and related disciplines for life sciences must be fast, reliable, and energy efficient. In miniature form, they provide minimally invasive and accurate diagnostic and therapeutic methods. Therapeutic ultrasound refers to the methods by which ultrasound is not only a means of diagnosis, but also a basic element of therapy, for example, tissue ablation, targeted drug delivery, ultrasound thrombolysis, and lithotripsy. Both focused and non-focused ultrasound can be used for these applications [111]. Piezoelectric materials are broadly used in medical appliances like ultrasound scanners for medical imaging. In pregnancy monitoring, an ultrasound scanner is used as shown in Fig. 2.22a. In an ultrasound scanner, the sonographer operating the scanner can view the condition of a fetus inside the womb. He can monitor and visualize the development of the baby. In an ultrasound scanner, a pulse is generated by means of a transducer at a frequency range of about 3–5 MHz. The pulse is sent by putting the transducer at the abdomen of the pregnant woman and the pulse of ultrasound is sent through the tissue and into the womb. The sound gets reflected by muscles and bones, and so on, and the reflected sound is received by a couple of piezotransducers. The transducers generate electrical pulses which are recorded to determine the size and condition of the baby inside the abdomen. Medical devices are generally equipped with a linear sequence containing, for example, 16, 32, or 64 individuals, individually controlled piezoelectric elements [112]. More detailed ceramics are preferred for their low-quality factor because they provide a high range. But they are relatively tiny and operate at low power and high frequency. Piezoelectric transducers are also used for ultrasonic dental scalers for removal of scaling, which is more efficient than traditional scale removal (Fig. 2.22b). In this scaling removal process, the piezoelectric actuator changes its size when induced with electric energy. The scalar vibrates against the tooth, creates sound energy, and in turn removes the scaling. Again, water flows to the tooth root by an external source. Water combined with vibration creates millions of tiny air bubbles, which collapse with high energy

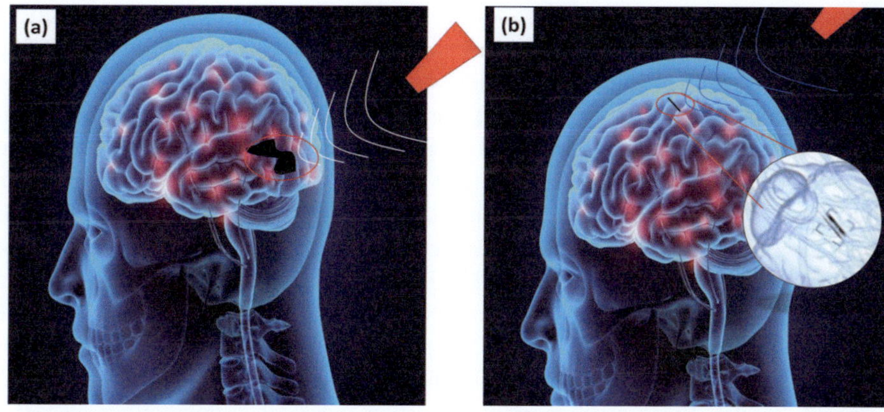

Fig. 2.23 (**a**) Noncontact tissue ablation and (**b**) targeted drug delivery

that removes the cell of bacteria. Piezoelectric scaler generally operates in the range 24–36 kHz with a peak power of 25 W [113]. The tool can be adapted to a variety of tasks, from cleaning to extraction, simply by changing the tip and power level. Such ultrasonic transducers are generally constructed with a ring-shaped piezoelectric monologue, which is glued to the coupling horn that allows the movement to be amplified. Heavy materials such as NCE80 are preferred to deliver high power without excessive heat production. The components of the ring can be replaced with a multi-layer piezo, which will drastically reduce the operating voltage, maintaining the same performance and, as such, reduce the required electrical safety at work in the human body [114, 115].

High-Intensity Focused piezoUltrasound (HIFU) used for noncontact tissue ablation is given in Fig. 2.23a. The ultrasound-based ablation of tissue, for example for the removal of tumors in the prostate or uterus, is carried out extracorporeal and is, therefore, noninvasive. For this therapy, HIFU is projected into the body with the help of piezoelectric elements [116]. To generate a directed wave, many piezoelectric elements are arranged on a spherical dome and excited synchronously to focus the energy in the shock wave. These acoustic waves have various effects in the body: In histotripsy, ultrasound produces cavitation bubbles whose mechanical energy causes cells to burst. Thermal ablation using ultrasound generates heat of over 42 °C by absorbing the wave energy so that the proteins of the cell structures become denaturated and entire tissue areas are coagulated in the sound focus [117]. HIFU can also be used to treat tremors or epilepsy without medication by targeting specific brain structures with heat. Monitoring in these cases is performed simultaneously by MRT. PI Ceramic offers a wide range of piezo components that can be used to build HIFU transducers: discs, plates, and focus bowls are manufactured in customer-specific geometries and electrode designs. The resulting focus elements are suitable for use in strong magnetic fields, for example, in MRT applications [118]. Targeted drug delivery using HIFU enables the release or activation of drugs at a specific site in the body (Fig. 2.23b). Various mechanisms for drug delivery are possible: the basic method is the injection of microspheres filled with medication or gas, which are distributed in the patient's bloodstream during treatment and are only used where they are in the sound field focus. Unstable cavitation caused by ultrasound causes the microspheres to burst and the medication to be released locally. Gas-filled microspheres can be injected to increase the uptake of therapeutic substances. The stable oscillation of these microspheres produces microcurrents which, in combination with the mechanical energy of the acoustic waves, increase the permeability of the blood vessel walls for the active substance in the blood. This sonoporation mechanism is used, for example, for the reversible opening of the blood–brain barrier—a process that can improve the medicinal therapy of neurodegenerative diseases such as Parkinson's or Alzheimer's significantly. In sonodynamic therapy, focused ultrasound waves induce the formation of reactive oxygen species which act cytotoxically on cells, for example, in cancer tumors [119]. As with tissue ablation, HIFU is generated for targeted drug delivery procedures using piezoelectric discs, plates, and focus bowls. Intravascular lithotripsy with miniaturized piezo components is used in a minimally invasive way to reduce atherosclerotic plaques in blood vessels or at the heart valves as well as life-threatening stenoses. Ultrasound waves increase the permeability of the blood vessel wall through sonoporation so that the medication can penetrate better and induce the dissolution of plaques. If a thrombus has already formed, its structure can be loosened with the help of mechanical ultrasound energy. This makes it easier for drugs to penetrate the thrombus and dissolve it, therefore, ensuring the flow of blood to vital organs and preventing embolisms. Miniaturized piezo tubes or tiny plates are suitable for intravascular lithotripsy and thrombectomy since they can generate ultrasound waves even when integrated into the finest catheters. PI ceramic manufactures miniaturized piezo elements with dimensions of less than one millimeter and refines them for specific applications, for example, by contacting them with strands or assembling them on substrates [120].

Some innovative medical practices include the use of ultrasonic actuators to break kidney stones inside the kidney and destroy malignant tissue. In addition, nowadays piezoelectric actuators are incorporated in surgical operation. Without the need for cauterization, the surgical process can be completed by means of an ultrasonic element. This process leads to less tissue damage, faster healing time, and less blood loss. A specific application for the detection of piezoelectric pressure is synthetic leather [121]. Biological quartz microbalance uses piezoelectric materials extensively. It can detect the odor of skin and diseases by changing its chemical composition. The microbalance can be used for various purposes like gas detection, chirality classification, composition analysis, viscosity determination, and many more. Quartz microbalance normally operates in the range of 5–10 MHz [122]. The quartz vibration frequency decreases with an increase in the mass of quartz. But the trend doesn't follow when the mass becomes more than 2%. The technique is also valid for the detection of bacteria.

In addition to bio implants, the piezoelectric materials are extensively used in those medical instruments where the vibratory property is more important. The best property of the piezoelectric transducer is that it is not needed to be biocompatible as it doesn't touch human cells while operating. Typical piezosurgical devices are ring-shaped stacked one upon the other, which is subjected to stress. The stacked actuators enhance the efficiency of the instrument as the electric field intensity is determined by applied voltage and thickness. The voltage is directly proportional to the electric field. As the thickness decreases a higher voltage can be generated inside the instrument [123]. In the instrument, the vibration is transduced toward the tip of the transducer. The instrument is designed in such a way that as the vibration progresses it gets amplified toward the tip. In surgery, piezoelectric devices generally made of ceramic materials are used for delicate operations to preserve the surrounding tissue. By controlling the vibration during surgical operation the surrounding tissues are secured and separation between interfaces is facilitated. Hard tissues like mineralized bones are damaged by frequencies of 25–39 kHz. However, neurovascular tissue is cut at frequencies above 50 kHz. There are no macro vibrations that can cause discomfort to the patient or alter the surrounding tissue. The tip oscillates in a linear direction and can cover a distance of 60–200 μm [124].

Initially, the piezo surgery was used in the dental industry only. In this treatment area many operations like dental implant removal, bone removal take place. Many of the operations require very small space facilitating the use of piezo surgery. In the implant removal process, the ultrasonic wave passes through the interfacial layer thus weakening the bond between tooth and implant. So, implant is removed easily as adhesive force is lessened [125]. Similarly, the collection of graft material is also a very demanding use of piezo surgery. After making preliminary cuts with a saw, ultrasonic vibrations reduce the need for cord blisters. In surgery performed on the lower jawline, protection of the lower alveolar nerve is important for patient recovery. As mentioned earlier, the use of piezo surgery prevents damage to this nerve tissue. By ultrasonic activity, the decomposition of particles occurs which enhances visibility. The piezo surgery has also applications in the field of orthopedic surgery and neurosurgery. But the disadvantage of piezo-surgery used in orthopedic surgery is that here more vibratory force is required for operation. The fragility and high cost of piezo equipments restrict its use in orthopedic surgery. The tip of the transducer may break creating the need for additional support. Again coolant supply should be there as the process lasts long and significant heat is generated [126]. Transdermal drug delivery occurs through the use of a needle which is painful, requires efficient practitioner, and has a chance of spread of disease. A jet injector device fulfills the limitation by providing greater consistency and reducing pain. One can use a new needless microjet injector powered by the rapid expansion of a piezoelectric actuator that provides electronic control of fluid displacement [127].

2.14.6 Energy Harvest

In a piezoelectric crystal, a little amount of electrical energy can be generated by pressing or deforming the crystal. Hence a large amount of electrical energy may be generated by pressing a large number of crystals at a time. Some industry demands electrical energy by the application of force, vibration, pressure, and so on. The electrical energy produced can be used or stored for future use. This is called energy harvesting. Energy harvesting can be done from any piezoelectric type. Scientists are working for energy harvesting from common materials that are available on earth. Inventors come up with new ideas of energy storage in piezoelectric devices, from shoes (Fig. 2.24a) that surround foot movements to heat to keep feet warm, and cell phones that charge outside of movements, from your body (Fig. 2.24b), highways that create street lights (Fig. 2.24c), contact lenses (Fig. 2.24d) that capture energy when you blink, and even devices that generate energy from the pressure of the falling rain [128–130]. The shoe with a built-in piezoelectric transducer has a spring system that vibrates when someone walks. It leads to the storage of energy in the battery. Similarly, in the case of a cellphone, each time it is charged excess energy is stored and at the need, it is discharged. The Footstep microcontroller-based power generation system is operated by a direct piezoelectric [131]. This is used to generate tension with the power of the steps. The energy harvesting process is very essential in public places like bus stops,

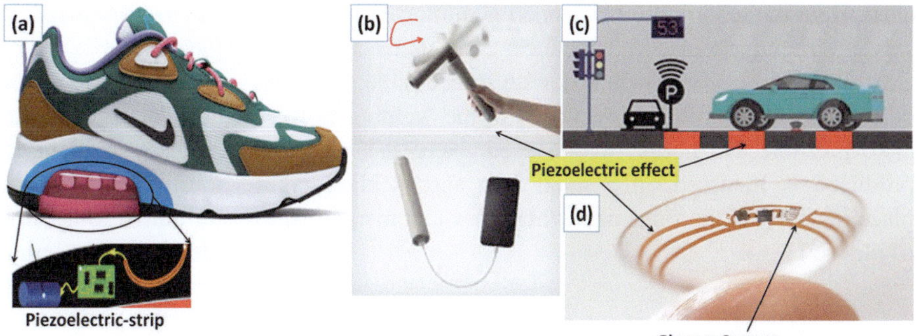

Fig. 2.24 Piezoelectric energy harvest in (**a**) Shoes, (**b**) Cellphones, (**c**) Roads that power streetlights, and (**d**) Contact lenses

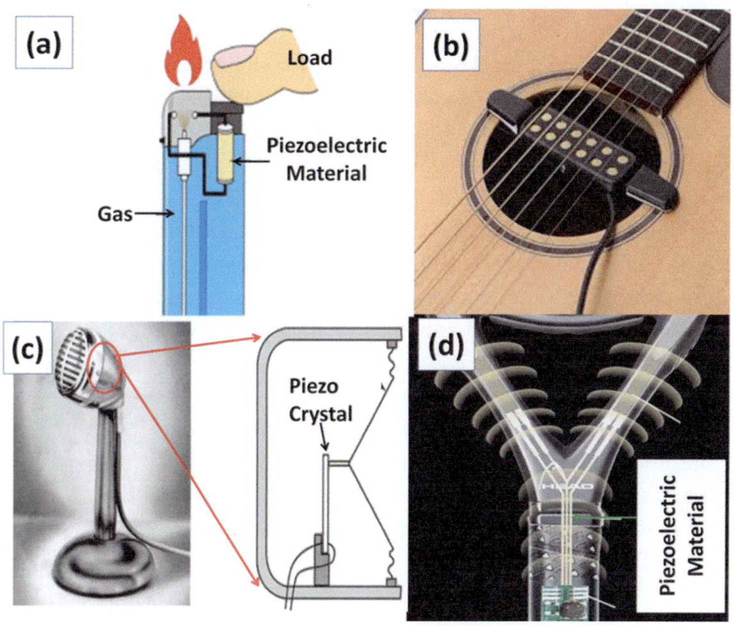

Fig. 2.25 (**a**) Piezoelectric igniter,(**b**) Instrument pickups, (**c**) Microphones, (**d**) Tennis Racquets

stations, theaters, shopping malls, and many more. Therefore, these systems are placed in public places where people walk and have to travel in this system to cross the entrance or exist [132]. These systems can then generate tension at every step of the foot. Hence a piezoelectric sensor is used to measure pressure, power, and acceleration through its change in electrical signals. This system uses a voltmeter to measure the scale, LED lights, weight measurement system, and a battery for a better demonstration of the system. Generally, the piezoelectric system is incorporated in the airport door or restaurant door. When a man steps near the door it opens up automatically without touching. Here, as a man steps a small pressure is exerted on the piezoelectric sensor on the door because of the man's weight. Hence it opens up automatically [133].

2.14.7 Household and Other Application

The piezoelectricity has extensive applications in various household appliances. It is used in music assets. In a piezoelectric igniter, a bottom is used to release a spring-loaded hammer, which in turn strikes a rod-shaped piezoelectric ceramic. Due to the high strain rate in piezoelectric material high voltage is created, which is enough to surpass the spark gap. Hence spark starts and fuel ignites (Fig. 2.25a) [134]. Many acoustic instruments (Fig. 2.25b) utilize piezoelectric pickups to convert acoustic vibration to electrical energy. Generally, the piezoelectric material is placed in between the instrument body and its support. In the case of the violin, as the strings vibrate, they generate electrical signals [135]. Some microphones use piezoelectric materials to convert sound vibrations into electrical output (Fig. 2.25c) [136]. These microphones usually have high output impedances, which have to happen when designing their respective preamplifiers. A somewhat unusual application of piezoelectricity integrates piezoelectric fibers into the neck of a tennis racket (Fig. 2.25d) along with a microcontroller in the handle. When the tennis player shoots the ball, it loses the frame of the racket and generates an electrical output that is powered, reversed, and returned to the fibers [137]. For ultrasonic cleaning piezoelectric actuators are used. Here the material to be cleaned is immersed inside the solvent. Then piezoelectric transducer controls the solvent to clean the material. Many objects with inaccessible areas can be cleaned using this methodology. In the welding process also, piezoelectric horns can be used. Here the vibration produced is focussed on a small area and welding takes place instantly at the same area where they can merge pieces of plastic [138].

2.15 Advantages of Piezoelectric Materials

The rise of piezoelectric technology is directly related to a set of inherent advantages. There is no need for an external force. It is easy to handle and use as it has small dimensions. High-frequency response means that the parameters change very rapidly. The high modulus of elasticity of many piezoelectric materials is comparable to that of many metals and goes up to 10^6 N/m^2. Although piezoelectric sensors are electromechanical systems that react to compression, the sensing elements show almost zero deflection [139]. This gives piezoelectric sensors ruggedness, an extremely high natural frequency, and excellent linearity over a wide amplitude range. Additionally, piezoelectric technology is insensitive to electromagnetic fields and radiation, enabling measurements under harsh conditions. Some materials used (especially gallium phos phate or tourmaline) are extremely stable at high temperatures, enabling sensors to have a working range of up to 1,000 °C [140].

2.16 Limitations of Piezoelectric Materials

Temperature and environmental conditions can affect the behavior of the transducer. They can only measure changing pressure hence they are useless while measuring static parameters. Piezoelectric sensors cannot be used for truly static measurements. A static force results in a fixed amount of charge on the piezoelectric material. In conventional read out electronics, imperfect insulating materials and reduction in internal sensor resistance cause a constant loss of electrons and yields a decreasing signal [141]. Elevated temperatures cause an additional drop in internal resistance and sensitivity. The main effect on the piezoelectric effect is that with increasing pressure loads and temperature, the sensitivity reduces due to twin formation. The output is low so some external circuit is attached to it. It is very difficult to give the desired shape to this material and also desired strength. Two main groups of materials are used for piezoelectric sensors: piezoelectric ceramics and single-crystal materials. The ceramic materials (such as PZT ceramic) have a piezoelectric constant/sensitivity that is roughly two orders of magnitude higher than those of the natural single crystal materials and can be produced by inexpensive sintering processes. The piezoeffect in piezoceramics is "trained," so their high sensitivity degrades over time. This degradation is highly correlated with increased temperature. The less-sensitive, natural, single-crystal materials (gallium phosphate, quartz, tourmaline), when carefully handled have a higher, almost unlimited long-term stability [142]. There are also new single-crystal materials commercially available such as Lead Magnesium Niobate-Lead Titanate (PMN-PT). These materials offer improved sensitivity over PZT but have a lower maximum operating temperature and are currently more expensive to manufacture.

2.17 Summary and Future Prospects

The most widespread and highly demanding material for sensors and actuators as per the current market scenario is a piezoelectric material. This chapter describes the basic mechanism involved in both direct and inverse piezoelectric materials. Various fabrication procedure of ceramics, polymer, metal, and composites used in a piezoelectric application has been explained. Process parameter (such as temperature, accuracy, linearity, resolution, stiffness, resonant frequency, mechanical amplification, quality factor, bandwidth, frequency constant, humidity, load ratings, vacuum), as well as the operational constant (piezoelectric voltage constant, permittivity constant, elastic and compliance, electromechanical coupling factor, Young's modulus, dielectric dissipation factor, piezoelectric frequency constant), has been discussed. Electric behavior, dielectric behavior, elasticity behavior, electromechanical behavior, coupling coefficient, material damping, mechanical loss, sound velocity, and acoustic impedance in piezoelectric materials have been discussed. This chapter describes a few of the potential applications, which can develop an interest in the researcher to further improvement of the materials as per the various requirement of future applications.

References

1. Campbell, S.: A short history of sonography in obstetrics and gynaecology. Facts. Views. Vis. Obgyn. **5**(3), 213–229 (2013)
2. Wudy, F., Stock, C., Gores, H.J.: Measurement methods, electrochemical: quartz microbalance. In: Garche, J. (ed.) Encyclopedia of Electrochemical Power Sources, pp. 660–672. Elsevier, Netherlands (2009). https://doi.org/10.1016/B978-044452745-5.00079-4
3. Dahiya, R.S., Valle, M.: Fundamentals of piezoelectricity. In: Robotic Tactile Sensing. Springer, Newyork. https://doi.org/10.1007/978-94-007-0579-1
4. https://www.americanpiezo.com/blog/top-uses-of-piezoelectricity-in-everyday-applications/. Accessed 25 Jan 2019

5. Kulkarni, H., Zohaib, K., Khusru, A., Aiyappa, K.S.: Application of piezoelectric technology in automotive systems. Mater. Today Proc. **5**, 21299–21304 (2018). https://doi.org/10.1016/j.matpr.2018.06.532
6. https://onscale.com/piezoelectricity/history-of-piezoelectricity/
7. Manjón-Sanz, A.M., Dolgos, M.R.: Applications of piezoelectrics: old and new. Chem. Mater. **30**(24), 8718–8726 (2018). https://doi.org/10.1021/acs.chemmater.8b03296
8. https://piezo.com/pages/history-of-piezoelectricity. Accessed 12 May 2019
9. https://en.wikipedia.org/wiki/Piezoelectricity. Accessed 12 May 2019
10. Whatmore, R.W.: Piezoelectric and pyroelectric materials and their applications. In: Miller, L.S., Mullin, J.B. (eds.) Electronic Materials. Springer, Boston, MA (1991). https://doi.org/10.1007/978-1-4615-3818-9_19
11. Uchino, K.: 1 - The development of piezoelectric materials and the new perspective. In: Uchino, K. (ed.) Woodhead Publishing Series in Electronic and Optical Materials, Advanced Piezoelectric Materials, pp. 1–85., ISBN. Woodhead Publishing, Sawston (2010). https://doi.org/10.1533/9781845699758.1. isbn:9781845695347
12. Zimmerman, D.: 'A more creditable way': The discovery of active sonar, the Langevin–Chilowsky patent dispute and the Royal Commission on Awards to Inventors. War in History. **25**(1), 48–68 (2018). https://doi.org/10.1177/0968344516651308
13. Hwang, W.-S., Park, H.C.: Finite element modeling of piezoelectric sensors and actuators. AIAAJ. **31**, 5 (1993). https://doi.org/10.2514/3.11707
14. Damjanovic, D.: Contributions to the piezoelectric effect in ferroelectric single crystals and ceramics. J. Am. Ceramic Soc. **88**(10), 2663–2676 (2005). https://doi.org/10.1111/j.1551-2916.2005.00671.x
15. Martin, R.M.: Piezoelectricity. Phys. Rev. B. **5**, 1607 (1972). https://doi.org/10.1103/PhysRevB.5.1607
16. Lefeuvre, E., Lallart, M., Richard, C., Guyomar, D.: Piezoelectric material-based energy harvesting devices: advance of SSH optimization techniques (1999–2009). In: Suaste-Gomez, E. (ed.) Piezoelectric Ceramics. InTech, London (2010) ISBN: 978-953-307-122-0, http://www.intechopen.com/books/piezoelectric-ceramics/piezoelectric-material-based-energy-harvestingdevices-advance-of-ssh-optimization-techniques-1999-2
17. MacCurdy, R.B., Reissmana, T., Garcia, E.: Energy management of multi-component power harvesting systems. Proc. SPIE. **6928**, 692809 (2008). 0277-786X/08/$18, http://robertmaccurdy.com/docs/2008_MacCurdy-Energy%20management%20of%20multi-component%20power%20harvesting%20systems.pdf). https://doi.org/10.1117/12.776545
18. Lamuta, C., Candamano, S., Crea, F., Pagnotta, L.: Direct piezoelectric effect in geopolymeric mortars. Mater. Des. **107**, 57–64 (2016). https://doi.org/10.1016/j.matdes.2016.05.108
19. Oshiki, M., Fukada, E.: Inverse piezoelectric effect and electrostrictive effect in polarized poly(vinylidene fluoride) films. J. Mater. Sci. **10**, 1–6 (1975). https://doi.org/10.1007/BF00541025
20. Bechmann, R.: Elastic, piezoelectric, and dielectric constants of polarized barium titanate ceramics and some applications of the piezoelectric equations. J. Acoust. Soc. Am. **28**, 347 (1956). https://doi.org/10.1121/1.1908324
21. Kargarnovin, M.H., Najafizadeh, M.M., Viliani, N.S.: Vibration control of a functionally graded material plate patched with piezoelectric actuators and sensors under a constant electric charge. Smart Mater. Struct. **16**(4). https://doi.org/10.1088/0964-1726/16/4/037
22. Ryu, J., Carazo, A.V., Uchino, K., Kim, H.-E.: Magnetoelectric properties in piezoelectric and magnetostrictive laminate composites. Jpn. J. Appl. Phys. **40**(Part 1), 8 (2001). https://doi.org/10.1143/JJAP.40.4948
23. Sihvola, A.H., Kong, J.A.: Effective permittivity of dielectric mixtures. IEEE Trans. Geosci. Remote Sens. **26**(4), 420–429 (1988). https://doi.org/10.1109/36.3045
24. Min, K.-B., Jing, L.: Numerical determination of the equivalent elastic compliance tensor for fractured rock masses using the distinct element method. Int. J. Rock Mech. Min. Sci. **40**(6), 795–816 (2003). https://doi.org/10.1016/S1365-1609(03)00038-8
25. Uchino, K.: Chapter 1 - The development of piezoelectric materials and the new perspective. In: Uchino, K. (ed.) Advanced Piezoelectric Materials Woodhead Publishing in Materials, 2nd edn, pp. 1–92., ISBN: 9780081021354. Woodhead Publishing (2017). https://doi.org/10.1016/B978-0-08-102135-4.00001-1
26. Yang, Z., Zhou, S., Jean, Z., Inman, D.: High-performance piezoelectric energy harvesters and their applications. Joule. **2**(4), 642–697 (2018). https://doi.org/10.1016/j.joule.2018.03.011
27. Fett, T., Munz, D., Thun, G.: Young's modulus of soft PZT from partial unloading tests. Ferroelectrics. **274**(1), 67–81 (2002). https://doi.org/10.1080/00150190213958
28. Egerton Dolores, L., Dillon, M.: Piezoelectric and dielectric properties of ceramics in the system potassium—sodium niobate. J. Am. Ceramic. Soc. **42**(9), 438–442 (1959). https://doi.org/10.1111/j.1151-2916.1959.tb12971.x
29. Kwok, K.W., Chan, H.L.W., Choy, C.L.: Evaluation of the material parameters of piezoelectric materials by various methods. IEEE Trans. Ultrason. Ferroelectr. Freq. Contr. **44**(4), 733–742 (1997). https://doi.org/10.1109/58.655188
30. Holden, A.N., Mason, W.P.: The elastic, dielectric and piezoelectric constants of heavy-water Rochelle salt. Phys. Rev. **57**, 54 (1940). https://doi.org/10.1103/PhysRev.57.54
31. Brown, C., Kell, R., Taylor, R., Thomas, L.: Piezoelectric materials, a review of progress. IRE Trans. Comp. Parts. **9**(4), 193–211 (1962). https://doi.org/10.1109/TCP.1962.1136768
32. Mitsuyu, T., Yamazaki, O., Ohji, K., Wasa, K.: Piezoelectric thin films of zinc oxide for saw devices. Ferroelectrics. **42**, 1 (1982). https://doi.org/10.1080/00150198208008116
33. Zhang, S., Fei, L.: High performance ferroelectric relaxor-PbTiO_3 single crystals: status and perspective. J. Appl. Phys. **111**, 031301 (2012). https://doi.org/10.1063/1.3679521
34. Tressler, J.F., Alkoy, S., Newnham, R.E.: Piezoelectric sensors and sensor materials. J. Electroceram. **2**, 257–272 (1998). https://doi.org/10.1023/A:1009926623551
35. Gerhard-Multhaupt, R.: Less can be more. Holes in polymers lead to a new paradigm of piezoelectric materials for electret transducers. IEEE Trans. Dielectr. Electr. Insul. **9**(5), 850–859 (2002). https://doi.org/10.1109/TDEI.2002.1038668
36. Lushcheĭkin, G.A.: New polymer-containing piezoelectric materials. Phys. Solid State. **48**, 1023–1025 (2006). https://doi.org/10.1134/S1063783406060011
37. Patel, I., Siores, E., Shah, T.: Utilisation of smart polymers and ceramic based piezoelectric materials for scavenging wasted energy. Sens. Actuator A Phys. **159**(2), 213–218 (2010). https://doi.org/10.1016/j.sna.2010.03.022
38. Furukawa, T.: Piezoelectricity and pyroelectricity in polymers. IEEE Trans. Electr. Insul. **24**(3), 375–394 (1989). https://doi.org/10.1109/14.30878
39. Ramadan, K.S., Sameoto, D., Evoy, S.: A review of piezoelectric polymers as functional materials for electromechanical transducers. Smart Mater. Struct. **23**, 3 (2014). https://doi.org/10.1088/0964-1726/23/3/033001
40. Dunn Martin, L., Minoru, T.: An analysis of piezoelectric composite materials containing ellipsoidal inhomogeneities. Proc. R. Soc. Lond. A. **443**, 265–287 (1993). https://doi.org/10.1098/rspa.1993.0145

41. Gururaja, T.R., Schulze, W.A., Cross, L.E., Newnham, R.E.: Piezoelectric composite materials for ultrasonic transducer applications. Part II: evaluation of ultrasonic medical applications. IEEE Trans. Son. Ultrason. **32**(4), 499–513 (1985). https://doi.org/10.1109/T-SU.1985.31624
42. Tiana, J., Han, P., Huang, X., Pan, H.: Improved stability for piezoelectric crystals grown in the lead indium niobate–lead magnesium niobate–lead titanate system. Appl. Phys. Lett. **91**, 222903 (2007). https://doi.org/10.1063/1.2817743
43. Shabbir, G., Kojima, S.: Anomalous variations in elastic properties of lead zirconate niobate-lead titanate single crystals in the vicinity of its ferroelectric phase transition. EPL (Europhysics Letters). **105** (5), 57001. https://doi.org/10.1209/0295-5075/105/57001
44. Gualtieri, J.G., Kosinski, J.A., Ballato, A.: Piezoelectric materials for acoustic wave applications. IEEE Trans. Ultrason. Ferroelectr. Freq. Contr. **41**(1), 53–59 (Jan. 1994). https://doi.org/10.1109/58.265820
45. Wang, Q.-M., Zhang, Q., Xu, B., Liu, R., Eric Cross, L.: Nonlinear piezoelectric behavior of ceramic bending mode actuators under strong electric fields. J. Appl. Phys. **86**, 3352 (1999). https://doi.org/10.1063/1.371213
46. Wasa, K., Matsushima, T., Adachi, H., Kanno, I., Kotera, H.: Thin-film piezoelectric materials for a better energy harvesting MEMS. J. Microelectromech. Syst. **21**(2), 451–457 (2012). https://doi.org/10.1109/JMEMS.2011.2181156
47. Yan, Z., Jiang, L.: Surface effects on the electromechanical coupling and bending behaviours of piezoelectric nanowires. J. Phys. D Appl. Phys. **44**(7). https://doi.org/10.1088/0022-3727/44/7/075404
48. Mahajan, R.P., Patankar, K.K., Kothale, M.B., et al.: Conductivity, dielectric behaviour and magnetoelectric effect in copper ferrite-barium titanate composites. Bull. Mater. Sci. **23**, 273–279 (2000). https://doi.org/10.1007/BF02720082
49. de Jong, M., Chen, W., Geerlings, H., et al.: A database to enable discovery and design of piezoelectric materials. Sci. Data. **2**, 150053 (2015). https://doi.org/10.1038/sdata.2015.53
50. Meguid, S.A., Deng, W.: Electro-elastic interaction between a screw dislocation and an elliptical inhomogeneity in piezoelectric materials. Int. J. Solids Struct. **35**(13), 1467–1482 (1998). https://doi.org/10.1016/S0020-7683(97)00116-9
51. Yang, F.: Electromechanical interaction of linear piezoelectric materials with a surface electrode. J. Mater. Sci. **39**, 2811–2820 (2004). https://doi.org/10.1023/B:JMSC.0000021458.32183.73
52. Eyvazian, A., Musharavati, F., Tarlochan, F., Pasharavesh, A., Rajak, D.K., Husain, M.B., Tran, T.N.: Free vibration of FG-GPLRC conical panel on elastic foundation. Struct. Eng. Mech. **75**(1), 1–18 (2020)
53. Qing, H.X.: Damping properties of epoxy-embedded piezoelectric composites. Key Eng. Mater. **512–515**., Trans Tech Publications Ltd., 1342–1346 (2012). https://doi.org/10.4028/www.scientific.net/kem.512-515.1342
54. Shekhani, H.N., Uchino, K.: Characterization of mechanical loss in piezoelectric materials using temperature and vibration measurements. J. Am. Ceram. Soc. **97**(9), 2810–2814 (2014). https://doi.org/10.1111/jace.12998
55. Bowen, C.R., Perry, A., Lewis, A.C.F., Kara, H.: Processing and properties of porous piezoelectric materials with high hydrostatic figures of merit. J. Eur. Ceram. Soc. **24**(2), 541–545 (2004). https://doi.org/10.1016/S0955-2219(03)00194-8
56. Alvarez-Arenas, T.E.G.: Acoustic impedance matching of piezoelectric transducers to the air. IEEE Trans. Ultrason. Ferroelectr. Freq. Contr. **51**(5), 624–633 (2004). https://doi.org/10.1109/TUFFC.2004.1320834
57. Sammoura, F., Shelton, S., Akhbari, S., Horsley, D., Lin, L.: A two-port piezoelectric micromachined ultrasonic transducer. In: 2014 Joint IEEE International Symposium on the Applications of Ferroelectric, pp. 1–4. International Workshop on Acoustic Transduction Materials and Devices & Workshop on Piezoresponse Force Microscopy, State College, PA (2014). https://doi.org/10.1109/ISAF.2014.6923004
58. Li, H., Tian, C., Daniel Deng, Z.: Energy harvesting from low frequency applications using piezoelectric materials. Appl. Phys. Rev. **1**, 041301 (2014). https://doi.org/10.1063/1.4900845
59. Liu, Y., Shan, J., Gabbert, U., Qi, N.: Hysteresis and creep modeling and compensation for a piezoelectric actuator using a fractional-order Maxwell resistive capacitor approach. Smart Mater. Struct. **22**, 11. https://doi.org/10.1088/0964-1726/22/11/115020
60. Zhang, S., Xia, R., Lebrun, L., Anderson, D., Shrout, T.R.: Piezoelectric materials for high power, high temperature applications. Mater. Lett. **59**(27), 3471–3475 (2005). https://doi.org/10.1016/j.matlet.2005.06.016
61. Wu, Z.: A wide linearity range current sensor based on piezoelectric effect. IEEE Sens. J. **17**(11), 3298–3301 (2017). https://doi.org/10.1109/JSEN.2017.2692258
62. Newcomb, C.V., Flinn, I.: Improving the linearity of piezoelectric ceramic actuators. Electron. Lett. **18**(11), 442–444 (1982). https://doi.org/10.1049/el:19820301
63. Kon, S., Horowitz, R.: A high-resolution MEMS piezoelectric strain sensor for structural vibration detection. IEEE Sens. J. **8** (12), 2027–2035 (2008). https://doi.org/10.1109/JSEN.2008.2006708
64. Clark, W.W.: Semi-active vibration control with piezoelectric materials as variable-stiffness actuators. In: Proceedings Volume SPIE 3672, Smart Structures and Materials 1999: Passive Damping and Isolation. SPIE, Newport Beach, CA (1999). https://doi.org/10.1117/12.349775
65. Morris, D.J., Youngsman, J.M., Anderson, M.J., Bahr, D.F.: A resonant frequency tunable, extensional mode piezoelectric vibration harvesting mechanism. Smart Mater. Struct. **17**, 6. https://doi.org/10.1088/0964-1726/17/6/065021
66. Shen, D., Choe, S.-Y., Kim, D.-J.: Analysis of piezoelectric materials for energy harvesting devices under high-g vibrations. Jpn. J. Appl. Phys. **46**(Part 1), 10A. https://doi.org/10.1143/JJAP.46.6755
67. Feenstra, J., Granstrom, J., Sodano, H.: Energy harvesting through a backpack employing a mechanically amplified piezoelectric stack. Mech. Syst. Signal Process. **22**(3), 721–734 (2008). https://doi.org/10.1016/j.ymssp.2007.09.015
68. Shekhani, H.N., Uchino, K.: Evaluation of the mechanical quality factor under high power conditions in piezoelectric ceramics from electrical power. J. Eur. Ceram. Soc. **35**(2), 541–544 (2015). https://doi.org/10.1016/j.jeurceramsoc.2014.08.038
69. Lobl, H.P., et al.: Piezoelectric materials for BAW resonators and filters. In: IEEE Ultrasonics Symposium Proceedings of an International Symposium (Cat. No.01CH37263), vol. 1, pp. 807–811. IEEE, Atlanta, GA (2001). https://doi.org/10.1109/ULTSYM.2001.991845
70. Sherrit, S., Mukherjee, B.K.: The use of complex material constants to model the dynamic response of piezoelectric materials. In: 1998 IEEE Ultrasonics Symposium. Proceedings (Cat. No. 98CH36102), vol. 1, pp. 633–640. IEEE, Sendai, Japan (1998). https://doi.org/10.1109/ULTSYM.1998.762229
71. Bur, A.J.: Measurements of the dynamic piezoelectric properties of bone as a function of temperature and humidity. J. Biomech. **9**(8), 495–507 (1976). https://doi.org/10.1016/0021-9290(76)90066-X
72. John, S., Sirohi, J., Wang, G., Wereley, N.M.: Comparison of piezoelectric, magnetostrictive, and electrostrictive hybrid hydraulic actuators. J. Intell. Mater. Syst. Struct. **18**(10), 1035–1048 (2007). https://doi.org/10.1177/1045389X06072355

73. Kuang, Z.-B., Yuan, X.-G.: Reflection and transmission of waves in pyroelectric and piezoelectric materials. J. Sound Vib. **330**(6), 1111–1120 (2011). https://doi.org/10.1016/j.jsv.2010.09.026
74. Neurgaonkar, R.R., Oliver, J.R., Nelson, J.G., Cross, L.E.: Piezoelectric and ferroelectric properties of La-modified and unmodified tungsten bronze Pb0.6Ba0.4Nb2O6 dense ceramics. Mater. Res. Bull. **26**(8), 771–777 (1991). https://doi.org/10.1016/0025-5408(91)90066-U
75. Amarande, L., Miclea, C., Tanasoiu, C.: Effect of excess PbO on the structure and piezoelectric properties of Bi-modified $PbTiO_3$ ceramics. J. Eur. Ceram. Soc. **22**(8), 1269–1275 (2002). https://doi.org/10.1016/S0955-2219(01)00442-3
76. Kang, M.-G., Jung, W.-S., Kang, C.-Y., Yoon, S.-J.: Recent progress on PZT based piezoelectric energy harvesting technologies. Actuators. **5**(1), 5 (2016). https://doi.org/10.3390/act5010005
77. Su, Y.M.: A determination of the rate of evaporation of zinc at atmospheric pressure. Masters Theses. (1967). https://scholarsmine.mst.edu/masters_theses/5141
78. Wang, Q.-M., Xiao-Hong, D., Xu, B., Cross, L.E.: Electromechanical coupling and output efficiency of piezoelectric bending actuators. IEEE Trans. Ultrason. Ferroelectr. Freq. Contr. **46**(3), 638–646 (1999). https://doi.org/10.1109/58.764850
79. Papila, M., Sheplak, M., Cattafesta, L.N.: Optimization of clamped circular piezoelectric composite actuators. Sens. Actuator A Phys. **147**(1), 310–323 (2008). https://doi.org/10.1016/j.sna.2008.05.018
80. Wang, Q.-M., Cross, L.E.: Constitutive equations of symmetrical triple layer piezoelectric benders. IEEE Trans. Ultrason. Ferroelectr. Freq. Contr. **46**(6), 1343–1351 (1999). https://doi.org/10.1109/58.808857
81. Randall, C.A., Kelnberger, A., Yang, G.Y., et al.: High strain piezoelectric multilayer actuators—a material science and engineering challenge. J. Electroceram. **14**, 177–191 (2005). https://doi.org/10.1007/s10832-005-0956-5
82. Juuti, J., Kordás, K., Lonnakko, R., Moilanen, V.-P., Leppävuori, S.: Mechanically amplified large displacement piezoelectric actuators. Sens. Actuator A Phys. **120**(1), 225–231 (2005). https://doi.org/10.1016/j.sna.2004.11.016
83. Sapper, E., Gassmann, A., Gjødvad, L., Jo, W., Granzow, T., Rödel, J.: Cycling stability of lead-free BNT–8BT and BNT–6BT–3KNN multilayer actuators and bulk ceramics. J. Eur. Ceram. Soc. **34**(3), 653–661 (2014). https://doi.org/10.1016/j.jeurceramsoc.2013.09.006
84. Dausch, D.E., Castellucci, J.B., Chou, D.R., von Ramm, O.T.: Theory and operation of 2-D array piezoelectric micromachined ultrasound transducers. IEEE Trans. Ultrason. Ferroelectr. Freq. Contr. **55**(11), 2484–2492 (2008). https://doi.org/10.1109/TUFFC.956
85. Fang, H.-B., Liu, J.-Q., Xu, Z.-Y., Lu, D., Wang, L., Chen, D., Cai, B.-C., Liu, Y.: Fabrication and performance of MEMS-based piezoelectric power generator for vibration energy harvesting. Microelectron. J. **37**(11), 1280–1284 (2006). https://doi.org/10.1016/j.mejo.2006.07.023
86. Guerrero, C., Sánchez, F., Ferrater, C., Roldán, J., García-Cuenca, M.V., Varela, M.: Pulsed laser deposition of epitaxial PbZrxTi1−xO3 ferroelectric capacitors with $LaNiO_3$ and $SrRuO_3$ electrodes. Appl. Surf. Sci. **168**(1–4), 219–222., ISSN 0169-4332 (2000). https://doi.org/10.1016/S0169-4332(00)00601-2
87. Damjanovic, D., Newnham, R.E.: Electrostrictive and piezoelectric materials for actuator applications. J. Intell. Mater. Syst. Struct. **3** (2), 190–208 (1992). https://doi.org/10.1177/1045389X9200300201
88. Leadbetter, J.: The design of ultrasonic lead magnesium niobate-lead titanate (PMN-PT) composite transducers for power and signal delivery to implanted hearing aids. Proc. Mtgs. Acoust. **19** (030029) (2013). https://doi.org/10.1121/1.4800367
89. Platt, S.R., Farritor, S., Garvin, K., Haider, H.: The use of piezoelectric ceramics for electric power generation within orthopedic implants. IEEE/ASME Trans. Mechatron. **10**(4), 455–461 (2005). https://doi.org/10.1109/TMECH.2005.852482
90. Zhang, Q.M., Pan, W.Y., Cross, L.E.: Laser interferometer for the study of piezoelectric and electrostrictive strains. J. Appl. Phys. **63**, 2492 (1988). https://doi.org/10.1063/1.341027
91. https://www.marketsandmarkets.com/Market-Reports/piezoelectric-devices-market-256019882.html. Accessed 15 May 2019
92. https://industry-experts.com/verticals/chemicals-and-materials/piezoelectric-materials-a-global-market-overview. Accessed 15 May 2019
93. https://www.globenewswire.com/news-release/2018/08/20/1553756/0/en/Global-Piezoelectric-Materials-Market-Report-2018-2014-2023-Market-Value-2017-2023-CAGRs.html. Accessed 15 May 2019
94. http://www.noliac.com/applications/aerospace-space/. Accessed 15 May 2019
95. https://www.piceramic.com/en/products/piezoelectric-materials/. Accessed 15 May 2019
96. Hoshyarmanesh, H., Ghodsi, M., Kim, M., Cho, H.H., Park, H.-H.: Temperature effects on electromechanical response of deposited piezoelectric sensors used in structural health monitoring of aerospace structures. Sensors. **19**(12), 2019 (2805). https://doi.org/10.3390/s19122805
97. Gluhihs, S., Kovalovs, A.: Reduction of the vibration in a helicopter, blade using piezoelectric actuators. Aviation. **10**(2), 3–6 (2006). https://doi.org/10.1080/16487788.2006.9635927
98. http://www.noliac.com/applications/item/show/active-vibration-damping/. Accessed 15 May 2019
99. Jbaily, A., Yeung, R.W.: Piezoelectric devices for ocean energy: a brief survey. J. Ocean Eng. Mar. Energy. **1**, 101–118 (2015). https://doi.org/10.1007/s40722-014-0008-9
100. Ishihara, C., et al.: Holographic 3-D imaging system using encoded wavefront: an underwater imaging system. In: Jones, J.P. (ed.) Acoustical Imaging, vol. 21. Springer, Boston, MA (1995). https://doi.org/10.1007/978-1-4615-1943-0_85
101. Wei, H., Wang, H., Xia, Y., Cui, D., Shi, Y., Dong, M., Liu, C., Ding, T., Zhang, J., Ma, Y., Wang, N., Wang, Z., Sun, Y., Wei, J. R., Guo, Z.: An overview of lead-free piezoelectric materials and devices. J. Mater. Chem. C. **6**, 12446–12467 (2018). https://doi.org/10.1039/C8TC04515A
102. Jones, D.J., et al.: Piezoelectric materials and their applications. In: Key Engineering Materials, vol. 122–124, pp. 71–144. Trans Tech Publications, Ltd. (1996). https://doi.org/10.4028/www.scientific.net/kem.122-124.71
103. Sanchez, A., Blanc, S., Yuste, P., Serrano, J.J.: A Low Cost and High Efficient Acoustic Modem for Underwater Sensor Networks, pp. 1–10. OCEANS 2011 IEEE, Santander, Spain (2011). https://doi.org/10.1109/Oceans-Spain.2011.6003428
104. Turner, R.C., Fuierer, P.A., Newnham, R.E., Shrout, T.R.: Materials for high temperature acoustic and vibration sensors: a review. Appl. Acoust. **41**(4), 299–324 (1994). https://doi.org/10.1016/0003-682X(94)90091-4
105. Powell, D.J., Hayward, G., Ting, R.Y.: Unidimensional modeling of multi-layered piezoelectric transducer structures. IEEE Trans. Ultrason. Ferroelectr. Freq. Contr. **45**(3), 667–677 (1998). https://doi.org/10.1109/58.677611
106. Ueno, T., Higuchi, T.: High sensitive and heat-resistant magnetic sensor using magnetostrictive/piezoelectric laminate composite. IEEE Trans. Magnet. **41**(10), 3670–3672 (2005). https://doi.org/10.1109/TMAG.2005.854795
107. Abe, S., Sakakibara, Y., Tomita, M., Shinoda, K.: Piezoelectric actuator and a piezoelectric pump injector incorporating the same. Patent no: US4553059A
108. Brewer, G.K., McInnes, J.C., Bayeh, D., Bennett, F.J., Taylor, R. K., Ballard, D.A., Barrett, G.A.: Ultrasonic toothbrushes employing an acoustic waveguide. Patent no: US7269873B2

109. Morita, T.: Miniature piezoelectric motors. Sens. Actuator A Phys. **103**(3), 291–300 (2003). https://doi.org/10.1016/S0924-4247(02)00405-3
110. Wang, Y., Cole, G.A., Su, H., Pilitsis, J.G., Fischer, G.S.: MRI compatibility evaluation of a piezoelectric actuator system for a neural interventional robot. In: 2009 Annual International Conference of the IEEE Engineering in Medicine and Biology Society, pp. 6072–6075. IEEE, Minneapolis, MN (2009). https://doi.org/10.1109/IEMBS.2009.5334206
111. Tandon, B., Blaker, J.J., Cartmell, S.H.: Piezoelectric materials as stimulatory biomedical materials and scaffolds for bone repair. Acta Biomater. **73**, 1–20 (2018). https://doi.org/10.1016/j.actbio.2018.04.026
112. Genovese, M.: Ultrasound transducers. J. Diagn. Med. Sonogr. **32**(1), 48–53 (2016). https://doi.org/10.1177/8756479315618207
113. http://www.noliac.com/news/item/show/piezos-in-health-care/. Accessed 18 May 2019
114. Xu, W.C., Lam, K.H., Choy, S.H., Chan, H.L.W.: A stepped horn transducer driven by lead-free piezoelectric ceramics. Integr. Ferroelectr. **89**, 1 (2007). https://doi.org/10.1080/10584580601077609
115. Stevenson, T., Quast, T., Bartl, G., Schmitz-Kempen, T., Weaver, P.M.: Surface mapping of field-induced piezoelectric strain at elevated temperature employing full-field interferometry. IEEE Trans. Ultrason. Ferroelectr. Freq. Contr. **62**(1), 88–96 (2015). https://doi.org/10.1109/TUFFC.2014.006683
116. Lai, N., Lai, Q., Liu, K., Xiong, G.: Quasi-self focusing high intensity and large power ultrasonic transducer. Patent no: US7602672B2
117. Li, Z., Zhang, T., Fan, F., Gao, F., Ji, H., Yang, L.: Piezoelectric materials as sonodynamic sensitizers to safely ablate tumors: a case study using black phosphorus. J. Phys. Chem. Lett. **11**(4), 1228–1238 (2020). https://doi.org/10.1021/acs.jpclett.9b03769
118. Bonomo, C., Brunetto, P., Fortuna, L., Giannone, P., Graziani, S., Strazzeri, S.: A tactile sensor for biomedical applications based on IPMCs. IEEE Sens. J. **8**(8), 1486–1493 (2008). https://doi.org/10.1109/JSEN.2008.920723
119. Gong, F., Cheng, L., Yang, N., Betzer, O., Feng, L., Zhou, Q., Li, Y., Chen, R., Popovtzer, R., Liu, Z.: Ultrasmall oxygen-deficient bimetallic oxide MnWOX nanoparticles for depletion of endogenous GSH and enhanced sonodynamic cancer therapy. Adv. Mater. **31**(23), 1900730 (2019). https://doi.org/10.1002/adma.201900730
120. Bazan-Peregrino, M., Arvanitis, C.D., Rifai, B., Seymour, L.W., Coussios, C.C.: Ultrasound-induced cavitation enhances the delivery and therapeutic efficacy of an oncolytic virus in an in vitro model. J. Control. Release. **157**(2), 235–242 (2012). https://doi.org/10.1016/j.jconrel.2011.09.086
121. Miller, N., Lingeman, J.: Treatment of kidney stones: current lithotripsy devices are proving less effective in some cases. Nat. Rev. Urol. **3**, 236–237 (2006). https://doi.org/10.1038/ncpuro0480
122. Olin, J.G., Sem, G.J.: Piezoelectric microbalance for monitoring the mass concentration of suspended particles. Atmos. Environ. (1967). **5**(8), 653–668 (1971). https://doi.org/10.1016/0004-6981(71)90123-5
123. Jaffe, H., Berlincourt, D.A.: Piezoelectric transducer materials. Proc. IEEE. **53**(10), 1372–1386 (1965). https://doi.org/10.1109/PROC.1965.4253
124. Peddigari, M., Kim, G.-Y., Park, C.H., Min, Y., Kim, J.-W., Ahn, C.-W., Choi, J.-J., Hahn, B.-D., Choi, J.-H., Park, D.-S., Hong, J.-K., Yeom, J.-T., Park, K.-I., Jeong, D.-Y., Yoon, W.-H., Ryu, J., Hwang, G.-T.: A comparison study of fatigue behavior of hard and soft piezoelectric single crystal macro-fiber composites for vibration energy harvesting. Sensors. **19**, 2196 (2019)
125. Chen-Glasser, M., Li, P., Ryu, J., Hong, S.: Piezoelectric Materials for Medical Applications. IntechOpen, London (2018). https://doi.org/10.5772/intechopen.76963
126. Gleizal, A., Bera, J., Lavandier, B., et al.: Piezoelectric osteotomy: a new technique for bone surgery—advantages in craniofacial surgery. Childs Nerv. Syst. **23**, 509–513 (2007). https://doi.org/10.1007/s00381-006-0250-0
127. Ma, B.: A PZT insulin pump integrated with a silicon micro needle array for transdermal drug delivery. In: 56th Electronic Components and Technology Conference 2006, p. 5. Springer, San Diego, CA (2006). https://doi.org/10.1109/ECTC.2006.1645724
128. Akaydin, H.D., Elvin, N., Andreopoulos, Y.: Energy harvesting from highly unsteady fluid flows using piezoelectric materials. J. Intell. Mater. Syst. Struct. **21**(13), 1263–1278 (2010). https://doi.org/10.1177/1045389X10366317
129. Rocha, J.G., Goncalves, L.M., Rocha, P.F., Silva, M.P., Lanceros-Mendez, S.: Energy harvesting from piezoelectric materials fully integrated in footwear. IEEE Transact. Industr. Electron. **57**(3), 813–819 (2010). https://doi.org/10.1109/TIE.2009.2028360
130. Kim, H.S., Kim, J., Kim, J.: A review of piezoelectric energy harvesting based on vibration. Int. J. Precis. Eng. Manuf. **12**, 1129–1141 (2011). https://doi.org/10.1007/s12541-011-0151-3
131. Veena, R.M., Reddy, B.H., Shyni, S.M.: Maximum energy harvesting from electromagnetic micro generators by footsteps using photo sensor. In: 2016 International Conference on Computation of Power, Energy Information and Commuincation (ICCPEIC), pp. 757–761. IEEE, Chennai (2016). https://doi.org/10.1109/ICCPEIC.2016.7557321
132. Li, X., Strezov, V.: Modelling piezoelectric energy harvesting potential in an educational building. Energy Convers. Manage. **85**, 435–442 (2014). https://doi.org/10.1016/j.enconman.2014.05.096
133. Solban, M.M., Moussa, R.R.: Piezoelectric tiles is a sustainable approach for designing interior spaces and creating self-sustain projects. In: IOP Conference Series: Earth and Environmental Science Simulation for Sustainable Built Environment, vol. 397, pp. 28–30. IOP Science, New Cairo, Egypt (2019). https://doi.org/10.1088/1755-1315/397/1/012020
134. Akkaya Oy, S., Özdemir, A.E.: Usage of piezoelectric material and generating electricity. In: 2016 IEEE International Conference on Renewable Energy Research and Applications (ICRERA), pp. 63–66. ICRERA, Birmingham (2016). https://doi.org/10.1109/ICRERA.2016.7884363
135. Barcus, L.M., Berry, J.F.: Musical instrument utilizing piezoelectric transducer. Patent no: US3325580A (1966)
136. Rahaman, A., Ishfaque, A., Jung, H., Kim, B.: Bio-inspired rectangular shaped piezoelectric MEMS directional microphone. IEEE Sens. J. **19**(1), 88–96 (2019). https://doi.org/10.1109/JSEN.2018.2873781
137. Lammer, H.: Racket with self-powered piezoelectric damping system. Patent no: US6974397B2 (2000)
138. Güney, M., Eşkinat, E.: Modeling of multilayered piezoelectric transducers with ultrasonic welding application. Smart Mater. Struct. **16**, 2. https://doi.org/10.1088/0964-1726/16/2/036. Accessed 20 May 2019
139. https://www.elprocus.com/what-is-a-piezoelectric-material-working/. Accessed 20 May 2019
140. https://www.semiconductorforu.com/advantages-disadvantages-piezoelectric-transducer/. Accessed 20 Aug 2019
141. Damjanovic, D.: Materials for high temperature piezoelectric transducers. Curr. Opinion Solid State Mater. Sci. **3**(5), 469–473 (1998). https://doi.org/10.1016/S1359-0286(98)80009-0
142. Schwarz, R.B., Vuorinen, J.F.: Resonant ultrasound spectroscopy: applications, current status and limitations. J. Alloys Comp. **310**(1–2), 243–250 (2000). https://doi.org/10.1016/S0925-8388(00)00925-7

3 Nanomaterials

Abstract

This chapter describes nanomaterials with respect to nanoscience and nanotechnology in the 1–100 nm range. The history gives a wonderful idea of how the gradual development of nanomaterials from ancient age to the current age. This chapter discusses the alteration in properties with the size reduction from bulk scale to nanoscale. Various classes of nanomaterials based on dimension and composition have been described. Different synthesis methods such as gas-, liquid- and solid-phase processes have been discussed for each category of nanomaterials. Nanotechnology network in various fields such as aero industries, automotive and naval industry, electronic industry, medical industries, energy harvest industries, food industries, textile industries, household application, and other industries has been discussed. Current problems and difficulties associated with nanomaterials have been discussed for future developments.

Keywords

Nanotechnology · Nanomaterials · Nano-scale · Size effects · Carbon-based materials · Graphene · Fullerene · Carbon nanotubes · Metal-based nanomaterials · Polymer-based nanomaterials · Nanocomposites

3.1 Introduction to Nanoscale World

The word "nano" is a Greek word meaning dwarf and is used to measure the one-billionth of any unit (denoted by 10^{-9} m). Nanomaterial is **defined as** the "material with any external dimension in the nanoscale or having internal structure or surface structure in the nanoscale, where nanoscale ranges approximately from 1 to 100 nm" [1–3]. Nanoscale material is defined as the material having at least one of its dimensions in the nano range (1–100 nm). Nanoscience is a multidisciplinary topic that deals with the investigation of the physical and chemical properties of materials in the nanoscale range. Nanoengineering or naotechnology deals with the design, manufacture, and application of structures, devices, and systems based on nanoscale. Nanoscale science and technology refers to the understanding and controlled manipulation of structures and phenomena with nanoscale dimensions, as illustrated in Fig. 3.1. Currently, scientists can observe nanoscaled structures using various types of equipment, such as scanning electron microscope (SEM), field emission scanning electron microscope (FESEM), transmission electron microscope (TEM), field ion microscope (FIM), scanning tunneling mircoscope (STM), and atomic force microscope (AFM) [4]. These instruments use electron beam to illuminate the structure and facilitate analysis along with the higher resolution and higher magnifications up to 1 million times. The size of the nanotechnology market is expected to reach $121.80 billion by 2025 after growing at an average annual growth rate of 14.3% in 2020–2025 [5]. The growth of the global market for nanotechnology is being driven by factors such as various increases in the introduction of nanoscience in a wide variety of materials. In addition, increasing government support and R&D funding as well as the advent of nanotechnology devices with their superior behavior offers lucrative opportunities for the nanotechnology market. Nanosciences and nanotechnology are the study of nanoparticles and devices that are used in all industrial sectors, including automobile, chemical, biomedical, mechanical, electronics, and almost all industries [6].

Nature is also a reservoir of design in nanoscaled structures. Day-by-day, scientists are continuously detecting many nanosystems in mineral, plants, and animals having superior activities, such as hydrophobic behavior, self-cleaning behavior of leaves, for example, a native Asian plant, lotus leaf; nanoscale optics in peacock feather (Fig. 3.2); zeolite mineral that have gas storage capacity, geckos can moves easily against gravity, upside down on a

A. Behera, *Advanced Materials*, https://doi.org/10.1007/978-3-030-80359-9_3

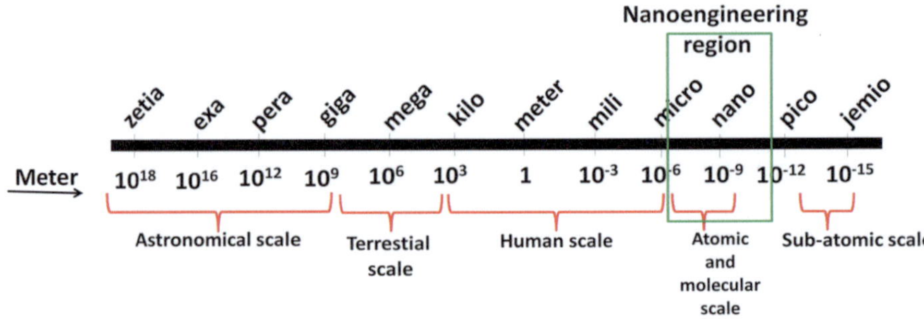

Fig. 3.1 Nanoscale range used in nanoengineering

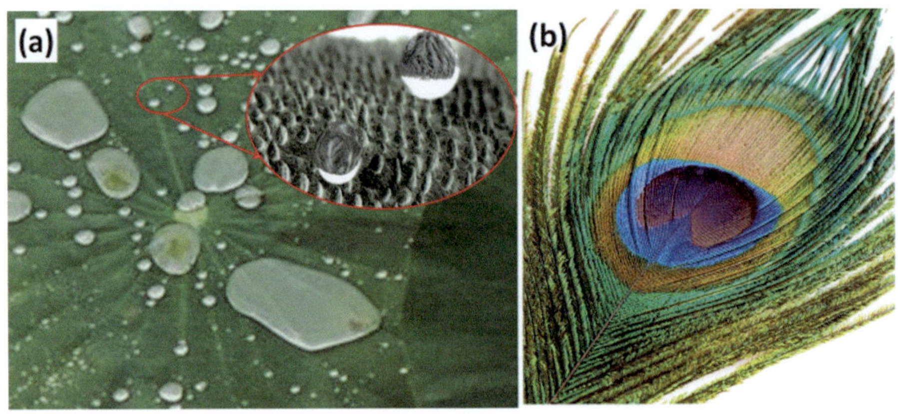

Fig. 3.2 Nanoscale in nature: (**a**) Nanobuds on lotus leaf responsible for superhydrophobicity and (**b**) nanoscale optics in peacock feather

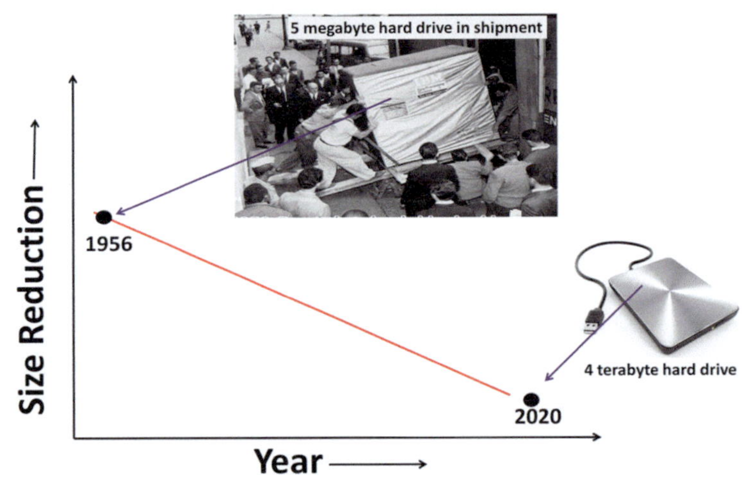

Fig. 3.3 Reduction in size from larger volume 1956s hard drive of 5 MB to nanotechnology product 4 TB hard drive

ceiling; glow of fireflies in the night; and iridescent colors in butterflies and many more. These findings are outstanding solutions and much beneficial for advanced technology [7, 8]. Figure 3.3 shows an advantage of nanotechnology that can reduce the volume of the hard drive from 1956 (5 MB memory, volume is of half of a general room) to 2000s (4 TB memory, volume is approximately 7 × 5 × 1 inches) [9]. Figure 3.4 shows a nanofly used in defense system which is also the beauty of nanotechnology [10].

3.2 History of Nanotechnology

Manipulating material at the atomic and molecular level to create new functions and properties depicts the advanced modern concept. Past artisans also controlled matter in the smallest of spaces. There are a number of relatively famous examples of ancient artifacts made with nanocomposites. The Lycurgus cup, for example, is an impressive decorative Roman treasure from around 400 AD; it is made of glass

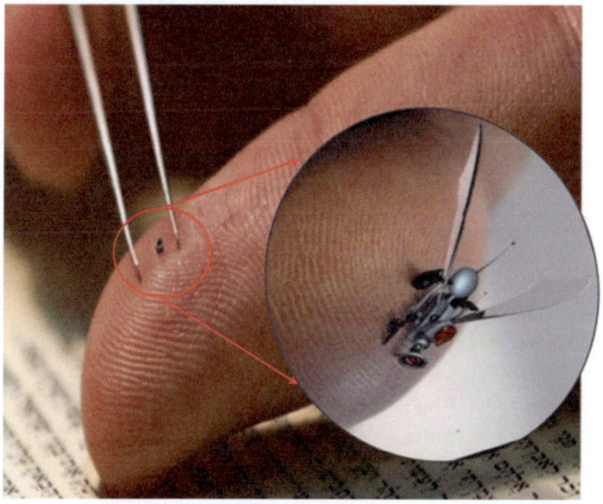

Fig. 3.4 A synthetic nanodevice: nanoflies

that changes color when light shines through it [11]. The glass contains gold and silver nanoparticles that are distributed in such a way that the glass appears green in the reflected light but shows a bright red when light passes through the glass. In the pre-Columbian Mayan city of Chichén Itzá, a corrosion-resistant blue pigment called Maya Blue was discovered, which was first produced in AD 800 [12]. It is a complex clay-containing material with nanopores in which the indigo dye has been chemically combined to produce an environmentally stable pigment. The Damascus steel swords from the middle east were manufactured between AD 300 and 1700 and are known for their impressive strength, tear resistance, and exceptionally sharp edge. The steel blades contain oriented tubular structures and nanoscale wires, which almost certainly improves the material properties. Ian Freestone from the University College London Institute of Archeology studied the Lycurgus Cup who were astonished about their high qualification on small scale [13]. It indicates they developed materials by trial and error similar to that of biological evolution. It is a mystery that how did they get such a homogenous distribution of nanoparticles? Historical structures are the result of centuries of trial and error with artisans who pass on their skills and knowledge from generation to generation along with the evolution in biology [14]. They did not know the processes in solids. "High-resolution microscopic analysis is used to uncover the nanostructure of these artifacts, but such an analysis cannot tell us how they were made." How did these metals dissolve in the glass? The great modern interest in nanotechnology is largely recognized by the brilliant American physicist Richard Feynman (1918–1988) [15]. In 1974, the Japanese engineering professor Norio Taniguchi called this field "nanotechnology". Nanotechnology really took off in the 1980s. That was when nanotechnology evangelist Dr. K. Eric Drexler first published his pioneering book *Engines of Creation: The Coming Era of Nanotechnology* [16]. In 1991, another Japanese scientist, Sumio Iijima, discovered carbon nanotubes that aroused great interest in new technical applications. Nanotechnology reached another important milestone this year with the awarding of the 2016 Nobel Prize in Chemistry to Jean-Pierre Sauvage, J. Fraser Stoddart, and Bernard Feringa, three scientists whose innovative work sparked the idea of converting molecules into machines [17].

3.3 Can We Make Small Devices?

Scientists are currently working with nanotechnology to send computers inside living body for a more accurate and specific diagnosis. Albert Swiston, a biomaterials scientist at MIT, tests a small pill that combines a microphone, a thermometer, and a battery to take multiple measurements simultaneously from the interior part of the body. It is the latest in a range of ingestible computers, such as a Proteus sensor that measures how patients take prescription medication, a Philips VitalSense that measures a patient's temperature, or a PillCam that allows people to skip colonoscopies [18]. The small pill device is so small about the size of a large multivitamin pill that you can take, and it can hear your heart and lungs. It has a built-in thermometer that shows the heart rate and respiratory rate and core temperature of the body. And then, this information can be sent outside the body to a smartphone or a laptop and tell them these vital functions without touching the body. It is not portable, it is ingestible.

When a computer takes apart and looks inside, there are many electronic parts, including a so-called microprocessor or, as it is commonly known, a chip. There are many, many small switches in this chip called transistors. There are around a 100 million switches on the latest chips. That is almost as many light switches as in the entire buildings in Rourkela. To make computers faster and faster, scientists and engineers had to make transistors smaller and smaller, so small that its average transistor is around 100 nanometers, which means that it could fit 1000 of them the width of a human hair [19]. The first transistor made in 1947 was about 2 in. wide. Today, these small transistors are manufactured using nanotechnology. With small transistors, one can take a laptop that is more powerful than computers that weighed a few kilograms 10 years ago. These tiny transistors also use very little power, so laptops run more on batteries.

In 1965, Gordon E. Moore, co-founder of Intel (NASDAQ: INTC), postulated that the number of transistors that can be packaged in a given space unit would double approximately every 2 years [20]. Nowadays, transistors installed on silicon chips are doubled every 18 months instead of every 2 years. G. E. Moore did not call this observation "Moore's Law" nor did he set about creating a

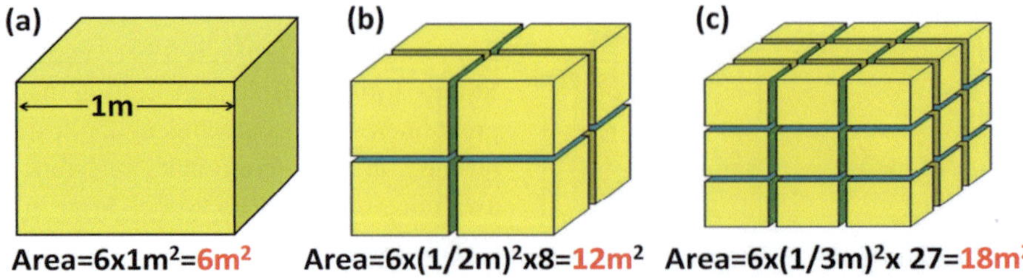

Fig. 3.5 Increasing trend of the relative surface area of a cube from (**a**) to (**c**) having initial 1 nm^3 volume. (**a**) It has a surface area of 6 m^2. (**b**) If this cube of the same volume is divided into eight small cubes, then the total surface area is 12 m^2. (**c**) If the same cube volume is divided into 27 small cubes, then the total surface area 18 m^2.

"law." Moore made this statement by noting emerging trends in chip manufacturing at Intel. Ultimately, Moore's vision became a prediction, which in turn became the golden rule known as Moore's Law. According to Moore's Law, one can assume that the speed and capacity of the computer will increase every 2 years, and there is a cost reduction. Another principle of Moore's law states that this growth is exponential. Experts agree that computers should reach the physical limits of Moore's law sometime in the 2020s [21]. Eventually, high transistor temperature would make it impossible to produce smaller circuits. This is because the cooling of the transistors requires more energy than the amount of energy that is already flowing through the transistors [22].

3.4 Size Effects

The properties of nanomaterials are very much different from the larger scale materials. Size reduction will be from bulk size to nano-size based on two principal factors.

- relative surface area and
- quantum confinement effect

When the particle size decreases, a greater portion of atoms is found at the surface compared to those inside. Therefore, nanomaterials have a greater surface area and surface energy per unit mass compared to larger particles (Fig. 3.5). The specific surface area of a system and surface to volume ratio are inversely proportional to particle size, and both increase sharply for particles <100 nm in diameter. *Quantum confinement* effect describes the physics of electron properties in solids with the reductions in particle size. This refers to the confinement of the motion of randomly moving electron and restricts its motion in specific energy levels (discreteness) [23]. This effect does not come into play by going from macro to micro-dimensions. However, it becomes dominant when the nanometer size range is reached. Quantum effects can begin to dominate the behavior of matter at the nanoscale—particularly at the lower end (single digit and lower than tens of nanometers)—affecting the optical, electrical, and magnetic behavior of materials. Materials reduced to the nanoscale can suddenly show very different properties compared to what they show on a macroscale. For instance, opaque substances become transparent (Cu); inert materials become catalysts (Pt); stable materials turn combustible (Al); solids turn into liquids at room temperature (Au); and insulators become conductors (Si) [24]. For nanosized objects, the concepts of "particles" and "waves" become mixed: light, being an electromagnetic wave (wave-like behavior), also behaves like it is made of particles in certain situations (particle behavior, also the same is true for what we traditionally think of as particles: electrons, protons, etc.; they sometimes behave like waves). Quantum mechanics is a way of describing this strange behavior and is a very powerful tool. Quantum mechanics is needed because otherwise, we would not understand many fundamental things, such as why are some materials metallic and others are insulating; where does magnetism come from; how are molecules formed; and why are different substances different colors. We should use the ideas of quantum mechanics whenever we are trying to understand the behavior of small objects—for example, why transistors cannot just keep on getting smaller? Quantum mechanics is more fundamental than Newton's laws which are a limiting form of quantum mechanics [25]. The acceptance by the general physics community of quantum mechanics is due to its accurate prediction of the physical behavior of systems, including systems where Newtonian mechanics fails.

3.5 Properties of Nanomaterials

The nano-scale is marked as the point at which the properties of a material change. Above this point, the properties of materials are treated as the bulk material effects, that is, related to what is the composition, what is the percentage of constituents, and how they are bound. Below a critical nano-point, the properties of a material change, although the type of atoms present and their relative orientations and geometry remain important. When we approach the nanoscale dimensions, we approach toward the atomic or molecular

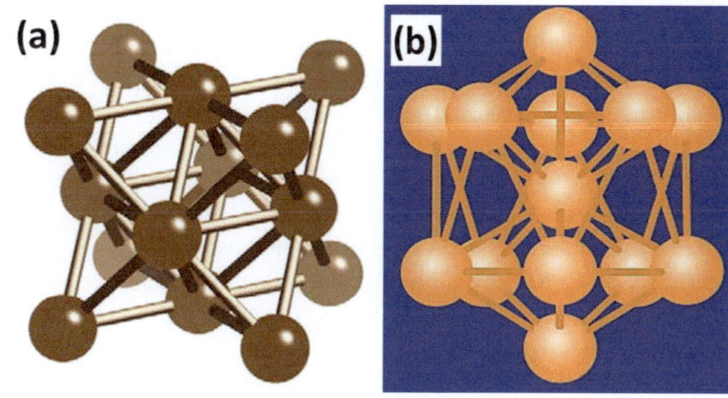

Fig. 3.6 Crystal structure of (**a**) Bulk (FCC) gold and (**b**) Nano (Icosahedral) Gold

scales. Atoms are the building blocks of all matter. They can be assembled in many ways to get the product per requirement. Both the chemistry and the geometric arrangement of the atoms can affect the property of the material. So if we have the ability to build matter atom by atom, we can do wonders. For example, we know that both graphite and diamond are made of pure carbon. So if we were able to convert the (carbon) atoms to graphite at our discretion, it would in principle be possible to make diamonds! Or if we could rearrange the atoms (silicon and oxygen) in the sand (and add some other trace elements), it would be possible to make a computer chip! Engineering at the nano level can significantly change the properties of products [26].

3.5.1 Structure Properties

To understand the novel properties of nanostructured materials, we need to understand the structure and its interrelationship properties. The microstructural features of importance in nanomaterials include grain size, morphology, nature of grain boundaries, interphase interfaces, nature of intragrain defects, composition profiles across grains and interfaces, and residual impurities from processing (such as point, linear, planar, and volume defects) and lead to the structure-sensitive properties of materials. Most nanomaterials are crystalline in nature, and they possess unique properties. The atomic structure of nanostructured material is unlike that seen in glass or conventional crystalline materials because of the large volume fraction of grain boundaries and interfaces. Hence, in nanocrystalline materials, a substantial fraction of atoms lies at the grain boundaries and interfaces. It is not surprising that the behavior of nanocrystalline materials is decided to a large extent by these defects, and as such, nanomaterials exhibit vastly different properties compared to bulk materials. The main defect types observed in nanocrystals are vacancies, grain boundaries, dislocations, twins, stacking faults, and triple junctions [27]. When going from a bulk material to a nanocrystal, the structure is affected, thus leading to changes in lattice parameters and consequently the possibility of having new phases associated with the residual strain on the particles arises. The crystal structure of nanomaterials may or may not be same as the bulk one but has different lattice parameters. For example, Au and Al nanoparticles are icosahedral in structure, but in bulk form, they are face-centered cubic (FCC) (Fig. 3.6). Indium is face-centered tetrahedral in bulk form but FCC when the size <6.5 nm [28]. In nanomaterials, the interatomic spacing decreases compared to bulk due to the long-range electrostatic forces and the short-range core–core repulsion. In Al, a decrease in separation from 2.86 Å to 2.81 Å results in the decrease in binding energy from 3.39 eV to 2.77 eV [29]. Nanoparticles can form suspensions. This is possible since the interaction of the particle surface with the solvent is strong enough to overcome density differences. In bulk materials, these interactions usually result in a material either sinking or floating in a liquid. Nanoparticles can be encountered as better aerosols (solids/liquids in air), suspensions (solids in liquids), or as emulsions (liquids in liquids) [30].

The changes in lattice parameters phenomenon have been observed for many materials either for spherical particles or for thin films. X-ray diffraction technique is one of the most used techniques for evaluating the structural properties of nanomaterials. In general, the diffraction pattern of crystalline nanoscale materials exhibits broadened and shifted peaks when compared to bulk, and these changes are associated with both size and strain. The effects of size and strain on the broadening of X-ray diffraction peaks in nanocrystals can be separated by constructing the so-called Williamson-Hall plots. The full width half maximum β_{total} can be written as originating from the broadening due to size β_{size} plus the contribution from strain β_{strain} as follows in Eq. 3.1:

$$\beta_{total} = \beta_{strain} + \beta_{size} = \frac{4\Delta d sin\Theta}{d cos\Theta} + \frac{k\lambda}{D cos\Theta}, \tag{3.1}$$

where Δd is the difference of d-spacing corresponding to a given diffraction peak, k is the Scherrer constant (in general assumed as 0.9 for spherical particles) is related to the shape of the nanocrystals, λ is the X-ray wavelength, and D is the average diffracted volume that is related to the nanocrystal size.

The second term of the Eq. 3.1 is the standard Scherrer's model applying for measuring the average crystal size [31] when considering a negligible microstrain at line broadening. The first term in the equation is related to microstrain that is defined as being the deviation of plane distance relative to their mean value, that is, $\Delta d/d$. By constructing a plot of $\beta_{total}cos\Theta$ vs. $4sin\Theta$, one obtains a line whereby the intercept is related to the inverse of crystal size and the slope yields the strain. The role of reduced coordination of surface atoms on the nanocrystals is clearly unveiled in the structural properties. The main effect is lattice contraction by up to 3% for some metallic nanocrystals [32]. The lattice contraction has been discussed in terms of a pressure that arises from the residual stress by using Laplace's law.

$$\Delta P = \frac{4\gamma}{D}, \tag{3.2}$$

where D is the diameter of a liquid droplet and γ is the surface tension.

For solid materials, the theory has been adapted and the strain can be expressed in terms of compressibility κ of bulk phase and surface tension γ. Following Wulff's theorem, the deviation of lattice parameter can be written as Eq. 3.3.

$$\frac{\Delta a}{a} = -\frac{4k\gamma}{3D}. \tag{3.3}$$

By assuming that κ does not change as size is reduced, the parameter γ can be estimated using the ($\Delta a/a$) vs. ($1/D$) plot. Even a small change in size can lead to strain at GPa levels which is enough for stabilizing high-pressure phase at ambient pressure. For example, internal stress is estimated at 2.5 GPa for 40 nm $BaTiO_3$ nanocrystals [33]. In thin films, similar properties are observed, but in this case, the substrate strain is due to lattice mismatch which promotes the thin film to be highly stressed such as observed for the ferroelectric $PbTiO_3$. By comparing the Raman frequencies observed in the thin film to those observed for bulk at high pressure, it was possible to estimate that a 92-nm-thick $PbTiO_3$ film has 1.7 GPa of strain. Other mechanisms are behind the lattice parameter changes in nanocrystals. For example, in the case of CeO_2, it is observed as a lattice expansion when the nanocrystal size decreases, and the phenomena have been identified as being due to the presence of oxygen vacancies [34].

3.5.2 Thermal Properties

The surface atoms play a significant role in determining the thermodynamic properties of nanostructured materials. The reduced coordination number of the surface atoms considerably increases the surface energy resulting in atomic diffusion at comparatively lower temperatures. The melting point of CdS nanoparticles of diameters below 3 nm is as low as 400 °C which is much lower than the bulk CdS melting point of 1600 °C [35]. Also, the gold nanoparticles with a diameter less than 3 nm experience a much lower melting temperature around 300 °C when compared to the bulk gold melting point of 1064 °C. This is due to the increase in surface-to-volume atoms ratio that provides a tremendous driving force for diffusion. Thus, the melting point goes on decreasing with a decrease in size. Moreover, the density of surface atoms varies considerably for different crystallographic planes, possibly leading to different thermodynamic properties. In general, increasing the number of grain boundaries will enhance phonon scattering at the disordered boundaries, resulting in lower thermal conductivity. Thus, nanocrystalline materials would be expected to have lower thermal conductivity compared to conventional materials. However, as the grain sizes assume nanodimensions, their size becomes comparable to the mean free paths of phonons that transport thermal energy [36]. Thus, nanomaterials can show widely different properties compared to coarse-grained materials due to the photon confinement and quantization effects of photon transport. It has been observed that in addition to the grain size, the shape also influences the thermal properties of nanomaterials. For example, one-dimensional nanowires may offer ultralow thermal conductivities, quite different from that of carbon nanotubes. In nanowires, the quantum confinement of phonons in one dimension can result in additional polarization modes compared to that observed in bulk solids. The strong phonon–phonon interactions and enhanced scattering at grain boundaries result in a significant reduction in the thermal conductivity of nanostructures [37]. Silicon nanowires are known to exhibit thermal conductivity at least about two orders of magnitude smaller than that of bulk silicon. In contrast, the tubular structures of carbon nanotubes result in an extremely high (~6600 W/mK) thermal conductivity along the axial direction. However, high anisotropy in their heat transport property is observed, making the thermal transport direction-dependent. Also, the presence of interface dislocations and defects can contribute to enhanced boundary scattering. All these factors can contribute to the lower thermal conductivity of multilayered nanostructured films. Due to higher surface energy, sintering of nanosize particles can take place at lower temperatures, over shorter time scales, than for larger particles. In theory, this does not affect the density of the final product, although flow difficulties and the tendency of nanoparticles to agglomerate complicates the matter. Also, the absorption of solar radiation is much higher in materials composed of nanoparticles than it is in thin films of continuous sheets of material. In both solar PV and solar thermal applications, by controlling the size, shape, and material of the particles, it is possible to control solar absorption. Zinc oxide particles have been found to have superior UV-blocking properties compared to its bulk substitute. This is one of the reasons why it is often used in the preparation of sunscreen lotions [38].

3.5.3 Mechanical Properties

The mechanical behavior such as tensile behavior, strain hardening, creep, fatigue, and deformation of nanocrystalline materials differs considerably from that of its bulk counterparts due to the complex interplay of defects in structures. Some of the special mechanical properties that were first observed for nanostructured materials are: The elastic modulus is 30–50% lower than that of conventional grain size materials and the hardness and strength were very high (e.g., the hardness values for pure nanocrystalline metals of ~10 nm grain size being up to seven times higher than for conventional coarse-grained metals of >1 μm) [39, 40]. In general, tensile and compressive strengths in almost all materials show significantly high values in the nanometer range. Nanoscale multilayers made of metal or ceramic showed an ultrahigh hardness. The Hall–Petch slope is negative below a critical grain size, which shows that the hardness decreases with decreasing grain size in the nanoscale range [41]. Ductility, possibly superplastic behavior, is observed at low temperature in brittle or intermetallic ceramics with a nanoscale range, possibly due to diffusion deformation mechanisms. Some novel nanostructures, which differ greatly from the bulk structure in terms of the atomic structure arrangement, have very different mechanical properties. For example, single or multi-walled carbon nanotubes have high mechanical strength and high elastic limit, which in turn lead to considerable flexibility. Bulk steel is much stronger than carbon (graphite), but in the nano range, carbon nanotube (CNT) is 100 times stronger than nanosteel and very flexible. Nanophase ceramics are of particular interest because they are more ductile at elevated temperatures compared to coarse-grained ceramics. Clay nanoparticles increase the reinforcement when installed in polymer matrices and lead to stronger plastics. These nanoparticles are hard and impart their properties to the polymer (plastic) [42].

The major factors contributing to the difference in the mechanical behavior of nanostructured materials are (1) very high fraction of atoms residing at grain boundaries and triple junctions exert a strong influence on the behavior of defects during application of stress. The influence of grain size on dislocation stability also has a significant influence on dislocation dynamics for material design and the mechanical properties of materials. Since dislocation interactions are the predominant deformation mechanisms that influence the mechanical property of coarse-grained materials, the plastic deformation mode of nanomaterials could be significantly different. The densities of dislocations, surface area to volume ratio, and grain size greatly influence the mechanical properties of the solid. The grain boundary sliding results in an enhancement in damping capacity (ability to absorb energy by converting mechanical energy into heat) of nanostructured materials. A decrease in grain size strongly affects strength and hardness. Single- and multi-walled carbon nanotubes exhibit high mechanical strengths and high elastic limits that result in considerable mechanical flexibility and reversible deformation. (2) Due to the abovementioned reasons, alternate deformation mechanisms like grain boundary migration/sliding, crack growth, and so on become important. (3) The porosity level in nanomaterials is strongly dependent on the processing technique and can also be significant, for example, in the case of mechanically alloyed specimens. Even after agglomeration, the pore size can be smaller than or equal to grain size. The presence of porosity can have a significant effect on the mechanical behavior of nanomaterials. (4) Finally, there can be segregation of different solutes at the grain boundaries in nanomaterials, which can influence the mechanical properties. The strength and toughness of both ceramics and metals can be greatly enhanced if they are made out of nanoscaled crystallites instead of the usual micron-sized grains [43]. This effect is already being employed extensively to make superior ceramics and tungsten carbide–cobalt composites that are used for industrial machinery, cutting tools, abrasives, bricks, pipes, floor tiles, and jewelry. Cutting tools made of nanomaterials, such as tungsten carbide, tantalum carbide, and titanium carbide, are much harder, much more wear-resistant, erosion-resistant, and last longer than their conventional (large-grained) counterparts [44].

3.5.4 Chemical Properties

Nanoscale structures have a very high surface area to volume ratio and potentially different crystallographic structures that may be lead to a radical alteration in chemical reactivity [45]. Significant factors influencing the catalytic activity and selectivity are surface structure, mobility of the active species as well as the mobility of the adsorbates on these active species. Nanomaterials have various advantages over its bulk counterpart such as short-range ordering, enhanced interaction with environments due to a large number of dangling bonds, great variety of the valence band electron structure, and self-structuring for optimum performance in chemisorptions. Catalysis using a finely divided nanoscale system can increase the rate, selectivity, and efficiency of chemical reactions. The electronic bands in metals become narrower when the size is reduced from bulk which changes the value of ionization potential. The ionization potentials at nanosize are higher than that for the bulk materials. Internal surface area can be increased by introducing atomic defects such as dislocations to enhance species diffusivity and chemical reactivity [46]. This enhances their use in the field of oxidation/reduction chemistry with many expected applications in fields like photocatalysis or photodegradation and detoxification of chemical waste and environmental pollutants. Nanoparticles can be employed to remove the contaminations in a medium through a chemical reaction to make them harmless. For example, groundwater can be purified by removing carbon tetrachloride from it using iron nanoparticles [47]. Gold-tipped carbon nanotubes can clean polluted water by trapping oil drops from polluted water. Light-activated nanoparticles like titanium dioxide are continued to be studied for their ability to remove contaminants from various media. Photoactive titanium dioxide (TiO_2) nanoparticles are used for cleaning polluted waters by removing various toxic metal ions such as mercury, cadmium, arsenic, chromium, and copper through reduction. Nanoparticles can also be used in treating water and contaminated air with various organic compounds, dyes, and pesticides [48]. The corrosion investigations of nanocrystalline materials showed a more intensive active anodic dissolution, compared to conventional materials. This difference in behavior can be explained by the small grain size, which results in a larger fraction of high-energy grain boundary defects in the material. Such high-energy grain boundary sites act as preferred anodic dissolution sites and can thus result in increased corrosion rates of nanocrystalline alloys [49].

3.5.5 Optical Properties

In bulk materials, optical emission and absorption depend on the transition between valence band and conduction band. Large changes in optical properties such as color are seen in low-dimensional semiconductors and metals, that is, nanoparticles often possess unexpected optical properties as they are small enough to confine their electrons and produce quantum effects [50]. In nanostructured systems, the effect of reduced dimensionality on electronic structure greatly affects the energies of the highest occupied molecular orbital and the lowest unoccupied molecular orbital. The transition between these states results in optical emission and absorption. Particularly, metals and semiconductors exhibit large changes in optical properties as a function of particle size. In semiconductor micro-crystallites, three-dimensional quantum confinement effects can be observed when the particle size approaches to exciton Bohr radius. This confinement results in novel optical properties. Due to these reasons, they have gained much attention and are being utilized for various applications in optoelectronic devices, such as optical data storage and high-speed optical communication. Optical properties such as fluorescence emission are also specifically dependent on the size of the nanocrystals. For example, it has been illustrated that with suitable size modifications, CdSe nanocrystals can fluoresce throughout the visible range of the electromagnetic spectrum. Size-dependent optical properties are exhibited by Au and Ag nanoparticles. Gold nanoparticles appear deep-red to black in solution as size reduces [51]. Colloidal solutions of Au nanoparticles have deep-red color which is progressively more yellow as the particle size increases as a result of surface plasmon resonance in low-dimensional materials. Semiconductor nanoparticles in the form of quantum dots show size-dependent behavior in the frequency and intensity of light emission as well as modified nonlinear optical properties and enhanced gain for certain emission energy or wavelength. Nanoparticles of gray silicon are red in color. Other properties that may be affected by reduced dimensionality include photocatalysis, photoconductivity, photoemission, and electroluminescence. The nanoparticles change their color with size due to surface plasmon resonance, and hence they are widely utilized for molecular sensing, diagnostics, and imaging [52].

3.5.6 Electrical Properties

The energy band structure and charge carrier density in the material can be modified from their bulk and in turn lead to changes in the associated electronic properties of the materials. These effects are normally termed as quantum confinement effect and relate to the structure and occupation of outermost electronic energy levels of the material. When the size of the system becomes comparable with the de-Broglie wavelength of electrons, the discrete nature of the energy state becomes observable once again, although a fully discrete energy spectrum is only observed in systems that are confined in all three dimensions. Discrete energy bands considerably change the transport properties of the system. In typical cases, the conducting materials become insulators below a critical length scale, as the energy band ceases to overlap. Nanoparticles made of semiconducting materials germanium, silicon, and cadmium are not semiconductors [53]. Electrical transport properties for the bulk system are determined by phonons scattering, impurities, and other carriers or scattering at rough surfaces. The electrical transport is diffusive, and the path of each electron represents a random walk. But in the nanostructured systems, electrons can travel through the system without randomization of the phase of their wave functions as the system dimensions are smaller than the electron mean free path for inelastic scattering. This results in additional localization phenomena that are basically related to phase interference. In highly confined structures such as quantum dots, conduction is mostly dependent on the presence of other charge carriers and hence the charge state of the quantum dot. In recent years, many advances are made in the field of molecular and nanoelectronics. Single molecules are expected to be able to control electron transport in molecular electronics. This offers the promise of exploring the vast variety of molecular functions for electronic devices. If the nanostructure has the same crystal structure as the bulk counterpart, the electronic states of the low-dimensional system are simply a subset of the electronic states of the bulk phase. Therefore, the wave vector components in the nanoscale directions can only take discrete values so that the wave vector components become quantized. The carbon nanotubes can act as conductors or semiconductors in behavior, but larger carbon structure (graphite) is a good conductor of electricity. Conductivity of carbon nanotubes changes with change in area of cross-section. Conductivity also changes when some shear force (or twist) has been given to the nanotube. The conductivity of a multiwalled carbon nanotube is different from that of a single nanotube with the same dimensions [54]. In high-energy density batteries, nanocrystalline materials are good materials for separator plates in batteries because they can hold considerably more energy than conventional ones. Nickel–metal hydride batteries made of nanocrystalline nickel and metal hydrides are envisioned to require far less frequent recharging and to last much longer.

3.5.7 Magnetic Properties

Magnetic properties are very important for various technologies and can be exploited in many aspects ranging from recording to drug delivery systems. The magnetic properties come from a genuine quantum effect related to spin interaction. In practice, the net magnetization is null for almost all materials because they are composed of magnetized regions called domains, and these domains have a random distribution of magnetization direction. The materials can be grouped into ferromagnetic, ferrimagnetic, anti-ferromagnetic, and paramagnetic. Each group has a different behavior when a magnetic field is applied [55]. A ferromagnetic material is characterized by the alignment of the magnetic moments leading to a spontaneous magnetization even in the absence of an external magnetic field. The antiferromagnetic and ferrimagnetic states arise when the magnetic moments' arrangement consists of two lattices with magnetic alignment in opposite directions. In the antiferromagnetism, these lattices have exactly the same magnetization values and the total magnetization is zero while for ferrimagnetic the lattices have different magnetization values thus resulting in a net magnetization. The paramagnetic states are characterized by the alignment of the magnetic moments when an external field is applied. Superparamagnetism is the main phenomenon observed at the nanoscale [56]. When going from a bulk ferromagnetic material to a nanoparticle, the typical hysteresis curve disappears and the system enters in a regime, whereby large reversible magnetization is observed. Superparamagnetism is observed in magnetic nanoparticles by which the magnetizations of the particles are randomly oriented, and they are aligned only under an applied magnetic field and the alignment disappears once the external field is withdrawn. This is due to the presence of only one domain in magnetic nanoparticles when compared to the multiple domains in bulk as shown in Fig. 3.7. By decreasing the size, the

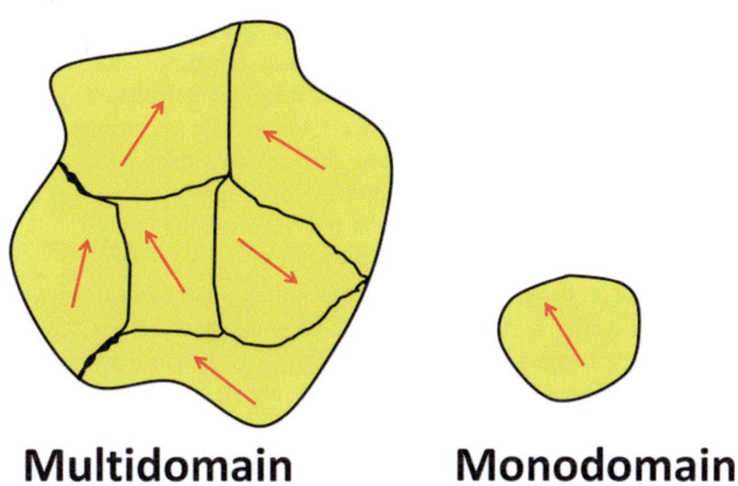

Fig. 3.7 Multidomain shown in macroscale and microscale materials, whereas monodomain shown by nanoscale material results super paramagnetism nature

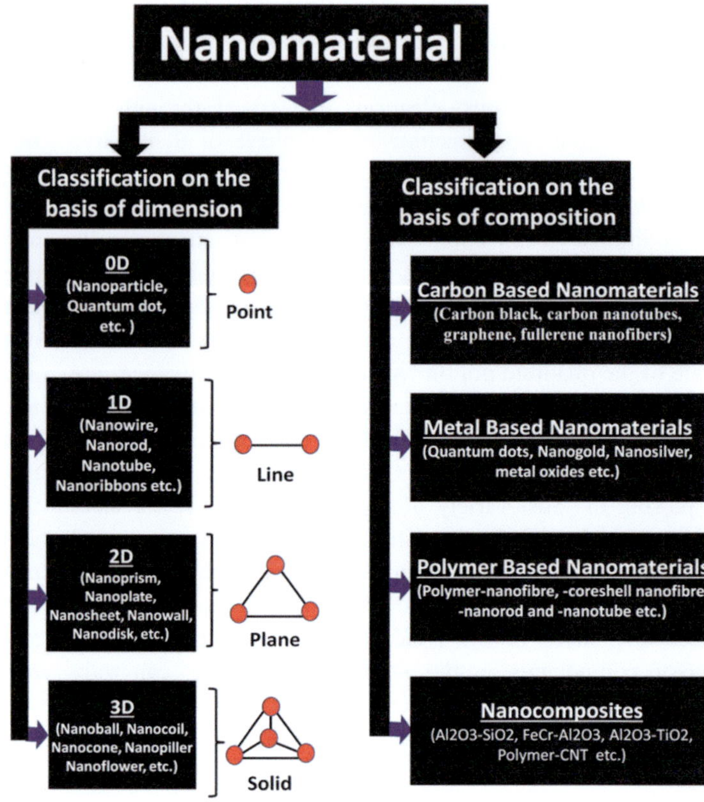

Fig. 3.8 Classification of nanomaterials

energy needed for rotating the magnetization out of the easy direction is reduced (the so-called anisotropy energy). Magnetic nanoparticles of palladium, platinum, and gold become nonmagnetic in their bulk form. Structural changes associated with size effects develop ferromagnetism in platinum and palladium. However, gold nanoparticles exhibit ferromagnetism when they are capped with appropriate molecules. The charge localized at the particle surface gives rise to ferromagnetic-like behavior. This observation indicated that the modifications of the d-band structure by chemical bonding can develop ferromagnetic-like behavior in metallic clusters. Magnetic nanoparticles have a variety of applications such as in nanoelectronics, biomedical sensors, drug delivery, magnetic resonance imaging, data storage, color imaging, bioprocessing, magnetic refrigeration, and ferrofluids [57]. The need to increase storage space on magnetic storage devices such as hard drives in computers resulted in the development of a new field of study called mesoscopic magnetism that involves the study of magnetic materials, specifically for films of nanomagnets. The principle behind the information storage mechanism includes the alignment of the magnetization in one direction of a very small region. Magnetic storage devices such as hard drives are based on tiny crystals of cobalt–chromium alloys [58].

3.6 Classification of Nanomaterials

The brief categorization of nanomaterials is given in Fig. 3.8.

3.6.1 Classification on the Basis of Dimension

Nanomaterials could be defined as materials with at least one of its dimensions in the range of 1–100 nm. Based on dimention, they can be classified as zero-dimensional (0-D), one-dimensional (1D), two-dimensional (2D), and three-dimensional (3D) nanomaterials.

3.6.1.1 Zero-Dimension (0-D)

In zero-dimensional (0-D) nanomaterials, all dimensions are measured in the nano range (no dimension is exceeding 100 nm). The most common 0-D nanomaterials are nanoparticles. Recently, several research groups synthesized 0-D nanomaterials, such as uniform particle sets (quantum dots), heterogeneous particle sets, core-shell quantum dots, onions, hollow spheres, and nanolenses [59]. A variety of physical and chemical methods have been developed to produce 0D-NMS with well-controlled dimensions. In addition, 0-D nanomaterials such as quantum dots are used in light-emitting diode (LED), solar cells, single-electron transistors, lasers, and so on.

3.6.1.2 One-Dimensional (1D)

In one-dimensional (1D) nanomaterials, only one dimension lies outside the nanoscale. This category includes nanotubes, nanorods, and nanowires. Nanotubes, nanowires, nanorods, nanoribbons, and nanoribbons are 1D nanomaterials [60, 61]. The 1D nanomaterial has a profound impact on nanoelectronics, nanodevices and systems, nanocomposite materials, alternative energy resources, and national security.

3.6.1.3 Two-Dimensional (2D)

In two-dimensional (2D) nanomaterials, there are two dimensions that lie outside the nanoscale. This class has plate-like shapes and includes graphene, nanofilms, nanolayers, nano-coatings, nanoprisms, nanoplates, nanosheets, nanowalls, and nanoplates [62]. In the past 10 years, a synthesis of 2D nanomaterials has become a focus of materials research due to its many different low-dimensional properties as volume properties. 2D nanomaterials with specific geometries have unique properties that depend on their shape and their later use as building blocks for the key components of nanodevices. These have new applications in sensors, photocatalysts, nanocontainers, nanoreactors, and templates for 2D structures of other materials [63].

3.6.1.4 Three-Dimensional (3D)

Three-dimensional (3D) nanomaterials are materials that are not limited to the nanoscale in any dimension. This class can contain bulk powder, nanoparticle dispersions, nanowire bundles, and nanotubes as well as multiple nanolayers [64]. In addition, 3D nanomaterials have recently drawn intense research interest because nanostructures have a larger surface area and provide sufficient absorption sites for all molecules involved in a small space. On the other hand, such materials with three-dimensional porosity could lead to better transport of the molecules.

3.7 Synthesis of Nanomaterials

There are two methods for nanoparticle synthesis: the top-down approach and the bottom-up approach described as shown in Fig. 3.9. The size reduction of bulk goods is from top (higher size) to bottom (lower size). A typical example of the process involved is ball milling that reduces the materials by heavy plastic deformations. The top-down process begins with large-scale materials that are then reduced to nanoscale with quicker production. The biggest problem with the top-down approach is the imperfection of the surface structure. Such an imperfection would have a significant impact on the physical chemistry and surface chemistry of nanostructures and nanomaterials. It is known that the conventional top-down technique can significantly damage the processed samples crystallographically. The bottom-up approach starts with atoms or molecules and builds up to nanostructures [65]. The manufacture is much cheaper. The bottom-up approach involves the accumulation of atom by atom, molecule by molecule, or group by group as given in the bottom part of Fig. 3.9. During self-assembly, the physical forces acting in the nano range are used to combine units into large, stable structures. This route is most commonly used to manufacture most nanoscale materials with the ability to create a uniform size, shape, and distribution. It effectively covers chemical synthesis and precisely controls the reaction to inhibit the growth of additional particles [66]. Typical examples are the quantum dot and the formation of nanoparticles from colloidal dispersion. The synthesis of nanoparticles involves better control of the particle size distribution, morphology, purity, quantity, and quality.

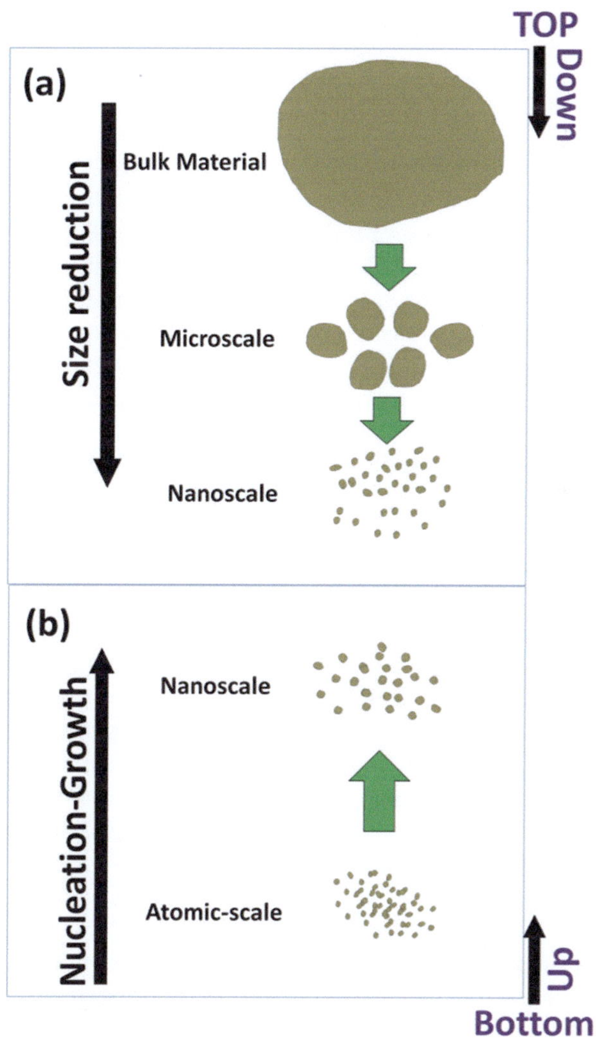

Fig. 3.9 (**a**) Top-down approach and (**b**) Bottom-up approach for the synthesis of nanomaterials

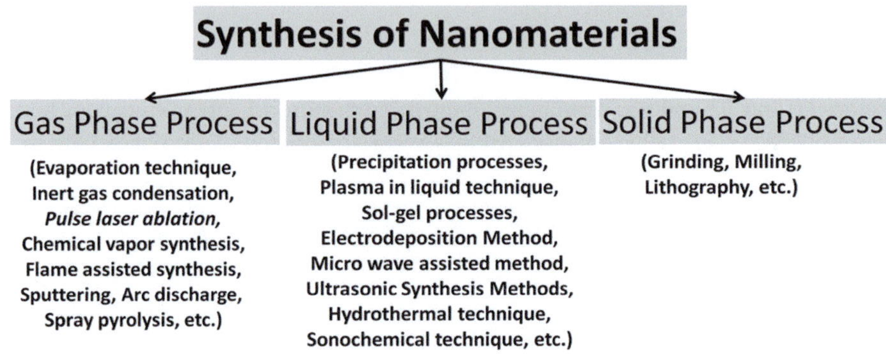

Fig. 3.10 Classification of synthesis of nanomaterials

Besides the abovementioned classification of synthesis, all the synthesis procedures can be divided into gas-phase process, liquid-phase process, and solid-phase process, as shown in Fig. 3.10.

3.7.1 Gas-Phase Processes

Processes for manufacturing gas-phase nanoparticles have attracted great interest over the years due to the variety of advantages they offer compared to other processes. These techniques are typically characterized by the ability to precisely control process parameters to adjust the shape, size, and chemical composition of nanostructures. Although the means and methods may differ, almost all manufacturing processes for gas-phase nanomaterials follow this order: (1) Energize the precursor materials in a gas phase, (2) transform the precursor material into small cluster, (3) force the cluster of these groups into nanoparticles, and (4) process for collecting manufactured nanoparticles. The growth of small nuclei into nanoparticles is called condensation and only occurs when the precursor vapor is oversaturated. The condensation process can be controlled by physical and chemical methods and is discussed subsequently [67].

Evaporation technique: This technique uses a laser/electron beam that vaporizes a target sample in an inert gas flow reactor. The starting material is locally heated to a high temperature, which enables evaporation. The vapor is cooled by collisions with the inert gas molecules and the resulting supersaturation induces the formation of nanoparticles [68]. A schematic of the evaporation technique has been given in Fig. 3.11.

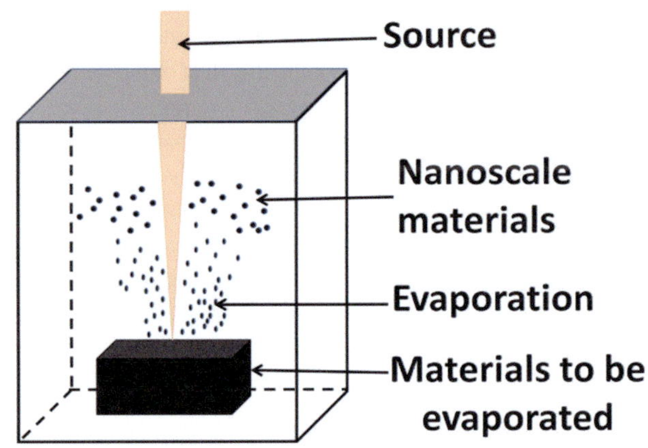

Fig. 3.11 Schematics of evaporation technique

Inert gas condensation: This method is the most rudimentary of all gas-phase manufacturing techniques. The process is as simple as heating material in an oven, usually under an inert gas such as nitrogen or helium. However, this method is only suitable for materials with low vapor pressure (Fig. 3.12). In other words, the material only evaporates at elevated temperatures, sometimes even at 2000 °C. This method is very useful in the manufacture of metallic nanoparticles because these materials have adequate evaporation rates at practical temperatures. Although the process is typically carried out in inert gas, reactive gases can also be introduced into the heated chamber to stimulate reactions [69].

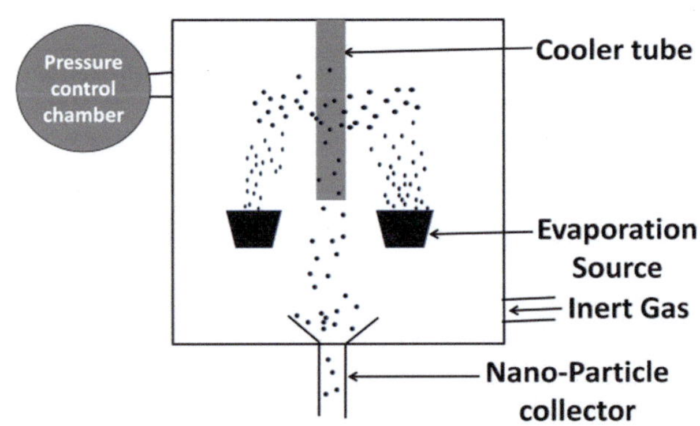

Fig. 3.12 Schematics of inert gas condensation technique

Pulse laser ablation: With this technique, instead of the entire sample, a narrower cloud of material is vaporized to generate steam. To achieve this, the high-energy laser focuses on a much localized position. The exposure with the laser is done in pulses and hence the name pulsed laser (Fig. 3.13).

This quickly heats a small patch of the material to a very high temperature at which the material is evaporated. Because of the small weight size of the sample to be evaporated, this can generally be used to produce a small amount of nanomaterials. However, the technique is very useful for the synthesis of nanomaterials from materials that otherwise cannot evaporate [70].

Chemical vapor synthesis: In chemical vapor synthesis, chemical vapors from precursor materials are converted in a reaction chamber. The reaction chamber is typically heated using Joule heating. The process is similar to chemical vapor deposition, but the chemical vapor synthesis process promotes the formation of nanoparticles instead of nanomaterial deposition as a thin film (Fig. 3.14). Therefore, the process parameters are set appropriately during synthesis to suppress film formation and promote particle nucleation in the gas phase. Typically, the precursor residence time in the reaction chamber is the most critical parameter in determining whether the film or parameter will form. This method is known for its flexibility as a nanomaterial synthesis process. The precursors can be in the solid, liquid, or gaseous phase at room temperature but are passed into the hot-wall reactor in the vapor phase. Chemical vapor synthesis is also considered a high-performance process because the production is continuous. Even a small reactor can produce a significant amount of nanomaterials compared to other manufacturing techniques [71].

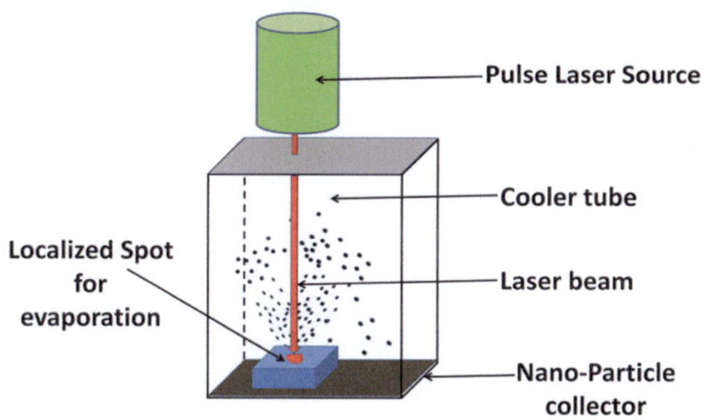

Fig. 3.13 Schematics of pulse laser ablation method

Flame-assisted synthesis: In the flame-assisted synthesis process, the energy required for particle nucleation is given by a flame, rather than being supplied externally from a secondary heat source (Fig. 3.15). With the flame-assisted synthesis method, particle nucleation and growth occur within the flame. This method is the most common of all synthetic processes and corresponds to the production of millions of tons per year in carbon black, fullerenes, and metal oxides. This method is particularly useful in the manufacture of metal oxides because the conditions within the flame and the environment are quite oxidizing [72].

Sputtering: Sputtering is a manufacturing process for gas-phase nanomaterials in which a solid precursor material is vaporized by high-speed ion bombardment of inert gas as shown in Fig. 3.16. This creates a cloud of atoms and groups

Fig. 3.14 Schematics of chemical vapor synthesis

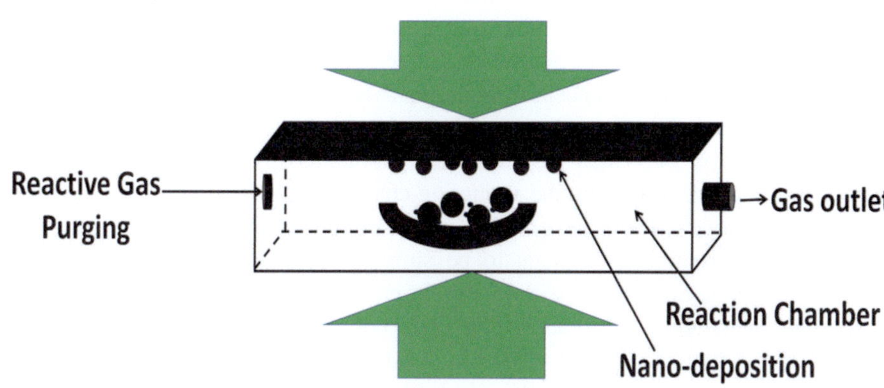

Fig. 3.15 Schematics of flame-assisted synthesis

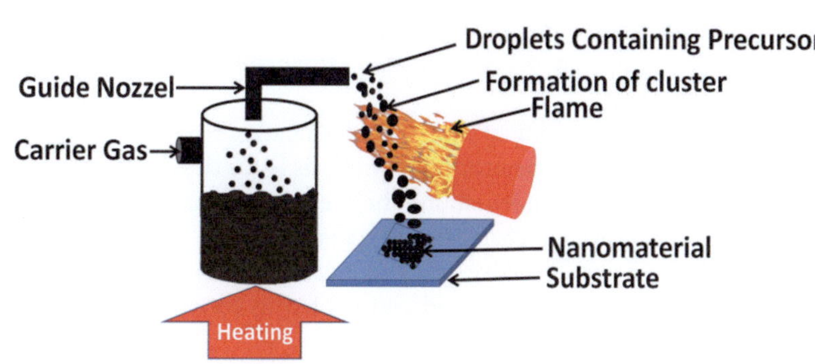

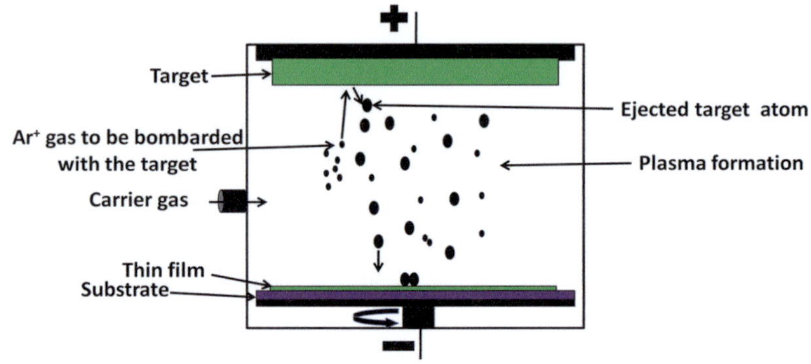

Fig. 3.16 Schematics of sputtering deposition

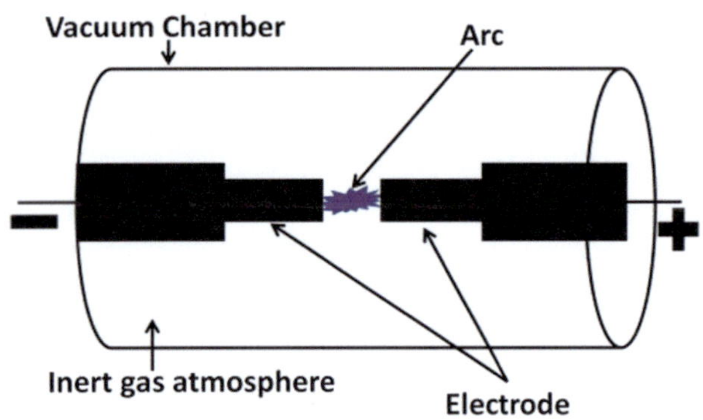

Fig. 3.17 Schematics of arc discharge technique

of atoms, which are then deposited on a substrate. The deposition is typically carried out in a vacuum environment because the mobility of the sprayed materials is hindered at higher pressures. The most common sputter sources are ion guns or hollow cathode plasma generators. The method offers advantages because the composition of the sprayed material is the same as that of the target material. Variation in chemical composition of deposited products can be obtained by introducing reactive gases into the sputtering chamber [73].

Arc discharge: The arc discharge technique for synthesizing nanoparticles appears to be simple, but it is difficult to get a high yield with a fine structure. These conditions are highly affected by the gap between the electrodes, the diameter of the electrodes, and the type of medium (liquid or gas) where the electrodes are placed (Fig. 3.17). Besides these factors, other factors play an important role to maximize the yield such as pressure, current, power supply type (ac or dc), voltage, temperature, and cathode shape [74].

Plasma vapor condensation: One of the outstanding strides in plasma processing for nanoparticles synthesis is the developed process of vapor condensation. The precursor material is put into the working chamber with a stable arc. The chamber is filled with reactive gas that becomes ionized; then molecular clusters are formed and cooled to produce nanoparticles (Fig. 3.18) [75].

Molecular beam epitaxy (MBE): In solid-source MBE, ultra-pure elements such as gallium and arsenic are heated in separate cells until they begin to slowly evaporate. The evaporated elements then condense on the wafer, where they may react with each other (Fig. 3.19) [76].

3.7.2 Liquid-Phase Processes

Some of the liquid-phase nanomaterial synthesis techniques are described subsequently:

Precipitation processes: Precipitation of solids from a solution containing metal ions is one of the most widely used production processes for nanomaterial synthesis. This approach can produce metal as well as metal oxide nanoparticles. The process is based on reactions of solute in solvents. A precipitant is added to produce the desired particle precipitation, and the precipitate is filtered and then subjected to heat treatment [77]. Schematics of the precipitation process are shown in Fig. 3.20.

Plasma generation in liquid technique: Many experimental setups of plasma in liquid generation have been reported, in which the liquid medium, electrode material, electrode configuration, electric power source, and other parameters were varied. Schematics of a typical plasma generation in liquid technique are given in Fig. 3.21. By examining electrode configurations and power sources, plasma in liquid generation can be subdivided into four main groups: (1) gas discharge between an electrode and the electrolyte surface, (2) direct discharge between two electrodes, (3) contact discharge between an electrode and the surface surrounding the electrolyte, and(4) radio-frequency and microwave generation. Similar processes include glow-discharge plasma, dielectric barrier discharge, gliding arc discharge techniques, gas-liquid interfacial plasma, plasma electrochemistry, dual plasma electrolysis, microplasma discharge, solution plasma, pulsed plasma in a liquid, discharge plasma in liquid, electric spark discharge, arc discharge, capillary discharge, streamer discharge, wire explosion process in water, DC diaphragm discharge, and AC capillary discharge. In contrast to gas discharge, two electrodes of similar size and shape are

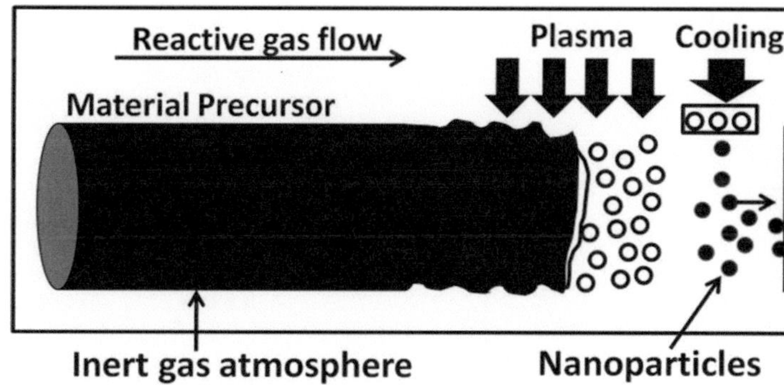

Fig. 3.18 Schematics of plasma vapor condensation

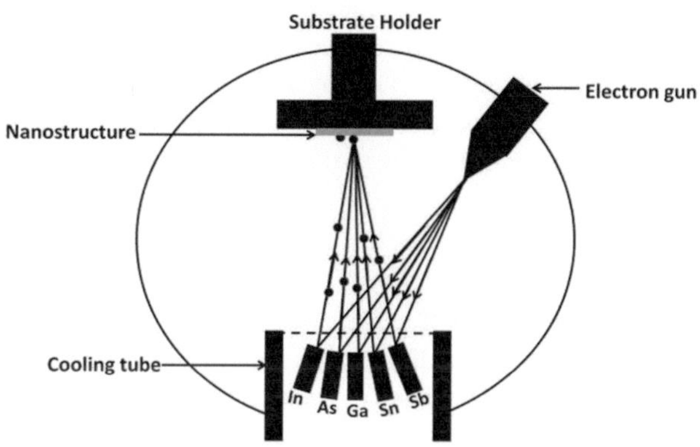

Fig. 3.19 Schematics of molecular beam epitaxy

immersed in the liquid at a short distance. Because of the direct discharge, most liquids containing conductive electrolytes such as deionized water, ethanol, and liquid nitrogen can be used in such systems. For nanoparticle production, both electrodes and ions in the liquid serve as raw materials for nanoparticle formation [78].

Sol-gel processes: The sol-gel technique is a long-established industrial process for the generation of colloidal nanoparticles from the liquid phase, which has been developed for the production of advanced nanomaterials and coatings. The main advantages of sol-gel techniques for the preparation of materials are low temperature of processing, versatility, and flexible rheology allowing easy shaping and embedding. They offer unique opportunities for accessing the organic–inorganic materials. The most commonly used precursors of oxides are alkoxides because of their commercial availability and the high liability of the M-OR bond allowing facile tailoring in situ during processing. Sol-gel syntheses (production of a gel from powdery materials) are wet chemical processes for the production of porous nanomaterials, nanostructured ceramic, polymers, and oxide nanoparticles. The synthesis takes place under relatively mild conditions and low temperatures. The term sol refers to dispersions of solid particles in the size range of 1–100 nm, which are finely distributed in water or organic solvents. In sol-gel processes, the production or deposition of the material takes place from a liquid sol state, which changes to a solid gel state through a sol-gel transformation (Fig. 3.22). The sol-gel transformation involves 3D crosslinking of the nanoparticles in the solvent, which gives the gel bulk properties. Controlled heat treatment in the air can convert gels to a ceramic oxide product. The addition of organic substances in the sol-gel process initially creates an organometallic compound from a solution containing an alkoxide (metallic compound of an alcohol, for example with silicon, titanium, or aluminum). The pH of the solution is adjusted with an acid or base, which also triggers the conversion of the alkoxide as a catalyst. Subsequent reactions are hydrolysis (division of a chemical bond by water), followed by condensation, and polymerization (reaction that leads from single-chain compounds to multichain or long-chain compounds). The particles or polymer oxide grow in the course of the reaction until a gel forms. In all cases, gel formation is followed by a drying step. The advantage is that the liquid phase enables the production of highly porous materials. This would normally not be possible in gas-phase reactors due to high temperatures. Due to the low process temperature, the substances can also be embedded in the gel during the synthesis stage. These can be saved or released in a controlled manner. The disadvantage of the sol-gel process lies in the difficult synthesis and drying steps, which make it difficult to extend the process. Organic impurities can also remain in the gel. The resulting necessary cleaning, drying, and heat post-treatment steps make this production process more complex than gas-phase synthesis [79].

Electrodeposition method: Electrodeposition method, also known as electroplating, is an electric current driving deposition method that gives precise control of coating the species epitaxially in the form of nanoparticles, nanowires, and so on onto a conductive target material. Electrodeposition is referred either to electroplating or to electrophoretic deposition. The electroplating process is usually based on an aqueous solution of ionic species, while electrophoretic deposition occurs in a suspension of particles (Fig. 3.23). In electroplating, there is a charge transfer during the deposition to produce the metal or oxide layer in the electrode. In general, the process of electrodepositing may be either

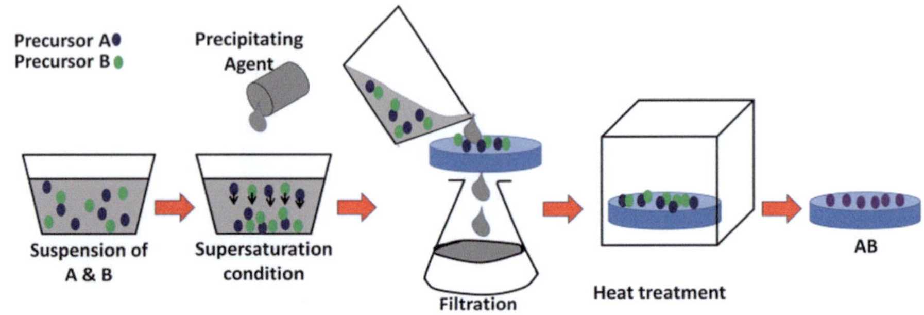

Fig. 3.20 Schematics of the precipitation process

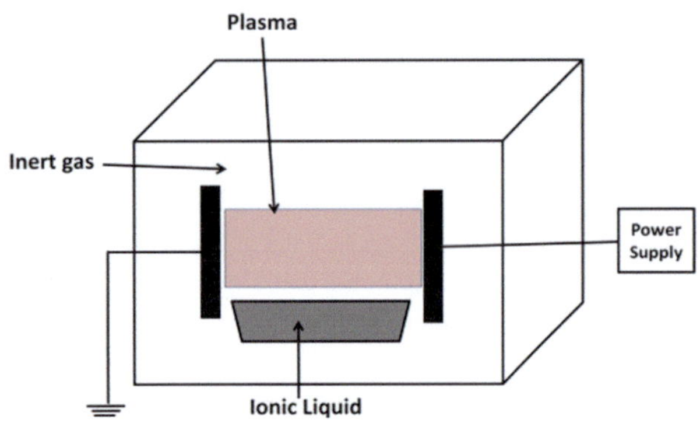

Fig. 3.21 Schematics of plasma generation in liquid technique

(1) an anodic process, where a metal anode is electrochemically oxidized in the solution, react together, and then deposit on the anode, or (2) a cathodic process, where components (ions, clusters, or nanoparticles) are deposited onto the cathode from solution precursors. Film thickness could be controlled by the electrodeposition time, since increasing the reaction time would cause more source material to be consumed and yield larger film thickness. The deposition rate can be followed by the variation of the current with time. Electrodeposition has three advantages over other liquid phase and gas-phase deposition techniques: (1) allow growing functional nanoparticles through complex 3D nanotemplates. Nanostructures and thin films can be deposited onto large specimen areas of complex shape, making the process highly suitable for industrial use. For example, electrodeposition can be performed within a nanoporous membrane, which serves to act as a template for nanoparticle growth; (2) applicable at room temperature from water-based electrolytes; and (3) suitable to be scaled down to the deposition of a few atoms or up to large dimensions, films thickness could range from a 1 nm to 10 μm [80].

Microwave-assisted method: Microwave-assisted methods have been widely applied in chemical reactions and the synthesis of nanoparticles and nanostructures. It is an attractive choice to promote reactions and is energy-effective heating compared to conventional heat conduction methods (such as an oil bath) due to the direct heating of the reaction mixture. Microwave heating offers many advantages over conventional heating. Microwaves are a form of electromagnetic radiation with frequencies between 300 MHz and 300 GHz and wavelengths between 1 m and 1 mm. Microwave ovens provide rapid heating compared to conventional conduction and convection heating systems. Microwave has a rapid processing time, 2 to 50 times faster than conventional heating. The rapid heating in this case is caused by the microwaves inducing the fast rotation of dipoles and the production of thermal energy, which heats the material. This means that a polar solvent is needed for microwave-assisted synthesis, and the higher polarity of solvents will result in the higher generation of heat upon microwave irradiation. Polar solvents heat easily throughout the whole volume. This is in direct contrast to the hot plate conventionally used in most liquid-phase synthesis methods, where the thermal energy is transferred by conduction and thermal convection. Nonpolar solvents mixed with small amounts of polar solvents could also be used in the microwave-assisted synthesis of nanoparticles due to the fact that energy transfer between the polar molecules and the nonpolar solvent is rapid. Furthermore, microwave-assisted synthesis is known to produce an increase in the boiling points of organic solvents, which was considered to be due to "nucleate boiling." Some examples of such microwave-assisted synthesis were reported for graphene-containing nanomaterials for lithium-ion batteries. Other examples of microwave-assisted synthesis of nanomaterials include nanoparticles of ZnO with different morphologies, Ag, Cu, Co_3O_4, composites of metal/metal oxide supported with CNT, and RE-doped YPO_4 (RE = Eu, Ce, and Tb) nanophosphors [81].

Ultrasonic synthesis method: Ultrasonic synthesis methods have emerged as a powerful synthesizing technique to prepare nanostructured materials. The ultrasound frequency range is above 20 kHz and can reach frequencies in the gigahertz level. By applying the ultrasonic radiation to the precursor solution, the chemical reaction is initiated. As a form of radiation, ultrasound can break chemical bonds and create products. The sudden implosion of the bubbles (about 100 μm in size) creates a local hot spot, which can reach temperatures as high as 5000 K and pressures of up to hundreds of atmospheres depending on the level of ultrasound power used. At the same time, the cooling rates are

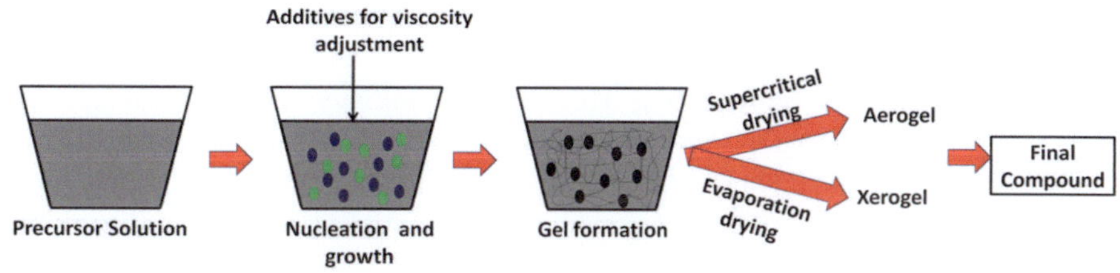

Fig. 3.22 Schematics of sol-gel technique

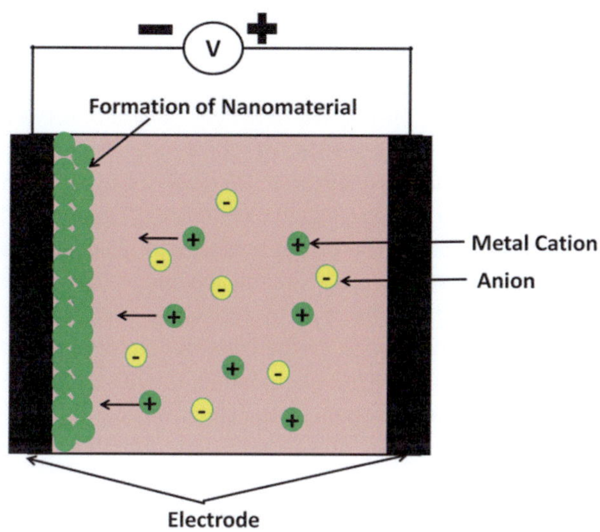

Fig. 3.23 Schematics of electrodeposition method

extremely high and can be as great as 1011 K/s. The capacity of ultrasound to create these extreme temperatures and pressures within a medium makes this technique attractive. The ultrasound leads to the creation, growth, and rapid collapse of small bubbles, which act as nucleation centers. The growth of the nucleus ends after bubble collapse. Size and shape selection in ultrasound synthesis is accomplished by controlling the precursor concentrations, pH, temperature, and surfactant concentration. Facile synthesis of fuel cell materials containing metallic nanoparticles was demonstrated via the ultrasonic synthesis method in the absence and presence of surfactants and alcohols. Recent studies on nanoparticles of Ag, carbon, silica, PdSn, and nanosized MOF also present ultrasonic synthesis method to be promising for the formation of various nanoparticle systems. Besides, ultrasound synthesis methods are used not only for the synthesis of nanomaterials but also for other purposes, such as loading nanoparticles inside nanotubes or surface functionalization of nanoparticles with polymers. Ag nanoparticle-loaded TiO_2 nanotubes and Pt nanoparticle-loaded CNTs were demonstrated for loading nanoparticles inside nanotubes via ultra-sonication [82].

Hydrothermal (solvothermal) process: Hydrothermal synthesis includes the various techniques of crystallizing substances. Hydrothermal phase equilibria have been studied by geochemists and mineralogists since the beginning of the twentieth century. Hydrothermal synthesis is the method of synthesis of single crystals that depends on the solubility of minerals in hot water under high pressure. The crystal growth is performed in an apparatus consisting of a steel pressure vessel called autoclave in which a nutrient is supplied along with water. A gradient of temperature is maintained at the opposite ends of the growth chamber so that the hotter end dissolves the nutrient and the cooler end causes seeds to take additional growth. The hydrothermal method has advantages over other types of crystal growth including the ability to create crystalline phases that are not stable at the melting point. Also, materials that have a high vapor pressure near their melting point can be grown by the hydrothermal method. The method is also particularly suitable for the growth of large good-quality crystals with good control over their composition. The particle size and morphology can be tuned by changing the experimental condition such as temperature, solvent, and so on. Hydrothermal method exploits the solubility of almost all inorganic substances in water at elevated temperature and pressure and subsequent crystallization of the dissolved material from the fluid. Water at elevated temperatures plays an essential role in the precursor material transformation because the vapor pressure is much higher and the structure of water at elevated temperatures is different from that at room temperature. Hydrothermal synthesis is commonly used to grow synthetic quartz, gems, emeralds, rubies, alexandrite, and other single crystals. The method has proved to be extremely efficient in the search for new compounds with specific physical properties. The techniques of crystallizing substances from high-temperature nonaqueous solutions at high vapor pressures are termed "solvothermal method." The solvothermal synthesis method uses organic solvents and supercritical CO_2 [83].

Sonochemical: Sonochemical synthesis is the process by which molecules undergo a chemical reaction using strong ultrasound radiation (20 kHz–10 MHz). Sonochemistry creates hot spots that can reach very high temperatures (5000 K–25,000 K), pressures of more than 1000 atmospheres, and heating and cooling rates of more than 10^{11} K/s. High-intensity ultrasound creates chemical and physical effects that can be used to manufacture or modify a variety of nanostructured materials. The principle that modifies nanostructures in the sonochemical process is acoustic cavitation. The advantages of sonochemical processes are high rates, controllable reaction conditions, higher ability to form uniform shapes, narrow particle size distributions, and high purity [84]. This method, initially proposed for the synthesis of iron nanoparticles, is nowadays used to synthesize different metals and metal oxides. The main advantages of the sonochemical method are its simplicity, operating conditions (in ambient conditions), and easy control of the size of nanoparticles by using precursors with different concentrations in the solution. Ultrasound power influences the chemical changes due to the cavitation phenomena involving the formation, growth, and collapse of bubbles in the liquid. The sonolysis technique involves passing sound waves of fixed frequency through a slurry or solution of carefully selected metal complex precursors. In a solvent with a vapor pressure of a certain threshold, the alternating waves of expansion and compression cause cavities to form, grow, and implode. Sonochemical reactions of volatile organometallics have been exploited as a general approach to the synthesis of various nanophase materials by changing the reaction medium [85].

3.7.3 Solid-Phase Processes

Several solid-phase processes adopted for nanomaterial fabrication are discussed subsequently.

Grinding and milling: Milling is a solid-state processing technique for the synthesis of nanoparticles. It is a top-down approach. Different types of mechanical mills are available which are commonly used for the synthesis of nanoparticles (Fig. 3.24). Due to mechanical limitations, it is very difficult to produce ultrafine particles using these techniques or it takes a very long time. However, simple operation, low cost of production of nanoparticles, and the possibility to scale it to produce large quantities are the main advantages of mechanical milling. The important factors affecting the quality of the final product are the type of mill, milling speed, container, time, temperature, atmosphere, size and size distribution of the grinding medium, process control agent, weight ratio of ball to powder, and extent of filling the vial [86]. Nanoalloying can be done using various predetermined elemental powder in ball mill.

Lithography: Lithography involves a number of manufacturing techniques that share the principle of transferring an image from a mask to a receiving substrate. A typical lithographic process consists of three successive steps (Fig. 3.25): (1) coating a substrate (Si wafer or glass) with a layer of sensitive polymer (called a resistor), (2) exposing the resistor to light, electrons, or ion beams; and (3) developing the resistance image with a suitable chemical (developer), which, depending on the type of resistance used, shows a positive or negative image on the substrate. In conventional microfabrication used in the semiconductor industry, the next step after lithography is pattern transfer from the resistor to the underlying substrate. This is achieved through a number of transmission techniques, such as chemical etching and dry plasma etching. Lithography techniques can be broadly divided into two main groups: (1) A method using a physical mask in which resistance is radiated through the mask that is in contact with or near the surface of the resistor. These methods are collectively referred to as mask lithography, of which photolithography is the most common. (2) Methods with the software mask. Here, a scanning beam irradiates the resistive surface successively point by point by a computer-controlled program in which the mask pattern is defined. These methods are collectively referred to as scanning

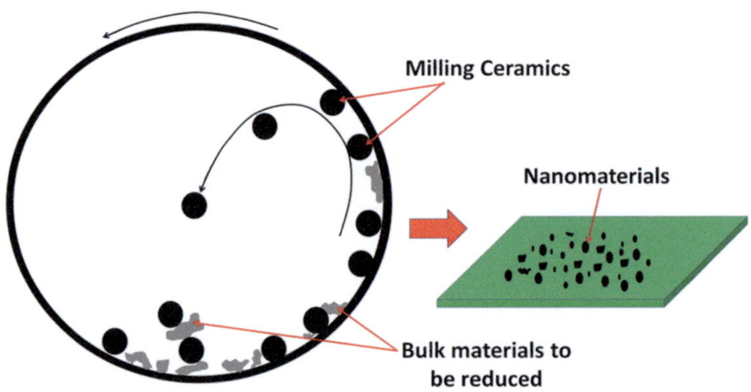

Fig. 3.24 Schematics of ball milling

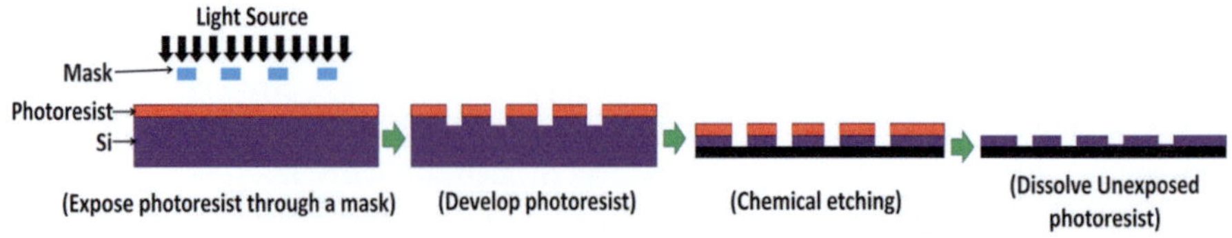

Fig. 3.25 Typical process sequence of lithography

lithography. The main difference between mask and scan lithography is speed: While mask lithography is a parallel and fast technique, scan lithography is a slow and serial technique. Another important difference is the resolution, which is generally higher with scanning methods. The price of higher resolution is the use of higher energy radiation sources, which imply expensive equipment. Photolithography uses light (UV, deep UV, extreme UV, or X-ray) to expose a layer of radiation-sensitive (photoresist) polymer through a mask. The mask is an almost optically flat glass plate (or quartz, depending on the light used), which contains the desired pattern: opaque areas (the pattern made of an absorbent metal) on a background transparent to UV rays. The image on the mask can be replicated as it is by placing the mask in physical contact with the resistor (photolithography in contact mode) or generally reducing it by a factor of 5 or 10 and projecting it onto the resistor layer by an optical system (Projection mode) [87].

3.8 Classification Based on Composition

Based on materials composition, they can be classified into carbon-based materials, metal-based materials, dendrimers, and nanocomposites.

3.8.1 Carbon-Based Materials

Generally, these nanomaterials (NM) contain carbon and have morphologies such as hollow tubes, ellipsoids, or spheres. Graphene, fullerenes (C_{60}), carbon nanotube (CNT), carbon nanofibers (CNF), carbon black, and carbon onions are included under the carbon-based NM category [88].

3.8.1.1 Graphene

Graphene is an allotrope of carbon in the form of a single layer of atoms in a two-dimensional hexagonal lattice in which one atom forms each vertex. It is the basic structural unit of other allotropes, including graphite, charcoal, carbon nanotubes, and fullerenes. Graphene is 1-atom-thick planar sheet of sp^2-bonded carbon atoms that are densely packed in a honeycomb crystal lattice. It can be viewed as an atomic-scale thick made up of carbon atoms and their bonds [89]. The carbon–carbon bond length in graphene is about 0.142 nm. Layers of graphene stacked on top of each other form graphite, with an interplanar spacing of 0.335 nm (Fig. 3.26) [90]. The separate layers of graphene in graphite are held together by van der Waals forces, which can be overcome during the exfoliation of graphene from graphite. Graphene is the strongest compound discovered that is 100–300 times stronger than steel with a tensile strength of 130 GPa and Young's modulus of 1 TPa to 150,000,000 psi. Graphene is the best conductor of heat at room temperature ($(4.84 \pm 0.44) \times 10^3$ to $(5.30 \pm 0.48) \times 10^3$ $Wm^{-1}\,K^{-1}$ and also the best conductor of electricity (electron mobility more than 200,000 $cm^2V^{-1}s^{-1}$) [91]. It conducts heat and electricity very efficiently and is nearly transparent. Other notable properties of graphene are its uniform absorption of light across the visible and near-infrared parts of the spectrum ($\pi\alpha \approx 2.3\%$) and its potential suitability for use in spin transport [92].

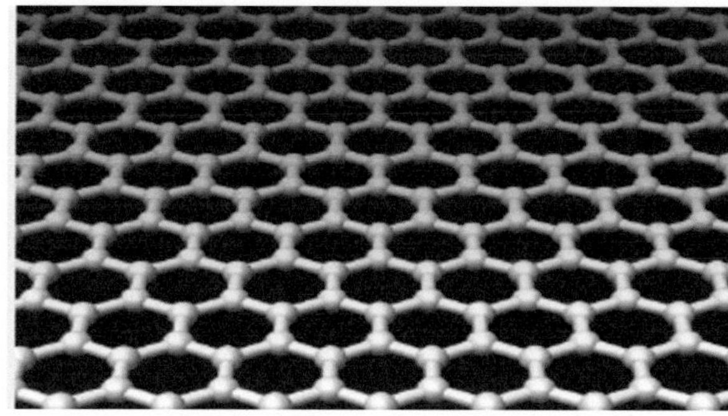

Fig. 3.26 Structure of graphene

3.8.1.2 Fullerene

A fullerene is an allotrope of carbon. The fullerene consists of carbon atoms that are connected by single and double bonds to form a closed or partially closed network with condensed rings of five to seven atoms. The molecule can be a hollow sphere, an ellipsoid, or many other shapes and sizes. Graphene (isolated atomic layers of graphite), a flat network of regular hexagonal rings, can be seen as a unit of the family. Fullerenes with a closed mesh topology are informally denoted by their empirical formula C_n, often written Cn, where n is the number of carbon atoms. However, for some values of n, there may be more than one isomer. The family Cn is named after buckminsterfullerene, the most famous member, which in turn is named after Buckminster Fuller. Closed fullerenes, especially the C_{60}, are also referred to informally as buckyballs due to their similarity to the standard soccer ball [93]. Nested closed fullerenes were called bucky onions. The solid mass form of pure or mixed fullerenes is called fullerite. In 1985, Harold Kroto of the University of Sussex and a member of the Rice University discovered fullerene in the soot residues that result from the evaporation of carbon in a helium atmosphere. Discrete points appeared in the product mass spectrum that corresponded to molecules with the exact mass of 60 or 70 or more carbon atoms (e.g., C_{60} and C_{70}) [94]. The ending "ene" in fullerene was chosen to indicate the carbons that are unsaturated and are only connected to three other atoms instead of the normal four atoms. The shortened "fullerene" was finally applied to the whole family. Kroto, Curl, and

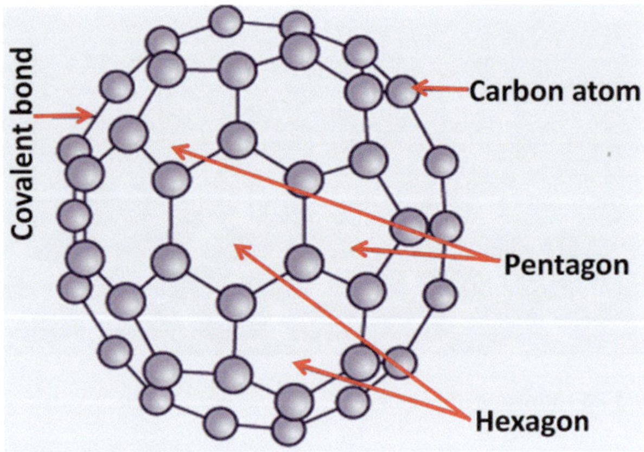

Fig. 3.27 Structure of fullerene

Smalley received the 1996 Nobel Prize in Chemistry [95] for their role in the discovery of this class of molecules.

3.8.1.2.1 Structure of Fullerene

Each carbon atom in the structure is covalently linked to three others. The 6:6 ring bonds (between two hexagons) can be considered as "double bonds" and are shorter than 6:5 bonds (between a hexagon and a pentagon) [96]. The C_{60} bucky ball consists of 60 carbons at 60 corners, which form a spherical structure (Fig. 3.27). It consists of 12 pentagonal and 20 hexagonal rings arranged side by side. These rings are conjugated with double bonds. The C—C bond length for hexagonal rings is 1.40 Å and 1.46 Å for pentagonal rings, with the average bond length being 1.44 Å [97]. Fullerenes have sp^2 and sp^3 hybridized carbon atoms. These molecules have an extremely high affinity for electrons and can be reversibly reduced to six electrons. Although this molecule consists of conjugated carbon rings, the electrons are not delocalized here, and therefore these molecules lack the property of superaromaticity. These molecules have a very high tensile strength and return to their original shape after more than 3000 atmospheric pressures [98].

3.8.1.2.2 Synthesis of Fullerene

Fullerenes were first synthesized through laser vaporization of carbon in an inert atmosphere, but this laser vaporization method produced very small amounts of fullerenes. However, large quantities of fullerene, C_{60}, were later synthesized through arc heating of graphite and laser irradiation of poly aromatic hydrocarbon [99].

Synthesis by laser vaporization of carbon: In this process, fullerenes are generated in a supersonic expansion nozzle by a pulsed laser that focuses on a graphite target in an inert atmosphere (helium). This process involves evaporating the carbon from a rotating solid graphite disc in a high-density helium flow using a focused pulsed laser.

Synthesis by electric arc heating of graphite: This method was designed in 1990 by Kratchmer and Huttman. An arc is created between graphite rods in an inert atmosphere, which creates a spongy condensate (soot). Part of this spongy condensate contains the removable fullerenes in toluene. The fullerenes in the carbon black are then extracted by solvation in a small amount of toluene, after extraction, the toluene (solvent) is removed using a rotary evaporator, and the solid mixture consisting mainly of C_{60} with a small amount of higher fullerenes is subjected to liquid chromatography procedures for the production of pure C_{60} [100].

Flame combustion method: The experimental setup for burning laminar low-pressure diffusion flames of benzene vapor, diluted with argon in oxygen, is shown in Fig. 3.28. The vertically adjustable burner consists of a fuel outlet pipe with an inner diameter of 10 mm (inner diameter), which is surrounded by a porous bronze plate with a diameter of 30.5 cm, through which oxygen is supplied. Sathish and Miyazawa recently reported the synthesis of fullerene nanowhiskers (FNW) by liquid–liquid interface precipitation in a carbon disulfide (CS_2) and isopropyl alcohol system. The high solubility of C_{60} in CS_2 is expected to lead to partial polymerization and the formation of the triclinic structure. By processing controlled flame combustion, superior fullerenes can be manufactured with sufficient precision to produce different blends of C_{60}, C_{70}, and superior fullerenes per ton, and the product can be tailored to the needs of a particular customer without the need for costly post-processing solvents. The method also has an advantage in doping the fullerene cluster by introducing the dopant into the benzene–oxygen mixture before combustion [101].

Synthesis by resistive arc heating of graphite: This process involves evaporating the carbon rods by resistive heating in a partial helium atmosphere. By resistively heating the carbon rods, the rod emits a weak gray–white plume, a soot-like substance made from fullerenes, which is deposited at the glass plates surrounding the carbon rods [102].

Synthesis by laser irradiation of polycyclic hydrocarbon (PAH): The direct synthesis of fullerenes was developed to obtain new homologs of the fullerene family that cannot be obtained in good quantities due to the uncontrolled process of graphite evaporation. This approach to fullerene synthesis is based on polycyclic aromatic hydrocarbon (PAH) with the necessary carbon skeletons. These PAH molecules wrap to fullerenes under immediate vacuum pyrolysis conditions. It has been reported that a polycyclic aromatic hydrocarbon consisting of 60 carbon atoms forms C_{60} fullerene when irradiated with a laser with a wavelength of 337 mn, as shown in Fig. 3.29 [99].

Purification of fullerenes during synthesis: In almost all of the processes described earlier, fullerenes are produced as a raw mixture of C_{60} together with other C_n species ($n \leq 60$) and conventional hydrocarbons. Individual fullerene clusters

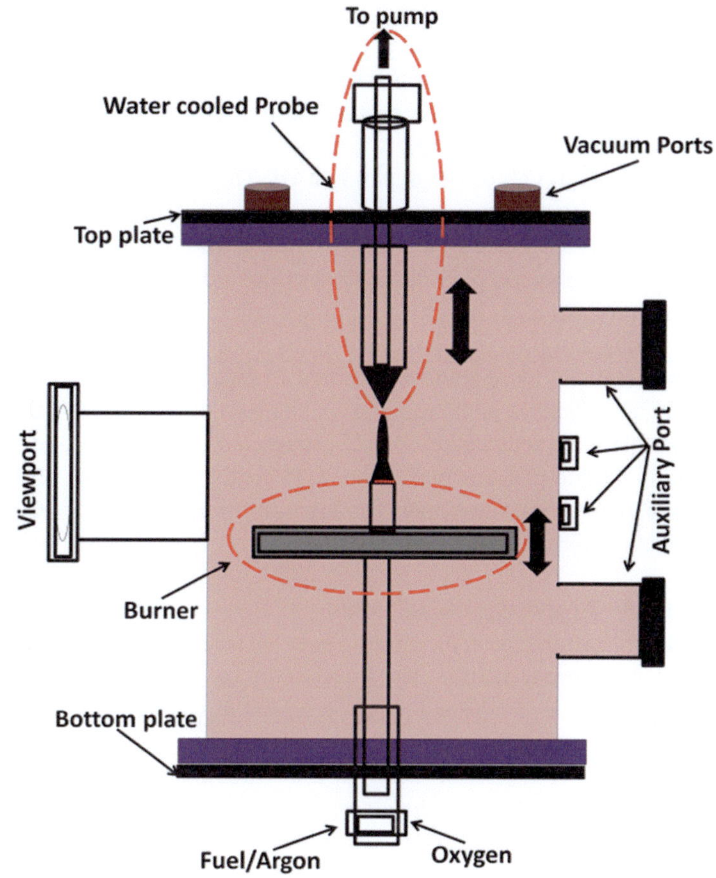

Fig. 3.28 Experimental setup for the flame combustion technique

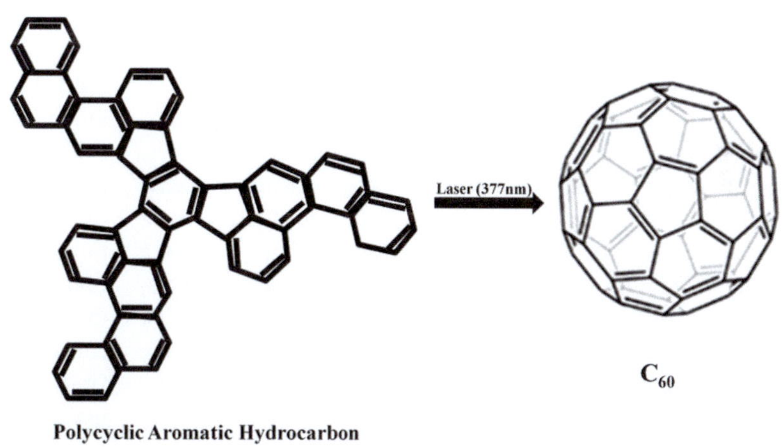

Fig. 3.29 Direct synthesis of fullerene by laser irradiation of polycyclic aromatic hydrocarbon

should therefore be isolated from raw soot. Kratschmer and Huffman produced fullerene with benzene. Current researchers used toluene as a better alternative to benzene because benzene is not a very friendly chemical. The fullerene extraction is carried out efficiently using various methods. The carbon black is first extracted continuously by boiling toluene in a Soxhlet solid extractor with one of the solvents such as chloroform, toluene, benzene, n-hexane, 1,2-dichlorobenzene, and so on. The second method for extracting fullerenes from soot involves sonicating the toluene-soot mixture at room temperature. The second basic fullerene cleaning process is an efficient alternative ultrasonic process for dispersing carbon black in tetrahydrofuran (THF) at a concentration of 0.1 gm/L. The experiment is carried out at room temperature, and it takes 20 min to complete the sonication. The solution is filtered and THF is evaporated by a rotary evaporator at 50 °C. The remaining impurities are mostly polyaromatic hydrocarbons and are washed in diethyl ether before being purified. A sublimation method called the nonsolvent-based method has been used successfully to extract fullerenes from the soot. C_{60} and C_{70} powders are known to sublime in vacuum at relatively low temperatures (between 350 °C and 450 °C). The advantage of producing fullerenes by sublimation process is that it does not contain any solvent. To isolate pure C_{60} and other individual fullerenes, liquid chromatography is used. The most widely used chromatographic separation employs neutral alumina (Al_2O_3) stationary phase and simple gravity feed. Chromatography on alumina provides a straightforward technology for isolating C_{60} and C_{70}. However, this method is very time-consuming and produces large quantities of organic wastes. A modified and more attractive method of separation of C_{60} and other fullerenes is the use of a special type of chromatogram where the activated charcoal is supported on silica gel, and the mobile phase is pure toluene. Recently, many solid phases have been developed for high-performance liquid chromatography of fullerenes [103].

3.8.1.2.3 Properties of Fullerene

Buckminsterfullerene is the largest observed nano-object that has wave-particle duality. The bonding of the fullerene is stable and can withstand high temperatures and high pressures. Since each carbon atom is only connected to three neighbors instead of the usual four carbon atoms, it is common to describe these bonds as a mixture of single and double covalent bonds. It is also known as an endohedral fullerene in which ions or small molecules are built into the atoms of the cage. The exposed surface of the structure can react selectively with other species while maintaining the spherical geometry. C_{60} tends to avoid double bonds in the pentagonal rings, which makes the electron delocalization band and does not make C_{60} "superaromatic" [104]. Fullerenes are stable but not completely unreactive. C_{60} behaves like an electron-poor alkene and reacts easily with electron-rich species. A carbon atom in the C_{60} molecule can be replaced by a nitrogen or boron atom that generates a C59N or C59B. The Bucky Balls collapse under high pressure and high temperature and form several one-, two- or three-dimensional carbon frames. Single-chain polymers are formed using the atom transfer radical addition polymerization route. "Ultra-hard fullerite" is a coined term that is often used to describe the material made by high temperature and high-pressure fullerite processing. Such treatment converts fullerite into a form of nanocrystalline diamond that has been reported to have remarkable mechanical properties. The sp^2-hybridized carbon atoms, which have their minimal energy in flat graphite, have to bend to form the sphere or the closed tube that creates angular stress [105]. The characteristic reaction of fullerenes is the electrophilic addition to 6,6 double bonds, which reduces the angular voltage by exchanging the hybridized carbons in sp^2 for the hybridized carbons in sp^3. The change in hybridized orbitals causes the bond angles to decrease from approximately 120° in sp^2 orbitals to approximately 109.5° in sp^3 orbitals [106]. This decrease in the bond angle allows the bonds to bend less when the ball or tube is closed, making the molecule more stable. Other atoms can be included in the fullerenes to form inclusion compounds known as endohedral fullerenes. An unusual example is the egg-shaped fullerene Tb_3NC_{84}, which violates the isolated pentagon rule.

C_{60} in extra virgin olive oil shows the characteristic purple color of flawless C_{60} solutions. Fullerenes are soluble in many organic solvents, such as toluene, chlorobenzene, and 1,2,3-trichloropropane. The solubilities are generally quite low, for example, 8 g/L for C_{60} in carbon disulfide. However, fullerenes are the only known allotrope of carbon that can dissolve in common solvents at room temperature. One of the best solvents is 1-chloronaphthalene, which dissolves 51 g/L C_{60}. Pure Buckminsterfullerene solutions have a deep purple color. C_{70} solution is reddish-brown. The top fullerenes C_{76} to C_{84} have different colors. Both pure and solvated C_{60} and C_{70} crystals with millimeter size can be grown from a benzene solution. Crystallization of C_{60} from a benzene solution below 30 ° C (if the solubility is maximum) results in a triclinic solidsolvateC_{60}•$4C_6H_6$ solvate. Above 30 °C, one obtains solvate-free FCC C_{60}. Fullerenes are usually electrically insulating, but when crystallized with alkali metals, the product can be conductive or even superconducting. Some fullerenes (e.g., C_{76}, C_{78}, C_{80}, and C_{84}) are inherently chiral because they are D2-symmetric and have been successfully resolved. Research efforts are underway to develop specific sensors for its enantiomers.

3.8.1.2.4 Applications

Fullerenes can produce excellent antioxidants. This property can be attributed to a large number of conjugated double bonds and a very high electronic affinity of these molecules (due to the low energy of the unoccupied molecular orbital). Fullerenes can react with various radicals before consumption. A single C_{60} molecule can interact with up to 34 methyl radicals before use [107]. For this reason, these molecules are called "the most efficient radical scavengers in the world" or "radical sponge." Perhaps, one of the major advantages of using these molecules as an antioxidant is that these can be localized within the cell. These molecules also act as effective cytoprotectors against UV-A radiation. These bind to reactive oxygen species and prevent damage to the cells. A water-soluble C_{60} derivative with polyvinyl pyrrolidone or a radical sponge is generally added to cosmetics. This prevents skin damage and premature aging without side effects. The main advantage of this molecule is that it is easily absorbed by intact skin. Fullerene molecules also prevent lipid peroxidation by eliminating peroxy radicals and thus prevent the associated cellular cytotoxicity. The greatest challenge for the scientists was their insolubility in an aqueous medium and their tendency to form aggregates. This has been overcome by various techniques, such as encapsulating fullerenes with hydrophilic molecules, suspending this molecule with other solvents, and conjugating it with other hydrophilic molecules. Fullerenes have received a lot of attention due to their potential as antiviral agents. Perhaps the most exciting aspect of this is its ability to suppress replication of the human immunodeficiency virus (HIV), thereby delaying the onset of AIDS [108].

Drug delivery is the proper transportation of a pharmaceutical compound to its site of action, while gene delivery is the introduction of foreign DNA into cells to achieve the desired effect. It is therefore extremely important to deliver these molecules safely and highly efficiently. Fullerenes are a class of inorganic carriers. These molecules are preferred because they have good biocompatibility, greater selectivity, retention of biological activity, and diffusion. The DNA sequences are linked to the amino acid derivatives of fullerene. These sequences become detached from the support when the amino groups are lost or denatured. Biochemical studies have shown that these derivatives have a greater protective capacity than the conventional vector. Fullerenes can be used in the administration of hydrophobic drugs. In fact, these carriers are used in the slow release of these hydrophobic drugs into the system. Significant anticancer activity was observed in the Paclitaxel C_{60} conjugate. An additional advantage is that they can easily diffuse through intact skin: A fullerene-based peptide has shown the ability to penetrate through the skin [109].

An organic polymer-based photovoltaic cell could be the answer to the search for a cheap and lightweight means of converting solar energy. Essentially, these solar cells work by transferring electrons from a material that is excited when exposed to light (known as a donor). In its excited state, this electron is picked up by an acceptor molecule, which is then transferred to the electrode. Fullerenes are excellent acceptors due to their high electronic affinity and their ability to transfer these electrons. These organic photovoltaic cells are composite of fullerenes and polymers and are called mass heterojunctions. Phenyl-C_{61}-butyric acid methyl ester (PCBM) is a common acceptor used in organic solar cells. It is generally used in conjunction with the polythiophene polymer (P3HT) as an electron donor. A recent study found that a hybrid system of fullerene and graphene molecules transferred a specific capacity of 135.36 Fg^{-1} when used as an electrode for supercapacitors, as opposed to the specific capacity of the pure graphene electrode (101.88 Fg^{-1}). According to the report, the fullerene/graphene connection showed a good retention time rate of 92.35% even after 1000 charge/discharge cycles. Fullerenes are used as anodes in lithium-ion batteries because carbon offers safety and longer life. A team of researchers examined a high-performance anode material based on hydrogenated fullerenes in rechargeable lithium-ion batteries. The results of the study show that the use of hydrogenated fullerenes as additives to commercial graphite led to an extraordinary decrease in the irreversible capacity of commercial carbon with an increase in its reversible capacity [110].

When an alkali metal is embedded in the cavities of the C_{60} fullerene, the newly formed compound becomes a superconductor. Accordingly, it has been reported that the compound K_3C_{60} has a high superconducting critical temperature with almost perfect three-dimensional superconductivity, current density, ductility, and high stability. Fullerenes have optical limiting properties. This refers to its ability to reduce the transmission of incident light. Therefore, these molecules can be used as an optical limiter that can be used in goggles and sensors. This optical limiter only lets light below a certain threshold and keeps the transmitted light at a constant level, far below the intensity that can damage the eye or the sensor. The one-of-a-kind molecular structure of fullerenes enables them to hydrogenate and dehydrogenate quite easily. The carbon rings in fullerene are conjugated with C=C double bonds. On hydrogenation, these bonds can be broken easily giving rise to C—C single bonds and C—H bonds. When heat is applied to these fullerene hydrides, the C—H bonds break easily to give back fullerene. This is because the bond strength of C—H is lower when compared to that of C—C. One fullerene molecule can hold up to 36 hydrogen atoms. The color of hydrogenated fullerenes changes from black to brown, red, orange, and, finally, to yellow as the hydrogen content increases. These molecules hold a promise of better, safer, and more efficient hydrogen storage devices than the ones that are currently being used. Fullerenes can be used as a hardening agent. Fullerenes can be the future of developing comparatively lightweight metals with greater tensile strength, without seriously compromising the ductility of the metal. This can be attributed to the small size and high reactivity due to the sp2 hybridization of the carbon. This enables dispersion strengthening metal matrix by interaction of the fullerenes and the metals. A 30% increase in the hardness of lightweight Ti-24.5AI-17N alloy has been observed with the addition of fullerenes [111].

3.8.1.3 Carbon Nanotube

Carbon nanotubes (CNTs) are cylindrical molecules that consist of rolled-up sheets of single-layer carbon atoms (graphene) [112]. They can be single-walled (SWCNT) with a diameter of <1 nm or multiwalled (MWCNT), consisting of several concentrically interlinked nanotubes, with diameters reaching more than 100 nm. Their length can reach several micrometers or even millimeters. Like their building block graphene, CNTs are chemically bonded with sp^2 bonds, an extremely strong form of molecular interaction. This feature combined with carbon nanotubes' natural inclination to rope together via van der Waals forces provides the opportunity to develop ultra-high-strength, low-weight materials that possess highly conductive electrical and thermal properties. This makes them highly attractive for numerous applications [113].

3.8.1.3.1 Synthesis of CNT

Fullerenes and carbon nanotubes are not necessarily products from high-tech laboratories. They usually form in places as banal as ordinary flames caused by burning methane, ethylene, and benzene and have been found in soot from indoor and outdoor air. However, these natural varieties can be very irregular in size and quality, as the environment in which they are produced is often very uncontrolled. Therefore, while they can be used in some applications, they may lack the high level of uniformity required to meet the many needs of research and industry. Recent efforts have focused on making a more uniform CNT. Such methods are promising for low-cost, large-scale nanotube synthesis based on theoretical models, although they have to compete with the rapid development of large-scale CVD production. The growth mechanism remains controversial, and more than one mechanism could be effective during the formation of CNT. A variety of techniques have been developed to produce CNT and MWNT with different structures and morphology in laboratory quantities. Some of the synthesis procedures of CNT are described subsequently.

Arc discharge: Nanotubes were observed in the carbon black of the graphite electrodes in 1991 during an arc discharge using a current of 100 amperes that was intended to generate fullerenes. However, the first macroscopic production of carbon nanotubes was carried out in 1992 by two researchers from the NEC Fundamental Research Laboratory. The procedure used was the same as in 1991. During this process, the carbon contained in the negative electrode is sublimed due to high discharge temperatures. The yield of this process is up to 30% by weight and produces single- or multi-walled nanotubes with a length of up to 50 μm with few structural defects. Arc discharge technology uses higher temperatures (over 1700 °C) for CNT synthesis, which generally causes CNT expansion with fewer structural defects compared to other processes [114].

Laser ablation: In laser ablation, a pulsed laser vaporizes a graphite target in a high-temperature reactor while an inert gas is purged in the chamber. Nanotubes form on the cooler surfaces of the reactor when the vaporized carbon condenses. A water-cooled surface can be included in the system to receive the nanotubes. This procedure was developed by Dr. Richard Smalley and his team at Rice University, who at the time of the discovery of carbon nanotubes lasered metals with a laser to make different metal molecules. When they found out about the existence of nanotubes, they replaced metals with graphite to produce multiwalled carbon nanotubes. Later that year, the team used a composite of graphite and metal catalyst particles (the best performance was a mixture of cobalt and nickel) to synthesize single-wall carbon nanotubes. The laser ablation process produces about 70% and mainly produces single-wall carbon nanotubes with a controllable diameter that is determined by the reaction temperature. However, it is more expensive than arcing or chemical vapor deposition [115].

Thermal plasma method: Single-wall carbon nanotubes (SW-CNT) can be synthesized using a thermal plasma method that Olivier Smiljanic first invented in 2000 at NIRS (National Institute for Scientific Research) in Varennes, Canada. The aim of this method is to reproduce the conditions that prevail in arc discharge and laser ablation approaches. Instead of graphite vapors, however, a carbon-containing gas is used to supply the required carbon. In this way, SW-CNT growth is more efficient (decomposition gas can use ten times less energy than graphite evaporation). The process is also continuous and inexpensive. A gaseous mixture of argon, ethylene, and ferrocene is placed in a microwave plasma torch, where it is atomized by the plasma at atmospheric pressure, which is in the form of an intense "flame." The fumes generated by the flame contain SW-CNT, carbon, and metal nanoparticles as well as amorphous carbon [116]. Another way to make SW-CNT with a plasma torch is to use the thermal plasma induction method, which was implemented in 2005 by groups at the University of Sherbrooke and the National Research Council of Canada. The process is similar to arcing since they both use ionized gas to reach the high temperature required to vaporize the carbonaceous substances and metal catalysts required for the subsequent growth of nanotubes. The thermal plasma is induced by high-frequency oscillating currents in a coil and kept in the inert gas stream. Typically, a soot and metal catalyst particle raw material is added to the plasma and then cooled to form SW-CNT. Different diameter distributions of SW-CNT can be synthesized. The induction thermoplasmic process can produce up to 2 g of nanotube material per minute, which is higher than the arc discharge or laser ablation process [117]. All these methods require the use of metals (e.g., iron, cobalt, and nickel) as catalysts. The basic elements for the formation of nanotubes are catalyst, a source of carbon, and sufficient energy. The common feature of these methods is the addition of energy to a carbon source to produce fragments (groups or single C atoms) that can recombine to generate CNT. The energy source may be electricity from an arc discharge, heat from a furnace (~900 °C) for CVD, or the high-intensity light from a laser (laser ablation). The CVD process currently holds the greatest promise, since it allows the production of larger quantities of CNTs under more easily controllable conditions and at a lower cost.

Chemical vapor deposition: During chemical vapor deposition (CVD), a substrate is made with a layer of metal catalyst particles, most commonly nickel, cobalt, iron, or a combination. Metal nanoparticles can also be synthesized in other ways, including oxide reduction or solid oxide solutions. CVD shows the most promise for industrial-scale deposition because of its price/unit ratio, and because CVD is capable of growing nanotubes directly on a desired substrate, whereas the nanotubes must be collected in the other growth techniques. The growth sites are controllable by careful deposition of the catalyst. In 2007, a team from Meijo University demonstrated a high-efficiency CVD technique for growing carbon nanotubes from camphor [118]. Researchers at Rice University, until recently led by the late Richard Smalley, have concentrated upon finding methods to produce large, pure amounts of particular types of nanotubes. Their approach grows long fibers from many small seeds cut from a single nanotube; all of the resulting fibers were found to be of the same diameter as the original nanotube and are expected to be of the same type as the original nanotube. The diameter of the nanotubes to be grown depends on the size of the metal particles. This can be controlled by stamped (or masked) deposition of the metal, annealing, or plasma etching of a metal layer. The substrate is heated to approximately 700 °C. To start the growth of nanotubes, two gases are released in the reactor: a process gas (such as ammonia, nitrogen, or hydrogen) and a carbon-containing gas (such as acetylene, ethylene, ethanol, or methane). Nanotubes grow at the locations of the metal catalyst. The carbon-containing gas is separated on the surface of the catalyst particle, and the carbon is transported to the edges of the particle, where it forms the nanotubes. This mechanism is still under investigation. The catalyst particles can remain at the tip of the growing nanotube, or at the base of the nanotube during growth, depending on the adhesion between the catalyst particle and the substrate. Thermal catalytic decomposition of hydrocarbons has become an active area of research and could be a promising path for CNT mass production [119]. For the production of CNT, the metal nanoparticles are mixed with a catalyst support such as MgO or Al_2O_3 to increase the surface area for a higher yield of the catalytic reaction of the carbon feedstock with the metal particles. One issue in this synthesis route is the removal of the catalyst support via an acid treatment, which sometimes could destroy the original structure of the CNT. However, alternative catalyst supports that are soluble in water have proven effective for nanotube growth.

In plasma-enhanced chemical vapor deposition, the nanotube growth will follow the direction of the electric field. By adjusting the geometry of the reactor, it is possible to synthesize vertically aligned carbon nanotubes (i.e., perpendicular to the substrate), a morphology that has been of interest to researchers interested in electron emission from nanotubes. Without the plasma, the resulting nanotubes are often randomly oriented. Under certain reaction conditions, even in the absence of a plasma, closely spaced nanotubes will maintain a vertical growth direction resulting in a dense array of tubes resembling a carpet or forest.

Super-growth CVD: Kenji Hata, Sumio Iijima, and her employees from AIST, Japan, developed super-growth CVD (chemical water-assisted vapor deposition). This process improves the activity and life of the catalyst by adding water to the CVD reactor. Dense arrays of vertically aligned nanotubes or "forests" were created that were normal to the substrate. The synthesis efficiency is about 100 times higher than for the laser ablation method. The time required to make SW-CNT forests of the height of 2.5 mm by this method was 10 min in 2004. Those SW-CNT forests can be easily separated from the catalyst, yielding clean SW-CNT material (purity >99.98%) without further purification. For comparison, the as-grown CNTs contain about 5–35% of metal impurities; it is therefore purified through dispersion and centrifugation that damages the nanotubes. Super-growth avoids this problem. Patterned highly organized SW-CNT structures were successfully fabricated using the super-growth technique. The mass density of super-growth CNT is about 0.037 g/cm^3. It is much lower than that of conventional CNT powders (~1.34 g/cm^3), probably because the latter contain metals and amorphous carbon. The super-growth method is basically a variation in CVD. Therefore, it is possible to grow material containing SW-CNT, DW-CNTs, and MW-CNTs and to alter their ratios by tuning the growth conditions. Their ratios change by the thinness of the catalyst. Many MW-CNTs are included so that the diameter of the tube is wide. The vertically aligned nanotube forests originate from a "zipping effect" when they are immersed in a solvent and dried. The zipping effect is caused by the surface tension of the solvent and the van der Waals forces between the CNT. It aligns the nanotubes into a dense material, which can be formed in various shapes, such as sheets and bars, by applying weak compression during the process. Densification increases the Vickers hardness by about 70 times and density is 0.55 g/cm^3. The packed CNTs are more than 1 mm long and have a carbon purity of 99.9% or higher; they also retain the desirable alignment properties of the nanotubes forest [120].

Liquid electrolysis method: In 2015, researchers from George Washington University discovered a new way to synthesize MW-CNT by electrolysis of molten carbonates. The mechanism is similar to CVD. Some metal ions were reduced to a metal form and bound to the cathode as a nucleation point for CNT growth [121]. The reaction on the cathode is

$$Li_2CO_3 \rightarrow Li_2O + CNTs + O_2 \tag{3.4}$$

The formed lithium oxide can in situ absorb carbon dioxide (if present) and form lithium carbonate as shown in the equation.

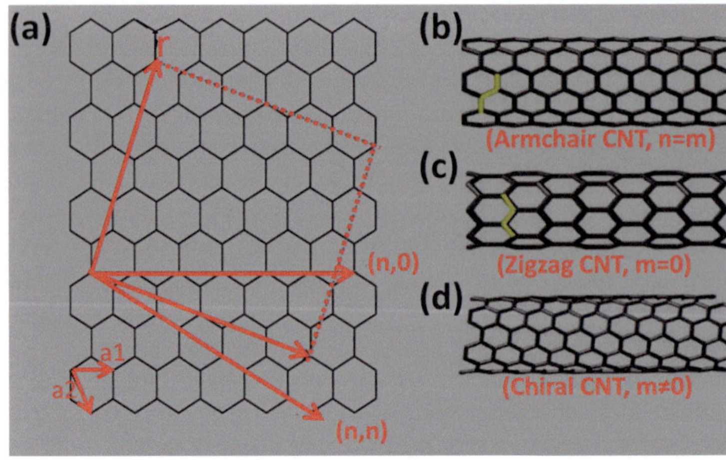

Fig. 3.30 (**a**) Graphene sheet that rolls to form single-walled and schematics of various classes of CNT: (**b**) Armchair, (**c**) Zigzag, and (**d**) Chiral CNT

$$Li_2O + CO_2 \rightarrow Li_2CO_3 \quad (3.5)$$

Thus, the net reaction is

$$CO_2 \rightarrow CNTs + O_2 \quad (3.6)$$

In other words, the reactant is only a greenhouse gas of carbon dioxide, while the product is high-valued CNT. This discovery was highlighted as a possible technology for carbon dioxide capture and conversions [122, 123].

Purification during synthesis of CNT: Catalyst removal, nanoscale metal catalysts are important components for the CVD synthesis of fluidized and fixed bed CNT. They make it possible to increase the growth efficiency of CNT and can control their structure and chirality. During synthesis, catalysts can convert carbon precursors into tubular carbon structures and can also form encapsulating carbon coatings. Together with metal oxide carriers, they can be bound to or incorporated into the CNT product. The presence of metallic contaminants can be problematic for many applications. In particular, catalyst metals such as nickel, cobalt, or yttrium can be toxicologically unsafe. Although the unencapsulated catalyst metals can be easily removed by washing with acid, the encapsulations require oxidative treatment to open their carbon shell. The effective removal of catalysts, especially encapsulation, while maintaining the structure of the CNT is a challenge. A new approach to breaking the encapsulation of carbonaceous catalysts is based on rapid thermal annealing [124].

3.8.1.3.2 Classification of CNT

Carbon nanotubes are classified into two types: single-walled carbon nanotubes (SW-CNT) and multiple-walled carbon nanotube (MW-CNT). During synthesis, a catalyst is required for SW-CNT, whereas MW-CNT can be produced without catalyst. Bulk synthesis for SW-CNT is difficult as it requires a suitable parametric condition and in MW-CNT bulk synthesis is easy. Purity is high in MW-CNT than in SW-CNT. For SW-CNT, CVD method yields 30–50 wt% purity and the arc discharge method yields 80% purity. For MW-CNT, CVD method yields 35–90 wt% purity. The twist ability is more in SW-CNT than MW-CNT.

In addition to the abovementioned classification, the entire CNT can be divided into three types: armchair CNT, zigzag CNT, and chiral CNT (Fig. 3.30b) [125]. The difference in these types of CNT is created depending on how the graphite is "rolled up" during its creation process. The choice of the rolling axis relative to the hexagonal network of the graphene sheet and the radius of the closing cylinder allows for different types of SW-CNT. The chiral vector is represented by a pair of indices, nand *m*, where these two integers correspond to the number of unit vectors along the two directions in the honeycomb crystal lattice of graphene. When $m = 0$, the nanotube is called "zigzag," when $n = m$ the nanotube is called "armchair," and all other configurations are designated as chiral [126].

The (n, m) notation

A cut and unwinded representation of a carbon nanotube as a strip of a graphene molecule superimposed on the diagram of the entire molecule (wispy background; Fig. 3.31). The arrow shows space A2 into which the A1 atom on one edge of the strip would fit when the strip was wrapped on the opposite edge. The zigzag and armchair configurations are not the only structures that a single-walled nanotube can have. To describe the structure of a general infinitely long tube, one should imagine it being sliced open by a cut parallel to its axis, that goes through some atom A, and then unrolled flat on the plane so that its atoms and bonds coincide with those of an imaginary graphene sheet- more precisely, with an infinitely long strip of that sheet. The two halves of the atom A will end up on opposite edges of the strip, over two

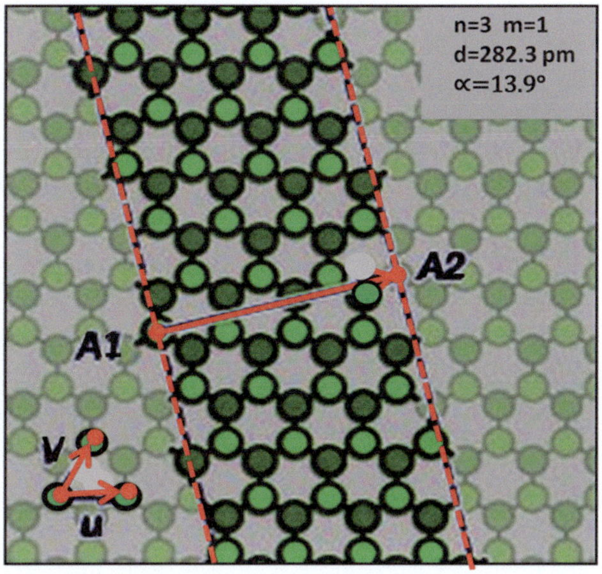

Fig. 3.31 (n,m) notation for the classification of CNT

atoms A1 and A2 of the graphene. The line from A1 to A2 will correspond to the circumference of the cylinder that went through atom A and will be perpendicular to the edges of the strip. In the graphene lattice, the atoms can be split into two classes, depending on the directions of their three bonds. Half the atoms have their three bonds directed the same way, and half have their three bonds rotated 180° relative to the first half. The atoms A1 and A2, which correspond to the same atom A on the cylinder, must be in the same class. It follows that the circumference of the tube and the angle of the strip are not arbitrary because they are constrained to the lengths and directions of the lines that connect pairs of graphene atoms in the same class [127].

Let **u** and **v** be two linearly independent vectors that connect the graphene atom A1 to two of its nearest atoms with the same bond directions. That is, if one number consecutively carbons around a graphene cell with C1 to C6, then **u** can be the vector from C1 to C3 and **v** be the vector from C1 to C5. Then, for any other atom A2 with the same class as A1, the vector from A1 to A2 can be written as a linear combination n**u**+m**v**, where n and m are integers. And, conversely, each pair of integers (n, m) defines a possible position for A2. Given n and m, one can reverse this theoretical operation by drawing the vector **w** on the graphene lattice, cutting a strip of the latter along lines perpendicular to **w** through its endpoints A1 and A2, and rolling the strip into a cylinder so as to bring those two points together. If this construction is applied to a pair (k, 0), the result is a zigzag nanotube, with closed zigzag paths of 2 k atoms. If it is applied to a pair (k, k), one obtains an armchair tube, with closed armchair paths of 4 k atoms [128].

3.8.1.3.3 Properties of CNT

CNTs are well suited for virtually any application requiring high strength, durability, electrical conductivity, thermal conductivity, and lightweight properties compared to conventional materials. The strength of the carbon–carbon bonds gives CNT amazing mechanical properties. No previous material has displayed the combination of superlative mechanical, thermal, and electronic properties attributed to them. Their densities can be as low as 1.3 g/cm^3(one-sixth of that of stainless steel). CNTs Young's moduli (measure of material stiffness) are superior to all carbon fibers with values >1 TPa which is approximately 5× higher than steel [129]. However, their strength is what really sets them apart. Carbon nanotubes are the strongest materials ever discovered by mankind. The highest measured tensile strength or breaking strain for a CNT was up to 63 GPa which is around 50 times higher than steel. Even the weakest types of CNT have strengths of several GPa. Besides that, CNTs have good chemical and environmental stability and high thermal conductivity (~3000 W/m/K, comparable to diamond). Currently, CNTs are mainly used as additives to synthetics. CNTs are commercially available as a powder, that is, in a highly tangled-up and agglomerated form. For CNT to unfold their particular properties, they need to be untangled and spread evenly in the substrate [130]. Another requirement is that CNT needs to be chemically bonded with the substrate, for example, a plastic material. For that purpose, CNTs are functionalized, that is, their surface is chemically adapted for optimal incorporation into different materials and for specific application. The electronic properties of CNT are also extraordinary. It has high electrical conductivity (comparable to copper). Especially notable is the fact that CNT can be metallic or semiconducting. The rolling action breaks the symmetry of the planar system and imposes a distinct direction with respect to the hexagonal lattice and the axial direction. Depending on the relationship between this axial direction and the unit vectors describing the hexagonal lattice, the nanotubes may behave electrically as either a metal or a semiconductor. Semiconducting nanotubes have bandgaps that scale inversely with diameter, ranging from approximately 1.8 eV for very small diameter tubes to 0.18 eV for the widest possible stable SW-CNT [131].

CNTs are not ideal structures; they contain defects formed during synthesis. Typically 1–3% of the C atoms are located at a defect site. The Stone–Wales defect (two pairs of five-membered and seven-membered rings, also known as 7-5-5-7) is one of the most common defects (Fig. 3.32). It leads to local deformation of the graphene sidewall, introducing an increased curvature. Higher the curvature, higher the reactivity. The strongest curvature is at the border between the two five-membered rings [132].

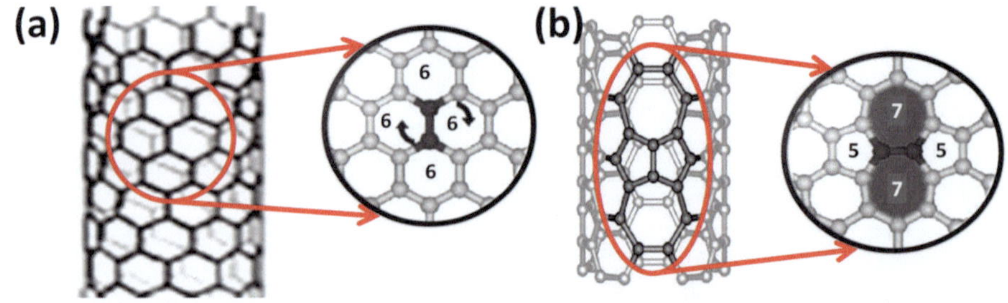

Fig. 3.32 (**a**) General configuration of CNT, and (**b**) The Stone-Wales defect present on CNT. *Red circle* highlights how the 6-6-6-6 general configuration dangles to form 7-5-5-7 configuration

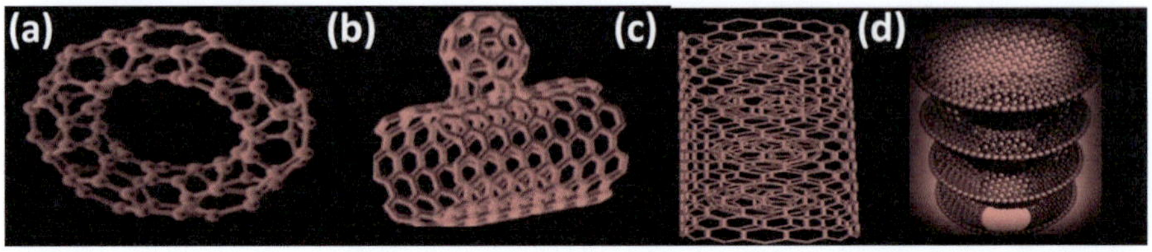

Fig. 3.33 Diffects (**a**) Torus, (**b**) Nanobud, (**c**) Peapod, and (**d**) Cup-stacked CNT

Torus: It is a carbon nanotube bent in a torus shape (i.e., doughnut shape;Fig. 3.33a) [133].

Nanobud: Carbon nanobuds are created by combining carbon nanotubes and fullerenes (Fig. 3.33b) [134].

Peapod: A carbon peapod is a novel hybrid carbon material that traps fullerene inside a carbon nanotube (Fig. 3.33c) [135].

Cup-stacked carbon nanotubes: Cup-stacked CNTs exhibit semiconducting behaviors due to the stacking microstructure of graphene layers (Fig. 3.33d) [136].

3.8.1.3.4 Application of CNTs

Currently, CNT is used in almost all structural and functional materials where high strength is essential. Semiconducting SW-CNTs are strong candidates for the next generation of high-performance, ultra-scaled, and thin-film transistors as well as for optoelectronic devices to replace silicon electronics. CNT transistors can offer performance advantages over silicon around sub-10 nm lengths [137]. While individual nanotubes generate discrete fine peaks in optical absorption and emission, macroscopic structures consisting of many CNT gathered together also demonstrate interesting optical behavior. For example, a millimeter-long bundle of aligned MW-CNTs emits polarized incandescent light by electrical current heating, and SW-CNT bundles are giving higher brightness emission at lower voltage compared to conventional tungsten filaments. Ink formulations based on CNT dispersions are attractive for printed electronic applications, such as transparent electrodes, RFID tags, thin-film transistors, light-emitting devices, and solar cells [138]. CNTs have been widely used as electrodes for chemical and biological sensing applications and many other electrochemical studies. With their unique one-dimensional molecular geometry of a large surface area coupled with their excellent electrical properties, CNTs have become important materials for the molecular engineering of electrode surfaces, where the development of electrochemical devices with region-specific electron transfer capabilities is of paramount importance. Given their high electrical conductivity, and the incredible sharpness of their tip (the smaller the tips' radius of curvature, the more concentrated the electric field, and the higher field emission), carbon nanotubes are considered the most promising material for field emitters and a practical example is CNT as electron emitters for field emission displays (FED).

Field emission display (FED) technology makes possible a new class of large-area, high-resolution, low-cost flat panel displays. However, FED manufacturing requires CNT to be grown in precise sizes and densities. Height, diameter, and tip sharpness affect voltage, while density affects the current. Buckypapers could find numerous applications; as one of the most thermally conductive materials known, Buckypaper could lead to the development of more efficient heat sinks for chips; a more energy-efficient and lighter background illumination material for displays; a protective material for electronic circuits from electromagnetic interference due to its unusually high current carrying capacity; or switchable surfaces. CNT is attractive for catalysis due to its exceptionally high surface area combined with the ability to attach essentially any chemical species to their sidewalls. Already, CNTs have been used as catalysts in many relevant chemical processes; however, controlling their catalytic activity is not easy. Initially, CNTs have been combined with molecules via very strong bonds (covalent bonds) that lead to very stable compounds. Such connection, however, implies a change in the structure of the nanotube and therefore in its properties. CNTs are used in sporting goods (high-strength bicycle frames, tennis rackets, hockey sticks, golf clubs and balls, skis, kayaks; sports arrows), Yachting (masts, hulls, and other parts of sailboats), Automotive, aeronautics, and space (lightweight, high-strength structural composites), industrial engineering (e.g., coating of wind-turbine rotor blades, industrial robot arms), electrostatic charge protection (for instance, researchers have a developed electrically conducting and flexible CNT film specifically for space applications) and radiation shielding with CNT-based nanofoams and aerogels, textiles (antistatic and electrically conducting textiles: smart textiles, bulletproof vests, water-resistant and flame-retardant textiles), and so on. [139].

3.8.1.4 Other Forms of Carbon-Based Nanomaterials

Graphene quantum dots (GQD) are graphene nanofragments with a size of <100 nm (1.0×10^{-7} m). The properties of GQD differ from "bulk" graphenes due to the quantum restriction effects, which can only be seen with a size of less than 100 nm [140]. Top-down methods cut bulk graphite in GQD with strong chemicals (e.g., mixed acids). Bottom-up methods assemble GQD from monomers of organic molecules (e.g., citric acid and glucose). Some examples of these processes are the microwave-assisted hydrothermal process, the soft template process, the hydrothermal process, the ultrasound peeling process, the electron beam lithography process, the chemical synthesis process, the electrochemical production process, graphene oxide reduction method (GO), catalytic C_{60} transformation method, and so on. Graphene nanoribbons ("nano-stripes" in the "zigzag" orientation) show spin-polarized metal edge currents at low temperatures, which also indicate applications in the new spintronics area. (In the "chair" orientation, the edges behave like semiconductors.) Graphene superlattices are periodically stacked graphene, and its insulating isomorphs provide a fascinating structural element in implementing highly functional superlattices at the atomic scale, which offers possibilities in designing nanoelectronic and photonic devices. Various types of superlattices can be obtained by stacking graphene and its related forms. The energy band in layer-stacked superlattices is found to be more sensitive to the barrier width than that in conventional III–V semiconductor superlattices. When adding more than one atomic layer to the barrier in each period, the coupling of electronic wavefunctions in neighboring potential wells can be significantly reduced, which leads to the degeneration of continuous sub-bands into quantized energy levels [141]. The bulk material spontaneously disperses in basic solutions or can be dispersed by sonication in polar solvents to yield monomolecular sheets, known as graphene oxide. Graphene oxide sheets have been used to prepare strong paper-like materials, membranes, thin films, and composite materials. Initially, graphene oxide attracted substantial interest as a possible intermediate for the manufacture of graphene. These sheets, called graphene oxide paper, have a measured tensile modulus of 32 GPa. The chemical property of graphite oxide is related to the functional groups attached to graphene sheets. These can change the polymerization pathway and similar chemical processes. Graphene oxide flakes in polymers display enhanced photo-conducting properties. Graphene is normally hydrophobic and impermeable to all gases and liquids (vacuum-tight). However, when formed into graphene oxide-based capillary membrane, both liquid water and water vapor flow through as quickly as if the membrane was not present. The graphene obtained by reduction of graphene oxide still has many chemical and structural defects which is a problem for some applications.

In 2011, researchers reported a simple approach to fabricate graphene fibers from CVD-grown graphene films. The method was scalable and controllable, delivering tunable morphology and pore structure by controlling the evaporation of solvents with suitable surface tension [142]. Flexible all-solid-state supercapacitors based on these graphene fibers were demonstrated in 2013. The fibers offer better thermal and electrical conductivity and mechanical strength. Thermal conductivity reached 1290 W/m/K, while tensile strength reached 1080 MPa. The researcher claimed that it is able to create a CNF that has a tensile strength of 80 GPa. Pillared graphene is a hybrid carbon structure consisting of an oriented array of CNT connected at each end to a sheet of graphene. It was first described theoretically by George Froudakis and colleagues of the University of Crete in Greece in 2008. Pillared graphene has not yet been synthesized in the laboratory, but it has been suggested that it may have useful electronic properties or as a hydrogen storage material [143]. In graphene Aerogel, an aerogel made of graphene layers separated by carbon nanotubes was measured at 0.16 mg/cm^3. A solution of graphene and carbon nanotubes in a mold is freeze-dried to dehydrate the solution, leaving the aerogel. The material has superior elasticity and absorption. It can recover completely after more than 90% compression and absorb up to 900 times its weight in oil at a rate of 68.8 g/sec. In 2015, a coiled form of graphene was discovered in graphitic carbon (coal) called Graphene nano coil [144]. The spiraling effect is produced by defects in the material's hexagonal grid that causes it to spiral along its edge, mimicking a Riemann surface, with the graphene surface approximately perpendicular to the axis. When voltage is applied to such a coil, current flows around the spiral, producing a magnetic field. The phenomenon applies to spirals with either zigzag or armchair patterns, although with different current distributions. A solenoid made with such a coil behaves as a quantum conductor whose current distribution between the core and exterior varies with applied voltage, resulting in nonlinear inductance. In hyperstage graphite, covalent modification of the π-electron basal planes of graphene enables the formation of new materials with enhanced functionality. An electrochemical method is reported for the formation of what is referred to as a hyperstage-1 graphite intercalation compound (GIC), which has a very large interlayer spacing $d_{001} > 15.3$ Å and contains disordered interstitial molecules/

ions. This material is highly activated and undergoes spontaneous exfoliation when reacted with diazonium ions to produce soluble graphenes with high functionalization densities of one pendant aromatic ring for every 12 graphene carbons. Critical to achieving high functionalization density is the hyperstage-1 GIC state, a weakening of the van der Waals coupling between adjacent graphene layers, and the ability of reactants to diffuse into the disordered intercalate phase between the layers [145]. The holey graphene, also called graphene nanomesh, is a structural derivative of graphene. Holey graphene is formed by removing a large number of atoms from the graphitic plane to produce holes distributed on and through the atomic thickness of the graphene sheets. These holes, sometimes with abundant functional groups around their edges, impart properties that are uncommon to intact graphene but advantageous toward various applications. In this review, strategies to prepare holey graphene and the related applications that take advantage of the unique structural motif of these materials. In 2013, a three-dimensional honeycomb of hexagonally arranged carbon was termed 3D graphene. Recently, self-supporting 3D graphene has also been produced. 3D structures of graphene can be fabricated by using either CVD or solution-based methods. A recent review by Khurram and Xu et al. has provided a summary of the state-of-the-art techniques for the fabrication of the 3D structure of graphene and other related two-dimensional materials. Recently, researchers at Stony Brook University have reported a novel radical-initiated crosslinking method to fabricate porous 3D free-standing architectures of graphene and CNT using nanomaterials as building blocks without any polymer matrix as support. These 3D graphene scaffolds/foams have applications in several fields such as energy storage, filtration, thermal management, and biomedical devices and implants [146]. Box-shaped graphene nanostructure appeared after mechanical cleavage of pyrolytic graphite has been reported recently. The discovered nanostructure is a multilayer system of parallel hollow nanochannels located along the surface and having quadrangular cross-section. The thickness of the channel walls is approximately equal to 1 nm. Potential fields of box-shaped graphene application include ultrasensitive detectors, high-performance catalytic cells, nanochannels for DNA sequencing and manipulation, high-performance heat sinking surfaces, rechargeable batteries of enhanced performance, nanomechanical resonators, electron multiplication channels in emission nanoelectronic devices, and high-capacity sorbents for safe hydrogen storage.

3.8.1.5 Metal-Based Nanomaterials

The metallic nanoparticle is a nano-sized metal with a size range from 10 to 100 nm. These are metal-based materials that we usually consider quantum dots, nanogold, nanosilver, and nano-metal-oxides. Metal nanoparticles have unique properties such as surface plasmon resonance and optical properties. Precious metals, especially silver and gold, have attracted the attention of researchers in various areas of science and technology, namely, catalysis, photography, and medicine as anticancer and antimicrobial agents [147]. Quantum dot is a very small particle in the nanometer range. They are made up of thousands of atoms. These semiconductor materials consist of silicon or germanium. The quantum dot can therefore be explained by semiconductor band theory. The bandgap relates to the energy difference between the valence band (top) and the conduction band (bottom). Metals that are electronically classified as semiconductors have a partially filled band that is separated from the empty conduction band by a bandgap. When it comes to interacting molecular orbitals, the two that interact are generally HOMO (Highest energy Occupied Molecular Orbital) and LUMO (Lowest energy Unoccupied Molecular Orbital). HOMO is the most energetic molecular orbital of a molecule, whereas LUMO is the vacant molecular orbital with the lowest energy of the other molecule. HOMO and LUMO are the pair that is close to the energy of an orbital pair in the two molecules and allows them to interact strongly. These orbitals are also called frontier orbitals because they are at the extreme limits of the electrons in the molecules. Therefore, quantum mechanics can be used to describe the energy of metallic nanoparticles.

Metal nanoparticles have been attracted to bulk metals by different industries due to their different physical and chemical properties. The catalysts used in metallic nanoparticles are selective and highly active, and they have a long lifespan for many chemical reactions. It was found that a DVD with a storage capacity of 10 TB corresponds to approximately 2000 films of conventional size [148]. This is only possible due to the optical properties of the gold nanorods embedded in the disc, which are randomly aligned. Zijlstra and his team used an optical spectrum and different polarization directions to store data. The optical properties of gold, silver, lead, and platinum nanoparticles result from the resonance vibration of their free electrons in the presence of light, also known as localized surface plasma resonance (LSPR). From a historical perspective, silver was considered a symbol of purity and was valuable compared to gold at that time because silver has many medicinal properties and heals many diseases. It has antibacterial and antiseptic properties. In ancient times, precious metals were used for stained glass to create the beautiful colors of beverage cups such as the Lycurgus cup. In 1890, bacteriologist Robert Koch discovered that potassium gold cyanide $K[Au(CN)_2]$, in low concentrations, has antimicrobial activity against the tubercle bacillus, from which gold is introduced into modern medicine [149].

3.8.1.5.1 Synthesis of Some Metal-Based Nanomaterials

There are two methods involved in the synthesis of metal nanoparticles: top-down approach and bottom-up approach. Some typical economical methods to synthesis of different metal nanoparticles such as gold, silver, etc. are discussed subsequently:

Chemical reduction method: Chemical reduction is the most common and widely used method for the preparation of gold nanoparticles. This method includes, in the presence of a reducing agent, gold salt is reduced. In 1857, Michael Faraday first studied gold colloids synthesis in the solution phase, wherein an aqueous medium gold chloride reduced with phosphorus. Citrate reduction method was reported in 1951 for the synthesis of gold nanoparticles (Au NP). Synthesis of Au NP was based on a single-phase reduction of gold tetrachloroauric acid by sodium citrate in an aqueous medium and produced particles about 20 nm. The major contribution for Au NP synthesis was published in 1994 and now it is popularly known as Burst–Schifrin method. This process uses two phases that exploit thiol ligands that strongly bind to gold due to the soft character of S and Au. With the help of a phase transfer agent such as tetraoctylammonium bromide, gold salt is transferred into an organic solvent and then organic thiol is added. At last, a strong reducing agent such as sodium borohydride (in excess) is added which gives thiolate-protected Au NP. The major benefit of synthesizing this process is ease of preparation, size-controlled, thermally stable nanoparticles, and reduced dispersity.

Silver nanoparticles are one of the most attractive inorganic materials because of its environment-free nature. Moreover, it has several applications in various fields like photography, diagnostics, catalysis, biosensor, and antimicrobial. There are two ways for reduction of silver nanoparticles: (1) by citrate anion and (2) by Gallic acid (GA). Reduction by citrate anion first is to stabilize the nanoparticles and to reduce the metal cation. To determine the growth of particles, this reactant played a major role. Citrate controls the size and shape of Ag Np. At different citrate concentrations, by using the boiling method, Ag Np with Plasmon maximum absorbance at 420 nm was produced. By increasing the concentration one to five times of sodium citrate to silver cation, that is, [citrate]/[Ag+], the elapsed time for Ag Np formation was 40 to 20 min reduced respectively, which indicates that a fraction of the Ag^+ was not reduced under equimolar conditions. At room temperature, reduction of Ag^+ in water can be achieved by using GA whose oxidation potential is 0.5 V. In benzoic acid structure, the hydroxyl group at determined positions plays an important role in the synthesis of metal nanoparticles. When hydroxyl groups are located at meta position, nanoparticle synthesis was not successful but it was achieved when hydroxyl groups are present at ortho and para positions. Here carboxylic group acts as a stabilizer and hydroxyl as the reactive part. To obtain silver colloids, NaOH addition is important. Then, the silver species reacting could be Ag_2O that has been reported as a good Ag Np precursor by thermal decomposition [150].

In the synthesis of platinum nanoparticles, the platinum metal precursor either in an ionic or a molecular state is taken. By the reducing agents, chemical changes are made to convert the precursor to platinum metal atoms. These metal atoms then combine into stabilizer or supported materials to form nanoparticles. For example, in chemical reduction, H_2PtCl_6 is reduced by NaBH4 or Zn to give rise to platinum nanoparticles. $H_2PtCl_6 + NaBH_4 = Pt +$ other reaction product H_2PtCl_6 is the common precursor used in the synthesis of platinum nanoparticles. H_2PtCl_6 is usually dissolved in the organic liquid phase or aqueous phase. By introducing decomposition, displacement, reducing agent, and electrochemical reactions, the dissolved metal precursor can be converted into solid metal. By these three methods, radiolytic, sonochemical, or electrochemical, the chemical step can be activated by physical mixing. Two different reactivites are generally used in the case of mixed metal nanoparticles for example, $RuCl_3$ and H_2PtCl_6 {Na6Pt $(So_3)_4$, Na_6Ru $(So_3)_4$}{$PtCl_2$ and $RuCl_3$)} various complex mixed precursors have been also used.

Physical method: To form Au NP, photochemical reduction of gold salts has been used. This formulation uses continuous-wave UV irradiation (250–400 nm), PVP as the capping agent, and ethylene glycol as the reducing agent. Glycol concentration and viscosity of the solvent mixture are the two factors, where Au NP formation is dependent. The process was further improved by the addition of Ag+ to the solution, leading to an increase in the production of Au nanoparticles. To synthesize Pt NP, irradiation and laser ablation techniques have also been used. In one method, irradiation was combined with ultrasonication. In this process, $H_2PtCl_6.6H_2O$ was added to a solution of 10 mm polypyrrole and SDS. Particle size is controlled by varying the length and time of ultrasonication and irradiation.

Biological method: Plant-mediated synthesis has gained more popularity due to being eco-friendly. Zingiber Officinale extract acts as a reducing agent as well as a stabilizer with particles ranging from 5 to 15 nm in diameter. To synthesize metallic NP, several bacteria and fungus-like prokaryotic bacteria and eukaryotic fungus are used. Plant extracts may have been employed for the reduction of aqueous metal ions. Biological methods may have a wide distribution in particle size but have a slow reaction rate. At room temperature, the extract is mixed with a metal salt solution, and within minutes the reaction is complete. By this method, gold and silver NP have been synthesized. The rate of nanoparticles production and their quantity can be affected by the plant extract concentration, its nature metal salt concentration, temperature, and the pH. By using a leaf extract of *Polyalthia longifolia*, Ag NP can be synthesized. Preparing metallic NP by the use of plant extract is environmentally friendly and economical. It brings controlled size and morphology of NP [151].

3.8.1.6 Polymer-Based Nanomaterials

Branched components that form polymers and whose surface exhibit chain ends suited for chemical manipulation as tools. Polymer-based nanomaterial is known as a dendrimer. Dendrimers are combinable to create hollow cavities or used as part of catalysis. Dendrimers represent a half step between molecular chemistry and polymer chemistry. The creation or preparation of dendrimers is either via divergence or convergence. Either the branches grow and then attach to the core or the branches grow from the core. In the human body, protein synthesis is the building of more complex structures from parts. It is the same concept for dendrimers. Dendrimers have an amazing capability, and their current application is through biomedicine with applications such as anticancer drugs, pain management, and timed released medications such as a transdermal patch or in gene therapy. In polymer-nanoparticle matrices, nanoparticles are those with at least one dimension of <100 nm. The transition from micro- to nano-particles leads to a change in its physical as well as chemical properties. Two of the major factors in this are the increase in the ratio of the surface-area-to-volume and the size of the particle. The increase in surface area-to-volume ratio, which increases as the particles get smaller, leads to increasing dominance of the behavior of atoms on the surface area of the particle over that of those interiors of the particle. This affects the properties of the particles when they are reacting with other particles. Because of the higher surface area of the NP, the interaction with the other particles within the mixture is more and this increases the strength and heat resistance, and many factors do change for the mixture [152].

3.8.1.6.1 Synthesis of Dendrimer

Different synthesis methods for polymer-based nanoparticles are discussed subsequently.

Ionotropic gelation method: Ionotropic gelation method is known as ionic cross-linked method. Chitosan nanoparticle prepared by this technique was first reported by Calvo and his group. This method is based on electrostatic interaction between the amine group of chitosan and negatively charged group of polyanion such as tripolyphosphate. First, chitosan can be dissolved in acetic acid in the absence or presence of the stabilizing agent, such as poloxamer, which can be added to the chitosan solution before or after the addition of polyanion. Polyanion or anionic polymer was then added, and nanoparticles were spontaneously formed under mechanical stirring at room temperature. The size and surface charge of particles can be modified by varying the ratio of chitosan and stabilizer. The size and surface charge of nanoparticles can be changed by changing the ratio of chitosan and stabilizer. For a high yield of nanoparticles, the critical processing parameter chitosan–tripolyphosphate (TPP) weight ratio should be controlled. The mean particle size of nanoparticle decreases with increase in solution temperature in ultrasonic radiation samples. With the change in physicochemical conditions, like p^H of the medium, a volume phase transition takes place. Structural changes can also be introduced by ionic strength variations such as the presence of KCl at low and moderate concentrations emphasizing swelling and weakness of chitosan–TPP ionic interactions and in turn particle disintegration. Using chitosan–TPP ionotropic gelation, the estradiol (E2)-loaded chitosan nanoparticles with an average size of 269.3–316 nm and a zeta potential of 25.4 mV have been reported with 64.7% entrapment efficiency. Besides this, the insulin-containing dilute alginate solution has been used by inducing an ionotropic pregel with calcium counter ions followed by polyelectrolyte complex coating with chitin. Using this method, proteins, and peptides, including insulin, ZDEVD-FMK, a caspase inhibitor peptide, cyclosporine small-interfering RNA (siRNA), and a basic fibroblast growth factor have been loaded. Shu and Zhu described the preparation of chitosan/TPP beads made up of a blend of chitosan and gelatin, coagulated in sesame oil, and cross-linked with TPP. These beads were very homogeneous and presented a high encapsulation efficiency of the model drug and were able to control the release for 30 h. Production of chitosan nanoparticles can be done by adding a volume of TPP solution of low concentration in acetic acid. A suspension of nanoparticles appeared when the concentrations of chitosan and TPP were in the range of 0.9–3.0 and 0.3–0.8 mg/mL, respectively. Here, the size of the nanoparticles can be controlled by varying the concentrations of both counter ions. Mesobuthuseupeus venom-loaded chitosan NP can be prepared via ionic gelation of TPP and chitosan. Particle size of NP was in the range of 370 nm with positive zeta potential. The in vitro release of nanoparticles showed an initial burst release of approximately 60% in the first 10 h, followed by a slow and much reduced additional release for about 60 h. Eudragit® L100 and Eudragit® L100-PLGA nanoparticles containing diclofenac sodium can be prepared. They demonstrated that Eudragit/PLGA nanoparticles show an initial burst release followed by a slower sustained release [153].

Microemulsion method: The microemulsion method is known as the covalent cross-linking method. Chitosan nanoparticle prepared by microemulsion technique was first developed by Maitra et al. [154]. This technique is based on the formation of covalent cross-link between the chitosan chain and a functional cross-linking agent. This method was first used to prepare chitosan nanoparticles by encapsulating 5-fluorouracil by cross-linking glutaraldehyde with amino groups in the molecular chain of chitosan. In the microemulsion method, a surfactant was dissolved in N-hexane. Thereafter, chitosan in acetic solution and glutaraldehyde was added to the surfactant/hexane mixture under continuous stirring at room temperature. Nanoparticles were formed in the presence of surfactants. The system was stirred to complete the cross-linking process, in which the free amine group of chitosan conjugates with glutaraldehyde. The organic solvent is then removed by evaporation under low pressure. The yield obtained was the cross-linked chitosan NP and excess surfactant. The excess surfactant was then removed by precipitate with $CaCl_2$ and then the precipitant was removed by centrifugation. The final nanoparticle suspension was dialyzed before lyophilization. This technique offers a narrow size distribution of <100 nm, and the particle size can be controlled by varying the amount of glutaraldehyde that alters the degree of cross-linking. Nevertheless, some disadvantages exist such as the use of organic solvent, time-consuming preparation process, and complexity in the washing step. Commonly applied cross-linking agents include polyethylene glycol (PEG), dicarboxylic acid, glutaraldehyde, or monofunctional agents such as epichlorohydrin [155].

Polyelectrolyte complex method: Polyelectrolyte complex or self-assembly polyelectrolyte is a term used to describe complexes formed by self-assembly of the cationic charged polymer and polyanions. There are several formulations of chitosan NPs that rely on polyelectrolyte complexation: it requires chitosan, a polycationic polymer, and a polyanionic polymer. No auxiliary molecules such as catalysts or initiators are needed and their action is generally performed in an aqueous solution. Chitosan is positively charged at low p^H values, and it spontaneously associates with negatively charged polyanions in solution to form polyelectrolyte complexes. Examples of polyanions include chondroitin sulfate and hyaluronate, dextran sulfate, carboxymethylcellulose, heparin, and DNA that have been utilized to prepare chitosan NPs. The preparation of polyelectrolyte complex (PEC) NP is quite simple and can be easily performed under mild conditions without the use of toxic organic reagents. In the case where DNA is used as a polyanion, the nanoparticles are spontaneously formed after the addition of DNA solution into chitosan dissolved in acetic acid solution, under mechanical stirring at or under room temperature. Chitosan–fucoidan complex NP can be prepared by polyelectrolyte complex method. Fucoidan is a negatively charged polysaccharide naturally occurring in many species of brown algae and many marine invertebrates. Fucoidan can increase the potency of NP for the delivery of anticoagulant agents on the site of application due to its antithrombotic activity. When fucoidan was dropped into chitosan solution, the inter- and intramolecular electrostatic attractions occurred between anionic sulfate groups from fucoidan and cationic amino groups of chitosan. These attractions could make the macromolecular chains of chitosan and fucoidan curl up, which leads to an insoluble chitosan-fucoidan complex formation. Therefore, at present, polyelectrolyte complexes are widely used as the carrier of drugs, nonviral vectors of transferred genes, biospecific sorbents, films, and gels [156, 157].

Emulsion solvent diffusion method: Chitosan NPs prepared by emulsion solvent diffusion method. This method is based on the partial miscibility of an organic solvent as an oil phase with water. An oil/water emulsion is obtained upon injection of an organic phase into chitosan solution containing a stabilizing agent (i.e., poloxamer) under mechanical stirring followed by high-pressure homogenization. The emulsion is then diluted with a large amount of water to overcome organic solvent miscibility in water. Due to the spontaneous diffusion of solvents, an interfacial turbulence is created between the two phases leading to the formation of small particles. As the concentration of water-miscible solvent increases, a decrease in the size of particles can be achieved. This method is suitable for hydrophobic as well as hydrophilic drugs. In the case of hydrophilic drug, a multiple water/oil/water emulsion needs to be formed with the drug dissolved in the internal aqueous phase. The major drawbacks of this method include harsh processing conditions (e.g., the use of organic solvents) and the high shear forces used during NP preparation.

Emulsion solvent evaporation method: This is one of the most popular methods for the encapsulation of drugs within water-insoluble polymer. It consists of preparing an emulsion, with different external phases depending on the nature of the polymer and the drug used for the encapsulation and evaporating the solvent with the subsequent formation of the microspheres/nanosphere. Alginate (a biodegradable and biocompatible copolymer of guluronic acid and mannuronic acid) has already acclaimed permission from the U.S. Food and Drug Administration for human use and is commonly administered orally for the treatment of reflux esophagitis and has been utilized as a nanoparticulate delivery system for frontline antitubercular drugs (ATDs; rifampicin, isoniazid, pyrazinamide, and ethambutol). The dosing frequency of ATDs by applying this methodology has been reduced, thus improving patient compliance. Alginate NPs were prepared by the cation-induced controlled gelification of alginate and have been used for encapsulating other antibiotics. Another example is the preparation of NP containing ketoprofen and Eudragit E 100 by emulsion solvent evaporation method [158].

Complex coacervation method: Coacervation techniques involve the separation of a polymer solution into two coexisting phases: a dense coacervate phase, which is rich in colloids, and a diluted equilibrium phase or supernatant, which is poor in colloids. Coacervation in aqueous systems is subdivided into simple and complex coacervation. In complex coacervation, there is a spontaneous liquid/liquid phase separation when oppositely charged colloids are mixed as a result of electrostatic attractions. The coacervation process occurs under mild conditions and consequently has a great potential for the microencapsulation of living cells and labile molecules, which are unable to withstand harsh conditions (heat and organic solvents) involved in other microencapsulation processes. By coacervation between the positively charged amine groups on the chitosan and negatively charged phosphate groups on the DNA, chitosan–DNA NPs have been reported to form readily. The weight ratio of the two polymers plays a major role in controlling particle size, surface charge, entrapment efficiency, and release characteristics of the nanoparticles to be formulated. The physicochemical and release characteristics of the chitosan (CS)–dextran sulfate (DS) NP can be varied by varying the ratios of two ionic polymers. The CS–DS NPs were developed for the delivery of water-soluble small and large molecules, including proteins, plasmid DNA, rhodamine-6 G, and bovineserum albumin under mild conditions. Lecithin/chitosan NP is formed by self-organizing interaction due to the negative lipid lecithin and the positively charged chitosan that can be loaded with drug-like progesterone. The encapsulation efficiency of these systems is lower when the drug is highly water-soluble [159].

Coprecipitation method: In coprecipitation method, chitosan-based NPs with a high degree of size uniformity were prepared by grafting D, L-lactic acid on chitosan to serve as a drug carrier for prolonged drug release. The lactic acid-grafted chitosan (LA-g-chitosan) was prepared by dehydrating the solvent cast thin film of chitosan-containing lactic acids. The LA-g-chitosan NPs were synthesized via a coprecipitation process by LA-g-chitosan in ammonium hydroxide to form coacervate drops. Spherical and uniformly dispersed chitosan and lactic acid-modified chitosan (LA-g-chitosan) NPs with a mean diameter of 10 nm were prepared. Albumin encapsulation efficiency as high as 92% and 96% was attained for chitosan and LA-g-chitosan NPs, respectively [160].

Self-assembly method: Amphiphilic compounds dispersed in water can form NPs with a core-shell structure by self-assembly. Amphiphilic NPs can be used as a carrier for the hydrophobic and hydrophilic drugs simultaneously due to the presence of hydrophobic core and hydrophilic shell. One benefit of using a hydrophilic shell is to reduce macrophage phagocytosis. Therefore, amphiphilic NPs have attracted increasing attention. For example, preparation of N-acylated chitosan NPs by hydrophobic self-assembly method to make the product-controlled release, self-assembled NPs containing hydrophobically modified chitosan for gene delivery, self-aggregates formed by modified glycol chitosan as a carrier for peptide drug, self-assembled NPs based on oleoyl-chitosan (OCH), and loaded doxorubicin in OCH NPs with an encapsulation efficiency of 52.6%, hydrophobic cholanic acid to modify glycol chitosan and subsequently form NPs by self-assembly in water solution [161].

3.8.1.6.2 Applications

The nanofiber, hollow nanofiber, core-shell nanofibers, and nanorods or nanotubes produced have a great potential for a broad range of applications including homogeneous and heterogeneous catalysis, sensors, filter applications, and optoelectronics. Tissue engineering preferably uses dendrimers [162]. This is mainly concerned with the replacement of tissues that have been destroyed by sickness or accidents or other artificial means. The examples are skin, bone, cartilage, blood vessels, and may be even organs. This technique involves providing a scaffold on which cells are added and the scaffold should provide favorable conditions for the growth of the same. Nanofibers have been found to provide very good conditions for the growth of such cells, one of the reasons being that fibrillar structures can be found on many tissues that allow the cells to attach strongly to the fibers and grow along them as shown. Nanoparticles such as graphene, CNT, molybdenum disulfide, and tungsten disulfide are being used as reinforcing agents to fabricate mechanically strong biodegradable polymeric nanocomposites for bone tissue engineering applications. The addition of these nanoparticles in the polymer matrix at low concentrations (~0.2 wt. %) leads to significant improvements in the compressive and flexural mechanical properties of polymeric nanocomposites. Potentially, these nanocomposites may be used to create novel, mechanically strong, lightweight composite bone implants. The results suggest that mechanical reinforcement is dependent on the nanostructure morphology, defects, dispersion of nanomaterials in the polymer matrix, and the cross-linking density of the polymer. In general, two-dimensional nanostructures can reinforce the polymer better than one-dimensional nanostructures, and inorganic nanomaterials are better reinforcing agents than carbon-based nanomaterials. Nano polymer tubes are also used for carrying drugs in general therapy and in tumor therapy in particular. The role of them is to protect the drugs from destruction in bloodstream, to control the delivery with a well-defined release kinetics, and in ideal cases, to provide vector-targeting properties or release mechanism by external or internal stimuli [163]. Rod or tube-like, rather than nearly spherical, nanocarriers may offer additional advantages in terms of drug delivery systems. Such drug carrier particles possess an additional choice of the axial ratio, the curvature, and the "all-sweeping" hydrodynamic-related rotation, and they can be modified chemically at the inner surface, the outer surface, and at the end planes in a very selective way. Nanotubes prepared with a responsive polymer attached to the tube opening allow the control of access to and release from the tube. Furthermore, nanotubes can also be prepared showing a gradient in its chemical composition along the length of the tube. Compartmented drug release systems were prepared based on nanotubes or nanofibers. Nanotubes and nanofibers, for instance, which contained fluorescent albumin with dog-fluorescein isothiocyanate, were prepared as a model drug as well as super paramagnetic NPs composed of iron oxide or nickel ferrite. The presence of the magnetic NPs allowed, first of all, the guiding of the nanotubes to specific locations in the body by external magnetic fields. Super paramagnetic particles are known to display strong interactions with external magnetic fields leading to large saturation magnetizations. In addition, by using periodically varying magnetic fields, the NPs were heated up to provide, thus, a trigger for drug release. The presence of the model drug was established by fluorescence spectroscopy and the same holds for the analysis of the model drug released from the nanotubes [164].

Core-shell fibers of nanoparticles with fluid cores and solid shells can be used to entrap biological objects such as proteins, viruses, or bacteria in conditions that do not affect their functions. This effect can be used among others for biosensor applications. For example, green fluorescent protein is immobilized in nanostructured fibers providing large surface areas and short distances for the analyte to approach the sensor protein. With respect to using such fibers for sensor applications, fluorescence of the core-shell fibers was found to decay rapidly, as the fibers were immersed into a solution containing urea: Urea permeates through the wall into the core where it causes denaturation of the green fluorescent protein. Biosensors, tissue engineering, drug delivery, or enzymatic catalysis are just a few of the possible examples. The incorporation of viruses and bacteria all the way up to microorganism should not really pose a problem, and the applications coming from such biohybrid systems should be tremendous [165].

3.8.1.6.3 Nanocomposites

As the word describes, this is a combination of nanoparticles or nanoparticles with other materials to form unique nanoproduct. Nanocomposite is a multiphase solid material where one of the phases has one, two, or three dimensions of <100 nm or structures having nano-scale repeat distances between the different phases that make up the material. In the broadest sense, this definition can include porous media, colloids, gels, and copolymers but is more usually taken to mean the solid combination of a bulk matrix and nano-dimensional phase differing in properties due to dissimilarities in structure and chemistry. Nanocomposites are found in nature, for example, in the structure of the abalone shell and bone [112]. In mechanical terms, nanocomposites differ from conventional composite materials due to the exceptionally high surface-to-volume ratio of the reinforcing phase and/or its exceptionally high aspect ratio. The reinforcing material can be made up of particles (e.g., minerals), sheets (e.g., exfoliated clay stacks), or fibers (e.g., carbon nanotubes or electrospun fibers). The area of the interface between the matrix and the reinforcement phases is typically an order of magnitude greater than for conventional composite materials. The matrix material properties are significantly affected in the vicinity of the reinforcement. This large amount of reinforcement surface area means that a relatively small amount of nanoscale reinforcement can have an observable effect on the macroscale properties of the composite. For example, adding carbon nanotubes improves electrical and thermal conductivity. Other kinds of nanoparticulates may result in enhanced optical properties, dielectric properties, heat resistance, or mechanical properties such as stiffness, strength, and resistance to wear and damage. In general, the nano reinforcement is dispersed into the matrix during processing. The percentage by weight (called mass fraction) of the nanoparticulates introduced can remain very low (on the order of 0.5–5%) due to the low filler percolation threshold, especially for the most commonly used nonspherical, high aspect ratio fillers (e.g., nanometer-thin platelets, such as clays, or nanometer-diameter cylinders, such as CNT).

Nanocomposites are of three types based on the matrix materials: (1) ceramic matrix nanocomposites (CMNC), (2) metal matrix nanocomposites (MMNC), and (3) polymer matrix nanocomposites (PMNC) are described subsequently.

3.8.1.6.4 Metal Matrix Nanocomposites (MMNC)

This type of composites can be classified as continuous and non-continuous reinforced materials. One of the more important nanocomposites is CNT metal matrix composites, which is an emerging new material that is being developed to take advantage of the high tensile strength and electrical conductivity of CNT materials. Critical to the realization of CNT-MMC possessing optimal properties in these areas is the development of synthetic techniques that are (a) economically producible, (b) provide for homogeneous dispersion of nanotubes in the metallic matrix, and (c) lead to strong interfacial adhesion between the metallic matrix and the CNT. In addition to CNT-MMC, boron nitride-reinforced MMC and carbon nitride MMC, Fe-Cr/Al_2O_3, Ni/Al_2O_3, Co/Cr, Fe/MgO, Al/CNT, and Mg/CNT are the new demanding areas on metal matrix nanocomposites [166].

Mechanical properties depend on the type of reinforcement. Comparing the mechanical properties (Young's modulus, compressive yield strength, flexural modulus, and flexural yield strength) of single- and multi-walled reinforced polymeric (polypropylene fumarate-PPF) nanocomposites to tungsten disulfide nanotubes-reinforced PPF nanocomposites suggest that tungsten disulfide nanotubes-reinforced PPF nanocomposites possess significantly higher mechanical properties, and tungsten disulfide nanotubes are better reinforcing agents than carbon nanotubes. Increases in the mechanical properties can be attributed to a uniform dispersion of inorganic nanotubes in the polymer matrix (compared to carbon nanotubes that exist as micron-sized aggregates) and increased cross-linking density of the polymer in the presence of tungsten disulfide nanotubes (increase in crosslinking density leads to an increase in the mechanical properties) [167].

3.8.1.6.5 Ceramic Matrix Nanocomposites (CMNC)

Ceramic matrix composites consist of ceramic fibers embedded in a ceramic matrix. The matrix and fibers can consist of any ceramic material, including carbon and carbon fibers. The ceramic occupying most of the volume is often from the group of oxides, such as nitrides, borides, and silicides, whereas the second component is often a metal. Ideally, both components are finely dispersed in each other to elicit particular optical, electrical, and magnetic properties as well as tribological, corrosion resistance, and other protective properties. Examples of CMCs are Al_2O_3/SiO_2, SiO_2/Ni, Al_2O_3/TiO_2, Al_2O_3/SiC, and Al_2O_3/CNT. The binary phase diagram of the mixture should be considered in designing ceramic-metal nanocomposites and measures have to be taken to avoid a chemical reaction between both components. The important point is for the metallic component that may easily react with the ceramic and thereby lose its metallic character. This is not an easily obeyed constraint because the preparation of the ceramic component generally requires high process temperatures. The safest measure thus is to carefully choose immiscible metal and ceramic phases [168]. The concept of CMNC was also applied to thin films that are solid layers of a few nanometer to some tens of micrometer thickness deposited upon an underlying substrate and that play an important role in the functionalization of technical surfaces. Gas flow sputtering by the hollow cathode technique turned out as a rather effective technique for the preparation of nanocomposite layers. The process operates as a vacuum-based deposition technique and is associated with high deposition rates up to some μm and the growth of nanoparticles in the gas phase.

3.8.1.6.6 Polymer Matrix Nanocomposites (PMNC)

In the simplest case, adding nanoparticulates to a polymer matrix can enhance its performance, often dramatically, by simply capitalizing on the nature and properties of the nanoscale filler (these materials are better described by the term nanofilled polymer composites). This strategy is particularly effective in yielding high-performance composites when uniform dispersion of the filler is achieved and the properties of the nanoscale filler are substantially different or better than those of the matrix. The uniformity of the dispersion is in all nanocomposites is counteracted by thermodynamically driven phase separation. Clustering of nanoscale fillers produces aggregates that serve as structural defects and result in failure. Layer-by-layer (LbL) assembly when nanometer-scale layers of nanoparticulates and polymers are added one by one. LbL composites display performance parameters 10–1000 times better than the traditional nanocomposites made by extrusion or batch-mixing [169]. Examples of various PMNC are thermoplastic/thermoset, polymer/layered silicate, polymer/Au, polyester/TiO_2, and polymer/CNT. Nanoscale dispersion of filler or controlled nanostructures in the composite can introduce new physical properties and novel behaviors that are absent in the unfilled matrices. Some examples of such new properties are fire resistance or flame retardancy and accelerated biodegradability. A range of polymeric nanocomposites is used for biomedical applications, such as tissue engineering, drug delivery, and cellular therapies. Due to unique interactions between polymer and nanoparticles, a range of property combinations can be engineered to mimic native tissue structure and properties. A range of natural and synthetic polymers are used to design polymeric nanocomposites for biomedical applications, including starch, cellulose, alginate, chitosan, collagen, gelatin, and fibrin, poly(vinyl alcohol) (PVA), poly(ethylene glycol) (PEG), poly(caprolactone) (PCL), poly(lactic-co-glycolic acid) (PLGA), and poly(glycerol sebacate) (PGS). A range of nanoparticles including ceramic, polymeric, metal oxide, and carbon-based nanomaterials are incorporated within polymeric network to obtain desired property combinations [170].

3.8.1.6.7 Synthesis of Nanocomposite

Most of these reinforcements are prepared by known techniques: chemical, mechanical (e.g., ball milling), vapor deposition, and so on. Surface modifications of reinforcements are carried out to give homogeneous distribution with less agglomeration and to improve interfacial bonding between the matrix and the nanosized reinforcements. In the case of CNTs, the use of surfactants, oxidation, or chemical functionalization of surfaces is some of the techniques employed. Chemical methods may be more effective, particularly for polymer and ceramic matrices. Physical blending and in situ polymerization are used for improving dispersion in the case of CNT-reinforced polymer composites, while alignment of CNT could be achieved by techniques such as ex situ techniques (filtration, template and plasma-enhanced chemical vapor deposition, force field-inducements, etc.). Despite their nano dimensions, most of the processing techniques of the three types of nanocomposites remain almost the same as in microcomposites [171]. Many methods that have been described for the preparation of CMNC are conventional powder method; polymer precursor route; spray pyrolysis; vapor techniques (CVD and PVD) and chemical methods, which include the sol-gel process, colloidal and precipitation approaches, and the template synthesis [172]. The most common techniques for the processing of MMNC are spray pyrolysis; liquid metal infiltration; rapid solidification; vapor techniques (PVD, CVD); electrodeposition; and chemical methods, which include colloidal and sol-gel processes. Difficulties have been encountered in preparing composites with very fine particles due to their induced agglomeration and non-homogeneous distribution. The use of ultrasound helped to improve the wettability between the matrix and the particles. Many methods have been described for the preparation of polymer nanocomposites, including layered materials and those containing CNTs. The most important synthesis processes are intercalation of the polymer or pre-polymer from solution, in situ intercalative polymerization, melt intercalation, direct mixture of polymer and particulates, template synthesis, in situ polymerization, and sol-gel process.

3.8.1.6.8 Application of Nanocomposite

Nanocomposites are currently being used in a number of fields, and new applications are being continuously developed. Applications of nanocomposites include thin-film capacitors for computer chips, solid polymer electrolytes for batteries, automotive engine parts and fuel cells, impellers and blades, oxygen and gas barrier coatings, food packaging, and so on. [173]. Other potential applications of nanocomposites are Fe/MgO nanocomposites used in catalyst and magnetic devices, Ni/PZT nanocomposites used in wear-resistant coatings and thermally graded coating, Ni/TiO_2 nanocomposites used in photoelectrochemical application, Al/SiC nanocomposites used in aerospace, naval and automotive structure, Cu/Al_2O_3 nanocomposites used in electronic packaging, Al/AlN nanocomposites used in microelectronic industries, Ni/TiN, Ni/ZrN nanocomposites are used in high-speed machinery, tooling, optical and magnetic storage materials, Nb/Cu nanocomposites used in structural materials for high-temperature application, Fe/$F_{23}C_6$/Fe_3B nanocomposites used in structural materials, Fe/TiN nanocomposites used in catalyst, Al/Al_2O_3 nanocomposites used in microelectronic industries, Au/Ag nanocomposites used in microelectronics, optical devices, and light energy conversion. Lightweight bodies made of metal- or polymer-based nanocomposites with suitable reinforcements are reported to exhibit low density and very high strength (e.g., carbon Bucky fibers, with a strength of 150 GPa and weight $\approx$1/fifth of steel). Also, two-phase heterogeneous nanodielectrics, generally termed dielectric nanocomposites, have wide applications in electric and electronic industries [174]. Metal and ceramic nanocomposites are expected to generate a great impact over a wide variety of industries, including the aerospace, electronic, and military, while polymer nanocomposites major impacts will probably appear in battery cathodes, microelectronics, nonlinear optics, sensors, and so on. Improved properties include significant enhancements in fracture strength (about 2 times) and toughness (about one-half time); no time-dependent wear transitions even at very low loads; higher high-temperature strength and creep resistance; increased hardness with increasing heat treatment temperature; hardness values higher than those of existing commercial steel and alloys; the possibility of synthesis of inexpensive materials; and a significant increase in Young's modulus (about 105%), shear modulus, and fracture strength (almost 3 times compared to microcomposites) [175].

CNT-polymer composites are reported to be potential candidates for data storage media, photovoltaic cells and photodiodes, optical limiting devices, and drums for printers. CNT-ceramic composites are reported to be potential candidates for aerospace and sports goods, composite mirrors, and automotive spares requiring electrostatic painting. Such materials have also been reported to be useful for flat panel displays, gas storage devices, toxic gas sensors, Li^+ batteries, robust but lightweight parts, and conducting paints [176]. The polymer-based nanocomposites are in the forefront of applications due to their more advanced development status compared to metal and ceramic counterparts, in addition to their unique properties. These include two- to threefold strength property increase, even with low reinforcement content (1–4 wt. (%)) (e.g., 102.7% in Young's modulus) with complete elimination of voids/holes, gas barrier properties (about 200,000 times over oriented PP and about 2000 times that of Nylon-6 with tenfold requirements of expensive organic modifiers), biodegradation and reduced flammability (about 60% reduction of heat release rate), and so on. [177]. Various types of polymer nanocomposites, containing insulating, semiconducting, or metallic nanoparticles, have been developed to meet the requirements of specific applications. Recently, some PLS nanocomposites have become commercially available, being applied as ablatives and as high-performance biodegradable composites as well as in electronic and food packaging industries. These include Nylon-6 (e.g., Durethan LDPU60 by Bayer Food Packages) and polypropylene for packaging and injection-molded articles, semi-crystalline nylon for ultrahigh barrier containers and fuel systems, epoxy electrocoating primers and high voltage insulation, unsaturated polyester for watercraft lay-ups and outdoor advertising panels, and polyolefin fire-retardant cables, electrical enclosures, and housings [178]. Heat-resistant polymer nanocomposites are used to make firefighter protective clothing and lightweight components suitable to work in situations of high temperature and stress. This includes hoods of automobiles and skins of jet aircrafts as opposed to heavier and costlier metal alloys.

They can also replace corrosion-prone metals in the building of bridges and other large structures with potentially lighter and stronger capabilities. Also, unsaturated polyester (UPE) nanocomposites can be employed in fiber-reinforced products used in marine, transportation, and construction industries. Currently, UPE/fiberglass nanocomposites are being used in boat accessories that are stronger and less prone to color fading [179]. Porous polymer nanocomposites can be employed for the development of pollution filters. Other promising technological application of polymer nanocomposites in the horizon is in airbag sensors, where nano-optical platelets are kept inside the polymer outer layer for transmitting signals at the speed of light gaining milliseconds to bring down the level of possible impact injuries [180]. Finally, polymer/inorganic nanocomposites with improved conductivity, permeability, water management, and interfacial resistance at the electrode are natural candidates for the replacement of traditional Nafion PEM in fuel cells and are currently under trial. Improvements in the mechanical properties of polymer nanocomposites have also resulted in their many general/industrial applications. These include impellers and blades for vacuum cleaners, power tool housings, and mower hoods and covers for portable electronic equipment, such as mobile phones and pagers. Another example is the use of polymer nanocomposites in glues for the manufacturing of pressure molds in the ceramic industry [181]. Polymer/clay nanocomposites are currently showing considerable promise for this application. The reduction of solvent transmission is another interesting aspect of polymer/clay nanocomposites. A study conducted by the UBE Industries has revealed significant reductions in fuel transmission through polyamide-6/66 polymers by incorporation of a nanoclay filler. As a result, these materials are very attractive for the development of improved fuel tanks and fuel line components for cars. In addition, the reduced fuel transmission means significant cost reductions. The presence of filler incorporation at nanolevels has also been shown to have significant effects on the transparency and haze characteristics of films in comparison to conventionally filled polymers. The ability to minimize the extent to which water is absorbed can be a major advantage for polymer materials that are degraded in moist environments.

3.8.1.7 Nanoporous Materials

Nanoporous materials consist of a regular organic or inorganic framework supporting a regular, porous structure. The size of the pores is generally 100 nm or smaller. Nanoporous materials are solid materials showing channels, pores, or cavities with at least one dimension in the nanoscale range. The most common and natural nanoporous material is zeolites. Zeolites are crystalline solids with well-defined structures that generally contain silicon, aluminum, and oxygen in their framework and cations, water, and/or other molecules within their pores. Because of their unique porous properties, these are used in a variety of applications in petrochemical cracking, ion-exchange (water softening and purification), and in the separation and removal of gases and solvents, gas storage, catalysis, structural materials, and delivery systems. They are often also referred to as molecular sieves. Most nanoporous materials can be classified as bulk materials or membranes. There are many natural nanoporous materials, but artificial materials can be manufactured. Nanoporous materials are omnipresent in our daily lives in air purifiers, water dispensers, catalytic converters, chemical industries, plant matter, and anywhere filtration is needed. They are also required in nanoreactors, low dielectric constant mediation, and guest–host interaction. Moreover, they are required in biomedical applications for slowing drug release, filtration in hemodialysis, inhibiting bacteria, and decontamination.

Nanoporous materials can be subdivided into three categories based on pore diameter set out by IUPAC (Interna tional Union of Pure and Applied Chemistry): Microporous materials (0.2–2 nm), mesoporous materials (2–50 nm), and macroporous materials (50–1000 nm). Examples of microporous materials include zeolites and metal-organic frameworks (MOF). Microporous adhesive tape is a surgical tape used to hold wound dressings and bandages in place, introduced in 1959 by 3M with the trade name Micropore. It can be used to hold gauze padding over small wounds, usually as a temporary measure until a suitable dressing is applied. Microporous tape is also used by some film and TV sound recordists to affix small radio microphones to actors' skin. Microporous material is also used as high-performance insulation material used from home applications up to metal furnaces requiring material that can withstand more than 1000 °C [182]. Typical mesoporous materials include some kinds of silica and alumina that have similarly-sized mesopores. Mesoporous carbon has direct applications in energy storage devices. Another very common mesoporous material is Activated Carbon which is typically composed of a carbon framework with both mesoporosity and microporosity depending on the conditions under which it was synthesized. The battery performance of mesoporous electroactive materials is significantly different from that of their bulk structure [183]. Macropores may be defined differently in other contexts. Within the context of porous solids (i.e., not porous aggregations such as soil), colloid and surface chemists define macropores as cavities that are larger than 50 nm. In soil, macropores are created by plant roots, soil cracks, soil fauna, and by aggregation of soil particles into peds [184].

3.8.1.7.1 Synthesis of Porous Materials

One method of synthesis is to combine polymers with different melting points so that upon heating one polymer degrades. A nanoporous material with consistently sized pores has the property of letting only certain substances pass through while blocking others. The most successful way to produce all sorts of nanoporous materials is the templating method. An organic (or sometimes inorganic) compound acts as a placeholder which later becomes a void space in nanoporous materials. The templating concept obviously allows control, mainly of the criterion of pore size but also pore shape. The creation of a porous material is always connected to the creation of a high surface area. The template structure can have sizes on many different length scales. The smallest pores are achieved for molecules as templates. The pores created have a shape and size similar to those of the molecules used. The pores can be spatially uncorrelated with each other. In this case, one speaks about molecular imprinting. Molecular imprinting is even successful for organic materials used to build networks and is therefore very often used for imprinting biomolecules to a matrix for biochemical application [185].

3.8.1.7.2 Applications of Nanoporous Materials

Aerogels are nanocrystalline porous and extremely lightweight materials and can withstand 100 times their weight. They are currently being used for insulation in buildings, household appliances, adsorbents and ion-exchangers, catalysis, nanoreactor quantum lines and dots, engineering polymers, guest–host interaction immobilized homogeneous catalysts, enzymes, nonlinear optics, low dielectric constant mediate, biomedical applications, decontamination and antibacterial agents, slow-release drugs, filter in hemodialysis potential, enzyme mimetics and biosensors, adjuvant in anticancer therapy, and so on. [186]

3.9 Emerging Application of Nanomaterials

Nanotechnology due to its enhanced quality has a significant impact on different engineering applications such as in the field of materials and manufacturing, households, agriculture, medicine, energy, communication, and transportation. The engineering field contributes most to the development and application of nanotechnology in the market. The nanotechnology market grows at a faster rate due to the very rapid adoption of nanotechnology in various fields. But the cost of nanotechnology and restriction in application in the extreme field offers a major barrier for the use of technology. Semiconductors including transistors and diodes were estimated as valued to be $119.02 billion in 2018 from a survey made by International Trade Center. This is also reflected in the boom in the semiconductor market which is estimated at $ 420 to $ 430 billion as of 2018, and semiconductor demand will observe a CAGR of 10–12% during the forecast period 2019–2025. Similar other electronic products using nanotechnology are experiencing steady growth in revenue growth. A key application of nanotechnology is found in the segment of electronic and semiconductor products, which is estimated to grow at a significant CAGR of 15.01% by 2025 [187].

3.9.1 Aero Industries

Nanotechnology plays an important role in the performance of aerospace materials. The lightweight of the aerospace vehicle incorporates nanotechnology so that it can lift up easily. Again by the use of nanotechnology, the aircraft may consume less fuel thus being cost-effective in traveling in space and reaching orbit. When nanomaterials are incorporated, nanosensors and nanorobots, it improves the performance of spaceships in exploring planets and moons thus being the "final frontier" [188]. Using material made of CNT reduces the weight of the spaceship and increases strength. The application of CNT in elevator cable leads to a reduction in cost for sending material into orbit. The outer layer of bio-nano robots used in space suits would minimize damage such as sealing punctures. The inner layer of the bio-nanosensor also helps astronauts by alerting and providing medication, thus helping astronauts when in trouble. A network of nanosensors is being employed in search of water on Mars. Thrusters for aircraft also use MEMS devices made of nanosensors, thus making the thrusting system of aircraft more effective. The main aim in deploying nanosensors is to reduce the weight of the aircraft [189]. CNTs produce light that uses sunlight pressure to reflect on mirrored solar cells to arrange a spacecraft. This solves the problem if there is a need to attract enough fuel into space in orbit, during a transplant mission, to create a spacecraft. Work with nanoseconds to track the level of chemicals on a spacecraft to monitor the performance of life support systems. Carbon nanotubes are currently produced for use in aircraft, making them ideal for use in, for example, electromagnetic shields and thermal controls. The game-changing outcome involved in nanotechnology uses high-power materials on almost any vehicle. For example, it has been established that a 20% reduction in weight can reduce fuel consumption by 15%. According to NASA, the total weight of a truck can be reduced by 63% by using advanced nanomaterials with twice the power of conventional composition [190]. This will not only save a significant amount of energy needed to orbit the spacecraft, but it will also make it possible to develop a separate phase to create an orbit in the orbit of the aircraft and further reduce running costs, increase mission reliability, and open the door to the alternative.

3.9.2 Automotive and Naval Industry

Nanostructured cemented carbide insulation materials are used on some marine vessels for increased durability. Nanomaterials used low weight cars, trucks, and boats can lead to significant fuel savings. Various nano-engineered materials such as high-powered rechargeable battery, a tire with low rolling resistance, fuel additives for cleaner exhaust, high-efficiency sensor and thermoelectric materials meant for temperature control, and solar panels are included under automotive products [191]. Nanostructured ceramics used in automotive parts have higher toughness leading to product longevity as wear and tear are minimized. The incorporation of nanotechnology in engineering products such as aluminum, concrete, steel improves the longevity, performance, and resilience of highway infrastructure components while reducing their cost. Again devices made of nanostructures such as nanoscale sensors and communications devices enhance the transportation infrastructure than can communicate to other vehicles so that it can help the driver to maintain safe distance, avoid collision, avoid congestion, and improve driver's ability to interface onboard electronics.

3.9.3 Electronic Industry

Nanotechnology has a profound contribution to the computer and electronics industry. Using this technique, larger information can be stored within a very short period of time. Transistors are the heart of computers, and their size is reduced sharply by adopting nanotechnology. By using magnetic random access memory (MRAM), the computer can be booted almost instantly. Nanometer-scale magnetic tunnel junctions are associated with MRAM. Hence data are saved almost quickly and effectively when the system shuts down. Flexible electronics is the gift of nanotechnology that finds application in aerospace, medical industries, Internet of things, and so on. Nanomembranes developed by flexible electronics are used in smartphones and e-reader displays. Wearable and "tattoo" sensors are manufactured by other flexible electronics by using nanomaterials such as grapheme and cellulosic nanomaterials. Lightweight, highly efficient electronics and flexible, flat electronics create door to the manufacture of countless smart products. Other electronics and computing products include cell phone casings, ultra-responsive hearing aids, conductive inks for printed electronics for RFID/smart cards/smart packaging, flash memory chips for smartphones and thumb drives, antimicrobial/antibacterial coatings on keyboards and cell phone casings, and flexible displays for e-book readers. Flexible display improves display quality in the electronic device. The thickness and weight of the screen can be increased, and power consumption can be decreased by incorporating smart electronics. It can also increase the density of the memory chip. It is believed that all the abilities of today's computers can be placed in one nanotechnology-based computer. Atomically thin indium-tin oxide sheets are demonstrated by researchers at the Royal Melbourne Institute of Technology that cost less and being flexible consume less power [192]. A plastic sheet with cadmium selenide nanocrystal deposition forms flexible electronic circuits. A simple fabrication process combined with flexibility and lower power requirement is desired nowadays. Integrating silicon nanophotonics components into CMOS-integrated circuits provides higher data transmission through the integrated circuit. Nanomagnet as switch consumes less power than transistors. By combining gold nanoparticles with organic molecules a transistor is produced known as a NOMFET (Nanoparticle Organic Memory Field-Effect Transistor) and uses very low power with high efficacy. Spintronic devices use magnetic quantum dots to be higher in density. It uses spin electronics technique [193]. Dense memory device is made by nanowires using an alloy of iron and nickel. When current is applied, the wire is magnetized along its length, and as magnetization proceeds, it is read by a stationary sensor thus putting it in memory. This method is called race track memory [194].

3.9.4 Medical Industries

The use of nanotechnology in the medical industry provides a precise solution for diagnostics, disease prevention, and treatment. Gold nanoparticles are vastly used in the treatment of cancer and other diseases [195]. The commercial applications have adopted the use of gold nanoparticles for use as probes in detecting sequence. The adoption of nanotechnology in imaging and diagnostic tools has opened the way for earlier diagnostics, better therapeutic success rates, and more individualized treatment options. For the treatment of atherosclerosis, nanotechnology has been extensively used. In a study, it is revealed that nanoparticles can mimic good cholesterol in our body called HDL. HDL helps to shrink plaque. Advanced nanopore materials allow or the development of novel gene sequencing technologies. It helps in single-molecule detection at low cost and high speed with minimal sample preparation and instrumentation [196]. Nanoparticle is considered to help deliver medicine directly to cancer cell without affecting other living healthy cells. This therapeutic medication process reduces the need for chemotherapy which often has a toxic effect on human cells. Customized nanoparticles are of molecular level. The technology employs nanoparticles to deliver drugs, light, and heat to specific cancer-affected cells. Particles are engineered so that they are attracted to diseased cells, which allow direct treatment of those cells. The healthy cells get unaffected. In North

Carolina State University, researchers delivered cardiac stem cells to damaged heart tissue. They use nanovesicles that differentiated between healthy cells and damaged cells and provided the stem cells to damaged cells only. In bone and neural tissue engineering, nanotechnology is widely used [197]. Researchers are working on tissue engineering in which complex tissue can be incorporated with nanotechnology and the organ associated can be grown in one day so that organ transplant can be much easier. Neurons can grow well on the conductive graphene surface. Graphene nanoribbons help in the repair of spinal cord injuries. It is expected that nanomedicines can replace vaccines, and it can be applied to the human body without needles. To combat the annual flu vaccine, researchers are in the process of inventing a universal vaccine scaffold that would cover more strains [198].

Cellulosic nanomaterials have potential applications in various engineering fields such as automotive, health care, defense, electronics, food, and construction industry. Cellulosic nanomaterials are less expensive, and it has high strength-to-weight ratio. Nano-bioengineering of enzymes aims at the extraction of cellulose from corn stalks, wood chips, and unfertilized perennial grasses. Antibodies are attached to carbon nanotubes in chips to detect cancer cells in the blood. Gold nanorods are used for the early detection of damage in the kidney. In this method, gold nanorods are functionalized to attach to the protein generated by kidneys. The color of the nanorod shifts when protein accumulates on the nanorod. The method is quick and inexpensive and a technique in which gold nanoparticles combined with infrared light can kill bacteria. This method can be used to sterilize surgical instruments. Quantum dots can also be used to treat antibiotic-resistant infections. A bandage fixed with nanogenerators can be pasted on the wound that can generate electric pulses regularly to recover the wound faster. Polymer nanoparticles produced in Chase Western Reserve University acts as a platelet. So it can be incorporated into the human body to reduce blood loss in trauma patients with internal bleeding. From the result, it is verified that the use of polymer nanoparticles as platelets greatly reduced blood loss. Nanorobots can also be manufactured to repair specific diseased cells. It can be programmed in such a way that it behaves as an antibody in our natural healing process [199].

3.9.5 Energy Harvest Industries

Constant effort is going on to produce clean, renewable, affordable energy sources that can meet the demand of the world, reduce consumption of energy, and lessen the environmental pollution. Nanotechnology greatly enhances the alternative energy approach to meet the world's energy demand. Nowadays, clean fuel can be produced from raw petroleum by using better catalysis made of nanomaterials. Nanotechnology can enhance the efficiency of vehicles by reducing fuel consumption in vehicles and power plants [200]. In power plants, scrubber and membrane can be made of nanotubes that will have less resistance than electric grids thus decreasing power consumption. Nanotechnology is also applicable in the oil and gas extraction industries. The valve used in offshore oil extraction can be made by nanoengineering and also nanotechnology can be applied to detect fractures in the lift pipe. Nanostructured solar cells are more efficient and can have higher conversion efficiency from sunlight to electricity. Solar cells can be printable with nanoparticles in flexible rolls rather than discrete panels. The study reveals that in the future, nanoparticles can also be paintable on solar cells thus causing less effort, less maintenance, less cost, and higher efficiency. Manufacturing solar cell by nanotechnology has various advantages. In this manufacturing process, a low-temperature printing-like process is conducted rather than a high-temperature diffusion process in conventional manufacturing. Thus, cost is highly reduced. Reduced installation costs are achieved by producing flexible rolls instead of rigid crystalline panels. Thin-film solar electric panels can be fitted to a computer and the same can also be woven into clothes to generate usable energy from other sources like light, body heat, and friction. [201].

It is found that nanotechnology can also be used in batteries for fast charging. Windmills use nanotubes for making strong wind blades. The wind blades can also be stronger, longer, and lightweight compared to other blades by incorporating nanomaterials. Likewise, various nano techniques are used to convert waste heat in computers, automobiles, power plants, and homes to usable power. The efficiency of energy-saving products is increased day by day. Nanotechnology enables more efficient lighting systems, lower energy consumption in advanced electronics, lighter and stronger vehicle chassis materials for the transportation sector, and light-responsive smart coatings for glass. Nanotechnology is used in fuel cells to reduce the cost of catalysts where hydrogen ion is produced from fuel. It also helps in improving the efficiency of membranes to separate hydrogen ions from other gases. Platinum is generally used as a catalyst in the process. As platinum is very expensive, researchers are using other materials instead of platinum, or in other ways, nanoparticles of platinum are used as catalysts in lesser amounts to reduce cost and increase efficiency. Fuel cells may act as membranes that selectively filter only hydrogen ions to pass through the cell but does not allow other atoms to pass. Nanotechnology is also incorporated to manufacture longer lasting and lighter weight fuel cells. Fuel cells can replace the battery in electric vehicles. Fuel cells generally use hydrogen as the fuel. But when hydrogen-based fuel cells are used in an electric vehicle, the storage of hydrogen is the main limiting factor as it has a chance to explode. A light and safe storage device is required for operation. Now nanostructured storage devices are manufactured that absorb hydrogen gas and supply whenever it is required [202].

By incorporation of nanotechnology, the battery can provide more power when desired and it should be recharged at a faster rate. The purpose is fulfilled by coating nanoparticles on the electrode. The coating increases the surface area of the electrode thus increasing current flow. This in turn provides more electricity to be flown from the electrode to chemicals inside the battery [202]. This technique increases the efficiency of hybrid vehicles by significantly reducing the weight of the batteries needed to provide adequate power. By using nanomaterials, the liquid is separated from solid electrodes in the battery. This separation prevents the low-level discharge that occurs in a conventional battery, which increases shelf-life [203]. Nanocatalysts decrease the requirement of temperature for the conversion of raw material into fuel. Catalysts made of nanoparticles have a larger surface area than conventional fuels. This facilitates the catalyst to interact with reacting chemicals more effectively. This effectiveness of catalyst can lead to extract diesel from coal. This technique is also implemented to extract oil from other sources such as crude oil.

3.9.6 Food Industries

Nanomaterials have an extensive impact on the food industry. Using nanomaterials in the food industry not only enhances the taste of food but also increases food safety and health benefits. Clay nanoparticles are used extensively in packages and bottles to restrict the passage of gasses like oxygen and carbon dioxide in lightweight bottles, cartons, and packaging films [204]. Storage bins also incorporate nanoparticles made of silver. Silver nanoparticles kill bacteria. Silver nanoparticles embedded with plastic kill bacteria from any material that was previously stored in the bins, minimizing health risks from harmful bacteria. People from Technische Universität München have invented a method in which carbon nanotubes can be sprayed to a plastic surface in which plastic surface acts as a sensor. The sensor can sense any rotten food present in the plastic wrapping. Researchers are using silicate nanoparticles to provide a barrier to gasses (for example oxygen) or moisture in a plastic film used for packaging. This can reduce the chance of spoiling food. Zinc oxide nanoparticles are also proved to be efficient for blocking UV rays when used on a plastic sheet. This improves the stability and strength of plastic and provides bacterial protection [205]. Nanosensors can also be used for identifying bacteria and other contaminants such as *Salmonella*. So with a very low cost, the food items can be checked for contamination. Hygenic quality checking using nanosensors can be done frequently without sending products frequently to the lab. If this process can be used efficiently in food processing, then contaminated foods can be eliminated from the market. Currently, nanocapsules are trending in the food industry. Nanocapsules are introduced inside the body, and it gets stored in the body as a super vitamin. Whenever deficiency of nutrients occurs in the body, the nanosensor can detect the deficiency of the vitamin. Hence, nanocapsule in form of super vitamin supplies the required nutrient to the body.

3.9.7 Textile Industries

Nanoscale addition or surface treatments of fabrics can provide slight ballistic energy deflection in the personal body wafer or help them with wrinkles, stains, and bacterial growth. Adding nanoparticles to fabrics improves the properties of fabric without increasing weight, stiffness, and thickness. It can be inferred that the inclusion of nanowhiskers into fabric used to make pants produces lightweight water and stain repellent material. Water-soaked nanowhiskers rise, making the fabric waterproof and stain-resistant. Silver nanoparticles in the fabric can kill the bacteria and make the clothes bacteria resistant [206]. Nanopores provide superior insulation for shoe inserts in cold weather. Nanoparticles that provide a "lotus plant" effect for used fabric tents and other materials that stay in the weather cause dirt in the rain. Protective clothing against hazardous chemicals using a polyurethane nanofiber honeycomb [207].

3.9.8 Household Application

Nano-engineered materials make superior household products such as air purifiers, degreasers and stain removers, antibacterial cleansers, environmental sensors, filters, antibacterial cleansers, and specialized paints and sealing products, such as self-cleaning house paints that resist dirt and marks. Nanotechnology helps in the enhancement of air quality. Nanotechnology may be used with catalysts that can transform toxic vapors that emerged from industrial plants and cars to harmless gases. This is possible due to the larger surface area of nanostructured materials. The larger surface area in the catalyst leads the catalyst to interact with more chemicals and convert more amount of hazardous air to harmless air. There are two ways by which air quality can be enhanced. One is by nanostructured catalysts and the other by using as membrane [208]. Coming to water treatment by nanotechnology, there are three major processes for water purification. One of the processes of water quality increment is groundwater purification by removal of industrial waste. Nanoparticles can be employed for the conversion of contaminating chemicals through reaction to make them harmless. It is revealed that the process can be used to remove contaminants from the pond that are dispersed at the bottom at a much lower cost. It avoids the expensive process of removal of water from the pond and treating it with conventional chemicals. Another method of water quality enhancement is the removal of salt from water. A deionization method is employed to convert salty or mineral-enriched water to freshwater. In this process, electrodes composed of nano-sized fibers are used to filter the minerals and salts present in water. The third process includes virus removal from water. In general, by filtration, the virus cannot be removed. When nanostructured fibers are used in the filtration process, it can filter viruses of nano diameter thus making water virus free. Researchers have found that a filter with nano-thin sheets of aluminum oxide can filter both oils and heavy metals from water [209].

Nanoscavengers are used in the water treatment process in which a core is used and outside of the core is applied with scavengers. The core is designed to be magnetized whenever required. The nanoparticles attach to the pollutants and kill them. After treatment, scavengers are removed by using magnetization. Nanostructured gold and palladium can be used to break down chlorinated contaminating groundwater. As palladium is very much expensive, it is made pellet structured so that each palladium atom can react with each pollutant thus making the process most cost-effective and efficient. Researchers at the University of Cincinnati have demonstrated a method of removing antibiotics from contaminated waterways. The method uses vesicle nanoparticles that absorb antibiotics.

It is found that graphene flakes can absorb radioactive ions in water. After absorption graphene oxide forms clumps and can be removed easily from water. As it is known that the grapheme with hole size in nanometer can be utilized to remove ions from water, so it is nowadays used to desalinate water at a lower cost than reverse osmosis. Water molecules can pass very easily through nanotubes than other types of pores; hence, the power requirement in desalination plants decreases. Now worldwide inexpensive, small water purification devices are used that uses carbon nanotubes. By adding boron atoms during the growth of carbon nanotubes causes the nanotubes to grow in a sponge-like material that can absorb many times its weight in oil. Hence, the nanotubes are used in oil spills for cleaning purposes. Nano hair can trap and measure mercury levels in water. The cost of desalinating seawater can be reduced by combining a nanomembrane with solar power. In addition, nanoscale additives in polymer composite materials are being used in baseball bats, tennis rackets, bicycles, motorcycle helmets, automobile parts, luggage, and power tool housings, making them lightweight, stiff, durable, and resilient [210].

3.9.9 Others

For example, cosmetics manufacturers often use carbonated titanium and zinc oxide dyes at the nanometer scale for skin in sunscreens and lotions. Nanoscale iron oxide powder is used as the base for blush and lipstick. Paintings with watercolor designs are inspired by diamond titanium alloys. Theaters with epoxide-containing carbon nanotubes reduce the risk of damaging the skin of the kay. Nanocomposite film is used to hold the air in long tennis balls. The strengthening of pharmaceuticals without increasing their weight by the addition of silica nanoparticles fills the gaps between carbon products [211]. It appears that porous and other nano-sized materials are found to have a much-needed appeal for electroluminescent silicon applications. The interest in luminescent Si is due to the need for integration of optoelectronic technologies, which are now based on III-V semiconductors that cannot be monolithically integrated with large-scale microelectronics. Therefore, the light-emitting Si creates a pathway for the use of optical and optical processing in a monolithic Si chip. There will be a variety of new features such as light-emitting, with light illuminated by lightning and some elements of optoelectronic integration. The loss of natural semiconductors is due to a change in the rate of contact between phonons (a large amount of energy associated with atomic vibes), which is not luminescent, and the change in radiative [212, 213].

3.10 Current Problems/Difficulties Associated With Nanomaterials

Some of the inherent difficulties of nanomaterials and their fabrications are discussed below [214–216]:

Instability of the particles: Nanoparticles are metastable. They are not stable at their size. They are more prone to attack and undergo transformation. Retaining stability is highly challenged as kinetics associated with nanoparticles are very rapid. So the nanoparticles are placed in some matrix to retain their size.

Strong Explosive: Nanoparticles in contact with oxygen causes rapid combustion. The larger surface area of nanoparticles is more prone to be in contact with oxygen. Rapid combustion leads to an explosion.

Impurity: As nanoparticles are highly reactive, there is a need for encapsulation of the nanoparticles. Encapsulation becomes necessary when they are produced in a solution. Hence, these secondary impurities become a part of the synthesized nanoparticles.

Biologically harmful: Nanomaterials are carcinogenic and cause irritation. Nanomaterials are transparent to cell dermis, and their high surface area makes them more sensitive to foreign materials. If inhaled, there is no chance of nanomaterials to come out of the body. In addition, their interaction with blood and the liver is found to be very harmful.

Difficulty in synthesis, isolation, and application: Free nanoparticles are hard to be used, as they react with foreign substances. Hence capsulation in a bigger, and stable molecule is a must for nanoparticles. Grain growth is inherently present in nanomaterials during their processing.

Recycling and disposal: Disposal of nanomaterials is still a challenge for researchers. Their toxicity restricts its casual disposal. The result of exposure to nanomaterials is still under study. Hence the uncertainty associated with the effects of nanomaterials is yet to be assessed to develop their disposal policies.

3.11 Opportunities and Challenges

Nanotechnology is a focus of interest in materials development by controlling the properties from their basic structural unit. This chapter described the understandings of nanotechnology and various developments of advanced materials. Restriction of nanosize concerning the nanodevice has been discussed. Various properties of nanomaterials discussed that are required before the application of nanotechnology in various dimensional (OD, 1D, 2D, and 3D) materials. Current potential application in aero-industries, automotive and naval industry, electronic industry, medical industries, energy harvest industries, food industries, textile industries, household application, and other industries has been discussed. This chapter helps the entire researchers to control the required properties with the development of various economic processes that can produce the new materials with dominating role in their working conditions.

References

1. Jeevanandam, J., Barhoum, A., Chan, Y.S., Dufresne, A., Danquah, M.K.: Review on nanoparticles and nanostructured materials: history, sources, toxicity and regulations. Beilstein J. Nanotechnol. **9**, 1050–1074 (2018). https://doi.org/10.3762/bjnano.9.98
2. https://www.sciencelearn.org.nz/resources/1651-nanometres-and-nanoscale
3. Behera, A., Mohapatra, S.S., Verma, D.K.: Nanomaterial: fundamental principle and application. In: Nanotechnology and Nanomaterial Applications in Food, Health and Biomedical Science, p. 343. Apple Academic Press & CRC Press, Palm Bay, FL., ISBN: 9781771887649 (2019)
4. https://justlearning.in/articles/meaning-applications-nanotechnology/2017/07/13/
5. https://www.industryarc.com/Report/15022/nanotechnology-market.html
6. I. Capek, Studies in Interface Science, Nanotechnology and nanomaterials, Elsevier, Netherland 23, 2006, Pages 1–69., ISSN 1383-7303, ISBN 9780444527165, See:doi: https://doi.org/10.1016/S1383-7303(06)80002-5
7. Xu, X., Wang, J., Long, Y.: Zeolite-based materials for gas sensors. Sensors. **6**, 1751–1764 (2006)
8. Megelski, S., Calzaferri, G.: Tuning the size and shape of zeolite L-based inorganic–organic host–guest composites for optical antenna systems. Adv. Funct. Mater. **11**, 277–286 (2001). https://doi.org/10.1002/1616-3028(200108)11:4<277::AID-ADFM277>3.0.CO;2-2
9. https://www.nano.gov/you/nanotechnology-benefits
10. https://www.defenceiq.com/defence-technology/articles/nano-drone-tech-is-advancing
11. https://www.smithsonianmag.com/history/this-1600-year-old-goblet-shows-that-the-romans-were-nanotechnology-pioneers-787224/
12. https://en.wikipedia.org/wiki/Maya_blue
13. Freestone, I., Meeks, N., Sax, M., et al.: The Lycurgus cup — a Roman nanotechnology. Gold Bull. **40**, 270–277 (2007). https://doi.org/10.1007/BF03215599
14. Woese, C.R.: A new biology for a new century. Microbiol Mol Biol Rev. **68**(2), 173–186 (2004). https://doi.org/10.1128/MMBR.68.2.173-186.2004
15. https://en.wikipedia.org/wiki/Richard_Feynman
16. Mohapatra, R.K.: Engineering Chemistry with Laboratory Experiments. PHI Publications, Delhi (2015) ISBN 978-81-203-5158-5
17. Barnes, J.C., Mirkin, C.A.: Profile of Jean-Pierre Sauvage, Sir J. Fraser Stoddart, and Bernard L. Feringa, 2016 Nobel Laureates in Chemistry. Proc. Natl. Acad. Sci. USA. **114**(4), 620–625 (2017). https://doi.org/10.1073/pnas.1619330114
18. https://www.npr.org/sections/alltechconsidered/2015/11/23/457129179/the-future-of-nanotechnology-and-computers-so-small-you-can-swallow-them
19. https://qz.com/852770/theres-a-limit-to-how-small-we-can-make-transistors-but-the-solution-is-photonic-chips/
20. https://en.wikipedia.org/wiki/Moore%27s_law
21. https://www.eetimes.com/moores-law-dead-by-2022-expert-says/
22. https://www.extremetech.com/extreme/210872-extremetech-explains-what-is-moores-law
23. http://nanoawesomeworld.blogspot.com/2012/09/quantum-confinement-in-nanoparticles.html
24. Macwan, D.P., Balasubramanian, C., Dave, P.N., Chaturvedi, S.: Thermal plasma synthesis of nanotitania and its characterization. J. Saudi Chem. Soc. **18**(3), 234–244., ISSN 1319-6103 (2014). https://doi.org/10.1016/j.jscs.2011.07.009
25. Ogborn, J., Taylor, E.: Quantum physics explains Newton's laws of motion. Phys. Educ. **40**, 26 (2004). https://doi.org/10.1088/0031-9120/40/1/001
26. Murty, B., Shankar, P., Baldev, R., Rath, B., James, M.: Unique properties of nanomaterials. Springer, Berlin, Heidelberg (2013). https://doi.org/10.1007/978-3-642-28030-6_2
27. Giallonardo, J.D., Avramovic-Cingara, G., Palumbo, G., et al.: Microstrain and growth fault structures in electrodeposited nanocrystalline Ni and Ni–Fe alloys. J. Mater. Sci. **48**, 6689–6699 (2013). https://doi.org/10.1007/s10853-013-7469-4
28. Loc, N., Howard, L., Giblin, S., Tanner, B.K., Terry, I., Hughes, A., Ross, I., Serres, A., Bürckstümmer, H., Evans, J.: Synthesis of monodisperse fcc and fct FePt/FePd nanoparticles by microwave irradiation. J. Mater. Chem. **15**, 5136–5143 (2005). https://doi.org/10.1039/b511850f
29. Makio Naito, Toyokazu Yokoyama, Kouhei Hosokawa, Kiyoshi Nogi, Nanoparticle Technology Handbook (3rd ed.), Elsevier Netherland, 2018, Pages 109–168, ISBN 9780444641106, doi: https://doi.org/10.1016/B978-0-444-64110-6.00003-2

30. Vafaei, S., Purkayastha, A., Jain, A., Ramanath, G., Borca-Tasciuc, T.: The effect of nanoparticles on the liquid–gas surface tension of bi 2 Te 3 nanofluids. Nanotechnology. **20**, 185702 (2009). https://doi.org/10.1088/0957-4484/20/18/185702
31. Bergeret, G., Gallezot, P.: Particle size and dispersion measurements. In: Ertl, G., Knözinger, H., Schüth, F., Weitkamp, J. (eds.) Handbook of Heterogeneous Catalysis. John Wiley & Sons, Inc., Hoboken, NJ (2008). https://doi.org/10.1002/9783527610044.hetcat0038
32. Diehm, P., Péter, Á., Karsten, A.: Size-dependent lattice expansion in nanoparticles: reality or anomaly? Chemphyschem. **13**, 2443–2454 (2012). https://doi.org/10.1002/cphc.201200257
33. Merrilea, M., Suresh, A., Wallace, P.: Thermodynamics for nanosystems: grain and particle-size dependent phase diagrams. Rev. Adv. Mater. Sci. **5**, 100–109 (2003)
34. Prieur, D., et al.: Size dependence of lattice parameter and electronic structure in CeO_2 nanoparticles. Inorg. Chem. **59**(8), 5760. Publication Date: April 1, 2020 (2020). https://doi.org/10.1021/acs.inorgchem.0c00506
35. Rodríguez-Castañeda, C.A., Moreno-Romero, P.M., Martínez-Alonso, C., Hailin, H.: Microwave synthesized monodisperse CdS spheres of different size and color for solar cell applications. J. Nanomater. **2015**., Article id: 424635, 10 (2015). https://doi.org/10.1155/2015/424635
36. Zhaojie, W., Joseph, A., Wanyoung, J., Javier, G., Chris, D.: Thermal conductivity of nanocrystalline silicon: importance of grain size and frequency-dependent mean free paths. Nano Lett. **11**, 2206–2213 (2011). https://doi.org/10.1021/nl1045395
37. Zhang, Z., Ouyang, Y., Cheng, Y., Chen, J., Li, N., Zhang, G.: Size-dependent phononic thermal transport in low-dimensional nanomaterials. Phys. Rep. **860**, 1–26., ISSN 0370-1573 (2020). https://doi.org/10.1016/j.physrep.2020.03.001
38. Smijs, T.G., Pavel, S.: Titanium dioxide and zinc oxide nanoparticles in sunscreens: focus on their safety and effectiveness. Nanotechnol. Sci. Appl. **4**, 95–112. Published 2011 Oct 13 (2011). https://doi.org/10.2147/NSA.S19419
39. Suryanarayana, C.: Mechanical behavior of emerging materials. Mater. Today. **15**(11), 486–498., ISSN 1369-7021 (2012). https://doi.org/10.1016/S1369-7021(12)70218-3
40. Tschopp, M.A., Murdoch, H.A., Kecskes, L.J., et al.: "Bulk" Nanocrystalline metals: review of the current state of the art and future opportunities for copper and copper alloys. JOM. **66**, 1000–1019 (2014). https://doi.org/10.1007/s11837-014-0978-z
41. Naik, S.N., Walley, S.M.: The Hall–Petch and inverse Hall–Petch relations and the hardness of nanocrystalline metals. J. Mater. Sci. **55**, 2661–2681 (2020). https://doi.org/10.1007/s10853-019-04160-w
42. Crosby, A.: Polymer nanocomposites: the "nano" effect on mechanical properties. Polym. Rev. **47**, 217–229 (2007). https://doi.org/10.1080/15583720701271278
43. Pelleg, J.: Mechanical Properties of Nanoscale Ceramics. Springer, New York (2014). https://doi.org/10.1007/978-3-319-04492-7_9
44. https://www.nanowerk.com/nanotechnology-applications.php
45. Sirelkhatim, A., Mahmud, S., Seeni, A., et al.: Review on zinc oxide nanoparticles: antibacterial activity and toxicity mechanism. Nano-Micro Lett. **7**, 219–242 (2015). https://doi.org/10.1007/s40820-015-0040-x
46. Pokropivny, V., Rünno, L., Nova, I., Alex, P., Sergei, V.: Introduction in Nanomaterials and Nanotechnology. Tartu University Press, Tartu (2007)
47. Wei-xian, Z., Daniel, E.: Applications of iron nanoparticles for groundwater remediation. Remed. J. **16**, 7–21 (2006). https://doi.org/10.1002/rem.20078
48. Guerra, F.D., Attia, M.F., Whitehead, D.C., Alexis, F.: Nanotechnology for environmental remediation: materials and applications. Molecules. **23**(7), 1760. Published 2018 Jul 18 (2018). https://doi.org/10.3390/molecules23071760
49. Mahesh, B.V., Raman, R.K.S.: Role of nanostructure in electrochemical corrosion and high temperature oxidation: a review. Metall. Mat. Trans. A. **45**, 5799–5822 (2014). https://doi.org/10.1007/s11661-014-2452-5
50. Ashrafi, A.: Quantum Confinement: an Ultimate Physics of Nanostructures. American Scientific Publishers, Los Angeles (2011)
51. https://www.sigmaaldrich.com/technical-documents/articles/materials-science/nanomaterials/gold-nanoparticles.html
52. Mafuné, F., Kohno, J.-y., Takeda, Y., Kondow, T.: Formation and size control of silver nanoparticles by laser ablation in aqueous solution. J. Phys. Chem. B. **104** (2000). https://doi.org/10.1021/jp001336y
53. Mohanty, D., Satpathy, S.K., Behera, B., Mohapatra, R.K.: Dielectric and frequency dependent transport properties in magnesium doped $CuFe_2O_4$ composite. Mater. Today Proc. (2020). https://doi.org/10.1016/j.matpr.2020.02.944
54. Stallard, J.C., Tan, W., Smail, F.R., Gspann, T.S., Boies, A.M., Fleck, N.A.: The mechanical and electrical properties of direct-spun carbon nanotube mats. Extr. Mech. Lett. **21**, 65–75., ISSN 2352-4316 (2018). https://doi.org/10.1016/j.eml.2018.03.003
55. Bhuyan, R.K., Mahapatra, R.K., Nath, G., Das, D., Sahoo, B.K., Pamu, D.: Effect of high energy ball milling on structural, microstrctural and optical properties of Mg_2TiO_4 nanoparticles. J. Mater. Sci. Mater. Electron. **31**, 1 (2020). https://doi.org/10.1007/s10854-019-02568-3.
56. Issa, B., Obaidat, I.M., Albiss, B.A., Haik, Y.: Magnetic nanoparticles: surface effects and properties related to biomedicine applications. Int. J. Mol. Sci. **14**(11), 21266–21305. Published 2013 Oct 25 (2013). https://doi.org/10.3390/ijms141121266
57. Barik, R., Satpathy, S.K., Behera, B., Biswal, S.K., Mohapatra, R. K.: Synthesis and spectral characterizations of nano sized lithium niobate ($LiNbO_3$) ceramic. Micro Nanosyst. **12**(2), 81–86 (2020). https://doi.org/10.2174/1876402911666190617114003
58. Chen, C., Ha, T., Hailu, A., Kim, T., Munteanu, M., Hwang, S.: Magnetic Recording Media Having Five Element Alloy Deposited Using Pulsed Direct Current Sputtering Utility. Seagate Technology LLC, California (2004) US20040253486A1
59. Jang, J., Kim, Y., Hwang, J., et al.: Biological responses of onion-shaped carbon nanoparticles. Nanomaterials (Basel). **9**(7), 1016. Published 2019 Jul 15 (2019). https://doi.org/10.3390/nano9071016
60. Toffoli, H., Erkoç, S., Toffoli, D.: Modeling of nanostructures. In: Leszczynski, J. (ed.) Handbook of Computational Chemistry. Springer, Dordrecht (2015). https://doi.org/10.1007/978-94-007-6169-8_27-2
61. Dong, B.X., Qiu, F., Li, Q., Shu, S.L., Yang, H.Y., Jiang, Q.C.: The synthesis, structure, morphology characterizations and evolution mechanisms of nanosized titanium carbides and their further applications. Nanomaterials (Basel). **9**(8), 1152. Published 2019 Aug 11 (2019). https://doi.org/10.3390/nano9081152
62. Hosseinidoust, Z., Basnet, M., van de Ven, T.G.M., Tufenkji, N.: One-pot green synthesis of anisotropic silver nanoparticles. Environ. Sci. Nano. **3**(6), 1259–1264 (2016)
63. Behera, A., Mallick, P., Mohapatra, S.S.: Nanocoatings for anti-corrosion: materials, fabrications and applications. In: Corrosion protection at the nanoscale, pp. 449–457. Elsevier Publisher, Netherland., ISBN No: 978-0-12-819359-4 (2020). https://doi.org/10.1016/B978-0-12-819904-6.00020-7
64. https://en.wikipedia.org/wiki/Nanoparticle
65. Wang, L., Sun, Y., Li, Z., Wu, A., Wei, G.: Bottom-up synthesis and sensor applications of biomimetic nanostructures. Materials (Basel). **9**(1), 53. Published 2016 Jan 18 (2016). https://doi.org/10.3390/ma9010053

66. https://www.nano.gov/nanotech-101/what/manufacturing
67. Yu, Y., Zhang, J., Zhong, H.: Heterogeneous condensation of water vapor on fine SiO_2 particles in two-section growth tube. Energy Fuel. **32**(12), 12750–12757 (2018)
68. Kruis, F., Fissan, H., Aaron, P.: Synthesis of nanoparticles in the gas phase for electronic, optical and magnetic applications - a review. J. Aerosol. Sci. **29**, 511–535 (1998). https://doi.org/10.1016/S0021-8502(97)10032-5
69. St Michael, J., Fee, P., Stillwater Hendrickson, W.A., St. Paul Pozarnsky, G.A., Inver Grobe Heights Walker, B.J.: Process for the Preparation of Reacted Aluminum or Copper Nanoparticles. DE10297575T5, Cima Nanotech Inc., German (2002)
70. Myungjoon, K., Saho, O., Taesung, K., Hidenori, H., Takafumi, S.: Synthesis of nanoparticles by laser ablation: a review. KONA Powder and Particle Journal. **37**, 3–18 (2017). https://doi.org/10.14356/kona.2017009
71. Charitidis, C.A., Georgiou, P., Koklioti, M.A., Trompeta, A.-F., Markakis, V.: Manufacturing nanomaterials: from research to industry. Manuf. Rev. **1**(11) (2014). https://doi.org/10.1051/mfreview/2014009
72. Teoh, W.Y.: A perspective on the flame spray synthesis of photocatalyst nanoparticles. Materials (Basel). **6**(8), 3194–3212. Published 2013 Jul 31 (2013). https://doi.org/10.3390/ma6083194
73. Muhl, S., Perez, A.: The use of hollow cathodes in deposition processes: a critical review. Thin Solid Films. **579**, 174–198 (2015). https://doi.org/10.1016/j.tsf.2015.02.066
74. Ahmed, M., Badawi, M., Ghatass, Z., Mohamed, M., Elkhatib, M.: Synthesize of silver nanoparticles by arc discharge method using two different rotational electrode shapes. J. Clust. Sci. **29**, 1169–1175 (2018). https://doi.org/10.1007/s10876-018-1430-2
75. https://www.science.gov/topicpages/v/vapor+condensation+method
76. https://en.wikipedia.org/wiki/Molecular-beam_epitaxy
77. Ali, A., Zafar, H., Zia, M., et al.: Synthesis, characterization, applications, and challenges of iron oxide nanoparticles. Nanotechnol. Sci. Appl. **9**, 49–67. Published 2016 Aug 19 (2016). https://doi.org/10.2147/NSA.S99986
78. Saito, G., Akiyama, T.: Nanomaterial synthesis using plasma generation in liquid. J. Nanomater. **2015**., Article Id: 123696, 21 (2015). https://doi.org/10.1155/2015/123696
79. Enas, M.: Ahmed, hydrogel: preparation, characterization, and applications: a review. J. Adv. Res. **6**(2), 105–121., ISSN 2090-1232 (2015). https://doi.org/10.1016/j.jare.2013.07.006
80. Ying, L., Wen-Zhi, J., Yan-Yan, S., Xing-Hua, X.: Superhydrophobicity of 3D porous copper films prepared using the hydrogen bubble dynamic template. Chem. Mater. **19**(23), 5758–5764 (2007). https://doi.org/10.1021/cm071738j
81. Song, J., Lim, S.: Effect of seed layer on the growth of ZnO nanorods. J. Phys. Chem. C. **119**(19), 10321–10328 (2015). https://doi.org/10.1021/jp0655017
82. Behera, A.: Shape-controlled metal nanoparticles for fuel cell applications. In: Nanotechnology in fuel cells. Elsevier Publisher, Netherland (2021)
83. Jianlin, L., Qingliu, W., Ji, W.: Synthesis of nanoparticles via solvothermal and hydrothermal methods. In: Handbook of Nanoparticles. Springer International Publishing, Switzerland (2016). https://doi.org/10.1007/978-3-319-15338-4_17
84. Srivastava, P., Ashokkumar, M.: Theoretical and Experimental Sonochemistry Involving Inorganic Systems. Springer, Netherlands (2011). https://doi.org/10.1007/978-90-481-3887-6
85. Le, A.K., Bender, J.A., Arias, D.H., Cotton, D.E., Johnson, J.C., Roberts, S.T.: Singlet fission involves an interplay between energetic driving force and electronic coupling in Perylenediimide films. J. Am. Chem. Soc. **140**(2), 814–826 (2018). https://doi.org/10.1021/jacs.7b11888
86. https://www.science.gov/topicpages/b/ball+milling+process
87. https://worldwidescience.org/topicpages/l/lithographic+masking+process.html
88. Casais-Molina, M.L., Cab, C., Canto, G., Medina, J., Tapia, A.: Carbon nanomaterials for breast cancer treatment. J. Nanomater. **2018**., Article Id: 2058613, 9 (2018). https://doi.org/10.1155/2018/2058613
89. Nasir, S., Hussein, M.Z., Zainal, Z., Yusof, N.A.: Carbon-based nanomaterials/allotropes: a glimpse of their synthesis, properties and some applications. Materials (Basel). **11**(2), 295. Published 2018 Feb 13 (2018). https://doi.org/10.3390/ma11020295
90. Pedro, D.A., Rafael, R., José, V.: Strong covalent bonding between two graphene layers. Phys. Rev. B. **77**, 045403 (2008). https://doi.org/10.1103/PhysRevB.77.045403
91. https://www.acsmaterial.com/graphene-facts
92. Haizhou, H., Shi, S., Wu, N., Wan, H., Wan, S., Hengchang, B., Litao, S.: Graphene-based sensors for human health monitoring. Front. Chem. **7** (2019). https://doi.org/10.3389/fchem.2019.00399
93. https://en.wikipedia.org/wiki/Buckminsterfullerene
94. https://en.wikipedia.org/wiki/Fullerene
95. https://www.nobelprize.org/prizes/chemistry/1996/press-release/
96. Chapleski Jr., R.C., Morris, J.R., Troya, D.: A theoretical study of the ozonolysis of C60: primary ozonide formation, dissociation, and multiple ozone additions. Phys. Chem. Chem. Phys. **16**, 5977–5986 (2014). https://doi.org/10.1039/C3CP55212H
97. http://www.cs.gordon.edu/courses/organic/dol/Fullerenes/
98. https://www.azonano.com/article.aspx?ArticleID=5158#:~:text=More%20importantly%2C%20buckminsterfullerene%20has%20a,proposed%20applications%20of%20the%20molecule
99. Nimibofa, A., Ebelegi, A., Abasi, C., Donbebe, W.: Fullerenes: synthesis and applications. J. Mater. Sci. Res. **7**, 3 (2018). https://doi.org/10.5539/jmsr.v7n3p22
100. Kwok, K.S., Chan, Y.C., Ng, K.M., Wibowo, C.: Separation of fullerenes C60 and C70 using a crystallization-based process. AICHE J. **56**, 1801–1812 (2010). https://doi.org/10.1002/aic.12105
101. Goel, A., Hebgen, P., Sande, J., Howard, J.: Combustion synthesis of fullerenes and fullerenic nanostructures. Carbon. **40**, 177–182 (2002). https://doi.org/10.1016/S0008-6223(01)00170-1
102. Kyesmen, P., Onoja, A., Amah, A.: Fullerenes synthesis by combined resistive heating and arc discharge techniques. Springerplus. **5** (2016). https://doi.org/10.1186/s40064-016-2994-7
103. https://en.wikipedia.org/wiki/Triple_point
104. Igor, K., Alexey, S., Alexey, P., Sotos, K., Thomas, D., Steven, S., Olga, B.: Seven-minute synthesis of PureCs-$C_{60}C_{16}$ from [60]Fullerene and Iodine Monochloride: First IR, raman, and mass spectra of 99 mol % $C_{60}C_{16}$. Chemistry (Weinheim an der Bergstrasse, Germany). **11**, 5426–5436 (2005). https://doi.org/10.1002/chem.200500185
105. Hu, Y.: Bending effect of sp-hybridized carbon (carbyne) chains on their structures and properties. J. Phys. Chem. C. **115**(5), 1843–1850 (2011). https://doi.org/10.1021/jp111851u
106. https://www.masterorganicchemistry.com/2017/10/10/orbital-hybridization-post/
107. Bakry, R., Vallant, R.M., Najam-ul-Haq, M., et al.: Medicinal applications of fullerenes. Int. J. Nanomedicine. **2**(4), 639–649 (2007)
108. Martinez, Z.S., Castro, E., Seong, C.S., Cerón, M.R., Echegoyen, L., Llano, M.: Fullerene derivatives strongly inhibit HIV-1 replication by affecting virus maturation without impairing protease activity. Antimicrob. Agents. Chemother. **60**(10), 5731–5741. Published 2016 Sep 23 (2016). https://doi.org/10.1128/AAC.00341-16
109. Yang, X., Ebrahimi, A., Li, J., Cui, Q.: Fullerene-biomolecule conjugates and their biomedicinal applications. Int. J. Nanomedicine. **9**, 77–92 (2014). https://doi.org/10.2147/IJN.S52829

110. Loutfy, R., Katagiri, S.: Fullerene materials for Lithium-ion battery applications. In: Perspectives of Fullerene Nanotechnology, pp. 357–367 (2002). https://doi.org/10.1007/0-306-47621-5_32
111. https://mstnano.com/applications-of-fullerenes/
112. https://en.wikipedia.org/wiki/Nanocomposite
113. Huang, J., Her, S.C., Yang, X., Zhi, M.: Synthesis and characterization of multi-walled carbon nanotube/graphene nanoplatelet hybrid film for flexible strain sensors. Nanomaterials (Basel). **8** (10), 786. Published 2018 Oct 4 (2018). https://doi.org/10.3390/nano8100786
114. Eatemadi, A., Daraee, H., Karimkhanloo, H., et al.: Carbon nanotubes: properties, synthesis, purification, and medical applications. Nanoscale. Res. Lett. **9**(1), 393. Published 2014 Aug 13 (2014). https://doi.org/10.1186/1556-276X-9-393
115. https://en.wikipedia.org/wiki/Synthesis_of_carbon_nanotubes
116. Ondrej, J., Marek, E., Lenka, Z., Vit, K., Martin, B., Jiřina, M., Antonín, R., Buršík, J., Magdaléna, K.: Carbon nanotubes synthesis in microwave plasma torch at atmospheric pressure. Mater. Sci. Eng. C. **26**, 1189–1193 (2006). https://doi.org/10.1016/j.msec.2005.09.024
117. https://www.azonano.com/article.aspx?ArticleID=1561
118. https://www.pharmatutor.org/articles/carbon-nano-tubes.
119. Zhang, Q., Huang, J.-Q., Zhao, M.-Q., Qian, W.Z.: Carbon nanotube mass production: principles and processes. ChemSusChem. **4**, 864–889 (2011). https://doi.org/10.1002/cssc.201100177review?page=2&quicktabs_latest_article_tabs=0
120. Rao, R., et al.: Carbon nanotubes and related nanomaterials: critical advances and challenges for synthesis toward mainstream commercial applications. ACS Nano. **12**(12), 11756–11784 (2018). https://doi.org/10.1021/acsnano.8b06511
121. Sharma, P., Pavelyev, V., Kumar, S., et al.: Analysis on the synthesis of vertically aligned carbon nanotubes: growth mechanism and techniques. J. Mater. Sci. Mater. Electron. **31**, 4399–4443 (2020). https://doi.org/10.1007/s10854-020-03021-6
122. https://www.bbc.com/future/article/20190806-how-vaccines-could-fix-our-problem-with-cow-emissions
123. Aqel, A., Kholoud, M.M., El-Nour, A., Ammar, R.A.A., Al-Warthan, A.: Carbon nanotubes, science and technology part (I) structure, synthesis and characterisation. Arab. J. Chem. **5**(1), 1–23., ISSN 1878-5352 (2012). https://doi.org/10.1016/j.arabjc.2010.08.022
124. Otor, H.O., Steiner, J.B., García-Sancho, C., Alba-Rubio, A.C.: Encapsulation methods for control of catalyst deactivation: a review. ACS Catal. **10**(14), 7630–7656 (2020). https://doi.org/10.1021/acscatal.0c01569
125. Zhang, M., Li, J.: Carbon nanotube in different shapes. Mater. Today. **12**(6), 12–18., ISSN 1369-7021 (2009). https://doi.org/10.1016/S1369-7021(09)70176-2
126. Vargas-Bernal, R., Herrera-Prez, G.: Carbon Nanotube- and Graphene Based Devices, Circuits and Sensors for VLSI Design. IntechOpen Limited, London (2012). https://doi.org/10.5772/38743
127. Yang, G., Li, L., Lee, W.B., Ng, M.C.: Structure of graphene and its disorders: a review. Sci. Technol. Adv. Mater. **19**(1), 613–648. Published 2018 Aug 29 (2018). https://doi.org/10.1080/14686996.2018.1494493
128. https://en.wikipedia.org/wiki/Carbon_nanotube
129. Radhamani, V., Lau, H.C., Ramakrishna, S.: CNT-reinforced metal and steel nanocomposites: A comprehensive assessment of progress and future directions. Comp. Part A Appl. Sci. Manuf. **114**, 170–187., ISSN 1359-835X (2018). https://doi.org/10.1016/j.compositesa.2018.08.010
130. https://www.nanowerk.com/nanotechnology/introduction/introduction_to_nanotechnology_22.php
131. James, E., Jan, S., Alan, W., Robert, Y., Milo, S.: Collapse of single-wall carbon nanotubes is diameter dependent. Phys. Rev. Lett. **92**, 095501 (2004). https://doi.org/10.1103/PhysRevLett.92.095501
132. https://en.wikipedia.org/wiki/Stone%E2%80%93Wales_defect
133. https://www.newworldencyclopedia.org/entry/Carbon_nanotube#:~:text=throughout%20the%20material.-,Torus,expected%20for%20certain%20specific%20radii
134. https://en.wikipedia.org/wiki/Carbon_nanobud
135. Simon, F., Monthioux, M.: Fullerenes inside carbon nanotubes: the peapods. In: Monthioux, M. (ed.) Carbon meta-nanotubes. John Wiley & Sons, Ltd., Hoboken, NJ (2011). https://doi.org/10.1002/9781119954743.ch5b
136. Liu, Q., Ren, W., Chen, Z.-G., Yin, L., Li, F., Cong, H., Cheng, H.-M.: Semiconducting properties of cup-stacked carbon nanotubes. Carbon. **47**(3), 731–736., ISSN 0008-6223 (2009). https://doi.org/10.1016/j.carbon.2008.11.005
137. Aaron, F., Mathieu, L., Shu-Jen, H., George, T., Breslin Christopher, M., Lynne, G., Lundstrom, M.S., Wilfried, H.: Sub-10 nm carbon nanotube transistor. Nano Lett. **12**, 758–762 (2012). https://doi.org/10.1021/nl203701g
138. He, L., Liao, C., Tjong, S.C.: Scalable fabrication of high-performance transparent conductors using graphene oxide-stabilized single-walled carbon nanotube inks. Nanomaterials (Basel). **8**(4), 224. Published 2018 Apr 7 (2018). https://doi.org/10.3390/nano8040224
139. https://www.science.gov/topicpages/l/lightweight+high+performance
140. Tian, P., Tang, L., Teng, K.S., Lau, S.P.: Graphene quantum dots from chemistry to applications. Mater. Today Chem. **10**, 221–258., ISSN 2468-5194 (2018). https://doi.org/10.1016/j.mtchem.2018.09.007
141. Yang, X., Yunlong, L., Huabin, C., Xiao, L., Shisheng, L., Bin, Y.: Ab initio study of energy-band modulation in graphene-based two-dimensional layered superlattices. J. Mater. Chem. **22**, 23821–23829 (2012). https://doi.org/10.1039/C2JM35652J
142. Xinming Li, Tianshuo Zhao, Kunlin Wang, Ying Yang, Jinquan Wei, Feiyu Kang, Dehai Wu, and Hongwei Zhu Directly drawing self-assembled, porous, and monolithic graphene fiber from chemical vapor deposition grown graphene film and its electrochemical properties, Langmuir, 2011 27 (19), 12164–12171, DOI: https://doi.org/10.1021/la202380g
143. Jain, V., Kandasubramanian, B.: Functionalized graphene materials for hydrogen storage. J. Mater. Sci. **55**, 1865–1903 (2020). https://doi.org/10.1007/s10853-019-04150-y
144. https://news.rice.edu/2015/10/16/graphene-nano-coils-are-natural-electromagnets-2/
145. Jeon, I., Yoon, B., He, M., Swager, T.M.: Hyperstage graphite: electrochemical synthesis and spontaneous reactive exfoliation. Adv. Mater. **30**, 3 (2018). https://doi.org/10.1002/adma.201704538
146. Kit, W.J., Gordon, X., Minmin, Z., Barbaros, Ö., Antonio, C.N., Soon, T.N., Cleo, C.: Polymer-enriched 3D graphene foams for biomedical applications. ACS Appl. Mater. Interfaces. **7**(15), 8275–8283 (2015). https://doi.org/10.1021/acsami.5b01440
147. Karolina Syrek, Joanna Grudzień, Aneta Sennik-Kubiec, Anna Brudzisz, Grzegorz D. Sulka, "Anodic titanium oxide layers modified with gold, silver, and copper nanoparticles", J. Nanomater., 2019, 9208734, 10 2019. doi: https://doi.org/10.1155/2019/9208734
148. Narayan, N., Meiyazhagan, A., Vajtai, R.: Metal nanoparticles as green catalysts. Materials (Basel). **12**(21), 3602. Published 2019 Nov 2 (2019). https://doi.org/10.3390/ma12213602
149. Benedek, T.: The history of gold therapy for tuberculosis. J. Hist. Med. Allied. Sci. **59**, 50–89 (2004). https://doi.org/10.1093/jhmas/59.1.50
150. Hosseinpour-Mashkani, S., Ramezani, M.: Silver and silver oxide nanoparticles: synthesis and characterization by thermal

decomposition. Mater. Lett. **130**, 259–262 (2014). https://doi.org/10.1016/j.matlet.2014.05.133
151. Khandel, P., Yadaw, R.K., Soni, D.K., et al.: Biogenesis of metal nanoparticles and their pharmacological applications: present status and application prospects. J. Nanostruct. Chem. **8**, 217–254 (2018). https://doi.org/10.1007/s40097-018-0267-4
152. Singh, S., Ahmed, R., Growcock, F.: Vital role of nanopolymers in drilling and stimulations fluid applications. Proc. SPE Ann. Techn. Conf. Exhibit., 1 (2010). https://doi.org/10.2118/130413-MS
153. Avadi, M.R., Sadeghi, A.M., Mohamadpour Dounighi, N., Dinarvand, R., Atyabi, F., Rafiee-Tehrani, M.: Ex vivo evaluation of insulin nanoparticles using chitosan and Arabic gum. ISRN Pharm. **2011**, 860109 (2011). https://doi.org/10.5402/2011/860109
154. Hembram, K.C., Prabha, S., Chandra, R., Ahmed, B., Nimesh, S.: Advances in preparation and characterization of chitosan nanoparticles for therapeutics. Artif. Cells Nanomed. Biotechnol. **44**(1), 305–314 (2016). https://doi.org/10.3109/21691401.2014.948548
155. Sahbaz, D.A., Yakar, A., Gündüz, U.: Magnetic Fe_3O_4-chitosan micro- and nanoparticles for wastewater treatment. Part. Sci. Technol. **37**(6), 732–740 (2019). https://doi.org/10.1080/02726351.2018.1438544
156. Lee, E.J., Khan, S., Lim, K.-H.: Chitosan-nanoparticle preparation by polyelectrolyte complexation. World J. Eng., 541–542 (2009)
157. Kumar, S., Dilbaghi, N., Rani, R., Bhanjana, G.: Nanotechnology as emerging tool for enhancing solubility of poorly water-soluble drugs. BioNanoScience. **2**, 227–250 (2012). https://doi.org/10.1007/s12668-012-0060-7
158. Hoa, L., Chi, N., Triet, N., Nhân, L., Chien, D.: Preparation of drug nanoparticles by emulsion evaporation method. J. Phys. Conf. Series, 187, 012047 (2009). https://doi.org/10.1088/1742-6596/187/1/012047
159. Sonvico, F., Cagnani, A., Rossi, A., et al.: Formation of self-organized nanoparticles by lecithin/chitosan ionic interaction. Int. J. Pharm. **324**(1), 67–73 (2006). https://doi.org/10.1016/j.ijpharm.2006.06.036
160. Bhattarai, N., Ramay, H.R., Chou, S.H., Zhang, M.: Chitosan and lactic acid-grafted chitosan nanoparticles as carriers for prolonged drug delivery. Int. J. Nanomedicine. **1**(2), 181–187 (2006). https://doi.org/10.2147/nano.2006.1.2.181
161. Jae, P., Seunglee, K., Ju-Ock, N., Rang-Woon, P., Hesson, C., Sang, S., In-San, K., Ick, K., Seo, J.: Self-assembled nanoparticles based on glycol chitosan bearing 5??-cholanic acid for RGD peptide delivery. J. Contr. Release. **95**, 579–588 (2004). https://doi.org/10.1016/j.jconrel.2003.12.020
162. Neel, J., Mark, G.: Applications of dendrimers in tissue engineering. Curr. Top. Med. Chem. **8**, 1225–1236 (2008). https://doi.org/10.2174/156802608785849067
163. Din, F.U., Aman, W., Ullah, I., et al.: Effective use of nanocarriers as drug delivery systems for the treatment of selected tumors. Int. J. Nanomedicine. **12**, 7291–7309. Published 2017 Oct 5 (2017). https://doi.org/10.2147/IJN.S146315
164. Wahajuddin, A.S.: Superparamagnetic iron oxide nanoparticles: magnetic nanoplatforms as drug carriers. Int. J. Nanomedicine. **7**, 3445–3471 (2012). https://doi.org/10.2147/IJN.S30320
165. Behera, A., Patel, S., Priyadarshini, M.: Fibre reinforced metal matrix nano-composites. In: Fiber-Reinforced Nanocomposites: Fundamentals and Applications, pp. 147–156. Elsevier Publisher, Netherland. ISBN: 978-0-12-819904-6 (2020). https://doi.org/10.1016/B978-0-12-819904-6.00007-4
166. Behera, A., Swain, B., Sahoo, D.K.: Fiber reinforced ceramic matrix nanocomposites. In: Fiber-Reinforced Nanocomposites: Fundamentals and Applications, pp. 359–368. Elsevier Publisher, Netherland., ISBN: 978-0-12-819904-6 (2020). https://doi.org/10.1016/B978-0-12-819904-6.00016-5
167. Lalwani, G., Henslee, A.M., Farshid, B., et al.: Tungsten disulfide nanotubes reinforced biodegradable polymers for bone tissue engineering. Acta Biomater. **9**(9), 8365–8373 (2013). https://doi.org/10.1016/j.actbio.2013.05.018
168. Khalid, N.R., Ejaz, A., Zhanglian, H., Mukhtar, A., Yuewei, Z., Sadia, K.: Cu-doped TiO_2 nanoparticles/graphene composites for efficient visible-light photocatalysis. Ceram. Int. **39**, 7107–7113 (2013). https://doi.org/10.1016/j.ceramint.2013.02.051
169. https://www.compositespress.com/wikicomposites/nanocomposite/
170. Camargo, P.H.C., Satyanarayana, K.G., Wypych, F.: Nanocomposites: synthesis, structure, properties and new application opportunities. Mat. Res. **12**, 1. São Carlos (2009). https://doi.org/10.1590/S1516-14392009000100002
171. Xiao-Lin, X., Mai, Y.-W., Xing-Ping, Z.: Dispersion and alignment of carbon nanotubes in polymer matrix: a review. Mater. Sci. Eng. R Rep. **49**, 89–112 (2005). https://doi.org/10.1016/j.mser.2005.04.002
172. Raman, N., Sudharsan, S., Pothiraj, K.: Synthesis and structural reactivity of inorganic–organic hybrid nanocomposites – a review. J. Saudi Chem. Soc. **16**(4), 339–352., ISSN 1319-6103 (2012). https://doi.org/10.1016/j.jscs.2011.01.012
173. Behera, A.: Fuel cell recycling. In: Nanotechnology in Fuel Cells. Elsevier Publisher, Netherland (2021)
174. Cao, Y., Irwin, P.C., Karim, Y.: The future of nanodielectrics in the electrical power industry. Dielectr. Electr. Insul. IEEE Trans. **11**, 797–807 (2004). https://doi.org/10.1109/TDEI.2004.1349785
175. Behera, A., Mallick, P.: Application of nanofibers in aerospace industry. In: Fiber-Reinforced Nanocomposites: Fundamentals and Applications. Elsevier Publisher, Netherland. ISBN: 978-0-12-819904-6 (2020). https://doi.org/10.1016/B978-0-12-819904-6.00020-7
176. Pandey, A., Patel, A.K., et al.: Enhanced tribological and bacterial resistance of carbon nanotube with ceria- and silver-incorporated hydroxyapatite biocoating. Nanomaterials (Basel). **8**(6), 363. Published 2018 May 24 (2018). https://doi.org/10.3390/nano8060363
177. Ramanathan, T., Ahmed, A., Sasha, S., Dikin, D., Herrera-Alonso, M., Richard, P., Douglas, A., Hannes, S., Chen, X., Rodney, R., Nguyen, S., Aksay, I., Robert, P.'h., Brinson, L.: Functionalized graphene sheets for polymer nanocomposites. Nat. Nanotechnol. **3**, 327–331 (2008). https://doi.org/10.1038/nnano.2008.96
178. Koo, J.: An overview of nanomaterials. In: Fundamentals, Properties, and Applications of Polymer Nanocomposites, pp. 22–108. Cambridge University Press, Cambridge (2016). https://doi.org/10.1017/CBO9781139342766.003
179. Thomason, J.L.: Glass fibre sizing: a review. Compos. A: Appl. Sci. Manuf. **127**, 105619., ISSN 1359-835X (2019). https://doi.org/10.1016/j.compositesa.2019.105619
180. Rathod, V.T., Kumar, J.S., Jain, A.: Polymer and ceramic nanocomposites for aerospace applications. Appl. Nanosci. **7**, 519–548 (2017). https://doi.org/10.1007/s13204-017-0592-9
181. https://wjpr.net/admin/assets/article_issue/1464659448.pdf
182. https://en.wikipedia.org/wiki/Microporous_material
183. https://www.wikiwand.com/en/Mesoporous_material
184. https://www.beilstein-journals.org/bjnano/series/78
185. Ertürk, G., Mattiasson, B.: Molecular imprinting techniques used for the preparation of biosensors. Sensors (Basel). **17**(2), 288. Published 2017 Feb 4 (2017). https://doi.org/10.3390/s17020288
186. Barrios, E., Fox, D., Li Sip, Y.Y., et al.: Nanomaterials in advanced, high-performance aerogel composites: a review. Polymers (Basel). **11**(4), 726. Published 2019 Apr 20 (2019). https://doi.org/10.3390/polym11040726
187. https://www.giiresearch.com/report/ina385591-nanotechnology-market-by-type-nanocomposites.html
188. https://www.nanotech-now.com/columns/?article=054

189. Sinha, A., Behera, A., Behera, A.: Nanotechnology in the space industry. In: Smart Remote Sensing Networks: From Nanotechnology to Space Technology Against Natural Disasters. CRC Publisher, Boca Raton, FL (2021)
190. https://www.nasa.gov/sites/default/files/atoms/files/spinoff2019_lres.pdf
191. https://sites.google.com/site/worldofbionics/benefits-and-application-of-nanotechnology
192. Zhang, H., Shuang, S., Wang, G., Guo, Y., Tong, X., Yang, P., Chen, A., Dong, C., Yong, Q.: TiO_2–graphene hybrid nanostructures by atomic layer deposition with enhanced electrochemical performance for Pb(II) and Cd(II) detection. RSC Adv. **5**, 4343–4349 (2015). https://doi.org/10.1039/C4RA09779C
193. Hirohata, A., Yamada, K., Nakatani, Y., Prejbeanu, I.-L., Diény, B., Pirro, P., Hillebrands, B.: Review on spintronics: principles and device applications. J. Magn. Magn. Mater. **509**, 166711., ISSN 0304-8853 (2020). https://doi.org/10.1016/j.jmmm.2020.166711
194. https://en.wikipedia.org/wiki/Racetrack_memory
195. Alharbi, K.K., Al-Sheikh, Y.A.: Role and implications of nanodiagnostics in the changing trends of clinical diagnosis. Saudi J. Biol. Sci. **21**(2), 109–117 (2014). https://doi.org/10.1016/j.sjbs.2013.11.001
196. Branton, D., Deamer, D.W., Marziali, A., et al.: The potential and challenges of nanopore sequencing. Nat. Biotechnol. **26**(10), 1146–1153 (2008). https://doi.org/10.1038/nbt.1495
197. Hofmann, M.C.: Stem cells and nanomaterials. Adv. Exp. Med. Biol. **811**, 255–275 (2014). https://doi.org/10.1007/978-94-017-8739-0_13
198. https://news.rice.edu/2016/09/19/graphene-nanoribbons-show-promise-for-healing-spinal-injuries/
199. Saha, M.: Nanomedicine: promising tiny machine for the healthcare in future-a review. Oman Med. J. **24**(4), 242–247 (2009). https://doi.org/10.5001/omj.2009.50
200. Shafique, M., Luo, X.: Nanotechnology in transportation vehicles: an overview of its applications, environmental, health and safety concerns. Materials (Basel). **12**(15), 2493. Published 2019 Aug 6 (2019). https://doi.org/10.3390/ma12152493
201. https://www.nbmcw.com/tech-articles/others-article/41632-era-of-nanotechnology.html
202. https://afdc.energy.gov/fuels/hydrogen_production.html
203. Behera, A.: 3D Print Battery. In: Nanobatteries and Nanogenarators, pp. 11–30. Elsevier Publisher, Netherland (2020) ISBN: ors/song/978-0-12-821548-7, https://www.elsevier.com/books/nanobatteries-and-nanogenerators/song/978-0-12-821548-7
204. Ameta, S.K., Rai, A.K., Hiran, D., Ameta, R., Ameta, S.C.: Use of nanomaterials in food science. In: Biogenic Nano-Particles and their Use in Agro-ecosystems, pp. 457–488. Published 2020 Mar 21. Springer, New York (2020). https://doi.org/10.1007/978-981-15-2985-6_24
205. Behera, A., Pan, J., Behera, A.: Temperature nanosensors for smart manufacturing. In: Nanosensors for smart manufacturing. Elsevier Publisher, Netherland (2020) ISBN: 978-0-12-823358-0
206. Mirjalili, M., Yaghmaei, N., Mirjalili, M.: Antibacterial properties of nano silver finish cellulose fabric. J. Nanostruct. Chem. **3**, 43 (2013). https://doi.org/10.1186/2193-8865-3-43
207. https://www.sciencedaily.com/releases/2008/02/080213133347.htm
208. Chaturvedi, S., Dave, P.N., Shah, N.K.: Applications of nano-catalyst in new era. J. Saudi Chem. Soc. **16**(3), 307–325., ISSN 1319-6103 (2012). https://doi.org/10.1016/j.jscs.2011.01.015
209. Haijiao, L., Wang, J., Stoller, M., Wang, T., Bao, Y., Hao, H.: An overview of nanomaterials for water and wastewater treatment. Adv. Mater. Sci. Eng. **2016**., Article Id: 4964828, 10 (2016). https://doi.org/10.1155/2016/4964828
210. http://powerelectronics.biz/benefits-and-applications/
211. Mishra, R., Militky, J., Baheti, V., Huang, J., Kale, B., Venkataraman, M., Bele, V., Arumugam, V., Zhu, G., Wang, Y.: The production, characterization and applications of nanoparticles in the textile industry. Text. Prog. **46** (2014). https://doi.org/10.1080/00405167.2014.964474
212. https://en.wikipedia.org/wiki/Photoluminescence
213. Takagi, H., Ogawa, H., Yamazaki, Y., et al.: Quantum size effects on photoluminescence in ultrafine Si particles. Appl. Phys. Lett. **56** (24), 2379 (1990)
214. Thangakani, J.A., Sheela, C.D., Dorothy, R., Renugadevi, N., Jeyasundari, J., Rajendran, S., Behera, A.: Copper nano-alloys. In: Nanotechnology in the Automotive Industry. Elsevier Publisher, Netherland (2021)
215. Behera, A., Chattergy, S.: Industrial Scale Up Applications of Nanomaterials Recycling. Elsevier Publisher, Netherland (2021) ISBN: 8779722122
216. Edelstein, A.S., Murday, J.S., Rath, B.B.: Challenges in nanomaterials design. Progr. Mater. Sci. **42**(1–4), 5–21., ISSN 0079-6425 (1997). https://doi.org/10.1016/S0079-6425(97)00005-4

Magnetostrictive Materials

4

Abstract

Noncontact sensors and actuators have a very demand in the advanced materials market. Use of magnetostrictive materials in advanced applications has been discussed in this chapter. A brief discussion on magnetostrictive materials has been made from the history to current development. Magnetostrictive sensors construction and working along with the electromagnetic properties and various associated magnetostrictive effects has been discussed here. Specific categories of materials such as iron-based alloys, Ni-based alloys, terfenol-D, metglas, ferromagnetic shape memory alloys, and so on used for magnetostrictive behavior have been discussed. All the potential applications in mechanical industries, aero-industries, automotive industries, biomedical industries, construction industries, energy-harvesting industries have been discussed. All the limitations described in this chapter, which can lead to the further development of magnetostrictive sensors and actuators in future research.

Keywords

Magnetostrictive materials · Magnetostrictive effect · Curie temperature · Hysteresis · Joul effect · Villari effect

4.1 Magnetostrictive Materials

Magnetostrictive materials are smart materials that are an essential part of applications going from dynamic vibration control, control surface deployment, and energy harvesting to stress and torque detecting. A primary application for magnetostrictive materials is vibration control of heavy structures [1]. Magnetostrictive material converts the magnetic energy to mechanical energy and vice versa as given in Fig. 4.1. In another it can explain, the magnetostrictive materials can convert magnetic energy into kinetic energy, or the reverse. Magnetostrictive materials are well-accepted build as noncontact actuators and sensors [2, 3]. Variations in magnetization result from magnetic moment rotations, which can be realized by the utilization of magnetic fields or stresses. Unlike piezoelectric, these materials can be used under static and low-frequency conditions with very low impedance but they offer high impedance at high frequencies (kHz range) due to the inductive nature of the materials [4]. Magnetostriction is the phenomenon of strong coupling between magnetic state and mechanical state. Most magnetostrictive materials have a higher Curie temperature than piezoelectric materials and hence can be operated at higher temperatures. Moreover, these materials do not need to be poled and hence there is no limitation in applying stress which can cause dipoling [5]. The small hysteresis shown by these materials translates to negligible self-heating and the need for less complicated control algorithms for actuation and precise sensing output. Figure 4.2 shows a magnetostrictive device used for actuation function. In addition, recent advances in materials science research have enabled more capable magnetostrictive materials in various forms, including amorphous or crystalline thin films, magnetostrictive particle-aligned polymer matrix composite, and sintered powder compacts suitable for mass production [6].

Magnetostriction is a property of magnetic materials that causes them to change their shape or dimensions during the process of magnetization. The variation of materials' magnetization due to the applied magnetic field changes the magnetostrictive strain until reaching its saturation value [7]. The effect was first identified in 1842 by James Joule when observing a sample of iron. The magnetization of magnetostrictive materials can be quantified by the magnetostrictive coefficient (Λ) which may be positive or negative and is defined as the fractional change in length as the magnetization of the material increases from zero to the saturation value [8]. The effect can be heard near transformers and high-power electric devices, produces a low-pitched humming sound "electric hum," where oscillating AC currents produce a changing magnetic field. This effect causes energy loss due

A. Behera, *Advanced Materials*, https://doi.org/10.1007/978-3-030-80359-9_4

Fig. 4.1 Illustrating magnetostriction: a cubic material distorts depending on the direction of its magnetic moment

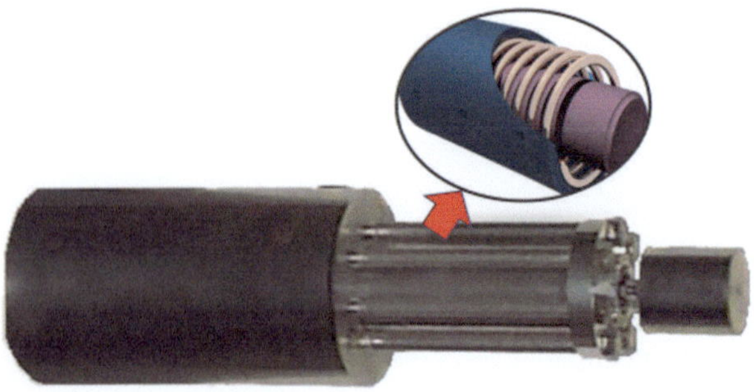

Fig. 4.2 Magnetostrictive transducer device. Cut-away of a transducer comprising: magnetostrictive material (inside), magnetizing coil, and magnetic enclosure completing the magnetic circuit (outside)

to frictional heating in susceptible ferromagnetic cores. A comparison between magnetostrictive material and a piezoelectric material has been given in Table 4.1.

4.2 History of Magnetostrictive Materials

Magnetostriction was initially reported by James P. Joule in the mid-1840s. He detected there is a change in the dimension of iron particles with a change in magnetization. He also discovered magnetostriction behavior in nickel in 1842 [11]. These effects are well accepted in stress/force sensor applications. In 1865, Villari discovered the reverse effect i.e. by the application of stress correspondingly change the magnetization in magnetostrictive material [12]. Anisotropy behavior by magnetostriction is discovered in single-crystal iron in 1926. In the mid of World War II, Ni-based magnetostrictive composites were used in transducers for sonar applications. Incorporation of rare earth metals such as terbium (Tb) and dysprosium (Dy) for magnetostrictive materials was reported in 1965 by Clark, due to their substantial magnetostriction at low temperatures [13]. The strains in these elements are of the order of $10{,}000 \times 10^{-6}$, or three orders of magnitude larger than those of nickel, but they are achieved at cryogenic temperatures. The temperature limitation and the fact that the field of piezoelectricity was gaining technical maturity hindered the development of magnetostrictive materials and led in the early 1970s to a search for a new class of transducer materials capable of high room-temperature strains. Highly magnetostrictive rare earths (R), principally samarium (Sm), terbium (Tb), and dysprosium (Dy), were combined with the magnetic transition metals nickel, cobalt, and iron by direct compound synthesis and by rapid sputtering into amorphous alloys [14, 15]. In contrast to the normal Curie temperature behavior of the R-Ni and R-Co compounds, the R-Fe compounds exhibit an increase in the Curie temperature with increasing rare earth concentration. This unusual property facilitates huge room temperature magnetostrictions, of up to 3000×10^{-6}, particularly in the $TbFe_2$ compound [16]. Both $TbFe_2$ and $DyFe_2$ alloys have extensive magneto-crystalline anisotropies and thus require substantial magnetic fields to drive them to immersion. Again, in 1975, Clark found out magnetostrictive effect in Terfenol-D ("Ter" for terbium, "Fe" for iron, "NOL" for Naval Ordnance Laboratory, and "D" for dysprosium) that shows high strain and blocked stress. Terfenol-D displays greater magnetostriction at room temperature [17]. Terfenol-D has extensively low magnetocrystalline anisotropy than both $TbFe_2$ and $DyFe_2$ [18]. One of the drawbacks of terfenol-D is that it is brittle in nature, which constrains its capacity to withstand shock loads or operate in tension. At the end of the 1900 and afterward many composite magnetostrictive materials have been developed such as polymer matrix and Terfenol-D particulate composite, Galfenol-a, oriented particulate composite, nanocrystalline hybrid metal-ferrites, etc. [19].

4.3 Mechanism of Magnetostrictive Effect

Internally, ferromagnetic materials have structures that are divided into domains, each of which is a region of uniform magnetic polarization. When a magnetic field is applied, the boundaries between the domains shift and the domains rotate [20]. The change in the magnetic domains of material results in a change in materials dimensions. This phenomenon is a consequence of magneto-crystalline anisotropy, which indicates it consume more energy to magnetize a crystalline material in one direction [21]. If a magnetic field is applied to the material at an angle to an easy axis of magnetization, the material will tend to rearrange its structure so that an easy axis is aligned with the field to minimize the free energy of the system. Since different crystal directions are associated with different lengths, this effect induces a strain in the material. The change in magnetic field gives the change in strain is known as Joule's effect. Most magnetic materials exhibit Joule magnetostrictions, only a small number of compounds containing rare earth elements provide strains in excess of 1000×10^{-6} [22]. These large strains are a direct consequence of a strong magnetoelastic coupling arising

Table 4.1 Comparative table between a magnetostrictive material with a piezoelectric material [9, 10]

	Actuation mechanism	Elongation (%)	Energy density	Bandwidth (kHz)	Hysteresis (%)	Approx. cost ($/cm^3)
Terfenol	Magnetostrictive material	0.2	20 J/m^3	10	2	400
PZT	Piezoelectric material	0.1	2.5 kJ/m^3	100	10	200

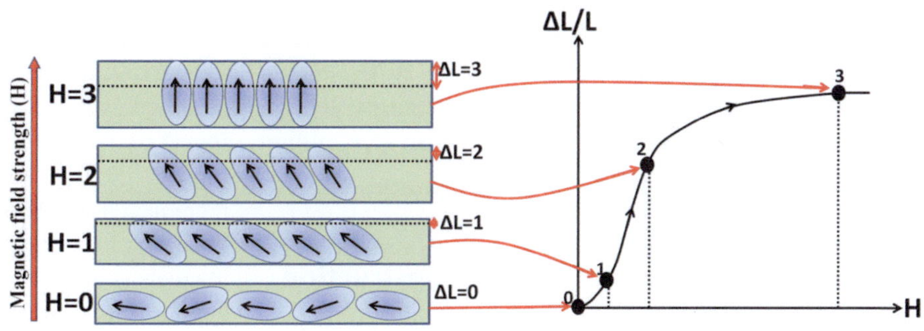

Fig. 4.3 Schematics of the magnetic domain orientation along with the strength of the magnetic line of force

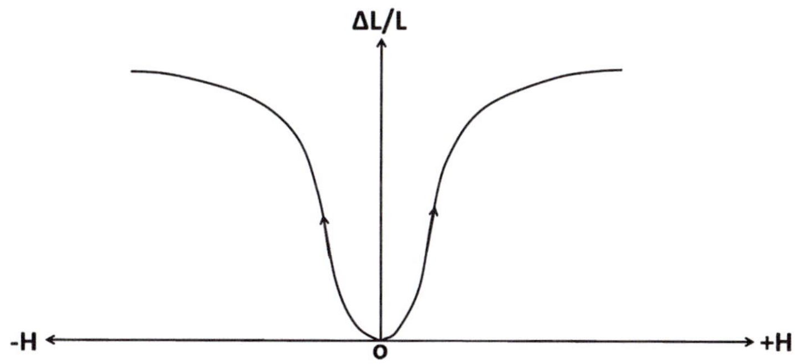

Fig. 4.4 Strain versus symmetric magnetic field

from the dependency of magnetic moment orientation with interatomic spacing. Whereas, the reciprocal effect, that is, by application of stress (mechanical energy) produces magnetization of the material is known as the Villari effect. Other effects such as ΔE-Effect, Matteucci effect, Wiedemann effect, Barret Effect, Nagaoka-Honda effect can be considered on the basis of their domain orientation [23–25].

For actuator applications, maximum rotation of magnetic moments leads to the highest possible magnetostriction output. This can be achieved by processing techniques such as stress annealing and field annealing. However, mechanical prestresses can also be applied to thin sheets to induce alignment perpendicular to actuation as long as the stress is below the buckling limit. These rotation and reorientation cause internal strains in the material structure. The strains in the structure lead to the stretching, in the case of positive magnetostriction of the material in the direction of the magnetic field. During this stretching process, the cross section is reduced in a way that the volume is kept nearly constant. The size of the volume change is so small that it can be neglected under normal operating conditions. Applying a stronger field leads to stronger and more definite reorientation of more and more domains in the direction of the magnetic field. When all the magnetic domains have become aligned with the magnetic field, the saturation point has been achieved [8].

Figure 4.3 will shows a strain versus magnetic field. Here the region between 0 and 1, where the applied magnetic field is small, the magnetic domains show almost no common orientation pattern. Depending on how the material was formed there may be a small amount of a common orientation pattern, which would show itself as a permanent magnet bias. The resulting strain depends very much on how homogeneous is the base structure of the magnetostrictive material and the material formulation. In region 1–2 ideally there should be an almost linear relationship between strain and magnetic field. Because the relationship is a simple one, it is easier to predict the behavior of the material and so most devices are designed to operate in this region. Beyond point 2, the relationship becomes nonlinear again as a result of the fact that most of the magnetic domains have become aligned with the magnetic field direction. At point 3, there is a saturation effect, which prevents the further increase in strain [26].

The idealized behavior of length change is in response to the applied magnetic field. When a magnetic field is established in the opposite direction, the field is understood to be negative, but the negative field produces the same elongation in the magnetostrictive material, as a positive field would. The shape of the curve is reminiscent of a butterfly and so the curves are referred as butterfly curves (Fig. 4.4).

The behavior of the magnetostrictive materials in various applications is complex because the changing conditions during operation cause changes in material properties [27]. A full understanding of the complexity will enable engineers to use the potential advantages of magnetostrictive materials and to optimize an actuator based on "giant" magnetostrictive materials. Giant magnetostrictive materials are a kind of functional materials developed since 1970s, known for their large magnetostrain and high energy density. It has a maximum strain usually in the hundreds part per million when saturated with a magnetic field.

4.4 Magnetostrictive Sensors Construction and Working

All ferromagnetic materials experience magnetostriction, but in some materials, the magnitude of the dimensional change is too small to be of practical use. For example, when a magnetostrictive bar or rod is placed in a magnetic field, the magnetic dipole is oriented parallel to the length of the bar co using the change in length of the bar. However, the length change for materials used in linear magnetostrictive sensors is very small, typically on the order of 10^{-6} m/m [28]. In magnetostrictive sensors, the wire, or bar, is referred to as a waveguide. It is typically made from an iron alloy and is mounted to a stationary part of the machine. The magnetic field is provided by a magnet, referred to as a position magnet, which is attached to the moving part being measured. Short pulses of current (1–3 μs) are applied to a conductor attached to the waveguide [29]. A typical magnetostrictive sensor with the waveguide, the position magnet, and a strain pulse converter has been given in Fig. 4.5.

In another case, torsional strain (twist) is induced in the waveguide, due to the interaction of the magnetic field caused by the current and the magnetic field caused by the position magnet. Because the current is applied as a pulse (referred to as an interrogation pulse), the twist travels down the wire as an ultrasonic wave, moving at approximately 2850 m/s [30]. This twist, or mechanical pulse, is detected by a signal converter (also referred to as a strain pulse converter), which relies on the Villari effect to create a voltage pulse indicating receipt of the mechanical strain wave [31]. The time between the initial current pulse and the detection of the mechanical pulse indicates the location of the position magnet, and therefore, the position of the moving part being measured. The interrogation rate, or update rate, can range from one time per second to over 4,000 times per second, with the maximum update rate determined by the length of the waveguide [32]. Magnetostrictive sensors provide absolute position information and, unlike incremental encoders, do not need to be re-homed when there is a loss of power. They can also use multiple position magnets with one waveguide, making them well-suited for applications that require position information for multiple components along the same axis, such as the knives on a slitting machine [33]. In addition to being noncontact, magnetostrictive designs enclose the waveguide in an extruded aluminum housing or stainless steel tube, so they are virtually impervious to contamination. They can also operate when there is a barrier between the position magnet and the waveguide- as long as the barrier is nonmagnetic, such as ceramic, plastic, aluminum, or stainless steel [34, 35].

4.5 Electromagnetic Properties

4.5.1 Permittivity

Permittivity (ε) is the property of the material to attract or repel electric fields as compared to free space (vacuum), given in Eq. (4.1.

$$\varepsilon = \varepsilon_0 \varepsilon_r \tag{4.1}$$

where ε_0 is the permittivity of free space $(1/(36\pi) \times 10^{-9})$ $F.m^{-1}$ and ε_r is the permittivity ratio of the material. Nickel is a solid conductive metal having a permittivity value of infinite.

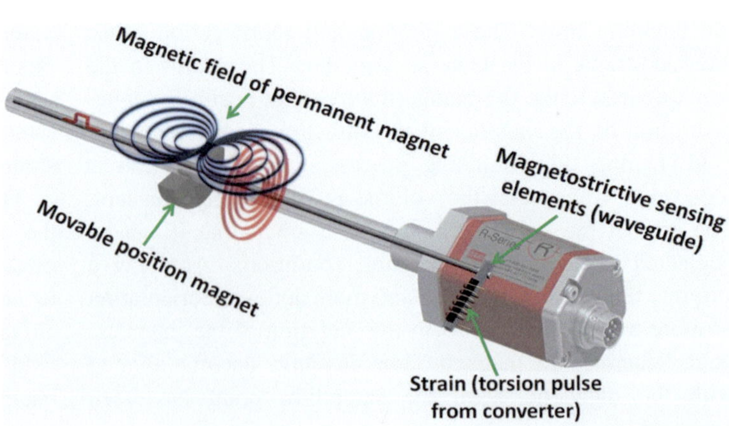

Fig. 4.5 The basic components of a magnetostrictive sensor include the waveguide, the position magnet, and a strain pulse converter.

4.5.2 Permeability

Permeability is the property of the material to attract or repel magnetic fields. An example of a permeable material can be found in the core of an electrical transformer. A magnetic field is produced by the primary winding and is conducted through the core to the secondary windings. The core material can attract or repel the magnetic field. In the case of an electrical transformer, it is desired to attract the magnetic field.

The value of attraction or repulsion of a magnetic field is known as permeability. Permeability is referenced to the magnetic field in free space (vacuum) by the Eq. 4.2:

$$\mu = \mu_0 \mu_r \tag{4.2}$$

where μ_0 is the permeability of free space ($4\pi \times 10^{-7}$) H. m^{-1} and μ_r is the permeability ratio of the material. The permeability of Ni is (753.98×10^{-6}) H. m^{-1}.

4.5.3 Magnetic Materials

This material has ability to attract or resist an induced magnetic field is the primary method that classifies the magnetic properties of the material. This is referred to as the permeability ratio of the material. The permeability ratio of the material is attributed to the spin and orbital spin of electrons around the nucleus of the atom and incomplete inner electron shells.

4.5.4 Diamagnetic Material

A diamagnetic material exhibits negative magnetism. A diamagnetic material repels an external magnetic field by producing a smaller internal magnetic field in opposite polarity. The external magnetic field will conduct around the diamagnetic material, rather than through the diamagnetic material. The permeability of diamagnetic material is less than one and negative.

4.5.5 Paramagnetic Material

A paramagnetic material has a permeability of less than 1. That is the magnetic field can conduct through the paramagnetic material although the magnetic field will experience less resistance by conducting around the paramagnetic material. An external magnetic field may tend to polarize the random moments by creating thermal agitations causing only a very small (partial) magnetism.

4.5.6 Ferromagnetic Material

Ferromagnetic materials are a refined material that has the ability to conduct higher magnetic flux compared to free space. Ferromagnetic materials have a permeability greater than one and therefore attract magnetic flux compared to free space. The three ferromagnetic elements are iron, cobalt, and nickel.

Soft ferromagnetic material has high permeability and will return to its natural state when the induced magnetic field is removed. Soft ferromagnetic materials are used in transformers where the magnetic flux changes direction every cycle. Nickel is classed as soft ferromagnetic material. Hard ferromagnetic materials have high permeability and will attempt to retain its induced magnetized state. Hard ferromagnetic materials are ideal permanent magnets.

4.5.7 Antiferromagnetic Material and Ferrimagnetic Material

Antiferromagnetic material relies on temperature to become permeable. The temperature that allows maximum permeability is known as the Neel Temperature. The permeability of antiferromagnetic material is usually less than 1. Naturally found materials, such as iron ore and nickel ore, that have an overall magnetic flux are known as ferrimagnetic.

4.5.8 Curie Temperature

The Curie temperature is the temperature where the stable magnetic properties of the material change due to the extra energy provided by heat. Ferromagnetic materials become paramagnetic when the temperature is above the Curie temperature. When the magnetic properties of the material change, this also changes the permeability of the material. A description of various types of magnetic materials with their inherent properties has been provided in Table 4.2 with examples of materials.

4.5.9 Generation of Magnetic Fields

A magnetic field is created by the flow of current. If the current is flowing in a coil the magnetic field is concentrated in the center of the coil. If the center of the coil is a ferromagnetic material, that attracts the magnetic field, then the magnetic field intensity and density can be increased. Figure 4.6 shows the concentration of the magnetic field in a coil of wire. The generation of the magnetic field in a magnetostrictive material is created by a wire coil or winding [36].

Table 4.2 Magnetic materials type and their properties

Magnetic material	Susceptibility to induced magnetic field	Permeability	Examples
Diamagnetic	Negative and small	< -1	Cu, Ag, Au
Paramagnetic	Positive and small	>1	Mg, Li
Ferromagnetic	Positive and very large	$>>1$	Fe, Ni
Anti-ferromagnetic	Positive and small	<1	NiO
Ferromagnetic	Positive and large	$>>1$	Fe_2O_3

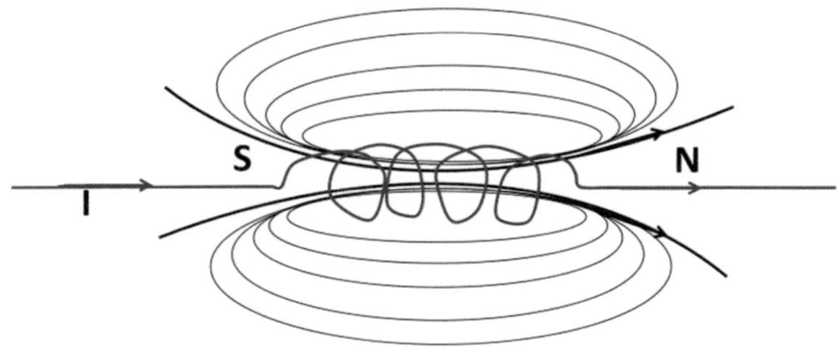

Fig. 4.6 Magnetic field created by a coil

A magnetic field is defined by two parameters, being magnetic field intensity (H) and magnetic flux density (B). The following equation states the relationship of magnetic field intensity and magnetic flux density,

$$B = \mu H = \frac{\varnothing}{A} \quad (4.3)$$

where H is the magnetic field intensity ($A.\ m^{-1}$), $\varnothing$ is the magnetic flux = BA (Wb) and A is the cross-sectional area (m^2) with the cross-sectional area being the area inside the windings. Due to the high intensity and density of magnetic flux in the center of the windings, miniscule currents are produced. These currents are known as eddy currents. To reduce the effect of eddy currents, the center core is usually made up of laminations of the ferromagnetic material. As the core is made of many laminations a stacking factor is usually incorporated in the cross-sectional area of the core material. A stacking factor takes into account the space between the laminations in the center core material [37].

4.5.10 Hysteresis

There are limits to the amount of magnetic flux density (B) that the material can withstand. If the magnetic field intensity (H) is increased beyond the saturation level (H_{sat}) the magnetic flux density will not increase beyond the saturation level for magnetic flux density. As the magnetic field intensity is determined by the current and the number of turns, there is a maximum current and a maximum number of turns in achieving the saturation level of magnetic flux density [38]. These limits can be shown as a hysteresis curve as shown in Fig. 4.7. Hysteresis also shows the different generation and degeneration of the magnetic flux as experienced in an alternating current source. Nickel has a saturation magnetic flux density (B_{sat}) of 0.617 T and a permeability of 7.547.54 × 10^{-7}H. m^{-1} leading to a saturation magnetic field intensity (H_{sat}) of 818.3 $A.\ m^{-1}$.

Like flux density, the magnetostriction also exhibits hysteresis versus the strength of the magnetizing field. The shape of this hysteresis loop (called "dragonfly loop") can be reproduced using the Jiles–Atherton model. The magnetostriction phenomenon is connected with the changes in the linear dimensions of the sample during the magnetization process. The magnetostrictive phenomenon has great technical importance. It can be used in the development of specialized position and level sensors, MEMS sensors, as well as ultrasonic transducers, and high accuracy actuators [39]. When a magnetic material is placed in an external magnetic field, its grains get oriented in the direction of the magnetic field. Now, even after the removal of an external magnetic field, some magnetization exists, which is called residual magnetism. This property of the material is called magnetic retentively of material. A hysteresis loop or B-H curve of a typical magnetic material is shown in Fig. 4.7. Magnetization B_r in the below hysteresis loop represents the residual magnetism of material [40].

Due to the retentivity of material, even after the removal of an external magnetic field some magnetization exists in the material. This magnetism is called residual magnetism of material. To remove this residual magnetization, we have to apply some external magnetic fields in opposite direction. This external magnetic motive force required to overcome the residual magnetism is called the "coercive force" of material. In the above hysteresis loop, $-H_c$ represents the coercive force. The material having a large value of residual

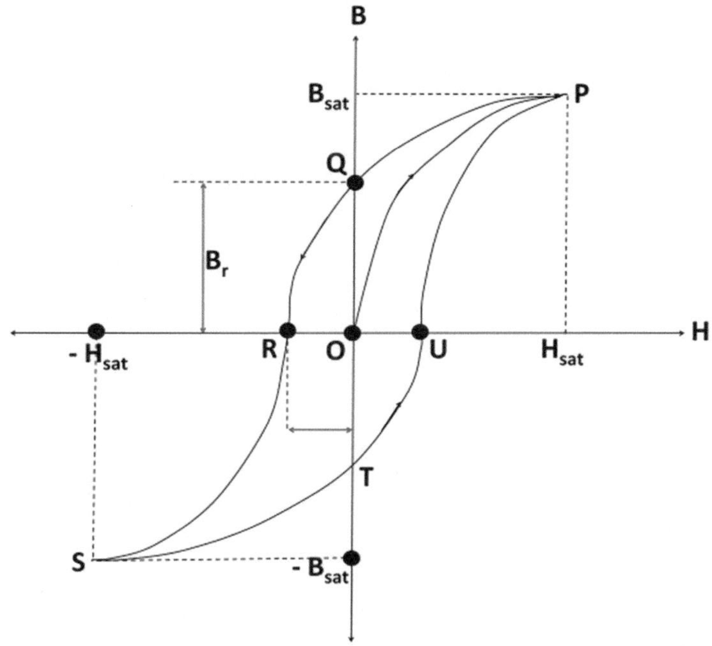

Fig. 4.7 Magnetic hysteresis loop in B–H curve of a typical magnetostrictive material

magnetization and coercive force is called magnetically hard materials. The material having very low value of residual magnetization and coercive force is called magnetically soft materials [41].

Magnetostriction in soft magnetic materials is caused by the changes of the total free magneto-mechanical energy of the material sample due to the transition from the paramagnetic to ferromagnetic state, during the cooling of the material and overturning of the Curie temperature. On the basis of magneto-crystalline anisotropy analyses, it may be stated that the dependence of magnetostriction (λ) on the magnetization (M) of magnetic material may be described by the fourth-order polynomial (Eq. 4.4):

$$\lambda(M) = a_1 M^2 + a_2 M^4 \quad (4.4)$$

This dependence is commonly reduced to the second-order polynomial [16]:

$$\lambda(M) = \frac{3}{2}\frac{\lambda_s}{{M_s}^2} M^2 \quad (4.5)$$

where λ_s is the saturation magnetostriction and M_s is saturation magnetization. In the case of soft magnetic materials, where relative permeability > > 1, Eq. 4.5 may be presented as:

$$\lambda(B) = \frac{3}{2}\frac{\lambda_s}{{M_s}^2} B^2 \quad (4.6)$$

where, B and B_s are flux density in the material and saturation flux density of the material, respectively. Such a form of the magnetostrictive characteristic model is more convenient for technical applications. For this reason, Eq. 4.6 was successfully used for the technical modeling of magnetostrictive actuators, especially with cores made of giant magnetostrictive materials, such as Terfenol-D. On the other hand, analysis of experimental results of measurements of $\lambda(B)$ hysteresis loop clearly indicates that accurate modeling of these characteristics requires consideration of $\lambda(B)$ hysteresis (which is different than B(H) hysteresis), as well as the so-called lift-off phenomenon [42]. The "lift-off" phenomenon is connected with the fact that, during the magnetization loop, magnetostriction never comes back to the value observed in the demagnetized state. The physical origins of hysteresis on the $\lambda(B)$ relation are connected with the interaction between residual stresses and magnetostrictive strain. In previous research, it was connected with the hysteretic magnetization equal to the difference between total magnetization (M) and hysteretic magnetization (M_{an}). Another approach to this hysteresis was based on the hyperbolic Bessel functions, connecting the magnetostriction with the efficient magnetizing field (H_e) in the magnetic material (eq. 4.7) [43]:

$$H_e = H + \alpha M \quad (4.7)$$

where H is the magnetizing field, whereas α is the interdomain coupling accordingly to the Bloch theory. The "liftoff" phenomenon is connected with the fact that, during the magnetization process, the $\lambda(B)$ characteristic never returns to zero. The physical origins of this effect are not clearly explained; however, it has been observed in experimental measurements of the magnetostrictive hysteresis loops $\lambda(B)$and $\lambda(H)$ of most ferromagnetic materials.

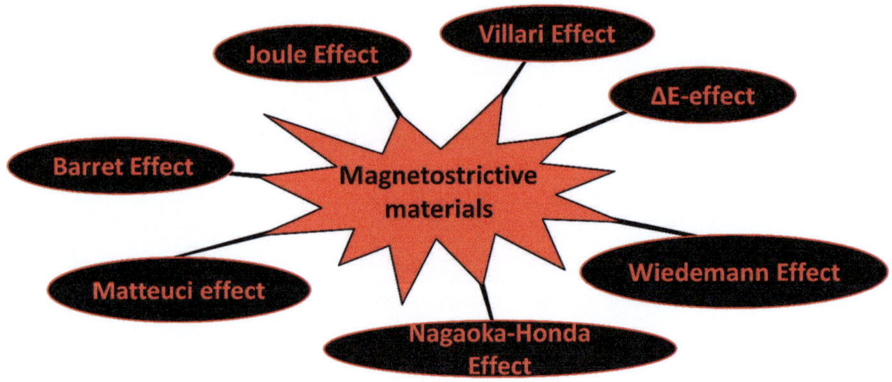

Fig. 4.8 Various effects associated with the magnetostrictive materials in the specific application

Known previous approaches to quantitative modeling of the "liftoff" phenomenon were focused on the reproduction of the shape of magnetostrictive hysteresis loops considering this phenomenon. The most important problem connected with modeling the magnetostrictive hysteresis loops $\lambda(B)$and $\lambda(H)$ is the fact, in the case of some soft magnetic materials, that local maxima occur on these dependencies. These local maxima were observed in experimental results; however, the quantitative model of such a magnetostrictive hysteresis loop was never presented before. Lack of such a model is a significant barrier for understanding the physical background of magneto-mechanical effects, as well as for the practical description of the behavior of ferromagnetic materials required for, for example, the development of transformers or nondestructive testing of elements made of constructional steels [44].

4.5.11 Inductance

Faraday's law states in Eq. 4.8:

$$E = -N \times \frac{d\varnothing}{dt} \tag{4.8}$$

where E is volts, N is the number of turns, $\varnothing$ is magnetic flux (Wb), and t is time. Inductance and current can be extracted from Faraday's law as Webers are equivalent to henry amps. Equation 4.9 is Faraday's law converted to henrys and amps.

$$V = L\frac{di}{dt} \tag{4.9}$$

4.6 Magnetostrictive Effects

Crystals of ferromagnetic materials change their shape when they are placed in a magnetic field. This phenomenon is called magnetostriction. It is related to various other physical effects. Magnetostriction is, in general, a reversible exchange of energy between the mechanical form and the magnetic form. The ability to convert an amount of energy from one form into another allows the use of magnetostrictive materials in actuator and sensor applications [45]. Figure 4.8 shows various physical effects that are related to magnetostrictive materials.

4.6.1 Joul Effect

The most understood effect which is related to magnetostriction is the Joule effect. As the applied magnetic field increases, magnetic domains tend to rotate toward the field direction, resulting in a dimensional change (Fig. 4.9) [46]. The field-induced strain, or magnetostriction, includes the longitudinal component and the transverse component. This is the expansion, positive magnetostriction, or contraction, negative magnetostriction, of a ferromagnetic rod in relation to a longitudinal magnetic field. This effect is mainly used in magnetostrictive actuators. Magnetostriction is a reversible material feature. In the absence of a magnetic field, the sample shape returns to its original dimensions. The ratio of DL/L in Terfenol-D is in the range of more than 1500 ppm and can be up to 4000 ppm at resonance frequency [47]. The increase in length (longitudinal strain) or the contraction of diameter (lateral strain) is roughly proportional to the applied magnetic field and this can be used for various purposes in an actuator mechanism.

The mechanical strain can be calculated as (Eq. 4.10),

$$S = C^H + d_\sigma H \tag{4.10}$$

where $c^H \rightarrow$ coefficient at constant field strength H and d_σ is the magnetostrictive constant at constant stress. The magnetic field strength, $H =$ Current (I) $\times$ Number of coilturns (N). Due to the fact that the axis of a typical bar-shaped sample is usually in line with the direction of magnetization only the

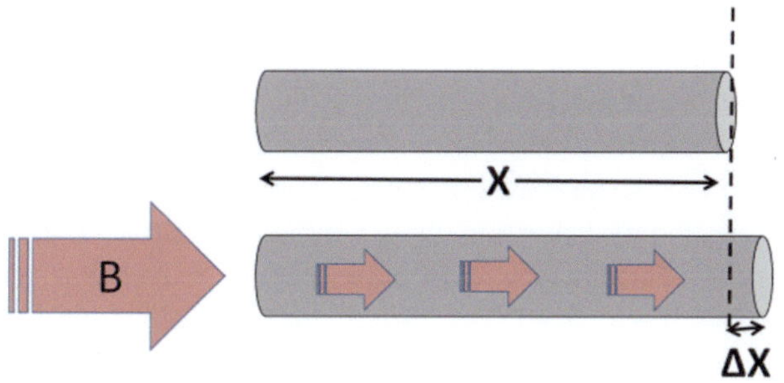

Fig. 4.9 Joul effects for actuation

axial component needs to be considered. Therefore d, μ, c can be treated as scalar quantities for simplification [48].

4.6.2 Villari Effect

Another widely utilized effect related to magnetostriction is the Villari Effect. As the mechanical load on the magnetostrictive material increases, all magnetic domains tend to be aligned in the basal plane perpendicular to the load and there is a change in the magnetic flux density (Fig. 4.10) [49]. The domain rotation induces magnetization change or known as the Villari effect. The stress-induced magnetization variation can eventually generate electrical energy on coils wound around the magnetostrictive materials following Faraday's law. Notable damping coefficients are available by dissipating the electrical energy on a shunt circuit. The Villari effect can also induce significant eddy current loss in electrically conductive magnetostrictive materials. The stress-induced eddy currents provide a compact damping mechanism that converts mechanical energy to Joule heat without introducing bulky shunt circuits [50]. The change in flux density by the mechanical load can be detected by a pickup coil and is proportional to the level of the applied stress. The Villari effect is reversible and is used in sensor applications. It can be analyzed using Eq. (4.11).

The magnetic induction,

$$B = d\sigma + \mu^{\sigma} \tag{4.11}$$

where d is magnetostrictive constant, σ is stress change, and μ^{σ} is permeability at constant mechanical stress σ. In a magnetostrictive application, the physical parameters described above do not remain constant during the operations, as such is the inverse of the Joule magnetostriction. Furthermore, the Villari effect exhibits many of the attributes of the direct magnetostrictive effect in as much as its physical origin also lies in the magnetoelastic coupling. The Villari effect has been the object of much study given its relevance in applications such as nondestructive evaluation and sensing [51].

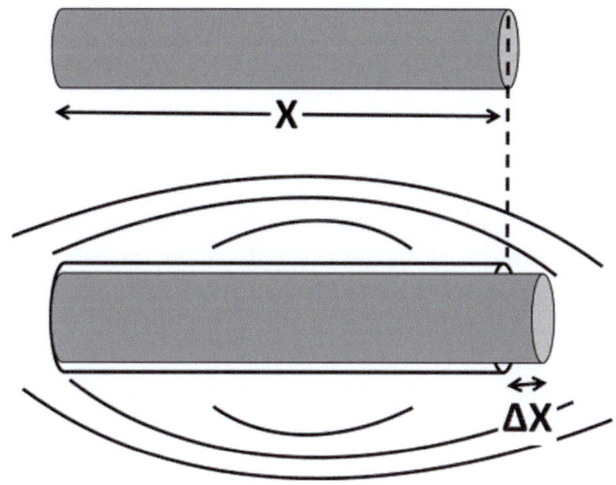

Fig. 4.10 Villari effect for sensing/harvesting

4.6.3 ΔE Effect

The **Δ**E-effect is another effect related to magnetostriction. It is the change of the Young's Modulus E due to a change in the magnetic field. Due to the change of Young's modulus, there is a change in the velocity of sound inside magnetostrictive materials, and this can be observed in tuneable vibration and broadband sonar systems [52]. The Young's modulus of magnetostrictive material is the inverse of the slope of the strain versus stress curve, which is estimated by piecewise linearization. The elasticity of magnetostrictive materials is composed of two separate but related attributes, namely the conventional stress-strain elasticity arising from interatomic forces and the magnetoelastic contribution due to the rotation of magnetic moments and ensuing strain which occur when stress is applied. This is known as the ΔE effect and is quantified by Eq. (4.12):

$$\Delta E = \frac{E_S - E_O}{E_O} \tag{4.12}$$

where E_0 is the minimum elastic modulus and E_S is the elastic modulus at magnetic saturation. Because the strain produced by magnetic moment rotation adds to the non-magnetic strain, the material becomes softer when the moments are free to rotate. Note that the material becomes increasingly stiffer as saturation is approached and magnetic moment mobility decreases. The ΔE effect is small in nickel ($\Delta E = 0.06$), but is quite large in Terfenol-D (ΔE up to 5) and certain transverse-field annealed $Fe_{81}B_{13.5}Si_{3.5}C_2$metglas ribbons ($\Delta E = 10$) [53].

4.6.4 Wiedemann Effect

Another effect related to magnetostriction is the Wiedemann effect. The physical background to this effect is similar to that of the Joule effect, but instead of a purely tensile or compressive strain forming as a result of the magnetic field, there is a shear strain which results in a torsional displacement of the ferromagnetic material (Fig. 4.11) [54]. A current-carrying ferromagnetic or amorphous wire will produce a circular magnetic field in a plane perpendicular to the wire and the moments will align predominantly in the circumferential direction. When an axial magnetic field is applied, some of the moments align in a helical fashion creating a helical magnetic field. The twist observed in the wire is called the Wiedemann effect [55].

The twisting of a ferromagnetic rod through which an electric current is flowing when the rod is placed in a longitudinal magnetic field. It was discovered by the German physicist Gustav Wiedemann in 1858 [56]. The Wiedemann effect is one of the manifestations of magnetostriction in a field formed by the combination of a longitudinal magnetic field and a circular magnetic field that is created by an electric current. If the electric current (or the magnetic field) is alternating, the rod will begin torsional oscillation. In linear approach angle of rod torsion α does not depend on its cross-section form and is defined only by current density and magnetoelastic properties of the rod (Eq. (4.13) [57]:

$$\alpha = j\frac{h_{15}}{2G} \tag{4.13}$$

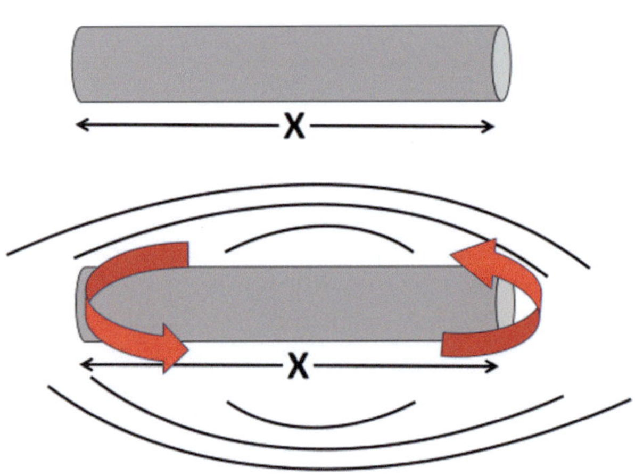

Fig. 4.11 Wiedemann effect for actuation/torque

where j is current density, h_{15} is magnetoelastic parameter that is proportional to the longitudinal magnetic field value, G is the shear modulus.

Magnetostrictive position sensors use the Wiedemann effect to excite an ultrasonic pulse. Typically, a small magnet is used to mark a position along a magnetostrictive wire. The magnetic field from a short current pulse in the wire combined with that from the position magnet excites the ultrasonic pulse. The time required for this pulse to travel from the point of excitation to a pickup at the end of the wire gives the position. Reflections from the other end of the wire could lead to disturbances. In order to avoid this, the wire is connected to a mechanical damper at that end [58].

4.6.5 Matteucci Effect

The inverse Wiedemann effect, known as the Matteucci effect, is the change in axial magnetization of a current-carrying wire when it is twisted [59]. It is observable in amorphous wires with helical domain structure, which can be obtained by twisting the wire, or annealing under twist. Alternating current fed to a coil creates a longitudinal magnetic field in a sample, and this, in turn, creates a magnetic flux density in the sample. The presence of the alternating magnetic flux can be detected by another coil, a pick-up coil which measures the rate of change in the magnetic flux density. Twisting the ferromagnetic sample induces a change in the magnetization of the sample, which results in a change in the rate of change of the magnetic flux density. By detecting the magnetization change using the pickup coil, the change in shear stress can be evaluated and as a result, the magnitudes of the applied torque can be calculated. The Matteucci effect is modified by introducing a permanent magnetic bias in the ferromagnetic sample and this is used in sensor applications [60].

4.6.6 Barret Effect

In certain extreme operational conditions, the volume of the material may change in response to a magnetic field. This effect is called Barret effect or volume magnetostriction. This volume change in response to a magnetic field and is so small that it can be neglected under normal operational conditions. The effect has little applicability in smart structure systems. For instance, while the magnetostriction curve of nickel rapidly reaches -35×10^{-6} at only 10 kA/m, the fractional volume change is only 0.1×10^{-6} at a much larger 7 field of 80 kA/m. In the alloy Invar (36% Ni–64%Fe) the fractional volume change at the Curie temperature, which is slightly above room temperature, compensates the intrinsic thermal expansion giving a compound with a nearly zero thermal expansion at room temperature [61].

4.6.7 Nagaoka–Honda Effect

The Nagaoka–Honda effect is the inverse of the Barret effect. This effect is the change of magnetic state caused by a change in the volume of a sample because of hydrostatic pressure. Due to the extreme operational conditions required to make it possible to detect these effects connected with the volume change, they have not found wide use in the industry [60].

4.7 Materials for Magnetostrictive Effects

The efficiency of a magnetostrictive vibrator as an electro-acoustic transducer depends not only upon the material used but also on such factors as the size and the kind of laminations used. Generally, the elements used in magnetostrictive materials are Iron, Nickel, Cobalt, Terbium, Dysprosium, Samarium, Gadolinium, Holmium, Erbium, Thulium, Metglas, etc. Some of the magnetostrictive materials have been discussed below:

4.7.1 Iron-Based Alloys

The effective combination of large spin-orbit coupling of heavy elements, the strong spin polarization of Fe, nonbonding states around the Fermi level, and an adjustable number of electrons, are of great importance to improve magnetostriction of transition-metal systems. As a potential magnetostrictive material, Fe-Al alloys exhibit excellent mechanical properties, low cost, and moderate magnetostriction, but the magnetostriction mechanism is still a mystery. The dependence of the room temperature magnetostriction is there on substituting Fe by different X elements (X = Al, Ga, Si, Ge) [62]. All Fe-X alloys exhibit at the Fe-rich side a similar phase diagram. The magnetostriction increases with increasing substitution exhibiting a first maximum close to 20 at.% for Ga and Al, 10 at. % for Ge, and 5 at. % for Si. The reported magnetostriction behavior of Fe-Ga is similar to that of Fe-Al alloys, while the magnetostriction behavior of Fe-Ge is similar to that Fe-Si alloys [63]. However, for all these alloys, the increase of magnetostriction is observed in disordered A_2 structures. Also with Tb doping into the Fe-Al alloy results in fivefold magnetostriction. Galfenol, Fe_xGa_{1-x}, and Alfenol, Fe_xAl_{1-x}, are newer alloys that exhibit 200–400 microstrains at lower applied fields (~200 Oe) and have enhanced mechanical properties from brittle Terfenol-D [64].

Fe-Co alloys, being magnetostrictive materials, are also suitable for energy harvesting applications due to their rich elements and lower cost compared with Terfenol-D and Galfenol. Fe-27% Co alloy is a good candidate for onboard power transformers core because of its very high magnetization saturation level and the size reduction of transformers that it can offer. This alloy exhibits unfortunately high magnitude magnetostriction constants leading to an unacceptable level of acoustic noise in operation when classical annealing operations are employed during the forming process. The cobalt ferrite is substituted with transition metals (such as Mn, Cr, Zn, and Cu or mixtures thereof), which provides mechanical properties for the materials effective for use as sensors and actuators. The substitution of transition metals lowers the Curie temperature of the material (as compared to cobalt ferrite) while maintaining suitable magnetostriction for stress-sensing applications [65].

An excellent giant-magnetostrictive material, Tb-Dy-Fe alloys (based on $Tb_{0.27-0.30}Dy_{0.73-0.70}Fe_{1.9-2}$ Laves compound) can be applied in sonar transducer systems, sensors, and micro-actuators. Nowadays, there are two different ways to substitute high-cost Tb and Dy alloying elements. One is to partially replace Tb or Dy with cheaper rare earth elements, such as Pr, Nd, Sm, and Ho; and the other is to use non-rare earth elements, such as Co, Al, Mn, Si, Ce, B, Be, and C, to substitute Fe to form single $MgCu_2$-type. Cobalt exhibits the largest room-temperature magnetostriction of a pure element at 60 microstrains. Cobalt ferrite ($CoO{\cdot}Fe_2O_3$) has high saturation magnetostriction (~200 parts per million) [66].

4.7.2 Ni-based Alloys

Ni-Fe alloys are among the most important soft magnetic alloy systems. Ni-Fe alloys are of particular interest because a broad variety of qualitatively different magnetic properties can be obtained by adjusting the composition and the preparation process. Ni-rich alloys are called Permalloys. The most exploited Ni-Fe alloy corresponds with 80% Ni and 20% Fe content. Both the crystal anisotropy and the magnetostriction cross the zero value for compositions around 80 wt.% Ni. The high magnetic permeability exhibited by this material makes it interesting for magnetic cores and magnetic shielding applications [67]. In addition, the low coercivity, negligible magnetostriction, and large anisotropic magnetoresistance, together with the possibility of reducing its size at the nanoscale, favor a wide range of applications. Treatments on materials with approximately this composition (and the deliberate introduction of other elements like Mo, Cu, and Cr) yield materials with extremely low hysteresis losses (coercive field less than 1 Am^{-1}), and a variety of hysteresis loop shapes controlled by induced anisotropy. The combination of low hysteresis losses and low classical losses makes these materials most suited to applications at high frequencies. Ni-Zn, Ni-Cu, Ni-Zn, Ni-Cu-Co, Ni-Cu-Zn-Co systems, and so on, also exhibit significant magnetostrictive properties [68].

4.7.3 Terfenol-D

Terfenol-D, an alloy of the formula $Tb_xDy_{1-x}Fe_2$ ($x \approx 0.3$), is a giant magnetostrictive material. Due to its material properties, Terfenol-D is excellent for use in the manufacturing of low frequency, high-powered underwater acoustics. Its initial application was in naval sonar systems. Among alloys, the highest known magnetostriction is exhibited by Terfenol-D, exhibits about 2000 microstrains in a field of 160 kA/m (2 kOe) at room temperature, and is the most commonly used engineering magnetostrictive material. Terfenol-D is used in active noise and vibration control systems, low-frequency underwater communications (sonar), linear and rotational motors, ultrasonic cleaning, machining and welding, micropositioning, and the detection of motion, force, and magnetic fields. Terfenol-D is currently available in a variety of forms, including monolithic rods, particle-aligned polymer matrix composites, and thin films [69]. The alloy has the highest magnetostriction of any alloy, up to 0.002 m/m at saturation. Terfenol-D has a large magnetostriction force, high energy density, low sound velocity, and a low Young's modulus. At its most pure form, it also has low ductility and low fracture resistance. Terfenol-D is a gray alloy that has different possible ratios of its elemental components that always follow a formula of $Tb_xDy_{1-x}Fe_2$. The addition of dysprosium made it easier to induce magnetostrictive responses by making the alloy require a lower level of magnetic fields. When the ratio of Tb and Dy is increased, the resulting alloy's magnetostrictive properties will operate at temperatures as low as −200 °C, and when decreased, it may operate at a maximum of 200 °C. The composition of Terfenol-D allows it to have a large magnetostriction and magnetic flux when a magnetic field is applied to it. This case exists for a large range of compressive stresses, with a trend of decreasing magnetostriction as the compressive stress increases. Magnetic heat treatment is shown to improve the magnetostrictive properties of Terfenol-D at low compressive stress for certain ratios of Tb and Dy [70].

4.7.4 Metglas

A new magnetostrictive material was introduced in 1978 which is based on amorphous metal, produced by rapid cooling of Fe, Ni, and Co alloys together with one or more of the elements Si, B, P. These alloys are known commercially as Metglas (metallic glass) and are commonly produced in thin-ribbon geometries [71]. Because of the extremely high coupling coefficients ($k > 0.92$), Metglas is a prime candidate for sensing applications in which a mechanical motion is converted into an electrical current or voltage. Metallic glass alloys such as metglas 2826 MB based amorphous alloys are the best-known candidates for magnetostrictive sensors because these amorphous alloys exhibit large saturation magnetostriction, high saturation magnetization, low anisotropy energies, and low coercivity. At present, these alloys are available only in the form of ribbons of thickness ranging from 10 to 50 μm. A series of post-treatment process, such as high-temperature annealing and epoxy treatment, is further required for amorphous alloy ribbons to be used as sensors. Therefore there are many difficulties in fabricating systems based on amorphous ribbons for micro-sensor applications. Magnetoelastic materials in the form of thin films are an alternative to ribbons and they can be integrated easily in micro/nano-electromechanical systems. This not only allows the miniaturization of sensor elements but also enables the same micro-fabrication technologies to be used in the production of both electronic and magnetic devices. In comparison with other MEMS technologies, for example, those incorporating piezoelectric materials, Magnetic MEMS offer a high power density, low-performance degradation, fast response times, and ease of fabrication. Thin films based on metallic glasses can be prepared by techniques such as thermal evaporation, flash evaporation, electrodeposition, molecular beam epitaxy, pulsed laser deposition, and sputtering. Vapor deposition offers a simple alternative to usual sputter deposition in obtaining thin films of supersaturated solid solutions and other metastable states [72]. The amorphous alloy $Fe_{81}Si_{3.5}B_{13.5}C_2$ has a high saturation-magnetostriction constant, λ, of about 20 microstrains and more, coupled with low magnetic-anisotropy field strength, H_A, of less than 1 kA/m (to reach magnetic saturation). It also exhibits a very strong ΔE-effect with reductions in the effective Young's modulus up to about 80% in bulk. This helps build energy-efficient magnetic MEMS [73].

4.7.5 Ferromagnetic Shape Memory Alloys

The ferromagnetic shape memory alloys are another class of smart materials that hold much promise due to the large strains that they can provide. While the Ni-Ti alloy commercially known as Nitinol features large recoverable strains in the order of $60{,}000 \times 10^{-6}$, it suffers from inferior dynamical response [79]. The possibility of combining the desirable aspects of shape memory with magnetostriction through actuating an SMA with a magnetic field is currently being investigated. The Heusler alloy, Ni_2MnGa, can show giant magnetostrictive effect as well as a shape memory effect. Shape memory alloys, NiFeGa, NiTiHf, FePd, and Fe_3Pt, and so on also exhibit a magnetostriction effect [74].

4.7.6 Other Materials

We report on the magnetostriction of hexagonal $HoMnO_3$ and $YMnO_3$ single crystals in a wide range of applied magnetic fields (up to $H = 14$ T) at all possible combinations of the mutual orientations of magnetic field and magnetostriction. It is found that the nonmonotonic behavior of magnetostriction of the $HoMnO_3$ crystal is caused by the Ho^{3+} ion; the magnetic moment of the Mn^{3+} ion parallel to the hexagonal crystal axis. The anomalies established from the magnetostriction measurements of $HoMnO_3$ are consistent with the phase diagram of these compounds. X_6Fe_{23}, $X =$ (Tb, Dy, Ho, and Er) compounds exhibited large magnetostrictions at 77 K. Among these compounds, the largest effect being obtained for Tb_6Fe_{23}. In Er_6Fe_{23}, the main source of magnetostriction in this compound is the Er ion. YCo_5, $SmCo_5$, $GdCo_5$, $LaNi_5$, and $GdNi_5$ systems are involved in the magnetostrictive effect. It is reported that the linear thermal expansion and magnetostriction effect are shown in $CeRu_2Si_2$ in magnetic fields up to 52.6 mT and at temperatures down to 1 Mk [75, 76].

4.8 Material Behavior

4.8.1 Magnetic Anisotropy

Magnetic anisotropy refers to the dependency of magnetic properties on the direction in which they are measured. It can be of several kinds, including crystal, stress, shape, and exchange anisotropy. Of these, however, only crystal anisotropy is a material property. In many crystalline materials, the magnetic moments do not rotate freely in response to applied fields, but rather they tend to orient along preferred crystallographic directions. This phenomenon is called the magnetocrystalline or crystal-anisotropy, and the associated anisotropy energy is the energy required to rotate the magnetic moments away from their preferred directions. Crystal anisotropy energy and magnetostriction are closely related effects; if the anisotropy was independent of the state of strain, there would be no magnetostriction. While accurate models for crystal anisotropy and its relation with the magnetization process exist for the simple cases of cubic and hexagonal crystals, models for complex crystal structures often rely on simplifying assumptions which reduce the analysis to the simpler cases [77].

4.8.2 Mechanical Behaviors

Mechanical impedance matching of the harvester is an important property of magnetostrictive materials for kinetic energy harvesting. Materials like Terfenol or Galfenol are quite rigid with a mechanical behavior near that of bulk iron [85]. The availability of softer magnetoelastic materials would enable vibration harvesting with lower stresses and higher strains, in the 0.1–1% range, with a rubber-like behavior. Metglas has been proposed for energy harvesting, has the advantage that it can be laminated to achieve a higher energy density. The material is a Fe-based amorphous ribbon with excellent magnetic softness and elastic response, and its production is in principle cheaper than Fe-Ga and Fe-Tb-Dy alloys [78]. Single crystal alloys exhibit superior microstrain but are vulnerable to yielding due to the anisotropic mechanical properties of most metals. It has been observed that for polycrystalline alloys with a high area coverage of preferential grains for microstrain, the mechanical properties (ductility) of magnetostrictive alloys can be significantly improved. Targeted metallurgical processing steps promote abnormal grain growth of {011} grains in Galfenol and Alfenol thin sheets, which contain two easy axes for magnetic domain alignment during magnetostriction. During subsequent hot rolling and recrystallization steps, particle strengthening occurs in which the particles introduce a "pinning" force at grain boundaries that hinders normal (stochastic) grain growth in an annealing step. Thus, single-crystal-like texture (~90% {011} grain coverage) is attainable, reducing the interference with magnetic domain alignment and increasing microstrain attainable for polycrystalline alloys as measured by semiconducting strain gauges [79].

4.9 Kinetics in Magnetostrictive Operation

Figure 4.12a showing an external magnetic field causes the rotation of magnetic domains in a ferromagnetic material, resulting in Joule effect or Villari effect. In a typical ferromagnetic material, the relation between the applied magnetic field H and the relative length change, that is, strain ($S = \Delta L/L$), looks like "S" curve as shown in Fig. 4.12b. One peculiar phenomenon observed in the curve in Fig. 4.12b is that S has the same sign regardless of the sign of H; furthermore, the curves for most ferromagnetic materials are usually highly nonlinear [80]. These issues must be properly addressed in designing a magnetostrictive transducer. To generate mechanical waves in a ferromagnetic material using the Joule effect, a dynamic magnetic field (H_D) should be applied to a ferromagnetic material. If we assume that H_D oscillates around O at a frequency of ω_0, the dominant frequency of the generated mechanical wave would be $2\omega_0$, because the sign of the generated wave is always the same. To achieve a linear response, an H_D value between $-\Delta H_D$ and $+\Delta H_D$ should be applied while the total field in a ferromagnetic material should be centered at point B that is located away from O; therefore, an appropriate static bias magnetic field H_S needs to be applied. To ensure linear behavior between the applied magnetic field and generated strain, ΔH_D must be sufficiently smaller than H_S. Optimal values of ΔH_D and H_S for efficient

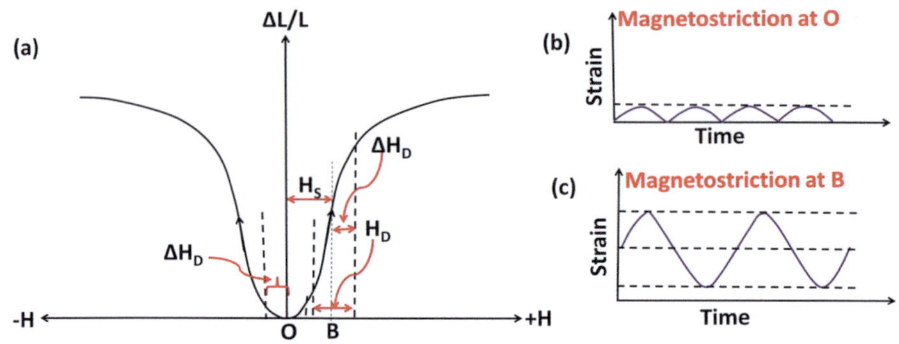

Fig. 4.12 (**a**) Schematics of typical magnetostrictive curve-related magnetic field (H) and mechanical strain (ΔL/L), the strain histories when time-vary magnetic fields (H_D) are applied at (**b**) O and (**c**) B, respectively

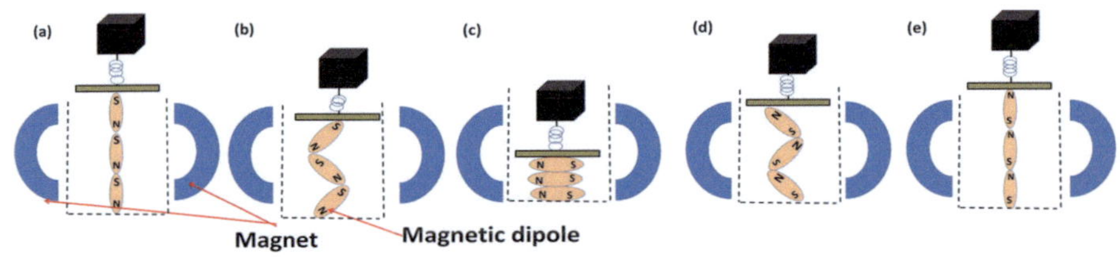

Fig. 4.13 (**a**) Schematics of magnetostrictive transducer operation

transduction may be found for selected magnetostrictive materials, desired wave modes, and transducer configurations. To increase transduction efficiency, compound materials such as Terfenol-D and Hyperco 50Aalloy (FeCoV soft magnetic alloy; saturation magnetostriction = 60–70 ppm) may be used, but Terfenol-D appears too brittle to be used as a base material for magnetostrictive patches in magnetostrictive patch transducers (MPTs) [81].

4.10 Potential Applications

We know that every substance is vulnerable to heat. When heat is applied, they tend to expand and when heat is removed, they tend to contract. In addition, magnetostrictive material can also expand or contract when the magnetic field is applied. Mostly ferrimagnetic and ferromagnetic materials exhibit magnetostrictive properties. Most of the ferromagnetic and ferrimagnetic materials can be magnetized under Curie point. The magnetostrictive property of the material is used in many devices such as hydrophones, torque meters, sonar scanning devices, telephone receivers, and so on, with the invent of "giant" magnetostrictive alloys nowadays it is being used in making devices like pumps, active vibration or noise control systems, high force linear motors, positioners for adaptive optics, sonar, chatter control of boring tools, medical and industrial ultrasonic, high-precision micropositioning, geological tomography, borehole seismic sources, hydraulic valves for fuel injection systems, hydraulic pumps, bone-conduction hearing aids, exoskeletal telemanipulators, deformable mirrors, self-sensing actuators, degassing in manufacturing processes such as rubber vulcanization, and industrial ultrasonic cleaning, etc. Ultrasonic magnetostrictive transducers have also been developed for making underwater sonar, surgical tools, chemical and material processing. Various emerging applications of magnetostrictive materials are described subsequently.

4.10.1 Magnetostriction in Mechanical Industries

Magnetostrictive transducers for lifting: Magnetostrictive transducers are used to convert mechanical energy into magnetic energy and vice versa. Due to bidirectional coupling between magnetic state and mechanical state of the material, the transducers can be utilized for sensor and actuator. The devices can also be called as electromechanical devices. Generally in magnetostrictive elements, current is passed through a conductive wire. The magnetic field is created by the passage of electric current. Moreover, in this mechanism, measuring current induced by a magnetic field indicates the magnetic field strength. The below Fig. 4.13 illustrates the working of magnetostrictive materials. The different figures explain the amount of strain produced from null magnetization to full magnetization [82]. The magnetostrictive effect is generated by the rotation of magnetic moments. In magnetostrictive elements, motion takes place only in one direction and it vibrates at twice the frequency of the drive mechanism. In majority, this transducer consists of a preload mechanism consisting of a bolt, a barrel-like permanent magnet that provides the bias magnetization, a spring washer, a cylindrical magnetostrictive rod, a surrounding copper-wire solenoid, and magnetic couplers [83].

Figure 4.13 depicts the magnetostrictive device where force may be applied. It has five subfigures. Figure 4.13 (c) depicts null magnetism when there is no external magnetic field is applied. Hence, there is no deformation in the magnet at all due to the null induction of magnetism. The magnetic field can vary from null to saturation value and the saturation value can vary from positive maximum saturation ($+H_{sat}$) to negative minimum saturation ($-H_{sat}$). Hence the value of magnetism also varies from $+B_{sat}$ to $-B_{sat}$ [96]. If we try to increase the magnetic field the magnetism cannot increase beyond saturation value. At maximum saturation magnetism $+B_{sat}$ the magnetic domains align in such a way that the material deforms the maximum as shown in Fig. 4.13 (e). Similarly at minimum saturation $-B_{sat}$ also the material deforms the maximum but the domains align themselves in the opposite direction as shown in Fig. 4.13(a). When the magnetism value lies in between zero to maximum or minimum, the magnetic dipoles are arranged in such a way that there is intermediate deformation in the material as shown in Fig. 4.13 (b) and (d). In other words, if we will apply force in the material as given in the figure then the magnetic dipole orientation may also change leading to a change in magnetism. This magnetism can be measured through the principle of Hall effect or by calculating the voltage produced in a conductor kept at the right angle to the flux produced. The resulting magnetism will be proportional to applied force [84].

Ultrasonic transducers: The magnetostrictive material changes its volume when subjected to magnetism as domain alignment takes place under the influence of the magnetic field. The volume change may increase or decrease depending upon the material (Fig. 4.14). Therefore, the magnetostrictive material finds applications in ultrasonic transducers. Ultrasonic material using magnetostriction is common nowadays which utilizes kinetic energy or mechanical energy as the source. Magnetostrictive ultrasonic transducers are used in ultrasonic cleaning and other processing applications [85].

Hybrid magnetostrictive/piezoelectric devices: Hybrid magnetostrictive materials are used in sonar and other applications. Hybrid devices consist of both piezoelectric and magnetostrictive materials. The piezoelectric materials are capacitive and magnetostrictive materials are inductive. So by combining both the effects superior quality can be obtained. By combining both a resonant electric circuit is formed. When driven at resonance, such a device behaves like a purely resistive load and only the energy effectively converted to mechanical motion or lost to inner losses needs to be supplied externally. This greatly simplifies amplifier design and helps for attaining high efficiencies [86]. Hybrid device has been introduced to overcome the difficulties involved in achieving motion at only one end of a Tonpilz piston-type sonar transducer. In the hybrid device, a quarter wavelength stack of piezoelectric Navy type I ceramic rings joined to a quarter wavelength Terfenol-D composite tube. The 90° shifting in the device between the piezoelectric process and the magnetostrictive process leads to motion at one end and cancellation at the other end. While the device is mechanically unidirectional, it becomes acoustically unidirectional only under array-baffled operation. The measured front-to-back pressure ratio is 5 dB for the device alone and 15 dB under array-loaded conditions. The concept of hybrid piezoelectric/magnetostrictive transduction has been also implemented for linear inchworm motors and rotational motors. The intrinsic 90° phase lag between the two types of elements provides natural drive timing for the inchworm, while the direction of motion can be easily reversed by changing the magnetic bias on the Terfenol-D elements. This motor achieves a zero-load speed of 25.4 mm/sec and a stall load of 115 N [87]. A hybrid magnetostrictive/piezoelectric rotational motor following the proof-of-concept transducer presented in. A piezoelectric stack clamps a piece of friction material onto the rotating disk, while two magnetostrictive rods move the clamp tangentially to the disk to produce the rotational motion. The sequence of the motion is, as indicated before, determined by the natural timing of the piezoelectric and magnetostrictive response. The device produces a speed of 4 rpm at excitation voltages of between 30 and 40 V and frequencies of between 650 and 750 Hz [16].

Sensor applications: The magnetostrictive mechanism can be incorporated into the design of sensors. The kind of sensor can detect the mechanical displacement, force, torque, and so on by considering electric or magnetic behavior, and so on. The sensor may produce electrical signals like current, voltage, and so on or may produce the magnetic field in response to mechanical activity. In magneto-mechanical sensors, the change in magnetoelastic property changes the anisotropy of the material. For completion of the sensing mechanism, a pickup coil is wrapped around the magnetostrictive material to detect the magnetization changes.

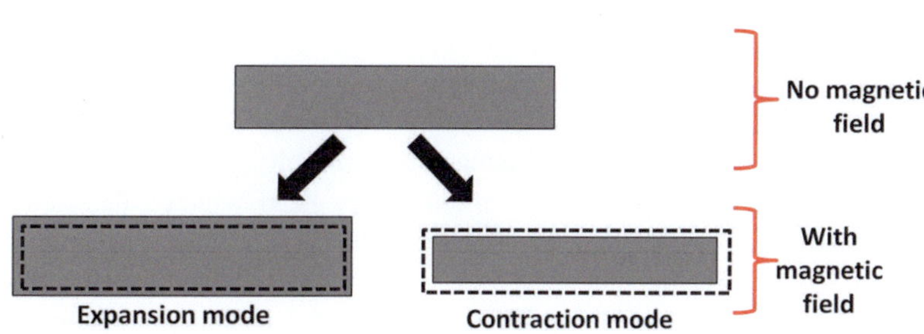

Fig. 4.14 Volumatic change of magnetostrictive material as the magnetic field applies to it

This in turn converts the magnetic field to the electrical signal. The principle that links the magnetization in the material with the voltage V generated across a pickup coil is the Faraday–Lenz law of electromagnetic induction, V = −N.A.Db.dt, in which N and A are, respectively, the number of turns and constant cross-sectional area of the coil, and B is the magnetic induction which quantifies the magnetization state through $B = \mu_0(H + M)$. In some fields, interferometry can also be used in order to detect the change in wave speed when magnetostrictive material changes its properties when external excitation is impended on it [88].

Torque sensors: Magnetostrictive torque sensors can be classified based on torque applied to it. We know that when torque is applied on the shaft, stress may be generated which may vary from $+\tau$ to $-\tau$ oriented at ±45° from the shaft axis. If any magnetostrictive ribbon will be attached to the shaft axis then due to stress magnetic field will be altered on the shaft depending upon the direction of stress. The differential property may be measured by the Hall effect or by any other available mechanism. This particular mechanism can be employed in fly-by-wire steering systems for the automotive industry. In another noncontact type torque meter, the permeability is changed due to variation in torque. The permeability is calibrated in terms of torque. Generally, this kind of mechanism is used in low-sensitive sensors. For example in one kind of sensor incorporated in the drill, the torque is measured by two sensing coils. The coils are connected in series. One coil is placed on the flute and the other on the shank. Due to the applied torque, the permeability of the drill bit changes. Therefore, the differential voltage generated in an excitation coil provides the AC magnetic field excitation [89].

Deformation and position sensors: Magnetostrictive ribbon, which is made up of a transverse annealed process, are very sensitive strain gauges. For example, a strain gauge made from Metglas 2605SC is annealed transversely for 10 mins at 390 °C and rapidly cooled in a saturation field. Magnetic permeability of the ribbon changes which in turn changes the permeability. Hence, the state of strain is changed. This sensor has an F equal to about 250,000, which compares extraordinarily well with resistive strain gauges ($F = 2$) and semiconductor gauges ($F = 250$). The local thermal expansion encounters a problem in magnetostriction as it can saturate the result. The said problem can be eliminated by placing the magnetostrictive with a highly viscous liquid [106]. A position detector can be attached with the magnetostrictive material that acts as an acoustic waveguide. Such a device runs through a waveguide and is connected to a target. There is a damper that prohibits reflection in waves and there is an emitter/receiver head which sends and receives either an acoustic or current pulse down the waveguide. The magnet interacts with the waveguide and changes the property of the material. The material property can be detected in two ways. In one process, the emitter sends a continuous current pulse down the waveguide, which produces a circumferential magnetic field that interacts with the axial field from the magnet. The resulting helical field produces a twist in the wire that travels back to the receiver head. In another method, the stiffness discontinuity produced by the magnet partially reflects back an acoustic pulse sent by the emitter. In both methods, the time consumed between the sent signal and the reflected signal is calibrated to know the location of the magnet along the waveguide. The sensor can measure fluid level by connecting the magnet to a float [90].

Magnetometers: Magnetic field can be reduced to one of the measuring lengths if magnetostriction of material is described as a function of magnetic field and the magnetic length can be measured by either of the device like optic fiber, capacitor, interferometer, and strain gauge. For example, in a magnetometer field annealed metallic glass, the ribbon can be bonded to a resonating PZT plate that is placed in a viscous fluid. An alternating voltage is applied to the PZT plate, which generates a longitudinal stress field. With proper bonding techniques, the dynamic stress in the metallic ribbon is congruent with that in the PZT while the static component is filtered out by the viscous fluid. In another type of magnetometer, both magnetostrictive and piezoelectric materials are bonded together. When the magnetic field is applied to magnetostrictive material, the material strains. Due to straining, the electric voltage is generated in the piezoelectric material. In another type of magnetometer magnetostrictive material is bonded to an optic fiber. By the application of a magnetic field, the magnetostrictive material deforms which in turn deforms the optical fiber. The optical length of the laser beam passing through the optical fiber changes that can be visible by the interferometer. Again, to detect the change in length of Terfenol-D rod a diode laser interferometer can be used when the magnetic field is applied. A maximum sensitivity of 160×10^{-6} A/m can be achieved [91].

Force sensors: Using the Villari effect, it is possible to make a simple and robust force transducer using crystalline or amorphous magnetostrictive materials. A magnetostrictive attribute that provides the principle of operation of a sensor is the dependence of the magnetization on the stress state of the material. The design consists of two magnetostrictive elements, one surrounded by an excitation coil and the other surrounded by a sensor coil and two rigid endplates. In a single mode of operation, an alternating voltage is applied to the excitation coil, which generates a magnetic flux in the sensor and a corresponding voltage in the sensor coil. When a force is applied, the magnetostriction in the elements causes a change in the magnetic flux, which is detected as a proportional change in the voltage in the sensing coil. In the second constant current operation mode, the excitation voltage can be varied to maintain a constant output voltage of the sensing coil. The change in excitation stress is

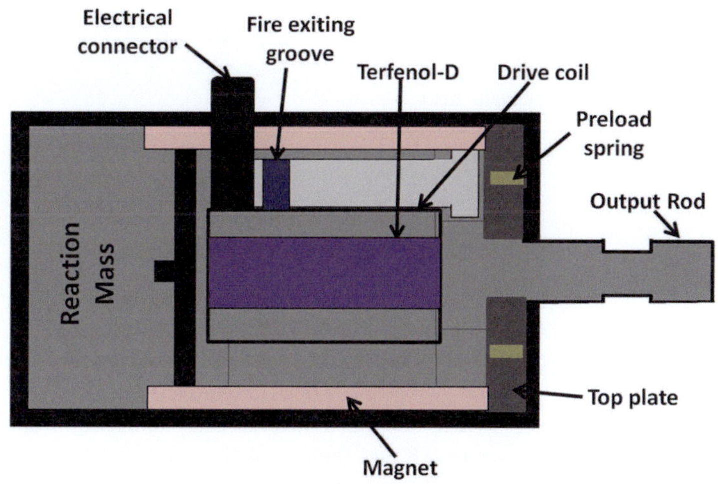

Fig. 4.15 Reaction mass actuator

related to the change in applied force. Compared to conventional force transducers, such as voltage gauge-based sensors, this sensor is simpler, more robust, and requires simpler electronics. A similar Villari effect sensor based on amorphous bands has been discussed in 4.6.2. Many other proposals have been discussed or patented, including impact sensors, pressure sensors, and force sensors based on magnetoelastic stress meters [92].

Reaction mass unit: Etrema has designed, built, and validated an actuator that is capable of generating useful forces, although it can operate over a wide range of frequencies. This particular reaction mass controller was designed to operate with a bandwidth of 150–2000 Hz. It was also able to generate a force of 4000 N or an acceleration of 30 g at a resonant frequency of 635 Hz. Seismic wave generation with this controller and reflection analysis provides an indication of underground and hidden structures and formations. Figure 4.15 shows a cross section of a reaction mass controller [93].

Linear motor based on Terfenol-D: Energen Inc. designed and manufactured a compact linear motor based on the intelligent Terfenol-D material. The central feature of this linear motor is the Terphenol-D rod surrounded by an electric coil which, when switched on, causes the rod to elongate. The drive is mounted between two terminals. By operating the controller and clamps in the correct order, the smart material bar moves forward or backward. Figure 4.16 shows the principle by which functionality is achieved. In the first step, it goes out and both ends are suppressed. Then release one of the clamps by force (step 2) and then, by feeding the actuator coil (step 3), the front end of the rod moves forward due to the Terphenol-D extension. During this extended state, the clamp is closed to hold the rod (step 4). Then the rear clamp is activated to allow movement (step 5). Then, the control spool is gradually turned off and the rear end of the rod moves forward (step 6). In the last sequence, the rear clamp is closed again to achieve full force. This arrangement achieved very precise position control over a few microns over a total length of 20 mm. The force capability was up to 3000 N [94].

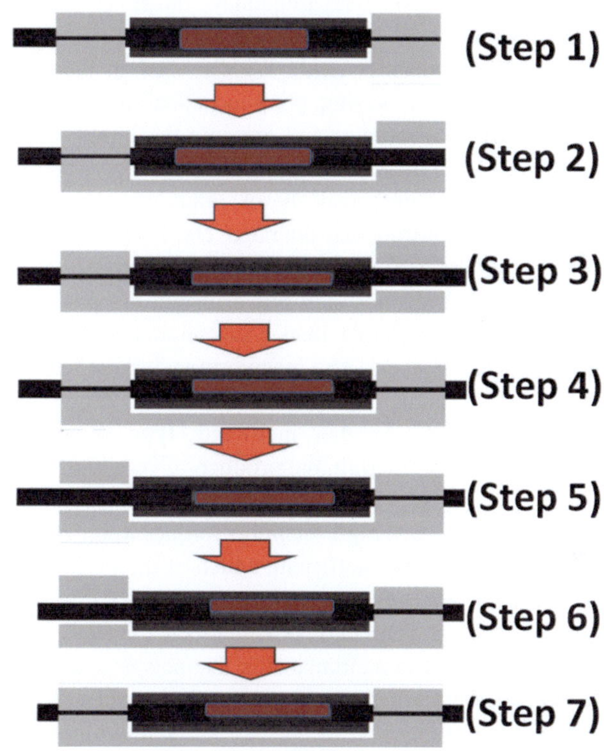

Fig. 4.16 Worm motor working steps

Terphenol-D in sonar transducers: A good sonar transducer should produce high mechanical force at low frequency. It is often necessary to reach another compromise between wide operating frequency bandwidth and low-quality factor Q [95]. The original, widely used Ni-based transducers with a magneto-mechanical coupling factor of 0.3 have been replaced by newer technology based on newer magnetostrictive materials. In fact, Terfenol-D sensors can operate at high mechanical forces and low

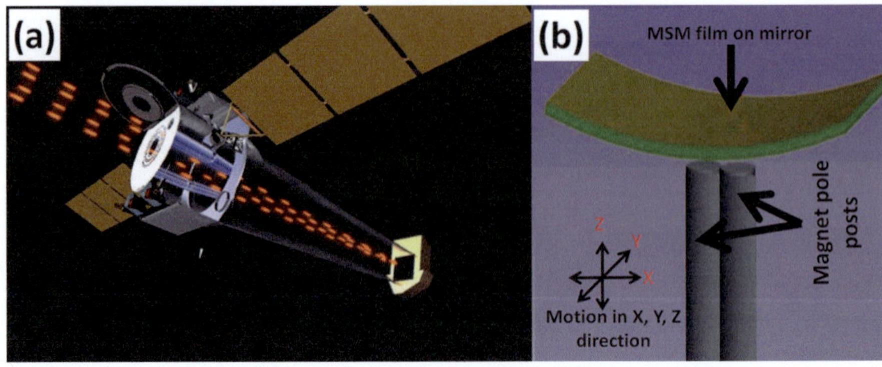

Fig. 4.17 Application of magnetostrictive film in (**a**) satellite, (**b**) the rectangular sample with local magnetic field, and (**c**) arrangement in the satellite

frequencies, as their magneto-mechanical coupling coefficients can be up to 0.8 and their quality coefficient Q is low. A typical Tonpilz sonar transducer is capable of operating in the 200 Hz frequency range at resonant frequencies [96].

Liquid-level measurement: Widely used in the petroleum, chemical, water-saving, pharmaceutical, food and beverage industries, to monitor and measure the liquid level in cans, as well as in air refueling systems and space, automotive refueling systems, and diesel refueling systems for hydrological monitoring and drainage, etc. [97].

Rotary motors: Smart hardware engines based on the principle of MRI are not only powerful but are potentially simpler and more reliable than conventional hydraulic or electromagnetic systems. The inchworm technique has been used in a rotary engine that produces 3 N.m of torque and a speed of 0.5 rpm. Another inchworm type device also provides a speed of 0.5 rpm, but produces a very high torque of 12 N.m and 800 μ_{rad} precision microsteps. Despite the high position accuracy and high retention torque, today's rotary inch motors tend to be inefficient. Much of the performance limit has been exceeded in the coordination of the rotary engine. Two linear Terphenol-D actuators are used to induce elliptical vibrations in a circular ring that acts as a stator and transmits the vibrations to the rotors pressed against the ring. The prototype is reported to provide a maximum torque of 2 N.m and a maximum speed of 17 rpm. There is a lot of research and commercial interest in the field of ultrasonic rotary engines. These engines are used in a wide range of applications, from autofocus camera lenses to robotic operators. Akuta has developed a torque activator that uses the Terphenol-D to achieve a relatively high speed of 13.1 rpm and a maximum torque of 0.29 N.m. This motor uses two Terphenol-D motor rods to cause axis rotation [98].

4.10.2 Magnetostriction in Aero-Industries

Application of magnetostrictive films in astronomical X-ray telescopes: Magnetostrictive material can have wide application in telescopes (Fig. 4.17). X-ray telescope Chandra can observe nearby galaxies by providing high angular resolution. The mirrors used in Chandra should be thick and formed by polishing and grinding. Therefore, the surface area of the mirror decreases by a considerable amount. For using mirrors in the telescope, the surface area of the mirror should be large being light in order to use it in launch vehicles as increment in surface area, increases the imaging capability of the telescope. However, the mirror is prone to deformation. For a purpose, the reflective side of the mirror is generally incorporated with smart materials that can detect various stimuli like the voltage, temperature change, and so on. One of the smart materials is magnetostrictive material that can counteract deformation in the mirror. The surface profile of the mirror is stabilized by using the magnetic field to magnetostrictive material which in turn can improve imaging quality [99]. It can be illustrated that the bending of the specimen can be measured by Zygo White Light Interferometer. It has a 2.21 μm resolution in the lateral direction and 0.1 nm resolution in the z-direction. The magnetic field is produced by a pseudo-magnetic head made up of two permanent magnet posts. The magnet posts can navigate through the glass samples. The system has 4 degrees of freedom which includes three translational motion and one rotational motion which ultimately change the orientation of magnetic field. The surface profile of the specimen is measured by White Light Interferometer [100].

Magnetostrictive materials for morphing aircraft: Wing structural morphing is used for increasing aerodynamic proficiency in a subsonic aerodynamic application. The use of various smart materials advances research on airfoil morphing. Conventional actuation methods that are replaced by smart actuation technology use pneumatic and hydraulic devices as well as ultrasonic motors. The sensitivity of smart actuators may be larger than that of conventional motors but energy consumption and life cycle are not much proficient for wider applications. Materials such as Terfenol-D and metallic ribbons have applications in actuators and sensors ranging from vibration controllers in aerospace to actuators in morphing structures. It provides a unique combination of strains, high forces, energy densities, and coupling coefficients. Moreover, different optimization techniques have resulted in various developments including hybrid giant magnetostrictive actuators [101].

Aerocraft wing pivot: Morphing technology can change the shape and efficiency of aerocraft. Here magnetostrictive actuator can be used to alter the geometry. In a two-spar wing of a Gulfstream III aircraft, the magnetostrictive actuator is used in wings. If a shaft is fabricated with magnetostrictive material such as Terfenol-D, the magnetic field along the shaft axle induces axial elongation that in turn changes the geometry. Here, the crystallographic direction of the structure should perfectly align the rod axis. In this technique, a small magnetic field creates high elongation (Fig. 4.18). In addition, hydraulic technology is also used for aerocraft for support in wing pivot. However, the hydraulic system can't be used for an unmanned aerial vehicle (UAV) and large aircrafts having longer wings as many pumps are required which makes the system heavy. So at this point, the magnetostrictive materials play a vital role to compensate the heavy hydraulic system. They replace the servo valve with torque motors. The hysteresis of the servo valve with the mechanically amplified actuator was observed to be less than that with the magnetically biased actuator. This difference is due to the additional magnetic hysteresis. The magnetostrictive actuator valve is faster than a conventional valve [102, 103].

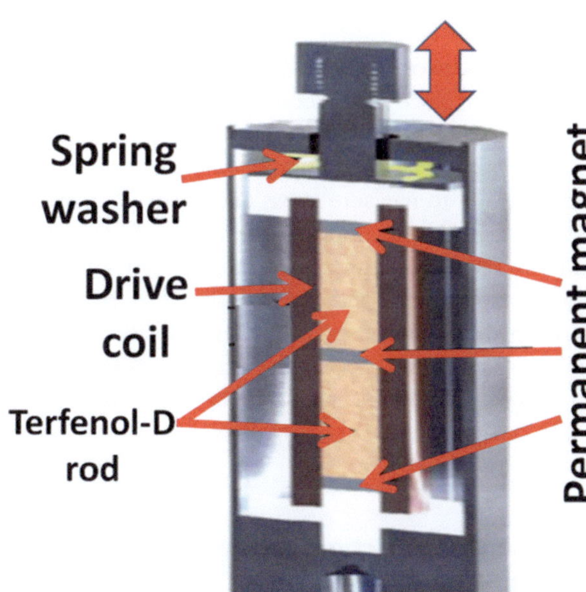

Fig. 4.18 (**a**) linear motor operated by MS, (**b**) operating device

4.10.3 Magnetostriction in Automotive Industries

Auto-drive magnetic guidance system: Autonomous driving or a self-driving is an emerging technology currently for automobile industries. Auto-drive technology means to carry passengers without the driver on a road. It is assisted by a powerful mechanism of an intelligent system. In a self-driving system, a magnetostrictive sensor is attached to the automated vehicle which can detect the magnetic field generated from the magnet set in the road. Road guidance systems are fitted with strong magnets (such as NdFeB, SrO, etc.) [104]. The magnetic guidance system works in three steps: (1) detection of road surface maker magnets and other larger magnetic disturbance of the road structure, quick response, and wide dynamic range, (2) selection of the marker magnetic field among all magnetic fields using bandpass filter signal processing electronic devices, and (3) determination of the marker position using the final signal processing software. The MI sensor effectively can sense a low magnitude of the magnetic field as well as a high magnitude of the magnetic field of 100 mG. Each MI sensor is designed as the differential type magnetic sensor, in which a pair of sensor head amorphous wires are aligned in parallel with a few intervals and set along the vehicle transport direction. An accurate running of a test bus of less deflection on the running line regardless of the time and the weather conditions are resulted. The magnetostrictive material assisted sensor array module has successfully worked covering the inside of the tunnel area, snow area, and the airport area regardless of the day and night times and weather conditions where the GPS is not available due to the electromagnetic wave problem (Fig. 4.19).

Magnetostrictive brake: Magnetostrictive brake is much more effective than any kind of brake available. It consumes very less power compared to the magnetic fail-safe brake in a robot. For example, for the same torque and same dimension magnetostrictive brake consumes only 2 W of power compared to 8 W in the fail-safe brake. In magnetostrictive brake, a stationary base is there upon which a hub is attached. Two solenoid assemblies are mounted on the recess of the hub diametrically. Solenoid core is made up of Terfenol-D or equivalent. The assembly is connected to the main shaft of the vehicle by means of a leaf spring. When no power is supplied to the solenoid the permanent magnet would be pulled on a stepped disc. Thus the ring is squeezed between the hub and stepped disc. The friction considers the squeezing of the ring to have a braking action. When power is supplied to the solenoid, the magnetostrictive element is pushed inward against a set of wedges that are in contact

Fig. 4.19 Self-driving magnetic guidance bus system

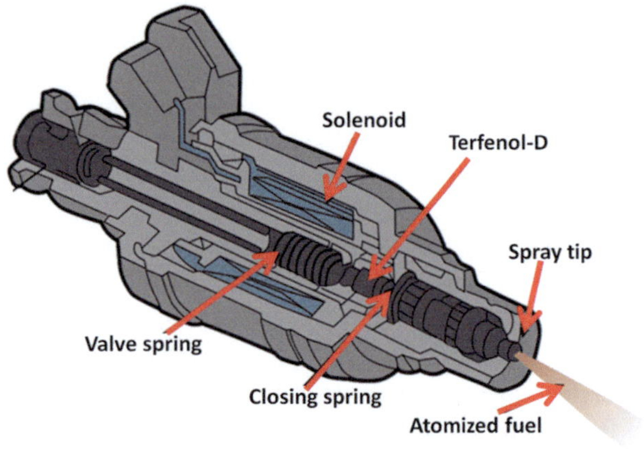

Fig. 4.20 A fuel injector

with the stepped disc. This axial displacement is sufficient to lift the stepped disk against a permanent magnetic field. Then as the ring is no longer squeezed hence now it is free to turn [105].

Fuel injector design concept: In fuel injector magnetostrictive material is used. For example, a prototype fuel injector assembly was manufactured which contains a Terfenol-D shaft as the core actuator. It contain 316 nonmagnetic stainless steel as the plunger material, 1020 soft magnetic steel as the injector housing material, and four different types of American wire gauge. The fuel injector design concept is illustrated in Fig. 4.20 [106].

Hydraulic lifts: Hydraulic cylinders are an essential component for many types of equipment, used to lift the heavy vehicle for work on the bottom side of a workshop. The operating principle of the feedback sensor is based on magnetostriction, a physical phenomenon that is observed in ferromagnetic materials. The basic components of a magnetostrictive transducer include the waveguide in a protective cover, electronic housing, and a position magnet that is connected to the piston. The waveguide is placed into a hydraulic cylinder, and the magnet moves along it. To register the position of a moving magnet as accurately as possible, the sensor only needs to measure the time of signal transmission from sending a current pulse to its return in a deformed form. Thus, you can get the reliable value of the object coordinates and the speed of its movement. Once this information is read, it is transmitted to the hardware controller [107].

4.10.4 Magnetostriction in Biomedical Industries

A contactless device operation mechanism is very much required in biomedical application, where magnetostrictive materials play a big role. For example, a copper coil coupled with a microscale Galfenol-silicon film is used to track the bone fracture healing. The load fracture decreases gradually from the osteosynthesis plate to the bone tissue as the bone healing process starts which generally creates stress on thin Galfenol film. Figure 4.21 showing the osteosynthesis plate with the Galfenol-based strain sensor and the geometry of the galfenol strain sensor. The stress generated on Galfenol film affects the inductance of the copper coil. The inductance in turn can be measured by the electrical network analyzer. The Galfenol film is very cheap and provides an accurate and wireless method to measure stress and strain. Magnetostrictive sensor can able to mimics the inner ear's cochlea and cilia. They produced high-aspect ratio Galfenol nanowires with length of 60 mm and diameters varying from 100 to 200 nm on different substrates. When acoustic pressure is created on Galfenol nanowires the magnetoresistive sensors detect it. This technique can also be used for medical scanning and sonar system [108].

4.10.5 Magnetostriction in Construction Industries

Suspender cables on bridges: In a typical magnetostrictive sensor, with the application of external pressures changes the length of the rod, which in turn creates a magnetic field that triggers a measurable current in the meter. Such sensors are often fixed to inspect spring cables on bridges [109].

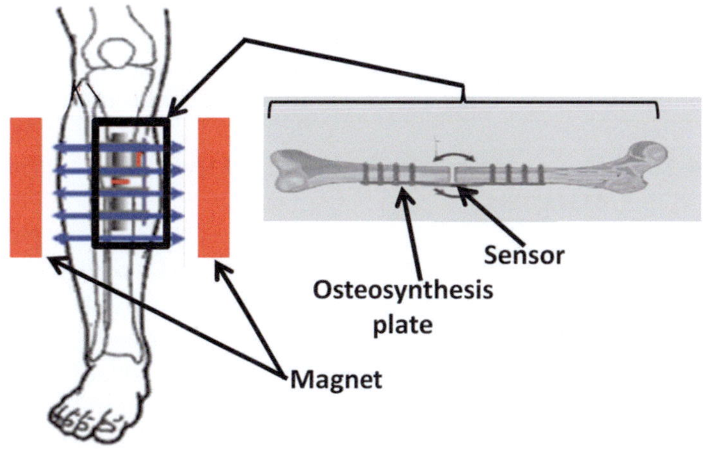

Fig. 4.21 Osteosynthesis plate with Galfenol-based strain sensor

Fig. 4.22 Suspender cables on bridges

Lift bridges: Lift bridges are built on busy waterways. They lift when the boat wants to travel on the waterway. As a result, part of the bridge is lifted by the hydraulic cylinders so far that the boat can easily pass under the raised bridge. Magnetostrictive sensors are installed inside the hydraulic cylinders. The sensors measure the stroke and transmit the values to the control unit. Hydraulic sensors ensure that part of the bridge reaches the minimum height for the boat to exit [110].

Liquid-level sensor: A magnetostrictive liquid level sensor (displacement sensor) accurately measures position through a voltage pulse signal from two different magnetic fields. It is a high-precision, long-distance absolute displacement measurement sensor based on magnetostriction. The measurement is a waveguide, the sensitive element of which is made of magnetostrictive material. When measured, the electron space of the sensor emits a current pulse that can be transmitted in a waveguide and generated around magnetic fields outside the waveguide. When this magnetic field encounters a moving magnetic ring that reflects a change in position, a magnetostrictive effect occurs. There will be a pulse of mechanical deformation within the waveguide, which is transmitted at a constant speed and will soon be detected in electronic space. The pulse time of a mechanical deformation transmitted in a waveguide is proportional to the distance between the magnetic ring of the moving position and the electron room, so it can accurately measure the distance by measuring time. Furthermore, the output signal is an absolute value, not a variable or gain signal, so there is no signal value drift or change and no regular scale change is required [111]. Figure 4.23 showing the magnetostrictive liquid level sensor. Since it does not come into contact with the moving position magnetic ring, it is wear-free, friction-free and has a long service life, good adaptation to the environment, reliable safety, and suitable for automation. Even in harsh environments (oil, dust, or other contaminated environments), it can usually work. Furthermore, it can withstand high temperatures, high pressure, and strong vibrations, and is widely used for displacement measurement and mechanical control. It has a stroke length of up to 3 m or more and a nominal accuracy of 0.05% F/s. The precision for those with a stroke length greater than 1 m can reach 0.02% F/S and the repeatability can be 0.002% F/S (the target sensor can be 0.002%). Therefore, it is widely used in the petrochemical industry, aviation, electricity and water saving, and so on [112]. Magnetostrictive materials are widely used in pharmacy, car refueling systems, water conservancies, food and beverage industries for controlling and measuring the liquid level of kinds of cans and water disposing, and so on.

4.10.6 Magnetostriction in Energy Harvesting Materials

Stirling thermoacoustic power converter and magnetostrictive alternator: By the innovators at NASA's Glenn Research center, two special technologies have been developed which make stirling engines more efficient and less costly. In this process, the Glennsthermos-acoustic power converter utilizes sound to convert heat into electric power [113]. This device can produce electricity with outstanding efficiencies; using heat-driven pressures and also using volume oscillations, starting from thermos-acoustic sources, power piezoelectric alternators or, from other power converter technologies. This device has got novelty that without moving parts this thermos-acoustic engine attains high thermal to electrical efficiencies; which is not like conventional stirling-based devices. Another improvement of Glenn's for stirling engines replaces the traditional linear alternator with a magnetostrictive alternator that converts the oscillating pressure wave into electric power. These new ideas offer a reliable and efficient way to generate power from any heat source benefiting applications like combined heat and power (CHP) system, distributer generation solar power generation, heat and cooling system [114]. The Glen magnetostrictive generator uses magnetostrictive materials stacked under biased and stress-induced magnetic compression. The acoustic energy of the motor travels through a resistance matching layer that is physically connected to the magnetostrictive mass. Compression screws hold the structure under compressive pressure, allowing the magnetostrictive material to be compressed to the micron scale and eliminating the need for bearings. Compression and expansion of the master-strict material change create a variable magnetic field, which is then induced by an electric current coil wrapped around the pile. This generator produces the electrical power of a sound pressure wave, and by adjusting the resonant frequency to match the motor, it can be a great replacement for a linear generator [115].

Vibration-driven generator: By magnetostrictive material, it is possible to create a vibrator generator that can generate several kilowatts of electricity and can be used for housing (Fig. 4.24). In these systems, one end of two long and thin tape recorders is fixed, but the weights are connected to the other ends (parallel beam structure). Fine electrical cables are wrapped around tape recorders for use as coils [116].

The magnetostatic effect is often used to generate seismic waves in sonar and electro-hydraulic actuators. An interesting feature of magnetostrictive materials is that when they are subjected to force, the relative deformation in them produces electricity, a property commonly used in energy collection methods. For example, Metglas magnetostrictive material ($Fe_{81}B_{13.5}Si_{3.5}C_2$) showed stimulating properties as a whole. The device for collecting energy vibration in the environment includes a giant Metglas connected to a copper substrate wrapped in a collecting coil. A magnetostrictive material mud gallphenol, which is a mixture of iron and French has a high piezomagnetic constant, good processing, and a large magnetic resonance feedback effect, with which

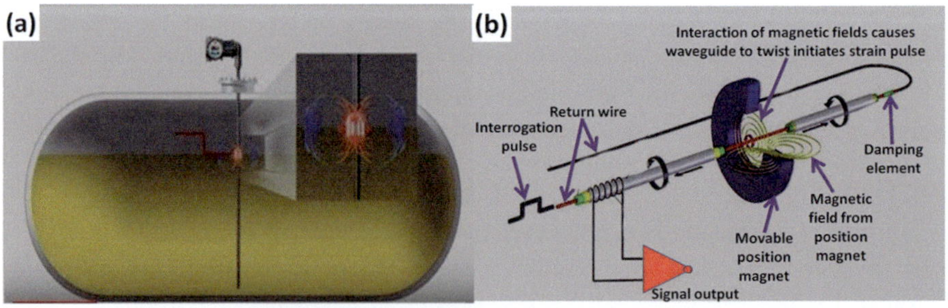

Fig. 4.23 (**a**) Tanker assisted with liquid level sensor, and (**b**) Levelling magnetostrictive device

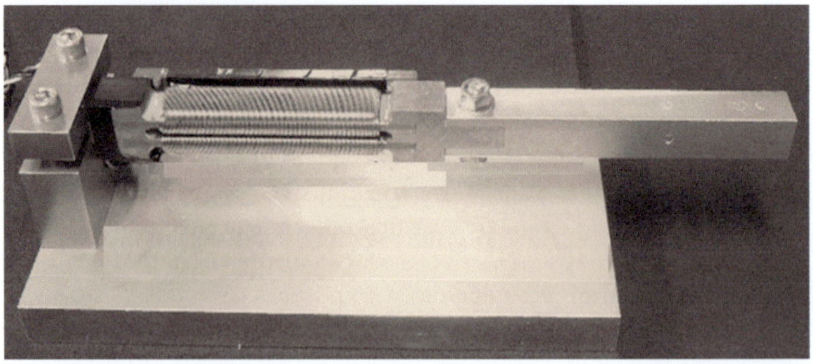

Fig. 4.24 Magnetostrictive vibration-driven generator that can generate several kilowatts of electricity and can be used for a residence

we can change the magnetism by mechanical pressure. These properties allow better energy collection from environmental vibrations than piezoelectric materials, which have limitations in terms of service life, power, and conversion efficiency. Magnetostrictive material connected to a copper substrate wrapped in a coil that can produce a maximum power of 3.5 mW at 395 Hz.

Band vibrators are used in various stages of ultrasonic application, such as ultrasonic transmitters, to generate and receive sound waves. Echo sounders, fish-finder, and SONAR use magnetic transmitters [117]. Noise from the liquid–solid interface is signaled to these devices. The time between transmitted and received pulse provides the distance to the target. Magnetic vibrators are used for ultrasonic energy in devices such as liquid, homogenization, dispersion, particle agglomeration, oxidation, emulsification, and high polymer decomposition, as well as cavitation mixers. More unusual applications are impact grinding, ultrasonic welding, impact welding, and painting of very thin wires [118].

4.10.7 Magnetostrictive Materials in Other Industries

Wireless thin film micro motor: Various types of ultrasonic motors have been developed using magnetostrictive materials. This is one of the uses of tape films. The linear micromotor is the result of a self-moving silicon wafer with a small band on the surface. The advantage of using smart materials such as Terfenol-D is that the connection without contact can be made by a magnetic field generated by an electric coil located at a certain distance from the moving parts. The magnetic field used creates a flexible form of resonance; which causes the plate to vibrate and cause movement at a speed of about 10–20 mm/s. A similar principle applies to rotary motors. In the 20 mT excitation field, the typical performance is 1.6 μN at 30 mph. Figure 4.25 shows the principle of operation of a thin film actuator based on a tape material [119].

Coating methods suitable for the use of terfenol-D and other magnetostrictive materials have been developed in some applications based on the magnetostrictive properties of the film. Sputter-stored tape films are used in simple, low-cost, noncontact arrangements with a high frequency of use in operational and microstructures. Depending on the composition and agglomeration conditions, it is possible to obtain a strain of 0.5 T for 700 ppm for TbDyFe. The deflection of the duct secured by the overlapping clamps may be convex or peeling. If only one side is coated, the wire can be placed under a magnetic field and controlled for various purposes.

Thin-film valve: The possible structure of the control valve is shown in Fig. 4.26. The area covered by a double-sided transverse substrate is shown in color. The design allows for less contact of the fluid flow with the moving parts through the magnetic field and has the advantage of the low mass of the moving parts with the help of this micro-valve structure [120].

Magnetic contact torque sensors: In these applications, Wideman, Villari, Δ*E*, and Matteucci effects are used to detect changes in magnetic quantities, providing numerical information on stress, force, or torque. Changes in mechanical properties, such as stress and strain, cause predictable changes in magnetic properties in a magnetostrictive material. The change in mechanical energy and the application of the sensor causes a change in magnetic energy. Typically, coils are used around a magnetostrictive material to detect

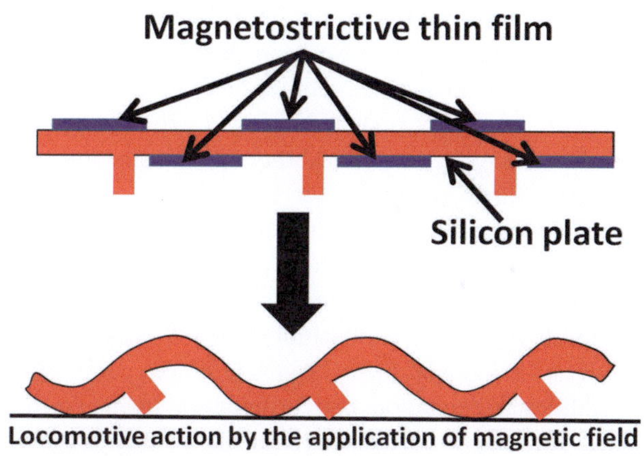

Fig. 4.25 Magnetostrictive thin-film actuator

Fig. 4.26 Magnetostrictive thin-film actuator: (**a**) Close state and (**b**) open state after application of magnetic field

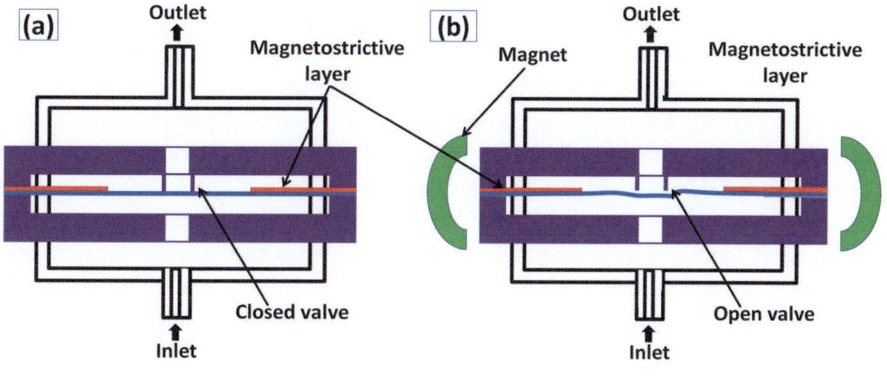

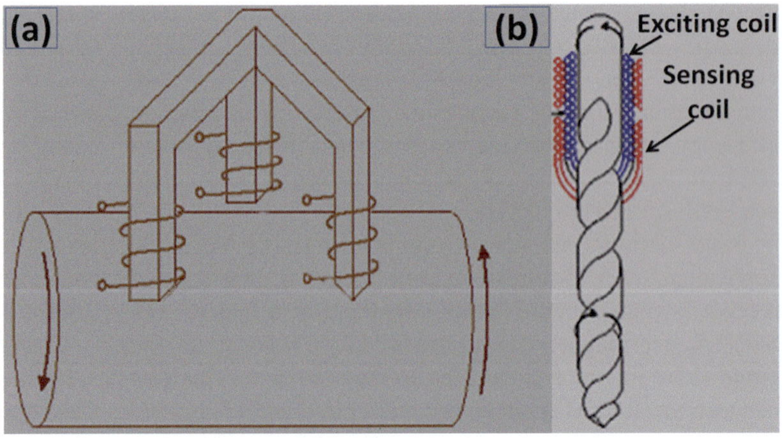

Fig. 4.27 Contactless sensor application based on magnetostrictive materials: (**a**) sensor based on tape technology and (**b**) another form of low sensitivity contact

changes in water permeability, while electronics are used to convert, filter, and amplify base data to generate strain and stress data that calculate load (pressure, force, and calculation). Figure 4.27a shows a sensor based on tape technology. During this application, torque is applied to the shaft and the shear stress is generated along the length of the vertical axis. It also causes tensile and compression stresses within ±45° of the axis. The two stress vectors are oriented at 90° to each other and have opposite properties (indicating tension and compression stresses). If the shaft contains a magnetostrictive material or a collar with a magnetostrictive material, the magnetic permeability measured in these directions changes. The change in magnetic flux can be measured without contact. Similar applications have been developed and can be used extensively in the automotive field. Another form of low-sensitivity contact is shown in Fig. 4.27b. The principle is based on changing the permeability due to transition stress. For less sensitive quantities, for example, cheap magnetic steel or alloys can be used to limit torque. In Fig. 4.27b, the rotation in the drill bit is measured by two arcs connected in series. One spiral is on the wing and the other on the shore. Changes in tangential permeability are less susceptible to changes at the moment than changes in limbic permeability. The additional excitation coil provides a magnetic field, whereas the output of the sensor is the voltage difference between the two pressure coils due to the different sensations of screw and glide permeability [121].

Magnetostrictive materials in remote tank meter: Another example of using the inverse magnetostrictive or magnetic impedance effect in sensor applications is the remote tank meter. Magnetostrictive materials or amorphous multilayer material are often used the inverse magnetostrictive effect. The instruments operate in the frequency range of 10 MHz to 8 GHz. The high frequency allows the sensor antenna to make small cuts, and a wide range of frequencies can be used for valid measurements. Based on changes in mechanical power (or deformation), there is a corresponding change in the AC conductivity of the stretched/stressed material. The use of a magnetostrictive material with high efficiency in converting mechanical energy into magnetic energy significantly increases the sensitivity to higher stress compared to conventional stress sensors [122].

Electromagnetically induced acoustic noise: Sound directly produces materials that vibrate when excited by electromagnetic forces. Some examples of this noise are the mains current, the transformer dagger, the jubilation of some rotating electric machines, or the sounds of headlamps. The snorting of high-voltage performance is due to corona discharge, not magnetism. This phenomenon depends on the application. The term electromagnetic noise is most commonly avoided because it is used in the field of electromagnetic immobility, which refers to radio frequencies. The term electrical noise refers to electrical disturbances that occur in electronic circuits rather than through sound. In the latter case, the terms electromagnetic oscillation or magnetic oscillation focused on a structural phenomenon [123].

Kinetic energy generating devices: The Villari effect is a design principle of any kinetic energy generating device based on magnetostrictive materials as shown in the Fig. 4.28. The full use of the process requires various functional components from the environment to the end user. A suitable mechanical frame is necessary to direct the environmental vibrations to the magnetostrictive material, which is why its design is extremely important to achieve maximum mechanical energy transfer. The conceptual implementations of the two mechanical parts are shown in Fig. 4.29a. The figure shows where the active material is used between the vibration source and the reference frame. The magnetostrictive rod (gray) is attached to a rigid frame. There is a compressive stress on the *Z*-axis and the rod produces a time-varying magnetization. Figure 4. 29b is suitable when a vibrating frame is available. Here, a magnetostrictive console beam is fixed to the vibrating frame at one end and the heavy mass at the other end. Due to the fluctuations of the mass, the material is subjected to a longitudinal voltage, which leads to time-varying magnetization. The two methods have some

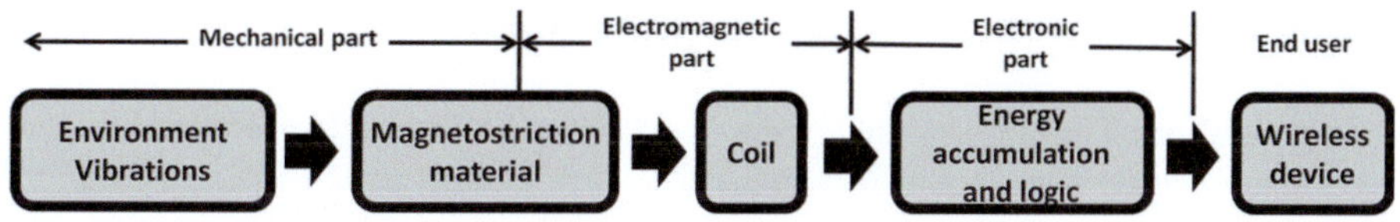

Fig. 4.28 Schematic representation of the functional process behind the kinetic energy harvesting by magnetostrictive materials

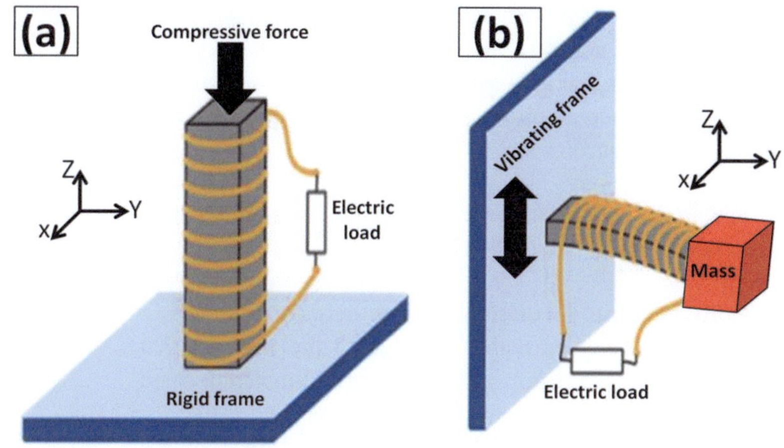

Fig. 4.29 Kinetic energy harvesting with magnetostrictive materials by (**a**) force driven and (**b**) velocity driven

common needs: a coil wrapped around a magnetostrictive material, and a magnetic circuit to transmit and close the magnetic flux lines [124].

The image Fig. 4.29a does not have a lower cutoff frequency, but it is usually larger than Fig. 4.29b structure. The latter has a bandwidth signal due to the characteristic resonant behavior of the console beam mass system and it works best with fluctuations around the resonant frequency of the system. Figure 4.29 shows that the electronics for optimizing and storing electrical energy are the "electric charge" of the harvesting device. Its development for magnetostrictive kinetic-energy harvesting (KEH) devices is still at an early stage, and many studies and many KEH applications have been proposed based on piezoelectrics. In general, all kinetic energy alerts are unregulated AC power supply using an electronic circuit with a suitable strategy to collect electrical energy.

The system can also be understood from an electromechanical point of view as a spring-mass damper system. Attenuation occurs through both mechanical friction and energy extraction, as the Villari effect also dissipates some of the mechanical energy. The material acts as a kind of electromechanical transformer, where we rely on the mechanical size (force and velocity) of the "primary" supplement, while " secondary" we have additional quantities (voltage and current) at the battery terminals. Using this use, the following relationship exists (Eq. 4.14):

$$\nu(t) \propto \frac{\partial B}{\partial t} = \frac{\partial B}{\partial \sigma}\frac{\partial \sigma}{\partial t} \quad (4.14)$$

The factor $\partial\sigma/\partial t$ is related to the oscillation frequency and is controlled by the oscillation source. The first factor is called the "piezomagnetic factor" and is a characteristic of the material [125].

4.11 Advantages/Disadvantages of MS Materials

The advantages of magnetostrictive materials are as follows [126, 127]:

1. **Noncontact easy installation:** Due to the involvement of magnetic setup, there is no need for additional actuation circuit connection (electric set up). Exterior magnet is enough to operate the materials.
2. **High reliability:** The magnetostrictive is based on the waveguide theory and without mechanical parts, so it has no friction and wears. The entire transducer was placed in a stainless steel tube without contact with the measuring medium. Therefore, it is reliable and durable.
3. **More stability:** The magnetostrictive structure has more stability, multi-output mode, and has overvoltage/high-frequency interference prevention.
4. **No insulation and maintenance:** No regular insulation and maintenance that requires an input power protection function that is available with opposite polarity.
5. **High accuracy:** The magnetostrictive fluid meter is based on a waveguide pulse, it measures the replacement time of the start and end pulses, so its accuracy is very high.
6. **Better resolution:** Resolution better than 0.01% full scale, higher than the resolution of other sensors.
7. **Safety:** The function is satisfactory without any precaution. In addition, with internally safe explosion protection,

the function of the magnetostrictive liquid meter is satisfactory.

8. **Suitable for system automation:** The output signal of the magnetostrictive measuring device (e.g., liquid meter) is standard, so the microcomputer is very suitable for information processing and easy entry into the network, which improves the automation of the measuring system.
9. **Can be operated with low impedance amplifiers:** The advantage of magnetostrictive sensors over other types of transducers is that they can be operated with conventional low impedance amplifiers, especially at frequencies well below resonance.

Few disadvantages involved with the magnetostrictive materials are as below [128, 129]

1. **Eddy currents formation:** at high frequency, there is a chance of eddy currents formation in the material that prevents the excitation of the core of the material. Such problems can be eliminated by using laminated materials.
2. **Current leakage and demagnetization:** Some of the challenges associated with the use of magnetostrictive materials are related to the effects of current leakage and demagnetization, which require the efficient design of the sensor's magnetic circuit.
3. **Higher product cost:** The production of advanced magnetostrictive joints is expensive because the conductors of advanced crystal converters must be fabricated using crystal growth techniques that provide directional hardening along the drive axis, combined with precision lamination, end diameter, and parallel machining of cut pieces.
4. **Significant nonlinearity and hysteresis:** In terms of device implementation, magnetostrictive materials have significant nonlinearity and hysteresis to the same extent as other intelligent materials, such as electrostrictive materials, in the case of a magnetostrictive fluid-level sensor.

4.12 Summary

The market for magnetostrictive materials is growing significantly due to the growing application of magnetostrictive materials in transducers, drives and motors, in building materials. In addition, market growth is driven by demand for magnetostrictive materials in various end-user sectors, such as the defense, aerospace, automotive, and consumer electronics industries. This chapter successfully described all the magnetostrictive materials along with their associated magnetostrictive effect, electromagnetic properties, applications, advantages, and limitations.

References

1. Deng, Z., Dapino, M.J.: Review of magnetostrictive materials for structural vibration control. Smart Mater. Struct. **27**, 11. https://doi.org/10.1088/1361-665X/aadff5
2. Squire, P.: Magnetostrictive materials for sensors and actuators. Ferroelectrics. **228**(1), 228, 305–319. https://doi.org/10.1080/00150199908226144
3. Apicella, V., Clemente, C.S., Davino, D., Leone, D., Visone, C.: Review of modeling and control of magnetostrictive actuators. Actuators. **8**(2), 45 (2019). https://doi.org/10.3390/act8020045.
4. Covaci, C., Gontean, A.: Piezoelectric energy harvesting solutions: a review. Sensors. **20**, 3512 (2020). https://doi.org/10.3390/s20123512
5. Mizuno, M.: Piezoelectric effects and materials. In: Encyclopedia of Thermal Stresses. Springer, New York. https://doi.org/10.1007/978-94-007-2739-7_326
6. Liang, X., Dong, C., Chen, H., Wang, J., Wei, Y., Zaeimbashi, M., He, Y., Matyushov, A., Sun, C., Sun, N.: A review of thin-film magnetoelastic materials for magnetoelectric applications. Sensors. **20**, 1532 (2020). https://doi.org/10.3390/s20051532
7. Zhan, Y.-S., Lin, C.-h.: A constitutive model of coupled magneto-thermo-mechanical hysteresis behavior for Giant Magnetostrictive materials. Mech. Mater. **148**, 103477 (2020). https://doi.org/10.1016/j.mechmat.2020.103477
8. Olabi, A.G., Grunwald, A.: Design and application of magnetostrictive materials. Mater. Des. **29**(2), 469–483 (2008). https://doi.org/10.1016/j.matdes.2006.12.016
9. Moffett, M.B., Clark, A.E., Wun-Fogle, M., Linberg, J.: Characterization of Terfenol-D for magnetostrictive transducers. J. Acoust. Soc. Am. **89**, 1448 (1991). https://doi.org/10.1121/1.400678
10. Shin, K.-H., Inoue, M., Arai, K.-I.: Elastically coupled magneto-electric elements with highly magnetostrictive amorphous films and PZT substrates. Smart Mater. Struct. **9**, 3. https://doi.org/10.1088/0964-1726/9/3/317
11. Calkins, F.T., Flatau, A.B., Dapino, M.J.: Overview of magnetostrictive sensor technology. J. Intell. Mater. Syst. Struct. **18**(10), 1057–1066 (2007). https://doi.org/10.1177/1045389X06072358
12. Kwun, H., Bartels, K.A.: Magnetostrictive sensor technology and its applications. Ultrasonics. **36**(1–5), 171–178 (1998). https://doi.org/10.1016/S0041-624X(97)00043-7
13. Hristoforou, E., Ktena, A.: Magnetostriction and magnetostrictive materials for sensing applications. J. Magn. Magn. Mater. **316**(2), 372–378 (2007). https://doi.org/10.1016/j.jmmm.2007.03.025
14. Jiles, D.C.: The development of highly magnetostrictive rare earth-iron alloys. J. Phys. D Appl. Phys. **27**, 1. https://doi.org/10.1088/0022-3727/27/1/001
15. Clark, A.E., DeSavage, B.F., Bozorth, R.: Anomalous thermal expansion and magnetostriction of single-crystal dysprosium. Phys. Rev. **138**, A216 (1965)
16. Narita, F., Fox, M.: A review on piezoelectric, magnetostrictive, and magnetoelectric materials and device technologies for energy harvesting applications. Adv. Eng. Mater. **20**(5), 1700743 (2018). https://doi.org/10.1002/adem.201700743
17. Grunwald, A., Olabi, A.G.: Design of a magnetostrictive (MS) actuator. Sensors Actuators A Phys. **144**(1), 161–175 (2008). https://doi.org/10.1016/j.sna.2007.12.034
18. Shih, J.-C., Hsu, S.-Y., Chao, L.-J., Chin, T.-S.: The magnetostriction of Tb(Fe0.9MnxAl0.1−x)2 alloys. J. Appl. Phys. **88**, 3541 (2000). https://doi.org/10.1063/1.1286469
19. Mathe, V.L., Sheikh, A.D.: Magnetostrictive properties of nanocrystalline Co–Ni ferrites. Phys. B Condens. Matter. **405**(17), 3594–3598 (2010). https://doi.org/10.1016/j.physb.2010.05.047

20. Parkin, S.S.P., Hayashi, M., Thomas, L.: Magnetic domain-wall racetrack memory. Science. **320**(5873), 190–194 (2008). https://doi.org/10.1126/science.1145799
21. Pandey, T., Parker, D.S.: Magnetic properties and magnetocrystalline anisotropy of Nd2Fe17, Nd2Fe17X3, and related compounds. Sci. Rep. **8**, 3601 (2018). https://doi.org/10.1038/s41598-018-21969-8
22. Dapino, M.J.: Magnetostrictive materials. In: Encyclopedia of Smart Materials. John Wiley & Sons, Inc., Hoboken, NJ (2002). https://doi.org/10.1002/0471216275.esm051
23. Sablik, M.J., Rubin, S.W.: Relationship of magnetostrictive hysteresis to the ΔE effect. J. Magn. Magn. Mater. **104–107**(1), 392–394 (1992). https://doi.org/10.1016/0304-8853(92)90847-H
24. Mohri, K., Humphrey, F., Yamasaki, J., Kinoshita, F.: Large Barkhausen effect and Matteucci effect in amorphous magnetostrictive wires for pulse generator elements. IEEE Trans. Magn. **21**(5), 2017–2019 (September 1985). https://doi.org/10.1109/TMAG.1985.1064026
25. Tzannes, N.S.: Joule and Wiedemann effects-the simultaneous generation of longitudinal and torsional stress pulses in magnetostrictive materials. IEEE Trans. Sonics Ultrasonics. **13**(2), 33–40 (1966). https://doi.org/10.1109/T-SU.1966.29373
26. Xiao, Z., Lo Conte, R., Chen, C., et al.: Bi-directional coupling in strain-mediated multiferroic heterostructures with magnetic domains and domain wall motion. Sci. Rep. **8**, 5207 (2018). https://doi.org/10.1038/s41598-018-23020-2
27. Newachecka, S., Youssef, G.: Magnetoelectricity beyond saturation. Mater. Horiz. **7**, 2124–2129 (2020). https://doi.org/10.1039/D0MH00595A
28. Pratt, J., Flatau, A.B.: Development and analysis of a self-sensing magnetostrictive actuator design. J. Intell. Mater. Syst. Struct. **6**(5), 639–648 (1995). https://doi.org/10.1177/1045389X9500600505
29. Barandiaran, J.M., Gutierrez, J.: Magnetoelastic sensors based on soft amorphous magnetic alloys. Sensors Actuators A Phys. **59** (1–3), 38–42 (1997). https://doi.org/10.1016/S0924-4247(97)80145-8
30. Zhang, L., Wang, B., Sun, Y., Weng, L., Wang, P.: Analysis of output characteristic model of magnetostrictive displacement sensor under a helical magnetic field and stress. IEEE Trans. Appl. Superconduct. **26**(4), 1–4 (2016., Art no. 0600904). https://doi.org/10.1109/TASC.2016.2520966
31. http://www.sanseer.com/index.php/news/zhihuiyuanqu/show_111.html?l=cn. Accessed 12 July 2012
32. https://www.linearmotiontips.com/how-do-magnetostrictive-sensors-work/. Accessed 12 July 2012
33. Zhang, Y., Liu, W., Zhang, H., Yang, J., Zhao, H.: Design and analysis of a differential waveguide structure to improve magnetostrictive linear position sensors. Sensors. **11**, 5508–5519 (2011). https://doi.org/10.3390/s110505508
34. Rueter, D., Morgenstern, T.: Ultrasound generation with high power and coil only EMAT concepts. Ultrasonics. **54**(8), 2141–2150 (2014). https://doi.org/10.1016/j.ultras.2014.06.012
35. Chen, Y., Snyder, J.E., Schwichtenberg, C.R., Dennis, K.W., McCallum, R.W., Jiles, D.C.: Metal-bonded co-ferrite composites for magnetostrictive torque sensor applications. IEEE Trans. Magn. **35**(5), 3652–3654 (1999). https://doi.org/10.1109/20.800620
36. Jenner, A.G., Smith, R.J.E., Wilkinson, A.J., Greenough, R.D.: Actuation and transduction by giant magnetostrictive alloys. Mechatronics. **10**(4–5), 457–466 (2000). https://doi.org/10.1016/S0957-4158(99)00065-3
37. Yamamoto, Y., Eda, H., Shimizu, J.: Application of giant magnetostrictive materials to positioning actuators. In: IEEE/ASME International Conference on Advanced Intelligent Mechatronics (Cat. No.99TH8399), pp. 215–220. IEEE, Atlanta, GA (1999). https://doi.org/10.1109/AIM.1999.803169
38. Deng, Z., Dapino, M.J.: Magnetic flux biasing of magnetostrictive sensors. Smart Mater. Struct. **26**, 5. https://doi.org/10.1088/1361-665X/aa688b
39. Szewczyk, R.: Model of the magnetostrictive hysteresis loop with local maximum. Materials (Basel). **12**(1), 105. Published 2018 (2018). https://doi.org/10.3390/ma12010105
40. Venkataraman, R., Krishnaprasad, P.S.: Approximate inversion of hysteresis: theory and numerical results [magnetostrictive actuator]. In: Proceedings of the 39th IEEE Conference on Decision and Control (Cat. No.00CH37187), vol. 5, pp. 4448–4454. IEEE, Sydney, NSW (2000). https://doi.org/10.1109/CDC.2001.914608
41. Hayashi, Y., Honda, T., Arai, K.I., Ishiyama, K., Yamaguchi, M.: Dependence of magnetostriction of sputtered Tb-Fe films on preparation conditions. IEEE Trans. Magn. **29**(6), 3129–3131 (1993). https://doi.org/10.1109/20.280877
42. Szewczyk, R.: Unified first order inertial element based model of magnetostrictive hysteresis and lift-off phenomenon. Materials. **12** (10), 1689 (2019). https://doi.org/10.3390/ma12101689
43. Callen, E.R., Clark, A.E., DeSavage, B., Coleman, W., Callen, H. B.: Magnetostriction in cubic Néel ferrimagnets, with application to YIG. Phys. Rev. **130**, 1735 (1963). https://doi.org/10.1103/PhysRev.130.1735
44. Augustyniak, M., Usarek, Z.: Finite element method applied in electromagnetic NDTE: a review. J. Nondestruct. Eval. **35**, 39 (2016). https://doi.org/10.1007/s10921-016-0356-6
45. Datta, S., Atulasimha, J., Flatau, A.B.: Figures of merit of magnetostrictive single crystal iron–gallium alloys for actuator and sensor applications. J. Magn. Magn. Mater. **321**(24), 4017–4031 (2009). https://doi.org/10.1016/j.jmmm.2009.07.067
46. Hathaway, K., Clark, A.: Magnetostrictive materials. MRS Bull. **18** (4), 34–41 (1993). https://doi.org/10.1557/S0883769400037337
47. Theodore, C.F.: Design, analysis, and modeling of giant magnetostrictive transducers. Retrospective Theses and Dissertations. 11832. https://lib.dr.iastate.edu/rtd/11832. (1997)
48. Zhang, J.X., Chen, L.Q.: Phase-field microelasticity theory and micromagnetic simulations of domain structures in giant magnetostrictive materials. Acta Mater. **53**(9), 2845–2855 (2005). https://doi.org/10.1016/j.actamat.2005.03.002
49. Dapino, M.J., Smith, R.C., Calkins, F.T., Flatau, A.B.: A coupled magnetomechanical model for magnetostrictive transducers and its application to Villari-effect sensors. J. Intell. Mater. Syst. Struct. **13** (11), 737–747 (2002). https://doi.org/10.1177/1045389X02013011005
50. X. Zhao, D. G. Lord. Application of the Villari effect to electric power harvesting. J. Appl. Phy.. vol. 99, 08M703 (2006); doi: https://doi.org/10.1063/1.2165133
51. Ren, L., Yu, K., Tan, Y.: Applications and advances of magnetoelastic sensors in biomedical engineering: a review. Materials (Basel). **12**(7), 1135. Published 2019 Apr 7 (2019). https://doi.org/10.3390/ma12071135
52. Dapino, M.J., Smith, R.C., Flatau, A.B.: Model for the ΔE effect in magnetostrictive transducers. In: Proc. SPIE 3985, Smart Structures and Materials 2000: Smart Structures and Integrated Systems. SPIE, Bellingham, WA (2000). https://doi.org/10.1117/12.388821
53. Saegusa, N., Morrish, A.H.: Crystallization and anisotropy in amorphous $Fe_{81}B_{13.5}Si_{3.5}C_2$. J. Magn. Magn. Mater. **31–34**(Part 3), 1555–1556 (1983). https://doi.org/10.1016/0304-8853(83)91012-0
54. Vinogradova, S., Cobbb, A., Light, G.: Review of magnetostrictive transducers (MsT) utilizing reversed Wiedemann effect. AIP Conf. Proc. **1806**(020008) (2017). https://doi.org/10.1063/1.4974549
55. Gianola, U.F.: Application of the Wiedemann effect to the magnetostrictive coupling of crossed coils. J. Appl. Phys. **26**, 1152 (1955). https://doi.org/10.1063/1.1722169

56. Karafi, M.R., Hojjat, Y., Sassani, F., Ghodsi, M.: A novel magnetostrictive torsional resonant transducer. Sensors Actuators A Phys. **195**, 71–78 (2013). https://doi.org/10.1016/j.sna.2013.03.015
57. Li, J.H., Gao, X.X., Xia, T., Cheng, L., Bao, X.Q., Zhu, J.: Textured Fe–Ga magnetostrictive wires with large Wiedemann twist. Scr. Mater. **63**(1), 28–31 (2010). https://doi.org/10.1016/j.scriptamat.2010.02.041
58. Park, J.S., Oh, O.K., Park, Y.W., Wereley, N.M.: A novel concept and proof of magnetostrictive motor. IEEE Trans. Magn. **49**(7), 3379–3382 (July 2013). https://doi.org/10.1109/TMAG.2013.2243132
59. Mohri, K., Humphrey, F.B., Kawashima, K., Kimura, K., Mizutani, M.: Large Barkhausen and Matteucci effects in FeCoSiB, FeCrSiB, and FeNiSiB amorphous wires. IEEE Trans. Magn. **26**(5), 1789–1791 (1990). https://doi.org/10.1109/20.104526
60. Mohri, K.: Sensitive bistable magnetic sensors using twisted amorphous magnetostrictive ribbons due to Matteucci effect. J. Appl. Phys. **53**, 8386 (1982). https://doi.org/10.1063/1.330369
61. Liu, J., Mo, P., Shi, M., Gao, D., Lu, J.: Multi-scale analysis of strain-dependent magnetocrystalline anisotropy and strain-induced Villari and Nagaoka-Honda effects in a two-dimensional ferromagnetic chromium tri-iodide monolayer. J. Appl. Phys. **124**, 044303 (2018). https://doi.org/10.1063/1.5036924
62. Kubota, T., Okazaki, T., Kimura, H., Watanabe, T., Wuttig, M., Furuya, Y.: Effect of rapid solidification on giant magnetostriction in ferromagnetic shape memory iron-based alloys. Sci. Technol. Adv. Mater. **3**, 2. https://doi.org/10.1016/S1468-6996(02)00020-7
63. Suzuki, S., Kawamata, T., Simura, R., Asano, S., Fujieda, S., Umetsu, R.Y., Fujita, M., Imafuku, M., Kumagai, T., Fukuda, T.: Anisotropy of magnetostriction of functional BCC iron-based alloys. Mater. Trans. **60**(11), 2235–2244 (2019). https://doi.org/10.2320/matertrans.MT-M2019146
64. Liu, H., Dong, B., Li, S., Meng, X., Gao, J., Jingping, Q., Li, Y.: Synthesis of Iron-based laves phase containing praseodymium magnetostrictive materials. J. Rare Earths. **25**(4), 449–451 (2007). https://doi.org/10.1016/S1002-0721(07)60454-2
65. Ito, S., Aso, K., Makino, Y., Uedaira, S.: Magnetostriction and magnetization of iron-based amorphous alloys. Appl. Phys. Lett. **37**, 665 (1980). https://doi.org/10.1063/1.92029
66. Ueno, T., Summers, E., Higuchi, T.: Machining of iron–gallium alloy for microactuator. Sensors Actuators A Phys. **137**(1), 134–140 (2007). https://doi.org/10.1016/j.sna.2007.02.026
67. Thomas, A.P., Gibbs, M.R.J., Vincent, J.H., Ritchie, S.J.: Technical magnetostriction parameters for application of metallic glasses. IEEE Trans. Magn. **27**(6), 5247–5249 (1991). https://doi.org/10.1109/20.278802
68. Willard, M.A., Claassen, J.C., Stroud, R.M., Francavilla, T.L., Harris, V.G.: (Ni,Fe,Co)-based nanocrystalline soft magnets with near-zero magnetostriction. IEEE Trans. Magn. **38**(5), 3045–3050 (2002). https://doi.org/10.1109/TMAG.2002.802435
69. Ryu, J., Priya, S., Vázquez, A., Kenji, C., Kim, U.H.-E.: Effect of the magnetostrictive layer on magnetoelectric properties in lead zirconate titanate/terfenol-D laminate composites. J. Am. Ceram. Soc. **84**(12), 2905–2908 (2004). https://doi.org/10.1111/j.1151-2916.2001.tb01113.x
70. Benbouzid, M.E.H., Reyne, G., Meunier, G., Kvarnsjo, L., Engdahl, G.: Dynamic modelling of giant magnetostriction in Terfenol-D rods by the finite element method. IEEE Trans. Magn. **31**(3), 1821–1824 (1995). https://doi.org/10.1109/20.376391
71. Senoy Thomas, Jinesh Mathew, P. Radhakrishnan, V.P.N. Nampoori, A.K. George, S.H. Al-Harthi, R.V. Ramanujan, M.R. Anantharaman. Metglas thin film based magnetostrictive transducers for use in long period fibre grating sensors. Sensors Actuators A Phys.,1611–2, 2010, 83–90, doi: https://doi.org/10.1016/j.sna.2010.05.006
72. Cugat, O., Delamare, J., Reyne, G.: Magnetic microsystems: MAG-MEMS. In: Azzerboni, B., Asti, G., Pareti, L., Ghidini, M. (eds.) Magnetic Nanostructures in Modern Technology. Springer, Dordrecht (2008). https://doi.org/10.1007/978-1-4020-6338-1_7
73. Gibbs, M.R.J., Hill, E.W., Wright, P.: Chapter 6: Magnetic microelectromechanical systems: MagMEMS. In: Buschow, K.H.J. (ed.) Handbook of Magnetic Materials, vol. 17, pp. 457–526., ISBN 9780444530226. Elsevier, Netherland (2007). https://doi.org/10.1016/S1567-2719(07)17006-2
74. Kokorin, V.V., Wuttig, M.: Magnetostriction in ferromagnetic shape memory alloys. J. Magn. Magn. Mater. **234**(1), 25–30 (2001). https://doi.org/10.1016/S0304-8853(01)00244-X
75. Mandal, S.K., Debnath, R., Singh, S., Nath, A., Dey, P., Nath, T. K.: Signature of magnetoelectric coupling of $xNiFe_2O_4 - (1\text{-}x) HoMnO_3$ (x=0.1 and 0.3) multiferroic nanocomposites. J. Magn. Magn. Mater. **443**, 222–232 (2017). https://doi.org/10.1016/j.jmmm.2017.07.081
76. Pavlovskii, N.S., Dubrovskii, A.A., Nikitin, S.E., Shaikhutdinov, K.A.: Magnetostriction of hexagonal $HoMnO_3$ and $YMnO_3$ single crystals. Phys. Solid State. **60**, 520–526 (2018). https://doi.org/10.1134/S1063783418030228
77. Schatz, F., Hirscher, M., Schnell, M.: Magnetic anisotropy and giant magnetostriction of amorphous TbDyFe films. J. Appl. Phys. **76**, 5380 (1994). https://doi.org/10.1063/1.357192
78. Zhou, Z., Li, J., Bao, X., Zhou, Y., Gao, X.: Improvement of bending strength via introduced (Dy,Tb)Cu phase at grain boundary on giant magnetostrictive Tb-Dy-Fe alloy by diffusing Dy–cu alloys. J. Alloys Compd. **826**, 153959 (2020). https://doi.org/10.1016/j.jallcom.2020.153959
79. Atulasimha, J., Flatau, A.B., Kellogg, R.A.: Sensing behavior of varied stoichiometry single crystal Fe-Ga. J. Intell. Mater. Syst. Struct. **17**(2), 97–105 (2006). https://doi.org/10.1177/1045389X06051075
80. Wan, Y., Fang, D., Hwang, K.-C.: Non-linear constitutive relations for magnetostrictive materials. Int. J. Non-Linear Mech. **38**(7), 1053–1065 (2003). https://doi.org/10.1016/S0020-7462(02)00052-5
81. Kim, Y.Y., Kwon, Y.E.: Review of magnetostrictive patch transducers and applications in ultrasonic nondestructive testing of waveguides. Ultrasonics. **62**, 3–19 (2015). https://doi.org/10.1016/j.ultras.2015.05.015
82. Wandass, J.H., Murday, J.S., Colton, R.J.: Magnetic field sensing with magnetostrictive materials using a tunneling tip detector. Sensors Actuators. **19**(3), 211–225 (1989). https://doi.org/10.1016/0250-6874(89)87074-X
83. Claeyssen, F., Lhermet, N., Le Letty, R., Bouchilloux, P.: Actuators, transducers and motors based on giant magnetostrictive materials. J. Alloys Comp. **258**(1–2), 61–73 (1997). https://doi.org/10.1016/S0925-8388(97)00070-4
84. https://www.waycon.biz/products/magnetostrictive-transducers/applications-magnetostrictive-transducer/. Accessed 19 July 2019
85. Fang, S., Zhang, Q., Zhao, H., Yu, J., Chu, Y.: The design of rare-earth giant magnetostrictive ultrasonic transducer and experimental Study on its application of ultrasonic surface strengthening. Micromachines. **9**, 98 (2018). https://doi.org/10.3390/mi9030098
86. Butler, S.C., Tito, F.A.: A broadband hybrid magnetostrictive/piezoelectric transducer array. In: OCEANS 2000 MTS/IEEE Conference and Exhibition Conference Proceedings (Cat. No.00CH37158), vol. 3, pp. 1469–1475. IEEE, Providence, RI, USA (2000). https://doi.org/10.1109/OCEANS.2000.881812
87. Goodfriend, M.J., Shoop, K.M.: Adaptive characteristics of the Magnetostrictive alloy, Terfenol-D, for active vibration

control. J. Intell. Mater. Syst. Struct. **3**(2), 245–254 (1992). https://doi.org/10.1177/1045389X9200300204
88. Paulsen, J.A., Ring, A.P., Lo, C.C.H., Snyder, J.E., Jiles, D.C.: Manganese-substituted cobalt ferrite magnetostrictive materials for magnetic stress sensor applications. J. Appl. Phys. **97**, 044502 (2005). https://doi.org/10.1063/1.1839633
89. Rao, G.S.N., Caltun, O.F., Rao, K.H., Subba Rao, P.S.V., Parvatheeswara Rao, B.: Improved magnetostrictive properties of Co–Mn ferrites for automobile torque sensor applications. J. Magn. Magn. Mater. **341**, 60–64 (2013). https://doi.org/10.1016/j.jmmm.2013.04.039
90. Chiriac, H., Marinescu, C.S.: New position sensor based on ultraacoustic standing waves in FeSiB amorphous wires. Sensors Actuators A Phys. **81**(1–3), 174–175 (2000). https://doi.org/10.1016/S0924-4247(99)00081-3
91. Koo, K., Dandridge, A., Tveten, A., Sigel, G.: A fiber-optic DC magnetometer. J. Lightwave Technol. **1**(3), 524–525 (1983). https://doi.org/10.1109/JLT.1983.1072141
92. Yang, Q., Yan, R., Fan, C., Chen, H., Liu, F., Liu, S.: A magneto-mechanical strongly coupled model for Giant Magnetostrictive force sensor. IEEE Trans. Magn. **43**(4), 1437–1440 (2007). https://doi.org/10.1109/TMAG.2007.891406
93. Stoyanov, P.G., Grimes, C.A.: A remote query magnetostrictive viscosity sensor. Sensors Actuators A Phys. **80**(1), 8–14 (2000). https://doi.org/10.1016/S0924-4247(99)00288-5
94. Teter, J.P., Sendaula, M.H., Vranish, J., Crawford, E.J.: Magnetostrictive linear motor development. IEEE Trans. Magn. **34**(4), 2081–2083 (1998). https://doi.org/10.1109/20.706805
95. Kim, Y., Kim, Y.Y.: A novel Terfenol-D transducer for guided-wave inspection of a rotating shaft. Sensors Actuators A Phys. **133** (2), 447–456 (2007). https://doi.org/10.1016/j.sna.2006.05.006
96. Taplin, L.B.: Microwave antenna and dielectric property change frequency compensation system in electrohydraulic servo with piston position control. Patent no: US4757745A (1988).
97. Li, Y., Sun, L., Jin, S., Sun, L.B.: Development of magnetostriction sensor for online liquid level and density measurement. In: 2006 6th World Congress on Intelligent Control and Automation, pp. 5162–5166. IEEE, Dalian (2006). https://doi.org/10.1109/WCICA.2006.1713375
98. Vranish, J.M., Naik, D.P., Restorff, J.B., Teter, J.P.: Magnetostrictive direct drive rotary motor development. IEEE Trans. Magn. **27** (6), 5355–5357 (1991). https://doi.org/10.1109/20.278837
99. Wang, X., Knapp, P., Vaynman, S., Graham, M.E., Cao, J., Ulmer, M.P.: Experimental study and analytical model of deformation of magnetostrictive films as applied to mirrors for x-ray space telescopes. Appl. Optics. **53**(27), 6256–6267 (2014). https://doi.org/10.1364/AO.53.006256
100. Picon, L.L., Bright, V.M., Kolesar, E.S.: Detecting low-intensity magnetic fields with a magnetostrictive fiber optic sensor. Proc. Natl. Aerospace Electron. Conf. (NAECON'94), Dayton, OH, USA. **2**, 1034–1039 (1994). https://doi.org/10.1109/NAECON.1994.332929
101. Sun, J., Guan, Q., Liu, Y., Leng, J.: Morphing aircraft based on smart materials and structures: a state-of-the-art review. J. Intell. Mater. Syst. Struct. **27**(17), 2289–2312 (2016). https://doi.org/10.1177/1045389X16629569
102. Ashley, S.: Magnetostrictive actuators. ASME. Mech. Eng. **120** (06), 68–70 (1998). https://doi.org/10.1115/1.1998-JUN-3
103. Friedmann, P.P., Carman, G.P., Millott, T.A.: Magnetostrictively actuated control flaps for vibration reduction in helicopter rotors-design considerations for implementation. Math. Comp. Model. **33** (10–11), 1203–1217 (2001). https://doi.org/10.1016/S0895-7177(00)00307-1
104. Zheng, C.: Magnetoresistive sensor development roadmap (non-recording applications). IEEE Trans. Magn. **55**(4), 1–30 (2019., Art no. 0800130). https://doi.org/10.1109/TMAG.2019.2896036
105. Pramod Kumar, K., Kadoli, R., Anil Kumar, M.V.: Mechanical and magnetic analysis of magnetostrictive disc brake system. In: 2010 5th International Conference on Industrial and Information Systems, pp. 644–649. IEEE, Mangalore (2010). https://doi.org/10.1109/ICIINFS.2010.5578627
106. Ghodsi, M.: Development of gasoline direct injector using giant magnetostrictive materials. IEEE Trans. Ind. Appl. **53**(1), 521–529 (2017). https://doi.org/10.1109/TIA.2016.2606591
107. Hartman, G.A., Sebastian, J.R.: Magnetostrictive actuator. Patent no: US5877432A (1997).
108. Zucca, M., Mei, P., Ferrara, E., Fiorillo, F.: Sensing dynamic forces by fe–ga in compression. IEEE Trans. Magn. **53**(11), 1–4 (2017., Art no. 2502704). https://doi.org/10.1109/TMAG.2017.2701859
109. Sun, Z., Ning, S., Shen, Y.: Failure investigation and replacement implementation of short suspenders in a suspension bridge. J. Bridge Eng. **22**(8) (2017). https://doi.org/10.1061/(ASCE)BE.1943-5592.0001089
110. Okada, N., Sasabuchi, T., Koike, K., Mineta, T.: MEMS magnetic sensor with bridge-type resonator and magnetostrictive thin film. Electron. Commun. Jpn. **101**(3), 90–95 (2018). https://doi.org/10.1002/ecj.12042
111. Barr, R., Gloden, M.L., Speecher, A.F.: Isolated magnetostrictive buffered liquid level sensor. Patent no: US7737684B2 (1997).
112. Zhang, L.: The output characteristics of galfenol magnetostrictive displacement sensor under the helical magnetic field and stress. IEEE Trans. Magn. **52**(7), 1–4 (2016., Art no. 4001104). https://doi.org/10.1109/TMAG.2016.2529291
113. https://technology.nasa.gov/patent/LEW-TOPS-80. Accessed 20 July 2019
114. Papaelias, M., Liang, C., Kogia, M., Mohimi, A., Kappatos, V., Selcuk, C., Constantinou, L., Muñoz, C.Q.G., Marquez, F.P.G., Gan, T.-H.: Inspection and structural health monitoring techniques for concentrated solar power plants. Renew. Energy. **85**, 1178–1191 (2016). https://doi.org/10.1016/j.renene.2015.07.090
115. http://www.freepatentsonline.com/9871186.html. Accessed 20 July 2019
116. Marciello, M., Pellico, J., Fernandez-Barahona, I., Herranz, F., Ruiz-Cabello, J., Filice, M.: Recent advances in the preparation and application of multifunctional iron oxide and liposome-based nanosystems for multimodal diagnosis and therapy. Interf. Focus. **6** (6), 20160055 (2016). https://doi.org/10.1098/rsfs.2016.0055
117. Slaughter, J.C., Dapino, M.J., Smith, R.C., Flatau, A.B.: Modeling of a Terfenol-D ultrasonic transducer. In: Proc. SPIE 3985, Smart Structures and Materials 2000: Smart Structures and Integrated Systems. SPIE, Bellingham, WA (2000). https://doi.org/10.1117/12.388838
118. Zhang, Z.G., Ueno, T., Higuchi, T.: Magnetostrictive actuating device utilizing impact forces coupled with friction forces. In: 2010 IEEE International Symposium on Industrial Electronics, pp. 464–469. IEEE, Bari (2010). https://doi.org/10.1109/ISIE.2010.5637678
119. Z. G. Zhang, T. Ueno and T. Higuchi, Development of a magnetostrictive linear motor for microrobots using Fe-Ga (Galfenol) alloys. IEEE Trans. Magn.,vol. 45, no. 10, 4598–4600, 2009, doi: https://doi.org/10.1109/TMAG.2009.2022846
120. Liu, X., Zheng, X.: A nonlinear constitutive model for magnetostrictive materials. Acta Mech. Sinica. **21**, 278–285 (2005). https://doi.org/10.1007/s10409-005-0028-8
121. Yamasaki, J., Mohri, K., Manabe, T., Teshima, N., Fukuda, S.: Torque sensors using wire explosion magnetostrictive alloy layers. IEEE Trans. Magn. **22**(5), 403–405 (1986). https://doi.org/10.1109/TMAG.1986.1064494
122. García-Arribas, A., Gutiérrez, J., Kurlyandskaya, G.V., Barandiarán, J.M., Svalov, A., Fernández, E., Lasheras, A., De

Cos, D., Bravo-Imaz, I.: Sensor applications of soft magnetic materials based on magneto-impedance, magneto-elastic resonance and magneto-electricity. Sensors. **14**(5), 7602–7624 (2014). https://doi.org/10.3390/s140507602

123. Hilgert, T., Vandevelde, L., Melkebeek, J.: Application of magnetostriction measurements for the computation of deformation in electrical steel. J. Appl. Phys. **97**, 10E101 (2005). https://doi.org/10.1063/1.1847951
124. Hasegawa, K., Ueno, T., Kiwata, T.: Proposal of wind vibrational power generator using magnetostrictive material. IEEE Trans. Magn. **55**(7), 1–4 (2019., Art no. 8203104). https://doi.org/10.1109/TMAG.2019.2904538
125. Fang, Z.-W., Zhang, Y.-W., Li, X., Hu, D., Chen, L.-Q.: Integration of a nonlinear energy sink and a giant magnetostrictive energy harvester. J. Sound Vibr. **391**, 35–49 (2017). https://doi.org/10.1016/j.jsv.2016.12.019
126. http://www.itargetsensors.com/html_news/Magnetostrictive-Principle-and-Advantage-5.html#:~:text=Advantage%20of%20magnetostrictive%20liquid%20level%20sensor%3A&text=High%20Reliability%3A%20Magnetostrictive%20liquidometer%20is,non%2Dcontact%20with%20measuring%20medium. Accessed 21 July 2019
127. Ghalamestania, S.G., Vandeveldea, L., Dirckx, J., Melkebeek, J.A.A.: Magnetostriction and the advantages of using noncontact measurements. AIP Conf. Proc. **1253**, 171 (2010). https://doi.org/10.1063/1.3455455
128. Toshiyuki Ueno, Toshiro Higuchi. Novel composite of magnetostrictive material and piezoelectric actuator for coil-free magnetic force control. Sensors Actuators A Phys., 1291–2, 2006, 251–255, doi: https://doi.org/10.1016/j.sna.2005.09.061
129. Dapino, M.J.: On magnetostrictive materials and their use in adaptive structures. Struct. Eng. Mech. **17**(3(4)), 303–329 (2004). https://doi.org/10.12989/sem.2004.17.3_4.303

Chromogenic Materials

5

Abstract

Chromic materials on the surface of any materials are a focus of interest of the entire customer in the market nowadays. This chapter describes the fundamental of chromism in the various materials in various stimulated conditions. The history of chromogenic materials gives the idea of their nomenclature and materials development from ancient times to the current age. Categorization of the chromogenic system shows the various mechanism of chromism. Most applications of the materials such as photochromic, thermochromic, electrochromic materials, and so on, have been discussed with their materials behavior, mechanism, limitations. Hybrid chromic materials showing the integrity of various phenomenons in the materials to change their color are discussed briefly, which needs to be developed a lot in the current future.

Keywords

Chromogenic materials · Photochromic materials · Electrochromic materials · Gasochromic materials · Mechanochromic/piezochromic materials · Chemochromic materials · Biochromic materials · Magnetochromic materials · Phosphorescent materials

5.1 Introduction

Materials that change color are termed chromogenic materials and are described as chameleonic because they change their color reversibly as a response to changes in environmental conditions (such as change of temperature, brightness, stress, P^H, etc.) or by induced stimuli [1, 2]. These materials are one type of smart material. "Chromo" originates from the Greek language and means that something is colored [3]. A *chromogen* is a colored compound containing a *chromophore*—responsible for the observed color [4]. A *chromophore* is a group of atoms within a larger molecule that are principally responsible for the absorption of certain wavelengths of visible light, while transmitting wavelengths that are not absorbed [5]. It is the wavelengths that are not absorbed which are responsible for the color observed. The technical principle, by which these materials change color, can be explained by an alteration in the equilibrium of electrons caused by the stimulus, like cleavage of the chemical bonds or changes occurring inside the molecule, among electrons, with a consequent modification of optical properties, such as reflectance, absorption, emission, or transmission. When the stimulus ceases, the material returns to its original electronic state, regaining the original optical properties, thus the initial color or transparency. This process, named *chromism*, implies "pi" and "d" electron positions so that the phenomenon is induced by various external stimuli bearing the ability to alter electronic density of the compound or a substance [6]. Many natural compounds exhibit chromism and now a number of artificial compounds of specific chromic properties have been synthesized. Chromic materials have this interesting capability of reacting to external stimuli and exhibit modification in color. The markets for chromogenic smart materials cover automotive, architectural, aircraft, and information display. The global market size for chromogenic displays is currently estimated at $60.9 billion, climbing to $86 billion in 2007. The largest geographical producers of chromogenic materials are Asia (1.76 billion m^2), the Americas (972 million m^2), and Europe (906 million m^2), with the rest of the world producing 472 million m^2 [7]. The chromogenic family of materials is expanding with many subcategorized materials that cover any visibly switchable technology useful for glazing, mirrors, transparent displays, and a variety of other applications [8].

5.2 History of Chomogenic Materials

Photochromic materials are the first chromogenic materials observed from ancient times. Warriors painted photochromic materials on their bracelets to change the color under sunlight

A. Behera, *Advanced Materials*, https://doi.org/10.1007/978-3-030-80359-9_5

[9]. In the eighteenth century, photochromism was also the first color change property that was studied and reported [10]. Photochromism was discovered in the late 1880s, including work by Markwald, who studied the reversible change of color of 2,3,4,4-tetrachloronaphthalen-1(4H)-one in the solid state. He labeled this phenomenon “phototropy” [11]. The potential of oxidized developers in a color photographic process, however, was first realized by another German chemist, Rudolf Fischer, who, in 1912, filed a patent describing a chromogenic process to develop both positives and negatives using indoxyl, and thio-indoxyl-based color developers as dye couplers in a light-sensitive silver halide emulsion [12]. The following year he filed a patent listing various color developers and dye couplers, which have historically been used in Agfachrome and are still in use today in Fujichrome Velvia and Provia, and Ektachrome. In spite of this, Fischer never created a successful color print due to his inability to prevent the dye couplers from moving between the emulsion layers. This first solution to this problem, found by Agfa workers Gustav Wilmanns and Wilhelm Schneider, who created a print made of three layers of gelatin containing subtractive color dye couplers made of long hydrocarbon chains, and carboxylic or sulfonic acid. This turned the dye couplers into micelles, which can easily be scattered in the gelatin, while loosely tethering to it. Agfa patented both the developer for this print and its photographic process, and promptly developed and released in 1936 Agfacolor Neu, the first chromogenic print, which was a color print film that could be developed using a transparency. Agfa developed a chromogenic negative film by 1939, which could be developed directly on a companion paper to the film, although this film was never commercialized. Kodak too worked to solve the issue of the dye couplers movement and found a different solution. They used ionic insoluble carbon chains, which were shorter than Agfa’s for their dye couplers and were suspended within droplets of water in the gelatin layers of the print [13]. In 1942, Kodak released Kodacolor, the first published chromogenic color print film that could be developed from a negative. It became cheaper and simpler to develop counterparts to the alternatives at the time and could be used in the simplest of cameras. Around the year1950, the word “photochromism” was reported by Greek scientist Fritsch, German chemist Benno Homolka, and Israeli physical chemist Yehuda Hirshberg [14, 15]. After that, more words related to chromism were created to describe different new types of color change phenomena. In 1955, Kodak introduced a chromogenic paper named “Type C,” which was the first color negative paper Kodak sold to other labs and individual photographers. Although the paper’s name was changed to “Kodak Ektacolor Paper” in 1958, the terminology “Type-C Print” persisted, and has become a popular term for chromogenic prints made from negatives still in use today, with the name “Type-R Print” becoming its reversal film counterpart [16]. Notwithstanding the success of chromogenic prints in the amateur and professional market, it wasn’t considered a medium for fine art photography up to the 1970s [13]. The pioneers in the use of chromogenic prints and the use of color photography as a whole in fine art were photographers such as Ernst Haas, which was profiled by the Museum of Modern Art in its first exhibition of color photography in 1962. Other pioneering fine-art color photographers who printed their photographs on chromogenic prints include 7 and Stephen Shore. Their works, and those of many others, caused chromogenic prints to become the preferred medium for contemporary photography by the 1990s [17]. Today, chromogenic prints, like most color photographic prints, are developed using the RA-4 process. As of 2017, the major lines of professional chromogenic print paper are Kodak Endura and Fujifilm Crystal Archive. Plastic chromogenic “papers” such as Kodak Duratrans and Duraclear are used for producing backlit advertising and art [16]. The typical historic progress of chromogenic materials has been given in Fig. 5.1.

5.3 Concept of Chromogenic Materials

Chromogenic suggests a reversible change in optical characteristics due to microstructural changes such as phase separation, change of particle size, aggregation, or isotropic-anisotropic phase transitions within the material that occurs by the surrounding chemical or physical stimuli such as solvent, temperature, pressure, light, or electrons [18, 19]. The technical principle by which these materials function can be explained by an alteration in the equilibrium of electrons caused by the stimulus, with a consequent modification of optical properties, such as reflectance, absorption, emission, or transmission. This process involves a change in the microstructure or electronic state of substances. The majority of chromism in polymeric materials occurs in conjugated polymers. Conjugated polymers have loosely bound, freely moving electrons, similar to mobile electrons in

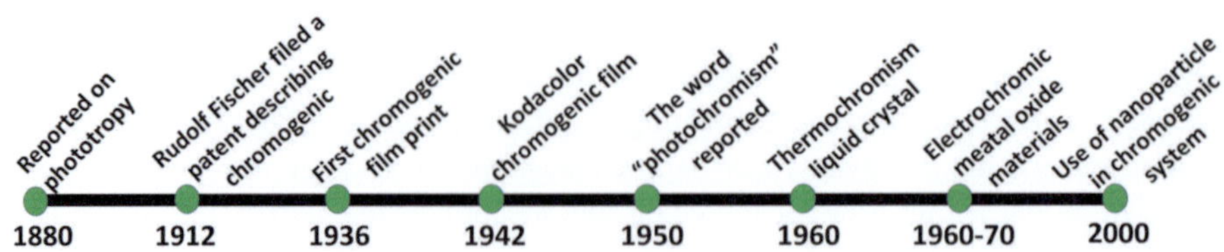

Fig. 5.1 Historic progress of chromogenic materials

metals. Chromic phenomena in such polymers are induced by various external stimuli which have the ability of altering the electronic configuration and energy transitions of electrons in the substance [20].

5.4 Classification of Chromogenic Materials

Those phenomena which involve the change in color of a chemical compound under an external stimulus fall under the generic term of chromisms. They take their individual names from the type of external influence, which can be either chemical or physical that is involved. Many of these phenomena are reversible. The following list includes all the classic chromisms [21, 22]:

Photochromism: Color change caused by light.
Thermochromism: Color change caused by heat.
Electrochromism: Color change caused by an electric current.
Gasochromism: Color change caused by a gas—hydrogen/oxygen redox.
Mechanochromism: Color change caused by mechanical energy.
Chemocromism: Color change caused by chemical energy
Halochromism: Color change caused by a change in pH.
Solvatochromism: Color change caused by solvent polarity.
Hydrochromism: Color change caused by interaction with bulk water or humidity.
Biochromism: Color change caused by a biochemical reaction.
Magnetochromism: Color change caused by a magnetic field.
Ionochromism: Color change caused by ions.
Vapochromism: Color change caused by the vapor of an organic compound due to chemical polarity/polarization.
Radiochromism: Color change caused by ionizing radiation.
Sorptiochromism: Color change when a species is surface adsorbed.
Aggregachromism: Color change on dimerization/aggregation of chromophores.
Chronochromism: Color change indirectly as a result of the passage of time.
Concentratochromism: Color change caused by changes in the concentration in the medium.
Cryochromism: Color change caused by lowering of temperature.
Rigidichromism: Color change caused by changes in the rigidity of the medium.
Metallochromism: Color change caused by metal ions.
Tribochromism: Color change caused by mechanical friction.
Cathodochromism: Color change caused by electron beam irradiation.
Amorphochromism: Color change caused by changes in crystalline habitat.
Crystallochromism: Color change due to changes in the crystal structure of a chromophore.

There are also chromisms that involve two or more stimuli. Examples include photoelectrochromism, photovoltachromism, bioelectrochromism, solvatophotochromism, thermosolvatochromism, halosolvatochromism, and electromechanochromism. Color changes are also observed in the interaction of metallic nanoparticles and their attached ligands with another stimulus. Examples include plasmonic solvatochromism, plasmonic ionochromism, plasmonic chronochromism, and plasmonic vapochromism [23, 24].

In some cases, the color will change on the basis of the energy that produces luminescence. The absorption of energy followed by the emission of light is often described by the term luminescence. The exact term for chromogenic system is based on the energy source, which are electrical (electroluminescence, galvanoluminescence, sonoluminescence), photons (light) (photoluminescence, fluorescence, phosphorescence, biofluorescence), chemical (chemiluminescence, bioluminescence, electrochemiluminescence), thermal (thermoluminescence, pyroluminescence, candololuminescence), electron beam (cathodoluminescence, anodoluminescence, radioluminescence), and mechanical (triboluminescence, fractoluminescence, mechanoluminescence, crystalloluminescence, lyoluminescence, elasticoluminescence) [25, 26]. Many of these phenomena are widely used in consumer products and other important outlets. Cathodoluminescence is used in cathode ray tubes, photoluminescence in fluorescent lighting and plasma display panels, phosphorescence in safety signs and low energy lighting, fluorescence in pigments, inks, optical brighteners, safety clothing, and biological and medicinal analysis and diagnostics, chemoluminescence and bioluminescence in analysis, diagnostics and sensors, and electroluminescence in the burgeoning areas of light-emitting diodes (LEDs/OLEDs), displays and panel lighting. Important new developments are taking place in the areas of quantum dots and metallic nanoparticles [27].

5.5 Photochromic Materials

Photochromic materials change color in response to change in light intensity. Photochromic materials respond to variations of incoming light intensity or the spectral distribution of light, modifying their color reversibly [28]. Owing to the induction effect of electromagnetic radiation by energetically rich photons of the near UV electromagnetic spectra, they are activated and change their molecular configuration and their light transmission coefficient, thus the chromatic spectrum at the exit, manifesting color and reduced transparency [29]. When the bright stimulus is removed, the color

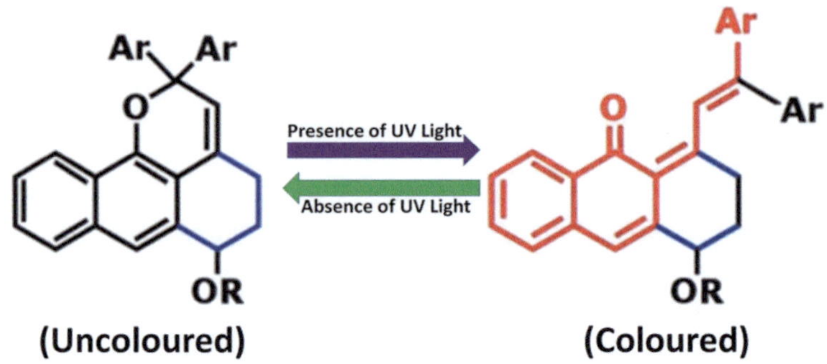

Fig. 5.2 Photochromic material changes its chemical structure to change color when exposed to UV light

disappears because the material returns to its original molecular configuration (e.g., Fig. 5.2). This color change is reversible, therefore has many useful applications. Color is produced when light interacts with materials in a variety of ways: (1) stimulated (reversible) color change, (2) the absorption and reflection of light, (3) the absorption of energy followed by the emission of light, (4) the absorption of light and energy transfer (or conversion), and (5) the manipulation of light [30].

Many biological systems exhibit photochromism. For example, rhodopsin is a natural photochromic substance, present in the retina of the eye. Rhodopsin is a pigment that is activated by light, producing a nerve stimulus transmitted to the cortex to start the process of visual perception. A similar protein was discovered in the bacterium. Many inorganic materials, such as copper, mercury, various metal oxides, and some minerals also exhibit photochromism. Inorganic photochromics are more appropriate for coatings on metal, glass, and ceramic surfaces while organic molecules such as spiropyrans, spirooxazines, and fulgides are suitable for use on textiles where they get a wide palette of colors across the visible range of light [31]. Photochromic materials in the form of pigments can be mixed with conventional materials and used in combination with other common pigments to obtain paints and inks with effects that vary from one color to another. Some devices are made up of photochromic substances placed between two layers of conventional materials having different energy absorption. These materials, based on the absorption of light, are usually employed in the context where the limitation of ultraviolet (UV) and infrared (IR) radiation is the principal objective, such as in the case of optical lenses for solar protection, which represents one of the principal applications [19]. Early photochromic glasses were usually made of glass and contained small crystals of silver halides (e.g., silver chloride) that darkened when exposed to light, just like old photographic films. Through the effect of silver halides and copper dopants contained in them, these glasses usually turn into a gray color when exposed to sunlight. In this mode, they function as a filter especially for infrared radiation, screening the heat emitted by the sun. The gray color is a result of silver cations from the silver halide, combining with copper electrons with the help of sunlight, forming metallic silver clusters. Photochromic glasses are also used for photographic reproduction and to realize "sensitive" frames and buildings. However, unlike those films, the darkening of photochromic lenses was reversible, that is, the lenses became clear again once the ambient light was lowered [32]. The crystals used in the glass of such lenses were minuscule, both in number and size. Less than 0.1% of silver halide crystals, which were 100 times thinner than human hair, were used in early photochromic glasses. Modern photochromic glasses, however, are usually made of plastic, rather than glass, and contain carbon-based (organic) molecules instead of silver compounds [33]. Such compounds are better, more efficient alternatives for achieving the quick-darkening and quick-clearing effect, as their molecular structure varies in accordance with the presence/absence of a certain type of light (usually UV light, as it's a component of sunlight). The use of photochromic windows has been limited until today due to certain technical and process-related problems such as obtaining a uniform dispersion of photochromic substances in the glass and the gradual loss of reversibility with time, also called fatigue. In some cases, experimental studies have already solved most of these technical difficulties, which will allow the increase of photochromic sheet dimension and stability with time, reducing production costs and increasing the number of cycles or useful life of the products, thus making this technology a feasible one [34].

The most commonly used photochromic molecules are naphthopyrans and oxanes, due to their ability to change their molecular structure reversibly upon exposure to UV light. Below is a generic example of such a reaction [35]:

5.5.1 Mechanism of Photochromic Materials

Molybdenum oxide is one of the most studied metal oxides regarding the application as a photochromic material when absorbing UV radiation [36]. The photochromism mechanism from MoO_3can be explained by the following Eqs. ((5.1), ((5.2), and ((5.3),

$$MoO_3 + h\lambda \rightarrow MoO_3* + e^- + h^+ \quad (5.1)$$

$$2h^+ + H_2O \rightarrow 2H^+ + O \quad (5.2)$$

$$MoO_3 + xH^+ + xe- \rightarrow H_xMo^{VI}{}_{1-x}Mo^V{}_xO_3 \quad (5.3)$$

When the film is irradiated with UV light (with excitation energy above the band gap), electrons and holes are formed (Eq. (5.1). The photogenerated electrons are then injected into the conduction band of the MoO_3 and the holes react with the adsorbed H_2O, originating protons (Eq. (5.2). These protons will then diffuse into the MoO_3 lattice forming the hydrogen molybdenum bronze (Eq. (5.3). The result will be a change in the optical absorption of MoO_3 that will change from colorless to blue. The photochromic properties of MoO_3 make it suitable for applications such as displays, erasable optical storage devices, and smart windows/glasses.

Tungsten oxide is another good candidate for photochromic applications. Exposure of amorphous WO_3 particles to UV radiation will induce a color change from colorless to an intense blue or brown. This color change is caused by the reduction of tungsten atoms followed by the formation of tungsten bronzes ($H_xW^V{}_xW^{VI}{}_{1-x}O_3$) (Eqs. (5.4 and (5.5). Electrons (e^-) and holes (h^+) are formed and the protons (H^+) necessary for coloration will be produced from the reaction of water absorbed on the surface or interior of the holes. Then the photogenerated electrons that were injected into the conductive band will react with H^+ and WO_3, producing hydrogen tungsten bronze ($H_xW^V{}_xW^{VI}{}_{1-x}O_3$). Tungsten oxide photochromism sensitivity is usually limited by its bandgap energy of 3.17 eV, corresponding to the near-UV range [37].

$$WO_3 + xe^- + xH^+ \rightarrow H_xW^V{}_xW^{VI}{}_{1-x}O_3 \quad (5.4)$$

$$W^{VI}{}_A + W^V{}_B + h\nu \rightarrow W^V{}_A + W^{VI}{}_B \quad (5.5)$$

For efficient use of solar radiation or laser, which covers the entire wavelength range, it is important to extend the coloration response to the visible light region. This is possible by the insertion of a small amount of cations in the crystallographic structure (like K^+ for WO_3 and $Li^+$for MoO_3) through cathodic polarization that will induce the formation of $K\times WO_3$ and this way a slight deformation in the oxide microstructure. Intermediate metastable trap states with energy levels lying within the bandgap region will be formed, which are accessible to visible light coloration [38].

The photochromic lenses were made from glass, in which the main photochromic elements were mainly silver compounds. These silver compounds were either silver chloride or silver halide. The compounds were embedded within the glass substrate. This substrate could vary in size and the most common size was about 5 nm. Normally, a number of manufacturers use borosilicate glass to produce these photochromic lenses [39]. The process involves adding silver chloride to the melt after which, cooling takes place. However, at this stage, the glass is bright blue and is not yet, photochromic. You must therefore subject it to heat treatment to a temperature of about 600 °C to obtain photochromic properties. The duration of the heat treatment and temperature will determine the sensitivity of the photochromic molecules. Now, when you subject glass photochromic lens to a shortwave light, it will begin to tint gradually. This implies that the silver chloride will begin to break into silver and chlorine. This is what leads to the darkening process. It can be seen that the chemical reactions given below show the oxidation and reduction processes [40].

$$Cl^- \xrightarrow{\text{Oxidation}} Cl + e^- \quad (5.6)$$

$$Ag^+ + e^- \xrightarrow{\text{Reduction}} Ag \quad (5.7)$$

Again, when you remove the UV light, silver will begin to combine with chlorine. It is at this stage that copper plays an integral role. That is, the presence of copper (I) chloride will help to reverse the process. It can be observed in the equation below [41]:

$$Cl + Cu^+ \xrightarrow{\text{Reduction}} Ag \quad (5.8)$$

$$\left[\overbrace{\text{oxidizing agent}}^{Cl} + \overbrace{\text{reducing agent}}^{Cu^+}\right] \rightarrow \left[\overbrace{\text{oxidizing species}}^{Cu^{2+}} + \overbrace{\text{reducing species}}^{Cl^-}\right] \quad (5.9)$$

The fact that this process is reversible makes it useful in making photochromic lenses. The commercialization of plastic photochromic lenses has seen a number of manufacturers demand polycarbonate sheets as a better alternative. By adopting polycarbonate materials, they can manufacture a range of photochromic lenses with varying designs and shapes. But more interestingly, the chemistry behind the working principle of plastic photochromic lenses is quite different from that of glass. These photochromic lenses depend on carbon-based compounds [42]. So far, there have been significant improvements in this industry with the aim of producing fast reacting and effective photochromic surfaces. For instance, initially photochromic lens manufacturers were using pyridobenzoxazines. This is basically an organic dye. Currently, they use naphthopyrans, with others opting for indenonaphthopyrans. The structure of these carbon-based compounds may vary broadly. This will depend on what the manufacturer demands to use for its products. However, how photochromic lens works will still remain the same. Whenever you expose these plastic photochromic lenses to a certain threshold of ultraviolet radiation, the chemical bonds in these compounds will be broken. This will result in a complete change in the molecular structure. What happens is that the weak bonds will break and the molecules will rearrange to form species that absorb light at stronger wavelengths [43].

5.5.2 Materials Used in Photochromic Materials

Photochromic materials are molybdenum oxide, tungsten oxide, silver chloride, silver halide, polycarbonate, pyridobenzoxazines, triarylmethanes, stilbenes, azastilbenes, nitrones, fulgides, spiropyrans, naphthopyrans, spirooxazines, quinones, and others. Furthermore, photochromic formulations of standard colors are possible, where for example, the yellow color changes to red and red into purple. Photochromic inksPhotochromic inks are applied on substrates via processes such as flexography, serigraphy, dry offset, and typography. It is possible to formulate inks, which combine photochromic pigments with thermochromic or other chromogenic pigments. To obtain the best results, the photochromic ink layer should be as compact as possible and in the range of several micrometers [44]. One of the oldest, and perhaps the most studied, families of photochromes is the spiropyrans. Very closely related to these are the spirooxazines. For example, the spiro form of an oxazine is a colorless leuco dye; the conjugated system of the oxazine and another aromatic part of the molecule is separated by an sp^3-hybridized "spiro" carbon. After irradiation with UV light, the bond between the spiro-carbon and the oxazine breaks, the ring opens, the spiro carbon achieves sp^2 hybridization and becomes planar, the aromatic group rotates, aligns its π-orbitals with the rest of the molecule, and a conjugated system forms with the ability to absorb photons of visible light, and therefore appear colorful [45]. When the UV source is removed, the molecules gradually relax to their ground state, the carbon-oxygen bond reforms, the spiro-carbon becomes sp^3 hybridized again, and the molecule returns to its colorless state. This class of photochromes in particular is thermodynamically unstable in one form and reverts to the stable form in the dark unless cooled to low temperatures. Their lifetime can also be affected by exposure to UV light. Like most organic dyes they are susceptible to degradation by oxygen and free radicals. Incorporation of the dyes into a polymer matrix, adding a stabilizer, or providing a barrier to oxygen and chemicals by other means prolongs their lifetime. The "diarylethenes" gained widespread interest, largely on account of their high thermodynamic stability [46]. They operate by means of a 6-pi electrocyclic reaction, the thermal analog of which is impossible due to steric hindrance. Pure photochromic dyes usually have the appearance of a crystalline powder, and to achieve the color change, they usually have to be dissolved in a solvent or dispersed in a suitable matrix. However, some diarylethenes have so little shape change upon isomerization that they can be converted while remaining in crystalline form. The photochromic trans-cis isomerization of *azobenzenes* has been used extensively in molecular switches, often taking advantage of its shape change upon isomerization to produce a supramolecular result. In particular, azobenzenes incorporated into crown ethers give switchable receptors and azobenzenes in monolayers can provide light-controlled changes in surface properties. Some *quinones* and phenoxynaphthacene quinone, in particular, have photochromicity resulting from the ability of the phenyl group to migrate from one oxygen atom to another. Quinones with good thermal stability have been prepared, and they also have the additional feature of redox activity, leading to the construction of many-state molecular switches that operate by a mixture of photonic and electronic stimuli [47].

Many inorganic substances exhibit photochromic properties, often with much better resistance to fatigue than organic photochromics. In particular, silver chloride is extensively used in the manufacture of photochromic lenses. Other silver and zinchalides are also photochromic. Yttrium oxyhydride is another inorganic material with photochromic properties. Photochromic coordination complexes are relatively rare in comparison to the organic compounds listed previously. There are two major classes of photochromic coordination compounds. Those are based on sodium nitroprusside and the ruthenium sulfoxide compounds. The mode of action is an excited state isomerization of a sulfoxide ligand on a ruthenium polypyridine fragment from S to O or O to S. The difference in bonding from between Ru and S or O leads to the dramatic color change and change in Ru(III/II) reduction potential. The ground state is always S-bonded and the metastable state is always O-bonded. Typically, absorption maxima changes of nearly 100 nm are observed. The metastable states (O-bonded isomers) of this class often revert thermally to their respective ground states (S-bonded isomers), although a number of examples exhibit two-color reversible photochromism. Ultrafast spectroscopy of these compounds has revealed exceptionally fast isomerization lifetimes ranging from 1.5 ns to 48 ps [48].

5.5.3 Limitations of Photochromic Glasses

One of the most commonly reported drawbacks of photochromic glasses is that they take somewhat longer to clear than to darken. For example, if a person walks indoors after being in the sun for a while, the glasses will allow only about 60% of light to pass through them for the first 5 minutes. Additionally, they could take as long as an hour to clear completely and become fully transparent again. Another major drawback is that since the photochromic compounds rely on the presence of heat and light to undergo changes, such glasses are found to be more effective in winter and less so during summer (when you need them the most) [49]. People also report that photochromic glasses are less effective inside cars and other automobiles. This is because the glass windows of these vehicles block UV rays to a great extent themselves, so the photochromic molecules present in the glasses do not receive enough UV rays to undergo changes and be effective.

5.5.4 Applications of Photochromic Materials

Changes from one color to another color are possible by mixing photochromic colors with base colors which can be used in paints, inks, and so on, for many different applications. The lenses of photochromic sunglasses get darker with increasing UV intensity and optimize the light that passes through them (Fig. 5.3). When the UV intensity is lower, for example in the interior of a building, *smart windows* become more transparent and make it easier for the user to see through while allowing more solar radiation to enter the building for balanced internal temperature. Photochromic materials in the form of pigments can be mixed with conventional. These are applied when driving or when skiing at a high altitude when the snow reflects extra light into your eyes (Fig. 5.4a) [50]. Currently, the *fashion industry* shows a good deal of interest in photochromic textiles, used to create colorful designs and effects, which change according to the incoming light intensity. The characteristics of these textiles depend on the substance used on them. The selection of the reflected waves depends on the type of molecules or liquid crystals used in microcapsules, which react in different ways to light. The most applied synthetic substances used as dyes in the textile fibers are spiropyrans, spironaphtoxazines, and chromenes. These compounds have been studied and show good resistance toward fatigue and photodegradation. Other compounds are fulgides and diarylethenes. Methods of application are various. Some of these methods involve embedding the photochromic dye in the polymer matrix during the spinning phase of the synthetic fibers such as polypropylene. Dyeing and screen printing processes are considered more appropriate than competing processes since the demand for photochromic textiles is limited to piece garments, rather than batch production. Photochromic textiles have been used to tailor bathing suits or for work wear in low light conditions for increased visibility of the wearer. Photochromic dye onto textile material, if

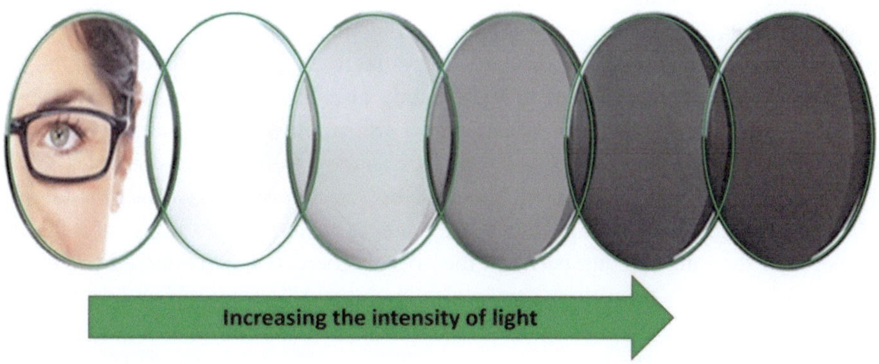

Fig. 5.3 Photochromic materials used in sun glass showing the color change with respect to the intensity of light

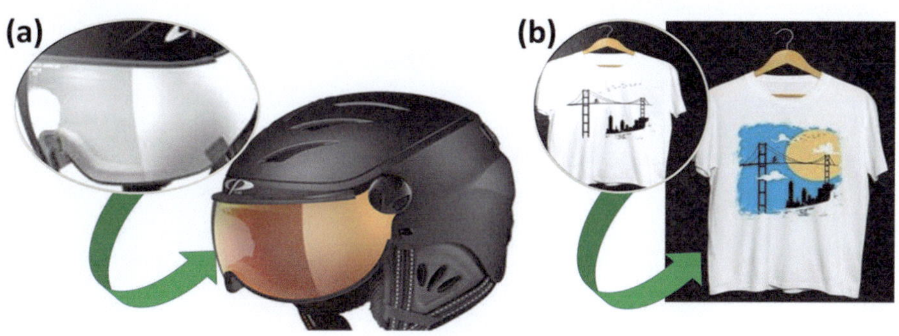

Fig. 5.4 Use of photochromic material in (**a**) headmet glass and (**b**) smart textile

appropriately applied, forms a photochromic system, which is as a matter of fact a sensor capable of detecting and reacting to UV light. This kind of smart textile can alert and protect against the negative influence of UV irradiation, showing the color at the moment in which such irradiation occurs (Fig. 5.4b) [51]. Photochromic systems have also been used for military applications, for example, in optical memory devices (photosensitive glass to register messages). In biology and chemistry, they have been used to create molecular switches and for optical data archiving in 3D. Light-sensitive photochromic materials are used in optical memory devices. For commercial use, for example, in consumer products (especially food) the photochromic inks must pass food container and packaging health and safety standards, and the producer must truthfully claim that they do not contain and potentially harmful or toxic substances [52]. Figure 5.5 shows the color change according to the intensity of light during morning to evening time.

5.6 Thermochromic and Thermotropic Materials

Thermochromic materials experience a change in color when subject to a change in temperature, that is, when the material is heated or cooled. They can be designed to change color at a specific transition temperature, which can be varied by doping the material with other compounds. Such materials have excellent practical applications, for example, a baby bath thermometer will change color to indicate the suitable temperature of the water. Thermochromic exists in two primary forms, liquid crystals and leucodyes [53]. Thermochromatic liquid crystals are structures that have many of the properties of a liquid while also having properties of solid crystals. When they are at lower temperatures these crystals appear to be black as they are in a more crystalline, solid form, which means that less light is reflected through them. While when they are at a higher temperature then the spacing between the crystals increases, which causes them to reflect light in different ways, and that causes different colors to be produced. The liquid crystals are used in products such as mood rings, thermometers, and inks. However, liquid crystals are very difficult to work with. Whereas leucodyes were more commonly used, for example, they were used to make heat-sensitive novelty mugs [54]. The development of the technology in creating these thermochromic materials is ongoing and as a result, the numerous applications are ever increasing. Leucodyes are combined with permanent-colored ink to produce a product, which changes from one color to another. The dyes are usually inside microcapsules where the dye is sealed inside with a solvent and an acid. At low temperatures, the solvent is a solid, which keeps the dyes and the acid close together, which means that they reflect light and produce a visible color but then as the temperature is increased the solvent melts, which allows the acid and the dye to be separated so no color is visible, which means that they become translucent so any color underneath is exposed. The applications of leuco dyes include labels and packaging to create a color change and in printing. This would result in an orange ink layer being printed, but once heated it would turn red since the leucodye would change from yellow to colorless [55].

Thermotropics, which exhibit large physical changes at certain temperatures, have also been studied and developed for glazing. These materials appear clear at lower temperatures but become opaque at higher temperatures. They can be used for skylights, inclined glazing, and upper windows where the view is not important. They can be totally passive, changing with ambient temperature for solar heating. Some designs use a resistive heating layer made from thin-film metals or transparent conductors, which enables electrical control of the physical change [56]. Thermotropics scatter light like polymer-dispersed liquid crystal (PDLC) materials, although PDLCs tend to be more isotropic scatterers. Most thermotropics are based on hydrogels, and polymer blends have been studied for higher temperature performance. One example of a thermotropic polymer gel is polyether/ethylene oxide/carboxyvinyl. A thermotropic called ThermoSEE has

Fig. 5.5 Photochromic glass in the building that changes color with respect to the intensity of the sunlight

been introduced by US Company Pleotint. The activation temperature can be set from −10 °C to 50 °C and the material is stable up to 85 °C [57].

5.6.1 Mechanism in Thermochromic Materials

The thermochromic effect can be abrupt at a certain temperature, or gradual, within a temperature range, depending on the material involved. This point (or range) is called the thermochromic switching or transition temperature. It is possible to adjust the switching temperature, T_g, by different approaches, depending on the type of material utilized (Fig. 5.6). Figure 5.7 has given a ray diagram of color change due to materials temperature. Metal oxide thermochromic materials such as V_2O_3, V_2O_5, and Ti_2O_3 are semiconductors below a critical temperature ($<T_c$), but behave as metals above T_c. In such systems, $T_g = T_c$ can be adjusted by doping [58]. For example, vanadium (IV) oxide, VO_2, is the transition metal oxide with the closest T_c ($T_c = 68$ °C). However, this temperature is still too high for many applications. Doping VO_2 using metal ions with larger atomic radii than the V^{4+} ion reduces the switching temperature. Tungsten (VI), niobium (V), and titanium (IV) ions are such dopants. For example, a 2 atm% addition of tungsten (VI) can reduce T_c to 25 °C. Similarly, it is possible to increase Tc by employing dopants with atomic radii smaller than the V^{4+} ion. The capacity of these materials to adopt different color states at different temperatures and to change color and return to their original color countless times in response to temperature fluctuations makes them particularly interesting. The thermochromic effect is based on a chemical equilibrium between two different forms of a molecule or between different crystalline phases. Two possible types of phase transformation causing thermochromism are termed first order and second order, depending on their thermodynamics. The color of transition metal oxides above the switching temperature can be affected by many factors. For example, heating of the metal oxide can change the oxidation state and this might result in errors of color interpretation because it may not behave as expected from a certain type of oxide [59]. Purity and material thickness are two other important factors to consider. Below the switching temperature of 68 °C, VO_2 single crystals are transparent both to visible and infrared light. The monoclinic crystal structure transforms to a tetragonal structure at T_c and the material becomes metallic. While still transparent in the visible spectrum, it is now reflective in the infrared region. This effect is useful for adaptive solar control applications. When thermochromic glazing reaches the switching temperature, it becomes reflective and prevents UV radiation to pass through the window, controlling the amount of heat caused by sunlight. When the temperature decreases below T_c, for example, in the afternoon or cooler months of the year, the window becomes transparent to UV

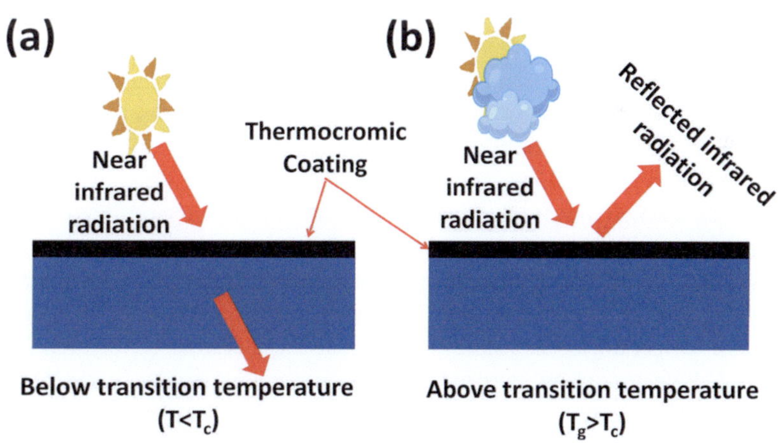

Fig. 5.6 Schematic of the mechanism of a thermochromic material when coating a surface for heat protection: (**a**) Below transition temperature and (**b**) above transition temperature

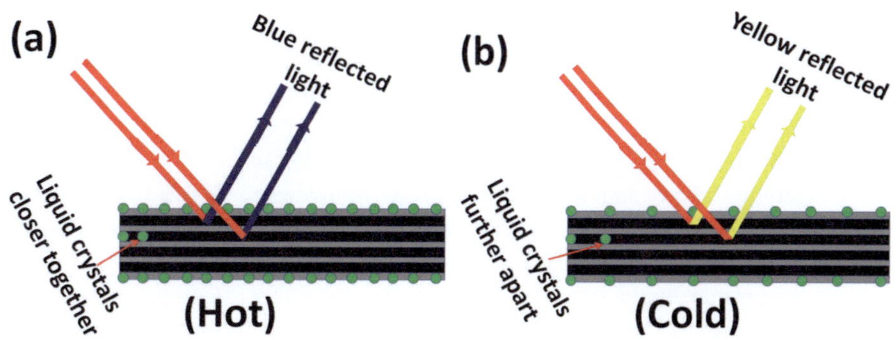

Fig. 5.7 Ray diagram of color change in thermochromic material at: (**a**) Hot condition and (**b**) Cold condition

radiation and allows making use of solar heating. Among inorganic systems, the thermochromism of mercury iodide (HgI_2) is notable. This compound displays a yellow–orange color at ambient temperature but turns red when heated to 200 °C. The color switching is due to a crystallographic transformation (a change in the position of iodine ions around mercury ions) at the switching temperature, similar to thermochromic metal oxides. Thermochromic colorants (pigments or dyes) are produced and commercialized by many chemical companies for applications in various sectors. Such colorants are used to produce paints and inks for superficial finishing or they are mixed with other materials such as polymers for bulk coloration. In the aerospace industry, thermochromic paints are used to change the emissivity of surfaces under a thermal effect [60].

Simply speaking, incoming light waves reflect off nearby crystals and add together by a process called interference, which produces the reflection. The color of the reflected light depends (in a very precise way) on how closely the crystals are together. Heat up or cool down your liquid crystals and changing the spacing between them, or push them into a different phase, altering the amount of interference and changing the color of the reflected light from black, through red and all the colors of the spectrum to violet and back to black again. In a nutshell, the liquid crystals look a different color depending on what temperature they are, because changes in temperature make them move closer together or further apart (depending on the material). In reality, the molecules in liquid crystals can form themselves into a number of different phases, and because they are not pointing randomly in all directions, they are generally anisotropic (they do different things to light when it hits them from different directions). In some thermochromic devices, the crystals start at low temperatures, in what is called the smectic phase, which means the molecules are organized in layers that slide easily past one another [61]. In this form, they happen to be completely transparent (they reflect little or no light, allowing virtually all light to pass through them). At higher temperatures, they shift to a different phase (known as chiral/cholesteric) and start to show shifting colors (sometimes called "color play") as they get hotter. At a certain higher temperature, known as the clearing point, the molecules stop behaving like liquid crystals altogether and shift to an entirely different form, known as an isotropic state (have the same optical properties in every direction). In this form, they are transparent once again [62]. Therefore, looking from the outside, you might see a material that changes from being transparent to colored, changes color as the temperature rises, and then goes back to transparent again; but on the inside, the molecules are doing radically different things at each temperature.

Thermochromic liquid crystals (TLCs) give a relatively accurate measurement of temperature within certain bands, so they are widely used in such things as strip thermometers (placed on a baby's forehead, perhaps, or stuck to the inside of a refrigerator or an aquarium tank) [63]. Typically, they are manufactured in the form of microscopic spheres (capsules) embedded in a plastic (polymer).

5.6.2 Materials used in Thermochromic Materials

Thermochromism is a phenomenon common in many chemical systems, both organic and inorganic, including metal oxides that transform into conductors at a specific temperature. Materials belonging to different groups such as polymers, solid-state semiconductors, or liquid crystals can show thermochromic behavior [64]. The thermochromic effect can be abrupt at a certain temperature, or gradual, within a temperature range, depending on the material involved. A number of *transition metal* compounds show temperature-dependent structural transformations with accompanying qualitative changes of their electronic properties at well-defined "critical" temperatures (τ_c). The most widely studied of these thermochromic materials is VO_2, which is monoclinic, semiconducting, and relatively infrared transparent at temperatures below τ_c, while it is tetragonal, metallic, and infrared reflecting above τ_c. Bulk-like VO_2 has $\tau_c \approx 68$ °C, which obviously is too high for practical windows. However, τ_c can be decreased to a comfortable temperature by doping with W and some other metals, while other dopants enhance τc. The decrease of τ_c in VO_2:W upon increasing W content has been reported to be from ~5 K/at.%W to ~24 K/at.%W. VO_2-based multilayer films with enhanced luminous transmittance and beneficial angular-dependent properties as well as on VO_2:Mg films for which the doping boosts the transmittance and simultaneously decreases τ_c [65].

VO_2, TiO_2, and multilayers based on these were produced by reactive DC magnetron sputtering. A basic problem with thermochromic fenestration using VO_2-based films (VO_2:Mg films) is the absorption at $\lambda < 0.5$ μm, and obviously, the optical properties would be much improved if the absorption edge could be moved toward the ultraviolet. The situation is similar to that in electrochromic NiO-based films, for which doping with Mg and some other elements known to form wide band gap oxides yielded improved transmittance. Similar band gap widening has been noted also in $Mg_xZn_{1-x}O$ nanoparticles [66].

5.6.3 Advantages and Limitations of Thermochromic Materials

Compared to other types of chromogenic windows, thermochromic windows have some advantages. Thermochromic materials are self-regulating with short response times; they reduce cooling and ventilation loads in an autonomous mode and eliminate problems of overheating by regulating the solar intake, contributing thus to energy savings; they diffuse the light in a constant and uniform manner both in the opaque as well as transparent state; they are simple to apply during building construction; they have low cost and long durability.

They also have some characteristics, which may be constituted as disadvantages: they are never completely transparent, thus they block beneficial solar rays in the winter; they can only be regulated using electrical circuits printed on the layers that cover the thermochromic film or layer, and some polymeric films have the tendency to turn yellow with time due to UV radiation (which can be solved by chemical stabilizers). A persistent problem with thermochromic VO_2 has been the difficulty to reach a sufficiently high luminous transmittance in thin films without resorting to so small thicknesses that the thermochromic effect in the infrared is essentially lost. The liquid crystals are more difficult to work with the leuco dyes as their performance decreases when they are repeatedly exposed to UV light and high temperatures. In addition, they are more expensive and are difficult to work with [67].

5.6.4 Applications

Thermochromic materials are used to make paints, inks or are mixed into molding or casting materials for different applications. A well-known product that makes use of this phenomenon is a ceramic mug (Fig. 5.8a), which changes color when a hot drink is poured inside. The transformation is reversible; thus, the color of the mug goes back to its original one when it cools down to room temperature. A less artistic approach is to have a mug, cup, or dish that changes to a color that indicates the food or drink is too hot or not [68]. Designers think these are great materials for creative experiments, for example, shirts that change color with temperature. Simple 'strip' thermometers that change color with a rise in temperature, so used as a color-temperature scale on a plastic strip that place on human skin. They are made using a mixture of pigments that become visible at different temperatures. Figure 5.8b is a picture of a simple household strip thermometer using a set of thermochromic pigments [58]. It is reading 18 °C with a hint of 21 °C, they are not very precise but give a quick and convenient indication of whether the room is warm or cool. You can design electric kettles incorporating thermochromic pigments in the plastic body that change color when the water boils. Other domestic uses of thermochromic pigments include food wrapping or containers for food stored in refrigerators or freezers to indicate they are cool enough [69]. Mood rings (Fig. 5.8c) use thermochromic pigments to change color as the temperature of your finger changes. Many baby products use thermochromic pigments as a safety feature to indicate if food/water is too hot to eat/bath a baby or young child. Thermochromic pigments are used in baby baths, baby toys, and baby feeding spoons to produce a color that indicates things are too hot or not (Fig. 5.9). If you include thermochromic pigments in acrylic paints, you can have painted walls or woodwork that change color at a temperature around room temperature, for example, the blue paint below 25 °C and looking yellow above 25 °C. Figure 5.10 shows the thermochromic painted car that changes color when in contact with the hot water.

Applied in buildings, thermochromic windows allow optimizing the energy consumption of the building. For window applications, thermochromism has been made possible by the use of metallic oxide coatings such as vanadium dioxide (VO_2) or tungsten trioxide (WO_3). Another approach

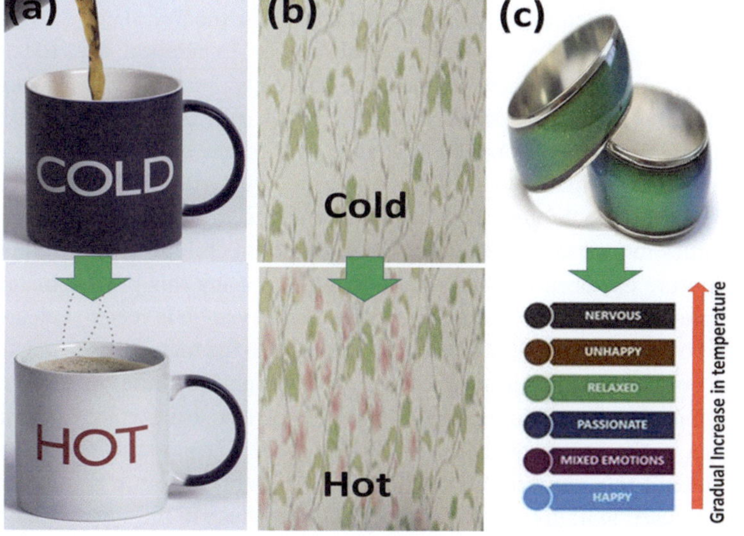

Fig. 5.8 (**a**) Ceramic mug, (**b**) wall design, and (**c**) mood ring

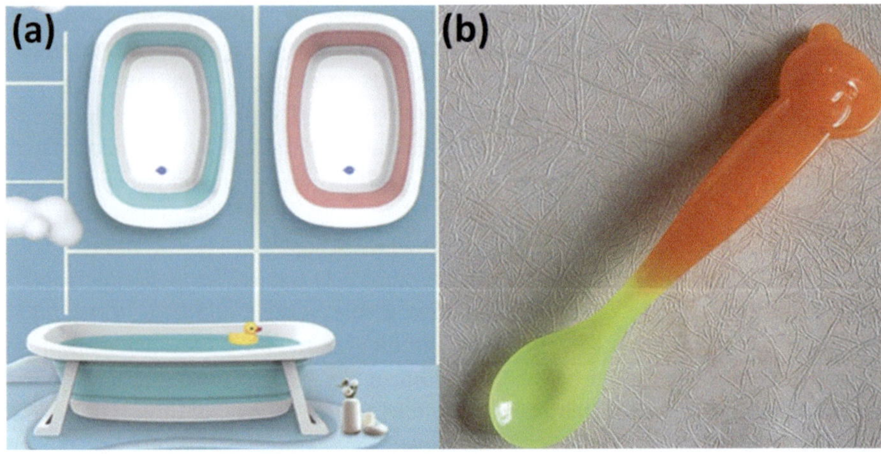

Fig. 5.9 Thermochromic materials in (**a**) baby bath and (**b**) baby feeding spoon

Fig. 5.10 Car design observable by hot water (**a**) Starting of pouring and (**b**) complete pouring

is to apply an intermediate gel layer between two layers of glass or plastic films. Thermochromic coatings based on VO_2 offer optimum performance with regard to interior comfort based on an ideal relationship between switching temperature and environmental temperature. Gel systems typically depend on a considerable change in the light scattering properties at the switching temperature. This may be considered as a variant of thermochromism, which is termed thermotropism. Thermotropism may occur due to different phenomena such as phase separation in a gel or phase transition between an isotropic and an anisotropic state [70]. Thermochromic windows automatically modify their transparency and thus their light absorption in relation to their external surface temperature. Initially transparent, above a critical temperature, which may vary from 10 to 90 °C, they become opaque. When the temperature decreases below the critical value, they return to being transparent [71]. Thermochromic inks are used on beverage cans or bottles to indicate whether or not the optimum chilling temperature has been reached. Many brands are using this principle to indicate to the consumer when the beverage is ready to be served. The ink can be directly applied to the metal cans or on a label, which is attached to the glass or plastic bottle. A common approach in the label or print design is to use a white area in the image, which turns to blue when the product is chilled to the desired temperature [72]. A thermochromic is even used in highways' road signs. The snowflake in the sign starts to change to blue below 2 °C and becomes completely blue below −1 °C to warn drivers about icy road conditions [73]. The fabric of the bag contains multiple layers of thermochromic ink printed in dots that are activated by heat patches positioned underneath. Many heat elements are mounted inside the bag and when a heating element is turned on (individually or combined in a group), the surface print changes color [60].

5.7 Electrochromic Materials

Electrochromic materials are characterized by an optical change upon the application of an electric field. This effect is used in liquid crystal displays [74]. Electrochromic materials change color or opacity reversibly upon application of an electric field or transfer of an electric charge. Electrochromism is due to the principle of oxidation-reduction (redox) reactions. This allows the electrochromic material to reversibly change color by gaining or losing an electron(s). The first discoveries on the phenomenon of electrochromism date back to 1953 when Kraus noted that upon application of an electric field, tungsten trioxide assumed an intense blue color [75]. The term electroluminescence is also used to describe this effect. Electroluminescent materials can produce brilliant colors if stimulated by an a.c. current. The difference between electrochromic materials

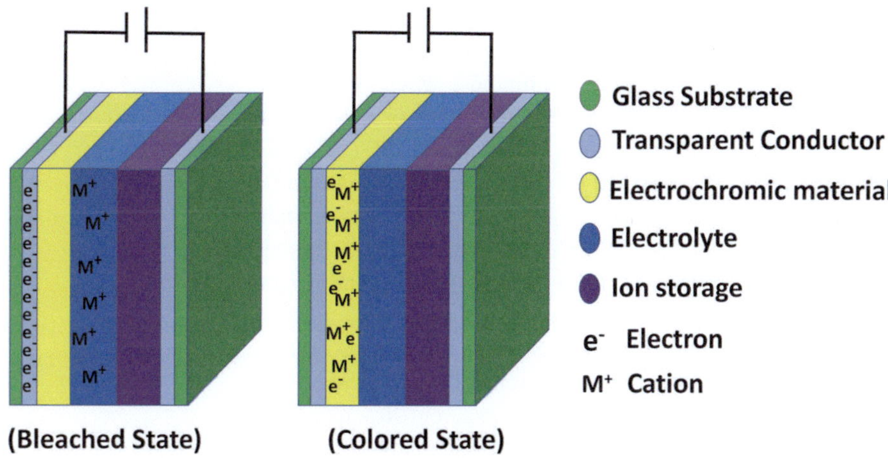

Fig. 5.11 Schematic of electrochemical redox and oxidation reactions observed in electrochromic materials

that can reversibly change color, and electro-optic materials that can reversibly change their optical characteristics (e.g., transparency or clarity), in response to electricity. Examples of electro-optic materials would include polymer dispersed liquid crystals, suspended particle devices, and films incorporating micro-blinds. Common examples of electrically powered technologies are electrochromics, suspended particle devices - also known as electrophoretic media, phase dispersed liquid crystals, and cholesteric liquid crystals. The market is divided into two major product areas, cathode ray tubes, and liquid crystals. These displays are seen in a variety of electronic products from televisions to personal data assistants. The major trend is toward thinner and lighter weight designs, based on flexible plastic substrates instead of glass. Cathode ray tube technology is likely to retain a good portion of the display market, but its share has been in decline since 2001. Three of the most dynamic areas of development are active matrix liquid crystal, plasma, and 'other' displays including OLED, electrochromic, and electroluminescent technologies [76].

5.7.1 Mechanism of Electrochromic Materials

An electrochromic device is essentially a rechargeable battery in which the electrochromic electrode is separated by a suitable solid or liquid electrolyte from a charge balancing counter electrode, and the color changes occur by charging and discharging the electrochemical cell with an applied potential of a few volts. After the resulting pulse of current has decayed and the color change has been effected, the new redox state persists, with little or no input of power, in the so-called memory effect [77]. Figure 5.11 illustrates an electrochromic device configuration. An electrochromic device conventionally consists of a "sandwich" configuration of electrodes, necessitating the use of at least one optically transparent electrode, such as indium tin oxide (ITO). This type of electrochromic, in its simpler configuration, is made of three layers of electrochromic materials deposited in the form of thin film on glass or polymer substrates. The middle layer is an ionic conductor (electrolyte); it loses ions when an electric current flows through it at relatively low voltage (1–5 V). The electrolyte layer is positioned between two layers: an electrochromic film, which acts as the electrode, and a layer for the accumulation of electrons (counter electrode). The counter electrode can be of any material with a suitable reversible redox reaction [78].

The electrochromic electrode of these devices, which can work either in the reflective or transmissive mode, is constituted by a conductive, transparent glass coated with electrochromic material. The electrolyte material in close proximity to the electrochromic layer, as well as transparent layers for setting up a distributed electric field that provides a reversible electrochemical reaction in devices operating in the reflective mode (like electrochromic displays); by contrast, in variable light transmission electrochromic devices (like electrochromic windows). Devices are designed to shuttle ions back and forth into the electrochromic layer with applied potential. Electrochemical stability can be increased by using interfacial layers [79].

In some cases, the ion storage layer is also an ionic conductor so efficient that it can substitute the electrolyte layer. Sometimes it also has electrochromic characteristics such that it can simultaneously substitute three layers (electrochromic, ionic conductor, and ion reservoir). In the case of devices based on organic semiliquid materials, two chemical species are mixed with a redox reaction on the surface of the electronic conductors, assuring optical modulation. The devices consist of electrochemical materials whose chemically active electrodes are deposited in the form of thin films. Following an electrical input at relatively low voltage (1–5 V), the electrochromic process takes place in which the material undergoes a phase transformation and a change in color and transparency. To guarantee the best adhesion between various layers and uniformity of surfaces, the assembly of the device is realized in an autoclave at a pressure of about 10 atm. Finally, the composite is sealed and rendered hermetic for common applications [80].

In the case of inorganic compounds, when an electric current is applied, the electrochromic effect occurs by the mutual injection or extraction of positive ions (M^+) and electrons (e^-). The passage of electric current provokes a change of electron density in the compound to maintain electrical neutrality [81]. This is the factor that modulates the optical behavior. In most of the solid metal oxide electrochromic materials, the atoms of these metals pass to a different valence from which they normally occur as oxides. These atoms at different valences noted as color centers can assign a certain color to a normally transparent structure. One of the most commonly used electrochromic materials is tungsten oxide (WO3), which occurs as a yellow, nontoxic, and solid powder at 20 °C. Its application temperature can range from −40 to120 °C. Used as a cathodic coloring compound in electrochromic films, it reacts to the influence of an electrical field (by electrochemical reduction) and changes color from transparent to blue. It can sustain more than 100 cycles of color change and shows a relatively good light fastness compared to organic electrochromic compounds based on polymers. This material is the principal chemical substance used for the production of smart glass for windows. It can be coupled with reflective metallic surfaces of Al, Ag, or Ni, to obtain optically active mirrors [82].

Equation 5.10 is an example of Ti metal, which is colorless in oxidized form (TiO_2) and blue-grey in reduced form (M_xTiO_2). In Eq. (5.11, the vanadium in oxidized form (V_2O_5) is brown/yellow and in reduced form ($M_xV_2O_5$) is pale blue. Niobium is colorless in oxidized form (Nb_2O_5) and in reduced form $M_xNb_2O_5$ is multicolor (Eq. (5.12). Tantalum is colorless in oxidized form (Ta_2O_5) and in reduced form $M_xTa_2O_5$ is deep blue (Eq. (5.13). Molebdnum is pale blue in oxidized form (MoO_3) and in reduced form M_xMoO_5 is very intense blue (Eq. (5.14). Tungsten is very pale yellowy in oxidized form (WO_3) and in reduced form M_xWO_3 is intense blue (Eq. (5.15). Manganese is brown in oxidized form (MnO_2) and in reduced form Mn_xO_2 is yellow (Eq. (5.16). Iron is brown in oxidized form (Fe_2O_3) and in reduced form Fe_3O_4 is black (Eq. (5.17). Cobalt is dark brown in oxidized form (Co_3O_4) and in reduced form $3CoO$ is pale yellow (Eq. (5.18). Nickel is grey/brown in oxidized form (NiO_x) and in reduced form M_yNiO_x is colorless (Eq. (5.19). Rhodium is yellow in oxidized form (Rh_2O_3. $5H_2O$) and in reduced form RhO_2. $2H_2O$ is green (Eq. (5.20). Iridium is brown in oxidized form (Ir_2O_4. H_2O) and in reduced form $M_xIr_2O_3$ is colorless (Eq. 5.21) [83–87].

$$TiO_2 + xM^+ + xe^- \leftrightarrow M_xTiO_2 \quad (5.10)$$

$$V_2O_5 + xM^+ + xe^- \leftrightarrow M_xV_2O_5 \quad (5.11)$$

$$Nb_2O_5 + xM^+ + xe^- \leftrightarrow M_xNb_2O_5 \quad (5.12)$$

$$Ta_2O_5 + xM^+ + xe^- \leftrightarrow M_xTa_2O_5 \quad (5.13)$$

$$MoO_3 + xM^+ + xe^- \leftrightarrow M_xMoO_5 \quad (5.14)$$

$$WO_3 + xM^+ + xe^- \leftrightarrow M_xWO_3 \quad (5.15)$$

$$M_XMnO_2 + xM^+ + e^- \leftrightarrow Mn_xO_2 \quad (5.16)$$

$$Fe_2O_3 + xM^+ + xe^- \leftrightarrow M_xFe_2O_3 \quad (5.17)$$

$$Co_3O_4 + xM^+ + e^- \leftrightarrow 3CoO + 2OH^- \quad (5.18)$$

$$NiO_x + yM^+ + e^- \leftrightarrow M_yNiO_x \quad (5.19)$$

$$Rh_2O_3.5H_2O + xM^+ + e^- \leftrightarrow RhO_2.2H_2O \quad (5.20)$$

$$Ir_2O_4.H_2O + xM^+ + e^- \leftrightarrow M_xIr_2O_3 \quad (5.21)$$

5.7.2 Materials Used

The most studied electrochromic materials are tungsten oxide (WO_3) and nickel oxide (NiO), as cationic and anodic electrochromic materials, respectively, but many other electrochromic metal oxides exist, such as V_2O_5, TiO_2

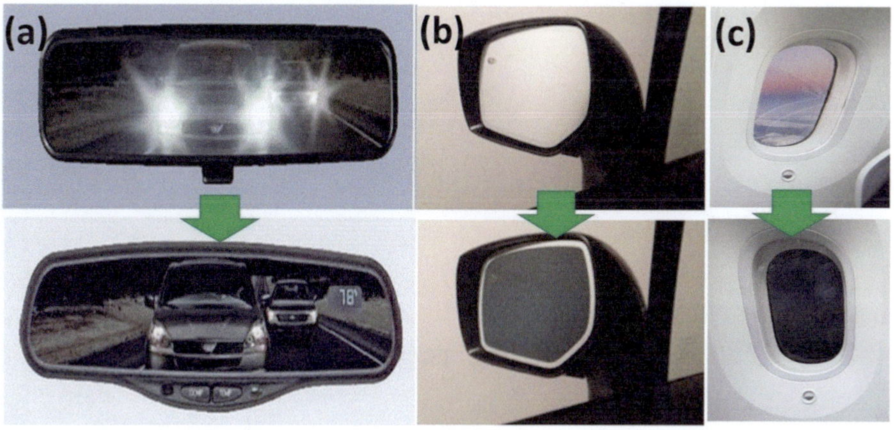

Fig. 5.12 Thermochromic materials in (**a**) rearview mirror, (**b**) side mirror, and (**c**) airplane window

MoO_3, Nb_2O_5, and Ta_2O_5 (cathodic electrochromic materials), and IrO_2, CoO_2, MnO_2, FeO_2, $Cr_2\,O_3$, and RhO_2 (anodic electrochromic materials) [88, 89]. These oxides form the basis for devices that can be used in a large variety of applications ranging from sunglasses to intelligent windows, from antifog rearview mirrors to computer displays. These films have been deposited by several techniques such as vacuum evaporation, sputtering, spray deposition, electrodeposition, electrochemical oxidation of tungsten metal, chemical vapor deposition, sol-gel method, and many more. The electrochromic effect also occurs in organic materials including various polymers (polyaniline, polypyrrole, polithiophene, polyisothiophene, phthalocyanines, viologens, and pyrazoline). Most of these materials, both organic and inorganic, are electrochromic at the solid or viscous liquid state while so far, no electrochromicity in the gas state has been demonstrated. Buckminsterfullerene also shows electrochromic behavior [78].

5.7.3 Applications

Electrochromic devices are one of the parts of smart material applications with huge commercial interest, with their controllable transmission, absorption, and reflectance. Most of the research in this field is concentrated on electrochromic windows of small- and large-area. Other possible applications such as flat panel display, building and automotive windows, solar cells, frozen food monitoring, mirror, and document authentication are also of great interest [90, 91]. The transparency of these windows changes depending on the voltage applied across the window. The less voltage that is applied across the window, the more transparent it is [92]. Using the rearview mirror and side mirror in Fig. 5.12a, b, a driver can dim the light according to his need and preference. Figure 5.12c shows the electrochromic coated airplane window for easy visibility without any hamper with the eye from external UV rays. Airbus A380 has fitted dimmable windows to the first-class cabin. Saint-Gobain has already shown a prototype electrochromic cabin window, which has a 40:1 contrast ratio with deeply colored visible transmittance of less than 1%. Electrochromic materials are very useful and have many applications including systems to reduce glare, thermal control, and as lenses in cameras and sunglasses (Fig. 5.13a) [93]. A big market for electrochromic materials today is dynamic antiglare mirrors that detect glare and automatically compensate for it, especially for nighttime driving safety. Chromogenic materials are finding application in specialized displays. Electrophoretic technology is being used in an electronic book, paper-like, and banner displays (Fig. 5.13b) that can store and display information and graphics on a flexible substrate, is becoming increasingly important [94]. Electrochromics can be found in low information content displays and indicators. Both phase dispersed liquid crystals and cholesteric liquid crystals are being used [95]. Many chromogenic materials can be fabricated on plastic substrates, which is an advantage for future display applications. Electrochromic materials can be used to light strips for decorating buildings and for industrial and public vehicles displays, for example, of safety notices. Electrochromism is the most versatile of all chromogenic technologies because it is the easiest to control and it can easily be used in combination with different stimuli such as stress or temperature [96].

Electronic paper (e-paper) is based on electronic ink (e-ink) and applied as a display in an e-book. The e-book display allows one to visualize text and images with effects very similar to printed paper, with even better contrast. It differs from a normal display because it does not use backlighting and reflects the ambient light as would do a piece of paper. Its main advantage compared to a backlight screen is the ability to read even under direct light conditions. According to the most widely used technique, the e-book display is produced by a multilayer polymeric composite

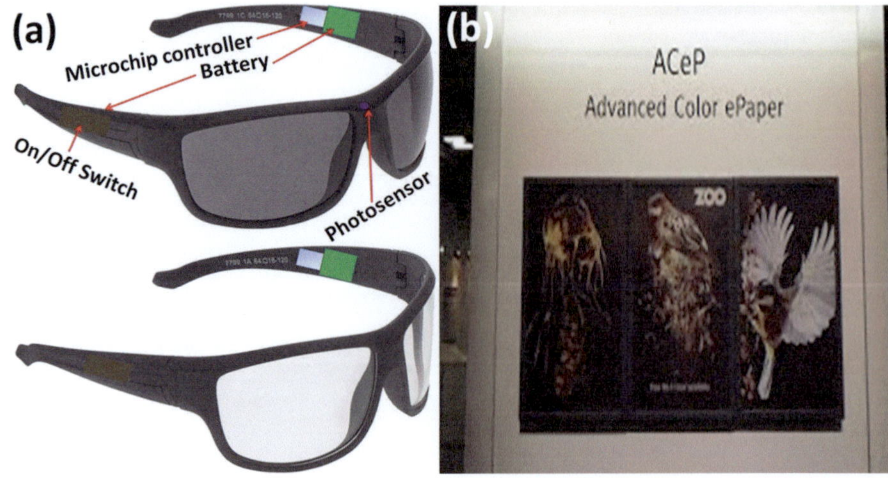

Fig. 5.13 Thermochromic materials in (**a**) electric spectacle and (**b**) banner display

that incorporates spherical microcapsules greatly reduced in size and which contain electrochromic pigments that react to electrical stimuli at low voltage (Fig. 5.14a). The spheres are electrically charged and have a positive part where the black pigments are located and a negative part where the white pigments are located (Fig. 5.14b). Through the electrical field, the spheres orient themselves to change the gray level in different parts of the screen (Fig. 5.14c). A Chinese firm Hanvon obtained e-papers capable of displaying various colors by placing a layer on the microspheres, which filters the light reflected from the spheres themselves. In 2013, e-Ink introduced spectra three colors electronic paper displays (white, red, and black) planned for electronic signage, announcing that other colors will be produced in the future [97].

Today, the most promising technology for a smart window is electrochromic (Fig. 5.15). The electrochromic glass changes its optical properties at the click of a button or is programmed to respond to changing sunlight and heat conditions in smart systems. Thus, it allows control over the amount of solar radiation, light, and heat passing through, providing visibility even in the darkened state and thus preserving visible contact with the outside environment. Electrochromic windows are useful in buildings (for windows and skylights), vehicles, and aircraft, according to a rational energy management scheme and environmental comfort. Electrochromic window multilayers are designed such that they allow the storage of ions and their movement back and forth for the insertion to or extraction from the electrochromic layer, through the application of an electric voltage. An electrochromic thin film stack with a thickness of about 1 μm is deposited on a glass substrate. The stack consists of ceramic metal oxide coating with three electrochromic layers sandwiched between two transparent electrical conductors. When a voltage is applied between the transparent electrical conductors, a distributed electrical field is set up. This field switches the glazing between a clear and transparent blue-gray translucent state, with no degradation in view, similar in appearance to photochromic sunglasses. The changing color can be modulated to intermediate states between clear and fully colored [98].

Furthermore, by using advances in transition metals, it was possible to develop hydride reflective electrochromic windows, which become reflective rather than absorbing, electrochromic antiglare glass that satisfies requirements of color upon request. The electrochromic windows with low-emittance coating and an insulating glass unit configuration, which can be used to reduce heat transfer from this absorptive glazing layer to the interior. The principal characteristic of electrochromic devices is the low energy consumption both during switching of the state as well as for storing the state of color in which the glass is located (this occurs without any energy consumption) [99].

Liquid crystals: The discovery of the liquid crystalline property possessed by certain organic compounds contributes a lot to the today world. While preparing cholesteryl benzoate, the substance seemed to present two distinct melting points with the formation of initially a rather cloudy liquid phase which successively turned perfectly clear at a higher temperature. It retained that the intermediate state was nothing but a new phase of the material, where in conditions of typical fluidity of the liquid state, there was a certain molecular order, which resembled the order of solid crystals. Because of the intermediate characteristics between those of isotropic liquids and those of crystals, the new phase was properly called liquid crystal. In a solid crystal, atoms (or molecules in the case of polymers) display a great degree of order both positional as well as orientational order, with very little freedom of displacement. On the contrary, in an isotropic liquid, molecules do not have any particular order while they have ample freedom of movement. The molecules of a liquid crystal have characteristics in between, conserving certain freedom of displacement while showing a tendency to assume preferential positions and orientations. Principally the

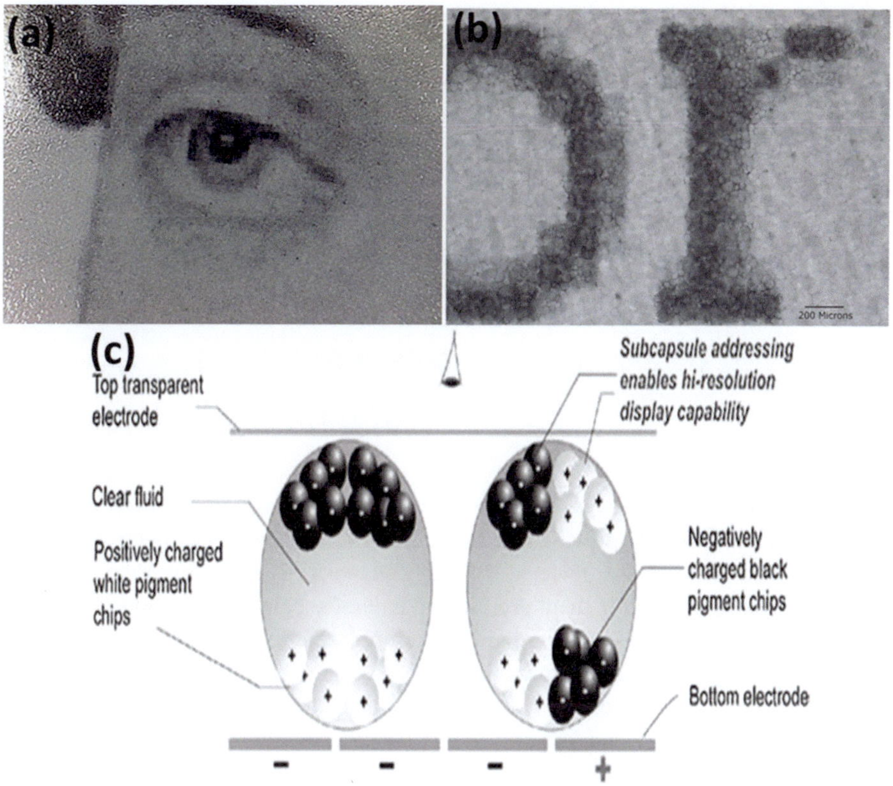

Fig. 5.14 (**a**) Image of eye on the e-ink screen of the Kindle 3 under direct light conditions, (**b**) View of an e-ink display under a microscope of 40× magnification, and (**c**) Cross section of electronic-ink microcapsules by e-Ink Corporation

Fig. 5.15 Sage electrochromic windows with tintable glass

anisotropy, which occurs due to the preferential orientation of the liquid crystal molecules, makes them useful in many commercial applications. For example, liquid crystal display (LCD) [100].

Liquid crystals can be processed into films, to produce electro-optic layers suitable for several applications. One example in today's market is Nanotec film by Unilux, a thin and flexible film with consists of a sandwich of two layers of PET that holds a polymer network ensuring constant and superfast conduction of electricity. This consists of a sponge-like structure that is filled with liquid crystals. Bistable liquid crystal devices have been extensively investigated because of their unique electro-optic properties, mainly the ability to maintain an image indefinitely without power consumption. The optical requirements of liquid crystals devices are light transmittance almost constant in two different states (on and off); a light transmittance factor of 60–80% in the active state, and 44–60% in the inactive state. Liquid crystals are commonly used on telephone screens, digital cameras, portable computers, and other electronic products. In these screens, the electrical current causes a reorientation of liquid crystal molecules parallel to the electric field, changing accordingly the mode in which light passing through the display becomes polarized [101].

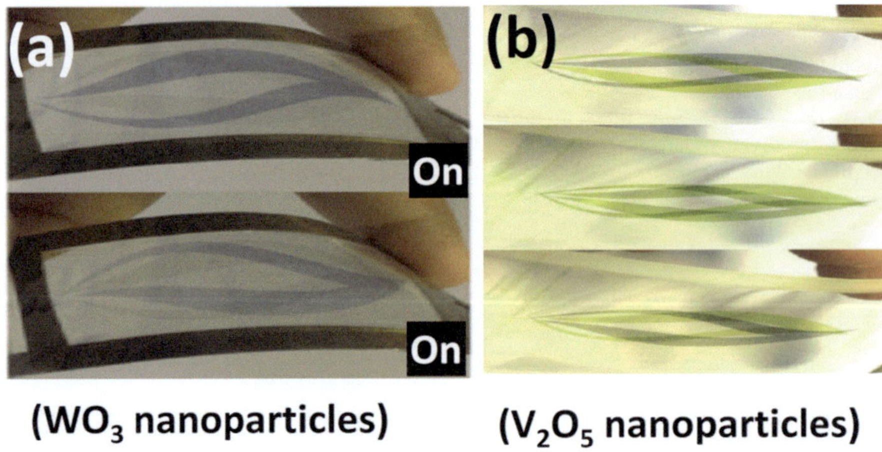

Fig. 5.16 Photograph of flexible electrochromic display with (**a**) WO_3 and (**b**) V_2O_5 nanoparticles, deposited by an inkjet printing technique

Electrochromic displays on flexible substrate: In the last decade, there has been a growing interest in the development of flexible devices. These devices bring the advantages of being less expensive with their production methods being easily accessible, such as sol-gel or inkjet printing. Several authors have reported the production of electrochromic devices, based on metaloxide nanostructures, on flexible substrates like polyethylene terephthalate (PET) or polyethylene naphthalate (PEN). 2D ultrathin WO_3 nanosheets obtained by exfoliation of tungstate-based inorganic-organic nanohybrids used in flexible devices (Fig. 5.16a). Figure 5.16b shows the V_2O_5 nanoparticles assisted flexible electrochromic display, developed by an inkjet printing technique. PET is used as a substrate, covered with Indium tin oxide (ITO) thin film as a working electrode and counter electrode. As an electrolyte, a solution containing lithium perchlorate is used in propylene carbonate [102].

5.8 Gasochromic Materials

Gasochromic materials change color in the presence of gas and they have the capacity to return to the initial state once the gas contact is removed [103]. The typical gasochromic materials are tin oxide (SnO_2), which reacts to flammable gases; palladium, which is sensitive to hydrogen (H_2); titanium tin oxide (Ti, Sn)O_2 which reacts with carbon monoxide (CO), propane (C_3H_8), ethanol (C_2H_5OH), and H_2; tungsten oxide (WO_3), which reacts with hydrogen disulfide (H_2S) [103, 104]. These gasochromic materials can be applied as a nanocoating by chemical vapor deposition processes. Optical coatings have been developed for applications in window systems with the function of reducing thermal energy loss or solar overheating of buildings, reducing heating, or cooling costs. These windows or other transparent building components, when exposed to the gas, become activated by the presence of a catalyst layer. Science fiction novels and movies are a great inspiration for innovation [105].

Unlike electrochromic devices, gasochromic devices do not need electrode layers. Gasochromic windows transmit more than 80% of solar radiation (Fig. 5.17a). Gasochromic windows involve a double pane construction with a gap between the two panes. One of the surfaces is coated with WO_3 and a thin layer of catalyst on top. The thin layers, invisible to the naked eye, transform regular glass into a gasochromic one [106].

5.8.1 Mechanism of Gasochromic Materials

Gasochromics are based on the materials' properties such as WO_3thin films to color in the presence of reducing gas with a suitable catalyst. Reducing gases, such as H_2, H_2S, and NH_3 will increase the surface conductivity of WO_3 by creating oxygen vacancies and will result in color changes. Oxidizing gases, such as O_3, NO_2, and also CO_2, will reduce the number of vacancies, resulting in a more transparent material, with a reduction in conductivity [107]. Gasochromic window construction follows a double pane model with a gap between the two panes. On one pane, a film of WO_3 is deposited with a thin layer of catalyst on top. Hydrogen gas is fed into the gap producing coloration of the WO_3 layer. This process is completely reversible by replacing the H_2 with O_2. When a small quantity of hydrogen is pumped into the gap between the panes, WO_3 obtains a blue tinge but remains transparent. The color change is attributed to H_2 intercalation with a WO_3 lattice that promotes the formation of small polarons as quasiparticle states [108]. The reaction involved is shown in Eq. (5.22:

$$WO_3(\text{colorless}) + xH_2 \rightarrow 2xH^+ + WO_3 \rightarrow H_{2x}WO_3(\text{blue}) \quad (5.22)$$

When oxygen is introduced, the reaction is reversed and the glass becomes clear again. A small electrolysis unit is hidden in the facade that produces the hydrogen gas and a

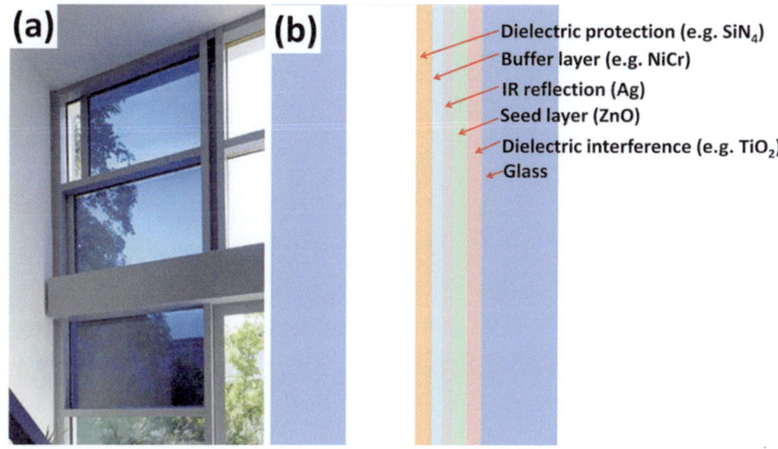

Fig. 5.17 (**a**) Darkening effect of gaschromic windows and (**b**) various layers that deposited on the glass

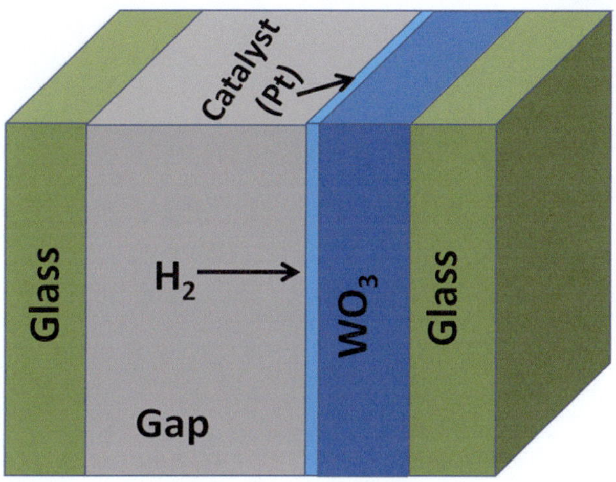

Fig. 5.18 Cross section of the gasochromic glazing

miniature pump diffuses the desired gas mixture into the cavity (Fig. 5.17b). The hydrogen concentration is very low and does not pose any risk. An example of a gasochromic window is shown in Fig. 5.18. Transmittance of 75–18% has been obtained, which is better than most electrochromic windows. The windows can color with 0.1–10% hydrogen, which is below the flammability concentration. The visual perception during the switching of color is fascinating: just like in a solar eclipse, the level of ambient light changes in a uniform fashion in a few minutes, in the total absence of noise.

Regarding MoO_3, the hydrogen atoms will intercalate into MoO_3, forming H_xMoO_3, which decomposes into MoO_2 + 1/ $2H_2O$ with a fast reduction of Mo^{6+} into Mo^{5+} states and slow and simultaneous formation of Mo^{4+} states. Oxygen has a larger electronegativity when compared to hydrogen. Therefore, most of the binary d-metal oxides are more stable than the corresponding hybrids. When exposed to gas mixtures with hydrogen, most transition metals form binary oxides. Hydrogen can intercalate into metal oxides acting as a donor or acceptor of electrons, inducing a variation in optical and electrical properties of the material, by forming molybdenum bronze, $H_xMo_x^{5+}Mo_{1-x}^{6+}O_3^{10}$. Hydrogen atoms occupy sites in the van der Waals gaps between double layers of MoO_6 octahedral as well as interlayer sites. This results in a small increase in the cell volume and distortion of the lattice changing the overall crystal symmetry from orthorhombic to monoclinic (Fig. 5.19) [109].

5.8.2 Applications of Gasochromic Materials

The group of gasochromic materials is mainly used in sensors and biosensors. An important application of MoO_3- and WO_3-based gasochromic devices is in the detection of hydrogen leakage, changing color from transparent to dark blue. A gasochromic WO_3/H_2 sensor can be used for the selection of specific mutant green algae that produce H_2 gas while tolerating high levels of O_2 presence. One example is the Chlamydomonas that are adapted to anaerobiosis in the dark and when illuminated they produce H_2 which is detected by a Pd/WO_3 system that turns blue. The selection of this algae will allow the development of biohydrogen production under atmospheric levels of O_2 [110].

5.9 Mechanochromic/Piezochromicmaterials

Mechanochromic/Piezochromicmaterials show a change in color when a mechanical stimulus, that is, stress, is applied [111]. These smart materials can respond to compressive, tensile, or more complex forms of stress, particularly for in situ failure monitoring due to fracture, corrosion, fatigue, or creep. Compared to other types of chromogenic materials such as thermochromic or photochromic materials, mechanochromic materials have not received much attention so far, because the related technology is not mature yet.

An interesting project is on flexible materials with stress-sensitive colors. Researchers from the University of

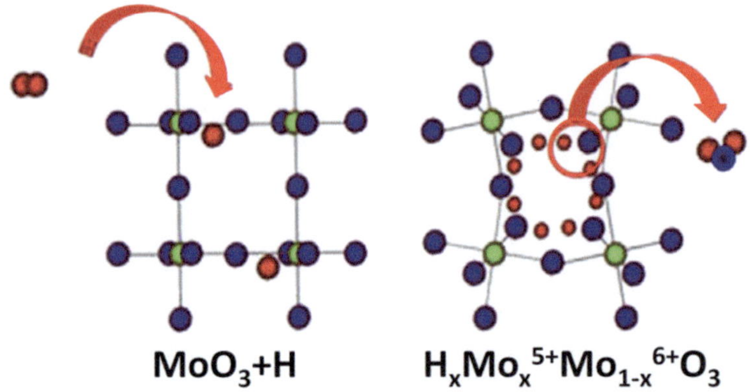

Fig. 5.19 Crystal structure of MoO_3 and hydrogen bronze

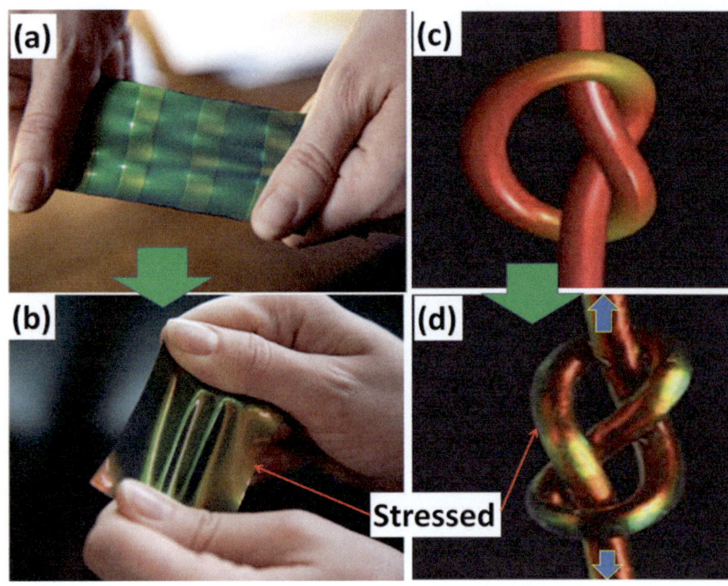

Fig. 5.20 (**a**, **b**) Stress in polymer opal, and (**c**, **d**) Stress in polymer rope

Cambridge and Fraunhofer Institute have developed what they call a "polymer opal" [112]. *Opal* is a hydrated amorphous form of silica ($SiO_2 \cdot nH_2O$); its water content may range from 3 to 21% by weight but is usually between 6% and 10%. Because of its amorphous character, it is classed as a mineraloid, unlike crystalline forms of silica, which are classed as minerals. While natural opals display a multitude of colors due to silica spheres that settle in different layers, polymer opals consist of one preferred layer of nanoparticles, resulting in uniform color [113]. A remarkable feature of polymer opal is the intense colors, which are based on the same physical principles found in nature, similar to that found in some beetles, butterflies, and peacocks. This type of color is known as structural color because it results from the diffraction of light due to the structural layout of nanoscale layers. An important advantage of structural color is that it is not dependent on time and temperature as long as the structure is preserved. This means that materials that display structural color would not fade over time, as do most painted or printed advertisements. Polymer opals can be stretched and twisted because the base material has a rubbery consistency. When stretched, the spacing between the nanospheres increases, resulting in a color change to the blue range of the spectrum. Compressing the material results in a color shift toward red, and when released, the material returns to its original color [114]. Figure 5.20a, b are showing the polymer opal gradually turns to blue when stretched and returns to its original color, in this case, green. Figure 5.20c, d showing a polymer rope with increasing the stress at both ends and change of color from red to yellowish.

5.9.1 Mechanism of Mechanochromism in Materials

There are three types of mechanisms generally involved in mechanochromic materials. By the application of pressure, the color changes generally by: (1) rearrangement of bonds, (2) forming or breaking dye aggregates, and (3) change of photonic pathlengths. Mechanically activated reactions that also change the optical properties of the mechanophore, before and after bond-breaking, give rise to mechanochromism. Remarkably, diacetylenes, a class of synthetically versatile conjugated molecules, have shown much utility in optical sensing for a large variety of biological systems and consumer product inventions. Representative structures of diacetylenic monomers, together with polymerization methodologies are shown in Fig. 5.21. Grinding or smearing small molecule dyes such as oligo-phenylene vinylene, and perylene, and straining elastomers containing a dispersion of these small molecule dyes are the example of mechanochromic process by forming or breaking dye aggregates. Alternating polymeric lamellae constructed from block copolymers is the third type of mechanochromic process by change of photonic pathlengths process [115].

5.9.2 Materials Used

Mechanochromic phenomena have been observed in various polymers and inorganic materials. Polymers such as poly(di-n-alkylsilanes), poly(3-alkylthiophene), poly (3-dodecylthiophene), and spiropyran were demonstrated to exhibit mechanochromicactivity. PMA-spiropyran solutions (PMA-SP-PMA) were subjected to ultrasound at 6–9 °C which caused the colorless solution to turn pink due to the mechanochemical ring opening of the spiropyran molecules. Exposure to ambient light and temperature for 40 min caused the color to disappear [116]. A polymeric mixture that involves cholesteryl derivatives can respond to very low levels of pressure (several bars) and switch reversibly from red to green color. The ability to obtain a reversible mechanochromic effect at several bars is an important achievement for practical and economically feasible applications. Chen et al. (2012) used computational analysis to develop new mechanochromic compounds using anthraquinone imide (AQI) derivatives with electron-donating substituents [117]. Three mechanochromic polymers with different color transformations were developed, that is, orange to deep red, dark purple to black, and green-yellow to red. A pressure of 900 MPa was applied to obtain the mechanochromic transformation. The color change was irreversible in these polymers. Mechanochromic luminescence was also observed in the synthesized compounds. Some inorganic materials also exhibit mechanochromism, including LiF and NaCl single crystals, $CuMoO_4$, and palladium complexes. The Ni (II), Pd (II), and Pt(II) dimethylglyoxime complexes change color under pressures of 6300–15,000 MPa. SmS undergoes a semiconductor to metal phase transformation under a pressure of 0.65 MPa, along with a mechanochromic response. The phase transition of $CuMoO_4$ requires a pressure of 250 MPa, reversibly modifying its color from green to brownish red. The pressure needed to induce the phase transition in $CuMoO_4$ was decreased significantly by substituting a small portion of Mo by W. A green to red transition occurs in the compound $CuMo_{0.9}W_{0.1}O_4$ at a pressure of 20 MPa. In other words, a simple push with a finger can cause a phase transition and the accompanying color change in the new material. Products employing mechanochromic materials are currently at the research and development stage in most cases [118].

5.9.3 Applications

A climbing rope was developed with the help of a mechanochromic dye at the Smart Structures Research Institute, Strathclyde University, which was able to detect critical levels of strain and signal the user, through a color change, that the rope is damaged and should be discarded [119]. One of the major driving forces for research in this field is the aerospace industry due to the high maintenance and inspection cost of air and space vehicles. An indirect approach involves microcapsules or hollow fibers, which contain a colored substance. Under high stress, the hollow fibers or microcapsules break up, releasing the colored ingredient to

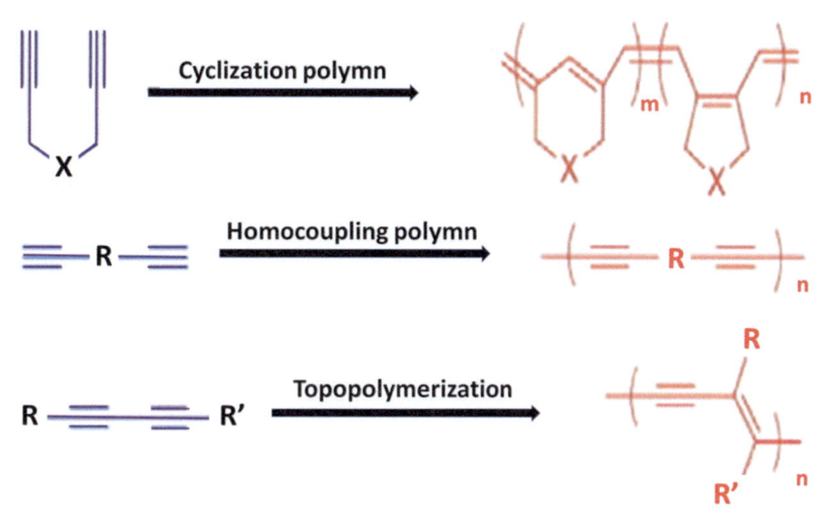

Fig. 5.21 Examples of synthetic methodologies for polydiacetylenes

Fig. 5.22 Tennis ball leaving a *red mark* on chromogenic paint

signal a critical condition of the related component. One difficulty with this approach is obtaining a homogeneous distribution of these fibers or microcapsules, which is necessary for it to function successfully. Two-part systems have also been developed where a dye and an activator, contained in separate microshells, mix and react to result in a colored substance when overload, cracking, or damage occurs. A direct approach is to use mechanochromic polymers such as diacetylene segmented copolymer or spiropyran for the same application, which is much simpler to manufacture and apply to the desired system [120]. The body of the smart toothbrush uses a mechanochromic liquid crystal cholesterol ester, which indicates to the user whether correct or excessive force is being applied during brushing. The product aims to prevent damage caused to teeth and gums due to brushing too hard [121].

Industrial roll covers are employed in papermaking, printing, and coating industries. Typically electronic sensors are used to prevent uneven or extreme pressures on these rolls, which are damaging them. Mechanochromic materials are proposed as a more feasible alternative to electronic sensors since electronic sensors typically require a power source, some communication equipment, data processing, and maintenance. Mechanochromic coatings would not require any of these. However, the use of mechanochromic coatings requires visibility: the whole roll surfaces should be visible during operation, which may not be possible for all such systems [122]. Mechanochromic Paint for Tennis, Volleyball, Basketball, and Similar Courts (Fig. 5.22) [123]. Those who play or watch tennis matches would appreciate how difficult it is to rule whether a ball served at 180 km/h that hits close to the service line or sideline is in or out. This problem was initially addressed by line judges and electronic sensors. An electronic line judge was first used in a tennis tournament in 1977. Because of the high cost and occasional breakdowns, other options have been also tried. Although this technology is more dependable than electronic line judges, it is costly and interruptions occur during the process. The court is painted with a paint that contains mechanochromic pigments with the ability to sense the force applied by the ball and reversibly change color from its normal color (green or blue) to a contrasting color such as red or orange. The mark left by the ball should fade away slow enough for the referee and players to decide whether it was in or out, for example, 60s. It is sufficient to use the mechanochromic paint for the lines and near them, since the decision only occurs in these regions and also because footprints due to the mechanochromic effect all over the court would cause confusion. The same type of paint could be used for other courts such as squash, volleyball, and basketball courts [124].

There are many types of scales in the market today, used for different weighing purposes. Mechanochromic materials scale is used in bathroom and kitchen scales [125]. A common and elegant design used for bathroom scales involves a glass platform and a digital display. Unlike this electronic scale, a mechanochromic scale would not need any batteries, mechanical or electronic parts. The product concept consists of a glass plate, under which there is a graphic design that involves mechanochromic ink. This image will show the corresponding weight of the person with the help of colors, graphics, or numbers. The mechanochromic image can be printed on a polymeric backing plate or it can be applied as a separate film. A similar approach can be used for a kitchen scale. However, in this case, a container is needed to contain the sugar, flour, or milk. Thus, a special container with a small base may be designed to provide more visibility for the mechanochromic graphic that must be seen by the user.

Overloading of packaged items could be a problem, especially with fragile, liquid, or gas contents. Some common situations where overload might occur are stacking of many boxes on top of each other in warehouses or supermarkets, which might cause damage to the lowermost box, or overloading of an individual box which, when lifted, may cause breakage or tearing of a box, and which may hurt the person who lifts it [126]. Mechanochromic labels or surfaces can be used to indicate an overload condition in the box or package. A related application would be tamper-evident closures. A mechanochromic film can be used in the closure area of a food or medicine container. Tampering or accidental damage to the closure area would be made visible through the color change of the film. A similar idea was mentioned specifically for pharmaceutical bottles and jars. Certain conditions in canned foods result in slight or obvious swelling or bulging of the container. For example, enzymatic action, nonenzymatic browning, and microbial growth cause carbon dioxide evolution and spoilage of the ingredients. A mechanochromic coating on the can could be designed and developed to warn the consumer about the possible spoilage of the canned ingredients [127].

Mechanochromic materials can also indicate a decrease of pressure if the packaging and the mechanochromic substance are designed accordingly. This effect could be used in pressurized tennis ball cans, or other pressurized packaging [128]. High-quality tennis balls are kept under 2 atm pressures so that the balls maintain their pressure while they are stored. A visible label on the packaging could be used to detect a fall in pressure which might be caused due to a defective package or because it was opened, to warn the customer about the condition of the product. Mechanochromic material used in mystery games intended for children. The game involves several cube blocks made of an elastomeric material such as synthetic rubber (polyisoprene or silicone rubber). Alternatively, wooden or plastic cubes can be covered with an elastomeric material. Each face of each block is printed with a mechanochromic image. When players rub two blocks to each other, a symbol appears temporarily on the faces which are being rubbed to each other, because of the pressure. One set of blocks is the key to the symbols. The players try to find out what types of symbols are written on the blocks and what they mean. When they put together all symbols on all sides of one or more blocks and when they find out which letter or number they correspond to, they solve the mystery. This game could be played by one or two players or by two or more teams [129].

5.10 Chemochromic Materials

Chemochromic materials are defined as materials that change color in the presence and by the effect of chemical agents with a change in color and/or transmission/reflection properties [130]. Chemochromic materials are available in a number of forms such as halochromic, solvatochromic, and hydrochromic materials. Gasochromic material is also a subgroup of chemocromic material. This is due to a chemical reaction taking place within the material. An electron that previously was a part of one chemical bond in one position may relocate to another position within the molecule, where its presence, or its absence, causes the molecule to absorb a certain color of light [131].

Halochromic materials can be considered a subgroup of chemochromic materials that change color as a response to pH changes in the environment [132]. Such materials are used in chemical indicators, such as Universal Indicator, which indicates over a pH range of 1–14 with a continuous color scale. Figure 5.23 showing the hybrid 3-glycidoxy-propyl-trimethoxy-silane (GPTMS) gel containing the chromophore molecule of the litmus dye, Resorufin (7-Hydroxy-3H-phenoxazin-3-one), was prepared by the sol-gel method through epoxy ring-opening reaction catalyzed by traces of 1-methylimidazole in acidic ethanol solutions [133].

Solvatochromic is the change in color activated by solvent polarity. Solvatochromic materials display the phenomenon called solvatochromism, typical of some chemical substances that are sensitive to a given solvent (liquid or gas) [134]. Solvatochromism is the phenomenon observed when the color due to a solute is different when that solute is dissolved in different solvents. Solvatochromism depends on the particular interaction between the molecules of the substance and the molecules of the solvent. The substance which contains chromophore groups is sensitive to the polarity of the solvent, which functions as a constant electrical field and determines the effects on the spectroscopic properties of the substance, hence a change in color. The solvatochromic effect is the way the spectrum of a substance (the solute) varies when the substance is dissolved in a variety of solvents. For example, Fig. 5.23 showing the color change of the wall in contact with urine before water spray. In this context, the dielectric constant and hydrogen bonding capacity are the most important properties of the solvent. With various solvents, there is a different effect on

Fig. 5.23 An example of a chromogenic material [136]

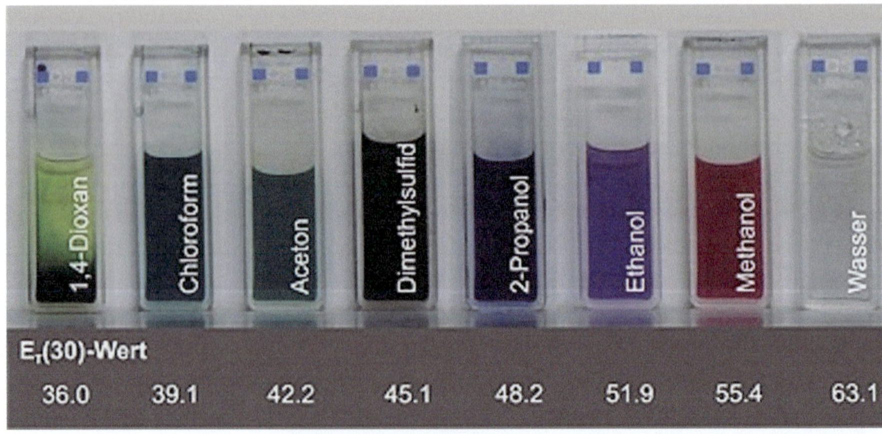

Fig. 5.24 Reichardt's dye dissolved in different solvents

the electronic ground state and excited state of the solute, so that the size of the energy gap between them changes as the solvent changes. This is reflected in the absorption or emission spectrum of the solute as differences in the position, intensity, and shape of the spectroscopic bands. When the spectroscopic band occurs in the visible part of the spectrum solvatochromism is observed as a change of color. This is illustrated by Reichardt's dye, as shown at the right (Fig. 5.24). Negative solvatochromism corresponds to a hypsochromic shift (or blue shift) with increasing solvent polarity. An example of negative solvatochromism is provided by4-(4′-hydroxystyryl)-*N*-methyl pyridinium iodide, which is red in 1-propanol, orange in methanol, and yellow in water. Positive solvatochromism corresponds to a bathochromic shift (or redshift) with increasing solvent polarity. An example of positive solvatochromism is provided by 4,4′-bis(dimethylamino)fuchsone, which is orange in toluene, red in acetone [135].

Hydrochromism is the change in color by interaction with bulk water or humidity. They form a category of a larger group that the materials which present the phenomenon called solvatochromism, typical of some chemical substances which are sensitive to a given solvent (liquid or gas) [137]. Figure 5.25 showing hydrochromic materials (polyamide/dye blend) changes color from green to orange when exposed to moisture.

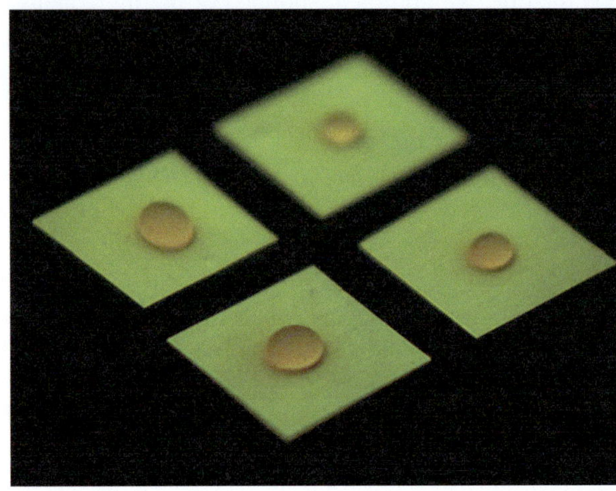

Fig. 5.25 *Green* colored hydrochromic materials changed to orange color in exposure to moisture

5.10.1 Applications

Chemochromic materials are found in a range of forms in a variety of applications. The most common form found is a dye. The smart material is used in litmus paper, which detects the acidity and alkalinity of chemicals [138]. The chemochromic chemical in the dye has low acidity and will change color depending on its p^H level. For example, methyl red is yellow if $p^H > 6$ and red if $p^H < 4$. In pregnancy tests, the chemicals used to detect and respond to HCG (human chorionic gonadotropin) traces found in a pregnant woman's urine. The more sensitive the chemical, the better the product is for the user. Chemochromic dyes are also used to show the ripeness of fruit, as the chemical reacts with gases released by the fruit as it ripens. The major application of chemochromic materials is their use in hydrogen leak detection in rocket engines. NASA has been using this technology to develop a chemochromic detector for sensing hydrogen gas leakage to protect humans from potential hydrogen-related incidents, such as detonation or frostbite from its high concentration [139].

Chemochromic material is creating paints that can change color to show corrosion of metal since the water in contact will change the pH. Halochromic materials change color if the ambient acidity changes, that is, in contact with an aqueous solution whose p^H may change over time. One possible application is to use paints that can change color in response to corrosion in the metal underneath them, for example, in the rusting of iron or steel the p^H of the water in contact with the metal changes in p^H [140]. For example, poly(3,4-ethylenedioxythiophene) which is purple red at a neutral state becomes blue or transparent in the oxidized state. Currently, many researchers are engaged in developing applications using halochromic dyes in textile products with the aim of realizing flexible p^H sensors, made with different techniques and for use in various fields. The coloration of halochromic nonwoven fabric made of nanofibers sensors is rapid due to the high porosity of the substrate made of nanofibers. Halochromic textiles pose unique opportunities for simple and smart solutions. For example, in the medical sector, halochromic bandages can be used for the monitoring of patients with burns. Since p^H changes during the healing process of the skin, with textile bandages that employ halochromic molecules, it is possible to monitor the level of healing without having to remove the bandage and cause pain or discomfort to the patient. Fabrics treated with halochromic dyes have also been used in geotextiles or protective clothing that measures pH alteration in the air in real time. A particular type of halochromic material is oxazines. These molecules are being studied to obtain paint with adjustable colors that change color after application so that their color corresponds to the color of adjacent areas or for decorative purposes. Colorants obtained with these molecules are characterized by high controllability of color. They change color upon the application of chemical agents and may maintain a specific color for a prolonged period such as several months. The color persists indefinitely unless a solution of basic p^H is applied [141].

Generally, solvatochromism is used for environmental sensors, in the analysis of probes with the capacity of determining the presence and the concentration of a solvent, and in molecular electronics for the construction of molecular switches. In industrial products, hydrochromic materials are used for special inks in the form of pigments, which in many cases, act similar to thermochromic ones: they are normally white and opaque but become transparent in the presence of water. In most of the applications, which were realized until today, these materials were applied on surfaces that already have a colored layer printed on them, forming a fine white film that rejects light waves, impeding them from reaching the printed image. At the moment in which the surface is wetted with water, the film which used to diffract the light acquires a viscosity such that it becomes permeable to electromagnetic frequencies and lets the light waves be filtered through, making visible the color image lying underneath. When the surface is dried, the ink film returns to its light impermeable condition, becoming white and opaque again. The best results with hydrochromic inks in terms of applicability and durability are obtained when the ink is applied on dry and smooth surfaces [142]. Various types of substrates can be used: soft sheet vinyl, paper, coated paper, styrene sheet, polyester fabric, soft PVC, and many more. The inks can be applied by screen printing (contact print) or spray coating, followed by passing it through a forced hot air tunnel, as in the production of printed shirts. Hydrochromic materials have been used in the textile sector to achieve dynamic patterns on textiles. Later, they were used to realize clothing and bathing suits, umbrellas, and table cloths. At first sight, under full appears like a traditional paisley patterned table cloth; monochromatic, decorated with floral designs with a shiny-opaque effect. However, when a cup of water accidentally tips over and the tablecloth becomes wet, the hidden graphic design reveals itself by becoming colored. Thus, an unpleasant incident at the table transforms into a positive experience as in the popular Italian saying "accidentally spilling wine on the table brings good luck" [143].

Solvatochromism of the photoluminescence/fluorescence of CNT has been identified and used for optical sensor applications. In one such application, the wavelength of the fluorescence of peptide-coated carbon nanotubes was found to change when exposed to explosives, facilitating detection. However, more recently the small chromophore solvatochromism hypothesis has been challenged for CNT in light of older and newer data showing electrochromic behavior. These and other observations regarding nonlinear processes on the semiconducting nanotube suggest colloidal models will require new interpretations that are in line with classic semiconductor optical processes, including electrochemical processes, rather than small molecule physical descriptions.

5.10.2 Limitations

One of the limitations of chemochromic chemicals is that the system of response would need a color chart to compare to, to understand what the stimulus is and what the chemical it detected. Another disadvantage of this material is that it depends on the subjective evaluation of the user and the sensibility of the human eye to colors. Furthermore, over time, the materials will begin to decay and react to other chemicals, lessening their sensitivity and effectiveness [144].

5.11 Biochromic Materials

Biochromism is a term used to define a color change due to a biochemical reaction. Biochromic materials were developed to detect and report the presence of pathogens with a color shift. Potential applications of biochromic materials include colorimetric detection of pathogens against food poisoning or bioterrorism. The target microbial enzymes release colored dyes upon hydrolysis [145]. This approach results in the formation of colored colonies that stand out against the background flora and allow the identification of pathogens easily. However, it requires the growth of a single cell into a colony and the detection procedures may last up to several days. In response to the disadvantages of conventional methods employed for pathogen detection, an effort is directed to the development of biosensors with a rapid detection capability of even small numbers of pathogens. Biochromic materials have been studied for their biosensor capabilities in the form of simple structures such as membranes, liposomes, and Langmuir-Blodgett (LB) monolayers. Biological membranes are critical to biological processes such as transport, signal transition, and molecular recognition. Cell membrane and related structures are mimicked to facilitate the design and development of biosensors. Conjugated polymers such as polydiacetylene (PDA) and polythiophene (PT) are commonly investigated for these purposes [146]. PDA has the ability to self-assemble into organized vesicles and films and to show a drastic color change from blue to red under heat, mechanical stress, or upon molecular recognition of certain pathogenic agents. PT is an organic polymer and shows chromic transitions upon excitation by heat, metals, chemicals, and proteins. PT-based materials functionalized with carbohydrates such as sialic acid and mannose have been produced to detect bacterial toxins. A redshift in UV absorbance was registered upon detection of toxins. A recent study investigated the direct application of surfactant functionalized PDA vesicles to detect common bacteria *Staphylococcus aureus* and *Escherichia coli* in a culture medium and apple juice. Bacterial detection of PDA vesicles coated on cellulose strips was also analyzed. Interestingly, apple juice alone could trigger a color shift in the films but the color shift was stronger in the presence of bacteria [147]. The colorimetric response of PDA films was shown to occur within several minutes upon contact with the solution containing selected bacteria. The color change is believed to occur due to the interaction of bacterially secreted membrane-active compounds such as bacteriocins and the receptors and/or (b)insertion of hydrophobic peptides and proteins into the PDA membrane. Cell surface receptors incorporated onto PDA vesicles or LB monolayers interact with specific viruses or toxins and a dramatic color change occurs from blue to red. The color change is irreversible in PDA supramolecular assemblies. The mechanism involved in the color change is not clear, although it is believed to be due to distortion of the backbone of the PDA polymer. The color intensity typically depends on the species, strains, growth rate, and population of pathogens. Biosensor signal amplification can be achieved by increasing the solution p^H or phospholipid content in the mixed lipid vesicles [148].

5.11.1 Application

Although biochromic materials are relatively new, some applications already emerge in the industry. One of the target markets of biochromic technology is the food industry, specifically in the detection of foodborne bacteria. The development of sensors suitable to be used on the packaging of any food product that may contain bacteria or toxins is an important target for health and quality concerns. There are many technologies being developed for such a purpose but biochromic devices have a high potential to offer important advantages such as easy adaptability to packaging and no external energy requirement. Researchers at Berkeley National Laboratory developed biochromic sensors (also called colorimetric biosensors) for the detection of viruses, bacteria, parasites, neurotoxins, and other pathogens. The engineered membranes consist of chemically or biologically specific ligands attached to a PDA backbone. These biochromic sensors have potential applications such as rapid test kits for influenza and other diseases, detection of foodborne bacteria such as *E. coli*, drug research and development, detection of DNA hybridization, and detection of pollutants [149, 150]. Figure 5.26 shows the application of biochromic material to the body part to detect the damaged cells.

5.12 Magnetochromic Materials

Magnetochromic materials possess the property of changing color when an external field is applied [151]. The spin-induced electric polarization can be magnetically controlled, creating a reorganization of the magnetic particles, and being a reversible process. The observed color will depend on the

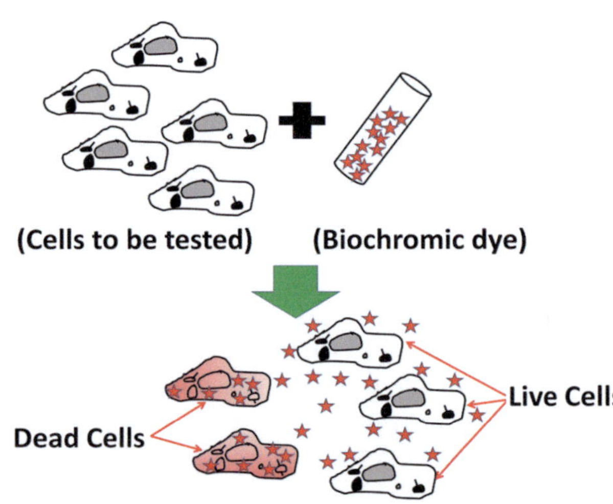

Fig. 5.26 Application of biochromic material to detect the damaged part of the body

particle size and the intensity of the applied magnetic field. Under an external magnetic field, the magnetic particles will form a long range ordered structure. Due to the chain-like distribution along the direction of the field, the particles can be viewed as a one-dimensional structure. By increasing the magnetic field intensity, the magnetic particles would become closer to one another, and at the same time, the magnetic attraction between the particles, the repulsion between solvent shells, and the electrostatic repulsion of the Stern layer will balance each other. Only when all these forces are in an equilibrium state, can the magnetic particles become an ordered structure, then being visible due to the phenomenon of magnetochromism [152]. The nanoparticle spacing dictates the wavelength of the light reflected by the material; according to Eq. ((5.23) (Braggs law):

$$m\lambda = 2nd\sin\theta \tag{5.23}$$

where m is the order of scattering, λ is the diffraction wavelength, n is the refraction index of water, d in the lattice plane spacing, and θ is the Bragg angle. When a stronger field is applied, reflections with a shorter wavelength will be observed, and by using Braggs law it is possible to make an estimation of the distance between adjacent nanoparticles along the direction of the magnetic field. Using a magnetic field as an external stimulus may improve the spectral tenability and the response rate of materials in chromogenic applications, and presents a facile integration into the existing photonic systems.

5.12.1 Applications

The most studied metal oxide nanoparticle with magnetochromic application is iron oxide, Fe_3O_4. This material can be used in applications, such as magnetic recording media, ferrofluid, and for biomedical purposes, and is considered to be a new platform of novel optoelectronic devices, sensors, and color displays [153]. Fe_3O_4 nanosheets can form one-dimensional photonic crystal under an external magnetic field. These magnetic nanoparticles can diffract visible light and display a variety of colors with changing intensity of the magnetic field. The shift of color is due to a decrease in the interparticle distance as the magnetic field increases, and as can be explained by Braggs law, the shorter the distance between the interparticles, the shorter the observed wavelengths. The behavior observed when using Fe_3O_4 monodisperse nanoclusters in water is very useful for applying magnetochromic material in displays [154]. By applying an external magnetic field to these suspensions, assembly of 1-D periodic photonic chains is induced, being a direct result of dipole–dipole interactions between magnetic particles. If the nanoparticles have a uniform size, the interparticle space becomes equivalent for all neighboring pairs and visible light will be diffracted with a certain color, depending on the interparticle space. The optical response of these magnetochromic materials to an external magnetic stimulus is fast, being a critical and important feature for potential applications in magnetic responsive color displays. Zhuan et al. developed a chip where they integrated Fe_3O_4 suspension into microwells, sealed with a PDMS (polydimethylsiloxane) microchannel to form a microfluidic system [155]. By applying an external magnetic field along the microchannel, a change in the microwells is observed. Also, Ge et al. were able to produce a small display using a suspension of Fe_3O_4 encapsulated in PDMS film. By applying an external magnetic field, the color change is induced and some letters appear [156]. Hu et al. describe how a magnetochromic effect can be applied in anticounter feiting [159]. By using Fe_3O_4 nanoparticles of different sizes and by applying an external magnetic field, these nanoparticles will present different colors depending on their size. Also, Ge et al. describe how it is possible to tune the diffraction wavelength by controlling the overall size of the nanoparticles: diffractions in red can be obtained with relatively large particles (150 nm), in the blue region they can be obtained with smaller particles (100 nm), while the medium-sized nanoparticles (100–150 nm) may present a diffraction wavelength tunable from blue to green, yellow, and red. In Fig. 5.27, the color change when the magnetic field of constant intensity is applied to different particle sizes of Fe_3O_4 nanoclusters can be observed [157].

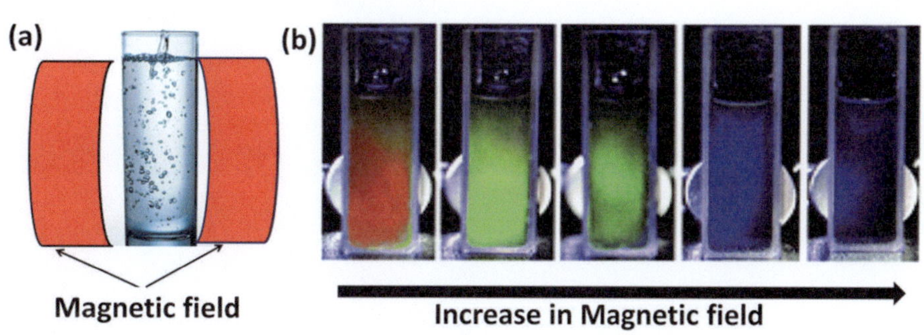

Fig. 5.27 (**a**) Fe_3O_4 nanosheet colloidal solution formed in response to an external magnetic field, and (**b**) Color changes as a response to an increase in the magnetic field intensity

5.13 Phosphorescent Materials

Phosphorescent smart materials are sometimes called 'after-glow' materials because they glow in the dark after being 'charged up' in the day. Phosphorescent pigments absorb natural or artificial visible light energy and store it in their molecules [158]. Phosphorescence is the delayed and slow emission of light from a material that was exposed to some other radiation, for example, UV or visible light. This energy is slowly released over a period which maybe just a few seconds or several hours. It is, therefore, a cheap and convenient "rechargeable" light source for particular applications, for example, Phosphorescent pigments have been used in emergency traffic signs, novelty decorations, evening lights in the garden (energized in daylight, toys, and glow-in-the-dark watches). The figures and hands of glow-in-the-night watches were painted with radioactive materials added to the pigment. These radioactive paints will glow for years without needing repainting or charging in light [159].

5.14 Ionochromic

Ionochromic is a term similar to chemochromic, indicating a reaction to the presence of ions in a medium by a color variation. A flow of ions through an ionochromic material results in a reaction/color change from the material. This material is in many ways similar to electrochromic materials which change color when electrons flow through them. Electrons, just like anions, carry a negative charge. Both electrochromic and ionochromic substances have their color change activated by the flow of charged particles. Ionochromic substances are suitable for the detection of charged particles. Some ionochromic substances can be used as indicators for complexometric titrations [160].

5.15 Vapochromism

Vapochromism strongly overlaps with solvatochromism since vapochromic systems are ones in which dyes change color in response to the vapor of an organic compound or gas. Vapochromic devices are the optical branch of electronic noses. The main applications are in sensors for detecting volatile organic compounds (VOCs) in a variety of environments, including industrial, domestic, and medical areas. An example of such a device is an array consisting of a metalloporphyrin (Lewis acid), a pH indicator dye, and a solvatochromic dye. The array is scanned with a flat-bed recorder, and the results are compared with a library of known VOCs. Vaporchromic materials are sometimes Pt or Au complexes, which undergo distinct color changes when exposed to VOCs [161].

5.16 Radiochromism

Radiochromic materials can change their color according to the radio wave. Radiochromic film is a type of self-developing film typically used in the testing and characterization of radiographic equipment such as CT scanners and radiotherapy linacs (Fig. 5.28). The film contains a dye that changes color when exposed to ionizing radiation, allowing the level of exposure and beam profile to be characterized. Unlike x-ray film, no developing process is required and results can be obtained almost instantly, while it is insensitive to visible light (making handling easier). Radiochromic films have been in general use since the late 1960s, although the general principle has been known since the nineteenth century. Radiochromic film can provide high spatial resolution information about the distribution of radiation. Depending on the scanning technique, submillimeter resolution can be

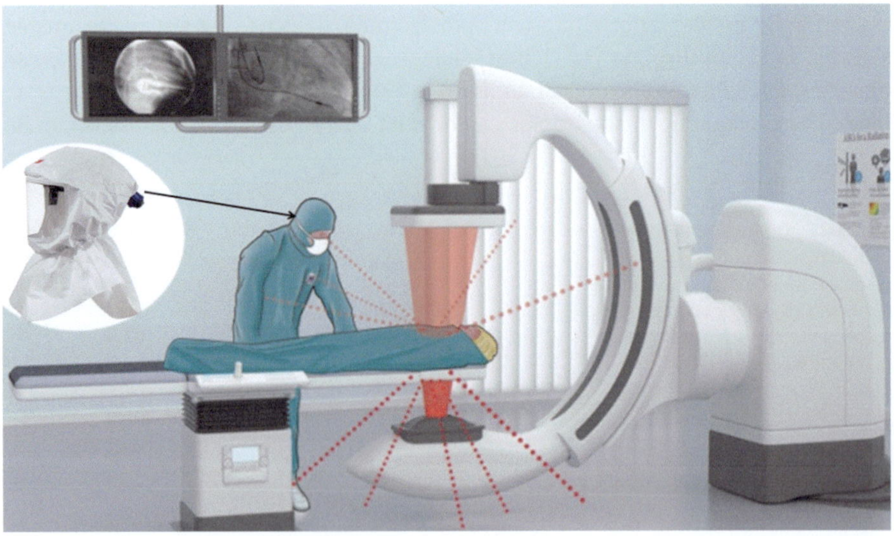

Fig. 5.28 Radiation protective mask on practicing interventional physicians

achieved. Unlike many other types of radiation detectors, radiochromic film can be used for absolute dosimetry where information about absorbed dose is obtained directly. It is typically scanned, for example, using a standard flatbed scanner, to provide accurate quantification of the optical density and therefore degree of exposure [162].

5.17 Sorptiochromism

Sorptiochromism refers to the phenomena of spectral shift resulting from certain species adsorbed (physically or chemically) on the metal nanoparticle surface. For those adsorbed molecules that are transparent around the plasmon resonance wavelengths, the sorptiochromism follows basically the same principles as in solvatochromism. Therefore, in this respect, sorptiochromism can be regarded as a particular subcategory of solvatochromism. This property can be used to detect certain species that possess a strong affinity to the metal particle surfaces, including thiols, as well as biomolecules like antigens and proteins. In certain cases, the adsorbates (i.e., dye molecules) are strongly absorptive near the localized surface plasmon resonance wavelengths. In this scenario, the interactions between metal nanoparticles and the adsorbed dyes would be more complex. As both the metal nanoparticle and the dyes under light excitation can be idealized as dipoles (plasmonic dipoles for the former, and molecular dipoles for the latter), and they have similar resonance frequencies, there may emerge strong plasmonic-molecular resonance coupling, which could significantly alter the absorption properties of both the metal nanoparticles and the dye molecules. Jensen et al. fabricated hybrid nanostructures using gold nanorods and another organic dye (2,2′-dimethyl-8-phenyl-5,6,5′,6′-dibenzoithia carbocyanine chloride). The gold nanorods and dye molecules had very similar resonance wavelengths (692 nm and 693 nm, respectively). They studied the effect on molecular resonance by systematically varying the concentration of the dye molecules and thus the nanorod-to-dye ratio, and found that the plasmon splitting in the resultant hybrid nanostructures increases in a linear manner with respect to the square root of the absorbance of the molecular layer. The plasmonic-molecular resonance coupling can be readily reconfigured into indirect sorptiochromic systems when stimuli-responsive chromophores are employed as the surface-bound molecular component [163].

5.18 Aggregachromism

The phenomena of aggregachromism have been well documented and discussed for dye-related systems. When dye molecules form aggregates (dimers, trimers, and higher oligomers), the electronic interaction between neighboring chromophores would alter the energy levels of molecular orbitals, resulting in spectral shifts. This can be interpreted using the dipole models. For example, two dye molecules, each idealized as a dipole, form dimers with two distinct configurations. When the two dipoles are aligned head-to-tail, this configuration stabilizes the energy levels owing to Coulombic attraction, and thus decreases the energy gap, resulting in a redshift of the absorption maximum. In contrast, when the two dipoles are coupled side by side, the energy levels become destabilized owing to Coulombic repulsion, increasing the energy gap and blue-shifting the absorption maximum. The cases for plasmonic nanostructures are somewhat similar, at least phenomenologically. If the two nanorods are paired in an end-to-end manner, meaning the aggregation axis is parallel to the longitudinal axis, the longitudinal plasmon would resonate to longer-wavelength excitations due to plasmonic coupling, leading to a redshift of the LSPR peak, just like that for J-aggregates of dye molecules; if the nanorods form a side by side dimer, like in H-aggregates, the localized surface plasmon resonance wavelength would shift to the blue end. Chen et al. presented a series of elegant studies regarding the monomers, dimers, trimers, and higher oligomers of gold nanospheres. Clusters assembled from gold nanospheres were encapsulated with polymers and subjected to density gradient centrifugation. Dimer and trimer samples were thus purified with good selectivity (95% and 81%, respectively), allowing for spectroscopic studies of nearly-pure dimers and trimers, rather than a mixed ensemble [164].

5.19 Chronochromism

The term "chronochromism" describes the processes where a system changes its color with respect to the passage of time. Clearly, a chronochromic system is designed and constructed on the basis of concerns on the kinetics of involved reactions, rather than the mechanism itself of the chromism. All chronochromic systems are examples of indirect chromisms; and technically, all chromic phenomena are chronochromic, because they all require a certain period (either short or long) to proceed. However, most of the reported chronochromic systems exploit slow reactions that cover a temporal duration typically of minutes, hours, days, and even months. This is because, as can be seen later, chronochromism is commonly used to translate time-related information into color change, in other words, to report time using a signal that is visual and intuitive. From the perspective of applications, apart from direct indication of time passage, chronochromic systems are usually employed for two other purposes: (a) self-destruction and

(b) real-time mimicking and an indication of some particular dynamic processes. In self-destruction purposes, disintegrate under predefined circumstances in a prescribed manner, exhibiting an irreversible evolution process along the free-energy gradient as dictated by the second law of thermodynamics. Therefore, self-destructive materials often find applications in security-related areas, in particular information security, for which chronochromic materials can often be employed. Self-erasing inks are classic examples of chronochromism. Generally, the self-erasing process could speed up on heat or visible light irradiation. It can be potentially used to store confidential and temporary information, or for self-expiring purposes like bus tickets, ATM receipts, and so on. In chronochromic materials for real-time mimicking and indication of dynamic processes, the usage of chronochromic materials for real-time mimicking and an indication of a variety of dynamic processes. There exists quite a number of dynamic processes in daily life that concerns people, like the spoilage of foods (owing to microbial growth or chemical deterioration), the corrosion of metals, the aging of rubber products, the wearing of machinery, degradation of chemical agents (owing to oxidation, deliquescence, decomposition, etc.), and the development of chronic diseases of human beings owing to long-term exposure to hazards. In these dynamic processes, time is a fundamental factor that determines how far these processes have evolved [165].

5.20 Concentratochromism

Concentratochromism is the change in color by changes in the concentration in the medium. For example, dilute solutions of dithizone are orange, while more concentrated solutions (not very soluble in methanol) are green. Another example is fullerenes mixed with methanol. Here, the hydrophobic cores tend to cluster together to minimize their exposures to the incompatible environment, whereas the hydrophilic amine pendants tend to molecularly dissolve into the polar solvent. The net outcome of these two antagonistic forces is the "dissolution" of the fullerene molecules into the solvent as nanoclusters. When the concentration of the aminated fullerene is increased, more buckyballs will be aggregated and larger nanoclusters will be formed. This increase in nanocluster size with concentration is probably the origin of the observed phenomenon of concentratochromism, because it is well known that the sizes of nanoclusters can affect their electronic and photonic properties to great extents. Both electronic absorption and light scattering are believed to be involved in the concentratochromic process [166].

5.21 Cryochromism

In fact, there are other forms of thermochromic materials, and some of these undergo color changes in very wide temperature ranges from well above ambient to cryogenic conditions. Cryochromism is the change in color caused by the lowering of temperature. Cryochromism still falls in the same scientific area of thermochromism, but the emphasis on cooling to below 0 °C may help draw research attention and effort to practical uses. For example, freezing indicator of water supply pipes in winter, vision guide of cooled beverages and ice creams, smart packaging for cryo-storage of medicines and living cells, and safety control for energy storage in liquid air or nitrogen. This communication describes the cryo-solvatochromism in solutions of chloro-nickel complexes in a 1-hydroxyalkyl-3-methylimidazolium cation-based ionic liquid containing added Cl^- ions. The color change of the ionic liquid solution from blue to green upon cooling from 22 to −13 °C [167].

5.22 Summary

In this chapter, variety of chromomeric system and their applications has been described with the involved mechanism. Most used photocromic, thermochromic, electrochromic, gasochromic, mechanochromic materials along with other chromic materials such as chemochromic, biochromic, magnetochromic materials, phosphorescent, ionochromic, vapochromic, radiochromic, sorptiochromic, aggregachromic, chronochromic, concentratochromic, and cryochromic materials has been included. Definitely, this chapter will enlighten researchers to enhance various materials' properties during optical characterization. This chapter throws a lot of direction to develop novel applications, which are yet to be discovered.

References

1. Lampert, C.M.: Chromogenic smart materials. Mater. Today. **7**(3), 28–35 (2004). https://doi.org/10.1016/S1369-7021(04)00123-3
2. Lotzsch, D., Eberhardt, V., Rabe, C., Materials, C.: Ullmann's encyclopedia of industrial. Chemistry. (2016). https://doi.org/10.1002/14356007.t07_t01
3. https://www.yourdictionary.com/chromo. Accessed 02 May 2019
4. https://en.wikipedia.org/wiki/Chromophore. Accessed 02 May 2019
5. He, G.S., Tan, L.-S., Zheng, Q., Prasad, P.N.: Multiphoton absorbing materials: molecular designs, characterizations, and applications. Chem. Rev. **108**(4), 1245–1330 (2008). https://doi.org/10.1021/cr050054x
6. Lang, F., Wang, H., Zhang, S., et al.: Review on variable emissivity Materials and devices based on smart Chromism. Int. J. Thermophys. **39**, 6 (2018). https://doi.org/10.1007/s10765-017-2329-0

7. https://www.marketwatch.com/press-release/smart-glass-market-global-share-size-trends-and-growth-analysis-forecast-to-2020-2024-2020-06-23?mod=mw_quote_news. Accessed 02 May 2019
8. https://www.marketsandmarkets.com/Market-Reports/electrochromic-glass-market-110534618.html. Accessed 02 May 2019
9. Huang, M.: Thesis: Encapsulation of Organic Thermochromic Materials with Silicon Dioxide. https://scholarcommons.usf.edu/cgi/viewcontent.cgi?article=9193&context=etd (2019)
10. Akita, M.: Photochromic organometallics, a stimuli-responsive system: an approach to smart chemical systems. Organometallics. **30**(1), 43–51 (2011). https://doi.org/10.1021/om100959h
11. Renzi-Hammond, L., Buch, J.R., Cannon, J., Hacker, L., Toubouti, Y., Hammond, B.R.: A contra-lateral comparison of the visual effects of a photochromic vs. non-photochromic contact lens. Cont. Lens Anterior Eye. **43**(3), 250–255 (2020). https://doi.org/10.1016/j.clae.2019.10.138
12. Fenech, A.; (2011) Lifetime of Ccolor Photographs in Mixed Archival Collections. Doctoral thesis, UCL (University College London)., https://discovery.ucl.ac.uk/id/eprint/1333217
13. https://en.wikipedia.org/wiki/Chromogenic_print. Accessed 06 May 2019
14. Periyasamy, A.P., Vikova, M., Vik, M.: A review of photochromism in textiles and its measurement. Text. Progr., 53–136 (2017). https://doi.org/10.1080/00405167.2017.1305833
15. Hadjoudisa, E., Mavridis, I.M.: Photochromism and thermochromism of Schiff bases in the solid state: structural aspects. Chem. Soc. Rev. **33**, 579–588 (2004). https://doi.org/10.1039/B303644H
16. http://citeseerx.ist.psu.edu/viewdoc/download?doi=10.1.1.369.2433&rep=rep1&type=pdf. Accessed 08 May 2019
17. https://www.loc.gov/preservation/outreach/tops/wilhelm/index.html. Accessed 08 May 2019
18. Mohr, G.J., Tirelli, N., Lohse, C., Spichiger-Keller, U.E.: Development of chromogenic copolymers for optical detection of amines. Adv. Mater. **10**(16), 1353–1357 (1998). https://doi.org/10.1002/(SICI)1521-4095(199811)10:16<1353::AID-ADMA1353>3.0.CO;2-X
19. Thummavichai, K., Xia, Y., Zhu, Y.: Recent progress in chromogenic research of tungsten oxides towards energy-related applications. Prog. Mater. Sci. **88**, 281–324 (2017). https://doi.org/10.1016/j.pmatsci.2017.04.003
20. Medeiros, S.F., Santos, A.M., Fessi, H., Elaissari, A.: Stimuli-responsive magnetic particles for biomedical applications. Int. J. Pharm. **403**(1–2), 139–161 (2011). https://doi.org/10.1016/j.ijpharm.2010.10.011
21. https://en.wikipedia.org/wiki/Chromism#:~:text=In%20chemistry%2C%20chromism%20is%20a,in%20the%20colors%20of%20compounds. Accessed 20 May 2019
22. He, T., Yao, J.: Photochromism of molybdenum oxide. J Photochem Photobiol C: Photochem Rev. **4**(2), 125–143 (2003). https://doi.org/10.1016/S1389-5567(03)00025-X
23. Vekshin, N., Savintsev, I., Kovalev, A., Yelemessov, R., Wadkins, R.M.: Solvatochromism of the excitation and emission spectra of 7-Aminoactinomycin D: implications for drug recognition of DNA secondary structures. J. Phys. Chem. B. **105**(35), 8461–8467 (2001). https://doi.org/10.1021/jp011168p
24. Bo, J., Zhang, J., Ma, J.-Q., Zheng, W., Chen, L.-J., Sun, B., Li, C., Hu, B.-W., Tan, H., Li, X., Yang, H.-B.: Vapochromic behavior of a chair-shaped supramolecular Metallacycle with ultra-stability. J. Am. Chem. Soc. **138**(3), 738–741 (2016). https://doi.org/10.1021/jacs.5b11409
25. Papaefthimiou, S.: Chromogenic technologies: towards the realization of smart electrochromic glazing for energy-saving applications in buildings. Adv. Build. Energy Res. **4**(1), 77–126 (2011). https://doi.org/10.3763/aber.2009.0404
26. Mooradian, A.: Photoluminescence of metals. Phys. Rev. Lett. **22**, 185 (1969). https://doi.org/10.1103/PhysRevLett.22.185
27. Chen, S., Hai, X., Chen, X.-W., Wang, J.-H.: In situ growth of silver nanoparticles on graphene quantum dots for ultrasensitive colorimetric detection of H_2O_2 and glucose. Anal. Chem. **86**(13), 6689–6694 (2014). https://doi.org/10.1021/ac501497d
28. Ferrara, M., Bengisu, M.: Materials that change color. In: SpringerBriefs in Applied Sciences and Technology. Cham, Springer (2014). https://doi.org/10.1007/978-3-319-00290-3_2
29. Webster, W.: Revelation and transparency in cColor vision refuted: a case Oo mind/brain identity and another bridge over the explanatory gap. Synthese. **133**, 419–439 (2002). https://doi.org/10.1023/A:1021294209237
30. Shang, M., Li, C., Lin, J.: How to produce white light in a single-phase host? Chem. Soc. Rev. **43**, 1372–1386 (2014). https://doi.org/10.1039/C3CS60314H
31. Faughnan, B.W., Staebler, D.L., Kiss, Z.J.: In: Wolfe, R. (ed.) Inorganic Photochromic Materials Applied Solid State Science, vol. 2, pp. 107–172., ISBN 9780120029020. Elsevier, Netherland (1971). https://doi.org/10.1016/B978-0-12-002902-0.50009-0
32. Crano, J.C.: Chromogenic materials, photochromic. In: Kirk-Othmer Encyclopedia of Chemical Technology. John Wiley & Sons, Inc., Hoboken, NJ (2000). https://doi.org/10.1002/0471238961.1608152003180114.a01
33. Granqvist, C.G.: Chromogenic materials for transmittance control of large-area windows. Crit. Rev. Solid State Mater. Sci. **16**(5), 291–308 (1990). https://doi.org/10.1080/10408439008242184
34. Štangar, U.L., Orel, B., Régis, A., et al.: Chromogenic WPA/TiO 2 hybrid gels and films. J. Sol-Gel Sci. Technol. **8**, 965–971 (1997). https://doi.org/10.1023/A:1018305501973
35. Sousa, C.M., Berthet, J., Delbaere, S., Polónia, A., Coelho, P.J.: Fast color change with photochromic fused Naphthopyrans. J. Org. Chem. **80**(24), 12177–12181 (2015). https://doi.org/10.1021/acs.joc.5b02116
36. Rouhani, M., Hobley, J., Subramanian, G.S., Phang, I.Y., Foo, Y. L., Gorelik, S.: The influence of initial stoichiometry on the mechanism of photochromism of molybdenum oxide amorphous films. Solar Energy Mater. Solar Cells. **126**, 26–35 (2014). https://doi.org/10.1016/j.solmat.2014.03.033
37. Shen, Y., Yan, P., Yang, Y., Hu, F., Xiao, Y., Pan, L., Li, Z.: Hydrothermal synthesis and studies on photochromic properties of Al doped WO3 powder. J. Alloys Comp. **629**, 27–31 (2015). https://doi.org/10.1016/j.jallcom.2014.11.218
38. Hea, T., Yao, J.: Photochromic materials based on tungsten oxide. J. Mater. Chem. **17**, 4547–4557 (2007). https://doi.org/10.1039/B709380B
39. Glebov, L.B.: Photochromic and Photo-Thermo-Refractive Glasses. In: Encyclopedia of Smart Materials. John Wiley & Sons, Inc., Hoboken, NJ (2002). https://doi.org/10.1002/0471216275.esm06
40. Gilroy, D., Conway, B.E.: Surface oxidation and reduction of platinum electrodes: coverage, kinetic and hysteresis studies. Can. J. Chem. **46**(6), 875–890 (1968). https://doi.org/10.1139/v68-149
41. Abazari, R., Sanati, S.: Perovskite $LaFeO_3$ nanoparticles synthesized by the reverse microemulsion nanoreactors in the presence of aerosol-OT: morphology, crystal structure, and their optical properties. Superlatt. Microstruct. **64**, 148–157 (2013). https://doi.org/10.1016/j.spmi.2013.09.017
42. Smith, G.P.: Photochromic glasses: properties and applications. J. Mater. Sci. **2**, 139–152 (1967). https://doi.org/10.1007/BF00549573
43. Li, S.-L., Han, M., Zhang, Y., Li, G.-P., Li, M., He, G., Zhang, X.-M.: X-ray and UV dual Photochromism, Thermochromism, Electrochromism, and amine-selective Chemochromism in an Anderson-like Zn7 cluster-based 7-fold interpenetrated

framework. J. Am. Chem. Soc. **141**(32), 12663–12672 (2019). https://doi.org/10.1021/jacs.9b04930
44. Yao, J., Hashimoto, K., Fujishima, A.: Photochromism induced in an electrolytically pretreated Mo03 thin film by visible light. Nature. **355**, 624–626 (1992). https://doi.org/10.1038/355624a0
45. Jiang, S., Ichihashi, J., Monobe, H., Fujihira, M., Ohtsu, M.: Highly localized photochemical processes in LB films of photo chromic material by using a photon scanning tunneling microscope. Opt. Commun. **106**(4–6), 173–177 (1994). https://doi.org/10.1016/0030-4018(94)90316-6
46. Masahiro Irie, Photochromic diarylethenes for optical data storage media, Molecular Crystals and Liquid Crystals Science and Technology. Section A. Molecular Crystals and Liquid Crystals, 227, 1993 - 1, 263–270, https://doi.org/10.1080/10587259308030979
47. Wang, L., Li, Q.: Stimuli-directing self-organized 3D liquid-crystalline nanostructures: from materials design to photonic applications. Adv. Funct. Mater. **26**(1), 10–28 (2016). https://doi.org/10.1002/adfm.201502071
48. Moritzen, P.A., El-Awady, A.A., Harris, G.M.: Kinetics and mechanisms of the formation, acid-catalyzed decomposition and intramolecular isomerization of oxygen-bonded (sulfito)-pentaaquochromium(III) perchlorate. Inorg. Chem. 1985. **24**(3), 313–318 (1985). https://doi.org/10.1021/ic00197a015
49. Ma, C., Taya, M., Xu, C.: Smart sunglasses based on electrochromic polymers. Polym. Eng. Sci. **48**(11), 2224–2228 (2008). https://doi.org/10.1002/pen.21169
50. Wang, L., Li, Q.: Photochromism into nanosystems: towards lighting up the future nanoworld. Chem. Soc. Rev. **47**, 1044–1097 (2018). https://doi.org/10.1039/C7CS00630F
51. Khattab, T.A., Rehan, M., Hamouda, T.: Smart textile framework: photochromic and fluorescent cellulosic fabric printed by strontium aluminate pigment. Carbohydr. Polym. **195**, 143–152 (2018). https://doi.org/10.1016/j.carbpol.2018.04.084
52. Wang, S., Liu, X., Yang, M., Zhang, Y., Xiang, K., Tang, R.: Review of time temperature indicators as quality monitors in food packaging. Packag. Technol. Sci. **28**(10), 839–867 (2015). https://doi.org/10.1002/pts.2148
53. White, M.A., LeBlanc, M.: Thermochromism in commercial products. J. Chem. Educ. **76**(9), 1201 (1999). https://doi.org/10.1021/ed076p1201
54. Garmaise, I., Garmaise, D.: Temperature sensing and indicating beverage mug. Patent no: US5678925A (1995)
55. Garshasbi, S., Santamouris, M.: Using advanced thermochromic technologies in the built environment: recent development and potential to decrease the energy consumption and fight urban overheating. Solar Energy Mater. Solar Cells. **191**, 21–32 (2019). https://doi.org/10.1016/j.solmat.2018.10.023
56. Seeboth, A., Ruhmann, R., Mühling, O.: Thermotropic and thermochromic polymer based materials for adaptive solar control. Materials. **3**(12), 5143–5168 (2010). https://doi.org/10.3390/ma3125143
57. Zhang, Z., Guo, L., Zhang, X., Hao, J.: Environmentally stable, photochromic and thermotropic organohydrogels for low cost on-demand optical devices. J. Colloid Interf. Sci. **578**, 315–325 (2020). https://doi.org/10.1016/j.jcis.2020.05.110
58. Cheng, Y., Zhang, X., Fang, C., Chen, J., Wang, Z.: Discoloration mechanism, structures and recent applications of thermochromic materials via different methods: a review. J. Mater. Sci. Technol. **34**(12), 2225–2234 (2018). https://doi.org/10.1016/j.jmst.2018.05.016
59. Lataste, E., Demourgues, A., Salmi, J., Naporea, C., Gaudon, M.: Thermochromic behavior (400<T°C<1200°C) of barium carbonate/binary metal oxide mixtures. Dyes Pigments. **91**(3), 396–403 (2011). https://doi.org/10.1016/j.dyepig.2011.05.016
60. Chowdhury, M.A., Joshi, M., Butola, B.S.: Photochromic and Thermochromic colorants in textile applications. J. Eng. Fibers Fabr. **9**(1), 107–123 (2014). https://doi.org/10.1177/155892501400900113
61. Chunhua, H., Englert, U.: Crystal-to-crystal transformation from a chain polymer to a two-dimensional network at low temperatures. Angewandte Chem. **44**(15), 2281–2283 (2005). https://doi.org/10.1002/anie.200462100
62. Guinneton, F., Sauques, L., Valmalette, J.-C., Cros, F., Gavarri, J.-R.: Role of surface defects and microstructure in infrared optical properties of thermochromic VO2 materials. J. Phys. Chem. Solids. **66**(1), 63–73 (2005). https://doi.org/10.1016/j.jpcs.2004.08.032
63. Sage, I.: Thermochromic liquid crystals. Liquid Cryst. **38**(11–12), 1551–1561 (2011). https://doi.org/10.1080/02678292.2011.631302
64. Sun, C., Xu, G., Jiang, X.-M., Wang, G.-E., Guo, P.-Y., Wang, M.-S., Guo, G.-C.: Design strategy for improving optical and electrical properties and stability of Lead-halide semiconductors. J. Am. Chem. Soc. **140**(8), 2805–2811 (2018). https://doi.org/10.1021/jacs.7b10101
65. Niklasson, G.A., Li, S.-Y., Granqvist, C.G.: Thermochromic vanadium oxide thin films: electronic and optical properties. J. Phys.: Conf. Ser. **559**, 012001 (2014). https://doi.org/10.1088/1742-6596/559/1/012001
66. Goran, G.C., et al.: Chromogenics for sustainable energy: some advances in thermochromics and electrochromics. Adv. Sci. Technol. **75**., Trans Tech Publications, Ltd., 55–64 (2010). https://doi.org/10.4028/www.scientific.net/ast.75.55
67. Khalid, M.W., Whitehouse, C., Ahmed, R., Hassan, M.U., Butt, H.: Remote thermal sensing by integration of corner-cube optics and thermochromic materials. Adv. Optic. Mater. **7**, 2 (2019). https://doi.org/10.1002/adom.201801013
68. Wilcox, I.: Beverage container having thermal indicator. Patent no: US20030000451A1 (2001).
69. Friend, C.M., Thorpe, C.: Smart consumer goods. In: Proc. SPIE 4763, European Workshop on Smart Structures in Engineering and Technology. SPIE, Bellingham, WA (2003). https://doi.org/10.1117/12.508658
70. Bastiani, M.D., Saidaminov, M.I., Dursun, I., Sinatra, L., Peng, W., Buttner, U., Mohammed, O.F., Bakr, O.M.: Thermochromic perovskite inks for reversible smart window applications. Chem. Mater. **29**(8), 3367–3370 (2017). https://doi.org/10.1021/acs.chemmater.6b05112
71. Salamati, M., Kamyabjou, G., Mohamadi, M., Taghizade, K., Kowsari, E.: Preparation of TiO_2@W-VO2 thermochromic thin film for the application of energy efficient smart windows and energy modeling studies of the produced glass. Constr. Build. Mater. **218**, 477–482 (2019). https://doi.org/10.1016/j.conbuildmat.2019.05.046
72. Kevin, A.T.: Container with thermochromic indicator. Patent no: US20120152781A1 (2011).
73. https://www.newscientist.com/article/dn13592-intelligent-paint-turns-roads-pink-in-icy-conditions/
74. Mortimer, R.J.: Electrochromic materials. Annu. Rev. Mater. Res. **41**, 241–268 (2011). https://doi.org/10.1146/annurev-matsci-062910-100344
75. Thrivikraman, G., Boda, S.K., Basu, B.: Unraveling the mechanistic effects of electric field stimulation towards directing stem cell fate and function: a tissue engineering perspective. Biomaterials. **150**, 60–86 (2018). https://doi.org/10.1016/j.biomaterials.2017.10.003
76. Cossari, P., Pugliese, M., Gambino, S., Cannavale, A., Maiorano, V., Gigliab, G., Mazzeo, M.: Fully integrated electrochromic-OLED devices for highly transparent smart glasses. J. Mater. Chem. C. **6**, 7274–7284 (2018). https://doi.org/10.1039/C8TC01665H

77. Svensson, J.S.E.M., Granqvist, C.G.: Electrochromic hydrated nickel oxide coatings for energy efficient windows: optical properties and coloration mechanism. Appl. Phys. Lett. **49**, 1566 (1986). https://doi.org/10.1063/1.97281
78. Somani, P.R., Radhakrishnan, S.: Electrochromic materials and devices: present and future. Mater. Chem. Phys. **77**(1), 117–133 (2003). https://doi.org/10.1016/S0254-0584(01)00575-2
79. Rowley, N.M., Mortimer, R.J.: New electrochromic Materials. Sci. Prog. **85**(3), 243–262 (2002). https://doi.org/10.3184/003685002783238816
80. Li, X., Perera, K., He, J., Gumyusenge, A., Mei, J.: Solution-processable electrochromic materials and devices: roadblocks and strategies towards large-scale applications. J. Mater. Chem. C. **7**, 12761–12789 (2019). https://doi.org/10.1039/C9TC02861G
81. Shah, K.W., Wang, S.-X., Soo, D.X.Y., Xu, J.: Viologen-based electrochromic materials: from small molecules, polymers and composites to their applications. Polymers. **11**(11), 1839 (2019). https://doi.org/10.3390/polym11111839
82. Lampert, C.M.: Large-area smart glass and integrated photovoltaics. Sol. Energy Mater. Sol. Cells. **76**(4), 489–499 (2003). https://doi.org/10.1016/S0927-0248(02)00259-3
83. Cantão, M.P., Cisneros, J.I., Torresi, R.M.: Electrochromic behavior of sputtered titanium oxide thin films. Thin Solid Films. **259**(1), 70–74 (1995). https://doi.org/10.1016/0040-6090(94)06401-6
84. Carpenter, M.K., Conell, R.S., Simko, S.J.: Electrochemistry and electrochromism of vanadium hexacyanoferrate. Inorg. Chem. **29** (4), 845–850 (1990). https://doi.org/10.1021/ic00329a054
85. Özer, N., Barreto, T., Büyüklimanli, T., Lampert, C.M.: Characterization of sol-gel deposited niobium pentoxide films for electrochromic devices. Sol. Energy Mater. Sol. Cells. **36**(4), 433–443 (1995). https://doi.org/10.1016/0927-0248(94)00197-9
86. Chigane, M., Ishikawa, M., Izaki, M.: Preparation of manganese oxide thin films by electrolysis/chemical deposition and electrochromism. J. Electrochem. Soc. **148**, 7. https://doi.org/10.1149/1.1376637
87. Burdukov, A.B., Vershinin, M.A., Pervukhina, N.V., Boguslvaskii, E.G., Eltsov, I.V., Shundrin, L.A., Selector, S.L., Shokurov, A.V., Voloshin, Y.Z.: Towards the clathrochelate-based electrochromic materials: the case study of the first iron(II) cage complex with an annelated quinoxaline fragment. Inorgan. Chem. Commun. **44**, 183–187 (2014). https://doi.org/10.1016/j.inoche.2014.03.032
88. Xi, T., Xiangkai, F., Jiang, Q., Liu, Z., Chen, G.: The synthesis and electrochemical properties of cathodic–anodic composite electrochromic materials. Dyes Pigments. **88**(1), 39–43 (2011). https://doi.org/10.1016/j.dyepig.2010.04.012
89. Dautremont-Smith, W.C.: Transition metal oxide electrochromic materials and displays: a review: part 2: oxides with anodic coloration. Displays. **3**(2), 67–80 (1982). https://doi.org/10.1016/0141-9382(82)90100-7
90. Kubo, T., Shinada, T., Kobayashi, Y., Imafuku, H., Toya, T., Akita, S., Nishikitani, Y., Watanabe, H.: Current state of the art for NOC-AGC electrochromic windows for architectural and automotive applications. Solid State Ion. **165**(1–4), 209–216 (2003). https://doi.org/10.1016/j.ssi.2003.08.043
91. Aegerter, M.A.: Sol–gel niobium pentoxide: a promising material for electrochromic coatings, batteries, nanocrystalline solar cells and catalysis. Sol. Energy Mater. Sol. Cells. **3–4**, 401–422 (2001). https://doi.org/10.1016/S0927-0248(00)00372-X
92. Cogan, S.F., Anderson, E.J., Plante, T.D., Rauh, R.D.: Electrochemical investigation of electrochromism in transparent conductive oxides. Appl. Opt. **24**(15), 2282–2283 (1985). https://doi.org/10.1364/AO.24.002282
93. Mortimer, R.J.: Switching colors with electricity: electrochromic materials can be used in glare reduction, energy conservation and chameleonic fabrics. Am. Scientist. **101**(1), 38 (2013) Gale Academic OneFile, Accessed 23 Aug 2020
94. Kobayashi, N., Miura, S., Nishimura, M., Urano, H.: Organic electrochromism for a new color electronic paper. Sol. Energy Mater. Sol. Cells. **92**(2), 136–139 (2008). https://doi.org/10.1016/j.solmat.2007.02.027
95. https://www.ainonline.com/sites/default/files/pdf/ain_2017_completions_0.pdf. Accessed 24 May 2019
96. http://citeseerx.ist.psu.edu/viewdoc/download?doi=10.1.1.202.3522&rep=rep1&type=pdf. Accessed 24 May 2019
97. Lakshmanan, R., Raja, P.P., Shivaprakash, N.C., Sindhu, S.: Fabrication of fast switching electrochromic window based on poly (3,4-(2,2-dimethylpropylenedioxy)thiophene) thin film. J. Mater. Sci: Mater. Electron. **27**, 6035–6042 (2016). https://doi.org/10.1007/s10854-016-4527-0
98. Granqvist, C.G.: Electrochromism and smart window design. Solid State Ion. **53–56**(Part 1), 479–489 (1992). https://doi.org/10.1016/0167-2738(92)90418-O
99. Ashrit, P.V., Bader, G., Girouard, F.E., Truong, V.-V.: Electrochromic materials for smart window applications. In: Proc. SPIE 1401, Optical Data Storage Technologies. SPIE, Bellingham, WA (1991). https://doi.org/10.1117/12.26121
100. Nicoletta, F.P., Chidichimo, G., Cupelli, D., De Filpo, G., De Benedittis, M., Gabriele, B., Salerno, G., Fazio, A.: Electrochromic polymer-dispersed liquid-crystal film: a new bifunctional device. Adv. Funct. Mater. **15**(6), 995–999 (2005). https://doi.org/10.1002/adfm.200400403
101. Goodman, L.A.: Passive liquid displays: liquid crystals, electrophoretics, and electrochromics. IEEE Trans. Consum. Electron. **CE-21**(3), 247–259 (1975). https://doi.org/10.1109/TCE.1975.266744.
102. Cao, X., Lau, C., Liu, Y., Wu, F., Gui, H., Liu, Q., Ma, Y., Wan, H., Amer, M.R., Zhou, C.: Fully screen-printed, large-area, and flexible active-matrix electrochromic displays using carbon nanotube thin-film transistors. ACS Nano. **10**(11), 9816–9822 (2016). https://doi.org/10.1021/acsnano.6b05368
103. Patil, P.S.: Gas-chromism in ultrasonic spray pyrolyzed tungsten oxide thin films. Bull. Mater. Sci. **23**, 309–312 (2000). https://doi.org/10.1007/BF02720088
104. Parthibavarman, M., Karthik, M., Prabhakaran, S.: Facile and one step synthesis of WO3 nanorods and nanosheets as an efficient photocatalyst and humidity sensing material. Vacuum. **155**, 224–232 (2018). https://doi.org/10.1016/j.vacuum.2018.06.021
105. van der Sluis, P.: Chemochromic optical switches based on metal hydrides. Electrochim. Acta. **44**(18), 3063–3066 (1999). https://doi.org/10.1016/S0013-4686(99)00021-3
106. Wittwer, V., Datz, M., Ell, J., Georg, A., Graf, W., Walze, G.: Gasochromic windows. Sol. Energy Mater. Sol. Cells. **84**(1–4), 305–314 (2004). https://doi.org/10.1016/j.solmat.2004.01.040
107. Tung, T.T., Duke III, C.B., Junker, C.S., O'Brien, C.M., Ross II, C. R., Barnes, C.E., Webster, C.E., Burkey, T.J.: Linkage isomerization as a mechanism for photochromic materials: cyclopentadienyl manganese tricarbonyl derivatives with chelatable functional groups. Organometallics. **27**(2), 289–296 (2008). https://doi.org/10.1021/om701101h
108. Lee, Y.-A., Kalanur, S.S., Shim, G., Park, J., Seo, H.: Highly sensitive gasochromic H2 sensing by nano-columnar WO3-Pd films with surface moisture. Sensors Actuators B Chem. **238**, 111–119 (2017). https://doi.org/10.1016/j.snb.2016.07.058
109. Hu, C.-W., Yamada, Y., Yoshimura, K.: Fabrication of nickel oxyhydroxide/palladium (NiOOH/Pd) nanocomposite for gasochromic application. Sol. Energy Mater. Sol. Cells. **177**, 120–127 (2018). https://doi.org/10.1016/j.solmat.2017.01.021
110. Takahashi, H., Okazaki, S., Nishijima, Y., Arakawa, T.: Optimization of hydrogen sensing performance of Pt/WO3 Gasochromic

film fabricated by sol–gel method. Sens. Mater. **29**(9), 1259–1268 (2017). https://doi.org/10.18494/SAM.2017.1585
111. Hupp, B., Nitsch, J., Schmitt, T., Bertermann, R., Edkins, K., Hirsch, F., Fischer, I., Auth, M., Sperlich, A., Doz, P., Steffen, A.: Stimulus-triggered formation of an anion–cation exciplex in Copper(I) complexes as a mechanism for mechanochromic phosphorescence. Angewandte Chem. **57**(41), 13671–13675 (2018). https://doi.org/10.1002/anie.201807768
112. https://www.aber.ac.uk/en/news/archive/2016/08/title-189046-en.html. Accessed 28 May 2019
113. https://www.polymersolutions.com/blog/polymer-opals/. Accessed 28 May 2019
114. Ruhl, T., Spahn, P., Hellmann, G.P.: Artificial opals prepared by melt compression. Polymer. **44**(25), 7625–7634 (2003). https://doi.org/10.1016/j.polymer.2003.09.047
115. Jiang, Y.: An outlook review: Mechanochromic materials and their potential for biological and healthcare applications. Mater. Sci. Eng. C. **45**, 682–689 (2014). https://doi.org/10.1016/j.msec.2014.08.027
116. Fleischmann, C., Lievenbrück, M., Ritter, H.: Polymers and dyes: developments and applications. Polymers. **7**(4), 717–746 (2015). https://doi.org/10.3390/polym7040717
117. Chen, F., Zhang, J., Wan, X.: Anthraquinone-imide-based dimers: synthesis, piezochromism, liquid crystalline and near-infrared electrochromic properties, Volume212, Issue17. September. **1**, 1836–1845 (2011). https://doi.org/10.1002/macp.201100065
118. Okazaki, M., Takeda, Y., Data, P., Pander, P., Higginbotham, H., Monkman, A.P., Minakata, S.: Thermally activated delayed fluorescent phenothiazine–dibenzo[a, j]phenazine–phenothiazine triads exhibiting tricolor-changing mechanochromic luminescence. Chem. Sci. **8**, 2677–2686 (2017). https://doi.org/10.1039/C6SC04863C
119. Klajn, R.: Spiropyran-based dynamic materials. Chem. Soc. Rev. **43**, 148–184 (2014). https://doi.org/10.1039/C3CS60181A
120. Stephanie, L., Potisek Douglas, A., DavisScott, R., White Nancy, R., Sottos Jeffrey, S.: Moore, self-assessing mechanochromic materials, patent no: US8236914B2 (2008).
121. Patel, M., Mintel, T.E., Kennedy, S., Gatzemeyer, J.J., Jimenez, E. J.: Toothbrush including a device for indicating brushing force. Patent no: US9578957B2 (2011).
122. Han, X., Liu, Y., Liu, G., Luo, J., Liu, S.H., Zhao, W., Yin, J.: A versatile naphthalimide–sulfonamide-coated tetraphenylethene: aggregation-induced emission behavior, mechanochromism, and tracking glutathione in living cells. Aggr. Induced Emission. **14** (6), 890–895 (2019). https://doi.org/10.1002/asia.201801854
123. https://technicalpaintservices.co.uk/catalogue/Tennis-Court-Paint-MUGAs-Line-and-Marking-Paints/
124. Qiu, W., Gurr, P.A., da Silva, G., Qiao, G.G.: Insights into the mechanochromism of spiropyran elastomers. Polym. Chem. **10**, 1650–1659 (2019). https://doi.org/10.1039/C9PY00017H
125. https://en.wikipedia.org/wiki/Weighing_scale
126. Walter, B.M.: Coextruded, heat-shrinkable, multi-layer, polyolefin packaging film. Patent no. US4352849A (1981).
127. Calvino, C., Neumann, L., Weder, C., Schrettl, S.: Approaches to polymeric mechanochromic materials. Ben Zhong Tang Trib. Spec. Iss. **55**(4), 640–652 (2016). https://doi.org/10.1002/pola.28445
128. https://www.pharmapproach.com/plastic-containers-for-pharmaceutical-use/
129. Ferrara, M., Bengisu, M.: Manufacturing and processes related to chromogenic materials and applications. In: Materials that Change Color. SpringerBriefs in Applied Sciences and Technology. Springer, Cham (2014). https://doi.org/10.1007/978-3-319-00290-3_3
130. Liu, P., Lee, S.-H., Tracy, C.E., Turner, J.A., Pitts, J.R., Deb, S.K.: Electrochromic and chemochromic performance of mesoporous thin-film vanadium oxide. Solid State Ion. **165**(1–4), 223–228 (2003). https://doi.org/10.1016/j.ssi.2003.08.044
131. Wang, M., Koski, K.J.: Reversible Chemochromic MoO3 nanoribbons through Zerovalent metal intercalation. ACS Nano. **9**(3), 3226–3233 (2015). https://doi.org/10.1021/acsnano.5b0033
132. Hariharan, P.S., Mothi, E.M., Moon, D., Anthony, S.P.: Halochromic isoquinoline with mechanochromic triphenylamine: smart fluorescent material for rewritable and self-erasable fluorescent platform. ACS Appl. Mater. Interf. **8**(48), 33034–33042 (2016). https://doi.org/10.1021/acsami.6b11939
133. Rosace, G., Guido, E., Colleoni, C., Brucale, M., Piperopoulos, E., Milone, C., Plutino, M.R.: Halochromic resorufin-GPTMS hybrid sol-gel: Chemical-physical properties and use as pH sensor fabric coating. Sens. Actuators B Chem. **241**, 85–95 (2017). https://doi.org/10.1016/j.snb.2016.10.038
134. Innocenzi, P., Brusatin, G., Abbotto, A., et al.: Entrapping of push-pull Zwitterionic chromophores in hybrid matrices for photonic applications. J. Sol-Gel Sci. Technol. **26**, 967–970 (2003). https://doi.org/10.1023/A:1020705312269
135. Inganäs, O., Salaneck, W.R., Österholm, J.-E., Laakso, J.: Thermochromic and solvatochromic effects in poly (3-hexylthiophene). Synth. Met. **22**(4), 395–406 (1988). https://doi.org/10.1016/0379-6779(88)90110-5
136. https://sites.google.com/site/smartmaterialswebsite/home/6-01-standard-one-foundations-for-leadership/standard-2-contextual-understanding/standard-3-planning-and-organization
137. Qi, S., Ren, X.-T., Dai, Y.-X., Wang, K., Li, W.-T., Gong, T., Fang, J.-J., Bo, Z., Gao, E.-Q., Wang, L.: Piezochromism and hydrochromism through electron transfer: new stories for viologen materials. Chem. Sci. **8**, 2758–2768 (2017). https://doi.org/10.1039/C6SC04579K
138. Hazra, A., Roy, A., Bhattacharjee, A., Barma, A., Roy, P.: Quinoline based chromogenic and fluorescence chemosensor for pH: effect of isomer. J. Mol. Struct. **1201**, 127173 (2020). https://doi.org/10.1016/j.molstruc.2019.127173
139. Mohajeri, N., T-Raissi, A., Bokerman, G., Captain, J.E., Peterson, B.V., Whitten, M., Trigwell, S., Berger, C., Brenner, J.: TEM–XRD analysis of PdO particles on TiO_2 support for chemochromic detection of hydrogen. Sensors Actuators B Chem. **144**(1), 208–214 (2010). https://doi.org/10.1016/j.snb.2009.10.064
140. Sakhri, A., Perrin, F.X., Aragon, E., Lamouric, S., Benaboura, A.: Chlorinated rubber paints for corrosion prevention of mild steel: a comparison between zinc phosphate and polyaniline pigments. Corros. Sci. **52**(3), 901–909 (2010). https://doi.org/10.1016/j.corsci.2009.11.010
141. Zhao, W., Monsur Ali, M., Aguirre, S.D., Brook, M.A., Li, Y.: Paper-based bioassays using gold nanoparticle colorimetric probes, anal. Chem. **80**(22), 8431–8437 (2008). https://doi.org/10.1021/ac801008q
142. Yan, X., Chang, Y., Qian, X.: Effect of concentration of thermochromic ink on performance of waterborne finish films for the surface of Cunninghamia Lanceolata. Polymers. **12**(3), 552 (2020). https://doi.org/10.3390/polym12030552
143. Lee, J., Pyo, M., Lee, S., et al.: Hydrochromic conjugated polymers for human sweat pore mapping. Nat. Commun. **5**, 3736 (2014). https://doi.org/10.1038/ncomms4736
144. https://ntrs.nasa.gov/search.jsp?R=20110003553
145. Chen, S., Liu, J., Zhang, S., Zhao, E., Yu, C.Y.Y., Hushiarian, R., Hong, Y., Tang, B.Z.: Biochromic silole derivatives: a single dye for differentiation, quantitation and imaging of live/dead cells. Mater. Horiz. **5**, 969–978 (2018). https://doi.org/10.1039/C8MH00799C
146. Bloor, D.: Experimental studies of polydiacetylene: model conjugated polymers. In: André, J.M., Bredas, J.L., Delhalle, J., Ladik, J., Leroy, G., Moser, C. (eds.) Recent Advances in the Quantum Theory of Polymers. Lecture Notes in Physics, vol. 113. Springer,

Berlin, Heidelberg (1980). https://doi.org/10.1007/3540097317_71

147. Kita, E., Marai, H., Michał, L., et al.: Mixed-ligand chromium(III)-oxalate-pirydinedicarboxylate complexes: potential biochromium sources: kinetic studies in NaOH solutions and effect on 3T3 fibroblasts proliferation. Transit. Met. Chem. **35**, 177–184 (2010). https://doi.org/10.1007/s11243-009-9311-z
148. Jelinek, R., Okada, S., Norvez, S., Charych, D.: Interfacial catalysis by phospholipases at conjugated lipid vesicles: colorimetric detection and NMR spectroscopy. Chem. Biol. **5**(11), 619–629 (1998). https://doi.org/10.1016/S1074-5521(98)90290-3
149. Page, T.H., Brown, A., Timms, E.M., Foxwell, B.M.J., Ray, K.P.: Inhibitors of p38 suppress cytokine production in rheumatoid arthritis synovial membranes: does variable inhibition of interleukin-6 production limit effectiveness in vivo? Arthritis Rheum. **62**(11), 3221–3231 (2010). https://doi.org/10.1002/art.27631
150. Zhao, B., Wang, X., Jin, H., Feng, H., Shen, G., Cao, Y., Yu, C., Lu, Z., Zhang, Q.: Spatiotemporal variation and potential risks of seven heavy metals in seawater, sediment, and seafood in Xiangshan Bay, China (2011–2016). Chemosphere. **212**, 1163–1171 (2018). https://doi.org/10.1016/j.chemosphere.2018.09.020
151. Yousef, E.S.S.: Linear and non-linear optical phenomena of glasses (photonics-photo chromic-electro and magneto optics): a review. Solid State Phenomena. **207**., Trans Tech Publications, Ltd., 1–35 (2013). https://doi.org/10.4028/www.scientific.net/ssp.207.1
152. Seevakan, K., Manikandan, A., Devendran, P., Baykal, A., Alagesan, T.: Electrochemical and magneto-optical properties of cobalt molybdate nano-catalyst as high-performance supercapacitor. Ceram. Int. **44**(15), 17735–17742 (2018). https://doi.org/10.1016/j.ceramint.2018.06.240
153. Tsujioka, T., Shimizu, Y., Irie, M.: Crosstalk in photon-mode photochromic multi-wavelength recording. Jpn. Soc. Appl. Phys. **33**(Part 1), 4A (1994). https://doi.org/10.1143/JJAP.33.1914
154. Lee, H., Jeon, S., Friedman, B., et al.: Simultaneous imaging of magnetic field and temperature distributions by magneto optical indicator microscopy. Sci. Rep. **7**, 43804 (2017). https://doi.org/10.1038/srep43804
155. Wang, F., Wang, G., Jiang, C., et al.: Shape-anisotropic enhanced damping in CoZr periodic arrays of nanohill structure. Nanoscale Res. Lett. **8**, 284 (2013). https://doi.org/10.1186/1556-276X-8-284
156. Sedó, J., Saiz-Poseu, J., Busqué, F., Ruiz-Molina, D.: Catechol-based biomimetic functional materials. Adv. Mater. **25**(5), 653–701 (2013). https://doi.org/10.1002/adma.201202343
157. Anuj, S., Kumar, P.M., Manoth, M., Sandhya, S., Kumar, S.V., Siddaramana, G.G., Raj, V.S., Narendra, K.: Preparation and characterization of biocompatible and water-dispersible Superparamagnetic Iron Oxide Nanoparticles (SPIONs). Adv. Sci. Lett. **3** (2), 161–167 (2010). https://doi.org/10.1166/asl.2010.1103
158. Sun, H., Liu, S., Lin, W., et al.: Smart responsive phosphorescent materials for data recording and security protection. Nat. Commun. **5**, 3601 (2014). https://doi.org/10.1038/ncomms4601
159. Flämmich, M., Frischeisen, J., Setz, D.S., Michaelis, D., Krummacher, B.C., Schmidt, T.D., Brütting, W., Danz, N.: Oriented phosphorescent emitters boost OLED efficiency. Org. Electron. **12**(10), 1663–1668 (2011). https://doi.org/10.1016/j.orgel.2011.06.011
160. Marsella, M.J., Swager, T.M.: Designing conducting polymer-based sensors: selective ionochromic response in crown ether-containing polythiophenes. J. Am. Chem. Soc. **115**(25), 12214–12215 (1993). https://doi.org/10.1021/ja00078a090
161. Dong, Y., Lam, J.W.Y., Li, Z., et al.: Vapochromism of Hexaphenylsilole. J. Inorg. Organomet. Polym. **15**, 287–291 (2005). https://doi.org/10.1007/s10904-005-5546-0
162. Jordan, K.: Review of recent advances in radiochromic materials for 3D dosimetry. J. Phys.: Conf. Ser. **250**, 012043 (2010). https://doi.org/10.1088/1742-6596/250/1/012043
163. https://en.wikipedia.org/wiki/Sorptiochromism
164. Cartwright, S.J.: Solvatochromic dyes detect the presence of homeopathic potencies. Homeopathy. **105**(1), 55–65 (2016). https://doi.org/10.1016/j.homp.2015.08.002
165. Zhang, C., Sun, L.-D., Yan, C.-H.: Noble metal plasmonic nanostructure related chromisms. Inorg. Chem. Front. **3**, 203–217 (2016). https://doi.org/10.1039/C5QI00222B
166. Peng, H., Leung, F.S.M., Wu, A.X., Dong, Y., Dong, Y., Yu, N.-T., Feng, X., Tang, B.Z.: Using Buckyballs to cut off light! Novel fullerene Materials with unique optical transmission characteristics. Chem. Mater. **16**(23), 4790–4798 (2004). https://doi.org/10.1021/cm049680l
167. Yu, L., George, Z.: Chen, Cryo-solvatochromism in ionic liquids. RSC Adv. **4**, 40281–40285 (2014). https://doi.org/10.1039/C4RA08116A

Smart Fluid

6

Abstract

This chapter describes the basics of rheology and various kind of rheological materials that are stimulated by electric field or magnetic field or light. Electro-rheological (ER) fluid, magneto-rheological (MR) fluid, ferrofluid, magneto-rheological elastomer, electro-conjugate liquid, photo-rheological fluid has been depicted here. Historical background, mechanism, materials used and synthesis procedure has been discussed in this chapter. Current potential applications based on various rheological materials in leading industries such as aero-industries, automobile and heavy machinery industries, military and defense industries biomedical industries, electronic industries, and other industries have been discussed.

Keywords

Electro-rheological (ER) fluid · Giant ER · Magneto-rheological (MR) fluid · Ferrofluid · Magneto-rheological elastomers · Electro-conjugate liquids · Photo-rheological fluid

6.1 Introduction

A smart fluid is a fluid whose properties (e.g., the viscosity) can be changed by applying the stimuli (electric field or a magnetic field). Rheological materials are one type of smart fluid. Smart fluids can be divided into five main classes: (1) electro-rheological (ER) fluids, (2) magneto-rheological (MR) fluids, (3) magneto-rheological elastomer (MRE) fluids, (4) electro-conjugate liquids, and (5) Photo-rheological fluid [1–3]. The properties of smart fluids have been known for the last 60 years, but were subject to only sporadic investigations up until the 1990s, when they were suddenly the subject of renewed interest, notably culminating with the use of an MR fluid on the suspension of the 2002 model of the Cadillac Seville STS automobile and more recently, on the suspension of the second-generation Audi TT [4]. Other applications include brakes and seismic dampers, which are used in buildings in seismically active zones to damp the oscillations occurring in an earthquake. The word **rheology** comes from greek words (rheo means flow and logia means study) [5]. It deals with the study of the flow of matter. It addresses the behavior of the materials with properties intermediate between those of ideal solids and ideal liquids. When subjected to external forces, solids (or truly elastic materials) will form, whereas liquids (or truly viscous materials) will flow [3]. The term was first used by Eugene C. Bingham and Crawford. It applies to substances which have a complex microstructure such as mud, sludge, suspensions, polymers, and other glass formers (e.g., silicates), many foods and additives, bodily fluids (e.g., blood), and other biological materials [6, 7]. Rheology properties are also observed in our daily life, for example, squeezing toothpaste tubes, pouring ketchup, kneading bread dough, skin lotion, hand wash gel, etc. (Fig. 6.1). Rheological materials are non-Newtonian fluids. Non-Newtonian fluid is a class of fluids whose viscosity changes with the strain rate unlike the Newtonian fluid, in which the viscosity changes with temperature but does not change with the strain rate [8].

Rheological materials follow Bingham's plastic behavior. Figure 6.2 shows the idealized shear stress versus shear rate characteristics of both the non-Newtonian and Newtonian fluid (Bingham plastic). Both the ER and MR fluids can be characterized by Bingham plastic behavior. When the applied field (electric or magnetic) is zero, then the smart fluid is Newtonian-like in behavior. The application of a suitable field and the subsequent formation of particle chains causes yield stress to develop. This yield stress must be overcome before flow can occur [9]. Thus the fluid behaves in a fashion similar to a Bingham plastic. Two regimes are then defined: (1) pre-yield, where the yield stress $\tau_b$is not exceeded and thus fluid flow does not occur (2) post-yield, where τ_b is exceeded and consequently flow occurs. The key difference between the performance of ER and MR fluids can be

A. Behera, *Advanced Materials*, https://doi.org/10.1007/978-3-030-80359-9_6

Fig. 6.1 (**a**) Tooth paste, (**b**) ketchup, (**c**) body lotion

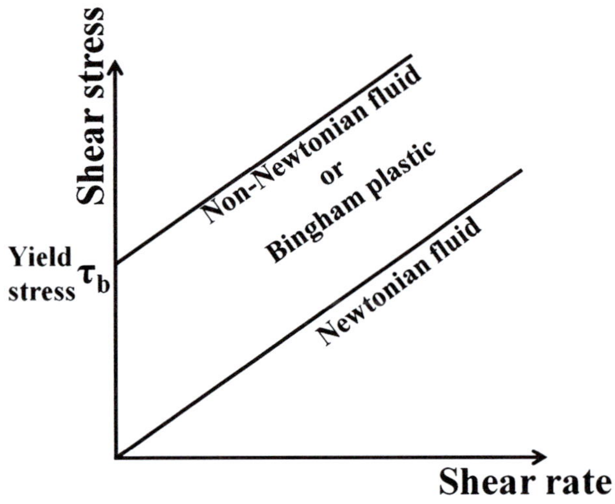

Fig. 6.2 Shear stress versus shear rate for Newtonian fluid and Bingham plastic

explained with reference to Fig. 6.2. In ER fluids, the maximum obtainable value of yield stress $\tau_b$is typically within the range of 3–5 kPa[3] [10]. In total contrast, dynamic yield strengths of 100 kPa are available from MR fluids. Moreover, modern ER fluids will operate within the temperature range from 15 to 90 °C whereas MR fluids will typically operate from −40 to 150 °C. From these factors, taken together with the electrical source required to excite the fluid up to 10 kV for ER fluids, 12 V for MR. It is clear why MR fluids are the clear favorite for commercial applications [11].

6.2 Electro-Rheological fluid

A major type of smart fluid is electro-rheological or ER fluids, whose resistance to flow can be quickly and dramatically altered by an applied electric field (note, the yield stress point is altered rather than the viscosity) [12]. Electro-rheological fluids, also called as electro-viscous fluids or electro-responsive fluids, are functional fluids, normally in the liquid state with good fluidity. Electro-rheological fluid is a liquid suspension of metals or zeolites which solidifies when a electric current is applied to it, becoming fluid again when the current is removed. Electrically active particles suspended in the fluid arrange themselves along the electric field lines to form rigid chains, resulting in increased viscosity, as shown in Fig. 6.3 [13]. ER fluids typically consist of a electrically active particle suspension in a hydrophobic liquid such as mineral or corn oil [14]. The particles usually have a size of 1–100 μm and a volume fraction between 5% and 50%. These fluids are capable of altering their viscosity in milliseconds. This phenomenon is reversible and repeatable. This makes the use of electro-rheological fluids highly competent in several applications such as clutches, dampers, brakes, shock absorbers, hydraulic valves, bulletproof vests where quick engagement and activation are important [15, 16]. The applied external field leads to the formation of particle chains that span the gap between the electrodes and thus increase the viscosity or even solidify the sample when the electric field is switched on. The particles rapidly orientate themselves in relatively regular chain-like columnar structures. The chain formation can be rationalized as being a consequence of the polarization of the particles in the electric field and it occurs when the polarizabilities (i.e., dielectric constants) of the medium and particles are different. Therefore, it is obvious that ER fluids are sensitive to the addition of small amounts of highly polarizable substances, such as water, such that moisture can strongly interfere with the performance of ER fluids. Since the solubility of water in the ER medium generally will be temperature-dependent, heating the ER fluid can have a further impact on the mechanical properties [17].

The apparent viscosity of these fluids changes reversibly by an order of up to 100,000 in response to an electric field [18]. For example, a typical ER fluid can go from the consistency of a liquid to that of a gel, and back, with response times on the order of milliseconds. The effect is sometimes called the Winslow effect after its discoverer, the American inventor Willis Winslow, who obtained a US patent on the effect in 1947 [19]. ER fluids, consisting of solid particles dispersed in an insulating liquid, display the special characteristic of electric-field-induced rheological variations. Nearly six decades after their discovery, ER fluids have emerged as materials of increasing scientific fascination and practical importance. In particular, ER fluids are divided into two different types, dielectric electro-rheological (DER) and giant electro-rheological (GER), which reflect the underlying electric susceptibility arising from the induced dielectric polarization and the orientational polarization of molecular dipoles, respectively [20]. The GER fluid was discovered in

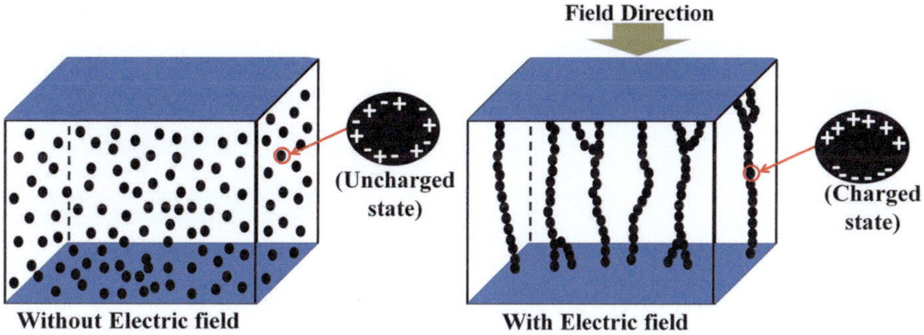

Fig. 6.3 Behavior of suspended particles in electro-rheological fluid under electric field

Table 6.1 Various carrier fluids and particles applied for ER fluid

Typical carrier fluids	Additive particles
Aldehydes, Aliphatic Esters, Aroclor, Carbon Tetrachloride, Caster Oil, Chlorobentzenediphenyl Alkanes, Chloroform, Cottonseed Oil, Di-2-Ethylhexyl Adipate, DibutylSebacate, Dielectric Oils, Different types of Ethers, Diphenyl Ethers, DiphenylSulphoxides, Grease, Hydrocarbon Oil, Ketones, Kerosene, Linseed Oil, Liquid Paraffin, Mineral Oil, Nitrobenzene, Olefins, Olive Oil, Orthochlorotoluene, Polychlorinated Biphenyls, Poly ChloroTrifluoroethylene, Polyalkylene Glycols, P-xylene, Polychlorinated biphenyls, Resin Oil, Silicone Oils, Transformer Oil	Alfa Silica, Alfa Methacrylate, Alginic Acid, Alumina, Alumina Silica Mixture, Aluminium Oleate, Aluminium Octoate, Aluminium Stearate, Boron, Azaporhin Systems, Barium Titanate, Cadmiumsulphide Phosphor, Calcium Stearate, Carbon, Cellulose, Ceramics, Charcoal, Chlorides, Colloidal Silica, Colloidal Kaolin Clay, Crystalline D Sorbitol, Diallylether, Diethylcarbocyanine Iodide, Diphenylthiazoleanthra Quinone, Divinylbenzene, Dyes, Gypsum, Iron Oxide, Mannitol, Metallic Semiconductors, Methyl Acrylate, Methyl Methacrylate, Microcell-C, Microcrystalline Cellulose, Micronized Mica, Monosaccharides, Molecular Sieves, N-Vinylpyrrolide, Nylon Powder, Olefins, Onyx Quartz, Polyvinyl Alcohol, Onyx Quartz, Pizo-ceramic, Phenolformaldehyde, Phthalocyanine, Polystyrene Polymer, Porhin, Phosphototungstomolybic Acid, Polymethacrylate Mixtures, Pyrogenic Silica, Quartz, Rotten Stone, Rubber, Silica Gel

2003 and is able to sustain higher yield strengths than many other ER fluids.

6.2.1 Materials Used in ER Fluid

ER particles are electrically active. They can be ferroelectric or, conducting material coated with an insulator, or electro-osmotically active particles. In the case of ferroelectric or conducting material, the particles would have a high dielectric constant [21]. There may be some confusion here as to the dielectric constant of a conductor, but "if a material with a high dielectric constant is placed in an electric field, the magnitude of that field will be measurably reduced within the volume of the dielectric," and since the electric field is zero in an ideal conductor, then in this context the dielectric constant of a conductor is infinite. A simple ER fluid can be made by mixing cornflour in light vegetable oil or (better) silicone oil [22]. In this fluid, the dispersions in an electrically insulating medium of 10–50% by volume of spherical silica particles prepared by hydrolyzing a silicon alkoxide of the general formula $Si(OR)_4$ in which R is an alkyl group in the presence of an alkali catalyst and drying at or below 500 °C, show excellent fluidity in the absence of an applied voltage, vary reversibly to the state of high viscosity or even to the state of gel on the application of a voltage, and have good storage stability [23]. The GER fluid consists of urea-coated nanoparticles of barium-titanium-oxalate suspended in silicone oil. The high yield strength is due to the high dielectric constant of the particles, the small size of the particles, and the urea coating. Another advantage of the GER is that the relationship between the electrical field strength and the yield strength is linear after the electric field reaches 1 kV/mm [24].

Some of the ER fluid constituents are dispersions of silica having a specified amount of silanol groups on its surface in oils; dispersions of ion exchange resins with adsorbed moisture in electrically insulating oils; dispersions of barium titanate and finely pulverized silica in electrically insulating oils; dispersions of water-containing phenolic resins in hydrophobic oils; dispersions of crystalline zeolites in electrically nonconducting fluids; and dispersions of cellulose, starch, soybean caseins, and the like in electrically insulating oils. Commercially available ER fluids having a very high cost, hence some carrier fluids and particles used to prepare ER fluids are given in Table 6.1 [25–27]. By various proportions of concentration, we can achieve many ER fluids according to our applications.

Another factor that influences the ER effect is the geometry of the electrodes. The introduction of parallel grooved electrodes showed a slight increase in the ER effect but perpendicular grooved electrodes doubled the ER effect. A much larger increase in ER effect can be obtained by coating

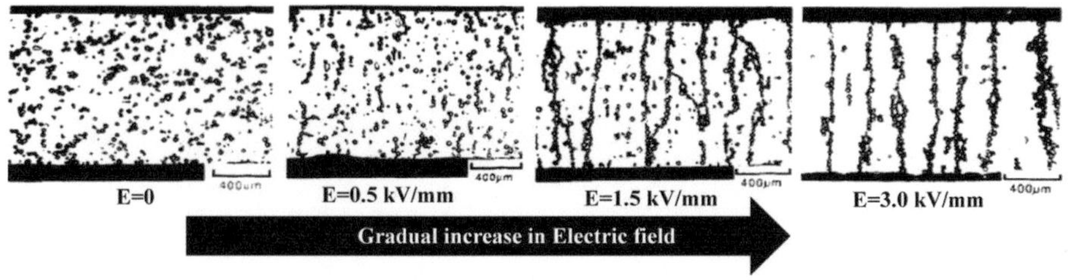

Fig. 6.4 Particles are gradually joining with each other to form chains as the gradual increase in the electric field

the electrodes with electrically polarisable materials [28]. This turns the usual disadvantage of dielectrophoresis into a useful effect. It also has the effect of reducing leakage currents in the ER fluid.

6.2.2 Preparation of ER Fluids

A typical procedure to prepare ER fluids is very simple. Initially, powder with different particles sizes will be prepared, with a precaution that all particles should be the same throughout. The powder will be passed through proper size by the sieve and weighed on an electronic weighing pan. This powder will be then mixed with a measured quantity of fluids which poured into the glass flask containing powder. While adding the oil into powder care should be taken to stir the mixture continuously so that powder gets mixed with oil throughout. This mixture will be stirred with the help of a glass rod continuously for half an hour for trough mixing, thus ER fluid will be prepared. After preparation of ER fluid, it will be passed through a vane pump 4–5 times for obtaining a homogenous mixture and good results [29]. The same process will be adopted for different combinations of the base fluid and powder particles to obtain different ER fluid samples. There are many different types of commercial ER fluid equipment, where ER fluid is composed of three parts [30, 31]: Base liquid, additive, and surfactant. Normally, the base liquid comprises mineral oils and other solutions, that are chosen for their electrical insulator properties The main property of a base liquid is its viscosity and electrical conductivity (10^{-9}–10^{-4} Ω/cm is normally required). Additives are highly polarizable suspended solid particles. These additives are responsible for the ER effects in an ER fluid. The additives typically consist of tiny particles or powders whose diameters are about 5–25 μm and can be suspended in the base oil for a long time. According to the theory behind the ER phenomenon, particles form many particle chains that can resist shear stress when an electric field is applied and can be easily polarized. The surfactant can prevent the degradation of particles from the ER fluid and activate the particles to cause the ER effect. The properties of ER fluids are not only related to the characteristics of the base liquid and particles, but also to their volume fraction or mass fraction in relation to the base liquid. After preparation, ER fluid undergo feasibility test with the help of viscosity test, temperature range test, sedimentation test, and break down test, etc.

6.2.3 Strengthening Mechanisms of Smart Fluid

Two basic phenomena have been proposed to explain the increase in shear strength of ER fluids, namely the interfacial tension or "water bridge" theory, and the electrostatic theory, polarization of the particles [32]. The water bridge theory assumes a three-phase system, the particles contain the third phase which is another liquid (e.g., water) immiscible with the main phase liquid (e.g., oil) [33]. With no applied electric field the third phase is strongly attracted to and held within the particles. This means the ER fluid is a suspension of particles, which behaves like a liquid. When an electric field is applied the third phase is driven to one side of the particles by electro-osmosis and binds adjacent particles together to form chains (Fig. 6.4). This bridge structure means the ER fluid has become solid. The interfacial tension between "the water and the ER carrier fluid then provides a source of shear resistance" [34]. The electrostatic theory assumes just a two-phase system, with dielectric particles forming chains aligned with an electric field [35]. An ER fluid can be constructed with the solid phase made from a conductor coated in an insulator. This ER fluid clearly cannot work by the water bridge model. However, although demonstrating that some ER fluids work by the electrostatic effect, it does not prove that all ER fluids do so. The advantage of having an ER fluid which operates on the electrostatic effect is the elimination of leakage current, i.e. potentially there is no direct current [36]. Of course, since ER devices behave electrically as capacitors, and the main advantage of the ER effect is the speed of response, an alternating current is to be expected.

6.2.4 Giant ER

The concept of a giant electro-rheological (GER) effect was first introduced by Wen et al. [37]. They examined nanoparticles of a urea-coated $BaTiO(C_2O_4)_2$-based ER fluid, which could reach yield stress orders of magnitude larger than the theoretical upper bound of the static yield stress in a conventional ER fluid [38]. The static yield stress of the GER fluid showed a linear relationship with the external applied electric field and varied by 1/R (R, radius of the GER particle), suggesting that smaller particles would generate a larger effect. The GER effect was only discovered in silicone oil according to the wetting effect between the nanoparticles and the continuous phase. The discovery of GER fluid expanded the possibility of practical applications of an ER fluid because the low yield behavior of conventional ER fluids is considered one of the main obstacles to its applications. To understand the GER effect, Lu et al. suggested that polar molecules (e.g., urea) adsorbed on the particles played a key role in generating the GER effect [39]. In this model, the attractive force between the particles consisted of dipole–dipole interaction between the oriented polar molecules and the dipole-charge interaction between the polar molecules and nanoparticles (Fig. 6.1). Moreover, Sheng and Wen recently reported a deep review of the mechanisms and dynamics in giant ER fluids compared to that of the dielectric ER fluids, where it was reiterated that the electric field could induce the alignment of urea dipolar filaments bridging the two boundaries of the nanoscale confinement using a molecular dynamics simulation [40].

The GER particles were fabricated by first dissolving barium chloride in distilled water at 50–70 °C. Separately, oxalic acid was dissolved in water at 65 °C in an ultrasonic tank, with titanium tetrachloride slowly added. The two solutions were mixed in an ultrasonic bath at 65 °C. Nanometresized barium-titanyl-oxalate particles were formed at this stage [41]. These bare particles, when mixed with silicone oil, have a maximum static ER yield stress of the order of 5 kPa. Addition of urea to the mixed solution led to the formation of a white colloid, which was cooled to room temperature. The precipitate was washed with water, filtered, and then dried to remove all trace water. The dried white powder consisted of the nanoparticles coated with urea, that is, ($BaTiO(C_2O_4)_2 + NH_2CONH_2$) [42]. In such a core/shell structure, urea serves as an ER promoter. Here the urea could be replaced with other chemicals, such as acrylamide. The ER fluids were prepared by mixing the powder with silicone oil and homogenizing it in a high-speed grinding mill for 2 h. The sample was then vacuum-dried at 105–120 °C for 1–3 h. The final suspension is stable, with no observable sedimentation after a few weeks on the shelf. The major concern is the use of oxalic acid for the preparation of the particles as it is a strong organic acid [43].

There was also a particle-size effect on the ER performance because of the difference in the surface area: doping with Rb reduced the size of the $BaTi(C_2O_4)_2$ particles coated with urea but achieved double the yield stress. Later, polar molecule-modification of SiO_2 and TiO_2 nanoparticles was applied to enhance the ER performance of these particles. In addition to the first applied polar molecule, urea, other polar molecules, such as triethanolamine and acetamide, were attempted as modification agents to inorganic particles. On the other hand, only the in situ added polar molecules could generate a GER effect rather than post modification of the synthesized inorganic particles. From the preparation of GER fluids, a high particle weight (or volume) fraction (nearly 50 wt% or even higher) was noticed. Therefore, it is essential to use suspension oils with low viscosity to reduce the off-state viscosity of the fluid. The wettability of the dispersed particles was also considered, upon which hydrophilic groups (hydroxyl)-terminated silicone oil was examined as a suspending liquid in the GER system [44]. A high yield stress is required for engineering devices. On the other hand, ER fluids generating high yield stress always have other problems, for example, serious sedimentation in an inorganics-based ER fluid and high zero-field viscosity of a GER fluid [45].

6.2.5 Microstructure and Properties

The widths of bright bands increase with increasing electric field. It is well known that when an electric field is applied, the ER fluid undergoes a transition to a gel-like state, the particles first form chains along the field direction, which then coalesces into columns, if the particle concentration is not too dilute [46]. An ER fluid is a substance which changes its apparent viscosity (rheological characteristic) according to the strength of the electric field immersing it, and its main features include a fast response speed and extremely low power consumption, among others, The ER fluids currently being developed may be classified into two types in terms of their characteristics. One is particle-type and another is homogeneous-type fluid [47].

ER fluids of particle-type are colloidal fluids which are solvents containing dispersed particles. It is clear from shearing speed versus shearing stress diagram that the particle-type ER fluids exhibit the characteristics of a Newtonian fluid when no electric field E is present but they reveal the characteristics of a Bingham fluid when an electric field is applied. In other words, when a voltage is applied, these ER fluids behave very much like Coulomb friction. Fluids of homogeneous type have been developed by using low molecule liquid crystal or macromolecular liquid crystal, and exhibit characteristics as shown in Fig. 6.5 [47]. Shearing stress nearly proportional to shearing speed is generated and its slope namely viscosity can be controlled by the electric field. As a result, it is possible to acquire a mechanical control force proportional to the speed under a constant electric field and to mechanically realize what is equivalent to the so-called differential control.

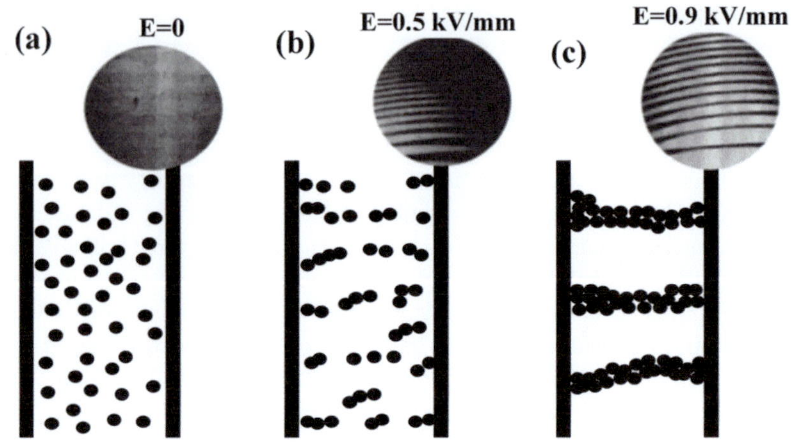

Fig. 6.5 Particle polarization and single sphere width chain formation with increasing field E. Gradual change of fluid properties observed as the changes in particle chaining

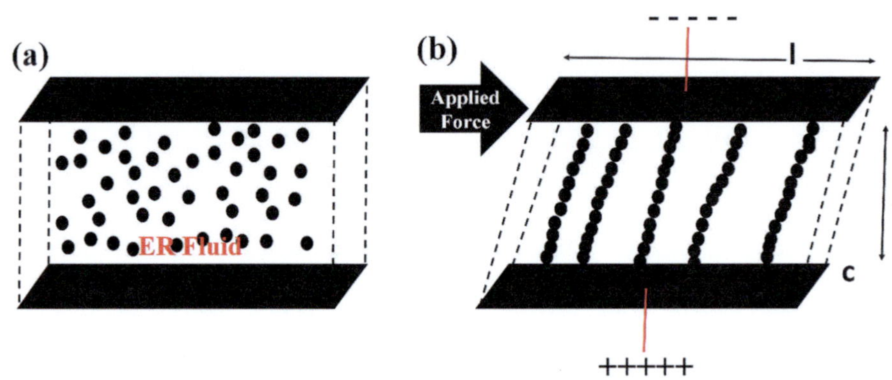

Fig. 6.6 Shear mode of ER fluid working

Any ER-based device or structure design depends on electromechanical characteristics in which shear stress and the applied electric field are used. Low electric field intensity leads to high stress. In ERF, the conducting properties are to know power consumption and heat dissipation. Low heat dissipation is expected (to avoid secondary/ancillary cooling systems). ER fluids should be operative over a broad temperature range and the use of anhydrous fluids gives a range of −20 to 700 °C [48]. ER fluid developed for commercial application must have (1) greater thermal stability, (2) density to particulate phase and continuous phase must be same to minimize sedimentation problems, (3) produces lower viscosity in the absence of electric field and higher viscosity in the presence of electric field, (4) dispersant used should have property like-low viscosity, low volatility, nontoxic, noncorrosive, nonflammable (for high-voltage applications where spark/arching possibilities are more) [49]. The dispersed phase should possess electrical attributes, has to be easily atomized from bulk state (to the surface area), must be easily dispersed with minimal use of additives and the solute must be nonabrasive. The charge migration mechanism for the dispersed phase is mainly surface phenomena (or) bulk phenomena. It depends on the columnar structure formed that depends on the transport mechanism, dipole moment, alignment of the dipoles. It depends on the nature of the liquid, porosity of the particle, characteristics of the surfactants, hydrous/anhydrous nature of the liquid, fluid density, properties of chemical activators present on the surface.

6.2.6 Modes of ER Fluid

There are three types of mode generally observed in ER fluid application, are discussed below:

Shear Mode: In this type of mode of ER damper, there are one or two parallel electrodes which can move parallel to each other and are always perpendicular to the electric field applied so that the fluid can have uniform shear and the ER fluid is present in between the two electrodes [50]. From Fig. 6.6, "c" and "l" are the breath and length of the electrode, respectively, and j is the electrode's gap. Here E is given voltage, "F" is net damping force, and "V" is the relative velocity of the electrodes. Two forces are acting in this ER damper (a) active force "F_c" because of ER effect and (b) passive force "F_y" due to the fluid viscosity. F_y, i.e., the passive force is always present and directly linked with the viscosity of the fluid as well as the damper geometric properties. During application of the electric field, a force F_c (because of creation of particles suspension lining up between the electrodes) i.e. static force which is needed to

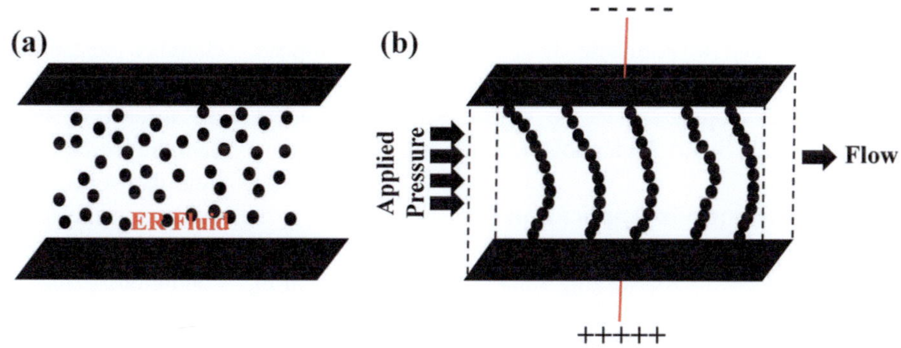

Fig. 6.7 Flow mode of ER fluid working

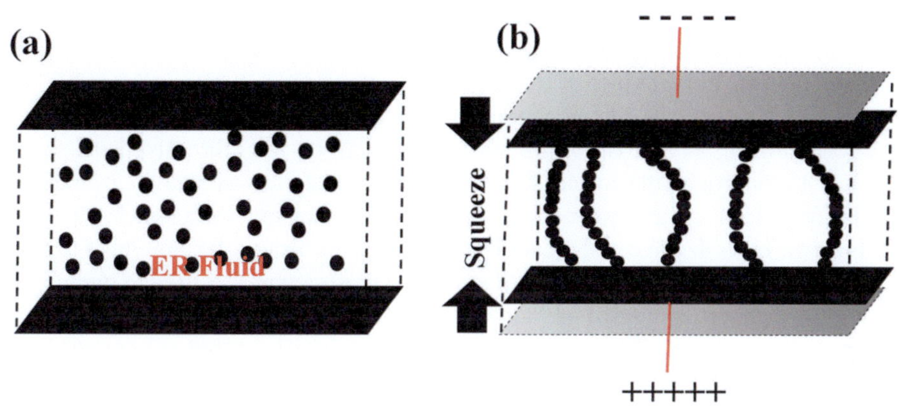

Fig. 6.8 Squeez mode of ER fluid working

overwhelmed so that the motion can occur [51]. The force F_c is a product of the area of the electrode and the yield strength of the fluid and does not depend on the electrode plate velocity. The net force F of damping of this ER damper is the sum of two components of force. The main aim of this ER damper is to give a large ratio of off-field to on-field damping by force ratios F_y and F_c. Because of this large ratio gives various responses by ER unit with changing voltage.

Flow Mode*:* In this type of mode the ER fluid is pressed between the two electrodes as given in Fig. 6.7. Because of this, the ER fluid is exposed to tensile, compression as well as shear. In the absence of the given electric field if the ER fluid is pressed it behaves like Newtonian fluid. There is a pressure drop that occurs at the flow rate. This pressure change in between the valve is because of the velocity of the ER fluid. Moreover, during the presence of the electric field, the yield stress is generated by the ER fluid which results in more pressure drop between the electrodes plate's length. The net damping force is the summation of two force components of this type of ER damper. In this type of mode, the device effectiveness is the across valves pressure drop with or without the effect of ER [52]. The electro-rheological fluids which are totally dependent on the applied electric fields are used in resistive force creation and damping. Examples of applications are active vibration suppression and motion control.

Squeeze Mode*:* In this mode, the gap between the electrodes is changed and the ER fluid is pressed or squeezed by the force acting normally. Figure 6.8 represents the squeezing mode of the ER fluid [53]. In squeeze mode operation, force obtained is at least one order of magnitude larger than in shear mode (plates moving parallel to each other). Here the damping, as well as the stiffness, are affected by the electrical field.

6.2.7 Applications

Depending on the end-user industry, the market of electro-rheological fluids may be segmented into automotive, electronics, defense, and other applications. In automobile, ER fluids are used for the suspension systems, dampers, and ER braking systems. Automotive and defense are the most important markets in ER applications. Due to the potential for the development of automotive technology, North America and Europe are important markets for electrorheology batteries. The emphasis on innovation and research and development in industries such as automotive and defense is one of the main reasons for the high demand for electro-rheological devices in both areas. The Asia Pacific is an important market for electro-rheological forces. Countries like China and Japan are making great progress in the use of smart fluids in electronics [54]. ER fluids can be used to develop tactile displays and flexible touch screens. The reason is intended to stimulate demand from the European league in the Asia Pacific segment in the coming years. Major players operating in the global electro-rheological market include Smart Technology Limited and Fludicon.

6.2.7.1 Automobile Application

Electro-rheological materials (Fig. 6.9) are used in vehicle suspension systems for steering strategy. The suspension in an automobile is a system of tires, air in the tires, springs, shock absorbers, and couplings that connect the vehicle to its wheels and allow movement between them. Suspension systems must be compatible with conflicting traction/handling and ride quality. Adjusting the suspension requires finding the right compromise. It is important that the suspension keeps the road wheel in maximum possible contact with the road surface, as all road or ground forces acting on the vehicle do so through the contact surfaces of the vehicle tires [55]. The suspension also protects the vehicle and any cargo or luggage from damage and wear. The design of the front and rear suspension of the vehicle may differ. The structure of the ER dampers is divided into upper and lower chambers by means of a piston. These chambers are completely filled with rheological fluid. As the piston moves, the ER fluid flows from one chamber to the other through an annular passage between the inner and outer cylinders. The inner cylinder is connected to the positive voltage generated by the high voltage supply unit represented by the positive (+) electrode. The outer cylinder is grounded, represented as a negative (−) electrode [57]. On the other hand, the gas chamber located outside the lower chamber acts as an accumulator of ER fluid caused by the movement of the piston. In the absence of an electric field, the ER damper generates a damping force caused only by the resistance of the viscous fluid. However, if a certain electric field is applied to the ER damper, it creates an additional damping force due to the limited load of the ER fluid. This damping force can be continuously regulated by regulating the intensity of the applied electric field [46].

The conventional application of ER fluids is in fast-acting hydraulic valves and clutches, with a distance between the plates of the order of 1 mm and the potential use of the order of 1 kV. Simply, when the electric field is applied, the ER hydraulic valve is closed or the ER clutch plates are connected, when the electric field is removed, the ER hydraulic valve is open, or the clutch plates are disconnected. Other common uses are in the case of ER brakes (think of the brake as a clutch with a fixed side) and shock absorbers (which can be thought of as closed hydraulic systems in which the impact is used to pump fluid through a valve) [56]. Several commercial applications have been examined and many more have not yet emerged, especially in recipes for electro-energy in the engine, in-seat dampers, the safety, and control system.

6.2.7.2 Electronic Industries

It has also been suggested that ER fluid has potential applications in flexible electronics, in which the fluid is embedded in elements such as keyboards and rollable

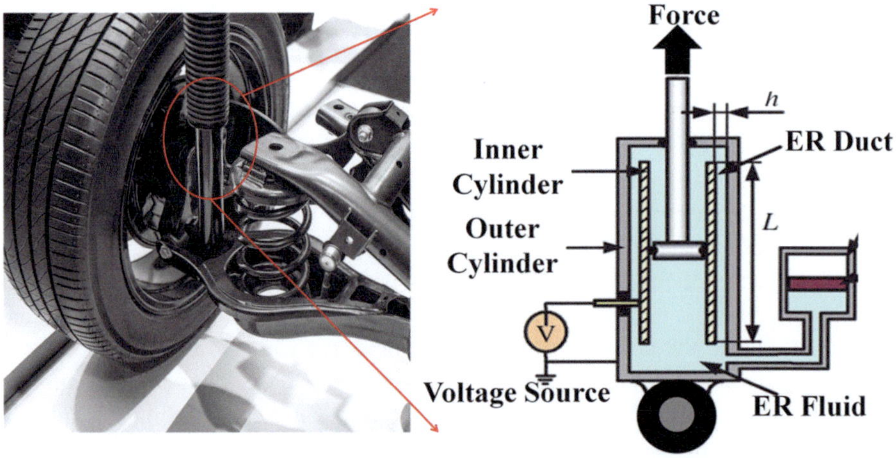

Fig. 6.9 ER damper on the car showing the magnified portions mechanism

Fig. 6.10 (**a**) Rollable keypad, (**b**) Screens, (**c**) Haptic device based on an ER fluid

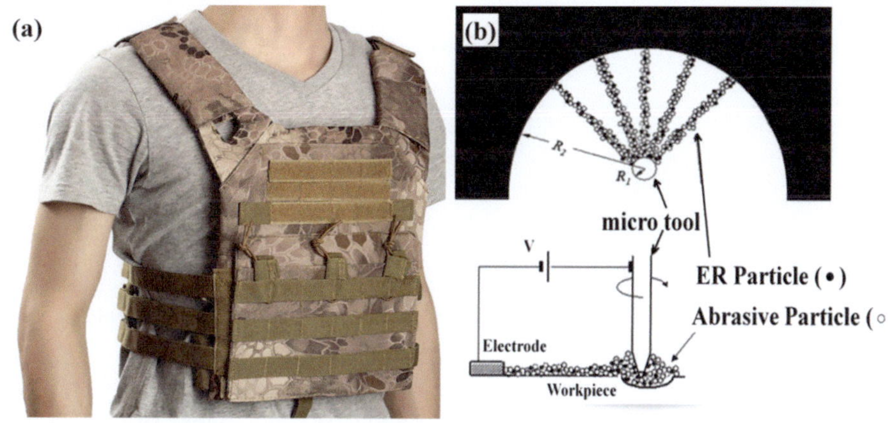

Fig. 6.11 (**a**) Bulletproof vests, (**b**) Principle of ER fluid-assisted polishing

screens, in which qualities to change the viscosity of the liquid allow the roll elements to become rigid for use and flexible to roll and retain for storage when not in use (Fig. 6.10a, b). Motorola applied for a patent for mobile applications in 2006 [57]. The property of ER liquids to be converted to a semisolid state is useful in the production of keyboards and roll-screens. When inactive, these devices can be turned on and when they are activated, the ER liquid hardens for use. These fluids are very fast with respect to reaction time.

The ER fluid is used in the haptic controller. Fig 6.10c shows a joystick-like haptic device that exerts resistance against lever movement. The lever position is detected by sensors and displayed on the computer screen. If the cursor enters special screens on the screen, the user sees a greater resistance force on the stick. The strength of the resistor is generated by ER fluid, based on zeolites, which lies between two concentric spheres where the inner sphere can rotate with the stick. In this way, the user feels the movement of the pointer as a haptic response. This response can support virtual reality (VR) applications or help blind people working with a computer [58].

ER fluid is used to control the flow of microfluidics in MEMS/NEMS. In microfluidic channels when there is no electric field applied to the ER fluid in thematic channels, it continues to flow. But when an electric field is applied, the ER fluid begins to form columns, with the effect of high-voltage performance to balance the pump pressure, leading to a blockage of the flow, if the electric field is strong enough [59]. In today's robotic industry, flexible joints are controlled by ER technology for fluid joints instead of hydroelectric controls which can perform much better than hydraulic-electric control. Engineers design and manufacture flexible gaskets that have less volume, quick reaction time, minimal wear, and can be easily controlled by microcomputers. ER fluids can provide all of these benefits over hydraulic-electric controls [60].

6.2.7.3 Other Applications

ER fluids are also likely to be used in the development of fireproof and bulletproof vests Fig. 6.11a. These liquids can be absorbed by the fabric and can become a hardcover by applying a small electric field almost immediately. Electro-rheological fluids can also be used to control fluid flow in variable flow controls. Other smart fluids change their surface tension in the presence of an electric field. This is used to produce very small controlled lenses: a drop of this liquid, trapped in a small cylinder and surrounded by oil, serves as a lens whose shape can be changed by applying an electric field [61]. The abrasion process of polishing using ER fluid, known as ER-fluid polishing, is potentially useful in the surface finish of three-dimensional micro or mesoscale devices. The polishing principle is schematically shown in Fig. 6.11b. An auxiliary circular electrode is constructed that surrounds the surface to be polished in the sample, and then the circular micro-tool is located in the middle region of the electrode, just above the surface. When the external electric field is applied after the supply of the liquid mixture ER and the abrasive particles, the particles are aligned with the electric field. Because the fields are concentrated and the field strength is greater near the tool, strong aggregation of a large amount of particles around the tool is maintained. As a result, the material under the tool is removed [62].

Fig 6.12 shows the ER fluid-assisted engine mount and hydrodynamic clutch. The pumps used to control pressure and the flow will be replaced by future ER technology. Due to the fact that the ER technology machines are less sophisticated, simple, low cost, long life, without wear and tear, etc. [63]. An ER engine mount has been shown in Fig. 6.12a. The ER mount was developed to carry a maximum weight of 200 kg. Its structure is similar to that of a normal hydraulic mount, which connects the upper and lower housing of the reduction unit. In this carrier, the inertia track is replaced by concentric annular drainage. Concentric rings that separate the flow streams, act as an electrode as they are electrically powered. Therefore, the internal flow of these components is controlled by the difference in the electric field between these

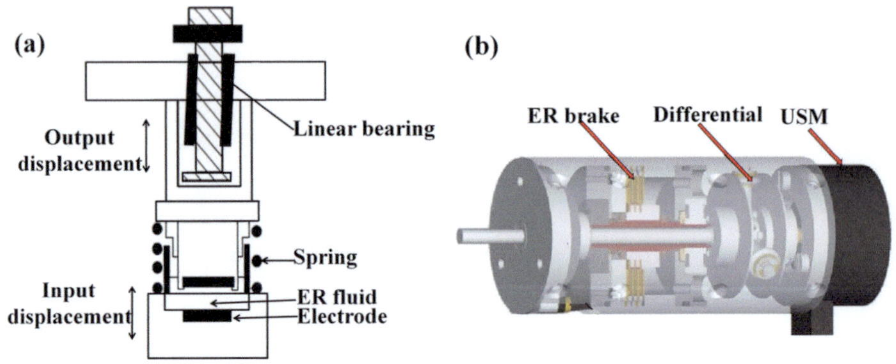

Fig. 6.12 (**a**) ER engine mount and (**b**) hydrodynamic clutch

electrons. Flow control affects the deflection of the mount [64].

The prototype hydrodynamic pump is driven by a continuous three-phase motor (2.5 kW), which is the speed of time transmitted by the transmitter. The electric motorcycle includes speed, start, stop, and return. The machine can be cooled due to airflow, especially at lower speeds. The torque meter is fitted to the tachometer between the electric motor and the test tube. The brake and tachometer are located near the pier. All joints of the machine parts of the test board shall be connected with simple screws. A special high-power mechanical section was built and manufactured with a high-power supply. The high-voltage system operates on the understanding of the flyback principle [65]. In order to protect the workers from bad electrical conductors during the tests, the outputs operate at 20 V DC volts, and their output power is limited to 18 W. Prior to the combination test, many pairs of fatty acids are attached to the promoter via a smooth contact ring [66]. These are places where ER drugs can be used.

6.2.8 Advantages/Disadvantages

An advantage is that an ER device can control considerably more mechanical power than the electrical power used to control the effect, i.e., it can act as a power amplifier. But the main advantage is the speed of response, there are few other effects able to control such large amounts of mechanical or hydraulic power so rapidly. Rise and decay times as short as a few milliseconds were observed at the highest shear rates. The inverse relationship between the characteristic rheological response times and the shear rate is consistent with the necessary roles of both field-induced structure formation and shear-induced strain growth in producing an ER shear stress. The ER fluid also has significant limitations such as settling and separation of the base fluid and particulates. This effect, called the sedimentation effect, may render the ER fluid entirely useless or ineffective changing its rheological properties [67]. Furthermore, the ER fluid has a limited range of operations and it cannot change its viscosity effectively in applications that require shear and flow modes. Collectively, these factors are hampering the ER fluids market [68]. A major problem is that ER fluids are suspensions, hence in time they tend to settle out, so advanced ER fluids tackle this problem by means such as matching the densities of the solid and liquid components, or by using nanoparticles, which brings ER fluids into line with the development of magneto-rheological fluids. Another problem is that the breakdown voltage of air is ~3 kV/mm, which is near the electric field needed for ER devices to operate. Sudden changes in the applied electric field in ER fluid were found to induce a variety of transient stress phenomena [69].

6.3 Magneto-Rheological Fluid

A magneto-rheological fluid (MR fluid, or MRF) is a type of smart fluid. Jacob Rabinow is credited with discovering MR fluid in the 1940s while working at the U.S. National Bureau of Standards. MR fluid, when subjected to a magnetic field (Fig. 6.13), the fluid greatly increases its apparent viscosity, to the point of becoming a viscoelastic solid. Importantly, the yield stress of the fluid when in its active ("on") state can be controlled very accurately by varying the magnetic field intensity. MRFs show a sharp variation of their non-Newtonian behavior, increasing their viscosity upon the application of an external magnetic field. The effect is reversible and the typical response time of commercial materials is in the 5 ms range [70]. MR fluid particles are primarily on the micrometer-scale and are too dense for Brownian motion to keep them suspended (in the lower density carrier fluid). The size of the MR particles is around 10 μm, while the nanosized particles produce a similar substance called ferrofluid[71].

6.3.1 Materials Used in MR fluid

MR fluid is quite similar to the ER fluid, except that the particle constituting the MR fluid must be ferromagnetic or

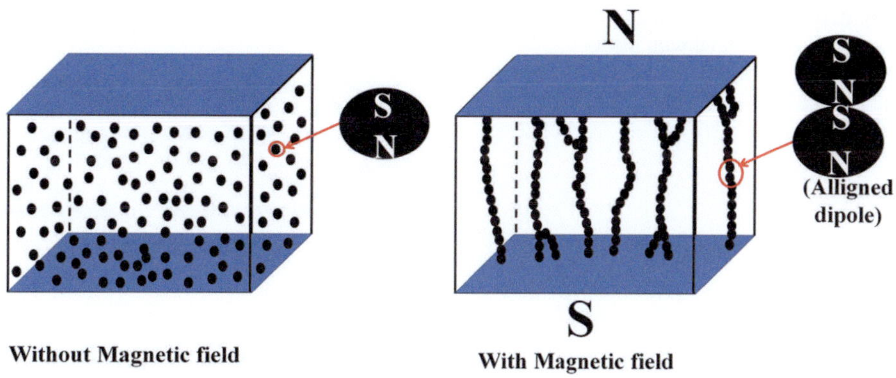

Fig. 6.13 Magneto-rheological fluid

Table 6.2 Materials for MR fluid

Magnetic particles	Carrier fluid
Iron, nickel, cobalt, Carbonyl iron particles, Iron-cobalt alloys, Ferrite etc.	Silicon oils, Mineral oils, Water etc.

a magnetic nonlinear particle. Iron, nickel, cobalt, and ceramic ferrites, etc., are good candidates for making MR fluids. Some of the magnetic particles and carrier fluids are given in Table 6.2. Ferrofluids, on the other hand, are stable dispersions made from superparamagnetic particles of size 5–10 nm, such as iron oxide [72].MR fluids, which comprise uniformly dispersed, noncolloidal magnetic material in a carrier fluid, are designed to change the properties (such as plasticity, elasticity, and apparent viscosity, etc.) when subjected to a magnetic field. A typical MR fluid consists of three parts: carbonyl iron particles, a carrier liquid, and proprietary additives [73]. Carbonyl iron particles are 20–40% of the fluid made of soft iron particles that are just 3–5 μm in diameter. A package of dry carbonyl iron particles looks like black flour because the particles are so fine. A Carrier Liquid is the iron particles are suspended in a liquid, usually hydrocarbon oil. Water is often used in demonstrating the fluid. Proprietary Additives is the third component of MR fluid, the additives are put into inhibit gravitational settling of the iron particles, promote particle suspension, enhance lubricity, modify viscosity, and inhibit wear. In the carbonyl iron particles, when a magnet is applied to the liquid, these tiny particles line up to make the fluid stiffen into a solid. This is caused by the dc magnetic field, making the particles lock into a uniform polarity. How hard the substance becomes depends on the strength of the magnetic field [74]. The particles unlock immediately, taking away the magnet.

6.3.2 Preparation of MR fluid

MRF is a noncolloidal mixture of ferromagnetic particles randomly dispersed in oil or water plus some surfactants useful to avoid the settling of the suspended particles. This material becomes suddenly smart and interesting as soon as a magnetic field passes through it. The ferromagnetic particles feel the induction field and acquire a magnetic bipole, then they move and redesign their arrangement start to flow and to form chains and linear structures. These microscopic chains have the macroscopic effect to change the apparent viscosity of the fluid [75]. Normally, the MR fluids are made of magnetizable (Table 6.2) particles dispersed in either insulating oil like mineral or silicone oils or conducting medium like water or glycol [76]. A variety of surfactants or lubricants are usually added to MR suspensions for preventing particle settling and enhancing the MR effect, as the typical diameter of the magnetic particle is about 3 to 5 microns. Various electrolytic and carbonyl iron powder-based MR fluids have been synthesized by mixing grease as a stabilizer, oleic acid as an antifriction additive, and gaur gum powder as a surface coating to reduce agglomeration of the MR fluid. MR fluid samples based on sunflower oil, which is bio-degradable, environmentally friendly, and abundantly available have also been synthesized. These MR fluid samples are characterized for the determination of magnetic, morphological, and rheological properties [77]. This study helps identify the most suitable localized MR fluid meant for the MR brake application.

In preparing the advanced MR fluid, carbonyl iron powder is used as magnetic particles in the suspended system. These carbonyl particles are coated with SiO2. This coating makes the MR fluid more stable as they prevent the aggregation of the magnetic particles. The average particle size and tap density were 7.0 μm and 4.3 g/cm^3, respectively. As a carrier fluid colorless silicone oil is used here. The essential rules of choosing carrier fluids, such as it are nonreactive towards the magnetic particles and toward the components/materials used in the device. Also boiling temperature, the vapor pressure at elevated temperatures, and freezing point are considered. The concentration of CI was fixed at 40 wt%. For reducing the sedimentation process of the CI particles, a very small amount of fumed silica that means Aerosil 200 was added as additional components [73].

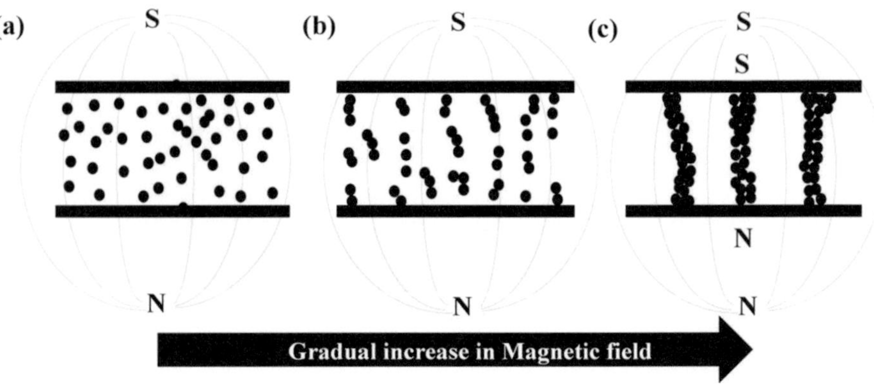

Fig. 6.14 Particulate chain formation in MR fluid with the application of magnetic field

Since the typical solid phase of MR fluids are metal or metal oxide ferrites and have a mass density larger than that of most suspending fluids, stabilizing a micrometer-sized particle against sedimentation is a challenging issue facing the technological application of these materials. Small magnetic particle sizes, high solid loadings, and high viscosities of dispersing oil are helpful in reducing the rate of sedimentation. Adding surfactant (stearates) or nano-silica particles into the MR fluids was found to be the best way to improve particle stability. There are two mechanisms proposed for explaining why silica particles can influence the stability of MRFs. First, silica–silica interparticle hydrogen bonding facilitates the formation of a thixotropic network that prevents the sedimentation and eventually reduces redispersion difficulties. Secondly, a silica particle may adhere to an iron particle by acid-base reaction in nonpolar media [78]. In the presence of the magnetic field, the shell of adsorbed silica will partially screen the magnetic interactions between iron particles, thus reducing the magneto-rheological effect.

6.3.3 Mechanism of Strengthening of MR Fluid

At the existence of the magnetic area, every particle has been converted to a dipole and creates a string having its particles which provides a structure that was semisolid and can withstand failure. The magnetic dipole and multipole moments induced on each particle interact with each other, leading to the formation of fibrillated particle structures along the direction of the magnetic field. These fibrillated chain structures induced by the external field and formed by the randomly dispersed particles are believed to be responsible for the dramatic rheological property increase, as in Fig. 6.14 [79]. When subjected to a magnetic field, the fluid greatly increases its apparent viscosity, to the point of becoming a viscoelastic solid. Importantly, the yield stress of the fluid when in its active ("on") state can be controlled very accurately by varying the magnetic field intensity. The upshot is that the fluid's ability to transmit force can be controlled with an electromagnet, which gives rise to its many possible control-based applications [86]. It is worth to remark that at low particle concentrations, the particle aggregates are mostly single strands, or 1D elongated structures, spanning the two walls. There are few interconnections between these strands and therefore it may be then hard to justify that these interconnections can be stronger than the intra-aggregate interactions unless other kinds of short-range attractions prevail [80]. At high particle volume fractions, the particles typically form thick columnar structures, whose sizes are larger than the structures formed at low particle concentrations. This is reflected when comparing the estimated fractal dimensions of these structures at high particle concentrations, and at low particle concentrations. In the absence of a magnetic field, the suspension behaves as a typical colloidal system. In the application of magnetic field, particles are aggregate and form a columnar chain-like structure in the field direction. The resulting field induces yield stress and viscoelasticity promotes rheological measurements in MR performance. The ratio between the material function in the presence of a magnetic field and the absence of a magnetic field during rheology measurements is defined as the MR effect [81]. The key property of MR fluid is yield stress and viscoelasticity. It is observed that by changing the microstructure of the fluid the yield stress of MRF can be enhanced. The conventional ferrofluids usually don't show yield stress and only give a rather small viscosity change in a magnetic field. There is no report of a negative MR effect to date, in parallel with the negative ER effect.

6.3.4 Microstructure and Properties of MR Fluid

Even though at first glance MRF does not look so impressive, the overall aspect is like a greasy quite heavy mud (Fig. 6.15), since MR fluids density is more than three times the density of water.

The yield shear stress is the main figure of merit of an MR fluid and derives from the non-Newtonian behavior of these fluids. The MR fluid behaves following a so-called Bingham law, which means that it exhibits a non-zero shear stress

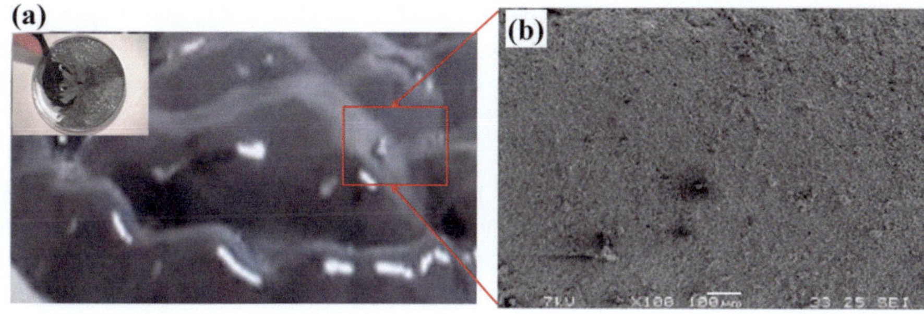

Fig. 6.15 (a) Prepared MR fluid and (b) magnified SEM micrograph of MR fluid

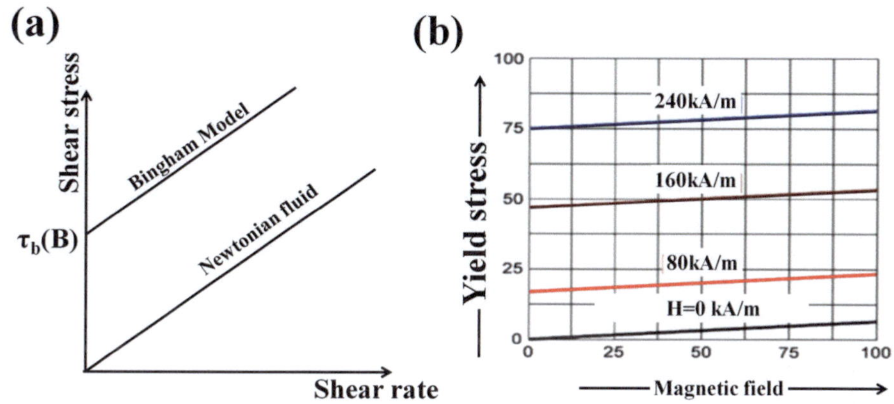

Fig. 6.16 (a) Bringham model of MR fluid and (b) yield stress versus magnetic field

value for a zero shear rate [90], behaving more like a solid than like a liquid, as shown in Fig. 6.16a. The value of the shear stress at no shear rate is called yield stress of the MR fluid and is controlled by the applied magnetic field as shown in Fig. 6.16b. The larger the field, the higher the yield stress. The higher the yield stress the higher the force, the material can withstand without flowing. Bearing a load is possible only because MR fluids can modify their aggregation states changing from a vicious free-flow liquid to a quasi-solid state [82].

However, they are limited by a low mechanical strength under a magnetic field, typically 5 kPa, comparing to more than 100 kPa for a micrometer-sized particle [83]. Fig 6.17 shows the yield stress of a carbonyl iron/silicone oil system versus the applied magnetic field. The average particle size is 5 μm and the particle volume fraction is 46 vol%. Without a normal compression pressure, this MR fluid can easily reach 120 kPa at H = 514 kA/m. The compression pressure can increase the yield stress to 600 kPa under a magnetic field of 514 kA/m. The mechanical strength of an MR fluid is controlled by the saturation magnetization of the suspended particle, and thus a particle with a large saturation magnetization is preferred [84].

A rheological study of the stabilization of microsized iron/silicone oil suspension by the addition of silica nanoparticle was made by Vicente. Fig 6.18 shows the normalized yield stress (σ_y/B^2) of an iron particle (~1 μm in diameter) dispersed in silicone oil containing nano-sized silica (~7 nm in diameter) versus silica concentration, where B is the magnetic flux density. When the silica concentration is low, below 20 g/L, the presence of nanosized silica particles hinders the MR effect, while once the silica concentration is larger than 20 g/L, the normalized yield stress seems to increase with the silica concentration [85]. As the authors claimed, this is a manifestation of the fact that in such high silica concentrations the magnetic interaction between silica-

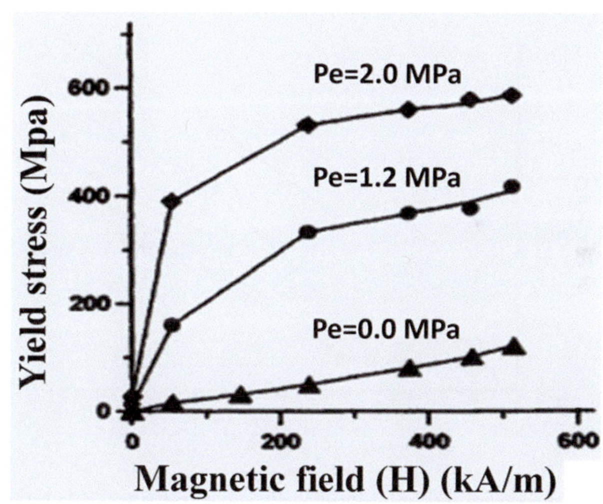

Fig. 6.17 The yield stress of carbonyl iron/silicone oil with particle volume fraction 46 vol% against the applied magnetic field. The average particle size is 5 μm

covered iron particles is almost fully shielded. The yield stress increase results from the crowding effect of the nanosilica particles, as at such a high silica concentration region the yield stress is independent of the iron concentration and the applied magnetic field.

Strain decreases with increasing the particle content. At these low concentrations, conventional micromechanical and macroscopic models for magneto-rheology apply. For larger concentrations, a two-step yielding process occurs and the suspensions behave either in the weak-link (large fields) or transition (low fields) regimes. These two processes are presumably associated first with the breaking of bonds between interconnected clusters and second to the breaking of bonds within clusters. For these high-particle concentrations, both the storage modulus and the first yield strain increase with particle loading. Actually, the stronger the secondary structure, the larger the linear viscoelastic region is. The secondary structure is due to the many-body interactions between particles or short-ranged interparticle interactions possibly coming from the remnant magnetization existing in the particles [86].

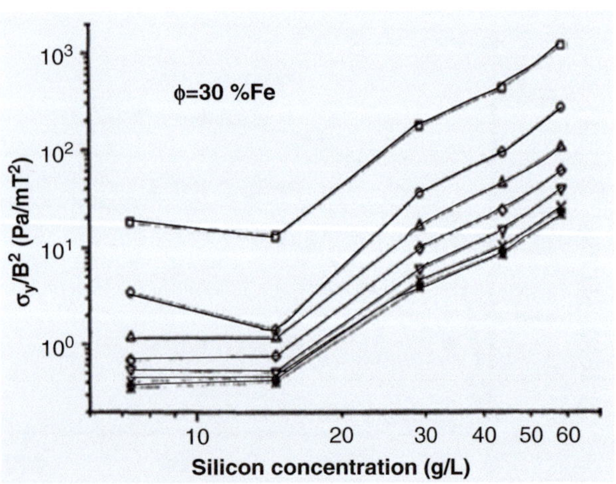

Fig. 6.18 Normalized yield stress (σ_y/B^2) of an iron particle (~1 μm in diameter) dispersed in silicone oil containing nano-sized silica (~7 nm in diameter) versus silica concentration. The symbols correspond to the following magnetic flux densities (mT): □ 0.345; ○ 0.725, Δ 1.085; ⋄ 1.42; ∇ 1.77; × 2.17, and * 2.38

6.3.5 Typical Modes of Application of MR Fluid

There are three main ways envisioned in current engineering applications: (a) flow mode, (b) shear mode, (c) squeeze mode (Fig. 6.19) [87]. These modes involve, respectively, fluid flowing as a result of the pressure gradient between two stationary plates; fluid between two plates moving relative to one another, and fluid between two plates moving in the direction perpendicular to their planes. In all cases, the magnetic field is perpendicular to the planes of the plates, so as to restrict fluid in the direction parallel to the plates.

Flow Mode: Flow mode, also called valve mode, exploits the fluid between two fixed walls, the magnetic field is normal to the flow directions and is typical for linear damper applications. Flow mode can be used in dampers and shock absorbers, by using the movement to be controlled to force the fluid through channels, across which a magnetic field is applied.

Shear Mode: A shear mode is mainly used in rotary applications such as brakes and clutches and the fluid is constrained between two walls which are in relative motion with the magnetic field normal to the wall direction. By shear mode, the rotational motion can be controlled.

Squeeze Mode: Squeeze mode is used mainly for bearing applications, is able to provide high forces and low displacements having the magnetic field normal to walls directions. Squeeze-flow mode is most suitable for applications controlling small, millimeter-order movements but involving large forces.

In all the abovementioned cases, the working principle is the same: the applied magnetic field regulates the yield stress of the fluid and changes its apparent viscosity. So the amount of dissipated energy of the system is simply controllable by acting on the coil current and the system can provide semi-active behavior.

6.3.6 Applications

Vibration is related to all aspects of life. This cannot be changed, but the negative effects can also be minimized. Magneto-rheological (MR) technology has been shown for many industrial applications such as shock absorbers,

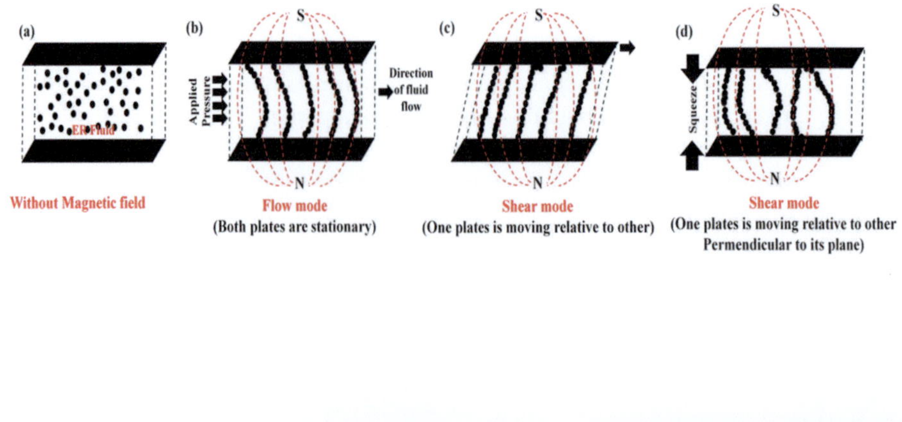

Fig. 6.19 (**a**) MR fluid without magnetic field, and different modes of work of MR fluid (**b**) Flow mode, (**c**) Shear mode, and (**d**) Squeeze mode

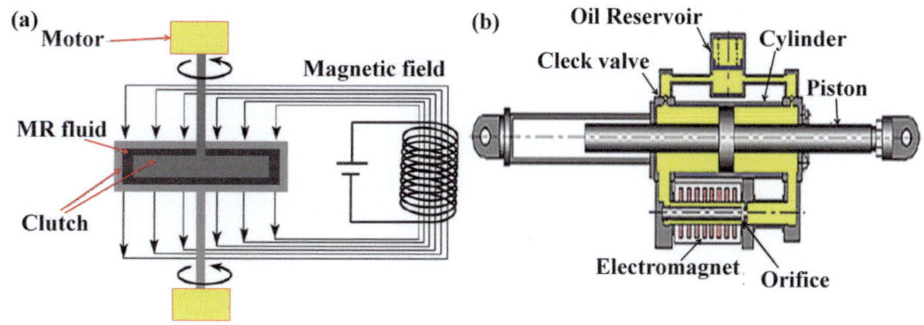

Fig. 6.20 (**a**) MR Clutch and (**b**) MR fluid suspension

actuators, and more. The range of applications is increasing due to the enormous advantages of magneto-rheological fluids. They are used in mechanics, aerospace, military and defense, electrical and electronics, manufacturing, and construction machinery [88, 89]. Demand ranges from nurses in the automotive engine category to anti-aging systems to systems based on military evidence. Structures based on future magneto-rheological circuits provide stability and dynamic vibration performance. An increase in the proportion of MR fractions in different companies should increase the demand in the forecast period in the market (2019–2027) [90]. Based on these trends, the global market for magneto-rheological fluid is expected to reach ~4.3 billion USD by the end of 2027 [91].

6.3.6.1 Automobile and Heavy Machinery Industries

If the shock absorbers of the vehicle suspension are filled with magneto-rheological fluid instead of simple oil or gas and the channels through which the damping fluid can flow between the two chambers are surrounded by electromagnets, the viscosity of the fluid and therefore critical shock absorber frequency may or it can be changed dynamically to allow stability control over very different road conditions. It is practically a magneto-rheological damper [92]. Figure 6.20 shows the MR clutch and MR fluid suspension. The magneride active suspension system allows the damping factor to adapt to the conditions once per millisecond. General Motors (in collaboration with Delphi Corporation) has developed this technology for automotive applications. He made his debut in Cadillac as "MagneRide" and in Chevrolet cars (all Corvettes manufactured since 2003). Other manufacturers have paid for use in their own vehicles, such as Audi and Ferrari, offering MagneRide for various models [93]. General Motors and other automotive companies are looking for a magneto-rheological clutch system for all-wheel drive systems using a button. This clutch system would use electromagnets to strengthen the fluid that would block the drive shaft in the drive system. Applications of smart fluids in couplings in which the torque is transmitted by the fluid without firm friction and corresponding wear. The transmission torque may be limited by the field strength. In the case of clutches and brakes, MR fluids are more suitable than ER fluids due to significantly higher shear stress. Because the MR fluid based on the iron particles in the silicone oil is located between the inner disk connected to the engine and the surrounding disks attached to the machine, a torque determined by the magnetic flux that penetrates the clutch is transmitted in another shaft [94].

In 2010, Porsche introduced the Porsche GT3 and GT2 models with a magneto-rheological engine mount. At high engine speeds, the magneto-rheological engine mounts become stiffer to achieve more accurate gear shifting by reducing the relative motion between the power train and the chassis/body. Since September 2007, Acura (Honda) has launched an advertising campaign that highlights the use of MR technology in passenger cars for the 2007 MDX model year [95]. Fig 6.21 showing the MR Fluid dampers in seat suspension. The semi-active vibration control system, first introduced in 1998, was used on the seats of 18-wheel trucks. The semi-active vibration control system includes an MR damper, a microprocessor, a sensor, a power driver, and auxiliary cables. Today, more than 5000 liquid-based vibration control systems are used in American trucks. These systems are highly valued by the drivers who experience them. They are effective for several hundred thousand kilometers and the number of errors recorded in this field is practically zero [96].

In MR fluid that is used in the automotive and other industries, the actuator consists of a disc bathing in an MR fluid surrounded by an electron. MR fluid is attached to the static structure and the discs are attached to the shaft. With the head rotation, the gas will rotate as well. Brake signals can be controlled by the direct transmission of electricity. In the presence of a magnetic field, the fluid transforms into semisolid by aligning magnetic particles in a chain-like structure, increasing the viscosity results in the yield strength. The components of the MR line are the disc, shaft, coil, fluid gap, and housing. Product selection is an important part of the analysis. Machine and waste materials are selected for structural reasons. Temperature affects the ferromagnetic apparatus, so the resulting heat must be released as soon as possible to prevent the detection of MR fluids. To remove heat from a magnetic field, the nonmagnetic materials having high

Fig. 6.21 MR fluid dampers in seat suspension

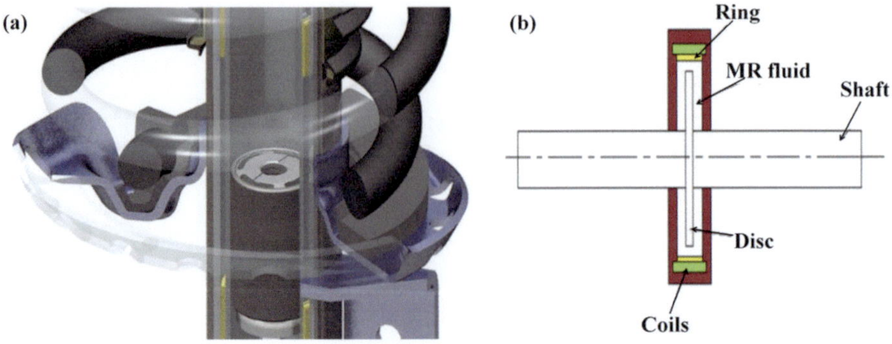

conductivity and high-convection coefficient are chosen. AISI 4340 has been selected as a nonmagnetic material and AISI 1010 is a major material used in steel applications, taking into account its cost, capacity, availability, and production strengths [97]. The damper is a device that transmits energy from a vibrating system in the form of heat or sound. MR dampers are an active source of energy used in system tubes. The output of the MR damper can vary depending on the magnetic field applied at the block. MR drugs have unique capabilities that make them effective in complex applications such as cars, tanks, recovery pistols, etc. Many MR vehicles have been manufactured and successfully used in dust transmission such as chassis systems, vehicle suspension systems, etc. [98]. Under the control of the MR axis, activators operating in the hydraulic system can use MR fluid as a functioning device. This system consists of MR blocks, the same size voltage with its component. The valve may consist of a metal body with an internal magnetic field, consisting of a ventilated and penetrating tube, located in the center of the body and outside. When you begin to enter MR fluids, magnetic fields are applied, thus increasing viscosity. Internal pressure is reinforced by an increase in the flow rate of the fluid, which reflects the increased viscosity [87, 99].

One of the most popular systems in the pipeline is dust accumulation. The MR mount uses a controlled fluid to disrupt the system's dust due to their support. Because of the rapid response time, it can change the appearance of dirt within milliseconds. In addition, it has a lower operating value compared to conventional liquid dampers [100, 101].

6.3.6.2 Military and Defense Industries

Magneto-rheological dampers are designed for use in military and commercial helicopters, as a safety device in the event of an accident. It is used to reduce trauma to the spine of passengers by reducing the level of injuries sustained during the accident. MR Fluid can be used as a gun recoil defense (Fig. 6.22) [102]. According to Newton's third law, as the firing is done the gun recoils. But recoil force after rifle shooting is very large. When a barrel or a rifle is carried on the shoulder, the retrieval unit is directed straight upwards and the wing exits for most of the reaction force. The cheek of the shooter can also absorb a bit of the buildup when the gun is firmly pressed against the panel on the cheek. The weapon comes on because of the spine and the hips but the muscles are usually restored to the shoulder. Many times it has a negative effect on the user. Side effects include broken bones, rib fractures, internal blood vessels, etc. The body takes a long time to recover from these injuries depending on the level of the injury. These injuries were common to soldiers in the war zone. These injuries can lead to death in rare cases if not properly reported. Nowadays, shoulder straps are made of thick foam, but they offer to a certain extent and prove to be used when not used properly. They are also tired after being used several times. To overcome this problem Looga foam can be replaced by liquid MR. MR fluid can be used as a vaporizer to reduce side effects. Compared to foam sheets, the longer ones last. MR fluid can work after thousands of cycles of "DOWN" and "ON" regions of phase change that prove extremely effective. MR fluids have no effect on atmospheric temperature (−50 to 150 °C) and therefore can be used in extreme cases, while the existing fluids are not efficient at extreme temperatures. MR fluid can be filled with bandages instead of foam that can be placed on the shoulders of the person using the gun as in Fig. 6.22. With time-controlled algorithms, MR fluid can be hardened quickly by applying direct magnetic to the field. The phase change of the MR fluid is in milliseconds, making it much

Fig. 6.22 Representation of using MR fluid to reduce gun recoil

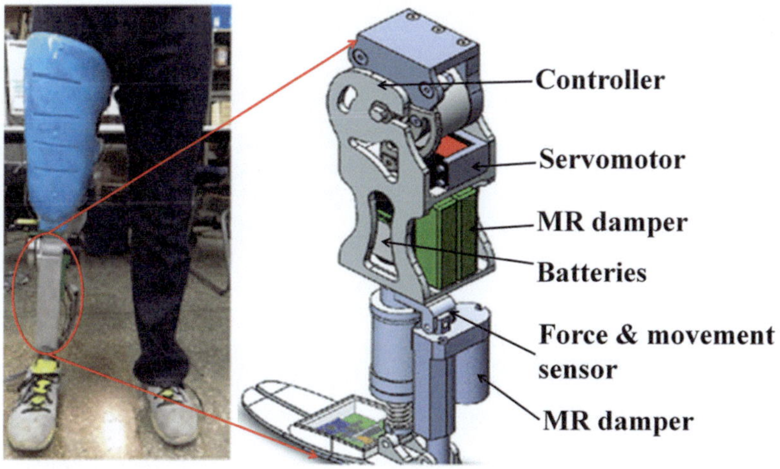

Fig. 6.23 MR dampers in prosthetics

easier to quickly change the phase of the fluid from the liquid to a solid substrate. This makes the user more secure from the risk of a recoil [103].

6.3.6.3 Biomedical Industries

Magneto-rheological dampers are used in the semi-active human leg muscles. Similar to those used in military and commercial helicopters, a prosthetic leg reduces the trauma to the patient's leg while jumping. This results in increased agility and mobility of patients. The MR device can be used with a prosthetic knee to provide a quick copy of the content and provide users with more natural footing. This is done with a small vapor of the MR fluid to absorb the shock due to the lower knee flexion movement (Fig. 6.23). This provides a gateway that meets all the requirements. Biedermann Motech developed HIP or Intelligence Prosthesis, suede over the knee. A group of sensors is attached with the MR equipment that determines the immediate state of the knee: knee angle, flexion strength, speed, and vibration speed [104].

MR Fluid is used to reduce the shock that occurs during the utilization of artificial muscles. Robots in human society need to communicate safely. So there is the first need for natural muscle based on pneumatics. This actuator is lightweight and has high power. However, artificial muscle has vibration problems. It is necessary to use the MR fluid. With this approach, the result of MR healing can be strong enough to give the desired results [105]. MR fluid was used in a ball-and-socket damper, as shown in Fig. 6.24. For example, the hand can be raised over the head, lower back, the whole body. In addition, the hand rotates in a 360° circle while holding on to the side. Thus an orthotic system for regeneration of the human body's limbs can be found to be similar to that of the solid. Therefore, the ball-and-socket configuration was chosen to design a new version of an MR damper, which allows for the functions to be simulated by the human shoulder. Providing the retention of the MR fluid content of filling the lip between the ball and socket of the joint, it should offer smart damping to compensate for the kinematics [106].

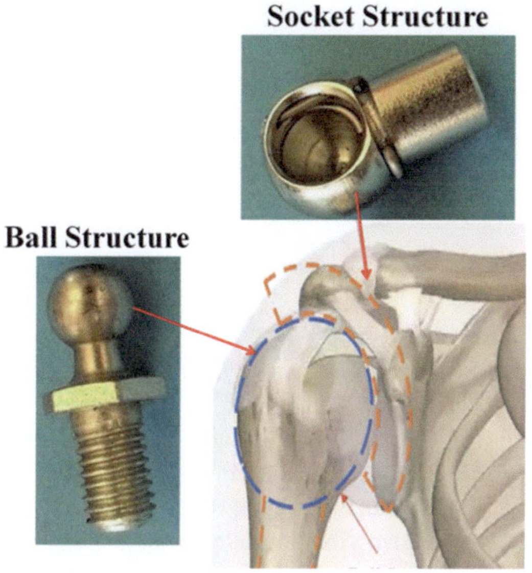

Fig. 6.24 Ball-and-socket damper in human shoulder joint

6.3.6.4 Other Industries

The MR fluid is developing seismic resistors, that when placed inside a building, will operate at the frequency of the building's resonance, absorbing harmful shock waves, and vibrations in the structure, allowing these strainers to withstand any building earthquake or at least earthquake resistance.MR dampers can be used as seismic dampers to operate at structural resonant frequencies, absorbing shock waves and vibrations that could have damaged the system. This allows dampers to absorb any building earthquake [107]. Figure 6.25a shows the MR seismic damper at the hinge and Fig. 6.25b shows the MR dampers in a bridge application.

MR fluid dampers are used to control the vibrations caused by the string wind. It is also used to prevent shock measures. In the developed world, the MR dampers used in the bridge under the cable as used to capture the moisture content of precipitation. The result is a significant reduction

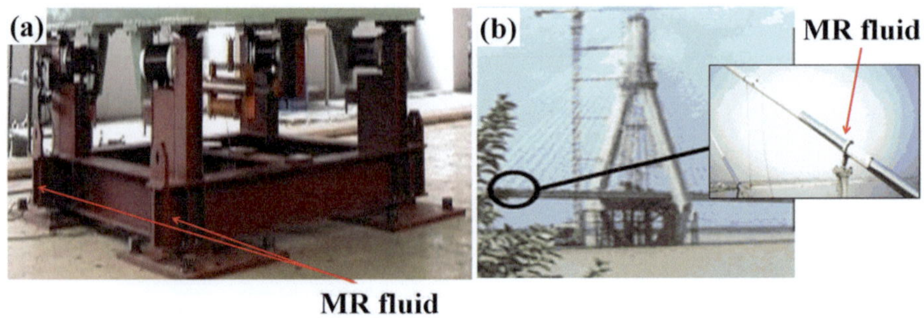

Fig. 6.25 (**a**) A MR seismic damper at a hinge, (**b**) Bridge with MR Dampers

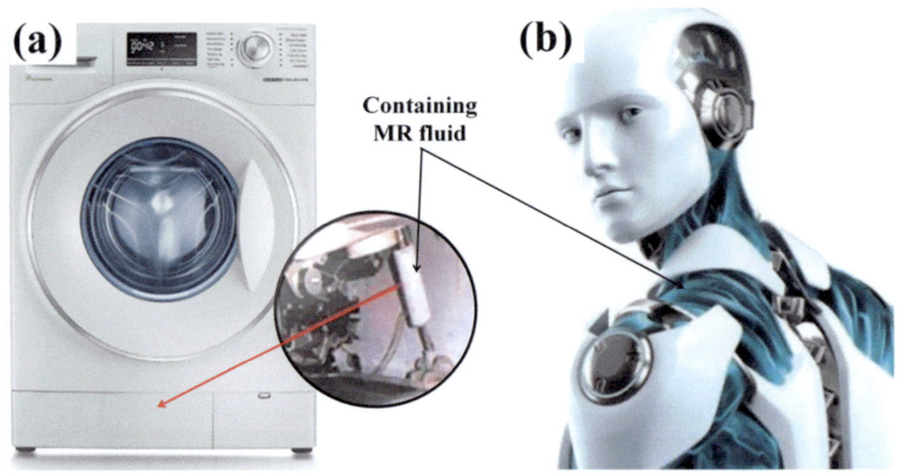

Fig. 6.26 (**a**) MR Fluid dampers in the washing machine, and (**b**) Robot movement parts containing MR fluid blood

in bridge structure vibrations. The MR fluid damper is applied to the washing machine to reduce vibrations during spinning (Fig 6.26a). Fig 6.26b shows a robot that the body contains MR fluid that can sustain the whole structure from the impact and fluctuating stress [108]. MR fluid systems are monitored to reduce vibration in air conditioners, satellites, etc., and also some electronic devices can be controlled by touching these MR fluids.

One of the most unexpected uses of MR fluid is its use as a polishing agent. Optical surfaces are polished in an MR finishing slurry, which has proven to be highly precise. Unlike the usual lap polishing, the shape and viscosity of the fluid can be controlled in varying the field. The computer algorithm predicts the final smoothness and form of the workpiece. It was used in the construction of the Hubble Space Telescope's corrective lens [109]. Since it can instantly and reversibly change shape, it could also be used to create scrolling Braille displays or reconfigurable molds [110].

6.3.7 Advantages and Disadvantages of MR Fluid

MR technology offers, continuous variable control of damping, motion and position control, have high dissipative force independent of velocity and greater energy density. It has a simple design (few or no moving parts) and quick response time (10 ms), consistent efficacy across extreme temperature variations, minimal power usage (typically 12 V, 1 Amp max current; fail safe to battery backup, which can fail safe to passive damping mode), inherent system stability (no active forces generated). MR fluids can be operated directly from low-voltage power supplies. MR technology can provide flexible, reliable control capabilities in designs. Comparing the two fluids, the operational temperature range is slightly wider for MR fluids (-40 to 147 °C) than for ER fluids (-20 to 121 °C), while the power demand is the same (1–50 W), even though to generate a high strength magnetic field and operate an MR fluid ($\approx$1 T). In the case of MR fluid, heavy equipment is required (either permanent magnets or copper coils, magnetically soft cores, and a high-power generator), while to operate an ER fluid system much lighter equipment is required (a voltage multiplier is a solid-state device quite simple and small).

Although smart fluids are rightly seen as having many potential applications, they are limited in commercial feasibility for the following reasons: (1) high density, due to the presence of iron, makes them heavy. However, operating volumes are small, so while this is a problem, it is not insurmountable, (2) high-quality fluids are expensive, (3) fluids are subject to thickening after prolonged use and need replacing, (4) settling of ferro particles can be a problem

Table 6.3 Comparison of ER and MR fluid

	ER Fluid	MR Fluid
Main properties	Base viscosity: 50–200 mPas Shear stress in field: 1–5 kPa Electric field of 1–5 kV/mm necessary High voltage(kV)–low current (mA)	Base viscosity: 200–1000 Shear stress in field: 10–50 kPa Magnetic field of 0.2–1 T necessary High voltage(kV)-low current (mA)
Advantages	Lower base viscosity Higher sedimentation stability Smaller and lighter actuators are possible	Much higher shear stress Lower temperature dependence Not sensitive to humidity
Possible applications	Shock absorber, vibration damper, clutches, brakes, force feedback, tactile devices, flexible grippers, hydraulic systems, positioning control, polishing	Shock absorber, vibration damper, clutches, brakes, force feedback, flexible grippers, polishing

for some applications, (5) cannot operate at extremely high/low temperatures [111, 112]. The properties, advantages, and disadvantages of both ER and MR fluid have been given briefly in Table 6.3.

6.4 Ferrofluid

A ferrofluid or ferromagnetic fluid is a liquid that becomes strongly magnetized in the presence of a magnetic field [113]. The difference between ferrofluid and MR fluid is the size of the particles. MR fluid particles primarily consist of micrometer-scale particles which are too heavy for Brownian motion to keep them suspended, and thus will settle over time because of the inherent density difference between the particle and its carrier fluid. As a result, both the MR fluid and ferrofluid have the same stimuli but have very different applications. A process for making a ferrofluid was invented in 1963 by NASA's Steve Papell to create liquid rocket fuel that could be drawn toward a pump inlet in a weightless environment by applying a magnetic field [114]. As the introduction of the name ferrofluid, the process improved, more highly magnetic liquids synthesized, additional carrier liquids discovered, and the physical chemistry elucidated by R. E. Rosensweig and colleagues; in addition, Rosensweig evolved a new branch of fluid mechanics termed ferrohydrodynamics [115]. Ferrofluids are colloidal liquids made of nanoscale ferromagnetic, or ferrimagnetic, particles suspended in a carrier fluid (usually an organic solvent or water). The magnetic particles in ferrofluid are around 10 nm in size. Particles of this size are known as colloids. Each tiny particle is thoroughly coated with a surfactant to inhibit clumping. Large ferromagnetic particles can be ripped out of the homogeneous colloidal mixture, forming a separate clump of magnetic dust when exposed to strong magnetic fields. The magnetic attraction of nanoparticles is weak enough that the surfactant's Vander Waals force is sufficient to prevent magnetic clumping or agglomeration. Ferrofluids usually do not retain magnetization in the absence of an externally applied field and thus are often classified as "superparamagnets" rather than ferromagnets. However, ferrofluids lose their magnetic properties at sufficiently high temperatures, known as the Curie temperature [116].

6.4.1 Mechanism

The material can act differently when it is nanometer sized. Nanoscale magnetite particles suspended in a liquid (ferrofluids) behave like paramagnets, meaning that it acts magnetically only in the presence of a magnet. But on the macroscale, magnetite is permanently magnetic. Nanotechnology takes advantage of special properties at the nanoscale such as paramagnetism to create new materials and devices. By applying a magnetic force to a ferrofluid, through a permanent magnet, the magnetic particles are attracted/repelled based on the magnetic field, all those little nanoparticles respond to the force and start moving. The surfactant used for ferrofluid remains ultra-slippery and refuses to let the nanoparticles attach to each other. These surfactants prevent the nanoparticles from clumping together, ensuring that the particles do not form aggregates that become too heavy to be held in suspension by Brownian motion. The magnetic particles in an ideal ferrofluid do not settle out, even when exposed to a strong magnetic, or gravitational field [117].

A surfactant has a polar head and non-polar tail (or vice versa), one of which adsorbs to a nanoparticle, while the non-polar tail (or polar head) sticks out into the carrier medium, forming an inverse or regular micelle, respectively, around the particle. Electrostatic repulsion then prevents the agglomeration of the particles. The addition of surfactants (or any other foreign particles) decreases the packing density of the ferro particles while in their activated state, thus decreasing the fluid's on-state viscosity, resulting in a "softer" activated fluid. The surfactants keep slipping away while nanoparticles move, forming quick bonds with the surfactant instead to become a special type of ligand or a coordinate bond with a metal atom. The surfactant's Vander Waals forces stop the magnetic nanoparticles aggregating in the solution. Different surfactants work in different ways but the general principle is that the surfactant creates a layer

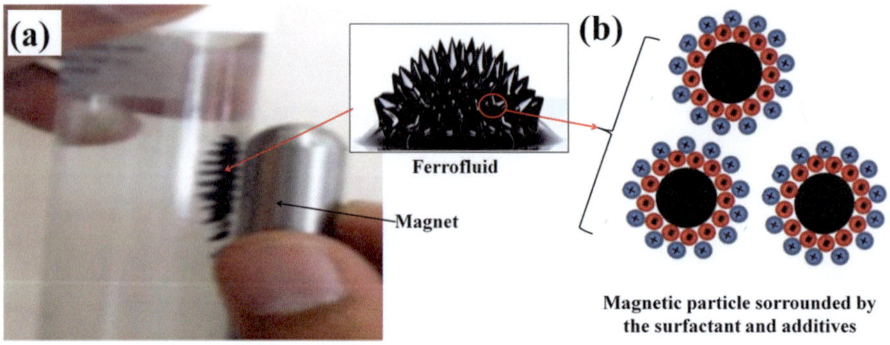

Fig. 6.27 (**a**) Ferrofluid influenced by a magnet (**b**) Presence and distribution of surfactant and additives in the magnetic particle

around the particle which will repel other coated nanoparticles. The surfactant ions form a layer of charge around the nanoparticle, repelling other charged, surfactant-coated particles. Whilst the addition of a surfactant is crucial, it has the negative effect of decreasing the viscosity of the fluid in the magnetized state and making it "softer." As most applications require a "hard" fluid in the magnetized form, this is an important factor to consider when choosing the ferrofluid composition. At the same time, the surface of the ferrofluid experiences a lot of surface tension, which allows the fluid to maintain shapes for long periods of time when it is drawn out. Meanwhile, Van der Waals forces are having a similar effect on the molecules within the mixture, allowing them to maintain a surprising amount of cohesion as the ferrofluid moves. As the ferrofluid is being pulled by the magnetic force, the heaviest parts are also being dragged back down by gravity at the same time. A dazzling display of spikes due to magnetic line of force, rivulets, and magical behavior observed as in Fig. 6.27a. Figure 6.27b showing the presence of surfactant around the magnetic particle. Unlike normal magnetite, it loses its charge every time and collapses back into a fluid, unable to hold a permanent, magnetically stabilized form and known as "superparamagnetic effect" [118].

When a ferrofluid layer is subjected to a uniform and vertically oriented magnetic field, an interfacial instability occurs, above a critical value of the magnetic field, giving rise to a hexagonal array (pattern) of peaks. A ferrofluid is a liquid which becomes highly magnetized in the presence of a magnetic field. Because ferrofluids are very easily magnetized (they have an incredibly high magnetic susceptibility), the peaks can be produced using a small bar magnet. The distinctive "spikey" shape of a magnetized ferrofluid is caused by the need to find the most stable shape in order to minimize the total energy of the system, an effect known as the normal-field instability [119]. The fluid is more easily magnetized than the surrounding air, so is drawn out along the magnetic field lines, resulting in the formation of peaks and troughs. When the magnetic force exceeds the stabilizing forces, this instability develops known as Rosensweig or normal-field instability. It can be explained by considering which shape of the fluid minimizes the total energy of the system [120].

6.4.2 Preparation of Ferrofluid

Ferrofluids are colloidal liquids made of nanoscale ferromagnetic, or ferrimagnetic, particles suspended in a carrier fluid (usually an organic solvent or water). Fluid responds to an applied magnetic field as one homogeneous system. A typical ferrofluid is comprised, by volume of about 5% solid component (magnetite, hematite, or some other compound containing iron), 85% liquid (carrier), and 10% surface-active agent (surfactant) [121]. Oleic acid, tetramethylammonium hydroxide, citric acid, soy lecithin, etc. are act as surfactants used to coat the nanoparticles. True ferrofluids are stable. This means that the solid particles do not agglomerate or phase separate even in extremely strong magnetic fields. However, the surfactant tends to break down over time (a few years), and eventually the nano-particles will agglomerate, and they will separate out and no longer contribute to the fluid's magnetic response. The fluids are produced either by the "top-down" or "bottom-up" approach. In the "top-down" method, a ferrofluid is prepared by grinding large particles to nano size. "bottom-up" methods are based on controlled nucleation and growth of nanoparticles such as vacuum evaporation, microemulsions, chemical coprecipitation, organic precursors, and sol-gel process. The coprecipitation synthesis route is simple and economical and typically utilized at the industrial level. When viewed as a colloid, a ferrofluid represents a complex system in which several competing forces need to be balanced to ensure dispersion stability. These forces are magnetic dipolar (attractive), Van der Waals (attractive), Brownian (random), steric (repulsive), and gravitational (attractive). The repulsive forces must outweigh attractive forces to avoid flocculation of the dispersed phase [122].

6.4.3 Applications

Ferrofluids have now found use in many applications from small electronic devices to space crafts to cancer treatments to art.

6.4.3.1 Aero-Industries

Ferrofluids can be created to self-assemble sharp tips similar to nanometer needles under the influence of a magnetic field. When they reach critical thinness, the needles begin to emit jets that in the future could be used as a conductor for small satellites like CubeSats [123]. When it comes to the future of space exploration, some really interesting concepts are evolving. Aim to get further and reduce the associated costs, one of the overall goals is to find a more effective and efficient way to send robotic spacecraft, satellites, and even crews to missions to their destinations. In that sense, all the ideas about nuclear power, ion engines, and even anti-matter. But this idea must be the strangest to date, known as ferrofluid propulsion, a new concept based on ionic liquids that are strongly magnetized and release ions when exposed to a magnetic field. According to new research by researchers at Michigan Tech's Ionic Space Laboratory, this concept could be the future of satellite propulsion. It is a completely new method of creating microthrusters: small nozzles that small satellites use to maneuver in orbit. Weighing less than 500 km (1100 pounds) can perform tasks that were once reserved for larger ones. It is therefore not surprising why researchers are looking at different types of microthrusters (Fig. 6.28a) to ensure that these satellites work efficiently. To solve this, ferrofluids may be a possible solution [124].

Ferrofluids have the ability to reduce friction. Applying a strong enough magnet to a surface, such as a neodymium one, can cause the magnet to slide on smooth surfaces with minimal resistance. Ferrofluids can also be used in semiactive shock absorbers in mechanical and aerospace applications (Fig. 6.28b). Although passive dampers are generally thicker and designed for a specific vibration source, active dampers consume more energy. Ferrofluid-based shock absorbers solve both of these problems and are becoming popular in the helicopter industry, which has to deal with large inertial and aerodynamic vibrations [125].

6.4.3.2 Electronics Engineering

Ferrofluid is used in rotary seals for computer hard drives and other rotating shaft motors and vibration damping speakers [126]. Ferrofluids are used to form liquid seals around drive shafts on hard drives. The rotating shaft is surrounded by magnets. A small amount of ferrofluid, located in the distance between the magnet and the shaft, will remain at the place of its attraction by the magnet. The fluid of the magnetic particle creates a barrier that prevents garbage from entering the interior of the hard drive (Fig. 6.29a). According to Ferrotec engineers, ferrofluid seals on rotating shafts typically support additional seals of 3–4 psi can be placed to form assemblies that can withstand higher pressures. Figure 6.29b shows the simple integration and separation of electronic equipment[127].

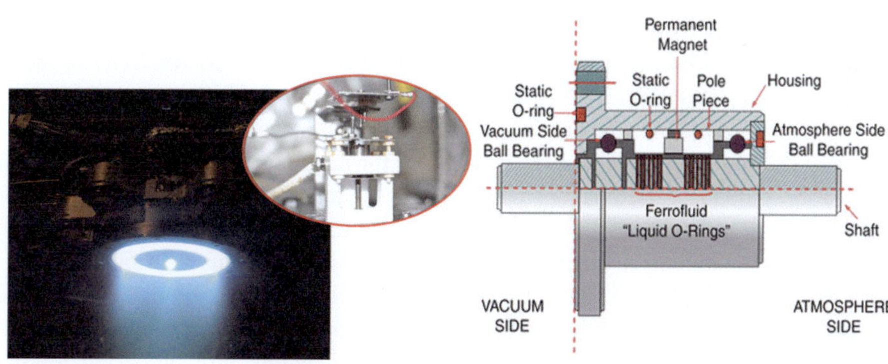

Fig. 6.28 (**a**) Ferrofluid thruster in satellite and (**b**) schematics of ferrofluid assisted shock absorbers

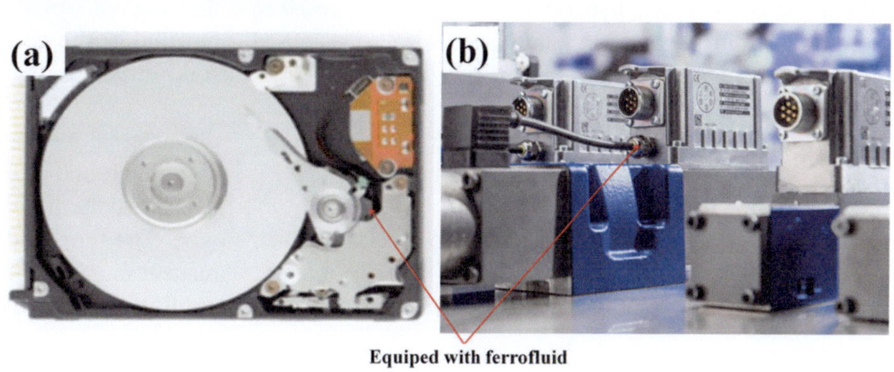

Fig. 6.29 Ferrofluid equipped in (**a**) Hard drive and (**b**) modern directional control valves offer integrated electronics or separate amplifiers for easy integration

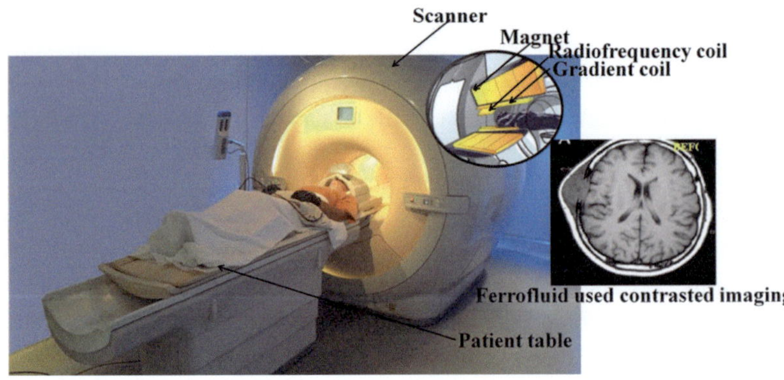

Fig. 6.30 Ferrofluid used in magnetic resonance imaging

Since 1973, ferrofluids have been used in speakers to remove heat from voice calls and passively suppress cone motion damping. They are located in what is usually the airgap around the voice coil that is held by the speaker magnet. Because ferrofluids are paramagnetic, they follow Curie's law and thus become less magnetic at higher temperatures. A strong magnet located near the voice coil (which generates heat) will attract cold ferrofluid more than hot ferrofluid, thus pulling the heated ferrofluid out of the electric voice coil and onto the radiator. This is a relatively efficient method of cooling that does not require additional energy input. The first Asian manufacturer released ferrofluids in commercial columns in 1979 [128]. This field grew rapidly in the early 1980s. Today, about 300 million audio transducers are produced with ferrofluid, including speakers installed on laptops, mobile phones, headphones, and earphones.

6.4.3.3 Medical Applications

In medicine, ferrofluid is used as a contrast agent for magnetic resonance imaging (Fig. 6.30). In the future, ferrofluids may also be used to carry medications to specific locations in the body [129]. Ferrofluids have been proposed for magnetic medication guidance. Medications can be encapsulated with ferrofluid and once injected into a specific area of the body that requires treatment, a magnetic field can be applied to hold the drug in that target area. Localization can limit the impact on the rest of the body and allow you to reduce the dosage level by reducing the side effects experienced by the patient. Target magnetic hyperthermia was also proposed to convert electromagnetic energy into heat. It is also proposed in the form of nanosurgery to separate one tissue from another, such as a tumor from the tissue in which ferrofluid plays a great role [130].

6.4.3.4 Other Industries

Ferrofluids have many optical applications because of their refractive properties; that is, every grain, a micro-magnet, reflects light. These examples include measuring the specific viscosity of a liquid exposed between a polarizer and an analyzer using a helium–neon laser. The research is underway to create an adaptive optical magnetic mirror made of ferrofluid for Earth-based astronomical telescopes. Optical filters are used to select different wavelengths of light. Replacing filters is cumbersome, especially when using wavelength constantly with adjustable lasers. Optical filters, which can be adjusted for different wavelengths by varying the magnetic field, can be built with a ferrofluid emulsion [131].

An external magnetic field overlying a liquid with different sensitivity (e.g., due to a temperature gradient) leads to an uneven force of the magnetic body, which leads to a form of heat transfer called thermomagnetic convection. This form of heat transfer can be useful if conventional convection of heat transfer is insufficient; for example, in miniature micro-scale devices or under reduced gravity conditions. Ferrofluids of suitable composition can show an extremely significant increase in thermal conductivity ($k = \sim 300\%$ of the thermal conductivity of the main liquid). Extensive ink gain is due to efficient heat transfer through the perforated nanoparticle pathway. As multifunctional "smart materials," it is possible to use special magnetic nanofluids with the reconstructed parity of thermal conductivity and viscosity that can remove heat, as well as stop vibrations (dampers). Such liquids can be used in microfluidic devices and MEMS [132].

Ferrofluids provide an interesting opportunity to collect vibrational energy from the environment. Existing methods for preparing low-frequency oscillations (<100 Hz) require the use of solid resonant structures. With the use of ferrofluids, the design of combinations no longer needed a solid structure. A simple example of the ferrofluid-based energy assembly is to place the ferrofluid in a container to use external mechanical vibrations to generate electricity in an orbit around a container surrounded by a permanent magnet. First, the liquid is placed in a container wrapped with wire. The ferrofluid is then externally magnetized by a permanent magnet. If external vibrations cause the ferrofluid to swell in the tank, then there is a change in the magnetic flux

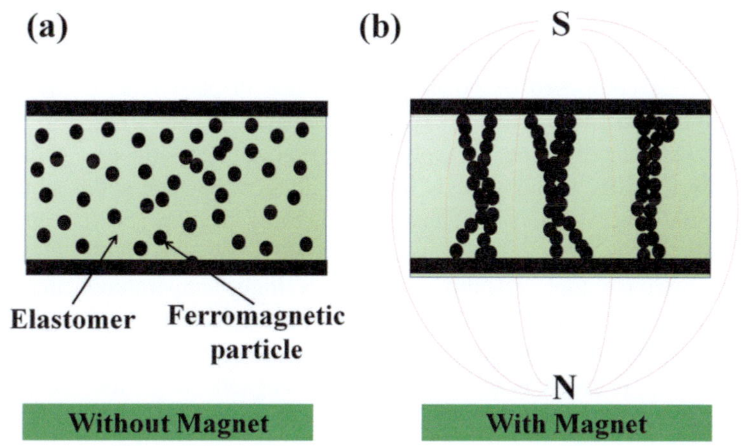

Fig. 6.31 Magneto-rheological elastomers effect (**a**) Before application of magnetic field and (**b**) after application of magnetic field

fields in relation to the wire coil. According to the Faraday law of electromagnetic induction, the voltage in a wire coil is caused by a change in magnetic flux [133].

6.5 Magneto-rheological Elastomers

Magneto-rheological elastomer (MRE) also known as magneto-sensitive elastomers is a class of solid that consists of a polymeric matrix embedded with micro- or nano-sized ferromagnetic particles such as carbonyl iron (Fig. 6.31) [134]. As a result of this composite microstructure, the mechanical properties of these materials can be controlled by the application of a magnetic field. Since the discovery in 1951 of the magneto-rheological phenomenon by Thomas Rabinow magneto-rheological fluids have proven to be commercially viable and suited to many applications. MR fluids have many advantages over the similar electro-rheological fluids, such as a larger change in modulus, lower power consumption, and the avoidance of the high voltages needed for ER fluids. However, there are still many concerns regarding the achievable yield stress, the stability, and durability of the fluid. Settling due to differences in specific gravities and wear of the magnetic particles can also lead to a reduction in the fluid's performance and eventual failure of the MR device. Although not suited to all the same applications magneto-rheological elastomers overcome many of these problems. MRE generally consists of a natural or synthetic rubber matrix interspersed with micron-sized (typically 3–5 μm) ferromagnetic particles [135]. This solid matrix for the particles avoids some common problems such as settling of particles normally associated with MR fluids and gels. Elastomers such as rubber are used as they are generally soft and/or deformable at room temperature, elastomers can have the ability to reversibly extend from 5 to 700%, depending on the specific material. Pure iron is generally used for the micron-sized particles in the rubber matrix, but some alloys of iron and cobalt can also be used to good effect, however, they are not as common. The ferromagnetic particles are added to the elastomer before it is cured. The configuration and rigidity of the chain structures depend on several factors including the strength and distribution of the applied magnetic field [136].

6.5.1 Materials Used

Among a variety of magnetic particle materials, micrometer-sized carbonyl iron (CI) powder is currently widely used as a magnetic particle for preparing MRE [174]. A basic requirement of elastic matrices for fabricating MRE is that the matrices have soft elastic property, meaning that the matrices can stably hold the magnetic particles under no magnetic field and have a finite deformation under a magnetic field. For the elastic matrices, there are lots of polymeric rubbers that can be considered as candidates, for example, silicone rubber, natural rubber, butadiene rubber, butyl rubber, polyurethane, polydimethylsiloxane, epoxy, etc. Besides the magnetic particles and the elastic matrices, additives are also key components for preparing MRE [137]. Silicone oil is usually used as an additive in the material fabrication of MRE. When the molecules of the silicone oil enter the matrix, the gaps between the matrix molecules are increased, and the conglutination of the molecules is decreased. Apart from increasing the plasticity and fluidity of the matrix, the additives can average the distribution of the internal stress in the materials, which makes a stable material property for MRE materials. The other additives include but are not limited to carbon black, carbon nanotubes, silver nanowire, Rochelle salt, gamma-ferrite additives, etc. [138, 139].

6.5.2 Preparation of MRE

MRE is composed of polarizable particles dispersed in a polymer medium. MREs are typically prepared by the curing process for polymers. The polymeric material (e.g., silicone rubber) in its liquid state is mixed with iron powder and several other additives to enhance its mechanical properties. Then the mixture vulcanizes at room temperature (called room-temperature vulcanizing) or at a high temperature higher than 120 °C (called high-temperature vulcanizing) [140]. If the elastomer, with suspended ferromagnetic particles, is cured in the presence of a magnetic field, the magnet is able particles will form chains along the direction of the magnetic field prior to the elastomer cross-linking process (curing) and resulting in an anisotropic MRE material. However, if the mixture is not cured in the presence of a magnetic field, and the particles are hence left randomly distributed, an anisotropic elastomer ferromagnet composite (EFC) is produced. Figure 6.32 shows the micrograph of ferromagnetic particles in a silicone polymeric elastomer matrix. An elastomer ferromagnet composite, randomly dispersed particles shown in Fig. 6.32a, and the magnetorheological elastomer, in which the aligned ferromagnetic particle chains have been shown in Fig. 6.32b. Recently, in 2017, an advanced technology, 3D printing has also been used to configure the magnetic particles inside the polymer matrix [141].

6.5.3 Application

Such properties are stimulated by their many promising applications, such as adaptively tuned vibration absorbers, adjustable mounting stiffness and suspension stiffness, and adjustable impedance surfaces. The vibration absorption unit consists of a magnetic conductor, a shear sleeve, a coil rod, an electromagnetic coil winding, and around cylindrical MRE that is vulcanized between the shear sleeve and the storage rod (Fig. 6.33a). The magnetic conductor, the coil rod, and the electromagnetic coil are supported by the MRE on the shear sleeve [142]. The magnetic conductor and the shell housing are bolted together, and the hip sleeve is secured to the lower housing. The outer surface of the sleeve rod is aligned with the inner surface of the magnetic conductor, and the magnetic conductor can move vertically along the high sleeve. The MRE operates in pure shear mode, and the magnetic conductor, the coil rod, and the electromagnetic coil form the dynamic mass of the MRE-based vibration absorber. The proposed MRE-based vibration absorber can absorb vibrational energy and thus reduce vibration [143].

Vibration isolators (Fig. 6.33b) are devices that can isolate an object, such as a part of the device, from a vibration source. Vibration isolators can be divided into two groups: base isolation and force isolation, and isolation modes have active and passive vibration isolation. The figure shows an example of the proposed arrangement and prototype of an

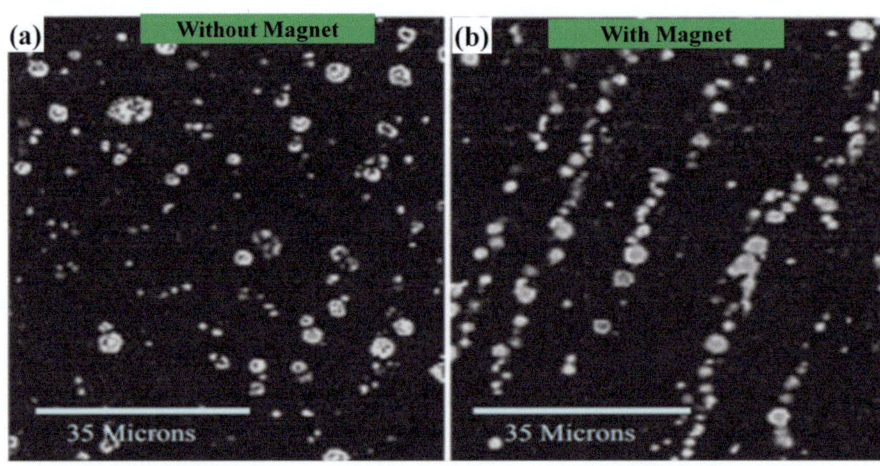

Fig. 6.32 Scanning electron micrographs of MRE (ferromagnetic particles+silicone polymeric elastomer matrix): (**a**) Without magnetic field and (**b**) With magnetic field

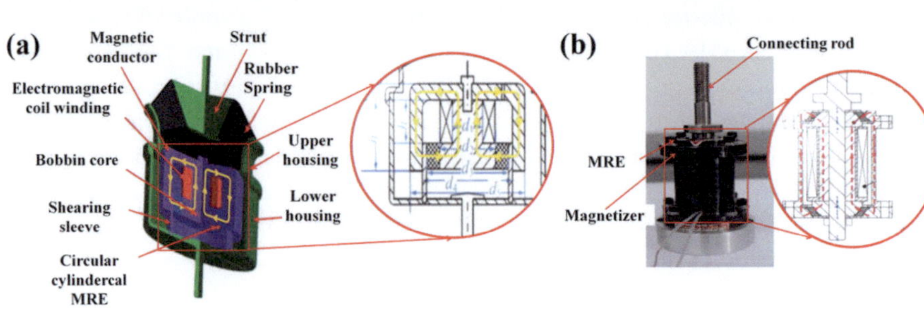

Fig. 6.33 (**a**) MRE-based dynamic vibration absorber and (**b**) MRE-based isolator

MRE-based insulator operating in compression/shear extension mode. This MRE insulator can be used as a device with controlled hardness and has great potential in the field of vibration suppression of heavy equipment [144].

Based on their field-sensitive properties, MREs have been developed for use as sensors and actuators, such as power sensors, magnetic sensors, magneto-sensitive voltage sensors, flexible triaxial touch sensors, three-sensor, self-propelled sensors, a magnetic and mechanical sensor, soft actuator, valve actuators, MEMS magnetometer, etc. In terms of absorption properties, the carbon iron filled in the composite has the ability to absorb microwaves and electromagnetic waves. The ability of the material that absorbs wool comes from the characteristics of the permissibility of carbon iron particles. As a sound absorption capability, conventional foam can only play a certain frequency absorption. The use of MRE foam promises a wider range of sound absorption coefficients. One of the potential applications of MRE in medical devices is the prosthetic foot. Conventional equipment often causes pain when walking, running, or jumping due to inelastic construction. A prosthetic foot equipped with an MRE can achieve an appropriate level of rigidity for the user based on the level of activity. Due to the low magnetic field, low hardness is suitable for comfort or accidental movement. High hardness is required for more selective activities such as running and jumping [145].

6.6 Electro-Conjugate Liquids

The electro-conjugate fluid (ECF) is a dielectric fluid, which works here as a smart/functional fluid. Applying a high voltage of several kilovolts between electrodes inserted into the fluid with an interelectrode gap of several hundred micrometers. This kind of fluid produces a jet flow when subjected to high DC voltage between the electrodes as shown in Fig. 6.34. This phenomenon is observed with the electro-conjugate fluid is known as ECF effect [146].

6.6.1 Application

This study introduces ECF into a flexible and novel robotic arm as an energy source. Figure 6.35 shows a conceptual view of the robot, consisting of flexible fingers filled with electro conjugated liquid, ECF jet generators, and an ECF (palm of the hand) tank. When high voltage is applied to each ECF nozzle generator, the ECF moves from the tank to each finger resulting in increased internal pressure in the fingers. As the fingers are properly reinforced, they bend when pressed. There are several similar fingers/hands that use pneumatic pressure, however, the proposed robot arm differs from these fingers/hands in the following points. First, the working fluid is completely closed in the robot's hand. Second, the actuator and pressure sources are integrated in the hand. Finally, the setup is quite simple compared to pneumatic robots, which require external compressors. Therefore, the ECF robotic arm has advantages for integration, reduction, and illumination [147].

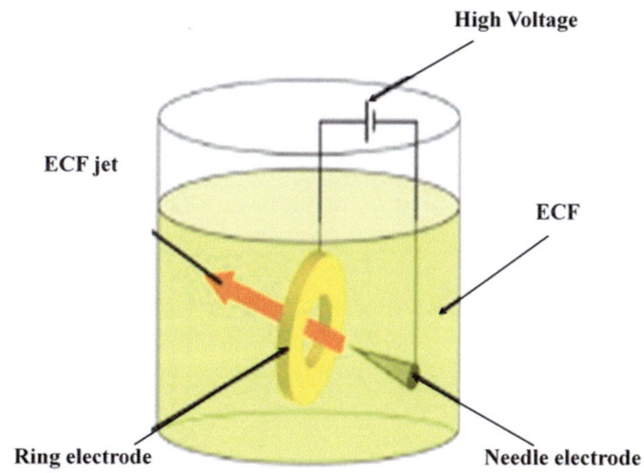

Fig. 6.34 Schematic diagram of ECF jet

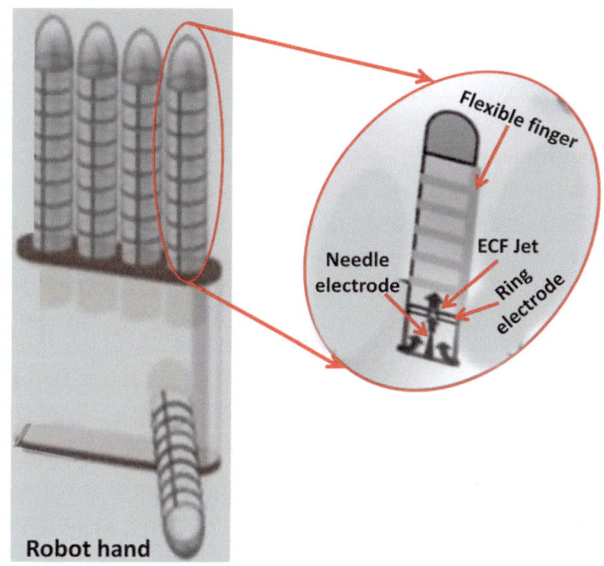

Fig. 6.35 ECF robot hand

6.7 Photo-Rheological Fluid

Photo-rheological (PR) fluids are fluids that exhibit changes in their viscosity upon ultraviolet (UV) light irradiation. Self-assembly of photoactive molecules in solution actuated using light [148]. Self-assembly refers to the spontaneous aggregation of molecules under a given set of conditions (concentration, temperature, type of solvent). Generally, the type of

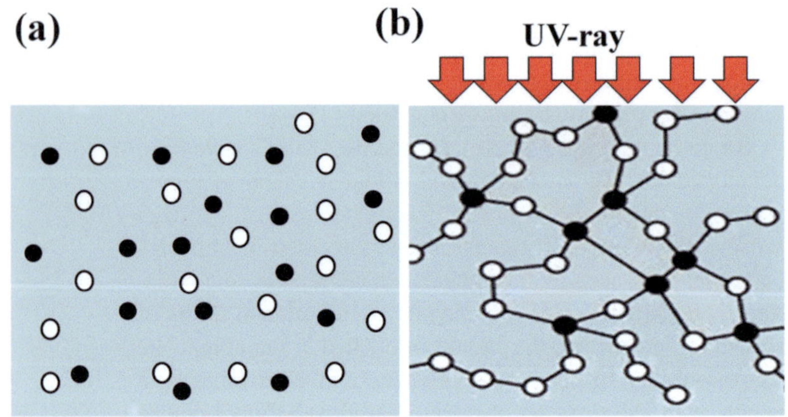

Fig. 6.36 PR fluid (**a**) without and (**b**) with UV ray

self-assembled structure formed is sensitive to the geometry when illuminated with light. For example, the molecules may initially assemble into a structure that yields a low viscosity (e.g., discrete spheres, Fig. 6.36a). Upon irradiation, however, the change in molecular shape facilitates assembly into a different type of structure that corresponds to a high viscosity (e.g., a network of chains, Fig. 6.36b). Thus, the nanoscale transformation of self-assembled structure is reflected as a dramatic change in viscosity, as shown by the fact that the solution on the left easily pours out of the vial whereas the one on the right is a thick, gel-like material that does not flow out of the vial [149]. Photo-responsive molecules undergo a chemical transformation upon absorption of light at certain wavelengths. One of the main types of photo-response is a *trans* to *cis* photo-isomerization about double bonds. Such a transformation occurs in molecules such as stilbenes, phenylalkenes, and azobenzenes. Stilbene (IUPAC name: 1,2-diphenylethylene) consists of a C-C double bond with phenyl groups on each carbon. This molecule undergoes a *trans* to *cis* photo-isomerization under UV light (313 nm). In the *trans*-isomer, the phenyl groups are on opposite sides of the double bond, whereas in the *cis*-isomer, they are on the same side [148].

6.7.1 PR Fluid Preparation

PR fluids are easy-to-make, cost-effective. These fluids have attracted increased attention since Lee et al. reported that they can be easily synthesized in the laboratory using commercially available chemicals and ingredients, such as spiropyran, instead of using conventional ingredients, such as *trans*-ortho-methoxycinnamic acid, alpha-cyclodextrin, and hydroxyethylcellulose [149].PR fluids are characterized by changes in their properties such as color and viscosity when irradiated by the UV light. Typically, UV sensors are implemented using semiconductors made of ZnO and ZnS, which have wide bandgaps and increased sensitivities to UV light. ZnO and ZnS can only detect UV wavelengths in the range of 320 nm to 400 nm because of the inherent limitations of UV-sensitive materials. Thus, nitride semiconductors, such as GaN, InN, and AlN, have been developed to detect shorter UV light wavelengths. In addition, many engineering studies have been conducted on UV sensors with various materials, such as titanium dioxide or diamond, to improve the performance of UV sensors. In particular, niobium pentoxide has a much higher sensitivity and external quantum efficiency than some other employed materials, and it has been extensively studied as a UV sensor. The aforementioned UV sensors can measure UV light exactly, they have fast responses, and the range of wavelengths that can be detected is broad. However, most of the aforementioned sensors require complicated electronic circuits and peripherals because they use semiconductors and respond to UV light electronically [148, 149].

6.7.2 Applications

Recently, research findings were presented on noncontact tunable damping using this new class of PR fluids. They can be successfully applied to perform amplitude adjustment of vibrational structure adjustment. PR fluids have gained considerable interest in their potential applications in microfluidic or MEMS devices due to their ability to control depending on the rheological conditions. PR fluids show only nonlinear and asymmetrically reversible responses [150]. This makes it difficult to standardize applications with photo-activation or photosensor, which is one of the key disadvantages of PR fluids for smart fluid applications. To solve this problem, new PR fluids are needed that can produce a linear and symmetrical response, regardless of whether the UV lamp is on or off. In this study, the photoluminescence (PL) nanoparticles (in powder form) of ZnS:Cu green phosphor were mixed with the synthesized PR fluids in order to change their fundamental photo-rheological characteristics [151]. Figure 6.37a shows the photo-rheological materials (cetyltrimethylammonium-bromide + *trans*-

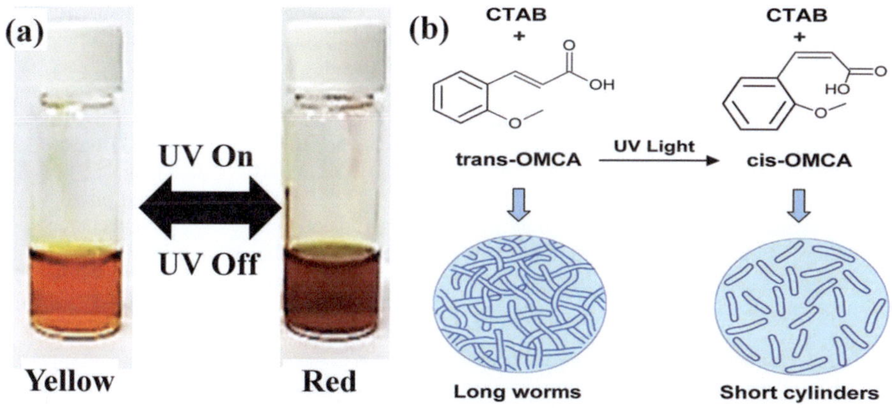

Fig. 6.37 (**a**) Photoresponsive activity of CTAB + OMCA, (**b**) Reduction in micellar lengthupon UV irradiation

ortho-methoxycinnamic acid) (i.e., CTAB + OMCA) that changes its viscosity on exposure to the UV-light. When OMCA is in its *trans*-form, its mixture with CTAB gives rise to long, entangled wormlike micelles and upon UV irradiation, *trans*-OMCA gets photoisomerized to *cis*-OMCA, and the corresponding change in molecular geometry causes a drastic reduction in micellar length (Fig. 6.37b).

6.8 Summary

Current interest on smart fluid technology is in demand for a new smart devices. Rheological materials on the smart application are given successfully in this chapter. Electric, magnetic, and UV light activated classified materials have been discussed along with their activation mechanism, and synthesis. For each categories, some of the materials have been focused on for giant effect which is the target materials in this current rheology scenario. Further study will be necessary to obtain a clear understanding of the giant effect both theoretically and experimentally. All studies related to the mechanisms and dynamic models should be verified using other materials, which is one of the reasons why material studies are being carried out continuously in the rheological community. All the stimuli moduli contact and noncontact based devices have been presented for micro application to heavy mass application. The materials and mechanism in this chapter will give a leading direction to grow the smart fluid application.

References

1. R. Stanway, Smart fluids: current and future developments, Materials Science and Technology, 20, 2004 - 8: Nanomaterials and Nanomanufacturing, 931–939, https://doi.org/10.1179/026708304225019867
2. Jong-Seok, O., Choi, S.-B.: State of the art of medical devices featuring smart electrorheological and magneto-rheological fluids. J. King Saud Univ. Sci. **29**(4), 390–400 (2017). https://doi.org/10.1016/j.jksus.2017.05.012
3. Choi, S.-B., Wereley, N.M., Li, W., Yu, M., Koo, J.-H.: Applications of controllable smart fluids to mechanical systems. In: Advances in Mechanical Engineering. SAGE Journals, Thousand Oaks, CA (2014). https://doi.org/10.1155/2014/254864
4. https://en.wikipedia.org/wiki/Smart_fluid. Accessed 12 Jul 2019
5. Olabi, A.G., Grunwald, A.: Design and application of magneto-rheological fluid. Mater. Des. **28**(10), 2658–2664 (2007). https://doi.org/10.1016/j.matdes.2006.10.009
6. Yang, F., Bick, A., Shandalov, S., Brenner, A., Oron, G.: Yield stress and rheological characteristics of activated sludge in an airlift membrane bioreactor. J. Membr. Sci. **334**(1–2), 83–90 (2009). https://doi.org/10.1016/j.memsci.2009.02.022
7. Kazemi-Beydokhti, A., Hajiabadi, S.H.: Rheological investigation of smart polymer/carbon nanotube complex on properties of water-based drilling fluids. Colloids Surf. A Physicochem. Eng. Asp. **556**, 23–29 (2018). https://doi.org/10.1016/j.colsurfa.2018.07.058
8. https://www.rheosense.com/applications/viscosity/newtonian-non-newtonian. Accessed 12 Jul 2019
9. Barnes, H.A.: The yield stress—a review or 'παντα ρει'-everything flows? J. Non-Newton. Fluid. **81**(1–2), 133–178 (1999). https://doi.org/10.1016/S0377-0257(98)00094-9
10. Zhen Dong, Y., Choi, H.J.: Electrorheological characteristics of poly(diphenylamine)/magnetite composite-based suspension. Materials. **2019**, 12 (2911). https://doi.org/10.3390/ma12182911
11. Lee, S.-R., Uhm, C.-H., Seong, M.-S., Oh, J.-S., Choi, S.-B.: Repulsive force control of minimally invasive surgery robot associated with three degrees of freedom electrorheological fluid-based haptic master. Proc. Inst. Mech. Eng. C J. Mech. Eng. Sci. **228**(9), 1606–1621 (2014). https://doi.org/10.1177/0954406213508935
12. Dimarogonas, A.D., Kollias, A.: Electrorheological fluid-controlled "smart" journal bearings. Tribol. Trans. **35**(4), 611–618 (1992). https://doi.org/10.1080/10402009208982163
13. Liua, Y.D., Choi, H.J.: Electrorheological fluids: smart soft matter and characteristics. Soft. Matter. **8**, 11961–11978 (2012). https://doi.org/10.1039/C2SM26179K
14. Leng, J.S., Liu, Y.J., Du, S.Y., et al.: Active vibration control of smart composites featuring electrorheological fluids. Appl. Compos. Mater. **2**, 59–65 (1995). https://doi.org/10.1007/BF00567378
15. Badam, A.P.R., Kumaraswamidhas, L.A., Rajak, D.K.: Modelling and simulation of semi-active hybrid control of vehicle suspension system with magneto – rheological dampers. Rom. J. Automotive Eng. **22**(1), 5–12 (2016)
16. Seo, Y.P., Choi, H.J., Seo, Y.: A simplified model for analyzing the flow behavior of electrorheological fluids containing silicananoparticle-decorated polyaniline nanofibers. Soft Matter. **8**, 4659–4663 (2012). https://doi.org/10.1039/C2SM07275K

17. Powell, J.A.: The mechanical properties of an electrorheological fluid under oscillatory dynamic loading. Smart Mater. Struct. **2**, 217 (1993). https://doi.org/10.1088/0964-1726/2/4/003
18. Wu, J., Liu, F., Guo, J., Cui, P., Xu, G., Cheng, Y.: Preparation and electrorheological characteristics of uniform core/shell structural particles with different polar molecules shells. Colloids Surf. A Physicochem. Eng. Asp. **410**, 136–143 (2012). https://doi.org/10.1016/j.colsurfa.2012.06.033
19. Monkman, G.J.: Addition of solid structures to electrorheological fluids. J. Rheol. **35**(7), 1385 (1991). https://doi.org/10.1122/1.550237
20. Ikazakidag, F., Kawaidag, A., Uchidadag, K., Kawakamiddag, T., Edamura, K., Sakurai, K., Anzai, H., Asakoddag, Y.: Mechanisms of electrorheology: the effect of the dielectric property. J. Phys. D: Appl. Phys. **31**, 3. https://doi.org/10.1088/0022-3727/31/3/014
21. Wu, J., Xu, G., Cheng, Y., Liu, F., Guo, J., Cui, P.: The influence of high dielectric constant core on the activity of core–shell structure electrorheological fluid. J. Colloid. Interf. Sci. **378**(1), 36–43 (2012). https://doi.org/10.1016/j.jcis.2012.04.044
22. Hong, Y., Wen, W.: Influence of carrier liquid on nanoparticle-based giant electrorheological fluid. J. Intell. Mater. Syst. Struct. **27** (7), 866–871 (2016). https://doi.org/10.1177/1045389X15596623
23. Yeh, J.-Y., Chen, L.-W.: Dynamic stability of a sandwich plate with a constraining layer and electrorheological fluid core. J. Sound Vib. **285**(3), 637–652 (2005). https://doi.org/10.1016/j.jsv.2004.08.033
24. Zhang, Y., Lu, K., Rao, G.: Electrorheological fluid with an extraordinarily high yield stress. Appl. Phys. Lett. **80**, 888 (2002). https://doi.org/10.1063/1.1446999
25. Sung, J.H., Lee, Y.H., Jang, I.B., Choi, H.J., Jhon, M.S.: Synthesis and electrorheological characteristics of microencapsulated conducting polymer. Des. Monomers Polym. **7**(1–2), 101–110 (2004). https://doi.org/10.1163/156855504322890061
26. Goodwin, J.W., Markham, G.M., Vincent, B.: Studies on model electrorheological fluids. J. Phys. Chem. B. **101**(11), 1961–1967 (1997). https://doi.org/10.1021/jp962267j
27. Datta, S., Barua, R., Das, J.: A Review on ElectroRheological Fluid (ER) and Its Various Technological Applications. IntechOpen, London (2020). https://doi.org/10.5772/intechopen.90706
28. Zhen Dong, Y., Seo, Y., Choi, H.J.: Recent development of electro-responsive smart electrorheological fluids. Soft Matter. **15**, 3473–3486 (2019). https://doi.org/10.1039/C9SM00210C
29. Yin, J., Zhao, X.: Titanate nano-whisker electrorheological fluid with high suspended stability and ER activity. Nanotechnology. **17**, 1. https://doi.org/10.1088/0957-4484/17/1/031
30. Hao, T.: Electrorheological suspensions. J. Colloid. Interf. Sci. **97** (1–3), 1–35 (2002). https://doi.org/10.1016/S0001-8686(01)00045-8
31. Park, J.K., Ryu, J.C., Kim, W.K., Kang, K.H.: Effect of electric field on electrical conductivity of dielectric liquids mixed with polar additives: DC conductivity. J. Phys. Chem. B. **113**(36), 12271–12276 (2009). https://doi.org/10.1021/jp9015189
32. Choi, Y., Sprecher, A.F., Conrad, H.: Vibration characteristics of a composite beam containing an electrorheological fluid. J. Intell. Mater. Syst. Struct. **1**(1), 91–104 (1990). https://doi.org/10.1177/1045389X9000100106
33. Stangroom, J.E.: Basic observations on electro-rheological fluids. J. Intell. Mater. Syst. Struct. **7**(5), 479–483 (1996). https://doi.org/10.1177/1045389X9600700501
34. Duan, X., Luo, W., Wu, W.: New theory for improving performance of electrorheological fluids by additives. J. Phys. D: Appl. Phys. **33**, 23. https://doi.org/10.1088/0022-3727/33/23/314
35. Tam, W.Y., Yi, G.H., Wen, W., Ma, H., Loy, M.M.T., Sheng, P.: New electrorheological fluid: theory and experiment. Phys. Rev. Lett. **78**, 2987 (1997). https://doi.org/10.1103/PhysRevLett.78.2987
36. Anderson, R.A.: Electrostatic forces in an ideal spherical-particle electrorheological fluid. Langmuir. **10**(9), 2917–2928 (1994). https://doi.org/10.1021/la00021a013
37. Wen, W., Huang, X., Yang, S., et al.: The giant electrorheological effect in suspensions of nanoparticles. Nat. Mater. **2**, 727–730 (2003). https://doi.org/10.1038/nmat993
38. Gong, X., Wu, J., Huang, X.X., Wen, W., Sheng, P.: Influence of liquid phase on nanoparticle-based giant electrorheological fluid. Nanotechnology. **19**, 16. https://doi.org/10.1088/0957-4484/19/16/165602
39. Jiaxing Li, Xiuqing Gong, Shuyu Chen, Weijia Wena), and Ping Sheng, Giant electrorheological fluid comprising nanoparticles: carbon nanotube composite, J. Appl. Phys., 107, 9 (2010), doi: https://doi.org/10.1063/1.3407503
40. Liu, L., Chen, X., Niu, X., Wena, W., Sheng, P.: Electrorheological fluid-actuated microfluidic pump. Appl. Phys. Lett. **89**, 8 (2006). https://doi.org/10.1063/1.2337877
41. Choi, S.-B., Seong, M.-S., Kim, K.-S.: Vibration control of an electrorheological fluid-based suspension system with an energy regenerative mechanism. Proc. Inst. Mech. Eng. D J. Automobile Eng. **223**(4), 459–469 (2009). https://doi.org/10.1243/09544070JAUTO968
42. Hou, J., Shi, L., Zhu, Q.: Electrorheological properties and structure of ($BaTiO(C_2O_4)_2/NH_2CONH_2$). J. Solid State Chem. **179**(6), 1874–1878 (2006). https://doi.org/10.1016/j.jssc.2006.02.026
43. Cheng, Y., Wu, K., Liu, F., Guo, J., Liu, X., Xu, G., Cui, P.: Facile approach to large-scale synthesis of 1D calcium and titanium precipitate (CTP) with high electrorheological activity. ACS Appl. Mater. Interfaces. **2**(3), 621–625 (2010). https://doi.org/10.1021/am900841m
44. Ma, N., Dong, X., Niu, C., Han, B.: A facile electrostatic spraying method to prepare polyvinylpyrrolidone modified TiO_2 particles with improved electrorheological effect. Soft Mater. **15**, 4 (2017). https://doi.org/10.1080/1539445X.2017.1356734
45. Unal, H.I., Sahan, B., Erol, O.: Effect of surfactant on electrokinetic properties of polyindole/TiO_2-conducting nanocomposites in aqueous and nonaqueous media. Colloid Polym. Sci. **292**, 499–509 (2014). https://doi.org/10.1007/s00396-013-3094-7
46. Vance, J.M., Ying, D.: Experimental measurements of actively controlled bearing damping with an electrorheological fluid. J. Eng. Gas Turbines Power. **122**(2), 337–344 (2000). https://doi.org/10.1115/1.483212
47. Guoguang, Z., Furusho, J., Sakaguchi, M.: Vibration suppression control of robot arms using a homogeneous-type electrorheological fluid. IEEE/ASME Trans. Mechatron. **5**(3), 302–309 (Sept. 2000). https://doi.org/10.1109/3516.868922
48. Liu, Y., Yuan, J., Dong, Y., Zhao, X., Yin, J.: Enhanced temperature effect of electrorheological fluid based on cross-linked poly (ionic liquid) particles: rheological and dielectric relaxation studies. Soft Matter. **13**, 1027–1039 (2017). https://doi.org/10.1039/C6SM02480G
49. Hong, J.-Y., Kwon, E., Jang, J.: Fabrication of silica/polythiophene core/shell nanospheres and their electrorheological fluid application. Soft Matter. **5**, 951–953 (2009). https://doi.org/10.1039/B821291K
50. Hosseini-Sianaki, A., Bullough, W.A., Firoozian, R., Makin, J., Tozer, R.C.: Experimental measurements of the dynamic torque response of an electrorheological fluid in the shear mode. Int. J. Mod. Phys. B. **06**(15–16), 2667–2681 (1992). https://doi.org/10.1142/S0217979292001365
51. Gast, A.P., Zukoski, C.F.: Electrorheological fluids as colloidal suspensions. J. Colloid. Interf. Sci. **30**, 153–202., ISSN 0001-8686 (1989). https://doi.org/10.1016/0001-8686(89)80006-5
52. Goldasz, J., Sapinski, B.: Nondimensional characterization of flow-mode magnetorheological/electrorheological fluid

dampers. J. Intell. Mater. Syst. Struct. **23**(14), 1545–1562 (2012). https://doi.org/10.1177/1045389X12447293

53. Jung, W.J., Jeong, W.B., Hong, S.R., Choi, S.-B.: Vibration control of a flexible beam structure using squeeze-mode ER mount. J. Sound Vib. **273**(1–2), 185–199 (2004). https://doi.org/10.1016/S0022-460X(03)00478-4
54. https://www.transparencymarketresearch.com/electrorheological-fluid-market.html. Accessed 25 July 2019
55. Sassi, S., Cherif, K.: Combination of homogeneous electrorheological fluid and multi-electrodes damper for a better control of car suspension motion. Int. J. Mech. Syst. Eng. **2**, 112 (2016). https://doi.org/10.15344/2455-7412/2016/112
56. Coulter, J.P., Weiss, K.D., Carlson, J.D.: Engineering applications of electrorheological materials. J. Intell. Mater. Syst. Struct. **4**(2), 248–259 (1993). https://doi.org/10.1177/1045389X9300400215
57. Choi, K., Gao, C.Y., Do Nam, J., Choi, H.J.: Cellulose-based smart fluids under applied electric fields. Materials (Basel). **10**(9), 1060 (2017). https://doi.org/10.3390/ma10091060
58. Bose, H., Berkemeier, H.-J.: Haptic device working with an electrorheological fluid. J. Intell. Mater. Syst. Struct. **10**(9), 714–717 (1999). https://doi.org/10.1106/M9QQ-FBPV-5MGD-FGE3
59. Das, S., Karthik, V., Pabi, S.K., Behera, A., Patel, S.K., Swain, B., Roshan, R., Behera, A.: Enhancement of thermal conductivity of Cu-Cr dispersed nanofluids according to multiscale modelling. Mater. Today Proc. **33**(Part 8), 5514–5520 (2020). https://doi.org/10.1016/j.matpr.2020.03.330
60. Li, J., Jin, D., Zhang, X., Zhang, J., Gruver, W.A.: An electrorheological fluid damper for robots. In: Proceedings of 1995 IEEE International Conference on Robotics and Automation, vol. 3, pp. 2631–2636. IEEE, Nagoya, Japan (1995). https://doi.org/10.1109/ROBOT.1995.525654.
61. https://www.sciencealert.com/liquid-armour-is-now-a-thing-and-it-stops-bullets-better-than-kevlar. Accessed 25 July 2019
62. Yunwei, Z., Dexu, G., Xiaomin, L., Jintao, Z.: A study on electrorheological fluid-assisted polishing of tungsten carbide. In: 2011 International Conference on Mechatronic Science, Electric Engineering and Computer (MEC), pp. 380–383. IEEE, Jilin (2011). https://doi.org/10.1109/MEC.2011.6025480
63. Liu, L., Chen, X., Niu, X., Wena, W., Sheng, P.: Electrorheological fluid-actuated microfluidic pump. Appl. Phys. Lett. **89**(083505) (2006). https://doi.org/10.1063/1.2337877
64. Choi, S.-B., Hong, S.-R.: Graduate assistant, dynamic modeling and vibration control of electrorheological mounts. J. Vib. Acoust. **126**(4), 537–541 (2004). https://doi.org/10.1115/1.1805006
65. Choi, S.-B., Yook, J.-Y., Choi, M.-K., Nguyen, Q.H., Lee, Y.-S., Han, M.-S.: Speed control of DC motor using electrorheological brake system. J. Intell. Mater. Syst. Struct. **18**(12), 1191–1196 (2007). https://doi.org/10.1177/1045389X07083135
66. Davidson, J.R., Krebs, H.I.: An electrorheological fluid actuator for rehabilitation robotics. IEEE/ASME Trans. Mechatron. **23**(5), 2156–2167 (Oct. 2018). https://doi.org/10.1109/TMECH.2018.2869126
67. Wena, W., Huang, X., Sheng, P.: Particle size scaling of the giant electrorheological effect. Appl. Phys. Lett. **85**, 299 (2004). https://doi.org/10.1063/1.1772859
68. Sturk, M., Wu, X.M., Wong, J.Y.: Development and evaluation of a high voltage supply unit for electrorheological fluid dampers. Int. J. Vehicle Mech. Mob. **24**, 2 (1995). https://doi.org/10.1080/00423119508969083
69. Liu, L., Niu, X., Wena, W., Sheng, P.: Electrorheological fluid-actuated flexible platform. Appl. Phys. Lett. **88**(173505) (2006). https://doi.org/10.1063/1.2196847
70. Zhu, X., Jing, X., Cheng, L.: Magnetorheological fluid dampers: a review on structure design and analysis. J. Intell. Mater. Syst. Struct. **23**(8), 839–873 (2012). https://doi.org/10.1177/1045389X12436735
71. Ashtiani, M., Hashemabadi, S.H., Ghaffari, A.: A review on the magnetorheological fluid preparation and stabilization. J. Magn. Magn. Mater. **374**, 716–730 (2015). https://doi.org/10.1016/j.jmmm.2014.09.020
72. Jolly, M.R., Bender, J.W., Carlson, J.D.: Properties and applications of commercial magnetorheological fluids. J. Intell. Mater. Syst. Struct. **10**(1), 5–13 (1999). https://doi.org/10.1177/1045389X9901000102
73. Miao, C., Shen, R., Wang, M., Shafrir, S.N., Yang, H., Jacobs, S. D.: Rheology of aqueous magnetorheological fluid using dual oxide-coated carbonyl iron particles. J. Am. Ceram. Soc. **94**(8), 2386–2392 (2011). https://doi.org/10.1111/j.1551-2916.2011.04423.x
74. Kolekar, S.: Preparation of magnetorheological fluid and study on its rheological properties. Int. J. Nanosci. **13**(02), 1450009 (2014). https://doi.org/10.1142/S0219581X14500094
75. Goncalves, F.D.: A review of the state of the art in magnetorheological fluid technologies--Part I: MR fluid and MR fluid models. Shock Vibr. Dig. **38**(3), 203 (2020)
76. Wang, G., Zhao, D., Ma, Y., Zhang, Z., Che, H., Jingbo, M., Zhang, X., Yu, T., Dong, X.: Synthesis of calcium ferrite nanocrystal clusters for magnetorheological fluid with enhanced sedimentation stability. Powder Technol. **322**, 47–53 (2017). https://doi.org/10.1016/j.powtec.2017.08.065
77. Shafrir, S.N., Romanofsky, H.J., Skarlinski, M., Wang, M., Miao, C., Salzman, S., Chartier, T., Mici, J., Lambropoulos, J.C., Shen, R., Yang, H., Jacobs, S.D.: Zirconia-coated carbonyl-iron-particle-based magnetorheological fluid for polishing optical glasses and ceramics. Appl. Opt. **48**, 6797–6810 (2009). https://doi.org/10.1364/AO.48.006797
78. Esmaeilnezhad, E., Choi, H.J., Schaffie, M., Gholizadeh, M., Ranjbar, M., Kwon, S.H.: Rheological analysis of magnetite added carbonyl iron based magnetorheological fluid. J. Magnet. Magn. Mater. **444**, 161–167 (2017). https://doi.org/10.1016/j.jmmm.2017.08.023
79. Popplewell, J., Rosensweig, R.E.: Magnetorheological fluid composites. J. Phys. D: Appl. Phys. **29**, 9. https://doi.org/10.1088/0022-3727/29/9/011
80. Fernández-Toledano, J.C., Rodríguez-López, J.: Two-step yielding in magnetorheology. J. Rheol. **58**, 5 (2014). https://doi.org/10.1122/1.4880675
81. Rabbani, Y., Ashtiania, A.M., Hashemabadi, S.H.: An experimental study on the effects of temperature and magnetic field strength on the magnetorheological fluid stability and MR effect. Soft Matter. **11**, 4453–4460 (2015). https://doi.org/10.1039/C5SM00625B
82. Dai, J., Chang, H., Zhao, R., Huang, J., Li, K., Xie, S.: Investigation of the relationship among the microstructure, rheological properties of MR grease and the speed reduction performance of a rotary micro-brake. Mech. Syst. Sign. Process. **116**, 741–750 (2019). https://doi.org/10.1016/j.ymssp.2018.07.004
83. Xu, Y., Liao, G., Liu, T.: Magneto-Sensitive Smart Materials and Magnetorheological Mechanism. IntechOpen, London (2019). https://doi.org/10.5772/intechopen.84742
84. Kumbhar, B.K., Patil, S.R., Sawant, S.M.: Synthesis and characterization of magneto-rheological (MR) fluids for MR brake application. Eng. Sci. Technol. Int. J. **18**(3), 432–438 (2015). https://doi.org/10.1016/j.jestch.2015.03.002
85. Hao, T.: Chapter 3 - The positive, negative, photo-ER, and electromagnetorheological (EMR) effects. In: Studies in Interface Science, vol. 22, pp. 83–113., ISBN 9780444521804. Elsevier, Netherland (2005). https://doi.org/10.1016/S1383-7303(05)80018-3

86. Lokander, M., Stenberg, B.: Improving the magnetorheological effect in isotropic magnetorheological rubber materials. Polym. Test. **22**(6), 677–680 (2003). https://doi.org/10.1016/S0142-9418(02)00175-7
87. Wang, J., Meng, G.: Magnetorheological fluid devices: principles, characteristics and applications in mechanical engineering. Proc. Inst. Mech. Eng. Part L J. Mater. Des. Appl. **215**(3), 165–174 (2001). https://doi.org/10.1243/1464420011545012
88. Carlson, J.D., Jolly, M.R.: MR fluid, foam and elastomer devices. Mechatronics. **10**(4–5), 555–569 (2000). https://doi.org/10.1016/S0957-4158(99)00064-1
89. Badam, A.P.R., Kumaraswamidhas, L.A., Rajak, D.K.: Modelling and simulation of semi-active hybrid control of vehicle suspension system with magneto - rheological dampers. Roman. J. Automot. Eng. **22**(1), 5–12 (2016)
90. Yao, G.Z., Yap, F.F., Chen, G., Li, W.H., Yeo, S.H.: MR damper and its application for semi-active control of vehicle suspension system. Mechatronics. **12**(7), 963–973 (2002). https://doi.org/10.1016/S0957-4158(01)00032-0
91. Milecki, A., Hauke, M.: Application of magnetorheological fluid in industrial shock absorbers. Mech. Syst. Signal Process. **28**, 528–541 (2012). https://doi.org/10.1016/j.ymssp.2011.11.008
92. Chen, L.: Using magnetorheological(MR) fluid as distributed actuators for smart structures. In: 2009 4th IEEE Conference on Industrial Electronics and Applications, pp. 1203–1208. IEEE, Xi'an (2009). https://doi.org/10.1109/ICIEA.2009.5138393
93. http://www.formula1-dictionary.net/damper_magnetorheological.html. Accessed 27 July 2019
94. Sassi, S., Cherif, K., Mezghani, L., Thomas, M., Kotrane, A.: An innovative magnetorheological damper for automotive suspension: from design to experimental characterization. Smart Mater. Struct. **14**, 4. https://doi.org/10.1088/0964-1726/14/4/041
95. file:///C:/Users/Ajit%20Behera/Desktop/PCNA18_0106_us.pdf. Accessed 28 July 2019
96. Kasemi, B., Muthalif, A.G.A., Rashid, M.M., Fathima, S.: Fuzzy-PID controller for semi-active vibration control using magnetorheological fluid damper. Proc. Eng. **41**, 1221–1227 (2012). https://doi.org/10.1016/j.proeng.2012.07.304
97. Chen, S., Huang, J., Jian, K., Ding, J.: Analysis of influence of temperature on magnetorheological fluid and transmission performance. Adv. Mater. Sci. Eng. **2015**, 583076 (2015). https://doi.org/10.1155/2015/583076
98. Kwok, N.M., Ha, Q.P., Nguyen, M.T., Li, J., Samali, B.: Bouc–Wen model parameter identification for a MR fluid damper using computationally efficient GA. ISA Trans. **46**(2), 167–179 (2007). https://doi.org/10.1016/j.isatra.2006.08.005
99. Jung, H.-J., Spencer, B.F., Lee, I.-W.: Control of seismically excited cable-stayed bridge employing magnetorheological fluid dampers. J. Struct. Eng. **129**, 7 (2003). https://doi.org/10.1061/(ASCE)0733-9445
100. Ryoji, I., Dodbiba, G., Fujita, T.: MR fluid of liquid gallium dispersing magnetic particles. Int. J. Mod. Phys. B. **19**(07–09), 1430–1436 (2005). Structures and Properties). https://doi.org/10.1142/S0217979205030402
101. Guo, H.T., Liao, W.H.: A novel multifunctional rotary actuator with magnetorheological fluid. Smart Mater. Struct. **21**, 6. https://doi.org/10.1088/0964-1726/21/6/065012
102. Li, Z.C., Wang, J.: A gun recoil system employing a magnetorheological fluid damper. Smart Mater. Struct. **21**, 10. https://doi.org/10.1088/0964-1726/21/10/105003
103. Kumar, J.S., Paul, P.S., Raghunathan, G., et al.: A review of challenges and solutions in the preparation and use of magnetorheological fluids. Int. J. Mech. Mater. Eng. **14**, 13 (2019). https://doi.org/10.1186/s40712-019-0109-2
104. Park, J., Yoon, G.-H., Kang, J.-W., Choi, S.-B.: Design and control of a prosthetic leg for above-knee amputees operated in semi-active and active modes. Smart Mater. Struct. **25**, 8. https://doi.org/10.1088/0964-1726/25/8/085009
105. Senkal, D., Gurocak, H.: Haptic joystick with hybrid actuator using air muscles and spherical MR-brake. Mechatronics. **21**(6), 951–960 (2011). https://doi.org/10.1016/j.mechatronics.2011.03.001
106. El Wahed Ali, K., Chen, W.H.: Performance evaluation of a magnetorheological fluid damper using numerical and theoretical methods with experimental validation. Front. Mater. **6**, 27 (2019). https://doi.org/10.3389/fmats.2019.00027
107. Li, D.D., Keogh, D.F., Huang, K., Chan, Q.N., Yuen, A.C.Y., Menictas, C., Timchenko, V., Yeoh, G.H.: Modeling the response of magnetorheological fluid dampers under seismic conditions. Appl. Sci. **9**, 4189 (2019). https://doi.org/10.3390/app9194189
108. Chrzan, M.J., Carlson, J.D.: MR fluid sponge devices and their use in vibration control of washing machines, Proc. SPIE 4331. In: Smart Structures and Materials 2001: Damping and Isolation. SPIE, Bellingham, WA (2001). https://doi.org/10.1117/12.432719
109. https://en.wikipedia.org/wiki/Magnetorheological_fluid. Accessed 29 July 2019
110. Zhang, L., Guo, S., Yu, H., et al.: Performance evaluation of a strain-gauge force sensor for a haptic robot-assisted catheter operating system. Microsyst. Technol. **23**, 5041–5050 (2017). https://doi.org/10.1007/s00542-017-3380-2
111. https://digitalcommons.linfield.edu/cgi/viewcontent.cgi?article=1092&context=studsymp_sci. Accessed 29 July 2019
112. https://www.seminarsonly.com/mech%20&%20auto/magneto-rheological-damper.php. Accessed 29 July 2019
113. Lahaye, T., Koch, T., Fröhlich, B., et al.: Strong dipolar effects in a quantum ferrofluid. Nature. **448**, 672–675 (2007). https://doi.org/10.1038/nature06036
114. https://www.reed-sensor.com/glossary/ferrofluid/. Accessed 29 July 2019
115. https://www.sfxc.co.uk/products/ferrofluid-magnetic-fluid. Accessed 29 July 2019
116. https://en.formulasearchengine.com/wiki/Ferrofluid. Accessed 29 July 2019
117. Assadsangabi, B., Tee, M.H., Takahata, K.: Electromagnetic microactuator realized by ferrofluid-assisted levitation mechanism. J. Microelectromech. Syst. **23**(5), 1112–1120 (2014). https://doi.org/10.1109/JMEMS.2014.2305112
118. Assadsangabi, B., Tee, M.H., Takahata, K.: Ferrofluid-assisted levitation mechanism for micromotor applications. In: 2013 Transducers & Eurosensors XXVII: The 17th International Conference on Solid-State Sensors, Actuators and Microsystems (Transducers & Eurosensors xxvii), pp. 2720–2723. IEEE, Barcelona (2013). https://doi.org/10.1109/Transducers.2013.6627368
119. Banerjee, S., Widom, M.: Shapes and textures of ferromagnetic liquid droplets. Braz. J. Phys. **31**(3), 360–365 (2001). https://doi.org/10.1590/S0103-97332001000300005
120. Renardy, Y., Sun, S.M.: Stability of a layer of viscous magnetic fluid flow down an inclined plane. Phys. Fluids. **6**, 10. https://doi.org/10.1063/1.868056
121. Khairul, M.A., Doroodchi, E., Azizian, R., Moghtaderi, B.: Experimental study on fundamental mechanisms of Ferro-fluidics for an electromagnetic energy harvester. Ind. Eng. Chem. Res. **55**(48), 12491–12501 (2016). https://doi.org/10.1021/acs.iecr.6b03161
122. Bocanegra-Diaz, A., Mohallem, N.D.S., Novak, M.A., Sinisterra, R.D.: Preparation of ferrofluid from cyclodextrin and magnetite. J. Magnet. Magn. Mater. **272–276**, 2395–2397 (2004). https://doi.org/10.1016/j.jmmm.2003.12.975
123. https://en.uj.edu.pl/en_GB/news/-/journal_content/56_INSTANCE_SxA5QO0R5BDs/81541894/143116349. Accessed 30 July 2019

124. https://www.imeche.org/news/news-article/bizarre-magnetic-ferrofluids-could-propel-cheaper-more-efficient-satellites. Accessed 30 July 2019
125. Mark R., JollyAndrew D.: Meyers, helicopter vibration control system and rotating assembly rotary forces generators for canceling vibrations. Patent no: US8435002B2 (2009)
126. Kuldip R.: Ferrofluid rotary-shaft seal apparatus and method. Patent no: US4357024A (1980)
127. Jibin, Z., Lu, Y.: Numerical calculations for ferrofluid seals. IEEE Trans. Magn. **28**(6), 3367–3371 (1992). https://doi.org/10.1109/20.179812
128. https://www.wikiwand.com/en/Ferrofluid
129. Casula, M.F., Floris, P., Innocenti, C., Lascialfari, A., Marinone, M., Corti, M., Sperling, R.A., Parak, W.J., Sangregorio, C.: Magnetic resonance imaging contrast agents based on iron oxide superparamagnetic ferrofluids. Chem. Mater. **22**(5), 1739–1748 (2010). https://doi.org/10.1021/cm9031557
130. Lübbe, A.S., Alexiou, C., Bergemann, C.: Clinical applications of magnetic drug targeting. J. Surg. Res. **95**(2), 200–206 (2001). https://doi.org/10.1006/jsre.2000.6030
131. John, P., Laskar Junaid, M.: Optical properties and applications of ferrofluids—a review. J. Nanofluids. **1**(18), 3–20 (2012). https://doi.org/10.1166/jon.2012.1002
132. Zhu, T., Cheng, R., Lee, S.A., et al.: Continuous-flow ferrohydrodynamic sorting of particles and cells in microfluidic devices. Microfluid. Nanofluid. **13**, 645–654 (2012). https://doi.org/10.1007/s10404-012-1004-9
133. https://www.magcraft.com/blog/what-is-a-ferrofluid. Accessed 30 July 2019
134. J. M. Ginder, S. M. Clark, W. F. Schlotter And M. E. Nichols, Magnetostrictive phenomena in magnetorheological elastomers, , Int. J. Mod. Phys. B 16, 17–18, 2412–2418 (2002), doi: https://doi.org/10.1142/S021797920201244X
135. Ginder, J.M., Nichols, M.E., Elie, L.D., Tardiff, J.L.: Magnetorheological elastomers: properties and applications. In: Proc. SPIE 3675, Smart Structures and Materials Smart Materials Technologies. SPIE, Bellingham, WA (1999). https://doi.org/10.1117/12.352787
136. Boczkowska, A., Awietjan, S.: Microstructure and Properties of Magnetorheological Elastomers. IntechOpen, London (2012). https://doi.org/10.5772/50430
137. Liu, T., Xu, Y.: Magnetorheological Elastomers: Materials and Applications. IntechOpen, London (2019). https://doi.org/10.5772/intechopen.85083
138. Chen, L., Gong, X.L., Li, W.H.: Effect of carbon black on the mechanical performances of magnetorheological elastomers. Polym. Test. **27**(3), 340–345 (2008). https://doi.org/10.1016/j.polymertesting.2007.12.003
139. Hu, T., Xuan, S., Ding, L., Gong, X.: Stretchable and magneto-sensitive strain sensor based on silver nanowire-polyurethane sponge enhanced magnetorheological elastomer. Mater. Des. **156**, 528–537 (2018). https://doi.org/10.1016/j.matdes.2018.07.024
140. Zhang, W., Gong, X.-l., Li, J.-f., Zhu, H., Jiang, W.-q.: Radiation vulcanization of magnetorheological elastomers based on silicone rubber. Chin. J. Chem. Phys. **22**, 5. https://doi.org/10.1088/1674-0068/22/05/535-540
141. Chokkalingam, R., Senthur, R., Mahendran, M.: Magnetomechanical behavior of Fe/PU magnetorheological elastomers. J. Compos. Mater. **45**(15), 1545–1552 (2011). https://doi.org/10.1177/0021998310383733
142. Xin, F.-L., Bai, X.-X., Qian, L.J.: Principle, modeling, and control of a magnetorheological elastomer dynamic vibration absorber for powertrain mount systems of automobiles. J. Intell. Mater. Syst. Struct. **28**(16), 2239–2254 (2017). https://doi.org/10.1177/1045389X16672731
143. Deng, H.X., Gong, X.L.: Adaptive tuned vibration absorber based on magnetorheological elastomer. J. Intell. Mater. Syst. Struct. **18**(12), 1205–1210 (2007). https://doi.org/10.1177/1045389X07083128
144. Xiaoyu, G., Yu, Y., Li, J., Li, Y.: Semi-active control of magnetorheological elastomer base isolation system utilising learning-based inverse model. J. Sound Vib. **406**, 346–362 (2017). https://doi.org/10.1016/j.jsv.2017.06.023
145. Ubaidillah, et al.: Potential applications of magnetorheological elastomers. Appl. Mech. Mater. **663**., Trans Tech Publications, Ltd., 695–699 (2014). https://doi.org/10.4028/www.scientific.net/amm.663.695
146. Abe, R., Takemura, K., Edamura, K., Yokota, S.: Concept of a micro finger using electro-conjugate fluid and fabrication of a large model prototype. Sensors Actuators A Phys. **136**(2), 629–637 (2007). https://doi.org/10.1016/j.sna.2006.10.046
147. Takemura, K., Yokota, S., Edamura, K.: A Micro Artificial Muscle Actuator Using Electro-Conjugate Fluid. In: Proceedings of the 2005 IEEE International Conference on Robotics and Automation, pp. 532–537. IEEE, Barcelona, Spain (2005). https://doi.org/10.1109/ROBOT.2005.1570173
148. Min, K.P., Kim, G.-W.: Photo-rheological fluid-based colorimetric ultraviolet light intensity sensor. Sensors. **19**, 1128 (2019)
149. Wang, Y., Shi, H., Fang, B., Zakin, J., Yu, B.: Heat transfer enhancement for drag-reducing surfactant fluid using photo-rheological counterion. Exp. Heat Transf. A J. Therm. Ener. Gen. Trans. Storage Conv. **25**(3), 139–150 (2012). https://doi.org/10.1080/08916152.2011.582569
150. Cho, M.-Y., Kim, J.-S., Choi, S.-B., Kim, G.-W.: Non-contact tunable damping of a cantilever beam structure embedded with photo-rheological fluids. Smart Mater. Struct. **25**, 2. https://doi.org/10.1088/0964-1726/25/2/025022
151. Smet, P.F., Moreels, I., Hens, Z., Poelman, D.: Luminescence in sulfides: a rich history and a bright future. Materials. **3**(4), 2834–2883 (2010). https://doi.org/10.3390/ma3042834

Superalloys

7

Abstract

Superalloys are a group of high-temperature advanced material. Many high-end machines and engines rely on superalloys to increase efficiency. These materials are used in high-speed missile propulsion engines to high-temperature chemical industry. This chapter provides a simple historical idea of a gradual increase in the operating temperature of a material with respect to their gradual development in processing routes. The increase in mechanical strength with temperature has been discussed for various superalloys such as Ni-based superalloys, Co-based superalloys, and Fe-based superalloys. Various methods for producing single crystal and polycrystalline superalloys have been described in this chapter. Various applications have been discussed in which superalloys are the only materials involved in improving efficiency.

Keywords

Superalloy · Ni-based superalloys · Co-based superalloys · Fe-based superalloys · Solid solution strengthening · Precipitation strengthening · Oxide dispersion strengthening · Grain boundary strengthening · Anti-phase boundary strengthening · Single-crystal superalloys · Superalloy coating · Bond coats

7.1 Superalloy

Superalloys are high-performance alloys, which have gained demand for its high strength and durability. Superalloys helped us in conquering the world from the space to the deep of ocean and earth. These alloys are mostly used in high-temperature equipment like an aircraft engine, jet engine, marine turbine engines, gas turbines engines, coal conversion equipment, high-temperature chemical processing industries, etc. (Fig. 7.1) [1–3]. Superalloy can be defined as a heat-resisting alloy capable of good mechanical strength and resistance to surface degradation at temperatures above 650 °C (1200 °F) [4, 5]. A designed alloy can be considered as a superalloy if its operating temperature can be more than 0.7 times of its absolute melting temperature (0.7 T_m) [6]. Figure 7.2 shows different parts of aeroengine using various superalloys. Generally, Ni-based superalloy has a service temperature of 0.9 times its absolute melting temperature (0.9 T_m) [7]. For high-temperature materials, oxidation and creep resistance are the major design criteria. Ni, Co, or Fe, which offer good resistance to thermal creep deformation, corrosion, or oxidation with good surface stability are the base alloying elements of superalloy [8]. Typically, superalloys have austenitic face-centered cubic crystal structure matrix. Over two decades, the thrust of jet engines and their fuel consumption have improved by more than 65% and 18–25%, respectively. These improvements are possible because of the advances in the high-temperature properties of superalloys. Ni-superalloys have very long life at service temperatures of 800–1000 °C, which makes them the most desired material for the hottest parts of gas turbine engines [9]. Figure 7.3 showing temperature with pressure profiles along the engine of a jet. The most common applications of superalloys in engine components are high-pressure turbine blades, discs and combustion chamber, thrust reversers, and afterburners [10, 11]. High cost is the primary concern along with the increased demand of superalloy. Therefore, the evolution of superalloys concern to the lowering of production cost and the invention of new production techniques.

7.2 History of Superalloys

The first successful flight was made by Wright brothers on December 17th, 1903 [12]. The race to create the most efficient flight machines encouraged by both commercial and defense demand began and eventually led to the development of the jet engine. High-temperature resistant materials with a balance of good strength and toughness are

A. Behera, *Advanced Materials*, https://doi.org/10.1007/978-3-030-80359-9_7

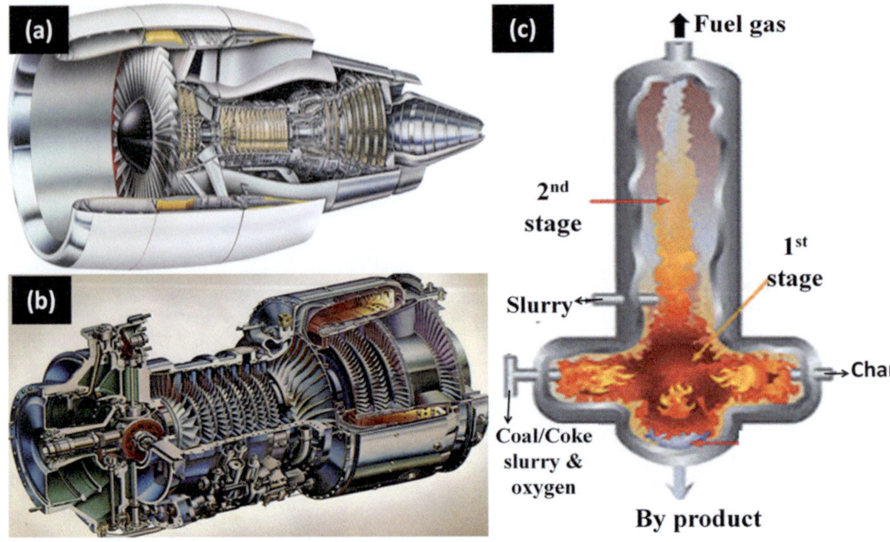

Fig. 7.1 (**a**) Jet engine, (**b**) marine turbine engines, and (**c**) coal conversion equipments

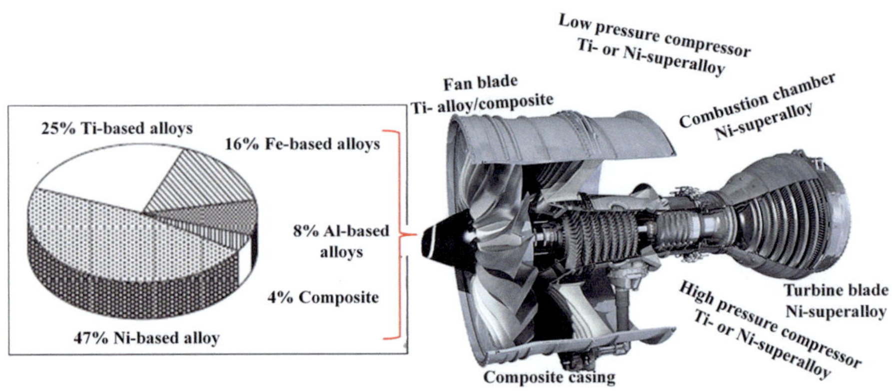

Fig. 7.2 Use of various superalloys in aeroengine (minimum percentage of materials usage in any engine is given here)

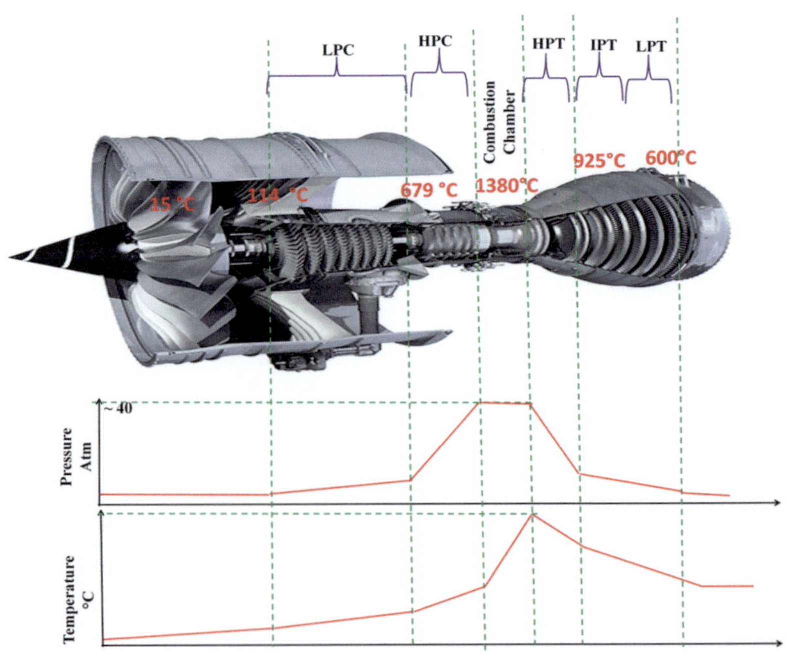

Fig. 7.3 A typical jet engine, showing the different stages: IPC (intermediate pressure compressor), HPC (high-pressure compressor), HPT (high-pressure turbine), IPT (intermediate pressure turbine), LPT (low-pressure turbine), and the pressure and temperature profiles along the engine

needed for jet engine construction. In the mid-1940s, the term "superalloy" was first used to describe high-temperature alloys [13]. At the same time, vacuum melting was invented. Its motive was to improve the reliability of the steels of that era. The origin of the name was that "superalloys" of a stainless variety led to improved iron-based alloys, whose name became superalloys with the hyphen dropped. After World War II, a group of these alloys developed for use in turbosuperchargers and aircraft turbine engines that required high performance at elevated temperatures [14]. As in film industries, the term "superhero" is used because of their patronizing activity; same as in case of metal, the patronizing property that is high-temperature properties obtained which named the "superalloy." The applications of superalloys have expanded to several other areas, which include aircraft and land-based gas turbines, rocket engines, chemical, and petroleum plants. Their ability to retain their strength after prolonged exposure to the working environment above 650 °C makes them well suited for these applications [15]. In the 1970s, directional solidification casting was invented. The advancement in that method gives rise to the single crystal (SC) superalloys. The process of production of single-crystal castings is similar to directional solidification by selecting a single grain. Application of gas turbine engines in aeroplanes drives the demand by advancing and expanding of the superalloy technology [16]. Table 7.1 gives a list of patents representing the gradual development of superalloy. This timescale shows the evolution of alloy composition as per different applications.

7.3 Basic Metallurgy of Superalloys

Above 540 °C service temperature, ordinary steels and titanium alloys are no longer strong enough for structural applications. In the case of steels, there is a possibility of enhanced corrosion attack. Where highest service temperatures (below the melting temperatures, about 1200 °C to 1400 °C for most alloys) are desired along with strength, Ni-base superalloys are the most suitable materials. Nickel-base superalloys can be used for service temperatures of a higher fraction of their melting points compared to other commercially available materials. Though refractory metals have higher melting points than superalloys, they do not have desirable properties such as superalloys. It reduced the application range of such materials. Strength of superalloy is directly related to the chemistry of the alloys and heat treatment techniques followed by their melting, casting, forging, and working processes. Though Fe and Co are not in FCC at normal room temperature, generally they have face-centered cubic (FCC-austenitic) crystal structure when iron, nickel, and cobalt are the basis for superalloys. Both Fe and Co

Table 7.1 Historical development from high-temperature alloys to the term "superalloy"

Year	Gradual development of materials according to the paper and patent
1905	W. von Bolton consumable electrode arc melted tantalum in a cooled copper crucible under a low pressure of argon
1906	Ni–Cr binary alloy for electrical apparatus
1907	Co–Cr binary alloy for cutting tools used to machining automotive engine components
1907	Ni–Cr binary alloy for cutting tools used to machining automotive engine components
1910–1915	Austenitic (γ-phase) stainless steels were developed for the high temperatures in gas turbines
1917	Fe (47%), Cr (23%), Ni (30%, Co may be substituted for Ni alloy for heating elements
1917	W. Rohn first melted nickel alloy in a vacuum resistance heated furnace
1924	Fe (balance), Ni (3.75–4.75%), Al (5.75–6.25%), Si (1.75–2.25%), C (2.4–2.8%) alloy for internal combustion engine parts, and valve head
1924	Fe (balance), Cr (10–15%), Ni (25–40%), Co (<10%), W (2–5%), Nb (1–3%), Ti (0.1–0.2%), Mn (0.5–1.0%), B (0.2–1.0%), C (0.3–1.0%) alloy for blades for steam turbines
1924	Fe (balance), Cr (5–9%), Ni (5%), W (0.25–1%), Si (1–5%), C (0.05–0.6%) alloy for turbine blades and electrical heating elements
1925	Fe (balance), B (0.75–4%) alloy for pistons, piston rings, and valves for internal combustion engines
1925	Fe (balance), Cr (2.8–7%), Co (0.2–5%), W (0.2–7%), Si (<3%), C (0.2–5%) alloy for forgeable engine valves, motor parts and such
1926	Two VIM furnaces in operation melting 80Ni-20Cr and 65Ni-15Fe-20Cr for thermocouples and denture alloys
1929	80Ni-20Cr alloy was the norm, with small additions of Ti and Al.
1930	Fe (balance), Ni (33–48%) alloy for internal combustion engine valves
1931	Fe (balance), Cr (1–20%), Al (1.5–4.5%), Si (<4%) alloy for engine valves
1950	Dr. Mohling melts first large heat, ten tons, vacuum induction melt of aluminum and titanium containing strengthened superalloy at Allegheny Ludlum Steel laboratory in Watervliet, NY.
1952	Special Metals Co., New Hartford, N.Y. produces the first production heat of Waspaloy in a 6-lb furnace for Pratt & Whitney J48 turbine engine blades
1960	Allvac Metals Co., Monroe, N.C. exclusively produces double vacuum melt (VIM/VAR) superalloys
1962	World's largest vacuum induction furnace (12,000 lb) installed by Allvac Metals Co. in anticipation of market acceptance and growth of vacuum melted superalloys

transform and become FCC at high temperatures or due to the presence of alloying elements. On the other hand, Ni is FCC at all temperatures and a preferable element for FCC stabilizer [17]. In Fe- and Co-based superalloys, the formed FCC is generally stabilized by the alloying elements. Notably, the addition of Ni provides the best properties. Superalloys are strengthened because of the basic nature of their FCC matrix (high Schmid's factor with low slip system) and its chemistry and also by the presence of special strengthening phases (usually precipitates) [18]. The phases of superalloys at high temperatures have a tendency to transform into stable lower temperature phases. The austenitic FCC matrices of superalloys have extended solubility for some alloying additions, excellent ductility, and favorable characteristics for precipitation of uniquely effective strengthening phases [19]. The density of pure Fe is of 7.87 g/cm^3 and pure Ni and Co is of about 8.9 g/cm^3. Densities of FeNi-base superalloys are about 7.9–8.3 g/cm^3, Co-based superalloys are about 8.3–9.4 g/cm^3, and Ni-based superalloys are about 7.8–8.9 g/cm^3. Alloying additions influence superalloy densities (Al, Ti, and Cr reduce density, whereas W, Re, and Ta increase density). The corrosion resistance of superalloys depends on the alloying elements (like Cr and Al) along with the operating environment. Cr and Al protect the surface by forming oxide layers that envelope the superalloy, and further oxidation can be stopped. Cr, Fe, Co, Mo, and Re help in the separation to the γ matrix, while Al, Ti, Nb, Ta, and V help in the separation to the γ' precipitates [20]. B and Y improve the adhesive force of the oxide layer to the substrate. B and Zr help in grain boundary strengthening (segregation of grain boundaries reduces the energy of grain boundary and results in better cohesion and ductility of grain boundary). The C and carbide former (e.g., Cr, Mo, W, Nb, Ta, Ti, or Hf) also improve grain boundary strengthening. These can form precipitates of carbides at grain boundaries, which results in reduction of grain boundary sliding. The melting temperature of the pure Ni is 1453 °C, pure Co is 1495 °C, and pre Fe is 1537 °C [21]. Melting temperatures of Ni-base single-crystal superalloys are higher compared to Co-based superalloys. Coating technology can be considered as an integral part of the development and application of superalloys. Table 7.2 gives the information on the categorization in gradual improvement in superalloy generation.

7.4 Strengthening Mechanisms of Superalloys

Generally, the wrought variety has somewhat less volume fraction of intermetallic phases than that of the cast variety. The fundamental point of the design of these alloys is to ensure that the compatibility of matrix solid solution phase and the dispersed strengthening phase in it. Strengthening is achieved by solid solution and hard particle dispersion and the microstructural stability at elevated operating temperature arises due to the very close matching of the lattices of the matrix and precipitate phases. As high-tech technologies demand materials for use at still higher temperatures, the first choice falls on intermetallic compounds, which are characterized by stronger bonding and consequent higher melting temperatures. The main limitation of intermetallics is their low ductility at lower temperatures. The choice of intermetallic structures having multiple slip systems and being amenable to plastic accommodation at grain boundaries, essential to impart ductility in polycrystalline materials, has led to the development of only a few structural intermetallics with limited range of applications. Meeting contradictory requirements such as low-temperature ductility, high- temperature strength, creep, and oxidation resistance has remained elusive in single-phase materials [25]. There has been some success in two-phase or multiphase materials and composites to achieve such combination of properties. The $\gamma + \gamma'$ structure in Ni(Ti/Al) system is the best example to illustrate this point. The product intermetallic phases in superalloy are coherent with each other and also resistant to coarsening of the dispersed phase at high operating temperatures [26]. All the major and minor strengthening mechanisms observed in superalloys are discussed subsequently.

7.4.1 Solid Solution Strengthening

Solid solution strengthening is a type of strengthening in which atoms of an alloying element are added to the crystalline lattice of the base metal that form a solid solution. Solid solution strengthening of the γ matrix in single crystal Ni-base superalloys helps in improving the creep strength at high temperatures [27]. Therefore, hardening elements of strong partition ability inside solid solution of the matrix are beneficial for creep strength at elevated temperature. Ti andW can cause more reliable partitioning of the solid solution of γ phase results in improved creep resistance. The requirements for a useful solid-solution strengthening are [28]:

(a) Additives should form a large portion of solid solution in the matrix.
(b) They should have the highest possible dissimilarity in atomic size compared to the matrix.
(c) They should have high melting points. Diffusion and dislocation cross-slip at higher temperature is essential in maintaining the strength of an superalloy, which can be achieved by the addition of Mo and W. They slowly diffuse at higher temperature and decrease the stacking fault energy between partial dislocations, such as Co.

Table 7.2 Various element addition for further improvement of the commercial properties in the superalloys [22–24]

Significant generations	Additive elements to superalloy
First-generation superalloys	• These alloys were derived from wrought superalloys to take advantage of vacuum melting and casting introduced in the early 1950s. • Increased Al, Ti, Ta, Nb content in order to increase the γ' volume fraction in these alloys. The volume fraction of the γ' precipitates increased to about 50–70% with the advent of SC, or monocrystal, solidification techniques for superalloys that enable grain boundaries to be entirely eliminated from a casting. Because the material contained no grain boundaries, carbides were unnecessary as grain boundary strengthers and were thus eliminated. • Improved castability through additions of Ta, increased solid solution strengtheners by adding W and Mo to impart higher temperature capability, and addition of Hf to improve stress rupture ductility and minimize porosity via increased amounts of low melting point eutectic. Subsequently, elimination of transverse grain boundaries by direct solidification (DS) of blades and vanes led to a further improvement in temperature capability. Although transverse grain boundary elimination was the original motivation for DS, it was the favored $\langle 100 \rangle$ growth directions and improved heat treatment characteristics from which most benefits occurred. Alignment of a $\langle 100 \rangle$ direction along the blade span not only provided intrinsically high creep resistance but also greatly improved the thermal fatigue life, since this is a low Young's modulus direction. DS turbine blades and vanes began appearing in military and commercial aeroengines in the 1960s. A logical extension of eliminating transverse grain boundaries in DS technology was to remove grain boundaries altogether to produce SC blades and vanes. • V and Nb also incorporated in this generation.
Second-generation superalloys	• Introduction of Rhenium (Re) about 3 wt % was marked by increased additions of refractory alloying elements, while maintaining a γ' volume fraction above 60%. Re is known to partition mainly to the γ-matrix, to retard γ' coarsening, and to increase the γ'/γ misfit. It is argued that some of the enhanced resistance to creep comes from the promotion of γ' "rafting" by Re. It is also claimed that Re reduces the overall diffusion rate in Ni-based superalloys. • Re promotes the formation of brittle topological-closed-pack phases, which has led to the strategy of reducing Co, W, Mo, and particularly Cr. • The second-generation SC alloys provide about a 30 °C increase in metal operating temperature, together with adequate resistance to oxidation, thermal fatigue, and deleterious phase formation relative to the first-generation SC alloys.
Third-generation superalloys	• The third-generation superalloys are marked by higher additions of Re (6 wt%) due to its dual role of improving creep strength as well as environmental resistance in spite of a low Cr level. • The second-generation SC alloys provide about a 60 °C metal operating temperature improvement.
Fourth-generation superalloys	• Additions of Pt group metals, especially ruthenium (Ru), to third-generation superalloys to further improve the phase stability in the presence of higher refractory element contents, thereby improving the strength. Further, the higher negative γ/γ' misfit resulted in γ' rafting perpendicular to the applied stress, thereby acting as a barrier to deformation under high temperature and low stress conditions.
Fifth-generation superalloys	• Addition of higher Ru contents ($\geq$6 wt%). • Higher lattice misfit and further solid solution strengthening by Mo and Re as well as Ru, and improved phase stability by Ru can all contribute to superior creep resistance over the previous generations. • The lower Cr content of these two alloys has resulted in poorer oxidation resistance at elevated temperatures, owing to vaporization of Ru and Re oxides.
Sixth-generation superalloys	• Use of both Re and Ru. • The tensile properties of this generation superalloy are claimed that these properties are superior to those of all other generations of superalloys. • The effect of Ru on the promotion of TCP phases is not well-determined. Early reports determined that Ru decreased the supersaturation of Re in the matrix and thereby diminished the susceptibility to TCP phase formation.

7.4.2 Precipitation Strengthening

Precipitation strengthening/hardening is also known as age hardening or particle hardening. It incorporates heat treatment processes to increase the yield strength. In the case of superalloys, it provides excellent high-temperature strength. Precipitation strengthening depends on changes in solid solubility with temperature. They produce fine particles (impurity phase), which hinder the movement of dislocations. Increase in the temperature of Ni-based superalloys develops ordered $Ni_3(Al, Ti)$ precipitates (γ' phase). This precipitation helps in increasing the strength with increase in service temperature. The γ' phase is an ordered structure of $Ni_3(Ti, Al)$ and can have a variety of compositions depending on the Al and Ti percentage in the alloy [29].

7.4.3 Oxide Dispersion Strengthening

Oxide dispersion strengthening consists of a small amount of dispersed oxide particles within the matrix. They are very useful for applications like high-temperature turbine blades and heat exchanger tubing. The incoherency of the oxide particles within the lattice of the material hinders the movement of dislocations within the material and reduces creep. Because of the incoherency of oxide particles, dislocations can only move by climb [30].

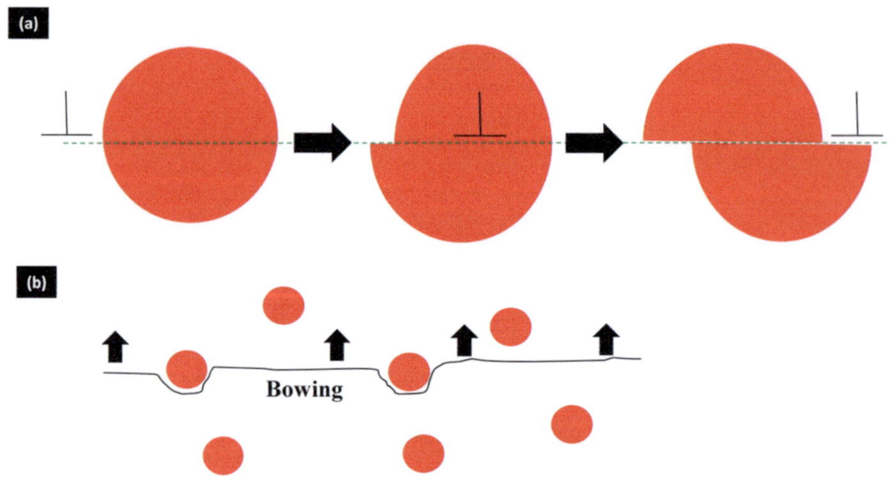

Fig. 7.4 (**a**) Antiphase boundary strengthening, (**b**) Orowan bowing around the precipitate

7.4.4 Grain Boundary Strengthening

Grain boundaries are invincible borders for dislocations inside a grain. The number of dislocations within a grain affects the stress buildup, which activates dislocation sources in the adjacent grains and enables deformations in the neighboring grain. In superalloys, stable carbide phases form by carbide former elements like C, Cr, Mo, W, C, Nb, Ta, Ti, and Hf. These phases are located at the grain boundaries and reduce sliding of grain boundaries at elevated temperature. B and Zr separate at the grain boundaries. As a result, grain boundary energy is minimized, and better creep strength and ductility can be achieved [31].

7.4.5 Antiphase Boundary Strengthening

Antiphase boundary strengthening is also known as order strengthening. Order strengthening brought about from the interaction of dislocations with ordered precipitates, forming anti-phase boundaries as dislocations move throughout the crystal, can lead to significant increases in strength and creep resistance (Fig. 7.4a). For this reason, order strengthening is often exploited for high-temperature creep-resistant superalloys used in turbine blades. Antiphase domains carry with them a surface energy penalty when compared to the perfect lattice due to their chemical disorder, and the presence of these boundaries impedes dislocation motion throughout the crystal, leading to increased strength under shear stress [32]. Figure 7.4 shows the process of an edge dislocation propagating through an ordered particle. As the dislocation moves throughout the particle, lattice planes are displaced from their equilibrium configuration. This forms a higher energy state than when compared to the equilibrium configuration, and the change in energy is called the antiphase boundary energy [33]. This can increase the degree of strengthening created from precipitation hardening, making it more difficult for cutting, and instead increasing the likelihood of Orowan bowing around the precipitate (Fig. 7.4b). Precipitates act as pinning points for dislocations. Bowing leads to unpinning leaving behind dislocation loops around the particles [34].

7.5 Types of Superalloys

Superalloys are generally classified on the basis of three major constituent elements with respect to commercial applications are Ni-based, Co-based, and Fe-based superalloys.

7.5.1 Ni-based Superalloys

Nickel-based superalloys are mostly strengthen by solid solution and precipitation strengthening. Alloying element compositions in Ni-based superalloys are 38–76% Ni, 10–20% Cr, up to 8% Al and Ti, 5–10% Co along with some minor alloying elements like Mo, W, Ta, Hf, and Nb [35]. Those elemental additions in these superalloys can be categorized as:

(a) γ formers (elements that tend to partition to the matrix)
(b) γ' formers (elements that partition to the γ' precipitate)
(c) carbide formers, and
(d) elements that segregate to the grain boundaries.

The γ formers are elements of groups V, VI, and VII (such as Co, Cr, Mo, W, and Fe). The diameters of atoms of these alloying elements vary from 3% to 13% of Ni atom. The γ' formers are the elements of groups III, IV, and V (such as Al, Ti, Nb, Ta, Hf). The diameters of atoms of these alloying elements vary from 6% to 18% of Ni atom [36]. Each element have some special characteristic in superalloys, which are given in Table 7.3.

Table 7.3 Contribution of various elements present in Ni-based superalloy

Elements	Composition range (wt %)	Properties
Ni	50–70%	• Stabilizes FCC matrix. • It forms base matrix (γ phase). • Helps to develop γ′ phase like $Ni_3(Ti/Al)$. • Inhibits formation of deleterious phases.
Co	30–60%	• Co has higher melting points than Ni. • They offer solid solution strengthening. • Raises solvus temperature of γ′ and lowers stacking fault energy (thereby making cross-slip of screw dislocations more difficult).
Fe	30–60%	• Fe and Co have higher melting points than Ni. They offer solid solution strengthening. • Fe is cheaper than Ni and Co, so can be replaced by both as per requirement.
Al	0.5–6%	• It is a γ′ former. • Also forms a protective oxide Al_2O_3, which provides oxidation resistance at higher temperature than Cr_2O_3.
Ti	1–4%	• It is a γ′ former.
Cr	5–20%	• Cr is used for oxidation and corrosion resistance. it forms a protective oxide Cr_2O_3. • Imparts sulfidation resistance as well as solid solution strengthening, and forms grain boundary carbides.
B	0–0.1%	• Provide strength to grain boundaries. • Improves stress rupture properties.
Zr	0–0.1%	• Provide strength to grain boundaries. • Improves stress rupture properties.
Nb	0–5%	• form γ″, a strengthening phase at lower (below 700 °C) temperature,
V	0–5%	• Form γ″.
Ta	0–0.1%	• Form solid solution strengthening and carbide formation. They are heavy, but have extremely high melting points.
C	0.05–0.2%	• Form MC and $M_{23}C_6$ (M = metal) carbides that are the strengthening phase in the absence of γ′.
W	0–1%	• Solid solution strengthening. • Form MC-type carbide.
Mo	1–10%	• Form solid solution strengthening and carbide formation. • They are heavy, but have extremely high melting points.
Re, W, Hf	1–10%	• Refractory metals, added in small amounts for solid solution strengthening (and carbide formation). • They are heavy, but have extremely high melting points.
La and Y	0–0.1%	• Improve oxidation resistance.

7.5.1.1 Phases of Ni-based Superalloys

Different phases of Ni-based superalloys are described subsequently.

Gamma (γ): The gamma phase, a continuous matrix (Fig. 7.5), is a disordered FCC crystal structure of Ni-based austenitic phase that usually contains a high percentage of Co, Cr, Mo, and W, which helps in the formation of the solid solution. This provides ductility and structure of the precipitates [37]. C, Nb, Fe, Ti, Al, V, and Ta are the other alloying elements found in most commercial Ni-based superalloys.

Gamma Prime (γ′): The primary strengthening of Ni-based superalloys is because of the ordered FCC structured Ni_3Al or/and Ni_3Ti intermetallics, also known as γ′ phase. It is an ordered $L1_2$ (pronounced L-one-two) means it has a particular atom on the face of the unit cell, and one specific atom on the corners of the unit cell (Fig. 7.6) [38]. For Ni-based superalloys, Ni atoms are on the faces, while Ti or Al atoms are on the corners. The close match in matrix/precipitate lattice parameter (~0–1%) combined with the chemical compatability allows the γ′ to precipitate homogeneously throughout the matrix and have long-time stability. Importantly the flow stress of the γ′ phase increases with increase in temperature up to 650 °C [39]. Al and Ti are the major alloying element and are added in mutual proportions so that precipitation occurs for a high volume percentage in the matrix. The volume fraction of the γ′ precipitate can be 70% for the high-performance superalloys. So many factors contribute to the hardening of superalloys due to γ′ phase are γ′ fault energy, γ′ strength, coherency strains, volume fraction of γ′, and γ′ particle size. Extremely small γ′ phase always precipitates as spheres. A given volume of a sphere has 1.24 times less surface area compared to a cube of the same volume, thus the precipitate shapes as a sphere to minimize surface energy. However, with coherent particles, the interfacial energy can be minimized and forms cubes to allow the crystalographic planes of the cubic matrix and precipitates to be continuous. Thus the morphology can change from spheres to cubes with the growth of γ′ (as shown in Fig. 7.7) or can be in the form of plates which depend on the matrix/precipitate lattice mismatch. For more significant mismatch, the critical particle size and morphological change

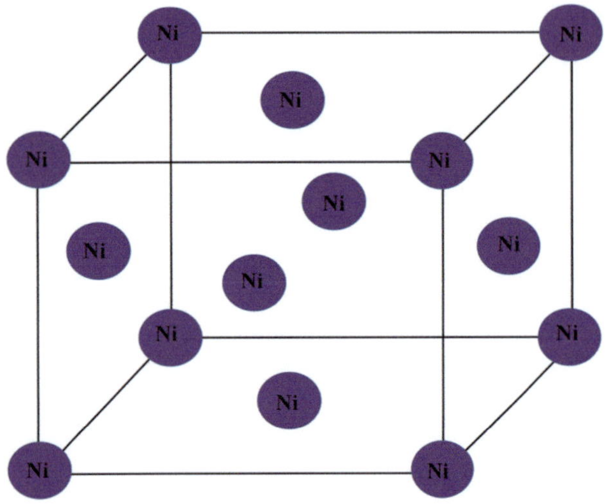

Fig. 7.5 Crystal structure for γ

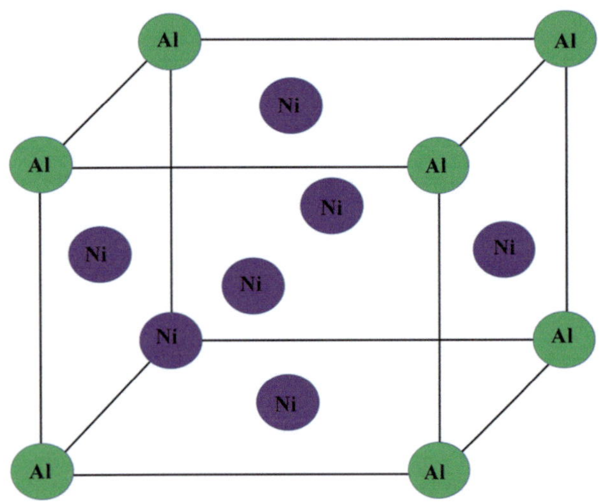

Fig. 7.6 Crystal structure for γ' shows the distribution of elements

of spheres to cubes (or plates) are reduced. Coherency can be lost by overaging. Directional coarsening (aspect ratio) and rounding of the cube edges are the indications of loss of coherency. Increase in directional coarsening for the increasing (positive or negative) mismatch can be expected [40].

However, as it can be seen from the ($\gamma + \gamma'$)/γ' phase boundary on the ternary sections of the Ni, Al, Ti phase diagram, the phase is not strictly stoichiometric. There may exist an excess of vacancies on one of the sublattices, which leads to deviations from stoichiometry; alternatively, some of the Ni atoms might occupy the Al sites and vice versa. In addition to Al and Ti, Nb, Hf, and Ta partition preferentially observed into γ' [41].

Furthermore, due to their similar lattice parameters, the γ' is coherent with the γ when the precipitate size is small. The γ' phase is an atomically ordered phase, so dislocations in the γ have difficulties in penetrating γ'. The ordering interferes with dislocation movement and strengthens the alloy [42].

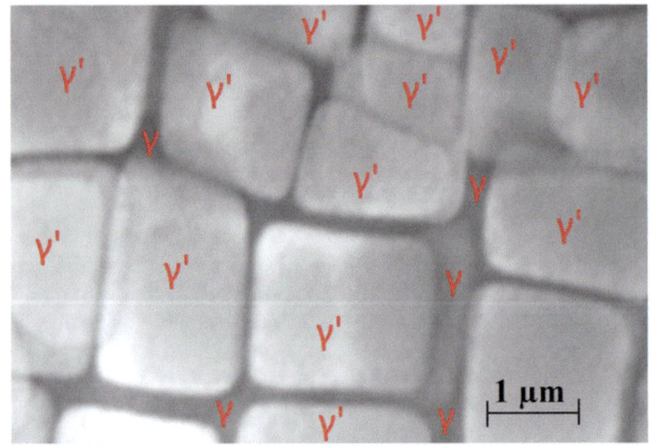

Fig. 7.7 Transmission electron micrograph showing a large fraction of cuboidal γ' particles in a γ matrix

The small misfit between the γ and γ' lattices is important for two reasons. First, when combined with the cube–cube orientation relationship, it ensures a low γ/γ' interfacial energy. The ordinary mechanism of precipitate coarsening is driven entirely by the minimization of total interfacial energy. A coherent or semi-coherent interface therefore makes the microstructure stable, a property that is useful for elevated temperature applications. The magnitude and sign of the misfit also influence the development of microstructure under the influence of a stress at elevated temperatures. The misfit is said to be positive when the γ' has a larger lattice parameter than γ. The misfit can be controlled by altering the chemical composition, particularly the aluminum to titanium ratio. A negative misfit stimulates the formation of rafts of γ', essentially layers of the phase in a direction normal to the applied stress. This can help reduce the creep rate if the mechanism involves the climb of dislocations across the precipitate rafts [43]. The graphical representation of the mechanical resistance of both the soft and hard phase of a typical superalloy is given in Fig. 7.8.

Gamma Double Prime (γ''): The γ'' phase is a composition of Ni_3Nb or Ni_3V and helps in strengthening Ni-based superalloys at lower temperatures (<650 °C). It is coherent with γ matrix phase, but at high temperatures, it dissolves in the matrix. This phase exists in the form of very small disks. Its structure is $D0_{22}$ (ordered BCT) (Fig. 7.9). This precipitate is coherent with γ'. It is the main strengthening phase in IN-718. These anisotropic discs form due to the lattice mismatch between the FCC matrix and the BCT precipitate. This lattice mismatch leads to high coherency strains which, together with order hardening, comprise the primary strengthening mechanisms. Because of its dissolving nature above 650 °C, it is very unstable for high-temperature applications [44].

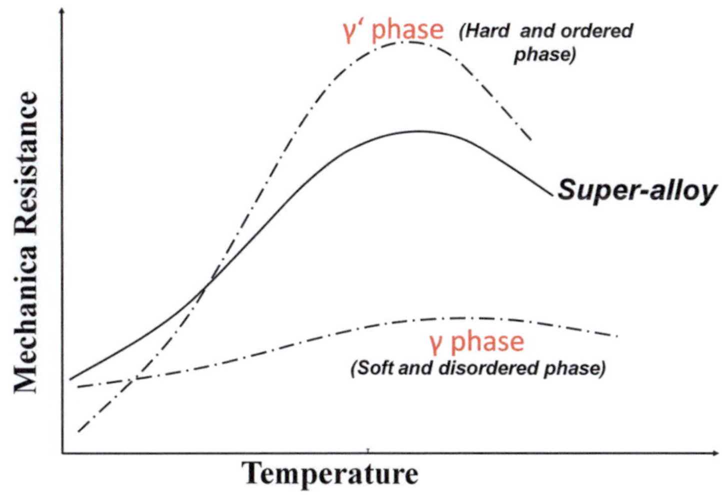

Fig. 7.8 Mechanical resistance of both γ matrix and γ′ hard phase constituting superalloy

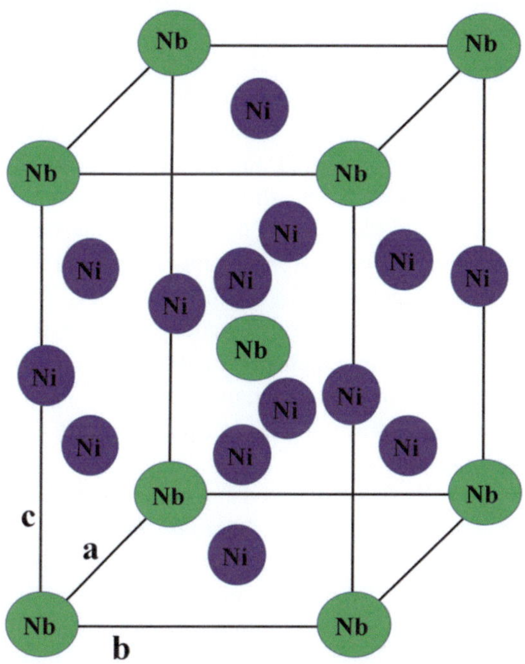

Fig. 7.9 Crystal structure for γ″ (Ni_3Nb) is body-centered tetragonal

Carbides: About 0.05–0.2% carbon is added to the alloys. It combines with highly reactive and refractory elements like Ti, Ta, and Hf and forms different carbides (e.g., TiC, TaC, or HfC). During the heat treatment process and the service, these carbides decompose and lower carbides like $M_{23}C_6$ and M_6C form. Generally, these carbides form on the grain boundaries (Fig. 7.10). All these carbides have FCC crystal structure. Their properties may vary upon their stability during the service. In multigrain superalloys, carbides are beneficial as they increase rupture strength at high service temperature. The dominating reactions for the formation of these mentioned carbides are given in Eqs. ((7.1), ((7.2), and ((7.3) [45]:

$$MC + \gamma \rightarrow \mathrm{M_6C} + \gamma' \tag{7.1}$$

$$MC + \gamma \rightarrow \mathrm{M_{23}C_6} + \gamma' \tag{7.2}$$

$$\mathrm{M_6}C + M' \rightarrow \mathrm{M_{23}C_6} + \mathrm{M''} \tag{7.3}$$

where M' and M'' can be referred as metals like Cr, Co, Ni, or Mo. Mostly carbides are precipitated at the grain boundaries of Ni superalloys while settled at intergranular sites of cobalt and iron superalloys. Generally, carbides formin solid-solution alloys due to a prolonged service exposures.

MC Carbide: MC carbides are a major source of carbon for subsequent phase reactions during heat treatment and service (usually forms in superalloys during freezing). MC usually exhibits a coarse, random, cubic, or script morphology, which tends to generate on the grain boundaries. ("M" may be Cr, Fe, W, and Mo) [46]. They are distributed heterogeneously through the alloy, both in intergranular and transgranular positions, often interdendritically. It reduces ductility. The preferred order of formation (in order of decreasing stability) in superalloys for these carbides is HfC, TaC, NbC, and TiC. In these carbides, M atoms can readily substitute for each other, as in (Ti, Nb)C [47]. However, the less reactive elements, principally Mo and W, can also substitute in these carbides. For example, (Ti, Mo)C is found. The change in stability order is due to the Mo or W substitution, which weakens the binding forces in MC carbides. This typically leads to the formation of the more stable compounds $M_{23}C_6$ and M_6C-type carbides in the alloys during processing or after heat treatment or service (Eqs. (7.1, (7.2, and (7.3). Alloys with high Nb and Ta contents contain MC carbides that do not break down easily during processing or solution treatment at temp ≈1200 °C [48].

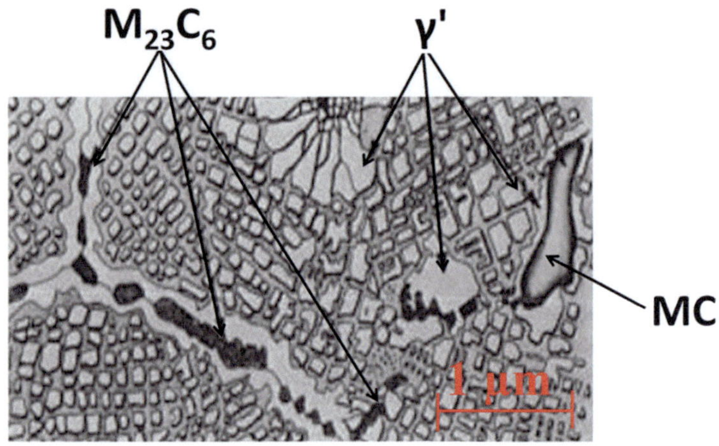

Fig. 7.10 Various carbide formation in Ni-based superalloys, which are present at grain boundary

$M_{23}C_6$ Carbide: $M_{23}C_6$is readily form in alloys with moderate to high Cr content. They form during lower temperature heat treatment and service (i.e., 760–980 °C) both from the degeneration of MC carbide and form soluble residual carbon in the alloy matrix. When W or Mo is present, the approximate composition of $M_{23}C_6$ is $Cr_{21}(Mo, W)_2C_6$, although it also has been shown that appreciable Ni can substitute in the carbide [49]. It is also possible for small amounts of Co or Fe to substitute for Cr. Rupture strength is improved by the presence of $M_{23}C_6$ particles, apparently through the inhibition of grain-boundary sliding. The carbide $M_{23}C_6$ is found primarily at grain boundaries and usually occurs as irregular, discontinuous, blocky particles, although plates and regular geometric forms have been observed. In some alloys, cellular structures of $M_{23}C_6$ have been observed. These can cause premature failures, but can be avoided by proper processing and/or heat treatment.

M_6C Carbide: Typical formulas for M_6C are (Ni, Co)$_3$Mo$_3$C and (Ni, Co)$_2$W$_4$C. M_6C carbides have a complex cubic structure. M_6C forms when Mo and/or W content is more than 6–8 at.%, typically in the range of 815–980 °C. M_6C carbides are formed when Mo or W acts to replace Cr in other carbides; unlike the more rigid $M_{23}C_6$, the compositions can vary widely. M_6C carbides are stable at higher temperature than $M_{23}C_6$ carbides. M_6C is more commercially important as a grain-boundary precipitate for controlling grain size during the processing of wrought alloys. The M_6C carbides also can precipitate in blocky form in grain boundaries and less often in a Widmanstatten intragranular morphology [50].

Oxide and Nitrides: Though all of the advanced Ni-based superalloys are vacuum-melted, the gaseous impurities O, and N cannot be completely removed. As a result, these gases usually combine with the more active elements present to form relatively stable second-phase particles. Oxide precipitates can be seen in processed superalloys, but they are extremely scarce, probably as a result of the efficient use of vacuum-melting and vacuum-casting techniques. Nitrogen is not removed as efficiently by use of vacuum-processing and nitrides are sometimes observed, usually as bulky TiN. At the stoichiometric composition, TiN is FCC [51].

Borides: The presence of B in Ni-based superalloys has been shown to increase stress-rupture properties. The optimum property occurs when small amounts of B and Zr are present. It is believed that B and Zr occupy positions at or near the grain boundaries, thus impeding the process of dislocation annihilation [83]. Borides, when present, appear as a blocky grain-boundary phase, thus supporting the postulate that B segregates to the grain boundaries. This could possibly be caused by the formation of the low-melting nickel–boron eutectic which has a melting point of approximately 1140 °C. For this reason, the B levels are intentionally kept below the solubility limits, thus restricting the formation of borides [52].

Topologically Close-Packed Phases (TCP): Adding new elements is usually good because of solid solution strengthening, but engineers need to be careful about which phases precipitate. Precipitates can be classified as geometrically close-packed (GCP), topologically close-packed (TCP), or carbides. GCP phases are usually good for mechanical properties, but TCP phases are often deleterious [53]. These are generally undesirable, brittle phases that can form during heat treatment or service. The cell structure of these phases has close-packed atoms in layers separated by relatively large interatomic distances. The layers of close packed atoms are displaced from one another by sandwiched larger atoms, developing a characteristic "topology." These compounds have been characterized as possessing a topologically close-packed structure. Conversely, γ' is close-packed in all directions and is called geometrically close-packed. **TCP** (Sigma, μ, Laves, etc.) usually forms as plate (which appears as needles on a single-plane microstructure) (Fig. 7.11) [54]. The plate-like structure negatively affects mechanical properties that are on ductility and creep rupture. Sigma appears to be the most deleterious, while strength retention has been observed in some alloys containing mu and Laves.

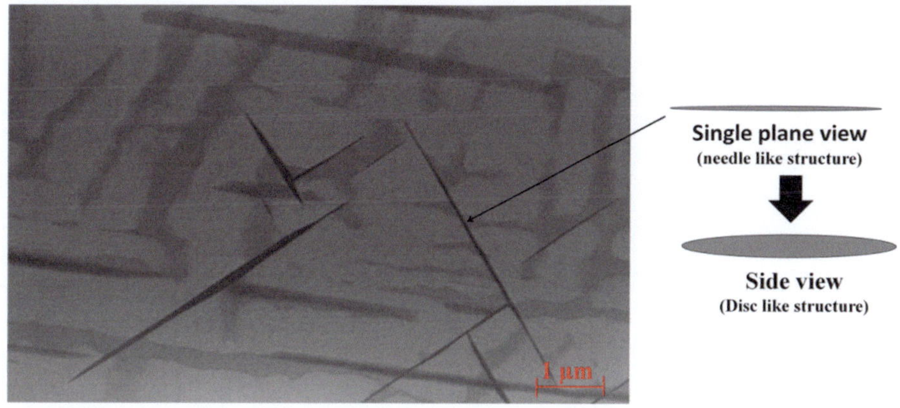

Fig. 7.11 Morphology of TCP phase

TCPs are potentially damaging for two reasons: they tie up γ and γ′ strengthening elements in a non-useful form, thus reducing creep strength, and they can act as crack initiators because of their brittle nature. Because TCP phases are not truly close packed, they have few slip systems and are very brittle. They are additionally bad because they "scavenge" elements away from GCP phases. Many elements that are good for forming γ′ or have great solid solution strengthening may precipitate TCP. Engineers need to find the balance that promotes GCPs while avoiding TCPs. The TCP phase is incoherent with the γ matrix. The TCP phase is surrounded by a "depletion zone" where there is no γ′. TCP phases frequently cause embrittlement, and although such effects can be minimized by adjusting other structural features (e.g., grain size) which influence sensitivity of properties [55].

Delta (δ) Phase is not close-packed. The structure is orthorhombic (Ni_3Nb), acicular (needle-like). The main issue with this phase is that it is not coherent with γ, but it is not inherently weak. It typically forms from decomposing γ″, but sometimes it is intentionally added in small amounts for grain boundary refinement [56]. **Phase η (Ni_3X)** that forms with high content of Ti, Ta, or Nb after long exposure to high temperatures. Ni_3Ti is a GCP having the structure DO_{24} (ordered HCP). The phase is not the worst, but it is not as good as γ′. It can be useful in controlling grain boundaries. When precipitates on the grain boundaries, it leads to a decrease in creep characteristics (time to fracture), whereas if precipitates inside the grains with Widmannstatten morphology, which decreases strength [57]. **Sigma (σ)phase** is a packed phase with tetragonal lattice and needle or block morphology. This brittle phase that can be chemically composed by FeCr, FeCrMo, CrFeMoNi, CrCo, or CrNiMo, is formed through a destabilization of austenitic γ matrix after a long-term exposure to high temperatures in the <540; 980 > °C range. Decreasing the amount of Ni in favor of elements, such as Fe, Cr, or Mo, which promotes the formation of this phase leads to a decrease in creep strength and decrease of ductility at normal temperatures [58]. Laves phases are electron compounds of elements in which the difference of atomic radii is 30%. It is possible to describe them by a general formula $(Fe,Cr,Mn,Si)_2(Mo,Ti,Nb)$ and they exhibit rhombohedral crystal structure and acicular morphology. They can precipitate in Ni superalloys containing Fe in a temperature range between 650 and 1100 °C either inside the grains or intercrystallically. They form coarse Widmanstatten platelets. Their precipitation occurs if the amount of Mo or W (or possibly their common volume) is >3% or there is an increased amount of Ti, Nb, Mo, and decreased amount of Ni. Due to the influence of these phases, decrease in ductility and yield strength occurs at normal temperature, the influence on creep characteristics is negligible [59]. Fe_2Nb, Co_2Ti, Fe_2Ti phases are considered as mu (μ) phase. These are the TCP phase having hexagonal structure. Morphology of mu phase is globules or platelets [60].

7.5.1.2 Properties of Ni-based Superalloys

Nickel-based superalloys used in jet engines have a high concentration of alloying elements (up to about 50% by weight) to provide strength, creep resistance, fatigue endurance, and corrosion resistance at high temperature. The types and concentration of alloying elements determine whether the superalloy is a solid solution-hardened or precipitation-hardened material. Precipitation-hardened superalloys are used in the hottest engine components, with their high-temperature strength and creep resistance improved by the presence of γ′ [Ni_3Al, Ni_3Ti, $Ni_3(Al,Ti)$] and other precipitates that have high thermal stability. Typical composition of various alloys is given in Table 7.4 [35, 61, 62]. Creep resistance is dependent, in part, on slowing the speed of dislocation motion within a crystal structure. In modern Ni-based superalloys, the γ′-$Ni_3(Al,Ti)$ phase present acts as a barrier to dislocation motion. For this reason, this γ′ intermetallic phase, when present in high volume fractions, drastically increases the strength of these alloys due to its ordered nature and high coherency with the γ matrix [63].

Ni-based superalloy products, such as Inconel 625, exhibit a tensile strength of 690 MPa and a yield strength of 275 MPa. Some Ni-based superalloys can withstand

Table 7.4 Nominal chemical compositions of some Ni-based superalloys (wt.%)

Common name	Ni	Cr	Co	Fe	Mo	Al	Ti	Si	Mn	C	W	Nb	Ta	B	Zr	Other
Alloy X	47.3	22.0	1.5	18.5	9.0			0.5	0.5	0.10	0.6					
Alloy 617	54.0	22.0	12.5		9.0	1.0				0.07						
Alloy 706	41.5	16.0		40.0		0.2	1.8	0.2	0.2	0.03		2.9				
Alloy 718	52.4	19.0		18.5	3.0	0.5	0.9	0.2	0.2	0.04		5.1				
Alloy 783	28.5	3.0	34.0	26.0		5.4	0.1					3.0				
Alloy 901	42.5	12.5		36.0	5.8		2.9									
Waspaloy	58.0	19.5	13.5		4.3	1.3	3.0			0.08				0.006	0.06	
UDIMET alloy 500	54.0	18.0	18.5		4.0	2.9	2.9			0.08				0.006	0.05	
UDIMET alloy 700	53.0	15.0	18.5		5.2	4.3	3.5			0.08				0.030		
NIMONIC alloy 75	73.0	18.0	2.0	5.0	0.3	0.3	0.2	1.0	1.0	0.08				0.001	0.05	
NIMONIC alloy 80A	76.0	19.5				1.4	2.4	0.3	0.3	0.06				0.003	0.06	
NIMONIC alloy 81		30.0	2.0	1.0	0.3	0.9	1.8	0.5	0.5	0.05				0.002	0.06	
NIMONIC alloy 90		18.0	15.0	1.0	0.3	1.1	2.0	1.5	1.0	0.05				0.001	0.04	
NIMONIC alloy 105		13.0	18.0	1.0	4.5	4.5	0.9	1.0	1.0	0.12				0.004	0.07	
NIMONIC alloy 115	60.0	14.2	13.2		3.3	4.9	3.8			0.15				0.160	0.04	
NIMONIC alloy 263	54	19.0	19.0	0.7	5.6	0.3	1.9	0.4	0.2	0.04				0.001	0.02	
Alloy B-1900	64.0	8.0	10.0		6.0	6.0	1.0			0.1			4.0	0.015	0.10	
INCONEL 600		15.0		8.0				0.2	0.5	0.08						
INCONEL 718		19.0	1.0	BAL	3.0	0.6	0.9	0.3	0.3	0.06				0.003		Nb + Ta= 5.25
INCONEL 738	61.0	16.0	8.5		1.7	3.4	3.4		0.5	0.17	2.6	0.9	1.7	0.010	0.10	
INCONEL 751		15.0		7.0		1.2	2.3	0.2		0.05						Nb = 0.95
INCONEL 939	48.0	22.0	19.0			1.9	3.7			0.15	2.0		1.4	0.009	0.09	
INCOLOY 904		0.1	14.0	Bal.	0.2		1.4	0.2	0.4	0.05						
MERL 76	54.5	12.5	18.5		3.2	5.0	4.4			0.02		1.4		0.020	0.06	0.4 Hf
René 41	55.0	19.0	11.0		10.0	1.5	3.1			0.09				0.005		
TD NiCr	78.0	20.0														2.0 ThO_2
Alloy MA754	78.0	20.0				0.3	0.5			0.05						0.6 Y_2O_3
PWA 1480	62.0	10.0	5.0				5.0	1.5			4.0		12.0			
DS Mar M 200 + Hf	60.5	8.0	9.0			5.0	1.9			0.13	12.0	1.0		0.015	0.03	2.0 Hf
Rene N4	62.0	9.8	7.5		7.5	4.2	3.5			0.06	6.0	0.5	4.8	0.004		0.15 Hf
CMSX-4	62.0	7.0	10.0		0.6	5.6	1.0				6.0		6.0			3.0 Re, 0.1 Hf
Waspaloy	58.0	19.5	13.5		4.3	1.3	3.0			0.08				0.006	0.06	

temperatures beyond 1200 °C, depending on the composition of the alloy. When exposed to oxidation, superalloys of this kind naturally create a protective oxide phase layer on the surface. As a result, the material is able to resist oxidative corrosion. This is the reason why some Ni superalloys are also used in industries where the equipment needs to be submerged in seawater. Various Ni-superalloys are Hastelloy, Inconel, Waspaloy, Rene alloys, Incoloy, Osprey® alloy [64]. The weldability of Ni-based superalloys is significantly better than that of other metals and alloys. This may be attributed to the material's creep resistance and high yield strength. For example, Osprey® Alloy-718 Ni-based superalloys have 600–980 MPa yield strength, 19%–71% elongation, and 29–108 J impact toughness [65]. The range is due to varying the fabrication procedure (from conventional vacuum melting to additive manufacturing technique) and varying the testing direction (vertical to horizontal testing). Osprey® Alloy 625 has attractive mechanical strength in combination with excellent corrosion resistance, performs well in ambient to elevated temperatures up to ~1000 °C [66].

Trace elements include residual gasses, such as N, O, and H, and nonmetallic impurities (S and P) can affect mechanical properties, alloy performance, castability, and weldability. The environment of a gas turbine blade contains oxygen and other elements, such as sulfur, that react readily with metals at operating temperatures. Hot corrosion is a severely accelerated form of environmental attack, which occurs by a combination of oxidation and sulfidation in contaminated operating environments [67]. The most common contaminants, which lead to hot corrosion, are Na and S. However, other metallic impurities, such as K, V, Pb, and Mo, can lead to accelerate attack. Two distinct hot corrosion mechanisms have been recognized: Type 1 hot corrosion occurs between 800 °C and 925 °C, while Type 2 hot corrosion occurs between 600 °C and 750 °C. In both cases, the protective oxide layer is melted by a fluxing reaction, allowing greatly accelerated oxidation and sulfidation to occur. Residual oxygen results in nonmetallic oxides (dross inclusions), which reduce mechanical properties and weldability and can initiate fatigue cracking. Residual nitrogen produces microporosity, alloy/crucible wetting, and formation of carbonitrides, which also can serve as fatigue initiation sites. N and O can affect DS casting columnar grain control and initiate grain defects in SX castings. Nonmetallic impurities, such as S and P, reduce grain boundary ductility resulting in DS grain boundary cracking and weld solidification cracking. Sulfur also increases alloy/crucible wetting and reduces oxidation resistance and coating/TBC adherence [68].

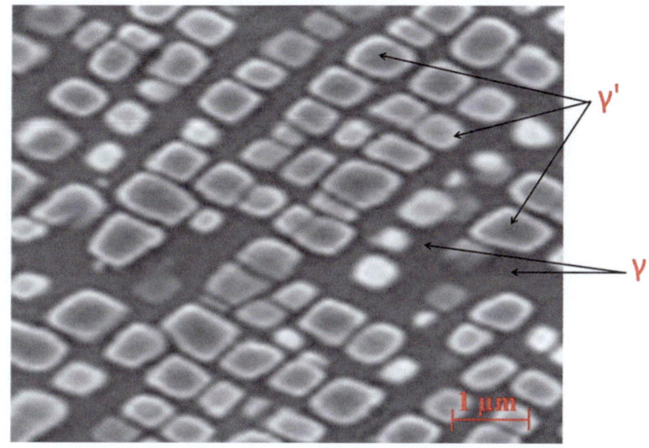

Fig. 7.12 Morphology of Co–Al–W superalloys showing γ and γ′ phases

7.5.2 Co-based Superalloys

The major composition of Co-based superalloys is Co, Cr, and Ni. The base combination of most cobalt alloys is **Co–Cr**, with Cr acting as a strengthening alloy. Adding W and/or Mo can affect strengthening even more [69]. Cobalt crystallizes in the HCP structure below 417 °C and at higher temperatures, it transforms to FCC. To avoid this transformation during service, Co-base alloys are alloyed with Ni in order to stabilize the FCC structure between room temperature and the melting point. Nickel is even more heat resistant than cobalt and is added to widen components' temperature resistance range. Cobalt has a higher melting point than Ni-based superalloys and has superior hot corrosion resistance, wear resistance, and thermal fatigue, whereas their strengthening mechanisms are inferior to γ′ precipitation strengthening. Co-superalloy contains 30–65% Co, 19–30% Cr, up to 35% Ni, and other minor elements that can provide good resistance against lead oxides, sulfur oxides, and other corrosive compounds in the combustion gas [70]. Co-superalloys are used in jet engine components that require excellent corrosion resistance. Co-based superalloys have depended on carbide precipitation and solid solution strengthening for mechanical properties. Carbide-strengthened Co-based superalloys are used in lower stress, higher temperature applications such as stationary vanes in gas turbines. Despite that, traditional Co-based superalloys have not found widespread usage because they have a lower strength at high temperature than Ni-based superalloys. A 2006 report on metastable γ′-$Co_3(Al,W)$ intermetallic compound with the $L1_2$ structure suggests Co-based alloys as alternative to traditional Ni-based superalloys. It was found that Mo, Ti, Nb, V, and Ta partition to the γ′ phase, while Fe, Mn, and Cr partition to the matrix γ in Co-based superalloys [71]. The two-phase microstructure consists of cuboidal γ′ precipitates embedded in a continuous γ matrix and is therefore morphologically identical to the microstructure observed in Ni-based superalloys (Fig. 7.12). The role of various alloying elements in Co-superalloys has been given in Table 7.5. The composition of cast and wrought Co-based superalloys has been given in Table 7.6.

High degree of coherency between the two phases results in the superior strength at high temperatures, in Ni superalloys. This provides a pathway for the development of a new class of load-bearing Co-based superalloys for application in severe environments. In these alloys, W is the crucial addition for forming the γ′ intermetallic compound; this makes them much denser (>9.6 g/cm^3) compared to Ni-based superalloys. Recently a new class of γ + γ′-cobalt-based superalloys has been developed that are W-free and have much lower density comparable to Ni superalloys. The tungsten-free γ′ phase has the composition $Co_3(Nb,V)$ and $Co_3(Ta,V)$. In addition to the fact that many of the properties of these new Co-based superalloys could be better than those of the more traditional Ni-based superalloys. Therefore, if the high-temperature strength could be improved, the development of novel Co-based superalloys could allow for an increase in jet engine operation temperature resulting in an increased efficiency [72, 73].

Oxygen presence can weaken the structure. Vacuum melting is therefore commonly used to give strict control over the elemental makeup of the superalloy. It is also critical that the specified compositions are adhered to, as excess soluble metals could form unwanted and deleterious phases [74].

7.5.2.1 Phases of Co-based Superalloys

Co superalloys possess various phases and intermetallics similar to Ni superalloys. **Gamma (γ) Phase** is the matrix of Co superalloy. Alloying elements found in Co-based alloys are C, Cr, W, Ni, Ti, Al, Ir, and Ta. Chromium is used in cobalt-based superalloys as it provides oxidation and corrosion resistance, critical for material use in gas turbines [75]. **Gamma prime (γ′)** phase constitutes the precipitate used to strengthen the alloy. In this case, it is usually close packed with an $L1_2$ structure of Co_3Ti or FCC Co_3Ta, though

Table 7.5 Role of various alloying elements in Co superalloys

Elements	Properties
Cr	• Improves oxidation and hot corrosion resistance. • Produce strengthening by formation of M_7C_3, and $M_{23}C_6$ carbides.
Ni	• Act as FCC stabilizer in matrix. • Produces strengthening by formation of intermetallic compound Ni_3Ti. • Improve forgeability.
Mo and W	• Solid solution strengtheners. • Strengthening by forming intermetallic compounds Co_3M and M_6C carbide.
Ti and Nb	• Solid solution strengtheners. • Strengthening by forming intermetallic compounds Co_3M, MC, and M_6C carbide.
C	• Strengthening by forming carbides MC, M_7C_3, $M_{23}C_6$, and M_6C carbide.
Al	• Improves oxidation resistance. • Formation of intermetallic compounds CoAl.
Ti	• Produces strengthening by forming MC carbide and intermetallic compound Co_3Ti with sufficient Ni produces strengthening by formation of intermetallic compound Ni_3Ti.
B and Zr	• Produces strengthening by effect on grain boundaries and by precipitate formation. • Zr produces strengthening by formation of MC carbides.
Y and La	• Increases oxidation resistance.

Table 7.6 Composition of cast and wrought Co-based superalloys

	Elemental composition in wt%								
Alloys	Cr	Ni	W	C	B	Ta	Zr	others	Co
X-40	26.5	10.5	7.5	0.5	0.01				Balanced
X-45	26.5	10.5	7.5	0.75	0.01				Balanced
FSX-414	29.5	10.5	7	0.35	0.01				Balanced
FSX-418	29.5	10.5	7	0.35	0.01			0.15 Y	Balanced
MAR-M 302	21.5	–	10	0.85	0.005	9	0.2	–	Balanced
MAR-M 509	21.5	10	7	0.60		3.5	0.5	0.2 Ti	Balanced
WI-52	21	–	11	0.45				2.0 Nb	Balanced
AiResist 13	21	1	11	0.45				3.5 Al	–
HA-188	25.5	22	15	0.08				0.2 Si, 0.08 La	Balanced
HA-25	20	10	15	0.1				1.5 Mn	–

both W and Al have been found to integrate into these cuboidal precipitates. The elements Ta, Nb, and Ti integrate into the γ' phase and are quite effective at stabilizing it at high temperatures [76]. This stabilization is quite important as the lack of stability is one of the key factors that makes Co superalloys weaker than their Ni superalloys at elevated temperatures. The metastable ternary γ'-Co_3(Al,W) phase becomes a stable phase by additions of γ'-stabilizing elements, such as Ti, Ta, and Hf [77]. The effectiveness of alloying elements in stabilizing the γ' phase was compared using the change in the γ'-solvus temperature. In the Co–Al–W ternary system, the metastable γ'-solvus temperature is not high compared with the γ'-solvus temperatures of Ni-base superalloys: only 1033 °C in a Co-9Al-11 W alloy that has a high volume fraction of the γ' phase (79% at 900 °C). The increase in the γ'-solvus temperatures per atomic percent quaternary elemental additions is there when alloying additions were made by substituting for Co in Co–Al–W ternary alloys. The most effective element is Ta, followed by Ti, Nb, W, and Hf. An addition of 1 at% Ta increases the γ' solvus temperature by 45 °C. Additions of V and Ni have a slightly positive effect, whereas additions of Cr, Mn, and Fe have a negative effect. According to the elemental partitioning study, the γ'-stabilizing elements tend to partition to the γ' phase against the γ phase, and the γ'-destabilizing elements show the opposite partitioning behaviors [78]. The γ'-solvus temperatures with additions of Ni, Cr, and Fe, by substituting for Co. Both Cr and Fe destabilize the γ' phase, and the γ'-solvus temperature is reduced by more than 100 °C by adding 20 at% of either element. Also noteworthy is that the $\gamma + \gamma'$ two-phase field is continuous between the Co–Al–W and Ni–Al–W ternary systems. The γ'-solvus temperature is increased with additions of Ni. The increment by additions of Ni is approximately 200 °C due to the large solubility of Ni, although its contribution is small on a per-atomic-percent basis [79]. Carbide phases appearance in Co-based superalloys does provide precipitation hardening, but decreases low-temperature ductility. They are hardened by carbide precipitation; carbon content is thus critical. These include $M_{23}C_6$, M_6C, and MC carbides. Transforms into

$M_{23}C_6$ during exposure to temperatures in the range of 816–927 °C for 3000 h [80].

Co-base alloys are more likely to precipitate undesirable topologically close-packed (TCP) structures such as σ. On the contrary, Co_7W_6 and Co_2(Ta, Nb, Ti) are TCP phases that are likely to cause the deterioration of mechanical properties. These phases may also appear in some developmental Co-based superalloys, but a major point of engineering these alloys is avoiding TCPs [81].

7.5.3 Fe-based Superalloys

Iron-based superalloys are used in jet engines for their high-temperature properties and low thermal expansion. These superalloys contain 15–60% Fe and 25–45% Ni, with minor elements [82]. Fe-superalloys are also known as "super chrome steels," these metals are at the highest end of the range of high-temperature, high-strength steels. Besides chrome, other additives can be Ni, Ti, Mn, Mo, V, Si, and C. Fe is added to replace some of Ni as it has lower cost. Therefore, Fe superalloys are less expensive than Co- or Ni-superalloys. Microstructure consists of austenitic FCC matrix and can be strengthened by solid solution strengthening (Mo, Cr) and precipitation hardening (Ti, Nb, Al) by forming intermetallic phases. These are used in blades, discs, and engine casings, which require low thermal expansion properties at high temperature [83]. Fe-superalloys expand less than Ni- or Co- superalloys at high temperature, which is an important material property for engine components requiring closely controlled clearances between rotating parts [84]. The composition of several Fe-superalloys used in jet engines is given in Table 7.7. Fe-superalloy is hardened by solid solution strengthening and precipitation strengthening. Al, Nb, and C are used as alloying elements to promote the formation of hard intermetallic precipitates or carbides that are stable at high temperature. The precipitates are similar to those present in Ni-based superalloys, and include γ' Ni_3(Al,Ti), γ''(Ni_3Nb) and various types of carbides. Precipitates provide Fe-superalloy with good resistance against creep and stress rupture at elevated temperature. Chromium is used to form an oxide surface layer to protect the metal from hot corrosive gases and oxidation [85].

The Fe-superalloys are of three types: alloys that can be strengthened by a martensitic type of transformation, alloys that are austenitic and are strengthened by a sequence of hot and cold working (usually, forging at 1093 °C to 1149 °C followed by finishing at 649–871 °C), and austenitic alloys strengthened by precipitation hardening. Table 7.8 shows various categories of Fe-based AISI 600 series alloys. The last group can be considered as superalloys, the others being categorized as high-temperature, high-strength alloys. In general, the martensitic types are used at temperatures below 538 °C; the austenitic types, above 538 °C [86]. Wear resistance increases with carbon content. Maximum wear resistance is obtained in alloys 611, 612, and 613, which are used in high-temperature aircraft bearings and machinery parts subjected to sliding contact. Oxidation resistance increases with chromium content. The martensitic chromium steels, particularly alloy 616, are used for steam-turbine blades. The superalloys are available in all conventional mill forms—billet, bar, sheet, and forgings and special shapes are available for most alloys. In general, austenitic alloys are more difficult to machining than martensitic types, which machine best in the annealed condition. Austenitic alloys are usually "gummy" in the solution-treated condition and machine best after being partially aged or fully hardened. Crack sensitivity makes most of the martensitic steels difficult to weld by conventional methods. These alloys should be annealed or tempered prior to welding; even then, preheating and postheating are recommended. Welding drastically lowers the mechanical properties of alloys that depend on hot/cold work for strength [87]. All of the martensitic low-alloy steels machine are satisfactorily and are readily fabricated by hot working and cold working. The martensitic secondary-hardening and chromium alloys are all hot worked by preheating and hot forging. Austenitic alloys are more difficult to forge than the martensitic grades.

7.5.3.1 Phases of Fe-based Superalloys

The use of steels in superalloy applications is of interest because certain steel alloys have shown creep and oxidation resistance similar to that of Ni-based superalloys, while being far less expensive to produce. Like the phases found in Ni-based superalloys, Fe-based alloys feature a matrix phase (**γ phase**) of austenite iron (FCC). Alloying elements that are commonly found in these superalloys include Al, B, C, Co, Cr, Mo, Ni, Nb, Si, Ti, W, and Y. While Al is introduced for its oxidation benefits, Al additions must be kept at low weight fractions (wt.%), because Al stabilizes a ferritic (BCC) primary phase matrix, which is an undesirable phase in superalloy microstructures, as it is inferior to the high-temperature strength exhibited by an austenitic (FCC) primary phase matrix [88]. γ' phase is introduced as precipitates to strengthen the alloy. Like in Ni-based alloys, γ'-Ni_3Al precipitates can be introduced with the proper balance of Al, Ni, Nb, and Ti additions [89].

Two major types of austenitic stainless steels exist and are characterized by the oxide layer that forms at the surface of the steel: chromia-forming or alumina-forming stainless steel. Chromia-forming stainless steel is the most common type of stainless steel produced. However, chromia-forming steels do not exhibit high creep resistance at high operating temperatures, especially in environments with water vapor, when compared to Ni-based superalloys. Alumina-forming

Table 7.7 Composition of various commercial Fe superalloys

Alloys	Composition (wt%)									
	Fe	Ni	Cr	Mo	W	Co	Nb	Al	C	Ta	V
Haynes 556	29.0	21.0	22.0	3.0	2.5	20.0	0.1	0.3	0.1	0.5	–
Incoloy	44.8	32.5	21.0	–	–	–	–	0.6	0.36	–	–
A-286	55.2	26.0	15.0	1.25	–	–	–	0.2	0.04	–	0.3
Incoloy 903	41.0	38.0	0.2	0.2	–	14.5	3.0	0.7	0.04	–	

Table 7.8 Categories of Fe-based AISI 600 series alloys

Category	AISI 600 series	
I	601–604	Martensitic low-alloy steels
II	610–613	Martensitic secondary hardening steels
III	614–619	Martensitic chromium steels
IV	630 through 635	Semiaustenitic and martensitic precipitation-hardening stainless steels
V	650 through 653	Austenitic steels strengthened by hot/cold work
VI	660 through 665	Austenitic superalloys; all grades except alloy 661 are strengthened by second-phase precipitation

austenitic stainless steels feature a single-phase matrix of austenite iron (FCC) with an alumina oxide at the surface of the steel. Alumina is more thermodynamically stable in oxygen than chromia. More commonly, however, precipitate phases are introduced to increase strength and creep resistance. In alumina-forming steels, NiAl precipitates are introduced to act as Al reservoirs to maintain the protective alumina layer. In addition, Nb and Cr additions help form and stabilize alumina by increasing precipitate volume fractions of NiAl [90].

7.6 Single-crystal Superalloys

It has become more common to use single-crystal rather than poly-crystal turbine blades. The reason for this can be attributed to two things: enhanced creep and fatigue properties. Good creep and fatigue properties are two of the most important factors for gas turbine blades. During creep, grain boundary sliding is a major concern. By using single-crystal instead of polycrystal material, grain boundary sliding is avoided since no grain boundaries are present in single crystals [91]. Single crystals are also anisotropic, which means that they have different properties in different directions, for example, different stiffnesses in different crystallographic directions. Fatigue life is enhanced by a low Young's modulus, since the stresses will be lower for a crystal orientation with low stiffness compared to a direction with a higher stiffness when a constant strain is considered [92].

Single-crystal superalloys can be prepared by directional solidification technique. Since introduction of single crystal casting technology, single-crystal superalloys development has focused on increased temperature capability, and major improvements in alloy performance have been associated with the introduction of new alloying elements, including Re and Ru [93]. Single crystal superalloys have wide application in the high-pressure turbine section of aero and industrial gas turbine engines due to the unique combination of properties and performance. With increasing turbine entry temperature, it is important to gain a fundamental understanding of the physical phenomena occurring during creep deformation of single crystal superalloys under such extreme condition (i.e., high temperature and high stress). The creep deformation behavior of superalloy single crystal is strongly temperature, stress, orientation, and alloy dependent. For a single-crystal superalloy, there are three different modes of creep deformation under regimes of different temperature and stress: primary, tertiary, and rafting [94]. At low temperature (~750 °C), SX alloys exhibit mostly primary creep behavior. The extent of primary creep deformation depends strongly on the angle between the tensile axis and the <001>/<011 > symmetry boundary. At temperature above 850 °C, tertiary creep dominates and promotes strain-softening behavior. When temperature exceeds 1000 °C, the rafting effect is prevalent where cubic particles transform into flat shapes under tensile stress. The rafts would also form perpendicular to the tensile axis, since γ phase was transported out of the vertical channels and into the horizontal ones. After conducting unaxial creep deformation of <001> orientated CMSX-4 single crystal superalloy at 1105 °C and 100 MPa, it is observed that the rafting is beneficial to creep life since it delays evolution of creep strain. In addition, rafting would occur quickly and suppress the accumulation of creep strain until a critical strain is reached [95].

In the case of cast superalloys, the introduction of directionally solidified (DS) columnar grained blades led to considerable increases in creep strength owing to the elimination of transverse grain boundaries. During solidification, this single grain grows to encompass the entire part. Even more spectacular improvements in creep strength became possible for SC castings, since besides the total elimination of grain boundaries, the chemistries could be altered (rebalanced) to increase the solidus temperature beyond the γ' solvus. This avoided the formation of eutectic nodules and made it possible to use high-temperature homogenization, with complete solutionizing of the γ' and subsequent formation of a uniform distribution of fine precipitates [96]. In general, the lack of transverse grain boundaries coupled with the lower modulus can result in three to five times improvement in creep rupture life. Many of the early DS blades produced from the late 1960s were hollow, with air cooling channels about half a millimeter in diameter. Today's SC blades often contain even more intricate cooling circuits, and the wall thickness can be as small as a few tenths of a millimeter. These developments have resulted in progressively higher operating temperatures [97].

7.7 Processing of Superalloys

The historical developments in superalloy processing have brought about considerable increases in superalloy operating temperatures. Superalloys were originally iron based and cold wrought prior to the 1940s. In the 1940s, investment casting of cobalt-based alloys significantly raised operating temperatures. The development of vacuum melting in the 1950s allowed for very fine control of the chemical composition of superalloys and reduction in contamination and in turn led to a revolution in processing techniques such as directional solidification of alloys and single crystal superalloys [98]. Superalloy processing begins with the fabrication of large ingots that are subsequently used for one of three major processing routes: (1) remelting and subsequent investment casting, (2) remelting followed by wrought processing, or (3) remelting to form superalloy powder that is subsequently consolidated and subjected to wrought processing operations. Ingots are fabricated by vacuum induction melting (VIM) in a refractory crucible to consolidate elemental and/or revert materials to form a base alloy. Although selected alloys can potentially be melted in air/slag environments using electric arc furnaces, VIM melting of superalloys is much more effective in the removal of low-melting-point trace contaminants. Following the vaporization of the contaminants, the carbon boil reaction is used to deoxidize the melt before the addition of the reactive forming elements such as Ti, Al, and Hf. Once the desired alloy composition of the VIM ingot is attained, the solidified ingot is then subsequently subjected to additional melting or consolidation processes that are dependent upon the final application of the material. Charge weights of VIM ingots may range from $\sim$2500 kg to in excess of 27,500 kg [99]. Considering the stringent requirements for minimizing defects in gas industry and turbines, etc., a detailed understanding of structure evolution in each of these processing paths is essential. In the following sections, we briefly discuss the well-known processing approaches.

7.7.1 Casting and Forging

The casting process is important in the production of heat-resistant superalloy engine components. Casting and forging are traditional metallurgical processing techniques that can be used to generate both polycrystalline and monocrystalline products. Polycrystalline casts tend to have higher fracture resistance, while monocrystalline casts have higher creep resistance. Jet turbine engines employ both poly and mono crystalline components to take advantage of their individual strengths. The disks of the high-pressure turbine, which are near the central hub of the engine are polycrystalline. The turbine blades, which extend radially into the engine housing, experience a much greater centripetal force, necessitating creep resistance. As a result, turbine blades are typically monocrystalline or polycrystalline with a preferred crystal orientation [100].

The investment casting method (Fig. 7.13a) is an old technique to produce very intricate shape. The shape of a pattern is made by low melting temperature materials like wax. Then the wax pattern has to dip inside a ceramic slurry and allowed to cool. After cooling the hard shell of ceramic can be formed as a mold. For the mold cavity, we have to heat the mold and remove the wax. After we get a finished mold, the common process of casting can be followed to get the finished product. After cooling, to remove the cast product, we have to break the mold completely. The one-time usability of the mold can be considered as a major disadvantage of this process. Though we can cast several small parts at once. Due to the high thermal resistivity of the mold, it is very suitable to produce materials with high melting temperature like superalloys. Investment casting leads to a polycrystalline final product, as nucleation and growth of crystal grains occur at numerous locations throughout the solid matrix. Generally, the polycrystalline product has no preferred grain orientation [101].

In vacuum induction melting (Fig. 7.13b), the metals melt with the help of induction heating within a vacuum chamber. This process was discovered in the early twentieth century and the first prototype was made in 1920. It is generally used for the materials which have more affinity to make oxides and nitrides. As this process takes place in a vacuum chamber, it has a higher tolerance for producing high purity metals and alloys. Ni and Ti materials are more commonly used in this process [102]. With the help of induction coils made up of copper, the supplied alternating current generates induced eddy current and helps in heating the crucible. Then the

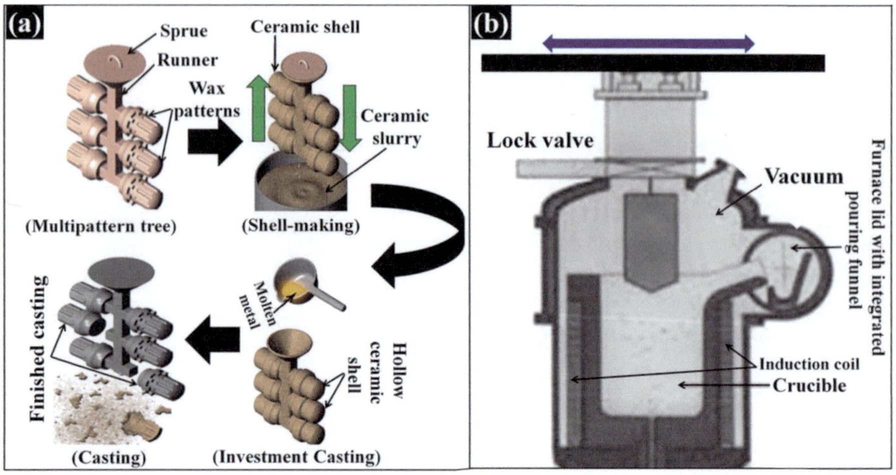

Fig. 7.13 Schematic diagram of (**a**) investment casting process and (**b**) vacuum induction melting process

Table 7.9 Common refractory materials used for crucible liner

Refractory material	Melting temperature (°C)	Thermal shock resistance	Application
Graphite	2300	Excellent	Copper and its alloys
ZrO_2	2300	Poor	Steels and superalloys
Al_2O_3-spinel	1900	Relatively good	Steels and superalloys
MgO-spinel	1900	Poor	Steels and superalloys
Al_2O_3	1900	Good	Steels and superalloys
MgO	1600	Good	Steels and superalloys

molten metal goes to the mold. If we use graphite as the crucible material for Ti then it may form TiC in superalloys. Some common use of crucible is given in Table 7.9.

A method of forming a fine-grained equiaxed casting is by melting metal and removing most of the superheat of the molten metal. The molten metal is placed in a mold and optionally subjected to turbulence whereupon it solidifies to form the casting of the desired microstructure [103]. To get more purity in superalloys, i.e., with very low content of O_2 and N_2, electron beam cold hearth refining is preferred over vacuum induction melting process. In case of vacuum induction melting ceramic mold is used so entrapped ceramic particle may present in the finished cast bars. A drawback to electron beam cold hearth refining is the loss of volatile elements such as chromium. However, this loss can be compensated for by enriching the input material or by in-process additions [104]. Vacuum arc remelting (Fig. 7.14a) is a secondary process to improve the quality of the material by removing the contaminated materials from the alloys. This process mostly produces highly reactive materials like steels, Ti-based alloys, and superalloys. According to the purity to achieve, this process can be repeated for several times. This process can be eliminated. Most of the commercial products avoid such processes because firstly this is a secondary process and secondly due to their time consumption and cost. In this process, due to a very small gap between the one end of the metal and the electrode, electric arc generates and helps in melting the metal. The setup is completely vacuumed or an inert gas may be used depending on the material used to melt. The water flow arrangement is there to regulate the temperature of the furnace and can maintain the cooling rate of the material so that we can avoid the shrinkage defects and the segregation microstructures [105].

Electroslag remelting (Fig. 7.14b) is also a secondary process carried out after vacuum induction melting. The other name of this process is electro-flux remelting. The purpose is to get the refined form of the as-cast alloys. The lower part of the crucible is made up of copper and water-cooled. From the top surface of the cast product which is closer to the electrode melts first and gradually melts the bottom surface at last. Metal particles are heavier than the impurities those form slag, so metal particles fall down and finally, all pure metals are at the bottom and slags are at the top. The cooling is directional and starts from the bottom to the top surface. In this way, the top surface having slags is separated from the pure metal residing at the bottom of the ingot made after cooling. These are very useful for manufacturing alloys with high service temperature and used for critical applications like steels and superalloys [106].

7.7.2 Powder Metallurgy Process

Powder metallurgy is a route different from that of liquid metallurgy process. The most crucial problem in liquid metallurgy route during the manufacturing of alloys is the difference between the melting temperature of the different

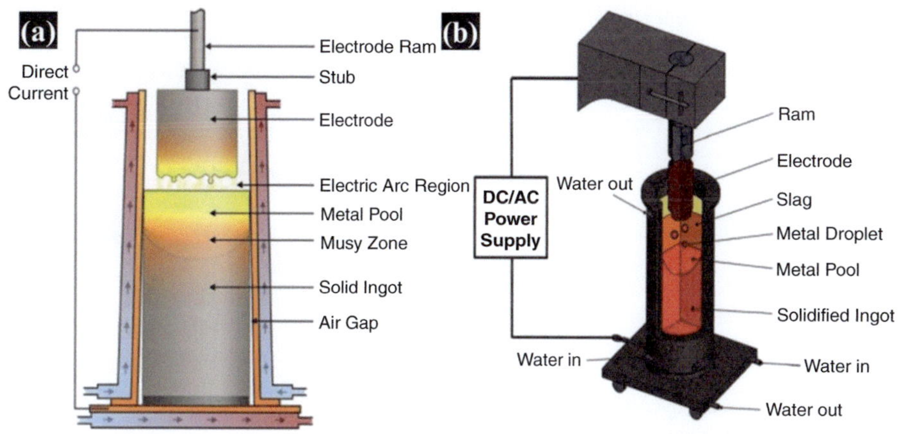

Fig. 7.14 Schematic diagram of (**a**) vacuum arc remelting and (**b**) electroslag remelting

elements. If the temperature difference is much higher than the problem of homogeneity may occur. Also, the difference in density of elements may have a higher impact on homogeneity in the system. In the case of powder metallurgy, we can get the finished product directly. There is no requirement of post-processing of finished products. Also, we do not have to melt the elements and can obtain better homogeneity by proper mixing of the elemental powders [107]. The steps for powder metallurgy process are given in Fig. 7.15.

In this process, the homogenized mixture is compacted in a predetermined structured mold to form the green pellets several types of hydraulic cold/hot compaction machine (such as equiaxial pressing, hot isostatic pressing, spark plasma sintering machine, etc.) is generally used for the making of compacts. The green pellets then follow sintering process. For sintering, we have to heat the material with a low heating rate up to a particular temperature and hold for some time. During this holding, all the elemental powders react with each other and form the desired phase. If we use some powders like oxides, then those oxide particles are entrapped in the phases formed by other metallic powders. This can strengthen the material by oxide dispersion strengthening. This process can also be called as oxide dispersion processing [108]. It does not need much post-processing like liquid metallurgy route. But porosity will be present in the material and very difficult to avoid in powder metallurgy processes. Unlike the sand castings and investment castings, we can reuse the dies for many times. Some metals like Ti may react with carbon and make carbides. In that case, we can avoid these processes which use graphite crucible. The environment for all types of sintering may be inert or vacuum so that we can avoid the formation of unwanted phases like oxides, nitrides, etc. [109].

7.7.3 Additive Manufacturing

As advanced production technology, additive manufacturing is increasingly being used to produce complex 3D objects through digitally regulated phase change deposition and solvent-based reactive materials and inks. This fabrication method generally begins with layer-by-layer processing as shown in Fig. 7.16 to fabricate the superalloy. Among various types of 3D printing, some industrial practices superalloy production are selective laser melting, electron beam melting, etc. [110].

Powder bed fusion is the 3D printing technology in which particles are fused together in a powder bed. Various sources, such as lasers or electron beams, are used to fuse particles from a powder bed. A layer of ultrafine powder is spread over a continuous layer with a roller. After the desired shape is completed, the non-molten powder is removed for reuse in the next production process. The powder bed fusion process includes direct metal laser sintering (DMLS), electron beam fusion (EBM), selective thermal sintering (SHS), selective laser fusion (SLM), and selective laser sintering (SLS) [110, 111]. Directed energy deposition (DED) processes include laser-engineered net shaping (LENS), direct laser fabrication (DLF), direct metal deposition (DMD), and 3D laser coating (LC) [112]. DED is a more complex printing process that is often used for repair or adjustment. There are components to add other materials. A typical DED machine consists of a nozzle mounted on a multi-axis arm that deposits molten material on a specific surface and solidifies on that surface. The process is basically similar to extrusion of material, but the nozzle can move in multiple directions and is not mounted on a certain axis. Materials that can be deposited from any angle with the aid of machines with four and five axes are fused during deposition with a laser or electron beam. This process can be used for polymers, ceramics, but is generally used with metals in powder or wire form [113]. In binder jetting process (BJP), both powder-based material and binder are used. The glue acts as a binder between the layers of dust. Binders are generally in liquid form and building materials are in powder form. The print head moves horizontally along the X and Y axes of the machine and alternately deposits layers of construction and binding material. After each layer, the objects to be printed

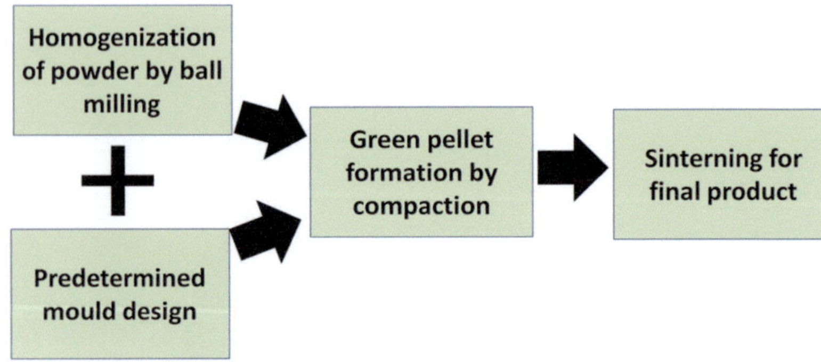

Fig. 7.15 Powder metallurgy processing steps

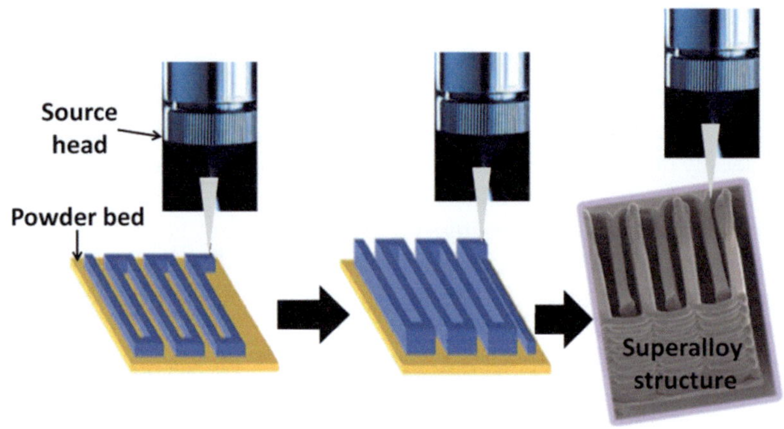

Fig. 7.16 3D-printed layer-by-layer to create superalloy

are placed on their construction platform. Due to the binding method used, the properties of the material are not always suitable for structural parts, and although the printing speed is relatively high, further post-processing can extend the overall course of time [114].

7.7.4 Directional Solidification Process

Directional solidification uses a thermal gradient to promote nucleation of metal grains on a low temperature surface as well as to promote their growth along the temperature gradient (Fig. 7.17). This leads to grains elongated along the temperature gradient and significantly greater creep resistance parallel to the long grain direction. In polycrystalline turbine blades, directional solidification is used to orient the grains parallel to the centripetal force. It is also known as dendritic solidification [115].

7.7.5 Single Crystal Growth

Single crystal growth starts with a seed crystal which is used to template growth of a larger crystal. The overall process is lengthy, and additional processing via machining is necessary after the single crystal is grown [116]. Single-crystal objects are outside of most people's experience. Forming single-crystal metal objects requires both special alloys and special casting techniques. The alloys are almost always Ni-based, with as many as nine minor metal components including five or more percent of Cr, Co, W, Ta, Al, and/or Re. The casting method is known as "directional solidification," and involves carefully cooling a cast metal part starting at one end to guarantee a particular orientation of its crystal structure. That orientation is chosen, naturally, based on expected stresses in the finished part [117]. Figure 7.18 shows the different microstructure of equiaxed, directional solidified, and single crystal structure with their gradual increase in creep resistance.

The primary application for single crystal superalloys is the manufacture of jet engine turbine blades, which must endure tremendous forces at extremely high temperatures for prolonged periods of time. Under such conditions, metals with a grain structure tend to slowly deform, along grain boundaries [118].

7.7.6 Post-fabrication Processing

The main purposes of heat treatment of superalloys are: (1) to give precipitation hardening, (2) to achieve desired precipitation of carbide, (3) to relieve the embrittling effects of mechanical working processes in wrought alloys through recrystallization and grain growth, (4) to create optimum grain size through grain growth (in cast and wrought alloys), and through recrystallization and grain growth in conjunction

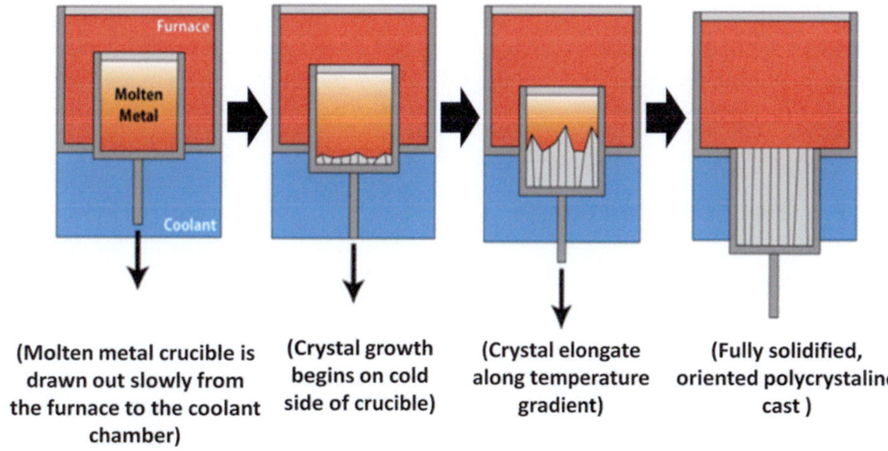

Fig. 7.17 Schematic of directional solidification

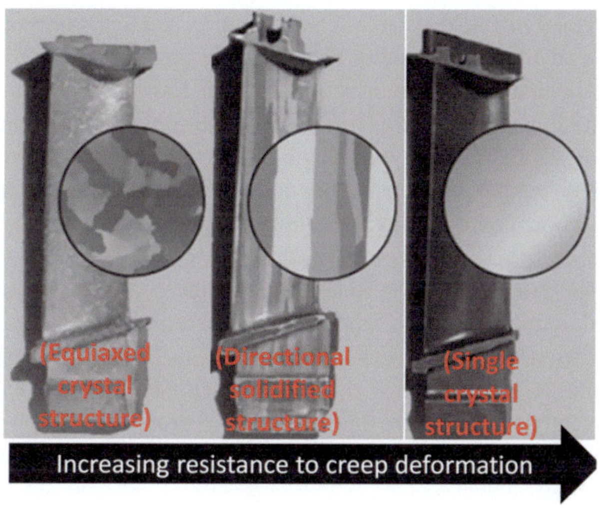

Fig. 7.18 Microstructure of (**a**) equiaxed, (**b**) directional, and (**c**) single crystal

with mechanical deformation—so-called thermomechanical processing [119]. The large grain size gives improved creep strength, reduced creep extension to failure, and reduced short-term strength and failure strength. The mechanical working of wrought superalloys results in shaping the component by forging, rolling and extrusion, etc. Post cast processing can homogenize the microstructures, for example, eliminating segregation of alloying additions after casting, and distributing MC carbides. Normally the superalloys are seldom used in the as-work state, because of reduced ductility (residual stress) and avoid the worked structure, that is always unstable in high-temperature situations. To optimize properties (often of an alloy/coating system), solution-treated Ni-based superalloys are aged at two different temperatures within the γ/γ. The higher temperature heat treatment precipitates coarser particles of γ. The lower temperature heat treatment leads to a finer, secondary dispersion of γ. The net result is a bimodal distribution of γ. The solution heat treatment temperature determines not only the amount of γ' that dissolves but also the γ grain size. This coarsens if all the γ' are dissolved, since the grain boundaries are no longer pinned by precipitates [120]. The steps to perform precipitation-hardening treatment are:

Solution Treatment: Heated to the single-phase region. Precipitation of grain boundary carbides with suitable morphology often requires a higher temperature (1100–1200 °C).

Quenching: Rapid cooling to room temperature to form a supersaturated solid solution.

Aging: Decomposition of the supersaturated solid solution in the two-phase field to form the fine precipitates supersaturated solid solution, an unstable condition and easy to form metastable phases to lower the energy of the system.

7.8 Problem Persist on Prepared Superalloy

7.8.1 Oxidation Effects

Along with hot corrosion, oxidation is one of the two basic types of deteriorative effects that superalloy coatings are designed to prevent. Oxygen reactions are the most common environmental threats for alloy materials, and although superalloys continue functioning with low risk at temperatures below 871 °C, at higher temperatures oxidation can seriously degrade material quality and performance, particularly for Ni- and Co-superalloys [121]. At lower temperature levels, Cr content is the primary defense against oxygen reactions, while at higher temperatures (>982 °C), Al becomes more important because it is the base in a compound that forms a protective oxide scale. When a superalloy's Al content is insufficient for preventing oxidation, a protective coating can be used to maintain service life by preventing selective oxidation at the grain boundaries and surface carbides, and by inhibiting internal oxidation. Chromium can sometimes decrease high-temperature mechanical strength, so many superalloys include a lower chromium content to maintain tensile strength, but this lowers their hot corrosion resistance [122].

7.8.2 Hot Corrosion Effects

Hot corrosion is one of the most common accelerated oxidation processes, and is usually caused by environmental contaminants, such as salt, or the sulfur found in fuel. Although the Cr content in most Ni- and Co-superalloys is usually sufficient for impeding hot corrosion, some Ni base varieties are prone to this deteriorative effect, particularly those with improved rupture strength at higher temperatures. Protective coatings, especially overlay coatings, and sometimes environmental inhibitors may be needed to shield high-strength Ni-base superalloys. Coating material can diffuse inward into the base metal, which may cause a decrease in the metal's incipient melting temperature. Once a superalloy has been heated past its incipient melting point, it causes a drop in grain-boundary strength and ductility and these properties cannot be restored via heat treatment. The degree of oxidation and hot corrosion usually determines the upper temperature limit for a superalloy, while wrought alloys used in lower temperature applications, such as rotating seals or turbine disks, can generally function without protective coatings [123].

7.9 Coating for Superalloy

Superalloys are operated in industrial environments containing corrosive species such as sulfur, chlorine, or carbon-containing compounds, water vapor, alkali, and alkaline earth metal salts, or ashes such as vanadates. At elevated temperatures, such compounds cause a wide range of attack types on most metallic alloys such as oxidation, carburization, sulfidation, hot corrosion, or a combination of their mechanisms. Table 7.10 shows effect of properties of superalloys in various engines [124–126]. But even in air, the oxidation resistance of Ni- and Co-based superalloys is not sufficient for continuous operation above 1000 °C. Oxide scales grow too fast and the subsurface zone of the alloys is changed and loses its mechanical strength. Despite these limitations, advances and improvements of industrial technology have led to more efficient processes and more powerful engines with increased operation temperatures, and the alloys used for their construction are pushed toward the applicable limit even as far as mechanical properties are concerned, such as fatigue, tensile strength, and creep resistance. Therefore, higher additions of alloying elements for corrosion protection such as Cr, Al, or Si cannot be used as they either lead to embrittlement or lower the melting point and therefore the creep strength at the high target temperatures. The only way is to apply high-temperature coatings to face aggressive corrosive high-temperature atmospheres and make processes possible which could not be operated efficiently and reliably without such coatings [127]. The requirements from coatings are: (1) slow-growing, thermodynamically stable scale formation and high concentration of the scale-forming elements; (2) thermal stability, no detrimental phase changes; (3) erosion resistance; (4) good adhesion; (5) low interdiffusion of the substrate; (6) mechanical compatibility: similar modulus and thermal expansion to avoid stresses at the interface; (7) no crack initiation, not too brittle; (8) good processability; (9) low price. The first purpose of coatings was therefore to protect the poor oxidation resistance of the base alloy (aluminide, Pt-aluminide, MCrAlY). A second type of coatings applied to high-temperature parts is known as thermal barrier coatings (TBCs). These are ceramic coatings with very low thermal conductivity. Despite being typically 1/5 mm thin, this allows for a drop of 100–300 °C between the gas and metal surface temperatures [9, 128]. Such coatings, however, are "oxygen transparent" and do not prevent oxidation of the underlying substrate.

There are three coatings that have been developed: diffusion coatings, overlay coatings, and thermal barrier coatings. Diffusion coatings, mainly constituted with aluminide or platinum-aluminide, are still the most common form of surface protection. To further enhance resistance to corrosion and oxidation, MCrAlX-based overlay coatings (M = Ni or Co, X = Y, Hf, Si) are deposited to surface of superalloys. Compared to diffusion coatings, overlay coatings are less dependent on the composition of the substrate, but also more expensive, since they must be carried out by air or vacuum plasma spraying or else electron beam physical vapor deposition. Thermal barrier coatings provide by far the best enhancement in working temperature and coating life.

Table 7.10 Effect of superalloys properties in various engine

Effects	Aircraft engines	Land-based power generation	Marine engines
Oxidation	Severe	Moderate	Moderate
Hot corrosion	Moderate	Severe	Severe
Interdiffusion	Severe	Moderate	Light
Thermal fatigue	Severe	Light	Moderate

7.9.1 Thermal Barrier Coatings

Thermal barrier coatings (TBCs) are a ceramic multilayer film applied to the superalloy surface to increase the operating temperature of the engine. It is estimated that modern TBC of thickness 300 μm, if used in conjunction with a hollow component and cooling air, has the potential to lower metal surface temperatures by a few hundred degrees [129]. The oxidation-resistant coating is also called bond coat because it provides a layer on which the ceramic TBC can adhere. Figure 7.19 showing bond-coat+TBC [130]. The coating is an insulating layer that reduces the heat conducted into the superalloy. Yttria-stabilized zirconia (YSZ) is the most common coating material and is used on engine components in the combustor chamber and turbine sections, including high-pressure blades and nozzle guide vanes [131]. YSZ is used as a TBC on components in the hot section, i.e., turbine section, of aircraft engines and industrial gas turbines. The low thermal conductivity of TBC materials facilitates increased performance by lowering component temperatures in the presence of high-temperature combustion. YSZ has a relatively high coefficient of thermal expansion (CTE) and is near that of the Ni- and Co-based superalloys used for turbine components. This coefficient of thermal expansion match minimizes stress induced by differential expansion between the coating and its substrate. To a lesser extent, other materials are used as stabilizers, including calcia (CaO), ceria (CeO_2), and magnesia (MgO) [132]. Stabilization is necessary to eliminate the volumetric change associated with the phase transformation from monoclinic to tetragonal during thermal excursions, which occurs as the turbine ramps from ambient to operational temperatures. TBC coating systems include a bond coat of an oxidation- and corrosion-resistant MCrAlY alloy. The "M" represents one or more of the elements Fe, Ni, or Co. These alloys form complex oxides from the Al, Cr, and Y, which serve as oxygen and corrosion barriers to protect the less resistant substrate.

Fig. 7.19 TBC-coated high-pressure turbine blade

Failure of thermal barrier coating usually manifests as delamination, which arises from the temperature gradient during thermal cycling between ambient temperature and working conditions coupled with the difference in thermal expansion coefficient of the substrate and coating. There are various degradation mechanisms for thermal barrier coating, and some or all of these must operate before failure finally occurs, involve in [133, 134]: (1) oxidation at the interface of thermal barrier coating and underlying bond coat; (2) the depletion of aluminum in bond coat due to oxidation and diffusion with substrate; (3) thermal stresses from mismatch in thermal expansion coefficient and growth stress due to the formation of thermally grown oxide layer, imperfections near thermally grown oxide layer; (4) various other complicating factors during engine operation; (5) additionally, TBC life is very dependent upon the combination of materials (substrate, bond coat, ceramic) and processes (EB-PVD, plasma spraying) used. Table 7.11 describes various the damage of hot section components of combustion chamber, nozzle or vane, and blade.

Thermal spraying consists of melting a consumable (most often powder or wire) and projects it as molten particles onto the substrate. Upon impact with the substrate, the molten particle flattens and solidifies very rapidly. The adhesion is primarily mechanical (for this reason, sprayed MCrAlY coatings are often given a diffusion heat treatment to obtain good adherence). A sprayed coating will typically include voids (i.e., have some percentage of porosity), and oxide particles. The amount of both depends on the processing parameters. The adhesion of the coating depends on the cleanliness of the substrate surface, and its area (a high surface roughness is desirable for good adhesion),

Table 7.11 Typical damage of hot section components

Sections	Primary damages
Combustion chamber	Thermal fatigue cracking, wear, coating spallation, oxidation, corrosion, microstructural degradation
Nozzle or vane	Thermal fatigue cracking, wear, coating spallation, oxidation, corrosion, erosion, microstructural degradation
Blade	Creep, high cycle fatigue, thermal fatigue cracking, wear, coating spallation Oxidation, corrosion, erosion, microstructural degradation

on the velocity of the particles, etc. Abradable coatings come under thermal spraying processes. Abradable coatings are used to reduce the clearance between the tip of the turbine blade and the inner diameter of the casing, to an extent that precision machining cannot achieve. MCrAlY or TBC coatings are used depending on the component temperature. For this purpose, a high porosity is usually desirable and thermal spraying is therefore ideal. The use of abradable coatings can result in up to 1% improvement in efficiency [135]. There are various types of thermal spraying methods, such as flame spraying, plasma spraying, electron beam physical vapor deposition, high-velocity oxy-fuel, etc.

Flame Spraying: Flame spraying is widely used where a cost-effective thermal spray coating is desired and a lower quality can be tolerated. Flame spraying offers a cost-effective alternative for applying metallic and ceramic coatings in a less demanding environment. The utilization of the flame spraying surface treatment allows the spraying of a wide variety of metallic or ceramic coatings onto a large range of component materials where good wear resistance and excellent impact resistance are required. The flame spraying process is fueled by a heat source that is created by a chemical reaction between a fuel of combustion and oxygen to create a stream of gas. The thermal spray material is then put into the flame in the form of a wire and compressed air is used to atomize the molten particles before propelling them onto the substrate [136].

Plasma Spraying: Of the various thermal spray methods, one of the more ideal and commonly used techniques for coating superalloys is plasma spraying. The coating material is in powder form and injected into a high-temperature plasma jet which heats the powder before accelerating it out toward the substrate. Plasma spraying can accommodate a very wide range of materials, much more than the other techniques. As long as the difference between melting and decomposition temperatures is greater than 27 °C, a material can be melted and applied as a coating via plasma spraying. APS and LPPS are widely used to apply MCrAlY coatings [137].

Electron Beam Physical Vapor Deposition (EBPVD): In these methods, film growth is obtained by condensation of a vapor on the substrate. The vapor can be produced by heating the consumable enough to obtain evaporation, or by mechanically knocking the atoms off (e.g., sputtering). In EBPVD, the evaporation is obtained with a focused electron beam under a vacuum. EBPVD TBCs offer vastly longer lives in jet engine operating conditions, and EBPVD is therefore the preferred process for application of these coatings. In the case of jet engine operation conditions, it has been shown that the use of EBPVD increases by 7–13 times the life of the coating, therefore easily compensating for the higher cost of the latter process [138].

High-velocity Oxyfuel (HVOF): The high-velocity oxy-fuel coating is coming under the thermal spray coating process. In HVOF processes, the kinetic energy of the coating material is increased by producing high-velocity gas. Due to the high-velocity gas, the coating material will travel at supersonic speed with high-impact energy and strike the base material. By this principle, HVOF will be able to produce dense coatings at relatively low temperatures. In HVOF coating process, kinetic energy is more important when compared with the temperature. Due to the higher kinetic energy and impact force, coating particles are responsible for the bonding effects with the substrate. Normally the bonding is due to the adhesion, adherence, diffusion, chemical reactions, and sometimes partial fusion between the contact surfaces. During the rapid solidification of the spray, particles become in close contact with the substrate surface. These bonding effects permit the formation of a continuous coating layer. HVOF spraying has advantages over other thermal spray processes, which are [139, 140]: (1) higher density (lower porosity) due to greater particle impact velocities, (2) higher strength bond to the underlying substrate and improved cohesive strength within the coating, (3) lower oxide content due to less in-flight exposure time, (4) retention of powder chemistry due to reduced time at temperature, (5) smoother as-sprayed surface due to higher impact velocities and smaller powder sizes, (6) better wear resistance due to harder, tougher coatings.

7.9.2 Pack Cementation Process

Pack cementation is a widely used chemical vapor deposition technique which consists of immersing the components to be coated in a metal powder mixture and ammonium halide activators and sealing them in a retort. The entire apparatus is placed inside a furnace and heated in a protective atmosphere to a lower than normal temperature for diffusion to take place, due to the halide salts chemical reaction which causes an eutectic bond between the two metals. The new surface alloy that is formed due to thermal diffused ion migration has a metallurgical bond to the surface substrate and a true intermetallic layer found in the gamma layer of the new surface alloys. This process includes aluminizing, chromizing, siliconizing, sherardizing, boronizing, titaniumizing. Pack cementation has had a revival in the last 10 years as it is being combined with other chemical processes to lower the temperatures of metal combinations even further and give intermetallic properties to different alloy combinations for surface treatments [141].

7.9.3 Bond Coats

Bond coat help to adhere the thermal barrier coating to the superalloy substrate. Additionally, the bond coat provides oxidation protection and functions as a diffusion barrier against the motion of substrate atoms toward the environment. Bond coat (the bond coat belongs to the class of diffusion or overlay coatings) in combination with a ceramic layer on top to form a thermal barrier coating (TBC) system [142]. There are five major types of bond coats, the aluminides, the platinum–aluminides, MCrAlY, cobalt–cermets, and nickel–chromium [143]. For the aluminide bond coatings, the final composition and structure of the coating depend on the composition of the substrate. Aluminides also lack ductility below 750 °C and exhibit a limited by thermomechanical fatigue strength. The Pt-aluminides are very similar to the aluminide bond coats except for a layer of Pt (5–10 μm) deposited on the turbine blade. The Pt is believed to aid in oxide adhesion and contributes to hot corrosion. The cost of Pt plating is justified by the increased blade life span. The MCrAlY is the latest generation of bond coat and does not strongly interact with the substrate. Normally applied by plasma spraying, MCrAlY coatings are secondary aluminum oxide formers. This means that the coatings form an outer layer of chromium oxide (chromia), and a secondary aluminum oxide (alumina) layer underneath. Yttrium enhances the oxide adherence to the substrate and limits the growth of grain boundaries. Investigation indicates that addition of Re and Ta increases oxidation resistance [144]. Various types of bond coat have been discussed subsequently.

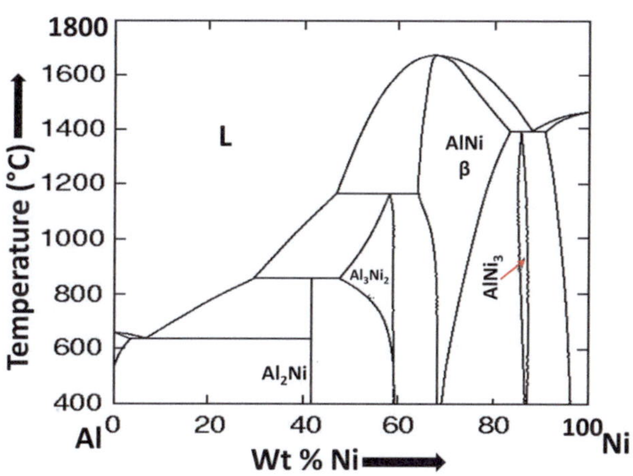

Fig. 7.20 The Al–Ni phase diagram

7.9.3.1 Auminides Bond Coats

In modern applications involving very high temperatures or severe hot corrosion problems, aluminide coatings provide significant protection. Diffusion aluminide coatings are based on the intermetallic compound β-NiAl (Fig. 7.20) [145]. Although different processes exist to form them, pack cementation is the most widely used as it is inexpensive and well adapted to coating of small parts. Pack cementation falls in the category of chemical vapor deposition. In this process, the components to be coated are immersed in a powder mixture containing Al_2O_3 and aluminum particles. About 1–2 wt% of ammonium halide activators are added to this "pack." This is then heated to temperatures around 800–1000 °C in argon or H_2 atmosphere. At these temperatures, aluminum halides form which diffuses through the pack and reacts on the substrate to deposit Al metal. The activity of Al maintained at the surface of the substrate defines two categories of deposition methods: low and high activity, also referred to as outward and inward diffusion respectively. In cements with low aluminum contents (low activity/outward), the formation of the coating occurs mainly by Ni diffusion and results in the direct formation of a Ni-rich NiAl layer (Fig. 7.21). The process requires high temperature (1000–1100 °C). In service, the interdiffusion with the substrate is very limited and the gradient of Al in β is low [146]. In cements with high aluminum contents (high activity/inward), the coating forms mainly by inward diffusion of aluminum and results in the formation of Ni_2Al_3 and possibly β-NiAl. Aluminizing temperatures can be lower

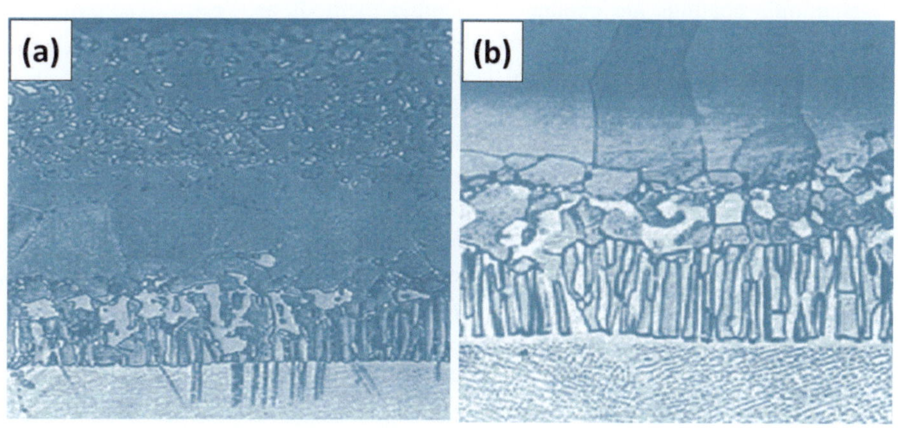

Fig. 7.21 Microstructures of two types of aluminides coating on superalloy, (**a**) high activity/inward diffusion and (**b**) low activity/outward diffusion

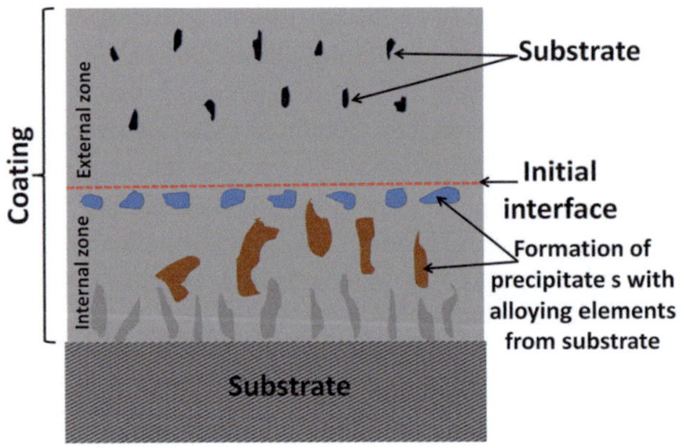

Fig. 7.22 Schematic illustration of aluminide coating obtained by low activity pack cementation

(700–950 °C). There can be a high Al concentration gradient in the coating and also significant interdiffusion with the substrate during service. For these reasons, a diffusion heat treatment is generally given at 1050–1100 °C to obtain a fully β layer [147].

The substrate composition strongly influences the final microstructure of the system, in a manner which also depends on the process. In low activity/outward diffusion coatings, the alloying elements present in the substrate will also tend to diffuse into the coating layer, to an extent limited by their solubility (Fig. 7.22). In high activity/inward diffusion coatings, they enter in solution the compound layer in formation, or as precipitates potentially forming during the treatment. Of the elements present in the alloy, Ti is thought to be particularly detrimental to the oxidation resistance of such coatings, as it leads to the formation of TiO_2 crystals which break the alumina layer. An typical microstructure of low activity aluminide coating is illustrated below. The external zone is typically Al-rich β-NiAl, while the internal one is Ni rich [148].

7.9.3.2 Pt–Aluminides Bond Coats

NiAl coatings develop strongly from interdiffusion with the substrate, which results in formation of γ′ at the expense of β. Pt–aluminide coatings are obtained by similar methods as conventional aluminide. The layer of Pt deposited is typically of 5–10 μm. The plating can increase the life of the blades by up to three times and here, the cost of Pt plating is easily compensated. Pt additions not only did not provide a diffusion barrier but also enhanced Al diffusion. When applied on the second-generation superalloy CMSX4, Pt formed TCP phases with some elements of the substrate (Re, W, Mo, Cr). Pt improves oxide adherence and also contributes to better hot corrosion resistance. Pt appears to partially substitute Ni in β-NiAl and also to form $PtAl_2$ which is believed to act as an Al reservoir. It has also been proposed that Pt acts in a similar manner as Y in MCrAlY coatings [149].

7.9.3.3 MCrAlY Bond Coats

Typical MCrAlY bond coats (M = Fe, Co, or Ni) contain at least four elements, which means that coating methods such as pack cementation are considerably more difficult to use, as the activity of each element in the pack would have to be controlled carefully so as to obtain a coating of required composition. The presence of a significant amount of Cr gives these coatings excellent corrosion resistance combined with good oxidation resistance. Alternative methods are therefore preferred, such as air plasma spray, low-pressure plasma spray, or electron beam physical vapor deposition. These are detailed in other sections. Deposition is followed by a high-temperature heat treatment in vacuum to allow interdiffusion and therefore improve adhesion. MCrAlY coatings typically exhibit a two-phase microstructure β + γ. The presence of γ increases the ductility of the coating thereby improving thermal fatigue resistance. As for β-NiAl coatings, high temperature exposure results in depletion of the Al both to the thermally grown oxide and to the substrate by interdiffusion. As the amount of Al decreases, the β phase tends to dissolve. For this reason, it is often described as an aluminum reservoir, and coating life often measures in terms of depletion of β [150].

The **M** of MCrAlY stands for either Ni or Co or a combination of both (when applied to steels, it can also be Fe), depending on the type of superalloy. Co based appears to have superior resistance to corrosion. Cr provides hot-corrosion resistance, but the amount that can be added is limited by the effect it is expected to have on the substrate, and the formation of Cr-rich phases in the coating. Al content is typically around 10–12 wt%. Since oxidation life is essentially controlled by the availability of Al, it would be tempting to increase the aluminium content. However, this results in significant reduction of ductility. MCrAlY also typically contains 1 wt% Y, which enhances adherence of the oxide layer. It was initially thought that Y helped the formation of oxide pegs which helped anchor the oxide layer to the coating. The main role of Y is to combine with sulfur and prevent its segregation to the oxide layer, which is otherwise detrimental to its adhesion. Additions of Hf play a similar role. Silicon significantly improved cyclic oxidation resistance; however, it also decreases the melting point of the coating; 5 wt% are enough to lower the melting temperature to about 1140 °C. Additions of **Re** improve isothermal or cyclic oxidation resistance and thermal cycle fatigue. Additions of Ta can also increase the oxidation resistance [151].

7.10 Applications of Superalloys

Ni-based superalloys have extensive applications in various industries where high temperature application is required. Ni-based superalloys are used in industries like submarine factory, heat exchanger tubing, industrial gas turbines, aerospace (turbine blade and jet/rocket engine), nuclear reactor, etc. Early aircraft gas turbines in the United States employed nickel-base and cobalt-base alloy turbine blades and vanes, iron-base alloy disks, and stainless or nickel-base alloy combustion cans. However, superalloys then accounted for less than 10% of engine weight and were limited to use below 815 °C. Now, in modern aircraft engines, superalloys account for as more as 70% of engine weight. Some of these alloys can withstand temperatures as high as 1040–1100 °C. Engines use Ni superalloys almost entirely for turbine blades, Ni- or Fe-superalloys for turbine wheels, and Ni- or Co-superalloys for vanes and combustion cans. Wider use of these improved superalloys has made it possible to increase turbine inlet temperatures from less than 815 °C to well over 1100 °C with the result that engine performance has increased dramatically. Some of the potential applications are discussed subsequently [152].

The bulk of tonnage of superalloy used is: aircraft gas turbines (disks, combustion chambers, bolts, casings, shafts, exhaust systems, cases, blades, vanes, burner cans, afterburners, thrust reversers), steam turbine power plants (bolts, blades, stack gas reheaters), reciprocating engines (turbochargers, exhaust valves, hot plugs, valve seat inserts), metal processing (hot-work tools and dies, casting dies), medical applications (dentistry uses, prosthetic devices), space vehicles (aerodynamically heated skins, rocket engine parts), heat-treating equipment (trays, fixtures, conveyor belts, baskets, fans, furnace mufflers), nuclear power systems (control rod drive mechanisms, valve stems, springs, ducting), chemical and petrochemical industries (bolts, fans, valves, reaction vessels, piping, pumps), pollution control equipment (scrubbers etc.), metals processing mills (ovens, afterburners, exhaust fans), and coal gasification and liquefaction systems (heat exchangers, reheaters, piping) [153, 154]. Again, some of these applications are described subsequently.

7.10.1 Gas Turbine Engines

In gas turbines, superalloys are used in those areas where operating temperature is very high. It may also be used in those areas where strength, creep resistance, corrosion resistance, and oxidation resistance are needed. New jet engines are considered to be more efficient at higher operating temperature. The use of superalloy in gas turbine increases the operating temperature range from 649 °C to 704 °C [155]. The high-temperature operation not only increases power output but also decreases emissions extensively.

7.10.2 Turbine Blades

Turbine blade is a special part of gas turbines and aeroengines, where superalloys are most preferable material. At higher temperature of about 400 °C and above, Ti cannot sustain its mechanical strength. So Ti can be used in comparatively colder regions. But for very high-temperature region, superalloys are needed. Turbine blade airfoils generally experience longitudinal stresses to approximately 138 MPa and temperatures of 650–980 °C in their airfoil section. The blade root, which attaches to the disk, is outside the hot gas path. It is exposed to a maximum temperature of 760 °C but is subjected to tensile stresses of 276–552 MPa. In addition to these demanding strength requirements, turbine bucket materials must also have adequate ductility to tolerate creep deformation, to resist low-cycle fatigue deformation, and to allow seating of the blade in the disk slot. Since these parts are in contact with high-temperature combustion products, good oxidation resistance is mandatory [2]. During cyclic operation, thermal gradients from heating and cooling of turbine sections of varying sizes induce thermal stresses, which subject the blade airfoils to thermal fatigue. Materials with high strength, ductility, high thermal conductivity, and low thermal expansion provide the best resistance to thermal fatigue. In addition, blade materials must be stable microstructurally so that their properties are retained for longer time. They should also have good impact strength, good castability, and low density. In aeroengine turbine blades and steam turbine blades, Ni-superalloys are mostly used. In these applications, generally single crystal-based blades (Fig. 7.23) are used as single crystal material is free

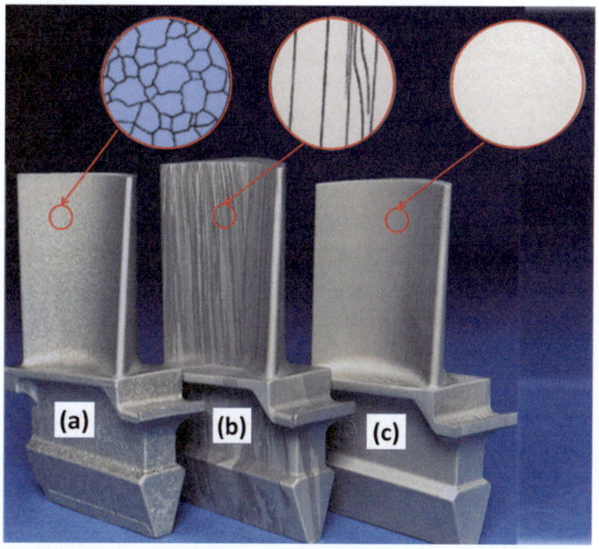

Fig. 7.23 Grain structures of (**a**) conventionally cast, (**b**) directionally solidified, and (**c**) single crystal turbine blades

from γ/γ' grain boundaries. Through γ/γ' grain boundaries, easy diffusion is possible which facilitates to reduce the resistance of material to creep. Again we know that at high temperature, grain boundary has less melting point than grains. So there is chance for crack formation in grain boundary. Hence effort is made to manufacture turbine blades with single crystal material. But it is difficult to manufacture single-crystal materials. In many cases, directionally solidified columnar grain structure finds the alternate application. It has many γ grains, but the boundaries are mostly parallel to the major stress axis. This structure is not very efficient than single crystal but very useful compared to blade with the equiaxed grain structure [156]. In single-crystal alloys, the grain boundary solutes are removed which facilitates to increase melting temperature of alloy due to localized melting with chemical segregation. It is also a reason for which single crystal superalloys are mostly demanded compared to conventionally cast polycrystalline superalloys, as in Fig. 7.23. The single-crystal alloy can be utilized for heat treatment in the range 1240–1330 °C which allows dissolution of coarse γ'. Coarse γ' was present due to solidification process. So the heat treatment process leads the coarse γ' material to be precipitated in fine structure. Rather than the property of removing grain boundary, the possibility of application of heat treatment process makes the alloy to be used at high-temperature applications. High heat treatment temperature leads the γ' to precipitate in a finer form [157].

Most severe environment is observed in high-pressure turbine jet engines. Here in addition to very high temperature close to melting temperature, high pressure is also developed due to rotation at more than 10,000 rpm. But use of thermal barrier coating in blade restricts the melting due to high temperature. Before reaching to critical condition, blades suffer accelerated oxidation and hot corrosion. Good coating can improve hot corrosion resistance or oxidation of the components [158].

7.10.3 Turbine Discs

In turbine system, the turbine blade is connected to the turbine disc which in turn connected to the shaft. But turbine disc is of different property than turbine. Turbine disc experiences low temperature than turbine but it should be highly fatigue resistant. Generally turbine disc (Fig. 7.24) is casted and forged so that polycrystalline microstructure is developed throughout. But as we know large columnar grain structure and chemical segregation take place in casted material. The latter leads to nonuniformity in mechanical property. Hence to eliminate the difficulty clean fine powdered material is used for manufacturing of disc by powder metallurgy route [159]. Turbine disc operating temperatures are generally around 760 °C. This temperature is experienced only on the outer rim in the area of blade attachment. In the hub portion, toward the center of the disk, temperatures are much lower. Operating stresses due to centrifugal loads in these rotating components are high at the rim and still higher toward the center, >483 MPa. Materials, therefore, require very high tensile strength at hub operating temperatures for burst protection and high creep strength at rim operating conditions. Also, good fatigue resistance, both low cycle and high cycle, is the primary disc requirements. For many small gas turbines, such as small aircraft engines, the turbine disc and blades are cast as a single piece, generally known as an integral wheel, and are called as BLISKs. Materials for integral wheels must possess the combination of blade and disk, in addition to very good castability [160].

Fig. 7.24 Powder metallurgical aeroengine disc

7.10.4 Turbine Nozzle Guide Vanes

These are static components, turbine nozzle guide vanes are subjected to relatively low stresses, generally below 69 MPa. Operating temperatures, however, can be very high as the hot gases entering the turbine section impinge directly upon the first-stage vane. Metal temperatures of approximately 1100 °C are often encountered in advanced turbine vanes so that there is the use of superalloy. The primary material requirements are creep strength at very high temperatures, thermal fatigue resistance, and resistance to surface degradation by oxidation, hot corrosion, and erosion. As with turbine buckets, secondary requirements are impact strength, castability, microstructural stability, and weldability in some designs [161]

7.10.5 Turbochargers

It is known that the internal combustion engine uses both air and fuel to initiate combustion. For boosting of power, a

Fig. 7.25 Turbocharger of Ni-superalloy, Inconel

turbocharger is used. Figure 7.25 shows a Ni-based (Inconel) turbocharger. A turbocharger forces more air to the combustion chamber for combustion; thus, increases power output. The turbocharger has two components. One turbine that operates by the exhaust gas drives an air pump, which pushes more air into the engine. As it is driven by exhaust gasses and rate of spin is very large so it gets very hot. Hence it should be strong and oxidation resistant [162].

7.10.6 Combustion Cans

The required strength of combustion can (Fig. 7.26) materials are relatively low and their strength must be maintained to operate at a temperature around 1100 °C. Oxidation resistance is the prime requisite, along with resistance to thermal fatigue and buckling. The material must also have good weldability and formability to facilitate fabrication. Combustion vessels are expected to experience extremely high temperatures and this is where Ni-based alloys are best used [163].

7.10.7 Steam Turbines and Nuclear Application

In steam power plant, the turbine experiences very high temperature. The turbine requires to be made up of single-crystal high-temperature resistant materials. Hence generally Ni-superalloys are used in turbine. The novel properties, like temperature resistance and strength of Ni-superalloys, make

Fig. 7.26 Combustion cans

its use in nuclear industry. Ni-superalloys are also used in reactor core and control rod [164].

7.10.8 Aero and Land Turbines

Cobalt superalloys are creep resistant and fatigue resistant. These are used in nonrotating applications where stress produced is very less. Hence static nonrotating components and turbine vanes of aero turbines are manufactured with Co superalloys. Cobalt superalloy has better thermal conductivity and lower coefficient of thermal expansion. Hence it acts suitably where thermal fatigue is an issue [165].

7.10.9 Oil and Gas Industry

In oil and gas industry, Ni-superalloys are increasingly used. In this industry, material is supposed to be corroded. So the use of suitable material is very challenging. Hydrogen sulfide, chlorides, free sulfur, and carbon dioxide are present in the working environment. In some environments also, high temperature and pressure are also exerted. So special materials are required for processing of natural gas and oil. In oil and natural gas production, Nickel superalloys 718, 725, and 925 are generally used. Molybdenum and chromium are major constituents in these alloys, which prohibit the corrosion. Alloy 718 was initially developed for use in aerospace and gas turbines, but has become the preferred material for the manufacture of wellhead components, auxiliary and down-hole tools, and subsurface safety valves [166].

7.10.10 Engine of Y2K Superbike

The bike uses a Rolls Royce Allison 250 gas turbine engine, which has been shown in Fig. 7.27. In superbike, very high temperature is generated in the engine. Hence Ni based

Fig. 7.27 Gas turbine engine of Y2K superbike

Fig. 7.28 Pressurized water reactor vessel head

superalloy is used in its blades so as to withstand high temperature. In engine, fuel is atomized in a combustion chamber located in front of turbine wheels. As the ignition takes place gas is emitted from the combustion process. It expands and rotates the fan resulting in a very large amount of power in the form of thrust [167].

7.10.11 Pressurized Water Reactor Vessel Head

The mechanisms for environmentally sensitive cracking in water-cooled reactors have been observed due to intergranular stress corrosion cracking, irradiation-assisted stress corrosion cracking, and corrosion fatigue. These mechanisms are affected by several variables including metallurgical structure, irradiation effects on grain boundary, and the aqueous environment. Figure 7.28 has shown a pressurized water reactor vessel head. Here, Ni superalloy is well fit for the application, but is sensitive to the presence of impurities including phosphorous, silicon, boron, and sulfur [168].

7.10.12 Reactor Vessel

Ti-lined reactor vessel (Fig. 7.29) is abundantly used in mining operation. Reactor vessels and many chemical industries use Ti lining for protection from corrosion and abrasion. These lining systems are required for processes that are corrosive and can operate at elevated temperatures. When used in temperature-exposed chemical process, Ni- and Co-superalloy lined reactor vessels have been found to be suitable for resisting the wear and corrosion [169].

7.10.13 Tube Exchanger

Straight heat tube exchanger (Fig. 7.30) uses Ni-superalloy for high-temperature operation. Initially, high-temperature heat exchangers are used for gas turbines made of ceramics. But nowadays Ni-based superalloy replaces ceramics for its enhanced properties [170].

7.10.14 Ti-Tubed Salt Evaporator for Table Salt

The process of obtaining table salt from seawater makes use of "Ti-based salt evaporator" (Fig. 7.31). The process includes obtaining salt by evaporation of seawater. This process is facilitated by Ti-based evaporator. Due to the involvement of high temperature, Ti alloys can be replaced by the superalloy to enhance the erosion resistance [171].

7.10.15 Casting Shell

Casting shell or mold is often made up of superalloys. When molten metal is poured to the mold, the casting shell experiences a large amount of heat. So, in order to sustain the high temperature, the shell material is usually made up of Ni-based or Ti-based superalloys depending on the temperature generated. Again by using superalloys, the surface finish and quality of casting material are enhanced as shown in below figures [172]. Figure 7.32 is showing the pattern of

Fig. 7.29 Reactor vessels

Fig. 7.30 Tube exchanger

Fig. 7.31 Salt evaporator

turbocharger rotor, the investment mold, and the Ni-based final product.

Cobalt-based superalloys have also applications in various fields like rocket motors, power plants, space vehicles, gas turbines, etc. It is also found out that for fabrication of alkali metal thermal to electrical conversion cells, cobalt-based superalloys are used as structural members. It is also extensively used in space applications and other heat source applications.

7.11 Summary

Superalloys have excellent heat-resistant properties and retain their stiffness, strength, toughness, and dimensional stability at temperatures much higher than the conventional structural materials. This chapter discussed various Ni-based, Co-based, and Fe-based superalloys. Constituting elements and effect on the mechanical properties of the superalloys at high temperature is the focus of interest of this chapter. All the high-end application at different condition at high temperature has been discussed. Still a lot of mystery to be found out in superalloy posttreatment: on microstructure variation with respect to environmental parameter, which are the future prospective and interest.

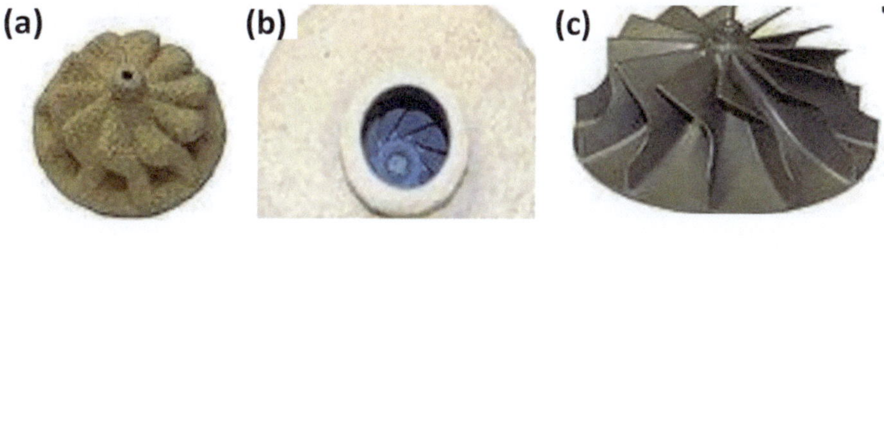

Fig. 7.32 (**a**) The investment shell for casting a turbocharger rotor, (**b**) a view of the interior investment shows the smooth surface finish, and (**c**) the completed workpiece of turbocharger rotor

References

1. Furrer, D., Fecht, H.: Ni-based superalloys for turbine discs. JOM. **51**, 14–17 (1999). https://doi.org/10.1007/s11837-999-0005-y
2. Arakere, N.K., Swanson, G.: Effect of crystal orientation on fatigue failure of single crystal nickel base turbine blade superalloys. ASME. J. Eng. Gas Turbines Power. **124**(1), 161–176 (2002). https://doi.org/10.1115/1.1413767
3. Grundy, E.: Other applications of superalloys. Mater. Sci. Technol. **3**(9), 782–790 (2013). https://doi.org/10.1179/mst.1987.3.9.782
4. Hardy, M.C., Detrois, M., McDevitt, E.T., et al.: Solving recent challenges for wrought Ni-base superalloys. Metall. Mat. Trans. A. **51**, 2626–2650 (2020). https://doi.org/10.1007/s11661-020-05773-6
5. Meetham, G.W.: High-temperature materials. Ullmann's Encycl. Indus. Chem. (2001). https://doi.org/10.1002/14356007.a13_025.
6. Nowotnik, A.: Nickel-Based Superalloys, Reference Module in Materials Science and Materials Engineering. Elsevier, Netherland (2016). https://doi.org/10.1016/B978-0-12-803581-8.02574-1
7. Gialanella, S., Malandruccolo, A.: Superalloys Aerospace Alloys. In: Topics in Mining, Metallurgy and Materials Engineering. Springer, Cham (2020). https://doi.org/10.1007/978-3-030-24440-8_6
8. Brady, M.P., Magee, J., Yamamoto, Y., Helmick, D., Wang, L.: Co-optimization of wrought alumina-forming austenitic stainless steel composition ranges for high-temperature creep and oxidation/corrosion resistance. Mater. Sci. Eng. A. **590**, 101–115 (2014). https://doi.org/10.1016/j.msea.2013.10.014
9. Swain, B., Mallick, P., Patel, S., Roshan, R., Mohapatra, S.S., Bhuyan, S., Priyadarshini, M., Behera, B., Samal, S., Behera, A.: Failure analysis and materials development of gas turbine blades. Mater. Today Proc. **33**(Part 8), 5143–5146 (2020). https://doi.org/10.1016/j.matpr.2020.02.859
10. Sohn, Y.H., Lee, E.Y., Nagaraj, B.A., Biederman, R.R., Sisson, R. D.: Microstructural characterization of thermal barrier coatings on high pressure turbine blades. Surf. Coat. Technol. **146–147**, 132–139 (2001). https://doi.org/10.1016/S0257-8972(01)01369-X
11. Leme, S., Domingos, A.: Finite Element Analysis to Verify the Structural Integrity of an Aeronautical Gas Turbine Disc Made from Inconel 713LC Superalloy. In: Advanced Engineering Forum, vol. 32, pp. 15–26. Trans Tech Publications, Ltd., Bäch SZ (2019). https://doi.org/10.4028/www.scientific.net/aef.32.15
12. https://airandspace.si.edu/exhibitions/wright-brothers/online/fly/1903/. Accessed 05 Aug 2019
13. Kracke, A.: Superalloys, the most successful alloy system of modern times - past, present and future. https://www.tms.org/Superalloys/10.7449/2010/Superalloys_2010_13_50.pdf.
14. http://megamex.com/superalloys.htm. Accessed 05 Aug 2019.
15. https://www.tms.org/Meetings/Specialty/Superalloys2000/SuperalloysHistory.html. Accessed 05 Aug 2019.
16. https://www.nap.edu/read/23490/chapter/6. Accessed 05 Aug 2019.
17. Tawancy, H.M.: On the precipitation of intermetallic compounds in selected solid-solution-strengthened Ni-base alloys and their effects on mechanical properties. Metallogr. Microstruct. Anal. **6**, 200–215 (2017). https://doi.org/10.1007/s13632-017-0352-y
18. Tsao, T., Yeh, A., Kuo, C., et al.: The high temperature tensile and creep behaviors of high entropy superalloy. Sci. Rep. **7**, 12658 (2017). https://doi.org/10.1038/s41598-017-13026-7
19. Huang, S., Gao, Y., An, K., Zheng, L., Wu, W., Teng, Z., Liaw, P. K.: Deformation mechanisms in a precipitation-strengthened ferritic superalloy revealed by in situ neutron diffraction studies at elevated temperatures. Acta Mater. **83**, 137–148 (2015). https://doi.org/10.1016/j.actamat.2014.09.053
20. Liu, X., Pan, Y., Chen, Y., Han, J., Yang, S., Ruan, J., Wang, C., Yang, Y., Li, Y.: Effects of Nb and W additions on the microstructures and mechanical properties of Novel γ/γ' Co-V-Ti-based superalloys. Metals. **8**(7), 563 (2018). https://doi.org/10.3390/met8070563
21. https://www.phase-trans.msm.cam.ac.uk/2003/Superalloys/SX/SX.html
22. Kawagishi, K., Harada, H., Sato, A., et al.: The oxidation properties of fourth generation single-crystal nickel-based superalloys. JOM. **58**, 43–46 (2006). https://doi.org/10.1007/s11837-006-0067-z
23. Li, Y., Wang, L., Zhang, G., Zhang, J., Lou, L.: Creep anisotropy of a 3rd generation nickel-base single crystal superalloy at 850 °C. Mater. Sci. Eng. A. **760**, 26–36 (2019). https://doi.org/10.1016/j.msea.2019.05.075
24. Yeh, A.C., Tsao, T.K., Chang, Y.J., Chang, K.C., Yeh, J.W., Chiou, M.S., Jian, S.R., Kuo, C.M., Wang, W.R., Murakami, H.: Developing new type of high temperature alloys–high entropy superalloys. Int. J. Metall. Mater. Eng. **1**, 107 (2015). https://doi.org/10.15344/2455-2372/2015/107
25. Zheng, L., Schmitz, G., Meng, Y., Chellali, R., Schlesiger, R.: Mechanism of intermediate temperature embrittlement of Ni and Ni-based superalloys. Crit. Rev. Solid State Mater. Sci. **37** (3) (2012). https://doi.org/10.1080/10408436.2011.613492
26. Lv, D.C., McAllister, D., Mills, M.J., Wang, Y.: Deformation mechanisms of D022 ordered intermetallic phase in superalloys. Acta Mater. **118**, 350–361 (2016). https://doi.org/10.1016/j.actamat.2016.07.055
27. Heckl, A., Neumeier, S., Göken, M., Singer, R.F.: The effect of Re and Ru on γ/γ' microstructure, γ-solid solution strengthening and creep strength in nickel-base superalloys. Mater. Sci. Eng. A. **528** (9), 3435–3444 (2011). https://doi.org/10.1016/j.msea.2011.01.023
28. Pouranvari, M.: Solid solution strengthening of transient liquid phase bonded nickel based superalloy. Mater. Sci. Technol. **31** (14), 1773–1780 (2015). https://doi.org/10.1179/1743284715Y.0000000005
29. Benjamin, J.S.: Dispersion strengthened superalloys by mechanical alloying. Metall. Trans. **1**, 2943–2951 (1970). https://doi.org/10.1007/BF03037835
30. Capdevila, C., Bhadeshia, H.K.D.H.: Manufacturing and microstructural evolution of mechanuically alloyed oxide dispersion strengthened superalloys. Adv. Eng. Mater. **3**(9), 647–656 (2001). https://doi.org/10.1002/1527-2648(200109)3:9<647::AID-ADEM647>3.0.CO;2-4
31. Kolb, M., Freund, L.P., Fischer, F., Povstugar, I., Makineni, S.K., Gault, B., Raabe, D., Müller, J., Spiecker, E., Neumeier, S., Göken, M.: On the grain boundary strengthening effect of boron in γ/γ' Cobalt-base superalloys. Acta Mater. **145**, 247–254 (2018). https://doi.org/10.1016/j.actamat.2017.12.020
32. Tawancy, H.M.: Order-Strengthening in a nickel-base superalloy (hastelloy alloy S). Metall. Trans. A. **11**, 1764–1765 (1980). https://doi.org/10.1007/BF02660534
33. Long, H., Liu, Y., Kong, D., Wei, H., Chen, Y., Mao, S.: Shearing mechanisms of stacking fault and anti-phase-boundary forming dislocation pairs in the γ' phase in Ni-based single crystal superalloy. J. Alloys Comp. **724**, 287–295 (2017). https://doi.org/10.1016/j.jallcom.2017.07.020
34. Friák, M., Buršíková, V., Pizúrová, N., Pavlů, J., Jirásková, Y., Homola, V., Miháliková, I., Slávik, A., Holec, D., Všianská, M., Koutná, N., Fikar, J., Janičkovič, D., Šob, M., Neugebauer, J.: Elasticity of phases in Fe-Al-Ti superalloys: impact of atomic order and anti-phase boundaries. Crystals. **9**, 299 (2019). https://doi.org/10.3390/cryst9060299
35. Thakur, A., Gangopadhyay, S.: State-of-the-art in surface integrity in machining of nickel-based super alloys. Int. J. Mach. Tools

Manuf. **100**, 25–54 (2016). https://doi.org/10.1016/j.ijmachtools.2015.10.001
36. Sharman, A.R.C., Hughes, J.I., Ridgway, K.: Workpiece surface integrity and tool life issues when turning inconel 718™ nickel based superalloy. Mach. Sci. Technol. Int. J. **8**(3) (2004). https://doi.org/10.1081/MST-200039865
37. Ojo, O.A., Chaturvedi, M.C.: On the role of liquated γ' precipitates in weld heat affected zone microfissuring of a nickel-based superalloy. Mater. Sci. Eng. A. **403**(1–2), 77–86 (2005). https://doi.org/10.1016/j.msea.2005.04.034
38. Nitz, A., Nembach, E.: Critical resolved shear stress anomalies of the L12-long-range ordered γ'-phase of the superalloy NIMONIC 105. Mater. Sci. Eng. A. **263**(1), 15–22 (1999). https://doi.org/10.1016/S0921-5093(98)01025-9
39. Bocchini, P.J., Sudbrack, C.K., Noebe, R.D., Seidman, D.N.: Temporal evolution of a model Co-Al-W superalloy aged at 650 °C and 750 °C. Acta Mater. **159**, 197–208 (2018). https://doi.org/10.1016/j.actamat.2018.08.014
40. Véron, M., Bréchet, Y., Louchet, F.: Directional coarsening of Ni-based superalloys: Computer simulation at the mesoscopic level. Acta Mater. **44**(9), 3633–3641 (1996). https://doi.org/10.1016/1359-6454(96)00011-0
41. Strondl, A., Fischer, R., Frommeyer, G., Schneider, A.: Investigations of MX and γ'/γ'' precipitates in the nickel-based superalloy 718 produced by electron beam melting. Mater. Sci. Eng. A. **480**(1–2), 138–147 (2008). https://doi.org/10.1016/j.msea.2007.07.012
42. Charpagne, M.-A., Vennéguès, P., Billot, T., Franchet, J.-M., Bozzolo, N.: Evidence of multimicrometric coherent γ' precipitates in a hot-forged γ–γ' nickel-based superalloy. J. Microsc. **263**(1), 106–112 (2016). https://doi.org/10.1111/jmi.12380
43. le Graverend, J.-B., Cormier, J., Jouiad, M., Gallerneau, F., Paulmier, P., Hamon, F.: Effect of fine γ' precipitation on non-isothermal creep and creep-fatigue behaviour of nickel base superalloy MC2. Mater. Sci. Eng. A. **527**(20), 5295–5302 (2010). https://doi.org/10.1016/j.msea.2010.04.097
44. Zhang, H., Li, C., Guo, Q., Ma, Z., Li, H., Liu, Y.: Improving creep resistance of nickel-based superalloy Inconel 718 by tailoring gamma double prime variants. Scripta Mater. **164**, 66–70 (2019). https://doi.org/10.1016/j.scriptamat.2019.01.041
45. Dong, X., Zhang, X., Kui, D., Zhou, Y., Jin, T., Ye, H.: Microstructure of carbides at grain boundaries in nickel based superalloys. J. Mater. Sci. Technol. **28**(11), 1031–1038 (2012). https://doi.org/10.1016/S1005-0302(12)60169-8
46. Al-Jarba, K.A., Fuchs, G.E.: Carbon-containing single-crystal nickel-based superalloys: Segregation behavior and carbide formation. JOM. **56**, 50–55 (2004). https://doi.org/10.1007/s11837-004-0201-8
47. Kontis, P., Li, Z., Collins, D.M., Cormier, J., Raabe, D., Gault, B.: The effect of chromium and cobalt segregation at dislocations on nickel-based superalloys. Scripta Mater. **145**, 76–80 (2018). https://doi.org/10.1016/j.scriptamat.2017.10.005
48. Peng-jie, Z., Jin-jiang, Y., Xiao-feng, S., Heng-rong, G., Zhuang-qi, H.: Roles of Zr and Y in cast microstructure of M951 nickel-based superalloy. Trans. Nonferrous Metals Soc. China. **22**(7), 1594–1598 (2012). https://doi.org/10.1016/S1003-6326(11)61361-7
49. Szczotok, A., Rodak, K.: Microstructural studies of carbides in MAR-M247 nickel-based superalloy. IOP Conf. Ser.: Mater. Sci. Eng. **35**, 012006. https://doi.org/10.1088/1757-899X/35/1/012006
50. Pigrova, G.D., Rybnikov, A.I.: Carbide phases in a multicomponent superalloy Ni-Co-W-Cr-Ta-Re. Phys. Metals Metallogr. **114**, 593–595 (2013). https://doi.org/10.1134/S0031918X13070089
51. Pérez-González, F.A., Garza-Montes-de Oca, N.F., Colás, R.: High temperature oxidation of the Haynes 282© Nickel-Based superalloy. Oxid. Met. **82**, 145–161 (2014). https://doi.org/10.1007/s11085-014-9483-6
52. Ghasemi, A., Pouranvari, M.: Thermal processing strategies enabling boride dissolution and gamma prime precipitation in dissimilar nickel-based superalloys transient liquid phase bond. Mater. Des. **182**, 108008 (2019). https://doi.org/10.1016/j.matdes.2019.108008.
53. Rae, C.M.F., Reed, R.C.: The precipitation of topologically close-packed phases in rhenium-containing superalloys. Acta Mater. **49** (19), 4113–4125 (2001). https://doi.org/10.1016/S1359-6454(01)00265-8
54. Krakow, R., Johnstone, D.N., Eggeman, A.S., Hünert, D., Hardy, M.C., Rae, C.M.F., Midgley, P.A.: On the crystallography and composition of topologically close-packed phases in ATI 718Plus®. Acta Mater. **130**, 271–280 (2017). https://doi.org/10.1016/j.actamat.2017.03.038
55. Robert, J.M.: Polycrystalline nickel-based superalloys: processing, performance, and application. In: Encyclopedia of Aerospace Engineering. John Wiley & Sons, Inc., Hoboken, NJ (2010). https://doi.org/10.1002/9780470686652.eae217
56. Chen, M., Wang, G., Li, H., et al.: Precipitation and dissolution behaviors of δ phase inside a deformed nickel-based superalloy during annealing treatment. Appl. Phys. A Mater. Sci. Process. **125**, 447 (2019). https://doi.org/10.1007/s00339-019-2741-3
57. Kulo, N., He, S., Ecker, W., Pippan, R., Antretter, T., Razumovskiy, V.I.: Thermodynamic and mechanical stability of Ni3X-type intermetallic compounds. Intermetallics. **114**, 106604 (2019). https://doi.org/10.1016/j.intermet.2019.106604.
58. Proult, A., Donnadieu, P.: Identification of the fault vectors of planar defects in the sigma phase of a nickel-based superalloy. Philos. Mag. A. **72**(2), 403–414 (2006). https://doi.org/10.1080/01418619508239932
59. Xiao, H., Li, S.M., Xiao, W.J., Li, Y.Q., Cha, L.M., Mazumder, J., Song, L.J.: Effects of laser modes on Nb segregation and Laves phase formation during laser additive manufacturing of nickel-based superalloy. Mater. Lett. **188**, 260–262 (2017). https://doi.org/10.1016/j.matlet.2016.10.118
60. Zhao, J., Henry, M.F.: The thermodynamic prediction of phase stability in multicomponent superalloys. JOM. **54**, 37–41 (2002). https://doi.org/10.1007/BF02822603
61. Shekhar, R., Arunachalam, J., Das, N., Murthy, A.M.S.: Chemical and structural characterisation of nickel based superalloys doped with minor and trace elements. Mater. Sci. Eng. A. **435–436**, 491–498 (2006). https://doi.org/10.1016/j.msea.2006.07.033
62. Darolia, R.: Development of strong, oxidation and corrosion resistant nickel-based superalloys: critical review of challenges, progress and prospects. Int. Mater. Rev. **64**(6), 355–380 (2019). https://doi.org/10.1080/09506608.2018.1516713
63. Wenderoth, M., Völkl, R., Vorberg, S., Yamabe-Mitarai, Y., Harada, H., Glatzel, U.: Microstructure, oxidation resistance and high-temperature strength of γ' hardened Pt-base alloys. Intermetallics. **15**(4), 539–549 (2007). https://doi.org/10.1016/j.intermet.2006.09.009
64. https://www.reade.com/products/superalloy-metal-and-powders. Accessed 09 Aug 2019.
65. https://www.additive.sandvik/en/products-services-am/metal-powder/nickel-based-superalloys/. Accessed 09 Aug 2019.
66. https://integration.metalpowder.sandvik/en/products/metal-powder-alloys/superalloys/ni-based-superalloys-for-additive-manufacturing/. Accessed 09 Aug 2019.
67. Gurrappa, I., Yashwanth, I.V.S., Mounika, I., Murakami, H., Kuroda, S.: The Importance of Hot Corrosion and Its Effective Prevention for Enhanced Efficiency of Gas Turbines. IntechOpen, London (2015). https://doi.org/10.5772/59124
68. Harris, K., Wahl, J.B.: Developments in superalloy castability and new applications for advanced superalloys. Mater. Sci. Technol. **25**

(2) Malcolm McLean Memorial Symposium: The superalloys: from processing to performance', 147–153 (2009). https://doi.org/10.1179/174328408X355442
69. Rajan, K., Vander Sande, J.B.: Room temperature strengthening mechanisms in a Co-Cr-Mo-C alloy. J. Mater. Sci. **17**, 769–778 (1982). https://doi.org/10.1007/BF00540374
70. Mudgal, D., Singh, S., Prakash, S.: Hot corrosion behavior of some superalloys in a simulated incinerator environment at 900 °C. J. Mater. Eng. Perform. **23**, 238–249 (2014). https://doi.org/10.1007/s11665-013-0721-x
71. Nithin, B., Samanta, A., Makineni, S.K., et al.: Effect of Cr addition on γ–γ′ cobalt-based Co–Mo–Al–Ta class of superalloys: a combined experimental and computational study. J. Mater. Sci. **52**, 11036–11047 (2017). https://doi.org/10.1007/s10853-017-1159-6.
72. Reyes Tirado, F.L., Taylor, S., Dunand, D.C.: Effect of Al, Ti and Cr additions on the γ-γ′ microstructure of W-free Co-Ta-V-based superalloys. Acta Mater. **172**, 44–54 (2019). https://doi.org/10.1016/j.actamat.2019.04.031
73. Jin, M., Miao, N., Zhao, W., Zhou, J., Qiang, D., Sun, Z.: Structural stability and mechanical properties of Co_3(Al, M) (M = Ti, V, Cr, Zr, Nb, Mo, Hf, Ta, W) compounds. Comp. Mater. Sci. **148**, 27–37 (2018). https://doi.org/10.1016/j.commatsci.2018.02.015
74. Klein, L., Bauer, A., Neumeier, S., Göken, M., Virtanen, S.: High temperature oxidation of γ/γ′-strengthened Co-base superalloys. Corros. Sci. **53**(5), 2027–2034 (2011). https://doi.org/10.1016/j.corsci.2011.02.033
75. Murray, S.P., Stinville, J., Callahan, P.G., et al.: Low cycle fatigue of single crystal γ′-containing co-based superalloys at 750 °C. Metall. Mat. Trans. A. **51**, 200–213 (2020). https://doi.org/10.1007/s11661-019-05508-2
76. Lass, E.A., Sauza, D.J., Dunand, D.C., Seidman, D.N.: Multicomponent γ′-strengthened Co-based superalloys with increased solvus temperatures and reduced mass densities. Acta Mater. **147**, 284–295 (2018). https://doi.org/10.1016/j.actamat.2018.01.034
77. Jiang, C.: First-principles study of Co_3(Al,W) alloys using special quasi-random structures. Scripta Mater. **59**(10), 1075–1078 (2008). https://doi.org/10.1016/j.scriptamat.2008.07.021
78. Ooshima, M., Tanaka, K., Okamoto, N.L., Kishida, K., Inui, H.: Effects of quaternary alloying elements on the γ′ solvus temperature of Co–Al–W based alloys with fcc/L12 two-phase microstructures. J. Alloys Compd. **508**(1), 71–78 (2010). https://doi.org/10.1016/j.jallcom.2010.08.050
79. Zhu, J., Titus, M.S., Pollock, T.M.: Experimental investigation and thermodynamic modeling of the co-rich region in the Co-Al-Ni-W quaternary system. J. Phase Equilib. Diffus. **35**, 595–611 (2014). https://doi.org/10.1007/s11669-014-0327-5
80. Cartón-Cordero, M., Campos, M., Freund, L.P., Kolb, M., Neumeier, S., Göken, M., Torralba, J.M.: Microstructure and compression strength of Co-based superalloys hardened by γ′ and carbide precipitation. Mater. Sci. Eng. A. **734**, 437–444 (2018). https://doi.org/10.1016/j.msea.2018.08.007
81. Lopez-Galilea, I., Koßmann, J., Kostka, A., et al.: The thermal stability of topologically close-packed phases in the single-crystal Ni-base superalloy ERBO/1. J. Mater. Sci. **51**, 2653–2664 (2016). https://doi.org/10.1007/s10853-015-9579-7
82. https://shodhganga.inflibnet.ac.in/bitstream/10603/79593/7/07_chapter%201.pdf.
83. Ates, H., Turker, M., Kurt, A.: Effect of friction pressure on the properties of friction welded MA956 iron-based superalloy. Mater. Des. **28**(3), 948–953 (2007). https://doi.org/10.1016/j.matdes.2005.09.015
84. Yeh, T.-K., Chang, H.-P., Wang, M.Y., Kai, J.J., Yuan, T.: Corrosion of nickel and iron based superalloys in high temperature gas environments. ECS Trans. **50**, 31 (2013). https://doi.org/10.1149/05031.0033ecst
85. Sidhu, T.S., Malik, A., Prakash, S., et al.: Oxidation and hot corrosion resistance of HVOF WC-NiCrFeSiB coating on Ni- and Fe-based superalloys at 800 °C. J. Therm. Spray Tech. **16**, 844–849 (2007). https://doi.org/10.1007/s11666-007-9106-8
86. Zhang, B.B., Yan, F.K., Zhao, M.J., Tao, N.R., Lu, K.: Combined strengthening from nanotwins and nanoprecipitates in an iron-based superalloy. Acta Mater. **151**, 310–320 (2018). https://doi.org/10.1016/j.actamat.2018.04.001
87. Özgür Özgün, H., Gülsoy, Ö., Yilmaz, R., Findik, F.: Injection molding of nickel based 625 superalloy: sintering, heat treatment, microstructure and mechanical properties. J. Alloys Compd. **546**, 192–207 (2013). https://doi.org/10.1016/j.jallcom.2012.08.069
88. Moody, N.R., Baskes, M.I., Robinson, S.L., Perra, M.W.: Temperature effects on hydrogen-induced crack growth susceptibility of iron-based superalloys. Eng. Fract. Mech. **68**(6), 731–750 (2001). https://doi.org/10.1016/S0013-7944(00)00122-3
89. Gopala Rao Thellaputta, Pulcharu Subhash Chandra, C.S.P. Rao, Machinability of nickel based superalloys: a review, Mater. Today Proc., 4, 2, Part A, 2017, 3712–3721, doi: 10.1016/j.matpr.2017.02.266.
90. Miller, M.K., Hetherington, E.M.G.: Atom probe analysis of B' precipitation in a model iron-based Fe-Ni-Al-Mo superalloy. J. Phys. Colloques. **50**, C8.425–C8-428 (1989). https://doi.org/10.1051/jphyscol:1989872
91. Murakumo, T., Kobayashi, T., Koizumi, Y., Harada, H.: Creep behaviour of Ni-base single-crystal superalloys with various γ′ volume fraction. Acta Mater. **52**(12), 3737–3744 (2004). https://doi.org/10.1016/j.actamat.2004.04.028
92. Reed, R.C., Tao, T., Warnken, N.: Alloys-By-Design: Application to nickel-based single crystal superalloys. Acta Mater. **57**(19), 5898–5913 (2009). https://doi.org/10.1016/j.actamat.2009.08.018
93. Yeh, A.C., Tin, S.: Effects of Ru and Re additions on the high temperature flow stresses of Ni-base single crystal superalloys. Scr. Mater. **52**(6), 519–524 (2005). https://doi.org/10.1016/j.scriptamat.2004.10.039
94. Ma, A., Dye, D., Reed, R.C.: A model for the creep deformation behaviour of single-crystal superalloy CMSX-4. Acta Mater. **56**(8), 1657–1670 (2008). https://doi.org/10.1016/j.actamat.2007.11.031
95. Hong, H.U., Kang, J.G., Choi, B.G., Kim, I.S., Yoo, Y.S., Jo, C.Y.: A comparative study on thermomechanical and low cycle fatigue failures of a single crystal nickel-based superalloy. Int. J. Fatigue. **33**(12), 1592–1599 (2011). https://doi.org/10.1016/j.ijfatigue.2011.07.009
96. Woodford, D.A., Mowbray, D.F.: Effect of material characteristics and test variables on thermal fatigue of cast superalloys. A review. Mater. Sci. Eng. A. **16**(1–2), 5–43 (1974). https://doi.org/10.1016/0025-5416(74)90135-9
97. Woodford, D.A., Frawley, J.J.: The effect of grain boundary orientation on creep and rupture of IN-738 and nichrome. Metall. Trans. **5**, 2005–2013 (1974). https://doi.org/10.1007/BF02644493
98. Khan, T., Caron, P., Nakagawa, Y.G.: Mechanical behavior and processing of DS and single crystal superalloys. JOM. **38**, 16–19 (1986). https://doi.org/10.1007/BF03258708
99. Satyanarayana, D.V.V., Eswara, P.N.: Nickel-Based Superalloys. In: Prasad, N., Wanhill, R. (eds.) Aerospace Materials and Material Technologies Indian Institute of Metals Series. Springer, Singapore (2017). https://doi.org/10.1007/978-981-10-2134-3_9
100. Erickson, G.L.: A new, third-generation, single-crystal, casting superalloy. JOM. **47**, 36–39 (1995). https://doi.org/10.1007/BF03221147
101. Li, Z., Xiong, J., Xu, Q., Li, J., Liu, B.: Deformation and recrystallization of single crystal nickel-based superalloys during investment casting. J. Mater. Process. Technol. **217**, 1–12 (2015). https://doi.org/10.1016/j.jmatprotec.2014.10.019
102. Matysiak, H., Zagorska, M., Andersson, J., Balkowiec, A., Cygan, R., Rasinski, M., Pisarek, M., Andrzejczuk, M., Kubiak, K.,

Kurzydlowski, K.J.: Microstructure of Haynes® 282® superalloy after vacuum induction melting and investment casting of thin-walled components. Materials. **6**, 5016–5037 (2013)

103. Niu, J.P., Sun, X.F., Jin, T., Yang, K.N., Guan, H.R., Hu, Z.Q.: Investigation into deoxidation during vacuum induction melting refining of nickel base superalloy using CaO crucible. Mater. Sci. Technol. **19**(4), 435–439 (2003). https://doi.org/10.1179/026708303225010704
104. Harker, H.R.: Experience with large scale electron beam cold hearth melting (EBCHM). Vacuum. **41**(7–9), 2154–2156 (1990). https://doi.org/10.1016/0042-207X(90)94211-8
105. Nastac, L.: Multiscale modelling approach for predicting solidification structure evolution in vacuum arc remelted superalloy ingots. Mater. Sci. Technol. **28**(8), 1006–1013 (2013). https://doi.org/10.1179/1743284712Y.0000000010
106. Jiang, Z., Hou, D., Dong, Y., et al.: Effect of slag on titanium, silicon, and aluminum contents in superalloy during electroslag remelting. Metall. Mat. Trans. B. **47**, 1465–1474 (2016). https://doi.org/10.1007/s11663-015-0530-8
107. Gessinger, G.H., Bomford, M.J.: Powder metallurgy of superalloys. Int. Metall. Rev. 1974. **19**(1), 51–76 (2013). https://doi.org/10.1179/imtlr.1974.19.1.51
108. Hack, G.A.J.: Developments in the production of oxide dispersion strengthened superalloys. Powder Metall. **27**(2), 73–79 (2013). https://doi.org/10.1179/pom.1984.27.2.73
109. Tang, C.F., Pan, F., Qu, X.H., Duan, B.H., He, X.B.: Nickel base superalloy GH4049 prepared by powder metallurgy. J. Alloys Comp. **474**(1–2), 201–205 (2009). https://doi.org/10.1016/j.jallcom.2008.06.038
110. Jia, Q., Dongdong, G.: Selective laser melting additive manufacturing of Inconel 718 superalloy parts: densification, microstructure and properties. J. Alloys Comp. **585**, 713–721 (2014). https://doi.org/10.1016/j.jallcom.2013.09.171
111. Zhang, F., Levine, L.E., Allen, A.J., Stoudt, M.R., Lindwall, G., Lass, E.A., Williams, M.E., Idell, Y., Campbell, C.E.: Effect of heat treatment on the microstructural evolution of a nickel-based superalloy additive-manufactured by laser powder bed fusion. Acta Mater. **152**, 200–214 (2018). https://doi.org/10.1016/j.actamat.2018.03.017
112. Jinoop, A., Paul, C., Bindra, K.: Laser-assisted directed energy deposition of nickel super alloys: a review. Proc. Inst. Mech. Eng. Part L J. Mater. Des. Appl. **233**(11), 2376–2400 (2019). https://doi.org/10.1177/1464420719852658
113. Lu, N., Lei, Z., Hu, K., Yu, X., Li, P., Bi, J., Wu, S., Chen, Y.: Hot cracking behavior and mechanism of a third-generation Ni-based single-crystal superalloy during directed energy deposition. Addit. Manuf. **34**, 101228 (2020). https://doi.org/10.1016/j.addma.2020.101228.
114. Amir Mostafaei, S., Neelapu, H.V.R., Kisailus, C., Nath, L.M., Jacobs, T.D.B., Chmielus, M.: Characterizing surface finish and fatigue behavior in binder-jet 3D-printed nickel-based superalloy 625. Addit. Manuf. **24**, 200–209 (2018). https://doi.org/10.1016/j.addma.2018.09.012
115. Liu, L., Huang, T.W., Zhang, J., Fu, H.Z.: Microstructure and stress rupture properties of single crystal superalloy CMSX-2 under high thermal gradient directional solidification. Mater. Lett. **61**(1), 227–230 (2007). https://doi.org/10.1016/j.matlet.2006.04.037
116. Liu, W., DuPont, J.N.: Effects of melt-pool geometry on crystal growth and microstructure development in laser surface-melted superalloy single crystals: mathematical modeling of single-crystal growth in a melt pool (part I). Acta Mater. **52**(16), 4833–4847 (2004). https://doi.org/10.1016/j.actamat.2004.06.041
117. Zhi, F.H., Liu, L.: Progress of directional solidification in processing of advanced materials. Mater. Sci. Forum. **475–479**., Trans Tech Publications, Ltd., 607–612 (2005). https://doi.org/10.4028/www.scientific.net/msf.475-479.607
118. Rae, C.M.F., Reed, R.C.: Primary creep in single crystal superalloys: origins, mechanisms and effects. Acta Mater. **55**(3), 1067–1081 (2007). https://doi.org/10.1016/j.actamat.2006.09.026
119. Deshpande, A., Deb Nath, S., Atre, S., Hsu, K.: Effect of post processing heat treatment routes on microstructure and mechanical property evolution of Haynes 282 Ni-based superalloy fabricated with Selective Laser Melting (SLM). Metals. **10**, 629 (2020)
120. Wegener, T., GÃnther, J., Brenne, F., Brenne, F., Niendorf, T.: Role of Post-Fabrication Heat Treatment on the Low-Cycle Fatigue Behavior of Electron Beam Melted Inconel 718 Superalloy. In: Shamsaei, N., Daniewicz, S., Hrabe, N., Beretta, S., Waller, J., Seifi, M. (eds.) Structural Integrity of Additive Manufactured Parts, pp. 465–483. ASTM International, West Conshohocken, PA (2020). https://doi.org/10.1520/STP162020180108
121. Smialek, J.L., Jayne, D.T., Schaeffer, J.C., Murphy, W.H.: Effects of hydrogen annealing, sulfur segregation and diffusion on the cyclic oxidation resistance of superalloys: a review. Thin Solid Films. **253**(1–2), 285–292 (1994). https://doi.org/10.1016/0040-6090(94)90335-2
122. Andrieu, E., Molins, R., Ghonem, H., Pineau, A.: Intergranular crack tip oxidation mechanism in a nickel-based superalloy. Mater. Sci. Eng. A. **154**(1), 21–28 (1992). https://doi.org/10.1016/0921-5093(92)90358-8
123. Mahobia, G.S., Neeta Paulose, S.L., Mannan, R.G., Sudhakar, K., Chattopadhyay, N.C.S.S., Singh, V.: Effect of hot corrosion on low cycle fatigue behavior of superalloy IN718. Int. J. Fatigue. **59**, 272–281 (2014). https://doi.org/10.1016/j.ijfatigue.2013.08.009
124. Goward, G.W., Boone, D.H.: Mechanisms of formation of diffusion aluminide coatings on nickel-base superalloys. Oxid. Met. **3**, 475–495 (1971). https://doi.org/10.1007/BF00604047
125. An, T.F., Guan, H.R., Sun, X.F., et al.: Effect of the θ–α-Al_2O_3 transformation in scales on the oxidation behavior of a Nickel-base superalloy with an aluminide diffusion coating. Oxid. Met. **54**, 301–316 (2000). https://doi.org/10.1023/A:1004654429687
126. Bezençon, C., Schnell, A., Kurz, W.: Epitaxial deposition of MCrAlY coatings on a Ni-base superalloy by laser cladding. Scr. Mater. **49**(7), 705–709 (2003). https://doi.org/10.1016/S1359-6462(03)00369-5
127. Strang, A., Lang, E.: Effect of Coatings on the Mechanical Properties of Superalloys. In: Brunetaud, R., Coutsouradis, D., Gibbons, T.B., Lindblom, Y., Meadowcroft, D.B., Stickler, R. (eds.) High Temperature Alloys for Gas Turbines 1982. Springer, Dordrecht (1982). https://doi.org/10.1007/978-94-009-7907-9_19
128. Sidhu, T.S., Agrawal, R.D., Prakash, S.: Hot corrosion of some superalloys and role of high-velocity oxy-fuel spray coatings-a review. Surf. Coat. Technol. **198**(1–3), 441–446 (2005). https://doi.org/10.1016/j.surfcoat.2004.10.056
129. Behera, A., Mishra, S.C.: Prediction and analysis of deposition efficiency of plasma spray coating using artificial intelligence method. Open J. Comp. Mater. **2**, 54–60 (2012). https://doi.org/10.4236/ojcm.2012.22008
130. Ali, M.Y., Nusier, S.Q., Newaz, G.M.: Mechanics of damage initiation and growth in a TBC/superalloy system. Int. J. Solids Struct. **38**(19), 3329–3340 (2001). https://doi.org/10.1016/S0020-7683(00)00261-4
131. D'Ans, P., Dille, J., Degrez, M.: Thermal fatigue resistance of plasma sprayed yttria-stabilised zirconia onto borided hot work tool steel, bonded with a NiCrAlY coating: experiments and modelling. Surf. Coat. Technol. **205**(11), 3378–3386 (2011). https://doi.org/10.1016/j.surfcoat.2010.11.054
132. Gurrappa, I.: Thermal barrier coatings for hot corrosion resistance of CM 247 LC superalloy. J. Mater. Sci. Lett. **17**, 1267–1269 (1998). https://doi.org/10.1023/A:1006547306416

133. Tawancy, H.M., Sridhar, N., Abbas, N.M., et al.: Failure mechanism of a thermal barrier coating system on a nickel-base superalloy. J. Mater. Sci. **33**, 681–686 (1998). https://doi.org/10.1023/A:1004333627312
134. Clarke, D.R., Christensen, R.J., Tolpygo, V.: The evolution of oxidation stresses in zirconia thermal barrier coated superalloy leading to spalling failure. Surf. Coat. Technol. **94–95**, 89–93 (1997). https://doi.org/10.1016/S0257-8972(97)00483-0
135. Sidhu, T.S., Prakash, S., Agrawal, R.D.: Studies of the metallurgical and mechanical properties of high velocity oxy-fuel sprayed stellite-6 coatings on Ni- and Fe-based superalloys. Surf. Coat. Technol. **201**(1–2), 273–281 (2006). https://doi.org/10.1016/j.surfcoat.2005.11.108
136. Kamal, S., Jayaganthan, R., Prakash, S.: Mechanical and microstructural characteristics of detonation gun sprayed NiCrAlY+0.4wt% CeO_2 coatings on superalloys. Mater. Chem. Phys. **122**(1), 262–268 (2010). https://doi.org/10.1016/j.matchemphys.2010.02.046
137. Sun, J., Qian-Gang, F., Liu, G.-N., Li, H.-J., Shu, Y.-C., Gao, F.: Thermal shock resistance of thermal barrier coatings for nickel-based superalloy by supersonic plasma spraying. Ceram. Int. **41**(8), 9972–9979 (2015). https://doi.org/10.1016/j.ceramint.2015.04.077
138. Unal, O., Mitchell, T.E., Heuer, A.H.: Microstructures of Y_2O_3-stabilized ZrO_2 electron beam-physical vapor deposition coatings on Ni-base superalloys. **77**(4), 984–992 (1994). https://doi.org/10.1111/j.1151-2916.1994.tb07256.x
139. Sidhu, T.S., Prakash, S., Agrawal, R.D.: Hot corrosion resistance of high-velocity oxyfuel sprayed coatings on a nickel-base superalloy in molten salt environment. J. Therm. Spray. Tech. **15**, 387–399 (2006). https://doi.org/10.1361/105996306X124392
140. Sidhu, T.S., Prakash, S., Agrawal, R.D.: Performance of high-velocity oxyfuel-sprayed coatings on an Fe-based superalloy in Na_2SO_4-60%V_2O_5 environment at 900 °C part II: Hot corrosion behavior of the coatings. J. Mater. Eng. Perform. **15**, 130–138 (2006). https://doi.org/10.1361/105994906X83411
141. Xiang, Z.D., Burnell-Gray, J.S., Datta, P.K.: Aluminide coating formation on nickel-base superalloys by pack cementation process. J. Mater. Sci. **36**, 5673–5682 (2001). https://doi.org/10.1023/A:1012534220165.
142. Peng, H., Guo, H., Yao, R., He, J., Gong, S.: Improved oxidation resistance and diffusion barrier behaviors of gradient oxide dispersed NiCoCrAlY coatings on superalloy. Vacuum. **85**(5), 627–633 (2010). https://doi.org/10.1016/j.vacuum.2010.09.006
143. Ito, K., Ogawa, K.: Effects of spark-plasma sintering treatment on cold-sprayed copper coatings. J. Therm. Spray. Tech. **23**, 104–113 (2014). https://doi.org/10.1007/s11666-013-0047-0
144. Singh, H., Puri, D., Prakash, S., Maiti, R.: Characterization of oxide scales to evaluate high temperature oxidation behavior of Ni–20Cr coated superalloys. Mater. Sci. Eng. A. **464**(1–2), 110–116 (2007). https://doi.org/10.1016/j.msea.2007.01.088
145. Haynes, J.A., Pint, B.A., Zhang, Y., Wright, I.G.: Comparison of the cyclic oxidation behavior of β-NiAl, β-NiPtAl and γ–γ′ NiPtAl coatings on various superalloys. Surf. Coat. Technol. **202**(4–7), 730–734 (2007). https://doi.org/10.1016/j.surfcoat.2007.06.039
146. Yu, Z., Hass, D.D., Wadley, H.N.G.: NiAl bond coats made by a directed vapor deposition approach. Mater. Sci. Eng. A. **394**(1–2), 43–52 (2005). https://doi.org/10.1016/j.msea.2004.11.017
147. Kiruthika, P., Makineni, S.K., Srivastava, C., Chattopadhyay, K., Paul, A.: Growth mechanism of the interdiffusion zone between platinum modified bond coats and single crystal superalloys. Acta Mater. **105**, 438–448 (2016). https://doi.org/10.1016/j.actamat.2015.12.014
148. Tolpygo, V.K., Clarke, D.R.: Microstructural evidence for counter-diffusion of aluminum and oxygen during the growth of alumina scales. Mater. High Temp. **20**(3), 261–271 (2003). https://doi.org/10.1179/mht.2003.030
149. Benoist, J., Badawi, K.F., Malié, A., Ramade, C.: Microstructure of Pt-modified aluminide coatings on Ni-based superalloys. Surf. Coat. Technol. **182**(1), 14–23 (2004). https://doi.org/10.1016/S0257-8972(03)00871-5
150. Toscano, J., Vaβen, R., Gil, A., Subanovic, M., Naumenko, D., Singheiser, L., Quadakkers, W.J.: Parameters affecting TGO growth and adherence on MCrAlY-bond coats for TBC's. Surf. Coat. Technol. **201**(7), 3906–3910 (2006). https://doi.org/10.1016/j.surfcoat.2006.07.247
151. Meng, G.-H., Liu, H., Liu, M.-J., Xu, T., Yang, G.-J., Li, C.-X., Li, C.-J.: Highly oxidation resistant MCrAlY bond coats prepared by heat treatment under low oxygen content. Surf. Coat. Technol. **368**, 192–201 (2019). https://doi.org/10.1016/j.surfcoat.2019.04.046
152. Goward, G.W.: Current research on the surface protection of superalloys for gas turbine engines. JOM. **22**, 31–39 (1970). https://doi.org/10.1007/BF03355665
153. https://www.coursehero.com/file/p2ab6i8u/Boron-in-quantities-of-0003-to-003-wt-and-less-frequently-small-additions-of/. Accessed 15 Aug 2019.
154. https://super-metals.com/industrial-resources/nickel-based-alloys-phases-and-corrosion-resistance/. Accessed 15 Aug 2019.
155. Pollock, T.M., Tin, S.: Nickel-based superalloys for advanced turbine engines: chemistry, microstructure and properties. J. Propul. Power. **22**(2) (2006). https://doi.org/10.2514/1.18239
156. Lvova, E., Norsworthy, D.: Influence of service-induced microstructural changes on the aging kinetics of rejuvenated Ni-based superalloy gas turbine blades. J. Mater. Eng. Perform. **10**, 299–312 (2001). https://doi.org/10.1361/105994901770345015
157. Broomfield, R.W., Ford, D.A., Bhangu, J.K., Thomas, M.C., Frasier, D.J., Burkholder, P.S., Harris, K., Erickson, G.L., Wahl, J.B.: Development and turbine engine performance of three advanced rhenium containing superalloys for single crystal and directionally solidified blades and vanes. ASME. J. Eng. Gas Turbines Power. **120**(3), 595–608 (1998). https://doi.org/10.1115/1.2818188
158. Yao, H., Bao, Z., Shen, M., Zhu, S., Wang, F.: A magnetron sputtered microcrystalline β-NiAl coating for SC superalloys. Part II. Effects of a NiCrO diffusion barrier on oxidation behavior at 1100°C. Appl. Surf. Sci. **407**, 485–494 (2017). https://doi.org/10.1016/j.apsusc.2017.02.245
159. Fecht, H., Furrer, D.: Processing of Nickel-base superalloys for turbine engine disc applications. Adv. Eng. Mater. **2**(12), 777–787 (2000). https://doi.org/10.1002/1527-2648(200012)2:12<777::AID-ADEM777>3.0.CO;2-R
160. Hu, D., Mao, J., Song, J., Meng, F., Shan, X., Wang, R.: Experimental investigation of grain size effect on fatigue crack growth rate in turbine disc superalloy GH4169 under different temperatures. Mater. Sci. Eng. A. **669**, 318–331 (2016). https://doi.org/10.1016/j.msea.2016.05.063
161. Torroba, A.J., Koeser, O., Calba, L., et al.: Investment casting of nozzle guide vanes from nickel-based superalloys: part I – thermal calibration and porosity prediction. Integr. Mater. **3**, 344–368 (2014). https://doi.org/10.1186/s40192-014-0025-5
162. Shi, Z., Dong, J., Zhang, M., Zheng, L.: Solidification characteristics and segregation behavior of Ni-based superalloy K418 for auto turbocharger turbine. J. Alloys Comp. **571**, 168–177 (2013). https://doi.org/10.1016/j.jallcom.2013.03.241
163. Saegusa, F., Shores, D.A.: Corrosion resistance of superalloys in the temperature range 800–1300 °F (430–700 °C). JMES. **4**, 16–27 (1982). https://doi.org/10.1007/BF02833378
164. Baldridge, T., Poling, G., Foroozmehr, E., Kovacevic, R., Metz, T., Kadekar, V., Gupta, M.C.: Laser cladding of Inconel 690 on Inconel 600 superalloy for corrosion protection in nuclear

applications. Opt. Lasers Eng. **51**(2), 180–184 (2013). https://doi.org/10.1016/j.optlaseng.2012.08.006
165. Patel, S.K., Swain, B., Behera, A.: Advanced processing of superalloys for aerospace industries. In: Functional and Smart Materials. CRC Press (2020). https://doi.org/10.1201/9780429298035. eISBN: 9780429298035. https://www.taylorfrancis.com/books/e/9780429298035/chapters/10.1201/9780429298035-6
166. https://www.tms.org/superalloys/10.7449/2008/Superalloys_2008_31_39.pdf. Accessed 18 Aug 2019.
167. Broomfield, R.W., Ford, D.A., Bhangu, H.K., Thomas, M.C., Frasier, D.J., Burkholder, P.S., Harris, K., Erickson, G.L., Wahl, J.B.: Development and Turbine Engine Performance of Three Advanced Rhenium Containing Superalloys for Single Crystal and Directionally Solidified Blades and Vanes. In: Proceedings of the ASME 1997 International Gas Turbine and Aeroengine Congress and Exhibition Manufacturing Materials and Metallurgy; Ceramics; Structures and Dynamics; Controls, Diagnostics and Instrumentation; Education; IGTI Scholar Award, vol. 4. ASME, Orlando, FL, USA. V004T12A001. https://doi.org/10.1115/97-GT-117
168. Dong, L., Zhang, X., Han, Y., Peng, Q., Deng, P., Wang, S.: Effect of surface treatments on microstructure and stress corrosion cracking behavior of 308L weld metal in a primary pressurized water reactor environment. Corros. Sci. **166**, 108465 (2020). https://doi.org/10.1016/j.corsci.2020.108465
169. Sehoana, M.P., Fazluddin, S.: Development of lining materials for reactor vessel used in the CSIR titanium process. IOP Conf. Ser.: Mater. Sci. Eng. **655**, 012007 (2019). https://doi.org/10.1088/1757-899X/655/1/012007
170. Matsuo, T., Ueki, M., Takeyama, M., et al.: Strengthening of nickel -base superalloys for nuclear heat exchanger application. J. Mater. Sci. **22**, 1901–1907 (1987). https://doi.org/10.1007/BF01132914
171. Hofmeister, M., Klein, L., Miran, H., Rettig, R., Virtanen, S., Singer, R.F.: Corrosion behaviour of stainless steels and a single crystal superalloy in a ternary LiCl–KCl–CsCl molten salt. Corros. Sci. **90**, 46–53 (2015). https://doi.org/10.1016/j.corsci.2014.09.009
172. Szeliga, D., Kubiak, K., Ziaja, W., Cygan, R., Suchy, J.S.Z., Burbelko, A., Nowak, W.J., Sieniawski, J.: Investigation of casting–ceramic shell mold interface thermal resistance during solidification process of nickel based superalloy. Exp. Therm. Fluid Sci. **87**, 149–160 (2017). https://doi.org/10.1016/j.expthermflusci.2017.04.024

Bulk Metallic Glass

8

Abstract

This chapter gives a brief introduction of one category of high-strength advanced material that is bulk metallic glass. Bulk metallic glass-forming liquids are alloys with typically three to five metallic components that have a large atomic size mismatch and a composition close to a deep eutectic. They are dense liquids with small free volumes and high viscosities that are several orders of magnitude higher than those in pure metals or previously known alloys. All the associated mechanism with the bulk amorphization and their strength at room temperature as well as the elevated temperature has been discussed. Thermodynamics and kinetics of the liquid metal during the formation of the glass phase are clearly depicted here. Various bulk metallic glass, such as metal–metal-type alloys, metal–metalloid-type alloys, pd–metalloid-type alloys, has been discussed along with a different kind of fabrication technique. The application of different kind of bulk metallic glass along with their in-developed properties has been discussed.

Keywords

Bulk metallic glass (BMG) · BMG structure · Plasticity · Brittleness · Metal–metal-type alloys · Metal–metalloid-type alloys · Pd–metalloid-type alloys

8.1 Introduction on BMG

The atoms in an ordinary metal arrange themselves into a repeating pattern of crystals or grains with different sizes and shapes upon cooling from the liquid state. Metals typically do not solidify into single crystals, they have inherent weaknesses such as weak strength along the grain boundary. Therefore, at higher stress and temperature, grains slide past each other resulting in metal deformation. In addition, extra atoms are often present in grains causing planes of distortion called dislocations. Dislocations easily move through metal that is under stress, again causing deformation [1]. Grain boundaries and dislocations greatly lower a metal's strength compared to its theoretical maximum. Common metals have approximately one dislocation in one billion molecules or 1014 dislocations in a cubic centimeter [2]. It is these dislocations, rather than the natural ordering of metals, that ultimately define the properties of metals. The structure of metallic glass is very different from that of conventional metals. Rather than arranging themselves into repeating patterns of grains, the atoms of metallic glasses are "frozen" at very high cooling rates in order to suppress the nucleation of crystalline phases results in a random, disordered structure, similar to glass structure [3]. Metallic glasses having a diameter or section thickness of at least 1 mm are classified as "bulk" metallic glasses [4]. Figure 8.1 is showing the typical strength of a general glass and bulk metallic glass after a ballistic impact.

Bulk metallic glasses (BMG) are generally multicomponent systems, that is, there are three or more elements are present. All the elements constitute noncrystalline phase. They can be produced at low solidification rates ~103 K/s or less [5]. There is a large difference between the glass transition temperature and the crystalline temperature hereby resulting in a large supercooled liquid region. Figure 8.2 represents the difference between crystalline, polycrystalline, amorphous, and BMG atomic arrangement. BMG has advantages over crystalline metals including high hardness, high elastic energy storage, and high corrosion resistivity [6]. Bulk metallic glass is twice as strong as titanium, tougher and more elastic than ceramics, and having excellent wear and corrosion resistance makes them attractive for a variety of applications. It can even be cast in a mold to near net shapes.

8.2 History on BMG

The first BMG was discovered by P. Duwez in 1960 by rapid quenching of an $Au_{80}Si_{20}$ liquid [7]. He used a high cooling rate and produced a thickness less than 100 μm. The molten

A. Behera, *Advanced Materials*, https://doi.org/10.1007/978-3-030-80359-9_8

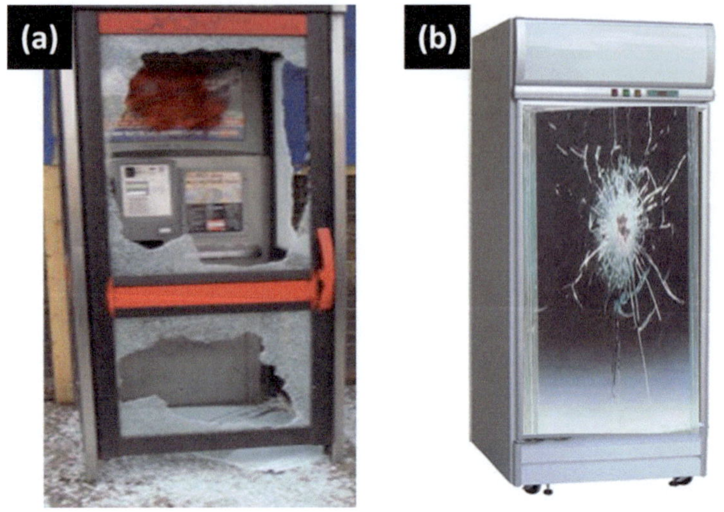

Fig. 8.1 A typical effect of (**a**) General glass and (**b**) Bulk metallic glass after a ballistic impact

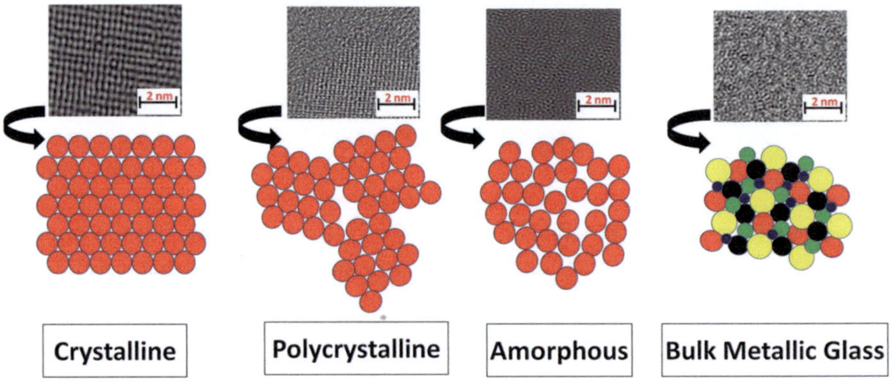

Fig. 8.2 Representation of crystalline, polycrystalline, amorphous, and bulk metallic glass atomic arrangement (TEM morphology at top indicates for respective materials)

metal reaches its glass transition temperature without enough time or energy to crystallize and instead solidifies as metallic glass. Because the material did not conduct heat well, only thin ribbons of metallic glass could be created because of the uniformity and speed of cooling that was required. A few years later, Chen and Turnbull (1969) were able to make amorphous spheres of ternary Pd–X–Si with X = Ag, Cu, or Au [8]. The alloy $Pd_{77.5}Cu_6Si_{16.5}$ could be made glassy with a diameter of 0.5 mm and the existence of a glass transition was demonstrated. In some ternary Pd–Cu–Si and Pd–Ag–Si alloys, the supercooled liquid range (the difference between the first crystallization temperature and the glass transition) extended to 40 K [9]. In addition, Chen (1974) made systematic investigations on Pd–Y–P alloys (Y = Ni, Co, Fe) and obtained a critical casting thickness on the order of 1 mm in these alloys. The research went on and in 1976H [10]. Liebermann and C. Graham developed a new method for making thin strips of amorphous metal with an alloy of Fe, Ni, P, and B, whose name was Metglas. Lee et al. (1982) obtained a slightly higher critical casting thickness in $Au_{55}Pb22._5Sb_{22.5}$. At the beginning of the 1980s, the Turnbull group revisited Pd–Ni–P alloys [11]. By subjecting the specimens to surface etching followed by a succession of heating and cooling cycles, they decreased heterogeneous nucleation and were thus able to make glassy ingots of $Pd_{40}Ni_{40}P_{20}$ with a diameter of 5 mm. Two years later, the same group could extend the critical casting thickness to 1 cm by processing the Pd–Ni–P melt in a boron oxide flux [12]. Pd–Ni–P may be considered as the first BMG to be developed. At this stage, however, people did not recognize the potential for BMG compositions to become widespread or their possible applications. During the late 1980s, the Inoue group in Sendai, Japan investigated rare-earth materials with Al and ferrous metals [13]. While studying rapid solidification in these systems, they found exceptional glass-forming ability in La–Al–Ni and La–Al–Cu alloys [14]. Cylindrical samples with diameters of up to 5 mm or sheets with similar thicknesses were made fully glassy by casting $La_{55}Al_{25}Ni_{20}$ (or later $La_{55}Al_{25}Ni_{10}Cu_{10}$ up to 9 mm) into Cu molds [15]. In the early 1990s, the same group developed glassy Mg–Cu–Y and Mg–Ni–Y alloys with the largest glass-forming ability obtained in $Mg_{65}Cu_{25}Y_{10}$ (Inoue et al. 1991) [16]. At the same time, the Inoue group developed a family of Zr-based Zr–Al–Ni–Cu alloys (Zhang et al.

Table 8.1 Year-wise various BMG development and investigations

Year	BMG
1974	Pd–Cu–Si
1975	Pt–Ni–P, Au–Si–Ge
1982	Pd–Ni–P
1988	Mg–Ln–M (Ln = Lanthanide metal, M = Ni, Cu, or Zn)
1989	Ln–Al–TM (TM-group transition metal), Ln–Ga–TM
1990	Zr–Ti–Al–TM
1993	Ti–Zr–TM, Zr–Ti–TM–Be, Zr–Ti–Cu–Ni–Be
1994	Nd(Pr)–Al–Fe–Co
1995	Zr–(Nb, Pd)–Al–TM, Cu–Zr–Ni–Ti, Fe-(Nb, Mo)-(Al, Ga)-(P, C, B, Si, Ge)
1996	Pd–Cu(Fe)–Ni–P, Co-(Al, Ga)-(P, B, Si), Fe-(Zr, Hf, Nb)-B, Co-Fe-(Zr, Hf, Nb)-B, Ni-(Zr, Hf, Nb)-(Cr, Mo)-B
1997	Pd–Cu–B–Si
1998	Ti–Ni–Cu–Sn, La–Al–Ni–Cu–Co
1999	$La_{55}Al_{25}Ni$, Zr-10Al-5Ti-17.9Cu-14.6Ni, Ni-(Nb,Cr,Mo)-(P,B)
2000	Zr–Nb–Cu–Fe–Be, Zr–Ti–Cu–Ni–Be, Pd–Ni–P
2001	Zr–Al–Ni–Cu, Zr–Ti–Ni–Cu–Be, Zr–Ta–Cu–Ni–Al
2002	Fe–Mn–Mo–Cr–C–B
2003	Ni–Nb–Sn–X (X = B, Fe, Cu), Pr–Cu–Ni–Al, Nd–Al–Fe–Co
2004	Pt–Cu–Ni–P, Cu–Zr–Al–Y
2005	Zr–Cu–Ti
2006	Fe–Co–Nd–Dy–B, Mg–Cu–Y, Fe–Ni–P–B
2007	Mg–Cu–Y, Cu–Ag–Zr–Al, Pr–Ni–Al
2008	Ni–Nb–Cr–Mo–P–B, Pt–Cu–Ni–P, Mg–Cu–Y, **Mg–Cu–Ni–Ag–Zn–Gd–Y**
2009	Zr–Ti–Cu–Ni–Be, Zr–Ti–Be–Cu, Zr–Al–Cu–Ag, (Zr–Cu–Ni–Al)Si
2010	Zr–Ti–Cu–Ni–Al, Fe–Mo–P–C–B, Ti–Zr–Ni–Be–Cu
2011	Fe–Cr–Mo–C–B–Y, La–Al–Ni
2012	Zr–Cu–Al, Zr–Cu–Al, Gd–Ni–Al
2013	Fe–Cr–Mo–C–B, **Zr–Al–Cu–Ni, Zr–Ni–Al**
2014	Gd–Al–Co–Ni
2015	**Zr–Cu–Ni–Al**
2016	Zr–Ti–Cu–Ni–Be
2017	Ti–Zr–Be–Ni
2018	**Zr–Nb–Cu–Ni–Al,** Fe–Si–B, **Fe–Cr(Mo)–Zr–B, Cu–Mn–Al,** Fe–Si–B–Cr
2019	Ti–Zr–Be–Ni, Ti–Zr–Be–Ni–Cu
2020	Zr–Cu–Ni–Al, Ti–Zr–Be-(Ni/Fe), Ti–Zr–Be–Ni–Cu

1991) with high glass-forming ability and thermal stability [17]. The critical casting thickness in these alloys ranged up to 15 mm and the supercooled liquid region was extended to 127 K for the alloy $Zr_{65}Al_{7.5}Ni_{10}Cu_{17.5}$ [18]. The development of these alloys demonstrated that BMG compositions were not a laboratory curiosity and could be quite interesting for engineering applications. The significance of Inoue's work was quickly recognized by Johnson and others from Caltech [19]. The Johnson group started the search for BMG compositions in the early. In 1992, the amorphous commercial alloy, Vitreloy was developed at Caltech, as part of NASA's Department of Energy and Research for new aerospace materials [20].

Between 1988 and 1992, other studies discovered multiple glass alloys with a glass transition and a supercooled liquid region. From these studies, loose glass alloys of La, Mg, and Zr were made and these alloys showed plasticity even when their web thickness was increased from 20 to 50 μm [21]. Plasticity was a clear difference compared to the amorphous metals of the past that became brittle at those thicknesses. In 2018, a team from the SLAC National Accelerator Laboratory, the National Institute of Standards and Technology (NIST), and Northwestern University reported on the use of artificial intelligence to predict and evaluate samples of 20,000 different metallic glass alloys in a year. Their methods promise to accelerate research and time to market for new amorphous metal alloys. Table 8.1 lists the typical BMG systems and the year in which they were first reported.

8.3 Mechanism of BMG Formation

If the quench rate of liquid metal is slow enough, the system kinetics allows the crystalline phase to nucleate below the melting temperature (T_m), and the liquid is transformed to a crystalline solid. On the other hand, if the quench rate of the liquid is sufficiently high, the nucleation of the crystalline phase will be impeded. Consequently, the liquid can be cooled below T_m without crystallization. In this case, the liquid is in the supercooled state, which is a metastable state as compared with the crystalline state. Upon further cooling of the supercooled liquid, its viscosity increases. Eventually, at a certain temperature below T_m, the viscosity of the supercooled liquid reaches a value around 1012 Pa.s, i.e., 1013 Poise, which is almost 14–15 orders of magnitude higher than the viscosity of liquid at T_m. By definition, this temperature is called the glass transition temperature (T_g) at which the atomic structure of the supercooled liquid remains almost stationary (Fig. 8.3). In other words, at Tg, the mobile melt turns to a rigid glass. In the rapid quenching processes, the critical cooling rate (R_c) is the slowest cooling rate that allows the material to avoid crystallization. Indeed, R_c is a measure of the glass-forming ability (**GFA**) [22]. GFA of metallic liquids is the ratio between the glass transition temperature (T_g) and the melting temperature (T_m), which is commonly referred to as the reduced glass transition (T_{rg}). That is, an alloy with a lower R_c has a better GFA than an alloy having higher R_c. On the other hand, the cooling rate and sample dimensions (for example, its thickness) are closely related. Hence, R_c also corresponds to the critical thickness (T_c) which is the maximum thickness that a glassy sample could be obtained from a given alloy. Compared to the traditional mineral (silicate) or organic glass-forming materials, pure metals, and ordinary metallic alloys readily crystallize upon quenching. Therefore, the formation of glassy metals is rather difficult compared to traditional non-metallic glasses. This is why the formation of the first metallic glass from Au–Si alloy required a very fast cooling rate of 10^6 K/s and very small dimensions of 50 μm. For many alloys and pure metals, a fast cooling rate on the order of 10^5 to 10^{10} K/s is required to generate a glass. Such metallic glasses are called conventional metallic glass [23]. In contrast with conventional metallic glass which has critical thicknesses in the submillimeter range, alloys with good GFA, which allows them to be casted in samples with at least 1 mm thickness, are called bulk metallic glasses (BMG) [24].

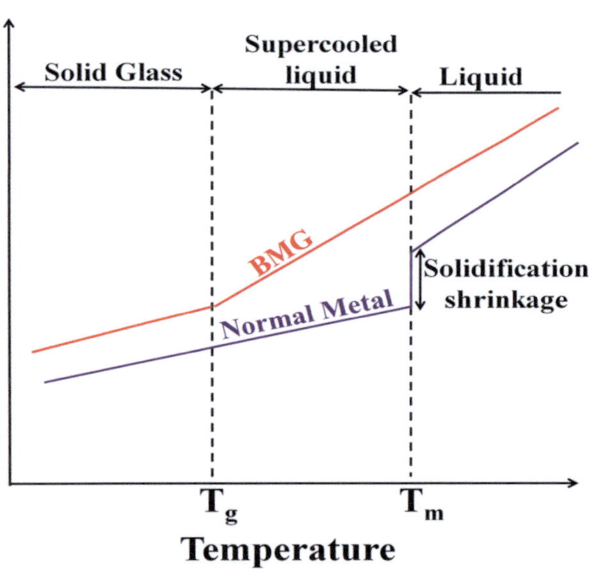

Fig. 8.3 Representation of the relationship between the temperature and volume of BMG in comparison with normal solidification of metal

The term "bulk" was arbitrarily defined as the millimeter scale. Later on, Turnbull et al. obtained a Pd–Ni–P BMG with a centimeter thickness and cooling rate of only 10 K/s. They had highly purified the melt to minimize the possibility of heterogeneous nucleation. However, the expensive Pd-based ternary alloys mainly remained as laboratory materials. Inoue et al., in the late 1980s, discovered new multicomponent alloy systems of common metallic elements having low critical cooling rates. By systematic investigation of the GFA in ternary systems of Fe and Al metals with rare-earth elements, they obtained exceptional BMGs such as La–Cu–Al samples with thicknesses of several millimeters. Based on these findings, quaternary and quinary BMGs, such as La–Al–Cu–Ni and La–Al–Cu–Ni–Coat, were synthesized with low cooling rates of less than 100 K/s. The largest diameter of a metallic glass rod produced to date is 80 mm in a $Pd_{42}.5Cu_{30}Ni_{7.5}P_{20}$ alloy [25]. Further research revealed similar BMG forming alloys designed by partially replacing the rare-earth metals with Mg (an alkaline metal), such as Mg–Cu–Y, Mg–Ni–Y, etc., along with a family of multicomponent Zr-based BMGs, such as Zr–Cu–Ni, Zr–Cu–Ni–Al. It is noteworthy that Inoue and Takeuchi et al. have reviewed and provided a sophisticated classification of BMGs, and suggested several empirical rules to design new BMG forming alloys [26], as discussed below.

In the microstructure, BMG has unique atomic configurations that are significantly different from those for conventional metal glasses. Thermodynamically, these melts are energetically closer to the crystalline state than other metallic melts due to their high packing density in conjunction with a tendency to develop short-range order. Density measurements show that the density difference between BMG and fully crystallized state is in the range 0.3–1.0%, which is much smaller than the previously reported range of about 2% for ordinary amorphous alloys. Such small differences in values indicate that the BMGs have higher dense randomly packed atomic configurations [27].

An important consequence of this was that metallic glasses could only be produced in a limited number of forms (typically ribbons, foils, or wires) in which one dimension was small so that heat could be extracted quickly enough to achieve the necessary cooling rate. As a result, metallic glass specimens (with a few exceptions) were limited to the thickness of less than 100 μm [28].

8.4 Thermodynamic and Kinetic Aspects of Glass Formation in Metallic Liquids

Understanding glass formation is equivalent to understanding both the thermodynamic and kinetic aspects of crystallization. Figure 8.4 shows a time-temperature-transformation diagram of metallic liquid with a higher cooling rate than the critical cooling rate producing BMG. If a liquid is cooled below the T_m, the free energy difference between the liquid and a crystal provides a driving force for crystal nucleation, while the creation of the liquid–crystal interface creates a positive interfacial energy that disfavors nucleation. This results in an energy barrier that a local composition fluctuation needs to overcome in order to form a nucleus. To enable the growth of such a nucleus, atoms within the liquid need to be rearranged. The rate of such atomic transport is described by the atomic diffusivity or viscosity [29, 30]. The resulting crystal nucleation rate (I_V) per unit volume is the product of a thermodynamic term, which depends on the probability of a fluctuation to overcome the nucleation barrier, and a kinetic term, which depends on atomic diffusion (or viscosity):

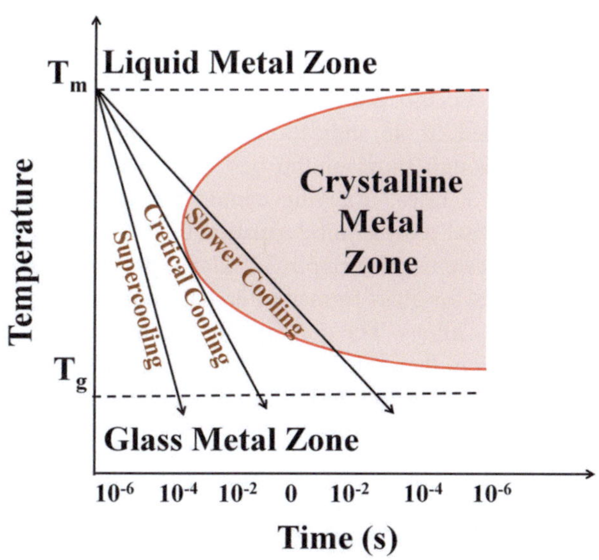

Fig. 8.4 Time–temperature–transformation diagram of metallic liquid, showing the higher cooling rate than the critical cooling rate producing BMG

From Turnbull theory, Eq. ((8.1) is,

$$I_V = \frac{A_v}{\eta(T)} \exp\left(-\frac{\Delta G^*}{k_B T}\right) \tag{8.1}$$

where A_v is a constant on the order of 10^{32} Pa.s/(m^{3}s), k_B is the Boltzmann constant, and ΔG^* is the nucleation barrier for forming a spherical nucleus represented as,

$$\Delta G^* = \frac{\left(\frac{16}{3}\right)\pi\sigma^3}{[\Delta g(T)]^2} \tag{8.2}$$

where Δg is the difference in Gibbs free energy (per unit volume), and σ is the interfacial energy between the liquid and the nuclei.

The temperature dependence of the Gibbs free-energy difference can be approximated to be a function of undercooling, i.e.

$$\Delta g = \alpha(T)\Delta s_f(T_m - T) \tag{8.3}$$

where $\alpha(T)$ is a temperature-dependent correction factor (which decreases slowly from 1 at T_m to 0.7 at T_g) and Δs_f is the entropy of fusion per volume of the liquid.

The kinetic parameter influencing glass formation is viscosity (or diffusion). Viscosity and diffusivity are often related by the Stokes–Einstein relation,

$$\text{The atomic diffusivity}, D = \frac{k_B T}{(3\pi\eta l)} \tag{8.4}$$

where ⍰ is the viscosity, $l$is the (average) atomic diameter. The viscosity of liquids is commonly described by a modification of the Vogel–Fulcher–Tamman (VFT) relation,

$$\eta(T) = \eta_0 \exp\left(\frac{D^* T_0}{T - T_0}\right) \tag{8.5}$$

where D^* is the fragility parameter ($1 \leq D^* \leq 100$), T_0 is the VFT temperature and ⍰$_0$ is a constant inversely proportional to the molar volume of the liquid. Physically, T_0 is the temperature where the barrier with respect to flow would become infinite. The fragility describes the degree to which the viscosity of a supercooled liquid deviates from an Arrhenius behavior. Liquids are commonly referred to as "fragile" when $D^* < 10$ and "strong" when $D^* > 20$. Strong liquids have a high equilibrium melt viscosity and show a more Arrhenius-like temperature dependency of viscosity than fragile liquids. For pure metals, $D^* \sim 1$, while for the network glass SiO_2, the classical example for a strong glass former, $D^* = 100$. The viscosity of SiO_2 strictly follows an Arrhenius-like temperature dependence.

Apparently, the liquidus temperature, the VFT temperature, and the fragility significantly influence the crystal nucleation rate.

$$T_r = \frac{T}{T_m} \tag{8.6}$$

The reduced VFT temperature,

$$T_{r0} = \frac{T_0}{T_m} \tag{8.7}$$

$$A = \frac{16\pi\sigma^3}{\left(3k_B\alpha^2 T_{liq}{}^3 \Delta S_f{}^2\right)} \tag{8.8}$$

By taking Eqs. ((8.6), ((8.7), and ((8.8), the homogeneous crystal nucleation becomes,

$$I_V = \frac{A_v}{\eta} \exp\left(-\frac{D^* T_{r0}}{T_r - T_{r0}}\right) \exp\left(-\frac{A}{T_r(1 - T_r)^2}\right) \tag{8.9}$$

The first exponential term decreases rapidly with undercooling in the range $T_{r0} < T_r < 1$, while the second exponential term increases rapidly in the same range. Thus, the maximum nucleation rate occurs at intermediate undercooling. From Eq. (8.9, it becomes immediately apparent that the nucleation rate is suppressed more for higher reduced VFT temperatures. Also, the second exponential term increases less rapidly with undercooling for lower liquidus temperatures, because of an increased value of A. However, since it was believed at this time that the VFT temperature equals the T_g, and replaces the reduced VFT temperature T_{r0} with the reduced glass transition T_{rg}. It is found that homogeneous nucleation was essentially suppressed for $T_{rg} \geq \frac{2}{3}$.

BMG exhibit a large supercooled liquid region. The difference between the glass transition temperature, T_g, and the crystallization temperature, T_x, that is, $\Delta T_x = T_x - T_g$, is large, usually, a few tens of degrees, and the highest reported value so far is 131 K in a $Pd_{43}Ni_{10}Cu_{27}P_{20}$ alloy [31]. Bulk metallic glasses are multicomponent alloys since specific element additions decrease the liquidus temperature and thus improve glass formation by increasing T_{rg}. Furthermore, the addition of elements that are chemically and topologically different from the other species (in atomic size and valence) frustrates crystal formation. Consequently, a liquid may not crystallize at all and form a BMG, if sufficient elements of different atomic sizes and valences are combined.

8.5 Empirical Rules

The excellent glass-forming ability of the new alloys has been generally attributed to the increased atomic packing density in the multicomponent system, as there are more atoms of the right size to fill free space in the randomly packed glass structure. This appears to be true since the total energy of alloys with directionless metallic bonding depends on the packing density; denser packing leads to lower energy and thereby higher stability. It is difficult to locate the compositions for the best glass formers in multicomponent alloy systems. Most multicomponent metallic glasses found so far have been identified by trial and error, and the development of new BMGs requires considerable experience and involves major commitments in time and resources. There is thus a compelling demand for research that uncovers the underlying mechanisms for the formation of BMGs.

Besides consideration of the packing density, the improved GFA of the multicomponent systems has also been nominally understood by Inoue empirical rules for the formation of BMG, as below [32, 33]:

1. Alloys should be multicomponent systems consisting of more than three elements. More the elements involved, lower the chance of crystal structure formation, and hence greater the chance of glass formation.
2. There should be a significant difference in the atomic size ratios, greater than 12%, among the three main constituent elements. For example, Be atom is much smaller than a Zr atom, in BeZr-based BMG.
3. The three main constituent elements should have negative heats of mixing.

Deep eutectic mixture of each binary and ternary alloy system is used as an indication for seeking good glass formers. Although most of the best glass formers follow those empirical rules, implying certain physical principles indeed play vital roles in the formation of BMGs. In multicomponent systems, the empirical rules represent only the bare essentials for glass formation and are not sufficient for designing new alloys. The definite physical mechanisms for BMG formation, therefore, remain unclear, and the laws for quantitative composition design of bulk metallic glasses are still unknown.

8.6 BMG Structure

In practical materials, knowledge of short-range order is insufficient to determine the overall structure of a disordered solid, so that metallic glasses must be extended to the macroscopic scale. Defining the structure of metallic glasses beyond the nearest-neighbor short-range order has remained an outstanding issue in metallic glass research [34]. Recently, Miracle suggested a scheme for modeling medium-range order in multicomponent metallic glasses. In his model, efficiently packed solute-centered atomic clusters are retained as local structural units. An extended structure is produced by idealizing these clusters as spheres and efficiently packing these sphere-like clusters in FCC and HCP configurations to fill three-dimensional space [35]. Because of internal strains and topological blocking, the order of the cluster-forming solutes cannot extend beyond a few cluster diameters, and thus the disordered nature of metallic glasses can be retained beyond the nanoscale. Based on experimental measurements and computational simulations, Sheng and coworkers proposed an alternative cluster packing scheme to resolve the atomic-level structure of amorphous alloys. By analyzing a range of model binary alloys involving different chemistries and atomic size ratios, they elucidated the different types of short-range order as well as the structure of medium-range order [36]. Their results suggest that icosahedral fivefold packing is a more favorable ordering pattern for short-range ordered cluster–cluster connection in metallic glasses than the FCC or HCP packing schemes. With consideration of the chemical effect, recent experimental, and theoretical studies show more complicated cluster packing schemes in real multicomponent alloys. For example, chemical heterogeneity can lead to the coexistence of multiple cluster packing schemes giving rise to medium-range order in the same alloy [37]. However, these experiments give only average and one-dimensional structural information, although plausible three-dimensional structural models can be reconstructed by trial and error using reverse Monte Carlo and ab initio molecular dynamics simulations [38]. The main problem is that these methods cannot give unique atomic configurations, particularly for multicomponent alloys. In this sense, experimental observations of the local atomic structure of disordered metallic glasses are still lacking, and definite evidence of the local atomic order suggested by various theoretical models remains inconclusive [39]. Figure 8.5 shows the scanning electron micrograph of ZrTiNbCuNiBe BMG.

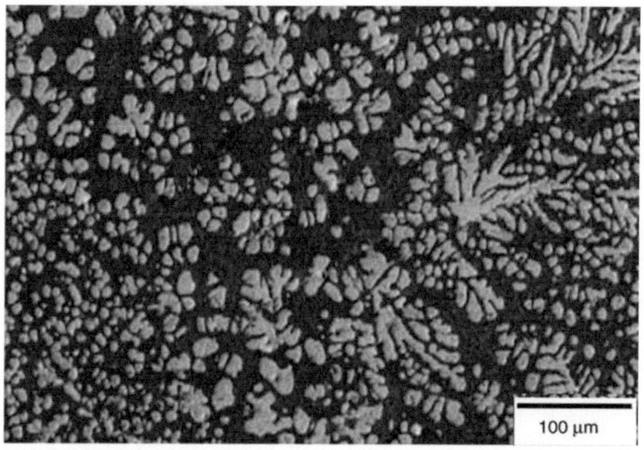

Fig. 8.5 SEM backscattered electron image of in situ ZrTiNbCuNiBe composite microstructure

8.7 Dynamics of BMG Structure Formation

Understanding glass formation is equivalent to understanding both the thermodynamic and kinetic aspects of crystallization. In view of the thermodynamic relationship between structure and phase stability in crystalline materials, the atomic origins of BMG formation have been discussed intensively from geometrical and topological perspectives of the dense atomic packing as introduced above. In principle, the formation of metallic glasses is a competition between the stability of supercooled liquids and the formation kinetics of rival crystalline phases [40]. As both liquid stability and crystallization kinetics are time-related, and as metallic glasses are essentially out-of-equilibrium systems, the formation of BMG involves structural evolution in time and thus may not be studied in terms of thermodynamics alone. Therefore, it seems to be more appropriate to explore the glass-forming mechanism and glass-forming ability from the perspective of the dynamics of supercooled liquids [41].

Several kinds of temperature-dependent relaxations have been observed experimentally in metallic glasses. (1) In the supercooled liquid state, α- or structural relaxation corresponds to an increase in shear viscosity and shear modulus during cooling, resulting in the change of the glass formers from liquid behavior to viscoelastic behavior [42]. In general, superior glass formers exhibit slower dynamics and a longer α-relaxation time at temperatures above the glass transition point. This is simply because the slow dynamics offers a low critical cooling rate for glass formation and, thus, has been used to explain the effect of alloying on the improved glass-forming ability of BMGs empirically [43]. (2) Nevertheless, the intrinsic correlation of the dynamic process with the atomic structure and chemistry of BMGs has not been well elucidated. It has been suggested that the development of icosahedral short-range order in the supercooled liquid regions may play an important role in glass formation because the densely packed atomic structure triggers slow dynamics near the glass transition point, a phenomenon known as a dynamic arrest. (3) Recently, MD simulations have suggested that the slow dynamic process may not be the only origin of the high stability of supercooled liquids and, instead, dynamic heterogeneity may play an important role in the excellent glass-forming ability of BMGs [44]. Significant dynamic heterogeneity, coupled with structural and chemical inhomogeneity, has been observed in a $Cu_{45}Zr_{45}Ag_{10}$ alloy. Interestingly, a high icosahedral cluster population and silver-poor environments are responsible for slow dynamics. In contrast, a low icosahedral cluster population and silver-rich environments correspond to fast dynamics [45].

The strong coupling between chemical and dynamic heterogeneity offers an alternative way to stabilize the supercooled liquid by partitioning the slow and fast dynamics regions, which can effectively prevent the nucleation of crystallites. As more or less chemical heterogeneity widely exists in multicomponent alloys, chemical and dynamic coupling seems to be a universal phenomenon in BMGs, which may suggest a new scheme that could be to elucidate the correlation between dynamic heterogeneity and glass-forming ability in multicomponent alloys and provide new insight into the dynamic origins of BMG formation.

8.8 Plasticity or Brittleness

The plastic flow stress in shear is proportional to the elastic shear modulus, thus the shear modulus (μ) is a measure of the difficulty of plastic flow. Similarly, the bulk modulus (B) is a measure of the difficulty of cracking. Thus high values of the shear-to-bulk modulus ratio (μ/B) should favor brittleness and vice versa [46]. For polycrystalline metals, there is a scale from ductile, low μ/B (e.g., Ag, Au, Cd, Cu) to brittle, high μ/B (e.g., Be, Ir). The critical μ/B ratio for different crystal structure metals is given in Eqs. ((8.10), ((8.11), and ((8.12) [47].

$$\left(\frac{\mu}{B}\right)_{crit} = 0.32 - 0.57 \text{ (for fcc metals)} \tag{8.10}$$

$$\left(\frac{\mu}{B}\right)_{crit} = 0.60 - 0.63 \text{ (for hcp metals)} \tag{8.11}$$

$$\left(\frac{\mu}{B}\right)_{crit} = 0.35 - 0.68 \text{ (for bcc metals)} \tag{8.12}$$

Thus critical modulus ratio $(\mu/B)_{crit}$ is not very well defined even for one structure type and $(\mu/B)_{crit}$ is affected by anisotropy.

8.9 Classification of BMG

Inoue classified the BMG into three types, that is (1) metal–metal-type alloys, (2) metal–metalloid type alloys, and (3) Pd–metalloid-type alloys. The structural configurations of these categories of BMG have been given in Fig. 8.6.

8.9.1 Metal–Metal-Type Alloys

In the metal–metal alloy, the glass consists of icosahedral clusters. The critical size for a transition from icosahedral cluster to icosahedral phase is around 8 nm. When the BMG is annealed in supercooled liquid region, the icosahedral quasicrystalline phase (I-phase) precipitates in the primary crystallization step, and the I-phase transforms to stable crystalline phases at higher temperatures. The precipitation of I-phase is structural heredity of the local structure of the BMG [48]. The existence of icosahedral clusters provides seeds for the precipitation of the I-phase and indicates the importance of the icosahedral clusters as the fundamental structural unit. Based on nucleation theory revealed that the activation energy for nucleation of I-phase is smaller than that for nucleation of crystals in the undercooled alloy melt. The icosahedral clusters (or icosahedral short-range order) in the amorphous state would provide an additional barrier for the nucleation of the crystalline phases. Since the I-phase with fivefold rotational symmetry would be incompatible with the translational symmetry of normal crystalline phases, therefore it has to be dissociated before the formation of the crystalline phases could occur. From kinetics point of view, the crystallization of BMG requires a substantial redistribution of the component elements across the icosahedral liquid [49]. The highly dense, randomly packed structure of the BMG in its supercooled state results in extremely slow atomic mobility, thus making the redistribution of atoms on a large scale very difficult. This fundamental structural discontinuity between the crystalline and the amorphous state suppresses the nucleation and growth of the crystalline phase from the supercooled liquid and results in an excellent GFA [50].

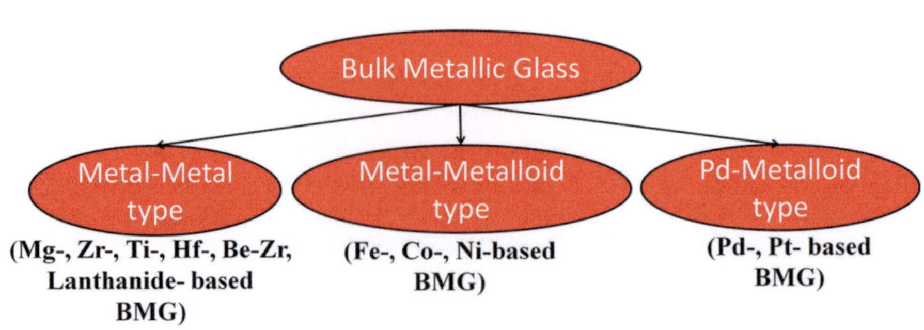

Fig. 8.6 Classification of BMGs

8.9.2 Metal–Metalloid-Type Alloys

For the metal–metalloid-type glassy alloys, for instance, Fe (Co)–Nb–B, a network atomic configurations consisting of trigonal prisms which are connected with each other through glue atoms comprising Zr, Nb, Ta, or lanthanide metal are commonly found. Fe-based BMGs form primary crystals of complex FCC, Fe23B6 phase with a large lattice parameter of 1.1 nm and a unit volume consisting of 96 atoms [51].

8.9.3 Pd–Metalloid-Type Alloys

Pd-based BMG does not satisfy the three empirical rules proposed by Inoue, and the structural investigation shows that Pd–Cu–Ni–P BMG consists of two large clustered units of a trigonal prism capped with three half-octahedra for the Pd–Ni–P and a tetragonal dodecahedron for the Pd–Cu–P region. As is evident from the distinctly different GFA between Pd–Ni–Cu–P and Pd–Ni–P, the coexistence of the two large different clustered units seems to play an important role in the stabilization of the supercooled liquid for the Pd-based alloy. This is in turn attributed to the strong bonding nature of metal–metalloid atomic pairs in the clustered units and the difficulty of rearrangement among the two kinds of clustered units [52].

8.10 Processing of Metallic Glasses

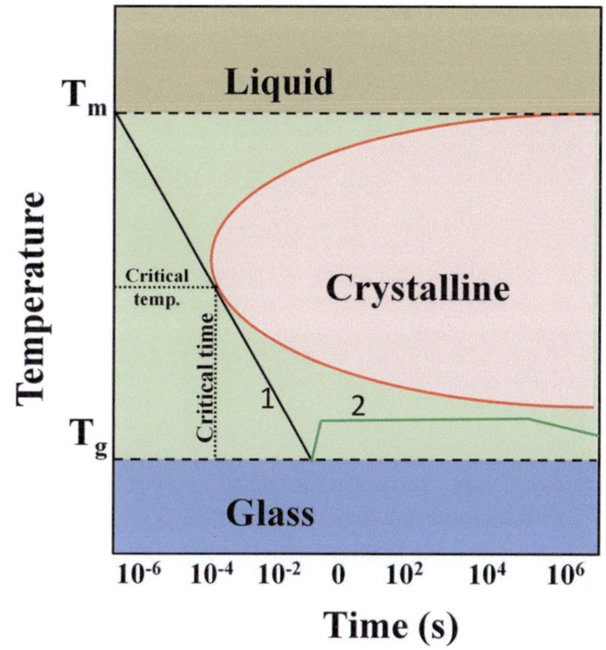

Fig. 8.7 Schematic TTT diagram illustrating the processing methods of BMG formers

Figure 8.7 shows the schematic TTT diagram illustrating the processing methods of BMG formers. During direct casting, forming takes place simultaneously with the required fast cooling to avoid crystallization and during thermoplastic forming (TPF), the required fast cooling and forming are decoupled. The amorphous BMG is reheated, where the available processing window is much larger than during direct casting, resulting in better controllability of the process. After the forming processing step, no fast cooling is required to avoid crystallization during cooling. Crystallization occurs when the cooling or heating path intersects with the crystallization curve of the TTT diagram (Fig. 8.7) [53]. To circumvent such an intersection, the unique crystallization behavior among BMG formers permits two principally different processing paths. Path 1, in Fig. 8.7, indicates the critical cooling rate for glass formation, the slowest cooling rate that avoids crystallization, and shows that a cooling rate such as path 1 or faster can be used for a casting process. During direct casting, the forming takes place simultaneously with required fast cooling to avoid the "crystallization nose." Path 2 represents the processing path for TPF [54]. Here, the required fast cooling and forming are decoupled, which will be shown to lift the geometrical limitations imposed by the critical casting thickness, to provide better control, and to yield the highest dimensional accuracy.

Various fabrication routes have been deployed as given in Fig. 8.8 for the synthesis of BMGs.

8.10.1 Liquid State Processes

8.10.1.1 Direct Casting

Casting methods such as suction casting and die casting have been used as BMG net-shape fabrication processes. The suction casting method yields high-quality parts with particularly low porosity compare to samples prepared by die casting. Currently, liquid metal technologies are using die casting, which is adopted from conventional aluminum die casting for high-volume manufacturing of Zr-based BMG [55]. The typically low melting temperature BMG formers are beneficial in a direct casting process because this process reduces tool costs and wear, lowers energy consumptions, shortens cycle time, homogeneous microstructure, and less solidification shrinkage. However, the critical cooling rate, T_c, and the TTT diagram are significantly affected by the maximum temperature prior to cooling. During suction casting as well as die-casting shrinkage has to be taken into consideration. Fortunately, the absence of a first-order phase transition during solidification reduces solidification shrinkage in BMG formers drastically. Phase-transition shrinkage of about 5% is typical for casting alloys. The

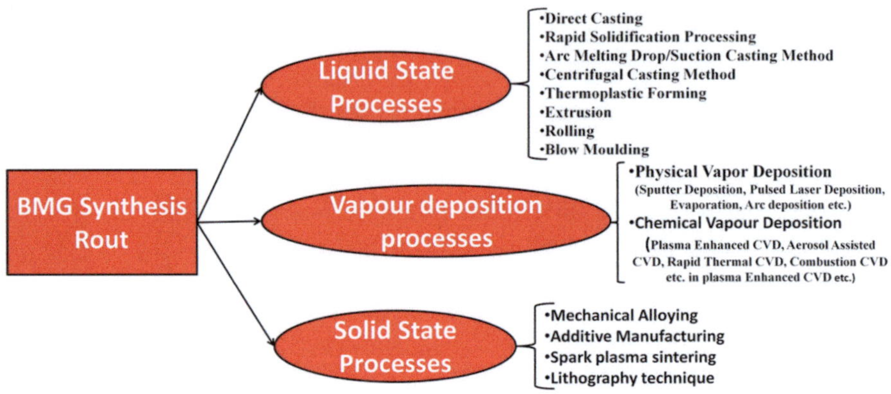

Fig. 8.8 Fabrication routs of BMG materials

solidification shrinkage during casting of a Zr-based BMG former, in comparison, is approximately only 0.5%. The one order of magnitude reduction in solidification shrinkage of BMG formers over conventional casting alloys suggests a much higher dimensional accuracy in the casting of BMG. It has also been observed that the casting environment is crucial to the cooling process. Even the low solidification shrinkage in BMG formers creates a gap between the mold and methods such as die casting has been used for near-net-shape processing of BMG. The cooling rate for the casting process is such that crystallization of the glass is narrowly avoided. The disadvantages of this process are: high viscosity of BMGs leads to low fluidity which makes casting difficult, internal stresses are developed due to rapid cooling, vacuum environment is necessary to avoid the reaction with atmospheric gases [56].

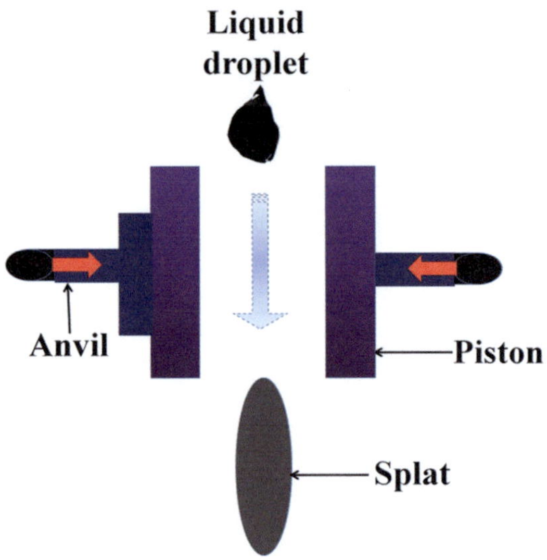

Fig. 8.9 Drop smasher (Piston and anvil technique)

8.10.1.2 Rapid Solidification Processing

The first amorphous metallic alloy, $Au_{75}Si_{25}$, was produced by Duwez et al. by applying a rapid quenching method with a cooling rate of ~10^8 °C/s [57]. This method was discovered by them and known as splat quenching or piston and anvil technique shown in Fig. 8.9. To understand the mechanical properties of metallic glass, the first amorphous metallic alloys were not proper on account of shape and size. In the following years, new rapid quenching methods such as melt spinning, twin roller, and drop melt extraction techniques were developed and achieved restrictions on dimensions partly. Especially, the melt spinning method enabled mass production of ribbon metallic glass. All these rapid quenching techniques used in the production of metallic glass were based on rapid quenching with a cooling rate range of 10^4–10^8 °C/s [58].

Melt spinning (Fig. 8.10) is a technique which involves rapid solidification of the molten glassy melt at cooling rates ~10^6 K/s to attain a metallic glass structure. This requires that the heat be removed from the melt at a very high rate due to which the section thickness of the final product is limited to micron ranges. The traditional methods of achieving such high cooling rates are, molten metal is atomized into small droplets and then these droplets are solidified either by being exposed to a stream of cold air or inert gas. Another method of solidification is by impinging these droplets, i.e., splatting on a good heat-conducting surface. In this method, a flowing stream of molten metal is continuously solidified by moving it in contact with a moving chill surface. The metallic glasses that are formed are in the shape of ribbons or sheets or wires. This technique involves rapid melting at the surface of a bulk melt and then solidification via rapid heat extraction into the unmelted region. A way to do this is via laser treatment [59].

Splat quenching typically involves rapid solidification of molten metal in between twin rollers which are cooled continuously to remove heat, as shown in Fig. 8.11. A thin sheet of metallic glass is formed with a low volume to area ratio. It is basically a liquid rolling technique. A better technique is Duwez and Willen's gun technique [60]. Here, the molten metal is thrown toward a quencher plate due to which its area rapidly increases and it rapidly cools to form a thin sheet. A

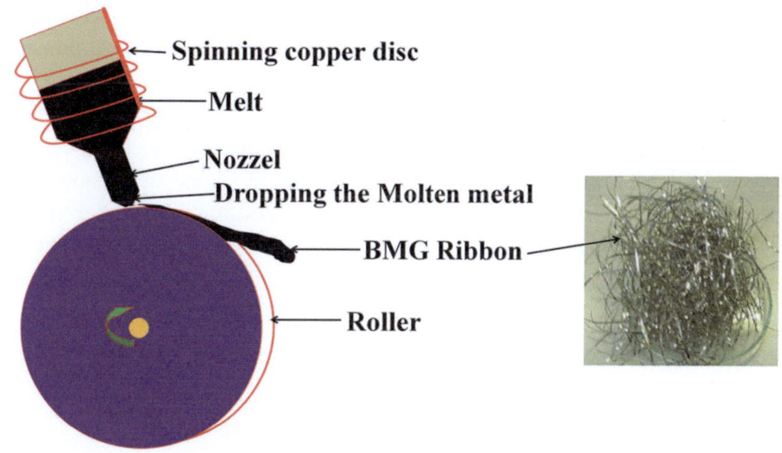

Fig. 8.10 Melt spinning method to obtain the BMG strips

wider range of near amorphous metallic glasses can be produced.

The two important process parameters for splat quenching are the velocity of the droplets and their volume. If the volume is too large or the velocity is too low, the droplet doesn't completely solidify. Hence, it is experimentally determined as to what is the optimum droplet size and velocity that is suitable for forming a thin sheet having uniform thickness, composition, and good mechanical properties. Products that are produced using splat quenching generally have a "near amorphous" structure and excellent paramagnetism due to which this technique is useful for applications related to magnetic shielding, etc. [61].

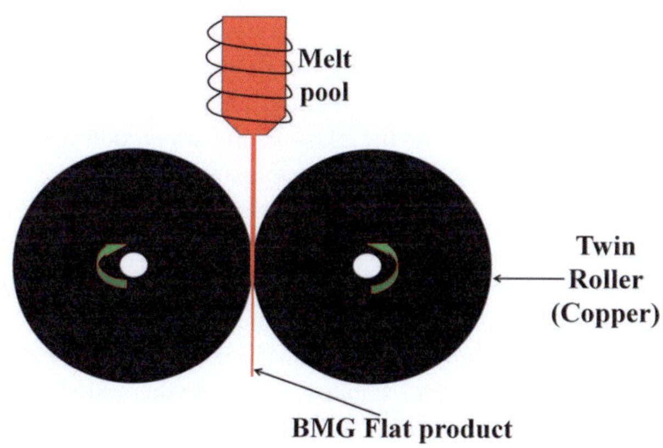

Fig. 8.11 Twin rolling system

8.10.1.3 Arc Melting and Drop/Suction Casting

Arc melting and drop-casting methods are well-established and known processes for BMG alloys. Drop casting is typically used to process materials with diameters larger than 6 mm, while suction casting is used for casting of materials with diameters less than 6 mm. This method is to combine and optimize these technologies to create an integrated drop and suction-casting, which is used to quickly and efficiently fabricate BMG alloys in various sizes and geometries [62]. A schematic picture of arc melting and drop/suction casting method is given in Fig. 8.12. Arc melting method is mostly used for a low critical cooling rate required alloys. The arc between a tungsten tip and the copper heart melts alloy under extremely pure environment and inert gas atmosphere. After alloy preparation, molten metal is transferred into water-cooled drop/suction casting molds, which sit under a vacuum bell. Arc melting drop/suction casting system consists of a water-cooled heart system and various types of mold, preferentially copper.

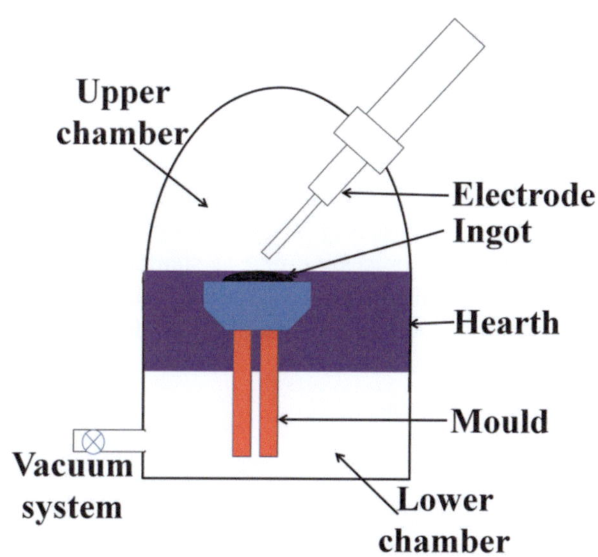

Fig. 8.12 Arc melt and drop/suction casting

8.10.1.4 Centrifugal Casting Method

Centrifugal casting system consists of two main parts: casting apparatus and injection system of molten alloy. The casting apparatus includes a cylindrical copper mold, which is rotated by a motor with belt transmission. The transmission allows to change the speed of rotating mold [63]. Moreover, the injection system consists of a quartz container where the alloy ingot is melted by inducing melting. The process of ingot melting is realized by using high-frequency power supply. Figure 8.13 shows a schematic system diagram of the centrifugal casting method for the production of ring shape BMG parts. The process parameters about the solidification of BMG alloys depend on many parameters such as duration, metal density, thermal conductivity, cooling rate, and melting temperature.

8.10.1.5 Thermoplastic Forming

Thermoplastic forming is a technique that takes advantage of the drastic softening of the BMG on heating above the glass transition temperature in order to form the glass into complex shapes. The glass is kept in the supercooled liquid region (SCLR) where it exists in a metastable state and then it is crystallized. The process is also known as hot forming, hot pressing, super plastic forming, viscous flow working, and viscous flow forming. The two important parameters to maximize the formability of the glass in the supercooled liquid region are the viscosity and the processing time. The viscosity has to be low and the processing time has to be high. The extent to which a BMG can be formed in its supercooled liquid state is dependent on the variation of viscosity with temperature and the crystallization time. TPF takes advantage of the drastic softening of the BMG former upon heating above T_g and its thermal stability [64]. The thermal stability is a measure of the ability of BMG formers to retain an amorphous structure when heated above its glass-transition temperature and is quantified by the width of the SCLR. During TPF, the BMG is reheated into the SCLR, where it relaxes into a supercooled and metastable liquid before it eventually crystallizes. An example of the temperature dependence of the crystallization time and viscosity for $Zr_{44}Ti_{11}Cu_{10}Ni_{10}Be_{25}$. At low temperatures, a long processing time is available for $Zr_{44}Ti_{11}Cu_{10}Ni_{10}Be_{25}$, accompanied by high viscosity. At high temperatures, on the other hand, the viscosity is significantly reduced but, at the same time, the processing time is shortened [65]. The advantages of thermoplastic forming methods are, (1) decoupling of forming and fast cooling due to which a wide range of complex shapes can be produced, (2) higher dimensional accuracy as compared to other techniques due to very low solidification shrinkage, (3) can be processed in air and unaffected by heterogeneous influences, and (4) low processing temperature and pressure due to which no significant investment required. The limitations are: (1) more number of steps as compared to casting due to decoupling of forming and rapid cooling and (2) requires skilled manpower due to novel methods of processing.

8.10.1.6 Extrusion

Extrusion is a thermoplastic formability process as shown in Fig. 8.14. Extrusion produces the highest aspect ratio shapes of constant cross section [66]. Early work on extrusion of BMG focused on consolidating BMG powder and processing in the supercooled liquid region. Kawamura and coworkers pioneered this method and achieved full strength compacts using Zr–Al–Ni–Cu powders. For the same alloy, the relationship between extrusion temperature, pressure, and ram speed was determined and summarized in an extrusion map.

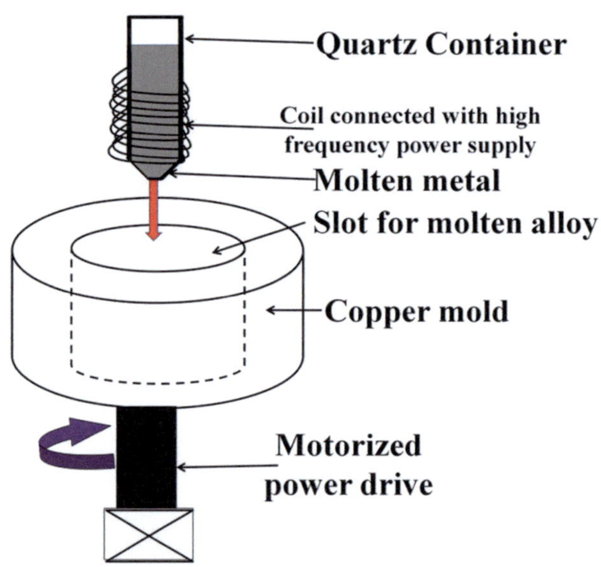

Fig. 8.13 Centrifugal casting setup with injection system of molten alloy

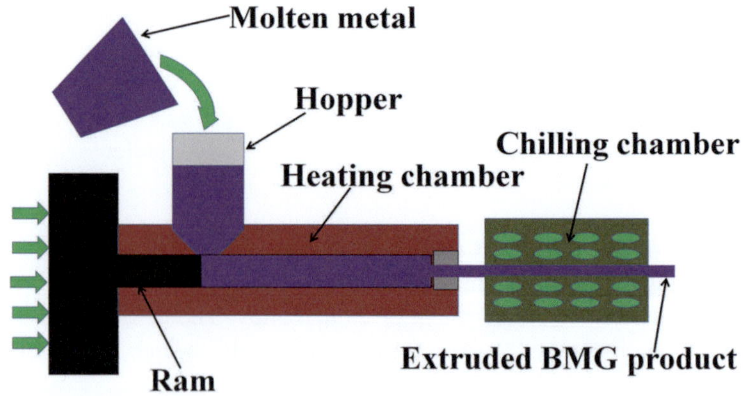

Fig. 8.14 BMG extrusion process

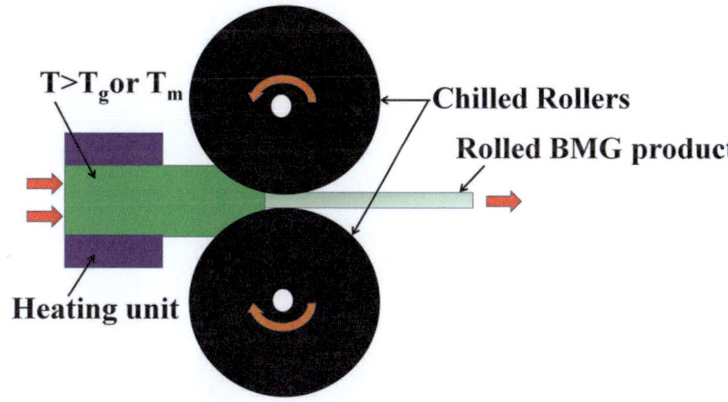

Fig. 8.15 Rolling process of BMG

Based on these findings and theoretical considerations, the authors concluded extrusion length of up to 1 m for a 3-mm diameter may be achieved under practical pressures. Various BMG formers can be used for extrusion, resulting in different extrusion ratios. An extrusion ratio of up to five was achieved with minimal crystallization for Cu–Ti–Zr–Ni. Similar results were obtained for $Cu_{54}Ni_6Zr_{22}Ti_{18}$, $Cu_{50}Ti_{32}Zr_{12}Ni_5Si_1$, and $Zr_{58.5}Nb_{2.8}Cu_{15.6}Ni_{12.8}Al_{10.3}$ [67, 68]. TPF-based extrusion of BMG formers offers a valid fabrication method for constant cross-sectional diameter profiles. BMG former with high formability should be chosen for TPF-based extrusion since the required forming pressure scales directly with the BMG formability. The dimensional accuracy of this process is, however, limited due to the unavoidable but predictable swelling.

8.10.1.7 Rolling

In rolling, the metal is most commonly obtained in the form of sheets. Sheets are either used directly as an end product or are processed further using one of the many sheet-forming technologies, as shown in Fig. 8.15. About 90% of sheet metal is produced by rolling. Rolling for metallic glasses can be categorized into two processes; one is based on liquid processing and the other on thermoplastic forming [69]. Melt spinning is one of the earliest examples of liquid processing-based rolling. Within melt spinning, the liquid sample is quenched by injecting on a single, fast-spinning copper roll. To improve the control over sheet dimensions, in particular the thickness, twin-roll casting has been employed. Simulations suggest that twin-roll casting can be extended into a continuous process. Recently, sheets have been produced from Fe-based BMGs under adequate control over sheet dimensions processed in a continuous way [70]. During liquid-phase rolling even the highest practical strain rates result in homogeneous deformation. On the other extreme is cold rolling, where the amorphous BMG is rolled at room temperature. Due to shear localization in the deformation region (Fig. 8.15b), cold rolling is limited to already initially thin, <1 mm, samples to prevent catastrophic failure along with a dominant shear band. In addition, the strain per pass is very limited and consequently, a large number of passes are required to achieve significant overall strains. It has been found that cold rolling affects the mechanical properties of BMGs. For $Zr_{55}Cu_{30}Al_{10}Ni_5$, plasticity, even in tension, increases with cold rolling through the introduction of shear bands. Similar observations were made when cold rolling $Zr_{41}Ti_{14}Cu_{12}Ni_{10}Be_{23}$ BMG. Cold rolling of elemental films is also a method to create amorphous material through a solid-state reaction.

8.10.1.8 Blow Molding

Although TPF-based hot rolling and extrusion provide methods to achieve high aspect ratios, the geometric range that can be achieved with these techniques is limited to rolling into planar shapes and to extrusion to a constant cross section [71]. Net shaping of complex 3D shapes, which are large compare to their wall thickness, using TPF-based techniques such as compression molding is challenging due to the friction between the mold and BMG, which often requires impractically high pressures. For example, to deform a $Zr_{44}Ti_{11}Cu_{10}Ni_{10}Be_{25}$ rod of 10 mm diameter and 30 mm height into a disc of 60 mm diameter and 0.8 mm height a forming pressure of 30 MPa is required, about 100 times higher than the flow stress of the alloy at this temperature and strain rate. Friction between the forming BMG and the mold can be reduced to some degree by using lubricants, which result in some slippage. The improvement is quite limited and the use of lubricants also sacrifices the otherwise excellent achievable surface finish. A more effective strategy to reduce forming pressure is to reduce the contact area between BMG and mold. This is the case for the previously discussed hot-rolling where the contact area reduces to a 2D-line contact. A similar effect is observed in an extrusion process where the contact area is reduced to the effective die length. Friction can be eliminated altogether during free expansion in blow molding. Even when blow molding into a mold, the vast majority of lateral straining is achieved before the deforming BMG comes into contact with

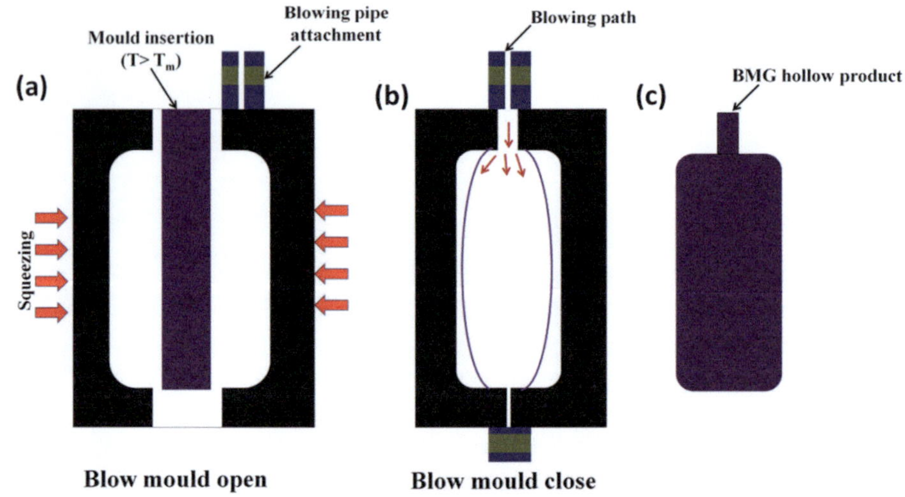

Fig. 8.16 BMG product by blow molding (**a**) Open mold, (**b**) Close mold and blow start, and (**c**) Final hollow product

the mold (Fig. 8.16). From a practical point of view, a pressure difference of 1 atm is desired for blow molding of BMGs. This pressure difference can be accomplished by creating a low vacuum on one side. We estimate the maximum strain that can be achieved with a $Zr_{44}Ti_{11}Cu_{10}Ni_{10}Be_{25}$ BMG former, when blow molding under such a pressure difference of 105 Pa, for a membrane of radius 20 mm and thickness 1 mm [72].

8.10.2 Vapor Deposition Process

8.10.2.1 Physical Vapor Deposition (PVD)

Figure 8.17 showing the PVD processes. Here the material to be deposited is vaporized by physically heating or by sputtering and then deposited onto a surface [73]. Various PVD techniques adopted for BMG are sputter deposition, pulsed laser deposition, thermal evaporation and electron beam evaporation, cathode arc deposition, etc.

Sputter Deposition: In the sputtering process, the magnetic field and the electric field are supplied in such a way to produce plasma in the vacuum chamber. A carrier gas is purged in the vacuum and energized that is bombarded with the target atom. The target atom escaped from the surface and deposited on the substrate surface. Ye et al. performed an RF sputtering using the same target but different at wt% of $Zr_{53}Cu_{29}Al_{12}Ni_6$ was deposited on the silicon (100) substrate. The target atom may be a premixed aggregate or the elemental exposed for co-sputtering. Recently, as-formed bulk metallic glass ZrCuNiAl was chosen as the target and direct current was used for sputtering. During sputtering Argon ions are bombarded towards the ZrCuNiAl bulk metallic glass target to remove surface atomic species which were then deposited on a 304 stainless steel substrate to form a ZrCuNiAl thin film which revealed greater adhesion with an increase in sputtering power. Conventionally, sputter deposition is used for the processing of thin-film metallic glasses with a single alloy target or multiple elemental targets [74].

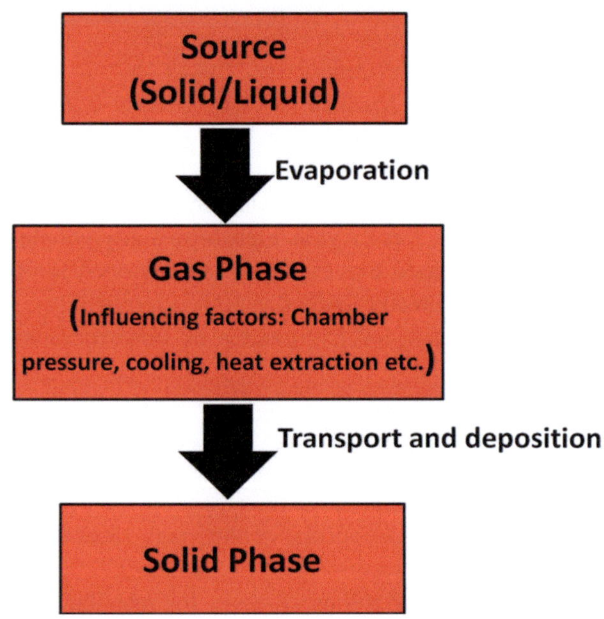

Fig. 8.17 General PVD technique flow diagram

Pulsed Laser Deposition: Pulsed laser deposition (PLD) is ideal for the formation of thin films with complex compositions. Here, the high-powered laser is used to vaporize the material. Here, photon energy is transferred to the bulk which is used to ablate the surface region into plasma. This plasma is then deposited on a substrate to form a thin film. The quality of the thin film formed is affected by various parameters like substrate temperature, the ambient gas pressure of chamber, and substrate target distance. Ag–Cu metallic glass was deposited on Ni foam by the PLD technique for application in zinc-air batteries. $Ti_{64}Cu_{12}Zr_{11}Co_5$ (Mo, Nb) was deposited by PLD technique on 316 L stainless steel and the optimum amorphicity was achieved at laser power 8 W [75].

Evaporation: Thermal evaporation and electron beam evaporation processes are also employed for TFMG production. Thermal evaporation is one of the evaporation methods. It consists of the sample holder, which is a container with a small orifice (Knudsen cell). In this orifice, the solid comes to quasi-equilibrium with its vapor. To raise the vapor pressure of a target material, a DC current is passed through an electric resistance heater such that the target melts and its vapor pressure is increased to a useful range. The particles of the material escape the cell and arrive at the substrate and form a solid layer. For example, $Co_{75}Fe_{14}Ni_4Si_5B_2$ were coated on naturally oxidized silicon substrate by thermal evaporation. Metglass amorphous magnetic thin films were coated using the thermal evaporation technique for applications in magnetostrictive sensor devices [76].

Cathode Arc Deposition: in this method, the high-powered electric arc is discharged at the target material and the vaporized materials deposited at the cathode.

8.10.2.2 Chemical Vapor Deposition (CVD)

In this method, the target material is exposed to some chemical reagent which decomposes on/reacts with it and produces the desired coating on it. CVD can be classified according to the operating pressure viz. atmospheric pressure or low pressure or ultrahigh vacuum pressure. The latter two are more prevalent nowadays. A few important CVD techniques are: plasma-enhanced CVD, aerosol-assisted CVD, rapid thermal CVD, combustion CVD, etc., in plasma enhanced CVD, plasma is used to increase the affinity of the chemical precursor for the target. It allows for low-temperature deposition and is useful especially for the semiconductor industry. In aerosol-assisted CVD, the precursors are transported to the target via an aerosol gas which can be generated ultrasonically. In rapid thermal CVD, a heating lamp is used to heat the target to allow easy deposition. In combustion CVD, a flame is used in an open atmosphere to deposit thin film coatings [77].

8.10.3 Solid-State Processes

8.10.3.1 Mechanical Alloying

Mechanical alloying is a powder metallurgy technique. The different steps involved in mechanical alloying (Fig. 8.18) for BMG are [78]:

Step I: The blended elemental powder particles and the grinding medium (Tungsten carbide or stainless steel balls) are kept in a container.

Step II: The container is agitated at high speed for some predetermined time duration.

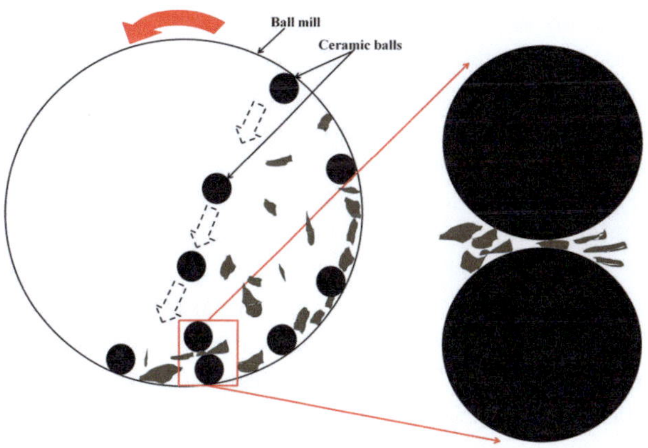

Fig. 8.18 Mechanical alloying by ball milling that prepared for hot pressing and cooling

Step III: The soft powders of each metal get crushed and assumed flat shape with a thin cross section. These flat structures have a layered arrangement.

Step IV: Due to heavy plastic deformation, crystal defects such as dislocations, vacancies, grain boundaries, etc., are introduced in the glass. Temperature also rises.

Step V: Due to the rise in temperature, diffusion is facilitated resulting in the mixing of the metallic powders to form alloys.

Step VI: These alloys can then be molded to the desired shape using techniques such as hot isostatic processing, hot extrusion, vacuum hot pressing, etc.

8.10.3.2 Additive Manufacturing

Additive manufacturing of metallic glass has recently received considerable attention as a promising technique to manufacture complex geometries with a desirable thickness. Powder bed fusion (PBF) techniques like selective laser melting (SLM) or electron beam melting (EBM), in addition to directed energy deposition (DED) methods, such as laser engineered net shaping (LENS) have high potential to overcome the limitations of traditional methods for metallic glass processing [79, 80]. Multiple factors, such as processing parameters (source power, scan speed, hatch spacing, laser energy density, laser spot size, etc.), powder/substrate microstructure (crystalline or amorphous), and chamber/substrate temperature influence the quality of the resulting amorphous phase. Importantly, the effective cooling rate depends on the scanning speed of the source beam, as well as several other processing factors such as the powder composition and temperature gradient between the melt pool and the substrate. In particular, by adjusting the process parameters, it would be possible to achieve an effective cooling rate in the range of 10^3 to 10^8 K/s during the SLM process. It is noteworthy that an effective cooling rate of ~10^3 K/s is plausible for a Zr-based BMG, which has a critical cooling rate of ~10^2 K/

s or lower. A proper processing regime and conditions can produce a desirable high density. Moreover, it has been demonstrated that the amorphous phase is achievable on both amorphous and crystalline substrates owing to the inherent high cooling rate of the laser deposition process. Powder size can also significantly affect the quality and uniformity of the final amorphous phase. A larger powder size could induce low thermal stability leading to undesirable severe crystallization, while a smaller powder size could offer a monolithic amorphous structure. DMLS use to produce a fully amorphous FeCrMoCB BMG whose thickness is more than 15 times the critical casting thickness in all dimensions, and larger than any traditionally produced Fe-based BMG recorded. The DMLS-fabricated BMG is 97% dense and has no major external cracks. Due to the brittleness associated with BMG and the FeCrMoCB alloy, DMLS process parameter alteration alone was insufficient to eliminate all microcracking. By eliminating microcracks through improved alloy design, stress-induced nucleation should subside [81].

8.10.3.3 Spark Plasma Sintering

Spark plasma sintering (SPS) is a powder metallurgy technique that has the capability to sinter the compacted material in the shortest possible time to ensure that no crystallization commences anywhere during sintering. The densification, an important aspect during the powder metallurgy route, can be additionally improved by the introduction of dynamic pressing and high stresses along with rapid heating. This strategy even guarantees the glass formation of marginal glass formers. Moreover, the SPS technique enables sintering at comparatively lower temperatures and shorter times which are the desired processing attributes for BMG formation [82].

8.10.3.4 Lithography Technique

Advancements in technologies such as microelectromechanical systems (MEMS), electronics devices, and medical devices have created an increasing demand for miniature parts and components on the length scale of micrometers to centimeters. A currently used miniature-fabrication method is lithography, electroplating, and molding [83]. BMGs have many attractive characteristics for top-down nanofabrication. Among them, there are homogeneous structures on the nanorange scale, the low required forming pressure and temperature, and their superior mechanical properties. For example, TPF-based processing techniques can be used to imprint $Pd_{40}Ni_{20}Cu_{20}P_{20}$ BMG into V-groove shapes, which are prepared by electron-beam lithography and anisotropic etching with a minimal width of 200 nm [84]. Using the flowed area into the groove, the ability of $Pd_{40}Ni_{20}Cu_{20}P_{20}$ to fill submicrometer features and also the effect of processing parameters, such as the filling time on the formability, can be determined; 500-nm features on Mg-based BMG using a patterned nickel mold were formed and the nanoformability of Zr–Cu–Ni–Al BMG using nanocrystals of Zn–Mn–O as a die material was explored. Despite a large number of studies on the formability of BMGs on the submicrometer length scale, evidence for nanoimprinting with BMGs has been rare. Very recently, it has been suggested that on the nanometer scale the forming of BMGs is not primarily controlled by the viscosity but, instead, by the capillary forces [85].

8.11 Fundamental Characteristics of BMG Alloys

It is well known that BMG exhibits excellent properties such as ultrahigh strength, large elastic elongation, low Young's modulus, high corrosion resistance, good soft magnetic properties, super-smooth surface, and viscous deformability due to their unique structures of dense random atomic configurations, freedom from lattice defects and an excess free volume.

Metallic glasses, with distinct mechanical, physical, and chemical properties, are attractive materials in fields like MEMs, NEMs devices where the fusion of electrical, magnetic, and optical components, are required.

8.11.1 Mechanical Properties

One of the most notable properties of BMG is their extreme hardness and strength, which makes them attractive for applications where strength is critical. Figure 8.19 shows the higher mechanical strength of BMG in comparison to other structural materials. For crystalline metals and alloys, the factors controlling their mechanical properties have been well investigated through the development of dislocation and electronic theories, which can explain, in general, the atomic and electronic origins of the strength and ductility of

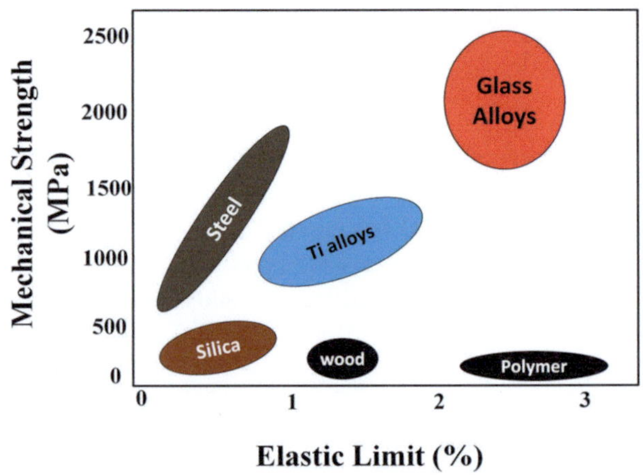

Fig. 8.19 Comparison of mechanical strength of BMG with other structural materials

crystalline materials [86]. For disordered materials such as metallic glasses, however, the definite correlation between mechanical behavior and atomic and electronic structures has not been properly established. It has long been recognized that the mechanical properties of metallic glasses are closely related to the chemical and physical properties of their component elements. The significant differences in mechanical performance, such as strength and ductility that are observed with changes in the chemical composition of metallic glasses indicate the existence of an intrinsic relationship between the mechanical properties and the atomic and electronic structures of metallic glasses [87]. BMG possesses very high strength compared to their crystalline counterparts, they generally suffer from low ductility at low temperatures. Room-temperature plastic deformation of metallic glasses has been known to occur through shear band movement, in which plastic flow driven mainly by shear stresses is localized within a nano-scale zone. Due to structural and/or thermal softening, these bands are preferential sites for further plastic flow and lead to final failure in a brittle manner that typically breaks a sample along with a single shear band without detectable plastic strains. Therefore, highly localized shearing and resultant strain softening are responsible for the poor ductility of BMGs. It has been recognized that the ductility improvement of BMG largely depends on the suppression of the localized strain softening caused by shear bands [88].

Ti-based BMGs have relatively low Young's modulus (80–120 GPa), high fracture strength (1700–2500 MPa), and excellent specific strength. Zr-based BMG has been featured with a high hardness that is about twice to three times that for conventional crystalline 316 L SS, Ti alloys, and Zr alloys, and high yield strength that is considerably higher than that of those above mentioned conventional crystalline metallic biomaterials. Plastic deformation in metallic glasses is associated with inhomogeneous flow in highly localized shear bands at low temperatures. The degree of plastic deformation before failure varies greatly from composition to composition among metallic glasses. Zr-, Pt-, and Pd-based BMG have all shown plasticity in compression and fracture toughness as high as 70 $MPa.m^{1/2}$. In contrast, Mg- and Fe-based BMG are both locally and globally brittle, showing significantly different fracture surfaces and fracture toughness less than 5 $MPa.m^{1/2}$ [89].

BMGs have much higher tensile strengths and much lower moduli than crystalline alloys of relatively similar compositions. The difference in these values between BMG and crystalline alloys is as large as 60%. The significant difference in mechanical properties is a reflection of the differences in deformation and fracture mechanisms between BMG and crystalline materials. For the same reasons that BMG exhibits exceptional mechanical strength, namely high elastic limit and the lack of long-range order, they also exhibit high mechanical hardness. Pd- and Zr-based alloys have reported hardness in excess of 500 Hv, while Fe-based metallic glass alloys have reported hardness in excess of 1000 Hv. The size of the pileup, or the distance from the edge of the indent to the outermost shear band, increases with increasing load. Furthermore, the size of the pileup and the density of the shear bands are compositionally dependent. Zr-based BMGs have smaller pileup zones than Pd-based alloys, indicating stronger resistance to deformation [90].

Fatigue in metallic glasses is characterized by limited plastic deformation. Crack growth thresholds are on the lower end of most structural metallic alloys, typically in the vicinity of 1.1–1.3 $MPa.m^{1/2}$. The endurance limit in most metallic glasses is only about 10% of the ultimate tensile stress. In some Zr-based metallic glasses, endurance limits as high as 40% of the ultimate tensile [91]. Prolonged exposure of BMG to temperatures near or below the glass transition temperature, T_g, may lead to a considerable change in their properties, even if there is no structural change from amorphous to crystalline state. They exhibit a marked loss in ductility upon annealing below the crystallization temperature.

8.11.2 Tribological Properties

Excellent mechanical properties are a sign of promising tribological properties. Blau showed no amorphous to crystalline (or vice versa) transformation evidence on a worn surface in case of ZrCuNiTiAl metallic glasses while in Fu and coworker reported that in the Zr-based BMG not only crystallization occurred on amorphous metallic glasses during tribological contact but also crystalline metallic glasses are re-amorphized. Zeynep and coworkers have compared the coefficient of friction of Zr52.5Ti5Cu17.9Ni14.6Al10 with aluminum 6061 and Stainless Steel AISI 304 [92]. Here block on a ring setup was used to study friction and wear properties, Zr-based BMG has a lower coefficient of friction than that of aluminum and stainless steel. During sliding, the wear rate decrease is controlled through the material transfer phenomenon between the counter surface material on the block on a ring setup and the formation of protective oxide layers on the Zr-based BMG surface. The coefficient of friction first rises due to initial damage on major surface asperities at lower loads and then decreases due to compacted and flattened debris at higher loads. High wear rate of the Zr-based BMG is explained by their poor ductility in tension because the BMG experiences high tensile stress during abrasive wear processes. The adhesion properties of TF MG show wide variations with composition, deposition technique, and substrate. In the case of Zr47Cu31Al13Ni9 metallic glass films on 316 L, stainless steel, a high density of shear bands is formed along the sliding direction with ductile deformation mode with good film plasticity. Zr58Cu25Al11Ni6 deposited on Si (001) by pulsed DC magnetron sputtering, the mean COF value was around 0.3–0.4 and no apparent crack or delamination was found adjacent to the scratch track of the coating, suggesting good adhesion [93].

8.11.3 Magnetic Properties

Metallic glasses, with magnetic properties, are required for various multidirectional applications. For example, an amorphous high Curie temperature magnetic semiconductor is worked with combined functionalities: negative temperature dependence of resistivity, ferromagnetism, and high transparency, all in one system fabricated by the introduction of oxygen into the ferromagnetic metallic glass (Co–Fe–Ta–B). Other than combined functionalities, another distinct feature of amorphous magnetic systems is that they allow direct one-step manufacturing of final shape products. Moreover, properties can be easily tailored by exploiting the flexibility offered by BMGs in terms of composition, size, and shape. Commercial applications in the area of magnetic recording, hard magnets, high-density rewritable devices, magnetic refrigerants, and magnetostrictive devices are highly anticipated [94]. BMGs, usually, exhibit soft magnetic properties because of the absence of structural inhomogeneities such as grain boundaries and crystal defects that hinder the domain wall motion during magnetization. Fe-, Ni-, and Co-based metallic glasses have been extensively studied having excellent soft magnetic properties in the form of low coercive force and high magnetic permeability. Soft magnetic properties have been exploited for applications in different fields such as cores, transformers, cantilevers, and sensors. As opposed to BMG characteristic nature of soft magnetic properties, the appearance of hard magnetic properties—primarily because of the presence of nanocrystalline phases, precipitates, or other defects is, also, very appealing and therefore, has attracted wide-scale research. BMGs with hard magnetic properties have potential applications in recording disks, actuators, and permanent magnetic motors, etc. Nd–Fe–Al, Pr–Fe–Al, and Sm–Fe–Al are some of the amorphous systems with decent combination of both appreciable GFAs and hard magnetic properties [95].

BMGs possess magnetostriction behavior that expands their scope in the field of sensors and actuators. $Fe_{64}Co_{17}Si_{7}B_{12}$ has been used as a magnetostrictive material along with a piezoelectric element to fabricate a multiferroic magnetoelectric composite for energy harvesting. Even a weak electromagnetic signal can deform the metallic glass ribbon because of its high magnetostriction value. Ni–Mn–Ga has also been used as a magnetostrictive material in another composite. The same effect of magnetostriction coupled with soft magnetic properties of BMGs has been exploited in the fabrication of a high-performance biosensor. Magnetostriction can be tuned using various approaches, for instance, Fe–Co–B–Si–Nb is demonstrated to improve with the addition of carbon showing a linear correlation between the applied magnetic field and resulting displacement which is highly important for sensing applications [96]. Magnetocaloric effect, a reversible temperature changing phenomenon, that occurs when a material is imposed to a changing magnetic field, usually accompanied by a change in magnetic entropy, is also displayed by various BMG. This effect can be employed for magnetic refrigeration, an energy-efficient and environmentally friendly approach. Although attractive candidate materials such as $Gd_5Si_2Ge_2$ compound or other rare earth-based systems showing first-order or second-order magnetic transition exist but they have major shortcomings such as low corrosion resistance, hysteresis losses, and poor mechanical properties. Fe-based amorphous systems, on the other hand, possess a captivating combination of mechanical properties, soft magnetic properties—with magnetic hysteresis approaching zero-, tunable curie temperatures, and excellent corrosion resistance. Li et al. demonstrated Fe–Mn–P–B–C metallic glass exhibiting both tunable magnetic entropy change and Curie temperature by varying Mn content. This system reaches the highest refrigerant value of 147.09 Jkg^{-1} and can be exploited for low-cost and near room temperature magnetic refrigerant. Likewise, Liet al. studied $(Fe_{0.76}1\text{-}xTmxB_{0.24})_{96}Nb_4$ BMG by doping the system with Thulium. The doping enhances the magnetocaloric response as well as makes the Curie temperature tunable. This system, too, has the potential to be used as room-temperature magnetic refrigerant. Additionally, Re-based BMG is also opted out as excellent candidate materials for magnetic refrigerants [97].

In 1996, rare earth Nd-based BMG (REBMG) with hard magnetic properties at room temperature (RT) have been developed. These hard magnetic properties are ascribed to a unique clustered amorphous structure. Y–Sc–Al–Co and Pr (Nd)-Al–Ni–Cu are REBMGs, developed in 2003. With Fe addition, it is found that the Pr-based REBMGs have an inhomogeneous cluster structure (like Nd-based REBMG) exhibiting complex magnetic properties. In 2004, the Ce-based REBMG with ultralow low elastic modulus and glass transition temperature Tg was reported, and very soon the Ce–Al–Cu (Ni) alloys with diameters up to 12 mm were developed. The glass formation in the RE-based alloys was investigated according to the elastic modulus rule in combination with the classic glass-forming criteria. The work result in the emergence of a series of new Sc-, Gd-, Sm-, Tb-, Dy-, Ho-, Er-, Yb-, Tm- and Lu-based REBMGs with good GFA, high thermal stability, and tunable physical and mechanical properties. For REBMGs, their GFA and properties are sensitive to minor element additions. For instance, in Ce-based

REBMGs, in addition to their low T_g, their formation is very sensitive to minor addition of Co. The main commercial prospects of the REBMGs lie in their high performance in applications as promising microstructures, hard magnets, and magnetic refrigerants, and so on [98].

Fe-, Ni-, Co-based BMG has soft magnetic properties applied in dust core applications, room temperature magnetic semiconductors, current sensors, targeted delivery. Nd–Fe–Al, Pr–Fe–Al BMG has hard magnet properties applied in magnetic recording disc, solenoid actuators, permanent magnet motors for a modern car, cosmetics. Fe–Co–Si–B, Ni–Mn–Ga BMG has magnetostriction property applied in sensors, actuators, multiferroic magnetocaloric composite. Tb–Co–Al, Ce–La–Cu–Co BMG has spin dynamics and heavy fermion behavior. Gd–Ni–Al, Fe–Mn–P–B–C BMG has a magnetocaloric effect used in the magnetic refrigerators. Fe–Cu–Nb–Si–B–V BMG has piezomagnetism used in highly sensitive, noncontact, fast response force sensitive sensors.

8.11.4 Chemical Properties

The study of the corrosion behavior of BMG is of great importance to understand their chemical and environmental stability. All metallic glasses have inherently high reactivity because they are highly metastable materials. This increased reactivity allows for the rapid production of a passivation layer. Therefore, corrosion resistance is strongly dependent on the composition of the glass and its environment [99]. Ti-based BMG exhibited passive behavior at the open-circuit potential with a low corrosion rate; a susceptibility to localized corrosion in the form of pitting corrosion; the localized corrosion resistance was statistically equivalent to, or better than, the conventional crystalline biomedical alloys, including 316 L stainless steel, pure Ti and Ti-based biomedical alloys (such as Ti–6Al–4 V) [100]. Stainless steel in the solution containing chlorine ion is generally going to pitting corrosion, intergranular corrosion and stress corrosion and hydrogen embrittlement, the amorphous FeCr alloy–stainless steel can make up for these deficiencies, this makes the iron-based amorphous alloys widely used in the oil, coal, steel and shipbuilding and other fields [101].

8.11.5 Electrical Property

Metal glass has special electrical properties. Compared with crystalline alloy, the resistivity of metal glass is obviously improved, and the resistance temperature coefficient of metal glass is smaller than that of crystalline alloy. The low resistance temperature coefficient is helpful to eliminate thermal drift and integrate with the temperature sensors. Therefore, metal glass becomes the best choice for developing multifunctional electronic skin, which can give the intelligent robots more accurate "perception."

8.12 Forming and Jointing of BMG

In most of the commercially available BMG products, vacuum die casting is used for producing the desirable shapes of the components. Unlike die casting of conventional crystalline materials where the flow of molten metal can be improved by heating the mould, die casting of BMG usually requires faster cooling for retaining the amorphous structure. Therefore, the casting of very thin and intricate shapes, as well as very large workpieces, is difficult. In addition to casting, forming the BMG parts through viscous flow in the supercooled liquid state is also a viable process. In order to better integrate the BMG components into the structure of machines or other products, the possibility of joining BMG with BMG or crystalline materials is desirable. The major contribution in this aspect was made by Kawamura et al. who demonstrated the success of several welding processes, namely, friction welding, spark welding, and electron-beam welding. The major consideration while welding BMG is whether the glassy state can be retained after the processes, which usually involve the supply of heat for softening the metallic glasses. It confirmed that with the careful control of process parameters, the joints remain amorphous and the strength is comparable to the bulk of the metallic glass [102–104].

8.13 Metallic Glass Foam

BMG foam offers unique properties in biomedical applications. The lack of macroscopic plasticity has prevented the widespread proliferation of BMG. The porous structures with suitable pore sizes (200–500 μm) are favorable for cell attachment, proliferation, and differentiation. The strength and Young's modulus of the porous materials can be adjusted through the adjustment of the pore size and the porosity in order to match the requirement of bone repairs as orthopedic implants. Porous BMG $Zr_{57}Nb_5Cu_{15.4}Ni_{12.6}Al_{10}$ has been fabricated by low-pressure infiltration of carbon microspheres and by casting the alloy into a bed of leachable BaF_2 salt that is subsequently dissolved after solidification. And the obtained foam structure is macroscopically homogeneous with a uniform pore size of 212–250 μm and the strength is 100 MPa at 50% strain. In addition, the BMG foams with honeycomb structures have been fabricated by a combination of lithography and thermoplastic forming fabrication, and the deformation made of a BMG honeycomb can be altered from brittle to ductile through varying the ratio of ligament length to ligament thickness and a 0.2% increase in density doubles strength and energy absorption [105, 106].

8.14 Metallic Glass Coatings

Considering their attractive mechanical, physical, and chemical properties of BMG, there has been a substantial and long-standing interest in producing amorphous alloy layers on conventional crystalline metallic substrates since the 1970s [107]. Laser cladding works by using a focused beam of photons to melt a pre-deposited powder onto the surface of substrate material. The rapid heating and cooling induced during laser processing results in minimal thermal effects on the base material during laser cladding. To create an optimal clad layer using a laser there are many factors to consider such as the effect of material properties on the absorption and melt pool geometry as well as the laser parameters which can affect the clad dimensions and microstructure. BMG thin-film coating can be used as a hydrophobic surface. The measured water contact angles of thin-film metallic glass are around 101°. The surface free energy value $Zr_{53}Cu_{33}Al_9Ta_5$ thin-film metallic glass is 26.9 mN/m. The two main factors affecting the contact angle are, namely, the surface tension of the liquid itself and the surface free energy of the object surface. Thus, materials with low surface free energy will have a high contact angle. Commercialization studies are ongoing to investigate the wetting angle of thin-film metallic glass on different materials and the influence of substrate roughness to wettability [108].

8.15 Application

Metallic glasses provide entirely new possibilities and enhanced performance to a wide variety of applications everything from commercial applications such as high-temperature structural materials, electronic materials to sports equipment.

8.15.1 Aerospace Industries

BMG is used in spacecraft shielding (Fig. 8.20a). Spacecraft shielding is defined as the outer layer of a satellite or spacecraft that protects it against micrometeorite and orbital debris, radiation damage, and re-entry temperatures [109]. There are several problems with the design and implementation of shields, particularly in the area of micrometeorite and orbital debris shielding. Spacecraft and satellites need to have the lowest possible mass due to the enormous cost per pound of putting them into orbit or deep space. However, low Earth orbit is currently littered (and increasingly so) with orbital debris, primarily remnants of rocket upper stages, satellites, and pieces of spacecraft that have broken away or have collided with other objects. The major threat is that this debris is traveling at 8–18 km/s, and any piece larger than a few centimeters has the kinetic energy to potentially become a "spacecraft killer." Large debris is tracked with radar, but the smaller debris (below a centimeter or so) is too small to track and must be mitigated by shields in the event of a collision. The International Space Station, for example, employs over 500 different shield designs into its outer skin, which are

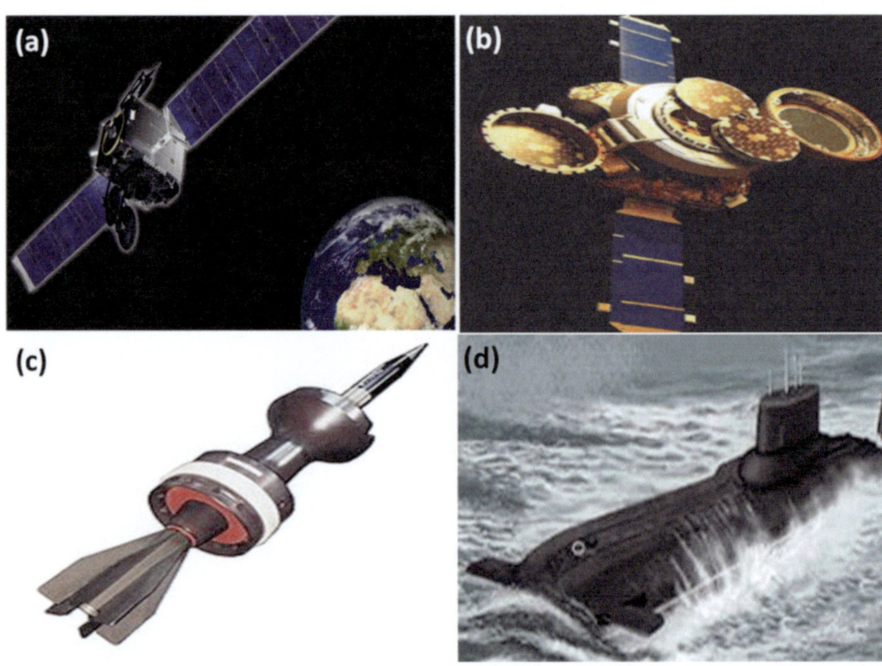

Fig. 8.20 (**a**) Spacecraft shielding, (**b**) Solar wind collector, (**c**) Antitank armor-piercing, (**d**) Magnetic mass hull materials

BMG designed to protect a variety of vital components. Metallic glasses have even been used to collect solar winds on NASA spacecraft. Genesis has received its final piece of science equipment: a solar wind collector (Fig. 8.20b) made of a new formula of BMG [110]. This metallic glass collector and other solar wind collector tiles on the spacecraft will collect the first-ever samples of the solar wind as the spacecraft floats in the oncoming solar stream. Genesis will have collected elements of the solar wind such as isotopes of oxygen and nitrogen. The critical properties for gear materials are high wear and fatigue resistance, which may be correlated with a high hardness and toughness that used BMG [111]. BMG technologies are applied in military materials that are stronger, lighter, and more effective at high temperatures and stress. These can replace depleted uranium penetrators in antitank armor-piercing projectiles because of their similar density and self-sharpening behavior (Fig. 8.20c). Unlike most crystalline metal projectiles, which flatten on impact, the sides of BMG-composite sheer away under dynamic loading, explained by Todd C. Hufnagel of The Johns Hopkins University. Technologies are working with Lockheed Martin Missiles and Fire Control to develop lighter and stronger ceramic–BMG composite armor tiles [112].

Wear-resistant gears are produced from BMG. A gear with superior wear resistance and excellent surface finish, as compared to EDMed gears, is fabricated in centimeter scale after carefully selecting from the tested BMG compositions. The gear-on-gear testing reveals that Cu–Zr-based BMG shows a 60% improvement in wear resistance properties in comparison to Viscomax C300, which is a very special high-performance steel used by NASA for Mars rover Curiosity.

BMG is used in reduced magnetic mass hull materials (Fig. 8.20d), which is very difficult to detect by satellite; moderate temperature, lightweight alloys for aircraft and rocket propulsion; and wear-resistant machinery components for ground, marine, and air vehicles. "In the future a ship made of BMG could be five times larger, or weigh five times less," [113]. Other BMG assisted equipment can be observed in the Mars rover, planetary gear of spacecraft, strain wave gears, etc. as shown in Fig. 8.21.

8.15.2 Automobiles Industries

BMG is spreading a great interest in the automobile industries. With the advancement of process control in the automotive and other industries, there is a need for advanced sensors. BMG is used in automobile pressure sensors (Fig. 8.22a). By using BMG for the diaphragm of a pressure sensor, the diaphragm is more flexible, resulting in a high strength, high sensitivity sensor [114]. Some metal glass also has good hydrogen storage properties, which can be used as a hydrogen storage material (Fig. 8.22b) for vehicles with hydrogen as fuel. The use of hydrogen-fueled cars can reduce air pollution [115]. Hydrogen energy is a new energy source that is both efficient and clean, and the storage of hydrogen energy is the main bottleneck restricting the arrival of hydrogen energy society. The hydrogen storage capacity of

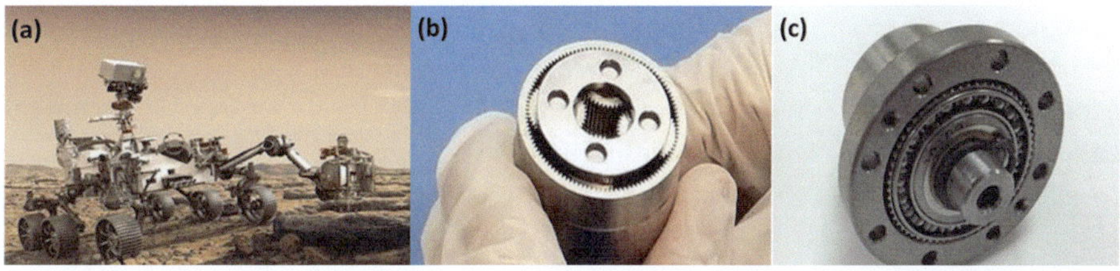

Fig. 8.21 BMG in (**a**) Mars rover, (**b**) Planetary gear of spacecraft, and (**c**) Strain wave gears

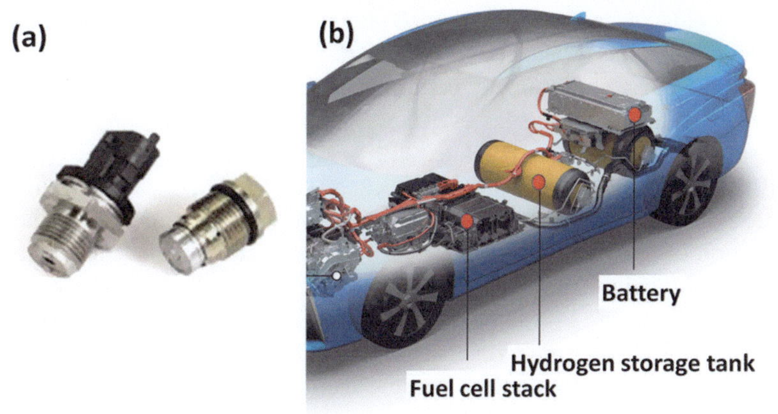

Fig. 8.22 (**a**) Pressure sensor and (**b**) hydrogen storage material

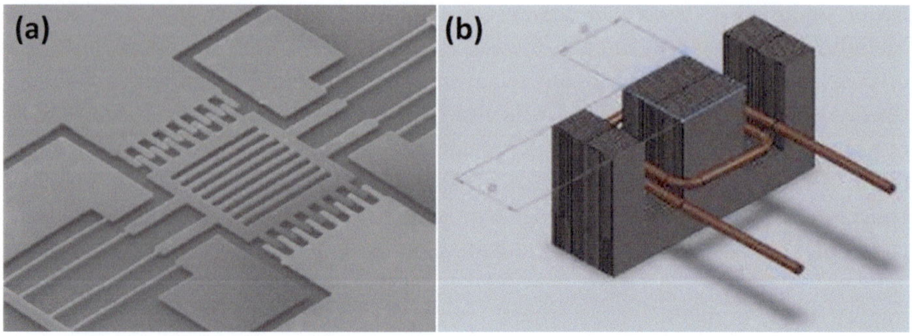

Fig. 8.23 BMG used in (**a**) Microcantilever, and (**b**) Transformer core material

amorphous alloys has attracted widespread attention. Zander et al.'s research shows that the hydrogen storage capacity of zr-based amorphous materials can be close to that of the best crystalline hydrogen storage materials. However, amorphous alloy hydrogen storage materials are more concerned than crystalline hydrogen storage materials. There are many "defects" in amorphous alloys that resemble crystalline materials, which may provide more hydrogen sites and have higher hydrogen storage capacity. Amorphous atoms have a variety of glassy transitions that can provide more potential for hydrogen occupancy [116].

8.15.3 Electrical and Electronic Industries

The main challenge in the designing of fuel cells is the availability of low-cost, effective materials along with their appropriate manufacturing routes. Silicon is a common choice for most of the recently developed fuel cells but it possesses low shock resistance and poor electrical conductivity. Additionally, the Si-based micro fuel cells (MFCs) demand expensive and complex processing techniques. The thermoplastic formability of BMG, which is a low cost and economical fabrication method and which enables the fabrication of hierarchical structures with length scales ranging from centimeters down to nanometers, is exploited to make a BMG micro fuel cell [117]. In this cell, the crucial components such as the catalyst layer, diffusion layer, and flow fields are made of bulk metallic glasses, using TPF. The flow fields are made of Zr-based BMG embossed via TPF while the catalyst layer is made of Pt-based BMG that exhibits remarkable catalytic performance. Zr-based current collectors show advantages over conventionally used stainless steel, silicon, and graphite. TPF-based embossing on BMGs that are durable and electrochemically active thus provides a versatile, fast, and economical route for the production of novel micro fuel cells [118]. Due to outstanding formability, low processing temperature, and inherent functionalization, Pt–Cu–Ni–P BMG has been employed for the fabrication of a microcantilever (Fig. 8.23a). The BMG microresonators offer tunable performance and a wide range of sensing. The said system is tested at different annealing temperatures to investigate the effect of structural relaxation. Annealing results in an increase in quality factor and resonant frequency with the former approaching a peak value of 8100 in vacuum. Quality factors approaching 8100 in a vacuum and 2000 in ambient conditions are achieved, which are the highest reported values for any system having this geometry. The most convincing attribute of this system is the performance tunability which is achieved by varying annealing temperature and processing conditions to optimize the cantilever for different circumstances. In a different work, twisted microcantilevers have been fabricated via surface micromachining of thin-film metallic glasses. The silicon-based microscanners used in MEMS possess certain limitations since silicon has a stiff, brittle nature, and contain defects when microprocessed [119]. Such undesirable characteristics create problems during torsional stresses in microscanners which subsequently limits the resulting scanning performance. Again, BMGs come to the rescue by offering desired properties such as low Young's modulus, high fracture toughness, and strength. A microscanner is developed in which the moving torsion bars are made of metallic glass. A large rotating angle with diminished power consumption, better sensing, and actuation performances, as compared to both single and polycrystalline silicon, have been noted. So far, the lowest values for power consumption, and the highest for rotating angle, are reported using this MG-based microscanner. Fe–B–Nd–Nb BMG system that exhibits good fracture toughness, vast super-cooled liquid region, and excellent magnetic properties, is used for the fabrication of a high-yield current sensor (cantilever) [120].

With the miniaturization of semiconductor chip parts and the acceleration of assembly, there is a corresponding demand for semiconductor die collets with higher strength and fine nozzles [121]. By combining our BMG forming technology and fine hole processing technology, we can produce a collet for chip transfer that has a fine nozzle and excellent abrasion resistance. In addition, using BMG can solve material-based issues, such as malfunctions caused by static electricity with resin collets, or chipping problems with ceramic collets [122]. Compared to conventional transformers, the power transformers with metallic glasses as a core are found to help in reducing the core loss (Fig. 8.23b). In addition, the size and weight are also very less. The loss is less in the latter case because of its ferromagnetic property with low hysteresis loss and high electrical resistivity property. BMG in transformers is found to improve the efficiency of power distribution in transformers. These transformers are used to convert high-voltage current into the low-voltage current to be used for domestic appliances (120 V and 240 V). Sputtered BMG films have also been used to fabricate actuators and microcantilevers for MEMS applications. Metallic glass nanowires fabricated by these methods show interesting multi-harmonic oscillations with possible applications in mechanical and magnetic sensors. The development of such nano-sized BMG structures may lead to a host of applications of BMGs in nanotechnology and bio-nanotechnology [64].

8.15.4 Biomedical Industries

With a continuously increasing aging population and the improvement of living standards, large demands of biomaterials are expected for a long time to come. Further development of novel biomaterials, that are much safer and of much higher quality, in terms of biomedical along with mechanical properties, is therefore of great interest for both the research scientists and clinical surgeons. Metals and alloys have been widely used as biomedical materials and are indispensable in the medical field. For example, about 70% of the structural materials employed in implants are metallic materials that are used to replace hard tissues or as hard-tissue prostheses, including bone screws, bone plates, artificial hip joints, knee joints, and dental implants. Currently, titanium, which has excellent biocompatibility, is the main material used for these, but there are problems such as poor processability and limited shape. BMG has high strength, flexibility, and a high degree of freedom in shape, and is expected to be a next-generation material to replace titanium. The advantages of BMGs compared with conventional crystalline metals and alloys include superior strength, high elasticity, and excellent wear and corrosion resistance, attributable to the lack of grain boundaries and crystal defects that usually lead to the weakening of material strength, intergranular corrosion, and stress-corrosion cracking in biological environments. BMGs have much lower Young modulus than crystalline metals, allowing BMGs to be more compatible with bone, reducing the stress shielding effect that can cause disuse atrophy of cortical bone. Therefore, BMGs possess attractive properties for load-bearing biomedical-implant applications. Small, high-precision orthodontic brackets can be manufactured using BMG [123]. High precision slot formation minimizes movement of the corrective wire and facilitates treatment. Additionally, the reduction in size can ease the burden on the patient during correction as well as discomfort in the mouth. [124].

BMG also applied as soft tissue stents due to their high flexibility, providing enhanced compliance with blood vessel biomechanics and thereby minimal damage to vessels. For example, Mg-based BMG has been developed for application as biodegradable implants. In addition to their higher strength and elasticity in comparison with the crystalline biodegradable magnesium-based alloys, biodegradable metallic glasses also show a distinct reduction in hydrogen evolution in vivo which is a general problem that occurs as a result of biocorrosion in physiological solutions and forms an unwanted gas pocket around the implants. Owing to the broad solubility of the glassy alloys, the chemical composition of BMGs can also be optimized to match the biocorrosion characteristics of the environment, providing control over degradation and hydrogen evolution rates, which is very difficult to achieve using crystalline alloys [125].

8.15.5 Other Applications

The first application of BMG in sports to be found was as golf club heads, for making golf plates [126]. Low density and high strength-to-weight ratio, other properties such as low elastic modulus and lower vibrational response provide a softer, more solid feel for better control when a golfer strikes the ball. The negligible hysteresis loss of BMG means that less energy is absorbed by the clubhead at impact, so more energy is transferred to the ball. Steel club heads transfer about 60% of the input energy to the ball and titanium transfers 70%, whereas the BMG transfers 99%. Sports BMG sports equipment include tennis rackets, baseball bats, skis, and snowboards, bicycle frames, hunting bows, and even edged tools such as axes. As the magnetic properties of the metallic glasses are not affected by irradiation they are used in making containers for nuclear waste disposal. Because of their high electrical resistivity and nearly zero temperature coefficient of resistance, these materials are used in making cryo thermometers, magnetoresistance sensors, and computer memories [127]. Figure 8.24 shows the BMG nuclear shielding, and BMG magnets for the levitated trains.

Au-based BMG is used in jewelry applications. From a processing point of view, it is desirable to possess a large supercooled liquid region which gives access to a low forming viscosity, which in turn facilitates superplastic forming. Currently, BMG daily used materials are cell phones casing (Fig. 8.25a), watch frame (Fig. 8.25b), and connecting part for optical fibers (Fig. 8.25c), which are

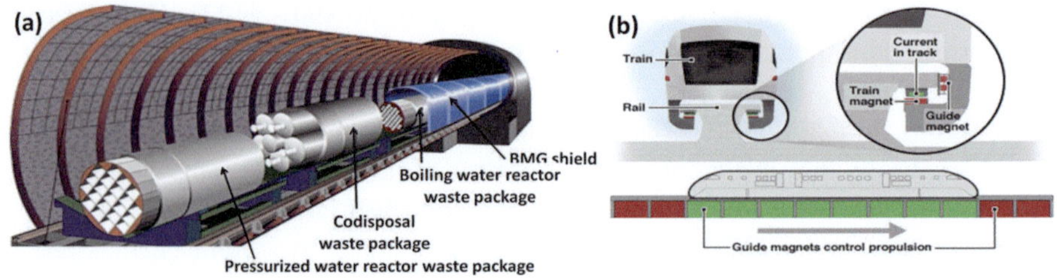

Fig. 8.24 BMG (**a**) nuclear shielding, and (**b**) magnets for levitated train

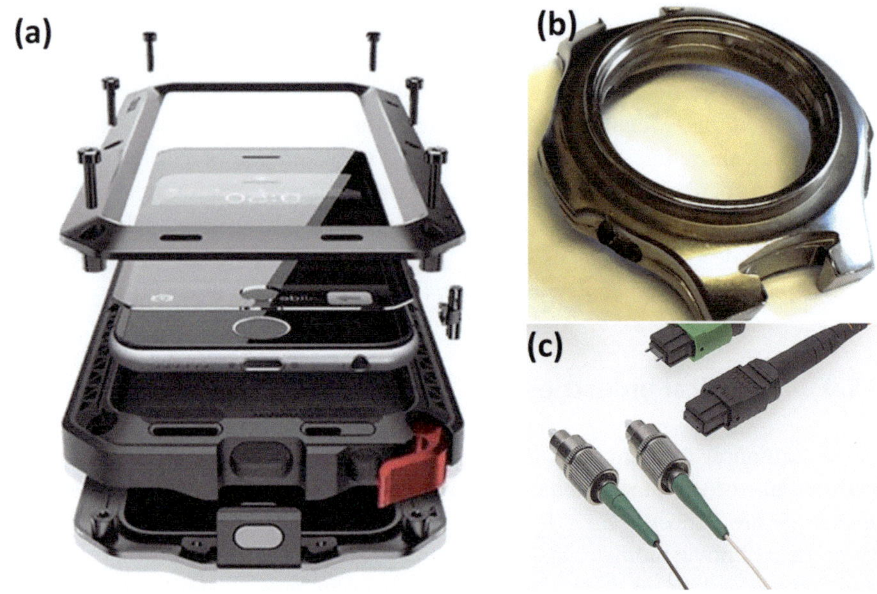

Fig. 8.25 BMG used (**a**) Cell phone casing, (**b**) Watch frame, and (**c**) Connecting part for optical fibers

almost indestructible. BMG razor blade stays super sharp for a year. Various shapes of optical mirrors, casing in electromagnetic instruments, connecting part for optical fibers, shot penning balls, electromagnetic shielding plates, soft magnetic choke coils, soft magnetic high-frequency power coils, high torque geared motor parts, high corrosion resistant coating plates, machinery structural materials, tool materials, die materials, vessels for lead-free soldering, in-printing plate, high density of information-storage material, yoke material for the linear actuator, high-frequency type antenna material, magnetic iron core for high rotation speed motor are made up of BMG [128, 129].

8.16 Summary

The atomic structure of metallic glasses is a current topic of intense discussion to produce a dimension more than 30 mm. This chapter intensifies the mechanism of advantages that in builds in the bulk metallic structure in various functional components. Most of the atomic configurations, particularly in multicomponent alloys, remain an unsolved mystery, and thus it is still a challenge to design BMGs based on atomic packing laws. Various production routes in solid-, liquid- and gas phases are described in this chapter. Further improvement in production technique, as well as the increase in product dimension, is a vast gap in these new materials, attracting to be solved by the research.

References

1. Diao, H.Y., Feng, R., Dahmen, K.A., Liaw, P.K.: Fundamental deformation behavior in high-entropy alloys: an overview. Curr. Opinion Solid State Mater. Sci. **21**(5), 252–266 (2017). https://doi.org/10.1016/j.cossms.2017.08.003
2. https://www.yalescientific.org/2012/03/bulk-metallic-glasses-constructing-the-future/. Accessed 01 Sept 2019
3. Cheng, Y.Q., Sheng, H.W., Ma, E.: Relationship between structure, dynamics, and mechanical properties in metallic glass-forming alloys. Phys. Rev. B. **78**, 014207 (2008). https://doi.org/10.1103/PhysRevB.78.014207.
4. Telford, M.: The case for bulk metallic glass. Mater. Today. **7**(3), 36–43 (2004). https://doi.org/10.1016/S1369-7021(04)00124-5
5. Chatterjee, R.: Manufacturing of metallic glasses. Adv. Mater. Manuf. Charact. **7**, 1 (2017). https://doi.org/10.11127/ijammc2017.04.05
6. Schroers, J., Johnson, W.L.: Ductile bulk metallic glass. Phys. Rev. Lett. **93**, 255506.– Published 16 December 2004 (2004). https://doi.org/10.1103/PhysRevLett.93.255506
7. Tasci, E.S., Sluiter, M.H.F., Pasturel, A., Villars, P.: Liquid structure as a guide for phase stability in the solid state: discovery of a stable compound in the Au–Si alloy system. Acta Mater. **58**(2), 449–456 (2010). https://doi.org/10.1016/j.actamat.2009.09.023

8. Pampillo, C.A., Chen, H.S.: Comprehensive plastic deformation of a bulk metallic glass. Mater. Sci. Eng. **13**(2), 181–188 (1974). https://doi.org/10.1016/0025-5416(74)90185-2
9. Fiore, G., Battezzati, L.: Engraving of a Pd77.5Cu6Si16.5 bulk metallic glass. Bulk Metall. Glasses. **9**(6), 509–511 (2007). https://doi.org/10.1002/adem.200700049
10. Assor, A., Tzelgov, J., Thein, R., Ilardi, B.C., Connell, J.P.: Assessing the correlates of over and underrating of academic competence: a conceptual clarification and a methodological proposal. **61**(6), 2085–2097., First Published: December 1990 (1990). https://doi.org/10.1111/j.1467-8624.1990.tb03590.x
11. Shinya, H.: Phonon excitations in Pd40Ni40P20 bulk metallic glass by inelastic X-Ray scattering. Mater. Sci. Forum. **879**., Trans Tech Publications, Ltd., 767–772 (2016). https://doi.org/10.4028/www.scientific.net/msf.879.767
12. Mukai, T., Nieh, T.G., Kawamura, Y., Inoue, A., Higashi, K.: Effect of strain rate on compressive behavior of a $Pd_{40}Ni_{40}P_{20}$ bulk metallic glass. Intermetallics. **10**(11–12), 1071–1077 (2002). https://doi.org/10.1016/S0966-9795(02)00137-1
13. Jörg, F.L.: Bulk metallic glasses. Intermetallics. **11**(6), 529–540 (2003). https://doi.org/10.1016/S0966-9795(03)00046-3
14. Lu, Z.P., Hu, X., Li, Y., Ng, S.C.: Glass forming ability of La–Al–Ni–Cu and Pd–Si–Cu bulk metallic glasses. Mater. Sci. Eng. A. **304–306**, 679–682 (2001). https://doi.org/10.1016/S0921-5093(00)01563-X
15. Zhang, T., Ye, F., Wang, Y., et al.: Structural relaxation of $La_{55}Al_{125}Ni_{10}Cu_{10}$ bulk metallic glass. Metall. Mat. Trans. A. **39**, 1953–1957 (2008). https://doi.org/10.1007/s11661-007-9369-1
16. Gao, R., Hui, X., Fang, H.Z., Liu, X.J., Chen, G.L., Liu, Z.K.: Structural characterization of $Mg_{65}Cu_{25}Y_{10}$ metallic glass from ab initio molecular dynamics. Comput. Mater. Sci. **44**(2), 802–806 (2008). https://doi.org/10.1016/j.commatsci.2008.05.031
17. Chen, W., Wang, Y., Qiang, J., Dong, C.: Bulk metallic glasses in the Zr-Al-Ni-Cu system. Acta Mater. **51**(7), 1899–1907 (2003). https://doi.org/10.1016/S1359-6454(02)00596-7
18. Shan, G.B., Li, J.X., Yang, Y.Z., Qiao, L.J., Chu, W.Y.: Hydrogen-enhanced plastic deformation during indentation for bulk metallic glass of $Zr_{65}Al_{7.5}Ni_{10}Cu_{17.5}$. Mater. Lett. **61**(8–9), 1625–1628 (2007). https://doi.org/10.1016/j.matlet.2006.07.181
19. H. Ma, Q. Zheng, J. Xu, Y. Li, Doubling the Critical Size for Bulk Metallic Glass Formation in the Mg–Cu–Y Ternary System, 20, 9 2005 , pp. 2252-2255, https://doi.org/10.1557/jmr.2005.0307.
20. https://spinoff.nasa.gov/Spinoff2004/ch_7.html. Accessed 08 Sept. 2019.
21. https://www.bulkmetallicglass.it/en/index.php. Accessed 08 Sept. 2019.
22. Mukherjee, S., Schroers, J., Zhou, Z., Johnson, W.L., Rhim, W.-K.: Viscosity and specific volume of bulk metallic glass-forming alloys and their correlation with glass forming ability. Acta Mater. **52**(12), 3689–3695 (2004). https://doi.org/10.1016/j.actamat.2004.04.023
23. Schroers, J.: Processing of bulk metallic glass. Adv. Mater. **22**(14), 1566–1597 (2010). https://doi.org/10.1002/adma.200902776
24. Lu, Z.P., Liu, C.T.: A new glass-forming ability criterion for bulk metallic glasses. Acta Mater. **50**(13), 3501–3512 (2002). https://doi.org/10.1016/S1359-6454(02)00166-0
25. Wang, W.H., Dong, C., Shek, C.H.: Bulk metallic glasses. Mater. Sci. Eng. R. Rep. **44**(2–3), 45–89 (2004). https://doi.org/10.1016/j.mser.2004.03.001
26. Suryanarayana, C., Inoue, A.: Iron-based bulk metallic glasses. Int. Mater. Rev. **58**(3), 131–166 (2013). https://doi.org/10.1179/1743280412Y.0000000007
27. Ding, S., Liu, Y., Li, Y., et al.: Combinatorial development of bulk metallic glasses. Nat. Mater. **13**, 494–500 (2014). https://doi.org/10.1038/nmat3939
28. Khan, M.M., Nemati, A., Rahman, Z.U., Shah, U.H., Asgar, H., Haider, W.: Recent advancements in bulk metallic glasses and their applications: a review. Crit. Rev. Solid State Mater. Sci. **43**(3), 233–268 (2018). https://doi.org/10.1080/10408436.2017.1358149
29. Fecht, H.-J., Johnson, W.L.: Thermodynamic properties and metastability of bulk metallic glasses. Mater. Sci. Eng. A. **375–377**, 2–8 (2004). https://doi.org/10.1016/j.msea.2003.10.254
30. Fecht, H.J.: Thermodynamic properties of amorphous solids-glass formation and glass transition-(overview). Mater. Transact. **36**(7), 777–793 (1995). https://doi.org/10.2320/matertrans1989.36.777
31. Gallino, I., Schroers, J., Busch, R.: Kinetic and thermodynamic studies of the fragility of bulk metallic glass forming liquids. J. Appl. Phys. **108**(6). https://doi.org/10.1063/1.3480805
32. Patel, S.K., Swain, B.K., Behera, A., Mohapatra, S.S.: Metallic glasses: a revolution in material science. In: Metallic Glasses., ISBN: 978-1-78985-488-6. InTech Press, London (2020). https://doi.org/10.5772/intechopen.90165
33. Tang, C., Zhou, H.: Thermodynamics and the Glass Forming Ability of Alloys. IntechOpen, London (2011). https://doi.org/10.5772/20803
34. Joannopoulos, J.D., Cohen, M.L.: Theory of short-range order and disorder in tetrahedrally bonded semiconductors. In: Ehrenreich, H., Seitz, F., Turnbull, D. (eds.) Solid State Physics, vol. 31, pp. 71–148., ISBN 9780126077315. Academic Press, Cambridge, MA (1976). https://doi.org/10.1016/S0081-1947(08)60542-1
35. Park, E.S., Kim, D.H.: Design of Bulk metallic glasses with high glass forming ability and enhancement of plasticity in metallic glass matrix composites: a review. Met. Mater. Int. **11**, 19–27 (2005). https://doi.org/10.1007/BF03027480
36. Espallargas, N., Aune, R.E., Torres, C., Papageorgiou, N., Muñoz, A.I.: Bulk metallic glasses (BMG) for biomedical applications-A tribocorrosion investigation of $Zr_{55}Cu_{30}Ni_5Al_{10}$ in simulated body fluid. Wear. **301**(1–2), 271–279 (2013). https://doi.org/10.1016/j.wear.2012.12.053
37. Kruzic, J.J.: Bulk metallic glasses as structural materials: a review. Adv. Eng. Mater. **18**(8), 1308–1331 (2016). https://doi.org/10.1002/adem.201600066
38. Chen, M.: A brief overview of bulk metallic glasses. NPG Asia Mater. **3**, 82–90 (2011). https://doi.org/10.1038/asiamat.2011.30
39. Das, J., Tang, M.B., Kim, K.B., Theissmann, R., Baier, F., Wang, W.H., Eckert, J.: Work-hardenable. Ductile bulk metallic glass. Phys. Rev. Lett. **94**, 205501.–Published 23 May 2005. https://doi.org/10.1103/PhysRevLett.94.205501
40. Şopu, D., Ritter, Y., Gleiter, H., Albe, K.: Deformation behavior of bulk and nanostructured metallic glasses studied via molecular dynamics simulations. Phys. Rev. B. **83**, 100202. Published 23 March 2011. https://doi.org/10.1103/PhysRevB.83.100202
41. Celtek, M., Sengul, S., Domekeli, U.: Glass formation and structural properties of $Zr_{50}Cu_{50}$-xAlx bulk metallic glasses investigated by molecular dynamics simulations. Intermetallics. **84**, 62–73 (2017). https://doi.org/10.1016/j.intermet.2017.01.001
42. Kong, J., Ye, Z., Chen, W., Shao, X., Wang, K., Zhou, Q.: Dynamic mechanical behavior of a Zr-based bulk metallic glass composite. Mater. Des. **88**, 69–74 (2015). https://doi.org/10.1016/j.matdes.2015.08.132
43. Ju, S.-P., Huang, H.-H., Wu, T.-Y.: Investigation of the local structural rearrangement of $Mg_{67}Zn_{28}Ca_5$ bulk metallic glasses during tensile deformation: a molecular dynamics study. Comp. Mater. Sci. **96**(Part A), 56–62 (2015). https://doi.org/10.1016/j.commatsci.2014.09.005
44. Wang, Q., Pelletier, J.M., Xia, L., Xu, H., Dong, Y.D.: The viscoelastic properties of bulk $Zr_{55}Cu_{25}Ni_5Al_{10}Nb_5$ metallic glass. J. Alloys Comp. **413**(1–2), 181–187 (2006). https://doi.org/10.1016/j.jallcom.2005.02.109
45. Fujita, T., Konno, K., Zhang, W., Kumar, V., Matsuura, M., Inoue, A., Sakurai, T., Chen, M.W.: Atomic-scale heterogeneity of a

multicomponent bulk metallic glass with excellent glass forming ability. Phys. Rev. Lett. **103**, 075502.–Published 12 August 2009. https://doi.org/10.1103/PhysRevLett.103.075502
46. Plummer, J.D., Figueroa, I.A., Hand, R.J., Davies, H.A., Todd, I.: Elastic properties of some bulk metallic glasses. J. Non-Cryst. Solids. **355**(6), 335–339 (2009). https://doi.org/10.1016/j.jnoncrysol.2008.12.011
47. Corteen, J.: Shear Banding in Metallic Glasses: a Mathematical Perspective Inspired by Soil Mechanics. University of Sheffield, PhD thesis (2014)
48. Ding, H.Y., Shao, Y., Gong, P., Li, J.F., Yao, K.F.: A senary TiZrHfCuNiBe high entropy bulk metallic glass with large glass-forming ability. Mater. Lett. **125**, 151–153 (2014). https://doi.org/10.1016/j.matlet.2014.03.185
49. Fang, H.Z., Hui, X., Chen, G.L., Liu, Z.K.: Al-centered icosahedral ordering in $Cu_{46}Zr_{46}A_{18}$ bulk metallic glass. Appl. Phys. Lett. **94**, 9 (2009). https://doi.org/10.1063/1.3086885
50. S. Mechlera), G. Schumacher, I. Zizak, M.-P. Macht, and N. Wanderka, Correlation between icosahedral short range order, glass forming ability, and thermal stability of Zr–Ti–Ni–Cu–(Be) glasses, Appl. Phys. Lett. 91, 2, https://doi.org/10.1063/1.2755924, (2007)
51. Naz, G.J., Dong, D., Geng, Y., et al.: Composition formulas of Fe-based transition metals-metalloid bulk metallic glasses derived from dual-cluster model of binary eutectics. Sci. Rep. **7**, 9150 (2017). https://doi.org/10.1038/s41598-017-09100-9
52. Takeuchi, A., Chen, N., Wada, T., Yokoyama, Y., Kato, H., Inoue, A., Yeh, J.W.: $Pd_{20}Pt_{20}Cu_{20}Ni_{20}P_{20}$ high-entropy alloy as a bulk metallic glass in the centimeter. Intermetallics. **19**(10), 1546–1554 (2011). https://doi.org/10.1016/j.intermet.2011.05.030
53. Hays, C.C., Schroers, J., Johnson, W.L.: Vitrification and determination of the crystallization time scales of the bulk-metallic-glass-forming liquid $Zr_{58.5}Nb_{2.8}Cu_{15.6}Ni_{12.8}Al_{10.3}$. Appl. Phys. Lett. **79**, 11 (2001). https://doi.org/10.1063/1.1398605
54. Kumar, G., Desai, A., Schroers, J.: Bulk metallic glass: the smaller the better. Adv. Mater. **23**(4), 461–476 (2011). https://doi.org/10.1002/adma.201002148
55. Qiu, C.L., Chen, Q., Liu, L., Chan, K.C., Zhou, J.X., Chen, P.P., Zhang, S.M.: A novel Ni-free Zr-based bulk metallic glass with enhanced plasticity and good biocompatibility. Scr. Mater. **55**(7), 605–608 (2006). https://doi.org/10.1016/j.scriptamat.2006.06.018
56. Li, W., Gao, Y., Bei, H.: On the correlation between microscopic structural heterogeneity and embrittlement behavior in metallic glasses. Sci. Rep. **5**, 14786 (2015). https://doi.org/10.1038/srep14786.
57. Johnson, W.L., Peker, A.: Synthesis and properties of bulk metallic glasses. In: Otooni, M.A. (ed.) Science and Technology of Rapid Solidification and Processing NATO ASI Series (Series E: Applied Sciences), vol. 278. Springer, Dordrecht (1995). https://doi.org/10.1007/978-94-011-0223-0_3.
58. Wu, Y., Zhang, L., Chen, S., Li, W., Zhang, H.A.: Multiple twin-roller casting technique for producing metallic glass and metallic glass composite strips. Materials (Basel). **12**(23), 3842. Published 2019 Nov 21 (2019). https://doi.org/10.3390/ma12233842
59. Aliaga, L.C.R., Beringues, J.F., Suriñach, S., Baró, M.D.: Comparative study of nanoindentation on melt-spun ribbon and bulk metallic glass with $Ni_{60}Nb_{37}B_3$ composition. **28**, 2740–2746 (2013). https://doi.org/10.1557/jmr.2013.260
60. Aqida, S.N., Shah, L.H., Naher, S., Brabazon, D.: Rapid solidification processing and bulk metallic glass casting. In: Hashmi, S., Batalha, G.F., Van Tyne, C.J., Yilbas, B. (eds.) Comprehensive Materials Processing, pp. 69–88. Elsevier, Netherland (2014). https://doi.org/10.1016/B978-0-08-096532-1.00506-9
61. Bordeenithikasem, P., Liu, J., Kube, S.A., et al.: Determination of critical cooling rates in metallic glass forming alloy libraries through laser spike annealing. Sci. Rep. **7**, 7155 (2017). https://doi.org/10.1038/s41598-017-07719-2.
62. Bakkal, M., Karagüzel, U., Kuzu, A.T.: In: Koç, M., Özel, T. (eds.) Manufacturing Techniques of Bulk Metallic Glasses, Modern Manufacturing Processes (2019). https://doi.org/10.1002/9781119120384.ch6
63. Nowosielski, R., Babilas, R.: Fabrication of bulk metallic glasses by centrifugal casting method. J. Achiev. Mater. Manuf. Eng. **20**, 1 (2007) Materials Science, http://jamme.acmsse.h2.pl/papers_vol20/1484S.pdf
64. Schroers, J., Pham, Q., Desai, A.: Thermoplastic forming of bulk metallic glass— a technology for MEMS and microstructure fabrication. J. Microelectromech. Syst. **16**(2), 240–247 (April 2007). https://doi.org/10.1109/JMEMS.0007.892889
65. Zhou, W., Li, H., Li, Z., Zhu, X., He, L.: Bonding of $Zr_{44}Ti_{11}Cu_{10}Ni_{10}Be_{25}$ bulk metallic glass and $AZ_{31}B$ magnesium alloy by hot staking extrusion. Vacuum. **166**, 240–247 (2019). https://doi.org/10.1016/j.vacuum.2019.05.020
66. Chiu, H.M., Kumar, G., Blawzdziewicz, J., Schroers, J.: Thermoplastic extrusion of bulk metallic glass. Scripta Mater. **61**(1), 28–31 (2009). https://doi.org/10.1016/j.scriptamat.2009.02.052
67. Kim, T.-S., Lee, J.-K., Kim, H.-J., Bae, J.-C.: Consolidation of $Cu_{54}Ni_6Zr_{22}Ti_{18}$ bulk amorphous alloy powders. Mater. Sci. Eng. A. **402**(1–2), 228–233 (2005). https://doi.org/10.1016/j.msea.2005.04.044
68. Hays, C.C.: Glass forming ability in the Zr-Nb-Ni-Cu-Al bulk metallic glasses. Mater. Sci. Forum. **343–346**., Trans Tech Publications, Ltd., 103–108 (2000). https://doi.org/10.4028/www.scientific.net/msf.343-346.103
69. Lee, M.H., Lee, K.S., Das, J., Thomas, J., Kühn, U., Eckert, J.: Improved plasticity of bulk metallic glasses upon cold rolling. Scr. Mater. **62**(9), 678–681 (2010). https://doi.org/10.1016/j.scriptamat.2010.01.024
70. Akihiro Makino, Takeshi Kubota, Chuntao Chang, Masahiro Makabe; Akihisa Inoue, FeSiBP bulk metallic glasses with unusual combination of high magnetization and high glass-forming ability, Mater. Trans., 48, 11 (2007) 3024 to 3027.
71. Schroers, J., Pham, Q., Peker, A., Paton, N., Curtis, R.V.: Blow molding of bulk metallic glass. Scr. Mater. **57**(4), 341–344 (2007). https://doi.org/10.1016/j.scriptamat.2007.04.033
72. Schroers, J., Hodges, T.M., Kumar, G., Raman, H., Barnes, A.J., Pham, Q., Waniuk, T.A.: Thermoplastic blow molding of metals. Mater. Today. **14**(1–2), 14–19 (2011). https://doi.org/10.1016/S1369-7021(11)70018-9
73. Ketov, S.V., Joksimovic, R., Xie, G., Trifonov, A., Kurihara, K., Louzguine-Luzgin, D.V.: Formation of nanostructured metallic glass thin films upon sputtering. Heliyon. **3**(1), e00228 (2017). https://doi.org/10.1016/j.heliyon.2016.e00228.
74. Wang, C., Yiu, P., Chu, J.P.: Zr–Ti–Ni thin film metallic glass as a diffusion barrier between copper and silicon. J. Mater. Sci. **50**, 2085–2092 (2015). https://doi.org/10.1007/s10853-014-8770-6.
75. Saraf Bidabad, M., Saniei, S.Z.: Feasibility of Ti-based metallic glass coating in biomedical applications. In: 2013 20th Iranian Conference on Biomedical Engineering (ICBME), pp. 250–254. ICBME (2013). https://doi.org/10.1109/ICBME.2013.6782229
76. Yang, B.J., Lu, W.Y., Zhang, J.L., et al.: Melt fluxing to elevate the forming ability of Al-based bulk metallic glasses. Sci. Rep. **7**, 11053 (2017). https://doi.org/10.1038/s41598-017-11504-6.
77. Blanquet, E., Mantoux, A., Pons, M., Vahlas, C.: Chemical vapour deposition and atomic layer deposition of amorphous and nanocrystalline metallic coatings: towards deposition of multimetallic films. J. Alloys Comp. **504**(Suppl. 1), S422–S424 (2010). https://doi.org/10.1016/j.jallcom.2010.03.205
78. Eckert, J.: Mechanical alloying of bulk metallic glass forming systems. Mater. Sci. Forum. **312–314**., Trans Tech Publications,

Ltd., 3–12 (1999). https://doi.org/10.4028/www.scientific.net/msf.312-314.3
79. Lindwall, J., Malmelöv, A., Lundbäck, A., et al.: Efficiency and accuracy in thermal simulation of powder bed fusion of bulk metallic glass. JOM. **70**, 1598–1603 (2018). https://doi.org/10.1007/s11837-018-2919-8
80. Zheng, B., Zhou, Y., Smugeresky, J.E., et al.: Processing and behavior of Fe-based metallic glass components via laser-engineered net shaping. Metall. Mat. Trans. A. **40**, 1235–1245 (2009). https://doi.org/10.1007/s11661-009-9828-y
81. Ibrahim, M.Z., Sarhan, A.A.D., Kuo, T.Y., Hamdi, M., Yusof, F., Chien, C.S., Chang, C.P., Lee, T.M.: Advancement of the artificial amorphous-crystalline structure of laser cladded FeCrMoCB on nickel-free stainless-steel for bone-implants. Mater. Chem. Phys. **227**, 358–367 (2019). https://doi.org/10.1016/j.matchemphys.2018.12.104
82. Xie, G., Zhang, W., Louzguine-Luzgin, D.V., Kimura, H., Inoue, A.: Fabrication of porous Zr–Cu–Al–Ni bulk metallic glass by spark plasma sintering process. Scr. Mater. **55**(8), 687–690 (2006). https://doi.org/10.1016/j.scriptamat.2006.06.034
83. Singer, J., Pelligra, C., Kornblum, N., et al.: Multiscale patterning of a metallic glass using sacrificial imprint lithography. Microsyst. Nanoeng. **1**, 15040 (2015). https://doi.org/10.1038/micronano.2015.40
84. Chua, J.P., Wijaya, H.: Nanoimprint of gratings on a bulk metallic glass. Appl. Phys. Lett. **90**(3). https://doi.org/10.1063/1.2431710
85. Li, B.S., Xie, S., Kruzic, J.J.: Toughness enhancement and heterogeneous softening of a cryogenically cycled Zr–Cu–Ni–Al–Nb bulk metallic glass. Acta Mater. **176**, 278–288 (2019). https://doi.org/10.1016/j.actamat.2019.07.012
86. Xu, D., Lohwongwatana, B., Duan, G., Johnson, W.L., Garland, C.: Bulk metallic glass formation in binary Cu-rich alloy series – $Cu_{100}-xZrx$ (x=34, 36, 38.2, 40 at.%) and mechanical properties of bulk $Cu_{64}Zr_{36}$ glass. Acta Mater. **52**(9), 2621–2624 (2004). https://doi.org/10.1016/j.actamat.2004.02.009
87. Szuecs, F., Kim, C.P., Johnson, W.L.: Mechanical properties of $Zr_{56.2}Ti_{13.8}Nb_{5.0}Cu_{6.9}Ni_{5.6}Be_{12.5}$ ductile phase reinforced bulk metallic glass composite. Acta Mater. **49**(9), 1507–1513 (2001). https://doi.org/10.1016/S1359-6454(01)00068-4
88. Wang, J., Li, R., Hua, N., Lu, H., Zhang, T.: Ternary Fe–P–C bulk metallic glass with good soft-magnetic and mechanical properties. Scr. Mater. **65**(6), 536–539 (2011). https://doi.org/10.1016/j.scriptamat.2011.06.020
89. Calin, M., Zhang, L.C., Eckert, J.: Tailoring of microstructure and mechanical properties of a Ti-based bulk metallic glass-forming alloy. Scr. Mater. **57**(12), 1101–1104 (2007). https://doi.org/10.1016/j.scriptamat.2007.08.018
90. Liu, C.T., Chisholm, M.F., Miller, M.K.: Oxygen impurity and microalloying effect in a Zr-based bulk metallic glass alloy. Intermetallics. **10**(11–12), 1105–1112 (2002). https://doi.org/10.1016/S0966-9795(02)00131-0
91. Li, Y., Shen, Y., Leu, M.C., Tsai, H.-L.: Mechanical properties of Zr-based bulk metallic glass parts fabricated by laser-foil-printing additive manufacturing. Mater. Sci. Eng. A. **743**, 404–411 (2019). https://doi.org/10.1016/j.msea.2018.11.056
92. Zhou, K., Chen, C., Liu, Y., Pang, S., Hua, N., Yang, W., Zhang, T.: Effects of lutetium addition on formation, oxidation and tribological properties of a Zr-based bulk metallic glass. Intermetallics. **90**, 81–89 (2017). https://doi.org/10.1016/j.intermet.2017.07.007
93. Tam, C.Y., Shek, C.H.: Abrasive wear of $Cu_{60}Zr_{30}Ti_{10}$ bulk metallic glass. Mater. Sci. Eng. A. **384**(1–2), 138–142 (2004). https://doi.org/10.1016/j.msea.2004.05.073
94. Stoica, M., Kumar, S., Roth, S., Ram, S., Eckert, J., Vaughan, G., Yavari, A.R.: Crystallization kinetics and magnetic properties of $Fe_{66}Nb_4B_{30}$ bulk metallic glass. J. Alloys Comp. **483**(1–2), 632–637 (2009). https://doi.org/10.1016/j.jallcom.2007.11.150
95. McCallum, R.W., Lewis, L.H., Kramer, M.J., Dennis, K.W.: Magnetic aspects of the ferromagnetic "bulk metallic glass" alloy system Nd–Fe–Al. J. Magn. Magn. Mater. **299**(2), 265–280 (2006). https://doi.org/10.1016/j.jmmm.2005.04.013
96. Fornell, J., González, S., Rossinyol, E., Suriñach, S., Baró, M.D., Louzguine-Luzgin, D.V., Perepezko, J.H., Sort, J., Inoue, A.: Enhanced mechanical properties due to structural changes induced by devitrification in Fe–Co–B–Si–Nb bulk metallic glass. Acta Mater. **58**(19), 6256–6266 (2010). https://doi.org/10.1016/j.actamat.2010.07.047
97. Zhang, H., Li, R., Xu, T., Liu, F., Zhang, T.: Near room-temperature magnetocaloric effect in FeMnPBC metallic glasses with tunable curie temperature. J. Magnet. Magn. Mater. **347**, 131–135 (2013). https://doi.org/10.1016/j.jmmm.2013.07.020
98. Li, S., Zhao, D.Q., Pan, M.X., Wang, W.H.: A bulk metallic glass based on heavy rare earth gadolinium. J. Non-Cryst. Solids. **351** (30–32), 2568–2571 (2005). https://doi.org/10.1016/j.jnoncrysol.2005.07.005
99. Peter, W.H., Buchanan, R.A., Liu, C.T., Liaw, P.K., Morrison, M. L., Horton, J.A., Carmichael, C.A., Wright, J.L.: Localized corrosion behavior of a zirconium-based bulk metallic glass relative to its crystalline state. Intermetallics. **10**(11–12), 1157–1162 (2002). https://doi.org/10.1016/S0966-9795(02)00130-9
100. Liu, L., Qiu, C.L., Chen, Q., Zhang, S.M.: Corrosion behavior of Zr-based bulk metallic glasses in different artificial body fluids. J. Alloys Comp. **425**(1–2), 268–273 (2006). https://doi.org/10.1016/j.jallcom.2006.01.048
101. Wang, R., Wang, Y., Yang, J., Sun, J., Xiong, L.: Influence of heat treatment on the mechanical properties, corrosion behavior, and biocompatibility of $Zr_{56}Al_{16}Co_{28}$ bulk metallic glass. J. Non-Cryst. Solids. **411**, 45–52 (2015). https://doi.org/10.1016/j.jnoncrysol.2014.12.018
102. Wong, C.H., Shek, C.H.: Friction welding of $Zr_{41}Ti_{14}Cu_{12.5}Ni_{10}Be_{22.5}$ bulk metallic glass. Scripta Materialia. **49**(5), 393–397 (2003). https://doi.org/10.1016/S1359-6462(03)00306-3
103. Kawamura, Y., Ohno, Y.: Spark welding of $Zr_{55}Al_{10}Ni_5Cu_{30}$ bulk metallic glasses. Scr. Mater. **45**(2), 127–132 (2001). https://doi.org/10.1016/S1359-6462(01)01003-X
104. Kim, J., Kawamura, Y.: Electron beam welding of the dissimilar Zr-based bulk metallic glass and Ti metal. Scripta Mater. **56**(8), 709–712 (2007). https://doi.org/10.1016/j.scriptamat.2006.12.046
105. Brothers, A.H., Dunand, D.C.: Syntactic bulk metallic glass foam. Appl. Phys. Lett. **84**(7) (2004). https://doi.org/10.1063/1.1646467
106. Lee, M.H., Sordelet, D.J.: Synthesis of bulk metallic glass foam by powder extrusion with a fugitive second phase. Appl. Phys. Lett. **89**(2) (2006). https://doi.org/10.1063/1.2221882
107. Yoon, S., Bae, G., Xiong, Y., Kumar, S., Kang, K., Kim, J.-J., Lee, C.: Strain-enhanced nanocrystallization of a CuNiTiZr bulk metallic glass coating by a kinetic spraying process. Acta Mater. **57**(20), 6191–6199 (2009). https://doi.org/10.1016/j.actamat.2009.08.045
108. Liu, W.D., Liu, K.X., Chen, Q.Y., Wang, J.T., Yan, H.H., Li, X.J.: Metallic glass coating on metals plate by adjusted explosive welding technique. Appl. Surf. Sci. **255**(23), 9343–9347 (2009). https://doi.org/10.1016/j.apsusc.2009.07.033
109. Davidson, M., Roberts, S., Castro, G., Dillon, R.P., Kunz, A., Kozachkov, H., Demetriou, M.D., Johnson, W.L., Nutt, S., Hofmann, D.C.: Investigating amorphous metal composite architectures as spacecraft shielding. Adv. Eng. Mater. **15**(1–2), 27–33 (2013). https://doi.org/10.1002/adem.201200313
110. Jurewicz, A.J.G., et al.: The genesis solar-wind collector materials. In: Russell, C.T. (ed.) The Genesis Mission. Springer, Dordrecht (2003). https://doi.org/10.1007/978-94-010-0241-7_2
111. Hofmann, D., Polit-Casillas, R., Roberts, S., et al.: Castable bulk metallic glass strain wave gears: towards decreasing the cost of

high-performance robotics. Sci. Rep. **6**, 37773 (2016). https://doi.org/10.1038/srep37773
112. Schuster, B.E., Wei, Q., Ervin, M.H., Hruszkewycz, S.O., Miller, M.K., Hufnagel, T.C., Ramesh, K.T.: Bulk and microscale compressive properties of a Pd-based metallic glass. Scr. Mater. **57**(6), 517–520 (2007). https://doi.org/10.1016/j.scriptamat.2007.05.025
113. Pang, S.J., Zhang, T., Asami, K., Inoue, A.: Synthesis of Fe–Cr–Mo–C–B–P bulk metallic glasses with high corrosion resistance. Acta Mater. **50**(3), 489–497 (2002). https://doi.org/10.1016/S1359-6454(01)00366-4
114. Nishiyama, N., Amiya, K., Inoue, A.: Recent progress of bulk metallic glasses for strain-sensing devices. Mater. Sci. Eng. A. **449–451**, 79–83 (2007). https://doi.org/10.1016/j.msea.2006.02.384
115. Yulei, D., Xu, L., Shen, Y., Zhuang, W., Zhang, S., Chen, G.: Hydrogen absorption/desorption behavior of $Mg_{50}La_{20}Ni_{30}$ bulk metallic glass. Int. J. Hydrog. Energy. **38**(11), 4670–4674 (2013). https://doi.org/10.1016/j.ijhydene.2013.02.043
116. Dong, F., He, M., Zhang, Y., Luo, L., Yanqing, S., Wang, B., Huang, H., Xiang, Q., Yuan, X., Zuo, X., Han, B., Xu, Y.: Effects of hydrogen on the nanomechanical properties of a bulk metallic glass during nanoindentation. Int. J. Hydrog. Energy. **42**(40), 25436–25445 (2017). https://doi.org/10.1016/j.ijhydene.2017.08.141
117. Sekol, R.C., Kumar, G., Carmo, M., Gittleson, F., Hardesty-Dyck, N., Mukherjee, S., Schroers, J., Taylor, A.D.: Bulk metallic glass micro fuel cell. Small. **9**(12), 2081–2085 (2013). https://doi.org/10.1002/smll.201201647
118. He, J.J., Li, N., Tang, N., Wang, X.Y., Zhang, C., Liu, L.: The precision replication of a microchannel mould by hot-embossing a Zr-based bulk metallic glass. Intermetallics. **21**(1), 50–55 (2012). https://doi.org/10.1016/j.intermet.2011.10.001
119. Li, D.-F., Shen, Y., Xu, J.: Bending proof strength of $Zr_{61}Ti_2Cu_{25}Al_{12}$ bulk metallic glass and its correlation with shear-banding initiation. Intermetallics. **126**, 106915 (2020). https://doi.org/10.1016/j.intermet.2020.106915.
120. Kanik, M., Bordeenithikasem, P., Kumar, G., Kinser, E., Schroers, J.: High quality factor metallic glass cantilevers with tunable mechanical properties. Appl. Phys. Lett. **105**(13). https://doi.org/10.1063/1.4897305
121. https://www.ad-na.com/en/product/jewel/material/bmg.html. Accessed 01 Sept 2019.
122. Bishop, D., Pardo, F., Bolle, C., et al.: Silicon micro-machines for fun and profit. J. Low Temp. Phys. **169**, 386–399 (2012). https://doi.org/10.1007/s10909-012-0654-z.
123. Sfondrini, M.F., Cacciafesta, V., Maffia, E., Massironi, S., Scribante, A., Alberti, G., Biesuz, R., Klersy, C.: Chromium release from new stainless steel, recycled and nickel-free orthodontic brackets. Angle Orthod. **79**(2), 361–367 (2009). https://doi.org/10.2319/042108-223.1
124. Qin, F.X., Wang, X.M., Inoue, A.: Observation of bone-like apatite on Ti-coated $Zr_{55}Al_{10}Ni_5Cu_{30}$ bulk metallic glass after alkali treatment. Intermetallics. **16**(7), 917–922 (2008). https://doi.org/10.1016/j.intermet.2008.04.008
125. Huang, L., Pu, C., Fisher, R.K., Mountain, D.J.H., Gao, Y., Liaw, P.K., Zhang, W., He, W.: A Zr-based bulk metallic glass for future stent applications: materials properties, finite element modeling, and in vitro human vascular cell response. Acta Mater. **25**, 356–368 (2015). https://doi.org/10.1016/j.actbio.2015.07.012
126. Teng-ho C.: Golf club head. Patent no: US5564994A (1996).
127. Ashley, S.: Metallic glasses bulk up. ASME Mech. Eng. **120**(06), 72–74 (1998). https://doi.org/10.1115/1.1998-JUN-4
128. Schroers, J., Lohwongwatana, B., Johnson, W.L., Peker, A.: Precious bulk metallic glasses for jewelry applications. Mater. Sci. Eng. A. **449–451**, 235–238 (2007). https://doi.org/10.1016/j.msea.2006.02.301
129. Kim, J.H., Lee, C., Lee, D.M., Sun, J.H., Shin, S.Y., Bae, J.C.: Pulsed Nd:YAG laser welding of $Cu_{54}Ni_6Zr_{22}Ti_{18}$ bulk metallic glass. Mater. Sci. Eng. A. **449–451**, 872–875 (2007). https://doi.org/10.1016/j.msea.2006.02.323

High Entropy Materials

9

Abstract

For more than 5000 years, metals and alloys have been formed in roughly the same way-propelling civilization from the Bronze Age to the industrial revolution and the advanced materials age. Now there is a new technique on the horizon that could help us take another big leap forward. This chapter describes high entropy materials (HEM) such as high entropy alloys (HEA), high entropy ceramics (HEC), high entropy polymer (HEP), and high entropy hybrids (HEH). These are the new class of materials that can extract strength from its single phase. For better understanding, an extensive description of various associated key concepts and mechanisms of high entropy alloys are described here. Various core effects such as high entropy effect, lattice distortion effect, sluggish diffusion effect, cocktail effect describe the strengthening process of the materials. Liquid-, solid- and gas-state route, electrochemical process, and additive manufacturing provide direction to synthesis the high entropy materials. High-entropy superalloys, high-entropy bulk metallic glasses, high-entropy flexible materials, and high-entropy coatings have been discussed. Various properties of HEA have been discussed with the example of current fabricated multi-materials. Emerging applications, of HEA such as in automobile, aero-vehicle, machinery, nuclear reactors, electrical and electronics, biomedical, and other applications, have been discussed here.

Keywords

High entropy alloys (HEA) · High entropy effect · Lattice distortion effect · Sluggish diffusion effect · Cocktail effect · Strain hardening · Grain-boundary hardening · Solid-solution hardening · Precipitation hardening · High-entropy superalloys · High-entropy bulk metallic glasses · Light materials HEA · High-entropy flexible materials · High-entropy coatings · Complex concentrated alloys

9.1 Introduction

High-entropy material (HEM) is currently the focus of significant attention in materials science and engineering due to their in situ potential properties. Furthermore, research indicates that HEM has considerably better strength-to-weight ratios, with a higher degree of fracture resistance, tensile strength, creep strength, as well as corrosion and oxidation resistance than conventional alloys [1–3]. Various HEM such as high entropy alloy (HEA), high entropy ceramics (HEC), high entropy polymers (HEP), and high entropy hybrids (HEH); in comparison to other materials have been given in Fig. 9.1.

9.2 High Entropy Alloys

Alloys like superalloys, steel, and so on, are based on one principal element with alloying additions done to improve their structural and functional properties. In contrast, HEAs are multicomponent alloys having constituents in equiatomic or near equiatomic ratios. HEAs are alloys that are formed by mixing equal or relatively large proportions of (usually) five or more elements [4]. Prior to the synthesis of these substances, typical alloys comprised one or two major components with smaller amounts of other elements. The term "high-entropy alloys" was coined because the entropy of mixing is substantially higher when there is a larger number of an element in the mix, and their proportions are more nearly equal. Whereas the content of the main elements is in the range of 5–35%, while the content of minor elements is <5% [5]. HEAs may also contain minor elements to modify the properties of the base alloy, further expanding the number of elements in HEAs. This composition-based definition prescribes elemental concentrations only and places no bounds on the magnitude of entropy. HEAs predominantly consist of a simple FCC, BCC, or FCC + BCC structure solid

A. Behera, *Advanced Materials*, https://doi.org/10.1007/978-3-030-80359-9_9

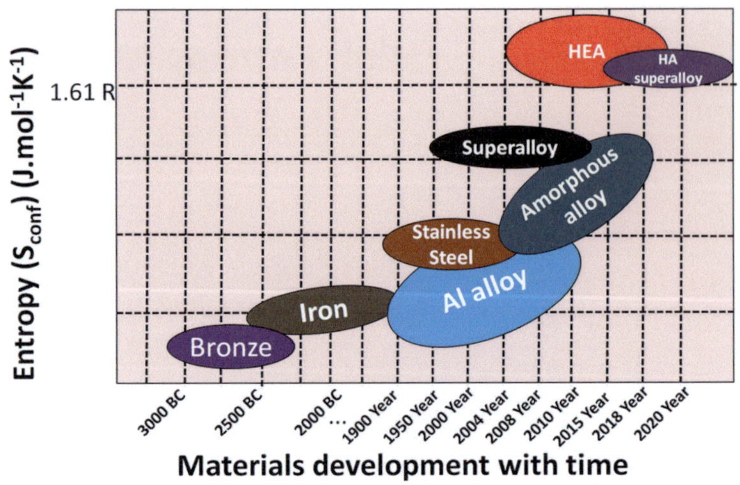

Fig. 9.1 Year-wise materials development and their corresponding entropy

solution phase owing to the high entropy of mixing, instead of many intermetallic phases or other complex phases [6]. HEA can be defined as the alloys that form solid solution with no intermetallic phases, as the formation of ordered phases decreases the entropy of the system. They exhibit simple solid solution structures owing to their high configurational entropy. The entropy-based definition assumes the alloy can be represented by the "liquid solution and high-temperature solid solution states where the thermal energy is sufficiently high to cause different elements to have random positions within the structure" [7].

9.3 Historical Development of High Entropy Alloy

Over 5000 years ago, the first alloy developer diluted elemental copper with a small amount of tin to make bronze. An immediate success, this simple alloying approach changed the course of civilizations—and humanity has used the same approach to develop alloys ever since. The idea is simple: an element with some attractive properties is used as a base, and minor amounts of other elements are added to improve the properties. While fewer than a dozen metallic elements were known before the eighteenth century, limiting options for early alloy developers, the remaining elements were subsequently discovered in quick succession and the best have now all been exploited. HEAs were considered from a theoretical standpoint as early as 1980 [8]. The first detailed study of multicomponent alloys consisting of a large number of constituents in equal or near-equal proportions was undertaken in 1981 by Vincent [9]. Prof. Jien-Wei Yeh from National Tsing Hua University came up with his idea for ways of actually creating high-entropy alloys in 1995 [10]. Jien-Wei Yeh and his team of Taiwanese scientists invented and built the world's first high-entropy alloys in 2004 that can withstand extremely high temperatures and pressures [11]. Ranganathan discussed similar types of alloys in 2003, describing them as multi-material cocktails, and Yeh et al. did likewise in 2004, describing them as high entropy alloys. In the twentieth century, HEA base elements provided lightweight and high-temperature alloys, essential for aerospace industries, nuclear reactors, jet aircraft, nuclear weapons, long-range hypersonic missiles, communication, and navigation via satellites and GPS. Recently, there have been a growing number of studies of multicomponent and high entropy alloys, as investigators have realized the potential for discovering new materials with valuable properties in the uncharted phase space in the middle of multicomponent systems [12].

9.4 The Key Concept of Multicomponent HEA

Yeh et al.'s term "high entropy alloys" has become accepted to refer to multicomponent alloys consisting of a large number of constituents in equal or near-equal proportions. Multicomponent alloys, however, do not in general exhibit particularly high entropy, or even high configurational entropy (Yeh et al.'s original concept) [9]. They often exhibit a surprisingly small number of phases with wide solubility ranges, caused partly by high entropy effects. In fact, there are many different kinds of multicomponent alloys, whether or not they have a large number of components in equal or near-equal proportions, depending on their development method (conventional alloying, extended alloying, ternary quaternary equiatomic substitution, etc.), constituent species (metallic, nonmetallic, semimetal, compound, etc.), manufacturing method (casting, rapid solidification, vapor deposition, co-sputtering, etc.), and resulting phase structure and microstructure (single-phase solid solution, dendritic, duplex eutectic or peritectic, multiphase, amorphous, etc.) [13]. Based on previous research, it can be noted that the multi-principal-elemental mixtures of HEAs result in:

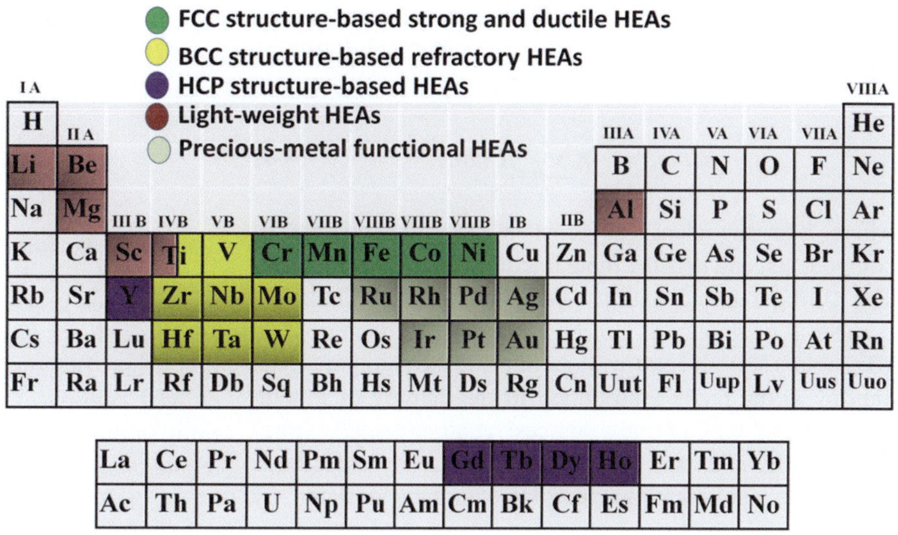

Fig. 9.2 Periodic table showing the main elements that take part in the synthesis in the five most typical HEA groups including FCC structure-based strong and ductile HEAs, BCC structure-based refractory HEAs, HCP structure-based HEAs, light-weight HEAs, and precious-metal functional HEAs. Minor elements of HEA systems including substitutional elements/interstitial elements (B, C, N, O, etc.) are not marked here

(1) high-entropy, (2) lattice distortion, (3) sluggish diffusion, and (4) cocktail effects. In brief, the high-entropy plays an important role in simplifying the microstructures so that they principally consist of simple solid solution phases with FCC and BCC structures. Lattice distortion further influences mechanical, physical, and chemical properties. Sluggish diffusion leads to the alloys developing nanocrystalline or even amorphous structures. Finally, the cocktail effects result in a composite effect on properties, wherein the interactions among the different elements themselves play an important role [14].

This design principle was originally aimed at the stabilization of single-phase massive solid solutions consisting of four or more elements through high configurational entropy. Beyond the original concept of single-phase solid solutions, several variants of this new design approach have been suggested including non-equiatomic, multiphase, interstitial, duplex, precipitate containing, and metastable HEAs. Several typical HEA systems have been developed, including (1) FCC structure-based strong and ductile HEAs, (2) BCC structure-based refractory HEAs, (3) HCP structure-based HEAs, (4) light-weight HEAs, and (5) precious-metal functional HEAs (Fig. 9.2). The number of HEA compositions explored is increasing rapidly due to the growing research efforts in this field, driven by promising properties that have been observed for some of these materials such as good cryogenic toughness and high strain hardening [15].

9.5 Thermodynamics of Solid Solution in HEA

Gibbs' phase rule (Eq. (9.1) can be used to determine an upper bound on the number of phases that will form in an equilibrium system.

$$F = C - P + 2 \tag{9.1}$$

In 2004, Cantor synthesized a 20-component alloy containing 5 at% of Mn, Cr, Fe, Co, Ni, Cu, Ag, W, Mo, Nb, Al, Cd, Sn, Pb, Bi, Zn, Ge, Si, Sb, and Mg. At constant pressure, the phase rule would allow for up to 21 phases at equilibrium, but far fewer actually formed. The predominant phase was an FCC solid solution phase, containing mainly Fe, Ni, Cr, Co, and Mn. From that result, the FeCrMnNiCo alloy, which forms only a solid solution phase, was developed. The Hume–Rothery rules have historically been applied to determine whether a mixture will form a solid solution. Research into high-entropy alloys has found that in multicomponent systems, these rules tend to be relaxed slightly [16]. Yeh also developed multicomponent alloys that consisted mostly or entirely of solid solution phases. He attributed this result to the high configurational, or mixing, entropy of a random solid solution containing numerous elements. The equilibrium phase will be achieved when the Gibbs free energy of formation (ΔG) will be lowest as in Eq. ((9.1).

$$\Delta G = \Delta H - T\Delta S \tag{9.2}$$

For predicting the equilibrium state of an alloy, the free energy changes from the elemental state to other states are often compared so that the state with the lowest mixing free energy (ΔG_{mix}) can be determined. From Eq. (9.2, it follows that the differences in free energy (ΔG_{mix}), enthalpy (ΔH_{mix}) and entropy (ΔS_{mix}) between the elemental and mixed states are related by:

$$\Delta G_{\text{mix}} = \Delta H_{\text{mix}} - T\Delta S_{\text{mix}} \tag{9.3}$$

The entropy of mixing or the configurational entropy for a random ideal solid solution can be calculated by Eq. (9.4).

$$\Delta S_{\text{mix}} = -R\sum\nolimits_{i=1}^{N} C_i ln C_i \tag{9.4}$$

where R is the ideal gas constant, N is the number of components, and c_i is the mole fraction of the ith component. The configurational entropy increases with the increased number of components. Notably, the term "high-entropy" in HEA refers to high entropy of mixing and the configurational entropy is the main contribution to the entropy of mixing amongst the four entropy sources: configurational, vibrational, electronic, and magnetic entropy. For equiatomic alloys, the configurational entropy can be calculated by Eq. (9.5.

$$\Delta S_{\text{mix}} = R \ln N \tag{9.5}$$

From this, it can be seen that the alloys in which the components are present in equal proportions will have the highest entropy, and adding additional elements will increase the entropy. A five-component, equiatomic alloy will have an entropy of mixing of 1.61R. However, entropy alone is not sufficient to stabilize the solid solution phase in every system. The enthalpy of mixing (ΔH), must also be taken into account. This can be calculated using:

$$\Delta H_{\text{mix}} = \sum_{i=1, I \neq j}^{N} 4\Delta H_{AB}{}^{\text{mix}} C_i C_j \tag{9.6}$$

where, $\Delta H_{AB}{}^{\text{mix}}$ is the binary enthalpy of mixing for A and B. To form a complete solid solution, ΔH_{mix} should be between −10 and 5 kJ/mol. In addition, if the alloy contains any pair of elements that tend to form ordered compounds in their binary system, a multicomponent alloy containing them is also likely to form ordered compounds. Both of the thermodynamic parameters can be combined into a single, unitless parameter Ω:

$$\Omega = \frac{T_m \Delta S_{\text{mix}}}{|\Delta H_{\text{mix}}|} \tag{9.7}$$

where, T_m is the average melting point of the elements in the alloy. Ω should be greater than or equal to 1.1 to promote solid solution development. Again, the atomic radii of the components must also be similar to form a solid solution following Eq. ((9.8). The difference in atomic radii in solid solution can be represented by δ,

$$\delta = \sqrt{\sum_{i=1}^{N} C_i \left(1 - \frac{r_i}{\bar{r}}\right)^2} \tag{9.8}$$

where r_i is the atomic radius of element i and $\bar{r} = \sum_{i=1}^{N} C_i r_i$.

Formation of a solid solution phase requires a $\delta \leq 6.6\%$, but some alloys with $4\% < \delta \leq 6.6\%$ do form intermetallics. Following Boltzmann's hypothesis, the mixing configurational entropy of an n-element equimolar alloy changing from an elemental to a random solution state (ideal state or regular state) can be calculated from:

$$\Delta S_{\text{mix}} = \text{R} \ln(\text{n}) \tag{9.9}$$

where R is Boltzmann's constant (gas constant) = 8.31 J/K mol. Figure 9.3 shows the entropy of mixing, calculated by Eq. (9.8, as a function of the number of elements in the equimolar alloys. Thus, binary and five-element equimolar alloys have solution states with an entropy of mixing of 5.76 and 13.37 J/K mol, respectively. The entropy of mixing for terminal solution or ordered compound states is expected to be smaller due to the limited number of ways they can mix.

Based on the characteristics of Fig. 9.2, HEAs have been preferentially designated to be alloys that comprise 5–13 major metallic elements. The lower limit of five elements is imposed because it is considered to be the point at which the mixing entropy is high enough to counterbalance the mixing enthalpy in most alloy systems and thus ensure the formation of solid solution phases [17]. Beyond 13 elements, there is a leveling off of the curve in Fig. 9.3, thus suggesting that little

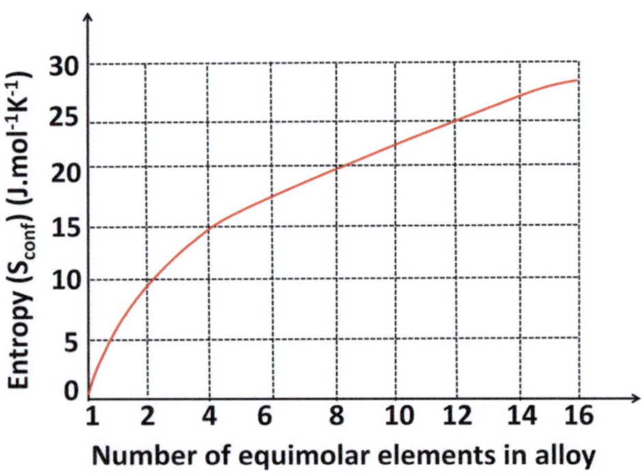

Fig. 9.3 The entropy of mixing with respect to the number of elements for equimolar alloys

further benefit will be brought about by composing alloys of a greater number of elements. The concentration of each element need not be equimolar but can be between 5 and 35 at%, therefore broadening the number of possible HEA systems [18]. Thus, HEAs do not contain any element whose concentration exceeds 50 at%, as is the case in the traditional alloys. Numerous alloys can therefore be generated that satisfy the HEA criteria. For instance, if 13 arbitrary elements are selected from the periodic table, then a total of 7099 five to 13 element alloy systems could be obtained, as determined by the following [19]:

$$C^{13}{}_5 + C^{13}{}_6 + C^{13}{}_7 + C^{13}{}_8 + C^{13}{}_9 + C^{13}{}_{10} + C^{13}{}_{11} + C^{13}{}_{12} + C^{13}{}_{13} = 7099 \quad (9.10)$$

The number of possible alloys is further increased by the fact that the alloys may or may not be equimolar, and other minor elements could be added to modify their properties. Examples of these three different types of HEAs are AlCoCrCuFeNi, $AlCo_{0.5}CrCuFe_{1.5}Ni_{1.2}$ and $AlCo_{0.5}CrCuFe_{1.5}Ni_{1.2}B_{0.1}C_{0.15}$. Based on the HEA definition, it is possible for alloys to be grouped roughly into three categories according to their mixing entropy in the random solution state, namely (1) low-entropy alloys (traditional alloys) with one or two major elements, (2) medium-entropy alloys with two to four major elements, and (3) high-entropy alloys with at least five major elements, as shown in Fig. 9.4. It should be noted that the random solution states are defined as liquid solution and high-temperature solid solution states where the thermal energy is sufficiently high to cause different elements to have random positions within the structure. Thus, high-entropy alloys are defined by the high entropy of the random solution state of multi-principal-element alloys [20].

From Hume–Rothery rules, we recognize the factors that affect the formation of binary solid solutions, which include atomic size difference, valence electron concentration (VEC), the crystal structure of the solute and solvent atoms, and difference in electronegativity. If the average VEC of the alloy is ≥ 8, the alloy will form an FCC lattice. If the average VEC is <6.87, it will form a BCC lattice. For values in between ($6.87 \leq VEC < 8.0$), it will form a mixture of FCC and BCC. VEC has also been used to predict the formation of σ-phase intermetallics (which are generally brittle and undesirable) in Cr and V containing HEAs [21]. Besides these factors, enthalpy and entropy of mixing are the most important phase formation parameters for HEAs. Higher entropy of mixing will lead to a lower Gibbs free energy ($\Delta G = \Delta H_{mix} - T\Delta S_{mix}$) which tends to stabilize the formation of solid-solution phases, rather than intermetallic phases [22]. For multicomponent systems, the ratio entropy and enthalpy of mixing ($T\Delta S_{mix}/\Delta H_{mix}$) would be more important to predict the formation of solid solution phases, and thus, a parameter can be defined Ω,

$$\Omega = T_m \frac{T\Delta S}{\Delta H_{mix}} \quad (9.11)$$

where, T_m is the average melting point of the alloy system. Solid solution phase tends to form as long as $\Omega > 1$ is satisfied which means that the effect of the entropy of mixing is greater than that of the enthalpy of mixing at the melting temperature.

9.6 Core Effects of HEA

The multi-principal element character in HEAs leads to some important effects that are much less pronounced in conventional alloys. These can be considered as four "core effects" are discussed subsequently.

9.6.1 The High Entropy Effect

The high-entropy effect states that the higher entropy of mixing (mainly configurational) in HEAs lowers the free energy of solid solution phases and facilitates their formation, particularly at higher temperatures. Due to this enhanced mutual solubility among constituent elements, the number of phases present in HEAs can be evidently reduced. HEAs

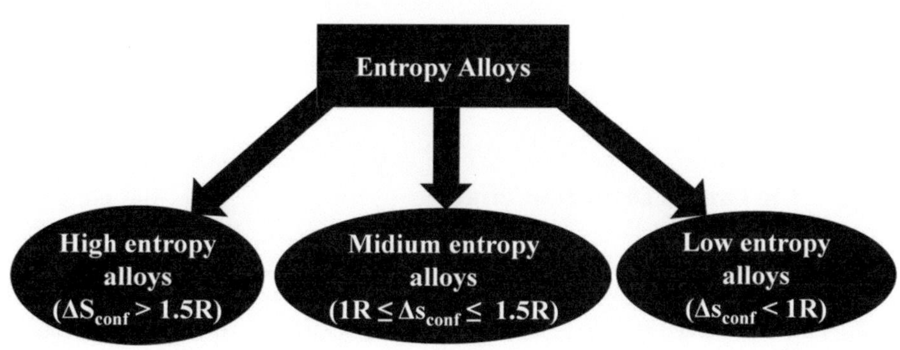

Fig. 9.4 The alloy world divided by the mixing entropy of their random solution states

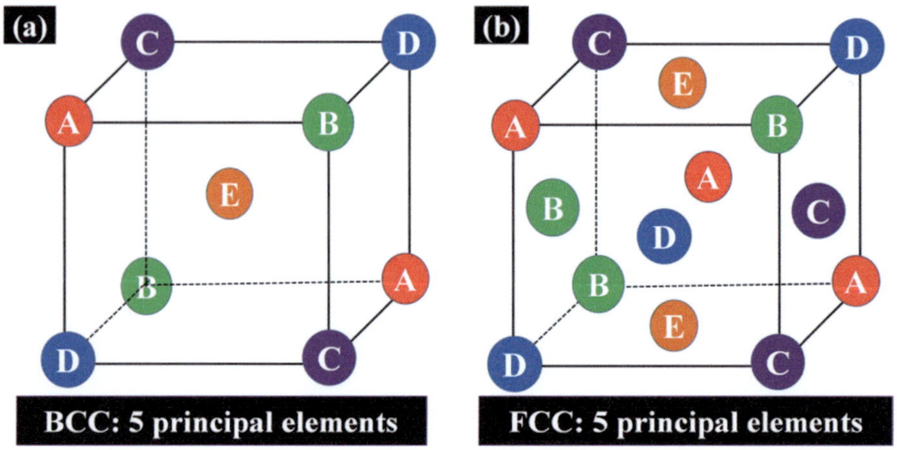

Fig. 9.5 Typical (**a**) BCC- and (**b**) FCC-HEA structures with five principal elements

composed of chemically compatible elements are composed of only a few solid solution phases or even one single phase, which is attributed to their high entropies of mixing. The high entropy of mixing enhances the mutual solubility among elements and prevents phase separation into terminal solution phases or intermetallic compounds. Although intermetallic phases can form in some HEAs as a result of the strong bonding between some metallic elements, even these phases tend to include a lot of other elements and have a significant reduction in their degree of ordering. In solid-state thermodynamics, the entropy of mixing is a well-known factor to account for the temperature effect on the increasing concentration of vacancies and the increased solubility of solutes that have weak bonds with the solvent atoms [23]. Moreover, for intermetallic compounds, the entropy of mixing is known to broaden the solubility of other elements. All these phenomena can be explained with reference to Eq. (9.3, in which the entropy of mixing competes with the enthalpy of mixing. Moreover, the $T\Delta S_{\text{mix}}$ term means that the mixing entropy becomes more dominant at higher temperatures. Therefore, the significantly higher entropy of mixing of HEAs for the random solution state is expected to significantly extend the solubility range for either terminal solutions or intermetallic compounds and form simple multielement solution phases, especially at higher temperatures. This competition between the enthalpy of mixing and entropy of mixing is thus a good alloy design parameter to predict the mutual solubility in solid solution phases. At elevated temperatures the higher the number of elements in a disordered state, the less the possibility of forming intermetallic phases. This uses the Boltzmann Equation (Eq. (9.9) [24].

9.6.2 The Lattice Distortion Effect

A multielement lattice is highly distorted because all atoms are solute atoms and their atomic sizes are all different from one another, as shown in Fig. 9.5. It might be expected that for sufficiently large atomic size differences, the distorted lattice will collapse into an amorphous structure since lattice distortion energy would be very high for retaining a crystalline configuration. Larger atoms push away their neighbors and small ones have extra space around. The strain energy associated with lattice distortion raises the overall free energy of the HEA lattice. It also affects the characteristic of HEAs. The distortion in either crystalline or amorphous structures influences the mechanical, thermal, electrical, optical, and chemical behavior of materials, by so-called lattice distortion effects [25]. For example, lattice distortion effects in HEA cause high solid solution hardening, thermal resistance, electrical resistance, and X-ray diffuse scattering, which is similar to the temperature effect on diffuse scattering.

9.6.3 The Sluggish Diffusion Effect

Sluggish diffusion involves the kinetics of the system; low diffusion rate, increases thermal stability, increase in recrystallization temperature, slows down grain growth, slows down phase separation, and improves creep resistance, which might benefit the microstructure. Phase transformations that depend on atomic diffusion require the cooperative diffusion of elements to attain the equilibrium partitioning among the phases. This, in combination with the lattice distortion which hinders atomic movement, will limit the effective diffusion rate in HEAs. In conventional casting of HEAs, the phase separation during cooling is often inhibited at higher temperatures and therefore delayed until lower temperatures. This is the reason why the as-cast structures of HEAs often have nano-precipitates in the matrix. This is also the reason for the higher recrystallization temperatures and activation energies of deformed HEAs [26]. The presence of nanocrystals in as-cast material and amorphous materials in sputter-deposited thin films and high recrystallization temperatures enlivened the sluggish diffusion theory. Cross-diffusion happens when the focus angle of one element prompts or changes the motion of another element. This occurs when one component changes the chemical attributes of other elements in the alloy system. Increasing the number of the composition of the elements in an alloy may make the diffusion become sluggish and reduce the temperature of the system.

The diffusion and phase transformation kinetics in HEA is slower than their conventional counterparts. This can be understood from two aspects. Firstly, in HEA the neighboring atoms of each lattice site are somewhat different. Thus, the neighbors before and after an atom jump into a vacancy are different. The difference in local atomic configuration leads to different bonding and therefore different local energies for each site. When an atom jumps into a low-energy site, it becomes "trapped" and the chance to jump out of that site will be lower. In contrast, if the site is a high-energy site, then the atom has a higher chance to drop back to its original site. Either of these scenarios slows down the diffusion process. Note that in conventional alloys with a low solute concentration, the local atomic configuration before and after jumping into a vacancy is, most of the time, identical [27]. Secondly, the diffusion rate of each element in a HEA is different. Some elements are less active (e.g., elements with high melting points) than others so these elements have lower success rates for jumping into vacancies. However, phase transformations typically require the coordinated diffusion of many kinds of elements. For example, the nucleation and growth of a new phase require the redistribution of all elements to reach the desired composition. Grain growth also requires the cooperation of all elements so that grain boundaries can successfully migrate. In these scenarios, the slow-moving elements become the rate-limiting factor that impedes the transformation. The slow kinetics in HEAs allows readily attainable supersaturated states and nano-sized precipitates, even in the cast state.

9.6.4 The 'Cocktail' Effect

The cocktail effect affects the properties of the predetermined system. It includes the compositional mixtures of elements where the consequence of the blend is both unpredictable and unexpected because of the distinctive properties the individual element provides. Adding an element to the mixture with properties realized will enhance the combination through the cocktail effect. High entropy alloys may exhibit properties dependent on the reaction between elements in the system. They exhibit a composite effect coming from the basic features and interactions among all the elements themselves, in addition to the indirect effects of the various elements on the microstructure [28]. For example, if more light elements are used, the overall density will be reduced. If more oxidation-resistant elements are used, such as Al, Cr, and Si, the oxidation resistance at high temperatures can be improved. If an element such as Al is added, which has strong bonding with the other elements present, such as Co, Cr, Cu, Fe, and Ni, and promotes the formation of a BCC phase, the strength will be increased [29].

9.7 Transformations in HEA

A liquid phase undercooled below the melting temperature is a simple example of a phase transformation. Phase transformations involve a change in structure and composition (for multiphase systems) that rearrange and redistribute atoms via diffusion. The process of phase transformation involves nucleation and growth. Nucleation of the new phases is the formation of stable nuclei of the new phases. Nuclei are often formed at grain boundaries and other defect sites. The growth of the new phases is at the expense of the original phases. Figure 9.6 is showing the nucleation and growth of HEAs. The nucleation can be heterogeneous and homogeneous.

Several types of transformation were observed in high entropy alloys such as precipitation transformations, allotropic transformation, eutectic transformation, massive transformation, continuous transformation, spinodal decomposition, order-disorder transformations during nucleation, and growth of HEA. The precipitation of phases constitutes the main mechanism of strengthening in HEA [30]. The high solvus temperature of the precipitation phase enables its stability in high temperatures and enforces the strengthening of the alloy. For example, the formation of an ordered intermetallic compound in non-equimolar 8.9Al-17.2Co-9.2Cr-8.2Fe-6Ni-50.5Ti (at%) system. In allotropic transformation in HEA, the formation of simple solid solutions with FCC or BCC structure was observed. In addition, similar disordered and ordered versions of the same base structure often coexist. For example, the coexistence of BCC

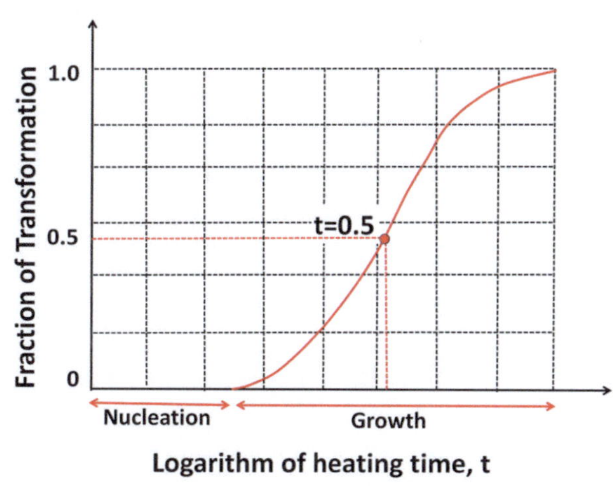

Fig. 9.6 Nucleation and growth curve showing the single-phase HEA transformation

and ordered B2 phases or the coexistence of FCC and ordered L12 phases were frequently reported. The exploitation of the eutectic transformations has been considered in HEAs, to enhance the strengthening and ductility of the alloys. HEAs with FCC structure have good ductility and poor strength, whereas HEAs with BCC structure have great strength but they lack plasticity [31]. Hence, aiming to design an alloy with combined strength and ductility, eutectic high entropy alloys (EHEA) have been developed. In continuous transformation the main characteristics are the diffusive interfaces between two phases with gradually changing in composition and that the decomposition proceeds with uphill diffusion, against the concentration gradient. In HEAs, the spinodal decomposition and the order-disorder transformation have been observed. An interesting fact in HEA is that spinodal decomposition and ordering take place simultaneously, which is thermodynamically contrasting. This phenomenon was explained due to the atomic size difference and the high elastic interactions. Spinodal decomposition and ordering in the dendritic constituent observed in AlxCoCrCuFeNi alloys with high Al content ($x>=1$), leading to the formation of the coherent A2 and B2 phases, accompanied by precipitation of Cu-rich particles. The $Al_{0.5}CoCrCuFeNi$ system contains two FCC phases and an ordered L_{12} phase. The L_{12} phase was rich in Ni and Al, with some solubility in Cu. The first FCC_1 phase was rich in Ni, Fe, Cr, and Co while the second FCC_2 phase was only Cu-rich. All phases were present in both dendritic and interdendritic structures. In the case of the spinodal decomposition, the enthalpy of mixing is positive ($\Delta H_{mix}>0$), whereas the ordered phases exhibit with negative enthalpy of mixing ($\Delta H_{mix}<0$). The bond energy and the atomic interactions between two dissimilar atoms explain the short-range order (SRO) interactions but fail to explain the long-range order (LRO) interactions and the spinodal decomposition (continuous transformations) [11]. The elastic interactions were responsible for this phenomenon and explain the LRO interactions. The high atomic difference causes extreme lattice deformations and strains, resulting in the increase of the elastic energy and promoting ordering and spinodal decomposition concurrently. In order-disorder transformations the enthalpy of mixing is negative and the formation of solid solutions is exhibited. In $CoCrFeNiAl_{0.3}$, $CoCrFeNiTi_{0.3}$, $CoCrFeNiAl_{0.3}Mo_{0.1}$ and $CoCrFeNiAl_{0.3}Mo_{0.3}$ systems, the formation of ordered L12 and disordered FCC structures has been found [32].

9.8 Phase Selection Approach in HEA

Hume–Rothery rules are perhaps the earliest guide to the formation of solid solution alloys. These rules state that extended solid solutions are favored in alloys whose elements have similar atom sizes, crystal structures, electronegativities, and valencies [33]. To apply these concepts to predict solid solution phase formation in complex alloys, the composition-weighted terms for differences in atom radii (δr) and electronegativity ($\delta\chi$), and an average valence electron concentration (VEC) has been developed. Thermodynamic considerations are reflected through the enthalpy of mixing (H^{SS}) and through a Ω term that combines H^{SS}, entropy of mixing (S^{SS}), and melting temperature, T_m [34].

Equations for these terms are:

$$\delta_r = \sqrt{\sum c_i\left(1-\frac{r_i}{\overline{r}}\right)^2} \tag{9.12}$$

$$\delta_\chi = \sqrt{\sum c_i\left(1-\frac{\chi_i}{\overline{\chi}}\right)^2} \tag{9.13}$$

$$VEC = \sum C_i VEC_i \tag{9.14}$$

$$H^{SS} = \sum_{i<j} 4H_{ij}c_i c_j \tag{9.15}$$

$$\Omega = \left(\sum C_i T_{m,i}\right)\left(\left|H^{SS}\right|\right) \tag{9.16}$$

Here r_i is atomic radius, χ_i is electronegativity, VEC_i is valency electron concentration, $T_{m,\,i}$ is melting temperature of elements i, c_i *and* c_j are the atom fractions of atoms i and j, $\overline{r}=\sum c_i r_i$ and $\overline{\chi}=\sum c_i\chi_i$ are the average atomic radius and electronegativity; and H_{ij} is the enthalpy of mixing of elements i and j at the equimolar concentration in regular binary solutions. The term $\delta_r$and δ_χ are often reported as percentages as follows (Eqs. 9.17 and 9.18):

$$\delta_r = \left(\sqrt{\sum c_i\left(1-\frac{r_i}{\overline{r}}\right)^2}\right)\times 100 \tag{9.17}$$

$$\delta_\chi = \left(\sqrt{\sum c_i\left(1-\frac{\chi_i}{\overline{\chi}}\right)^2}\right)\times 100 \tag{9.18}$$

No correlations are found between phases formed and $\delta\chi$ or VEC when a large number of alloys and a range of alloy families are considered – solid solution (SS), intermetallics (IM), and (SS + IM) multi-principal element alloys (MPEAs) all have similar ranges in $\delta\chi$ and VEC values. However, VEC can separate phases when a limited number of alloys are considered within a given alloy family. Compositions for BCC and FCC phases are separated in as-cast

$Al_xCoCrCuFeNi$ and $Al_xCoCrFeNi_2$ alloys ($0 \leq x \leq 2$), and σ-phase formation is predicted in annealed alloys containing Cr and Fe along with Al, Co, Mn, Ni, Ti, and/or V. These correlations become unreliable as more elements are added, for example, the same VEC range predicts BCC + FCC or σ phase formation. The addition of Mn is suggested to make these predictions unreliable. Most empirical approaches to predict SS or IM phases in HEAs use δr and H^{SS} or Ω. Atom size mismatch and H^{SS} are well-known empirical criteria for amorphous alloys. These parameters separate SS and amorphous phases in HEAs, but IM phases overlap with both of these fields [35]. The ability to separate SS and amorphous phases can be understood since H^{SS} is a property of disordered solution phases, to which SS and amorphous phases belong. In all of these analyses, negative H^{SS} values are claimed to stabilize IM phases or to destabilize SS phases by competing with S^{SS}. These statements are incorrect since negative H^{SS} values work together with S^{SS} to stabilize solid solutions. Large, negative values of H^{IM} will destabilize SS phases by competing with S^{SS}. H^{SS} may be a proxy for H^{IM}-systems with large, negative H^{SS} will generally have even more negative values of H^{IM} and thus be prone to IM phase formation [36].

9.9 Fabrication Routes of HEA

High-entropy alloys fabrication is very difficult, generally requires both expensive materials and specialty processing techniques. Production of HEA is classified into liquid-state route, solid-state route, gaseous-state route, electrochemical process, and additive manufacturing process (Fig. 9.7) [37]. Most HEAs have been produced using liquid-phase methods includes arc melting, electric resistance melting, inductive melting, Bridgman solidification, and laser cladding manufacturing. Thesis liquid-mix processes are always equipped with a vacuum chamber to avoid any contamination. Solid-state processing is generally done by mechanical alloying using a high-energy ball mill. This method produces powders that can then be processed using conventional powder metallurgy methods or spark plasma sintering. This method allows for alloys to be produced that would be difficult or impossible to produce using casting, such as AlLiMgScTi. The solid-state route involves mechanical alloying and subsequent consolidation processes. Some studies have shown that mechanical alloying produces homogenous and stable nanocrystalline microstructure. Gas-phase processing includes processes such as sputtering deposition, molecular beam epitaxy (MBE), pulse-laser deposition (PLD), vapor phase deposition, and atomic layer deposition which can be used to carefully control different elemental compositions to get high-entropy metallic or ceramic films. Other routes for processing the HEAs are electrodeposition, and additive manufacturing [38].

9.9.1 HEA Preparation by Liquid-State Route

Vacuum arc melting: A popular liquid processing method is arc melting. The torch temperature of the arc-melting furnace can be very high (>3000 °C) and can be controlled by adjusting the electrical power. Hence, most of the high-melting elements can be mixed in their liquid state by this kind of furnace. However, for elements with a low melting point, which are easy to evaporate, for example, Mg, Zn, and Mn, the arc-melting process may not be the best choice, because the composition cannot be precisely controlled. In this case, resistance heating or induction heating may be much more appropriate. It can be found that high cooling rates favored the formation of polycrystalline phases with a size of few nanometers. However, relatively low cooling rates led to the formation of typical dendritic and interdendritic microstructures due to elemental segregation. This work clearly demonstrated that for certain "HEA", a single solid-solution phase can only form at relatively high cooling rates, and their high-entropy state is valid in a metastable condition. Annealing at elevated temperatures or using slow cooling rates induces the formation of multiple phases, resulting in a dramatic reduction in the configurational entropy of mixing due to elemental partitioning among these phases [39].

Bridgman solidification: Another liquid technique is Bridgman solidification, which is also called the Bridgman–Stockbarger method. This method is named after the Harvard physicist, Percy Williams Bridgman, and the Massachusetts Institute Technology (MIT) physicist, Donald C. Stockbarger [40]. This technique is primarily used for growing single-crystal ingots and involves heating the polycrystalline material above its melting point and slowly cooling it from one end of its container, where a seed crystal is located. A single crystal of the same crystallographic orientation as the seed material is grown on the seed, and progressively formed along the length of the container. Such a process can be carried out in a horizontal or vertical geometry. The Bridgman method is a popular way for producing certain semiconductor crystals for which the Czochralski process is more difficult, such as gallium arsenide [41].

Thermal spray: Thermal-spray (TS) technology is used to fabricate various HEA coatings. The first work done in this regard is thermally sprayed $AlCrFeMo_{0.5}NiSiTi$ and $AlCoCrFeMo_{0.5}NiSiTi$ alloys for producing a protective layer with a thickness of around 200 μm on a substrate. In the thermal-spray process, finely divided HEA powders are

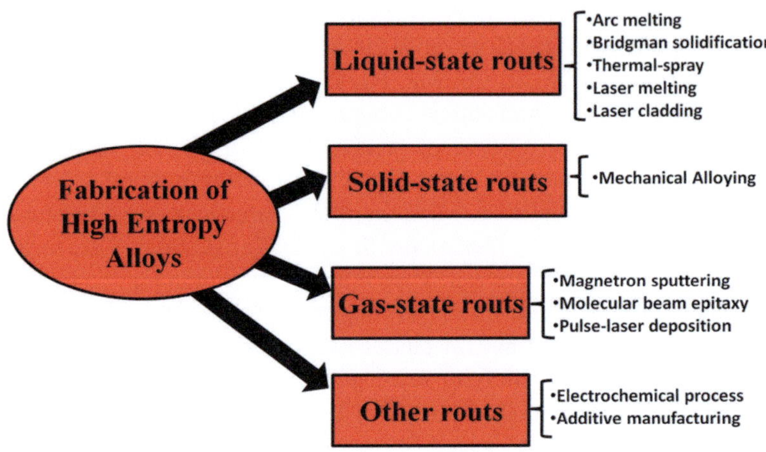

Fig. 9.7 Fabrication routes of HEAs

initially melted on prepared substrates to form spray deposits. The powders for the spray process were prepared by crushing arc-melted alloying and then ball mill the crushed particles into finer powders. Both coatings are composed of a primary BCC phase, but there are also other unknown phases present. The coatings exhibit a lamellar structure typical of thermal-sprayed coatings. The required heat is generated by combustible gases, electric arcs in the thermal-spraying gun. As the target material is gradually heated up, it is converted to a molten state and will be accelerated by the compressed gas [42]. The confined stream of particles is carried to the substrate and strikes the surface to flatten and forms a thin splat. These splats are compatible with the irregularities of the prepared surface and with each other. Moreover, these sprayed particles are accumulated on the substrate by cooling and building up one by one into a cohesive structure. Thus, coatings are formed. The hardness of the as-sprayed $AlCrFeMo_{0.5}NiSiTi$ and $AlCoCrFe\text{-}Mo_{0.5}NiSiTi$ coatings is around 525 and 485 H_V, respectively, because of the super-saturated concentration of solutes in the matrix. Both coatings harden significantly at high temperatures because of the precipitation of the Cr-silicide phase. For example, the hardness of both coatings is around 925 H_V after heat treatment at 800 °C for 1 h. Despite the apparently lower hardness of the as-sprayed coatings, they show wear resistance similar to AISI 52100 bearing steel. Annealed coatings, owing to their higher hardness, are evidently more wear resistant than AISI 52100 and AISI H13 tool steel [43].

Cladding: Laser cladding is used to prepare the low-cost HEA coating. This technology is similar to the thermal spray method in that it has an energy source to melt the feedstock that is being applied to a substrate. What differs is that it uses a concentrated laser beam as the heat source, and it melts the substrate that the feed stock is being applied to. This technique normally results in a metallurgical bond that has superior bond strength over thermal spray. The resultant coating is dense with no voids or porosity. One of the advantages of the laser-cladding process is the laser beam that can be focused and concentrated to a very small area and keeps the heat-affected zone of the substrate very shallow. This feature minimizes the chance of cracking, distorting, or changing the metallurgy of the substrate. Additionally, the lower total heat minimizes the dilution of the coating with materials from the substrate. The coating prepared by laser cladding has a simple BCC solid solution with high microhardness, high resistance to softening, and large electrical resistivity [44]. Zhang et al. fabricated AlCoCrFe6NiSiTi coatings about 1.2 μm in thickness via laser cladding [45]. The coating is composed of equiaxed polygonal grains surrounded by interdendritic regions. The composition of the BCC structure with a hardness of the coating is about 780 H_V. Huang et al. tested the thermal stability and oxidation resistance of 2-mm-thick AlCrSiTiV HEA coating on Ti-6Al-4 V substrate prepared via laser cladding. An FCC-structured CoCrCuFeNi HEA was laser cladded onto a steel substrate. It was found that both the phase constitution and the dendritic morphology of the alloy coating showed impressive stability up to 0.7 T_m (750 C), and a high level of hardness could be retained even after annealing at 1000 C for 5 h. Chen et al. prepared Al-Co-Cr-Fe-Mo-Ni layer claddings via tungsten inert gas (TIG) welding technique [46]. Hardness of the single-layer coating is about 500 H_V, and that of the double-layer coating is about 800 H_V. The wear resistance of the double layer coating is more than three times better than the steel substrate [47].

9.9.2 HEA Preparation by Solid-State Route

Mechanical alloying (MA) is a solid-state powder-processing technique involving repeated cold welding, fracturing, and re-welding of powder particles in a high-energy ball mill. Originally developed to produce oxide-dispersion strengthened Ni- and Fe-base superalloys for applications in the aerospace industry, MA has now been shown to be capable of synthesizing a variety of equilibrium and nonequilibrium alloys starting from blended elemental or prealloyed powders. By MA metals may be mixed to produce superalloys [48]. Mechanical alloying occurs in three steps. First, the alloy materials are combined in a ball mill and ground into fine powders. A hot-isostatic pressing (HIP) process is then applied to simultaneously compress and sinter the powders. A final heat-treatment stage helps remove existing internal stresses produced during any cold compaction, which may have been used. This MA process has successfully produced alloys suitable for high-heat turbine blades and other aerospace components [49].

9.9.3 HEA Preparation by Gas-State Route

Gas-state route involves magnetron sputtering (MS), molecular beam epitaxy (MBE), pulse-laser deposition (PLD). MS uses certain high-energy particles to bombard the surface of a specific target material with the help of Ar carrier gas. Under the action of a strong electric field, Ar will ionize to Ar^+ ion and e^- [50]. Then, ions are accelerated toward the cathode in the electric field to bombard the target surface with high energy, resulting in the occurrence of a sputtering target. The target will be sputter deposition to the substrate surface to form a thin film. The as-deposited film is amorphous and aging is required at low temperature to reduce the retained stress and to enhance the mechanical properties of the films. MBE is a process for growing thin, epitaxial films of a wide variety of materials, ranging from oxides to semiconductors to metals. It was first applied to the growth of compound semiconductors. The growth process is performed in an ultrahigh-vacuum environment created by continuous pumping by a turbopump. A sample holder with a substrate is attached to a magnetic manipulator which can be rotated during growth. K-cells are attached to the bottom of the chamber, pointing toward the substrate. Selective opening of shutters on each K-cell ensures the growth using only specific elements [51]. In PLD, a high-power pulsed laser beam is focused inside a vacuum chamber to strike a target of the material that is to be deposited. This material is vaporized from the target (in a plasma plume), which deposits it as a thin film on a substrate. This process can occur in ultrahigh vacuum or in the presence of a background gas, such as oxygen which is commonly used when depositing oxides to fully oxygenate the deposited films. When the laser pulse is absorbed by the target, energy is first converted to electronic excitation and then into thermal, chemical, and mechanical energy resulting in evaporation, ablation, plasma formation, and even exfoliation. The ejected species expand into the surrounding vacuum in the form of a plume containing many energetic species including atoms, molecules, electrons, ions, clusters, particulates, and molten globules, before depositing on the typically hot substrate [52].

9.9.4 HEA Preparation by Electrochemical Process

Owing to its unique advantages of cost-effectiveness, reliability, and also atom-by-atom replication on the given substrate surface profile, the electrochemical process has been widely used for high-quality coatings. The electrochemical process in which cathodic electrode position takes place through a turbulent flow of electrolyte with an electric potential difference between the electrodes. For example, HEAs such as TiNbTaZr and TiNbTaZrHf were produced through an electrochemical reduction in molten salt. The structural investigations revealed that the TiNbTaZr alloy comprises two BCC phases while the TiNbTaZrHf alloy has one disordered BCC phase.

9.9.5 Additive Manufacturing Process

Owing to the reduced defects, low cost, and high efficiency, the additive manufacturing (AM) technique has attracted increasing attention and has been applied in HEA in recent years. It was found that AM-processed HEAs possess an optimized microstructure and improved mechanical properties. Compared with the casting counterparts, AM-HEAs were found to have a superior yield strength and ductility as a consequence of the fine microstructure formed during the rapid solidification in the fabrication process. The posttreatment, such as high isostatic pressing (HIP), can further enhance their properties by removing the existing fabrication defects and residual stress in the AM-HEA [53]. Furthermore, the mechanical properties can be tuned by either reducing the preheating temperature to hinder the phase partitioning or modifying the composition of the HEA to stabilize the solid-solution phase or ductile intermetallic phase in AM materials. Moreover, the processing parameters, fabrication orientation, and scanning method can be optimized to further improve the mechanical performance of the as-built-HEAs.

9.10 Strengthening Mechanisms

Strengthening (or hardening) of crystalline solids occurs when moving dislocations interact with crystalline defects. Depending upon the type of defects, strengthening mechanisms are traditionally divided into four categories:

(1) solid-solution strengthening, which is associated with point defects in the crystal; (2) strain hardening, which is associated with line defects in the crystal; (3) grain boundary hardening, which is associated with planar defects in the crystal; and (4) precipitation and/or dispersion hardening, which is associated with volumetric defects in the crystal. Since each of the above hardening mechanisms operates independently, they contribute individually and simultaneously to the overall strength of a crystal.

The resultant strength of the crystal, σ_t is, therefore,

$$\sigma_t = \sigma_{\text{intrinsic or frictional strength of the crystal}} + \sigma_{\text{solid solutions}} + \sigma_{\text{strain hardening}} + \sigma_{\text{precipitation/dispersion}} + \sigma_{\text{grain boundary}} \quad (9.19)$$

For high-entropy alloys, there have been limited systematic studies on the above four-component mechanisms.

9.10.1 Strain Hardening

Single-phase FCC FeCrNiCoMn was relatively soft in the as-cast state whereas, cold working significantly increased its strength. The result shows that 50% cold working can enhance the yield strength by a factor of five (from 200 MPa to 1000 MPa). However, it is also noted that plastic instability develops immediately after yielding in samples cold-rolled over 50%, indicating it is at the plastic strain limit of the alloy. The single-phase FCC-$Fe_{40}Mn_{27}Ni_{26}Co_5Cr_2$ also exhibited extended ductility and significant strain hardening. For this alloy, plastic instability also occurs right after the yielding when it has been cold-worked 60% [54].

9.10.2 Grain-Boundary Hardening

There have been limited investigations yet that to examine the effect of grain size on the yield strength of HEAs. For example, cold rolling of an FCC-FeCrNiCoMn alloy to 70% and then, annealed the cold-rolled samples at 850–950 C to produce various grain sizes. Microhardness tests were subsequently carried out on the annealed samples and it was found that $k_{HP} = 677$ MPa-(μm)$-^{1/2}$, where k_{HP} is the Hall–Petch coefficient in the classical (Eq. (9.20 [55].

$$H = H_0 + k_{HP} d^{-1/2} \quad (9.20)$$

It has been shown that, in the conventional metallic alloys, the upper bound value of $k^{fcc}{}_{HP}$ is about 600 MPa-(μm)$^{-1/2}$ and $k^{bcc}{}_{HP} \approx 3k^{fcc}{}_{HP}$. In comparison, the k_{HP} value for the above FCC FeCrNiCoMn alloy is in line with and slightly higher than that for the conventional metallic alloys. Therefore, we may conclude that grain refinement is an effective strengthening mechanism in HEA, comparable to and probably slightly stronger than that observed in conventional FCC metals [56].

9.10.3 Solid-Solution Hardening

As discussed earlier, it is impossible to differentiate solvent and solute in a HEA. It is noted that the "intrinsic" strength of a HEA is usually higher than that of the individual constituent elements. Since in a single-phase HEA, atoms are homogeneously spread throughout the alloy and the lattice distortion on average would be uniform, dislocations are anticipated to experience a uniform viscous resistance while moving through the lattice. The traditional picture for solid-solution strengthening, in which a distinct local lattice distortion caused by the individual solute, is difficult to visualize. It is worth noting that the presence of interstitial atoms in a BCC-HEA caused a dramatic hardening. Based on the traditional wisdom, interstitial solid-solution strengthening is generally much more significant as compared to substitutional strengthening [57].

9.10.4 Precipitation Hardening

The strengthening phases were identified to be tetragonal $Cr_5Fe_6Mn_8$ phase in BCC-$Al_{0.5}CrFe_{1.5}MnNi_{0.5}$ alloys and $Cr_5Fe_6Mn_8$ (R phase) in $CuCr_2Fe_2NiMn$. These precipitates were relatively bulky (>micron size), quite different than the fine precipitates (~nm) found in the traditional precipitation-hardened alloys. Furthermore, microstructural kinetics such as precipitate size, shape and distribution, and the feature of precipitate matrix interfaces were not carefully characterized; in other words, the nature of age hardening was largely unclear. He and Lu reported the observation of significant hardening of a quaternary Fe1-x-y-zCoxNiyCrzHEA by adding Ti and Al, which leads to phase separation and subsequent precipitation of fine precipitates (~10 nm). The base alloy is relatively soft in the homogenized state (yield strength < 200 MPa), similar to a regular FCC-alloy. However, after aging at 650 C, the alloy exhibits yield strength of over 1 GPa and still keeps a respectable tensile ductility of over 15% [58].

9.11 High-Entropy Superalloys (HESA)

High-entropy superalloys were proposed in 2015. Conventionally, the most important family of superalloys is the Ni-based ones. The replacement of Ni by other elements could reduce cost and/or density. The high structural stability and slow diffusion kinetics might be helpful for the thermal stability of the gamma prime phase. Yeh et al. developed a series of HESAs based on the Al-Co-Cr-Fe-Ni-Ti system [59]. These alloys do have reduced densities (below 8 g/cm^3) and lower raw material costs (about 20% lower than common Ni-based superalloys). Among these alloys, $Al_{0.1}Co_{1.5}CrFeNi_{1.5}Ti_{0.4}$ has good oxidation resistance, high hot hardness (~350 Hv at 800 C), and does not contain the detrimental ⍰ phase [60]. $Ni_{44}Al_{3.9}Co_{22.3}Cr_{11.7}Fe_{11.8}Ti_{6.3}$ (in wt.%) and $Ni_{51}Al_5Co_{18}Cr_7Fe_9Ti_5Ta_2Mo_{1.5}W_{1.5}$ (in wt. %) show exceptional oxidation resistance and hot corrosion resistance that is superior to commercial superalloy CM247LC. $Al_{10}Co_{25}Cr_8Fe_{15}Ni_{36}Ti_6$ shows the γ/γ' microstructure characteristic of superalloys. This alloy has a tensile strength of 650 MPa at 800 C, which is higher than that of Inconel 617 (<500 MPa). Senkov et al. used a very different approach to obtain HESAs by using mostly refractory elements. They synthesized an alloy $AlMo_{0.5}NbTa_{0.5}TiZr$ that contains high-density cubic/plate-like precipitates embedded in a matrix, which looks similar to the structures of Ni-based superalloys [61]. The precipitates have a disordered BCC structure, while the matrix has an ordered B2 structure. The strength of this alloy is significantly higher than Ni-based superalloys in the temperature range of 20 to 1200 C [62].

9.12 High-Entropy Bulk Metallic Glasses

Although HEAs and bulk metallic glasses (BMGs) are both multicomponent and their compositions look similar at first glance, these two classes of materials are based on completely different concepts. For HEAs, more than five principal elements are required. In contrast, most BMGs are based on one or two principal elements. More importantly, BMGs are characterized by their amorphous structure, but there is no structural requirement for HEAs. However, HEAs do not have a solvent element, and the configurational entropy is higher. It has been pointed out that higher configuration entropy could be beneficial for the glass-forming ability (GFA) of BMGs. Thus, it is reasonable to expect that some HEAs can be good glass formers. However, since most reported HEAs are crystalline, high configurational entropy apparently does not directly translate to glass formation. Inoue's empirical of BMG rules state that (1) more than three composing elements, (2) large atomic size differences, and (3) large negative mixing enthalpies favor the formation of BMGs. The first rule is in line with the design concept of HEAs. Indeed, high configurational entropy in HEAs was found to be beneficial for glass formation. This was evidenced by lower critical cooling rates in BMGs with higher ΔS_{mix}. The second and third rules for BMG formation are exactly opposite to the requirements to form simple phases in HEAs. This suggests that one should choose HEAs that are predicted to form intermetallic phases rather than simple phases as candidates for HEBMGs [11].

Other factors also play critical roles: Guo et al. showed that HEAs do form metallic glasses when they have a large negative enthalpy of mixing and large atomic size difference. These conditions are in line with the empirical rules for BMG formation. The number of reported high entropy BMGs is still not large. Indeed, very recently Guo et al. analyzed the ΔH_{mix} and δ of some HEAs and BMGs and found that simple-solution-type HEAs and BMGs fall in opposite corners of the δ-ΔH_{mix} plot. They further pointed out that as long as the competition from intermetallic phases can be ruled out by composition adjustment, HEAs that locate in the amorphous phase region of the δ-ΔH_{mix} plot indeed solidify into metallic glasses. This strategy has been successfully demonstrated in a series of Zr-containing alloys prepared via melt spinning. Most of these HE-BMGs have diameters smaller than 3 mm. Takeuchi et al. successfully prepared $Pd_{20}Pt_{20}Cu_{20}Ni_{20}P_{20}$ HE-BMG with a maximum diameter of 10 mm [63]. But they suggested that conventional GFA assessment parameters such as T_{rg} cannot evaluate the GFA of HE-BMG effectively. Other factors, for example, Gibbs free energy assessments, need to be taken into account. The distinct composition design in HEAs brings with their amorphous structure other special properties. For example, an extremely high crystallization temperature (probably the highest among reported BMG) of over 800 C and good performance as a diffusion barrier between Cu and Si was observed in 20 nm thick NbSiTaTiZr alloy film.

9.13 Light Materials HEAs

Because of the strong demand for energy conservation, the development of advanced light metals has been an important issue. However, conventional light metals often have mediocre strength and limited high-temperature applications. In contrast, HEAs are known for high strength, good resistance to thermal softening, and good structural thermal stability [5]. Thus, light HEAs are a direction with great potential. Candidate elements include all low-density elements (e.g., Al, Mg, Ti, Si, V, Li, Ca, Sc, etc.) and other elements to optimize properties. For example, Juan et al. designed $Al_{20}Be_{20}Fe_{10}Si_{15}Ti_{35}$, which has a density of 3.91 g/cm^3 [33]. Li et al. developed the $Mg_x(MnAlZnCu)_{100-x}$ alloys with densities ranging from 2.20 to 4.29 g/cm^3 and compressive strength around 400–500 MPa [64]. Stepanov et al. prepared the AlNbTiV alloy, which has a density of 5.59 g/

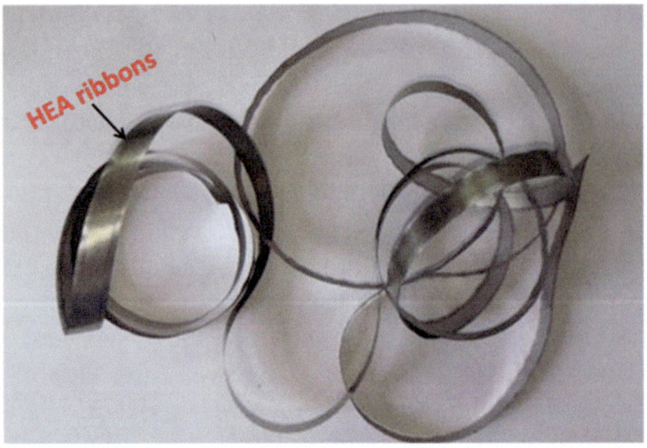

Fig. 9.8 High-entropy alloy ribbons by vacuum suspension quenching

cm^3, a high yield strength of 1029 MPa, and ductility of 5% at room temperature [65]. Youssef et al. reported very high hardness (5.9 GPa) in $Al_{20}Li_{20}Mg_{10}Sc_{20}Ti_{30}$, which also has a very low density (2.67 g/cm^3) [66]. According to these studies, light HEAs do have high strength. As long as reasonable ductility can be achieved, these alloys will have great potential in the industry.

9.14 High-Entropy Flexible Materials

HEA tends to have a solid solution structure, which means that these alloys had good plastic deformation capacity. The FCC-HEAs, such as CoCrFeMnNi, $Al_{0.3}CoCrFeNi$, CoCrFeNi, and many more, show excellent tensile plasticity that can exceed 50% at room temperature. Therefore, these alloys can be deformed by plastic deformation such as rolling, extrusion deformation, drawing deformation, and so on, which can be made into foils, ribbon filaments, and so on; such materials tend to have the characteristics of flexible materials. In addition, further methods to obtain an alloy of flexible material are the use of melt spinning method or coating. The single-crystal structure, the $Al0._3CoCrFeNi$ alloy can be made with the Bridgman solidification, which found that the elongation of this alloy is ~80%, the alloy shows an excellent plastic deformation capacity. $Al0._3CoCrFeNi$ HEA shows the elongation of more than ~60% with forging that formed the fibers with this alloy. In addition, the high-entropy alloy ribbons and fibers can also be prepared by a vacuum suspension quenching system, which prepared the $CoFeNi(AlBSi)_x$ ribbons. HEA films can be prepared by CVD and PVD, the thickness of these films tends to be 0.5–2 μm, which become two-dimensional materials, also such films after separation from the substrate will be flexible materials. Figure 9.8 shows the $CoFeNi(AlBSi)_x$ high-entropy alloy ribbons by vacuum suspension quenching.

9.15 High-Entropy Coatings

HEA coatings have been fabricated by different techniques including thermal spraying, gas tungsten arc welding (GTAW), laser cladding, and sputtering. These coatings have been tested for applications such as protective surface coatings and diffusion barriers. Because HEA coatings have a small thickness and are formed via rapid solidification that increases the nucleation rate to yield a small grain size and tends to inhibit phase separation to cause a supersaturated state [35]. Sluggish diffusion of HEA coatings generally reduces the coarsening rate of grains and phases and retards the crystallization of amorphous structures at high temperatures. Besides these, high entropy and severe lattice distortion effects also affect the microstructure and properties of HEA coatings in different aspects, which could bring advantages for special applications.

9.16 Typical Properties of HEA

High-performance HEA should simultaneously possess several key properties: excellent mechanical strength, resistance to thermal creep deformation, good surface stability, and resistance to corrosion or oxidation. Some of these properties on HEA materials are discussed below.

9.16.1 Strength and Hardness

The crystal structure of HEAs has been found to be the dominant factor in determining mechanical properties. BCC HEAs typically have high yield strength and low ductility and vice versa for FCC HEAs. HEAs have four intrinsic core effects and are favor high-temperature applications. For example, high entropy enhances phase and microstructural stability, especially at high temperatures. Slow diffusion retards structure recovery, and thus improves high-temperature resistance. Lattice distortion leads to high friction for dislocation motion. On the other hand, the "cocktail effect" of high entropy alloys intensified the interaction of various alloying elements. Therefore, the high entropy alloy reflects the composite effect improving the strength and hardness.

The strength of the AlCoCrCuFeNi-system is one of the more studied sets of properties for HEAs. Varying the content of Al will change the crystal structure of the alloy and result in a change in strength. Ordinarily, an FCC phase is softer and more resistant to change under elevated temperatures while a BCC phase is harder and more sensitive to high temperatures. For lower Al content there is a rather ductile FCC phase while increasing it will introduce an ordered BCC phase, which will increase in volume with more Al and thus increase the strength of the alloy while lowering its ductility [67]. The hardness ranges from ≈100 to 600 H_V with lower values for higher volumes of FCC phase and vice versa. The strengthening is attributed to solid solution strengthening with the larger Al-atoms, precipitation strengthening of nanophases as well as increasing the ratio of the stronger BCC phase. The alloy has high strength at elevated temperatures up to 800 °C. Heat-treatments such as multistep forging, annealing, and aging at different temperatures can alter the properties. As for annealing, it increases the FCC phase as well, making it a possible control tool for the volume ratio between the FCC and BCC phases. Heat-treatments will also affect the modulus of elasticity E, which in the as-cast state has been shown to be 180 GPa, which is lowered to 95 GPa after 0.5 h of annealing at 550 °C. Extending the annealing to 2 h again increases E to 115 GPa. The lowest E was achieved for the splat-quenched alloy, 40 GPa. Similar characteristics are seen for the Cu-free version, AlCoCrFeNi which is a bit stronger than AlCoCrCuFeNi since copper often segregates to the interdendrite regions. Also, the addition of Ti in either the AlCoCrCuFeNi or the AlCoCrFeNi system affects. For low amounts of Ti in the system without Cu, there are two BCC phases. For higher concentrations of Ti Laves phase forms [68]. It shows a high strength. Particularly, the $AlCoCrFeNiTi_{0.5}$ alloy has a compressive yield strength of 2.26 GPa, a fracture strength of 3.14 GPa, and a plastic strain of 23.3%. This system gets its strength from among other factors BCC solid solution strengthening, precipitation strengthening, and nanocomposite effects. If the concentration of Co is raised one of the two BCC phases transforms to an FCC, which lowers the strength but also the ductility. Mn addition can contribute to great age hardening, such as with $Al_{0.3}CrFe_{1.5}MnNi_{0.5}$ and $Al_{0.5}CrFe_{1.5}MnNi_{0.5}$ where hardness of 850 H_V is reached due to the BCC matrix transforming into a Cr-Mn-Fe σ-phase with the same composition. The addition of Mn to the system without Al and Cu has also been studied, and it was found that the tensile yield strength is slightly increased from 140 to 210 MPa, but with lowered ductility. The increase in strength is attributed to the lattice strain effect of having multiple principal elements. All in all, the alloy is rather soft and ductile, but tensile strength can be increased by treatments based on lowering the grain size. Mo has been added to both AlCrCoCuFeNi and AlCrCoFeNi. The effects were similar for both alloys. For low Mo contents the original phase structure is maintained, while for a bit higher Mo addition, a secondary phase that is α-phase will form, which then grew in volume with increasing amounts of Mo [69]. The alloy without Cu exhibited the best mechanical properties, both for tensile and compressive tests. Yield strength ranged from 1051 to 2757 MPa and compressive fracture strength ranging from 2280 to 3036 MPa as $AlCrCoFeNiMo_x$ had x going from 0 to 0.5. The strengthening is attributed to solid solution strengthening by the Mo-atoms and precipitation strengthening of the α-phase. Nb added to AlCrCoFeNi leads to a Laves phase beside the BCC solid solution [70]. Taking the Nb-content from 0 to 0.5 increases the yield strength from 1373 to 2473 MPa while decreasing the plastic strain limits from 24.5 to 4.1%. The hardness is found to increase linearly with the Nb content. Addition of Si from 0 to 1 in AlCrCoFeNiSi, increases the strength and lowers the plastic strain. The best combination is seen for $x = 0.4$ which reaches a compressive yield strength of 1481 MPa, fracture strength of 2444 MPa, and plastic strain 13.38%. The phase of the alloy system is mainly a BCC solid solution, with increasing Si concentration leading to a secondary phase precipitating at the grain boundaries which in turn changes the fracture type from plastic to brittle [71].

The compressive properties of a few BCC HEAs including TaNbHfZrTi, NbMoTaW, and VNbMoTaW, have been calculated at elevated temperatures. The yield strength of the NbMoTaW and VNbMoTaW HEAs above 800 C was around 600–800 MPa and 400–450 MPa at 1600 C, much superior to that of the traditional superalloys Inconel 718 and Haynes 230, confirming that these refractory BCC HEAs have a strong resistance to high-temperature softening. For FCC FeCoNiCrMn HEAs, even after 1 month of aging at 950 C, the typical FeCoNiCrMn alloy remained a single phase with no precipitation of any second phase, indicating good microstructural and phase stability [72]. A single-phase nanocrystalline $Al_{20}Li_{20}Mg_{10}Sc_{20}Ti_{30}$ alloy was developed with a density of 2.67 g cm^{-3} and microhardness of 4.9–5.8 GPa, which would give it an estimated strength-to-weight ratio comparable to ceramic materials such as silicon carbide, although the high cost of scandium limits the possible uses. Rather than bulk HEAs, small-scale HEA samples (e.g., NbTaMoW micro-pillars) exhibit extraordinarily high yield strengths of 4–10 GPa, one order of magnitude higher than that of its bulk form, and their ductility is considerably improved [73]. Additionally, such HEA films show substantially enhanced stability for high-temperature, long-duration conditions (at 1100 C for 3 days). In particular, alloys of Ti, Hf, and Zr have been shown to have enhanced work hardening and ductility characteristics. $Al_5Ti_5Co_{35}Ni_{35}Fe_{20}$ HEA after hot rolling and air-quenching, with exposure to a temperature range of 650–900 °C for 7 days [74] and subsequent air-quenching caused γ' precipitation distributed uniformly throughout the microstructure. The high-temperature exposure resulted in the growth of the γ' particles and at temperatures higher than 700 C, additional precipitation of γ' was observed. The highest mechanical properties were obtained after exposure to 650 °C with a yield strength of 1050 MPa and ultimate tensile yield strength of 1370 MPa. Increasing the temperature further decreased the mechanical properties. A series of quaternary non-equimolar HEA $Al_xCo_{15}Cr_{15}Ni_{70-x}$ with x ranging from 0 to 35% has been studied. In these HEAs, the lattice structure changed from FCC to BCC as Al content increased, and with Al content in the range of 12.5 to 19.3 at%, the γ' phase formed and strengthened the alloy at both room and elevated temperatures. With Al content at 19.3 at%, a lamellar eutectic structure formed composed of γ' and B_2 phases. Due to high γ' phase fraction of 70 vol%, the alloy had a compressive yield strength of 925 MPa and fracture strain of 29% at room temperature and high yield strength at high temperatures as well with values of 789, 546, and 129 MPa at the temperatures of 973, 1123, and 1273 K. In general, refractory HEA have exceptional strength at elevated temperatures but are brittle at room temperature. The HfNbTaTiZr alloy is an exception with a plasticity of over 50% at room temperature. However, its strength at high temperatures is insufficient. The mechanical properties of the refractory HEAs HfMoTaTiZr and HfMoNbTaTiZr have been studied. Both alloys have a simple BCC structure. The yield strength of HfMoNbTaTiZr had a yield strength 6 times greater than HfNbTaTiZr at 1200 °C with a fracture strain of 12% retained in the alloy at room temperature [75].

9.16.2 Wear Resistance

Refractory metals, with their high wear resistance and high melting point, have been used as components in several high-entropy systems. Most of the refractory alloys show high hardness and strength. Compressive yield strengths range from 900 to 2000 MPa, and the hardness ranges from 3 to 5.8 GPa respectively [76]. All alloys show mainly a BCC solid solution, but for some, there are also intermetallics and other BCC phases. The strongest alloy, for instance, $AlMo_{0.5}NbTa_{0.5}TiZr$ has two BCC phases with similar lattice parameters but different compositions. Most of the wear resistance study has been done on the AlCoCrCuFeNi system. Here, HEAs with comparatively low hardness having high wear resistance and vice versa. For instance, $Al_{0.5}CoCrCuFeNi$ with a hardness of 223 H_V has a wear resistance similar to that of SKD61 steel, which has a hardness of 567 H_V. This high wear resistance is attributed to surface hardening of the ductile FCC phase during the wear test. The type of wear has been shown to be dependent on Al and Ti concentration, where for low Al and Ti concentrations it is delamination wear and for higher concentrations, it is oxidation wear (this is also related to the phases, with delamination in FCC phases and oxidation in BCC). Some V can also increase the wear resistance while increasing the Fe concentration lowers both hardness and wear resistance. $Al_{0.2}Co_{1.5}CrFeNi_{1.5}Ti_{1.0}$ is an example of a HEA with extremely good wear resistance, twice the wear resistance of SKH51 and 3.6 times the resistance of SUJ2 while having a hardness similar to that of SUJ2. This alloy also shows superior hot hardness. The friction and wear behaviors of MoNbTaVW, MoNbTaTiZr, and HfNbTiZr RHEAs and (HfNbTaTiVZr)N, (HfNbTaTiZr)N, and (HfNbTaTiZr) have been studied. All these materials show very high hardness, giving wear resistance superior to a commercial superalloy, refractory metals and alloys, and tool steel. Wear resistance of MoNbTaVW has been found relative to Inconel 718 (well known for its high wear resistance). The coefficient of friction of MoNbTaVW-100Cr6 and MoNbTaVW-Al2O3 pairs was; 0.7 and 0.5, respectively. The MoNbTaTiZr wear resistance is better than that of MoNbTaVW, although the alloys have the same hardness. The better wear resistance of MoNbTaTiZr was by the presence of a harder secondary HCP phase and easier formation of lubricating oxides based on Ti and Zr [77].

9.16.3 Fatigue

Fatigue behavior of HEAs $Al_{0.5}CoCrCuFeNi$ and $Al_{7.5}Cr_{22.5}Fe_{35}Mn_{20}Ni_{15}$ has been studied. $Al_{0.5}CoCrCuFeNi$, consisting of two FCC phases, has a fatigue endurance limit between 540 and 945 MPa and a ratio between the fatigue endurance limit and the UTS calculated to between 0.402 and 0.703, while $Al_{7.5}Cr_{22.5}Fe_{35}Mn_{20}Ni_{15}$, with dendrite BCC and interdendrite FCC along with a dispersion of solution B2 Ni-Al particles, has a fatigue endurance limit between 540 and 630 MPa but due to the lack of UTS data the ratio is not calculated. Some scatter in the fatigue life was observed, and this is attributed to microstructural defects, and it is believed that reducing these will lead to a fatigue behavior better than that of conventional alloys. The FCC-type $Al_{0.5}CoCrCuFeNi$ shows better fatigue resistance than the BCC-type $Al_{7.5}Cr_{22.5}Fe_{35}Mn_{20}Ni_{15}$, which still shows a good result compared to conventional alloys, showing the promise for the fatigue resistance of HEAs as compared to conventional alloys in general [78].

9.16.4 Chemical Properties

High corrosion resistance of high-entropy alloys can be achieved by appropriately choosing the chemical composition. The possibility of variation of the constituent elements and the ability to form an amorphous phase are key advantages of HEA that allow them to possess good electrochemical properties. Some of the HEAs exhibit better oxidation resistance than ordinary alloys. For instance, $Al_xCo1.5CrFeNi_{1.5}Ti_y$ exhibits greater oxidation properties than both SUJ2 and SKH51 at both 600 and 800 C and the refractory $NbCrMo_{0.5}Ta_{0.5}TiZr$ have better oxidation properties than commercial Nb-alloys [79]. A laser cladding of Ti-6Al-4 V by TiVCrAlSi is shown to improve its oxidation properties and $(Al_{23.1}Cr_{30.8}Nb_{7.7}Si_{7.7}Ti_{30.7})N_{50}$ and $(Al_{29.1}Cr_{30.8}\text{-}Nb_{11.2}Si_{7.7}Ti_{21.3})N_{50}$ has both been shown to have good resistance to oxidation. The high-temperature oxidation properties of (Ti-Zr-Hf-V-Nb)N coatings have been shown to improve if deposited under increased stress strain. In general Al and for refractory alloys Ti and Si have been shown to improve oxidation properties, while Fe (in general) and V (at least for refractory alloys) have been shown to decrease oxidation properties. In general, the alloys containing both Al and Cr in rather high concentration, conventionally gives good resistance in other alloys. Al and Cr resist by creating a protective Al_2O_3 and Cr_2O_3 layer respectively on the surface. The corrosion properties of the alloy $Cu_{0.5}$NiAlCoCrFeSi were thoroughly compared with those of 304 AISI stainless steel. As can be seen, in NaCl and H_2SO_4 the room-temperature corrosion resistance of the alloy $Cu_{0.5}$NiAlCoCrFeSi is higher than that of stainless steel. This is first of all indicated by lower corrosion rate which is determined using the corrosion current density, ion valences, density of alloys, and many more. It is clear that better corrosion properties can be achieved if the HEA contains almost no grain boundaries and is in a nearly-amorphous state. However, the alloy $Cu_{0.5}$NiAlCoCrFeSi exhibits a trend to degradation of certain surface regions and pitting due to a narrower potential range in the passive region and lower pitting-formation potential in the case of degradation of a passive film [80].

In the case of CuCrFeNiMn HEA, high corrosion resistance is mainly due to the Cu concentration, degree of Cu segregation, and the structure of the alloy. It is found that better corrosion property of $CuCr_2Fe_2Ni_2Mn_2$ HEA is due to the low concentration of Cu, whereas, the poorest corrosion properties were determined for the alloy $Cu_2CrFe_2NiMn_2$. The negative influence of Cu was explained by the tendency of this element to form the FCC structure, which leads to a decrease in corrosion resistance [81]. The electrochemical behavior of the alloy $Cu_{0.5}$NiAlCoCrFeSi at elevated temperatures has been studied. It was shown that the corrosion resistance of the samples decreases with increasing temperature, the decrease being more pronounced in 1 M H_2SO_4 solution than in 1 M NaCl solution. It should be noted that the HEA is characterized by higher corrosion resistance than the 304 AISI stainless steel in the acid medium in the whole temperature range, but the lack of the passive film leads to more intense corrosion. In 1 M NaCl solution, the HEA is less corrosion resistant than the 304 AISI stainless steel at elevated temperatures owing to the lower activation energy. The corrosion behavior of HfNbTaTiZr has been studied by taking nitric and fluorinated nitric acids at 27 and 120 °C. The alloy passivated spontaneously during potentiodynamic polarization in 11.5 M HNO_3 and 11.5 M HNO_3 10.05 M NaF at 27 °C. The corrosion rate was also negligible during exposure in boiling at 120 °C, 11.5 M HNO_3 for 240 h. The formation of a protective passive film consisting of Ta_2O_5 in contrast to the air-formed native film consisting of ZrO_2 and HfO_2. HfNbTaTiZr was severely corroded in boiling fluorinated nitric acid at 120 °C. The nonprotective surface film consisted of ZrF_4, $ZrOF_2$, and HfF_4 along with oxides containing Ta, Nb, and Ti. The HEA MoNbTaTiZr showed excellent corrosion resistance comparable to Ti6Al4V, with no pitting corrosion effect in the use of biomedical applications. The corrosion behavior was superior to the 316 L SS and CoCrMo alloys [82].

9.16.5 Electrical Properties

Electrical resistivity typically ranges from 100 to 200 μΩ-cm in Al_xCoCrFeNi alloys ($0 \leq x \leq 2$). For each of these alloys, electrical resistivity increases linearly with temperature. Increasing the Al content transforms the microstructures from FCC to BCC + FCC to BCC, giving a non-monotonic dependence of electrical resistivity. The electrical resistivity of the FCC phase is higher than that of the BCC phase at the same composition, and the resistivity in the two-phase field follows a linear average of the volume fraction of the BCC and FCC phases. A non-monotonic dependence of electrical resistivity on Al content is also likely due to the transition from the BCC to FCC microstructures [83]. Cold rolled alloys have higher resistivities than homogenized material. Using AlxCoCrFeNi alloys as a baseline, other composition effects can be inferred but are difficult to determine precisely since the phases also change. Titanium additions give AlxCoCrFeNiTi alloys ($0 \leq x \leq 2$) that consist of BCC, B2, and other intermetallic phases and show a non-monotonic dependence of resistivity on Al content. The as-cast resistivity ranges from 60 to 114 μΩ-cm and increases to 132–396 μΩ-cm after annealing due to increasing volume fraction of intermetallic precipitates. Removing Cr from the baseline gives AlxCoFeNi and CoFeNiSix alloys ($0 \leq x \leq 1$). Increasing Al transforms CoFeNi from FCC to BCC + B2, while increasing Si forms silicide phases. Room temperature resistivities range from 17 to 83 μΩ-cm for this family of alloys. Removing Cr and adding both Al and Si in $(AlSi)_x$CoFeNi ($0 \leq x \leq 0.8$) gives resistivities ≤80 μΩ-cm except at the highest Al and Si contents, where the resistivity is 265 μΩ-cm [84]. These alloys are studied in the as-cast condition, and elemental segregation is evident in most of the microstructures. In a zero magnetic field, the electrical resistivity of single-phase BCC $Hf_8Nb_{33}Ta_{34}Ti_{11}Zr_{14}$ decreases from 46 μΩ-cm to ~36 μΩ-cm with a decrease in temperature from 300 K to ~8 K, and then sharply drops to zero at

$T_c \approx 7.3$ K. The lattice parameter and Debye temperature of the alloy obey a rule of mixtures of the pure elements so that the elements are concluded to distribute randomly in the lattice. Electronic-dependent properties such as the normal-state electronic specific heat coefficient and the superconducting transition temperature do not obey this rule of mixtures. The resistivities of other refractory alloys, CrTiV and CrTiVYZr, range from 80 to 100 μΩ-cm. These alloys were produced by magnetron sputtering [85].

9.16.6 Thermal Properties

Thermal conductivity of HEA $Al_xCoCrFeNi$ ($0 \leq x \leq 2$) and $Al_xCrFe_{1.5}MnNi_{0.5}Mo_y$ ($x = 0.3, 0.5$; $y = 0, 0.1$) systems has been studied. Alloys of the first system were annealed at 1273 K and water quenched, while alloys of the second system were studied in the as-cast condition. Thermal conductivity and thermal diffusivity values for these alloys increase with increasing temperature from 293 to 573 K and are in the range of 10–27 Wm^{-1} K^{-1} and 2.8–3.5 mm^{-2} s^{-1}, respectively [86]. These values are lower than for pure metallic elements but are similar to highly alloyed steels and Ni superalloys. The temperature influence is opposite that typically observed for pure metals, but is similar to that of alloys such as Inconel and stainless steels. Single-phase FCC alloys (low Al content) have almost half the thermal conductivity of single-phase BCC alloys (high Al content). Within single-phase regions, the thermal conductivity decreases with increasing Al content. These behaviors are analyzed in terms of lattice distortions and an increased phonon mean free path due to lattice thermal expansion at higher temperatures [87].

9.16.7 Magnetic Properties

$CoCrCuFeNiTi_x$ HEA depends on the T_i concentration showed that two alloys with the FCC structure, CoCrCuFeNi and $CoCrCuFeNiTi_{0.5}$, are typical paramagnets. As the T_i concentration increases, a transition to the superparamagnetic state occurs due to the formation of an amorphous phase containing nano-particle inclusions. The influence of the decomposition of phases formed on the magnetic properties of the HEA AlCoCrCuFeNi was investigated [88]. The magnetic properties (magnetization) of the alloy after thermal aging appeared to be better than those of the same as-cast alloy. It should be emphasized that the samples exhibit the properties of ferromagnets. It is believed that thermal aging causes improvement of magnetic properties due to the decomposition of the Cr_7Fe domain into ferromagnetic (Fe_7Co) and antiferromagnetic (Cr) subdomains. Cooling by sputtering was followed by the appearance of the magnetic properties typical of soft magnetic materials. At room temperature, the magnetization of the alloy was maximum and abruptly decreased with increasing temperature. The saturation magnetization (Ms) of an alloy $FeCoNi(AlSi)_x$, first of all, depends on the chemical composition of the alloy and monotonically decreases as x increases. The saturation magnetization of $FeCoNi(AlSi)_x$ is higher than that of the alloy AlxCoCrFeNi (1.151 T vs. 0.106 T, respectively), which makes the former alloy more promising for application as soft magnetic material. It is found that the low saturation magnetization of AlxCoCrFeNi is due to the presence of Cr atoms whose magnetic moments are antiparallel to those of other elements and the BCC type of the crystal lattice. HEA with the FCC lattice is characterized by high M_s in contrast to the alloys with the BCC lattice. This feature originates from the higher density of atomic packing in the FCC lattice and the high content of ferromagnetic elements (Fe, Co, Ni) [89].

Nearly all complex, concentrated alloys (CCAs) have been studied for magnetic properties that contain Co, Fe, and Ni. CoFeNi is a single-phase solid solution alloy with an FCC crystal structure and is ferromagnetic with a saturation magnetization (M_s) of 151 emu/g. For reference, M_s for pure Fe is 218 emu/g and for pure Ni is 55 emu/g. The FCC structure transforms to FCC + BCC/B_2 with Al additions in $Al_xCoFeNi$ or to FCC + silicides in $CoFeNiSi_x$. All alloys are ferromagnetic, and M_s decreases to 102 emu/g as Al increases from $x = 0$ to 1, or 80.5 emu/g as Si increases from $x = 0$ to 0.75. The magnetostriction effect is very small, which is essential to ensure that materials are not stressed in an external magnetic field. Adding Al and Si together in $(AlSi)_xCoFeNi$ ($0 \leq x \leq 0.8$), the properties at $x = 0.2$ give M_s, coercivity, electrical resistivity, yield strength, and strain without fracture that make the alloy an attractive soft magnetic material. Adding Al and Cr to CoFeNi in homogenized $Al_xCoCrFeNi$ ($0 \leq x \leq 2$) alloys gives ferromagnetic behavior at 5 K and 50 K but paramagnetic properties at 300 K due to changing alloy phases [90]. Combining AlCrCu and CoFeNi gives AlCoCrCuFeNi that is ferromagnetic in the as-cast state. It contains both paramagnetic (Cu-based FCC and Al-Ni-based B_2) and ferromagnetic (CoCrFe-based FCC and CrFe-based BCC) phases. Aging increases M_s and coercivity by decomposing CoCrFe-rich regions into ferromagnetic CoFe-rich and antiferromagnetic Cr-rich domains. Adding Al_xCrTi to CoFeNi produces as-cast $Al_xCoCrFeNiTi$ alloys ($0 \leq x \leq 2$) that are ferromagnetic at room temperature and consist of BCC (Cr + Fe-rich), B2 (Al + Ni-rich), and other intermetallic phases. M_s is low, <16 emu/g, for both as-cast and annealed $Al_xCoCrFeNiTi$ alloys. CoFeNi has also been combined with AlCrCuW, AlCrCuWZr, and Bi-Mn. Other studied alloys include $CuFeMnNiTiSn_x$ ($0 \leq x \leq 1$), AlBFeNiSi and AlBFeNiSiNb. The magnetic properties in these alloys are influenced by composition through the fraction of magnetic elements present. Processing and thermal history also influence magnetic properties through the phases formed. As-processed material typically has different microstructures and magnetic properties compared to annealed materials [91].

9.16.8 Hydrogen Storage Properties

Currently, hydrogen is considered a renewable and sustainable solution for reducing worldwide fossil fuel consumption. However, the difficulty of hydrogen storage is still an important practical drawback. Among the various storage materials, alloys and intermetallics are very important for practical applications due to their high-volume density, safety, and reversibility. The BCC and Laves phase alloys exhibit high reactivity with hydrogen at room temperature, and these alloys are considered promising hydrogen storage materials for stationary applications. The HEA, CoFeMnTiVZr with C_{14} Laves phase structures can absorb and desorbs up to 1.6 wt.% of hydrogen at room temperature. The effect of the alloy's chemical composition on the hydrogen storage properties was elucidated in terms of the lattice constant, element segregation, and hydride formation enthalpies. The multicomponent HEAs with BCC and Laves phase structures present countless opportunities for discovering new metal hydrides with superior hydrogen storage properties related to practical applications. The ZrTiVCrFeNi HEA synthesized from a feedstock composed of elemental powders via the LENS process gives better hydrogen storage properties. The prepared multicomponent ZrTiVCrFeNi alloy can absorb and desorbs hydrogen at temperatures as low as 50 °C. A thorough understanding of the structural and accompanying hydrogen storage property changes is of fundamental importance for the future development of HEAs as potential materials for hydrogen storage [92]. ScTiVCrCo, ScTiVCrNi, ScTiVCrCu HEAs shows the maximum hydrogen absorption capacity of 1.76%, 2.19%, and 2.56% respectively.

9.16.9 Irradiation Properties

The irradiation can produce a wide variety of microstructural changes in metals and alloys, including the formation of point defects such as self-interstitial atoms and vacancies, defect clusters such as dislocation loops and stacking fault tetrahedrons, and cavities such as voids and the gas-filled bubbles, causing the properties of nuclear materials degradation [93]. Thus, some strategies have been proposed to explore and design novel structural materials with high radiation resistance. In many cases, the general strategy is based on the introduction of a high density of defect sinks (grain or phase boundary), which can enhance the point-defect recombination and decrease the defect density, such as nanocrystalline steels, ODS steels, and so on [94]. It was found that HEAs possess excellent mechanical properties at high temperature, high corrosion resistance, and no grain coarsening and self-healing ability under irradiation, especially, the exceptional structural stability and lower irradiation-induced volume swelling, compared with other conventional materials. Thus, HEAs have been considered as the potential nuclear materials used for future fission or fusion reactors, which are designed to operate at higher temperatures and higher radiation doses up to several hundreds of displacement per atom (dpa) [95].

9.16.10 Diffusion Barrier Properties

The high concentrations of multiple elements in HEAs lead to slow diffusion. The activation energy for diffusion was found to be higher for several elements in CoCrFeMnNi than in pure metals and stainless steels, leading to lower diffusion coefficients. The accurate measurement of diffusion coefficients is challenging and viewed broadly, differences within one order of magnitude are not considered significant. We conclude that the measured diffusion coefficients in $CoCrFeMn_{0.5}Ni$ are not essentially different from diffusion in elements and conventional alloys. We base this conclusion on a comparison with a wide range of measured diffusion coefficients for FCC metals and alloys at melting temperature. Appearance of nanocrystals in as-cast AlxCoCrCuFeNi and retention of nanocrystals in AlCrMoSiTi after annealing was interpreted to signify slow diffusion. Furnace cooling of$Al_{0.5}CoCrCuFeNi$ avoids the formation of low-temperature phases, and AlMoNbSiTaTiVZr is a better diffusion barrier than TaN/TiN or Ru/TaN and these support the sluggish diffusion hypothesis. The diffusion coefficients in $CoCrFeMn_{0.5}Ni$ are generally higher than in conventional materials when compared at the same temperature. For example, diffusion coefficients of Ni (D_{Ni}) in $CoCrFeMn_{0.5}Ni$, Fe-15Cr-20Ni steel, and pure Fe at 1173 K are 14.3 × 10–18 m^2/s, 6.56 × 10^{-18} m^2/s, and 3.12 × 10^{-18} m^2/s, respectively, and at 1323 K (the highest measurement temperature) are 5.74 × 10^{-16} m^2/s, 2.15 × 10^{-16} m^2/s and 1.2 × 10^{-16} m^2/s, respectively. This shows that diffusion coefficients of Ni are higher in $CoCrFeMn_{0.5}Ni$ than in austenitic steels and pure metals when compared at the same temperature, in the range of 873–1323 K. Diffusion coefficients appear lower in $CoCrFeMn_{0.5}Ni$ than in selected FCC elements and conventional alloys after normalizing activation energies by the melting temperature (or solidus temperature for alloys). This is based on the empirical observation that diffusion coefficients at T_m, DT_m, are roughly equal for metals and alloys with the same crystal structure and the same bonding type, and on the assumption that the preexponential term, D0, is constant at T_m. DTm values range from 4.5 × 10^{-14} to 6.6 × 10^{-12} m^2s^{-1} for a wide range of FCC elements and 6.4 × 10^{-14} to 4.9 × 10^{-12} m^2s^{-1} for a wide range of binary alloys. All of the DT_m data for $CoCrFeMn_{0.5}Ni$ fall in this range. This shows that diffusion in $CoCrFeMn_{0.5}Ni$ is not significantly different from a wide range of other FCC metals and alloys when the data are normalized at T_m [96].

9.17 Difference between BMG and HEA

BMG is to avoid crystallization in bulk samples, whereas in HEA, using the complex thermodynamics of very concentrated alloys to obtain unusual properties in crystalline samples (in fact, solid solutions). Number of elements in BMGs is not always 5 (and in HEAs should be 5 or more) and not in near equimolar quantities (HEAs are usually near equimolar), BMGs are amorphous and HEAs are crystalline. Both should be single-phase, BMGs amorphous, HEAs solid solution [97]. BMG is a class of multicomponent metallic materials, at which crystallization is kinetically suppressed by a number of factors like atomic diameter difference, high natural supercooling (due to near-eutectic composition), negative heat of constituent elements mixing, and so on Materials systems with high GFA a thermodynamically unstable, and then heating ("turning on" of diffusion) may cause their crystallization [98]. On the other hand, HEA exists in single-phase solid solution crystalline state characterized by high configurational entropy. They just have a strong thermodynamically dictated tendency to avoid the formation of ordered solid solutions as well as intermetallic compounds. The main term responsible for their stability is entropy term, so their heating improves their stability.

9.18 Complex Concentrated Alloys (CCAs), Multi-Principal Element Alloys (MPEAs)

HEA field was dominated by configurational entropy and the search for single-phase, solid solution alloys. The HEA definition further restricted studies to alloys with 5 or more principal elements, even though interesting results were being obtained in concentrated alloys with only 3 or 4 principal elements. This restriction of thought began to evolve in unproductive ways by excluding new results in new alloy systems based solely on the number of elements used or the number and types of phases formed. This early exclusion seems strange, especially in a field founded on the concept of unbounded expansiveness of alloy compositions and microstructures [99]. It was concluded that the HEA field is too broad to be adequately described by a single definition or microstructure, and so new terms were introduced to help remove these barriers. These new terms include complex concentrated alloys (CCAs), multi-principal element alloys (MPEAs), or simply "baseless" alloys. These more inclusive terms retain focus on concentrated alloys that are also compositionally complex and thus have no single, dominant element. Definitions ultimately set boundaries, and so these new terms intentionally avoid specific definitions based on the number or concentration ranges of elements used or the number or types of phases formed. These new terms also have no implications regarding the magnitude or importance of configurational entropy [100]. These new terms include every alloy that satisfies HEA definitions, and they also include many alloys that are excluded by HEA definitions. They thus expand the HEA field by including concentrated ternary and quaternary alloys, by allowing elemental concentrations in more than 35 at.%, and by including single-phase intermetallic alloys and alloys with any number of solid solution and intermetallic phases. Finally, the early years of the HEA field were dominated by studies of metallic alloys for structural applications, but these new terms specifically include nonmetallic alloys and alloys developed for functional applications. CCAs are used to include all alloys in the HEA field, as well as alloys that satisfy the motivation of studying complex, concentrated alloys or the vast number of compositions and microstructures in the central regions of multicomponent phase diagrams.

9.19 Application of HEA

High entropy alloys currently spreading their demand among high strength materials. Some of the emerging application in various sectors has been discussed below.

9.19.1 Automobile Industries

Weight reduction remains the most cost-effective way to reduce fuel consumption and greenhouse gas emissions in the transport sector. It is estimated that for every 10% of vehicle weight loss, there is a 7% reduction in fuel consumption, which means that there is an approximately 20 kg reduction in carbon dioxide per kilogram of vehicle weight [101]. The main disadvantage of existing lightweight materials is their limited properties or high production cost. Therefore, researchers are making continuous efforts to develop new types of lightweight alloys that are low in price and capable of meeting the requirements of the properties. In the past three decades, efforts have been made to replace steel and iron with Al, Mg, and composite materials. HEA multicomponent alloys have shown promising materials for automobile applications. HEAs are used in bearings, bushings, and motor equipment. Low cost, high entropy mechanisms with their unique Al composition are up to 20% cheaper to manufacture and have the potential to change the market landscape for standard brass bearings, bushings, gears, and housings (Fig. 9.9). These alloys consist of various mixtures of Cu, Ni, Mn, Zn, Al, and Sn, which allow for the specific cost adjustment, strength, resistance to wear and corrosion, and exhibit the same self-lubricating properties as typical brass [102].

9.19.2 Aero-Vehicle Industries

HEA has great potential for relatively low density. Maybe not lower than Ti-alloys, but even lower than Ni-based superalloys. They show great promise of stability at high temperatures. These properties suggest that HEA could be potential replacement materials for high-temperature components behind the combustion chamber, such as the turbine exhaust chamber (Fig. 9.10a) and the mid turbine frame, but also the turbine outlet (nozzle and cone), as these parts are now mostly made of relatively heavy Ni-alloys, while cold components are made of lighter elements such as Ti-alloys. Friction heating can raise the surface temperature of supersonic aircraft to more than 300 C and temperatures in the compressor sections of first- and second-stage turbine engines can exceed 400 °C. HEAs are good candidates for replacing steel and Ti-alloys. Applications may include aerodynamic engine compressor blades, which are often made of Ti-based alloys [103].

A jet engine consists of turbine exhaust housing, also known as a turbine rear frame or a tail bearing housing. The purpose of this part is to support the end of the low-pressure rotor, mount the motor on the body of the aircraft, and remove the angular component of the output current. Although the gas temperatures in the turbine exhaust housing are not as high as in the turbines due to the gas removal work, they must still face relatively high temperatures and must therefore be able to maintain a number of their mechanical properties to continue their engine function. It must also be resistant to corrosion and oxidation at these elevated temperatures and must be weldable, moldable, and generally malleable. Unfortunately, having these qualities is not enough. The materials used in the turbine exhaust housing must also be fatigue resistant (mainly LCF), not too heavy, weldable, flexible, and generally capable of forming and gluing. Currently, Inconel 718 is the most widely used alloy for turbine exhaust and can withstand temperatures of $\approx$ 650 C [104]. In addition, load-bearing structure (LCS) requires a refractive elongation of at least 5%. Covering materials for thermal protection is limited in particular their thermal capacity (e.g., thermal stability, creep strain), followed by their formability and oxidation characteristics. Operating temperatures for heat shields are expected to be $\approx$670 °C with maximum temperatures $\approx$760 °C. However, it is expected that in the future these temperatures may reach up to $\approx$900 C with average temperatures $\approx$800 °C, using HEAs.

Ni-based HEAs are suitably used in the mid turbine frame (MTF) (Fig. 9.10b). In front of the turbine exhaust case (TEC) in the engine between the high- and low-pressure turbine, there is a turbine system MTF. They make the center of the center aisle, which carries HP and LP rotors, and transfers their cargo to the outer cavity. The MTF requirement is the same as the TEC requirement, but because the MTF is located in front of the lower pressure turbine, the gas temperature is slightly higher [105]. It must have some stiffness to make sure that the blade tips can rotate without being restrained by the casings, which must be beaten during normal flights even during extreme movements. The MTF often divides the workings, a curve in which the LCS is protected by a heat shield. The fixed assets for the LCS are the same as

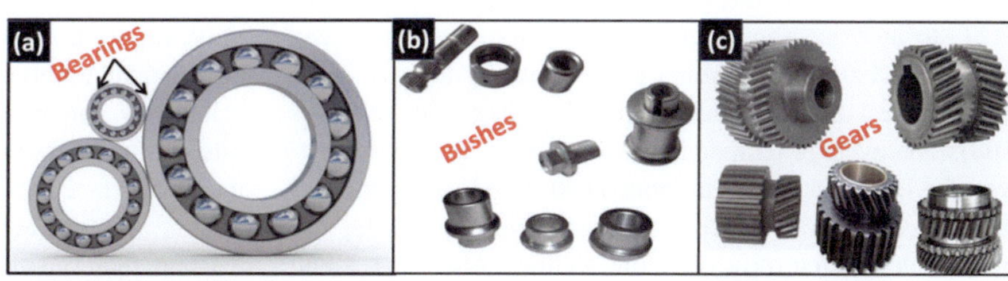

Fig. 9.9 Automobile (**a**) Bearings, (**b**) Bushes, and (**c**) Gear Motors

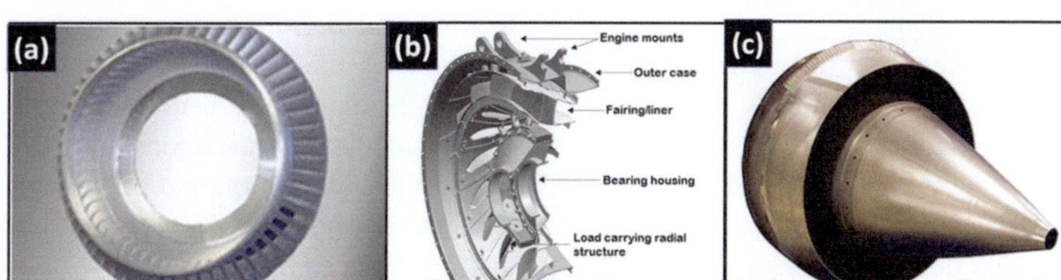

Fig. 9.10 (**a**) Turbine exhaust case, (**b**) Mid Turbine Frame, with supporting material inside heat shielding fairings in the flow path, (**c**) Exhaust cone and nozzle attached to an airplane

Fig. 9.11 The main landing gear of an aero-vehicle

the LCS assets for the TEC, with the initial assets for the LCF followed by strength and rigor. The most interesting alloy, still available today, is Inconel 718. The choice of alloys for decoration is limited by their powerful properties, followed by their properties and retail-functional LCF/thermo-mechanics. The alloy currently being considered is Mar-M-247 based on Ni-Mar-M-509 [106]. Behind the TEC we find the latest jet engine parts (civilian), nozzles, and tires (in jet engines for military applications with frequent burns). Figure 9.10c shows the HEA exhaust cone and nozzle attached to an airplane. Modern tires are usually made of Ti-alloys, but cones for new motors are made from superalloys made of stainless steel. Future research on cones is currently focused on CMCs based on Al_2O_3 and SiC. Effective properties include antimicrobial properties, hot springs, high stability, and density. If nozzle and cone can be synthesized with TEC rather than separately, care can be taken to improve the communication properties that arise when the three parts are combined [107].

Superalloys are generally used in high-temperature applications such as on the hot side of the turbine engine. As engine heat efficiency can only be increased by increasing the heat capacity of the turbine blades and an increase of 40 °C to increase engine efficiency by about 1%, we need equipment that can withstand high operating temperatures. The historical development of superalloys began in the 1940s [108]. The average temperature rises to 6 °C per year; however, the limit since 2010 is about 1100 °C. Most superalloys are MEAs. Since 1940, configurational entropy has rapidly grown into the entropy medium, falling between 1.2 and 1.4R. It does not go beyond 1.5R [109]. For creep resistance, the grain boundary is a negative factor. We find that the general trend is favorable for alloy series, alloys, and single-crystal series. This means that the total temperature range increases with the entropy gain in each category and also in the same series (N, IN, MarM, TM, PWA, CMSX, and TMS headphones, respectively). We also note that in each category there is a different degree of "alloy" that can be rated higher than the alloy or equivalent grade [110]. This means that component selection and focus is also an important factor. So, we must design HESA with matching elements, not just increase entropy, to improve the temperature potential of each category. With a good strength-to-weight ratio, good oxidation resistance, medium resistance, consumption resistance, high power, light and wear resistance, the HEA is a good material for compressors, mixing chambers, nozzles, and gas turbine engines. Currently, HEA landing gear (Fig. 9.11) is a focus of point. A list of general-purpose aircraft components contains axil, shock, and drag strut, walking beam, and torsion links forming on Ti-10 V-2Fe-3Al. Ti-10 V-2Fe-3Al in wing/flap systems has been successfully used with alloy forging [111]. Tracks and dynamic body allow movement between the wings and the moving slits. They are made of stainless steel. Recently, the Ti-10 V-2Fe-3Al was made with a wire on the side of the wing.

9.19.3 Machineries

HEAs can be used to protect the surface of machine components and tools because of their high hardness, durability, high temperature softening resistance, corrosion protection, and combinations of these properties. Hard-facing technology can be used where HEAs are manufactured in rods and powders and then plasma arc or thermally sprayed onto the surface of tools and other components. The hard face process involves adding a thick layer of wear or corrosion-resistant material through welding, thermal spray welding. Common industrial uses include molds, dies, tools, nozzles, and many more. An important application of HEA is hot working mold and dies (Fig. 9.12a), which can maintain the extrusion temperature of 1200 °C. For alloy extrusion, the maximum extrusion temperature is 500 °C. The steel can hold the heater up to 550 °C but becomes considerably soft at temperatures above 600 °C. This means that the SKD 61 steel nozzle may require extrusion of Al alloys but is far

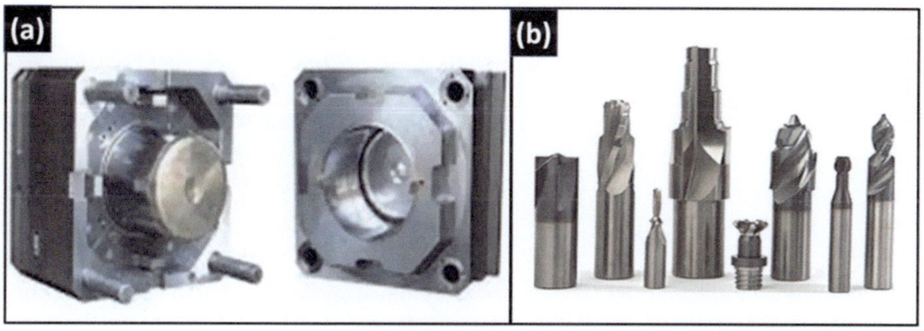

Fig. 9.12 (**a**) Molds and dies and (**b**) high-speed cutting tool

from the requirement for steel extrusion. Thus, the success of HEA for this use would make Steele's extrusion more popular for producing different cross sections for many structural and architectural forms. It can be seen that three HEA candidates (commercial alloys IN718, IN718H, and T-800) hold their hardness above HV 400 at 1000 °C. This shows the potential applications of HEA for such high-temperature extrusion matrices in steel [112]. HEA is used in high-speed cutting tools, showing better friction and wear resistance with respect to ordinary high-speed steel. Figure 9.12b is showing various HEA high-speed cutting tools. When the outer load and rotational speed are fixed, the friction of the HEA coated cutting tool is always lower than that of the conventional high-speed steel. If the tool rotates at a high speed, the higher the coefficient of friction, the higher the frictional resistance. Currently, the chip will easily hold in a knife and it will scratch the surface of the material, reducing the surface finish of the material. Accordingly, the HEA coating on the surface of the cutting edge can reduce the frictional force of the cutting process, which extends the life of the tool and improves the surface quality of the machined surface [113]. High entropic carbides have an important application for cutting tools for their combined effect of higher hardness and fracture toughness that results in better cutting performance [114]. High entropy nitride (HEN) coating can improve the life of coating. Hard coating in addition increases its hardness and toughness. Higher oxidation resistance will improve cutting performance. For example, HEN $(Al_{0.34}Cr_{0.22}Nb_{0.11}Si_{0.11}Ti_{0.22})_{50}N_{50}$ [115].

Hard-facing materials can extend the life of durable plates, cylinders, excavators, and so on, using HEA. It shows the wear resistance of the friction against the hardness of various alloys. High Cr cast iron is the typical durable alloy for hard turning. The PDA HEA, which is a HEA with many carbides and north winds, can withstand wear seven times greater than high chromium cast iron. In addition, the 17.5Cr HEA, which is further modified with more Cr, has 12 times the resistance. It contains many hard borides and carbides. Due to its better abrasion resistance, HEA is also used in functional coating materials in nonstick, anti-fingerprint, antibacterial, and aesthetic equipment. The protective layer can improve corrosion, oxidation, thermal, and wear resistance of structural

Fig. 9.13 Nuclear fuels and high-pressure vessels

parts. A welding coating is the necessary welding layer between the substrate overlay and the upper layer of thermal barrier Y-stabilized zirconia, which provides good welding and oxidation resistance. Compared to other adhesives, HEA provides better adhesive coating properties, for example, HEA $NiCo_{0.6}Fe_{0.2}Cr_{1.5}$-$SiAlTi_{0.2}$ adhesive coatings [116].

9.19.4 Nuclear Application

No suitable material has good high-temperature properties and radiation resistance. New and modern nuclear materials are needed to ensure the safe operation of nuclear reactors and to maintain a sustainable energy supply. HEAs are used successfully in the nuclear industry. Improved radiation and high corrosion resistance make HEA potential candidates for investment in materials used in nuclear fuel and high-pressure vessels (Fig. 9.13). The fourth-generation nuclear reactor emphasizes the very economical operation, increased safety, minimal waste, and reproduction. For example, a high-speed gas-cooled reactor is a candidate for fourth-generation reactors. Operating temperature is about 850 C [117]. The Oak Ridge National Laboratory has published a series of documents on radiation damage from concentrated

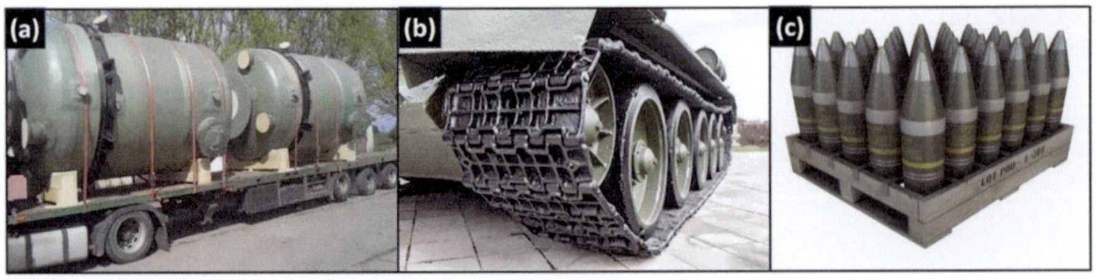

Fig. 9.14 (**a**) Chemical vessels, (**b**) track of tanks, and (**c**) artillery shells

solid solutions containing Ni. They found an interesting trend that resistance to radiation damage generally increases with the number of components in a Ni-containing FiC system. In addition, FCC Fe-28% Ni-27% Mn-18% Cr HEA, radiation damage at elevated temperatures (400–700 °C) generally showed less radiation degradation as compared to conventional ossifying alloys Fe-Cr-Ni or Fe-Cr-Mn. HEA radiation shows signs of relatively slow diffusion of solute with limited depletion of solute or enrichment of grain boundaries. No leaks were observed in any radiation situation. Due to high temperature and stability, high oxidation resistance, and low neutron cross-sectional absorption, such HEAs will be more widely used in fourth-generation nuclear power plants [118].

9.19.5 Electrical and Electronics

Superionic conductivity for the solid-state battery such as HEA $(MgCoNiCuZn)_{1-x}Li_xO$ showing ionic conductivity at least two orders larger than glassy lithium-phosphorus-oxynitride electrolyte (LiPON) and among the largest ever observed in oxides [119]. HEA is used in superconductor materials with higher critical temperatures and critical current. Using lanthanum, cerium, praseodymium, neodymium, and samarium, they constructed a high-entropy alloy capable of withstanding fatigue and ductility. This allows researchers to create superconductors that continue to function under high temperatures. They can be used to suppress electromagnetic interference, especially in electronics. For example, a coating of 1 μm at 13,000 MHz can be effective as a screen in commercial applications [120]. The HEA $Al_{2.08}CoCrFeNi$ with near-constant resistivity is useful for electronic applications.

9.19.6 Biomedical Applications

With excellent mechanical, electrical, electrochemical, and anticorrosion properties, high entropy alloys are alternative alloys for biomedical applications. In recent years, Nb, Ta, and Zr are considered the most favorable nontoxic and nontoxic allergenic alloying elements for titanium alloys for biomedical applications. Of the Ti-type alloy systems, the most reported biomaterials are TiNbTaMo, TiNbTaSn, TiNbTaZr, TiNbSn, TiZrCrMo, TiMoNb, TiMoTa, TiZrMo, and TiNbZrTa. Many of these alloys contain different percentages of Nb, Ta, Zr, Mo, Sn, or Cr. These alloys are generally made by melting and casting, followed by heat treatment or sintering, followed by hot forging and cold rolling. They have a modulus of elasticity between 42 and 93 GPa, tensile strength between 597 and 1037 MPa, elastic limit 547–908 MPa, and elongations 13–19%. A new Ti-based alloy (TiNbZrTaFe) with a yield strength of 2425 MPa and a Young modulus of 52 GPa was recently reported. The materials were obtained after heat treatment at 960 C, with a double structure and an average grain size of 200–300 nm. A high entropy multicomponent TiZrNbTaFe alloy shows remarkable corrosion resistance in simulated body fluids and proved to be a promising alternative for biomedical applications [121].

9.19.7 Other Applications

HEA can be used in coatings used for food preservation and cookware due to corrosion, antioxidants, and wear resistance. Bacteria like E-coli and colonies can be prevented from multiplying [122]. Figure 9.14 showing the HEA used chemical vessel, tank moving belt, and artillery shells. The CrMnFeCoNi alloy can be used in cryogenic applications such as liquefied gas storage and can retain mechanical properties at temperatures up to 77 K [123]. HEAs are used as solder and solder filler to weld pure Ti and Cr-Ni-Ti stainless steel, cemented carbide, and steel, respectively. Due to the pronounced corrosion resistance, the HEA can be applied in all kinds of corrosive working conditions, such as the ocean floor, cases prone to electrochemical reactions and various chemical containers, and so on [124]. Eutectic HEA $AlCoCrFeNi_{2.1}$, with excellent resistance to casting and corrosion for marine propellers and in the forged state with ultrahigh resistance, equivalent to steel for structural applications (for commercialization). HEAs can be used to produce high-quality golf surfaces, high hardness steel tubes, and turbine blades that operate under high loads and high circumstances, technology for national defense, such as tank tracks, artillery shells, and so on.

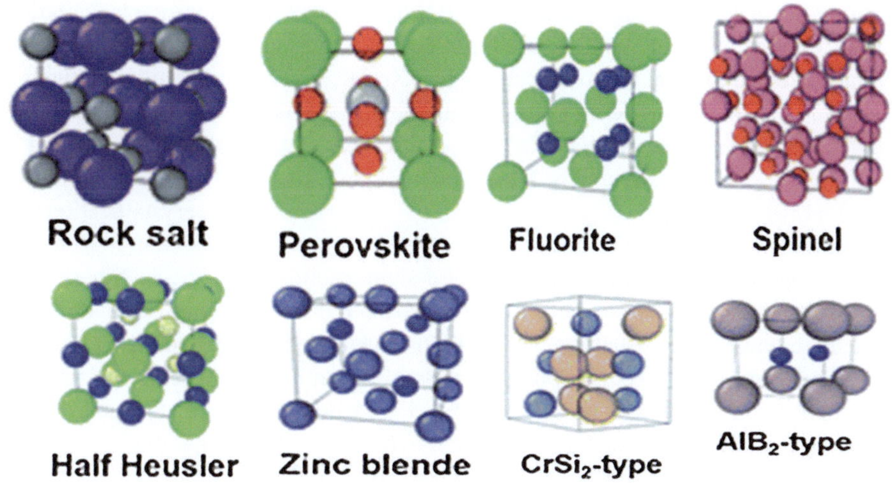

Fig. 9.15 Structural diversity of the HEC family, anions (*dark gray*), and randomly distributed cations

9.20 High Entropy Ceramics

The development of materials for engineering applications exceeding 2000°C in oxidizing atmospheres, such as hypersonic vehicles and spacecraft. To date, ultrahigh temperature ceramics and high entropy ceramics are the limited groups of materials that can withstand such extreme environments. They are based on the refractory borides, carbides, and nitrides of the group of IV and V transition metals and are typically defined as having melting temperatures higher than 3000 C. High entropy ceramic (HEC) or single-phase ceramic is novel material with no less than four different cations or anions. The development of HEC follows the "configurational entropy stabilized single phase" concept. HECs consist of multicomponent ceramic oxides, carbide, borides, and nitrides. For example, Bulk (HfTaZrNb)C, (Zr, V,Ti,Ta,Hf)B_2 and B_4(HfMo_2TaTi)C [125, 126]. The HECs exhibit homogeneous crystalline single-phase despite the complex structure of ceramic compounds and offer superior properties over the constituents and conventional ceramics. The advantages of HECs are their compositional and structural diversity, and many of them have a bandgap, which makes them potential functional materials for a wide range of applications. Another advantage of HECs is that they have ample structural diversity (Fig. 9.15). The members of the HEC family are increasing rapidly. In 2015, Rost et al. reported on entropy-stabilized HEC oxides. This work stimulated interest in the properties of these materials, such as dielectric, Li-battery, and low thermal conductivity [127]. Currently, researchers extend the family members to include diborides, chalcogenides, silicides, and intermetallics such as half Heusler, for a broad range of applications from structural to functional. The earlier reported HECs all had rock salt structure, including nitrides, carbides, and the well-studied $Mg_{0.2}Co_{0.2}Ni_{0.2}Cu_{0.2}Zn_{0.2}O$, cubic perovskite oxides, fluorite oxides, and hexagonal diborides. To measure the ceramic strength, micropillar compressions were carried out at room temperature on an HEC and generally possess high hardness (>20 GPa) and strength (>500 MPa) with excellent oxidation resistance and good resistance to thermal shock even at temperatures exceeding 2000 °C [128].

Jiang et al. fabricated several high-entropy perovskite oxides and concluded that the Goldschmidt's tolerance factor, instead of cation-size differences on a specific lattice site, influences the formation and temperature-stability of single cubic perovskite solid solutions [129]. They found that the calculated tolerance factors that form single-phase high-entropy perovskite phases were close to unity ($0.97 \leq t \leq 1.03$). They suggested that a Goldschmidt tolerance factor close to unity ($t \approx 1.00$) is important, but not a solely sufficient criterion to form a single high-entropy perovskite phase. The above models illustrate the use of descriptors related to mixing enthalpy. There has also been an attempt to develop an "entropy descriptor." Vecchio et al. proposed "entropy forming ability (EFA)" as a descriptor to address the synthesizability of HECs from first principles. The EFA was estimated as the inverse of the standard deviation of the energy distribution spectrum of the enthalpies of all of the possible atom configurations of 10-atom supercells

$$\mathrm{EFA} = \left(\sqrt{\frac{\sum_{i=1}^{n} g_i (H_i - H_{\mathrm{mix}})^2}{\left(\sum_{i=1}^{n} g_i\right) - 1}} \right)^{-1} \tag{9.21}$$

where n is the total number of sampled geometrical configurations, and g_i is their degeneracies. H_i is the enthalpy of formation at zero temperature of a supercell, and H_{mix} is the average of the enthalpies H_i of all of the possible supercell configurations. A larger EFA value corresponds to larger entropy for a given composition. The EFA descriptor was used to screen all of the possible compositions of five-metal carbides selecting from 8 chosen cations (Ti, Zr, Hf, V, Nb, Ta, Mo, W). Then the six compositions with the highest EFA values (the highest one is $MoNbTaVWC_5$ with EFA = 125 atom e/V) were experimentally fabricated as single-phase HECs [130].

Solid-state reaction, wet chemical, and epitaxial growth methods are used to synthesize HECs. Solid-state reaction is the predominant synthesis route used to prepare HECs, in which the powders are usually produced by mechanochemistry or mixing of the precursor powders during ball milling, and then the resultant powders are sintered. HECs prepared from low melting point precursors, such as chalcogenides, are synthesized by mechanochemistry during ball milling. Another HEC synthesis routine is the epitaxial growth of films. So far in the literature, only pulsed laser deposition (PLD) has been used to prepare HEC oxides. One advantage of this method is that single crystals can be grown, therefore grain boundary segregation is avoided and precise elemental ratios can be achieved. Epitaxial growth is also useful for preparing structures that are not possible to produce by solid-state reaction and wet chemical methods, such as superlattices and heterostructures [131].

An ideal HEC should have a lattice with long-range periodicity but compositional disorder, that is, the periodic lattice sites are randomly occupied by the constituent atoms. Therefore, the structural characterization of HECs is mainly concerned with the determination of single phase (i.e., long-range periodic lattice) and the homogeneity of the constituent elements (i.e., truly random occupation). It is worth noting that although in most HEC development the achievement of a homogeneous single phase is emphasized, sometimes second phases and phase separation can be beneficial to the properties. HECs are applied in low thermal conductivity, thermoelectric, structural materials, catalyst, dielectrics, and magnetic materials [132].

9.21 High Entropy Polymer

The cocrystallization of multiple principal elements with a near-equimolar ratio has been widely pursued as an approach to create high-entropy alloys, as plentiful spatial distribution patterns of constituent elements within crystal lattices are attainable. In the field of organic materials, the cocrystallization of two disparate organic compounds is also feasible, which broadens the features of organic crystals as well. However, different from metal alloys, the spatial packing scheme of molecular motifs (structural units) is fixed principally via lattice interactions, and therefore the entropy of organic co-crystals mainly comes from the acceptable variation of molecular conformation. The axial rotation of structural units or molecular constituents within crystalline lattices can yield a significant contribution to crystal entropy as well and nevertheless is less feasible, which is clarified as the major feature of plastic crystals. In addition to the crystallization of copolymers, the cocrystallization of homopolymers and copolymers is another way to include disparate motifs into crystalline lattices. Nevertheless, the cocrystallization of two disparate long-chain molecules is rarely found so far because two dissimilar kinds of polymers are less likely to be fully miscible in the solid state [133]. The vast majority of organic binary blends able to exhibit eutectic behavior are classified to be the mixtures of crystalline polymers and monomers or the mixtures of low-MW organics. Nevertheless, for the mixture of poly(vinylidene fluoride) and poly (vinyl fluoride) homopolymers, the rare isomorphic behavior has been identified. Moreover, the cocrystallization of two selected random copolymers with various ratios of trifluoroethylene comonomers has been verified feasible as well. The solvent evaporation has been identified as able to induce the cocrystallization of dissolved polyvinylidene fluoride homopolymer and poly (vinylidene fluoride trifluoroethylene) copolymer solutes [134].

9.22 High Entropy Hybrid

Undoubtedly, entropy has a tremendous impact on the crystal chemistry of hybrid inorganic-organic materials and includes all aspects of dynamic effects and structure-property relationships. In hybrid materials in general, it is well established that long organic linkers and porous materials lead to high vibrational entropies, while, at the same time, substitutional defects seem to be relatively easy to introduce in hybrid materials. For the impact of rotational entropy, the field of hybrid materials can learn from related research areas. For instance, the structure of helical polymers is mainly influenced by salt-bridge like interactions, which have been designed accordingly. It is worth remembering here that the nearly unlimited variety of organic linker materials allow for the specific design of hydrogen bonds, van der Waals forces, π-π interactions, and so on. The solubility of pristine fullerenes can be enhanced by mixing C_{60} and C_{70} due to the associated increase in configurational entropy. This “entropic dissolution” allows the preparation of field-effect transistors with electron mobility of 1 $cm^2V^{-1}\ s^{-1}$ and polymer solar cells with a highly reproducible power-conversion efficiency of 6%, as well as a thermally stable active layer. High configurational entropy is associated with C_{60}:C_{70} mixtures, which in addition facilitates the formation of amorphous thin films [135].

9.23 Summary

This chapter successfully described high entropy material, the newly developed material of this current era. Various high entropy materials, their synthesis route, and application have been given in this chapter. Why there is a requirement of single-phase with multicomponent system mechanisms have been well depicted here. Increment of strength at high temperature by strain hardening, grain-boundary hardening, solid-solution hardening, precipitation hardening for HEA and HEC has been given, whereas coexistence of copolymers has been given here. For about 15 years, engineers around the world have been trying to make these innovative materials ready for series production. But high-entropy alloys are still too expensive and difficult to process. More research is required to develop high strength in these materials with economical processing route.

References

1. https://en.wikipedia.org/wiki/High_entropy_alloys. Accessed 08 Sep 2019
2. Miracle, D.B., Senkov, O.N.: A critical review of high entropy alloys and related concepts. Acta Mater. **122**, 448–511., ISSN 1359-6454 (2017). https://doi.org/10.1016/j.actamat.2016.08.081
3. Daoud, H., Manzoni, A., Wanderka, N., Glatzel, U.: High-temperature tensile strength of $Al_{10}Co_{25}Cr_8Fe_{15}Ni_{36}Ti_6$ compositionally complex alloy (high-entropy alloy). JOM. **67**, 2271–2277 (2015). https://doi.org/10.1007/s11837-015-1484-7
4. https://www.army.mil/article/234809/researchers_create_materials_by_design_for_future_army. Accessed 08 Sep 2019
5. He, J.Y., Zhu, C., Zhou, D.Q., Liu, W.H., Nieh, T.G., Lu, Z.P.: Steady state flow of the FeCoNiCrMn high entropy alloy at elevated temperatures. Intermetallics. **55**, 9–14 (2014). https://doi.org/10.1016/j.intermet.2014.06.015
6. Shabani, A., Toroghinejad, M.R., Shafyei, A., et al.: Microstructure and mechanical properties of a multiphase FeCrCuMnNi high-entropy alloy. J. Mater. Eng. Perform. **28**, 2388–2398 (2019). https://doi.org/10.1007/s11665-019-04003-4
7. Ikeda, Y., Grabowski, B., Körmann, F.: Ab initio phase stabilities and mechanical properties of multicomponent alloys: a comprehensive review for high entropy alloys and compositionally complex alloys. Mater. Charact. **147**, 464–511., ISSN: 1044-5803 (2019). https://doi.org/10.1016/j.matchar.2018.06.019
8. https://www.tandfonline.com/doi/full/10.1179/1743284714Y.0000000749. 08 Sept 2019
9. Cantor, B.: Multicomponent and high entropy alloys. Entropy. **16** (9), 4749–4768 (2014)
10. https://materials.typepad.com/mrs_meeting_scene/2016/11/high-entropy-alloys-an-interview-with-jien-wei-yeh.html. 08 Sept 2019
11. Tsai, M.-H., Yeh, J.-W.: High-entropy alloys: a critical review. Mater. Res. Lett. **2**(3), 107–123 (2014). https://doi.org/10.1080/21663831.2014.912690
12. George, E.P., Raabe, D., Ritchie, R.O.: High-entropy alloys. Nat. Rev. Mater. **4**, 515–534 (2019). https://doi.org/10.1038/s41578-019-0121-4
13. https://www.mdpi.com/1099-4300/16/9/4749/pdf. Accessed 10 Sept 2019
14. Bee, M.A., Micheyl, C.: The cocktail party problem: what is it? How can it be solved? And why should animal behaviorists study it? J. Comp. Psychol. **122**(3), 235–251 (2008). https://doi.org/10.1037/0735-7036.122.3.235
15. http://www.dierk-raabe.com/combinatorial-discovery-of-high-entropy-alloys/. Accessed 10 Sept 2019
16. https://en.wikipedia.org/wiki/Hume-Rothery_rules. Accessed 10 Sept 2019
17. Yeh, J.-W.: Recent progress in high-entropy alloys. Eur. J. Control. **31**, 633–648 (2006). https://doi.org/10.3166/acsm.31.633-648
18. Senkov, O.N., Miller, J.D., Miracle, D.B., Woodward, C.: Accelerated exploration of multi-principal element alloys with solid solution phases. Nat. Commun. **6**, 6529 (2015). https://doi.org/10.1038/ncomms7529
19. https://www.angelo.edu/faculty/kboudrea/periodic/trans_transition.htm. Accessed 10 Sept 2019
20. Ye, Y.F., Wang, Q., Lu, J., Liu, C.T., Yang, Y.: High-entropy alloy: challenges and prospects. Mater. Today. **19**(6), 349–362., ISSN: 1369-7021 (2016). https://doi.org/10.1016/j.mattod.2015.11.026
21. Jiang, H., Huang, T.-D., Chao, S., Zhang, H.-B., Han, K.-M., Qin, S.-X.: Microstructure and mechanical behavior of $CrFeNi_2V_{0.5}W_x$ ($x = 0, 0.25$) high-entropy alloys. Acta Metallurgica Sinica (Eng. Lett.). **33**(8), 1117–1123 (2020). https://doi.org/10.1007/s40195-020-01029-9
22. https://courses.lumenlearning.com/cheminter/chapter/entropy-and-free-energy/. 10 Sept 2019
23. http://www.dierk-raabe.com/high-entropy-alloys/. 10 Sept 2019
24. https://en.wikipedia.org/wiki/Boltzmann_equation#:~:text=The%20Boltzmann%20equation%20or%20Boltzmann,by%20Ludwig%20Boltzmann%20in%201872. Accessed 10 Sept 2019
25. He, Q., Yang, Y.: On lattice distortion in high entropy alloys. Front. Mater. **5** (2008). https://doi.org/10.3389/fmats.2018.00042. https://www.frontiersin.org/article/10.3389/fmats.2018.00042
26. Stepanov, N., Shaysultanov, D.G., Nikita, Y., Sergey, Z., Ladygin, A.N., Gennady, S., Mikhail, T.: High temperature deformation behavior and dynamic recrystallization in CoCrFeNiMn high entropy alloy. Mater. Sci. Eng. A. **636**, 188–195 (2015). https://doi.org/10.1016/j.msea.2015.03.097
27. Kottke, J., Utt, D., Laurent-Brocq, M., Fareed, A., Gaertner, D., Perrière, L., Rogal, Ł., Stukowski, A., Albe, K., Divinski, S.V., Wilde, G.: Experimental and theoretical study of tracer diffusion in a series of (CoCrFeMn)100−xNix alloys. Acta Mater. **194**, 236–248., ISSN 1359-6454 (2020). https://doi.org/10.1016/j.actamat.2020.05.037
28. Zhang, D.C., Mao, Y.F., Li, Y.L., Li, J.J., Yuan, M., Lin, J.: Effect of ternary alloying elements on microstructure and superelastictity of Ti–Nb alloys. Mater. Sci. Eng. A. **559**, 706–710 (2013). https://doi.org/10.1016/j.msea.2012.09.012
29. Barella, S., Ciuffini, A.F., Gruttadauria, A., Mapelli, C., Mombelli, D., Longaretti, E.: Corrosion and oxidation behavior of a Fe-Al-Mn-C duplex alloy. Materials (Basel). **12**(16), 2572. Published 2019 Aug 12 (2019). https://doi.org/10.3390/ma12162572
30. Basu, I., De Hosson, J.T.M.: Strengthening mechanisms in high entropy alloys: fundamental issues. Script. Mater. **187**, 148–156., ISSN 1359-6462 (2020). https://doi.org/10.1016/j.scriptamat.2020.06.019
31. Nikita, Y., Stepanov, N., Sergey, Z., Mikhail, T., Gennady, S.: Structure and mechanical properties of B2 ordered refractory AlNbTiVZr x (x = 0–1.5) high-entropy alloys. Mater. Sci. Eng. A. **704** (2017). https://doi.org/10.1016/j.msea.2017.08.019
32. http://www.mie.uth.gr/ekp_yliko/HighEntropyAlloys.pdf. Accessed 13 Sept 2019
33. Tseng, K.-K., Yang, Y.C., Juan, C.C., Chin, T.S., Tsai, C.W., Yeh, J.-W.: A light-weight high-entropy alloy $Al_{20}Be_{20}Fe_{10}Si_{15}Ti_{35}$. Sci.

China Technol. Sci. **61** (2017). https://doi.org/10.1007/s11431-017-9073-0
34. Wang, Z., Huang, Y., Yang, Y., Wang, J., Liu, C.T.: Atomic-size effect and solid solubility of multicomponent alloys. Scr. Mater. **94** (2015). https://doi.org/10.1016/j.scriptamat.2014.09.010
35. Pickering, E.J., Jones, N.G.: High-entropy alloys: a critical assessment of their founding principles and future prospects. Int. Mater. Rev. **61**(3), 183–202 (2016). https://doi.org/10.1080/09506608.2016.1180020
36. https://www.isda.org/2018/09/26/joint-trades-final-stages-of-initial-margin-phase-in-comment-letter/. Accessed 13 Sept 2019
37. Alshataif, Y.A., Sivasankaran, S., Al-Mufadi, F.A., et al.: Manufacturing methods, microstructural and mechanical properties evolutions of high-entropy alloys: a review. Met. Mater. Int. **26**, 1099–1133 (2020). https://doi.org/10.1007/s12540-019-00565-z
38. Willmott, P., Huber, J.: Pulsed laser vaporization and deposition. Rev. Modern Phys. **72**, 315 (2000). https://doi.org/10.1103/RevModPhys.72.315
39. Zhang, Y.: Microstructures and properties of high entropy alloys. Progr. Mater. Sci. **61**, 1–93 (2014)
40. Distanov, V., Nenashev, B.G., Kirdyashkin, A.G., Serboulenko, M.G.: Proustite single-crystal growth by the Bridgman–Stockbarger method using ACRT. J. Cryst. Growth. **235**, 457–464 (2002). https://doi.org/10.1016/S0022-0248(01)01933-9
41. https://en.wikipedia.org/wiki/Czochralski_method. Accessed 15 Sept 2019
42. Herman, H., Sampath, S., Mccune, R.: Thermal spray: current status and future trends. MRS Bull. **25**, 119 (2000). https://doi.org/10.1557/mrs2000.119
43. https://www.ahssinsights.org/news/tool-materials-and-die-wear/. Accessed 15 Sept 2019
44. Hushchyk, D.V., Yurkova, A.I., Cherniavsky, V.V.: Nanostructured AlNiCoFeCrTi high-entropy coating performed by cold spray. Appl. Nanosci. (2020). https://doi.org/10.1007/s13204-020-01364-4
45. He, X., Kong, D., Song, R.: Microstructures and properties of laser cladding Al-TiC-CeO_2 composite coatings. Materials (Basel). **11** (2), 198. Published 2018 Jan 26 (2018). https://doi.org/10.3390/ma11020198
46. Jie, C., Pei, C., Pei, H., Ming, C., Yin-Yu, C., Weite, W.: Deposition of multicomponent alloys on low-carbon steel using Gas Tungsten Arc Welding (GTAW) cladding process. Mater. Transact. **50**, 689–694 (2009). https://doi.org/10.2320/matertrans.MRP2008276
47. Lin, Y.C., Cho, Y.H.: Elucidating the microstructure and wear behavior for multicomponent alloy clad layers by in situ synthesis. Surf. Coat. Technol. **202**, 4666–4672 (2008). https://doi.org/10.1016/j.surfcoat.2008.03.033
48. Mu, Y.K., Jia, Y.D., Xu, L., Jia, Y.F., Tan, X.H., Yi, J., Wang, G., Liaw, P.K.: Nano oxides reinforced high-entropy alloy coatings synthesized by atmospheric plasma spraying. Mater. Res. Lett. **7** (8), 312–319 (2019)
49. Suryanarayana, C.: Mechanical alloying: a novel technique to synthesize advanced materials. Research (Wash D C). **2019**, 4219812. Published 2019 May 30 (2019). https://doi.org/10.34133/2019/4219812
50. https://angstromengineering.com/tech/magnetron-sputtering/. Accessed 18 Sept 2019
51. Capper, P., Irvine, S., Joyce, T.: Epitaxial crystal growth: methods and materials. In: Kasap, S., Capper, P. (eds.) Springer Handbook of Electronic and Photonic Materials. Springer Handbooks. Springer, Cham (2017). https://doi.org/10.1007/978-3-319-48933-9_14
52. https://en.wikipedia.org/wiki/Pulsed_laser_deposition. Accessed 18 Sept 2019
53. Chen, S., Yang, T., Liaw, P.: Additive manufacturing of high-entropy alloys: a review. Entropy. **20**, 937 (2018). https://doi.org/10.3390/e20120937
54. Sun, J., Han, J., Yang, Z., et al.: Rebuilding the strain hardening at a large strain in twinned Au nanowires. Nanomaterials (Basel). **8** (10), 848. Published 2018 Oct 18 (2018). https://doi.org/10.3390/nano8100848
55. https://royalsocietypublishing.org/doi/10.1098/rspa.2015.0890. Accessed 18 Sept 2019
56. Yuntian, Z., Terry, L.: Observations and issues on mechanisms of grain refinement during ECAP process. Mater. Sci. Eng. A. **291**, 46–53 (2000). https://doi.org/10.1016/S0921-5093(00)00978-3
57. Li, Z., Tasan, C.C., Springer, H., Gault, B., Raabe, D.: Interstitial atoms enable joint twinning and transformation induced plasticity in strong and ductile high-entropy alloys. Sci. Rep. **7**, 40704. Published 2017 Jan 12 (2017). https://doi.org/10.1038/srep40704
58. Sohn, S.S., Song, H., Jo, M.C., Song, T., Kim, H.S., Lee, S.: Novel 1.5 GPa-strength with 50%-ductility by transformation-induced plasticity of non-recrystallized austenite in duplex steels. Sci Rep. **7**(1), 1255. Published 2017 Apr 28 (2017). https://doi.org/10.1038/s41598-017-01514-9
59. Oleg, S., Witusiewicz Victor, T., Sergej, G., Daniel, R., Ulrike, H.: Phase equilibria in the Al–Co–Cr–Fe–Ni high entropy alloy system: thermodynamic description and experimental study. Front. Mater. **7** (2020) https://www.frontiersin.org/article/10.3389/fmats.2020.00270, https://doi.org/10.3389/fmats.2020.00270
60. https://www.sandvik.coromant.com/en-gb/knowledge/materials/pages/workpiece-materials.aspx. Accessed 19 Sept 2019
61. Senkov, O.N., Wilks, G.B., Miracle, D.B., Chuang, C.P., Liaw, P. K.: Refractory high-entropy alloys. Intermetallics, 18, 1758–1765 (2010)
62. Izabela, K., Marek, P., Jerzy, B.: Structure and hydrogen storage properties of a high entropy ZrTiVCrFeNi alloy synthesized using Laser Engineered Net Shaping (LENS). Int. J. Hydrog. Ener. **38** (2013). https://doi.org/10.1016/j.ijhydene.2013.05.071
63. Takeuchi, A., Chen, N., Wada, T., Yokoyama, Y., Kato, H., Inoue, A., Yeh, J.W.: $Pd_{20}Pt_{20}Cu_{20}Ni_{20}P_{20}$ high-entropy alloy as a bulk metallic glass in the centimeter. Intermetallics. **19**(10), 1546–1554 (2011). https://doi.org/10.1016/j.intermet.2011.05.030
64. Rui, L., Jia, G., Ke, F.: Microstructure and mechanical properties of MgMnAlZnCu high entropy alloy cooling in three conditions. Mater. Sci. Forum. **686**, 235–241 (2011). https://doi.org/10.4028/www.scientific.net/MSF.686.235
65. Stepanov, N.D., Shaysultanov, D.G., Salishchev, G.A., Tikhonovsky, M.A.: Structure and mechanical properties of a light-weight AlNbTiV high entropy alloy. Mater. Lett. **142**, 153–155 (2015). https://doi.org/10.1016/j.matlet.2014.11.162
66. Khaled, Y., Alexander, Z., Changning, N., Douglas, I., Carl, K.: A novel low-density, high-hardness, high-entropy alloy with close-packed single-phase nanocrystalline structures. Mater. Res. Lett. **2**, 1 (2014). https://doi.org/10.1080/21663831.2014.985855
67. Soni, V., Senkov, O.N., Gwalani, B., et al.: Microstructural design for improving ductility of an initially brittle refractory high entropy alloy. Sci. Rep. **8**, 8816 (2018). https://doi.org/10.1038/s41598-018-27144-3
68. Rabadia, C.D., Liu, Y.J., Chen, L.Y., Jawed, S.F., Wang, L.Q., Sun, H., Zhang, L.C.: Deformation and strength characteristics of laves phases in titanium alloys. Mater. Des. **179**, 107891., ISSN 0264-1275 (2019). https://doi.org/10.1016/j.matdes.2019.107891
69. Chan, K.W., Tjong, S.C.: Effect of secondary phase precipitation on the corrosion behavior of duplex stainless steels. Materials (Basel). **7**(7), 5268–5304. Published 2014 Jul 22 (2014). https://doi.org/10.3390/ma7075268
70. Woei-Ren, W., Wei-Lin, W., Jien-Wei, Y.: Phases, microstructure and mechanical properties of AlxCoCrFeNi high-entropy alloys at

elevated temperatures. J. Alloys Comp. **589**, 143–152 (2014). https://doi.org/10.1016/j.jallcom.2013.11.084
71. Stepanov, N., Nikita, Y., Skibin, D.V., Mikhail, T., Gennady, S.: Structure and mechanical properties of the AlCrxNbTiV (x = 0, 0.5, 1, 1.5) high entropy alloys. J. Alloys Comp. **652**, 266–280 (2015). https://doi.org/10.1016/j.jallcom.2015.08.224
72. Gludovatz, B., Hohenwarter, A., Thurston, K.V., et al.: Exceptional damage-tolerance of a medium-entropy alloy CrCoNi at cryogenic temperatures. Nat Commun. **7**, 10602. Published 2016 Feb 2 (2016). https://doi.org/10.1038/ncomms10602
73. Zou, Y., Ma, H., Spolenak, R.: Ultrastrong ductile and stable high-entropy alloys at small scales. Nat. Commun. **6**, 7748. Published 2015 Jul 10 (2015). https://doi.org/10.1038/ncomms8748
74. Bała, P., Górecki, K., Bednarczyk, W., Wątroba, M., Lech, S., Kawałko, J.: Effect of high-temperature exposure on the microstructure and mechanical properties of the $Al_5Ti_5Co_{35}Ni_{35}Fe_{20}$ high-entropy alloy. J. Mater. Res. Technol. **9**(1), 551–559 (2020). https://doi.org/10.1016/j.jmrt.2019.10.084
75. Tseng, K.-K., Juan, C.-C., Tso, S., Chen, H.-C., Tsai, C.-W., Yeh, J.-W.: Effects of Mo, Nb, ta, Ti, and Zr on mechanical properties of Equiatomic Hf-Mo-Nb-ta-Ti-Zr alloys. Entropy. **21**(1), 15 (2019). https://doi.org/10.3390/e21010015
76. https://fdocuments.in/document/high-entropy-alloys-chalmers-publication-library-cpl-entropy-alloys-pm.html. Accessed 21 Sept 2019
77. Mathiou, C., Poulia, A., Georgatis, E., Karantzalis, A.: Microstructural features and dry - sliding wear response of MoTaNbZrTi high entropy alloy. Mater. Chem. Phys. (2017). https://doi.org/10.1016/j.matchemphys.2017.08.036
78. Capar, I.D., Kaval, M.E., Ertas, H., Sen, B.H.: Comparison of the cyclic fatigue resistance of 5 different rotary pathfinding instruments made of conventional nickel-titanium wire, M-wire, and controlled memory wire. J. Endod. **41**(4), 535–538 (2015). https://doi.org/10.1016/j.joen.2014.11.008
79. Chang, C.-H., Titus, M.S., Yeh, J.-W.: Oxidation behavior between 700 and 1300 °C of refractory TiZrNbHfTa high-entropy alloys containing aluminum. Adv. Eng. Mater. **20**, 1700948 (2018). https://doi.org/10.1002/adem.201700948
80. Ha, H.Y., Jang, J.H., Lee, T.H., et al.: Investigation of the localized corrosion and passive behavior of type 304 stainless steels with 0.2–1.8 wt % B. Materials (Basel). **11**(11), 2097. Published 2018 Oct 25 (2018). https://doi.org/10.3390/ma11112097
81. Eliaz, N.: Corrosion of metallic biomaterials: a review. Materials (Basel). **12**(3), 407. Published 2019 Jan 28 (2019). https://doi.org/10.3390/ma12030407
82. Jayaraj, J., Thinaharan, C., Ningshen, S., Mallika, C., Kamachi, M.: Corrosion behavior and surface film characterization of TaNbHfZrTi high entropy alloy in aggressive nitric acid medium. Intermetallics. **89**, 123–132 (2017). https://doi.org/10.1016/j.intermet.2017.06.002
83. Chou Hsuan-Ping, Chang Yee-Shyi, Lee Po-Han, Yeh Jien-Wei (2009) Microstructure, thermophysical and electrical properties in AlxCoCrFeNi ($0 \leq x \leq 2$) high-entropy alloys. Mater. Sci. Eng. B. 163. 184–189. https://doi.org/10.1016/j.mseb.2009.05.024
84. Du-Cheng, T., Fuh-Sheng, S., Shou-Yi, C., Hsiao-Chiang, Y., Min-Jen, D.: Structures and characterizations of TiVCr and TiVCrZrY films deposited by magnetron sputtering under different bias powers. J. Electrochem. Soc. **157**, K52 (2010). https://doi.org/10.1149/1.3285047
85. Bauer, E., Andrzej, Ś., Freeman, E., Sirvent, C., Maple, M.: Kondo insulating behaviour in the filled skutterudite compound $CeOs_4Sb_{12}$. J. Phys. Conden. Matter. **13**, 4495 (2001). https://doi.org/10.1088/0953-8984/13/20/310
86. Zhao, M., Pan, W.: Effect of lattice defects on thermal conductivity of Ti-doped, Y_2O_3-stabilized ZrO_2. Acta Mater. **61**, 5496–5503 (2013). https://doi.org/10.1016/j.actamat.2013.05.038
87. https://physics.info/expansion/. Accessed 23 Sept 2019
88. Singh, S., Wanderka, N., Kiefer, K., Siemensmeyer, K., Banhart, J.: Effect of decomposition of the Cr-Fe-co rich phase of AlCoCrCuFeNi high entropy alloy on magnetic properties. Ultramicroscopy. **111**, 619–622 (2010). https://doi.org/10.1016/j.ultramic.2010.12.001
89. Oxley, P., Goodell, J., Molt, R.: Magnetic properties of stainless steels at room and cryogenic temperatures. J. Magn. Magnet. Mater. **321**, 2107–2114 (2009)
90. Yih-Farn, K., Po-Han, L., Ting-Jie, C., Po-Chou, C., Jien-Wei, Y., Su-Jien, L.: Electrical, magnetic, and Hall properties of Al x CoCrFeNi high-entropy alloys. J. Alloys Comp. **509**, 1607–1614 (2011). https://doi.org/10.1016/j.jallcom.2010.10.210
91. https://en.wikipedia.org/wiki/Annealing_(metallurgy). Accessed 24 Sept 2019
92. https://www1.eere.energy.gov/hydrogenandfuelcells/pdfs/executive_summaries_h2_storage_coes.pdf. Accessed 24 Sept 2019
93. Zinkle, S.J.: Radiation-induced effects on microstructure. Comprehens. Nucl. Mater. **1**, 65–98 (2012). https://doi.org/10.1016/B978-0-08-056033-5.00003-3
94. Zhang, Y.: Irradiation Behavior in Entropic Materials. Springer, New York (2019). https://doi.org/10.1007/978-981-13-8526-1_6
95. Song-qin, X., Zhen, W., Teng-fei, Y., Yong, Z.: Irradiation behavior in high entropy alloys. J. Iron Steel Res. Int. **22**(10), 879–884., ISSN 1006-706X (2015). https://doi.org/10.1016/S1006-706X(15)30084-4
96. https://www.science.gov/topicpages/h/high+ni+excess. Accessed 24 Sept 2019
97. Wang, W.H.: High entropy metallic glasses. JOM. **66** (2014). https://doi.org/10.1007/s11837-014-1002-3
98. https://www.intechopen.com/books/metallic-glasses/metallic-glasses-a-revolution-in-material-science. Accessed 24 Sept 2019
99. Gorsse, S., Couzinié, J.-P., Miracle, D.B.: From high-entropy alloys to complex concentrated alloys. Comptes Rendus Physique. **19**(8), 721–736., ISSN 1631-0705 (2018). https://doi.org/10.1016/j.crhy.2018.09.004
100. https://www.nap.edu/read/1792/chapter/6. Accessed 24 Sept 2019
101. Cheah, L.: Cars on a diet: the material and energy impacts of passenger vehicle weight reduction in the US. MIT, Cambridge, MA (2010)
102. Poletti, M.G., Fiore, G., Gili, F., Mangherini, D., Battezzati, L.: Development of a new high entropy alloy for wear resistance: FeCoCrNiW0.3 and FeCoCrNiW0.3+5at.% of C. Mater. Des. **115**, 247–254., ISSN 0264–1275 (2017). https://doi.org/10.1016/j.matdes.2016.11.027
103. Zhang, L., Chen, L., Wang, L.: Surface modification of titanium and titanium alloys: technologies, developments, and future interests. Adv. Eng. Mater. **22**, 1901258 (2020). https://doi.org/10.1002/adem.201901258
104. https://www.farinia.com/additive-manufacturing/3d-materials/inconel-718-aerospace-additive-manufacturing. 25 Sept 2019
105. https://kraftwerkforschung.info/en/higher-temperatures-in-turbines/. 25 Sept 2019
106. https://www.asminternational.org/documents/10192/1849770/ACFAAD4.pdf/35831243-7132-4243-91a1-705e67b1cf4b. Accessed 25 Sept 2019
107. Svensson, D.: High Entropy Alloys: Breakthrough Materials for Aero Engine Applications, Master Theses (IMS). https://hdl.handle.net/20.500.12380/214141 (2015).
108. Molland, A.: The maritime engineering reference book. In: Butterworth-Heinemann, pp. 636–727., ISBN 9780750689878. Elsevier, Netherland (2008). https://doi.org/10.1016/B978-0-7506-8987-8.00009-3
109. https://en.wikipedia.org/wiki/Superalloy. Accessed 26 Sept 2019
110. https://www.esabna.com/us/en/education/blog/understanding-the-aluminum-alloy-designation-system.cfm. Accessed 26 Sept 2019

111. https://en.wikipedia.org/wiki/Components_of_jet_engines. Accessed 26 Sept 2019
112. https://matmatch.com/learn/material/inconel-718. Accessed 26 Sept 2019
113. Xiang, Y., Shuqiong, X.: Properties and preparation of high entropy alloys. MATEC Web Conf. **142**, 03003 (2018). https://doi.org/10.1051/matecconf/201714203003
114. Holmström, E., Lizárraga, R., Linder, D., Salmasi, A., Wang, W., Kaplan, B., Mao, H., Larsson, H., Vitos, L.: High entropy alloys: substituting for cobalt in cutting edge technology. Appl. Mater. Today. **12**, 322–329., ISSN 2352-9407 (2018). https://doi.org/10.1016/j.apmt.2018.07.001
115. Chim, Y.C., Xiang, D., Zeng, X.T., Zhang, S.: Oxidation resistance of TiN, CrN, TiAlN and CrAlN coatings deposited by lateral rotating cathode arc. Thin Solid Films. **517** (2009). https://doi.org/10.1016/j.tsf.2009.03.038
116. https://www.phase-trans.msm.cam.ac.uk/2003/Superalloys/coatings/index.html. Accessed 26 Sept 2019
117. https://en.wikipedia.org/wiki/Generation_IV_reactor. Accessed 26 Sept 2019
118. Teuchert, E., Axmann, J.K., Klüver, B., Meyer, P.-J., Hollinger, H., Kausz, I., Debray, W., Erve, M., Riess, R., Nieder, R., Kotthoff, K., Budnick, D., Dörfler, U., Dreyer, S., Schillings, K.-L., Luster, V.P., Essig, C., Koch, G., Träger, S., Max, A., Krebs, W.-D., Stoll, W., Heit, W., Warnecke, E., Brennecke, P., Merz, E., Schumacher, U., Herold, H.: Nuclear technology. In: Ullmann's Encyclopedia of Industrial Chemistry. John Wiley & Son Ltd., Hoboken, NJ (2000). https://doi.org/10.1002/14356007.a17_589
119. Yibo, S., Ye, L., Fitzhugh, W., Wang, Y., Gil-González, E., Kim, I., Li, X.: A more stable lithium anode by mechanical constriction for solid state batteries. Energy Environ. Sci. **3**, 908–916 (2020)
120. Kumar, S., Kumar, S., Singh, V., Kumar, J.J.: High-entropy alloys, a review. Int. Res. J. Eng. Technol. **3**, 9 (2016)
121. Gabriela, P., Ghiban, B., Popescu, C., Rosu, L., Roxana, T., Ioan, C., Vasile, S., Daniela, D., Ionut, C., Mihai, O., Beatrice, C.: New TiZrNbTaFe high entropy alloy used for medical applications. IOP Conf. Series Mater. Sci. Eng. **400**, 022049 (2018). https://doi.org/10.1088/1757-899X/400/2/022049
122. Murty, K., Indrajit, C.: Structural materials for Gen-IV nuclear reactors: challenges and opportunities. J. Nucl. Mater. **383**, 189–195 (2008). https://doi.org/10.1016/j.jnucmat.2008.08.044
123. Otto, F., Dlouhy, A., Pradeep, K.G., Kubenova, M., Raabe, D., Eggeler, G., George, E.P.: Decomposition of the single-phase high-entropy alloy CrMnFeCoNi after prolonged anneals at intermediate temperatures. Acta Mater. **112** (2016). https://doi.org/10.1016/j.actamat.2016.04.005
124. Shi, Y., Yang, B., Liaw, P.K.: Corrosion-resistant high-entropy alloys: a review. Metals. **7**, 43 (2017)
125. Castle, E., Csanádi, T., Grasso, S., Dusza, J., Reece, M.: Processing and properties of high-entropy ultra-high temperature carbides. Sci. Rep. **8**(1), 8609. Published 2018 Jun 5 (2018). https://doi.org/10.1038/s41598-018-26827-1
126. Sajid, A., Hanzhu, Z., Farid, A.: High-entropy ceramics. IntechOpen, London (2019). https://doi.org/10.5772/intechopen.89527
127. Christina, R., Edward, S., Trent, B., Ali, M., Elizabeth, D., Dong, H., Jacob, J., Stefano, C., Jon-Paul, M.: Entropy-stabilized oxides. Nat. Commun. **6**, 8485 (2015). https://doi.org/10.1038/ncomms9485
128. https://en.wikipedia.org/wiki/Ultra-high-temperature_ceramics. Accessed 26 Sept 2019
129. Sicong, J., Tao, H., Joshua, G., Naixie, Z., Jiuyuan, N., Mingde, Q., Tyler, H., Kenneth, V., Jian, L.: A new class of high-entropy perovskite oxides. Script. Mater. **142**, 116–120 (2018)
130. Sarker, P., Harrington, T., Toher, C., et al.: High-entropy high-hardness metal carbides discovered by entropy descriptors. Nat. Commun. **9**, 4980 (2018). https://doi.org/10.1038/s41467-018-07160-7
131. Gerstl, M., Navickas, E., Friedbacher, G., Kubel, F., Ahrens, M., Fleig, J.: The separation of grain and grain boundary impedance in thin yttria stabilized zirconia (YSZ) layers. Solid State Ionics. **185** (1), 32–41 (2011). https://doi.org/10.1016/j.ssi.2011.01.008
132. Oses, C., Toher, C., Curtarolo, S.: High-entropy ceramics. Nat. Rev. Mater. **5**, 295–309 (2020). https://doi.org/10.1038/s41578-019-0170-8
133. Yadav, A.V., Shete, A.S., Dabke, A.P., Kulkarni, P.V., Sakhare, S.S.: Co-crystals: a novel approach to modify physicochemical properties of active pharmaceutical ingredients. Indian J. Pharm. Sci. **71**(4), 359–370 (2009). https://doi.org/10.4103/0250-474X.57283
134. Yin, Z., Tian, B., Zhu, Q., Duan, C.: Characterization and application of PVDF and its copolymer films prepared by spin-coating and Langmuir-Blodgett method. Polymers (Basel). **11**(12), 2033 (2019). https://doi.org/10.3390/polym11122033
135. Zerio Mendaza, A.D., Melianas, A., Rossbauer, S., Bäcke, O., Nordstierna, L., Erhart, P., Olsson, E., Anthopoulos, T.D., Inganäs, O., Müller, C.: High-entropy mixtures of pristine fullerenes for solution processed transistors and solar cells. Adv. Mater. **27**, 7325–7331 (2015). https://doi.org/10.1002/adma.201503530

10 Self-Healing Materials

Abstract

In the current era, the self-healing properties found in some advanced materials have a resemblance with the biological wound healing process. This chapter describes the gradual development of metal, ceramic, polymer, and others in the direction of self-healing at room temperature as well as in high-temperature applications. Various classifications of materials such as autonomic, nonautonomic, intrinsic, and extrinsic processes of self-healing are described with their mechanism. Self-healing polymer is a step ahead of other materials in cent percent healing. Different concepts and thermodynamic approaches of thermoplastic, as well as thermosetting polymer, have been discussed broadly. Applications in aero industries, automobile, electrical and electronics, mechanical, medical, textile, and other industries have been discussed extensively.

Keywords

Self-healing materials · Autonomic self-repair materials · Non-autonomic self-repair materials · Self-healing metal · Self-healing ceramics · Self-healing polymers · Self-healing coatings · Self-healing hydrogels

10.1 Introduction and Overview

Things break in our daily life such as scratching of car body due to misjudgment of parking width, scratch or crack on the tablet screen when sudden fall down, and scratching of badminton racket with a little force contact with the floor. It is frustrating when it happens in expensive things. The formation of a microscopic crack in spacecraft or in concrete dam makes a huge sense, which is the importance of the development of self-healing materials. Self-healing materials are classified as smart structures and can adapt to various environmental conditions according to their sensing and actuation properties. The name of the material was publicly awarded by a movie "Mission Mangan" in 2019, showing the solar panel surface healing after an asteroid strike. Fortunately, we are amid a fascinating rise in self-healing materials that promise to do just that. But what does "self-healing" mean in this context? The term brings to mind something biological, and indeed, self-repair is a hallmark of living things, from DNA and cell wall repair in single-celled organisms all the way to wound and bone repair in our own bodies. Yet neither your tablet screen nor your car frame is alive but able to heal which is the beauty of self-healing materials. Self-healing material can be defined as "the materials when damaged through various mode (thermal, mechanical, ballistic or other means) have the ability to heal and restore the material to its original set of properties" [1, 2]. Common terms such as self-repairing, autonomic healing, and autonomic repairing are used to define such a property in materials. Self-healing materials may be metal or ceramic or polymer cementitious materials and their composites (Fig. 10.1) [3, 4]. For example, self-healing polymers activate in response to an external stimulus (light, temperature change, etc.) to initiate the healing processes. Damage on a microscopic level has been shown to change the thermal, electrical, and acoustical properties of materials, and the propagation of cracks can lead to eventual failure of the material. A material that can intrinsically correct damage caused by normal usage could prevent costs incurred by material failure and lower costs of some different industrial processes through a longer part life time and reduction of inefficiency caused by degradation over time. Healing mechanisms vary from an intrinsic repair of the material to the addition of a repair agent contained in a microscopic vessel [5].

The market size of the global self-healing materials was estimated at USD 291.4 million in 2018, growing at a CAGR of 46.1% from 2019 to 2025 [6]. Growing demand for advanced self-repairing polymers from end-use industries and the ability of materials to repair damages caused due to mechanical friction are the factors estimated to drive overall

A. Behera, *Advanced Materials*, https://doi.org/10.1007/978-3-030-80359-9_10

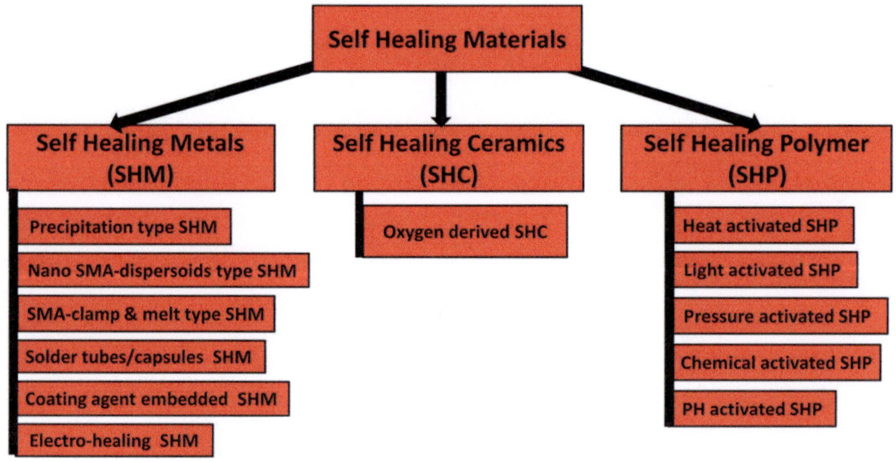

Fig. 10.1 Classification of self-healing materials

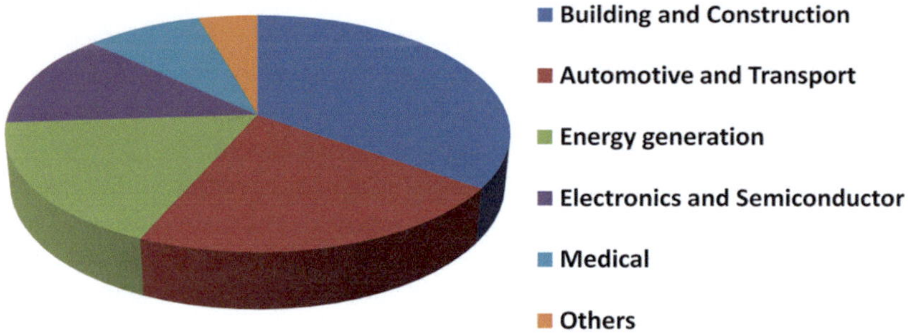

Fig. 10.2 Current market status of global self-healing materials

market growth in the forecast period (Fig. 10.2). Rising demand for advanced self-repairing polymers improve productivity and minimize losses, related to product damages, particularly from automotive and electronic devices manufacturing industries is expected to drive growth [7]. In addition, a rise in the application of such materials in paints and coatings in the aforementioned industries is expected to promote growth. Factors such as high cost of such products coupled with the presence of a number of substitutes are estimated to restrain the overall market growth. In addition, limited awareness of the products primarily across the less developed economies is expected to limit the market growth. The building and construction segment dominated the market accounting for 33.4% of the total demand in 2018 [8]. The growing adoption of such materials by the major players in the electronics and semiconductors markets is expected to drive the industry growth over the forecast period. In addition, the usage of such products in patented applications is expected to benefit the industry growth over the forecast period. The automotive and transportation sector emerged as one of the major applications for the products on account of the rising affinity of the consumers toward the lowering of the maintenance cost of the paints. For instance, Ford developed a new technique of self-healing film that protects the car from damage and scratches, which is expected to be implemented on a commercial scale. The self-healing materials are exclusively used in the energy generation industry in ionic conductors, insulators, and electrical conductors; in power generation; and in oil/gas refineries. Moreover, electronics companies such as Samsung and Apple are using self-healing materials in mobiles, laptops, and desktops that are expected to drive their demand over the forecast period.

10.2 History of Self-Healing Materials

Self-healing materials are mimicked from nature. Plants and animals have the capacity to seal and heal wounds. In plants, rapid self-sealing prevents the plants from desiccation and infection by pathogenic germs. This gives time for the subsequent self-healing of the injury which in addition to wound closure also results in the (partly) restoration of mechanical properties of the plant organ. Based on a variety of self-sealing and self-healing processes in plants, different functional principles were transferred into bio-inspired self-repairing materials. The connecting link between the biological model and the technical application is an abstraction describing the underlying functional principle of the biological model that is mainly influenced by the physical–chemical processes. The ancient Romans used a form of lime mortar that has been found to have self-healing properties. By

2014, geologist Marie Jackson and her colleagues had recreated the type of mortar used in Trajan's Market and other Roman structures, such as the Pantheon and the Colosseum, and studied its response to cracking [9]. The Romans mixed a particular type of volcanic ash called Pozzolana Rosse, from the Alban Hills volcano, with quicklime and water [16]. They used it to bind together decimeter-sized chunks of tuff, an aggregate of volcanic rock. As a result of Pozzolanic activity as the material cured, the lime interacted with other chemicals in the mix and was replaced by crystals of a calcium aluminosilicate mineral called Stratlingite. Crystals of plate Stratlingite grow in the cementitious matrix of the material, including the interfacial zones where cracks would tend to develop. This ongoing crystal formation holds together the mortar and the coarse aggregate, countering crack formation and resulting in a material that has lasted for 1900 years. Related processes in concrete have been studied microscopically since the nineteenth century. Self-healing materials only emerged as a widely recognized field of study in the twenty-first century. The first international conference on self-healing materials was held in 2007 [10]. Toohey et al. did this with an epoxy substrate containing a grid of microchannels containing dicyclopentadiene (DCPD) and incorporated Grubbs' catalyst to the surface [11]. This showed partial recovery of toughness after fracture and could be repeated several times because of the ability to replenish the channels after use. The process is not repeatable forever because the polymer in the crack plane from previous healings would build up over time. Inspired by rapid self-sealing processes in the twining liana Aristolochiamacrophylla and related species (pipevines), a biomimetic PU-foam coating for pneumatic structures was developed. With respect to the low coating weight and thickness of the foam layer, maximum repair efficiencies of 99.9% and more have been obtained [12]. Other role models are latex bearing plants as the weeping fig (Ficus benjamina), the rubber tree (Heveabrasiliensis), and spurges (Euphorbia spp.) in which the coagulation of latex is involved in the sealing of lesions. Different self-sealing strategies for elastomeric materials were developed showing significant mechanical restoration after a macroscopic lesion [13]. The year-wise gradual development of self-healing materials has been given in Table 10.1.

10.3 Types of Self-Healing Processes

Incorporation of self-healing properties in manmade materials very often cannot perform the self-healing action without an external trigger. As a result of the multitude of different approaches used and materials studied, self-healing materials can be divided into two different classes, depending on the required trigger and the nature of the self-healing process: autonomic (without any intervention) and non-autonomic (needs human intervention/external triggering) as shown in Fig. 10.3. In autonomic self-healing materials, the harnessed chemical potential is automatically released and facilitates the repair/healing in response to damage/rupture. On the other hand, non-autonomic self-healing materials need external intervention. These materials are also of various types depending on their chemistry and applications [14, 15]. Still another property of the respective self-healing process could be used to distinguish material subclasses, leading to the terms "extrinsic" and "intrinsic" self-healing. Extrinsic self-healing materials themselves do not owe a hidden intrinsic capability for self-healing; rather, the healing process is based on external healing components, such as micro- or nano-capsules, intentionally embedded into the matrix materials to make them self-healing [16]. The content of these capsules becomes the mobile phase upon damage. The capsules or networks are made of a polymer, such as polyurethane (PU) and poly(urea-formaldehyde) (PUF). However, there are also some shortcomings of the microcapsule-mediated extrinsic self-healing approach, as multiple times, healing is not possible and catalysts are usually expensive [17]. On the other hand, intrinsic self-healing requires no separate healing agents, which is to be preferred but, depending on the material class and healing mechanism, not always feasible. Formation of (secondary or primary) chemical bonds and physical interactions between the interfaces of the crack (adhesion, wetting) are successful examples to achieve self-healing via this route, provided that the crack width is below a certain limit. An intrinsic self-healing system utilizes noncovalent interactions or dynamic covalent chemistry. The noncovalent interactions used are hydrogen bonding, host–guest complex, metal–ligand interaction, and so on. [18]. The reversible dynamic covalent chemistry (DCC) involves Diels–Alder reaction, imine chemistry, radical exchange, and transesterification. The healing mechanism may be triggered by some external stimuli. Due to the reversible nature of the supramolecular interactions or DCC employed in the intrinsic self-healing system, healing of damaged area is possible multiple times. Thus, the current research goal is focused toward designing next-generation materials capable of exhibiting intrinsic self-healing behavior.

To quantify healing, researchers have proposed multiple definitions of healing efficiencies. For convenience, healing efficiency (η) can be defined as a ratio of changes in material properties as in Eq. ((10.1) [19]

$$\eta = \frac{f_{\text{HEALED}} - f_{\text{DAMAGED}}}{f_{\text{VERGIN}} - f_{\text{DAMAGED}}} \tag{10.1}$$

where f is the property of interest.

Table 10.1 The development of different self-healing materials and applications.

Development of various self-healing materials or materials in various application	Year
Self-curable resinous compositions containing *N*-methylol amide groups	1980
Heat-curable surface-coating agents	1985
Heat-curable resin-coating composition	1985
Self-repairing, reinforced matrix materials	1990
Microencapsulated adhesive	1992
Crack repair using active and passive modes for smart cement matrices	1994
Microcapsule-type curing agent fiber-reinforced composite material	1994
Polymerization of dicyclopentadiene	1994
Process for the preparation of microencapsulated polymers	1996
Self-healing alloy composite	1997
Feasibility study of a passive smart self-healing cementitious composite	1998
Semiconductor encapsulating epoxy resin composition	1999
Autogenous healing of cracks in concrete	1999
Self-healing SiC_f/SiC ceramic composite	2001
Autonomic healing of polymer composites	2001
Self-healing polymers in sports equipment	2001
Thermally re-mendable cross-linked polymers	2002
In situ (urea-formaldehyde) microencapsulation of dicyclopentadiene	2003
Self-healing cracks in underfill material between an I/C chip	2003
Self-healing coating system	2006
Polydimethylsiloxane-based self-healing materials	2006
Polyalkyleneimines Crosslinked gels used as sealants used as medical devices	2007
Self-healing poly (dimethyl siloxane) elastomer	2007
Liquid dicyclopentadiene embedded in an epoxy matrix	2007
Self-healing carbon fiber-reinforced polymer for aerospace applications	2007
Self-healing photoreceptor	2008
Composite laminate with self-healing layer	2008
Phase separated self-healing polymer coatings	2009
Self-repairing IC package design by microencapsulation process	2009
Packaged electronic device having metal comprising self-healing die attach material	2009
A biomimic shape memory polymer-based self-healing particulate composite	2010
Microencapsulation of liquid-phase amine for self-healing epoxy composites	2010
Information medium or paper comprising a self-repairing material	2011
Self-healing electrical insulation	2011
Radio frequency magnetic field responsive polymer composites for self-crack healing	2011
Activation and deactivation of self-healing in supramolecular rubbers	2012
Self-healing polymeric materials via unsaturated polyesters	2013
Renewable self-healing capsule system	2013
Puncture-healing thermoplastic resin carbon fiber-reinforced composites	2013
Self-healing polymers based on thermally reversible Diels–Alder chemistry	2013
Self-healing Cr_2AlC ceramic	2013
Alkoxyamine in self-healing of epoxy	2014
Al-shape memory alloy self-healing metal matrix composite	2014
Self-healing nanofiber-reinforced polymer composites	2015
Self-healing carbon fiber composites	2015
Self-healing metal capacitor	2015
Self-healing coatings for oil and gas applications	2015
Zinc alloy ZA-8/shape memory alloy self-healing metal matrix composite	2015
Self-healing hydrogels	2016
Highly stretchable autonomous self-healing elastomer	2016
pH-responsive self-healing injectable hydrogel based on *N*-carboxyethyl chitosan	2017
Graphene oxide microcapsules for self-healing waterborne polyurethane coatings	2017
Self-healing of poly (*p*-phenylenebenzobisoxazole) fiber composites	2018

(continued)

Table 10.1 (continued)

Development of various self-healing materials or materials in various application	Year
Self-healing electronic skins for aquatic environments	2019
Self-healing energy storage device	2019
Self-Healing Zn-δ-MnO_2 Battery	2019
Self-Healing, Anti-Freezing, and Non-Drying Strain Sensor	2020
Self-Healing TPU/Vitrimer Polymer	2020

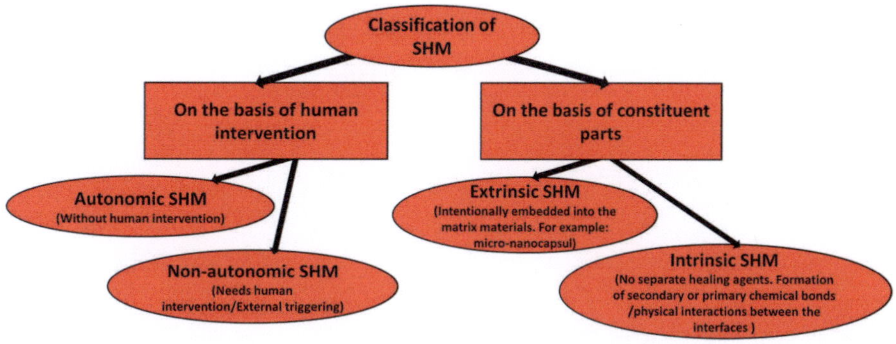

Fig. 10.3 Classification of self-healing materials

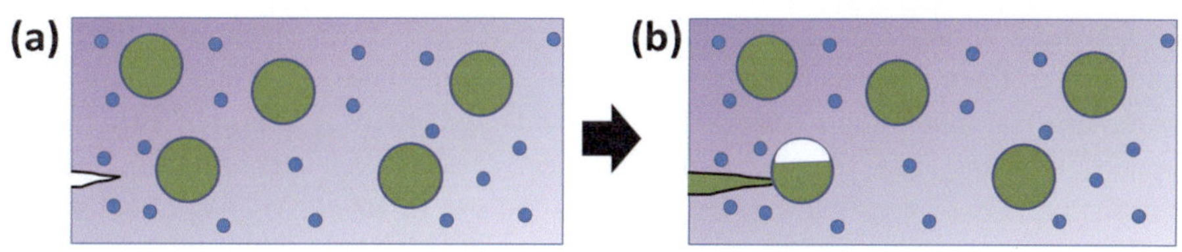

Fig. 10.4 The autonomic self-healing system

10.4 Autonomic Self-Repair Materials

An interesting variant of the passive self-healing method was developed by White and Sottos in 2001, where spheres-shaped healing fluid is used as shown in Fig. 10.4. They describe the autonomic healing concept as follows: A microencapsulated healing agent is embedded in a structural epoxy matrix containing a catalyst (Grubb's) capable of polymerizing the healing fluid [20]: (a) cracks form in the matrix wherever damage occurs [21], (b) the crack ruptures the microcapsules releasing the healing agent into the crack plane through capillary action [22], and (c) the healing fluid contacts the catalyst, which is randomly distributed in the matrix and triggers the polymerization reaction that helps to heal the damage by bonding the crack surfaces closed [23]. For example, the propagating crack ruptured the urea–formaldehyde spheres, which released the dicyclopentadiene fluid. In the rest state, the released fluid contacted the ruthenium (Grubb's) catalyst embedded at random in the epoxy matrix, which initiated the polymerization of the fluid and repaired the crack [24].

10.5 Non-autonomic Self-Repair Materials

An alternative approach toward self-healing materials is one in which the material heals in response to externally applied stimuli, such as heat or light. This process requires human intervention, but, at the same time, it allows the healing process to be performed in a controlled fashion at the user's discretion and, in some cases, multiple times [25]. As a result, such non-autonomic self-healing materials often utilize chemistries that are very different from those used in autonomic variants. The characteristic and necessary trait shared by each of these systems is that the structural framework of the material contains a reversible, or dynamic, chemical bond—usually weak covalent bonds, metal–ligand coordination bonds, or directed electrostatic interactions. Additionally, hydrogen bonding has served as the basis for a recently reported room-temperature self-healing rubber as well as other systems that display self-healing potential. The basic concept behind these materials is that they can exist in either polymeric or monomeric states, depending on their environments. Hence, the material can be toggled between exhibiting good mechanical properties (polymer) and good properties for facilitating mass transport to damaged areas (monomer) [26].

10.6 Materials for Self-Healing Purposes

Here, we will discuss various materials such as metal, ceramic, and polymer. The most market-dominating material is the polymer and will be discussed broadly in the end part of this section.

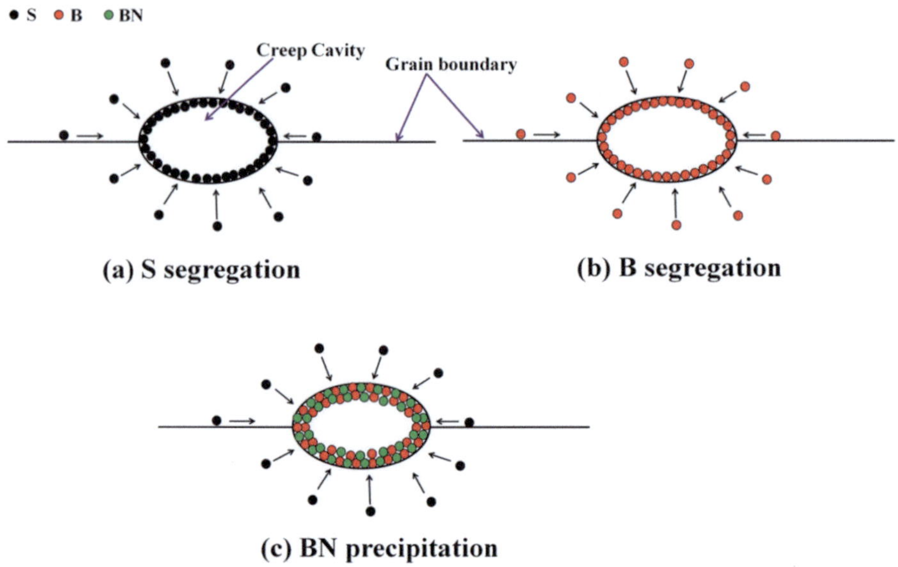

Fig. 10.5 Schematics of S and B segregation and BN precipitation on the creep effect

10.6.1 Self-Healing in Metals

Self-healing in metallic materials has been explored with the adoption of three basic techniques: (1) the use of precipitates from supersaturated solid solutions to fill voids and micro-cracks, (2) the use of shape memory alloy wires as reinforcements of metallic matrices, and (3) the use of low melting temperature metals as reinforcement of metallic matrices.

10.6.1.1 Precipitation From Supersaturated Solid Solutions

A precipitation reaction is prevalent in metallic systems characterized by decreasing solubility for solute atoms with a decrease in temperature or vice versa [27]. Usually, at lower temperatures, the equilibrium amount of solute atoms, which can be accommodated within the matrix phase, is lower compared to the higher solubility at higher temperatures. Adoption of slow cooling from a single-phase region will result in the maintenance of the equilibrium amount of solutes in the matrix with a temperature drop and the excess solute atoms precipitated as a second phase, which is usually an intermetallic compound. However, rapid cooling from the high-temperature, single-phase region will result in little time for the excess solute atoms to diffuse out from the matrix resulting in a state of supersaturation of solute atoms in the matrix. The excess solute atoms "frozen" within the matrix change the thermodynamic state of the metallic system from one that is stable to one that is metastable. The metastability raises the overall energy in the metallic system, which has a natural tendency to reduce its excess energy to regain stability under the prevailing condition. This quest drives the process of precipitation, which involves the migration of excess solute atoms to nucleation sites, the arrangement of the solute atoms to a lower energy configuration, and the formation of cluster, before eventually nucleating the precipitate phase [28].

Self-healing by precipitation of solute atoms has been reported in a few metallic systems. Lumley and Polmear reported healing attributes of underaged wrought Al–Cu–Mg–Ag alloy when subjected to creep tests [29]. The healing behavior was attributed to the facilitation of dynamic precipitation by retained solute atoms in the matrix which inhibits dislocation motion during creep. A similar mechanism was reported to have caused the healing of incipient fatigue cracks in the same alloy. Laha et al. and Shinya et al. have shown that Boron Nitride (BN) imparted self-healing behavior in 304 stainless steel modified with B, Ce, and Ti. It was reported that the precipitation of soluble sulfur in the creep cavity of 314 stainless [30] steel promotes creep failure in this material. The modification of this alloy with B, Ce, and Ti resulted in the preferential precipitation of BN at the creep cavity sites (Fig. 10.5). The presence of BN, which is stable at high temperature in the creep cavity surface, led to improved creep resistance of the modified 314 stainless steel. The replacement of S with BN at the creep cavity site was assisted by the formation of the sulfides of Ce and Ti. This process has been harnessed to achieve self-healing at heterogeneous sites, such as voids, defects, grain boundaries, and free surfaces, which can serve as nucleation sites. Volumetric defects such as voids and micro-cracks can be healed through the formation of precipitates, which fill up the voids with atomic species. This process has been reported to result in the restoration of mechanical strength and toughness in a number of metallic materials. Also, fatigue and creep life improvements have been reported in Al–Cu alloys subjected to this self-healing treatment. Although there exists satisfactory evidence of achieving self-healing in metallic alloys via this approach, it is important to determine the optimum conditions (such as critical size and morphology of the solute precipitates, secondary aging temperature, and time) for obtaining high self-healing efficiency under static and dynamic loading conditions. The advantage of this technique lies in the fact that self-healing is achieved in situ based on microstructural modifications of the metallic alloys without any macroscopic modification [31].

10.6.1.2 Reinforcement of Metallic Matrices With Shape Memory Alloy Wires

The use of shape memory alloy (SMA) as reinforcement in metals has also been explored as a design strategy to achieve self-healing in metallic systems. This approach involves the embedding of SMA wires into a metallic matrix [32]. The SMAs are known to have two phases: a low-temperature martensitic phase, which is soft and deformable, and a high-temperature austenitic phase, which is relatively stronger. The sort of deformation that occurs in SMA is quite unlike slip, which involves permanent atomic-scale displacement. The deformation that occurs is by twinning, which is a reversible deformation process essentially by reorienting and self-accommodating the domain boundaries created by the deformation process. If a metallic material containing SMA wires undergoes damage in the form of crack formation, for instance, the crack surface will seek outright separation, which will stretch the wire. If the damaged material is heated above the transition temperature for the SMA wire, the austenitic phase forms and tries to revert back to the original size of the wire [33]. The contraction of the wire will induce compressive stresses on the cracked surfaces, which will result in some degree of crack closure, hence affecting some level of damage control (Fig. 10.6). This approach has been explored by several researchers. Recently, Rohatgi reinforced Al-A380 alloy with NiTi SMA wire; it was found that crack width reduced after self-healing treatment [34]. However, the crack could not be closed completely, and no strength was recovered due to poor interfacial bonding between the SMA wire and the matrix and the oxidation of the crack surface. It was also reported that the rate of crack closure was dependent on the dimension of the wire and not the number of the wires. Ferguson et al. reported improved healing in ZA-8 alloy reinforced with NiTi SMA wire. The damaged crack was completely closed, and about 30% strength and ductility were recovered after subjecting the damaged samples to healing treatment. The advantage of this design strategy is that macroscopic damage can be repaired without outright replacement of the damaged component, but there is a need for enhanced interfacial bonding between the SMA wires and the matrix. Also, this process is somewhat expensive considering the cost of SMA wires; hence, it becomes imperative to study the viability of low-cost SMA wires as self-healing reinforcements in metallic systems [35].

10.6.1.3 Reinforcement of Metallic Matrices with Low Melting Temperature Alloy

A low melting temperature alloy can be used as reinforcement in a metallic material with the low melting temperature alloy serving as a healing agent [36]. The incorporation of the low melting alloy can be tricky but experimentally has been achieved by infiltrating a low melting point alloy into hollow ceramic tubes. The impregnated tubes are then used to reinforce the metallic material. If a crack develops within the material, the crack path will most likely cut through the infiltrated tubes giving space for the low melting temperature material to flow by capillary pressure and surface tension when heated to the healing temperature (which is typically the melting point of the low melting point material) [37]. Once the healing temperature is attained, the solder melts and flows to fill up the damaged site and on solidification fills up the crack. Alaneme and Omusule, Oladijo et al., and a host of other researchers have demonstrated that the healing efficiency from this healing process is promising, and strength restoration of at least 60% of the original pre-damage strength is achieved via this process. This healing process must be acknowledged success based on wettability between the metallic material and the low melting temperature alloy. Aluminum alloy with $Sn_{60}Pb_{40}$ solder is one among the few metallic systems where this self-healing approach has been explored. In the work of Alaneme and Omosule, self-healing efficiency of 91% was achieved when $Sn_{60}Pb_{40}$ was used as

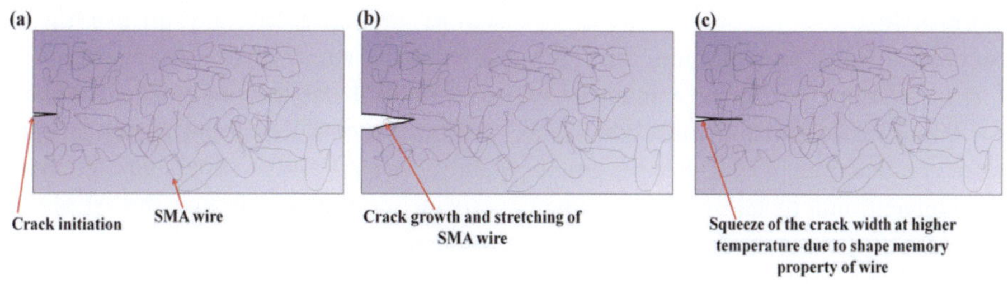

Fig. 10.6 (**a**) Crack initiation, (**b**) Crack growth and stretching of SMA wire (Deformation of wire at low temperature that is at martensite phase), and (**c**) Squeeze of the crack width at higher temperature due to shape memory property of wire (Recovery of wire at high temperature that is at austenitic phase)

low melting point healing agent [38]. However, a much lower self-healing efficiency of 61% was achieved in the work of Oladijo et al. Additionally, the healing efficiencies were obtained under different loading conditions (tensile and impact), and this would have contributed significantly to the disparities in the healing efficiencies of the different materials despite having the same healing agent [39].

10.6.2 Classification of Self-Healing Metals

Self-healing metals are of two groups defined by (1) healing of nanoscale voids in the nanometer range (precipitation and nano SMA-dispersoids concepts) and (2) healing of macro-scale cracks in the millimeter range (SMA-clamp & melt, solder tubes/capsules, coating agent, and electro-healing concepts) [40, 41]. The healing within each group occurs exclusively on the respective scale. Within the first group, only the nanoscale is accessible, and, thus, if macrocracks appear, they cannot be healed and lead eventually to fracture. That stated, because the crack coalescence process involves the interaction of spatially dispersed cracks, it can be safely proposed that these concepts would also be effective in slowing down the overall failure process. Within the second group, only the macroscale is observable and, thus, nanovoids are not healed until they grow or coalesce to form macrocracks [42]. This could be seen as a disadvantage for the service life because the presence of unhealed nanovoids would cause fast growth of secondary macrocracks, even if the first macrocrack is healed. The self-healing concepts in the nanoscale group pursue the management of nanoscale damage to prevent macroscale damage. Three of the macroscale concepts (below section) utilize liquid-assisted healing in the sense that the healing temperature must be close to the melting temperature of the solder (i.e., of the low-melting material used to "glue" the cracks together).

10.6.3 Proposed Self-Healing Concepts in Metals

10.6.3.1 High-T Precipitation

The self-healing concept using precipitation at high temperatures (high-T precipitation) is the most intensively studied concept. The original microstructure must contain a supersaturated amount of solute atoms [43]. The supersaturation can be achieved by conventional metallurgical treatment given that the phase diagram of the constituent atoms shows sufficient solubility of the solute atoms at high temperatures. Quenching from these temperatures produces a metastable, supersaturated solid solution, which upon additional aging tends to precipitate secondary phases. However, for the high-T precipitation concept to work properly, precipitation should not happen spontaneously throughout the microstructure but only in localized regions where nanovoids are present. These nanovoid forms at a grain boundary during the damage phase and acts as an attractor for the solute atoms during the healing stage and eventually as a nucleation site for the precipitation process. An important feature of the high-T precipitation concept is that the temperature during the service life must be sufficiently high [44]. An elevated temperature is required to enable lattice diffusion of the solutes toward the nanovoids as indicated by the arrows in the healing stage. As mentioned earlier, the temperature must be not too high to prevent the nucleation of precipitates at sites other than the nanovoids [45]. For example, He et al. found that the addition of B and N significantly accelerates Cu precipitation in a Fe–Cu alloy and that most open-volume defects (nanovoids) can be closed. This result was corroborated by He et al. that Cu precipitation occurs in the form of spherical nanoscale precipitates inside the grains and in the form of dislocation and interface decoration [63]. More recent investigations by the Delft group focused on replacing Cu by Au (i.e., FeAuBN). The choice of Au was motivated by an atomic-scale analysis showing that homogenous nucleation can be prevented while enhancing nucleation at the damaged nanovoids sites. It is confirmed that Au strongly precipitates at nanovoids, and creep tests showed an improved creep lifetime [46].

10.6.3.2 Low-T Precipitation

The low-T precipitation concept is closely related to the high-T precipitation concept, with the main difference being that at all stages during service the temperature can be lower. As in the case of the high-T precipitation concept, the microstructure required for the low-T precipitation concept must contain supersaturated solute atoms [47]. A distinctive and necessary feature for the healing process is that the solute atoms tend to segregate to dislocation cores. In particular, the solute atoms within the dislocation cores can be considered mobile even at low temperatures, which is a consequence of pipe diffusion. When localization of dislocations in pile-ups leads to stress concentration and formation of nanovoids, the mobile solute atoms are attracted and diffuse through the dislocations toward the stress region. Precipitation within the nanovoids eventually leads to void closure and healing of the damaged region. Potential material systems with such a self-healing ability are Al alloys supersaturated with Cu solutes [48]. A crucial requirement is that the Al alloys are prepared in an under-aged condition, that is, aging should be aborted before the peak strength of the material is reached. This condition guarantees that there are enough Cu atoms left in the solution for performing the self-healing process in the form of precipitation. Cu atoms are assumed to diffuse through dislocations

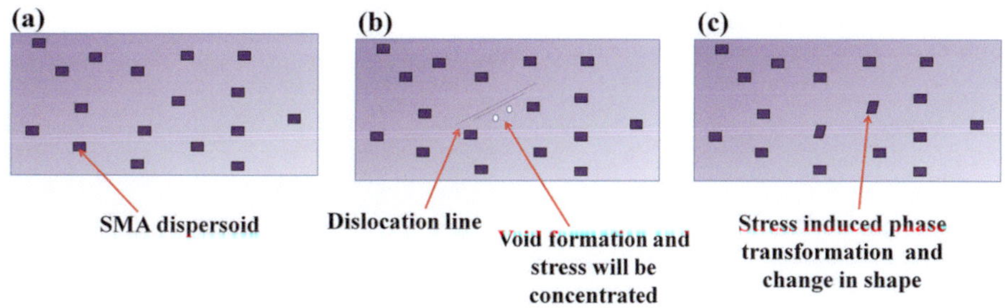

Fig. 10.7 SMA dispersoid self-healing materials

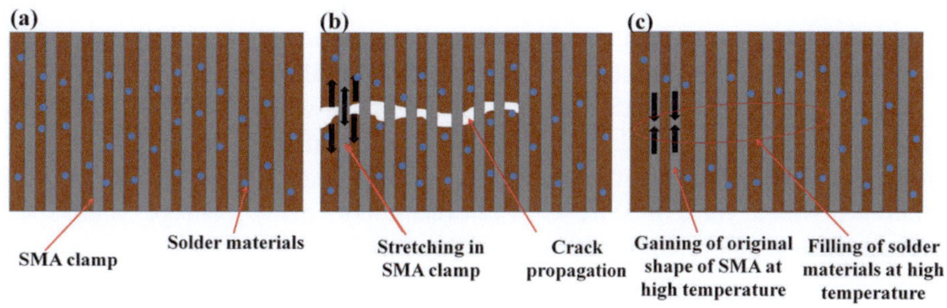

Fig. 10.8 SMA-clamp and solder materials melt self-healing method (**a**) before crack, (**b**) crack formation and SMA clamp deformation (stretching), and (**c**) after heating, shape recovery of clamp and filling of the gap by solder melt

by pipe diffusion toward the nanovoids and precipitate there, leading to void closure [49].

10.6.3.3 Nano SMA Dispersoids

Crack closure is a well-known phenomenon under cyclic loading conditions, and phase transformation is often discussed as one of the mechanisms responsible for it. The original microstructure consists of a host matrix with embedded coherent SMA nanoparticles. The nanoparticles are stabilized by the host matrix in their high-temperature phase (austenite phase) [50]. When damage is initiated in the form of dislocation localization and nanovoid formation (Fig. 10.7), the nanoparticles are activated. In particular, the stress field of the nanovoid is thought to trigger phase transformation of the SMA nanoparticle from the austenite into the martensite phase. This phase transformation is accompanied by a strong change in the shape of the particle that induces local strain fields on the host matrix and eventually leads to crack closure. A crucial stage of the described self-healing process is crack closure by the phase-transforming nanoparticle. Xu and Demkowicz investigated the migration of a stress-driven grain boundary [51]. When a stress-driven grain boundary moves toward the crack, the size of the crack is reduced, and, depending on the stress magnitude, the crack can be completely closed leaving behind only geometrically necessary dislocations. In the nano SMA dispersoids self-healing concept, the moving grain boundary is replaced by the moving interface between the host matrix and nanoparticle during its phase transformation.

10.6.3.4 SMA-Clamp and Melt

The SMA "clamp and melt" concept is possibly the oldest self-healing concept for bulk metals. Because of the wide interest, the SMA-clamp and melt concept is currently the best-investigated macroscale concept [52]. The crucial difference from the self-healing concepts discussed so far relates to the characteristic length scale that is the relevant microstructural features as well as the cracks to be healed are in the millimeter regime. In particular, a composite microstructure is required that consists of SMA reinforcement wires embedded in a solder matrix material (Fig. 10.8). The solder material is the "glue", and for that purpose, it must have a melting point that is considerably lower than that of the SMA wires [53]. When the stress applied to an SMA solder composite exceeds the ultimate tensile strength of the solder material, a crack is produced in the solder material. The SMA wires have a higher ultimate tensile strength, and they respond to the applied stress by a transformation into the martensite phase. To achieve self-healing, the composite sample needs to be externally heated to a temperature above the austenite transition temperature. The increase in temperature leads to phase transformation of the SMA wires from the martensite to austenite phase. The transition is accompanied by compressive stress that contracts the composite sample, bringing the cracked surfaces together. The temperature needs to be further increased toward the melting point of the solder. Once the cracked surfaces have started to melt, they can rejoin and the temperature can be reduced to the original, low value [54]. A unique advantage of this self-healing concept is that it can be repeated limitlessly. This is indicated by the fact that the original structure is exactly the same as the healed structure. A marked disadvantage is the strong anisotropic response and requirement of the externally applied triggering process. In the case of the SMA-clamp and melt concept, the external trigger is the heat transfer required to transform the SMA wires and melt the solder. This renders

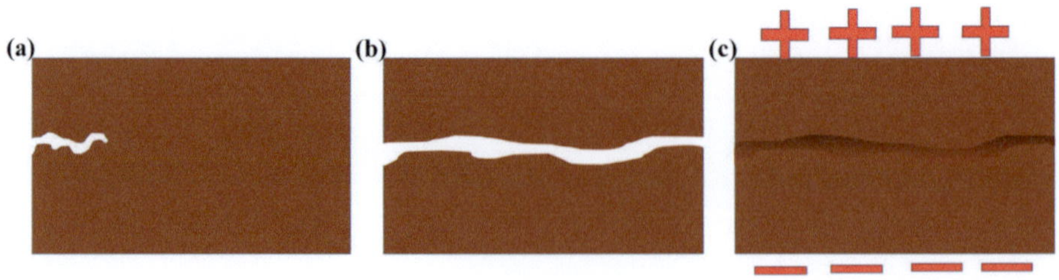

Fig. 10.9 Electro healing process of self-healing materials

the process non-autonomous such that possible industrial applications would require routine service intervals [55]. Ferguson et al. investigated the self-healing properties of a commercial ZnAlCu alloy enforced with TiNi SMA wires. In another study, NASA tested Al-based alloys with the focus on improving the damage tolerance of aeronautical structures. The authors could show that an AlSi matrix reinforced with 2 vol% of SMA wires can retain more than 90% ultimate tensile strength after self-healing [56].

10.6.3.5 Solder Tubes/Capsules

The solder tubes/capsules concept is to encapsulate a solder material inside ceramic capsules or ceramic tubes, inside a host matrix of a higher melting material. The role of the solder is critically modified in the solder tubes/capsules concept [57]. For the SMA-clamp and melt concept, the solder constitutes the host matrix, and the crack to be healed appears in the solder itself (i.e., it is the ultimate tensile strength of the solder that determines crack initiation). For the solder tubes/capsules concept, the host matrix is composed of a different, high-melting material with an ultimate tensile strength that can be significantly larger than that of the solder. The solder is activated only when a crack in the host matrix has formed. Activation of the solder is achieved by increasing the temperature above the melting temperature of the solder. After activation, the solder wets the crack surfaces, fills the crack as a result of capillary pressure and surface tension, and solidifies, thereby closing the crack. Despite the conceptual analogy to the successful self-healing concept in polymers, the solder tubes/capsules concept applied to metals involves many complications. Designing the original microstructure is difficult when solder capsules are used. The ceramic capsules need to contain holes so that the solder can be filled in. Three or even more critical problems with the solder tubes/capsules concept occur during the damage and healing phases. First, for the solder to have any effect, the crack must not only hit a capsule but must also destroy the ceramic shell such that the solder can escape. This condition is not easily fulfilled because the crack can spread along the interface of the host matrix and the ceramic shell. Second, even if the first condition is fulfilled and the solder can be activated to flow into the crack, the solder must wet the crack properly and, more importantly, bind strongly to the crack surface. Third, after the healing process, new voids are created inside the bulk, possibly leading to additional weakening of the sample. Practically, the described conditions appear too difficult to be fulfilled in metallic systems [58].

10.6.3.6 Coating Agent

The main concept of a coating for self-healing is the application of solder as the self-healing agent. It provides the present concept similar to the previously discussed macroscale concepts. To trigger self-healing with the solder, a heat treatment is required. After the application of the heat treatment, the crack should be filled with solder as in the case of the solder tubes/capsules concept. A major requirement is to develop good wetting of the crack surfaces and strong bonding between the host matrix and solder. An advantage of the coating agent concept over the solder tubes/capsules concept is that no voids occur after healing. It is only the solder coating that is locally reduced. Leser et al. chose to carry out a proof-of-principle study of the coating agent concept on a Ti alloy with a $In_{60}Sn_{40}$ (composition in wt%) coating. The melting temperature of this coating material is 124 °C. Here, a low melting temperature is not only crucial for the healing heat treatment but also for avoiding over-aging of the base material in the process [59].

10.6.3.7 Electro-Healing

The electro-healing concept has been developed by Zheng et al. who were inspired by a partially successful, electro-pulsing-based healing of microcracks, as shown in Fig. 10.9. The electro-healing concept is a macroscale concept that differs quite significantly from the other macroscale concepts [59]. In particular, no composite matrix material and no solder as healing agents are required. Instead, the cracked sample needs to be immersed in an electrolyte solution and a voltage needs to be applied. The subsequent electrochemical reaction leads to the deposition of material inside the crack, eventually closing it. The electro-healing concept is very appealing for several reasons: (1) the original microstructure needs no modification, (2) experience from the field of electrochemical metallic coatings can be utilized for design and optimization purposes, and (3) very strong bonding between the crack surfaces and the newly deposited material can be achieved, enabling good material properties in the healed stage [60]. A present limitation is that a through-thickness crack is required for the self-healing process to work.

10.7 Self-Healing Ceramics

In self-healing ceramics, the ceramic can repair due to the reaction between oxygen molecules, which enter a crack, and silicon carbide in the ceramic [61]. This reaction forms silicon dioxide, which reacts with another ceramic component, alumina, to form a material that fills the gap and crystallizes into a hardened form. Research conducted in 2018 found that adding trace amounts of manganese oxide to alumina grains allowed the ceramic to heal in under 60 s at 1000 °C [62]. This was noted as the operating temperature of the aircraft engines where this ceramic might be used as turbine material. It is a non-autonomic healing process.

Recently, the autogenous self-healing of hydraulic concrete materials has been the object of intensive scientific and application-oriented research. In damaged concrete, water permeates through the cracks and induces the local formation of hydrates which lead, depending on water pressure, pH-value, crack morphology and crack width, to a detectable crack healing [88]. The crack width plays an important role in the self-healing process. For this reason, Engineered Cementitious Composites (ECC) have been developed that control the crack width to values below 60 μm even for very high strains well above the strain leading to the first observable crack. In addition, a bio-inspired approach to self-healing concrete is being investigated: Bacteria are immobilized in the concrete and will be activated if water permeates into fresh cracks, where they start to precipitate minerals. In contrast, little is known about the regeneration processes in cracked engineered ceramic materials. Thus, the maximum crack length and, more importantly, the width of the mendable cracks, the optimal conditions (preferably low temperatures) as well as the local heat transport properties which change with crack growth and shrinkage, still remain as open questions. With most current ceramics, very high local temperatures are required for repair due to the high activation energies of the diffusive mass transport in the covalent or ionic structures of ceramics (e.g., $SiC/Al_6Si_2O_{13}$ composite at 1300 °C). On the other hand, it could be shown that Al_2O_3- and $Al_6Si_2O_{13}$ reinforced by SiC-whiskers can completely heal cracks up to a length ~ 100 μm [90]. The extensive regeneration of the original strength could be demonstrated in SiC- and Al_2O_3 (sintering temperature 1400 to 1600 °C). Considerably lower reaction temperatures can be expected if the grain boundary is wetted by an intergranular glass phase. Depending on the glass transition temperature T_g (Na-Ca silicate glass 500 °C, and SiO_2 1100 °C), stress relaxation processes, viscoelastic flow processes, and diffusion can be initiated at considerably lower activation temperatures [91]. The healing reaction starts with grain boundary and surface diffusion at contact points in the region of crack faces. Further energy supply results in the formation of spherical and cylindrical pores that can collapse at high temperatures provided they have small radii. The driving force is the reduction in surface energy (*Rayleigh instability*). In addition, the oxidation of nonoxidic phases in the oxidizing atmosphere is the another effective healing mechanism [63].

Furthermore, in multi-component and multi-phase ceramic materials, the formation of a eutectic melt, as well as the local particle rearrangement induced by a phase transition, is considered a promising healing mechanism. While most of the work in the field of ceramics focuses on crack healing processes at temperatures above the working temperature, few efforts have been made on healing below these working conditions. As an example, a glass phase containing Si_3N_4/SiC already features significant healing at 1000 °C (also under cyclic stress), which leads to a distinct increase in the static and dynamic fatigue strength. Further investigations on this material have demonstrated a significant influence on the stress frequency [64].

10.8 Self-Healing Polymers

In the last century, polymers became a base material in everyday life for products such as plastics, rubbers, films, fibers, or paints. This huge demand has forced them to extend their reliability and maximum lifetime, and a new design class of polymeric materials that are able to restore their functionality after damage or fatigue was envisaged [65]. These polymer materials can be divided into two different groups based on the approach to the self-healing mechanism: intrinsic or extrinsic. Autonomous self-healing polymers follow a three-step process very similar to that of a biological response. In the event of damage, the first response is triggering or actuation, which happens almost immediately after damage is sustained. The second response is the transport of materials to the affected area, which also happens very quickly. The third response is the chemical repair process. This process differs depending on the type of healing mechanism that is in place (e.g., polymerization, entanglement, and reversible cross-linking) [66]. These self-healing materials can be classified in three different ways: capsule-based, vascular, and intrinsic [67]. Various self-healing polymer processes are described subsequently:

10.8.1 Mechanically Triggered Self-Healing

These self-healing materials were those healed in response to a mechanical trigger. Mechanically triggered healing materials contain (1) a resin (e.g., epoxy) and a hardener (e.g., diamine) that are contained within adjacent fibers or

(2) one component rests within the fibers and the other component is microencapsulated and/or embedded within the polymer matrix [68]. White et al. developed self-healing polymers containing a catalyst, originally **Grubbs' catalyst**, which in later systems was encapsulated, and microcapsules that were filled with a healing agent, typically dicyclopentadiene. Microcapsule-polymer composite system parameters, such as fiber or capsule wall thickness, fiber diameter, capsule size and concentration, healing agent, catalyst type, and catalyst incorporation method, have all been well studied to optimize the release of the healing agent, the mixing of the healing agent with the catalyst (where relevant), and the rate of polymerization/curing following a mechanical trigger. In general, these systems have been moving away from air-sensitive and expensive components, such as **Grubbs' catalyst**. For instance, a relatively inexpensive, stable, tin-based catalyst has been utilized to catalyze the polycondensation of siloxane-based macromonomers following the mechanically triggered release of these monomers from microspheres. A careful selection of microcapsule material, size, and concentration has permitted the preparation of materials with increased toughness of the pristine/undamaged polymer composite and enhanced healing efficiency and, therefore, increased resistance to bulk material crack formation. Compared to many other self-healing material designs, those that employ a microencapsulated repairing agent are attractive because they allow for autonomous healing, they can be adapted to a number of different polymers (with differing material properties), and they have the potential for quantitative healing [69].

10.8.2 Ballistic Impact Self-Healing

Perhaps the most provocative self-repair experiment was the observation by R. Fall et al. and Kalista et al. that bullet holes in a plastic plate would heal up instantly. The high-velocity projectile could penetrate the polymer plate and the holes would reseal faster than the eye could see [70]. In several poly(ethylene-co-methacrylic acid) copolymers (EMMA), projectile puncture has been investigated [71]. The polymer contained about 5.4 mol% methacrylic acid groups, which had been partially (60%) neutralized by sodium and had a melting point of 93 °C. The central damage zone is about the size of the bullet cross-section but has been healed. It is found that the ionic content of the polymer was not important for healing and that self-repair occurred by a two-stage process: Stage-1 involved melt elastic recovery followed by Stage-2 with sealing and polymer-chain interdiffusion [72]. Kalista et al. conclude that the impact energy of the projectile was sufficient to melt the polymer in the damage zone and melt recoil followed by interdiffusion promoted self-repair. A minimum balance analysis of this process would give the temperature rise, $\Delta T = T(\text{impact}) - T(\text{sample})$, in the damage zone of radius R and mass m as;

$$\Delta T = \frac{\Delta U - m\Delta H_{\mathrm{f}}}{mC_p}, \qquad (10.2)$$

where $\Delta U = \frac{1}{2}m\left({V_1}^2 - {V_2}^2\right)$ is the energy dissipated with impact velocity V_1 and exit velocity V_2, ΔH_f is the heat of fusion to melt the polymer, and C_p is the heat capacity. The mass of the damage zone is $m = \rho\pi R^2\text{h}$. The impact velocity gives energy of $U_1 = 0.9\ J$, $R = 2$ mm, $h = 1$ mm, $C_p \approx 2.7\text{JK}^{-1}\text{g}^{-1}$, $\Delta H_{\mathrm{f}} = 429\ \text{Jg}^{-1}$(PE value) and $\rho \approx 1\ gcc^{-1}$. These values give $\Delta T \approx 100^{\circ}\text{C}$ in the damage zone, which is sufficient for melting and rehealing with $T_m = 92^{\circ}\text{C}$ and $T(\text{sample}) = 22\ ^{\circ}\text{C}$. However, at $-30\ ^{\circ}\text{C}$, this is not sufficient energy for melting [106]. Also, if the mass of the damage zone increases with increasing temperature, the numerator in the above relation goes rapidly to zero and healing will not occur, as observed by Kalista et al. In related experiments, Kalista et al. demonstrated that the same polymer plates also self-repair when cut with a saw, since the friction generated by the saw is sufficient to thermally weld the surfaces together. However, the polymer plate when cut with a very sharp razor blade did not self-heal due to insufficient mechanical energy dissipation at the crack interface. Room temperature projectile testing with LDPE films showed no tendency for self-repair. In general, one should be able to design most thermoplastics for ballistic self-healing. These could find unique applications in space-capsule protection against micrometeorites. In the Kalista ballistic-healing experiments, self-healing is obtained by transforming "hard matter" to "soft matter" due to mechanical action [108]. Decker et al. have shown that ballistic impact resistance can be obtained by transforming a liquid into a solid through the use of shear-thickening fluids (STF). The STF consists of a colloidal particle suspension that percolates rigidity under high deformation rates and has become the basis for the invention of "liquid armor." The advantage of this system is that after impact, the instantly rigidized matter returns to the liquid state and rapidly heals itself by restoring the local concentration of particles. The STF suspensions when mixed with Kevlar fibers have shown remarkable stab resistance when subjected to repeated high-velocity trauma from sharp objects such as knives and ice picks [73].

10.8.3 Thermally Triggered Self-Healing

There are many types of dynamic bonds that have been incorporated into materials to render them intrinsically self-healing. With only a few exceptions, materials based on non-covalent reversibly formed bonds tend to heal without any addition of energy following mechanical damage/stress, whereas ruptured covalent dynamic bonds require the input of some form of energy for healing (bond reformation) to occur [74]. Although the self-healing systems based on covalent bonds would no longer be capable of autonomous healing, these materials would benefit from improved mechanical properties, which may be necessary for certain applications. For this reason, many material chemists have invested time into designing self-healing materials from thermally controllable covalent dynamic bonds, such as bonds formed from Diels–Alder (DA) cycloaddition, disulfide bonds, alkoxyamine bonds, and imine bonds. DA chemistry is a highly promising technique for generating polymeric systems with thermally triggered self-healing. A few different strategies for generating self-healing materials from DA polymer networks have been investigated. One strategy consists of polymerizing di- or multifunctional monomers (e.g., a bis- or multifuran functionalized monomer with a bis- or multimaleimide functionalized monomer) [75]. Chen et al. were among the first to utilize this technique to generate a cross-linked material from a furan tetra-linker (diene) and a maleimide trilinker (dienophile), which healed above 120 °C. Another strategy that has been explored for creating self-healing DA-cross-linked polymers is that which involves cross-linking a linear thermoplastic bearing pendant furan or maleimide groups with either a di-linker or another linear polymer with pendant functional groups [76].

10.8.4 Optically Triggered Healing

Light can act as external stimuli to control dynamic bond alteration to trigger the healing process in the materials. The first optically triggered self-healing systems were based on reversible cycloaddition reactions, which could be cycled back and forth to open and close cross-links in a polymer network [77]. Typical examples for such optically triggered, reversible reactions are the [2 + 2] cycloaddition of coumarin and cinnamoyl groups as well as the [4 + 4] cycloaddition of anthracene. The latter type of pericyclic reaction has become more important to chemists due to the longer wavelength at which the forward and backward dynamic bond reactions proceed. Anthracene cycloaddition occurs under irradiation at 340 nm, which is a wavelength that is tolerated by most functional groups and is even close to the wavelength of sunlight. As a result, an anthracene-based system is potentially attractive for its application to environmentally exposed surfaces. Although the linkages formed by the cycloaddition of cumarins, cinnamic acids, or anthracenes are hypothesized to be the first to cleave under mechanical stress/damage (in which case material healing can be achieved after the material is exposed to the relevant wavelength), if one desires to controllably cleave the network to allow for polymer network rearrangement, a different wavelength is required than that needed to reform/heal the network [78]. Metathesis is the key mechanism involved in the healing of this type of system [79]. Exposing bonds that are susceptible to metathesis to irradiation results in the homolytic cleavage of these bonds, yielding highly reactive radicals, which subsequently recombine. Through this dynamic bond exchange, the network is able to accommodate and repair damage and, therefore, mend itself. Examples of bonds that can undergo light-triggered metathesis are disulfide bonds and allyl sulfide bonds. Fortunately, allyl sulfide groups are preserved in the polymer network regardless of whether the photoinitator has been consumed, which means that after adding more photoinitiators, self-healing is again possible [80, 81].

10.8.5 Other Methods for Triggering Healing

Many intrinsically self-healing materials are designed to undergo reversible bond formation in the presence of a chemical trigger. Specifically, most of these reversible systems are sensitive to pH value or the presence of redox agents. Attractive examples of pH-sensitive systems are those based on boronic-acid-salicylhydroxy complexation, hydrazone formation of acylhydrazines, and crown ether–benzylammonium salt ionic interactions [82, 83]. A typical example of redox agent-triggered self-healing systems is that which incorporates disulfide linkages. It is no surprise that polymers containing disulfide linkages have become increasingly attractive for self-healing material applications, given the fact that healing can be stimulated and controlled by redox agents, in addition to heat and irradiation [84]. Some self-healing materials have been designed to heal when electrochemically stimulated. For instance, conducting polymers have been used to coat metal surfaces to protect them from corrosion by releasing healing ions and reforming the damaged coating at the surface following damage. In addition, another electrically conductive material, N-heterocyclic carbenes (NHCs), has been found to heal when a crack is formed after a current is passed through the system [85].

10.8.6 Stages of Passive Self-Healing in Polymer

Repair of cracks and microscopic damage has been explained based on the following stages: (a) surface rearrangement; (b) surface approach; (c) wetting; (d) diffusion; and (e) randomization. These are discussed with respect to self-healing materials in the following sections [86].

Surface rearrangement: When the freshly damaged surfaces or microvoids are created in fracture or fatigue, one should consider the roughness or topography of the surface and how it changes with time, temperature, and pressure following contact with the healing fluid. In fractured polymers, rearrangement of fibrillar morphology and other factors affect the rate of crack healing. Chain-end distributions near the surface can change as molecules diffuse back into the bulk. If the chain ends are needed for reaction with the fluid, they could be designed to preferentially migrate to the surface using lower surface tension moieties on the chain ends. Spatial changes in the molecular weight distribution can also occur, for example, where the low-molecular-weight species preferentially migrate to the surface. Nanoparticles in the bulk could preferentially migrate into nanovoids. In time-release solvents or adhesives, surface rearrangement is affected by the polymer–solvent interaction. Chemical reactions, for example, oxidation and cross-linking, can occur on the surface and complicate the dynamics of diffusion. Solvents used in the passive healing experiments could also cause additional damage, for example, by causing crazes and microvoids to swell and allow them to propagate further. Each material and self-repair release technique possess unique surface rearrangement processes that may need to be quantified. The use of solvent systems that promote surface segregation of chain ends, for example, would be highly conducive to rapid healing of damage. The relation between solvent concentration and surface rearrangement dynamics needs to be quantified, primarily through the effect of solvent, on both the glass transition temperature T_g and the relaxation times of the surface molecules [87].

Surface approach: This stage is the most critical for self-healing materials. Simply, no healing occurs if the surfaces are not brought together or the gap is not filled with the healing fluid. Thus, any debris left over from the damage process could pry the surface apart to prevent surface approach and terminate the self-healing process. The surface approach applies to crack surfaces that are brought together to heal alone or in the presence of the healing fluid. This stage of healing considers the time-dependent contact of the different parts of the surfaces to create the interface. The surface approach may be especially important in composites where the nature of the damage can involve the polymer matrix, fibers, and the matrix–fiber interface. If the healing fluid only bonds to one surface, then little or no healing will occur. The healing solvent may also force the surfaces together by the pressure of swelling [88].

Wetting: When the damaged surfaces approach, they need to wet each other and form an interface before the healing process can continue. With self-repair fluids, the wetting and compatibility of the damaged surface by the fluid must also be considered. The capillary action is required to fill the voids in wetting and spreading of a fluid on a surface. Wettability on the surfaces by the fluid can be determined as part of the self-healing design. Wetting can occur in a time-dependent fashion at the interface. Typically, wetted pools are nucleated at random locations at the interface and propagate radially until coalescence and complete wetting are obtained [89].

Diffusion stage: The diffusion stage is the most critical stage of strength development for both crack healing and self-healing systems at polymer–polymer interfaces. It is a dominant stage in the ballistic self-healing experiments when the elastically recoiling melt from the penetration hole rapidly wets and interdiffuses to make a good seal across the bullet hole. When the local stress exceeds the yield stress, the deformation zone forms, and the oriented craze fibrils consist of mixtures of fully entangled matrix chains and partially interpenetrated minor chains. Fracture of the weld occurs by disentanglement of the minor chains or by bond rupture. It is interesting to note that if the stress rises to the point where random bond rupture in the network begins to dominate the deformation mechanism, instead of disentanglement, then the weld will appear to be fully healed, regardless of the extent of interdiffusion [90].

10.8.7 Damage and Healing Theories

10.8.7.1 Percolation Theory of Damage and Healing

This approach evaluating the fracture energy (G_{IC}) of a healing interface of A-B polymers is represented in Fig. 10.10. In this double cantilever beam (DCB) system, a crack propagates through the interface region preceded by a deformation zone at the crack tip [91]. For cohesive failure, the fracture energy can be determined by the J-integral method, as described by Hutchinson and Tvergaard [92], where G_{IC} is the integral of the traction stresses with crack opening displacements (d), in the cohesive zone, following yielding at a local yield or craze stress σ_Y. The cohesive zone at the crack tip breaks down by a percolation process, as described herein, at a maximum stress value, $\sigma_m > \sigma_Y$. Typical ratios of σ_m/σ_Y are about 4–5. The yield stress dominates the fracture process for non-crazing matrices such as thermosets, and this is determined by the twinkling fractal theory (TFT) [93]. Both σ_m and δ_m are rate dependent and in the simplest case, the fracture energy is determined by:

$$G_{IC} = \sigma_m \delta_m \tag{10.3}$$

where δ_m is the critical crack opening displacement. Both σ_m and δ_m depend on the damage-zone structure and the microscopic deformation mechanisms controlling the percolation fracture process via disentanglement and bond rupture. To convert these percolation concepts into quantitative fracture terms for healing processes. This experiment can be used to interpret fracture and healing on any 2D or 3D lattice with initial tensile modulus E and subject to random bond fracture. Random bond scission causes the formation of microvoids, which coalesce into larger voids and facilitate a macroscopic crack propagating through the net at the percolation threshold. The Hamiltonian for the stored elastic energy can be formulated using the Kantor and Webman approach for specific lattices, the Born and Huang method, or using the simple engineering strain energy density approach as follows [94]. The stored elastic strain energy density U, (energy per unit volume), in the lattice due to an applied uniaxial stress σ is determined by Eq. ((10.4).

$$U = \frac{\sigma^2}{2E} \tag{10.4}$$

The stored strain energy dissipation per unit volume U_r to fracture a network consisting of ν bonds per unit volume is

$$U_f = \nu D_0[p - P_c] \tag{10.5}$$

where D_0 is the bond fracture energy and $[p - P_c]$ is the percolation fraction of bonds that must be broken to cause a fracture in the network. In this approach, the strain energy U is first stored in the net and inquire if this energy is sufficient to break $\nu[p - P_c]$ bonds per unit volume when it releases at a certain strain energy density $U^* = \frac{\sigma^{*2}}{2E}$ such that at the critical condition,

$$U^* \geq U_f \tag{10.6}$$

Substituting for U^**and* U_f in Eq. ((10.7) and solving for the critical stress σ^*, we obtain the net solution for the critical fracture stress as,

$$\sigma^* = \{2E\nu D_0[p - P_c]\}^{1/2} \tag{10.7}$$

This relation was found to be applicable to damage and healing events in carbon nanotubes. Eq. ((10.7) is applied to a range of polymer materials where the entanglement density is $\nu = \frac{\rho}{M_c}$. Equation ((10.7) predicts that the fracture stress increases with the square root of the bond density $\nu \tilde{\frac{1}{M_c}}$. The percolation parameter p, is in effect, the normalized bond density such that for a perfect net without defects, $p = 1$, and for a net that is damaged or contains missing bonds, then $p < 1$. Obviously, as p approaches P_c, the fracture stress decreases toward zero and we have a very fragile material. The fatigue lifetime is also calculated using this concept and is presented in a later section. This fracture relation could therefore be used to evaluate durability, fatigue damage accumulation, healing processes, or retention strength of a material by tracking damage through a single parameter p. For thermosets, p is related to the extent of reaction of the cross-link groups and this could be critical in the fiber-matrix interface of composites.

The time dependence of healing $R(t)$ can be described from Eq. ((10.7) as:

$$R(t) = \frac{\sigma(t)}{\sigma_\infty} = \left[\frac{(p(t) - p_c)}{(1 - p_c)}\right]^{1/2} \tag{10.8}$$

where σ_∞ is the virgin strength with $p = 1$ and $p(t)$ is the time-dependent bond fraction recovery. For example, if a diffusion process controls $p \sim t^{1/2}$, then $R \sim t^{1/4}$, as observed for crack healing. The latter equation is similar in some respects to the generic kinetic healing rate equation we proposed for mechanical recovery in solid rocket propellants, filled elastomers, block copolymers, and hard elastic polypropylene:

$$R(t) = 1 - \frac{[1 - R_0]}{[1 + K_t]^\alpha} \tag{10.9}$$

Here R_0 is an instantaneous recovery fraction, which is dependent on the initial extent of the damage, K is a temperature-dependent parameter, and α is a rate constant. This empirical relation gives $R = R_0$ at $t = 0$ and $R = 1$ as t goes to infinity and was found to adequately describe many healing processes.

10.8.7.2 Fracture and Healing by Bond Rupture and Repair

The percolation theory predicts that fracture by bond rupture of linear polymers is in accord with Flory's suggestion of the chain-end effect, via, $P = 1 - \frac{M_c}{M}$ and $P_c = 1 - \frac{M_e}{M_c}$ such that:

$$\frac{G}{G_\infty} = \frac{\left[1 - \frac{M_c}{M(t)}\right]}{\left[1 - \frac{M_c}{M_\infty}\right]} \tag{10.10}$$

Thus, healing in this system involves repairing the molecular weight from Mc back to its original value M_∞. This could be done using a liquid monomer, which reacts with the catalyst to make new chains in a crack zone, as suggested by White et al. and Wudl et al. [95].

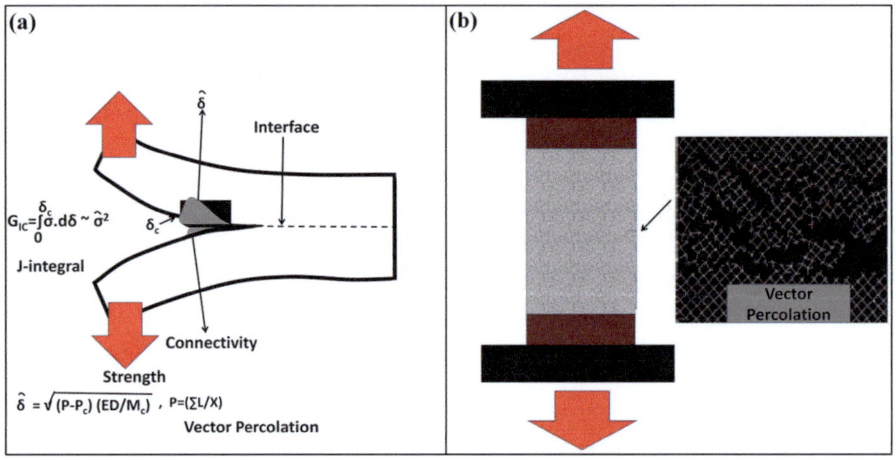

Fig. 10.10 The microscopic entanglement structure, e.g., at an interface or in the bulk, is related to the measured macroscopic fracture energy G_{Ic} via the RP theory of breaking connectivity in the embedded plastic zone (EPZ) at the crack tip. The RP theory determines σ_{max} in the EPZ, which is related to G_{Ic} via Hutchinson's J-integral theory. The percolation parameter p, which is a measure of damage and healing, is related to the interface molecular structure, where S is the number of chains of length L in an interface of width X. (**a**) The net of initial modulus, E, is stressed in uniaxial tension to a stress σ. (**b**) Release of stored strain energy by a random bond fracture in the net results in a percolating system near the fracture threshold and a very broad distribution of stress on the bonds

10.8.7.3 Fracture and Healing of an Ideal Rubber

For an ideal rubber, $p = 1$, the modulus $E = \nu kT$ where ν is the cross-link density. Thus, Eq. ((10.7) predicts:

$$\frac{\sigma(t)}{\sigma_\infty} = \frac{E(t)}{E_\infty} \tag{10.11}$$

$$\frac{G(t)}{G_\infty} = \frac{\nu(t)}{\nu_\infty} \tag{10.12}$$

An ideal rubber with a perfect lattice was essentially unbreakable at low strains ($\lambda < 4$) since insufficient strain energy could be stored to break the requisite percolation number of bonds, but fracture could occur by strain hardening to produce a tenfold increase in modulus E, by introducing defects ($P = P_c$) via fatigue or radiation, by decreasing the bond energy D_0 an order of magnitude, or by raising the temperature to 3000 K. Alternatively, by making the cross-links through the use of partially intercalated nanoclays, damage occurs by opening the nanobeams of the clay and reducing the cross-link density. However, this system can self-repair through diffusion and intercalation to restore the original cross-link density, as noted by Zhu and Wool [96].

10.8.7.4 Fracture and Healing of Thermosets

With highly cross-linked materials such as typical unsaturated polyesters and epoxies used in composite materials, the percolation theory predicts:

$$\frac{\sigma(t)}{\sigma_\infty} \sim \left[\frac{\nu(t)}{\nu_\infty}\right]^{1/2} \tag{10.13}$$

The repair of the cross-link density $\nu(t)$ can be done by chemical means, as suggested by White et al., Dry and Sottos, and Chen et al. for thermally remendable cross-linked materials. Raghavan and Wool examined a matrix of possibilities for healing and repair in thermoset polymers. Here we examined a compact tension (CT) fracture mechanics specimen made with bottom part A and top part B with the A–B interface along the crack plane. The A–B interfaces considered were as follows [97, 98]:

1. **Liquid–liquid thermoset (L-L):** This is the virgin state for reference after curing the vinyl ester (Dow Derakane 411-C50) by free-radical polymerization using an organic peroxide initiator (USP 245). For reference, the virgin L-L thermoset had a fracture energy $G_{IC} \approx 500$ J/m^2.
2. **Liquid–solid virgin interface (L-S$_v$):** The liquid was cured on a virgin solid made from the same liquid. This is important in composite repair and recycling processes. This interface surprisingly only achieved values of $G_{IC} \approx 15 - 200$ J/m^2, which is substantially less than the virgin state. Many different forms of cure were attempted, but the result was always the same, a procured solid sample does not bond with 100% strength to its own liquid, which could be a major problem in the design of autonomic self-healing materials. Perhaps a chain-transfer agent would have helped to promote additional connectivity between the polymerizing liquid and the solid cross-linked material.

3. **Solid virgin-solid virgin interface (S_v-S_v):** This was a welding experiment with the two solid virgin surfaces. This interface showed some strength development when welded above T_g and $G_{IC} = 66 - 107$ J/m^2. The best results were obtained with the lowest cross-link density. This interface clearly can benefit from the release of self-healing fluids.
4. **Solid virgin-solid fracture interface (S_v-S_F)**: The fractured surface was welded with a virgin surface. No strength was developed after welding above T_g, as might be expected from the very difficult wetting stage involving the rough fracture surface and the smooth virgin cured surface.
5. **Solid fracture-solid fracture interface (SF-SF):** This is a crack healing experiment, and little strength was developed at healing temperatures above T_g with $G_{IC} = 8$ J/m^2. Below T_g, healing fluids would have to be released into the interface.
6. **Liquid-solid fracture interface (L-SF):** The fractured solid is cured in contact with its own liquid. This is important for autonomic healing processes, where the healing liquid in the hollow fibers or UF spheres is similar to the matrix. This interface behaved like the L-S_v interface and only a small fraction of the strength was recovered, again raising a warning signal for autonomic self-healing design in highly cross-linked thermosets. Again, it should be possible to improve on this healing mechanism with free-radical transfer agents.
7. **Liquid-solid fracture, chemically treated interface (L-S_{FC}):** This interface involves first treating the fracture surface with a chemical treatment, such as styrene monomer which would polymerize along with the liquid. This is important in repair and recycling. The in situ polymerization of styrene gave the best results with $G_{IC} = 320$ J/m^2, which is still considerably less than the virgin state (500 J/m^2). With increasing molecular weight of the PS samples, the repair strength decreased.
8. **Solid virgin with solid fractured chemically treated interface (Sv-SFC):** This interface is relevant to welding and repair. Superglue gave a repair strength of 107 J/m^2, which is 22% of the virgin state.
9. **Solid fracture-solid fracture chemically treated interface (SF-SFC):** This interface is important in wear and repair. Typical values with adhesives (superglue) gave strengths $\approx$100 J/m^2. This interface resembles that used in the autonomic healing experiments of White and Sottos where the healing fluids are contained in the UF spheres and released to the interface when damage occurs.
10. **Solid fracture chemically treated–solid fracture chemically treated interface (SFC-SFC):** The results were similar to the chemical treatments giving some of the best results with fracture energies near 300 J/m^2. One of the chemical treatments consisted of using 2 wt% solutions of polystyrene in toluene. The PS molecular weights were 220, 596, and 4340 kDa. The idea was that connectivity between the solid interfaces could be achieved by diffusion of the PS chains into the cross-linked network, which would have been swollen by the toluene solvent. This result is encouraging for the design of solvent-healing systems.

10.8.8 Healing of Polymer–Polymer Interfaces

Self-healing of thermoplastic polymer–polymer interfaces occurs during ballistic impact, solvent bonding, and sub-T_g surface welding. For healing of symmetric A–A interfaces, the gradient percolation width X is the average monomer interdiffusion distance, the percolation parameter $P = \frac{\sum L}{X}$, where $\sum$is the areal density of minor chains of interdiffused contour length L, and the percolation threshold $P \approx \frac{L_c}{M} \approx 0$. In the general healing case where the chain ends are randomly distributed in space near the interface, $\frac{\sum L}{X} \frac{\tilde{1}}{M}$, L~ (t/M)$^{1/2}$, then the fracture energy increases as $G \approx L$

$$\frac{G(t)}{G_\infty} = \left[\frac{t}{\tau^*}\right]^{1/2} \tag{10.14}$$

where the weld time τ^*~M, when $M > M^*$, and τ~M^3, and when $M < M^*$. Typically, $M^* \approx 8M_c$. When the chain ends are segregated to the surface, which could occur due to fracture or the presence of low surface energy chain ends, $\sum$is constant with time since all the chains begin diffusing at the same time and we obtain:

$$\frac{G(t)}{G_\infty} = \left[\frac{t}{\tau^*}\right]^{1/4} \tag{10.15}$$

where again, τ^*~M. This allows healing to occur faster compared to Eq. (10.14. In balastic self-repair experiment, the time τ^* is temperature dependent and the needed time for healing competes with the cool-down process [99]. When using healing fluids with monomers that are different than the matrix monomers, one can encounter an asymmetric, potentially incompatible interface. For asymmetric incompatible A–B interfaces of width d~$\chi^{-1/2}$, where χ is the Flory–Huggins interaction parameter, we have again the percolation parameter for the diffuse interface as P=$\sum$$L/X$. In this case, X~d, L~d^2 and $\sum$is constant such that P~d. Since the fracture energy G ~[d-dc], where dc is the tube diameter, then the fracture energy depends on the normalized width w = d/dc as,

$$\frac{G(t)}{G_\infty} = \left[\frac{w(t) - 1}{w_\infty - 1}\right] \quad (10.16)$$

The latter relation is supported by experimental data on a wide range of A–B interfaces, which were analyzed by Benkoski et al. and Cole et al.

Incompatible A–B interfaces are typically quite weak compared to welded homopolymers. To make such interfaces stronger, they can be reinforced with A–B co-polymer compatiblizer chains, as demonstrated by Brown and co-workers [100]. Thus, one could incorporate compatibilizers into the healing fluid, which when released would have sufficient mobility, for example, in a solvent, to heal the interface.

10.8.9 Fatigue Healing

For healing of thermoplastic interfaces, the total interpenetration of chains (X approaches Rg) is not necessary to achieve complete strength when $M > M^*$ and $t^* < T_r$. It is only necessary to diffuse a distance equivalent to the radius of gyration of M^*. However, a word of caution: while complete strength may be obtained in terms of critical fracture measures such as G_{1c} and K_{1c}, the durability, measured in sub-critical fracture terms, such as the fatigue crack propagation rate da/dN, may be very far from its fully healed state at t^*. We have shown that while the weld toughness K_{1c} increases linearly with interdiffusion depth X, as $K_{Ic} \sim X$, the fatigue crack propagation behavior of partially healed welds behaves as [101]:

$$\frac{da}{dN} \sim X^{-5} \quad (10.17)$$

It is a very strong function of interdiffusion and underscores the penalty to pay for partial welding. Thus, the weld strength may be near, or at the virgin strength, but the fatigue strength may be dramatically reduced below its maximum value. Thus, one should always design a healing time with respect to T_r to achieve maximum durability of welds and interfaces. This is rather a subtle processing point, which is often not appreciated by the manufacturing industry and is important for self-healing design, namely, that fatigue and strength are related but not similar in terms of healing parameters.

In fatigue of materials in general with applied stress $\sigma_{app} \sim \sigma_c$, the percolation theory suggests a new approach as follows: The lifetime τ occurs when the initial fraction of bonds P_i is reduced to P_f such that fracture occurs at the applied stress as:

$$\sigma_{app} = \{2ED_0\nu[P_f - P_c]\}^{1/2} \quad (10.18)$$

This gives the critical bond fraction P_f for the applied stress as

$$P_f = P_c + \frac{{\sigma_{app}}^2}{2ED_0\nu} \quad (10.19)$$

The time dependence of P can be deduced from a steady-state bond fracture concept via

$$P_f = P_i - \tau\left(\frac{dp}{dt}\right) \quad (10.20)$$

The failure time τ is the time required to reduce P_i to P_f at the prevailing breakage rate *dp/dt*. The rate of bond rupture *dp/dt* can be given by a thermally activated state theory as:

$$\frac{dp}{dt} = \left(\frac{1}{\tau_0}\right)\exp\left(-D_0\left[1 - \sigma_{app}/\sigma_c\right]/kT\right) \quad (10.21)$$

where the energy for bond fracture D_0 is linearly reduced by the applied stress. Substituting for $\frac{dp}{dt}$ and P_f and solving for τ, we obtain the lifetime of the material as:

$$\tau = \tau_0\left\{[P_i - P_c] - \frac{{\sigma_{app}}^2}{2ED_0\nu}\right\}\exp\left(\frac{D_0\left[1 - \frac{\sigma_{app}}{\sigma_c}\right]}{kT}\right) \quad (10.22)$$

where $\frac{D_0}{kT} \approx 133$ at room temperature and $\tau_0 \approx 10^{-12}$ s is the vibrational period for the bonds being broken. For a material without defects, $P_i = 1$ and σ_c is determined from Eq. ((10.20) using $P_f = 1$. Note that the applied stress enters in both the front factor and exponential factor for *t*. When the applied stresses are small, the exponential factors dominate and the ratio of two lifetimes at applied stresses σ_1 and σ_2 would be:

$$\frac{\tau_1}{\tau_2} = \frac{\exp\left[\frac{\sigma_2 - \sigma_1}{\sigma_c}\right]D_0}{kT} \quad (10.23)$$

When the damaged bond fraction is restored by self-repair processes such that $P_i \rightarrow 1$ in Eq. ((10.22), the material is rejuvenated and its lifetime is considerably extended. Thus, the healing fluids allowed the initial pre-crack to be completely healed ($P_i \rightarrow 1$) and that during fatigue, the self-healing fluids retarded the crack advance compared to the control as the rate of bond rupture *dp*/*dt* was reduced.

10.8.10 The Hard-to-Soft Matter Transition

10.8.10.1 Twinkling Fractal Theory of T_g

A topic that is most relevant to this discussion on self-healing is the fundamental understanding of the hard-to-soft and soft-to-hard matter transition, typically called the glass transition. The glass transition temperature T_g is arrived at by heating a hard cold glassy material to a suitable temperature T_g, whereby it becomes soft; or by providing sufficient mechanical energy to make the material flow by yielding; or by making the sample dimensions sufficiently small, typically nanoscale where the T_g drops considerably; or by examining the top layer of a solid sample at $T < T_g$, where it appears to have a mobile layer; or by increasing the cross-link density; or by changing molecular weight or by removing a solvent, etc. [102, 103]. The classical onset of the glass transition temperature T_g is shown in Fig. 10.11 where the slope of volume *V* vs. *T* changes at T_g. The slopes are the thermal expansion coefficients α_g and α_L for the glass and liquid, respectively. It considers the anharmonic interaction potential U(x) of quasiharmonic diatomic oscillators (Fig. 10.12) in the liquid state as a function of temperature. The anharmonicity controls the vibrational frequencies ω , with temperature *T*, pressure *P*, and stress σ, and the extent of the anharmonicity controls the magnitude of the thermal expansion coefficient α. The vibrational frequencies, degree of freedom ($d = 1$, 2 or 3), and their density of states $G(\omega)$ determine the thermal properties, such as the specific heat capacity *Cp* and thermal conductivity *K*. As T_g is approached from above, the average bond distance between the oscillators contracts, the volume T_g decreases, and the bond expansion factor x approaches its critical value x_c such that when $x > x_c$ the molecules are in the ergodic liquid state and when $x < x_c$, the molecules on average exist in the non-ergodic solid state. The latter idea parallels the Lindemann (1910) theory of melting, which states that when the vibrational amplitudes exceed a critical value of the bond length ($x = 0.11$), melting occurs [103]. The Boltzmann distribution of oscillators at various quantized energy levels in the anharmonic potential energy function of the atoms such that at any temperature *T*, there exists a fraction P_s of solid atoms ($x < x_c$) and a fraction P_L of liquid atoms. The solid fraction P_S at any temperature *T* is determined from the integral (0 to x_c) of the Boltzmann energy populations $\varphi(x)$:

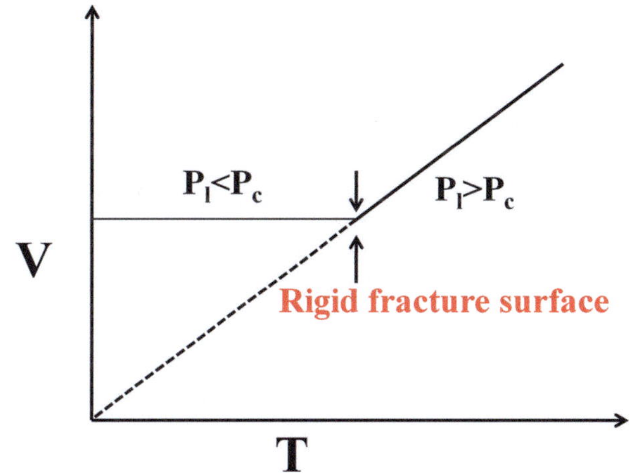

Fig. 10.11 Volume–temperature curve for a glass former. T_g occurs at $P_s = P_c$ when the fractal structure form

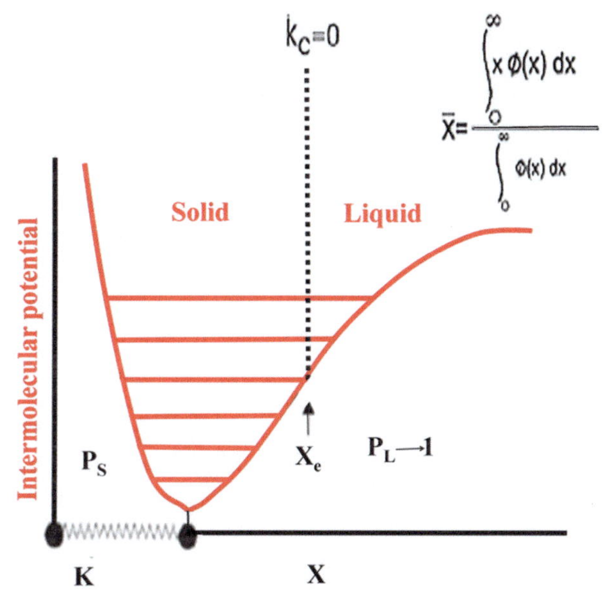

Fig. 10.12 Diatomic anharmonic Morse oscillator $U(x) = D_0[1 - E^{-\Delta X}]^2$ with Boltzmann energy levels and populations $\varphi(X)\,\tilde{\exp}\left[\frac{U(x)}{kT}\right]$ show the liquid $X > X_c$ and solid $X < X_c$ phase diagram

$$P_s(T) = \int_0^{x_c} \varphi(x)dx \qquad (10.24)$$

In Eq. ((10.24), $\varphi(x)\,\tilde{\exp}\,\frac{-U(X)}{kT}$. As T_g is approached from above, example in Fig. 10.12, the fraction of solid atoms P_s grows with a cluster size distribution function determined by percolation theory and eventually percolates at the critical percolation threshold P_c.

The rigid (black) percolation cluster has a fractal dimension D_f and is not stationary but is quite dynamic, since locally the solid and liquid atoms are in dynamic equilibrium. Thus at T_g, to the observer, there appears to exist a fractal structure which "twinkles" with a frequency spectrum $F(\omega)$ as liquid and solid atoms exchange. This "twinkling" fractal structure is invisible to the usual scattering experiments, since the liquid and solid are essentially indistinguishable to such scattering experiments.

The existence of the twinkling fractal near T_g has a profound effect on many properties, such as the mechanical loss peak. It suggests that the onset of T_g is purely kinetic rather than thermodynamic such that at very slow rates, the fractal appears liquid and at very high rates approaching u_0, the fractal appears quite rigid. This effect dominates the physics of the rate dependence of yield in amorphous thermoplastic polymers and the physics of liquid armor. The change in heat capacity ΔC_p which appears as a pseudo-second-order phase transition near T_g should be determined predictably by the changes in the degree of freedom of the oscillators from $d = 3$ to $d = D_f$ as well as the density of states $G(\omega) \approx \omega^{df}$, where df is the fracton dimension. The T_g value of thin films will be affected by finite-size percolation effects. The existence of the twinkling fractal will also allow welding of glassy polymers below their T_g via gradient percolation effects, where the fractal dances while twinkling. When the temperature drops below T_g, then the fractal has its greatest manifestation on the evolution of the glassy structure. With decreasing temperature, the low-frequency components of the twinkling fractal spectrum slow down or become negligible and the fractal becomes quite rigid such that the normal volumetric contraction experienced in the liquid state deviates from the extrapolated liquidus line V_∞ and a non-equilibrium fractal-cavitation process commences resulting in the usual ΔV noted between V_g and V_∞ (Fig. 10.11) Thus, the thermal expansion coefficient in the glass α_g is less than in the liquid state α_L approximately as $\alpha_g \approx \rho_c \alpha_L$. For the Morse oscillator of equilibrium interatomic distance R_0, anharmonicity factor a, bond energy D_0, the critical distance $x_c \approx 1/3a$ andT_g is determined from Eq. ((10.25), at $\rho_c = \rho_s$, by:

$$T_g \approx \frac{(1-\rho_c)4D_0}{9k} \approx \frac{2D_0}{9k} \tag{10.25}$$

The linear thermal expansion coefficient in the liquid is obtained from the average position of the bond length using the integral of the Boltzmann populations (Fig. 10.12) and is obtained in the quasi-anharmonic approximation as:

$$\alpha_L = \frac{3k}{4D_0R_0a} \tag{10.26}$$

Equation ((10.26) is related to T_g via Eq. ((10.25). For example, if $D_0 = 3.5$ kcal mol^{-1}, Eq. ((10.26) gives $T_g = 118$ °C; using R_0= 3A and $a = 2/$A, then $\alpha_L \approx$ z 70 ppmK^{-1} and $\alpha_g \approx$ 35 ppmK^{-1}, which are typical values for engineering thermoplastics. It is interesting to note that the relation $\alpha_L T_g = \frac{1}{6aR_0} \approx 0.03$, which was found to be true for many polymers and composite resins. Since the modulus $E \sim D_0$, one also expects that $E \tilde{\frac{1}{\alpha_{L0}}}$.

Physical aging occurs below T_g through the relaxation of ΔV via the twinkling frequencies and is complex. Even well below T_g, there is a non-zero predictable fraction of liquid atoms remaining, which will cause the twinkling process to continue at an ever-slowing pace but allow the non-equilibrium structure to eventually approach a new equilibrium value near V_∞. A near-equilibrium glass can be made by removing the fractal constraints in 3d at T_g by forming the material using 2d vapor deposition, as recently observed by Ediger et al. [104]. They used vapor deposition of indometracin to make films, which were reported to have maximum density and possess exceptional kinetic and thermodynamic stability, consistent with the liquid structure extrapolated into the glass. We have suggested that in this case with $\Delta V = 0$, a critical observation would be that $\alpha_g = \alpha_L$.

10.8.10.2 Healing below the Glass Transition Temperature

Healing of polymer–polymer interfaces below T_g, as demonstrated by Boiko et al. can occur due to softening of the surface layer. We have treated the surface layer softening as a gradient rigidity percolation issue [105]. The surface rubbery layer concept in thick films is interesting, and this percolation theory suggests that for free surfaces there is a gradient of $p(x)$ near the surface, where $x < \xi$ (cluster size correlation length) and hence a gradient in both T_g and modulus E. If the gradient of p is given by $p(x) = (1 - x/\xi)$ then the value of x_c for which the gradient percolation threshold p_c occurs, and which defines the thickness of the surface mobile layer, is given by the percolation theory as

$$x_c = \frac{b(1-P_c)}{\left\{P_c^{\ v}\left[1-\frac{T}{T_g}\right]^v\right\}} \tag{10.27}$$

Here b is the bond length and v is the critical exponent for the cluster correlation length $\xi \sim (P - P_c)^{-v}$. For example with polystyrene, when $T = T_g - 10$ K, $T_g = 373$ K,and using b = 0.154 nm, P_c= 0.4, v= 0.82, then the thickness of the surface mobile layer x_c= 3.8 nm. This could allow for healing to occur below T_g assuming that the dynamics are fast enough.

If $G_{IC} \sim X^2$ for entangled polymers, then we could deduce that for sub T_g healing at $\Delta T = T_g - T$, as

$$G_{IC} \sim \left[\frac{1}{\Delta T}\right]^{2v} \qquad (10.28)$$

This appears to be in qualitative agreement with Boiko et al.'s data who examined the fracture energy of polystyrene interfaces during welding at temperatures up to 40 K below T_g. The TFT also suggests the presence of some interesting dynamics in the mobile layer since the fractal clusters are essentially "dancing" with a spectrum of frequencies related to the density of states $G_\omega \sim \omega^{df}$, where the Orbach fracton dimension $df = 1.33$.

10.8.10.3 Twinkling Fractal Theory of Yield Stress

The following analysis is important for the design of self-repair systems subject to ballistic impact and for the basic understanding of rate effects on yield stress of solids and shear-thickening fluids. For an amorphous solid below T_g, subjected to triaxial stresses $\sigma_{1,}$, σ_2, σ_3,the distortional strain energy function U_d is determined by (Von Mises) as [166]:

$$U_d = \frac{(1-v)\left[(\sigma_{1,} - \sigma_{2,})^2 + (\sigma_{2,} - \sigma_{3,})^2 + (\sigma_2 - \sigma_{3,})^2\right]}{6E} \qquad (10.29)$$

Here, v is the Poisson ratio and E is the isotropic elastic Young's modulus. In the simple uniaxial case, we obtain the stored strain energy from Eq. ((10.29) with $v = 0.5$ as $U = \sigma 2/2E$. This stored energy must be utilized to overcome a percolation number of interatomic anharmonic oscillator bonds of energy U(xc). For a Morse oscillator U(xc) = 0.08 D_0, which is the energy necessary to reach the Lindemann bond expansion at xc. The number of such oscillators per unit volume is $1/V_m$, where V_m is the molar volume. At the critical stored energy, there is yielding when the amorphous solid is raised to the twinkling fractal state by the release of the mechanical energy and flow begins in accord with:

$$\frac{\sigma_y{}^2}{2E} \geq 0.08 D_0 \frac{[p_s - p_c]}{V_m} \qquad (10.30)$$

In which the solid fraction $p_s$is given by Eq. ((10.24) and is both temperature and rate dependent. Thus, the yield stress is obtained as:

$$\sigma_y = \{0.16E[p_s - p_c]D_o/V_m\}^{1/2} \qquad (10.31)$$

Note that the term D_o/V_m corresponds to the traditional cohesive energy density. For example, if E = 1 GPa, p = 1 (high rate of deformation), p_c=1/2, D_0= 3 kcal mol^{-1}, $V_m = \frac{2M_0}{\rho}$, $\rho = 1 g/cc$, the monomer molecular weight $M_0 = 100$ g/mol, then Eq. ((10.31) predicts that σ_y=71 MPa. This value is typical for high-performance amorphous polymers with $T_g \approx 63$ °C.

The magnitude of p_s is also rate- and temperature-dependent in a manner determined by the twinkling fractal density of states. As T approaches T_g from below, p_s decreases toward p_c and σ_y decreases accordingly. At high rates of deformation, p increases and σ_y increases. At low rates of deformation, p decreases toward pc, and this is the basis for designing liquid armor with shear-thickening fluids, which are liquid layers with particles at $p > p_c$ that turn into solids at high rates. Thus, $p_s(\gamma)\left(\frac{\gamma}{\omega_0}\right)^{\tilde{df}}$ such that we can express Eq. (10.31 as

$$\frac{\sigma_y}{\sigma_\infty} = \left\{\left[\left(\frac{\gamma}{\omega_0}\right)^{df}\right] - p_c[1-p_c]\right\}^{1/2} \qquad (10.32)$$

Here, σ_∞ is the yield stress at $\gamma = \omega_0$. There exists a critical deformation rate given by

$$\gamma_c = \omega_0 P_c \qquad (10.33)$$

when $\gamma < \gamma_c$, then the material is liquid-like and when $\gamma > \gamma_c$, then the yield stress increases from zero to its maximum value at u_0. Thus, the yield stress increases with the rate of testing, and the higher the rate, the more stored energy is required, which facilitates ballistic healing at high rates.

The TFT concept of T_g suggests that twinkling fractal cluster at T_g is quite soft, especially when sensed at low rates of deformation. One can immediately understand why the loss tangent tan$\delta = E''/E'$ (loss/storage modulus) reaches its maximum value near T_g. The twinkling fractal frequencies are given by

$$F(\omega)\omega^{\tilde{df}} \exp \frac{-[|U(T) - U_c|]}{kT} \qquad (10.34)$$

Here, the first term ω^{df} is the vibrational density of states for a cluster of frequency ω and the second exponential term is the probability that the vibration will cause a "twinkle" or change from solid to liquid or vice versa. Note that this energy difference is always positive and behaves as $|U(T) - U_c| \sim [T^2 - T_c{}^2]$. In a single mechanical cycle, the stored energy is released or dissipated by the twinkling process as liquid and solid clusters of frequency u exchange at the fastest rates in accord with the density of states $G(\omega) = \omega^{df}$. As the temperature drops below T_g, the percolating cluster increases in mass and the stored energy increases. However, the twinkling frequencies decrease their rate due to the increased energy barrier $\Delta E = |U(T) - Uc|$ for the solid-to-liquid transition, and the energy dissipation decreases. Near T_g, we can approximate the temperature dependence of $U(x)$ for a Morse oscillator as $U(T) \approx \alpha^2 D_0 a^2 T^2$, such that the temperature dependence of $F(\omega)$ is:

$$F(\omega, T) = \omega^{df}\, exp\, \frac{-\left[\beta\left(T^2 - {T_g}^2\right)\right]}{kT} \tag{10.35}$$

Here, the constant $\beta = \alpha^2 D_0 a^2$. Thus, the activation energy ΔE is temperature-dependent approximately as $\Delta E \sim [(T/T_g)^2 - 1]$ and changes rapidly with T. When $T > T_g$, the rigid clusters decrease, the stored energy decreases, and the rate of liquid-to-solid transitions decreases. Thus, typically T will reach a maximum value near T_g. The twinkling fractal, although invisible to scattering experiments, could be the dynamics engine of the amorphous state and plays an important role in complex processes, such as physical aging and self-healing.

10.8.11 Fracture Mechanics of Polymeric Materials

Damage in polymers and composites often involves some form of fracture, and for structural applications, the recovery of fracture properties is an important research area. The primary fracture loading conditions for self-healing specimens have been quasi-static fracture, fatigue, and impact as shown in Fig. 10.13. These loading conditions cause several types of fracture, including Mode I crack opening, Mode III tearing, mixed-mode cutting, matrix–fiber delamination, and transverse (shear) cracking. [67].

10.8.12 Self-Healing of Thermoplastic Materials

Self-healing of thermoplastic polymers can be achieved via a number of different mechanisms given below.

10.8.12.1 Healing by Molecular Interdiffusion Approach

The polymers investigated cover amorphous, semi-crystalline, block copolymers, and fiber-reinforced composites. It has been discovered that when two pieces of the same polymer are brought into contact at a temperature above its glass transition (T_g), the interface gradually disappears and the mechanical strength at the polymer–polymer interface

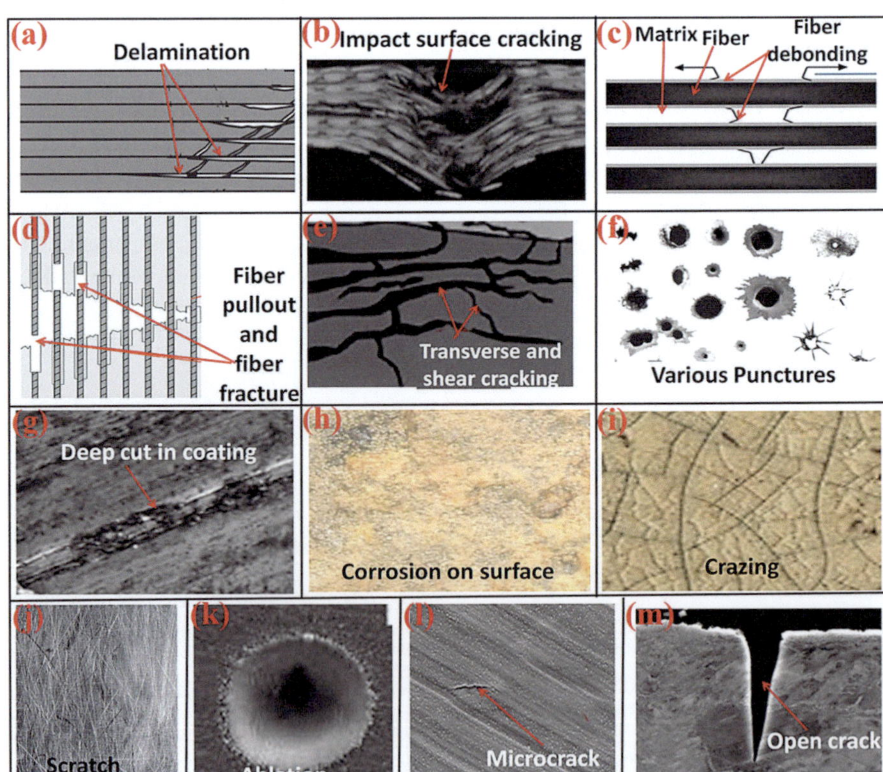

Fig. 10.13 Damage modes in polymer composites: (**a**) Delamination, (**b**) Impact/indentation surface cracking, (**c**) Fiber debonding, (**d**) Fiber rupture and pullout, (**e**) Transverse and shear cracking, (**f**) Puncture, (**g**) Deep cut in coating, (**h**) Corrosion in protected metal, (**i**) Crazing, (**j**) Scratch, (**k**) Ablation, (**l**) Microcracking, and (**m**) Opening crack

increases as the crack heals due to molecular diffusion across the interface [106]. The healing process was examined at atmospheric pressure or in vacuum for healing times ranging from minutes to years and at healing temperatures above the T_g of the polymers that typically varied from −50 to +100 °C [107]. Jud and Kausch studied the effect of molecular weight and degree of copolymerization on the crack healing behavior of poly(methylmethacrylate) (PMMA) and PMMA-poly (methoxyethylacrylate) (PMEA) copolymers. The self-healing ability of the copolymers was tested by clamping and heating these samples in which the fractured surfaces (of single-edge notched and compact tension specimens) were brought together and held for set periods of time. Various experimental parameters were investigated, which included the time between fracturing and joining of the fractured surfaces, the healing time, the healing temperature, and the clamping pressure. It appeared that a temperature of 5 °C higher than the T_g and a healing time of more than 1 min were required to produce healing greater than that could be attributed to simple surface adhesion. An increase in the time between fracture initiation and self-healing of the fractured surfaces was found to significantly inhibit healing, dropping optimum property recovery from 20% to 80%. Visual healing of the fracture surfaces was found to occur before a significant recovery in strength was achieved, with the interdiffusion of numerous chain segments (rather than entire chains) being reported as the most likely healing mechanism. In particular, Wool and O'Connor suggested a five-stage model to explain the crack healing process in terms of surface rearrangement, surface approach, wetting, diffusion, and randomization (Fig. 10.14). The phenomena of crack healing in the thermoplastics occur most effectively at or above the T_g of these materials [108].

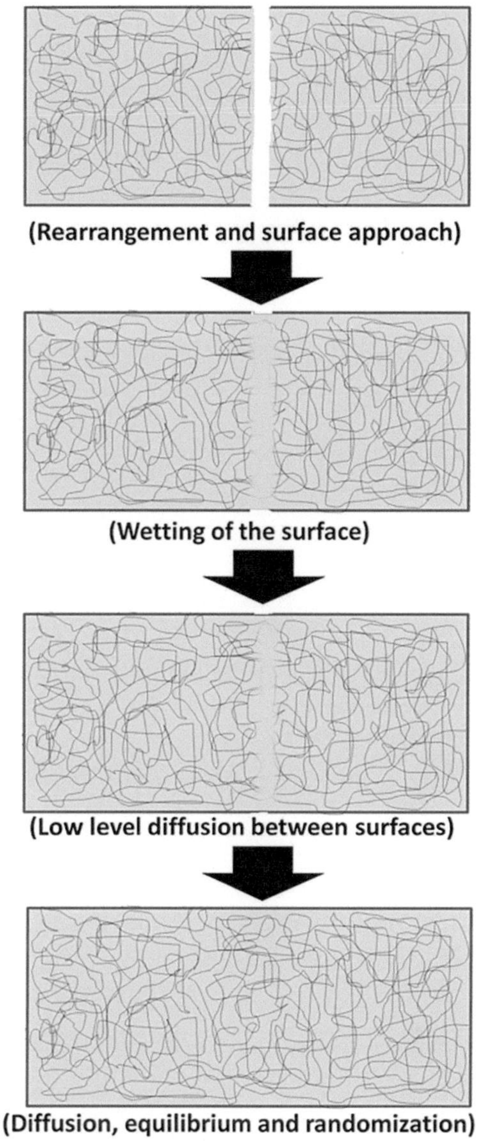

Fig. 10.14 Mechanisms involved in self-healing via molecular interdiffusion

10.8.12.2 Healing by Recombination of Chain-Ends Approach

Recombination of chain ends is a relatively new technique proposed to heal structural (strength loss) and molecular (chain scission) damages in certain thermoplastics. This approach does not rely on the constrained chain confirmations to promote site-specific chain scission (Fig. 10.15). Some engineering thermoplastics prepared by condensation reactions, such as polycarbonate (PC), polybutyleneterephthalate (PBT), polyether-ketone (PEK), Polyphenylene ether (PPE), and PEEK, can be healed by a simple reaction that reverses the chain scission. The authors observed that the self-healing reaction of this polymer did occur in the solid state, and a series of events was identified prior to and during the healing process. These events include (1) occurrence of chain cleavage due to degradation; (2) diffusion of oxygen into the polymer materials; (3) recombination of the cleaved chain ends by the catalytic redox reaction under oxygen atmosphere and in the presence of copper/amine catalyst; and (4) water discharge as a result of the self-healing reaction. Figure 10.16 shows the chain scission, healing initiation, and healing completion reactions in self-healing PC.

As such, the kinetics of the self-healing reaction was found to depend on factors such as oxygen concentration and mobility of the polymer chain (affected by the concentration of the plasticizer). It was also observed that the speed of the healing reaction decreases with an increase in the reaction time due to a reduction in the polymer chain mobility with increasing molecular weight as the reaction progresses and a gradual decrease in available hydroxyl (OH) end groups as they are consumed by the recombination reaction. The feasibility of the healing process was found to depend on the type of end groups present, which is in turn affected by

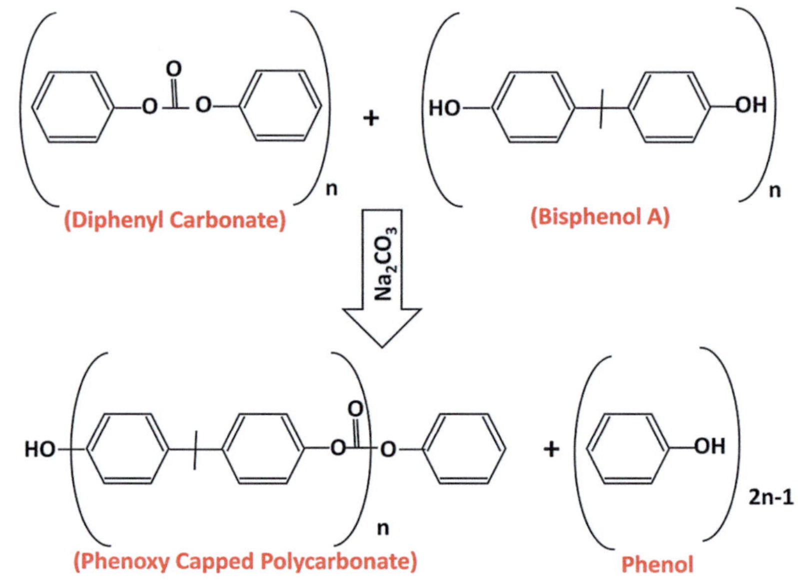

Fig. 10.15 Reaction mechanism for self-healing PC production

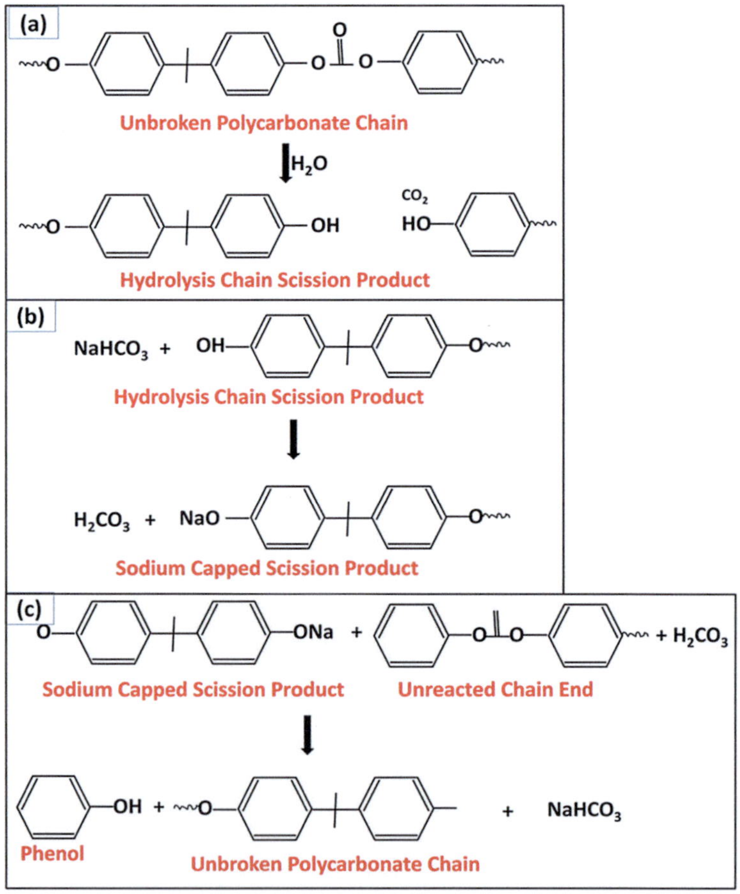

Fig. 10.16 (**a**) Chain scission, (**b**) Healing initiation, and (**c**) Healing completion reactions in self-healing PC

the synthesis method of the PC. It has been reported that although the repair of the standard PC prepared by bisphenol-A and phosgene was not feasible, the use of sodium carbonate (Na_2CO_3) as a healing agent for the PC prepared by ester exchange of a diester carbonate and a hydroxyl compound was successful. Healing efficiencies up to 98% in tensile strength and molecular weight recovery were achieved after a healing period of more than 600 h. Self-healing of

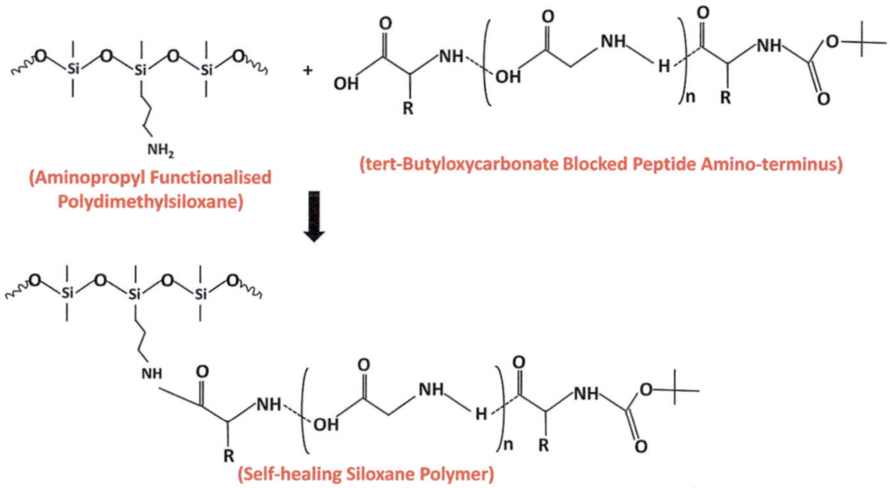

Fig. 10.17 A production route for self-healing organo-siloxane polymers

hydrolysis scissored chains in the PC occurred through recombination of the phenolic end groups and the phenyl end groups and was accelerated by the presence of a small amount (0.1 ppm) of Na_2CO_3. This healing mechanism is only applicable to certain types of thermoplastics capable of recombining chain ends via a specific reaction mechanism [109].

10.8.12.3 Self-Healing Via Reversible Bond Formation

The chain mobility in thermoplastics can also be used to heal fractures at ambient temperatures by the inclusion of reversible bonds in the polymer matrix. This provides an alternative approach to the UV light or catalyst-initiated healing of the covalent bonds as discussed in the previous sections and utilizes hydrogen or ionic bonds to heal damaged polymer networks [110].

Organo-siloxane: The self-healing materials described were relating to the production of polypeptide-polydimethylsiloxane copolymers (Fig. 10.17) in which the silicon-based primary polymeric networks were grafted or block copolymerized with a secondary network of crosslinking agents (such as peptides). The secondary crosslinking components comprise polymer domains with intermediate-strength crosslinks formed via hydrogen and/or ionic bonding. The intermediate-strength crosslinks provide a good overall toughness to the material while allowing for self-healing due to the possibility of reversible crosslinking. Healing was initiated when the fractured surfaces came in contact either through physical closure or via solvent-induced chain mobility. It was claimed that the healing times could be adjusted by varying the structure of the polymer, the degree of crosslinking, or the strength of the crosslinks [111].

Ionomers: Ionomers are defined as polymers comprising less than 15 mol% ionic groups along the polymer backbone. In particular, the self-healing ability of poly(ethyleneco-methacrylic acid) (EMAA)-based ionomers following high-speed impact was investigated along with proposals of possible healing mechanisms. While it is recognized that the existing EMAA ionomers with self-healing properties are not suitable for some applications, the suitable ionomers could be synthesized or modified by fillers or fibers based on a better understanding of the associated healing phenomenon. For example, EMAA polymer and methacrylic acid (MA) with sodium cation and an ionomer based on Surlyns 8940 and is marketed for its ability to self-heal upon high-speed impact. The healing was reported to occur almost instantaneously following balastic impact. Fall proposed that the ionic content and its order–disorder transition was the driving force behind the healing process. It has been hypothesized that the self-healing response was related to ionic aggregation and melt flow behavior of these copolymers. Healing was expected to occur if sufficient energy was transferred to the polymer upon impact, heating the material above its order–disorder transition resulting in disordering of the aggregates. During the post-puncture period, the ionic aggregates have the tendency to reorder and patch the hole [112].

10.8.12.4 Healing by Photo-Induced Approach

The first example of photo-induced self-healing in PMMA was reported by Chung et al. The photochemical [2 + 2] cycloaddition of cinnamoyl groups was chosen as the healing mechanism since photo-cycloaddition produced cyclobutane structure and the reversion of cyclobutane to the original cinnamoyl structure readily occurs in a solid state upon crack formation and propagation. The feasibility of this concept was tested by blending a photo-cross-linkable cinnamate monomer, 1,1,1-tris-(cinnamoyloxymethyl) ethane (TCE), with urethane dimethacrylate (UDME), triethyleneglycoldimethylacrylate (TEGDMA)-based monomers, and a visible-light photoinitiator camphor quinone (CQ) (Fig. 10.18). The mixture was polymerized into a very hard and transparent film after irradiation for 10 min with a 280 nm light source. Healing of the fractures in these films was achieved by re-irradiation for 10 min with a light of l4280 nm. The healing was shown to only occur upon exposure to the light of the correct wavelength, proving that the healing was light initiated. Healing efficiencies in flexural strength up to 14% and 26% were reported using light or a combination of light and heat (100 °C). A mechanism of fracturing and healing was shown in Fig. 10.19. In this particular system, however, healing was limited to the surfaces being exposed to light, meaning that internal cracks or thick substrates are unlikely to heal [109].

10.8.12.5 Living Polymer Approach

For the purpose of providing protection against damage mechanisms unique to space applications such as ionizing radiation damage, the development of self-healing polymeric materials using living polymers as the matrix resins has been proposed. It suggested preparing living polymers with a number of macro radicals (polymer chains capped with radicals). The living polymers can be theoretically synthesized by either ionic polymerization or free-radical polymerization during which the polymer chains grow without chain transfer and termination. As a consequence, the chain ends of the living polymers are equipped with active groups capable of resuming polymerization if additional monomer is added to the system. The free radical living polymerization is likely more suitable for this purpose considering the high reactivity and stringent conditions required for the ionic living polymerization. In this approach, the degradation of the material upon exposure to ionization or UV radiation is potentially prevented because of possible recombination reactions between the new free radicals generated and the macroradicals on the chain ends. Such a molecular scale healing process is controlled by the diffusion rate of the macroradicals, which is in turn affected by the T_g of the polymer. Below T_g, the diffusion rate of the macroradicals in the condensed state is low, resulting in a slow healing process. The electron spin resonance data indicated that such polymers should be capable of providing self-healing capabilities at temperatures up to 127 °C.

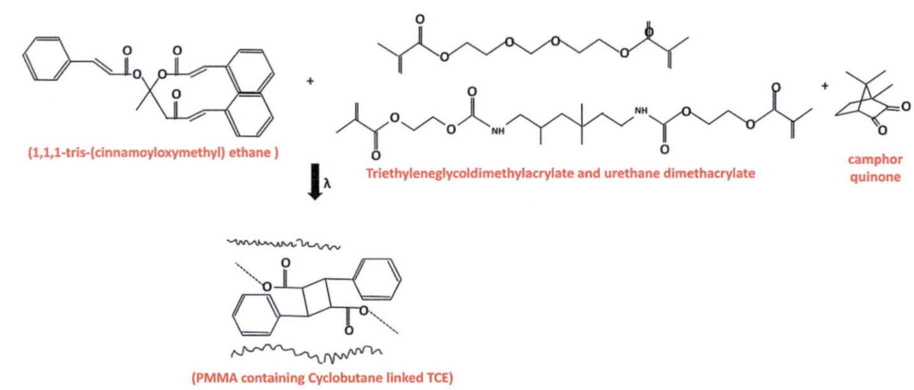

Fig. 10.18 Route to producing photo-healable PMMA

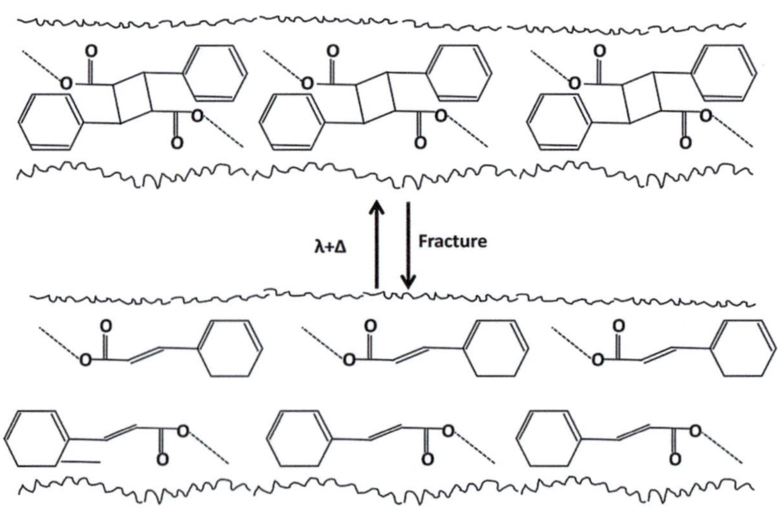

Fig. 10.19 Mechanism of fracture and repair of photo-induced healing in PMMA

Although Chipara and Wooley demonstrated the living polymer approach in a PS matrix, it may also be applicable to thermosets. Such a self-healing system does not require the addition of catalysts in the polymer and may provide protections for space materials against various degradation environments. However, the concept requires further investigation in terms of working conditions required to prevent premature deactivation of the living radicals and the applicability of the concept to different polymer matrices. It is proposed that such a molecular healing process can be combined with the inclusion of microencapsulated monomers to provide a multiscale self-healing system. As the polymer chains remain active, the release of the monomer in the event of a crack is expected to restart the polymerization process and heal the microcracks [113].

10.8.12.6 Self-Healing by Nanoparticles Approach

This type of polymer–nanoparticle composite actively responds to the damage and can potentially heal itself multiple times as long as the nanoparticles remain available within the material. The nanoparticles have a tendency to be driven toward the damaged area by a polymer-induced depletion attraction and that larger particles are more effective than small particles for migrating to the damaged region at relatively short time scales. Once particle migration has occurred, the system can then be cooled down so that the coating forms a solid nanocomposite layer that effectively repairs the flaws in the damaged surface. The example involves a 50-nm-thick silicon oxide (SiO_2) layer deposited on top of a 300-nm-thick PMMA film embedded with 3.8 nm CdSe/ZnS nanoparticles. The migration of the nanoparticles toward the cracks in the brittle SiO_2 layer is dependent on the enthalpy and entropic interactions between the PMMA matrix and the nanoparticles. The phenomenon of self-healing by nanoparticles has been explained by the polymer chains close to the nanoparticles being stretched and extended, driven by the tendency to minimize nanoparticles–polymer interactions via segregation of the nanoparticles in the crack and pre-crack regions. The nanoparticles are more effective than the larger particles for healing because they diffuse faster than the larger ones. One of the key enabling requirements for this type of auto-responsive healing technique relies on the ability to functionalize the surface of the nanoparticles with suitable ligands [114].

10.8.13 Self-Healing of Thermoset Materials

Most of the structural applications generally require rigid materials with a thermal stability that most thermoplastics do not possess. The rigidity and thermal stability of thermosets come from their cross-linked molecular structure, meaning that they do not possess the chain mobility so heavily utilized in the self-healing of thermoplastics. As a result of their different chemistry and molecular structure, the development of self-healing thermosets has followed distinctly different routes. The most common approaches for autonomic self-healing of thermoset-based materials involve the incorporation of self-healing agents within a brittle vessel prior to the addition of the vessels into the polymeric matrix. These vessels fracture upon loading of the polymer, releasing the low-viscosity self-healing agents to the damaged sites for subsequent curing and filling of the microcracks. The exact nature of the self-healing approach depends on (1) the nature and location of the damage; (2) the type of self-healing resins; and (3) the influence of the operational environment [115].

10.8.13.1 Hollow Glass Fiber Systems

Hollow glass fibers have already been shown to improve structural performance of materials without creating sites of weakness within the composite. These hollow fibers offer increased flexural rigidity and allow for greater custom tailoring of performance, by adjusting, for example, both the thickness of the walls and the degree of hollowness. By using hollow glass fibers in the polymer composites, alone or in conjunction with other reinforcing fibers, it would be possible to not only gain the desired structural improvements but also to introduce a reservoir suitable for the containment of a healing agent. The healing agents contained within the glass fibers have been either a one-part adhesive, such as cyanoacrylate, or a two-part epoxy system, containing both a resin and a hardener, where either both are loaded in perpendicular fibers or one embedded into the matrix and the other inside fibers. One of the initial challenges encountered when creating this type of self-healing systems is the development of a practical technique for filling the hollow glass fibers with a repair agent (Fig. 10.20). When approaching this problem, the dimensions of the glass fiber itself must be considered, including diameter, wall thickness, and fiber hollowness as well as the viscosity and healing kinetics of the repair agent. The fiber filling method involving "capillary action" is assisted by vacuum, which is now the main commonly used process. The chosen glass fiber should also be evaluated for its capacity to survive the composite manufacturing process without breakage, while still possessing its ability to rupture during a damage event to release the required healing agent. This mechanism will obviously depend upon the viscosity of the healing material as well as the kinetics of the repair process. Various techniques have then used liquid dyes inside the composites in order to serve as a damage detection mechanism, providing hence a visible indication of the damage site, while allowing a clear evaluation of the flow of healing agents to those sites. The concentration of healing fibers within the matrix depends on their special distribution, and the final dimensions of the specimen, which have direct effects on the mechanical properties of the resulting

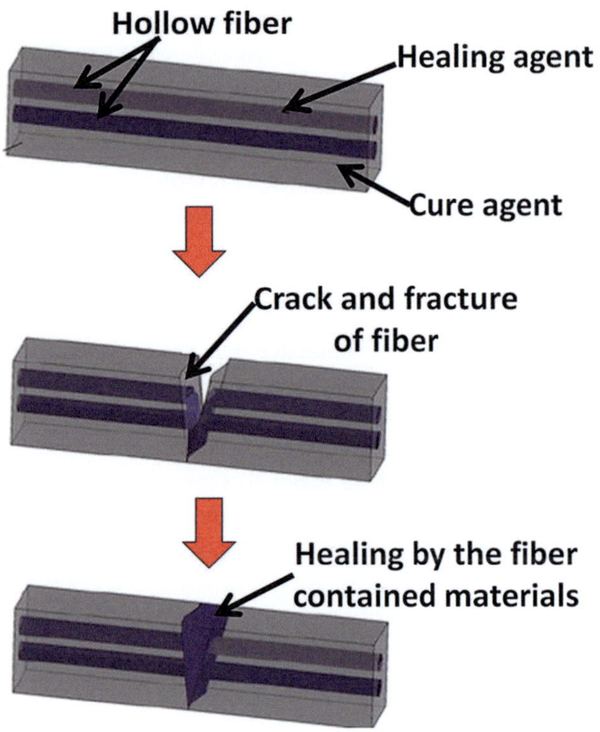

Fig. 10.20 Hollow fibers self-healing

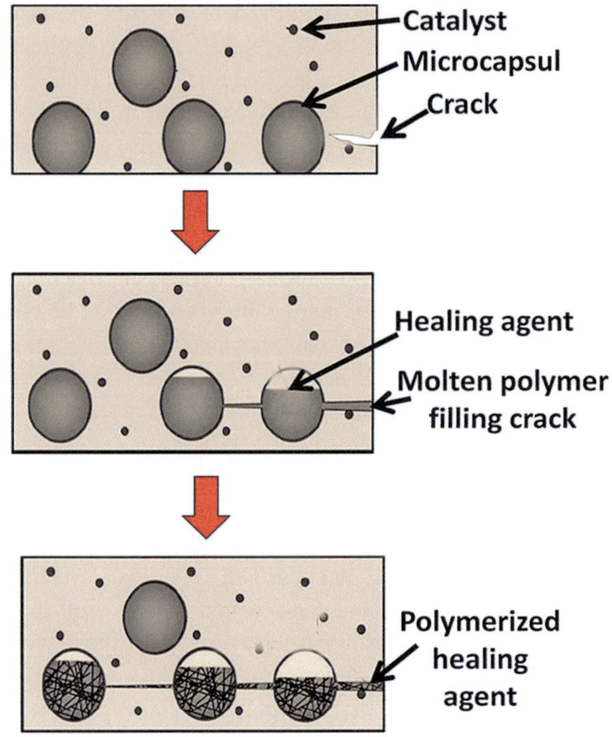

Fig. 10.21 Microencapsulated Healing system

composite material. The stacking sequence of the fibers within the composite plays a role in inhibiting plastic deformation and delamination and will also affect the response to an impact damage event. In order to maintain high mechanical properties, repair fibers need to be adequately spaced within the composite. The work of Williams et al. has considered the development of autonomic self-healing within a carbon fiber-reinforced polymer (CFRP) and has demonstrated the significant strength recovery (>90%), which was possible when a resin-filled hollow glass fiber system was distributed at specific interfaces within a laminate, thereby minimizing the reduction in mechanical properties while maximizing the efficiency of the healing event [116].

10.8.13.2 Based on Microencapsulated Healing System

The microencapsulation approach is by far the most studied self-healing concept in recent years. Polymer adopts by embedding a microencapsulated liquid healing agent and solid catalytic chemical materials within a polymer matrix (Fig. 10.21). Hence, upon damage-induced cracking in the matrix, microcapsules are supposed to release their encapsulated liquid healing agent into the crack planes. At the same time, the microcapsule walls must be resistant enough to processing conditions of the host composite while maintaining excellent adhesion with the cured polymer matrix to ensure that the capsules rupture upon composite fracture. This particular approach involves the incorporation of a microencapsulated healing agent and a dispersed catalyst within a polymer matrix. Upon damage-induced cracking, the microcapsules are ruptured by the propagating crack fronts resulting in the release of the healing agent into the cracks by capillary action. The subsequent chemical reaction between the healing agent and the embedded catalyst heals the material and prevents further crack growth. There are some obvious similarities between the microencapsulation and hollow fiber approaches, but the use of microcapsules alleviates the manufacturing problems experienced in the hollow fiber approach. The microencapsulation approach is also potentially applicable to other brittle material systems such as ceramics and glasses [117].

The size of the microcapsule also plays a role in the performance of the system, in terms of the effect on toughness of the composite, and the nature of the interface between microcapsule and polymer matrix. Based on these relationships, the size and weight fraction of microcapsules can be rationally chosen to give optimal healing of a given crack size. Generally, the shell wall thickness is largely independent of manufacturing parameters and is typically between 160 and 220 nm thick; nevertheless, slight adjustments can be made during the encapsulation procedure to alter the resulting microcapsules. The capsules were found to possess a uniform urea-formaldehyde (UF) shell wall (77 nm average thicknesses) and display good thermal stability. However, there are drawbacks with UF microcapsules:

first, (1) the formation of agglomerated nanoparticles debris that could act as crack initiation sites within the host matrix; second, (2) rough and porous wall surfaces formed by agglomerated nanoparticles that may reduce the adhesion between the microcapsules and matrix; and, finally, (3) rubbery and thin capsule walls (160–220 nm) that lead to the loss of core material during storage and cause handling difficulties during the processing of the composites. In addition to UF microcapsules, melamine-formaldehyde and polyurethane shell wall materials were successfully used to prepare microcapsules of various healing materials. A suitable self-healing system should (1) be easily encapsulated; (2) remain stable and reactive over the service life of the polymeric components under various environmental conditions; and (3) respond quickly to repair damage once triggered. The resulting microcapsules need to possess sufficient strength to remain intact during the processing of the polymer matrix, rupture (rather than de-bond) in the event of the crack, capable of releasing the healing agent or catalyst into the crack, and have minimal adverse affects on the properties of the neat polymer resin or reinforced composite. The availability of active catalysts for crack healing was affected by factors, such as the order of mixing, the type of matrix resin, type of curing agent, the catalyst particle size, and the amount of catalyst added.

10.8.13.3 Based on Fatigue Cracks Retardation Self-Healing System

To retard the growth of fatigue cracks, shape-memory alloy (SMA) wires are well suited to this application since they exhibit a thermoelastic martensitic phase transformation, contracting above their transformation temperature and exerting large recovery stresses of up to 800 MPa, when constrained at both ends. When an SMA wire is embedded within an epoxy matrix, the full recovery force acts at the free edges of the component. Therefore, an SMA wire bridging a crack should induce a large closure force on the crack. The self-healing polymers with embedded SMA wires, where the addition of SMA wires shows improvements of healed peak fracture loads by up to 160% (comparatively with specimen without SMA), approaching the performance of the virgin material. Moreover, the repairs can be achieved with reduced amounts of healing agent. The improvements in performance were attributed mainly to the crack closure, which reduces the total crack volume and increases the crack fill factor for a given amount of healing agent and the heating of the healing agent during polymerization, which increases the degree of cure of the polymerized healing agent. Although the examples described are restricted to NiTi alloys, shape memory polymers can also be used. This technology is likely to require heating for the healing to occur and is applicable to repair surface scratches. Other self-healing technologies involving SMAs were either involved non-polymeric systems or were used to assist other repair processes rather than complete the repair itself [118].

10.8.13.4 Three-Dimensional Microchannel Structure Self-Healing Systems

Complex microvascular networks are widely observed in biological systems, such as leaf venation and blood vascularisation. Indeed, in the latter case, the human circulatory system is comprised of vessels of varying diameter and length: arteries, veins, and capillaries. These vessels function together in a branched system to supply blood to all points in the body simultaneously. However, due to their complex architecture, replication of these microvascular systems remains a significant challenge for those pursuing synthetic analogs. These microvascular networks can be created via soft lithographic methods, in which all microchannels can be fabricated at the same time, laser ablation, or direct write methods, which are more suited for building three-dimensional (3D) micro-channel structures. One of the main advantages of those systems compared to both the hollow fiber and microcapsule systems is their ability to heal the same location in the material more than once [119].

10.8.13.5 Inclusion of Thermoplastic Additives System

The use of thermoplastic additive as a self-healing agent for thermoset matrices was first reported by Zako and Takano in 1999. Using thermoplastic additives instead of thermally reversible crosslinks enables the original polymer matrix to remain unaltered during incorporation of the healing capability as well as providing solidifiable crack filler capable of rebonding fracture surfaces. The feasibility of this technology was demonstrated using up to 40 vol% of thermoplastic epoxy particles (average diameter of 105 mm) in a glass fiber-reinforced epoxy composite. Upon heating, the particles melted, flowed into internal cracks or flaws, and healed them (Fig. 10.21). The crack was healed by the application of heat, which triggered flow and subsequent polymerization of the embedded particles.

Various thermoset polymer self-healing approaches are described below:

10.8.13.6 Thermally Reversible Cross-Linked Approach

This self-healing concept involves the development of a new class of cross-linked polymer capable of healing internal cracks through thermo-reversible covalent bonds (Fig. 10.22). The mechanical properties of this type of polymers were comparable to those of the epoxy resins and the other thermoset resins commonly used in fiber-reinforced composites. Therefore, this type of polymer may be used to fabricate fiber-reinforced polymer composites for structural applications. The use of thermally reversible crosslinks to

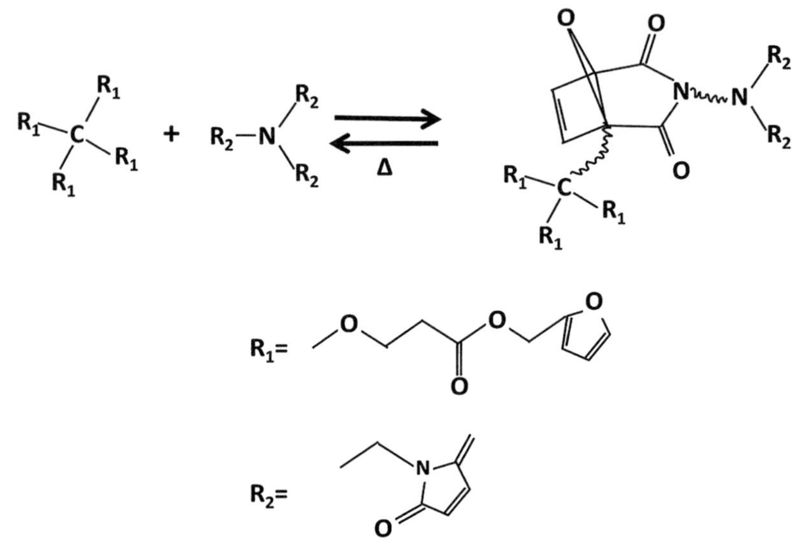

Fig. 10.22 Cross-linking agents and thermally reversible crosslinking mechanism in self-healing polymers

heal thermosets eliminates the need to incorporate healing agent vessels or catalysts in the polymeric matrix although heat is now needed to initiate the healing. The exploration of a thermally reversible reaction such as the Diels–Alder (DA) reaction for self-healing application has been pioneered by Chen et al. They described a "remendable" material capable of offering multiple cycles of crack healing. This approach also offers advantages over the popular microencapsulation approach because it eliminates the need for additional ingredients, such as catalyst, monomer, or special treatment of the fracture interface. Solid-state reversibility of the cross-linking structure via DA and retro-DA reactions was tested and confirmed by subjecting the polymerized films to different heating and quenching cycles and analyzing the corresponding chemical structure by solid-state NMR. Healing in the thermally reversible crosslinked polymers depends upon the fracture and repair of the specific covalent bonds. It is proposed that the bond strength between the furan and maleimide moieties is much lower than that of the other covalent bonds, meaning the retro-DA reaction should be the main pathway for crack propagation. Since the inter-monomer linkages formed by the DA cycloaddition are disconnected upon heating to 120 °C then reconnected upon cooling, the self-healing process does not occur at a temperature lower than 120 °C. Quantification of the healing efficiency by fracture tests shows that it was about 50% at 150 °C, and 41% at 120 °C. Multiple healing at or near the same interface was also observed although the critical load at fracture of the third cracking was about 80% of the second. This drop in mechanical properties from the second to the third healing process was attributed to the healed region. There is potential to use this type of thermally reversible crosslink instead of the DA reagents for self-healing purposes. It should be noted that the practicability of the current embodiment of the technology needs to be improved, since the healing reaction only takes place under extreme conditions (anisole solution maintained at 100 °C for 24 h) [120].

10.8.13.7 Chain Rearrangement Approach

Healing of thermosets has also been shown to achieve by rearranging polymer chains at ambient or elevated temperatures. Similarities exist between this technology and thermoplastic molecular interdiffusion technologies. Chain rearrangement occurring at ambient temperature heals cracks or scratches via interdiffusion of dangling chains or chains lippage in the polymer network. These two ambient temperature modes of healing eliminate the need for heating cycles during healing that were required for the thermoplastic additives or the thermally reversible crosslinks approach. Fractured epoxy resins made from diglycidyl ether of bisphenol-A (DGEBA), nadic methyl anhydride (NMA), and benzyl dimethylamine (BDMA) were shown to repeatedly heal when heated to above150 °C. Healing was assessed visually and by double torsion fracture testing; each resulted in a 100% healing efficiency over multiple fracture events. When subjecting to different thermal treatments, the healing process was independent of the healing temperature or the presence of unreacted monomer but only occurred when the epoxy was heated above its T_g (120 °C). Healing was attributed to Micro-Brownian motion of the polymer chains with local flow enabling good interfacial bonding and the restoration of the original surface contours. In 2007, Yamaguchi et al. reported the first self-healing thermoset based on molecular interdiffusion of dangling chains. These self-healing polymers consisted of a polyurethane network made using a trifunctional polyisocyanate, polyester-diol, and a dibutyl-tin-dilaurate catalyst. The authors varied reagent ratios to manipulate the crosslink density and therefore the number of dangling chain ends. Healing was assessed visually by checking slit closure of cut specimens over time. Using the correct reagent ratios enabled healing to occur rapidly (10 min) once the cut surfaces were brought in contact with each other. The interdiffusion of dangling chains was also found to contribute to healing in epoxies and in polyurethane with initiator residues forming loose chain ends [121].

10.8.13.8 Metal-Ion-Mediated Healing Approach

Self-healing via metal-ion-mediated reactions was developed for the repair of lightly cross-linked hydrophilic polymer gels. This technology involves rearrangement of cross-linked networks; however, additionally, change occurs as metal ions are absorbed from an aqueous solution and then incorporated into the hydrogel. The metal-ion-mediated healing of hydrogels is distinct from self-healing systems discussed earlier because the "healed" material has an entirely different structure and set physical properties from the "unhealed" material, making comparisons between the systems difficult. The self-healing hydrogels contain flexible hydrophobic side chains with a terminal carboxyl group and undergo healing at ambient temperature through the formation of coordination complexes mediated by transition-metal ions. A series of monomers made by reacting amino acids with acryloyl chloride were tested, but a gel based on acryloyl-6-amino caproic acid (A6ACA) was studied extensively (Fig. 10.23). The factors affecting the healing ability include the metal-binding capacity of the gel, the nature of the complexation, and the ability to deform under stress [122].

10.8.13.9 Other Approaches of Thermoset Self-Healing Approach

A number of other approaches have been undertaken during the development of self-healing thermosets including the use of SMA, passivating additives, and water-absorbent matrices. These approaches can be separated from those discussed in previous sections, as they do not repair structural defects but address other properties such as surface smoothness or permeability. The focus of these technologies on nonstructural repairs makes comparison with traditional self-healing polymers difficult. However, these approaches represent novel developments opening avenues to alternative applications of self-healing polymeric systems.

10.9 Self-Healing Coatings

Coatings allow the retention and improvement of the bulk properties of a material. They can protect a substrate from environmental exposure. Thus, when damage occurs (often in the form of microcracks), environmental elements like water and oxygen can diffuse through the coating and may cause material damage or failure. Microcracking in coatings can result in mechanical degradation or delamination of the coating or in electrical failure in fiber-reinforced composites and micro-electronics, respectively. As the damage is on such a small scale, repair, if possible, is often difficult and costly. Therefore, a coating that can automatically heal itself (self-healing coating) could prove beneficial by automatic recovering properties (such as mechanical, electrical, and aesthetic properties) and thus extending the lifetime of the coating. Self-healing coating includes microencapsulation and the introduction of reversible physical bonds, such as hydrogen bonding, ionomers, and chemical bonds (Diels–Alder chemistry). Microencapsulation is the most common method to develop self-healing coatings. To prove the effectiveness of microcapsules in polymer coatings for corrosion protection, researchers have encapsulated a number of materials. These materials include isocyanates monomers such as DCPD, GMA, epoxy resin, linseed oil, and tung oil. For encapsulation of core like as mentioned earlier, number of shell materials have been utilized, such as phenol formaldehyde, urea formaldehyde, dendritic or PAMAM, melamine formaldehyde, and so on. Each shell material has its own merits and demerits. Even these shell materials extended their applications in controlling the delivery of pesticides and

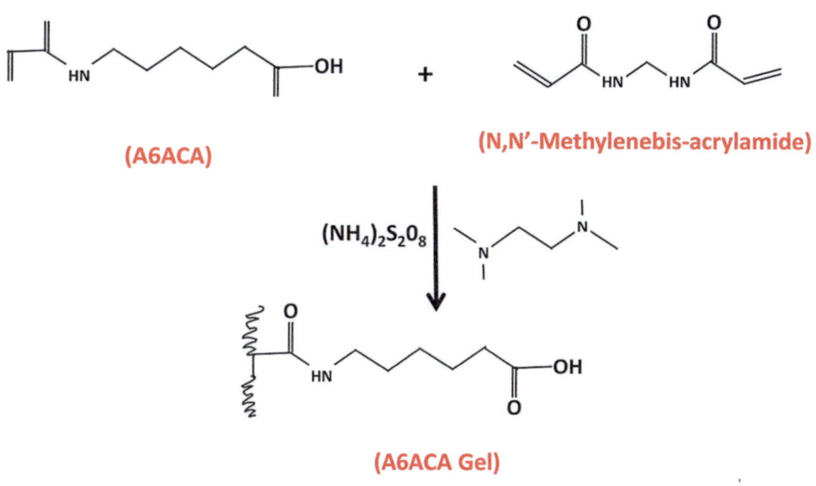

Fig. 10.23 Reaction used to produce self-healing hydrogels

drugs. By using the aforementioned materials for self-healing in coatings, it was proven that microencapsulation effectively protects the metal against corrosion and extends the lifetime of a coating [123].

10.10 Self-Healing Hydrogels

Hydrogels are constructed by the crosslinked polymer networks as water-swollen gels. Hydrogels have received significant attention as the extracellular matrix mimics for biomedical applications because of their water-retention abilities, appropriate elasticities, and network structures. The self-healing properties, originated from phenomena of wound healing in organisms, are used to describe materials with the ability to restore the morphology and mechanical properties after repeated damages. The microcapsule-laden hydrogels were developed that released healing agents at damage sites. However, the irreversible healing process and potential interference of fillers limited their applications. Besides, many dynamic hydrogels typically relied on external stimuli, such as high temperature, low pH, and light, to trigger dynamic crosslinks. The external stimuli would have adverse effects on the cells and living tissues. Self-healing hydrogels are referred that automatically and reversibly repair the damages and recover the functions. Self-healing hydrogels can be prepared through dynamic covalent bonds and noncovalent interactions. The dynamic equilibrium between dissociation and recombination of various interactions leads the hydrogel to heal damages and reform shapes. Commonly, dynamic covalent bonds exhibit stable and slow dynamic equilibriums, while noncovalent interactions show fragile and rapid dynamic equilibriums. With versatile mechanical properties, self-healing hydrogels can be manufactured with robust, shear-thinning, or cell-adaptable properties for a broad range of applications, such as soft robots, 3D printing, and drug/cell delivery. Current applications in biomedical applications of self-healing hydrogels are described [124].

10.11 Applications

It is not difficult to imagine all kinds of applications for self-healing materials, from bridges and buildings that repair their own cracks to car fenders. Matrix self-healing composite used in jet turbine engines, land-based power generation, nuclear fission and fusion reactors, heat exchangers, and furnace components. A microscopic breach in an airplane's fuselage grows and ruptures. Nickel-based superalloys have been used in high-temperature, high-stress applications (such as airplane turbine blades and vanes) for years. An increased effort to develop lightweight, high-temperature, creep-resistant substitute materials has led to the formulation and implementation of silicon carbide ceramic matrix composites (SiC/SiC CMCs) [125]. In current-generation SiC/SiC composites, the matrix tends to crack prematurely when stresses are applied, thus requiring the SiC fibers to carry almost the entire load. To remedy this deficiency, innovators at NASA's Glenn Research Center have successfully created an engineered matrix composite that offers increased resistance to cracking and "self-heals" the cracks that do occur, by converting oxygen into a low-viscosity oxide that acts as a sealant. This groundbreaking concept provides considerable flexibility in designing the composite matrix for a potentially wide variety of high-temperature applications

Mechano-responsive healing polymers are used in aircraft, rotocraft, and spacecraft. NASA Langley Research Center is developing an innovative self-healing resin that automatically reacts to mechanical stimuli. Current structural materials are not self-healing, making it necessary to depend on complicated and potentially destructive repair methods and long downtimes. Unlike other proposed self-healing materials that use microencapsulated healing agents, this technology utilizes viscoelastic properties from inherent structural properties. The resulting technology is a self-healing material with rapid rates of healing and a wide range of use temperatures [126]. NASA inventors have developed a set of puncture self-healing materials comprised of commercially available, known self-healing polymer resin and additive blends. A range of puncture healing blends was developed by melt blending self-healing polymers with non-self-healing polymeric materials [127]. Self-healing low-melt polyimides are used in aerospace (aircraft, helicopters, rockets, expandable or inflatable structures/architectures), defense (missiles, ground vehicles, habitation tents, ships, submarines, unmanned aerial vehicles, automotive, power production (wiring, solar panels), textiles. NASA Kennedy Space Center seeks partners interested in the commercial application of a thin-film, high-performance, polyimide and self-healing/sealing systems for nearly a decade. The self-healing/sealing properties of the materials are provided as a self-sealing polyimide film, a layered composite, as a healant in embedded microcapsules, or a combination thereof. When cut or otherwise damaged, the self-sealing film and/or microcapsule healant will result in a repair of the damaged area. The capability to heal or self-repair in applications such as wire insulation, inflatable structure inner linings, spacesuits, and solar panels is a key technology area for NASA, and developmental testing of these novel materials has shown great promise [128]. Use of microencapsulated dicyclopentadiene (DCPD) monomer and Grubbs' catalyst to self-heal epoxy polymer coating was later adapted to epoxy adhesive films that are commonly used in the aerospace and automotive industries for bonding metallic and composite substrates. Recently, microencapsulated liquid suspensions of metal or carbon black were used to restore electrical conductivity in a multilayer microelectronic device

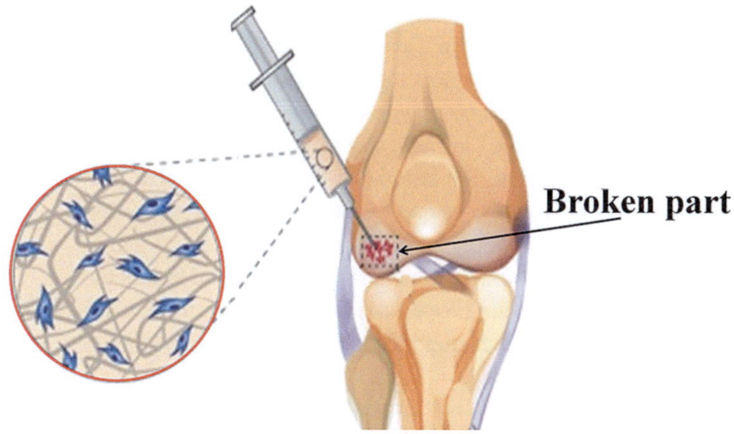

Fig. 10.24 Self-healable hydrogel at knee broken part

and battery electrodes respectively; however, the use of microencapsulation for restoration of electrical properties in coatings is limited. Liquid metal microdroplets have also been suspended within silicone elastomer to create stretchable electrical conductors that maintain electrical conductivity when damaged, mimicking the resilience of soft biological tissue. The most common application of this technique is proven in polymer coatings for corrosion protection. Corrosion protection of metallic materials is of significant importance on an economical and ecological scale [129, 130]. Puncture-healing engineered polymer blends are used in radiation shielding, fuel tank liners, healing layers in ballistic protection for armor, helmets and other personal protective equipment, packaging material, human prosthetics, wire insulation material, space habitats and structures, micrometeoroid, and orbital debris protective liners. Lamborghini and MIT's Terzo Millennio is a self-healing future supercar. That technology can also be used to monitor the car's carbon-fiber structure. If small cracks develop—say, from a collision—the charge may move through the body differently, which can kick-start a "self-repairing" process in order to prevent the cracks from growing.

The electronics industry has been a major driving force in the development of novel materials. The demand for flexible and thinner displays has led to the need for improved materials, ensuring added features in the newest electronics do not compromise on the lifetime of the product. Self-healing materials have the potential of becoming very useful within the electronics industry in the construction of tiny interior components and external materials and displays. This will allow more durable, functional products that can last longer, even when damaged or dropped [131]. Figure 10.24 shows a self-healable hydrogel doped for repair of the broken part [132].

Taking the textile industry a step further to the future of fabrics is a mind-blogging invention of different self-healing materials. There are several strategies used in the industry to achieve such results on fabrics such as reversible cross-links,

Fig. 10.25 Self-healable concrete

using nanotechnology that releases healing agents, and employing technologies like shape memory effect, nanoparticle migration, electro hydrodynamics, conductivity, and co-deposition. The textile industry is progressing toward creating more functional products and smart fabrics for sustainable and greener future self-repairing garments [133].

Self-healing material application has demand in concrete, coatings, polymers, asphalt, ceramic, metals, and end-use industry such as building and construction, transportation, and mobile devices. [134–136]. Figure 10.25 shows a self-healable concrete. Smart coating for corrosion detection and protection is used in bridges, automobiles, ships, pipes, and other infrastructure, machinery, airplanes. NASA Kennedy Space Center seeks partners interested in the commercial application of a smart, environmentally friendly coating system for early detection and inhibition of corrosion and self-healing of mechanical damage without external intervention. This coating will have the inherent ability to detect the onset of corrosion in the coated substrate and respond autonomously to control it. Microcapsules in polymer for self-healing purposes have been used in the paper industry for a range of different purposes, for example, in self-copying carbonless copy paper, and in the food and packaging industries for applications, such as control of aroma release and temperature or humidity indicators [137]. Other possible applications might include encapsulation of antimicrobial agents or scavengers in active packaging. Recently, Andersson et al. have developed microcapsules with a hydrophobic core surrounded by a hydrophobically modified polysaccharide membrane in aqueous suspension to obtain capsules fulfilling both the criteria of small capsule size and reasonably high solids content to match the requirements set on surface treatment of paperboard for enhancement of packaging functionality, and they have shown a reduced tendency for deteriorated barrier properties and local termination of cracks formed upon creasing [138].

10.12 Summary

Self-healing materials are successfully described in this chapter. Different mechanisms associated with the self-healing metal, self-healing polymer, self-healing ceramic, and other materials of self-healing will give a new direction to high-end industries like space industries, where most of the equipment is affected by high-energy particles. One day, we might even have implant parts for the human body that can heal themselves as well as their natural equivalents.

References

1. Wu, M., Johannesson, B., Geiker, M.: A review: self-healing in cementitious materials and engineered cementitious composite as a self-healing material. Constr. Build. Mater. **28**(1), 571–583., ISSN: 0950-0618 (2012). https://doi.org/10.1016/j.conbuildmat.2011.08.086
2. Wool, R.P.: Self-healing materials: a review. Soft Matter. **4**, 400–418 (2008). https://doi.org/10.1039/B711716G
3. Shchukin, D.G., Möhwald, H.: Self-repairing coatings containing active nanoreservoirs. Nano Micro Small. **3**(6) Complex Materials, 926–943 (2007). https://doi.org/10.1002/smll.200700064
4. White, S., Sottos, N., Geubelle, P., et al.: Autonomic healing of polymer composites. Nature. **409**, 794–797 (2001). https://doi.org/10.1038/35057232
5. Van Tittelboom, K., De Belie, N.: Self-healing in cementitious materials-a review. Materials. **6**, 2182–2217 (2013). https://doi.org/10.3390/ma6062182
6. https://www.marketsandmarkets.com/Market-Reports/self-healing-material-market-46412119.html#:~:text=%5B130%20Pages%20Report%5D%20The%20global,95.0%25%20between%202016%20and%202021. Accessed 02 Jan 2020
7. https://www.grandviewresearch.com/industry-analysis/self-healing-materials. Accessed 02 Jan 2020
8. https://www.marketresearchfuture.com/reports/self-healing-materials-market-5503. Accessed 02 Jan 2020
9. Savatin, D.V., Gramegna, G., Modesti, V., Cervone, F.: Wounding in the plant tissue: the defense of a dangerous passage. Front. Plant Sci. **5**, 470 (2014). https://doi.org/10.3389/fpls.2014.00470
10. https://www.wikiwand.com/en/Self-healing_material. Accessed 02 Jan 2020
11. Bekas, D.G., Tsirka, K., Baltzis, D., Paipetis, A.S.: Self-healing materials: a review of advances in materials, evaluation, characterization and monitoring techniques. Compos. Part B. **87**, 92–119 (2016). https://doi.org/10.1016/j.compositesb.2015.09.057
12. Toohey, K., Sottos, N., Lewis, J., et al.: Self-healing materials with microvascular networks. Nat. Mater. **6**, 581–585 (2007). https://doi.org/10.1038/nmat1934
13. Harrington, M.J., Speck, O., Speck, T., Wagner, S., Weinkamer, R.: Biological archetypes for self-healing materials. In: Hager, M., van der Zwaag, S., Schubert, U. (eds.) Self-healing Materials Advances in Polymer Science, vol. 273. Springer, Cham (2015). https://doi.org/10.1007/12_2015_334
14. Yang, Y., Ding, X., Urban, M.W.: Chemical and physical aspects of self-healing materials. Prog. Polym. Sci. **49–50**, 34–59 (2015). https://doi.org/10.1016/j.progpolymsci.2015.06.001
15. Huynh, T.-P., Sonar, P., Haick, H.: Advanced materials for use in soft self-healing devices. Adv. Mater. **29**(19), 1604973 (2017). https://doi.org/10.1002/adma.201604973
16. AbdolahZadeh, M., van der Zwaag, S., Garcia, S.J.: Routes to extrinsic and intrinsic self-healing corrosion protective sol-gel coatings: a review. Self-Healing Mater. **1**, 1–18 (2013). https://doi.org/10.2478/shm-2013-0001
17. Billiet, S., Hillewaere, X.K.D., Teixeira, R.F.A., Du Prez, F.E.: Chemistry of crosslinking processes for self-healing polymers. Macromolecular. **34**(4), 290–309 (2013). https://doi.org/10.1002/marc.201200689
18. Garcia, S.J.: Effect of polymer architecture on the intrinsic self-healing character of polymers. Eur. Polym. J. **53**, 118–125 (2014). https://doi.org/10.1016/j.eurpolymj.2014.01.026
19. Rule, J.D., Brown, E.N., Sottos, N.R., White, S.R., Moore, J.S.: Wax-protected catalyst microspheres for efficient self-healing materials. Adv. Mater. **17**(2), 205–208 (2005). https://doi.org/10.1002/adma.200400607
20. Guadagno, L., Longo, P., Raimondo, M., Naddeo, C., Mariconda, A., Vittoria, V., Iannuzzo, G., Russo, S.: Use of Hoveyda-Grubbs' second generation catalyst in self-healing epoxy mixtures. Compos. Part B. **42**(2), 296–301 (2011). https://doi.org/10.1016/j.compositesb.2010.10.011
21. Kessler, M.R., Sottos, N.R., White, S.R.: Self-healing structural composite materials. Compos. A: Appl. Sci. Manuf. **34**(8), 743–753 (2003). https://doi.org/10.1016/S1359-835X(03)00138-6
22. Brown, E.N., White, S.R., Sottos, N.R.: Retardation and repair of fatigue cracks in a microcapsule toughened epoxy composite-part II: in situ self-healing. Compos. Sci. Technol. **65**(15–16), 2474–2480 (2005). https://doi.org/10.1016/j.compscitech.2005.04.053
23. Samadzadeh, M., HatamiBoura, S., Peikari, M., Kasiriha, S.M., Ashrafi, A.: A review on self-healing coatings based on micro/nanocapsules. Prog. Org. Coat. **68**(3), 159–164 (2010). https://doi.org/10.1016/j.porgcoat.2010.01.006
24. Rule, J.D., Sottos, N.R., White, S.R.: Effect of microcapsule size on the performance of self-healing polymers. Polymer. **48**(12), 3520–3529 (2007). https://doi.org/10.1016/j.polymer.2007.04.008
25. Thakur, V.K., Kessler, M.R.: Self-healing polymer nanocomposite materials: a review. Polymer. **69**, 369–383 (2015). https://doi.org/10.1016/j.polymer.2015.04.086
26. Jun Tan, Y., Jiake, W., Li, H., Tee, B.C.K.: Self-healing electronic materials for a smart and sustainable future. ACS Appl. Mater. Interf. **10**(18), 15331–15345 (2018). https://doi.org/10.1021/acsami.7b19511
27. Lifshitz, I.M., Slyozov, V.V.: The kinetics of precipitation from supersaturated solid solutions. J. Phys. Chem. Solids. **19**(1–2), 35–50 (1961). https://doi.org/10.1016/0022-3697(61)90054-3
28. Alaneme, K.K., Bodunrin, M.O.: Self-healing using metallic material systems – a review. Appl. Mater. Today. **6**, 9–15 (2017). https://doi.org/10.1016/j.apmt.2016.11.002
29. Ferguson, J.B., Schultz, B.F., Rohatgi, P.K.: Self-healing metals and metal matrix composites. JOM. **66**, 866–871 (2014). https://doi.org/10.1007/s11837-014-0912-4

30. Kilicli, V., Yan, X., Salowitz, N., et al.: Recent advancements in self-healing metallic materials and self-healing metal matrix composites. JOM. **70**, 846–854 (2018). https://doi.org/10.1007/s11837-018-2835-y
31. Greil, P.: Generic principles of crack-healing ceramics. J. Adv. Ceram. **1**, 249–267 (2012). https://doi.org/10.1007/s40145-012-0020-2
32. Ferguson, J.B., Schultz, B.F., Rohatgi, P.K.: Zinc alloy ZA-8/shape memory alloy self-healing metal matrix composite. Mater. Sci. Eng. A. **620**, 85–88 (2015). https://doi.org/10.1016/j.msea.2014.10.002
33. Zhu, P., Cui, Z., Kesler, M.S., Newman, J.A., Manuel, M.V., Wright, M.C., Brinson, L.C.: Characterization and modeling of three-dimensional self-healing shape memory alloy-reinforced metal-matrix composites. Mech. Mater. **103**, 1–10 (2016). https://doi.org/10.1016/j.mechmat.2016.09.005
34. Poormir, M.A., Khalili, S.M.R., Eslami-Farsani, R.: Optimal design of a bio-inspired self-healing metal matrix composite reinforced with NiTi shape memory alloy strips. J. Intell. Mater. Syst. Struct. **29**(20), 3972–3982 (2018). https://doi.org/10.1177/1045389X18803448
35. Poormir, M.A., Khalili, S.M.R., Eslami-Farsani, R.: Investigation of the self-healing behavior of Sn-Bi metal matrix composite reinforced with NiTi shape memory alloy strips under flexural loading. JOM. **70**, 806–810 (2018). https://doi.org/10.1007/s11837-018-2826-z
36. Dorri Moghadam, A., Schultz, B.F., Ferguson, J.B., et al.: Functional metal matrix composites: self-lubricating, self-healing, and nanocomposites-an outlook. JOM. **66**, 872–881 (2014). https://doi.org/10.1007/s11837-014-0948-5
37. Rohatgi, P.K.: Al-shape memory alloy self-healing metal matrix composite. Mater. Sci. Eng. A. **619**, 73–76 (2014). https://doi.org/10.1016/j.msea.2014.09.050
38. Gupta, N.K., Kumar, M., Thakre, G.D.: Mechanical characterization of 60Pb40Sn reinforced Al6061 self-healing composite. In: Prakash, C., Singh, S., Krolczyk, G., Pabla, B. (eds.) Advances in Materials Science and Engineering Lecture Notes in Mechanical Engineering. Springer, Singapore (2020). https://doi.org/10.1007/978-981-15-4059-2_4
39. Oladijo, O.P., Bodunrin, M.O., Sobiyi, K., Maledi, N.B., Alaneme, K.K.: Investigating the self-healing behaviour of under-aged and 60Sn-40Pb alloy reinforced aluminium hybrid composites. Thin Solid Films. **620**, 201–205 (2016). https://doi.org/10.1016/j.tsf.2016.08.071
40. Cheng, Y., Xiao, X., Pan, K., Pang, H.: Development and application of self-healing materials in smart batteries and supercapacitors. Chem. Eng. J. **380**, 122565 (2020). https://doi.org/10.1016/j.cej.2019.122565
41. Nakahata, M., Takashima, Y., Yamaguchi, H., et al.: Redox-responsive self-healing materials formed from host–guest polymers. Nat. Commun. **2**, 511 (2011). https://doi.org/10.1038/ncomms1521
42. Guoqiang, L., Amir, S.: A viscoplastic theory of shape memory polymer fibres with application to self-healing materials. Proc. R. Soc. A. **468**, 2319–2346 (2012). https://doi.org/10.1098/rspa.2011.0628
43. Shinya, N., Kyono, J., Laha, K., Masuda, C.: Self healing of creep damage through autonomous boron segregation and boron nitride precipitation during high temperature use of austenitic stainless steels. In: Proceedings of the First International Conference on Self Healing Materials, pp. 18–20. Noordwijkaan Zee, The Netherlands (2007) file:///C:/Users/Ajit%20Behera/Desktop/documents_35.pdf
44. Rajak, P., Kalia, R.K., Nakano, A., Vashishta, P.: Faceting, grain growth, and crack healing in alumina. ACS Nano. **12**(9), 9005–9010 (2018). https://doi.org/10.1021/acsnano.8b02484
45. Laha, K., Kyono, J., Sasaki, T., et al.: Improved creep strength and creep ductility of type 347 austenitic stainless steel through the self-healing effect of boron for creep cavitation. Metall. Mat. Trans. A. **36**, 399–409 (2005). https://doi.org/10.1007/s11661-005-0311-0
46. Zhang, S., Cizek, J., Yao, Z., Oleksandr, M., Kong, X., Liu, C., van Dijk, N., van der Zwaag, S.: Self healing of radiation-induced damage in Fe-Au and Fe-cu alloys: combining positron annihilation spectroscopy with TEM and ab initio calculations. J. Alloys Compd. **817**, 152765 (2020). https://doi.org/10.1016/j.jallcom.2019.152765
47. Grabowski, B., Tasan, C.C.: Self-healing metals. In: Hager, M., van der Zwaag, S., Schubert, U. (eds.) Self-healing Materials Advances in Polymer Science, vol. 273. Springer, Cham (2016). https://doi.org/10.1007/12_2015_337
48. He, S.M., Brandhoff, P.N., Schut, H., et al.: Positron annihilation study on repeated deformation/precipitation aging in Fe–Cu–B–N alloys. J. Mater. Sci. **48**, 6150–6156 (2013). https://doi.org/10.1007/s10853-013-7411-9
49. Lumley, R.: Self healing in aluminium alloys. In: van der Zwaag, S. (ed.) Self Healing Materials Springer Series in Materials Science, vol. 100. Springer, Dordrecht (2007). https://doi.org/10.1007/978-1-4020-6250-6_11
50. Rabeeh, B.M., Fouad, Y.: The synthesis and processing of self-healing materials: a lamellar shape memory alloy in composite structure. In: Sano, T., Srivatsan, T.S. (eds.) Advanced Composites for Aerospace, Marine, and Land Applications II. Springer, Cham (2015). https://doi.org/10.1007/978-3-319-48141-8_22
51. Xu, G., Demkowicz, M.J.: Crack healing in nanocrystalline palladium. Extreme Mech. Lett. **8**, 208–212 (2016). https://doi.org/10.1016/j.eml.2016.03.011
52. Hossain, S., Ravindra, N.M.: Self-healing in materials: an overview. In: TMS 2019 148th Annual Meeting & Exhibition Supplemental Proceedings The Minerals, Metals & Materials Series. Springer, Cham (2019). https://doi.org/10.1007/978-3-030-05861-6_153
53. Neuser, S., Michaud, V., White, S.R.: Improving solvent-based self-healing materials through shape memory alloys. Polymer. **53**(2), 370–378 (2012). https://doi.org/10.1016/j.polymer.2011.12.020
54. Li, G., Zhang, P.: A self-healing particulate composite reinforced with strain hardened short shape memory polymer fibers. Polymer. **54**(18), 5075–5086 (2013). https://doi.org/10.1016/j.polymer.2013.07.010
55. Nosonovsky, M., Rohatgi, P.K.: Development of metallic and metal matrix composite self-healing materials. In: Biomimetics in Materials Science Springer Series in Materials Science, vol. 152. Springer, New York, NY (2011). https://doi.org/10.1007/978-1-4614-0926-7_5
56. https://ntrs.nasa.gov/archive/nasa/casi.ntrs.nasa.gov/20140013299.pdf. Accessed 08 Jan 2020
57. Narumi, K., Qin, F., Liu, S., Cheng, H., Gu, J., Kawahara, Y., Islam, M., Yao, L.: Self-healing UI: mechanically and electrically self-healing materials for sensing and actuation interfaces. In: UIST '19: Proceedings of the 32nd Annual ACM Symposium on User Interface Software and Technology, pp. 293–306. ACM, New Orleans (2019). https://doi.org/10.1145/3332165.3347901
58. Take, N., Joshi, S.N., Dede, E.M.: Self-healing technology for metallic die attach materials in electronics. Adv. Mater. **21**(8), 1900245 (2019). https://doi.org/10.1002/adem.201900245
59. Zhang, S., van Dijk, N., van der Zwaag, S.: A review of self-healing metals: fundamentals, design principles and performance. Acta Metall. Sin. (Engl. Lett.). **33**, 1167–1179 (2020). https://doi.org/10.1007/s40195-020-01102-3
60. Zheng, C., Yue, Y., Gan, L., Xu, X., Mei, C., Han, J.: Highly stretchable and self-healing strain sensors based on Nanocellulose-

supported graphene dispersed in electro-conductive hydrogels. Nano. **9**(7), 937 (2019). https://doi.org/10.3390/nano9070937
61. Liu, W., Sun, X., Khaleel, M.A.: Predicting Young's modulus of glass/ceramic sealant for solid oxide fuel cell considering the combined effects of aging, micro-voids and self-healing. J. Power Sources. **185**(2), 1193–1200 (2008). https://doi.org/10.1016/j.jpowsour.2008.07.017
62. Cai, D., Jia, D., Yang, Z., Zhu, Q., Ocelik, V., Vainchtein, I.D., De Hosson, J.T.M., Zhou, Y.: Effect of magnesium aluminum silicate glass on the thermal shock resistance of BN matrix composite ceramics. J. Am. Ceramic. Soc. **100**(6), 2669–2678 (2017). https://doi.org/10.1111/jace.14795
63. Tavangarian, F., Li, G.: Bio-inspired crack self-healing of SiC/spinel nanocomposite. Ceram. Int. **41**(2(Part B)), 2828–2835 (2015). https://doi.org/10.1016/j.ceramint.2014.10.103
64. Takahashi, K., Jung, Y.-S., Nagoshi, Y., Ando, K.: Crack-healing behavior of Si3N4/SiC composite under stress and low oxygen pressure. Mater. Sci. Eng. A. **527**(15), 3343–3348 (2010). https://doi.org/10.1016/j.msea.2010.02.060
65. Benight, S.J., Wang, C., Tok, J.B.H., Bao, Z.: Stretchable and self-healing polymers and devices for electronic skin. Prog. Polym. Sci. **38**(12), 1961–1977 (2013). https://doi.org/10.1016/j.progpolymsci.2013.08.001
66. Trask, R.S., Williams, H.R., Bond, I.P.: Self-healing polymer composites: mimicking nature to enhance performance. Bioinspir. Biomim. **2**(1) (2007). https://doi.org/10.1088/1748-3182/2/1/P01
67. Blaiszik, B.J., Kramer, S.L.B., Olugebefola, S.C., Moore, J.S., Sottos, N.R., White, S.R.: Self-healing polymers and composites. Annu. Rev. Mater. Res. **40**, 179–211 (2010). https://doi.org/10.1146/annurev-matsci-070909-104532
68. Lv, L., Yang, Z., Chen, G., Zhu, G., Han, N., Schlangen, E., Xing, F.: Synthesis and characterization of a new polymeric microcapsule and feasibility investigation in self-healing cementitious materials. Constr. Build. Mater. **105**, 487–495 (2016). https://doi.org/10.1016/j.conbuildmat.2015.12.185
69. Brown, E.N., Sottos, N.R., White, S.R.: Fracture testing of a self-healing polymer composite. Exp. Mech. **42**, 372–379 (2002). https://doi.org/10.1007/BF02412141
70. Varley, R.J., van der Zwaag, S.: Development of a quasi-static test method to investigate the origin of self-healing in ionomers under ballistic conditions. Polym. Test. **27**(1), 11–19 (2008). https://doi.org/10.1016/j.polymertesting.2007.07.013
71. Kalista Stephen, J., Ward Thomas, C.: Thermal characteristics of the self-healing response in poly(ethylene-co-methacrylic acid) copolymers. J. R. Soc. Interf. **4**, 405–411 (2007). https://doi.org/10.1098/rsif.2006.0169
72. Peterson, A.M., Kotthapalli, H., Aflal, M., Rahmathullah, M., Palmese, G.R.: Investigation of interpenetrating polymer networks for self-healing applications. Compos. Sci. Technol. **72**(2), 330–336 (2012). https://doi.org/10.1016/j.compscitech.2011.11.022
73. Wang, Y., Ding, L., Zhao, C., Wang, S., Xuan, S., Jiang, H., Gong, X.: A novel magnetorheological shear-stiffening elastomer with self-healing ability. Compos. Sci. Technol. **168**, 303–311 (2018). https://doi.org/10.1016/j.compscitech.2018.10.019
74. Liu, Y.-L., Chuo, T.-W.: Self-healing polymers based on thermally reversible Diels–Alder chemistry. Polym. Chem. **4**, 2194–2205 (2013). https://doi.org/10.1039/C2PY20957H
75. Fortunato, G., Marroccoli, V., Corsini, F., Turri, S., Griffini, G.: A facile approach to durable, transparent and self-healing coatings with enhanced hardness based on Diels-Alder polymer networks. Prog. Org. Coat. **147**, 105840 (2020). https://doi.org/10.1016/j.porgcoat.2020.105840
76. Gao, D., Zhang, J., Lyu, B., Ma, J., Yang, Z.: Polyacrylate crosslinked with furyl alcohol grafting bismaleimide: a self-healing polymer coating. Prog. Org. Coat. **139**, 105475 (2020). https://doi.org/10.1016/j.porgcoat.2019.105475
77. Habault, D., Zhanga, H., Zhao, Y.: Light-triggered self-healing and shape-memory polymers. Chem. Soc. Rev. **42**, 7244–7256 (2013). https://doi.org/10.1039/C3CS35489J
78. Zhang, H., Zhao, Y.: Polymers with dual light-triggered functions of shape memory and healing using gold nanoparticles. ACS Appl. Mater. Interf. **5**(24), 13069–13075 (2013). https://doi.org/10.1021/am404087q
79. Ahner, J., Micheel, M., Geitner, R., Schmitt, M., Popp, J., Dietzek, B., Hager, M.D.: Self-healing functional polymers: optical property recovery of conjugated polymer films by uncatalyzed imine metathesis. Macromolecules. **50**(10), 3789–3795 (2017). https://doi.org/10.1021/acs.macromol.6b02766
80. Arslan, M., Kiskan, B., Yagci, Y.: Recycling and self-healing of Polybenzoxazines with dynamic sulfide linkages. Sci. Rep. **7**, 5207 (2017). https://doi.org/10.1038/s41598-017-05608-2
81. Ahner, J., Pretzel, D., Enke, M., Geitner, R., Zechel, S., Popp, J., Schubert, U.S., Hager, M.D.: Conjugated oligomers as fluorescence marker for the determination of the self-healing efficiency in mussel-inspired polymers. Chem. Mater. **30**(8), 2791–2799 (2018). https://doi.org/10.1021/acs.chemmater.8b00623
82. Guimard, N.K., Oehlenschlaeger, K.K., Zhou, J., Hilf, S., Schmidt, F.G., Barner-Kowollik, C.: Current trends in the field of self-healing materials. Switchable Macromol. Syst. **213**(2), 131–143 (2012). https://doi.org/10.1002/macp.201100442
83. Tu, Y., Chen, N., Li, C., Liu, H., Zhu, R., Chen, S., Xiao, Q., Liu, J., Ramakrishna, S., He, L.: Advances in injectable self-healing biomedical hydrogels. Acta Biomater. **90**, 1–20 (2019). https://doi.org/10.1016/j.actbio.2019.03.057
84. Hoogenboom, R.: Hard autonomous self-healing supramolecular materials—a contradiction in terms? Angewandte Chemie. **51**(48), 11942–11944 (2012). https://doi.org/10.1002/anie.201205226
85. Neilson, B.M., Tennyson, A.G., Bielawski, C.W.: Advances in bis (N-heterocyclic carbene) chemistry: new classes of structurally dynamic materials. J. Phys. Org. Chem. **25**(7), 531–543 (2012). https://doi.org/10.1002/poc.1961
86. Urdl, K., Kandelbauer, A., Kern, W., Müller, U., Thebault, M., Zikulnig-Rusch, E.: Self-healing of densely crosslinked thermoset polymers—a critical review. Prog. Org. Coat. **104**, 232–249 (2017). https://doi.org/10.1016/j.porgcoat.2016.11.010
87. Skorb, E.V., Andreeva, D.V.: Layer-by-layer approaches for formation of smart self-healing materials. Polym. Chem. **4**, 4834–4845 (2013). https://doi.org/10.1039/C3PY00088E
88. Jonkers, H.M.: Self healing concrete: a biological approach. In: van der Zwaag, S. (ed.) Self Healing Materials Springer Series in Materials Science, vol. 100. Springer, Dordrecht (2007). https://doi.org/10.1007/978-1-4020-6250-6_9
89. Kazakli, M., Mutch, G.A., Qu, L., Triantafyllou, G., Metcalfe, I.S.: Autonomous and intrinsic self-healing Al_2O_3 membrane employing highly-wetting and CO_2-selective molten salts. J. Membr. Sci. **600**, 117855 (2020). https://doi.org/10.1016/j.memsci.2020.117855
90. Sun, D., Sun, G., Zhu, X., Pang, Q., Yu, F., Lin, T.: Identification of wetting and molecular diffusion stages during self-healing process of asphalt binder via fluorescence microscope. Constr. Build. Mater. **132**, 230–239 (2017). https://doi.org/10.1016/j.conbuildmat.2016.11.137
91. Yang, Y., Zhu, B., Yin, D., Wei, J., Wang, Z., Xiong, R., Shi, J., Liu, Z., Lei, Q.: Flexible self-healing nanocomposites for recoverable motion sensor. Nano Energy. **17**, 1–9 (2015). https://doi.org/10.1016/j.nanoen.2015.07.023
92. Luo, F., Sun, T.L., Nakajima, T., Kurokawa, T., Zhao, Y., Ihsan, A.B., Guo, H.L., Li, X.F., Gong, J.P.: Crack blunting and advancing behaviors of tough and self-healing Polyampholyte hydrogel.

Macromolecules. **47**(17), 6037–6046 (2014). https://doi.org/10.1021/ma5009447

93. Wool, R.P.: Twinkling fractal theory of the glass transition. Am. Phys. Soc. Div. Polym. Phys. **46**(24), 2765–2778 (2008). https://doi.org/10.1002/polb.21596
94. Lee, M.W.: Prospects and future directions of self-healing fiber-reinforced composite materials. Polymers. **12**(2), 379 (2020). https://doi.org/10.3390/polym12020379
95. McDonald, S.A., Coban, S.B., Sottos, N.R., et al.: Tracking capsule activation and crack healing in a microcapsule-based self-healing polymer. Sci. Rep. **9**, 17773 (2019). https://doi.org/10.1038/s41598-019-54242-7
96. Utrera-Barrios, S., Santana, M.H., Verdejo, R., López-Manchado, M.A.: Design of rubber composites with autonomous self-healing capability. ACS Omega. **5**(4), 1902–1910 (2020). https://doi.org/10.1021/acsomega.9b03516
97. Hayes, S.A., Jones, F.R., Marshiya, K., Zhang, W.: A self-healing thermosetting composite material. Compos. A: Appl. Sci. Manuf. **38**(4), 1116–1120 (2007). https://doi.org/10.1016/j.compositesa.2006.06.008
98. Jony, B., Thapa, M., Mulani, S.B., Roy, S.: Repeatable self-healing of thermosetting fiber reinforced polymer composites with thermoplastic healant. Smart Mater. Struct. **28**, 2. https://doi.org/10.1088/1361-665X/aaf833
99. Kim, Y.H., Wool, R.P.: A theory of healing at a polymer-polymer interface. Macromolecules. **16**(7), 1115–1120 (1983). https://doi.org/10.1021/ma00241a013
100. Imato, K., Takahara, A., Otsuka, H.: Self-healing of a cross-linked polymer with dynamic covalent linkages at mild temperature and evaluation at macroscopic and molecular levels. Macromolecules. **48**(16), 5632–5639 (2015). https://doi.org/10.1021/acs.macromol.5b00809
101. Brown, E.N., Moore, J.S., White, S.R., Sottos, N.R.: Fracture and Fatigue Behavior of a Self-Healing Polymer Composite. Cambridge University Press, Cambridge. https://doi.org/10.1557/PROC-735-C11.22
102. Banshiwal, J.K., Tripathi, D.N.: Self-Healing Polymer Composites for Structural Application. IntechOpen, London (2019). https://doi.org/10.5772/intechopen.82420
103. Williams, M.L., Landel, R.F., Ferry, J.D.: The temperature dependence of relaxation mechanisms in amorphous polymers and other glass-forming liquids. J. Am. Chem. Soc. **77**(14), 3701–3707 (1955). https://doi.org/10.1021/ja01619a008
104. Bartlett, M.D., Dickey, M.D., Majidi, C.: Self-healing materials for soft-matter machines and electronics. NPG Asia Mate. **11**(1), 21 (2019). https://doi.org/10.1038/s41427-019-0122-1
105. Sung Hwan, J., Kim, J.C., Noh, S.M., Cheong, I.W.: Environmentally adaptable and temperature-selective self-healing polymers. Macro Molecul. Rapid. Commun. **39**(24), 1800689 (2018). https://doi.org/10.1002/marc.201800689
106. Zhu, D.Y., Rong, M.Z., Zhang, M.Q.: Self-healing polymeric materials based on microencapsulated healing agents: from design to preparation. Prog. Polym. Sci. **49–50**, 175–220 (2015). https://doi.org/10.1016/j.progpolymsci.2015.07.002
107. Wang, X., Liu, F., Zheng, X., Sun, J.: Water-enabled self-healing of polyelectrolyte multilayer coatings. Angewandte Chemie. **50** (48), 11378–11381 (2011). https://doi.org/10.1002/anie.201105822
108. Wu, D.Y., Meure, S., Solomon, D.: Self-healing polymeric materials: a review of recent developments. Prog. Polym. Sci. **33**, 479–522 (2008). https://doi.org/10.1016/j.progpolymsci.2008.02.001
109. Aïssa, B., Therriault, D., Haddad, E., Jamroz, W.: Self-healing materials systems: overview of major approaches and recent developed technologies. Adv. Mater. Sci. Eng. **2012**, 854203., 17 pages (2012). https://doi.org/10.1155/2012/854203
110. Barthel, M.J., Rudolph, T., Teichler, A., Paulus, R.M., Vitz, J., Stephanie, H., Hager, M.D., Schacher, F.H., Schubert, U.S.: Self-healing materials via reversible crosslinking of poly(ethylene oxide)-block-poly(furfurylglycidyl ether) (PEO-b-PFGE) block copolymer films. Adv. Funct. Mater. **23**(39), 4921–4932 (2013). https://doi.org/10.1002/adfm.201300469
111. Zhang, M., Rong, M.: Design and synthesis of self-healing polymers. Sci. China Chem. **55**, 648–676 (2012). https://doi.org/10.1007/s11426-012-4511-3
112. Varley, R.: Ionomers as self healing polymers. In: van der Zwaag, S. (ed.) Self Healing Materials Springer Series in Materials Science, vol. 100. Springer, Dordrecht (2007). https://doi.org/10.1007/978-1-4020-6250-6_5
113. Fainleib, A.M., Purikova, O.H.: Self-healing polymers: approaches of healing and their application. Polym. J. **41**(1), 4–18 (2019). https://doi.org/10.15407/polymerj.41.01.004
114. Zhang, F., Pengfei, J., Pan, M., Zhang, D., Huang, Y., Li, G., Li, X.: Self-healing mechanisms in smart protective coatings: a review. Corr. Sci. **144**, 74–88 (2018). https://doi.org/10.1016/j.corsci.2018.08.005
115. Neusera, S., Michaud, V.: Effect of aging on the performance of solvent-based self-healing materials. Polym. Chem. **4**, 4993–4999 (2013). https://doi.org/10.1039/C3PY00064H
116. Trask, R.S., Bond, I.P.: Biomimetic self-healing of advanced composite structures using hollow glass fibres. Smart Mater. Struct. **15**, 3. https://doi.org/10.1088/0964-1726/15/3/005
117. Ullah, H., Azizli, K.A.M., Man, Z.B., Che Ismail, M.B., Khan, M. I.: The potential of microencapsulated self-healing materials for microcracks recovery in self-healing composite systems: a review. Polym. Rev. **56**(3), 429–485 (2016). https://doi.org/10.1080/15583724.2015.1107098
118. Jones, A., Rule, J.D., White, S.: Life extension of self-healing polymers with rapidly growing fatigue cracks. J. R. Soc. Interface. **4**(13), 395–403. https://doi.org/10.1098/rsif.2006.0199
119. Hia, I.L., Vahedi, V., Pasbakhsh, P.: Self-healing polymer composites: prospects, challenges, and applications. Polym. Rev. **56**(2), 225–261 (2016). https://doi.org/10.1080/15583724.2015.1106555
120. An, S.Y., Arunbabu, D., Noh, S.M., Song, Y.K., Oh, J.K.: Recent strategies to develop self-healable crosslinked polymeric networks. Chem. Commun. **51**, 13058–13070 (2015). https://doi.org/10.1039/C5CC04531B
121. Javierre, E.: Modeling self-healing mechanisms in coatings: approaches and perspectives. Coatings. **9**, 122 (2019). https://doi.org/10.3390/coatings9020122
122. Wang, C.H., Sidhu, K., Yang, T., Zhang, J., Shanks, R.: Interlayer self-healing and toughening of carbon fibre/epoxy composites using copolymer films. Comp. A: Appl. Sci. Manuf. **43**(3), 512–518 (2012). https://doi.org/10.1016/j.compositesa.2011.11.020
123. Sauvant-Moynot, V., Gonzalez, S., Kittel, J.: Self-healing coatings: an alternative route for anticorrosion protection. Prog. Org. Coat. **63**(3), 307–315 (2008). https://doi.org/10.1016/j.porgcoat.2008.03.004
124. Taylor, D.L., In Het Panhuis, M.: Self-healing hydrogels. Adv. Mater. **28**(41), 9060–9093 (2016). https://doi.org/10.1002/adma.201601613
125. Kaya, H.: The application of ceramic-matrix composites to the automotive ceramic gas turbine. Compos. Sci. Technol. **59**(6), 861–872 (1999). https://doi.org/10.1016/S0266-3538(99)00016-0
126. Davis, D., Hamilton, A., Yang, J., et al.: Force-induced activation of covalent bonds in mechanoresponsive polymeric materials. Nature. **459**, 68–72 (2009). https://doi.org/10.1038/nature07970
127. https://www.forbes.com/2009/11/30/battle-jacket-goodyear-technology-breakthroughs-materials.html#35e5b7ad331a. Accessed 15 Jan 2020

128. https://ntrs.nasa.gov/citations/20190000778. Accessed 15 Jan 2020
129. Ullah, H., Azizli, K.A.M., Man, Z.B., Che Ismail, M.B., Khan, M. I.: The potential of microencapsulated self-healing materials for microcracks recovery in self-healing composite systems: a review. Polym. Rev. **56**, 3 (2016). https://doi.org/10.1080/15583724.2015.1107098
130. https://www.cnet.com/roadshow/news/lamborghini-mit-terzo-millennio-self-healing-supercar/. Accessed 21 Jan 20
131. Tee, B., Wang, C., Allen, R., et al.: An electrically and mechanically self-healing composite with pressure- and flexion-sensitive properties for electronic skin applications. Nat. Nanotech. **7**, 825–832 (2012). https://doi.org/10.1038/nnano.2012.192
132. Frei, R., McWilliam, R., Derrick, B., et al.: Self-healing and self-repairing technologies. Int. J. Adv. Manuf. Technol. **69**, 1033–1061 (2013). https://doi.org/10.1007/s00170-013-5070-2
133. Lahiri, S.K., Zhang, P., Zhang, C., Liu, L.: Robust Fluorine-free and self-healing superhydrophobic coatings by H3BO3 incorporation with SiO_2-Alkyl-Silane@PDMS on cotton fabric. ACS Appl. Mater. Interf. **11**(10), 10262–10275 (2019). https://doi.org/10.1021/acsami.8b20651
134. Wiktor, V., Jonkers, H.M.: Quantification of crack-healing in novel bacteria-based self-healing concrete. Cem. Concr. Compos. **33**(7), 763–770 (2011). https://doi.org/10.1016/j.cemconcomp.2011.03.012
135. Schlangen, E., Sangadji, S.: Addressing infrastructure durability and sustainability by self healing mechanisms - recent advances in self healing concrete and asphalt. Proc. Eng. **54**, 39–57 (2013). https://doi.org/10.1016/j.proeng.2013.03.005
136. https://www.bbc.com/news/business-45357786. Accessed 22 Jan 2020
137. Mlalila, N., Kadam, D.M., Swai, H., Hilonga, A.: Transformation of food packaging from passive to innovative via nanotechnology: concepts and critiques. J. Food Sci. Technol. **53**(9), 3395–3407 (2016). https://doi.org/10.1007/s13197-016-2325-6
138. Hu, B., Chen, L., Lan, S., Ren, P., Wu, S., Liu, X., Shi, X., Li, H., Du, Y., Ding, F.: Layer-by-layer assembly of polysaccharide films with self-healing and antifogging properties for food packaging applications. ACS Appl. Nano Mater. **1**(7), 3733–3740 (2018). https://doi.org/10.1021/acsanm.8b01009

11 Self-Cleaning Materials

Abstract

Self-cleaning surfaces have attracted significant attention in the scientific and industrial community for their peculiar fundamental aspects. In the current chapter, the fundamental principles of self-cleaning materials are briefly discussed. Fabrication strategies of various self-cleaning materials are described here in detail. The low surface energy materials (silicones, fluorocarbons, organic materials, and inorganic materials) required for superhydrophobic application are discussed extensively with various fabrication techniques, such as electrospinning technique, wet chemical reaction, electrochemical deposition, hydrothermal reaction, lithography, layer-by-layer assembly, plasma treatment, and 3D printing technique. The applications of self-cleaning surfaces are discussed in terms of the equipment used in aero-industries, maritime industry, automobile industries, electronic industries, medical industries, textile industries, and other industries.

Keywords

Self-cleaning materials · Wettability · Surface tension · Surface energy · Surface roughness · Air pockets · Hydrophobic surfaces · Superhydrophobic surfaces · Hydrophilic surfaces · Super-hydrophilic surfaces · Photocatalysis self-cleaning materials

11.1 What Is Self-Cleaning Property of Materials?

Everyone wants a dirt-free clean surface. In the current decade, self-cleaning material is going to fulfill this dream. This type of surface has the ability to wash out any surface contaminants (dirt, pollutants, and bacteria) that lead to labor-saving and offers a healthy environment. The self-cleaning quality of the surfaces is mainly mimicked from natural occurrences detected in *lotus leaves, gecko feet, water striders, and in many creatures* in this world [1–3]. Researchers have been inspired by this type of surface, where elements of nature help in solving realistic problems. The self-cleaning characteristics can be achieved on any synthetic surface by wettability control. Self-cleaning activity on the surfaces is preferentially applied in many industries, domestic, agriculture, and military sector [4–7].

11.2 History of Self-Cleaning Materials

Paz et al. first attempted self-cleaning surface fabrication in 1995 by the formation of a transparent TiO_2-film on a glass substrate [8]. Gradual research on this topic gave rise to the first commercial production in 2001, that is, Pilkington glass-coated surface. After the commercial production of self-cleaning glasses, TiO_2 nanoparticles are incorporated into other material surfaces to facilitate the self-cleaning property [9]. Gradual detection of different self-cleaning objects in various creatures and materials is presented in Fig. 11.1.

11.3 Classification of Self-Cleaning Materials

According to the surface activities, the self-cleaning surfaces have been divided mainly into three groups: (1) hydrophobic and superhydrophobic surface, (2) hydrophilic and superhydrophilic surface, and (3) photocatalytic surface. The self-cleaning activity in superhydrophilic surfaces was performed by the washing of dust by spreading the phenomenon of the water droplet. At the same time, the superhydrophobic surfaces remove the dirt by rolling water droplets on the surfaces. Moreover, photocatalytic surfaces were formed by the deposition of metal oxides (WO_3, TiO_2, ZrO_2, ZnO, etc.) on the substrate surface. This surface stops dirt deposition by photocatalytic effect or by UV-assisted cleaning methodology [10]. Figure 11.2 shows hydrophobic,

A. Behera, *Advanced Materials*, https://doi.org/10.1007/978-3-030-80359-9_11

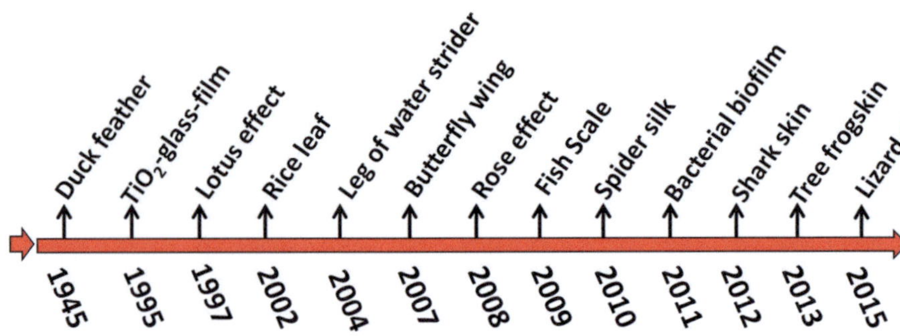

Fig. 11.1 Year-wise development of the self-cleaning materials

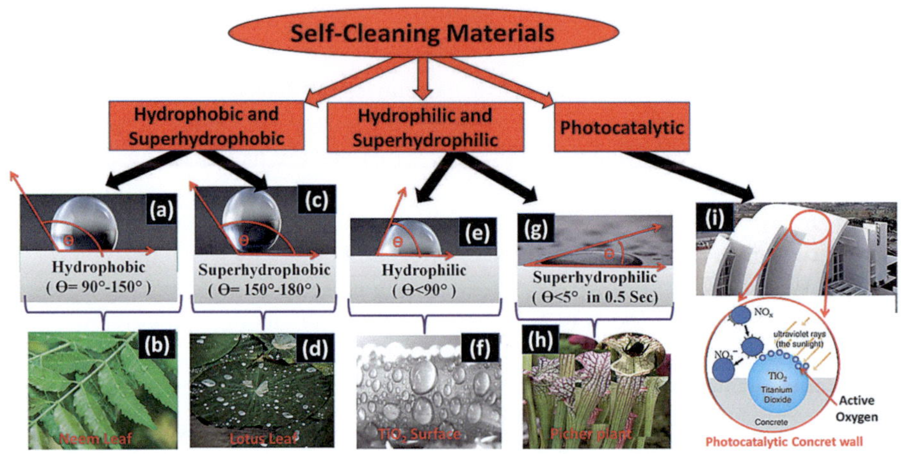

Fig. 11.2 Classification of self-cleaning materials showing (**a**) Hydrophobic contact angle range; (**b**) Hydrophobic Neem plant leaf; (**c**) Superhydrophobic contact angle range; (**d**) Superhydrophobic Lotus Leaf; (**e**) Hydrophilic contact angle range; (**f**) Hydrophilic surface; (**g**) Superhydrophilic contact angle range; (**h**) Superhydrophilic pitcher plant; and (**i**) Photocatalytic concrete material and Photocatalytic TO_2 effect on the surface

superhydrophobic, hydrophilic, superhydrophilic, and photocatalytic materials. The main factor responsible for categorization of these materials, wettability, is given in the figure mentioning their angle condition.

11.4 Surface Characteristics of Self-Cleaning Materials

Hydrophobicity or hydrophilicity of a surface decides the self-cleaning characteristic of a material. In the cleaning of aqueous or organic material from the surface, water plays a vital role. The self-cleaning property of a surface is determined by the water contact angle. This angle is also influenced by the micro/nano-roughness of the surface that is well described in the following models with respect to the wettability (or adhesion) of a self-cleaning surface.

11.4.1 Wettability

The wettability of a surface was measured by static contact angle (θ) [11]. Several factors, such as surface roughness, surface preparation, and cleanliness, decide the contact angle. In hydrophilic surfaces ($0 \leq \theta \leq 90°$), the liquid wets the surface, whereas in hydrophobic surfaces ($90° < \theta \leq 180°$), the liquid cannot wet the surface. The surfaces having water contact angles in the range of 150° and 180° are called superhydrophobic surfaces. The co-relationship of θ with wettability is illustrated in Fig. 11.3a. All the theories of wetting phenomena originate from the equation of Thomas Young [12].

11.4.1.1 Young's Model of Wetting

Young's model of wetting is used to describe the relationship between the water droplet contact angle θ with the surface energies of the water, the self-cleaning surface, and the surrounding air (Fig. 11.3a). This model is generally used to explain the self-cleaning on an ideally flat surface as seen in the lotus leaves mechanism. Young's model is presented in Eq. (11.1) [13].

$$Y_{SG} - Y_{SL} = Y_{LG} \cos \Theta \tag{11.1}$$

where,

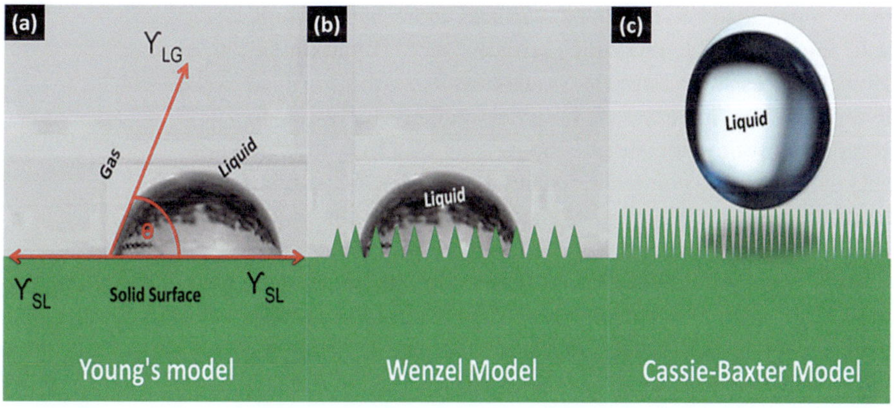

Fig. 11.3 Schematic of (**a**) Young's Model, (**b**) Wenzel, and (**c**) Cassie-Baxter models for wettability

θ = Water contact angle,
Y_{SG} = Surface energy of the surface-air interface.
Y_{SL} = Surface energy of surface-liquid interface.
Y_{LG} = Surface energy of liquid-air interface.

Equation 11.1 simplified as,

$$\cos(\Theta) = \frac{Y_{SG} - Y_{SL}}{Y_{LG}} \tag{11.2}$$

This equation is only reasonable for an ideal solid surface that is surface is smooth, inert, chemically homogeneous, and rigid.

11.4.1.2 Wenzel's Model of Wetting

The dependency of wettability on surface roughness was discovered by Wenzel in 1936. Wenzel's model of wettability dominants over Young's model when a water droplet impinges on a non-flat surface, and the surface morphology results in a larger surface area than that of a surface of perfectly flat texture (Fig. 11.3b). Wenzel's model is presented in Eq. (11.3) [14]:

$$\cos(\Theta) = R_f \cos(\Theta_0) \tag{11.3}$$

where,

Θ = Water contact angle predicted by Wenzel's model.
R_f= Ratio of surface areas of rough surface and flatten surface.

Thus, Wenzel's wetting model describes the droplet and rough surface interface.

11.4.1.3 Cassie-Baxter's Model of Wetting

The effect of surface roughness on wettability was discussed by Cassie and Baxter in 1944. This model is a modification of Wenzel's model and is applicable for complex systems that are representative of water–surface interactions in nature. This model shows water droplet that traps air in micro/nano-rough surface just below it (Fig. 11.3c). The Cassie–Baxter model is presented in Eq. (11.4) [15]:

$$\cos(\Theta_{CB}) = R_f \cos(\Theta_0) - f_{LA}\left(R_f \cos(\Theta_0) + 1\right) \tag{11.4}$$

where,

Θ_{CB} = Contact angle of water droplet predicted by Cassie-Baxter's model.
f_{LA} = Liquid–air fraction, the fraction of the liquid droplet that is in contact with air.

Cassie Baxter's wettability model describes the interfacial phenomena, that is, the air pocket formation between the droplet and the surface topography. According to Wenzel, in the hydrophilic surface, the water droplet spreads more when compared to the hydrophobic surface due to lesser effective surface area in the hydrophobic surface. Both the phenomena are thermodynamically favorable. Here, it can be concluded that the water droplet remains in complete contact with the solid surface, and this state is called the Wenzel state of wettability (Fig. 11.3a). The relationship between the apparent contact angle on a rough surface (Θ_{rough}) and its intrinsic contact angle (Θ_{flat}) has been described by the Wenzel equation:

$$\Theta_{rough} = \Theta_{flat} \tag{11.5}$$

where r is the roughness factor, defined as the ratio of the actual surface area to its horizontal projection. Since r is always >1 for a rough surface, this equation predicts that if Θ_{flat}>90°, $\Theta_{rough} > \Theta_{flat}$, and if $\Theta_{flat} < 90^\circ$, $\Theta_{rough} < \Theta_{flat}$, Therefore, in the Wenzel state, the surface roughness will make intrinsically hydrophobic surfaces more hydrophobic and intrinsically hydrophilic surfaces more hydrophilic. However, as the surface roughness or the surface hydrophobicity increases, it becomes unlikely for water to completely follow the surface topography of a hydrophobic substrate to have complete contact with the solid surface since the system is in a higher energy state when the hydrophobic substrate is in complete contact with water than when it is only in partial contact with water. Instead, air may be trapped between the water and the surface texture. As a result, water is in contact with a composite surface of solid and air and forms droplets (known as fakir droplets) [16]. Such a state is called Cassie State (Figure 11.3b). The apparent contact angle, in this case, has been described by the Cassie-Baxter equation,

$$\cos \Theta_{\text{rough}} = \phi_s \cos \Theta_{\text{flat}} + \phi_G \cos \Theta_{\text{LG}} = \phi_s \cos \Theta_{\text{flat}} - (1 - \phi_s) \tag{11.6}$$

where ϕ_s and ϕ_G are the fractions of solid and air contacting the water $\phi_s + \phi_G = 1$. Since the contact angle of water on air ϕ_{LG} is 180°cos $\Theta_{\text{LG}} = -1$, air entrapment will remarkably increase the apparent surface hydrophobicity. Based on Eq. (11.6), monotonic decrease in Θ_{rough} results in an increase of Θ_{rough} and eventually leads to a superhydrophobic state.

Wenzel state or Cassie state exists when a water droplet is in contact with a rough solid surface. The relationship between cos Θ_{rough} and cos Θ_{flat} for the above-mentioned states is represented in Fig. 11.4, according to Eqs. 11.5 and 11.6, as previously demonstrated [10]. The two lines corresponding to the two states intersect at one critical angle Θ_c

$$\cos \Theta_c = \frac{1 - \phi_s}{r - \phi_s} \tag{11.7}$$

It was proved that with increasing cos Θ_{rough}, the energy of the system decreases monotonically [11]. Therefore, when water contacts a solid surface with $\Theta_{\text{flat}} < \Theta_c$, the Wenzel state is thermodynamically more favorable (with smaller Θ_{rough} and greater cos Θ_{rough} than those in the Cassie state) and hence should be preferred by the system from a thermodynamic point of view; when water contacts a solid surface with $\Theta_{\text{flat}} > \Theta_c$, the Cassie state is thermodynamically more favorable. It should be noted that, according to Eq. (11.5), Θ_c is always >90°, since $\Theta_s < 1$ and $r > 1$.

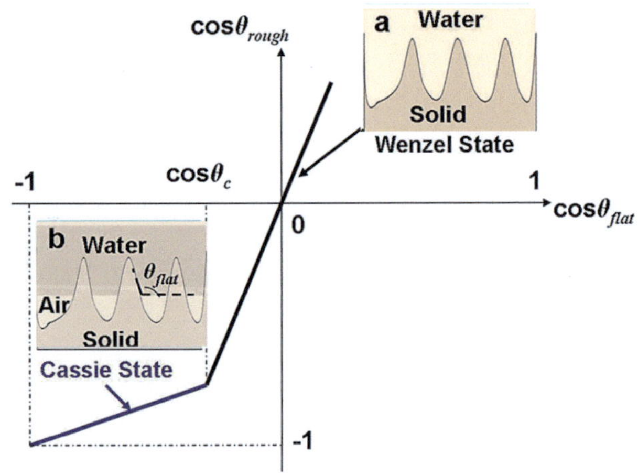

Fig. 11.4 Relationship of cos Θ_{rough} with cos Θ_{flat}. The black solid and blue solid lines correspond to the Wenzel state and the Cassie state, respectively

11.4.1.4 Transition between Cassie and Wenzel States

In their energy profile, the Wenzel and Cassie Baxter states usually correspond to separated energy minima. The Cassie state has higher energy than the Wenzel state. It was observed that there is always an energy barrier when the transition happens. The energy barrier can be influenced by surface features; thus, this energy barrier can be sufficiently large to prevent spontaneous transition into the other primary minimum. This energy barrier can be got overcome by dynamic effects [17].

Cassie–Wenzel transition: The application of force/pressure can create the Cassie–Wenzel transition, for example, impingement of the droplet from a significant height, droplet bouncing, droplets evaporation, application of voltage, or substrate vibration. The critical pressure required for the transition of Cassie–Wenzel is ~300 Pa. The dynamic pressure of rain droplets can reach up to 104–105 Pa, which is much larger than 300 Pa. Therefore, the superhydrophobic surface should be more stable, which is of major importance for a large amount of outdoor applications such as photovoltaic devices covering, windows, coating of building, and so on. Obviously, transitions (Cassie–Wenzel or Wenzel–Cassie) decide the life and death of fakir droplets (droplets in CB state). Generally, a low contact angle hysteresis (CAH) and sliding angle (SA) are used to identify the Cassie wetting. However, at the time of Cassie–Wenzel transition, the water droplets penetrate into the grooves between the pillars, and CAH increases drastically [18].

To elucidate the transition mechanism, Patankar proposed a "sag" transition by virtue of the pillars' modified surfaces. He observed that at the time of droplet impingement on the micro-patterned surface, a meniscus formation is identified due to the action of different pressures. With the increase in applied pressure to the droplet, the movement of the meniscus toward the relief bottom results in the final collapse transition from the Cassie state to the Wenzel state. According to this theory, the pillars that form the relief must be long enough to enhance the stability of the CB state and avoid the collapse transition. By increasing the pillars above a critical height, the "sag" transition becomes impossible because, with the increase in height of the pillars to a certain value, the critical pressure in a droplet will be prevailed over, which would result in the de-pinning of the three-phase contact line. A decrease in the space between two pillars of the modified surface will enhance the pressure. It was also observed that exceeding the critical pressure is essential but not enough for the wetting transition. However, after reaching the critical pressure, the three-phase contact line may impulsively slide along surfaces of the adjacent pillars once this surface is wetted. For this situation, the transition component of hydrophobic and hydrophilic surfaces is very extraordinary. For hydrophobic materials, the multi-scaled structure of the surface builds the possible boundary between the CB and Wenzel states. Additionally, the wetting progress of their side surface is thermodynamically ominous. Be that as it may, contrasted with the Wenzel state, the Cassie state consistently has a higher vitality state for hydrophilic materials, yet it is settled by the energy barrier. The expansion of the fluid air interface over the span of infiltration affects the energy barrier. These modified superhydrophobic surfaces dependent on inalienably hydrophilic materials can be created along these lines. In this way, the apparent hydrophobicity of the surface morphologies dependent on innately hydrophilic materials can be comprehended [19].

Wenzel–Cassie transition: The change from the CB state to the Wenzel state was considered as an irreversible procedure taking into account the way that the Wenzel droplet is at the global minimum energy state. Presently that the Cassie-Wenzel transition has been all around contemplated and can without much of a stretch be accomplished in various manners, what about the reverse transition? Sometimes, water droplets can travel from the Wenzel state to the CB state despite the fact that the Wenzel state is the enthusiastically increasingly positive one for the surfaces utilized. At the point when the CB droplet is the thermodynamically preferred state and the Wenzel droplet is in a metastable express, the change can be accomplished [20].

Effectively, the transition can be activated easily by heating the substrate.

11.4.2 Drag Reduction

As the Cassie–Baxter model shows, superhydrophobic surfaces have an exceptionally convex microstructure, which keeps up a large amount of air between the solid surface, for example, the genuine contact surface between the fluid and the superhydrophobic surface comprises of two interfaces: liquid–solid interface and liquid–air interface [21]. On a superhydrophobic surface, the droplet can just come into contact with the isolated convex microstructure, which incredibly diminishes the contact surface of the water and the solid surface, and the coefficient of friction of the fluid gas interface is a lot lower than the liquid–solid interface so that the superhydrophobic surface will accomplish the objective of decreasing liquid opposition. Ke et al. established that the static contact edge had little impact in decreasing opposition, and the dynamic contact angle was a significant factor in lessening liquid resistance. Applying a superhydrophobic surface to water funnels and pipelines can lessen the friction resistance of the transport medium in the pipeline, in this manner diminishing the expense of pipeline transport. Again, making a superhydrophobic surface on the surface of the boat can diminish the obstruction of the boat to liquids, spare vitality, and lessen the utilization of energy [22].

11.4.3 Surface Tension and Surface Energy

Cohesion forces hold fluid atoms together, and since the particles on a superficial level are not encircled every way by other fluid particles, they have to apply a more grounded force on their neighbors. This more grounded force at the surface is the surface pressure and is the thing that gives protection from objects being squeezed into the fluid. Surface pressure is communicated as an estimation of force required to break a film of fluid that is 1 cm long. The force esteem is communicated in dynes (dyn), where 1 dyne is equivalent to 10 μN. The surface pressure esteems at 20°C for benzene, water, and mercury are 29, 73, and 487 dyn/cm individually [23]. On account of solids, the particles on a superficial level are additionally not as unequivocally reinforced as those inside, and the subsequent vitality is alluded to as surface vitality. The higher the surface vitality of a material, the higher its holding potential with different materials. On account of a fluid laying on a solid surface in a vaporous domain, the contact angle can be characterized as the angle framed by the fluid at the gas, fluid, and solid boundary. Because of the high surface pressure of water, it will in general structure circular droplets on surfaces to diminish its zone and in this manner vitality. The contact angle is thus commonly enormous, although this relies upon the surface material. Because of its high surface tension, water has a high limit with respect to holding if the surface is correct. Hydrophobic surfaces are ones with a low surface vitality that in this way do not draw in water to them. For these materials, the contact angle is more prominent than 90°. Hydrophilic surfaces, for example, glass, have high surface energy and water spreads out over them. For these materials, the contact angle is under 90°. We can change the surface qualities by applying low surface energy materials on any materials. There are additionally some other low surface energy materials, for example, low-thickness polyethylene, dimethylformamide, alkylketene, polycarbonate, polyamide, ZnO, and TiO_2 [24].

11.4.4 Surface Roughness and Air Pockets

The function of the rough textured surface is to make air pockets. A water droplet sits over the entrapped air and this definitely decreases the contact between solid and liquid, permitting the droplets to shape close to the sphere which effectively slips off. Scanning probe microscopy shows the lotus leaves are secured with 1–5 μm knocks called papillae underneath a waxy crystalline top layer. This implies raindrops will move off, alongside any surface dust. A macroscopically smooth surface ordinarily displays a microscopic roughness on various length scales, and the surface microscale and nanostructures along with hydrophobic epicuticular wax crystalloids bring about superhydrophobicity. Similarly, as with the lotus leaf, the flower petal has a hydrophobic wax layer, covering a progressively finished surface framed from nanofolds in its fingernail skin and tapered formed papillae. In any case, the thing that matters are the size of these structures—16 μm in measurement contrasted with 11 μm for the lotus leaf. This implies water can infiltrate the organized surface, making a more noteworthy solid–fluid interface and therefore the stickiness. The sticking is known as the Wenzel state, while the drops that move off the lotus leaf are in the Cassie-Baxter state [25].

11.5 Act of Self-Cleaning Surfaces

For liquid stream applications, not withstanding the high contact angle, superhydrophobic surfaces ought to likewise have extremely low water CAH. On account of these surfaces, water droplets move off (with some slip) the surface and take contaminant with them giving self-cleaning capacity, known as the Lotus effect. The CAH is the distinction between the advancing and receding contact angles, which are two stable qualities. It happens because of roughness and surface heterogeneity. Contact angle hysteresis mirrors a major asymmetry of wetting and dewetting and the irreversibility of the wetting/dewetting cycle. It is a proportion of energy dissipation during the flow of a droplet along a solid surface. For a droplet moving along the solid surface (for instance, if the surface is inclined), the contact angle at the front of the drop (advancing contact edge, ⍰$_{adv}$) is more noteworthy than that at the rear of the drop (receding contact edge, ⍰$_{rec}$), because of roughness, bringing about the contact angle hysteresis otherwise called a sliding angle and roll off-angle [26]. Surfaces with low CAH have low water roll-off-angle, which means the angle to which a surface must be inclined to roll off of water drops (i.e., extremely low water CAH). It is comprehended that during the move off of water droplets, some slip is related [27]. Schematics of self-cleaning surfaces, for example, superhydrophobic, superhydrophilic, and photocatalytic are shown in Fig. 11.5.

Normally, the CAH is estimated during the development and compression of the droplets incited by setting a needle in the water droplets and ceaselessly providing and pulling back the water through the needle. The advancing contact angle θ_A is recorded when the liquid volume arrives at the most extreme before the liquid–solid interfacial region begins to increase; the receding contact angle is recorded when the

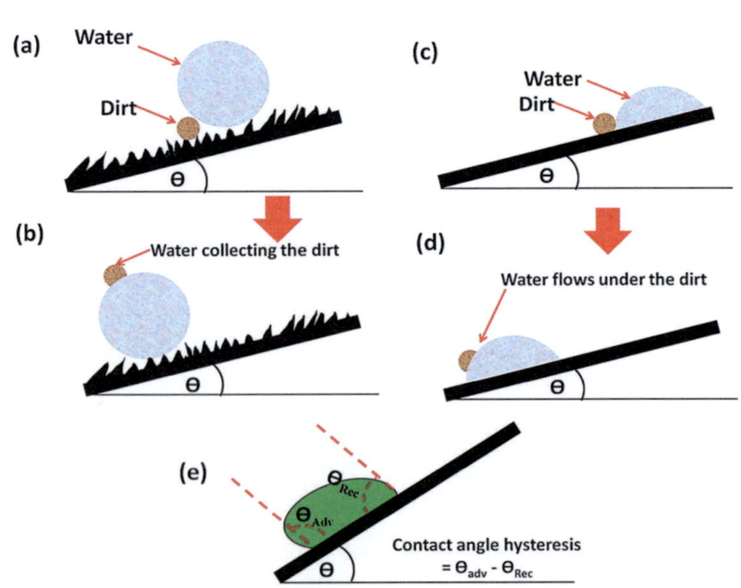

Fig. 11.5 Superhydrophobic surface at (**a**) point of touching of water droplet with the dirt, (**b**) dirt collecting by the water droplet; Superhydrophilic surface at (**c**) point of touching of a water droplet with the dirt, (**d**) dirt collecting by the water droplet, and (**e**) shows the CAH

liquid volume arrives at least before the interfacial area begins to diminish. Furthermore, the SA, which is characterized as the minimum angle that the substrate should be inclined before a droplet begins to roll off, is additionally used to describe the wettability of a surface. The relation between the CAH and the SA can be depicted in the below equation [28]:

$$\mathrm{mg}(\sin\theta)/w = \gamma_{\mathrm{lG}}(\cos_{\mathrm{Rec}} - \cos_{\mathrm{Adv}}) \quad (11.8)$$

where θ is the SA, g is the acceleration due to gravity, m is the mass of the droplet, and w is the diameter of the wetting area. From Eq. (11.8), it can be seen that the SA depends on the mass or the size of the droplet, and for the same sized droplet, small CAH will lead to a small SA. Therefore, SA can be used to compare the wettability of two surfaces only when the same sized droplet is used, and the contact CAH is a better parameter for the purpose of characterizing the surface wettability. Typical superhydrophobic surfaces should have a static water contact angle of >150° and CAH of <10° [29].

11.6 Hydrophobic and Superhydrophobic Surfaces

The word hydrophobic originates from the Greek word (hydro implies: water and phobia imply: dreading or abhorring or fear). In our day-to-day life, organofluorine compounds are utilized for nonstick coatings. Teflon, polytetrafluoroethylene, has been protecting our broiling pan since the 1940s. Be that as it may, with expanding proof of organofluorines' ecological persistence, bioaccumulation, and toxicity, the chase is on for new nonstick solutions [30]. Water on hydrophobic surfaces will show a high contact angle. Examples of hydrophobic molecules incorporate alkanes, oils, fats, and oily substances in normal conditions. If the contact angle falls within 90° to 150°, that surface is called a hydrophobic surface. If the contact angle is more than 150°, and the sliding angle is under 10°, at that point, the water droplets will fall effectively with the dust (Fig. 11.4a, b). In this phenomenon, there is no reverse force included; it is a nonappearance of attraction. Hydrophobic molecules tend to be in general be nonpolar and, subsequently, incline toward other neutral molecules and nonpolar solvents [31]. Since water molecules are polar, hydrophobes do not dissolve well among them. Hydrophobic molecules in water regularly cluster together, shaping micelles. This property mimicked from lotus leaves (Nelumbo nucifera) is water repellent, and adhesive resistance keeps them free from contamination even when rinsed with polluted water [32]. The leave cells have papillae or microasperties that results in very rough surface. Just above the microscale roughness, the papillae surface overlying with nanoscale asperities contains epicuticular wax (hydrophobic hydrocarbons). Basically, the plant cuticle is a composite material consisting of spreaded cutin with low surface energy waxes, growing at different hierarchical levels [33]. The different leveled surfaces of the lotus leaves are made up of convex cells (which look like bumps) and a much smaller layer of waxy tubules. The water pearls on the leaves of the plants rest on top of the nanofunctions because air is trapped in the valley of the convex cells, which minimizes the contact area of the water droplets. Therefore, the lotus leaves indicate a remarkable superhydrophobia. The static contact angle and the hysteresis of the contact angle of the lotus leaf are determined, respectively, around 164° and 3° [34]. At small tilt angles, the water droplets on the sheet roll up and cause dirt or contaminants, leading to self-cleaning. The ability to form droplets and roll depends not only on hydrophobicity but also on contact angle hysteresis. In the plant world, the other natural superhydrophobic surfaces are taro leaves (*Colocasia esculenta*), Indian canna leaves (Cannageneralis bailey), and rice leaves (whatever the type of rice) which represent superhydrophobicity. In the animal world, some examples are the butterfly wings, the leg of the mosquitoes (*Gerris remigis*), the feet of Gecko, and so on [35, 36]. SEM micrographs of butterfly wings show a hierarchy, resulting from aligned micro-grooves, covered with fine strip-stacked nano-rays. The legs of spiders (*G. remigis*) are highly water repellent due to their hierarchical morphology. They are made up of hydrophobic waxy micro hairs, and each hair is covered with nano grooves. As a result, the air is trapped between micro and nano hairs, which repel water. In Gecko's strong grip, when studying the surface characteristics of the toes, there is a hierarchical morphology of each foot that is made up of millions of tiny hairs called bristles. In addition, each dental floss is made up of a smaller pile, and each pile is covered with a flat spatula. In addition to strong adhesion to adhere to the surface, the gecko foot has a unique self-cleaning property that does not require water like the lotus leaf [37].

Hence, superhydrophobicity requires two things:

1. Hydrophobic (non-polar) surface chemistry.
2. Micro- or nanostructured surface texture.

11.6.1 History of Hydrophobic Materials

A few occurrences of superhydrophobicity were distinguished before the term was actually utilized in the literature. Around the 1900s and early 1920s, contact edges (CA) around 160° or more were seen on surfaces coated with soot, and another with galena altered with steric acids shows that research related to superhydrophobic materials goes back to over a century prior [38]. A couple of decades later, Wenzel, Cassie, and Baxter established the theory behind superhydrophobic characteristics. Between the 1940s and 1990s, research advanced gradually particularly regarding geometry impacts on wetting behavior on surfaces. Structural spatial surface investigation was—as is often the situation in science—a consequence of another method: The introduction of the scanning electron microscope (SEMs) in the late 1960s. The properties of man-made superhydrophobic materials were reported in the 1970s and 1980s, and research has accelerated since the 1990s [39]. Later, the theory "lotus impact," which portrays the characteristic of superhydrophobic property in lotus leaves, was proposed by Neihnus and Barthlott, which catalyzed research in the field as exemplified by the first critical review on superhydrophobicity by McCarthy et al. in 1999 [40]. Over the most recent few decades, huge research endeavors have been directed toward the formation of superhydrophobic materials utilizing moderately basic and low-cost methods. A large number of the methods reported in the literature include microfabrication and compound procedures. Some particular procedures incorporate etching, chemical vapor deposition, thermal through-air bonding, and two-beam laser interference. Currently, superhydrophobic materials have been sought for applications, including anti-fog and anti-corrosion surfaces, self-cleaning, deicing, drag decrease, and medical services [41, 42].

11.6.1.1 Direction of Hydrophobicity From Nature

If you observe nature you can find various instances of normally happening superhydrophobic surfaces, for example, duck feathers, butterfly wings, and lotus leaves. Certain plant leaves, such as lotus leaves, are known to be superhydrophobic because of their roughness and the presence of thin wax film on the leaf surface, and the phenomenon is known as the "Lotus effect" [43]. Amazingly, water-repellent superhydrophobic surfaces can be created by utilizing roughness combined with hydrophobic coatings. Other cases of self-cleaning and water-repellent natures are peanut leaf, rose flower petals, poplar leaf, *Salvinia molesta* floating leaves, butterfly wing, fish scale, water strider, compound eyes of mosquito, gecko feet, desert beetles, spider silk, cactus, and many more. These surfaces effectively repulse water, making it cluster together and structure little beads, since it is more attracted to itself than the surface. Artificial superhydrophobic surfaces are manufactured based on the motivations of the lotus leaf by utilizing two sorts of approaches: making hierarchical structures (small scale and nanostructures) on hydrophobic substrates or artificially adjusting various leveled organized surfaces with a low surface energy material [44, 45]. Various investigations have affirmed that the combination of hierarchical roughness, alongside a low surface free energy material, prompts the manufacture of a superhydrophobic surface with apparent water CA $>$ 150, a low sliding point (SA) $<$ 10, demonstrating self-cleaning effect.

Indeed, the superhydrophobic property is often indicated as the lotus effect. The leaves of the lotus comprise micro-scale and nano-scale papillae that are protected in a hydrophobic wax. This double structure makes the leaves superhydrophobic and water reaches a contact angle up to 170° [46]. The subsequent self-cleaning effect implies that lotus leaves are liberated from dust and bacteria, regardless of growing in dirty ponds. Additionally, the manners in which geckos can walk rapidly while upside down fascinates lay-people and researchers. The pads of a gecko's feet are secured with little fibers made of the protein keratin. These enable the gecko to adhere tightly to a surface and yet permit it to lift its feet rapidly so it can walk along a surface at speed without tumbling off. If these fibers become dirty, the capacity would be lost. Presently, it has been found that the gecko secretes oil that bestows superhydrophobic usefulness to keep the toe filaments clean and therefore in great working condition [47].

Another model from nature is the rose petal. Its surface is likewise hydrophobic; however, it behaves in an unexpected way. "At the point when you put a drop of water on a rose petal, you more likely than not saw that the water sticks on the surface—it does not roll off. It is a sticky hydrophobicity. The mystery of life was to imagine a complex, profoundly refined hierarchical structuring of the surface." The surfaces from the crystal structure, for example, tubules, platelets, or strings, however, cover another finished layer of cells or hairs making a few superimposed degrees of "hierarchical sculpturing" at the microscale and nanoscales [48].

A numerous example of plant and animal is given below those showing superhydrophobicity phenomenons:

Lotus Leaf: The lotus leaf, a regular model, is a species that normally develops in swamps and shallow waters in eastern Asia and eastern North America. Lotus is an image of purity and sacredness in numerous Asian nations, particularly in India, China, and so on. The anti-sticking and dirt-cleaning capacity of the lotus was portrayed by an old renowned artist as "Live in the silt but not imbrued." The mechanism behind this reality stayed a mystery till the advancement of the SEM presented in the mid-1960s. There was a great deal of papillae on the lotus surface in an arbitrary dispersion with a diameter ranging from 5 to 9 μm, illustrated in Fig. 11.6. A branch-like nanostructure was found on the microstructure, shaping a hierarchical structure. It was seen that the average diameter of nanosticks was 124.3 ± 3.2 nm on each micropapillae [49]. These hill-and-valley like papillae with branch-like nanostructures spread on the lotus haphazardly and uniformly, which reached among surface and

Fig. 11.6 Lotus effect and roughness effect on superhydrophobicity

Fig. 11.7 Rice leaf effect

water was small, and the hydrophobic nanostructure restricts penetration of water into valleys simultaneously. Superhydrophobicity and low adhesion could be created by such a sort of various hierarchical structures. Not exclusively were the nanostructures found on the top surface, yet in addition, the lower surface represented in the figure and the lower surface of the lotus likewise indicated water repellency [50].

Rice Leaf: Rice is believed to be the major food of some of the nations like India. The droplet on the rice leaf is almost spherical and could not wet the surface, as illustrated in Fig. 11.7. In this figure, we can see a lot of the lengthwise grooves at the macro-level on the rice leaf surface. Through the images, the layered double structure is seen, which is like a lotus leaf. The average diameter of protrusion is between 5 and 8 μm, demonstrating a water contact edge of about $157 \pm 2°$ [51]. The most extraordinary attribute of surface structure between lotus leaf and rice leaf was laid in the arrangement mode of prominence. Anisotropic and hierarchical structures were found on rice leaves. The papillae were organized in one dimension which was nearly parallel to the edge direction of the rice leaf, as shown in the figure. Investigations have been done to demonstrate that this arrangement of surface structure could have an extraordinary impact on the wettability of material. It shows extraordinary superhydrophobic properties as well as an anisotropic sliding property. It is demonstrated that any water droplet can easily roll off along the longitudinal direction in Fig. 11.7; however, it is harder to move along the vertical direction. The sliding angle along the longitudinal direction is 4°, yet 12° along the vertical direction. These anisotropic disseminated papillae in the figure are outwardly expressed as grooves. It is easy for droplets to roll off along the direction of groves [52].

Taro (Colocasia esculenta): It is a wetland herbaceous perennial, which is found throughout the tropic sand of much of the subtropics. Taro has a good growth rate, and it spreads fast over the water bodies and colonizes the surrounding marshy land areas too. Leaf lamina is adaxially matte waxy glaucous and water-shedding, infructescence is not erect, and the central corm is large, cylindrical, up to 35 cm long and 15 cm in diameter, with small side cormels. The most recent researches about taro have been concentrated on pollution cleaning, ecological restoration abilities, and nutritive/medical properties [53]. Taro leaf is indicated by the surface micro- and nanoscale structures and the superhydrophobic surface due to the reduced contact area between the water and solid interface, which is similar to lotus leaves. The taro leaf has its own surface composite structure with a combination of micro−/nanosized structures, as shown in Fig. 11.8. Microscale hexagonal shape prominence, protruding round shapes,

Fig. 11.8 Taro leaf texture

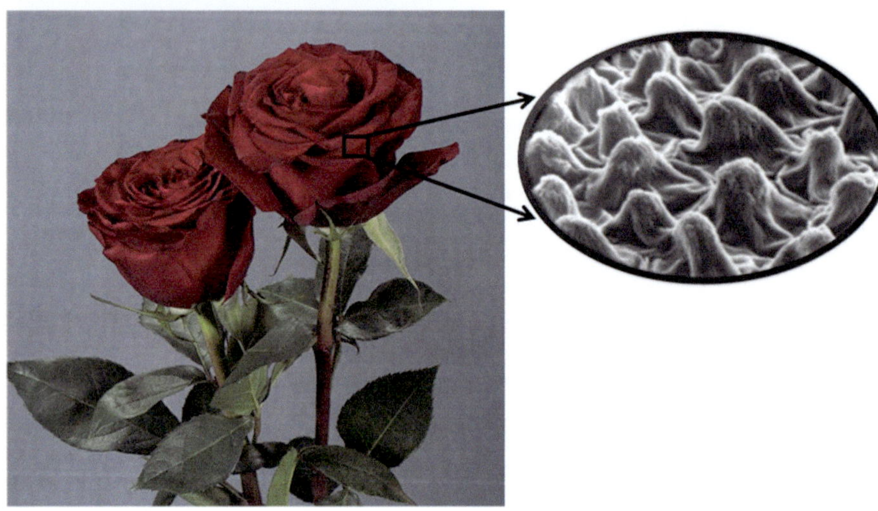

Fig. 11.9 Rose petal

and nanosized contour lines near the hexagonal shapes with petal-like nanostructures were observed on the surface. The taro leaf has a unique microstructure that is owned by many elliptic protrusions with an average diameter of about 10 μm on its surface. These elliptic protrusions were uniformly distributed in their corresponding nest-like caves that many nanoscale pins were also found harmoniously distributed on the surface under a higher SEM magnification. The hierarchical structure on its surface together with the formed microstructure has aroused the surface superhydrophobicity, which is confirmed by a higher CA of about $159 \pm 2°$, and a lower SA of about 3° [54].

Rose Petal: The abovementioned lotus leaf, taro, and rice leaf are several typical plant leaf representatives owning high water CA and low SA. Water could be formed almost as spherical droplets on the leaf surface and easily be moved away. However, another kind of plant surface with higher water CA but higher SA is existing as well, such as red rose petal. The droplet is no more like that of lotus leaf but is hard to be moved away freely in this situation. This kind of special phenomenon is called as "rose-petal effect." [55]. A huge difference exists between the leaf surface and the red rose petal. The red rose petal shows not only higher superhydrophobicity but also a great adhesive power compared to the leaf. Compared to the researches about lotus and rice, researches on red rose petal are few, but it is still successful. Orderly arrayed microtubercle structures are present on red rose petal surface, and nanoscale covering is folded on it. As shown in Fig. 11.9, even if the petal is overturned for 180°, the surface with regularly arrayed papillae still has good adhesion properties for droplets, and it will not drop out from the petal. The fresh red rose surface is composed of an array of plump papillae. A periodic array of micropapillae with an average diameter of 16 μm and height of 7 μm is observed. The measured water CA is 152.4°. According to the theories of two typical models, including Wenzel and Cassie state, the contact area between water and surface in Wenzel state is larger than the one in Cassie state; therefore, the water in Wenzel model owned a higher CA [56].

Fig. 11.10 *Salvinia molesta* leaf

Fig. 11.11 (**a**) *Callistephus chinensis*, (**b**) *Nasturtium Leaves*, and (**c**) *Brassica oleracea*

Other petals such as sunflower petals, Chinese kaffir lily, and so on also exhibit adhesion superhydrophobicity, petals like white orchids and poinsettias show high CA, and another special calla petal with a CA $\approx$ 0° exhibits extreme superhydrophilicity.

Salvinia molesta fern: In many instances, nature has used superhydrophobicity to allow plants and insects to survive underwater for long periods of time. One example is *Salvinia molesta*, an extremely invasive fern that can survive underwater. Its water-repellent surface holds a protective air layer via an array of whisk-shaped hairs (called trichomes) that make up the surface, shown in Fig. 11.10. The tips of the whisks are chemically distinct, being hydrophilic, and this firmly pins a water layer to the surface with air trapped underneath. The pinning effect keeps the air layer as large as 3.5 mm under negative pressure in small individual pockets [57].

Callistephus chinensis: This plant grows easily in rich, moist, well-drained soils in full sun to part shade. This is an annual or biennial herb with one erect, mostly unbranched stem growing 20–100 cm tall. The adaxial surface of leaves of *Callistephus chinensis*, having a CA of 139°, was densely covered with conoid trichomes, shown in Fig. 11.11a [58].

Nasturtium Leaves: The bumpy texture traps air between leaf and water, forcing the water to bead and perch on the peaks rather than sit and spread in the valleys. With nothing to anchor them, the droplets roll around with little contact resistance until they fall off the edge of the leaves, shown in Fig. 11.11b [59].

Brassica oleracea: *Brassica oleracea* is a plant species that includes many common foods as cultivars, including cabbage, broccoli, cauliflower, kale, Brussels sprouts, collard greens, savoy cabbage, kohlrabi, and gailan, as shown in Fig. 11.11c. In its uncultivated form, it is called wild cabbage. *Brassica oleracea* exhibits CA more than 160° [60].

Water Strider Leg: The water strider has six legs, with the short propodium is used to catch prey, and the other four legs are the main apparatus to make it stand effortlessly and move quickly on the water, exhibiting nonwetting property. This antiwetting feature was believed to be due to a surface–tension effect caused by secreted wax at first. There is a hierarchical micro/nanostructure of the water strider leg. Biomimetic researches revealed that the hierarchical structure of a water spider's leg is covered by innumerable oriented setae that have diameters in a few to tens of micrometers. As is shown in Fig. 11.12, the setae are arranged at an inclined angle of about 20° from the surface, and the set surface has a groove structure with depth and width in about 100 nm, in which tiny air bubbles could be trapped. A high-sensitivity balance system was used to construct force-displacement curves for water striders legs when pressing on the water surface, and as a result, the maximal supporting force exerted by a water strider on the water surface reaches at least 750 dynes, which is more than 60 times the weight of its body, exhibiting striking flotation ability. Inspired by the water strider legs' antiwetting property [61],

Shark Skin: Fish scale is considered as owning the ability to reduce the friction between water and fish body. Shark skin is famous for its superior drag reduction and antifouling performance which is induced by its unique surface morphology. Actually, shark skin is composed of its riblet microtexture, flexion of scales, and a mucous layer. Micrometer-sized grooved scales are grown on shark skin, which is called dermal denticles, interlocking to form a natural nonsmooth surface. The dermal denticles are covered by spaced microriblets, and its distribution arrangement is

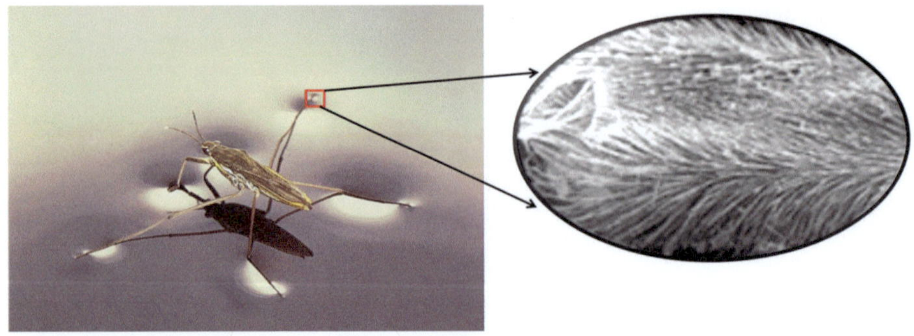

Fig. 11.12 Water strider magnified leg

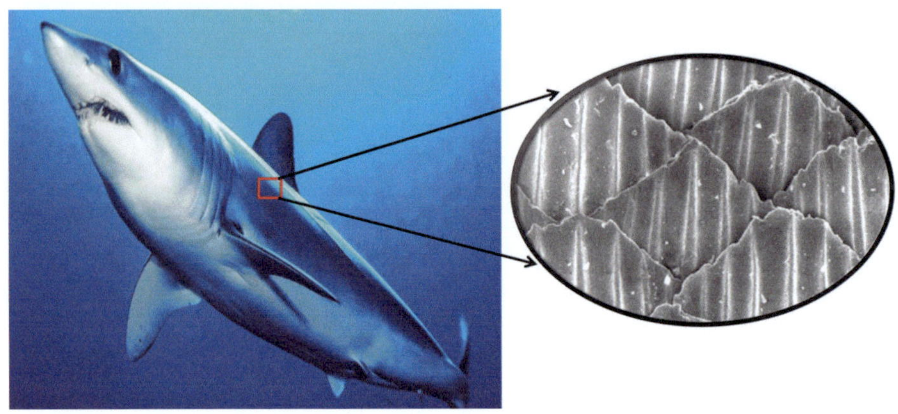

Fig. 11.13 Shark skin riblet microtexture

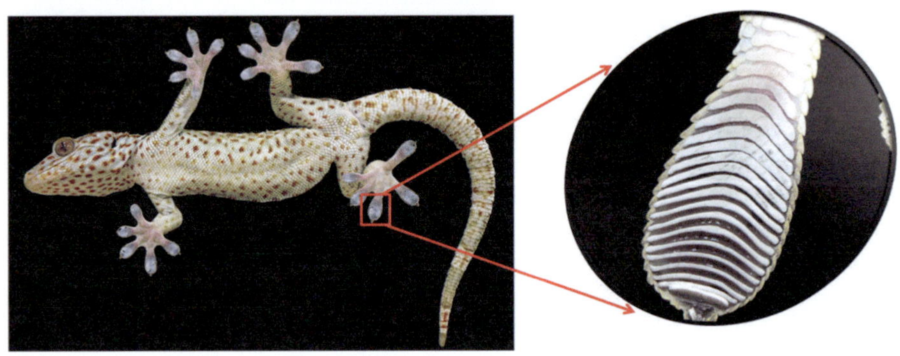

Fig. 11.14 Gecko toe and its magnified image

paralleled to the moving direction, as shown in Fig. 11.13. This surface could reduce vortices formation, and the riblets are considered to lift the vortices off the surface, which results in moving easily in the water field [62].

Gecko Toe: The gecko is famous for its extraordinary ability to climb rapidly and steadily up smooth vertical surfaces. The gecko toe exhibits high CA and adhesive force. And this mysterious ability is attributed to its toe tissue, which is composed of nearly 500,000 keratinous hairs called setae. The microscale of the setae is ranged from 30 to 130 μm. The setae that contain hundreds of smaller hairs are called spatulas, and its diameter is about 7 nm. These hierarchical structures on the soles of gecko feet, shown in Fig. 11.14, are relatively large compared to the structures found on superhydrophobic plants and insects, and as a result, any water coming into contact with them strongly adheres to the surface. Each spatula hair produces a minuscule force of about 10^{-7} N through van der Waals' interactions, but millions of hairs create a formidable adhesion of about 10 N/cm^2 collectively. This is exactly what makes geckos freely walk on smooth and vertical even turned upside down surfaces [63].

Springtail insect body: Spring tails are small soil-dwelling arthropods that have remarkable cuticles with intrinsically omniphobic surfaces displaying both static repellency and pressure resistance to impacts of drops such as raindrops. The springtail insect's body is hierarchically structured and able to hold air bubbles to repel water, shown in Fig. 11.15. The springtail surface has highly ordered rhombic or hexagonal honeycomb-like patterns, composed of three different hierarchical layers. The resulting structure forms nanocavities (0.3–1 μm) covering the entire body, but in

Fig. 11.15 Springtail insect's body is a hierarchically structure

Fig. 11.16 Watermelon outer surface

cross section, there are characteristic, mushroom-shaped overhangs. This particular cross-sectional structure keeps nanosized air bubbles in the structure, and this retention of the gas avoids wetting even of non-polar liquids [64].

Watermelon Surface (*Citrullus lanatus*): *Watermelon* is a plant species in the family Cucurbitaceae, a vine-like flowering plant originally domesticated in West Africa. It is a highly cultivated fruit worldwide, having more than 1000 varieties. Watermelon is a scrambling and trailing vine in the flowering plant family Cucurbitaceae. The hydrophobic surface of the watermelon can be observed in Fig. 11.16 [65].

11.6.2 Type of Superhydrophobic Surface in Plant Leaves

Surface hydrophobicity could be affected by its surface morphology, characters, and chemical structures of surface constituents. Hydrophobic surface morphology is classified according to the shape of its units, including convex hull type, scallops type, stripe type, ripple type, taper type, and so on. [66]. Basically, we can observe two micro-texture as given subsequently.

With Hierarchical Structure: The surface morphology is similar to that of the lotus leaf of small papillate hills, with an average diameter of about 6 μm. In addition, these micrometer-scaled hills and the valleys between the hills are textured by intricate nanostructures. The obtained surface structure is also very similar to that of the lotus leaf. Besides the lotus leaf as a template, other superhydrophobic plant leaves were also used as templates are shown in Fig. 11.17 [67].

With Unitary Structure: Fig. 11.18 shows the natural superhydrophobic surfaces with unitary structure. It shows typical images of the rear face of the ramee leaf with superhydrophobicity (CA of water is 164°). It is clearly seen that many quite slick fibers with diameters from 1 to 2 μm were uniformly distributed on the surface, forming a unitary structure different from the surface with hierarchical structures of the plant leaves referred to earlier. Interestingly, this unique surface structure is also found in the watermelon. The surface structure of Chinese watermelon is surprisingly similar to that of ramee leaf with a CA of water of 159°, indicating that the hierarchical structure is not a necessary condition to form the superhydrophobic surface in nature. Importantly, their mechanical properties for the only

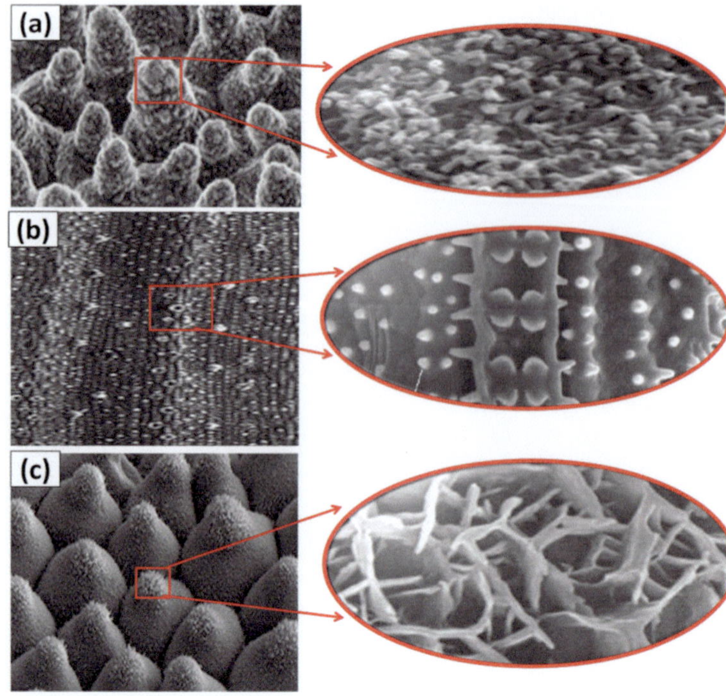

Fig. 11.17 Magnified images of natural superhydrophobic surfaces with hierarchical structures. (**a**) Lotus leaf hierarchical pattern, (**b**) Rice leaf hierarchical pattern, and (**c**) Taro leaf hierarchical pattern that can produce contact angle of 162°, 157°, and 159°, respectively

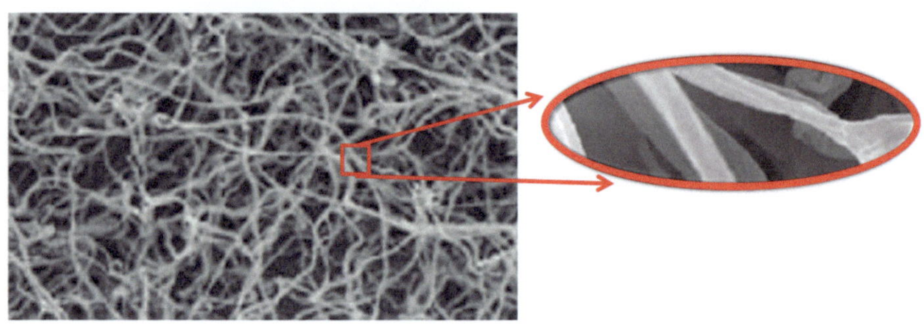

Fig. 11.18 Unitary superhydrophobic surfaces in ramee/watermelon surface. It will produce water CA of 164° for ramee and 159° for watermelon

microstructure are better than that of the hierarchical structures [68].

11.7 Hydrophilic and Superhydrophilic Self-Cleaning Surfaces

Hydrophilic, meaning to love water (comes from the Greek words: hydro meaning water and philos meaning love), is generally defined as having a CA of <90° [69]. Superhydrophilicity allows surfaces to clean a wide variety of dirt or debris, and here the CA is <5°. On such a surface, water droplets spread very rapidly and water flows from the surface at a considerable speed. For self-cleaning superhydrophilic surfaces, cleaning occurs because the water on the surface can act to a great extent (extremely small CA with water) to position itself between contaminating debris and the surface to be washed. From nature, pitcher plant and shark skin are examples of self-cleaning superhydrophilic surfaces. Pitcher plant is a carnivorous plant. Insects caught on the edge of the petal jar (peristome) fall prey to the bottom of the jar [70]. One of the most important characteristics of the jar to catch insects is its slippery surface (Fig. 11.19). The surface of the peristome shows superhydrophilicity due to microtopography of the surface and the secretion of hygroscopic nectar. As a result, stable water films are formed in wet conditions. This water film causes insects that walk above the rim to slide down from the jar and repelling the oils on their feet. Another example of anti-fouling, self-cleaning, and low-adhesion surfaces is shark skin. This hydrophilic surface allows sharks to quickly maneuver in the water. Shark skin is made up of periodically arranged diamond-shaped dermal denticles superimposed on triangular triangle riblets. The wettability of these surfaces is normally high, so the water CA tends to be approximately 0°. Despite such cleaning ability, it is still in a mature state with respect to hydrophobic coatings, and research in this field is still underway to innovate effective cleaning ability for such coatings by varying material compositions. These are high surface energy substrates that attract water and allow surface wetting.

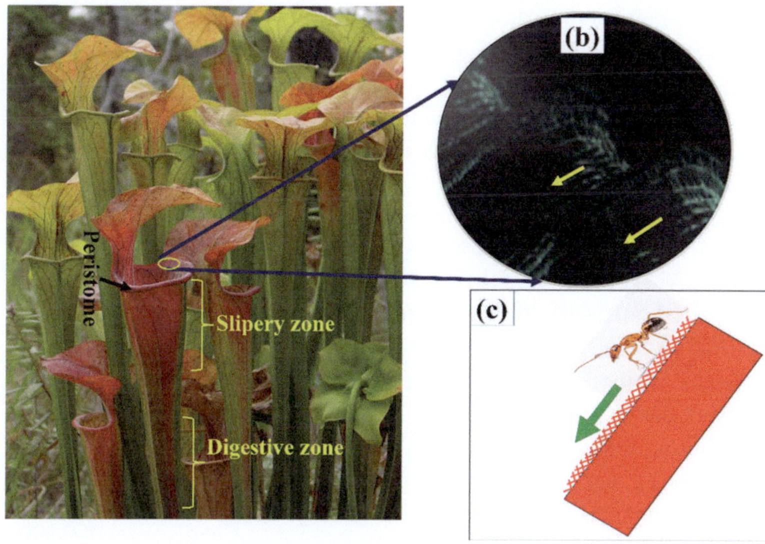

Fig. 11.19 (**a**) Superhydrophilic pitcher plant; (**b**) Topography of peristome, arrow mark indicates the direction toward inside of pitcher; and (**c**) Slippery mechanism: When peristome is wet by nector or rain water, insects aquaplane across the surface and fall into the inside of the pitcher, where they are digested

Many surfaces tend to be more water-friendly, including glass, steel, or stainless steel and many coatings and paints. If the droplets extend almost flat, with a CA of less than about 5°, they will show superhydrophilicity [71].

11.8 Photocatalysis Self-Cleaning Materials

One of the most widely used self-cleaning products, TiO_2, uses a unique self-cleaning mechanism that combines an initial photocatalytic step and subsequent superhydrophilicity. TiO_2 is widely used because of its nontoxicity, availability, cost-effectiveness, chemical stability, and favorable physical and chemical properties [72]. A TiO_2 layer, usually on glass windows, when exposed to UV light, will generate free electrons that will interact with oxygen and water in the air to create free radicals. These free radicals in turn decompose any embedded organic matter deposited on the glass surface. TiO_2 also transforms normally hydrophobic glass into a superhydrophilic surface (CA of approximately 0°) [73]. Therefore, when precipitation occurs, instead of the water dripping onto the window surface and instantly falling onto the glass, the raindrops spread quickly on the hydrophilic surface. The water will descend along the surface of the window, in the form of a film rather than a drop, essentially acting as a squeegee to remove debris from the surface. Studies have shown that the self-cleaning effect of TiO_2 can be enhanced by water flow or precipitation. Therefore, one of the best applications of TiO_2 self-cleaning surfaces is construction materials, as these could be exposed to abundant sunlight and precipitation. Later, in the 1990s, a wide range of self-cleaning hydrophilic coatings such as cement, tile, marquee materials, glass, aluminum cladding, and plastic films, were commercialized. Hydrotech™, a photo-induced superhydrophilic technology was introduced by TOTO Ltd. (a Japanese company). This technology uses sunlight to break down washable dirt/contaminants. It could be effectively used in building materials, coatings, and construction paints. High processing temperatures are required to manufacture TiO_2-coated, self-cleaning photocatalytic glazing products in commercial applications. There are several forms of TiO_2, among which the primary phases are the anatase, rutile, and brookite phases [74]. The most common form of TiO_2 is the rutile phase, which is densely packed and used in pigments such as sunscreens and paints. The anatase phase is rare and has an open crystal structure, making it highly photocatalytic. The anatase and rutile phases have a tetragonal structure. The brookite phase is orthorhombic, and it is extremely rare. The TiO_2 anatase phase when heated to more than 400 °C becomes the rutile phase. Photocatalysis can generally be classified into two kinds of processes. The process in which the adsorbate molecule is photoexcited and interacts with the catalyst substrate in the ground state is known as catalyzed photoreaction. Furthermore, if the initial photoexcitation takes place on the catalytic substrate and transfers an electron or energy to a molecule in the ground state, the process is called sensitized photoreaction. The quantum yield (number of events that occur per absorbed photon) determines the efficiency of a photocatalytic process. By analyzing all possible paths for electrons and holes, quantum efficiency or yield is calculated. As TiO_2 is a semiconductor, during the absorption of light greater than or equal to its prohibited band, it is excited to produce electrons and holes. Most of these charge carriers undergo recombination and some migrate to the surface. The electrons produced move from the valence band to the conduction band, where it reacts with atmospheric oxygen to produce superoxide radicals. These superoxide radicals are very energetic, breaking down organic dirt into carbon dioxide and water, which is called the cold combustion process. The decomposition of stearic acid into carbon dioxide and water vapor occurs in the presence of atmospheric oxygen. TiO_2 is even used in paints and cosmetics as a pigment and as a food additive. The material is also used in antipollution applications and for water purification (the membrane is covered with TiO_2 that kills bacteria in the water) [75].

11.9 Materials Used for Synthesis of Superhydrophobic Surfaces

Different methods adopted to achieve superhydrophobic nanosurfaces are introduction of (1) some nanomaterials; (2) low surface energy materials such as fluorochemicals and silicones; and (3) hybrid methodology such as low surface energy materials along with nanofillers [76].

The roughness can be attained either by using various nanoparticles or by growing nano-morphologies using various metals and metal oxides. The transition metals of d-orbitals in the periodic table (group III to XII) have peculiarities such as variable oxidation states, electrical properties, and colored compound formation due to loosely bound valence shell electrons. The oxidation number increases from left to right and decreases at the end of d-orbitals in the periodic table. Due to this variation in oxidation states, most of the transition metals can create various oxide complexes by forming coordination bonds with ligands, such as water, ammonia, and chloride ions. Hence, these transition metals are being utilized as substrate and their oxidation/anodization are essential for the formation of a rough surface. The surface wettability is mainly governed by multiscale roughness (from micro- to nanoscale) and chemical modification done over the roughened surface. Materials employed for chemical modification possess intrinsically low surface energy due to nonpolar chemistries and closely packed, stable atomic structures can exhibit water repellency. An example includes polysiloxanes (-Si-O-Si- groups), fluorocarbons (CF_2/CF_3), non-polar materials (with bulky CH_2/CH_3 groups), or polymers with combined chemistry. These low surface energy materials are coated over an already rough surface. The transition metal oxides and nanomaterials in group VIII and IB (e.g. Fe, Co, Ni, Cu, and Ag) are successfully utilized to grow micro- or nanoscale morphologies by using neutralization and oxidation–reduction reaction. Solubility of organometallic precursors, crystal size, and shapes are crucial parameters for nucleation. Moreover, they show strong interaction with O, N, and S ligands that effectively modify nanocoating with any low surface energy material. Most of the transition metals and their oxide nanoparticles possess multifunctional characteristics, such as corrosion-resistant, antibacterial behavior, photocatalytic activity, magnetic properties, electrical properties, optical properties, and so on. Other properties considered for material selection include hardness, stiffness, elastic modulus, durability, corrosion resistance, sustainability, and cost [77].

Various materials that can develop the superhydrophobicity on the surface are discussed subsequently:

ZnO-based Surfaces: Zinc oxide (ZnO) is an n-type semiconductor with large exciting binding energy (60 meV) at room temperature with unique optoelectronic, piezoelectric, and optical transparency. Additionally, it possesses several advantages, such as environmentally safe, low cost, corrosion-resistant, and antibacterial behavior, which enable its application in chemical/biological sensors, solar cells, UV detectors, piezoelectric materials, and low friction/wear coatings for wide temperature range. Generally, the surface of ZnO is known to be hydrophilic, and it can be transformed into superhydrophobic by surface modification with low surface energy materials. The phenomenon of achieving superhydrophobicity by incorporation of ZnO is mainly attributed to low surface energy and the ability of the surface to hold air within the micro- and nano-structures developed. However, zinc oxide-based surfaces are reversible in nature and can be converted to superhydrophilic surfaces once subjected to UV irradiation. Zinc oxide possesses excellent photocatalytic activity for organic compound degradation. Once ZnO surface gets subjected to UV light, the electrons in the valence band get excited to conduction band and simultaneously create holes in the valence band. These electrons and holes move around the surface of ZnO and begin redox reaction with oxygen and water which lead to disintegration of low surface energy material coatings over the surface and transforming it to superhydrophilic type. The rate of photocatalytic degradation by ZnO has been quantified by analyzing decolorization of methylene blue dye using UV–visible spectrophotometer [78].

TiO_2-based Surfaces: Titanium dioxide (TiO_2) nanoparticles offered excellent chemical and physical stability, high surface area, low cost, and reduced toxicity. Photocatalytic activity of TiO_2 received tremendous attention due to its unique electronic and optical properties, that is, high ability of UV absorption and photocatalytic decomposition of organic molecules. These singularities are based on surface interaction phenomenon, which gets triggered by factors such as light, temperature, electric potential, solvent, and mechanical actions. UV radiations with energy higher than TiO_2 bandgap (3.2 eV) are engrossed by TiO_2 which resulted in the formation of electron–hole pairs. These holes and electrons react with H_2O and O_2, which produced highly oxidizing OH and O_2 radicals. These oxidizing agents produced by the photocatalytic process can mutilate molecules into H_2O and CO_2. When TiO_2 is used for UV protection of organic materials, the photocatalytic activity is inhibited by the surface reformation of TiO_2. This implies that photocatalytic oxidation and photoinduced switching wettability proceed on the TiO_2 surface under UV irradiation [79].

Clay-based surfaces: Clay is the cheapest material available in various forms of hydrated aluminosilicate. Unique properties of clay minerals include ion exchange capacity, nano-layered construction, high specific surface area for adsorption of stable colloidal dispersion forming tendency, and interlayer chemical modification made by investigation of clay chemistry an important field of research. Thin platelets of clay exhibit an extreme tendency to self-assemble into unique geometrical micro-, nano-scale structures that create roughness and subsequent chemical treatment that lead to superhydrophobicity. Due to isomorphous substitution in the lattice, large negative surface charge is developed over clay crystallite. Thus, the clay basal surface charge containing silicate can be modified either by adding functional coupling agents or by adsorption of H^+ and OH^- [80].

Silicones: PDMS (polydimethylsiloxane) belongs to a group of organosilicon compounds, commonly called silicones. The intrinsic deformability and hydrophobic properties of PDMS make it a very suitable material for the production of superhydrophobic surfaces. Various methods are practiced to produce superhydrophobic surfaces using PDMS. The modification of the surface in the PDMS can be performed using a pulsed CO_2 laser as the excitation source to introduce peroxide groups on the surface of the PDMS. These peroxides are capable of initiating graft polymerization of 2-hydroxyethyl methacrylate in PDMS. The water contact angle of the treated PDMS was measured at 175°. The reason for such an increase in the angle of contact with water was due to the porosity and order of the chains on the PDMS surface. The PDMS elastomer containing microcomposite and nanocomposite structures is also used to produce superhydrophobic surfaces by laser engraving to induce roughness on the PDMS surface. The surface produced by this technique had a water CA of up to 160° and an SA of < 5°. Electrospinning technique is used to produce superhydrophobic membranes. Here, the electrospun fibers formed by a PS-PDMS block mixed with a PS homopolymer (polystyrene) have reached a water CA of approximately 163°. The large angle of contact with water is due to the combined effect of the enrichment of the fiber surfaces by the PDMS component and the roughness of the surface due to the small diameter of the fiber (150–400 nm). The superhydrophobic surface can be synthesized using a casting technique. In the casting process, a PS-PDMS micellar solution in the presence of moist air gave a superhydrophobic surface with a water contact angle of about 163° [81].

Fluorocarbons: Fluoropolymers are attracting a lot of interest these days due to their extremely low surface energies. Curing these polymers will result in superhydrophobic surfaces. Fluorine is one of the most electronegative elements with the highest ionization potential element among all. Its relatively smaller atom size and low polarizability strongly influence bonding and electronic properties. Van der Waals interactions between fluorinated chains are weak because of its limited polarizability, which results in low cohesive energy of fluorocarbons. The low cohesive energy attributes low surface tension, high vapor pressure, low dielectric constant, high gas solubility, and high compressibility. The incorporation of fluorine-based functional monomers in acrylic copolymers have emerged as a rapid research topic in materials science, which possesses various striking features that play a crucial role in water-repellent behaviors. The examples of acrylic copolymers include poly[(triisopropyloxysilyl) propyl methacrylate]-*block*-poly [2-(perfluorohexyl) ethyl acrylate] and poly(glycidyl methacrylate)-*block*-poly[2-(perfluorohexyl) ethyl acrylate], poly [2-(perfluorohexyl) ethyl acrylate-*co*-methyl methacrylate-*co*-methyl butyl acrylate], and poly[1H, 1H, 2H, 2H-perfluorodecyl acrylate-*co*-methyl methacrylate-*co*-butyl acrylate-*co*methacrylic acid]. Superhydrophobicity can be obtained by stretching a Teflon film (polytetrafluoroethylene). The superhydrophobic property obtained is due to the presence of fibrous crystals with large fractions of space on the surface. The rough surface was produced by treatment with oxygen plasma on the Teflon surface and showing a superhydrophobic nature with a water CA of 168°. Due to limited solubility, many fluorinated materials have not been used directly but have been bonded with other raw materials to create superhydrophobic surfaces. A microporous transparent honeycomb polymer film with microsized pores can be produced by pouring a polymer solution under wet conditions. In this process, the fluorinated glass substrate was placed on a substrate support. A metal blade was attached perpendicular to the substrate and the space between the blade and the substrate was adjusted. A fluorinated copolymer solution was provided between the slide and the substrate. Moist air (relative humidity of approximately 60% at room temperature) was supplied to the surface of the solution at a flow rate of 10 per min. Transparent honeycomb-patterned films produced by this method exhibited superhydrophobic properties with a water CA of approximately 160° [82].

CNT-based Surfaces: Carbon nanotubes (CNT) attracted tremendous interest due to their one-dimensional nanostructure, excellent mechanical strength, low density, high porosity, and hydrophobic nature. It is a hollow-tubular nanostructure with a large aspect ratio. It has been reported that these CNT forests, that is, vertically aligned CNT arrays structured at micro- and nanoscale can show superhydrophobic behavior. In the fabrication of superhydrophobic coatings using CNT, the most appealing task is to stabilize water droplets on aligned CNT surfaces for a prolonged time due to its high surface energy. To surmount this problem, the coating of low surface energy polymers is used to convalesce the water repellency [83].

Graphene-based Surfaces: Graphene is a 2D, single-atom nanosheet consisting of sp^2-hybridized carbon atoms exhibiting several distinct properties, such as excellent mechanical modulus, high thermal conductivity, and chemical stability. Graphene sheet modified with epoxy, phenol, hydroxyl, and carbonyl functionalities is considered an outstanding manipulatable precursor. Many efforts were made to align these carbon allotrophs at micro/nanoscale to achieve superhydrophobic effect. However, graphene suffers from some drawbacks such as high cost, low dispersibility, and surface preparation. Thus, graphene oxide (GO) is generally considered as the best precursor in place of graphene. However, it is hard to attain a structural hierarchy of 2D graphene with superhydrophobicity. A link of this hitch was provided by Choi et al. who achieved petal-like, porous structural hierarchy by regulating interpenetrating network and compactly interlocked structure of graphene/Nafion nanohybrid films [84].

Bio-based Materials: Most of the previously reported synthesis processes are having various environmental and safety issues. Especially with spraying methods, the coating prepared by using organic solvents such as ethanol, acetone, tetrahydrofuran, xylene which are flammable, toxic, and explosive causes health hazards for people and the environment. Additionally, nonbiodegradable inorganic particles such as silica, titanium dioxide, calcium carbonate, and ferrous oxide are frequently used to build a rough surface structure. Till date, limited applications of bio-based materials in superhydrophobic coatings have been reported. In 2018, Wang et al. took full advantage of castor oil and abundant renewable resources and prepared a superhydrophobic surface using biomass-based polyurethane and SiO_2 nanoparticles. Here, bio-based material castor oil which contained three long tridecyl side chains was used as film-forming resin and low surface energy material as well. Results showed a substantial rise in water CA and abrasion resistance with an increase in the proportion of castor oil and SiO_2 along with improved transparency [85].

More recently, polylactide has become an important research topic and has drawn increasing attention in the past few years. It is one of the most promising polymers with high biodegradability, mechanical properties, and recyclability. Giuntoli et al. synthesized fluoro-functionalized polylactic acid (PLA) utilizing two alcohols (3,3,4,4,5,5,6,6,7,7,8,8,8-tridecafluorooctyl) benzyl alcohol and 3,3,4,4,5,5,6,6,7,7,8,8,8-tridecafluoro-1-octanol as a co-initiator for ring-opening polymerization of lactides. The coating was stable to artificial aging (750 h), and the fluorinated functional group showed the water-repellent effect with respect to the nonfluorinated PLAs. Bio-based materials, such as lignin-coated cellulose nanocrystal and cellulose nanofibril, were investigated as potential materials to enhance the mechanical properties of the coating. Due to non-toxicity, biodegradable and simplified fabrication has drawn significant research interest as superhydrophobic materials. Recently, various low surface energy materials were coated onto cellulose nanocrystals to fabricate superhydrophobic surfaces. For example, Huang et al. fabricated biodegradable, hydrophobic-modified, lignin-coated cellulose nanocrystal particles for self-cleaning application. Besides nanocrystals, cellulose nanofibrils also offer new insights into the development of binders used in the superhydrophobic coating. Coating was composed of sodium oleate and biobased cellulose nanofibrils hydrophobized with either alkyl ketene dimer or amino propyl trimethoxy silane. Such coatings have very promising application in areas, such as in bio-based building materials, microfluidic devices, and food packaging [86].

11.10 Synthesis of Self-Cleaning Surfaces

Self-cleaning surface preparation methods to produce micro/nanostructured textures include the following: lithography, chemical vapor deposition, physical vapor deposition, electrochemical deposition, electrospinning, wet chemical reaction, hydrothermal reaction, solution immersion, templating, layer-by-layer self-assembly technique, plasma treatment, sol-gel polymerization reaction, flame treatment, nanocasting, 3D printing, and so on. These techniques are discussed subsequently:

11.10.1 Microlithography and Nanolithography

The lithography method has been widely used in fabricating hydrophobic self-cleaning surfaces in the past years. This method is a kind of technology using a series of produce procedures to deprive specific sections on target surfaces, forming micro-/nano-hierarchical structures. Microlithography and nanolithography refer specifically to lithographic patterning methods capable of structuring material on a fine scale. Typically, features smaller than 10 μm are considered microlithographic, and features smaller than 100 nm are considered nanolithographic [87]. Photolithography is one of these methods often applied to semiconductor device fabrication. Photolithography is also commonly used for fabricating micro-electro-mechanical systems (MEMS) devices. *Photolithography* generally uses a prefabricated photomask or reticle as a master from which the final pattern is derived. Although photolithographic technology is the most commercially advanced form of nanolithography, other techniques are also used. Some, for example, electron beam lithography, are capable of much greater patterning resolution (sometimes as small as a few nm). Electron beam lithography is also important commercially, primarily for its use in the manufacture of photo-masks. Electron beam lithography as it is usually practiced is a form of maskless lithography, in that a mask is not required to generate the final pattern. Instead, the final pattern is created directly from a digital representation on a computer, by controlling an electron beam as it scans across a resist-coated substrate. Electron beam lithography has the disadvantage of being much slower than photolithography. In addition to these commercially well-established techniques, a large number of promising microlithographic and nanolithographic technologies exist or are being developed, including nanoimprint lithography, interference lithography, X-ray lithography, extreme ultraviolet lithography, magnetolithography, and scanning probe lithography [88, 89].

11.10.2 Chemical Vapor Deposition

Chemical vapor deposition (CVD) method indicates that the process of the vapor of gas state reactant or liquid state reactant with film component element and other needed gas is introduced into the reaction chamber, and the film is formed on the substrate surface after producing a chemical reaction. CVD is a well-established technique for the deposition of a wide variety of films with different compositions and thicknesses down to a single layer of atoms. Microfabrication processes widely use CVD to deposit materials in various forms, including monocrystalline, polycrystalline, amorphous, and epitaxial. These materials include silicon (dioxide, carbide, nitride, and oxynitride), carbon (fiber, nanofibers, nanotubes, diamond, and graphene), fluorocarbons, filaments, tungsten, titanium nitride, and various high-k dielectrics. Many films are obtained via the CVD method in the fabrication of super large-scale integration. This method could improve film adherence by 30%. For example, (1) aligned CNT film can be fabricated with lotus-like morphology on a silica substrate; (2) plasma CVD was used to modify the titanium surface that exhibited WCA of 145° and could be switched to superhydrophilic, with contact angle of 0°; (3) plasma enhanced-CVD was used to deposit the silicon doped hydrocarbon films; and (4) highly hydrophobic and superhydrophobic fluorocarbon surfaces via combining deposition of nanoparticle films and overcoating of such films by a magnetron sputtered polytetrafluoroethylene layer [90].

11.10.3 Physical Vapor Deposition (PVD)

Physical vapor deposition (PVD) is a vaporization coating technique that involves the transfer of material at the atomic level. The process can be described according to the following sequence of steps: (1) the material to be deposited is converted into a vapor by physical means (high-temperature vacuum or gaseous plasma), (2) the vapor is transported to a region of low pressure from its source to the substrate, and (3) the vapor undergoes condensation on the substrate to form a thin film. Typically, PVD processes are used to deposit films with thicknesses in the range of a few nanometers to thousands of nanometers. PVD thin-film technology covers a rather broad range of deposition techniques, including electron-beam or hot-boat evaporation, reactive evaporation, and ion plating. PVD techniques also include processes based on sputtering, whether by plasma or by an ion beam. PVD is also used to describe the deposition from arc sources that may or may not be filtered. In general, this process can be divided into two groups: evaporation and sputtering. Evaporation refers to thin films being deposited by thermal means, whereas in the sputtering mode, the atoms or molecules are dislodged from the solid target through the impact of plasma gaseous ions [91].

11.10.4 Electrochemical Deposition

Electrochemical deposition has been extensively employed to construct biomimetic superhydrophobic surfaces since it is a versatile technique to prepare microscale and nanoscale structures. Electrochemical deposition is a process by which a thin and tightly adherent desired coating of metal, oxide, or salt can be deposited onto the surface of a conductor substrate by simple electrolysis of a solution containing the desired metal ion or its chemical complex. Often, coatings are used for a variety of reasons, including creating reflecting surfaces, conducting paths in printed circuits, magnetic layers, surfaces with specific friction parameters in sliding bearings, and restoration of the surface of worn parts, among others applications. Electrochemical method is a way to utilize chemical reactions that take place in a solution at the interface of an electron conductor. These reactions involve electron transfer between the electrode and the electrolyte or species in solution. Synthesis of ZnO thin films with diverse nanostructure, including nanodot, nanowire- and nanoflowers on zinc foils by a simple and rapid electrochemical anodization method. Under the DC or AC electric field, the electro-induced surface wettability conversion from the superhydrophobic to hydrophilic state was observed, and the generation of surface defective sites on ZnO films under electric field was used to explain the transition mechanism. It is found that the reversibility of wettability conversion is dependent on the structure types. Synthesis of superhydrophobic surface on metals using a very simple galvanic deposition of metallic salt solution with a high water CA of 173°. The resulted surfaces can effortlessly float on a water surface similar to pond skaters, which can be supported on water by superhydrophobic legs with a hierarchical structure based on an oriented microscale with nanogrooves. Jiang et al. reported a new microscale and nanoscale hierarchical structured copper mesh films with superhydrophobic and superoleophilic properties by a simple electrochemical deposition induced by long-chain fatty acids. Interestingly, the as-prepared mesh film works well as a separator of diesel oil, which is very useful for practical separation applications. Since the films were electrodeposited and grew within the interstitial spaces between the hydrogen bubbles, the pore diameter and wall thickness of the porous copper films were successfully tailored by adjusting the concentration of the electrodeposition electrolyte [92].

11.10.5 Electrospinning Method

Electrospinning involves three main components: the polymer fluid usually contained in a capillary such as a hypodermic needle, an electric field source, and a collector; however, each setup has different variations. When a sufficiently high voltage is applied to a liquid droplet, the body of the liquid becomes charged, and electrostatic repulsion counteracts the surface tension and the droplet is stretched; at a critical point, a stream of liquid erupts from the surface. This point of eruption is known as the Taylor cone. If the molecular cohesion of the liquid is sufficiently high, stream breakup does not occur (if it does, droplets are electrosprayed) and a charged liquid jet is formed. As the jet dries in flight, the mode of current flow changes from ohmic to convective as the charge migrates to the surface of the fiber. The jet is then elongated by a whipping process caused by electrostatic repulsion initiated at small bends in the fiber until it is finally deposited on the grounded collector. The elongation and thinning of the fiber resulting from this bending instability lead to the formation of uniform fibers with nanometer-scale diameters [93]. Once the Taylor cone is formed, the charged liquid jet is ejected toward the collector. Collectors can be but are not limited to, stationary flat plates, rotating drums, mandrels, and disks. Depending on the solution viscosity, solid fibers will be formed as the solvent evaporates from the whipping motion that occurs during its flight time from the Taylor cone to the collector. The result is a non-woven fiber mat that is deposited on the collector. This jet travels in air causing the solvents to evaporate as the jet dries and is deposited on the collecting surface as a fiber. Instability of the jet due to electric charge results in a random arrangement of fibers. However, various methods including the use of an alternating current (AC) field source, a negative ion source such as a corona assembly, or a rotating collector help distribute charge thereby reducing instability and producing aligned fibers. It is important to know that completely or perfectly aligned fibers are difficult to obtain [94].

11.10.6 Wet Chemical Reaction

A wet chemical reaction, a straightforward and easy way to make rough surfaces, is effectively applied to control the size, shape, and dimensionality of micro- and nanometer-scale materials, such as nanoparticles, nanowires, and mesoporous inorganics. In the past years, especially in the last 4 years, this method was often employed in tailoring surface morphology for the fabrication of biomimetic superhydrophobic surfaces on various metal substrates, such as steel, copper, and aluminum. Wet-chemical synthesis routes deal with chemical reactions in the solution phase using precursors at proper experimental conditions. Each wet-chemical synthesis method differs from the others, meaning that one cannot find a general rule for these kinds of synthesis approaches. These synthesis strategies have been used for the preparation of 2D nanomaterials that are unable to be prepared by top-down approaches. Wet-chemical synthesis routes offer a high degree of controllability and reproducibility for 2D nanomaterial fabrication. Solvothermal synthesis, template synthesis, self-assembly, oriented attachment, hot injection, and interface-mediated synthesis are the main wet-chemical synthesis routes for 2D nanomaterials. For example, Jiang et al. have fabricated superhydrophobic surfaces on a copper substrate with a simple chemical composition method by immersing the substrate into the n-tetradecanoic acid solution for about 1 week without any further modification. Zhang et al. described a simple surface roughness generation method by chemical etching on polycrystalline metals with an acidic or basic solution. After treatment with fluoroalkyl silane, the etched metallic surfaces exhibited superhydrophobicity [95].

11.10.7 Templating

Template-based methods prepare superhydrophobic surfaces using an intermediate surface termed as a master template. Master templates can be divided into two major groups: polymers and metals. Natural surfaces, such as leaves, insect wings, and reptile skins, are typically used as prototypes to produce polymeric master templates. The desired micro- and nano-scale surface roughness of the metal's master templates can be created through other methods, such as etching. The templating process includes the use of a master template having the desired specifications and then molding to replicate these specifications. Templates are subsequently removed by either lifting the replica off the template or by dissolving the templates. This is a widely used method for preparing polymeric surfaces due to it being a low cost, rapid, and repeatable procedure. The main advantage of this method is that natural surfaces can be replicated. Peng et al. prepared a PDMS master template by casting the liquid PDMS directly onto a natural, fresh *Xanthosoma sagittifolium* leaf and curing it at 50°C for 4 h, then removing the hardened PDMS template. Subsequently, this master template was used for the preparation of a replica composed of polyaniline on cold-rolled steel. The superhydrophobic polyaniline film had almost the identical micro- and nano-scale surface morphology as the natural Xanthosoma leaf.

Template-based methods are another imprint-related way to prepare superhydrophobic surfaces. Templates can be fabricated using lithography. The original prototypes of the templates can be filter paper, insect wings, reptile skins, and plant leaves. From a chemical and morphological point of view, the template can even be molecules and polymers. A high-quality template can be repeatedly used to make a large number of samples. However, when it comes to micro- and nano-scaled structures, the operators are required to be essentially careful to avoid damaging both the samples and the templates. Also, the templating methods might not be fit to produce irregular surface patterns with excessively complex structures due to the peeling off [96].

11.10.8 Solution Immersion Process

The solution immersion process is a simple and universal one-step process to render substrates superhydrophobic by immersion of the substrates in a solution that generally contains low-surface energy agents. However, the defects of this method are unstable coatings, and it is time-consuming. Control of the nanoroughness is also difficult using this method. With the increase in immersion time, clusters formed became denser, covered the surface more, and the skeleton of 3D porous structure got thicker and rougher while pore size decreased. Because of the coexistence of advantages and disadvantages mentioned earlier, many researchers still attempt to make progress in this aspect. One recently reported example was based on the use of pristine multiwall carbon nanotube (MW-CNT) and polybenzoxazine in a solution immersion process to construct nanocomposite coating on ramie fabric. A complete cycle is as follows: First, the ramie fabric was immersed into the MW-CNT/benzoxazine monomer (3,3-′-(((2,2-dimethylpropane-1,3-diyl)bis(oxy))bis (4,1-phenylene))bis(3,4-dihydro-2H-benzo[*e*][1,3]oxazine-6-carbonitrile), BOZ) mixture suspension, then washed several times with deionized water and ethanol, dried under vacuum, and finally heated at 130°C. The water CA value of MW-CNT-BOZ mixture suspension (MW-CNTs: 1.0 mg m/L, BOZ: 1.0 mg m/L) for 20 cycles can reach 152°. Finally, the relationship of superhydrophobicity and conductivity with the number of repeated cycles and the concentration of MW-CNT suspension is highlighted. Clean commercially obtained copper foam was used as a substrate to prepare non-flaking superhydrophobic coating through a one-step solution-immersion process. Copper stearate with micro- and nanoscale hierarchical surface morphology can be obtained after copper foams are immersed in ethanolic stearic acid solution for several days at room temperature. Wang et al. prepared a series of superhydrophobic light alloys (including AZ91D Mg alloy, 5083 Al alloy, and TC4 Ti alloy) by immersing the substrates in 1H,1H,2H,2H-perfluorooctyltrichlorosilane (PFOTS), ethanol, and H_2O–H_2O_2 mixed solution. Through this one-step method, the light alloys have micro/nano surface structures and low surface energy at the same time [97].

11.10.9 Self-Assembly and Layer-by-Layer Methods

Self-assembly and layer-by-layer are other simple and inexpensive methods to prepare micro-, nano-, and dual-scale superhydrophobic surface structures as well as the obtained structures can be finely controlled. For example, Jiang et al. prepared a conducting and superhydrophobic rambutan-like hollow spheres of polyaniline using a self-assembly method in the presence of perfluoro-octane sulfonic acid (PFOSA), which is employed as a dopant and soft template. An artificial lotus leaf structure on cotton substrates was constructed via the controlled assembly of carbon nanotubes. To control the assembly of carbon nanotubes on cotton fibers, the cotton fibers were modified using treated carbon nanotubes as macro-initiators. Unlike most synthetic textiles, cotton textiles have perfect absorption abilities for most liquids. So the generation of superhydrophobicity on cotton is quite different from these synthetic textiles. Similar results were also obtained by introducing a new strategy to prepare hydrophobic to superhydrophobic surfaces by depositing about 2.5 nm FePt nanoparticles with varying degrees of fluorinated ligands on their surface. It is the introduction of magnetic materials on superhydrophobic surfaces, which might help in preventing oxidation and prolonging the lifetime of magnetic media. Cohen et al. indicated that the layer-by-layer assembly method can allow controlling the placement and level of aggregation of nanoparticles with different sizes within the resultant multilayer thin film. A transparent and antireflective superhydrophobic film based on SiO_2 nanoparticles of various sizes was obtained, presenting important respective in practical applications [98].

11.10.10 Plasma Treatment

Plasma treatments have been widely applied to polymers, metals, or elastomers for the fabrication of engineered materials, including process for tailoring the wetting properties to either superhydrophilic or superhydrophobic surfaces. For example, a hydrophobic coating of CNT by methane glow discharges plasma at low pressure. The surface free energy of the plasma-treated CNT decreases drastically, revealing superhydrophobic modification of CNT by the CH_4 plasma. An engineered surface with selected hydrophilic/hydrophobic character was obtained by employing simple plasma-based techniques to independently tailor the substrate topography and the surface chemistry. Badyal et al. studied a Stenocara Beetle's back for micro-condensation using plasma chemical patterned superhydrophobic and superhydrophilic surfaces. They found that the micro-condensation efficiency is mainly dependent on the chemical nature of the hydrophilic pixels and their dimensions [99].

11.10.11 Sol-Gel Method

The best advantage to employ a sol-gel method in fabricating biomimetic superhydrophobic surfaces is for all kinds of solid substrates, such as metals, glass, silicon wafer, polymers, and textiles. For example, a facile sol-gel method for finely controlling the wettability transition temperature of colloidal-crystal films from superhydrophilic to superhydrophobic was obtained. Superhydrophobic cotton textiles can be prepared by in situ growing microsized silica particles on hydrophilic cotton textiles followed by the hydrophobization step. It is interestingly noted that, on the perfluorooctyl-modified cotton sample, the static CA for a 15 mg sunflower drop is about 140° and the roll-off angle for the same drop is about 24°, which is a breakthrough in fabricating superhydrophobic surface to be applied in practice. Another example is biomimetic superhydrophobic surfaces on engineering materials of copper alloy prepared using hexamethylenetetramine and ethylene glycol, a strong bidentate chelating agent to Cu^{2+} and Fe^{2+} ions with a high stability constant, as the capping reagents. An ordered pore indium oxide array film with controllable superhydrophobic and lipophobic properties was fabricated by sol-dipping method using the polystyrene colloidal monolayers. It is found that, with the increase of pore size in the obtained films, the superhydrophobicity could be controlled and was gradually enhanced due to the corresponding increase in roughness caused by nanogaps produced by the thermal stress in the annealing process with the increase in film thickness. Polymerization reaction is a promising route in changing the surface free energy and in tailoring surface morphology for the fabrication of biomimetic superhydrophobic surfaces in recent years. Jon and Choi et al. presented a facile and effective method for water-repellent coating with a new random copolymer, poly (TMSMA-r-fluoroMA) [3-(trimethoxysilyl)propylmethacrylate] (TMSMA), methacylate (MA)] on oxide-based substrates. The water CA is almost 163°, and SA is <2°. A superhydrophobic surface originated from quincunx-shape composite particles was constructed by utilizing the encapsulation and graft of silica particles to control the surface chemistry and morphology [100].

11.10.12 Flame Treatment

Flame treatment is a special and simple method of making a superhydrophobic surface. Based on the difference in thermal properties of organic and inorganic materials, the micro/nanostructure and morphology of the superhydrophobic biomimetic surfaces were fabricated. An organic–inorganic compound consisting of polydimethylsiloxane particles and SiO_2 can be used as a powerful hydrophobic material. In this compound, these two components showed different thermal responses in the flame treatment process and then built the expected superhydrophobic micro/nanostructure. The roughness of the superhydrophobic surface could be controlled by modifying the flame treatment time to obtain an ultra-low slip angle that reaches a limit value of 1° and considerably eliminates anisotropic wettability. Compared to other conventional methods, this method has the advantages of simplicity, high efficiency, and low cost [101].

11.10.13 Nanocasting

Nanocasting is a method based on smooth lithography that uses elastomeric molds to produce nanostructured surfaces. For example, polydimethylsiloxane was poured onto the lotus leaf and used to create a negative PDMS matrix. The PDMS was then coated with a trimethylchlorosilane non-stick monolayer and used to make a positive PDMS matrix from the former. Since the natural structure of the lotus leaf allows a pronounced self-cleaning ability, this modeling technique was able to reproduce the nanostructure, resulting in surface wettability similar to that of the lotus leaf. Furthermore, the ease of this methodology allows for the translation into massive replication of nanostructured surfaces [102].

11.10.14 3D Printing

Developments in the field of 3D printing techniques have helped in the generation of topographical structures with the desired superhydrophobic properties. Single-step 3D printing process has been developed, wherein an ordered mesh is created using a viscoelastic ink with low surface energy materials added. Juan et al. generated nanoporous membrane where the ink contains hydrophobic nanosilica-filled PDMS. A membrane can be used for the process of oil-water separation. Multistep processes include first 3D printing the meshed surface with the structures and then dip-coating it with low surface energy coatings [103].

11.10.15 Fabrication of Hydrophilic Materials

In general, improving the hydrophilicity of surfaces can be accomplished in two ways: by depositing a molecular film of new material more hydrophilic than the substrate or by modifying the chemistry of the substrate surface. Deposition of surface coatings is the most common method for inorganic substrates. However, modification of surface chemistry is used in the case of polymeric materials. Several techniques for synthesizing surface hydrophilicity are discussed subsequently:

11.10.16 Deposited Molecular Structures

Monolayers can be formed from organic molecules that adsorb to the solid surface. Organic molecules can come from a solution or a vapor phase. This ultimately changes the wetting characteristics of the surface. For example, alkanethiols have been generally used in Au, Ag, Cu, and Pd. Chlorosilanes have been used in silicon oxide, Al, and Ti. Phosphoric acids can also be used in Ti and Al. The Langmuir-Blodgett film technique is used to mechanically deposit monolayers and multilayers. However, a major disadvantage of using this method is that it suffers from poor stability when multiple layers are contacted with liquids. The hydrophilic surface is obtained if the end of the deposited organic layers is polar. If a saturated hydrocarbon group or a fluorinated group is at the end of the layer, the water will not be attracted to the surface, which will cause hydrophobic conditions. The hydrogen bond is how the water molecules will be attracted to the surface if the surface has chemical groups like –OH, –COOH, and POOH [104]. Biomacromolecules, such as albumin and heparin, have been widely used in the health sector to provide hydrophilic characteristics that complement the body's needs. Synthetic polymers such as polyethylene glycol and phospholipid macromolecules have been extensively studied. The presence of a hydrophilic coating in bioengineering applications can cause the adsorption of proteins on the surface, which is highly undesirable. Therefore, having such protective coatings prevents protein adsorption when the materials come into contact with biological fluids [105].

11.10.17 Modification of Surface Chemistry

The modification of surface chemistry has been studied in the last decades. Plasma, corona, flame, photons, electrons, ions, X-rays, UV-rays, and ozone are methods that have been studied to modify the chemistry of polymer surfaces without affecting their volume properties. Oxidation of polymer surfaces can be accomplished by plasma treatment in an air or oxygen, corona, and flame environment. In plasma and corona treatments, the electrons are accelerated and then bombard the polymer with energies two to three times higher than necessary to break the molecular bonds, producing free radicals. This generates cross-links and reacts with the surrounding oxygen to produce oxygen-based functionality. Hydroxyl, peroxy, carbonyl, ether, and ester groups are typical polar groups that are created on the surface. In flame treatment, the surface combustion of the polymer takes place with the formation of radicals such as hydroperoxide and hydroxyl. The oxidation depth by flame treatment is approximately 5–10 nm. This increases to more than 10 nm for air plasma treatment. Plasma, corona, and flame treatments cause significant surface oxidation and ultimately highly wettable surfaces. Due to environmental chemistry, the polar groups produced during surface oxidation tend to be buried in the mass when in contact with air for an extended period. However, its presence remains on the surface in contact with water or any other polar environment. Many polymers undergo oxidation and degradation under UV light. For example, outdoor consumer products made from polymers require the incorporation of UV absorbers when exposed to the sun to prevent discoloration and cracking. The wavelength of light varies from 10 nm to 400 nm, and the incident photons have enough energy to break the intermolecular bonds of most polymers. This allows for structural and chemical changes in the macromolecules. Exposure of the polymer to UV radiation has the result of making the surface more hydrophilic due to chain cleavage, crosslinking, and increased density of oxygen-based polar groups. Surface hydrophilicity can be improved by alkaline treatment of polymers at elevated temperatures. The surface of the polymer, such as polyethylene terephthalate, contains groups such as hydroxyl and carboxyl groups that contribute to the hydrophilicity of the interaction during etching with concentrated bases. Oxide conducting surfaces can be electrochemically treated using an anode potential to control their wetting characteristics [106].

11.11 Properties of Superhydrophobic Materials

Weak mechanical durability is the main issue limiting the widespread application of superhydrophobic coatings because their micro- or nano-scale roughness structures are mechanically weak and readily abraded. The static contact and roll-off angle remained unchanged generally after heating up to 400°C for 1 h. Heating the coatings at even higher temperatures decomposed the roughness, causing that the coating lost its super amphiphobicity. However, some surfaces exhibit resistance toward heat up to a certain temperature and then lose the superhydrophobicity nature of the surface. For example, porous inorganic silica films on glass substrates, by the sol-gel process, using tetramethoxysilane and isobutyltrimethoxysilane as precursors. At elevated temperatures (>200°C), the hydrophobic isobutyl group in the silica surface decomposes and leads to the loss of superhydrophobicity. The contact angle decreased with increasing temperature and the CAH increased. However, when the as-prepared silica films were modified with fluoroalkyl silanes, the thermal stability was significantly improved: at a temperature of 400°C, the surface remained stable. Further heating resulted in a decrease of contact angle and when the temperature reached 500°C, the film was no longer hydrophobic (contact angle almost became 0°) [107]. In addition to mechanical damage, the wettability of superhydrophobic surfaces tends to decrease in harsh environments due to chemical degradation. The chemical stability of the composite coating was also evaluated by measuring the changes in the CA and SA after immersion in a corrosive acid, base, and organic solvents. The treated coatings did not show any obvious changes and appeared to be as dry as the non-treated samples. Furthermore, further research was conducted to study the chemical stability of the composite coating in a strong acid environment. The electrochemical tests for the as-prepared composite coatings on tin plate electrodes found that there is no effect of acid UV photodegradation that conducted up to 96 h. The result indicates the composite coating is durable and chemically stable [108]. In general, the superhydrophobicity of a superhydrophobic coating deteriorates due to the degradation of hydrophobic organic molecules at high temperatures. When the heating duration was further increased for a long period of time, the coating finally became hydrophilic, and this is due to the decomposition of the hydrophobic groups on the surface [109].

Many works have been carried out to study the stability and durability of superhydrophobic coatings on different metallic substrates, in water and various organic solvents. TiO_2/SiO_2-coated steel substrates investigated with the effect of water, chloroform, and decade on the CA after every 24 h for a period of 8 days. It was shown that the CA remained constant, and the surfaces were stable even after 8 days. In another work, coated Zn-substrates were immersed in toluene for 24 h at room temperature. The resulting contact angles had no variation, suggesting a good stability of the surface. The stability of superhydrophobic surfaces on different substrates in seawater was also intensively studied and revealed the conservation of superhydrophobicity even after a long duration exposure created a superhydrophobic film by using electroless Ni-P composite coating on carbon steels. The superhydrophobic film has good stability in the air at room temperature and good corrosion resistance in 5 wt% NaCl solution, neutral salt spray test, and water erosion test. Nanocomposite superhydrophobic layer coated low C-steel steel under atmospheric conditions and aggressive media results in the most durable protection against corrosion. Nanocomposite superhydrophobic coated MgMnCe alloy shows that preliminary plasma electrolytic oxidation of alloy improves the adhesion and water-repelling properties of the coatings and consequently increases the corrosion resistance of Mg alloy [110]. The stability of superhydrophobic surfaces over a wide pH range is crucial for their use as engineering materials in several industrial applications. Superhydrophobic coated Al-alloy (CA $\approx$ 160°–162°) and Mg-alloys (>145°) were investigated for the stability over the pH range from 1 to 14. It concluded no effect of the pH of the aqueous solution [111]. The stability of superhydrophobic films against UV irradiation is extremely important, especially for outdoor surfaces that are exposed to UV light. The TiO_2/SiO_2-coated steel surfaces strongly resist UV radiation: the CA remains constant even after a period of 5 h UV exposure. This was explained by the fact that in the presence of SiO_2 nanoparticles, the high-energy electrons generated by TiO_2 under UV cannot diffuse to the surface, and thus, no hydroxyl radicals can be formed and no oxidation will occur. While superhydrophobic ZnO nanorod films that lost their superhydrophobicity under UV illumination for 2 h, the water contact angle dropped from 160° to 0°. The most stable superhydrophobic silica surfaces treated with fluoroalkyl silanes show resistance after 5500 h UV irradiation without degradation of either CA or CAH [112].

11.12 Other Terminology with Phobic and Philic

With the rapid development of the research field related to wettability, thousands of surfaces showing special wetting properties have been developed. Meanwhile, many terms are created and used to describe those special wetting phenomena. In addition to the "classical" terms, "hydrophobic" and "hydrophilic," other terms such as "oleophobic/oleophilic,"

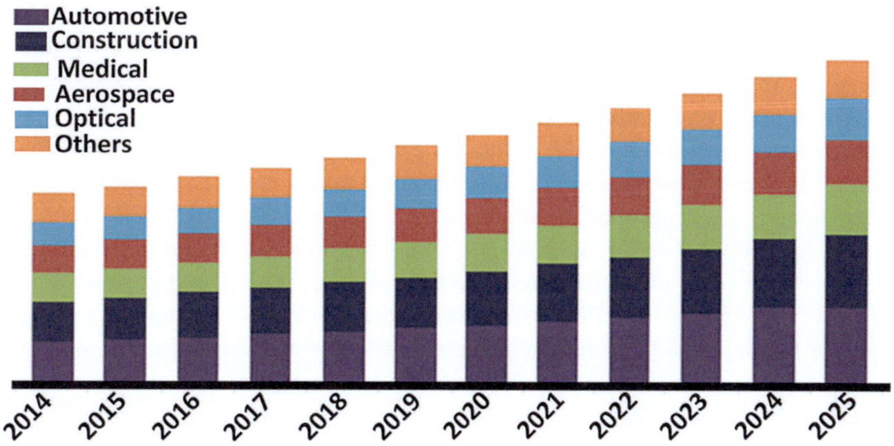

Fig. 11.20 Year wise growth rate of self-cleaning materials in different industries in the running market

"hygrophobic/hygrophilic," "lyophobic/lyophilic," "amphiphobic," and "omniphobic" have also appeared in the many recent publications [113–115]. They are preceded by a prefix "super" to express their extreme cases. "Superomniphobicity" is used by some people to describe the surfaces that are repellent to both high and low surface tension liquids (oils), which has the same meaning as that of "superamphiphobicity." As is well know, "phobic" and "philic" come from the Greek words for "fear" and "love," respectively. In some terminology, "hydro" (Greek word), "oleo" (Latin word), and "hygro" (Greek word) refer to "water", "oils or fats," and "liquid", respectively. It is also worth noting that "oleo" is a Latin word, while "lipo" is the Greek word for oils or fats. When the probe liquid is not certain, we can consider temporarily using "hygrophobic" and "hygrophilic, the liquid droplets in those models can be water or it can be oils, and it can also be other kinds of liquids, so "hygrophobic" and "hygrophilic" are temporarily adopted for liquid-repelling and liquid-loving. The surfaces simultaneously showing superhydrophobic and superoleophobic properties are generally known as the "superamphiphobic" interfaces. Sometimes "superomniphobic" is also used to describe a surface that repels various liquids from high surface tension water to low surface tension oils, since "amphi" comes from the Greek word for "both" and "omni" comes from the Latin word for "all" [116, 117].

11.13 Applications of Self-Cleaning Materials

Self-cleaning coatings can be implemented in various surfaces irrespective of their complex geometry and composition. Self-cleaning glass is a specialized category of glass that has hydrophilic and photolytic properties and can keep its surface free of dirt and grime. The global hydrophobic coatings market size was USD 114 million in 2015 and is expected to experience a significant rise due to its increasing reach in the automotive, aerospace, and construction industries [118]. Self-cleaning glass is applied in areas such as glass facades, outdoor store fountains, display cases, terraces, balconies, suspended glazing, windows, and doors. Hydrophobic coatings for glass components provide high resistance to water in heavy rain. As a result, high demand for protective window and window manufacturing is expected. The aerospace industry is expected to gain ground due to the increasing use of anti-corrosion and anti-ice or anti-wetting coatings. Demand for the product is expected to increase significantly in the medical industry and optical industry. The global self-cleaning glass market was valued at USD 118 million in 2018 and is projected to reach USD 162.9 million by 2026, at a compound annual rate of 4.1% [119]. Figure 11.20 illustrates the significant growth rates of self-cleaning materials in the automotive, aerospace, medical, construction, optical, and other sectors. North America and Europe account for more than 50% of the total market share due to high production volumes from the end-user segments, in particular automobiles, aircraft, and medical devices. The Middle East and Africa are expected to experience significant growth due to growth in the automotive sector in the region. The main players in the world market for self-cleaning glass are Saint-Gobain Glass, Nippon Sheet Glass Co. Ltd., Cardinal Glass Industries, Inc., Asahi Glass Co., Atis Group, Guardian Industries, Roof-Maker Limited, Wuxi Yaopi Glass Engineering Co., Viridian Glass, Kneer-Südfenster, Clear Glass Solutions, Dongguan City of East Pearl River Glass Company Limited, Foshan Qunli Glass Company Limited, Gevergel (FoShan) Engineering Glass Company Limited, ITC International Trading & Consulting Pty Limited, PPG Industries, Inc., Saint-Gobain SA., Shanghai HuZheng Nanotechnology Company Limited, and others [120, 121].

The self-cleaning surface prevents or reduces the accumulation of contaminants such as water droplets (rain and condensation), dust, stains, and organic matters which washes away the surface contaminants with the help of gravity. Some of the basic functions of these materials in various industries are given below:

Antifogging and Anti-ice Materials: Accumulation of fog over surfaces may cause detrimental side effects such as inferior optical properties, layer of fog influences on temperature, humidity, and convection, adherence on surface causes instant formation of dense ice layer, and causes inconsistent supply of electrical and telecommunication network. Antifogging coatings are an effective approach to resolve the issue of fogging over surfaces and are capable of handling a wide range of environmental challenges. Ice build-up on solid materials, especially accumulation on high-voltage electric power lines and conductors in some cold regions, may lead to mechanical line failure or insulator flashovers. Ice deposition also results in serious traffic disruption in some main highways and railway lines. It was a challenge to construct solid surfaces with anti-icing properties. The superhydrophobic surfaces have shown their speculated capability to reduce the accumulation of snow and ice and to even completely prevent the formation of ice on solid surfaces. Several different companies have made self-cleaning superhydrophobic coatings spray that contained polytetrafluroethylene particles and a binder was specifically designed for antennas and other transmission systems, particularly located at a high elevation, to reduce the accumulation of snow [122].

Anticorrosion: The concept of preparing surfaces that repel water creates huge opportunities in the area of corrosion inhibition for metals and alloys. Given their water repellency, superhydrophobic coatings form an important and successful method to slow down the breaking of the oxide layer of metals and thus prevent the metal surface underneath from further corrosion. Several works have been carried out to study the corrosion resistance of metals coated with superhydrophobic surfaces. It has been investigated that superhydrophobic coatings have the potential to replace conventional techniques such as chromatin, cathodic protection, sacrificial anodic coatings, and chemical conversion coatings. Its synergistic effect of low surface energy and air valleys hinders the direct contact of aggressive ions and metal surface by forming slippery surfaces and large peaks that effectively reduced the metal corrosion. The concept of air valleys works on "capillarity effect" which forces out corrosive ions from the pores of superhydrophobic surface by the action of Laplace pressure [123].

Antifouling: Any natural and artificial surfaces submerged in the water are subjected to fouling, and this process of microorganism deposition is known as biofouling, and it is detrimental for marine transportation, the major sector that is severely suffering from the issue. This unceasing accumulation of marine organisms resulting in a drastic increase in net weight of the ship, increase in drag when voyaging, and increase in fuel consumption. Anti-fouling coatings are routinely used to prevent fouling. According to the action of mechanism, antifouling coatings employs two approaches: (1) prevention from attachment and (2) detachment of biofoulant. The first approach is preferred to use materials with maximum critical surface tension (hydrophilic materials), whereas the second approach seeks coatings with micro-nanoscopic roughness and low surface energy values such as fluorocarbons, siloxanes, and hydrocarbons. The biofouling mechanism cannot be quantified using apparent contact angle, sliding angle, or contact angle hysteresis as the volume of water is infinite. Thus, the only available measure of biofouling is the fraction of wetted contact area. The reduction of the wetted area minimizes the probability that biological organisms encounter a solid surface. The design of such surfaces should involve optimization between mechanical stability and minimal wetted area [124].

Antifriction: As is well known, the water strider shows an excellent drag-reducing ability on water. To mimic the legs of water striders, superhydrophobic drag reduction parts were fabricated. Many techniques are applied to the water vehicle to reduce the friction force on its sidewall with the water. For example, modifying the sandblasting pressure, followed by decorating the surface with polydimethylsiloxane (PDMS) to obtain a superhydrophobic surface for friction reduction [125].

The growth of these self-cleaning surface materials with their above-mentioned properties has been used in many sectors such as automobile industries, aero-industries, chemical industries, textiles industries, and electronics industries are discussed below:

11.13.1 Aero-Industries

Between the years 2000 and 2005, there is a focus on fixed-wing aircraft by hydrophobic and self-cleaning structural aircraft parts. Coating for de-icing an aircraft, which has other potential applications in the aerospace industry, such as coating of helicopter rotor blades and an aircraft maintenance system. Self-cleaning superhydrophobic coating for aircraft structures (especially Ti or Ti-based alloy structures) that are sensitive to the accumulation of ice, water, and other contaminants [126]. Hard, wear-resistant, and phobic coating for an aerodynamic surface to improve de-icing properties through low-pressure plasma vapor deposition technologies, such as plasma-assisted CVD, sputtering, and reactive spraying. The liner includes a functional top layer that is harder than the surface of the airfoil and has a high angle of contact with water. Hydrophobic coatings with anticorrosive and self-cleaning properties for spacecraft survival systems and self-cleaning of exterior walls [127]. Self-cleaning coating with antimicrobial properties for use in aircraft cabins.

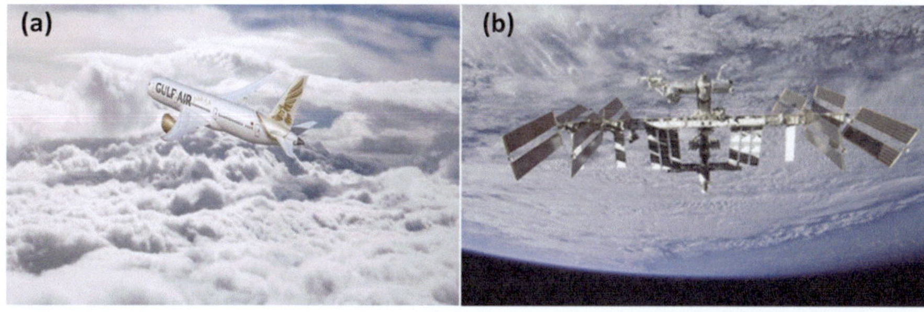

Fig. 11.21 Self-cleaning coating is used in (**a**) Aeroplane outer surface to avoid the galvanic activities, (**b**) Satellite to avoid the clogging effect

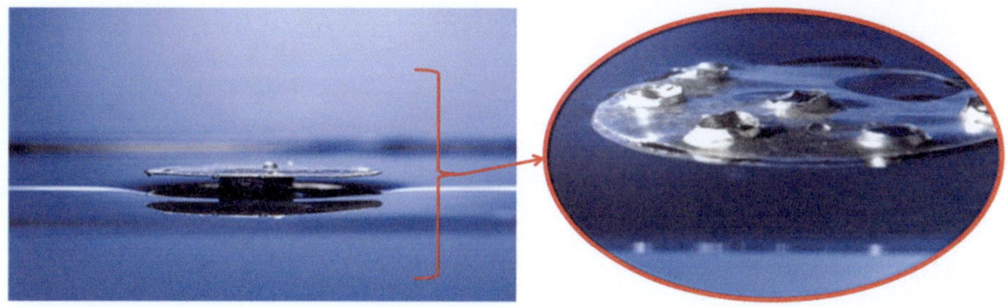

Fig. 11.22 The superhydrophobic structure remains afloat

Figure 11.21 shows the self-cleaning exterior surface used on aeroplanes and satellites.

11.13.2 Maritime Industry

Several boat and water-running robots developed based on the geometrical and movement characteristics of a water strider. Superhydrophobic coatings have important applications in the maritime industry. They can yield skin friction drag reduction for ships' hulls, thus increasing fuel efficiency. Such a coating would allow ships to increase their speed or range while reducing fuel costs. They can also reduce corrosion and prevent marine organisms from growing on a ship's hull. The coatings also make removal of salt deposits possible without using freshwater. Objects subject to constant friction like boat hulls would require constant reapplication of such a coating to maintain a high degree of performance [128]. Superhydrophobic materials can be used as a Buoyant devices. Multifaceted superhydrophobic surfaces can trap a large air volume, which points toward the possibility of using superhydrophobic surfaces to create buoyant devices (Fig. 11.22). Guo's lab created a structure in which the treated surfaces on two parallel aluminum plates face inward, not outward, so they are enclosed and free from external wear and abrasion. The surfaces are separated by just the right distance to trap and hold enough air to keep the structure floating—in essence creating a waterproof compartment. The superhydrophobic surfaces will keep water from entering the compartment even when the structure is forced to submerge in water. Even after being forced to submerge for 2 months, the structures immediately bounced back to the surface after the load was released. The structures also retained this ability even after being punctured multiple times because the air remains trapped in the remaining parts of the compartment or adjoining structures. Between the treated plates to trap and hold enough air to keep the structure floating [129].

11.13.3 Automobile Industries

The self-cleaning coating was developed at the Nissan Technical Center in the United Kingdom in conjunction with Nanotechnology firm, Nano labs, in the hope that Nissan owners would never have to wash their cars again. Nissan is the first car maker to apply this technology, called "Ultra-Ever Dry," on automotive bodywork [130]. It has responded well to common use cases including rain, spray, frost, sleet, and standing water. Figure 11.23a shows the effect of mud on the car body after and before being coated with the self-cleaning materials. Self-cleaned materials were also used on the front glass and mirror as shown in Fig. 11.23b, c. Self-cleaning water resistance plastics are also used in automotive industries (Fig. 11.23d).

11.13.4 Electronic Industries

Hydrophobic electronic treatments can be used to increase reliability, improve performance, and add value to consumer electronics by repelling oil and water and allowing easy cleaning of fingerprint-resistant surfaces. Hydrophobic applications include anti-wetting applications, inkjet printer nozzles, microfluidic channels, inkjet repellency, hard drives, stainless steel components, needles and syringes, and many

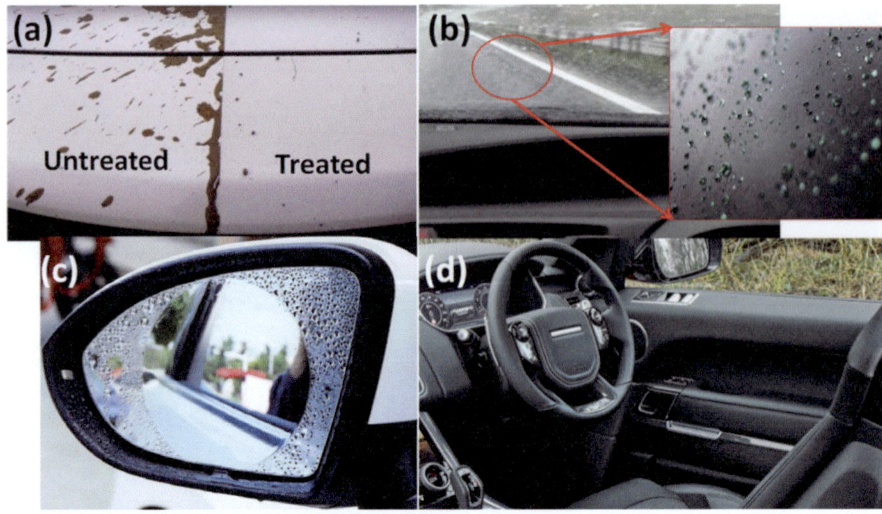

Fig. 11.23 Self-cleaned surface at (**a**) Car body, (**b**) Front glass, (**c**) Mirror, and (**d**) Cabin polymer

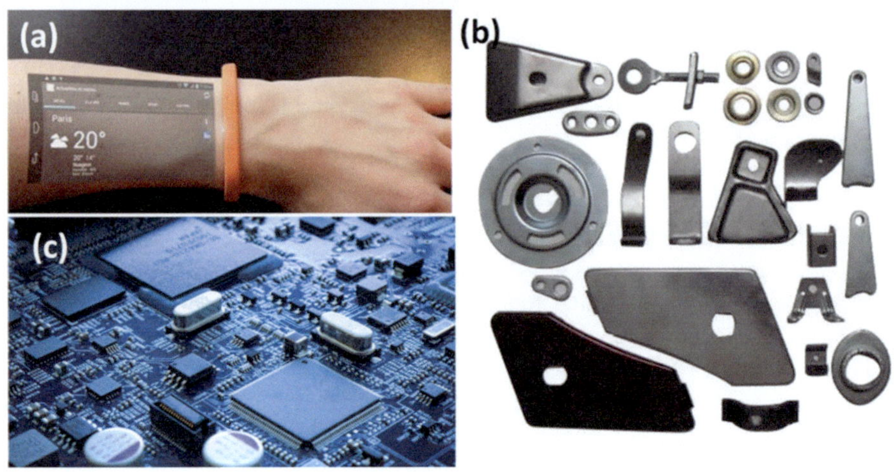

Fig. 11.24 (**a**) Inflexible and the wearable electronic device, (**b**) Stainless steel components, and (**c**) Electronic circuits

others hydrophobic electronic repellency treatments for electronics compatible with the following types of materials: metals, polymers/fabrics, glass/ceramics, semiconductors, and many others. Figure 11.24a–c shows some self-cleaning surface coating observed in a wearable electronic device, various stainless steel electronics components, and electronic circuits. High-sensitive super-hydrophobic self-cleaning infrared nano-sensor is also synthesized by carbon nanoparticles or TiO_2 nanotube layers or ZnO nano-rod. Now, most of the glass materials in electronics have a high demand for transparent or colored superhydrophobic material coating [131]. Micro/nanochannels applications requiring a reduction in fluid flow. Superhydrophobic surfaces prevent the formation of menisci at contacting interfaces and can be used to minimize high adhesion and stiction. It is found that by creating the passage similar to the superhydrophobic surface of rice leaf, there is no loss during liquid transportation and also better control to flow the fluid in a particular direction in Microfluidic passage [132]. The coatings are mostly used on things such as electronic components, which are not prone to wear. These very tiny structures are by their nature very delicate and very easily damaged by wear, cleaning, and lose its superhydrophobic properties. It is used mainly in sealed environments that are not exposed to wear or cleaning, such as electronic components (like the inside of smartphones) and air conditioning heat transfer fins to protect from moisture and prevent corrosion.

11.13.5 Medical Industries

In healthcare industries, it is necessary to equip materials and surfaces with a high level of hygiene using antimicrobial agents to protect them against bacteria and other microorganisms to prevent infections caused by bacteria and contribute significantly to reducing health costs. Thanks to innovative hydrophilic coatings, invasive operations that were previously impossible can now be carried out safely through the cardiovascular, circulatory, neurological, urological, and vascular systems, allowing for faster recoveries and fewer complications. This growing and exciting field stimulates growth and demand for new tools and innovations.

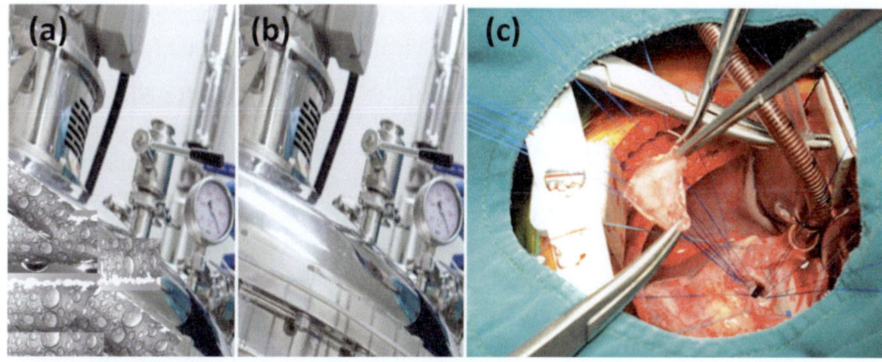

Fig. 11.25 Liquid nitrogen equipped cylinder (**a**) without self-cleaning coating, (**b**) with self-cleaning coating, and (**c**) self-cleaning material coated surgical instruments for easily visible surgery

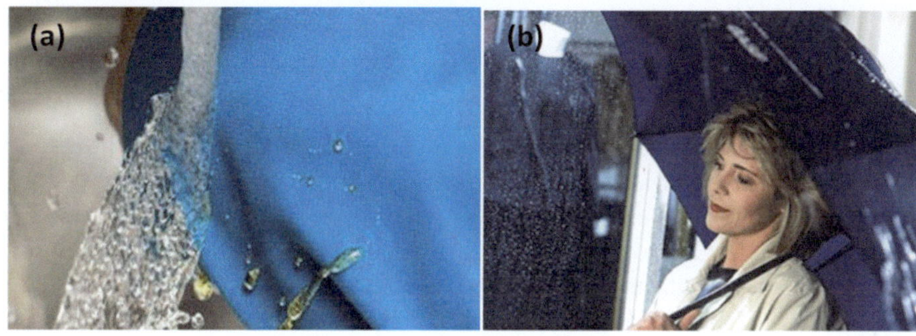

Fig. 11.26 Self-cleaning coated materials in (**a**) Wear materials, and (**b**) Umbrella

Our hydrophilic coatings are formulated to attract and retain water, giving them lubricating qualities that reduce potential trauma and improve performance during these invasive procedures. Hydromer medical device coatings provide lubricity, some of which are low-particle coatings, and help improve the performance of these medical devices [133]. Additional self-cleaning products included reusable tissue box covers, mousepads, reception mats, and placemats that provide a cleaner rest area for personal effects and medical equipment. And the newest product the team is about to launch is a transparent, self-cleaning film for kiosks and tablets [134]. Figure 11.25a shows the nitrogen cylinder that is equipped in medical industries which clearly shows the effect of the self-cleaning material coating. Figure 11.25b shows the surgical instruments that are coated with the self-cleaning materials to avoid any contamination in precise surgery. In biomedical devices, the patterned superhydrophobic surfaces also have promise for lab-on-a-chip microfluidic devices and can drastically improve surface-based bioanalysis in biomedical analysis. Due to the extreme repellence and in some cases bacterial resistance of hydrophobic coatings, there is much enthusiasm for their wide potential uses with surgical tools and medical equipment. Newer engineered surface textures on stainless steel are extremely durable and permanently hydrophobic. Optically, these surfaces appear as a uniform matte surface, but microscopically they consist of rounded depressions 1 to 2 microns deep over 25–50% of the surface. These surfaces are produced for buildings that will never need cleaning [135]. One of the main concerns in many industries including the medical industry is the environmental pollution of materials and equipment. The growth of biofilms as a natural spontaneous phenomenon can become a problem in the case of hospital infections. To kill the biofilm or reduce their adhesion, two strategies of bactericidal surfaces and anti-bioadhesion surfaces are considered, and the SHS is also used for this purpose [136].

11.13.6 Textile Industries

These materials are based on a lotus leaf. The surface of the prepared textile has a hierarchical structure comprising both micrometer and submicrometer features that make it difficult for water droplets to spread. As a result, water droplets form tight spheres that easily roll off, also can pick up dirt particles en route. The important application of TiO_2 is the self-cleaning of fabrics. Coffee and red wine stains on cotton fabrics can be easily removed with TiO_2 nanosols under UV radiation. Furthermore, the TiO_2 nanosol coating has demonstrated excellent UV protection against the cotton fabric. Fabrics coated with cellulose acetate and TiO_2 could retain their self-cleaning activity after several stains and washes. The TiO_2-SiO_2–coated fabrics have been tested for their self-cleaning ability against fading of red wine stains. TiO_2-SiO_2 with a molar ratio of 30:70 has shown optimal efficacy in stain removal. Fabrics coated with porous $Au/TiO_2/SiO_2$ nanocomposites manufactured to enhance self-cleaning performance assisted by visible light. Red wine and brown stains on the nanocomposite-coated tissues were completely removed after 20 h. of visible light irradiation [137]. Figure 11.26 shows some self-cleaning coated (TiO_2, TiO_2-SiO_2) fabric for textile and umbrella cloth. Cotton fabric with silica or titania particles coating by sol-gel technique protects the fabric from UV light and makes it superhydrophobic. An efficient routine has been reported for making polyethylene superhydrophobic and thus self-cleaning. 99% of dirt on such a surface is easily washed away. Superhydrophobic surfaces with self-cleaning properties also have many applications in the textile industry. For example, self-cleaning textile shirts, blouses, skirts, and trousers, which are stain-proof, have already been synthesized [138].

11.13.7 Other Industries

Nowadays, self-cleaning materials are growing their importance from larger industries to smaller retailer industries. In energy harvesting equipment like solar cells, more than 30% of efficiency are lost due to the accumulation of dust particles and other organic pollutants. This problem can be rectified by the application of self-cleaning materials on the upper surface of the solar panel [139]. A typical solar cell (silicon micromorph) can be designed on a glass substrate via plasma-enhanced CVD method to develop the self-cleaning and antireflective properties. Superhydrophobic coatings are used in ultra-dry surface applications. Some of the well-developed activities are discussed below:

Oil–Water Separator: Hydrophobic materials are used for oil removal from water, the management of oil spills, and chemical separation processes to remove non-polar substances from polar compounds. Large amounts of crude oil are now extracted from reservoirs using injected steam, but separating the resulting emulsion is difficult and requires materials that can function above 130°C. A new filtration process has been developed based on a stainless steel mesh coated with Zn-oxide nanotetrapods. By itself, it is quite hydrophobic because of the layer of air on the surface, but it is also oleophilic. The mesh membrane forms an interconnected porous plastron network that lets oil through. But it keeps water droplets suspended above the air pockets formed between the protruding nanotetrapods. The water droplets are in the Cassie-Baxter state, in contrast to oil drops that are in Wenzel mode and permeate the mesh. The filter can reduce the water content of viscous oil [140]. Due to the increase in industrial water laden with oil, the separation of oil and water mixture has become one of the global challenges. The combination of "superhydrophobic and superoleophilic," "superhydrophilic and under water superoleophobic," and "superhydrophilic and superoleophobic" are used to separate oil and water. By the use of different strategies, a series of materials such as membranes, tissues, foams, and sorbents have been developed for separating water and oil [141].

Ultra-Slippery Surface: It is also called slippery liquid-infused porous surfaces. This surface developed on the basis of the pitcher plant principle. Here, a textured or porous sponge-like surface is present in which the pores are filled by lubricating fluid that has a strong chemical affinity to the underlying textured surface. With this combination, a slip can repel anything that is immiscible with the lubricant. If there is a requirement to repel oil-based fluid, then it can engineer the lubricant to either be aqueous or it can use a perfluorinated fluid that is immiscible with water and an oil phase. Based on these design criteria, it can explore all kinds of liquids as a lubricant. These materials make them ideal for coating oil storage and transportation equipment [142].

Water Harvest: This is based on the hybridization of the hydrophilic surface with the hydrophobic surface like rice leaf ridges. The lubricated surfaces could also be useful for water collection in parts of the world where there are water shortages. Only a hydrophobic slip did not provide the most efficient way to harvest water vapor or fog as it limited water drop nucleation. And simply using a hierarchical textured surface led to droplets being pinned as with the rose petal, rather than rolling off to be collected. In combination with the rice leaf phenomenon, it now attracts the water vapor or water droplets from the air; once the water is in contact with the surface, it can slide away easily. A set of competing effects all contribute to a surface that repels liquids regardless of how they wet the surface. The hydrophilic lubricant helps water droplets nucleate, and the rice leaf ridges then allow drops to roll off. Roughly 500 mg of water per cm^2 in an hour can be collected by these materials, which is close to 10 times more water than the typical fog-harvesting material [143].

Self-cleaning Windows, Windshields, Roof tile, Exterior Paints for Buildings: This material is based on the lotus leaf principle. It has micrometer and submicrometer hierarchical structure to roll off the water droplet and also can clean during water roll-off time. Superhydrophobic surfaces have low drag for fluid flow and low tilt angle. Polysiloxane emulsion with particles of TiO_2 and SiO_2 is used as exterior paint and can be used on mineral (concrete) and organic (non-elastic such as shielding) substrates. Dirt particles have a reduced adhesion to the painted surface and are easily cleaned off by the rain. The droplets of car window, glasses, and tall-building glass surface slide to take away the dirt, but with the evaporation of droplets, the dirt will be stuck on the surface. The solution to this problem is found out that the droplets on the flat surface can remove the dirt by sliding. This removal is due to the microstructure of the superhydrophobic surface for which the droplets could roll to remove dirt with a small tilt [144].

Agriculture Fields: In agricultural fields, it is found that the bionic convex triangular trough ridging shovel and the bionic convex arc-shaped trough ridging shovel had more obvious resistance reduction effects than other types of the shovel. This works on the basic principle of superhydrpphobic shark skin [145].

Nonshrinkable Materials: Spiders and fire ants can survive long periods under or on the surface of the water. It is possible by trapping air in an enclosed area. *Argyroneta aquatic* spiders, for example, create an underwater dome-shaped web—a so-called diving bell—which they fill with air carried from the surface between their super-hydrophobic legs and abdomens. Similarly, fire ants can form a raft by trapping air among their superhydrophobic bodies [146]. By taking this phenomenon, non-shrinkable textile has been developed.

Also use of self-cleaning materials in the exterior paints for buildings, hydrophobic concrete, nonstick utensils, optical applications, such as solar panels, lenses, and mirrors, shower screens, solar cells and panels, toilet urinals to eliminate the need for flushing water has been developed.

11.14 Limitations of Self-Cleaning Materials

Although there have been major developments in the field of superhydrophobic materials and surfaces, still not many commercial products are being developed with these properties. Some of the major reasons and limitations are listed subsequently [147–149]:

Mechanical Stability: These very tiny structures are by nature very delicate and very easily damaged by wear, cleaning, or any sort of friction; if the structure is damaged even slightly it loses its superhydrophobic properties. This technology is based on the microstructure of the hairs of a lily pad that makes water just roll off. Rub a lily leaf a little and it will no longer be superhydrophobic. Unlike a lily leaf, which can heal and grow new hairs, a coating will not do this in manmade structural materials.

Materials Stability and Durability: Any superhydrophobic surface requires stable micro- and nanotopography with hydrophobic surface chemistry. At nanolevels, especially in case of polymers, this is usually not easy as the strands act like wet noodles and can easily wind and matte down losing the superhydrophobic properties. Also, there is a standoff between durability and superhydrophobic behavior, as it is not easy to bond any superhydrophobic nanoparticle without degrading or destroying its superhydrophobic nature.

Cost Issues: Most of the fabrication methods indicated in the aforementioned section are costly due to the higher overall cost of raw materials or other expensive process requirements. Also, most of the processes like lithography and template method cannot be used to generate large area substrates with nanostructured surfaces. So, producing small chip dies and stitching them together may increase the cost.

Nonuniform Coatings and Wastage of Reactants: Coating of low-energy materials using spin or dip or spray coating on the surfaces to provide superhydrophobic properties is also not foolproof. A major problem is the nonuniform application of the material on the surface. Also, there is wastage of reactant materials due to spraying or spillage in most of the coating techniques.

Oil and Surfactant Wetting Issues: Although the surfaces have anti-wetting properties due to amplifying the surface tension of water, not much research has been done to study the effect of lower surface tension developed in water due to mixing with surfactants or with oil. Since surface tension will reduce, the complete wetting of surface will occur, and the superhydrophobic behavior will be reduced or completely removed.

Coating Impingement Issues: Superhydrophobic coatings have usually been thin layers on material surfaces or have been surface effects. Such coatings or effects can be destroyed by simple rubbing or localized high-pressure stream of water.

Vapor Condensation Issues: Although all coated surfaces repel water, water vapor is not repelled. If the temperature of the coating is below the dew point, then condensation will happen and the surface will get wetted substantially losing its superhydrophobic properties.

Selective Base Material and Temperature Ranges for Each Process: Each of the processes for fabricating a superhydrophobic surface is developed for only certain base material and temperature of working. The fabrication method fails if the material is substituted. Also, not much research has been done to check the properties of the surfaces at varying temperature and environmental conditions.

Health and Environmental Risks: Most of the superhydrophobic coatings are fluorine-based. Although small quantities of fluorine are good, continuous exposure to fluorine-based products having high concentrations can cause health issues, such as teeth and bone decay and harm to kidney, muscles, and nerves. Also, exposure to fluorine gas can cause eyes and nose irritations. The use of inorganic agents and polymers as coatings, that are nonbiodegradable, can also cause a risk of environmental pollution during disposal.

11.15 Summary

This chapter described the fundamental knowledge of self-cleaning materials. From the historical development of this material, it is clear about the great amount of effort that has been put into the research to understand the mechanisms. The

Young equation, Wenzel equation, and Cassie-Baxter equation are the fundamental theories to describe the wetting phenomena on self-cleaning surfaces. Although superhydrophobicity is only a recently developed concept, it has already become important to a lot of research and will be potentially important to people's life.

References

1. Liu, M., Wang, S., Jiang, L.: Nature-inspired super wettability systems. Nat. Rev. Mater. **2**, 17036 (2017). https://doi.org/10.1038/natrevmats.2017.36
2. Behera, A., Rajak, D.K., Jeyasubramanian, K.: Fabrication of nanostructure with excellent self-cleaning property. In: Design, fabrication and characterization of multifunctional nanomaterials. Elsevier Publisher. ISBN: 9780128205587. (In Press)
3. Nishimotoab, S., Bhushan, B.: Bioinspired self-cleaning surfaces with superhydrophobicity, superoleophobicity, and superhydrophilicity. RSC Adv. **3**, 671–690 (2013) https://doi.org/10.1039/C2RA21260A
4. Li, Y., Zhang, H., Yao, Y., Li, T., Zhang, Y., Lic, Q., Dai, Z.: Transfer of vertically aligned carbon nanotube arrays onto flexible substrates for gecko-inspired dry adhesive application. RSC Adv. **5**, 46749–46759 (2015). https://doi.org/10.1039/C5RA06206C
5. Andersen, N.M.: The evolution of dispersal dimorphism and other life history traits in water striders (Hemiptera: Gerridae). Entomol. Sci. **3**(1), 187–199 (2000) ref.3 pp. of Conference Title : Proceedings of the International Japanese-Czech New-Year Seminar of Entomology, 8–10 January, Kochi, Japan. https://www.cabdirect.org/cabdirect/abstract/20000507422
6. Stan, M.S., Badea, M.A., Pircalabioru, G.G., Chifiriuc, M.C., Diamandescu, L., Dumitrescu, I., Trica, B., Lambert, C., Dinischiotu, A.: Designing cotton fibers impregnated with photocatalytic graphene oxide/Fe, N-doped TiO_2 particles as prospective industrial self-cleaning and biocompatible textiles. Mater. Sci. Eng. C. **94**, 318–332 (2019). https://doi.org/10.1016/j.msec.2018.09.046
7. Cedillo-González, E.I., Barbieri, V., Falcaro, P., Torres-Martínez, L.M., Juárez-Ramírez, I., Villanova, L., Montecchi, M., Pasquali, L., Siligardi, C.: Influence of domestic and environmental weathering in the self-cleaning performance and durability of TiO_2 photocatalytic coatings. Build. Environ. **132**, 96–103 (2018). https://doi.org/10.1016/j.buildenv.2018.01.028
8. Paz, Y., Luo, Z., Rabenberg, L., Heller, A.: Photooxidative self-cleaning transparent titanium dioxide films on glass. J. Mater. Res. **10**(11), 2842–2848. https://doi.org/10.1557/JMR.1995.2842
9. Banerjee, S., Dionysiou, D.D., Pillai, S.C.: Self-cleaning applications of TiO2 by photo-induced hydrophilicity and photocatalysis. Appl. Catal. B Environ. **176–177**, 396–428 (2015). https://doi.org/10.1016/j.apcatb.2015.03.058
10. Dong, H., Zeng, G., Lin, T., Fan, C., Zhang, C., He, X., He, Y.: An overview on limitations of TiO_2-based particles for photocatalytic degradation of organic pollutants and the corresponding countermeasures. Water Res. **79**, 128–146 (2015). https://doi.org/10.1016/j.watres.2015.04.038
11. Blossey, R.: Self-cleaning surfaces-virtual realities. Nat. Mater. **2**, 301–306 (2003). https://doi.org/10.1038/nmat856
12. Gao, L., McCarthy, T.J.: Wetting 101°. Langmuir. **25**(24), 14105–14115 (2009). https://doi.org/10.1021/la902206c
13. Kim, S., Cheung, E., Sitti, M.: Wet self-cleaning of biologically inspired elastomer mushroom shaped microfibrillar adhesives. Langmuir. **25**(13), 7196–7199 (2009). https://doi.org/10.1021/la900732h
14. Nosonovsky, M.: Slippery when wetted. Nature. **477**, 412–413 (2011). https://doi.org/10.1038/477412a
15. Kamegawa, T., Irikawa, K., Yamashita, H.: Multifunctional surface designed by nanocomposite coating of polytetrafluoroethylene and TiO_2 photocatalyst: self-cleaning and super hydrophobicity. Sci. Rep. **7**, 13628 (2017). https://doi.org/10.1038/s41598-017-14058-9
16. Bartolo, D., Bouamrirene, F., Verneuil, É., Buguin, A., Silberzan, P., Moulinet, S.: Bouncing or sticky droplets: impalement transitions on superhydrophobic micropatterned surfaces. Europhy. Lett. **74**(2), 299 (2005). https://doi.org/10.1209/epl/i2005-10522-3
17. Koishi, T., Yasuoka, K., Fujikawa, S., Ebisuzaki, T., Zeng, X.C.: Coexistence and transition between Cassie and Wenzel state on pillared hydrophobic surface. PNAS. **106**(21), 8435–8440 (2009). https://doi.org/10.1073/pnas.0902027106
18. Zhang, K., Li, Z., Maxey, M., Chen, S., Karniadakis, G.E.: Self-cleaning of hydrophobic rough surfaces by coalescence-induced wetting transition. Langmuir. **35**(6), 2431–2442 (2019). https://doi.org/10.1021/acs.langmuir.8b03664
19. Cai, T., Jia, Z., Yang, H., et al.: Investigation of Cassie-Wenzel wetting transitions on microstructured surfaces. Colloids Polym. Sci. **294**, 833–840 (2016). https://doi.org/10.1007/s00396-016-3836-4
20. Fang, W., Guo, H.-Y., Li, B., Li, Q., Feng, X.-Q.: Revisiting the critical condition for the Cassie–Wenzel transition on micropillar-structured surfaces. Langmuir. **34**(13), 3838–3844 (2018). https://doi.org/10.1021/acs.langmuir.8b00121
21. Bhushan, B., Jung, Y.C.: Natural and biomimetic artificial surfaces for superhydrophobicity, self-cleaning, low adhesion, and drag reduction. Progress. Mater. Sci. **56**(1), 1–108 (2011). https://doi.org/10.1016/j.pmatsci.2010.04.003
22. Bixlera, G.D., Bhushan, B.: fluid drag reduction and efficient self-cleaning with rice leaf and butterfly wing bioinspired surfaces. Nanoscale. **5**, 7685–7710 (2013). https://doi.org/10.1039/C3NR01710A
23. Pan, S., Guo, R., Björnmalm, M., et al.: Coatings super-repellent to ultralow surface tension liquids. Nat. Mater. **17**, 1040–1047 (2018). https://doi.org/10.1038/s41563-018-0178-2
24. Jerman, I., Koželj, M., Orel, B.: The effect of polyhedral oligomeric silsesquioxane dispersant and low surface energy additives on spectrally selective paint coatings with self-cleaning properties. Sol. Energy Mater. Sol. Cells. **94**(2), 232–245 (2010). https://doi.org/10.1016/j.solmat.2009.09.008
25. Wang, Z., Koratkar, N.: Combined micro−/nanoscale surface roughness for enhanced hydrophobic stability in carbon nanotube arrays. Appl. Phys. Lett. **90**(14), 143117 (2007). https://doi.org/10.1063/1.2720761
26. He, B., Lee, J., Patankar, N.A.: Contact angle hysteresis on rough hydrophobic surfaces. Colloids Surf A. **248**(1–3), 101–104 (2004). https://doi.org/10.1016/j.colsurfa.2004.09.006
27. Yilbas, B.S., Al-Sharafia, A., Alia, H., Al-Aqeeli, N.: Dynamics of a water droplet on a hydrophobic inclined surface: influence of droplet size and surface inclination angle on droplet rolling. RSC Adv. **7**, 48806–48818 (2017). https://doi.org/10.1039/C7RA09345D. (Paper)
28. Li, S., Page, K., Sathasivam, S., Heale, F., He, G., Lu, Y., Lai, Y., Chen, G., Carmalt, C.J., Parkin, I.P.: Efficiently texturing hierarchical superhydrophobic fluoride-free translucent films by AACVD with excellent durability and self-cleaning ability. J. Mater. Chem. A. **6**, 17633–17641 (2018). https://doi.org/10.1039/C8TA05402A

29. Zhu, J., Dai, X.: A new model for contact angle hysteresis of superhydrophobic surface. AIP Adv. **9**(6), 065309 (2019). https://doi.org/10.1063/1.5100548
30. https://www.chemistryworld.com/features/superhydrophobic-materials-from-nature/3010321.article#/
31. Schnell, G., Polley, C., Bartling, S., Seitz, H.: Effect of chemical solvents on the wetting behavior over time of femtosecond laser structured Ti6Al4V surfaces. Nanomaterials. **10**, 1241 (2020). https://doi.org/10.3390/nano10061241
32. Jung, Y.C., Bhushan, B.: Mechanically durable carbon nanotube−composite hierarchical structures with superhydrophobicity, self-cleaning, and low-drag. ACS Nano. **3**(12), 4155–4163 (2009). https://doi.org/10.1021/nn901509r
33. Ming, R., VanBuren, R., Liu, Y.: Genome of the long-living sacred lotus (Nelumbo nucifera Gaertn.). Genome. Biol. **14**, R41 (2013). https://doi.org/10.1186/gb-2013-14-5-r41
34. Shang, Q., Zhou, Y.: Fabrication of transparent superhydrophobic porous silica coating for self-cleaning and anti-fogging. Ceram. Int. **42**(7), 8706–8712 (2016). https://doi.org/10.1016/j.ceramint.2016.02.105
35. Zhu, H., Wu, L., Meng, X., Wang, Y., Huang, Y., Lin, M., Xia, F.: An anti-UV superhydrophobic material with photocatalysis, self-cleaning, self-healing and oil/water separation functions. Nanoscale. **12**, 11455–11459 (2020). https://doi.org/10.1039/D0NR01038C
36. Xu, Q., Wan, Y., Hu, T., et al.: Robust self-cleaning and micromanipulation capabilities of gecko spatulae and their bio-mimics. Nat. Commun. **6**, 8949 (2015). https://doi.org/10.1038/ncomms9949
37. Xu, Q., Zhang, W., Dong, C., Sreeprasad, T.S., Xia, Z.: Biomimetic self-cleaning surfaces: synthesis, mechanism and applications. J. R. Soc. Interface. **13**(122), 20160300 (2016). https://doi.org/10.1098/rsif.2016.0300
38. Cunha, A.G., Gandini, A.: Turning polysaccharides into hydrophobic materials: a critical review. Part 1. Cellulose. Cellulose. **17**, 875–889 (2010). https://doi.org/10.1007/s10570-010-9434-6
39. Cunha, A.G.: Gandini, A. turning polysaccharides into hydrophobic materials: a critical review. Part 2. Hemicelluloses, chitin/chitosan, starch, pectin and alginates. Cellulose. **17**, 1045–1065 (2010). https://doi.org/10.1007/s10570-010-9435-5
40. David, Q., Mathilde, R.: Non-adhesive lotus and other hydrophobic materials. Philos. Trans. R. Soc. A. **366**, 1539–1556 (2008). https://doi.org/10.1098/rsta.2007.2171
41. Howarter, J.A., Jeffrey, P.: Youngblood, self-cleaning and next generation anti-fog. Surf. Coat. **26**(6), 455–466 (2008). https://doi.org/10.1002/marc.200700733
42. Syed, J., Tang, S., Meng, X.: Super-hydrophobic multilayer coatings with layer number tuned swapping in surface wettability and redox catalytic anti-corrosion application. Sci. Rep. **7**, 4403 (2017). https://doi.org/10.1038/s41598-017-04651-3
43. Zhong-Ze, G., Uetsuka, H., Takahashi, K., Nakajima, R., Onishi, H., Fujishima, A., Sato, O.: Structural color and the Lotus effect. Angewandte. Chemie. **115**(8), 922–925 (2003). https://doi.org/10.1002/ange.200390204
44. Bellanger, H., Darmanin, T., de Givenchy, E.T., Guittard, F.: Chemical and physical pathways for the preparation of superoleophobic surfaces and related wetting theories. Chem. Rev. **114**(5), 2694–2716 (2014). https://doi.org/10.1021/cr400169m
45. Avrămescu, R.-E., Ghica, M.V., Dinu-Pîrvu, C., Prisada, R., Popa, L.: Superhydrophobic natural and artificial Surfaces—a structural approach. Materials. **11**, 866 (2018). https://doi.org/10.3390/ma11050866
46. Wang, Z., Tang, Y., Li, B.: Excellent wetting resistance and anti-fouling performance of PVDF membrane modified with superhydrophobic papillae-like surfaces. J. Membr. Sci. **540**, 401–410 (2017). https://doi.org/10.1016/j.memsci.2017.06.073
47. Sethi, S.K., Soni, L., Manik, G.: Component compatibility study of poly(dimethyl siloxane) with poly(vinyl acetate) of varying hydrolysis content: an atomistic and mesoscale simulation approach. J. Mol. Liq. **272**, 73–83 (2018)
48. Sethi, S.K., Manik, G.: Recent progress in super hydrophobic/hydrophilic self-cleaning surfaces for various industrial applications: a review. Polym. Plast. Technol. Eng. **57**(18), 1932–1952 (2018). https://doi.org/10.1080/03602559.2018.1447128
49. Yuan, Z., Chen, H., Zhang, J.: Facile method to prepare lotus-leaf-like super-hydrophobic poly(vinyl chloride) film. Appl. Surf. Sci. **254**(6), 1593–1598 (2008). https://doi.org/10.1016/j.apsusc.2007.07.140
50. Xi, W., Qiao, Z., Zhu, C., Jia, A., Li, M.: The preparation of lotus-like super-hydrophobic copper surfaces by electroplating. Appl. Surf. Sci. **255**(9), 4836–4839 (2009). https://doi.org/10.1016/j.apsusc.2008.12.012
51. Kurokawa, Y., Nagai, K., Huan, P.D., Shimazaki, K., Qu, H., Mori, Y., Toda, Y., Kuroha, T., Hayashi, N., Aiga, S., Itoh, J.-i.: Rice leaf hydrophobicity and gas films are conferred by a wax synthesis gene (LGF1) and contribute to flood tolerance. New Phytol. **218**(4), 1558–1569 (2018). https://doi.org/10.1111/nph.15070
52. Lee, S.G., Lim, H.S., Lee, D.Y., Kwak, D., Cho, K.: Tunable anisotropic wettability of rice leaf-like wavy surfaces. Adv. Funct. Mater. **23**(5), 547–553 (2013). https://doi.org/10.1002/adfm.201201541
53. Ferguson, L.R., Roberton, A.M., McKenzie, R.J., Watson, M.E., Harris, P.J.: Adsorption of a hydrophobic mutagen to dietary fiber from taro (Colocasia esculenta), an important food plant of the south pacific. Nutr. Cancer. **17**(1), 85–95 (1992). https://doi.org/10.1080/01635589209514175. 2009
54. Kumar, M., Bhardwaj, R.: Wetting characteristics of Colocasia esculenta (Taro) leaf and a bioinspired surface thereof. Sci. Rep. **10**, 935 (2020). https://doi.org/10.1038/s41598-020-57410-2
55. Ebert, D., Bhushan, B.: Wear-resistant rose petal-effect surfaces with superhydrophobicity and high droplet adhesion using hydrophobic and hydrophilic nanoparticles. J. Colloid Interface Sci. **384**(1), 182–188 (2012). https://doi.org/10.1016/j.jcis.2012.06.070
56. Huang, H.-c., Zacharia, N.S.: Layer-by-layer rose petal mimic surface with oleophilicity and underwater oleophobicity. Langmuir. **31**(2), 714–720 (2015). https://doi.org/10.1021/la504095k
57. Tricinci, O., Terencio, T., Mazzolai, B., Pugno, N.M., Greco, F., Mattoli, V.: 3D micropatterned surface inspired by Salvinia molesta via direct laser lithography. ACS Appl. Mater. Interfaces. **7**(46), 25560–25567 (2015). https://doi.org/10.1021/acsami.5b07722
58. Cheng, L., Xu, Y., Grotewold, E., et al.: Characterization of Anthocyanidin synthase (ANS) gene and anthocyanidin in rare medicinal plant-Saussurea medusa. Plant. Cell Tissue Organ Cult. **89**, 63–73 (2007). https://doi.org/10.1007/s11240-007-9211-x
59. Koch, K., Dommisse, A., Barthlott, W.: Chemistry and crystal growth of plant wax tubules of lotus (Nelumbo nucifera) and nasturtium (Tropaeolum majus) leaves on technical substrates. Cryst. Growth Des. **6**(11), 2571–2578 (2006). https://doi.org/10.1021/cg060035w
60. Thongsook, T., Barrett, D.M.: Purification and partial characterization of broccoli (Brassica oleracea Var. Italica) peroxidases, J Agric Food Chem. **53**(8), 3206–3214 (2005). https://doi.org/10.1021/jf048162s
61. Feng, X.-Q., Gao, X., Wu, Z., Jiang, L., Zheng, Q.-S.: Superior water repellency of water strider legs with hierarchical structures: experiments and analysis. Langmuir. **23**(9), 4892–4896 (2007). https://doi.org/10.1021/la063039b

62. Fu, Y.F., Yuan, C.Q., Bai, X.Q.: Marine drag reduction of shark skin inspired riblet surfaces. Biosurf. Biotribol. **3**(1), 11–24 (2017). https://doi.org/10.1016/j.bsbt.2017.02.001
63. Badge, I., Stark, A., Paoloni, E., et al.: The role of surface chemistry in adhesion and wetting of gecko toe pads. Sci. Rep. **4**, 6643 (2014). https://doi.org/10.1038/srep06643
64. Hensel, R., Helbig, R., Aland, S., et al.: Tunable nano-replication to explore the omniphobic characteristics of springtail skin. NPG Asia Mater. **5**, e37 (2013). https://doi.org/10.1038/am.2012.66
65. Wani, A.A., Sogi, D.S., Singh, P., Wani, I.A., Shivhare, U.S.: Characterisation and functional properties of watermelon (Citrullus lanatus) seed proteins. J. Sci. Food Agric. **91**(1), 113–121 (2011). https://doi.org/10.1002/jsfa.4160
66. Paul Roach, A., Shirtcliffe, N.J., Newton, M.I.: Progess in superhydrophobic surface development. Soft. Matter. **4**, 224–240 (2008). https://doi.org/10.1039/B712575P
67. Zhang, D., Li, L., Wu, Y., Zhu, B., Song, H.: One-step method for fabrication of bioinspired hierarchical superhydrophobic surface with robust stability. Appl. Surf. Sci. **473**, 493–499 (2019). https://doi.org/10.1016/j.apsusc.2018.12.174
68. Guo, Z., Liu, W.: Biomimic from the superhydrophobic plant leaves in nature: binary structure and unitary structure. Plant Sci. **172**(6), 1103–1112 (2007). https://doi.org/10.1016/j.plantsci.2007.03.005
69. Zhang, L., Dillert, R., Bahnemann, D., Vormoor, M.: Photo-induced hydrophilicity and self-cleaning: models and reality. Energy Environ. Sci. **5**, 7491–7507 (2012). https://doi.org/10.1039/C2EE03390A
70. Jesus, M.A.M.L., Neto, J.T.D.S., Timò, G., et al.: Superhydrophilic self-cleaning surfaces based on TiO_2 and TiO_2/SiO_2 composite films for photovoltaic module cover glass. Appl. Adhes. Sci. **3**, 5 (2015). https://doi.org/10.1186/s40563-015-0034-4
71. Park, E.J., Yoon, H.S., Kim, D.H., Kim, Y.H., Kim, Y.D.: Preparation of self-cleaning surfaces with a dual functionality of superhydrophobicity and photocatalytic activity. Appl. Surf. Sci. **319**, 367–371 (2014). https://doi.org/10.1016/j.apsusc.2014.07.122
72. Son, J., Kundu, S., Verma, L.K., Sakhuja, M., Danner, A.J., Bhatia, C.S., Yang, H.: A practical superhydrophilic self cleaning and antireflective surface for outdoor photovoltaic applications. Sol. Energy Mater. Sol. Cells. **98**, 46–51 (2012). https://doi.org/10.1016/j.solmat.2011.10.011
73. Kamegawa, T., Shimizu, Y., Yamashita, H.: Superhydrophobic surfaces with photocatalytic self-cleaning properties by nanocomposite coating of TiO_2 and polytetrafluoroethylene. Adv. Mater. **24**(27), 3697–3700 (2012). https://doi.org/10.1002/adma.201201037
74. Yuan, S., Chen, C., Raza, A., Song, R., Zhang, T.-J., Pehkonen, S. O., Liang, B.: Nanostructured TiO_2/CuO dual-coated copper meshes with superhydrophilic, underwater superoleophobic and self-cleaning properties for highly efficient oil/water separation. Chem. Eng. J. **328**, 497–510 (2017). https://doi.org/10.1016/j.cej.2017.07.075
75. Adachi, T., Latthe, S.S., Gosavi, S.W., Roy, N., Suzuki, N., Ikari, H., Kato, K., Katsumata, K.-i., Nakata, K., Furudate, M., Inoue, T., Kondo, T., Yuasa, M., Fujishima, A., Terashima, C.: Photocatalytic, superhydrophilic, self-cleaning TiO_2 coating on cheap, light-weight, flexible polycarbonate substrates. Appl. Surf. Sci. **458**, 917–923 (2018). https://doi.org/10.1016/j.apsusc.2018.07.172
76. Yang, Z., Wang, L., Sun, W., Li, S., Zhu, T., Liu, W., Liu, G.: Superhydrophobic epoxy coating modified by fluorographene used for anti-corrosion and self-cleaning. Appl. Surf. Sci. **401**, 146–155 (2017). https://doi.org/10.1016/j.apsusc.2017.01.009
77. Zhang, X., Guo, Y., Zhang, Z., Zhang, P.: Self-cleaning superhydrophobic surface based on titanium dioxide nanowires combined with polydimethylsiloxane. Appl. Surf. Sci. **284**, 319–323 (2013). https://doi.org/10.1016/j.apsusc.2013.07.100
78. Wu, X., Zheng, L., Wu, D.: Fabrication of superhydrophobic surfaces from microstructured ZnO-based surfaces via a wet-chemical route. Langmuir. **21**(7), 2665–2667 (2005). https://doi.org/10.1021/la050275y
79. Nishimoto, S., Kubo, A., Nohara, K., Zhang, X., Taneichi, N., Okui, T., Liu, Z., Nakata, K., Sakai, H., Murakami, T., Abe, M., Komine, T., Fujishima, A.: TiO2-based superhydrophobic–superhydrophilic patterns: fabrication via an ink-jet technique and application in offset printing. Appl. Surf. Sci. **255**(12), 6221–6225 (2009). https://doi.org/10.1016/j.apsusc.2009.01.084
80. Peng, C.-W., Chang, K.-C., Weng, C.-J., Lai, M.-C., Hsu, C.-H., Hsu, S.-C., Hsu, Y.-Y., Hung, W.-I., Wei, Y., Yeh, J.-M.: Nano-casting technique to prepare polyaniline surface with biomimetic superhydrophobic structures for anticorrosion application. Electrochimi. Acta. **95**, 192–199 (2013). https://doi.org/10.1016/j.electacta.2013.02.016
81. Zhang, J., Seeger, S.: Silica/silicone nanofilament hybrid coatings with almost perfect superhydrophobicity. ChemPhysChem. **14**(8), 1646–1651 (2013). https://doi.org/10.1002/cphc.201200995
82. Zhu, L., Xiu, Y., Xu, J., Tamirisa, P.A., Hess, D.W., Wong, C.-P.: Superhydrophobicity on two-tier rough surfaces fabricated by controlled growth of aligned carbon nanotube arrays coated with fluorocarbon. Langmuir. **21**(24), 11208–11212 (2005). https://doi.org/10.1021/la051410
83. Lau, K.K.S., Bico, J., Teo, K.B.K., Chhowalla, M., Amaratunga, G.A.J., Milne, W.I., McKinley, G.H., Gleason, K.K.: Superhydrophobic carbon nanotube forests. Nano Lett. **3**(12), 1701–1705 (2003). https://doi.org/10.1021/nl034704t
84. Nguyen, D.D., Tai, N.-H., Leea, S.-B., Kuo, W.-S.: Superhydrophobic and superoleophilic properties of graphene-based sponges fabricated using a facile dip coating method. Energy Environ. Sci. **5**, 7908–7912 (2012). https://doi.org/10.1039/C2EE21848H
85. Doshi, B., Sillanpää, M., Kalliola, S.: A review of bio-based materials for oil spill treatment. Water Res. **135**, 262–277 (2018). https://doi.org/10.1016/j.watres.2018.02.034
86. Parvate, S., Dixit, P., Chattopadhyay, S.: Superhydrophobic surfaces: insights from theory and experiment. J. Phys. Chem. B. **124**(8), 1323–1360 (2020). https://doi.org/10.1021/acs.jpcb.9b08567
87. Sung, Y.H., Kim, Y.D., Choi, H.-J., Shin, R., Kang, S., Lee, H.: Fabrication of superhydrophobic surfaces with nano-in-micro structures using UV-nanoimprint lithography and thermal shrinkage films. Appl. Surf. Sci. **349**, 169–173 (2015). https://doi.org/10.1016/j.apsusc.2015.04.141
88. Amalathas, A.P., Alkaisi, M.M.: Efficient light trapping nanopyramid structures for solar cells patterned using UV nanoimprint lithography. Mater. Sci. Semicond. Process. **57**, 54–58 (2017). https://doi.org/10.1016/j.mssp.2016.09.032
89. Liu, R., Chi, Z., Liang, C., Weng, Z., Lu, W., Li, L., Saeed, S., Lian, Z., Wang, Z.: Fabrication of biomimetic superhydrophobic and anti-icing Ti6Al4V alloy surfaces by direct laser interference lithography and hydrothermal treatment. Appl. Surf. Sci. **534**, 147576 (2020). https://doi.org/10.1016/j.apsusc.2020.147576
90. Sun, H., Wang, C., Pang, S., Li, X., Tao, Y., Tang, H., Liu, M.: Photocatalytic TiO2 films prepared by chemical vapor deposition at atmosphere pressure. J. Non-Crystal. Solids. **354**(12–13), 1440–1443 (2008). https://doi.org/10.1016/j.jnoncrysol.2007.01.108
91. Weng, K.-W., Huang, Y.-P.: Preparation of TiO_2 thin films on glass surfaces with self-cleaning characteristics for solar concentrators. Surf. Coat. Technol. **231**, 201–204 (2013). https://doi.org/10.1016/j.surfcoat.2012.06.058

92. Qing, Y., Hu, C., Yang, C., An, K., Tang, F., Tan, J., Liu, C.: Rough structure of electrodeposition as a template for an ultrarobust self-cleaning surface. ACS Appl. Mater. Interfaces. (9, 19), 16571–16580 (2017). https://doi.org/10.1021/acsami.6b15745
93. Sas, I., Gorga, R.E., Joines, J.A., Thoney, K.A.: Literature review on superhydrophobic self-cleaning surfaces produced by electrospinning. J. Polym. Sci. Part B Polym. Phys. **50**(12), 824–845 (2012). https://doi.org/10.1002/polb.23070
94. Zheng, G., Jiang, J., Wang, X., et al.: Self-cleaning threaded rod spinneret for high-efficiency needleless electrospinning. Appl. Phys. A Mater. Sci. Process. **124**, 473 (2018). https://doi.org/10.1007/s00339-018-1892-y
95. Verhovšek, D., Veronovski, N., Štangar, U.L., Kete, M., Žagar, K., Čeh, M.: The synthesis of anatase nanoparticles and the preparation of photocatalytically active coatings based on wet chemical methods for self-cleaning applications. **2012**, 329796 (2012). https://doi.org/10.1155/2012/329796
96. Dong, Z., Xia, Y., Chen, X., Shi, S., Lei, L.: PDMS-infused poly (high internal phase emulsion) templates for the construction of slippery liquid-infused porous surfaces with self-cleaning and self-repairing properties. Langmuir. **35**(25), 8276–8284 (2019). https://doi.org/10.1021/acs.langmuir.9b01115
97. Wang, S., Feng, L., Jiang, L.: One-step solution-immersion process for the fabrication of stable bionic Superhydrophobic Surfaces. Adv. Mater. **18**(6), 767–770 (2006). https://doi.org/10.1002/adma.200501794
98. Nakamura, C., Manabe, K., Tenjimbayashi, M., Tokura, Y., Kyung, K.-H., Shiratori, S.: Heat-shielding and self-cleaning smart windows: near-infrared reflective photonic crystals with self-healing omniphobicity via layer-by-layer self-assembly. ACS Appl. Mater. Interfaces. **10**(26), 22731–22738 (2018). https://doi.org/10.1021/acsami.8b05887
99. Swain, B., Pati, A.R., Mallick, P., Mohapatra, S.S., Behera, A.: Development of highly durable superhydrophobic coatings by one-step plasma spray methodology. J. Therm. Spray Tech. **30**, 405–423 (2021). https://doi.org/10.1007/s11666-020-01132-4
100. Kumar, D., Wu, X., Qitao, F., Ho, J.W.C., Kanhere, P.D., Lin, L., Chen, Z.: Development of durable self-cleaning coatings using organic–inorganic hybrid sol–gel method. Appl. Surf. Sci. **344**, 205–212 (2015). https://doi.org/10.1016/j.apsusc.2015.03.105
101. Liu, P.-F., Miao, L., Deng, Z., Zhou, J., Yufei, G., Chen, S., Cai, H., Sun, L., Tanemura, S.: Flame-treated and fast-assembled foam system for direct solar steam generation and non-plugging high salinity desalination with self-cleaning effect. Appl. Energy. **241**, 652–659 (2019). https://doi.org/10.1016/j.apenergy.2019.02.030
102. Zhao, F., Ma, Z., Xiao, K., Xiang, C., Wang, H., Huang, X., Liang, S.: Hierarchically textured superhydrophobic polyvinylidene fluoride membrane fabricated via nanocasting for enhanced membrane distillation performance. Desalination. **443**, 228–236 (2018). https://doi.org/10.1016/j.desal.2018.06.003
103. Davoudinejad, A., Ribo, M.M., Pedersen, D.B., Islam, A., Tosello, G.: Direct fabrication of bio-inspired gecko-like geometries with vat polymerization additive manufacturing method. J. Micromech. Microeng. **28**(8), 085009. https://doi.org/10.1088/1361-6439/aabf17
104. Li, M., Zhai, J., Liu, H., Song, Y., Jiang, L., Zhu, D.: Electrochemical deposition of conductive superhydrophobic zinc oxide thin films. J. Phys. Chem. B. **107**(37), 9954–9957 (2003). https://doi.org/10.1021/jp035562u
105. Zhang, X., Shi, F., Yu, X., Huan Liu, Y.F., Wang, Z., Jiang, L., Li, X.: Polyelectrolyte multilayer as matrix for electrochemical deposition of gold clusters: toward super-hydrophobic. J. Am. Chem. Soc. **126**(10), 3064–3065 (2004). https://doi.org/10.1021/ja0398722
106. Satapathy, M., Varshney, P., Nanda, D., Mohapatra, S.S., Behera, A., Kumar, A.: Fabrication of durable porous and non-porous superhydrophobic polymer composite coatings with excellent self-cleaning property. Surf. Coat. Technol. **341**, 31–39 (2018). https://doi.org/10.1016/j.surfcoat.2017.07.025
107. Wang, F.J., Lei, S., Ou, J.F., Xue, M.S., Li, W.: Superhydrophobic surfaces with excellent mechanical durability and easy repairability. Appl. Surf. Sci. **276**, 397–400 (2013). https://doi.org/10.1016/j.apsusc.2013.03.104
108. Mechanically durable superhydrophobic surfaces. Adv. Mater. **23** (5), 673–678 (2011). https://doi.org/10.1002/adma.201003129
109. Jia, S., Lu, Y., Luo, S., Qing, Y., Wu, Y., Parkin, I.P.: Thermally-induced all-damage-healable superhydrophobic surface with photocatalytic performance from hierarchical BiOCl. Chem. Eng. J. **366**, 439–448 (2019). https://doi.org/10.1016/j.cej.2019.02.104
110. Ishizaki, T., Masuda, Y., Sakamoto, M.: Corrosion resistance and durability of superhydrophobic surface formed on magnesium alloy coated with nanostructured cerium oxide film and fluoroalkylsilane molecules in corrosive NaCl aqueous solution. Langmuir. **27**(8), 4780–4788 (2011). https://doi.org/10.1021/la2002783
111. Yeganeh, M., Mohammadi, N.: Superhydrophobic surface of mg alloys: a review. J. Magnesium Alloys. **6**(1), 59–70., ISSN 2213-9567. (2018). https://doi.org/10.1016/j.jma.2018.02.001
112. Deng, Z.-Y., Wang, W., Mao, L.-H., Wanga, C.-F., Chen, S.: Versatile superhydrophobic and photocatalytic films generated from TiO_2–SiO_2@PDMS and their applications on fabrics. J. Mater. Chem. A. **2**, 4178–4184 (2014). https://doi.org/10.1039/C3TA14942K
113. Palamà, I.E., D'Amone, S., Arcadio, V., Caschera, D., Toro, R.G., Gigliacd, G., Cortese, B.: Underwater Wenzel and Cassie oleophobic behavior. J. Mater. Chem. A. **3**, 3854–3861 (2015). https://doi.org/10.1039/C4TA06787H
114. Yong, J., Chen, F., Yang, Q., Huo, J., Hou, X.: Superoleophobic surfaces. Chem. Soc. Rev. **46**, 4168–4217 (2017). https://doi.org/10.1039/C6CS00751A
115. Zeira, A., Chowdhury, D., Hoeppener, S., Liu, S., Berson, J., Cohen, S.R., Maoz, R., Sagiv, J.: Patterned organosilane monolayers as lyophobic−lyophilic guiding templates in surface self-assembly: monolayer self-assembly versus wetting-driven self-assembly. Langmuir. **25**(24), 13984–14001 (2009). https://doi.org/10.1021/la902107u
116. Chua, Z., Seeger, S.: Superamphiphobic surfaces. Chem. Soc. Rev. **43**, 2784–2798 (2014). https://doi.org/10.1039/C3CS60415B
117. Kota, A., Kwon, G., Tuteja, A.: The design and applications of superomniphobic surfaces. NPG Asia Mater. **6**, e109 (2014). https://doi.org/10.1038/am.2014.34
118. Zhao, X., Zhao, Q., Yu, J., Liu, B.: Development of multifunctional photoactive self-cleaning glasses. J. Non-Cryst. Solids. **354** (12–13), 1424–1430 (2008). https://doi.org/10.1016/j.jnoncrysol.2006.10.093
119. Powell, M.J., Quesada-Cabrera, R., Taylor, A., Teixeira, D., Papakonstantinou, I., Palgrave, R.G., Sankar, G., Parkin, I.P.: Intelligent multifunctional $VO_2/SiO_2/TiO_2$ coatings for self-cleaning, energy-saving window panels. Chem. Mater. **28**(5), 1369–1376 (2016). https://doi.org/10.1021/acs.chemmater.5b04419
120. https://www.mordorintelligence.com/industry-reports/superhydrophobic-coatings-market
121. https://www.industryarc.com/Report/16278/superhydrophobic-coatings-market.html
122. Caldona, E.B., De Leon, A.C.C., Thomas, P.G., Naylor, D.F., Pajarito, B.B., Advincula, R.C.: Superhydrophobic rubber-modified polybenzoxazine/SiO_2 nanocomposite coating with anticorrosion, anti-ice, and superoleophilicity properties. Ind.

Eng. Chem. Res. **56**(6), 1485–1497 (2017). https://doi.org/10.1021/acs.iecr.6b04382

123. Zhang, H., Yang, J., Chen, B., Liu, C., Zhang, M., Li, C.: Fabrication of superhydrophobic textured steel surface for anti-corrosion and tribological properties. Appl. Surf. Sci. **359**, 905–910 (2015). https://doi.org/10.1016/j.apsusc.2015.10.191
124. Goetz, L.A., Jalvo, B., Rosal, R., Mathew, A.P.: Superhydrophilic anti-fouling electrospun cellulose acetate membranes coated with chitin nanocrystals for water filtration. J. Membrane Sci. **510**, 238–248 (2016). https://doi.org/10.1016/j.memsci.2016.02.069
125. Conradi, M., Drnovšek, A., Gregorčič, P.: Wettability and friction control of a stainless steel surface by combining nanosecond laser texturing and adsorption of superhydrophobic nanosilica particles. Sci. Rep. **8**, 7457 (2018). https://doi.org/10.1038/s41598-018-25850-6
126. Tarquini, S., Antonini, C., Amirfazli, A., Marengo, M., Palacios, J.: Investigation of ice shedding properties of superhydrophobic coatings on helicopter blades. Cold Reg. Sci. Technol. **100**, 50–58 (2014). https://doi.org/10.1016/j.coldregions.2013.12.009
127. Wang, F., Xie, T., Junfei, O., Xue, M., Li, W.: Cement based superhydrophobic coating with excellent robustness and solar reflective ability. J. Alloys Compd. **823**, 153702 (2020). https://doi.org/10.1016/j.jallcom.2020.153702
128. Eseev, M., Goshev, A., Kapustin, S., Tsykareva, Y.: Creation of Superhydrophobic coatings based on MWCNTs Xerogel. Nano. **9** (11), 1584 (2019). https://doi.org/10.3390/nano9111584
129. Hwang, G.B., Patir, A., Page, K., Lu, Y., Allanb, E., Ivan, P.: Parkin, buoyancy increase and drag-reduction through a simple superhydrophobic coating. Nanoscale. **9**, 7588–7594 (2017). https://doi.org/10.1039/C7NR00950J
130. https://www.deccanchronicle.com/140427/business-autos/article/nissan-develops-%E2%80%98self-cleaning%E2%80%99-car-prototype
131. https://www.businesswire.com/news/home/20190902005092/en/Global-Market-Hydrophobic-Superhydrophobic-Oleophobic-Omniphobic-Coatings
132. Zhao, X., Park, D.S., Soper, S.A., Murphy, M.C.: Microfluidic gasketless interconnects sealed by superhydrophobic surfaces. J. Microelectromech. Syst. **29**, 894–899. https://doi.org/10.1109/JMEMS.2020.3000325
133. https://www.vendop.com/vendor/hydromer-inc/
134. Mohan Raj, R., Raj, V.: Fabrication of superhydrophobic coatings for combating bacterial colonization on Al with relevance to marine and medical applications. J. Coat. Technol. Res. **15**, 51–64 (2018). https://doi.org/10.1007/s11998-017-9945-2
135. Han, K., Park, T.Y., Yong, K., Cha, H.J.: Combinational biomimicking of lotus leaf, mussel, and sandcastle worm for robust superhydrophobic surfaces with biomedical multifunctionality: antithrombotic, antibiofouling, and tissue closure capabilities. ACS Appl. Mater. Interfaces. **11**(10), 9777–9785 (2019). https://doi.org/10.1021/acsami.8b21122
136. Chauhan, P., Kumar, A., Bhushan, B.: Self-cleaning, stain-resistant and anti-bacterial superhydrophobic cotton fabric prepared by simple immersion technique. J. Colloid Interface Sci. **535**, 66–74 (2019). https://doi.org/10.1016/j.jcis.2018.09.087
137. Xue, C.H.: UV-durable superhydrophobic textiles with UV-shielding property by coating fibers with ZnO/SiO_2 core/shell particles. Adv. Mater. Res. **441**., Trans Tech Publications, Ltd., 351–355 (2012). https://doi.org/10.4028/www.scientific.net/amr.441.351
138. Xue, C.-H., Jia, S.-T., Chen, H.-Z., Wang, M.: Superhydrophobic cotton fabrics prepared by sol–gel coating of TiO_2 and surface hydrophobization. Sci. Technol. Adv. Mater. **9**(3), 035001 (2008). https://doi.org/10.1088/1468-6996/9/3/035001
139. Sutha, S., Suresh, S., Raj, B., Ravi, K.R.: Transparent alumina based superhydrophobic self–cleaning coatings for solar cell cover glass applications. Sol. Energy Mater. Sol. Cells. **165**, 128–137 (2017). https://doi.org/10.1016/j.solmat.2017.02.027
140. Zhou, X., Zhang, Z., Xu, X., Guo, F., Zhu, X., Men, X., Ge, B.: Robust and durable superhydrophobic cotton fabrics for oil/water separation. ACS Appl. Mater. Interfaces. **5**(15), 7208–7214 (2013). https://doi.org/10.1021/am4015346
141. Lin, Z., Li, H., Lai, X., Xiaojing, S., Liang, T., Zeng, X.: Thiolated graphene-based superhydrophobic sponges for oil-water separation. Chem. Eng. J. **316**, 736–743 (2017). https://doi.org/10.1016/j.cej.2017.02.030
142. Tian, Z., Lei, Z., Chen, Y., Chen, C., Zhang, R., Chen, X., Bi, J., Sun, H.: Inhibition effectiveness of laser-cleaned nanostructured aluminum alloys to sulfate-reducing bacteria based on superwetting and ultra-slippery surfaces. ACS Appl. Bio Mater. **3**, 6131–6144 (2020). https://doi.org/10.1021/acsabm.0c00714
143. Yang, X., Song, J., Liu, J., et al.: A twice electrochemical-etching method to fabricate Superhydrophobic-Superhydrophilic patterns for biomimetic fog harvest. Sci. Rep. **7**, 8816 (2017). https://doi.org/10.1038/s41598-017-09108-1
144. Bhushan, B., Jung, Y. C., Koch, K.: Micro-, nano- and hierarchical structures for superhydrophobicity, self-cleaning and low adhesion. Philos. Trans. R. Soc. A **367**(1894) (2009). https://doi.org/10.1098/rsta.2009.0014
145. Yuan, Z., Chen, H., Zhang, J., Zhao, D., Liu, Y., Zhou, X., Li, S., Shi, P., Tang, J., Chen, X.: Preparation and characterization of self-cleaning stable superhydrophobic linear low-density polyethylene. Sci. Technol. Adv. Mater. **9**(4), 045007. https://doi.org/10.1088/1468-6996/9/4/045007
146. Lin, S., Lee, E.K., Nguyen, N., Khine, M.: Thermally-induced miniaturization for micro- and nanofabrication: progress and updates. Lab Chip. **14**, 3475–3488 (2014). https://doi.org/10.1039/C4LC00528G
147. Zhong, L., Haghighat, F.: Photocatalytic air cleaners and materials technologies – abilities and limitations. Build. Environ. **91**, 191–203 (2015). https://doi.org/10.1016/j.buildenv.2015.01.033
148. Zhang, X., Li, Z., Liu, K., Jiang, L.: Bioinspired multifunctional foam with self-cleaning and oil/water separation. Adv. Fuct. Mater. **23**(22), 2881–2886 (2013). https://doi.org/10.1002/adfm.201202662
149. Anand Ganesh, V., Raut, H.K., Naira, A.S., Ramakrishna, S.: A review on self-cleaning coatings. J. Mater. Chem. **21**, 16304–16322 (2011). https://doi.org/10.1039/C1JM12523K

Ultralight Materials

12

Abstract

Everyone wants the lightest materials to their use. Lighter materials are the most important parameter for a dynamic system. This chapter describes all the types of latest ultralight materials having a density of <10 mg/cm^3. Aerogel, aerographite, aerographene, 3D graphene, carbyne, microlattice, and foam come under ultralight material. Classification, fabrication, properties, and applications of each of the ultralight material has been described here extensively.

Keywords

Ultralight materials · Aerographite · Aerographene · 3D graphene · Carbyne · Micro-lattice materials · Metallic micro-lattice · Polymer microlattice · Ceramic microlattice · Composite microlattice · Metallic foams · Ceramic foam · Polymeric foam

12.1 Introduction of Ultralight Materials

Ultralight materials are solids with a density of <10 mg/cm^3 [1]. Ultralight material is defined by its cellular arrangement and its stiffness and strength that make up its solid constituent. Ultralight materials are produced to have the strength of bulk-scaled properties at a micro-size. Also, they are designed to not compress even under extreme pressure, which shows that they are stiff and strong. Ultralight materials also have elastic properties. Some ultralight materials are designed with more pores to allow the structure to have better heat transfer, which is needed for many materials, for example, pipes. In compression experiments, ultralight materials almost always show a complete recovery from strains exceeding 50% [2]. They include 3D graphene materials, carbyne materials, aerographite materials, aerographene materials, various aerogel materials, foams, and microlattice materials [3]. The density of air is about 1.275 mg/cm^3, which means that the air in the pores contributes significantly to the density of these materials in atmospheric conditions. Several ultralight materials have been shown in Fig. 12.1.

12.2 Aerogel

Aerogel is a synthetic porous ultralight material derived from a gel in which the liquid component for the gel has been replaced with a gas. Despite the name, aerogels are solid, rigid, and dry materials that do not resemble a gel in their physical properties: The name comes from the fact that they are made *from* gels. The high air content (99.98% air by volume) makes it one of the world's lightest solid materials as shown in Fig. 12.2. Aerogels can be made from a variety of chemical compounds and are a diverse class of materials with unique properties. They are known as excellent insulators and usually have low density and low thermal conductivity. The result is a solid with extremely low density and extremely low thermal conductivity. Other names of aerogel include *frozen smoke*, *solid smoke*, *solid air*, *solid cloud*, and *blue smoke* owing to its translucent nature and the way light scatters in the material. Silica aerogels feel like fragile expanded polystyrene to the touch, while some polymer-based aerogels feel like rigid foams. Aerogels can be made from a variety of chemical compounds. Aerogel was first created by Samuel Stephens Kistler in 1931 as a result of a bet with Charles Learned over who could replace the liquid in "jellies" with gas without causing shrinkage [4]. Aerogels are produced by extracting the liquid component of a gel through supercritical drying. This allows the liquid to be slowly dried off without causing the solid matrix in the gel to collapse from capillary action as would happen with conventional evaporation. The first aerogels were produced from silica gels. Kistler's later work involved aerogels based on alumina, chromia, and tin dioxide. Carbon aerogels were first developed in the late 1980s. Aerogel is not a single material with a

A. Behera, *Advanced Materials*, https://doi.org/10.1007/978-3-030-80359-9_12

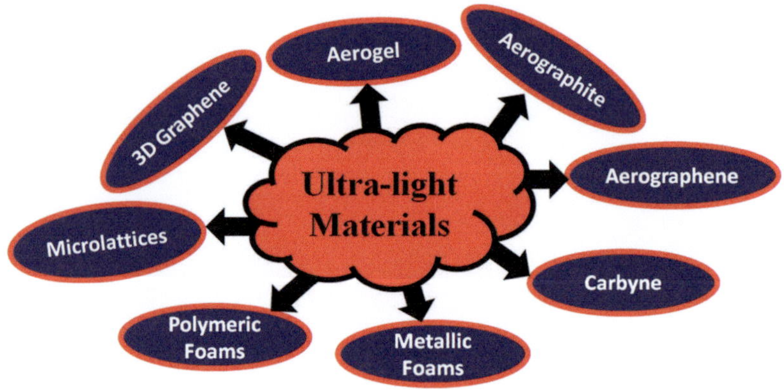

Fig. 12.1 Various types of ultralight materials

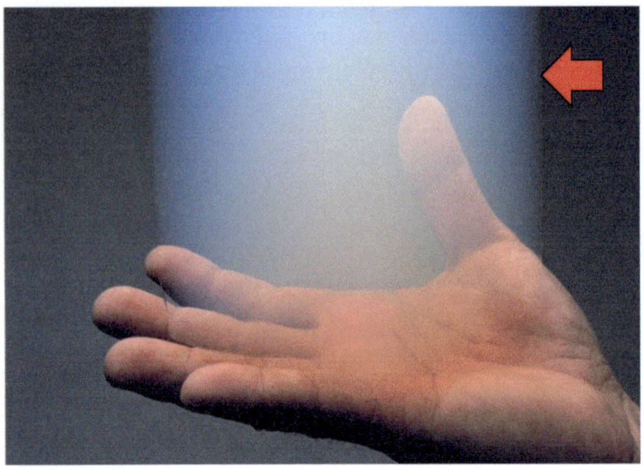

Fig. 12.2 Aerogel

set chemical formula; instead, the term is used to group all materials with a certain geometric structure. The concept of aerogels was introduced in the 1930s [5].

12.2.1 Classification of Aerogel

Based on different microstructures there are of three types of aerogels: (1) microporous aerogel, (2) mesoporous aerogel, and (3) mixed porous aerogel (Fig. 12.3). Based on chemical structure, these are of five types of aerogels: (1) oxides, (2) polymers, (3) mixed, and (4) composite. Various aerogel materials are discussed below:

Silica Aerogel: Silica aerogel is the most common type of aerogel and the most extensively studied and used. It is silica-based and can be derived from silica gel or by a modified Stober process. The lowest density silica nanofoam weighs 1,000 g/m^3, which is the evacuated version of the record aerogel of 1,900 g/m^3. The density of air is 1,200 g/m^3 (at 20 °C and 1 atm) [6]. The silica solidifies into three-dimensional, intertwined clusters that make up only 3% of the volume. Conduction through the solid is therefore very low. The remaining 97% of the volume is composed of air in extremely small nanopores. The air has little room to move, inhibiting both convection and gas-phase conduction. Silica aerogel also has a high optical transmission of ~99% and a low refractive index of ~1.05. This aerogel has remarkable thermal insulative properties, having an extremely low thermal conductivity: from 0.03 W/(mK) in atmospheric pressure down to 0.004 W/(mK) in a modest vacuum, which corresponds to *R*-values of 14 to 105 (US customary) or 3.0 to 22.2 (metric) for 3.5 in (89 mm) thickness. For comparison, typical wall insulation is 13 (US customary) or 2.7 (metric) for the same thickness. Its melting point is 1,473 K (1200 °C; 2192 °F) [7].

Carbon Aerogel: Carbon aerogels are composed of particles with sizes in the nanometer range covalently bonded together. They have very high porosity (over 50%, with pore diameter under 100 nm) and surface areas ranging between 400 and 1,000 m^2/g [8]. They are often manufactured as composite paper: nonwoven paper made of carbon fibers, impregnated with resorcinol–formaldehyde aerogel, and pyrolyzed. Depending on the density, carbon aerogels may be electrically conductive, making composite aerogel paper useful for electrodes in capacitors or deionization electrodes. Due to their extremely high surface area, carbon aerogels are used to create supercapacitors, with values ranging up to thousands of farads based on a capacitance density of 104 F/g and 77 F/cm^3. Carbon aerogels are also extremely "black" in the infrared spectrum, reflecting only 0.3% of radiation between 250 nm and 14.3 μm, making them efficient for solar energy collectors.

Metal oxide aerogel: Metal oxide aerogels are used as catalysts in various chemical reactions/transformations or as precursors for other materials. Aerogels made with aluminum oxide are known as alumina aerogels. These aerogels are used as catalysts, especially when "doped" with a metal other than aluminum. Nickel–alumina aerogel is the most common combination. Alumina aerogels are also being considered by NASA for capturing hypervelocity particles; a formulation doped with gadolinium and terbium could fluoresce at the particle impact site, with the amount of

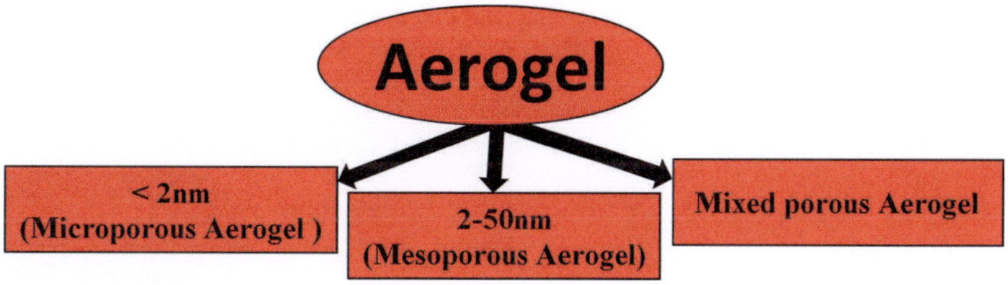

Fig. 12.3 Classification of aerogel based on pore size

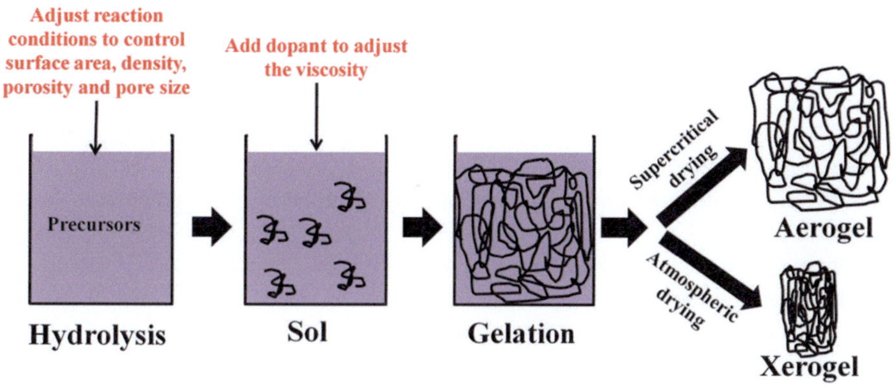

Fig. 12.4 Schematics of the sol-gel process

fluorescence dependent on impact energy [9]. One of the most notable differences between silica aerogels and metal oxide aerogel is that metal oxide aerogels are often variedly colored.

Other Aerogels: Organic polymers can be used to create aerogels. SEA gel is made of agar. Cellulose from plants can be used to create a flexible aerogel. GraPhage is the first graphene-based aerogel assembled using graphene oxide and the M13 bacteriophage. Chalcogel is an aerogel made of chalcogens (the column of elements on the periodic table beginning with oxygen) such as sulfur, selenium, and other elements. Metals less expensive than platinum have been used in their creation. Aerogels made of cadmium selenide quantum dots in a porous 3D network have been developed for use in the semiconductor industry. Aerogel performance may be augmented for a specific application by the addition of dopants, reinforcing structures, and hybridizing compounds.

12.2.2 Fabrication of Aerogel

Generally, aerogels are prepared based on sol-gel formulation. Some of the processes are discussed subsequently:

12.2.2.1 Sol-gel Process

Aerogels are synthesized via a sol-gel process consisting of three main steps: gel preparation, aging of the gel, and drying (Fig. 12.4). Typical silica aerogel source materials, precursors, and preparation steps are given in Fig. 12.5.

Gel preparation: Solid nanoparticles grow crosslink and finally form a three-dimensional solid network with solvent-filled pores. To begin with, a gel is created in solution and then the liquid is carefully removed to leave the aerogel intact; initially, the creation of a colloidal suspension of solid particles known as a "sol" takes place; for example, silica gels are synthesized by hydrolyzing monomeric tetrafunctional and trifunctional silicon alkoxide precursors using a mineral acid or a base as a catalyst [10]. There are many ways to create silica-based sol-gels. One is by mixing with a silicon alkoxide, such as tetraethoxysilane (Si $(OC_2H_5)_4$), tetramethoxysilane (TEOS), and polyethoxydisiloxane with a catalyst such as ethanol and water to make it polymerize and thus produce a water-based silica gel as shown in Eq. (12.1). A solvent, such as methanol, is used to extract and replace the water:

$$\begin{aligned} &Si(OCH_2C_3)_4(liq) + 2H_2(liq) \\ &\rightarrow SiO_2 + CH_2CH_3OH(liq) \end{aligned} \quad (12.1)$$

The oxide suspension begins to undergo condensation reactions which result in the creation of metal oxide bridges (either M-O-M, "oxo" bridges, or M-OH-M, "ol" bridges) linking the dispersed colloidal particles. These reactions generally have moderately slow reaction rates, and as a result, either acidic or basic catalysts are used to improve the processing speed [11].

Aging of the Gel: It provides strength to the structure of the gel. The gel prepared earlier is aged in its mother solution. This aging process strengthens the gel so that minimum shrinkage occurs during the drying step. After gelification, the gel is left undisturbed in the solvent to complete the reaction. After completion of the reaction, the aerogel product is formed. Inorganic aerogels can be prepared via sol-gel

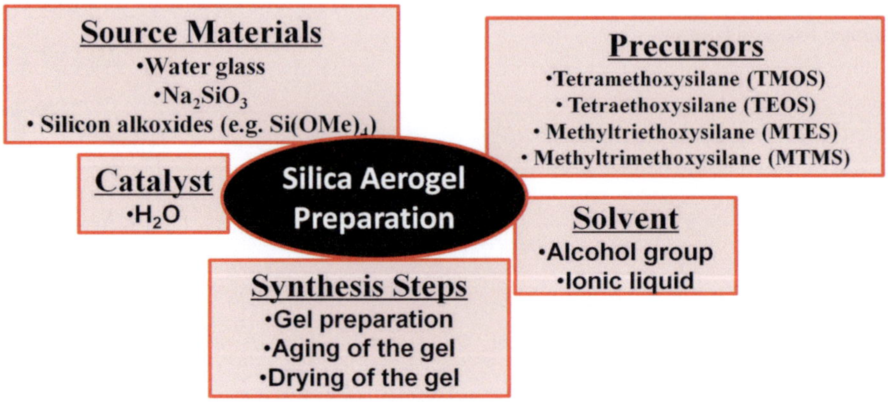

Fig. 12.5 Constituents of silica aerogel and the associated steps

processing, a technique that requires alkoxides or metal salts in alcoholic or aqueous solutions and subjecting to supercritical drying [12].

Drying: The solvent has to be removed while preserving the solid aerogel network. This can be done either by supercritical drying or at ambient conditions. Aerogel materials are typically prepared by removing the solvent contained in a gel matrix by extraction in a supercritical fluid medium. This can be accomplished by bringing the gel solvent system above its critical temperature and pressure and subsequently relieving pressure above the critical temperature until only vapor remains. Alternatively, the gel solvent system can be extracted from the wet gel with an appropriate solvent. Liquid carbon dioxide is the most popular extraction solvent because it is inexpensive and has a relatively low critical temperature and critical pressure. Crack-free silica aerogels can also be obtained via solvent exchange and resulting surface modification of wet gels using isopropyl alcohol, trimethylchlorosilane, or n-hexane solution.

For dense silica, solid conductivity is relatively high (a single pane window transmits a large amount of thermal energy). However, silica aerogels possess a very small (~1–10%) fraction of solid silica. Additionally, the solids that are present consist of very small particles linked in a three-dimensional network with many "dead ends." Therefore, thermal transport through the solid portion of silica aerogel occurs through a very tortuous path and is not particularly effective. The use of methyltrimethoxysilane coprecursor makes the aerogel hydrophobic and makes it able to hold water droplets on the surface. The porosity of silica aerogels was determined by helium pycnometry using the following formula and was found to be 1,900 kg/m^3 [13]:

$$\text{Porosity}\ (\%) = \left(1 - \frac{\rho_b}{\rho_s}\right), \tag{12.2}$$

where ρ_b is the bulk density and ρ_s is the skeleton density.

The hydrophobic aerogels have also been obtained via the co-recursor method pioneered by Schmidt and Schwertfeger [14]. Hydrolysis and condensation rates of all the co-precursors were observed to be slower than that of TEOS because the former contains one or more nonreactive alkyl/aryl groups, which are nonhydrolyzable, and a three-dimensional solid network is achieved as per the following chemical reaction:

$$\mathrm{nSi(OC_2H_5)_4 + 2H_2O \xrightarrow[\text{Oxalic Acid}]{\mathrm{C_2H_5OH}} SiO_2 + 4nC_2H_5OH.} \tag{12.3}$$

When a sufficient amount of TEOS has hydrolyzed, the silyl groups of the co-precursor get attached to the silica clusters as per the following chemical reactions:

$$\mathrm{-Si \equiv (OH)_3 + (OCH_3)_3Si - CH_3 \rightarrow (-Si \equiv O \equiv Si) - CH_3 + 3CH_3OH} \tag{12.4}$$

$$\mathrm{= Si - OH + Cl - \left(H - Si - (CH_3)_2\right) + HCl} \tag{12.5}$$

$$\mathrm{-O - \left(Si - (OH)_3\right) + (C_2H_5O)_3 - (Si - (C_6H_6)) \rightarrow -O - Si \equiv Si - (C_6H_6) + 3C_2H_5OH} \tag{12.6}$$

As the silica clusters get attached to nonhydrolyzable organic groups (silyl) on their surfaces, the aerogels become hydrophobic. The hydrophobicity of aerogels will increase with the numbers of alkyl/aryl groups attached to the surface.

Supercritical Drying: Supercritical drying (SCD) is the first and most commonly used method for silica aerogels. Obviously if one wishes to produce an aerogel, he must replace the liquid with air by some means in which the surface of the liquid is never permitted to recede within the gel. If a liquid is held under pressure always greater than the vapor pressure, and the temperature is raised, it will be transformed at the critical temperature into a gas without two phases having been present at any time. There are two types of SCD methods: high-temperature supercritical drying (HT-SCD) and low-temperature supercritical drying (LT-SCD) from CO_2 or the Hunt Process. HT-SCD is carried out in three steps: First, the aged gel is placed in an autoclave filled halfway with the same solvent held in the gel's pores. The vessel is then sealed and heated slowly past the solvent's critical temperature and pressure (i.e. most-used organic solvents have a relatively higher critical temperature or 300–600 K with a critical pressure of 30–80 atm) [15]. Second, the fluid is isothermally depressurized. Finally, at ambient pressure, the autoclave is cooled to room temperature. In the case of silica aerogels, methanol is most frequently used as a solvent for HT-SCD. At its critical point (512.6 K and 79.783 atm), methanol can react with OH groups on the surface of the gel backbone to form CH_3O groups, which make the silica aerogels partially hydrophobic. HT-SCD has been found the best way to minimize shrinkage of the gel. For each possible solvent, drying pressures are known for which shrinkage of the aerogels stays below 5%. The process of LT-SCD is similar to that of HT-SCD and is also carried out in three steps: Firstly, the aged gel is placed in an autoclave but now filled with the saver, nonflammable liquid CO_2 (with critical temperature and pressure of 304.2 K and 72.786 atm, respectively) at 4–10 °C until 100 bar to replace the solvent in the pores of the gel. When all solvents are replaced, the autoclave is heated to 313 K while maintaining 100 bar. Second, the fluid is isothermally depressurized. Finally, at ambient pressure, the autoclave is cooled to room temperature. Also, aerogels dried by LT-SCD show shrinkage but, compared to HT-SCD, not caused by the SCD process but by replacement of the original solvent with liquid CO_2 [16].

Ambient pressure drying: Ambient pressure drying (APD) is of most interest to lower the costs compared to the expensive drying processes of HT-SCD or LT-SCD. APD is generally carried out in two steps: First, silylation of all OH groups must take place for preventing adsorption of water resulting by the formation of hydrophobic aerogels. This is carried out by replacing the present solvent with a water-free solvent and a silylating agent (e.g. hexamethyldisilazane HMDS), resulting in a replacement of H from OH groups by alkyl such as CH_3 [17]. Second, drying is carried out by ambient pressure evaporation and consists of three steps: After a warming period, the first drying period occurs where the volume loss of the gel balances those of the evaporated liquid as free water moves continuously to the external surface by capillary forces. In the second drying period or falling rate period, diffusive vapor transport will dominate allowing liquid to escape slowly to the exterior.

In the case of organic aerogels derived from the sol-gel polymerization of resorcinol with formaldehyde, thermal conductivity components are clearly correlated with the aerogel structure; that is, the solid conductivity can be determined by the porosity and connectivity between the particles, while the gaseous conductivity can be influenced by pore size and mass-specific infrared absorption of the building units influences radiative transport. Polymer aerogels were prepared from mixtures containing a fixed stoichiometric amount of formaldehyde and varying proportions of resorcinol (RF) and 2,4-dihydroxybenzoic acid (DHBAF) with the objective of combining the advantages of high mesopore volume and solid content of RF aerogels with the ion exchange capacity of DHBAF aerogels, and results show that the aerogel properties vary systematically as the synthesis conditions are changed. It was found that the addition of R to the synthesis mixture resulted in increased values of surface area, mesopore volume, and mean diameter while simultaneously maintaining the ion exchange capacity of the wet gel. In the TG-DTA of some of the silica aerogel samples, there is a rapid increase in the weight loss of hydrophilic silica aerogels at 50–100 °C due to evaporation of trapped H_2O and alcoholic groups from hydrophilic silica aerogels that were produced by the condensation reactions of Si-OH and $Si(OC_2H_5)$ groups, whereas the percentage of weight loss is negligible up to the temperature of thermal stability in case of hydrophobic aerogels [18]. The effect of heat treatment on hydrophobicity and specific surface area has also been investigated by several researchers. The results of these studies indicate that the hydrophobicity of silica aerogel decreased with increasing the heating temperature to 350 °C. On further increasing the heating temperature to 500 °C, silica aerogel becomes completely hydrophilic. Some results for methyltriethoxysilaneco precursor-based aerogels show that the hydrophobicity of the silica around 573K corresponding to oxidation of aerogel could be maintained up to 350 °C [19].

12.2.2.2 3D Printing

The 3D printing of aerogels is revolutionizing the field by enabling a fast and accurate fabrication of complex 3D porous structures, thereby introducing new functionalities, lower costs, and higher reliability in aerogel manufacturing. 3D printing, in general, is a type of additive manufacturing technique that builds 3D objects through a layer-by-layer growth process. This technique makes it possible to fabricate highly customizable and complex structures for many industrial sectors in significantly reduced times while using a variety of materials, such as polymers, ceramics, and metals. The 3D printing of aerogels is considered a hybrid fabrication technique to produce extremely lightweight 3D structures, employing new depositional strategies for the creation of the 3D gel constructs while utilizing the common drying methods of supercritical drying and freeze-drying, as discussed previously. 3DP of aerogel techniques are categorized depending on the sol-gel transitions during the printing process. These categories include: (1) direct ink writing (DIW), where a gel is formed prior to printing; (2) stereolithography (SLA), where the sol-gel transition occurs during printing; and (3) inkjet printing (IJP), where the sol-gel transition occurs after printing.

12.2.2.3 Properties of Aerogel

Aerogel is a material that is 99.8% air. Aerogels have a porous solid network that contains air pockets, with the air pockets taking up the majority of space within the material. The lack of solid material allows aerogel to be almost weightless. Aerogels are good thermal insulators because they almost nullify two of the three methods of heat transfer-conduction (they are mostly composed of insulating gas) and convection (the microstructure prevents net gas movement). They are good conductive insulators because they are composed almost entirely of gases, which are very poor heat conductors. (Silica aerogel is an especially good insulator because silica is also a poor conductor of heat; a metallic or carbon aerogel, on the other hand, would be less effective.) They are good convective inhibitors because air cannot circulate through the lattice. Aerogels are poor radiative insulators because infrared radiation (which transfers heat) passes through them. For example, the thermal conductivity of air is about 25 mW/mK at STP and in a large container but decreases to about 5 mW/m·K in a pore 30 nm in diameter [20]. Owing to its hygroscopic nature, aerogel feels dry and acts as a strong desiccant. Aerogels by themselves are hydrophilic, and if they absorb moisture they usually suffer a structural change, such as contraction, and deteriorate, but degradation can be prevented by making them hydrophobic by using a chemical treatment. One way to waterproof the hydrophilic aerogel is by soaking the aerogel with some chemical base that will replace the surface hydroxyl groups (–OH) with non-polar groups (–O*R*), a process that is most effective when *R* is an aliphatic group. Aerogels with hydrophobic interiors are less susceptible to degradation than aerogels with only an outer hydrophobic layer, even if a crack penetrates the surface [21].

Silica aerogels have interesting optical properties with high transmittance of radiation within the range of visible light (i.e. radiation with a wavelength between 380 and 780 nm). Monolith translucent silica aerogel in a 10-mm thick packed bed has a solar transmittance T_{SOL} of 0.88. Heat treatment of the aerogels can increase their transparency further, that is, currently by up to 6%, because of water desorption and burning of organic components. The optical properties can be influenced furthermore by parameters of the sol-gel process, that is, by selecting optimal synthesis parameters. Light reflected by (silica) aerogels appears bluish, and transmitted light appears slightly reddened. This scattering of the light can be explained by bulk or Rayleigh scattering and by exterior surface scattering. Rayleigh scattering is caused by the interaction with inhomogeneities in solids, liquids, or gases, such as dust particles in the atmosphere, and becomes more effective when the size of the particles is similar to the wavelength of the incident light. The presence of a certain number of pores within this range in aerogels can act as so-called "scattering centers." The efficiency of scattering will depend on the size of the scattering centers, while different wavelengths of radiation will scatter with different magnitudes. Silica aerogels can also have high transparency in the infrared spectrum, that is, a TIR of 0.85 [22]. Monolith silica aerogels have a lower speed of sound than air. Sound velocities down to 40 m/s have been measured, whereas (non-monolith) commercial products claim to have a sound velocity of ~100 m/s through the structure. Granular aerogels are exceptional reflectors of audible sound, making excellent barrier materials. By combining multiple layers with different granular sizes, average attenuations of 60 dB have been found for a total thickness of only 7 cm [23].

12.2.2.4 Applications of Aerogel

Based on application, aerogels are of three types: Cryogel (for low-temperature applications), Spaceloft (for mid-temperature applications), and Pyrogel (for high-temperature applications) as presented in Fig. 12.6. Cryogel is generally found in sub-ambient piping and equipment, cryogenic storage, sea transport, industrial gases, liquefied natural gas import/export pipelines, chilled water systems,

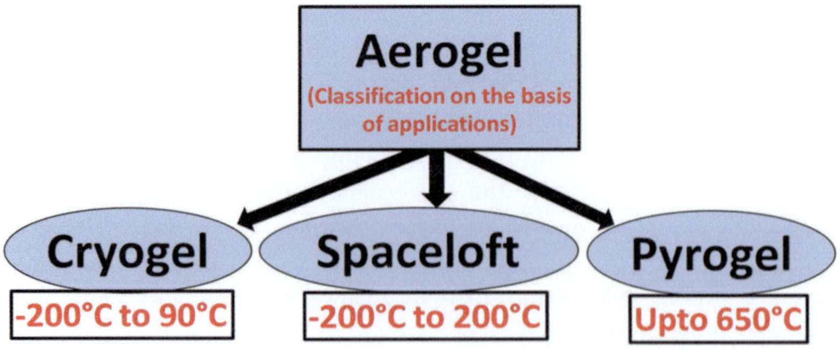

Fig. 12.6 Classification of aerogel on the basis of applications

gloves, jackets, sleeping bags, and boots. Spaceloft is found in building and construction, heating, water and conditioning systems, and household appliances. Pyrogel is found in industrial and petrochemical equipment, power generation equipment, fire protection equipment, aerospace and aeronautic transportation, and so on. Common applications of aerogel include enhancing the thermal performance of energy-saving materials and sustainable products for buildings, acting as a high-performance additive to coatings, prevention of corrosion under insulation, uses in imaging devices, optics, and light guides, thermal breaks and condensation control, architectural lighting panels, outdoor and sports gear and clothing, and more.

In space vehicle: Aerogel is a good dust collector that is deployed in space because the aerogel can capture particles at highspeed impacts. During the comet encounter, the aerogel grid was deployed such that particles from the coma of the comet impacted with the aerogel and embedded themselves in the porous network of the aerogel. Having captured the cometary particles, the aerogel was retracted into the spacecraft and returned to earth. Upon returning to earth, the spacecraft will release the sample return capsule, containing the aerogel grid, which will be parachuted down to the testing and training range [24]. Two grids of aerogel (Fig. 12.7) were assembled for the collection of the cometary and interstellar particles. These grids were mounted back-to-back so that one grid would capture interstellar particles, while the other would be used to capture cometary particles. Each grid was composed of one 130 cells of gradient density, silica aerogel. Aerogel is an excellent hypervelocity particle capture medium because it is a highly porous material whose microstructure is made up of nanoscale filaments. Since the particles impacting the aerogel are micron in size, while the filaments of the aerogel are nanometer in size, the filaments yield to the force exerted by the particles. As the individual filaments are broken and destroyed, the kinetic energy of the particles is gradually transferred to thermal and mechanical energy, the particle is slowed and finally stopped, and largely intact. Gradient density silica aerogel was developed and produced for the Stardust particle capture grids, since it was considered to have superior capture properties [26].

The 2003 Mars Exploration Rover, named Spirit and Opportunity, was launched in May and June 2003. They were successfully landed and deployed on the surface of Mars in January 2004. Their exploration missions began in February 2004, and they are both still actively exploring the Martian landscape (Fig. 12.8). High strength and low weight properties make aerogel a suitable material for space vehicle

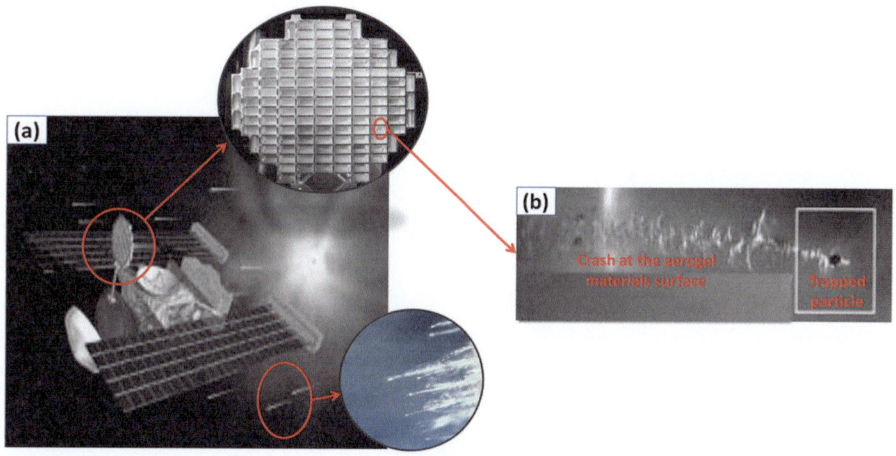

Fig. 12.7 (**a**) Spacecraft collecting the high-velocity cosmic dust and (**b**) the particle trapped aerogel holder surface [25]

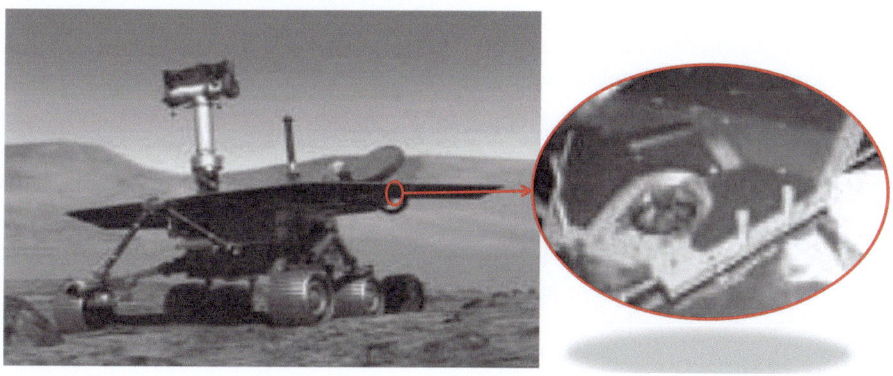

Fig. 12.8 The Mars Exploration Rover on the surface of Mars and magnified aerogel placed on one of the WEB walls of a Mars rover [27]

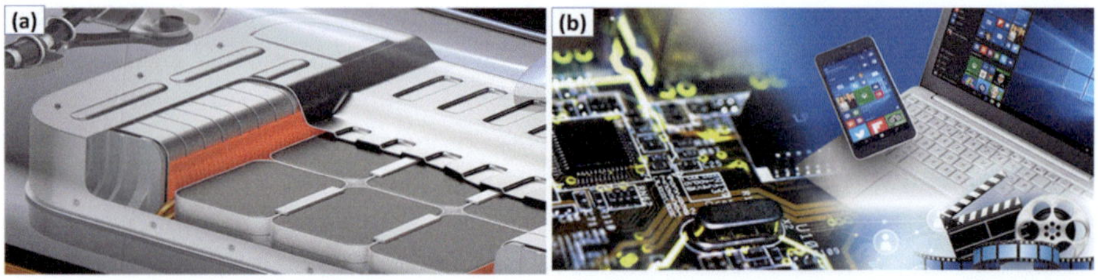

Fig. 12.9 Aerogel used in (a) Electric vehicles (Ultra-thin, flexible PyroThin aerogel sheets are ideal solutions for use between LiB cells and modules for mitigating thermal runaway propagation), and (b) portable electronics

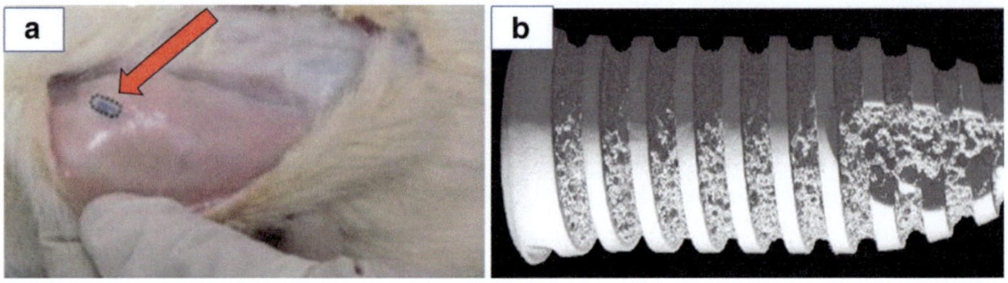

Fig. 12.10 (a) Pigmented aerogel implant in domestic rat and (b) Ti-based aerogel dental implant

outer frames. In aircraft de-icing, a new proposal uses a CNT aerogel. A thin filament is spun on a winder to create a 10-μm thick film. The amount of material needed to cover the wings of a jumbo jet weighs 80 g (2.8 oz). Aerogel heaters could be left on continuously at low power to prevent ice formation [28].

Energy Storage: Over the past decades, great research efforts have been devoted to building high-performance, cost-effective, renewable, and sustainable energy storage devices to meet the need for renewable energy, electric vehicles, and portable electronics (Fig. 12.9). Among many devices developed to address such needs, advanced batteries and supercapacitors are leading this quest. Batteries, owing to their high energy density and ability to supply a constant source of electrical power, dominate the worlds of portable electronics and electric vehicles. For batteries, the aerogel structures create space within the electrode to accommodate any volume changes that occur during the charge–discharge cycles that are unavailable in traditional electrode materials. The chemistry, fabrication, and properties of a variety of carbon-based aerogels have been extensively investigated for energy storage applications. While CNT, graphene, and graphene derivatives have been used extensively in producing aerogels for energy storage applications, bio-based carbon materials are abundant and extremely cost-effective, while offering biodegradability and biocompatibility, offering flexible, fully disposable electrodes [29].

Biomedical Applications: The typical low density, high porosity, and high surface area of aerogels make them suitable for many biomedical applications, including drug delivery, tissue engineering, implantable devices, biomedical imaging, and biosensing (Fig. 12.10) [30]. Previously, only silica-based aerogels had been investigated for use in these applications; however, their poor biodegradability inhibited their use in pharmaceutical and biomedical applications. At present, polysaccharides such as starch, alginate, and chitosan are being investigated as biomedical aerogels, as they provide low toxicity, biocompatibility, and biodegradability. Aerogels can be used as a drug delivery system owing to its biocompatibility. Due to its high surface area and porous structure, drugs can be adsorbed from supercritical CO. The release rate of the drugs can be tailored by varying the properties of the aerogel.

Electrical Industries: By filling pores of silica aerogel with ferric oxide, a nanocomposition with a soft magnetic behavior, low density, and high electric resistance can be formed. Conducting carbon aerogels is especially promising for electrical engineering. The large surface of the electrodes ensures the correspondingly large trapping capacity of the device. The most striking application of carbon aerogel electrodes is the development of accumulators with a very high capacity on their basis. For this purpose, two electrodes made of carbon aerogel were placed in an electrolyte and connected to standard electrodes. The electrolyte penetrates into the aerogel pores, which causes the area of contact between the electrolyte and electrodes to become many times larger, and the capacitance of this device correspondingly increases up to 45 F [31]. Such an accumulator was developed for a car with an electric motor. Carbon aerogels are used in the construction of small electrochemical double-layer supercapacitors (Fig. 12.11a) and electromagnetic shielding (Fig. 12.11b). Due to the high surface area of the aerogel, these capacitors can be 1/2000th to 1/5000th the size of similarly rated electrolytic capacitors. Aerogel supercapacitors can have a very low impedance compared to normal supercapacitors and can absorb or produce very high peak currents. At present, such capacitors are polarity sensitive and need to be wired in series to achieve a working voltage of greater than about 2.75 V [32]. Aerogels are the materials with the lowest dielectric constants, and the use of such materials in microelectronics

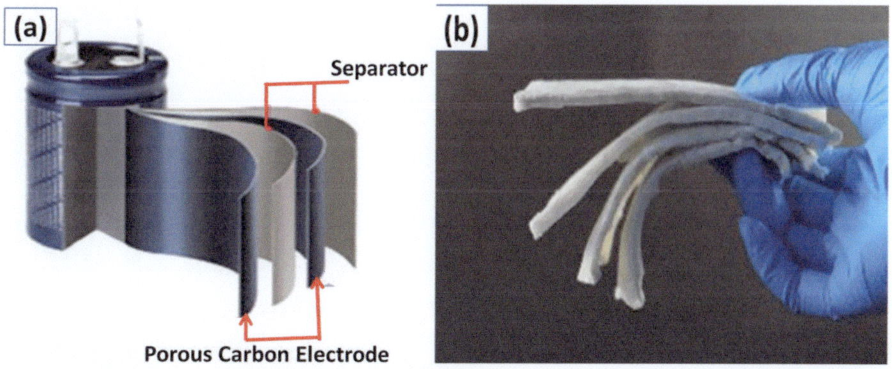

Fig. 12.11 Aerogel used (**a**) Supercapacitor and (**b**) Electromagnetic shielding

Fig. 12.12 Aerogel used in (**a**) high-temperature hot gas transport system, (**b**) home chimney, and (**c**) cryogenic insulation of the pipe

makes it possible to substantially reduce the appropriate parasitic capacitances and thus increase the response speed. This primarily refers to interlayer dielectrics in multilayer printed circuit boards. For this purpose, aerogel is manufactured in the form of thin films that can be applied with good enough adhesion onto silicon or glass substrates at the gel stage. The dielectric constant of aerogel depends on its porosity.

High-temperature Application: There are a number of industrial processes that require pumping hot fluids from one place to another (Figs. 12.12a, b). Many older plants were designed and built with little or no insulation, especially at times where energy was all abundant and cheap. These days, significant savings can be generated by insulating the piping systems of steam cycles, chemical processes, or oil and gas processing refineries. This opens up a large potential for high-performance insulation systems, particularly for older or already existing plants with restricted access and tightly spaced arrangements of individual pipes. Aspen's Pyrogel is a product that was developed especially for high-temperature applications [33]. With a low thermal conductivity (~13 mW/mK), they show remarkable characteristics compared to traditional thermal insulation materials. Its low density and thermal conductivity seem to hint that graphene aerogel could be used as a suitable insulator for any environmental condition.

Cryogenic Applications: In the field of cryotechnology, which includes, for example, transport and storage of liquefied gases or frozen biomedical specimens, aerogel insulation offers numerous advantages (Fig. 12.12c). Besides its superinsulation properties, aerogels tend to embrittle less with decreasing temperature than, for example, polymer foam insulation. Aspen aerogels have developed a special low-temperature aerogel blanket insulation product, the Cryogel Z. Granular products such as Cabot's Nanogel offer great versatility for lining nontrivial vessel and piping geometries, which are part of more complex cryogenic systems. Cryogenic aerogel insulation is also investigated by NASA's rocket engine research teams for liquid hydrogen and liquid oxygen-propelled drive systems [34].

Transparent Walls: Aerogel strongly absorbs IR radiation and has the lowest thermal conductivity among solids. This makes it attractive for use as a transparent heat insulator that transmits solar light and efficiently suppresses the heat transfer. As was pointed out in [35], 91% of the solar radiation at the ground level lies in the wavelength range of 400–2000 nm, in which the transparency of aerogel is not lower than that of ordinary glass. However, compared to glass, aerogel, being a unique heat insulator, allows for a more efficient accumulation of the solar heat and preservation of the heat delivered from heating systems; that is, it can be used as transparent walls in solar collectors and living quarters (Fig. 12.13). Aerogel is placed between two thin windowpanes hermetically attached to a steel frame with butyl cement. A similar structure but without an absorber of solar rays was developed for constructing a transparent wall in living quarters. Aerogel's transparency for insulation applications, which will allow their use in windows and skylights, gives architects and engineers the opportunity of reinventing architectural solutions [36].

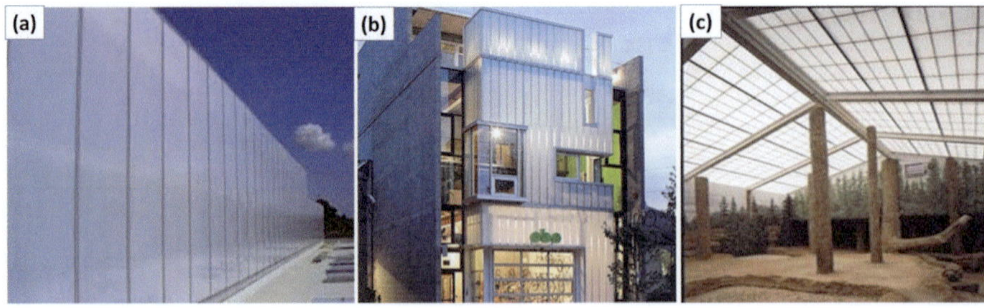

Fig. 12.13 Application of aerogel in transparent (**a**) wall, (**b**) home, and (**c**) roof

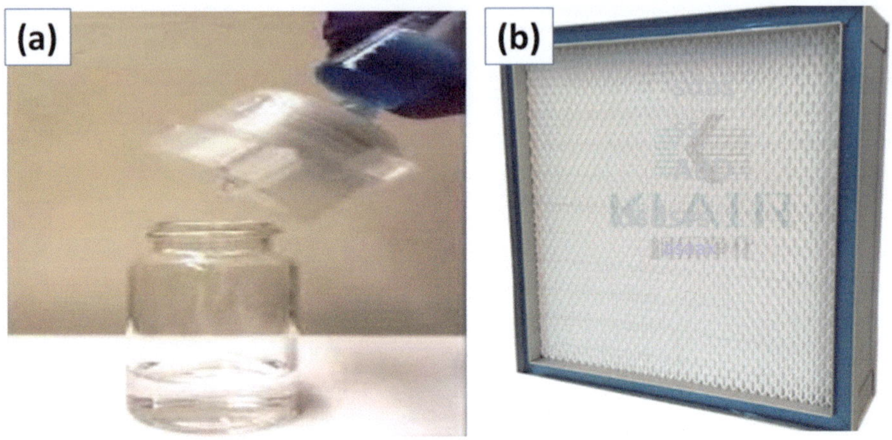

Fig. 12.14 (**a**) Water purifier siring neck and (**b**) mini pleat silica gel air filter

Purification of Liquids and Gases: Due to their large surface area, aerogels can be used as filters for various purposes. Hydrophobic aerogels are efficient absorbers of substances that are soluble and insoluble in water. The absorption capacity of aerogel is two orders of magnitude larger than that of granular activated carbon of the same mass. Aerogels are used for absorbing long-lived nuclear wastes, which are contained in liquid radioactive wastes. In water purification, chalcogels have shown promise in absorbing the heavy metal pollutants, mercury, lead, and cadmium, from water (Fig. 12.14a). Aerogel was used for purifying the air of dangerous aerosol particles (viruses and bacteria) with dimensions of 20–2000 nm. Aerogel can also be used in the purification of gases emerging from the exhaust pipe of a car (Fig. 12.14b) [31]. Our indoor environments are polluted by releasing many pollutants such as chloride from tap-water, VOCs from organic solvents, formalin from furniture and paints, SO? and NO? from incomplete combustion of gasses and many hydrocarbons, and so on [37].

Sorbent Super-material: Owing to aerogel properties and extra-low density, graphene aerogel can absorb up to 900 times its weight in oil. The projection of it may be used to clean up oil spills or leaks of dangerous chemicals. Furthermore, graphene aerogel can return to its shape even after 90%. This means it can absorb the hazardous liquid, be wrung or squeezed into a container, and used again.

Sports Goods: Aerogel is used in many lightweight, high-strength sports goods such as racquets for tennis, squash, and badminton. Figure 12.15 shows aerogel-assisted badminton racquets, women gloves, sports ultralight shoes, and aerogel balaclava for high-altitude mountaineering.

12.3 Aerographite

Aerographite was developed jointly by a team of researchers at the University of Kiel and the Technical University of Hamburg in Germany and was first reported in a scientific journal in June 2012. Aerographite has a density of 0.2 mg/ cm^3 [38]. Aerographite is synthetic foam consisting of a porous interconnected network of tubular carbon, grown at nanoscale, and contains 99.99% air. Figure 12.16 shows an aerographite that is placed on the tip of the grass. Its material properties include electrical conductivity and the ability to be compressed by a factor of 1,000 before springing back to its original shape. The material is robust enough to support 40,000 times its own weight [39]. If you were to squash aerographite like a sponge, it would just spring back into its original shape. The scientist tested this by compressing a three-millimeter-tall piece of aerographite down to a few hundred micrometers. Because aerographite is electrically conductive and chemical resistant, the scientists believe that the material could be used in ultra-lightweight batteries or

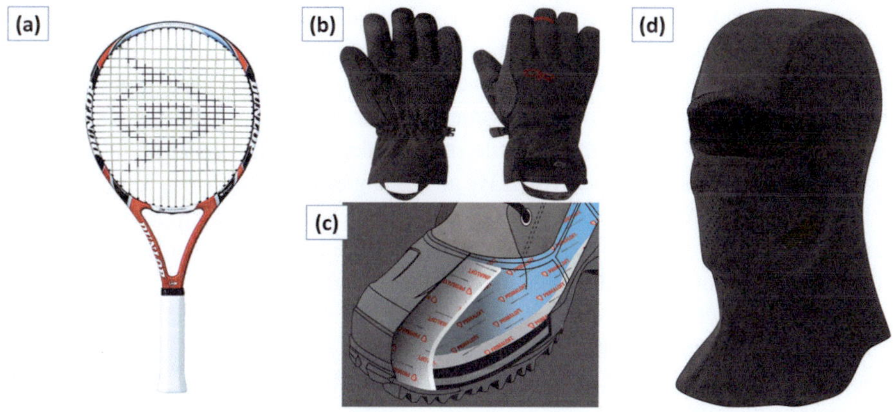

Fig. 12.15 (**a**) Badminton racquets, (**b**) Women gloves, (**c**) Sports shoes, and (**d**) Balaclava

Fig. 12.16 Aerographite on the grass

supercapacitors. The material also has a few other interesting properties that make it hydrophobic, a good insulator, readily absorb visible light, and opaque to X-rays [40].

12.3.1 Synthesis of Aerographite

Synthesis of aerographite has been reported in 2012 by CVD process using ZnO template. Aerographite is synthesized in a two-step process, with the first being the manufacturing of the template structure (ZnO) which is followed by its replication in the CVD process. The template consists of micron-thick rods, often in the shape of multipods, that can be synthesized by mixing comparable amounts of Zn and polyvinyl butyral powders and heating the mixture at 900 °C. The samples are placed in a defined area in the reactor on a silicon wafer. The synthesis starts with heating the reactor up to 760 °C under an argon flow of 0.2 L/min. The toluene vapors are injected as a carbon source. Finally, the reactor is heated up to 900 °C for 1 h without further injection of the carbon source. The synthesis ends with a cooling down phase without the supply of hydrogen but an argon flow of 0.4 L/min. A thin (~15 nm), discontinuous layer of carbon is deposited on ZnO which is then etched away by adding hydrogen gas to the reaction chamber. Thus, the remaining carbon network closely follows the morphology of the original ZnO template. In particular, the nodes of the aerographite network originate from the joints of the ZnO multipods [41].

Requirements for the metal oxides in the synthesis are low activation energy for chemical reduction, a metal phase that can nucleate graphite, and a low evaporation point of the metal phase (ZnO, SnO). From the engineering perspective, the developed CVD process enables the use of ceramic powder processing (use of custom particles and sintering bridges) for the creation of templates for 3D carbon via CVD. Key advantages compared to commonly use metal templates are shape variety of particle shapes, the creation of sintering bridges, and removal without acids. Originally demonstrated on just µm-sized meshed graphite networks, the CVD mechanism had been adopted after 2014 by other scientists to create nm-sized carbon structures.

12.3.2 Properties of Aerographite

Aerographite is a black freestanding material that can be produced in various shapes occupying a volume of up to several cubic centimeters. It consists of a seamless interconnected network of carbon tubes that have micron-scale diameters and a wall thickness of about 15 nm. Because of the relatively lower curvature and larger wall thickness, these walls differ from the graphene-like shells of carbon nanotubes and resemble vitreous carbon in their properties. These walls are often discontinuous and contain wrinkled areas that improve the elastic properties of aerographite. The carbon bonding in aerographite has an sp^2 character as confirmed by electron energy loss spectroscopy and electrical conductivity measurements. Upon external compression, the conductivity increases, along with material density, from ~0.2 S/m at 0.18 mg/cm^3 to 0.8 S/m at 0.2 mg/cm^3. The conductivity is higher for a denser material, 37 S/m at 50 mg/cm^3. Aerographite is superhydrophobic and thus it repels water [42]. Owing to its interconnected tubular network structure, aerographite resists tensile forces much better than other carbon foams as well as silica aerogels. It sustains extensive elastic deformations and has a very low Poisson's ratio [43]. Complete shape recovery of a 3-mm-tall sample after it was compressed down to 0.1 mm is possible. Loading and unloading stress–strain response of aerographite showing the elastic behavior under compressive and tensile stress as shown in Fig. 12.17. Its ultimate tensile strength depends on material density and is about 160 kPa at 8.5 mg/cm^3 and 1 kPa at 0.18 mg/cm^3; in comparison, the strongest silica aerogels have an ultimate tensile strength of 16 kPa at 100 mg/cm^3. Young's modulus is 15 kPa at 0.2 mg/cm^3 in tension but is much lower in compression, increasing from 1 kPa at 0.2 mg/cm^3 to 7 kPa at 15 mg/cm^3. Their specific energy of 1.25 Wh/kg is comparable to that of CNT electrodes (~2.3 Wh/kg) [45].

12.3.3 Applications of Aerographite

Aerographite is electrically conductive. Aerographite electrodes have been tested in an electric double-layer capacitor (EDLC, also known as supercapacitor) and endured the mechanical shocks related to loading–unloading cycles and crystallization of the electrolyte that occurs upon evaporation of the solvent. It gives excellent potential for the creation of smaller, lighter batteries. These lightweight batteries could be used to further develop green transportation by increasing the miles-per-charge in electric vehicles.

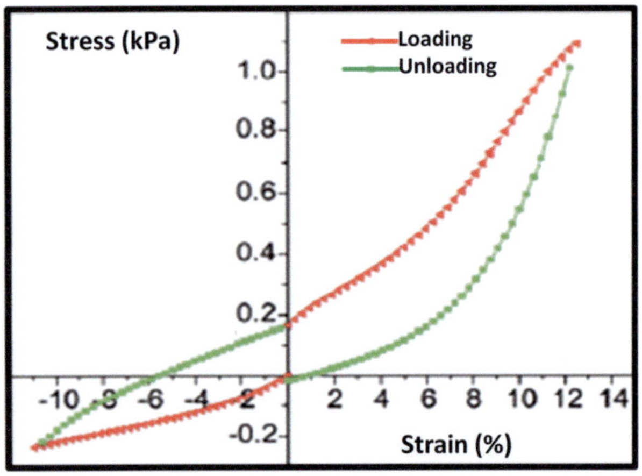

Fig. 12.17 Stress–strain response of aerographite [44]

12.4 Aerographene

Aerographene or graphene aerogel is, as of March 2013, the least dense solid known at 0.16 mg/cm^3, less than the Helium. It is approximately 7.5 times less dense than air. Aerographene was discovered at Zhejiang University by a team of scientists led by Gao Chao [46]. He and his team had already successfully created macroscopic materials made out of graphene. Its density is such that blocks of it can be balanced on small plants and plant structures (e.g., flowers and grasses). These materials were one-dimensional and two-dimensional. The synthesis was accomplished by the freeze-drying of carbon nanotube solutions and large amounts of graphene oxide. Residual oxygen was then removed chemically.

12.4.1 Synthesis

The synthesis of aerographene is achieved by freeze-drying that can yield graphene sponges of arbitrary size. Freeze-drying is a process that works by initially freezing the material and then later reducing the surrounding pressure to allow the frozen water present in the material to sublimate. In aerographene synthesis, a solution of graphene and CNT is created and later it is poured into a mold, which is then freeze-dried [47]. Freeze-drying helps to dehydrate the solution, leaving single-atom-thick layers of graphene, supported by CNT. Residual oxygen is finally removed chemically. With no need for templates, its size only depends on that of the container. A bigger container helps to produce aerographene in bigger size, even as much as thousands of cubic centimeters or larger. Another process is based on the sol-gel polymerization of resorcinol (R) and formaldehyde (F) with sodium carbonate as a catalyst (C) in an aqueous suspension of graphene oxide (Fig. 12.18). The graphene oxide is produced by the Hummers method, and the suspension is prepared by ultra sonification. The molar ratio of resorcinol–formaldehyde is 1:2, the reactant concentration in the starting mixture is 4 wt% resorcinol–formaldehyde solids, and the concentration of graphene oxide in suspension is 1 wt% [48]. The molar ratio of resorcinol–catalyst is 200:1. The sol-gel mixture is cured in sealed glass vials at 85 °C. After gelation, the wet graphene oxide–resorcinol–formaldehyde gels are removed from the glass vials and washed in

Fig. 12.18 Synthesis of aerographene

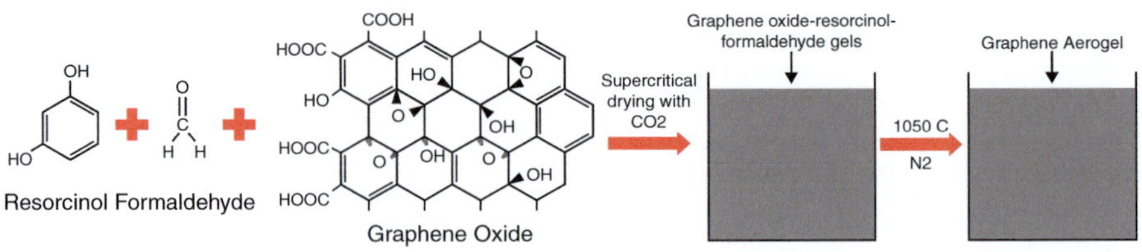

acetone to remove water from the pores. Supercritical CO_2 is used to dry the graphene oxide–resorcinol–formaldehyde gels and pyrolysis at 1050 °C under nitrogen yielded the final graphene aerogel.

12.4.2 Properties

Aerographite has truly superb elasticity and absorption. The aerographene can recover completely after more than 90% compression and absorb up to 900 times its own weight in oil, at a rate of approximately 68.8 g/sec. It shows hydrophobic nature with a water contact angle of more than 136° (Fig. 12.19) [49]. In contrast to hydrophobic wettability to water, the aerographite shows excellent oleophilic properties. It also remains afloat when saturated with oil for months. With these two features combined, the material has massive potential to be used to mop up oil spills and squeezed to reclaim the oil. Due to its elasticity, both the oil absorbed and graphene aerogel can be recycled. Aerographene is robust, highly flexible, can be rolled for storage, and capable of withstanding a number of large-strain compressive cycles without affecting their properties. Aerographene works efficiently under harsh conditions, such as in high or low temperatures. Traditional sorbents (such as polyurethane-based and polyethylene-based materials) cannot be used above 200 °C, and others will be very brittle at low temperature. In this regard, aerographene exihibits exceptional features [50]. When exposed to an ethanol flame, the graphene aerogel did not support any burning and remains inert. With the sudden change in environmental temperature from extremely higher to extremely lower temperature, aerographene is the best material to be used. This helps to recycle and reuse aerographene multiple times without worrying about its decomposition.

12.4.3 Applications

Aerographene has potential applications for cleaning up oil spills due to its ability to absorb 900 times its weight in oil. It can also be used to gather dust from the tails of comets. After bulk production, it is highly expected to cover all the high-end applications.

12.5 3D Graphene

Graphene is thought to be the strongest of all known materials on the planet made from pure carbon, ultra-thin graphene. The 3D architecture of 2D graphene is an attempt to extend the successful utilization of graphene. Generally, graphene is the 2D planar sheet of carbon atoms packed in the honeycomb-like lattice, whereas 3D graphene is the porous network of the interconnected graphene sheet. Researchers at MIT found a way to turn 2D graphene into a 3D structure by designing a new material with a sponge-like configuration through a process that compressed and fused graphene flakes into a porous, sponge-like 3D form using a combination of heat and pressure that is 5% the density of steel and about 10 times as strong. This process produced a strong, stable structure of which the form resembles that of some corals as shown in Fig. 12.20. These shapes, which have an enormous surface area in proportion to their volume, proved to be remarkably strong. The superstrong and lightweight 3D graphene has been shown to be stronger than its 2D counterpart and offers greater potential uses. 3D graphene foam is a porous, high-surface-area form of graphene. The 3D graphene material is composed of curved surfaces under deformation. It resembles what would happen with sheets of paper. Paper has little strength along its length and width and can be easily crumpled up. But when made into certain shapes, for example, rolled into a tube, suddenly the strength

Fig. 12.19 More than 150° WCA signifies the hydrophobicity nature of aerographene

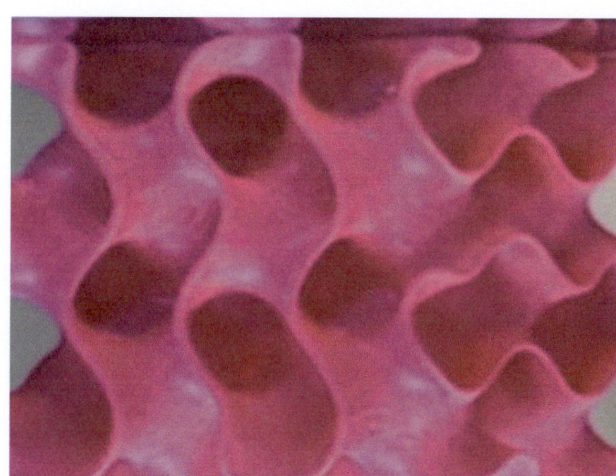

Fig. 12.20 3D graphene [10]

along the length of the tube is much greater and can support substantial weight. Similarly, the geometric arrangement of the graphene flakes after treatment naturally forms a very strong configuration. Researchers say the ability to tune the mechanics of materials simply by adjusting its geometry opens the door to a wide variety of practical applications where you need strong, lightweight materials like those for cars and buildings. Moreover, because of the porous nature of the structures, the process and material are also well suited to filtration and energy storage applications [51].

Particularly, 3D graphene structures have been produced on various substrates, such as nanoparticles, metal foams, non-metal porous structures, and without any template. Due to these facile synthesis approaches, 3D graphene has significant importance in emerging applications in electrochemistry energy storage, environmental science, and biomedical engineering. Specifically, 3D nanostructured graphene is in general a porous structure and does not consist of planar graphene sheets. 3D graphene tends to pose good electrical conductivity and mechanical integrity, which makes it well-suited as anode material in lithium and sodium-ion batteries [52, 53].

12.5.1 Synthesis of 3D graphene

Freestanding 3D graphene or porous networks of graphene can be attained by various methods including template assembly and self-assembled graphene nanosheets (given in Fig. 12.21). The template strategy was commonly applied in several methods including chemical vapor deposition, lithography, direct electrochemical methods, hydrothermal, solvothermal, sol-gel method, and so on. [54]. Various templates such as Ni foam, Cu foam, polystyrene spheres, and so on were used for the fabrication of 3D graphene. The template-based synthesis of the 3D porous network generally consists of three steps. In the first step, the reaction precursors are attached to the template either by incorporation or by impregnation. In the second step of the reaction, growth or nucleation is continued to form the solid species either in or on the template surface. In the last step, the template was removed in several ways to form a 3D porous structure. The assembly of graphene into 3D hierarchical architectures has been recognized as one of the most promising strategies for "bottom-up" nanotechnology. With the tremendous ascent, 3D assembled macrostructures possess new collective physiochemical properties that are remarkably different from both the individual building blocks and their bulk materials, which further extend their application capabilities.

12.5.2 Template-Assisted Processes

12.5.2.1 Chemical Vapor Deposition (CVD)

Graphene synthesis by CVD is a process where a carbon source is decomposed on a metal substrate to form a single layer or few layers of the graphene sheet. 3D graphene structure with controlled layers and sizes can be produced by CVD on prefabricated 3D metal framework. Nickel and copper substrates have been applied for the synthesis of 3D graphene structures [55]. It was reported that the primary growth mechanisms of graphene on nickel substrates were adsorption and incorporation of carbon atoms within the substrate and precipitation of carbon atoms from the substrate upon cooling (out-diffusion). Although surface growth is one of the growth mechanisms on a nickel substrate, the carbon dissolution–precipitation is more dominant, and thus several layers of graphene structure can be deposited on the metal framework, thereby strengthening the stability of the structure. Hence, nickel substrate is the most preferred substrate for synthesizing 3D graphene foam. On the copper substrate, the formation of graphene was governed purely by surface growth which resulted in lower carbon solubility. Hence, only a single layer of graphene is deposited on the template making it difficult to support the interconnected 3D network

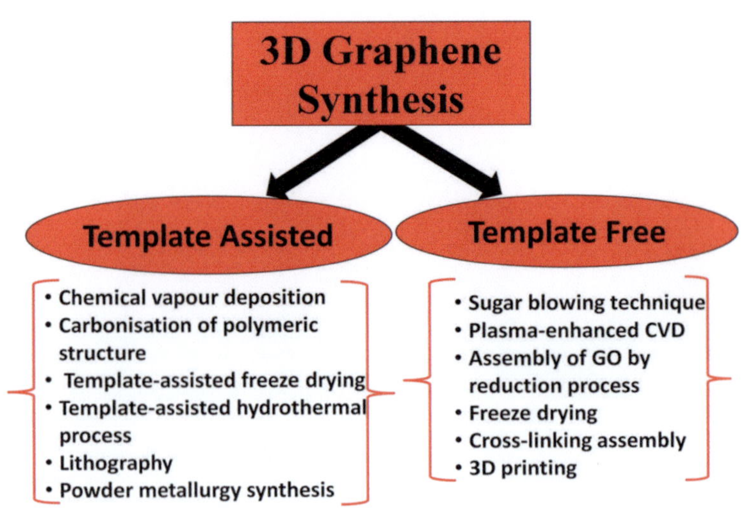

Fig. 12.21 Classification of 3D graphene synthesis

of graphene. This type of structure breaks down easily during handling. Apart from using metal substrates, other materials such as metal oxides, metal nanostructures, and metallic salts were used as templates for the synthesis of 3D graphene structures. The first 3D graphene foam was made by CVD on Ni-based template. Here, the Ni substrate was pyrolyzed with (CH_4) as the carbon source at 1000°C and atmospheric pressure. Polymethyl methacrylate (PMMA) film was formed on the surface of the 3D graphene foam to support the structure and prevent it from disintegrating when Ni was etched with HCl. Finally, the PMMA was dissolved with hot acetone to produce free-standing graphene foam. The graphene foam produced had a specific surface area of ~850 m^2/g, interlayer spacing of 0.34 nm, and porosity of ~99.7% [56].

12.5.2.2 Carbonization of Polymeric Structure

Carbonization of the organic-based framework is a relatively simple technique for the synthesis of carbon-rich materials. In this process, 3D polymeric structures such as highly cross-linked framework, hydrogel, or aerogel are subjected to thermal annealing (carbonization) to synthesize a 3D graphene-based structure. This approach is for the development of 3D carbon foam from poly-HIPE polymeric structure for supercapacitor application. In their work, the poly-HIPE was sulfonated and pyrolyzed at 850 °C to produce a 3D graphitic configuration with a specific surface area of 389 m^2/g and open voids of 14 μm interconnected by 4-μm windows in a monolithic structure. A novel elastic carbon foam with ultrahigh porosity (99.6%), high specific surface area (268 m^2/g), and ultralight weighted or low density (5 mg/cm^3) was developed by direct carbonizing of melamine foam at 800 °C. Their results demonstrated that the elastic carbon foam was able to absorb 148–411 times of its own weight, depending on the density of the specific organic solvents. This direct carbonization technique enabled the fabrication of 3D porous graphene monolithic structure with a high specific area, large void, lightweight, and excellent adsorption properties.

12.5.2.3 Lithography

Lithography has emerged as a powerful tool for attaining 3D architectures. 3D graphene can be achieved in lithography using photoresist structures. The thin nickel film was sputtered uniformly on the lithographically defined conductive 3D carbon networks. Further annealing of the Ni-coated carbon resulted in the fabrication of the multilayer graphene-coated 3D Ni electrode. The multilayer graphene hollow structure can be formed by removing the Ni through acid etching. Overall, three steps are required to convert the predefined 3D pyrolyzed photoresist films into 3D porous graphene. In the first step, the Ni was sputtered on the 3D amorphous carbon structure, while the second step consisted of annealing. In the last step, the Ni was removed by acid etching to give a multilayer graphene hollow structure. One of the major advantages of using the porous network is the huge available surface area and availability of the interior surface for the electrochemical reaction. Sometimes, the high density of nanopores and the wide size distribution of the pores may lead to overlapping diffusion regions that may limit the mass transport to the interior region of the porous network. The benefit of the high surface area does not remain much effective. The highly ordered fabrication of the mesoporous material might be a possible solution [57]. Lithography provided more controlled fabrication of 3D interconnected porous electrodes. The controlled morphology provided several pathways for the analyte to diffuse, and the electrochemical reaction takes place throughout the porous electrode surface. The nanofabrication through conventional lithography is a complex, expensive, and time-consuming process.

12.5.2.4 Template-assisted Freeze-Drying

Template-assisted freeze-drying or ice-template is the process that utilizes the rapid freezing of aqueous dispersion containing amphiphilic polymers to construct a porous 3D structure. This process is also termed the ice-segregation-induced-self-assembly (ISISA). GO was first incorporated into the aqueous media either with or without additional chemical binders before freeze-drying. In this process, the formation of ice crystals provided the structural support for the 3D graphene structure, and the morphological properties and porosity of the structure were dependent on the freezing temperature and temperature gradient. During freeze-drying, the GO sheets were ejected and entrapped between the neighboring ice crystals, yielding highly oriented interconnected GO-composite layers that formed a solid 3D graphene-based structure. However, the pore structure of the graphene configuration fabricated by ISISA was inconsistent throughout the structure which was attributed to the local differences in freezing time and the temperature gradient applied onto the template during the synthesis process [58].

Many chemical additives were used to fabricate 3D graphene-based structures by ISISA, and these included chitosan, polyvinyl alcohol (PVA), carboxymethyl cellulose, and gum. Typically, the process involved mixing of GO suspension with the chemical additive solution. The homogeneous mixture of GO-composite solution was frozen overnight prior to freeze-drying or drying by super critical CO_2 to form 3D graphene aerogel. For instance, the green development of xanthan gum–GO hybrid aerogel has been demonstrated by mixing 5 mL GO dispersion (0.002 g/mL) with 5 mL of xanthan gum solution (0.01 g/mL). The resultant mixture was subjected to freezing at −40 °C for 1 h before freeze-drying for 8 h. From their findings, the freezing temperature played an important role in the formation of the porous structure. The aerogel that was frozen at a lower temperature (−40 °C) displayed an oriented hierarchical thin sheet structure, and the formation of networks connecting the co-aligned and pores channel was also observed. Conversely, for the sample that was frozen at higher temperature (−10 °C), the thickness of the GO-composite sheets increased and the aerogel had a disoriented structure. The orientation of the aerogel also affected the adsorption performance as the reported adsorption efficiency for rhodamine B and methylene blue dyes of the aerogel frozen at −40 °C was higher than that at −10 °C. The volume per unit mass of the developed aerogel was reported to range from 52.7 to 63.2 cm^3/g, indicating the XG/GO aerogel had a tunable porosity [59]. The ISISA or template-assisted freeze-drying is one of the most commonly used techniques for the synthesis of 3D graphene structures as the process has relatively straightforward steps.

12.5.2.5 Template-assisted Hydrothermal Process

Template-assisted hydrothermal synthesis involves the use of a porous foam-like template, and the 3D graphene structure is obtained through the hydrothermal process. The foam template used can be either metallic or polymeric foam. In the fabrication process, the templates were soaked in GO dispersion, and the soaked template was subjected to hydrothermal process for deposition of GO sheets on the templates wall. After chemically reducing the GO, the template skeleton was removed to obtain the graphene-based structure. Typically, the removal of the metallic template was achieved by chemical etching, whereas the polymeric template was eliminated by solvent dissolution or thermal decomposition. In this synthesis method, the quality of 3D graphene foam is poorer than that fabricated by CVD due to the stacking of multilayered GO sheets. Furthermore, this method is associated with higher manufacturing costs due to the sacrifice of the template support at the end of the process and the higher difficulty in controlling the morphological properties of the final product.

12.5.2.6 Powder Metallurgy Synthesis

Powder metallurgy can provide both growth templates and carbon sources for producing the 3D graphene foams in thermal annealing. This unique method consists of two steps, first is the powder metallurgy method for blending Ni and carbon precursors and the second is the thermal decomposition of carbon precursors over Ni for graphene formation. To start with, Ni powders (as templates) and sucrose (as carbon precursors) are mixed into a composite in deionized water [60]. Next, water is removed from the mixture via mechanical stirring at 120 °C. The nickel/sucrose powders are dried at 80 °C, and the Ni carbon blends are readily processed into pellets under hot pressing with the assistance of a pestle and mortar. The pellets embedded with both Ni and carbon precursors are loaded onto a quartz tube furnace for a CVD process at a heating rate of 10–1000 °C/min for 30 min. After this stage, the pellets (coated with graphene) are moved to the colder zone of the furnace for cooling. The graphene–Ni pellets are then etched in an aqueous solution of iron chloride to remove the Ni. Finally, the freestanding 3D graphene foams form after the drying process. Such 3D graphene possesses good electrical and mechanical properties. This method is simple and allows for the customization of 3D graphene architecture to change the shape of the press dye and introduce various types of additives.

12.5.3 Template-Free Processes

12.5.3.1 Sugar Blowing Technique

The sugar blowing method has been reported in the literature for the preparation of a 3D graphene structure. This technique was inspired by the traditional food art of "blown sugar" in which a mixture of glucose and NH_4Cl was heated. The molten syrup was gradually polymerized, and the gas released chemically by NH_4Cl blew the glucose-derived polymer into a number of bubbles. The generation of the gas caused the bubble walls to become thinner, and the polymeric fluid and small molecules were removed from the polymer walls. The ultrathin graphitic bubble networks were formed from the carbonization of the polymeric wall at high temperatures. The 3D structure produced by this method is known as 3D strutted graphene. The strutted graphene with an average width of 3.5 μm graphitic skeleton, density of 3 mg/cm^3, and specific surface area of 1005 m^2/g was used for high power density supercapacitors application [61].

12.5.3.2 Plasma-enhanced CVD (PE-CVD)

A direct PE-CVD approach without sacrificial templates has been used to construct 3D graphene-based structure. This method utilized nonthermal radio frequency plasma to radicalize a monomer, generating chemical reactive species to form uniform coatings and depositing the coated species vertically to construct a 3D structure. In 3D graphene structure development, conductive substrates such as Ni, Cu, Au, or stainless steel were often used together with CH_4 as the carbon source to produce graphene foam without the assistance of a template. In general, the 3D graphene structure developed by this technique is highly cross-linked with uniform and robust graphene films. The graphene layers are firmly adhered to the substrates with many vacant active sites on the edges of the graphene sheets, which render them suitable for surface-related applications, such as adsorption, bio-sensing, and catalysis [62].

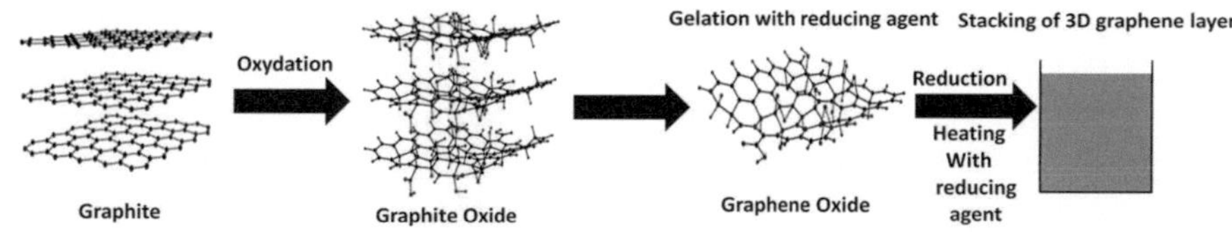

Fig. 12.22 Steps of chemical reduction process of graphite

12.5.3.3 Assembly of GO by Reduction Process

The assembly of 2D GO sheets into 3D graphene-based structures is initiated from the dispersion of GO in aqueous media. Due to the strong electrostatic repulsion force in GO dispersion, the 2D GO sheets form a stable colloidal system hindering the self-assembly of GO into a 3D structure. Several methods have been proposed to overcome this problem, and one of them is the reduction of GO either by chemical or by thermal treatment. In situ reduction of GO by chemical additives is a potential approach for the fabrication of 3D graphene structures (Fig. 12.22). The mechanism for the chemical reduction of GO is governed by sol-gel chemistry. The process involved the elimination of the oxygenated functional groups from the GO, triggering gelation of the GO sheets into a 3D form known as rGO hydrogel (Fig. 12.22). A large variety of reducing agents have been tested, and these included acids, hydrazine, metal or metal oxides, reducing salts, and plant extracts. In a typical chemical reduction process, the reducing agent was introduced to the GO dispersion and subsequently heated to complete the reduction process. During reduction, the reaction vessel was left to stand without stirring to enable stacking of graphene layers in an orderly arrangement. The properties of the developed 3D graphene structure were dependent on the type of reducing agent used. It was reported that mild reducing agents, such as l-ascorbic acid, could assist in the construction of more uniform and stable graphene structures in water. The stability of the rGO hydrogel could be attributed to the formation of hydrogen bonds between the oxidized ascorbic acid and residual oxygen functional groups as well as the disruption of π-π stacking between the rGO hydrogel which prevented agglomeration. Recently, green reducing agents such as plant-derived extracts (tea leaves, spinach leaves, and Roselle flower) were investigated as alternative reagents for GO reduction. It was reported that the plant extracts could effectively reduce GO [38].

The hydrothermal reduction process is commonly used to reduce GO to initiate the self-assembly of GO under a high temperature and pressure environment. In this method, a fixed concentration of GO dispersion was heated in a Teflon-lined hydrothermal reactor to form the hydrogel. During the hydrothermal process, the reduction of GO was achieved through the elimination of carboxylic groups, which significantly reduced the electrostatic repulsion in the GO dispersion. This phenomenon triggered the self-assembly of graphene-based hydrogel through the π-π deposition of GO sheets at the graphene basal planar. The feasibility of adjusting the pores morphology of hydrogel through the hydrothermal process was demonstrated by altering GO concentration, hydrothermal reaction time, or addition of metallic catalyst. The hydrothermal reduction also offers a possibility for elemental functionalization onto the graphene lattice through doping. Du et al. (2014) have successfully synthesized nitrogen-doped rGO structures for high-performance lithium-ion battery anode by hydrothermal process. In their work, the nitrogen-doped rGO was formed by hydrothermal treatment of a mixture of melamine and GO suspension in a Teflon-lined autoclave at 180 °C for 6 hours. From their work, the nitrogen-doped rGO exhibited Type IV adsorption isotherm indicating the existence of mesopores within the structure with a Brunauer-Emmett-Teller (BET) specific surface area of 146.0 m^2/g [39]. In addition, the hydrothermal process can be extended to the solvothermal process by replacing the water with organic solvents. This method enables the use of chemicals that are soluble in organic solvents for the functionalization of the 3D graphene structures.

12.5.3.4 Freeze-Drying

Freeze-drying is commonly practiced to eliminate the water molecules trapped inside the structure of rGO hydrogel to produce an orderly configured aerogel. The preparation of rGO aerogel from GO sheets involved three key steps. Generally, the GO sheets are transformed into rGO hydrogel through a reduction process, followed by freezing of the hydrogel to crystallize the water molecules. The water crystals act as the pore-forming agent. Finally, the solidified water molecules are sublimated under vacuum conditions to form a 3D porous rGO aerogel [63]. The porous microstructure of the 3D graphene-based aerogel is highly dependent on the nature of hydrogel, freezing conditions, and temperature gradient. 3D graphene aerogel was produced by adding L-ascorbic acid into an aqueous GO suspension. Upon stirring and heating, the mixture was allowed to stand for at least 16 h for construction of rGO hydrogel. The hydrogel was then purified and subjected to freezing and freeze-drying to form the graphene-based aerogel. Two types of sublimation systems, that is freeze-drying and supercritical CO_2 drying, have been investigated [64]. It is found that the graphene aerogel made by supercritical CO_2 drying exhibited remarkable mechanical properties, as it could withstand 14,000 times its own weight than that prepared by freeze-drying

(3300 times its own weight). The aerogel prepared by supercritical CO_2 showed a BET surface of 512 m^2/g which was higher than that of aerogel prepared by freeze-drying (11.8 m^2/g). Another hydrothermal technique is using reducing agents (vitamin C (VC), ethylenediamine (EDA), and ammonia for the formation of graphene aerogel [65]. The specific surface areas for graphene aerogels reduced by VC, EDA, and ammonia were reported to be 661, 440, and 1089 m^2/g, respectively. However, the highest mechanical strength was demonstrated by the VC-reduced graphene aerogel. The researchers further reported that the surface roughness could be governed by the pH values that were determined by the functional groups of reducing agents. The pH of the mixed solutions of VC-GO, EDA-GO, and ammonia-GO before hydrothermal reduction were 3.4, 11, and 10.4, respectively. The acidic condition promoted agglomeration and fragmentation of graphene sheets during the reduction process, whereas the basic condition favored the production of larger graphene sheets with thin morphology.

12.5.3.5 Cross-linking Assembly

Cross-linking-induced assembly is based on the addition of different chemicals to promote gelation of the GO dispersion, construction, and strengthening of the 3D structures through chemical or physical cross-linking. In this approach, the chemical additives act as pore-forming agents and introduce new functional groups to the 3D graphene structure. The first cross-linking agent reported in the literature was polyvinyl alcohol (PVA) by Bai et al. (2010) [66]. It was found that the gelation between GO and PVA was governed by hydrogen bonding between the functional groups (hydroxyl, epoxy, and carboxyl groups) on the GO surface and the hydroxyl-rich PVA chains. One notable finding from their work was that gelation of GO-PVA hydrogel was a reversible process made possible by tuning the pH of the system. In their research, the hydrogel was formed at pH <7 but remained in the aqueous form at pH >7. This sol-gel transition of GO-PVA hydrogel occurred due to insufficient binding force between the molecules under high pH conditions. The hydrogel produced under acidic condition was found to be much stronger than that prepared under elevated pH conditions. This was attributed to a reduction of GO sheets through acidification that facilitated the π-π stacking of graphene sheets. The development of GO-PVA hydrogel was also reported by other researchers for applications, such as adsorbent, material reinforcement, and wound dressing. Apart from PVA, various chemicals were reported to be effective cross-linkers for the construction of 3D graphene-based structures, such as polymers (poly (N-isopropylacrylamide), polypyrrole and polyaniline), polysaccharide-based materials (chitosan, sodium alginate, carboxymethyl cellulose, and cyclodextrin), bivalent, and trivalent metal ions (Ni^{2+}, Mg^{2+}, and Fe^{3+}). The type of cross-linking mechanisms formed between the GO and cross-linker depends on the cross-linking chemical used. For example, the cross-linking mechanism between chitosan and GO was reported to be a nucleophilic addition mechanism, as the epoxy groups in GO were cross-linked with the amino groups of chitosan. Huang et al. (2015) hypothesized that the GO sheets and polyethyleneimine (PEI) films were cross-linked through electrostatic interaction and covalent bonding between the epoxy groups of GO and the primary amine groups of PEI. The cross-linked GO-PEI film was more mechanically stable in water [67, 68].

12.5.3.6 3D Printing

3D printing is another effective strategy to prepare 3D graphene structures. The 3D graphene structure that is synthesized by 3D printing can achieve higher mass loading and other tailored properties for energy storage applications. 3D printing refers to the process of connecting or solidifying materials to create 3D objects under the control of the computer. The key challenges in this method are (1) developing printable GO-based or graphene-based inks with suitable viscous and thixotropic properties and (2) maintaining the intrinsic properties of single graphene sheets, such as large surface area and excellent mechanical and electrical performance. The GO inks were prepared by combining highly concentrated GO suspensions with silica powder to form a homogenous, highly viscous, and thixotropic ink for printing. Then, the inks were transferred into a spring barrel and extruded through a small-size nozzle to pattern 3D architectures in an iso-octane bath (to avoid drying of the inks in air). Afterward, the printed microlattice architecture was supercritically dried to remove the solution before being heat-treated at 1050 °C under N_2 atmosphere for carbonization, while the GO was reduced to graphene at high temperature. Finally, the silica powder was washed with hydrogen fluoride (HF) solution. Sha et al. [69] reported the synthesis of grapheme foam via the 3D printing method. Nickel particles with a diameter of 2–3 μm were used as the template and catalyst and sucrose coated on nickel particles was used as the carbon source. A thin layer of powder containing sucrose-coated nickel particles was first set on the platform, followed by laser radiation to produce graphene layers. After first rastering, another thin layer of powder was manually added onto the top, followed by the same laser radiation process. After repeating 20 times and template removal, graphene foam was obtained. The obtained materials owned excellent physical properties, such as a high porosity of 99.3 %, a low density of 15 mg/cm^3, high electrical conductivity of 8.7 S/cm, and remarkable storage modulus of 11 kPa.

12.5.4 Factors Influencing the Synthesis

The formation of 3D graphene structures will be greatly influenced by various conditions. According to the choice of 3D graphene structure formation method and the relevant parameters, researchers can tailor the fabrication process and obtain specified 3D porous architectures to satisfy the needs of the corresponding application. It is well established that the pore morphologies have affinities to some physical properties of the 3D architectures. For instance, the mechanical stability of the as-prepared 3D graphene structure is determined by the interconnectivity, degree of crosslinking, and so on. Meanwhile, some properties, such as the pore size distribution and density of the 3D graphene structures, can be tailored by controlling the synthesis parameters. The assembly behavior of 3D graphene structures governs by adjusting the concentration of the initial GO and the reaction time of the hydrothermal process. The results demonstrated that the concentration of GO could affect the size and strength of the 3D graphene structures, while the reduction time could be used to tailor the electrical conductivity and size of the resultant 3D graphene structure. The density and pore size distribution of the 3D graphene structures were also determined by the size of the selected GO precursors. Compared to the large-size GO nanosheets, smaller ones would result in a higher 3D graphene structure density and smaller pore size. Some other properties, such as the pore wall thickness of a hydrogel, would be affected by introducing site-specific crosslinkers. The orientation of the pores in the 3D construction could be controlled by adjusting the orientation of the 2D GO precursors [70]. The orientation-controlled 3D graphene structures could exhibit anisotropic conductivity and acoustic vibrations that were guided in desired directions. After the formation of the graphene hydrogel, the drying method could help to adjust the volume and density of the porous framework.

12.5.5 Properties of 3D Graphene

3D graphene films have many interesting and sometimes advantageous properties when compared to 2D graphene films. They are often combined with other components to improve their properties and provide them with the opportunity for use, for example, in the field of energy storage on a large scale. In general, 3D graphene structures with chemically bonded structures have better properties than physically assembled structures, such as lower contact resistance, better conductivity, and they are stronger, tougher, and more flexible. Changes in the fabrication conditions caused differences in the structural features, such as the orientation and arrangement of the graphene sheets, the physical or chemical links between the graphene sheets, the pore size and porosity, and the number of layers of graphene sheets, changing their properties. The orientations of the graphene sheets in graphene foam (GFs) and graphene sponge (GSs) are different [71]. In GFs, the graphene sheets have no orientation. Thus, GFs have isotropic structures and properties. Conversely, the graphene nanosheets in GSs are stacked in highly ordered film-like structures and these large graphene films are aligned nearly parallel with each other, creating an anisotropic structure. The preparation process for GFs is very versatile and can control both the macrostructure and the microstructure. The pore size and porosity can be tuned by changing the pore structure of the Ni foam. The average number of graphene layers, specific surface area, and density of GFs can be controlled by changing the CH_4 concentration. A higher CH_4 concentration led to an increase in the number of graphene layers and consequently large changes in the specific surface area, density, and electrical conductivity of the GFs. The thickness and mass of the GFs increased as the number of graphene layers increased, while the density and specific surface area decreased with an increasing number of layers. The reason for the decrease in the density with an increase in the number of graphene layers may have caused the framework to shrink and the HCl and acetone etching reduced with an increase in the thickness of the GFs. The microstructure and properties of GSs can be tuned by adjusting the synthesis conditions. When producing GSs by freeze-drying graphene hydrogel (GH), the freezing temperature is critical [72].

The mechanical properties of 3D graphene materials usually depend on their shapes, which are determined by the supporting templates. Samad et al. investigated the morphology of nickel foam and the graphene foam formed by dip-coating. The stress–strain curve of the graphene foam shows the division for three regions, that is, elasticity, plateau, and densification. The elastic modulus is 89 ± 6 kPa [73]. After release at 40% strain, it could recover from $20 \pm 5\%$ of the strain, termed post-compaction strain recovery. The ultralight graphene aerogel also presented excellent stretchable capability. Produced by the functionalization of graphene oxides in an ethylenediamine aqueous solution, freeze-drying, and microwave treatment, the pristine graphene aerogel features a complete recovery of its original volume under external pressure. However, the functionalized graphene hydrogel, a product without the freeze-drying process, shows only partial recovery. CVD-grown graphene foam connected to the PDMS can derive a composite, which has shown good flexibility, the ability to bend, stretch, and twist without causing any distortion in the material. The addition of graphene foam in such composite indicated an improvement in mechanical strength by 30%. Indeed, when PDMS is combined with graphene foam to form a composite, an almost complete recovery of the strain is achieved, suggesting that the addition of polymers improves flexibility.

In addition, the self-assembled graphene hydrogel prepared in a one-step hydrothermal method exhibits good mechanical strength.

The vacuum infusion of graphene foam into PDMS provides a flexible composite with electromechanical integrity for repeatability and stability, and it retains its electrical resistance under shear stress. Gradually increasing pressure above 900 kPa, the resistance increases rapidly and becomes an insulator at 1300 kPa [74]. A compressive strain results in an increase in the resistance of the composite to 120% of the initial value. Such 3D graphene PDMS composite presents sound electrochemical integrity with good repeatability and stability under compression. The correlation of electrical resistance with stretching force shows that resistance increases under increasing maximum tensile strain, and it remains stable under a higher number of cycles [75]. The bulk electrical conductivity of a chemically bonded GA is about 1 S/cm, more than two orders of magnitude greater than those reported for the macroscopic 3D graphene structures prepared with either physical crosslinkers alone or partial chemical bonding. GO hydrogel has a conductivity of 5×10^{-3} S/cm and 3D RGO has a conductivity of 2.5×10^{-3} S/cm. In comparison, graphene foam (GFs) has a much higher electrical conductivity owing to its unique continuous interconnected networks. The GAs had similar high electric conductivities of around 1 S/cm, both with and without an organic crosslinker [76]. This may have been caused by crosslinking between the functional groups on the surfaces and edges of the GO during the sol-gel process. 3D graphene foam shows great biocompatibility as a platform for supporting the growth of bone tissues and neural stem cells. Graphene itself is not sufficient to form the calcium phosphate compound, which is a prerequisite for producing bones. However, graphene foam is highly biocompatible when connected with osteo-compatible polymers, such as poly(vinylidene fluoride) and polycaprolactone. Indeed, the graphene biopolymer composite allows the stable growth of $CaHPO_4$ due to good osteoconductivity, which can be used in scaffolding [77].

12.5.6 Application

3D porous materials with interconnected ordered structures are gaining technological importance in various applications. 3D architecture substantially improved the performance of certain devices. The 3D graphene design has a good future in many challenging applications, such as stretchable electronics, supercapacitor, catalyst support, stretchable conductors, environmental protection, biofuel cells, gas sensors, and electrochemical sensors. 3D graphene structures are considered an attractive and competent material for application in the fields of energy storage and conversion, including fuel cells, batteries, solar cells, and supercapacitors. Some emerging applications are discussed subsequently.

Stretchable electronics: The preparation of rubbery and stretchy conducting materials has attracted increasing interest over the last decade due to its potential application in stretchable electronics. Recently, Cheng and coworkers demonstrated that the composites of graphene foams (GFs)-poly(dimethyl siloxane) (PDMS) fabricated by infiltrating GFs with PDMS are a good stretchable conductor. The 3D continuous and interconnected network ofGFs carries the electricity, and PDMS provides flexibility. The GF-PDMS composites can be bent, stretched, and twisted without breaking and also easily recover to their initial form after the strain is released, indicating their good mechanical robustness and elasticity. Moreover, the seamless interconnected network of GF facilitates fast transport of charge carriers as confirmed by a high electrical conductivity with an ultralow graphene loading. The incorporation of excellent mechanical robustness and electronic performance endows the composites with great potential for use as stretchable conductors. Bending and stretching tests were carried out to quantify the mechanical stability of the GF-PDMS composites.

Supercapacitors: Supercapacitors, also known as ultracapacitors and electric double-layer capacitors (EDLCs), are energy storage devices that possess high power densities, long life, and high rate capability. Supercapacitor electrodes based on 3D hierarchical graphene/polypyrrole aerogels exhibit excellent electrochemical performance, including a high specific capacitance up to 253 F/g, a good rate performance, and outstanding cycle stability. It is reported that a CMG-based supercapacitor with specific capacitances of 135 and 99 F/g in aqueous and organic electrolytes, respectively, demonstrate that CMGs are promising electrode materials for supercapacitors. However, this specific capacitance is far below the theoretical value of 550 F/g calculated for single-layer graphene [78]. The capacitance of an EDLC is generated by electrostatic charge accumulation at the electrode/electrolyte interfaces and is usually proportional to the effective specific surface area of the electrode material. CMGs with 3D interpenetrating microstructures can provide large accessible surface areas for forming electric double layers and facile routes for electron and electrolyte transportations in interconnected conductive networks. Thus, supercapacitors with 3D porous graphene-based materials as active electrodes are expected to have good performances.

Catalysis: Graphene has been widely explored either as inherent catalysts or as supports for other catalytic components. The porous interconnected network is beneficial to ion diffusion and transfer kinetics and provides a special reaction microenvironment and conductively multiplexed pathways for rapid charge transfer and conduction. 2D GO sheets can self-assemble into a 3D porous composite hydrogel upon the addition of an enzyme, hemoglobin (Hb), which exhibits improved activity and stability for catalyzing a peroxidation reaction of pyrogallol in organic media in comparison with free Hb or GO [79]. It is believed that the aqueous microenvironment of the GO hydrogel can efficiently protect Hb from deactivation in organic media. Moreover, the substrate pyrogallol is more hydrophilic than the product of purpurogallin; thus, the composite hydrogel in organic media can act as a transit station to enrich the substrate and exclude the product, promoting the enzymatic reaction in the hydrogel matrix. This pioneering work indicated that 3D porous GO hydrogels have great potential in enzymatic biotransformations. GO/graphene has been found to have potential catalytic properties in some reactions, including the hydration of alkynes, oxidation, oxidative coupling, Friedel-Crafts addition, aza-Michael addition, polymerization, and photo-oxidation. Also, GO/graphene can be used as catalyst support in many reactions, such as Suzuki-Miyaura coupling and photocatalysis, owing to its two-dimensional structure, large surface area, extraordinary electronic and mechanical properties, and the abundant functional groups on the GO surfaces that provide many favorable sites where functional nanocomponents can anchor [80]. Graphene-based catalysts can be used in organic synthesis, sensors, environmental protection, and energy-related systems. However, the large resistance from structural defects and the strong planar stacking of graphene sheets lead to drastic deterioration in the properties. These shortcomings can be overcome by forming 3D graphene skeletons [81].

Fuel Cells/Batteries: 3D graphene, either grown on the catalyst substrate or reduced from graphene oxide, possesses a large surface area, high electrical conductivity, and good electrochemical stability. In conjunction with various other materials, including, but not limited to, metal/metal oxides, polymers, and carbon allotropes, the 3D graphene-based composites are promising electrode materials for high-performance supercapacitors and lithium-ion batteries. However, there are still various challenges in fabricating perfect 3D graphene materials for practical applications in energy storage devices. First, the efficient surface area is still limited due to difficulties in precise tune and control, the porosity, and structure in the 3D graphene architecture during the synthesis, and hence the obtained specific surface area is much lower than the theoretical value. Second, the electrical conductivity is high but far lower than the theoretical conductivity of ideal single-layer graphene. Third, the ion and electron diffusion in the intrinsic microstructure should be further investigated to further improve electrochemical performance. Last, 3D graphene-based energy storage devices should be endowed with more functions, such as flexibility, stretchability, and wearability, to improve their practical applications. In fuel cells, the role of 3D grapheme structure is generally a part of the catalyst. Yan's group used 3D graphene as the anode in microbial fuel cells. The maximum power density reached was 427.0 W/m^3, which is higher than that of the microbial fuel cells fabricated using carbon felt as the anode. The macroporous structure of the graphene structure ensured that the microbes could easily diffuse and propagate inside the materials, resulting in a higher MFC performance. Similar to 2D GO/graphene, 3D graphene has been used as active electrode materials in lithium-ion and Li-S batteries. The 3D electrode of sulfur embedded in porous graphene structure is used in lithium-sulfur batteries. The grapheme structures worked as a framework that could provide a high electronic conductive network, the ability to absorb intermediate polysulfides, and mechanical support to accommodate the volume changes during the charge and discharge processes. As a result, the S-GS electrodes with 80 wt.% sulfur could deliver a high areal specific capacity (4.53 mAh/cm^2 after 300 cycles) and a slow decay rate at 0.1 C (0.08% per cycle after 300 cycles). This is a significant step toward the application of Li-S batteries [82, 83].

Electrochemical Sensor: 3D graphene provides a promising platform for the sensing of various electrochemical analytes. The 3D graphene network provides multiplexed conductive pathways and offers rapid charge transfer. These factors contribute to enhancing the sensitivity of the electrochemical sensor. Highly conductive and monolithic 3D graphene foam was synthesized by CVD and was applied for the sensing of dopamine. The sensing of dopamine is challenging due to its close electro-oxidation peak with uric acid. The 3D graphene electrodes demonstrate the tendency to distinguish between the dopamine and the uric acid peaks [84]. The sensor displayed a wide linear range of 25 nM to 25 μM. The excellent sensitivity, selectivity, and wide linear range of the sensor are attributed to high charge transfer, availability of the high active surface area, and π–interactions of the dopamine with graphene. Similarly, electrochemically synthesized 3D ErGO on AuE displays high conductive channels and strong electrocatalytic behavior for the sensing of dopamine compared to compact 2D ErGO. Compact 2D ErGO provided a limited excess to an electrolyte containing dopamine, and fewer graphene edges are exposed for the electrocatalysis which is responsible for lower sensitivity [85].

Environmental Remediation: There has been a growing need for recyclable absorbents that can remove organic pollutants or oil spills from water. One of the important features of 3D GO/graphene structures is their large accessible specific surface areas, as they can effectively absorb organic and inorganic contaminants. Furthermore, 3DGNs can be easily separated from the solution, making them convenient to collect and recycle. Thus, GO/graphene foams and sponges can be used as super-absorbers in environmental remediation to remove a range of organic contaminants and heavy metal ions with high efficiencies. A GO/DNA composite hydrogel can efficiently extract and remove the model dye safranine O from water. After incubating 0.6 mL of aqueous safranine O solution (0.1 mg/mL) with 0.2 mL of GO/DNA hydrogel for 24 h, nearly 100% model dye was absorbed by the hydrogel, as confirmed by absorption spectra. The loading capacity for safranine O was estimated to be as high as 960 mg/g GO. The high loading capacity is believed to result from the synergistic effect of electrostatic and π-π interactions between dye and GO as well as DNA. The large specific surface area of the 3D porous structure of hydrogel would also give a positive contribution to the high loading capacity. The hydrophobic porous rGO films are super wetting for organic solvents, exhibiting great potential for use as selective super absorbants. The absorption capacities of an rGO foam with a weight density of 0.03 g/cm^3 for a selection of motor oils and organic solvents in terms of its weight gain have been investigated. It is clear that its uptake capacity is up to 37 times its weight for motor oils and up to 26 times for organic solvents, which is much larger than those of compact rGO films and graphite. Moreover, the porous rGO films are very stable and can be regenerated by removing the adsorbed [86]. Heavy metal ions that cause water pollution are currently attracting much attention owing to their toxic effects on the health of humans and other organisms in the environment. GO is a suitable adsorbent to remove a wide range of heavy metal ions. 3D GO foam/Fe_3O_4 nanocomposites are used for the removal of Cr (IV) [87].

Other Applications: Concrete for a large-scale structure such as a bridge or building might be made with this porous geometry, with the new 3D material providing the compatible strength needed with only a fraction of the weight. The new material might also be used in some filtration systems for water or chemical processing, due to its shape featuring many tiny pore spaces. 3D printed graphene objects are highly coveted in certain industries, including aerospace, separation, batteries, heat management, sensors, and catalysis [88–90].

12.6 Carbyne

Scientists in Vienna have successfully created a stable form of carbyne, the world's strongest material. Carbyne is a linear acetylenic carbon (LAC)—an infinitely long carbon chain. It can be considered as a one-dimensional allotrope of carbon. Carbyne has a chemical structure with alternating single and triple bonds: $(-C\equiv C-)_n$. [91]. It is one of the allotropes of carbon with purely *sp* hybridization. Its structure makes it highly reactive, which means that as quickly as it is manufactured, it is destroyed. However, carbyne is an extremely unstable form of carbon, so its existence could not be confirmed until it was isolated recently. Previously, a record of a chain of 100 atoms existed (which due to the instability of the allotrope, was soon destroyed). Earlier this year, this record was beaten by producing a chain of 6400 carbon atoms, and the chain continues to be stable. It was proposed theoretically that carbyne may be stable at high temperatures. It would thus be the ultimate member of the polyyne family. This polymeric carbyne is of considerable interest to nanotechnology as its Young's modulus is 32.7 TPa, 40 times that of diamond and 30 times that of carbon nanotube. It has also been identified in interstellar space; however, its existence in condensed phases has been contested recently, as such chains would crosslink exothermically (and perhaps explosively) if they approached each other. It has been theoretically predicted that carbyne may be stable at high temperatures (3000 K). Indications of naturally formed carbyne were observed in environments such as shock-compressed graphite, interstellar dust, and meteorites. The carbyne ring structure is the ground state for small (up to about 20 atoms) carbon clusters. Experimentally, many different methods of fabrication of finite-length carbon chains have been demonstrated, including gas-phase deposition, epitaxial growth, electrochemical synthesis, or "pulling" the atomic chains from graphene or CNTs. Carbyne's electrical properties increase with chain length, which could make it very useful in nanoscale electronics. Further investigations into carbyne have found that it is stable under tension (otherwise it would already be falling apart), and its bandgap (energy difference between the top of the valence band and the bottom of the conduction band of the atoms) is very sensitive to twisting, so it could be useful as a sensor for torsion or magnetic fields (if attached to something that will make it twist). The fact that it is so incredibly strong and lightweight makes the revolution in materials application.

12.6.1 History of Development

The first claims of detection of this allotrope were made in 1960 and repeated in 1978. A 1982 reexamination of samples from several previous reports determined that the signals originally attributed to carbyne were in fact due to silicate impurities in the samples. Carbyne was first proposed in 1885 by Adolf von Baeyer who described the existence of linear acetylenic carbon. Astronomers believe that they have detected carbyne in meteorites and interstellar dust. The absence of carbyne crystalline rendered the direct observation of a pure carbyne-assembled solid still a major challenge because carbyne crystals with well-defined structures and sufficient sizes are not available to date. This is indeed the major obstacle to the general acceptance of carbyne as a true carbon allotrope. In 1991, carbyne was allegedly detected among various other allotropes of carbon in samples of amorphous carbon black vaporized and quenched by shock waves produced by shaped explosive charges. In 1995, the preparation of carbyne chains with over 300 carbons was reported [92]. They were claimed to be reasonably stable, even against moisture and oxygen, as long as the terminal alkynes on the chain are capped with inert groups (such as tert-butyl or trifluoromethyl) rather than hydrogen atoms. The study claimed that the data specifically indicated carbyne-like structures rather than fullerene-like ones. The 1995 report claimed detection of carbyne chains of indeterminate length in a layer of carbonized material, about 180 nm thick, resulting from the reaction of solid polytetrafluoroethylene (PTFE, Teflon) immersed in alkali metal amalgam at ambient temperature (with no hydrogen-bearing species present) [93]. In 2004, an analysis of a synthesized linear carbon allotrope found it to have cumulene electronic structure-sequential double bonds along asp-hybridized carbon chain rather than the alternating triple-single pattern of linear carbyne. In 2016, the synthesis of linear chains of up to 6,000 sp-hybridized carbon atoms was reported. The chains were grown inside double-walled carbon nanotubes and are highly stable protected by their hosts. While the existence of "carbyne" chains in pure neutral carbon material is still disputed, short $(-C{\equiv}C-)_n$ chains are well established as substructures of larger molecules (polyynes). As of 2010, the longest such chain in a stable molecule had 22 acetylenic units (44 atoms), stabilized by rather bulky end groups. In 2016, it is reported that Austrian researchers have found a way to make it avoiding such destruction during the fabrication of carbine. They took two sheets of graphene, laid them on top of each other, and then rolled the whole thing up to create a double-walled tube. Then, they synthesized the Carbyne inside the tube providing a protective casing allowed the material to remain intact and can put 6400 atoms together to remain in a chain for as long as they want [94]. It remains to be seen how useful Carbyne will be while wrapped up, but for now, it is the best that researchers can achieve.

12.6.2 Synthesis of Carbyne

12.6.2.1 Polycondensation of Carbon Suboxide with Bis(Bromomagnesium) Acetylide

The cumulene modification of carbyne was synthesized for the first time in 1968. For this purpose, an original two-step synthetic method was developed. In the first step, the polycondensation of carbon suboxide (C_3O_2) with bis (bromomagnesium)acetylide by means of a Grignard reaction was carried out to form a polymeric acetylene–allene glycol.

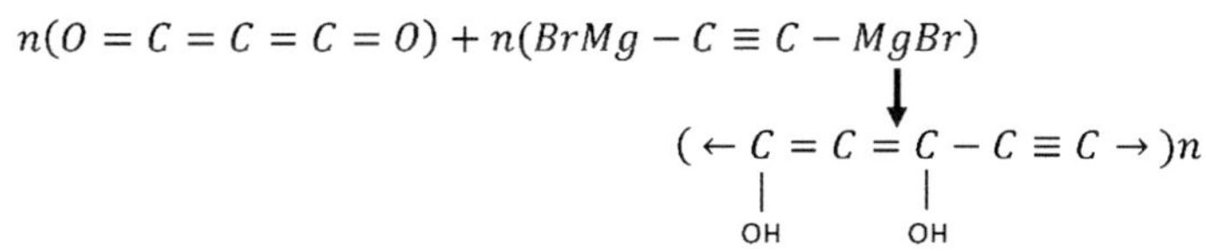

In the second step, polymeric glycol obtained was then reduced by the action of $SnCl_2$ in an acidic medium

$$\text{equation } (12.7) \xrightarrow{SnCl_2} (= c = c =)_n \qquad (12.8)$$

The high-molecular-(equation 12.8) weight cumulene is an insoluble dark brown powder with a highly developed specific surface (200–300 m^2/cm^3). When the polycumulene obtained is annealed for many hours at 1000 °C and reduced pressure, it partially crystallizes. A study of the annealed product using transmission electron microscopy revealed two types of single crystals corresponding to the α and β modification of carbyne [95].

12.6.2.2 Dehydrohalogenation of Polymers

One of the most convenient and accessible methods for synthesizing carbyne is the chemical dehydrohalogenation of some halogen-containing polymer (HCP). The main difference between this method and the polycondensation processes is that the carbon backbone is formed beforehand by polymerizing monomers, and when the carbyne is synthesized, the task is confined to the complete elimination of the hydrogen halide with retention of the linear structure. For the formation of carbyne through dehydrohalogenation, an equal number of the halogen and hydrogen atoms must be attached to neighboring carbon atoms in the original polymer chain. This allows complete dehydrohalogenation, that is, chemical carbonization of the original polymer. This condition is satisfied in poly(vinylidenehalides) (PVDH) and poly (1,2-dibromoethylene) as well as in poly(1,1,2- and 1,2,3-trichlorobutadiene). During the studying of the dehydrohalogenation of HCP under the action of alcoholic solutions of bases, it was noted that the addition of some polar solvents accelerates the reaction significantly. A study of the influence of numerous solvents on the dehydrohalogenation reaction resulted in the development of an efficient dehydrohalogenating system, which consists of a mixture of a saturated solution of KOH in ethanol with THF in a 4:6 ratio. The use of THF allowed the performance of the dehydrohalogenation of HCP under mild conditions, that is, at room and lower temperatures, due to its specific solvating nature and complex-forming ability, and thereby made it possible to avoid secondary cross-linking reactions and the formation of graphite-like defect structures. A number of PVDH, namely, poly(vinylidene bromide), poly(vinylidene chloride), and poly(vinylidenefluoride) (PVDF) have been used as the starting polymers for the synthesis of carbyne. Among the PVDH, the most promising polymer from the viewpoint of obtaining carbyne is PVDF owing to its better solubility. However, the dehydrofluorination of PVDF proceeds slowly because of the great strength of the halogen–carbon bond [96].

12.6.2.3 Dehydrogenation of Polyacetylene

A recent study of the interaction of polyacetylene, (CH), with metallic potassium under high-pressure conditions revealed that the compression of polyacetylene in the presence of potassium up to 4 GPa results in complete dehydrogenation of polyacetylene to form potassium hydride and a carbon matrix intercalated with potassium. The intercalation compound thus obtained reacts vigorously with water and mineral acids. After the decomposition of this compound by treatment with hydrochloric or nitric acids, brown plate-like crystals of hexagonal habit (measuring ~1 mm and up to 1 g) were isolated from the reaction products. The process as a whole may be described by the following scheme

$$(CH)_n + 1.7\,nK \xrightarrow[800\,^{\circ}C]{4GPa} (CK_{0.7})_n + nKH \quad \xrightarrow{HCl} \quad (=C=)_n$$

The selection of diffraction pattern from a crystal and the distribution of the intensities of the reflections correspond to the hexagonal crystal lattice of carbyne with the unit cell parameters a = 8.86 Å and c = 16.000 Å. The results obtained indicate that carbyne is capable of forming intercalation compounds with alkali metals.

12.6.2.4 Synthesis of Carbyne in Plasma

During the investigation of thermal decomposition of hydrocarbons (acetylene, propane, heptane, and benzene), as well as CC1441 and acetone, in the stream of the nitrogen plasma of a UHF discharge with subsequent quenching of the reaction products by diluting the plasma jet with an inert gas (N_2 or Ar), dispersed carbon powders were obtained containing 97–99% carbon and minor (<1%) impurities of N_2 and O_2 [97]. The concentration of other impurity elements such as Si, B, Cu, Fe, Mo, Pb did not exceed 0.003%. Plasma chemical carbon had a predominantly amorphous structure and contained separate single and polycrystals of carbyne. To isolate the crystalline phase, the samples were treated with a boiling mixture of acids (H_2SO_4 and HNO_3). Owing to the selective oxidation of aromatic carbon, the polycrystals of white and brown color, as well as single crystals of white color, were obtained. The single crystals generally had a lamellar habit and hexagonal shape. The value of the hexagonal unit cell parameter was found to be 8.94 Å, which is close to that of carbyne obtained by the oxidative dehydropolycondensation of acetylene. A study of the effect of the plasma-chemical synthesis conditions (plasma parameters, temperature, and concentration of the gas mixture) allowed optimization of this process. It was established that a moderate plasma temperature and a low concentration of reagents, which was achieved by diluting them with an inert gas (Ar or He) during injection into the plasma, facilitate the formation of carbyne, and the process of carbyne formation does not depend on the nature of the original organic compounds and the type of bonds in them [98].

12.6.2.5 Laser-induced Sublimation of Carbon

In 1971, carbyne was obtained for the first time by condensing carbon vapor formed through the laser vaporization of pyrographite. During the irradiation of a pyrographite plate by a laser beam (Nd-laser with a pulse energy of 500 J), intensive vaporization and melting of the carbon were observed with the formation of crater at the point of beam incidence. The carbon vapor rushed as a diverging flow out of the crater formed and condensed as a silvery–white layer on a flat substrate. The density of the white carbon measured by weighing the particles in heavy liquids was found to be 2.48 g/cm^3. The material obtained consisted of amorphous and crystalline particles of carbyne. An analysis of the point electron microdiffraction patterns from several single crystals selected after mechanical disintegration of the carbyne samples revealed that different sets of interplanar spacings correspond to these crystals. The following values of the unit cell parameter a were determined for different single-crystal particles: 8.24 Å, 8.80 Å, 8.94 Å (which is close to the value for carbyne obtained by the oxidative dehydropolycondensation of acetylene and by the decomposition of hydrocarbons in a plasma), 9.02, and 9.11 Å. It was concluded based on these results that there are numerous polymorphous modifications (polytypes) of carbyne with different unit cell parameters [99].

12.6.2.6 Deposition of Carbyne from an Electric Arc

The vaporization of carbon in an electric arc with subsequent condensation of the vapor is widely employed for obtaining carbon films of various structures. An electron microscopic study of carbon deposited on a crystalline substrate by vaporizing spectroscopically pure carbon blacks in an arc revealed that it consists predominantly of carbyne and graphite. When polymerization and subsequent crystallization of the carbon vapor on the cold substrate are sufficiently slow (heterogeneous), the carbyne forms of carbon predominate in the condensation product formed. The deposit is an aggregate of structural fragments with a plate-like (lamellar) morphology. The electron diffraction pattern from the single-crystal particles corresponds to a hexagonal lattice of carbyne with a unit cell parameter a = 8.94 Å. During the condensation of carbon from an arc discharge produced between two graphite electrodes, the carbon films were prepared. After annealing apart from amorphous regions corresponding to graphite clusters, the crystalline fragments of carbyne were also observed with a lattice of hexagonal type and a unit cell parameter a = 8.70 Å.

12.6.2.7 Ion-assisted Condensation of Carbyne

The method of ion-assisted condensation occupies a special position among the methods for obtaining carbyne and follows three steps: (1) carbyne films with a controllable degree of ordering from amorphous to single-crystal layers; (2) carbynes of a required modification, that is, with controlled crystal lattice parameters of carbyne; and (3) films of other, including metastable forms, which are obtained by varying the ion-irradiation parameters [100]. Prospects for creating the layered structures, such as superlattices, based on various carbon modifications in a single technological process are thereby opened up. This method was developed for the preparation of carbyne at the M. V. Lomonosov Moscow State University based on theoretical and experimental investigations in the field of the ion stimulation of processes on the surface of a solid. It was shown that during the vacuum condensation of thin carbon films under the conditions of either simultaneous (with condensation) or alternate irradiation by slow ions of inert gases, the stimulation of phase transitions in the films can be observed with the formation of metastable phases. The yield of this process is highest under conditions conducive to the resonance charge exchange of ions on the amorphous condensate. The resonance charge exchange of the levels of an ion (ionized) and a surface (filled with electrons) is observed in selective ranges of ion energies. During ion-assisted condensation in a high vacuum (10^{-7} Torr), the streams of carbon and inert gas ions (Ar^{+}) impinge on the substrate either simultaneously or alternately. The carbon flow is formed by thermal vaporization or ion sputtering of a graphite sample. The energy of the ions impinging on the substrate varies within the range 0–200 eV, the ion current density (j) at the substrate is 10–1000 gA/cm^2, and the rate of film growth is 10–1000 A/min. The thickness of the films formed is 200–1000 A. The carbon films prepared by alternate condensation and ion irradiation are generally amorphous, but their structure and properties depend greatly on the irradiation conditions [101].

12.6.3 Properties

Carbyne (Linear acetylenic carbon) is just a single strand of molecules often referred to as a "one-dimensional" structure. Carbyne is one atomic thick and has relatively massive surface area. A single gram of graphene, for example, has a surface area of about five tennis courts. This could be very important in areas such as energy storage (batteries, supercapacitors), where the surface area of the electrode is directly proportional to the energy density of the device. In the aforementioned hydrogen storage sponge, too, the huge surface area of carbyne could result in a lot of hydrogen being mopped up relative to the size of the device. The effective surface of a carbyne network was estimated to exceed 13,000 m^2/g, four times larger than the theoretical value of graphene [102]. A further distinction for carbyne is based on the two possible forms: The term α-carbyne has been used to describe a bond length alternated structure characterize by alternating conjugated single and triple CC bonds; on the other hand, β-carbyne denotes an infinite chain characterized by all-equal double bonds. The former is also called "polyyne" in the literature while the latter "poly-cumulene" or just "cumulene" as sketched. The first case can be described as an infinite 1D crystal with a basis of two carbon atoms showing a bond length alternation. The second case can be described as a monoatomic sp-carbon chain, where all the CC bonds are equal having a double character (=C=C=), thus corresponding to bond length alternation (BLA) = 0. These two structures would correspond to an insulating/semiconducting system for BLA≠0 (polyyne) or truly metallic systems when BLA = 0 (polycumulene), revealing the usual connection between molecular structure (BLA) and electronic properties (bandgap) which is peculiar of polyconjugated molecules. Due to Peiers distortion and indirect connection to other polyconjugated molecules, the only stable form possible is α-carbyne for an infinite chain. It is shown that amorphous carbyne tends to crystallize under the influence of shock pressure. Thick films of carbyne transform into the mixture of an amorphous carbon phase and

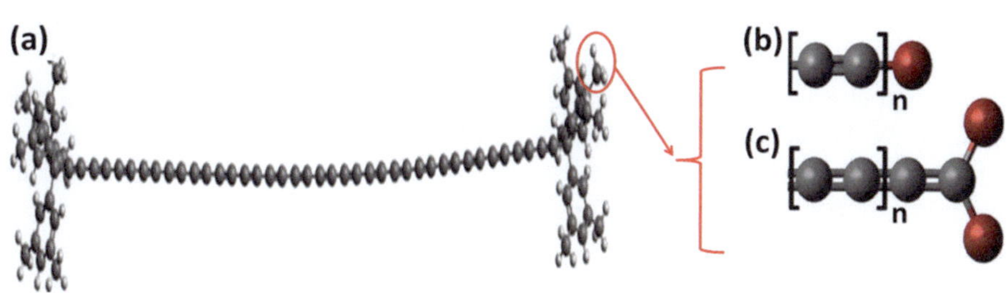

Fig. 12.23 (**a**) A typical carbyne structure, (**b**) reaction of the head with one element, and (**c**) reaction of the head with multiple elements [104]

nanocrystalline cubic diamond. Thin films of carbyne transform to the material that contains a significant amount of the crystalline carbyne phase [103]. Figure 12.23(a) shows the typical carbyne structure and the reaction of the carbine head with one element (Fig. 12.23b), and the reaction of head with multiple elements (Fig. 12.23c).

A new form of carbon, dubbed carbyne, is stronger and stiffer than any known material. In fact, carbyne is about two times stronger than graphene and carbon nanotubes, which until now were the strongest materials by some margin. Carbyne chains have been claimed to be the strongest material known per density. Calculations indicate that carbyne's specific tensile strength (strength divided by density) of 6.0–7.5 $\times$ 10^7 Nm/kg beats graphene (4.7–5.5 $\times$ 10^7 Nm/kg), carbon nanotubes (4.3–5.0 $\times$ 10^7 Nm/kg), and diamond (2.5–6.5 $\times$ 10^7 Nm/kg). Its specific modulus (Young's Modulus divided by density) of around 10^9 Nm/kg is also double that of graphene, which is around 4.5 $\times$ 10^8 Nm/kg. Carbyne has a very high value of the Young modulus (up to 32 TPa) and specific stiffness of about 10^9 Nm/kg, which is much higher than all the other materials, including CNT, graphene, and diamond. The strength of carbyne depends on the breaking strain and force, which can be estimated by phonon instability or the activation barrier of one bond breaking. The breaking strain of carbyne is around 18–19% with a breaking force 12 nN. The specific strength of carbyne is 6.0–7.5 $\times$ 10^7 Nm/kg, again significantly outperforming known materials including graphene, CNTs, and diamond [105].

Stretching carbyne 10% alters its electronic bandgap from 3.2 to 4.4 eV. Outfitted with molecular handles at the chain's ends, it can also be twisted to alter its bandgap. With a 90° end-to-end twist, carbyne turns into a magnetic semiconductor. In 2017, the band gaps of confined linear carbon chains (LCC) inside double-walled carbon nanotubes with lengths ranging from 36 up to 6000 carbon atoms were determined for the first time ranging from 2.253 to 1.848 eV, following a linear relation with Raman frequency [106]. This lower bound is the smallest bandgap of linear carbon chains observed so far. The comparison with experimental data obtained for short chains in the gas phase or in solution demonstrates the effect of the DW-CNT encapsulation, leading to an essential downshift of the bandgap. The LCCs inside double-walled carbon nanotubes lead to an increase of the photoluminescence (PL) signal of the inner tubes up to a factor of 6 for tubes with (8,3) chirality. This behavior can be attributed to a local charge transfer from the inner tubes to the carbon chains, counterbalancing quenching mechanisms induced by the outer tubes. Actually, the carbyne cannot be stretched, it can be bent into an arc or circle (Fig. 12.24), and by doing so, the additional strain between the carbon atoms alters the electrical bandgap. This property could lead to some interesting uses in micro-electro-mechanical systems.

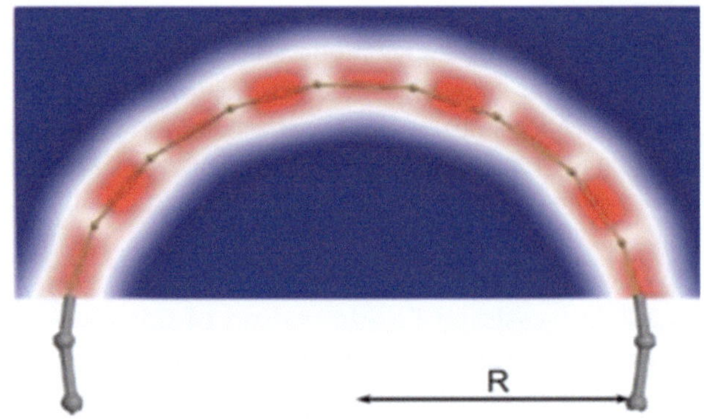

Fig. 12.24 Bending of carbine [185]

By adding different molecules to the end of a carbyne chain, such as a methylene (CH_2) group, carbyne can also be twisted much like a strand of DNA—again adding strain and modifying the electrical bandgap. By "decorating" carbyne chains with different molecules, other properties can be added. For example, by taking some calcium atoms on the end, which like to mop up spare hydrogen molecules, and suddenly it has a high-density, reversible hydrogen storage sponge [107].

Thermal conductivity has been recently estimated to reach extremely high values (200–80 kW/m/K for cumulenes and polyynes at room temperature) well above the best values of graphene (5 kW/m/K) and nanotubes (3.5 kW/m/K). Such super-high thermal conductivity is attributed to high phonon frequencies, and long phonon means free path allowing ballistic thermal transport up to the micron-scale. Carbyne, especially in the crystalline state, exhibits remarkable inertness toward various oxidizing agents. Only the reaction of ozone with carbyne results in its complete degradation. With respect to boiling in a mixture of HNO_3 with H_2SO_4 (I:1), as well as other chemical agents, carbyne displays a high chemical inertness, approaching that of a diamond. It reacts with chlorine only at temperatures above 800 °C with simultaneous degradation of the original polymer and the formation of polychlorides of undetermined structure. No hydrogenation of a suspension of carbyne in alcohol at 200 °C and 20 MPa is observed over Raney nickel or with the use of homogeneous hydrogenation catalysts. Carbyne exhibits a catalytic activity in dehydrogenation and dehydration reactions [108].

12.6.4 Applications of Carbyne

Carbyne has a long list of unusual and highly desirable properties that make it an interesting material for a wide range of applications, from nanoelectronic/spintronic devices to hydrogen storage to higher-density batteries. Carbyne chains can take on side molecules that may make the chains suitable for energy and hydrogen storage [109].

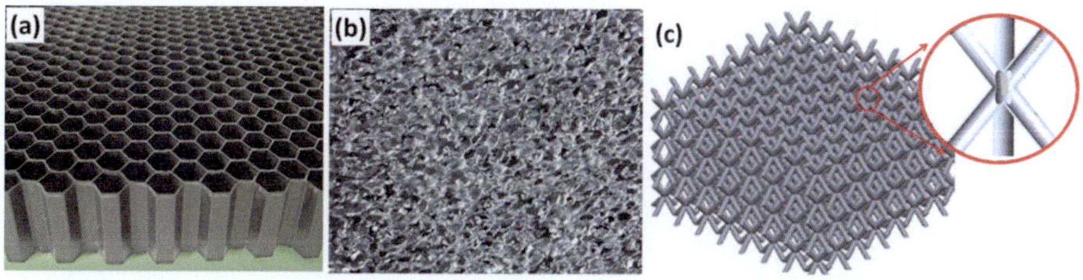

Fig. 12.25 Different physical appearance of cellular materials, (**a**) Honeycomb, (**b**) Metal foam, and (**c**) Microlattice structure

12.7 Microlattice Materials

The materials used in this microlattice structure are hollow, allowing for this structure to be ultralight. Figure 12.25 distinguishes microlattice when compared to other hollow and porous materials. The microlattice can exceed strains of above 50% and can restore its original dimensions. This remarkable structure does obtain some microscopic fractures that decrease its energy absorption during the few times of compression, but then it maintains relatively constant energy absorption. It is these fractures or cracks that occur near junction points of the cell that cause it to lose 2% of its height after it is compressed. After the third compression, stable fractures are formed, and the material has a relatively constant superelastic behavior.

Based on materials used in microlattice, these are of four types: metallic microlattice, polymer microlattice, ceramic microlattice, and composite microlattice.

12.7.1 Metallic Microlattice

A metallic microlattice is a synthetic porous metallic material having a density as low as 0.9 mg/cm^3 [110]. It is one of the lightest structural materials. It was developed by a team of scientists from California-based HRL Laboratories, in collaboration with researchers at the University of California, Irvine, and Caltech, and was first announced in November 2011. The prototype samples were made from a nickel–phosphorus alloy. In 2012, the microlattice prototype was declared 1 of 10 World-Changing Innovations by Popular Mechanics. Metallic microlattice technology has numerous potential applications in automotive and aeronautical engineering. A metallic microlattice is composed of a network of interconnecting hollow struts. In the least-dense microlattice sample reported, each strut is about 100 micrometers in diameter, with a 100 nm thick wall. The completed structure is about 99.99% air by volume, and by convention, the mass of air is excluded when the microlattice density is calculated [111].

To produce metallic microlattice, the HRL/UCI/Caltech team first prepared a polymer template using a technique based on self-propagating waveguide formation, although it was noted that other methods can be used to fabricate the template. The process passed UV light through a perforated mask into a reservoir of UV-curable resin. Fiber-optic-like "self-trapping" of the light occurred as the resin cured under each hole in the mask, forming a thin polymer fiber along the path of the light. By using multiple light beams, multiple fibers could then interconnect to form a lattice. The process was similar to photolithography in that it used a two-dimensional mask to define the starting template structure, but differed in the rate of formation: where stereolithography might take hours to make a full structure, the self-forming waveguide process allowed templates to be formed in 10–100 s. In this way, the process enables large free-standing 3D lattice materials to be formed quickly and scalably. The template was then coated with a thin layer of metal by electroless nickel plating, and the template is etched away, leaving a free-standing, periodic porous metallic structure. Nickel was used as the microlattice metal in the original report. Owing to the electrodeposition process, 7% of the material consisted of dissolved phosphorus atoms, and it contained no precipitates.

12.7.1.1 Manufacturing of Metallic Lattice Structure

There are several manufacturing processes of metallic microlattice structures that are discussed subsequently.

Investment Casting: Minimum relative density of approximately 0.2% can be achieved by this method. Investment casting is one of the conventional methods to create cellular structures by injection molding or rapid prototyping methods, where sacrificial truss patterns with attached face sheets are produced from a volatile wax or polymer such as polyurethane. In this process, a pattern is coated with ceramic casting slurry and dried with the help of a system of gating and risers. The wax or polymer is later removed by melting or vaporization and then the lattice material is produced by filling the empty mold with liquid metal. A range of cell topologies is possible with this method, such as pyramidal, tetrahedral and 3D kagome. Fabrication of complex, nonplanar shapes featuring trusses with high nodal connectivity is possible with this approach. However, it is difficult to fabricate structures with near-optimal, low relative density cores because the metal pathways in the molds become prohibitively small and complex and subsequently suffer

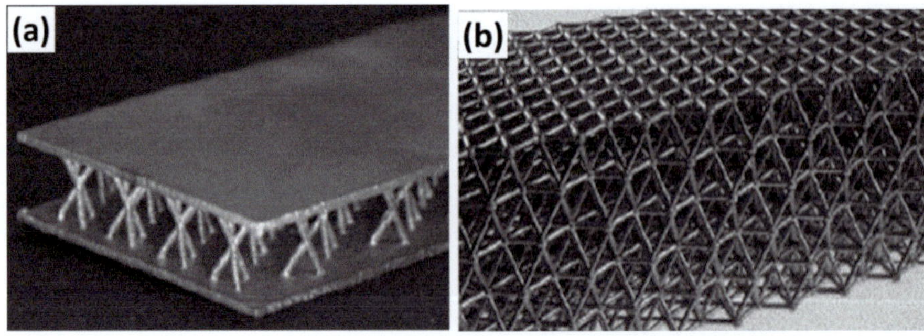

Fig. 12.26 Aluminum alloy investment casted microlattice (**a**) Sandwich structure, and (**b**) Octet-truss lattice structure

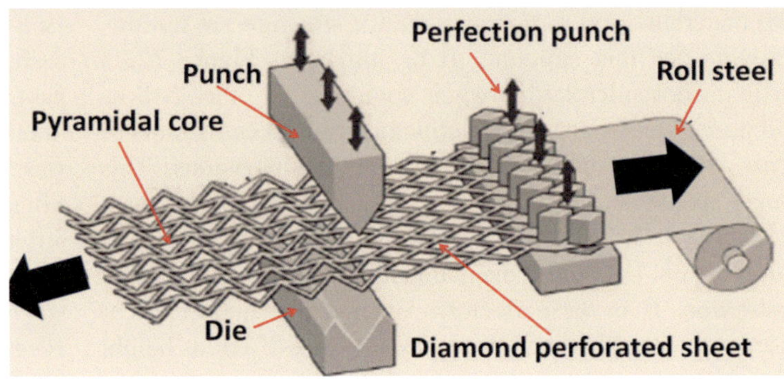

Fig. 12.27 Aluminum deformation forming to get the microlattice patterns

from increased susceptibility to casting defects. Alloys with high fluidity must be used which limits material choice. This method is expensive and time-consuming, and the produced structures contained significant porosity. Al/Si and Si/brass sandwich beam manufactured with tetrahedral cores and octet-truss lattice material from Al-alloy, both using investment casting with injection-molded polystyrene pre-forms. Wadley et al. and Wang et al. used rapid-prototyped Acrylonitrile Butadiene Styreneto to manufacture a sacrificial pattern for investment casting using Cu-Be alloy, shown in Fig. 12.26 [112].

Deformation Forming: Minimum 1.7% relative density was achieved by this process. Deformation forming is another method of producing periodic open-cell lattice structures by press forming operation. Using the forming and subsequent assembly process, cell sizes of millimeters to several centimeters can be obtained. It utilizes sheet perforation and shaping techniques. Perforated metal such as stainless steel sheets with hexagonal or diamond-shaped holes can be deformed at the nodes to produce sheets of tetrahedrons or pyramidal structure, as shown in Fig. 12.27. The processed material requires annealing treatment to soften the strain-hardened struts. Lattice structure manufactured using deformation forming showed greater ductility than the investment casting process. Relative densities between 1.7% and 8% can be achieved by varying the sheet thickness and the dimensions of the holes [113].

Another technique adopted in the deformation forming process involves shearing and expanding metal sheets. Low carbon steel sheet can be cut by laser and expanded widthwise to form a metal mesh. The metal mesh was later bent along the lines connecting the longer ends of the diamond shapes, forming a corrugated sheet. Then, the shorter struts were rotated by a 120° angle, and a quasi kagome truss was produced. Corrugated shapes or egg box topologies can be formed by a simple press forming operation on solid sheets, made from high formability alloys. Corrugated and prismatic structures can also be manufactured using the lotting technique. This method has been used to produce square honeycomb cores and diamond prismatic cores.

In woven metal textiles, the minimum relative density achieved in this method is 10%. The woven metal textile approach is a simple method of weaving, braiding, and sewing wire drawn from metal alloy to produce an open-cell woven structure. The wire orientation is possible to be arranged at any angle. Fig. 12.28a shows 0°/90° orientation and Fig. 12.28b shows 45° orientation where plain weave structure and pyramidal truss structure is shown at the top and bottom respectively. Multifunctional uses are limited, as the wires are not bonded together in normal practice. This process offers a host of options as virtually all metals can be used to produce wires and a variety of truss arrangements available. Relative densities of around 10% can be achieved with this method [114].

Nonwoven metal textile approach produces textiles by layering wires and tubes made of metal such as stainless steel and subsequently joined together by brazing. Square and diamond cell structures with relative densities between 3% and 23% can be produced by this method. The structures can be processed further by bending the layers to form

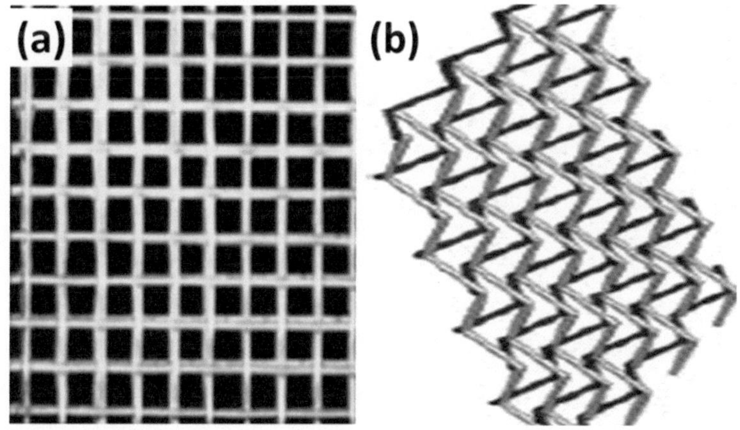

Fig. 12.28 Woven metal textile (**a**) 0°/90° orientation and (**b**) Pyramidal truss

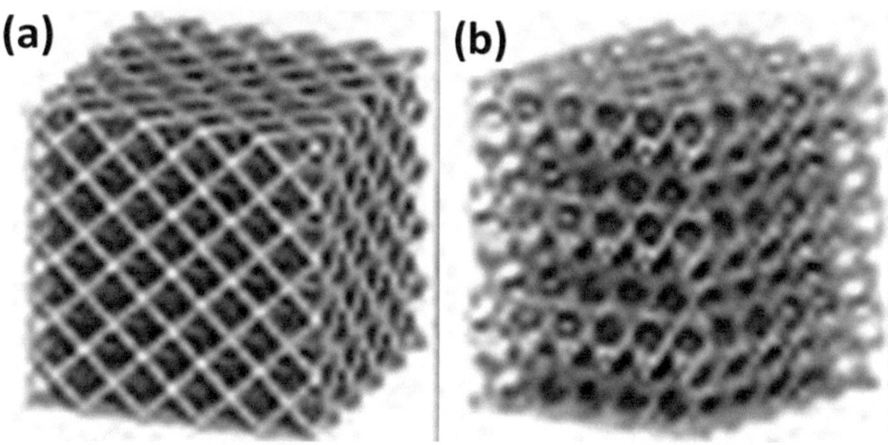

Fig. 12.29 Nonwoven metal textiles, solid and hollow micro truss (**a**) Octate truss, and (**b**) 3D Kagome

pyramidal structures. Examples of nonwoven metal textiles are shown in Fig. 12.29 [115].

Additive Manufacturing: Selective Laser Melting (SLM) belongs to the group of additive manufacturing techniques. The principle of SLM process is based on the fact that the metal powder is applied in very thin layers on a building platform, which is later completely melted using thermal energy induced by a laser beam. The cross-section area of a part is built by melting and re-solidifying metal powder in each layer, and then a new layer of powder is deposited and leveled by a wiper after the building platform is lowered. The laser beam can be redirected and focused across the powder bed following a computer-generated pattern by scanner optics in such a way that the powder particles are possible to selectively melt where desired. Schematic of SLM process is shown in Chap. 20, Fig. 20.13. Various types of metal powders can be used in SLM process including stainless steel, Cu, Ni, Cr, Ti, and super-alloys. Although freeform fabrication processes are capable of building any arbitrary shape, the SLM process has some limitations. It is difficult to produce overhanging geometries because of poor heat conduction in the powder bed below the newly laid exposed powders. The build angle of the truss has a significant effect on the mechanical properties. The most acute build angle possible is approximately 25° to the horizontal.

There is an advanced process similar to SLM but instead of using laser, the electron beam is used as the energy source in this method to melt layers of metal powders in vacuum, and the process is called electron beam melting (EBM). First, a tungsten filament is heated to generate the electron beam, and the electrons are accelerated to the build table onto the metal powder using an accelerating voltage of 60 kV. Electromagnetic coils are used to focus and deflect the electron beam for controlling purposes. Similar to SLM, the EBM also manufactures 3D objects by following layer-by-layer build until the structure is completed. But unlike SLM, the base metal plate and the powder bed need to be preheated prior to electron scanning in the EBM [116].

Self-propagating Photopolymer Waveguide Technique: Metallic microlattices have been realized based on thiolene polymer templates. This technique was used to produce microlattice with a hollow tube, allowing for this structure to be ultralight, 0.9 mg/cm^3. Fig. 12.30 shows the steps involved in this technique. A template using a polymer had been created with the required repeating cell structure. The

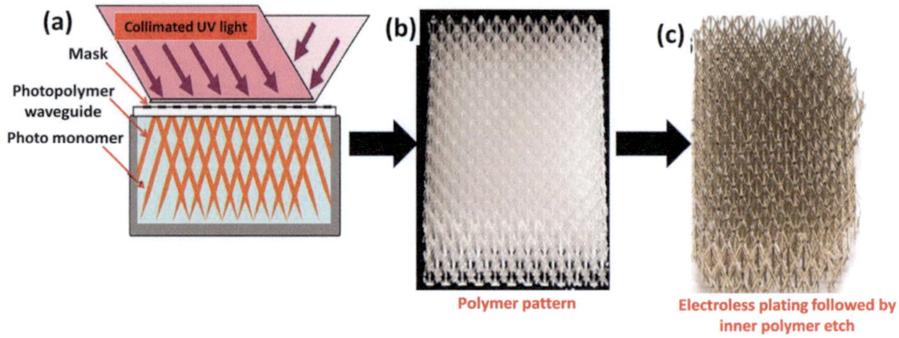

Fig. 12.30 Self-Propagating photopolymer waveguide technique for preparing the microlattices, (**a**) Polymer patterning by UV ray exposure, (**b**) Polymer microlattice templates, (**c**) Electroless plating and by etch removal of the template

polymer starts as a liquid that hardens when ultraviolet light shines on it. A "patterned mask" that is similar to a stencil was used to shine the ultraviolet light through the open areas of the mask onto the liquid polymer, which results in hardening of the areas of liquid polymer that are exposed to the UV light. A 3D array of repeating cells were created using this technique. Later, the polymer template was coated with metal, such as nickel phosphorous, and finally, the polymer template is removed by etching it out. Microlattices with relative densities from 0.01% to 8.4% produced by this approach [117].

12.7.1.2 Properties

The mechanical properties of metallic microlattices depend on various factors, such as the mechanical properties of parent material, size and shape of the cell, periodicity and connectivity between cell walls or struts, type of strut, that is, solid or hollow, type of porosity, relative density of the materials, and so on. The ratio of the lattice density to the density of the parent material is defined as the relative density of the lattice structure (Eq. 12.10). The properties of lattice structures are strongly dependent on the manufacturing method used. The material behavior in microlattice structure can be bending-dominated or stretch-dominated. Bending-dominated behavior is found in open-cell or stochastic materials, while stretch-dominated behavior is common in closed-cell or sometimes open-cell periodic materials. The modulus and initial yield strength of stretching-dominated structures are much greater than those of bending-dominated structures of the same relative density due to their different collapse modes and, hence, are more weight-efficient for structural applications [118]. Both types of structure often experience an initial settling period occurring due to broken cell edges from post-processing, followed by a linear elastic region represented by the solid black lines. On the other hand, the Pyramid cell type is a desirable core material for sandwich construction due to its larger shearing strength than the uniaxial strength. It was suggested that the optimum design of lattice materials depends on two factors, the relative density (ρ*) and the number and stacking order of struts. The relative density (Eq. 12.10) must be greater than a certain critical value for that lattice structure; otherwise, the lattice structure will collapse early [119].

$$\text{Relative density } (\rho^*) = \frac{\text{Density of lattice material}}{\text{Density of parent materials}} \tag{12.10}$$

12.7.1.3 Applications of Metallic Microlattice

Metallic microlattice may find potential applications as thermal and vibration insulators such as shock absorbers and may also prove useful as battery electrodes and catalyst supports. Additionally, the ability of microlattice to return to their original state after being compressed may make them suitable for use in spring-like energy storage devices. Automotive and aeronautical manufacturers are using microlattice technology to develop extremely lightweight and efficient structures that combine multiple functions, such as structural reinforcement and heat transfer, into single components for high-performance vehicles [120]. Metallic microlattice conducts electricity and can handle temperatures exceeding 400 °C and could be applied in aerospace structures, such as satellites, space telescopes, and airplanes. Aerospace industries have a strong interest in lightweight structural concepts that can absorb acoustic, shock, and vibration energy. Boeing 360 helicopter was partly manufactured using sandwich materials which resulted in weight-saving, number of parts, tooling costs, and manufacturing time reduction. Microlattices can be used in a sandwich construction of future aircraft fuselages and wing structures, offering higher performance per unit cost, as microlattice materials are an excellent candidate to use as core material of sandwich panel construction, resulting in more weight efficiency. In general, stretching and compression without bending sandwich core topologies are the preferred types. Miller et al. proposed a new protection system for flight recorders where microlattice material layer protects the memory device against crash [121]. Microlattices have excellent energy absorption capacity, and this particular property is of special interest to automotive industry as it is mandatory to use energy-absorbing materials for protecting passengers from impact when designing a car or motor vehicle. It is important to keep the peak force transmitted through the structure below the limits that a human can withstand. The energy-absorbing behavior of microlattices can be influenced within a certain range by varying the cell topology, alloy, and relative density.

12.7.2 Polymer Microlattice

HRL Laboratories (Malibu, California) invented a unique fabrication method to form polymer microlattice materials. This was made possible due to a breakthrough in fabrication technique based on an interconnected pattern of self-propagating polymer waveguides. This process is capable of rapidly fabricating ordered open cellular materials, with feature sizes comparable to stochastic microscale foams. It was formed in a photosensitive monomer from a single point exposure of light. The compressive behavior of these micro-truss materials has been correlated with variations in density, cell size, truss angle, and the properties of the solid polymer. The shear strength of the micro-truss materials also depends on the tensile and compressive strength of the polymer as well as on the failure mode of the structure. For instance, although the general 3D Kagome material has been proved to be mechanically efficient, fabrication techniques for these or similar materials have been limited until recently. Based on static compression experimental results, it was found that the compressive strength of the microlattice materials with six-fold symmetry was greater than the materials produced with threefold symmetry. This was attributed to the secondary nodes, in the six-intersecting waveguide materials, reducing the slenderness ratio of the truss members, without increasing the relative density. More recently, the researchers from HRL Laboratories changed the parent materials of microlattice materials, and carbon microlattice materials were fabricated by pyrolyzing a polymer precursor template formed from an interconnected 3D array of self-propagating photopolymer waveguides. In this, the polymer microlattice precursor is impregnated with polyarcylonitrile and this is heated to 1000 °C for 1 h [122]. The microlattice materials made by carbon can be optimized for multifunctional applications, such as internal fluid transport, while carrying mechanical loads. Such functionality is highly desirable for heat transfer media or electrodes in electromechanical devices. Flexible microlattices were produced using an interconnected pattern of self-propagating polymer waveguides developed at HRL Laboratories.

12.7.2.1 Applications of Polymer Microlattice

Microlattices have been proposed for a number of applications, ranging from electrodes for batteries, photonic crystals for strain sensing, and for use as ultrasonic filters exploiting local resonances. Microlattices are also promising as energy and sound-absorbing materials, for instance, in helmets or other protective equipments. Foam materials are traditionally used in these applications, as they can be produced relatively easily and their properties can be adjusted over a wide range. The damping of structural vibrations is important in many engineering systems, including machinery, automotive, aircraft, and satellites. As structural materials typically show a very low amount of intrinsic damping, they are often combined with viscoelastic materials, such as foams, to deliver a sufficient amount of vibration mitigation [123].

12.7.3 Ceramic MicroLattice

Ceramic materials allow reducing the scale of lattice structures to nanoscale. Bauer et al. manufactured ceramic lattice structures using 3D laser lithography, in which polymeric structures were printed and then coated with Al_2O_3. They achieve a variety of microlattice structures, with strut size of the order of 2 mm and a cell size of the order of 10 mm. Periodic nanolattice ceramics from hollow titanium nitride (TiN) nanolattices were successfully fabricated using a multistep negative pattern process, including digital design by additive manufacturing (direct laser printing), deposition of TiN, and etching out of polymer. The octahedral nanolattice was designed using a series of tessellated regular octahedra connected at their vertices. The resulting nanolattice material was approximately 100 mm in each direction and it is one of the smallest periodic lattice materials. These materials do not fracture under the applied load, even after multiple loading cycles, and their sustained tensile stresses are 1.75 GPa, which is one of the highest amongst other cellular materials [124].

Ultralight hollow ceramic nanolattices absorb energy, recover after significant compression, and reach an untapped strength and stiffness material property space. This is achieved using high-strength atomic layer deposition Al_2O_3 engineered into a thin-walled nanolattice that is capable of deforming elastically via shell buckling. The ultralight ceramic nanolattices represent the concept of materials by design, where it is possible to transform a strong and dense brittle ceramic into a strong, ultralight, energy-absorbing, and recoverable metamaterial. These results serve to emphasize the critical connection between material microstructure, hierarchical architecture, and mechanical properties at relevant length scales [125].

Some of the ceramic synthesized materials are discussed below:

TiO_2 Sponge: The precursor solution was prepared through the solution blow-spinning technique. Tetrabutyl titanate [$Ti(OBu)_4$] and polyvinylpyrrolidone [PVP

(M_w = 1,300,000)] with a 2:1 mass ratio were mixed with ethanol and acetic acid at a mass ratio of 3:1. The solution was magnetically stirred for ~6 h in a capped bottle at room temperature. The mixture was loaded into a 1-ml syringe with a specific coaxial needle. The precursor solution with 7 wt % of PVP was injected from the axle with a speed of 3 ml/h toward a porous and air-permeable cage collector placed at a distance of 20 cm, under gas pressure of 69 kPa (airflow velocity is about 21 m/s at the exit) from the outer shaft. After spinning for 20 min at room temperature, a sponge of Ti $(OBu)_4$/PVP was obtained in the cage. The $Ti(OBu)_4$/PVP sponge was immediately treated at a heating rate of 2 °C/min and then held at 450 °C in the air for 200 min before being cooled down in the furnace. After sufficient heating time, the grain size of the ceramic nanofiber only showed slight changes at the experimental temperatures, which were lower than the heating temperature during fabrication. This avoids the influence of grain growth on the deformation of sponges at high temperatures [126].

ZrO_2 Sponge: Zirconium oxychloride ($ZrOCl_2 \cdot 8H_2O$) and PVP with a 2:1 mass ratio were mixed with ethanol and deionized water at a mass ratio of 1:1. The solution was then magnetically stirred for ~6 h in a capped bottle at room temperature. The spinning process was the same as that for the TiO_2 sponge. The sponge was then treated at a heating rate of 2 °C/min with a holding time of 200 min at a temperature of 800 °C in air and then cooled in the furnace. The precursor solution of YSZ was fabricated by zirconium *n*-propoxide, yttrium nitrate, PVP, diacetone, and ethanol [126].

$BaTiO_3$ Sponge: Barium acetate [$Ba(Ac)_2$] was dissolved in acetic acid, and $Ti(OBu)_4$ was then added drop by drop with continuous stirring. After dissolution, 7 wt% of PVP and ethanol was added into the solution (the molar ratio of Ba $(Ac)_2$ and $Ti(OBu)_4$ is 1:1; the mass ratio of PVP and Ti $(OBu)_4$ is 1:2). The solution was stirred for ~6 h at room temperature. The spinning process was the same as that for the TiO_2 sponge. The obtained sponge was then treated at a heating rate of 2 °C/min and a holding time of 1 h at 750 °C in air and then cooled in the furnace [127].

12.7.4 Composite Microlattice

Fiber-reinforced composite cellular materials with microlattice truss topologies have been recently shown to fulfill the apparent gap between existing materials and the limit of the unattainable material in the low-density region. These periodic spatial bar lattice structures often find applications in reinforcing components for extremely lightweight construction on one hand and for explosion protection and multifunctional applications on the other. These lattice truss materials are produced by adapting relevant technologies from the hot press molding, interlocking, and textile techniques. Xiong et al. introduced a novel and practical methodology to fabricate composite pyramidal truss cores with unidirectional carbon/epoxy prepreg by the molding hot-press method [128]. In that method, all the continuous fibers of the composite were aligned in the direction of the fabricated truss. These results show that the fabricated low-density truss cores have superior compressive strength and thus could be used in the development of novel lightweight multifunctional structures.

12.8 Foams

Foams are lightweight cellular materials inspired by nature. Wood, bones, and sea sponges are some well-known examples of these types of structures. Currently, inorganic foam plays a dominant role in various functional and structural applications. Foams are special cases of porous metals. Solid foam originates from liquid foam in which gas bubbles are finely dispersed in a liquid. Generally, the ultralight foams are categorized into two parts that is metallic- and polymeric-foam. Again, this foam is generally found in two forms, that is, open pore foam and close pore foam [129].

12.8.1 Metallic Foams

Metal foams with high levels of controlled porosity are an emerging class of ultra-lightweight materials that are receiving increased attention for both commercial and military applications. Aluminum metal foam materials, which can be fabricated into a variety of functional geometries, offer significant performance advantages for weight-sensitive applications. Metal foams exhibit high stiffness-to-weight and strength-to-weight ratios and thus offer potential weight savings. They also have the ability to absorb high amounts of energy during compressive deformation for efficient crash energy management [130].

12.8.1.1 Classification of Metallic Foam

Among the close and open metallic foam, the open variety is ideal for vibration and sound absorption, filtration and catalysis at high temperatures, and in medical devices. Open celled metal foams, also called metal sponges, have a wide variety of applications including heat exchangers (compact electronics cooling, cryogen tanks, PCM heat exchangers), energy absorption, flow diffusion, and lightweight optics. Extremely fine-scale open-cell foams, consisting of cells smaller than visible to the naked eye, are used as high-temperature filters in the chemical industry. Metal sponges have a very large surface area for their weight, so catalysts are often metal sponges, such as palladium black, platinum

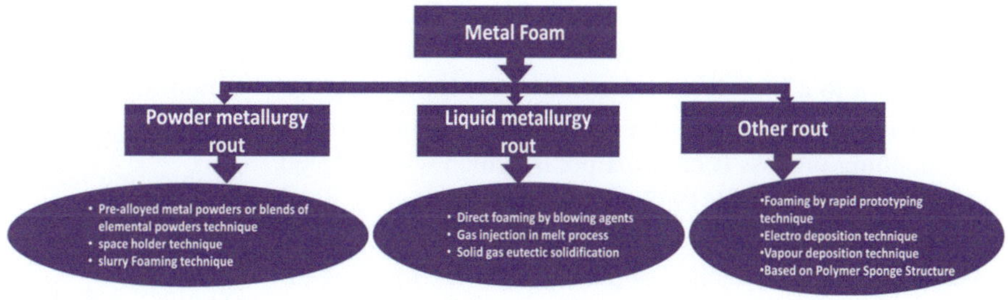

Fig. 12.31 Schematic grouping of the commercially most relevant production methods of metal foams and sponges

sponge, spongy nickel, and so on. Sometimes, a metal such as osmium or palladium hydride is metaphorically called a "metal sponge" because it soaks up hydrogen "like a sponge soaks up water". The closed variety is used for structural applications requiring load-bearing features and for weight-saving and impact-absorbing structures in vehicles. Closed-cell metal foams are primarily used as an impact-absorbing material, similar to the polymer foams in a bicycle helmet but for higher impact loads. Unlike many metals, foams remain deformed after impact and can therefore only be used once. Closed-cell foams retain the fire-resistant and recycling capability of other metallic foams but add the ability to float in water.

12.8.1.2 Synthesis of Metallic Foam

Syntheses of metallic foam are discussed below with respect to the classification given in Fig. 12.31.

12.8.1.2.1 Powder Metallurgy (P/M) Rout

The overall Fraunhofer powder metallurgy process is based on blending metal powders with a foaming agent, compacting the powder mixture to high density and then heating the compact to near the melting point of the metal. In this process either pre-alloyed metal powders or blends of elemental powders mix with a small amount of foaming agent (if metal hydrides are used, content of less than 1% is sufficient in most cases) [131]. After the foaming agent is uniformly distributed within the matrix powders, the mixture is compacted to yield a dense, semi-finished product without any residual open porosity. Typical examples of such compaction methods include uniaxial pressing, extrusion, and powder rolling. Further shaping of the foamable material can be achieved through subsequent metal working processes such as rolling, swaging, or extrusion. Examples of the semi-finished products include billets, plates, and rods. Following the powder consolidation and metalworking steps, the foamable precursor material is heated to the process temperature. At this stage, the foaming agent decomposes, forming a gas that is trapped inside the compacted powder body. Gas bubble voids form within the expanding body of semi-solid metal and are retained during solidification. This process results in a lightweight structure with a high degree of closed-cell porosity. Metal foam parts can be produced to net shape in several different configurations. The density of the metal foams can be controlled by adjusting the content of the foaming agent and several other foaming parameters, such as temperature and heating rate [132].

Metallic Foam can be synthesized with uniform pore diameter using the space holder technique. In this method, metal powder is alloyed with materials such as carbamide, ammonium bicarbonate, dolomite, titanium hydride, and so on. Pellets can be made by compaction of powder without using binder material. These pellets are heated upto melting temperature of space holder material. During the sintering process, blowing agents are removed leaving behind space holder particles from the pores. So, the highly porous material can be fabricated. From this method, 90% porosity can be achieved with good strength. The advantage of this method is that complex shapes can be fabricated and it is economical. This method is generally used for the fabrication of porous steel [133]. Another powder metallurgy technique is the slurry Foaming technique. Slurry of metal powder is first made and blowing agents are added. Some additives can also add to improve the flow properties of slurry. This slurry is poured into a mold and heated at high temperatures. At high temperatures, slurry expands and releases the gas due to the decomposition of blowing agents. So, porous structure can be obtained [134]. The most commonly used aluminum alloys for foaming are Al-Si, 2xxx series, and 6xxx series Al-alloys. The powder metallurgy approach as developed by Fraunhofer, however, can also be used for other metals, including steel, lead, tin, zinc, brass, and bronze.

12.8.1.2.2 Liquid Metallurgy Route

Foam can be manufactured from liquid metal by direct foaming or by indirect foaming with the help of casting around solid space holding fillers. Some methods of foam manufacturing in the liquid metallurgy route have been discussed below.

Direct Foaming by Blowing Agents: Shinko Wire, Amagasaki City, Japan, provided the first commercial production of metal foams in the late 1980s. In this process, calcium is first added to an aluminum melt and stirred in the air to produce oxides and raise their viscosity. Subsequently, powder of the blowing agent (TiH_2) is dispersed quickly into the melt by stirring, decomposing into gaseous hydrogen and titanium at the melt temperature. The melt starts then foaming inside the crucible or inside a dedicated mold. Finally, it is cooled down to a big block and usually sliced into plates of the desired thickness. This type of metallic foam and the corresponding process is called Alporas, which was patented in the United States in 1987 [135]. This method is most explored in literature among all methods of foam manufacturing. This method is mainly used for foaming Al and its alloys, although other metals can be foamed by this method. In this method, metal is melted and suitable viscosity-enhancing material is added to increase the viscosity of the melt. Graphite or stainless steel stirrer can be used for uniform mixing. Then, blowing agents are added for blowing of gas for foaming. Metal hydrides, metal carbonates, or metal oxides are generally used as blowing agents. Blowing agent with decomposition temperature matching with a melting point of the metal is generally chosen for a better quality of resultant foam. After uniform mixing of blowing agent with molten metal, the temperature of the mixture is increased. As soon as the temperature is reached to the decomposition temperature of blowing agent, it starts to blow bubbles of gas that are trapped inside melt. After gas entrapment in the liquid melt, it is suddenly cooled to avoid any effect of gravity during solidification.

Various parameters that can affect foam quality in the direct foaming method are the amount of viscosity-enhancing agent, amount of blowing agent, stirring time, holding time after blowing agent addition, the temperature at which viscosity enhancement agents and blowing agents are added, solidification method and time, and so on. Titanium hydride powder is generally used for the fabrication of porous aluminum and other alloy and it is the most popular process for melt processing routes and powder metallurgy routes. By adding titanium hydride (1.6%wt) at 680 °C in Al-alloy melt, decomposition of blowing agent will occur at high temperature, and it releases the hydrogen. During decomposition of titanium hydride, chemical reactions as per Eq. (12.11) may occur. Yang et. al. carried out the fabrication of Mg foam by use of calcium carbonate as a blowing agent. Exothermic reaction as per Eq. (12.12) occurs within the temperature range of 575–750 °C.

$$TiH_2 = Ti + H_2 \tag{12.11}$$

$$Mg + CaCo_3 = MgO + CaO + CO \tag{12.12}$$

Zirconium hydride, magnesium hydroxide, and dolomite can also be used as a blowing agents. Chromium nitride and strontium carbonate are also used as foaming agents in the manufacturing of porous steel. Foams manufactured by direct foaming technique have superior energy absorbing capacity and have a highly isotropic and homogeneous cellular structure. So it can be concluded that direct foaming technique is an excellent candidate for studying mechanical variability [136].

Gas Injection in Melt Process: Foam can be manufactured by injecting pressurized gas in melt directly. Control of the viscosity of melt in this process is critical. Viscosity can be controlled by varying the temperature of melt or adding ceramics. Gas injection method is extensively used for the foaming of Al and its alloys. This method is not suitable for foaming of materials that oxidize rapidly like Ti and Mg. Gases that can be used in this method are air, nitrogen, carbon dioxide, inert gases, oxygen, and even water. Advantages of this method are that large volumes of foam can be produced continuously and complex shapes can be produced by selecting proper mold [137].

Solid gas eutectic solidification: Some liquid metals make a eutectic system in hydrogen gas. When material melts into hydrogen atmosphere under high pressure around 50 atm, homogeneous melt charged with hydrogen is obtained [138]. When liquid metal is cooled by reducing temperature and pressure, the melt will be transformed into two-phase region, that is, the mixture of solid and gas. Solidification will start by the removal of heat. This method can be used for Ni, Cu, Al, and Mg. The main disadvantage of this method is poor homogeneity.

12.8.1.3 Foaming by Rapid Prototyping Technique

Foaming by rapid prototyping includes a 3D printing process for producing sacrificial templates. This template can be used for metallic slurry to produce a porous metal structure. Sacrificial template is made layer by layer with a binder. After one layer of binder, another layer of metallic/ceramic power is created by spray. In this way, the entire part is completed by layers. For example, porous Mg can be synthesized by this technique. Open cellular Mg can be fabricated by this method. The main disadvantage of this method is the higher cost [139].

12.8.1.4 Electro-Deposition Technique

In this method, the metal used is in form of ions in an electrolyte. The metal is galvanically deposited on polymer foam by electro-deposition technique. Then, it is removed by thermal treatments. Foamed polymer is replaced by metallic material. So, actual foaming is not occurring in a solid state. Also, polymer foam requires some electrical conductivity for galvanic deposition initially. From this method, Ni-Cr foams can be fabricated with a density of 0.4–0.65 g/cm^3 [140].

12.8.1.5 Vapor Deposition Technique

In this method, gaseous metals are used. Polymer precursor is used to define the geometry of metallic foam. Metal vapors are produced in a vacuum chamber and allow condensing in precursor. Nickel foams and nickel carbonyl foams can be fabricated by this method. Densities obtained are in the range of 0.2–0.6 g/cm^3. There are different methods for fabricating metallic porous material structures, including metallic foam from liquid metal, powdered metal and metallic ions in electrolyte, and so on. Metallic foam can replace conventional parts with respect to its physical and mechanical properties and also to its economical factors by selecting suitable production methods [141].

12.8.1.6 Based on Polymer Sponge Structure

Metal sponges can be manufactured by replicating an open porous polymeric structure. Their morphology is based obviously on the structure of the polymeric foam, for example, polyurethane foam. This implies certain advantages, such as their quality in the sense of pore homogeneity and distribution is superior and polymeric foam technology is quite advanced and established compared to the metallic foam one. However, it also has several disadvantages concerning the number of production steps, dimensions, costs, and so on. An example is given by ERG (Oakland, CA, USA) which produces sponges of different metals, such as Al, Cu known under the tradename Duocel®, following a foam replication or investment casting method. Here, the polymer sponge is filled with a slurry of heat-resistant material, for example, a mixture of mullite, phenolic resin, and calcium carbonate. After drying at a certain temperature, the polymer is decomposed and the form is stable enough for casting with the corresponding metal. The different grades are given in pores per inch (ppi), and porosities in the range of 80–97% are achieved [142].

A similar method, but with an additional wax covering of the polymeric foam for increasing stiffness, is used by Mayser GmbH, formerly M-pore (Dresden, Germany). The drying step with temperatures up to 350 °C allows one to remove the wax and the polymer, and metallic sponges with large and regular pores up to 10 mm in diameter can be produced. Another method consists of a metallic coating of the polymeric sponge with subsequent sintering and removal of the polymer. An example is the deposition onto polyurethane foam by, for example, chemical vapor deposition of nickel tetracarbonyl ($Ni(CO)_4$), which decomposes to elemental nickel and carbon monoxide at a heating temperature of 150–200 °C. This method was developed and commercialized by Inco (Mississauga, ON, Canada). Now, this technology is used by Alantum (Seongnam-City, Korea), with a large production plant in China, where they also produce Fe and Cu foams. Recemat (Dodewaard, the Netherlands) produces sponges of Ni, stainless steel, Ti, and so on., being metalized with the subsequent removal of the polyurethane sponge by pyrolysis. Generally, the initial metallization is done with Ni metal. The resulting Ni metal foam, after being heat-treated, is used in applications, where high electrical and thermal conductivity are of importance [143].

Spray Forming: In the spray forming method, liquid metal is first atomized. Spray of fast-flying atomized metal is created which is sprayed on the substrate of required shape to obtain the dense solid form of metal. If blowing agent powders are added in atomized spray, it results in the release of gas which upon solidification becomes the reason of porosity in the resultant foam structure. The main advantages of the spray forming method are low oxide content and fine grain size. This method can be used for steel and copper alloy. Porosities obtain from this method are around 60%. The main disadvantage of this method is less uniformity in porosity [144].

12.8.1.7 Properties

One of the most significant attributes of metal foams is their characteristic nonlinear deformation behavior. This attribute lends itself to applications in which both lightweight construction and efficient absorption of deformation energy are important. In automotive applications, the crush energy absorption behavior of metal foams due to high strain rate deformation is important in designing vehicles for optimum crashworthiness. When subjected to uniaxial compressive loading, metal foams exhibit deformation behavior very similar to that of polymer foams. Three steps of compressive deformation are observed with metal foams [145]: (1) metal foam shows linear elastic deformation related to bending of the cell walls and (2) the foam begins to undergo plastic deformation due to cell wall buckling. The initial wall buckling events result in the slight drop in stress observed at the transition between Step 1 and Step 2. In this region, the compressive stress increases slightly as a function of strain, (3) this step is characterized by progressive collapse and crushing of cell walls. Significant densification of the foam occurs, resulting in a strong increase in stress with increasing compressive strain.

Mechanical properties depend mainly on the density but are also influenced by the quality of the cellular structure in the sense of cell connectivity, cell roundness and diameter distribution, the fraction of the solid contained in the cell nodes, edges, or the cell faces, and so on. There is a wide range of functional properties of metal foams originating from their cellular character. This structure implies a large surface area and a number of cells. Very often, the combination of functional properties, such as acoustic, thermal, electrical, or chemical resistance, with mechanical properties, like strength or stiffness, allows interesting new applications. Foams in general are known for their thermal insulation and crash properties [146].

12.8.1.8 Application

Lightweight materials with high stiffness are often desired for various applications are discussed below.

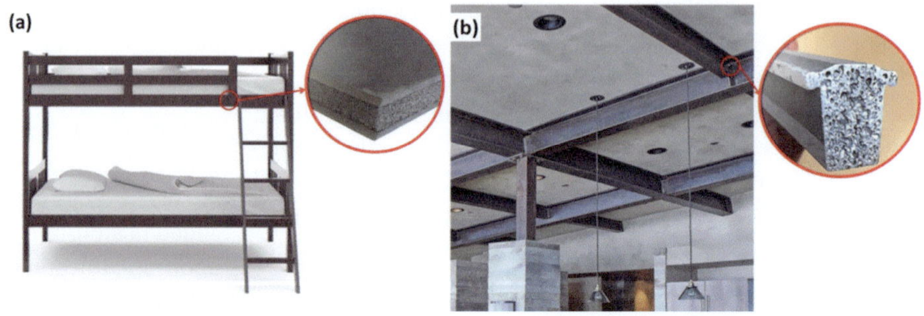

Fig. 12.32 (**a**) Folding bed and (**b**) Supporting platform for large buildings

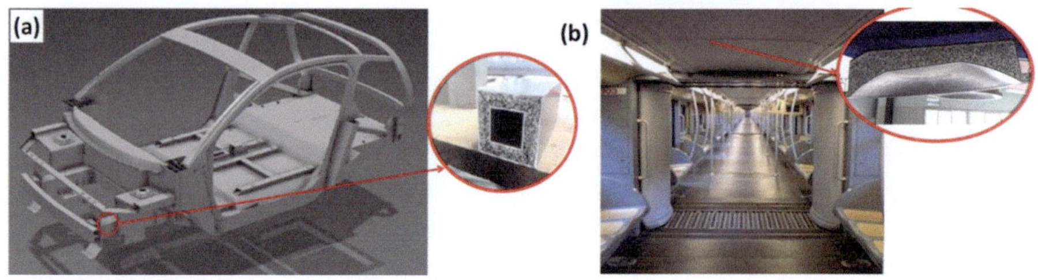

Fig. 12.33 (**a**) Small crash-absorbing elements for Audi Q7, and (**b**) High-velocity train interior

Figure 12.34 (**a**) Ceiling and wall of the large audience hall, and (**b**) Restaurant roof and wall

Structural applications: Aluminum foam sandwiches (AFS) and steel aluminum foam sandwiches (SAS) are promising products for structural applications and are already on the market. AFS panels are used as support frames, for example, for solar panels, mirrors, and so on and everywhere where light and rigid metallic panels are needed. The sliding bed (Fig. 12.32a) is made of welded AFS parts, and the construction is 28% lighter than the cast part with the same stiffness but improving vibration damping. Many household things like folding bed, table, beam, and so on (Fig. 12.32b) are made up of aluminum foam in its sandwiched core [147].

The closed metallic foam variety is used for structural applications requiring load-bearing features and for weight-saving and impact-absorbing structures in vehicles. Advanced automobile industries like Audi, Ferrari, and so on are using Al-foam for their part for crash-resistance materials as shown in Fig. 12.33a [148]. Figure 12.33 shows the interior metallic foam applications in high-velocity trains.

The open variety is ideal for vibration and sound absorption, filtration, and catalysis at high temperatures, for heat exchange, and in medical devices [149, 150]. Figure 12.34 shows the vibration and sound absorption foam in the ceiling and wall of the large audience hall and restaurant.

12.8.2 Ceramic Foam

Porous ceramics are of great interest due to their numerous potential applications in catalysis, adsorption and separation, filtration of molten metals or hot gases, refractory insulation of furnaces, and hard tissue repair and engineering [151].

12.8.2.1 Synthesis

The three main processing routes for the fabrication of macroporous ceramics are the direct foaming technique, the replica technique, and the sacrificial template method. The processing route ultimately determines the microstructure of the final macroporous ceramic. Therefore, the selection of a given processing method depends strongly on the microstructure needed in the end application as well as on the inherent features of the process such as cost, simplicity, and versatility [152].

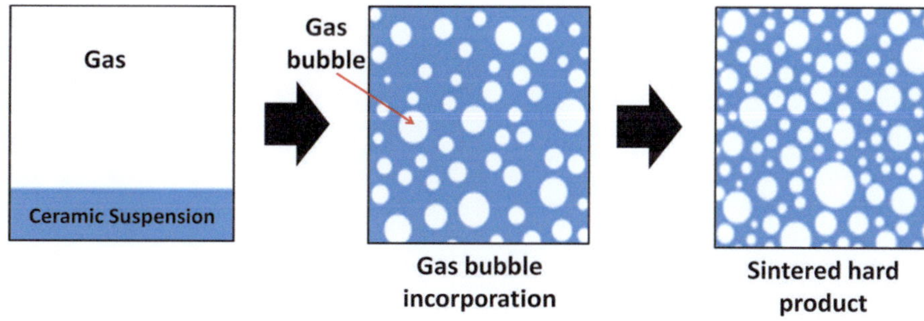

Fig. 12.35 Direct foaming processing routes used for the production of macroporous ceramics

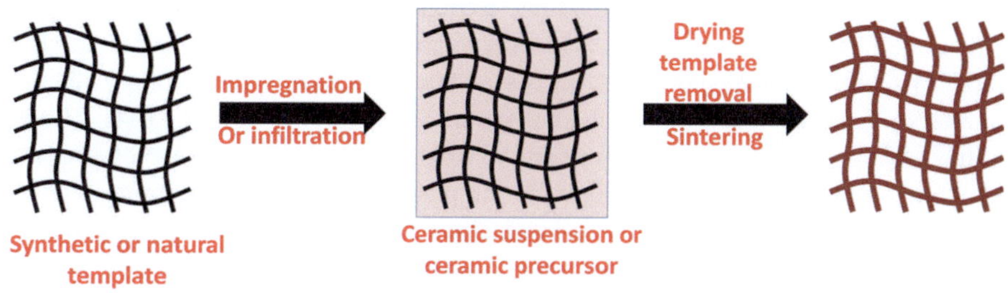

Fig. 12.36 Replica processing routes used for the production of macroporous ceramics

12.8.2.1.1 Direct Foaming Technique

The direct foaming technique is particularly suitable for the fabrication of open and closed porous structures with porosities ranging from 45% to 97% and cell sizes between 30 mm and 1 mm. Direct foaming methods involve (Fig. 12.35) the incorporation of a gaseous phase into a ceramic suspension consisting of ceramic powder, solvent, dispersants, surfactants, polymeric binder, and gelling agents. The incorporation of the gaseous phase is carried out either by mechanical frothing, injection of a gas stream, gas releasing chemical reactions, or solvent evaporation. The liquid foams obtained upon gas incorporation are thermodynamically unstable due to their large interfacial area. Therefore, gas bubbles that initially nucleated as spheres grow as polyhedral cells. There are mainly three processes responsible for coarsening: drainage, coalescence, and Ostwald ripening [153].

Drainage of the liquid from the lamellas between the bubbles results in a close approach of bubble surfaces, which can lead to their coalescence and to foam collapse. Additionally, due to different Laplace pressures of bubbles with different sizes, gas diffusion occurs from smaller to larger bubbles. This migration of gas between bubbles leads to coarsening of the foam and to a broadening of the bubble size distribution. Eventually, the liquid foam collapses due to the combined action of these destabilization mechanisms. Additives are often used to avoid foam collapse by setting the foam structure shortly after air incorporation. This usually occurs by gelling or cross-linking organic compounds added to the suspension liquid medium. A drawback of these methods is the fact that they cannot avoid rapid bubble growth before the setting reaction takes place, which leads to a large average bubble and a wide bubble size distribution. We have recently shown that ultra-stable wet foams can be produced by using particles instead of surfactants as foam stabilizers in the direct-foaming method. The energy gain upon adsorption of a particle to the air–water interface of fresh gas bubbles can be as high as thousands of kT's, as opposed to the adsorption energy of a few kT's in the case of surfactants (k is the Boltzmann constant and T is the temperature). Therefore, particles can irreversibly adsorb at the surface of gas bubbles, in contrast to surfactants that adsorb and desorb at relatively short time scales at the interface.

12.8.2.1.2 Replica Technique

The replica method is based on the impregnation of a cellular structure with a ceramic suspension or precursor solution to produce a macroporous ceramic exhibiting the same morphology as the original porous material (Fig. 12.36). Many synthetic and natural cellular structures can be used as templates to fabricate macroporous ceramics through the replica technique. Here, we focus mainly on those processes that allow for the fabrication of bulk ceramic structures, while references are given for the recent studies describing the preparation of ceramic macroporous films and particles [154].

Synthetic templates: Since the sponge replica technique has become the most popular method to produce macroporous ceramics and is today extensively used in industry to prepare ceramic filters for molten metal filtration and other applications. This success is primarily attributed to the simplicity and flexibility of the method. In the polymer

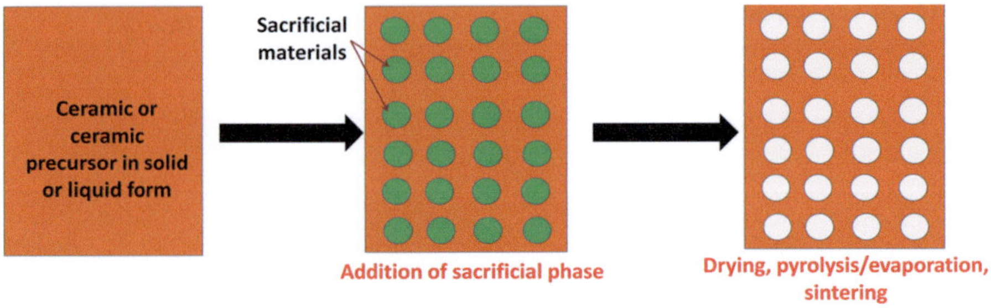

Fig. 12.37 Sacrificial template processing routes used for the production of macroporous ceramics

replica approach, a highly porous polymeric sponge (typically polyurethane) is initially soaked into a ceramic suspension until the internal pores are filled in with ceramic material [155]. The impregnated sponge is then passed through rollers to remove the excess suspension and enable the formation of a thin ceramic coating over the struts of the original cellular structure. The ceramic-coated polymeric template is subsequently dried and pyrolyzed through careful heating between 300 °C and 800 °C. Heating rates usually lower than 11 °C/min are required in this step to allow for the gradual decomposition and diffusion of the polymeric material, avoiding the build up of pressure within the coated struts. Binders and plasticizers are added to the initial suspension to provide ceramic coatings sufficiently strong to prevent cracking the struts during pyrolysis. Typical binders used are colloidal aluminum orthophosphate, potassium and sodium silicates, magnesium orthoborate, hydratable alumina, colloidal silica, polyvinyl butyral with polyethylene glycol as a plasticizer, and polymerizable monomers. After removal of the polymeric template, the ceramic coating is finally densified by sintering in an appropriate atmosphere at temperatures ranging from 1100 to 1700 °C depending on the material. Macroporous ceramics of many different chemical compositions have been prepared using the sponge replica technique. In contrast to ceramic suspension-derived reticulated structures, cellular materials obtained from preceramic polymers have crack-free struts due most likely to the improved wetting on the sponge and the partial melting of the cross-linked polymer during pyrolysis.

Natural templates: In addition to synthetic polymer foams, other cellular structures have been used as templates for the fabrication of macroporous ceramics through the replica approach. Cellular structures available in nature are particularly interesting as natural replica templates, due mainly to their special pore morphology and intricate microstructures, which might be difficult to produce artificially [156].

12.8.2.1.3 Sacrificial Template Method

The sacrificial template technique usually consists of the preparation of a biphasic composite comprising a continuous matrix of ceramic particles or ceramic precursors and a dispersed sacrificial phase that is initially homogeneously distributed throughout the matrix and is ultimately extracted to generate pores within the microstructure. This method leads to porous materials displaying a negative replica of the original sacrificial template, as opposed to the positive morphology obtained from the replica technique described above [154]. The biphasic composite is commonly prepared by (1) pressing a powder mixture of the two components, (2) forming a two-phase suspension that is subsequently processed by wet colloidal routes such as slip, tape or direct casting, or (3) impregnating previously consolidated preforms of the sacrificial material with a preceramic polymer or ceramic suspension (Fig. 12.37). The way that the sacrificial material is extracted from the consolidated composite depends primarily on the type of pore former employed. A wide variety of sacrificial materials have been used as pore formers, including natural and synthetic organics, salts, liquids, metals, and ceramic compounds. Synthetic and natural organics are often extracted through pyrolysis by applying long thermal treatments at temperatures between 200 and 600 °C. The long periods required for complete pyrolysis of the organic component and the extensive amount of gaseous by-products generated during this process are the main disadvantages of using organic materials as a sacrificial phase [157].

12.8.2.2 Polymeric Foam

Polymer foams have been widely used in a variety of applications: Insulation, cushion, absorbents, etc. Various polymers have been used for foam applications are polyurethane (PU), polystyrene (PS), polyethylene (PE), polypropylene (PP), poly(vinyl chloride) (PVC), polycarbonate (PC), and so on. They have low density so they are lightweight materials. Generally, polymer foams have low heat or sound transfer, making them optimal insulators. But they have inferior mechanical strength, low thermal and dimensional stability [158]

12.8.2.2.1 Classification of Polymer Foams

Polymer foams can also be defined as either closed cell or open-cell foams. In closed-cell foams, the foam cells are isolated from each other, and cavities are surrounded by complete cell walls. Generally, closed-cell foams have lower permeability, leading to better insulation properties and absorbing sounds, especially bass tones. Closed-cell

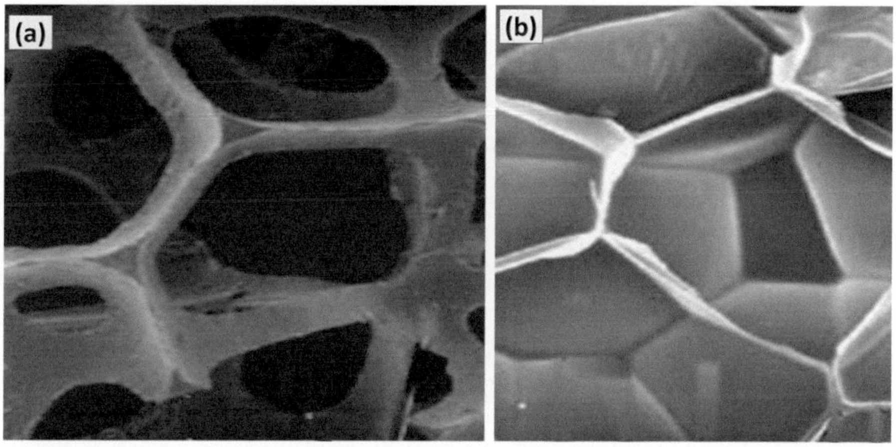

Fig. 12.38 Cellular solids: (**a**) Open-cell polyurethane foam, (**b**) Closed-cell polyethylene foam

foams are usually characterized by their rigidity and strength, in addition to the high *R*-value (Resistance to heat flow). Closed-cell polyurethane spray foam has among the highest *R*-values of any commercially available insulation. In open-cell foams, cells are connected with each other. They have a softer and spongier appearance. Open-cell foams are incredibly effective as a sound barrier in normal noise frequency ranges and provide better absorptive capability [159]. Figure 12.38 shows the SEM micrograph of open-cell and closed-cell polyethylene foam. The advantages of closed-cell foam compared to open-cell foam include its strength, higher R-value, and its greater resistance to the leakage of air or water vapor. The disadvantage of the closed-cell foam is that it is denser, requires more material, and, therefore, is more expensive.

Again, Polymer foams can be classified as *rigid or flexible foams*. Rigid foams are widely used in applications such as building insulation, appliances, transportation, packaging, furniture, and food and drink containers. Flexible foams are used as furniture, transportation, bedding, carpet underlay, textile, sports applications, and shock and sound attenuation. According to the size of the foam cells, polymer foams can be classified as: (1) macrocellular (>100 μm), (2) microcellular (1–100 μm), (3) ultramicrocellular (0.1–1 μm), and (4) nanocellular (0.1–100 nm). The thermal insulation performance of polyurethane rigid foam depends chiefly on the size of the foam pores. The smaller the diameter, the lower the thermal conductivity and the better the insulating effect [160].

12.8.2.2.2 Synthesis of Polymeric Foam

Foaming techniques of polymer foam are mechanical foaming, foaming with hollow glass beads, chemical foaming, and physical foaming. Foaming agents are substances that form the gaseous phase in the foams. For large-scale production, the direct utilization of foaming agents is the most commonly used method. Two types of foaming agents are often used: (1) chemical blowing agents and (2) physical blowing agents. Chemical blowing agents are usually reactive species that produce gases in the foaming process. Blowing agents that produce gas via chemical reactions include baking powder and isocyanates; when they react with water: $RNCO{+}H_2O \rightarrow RNH_2{+}CO_2$. Physical blowing agents are substances that gasify under foaming conditions. Typical physical foaming agents are volatile chemicals, such as chlorofluorocarbons(CFC), volatile hydrocarbons and alcohols, or inert gases such as nitrogen, carbon dioxide, argon, and water. In physical foaming, a blowing gas is first dissolved in the polymer to form a homogeneous mixture. This is usually done through pressurization. Subsequent pressure release or temperature increase results in a supersaturation state, and gas starts to form nuclei and expand. The traditional CFC physical blowing agents have good solubility in the polymer matrix, low diffusivity, and low thermal conductivity that result in foam products with good insulation and physical properties. However, their use has been greatly reduced globally because of their high ozone depletion effect. Other choices of foam blowing agents include HFCs; hydrocarbons (propane, butane, pentane, etc.); inert gases such as nitrogen, argon and CO_2; and water. HFCs do not destroy the ozone, but they have a negative impact on global warming (according to Montreal protocol), and their applications will most likely be regulated in the near future. Hydrocarbons present some serious problems, such as a greater fire hazard in closed-cell foams due to the entrapped blowing agent as well as adding to VOC emissions [161]. CO_2 is the most favorable choice because it is inexpensive, nontoxic, and environmentally benign (zero Ozone depletion potential and 100-year global warming potential compared to 1300 years for HFC). Many challenges must be met to enable the use of CO_2 as a blowing agent: (1) The low solubility of CO_2 in most polymer melts; (2) CO_2 has a high diffusivity in the polymer melt due to its small size, while this ensures a fast mixing process, it also results in quick escape of gas from the foam after processing; (3) CO_2 has a higher gas thermal conductivity in comparison to that of HFC blowing agents [162].

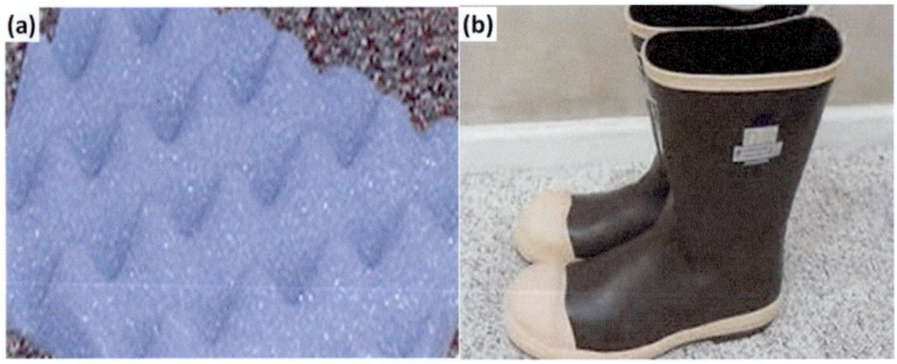

Fig. 12.39 (**a**) Foamed polymer foam base for transportation without jerking, (**b**) Polymer foam ultra lightweight footwear

12.8.2.3 Application

Applications based on polymeric foams are well established in the market and a large part of our daily life. Some applications based on other foamed materials, such as porous concrete or food foams, are also quite popular in a wide variety in our society [163]. Figure 12.39 shows the most used polymer foam base for transportation without jerking, and polymer foam ultra lightweight footwear.

12.9 Summary and Perspectives

This chapter covers the current status and the future progress of ultralight materials. Each material covers classification, synthesis procedure, properties, and applications. This chapter provides all the novel ideas on ultra-lightweight materials to be progressed in the near future and is the direction to strengthen the materials engineering area to be used in high-end applications even to be used for the travel from earth to mars.

References

1. Jiang, B., He, C., Zhao, N., et al.: Ultralight metal foams. Sci. Rep. **5**, 13825 (2015). https://doi.org/10.1038/srep13825
2. Woignier, T., Primera, J., Alaoui, A., Etienne, P., Despestis, F., Calas-Etienne, S.: Mechanical properties and brittle behavior of silica aerogels. Gels. **1**(2), 256–275 (2015). https://doi.org/10.3390/gels1020256
3. Mecklenburg, M., Schuchardt, A., Mishra, Y.K., Kaps, S., Adelung, R., Lotnyk, A., Kienle, L., Schulte, K.: Aerographite: ultra lightweight, flexible nanowall, carbon microtube material with outstanding mechanical performance. Adv. Mater. **24**, 3486–3490 (2012). https://doi.org/10.1002/adma.201200491
4. See: https://basalt.today/material/aerogel/. 02.2.2020
5. See: https://www.graphene-info.com/graphene-aerogel. 02.2.2020
6. Gurav, J.L., Jung, I.-K., Park, H.-H., Kang, E.S., Nadargi, D.Y.: Silica aerogel: synthesis and applications. J. Nanomaterials. **2010**, 11 (2010). https://doi.org/10.1155/2010/409310. 409310
7. See: https://en.wikipedia.org/wiki/Aerogel. 02.2.2020
8. Budtova, T.: Cellulose II aerogels: a review. Cellulose. **26**, 81–121 (2019). https://doi.org/10.1007/s10570-018-2189-1
9. See: https://www.newworldencyclopedia.org/entry/Aerogel. 02.2.2020
10. Thapliyal, P.C., Singh, K.: Aerogels as promising thermal insulating materials: an overview. J. Mater. **2014**, 10 (2014). https://doi.org/10.1155/2014/127049. 127049
11. Danks, A., Hall, S., Schnepp, Z.: The evolution of 'sol-gel' chemistry as a technique for materials synthesis. Mater. Horiz. **3** (2015). https://doi.org/10.1039/C5MH00260E
12. Dervin, S., Pillai, S.: An introduction to sol-gel processing for aerogels. Springer, New York (2017). https://doi.org/10.1007/978-3-319-50144-4_1
13. Rao, V., Kalesh, R.: Comparative studies of the physical and hydrophobic properties of TEOS based silica aerogels using different co-precursors. Sci. Technol. Adv. Mater. **4**, 6. 509–515 (2003). https://doi.org/10.1016/j.stam.2003.12.010
14. Yun, S., Luo, H., Gao, Y.: Superhydrophobic silica aerogel microspheres from methyltrimethoxysilane: rapid synthesis via ambient pressure drying and excellent absorption properties. RSCD Adv. **4**(9), 4535 (2014)
15. Jelle, B., Baetens, R., Gustavsen, A.: Aerogel insulation for building applications. Elsevier, Amsterdam (2015). https://doi.org/10.1002/9783527670819.ch45
16. Nur, Z., Ebrahim, A.-L., Mohamed, N., Ezzat, A.: Aerogel-based materials for adsorbent applications in material domains. E3S Web Conf. **90**, 01003 (2019). https://doi.org/10.1051/e3sconf/20199001003
17. Rao, A., Rao, A.: Modifying the surface energy and hydrophobicity of the low-density silica aerogels through the use of combinations of surface-modification agents. J. Mater. Sci. **45**, 51–63 (2010). https://doi.org/10.1007/s10853-009-3888-7
18. Rao, A., Sharad, B., Hiroshi, H., Gerard, P.: Synthesis of flexible silica aerogels using methyltrimethoxysilane (MTMS) precursor. J. Colloid Interface Sci. **300**, 279–285 (2006). https://doi.org/10.1016/j.jcis.2006.03.044
19. Shi, F., Wang, L., Liu, J., Zeng, M.: Effect of heat treatment on silica aerogels prepared via ambient drying. J. Mater. Sci. Technol. **23**, 402–406 (2007)
20. Stojanovic, A., Zhao, S., Angelica, E., Malfait, W., Koebel, M.: Three routes to superinsulating silica aerogel powder. J. Sol-Gel Sci. Technol. (2019). https://doi.org/10.1007/s10971-018-4879-4
21. Pratap, W., Ingale, S.: Comparison of some physico-chemical properties of hydrophilic and hydrophobic silica aerogels. Ceramics Int. **28**, 43–50 (2002). https://doi.org/10.1016/S0272-8842(01)00056-6
22. Bheekhun, N., Talib, A.R.A., Hassan, M.R.: Aerogels in aerospace: an overview. Adv. Mater. Sci. Eng. **2013**, 18 (2013). https://doi.org/10.1155/2013/406065. 406065
23. Riffat, S., Qiu, G.: A review of state-of-the-art aerogel applications in buildings. Int. J. Low-Carbon Technol. **8**, 1–6 (2012). https://doi.org/10.1093/ijlct/cts001
24. See: https://en.wikipedia.org/wiki/Stardust_(spacecraft). 08.2.2020

25. See: https://www.nasa.gov/mission_pages/stardust/mission/index-aerogel-rd.html. 08.2.2020
26. See: https://www.nasa.gov/vision/earth/technologies/aerogel.html. 08.2.2020
27. See: https://mars.nasa.gov/mer/mission/rover/temperature/, 08.2.2020
28. See: https://www.hisour.com/aerogel-40784/. 08.2.2020
29. Xuan, Y., Kaiyuan, S., Igor, Z., Emily, C.: Cellulose nanocrystal aerogels as universal 3D lightweight substrates for supercapacitor materials. Adv. Mater. (2015). https://doi.org/10.1002/adma.201502284
30. Stergar, J., Maver, U.: Review of aerogel-based materials in biomedical applications. J. Sol-Gel Sci. Technol. **77**, 738–752 (2016). https://doi.org/10.1007/s10971-016-3968-5
31. Akimov, Y.K.: Fields of Application of Aerogels (Review). Instrum. Exp. Tech. **46**, 287–299 (2003). https://doi.org/10.1023/A:1024401803057
32. Jyoti, G., Jung, I.-K., Hyung-Ho, P., Eul, K., Digambar, N.: Silica aerogel: synthesis and applications. J. Nanomaterials. **2010** (2010). https://doi.org/10.1155/2010/409310
33. See: https://www.aerogel.com/products-and-solutions/pyrogel-hps/. 12.2.2020
34. See: https://spinoff.nasa.gov/Spinoff2020/ip_2.html. 12.2.2020
35. Wang, H., Du, A., Ji, X., Zhang, C., Zhon, B., Zhang, Z., Shen, J.: Enhanced photothermal conversion by hot-electron effect in ultrablack carbon aerogel for solar steam generation. ACS Appl. Mater. Interfaces. **11**(45), 42057–42065 (2019)
36. Garnier, C., Muneer, T., McCauley, L.: Super insulated aerogel windows: Impact on daylighting and thermal performance. Build. Environ. **94** (2015). https://doi.org/10.1016/j.buildenv.2015.08.009
37. Jiang, X.Q., Mei, X.D., Feng, D.: Air pollution and chronic airway diseases: what should people know and do? J Thorac Dis. **8**(1), E31–E40 (2016). https://doi.org/10.3978/j.issn.2072-1439.2015.11.50
38. Hwei-Jay, C., Chi-Young, L., Nyan-Hwa, T.: Green reduction of graphene oxide by Hibiscus sabdariffa L. to fabricate flexible graphene electrode. Carbon. **80**, 725–733 (2014). https://doi.org/10.1016/j.carbon.2014.09.019
39. Meng, D., Sun, J., Chang, J., Yang, F., Shi, L., Gao, L.: Synthesis of nitrogen-doped reduced graphene oxide directly from nitrogen-doped graphene oxide as a high-performance lithium ion battery anode. RSC Adv. **4**, 42412–42417 (2014). https://doi.org/10.1039/C4RA05544F
40. Mishra, Y.K., Adelung, R.: ZnO tetrapod materials for functional applications. Mater Today. **21**(6), 631–651 (2018). https://doi.org/10.1016/j.mattod.2017.11.003. ISSN 1369-7021
41. Manawi, Y.M., Ihsanullah Samara, A., Al-Ansari, T., Atieh, M.A.: A review of carbon nanomaterials' synthesis via the chemical vapor deposition (CVD) method. Materials (Basel). **11**(5), 822 (2018). https://doi.org/10.3390/ma11050822
42. Qiu, B., Xing, M., Zhang, J.: Recent advances in three-dimensional graphene based materials for catalysis applications. Chem. Soc. Rev. **47**, 2165–2216 (2018). https://doi.org/10.1039/C7CS00904F
43. Svenja, G., Taro, F., Matthias, M., Daria, S., Yogendra, M., Rainer, A., Bodo, F., Karl, S.: Electro-mechanical piezoresistive properties of three dimensionally interconnected carbon aerogel (Aerographite)-epoxy composites. Composites Sci. Technol. **134**, 226–233 (2016). https://doi.org/10.1016/j.compscitech.2016.08.019
44. Yogendra, M., Sören, K., Arnim, S., Ingo, P., Xin, J., Dawit, G., Sebastian, W., Lupan, O., Rainer, A.: Versatile fabrication of complex shaped metal oxide nano-microstructures and their interconnected networks for multifunctional applications. Powder Part. **31**, 92–110 (2014). https://doi.org/10.14356/kona.2014015
45. See: https://www.hisour.com/aerographite-40791/. 15.2.2020
46. See: https://wikivisually.com/wiki/Aerographene. 15.2.2020
47. Weiwei, G., Zhao, N., Weiquan, Y., Zhen, X., Hao, B., Chao, G.: Effect of flake size on the mechanical properties of graphene aerogels prepared by freeze casting. RSC Adv. **7**, 33600–33605 (2017). https://doi.org/10.1039/C7RA05557A
48. Worsley, M., Baumann, T.: Carbon aerogels. Springer, New York (2016). https://doi.org/10.1007/978-3-319-19454-7_90-1
49. Meija, R., Signetti, S., Schuchardt, A., et al.: Nanomechanics of individual aerographite tetrapods. Nat. Commun. **8**, 14982 (2017). https://doi.org/10.1038/ncomms14982
50. Luo, Y., Jiang, S., Xiao, Q., Chen, C., Li, B.: Highly reusable and superhydrophobic spongy graphene aerogels for efficient oil/water separation [published correction appears in Sci Rep. 20185;8 (1):273]. Sci. Rep. **7**(1), 7162 (2017). https://doi.org/10.1038/s41598-017-07583-0
51. García-Moreno, F.: Commercial applications of metal foams: their properties and production. Materials (Basel). **9**(2), 85 (2016). https://doi.org/10.3390/ma9020085
52. Yang, Z., Chabi, S., Xia, Y., Zhu, Y.: Preparation of 3D graphene-based architectures and their applications in supercapacitors. Prog. Nat. Sci. Mater. Int. **25**(6), 554–562 (2015). https://doi.org/10.1016/j.pnsc.2015.11.010. ISSN 1002-0071
53. Ma, Y., Chen, Y.: Three-dimensional graphene networks: synthesis, properties and applications. Natl. Sci. Rev. **2** (2014). https://doi.org/10.1093/nsr/nwu072
54. Baig, N., Saleh, T.A.: Electrodes modified with 3D graphene composites: a review on methods for preparation, properties and sensing applications. Microchim Acta. **185**, 283 (2018). https://doi.org/10.1007/s00604-018-2809-3
55. Muñoz, R., Gómez-Aleixandre, C.: Review of CVD synthesis of graphene. Chem. Vap. Deposition. **19**, 297–322 (2013). https://doi.org/10.1002/cvde.201300051
56. Sim, Y., Kwak, J., Kim, S.-Y., Jo, Y., Kim, S., Kim, S.Y., Kim, J. H., Lee, C.-S., Jo, J.H., Kwon, S.-Y.: Formation of 3D graphene–Ni foam heterostructures with enhanced performance and durability for bipolar plates in a polymer electrolyte membrane fuel cell. J. Mater. Chem. A. **6**, 1504–1512 (2018)
57. Xiao, X., Beechem, T., David, W., Burckel, D., Ronen, P.: Lithographically defined porous Ni-carbon nanocomposite supercapacitors. Nanoscale. **6** (2013). https://doi.org/10.1039/c3nr05751h
58. Lei, Q., freezing, Z.H.C., drying, f.: A versatile route for porous and micro-/nano-structured materials. J Chem Technol Biotechnol. **86**, 172–184 (2011). https://doi.org/10.1002/jctb.2495
59. Yu, Z., Hu, C., Dichiara, A.B., Jiang, W., Gu, J.: Cellulose nanofibril/carbon nanomaterial hybrid aerogels for adsorption removal of cationic and anionic organic dyes. Nanomaterials. **10**, 169 (2020)
60. Sha, J., Lee, S.-K., Yilun, L., Zhao, N.Q., James, T.: Preparation of three-dimensional graphene foams using powder metallurgy templates. ACS Nano. **10** (2015). https://doi.org/10.1021/acsnano.5b06857
61. Wang, X., Zhang, Y., Zhi, C., et al.: Three-dimensional strutted graphene grown by substrate-free sugar blowing for high-power-density supercapacitors. Nat. Commun. **2013**, 4 (2905). https://doi.org/10.1038/ncomms3905
62. Sur, U.K.: Graphene: a rising star on the horizon of materials science. Int. J. Electrochem. **2012**, 12 (2012). https://doi.org/10.1155/2012/237689. 237689
63. Gao, X., Ma, W., Han, G., Chang, Y., Zhang, Y., Li, H.: The electrochemical properties of reduced graphene oxide film with capsular pores prepared by using oxalic acid as template. Int. J. Energy Res. **43**, 8177–8189 (2019). https://doi.org/10.1002/er.4813

64. Pan, Y., Cheng, X., Zhou, T., Gong, L., Zhang, H.: Spray freeze-dried monolithic silica aerogel based on water-glass with thermal superinsulating properties. Mater. Lett. **229**, 265–268 (2018)
65. Wan, W., Zhang, F., Shang, Y., Ruiyang, Z., Ying, Z.: Hydrothermal formation of graphene aerogel for oil sorption: The role of reducing agent, reaction time and temperature. New J. Chem. **40** (2016). https://doi.org/10.1039/C5NJ03086B
66. Xue, B., Zheng-Fang, Y., Yan-Feng, L., Lin-Cheng, Z., Liuqing, Y.: Preparation of crosslinked macroporous PVA foam carrier for immobilisation of microorganisms. Process Biochem. **45**, 60–66 (2010). https://doi.org/10.1016/j.procbio.2009.08.003
67. Schneider, A., Picart, C., Senger, B., Schaaf, P., Voegel, J.C., Frisch, B.: Layer-by-layer films from hyaluronan and amine-modified hyaluronan. Langmuir. **23**(5), 2655–2662 (2007). https://doi.org/10.1021/la062163s
68. Razieh, S., Zahra, S.P., Mousa, G.: Novel magnetic bio-sorbent hydrogel beads based on modified gum tragacanth/graphene oxide: Removal of heavy metals and dyes from water. J. Cleaner Prod. **142** (2016). https://doi.org/10.1016/j.jclepro.2016.10.170
69. Sha, J., Li, Y., Salvatierra, R.V., Wang, T., Dong, P., Ji, Y., Lee, S.-K., Zhang, C., Zhang, J., Smith, R.H., Ajayan, P.M., Lou, J., Zhao, N., Tour, J.M.: Three-dimensional printed graphene foams. ACS Nano. **11**(7), 6860–6867 (2017). https://doi.org/10.1021/acsnano.7b01987
70. Wang, Z., Gao, H., Zhang, Q., Liu, Y., Chen, J., Guo, Z.: Small. **15**, 1803858 (2019). https://doi.org/10.1002/smll.201803858
71. Ma, Y., Chen, Y.: Three-dimensional graphene networks: synthesis, properties and applications. Natl. Sci. Rev. **2**(1), 40–53 (2015). https://doi.org/10.1093/nsr/nwu072
72. Sun, C., Liu, Y., Sheng, J., Huang, Q., Lv, W., Zhou, G., Cheng, H.-M.: Status and prospects of porous graphene networks for lithium–sulfur batteries. Mater. Horiz. (2020). https://doi.org/10.1039/D0MH00815J. Advance Article
73. Bulakhe, R.N., Nguyen, V.H., Shim, J.-J.: Layer-structured nanohybrid MoS_2@rGO on 3D nickel foam for high performance energy storage applications. New J. Chem. **41**, 1473–1482 (2017). https://doi.org/10.1039/C6NJ02590K
74. Samad, A., Li, Y., Yuan-Qing, A.S., Kin, L.: Novel graphene foam composite with adjustable sensitivity for sensor applications. ACS Appl. Mater. Interfaces. **7** (2015). https://doi.org/10.1021/acsami.5b01608
75. Das, S., Rajak, D.K., Khanna, S., Mondal, D.P.: Energy absorption behavior of Al-SiC-graphene composite foam under a high strain rate. Materials. **13**, 783 (2020)
76. Zongping, C., Wencai, R., Libo, G., Bilu, L., Songfeng, P., Hui-Ming, C.: Three-dimensional flexible and conductive interconnected graphene networks grown by chemical vapour deposition. Nat. Mater. **10**, 424–428 (2011). https://doi.org/10.1038/nmat3001
77. Nieto, A., Dua, R., Zhang, C., Boesl, B., Ramaswamy, S., Agarwal, A.: Three dimensional graphene foam/polymer hybrid as a high strength biocompatible scaffold. Adv. Funct. Mater. **25**, 3916–3924 (2015). https://doi.org/10.1002/adfm.201500876
78. Zhang, X., Zhang, H., Li, C., Wang, K., Sun, X., Ma, Y.: Recent advances in porous graphene materials for supercapacitor applications. RSC Adv. **4**, 45862–45884 (2014). https://doi.org/10.1039/C4RA07869A
79. Cancan, H., Bai, H., Chun, L., Ge, S.: A graphene oxide/hemoglobin composite hydrogel for enzymatic catalysis in organic solvents. Chem Commun (Cambridge, England). **47**, 4962–4964 (2011). https://doi.org/10.1039/c1cc10412h
80. Gil, S., Luigi, R., Peter, S., Willi, B., Rolf, M.: ChemInform abstract: palladium nanoparticles on graphite oxide and its functionalized graphene derivatives as highly active catalysts for the suzuki-miyaura coupling reaction. Cheminform. **40** (2009). https://doi.org/10.1002/chin.200945090
81. Shen, Y., Fang, Q., Chen, B.: Environmental applications of three-dimensional graphene-based macrostructures: adsorption, transformation, and detection. Environ. Sci. Technol. **49**(1), 67–84 (2015). https://doi.org/10.1021/es504421y
82. Duan, L., Kong, W., Yan, W., Li, C.-H., Jin, Z., Zuo, J.-L.: Improving the capacity and cycling-stability of Lithium–sulfur batteries using self-healing binders containing dynamic disulfide bonds. Sustainable Energy Fuels. **4**, 2760–2767 (2020)
83. Liang, Y., Zhao, C.-Z., Yuan, H., et al.: A review of rechargeable batteries for portable electronic devices. InfoMat. **1**, 6–32 (2019). https://doi.org/10.1002/inf2.12000
84. Xiaochen, D., Xuewan, W., Lianhui, W., Hao, S., Hua, Z., Wei, H.: Chen Peng. 3D graphene foam as a monolithic and macroporous carbon electrode for electrochemical sensing. ACS Appl. Mater. Interfaces. **4** (2012). https://doi.org/10.1021/am300459m
85. Bo, Y., Da, K., Sen, L., Chang, L., Tong, Z.: Template-assisted self-assembly method to prepare three-dimensional reduced graphene oxide for dopamine sensing. Sens. Actuators B Chem. **205**, 120–126 (2014). https://doi.org/10.1016/j.snb.2014.08.038
86. Cheng, C., Li, S., Thomas, A., Kotov, N.A., Haag, R.: Functional graphene nanomaterials based architectures: biointeractions, fabrications, and emerging biological applications. Chem. Rev. **117**(3), 1826–1914 (2017). https://doi.org/10.1021/acs.chemrev.6b00520
87. Geetha Bai, R., Muthoosamy, K., Manickam, S., Hilal-Alnaqbi, A.: Graphene-based 3D scaffolds in tissue engineering: fabrication, applications, and future scope in liver tissue engineering. Int. J. Nanomedicine. **14**, 5753–5783 (2019). https://doi.org/10.2147/IJN.S192779
88. Michel, M.: Bioscaffold-induced brain tissue regeneration. Front. Neurosci. **13** (2019). https://doi.org/10.3389/fnins.2019.01156
89. See: https://www.gepowerconversion.com/press-releases/compact-mighty-n37-induction-motor-ge-among-highest-power-density. 18.2.2020
90. Yuxi, X., Qiong, W., Sun, Y., Bai, H., Shi, G.: Three-dimensional self-assembly of graphene oxide and dna into multifunctional hydrogels. ACS Nano. **4**(12), 7358–7362 (2010). https://doi.org/10.1021/nn1027104
91. See: https://en.wikipedia.org/wiki/Linear_acetylenic_carbon. 18.02.2020
92. See: http://everything.explained.today/Linear_acetylenic_carbon/. 18.2.2020
93. See: http://self.gutenberg.org/articles/eng/Linear_acetylenic_carbon. 18.2.2020
94. See: https://www.electrochem.org/ecs-blog/carbyne-is-stronger-stiffer-than-any-known-material. 18.2.2020
95. Casari, C.S., Tommasini, M., Tykwinski, R.R., Milani, A.: Carbon-atom wires: 1-D systems with tunable properties. Nanoscale. **8**, 4414–4435 (2016). https://doi.org/10.1039/C5NR06175J
96. Maclay, W.N., Fuoss, R.M.: Polyelectrolytes. VII. Viscosities of derivatives of poly-2-vinylpridine. J. Polym. Sci. **6**, 511–521 (1951). https://doi.org/10.1002/pol.1951.120060501
97. Tanzawa, T., Gardiner, W.C.: Reaction mechanism of the homogeneous thermal decomposition of acetylene. J. Phys. Chem. **84**(3), 236–239 (1980). https://doi.org/10.1021/j100440a002
98. Dalle, K.E., Warnan, J., Leung, J.J., Reuillard, B., Karmel, I.S., Reisner, E.: Electro- and solar-driven fuel synthesis with first row transition metal complexes. Chem Rev. **119**(4), 2752–2875 (2019). https://doi.org/10.1021/acs.chemrev.8b00392
99. Kudryavtsev Yu, P., Sergey, E., Babaev, V.G., Guseva, M.B., Valerii, K., Krechko, L.M.: Oriented carbyne layers. Carbon. **30**, 213–221 (1992). https://doi.org/10.1016/0008-6223(92)90082-8
100. Carlo, C., Alberto, M.I.: Carbyne: from the elusive allotrope to stable carbon atom wires. MRS Commun. (2018). https://doi.org/10.1557/mrc.2018.48

101. Pavel, S., Hoonkyung, L., Liubov, A., Abhishek, S., Boris, Y.: Calcium-decorated carbyne networks as hydrogen storage media. Nano Lett. **11**, 2660–2665 (2011). https://doi.org/10.1021/nl200721v
102. Yuan, Q., Ding, F.: Formation of carbyne and graphyne on transition metal surfaces. Nanoscale. **6**, 12727–12731 (2014). https://doi.org/10.1039/C4NR03757J
103. Piedade, A.P., Cangueiro, L.: Influence of carbyne content on the mechanical performance of nanothick amorphous carbon coatings. Nanomaterials (Basel). **10**(4), 780 (2020). https://doi.org/10.3390/nano10040780
104. Astruc, D.: Metal-carbene and -carbyne complexes and multiple bonds with transition metals. Springer, New York (2007). https://doi.org/10.1007/978-3-540-46129-6_11
105. Kotrechko, S., Mikhailovskij, I., Tatjana, M., Evgenij, S., Timoshevskii, A., Stetsenko, N., Yuriy, M.: Mechanical properties of carbyne: experiment and simulations. Nanoscale Res. Lett. **10**, 24 (2015). https://doi.org/10.1186/s11671-015-0761-2
106. Lei, S., Philip, R., Marius, W., Angel, R., Sören, W., Stephanie, R., Sofie, C., Wim, W., Paola, A., Thomas, P.: Electronic band gaps of confined linear carbon chains ranging from polyyne to carbyne. Phys. Rev. Mater. **1**, 075601 (2017). https://doi.org/10.1103/PhysRevMaterials.1.075601
107. See: https://www.extremetech.com/extreme/163997-carbyne-a-new-form-of-carbon-thats-stronger-than-graphene. 20.02.2020
108. Masaharu, T., Shingo, K., Toshinori, M., Takeshi, T.: Formation of hydrogen-capped polyynes by laser ablation of C60 particles suspended in solution. Carbon. **41**, 2141–2148 (2003). https://doi.org/10.1016/S0008-6223(03)00241-0
109. Pan, B., Xiao, J., Li, J., Liu, P., Wang, C., Yang, G.: Carbyne with finite length: The one-dimensional sp carbon. Sci. Adv. **1**(9), e1500857 (2015). https://doi.org/10.1126/sciadv.1500857
110. See: https://en.wikipedia.org/wiki/Metallic_microlattice. 20.02.2020
111. Anna, T., Tobias, S., Alan, J., William, C., Lorenzo, V.: Characterization of nickel-based microlattice materials with structural hierarchy from the nanometer to the millimeter scale. Acta Materialia. **60**, 3511–3523 (2012). https://doi.org/10.1016/j.actamat.2012.03.007
112. Deshpande, V.S., Norman, F., Mike, A.: Effective properties of the octet-truss lattice material. Journal of the Mechanics and Physics of Solids. **49**, 1747–1769 (2001). https://doi.org/10.1016/S0022-5096(01)00010-2
113. Rashed, M.G., Mahmud, A., Mines, R.A.W., Paul, H.: Metallic microlattice materials: a current state of the art on manufacturing, mechanical properties and applications. Mater. Des. **95**, 518–533 (2016). https://doi.org/10.1016/j.matdes.2016.01.146
114. Kiju, K.: Wire-woven cellular metals: the present and future. Prog. Mater. Sci. **69** (2014). https://doi.org/10.1016/j.pmatsci.2014.11.003
115. do Nascimento, R.M., Martinelli, A.E., Buschinelli, A.J.A.: Review article: recent advances in metal-ceramic brazing. Cerâmica. **49**(312), 178–198 (2003). https://doi.org/10.1590/S0366-69132003000400002
116. Manuela, G., Luca, I.: A literature review of powder-based electron beam melting focusing on numerical simulations. Addit. Manuf. **19** (2017). https://doi.org/10.1016/j.addma.2017.11.001
117. Rashed, M.G., Ashraf, M., Mines, R.A.W., Hazell, P.J.: Metallic microlattice materials: a current state of the art on manufacturing, mechanical properties and applications. Mater. Des. **95**, 518–533 (2016)
118. Deshpande, V.S., Mike, A., Norman, F.: Foam topology: bending versus stretching dominated architectures. Acta Materialia. **49**, 1035–1040 (2001). https://doi.org/10.1016/S1359-6454(00)00379-7
119. Maskery, I., Aremu, A.O., Parry, L., Wildman, R.D., Tuck, C.J., Ashcroft, I.A.: effective design and simulation of surface-based lattice structures featuring volume fraction and cell type grading. Mater Des. **155**, 220–232 (2018). https://doi.org/10.1016/j.matdes.2018.05.058. ISSN 0264-1275
120. Tobias, S., Christopher, R., Adam, S., Zak, E., Sophia, Y., William, C., Alan, J.: Designing metallic microlattices for energy absorber applications. Adv. Eng. Mater. **16** (2014). https://doi.org/10.1002/adem.201300206
121. Miller, D.L., Kersten G., Frost W.A.. Systems and methods for protecting a flight recorder, US8723057 (2014)
122. Alan, J., William, C., Steven, N.: Micro-scale truss structures with three-fold and six-fold symmetry formed from self-propagating polymer waveguides. Acta Materialia. **56**, 2540–2548 (2008). https://doi.org/10.1016/j.actamat.2008.01.051
123. Casper, A., Erik, A., Jakob, J., Ole, S.: On the realization of the bulk modulus bounds for two-phase viscoelastic composites. J. Mech. Phys. Solids. **63**, 228–241 (2014). https://doi.org/10.1016/j.jmps.2013.09.007
124. Meza, L., Greer, J.: Mechanical characterization of hollow ceramic nanolattices. J. Mater. Sci. **49** (2014). https://doi.org/10.1007/s10853-013-7945-x
125. Lucas, M., Satyajit, D., Julia, G.: Strong, lightweight, and recoverable three-dimensional ceramic nanolattices. Science (New York, N.Y.). **345**, 1322–1326 (2014). https://doi.org/10.1126/science.1255908
126. Wang, H., Zhang, X., Wang, N., et al.: Ultralight, scalable, and high-temperature-resilient ceramic nanofiber sponges. Sci. Adv. **3**(6), e1603170 (2017). https://doi.org/10.1126/sciadv.1603170
127. S Rezaee, G.R., Rashed, M.A.: Golozar. Electrochemical and oxidation behavior of yttria stabilized zirconia coating on zircaloy-4 synthesized via sol-gel process. Int. J. Corrosion. **2013**, 9 (2013). https://doi.org/10.1155/2013/453835. 453835
128. Liu, B., Sun, Y., Sun, Y., Zhu, Y.: Fabrication and compressive behavior of carbon-fiber-reinforced cylindrical foldcore sandwich structure. Composites A Appl. Sci. Manuf. **118**, 9–19 (2019). https://doi.org/10.1016/j.compositesa.2018.12.011. ISSN 1359-835X
129. See: https://www.azom.com/article.aspx?ArticleID=8097. 22.02.2020
130. Claar, T., Yu, C., Hall, I., Banhart, J., Baumeister, J., Seeliger, W.: Ultra-lightweight aluminum foam materials for automotive applications. SAE Trans. **109**, 98–106 (2000) Retrieved August 27, 2020, from http://www.jstor.org/stable/44643817
131. See: http://www.issp.ac.ru/ebooks/books/open/Powder_Metallurgy.pdf. 22.02.2020
132. John Banhart: Manufacturing routes for metallic foams. JOM. **52**(12), 22–27 (2000)
133. Rajak, D.K., Mahajan, N.N., SenthilKumaran, S.: Fabrication and experimental investigation on deformation behaviour of AlSi10Mg foam-filled mild steel tubes. Trans. Indian Inst. Metals. (2020). https://doi.org/10.1007/s12666-020-01879-y
134. Kennedy, A.: Porous metals and metal foams made from powders. IntechOpen, London (2012). https://doi.org/10.5772/33060
135. García-Moreno, F.: Commercial applications of metal foams: their properties and production. Materials. **9**, 85 (2016). https://doi.org/10.3390/ma9020085
136. CERAMIC ABSTRACTS. J. Am. Ceramic Soc. **47**, 209–242. https://doi.org/10.1111/j.1151-2916.1964.tb13830.x
137. Banhart, J., Baumeister, J. (1998). Production methods for metallic foams. https://doi.org/10.1557/PROC-521-121
138. Tiryakioğlu, M.: The effect of hydrogen on pore formation in aluminum alloy castings: myth versus reality. Metals. **10**(3), 368 (2020). https://doi.org/10.3390/met10030368
139. Dipen Kumar Rajak, L.A., Kumaraswamidhas, S., Das, S., Kumaran, S.: Characterization and analysis of compression load

behavior of aluminum alloy foam under the diverse strain rate. J. Alloys Compounds. **656**, 218–225 (2016)
140. John, B.: Manufacture, characterization and application of cellular metals and metal foams. Prog. Mater. Sci. **46**, 559–5U3 (2001). https://doi.org/10.1016/S0079-6425(00)00002-5
141. John, B.: Manufacturing routes for metallic foams. JOM. **52**, 22–27 (2012). https://doi.org/10.1007/s11837-000-0062-8
142. Matz, A.M., Mocker, B.S., Müller, D.W., Jost, N., Eggeler, G.: Mesostructural design and manufacturing of open-pore metal foams by investment casting. Adv. Mater. Sci. Eng. **2014**, 9 (2014). https://doi.org/10.1155/2014/421729. 421729
143. Ding, Y., Tao, X., Onyilagh, O., Fong, H., Zhu, Z.: Recent advances in flexible and wearable pressure sensors based on piezoresistive 3D monolithic conductive sponges. ACS Appl. Mater. Interfaces. **11**(7), 6685–6704 (2019). https://doi.org/10.1021/acsami.8b20929
144. Zhang, Q., Yunxia, G., Jing, L.: Atomized spraying of liquid metal droplets on desired substrate surfaces as a generalized way for ubiquitous printed electronics. Appl. Phys. A. **116** (2013). https://doi.org/10.1007/s00339-013-8191-4
145. Rajak, D.K., Gawande, S.A., Kumaraswamidhas, L.A.: Evaluation of mild steel hollow and foam filled circular tubes under axial loading. Adv. Mater. Lett. **9**(9), 660–664 (2018)
146. Yifeng, F., Josef, H., Ya, L., Shujing, C., Abdelhafid, Z., Majid, K. S., Nan, W., Yuxiang, N., Yan, Z., Zhi-Bin, Z., Qianlong, W., Mengxiong, L., Hongbin, L., Marianna, S., Clivia, M.S.T., Sebastian, V., Alexander, A.B., Xiangfan, X., Johan, L.: Graphene related materials for thermal management, 2d materials. **7**(1), 1–42 (2019)
147. Seeliger, H.W.: Aluminium foam sandwich (AFS) ready for market introduction. Adv. Eng. Mater. **6**, 448–451 (2004). https://doi.org/10.1002/adem.200405140
148. Pan, H.: Development and application of lightweight high-strength metal materials. MATEC Web Conf. **207**, 03010 (2018). https://doi.org/10.1051/matecconf/201820703010
149. Rajak, D.K., Kumaraswamidhas, L.A., Das, S.: An investigation on axial deformation behaviour of thin-wall unfilled and filled tube with aluminium alloy (Al-Si7Mg) foam-reinforced with SiC particles. J. Mater. Eng. Perform. **25**, 3430–3438 (2016)
150. Rajak, D.K., Nikhil, N.M., Das, S.: Fabrication and investigation of influence of CaCO3 as foaming agent on Al-SiCp foam. Mater. Manuf. Processes. **34**(4), 379–384 (2019)
151. Dipen Kumar Rajak, L.A., Kumaraswamidhas, S.: Das, Experimental fabrication and compression analysis characterization of LM30 Al-alloy foam with 5wt% SiCp at room temperature. Mater. Res. Expr. **5**, 066526 (2018)
152. André, S., Urs, G., Elena, T., Ludwig, G.: Processing routes to macroporous ceramics: a review. J. Am. Ceramic Soc. **89**, 1771–1789 (2006). https://doi.org/10.1111/j.1551-2916.2006.01044.x
153. Slavka, T., Zlatina, M., Konstantin, G., Nikolai, D., Martin, V., Ananthapadmanabhan, K.: Control of Ostwald ripening by using surfactants with high surface modulus. Langmuir ACS J. Surf. Colloids. **27**, 14807–14819 (2011)
154. Studart, A.R., Gonzenbach, U.T., Tervoort, E., Gauckler, L.J.: Processing routes to macroporous ceramics: a review. J. Am. Ceramic Soc. **89**, 1771–1789 (2006). https://doi.org/10.1111/j.1551-2916.2006.01044.x
155. Jan, L., Thijs, I., Vandermeulen, W., Steven, M., Wallaeys, B., Mortelmans, R.: Strong ceramic foams from polyurethane templates. Adv. Appl. Ceramics. **104**, 4–8 (2005). https://doi.org/10.1179/174367605225010990
156. Muhammad, N., Bao, X., Edirisinghe, J.: Processing of ceramic foams from polymeric precursor-alumina suspensions. Cell. Polym. **20**, 17–35 (2001). https://doi.org/10.1177/026248930102000102
157. Osayi, J.I., Iyuke, S., Ogbeide, S.E.: Biocrude production through pyrolysis of used tyres. J. Catalysts. **2014**, 9 (2014). https://doi.org/10.1155/2014/386371. 386371
158. Yanbin, F., Wang, X., Wang, L., Li, Y.: Foam concrete: a state-of-the-art and state-of-the-practice review. Adv. Mater. Sci. Eng. **2020**, 25 (2020). https://doi.org/10.1155/2020/6153602. 6153602
159. Leitao, C., Qiuxia, F., Yang, S., Bin, D., Jianyong, Y.: Porous materials for sound absorption. Composites Commun. **10**, 25–35 (2018). https://doi.org/10.1016/j.coco.2018.05.001
160. Roucher, A., Schmitt, V., Blin, J., et al.: Sol–gel process and complex fluids: sculpting porous matter at various lengths scales towards the Si(HIPE), Si(PHIPE), and SBA-15-Si(HIPE) series. J. Sol-Gel Sci. Technol. **90**, 95–104 (2019). https://doi.org/10.1007/s10971-018-4794-8
161. See: https://www.eea.europa.eu/themes/climate/ozone-depleting-substances-and-climate-change. 24.02.2020
162. Jason, L., Selina, Y., Guangming, L., Martin, J., Patrick, L.: Measurement methods for solubility and diffusivity of gases and supercritical fluids in polymers and its applications. Polym. Rev. **57** (2017). https://doi.org/10.1080/15583724.2017.1329209
163. Bo-Yeon, L., Jiyoon, K., Hyungjin, K., Chi-Woo, K., Sin-Doo, L.: Low-cost flexible pressure sensor based on dielectric elastomer film with micro-pores. Sens. Actuators A Phys. **240** (2016). https://doi.org/10.1016/j.sna.2016.01.037

Biomaterials

13

Abstract

In modern society, diseases have been increasing in humans as well as domestic animals because of pollution, accident, and lifestyle. The mutilation in the human body leads to expand the need for the replacement of tissues/organs where the availabilities of sources for tissues/organs is limited. Creating artificial tissues/organs for the replacement of damaged, dysfunctional tissues/organs becomes a big discipline in material science. This chapter describes the brief idea on the requirement of materials for implant inside or outside the body, material–body fluid interface, and interactions. The main governing factors associated with choosing the material as a biomaterial have been described. Typically, five classes of biomaterials such as metallic, ceramic, polymeric, composite, and natural biomaterials are discussed with their modification and application in various parts of the body. Synthesis processes and surface modifications have been presented to develop better biocompatible materials.

Keywords

Biomaterials · Biocompatibility · Wettability · Porosity · Stability · Metallic biomaterials · Ceramics biomaterials · Polymeric biomaterials · Biocomposite · Biologically derived biomaterials · Biocompatible coating · Surface treatment

13.1 Introduction

When humans face accident leading to damage of any part, the doctor looks for a material to replace the part or to treat the part. Those most wanted materials are the focus of interest to study biomaterials. Biomaterials are the bioactive materials used every day in biomedical applications, in dental implants, in case of surgery, and in drug delivery [1–3]. These biomaterials fulfill the purpose of medical science. A biomaterial is defined as any constituent that has been developed to cooperate with living systems for a healing purpose—either helpful in the augmenting, treatment, healing or in exchanging the biological role with human as well as domestic animal body and for the diagnostic purpose. The study and application of bioactive and biocompatible materials are known as biomaterials science and biomaterials engineering, respectively [4]. The word "bio" is related to the word "biological compatibility." The term "biomedical material" is used to signify a module of any biomedical device useful with or without intimate contact with tissue system of living body, whereas the definition of "biomaterials" was established to define the compound or composite used only in friendly commerce with active tissues system [5] and should be biocompatible with specific application [6]. In the laboratory, biomaterials are formulated smoothly by means of chemical approaches from polymers, utilizing metallic components as composite materials, ceramics, or extracted from natural resources. The market price for biomaterial is estimated to increase from USD 105 billion to USD 207 billion at a CAGR of 14.5% between 2019 and 2024 [7]. Better development of novel biomaterials, rising demand for medical implants, and the rising incidence of cardiovascular diseases, where factors like increasing funds and grants by government policies worldwide, are driving the growth of the market. Countries such as China, India, and Japan are rolling for the huge opportunity to merge in the biomaterial market.

13.2 History of Biomaterials

The main texts of medicine, "Charaka Samhita," "Sushruta Samhita," and "Astanga Hridaya," are obtained from the three old Indian books called The Great Triology which are about more than 2000 years old [8, 9]. These books are the nucleation point toward the development of biocompatible materials in the world. In the early stage of the nineteenth century, artificial threads, bands, and tubes were used in

A. Behera, *Advanced Materials*, https://doi.org/10.1007/978-3-030-80359-9_13

Table 13.1 Gradual development of various biocompatible materials

Materials developed	Patent filled year
Cobalt-based implant	1930
Artificial threads, bands, tubes, and the like for surgical and other purposes	1932
High molecular weight linear polyester-amides, Vitallium implant	1940
Individual tooth implant	1956
Artificial tooth, intra-osseous pins, and posts and their use	1965
Anchor screw for dental prosthesis	1967
Drug delivery device	1971
Dental implants	1972
Hip joint prosthesis sockets	1973
Microporous drug delivery device	1974
Implants for bones, joints, and tooth roots, Endosseous plastic implant	1975
Socket implant	1976
Ceramic endosseous implant	1977
Guide wire for catheters	1983
Hydroxylapatite or calcium triphosphate bone implant	1984
Bone screw, stent	1985
Ceramic-coated metal implants, PGA polymer implants	1986
Endosteal implant, Jaw-bone implant	1987
Adjustable compression bone screw, preloading spinal implants	1988
Cast dental implant abutment	1989
Load sharing femoral hip implant	1990
Vascular stent, Nitinol stent	1991
Fixation in skeletal bone	1992
Absorbable structures for ligament and tendon repair	1993
Injectable polyethylene oxide gel implant	1994
Implantable device with an internal electrode to promote tissue growth, Biocompatible Ti-implant	1995
Bioactive-gel compositions	1997
Multilayered metal stent	1997
Interlocking spinal inserts, Drug-releasing stent with a ceramic layer	1998
Bone anchoring system	1999
Orthopedic implant with sensors, Polyactive/polylactic acid coatings for an implantable device, Polymer film, Poly(ester amide) block copolymers	2004
Coatings on medical implants to guide soft tissue healing, Surgical scaffold, Polymeric stent patterns	2005
Nanoshells polymers	2006
Nano-patterned implant surfaces, Drug-Eluting Endoprosthesis	2007
Modular lag screw	2013
Triblock polymer coating	2017

surgical and other applications. Most medical implants are made up of metallic materials and can be sketched back to the middle of the nineteenth century, the era when the industrial revolution expanded the metal industry. After the nineteenth century, application of ceramics was seen in load-bearing bioimplant [10]. Stainless steel, Mg-based alloys, Ti-based alloys, Co-based alloys, NiTi alloys, and so on are the few recommended products that provide long-term success and biocompatibility despite a large number of metals and alloys produced in industries [11]. The gradual development of various biomaterials of polymer, ceramic, and composite is shown in Table 13.1. Currently, with the development of various materials, the number of long-lasting suitable biomaterials is increasing.

13.3 The Body Environment

The study of the body environment can evaluate the biocompatibility of any biomaterial. The human body contains 40–60% water of its total mass, where extracellular and intracellular fluids are the major fluid compartments [12]. Extracellular fluid (ECF) contains plasma that is found in the blood vessels and the interstitial fluid is the fluid that surrounds the cells, the lymph, and transcellular fluids (e.g., synovial fluids and cerebrospinal fluid), whereas Intracellular fluid (ICF) is a fluid part inside the cells. Homeostasis is the mechanism through which the distribution of body fluids and electrolytes are kept normal and constant. In these fluid

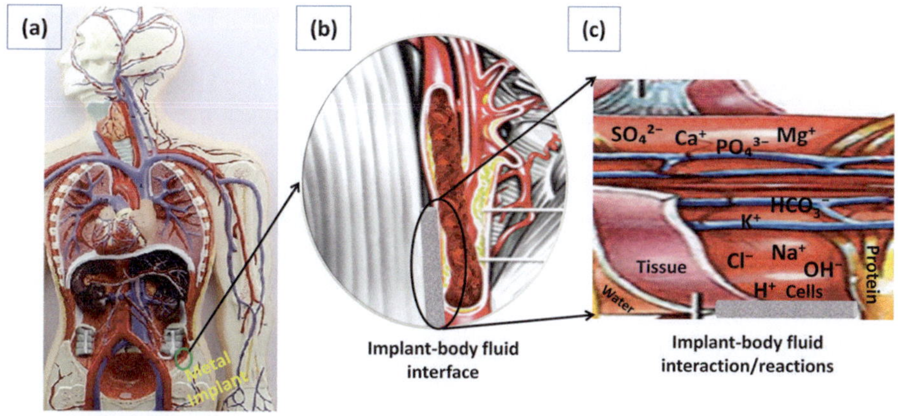

Fig. 13.1 (**a**) Metal implant at hip, (**b**) Magnified image of implant and body fluid (blood) interface, and (**c**) Cationic–anionic evolution at the interface

compartments, electrolyte plays a major role in organ function. The major metal ions take part in metabolism to regulate the osmolality and cell membrane potentials of body fluids and so forth as shown in Fig. 13.1. Major cations include H^+, Na^+, Ca^+, K^+, and Mg^+ ions. Major anions include HCO_3^-, OH^-, Cl^-, SO_4^{2-}, and PO_4^{3-} ions [13]. Soluble ions are probably the most significant components for implant deterioration in vivo. Chloride ions (and other halides) are the major ions that can corrode almost all metals and impede many protection methods. Hydrogen ion concentration and temperature are the two most important factors that affect the corrosion activities of materials with normal conditions, where body fluids have a temperature of 37 °C throughout the lifespan of an implanted material. Some pathogens during infection can cause electrochemical variation in the equilibrium when implants are placed in vivo and through it affect the healing process. In general and in the physiological condition, the two reduction reactions, hydrogen evolution reaction (HER) and oxygen evolution reaction (OER), are the two chief reduction reactions [14]. The hydrogen evolution reaction in acid and alkaline solutions is given in Eqs. (13.1) and (13.2), respectively. The oxygen evolution reaction with both acid and alkaline solutions is given in Eqs. (13.3) and (13.4), respectively.

$$2H^+ + 2e^- \rightarrow H_2 \tag{13.1}$$

$$2H_2O + 2e^- \rightarrow H_2 + 2(OH)^- \tag{13.2}$$

$$O_2 + 4H^+ + 4e^- \rightarrow 2H_2O \tag{13.3}$$

$$2(OH)^- \rightarrow O_2 + 4e^- + 2H_2O \tag{13.4}$$

The electrochemical reaction with concentration, temperature, and standard electrode potential of these chemical species undergoing the reaction is represented by the Nernst equation. The Nernst equation (E) for hydrogen evolution reaction and oxygen evolution reaction is given in Eqs. (13.5) and (13.6), respectively.

$$\begin{aligned} E &= 2.3026\frac{RT}{2F}\log\frac{[H^+]^2}{H_2} \\ &= -\frac{2.3026RT}{2F}\log[H_2] - \frac{2.3RT}{F}P^H \end{aligned} \tag{13.5}$$

$$E = +1.229 + \frac{2.3026RT}{4F}\log[O_2] - \frac{2.3026RT}{F}P^H \tag{13.6}$$

where $[H_2]$ and $[O_2]$ denote the partial pressures of hydrogen and oxygen, respectively.

In addition to the above, other possible reduction reactions at the implant are given in Eqs. (13.7)–(13.10).

$$O_2 + 2H_2O + 2e^- \rightarrow H_2O_2 + 2(OH)^- \tag{13.7}$$

$$H_2O_2 + 2e^- \rightarrow +2(OH)^- \tag{13.8}$$

$$O_2 + H_2O + e^- \rightarrow HO_2^+ + (OH)^- \tag{13.9}$$

$$HO_2^+ + H_2O \rightarrow H_2O_2 + (OH)^+ \tag{13.10}$$

From these equations, the resulting intermediate species in the form of ion or free radicals considerably affect the biological system by prompting oxidative stress in the cell. In reduction reaction, the species inhibit the process of oxidation in vivo, for example, reduction of disulfide bonds and other protein-like molecules [15]. The agents causing a local redux atmosphere around a metallic implant can impact the redox state of cells.

The hydrogen ion concentration $P^H = 7$ when when $H^+ = (OH)^-$ for neutrality at 25 °C is called neutral environment. However, P^H depends on temperature at neutrality. The P^H of the body fluid ranges from 7.4 to 5.5, and the hydrogen ion concentration could take 10 to 15 days to recuperate its usual value of approximately 6.81. The implant surface P^Hchanges from 4.0 to 9.0 due to some bacterial infection which is from acidic to alkaline, respectively. Enzymes play a major role in homeostasis, metabolism that can alter the acid–alkaline environment. In case of acidosis, the blood plasma P^H is nearly at 2.35, and P^H 7.45 regarded as alkalosis. Buffers can resist P^H change, where two major mechanisms control the body P^H, that is, chemical and physiological. Bicarbonate, phosphate, and protein buffers combine immediately with any added acid or alkali and interact rapidly which can prevent drastic changes in H^+ concentration in body fluids. In this rapid interacting situation, chemical buffers cannot stabilize the P^H, where respiratory and urinary response systems serve as secondary defense against accidental shift in P^H. The output of acids and bases of CO_2 is controlled by the physiological buffer system [16]. The urinary response system requires several hours and a greater quantity of buffers, whereas the respiratory response system buffers work rapidly. For arterial and venous blood, the partial pressure of oxygen in blood varies between 100 and 40 mm Hg, respectively. The corresponding value in air is 160 mm Hg where most biomaterials rely on oxygen to repassivate. The repassivation of metal surfaces is more difficult under low concentration of dissolved oxygen. Many food stuffs, basically carbohydrate, on degradation by amylase give glucose and further glucose is converted to gluconic acid due to oral pathogen. The gluconic acid and high chloride content are more corrosive than saliva, especially for teeth and other tissue systems. Fluoride-containing dental products and solutions are harmful to the passive layer, and some preparations with 2 wt% fluoride varnishes are rarely used by dentists [17]. The toxicity may be due to metal leaching out from nitinol, amalgams, and other dental materials that might cause oral cancer in case of repair or replacement of dental implants is the recent concern.

13.4 Governing Factors of Biomaterials

13.4.1 Biocompatibility

To replicate the biological performance of a specific body part, materials should be biocompatible in nature. It means compatibility or harmony of the materials with the living system is a must. Biocompatibility can be defined as "the capability of a material to achieve accomplishments with a proper host response in a precise application" [18]. Biomaterials should be nonallergenic, nonpyrogenic, noncarcinogenic, and nontoxic. Any discrepancy is considered as bio-incompatibility of the materials which form some effects as shown in Fig. 13.2. Bio-incompatible factors are inflammation, swelling, redness, immune system deactivation, warmth, infection, pain, blood clot, tumor formation, and implant calcification. Bio-functionality (B) is referred to as a set of characters where the material should be used safe and sound in biological organisms as given in Eq. (13.11). It must not be harmful to the local and systemic environment of the host (soft tissues, bone, ionic composition of plasma, as well as intra and extracellular fluids).

$$B = f(X_1, X_2 \ldots\ldots Xn) \tag{13.11}$$

Where X: material, design, stress distribution, articulation to allow movement, etc.

13.4.2 Wettability

Surface wettability is the process that can control protein adsorption and deals with cell attachment. During the implantation, wetting on the implant surface is due to the interaction

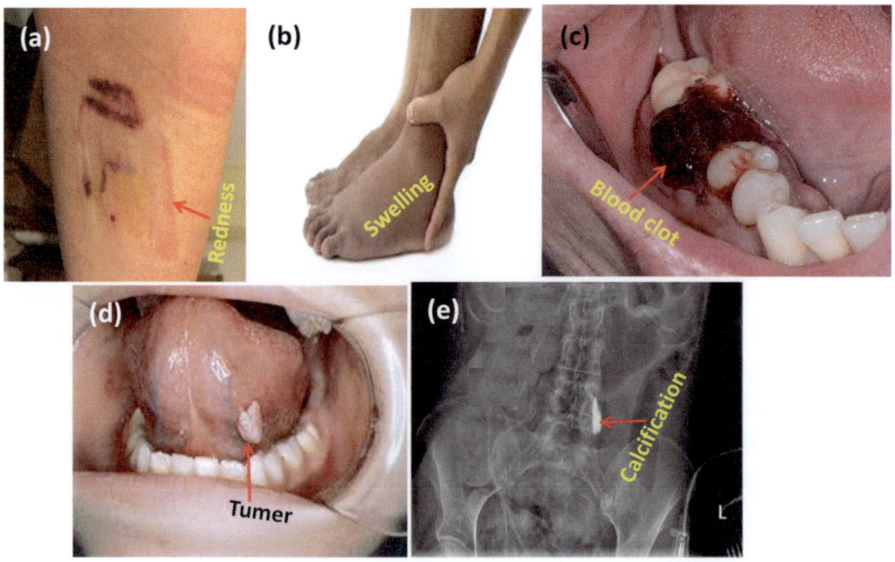

Fig. 13.2 (**a**) Redness, (**b**) Swelling, (**c**) Blood clot after dental surgery, (**d**) Tumor formation, and (**e**) Implant calcification

with the physiological fluid. Wetting is a measure of a fluid's ability to spread out on a solid substrate. Wettability mainly depends on the type of host materials, surroundings, and the type of implant materials. The cells composed of protein typically adhere selectively to the hydrophilic regions and less in a hydrophobic area, where the cell behavior is also dependent on the cell type and the material where few cells can adhere to superhydrophobic surfaces and want to adopt a round morphology upon attachment [19]. The decrease in contact angle is due to more hydrophilic –COOH groups whereas the increase in contact angle is due to hydrophobic –CH_3 groups on the surface [20].

13.4.3 Porosity

Suitable implant with macro/micro porosity leads to the successful growth of bone tissues. The implant material structure should be porous with about pore diameter of >100 μm [21]. Optimal pore diameter of 200–350 μm is preferred for good growth of bone tissue. It is very essential for the diffusion of essential nutrients, oxygen, and extracellular fluid with proper interconnectivity between the pores out of the cellular matrix. Other factors such as pore diameter, pore volume, pore size distribution, pore throat size, pore shape, and pore wall roughness are also taken into consideration for proper tissue growth [22].

Macro porosity of 150–900 μm satisfied the diffusion of waste and nutrient supply to the cellular network and microporosity having a pore size of less than 10 μm cares capillary, cell–matrix interactions, and vascularization where both macro- and micro-porosity play a vital role in tissue growth inside the scaffold environment. Uniform dimension and distributions of pores enhance the interaction and proper attachment of cells with applied scaffold [23].

13.4.4 Stability

Tensile strength, elastic modulus, yield strength, fatigue resistance, corrosion and surface finish, hardness, and creep-like properties should be wisely planned and evaluated in their applications. Physical properties such as mechanical strength of the organ to be fabricated may degrade or get resorbed in the body until a fully functional organ is formed. Higher mechanical strength, as well as higher corrosion resistance, wears behavior, and fatigue resistance should be studied for the stability of biomaterials [24].

13.5 Classification of Biomaterials

The invention of various biocompatible materials as biomaterials carried out by continuous research can be classified into two main groups: synthetic and natural biomaterials [25, 26]. Synthetic biomaterials are classified as metals, ceramics, biodegradable polymers, nonbiodegradable polymers, and composite materials (Fig. 13.3).

13.5.1 Metallic Biomaterials

Biomedical wires, pins, screws, and plates are the metallic preparations used almost exclusively for load-bearing implants, such as hip and knee prostheses and fracture fixation. In some cases, pure metals are used, whereas alloys commonly improve in material properties, such as strength and corrosion resistance and so on. Stainless steel, cobalt–chromium–molybdenum alloy, and titanium and titanium alloys are the three metal material groups governed by biomedical metals [27, 28]. These selected metals and alloys in biomedical applications are encouraged for their excellent

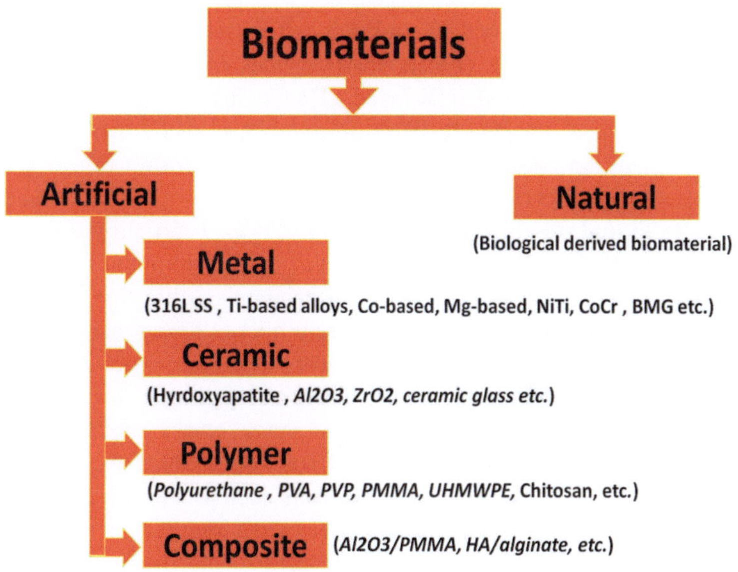

Fig. 13.3 Classification of biomaterials

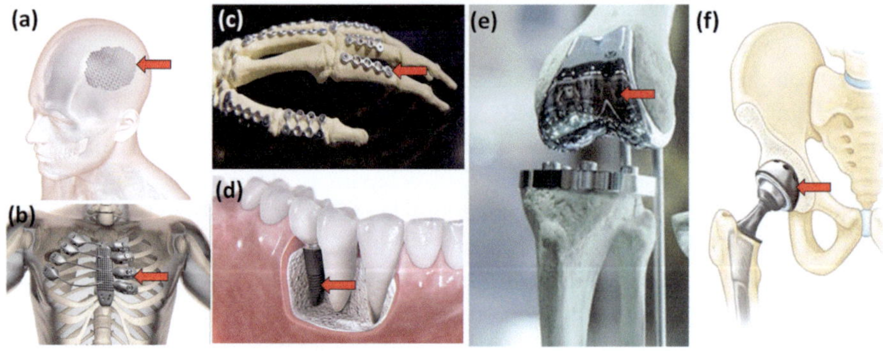

Fig. 13.4 Metal biomaterial implants used at marked position (**a**) Skull bone, (**b**) Chest rips, (**c**) Finger bone support, (**d**) Dental, (**e**) Knee, and (**f**) Hip

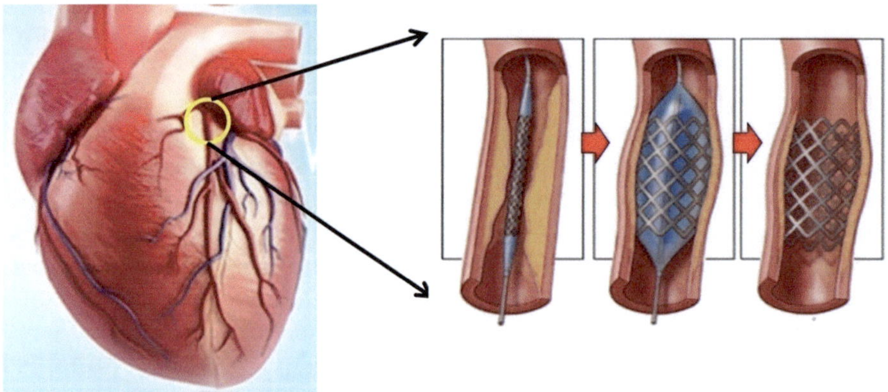

Fig. 13.5 Biocompatible NiTi smart alloy stent implanted in heart showing their transformation from the deformed shape to undeformed shape

thermal and electrical conductivity, biocompatibility, appropriate mechanical properties, corrosion resistance, and reasonable cost. Figure 13.4 shows some of the metallic implants at various places of human body.

13.5.1.1 Materials in Metallic Biomaterials

Stainless steel: One of the commonly investigated metals is 316L stainless steel (316L SS) in the surgical field due to its biocompatibility, attractive mechanical properties, corrosion resistance, cost-effectiveness, and easy availability. Metallic biomaterial in the form of plates, screws, and nails was suggested first for its implant application. The metal material contains lower carbon content that provides an equitable tendency of corrosion resistance. Under high stressed conditions and oxygen-depleted surroundings inside the body, there are enormous possibilities for the corrosion of 316L stainless steel [29].

CP Ti and Ti-alloys: Pure titanium (CP-Ti) is an alloy composed of titanium and oxygen. The oxygen content must be less than 0.5% to satisfy the British standard specification for use in surgical implants. The closely packed hexagonal structure of this alloy contains elements such as oxygen, nitrogen, and carbon having a greater solubility in the close-packed hexagonal structure at α-phase, where cubic form of β-phase and the oxygen in solution denotes the metal at single phase. Transition elements such as Mo, Nb, and V act as β-phase stabilizer, where elements form interstitial solid solutions with titanium to stabilize the α-phase [30]. There are three structural types of titanium alloys: Alpha (α), Alpha-Beta (α-β), or metastable β and Beta (β). The β-phase Ti alloys tends to exhibit a much lower modulus than α phase, which satisfies most of the other desires for orthopedic application. At 883 °C, Ti undergoes phase transformation which results from a hexagonal close-packed α-phase to a body-centered cubic β-phase also alloying elements added to stabilize the phase [31]. Aluminum and Vanadium are added for increasing the strength over that of CP-Ti to Ti alloy; while Aluminum is regarded as α-phase stabilizer and vanadium is a β-phase stabilizer. The stabilizing agents are added to minimize the formation of $TiAl_3$ to around 6% or less to decrease the susceptibility towards corrosion of the alloy. Both α- and β-forms can exist at room temperature, but the temperature depresses the α–β transition. Ti-6%Al-4%V has a two-phase structure of α and β grains. If more hardness is desired, addition of Mo, Nb, and Sn can be done. Both α and β-phase can be found out in Ti-6Al-7Nb and Ti-6Al-4Fe [32].

Cobalt-based alloy: In the 1930s, cobalt-based alloys were first used in medical implants. Cobalt-based biomaterial is one of the hardest and biocompatible alloys which shows its durability and versatility in the application of orthopedic implants for several years with high wear properties and corrosion resistance and the first used in dental implant. However, recently, the use has been broadened up to joint and fracture fixation. Numerous biocompatible Co-based composite exists, including CoCrMo, CoCr, and Ni-free

CoCrW alloys [33]. The basic classification of CoCr alloys are of two types: one is the CoCrMo alloy [Cr (27–30%), Mo (5–7%), Ni (2.5%)] which has been used for many decades in making artificial joints and dentistry, and the second one the CoNiCrMo alloy [Cr (19–21%), Ni (33–37%), and Mo (9–11%)] has been used for making the stems of prostheses for knee and hip which are heavily loaded joints. Due to the spontaneous formation of the passive oxide layer, chloride ions formed more, whereas Co-based alloys are highly resistant to corrosion in this environment within the human body. The microstructure modification of alloy alters the electrochemical and mechanical properties through thermal treatments to CoCrMo alloys and can be used as biomaterials. The corrosive products of CoCrMo are more toxic than those of stainless steel 316 L [34].

NiTi smart alloys (nitinol): NiTi implants are explored in various applications as it possess a mixture of novel properties such as shape memory effect, super plasticity enhanced biocompatibility, and high damping properties. Owing to these promising effects, it finds wide presentations in the industrial and medical fields, where the biomedical applications include intravascular stents (Fig. 13.5), orthodontic wires for dental, staples for foot surgery, bone fracture fixtures, and so on. The intramedullary nails and spinal implants are mainly made up of porous NiTi, where the elastic modulus of the porous nitinol implants is closer to that of the bone [35]. The porous behavior of this biomaterial permits tissue/bone cell penetration and integration. NiTi alloys implanted in soft tissues and in vitro experiments show brilliant biocompatibility in vivo. The released Ni from the implant induces allergic reaction, which leads to severe local tissue necrosis, irritation, and toxic reactions in an elevated concentration. Ni-rich NiTi with smart properties is used as a suitable implant [36].

Mg-alloys: Magnesium wires and their implants are used in biomedical applications such as draining vessels, cardiovascular, musculoskeletal, and general surgery toward the late nineteenth century. The unique composition of an alloy can affect the strength, ductility, and corrosion properties of the material. Generally, these alloys contain aluminum or rare earth elements. While these alloys are still being widely used in the biomedical field, another branch of magnesium research has concentrated on the production and assessment of Mg alloys containing arguably nontoxic elements such as Ca, Mn, Zn, and Zr [37]. These alloys can be binary, ternary, or more, with both components. The composition of the alloys is contributing significantly to different mechanical properties and corrosion behaviors. It pointedly decreases the stress shielding impact of bone repair. The requirement of secondary surgical treatment for removing the Mg implant is evitable because magnesium degrades in vivo, and it is excreted from urine [38].

Zinc Alloys: Zinc is another trace element that performs a significant role in the structure and function of proteins and is essential to catalytic functions in more than 300 enzymes, stabilizing protein subdomains, for example, DNA-binding domains of eukaryotic transcription factors, RNA polymerases, and accessory proteins engaged in nucleic acid replication inducing folding behavior. In this era, Zn alloys have bioresorbable metallic stent applications, where most tissue systems have good acceptance of excess Zn ions. The corrosion of Zn depends on anodic dissolution, cathodic reduction, and pH of the surrounding. ZnCl and ZnO begin to participate when the hydrogen ion concentration is 7.3 (p^H) in the corrosion process. The corrosion rate of pure Zn is lower than that of pure Mg and is not accompanying the evolution of hydrogen gas. Pure Zn wire provides compatibility and did not lead to inflammatory response or extensive thrombosis and shows considerable tissue multiplication within the partially degraded stent when used as stents in rat arteries for up to 6 months. The combination of Zn with Mg increases the corrosion rate with an improvement in mechanical properties of the applied alloy [39].

Iron Alloys: Like Zn implants, Fe alloys go through local corrosion facilitated by dissolved oxygen, a procedure that does not form any hydrogen gas. Iron has the lowest tendency to dissolve among biodegradable metals, where the degradation rates are comparable to that of arterial remodeling. Rapid degradation is prevented due to a low corrosion rate which is attributed to the formation of a protective surface oxide layer. Inflammatory response, excessive toxicity, or thrombosis is under control when studies using Fe bioresorbable stents in rabbits, pigs, and rats. High radial strength exhibited by Fe stent which allowed for extremely thin stent struts produces more ductile structure, making it easier to deploy into the artery. Superior mechanical properties and slow corrosion process are positive qualities but they may also produce some adverse impact [40]. The incomplete corrosion within the follow-up period has been described for Fe stents, proposing the necessity to increase the degradation rate through alloying. Production of iron oxide degradation by-products and the influx of metallic ions into the surrounding tissue also needed to be cautiously controlled. Iron overdose can include inflammation and increases in free radicals which can cause damage of lipid membranes, proteins, and DNA by long-term effects, where the daily intake of iron is 6–20 mg. When the Fe-based implant is exposed to strong magnetic fields, it can cause a heating effect to the implant. It can cause potential alteration in its shape or position, where the presence of iron in surgical implants interferes with magnetic resonance imaging (MRI) to visualize the anatomy and the physiological processes of the patient during diagnosis, healing, and follow-up observation. Iron nanoparticles at porous magnetic scaffolds could show potential stimulation of osseous tissue generation for enhancement of growth factors, hormones, and polypeptides, and promoting cell adhesion and proliferation. There is an associative pathway for bone formation, where Fe particles displayed upon the external magnetic field, the particles are displaced by applying compression and tensile forces on the attached cells and thus inducing cytoskeleton deformation and cell dragging, where it activates intracellular signaling pathways associated with natural processes of bone formation. The use of external magnetic field can also be used to initiate drug release, whereas localized heating may be used for thermal therapy of cancer and implant-associated infections [41].

Bulk metallic glasses: Bulk metallic glasses (BMGs) are the new class of metallic biomaterials, which show no long-range atomic order and display some smart mechanical, chemical, and physical properties when relating it to the traditional alloy [42]. Bio-inert BMG alloys can be contrived from Ti and Zr, which are accountable for a better match to cortical bone and reduce stress shielding when used in hard-tissue prostheses, and BMG devices are unaffected due to corrosion wear and have easy to deal with surface topography. Bulk metallic glasses could be used in orthopedic surgical scalpels, prostheses, and flexible vascular stents. Bioresorbable BMGs can be used as a temporary vascular stent, a bone screw, or surgical plate in case of intramedullary nails [19].

13.5.1.2 Advantages/Disadvantages of Metallic Biomaterials

Metallic biomaterials have better load-bearing capability along with better toughness, high mechanical resistance to wear, and shock. Besides these advantages, doctors have observed some symptoms with the metallic implant. Elemental release from some of the metallic biomaterials results in low biocompatibility. For example: (1) Ni, Cr, and Co releases from stainless steel and CoCr-based alloys and produces a toxic effect in body: (2) the long-term existence of Al and V ions in Ti alloys has been found to cause Alzheimer's disease, neuropathy, and osteomalacia, (3) the presence of Co has also been reported to have carcinogenic effects. A high friction coefficient and wear debris formation can produce an inflammatory reaction, leading to the loosening of implants due to osteolysis. Implant failure is associated with a high modulus of elasticity which leads to stress shielding. To hinder the elemental release, surface treatment of the biomaterials is necessary, which are possible nowadays by (1) heat treatment to form a passive layer and (2) dense surface coating preparation for example sputtering deposition technique [43].

13.5.2 Ceramic Biomaterials

Most ceramics are polycrystalline materials having affordable biomaterial properties such as resistance to corrosion, great strength and stiffness, hardness and brittleness and wear, and low density. Ceramic biomaterials are poor at tension force but effective at compression force [44]. Ceramics are characteristically thermal and electrical insulators and also used in the field of orthopedics, dentistry, and as medical sensors. These ceramic biomaterials have been used less extensively than either metals or polymers. Ceramics materials are a diverse class of biomaterials currently including three basic types: bioactive, bioinert, bioresorbable. Bioactive refers to a material that, upon being placed within the human body, interacts with the surrounding bone and in some cases even soft tissue. This occurs through a time-dependent kinetic modification of the surface, triggered by their implantation within the living bone. An ion-exchange reaction between the bioactive implant and surrounding body fluids results in the formation of a biologically active carbonate apatite (CHAp) layer on the implant that is chemically and crystallographically equivalent to the mineral phase in bone. Prime examples of these materials are synthetic hydroxyapatite [$Ca_{10}(PO_4)_6(OH)_2$], glass ceramic, and bioglass. Bioinert refers to any material

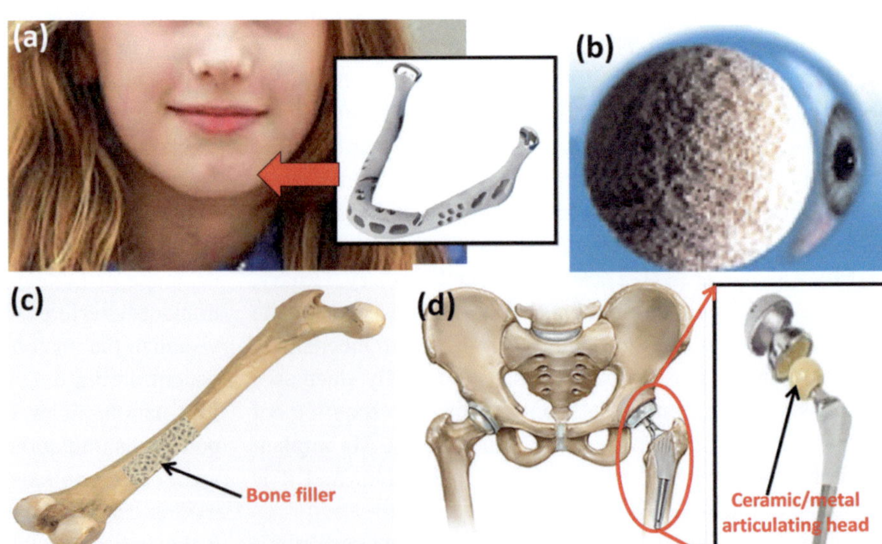

Fig. 13.6 Ceramic biomaterial implants as (**a**) lower jaw, (**b**) orbital implants, (**c**) bone cement (Ca-P based), (**d**) articulating head at the hip

that once placed in the human body has minimal interaction with its surrounding tissue, and examples of these are stainless steel, titanium, alumina, partially stabilized zirconia, and ultra-high-molecular-weight polyethylene. Generally, a fibrous capsule might form around bioinert implants, hence its biofunctionality relies on tissue integration through the implant. Silicon nitride (Si_3N_4), zirconia (ZrO_2), alumina (Al_2O_3), and pyrolytic carbon are termed as bioinert materials that are very compatible with the body parts. Bioresorbable refers to a material that upon placement within the human body starts to dissolve (resorbed) and slowly replaced by advancing tissue (such as bone). Common examples of bioresorbable materials are calcium phosphate, aluminum–calcium–phosphate (ALCAP) ceramics, tricalcium phosphate (TCP) ceramics, calcium aluminates ceramics, coralline, inc–sulfate–calcium–phosphate (ZSCAP) ceramics, zinc–calcium–phosphorous oxide (ZCAP) ceramics, and ferric–calcium–phosphorous–oxide (FECAP) ceramics. Calcium oxide, calcium carbonate, and gypsum are other common materials that have been utilized during the last three decades. Bioresorbable biomaterial can degrade itself by hydrolytic breakdown in the body while they are being replaced by rejuvenating natural tissue where chemical by-products produced by degradation of ceramic materials are absorbed and released via metabolic processes of the body. Most clinical applications of bioceramics relate to the repair of the joints and teeth, skeletal system, bone, and to augment both hard and soft tissues [45]. Figure 13.6 shows the ceramic biomaterial implants as the lower jaw, orbital implants, bone cement, and articulating head at the hip.

13.5.2.1 Materials in Ceramic Biomaterials

Hydroxyapatite: Hydroxyapatite (HA) is a ceramic composed of calcium and phosphate having high hardness. HA ceramic is widely used for creating scaffolds in tissue engineering which have a unique porous structure like bone [46]. Hydroxyapatite can be used for long-term bone replacement with delayed degradation. Hydroxyapatite ($Ca_{10}(PO_4)6$ $(OH)_2$) is a synthetic inorganic biomaterial that is similar to hard tissues such as the bone and teeth. HA ceramics are bioactive such that they stimulate hard tissue ingrowth and involves osseointegration when implanted within the human body. The porous scaffold of this material can be personalized to suit the interfacial surfaces of the implant. The porosity of the hydroxyapatite structure can be organized similar to the animal bone which is very ideal to be used in implants for artificial tooth, hip, and knee replacements. A coating of hydroxyapatite ceramics is applied to the core metallic implant using the PVD process (e.g., plasma spraying and sputter deposition). One common example is the femoral component of the hip implant that has been coated with hydroxyapatite ceramic to promote rapid ingrowth of bone structure. HA ceramic has also been applied in maxillofacial implants as bone filler and as orbital implants inside the eye socket [47].

Alumina (Al_2O_3): Alumina is a bioactive material having high density and high purity (>99.5%). Alumina (Al_2O_3) was the first ceramic material that was widely used clinically [48]. It is used in load-bearing hip prostheses and dental implants because of its grouping of high-strength good biocompatibility, excellent corrosion resistance, and high wear resistance. The outstanding wear and friction behaviors of (Al_2O_3) deal with the surface smoothness and surface energy of this ceramic. Among ceramic biomaterials, alumina has the maximum mechanical properties, but its tensile properties are still below compared to other metallic biomaterials [49].

Zirconia (ZrO_2): Zirconia bioceramics have several benefits over other ceramic materials due to the transformation toughening mechanisms operating in their microstructure and due to its high mechanical strength and fracture toughness. Zirconia bioceramics are mainly applied in total hip replacement ball heads. Very small traces of radio elements are present in Zirconia, which can be found even in fully refined ceramics, and have a negative effect on organs and tissues [50].

Pyrolytic carbon: Carbon exists in a variety of forms and is versatile in nature. Carbonaceous materials are very compatible with bone and other tissues, and the similarity of the mechanical characteristic of carbon to those of bone points that carbon is an exciting candidate for orthopedic implants. Unlike polymers, metals, and other ceramics, this carbonaceous material does not suffer from fatigue. Due to the intrinsic brittleness and low tensile strength parameters, these can be used in major load-bearing applications [51].

Bioglass and glass ceramics: Glass ceramic is a bioactive material where surface modification is carried out before implantation and can be widely used for filling bone defects [52]. Bioglass is porous in nature and hence is beneficial for bioactivity and resorption. Between the stable apatite crystals in the glass ceramic and the bone, the interface reaction was interpreted as a chemical process that involves the solubility of the glass ceramic and a solid state reaction. Brittleness and poor tensile strength are the drawbacks of glass ceramics in biomedical implants. When loaded in compression, the glass ceramics have outstanding strength, but these bioglasses fail at low stress when loaded in tension or bending [53].

Calcium phosphate ceramics: Calcium phosphate (CaP) ceramics are used as a resorbable or bioactive material in different physical forms. Porosity is one of the major properties of calcium phosphate bioceramic and is used as spongy bone. The CaP is very much bioactive in that the molecule can bond with living tissue, especially bone, and can form a bone-like apatite layer on its surface [54].

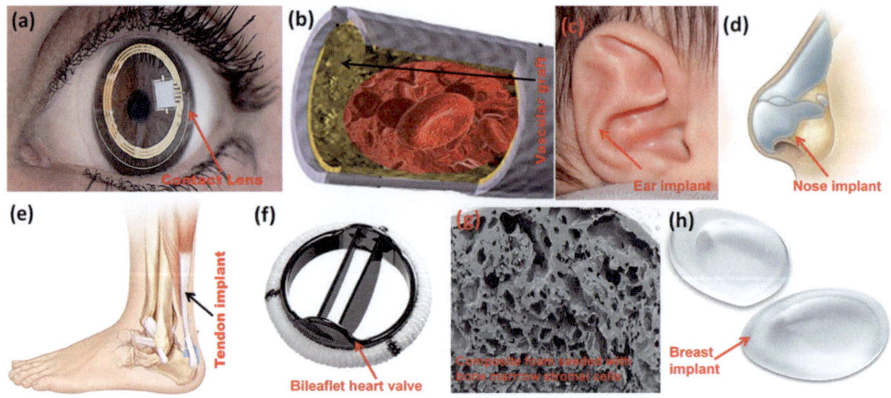

Fig. 13.7 Polymer biomaterial implants in (**a**) eye contact lens, (**b**) vascular graft, (**c**) ear, (**d**) nose, (**e**) leg tendon, (**f**) heart valve, (**g**) composite foam for bone, and (**h**) breast implant

13.5.2.2 Advantages and Disadvantages of Ceramic Biomaterials

The biomedical use of some ceramic biomaterials has some special properties higher compressive strength, high corrosion resistance, low thermal conductivity, and electrical conductivity. The limitations of ceramic biomaterials are that it has a very lower tensile strength, bending strength, and low impact resistance. The reproduction and processing are difficult for the fabrication of specific implants [55].

13.5.3 Polymeric Biomaterials

The use of natural polymers as biomaterials dates back thousands of years ago. Low strength and high flexibility are the necessary properties to enhance the use of polymeric biomaterial in the field of biomedical applications. Polymers are the most widely used materials in biomedical applications, where low strength with flexibility is necessary [56]. Polymers are mainly organic materials that can form large chains made up of a number of repeating units. The applications of polymeric biomaterials are more diverse than for metallic implants. The interchange ability of polymeric biomaterials is not that much abundant. The unique properties like flexibility, resistance to biochemical attack, good biocompatibility, and lightweight have no competitive value from other types of material used as biomaterials. The easy manufacturing of polymeric biomaterial depends on adequate physical and mechanical properties that can produce the variety according to the desired shape. The polymer biomaterials are mostly used in arteries, artificial tendons, veins, cements, teeth, valve, ear, nose, hearth, lenses, testicles, and breasts with desires structures [31, 57]. A few of the major applications of polymer biomaterial implant are shown in Fig. 13.7:

13.5.3.1 Materials in Polymeric Biomaterials

Polylactic acid: Polylactic acid, or polylactide (PLA), is a thermoplastic biodegradable polyester. The derivatives of PLA are L-PLA (PLLA), D-PLA (PDLA), and ceramic mixture of D and L-PLA (PDLLA). These polymers allow hydrolytic degradation through de-esterification. Once degraded of PLA, the monomeric components of each polymer are removed by natural biochemical pathways, where the living body system contains highly regulated mechanisms to remove monomeric units of lactic acids. Lactic acid is the degradation product of PLA and hence used as biomedical implants in the form of screws, anchors, pins, plates, rods, and as a mesh. The retention of PLA depends on its type, and the average retaining capacity is 6 months to 2 years inside the body. Upon degradation, for a support structure, this polymer gradually transfers the consignment to the body (e.g., bone) as that area heals [58].

Polyglycolic acid: The semicrystalline polymer polyglycolic acid (PGA) is insoluble in most organic solvents which have a high tensilemodulus of 12.5 GPa. PLA and PGA are approved by the U.S. FDA which are the first synthetic biodegradable polymers used in the production of resorbable stitches. Polyglycolic acid has an outstanding mechanical characteristic that degrades faster than PLA. PGA implants are biocompatible when implanted to the cortical and cancellous bone in case of defects of rabbits discovered in 1986. After 12 weeks of implantation, the maximum degradation at cancellous sites at implantation and partial degradation were seen in cortical bones without producing any inflammatory response. PGA on degradation harvests glycolic acid under physiological conditions and also some acidic by-products that could produce a strong inflammatory response. The PGA with HA composite scaffolds exhibited good resorption in vitro [33].

Poly(lactide-co-glycolide): Poly(lactide-co-glycolide) (PLGA) is widely used for its favorable mechanical properties and biocompatibility [59]. Limitation of both PLA and PGA are minimized by PLGA as biomaterial, where the premature degradation of PGLA by adjusting the ratio of the homopolymers can be allowed for some control over their degradation rates. Due to the presence of an amorphic PGA of about 75%, the PGA content of PGLA is hydrolytically unstable and shows faster degradation. Lactic acid and glycolic acid are the two by-products upon biodegradation of PGLA which can be excreted by urine as a usual human metabolism. PLGA biomaterial is applied as a scaffold for bone, skin, cartilage, and nerve regeneration apart from drug delivery [60].

Poly-para-dioxanone: The biodegradablepolyester poly-para-dioxanone (PDS) is used in bone fracture fixation, tissue engineering, and controlled drug delivery due to its mechanical flexibility, excellent biocompatibility, biodegradability, and bio-absorbability [61]. In the form of films, foams, nonwoven materials, molded products, laminates, adhesives, and coatings PDS possesses excellent potential for use in broad-spectrum medical implants. In the 1970s, after PLA and PGA, PDS was first developed. PDS is universally recognized as the first choice for the fracture internal fixation material due to its unique physical and biological properties as it is completely bioabsorbable in bone tissues. PDS vanishes completely within 180–200 days through control of its molecular weight, melting temperature, and crystallinity [35].

Polyvinyl alcohol: The water-soluble polymer polyvinyl alcohol (PVA) is nontoxic, biocompatible, fiber/film-forming ability, and chemical as well as mechanical resistance. PVA is widely used in the chemical and medical industries for the productions of fibers, films, coatings, cosmetics, pharmaceuticals, and so on. PVA hydrogels are easily produced by using the freezing-thawing and crosslinking technique. Freeze-thawed PVA hydrogels are nontoxic, biodegradable, biocompatible, noncarcinogenic, have good biocompatibility, smooth, biologically inactive, excellent transparency, rubbery in nature, and a high degree of swelling in water and thus broadly used in biomedical and pharmaceutical applications in the form of temporary skin covers or burn dressings [62].

Poly-methyl-methacrylate: The hard and brittle polymer polymethyl methacrylate (PMMA) appears to be unsuitable for most clinical applications, but it is used as a surgical aid. Under ambient environments, it can be manipulated in the operating theater or dental clinic, explaining its use in dentures and bone cement. PMMA cement with other polymer monomeric methylmethacrylate is used in many joint prostheses that form a dough, which can be placed in the bone according to its applicability [63].

Polyurethane*:* Polyurethane is used for dressing purposes, which is highly conformable, nonadherent, and semiocclusive. The foam is applicable to absorb exudates from the wound, thereby decreasing tissue soaking; concurrently, they maintain moist surroundings while with the sheet form exudation pools under the dressing [64]. These polyurethane polymeric biomaterials are used in dressing in the early inflammatory phase as well as in the proliferative period of repair since they do not stick to the regenerating tissue and leave it undisturbed at bandage changes. The dressings through polyurethane should be replaced frequently to increase comfort in heavily exuding wounds, whereas the rate of dressing change decreases as the healing progresses and less fluid is formed by the wound [37].

Ultrahigh-molecular-weight polyethylene: UHMWPE material is used as the bearing surface in total joint arthroplasty, and there is much research ongoing for the examination of its wear properties. Within 15 years, the use of UHMWPE with metal is of 90%. It is described that the submicron units found in periprosthetic tissues in the presence of polyethylene were present [65].

Polyvinyl pyrrolidone: Polyvinyl pyrrolidone (PVP) is a biocompatible synthetic polymer that is potentially used after UV treatment as a bioadhesive wound dressing matrix. PVP has been used broadly due to its lubricity and viscous properties to coat tissue-contacting surfaces and as a vitreous humor substitute [66].

Silicone rubbers: Silicone rubbers exhibit both advantages and disadvantages whether it is heat vulcanized through heat or vulcanizing silicones at room temperature. Heat-vulcanizing rubbers silicone is provided as a semisolid material that needs milling and packing under pressure [67].

Polypropylene fumarate: The unsaturated linear polyester polypropylene fumarate (PPF) is the degradation product of PPF (i.e. propylene glycol and fumaric acid) which is biocompatible that can be readily removed from the body [68]. The double bond along the backbone of the polymer can form a moldable composite to harden within 10–15 min. The variety of the PPF molecular weight provides its different mechanical properties and degradation time. During PPF synthesis, critical issues are the preservation of the double bonds and control of molecular weight. PPF scaffold is intended for guided tissue regeneration by providing injection for bone replacement as a composite material and sometimes used for osteoblast cultures [69].

Polyhydroxyalkanoates: Polyhydroxyalkanoate (PHA) is a polyester that is aliphatic and produced by microorganisms under uneven growth conditions. This is thermo-processable which makes them attractive as biomaterials for applications in medical devices and tissue engineering. Over the past years, PHA, particularly poly-3-hydroxybutyrate, poly-4-hydroxybutyrate, copolymers of 3-hydroxybutyrate, and 3-hydroxyhexanoate, copolymers of 3-hydroxybutyrate, 3-hydroxyvalerate, and poly-3-hydroxyoctanoate, were demonstrated to be suitable for tissue engineering, where PHA is biodegradable in the process of hydrolysis. PHA polymers can be surface modified, blended, or composed with other polymers, enzymes, and depending on their physical properties [70].

13.5.3.2 Advantages and Disadvantage of Polymeric Biomaterials

The main advantages of polymeric biomaterials are low density, low weight, easy to harvest, and flexibility, where easy degradability and low mechanical properties are the limitations in the field of biological implantation [71].

13.5.4 Biocomposite

The composite mixture is made up of different natural or synthetic polymers or a combination of both. Biocomposite materials contain biodegradable matrix and composite fiber reinforcement. Biocomposites are developed for their environmental benefit and enhanced performance. Bone composed of calcium phosphate ceramics in a highly organized collagen matrix achieves most of its mechanical properties as a natural composite material. In addition to a matrix material, obtaining certain properties of biomaterial composite with filler (reinforcement) can improve every one of the components [72]. During the fabrication of biocomposite, some matrix materials may be combined with different types of fillers where several phases can be obtained. The addition of particulate fillers with the polymer is called particulate composites. Plaster of Paris bandage is the first biocomposite used by an orthopedic surgeon, where it has been refined to fiber glass with a polymeric matrix at the current synthetic casting materials. The addition of chopped carbon to polyethylene components could improve the mechanical properties and could be applied to internal prostheses. Composite structures are typically produced from laminates where the laminate is a thin sheet of composite material in which all the fibers run in one direction and are held together by a thin coating of the polymer matrix material. Bulk composite is formed by the combination of laminate with other laminate, where the properties of this composite vary depending on the orientation of each layer of the laminate. There is the extensive use of composite biomaterials in the field of skin tissue engineering for wound healing and its repair as regenerative medicine. The common composite material used are chitosan–gelatin spongy mixtures, chitosan–alginate polyelectrolyte membranes, chitosan–cellulose–silver nanoparticle mixtures, chitosan gelatin–antibiotic mixtures chitosan gels containing EGF, tencel–chitosan–pectin composite, chitosan–fibrin nanocomposites like nanofibrous chitosan–silk fibroin composite, beads of chitosan–fibrin, chitosan–fibrin–sodium alginate, chitosan nanoparticles contained fibrin gels, thrombin receptor agonist peptide encapsulated in poly(N-vinyl caprolactam)–calcium alginate hydrogel film, biopolymeric matrices delivering angiogenic growth factors, or epidermal growth factor delivering micro- and nanoparticles for skin or dermal tissue regenerative applications [42, 73–75]. As anti-infective bandages with specific bactericidal activity against *S. aureus*, the most used composite is cellulose–chitosan and cellulose–poly(methylmethacrylate) fibers with the cell lytic enzyme lysostaphin. The composite of wound dressing material contains mainly different active molecules, such as enzymes, antioxidants, hormones, vitamins, and antimicrobial drugs [76]. Alginate-based biomaterials containing zinc alginate and silver alginate have the ability for tissue healing due to antimicrobial properties of zinc and silver and sometimes contain alginate films loaded with asiaticoside and alginate–chitosan membranes for better performance.

Composite of poly(N-vinyl-2-pyrrolidone) (PVP), kappa-carrageenan (KC), potassium chloride, and polyethylene glycol and PVP-KC hydrogels have been reported for their wide applications in tissue engineering. For their wound dressing application, hydrogel composites were prepared by exposure to higher doses of gamma radiation. Also, other hydrogel membranes that are made up of pectin and gelatin are widely used for wound dressing. Natural polymer derived from a fruit gum, fragrant manjack, snotty gobbles (Boraginaceae) with altered percentages of glycerin, extracellular polysaccharide of *Trametes versicolor*, a polymer of fungal origin, where the composites material with antibacterial components are also used in tissue engineering and has great potential for the treatment of wounds [77]. Most biodegradable matrix are based on aliphatic polyesters reinforced with various vegetable fillers (such as flax, jute, sisal, and kenaf). Biodegradable composites also carrying drug/genes are widely used in the field of biomedical engineering as in cosmetic orthodontics and tissue regenerative application [78]. These bioactive materials provide strengthening properties of the matrix as well as biocompatible. Applications of various composite biomaterials are shown in Fig. 13.8. Figure 13.9 depicts various materials used in polymer biocomposite.

13.5.4.1 Advantages and Disadvantages of Composite Biomaterials

Composite materials provide high biocompatibility, better corrosion resistance, and are inert, while the limitations are lack of consistency and difficulty to reproduce during fabrication.

13.5.5 Biologically Derived Biomaterials

The biomaterials that are derived from the biological source have the highest level of compatibility for the implant. Various biological-derived biomaterials are described subsequently.

13.5.5.1 Protein

Collagen*:* Collagen is the most abundant protein found in connective tissues in all animals (Fig. 13.10a). There are mainly 16 types of collagen secreted by fibroblasts and epithelial cells. The triple helix is the basic structural unit of collagen [79]. Collagen provides physical support for tissue due to its tensile strength and also influences cell proliferation, adhesion, migration, differentiation, and polarization. These properties support biological processes, such as development, tissue maintenance, regeneration, and repair, and in various pathological processes, such as tumor growth and

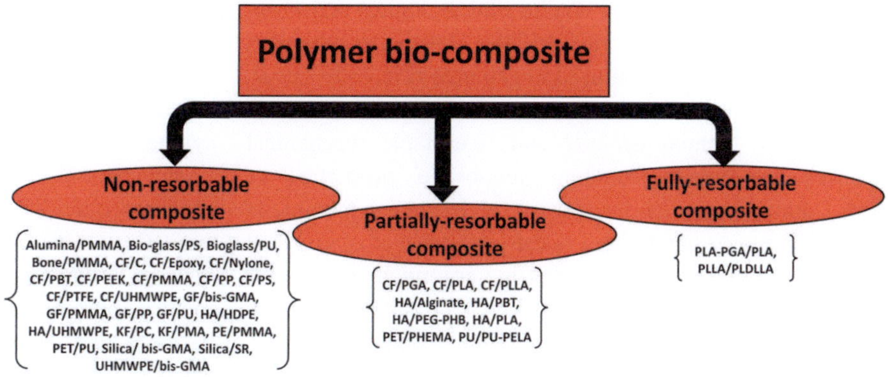

Fig. 13.9 Materials used in polymer biocomposite

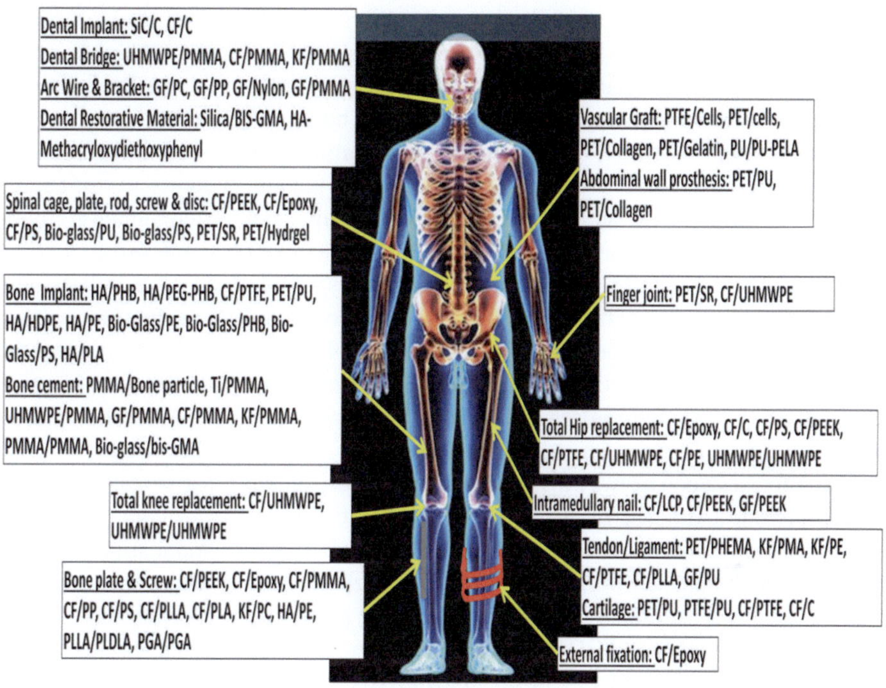

Fig. 13.8 Composite biomaterials used in various parts of the human body

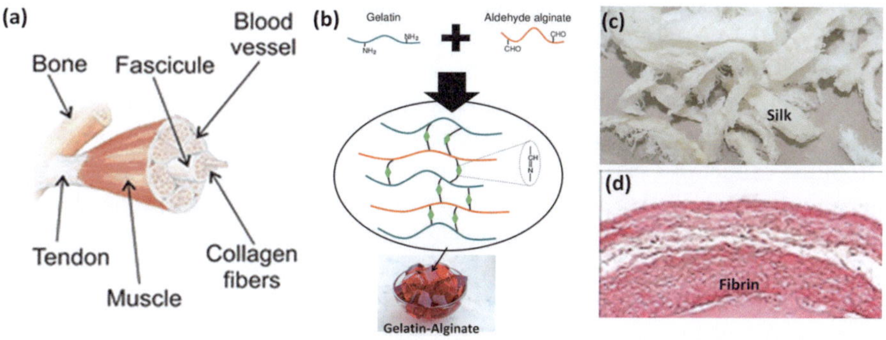

Fig. 13.10 (**a**) Collagen fiber, (**b**) Gelatine alginate, (**c**) Silk, and (**d**) Fibrin

metastasis. Collagen is generally obtained from various animal sources, where amniotic membrane (AM) is a very attractive source. An amniotic membrane is a thin membrane is filled with amniotic fluid surrounding the fetus. This membrane consists of an epithelial monolayer, a thick basement membrane, a compact layer, a fibroblast layer, and a spongy layer. The innermost layer is a monolayer of epithelial cells which is nearest to the fetus, anchored on the basement membrane. The basement membrane containing the collagen of the amniotic membrane includes types, that is, III, IV, V,

VII, and XVII, which have similar morphological and ultrastructural basement membrane of skin. Hence, the basement membrane of AM is often used to generate skin equivalents. Anti-inflammatory, antibacterial, antifibrosis, antiscaring as well as low immunogenicity and reasonable mechanical features are the outstanding supportive properties of AM. Amniotic membrane consists of basement membrane where the major component is the extracellular matrix and behaves as a natural scaffold that is very much suitable for cell seeding in tissue engineering. The major areas where the AM can be applied are eye, skin, cartilage, and nerve [80].

Gelatin–alginate: Gelatin is used as a plasma expander, wound dressing, adhesive, and absorbent pad for surgical application without antigenicity, while collagen is widely used in biomedical areas which expresses antigenicity in physiological conditions (Fig. 13.10b) [81]. By controlled hydrolysis of collagen, gelatin is obtained which is a fibrous insoluble protein that is a major component of skin, bone, and connective tissue. Glycine, proline, and hydroxyproline are the amino acids quantitatively high in content of gelatin. Glycine-XY triplets (X and Y are frequently proline and hydroxyproline) are the repeating units that are the content of gelatin. These sequences are triple-helical in the structure of gelatin and can form gels where helical regions form in the gelatin protein chains immobilizing water. In algae, the function of alginates is primarily skeletal. The gel located in the cell wall and intercellular matrix convening the strength and flexibility which is necessary to withstand the force of water in which the seaweed grows [48]. Alginate is a water-soluble polysaccharide extracted from marine brown algae such as *Laminariahyperborea* or soil bacteria such as *Azobactervinelandii* and composed of 1,4-linked β-D mannuronic acid (M) residues and 1,4-linked α-L-guluronic acid (G) in varying proportions, revealing carboxylic acid functionality at the C5 residue. The sponge consisting of gelatin from hydrolysis of collagen and alginate a polysaccharide from Phaeophyta was established by using as a crosslinking agent, that is, EDC. The sponges were composed of some natural substances which are traditionally used for burn treatment such as tamanu oil (from nuts of *Calophylluminophyllum*), cajeputi oil (from leaves of *Meulaleucaleucadendron*), madecassol (from the extract of *Centellaasiatica*), turmeric, and python fat. Sponges have a rather good ability in preventing infection and promoting wound healing compared to control. Cajeputi oil and madecassol-based sponges have the best potential for burn treatment. High density of the network of alginate matrix gels limits the cell growth and cell anchorage, while a strict requirement for survival is very narrow on alginate gels due to their hydrophilicity. By copolymerization with polyethylene glycol, the natural composite is used to improve their biocompatibility [82].

Silk: The unique combination of biocompatibility, biodegradability, self-assembly, mechanical stability, controllable structure, and morphology are the promising factors of silk (Fig. 13.10c) [83]. The proteinous polymer silk is spun into fibers by Lepidoptera larvae such as silkworms, spiders, scorpions, mites, and flies, which are fibrous proteins synthesized at epithelial cells that line glands in these organisms. Silk fibroin is a polymer that consists of repetitive protein sequences and can provide structural roles in cocoon formation, traps, web formation, nest building, safety lines, and egg protection. The silk biopolymer is applicable for treating burn victims, as a matrix of wound healing, and tissue regeneration. The peptides of silk fibroin are used in cosmetics due to their glossy, elastic coating power property, flexibility, easy spreading, and adhesion characters. Due to its crystalline nature, the fine silk powder is touted and relieves from sunburns which reflects UV radiation, and as a demulcent, it acts as a protective buffer between skin and environment. The ultra-micro silk powder is used in hair and massage oils, and water-dispersible finer grade silk powder is used in liquid cosmetic preparations [84]. The silk has the ability to fight against edema, impotence, cystitis, adenosine augmentation therapy, epididymitis, and cancer. Serratio peptidase is a silk protein derivative used in the market as an anti-inflammatory, anti-tumefacient for treating acute tonsiloctomy, sinusitis, tooth filling, oral surgery, cleaning, and extractions. Due to its nontoxic nature, silk fibroin is used in veterinary medication. Surgical sutures made up of silk fibers do not cause inflammatory reactions and are absorbed after wounds heal. Silk-based biodegradable microtubes for the repair of blood vessels and as molded inserts for bone, cartilage, and teeth reconstruction are other promising applications in the biomedical and bioengineered field, where the use of natural fiber mixed with biodegradable and bioresorbable polymers can produce joints and bone fixtures to alleviate pain for patients. The upcoming future applications of silk biomaterials in the field of biomedical research include new-generation soft contact lenses that enable greater oxygen permeability, skin grafts, artificial corneas, and epilepsy drug permeable devices [52].

Fibrin: The fibrin formation mechanism is elucidated primarily from the thrombin-mediated cleavage of fibrinogen. The precursor protein of fibrin is fibrinogen, which plays a great role in blood clotting (Fig. 13.10d) [85]. Fibrin gels were prepared by combining fibrinogen, NaCl, thrombin, and $CaCl_2$. Fibrin acts as hemostatic plug scaffold for cell proliferation, migration, and wound healing, and this potential is widely used in tissue engineering and medical application. Fibrin glue or fibrin sealant is formed in combination with fibrinogen and thrombin at very high amounts of calcium and FXIII, which is used as an adjunct to hemostasis in patients undergoing surgery. Tisseel (Immuno, Vienna, Austria), Beriplast (Behringwerke AG, Marburg/Lahn, FRG), and Biocol (CRTS, Lille, France) are the commercially available sealants and have been extensively used in clinical

Fig. 13.11 (**a**) Wood Cellulose fiber, (**b**) Chitin, (**c**) Agarose, (**d**) Carrageenan, (**e**) Fucoidan

application. Fibrin is natural and used as a biological scaffold for cell proliferation, migration, and differentiation applied in various tissue engineering in the field of skin grafts to burned areas. The use of fibrin glue at the immediate postoperative period enhances healing and minimizes scarring instead of sutures or pressure dressings implants. ARTISS fibrin sealant is used for burn treatment (Baxter International Inc., USA), is indicated to adhere autologous skin grafts, and to surgically prepared wound beds resulting from burns in case of adults and Pedi patients [86].

13.5.5.2 Polysaccharide

Cellulose: The most abundant polymer organic material on Earth is cellulose. Plants contain about 33% cellulose; however, wood contains 50% and cotton contains 90% [87]. Commercial paper production depends on cellulosic material (Fig. 13.11a). This provides the relative stiffness and rigidity of the cellulose. This intramolecular hydrogen bonding property of cellulose reflects its high viscosity in solution, a high tendency to crystallize, and its ability to form fibrillar strands. Natural cellulose spheres are often used in bioseparation, cell suspension culture, immobilized reaction, and as an adsorbent for sewage treatment [88]. Microbial cellulose synthesized from *Acetobacterxylinum* (AX) and its nanostructure show novelty in wound healing. Bacteria intake various carbon compounds, and during the process of biosynthesis, carbon compounds are polymerized into single, linear β-1,4-glucan chains secreted outside the cells through a linear row of pores located on their outer membrane. The cellulose derivatives produced by *Acetobacterxylinum* have also been explored as a potential scaffold material because of its unusual material properties and degradability. This bacterial cellulose has ultrafine network architecture, high hydrophilicity, and mouldability during fabrication. This cellulosic material is also suitable for use in micro nerve surgery and as an artificial blood vessel suitable for microsurgery [89].

Chitin–chitosan: Chitin is a nitrogenous polysaccharide that is white, hard, and inelastic in nature and widely found in the exoskeleton as well as in the internal structure of invertebrates. Chitin (Fig. 13.11b) is derived from many natural sources, including the exoskeleton of arthropods and insects, and is the second most abundant natural hydrophobic linear polysaccharide next to cellulose [90]. Chitosan is a semicrystalline linear polysaccharide polymer that is composed of randomly distributed β-(1→4)-linked D-glucosamine (deacetylated unit) and *N*-acetyl-D-glucosamine (acetylated unit), where the degree of crystallinity is a function of the degree of deacetylation [91]. Chitosan is extracted from chitin shells of shrimp and other crustaceans by the treatment of an alkaline substance, such as sodium hydroxide, where chitosan has a number of commercial and biomedical uses. Chitosan can deliver the drug in the form of bandages to reduce bleeding when applied to skin and also acts as an antibacterial agent.

Chitosan can decrease bleeding and be used within some wound dressings. Chitosan salts made from mixing chitosan with an organic acid (such as succinic or lactic acid) are used as hemostatic agents. The protonated chitosan (positive charge) leads to the engrossment of platelets and rapid thrombus formation where it acts as a hemostatic agent that works by an interaction between the cell membrane of erythrocytes (negative charge). Absorbent properties of chitosan can be increased when mixed with materials such as alginate with varying the rate of solubility and bio-absorbability of the chitosan salt. The chitosan salts are biocompatible and biodegradable by lysozyme in the body and the degraded molecules are glucosamine and the conjugate base of the acid such as lactate or succinate. It is widely used in self-healing polyurethane paint coating. Chitosan membranes are biologically used as artificial kidney membranes due to their suitable permeability and high tensile strength. Commercially production of cellulose and cuprophane is applicable for artificial kidneys as it acts as a semipermeable membrane [92]. Chitosan possesses all the acceptable properties such as optical clarity, mechanical stability, sufficient optical correction, gas permeability, wettability, and immunological compatibility, which are the requirements for the application of contact lenses. Partial depolymerization and purified squid pen chitosan produced by spin-casting technology are used to produce contact lenses that are clear, tough, and possess other required physical properties such as modulus, tensile strength, tear strength, elongation, water content, and oxygen permeability that are required for ideal contact lenses. Chitosan is suitable for the development of ocular bandage lens due to its antimicrobial and wound healing properties, where special attention has been paid to chitosan for the repair of articular cartilage. Articular cartilage is chiefly susceptible to injury, trauma, disease, or congenital abnormalities due to its avascular, alypmhatic, and aneural properties. Once damaged, it has little capacity for intrinsic repair [93].

Agarose: Agarose (Fig. 13.11c) is obtained from red seaweed, which exhibits many of the properties of alginate, emulating the extracellular matrix with high water uptake, but with the advantage of hydrogen bond-mediated self-gelation without the need for potentially toxic crosslinking agents such as genipin. It is adversely affected mechanically by the presence of cells: Cells within agarose weaken the gel strength due to the intervention of hydrogen bonding on behalf of crosslinking and gelation. With the combination of other polymers and proteins such as collagen, chitosan, and cellulose, the limitations can be hindered. Agarose composites have been explored for tissue engineering, mainly for neural, vascular, bone, and pancreatic tissues. Nowadays, agarose are required to produce bio-inks possessing the necessary properties for bio-printing purposes [94].

Carrageenan: Carrageenan (Fig. 13.11d) is a polygalacton derivative that is obtained from the Rhodophyceae members of red algae seaweeds. Carrageenan when combined with other materials, such as poly (oxyalkylene amine), methacrylic anhydride, and nanosilicates, produce a printable bio-ink capable of extrusion and shape retention, with elasticity, high fidelity, and stiffness to enable cross-linkable multilayered tissue constructs by both thermal and ionic gelation processes. Young's modulus of carrageenan is dependent on its concentration with values ranging from 0.10 MPa (1% carrageenan) to 0.66 MPa (3% carrageenan) and becoming increasingly unpredictable as water content increases. The average and usable tensile strength of carrageenan have been reported as 39.34 ± 0.51 MPa as well as elongation at break value of $19.5 \pm 0.4\%$. Bio-inks from carrageenan have good cell viability and attachment and an affinity for osteogenesis with collagen–hydroxyapatite-based biocomposite gel increases the strength for bone tissue engineering. The sulfated support in carrageenan mimics naturally occurring sulfated glycosaminoglycans in extracellular matrix of cartilage which is nontoxic, chondrogenic, and mechanically similar to that of native cartilage. The facile cross-linking and glycosaminoglycan-simulated carrageenan are the prospects for cartilage bio-inks in tissue regeneration [95].

Fucoidan: Fucoidan (Fig. 13.11e) is a water-soluble polysaccharide derived from marine brown algae and is anionically sulfated which shows bioactive properties such as antioxidant, anti-inflammatory, and angiogenicity and used for tissue engineering applications. Fucoidan having the rheological properties is conducive to bioprinting with shear-thinning behavior observed below 1.5% weight volume yet plastic behavior at 2% with a yield value of 2 Pa. The viscoelasticity property can be increased by increasing the concentration in solution and through the addition of NaCl and $CaCl_2$ and decreases at higher temperatures. Distinct rheological properties could be found for both linear and branched subtypes of fucoidan. Variable viscosity of fucoidan depends on different types of species of seaweed from which it is obtained which reflects differences in molecular weight, proportion of sulfates, and uronic acids [96].

13.6 Various Synthesis Techniques of Biomaterials

The fabrication of the porous scaffold with a network structure having precise mechanical strength and physicochemical features is required for successful cell seeding, and growth of tissue encompasses the best knowledge in the field of biomedical engineering and material science. Scientists are continuously working on various interdisciplinary fields for the development of scaffold fabrication techniques for biomedical purposes. There are some fabrications techniques described in the subsequent sections.

13.6.1 Solvent Casting

One of the simplest techniques of scaffold synthesis is solvent casting, where an organic solvent is used to solubilize the bioactive polymer and the solvent is allowed to evaporate to form the scaffold, sometimes slowly and sometimes rapidly with respect to the designed properties. Generally, there are two different ways of fabrication. In one of the processes, the mold of the desired shape is dipped in a solution of polymer and solvent and is left for a suitable time to draw the solution. In this process, there is the formation of a layer of polymer on the mold which is further processed for fabricating the scaffold. In the second process, the solution is poured into the mold and kept it for sufficient time for the solvent to evaporate which results in the mold filled with a layer of polymeric slurry, as shown in Fig. 13.12. Certain binders are added to the polymeric solution to enhance the strength and integrity of the scaffold structure. But the limitation is the risk of introducing toxic elements into the scaffold by the use of solvents; hence, proper care needs to be taken for the removal of toxic elements by proper drying of the scaffold under vacuum after fabrication, and the other drawback is the nonuniform porosity of the structure could be achieved [59].

13.6.2 Particulate Leaching

The leaching method is one of the most popular and widely used for scaffold synthesis. The limitations of the solvent casting technique are overcome by the leaching method. In the leaching method, certain porogen (porosity-inducing agents) are added for preparing the porous architecture of the scaffold (Fig. 13.13). Some salt, sugar, and wax are regarded as porogen by the researchers, where the porogen are first grounded into desired shape, size, and are introduced into the mold. The required quantity of polymeric solution is poured into the porogen which is filled with mold and is kept for 40–50 h for the solvent to evaporate. After the evaporation of water, the polymeric solution can be cast into a porous scaffold. After casting, the resulting scaffold is dried and washed in

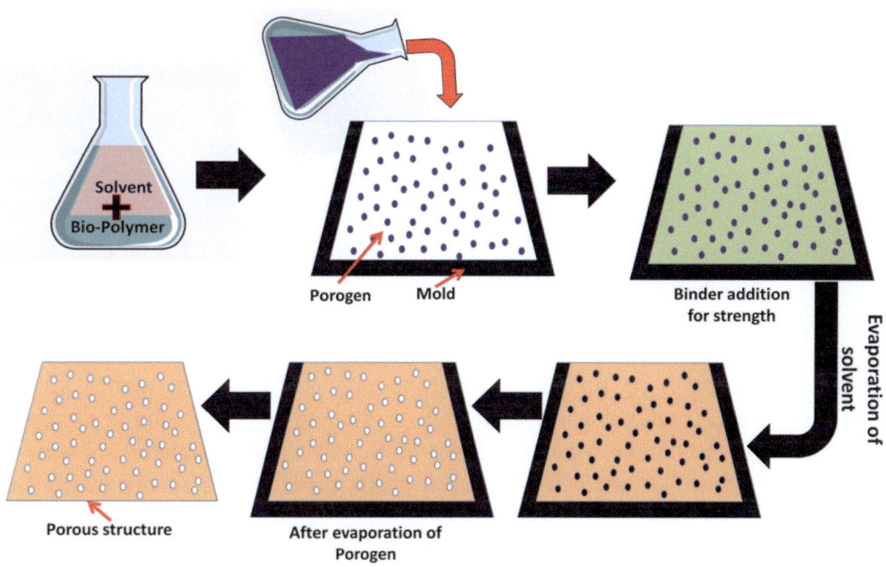

Fig. 13.12 Solvent casting process of PLA porous materials

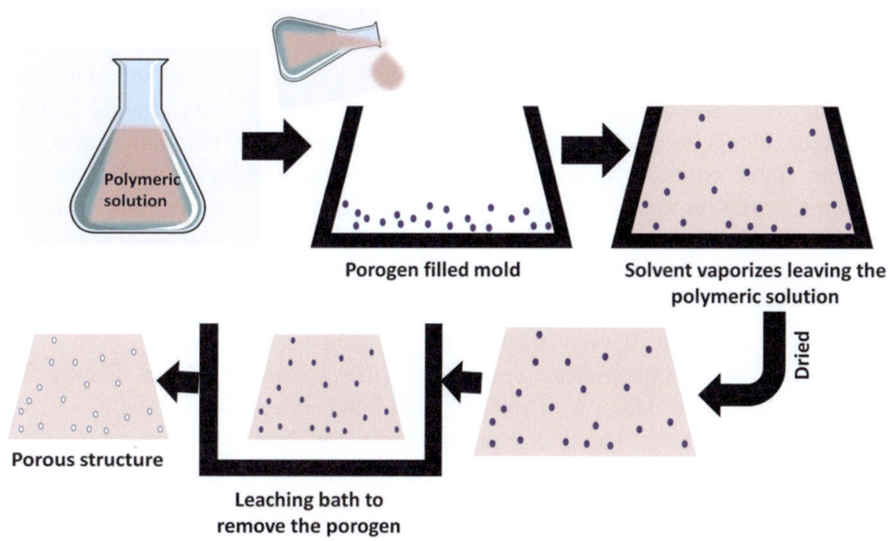

Fig. 13.13 Particulate leaching technique

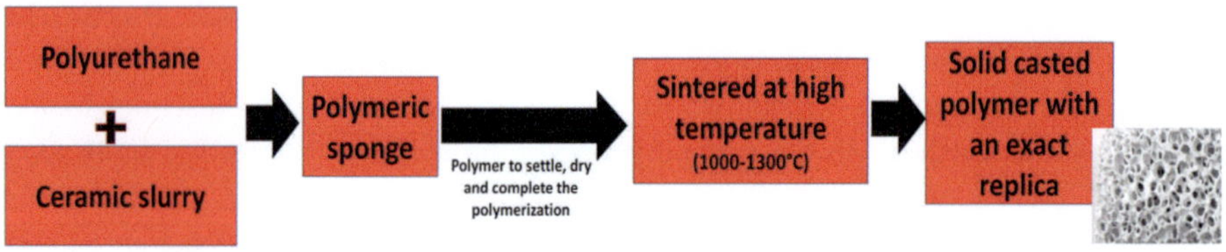

Fig. 13.14 Polymer sponge replication process

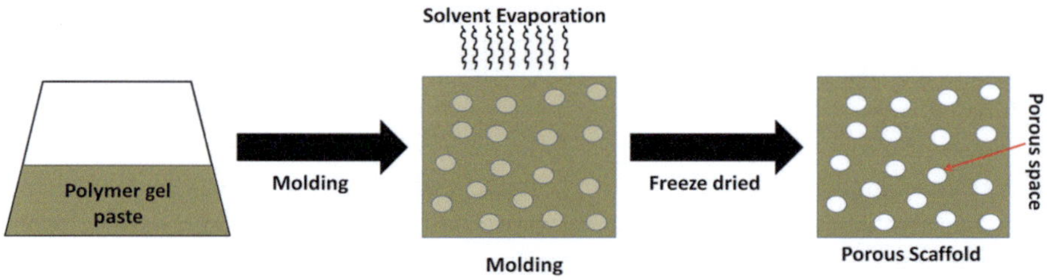

Fig. 13.15 Gas forming process

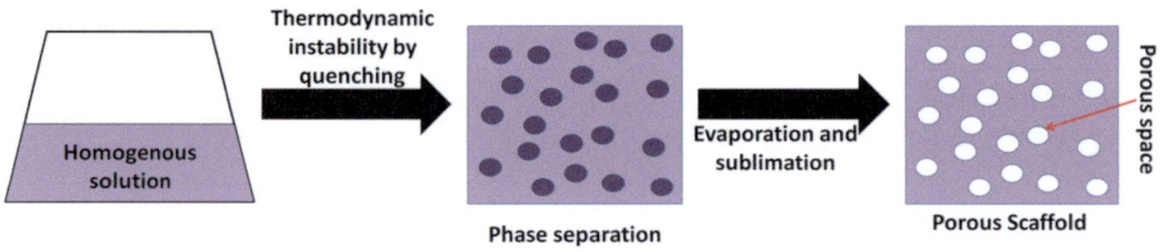

Fig. 13.16 Phase separation technique

a bath of deionized water to leach out the porogen. The wax-based porogen scaffold can be heat treated for melting the wax, and after the removal of wax, the scaffold structure gets porous. This leaching method is simple, and proper dimensional control on pore size could be obtained by controlling the quantity, shape, and size of the porogen selected. In this technique, a porosity of 90–95% with a pore diameter of about 500 μm had been achieved by inventors [97]. The lack of control over the interconnectivity of pores inside the scaffold structure is the major drawback of this method.

13.6.3 Polymer Sponge Replication Method

In the polymer sponge replication method (Fig. 13.14), the polymeric sponge made from polyurethane is dipped in the ceramic slurry which is then kept for 48 h to settle the polymer, complete the polymerization process, and then dry. The obtained polymeric sponge (dried) is then sintered at high temperature (1000–1300 °C) for the sponge to burn off and leave the imprint into the polymer which resulted in a solid casted polymer with a replica of the geometrical topographies of the polyurethane sponge [61]. The formed scaffold is porous in nature which shows high interconnectivity between the pores. The sponge prepared by the replication method acts as a prototype for the porous scaffold. In this method of fabrication, the polymeric sponge owes the high thermal and chemical stability and high hardness and gained widespread popularity among all other scaffold preparation processes.

13.6.4 Gas Foaming

Organic solvents (Fig. 13.15) are mostly used for the preparation of the ceramic slurry-made scaffold. Most scaffolds are heat treated after fabrication for the enhancement of their structural strength and stability. In some cases, the use of organic solvents sustains the incorporation of unwanted residues which can damage tissues and can affect cellular functions. There is no use of organic solvents in the case of the gas foaming technique, where the scaffolds are heat-treated post-processing. In this fabrication process, the high-pressure carbon dioxide (CO_2) gas (800 psi) is allowed to saturate the polymeric solution where the unstable CO_2 causes it to cluster inside the solution. The cluster of CO_2 gives rise to nucleation which introduces pores into the structure. For open-pore geometry, the gas foaming technique is sometimes combined with the particulate leaching technique. Sugar and salts, which are very soluble in water, act as porogen which can produce a porosity of around 90% with a pore diameter of 100 μm. The makeover scaffold is flexible for the interconnectivity between pores and tissue system [98].

13.6.5 Phase Separation

The phase separation technique is a very simple technique that involves a concentration gradient to separate two phases of polymeric solution for scaffold synthesis (Fig. 13.16). The polymeric solution has two phases: one with rich concentration and the other with lean concentration are dissolved with phenol or naphthalene where the concentration gradient

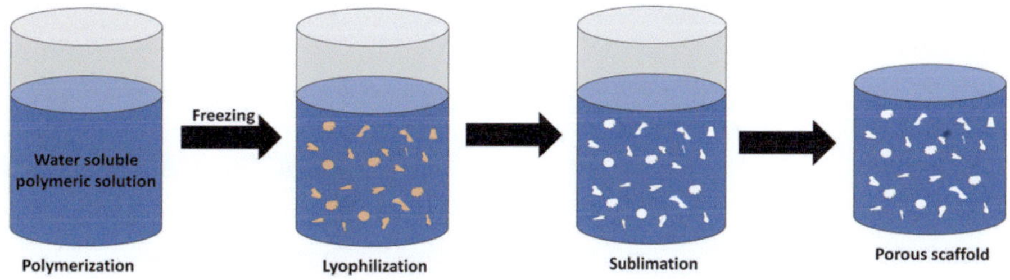

Fig. 13.17 Freeze-drying

depends on temperature. In the phase separation process, the temperature of the solution is then quickly dropped to separate the two phases of the solution followed by quenching. To separate the solvent from the scaffold, the evaporation and sublimation processes are carried out for the fabrication of a porous solid structure. Rapid prototyping and particulate leaching are the advanced phase separation methods, which can be combined with this synthesis method for preparing nanofibrous scaffolds [99].

13.6.6 Freeze Drying

Dehydration by freeze-drying (Fig. 13.17) is the process where a material is first frozen extremely at a very low temperature and the surrounding pressure is lowered for the sublimation of frozen water. The water-soluble polymeric solution of the desired polymer and suitable solvent is first prepared and then the solution is kept for polymerization. The solvent is separated from the resulting solution by lowering the temperature of the solution in a negative pressure environment which causes the solvent to sublime leaving behind a solid polymeric scaffold structure with porous networks. This freezing process under negative pressure conditions is called lyophilization. The p^H of the solution and the freezing rate of the porosity of the structure can be monitored. In this lyophilization process, there nothing is required for creating the porosity in the scaffold structure, But the drawback is that it is time-consuming and the pores are formed are quite smaller in size [100].

13.6.7 Electrodeposition

The electrospun machine is intended for the process of electrodeposition of polymeric as well as metallic and ceramic polymeric composite on a metallic plate. The deposits from electrodeposition are Ti and its alloys containing NiTi superelastic and shape memory alloy; the electrodeposited coating materials are ceramics, such as hydroxyapatite HA, octacalcium phosphate (OCP), brushite (DCPD), other calcium phosphates (Ca-P), DLC, and TiO_2; metals such as magnesium and tantalum; and polymers such as collagen, chitosan, and poly(ethylene glycol) (PEG) [101]. In the electrodeposited calcium phosphate, porous Ti alloys, Ti, Pt mesh, and supersaturated calcium phosphate solution were used as the cathode, electrode, and electrolyte, respectively. For the fulfillment, a desired Ca/P ratio range was 1.65. The calcium phosphates could homogenously coat and cover the entire titanium surface. The coating of HA whose thickness was approximately 25 μm exhibits a bond strength of 25 MPa. Adamek et al. prepared an HA-based bioactive surface on porous Ti6Al4V using electrochemical etching and cathode deposition technique [102]. Large pores and nanolamellas could be obtained on the HA layer. The cathodical deposition method was a favorable route for the enhancement of bioactivity on the surface.

13.6.8 Rapid Prototyping

Rapid prototyping (RP) is a computer-aided designing process, where the biomaterials can be fabricated using the layer-by-layer technique. RP technique can produce structures having similarities in composition, and reproducible architecture can also control the mechanical property and the biological effect of scaffolds. Selective laser melting (SLM), electron beam melting (EBM), selective laser sintering (SLS), and laser-engineered net shaping (LENS) are considered rapid techniques used to make scaffolds used for tissue engineering [64]. SLM process is laser-based powder bed fusion, where laser source permits full melting of metal powder and produces a homogenous part, and the used materials should be pure in case of alloyed metal powders. The direct metal laser sintering is practically used interchangeably with SLM, where EBM is one of the most established powder bed fusion techniques for metal. Like SLM, the arrangement consists of rake, build platform, powder hoppers, and an energy source. The electron beam energy is used rather than a laser in the process because of some inherent properties of electron beams, and it is essential to work under a high vacuum for EBM. The laser-based powder bed fusion technique is the process besides SLM and SLS. In between SLS and other powder bed fusion techniques, the difference is based on the material options and powder fusing mechanism. Unlike SLM, the raw powder processing during SLS is partially molten or heated to the sintering point. In LENS process, there is no possibility for making an item to

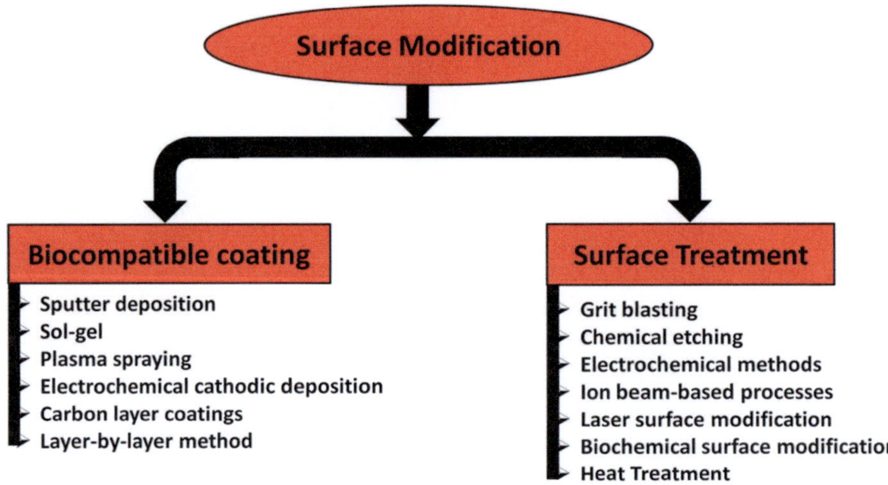

Fig. 13.18 Surface modification processes of implant

exact specifications without any secondary operations. The LENS process can go from metal and metal oxide powder to metal parts in many cases [103].

13.7 Surface Modification of Biomaterials

The formation of new bone is the process of bone healing named osseointegration. The number of biological reactions occurs in various stages while the implant is fixed into a body. First of all, there will be adsorption of water molecules and proteins and then one of the following processes will take place as described below [104, 105]:

1. New bone cell formation on the implant surface, bone cells proliferation, and differentiation leading to osseointegration.
2. Inflammatory response by the human body to reject the implant.
3. Micromotions of the implant leading to the formation of fibrous tissue instead of a bony interface that impedes osseointegration.

The whole process depends on the surface properties, such as surface chemistry, surface topography, surface roughness, and mainly the surface energy that changes with all the said properties. Typically, two types of techniques have been adopted for surface modification of biomaterials, that is, biocompatible surface coating and surface treatment as shown in Fig. 13.18.

13.7.1 Biocompatible Coating

The coating, which is assisting the early fixation of bone, improves prosthesis life despite biomechanical mismatch. Surface modification is the main objective of the metallic biomaterials for hard-tissue engineering for the improvement of bone conductivity through the formation of a bioactive layer [106]. Mostly, HA coating, bioglass coating, glass-ceramic coating, biocompatible nano-ceramic coating, and oxide coating are established by the following techniques. Surface coating of synthetic HA^- where a calcium phosphate compound is similar to the bone-promoted bone opposition to the surface. Bone cells can adhere to the surface of the apatite coating without any intervening layers when a metal implant is coated with HA. There is an excellent adhesion of the coated implant to the bone when the HA matrix of the bone cells later becomes integrated with the HA coating. The process of coating must be consistent with the substrate to be coated, and an effort should be made to keep the substrate intact during the coating process. Consequently, failure in the substrate-coating linkage is usually objectionable for any application because of its debris or ions that are released from implantation [68]. Some coating techniques are described below:

Sputtering: Sputter deposition is a PVD process. This process consists of the ejected material from the target and deposited onto a substrate. With a wide energy distribution, typically up to tens of eV (100,000 K), sputtered atoms are ejected from the target. The sputtered ions are typically only a small fraction of the ejected particles being ionized, and 1% can ballistically fly from the target in straight lines and impact energetically on the substrates or vacuum chamber. Where there are higher gas pressures, the ions collide with the gas atoms that act as a moderator and move diffusively, reaching the substrates or vacuum chamber wall and condensing after undergoing a random walk. The whole range from high-energy ballistic influence to low-energy thermalized motion is accessible by changing the background gas pressure where often the sputtering gas contains an inert gas such as argon. The atomic weight of the sputtering gas must be close to the atomic weight of the target for efficient momentum transfer, so for sputtering light elements, neon is preferable, while for heavy elements, krypton or xenon are used. Reactive gases are rarely used to sputter compounds if the reactive product is required [107].

Solgel: Solgel is one more popular process of HA coatings on Ti alloy substrate where CaP coatings are prepared by dipping the titanium in calcium (usually nitrate salt) and phosphorus gels for an appropriate time. HA coating obtained by sol–gel coating is generally better for biological performance than sputtering and plasma-spraying techniques. It is due to the formation of rough and porous coatings through this method which favors the circulation of physiological fluids and accelerates the transport of nutrients and tissue ingrowth when it is used for biomedical application. In the solgel method, the presence of hydroxyl groups promotes calcium and phosphate precipitation and hence improves the interactions with osteoblasts. The solgel method provides numerous advantages, such as the possibility of preparing homogeneous films, better control of the chemical composition of the coating, control of the film microstructure, and a reduction in the densification temperature of the ceramic layer. The solgel method requires less equipment than most available techniques and is less expensive. CaP coatings obtained through this method are porous, less dense, and fail to adhere properly to the substrate. The poor adhesion is an unfavorable factor that limits its use. The adhesion and density of the coatings can be improved by sintering at high temperatures and depending on the applied temperature where different CaP phases can be obtained [108].

Plasma spraying: The most widely used technique is plasma spraying to coat HA on Ti and its alloys substrate. The process of plasma spraying involves complex thermal changes between the plasma zone, powder particles, and the substrate. Particles are injected into the 10,000 °C plasma jet which undergoes extremely high heating within a few seconds. Some large-sized particles may not be fully molten because of their short duration in the plasma zone where the particles experience melting or even boiling [109]. The molten droplets are propelled and lastly impact the substrate surface, forming splats. When in contact with the surface, the melted particles condense and form a film of several tenths of microns with an average roughness of several micrometers, where the irregularity of surface associated with HA affords good bone-screw fixation. This process is a well-accepted coating method but exhibits some major drawbacks. A very rough surface will have more particulate concentration around the implant. The drawbacks include failure within the coating, discontinued dissolution of the coating after implantation, variation in bond strength at the coating metal interface, variations in coating density as a result of the process, poor adhesion between the coatings and substrates, microcracks on the coating surface, and tendency to delaminate. Therefore, the process is limited to thermally stable phases such as for HA, which makes it almost too difficult to immobilize bone morphogenetic proteins on the surface for the stimulation of bone healing. The plasma-spraying method can only be applied to areas that are in direct line-of-sight where only the parts of the substrate that are directly in the line of the spray are coated. Hence, it is problematic to uniformly coat implants of complex shapes and the internal surfaces of porous structures or scaffolds widely which are used in orthopedic and dental implants. Plasma-sprayed HA coatings exhibit suitable adhesion strength and can rapidly be deposited on the surface. Consequently, this method is broadly accepted for clinical applications [110].

Electrochemical cathodic deposition: Electrochemical Cathodic Deposition (ECD) is a potential technique where HA can be coated on titanium surfaces. In this process, an electric current is passed through two electrodes plunged in an electrolyte made up of a solution rich in the ions that are to be deposited onto the cathode substrate (Fig. 13.19) [111]. CaP is not directly electroplated onto the substrate;

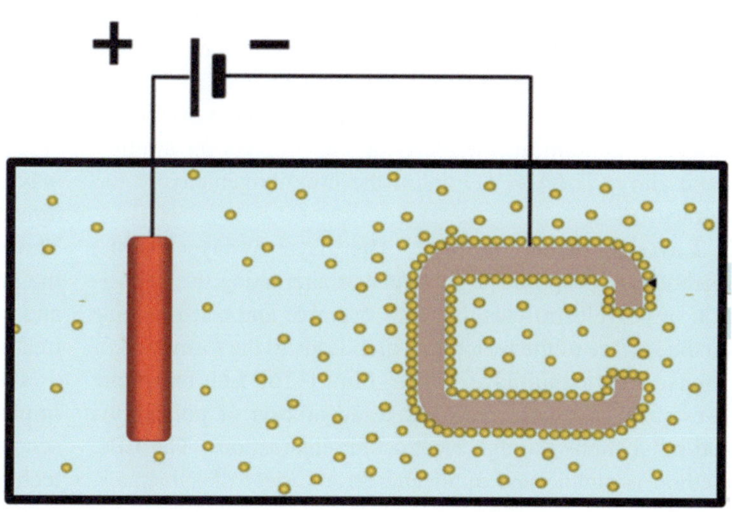

Fig. 13.19 Electrochemical cathodic deposition

instead, the electrochemical reaction that occurs is usually the reduction of water, as in Eq. (11.14);

$$H_2O + 2e^- \rightarrow H_2 + OH^- \quad (13.14)$$

In the reaction that results in a local increase in the pH at the surface of the electrode to a level where the CaP turns into insoluble, the heterogeneous nucleation occurs on the substrate which is more thermodynamically favorable than homogeneous nucleation in bulk solution and forming a film. The ECD process allows for an appropriate thickness of the coating to be accomplished on the metal substrate and has the benefit of being a non-line-of-sight method, enabling great irregularly designed objects to be coated uniformly [112].

Hydrogen gas is produced along with the hydroxide ion as described in the Eq. (11.14) which leads to gas bubbles that can adhere themselves to the surface of the titanium substrate, which can prevent further deposition of CaP and leading to low adherence of the coating to the substrate. Electrochemical Cathodic deposition sometimes results in the deposition of monetite ($CaHPO_4$) or brushite ($CaHPO_4.H_2O$) rather than HA [113].

Carbon layer coatings: Thin carbon layers are biocompatible and more resistant to fibrous encapsulation (biofilm) which is in the form of nanocrystalline diamond or diamond-like carbon (DLC). Diamond is chemically inert and uniquely biocompatible and has a number of applications in medicine [114]. The diamond coating on metallic biomaterials decreases the generation of macrophases and improves the wearability of devices in the field of orthopedic surgery. Carbon layer coating is a perfect coating material that protects the surfaces of surgical instruments against the aggressive biological environment and can solve tribological problems like friction, lubrication, and high wear-resistance characteristics along with chemical inertness corrosion resistance behavior. DLC is associated with high hardness and low friction coefficients than that of the sputtering process, where DLC has a possible alternative to solve tribological problems to protect the alloy surface using biocompatible titanium carbonitride (TiCN) coatings [115]. The TiCN coatings increase the biomaterial response where the preparation of titanium nitride (TiN) or TiCN layers on the titanium alloy (Ti-6Al-4V) control the thrombogenicity of the surface.

Layer-by-layer method: The layer-by-layer (LbL) method was proposed in the favor of alternating the deposition of polyelectrolytes that self-assemble and self-organize on the surface of the material, which leads to the formation of polyelectrolyte multilayer (PEM) films [116]. LbL technique is established on the consecutive adsorption of polyanions and polycations through electrostatic interactions. The LBL method is simple and applicable to a variety of substrates, where the electrostatic attraction among positively and negatively charged ions performs to be an appropriate choice as a driving force for multilayer build-up. The charged species of different types of films containing biological molecules like DNA, polypeptides, proteins, polysaccharides, viruses, and also various kinds of nanoparticles clay platelets, and carbon nanotubes have been successfully prepared. The LbL accumulated heparin/fibronectin biofunctional films were fabricated on titanium surfaces for the enhancement of blood anticoagulation and to accelerate the formation of endothelial tissue simultaneously. The LBL method assembled with chitosan/heparin coating on the surface of a coronary stainless steel stent could promote the formation of endothelial tissue which is safer than bare-metal stents because of the superior anticoagulative effect. The LbL self-assembly technique is based on the polyelectrolyte-mediated electrostatic adsorption of chitosan and gelatin and leads to the formation of multilayers on the titanium thin layers that can progress the surface biocompatibility of titanium oxide layers. Cells could stick to this modified surface with flattened morphology where controlled better cell–cell/cell–substrate interactions occur. Through enzymatic reaction, hydrogels containing heparin and dopamine prepared which has been used as another coating method to enhance the blood compatibility of metallic biomaterials. The surface with hydrogel not only offers a smooth and slippery surface but also prevents the formation of fibrous tissue and interaction with blood components. Heparin coating on the surfaces is an effective methodology to improve the blood compatibility of artificial materials [117].

13.7.2 Surface Treatment

Fibrin adhesion, blood vessel growth, and micromotions should be avoided to achieve the higher degree of osseointegration, and when the mechanical constancy is higher, the probability of implant loosening becomes smaller. By properly tailoring the surface of the implant, cell adhesion can be enhanced and micromotion can be reduced. The required interface can be developed by surface chemistry specifically by nanometer and micrometer scale topographies [118]. Titanium-based materials were experimented in many ways to improve bone integration. The surface roughness is directly related to impaction of cell morphology and growth. The physical placement of grooves and depression can alter the surface topography and also change the cell orientation and attachment. The most frequent physicochemical surface treatments used techniques are as follows:

Grit blasting: The formation of a porous layer on the implant surface can be obtained by the collision with microscopic particles which indicates the grit blasting is a physical technique. Grit blasting is the process where particles of aluminum oxide (Al_2O_3), titanium dioxide (TiO_2), or calcium carbonate ($CaCO_3$) are projected by a fluid carrier that may be of compressed air or liquid onto the surface [119]. This is the process of erosion which affects down to the micrometer scale that depends on the size of particles used for blasting and finally gives surfaces with average roughness less than 1 mm.

Chemical etching: Chemical etching is the process that acts on the dissolution of native oxide layer of metals. Chemical etching can be carried out in either acidic or alkaline conditions, where strong acids like H_2SO_4, HNO_3, HCl, H_3PO_4, and even HF are usually used for the etching of Ti and its alloys. The multiple combined strong acids serve as an effective way of creating a thin grid of nanopits on a titanium surface which is approximately 20–100 nm in diameter. Significantly acidic etching alters the oxide surface by introducing TiOH where the acidic hydroxyl groups are two times coordinated by titanium atoms. NaOH treatment promotes the density and stabilizes the basic hydroxyl groups that are singly coordinated to titanium atoms at the surface of the metallic oxide at a high pH value. To dissolve alumina particles implanted in the layer after a grit-blasting process, the acid etching progresses better. Depending on the material, an etched surface can also be achieved by revealing the surface to reactive plasma in a chamber apart from the electrochemical process or chemical wherein etching process reactive plasma produces reactive products. The sand/grit-blasting and chemical etching is a very random process that is hard to control the uniformity and distribution of nanostructures on the implant surfaces [120].

Electrochemical methods: In electrochemical methods, the implant is immersed in a solution of electrolytes containing acids, ions, or oxidants and connected to a pole of an electrical circuit. The electrochemical method splits into anodization, electropolishing, and electro erosion depending upon requirements. Anodization process is a powerful anodic oxidation process that can produce surfaces with improved oxide layer with an increased oxide layer thickness of up to micron size in comparison of 3–4 nm of the native oxide layer [121]. The anodization treatment renders the material to form topographically complex surfaces at the submicron level that produces porous coatings and/or leads to an increase in corrosion protection and reduced ion release which is due to increased oxide thickness. Anodization technique produces the nanosized texture that is an efficient means to enhance attachment and proliferation of osteoblast cells. For the thickening of the oxide layer throughout the implant, surface voltage is required in the form of direct current (galvanic current). At the anodization process, the titanium substrates serve as the anode, while an inert platinum sheet is regarded as a cathode. The parameters controlling the process variables, such as anode potential, electrolyte temperature, composition, and current, can alter the structural and chemical properties of anodic oxides on titanium. In this method, some diluted acids such as H_2SO_4, H_3PO_4, acetic acid, and others are commonly used as electrolytes [122]. Electropolishing is an industrial process where the process involves a controlled electrochemical dissolution of the amorphous native oxide to give a polycrystalline metallic surface that has wide application in the field of surface finishing of metal components. The new oxide layer has no defects, not dense, but exhibits a reduced roughness. The rate of material removal at its dimensional change will be in the range of 1–10 mm/min for titanium with defined process protocols [123]. In the electro-erosion process, the surrounding area of a Cu–Zn wire from its surface initiates a plasma zone that melts the surface of the titanium, implants locally, and grows a TiO_2 layer. The isotropic and smooth aspects of the relief are the main influential characteristics of the electro-eroded surfaces that cause surface fusion during the process which is somehow contaminated with copper and zinc. The temperature of fusion of materials conceals a thickened oxidized layer on the surface during the electro-erosion process, which would remove all surface inclusions and remove uniformly [124].

Ion beam-based processes: Ion-beam-based processes include ion implantation, ion-beam-assisted deposition (IBAD), and plasma treatment. This process is based on ionized particle bombardment which is associated with providing beneficial surface layers with desirable properties. Ion's energy has an impact on the material surface through attaching, sputtering, and implantation. Physical changes like the amorphic structure of metallic biomaterials can often show in the formation of highly disorder in implantation. Accurate dose and depth control, the versatility of ion species, low-temperature processing, nonequilibrium process, and reliability are the advantages of ion-beam-based processes. Yet, low penetration depth and its relatively high cost are the main drawbacks of this technique where the treatment is selective in the application towards specific areas of the surface. Thin-film formation and surface-modified layer formation are two categorized techniques under ion-beam technology for surface modification [125]. The thin film formation is meant to improve the specific surface properties of an implant and have a different chemistry from the substrate. For example, ion implantation of calcium and phosphorus on titanium implants (e.g., apatite film) leads to the construction of a thin film with dissimilar surface polarities and electrokinetic potentials in comparison to pure medical-grade titanium. The formation of thin films of TiO_2 and TiN is to achieve the improvement of surface properties, such as corrosion resistance, wear resistance, and bone conductivity. This process is effective to fabricate functionally graded materials. With the help of the film formation technique, the film composition can be easily modulated where the film adhesion to the substrate is weak.

The surface-modified layer is the second category of ion-beam technology which can be produced by ion implantation and ion mixing. The graded composition and an unintelligible interface between the surface layer and substrate, making surface delamination less of a problem are the advantages of the surface-modified layer [126].

Laser surface modification: Alteration of metallic surfaces and tailor the surface properties without unduly affecting the underlying properties of a material can be brought by laser-assisted surface modification. The energy beam from the laser can be focused accurately on the surface only where it is needed. Metals have the first atomic layers for opaque materials where the electromagnetic radiation of a laser beam is being absorbed. Hot gas jets and eddy currents are not needed wherein radiation spillage is outside the optically defined beam area. Transformation hardening, surface melting, and/or surface alloying of metallic biomaterials could cause by laser surface modification technology. Laser surface modification shows advantages in particular for biomedical applications and include, but not limited to controlled thermal penetration, chemical cleanliness, remote noncontact processing, and relative ease to automate. Lasers can improve bone cell–biomaterials interactions by fabricating the coat of titanium with tricalcium phosphate (TCP) ceramics to a coating thickness of 200–700 mm using optimized laser parameters. The thickness of the coating is improved by increasing laser power and/or powder feed rate while the coating thickness can be decreased with increasing laser scan speed [127].

Biochemical surface modification: Biologists, chemists, and material scientists put forth to design and develop biomaterials that are equipped with molecular cues simulating certain aspects of structure or function of natural electrocellular matrices in the area of cell and molecular biology. Biochemical modification of surface contributes to changing the biochemical properties of biomaterials by immobilizing bioactive molecules, such as peptides, proteins, enzymes, or growth factors that represent such molecular cues on the surface of biomaterials. For the expression of bio functions and to prompt desired biological responses, these hybrid materials are integrated with synthetic and biological components. There are three major mechanisms, that is, adsorption, entrapment, and covalent attachment can immobilize the biomolecules on the surface of a biomaterial. The binding proteins to polymers are moderately widespread, while the covalent binding of proteins to metallic surfaces has long been challenging and obscure. On the material surface, biomolecules are immobilized which is related to achieve bone formation on the surface of metallic biomaterials to be accelerated. There are several immobilization techniques to enhance bone formation on the biomaterial surface. As per requirement, the amount of proteins bound to metallic biomaterials can be controlled by surface treatment, immobilization chemistry, and depends on the type of material. In case the modified layer cannot make direct contact with the substrate of titanium and its alloys because of the presence of native oxide on titanium substrates, TiO_2 has an inert surface where only a few organic reagents, such as organophosphates, organosilanes, and photosensitive chemicals, can form strong chemical bonds [128]. Biomolecules are conjugated to the surface by reacting with the reactive groups followed by converting the metal surface into reactive groups as per example by covering a surface with organofunctional alkoxysilane molecules by self-assembly. This process is called silanization, in brief the surfaces are converted into amino groups through 3-aminopropyltriethoxysilane and further into aldehyde groups by the treatment of glutaraldehyde or maleimide groups by succinimidyl ester maleimide). Silane and thiols with different end groups are used as monolayer coatings of biomaterials that offer a wide variety of surface chemical properties. The hydrophilicity of the surface and certain chemical groups such as methyl, hydroxyl, or carboxyl groups can be introduced at the surface are controlled by coating. Through the amino-terminal group such as the formation of Schiff's base or by cysteine (i.e., the addition of HSe) Peptides/proteins can be conjugated to the surface. Arg–Gly–Asp (RGD) sequence which is peptides containing the ubiquitous has been immobilized which receives significant attention because RGD is the essential sequence mediating cell adhesion in many ECM proteins (e.g. vitronectin, fibronectin). Bone morphogenetic protein-4 (BMP-4) is introduced on the surface of Ti-6Al-4 V (Tie6Ale4V) alloy through lysozyme to improve hard-tissue response. In culture, collagen has positive properties that can affect cellular adhesion, proliferation, and differentiation of many cell types. Collagen induces the cellular response which is mainly mediated through the amino acid sequence of ArgeGlyeAsp (RGD) acting as the mediator, and the mediator is recognized by the integrin receptors located at the cell membrane. Due to this reason, collagen has been used to modify biomaterial surfaces successfully, which is made up of stainless steel and titanium [129].

It is necessary to reduce the surface protein adsorption in the case of some blood-contacting devices. For the reduction of protein adsorption, it is to passivate the surface with a layer of an inert protein such as albumin where cells cannot adhere to receptors for albumin, and due to this, albumin is used to passivate surfaces against cell adhesion. The anticoagulant and antithrombotic agents are used to immobilize on the biomaterial surface either for controlled release or for a sustained period. Anticoagulants can be incorporated into the chemical structure of a polymer in the case of polymeric materials which is an integral part of the biomaterial. Heparin can be immobilized on biomaterials surface where it has excellent anticoagulant and antithrombotic properties. Tropoelastin on the plasma coated stainless steel surface immobilize covalently have to improve the biocompatibility through promoting the endothelial cell attachment and proliferation relative to uncoated stainless steel controls. Various antithrombotic molecules (covalent, ionic, and physical adsorption) with specific chemical groups such as –COOH, $-NH_2$, and –OH) were immobilized by means of bonding covalently or by ionic bonding or by adsorbing physically onto the surface of clinical vascular devices promisingly in terms of blood biocompatibility [130].

Heat Treatment: In the presence of oxygen or in air, the appropriate heat treatment can change the composition of oxygen of the alloying elements present on the surface and shows biocompatibility. Low-temperature heat treatment of titanium alloy improves the surface with Ti and Al and promotes cell attachment. The heat treatment gives a better adhesion property at the interface for the surface-modified coating.

13.8 Summary

This chapter described the beauty of materials science and its development to replace the body organ. The basic concept of body environment and its interaction with implant materials has been given. Among several types of biomaterials, metallic biomaterials is for load-bearing application (with toughness); ceramic materials is for compressive load and wear application; and polymeric materials are for flexible body organ has been discussed clearly along with the composite and naturally derived biomaterials. Naturally derived biomaterials have been studied and applied in clinical applications such as artificial tissues/organs because they are capable of supporting cell proliferation, biodegradability, and remodeling tissues. Although the current results have not completely satisfied the clinical demand, the potential applications of naturally derived biomaterials are still highly considered; therefore, research on this field has now being taken place all over the world.

References

1. Cao, W., Larry, L.: Hench, bioactive materials. Ceramics Int. **22** (6), 493–507 (1996). https://doi.org/10.1016/0272-8842(95)00126-3
2. Wilson, C.J., Clegg, R.E., Leavesley, D.I., Pearcy, M.J.: Mediation of biomaterial-cell interactions by adsorbed proteins: a review. Tissue Eng. **11**(1-2) (2005). https://doi.org/10.1089/ten.2005.11.1
3. Wang, W., Yeung, K.W.K.: Bone grafts and biomaterials substitutes for bone defect repair: a review. Bioactive Mater. **2**(4), 224–247 (2017). https://doi.org/10.1016/j.bioactmat.2017.05.007
4. Silver, F.H., Christiansen, D.L.: Introduction to biomaterials science and biocompatibility. In: Biomaterials science and biocompatibility. Springer, New York, NY (1999). https://doi.org/10.1007/978-1-4612-0557-9_1
5. Lawrence, B.D., Marchant, J.K., Pindrus, M.A., Omenetto, F.G., Kaplan, D.L.: Silk film biomaterials for cornea tissue engineering. Biomaterials. **30**(7), 1299–1308 (2009). https://doi.org/10.1016/j.biomaterials.2008.11.018
6. Jandt, K.D.: Evolutions, revolutions and trends in biomaterials science – a perspective, special issue. Biomaterials. **9**(12), 1035–1050 (2007). https://doi.org/10.1002/adem.200700284
7. See: https://www.marketsandmarkets.com/Market-Reports/biomaterials-393.html?gclid=Cj0KCQjws536BRDTARIsANeUZ5_v618EenMXGC_dzOHnb33VlDM01YP0biqRY7fYElCD0uV_xSgGYREaAvcCEALw_wcB. 01.03.2020
8. See: https://en.wikipedia.org/wiki/Sushruta_Samhita. 01.03.2020
9. Shrivastava, S., Soundararajan, P., Agrawal, A.: Ayurvedic approach in chronic disease management. In: Noland, D., Drisko, J., Wagner, L. (eds.) Integrative and functional medical nutrition therapy. Humana, Cham (2020). https://doi.org/10.1007/978-3-030-30730-1_45
10. Dearnley, P.A.: A review of metallic, ceramic and surface-treated metals used for bearing surfaces in human joint replacements. Proc. Inst. Mech. Eng. H J. Eng. Med. **213**(2), 107–135 (1999). https://doi.org/10.1243/0954411991534843
11. Manam, N.S., Harun, W.S.W., Shri, D.N.A., Ghani, S.A.C., Kurniawan, T., Ismail, M.H., Ibrahim, M.H.I.: Study of corrosion in biocompatible metals for implants: a review. J. Alloys Compd. **701**, 698–715 (2017). https://doi.org/10.1016/j.jallcom.2017.01.196
12. Oliveira, A., et al.: In vitro studies of bioactive glass/polyhydroxybutyrate composites. Mat. Res. **9**(4), 417–423 (2006). https://doi.org/10.1590/S1516-14392006000400013
13. Roumelioti, M.E., Glew, R.H., Khitan, Z.J., et al.: Fluid balance concepts in medicine: principles and practice. World J. Nephrol. **7** (1), 1–28 (2018). https://doi.org/10.5527/wjn.v7.i1.1
14. Kim, J., Heo, J.N., Do, J.Y., Chava, R.K., Kang, M.: Electrochemical synergies of heterostructured Fe2O3-MnO catalyst for oxygen evolution reaction in alkaline water splitting. Nanomaterials (Basel). **9**(10), 1486 (2019). https://doi.org/10.3390/nano9101486. Published 2019 Oct 18
15. Vatansever, F., de Melo, W.C., Avci, P., et al.: Antimicrobial strategies centered around reactive oxygen species--bactericidal antibiotics, photodynamic therapy, and beyond. FEMS Microbiol. Rev. **37**(6), 955–989 (2013). https://doi.org/10.1111/1574-6976.12026
16. Frank, M., Gutowska, M.A., Martina, L., Dupont, S., Lucassen, M., Thorndyke, M.C., Bleich, M., Pörtner, H.-O.: Physiological basis for high CO2 tolerance in marine ectothermic animals: pre-adaptation through lifestyle and ontogeny? Open Access Biogeosciences (BG). **6**, 2313–2331 (2009). https://doi.org/10.5194/bg-6-2313-2009
17. See: https://qmro.qmul.ac.uk/xmlui/handle/123456789/36705. 02.03.2020
18. David, F.W.: On the mechanisms of biocompatibility. Biomaterials. **29**(20), 2941–2953 (2008). https://doi.org/10.1016/j.biomaterials.2008.04.023
19. Schroers, J., Kumar, G., Hodges, T.M., et al.: Bulk metallic glasses for biomedical applications. JOM. **61**, 21–29 (2009). https://doi.org/10.1007/s11837-009-0128-1
20. Naidich, J.V.: The wettability of solids by liquid metals. In: Cadenhead, D.A., Danielli, J.F. (eds.) Progress in surface and membrane science, vol. 14, pp. 353–484. Elsevier, Amsterdam (1981). https://doi.org/10.1016/B978-0-12-571814-1.50011-7. ISBN 9780125718141
21. Shuilin, W., Liu, X., Yeung, K.W.K., Liu, C., Yang, X.: Biomimetic porous scaffolds for bone tissue engineering. Mater. Sci. Eng. R Rep. **80**, 1–36 (2014). https://doi.org/10.1016/j.mser.2014.04.001

22. Tang, W., Lin, D., Yu, Y., Niu, H., Guo, H., Yuan, Y., Liu, C.: Bioinspired trimodal macro/micro/nano-porous scaffolds loading rhBMP-2 for complete regeneration of critical size bone defect. Acta Biomater. **32**, 309–323 (2016). https://doi.org/10.1016/j.actbio.2015.12.006
23. Murphy, C.M., O'Brien, F.J.: Understanding the effect of mean pore size on cell activity in collagen-glycosaminoglycan scaffolds. Cell Adhes. Migr. **4**(3), 377–381 (2010). https://doi.org/10.4161/cam.4.3.11747
24. Zhu, D., Cockerill, I., Su, Y., Zhang, Z., Fu, J., Lee, K.-W., Ma, J., Okpokwasili, C., Tang, L., Zheng, Y., Qin, Y.-X., Wang, Y.: Mechanical strength, biodegradation, and in vitro and in vivo biocompatibility of Zn biomaterials. ACS Appl. Mater. Interfaces. **11** (7), 6809–6819 (2019). https://doi.org/10.1021/acsami.8b20634
25. Amid, P.K.: Biomaterials - classification, technical and experimental aspects. In: Schumpelick, V., Kingsnorth, A.N. (eds.) Incisional Hernia. Springer, Berlin (1999). https://doi.org/10.1007/978-3-642-60123-1_13
26. Amid, P.K.: Classification of biomaterials and their related complications in abdominal wall hernia surgery. Hernia. **1**, 15–21 (1997). https://doi.org/10.1007/BF02426382
27. Mohapatra, R.K., El-ajaily, M.M., Alassbaly, F.S., Sarangi, A.K., Das, D., Maihub, A.A., Ben-Gweirif, S.F., Mahal, A., Suleiman, M., Perekhoda, L., Azam, M., Al-Noor, T.H.: DFT, anticancer, antioxidant and molecular docking investigations of some ternary Ni(II) complexes with 2-[(E)-[4-(dimethylamino)phenyl] methyleneamino]phenol. Chem. Papers. (2020). https://doi.org/10.1007/s11696-020-01342-8
28. Mohapatra, R.K., Mishra, U.K., Mishra, S.K., Mahapatra, A., Dash, D.C.: Synthesis and characterization of transition metal complexes with benzimidazolyl-2-hydrazones of o-anisaldehyde and furfural. J. Korean Chem. Soc. **55**(6), 926–931 (2011)
29. Mohapatra, R.K., Dash, M., Mishra, U.K., Mahapatra, A., Dash, D.C.: Synthesis and characterization of transition metal complexes with benzimidazolyl-2-hydrazones of glyoxal, diacetyl and benzil. Synth. React. Inorg. M. **44**(5), 642–648 (2014)
30. Radenković, G., Petković, D.: Metallic biomaterials. In: Zivic, F., Affatato, S., Trajanovic, M., Schnabelrauch, M., Grujovic, N., Choy, K. (eds.) Biomaterials in clinical practice. Springer, Cham (2018). https://doi.org/10.1007/978-3-319-68025-5_8
31. Marjanović-Balaban, Ž., Jelić, D.: Polymeric biomaterials in clinical practice. In: Zivic, F., Affatato, S., Trajanovic, M., Schnabelrauch, M., Grujovic, N., Choy, K. (eds.) Biomaterials in clinical practice. Springer, Cham (2018). https://doi.org/10.1007/978-3-319-68025-5_4
32. Mohan, P., Rajak, D.K., Catalin, P.I., Behera, A., Amigó-Borrása, V., Elshalakany, A.B.: Influence of β-phase stability in elemental blended Ti-Mo and Ti-Mo-Zr alloys. Micron. **142**, 102992 (2021). https://doi.org/10.1016/j.micron.2020.102992
33. Ma, P., Langer, R.: Degradation, structure and properties of fibrous nonwoven poly(glycolic acid) scaffolds for tissue engineering. MRS Proc. **394**, 99 (1995). https://doi.org/10.1557/PROC-394-99
34. Lemons, J.E., Lucas, L.C.: Properties of biomaterials. J. Arthroplasty. **1**(2), 143–147 (1986). https://doi.org/10.1016/S0883-5403(86)80053-5
35. Pezzin, A.P.T., Duek, E.A.R.: Hydrolytic degradation of poly (para-dioxanone) films prepared by casting or phase separation. Polym. Degradation Stability. **78**(3), 405–411 (2002). https://doi.org/10.1016/S0141-3910(02)00174-X
36. Pilliar, R.M.: Metallic Biomaterials. In: Narayan, R. (ed.) Biomedical materials. Springer, Boston, MA (2009). https://doi.org/10.1007/978-0-387-84872-3_2
37. Kathryne, S.B., Lai, B.F.L., Jayachandran, N.K., Paul Santerre, J.: Hemocompatibility of degrading polymeric biomaterials: degradable polar hydrophobic ionic polyurethane versus poly(lactic-co-glycolic) acid. Biomacromolecules. **18**(8), 2296–2305 (2017). https://doi.org/10.1021/acs.biomac.7b00456
38. Mueller, W.-D., Lucia Nascimento, M., de Mele, M.F.L.: Critical discussion of the results from different corrosion studies of Mg and Mg alloys for biomaterial applications. Acta Biomater. **6**(5), 1749–1755 (2010). https://doi.org/10.1016/j.actbio.2009.12.048
39. Yingchao, S., Cockerill, I., Wang, Y., Qin, Y.-X., Chang, L., Zheng, Y., Zhu, D.: Zinc-based biomaterials for regeneration and therapy. Trends Biotechnol. **37**(4) (2019). https://doi.org/10.1016/j.tibtech.2018.10.009
40. Hermawan, H., Alamdari, H., Mantovani, D., Dubé, D.: Iron–manganese: new class of metallic degradable biomaterials prepared by powder metallurgy. Powder Metallurgy. **51**(1), 38–45 (2008). https://doi.org/10.1179/174329008X284868
41. Wu, G., Li, P., Feng, H., Zhanga, X., Chu, P.K.: Engineering and functionalization of biomaterials via surface modification. J. Mater. Chem. B. **3**, 2024–2042 (2015). https://doi.org/10.1039/C4TB01934B
42. Parida, P., Mishra, S.C., Sahoo, S., Behera, A., Nayak, B.P.: Development and characterization of ethylcellulose based microsphere for sustained release of nifedipine. J. Pharm. Anal. **6**(5), 341–344 (2016). https://doi.org/10.1016/j.jpha.2014.02.001
43. Chen, Q., Thouas, G.A.: Metallic implant biomaterials. Mater. Sci. Eng. R Rep. **87**, 1–57 (2015). https://doi.org/10.1016/j.mser.2014.10.001
44. Huang, J., Best, S.M.: 1 - Ceramic biomaterials. In: Boccaccini, A. R., Gough, J.E. (eds.) Woodhead publishing series in biomaterials, tissue engineering using ceramics and polymers, pp. 3–31. Woodhead Publishing, Sawston (2007). https://doi.org/10.1533/9781845693817.1.3. ISBN 9781845691769
45. Jones, D.W.: Ceramic biomaterials. In: Key engineering materials, vol. 122–124, pp. 345–386. Trans Tech Publications, Ltd., Freienbach (1996). https://doi.org/10.4028/www.scientific.net/kem.122-124.345
46. Harun, W.S.W., Asri, R.I.M., Alias, J., Zulkifli, F.H., Kadirgama, K., Ghani, S.A.C., Shariffuddin, J.H.M.: A comprehensive review of hydroxyapatite-based coatings adhesion on metallic biomaterials. Ceramics Int. **44**(2), 1250–1268 (2018). https://doi.org/10.1016/j.ceramint.2017.10.162
47. Paul Ducheyne, W., Van Raemdonck, J.C., Heughebaert, M.: Heughebaert, Structural analysis of hydroxyapatite coatings on titanium. Biomaterials. **7**(2), 97–103 (1986). https://doi.org/10.1016/0142-9612(86)90063-3
48. Echave, M.C., Burgo, L.S., Pedraz, J.L., Orive, G.: Gelatin as biomaterial for tissue engineering. Curr. Pharm. Des. **23**(18), 3567–3584 (2017). https://doi.org/10.2174/0929867324666170511123101
49. Piconi, C., Maccauro, G., Muratori, F., Del Prever, E.B.: Alumina and zirconia ceramics in joint replacements. J. Appl. Biomater. Biomech. **1**(1), 19–32 (2003). https://doi.org/10.1177/228080000300100103
50. Liu, X., Huang, A., Ding, C., Chu, P.K.: Bioactivity and cytocompatibility of zirconia (ZrO2) films fabricated by cathodic arc deposition. Biomaterials. **27**(21), 3904–3911 (2006). https://doi.org/10.1016/j.biomaterials.2006.03.007
51. Ritchie, R.O., Dauskardt, R.H., Yu, W., Brendzel, A.M.: Cyclic fatigue-crack propagation, stress-corrosion, and fracture-toughness behavior in pyrolytic carbon-coated graphite for prosthetic heart valve applications. J. Biomed. Mater. Res. **24**(2), 189–206 (1990). https://doi.org/10.1002/jbm.820240206
52. Eleanor, M.P., Szybala, C., Boison, D., Kaplan, D.L.: Silk fibroin encapsulated powder reservoirs for sustained release of adenosine. J. Control. Release. **144**(2), 159–167 (2010). https://doi.org/10.1016/j.jconrel.2010.01.035
53. João, S.F., Gentile, P., Pires, R.A., Reis, R.L., Hatton, P.V.: Multifunctional bioactive glass and glass-ceramic biomaterials with

antibacterial properties for repair and regeneration of bone tissue. Acta Biomater. **59**, 2–11 (2017). https://doi.org/10.1016/j.actbio.2017.06.046
54. Kamitakahara, M., Ohtsuki, C., Miyazaki, T.: Review paper: behavior of ceramic biomaterials derived from tricalcium phosphate in physiological condition. J. Biomater. Appl. **23**(3), 197–212 (2008). https://doi.org/10.1177/0885328208096798
55. Jack, E.L.: Ceramics: past, present, and future. Bone. **9**(1 Supplement 1), 121–128 (1996). https://doi.org/10.1016/S8756-3282(96)00128-7
56. He, W., Benson, R.: 8 - Polymeric biomaterials. In: Kutz, M. (ed.) Plastics design library, applied plastics engineering handbook, 2nd edn, pp. 145–164. William Andrew Publishing, Norwich (2017). https://doi.org/10.1016/B978-0-323-39040-8.00008-0. ISBN 9780323390408
57. Kohane, D., Langer, R.: Polymeric Biomaterials in Tissue Engineering. Pediatr. Res. **63**, 487–491 (2008). https://doi.org/10.1203/01.pdr.0000305937.26105.e7
58. Athanasiou, K.A., Niederauer, G.G., Agrawal, C.M.: Sterilization, toxicity, biocompatibility and clinical applications of polylactic acid/ polyglycolic acid copolymers. Biomaterials. **17**(2), 93–102 (1996). https://doi.org/10.1016/0142-9612(96)85754-1. ISSN 0142–9612
59. Kirn, D., Takeno, M.M., Ratner, B.D., et al.: Glow discharge plasma deposition (GDPD) technique for the local controlled delivery of hirudin from biomaterials. Pharm. Res. **15**, 783–786 (1998). https://doi.org/10.1023/A:1011987423502
60. Calis, S., Jeyanthi, R., Tsai, T., et al.: Adsorption of salmon calcitonin to PLGA microspheres. Pharm. Res. **12**, 1072–1076 (1995). https://doi.org/10.1023/A:1016278902839
61. Wang, C., Chen, H., Zhu, X., Xiao, Z., Zhang, K., Zhang, X.: An improved polymeric sponge replication method for biomedical porous titanium scaffolds. Mater. Sci. Eng. C. **70**(Part 2), 1192–1199 (2017). https://doi.org/10.1016/j.msec.2016.03.037
62. Rachael, H.S., Masters, K.S., West, J.L.: Photocrosslinkable polyvinyl alcohol hydrogels that can be modified with cell adhesion peptides for use in tissue engineering. Biomaterials. **23**(22), 4325–4332 (2002). https://doi.org/10.1016/S0142-9612(02)00177-1
63. Hyun, J., Zhu, Y., Liebmann-Vinson, A., Thomas, P.B., Chilkoti, A.: Microstamping on an activated polymer surface: patterning biotin and streptavidin onto common polymeric biomaterials. Langmuir. **17**(20), 6358–6367 (2001). https://doi.org/10.1021/la010695x
64. Das, M., Balla, V.K., Kumar, T.S.S., Manna, I.: Fabrication of biomedical implants using laser engineered net shaping (LENS™). Trans. Indian Ceramic Soc. **72**(3), 169–174 (2013). https://doi.org/10.1080/0371750X.2013.851619
65. Edidin, A.A., Rimnac, C.M., Goldberg, V.M., Kurtz, S.M.: Mechanical behavior, wear surface morphology, and clinical performance of UHMWPE acetabular components after 10 years of implantation. Wear. **250**(1–12), 152–158 (2001). https://doi.org/10.1016/S0043-1648(01)00616-0
66. Jones, D.S., Djokic, J., Gorman, S.P.: The resistance of polyvinylpyrrolidone–Iodine–poly(ε-caprolactone) blends to adherence of Escherichia coli. Biomaterials. **26**(14), 2013–2020 (2005). https://doi.org/10.1016/j.biomaterials.2004.06.001
67. Cifková, I., Lopour, P., Vondráček, P., Jelínek, F.: Silicone rubber-hydrogel composites as polymeric biomaterials: I. Biological properties of the silicone rubber-p(HEMA) composite. Biomaterials. **11**(6), 393–396 (1990). https://doi.org/10.1016/0142-9612(90)90093-6
68. Xiao, L., Li, J., Brougham§, D.F., Fox§, E.K., Feliu⊥, N., Bushmelev, A., Schmidt, A., Mertens, N., Kiessling, F., Valldor, M., Fadeel, B., Mathur, S.: Water-soluble superparamagnetic magnetite nanoparticles with biocompatible coating for enhanced magnetic resonance imaging. ACS Nano. **5**(8), 6315–6324 (2011). https://doi.org/10.1021/nn201348s
69. Lee, K.-W., Wang, S., Fox, B.C., Ritman, E.L., Yaszemski, M.J., Lichun, L.: Poly(propylene fumarate) bone tissue engineering scaffold fabrication using stereolithography: effects of resin formulations and laser parameters. Biomacromolecules. **8**(4), 1077–1084 (2007). https://doi.org/10.1021/bm060834v
70. Chen, G.-Q., Qiong, W.: The application of polyhydroxyalkanoates as tissue engineering materials. Biomaterials. **26**(33), 6565–6578 (2005). https://doi.org/10.1016/j.biomaterials.2005.04.036
71. Boni, R., Ali, A., Shavandi, A., et al.: Current and novel polymeric biomaterials for neural tissue engineering. J. Biomed. Sci. **25**, 90 (2018). https://doi.org/10.1186/s12929-018-0491-8
72. Mohanty, A.K., Misra, M., Hinrichsen, G.: Biofibres biodegradable polymers and biocomposites: An overview. Macromol. Mater. Eng. **276, 277**(1), 1–24 (2000). https://doi.org/10.1002/(SICI)1439-2054(20000301)276:1<1::AID-MAME1>3.0.CO;2-W
73. Lan, G., Lu, B., Wang, T., Wang, L., Chen, J., Yu, K., Liu, J., Dai, F., Wu, D.: Chitosan/gelatin composite sponge is an absorbable surgical hemostatic agent. Colloids Surf. B: Biointerfaces. **136**, 1026–1034 (2015). https://doi.org/10.1016/j.colsurfb.2015.10.039
74. Park, S.-B., Lih, E., Park, K.-S., Joung, Y.K., Han, D.K.: Biopolymer-based functional composites for medical applications. Prog. Polym. Sci. **68**, 77–105 (2017). https://doi.org/10.1016/j.progpolymsci.2016.12.003
75. Tan, H.-L., Teow, S.-Y., Pushpamalar, J.: Application of metal nanoparticle–hydrogel composites in tissue regeneration. Bioengineering. **6**(1), 17 (2019). https://doi.org/10.3390/bioengineering6010017
76. Mogoşanu, G.D., Grumezescu, A.M.: Natural and synthetic polymers for wounds and burns dressing. Int. J. Pharm. **463**(2), 127–136 (2014). https://doi.org/10.1016/j.ijpharm.2013.12.015
77. Dinesh, M.: Pardhi, Didem Şen Karaman, Juri Timonen, Wei Wu, Qi Zhang, Saurabh Satija, Meenu Mehta, Nitin Charbe, Paul Mc Carron, Murtaza Tambuwala, Hamid A. Bakshi, Poonam Negi, Alaa AAljabali, Kamal Dua, Dinesh K Chaellappan, Ajit Behera, Kamla Pathak, Ritesh B. Wathar karo, Jessica M. Rosenholm. Anti-bacterial activity of inorganic nanomaterials and their antimicrobial peptide conjugates against resistant and non-resistant pathogens. Int. J. Pharm. **586**, 119531 (2020). https://doi.org/10.1016/j.ijpharm.2020.119531
78. Chabbaa, S., Matthewsb, G.F., Netravali, A.N.: Green' composites using cross-linked soy flour and flax yarns. Green Chem. **7**, 576–581 (2005). https://doi.org/10.1039/B410817E
79. Nishihara, T., Rubin, A.L., Stenzel, K.H.: Biologically derived collagen membranes. In: Stark, L., Agarwal, G. (eds.) Biomaterials. Springer, Boston, MA (1967). https://doi.org/10.1007/978-1-4615-6555-0_14
80. John, F., Cavallaro Paul, D., Kemp Karl, H.K.: Collagen fabrics as biomaterials. Biotechnol Bioeng. **43**(8), 781–791 (1994). https://doi.org/10.1002/bit.260430813
81. Choi, Y.S., Hong, S.R., Lee, Y.M., Song, K.W., Park, M.H., Nam, Y.S.: Study on gelatin-containing artificial skin: I. Preparation and characteristics of novel gelatin-alginate sponge. Biomaterials. **20**(5), 409–417 (1999). https://doi.org/10.1016/S0142-9612(98)00180-X
82. Ha, T.L.B., Quan, T.M., Vu, D.N., Si, D.M.: Naturally derived biomaterials: preparation and application. IntechOpen, London (2013). https://doi.org/10.5772/55668
83. Anshu, B.M., Gupta, V.: NANOMEDICINE, Silk fibroin-derived nanoparticles for biomedical applications. Nanomedicine. **5**(5) (2010). https://doi.org/10.2217/nnm.10.51
84. Nguyen, T.P., Nguyen, Q.V., Nguyen, V.H., et al.: Silk fibroin-based biomaterials for biomedical applications: a review. Polymers

(Basel). **11**(12), 1933 (2019). https://doi.org/10.3390/polym11121933
85. Ahmed, T.A.E., Dare, E.V., Hincke, M.: Fibrin: a versatile scaffold for tissue engineering applications. Tissue Eng. B Rev. **14** (2) (2008). https://doi.org/10.1089/ten.teb.2007.0435
86. Le Guéhennec, L., Layrolle, P., Daculsi, G.: A review of bioceramics and fibrin sealant. Eur. Cells Mater. **8**, 1–11 (2004). https://doi.org/10.22203/eCM.v008a01
87. Stanton, J., Xue, Y., Waters, J.C., Lewis, A., Cowan, D., Hu, X., la Cruz, D.S.-d.: Structure–property relationships of blended polysaccharide and protein biomaterials in ionic liquid. Cellulose. **24**, 1775–1789 (2017). https://doi.org/10.1007/s10570-017-1208-y
88. Park, T.-J., Murugesan, S., Linhardt, R.J.: Cellulose composites prepared using ionic liquids (ILs) - blood compatibility to batteries. In: Polysaccharide materials: performance by design, Chapter 7 ACS Symposium Series, vol. 1017, pp. 133–152. IntechOpen, London (2009). https://doi.org/10.1021/bk-2009-1017.ch007. ISBN13: 9780841269866eISBN: 9780841225343
89. Cheng, K., Catchmark, J.M., Demirci, A.: Effect of different additives on bacterial cellulose production by Acetobacter xylinum and analysis of material property. Cellulose. **16**, 1033–1045 (2009). https://doi.org/10.1007/s10570-009-9346-5
90. Shigemasa, Y., Minami, S.: Applications of chitin and chitosan for biomaterials. Biotechnol. Genetic Eng. Rev. **13**(1), 383–420 (1996). https://doi.org/10.1080/02648725.1996.10647935
91. Usami, Y., Minami, S., Okamoto, Y., Matsuhashi, A., Shigemasa, Y.: Influence of chain length of N-acetyl-d-glucosamine and d-glucosamine residues on direct and complement-mediated chemotactic activities for canine polymorphonuclear cells. Carbohydr. Polym. **32**(2), 115–122 (1997). https://doi.org/10.1016/S0144-8617(96)00153-1
92. Piskin, E.: Synthetic polymeric membranes: separation via membranes. In: Piskin, E., Hoffman, A.S. (eds.) Polymeric biomaterials. NATO ASI series (Series E: applied sciences), vol. 106. Springer, Dordrecht (1986). https://doi.org/10.1007/978-94-009-4390-2_8
93. Francis Suh, J.-K., Matthew, H.W.T.: Application of chitosan-based polysaccharide biomaterials in cartilage tissue engineering: a review. Biomaterials. **21**(24), 2589–2598 (2000). https://doi.org/10.1016/S0142-9612(00)00126-5
94. Hani, A.A., Wickham, M.Q., Leddy, H.A., Gimble, J.M., Guilak, F.: Chondrogenic differentiation of adipose-derived adult stem cells in agarose, alginate, and gelatin scaffolds. Biomaterials. **25** (16), 3211–3222 (2004). https://doi.org/10.1016/j.biomaterials.2003.10.045
95. Liu, J., Zhan, X., Wan, J., Wang, Y., Wang, C.: Review for carrageenan-based pharmaceutical biomaterials: Favourable physical features versus adverse biological effects. Carbohydr. Polym. **121**, 27–36 (2015). https://doi.org/10.1016/j.carbpol.2014.11.063
96. Park, S.-j., Lee, K.W., Lim, D.-S., Lee, S.: The sulfated polysaccharide fucoidan stimulates osteogenic differentiation of human adipose-derived stem cells. Stem Cells Dev. **21**(12) (2011). https://doi.org/10.1089/scd.2011.0521
97. I Rodrı́guez, M., Santamarina, M.H., Bollaı́n, M.C., Mejuto, R.C.: Speciation of organotin compounds in marine biomaterials after basic leaching in a non-focused microwave extractor equipped with pressurized vessels. J. Chromatogr. A. **774**(1–2), 379–387 (1997). https://doi.org/10.1016/S0021-9673(96)00912-0
98. Kim, T.K., Yoon, J.J., Lee, D.S., Park, T.G.: Gas foamed open porous biodegradable polymeric microspheres. Biomaterials. **27** (2), 152–159 (2006). https://doi.org/10.1016/j.biomaterials.2005.05.081
99. Liu, X., Ma, P.X.: Phase separation, pore structure, and properties of nanofibrous gelatin scaffolds. Biomaterials. **30**(25), 4094–4103 (2009). https://doi.org/10.1016/j.biomaterials.2009.04.024
100. Claire, M.B., Levingstone, T.J., Shen, N., Cooney, G.M., Jockenhoevel, S., Flanagan, T.C., O'Brien, F.J.: Freeze-drying as a novel biofabrication method for achieving a controlled microarchitecture within large, complex natural biomaterial scaffolds. Adv. Healthcare Mater. **6**(21), 1 (2017). https://doi.org/10.1002/adhm.201700598
101. Xue, J., Wu, T., Dai, Y., Xia, Y.: Electrospinning and electrospun nanofibers: methods, materials, and applications. Chem. Rev. **119** (8), 5298–5415 (2019). https://doi.org/10.1021/acs.chemrev.8b00593
102. Song, Y.W., Shan, D.Y., Han, E.H.: Electrodeposition of hydroxyapatite coating on AZ91D magnesium alloy for biomaterial application. Mater. Lett. **62**(17–18), 3276–3279 (2008). https://doi.org/10.1016/j.matlet.2008.02.048
103. Koike, M., Greer, P., Owen, K., Lilly, G., Murr, L.E., Gaytan, S. M., Martinez, E., Okabe, T.: Evaluation of titanium alloys fabricated using rapid prototyping technologies-electron beam melting and laser beam melting. Materials. **4**(10), 1776–1792 (2011). https://doi.org/10.3390/ma4101776
104. Chu, P.K., Chen, J.Y., Wang, L.P., Huang, N.: Plasma-surface modification of biomaterials. Mater. Sci. Eng. R Rep. **36**(5–6), 143–206 (2002). https://doi.org/10.1016/S0927-796X(02)00004-9
105. Rao, P.J., Pelletier, M.H., Walsh, W.R., Mobbs, R.J.: Spine interbody implants: material selection and modification, functionalization and bioactivation of surfaces to improve osseointegration. Orthoped. Surg. **6**(2), 81–89 (2014). https://doi.org/10.1111/os.12098
106. Lin, D.-J., Hung, F.-Y., Jakfar, S., Yeh, M.-L.: Tailored coating chemistry and interfacial properties for construction of bioactive ceramic coatings on magnesium biomaterial. Mater. Des. **89**, 235–244 (2016). https://doi.org/10.1016/j.matdes.2015.09.144
107. Behera, A., Aich, S.: Characterization and properties of magnetron sputtered nanoscale NiTi thin film and the effect of annealing temperature. Surf. Interface Anal. **47**, 805–814 (2015). https://doi.org/10.1002/sia.5777
108. Asri, R.I.M., Harun, W.S.W., Hassan, M.A., Ghani, S.A.C., Buyong, Z.: A review of hydroxyapatite-based coating techniques: sol-gel and electrochemical depositions on biocompatible metals. J. Mech. Behav. Biomed. Mater. **57**, 95–108 (2016). https://doi.org/10.1016/j.jmbbm.2015.11.031
109. Cheang, P., Khor, K.A.: Addressing processing problems associated with plasma spraying of hydroxyapatite coatings. Biomaterials. **17**(5), 537–544 (1996). https://doi.org/10.1016/0142-9612(96)82729-3
110. Lugscheider, E., Weber, T., Knepper, M., Vizethum, F.: Production of biocompatible coatings by atmospheric plasma spraying. Mater. Sci. Eng. A. **139**, 45–48 (1991). https://doi.org/10.1016/0921-5093(91)90594-D
111. Li, B., Hao, J., Min, Y., Xin, S., Guo, L., He, F., Liang, C., Wang, H., Li, H.: Biological properties of nanostructured Ti incorporated with Ca, P and Ag by electrochemical method. Mater. Sci. Eng. C. **51**, 80–86 (2015). https://doi.org/10.1016/j.msec.2015.02.036
112. Kuo, M.C., Yen, S.K.: The process of electrochemical deposited hydroxyapatite coatings on biomedical titanium at room temperature. Mater. Sci. Eng. C. **20**(1–2), 153–160 (2002). https://doi.org/10.1016/S0928-4931(02)00026-7
113. Prado Da Silva, M.H., Lima, J.H.C., Soares, G.A., Elias, C.N., de Andrade, M.C., Best, S.M., Gibson, I.R.: Transformation of monetite to hydroxyapatite in bioactive coatings on titanium. Surf. Coat. Technol. **137**(2–3), 270–276 (2001). https://doi.org/10.1016/S0257-8972(00)01125-7
114. Grill, A.: Diamond-like carbon coatings as biocompatible materials—an overview. Diamond Relat. Mater. **12**(2), 166–170 (2003). https://doi.org/10.1016/S0925-9635(03)00018-9

115. Ul-Hamid, A.: The effect of deposition conditions on the properties of Zr-carbide, Zr-nitride and Zr-carbonitride coatings– a review. Mater. Adv. **1**, 988–1011 (2020). https://doi.org/10.1039/D0MA00232A
116. Shenoy, D.B., Antipov, A.A., Sukhorukov, G.B., Möhwald, H.: Layer-by-layer engineering of biocompatible, decomposable core–shell structures. Biomacromolecules. **4**(2), 265–272 (2003). https://doi.org/10.1021/bm025661y
117. Hua Ai, Hongdi Meng, Izumi Ichinose, Steven A Jones, David K Mills, Yuri M Lvov, Xiaoxi QiaoBiocompatibility of layer-by-layer self-assembled nanofilm on silicone rubber for neurons, J. Neurosci. Methods 2003, 128, 1–2, 1–8. doi:https://doi.org/10.1016/S0165-0270(03)00191-2ISSN 0165–0270
118. Variola, F., Vetrone, F., Richert, L., Jedrzejowski, P., Yi, J.-H., Zalzal, S., Clair, S., Sarkissian, A., Perepichka, D.F., Wuest, J.D., Rosei, F., Nanci, A.: Improving biocompatibility of implantable metals by nanoscale modification of surfaces: an overview of strategies, fabrication methods, and challenges. Small. **5**(9), 996–1006 (2009). https://doi.org/10.1002/smll.200801186
119. Eckardt, A., Aberman, H.M., Cantwell, H.D., Heine, J.: Biological fixation of hydroxyapatite-coated versus grit-blasted titanium hip stems: a canine study. Arch. Orthop. Trauma Surg. **123**(1), 28–35 (2003). https://doi.org/10.1007/s00402-002-0451-2
120. Huynh, V., Ngo, N.K., Golden, T.D.: Surface activation and pretreatments for biocompatible metals and alloys used in biomedical applications. Int J Biomaterials. Volume. **2019**, 3806504 (2019). https://doi.org/10.1155/2019/3806504
121. Mohammadi, F., Golafshan, N., Kharaziha, M., Ashrafi, A.: Chitosan-heparin nanoparticle coating on anodized NiTi for improvement of blood compatibility and biocompatibility. Int. J. Biol. Macromol. **127**, 159–168 (2019). https://doi.org/10.1016/j.ijbiomac.2019.01.026
122. Lee, K., Choe, H.-C., Kim, B.-H., Ko, Y.-M.: The biocompatibility of HA thin films deposition on anodized titanium alloys. Surf. Coat. Technol. **205**, S267–S270 (2010). https://doi.org/10.1016/j.surfcoat.2010.08.015
123. Hryniewicz, T., Rokicki, R., Rokosz, K.: Surface characterization of AISI 316L biomaterials obtained by electropolishing in a magnetic field. Surf. Coat. Technol. **202**(9), 1668–1673 (2008). https://doi.org/10.1016/j.surfcoat.2007.07.067
124. Bigerelle, M., Anselme, K., Noël, B., Ruderman, I., Hardouin, P., Iost, A.: Improvement in the morphology of Ti-based surfaces: a new process to increase in vitro human osteoblast response. Biomaterials. **23**(7), 1563–1577 (2002). https://doi.org/10.1016/S0142-9612(01)00271-X
125. Cui, F.Z., Luo, Z.S.: Biomaterials modification by ion-beam processing. Surf. Coat. Technol. **112**(1–3), 278–285 (1999). https://doi.org/10.1016/S0257-8972(98)00763-4
126. Barnbauer, R., Mestres, P., Schiel, R., Klinkrnann, J., Sioshansi, P.: Surface-treated catheters with ion beam-based process evaluation in rats. Artif. Organ. **21**(9), 1039–1041 (1997). https://doi.org/10.1111/j.1525-1594.1997.tb00520.x
127. Kurella, A., Dahotre, N.B.: Review paper: surface modification for bioimplants: the role of laser surface engineering. J. Biomater. Appl. **20**(1), 5–50 (2005). https://doi.org/10.1177/0885328205052974
128. Xiao, Y., Martin, D.C., Cui, X., et al.: Surface modification of neural probes with conducting polymer poly(hydroxymethylated-3,4-ethylenedioxythiophene) and its biocompatibility. Appl. Biochem. Biotechnol. **128**, 117–129 (2006). https://doi.org/10.1385/ABAB:128:2:117
129. Priyadarshini, B., Rama, M., Chetan, U.: Vijayalakshmi. Bioactive coating as a surface modification technique for biocompatible metallic implants: a review. J. Asian Ceramic Soc. **7**(4), 397–406 (2019). https://doi.org/10.1080/21870764.2019.1669861
130. Zhang, Y.Z., Venugopal, J., Huang, Z.-M., Lim, C.T., Ramakrishna, S.: Characterization of the surface biocompatibility of the electrospun PCL-collagen nanofibers using fibroblasts. Biomacromolecules. **6**(5), 2583–2589 (2005). https://doi.org/10.1021/bm050314k

14 Advanced Plastic Materials

Abstract

From high-end sports cars to everyday minivan, plastics have revolutionized not only the automotive industry but also other industries. This chapter gives the idea of advanced plastic materials such as high-temperature plastic, conducting plastic, magnetic plastic, transparent plastic, and bioplastic. Each of these plastic materials is discussed with their subclassification, mechanism, constituting elements, and applications.

Keywords

Advanced plastic · High-temperature plastics · Conducting plastic · Magnetic plastic · Transparent plastic · Bioplastic

14.1 Introduction to Advanced Plastic

Advanced plastics or engineering plastics are the ones that have higher tenacity, as in they perform better than the normal plastics available in the market. The various type of advanced plastic is given in Fig. 14.1.

14.2 High-Temperature Plastics

The term "plastic" covers a host of any structure, from those used in kitchen utensils and plastic bags to those found in car parts and aircraft engines. Plastics are generally divided into three categories: standard plastic, engineering plastic, and high-performance plastic. Standard plastics are the most common due to their low cost, but they also have the weakest mechanical properties (e.g., easy deformation of a water bottle). Engineering plastics have a relatively higher strength to make engineering structures. High-performance plastics are must for the parts that require tight tolerances and extreme conditions. High-performance plastics are plastics that meet higher requirements than standard or engineering plastics [1, 2]. Some of their diverse applications include high-temperature fluid flow tubing, electrical wire insulators, architecture, and fiber optics. High-performance plastics can be defined as the plastic that has a continuous service temperature of more than 150 °C without any alternation of their mechanical properties, chemical and/or heat stability [3, 4]. There are many synonyms for the term high-performance plastics, such as high-temperature plastics, high-performance polymers, high-performance thermoplastics, or high-tech plastics. The term "polymers" is often used instead of "plastics" because both terms are used as synonyms in the field of engineering. If the term "high-performance thermoplastics" is used, it indicates for standard and engineering as well as high-performance plastics, whereas the thermosets and elastomers are outside of this classification that forms their own classes. Figure 14.2 depicts various standard plastics, engineering plastics, and high-performance plastics with respect to the temperature. However, the differentiation from less powerful plastics has varied over time; while nylon and poly(ethylene terephthal ate) were initially considered powerful plastics, they are now ordinary. High-heat thermoplastics exhibit good temperature capability as they possess high glass transition temperature, T_g as well as mechanical strength than those of thermosets. The improvement of mechanical properties and thermal stability has always been an important goal in the research of new plastics. Since the early 1960s, the development of high-performance plastics has been driven by corresponding needs in the aerospace and nuclear technology. Synthetic routes, for example, PPS, PES, and PSU were developed in the 1960s by Philips, ICI and Union Carbide. The market entry took place in the early 70s. A production of PEEK (ICI), PEK (ICI), and PEI (General Electric and GE) via polycondensation was developed in the 1970s. PEK was offered since 1972 by Raychem, however, made by an electrophilic synthesis. Since electrophilic synthesis has in general the disadvantage of a low selectivity to linear polymers and is using aggressive reactants, the product could hold only a short time on the

A. Behera, *Advanced Materials*, https://doi.org/10.1007/978-3-030-80359-9_14

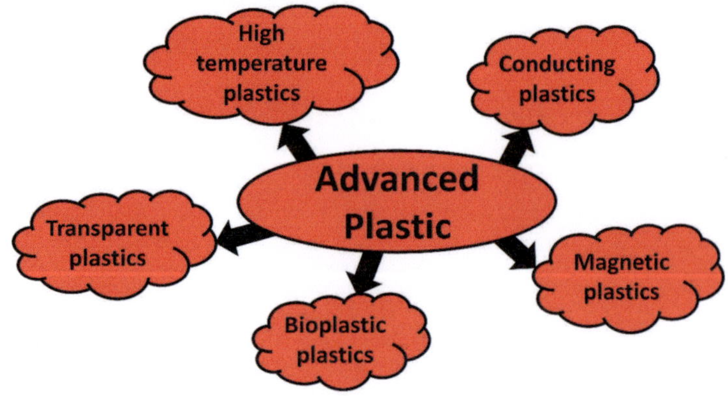

Fig. 14.1 Different types of advanced plastic

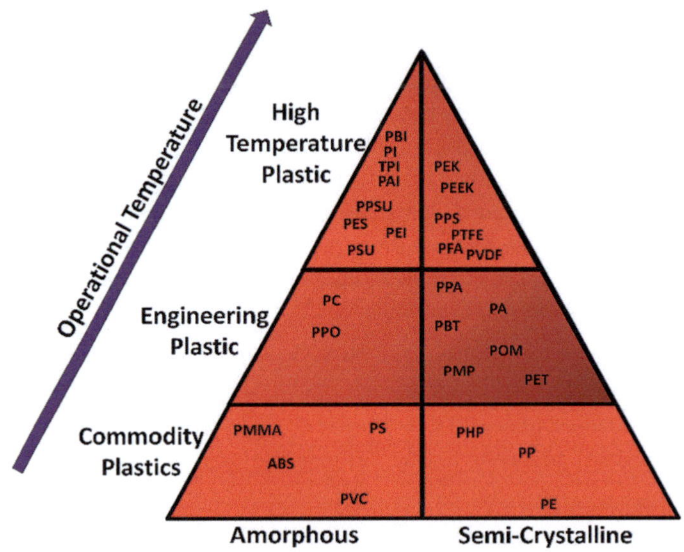

Fig. 14.2 A comparison of standard plastics, engineering plastics, and high-performance plastics

market. For this reason, the majority of high-performance plastics are nowadays produced by polycondensation processes. In manufacturing processes by polycondensation, a high purity of the starting materials is important. In addition, stereochemistry plays a role in achieving the desired properties in general. The development of new high-performance plastics is therefore closely linked to the development and economic production of the constituent monomers.

High-performance plastics are more expensive and used comparatively smaller quantites. The price per kilogram may be between \$5 (PA 46) and \$100 (PEEK) per kilo. The average value is slightly less than 15 USD/kg. High-performance plastics are thus about 3 to 20 times as expensive as engineering plastics. Also in future, there cannot be expected a significant price decline, since the investment costs for production equipment, the time-consuming development, and the high distribution costs are going to remain constant. Since production volumes are very limited with 20.000 t/year the high-performance plastics are holding a market share of just about 1%. Among the high-performance polymers, fluoropolymers have a 45% market share (main representatives: PTFE), sulfur-containing aromatic polymers 20% market share (mainly PPS), aromatic polyarylether and Polyketones 10% market share (mainly PEEK), and liquid crystal polymers (LCP) 6%. In the electrical and electronics industries are 41% and 24% of high-performance plastics used in the automotive industry, respectively. Thus, these are the largest consumers. All remaining industries (including chemical industry) have a share of 23%.

The global high-performance plastics market size was valued at around USD 14 billion in 2016 and is expected to cross USD 30 billion at a CAGR of approximately 7% by 2022. The market is highly application based. The factors contributing to the growth of the global high-performance plastics market are their properties which makes it more demanding over standard plastic. Asia-Pacific is the largest region for the high-performance plastics market, followed by Europe and North America. Emerging countries like China and India have a high demand for the automotive and electrical & electronics products due to factors such as the growth of industrialization [5].

14.2.1 High-Temperature Thermoplastics Structures and Stability

High-temperature thermoplastics comprise two molecular structures: amorphous (randomly ordered) and crystalline (highly ordered). For practical purposes, thermoplastics are either amorphous polymers or semicrystalline polymers. Polysulfones, Polyetherimides, Polyphenylsulfone are the amorphous structure which has a melting range rather than a defined melt point and broad glass transition temperature range. They have good strength, stiffness, isotropic dimensional stability, toughness and impact resistance, good surface appearance. Polyphenylene sulfide, liquid crystal polymer, polyetheretherketone, polyphthalamide are semicrystalline structure, which has sharply defined melting point and glass transition temperature. They have better chemical resistance, wear resistance, lower ductility and impact strength, opaque, low stiffness and creep resistance at high temperatures, poor dimensional stability. Both amorphous and crystalline high-temperature thermoplastics are used in the automotive, aerospace, medical, and electrical/electronic industries where demanding properties are required. High-temperature thermoplastics generally gain their temperature resistance from the introduction of rigid aromatic rings instead of aliphatic groups in their molecular structure [6]. This restricts the movement of the backbone chain and requires two chemical links to be broken (compared to one in aliphatic structures) for a chain break as shown in Fig. 14.3. Hence mechanical properties, high-temperature capability, and chemical resistance are greatly improved and can be often equivalent or even better than crosslinked, thermosetting polymers.

14.2.2 High-Temperature Plastic Materials

Various types of the high-temperature plastics that perform in a variety of applications have been discussed below.

Vespel: Vespel is well-known for its heat resistance. Vespel is the trademark of a range of durable high-performance polyimide-based plastics manufactured by DuPont. It is characterized by its ring-shaped molecular structure containing nitrogen (Fig. 14.4). It is a non-melting polyimide that can withstand heat up to 300 °C repeatedly without affecting its thermal or mechanical properties. Its popularity in large part is due to versatility in hostile and extreme environmental conditions. Unlike most plastics, Vespel does not produce significant out-gassing at high temperatures, which makes it ideal for lightweight heat shields and crucible support. Depending on the additive material (15% Graphite, 40% Graphite, 10% PTFE, and 15% Graphite, or 15% Moly), Vespel can withstand 350 h of 398 °C heat, losing only 50% of its initial tensile strength: 12,500 psi (unfilled base resin) reduces to 6000 psi. This loss is due to almost entirely oxidative degradation. It has flexural strength of 16,000 psi, high resistance to chemicals, low friction, and superior wear performance, excellent for electrical insulation. The parts will perform in inert environments, such as nitrogen or vacuum, with negligible loss of properties over time. Vespel is commonly used as a thermal

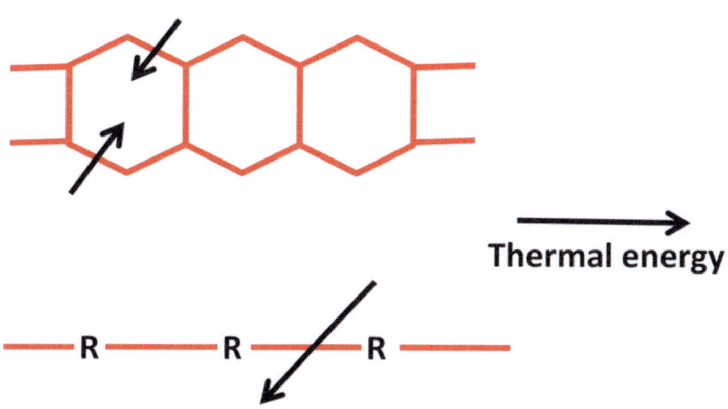

Fig. 14.3 Degradation of an aromatic and a straight-chain polymer due to thermal aging

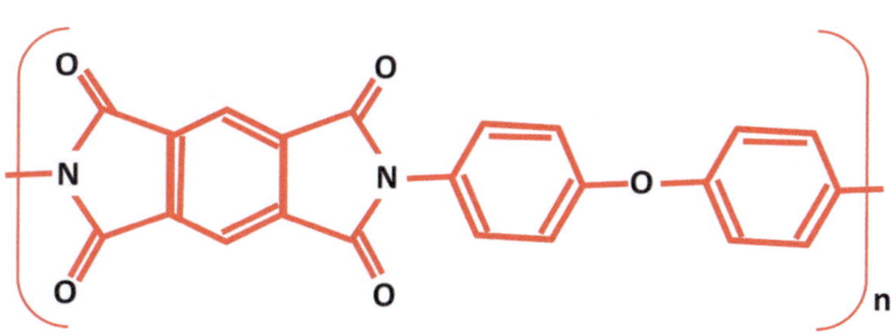

Fig. 14.4 Vespel structure

conductivity reference material for testing thermal insulators, because of the high reproducibility and consistency of its thermophysical properties. Unlike most plastics, it does not produce significant outgassing even at high temperatures, which makes it useful for lightweight heat shields and crucible support. It also performs well in vacuum applications, down to extremely low cryogenic temperatures. It does not become brittle even at subzero temperatures (−400 °F) [7]. However, Vespel tends to absorb a small amount of water, resulting in a longer pump time while placed in a vacuum. Vespel is used in high-resolution probes for NMR spectroscopy because its volume magnetic susceptibility ($-9.02 \pm 0.25 \times 10^{-6}$ for Vespel SP-1 at 21.8 °C) is close to that of water at room temperature (-9.03×10^{-6} at 20 °C). Negative values indicate that both substances are diamagnetic. Matching volume magnetic susceptibilities of materials surrounding the NMR sample to that of the solvent can reduce susceptibility broadening of magnetic resonance lines. Vespel can be processed by direct forming (DF) and isostatic molding to form basic shapes—plates, rods, and tubes. It combines the best properties of ceramics, metals, and plastics into one, offering unique strengths for major performance and cost benefits. It can be used as a lightweight alternative to metal and is often used in high-heat environments where thermoplastic materials lose their mechanical properties. It has a low coefficient of friction and superior wear performance—especially for applications where interfacial temperatures are high or require sliding contact with or without lubrication [8].

It is a popular choice for jet engines, industrial machinery, semiconductor, and transportation technology. Some aerospace valves must operate high in the atmosphere, in polar climates, or in space. Therefore the seals are exposed to extremely cold temperatures. PCTFE, which is a common valve seat material, will become brittle at temperatures below −123 °C. The compressive strength of the Vespel, SP21-grade in particular remains consistent when cooled to cryogenic temperatures. Therefore, the seals retain their ductility and are able to maintain a high-pressure seal without showing brittle fracture. Most commonly used in the aerospace industry are bearings, bushings, gaskets, insulators, piston rings, seal rings, thrust washers, valve disks, and washers. Vespel is not a biocompatible material, so can not use in medical applications, or prints parts [9].

Polyether Ether Ketones: Polyether ether ketone (PEEK) is a colorless organic thermoplastic polymer in the polyaryletherketone (PAEK) family (PEK, PEEK, PEEKK, PEKK, PEKEKK) and among them, PEEK is the most widely used and manufactured in large scale. It was originally introduced by Victrex PLC and then Imperial Chemical Industries (ICI) in the early 1980s. PEEKs are a family of polymers containing alternative ether and ketone groups in their repeating units. PEEK polymers are obtained by step-growth polymerization by the dialkylation of bisphenolate salts. Typically the reaction of 4,4′-difluorobenzophenone with the disodium salt of hydroquinone, which is generated in situ by deprotonation with sodium carbonate. The reaction is conducted around 300 °C in polar aprotic solvents-such as diphenyl sulphone (Fig. 14.5). Such kind of chemical structure imparts PEEKs semicrystalline nature with high-thermal resistance and excellent dielectric properties. PEEK has a glass transition temperature of around 143 °C and melts around 343 °C. Some grades have a useful operating temperature of up to 250 °C. The thermal conductivity increases nearly linearly with temperature between room temperature and solidus temperature. It is highly resistant to thermal degradation, as well as to attack by both organic and aqueous environments. It is attacked by halogens and strong Brønsted and Lewis acids, as well as some halogenated compounds and aliphatic hydrocarbons at high temperatures. It is soluble in concentrated sulfuric acid at room temperature, although dissolution can take a very long time unless the polymer is in a form with a high surface-area-to-volume ratio, such as a fine powder or thin film. It has high resistance to biodegradation [10].

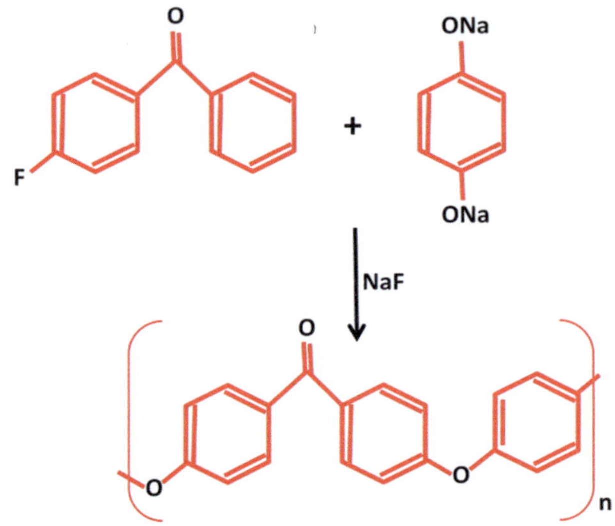

Fig. 14.5 Polyether ether ketone

It serves continuously at temperatures of 500 °F with the tensile strength of 16,000 psi, flexural strength of 25,000 psi, good wear and abrasion resistance, excellent electrical properties, and biocompatible with radiation resistance. Unlike other heteroaromatic polymers, PEEK can usually be winded with copper or aluminum wires directly at elevated temperatures without using any adhesives. This is quite beneficial for improving the reliability of the insulating products. At last, PEEK usually possesses very stable dielectric properties over a wide range of temperatures, frequencies, and humidity. Especially, the good resistance to steam makes PEEK unique dielectric for insulating applications because steam is a very harsh environment for insulating materials. It is a combination of water and high temperature and easily causes the hydrolysis of common HTPDs. It has been proven that PEEK can resist hydrolysis in temperatures up to 250 °C. PEEK can be used as HTPDs with both forms of engineering plastics and film. In its solid state PEEK is readily machinable, for example, by (CNC) milling machines, and is commonly used to produce high-quality plastic parts that are thermostable and both electrically and thermally insulating. Filled grades of PEEK can also be CNC machined, but special care must be taken to properly manage stresses in the material. PEEK has a flammability rating down to 1.45 mm and an LOI of 35%. Its smoke and toxic gas generation are extremely low. Crystallinity imparts excellent resistance to a wide range of liquids and superb fatigue performance. PEEK is insoluble in all common solvents. It does not undergo hydrolysis and can be used for thousands of hours in steam or high-pressure water without significant property degradation. More properties of PEEK include low friction, good dimensional stability, exception insulation properties, excellent sterilization resistance at high temperature, biocompatible, long life, and inherent purity. PEEK can be processed by conventional methods such as injection molding, extrusion, compression molding, etc. It is technically feasible to process granular PEEK into filament form and 3D printing parts from the filament material using fused deposition modeling (FDM)or fused filament fabrication (FFF) technology. PEEK has low resistance to UV light. PEEK has become a popular material for medical devices and suitable for contact with the human body. It is commonly used in the aerospace, automotive, oil and gas, and electronics industries as well. Because of its robustness, PEEK is used to fabricate items used in demanding applications, including bearings, piston parts, pumps, high-performance liquid chromatography (HPLC) columns, compressor plate valves, and electrical cable insulation [11]. In aircraft exterior parts, PEEK provides excellent resistance to rain erosion, while for aircraft interior components, its inherent flame retardancy and low smoke and toxic gas emission reduce hazard in the event of a fire (Fig. 14.6). In aircraft electrical systems, the polymer is used for the manufacture of convoluted tubing to protect wires and fiber optic filaments. In exterior parts of the aircraft, PEEK provides excellent resistance to rain erosion. In interior components, its inherent flame retardancy and low smoke and toxic gas emission reduce hazards in the event of a fire. Outstanding thermal properties enable PEEK polymer parts to withstand the elevated temperatures associated with soldering processes. PEEK can withstand exposure to pressures as high as 25,000 psi and temperatures up to 260 °C [12].

Torlon: Torlon is another thermoset polyamide with exceptional long-term strength and stiffness up to a continuous 260 °C. One of the first commercial Poly(amide-imide) (PAI) introduced by Amoco Chemical was synthesized via this reaction and marketed under the name of Torlon. Torlon is an amorphous material with a T_g of 280 °C. Torlon is considered an effective alternative to metal where high-temperature friction and wear applications are present. Torlon's chemical structure is given in Fig. 14.7. It has a tensile strength of 27,847 psi, flexural strength of 35,390 psi, high resistance to wear, creep, and chemicals, superior compressive strength, and impact resistance. The thermoset is also an ideal material for use in severe service environments. It has outstanding resistance to wear, creep, and chemicals, including strong acids and most organic chemicals, and is ideally suited for severe service environments. Torlon also has superior electrical and structural characteristics at high temperatures, excellent radiation resistance, an extremely low coefficient of linear thermal expansion, and exceptional dimensional stability. PAI is synthesized from either monomer containing amide groups or from monomers containing imide groups in their chemical structure. Poly(amide-imide)s can be also prepared by the reaction of trimellitic acid-anhydride or its chloride derivatives with various available diamines which proceeds via two-step routes, namely the formation of intermediate amic acid and subsequent thermal

Fig. 14.6 PEEK exterior and interior components in aircraft

Fig. 14.7 Torlon structure

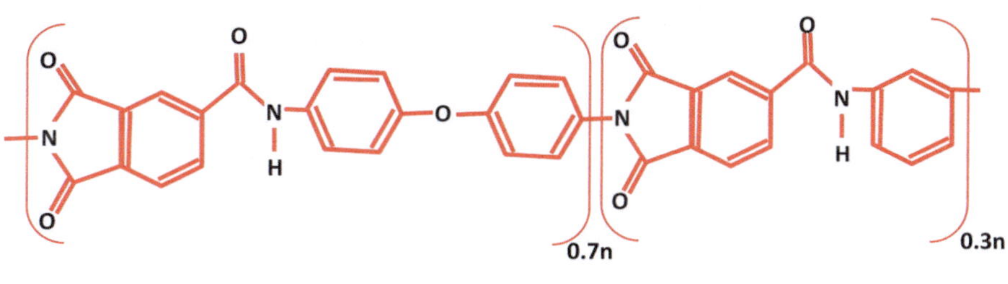

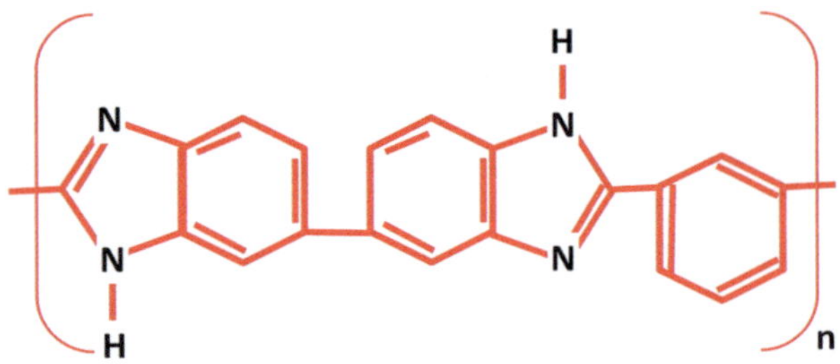

Fig. 14.8 Celazole PBI

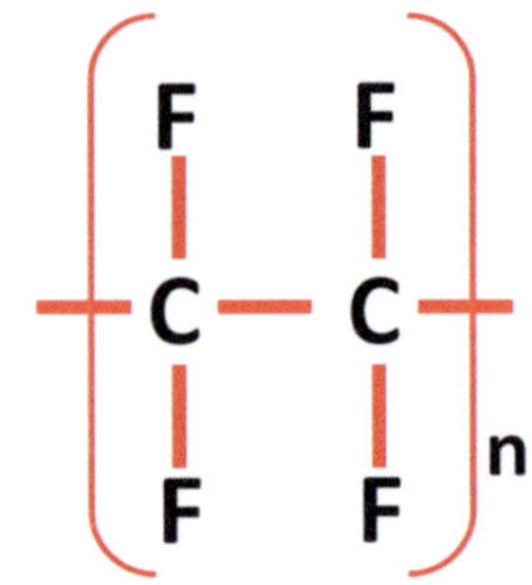

Fig. 14.9 PTFE plastics

cyclodehydration to form imide rings. Typical applications include wire coatings in electronic accessories, non-lubricating bearings, seals, valves, compressor, piston parts, bearing cages, bushings, and thrust washers. Torlon is typically used to make aircraft hardware and fasteners, mechanical and structural components, transmission and powertrain components, as well as coatings, composites, and additives. Common applications include mechanical and structural components, aircraft hardware, transmission and power train components, coatings, composites, and additives. Torlon is commercially available in the form of a solution in N-methyl-2-pyrrolidone and is mainly used as resins and adhesives needed in space vehicles, wire enamel, and decorative coating for kitchen appliances [13].

Celazole PBI: Celazole PBI polymers are uniquely suited for metal replacement and exhibit excellent thermal stability and wear resistance. Figure 14.8 shows the structure of Celazole PBI. It offers the highest heat resistance and mechanical property retention over 400 °C Celazole PBI of any unfilled plastic. It has a glass transition temperature of 427 °C. It has tensile strength of 16,000 psi, the flexural strength of 32,000 psi with higher chemical and wears resistance [14]. As an unreinforced material, Celazole PBI is very "clean" in terms of ionic impurity and it does not outgas (except water). These characteristics make this material very attractive to semiconductor manufacturers for vacuum chamber applications. Celazole PBI has excellent ultrasonic transparency which makes it an ideal choice for parts such as probe tip lenses in ultrasonic measuring equipment. Celazole PBI is also an excellent thermal insulator. Other plastics in melt do not stick to PBI. These characteristics make it ideal for contact seals and insulator bushings in plastic production and molding equipment. Celazole PBI is extremely hard and can be challenging to fabricate. Polycrystalline diamond tools are recommended when fabricating production quantities. Celazole tends to be notch sensitive [15].

Rulon: Rulon is the trade name for a family of PTFE plastics produced by Saint-Gobain Performance Plastics. Rulon plastics are known for their low coefficient of friction, excellent abrasion resistance, a wide range of operating temperatures, and chemical inertness. Common applications for Rulon include seals, piston rings, bearings, and electrical insulation. Rulon® is a proprietary, homogeneous material made primarily of PTFE-based resins engineered and developed for specific applications. The Rulon family of materials combines high compressive strength, a low coefficient of friction, and excellent abrasion and corrosion resistance running without lubrication. Figure 14.9 shows the structure of PTFE plastics. Rulon products are used in bearing and seal applications at temperatures from −240 °C to 260 °C with and without supplemental lubricants. Rulon® products are unique in that they do not cause stick-slip or erratic low-speed motion. They also withstand a variety of harsh environments such as extreme dryness, cryogenic temperatures, steam, and hydrocarbon fuel. Rulon® products primarily are used for mechanical, electrical, and chemical applications [16].

Rulon has several subgrades according to the application. Rulon® LR is marked by low deformation characteristics that increase the mechanical properties for slightly more load capacity with a corresponding decrease in flexibility. Rulon® LR is compatible with most hardened steel substrates. Mild steel is acceptable, though harder surfaces are better. It also has a virtually universal chemical inertness. Only molten sodium and fluorine at elevated temperatures and pressures show any signs of attack. This grade sometimes is used as a seal for piston rings but not in flexing lip-seal applications. Its color is dark maroon. Rulon® J is an all-plastic reinforced PTFE with lower friction and wears compared to other reinforced PTFE compounds. Rulon® J can be used against nonferrous and nonmetallic surfaces such as mild steel, 316 stainless steel, aluminum, brass, and other plastics. It has superb tribological properties and is suitable for use in bearing, seal, and wear-component applications. The mechanical strength of Rulon® J is slightly less than it is for other Rulon® grades. It is ideally suited for start/stop applications where stick-slip must be eliminated. Rulon® J should not be used in alkaline solutions, oxidizing acids, or steam environments. Color is dull gold. Rulon® W2 is a black PTFE-based material developed for use in fresh-water applications. It exhibits low friction and excellent wear characteristics (one of the lowest wear rates in freshwater) as well as good thermal dissipation, preventing shaft distress. Its properties are enhanced when wet. It is compatible with most metal substrates and soft mating surfaces. Rulon® W2 is a good alternative to Rulon® J when superior chemical resistance is needed. However, it should not be used on very soft mating surfaces or where electrical insulation is desired. Rulon® 142 was developed specifically for use as linear guideways on machine tools. It has a slightly lower deformation underload than Rulon® LR and is more thermally and electrically conductive. Typical characteristics include low wear, high thermal dissipation, and good dimensional stability. Rulon® 142 also is more abrasive and less chemically resistant than Rulon® LR; however, most other physical properties are similar. Strong acids and bases should be avoided, as they may attack the Rulon® 142 fillers. The color is bright turquoise. Rulon® 123 is a glossy black non-abrasive compound for softer mating surfaces, such as stainless steel. This material has excellent chemical resistance and is FDA and USDA compliant. It is less expensive than Rulon® J but is slightly less flexible and higher in wear. It has a high resistance to deformation, low coefficient of friction, and good thermal and electrostatic dissipation. This material has a maximum operating temperature of 288 °C. Rulon® 123 releases black wear debris over time and should not be used in ultra-dry, vacuum applications or where insulation is desired [17].

Ryton: Ryton is the commercial product of Polyphenylene sulfide (PPS). The first commercial process for PPS was developed by Edmonds and Hill in 1967 while working at Philips Petroleum under the brand name Ryton. PPS is a semicrystalline organic polymer consisting of aromatic rings linked by sulfides (Fig. 14.10). Maximum service temperature is 218 °C. PPS has not been found to dissolve in any solvent at temperatures below approximately 200 °C. It is a rigid and opaque polymer. PPS must be filled with fibers and fillers to overcome its inherent brittleness. High molecular weight (HMW) linear PPS has a molecular weight about double that of regular PPS. It offers an excellent balance of properties, including high-temperature resistance, chemical resistance, flowability, dimensional stability, and electrical characteristics. Also, it exhibits resistance to heat, acids, alkalies, mildew, bleaches, aging, sunlight, and abrasion. It absorbs only small amounts of solvents and resists dyeing. PPS can be molded, extruded, or machined to high tolerances. It is used to make filter fabric for coal boilers, papermaking felts, electrical insulation, specialty membranes, gaskets, and packings. PPS is produced by the reaction of sodium sulfide and dichlorobenzene in a polar solvent such as N-methylpyrrolidone and at a higher temperature at about 250 °C [18].

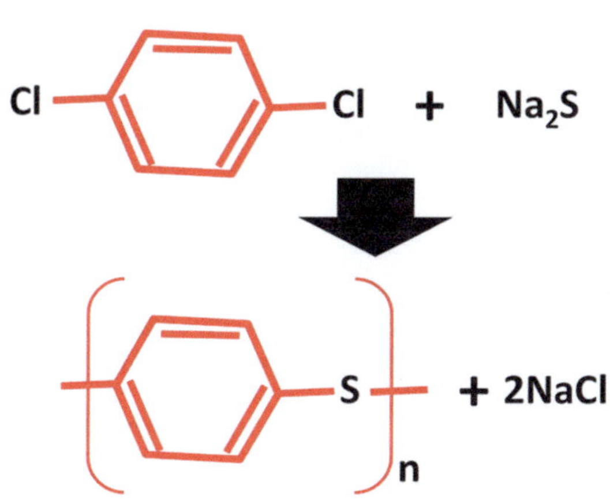

Fig. 14.10 Ryton formation

To produce molding grades, PPS is cured (chain extended or crosslinked) around the melting point of the polymer in the presence of a small amount of air. This curing process results in an increase in molecular weight, increased toughness, loss of solubility, decrease in melt flow, and decrease in crystallinity and darkening in color. PPS resin is generally reinforced with various reinforcing materials or blended with other thermoplastics in order to further improve its mechanical and thermal properties. PPS is more used when filled with glass fiber, carbon fiber, and PTFE. Regular PPS is an off-white, linear polymeric material of modest molecular weight and mechanical strength. When heated above its glass transition temperature (T_g ~85 °C), it crystallizes rapidly. The main three types of PPS include: linear PPS, cured PPS, and branched PPS. The molecular weight of Linear PPS polymer is nearly double as compared to regular PPS. The increased molecular chain length results in high tenacity, elongation, and impact strength. Cured PPS obtained from the heating of regular PPS in the presence of air (O_2). Curing results in molecular chain extension and formation of some molecular chain branches increases the molecular weight and provides some thermoset-like characteristics. Branched PPS has a higher molecular weight than regular PPS. The backbone of the extended molecule has extended polymer chin branched from it. Some of the key producers of the Ryton family includes PrimoSpire®, Tribocomp®, TORELINA®, TORAYCA®, RTP 1300, FORTRON®, CoolPoly®, Celstran®, DURAFIDE®, LNP™ LUBRICOMP™, LNP™ STAT-KON™, LNP™ THERMOCOMP ™, LUVOCOM® [19].

PPS has chosen alternatives to metals and thermosets for use in automotive parts, appliances, electronics, and several others applications. PPS is used in filter fabric for coal boilers, papermaking felts, electrical insulation, film capacitors, specialty membranes, gaskets, and packings. PPS is the precursor to a conductive polymer of the semi-flexible rod polymer family. The PPS, which is otherwise insulating, can be converted to the semiconducting form by oxidation or the use of dopants. These are used in automotive coolant, fuel, braking, transmission, engine, electrical and lighting components engine mounts; electrical: connectors, sockets, bobbins, relays, optical pickups, housings; industrial

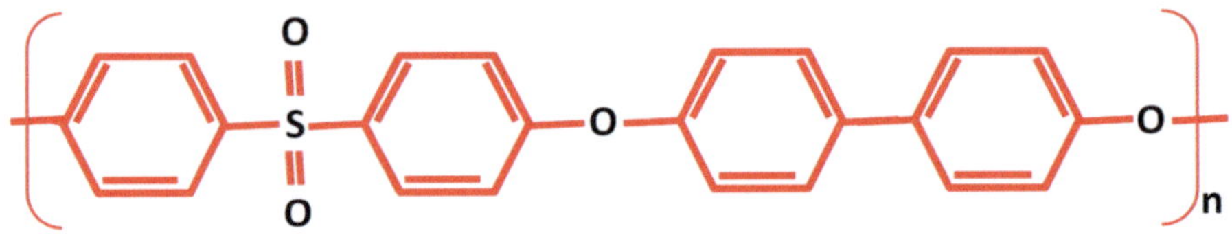

Fig. 14.11 Molecular structure of polyphenylsulfone

and consumer: hair straightener housings, hard disk drive components, chemical pumps, turbocharger air ducts, piping for downhole oil field applications, pump and motor parts, sensors, thermostats, blower housings, hot water manifolds, nonstick cookware coatings. PPS compounds (typically glass-reinforced grades) are used in medical application such as surgical instruments and device components and parts that require high dimensional stability, strength, and heat resistance [20].

Radel: Commercial name of Polyphenylsulfone (PPSU, **PPSF**) is Radel. It is transparent and rigid high-temperature engineering thermoplastic and belongs to the polysulfone family. Figure 14.11 shows the molecular structure of Polyphenylsulfone. It mainly consists of phenyl (aromatic) rings linked by sulfone (SO_2) groups. Polyphenylsulfone has a glass transition temperature of 288 °C and a heat-deflection temperature of 274 °C. The continuous use temperature has been suggested to be 260 °C. In these high-temperature properties, polyphenylsulfone is superior to other polysulfones such as standard polysulfone, polyethersulfone, or polyarylsulfone [21]. Processing requires a relative high mold temperature of about 365–395 °C whereas glass fiber reinforced grades require even higher temperatures. These are moldable plastics often used in rapid prototyping and rapid manufacturing applications. Polyphenylsulfone is heat and chemical-resistant suited for automotive, aerospace, and plumbing applications. Polyphenylsulfone has no melting point, reflecting its amorphous nature, and offers tensile strength up to 55 MPa. In plumbing applications, polyphenylsulfone fittings have been found to sometimes form cracks prematurely or to experience failure when improperly installed using non-manufacturer approved installation methods or systems [22]. Polyphenylsulfone uses in pipe fittings, battery containers, medical device parts, and sterilizable products for health care and nursing. Polyphenylsulfone is also used in the automotive and aerospace industries for applications where superior thermal and mechanical properties relative to conventional resins are required [23].

14.2.3 Application of High-Temperature Plastic

Various industrial high-temperature plastics have been given in Table 14.1 [24–28].

14.2.4 Advantages and Disadvantages of High-Temperature Plastics Over Metals

High-temperature plastics provide weight savings in many applications and as a result are often considered for metal replacement. Table 14.2 summarizes the advantages and disadvantages of high-temperature thermoplastics over metals.

14.3 Conducting Plastic

Polymers are insulating materials in general and they are used to make nonconductive coatings on wires. Polymers are organic macromolecules, a long carbonic chain, composed of structural repeat entities, called *mer*. These smallest units, for instance, are bonded by covalent bonds, repeating successively along a chain. A monomer, molecule composed of one *mer*, is the raw material to produce a polymer. The majority of polymers are insulators, due to unavailability of free electrons to create conductivity [29]. In a covalent bond, shown in Fig. 14.12, the electrons are locked in these strong and directional bonds, so when an electric field is applied, electrons can not drift. Therefore, these types of materials do not show a high conductivity. An electric current results in the orderly movement of charges in a material as a response to forces that act on them, when a voltage is applied. The positive charges flow in the direction of the electric field applied, whereas the negative charges move in the opposite direction. In the majority of materials, a current is resulted from the flux of electrons, known as electric conduction. The structure of these materials has conjugated chains that is an alternating single and double bond between the atoms. The process of doping of conductive polymers becomes easier due to these conjugated bonds. In this process, defects and deformations in the polymeric chain are formed. An electron-deformation pair, or also an electron-phonon cloud pair, is called polarons, which are responsible for the conductivity in polymers. Bipolarons and solitons, other types of quasi-particles, also participate in the conductivity mechanism. The type of soliton, bipolaron, or polaron formed depends on the dopant used [30].

The poor conductivity of polymers is also explained by the band theory shown in Fig. 14.12a. This theory says that the energy levels of electrons can occupy are grouped in

Table 14.1 Use of high-temperature plastic in various parts of specific industries

Industrial sector	Parts	Advantages
Automotive	Piston components, seals, washers, bearings, transmission components, transmission thrust washers, braking and air-conditioning systems, ABS brake systems engineer control systems, truck oil screens, starting disks in gears, etc.	• Greater design flexibility • Reduced development time • Lower assembly costs
Aerospace	Airbus interior components, bow-shaped luggage compartment retainers, cable conduits, cable clips, ventilation wheels inside aviation fans, suction manifold of aviation pumps, electrical wire harnesses, convoluted turbine, wire insulation, pump casings, impellers	• Excellent resistance to rain erosion • Inherent flame retardancy • Less smoke and toxic gas emission
Electrical and Electronics	Wire insulation for high-temperature applications, cable couplings and connectors, subsea connectors, coaxial connector jacks, subsea-controlled environment connectors, wafer transport carriers, surface-mounted trimming, potentiometers, appliance handles, electrical insulations, IC packaging trays	• Low warpage materials. • Design flexibility and design for manufacturing and assembly. • Miniaturization, which increases the temperature and mechanical requirements of plastic materials.
Machinery applications	Impeller wheels for regenerative pumps, pump rotors, multi-pin connectors, glue gun bushings, quick coupling systems, laundry system wheels, conductivity sensors and seals, compressor valve plates, heat exchanger parts, bearings	• Can withstand sliding friction, high temperatures and have good chemical resistance.
Medical Applications	• Polyphenylsulfone (PPS)—for the development of sterilizable containers. • Polyetherimide (PEI)—for both disposable and reusable medical devices • Polysulfones (PS) and polyethersulfones (PES)—for parts and membranes for dialyzers; instruments; parts for instruments; surgical theater luminaries; sterilizing boxes; infusion equipment; secretion bottles and reusable syringes. • Liquid crystal polymers (LCP)—for replacing metal in medical devices in techniques of minimally invasive surgery and microsystem technology. • Polyetheretherketones (PEEK)—for replacing glass, stainless steel, and other metals in a growing range of medical applications like dental instruments, endoscopes, dialyzers, handles on dental syringes, and sterile boxes that hold root canal files	• Preferable properties are biocompatible, bioresorbable etc.
Heat recovery application	• Water vapor condensation	• Corrosive resistant feature, polymer heat exchangers are also applied for flue gas recovery in the power generation process.

Table 14.2 Advantages and disadvantages of high-performance plastic over traditional metal

Advantages over metals	Disadvantages over metals
• Low density • Good noise and vibration damping • Electrical and thermal insulation or adjustable conductivity • Good chemical and corrosion resistance • Increased design freedom • Adaptable to high volume production processes • Adaptable to property modification for specific applications	• Greater thermal expansion • Poorer creep resistance • Lower thermal resistance • Susceptible to UV, moisture, and oxidation • Not considered to be a vapor barrier • Lower mechanical properties • Plastic parts generally must be redesigned over metal parts

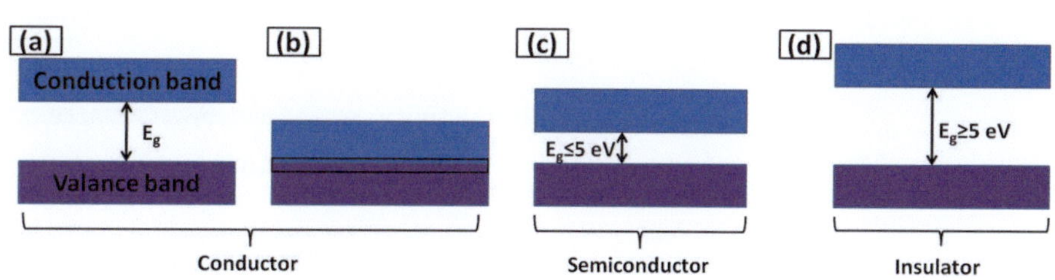

Fig. 14.12 Band structure for conductors, semiconductors, and insulators: (**a**) for monovalent metals; (**b**) divalent metal; (**c**) insulators; and (**d**) semiconductors

allowed bands and may have energy levels of the electron that are forbidden denominated bandgap. This theory is resulted of Schrodinger's equation applied in a periodic field of a crystal solid. The lowest bands are called valence bands and are inert from an electrical perspective. On the other hand, the highest bands, which participate in the electric conduction, are called conduction bands [31]. Some conductors have a partially filled valence band that is relatively easier to excite an electron to a higher energy level, as in Fig. 14.12a. Other conductors, such as divalent metals represented in Fig. 14.12b, can have an overlap of the empty conductive band with a totally filled valence band. For semiconductor and insulators, represented in Fig. 14.12 c, d, respectively, the valence electrons must cross the bandgap in order to result conduction. The difference is a semiconductor has a relatively smaller bandgap energy than insulators have. For a long period of time, polymers were considered as insulators. Until 1970, when the first intrinsic conductive polymer was produced by Shirakawa, Heeger, and Mac Diarmid, which resulted in the Nobel prize in 2000. The polymer was produced by the exposure of the polyacetylene to dopant compounds: oxidizing or reducing agents; electron-donor or electron-receptor of electrons [32]. The charges resulted from the doping process in conductive polymers are the reason for their great conductivity. The constant movement of the double bonds to stabilize the charge in the neighbor atoms causes, therefore, the movement of the charge, resulting in the conductivity. This movement of double bonds is called resonance and it describes the delocalized electrons within a molecule. A delocalized electron is an electron, presented in a π bond, which is shared by three or more atoms. Due to this process of the polaron formation, there is a change in the band structure of the conductive polymer [33]. It creates the polaronic conduction bands, allowed bands in the bandgap, reducing the bandgap energy, making the polymer able to conduct, as shown in Fig. 14.13.

Since the conductivity of a conductive polymer is due to the charge formed by the dopant, as the doping level increases, more charges are formed in the polymer and, thus, results in a greater conductivity. The conductivity of a conductive polymer is also temperature-dependent, because as the temperature increases the molecules become farther from each other. Thus, the doping effect is more effective and, consequently, the amount of charges, which is the doping level of the polymer is greater, increasing the conductivity. Moreover, as the temperature increases, the energy of an electron is related to the temperature by the Boltzmann relationship. Because of that, the greater the temperature, the greater is the energy of the electrons, and, consequently, the easier is to excite the electron to the conduction band [34].

Conductivity of conductive polymers is in the range of semiconductors' conductivity. Conductivity of polymers compared to those of other materials is presented in Fig. 14.14.

14.3.1 Historical Background

Somewhat surprisingly, a new class of polymers possessing high electronic conductivity (electronically conducting polymers) in the partially oxidized state was discovered. Alan J. Heeger, Alan G. Mac Diarmid, and Hideki Shirakawa played a major role in this breakthrough, and they received the Nobel Prize in Chemistry in 2000 "for the discovery and development of electronically conductive polymers" [35]. The most important representatives of these materials, polyaniline and polypyrrole, were already being prepared by chemical or electrochemical oxidation in the nineteenth century. Of course, for a long time, they were not called polymers, since the existence of macromolecules was not accepted until the 1920s. Therefore, it is somewhat interesting to review the story of polyaniline here, because it

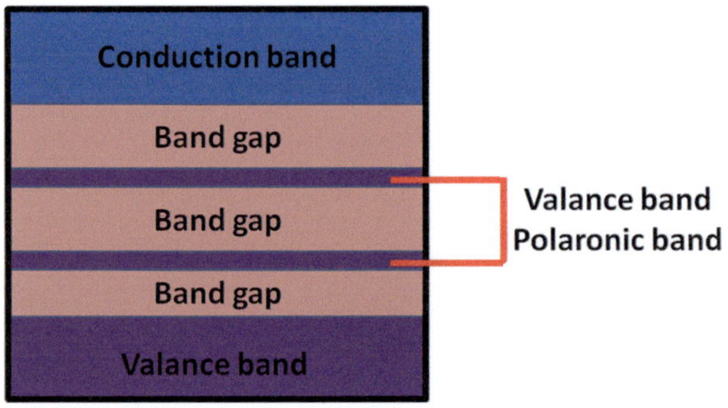

Fig. 14.13 Band structure of a conductive polymer

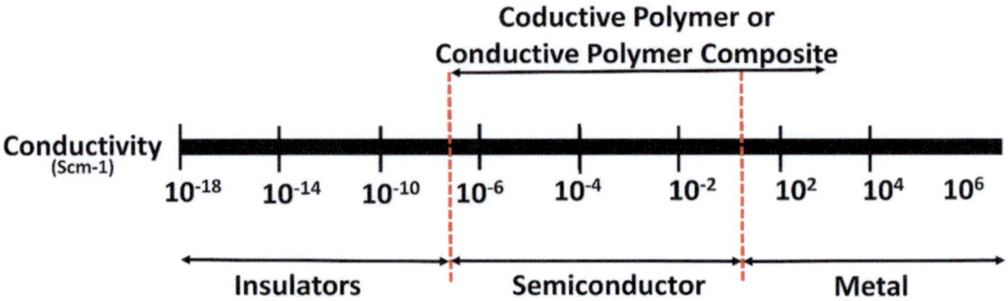

Fig. 14.14 Conductivity range of polymers compared to other materials conductivity

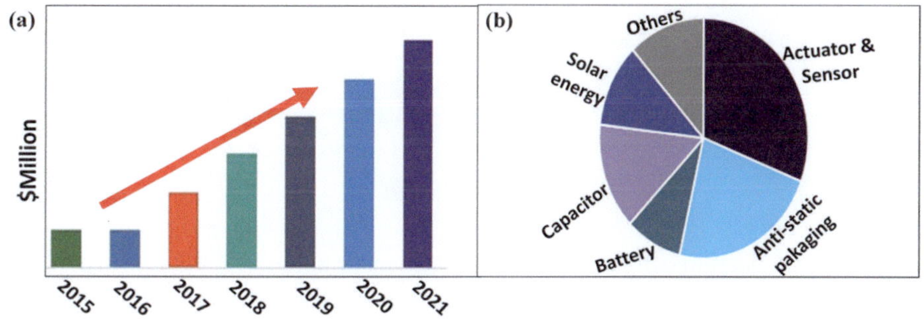

Fig. 14.15 Global conductive polymers market (**a**) Market size from the year 2015 to 2021, and (**b**) distribution of conductive polymer by various applications on 2015

provides an insight into the nature of the development of science. One may recall that aniline was prepared from the coal tar residues of the gas industry in the first half of the nineteenth century, and later played a fundamental role in the development of organic chemistry and the chemical industry. First, aniline dyes replaced dyes from natural sources. Then coal tar dyes were found use in medicine (to stain tissues), and the selective toxicity of these compounds was also discovered. This initiated the chemical production of medicines and the establishment of the pharmaceutical industry. In 1862 Dr. Henry Letheby, who was a physician and a member of the Board of Health in London, was interested in aniline because it was poisoning workers. Letheby observed that a bluish-green precipitate was formed at the anode during electrolysis, which became colorless when it was reduced and regained its blue color when it was oxidized again. In 1840, Fritzsche observed the appearance of blue color during the oxidation of aniline in acidic media. In 1950, Khomutov and Gorbachev discovered the autocatalytic nature of the electrooxidation of aniline. In 1962 Mohilner, Adams and Argersinger reinvestigated the mechanism of the electrooxidation of aniline in aqueous sulfuric acid solution at a platinum electrode. They proposed a free radical mechanism and wrote that "the final product of this electrode reaction is primarily the octamer emeraldine or a very similar compound" [36].

The first real breakthrough came in 1967 when Buvet delivered a lecture at the 18th Meeting of CITCE (later ISE), and this presentation appeared a year later in Electrochimica Acta. Polyanilines are particularly representative materials in the field of organic protolytic polyconjugated macromolecular semiconductors, because of their constitution and chemical properties. They also established that polyanilines also have redox properties and that the conductivity appears to be electronic. It was also shown that "polyanilines are also ion-exchangers." It was not until the 1970s that the development of conducting polymers gained traction and attention throughout the world of scientists and engineers. In 1975 polysulfur nitride was discovered to be superconducting at low temperatures, and this discovery paved the way for intensive research in the field of conductive polymers. Ultimately, the possibility of having organic materials with the conductive properties of nonorganic materials along with the intrinsic polymeric features of mechanical flexibility and relatively low-cost production gives rise to endless opportunities [37].

14.3.2 Industrial Market Status

The global conductive polymers market size was 330.7 k tons in 2015. Rising awareness towards shielding sensitive electronic devices medium from electrostatic discharge (ESD) and radio frequency interference (RFI) to electromagnetic interference (EMI) as a result of the increasing market size of sensitive electronic devices are expected to have a positive impact on market growth. The global conductive polymers market revenue has been given in Fig. 14.15a. Various applications of conductive polymers in 2015 have been given in Fig. 14.15b. Growing importance of electronics devices with reduced sound levels owing to stringent regulations imposed to control noise pollution is anticipated to result in buyers procuring conductive plastics including ABS and PPS in place of conventional metals and ceramics. Rising importance of ionic polymers in the electronics industry on account of providing excellent dimensional stability as well as good conducting characteristics is expected to remain a favorable factor for the growth of the conductive polymers market. These materials are electroactive in nature and are used for providing characteristics including good optical property, excellent mechanical flexibility, and high thermal stability. ABS, polycarbonates (PC), polyphenylene resin, nylon, polyethylene terephthalate (PPS), and polyvinylidene chloride (PVC) are the majorly used conductive polymer products [38].

Favorable outlook toward the electronics industry in China, Japan, South Korea, and Singapore as a result of new product launches including smartphones by market players such as Samsung and LG are expected to play a crucial role in upscaling the requirement of conductive polymers. In July 2015, the government of India launched the "Digital India Campaign" aimed at ensuring internet connectivity along with upscaling the production output in digital electronics on a domestic level. This initiative is expected to play a significant role in promoting the

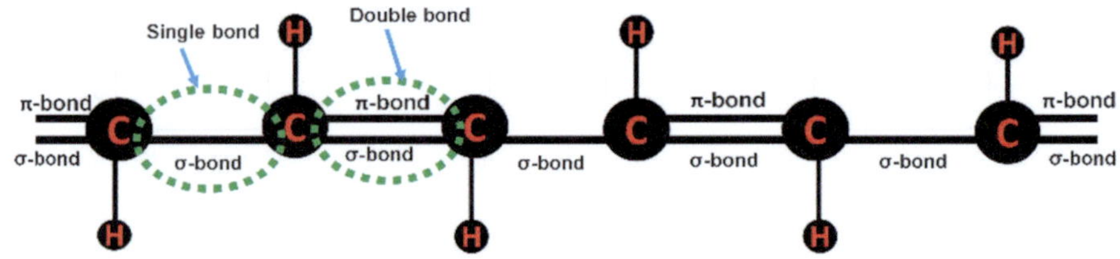

Fig. 14.16 Simplified schematic of a conjugated backbone: a chain containing alternating single and double bonds

production of electronic digital goods including smartphones, the web camera, and tablets which are likely to open new opportunities for the conductive polymers market over the next 8 years. Rising importance of lightweight materials in the automotive industry on account of increasing awareness towards improvement in fuel economy is expected to promote the use of ABS and PPS in wiring systems as conducting materials. Furthermore, increasing defense expenditure by governments in the Middle East and North America is expected to boost the production output of military aircraft on a domestic level which is supposed to promote the usage of conductive polymer over the forecast period [39].

14.3.3 Classification of Conducting Polymer

There are two subgroups of conductive polymers: (1) intrinsically conductive polymers, and (2) extrinsically conductive polymers. Intrinsically (or inherently) conductive polymers (ICPs), also known as conjugated polymers and synthetic metals, exhibit interesting electrical and optical properties previously found only in inorganic systems. Different types of ICPs can be prepared with a broad range of conductivities from 10^{-10} to 10^{+5} S/cm. The most attractive in a group of these polymers are polyaniline (PANI), polypyrrole (PPy), and poly (3, 4-ethylenedioxythiophene) (PEDOT) as one of the polythiophene (PTh) derivatives. They show high electrical conductivity and environmental stability, they are synthesized easily but have poor mechanical properties [40]. Extrinsically conductive polymers (ECPs), or conductive polymer composites (CPCs) are obtained by blending (melt mixing) an insulating polymer matrix, thermoplastic or thermosetting plastic, with conductive fillers. The three most important conductive fillers are carbon (carbon black (CB) and carbon nanotubes (CNTs)), metal powders and their compounds (indium tin oxide (ITO) and aluminium zinc oxide (AZO)), and ICPs (PPy, PANI). ECPs have special properties such as good electrical and thermal conductivity, corrosion resistance, and good mechanical properties. They are used as conductive and semi-conductive polymer fibers, corrosion-resistant coatings, ESD materials, electronics, and solar collectors. Their conductivity values are much lower than the conductivity of ICPs, mostly in the range between 10^{-5} and 10^{3} S/cm depending on the applications [41].

14.3.4 How Can Polymer Conduct Electricity?

There are two conditions for polymers to become conductive. The first condition is that conductive polymers consist of alternating single and double bonds, called conjugated double bonds. Such bonds contain localized sigma (σ) bonds which form strong chemical bonds. In addition, each double bond also contains a less strongly localized pi (Π) bond which is weaker (Fig. 14.16) [42]. However, bond conjugation is not enough to make the polymer material conductive. Hence, the second condition is that polymer structure has to be disturbed by removing electrons from (oxidation), or inserting them into it (reduction). The processes are termed as p-doping and n-doping. They can affect its surface and bulk structural properties (color, porosity, volume) [43].

Let us take a closer look at the molecular structure of polyaniline in Fig. 14.16 showing the molecular structure of different oxidation states of polyaniline. In terms of the conductivity of the polymer, we can observe the interchanging single and double bonds occurring throughout the backbone of the polymer. This is also known as the conjugated backbone, but what does it really mean when a polymer is conjugated? Conjugation in terms of linguistics refers to a link between two separate things. Essentially what is obtained in a conjugated backbone of a polymer is an overlapping of p-orbitals, which allows for the delocalization of electrons along with those orbitals [44]. The delocalized electrons thus may act as charge carriers as they are free to move throughout the whole system. Electron delocalization is also the mechanism responsible for the electrical conductivity of metals. To induce or enhance the electrical conductivity of the conjugated polymers, a process known as doping is carried out. The doping process is essentially like that of silicon semiconductors, in which there is a p-doping (oxidation, or addition of a delocalized electron-hole) and n-doping (reduction, or addition of a delocalized electron). By the doping process, the electrons are now free to move around, resulting in an electrical current as they pass along the conjugated polymer backbone [45]. The doping level of conducting polymer can be controlled by using different counter-ions (dopants) to modify the conductivity. The properties change of conductive polymers in response to dopant is given in Table 14.3. In addition, the conductivity of conducting polymers is not only determined by the type of monomers and dopants but also determined by the conditions under which the polymers are fabricated.

The factors that affect the polymer conductivity are the density of charge carriers, their mobility and direction, the presence of doping materials, and temperature. Conductive polymers are inappropriate oxidized or reduced state conductors due to their unique, stretched, pi-conjugation. Overlapped pi-orbitals from the valence band, and pi*-orbitals form a conductive band. Due to chemical or electrochemical oxidation of conductive polymers, the electrons are removed from the valence band, which leads to the presence of charge on the conductive polymer which is delocalized over several monomer units in the polymer and cause relaxation of the polymer geometry in the most stable form. Dopants can be incorporated into the polymer during synthesis or can be retrofitted. They can be anions or cations, for example, ClO^{4-}, Na^{+}, or larger polymer particles such as polyelectrolytes, poly(styrene sulfonic acid) (PSS), and poly (vinyl sulfonic acid) (PVS). The degree of doping represents the ratio of counterion and monomers in polymer. In 1977, it was discovered that the conductivity of polyacetylene can be increased by doping with iodine [46].

It is interesting to compare three simple carbon compounds, diamond, graphite, and polyacetylene. They may be regarded, as three-, two-, and one-dimensional forms of carbon materials (Fig. 14.17). Diamond and graphite are modifications of pure carbon, while in polyacetylene, one hydrogen atom is bound to each carbon atom. Diamond contains only s bonds and it is an insulator. High symmetry gives isotropic properties. Graphite and polyacetylene have mobile p electrons and for this reason, they are highly anisotropic conductors [47]. Polyacetylene as the simplest possible conjugated polymer obtained by polymerization of acetylene illustrates principles of conduction mechanisms in polymers. As a necessary consequence of the asymmetry of the polyacetylene ground sate, two equivalent polyene chains, L and R, are interconverted through the intervention of a soliton. The soliton as a mobile charge or neutral defect or a "kink" in the polyacetylene chain propagates down the chain and reduces the barrier for interconversion. The charge carrier in n-doped polyacetylene is a resonance-stabilized polyenyl anion of approximately 29-31 CH units in length, which the highest amplitude is at the center of the defect. Soliton moving from one end of the sample to another is explained by the bipolaron hopping mechanism. The instability of polyacetylene in the air (covalent bonds are formed between oxygen and carbon atoms and these bonds lower the conductivity of polyacetylene) intensified research towards the study of other conductive polymers.

14.3.5 Materials in Conducting Polymer

A variety of conducting polymers are polyacetylene (PA) [48], Polyaniline (PANI) [49], polypyrrole (PPY) [50], poly(phenylene)s(PPs) [51], Poly(p-phenylene)(PPP) [52], poly(p-phenylenevinylene) (PPV) [53], poly(3,4-ethylene dioxythiophene) (PEDOT) [54], polyfuran (PF) [55] and other polythiophene (PTh) derivatives [56], etc,

Polyacetylene: Polyacetylene (PA) is one of the polymers in the study that resulted in the Chemistry Nobel prize in 2000. Some of the useful features of PA include electrical conductivity, photoconductivity, gas permeability, supramolecular assemblies, chiral recognition, helical graphitic nanofiber formation, and liquid crystal. The chemical structure of PA is a linear polyene chain $[-(HC{=}CH)_n-]$ [57]. Its backbone provides an important opportunity for decoration

Table 14.3 Various properties by a redox reaction in addition of dopant in conducting polymer

Properties	By Oxidation	By Reduction
Conductivity	More	Less
Capacitance	More	Less
Transparent	Less	More
Hydrophobic	Less	More
Expanded/Contracted	Expanded	Contracted
Modulus	Higher	Lower

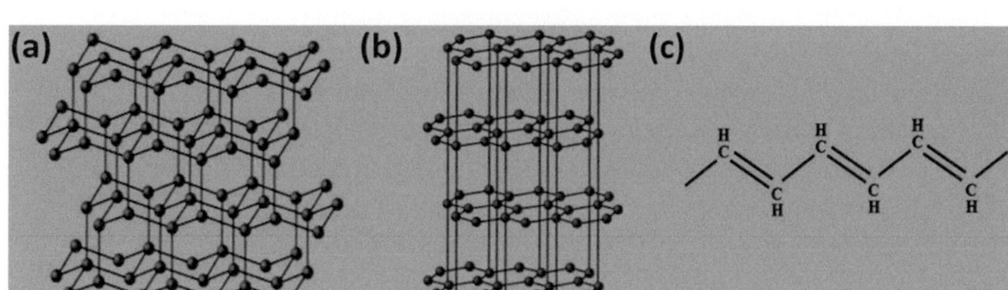

Fig. 14.17 Three-, two- and one-dimensional carbon materials: (**a**) diamond, (**b**) graphite, (**c**) polyacetylene chain

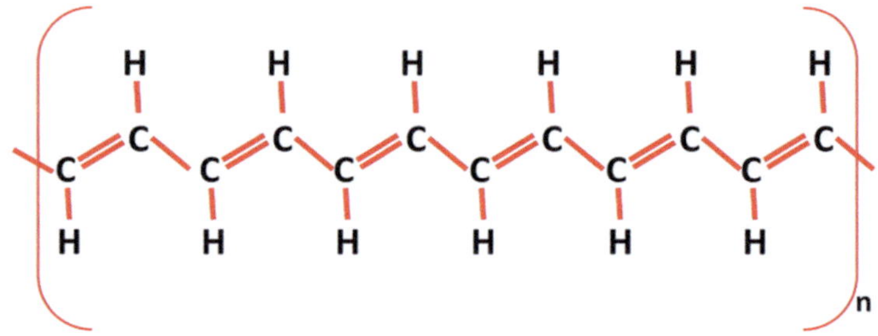

Fig. 14.18 Polyacetylene structure

with pendants due to the presence of repeated units of two hydrogen atoms. Each repeated unit of hydrogen could thus be replaced by one or two substitutes to yield monosubstituted or disubstituted PAs, respectively (Fig. 14.18).

Acetylene only or other monomers could be used in a number of methods to develop and synthesize polyacetylene. One of the methods is named Ziegler–Natta catalysis and involves the use of titanium and aluminum in the presence of gaseous acetylene. By changing the temperature and amount of catalyst, this method could be a beneficial way to develop polyacetylene while monitoring the structure and watching for the final polymer products. The polyacetylene could be synthesized by substituting the catalyst with $CoNO_3/NaBH_4$, and results show stabilities to oxygen and water. Developing and synthesizing polyacetylene could also be obtained by other methods of polymerization radiations such as glow discharge, ultraviolet, and ϒ-radiation. The use of radiation methods could be beneficial since they could avoid the use of catalysts and solvents. Compounds and composites of polyacetylene can be formed by hybridization with different materials to improve the conductivity such as dihexadecyl hydrogen phosphate, quaternized cellulose NPs, and Au NPs. Polyacetylenes are also called acetylene black (AB) or polyacetylene black depending on the preparation method. AB is a carbon nanomaterial, which is a specific subtype of carbon black and is produced by the controlled combustion of acetylene under pressurized air. ABs are usually used as substitutes for graphitic powder in the preparation of carbon paste electrodes, and are dispersed in chitosan solution to modify glassy carbon electrodes [58].

Polythiophene: Polythiophene (PTh) is interesting for its stable conductivity and high electrical conductivity ($10^3 Scm^{-1}$), and its conductivity varies with the type of dopant and polymerization. As CPs are based on conjugated systems, they are by nature nontransparent and refractory, polythiophene being a prime example. Polythiophene and its derivatives have been explored for electroactive scaffolds for cell culture, biosensors, and neural probes. Poly (3,4-ethylenedioxythiophene) (PEDOT) (Fig. 14.19) is considered the most successful PTh derivative due to its higher electrical conductivity and chemical stability which allows its use in biomedicine and biotechnology. The conductivity of

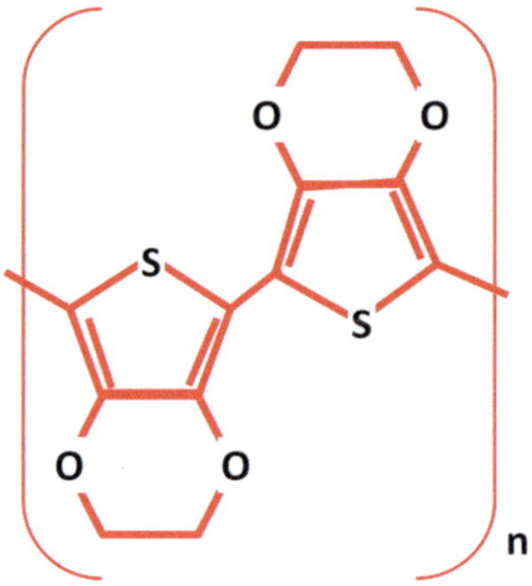

Fig. 14.19 Polythiophene

doped PEDOT is in the range of 0.4–400 S/cm. Oligomers consisting of 11 thiophene units have conductivity similar to that of higher molecular weight polythiophene. Consistent with this is the finding that short oligomers of thiophene have polymer properties, with conductivity and carrier mobility increasing as a function of the conjugation length up to the hexamer of thiophene. Transparency is one of the main characteristics based on the application where electrical conductivity is important, such as photographic films coated with antistatic coatings, which should be higher than 90%. CP transparency could be increased by dilution, which influences conductivity. There are different methods for polythiophene dilution, such as block copolymerization, alkyl side chain grafting onto the conjugated backbone, or blending with a transparent polymer, and producing composites via thiophene polymerization absorbed in an insulating polymer [59]. In addition, plasma polymerization, electrochemical procedures, and thin layer polythiophene deposition could be conducted. Polythiophene can be synthesized electrochemically and chemically. By obtaining a conductive film on the anode with a solution of thiophene and an electrolyte, the electrochemical polymerization of polythiophene results. Polythiophene polymerization's main benefit is that it does not require isolation and purification, but it could be irreversible and decompose. In chemical synthesis and development, there is a better monomer selection and better synthesis by using proper catalysts. PEDOT can be synthesized in various forms such as nanofilms, nanorod arrays, and nanofiber mats. The conductivity of polythiophene improved by making compound and composites with tosylate anion, poly(styrene sulfonate), sodium chloride, lithium perchlorate, sodium phosphate monobasic monohydrate, Br, poly(sodium 4-styrene sulfonate), poly(styrene sulfonic acid)/Au, methyl- or benzyl-capped diethylene glycol, tetraethylene glycol, alkyl side chains, and poly(ethylene oxide). In addition, PEDOT in the presence of EDOT-OH, C2-EDOT-COOH, C4-EDOT-COOH, C2-EDOT-NHS, and EDOT-N3 indicated better conductivity [60].

Polypyrrole: Polypyrrole (PPy) has the potential to be applied in biomedicine due to its impressive conductivity, outstanding redox properties, biocompatibility, easy synthesis, and environmental stability. It can be used for neural implants, biosensors, drug delivery, bioactuators, and molecular memory devices. Polypyrrole was first obtained by Weiss and colleagues in 1963 as a highly conductive polymer material from pyrolysis of tetraiodopyrrole. It is obtained by the oxidization of the pyrrole, resulting in the structure shown in Fig. 14.20[61]. PPy possesses an amorphous structure and is insoluble. PPy could modulate cellular activities, including deoxyribonucleic acid (DNA) synthesis by ES, migration, and cell proliferation in biological environments. PPy can easily be synthesized in large quantities at room temperature in a variety of common organic solvents and also in water. The conductivity of PPy films can be achieved up to 7.5×10^3 S/cm depending on the type and amount of dopant. Development, synthesis, and polymerization of PPy are achieved from many repetitions of pyrrole's oxidation of ferric chloride in methanol. Polymerization would be obtained by peeling the film from the anode. Myocytes, fibronectin, titanium, bovine leukemia virus (BLV) protein (gp51), poly(ethylene terephthalate) (PET), biotin, alginate, biotinylated GOx, silk fibroin, chlorpromazine (CPZ)-incorporated heparin (Hep), and tosylate ion are used to hybridize with polypyrrole to improve the conductivity. The PPy processability and functionality are significantly improved through copolymerization with other monomers with self-stabilized functional groups [62].

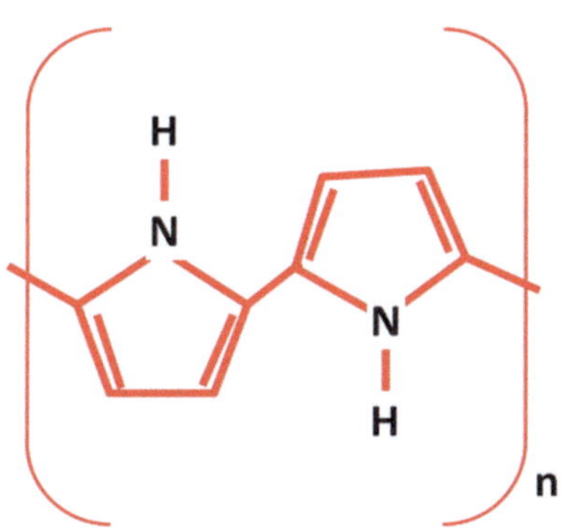

Fig. 14.20 Chemical structure of the Polypyrrole

Polyaniline: One of the most promising conjugated CPs is Polyaniline (PANi), due to its preparation simplicity, high electrical conductivity, and great environmental stability. Its production cost is low and it can be easily doped with inorganic and organic acids to prepare the conductive form. The thermal stability of PANI is superior to other ICPs. The conductivity of PANi is in the range of 30–200 S/cm depending on p/n doping. These properties make PANi suitable to be employed on different sensor applications, such as pH switching electrical conducting biomaterials, electrically active redox biopolymers, and matrixes for nanocomposite CP preparation. This polymer, in fact, is a family of polymers that are classified by aromatic rings bonded together by nitrogen atoms. Its structure is composed by x units of reduced species alternated with 1-x units of oxidized species, as observed in Fig. 14.21 [63]. It exists in various forms based on its oxidation level i.e. the fully oxidized pernigraniline base, half-oxidized emeraldine base, and fully reduced leucoemeraldine base. PANi emeraldine form is the most stable and conductive. PANi is also difficult to process due to its poor solubility in most of the available solvents. Tremendous development in PANi-based nanocomposite biopolymer preparation has been done. PANi is polymerized on the basis of the aniline monomer and can be found in one of three idealized oxidation states, which affect its conductivity. The partially oxidized structure is called emeraldine. After being doped with acid, it becomes a CP which is very stable at room temperature. Polyaniline can be obtained from a three-component-state physical mixture of leucoemeraldine [$(C_6H_4NH)_n$], emeraldine [([C6H4NH]$_2$[C6H4N]$_2$)$_n$], and per-nigraniline [$(C_6H_4N)_n$]. Emeraldine [([C_6H_4NH]$_2$-[C6H4N]$_2$)$_n$] is doped with acid and is in the most stable and conductive of the three states of the physical mixture used to obtain polyaniline.

Fig. 14.21 Chemical structure of the polyaniline

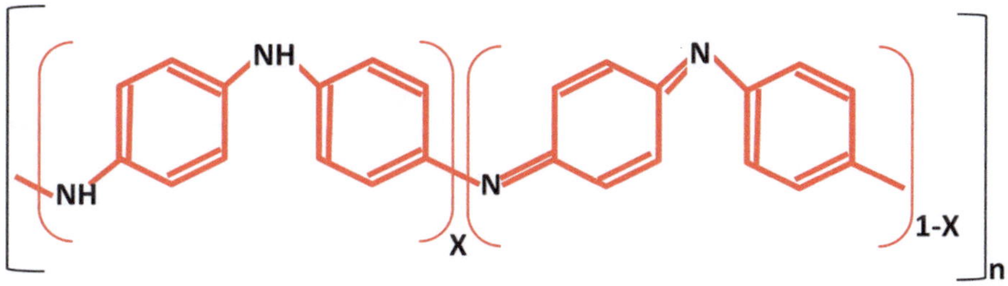

Fig. 14.22 Poly-(*p*-phenylene)

Different materials for hybridization with polyaniline to improve the conductivity are cyclodextrin, β-cyclodextrin ring, lyotropic liquid crystalline (LLC) materials, platinum, oxidant ammonium peroxydisulfate, ATQD, and methyl orange (MO) [64].

Poly-(*p*-phenylene): Poly-(p-phenylene) is one of the CP that contain an aromatic ring in its structure. The conductivity of Poly-(*p*-phenylene) doped with AsF5 is 1.5×10^4 S/cm. An insoluble powder, it has shown interesting thermal, chemical, and electrical properties. Different materials hybridization with polyphenylene to improve the conductivity, such as ZnO NPs, silicate platelets, polystyrene mixtures, amino or carboxyl groups, poly-L-lysine, and (diphenylamino)-*s*-triazine [65] (Fig. 14.22).

14.3.6 Preparation of Conducting Polymers

Organic conducting polymers can be prepared by pyrolysis and numerous methods such as chemical, electrochemical, photochemical, concentrated emulsion, inclusion, solid-state, and plasma polymerization. Among these preparation methods, there are three industrial ways to synthesize the conductive polymers: reactional chemistry, electrochemical, and photoelectrochemical, being the first one the most used, due to its high profitability and efficiency. The chemical process consists of the union of monomers by the addition of the oxidizing or reducing agents that makes the neutral polymer to a cationic or anionic ionic complex, ending the reaction by the bonding of this complex to the counter-ion of the redox agent. This process requires high control since the reaction is very exothermic and emits gases, requiring proper treatment and equipment of protection. Chemical synthesis is the oldest and the most popular route for the preparation of bulk quantities of CP on a batch scale. Chemical polymerization is typically carried out using relatively strong chemical oxidants such as KIO_3, $KMnO_4$, $FeCl_3$, K_2CrO_4, $KBrO_3$, $KClO_3$, $(NH_4)_2S_2O_8$. These oxidants are able to oxidize the monomers in solution, leading to the formation of cation radicals. These radicals further react with other monomers or n-mers, yielding oligomers or insoluble polymers. Chemical oxidative polymerization has some limitations related to the limited range of available chemical oxidants. The counterion of the oxidant ultimately will act as a dopant or co-dopant in the polymer. Therefore there is an obstacle to obtain CP with different dopants. The limited range of oxidants also, makes it difficult to oxidizing power in the reaction mixture and subsequently the degree of over oxidation during synthesis. Both the type of dopant and the level of doping are influential parameters affecting the final properties of CPs such as molecular weight, crosslinking, and conductivity. Moreover, a strong drawback in using these stoichiometric oxidants is the formation of a large amount of byproducts, in the case of ammonium persulfate (APS), the amount of resulting ammonium sulfate is about 1 kg per kg of the organic polymer [66].

The electrochemical method consists of the electronic deposition of the polymer in the electrode. The solution that the electrode is immersed has the monomers and the dopants. When a voltage or a current strong enough is applied, the monomers oxidize resulting in polymerization. This process results in polymers with the shape of the electrode, requiring posterior processing to get the shape desired. The electrochemical technique has received wider attention, because of the simplicity and the added advantage of obtaining simultaneously doped conductive polymers. In general chemical oxidation method produces CP as powders while electrochemical synthesis generally leads to deposition of conducting polymers films onto a supporting electrode surface by anodic oxidation (electropolymerization) of the corresponding monomer. When a positive potential is applied to the working electrode, the oxidation starts with the formation of the radical cation. The delocalized radical cations induce the radical-radical coupling to form firstly dimmers. The extended conjugation in the polymer results in a lowering of the oxidation potential compared to the monomer. The electrochemical oxidation and radical coupling process are repeated continuously and finally the deposited CP film is produced on the working electrode. The doping of the polymer is generally performed simultaneously by the incorporation of the doping anion into the polymer to ensure the electrical neutrality of the film [67]. The photoelectrochemical process is based on photoexcitation of the polymer or in compounds that have catalyst properties in presence of light, oxidizing the monomers resulting in polymerization. Even though this process is simple and environmentally friendly, the mechanical property of the resulting polymer is not good [68].

14.3.7 Applications of Conductive Polymer

Various applications on conductive plastics are discussed below:

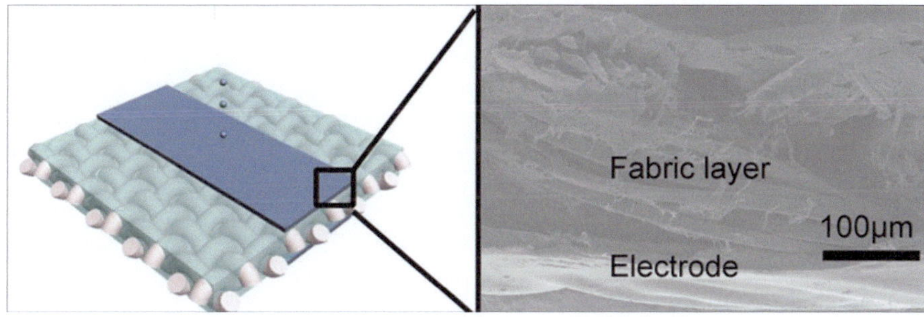

Fig. 14.23 Textile-based wearable devices

Biomedical Application: For the past few decades, conducting polymers have drawn much attention for their biomedical applications. Most biological cells are sensitive to electrical impulses; therefore, conducting polymers can be used in the field of tissue engineering to modulate cellular activities [69]. In terms of tissue engineering, 3D scaffolds utilizing biocompatible CP can be obtained by means of electrospinning. The conductive polymer is believed to also be applicable in areas such as wound healing, bone repair, and spinal cord regeneration. Medical industry used CP as artificial muscles by utilizing their response to an electrical field. The polymers can change their dimension in response to electrical stimuli and are known as electroactive polymers (EAP). Several groups have managed to produce an actuator device that can transform electrical pulses into mechanical movement, which is a process which mimics that of our own natural muscles. For example, PPy, PANI, polythiophene (PTh) and PEDOT, show biocompatibility, conductivity, reversible oxidation, redox stability, and excellent electrical and optical properties. These make them suitable for cell adhesion and tissue engineering application and smart textile application.PPy-coated electrospun poly(lactic-coglycolic acid) (PLGA) nanofibers (PPy-PLGA) were fabricated for neural tissue applications. Takamatsu et al. developed textile-based wearable devices inspired by the Japanese kimono dyeing technique (Fig. 14.23) to record high-quality EKG in a clinic and ambulatory conditions and to determine heart rate [70].

Conducting polymer used in drug delivery and protein purification [71]. To enhance the drug-targeting specificity and decrease systemic drug toxicity, many drug delivery system devices have emerged during the last few decades, which have been used for the treatment of different kinds of diseases. Various drug delivery systems include polymeric microspheres, polymer micelles, polymeric nanofibers, micro-nanogels, etc. One of the major drawbacks of the delivery system is to maintain strict control of ON=OFF state. To overcome this drawback, conducting polymers have been used as they show a reversible electrochemical response, i.e., they contract upon reduction and expand upon oxidation. Thus, this induced volume change is expected to favor the controlled release of various kinds of drugs. Conducting polymers are used in the field of tissue engineering [72]. In the case of living cells, electrical stimulation can modulate cell-to-cell attachment, cell proliferation, cell migration, and cell differentiation. The fabrication of polymer nanofibers (i.e., electrospinning of blended polymer solution consisting of conducting polymer and other kinds of biocompatible polymers) produces conductive composite nanofibers with high surface area, high porosity, good biocompatibility, and biodegradability, which have found use in tissue engineering. The natural protein gelatin and polygalacturonic acid (PGA)/gelatine are also used as a scaffold for tissue engineering. Some of the advantages of conducting polymer nanocomposites include easy loading, little influence on drug activity, and controlled release rate.

Automotive Applications: One-shot manufacturing of complex-shaped parts may be considered as one of the best solutions for the fast production of composite pieces dedicated to the automotive and railway industries. Replacement of metallic cross stiffeners with their composite parts made of GF/PP coming led yarn. The PEDOT: PSS sensors based on glass yarns were inserted into braided fabrics coupons (before thermal consolidation) and provided the ability to monitor the structural health of these composite parts in a real time in order to choose the appropriate composite architecture for the final automotive applications [73]. Figure 14.24 showing the conducting polymer in an automobile charging battery. CP-assisted electrorheological (ER) fluids have wide applications in clutch systems, brakes, hydraulic valves, and dampers for their adjustable properties of vibration control under an external stimulus. Particles of high dielectric constant and low conductivity dispersed in a non-conducting fluid medium are generally the component of ER fluids. This property under an electric field can be changed reversibly in a short period of time. Recently, conducting polymers have been mostly used as polarizable particles because of their superior physical properties such as better environmental stability and high polarizability [74].

Energy Storage Application: Energy has become the most important global concern because fossil fuels are going to be exhausted. Usually, conducting polymer nanostructures have higher specific capacitance values and can be an alternative in the development of the next-

Fig. 14.24 Conducting polymer in automobile charging battery

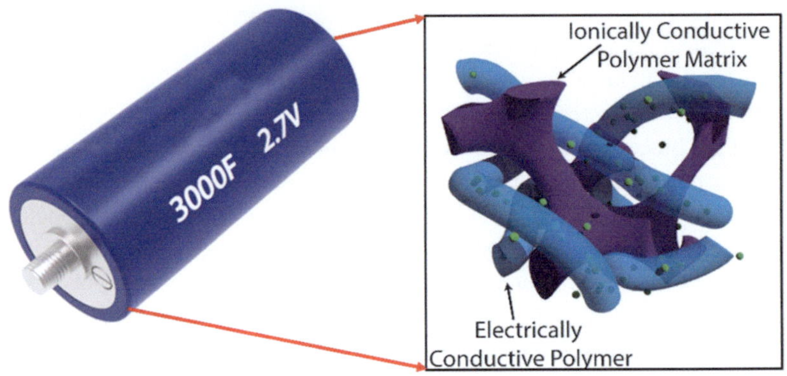

Fig. 14.25 Conducting polymer in energy storage application

generation energy storage devices [75]. Solar cells are energy conversion devices that convert sunlight to electric energy. Conducting polymers have unique properties of light absorbance and whole transporting when combined with metal oxide, which may contribute to the improvement of the photovoltaic efficiencies (Fig. 14.25). TiO_2 nanotube arrays are considered as good candidates for the construction of solar cells because they provide good pathways for electron migration [76]. Fuel cells convert the chemical energy directly into electricity by electrochemical reactions. In recent decades, fuel cells have attracted attention for their applications in electric vehicles. Due to high energy conversion efficiency, fuel portability and environment friendliness, direct methanol fuel cells (DMFC) have been a research focus in the field of energy application [77]. The effects of an electrocatalyst on the performance of DMFC have been investigated, and the CP with 1D-nanostructures have become good candidates as electrocatalyst supports. Conventional rechargeable Ni-Cd or Ni-metal hydride batteries are limited by their capacity and durability. On the other hand, lithium-ion batteries that are lighter and have much greater capacity have been considered as the most promising and practical rechargeable batteries. 1D-nanostructured materials have proved to be good candidates in Li-ion battery electrodes because of their high specific capacity and good cycle performance [78]. Supercapacitors are one of the most promising energy storage devices for a wide range of applications in electric vehicles, uninterruptible power supplies, etc. Compared to Li-ion batteries, supercapacitors exhibit higher specific power. There are mainly three kinds of electrode materials for supercapacitors, i.e., carbon, metal oxides, and conducting polymers. Conducting polymers have high specific capacitance, but their cyclic stability is poor. This drawback has been overcome by the fabrication of conducting polymer nanocomposites. Microwave absorption and electro-magnetic frequency interference (EMI) is a serious issue caused by the rapid proliferation of electronics, wireless systems, navigation, space technology, etc. EMI affects the performance of the electric device as well as various life forms, including human beings. Therefore shielding materials such as metals, carbon materials, and conducting polymers have been employed to prevent electromagnetic noise. The uses of conducting polymers as shielding materials have attracted increased attention due to their good electrical conductivity and processibility. Polyaniline microtubes, nanofibers, and polyaniline-multiwalled carbon nanotube nanocomposites can be used as microwave absorbers and electromagnetic interference-shielding materials [79]. Figure 14.26 showing the conducting polymer anode material.

Conducting polymer bulk films and nanostructures have potential applications as actuators or artificial muscles [80]. There are reports suggesting the movement of an all-polymeric, triple-layer artificial muscle based on polypyrrole. In a study conducted by Okamoto et al., it has been reported that an actuator based on doping/undoping-

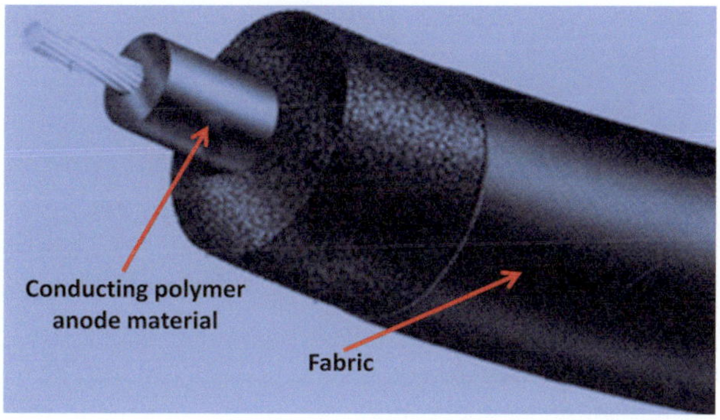

Fig. 14.26 Conducting polymer anode material

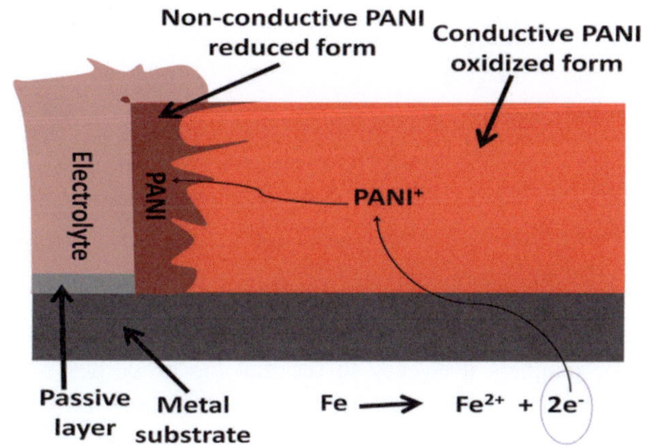

Fig. 14.27 Conducting polymer PANI released ion to prevent the corrosion

induced volume changes in anisotropic polypyrrole film. It has been reported that flash-welded polyaniline nanofiber actuators demonstrate unprecedented reversible rapid actuation upon doping. CP can be used as biomimetic dual sensing-actuator, nanofillers for the ferroelectric copolymer, actuator-driving, variable-focus lens system using PEDOT, multi-material hydrogel cantilevers to measure the contractile forces of cardiomyocyte cell sheets, and as an early prototype for the design of optimal cell-based biohybrid actuators. Polyaniline nanofibers when doped with gold can form a bi-stable electrical behavior that allows to be switched electrically between two states, which may be used in the fabrication of plastic digital nonvolatile memory devices [81].

Corrosion Protection Application: Conducting polymers of various forms will be electrodeposited onto oxidizable metals and electrochemical and environmental means will be used to access their application for corrosion protection. Polyaniline and its derivatives are among the most frequently studied CP used for corrosion protection. In addition, electro-active coatings of the PANI could provide adequate protection against corrosion of stainless steel and iron sheets. The extent of the corrosion protection offered by polyaniline coatings to mild steel was investigated in aqueous 3% NaCl solution, 0. 01 M Na_2SO_4 solution, and in aqueous solutions of $NaCl+Na_2SO_4$ with different concentrations by potentiodynamic polarization technique and electrochemical impedance spectroscopy. The results of these studies reveal that the corrosion resistance of the polyaniline-coated mild steel is significantly higher and the corrosion rate is considerably lower than that of uncoated steel [82]. Figure 14.27 showing the use of PANI conductive polymer to protect the metal substrate.

14.4 Magnetic Plastic

A plastic magnet is a non-metallic magnet made from an organic polymer. In 2002, researchers from Ohio State University and the University of Utah developed the world's first light-tunable plastic magnet [83]. The plastic material became 1.5 times more magnetic when blue light shines on it. Green laser light reversed the effect somewhat, by decreasing the material's magnetism to 60% of its normal level. The plastic magnet was made from a polymer made of tetracyanoethylene (T-CNE) combined with manganese (Mn) ions, atoms of the metal manganese with electrons removed. The magnet functioned up to a temperature of −198 °C. But PANiCNQ was the first magnetic polymer to function at room temperature. PANiCNQ is a combination of emeraldine-based polyaniline (PANi) and tetracyanoquinodimethane (TCNQ). It was created by scientist Naveed and colleagues at the University of Durham in 2004 [84]. PANiCNQ is a conductive polymer that is stable in air and is 1/100th the strength of conventional metal magnets. When combined with the free radical-forming TCNQ as an acceptor molecule, it can mimic the mechanism of metallic magnets. The magnetic properties arise from the fully pi-conjugated nitrogen-containing backbone combined with molecular charge transfer side groups. These properties cause the molecule to have a high density of localized spins that can give rise to the coupling of their magnetic fields.

Magnetic plastics have several advantages over their metallic counterparts, such as being less dense and are useful in applications where weight is a concern. Magnetic plastics have a superior cost-performance ratio and can be produced in a large variety of shapes. Sprayed magnets are typical composite materials. In magnetostrictive polymer, the magnetic powder is embedded in thermoplastics (polyamides), allowing the most diverse shapes to be created. Magnetic plastics are produced by embedding hard ferrite or rare earth magnet powder in plastic [85]. First, the magnet powder and the plastic are mixed in special equipment. Then the mixture is pressed or processed in modified injection molding machines. This process achieves very narrow tolerances, so that post-processing is normally unnecessary. This type of

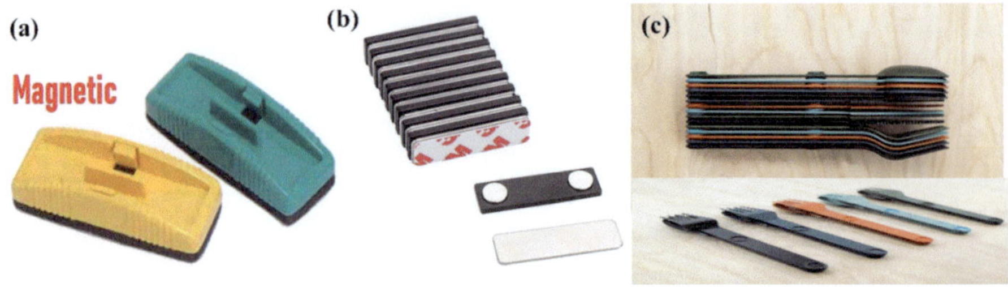

Fig. 14.28 Magnetic polymer assisted (**a**) duster, (**b**) Name Badge, and (**c**) Magware Magnetic Flatware

magnetostrictive polymer has been described in Chap. 4. For plastic magnets it is easy to see that the properties of the polymer matrices, especially viscoelasticity, may affect the processability as well as magnetic property. Furthermore, because identical poles in magnetic particles are orientated repulsively in the anisotropic plastic magnets.

14.4.1 Applications

Plastic magnets could have uses in computer hardware, for example as disc drives and in medical devices such as pacemakers and cochlear implants where the organic material is more likely to be biocompatible than its metallic counterpart [86]. The demand for magnetic plastics will be driven by growth in the consumer industry. As there is an increase in the production of personal computers, mobile phones, and generators, there will be increase in consumption of magnetic plastics. The Asia Pacific is the largest market for magnetic plastics. China is the dominant country market in the Asia Pacific, accounting for the majority of the magnetic plastics market in the Asia Pacific. Most of the larger manufacturing units are located outside the developed economies particularly in emerging economies, to take advantage of lower labor costs and overheads. Southeast Asian countries, India, South Africa, and Brazil are expected to have significant growth in near future. Some of the key players in this market are OM Group, ALL Magnetics Inc, Thyssen Krupp, Arnold Magnetic Technologies, and Kolektor Magnet Technology GmbH among others [87].

Magnetic plastics could lead to the development of inexpensive and flexible devices for quantum computing and superconducting electronics. Magnetic plastics are free from surface corrosion. Hence, they can be used in numerous application fields without additional coating. Magnetic plastics have applications in the food, electronics, and medical industry [88]. These can be used to manufacture magnetic switches or inductive assemblies. In the food industry, magnetic plastics are used to manufacture tanks, molds, scrapers, housings, and other parts. These components do not always come into direct contact with foodstuffs but may contaminate the food with parts of different sizes if damaged. This has given rise to the demand that all plastic parts, which may come into contact with food should be detectable and suitable for use in the food sector. Magnetic plastics can also be used to manufacture magnetic deflector beams, magnetic rollers for color television sets, rotating magnets in motors for office equipment, cooling fans, magnets for rotational or frequency sensor, and motors in the automotive industry among others [89, 90]. Figure 14.28 shows the magnetic polymer assisted duster, name badge, and magware magnetic flatware

14.5 Transparent Plastic

In everyday language, transparency is the effect of a sheet or film on the visibility of an object full of contrast, the specimen being interposed at a considerable distance from the object. Transparent plastics are used primarily when clarity of vision through the material is a concern. Transparent colored materials transmit the portion of the visible spectrum that allows the eye to see the desired color. Most plastic materials are not transparent: however, in this group acrylic transmits 92% of the light available (comparable to glass). Transparent plastics can be defined as the plastic that offers light transmission rates greater than 80% in thickness more than 1 mm in the visible light regime and has better impact resistance [91, 92]. Acrylics are 17 times more resistant to breakage than glass and are easily fabricated. PETG has 7 times the impact resistance of acrylic and Lexan has 17 times the impact resistance of acrylic. This group of materials allows their use in applications instead of glass. The average weight of this set of plastics is about one-half that of a comparable size of the glass. The Transparent Plastics market grows from USD 102.0 billion in 2017 to forecasted USD 137.7 billion by 2022, at a CAGR of 6.2%. The major factors that are expected to be driving the transparent plastics market are increasing adoption of transparent plastics by end-use industries, improved shelf-life of packaged food products, high demand in emerging regions, and ease of customization [93]. The regional consumption of this plastic is given in Fig. 14.29.

14.5.1 Major Factors of Transparency

Figure 14.30 shows the operation Transparent plastics which is at scattering state (voltage off) (Fig. 14.30a), and in

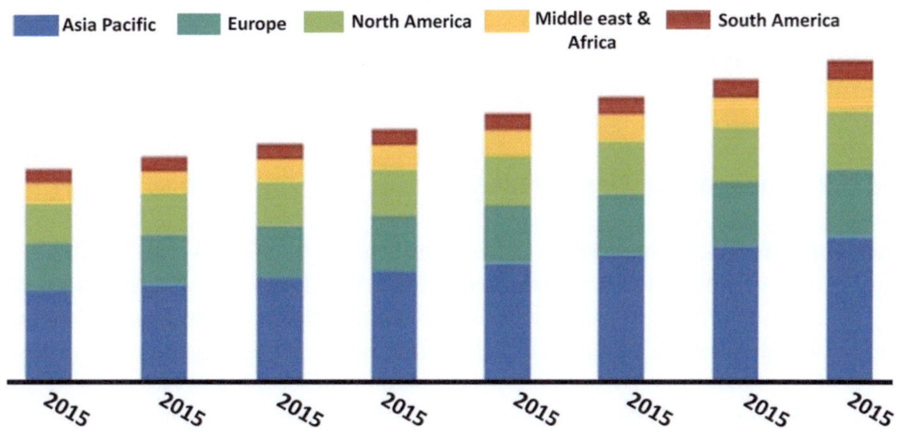

Fig. 14.29 The regional consumption of this transparent plastic

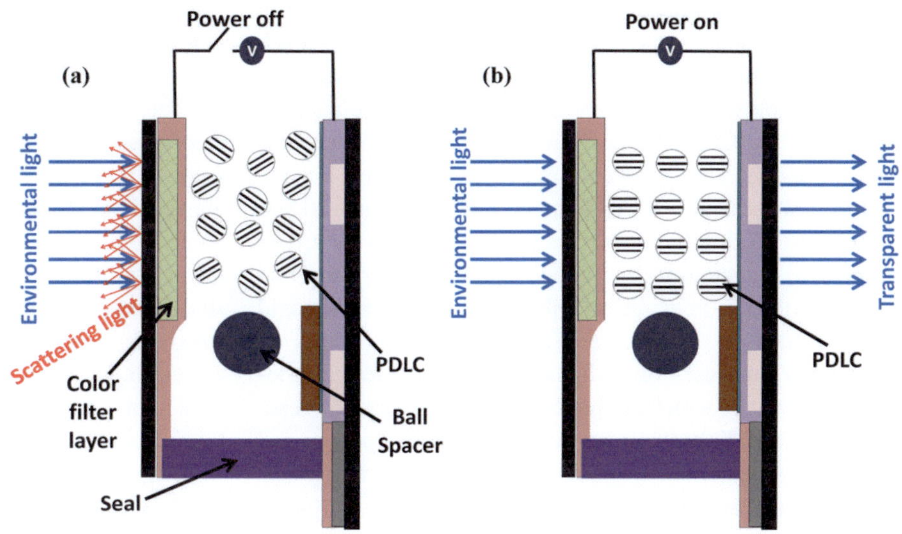

Fig. 14.30 (**a**) Scattering state and (**b**) transparent state

transparent state (voltage on) (Fig. 14.30b). The major factors are the presence of amorphous regions, low refractive index, low compressibility [94, 95]. Crystalline polymers have both crystallized and amorphous regions. The existence of both crystallized and amorphous regions causes large size heterogeneities with an order of several thousand angstroms, which seriously increases the light-scattering loss. In order to prohibit crystallization, homopolymers with branched side chains are preferred. But, the higher concentration of chain ends in the branched polymer increases the mobility of branched polymer segment and consequently lowers T_g. The important effect of the nature of the chain repeat units on T_g is a hindrance to free rotation along the polymer chain. To inhibit the lowering T_g and crystallization, the main chain which contains a cyclic structure should be considered. The fluorination reduces the refractive index, thus leads to less scattering, and also reduces the vibration absorption. The theoretical lower limit of the refractive index of organic polymers has been investigated by using the Lorentz-Lorenz equation. They concluded that the theoretical lower limit for an amorphous organic polymer is about 1.29. For example, the calculated refractive index of an amorphous poly (tetrafluoroethylene)(PTFE) from molecular structure using $K = 0.68$ is $n_0 = 1.293$. But -(CF2CF2)-the chain has a trend to be crystallized. The crystalline PTFE has a relatively high refractive index of 1.35 due to higher density. Therefore, amorphous polymer seems to be the lowest refractive index polymer and a most transparent polymer.

Perfluoro(butenyl vinyl ether) polymers (poly(FBVE)) polymer having a larger cyclic structure has larger compressibility than PMMA because of its larger molecular volume. The fluorination reduces the refractive index and thus leads to less scattering. But we must consider the opposite effects between isothermal compressibility and the refractive index on the isotropic scattering loss. The calculated isotropic light scattering loss of poly(FBVE) is about 4. 7 dB/km at 633 nm and is half of the value (9.7 dB/km) of PMMA**.** Light scattering loss is inversely proportional to the fourth power of wavelength. The inherent light scattering losses of poly (FBVE) at 1.3 μm are about 0.3 dB/km. The fluorination reduces the refractive index, thus leads to less scattering, and also reduces the vibration absorption. Amorphous perfluoropolymer is expected as a material for new low-loss polymer optical fiber and seems to be the most transparent polymer.

14.5.2 Transparent Plastic Materials

Acrylonitrile Butadiene Styrene (ABS): Acrylonitrile butadiene styrene (ABS) (chemical formula ($C_8H_8{\cdot}C_4H_6{\cdot}C_3H_3N$) n is a common thermoplastic polymer. ABS is a terpolymer made by polymerizing styrene and acrylonitrile in the presence of polybutadiene. The proportions can vary from 15 to 35% acrylonitrile, 5–30% butadiene, and 40–60% styrene. The result is a long chain of polybutadiene crisscrossed with shorter chains of poly(styrene-co-acrylonitrile). The nitrile groups from neighboring chains, being polar, attract each other and bind the chains together, making ABS stronger than pure polystyrene. The styrene gives the plastic a shiny, impervious surface. The polybutadiene and a rubbery substance provides toughness even at low temperatures. For the majority of applications, ABS can be used between −20 and 80 °C as its mechanical properties vary with temperature. The properties are created by rubber toughening, where fine particles of elastomer are distributed throughout the rigid matrix [96].

Polycarbonates: Polycarbonates (PC) are often thermoplastic polymers, which are produced in a condensation reaction from an aliphatic or aromatic dihydroxy species and phosgene. An alternative synthetic route provides transesterification with carbonic acid. Products are usually amorphous with small amounts of crystalline fractions. The preparation, purification, and in particular the achievement of high molecular weights can be difficult. In contrast, the thermal and mechanical resistance of polycarbonate is of outstanding quality. In the field of high temperature or impact-resistant POFs, Polycarbonates are a common material. For the fiber core bisphenol-A-type polycarbonates with a molecular weight $M_w = 23{,}000$ g/mol, a glass transition temperature $T_g = 145$ °C. Due to the polycondensation process, the purification of polycarbonates is challenging, which often results in a reduced transparency [97].

Poly(methyl Methacrylate) (PMMA): This transparent plastic is sometimes called acrylic glass. It has chemical formula $(C_5O_2H_8)_n$. For the glass/plastic laminate often called safety glass, laminated glass. PMMA can be produce hardened transparent in any shape, form a mold. PMMA is an economical alternative to polycarbonate (PC) when tensile strength, flexural strength, transparency, polishability, and UV tolerance are more important than impact strength, chemical resistance, and heat resistance. Additionally, PMMA does not contain the potentially harmful. PMMA is generally produced by emulsion polymerization, solution polymeriza tion, and bulk polymerization. PMMA produced by radical polymerization (all commercial PMMA) is atactic and completely amorphous. PMMA is suitable for processing by injection molding, extrusion, extrusion blow molding (impact modified acrylics only), thermoforming, and casting. Pre-drying is not necessary if a vented cylinder is used but if a normal cylinder is used then PMMA must be processed dry and it is advisable to pre-dry the granules for upto 8 h at 70–100 °C [98].

Polystyrene: Polystyrene (PS) is a synthetic aromatic hydrocarbon polymer made from the monomer known as styrene. The chemical structure of polystyrene is (C8H8)n. Polystyrene can be solid or foamed. General-purpose polystyrene is clear, hard, and rather brittle. It is an inexpensive resin per unit weight. It is a rather poor barrier to oxygen and water vapor and has a relatively low melting point. Polystyrene is one of the most widely used plastics, the scale of its production being several million tonnes per year. Polystyrene can be naturally transparent, but can be colored with colorants. As a thermoplastic polymer, polystyrene is in a solid (glassy) state at room temperature but flows if heated above about $T_g = 100$ °C and becomes rigid again when cooled. This temperature behavior is exploited for extrusion (as in Styrofoam) and also for molding and vacuum forming since it can be cast into molds with fine detail. Under ASTM standards, polystyrene is regarded as not biodegradable. It is accumulating as a form of litter in the outside environment, particularly along shores and waterways, especially in its foam form, and in the Pacific Ocean [99].

Cellulose Acetate Butyrate: Cellulose acetate butyrate (CAB) is a thermoplastic cellulose ester with 30–55 wt% butyryl content. It is the most commercially important mixed cellulose ester and has many attractive properties such as low viscosity, high transparency and surface gloss, improved resistance to moisture and ultraviolet light (UV), and good inter-coat adhesion. CAB is compatible with many resin systems, including most acrylics, polyesters, phenolics, ureas, and isocyanates. However, they are often incompatible with melamine, urea-formaldehyde, and many alkyds. Both the solubility and miscibility depend on the butyryl content; in general, with increasing butyryl content, the solubility in many resins and solvent increases. For example, CAB resins with high butyryl content are soluble in many esters, ketones, glycol ethers, blends of alcohols, and aromatic hydrocarbons. CAB is often used in coating formulations where low application viscosities at relatively high solid content are required. When added to coating formulations, it improves many coating properties like hardness, leveling, solvent resistance, clarity, and gloss while it reduces dry-to-touch time, cratering, orange peel, and blocking [100].

Glycol Modified Polyethylene Terephthalate: Polyethylene terephthalate, commonly abbreviated PET, or PETE is the most common thermoplastic polymer resin of the polyester family and is used in fibers for clothing, containers for liquids and foods, thermoforming for manufacturing, and in combination with glass fiber for engineering resins. The majority of the world's PET production is for synthetic fibers (in excess of 60%), with bottle production accounting for about 30% of global demand. In the context of textile applications, PET is referred to by its common name, polyester, whereas the acronym PET is generally used in relation to packaging. Polyester makes up about 18% of world polymer production and is the fourth-most-produced polymer after polyethylene (PE), polypropylene (PP), and polyvinyl chloride (PVC). PET consists of polymerized units of the monomer ethylene terephthalate, with repeating ($C_{10}H_8O_4$) units. PET exists both as an amorphous (transparent) and as a semicrystalline polymer. The semicrystalline material might appear transparent (particle size less than 500 nm) or opaque and white (particle size up to a few micrometers) depending on its crystal structure and particle size. The monomer bis(2-hydroxyethyl) terephthalate can be synthesized by the esterification reaction between terephthalic acid and ethylene glycol with water as a byproduct, or by transesterification reaction between ethylene glycol and dimethyl terephthalate (DMT) with methanol as a byproduct. Polymerization is through a polycondensation reaction of the monomers (done immediately after esterification/transesterification) with water as the byproduct [101].

Styrene Acrylonitrile: Styrene acrylonitrile (SAN) polymer has very high transparency and high gloss and can be colored with a variety of pigments. It maintains its gloss even at low temperatures. The chemical structure of SAN is $(C_8H_8)_n$-$(C_3H_3N)_m$. The copolymerization of acrylonitrile with styrene improves the heat and chemical resistance compared to styrene. The material is more rigid and harder than polystyrene and has higher scratch resistance. The polar acrylonitrile content increases the moisture absorption and lowers the electrical properties compared to polystyrene. SAN can have a yellow tint, which can be disguised with the use of blue tinting agents [102].

Acrylate Styrene Acrylonitrile: Acrylate styrene acrylonitrile (ASA) is structurally very similar to ABS. The spherical particles of slightly crosslinked acrylate rubber (instead of butadiene rubber), functioning as an impact modifier, are chemically grafted with styrene-acrylonitrile copolymer chains, and embedded in the styrene-acrylonitrile matrix. The acrylate rubber differs from the butadiene-based rubber by the absence of double bonds, which gives the material about ten times the weathering resistance and resistance to UV radiation of ABS, higher long-term heat resistance, and better chemical resistance. ASA is significantly more resistant to environmental stress cracking than ABS, especially to alcohols and many cleaning agents. n-Butyl acrylate rubber is usually used, but other esters can be encountered too, for example, ethyl hexyl acrylate. ASA has a lower glass transition temperature than ABS, 100 °C vs. 105 °C, providing better low-temperature properties to the material. ASA is a transparent polymer and has excellent resistance to UV light. It has excellent durability under a wide range of temperatures and environmental conditions, with minimal change in its gloss. ASA has the highest temperature resistance among the styrenics [103].

Methacrylate Acrylonitrile Butadiene Styrene: Methacrylate acrylonitrile butadiene styrene (MABS) is a clear, transparent material with thermal and mechanical properties equivalent to ABS. The transparency is achieved by matching the refractive indices of the matrix resin (the transparent acrylate-acrylonitrile-styrene polymer) with the polybutadiene rubber impact modifier. When the refractive indices match, light passes through the material. MABS is an amorphous thermoplastic with the same shrinkage as ABS and polycarbonate. They can be used in the same molds as these materials. MABS adheres easily to PVC by solvent bonding. MABS has the highest impact resistance of all the styrenics [104].

Styrene–Butadiene Copolymer: Styrene–butadiene copolymer (SBC) describes families of synthetic rubbers derived from styrene and butadiene. These materials have the good abrasion resistance and good aging stability when protected by additives. Engineering thermoplastic materials of styrene-butadiene copolymers are obtained when the styrene content is 70%. The materials are transparent, melt processable, and have excellent colorability. These polymers also have a good balance of stiffness, rigidity, and toughness. SBCs have the lowest densities (and hence the lightest parts) and the lowest moisture absorption compared to the other styrenics [105].

Cyclo Olefin Copolymers: Another interesting and growing class of transparent polymers is cyclo olefin polymers (COP) and cyclo olefin copolymers (COC). COC is a class of amorphous, transparent, and thermoplastic polymers and was introduced in 1994 by Hoechst. They were initially produced in a metallocene catalyzed reaction of cyclo olefins. Modern species are also made in a ring-opening metathesis polymerization (ROMP). Depending on the substituents, the properties of the obtained polymers vary widely. The initially developed type was the norbornene-type, which was based on the copolymerization of a (substituted) norbornene monomer or its higher homologs (e.g., tetracyclododecene) with a 1-olefine. This type of COC is commercialized by the name Topas by the German company Topas Advanced Polymers GmbH. A similar type of COC is known under the trade name Apel and is distributed through Mitsui Chemicals (Japan). The second group of polymers related to the COP was engineered by ZEON Corporation (Japan). They are made of cyclopentadiene derivates in a ROMP followed by hydrogenation. These polymers are sold as Zeonex (optical products) or Zeonor (molding products). Typical applications are lenses, packaging films, display foils, medical compounds, blister packaging, optical storage devices, capacitor films, or photovoltaic backsheet [106].

Thermoplastic Polyurethane Elastomers: Thermoplastic polyurethane elastomers (TPU) belong to the class of PUR, which represent a broad field of polymers covering applications like foams, paints, soles, pipes, gaskets, dashboards, or skis. PURs are usually made by a continuous polyaddition reaction of polyisocyanates and polyalcohols forming chains connected through the characteristic urethane group [107].

Silicones: Silicones are own class of synthetic polymeric compounds containing silicon. More precisely called polymerized siloxanes or polysiloxanes, silicones consist of an inorganic silicon-oxygen backbone chain (–SiO–Si–O–Si–O–)n with two organic groups attached to each silicon center. Unlike other polymers, this product family has silicon, not carbon, along the main chain. The pendant side groups can be aliphatic, aromatic, or fluorinated. Most commercially available silicones contain methyl groups and are called polydimethylsiloxanes. Silicones are also known as siloxanes, polyorganosiloxanes, or polysiloxanes. Silicones can be used from temperatures as low as 210 °C to as high as 250 °C. Silicones are transparent, hydrophobic, and resistant to UV and gamma radiation, have excellent electrical properties, a low dielectric constant, and high gas permeability, and are chemically inert and resistant to most chemicals. Silicone elastomers have relatively low tear strengths and abrasion resistance and are highly permeable to gases and hydrocarbons [108].

14.5.3 Influence of Nano-metal Oxides in Polymer Transparency

Titan Oxide: Titan oxide (TiO_2) on the microscale is often used as a white colorant for polymers because it scatters all visible wavelengths. Nano-TiO_2 in transparent polymers causes absorption primarily in the UV region and is therefore often used as a UV absorber for wavelengths below 400 nm. Additionally, influences on the refractive index and the mechanical and thermal properties have been reported. Nussbaumer et al. observed an increase in the refractive index in poly(vinyl alcohol) and Tao et al. raised the refractive index of PGMA (poly(glycidyl methacrylate)) and epoxy which also increased the Abbe number of the material [109].

Zinc Oxide and Cerium Oxide: These kinds of NPs are utilized to produce luminescent nanocomposites. For cerium ions, the relationships among blue shift, the valence state of ions, and particle size are described. Regarding other effects on the composite's properties, an increase of 26 °C in the glass transition temperature, an increase in the thermal stability, and an increase in the refractive index were observed with a PMMA matrix. The transmittance of a rather thin specimen (0.4 mm thickness, 0.5 wt% ZnO quantum dot (QD)) is high. On the other hand, thicker specimens have a low transmittance in the visual range (thickness: 3.5 mm, 1 wt% ZnO NP) or simply a transparent appearance (thickness: 3 mm, ZnO QD). A remarkable effect of the transmission of composites made of transparent polystyrene (PS) and CeO_2 NPs was observed by Parlak and Demir. The transmittance of the thin films showed a first-order exponential decay up to a particle concentration of 20 wt%, as the Rayleigh scattering theory proposes. Higher filling degrees do not impair further the transmission. This fact is explained as a result of interference in the multiple instances of light scattering by the quasi-ordered internal microstructure, developing at high filling degrees [110].

Tin Oxide: The conductivity and antistatic behavior of antimony-doped tin oxide (ATO) NPs dispersed in transparent polymers is described by Wakabayashi et al. and Sun et al. These examined films exhibit antistatic properties above a threshold value of only 0.2 vol%. This is explained by the forming of chain-like aggregates in the NPs. Modified indium tin oxide NPs affect the transparency such that both a UV and IR absorption occur in poly(urethane acrylate) while showing high transmittance values between 500 and 1000 nm (thickness: 0.1 mm) [111].

Iron Oxide and Cobalt: Iron oxide (Fe_3O_4) NPs and cobalt (Co) NPs can lead to a ferromagnetic behavior in a composite. Peluso et al. show that composites starting as iron (II) mercaptides show higher transmittance values in the range from 400 to 900 nm than composites starting as iron (III) mercaptides. High transparency in the composites is observed for wavelengths over 550 or 700 nm [112].

Aluminum Oxide and Zirconium Oxide: Aluminum oxide (Al_2O_3) NPs have a primarily positive effect on the mechanical properties of the composite. They increase the elastic modulus, the yield stress, and the impact strength. According to Sarwar et al., the optimum loading for the best mechanical properties is 2.5 wt% [113]. Zirconium oxide (ZrO_2) NPs diversely affect the composite properties. They can lead to improvements to the hardness and scratch resistance (for which a high filling degree of 15 wt% is needed), enhancements to the mechanical properties, adjustments to refractive index, and improvements to thermal properties [114].

Cadmium: Composites with NPs containing cadmium (Cd) are mainly prepared to achieve luminescent properties. The luminescent properties increase with good dispersion of the NP and primarily depend on the particle concentration and the particle size. The transmittance decreases strongly below a wavelength of 550 nm, whereas the composites have a good transmittance above [115].

Zinc Sulfide and Lead Sulfide: The main purpose behind using zinc sulfide (ZnS) particles is also luminescent

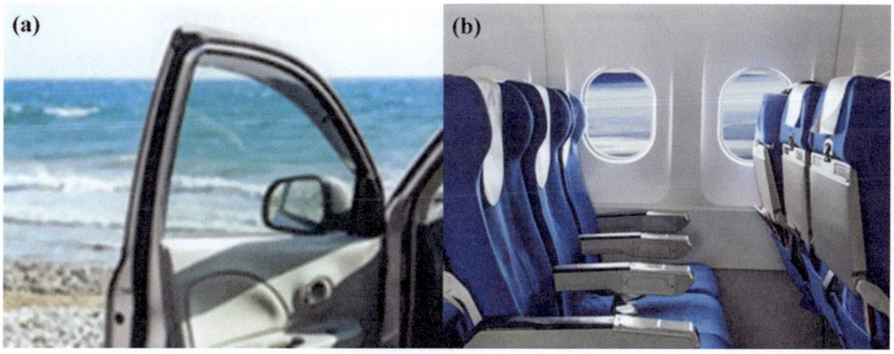

Fig. 14.31 Acrylonitrile butadiene styrene used in (**a**) Interior door material and (**b**) airplane seats

properties or high refractive index materials. Zhang et al. increased the refractive index of poly(vinylpyrrolidone) linearly from 1.5061 to 1.7523 with an increasing particle content up to 80 wt% while reducing the Abbe number. The transparency of thin films (thickness: 25 mm) above 500 nm was maintained [116]. As with ZnS, lead sulfide (PbS) is used to increase the refractive index. As such, Zimmermann et al. and Weibel et al. claim to have achieved the highest refractive index of 2.5 and 2.9 (600 nm), respectively, known for polymeric composite materials. Lu et al. offered insights into the transparence, indicating their nanocomposite films containing PbS particles exhibit good transparency (>90%) above 600 nm, although the PbS particles exhibit strong absorption below 600 nm in the UV-visible region [117].

14.5.4 Application

Vehicle Parts: ABS used in automotive trim components, automotive bumper bars. Thermoplastic polyurethane elastomer has applications including automotive instrument panels, caster wheels, and power tools. The cockpit canopy of the Lockheed Martin F-22 Raptor jet fighter is made from a piece of high optical quality polycarbonate, and is the largest piece of its type formed in the world. In vehicles, PMMA sheets are used in car windows, motorcycle wind shields, interior and exterior panels, fenders, etc. Also colored acrylic sheets are used in car indicator light covers, interior light covers, etc. It is also used for windows of a ship (salt resistance) and aviation purposes [118]. PMMA also opens new design possibilities for car manufacturers due to its pleasant acoustic properties, outstanding formability, and excellent surface hardness. Figure 14.31 shows the acrylonitrile butadiene styrene that used in interior door material and airplane seats.

Household Materials: ABS's lightweight and ability to be injection molded and extruded make it useful in manufacturing products such as drain-waste-vent pipe systems, buffer edging for furniture and joinery panels, luggage and protective carrying cases, pen housing, small kitchen appliances, and toys [119]. Polystyrene uses include protective packaging (such as packing peanuts, CD, and DVD cases), containers, lids, bottles, trays, tumblers, disposable cutlery, and in the making of models. Due to Cellulose acetate butyrates superior properties, it finds many uses in coatings including metal and wood coatings, gravure and flexographic printing inks, graphic arts, inkjet printing inks, and nail lacquer topcoats, automotive clear and base coats, and general industrial coatings. Thermoplastic poly-urethane elastomer has applications in drive belts, footwear, inflatable rafts, and a variety of extruded film, sheet, and profile applications [120]. TPU is well known for its applications in wire and cable jacketing, hose and tube, in adhesive and textile coating applications, as an impact modifier of other polymers. Transparent TPU is used in high-demanding transparent film applications like high-impact resistant glass structures. Thanks to its excellent impact and UV resistance, PMM is widely used in window and door profiles, canopies, panels, facade design, etc. It also facilitates light transmission and provides good heat insulation, hence a suitable choice for building greenhouses. PMMA is also used to build aquariums and marine centers. PMMA offers exceptional properties such as transparency, toughness, and aesthetics to produce chairs, tables, kitchen cabinets, bowls, table mats, etc, in any shape, color, or finishes. Cellulose acetate butyrate is resistant to household chemicals and can be used in the manufacture of numerous types of toys and sporting goods. Non-oriented PET sheets can be thermoformed to make packaging trays and blister packs. If crystallizable PET is used, the trays can be used for frozen dinners, since they withstand both freezing and oven baking temperatures [121]. Both amorphous PET and Bio-PET are transparent to the naked eye. Color-conferring dyes can easily be formulated into PET sheets. PET is also used as a waterproofing barrier in undersea cables. The combination of transparency and resistance to oils, fats, and cleaning agents make SAN very suitable for use in the kitchen as mixing bowls and basins and fittings for refrigerators. It is also used for the outer casings of thermally insulated jugs, for tableware, cutlery, coffee filters, jars, and beakers as well as storage containers for all kinds of foods. An additional application is in multi-trip tableware for the catering sector. Figures 14.32 and 14.33 show various household

Fig. 14.32 Applications of (**a**) polystyrene and (**b**) acrylonitrile butadiene styrene

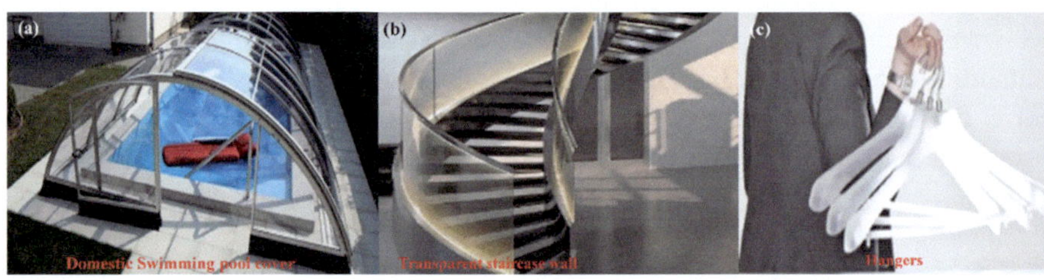

Fig. 14.33 Applications of (**a**) styrene acrylonitrile, (**b**) methacrylate acrylonitrile butadiene styrene, and (**c**) styrene-butadiene copolymer

applications of polystyrene, and acrylonitrile butadiene styrene, styrene acrylonitrile, methacrylate acrylonitrile butadiene styrene, and styrene-butadiene copolymer.

Medical Equipments: ABS used in inhalers, nebulizers, nonabsorbable sutures, tendon prostheses, drug-delivery systems tracheal tubes. Thermoplastic poly-urethane elastomer has applications in medical devices [122]. PMMA is a high purity and easy-to-clean material and hence is used to fabricate incubators, drug testing devices, storage cabinets in hospitals, and research labs. Due to its high bio-compatibility, PMMA is also applied as dental cavity fillings and bone cement. Cellulose acetate butyrate has a variety of applications in products such as textile fibers, wound dressings. The high purity, moisture barrier, clarity, and sterilization compatibility of COC resins make them an excellent alternative to glass in a wide range of medical products. Breakage prevention and weight reduction are common reasons for choosing COC in these applications. COC has very low energy and nonreactive surface, which can extend the shelf life and purity of medications such as insulin and other protein drugs in applications such as vials, syringes, and cartridges. The high UV transmission of COC also drives diagnostic applications such as cuvettes and microplates. COC plays an increasingly important role in microfluidics due to its chemical resistance, clarity, and unusually high mold detail replication which makes it possible to reliably mold submicron features. Most COC grades can undergo sterilization by gamma radiation, steam, or ethylene oxide.

Electronic Equipments: ABS used in enclosures for electrical and electronic assemblies, protective headgear. Particular forms of ABS filaments are ABS-ESD (electrostatic discharge) and ABS-FR (fire-resistant), which are used in particular for the production of electrostatically sensitive components. TPU is also a popular material found in outer cases of mobile electronic devices, such as mobile phones. Polycarbonate is mainly used for electronic applications that capitalize on its collective safety features. Being a good electrical insulator and having heat-resistant and flame-retardant properties, it is used in various products associated with electrical and telecommunications hardware. It can also serve as a dielectric in high-stability capacitors. A major application of polycarbonate is the production of compact discs, DVDs, and blu-ray discs. PMMA sheets are used for designing LED lights where it helps to maximize light-emitting potential. It is also used for the construction of the

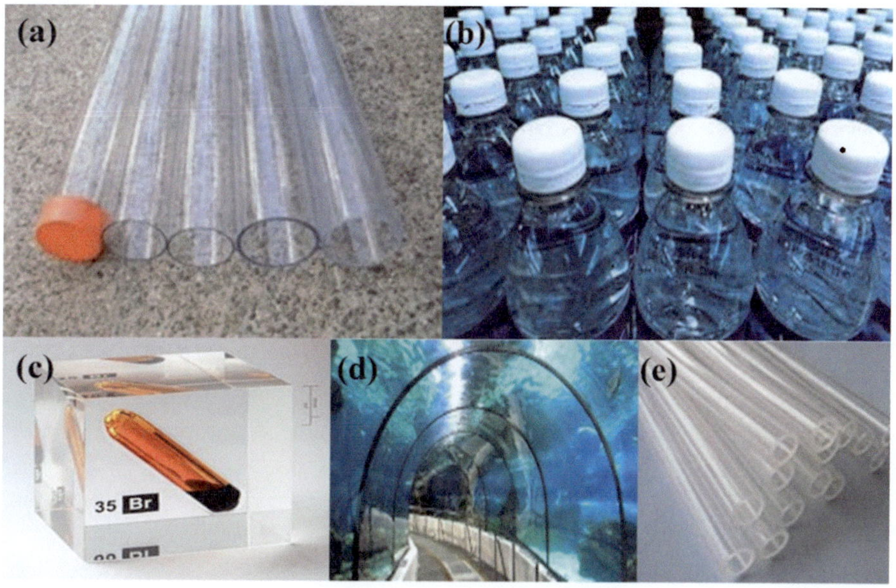

Fig. 14.34 Applications of (**a**) cellulose acetate butyrate tubes, (**b**) PET mineral water bottle, (c) PMMA shielded container holding the bromine chemical sample, (**d**) acrylate styrene acrylonitrile underwater transparent shielding, (**e**) silicone tubes

lamp. Due to its excellent optical clarity, high light transmission, and scratch resistance, PMMA is widely used in LCD/LED screens, laptops, smartphones display as well as electronic equipment displays [123]. PMMA is used in solar panels as cover materials whereas, PET is used as a substrate in thin-film solar cells.

Others: ABS used in musical instruments (recorders, plastic oboes and clarinets, and piano movements). Majority of cellulose acetate butyrate is consumed for making cigarette filters [124]. Plastic bottles made from PET are widely used for soft drinks. For certain specialty bottles, such as those designated for beer containment, PET sandwiches additional polyvinyl alcohol (PVOH) layer to further reduce its oxygen permeability. SAN is used for all types of outer covers, for example, printers, calculators, instruments, and lamps. COC is commonly extruded with cast or blown film equipment in the manufacture of packaging films. Most often, due to cost, COC is used as a modifier in monolayer or multilayer film to provide properties not delivered by base resins such as polyethylene. Grades of COC based on ethylene show a certain amount of compatibility with polyethylene and can be blended with PE via commercial dry blending equipment. These films are then used in consumer applications including food and healthcare packaging. Common applications of COC include shrink films and labels, twist films, protective or bubble packaging, and forming films. Another noted application which often relies on a high percentage of COC in the end product is pharmaceutical blister packaging [125]. These polymers are commercially used in optical films, lenses, touch screens, light guide panels, reflection films, and other components for mobile devices, displays, cameras, copiers, and other optical assemblies. Clear Silicone is a multipurpose adhesive and sealant which creates a waterproof, protective seal that is ideal for metal, glass, rubber, tile, and porcelain. It is designed to be used indoors and outdoors to repair everything from electrical connections to seams on boots to stop leaks in wet weather. ABS is used in golf club heads (because of its good shock absorbance). Thermoplastic poly-urethane elastomer has applications in sporting goods [126, 127]. Figure 14.34 shows the applications of cellulose acetate butyrate tubes, PET mineral water bottle, PMMA shielded container holding the bromine chemical sample, acrylate Styrene Acrylonitrile underwater transparent shielding, and Silicone tubes.

14.6 Bioplastic

The use of bioplastic is much more necessary in this current state of the art to go all out as a green technology to save the immunity of our world. Bioplastics possess a unique potential and thus, can help reduce greenhouse gas emissions effectively. Bioplastic can be defined as the plastic that is directly or indirectly derived from the biomass [128]. Bioplastics derived from agricultural byproducts (vegetable fats and oils, corn starch, straw, wood chips, sawdust, recycled food waste, etc), sugar derivatives (starch, cellulose, and lactic acid), microorganism (microorganism derivative plastic bottles and other containers), fossil-fuel (fossil-fuel plastics). Within the term bioplastics, it distinguishes (1) bio-based plastic and (2) biodegradable plastics, but bioplastic can also fulfill both of these criteria [129, 130]. Bio-based plastics are typically made from renewable sources by the action of living organisms. They can be polysaccharides (e.g., starches, such as thermoplastic starch (TPS); cellulose, such as regenerated cellulose; pectin, and chitin), proteins (e.g., wheat gluten, wool, silk, casein, and gelatine), lipids (e.g., animal fats, plant oils), or products of microorganisms (e.g., poly(hydroxyalkanoate)s (PHAs)

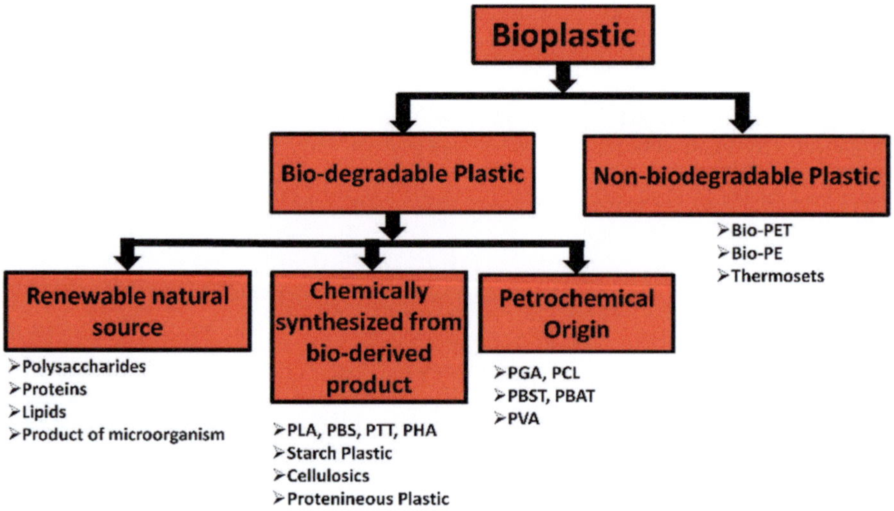

Fig. 14.35 Classification of bioplastic

such as poly(hydroxybutyrate) (PHB)). Furthermore, bio-based plastics can also be chemically synthesized from bio-derived products (e.g., poly(lactic acid) (PLA), poly (butylene succinate) (PBS), and poly(trimethylene terephthalate) (PTT)). Moreover, there is another fraction of plastics that is bio-based but not biodegradable that is called "drop-ins" (e.g., bio-PET, bio-PE, bio-PP) with identical features as their petrochemical ancestors. Moreover, biodegradable plastics, with a certain degree of biodegradability, can also be synthesized from petrochemical origins, such as poly(glycolic acid)(PGA), poly(caprolactone) (PCL), poly (butylene succinate-coterephthalate) (PBST), poly(butylene adipate-co-terephthalate) (PBAT) and poly(vinyl alcohol) (PVA). The classification of different bioplastics is illustrated in Fig. 14.35.

14.6.1 Market Growth of Bioplastic

While plastics based on organic materials were manufactured by chemical companies throughout the twentieth century, the first company solely focused on bioplastics—Marlborough Biopolymers—was founded in 1983. However, Marlborough and other ventures that followed failed to find commercial success, with the first such company to secure long-term financial success being the Italian company Novamont, founded in 1989. Currently, the global bioplastics market size was valued at USD 8.3 billion in 2019 and is expected to register a CAGR of 16.1% from 2020 to 2027 [131]. Growing population and urbanization coupled with the increasing awareness regarding health issues in emerging economies of Asia Pacific are likely to assist an end-use industry, which in turn is expected to escalate the demand for bioplastics over the forecast period. Moreover, rising product demand in the packaging industry will drive the market further. Bioplastics are also used in other application industries including agriculture, consumer goods, textile, automotive, and construction (Fig. 14.36). Non-biodegradable plastics segment led the market with a share of 54.96% in 2019 and will maintain its leading position over the forecast years. This growth is attributed to the high demand for electronic equipment, automotive housings, and consumer goods, such as carry bags, cups, films, and bottles. The packaging application led the market with a share of 62.4% in 2019. Increasing utilization of bioplastics to produce bags for compost, agricultural foils, horticultural products, nursery products, toys, and textiles is the key factor responsible for the segment growth. Moreover, bioplastics are used to produce disposable cups, bowls, plates, and food containers. The development of food-grade bioplastics has improved the shelf life of the food products, which drives their demand, thereby boosting the segment growth. Moreover, the majority of packaging products are

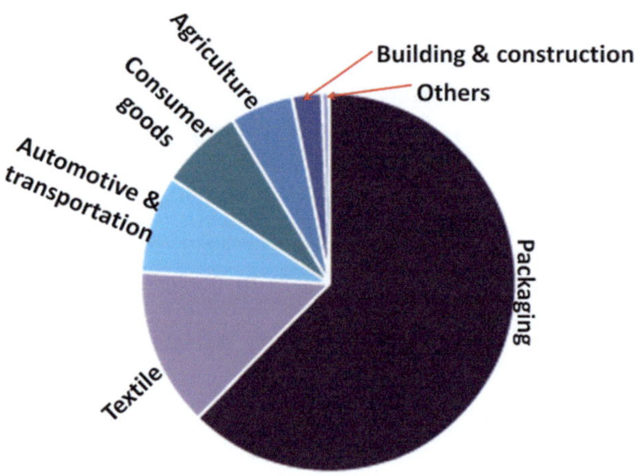

Fig. 14.36 Global bioplastics matrix share in the year 2019

currently produced from conventional plastics and end up in landfills, which is a situation that various governments intend to address. Bioplastics are also used to produce semiconductor casings, electronic device covers, connectors, and circuit boards. With the help of injection molding, bioplastics can be used to produce touch screen computer casings, headphones, loudspeakers, keyboards, laptops, games consoles, mobile phone casings, vacuum cleaners, etc. The growing demand for electrical and electronic products along with rising environmental concerns is expected to drive the demand in the consumer goods application segment. Some of the prominent companies in the bioplastics global market are Teijin Ltd., Toray Industries, Toyota Tsusho, M&G Chemicals, PTT Global Chemical Public Company Ltd., Showa Denko K. K., Nature Works LLC, SABIC, BASF SE, Futerro SA, E.I. du Pont de Nemours and Company, Braskem, Corbion N.V., Galactic, Solvay SA [131].

14.6.2 Biodegradation of Bioplastics

The categorization of bioplastic from the plastic is depending on the easiness of the biodegradation process. Biodegradation of any plastic is a process that happens at the solid/liquid interface whereby the enzymes in the liquid phase depolymerize the solid phase. Both bioplastics and conventional plastics containing additives are able to biodegrade. Bioplastics are able to biodegrade in different environments hence they are more acceptable than conventional plastics. Biodegradability of bioplastics occurs under various environmental conditions including soil, aquatic environments, and compost. Both the structure and composition of biopolymer or bio-composite have an effect on the biodegradation process, hence changing the composition and structure might increase biodegradability. Soil and compost as environmental conditions are more efficient in biodegradation due to their high microbial diversity. Composting not only biodegrades bioplastics efficiently but also significantly reduces the emission of greenhouse gases. Biodegradability of bioplastics in compost environments can be upgraded by adding more soluble sugar and increasing temperature. Soil environments on the other hand have a high diversity of microorganisms making it easier for biodegradation of bioplastics to occur. However, bioplastics in soil environments need higher temperatures and a longer time to biodegrade. Some bioplastics biodegrade more efficiently in water bodies and marine systems; however, this causes danger to marine ecosystems and freshwater. Hence it is accurate to conclude that biodegradation of bioplastics in water bodies which leads to the death of aquatic organisms and unhealthy water can be noted as one of the negative environmental impacts of bioplastics [132].

14.6.3 Types of Bioplastics

14.6.3.1 Polysaccharides Bioplastic

Polysaccharides are the most abundant macromolecules in the biosphere. These complex carbohydrates constituted by glycosidic bonds are often one of the main structural elements of plants (e.g., cellulose) and animal exoskeletons (e.g., chitin), or have an important role in plant energy storage (e.g., starch). A high variety of polysaccharides and their derivatives have been used to produce biodegradable films and thin membranes and used in several industries, such as food, medical, pharmaceutical, and specific industrial processes (e.g., pervaporation). Polysaccharide-based membranes have been widely used in the food industry in packaging and edible coatings [133]. Polysaccharide membranes are generally attractive due to their good barrier against oxygen and carbon dioxide (at low or moderate relative humidity) and good mechanical properties. However, their major drawback is related to their low barrier against water vapor due to their hydrophilic nature. Starch is the most abundant reserve polysaccharide in plants. As such, it is a renewable resource, biodegradable, produced in abundance at low cost, easy-to-handle, and can exhibit thermoplastic behavior. Starch granules are insoluble in cold water and are composed of two types of glucose polymers: amylose (the linear polymer which comprises approximately 20% of starch granules) and amylopectin (the branched polymer). Starch properties depend directly on the botanical source, granule size distribution and morphology, genotype, amylose/amylopectin ratio, and other factors such as composition, pH, and chemical modifications. Galactomannans are neutral polysaccharides obtained from the endosperm of dicotyledonous seeds of several plants, particularly the Leguminosae, where they function as carbohydrate reserves [134]. The three major galactomannans with interest in food and non-food industries are guar gum (Cyamopsis tetragonolobo), tara gum (Caesalpinia spinosa), and locust bean gum (Ceratonia siliqua). However, just locust bean gum and guar gum are considered commercially interesting due to their availability and price. These natural polysaccharides are commonly used in the food industry, mainly as stabilizers, thickeners, and emulsion stabilizers, as well as for the production of edible membranes and coatings. Cellulose is the most abundant occurring natural polymer on earth, being the predominant constituent in the cell walls of all plants. Cellulose is composed of a unique monomer: glucose under its β-D-glucopyranose form. Due to its regular structure and array of hydroxyl groups, it tends to form strong hydrogen-bonded crystalline microfibrils and fibers and is most familiar in the form of paper, paperboard, and corrugated paperboard in the packaging context [135]. Carrageenan is a naturally occurring hydrophilic, anionic sulfated linear polysaccharide extracted from red seaweeds, specifically from the

Rhodophyceae family (e.g., Chondrus crispus, Kappaphycus spp., Eucheuma spp., and Gigartina stellata). Carrageenan is approved as a food-grade additive, and it has been used mainly as an emulsifier and stabilizer in flavored milk, dairy products, pet food, dietetic formulas, and infant formulas [136]. Alginate is a linear polysaccharide that is abundant in nature and is synthesized by brown seaweeds (e.g., *Laminaria digitata* and *Ascophyllum nodosum*) and some soil bacteria. They are film-forming compounds because of their non-toxicity, biodegradability, biocompatibility, and low cost. In addition, other functional properties have been studied, such as thickening, stabilizing, suspending, and gel-producing among others [137]. Pullulan is a linear, water-soluble and neutral exopolysaccharide (EPS), can be used as a food additive, as a flocculent agent, or even as a blood plasma substitute, beyond film-forming agent. Pullulan membranes are edible, homogeneous, transparent, printable, heat sealable, flexible, and good barriers to oxygen. These properties and the fact of pullulan membranes inhibit fungal growth, make them a good material for food applications [138]. Gellan gum is an anionic water-soluble exopolysaccharide, produced by *Sphingomonas elodea*. Gellan gum was identified as a product with potential commercial value by Kelco during an extensive screening program of soil and water bacteria. In its original form (high acyl gellan), gellan gum has two acyl substituents (acetate and glycerate). Low acyl gellan gum is obtained with removal of acyl groups. High acyl gellan forms soft, elastic, non-brittle, thermo-reversible gels, and low acyl gellan tends to form firm, non-elastic brittle, and thermostable gels [139]. Xanthan gum is an exopolysaccharide produced by *Xanthomonas campestris* using glucose and sucrose as a sole carbon sources. It was the second microbial polysaccharide commercialized. FucoPol is a high molecular weight exopolysaccharide. This biodegradable, anionic, and water-soluble heteropolysaccharide is composed of fucose, galactose, glucose, glucuronic acid, and acyl groups, which account for 12-18 wt % of the FucoPol dry weight [140].

14.6.3.2 Proteins Bioplastic

A protein-based material could be defined as a stable 3-D macromolecular network stabilized and strengthened by hydrogen bonds, hydrophobic interactions, and disulfide bonds. However, as proteins themselves do not have sufficient plasticity to be handled, a plasticizer is required. Plasticizers are molecules with low molecular weight and volatility, which modify the 3-D structure of proteins by reducing the intermolecular forces and increasing the polymer chain's mobility. Bioplastics can be made from proteins from different sources such as wheat gluten and casein show promising properties as a raw material for different biodegradable polymers. Additionally, soy protein is being considered as another source of bioplastic. Soy proteins have been used in plastic production for over one hundred years. For example, body panels of an original Ford automobile were made of soy-based plastic. There are difficulties with using soy protein-based plastics due to their water sensitivity and relatively high cost. Therefore, producing blends of soy protein with some already available biodegradable polyester improves the water sensitivity and cost [141].

14.6.3.3 Poly(hydroxybutyrate) (PHB) Bioplastic

The biopolymer poly-3-hydroxybutyrate (PHB) is polyester produced by certain bacteria processing glucose, corn starch, or wastewater. Its characteristics are similar to those of petroplastic polypropylene. PHB production is increasing. The South American sugar industry, for example, has decided to expand PHB production to an industrial scale. PHB is distinguished primarily by its physical characteristics. It can be processed into a transparent film with a melting point higher than 130 °C, and is biodegradable without residue [142].

14.6.3.4 Poly(lactic acid) (PLA) Bioplastic

Polylactic acid (PLA) is transparent plastic produced from corn or dextrose. Superficially, it is similar to conventional petrochemical-based mass plastics like PS. It has the distinct advantage of degrading to nontoxic products. Unfortunately, it exhibits inferior impact strength, thermal robustness, and barrier properties (blocking air transport across the membrane). PLA and PLA blends generally come in the form of granulates with various properties, and are used in the plastics processing industry for the production of films, fibers, plastic containers, cups, and bottles. PLA is also the most common type of plastic filament used for home fused deposition modeling [143].

14.6.3.5 Poly(butylene succinate) (PBS) Bioplastic

Bio-PBS is biodegradable plastic that decomposes into water and carbon dioxide with the microorganism under the soil. PBS has a higher cost compared to traditionally used petroleum polymers that are nonbiodegradable. In the particular case of composite formulations, the higher cost of the polymers can be lowered with the addition of inexpensive fillers. PBS is a very promising biopolymer because its mechanical properties are comparable with those of widely used high-density polyethylene and isotactic polypropylene [144].

14.6.3.6 Poly(trimethylene terephthalate) (PTT) Bioplastic

Poly(trimethylene terephthalate) is one of the most important commercial polyesters. It is a semi-aromatic, semicrystalline thermoplastic that can be easily molded, thermoformed, and spun into fibers. It has mechanical and thermophysical properties similar to polyethylene terephthalate (PET) whereas its molding properties are comparable to polybutylene terepthalate (PBT). For example, it has good tensile and flexural strength, good dimensional stability, and excellent flow and surface finish. Like PBT, it also has good chemical resistance to a broad range of chemicals, including gasoline, carbon tetrachloride, oils, fat, alcohols, glycols, diluted acids, and bases. However, it is affected by hot water and steam. Polytrimethylene terephthalate is mainly used in fibers for carpets, and textiles known as Triexta fibers. Polyester fabrics and yarns made from this fiber are strong, very elastic (springs back into shape), and have high abrasion and wrinkle resistance. Because of its good dimensional stability and finishing qualities, it is also a good choice for engineering applications such as automotive parts, mobile phone housings, and many other industrial and consumer products. However, compared to PET and PBT it is used on a much smaller scale [145].

14.6.3.7 Polyhydroxyalkanoates (PHA) Bioplastic

Polyhydroxyalkanoates are linear polyesters produced in nature by bacterial fermentation of sugar or lipids. They are produced by the bacteria to store carbon and energy. In industrial production, the polyester is extracted and purified from the bacteria by optimizing the conditions for the fermentation of sugar. More than 150 different monomers can be combined within this family to give materials with extremely different properties. PHA is more ductile and less elastic than other plastics, and it is also biodegradable. These plastics are being widely used in the medical industry [146].

14.6.3.8 Poly(glycolic acid) (PGA) Bioplastic

Poly (glycolic acid) (PGA) has a similar chemical structure as PLA but exhibits very different characteristics. As a bio-degradable polymer, PGA can degrade quickly in the natural environment. It also has a high heat distortion temperature, good mechanical properties, and gas barrier properties due to high stereoregularity. PGA may be a beneficial supplement to PLA in certain single-use or relatively short-term applications, i.e. high gas barrier packaging. The combination of PGA and PLA can be achieved via co-polymerization, physical blending and multilayer lamination. Previous researches have shown that the degradability and the mechanical properties of PLA were significantly improved by combining PGA. However, PGA has been mainly used in bio-medical applications and has not been well produced at large scales for industrial (non-medical) applications due to its relatively higher cost compared with other biodegradable polymers. Traditionally, PGA is produced from petroleum resources and the production cost is prone to be varied by the supply and the price of crude oil. The development of new technology of PGA production is of vital importance for industrial scaling-up. For example, PGA can be made from the waste gases of coal chemical plants, which can dramatically reduce its production cost and carbon emissions [147].

14.6.3.9 Poly(caprolactone) (PCL) Bioplastic

Polycaprolactone (PCL) is a biodegradable polyester with a low melting point of around 60 °C and a glass transition temperature of about −60 °C. The most common use of polycaprolactone is in the production of specialty polyurethanes. Polycaprolactones impart good resistance to water, oil, solvent, and chlorine to the polyurethane produced. This polymer is often used as an additive for resins to improve their processing characteristics and their end-use properties (e.g., impact resistance). Being compatible with a range of other materials, PCL can be mixed with starch to lower its cost and increase biodegradability or it can be added as a polymeric plasticizer to polyvinyl chloride (PVC). PCL is degraded by hydrolysis of its ester linkages in physiological conditions (such as in the human body) and has therefore received a great deal of attention for use as an implantable biomaterial. In particular, it is especially interesting for the preparation of long-term implantable devices, owing to its degradation which is even slower than that of polylactide. PCL has been widely used in long-term implants and controlled drug release applications. However, when it comes to tissue engineering, PCL suffers from some shortcomings such as slow degradation rate, poor mechanical properties, and low cell adhesion. The incorporation of calcium phosphate-based ceramics and bioactive glasses into PCL has yielded a class of hybrid biomaterials with remarkably improved mechanical properties, controllable degradation rates, and enhanced bioactivity that are suitable for bone tissue engineering [148].

14.6.3.10 Poly(butylene succinate-co-terephthalate) (PBST) Bioplastic

Poly(butylene succinate-co-terephthalate) is known for being flexible and tough which makes it ideal for combination with other biodegradable polymers that have high modulus and strength, but are very brittle. It can be applied as garbage bags, wrapping films, disposable plastic products (lunch boxes, dishes, cups, etc.) [149].

14.6.3.11 Poly(butylene adipate-terephthalate) (PBAT) Bioplastic

Poly(butylene adipate-terephthalate) is a biodegradable random copolymer, specifically a copolyester of adipic acid, 1,4-butanediol, and terephthalic acid (from dimethyl terephthalate). PBAT is produced by many different manufacturers and may be known by the brand names ecoflex®, Wango, Ecoworld, Eastar Bio, and Origo-Bi. It is also called poly(butylene adipate-co-terephthalate) and sometimes polybutyrate-adipate-terephthalate (a misnomer) or even just "polybutyrate." It is generally marketed as fully biodegradable alternative to low-density polyethylene, having many similar properties including flexibility and resilience, allowing it to be used for many similar uses such as plastic bags and wraps. The structure of the PBAT polymer indicates the block copolymer due to the common synthetic method of first synthesizing two copolymer blocks and then combining them. PBAT is marketed commercially as fully biodegradable plastic, with BASF's ecoflex® showing 90% degradation after 80 days in testing. Particular applications that are highlighted by the manufacturers include cling wrap for food packaging, compostable plastic bags for gardening and agricultural use, and as water-resistant coatings for other materials, as in paper cups [18]. Due to its high flexibility and biodegradable nature, PBAT is also marketed as an additive for more rigid biodegradable plastics to impart flexibility while maintaining full biodegradability of the final blend [150].

14.6.3.12 Poly(vinyl alcohol) Bioplastic

Poly(vinyl alcohol) (PVA) is a water-soluble synthetic polymer, with a backbone composed only of carbon atoms and is biodegradable under both aerobic and anaerobic conditions. This polymer can be prepared by the hydrolysis of polyvinylacetate and is one of the most important synthetic polymers used in commercial, industrial, medical, and nutraceutical applications. The promising particularity of such synthetic polymer PVA is its potential as a biodegradable polymer which is considered as an unusual trait among the other synthetic carbon-chain polymers. Indeed, PVA has been found to be the only vinyl polymer which can be utilized by some microorganisms as a source of carbon and energy [151].

14.6.3.13 Bio-PET Bioplastic

Polyethylene terephthalate (PET) is known for its use in beverage bottles. Coca-Cola introduced plant bottle technology where the PET had been made from bio-based monoethylene glycol (from sugarcane) and terephthalic acid (from petrochemicals). Bio-PET has been around for some years now. It is a plastic-based on 30% renewable raw material and 70% oil-based raw material. Coca-Cola, for example, has been using Bio-PET since 2009 in the form of the "Plant Bottle." PET Power also offers its customers the possibility to use this type of plastic. The mechanical and thermal properties of Bio-PET are similar to other oil-based PET products. However, the color tone may slightly differ, depending on the type [152].

14.6.3.14 Bio-PE Bioplastic

The basic building block (monomer) of polyethylene is ethylene. Ethylene is chemically similar to and can be derived from ethanol, which can be produced by fermentation of agricultural feedstocks such as sugar cane or corn. Bio-derived polyethylene is chemically and physically identical to traditional polyethylene. It does not biodegrade but can be recycled. The Brazilian chemicals group Braskem claims that using this method of producing polyethylene from sugar cane ethanol captures (removes from the environment) 2.15 tonnes of CO_2 per tonne of Green Polyethylene produced [152].

14.6.4 Impact of Bioplastic on the Environmental

Materials such as starch, cellulose, wood, sugar, and biomass are used as a substitute for fossil fuel resources to produce bioplastics; this makes the production of bioplastics a more sustainable activity compared to conventional plastic production. The environmental impact of bioplastics is often debated, as there are many different metrics for "greenness" (e.g., water use, energy use, deforestation, biodegradation, etc.). Bioplastic production significantly reduces greenhouse gas emissions and decreases nonrenewable energy consumption. Firms worldwide would also be able to increase the environmental sustainability of their products by using bioplastics. Although bioplastics save more nonrenewable energy than conventional plastics and emit less greenhouse gas compared to conventional plastics, bioplastics also have negative environmental impacts such as eutrophication and acidification. Bioplastics induce higher eutrophication potentials than conventional plastics. Biomass production during industrial farming practices causes nitrate and phosphate to filtrate into water bodies; this causes eutrophication which is the richness of the nutrients in body waters. Eutrophication is a threat to water resources around the world since it kills aquatic organisms, creates dead zones, and causes harmful algal blooms. Bioplastics also increase acidification. The high increase in eutrophication and acidification caused by bioplastics is also caused by using chemical fertilizer in the cultivation of renewable raw materials to produce bioplastics [153].

Other environmental impacts of bioplastics include exerting lower human, terrestrial ecotoxicity, and carcinogenic potentials compared to conventional plastics. However, bioplastics exert higher aquatic ecotoxicity than conventional materials. Bioplastics and other bio-based materials increase stratospheric ozone depletion compared to conventional plastics; this is a result of nitrous oxide emissions during fertilizer application during industrial farming for biomass production. Artificial fertilizers increase nitrous oxide emissions especially when the crop does not need all the nitrogen. Minor environmental impacts of bioplastics include toxicity through using pesticides on the crops used to make bioplastics. Bioplastics also cause carbon dioxide emissions

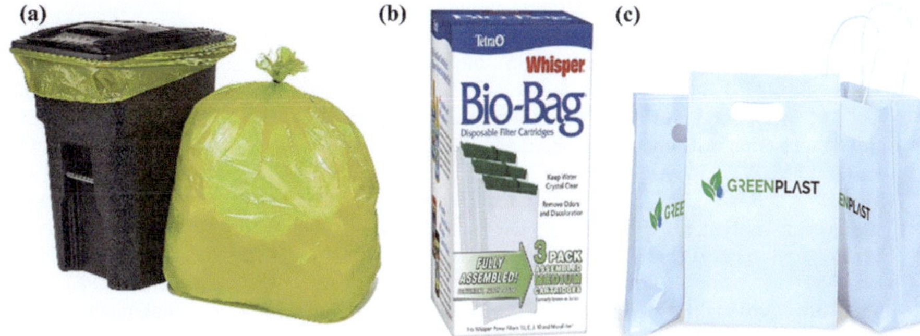

Figure 14.37 (**a**) Biodegradable garbage bag, (**b**) whisper bio-bag disposable, and (**c**) greenplast products

from harvesting vehicles. Other minor environmental impacts include high water consumption for biomass cultivation, soil erosion, soil carbon losses, and loss of biodiversity, and they are mainly are a result of land use associated with bioplastics.

14.6.5 Applications

Bioplastics are used for disposable items, such as packaging, crockery, cutlery, pots, bowls, and straws. Few commercial applications exist for bioplastics. In principle they could replace many applications for petroleum-derived plastics, however, cost and performance remain problematic. As a matter of fact, their usage is financially favorable only if supported by specific regulations limiting the usage of conventional plastics. Typical is the example of Italy, where biodegradable plastic bags and shoppers are compulsory since 2011 with the introduction of a specific law [154]. Beyond structural materials, electroactive bioplastics are being developed that promise to be used to carry electric current. Biopolymers are available as coatings for paper rather than the more common petrochemical coatings. There has been a marked increase in the use of biodegradable materials for a range of applications. Some of the applications include compostable bags, packaging, agriculture and horticulture, medical devices, consumer electronics, automotive. Figure 14.37 shows the bioplastic application in a biodegradable garbage bags, whisper bio-bag disposable, and green plant products.

Biomedical applications of vegetable oil-based polymeric materials are not restricted to the industrial arena and can also be extended to a variety of biomedical applications as surgical sealants and glues, pharmacological patches, wound healing devices, and drug carriers to scaffolds for tissue engineering. This is mainly because vegetable oil is a biobased raw material that can be metabolized in the human body, and therefore materials derived from them are potentially biocompatible. Moreover, the incorporation of vegetable oil moiety can enhance the biodegradation of the material. In fact, they have been attractive for several biomedical applications requiring materials that range from soft to hard. For example, poly(glycerol sebacate) has been used in several soft tissue engineering applications such as retinal, nerve, vascular and myocardial repair, as well as an adhesive sealant. Importantly, the two monomers, glycerol, and sebacic acid are both endogenous compounds found in human metabolism. In this line, poly(propylene sebacate)-based polyesters were found to be promising shape memory polymers with tunable switching temperatures. In vitro fibroblast response and degradation demonstrated that these polymers are potentially biocompatible and biodegradable, rendering them as a good candidates for biomedical devices such as different stents [155].

14.7 Summary and Future Prospects

Advanced plastic is also has taken a competitive position in the current edge technology. This chapter successfully discussed on various advanced plastic and their application in day-to-day life. All of these advanced plastic will lead the world with reducing the weight and increasing the efficiency of the system.

References

1. Kurdi, A., Li, C.: Recent advances in high-performance polymers-tribological aspects. Lubricants. **7**(1), 2 (2019). https://doi.org/10.3390/lubricants7010002
2. Coleman, J.N., Cadek, M., Blake, R., Nicolosi, V., Ryan, K.P., Belton, C., Fonseca, A., Nagy, J.B., Gun'ko, Y.K., Blau, W.J.: High-performance nanotube-reinforced plastics: understanding the mechanism of strength increase. Adv. Funct. Mater. **14**(8), 791–798 (2004). https://doi.org/10.1002/adfm.200305200
3. Sikder, A.K., Nirmala Sikder, A.: Review of advanced high performance, insensitive and thermally stable energetic materials emerging for military and space applications. J. Hazard. Mater. **112**(1–2), 1–15 (2004). https://doi.org/10.1016/j.jhazmat.2004.04.003
4. Bourbigot, S., Flambard, X.: Heat resistance and flammability of high-performance fibres: a review. Fire and Mater. **26**(4, 5, Special Issue: Fire behaviour of textiles), 155–168 (2002). https://doi.org/10.1002/fam.799
5. See: https://www.marketresearchfuture.com/reports/high-performance-plastics-market-1987. 02.02.2020

6. Subrahmanian, K.P.: High-temperature polymers and adhesives. In: Hartshorn, S.R. (ed.) Structural adhesives. Topics in applied chemistry. Springer, Boston, MA (1986). https://doi.org/10.1007/978-1-4684-7781-8_8
7. Iye, M., Moorwood, A.F.M.: Proceedings Volume 4008. Optical and IR telescope instrumentation and detectors. (2000). https://doi.org/10.1117/12.395459.
8. Murari, A., Barzon, A.: Ultra high vacuum properties of some engineering polymers. IEEE Trans. Dielectr. Electr. Insulation. **11**(4), 613–619 (2004). https://doi.org/10.1109/TDEI.2004.1324351
9. See: http://www.tkmna.com/wcm/idc/groups/public/documents/web_content/mdaw/mdi5/~edisp/pw022787.pdf. 02.02.2020
10. See: https://www.roncelli.com/peek-datasheet. 02.02.2020
11. Murari, A., Vinante, C., Monari, M.: Comparison of PEEK and VESPEL®SP1 characteristics as vacuum seals for fusion applications. Vacuum. **65**(2), 137–145 (2002). https://doi.org/10.1016/S0042-207X(01)00420-1
12. Rosato, D.V., Di Mattia, D.P., Rosato, D.V.: The properties of plastics. In: Designing with plastics and composites: a handbook. Springer, Boston, MA (1991). https://doi.org/10.1007/978-1-4615-9723-0_6
13. Wang, Y., Goh, S.H., Chung, T.S.: Miscibility study of Torlon® polyamide-imide with Matrimid 5218 polyimide and polybenzimidazole. Polymer. **48**(10), 2901–2909 (2007). https://doi.org/10.1016/j.polymer.2007.03.040
14. Sandor, R.B.: PBI (polybenzimidazole): synthesis, properties and applications. High Perform. Polym. **2**(1), 25–37 (1990). https://doi.org/10.1177/152483999000200103
15. Moon, J.D., Bridge, A.T., D'Ambra, C., Freeman, B.D., Paul, D.R.: Gas separation properties of polybenzimidazole/thermally-rearranged polymer blends. J. Membr. Sci. **582**, 182–193 (2019). https://doi.org/10.1016/j.memsci.2019.03.067
16. Zheng, J., et al.: The tribological properties of several PTFE based self-lubricating composites in vacuum, advanced materials research, vol. 535–537, pp. 144–148. Trans Tech Publications, Ltd., Freienbach (2012). https://doi.org/10.4028/www.scientific.net/amr.535-537.144
17. Robert, G.C., Pope, R.D.: A virtually ideal production system: specifying and estimating the VIPS model. Am. J. Agric. Econ. **76**(1), 105–113 (1994). https://doi.org/10.2307/1243925
18. Beever, W., Ryan, C., O'Connor, J., Lou, A.: Ryton®-PPS carbon fiber reinforced composites: the how, when, and why of molding. In: Johnston, N. (ed.) Toughened composites, pp. 319–327. ASTM International, West Conshohocken, PA (1987). https://doi.org/10.1520/STP24385S
19. See: https://www.solvay.jp/ja/binaries/Ryton-PPS-Design-Guide_EN-205287.pdf. 02.02.2020.
20. Lyu, M.-Y., Choi, T.G.: Research trends in polymer materials for use in lightweight vehicles. Int. J. Precision Eng. Manuf. **16**(1), 213–220 (2015). https://doi.org/10.1007/s12541-015-0029-x
21. Darvishmanesh, S., Tasselli, F., Jansen, J.C., Tocci, E., Bazzarelli, F., Bernardo, P., Luis, P., Degrève, J., Drioli, E., Van der Bruggen, B.: Preparation of solvent stable polyphenylsulfone hollow fiber nanofiltration membranes. J. Membr. Sci. **384**(1–2), 89–96 (2011). https://doi.org/10.1016/j.memsci.2011.09.003
22. Johannes, C.J., Darvishmanesh, S., Tasselli, F., Bazzarelli, F., Bernardo, P., Tocci, E., Friess, K., Randova, A., Drioli, E., Van der Bruggen, B.: Influence of the blend composition on the properties and separation performance of novel solvent resistant polyphenylsulfone/polyimide nanofiltration membranes. J. Membr. Sci. **447**, 107–118 (2013). https://doi.org/10.1016/j.memsci.2013.07.009
23. Wan, H., Wang, N., Yang, J., Si, Y., Chen, K., Ding, B., Sun, G., El-Newehy, M., Al-Deyab, S.S., Yu, J.: Hierarchically structured polysulfone/titania fibrous membranes with enhanced air filtration performance. J. Colloid Interface Sci. **417**, 18–26 (2014). https://doi.org/10.1016/j.jcis.2013.11.009
24. Puts, G.J., Crouse, P., Ameduri, B.M.: Polytetrafluoroethylene: synthesis and characterization of the original extreme polymer. Chem. Rev. **119**(3), 1763–1805 (2019). https://doi.org/10.1021/acs.chemrev.8b00458
25. Sperati, C.A., Starkweather, H.W.: Fluorine-containing polymers. II. Polytetrafluoroethylene. In: Fortschritte der hochpolymeren-forschung. Advances in polymer science, vol. 2/4. Springer, Berlin (1961). https://doi.org/10.1007/BFb0050504
26. Schmidt, M., Harmuth, S., Barth, E.R., Wurm, E., Fobbe, R., Sickmann, A., Krumm, C., Tiller, J.C.: Conjugation of ciprofloxacin with poly(2-oxazoline)s and polyethylene glycol via end groups. Bioconjugate Chem. **26**(9), 1950–1962 (2015). https://doi.org/10.1021/acs.bioconjchem.5b00393
27. Galante, A.M.S., Galante, O.L., Campos, L.L.: Study on application of PTFE, FEP and PFA fluoropolymers on radiation dosimetry. Nucl. Instrum. Methods Phys. Res. Sect. A. **619**(1–3), 177–180 (2010). https://doi.org/10.1016/j.nima.2009.10.103
28. See: https://miller-stephenson.com/product/ra-dry-film-dispersion/. 04.02.2020.
29. Ohashi, M., Kawakami, S., Yokogawa, Y., Lai, G.C.: Spherical aluminum nitride fillers for heat-conducting plastic packages. J. Am. Ceramics Soc. **88**(9), 2615–2618 (2005). https://doi.org/10.1111/j.1551-2916.2005.00456.x
30. Wilmer, D., Funke, K., Witschas, M., Banhatti, R.D., Jansen, M., Korus, G., Fitter, J., Lechner, R.E.: Anion reorientation in an ion conducting plastic crystal – coherent quasielastic neutron scattering from sodium ortho-phosphate. Phys. B Condensed Matter. **266** (1–2), 60–68 (1999). https://doi.org/10.1016/S0921-4526(98)01494-X
31. Bidan, G., Ehui, B., Lapkowski, M.: Conductive polymers with immobilised dopants: ionomer composites and auto-doped polymers-a review and recent advances. J. Phys. D Appl. Phys. **21**(7) (1988). https://doi.org/10.1088/0022-3727/21/7/001
32. Abu-Lebdeh, Y., Abouimrane, A., Alarco, P.J., Hammami, A., Ionescu-Vasii, L., Armand, M.: Ambient temperature proton conducting plastic crystal electrolytes. Electrochem. Commun. **6** (4), 432–434 (2004). https://doi.org/10.1016/j.elecom.2004.02.015
33. Zhong, S., Kazacos, M., Burford, R.P., Skyllas-Kazacos, M.: Fabrication and activation studies of conducting plastic composite electrodes for redox cells. J. Power Sources. **36**(1), 29–43 (1991). https://doi.org/10.1016/0378-7753(91)80042-V
34. Capacciolidag, S., Lucchesidag, M., Rolladag, P.A., Ruggeri, G.: Dielectric response analysis of a conducting polymer dominated by the hopping charge transport. J. Phys. Condensed Matter. **10** (25) (1998). https://doi.org/10.1088/0953-8984/10/25/011
35. See: https://www.sciencehistory.org/historical-profile/alan-g-macdiarmid-alan-j-heeger-and-hideki-shirakawa. 04.02.2020.
36. David, M.M., Adams, R.N., Argersinger, W.J.: Investigation of the kinetics and mechanism of the anodic oxidation of aniline in aqueous sulfuric acid solution at a platinum electrode. J. Am. Chem. Soc. **84**(19), 3618–3622 (1962). https://doi.org/10.1021/ja00878a003
37. Greene, R.L., Street, G.B., Suter, L.J.: Superconductivity in polysulfur nitride (SN)X. Phys. Rev. Lett. **34**, 577 (1975). https://doi.org/10.1103/PhysRevLett.34.577
38. See: https://www.gminsights.com/industry-analysis/conductive-polymers-market. 02.02.2020
39. See: https://www.transparencymarketresearch.com/conductive-polymers-market.html. 02.02.2020
40. Shacklette, L.W., Colaneri, N.F., Kulkarni, V.G., Wessling, B.: EMI shielding of intinsically conductive polymers. J. Viyl Technol. **14**(2), 118–122 (1992). https://doi.org/10.1002/vnl.730140214

41. Saleem, A., Frormann, L., Soever, A.: Fabrication of extrinsically conductive silicone rubbers with high elasticity and analysis of their mechanical and electrical characteristics. Polymers. **2**(3), 200–210 (2010). https://doi.org/10.3390/polym2030200
42. G Casalbore-Miceli, M.J., Yang, N., Camaioni, C.-M., Mari, Y., Li, H., Sun, M.L.: Investigations on the ion transport mechanism in conducting polymer films. Solid State Ionics. **131**(3–4), 311–321 (2000). https://doi.org/10.1016/S0167-2738(00)00688-3
43. Rault-Berthelot, J., Rozé, C., Granger, M.M.: Anodic oxidation of 2(9H-fluoren-9-ylidene) malononitrile and 2(9H-fluoren-9-ylidene)-2-phenylacetonitrile. Electrochemical behavior and physicochemical properties of the derived polymers. J. Electroanal. Chem. **1–2**(436), 85–101 (1997). https://doi.org/10.1016/S0022-0728(97)00244-1
44. Xia, L., Wei, Z., Wan, M.: Conducting polymer nanostructures and their application in biosensors. J. Colloid Interface Sci. **341**(1), 1–11 (2010). https://doi.org/10.1016/j.jcis.2009.09.029
45. Rudge, A., Davey, J., Raistrick, I., Gottesfeld, S., Ferraris, J.P.: Conducting polymers as active materials in electrochemical capacitors. J. Power Sources. **47**(1–2), 89–107 (1994). https://doi.org/10.1016/0378-7753(94)80053-7
46. Chiang, C.K., Fincher Jr., C.R., Park, Y.W., Heeger, A.J., Shirakawa, H., Louis, E.J., Gau, S.C., MacDiarmid, A.G.: Electrical conductivity in doped polyacetylene. Phys. Rev. Lett. **9**, 1098 (1997). https://doi.org/10.1103/PhysRevLett.39.1098. Erratum Phys. Rev. Lett. 40, 1472 (1978)
47. Grancarić, A.M., Jerković, I., Koncar, V., Cochrane, C., Kelly, F. M., Soult, D., Ligand, X.: Conductive polymers for smart textile applications. J. Ind. Textiles. **48**(3), 612–642 (2018). https://doi.org/10.1177/1528083717699368
48. Teemed, s., Heeger, A.J.: Polyacetylene, (CH) X: The Prototype Conducting Polymer. Ann. Rev. Phys. Chem. **33**, 443–469 (1982). https://doi.org/10.1146/annurev.pc.33.100182.002303
49. Mizoguchi, K., Nechtschein, M., Travers, J.-P., Menardo, C.: Spin dynamics in the conducting polymer, polyaniline. Phys Rev Lett. **63**, 66 (1989). https://doi.org/10.1103/PhysRevLett.63.66
50. Ramanavičius, A., Ramanavičienė, A., Malinauskas, A.: Electrochemical sensors based on conducting polymer-polypyrrole. Electrochim. Acta. **51**(27), 6025–6037 (2006). https://doi.org/10.1016/j.electacta.2005.11.052
51. Miyatake, K., Iyotani, H., Yamamoto, K., Tsuchida, E.: Synthesis of poly(phenylene sulfide sulfonic acid) via Poly(sulfonium cation) as a thermostable proton-conducting polymer. Macromolecules. **29** (21), 6969–6971 (1996). https://doi.org/10.1021/ma960768x
52. Clarke, T.C., Kanazawa, K.K., Lee, V.Y., Rabolt, J.F., Reynolds, J. R., Street, G.B.: Poly(phenylene sulfide) hexafluoroarsenate: a novel conducting polymer. J. Polym. Sci. **20**(1), 117–130 (1982). https://doi.org/10.1002/pol.1982.180200110
53. Lee, C.H., Yu, G., Moses, D., Heeger, A.J.: Picosecond transient photoconductivity in poly(p-phenylenevinylene). Phys. Rev. B. **49**, 2396 (1994). https://doi.org/10.1103/PhysRevB.49.2396
54. Boussoualem, M., King, R.C.Y., Brun, J.-F., Duponchel, B., Ismaili, M., Roussel, F.: Electro-optic and dielectric properties of optical switching devices based on liquid crystal dispersions and driven by conducting polymer [poly(3,4-ethylene dioxythiophene): polystyrene sulfonate (PEDOT:PSS)]-coated electrodes. J. Appl. Phys. **108**(11), 113526 (2010). https://doi.org/10.1063/1.3518041
55. González-Tejera, M.J., Sánchez de la Blanca, E., Carrillo, I.: Polyfuran conducting polymers: Synthesis, properties, and applications. Synth. Metals. **158**(5), 165–189 (2008). https://doi.org/10.1016/j.synthmet.2007.12.009
56. Yongjing, H., Shi, H., Song, H., Liu, C., Xu, J., Zhang, L., Jiang, Q.: Effects of a proton scavenger on the thermoelectric performance of free-standing polythiophene and its derivative films. Synth. Metals. **181**, 23–26 (2013). https://doi.org/10.1016/j.synthmet.2013.08.006
57. Tsukamoto, J., Takahashi, A., Kawasaki, K.: Structure and electrical properties of polyacetylene yielding a conductivity of 105 S/cm. Jpn. J. Appl. Phys. **29**(Part 1, Number 1) (1990). https://doi.org/10.1143/JJAP.29.125
58. Sailor, M., Klavetter, F., Grubbs, R., et al.: Electronic properties of junctions between silicon and organic conducting polymers. Nature. **346**, 155–157 (1990). https://doi.org/10.1038/346155a0
59. McCullough, D.R.: The chemistry of conducting polythiophenes. Adv. Mater. **10**(2), 93–116 (1999). https://doi.org/10.1002/(SICI)1521-4095(199801)10:2<93::AID-ADMA93>3.0.CO;2-F
60. Tsai, M.-H., Lin, Y.-K., Luo, S.-C.: Electrochemical SERS for in situ monitoring the redox states of PEDOT and its potential application in oxidant detection. ACS Appl. Mater. Interfaces. **11**(1), 1402–1410 (2019). https://doi.org/10.1021/acsami.8b16989
61. Wadhwa, R., Lagenaur, C.F., Cui, X.T.: Electrochemically controlled release of dexamethasone from conducting polymer polypyrrole coated electrode. J. Control. Release. **110**(3), 531–541 (2006). https://doi.org/10.1016/j.jconrel.2005.10.027
62. Balint, R., Cassidy, N.J., Cartmell, S.H.: Conductive polymers: towards a smart biomaterial for tissue engineering. Acta Biomaterialia. **10**(6), 2341–2353 (2014). https://doi.org/10.1016/j.actbio.2014.02.015
63. Huang, J.: Syntheses and applications of conducting polymer polyaniline nanofibers. Pure Appl. Chem. **78**(1) (2009). https://doi.org/10.1351/pac200678010015
64. Huang, J., Moore, J.A., Henry Acquaye, J., Kaner, R.B.: Mechanochemical route to the conducting polymer polyaniline. Macromolecules. **38**(2), 317–321 (2005). https://doi.org/10.1021/ma049711y
65. Murakami, M., Yasujima, H., Yumoto, Y., Mizogami, S., Yoshimura, S.: Electrical and structural properties of a new conducting polymer: pyrolytic poly (p-phenylene-1,3,4-oxadiazole). Solid State Communications. **45**(12), 1085–1088 (1983). https://doi.org/10.1016/0038-1098(83)91055-4
66. Toribio, F.O.: Biomimetic conducting polymers: synthesis, materials, properties, functions, and devices, polymer reviews. Conducting Polym. Biointerfaces Biomater. **53**(3), 311–351 (2013). https://doi.org/10.1080/15583724.2013.805772
67. Bargon, J., Mohmand, S., Waltman, R.J.: Electrochemical synthesis of electrically conducting polymers from aromatic compounds. IBM J. Res. Dev. **27**(4), 330–331 (1983). https://doi.org/10.1147/rd.274.0330
68. Gazotti, W.A., Nogueira, A.F., Girotto, E.M., Gallazzi, M.C., De Paoli, M.-A.: Flexible photoelectrochemical devices based on conducting polymers. Synth. Metals. **108**(2), 151–157 (2000). https://doi.org/10.1016/S0379-6779(99)00272-6
69. Guo, B., Glavas, L., Albertsson, A.-C.: Biodegradable and electrically conducting polymers for biomedical applications. Prog. Polym. Sci. **38**(9), 1263–1286 (2013). https://doi.org/10.1016/j.progpolymsci.2013.06.003
70. Heo, J.S., Eom, J., Kim, Y.-H., Park, S.K.: Recent progress of textile-based wearable electronics: a comprehensive review of materials, devices, and applications. Small. **14**(3), 1703034 (2018). https://doi.org/10.1002/smll.201703034
71. Svirskis, D., Travas-Sejdic, J., Rodgers, A., Garg, S.: Electrochemically controlled drug delivery based on intrinsically conducting polymers. J. Control. Release. **146**(1), 6–15 (2010). https://doi.org/10.1016/j.jconrel.2010.03.023
72. Guo, B., Ma, P.X.: Conducting polymers for tissue engineering. Biomacromolecules. **19**(6), 1764–1782 (2018). https://doi.org/10.1021/acs.biomac.8b00276
73. Morvan, M., Talou, T., Gaset, A., Beziau, J.F.: Electronic-nose systems for control quality applications in automotive industry. Sens. Actuators B Chem. **69**(3), 384–388 (2000). https://doi.org/10.1016/S0925-4005(00)00495-0

74. Lu, Q., Han, W.J., Choi, H.J.: Smart and functional conducting polymers: application to electrorheological fluids. Molecules. **23** (11), 2018 (2854). https://doi.org/10.3390/molecules23112854
75. Gurunathan, K., Murugan, A.V., Marimuthu, R., Mulik, U.P., Amalnerkar, D.P.: Electrochemically synthesised conducting polymeric materials for applications towards technology in electronics, optoelectronics and energy storage devices. Mater. Chem. Phys. **61** (3), 173–191 (1999). https://doi.org/10.1016/S0254-0584(99)00081-4
76. Koh, J.K., Kim, J., Kim, B., Kim, J.H., Kim, E.: Highly efficient, iodine-free dye-sensitized solar cells with solid-state synthesis of conducting polymers. Adv. Mater. **23**(14), 1641–1646 (2011). https://doi.org/10.1002/adma.201004715
77. Jinli Qiao, Takeo Hamaya, Tatsuhiro Okada: New highly proton-conducting membrane poly(vinylpyrrolidone)(PVP) modified poly (vinyl alcohol)/2-acrylamido-2-methyl-1-propanesulfonic acid (PVA–PAMPS) for low temperature direct methanol fuel cells (DMFCs). Polymer. **46**(24), 10809–10816 (2005). https://doi.org/10.1016/j.polymer.2005.09.007
78. Sengodu, P., Deshmukh, A.D.: Conducting polymers and their inorganic composites for advanced Li-ion batteries: a review. RSC Adv. **5**, 42109–42130 (2015). https://doi.org/10.1039/C4RA17254J
79. Graeme, A.S., Kao, P., Best, A.S.: Conducting-polymer-based supercapacitor devices and electrodes. J. Power Sources. **196**(1), 1–12 (2011). https://doi.org/10.1016/j.jpowsour.2010.06.084
80. Baughman, R.H.: Conducting polymer artificial muscles. Synth. Metals. **78**(3), 339–353 (1996). https://doi.org/10.1016/0379-6779(96)80158-5
81. Otero, T.F., Martinez, J.G., Arias-Pardilla, J.: Biomimetic electrochemistry from conducting polymers. A review: artificial muscles, smart membranes, smart drug delivery and computer/neuron interfaces. Electrochim. Acta. **84**, 112–128 (2012). https://doi.org/10.1016/j.electacta.2012.03.097
82. Sitaram, S.P., Stoffer, J.O., O'Keefe, T.J.: Application of conducting polymers in corrosion protection. J. Coat. Technol. **69**, 65–69 (1997). https://doi.org/10.1007/BF02696146
83. See: https://news.osu.edu/researchers-develop-worlds-first-light-tunable-plastic-magnet/. 08.02.2020
84. Kalia, S., Kango, S., Kumar, A., et al.: Magnetic polymer nanocomposites for environmental and biomedical applications. Colloid Polym. Sci. **292**, 2025–2052 (2014). https://doi.org/10.1007/s00396-014-3357-y
85. Strnat, K.: Permanent magnet on the basis of cobalt-rare earth alloys and method for its production (1975) patent no: US3998669A
86. Erwin, S.H.: MRI safety of Med-El C40/C40+ cochlear implants. Cochlear Implants Int. **2**(2), 98–114 (2013). https://doi.org/10.1179/cim.2001.2.2.98
87. See: https://www.persistencemarketresearch.com/market-research/magnetic-plastics-market.asp#:~:text=China%20is%20the%20dominant%20country,lower%20labor%20costs%20and%20overheads. 08.02.2020
88. Ansari, S.: Application of magnetic molecularly imprinted polymer as a versatile and highly selective tool in food and environmental analysis: recent developments and trends. TrAC Trends Anal. Chem. **90**, 89–106 (2017). https://doi.org/10.1016/j.trac.2017.03.001
89. Bernardi, D., Egeni, G.P., Parere, F., Pegoraro, M., Rossi, P., Rudello, V., Somacal, H., Vittone, E., Viviani, M.: Focused micro-beam single event with a scintillating foil trigger and magnetic blanking. Nucl. Instrum. Methods Phys. Res. B Beam Interactions Mater. Atoms. **152**(2–3), 377–385 (1999). https://doi.org/10.1016/S0168-583X(99)00018-X
90. Hamano, M.: Overview and outlook of bonded magnets in Japan. J. Alloys Compd. **222**(1–2), 8–12 (1995). https://doi.org/10.1016/0925-8388(94)04903-3
91. Motamedi, M., Warkiani, M.E., Taylor, R.A.: Transparent surfaces inspired by nature. Adv. Opt. Mater. **6**(14), 1800091 (2018). https://doi.org/10.1002/adom.201800091
92. Kaushika, N.D., Sumathy, K.: Solar transparent insulation materials: a review. Renew. Sustain. Energy Rev. **7**(4), 317–351 (2003). https://doi.org/10.1016/S1364-0321(03)00067-4
93. See: https://www.marketsandmarkets.com/Market-Reports/transparent-plastics-market-57341363.html. 10.02.2020
94. See: https://omnexus.specialchem.com/polymer-properties/properties/transparency. 10.02.2020
95. Koike, Y., Koike, K.: Progress in low-loss and high-bandwidth plastic optical fibers. J. Polym. Sci. B Plym. Phys. **49**(1), 2–17 (2011). https://doi.org/10.1002/polb.22170
96. Kulich, D.M., Gaggar, S.K., Lowry, V., Stepien, R.: Acrylonitrile–butadiene–styrene polymers. Encyclopedia Polym. Sci. Technol. https://doi.org/10.1002/0471440264.pst011
97. Wouters, M.E.L., Wolfs, D.P., van der Linde, M.C., Hovens, J.H. P., Tinnemans, A.H.A.: Transparent UV curable antistatic hybrid coatings on polycarbonate prepared by the sol–gel method. Prog. Org. Coat. **51**(4), 312–319 (2004). https://doi.org/10.1016/j.porgcoat.2004.07.020
98. Liu, H., Liu, D., Yao, F., Wu, Q.: Fabrication and properties of transparent polymethylmethacrylate/cellulose nanocrystals composites. Bioresour. Technol. **101**(14), 5685–5692 (2010). https://doi.org/10.1016/j.biortech.2010.02.045
99. Knoll, K., Nießner, N.: Styrolux+ and styroflex+ from transparent high impact polystyrene to new thermoplastic elastomers: syntheses, applications and blends with other styrene based polymers. Macromol. Symp. **132**(1), 231–243 (1998). https://doi.org/10.1002/masy.19981320122
100. Petersson, L., Mathew, A.P., Oksman, K.: Dispersion and properties of cellulose nanowhiskers and layered silicates in cellulose acetate butyrate nanocomposites. Appl. Polym. Sci. https://doi.org/10.1002/app.29661
101. Cui, F., Jafarishad, H., Zhou, Z., Chen, J., Shao, J., Wen, Q., Liu, Y., Zhou, H.S.: Batch fabrication of electrochemical sensors on a glycol-modified polyethylene terephthalate-based microfluidic device. Biosens. Bioelectr. **167**, 112521 (2020). https://doi.org/10.1016/j.bios.2020.112521
102. See: https://en.wikipedia.org/wiki/Styrene-acrylonitrile_resin. 12.02.2020
103. See: https://en.wikipedia.org/wiki/Acrylonitrile_styrene_acrylate. 12.02.2020
104. Xu, L., Liu, B., Zhang, M., Bai, Y., Zhang, J., Song, J.: Control of the particle microstructure during the synthesis of bulk-polymerized transparent methacrylate-acrylonitrile-butadiene-styrene resin. Polym. Eng. Sci. **60**(6), 1194–1201 (2020). https://doi.org/10.1002/pen.25373
105. Knoll, K., Nieβner, N.: Styroflex: a new transparent styrene-butadiene copolymer with high flexibility, synthesis, applications, and synergism with other styrene polymers. In: Applications of anionic polymerization research, Chapter 9 ACS Symposium Series, vol. 696, pp. 112–128. ACS, Washigton (1998). https://doi.org/10.1021/bk-1998-0696.ch009. ISBN13: 9780841235656eISBN: 9780841216686
106. Yamazaki, M.: Industrialization and application development of cyclo-olefin polymer. J. Mol. Catal. A Chem. **213**(1), 81–87 (2004). https://doi.org/10.1016/j.molcata.2003.10.058
107. Puszka, A., Kultys, A.: New thermoplastic polyurethane elastomers based on aliphatic diisocyanate: synthesis and characterization. J Therm Anal Calorim. **128**, 407–416 (2017). https://doi.org/10.1007/s10973-016-5923-7

108. Vincent, H.L., Kimball, D.J., Boundy, R.R.: Polysiloxane-silica hybrid resins as abrasion-resistant coatings for plastic substrates. In: Polymer wear and its control, Chapter 9 ACS Symposium Series, vol. 287, pp. 129–134. ACS, Washigton (1985). https://doi.org/10.1021/bk-1985-0287.ch009. ISBN13: 9780841209329eISBN: 9780841211193
109. Schilling, K., Bradford, B., Castelli, D., Dufour, E., Nash, J.F., Pape, W., Schulte, S., Tooley, I., van den Boschi, J., Schellauf, F.: Human safety review of "nano" titanium dioxide and zinc oxide. Photochem. Photobiol. Sci. **9**, 495–509 (2010). https://doi.org/10.1039/B9PP00180H
110. Eylem, D.T., Yazici, O.A., Sam Parmak, E.D., Gonultas, O.: Influence of tannin containing coatings on weathering resistance of wood: combination with zinc and cerium oxide nanoparticles. Polym. Degrad. Stabil. **152**, 289–296 (2018). https://doi.org/10.1016/j.polymdegradstab.2018.03.012
111. Angmo, D., Espinosa, N., Krebs, F.: Indium tin oxide-free polymer solar cells: toward commercial reality. In: Lin, Z., Wang, J. (eds.) Low-cost nanomaterials. Green energy and technology. Springer, London (2014). https://doi.org/10.1007/978-1-4471-6473-9_8
112. Mostafa, R.A., Adlii, A., Bakry, B.M.: Green fabrication of bentonite/chitosan@cobalt oxide composite (BE/CH@Co) of enhanced adsorption and advanced oxidation removal of Congo red dye and Cr (VI) from water. Int. J. Biol. Macromol. **126**, 402–413 (2019). https://doi.org/10.1016/j.ijbiomac.2018.12.225
113. Takashi 1, K., Kumar, P.B., Akira, T.: Synthesis of carbon nanotube composites in nanochannels of an anodic aluminum oxide film. Bull. Chem. Soc. Jpn. **72**(9) (1999). https://doi.org/10.1246/bcsj.72.1957
114. Yi, X., Zhang, C., Wang, J.-X., Wang, D., Zeng, X.-F., Chen, J.-F.: Synthesis of transparent aqueous zro2 nanodispersion with a controllable crystalline phase without modification for a highrefractive-index nanocomposite film. Langmuir. **34**, 6806–6813 (2018). https://doi.org/10.1021/acs.langmuir.8b00160
115. Balamurugan, S., Balu, A.R., Usharani, K., Suganya, M., Anitha, S., Prabha, D., Ilangovan, S.: Synthesis of CdO nanopowders by a simple soft chemical method and evaluation of their antimicrobial activities. Pac. Sci. Rev. A Nat. Sci. Eng. **18**(3), 228–232 (2016). https://doi.org/10.1016/j.psra.2016.10.003
116. Gross, S., Vittadini, A., Dengo, N.: Functionalisation of colloidal transition metal sulphides nanocrystals: a fascinating and challenging playground for the chemist. Crystals. **7**, 110 (2017). https://doi.org/10.3390/cryst7040110
117. Wattoo, M.H.S., Quddos, A., Wadood, A., Khan, M.B., Wattoo, F. H., Tirmizi, S.A., Mahmood, K.: Synthesis, characterization and impregnation of lead sulphide semiconductor nanoparticles on polymer matrix. J. Saudi Chem. Soc. **16**(3), 257–261 (2012). https://doi.org/10.1016/j.jscs.2011.01.006
118. Patil, A., Patel, A., Purohit, R.: An overview of polymeric materials for automotive applications. Mater Today Proc. **4**(2 Part A), 3807–3815 (2017). https://doi.org/10.1016/j.matpr.2017.02.278. ISSN 2214-7853
119. See: https://www.creativemechanisms.com/blog/eleven-most-important-plastics. 14.02.2020
120. See: https://polyurethane.americanchemistry.com/polyurethanes/Introduction-to-Polyurethanes/Applications/Thermoplastic-Polyurethane/. 14.02.2020
121. See: https://www.chemicalsafetyfacts.org/types-plastic-food-packaging-safety-close-look/. 18.02.2020
122. McKeen, L.W.: 3-Plastics used in medical devices. In: Modjarrad, K., Ebnesajjad, S. (eds.) Plastics design library, handbook of polymer applications in medicine and medical devices, pp. 21–53. William Andrew Publishing, Norwich (2014). https://doi.org/10.1016/B978-0-323-22805-3.00003-7. ISBN 9780323228053
123. See: https://www.materialstoday.com/polymers-soft-materials/news/transparent-displays-via-thin-plastic-coating/. 18.02.2020
124. See: https://www.marketsandmarkets.com/Market-Reports/cellulose-esters-market-173641209.html. 18.02.2020
125. Olds, W.J., Jaatinen, E., Fredericks, P., Cletus, B., Panayiotou, H., Izake, E.L.: Spatially offset Raman spectroscopy (SORS) for the analysis and detection of packaged pharmaceuticals and concealed drugs. Forensic Sci. Int. **212**(1-3), 69–77 (2011). https://doi.org/10.1016/j.forsciint.2011.05.016
126. Hosono, H.: Recent progress in transparent oxide semiconductors: materials and device application. Thin Solid Films. **515**(15), 6000–6014 (2007). https://doi.org/10.1016/j.tsf.2006.12.125
127. Douglas, W.: Sports helmet with transparent windows in the side walls (1990), patent no: US5101517A
128. Siracusa, V., Rocculi, P., Romani, S., Rosa, M.D.: Biodegradable polymers for food packaging: a review. Trends Food Sci. Technol. **19**(12), 634–643 (2008). https://doi.org/10.1016/j.tifs.2008.07.003
129. Lambert, S., Wagner, M.: Environmental performance of bio-based and biodegradable plastics: the road ahead. Chem. Soc. Rev. **46**, 6855–6871 (2017). https://doi.org/10.1039/C7CS00149E
130. Prieto, A.: To be, or not to be biodegradable... that is the question for the bio-based plastics. Microbial Biotechnol. **9**(5, Special Issue), 652–657 (2016). https://doi.org/10.1111/1751-7915.12393
131. See: https://www.grandviewresearch.com/industry-analysis/bioplastics-industry#:~:text=The%20global%20bioplastics%20market%20size%20was%20estimated%20at%20USD%208.3,USD%209.2%20billion%20in%202020.&text=The%20global%20bioplastics%20market%20is,USD%2026.0%20billion%20by%202027. 20.02.2020
132. Mehdi Emadian, S., Turgut, T.O., Demirel, B.: Biodegradation of bioplastics in natural environments. Waste Manag. **59**, 526–536 (2017). https://doi.org/10.1016/j.wasman.2016.10.006
133. Zárate-Ramírez, L.S., Romero, A., Bengoechea, C., Partal, P., Guerrero, A.: Thermo-mechanical and hydrophilic properties of polysaccharide/gluten-based bioplastics. Carbohydr. Polym. **112**, 24–31 (2014). https://doi.org/10.1016/j.carbpol.2014.05.055
134. Vipul, D.P., Jani, G.K., Moradiya, N.G., Randeria, N.P., Nagar, B. J., Naikwadi, N.N., Variya, B.C.: Galactomannan: a versatile biodegradable seed polysaccharide. Int. J. Biol. Macromol. **60**, 83–92 (2013). https://doi.org/10.1016/j.ijbiomac.2013.05.017
135. Park, H.-M., Misra, M., Drzal, L.T., Amar, K.M.: "Green" nanocomposites from cellulose acetate bioplastic and clay: effect of eco-friendly triethyl citrate plasticizer. Biomacromolecules. **5** (6), 2281–2288 (2004). https://doi.org/10.1021/bm049690f
136. Bakti, B.S., Cran, M.J., Bigger, S.W.: A review of property enhancement techniques for carrageenan-based films and coatings. Carbohydr. Polym. **216**, 287–302 (2019). https://doi.org/10.1016/j.carbpol.2019.04.021
137. Ward, A.M., Wyllie, G.R.A.: Bioplastics in the general chemistry laboratory: building a semester-long research experience. J. Chem. Educ. **96**(4), 668–676 (2019). https://doi.org/10.1021/acs.jchemed.8b00666
138. Rishi, V., Sandhu, A.K., Kaur, A., et al.: Utilization of kitchen waste for production of pullulan to develop biodegradable plastic. Appl. Microbiol. Biotechnol. **104**, 1307–1317 (2020). https://doi.org/10.1007/s00253-019-10167-9
139. Rukmanikrishnan, B., Ramalingam, S., Rajasekharan, S.K., Lee, J., Lee, J.: Binary and ternary sustainable composites of gellan gum, hydroxyethyl cellulose and lignin for food packaging applications: biocompatibility, antioxidant activity, UV and water barrier properties. Int. J. Biol. Macromol. **153**, 55–62 (2020). https://doi.org/10.1016/j.ijbiomac.2020.03.016
140. Niknezhad, S.V., Asadollahi, M.A., Zamani, A., et al.: Optimization of xanthan gum production using cheese whey and response surface methodology. Food Sci. Biotechnol. **24**, 453–460 (2015). https://doi.org/10.1007/s10068-015-0060-9
141. Jerez, A., Partal, P., Martínez, I., Gallegos, C., Guerrero, A.: Protein-based bioplastics: effect of thermo-mechanical processing.

Rheol Acta. **46**, 711–720 (2007). https://doi.org/10.1007/s00397-007-0165-z
142. Han, J., Hou, J., Zhang, F., Ai, G., Li, M., Cai, S., Liu, H., Wang, L., Wang, Z., Zhang, S., Cai, L., Zhao, D., Zhou, J., Xiang, H.: Multiple propionyl coenzyme a-supplying pathways for production of the bioplastic poly(3-Hydroxybutyrate-co-3-Hydroxyvalerate) in Haloferax mediterranei. Appl. Environ. Microbiol. https://doi.org/10.1128/AEM.03915-12
143. Zhao, X., Guerrero, F.R., Llorca, J., Wang, D.-Y.: New superefficiently flame-retardant bioplastic poly(lactic acid): flammability, thermal decomposition behavior, and tensile properties. ACS Sustain. Chem. Eng. **4**(1), 202–209 (2016). https://doi.org/10.1021/acssuschemeng.5b00980
144. Anstey, A., Muniyasamy, S., Reddy, M.M., et al.: Processability and biodegradability evaluation of composites from poly(butylene succinate) (PBS) bioplastic and biofuel Co-products from Ontario. J. Polym. Environ. **22**, 209–218 (2014). https://doi.org/10.1007/s10924-013-0633-8
145. Poulopoulou, N., Kasmi, N., Siampani, M., Terzopoulou, Z.N., Bikiaris, D.N., Achilias, D.S., Papageorgiou, D.G., Papageorgiou, G.Z.: Exploring next-generation engineering bioplastics: poly (alkylene furanoate)/poly(alkylene terephthalate) (PAF/PAT) blends. Polymers. **11**, 556 (2019)
146. Keshavarz, T., Roy, I.: Polyhydroxyalkanoates: bioplastics with a green agenda. Curr. Opin. Microbiol. **13**(3), 321–326 (2010). https://doi.org/10.1016/j.mib.2010.02.006
147. Samantaray, P.K., Little, A., Haddleton, D.M., McNally, T., Tan, B., Sun, Z., Huang, W., Jic, Y., Wanm, C.: Poly(glycolic acid) (PGA): a versatile building block expanding high-performance and sustainable bioplastic applications. Green Chem. **22**, 4055–4081 (2020). https://doi.org/10.1039/D0GC01394C
148. Sanchez-Garcia, M.D., Ocio, M.J., Gimenez, E., Lagaron, J.M.: Novel polycaprolactone nanocomposites containing thymol of interest in antimicrobial film and coating applications. J. Plast. Film Sheeting. **24**(3–4), 239–251 (2008). https://doi.org/10.1177/8756087908101539
149. Lee, S.-H., Kim, M.-N.: Isolation of bacteria degrading poly(butylene succinate-co-butylene adipate) and their lip A gene. Int. Biodeterioration Biodegrad. **64**(3), 184–190 (2010). https://doi.org/10.1016/j.ibiod.2010.01.002
150. Zhou, X., Mohanty, A., Misra, M.: A new biodegradable injection moulded bioplastic from modified soy meal and poly (butylene adipateco-terephthalate): effect of plasticizer and denaturant. J. Polym. Environ. **21**, 615–622 (2013). https://doi.org/10.1007/s10924-013-0578-y
151. Mandala, W.C.R., Saepudina), E., Nizardo, N.M.: Effect of addition of antibacterial compound from kelor leaves extract (Moringa oleifera Lam) to foodborne pathogen bacteria activity on crosslinked bioplastic poly(vinyl alcohol)/starch. AIP Conf. Proc. **2242**(1) (2020). https://doi.org/10.1063/5.0010360
152. Lackner, M.: Bioplastics, Kirk-Othmer Encyclopedia of Chemical Technology. https://doi.org/10.1002/0471238961.koe00006
153. Iles, A., Martin, A.N.: Expanding bioplastics production: sustainable business innovation in the chemical industry. J. Cleaner Prod. **45**, 38–49 (2013). https://doi.org/10.1016/j.jclepro.2012.05.008
154. See: https://en.wikipedia.org/wiki/Biodegradable_plastic. 24.02.2020
155. Xia, Y., Larock, R.C.: Vegetable oil-based polymeric materials: synthesis, properties, and applications. Green Chem. **12**, 1893–1909 (2010). https://doi.org/10.1039/C0GC00264J

15 Energy Harvesting and Storing Materials

Abstract

Energy is always required in our daily work, from our body movement to moving big industries. Each energy is convertible, that is, energy can change from one form to another while work. This chapter focused on how economically one energy can be converted to electric energy with the utilization of various materials. This chapter gives the basic knowledge on renewable energy and its requirement in the green world. How to harvest energies and how to store these energies have been clearly explained in this chapter. The role of various types of currently used batteries has been discussed here.

Keywords

Ambient energy sources · Photo-energy harvest · Thermal energy harvest · Mechanical energy/vibrational energy harvest · Electromagnetic energy harvesting · Electrostrictive energy harvesting · Magnetostrictive energy harvesters · Chemical energy · Wind energy harvest · Tide energy · Energy storage · Batteries

15.1 Introduction

Energy harvesting is one of the key emerging technologies of the twenty-first century. The field of power harvesting has experienced significant growth in the past few years due to the ever-increasing desire to produce portable and wireless electronics with an extended life span. This chapter focuses on the collection of energy from the environment, more generally from renewable energy sources. Most often, it involves small systems with tiny amounts of power, ranging from nanowatts to hundreds of milliwatts [1]. The main category of applications of these power levels is wireless devices. The applicability of energy harvesting to particular devices depends on the type and amount of the available ambient energy as well as on size limitations. It has been shown, for example, that energy harvesting from human motion is generally not enough to power laptops or mobile phones but is viable for many types of wireless sensors. Another important factor that has to be regarded when considering energy harvesting as a solution, for example, batteries, the currently dominant wireless power source, is the ability to satisfy the power, size, weight, lifetime, and ecological demands of the specific application. Motion, temperature gradients, light, electromagnetic radiation, and chemical energy can all be used as sources for energy harvesting [2]. Energy harvesting technology can be defined as the process by which energy from the physical environment is captured and converted into usable electrical energy in real-time and used immediately so that energy only ever needs to be stored temporarily. Energy harvesting is also known as power harvesting or energy scavenging or ambient power [3, 4]. External sources are, for example, solar power, thermal energy, wind energy, salinity gradients, and kinetic energy, also known as ambient energy. To provide a reliable source of energy for a wireless sensor system, one can consider extracting energy from the environment to complement the battery energy storage or even replace it [5].

The most common, for which promising results have already been achieved, is the extraction of power from various sources, such as light energy (captured from sunlight or room light via photo sensors, or solar panels), thermal energy (from furnaces, domestic radiators, human skin, vehicle exhausts, and friction sources), mechanical vibrations (from sources such as car engine compartment, trains, ships, helicopters, bridges, office floors, train stations floors, nightclubs floors, speakers, window panes, walls, fridges, washing machines, microwave ovens, pumps, motors, compressors, chillers, conveyors), radio frequency (from microwaves, infrared, cell phones, and high power line emissions), and chemical energy (from the surface of electrodes). Light harvesters use the photoelectric effect. Thermal harvesters use the thermoelectric effect (also

A. Behera, *Advanced Materials*, https://doi.org/10.1007/978-3-030-80359-9_15

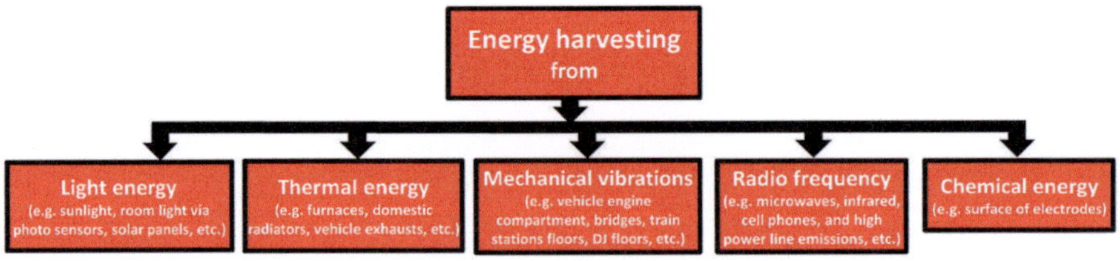

Fig. 15.1 Various sources from which energy can be harvested

known as the Seebeck effect) [6–10]. There are three main mechanisms by which vibrations can be converted into electrical energy: electromagnetic, electrostatic, and piezoelectric. Among them, piezoelectric vibration-to-electricity converters have received much attention, as they have high electromechanical coupling and no external voltage source requirement, and they are particularly attractive for use in MEMS. Energy harvesting using piezoelectric materials allows for a device that is self-contained, that is, does not require any external supporting source. Furthermore, piezoelectric energy-harvesting devices have a minimum of moving parts and are capable of generating power with voltage levels that can be easily conditioned (e.g., converted to DC or boosted) [11]. The categorization of various energy-harvesting sources is illustrated in Fig. 15.1.

15.2 Types of Ambient Energy Sources

Various energy-harvesting technologies have been explained subsequently.

15.2.1 Photo-Energy Harvest

The sun is a virtually infinite reservoir of renewable energy. The size of the solar energy reservoir is considerably larger than all other energy sources combined, including both renewable and nonrenewable sources. It has been estimated that if only 2% of the solar energy hitting the earth's surface each day was captured, all of humanity's current energy needs would be met and even exceeded. In 2013, the global solar power generating capacity increased by 26%, from 31 Gigawatts (GW) in 2012 to 39 GW in 2013 [12]. The importance of solar power in the future of the energy industry prompts a closer look at four ways to utilize sun light sources and are discussed below:

Photovoltaic Modules: Photovoltaic (PV) modules, often referred to as solar panels, are the most ubiquitous solar power harvesting technology. PV modules have been applied for residential and commercial power generation in buildings. PV modules generate electricity by energizing electrons inside their material system. PV modules are made from semiconductor materials, which have the ability to absorb sunlight and create electric energy. The solar cell consists of a p-type semiconductor and an n-type semiconductor. When sunlight reaches the semiconductor materials of PV cells, free electrons are forced to flow in a certain direction [13]. The negatively charged electrons move toward the n-type semiconductor, while the positively charged electrons move toward the p-type semiconductor. The flow of moving electrons creates an electrical current when connected to the electrical load. A major reason why semiconductor materials are used for PV modules is their ability to separate electrons into high- and low-energy states, which are separated by an energy gap that is inherent to the material. When the sunlight is absorbed in the semiconductor, the absorbed energy promotes a low-energy electron to a high-energy electron, which then flows into the external circuit. Positively charged holes are also generated in this process and maintain charge balance in the overall system. As the amount of absorbed sunlight increases, so does the number of high-energy electrons, which then creates a flowing electric current. The energy difference between low-energy electrons and high-electrons, that is, the size of the energy gap, determines the voltage output by the PV module [14].

Solar Thermal: Solar thermal, also know as concentrating solar power (CSP), is a lesser known method to extract solar power. Unlike PV modules, solar thermal is not applied commonly for residential or commercial building power generation. Solar thermal, however, has a tremendous potential to produce solar power on a utility scale. The 377 MW Ivanpah solar plant in California's Mojave Desert is the world's largest solar energy facility. The energy-harvesting process of CSP is inherently different from solar photovoltaics. In CSP, sunlight is directed using mirrors and then concentrated at a single point [15]. The concentrated optical energy is then used to boil water and generate high-energy steam. The high-energy steam then turns a turbine generator to produce electricity. Steam-powered generators are a very established technology to generate electricity that is used by virtually all traditional power plants, including coal, natural gas, and nuclear power plants. The inherent difference between solar thermal and traditional power plants is that CSP uses renewable energy input to generate steam [16].

Biofuels (Natural Photosynthesis): A major challenge of solar photovoltaic and solar thermal energy generation is their inability to store energy effectively. This creates a demand–response issue, as the generated electricity needs to be consumed instantly. Plants, however, have the ability to store solar energy using photosynthesis. In the process of photosynthesis, plants take in CO_2 and water to produce oxygen and glucose [17]. Photosynthesis has been extensively studied and the abundance of plants on the earth's surface speaks for the scalability of natural photosynthesis. The products of the photosynthetic process of certain plants can then be further refined to create chemical fuels that have sufficient energy density for practical applications. This process is already being applied in making ethanol fuels from corn feedstock. Given the controversy of using corn to generate fuels, researchers are investigating additional crops that can be used for biofuel generation. Their aim is to find biofuel crops that grow quickly and efficiently, have a high energy density, and mitigate the food versus fuel conflict. Despite the immense scale of biofuels, one of the main drawbacks of natural photosynthesis remains its inherently low efficiency. The energy conversion efficiency of natural photosynthesis is <0.5%, which is a major reason why many biofuel crops require significant refining steps to be converted into usable biofuels [18].

Artificial Photosynthesis: Artificial photosynthesis aims to address the inherent flaw of natural photosynthesis by engineering a chemical scheme that converts sunlight into usable chemical fuels. Artificial photosynthesis intends to mimic the process of natural photosynthesis and to increase the energy conversion efficiency by replacing the biological agents in the natural scheme with solid-state materials that drive chemical reactions to produce the desired fuels. The principle of artificial photosynthesis is based on electrochemistry [19]. Many naturally occurring chemical reactions generate electricity with a distinct voltage. This phenomenon was the scientific basis used to engineer some of the world's first battery systems. Like many phenomena in the chemical world, electrochemical reactions are reversible. This means that if one applies electricity with a certain voltage to a chemical system, one can initiate a chemical reaction that does not occur naturally. A common example of this concept is the production of aluminum: In its natural state, aluminum is found in various compounds, including aluminum oxide, which is more chemically stable than pure aluminum in nature [20]. In this case, the natural electrochemical reaction is for oxygen to react with pure aluminum to form aluminum oxide. The reverse reaction can be initiated by flowing electricity through the aluminum oxide to create pure aluminum. This process of making aluminum is known as the Hall-Heroult process and is still a common industrial practice today (Fig. 15.2) [21].

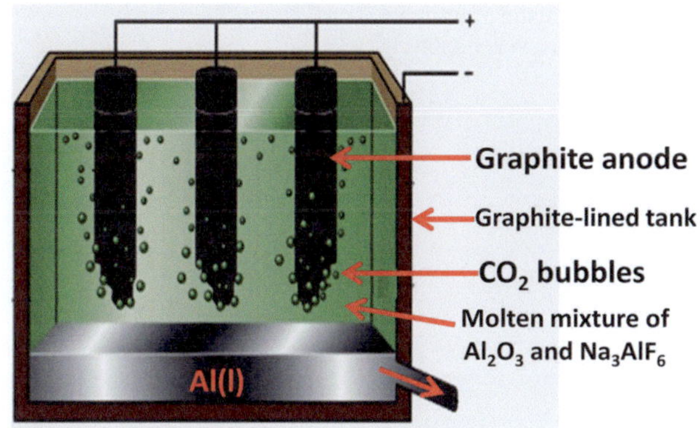

Fig. 15.2 Schematic of the Hall–Heroult process where electricity is applied through a graphite anode

Unlike PV modules, the electricity is not extracted into an external circuit but used to initiate chemical reactions at the surface of the material. The chemical reactions can be accelerated with certain catalyst materials, which increase the energy conversion of the overall system when engineered effectively. Various systems have the ability to generate hydrogen via water splitting (Fig. 15.3). The solar-generated hydrogen is a usable chemical fuel that can be used for a variety of applications. Current artificial photosynthesis systems fabricated by National Renewable Energy Laboratory (NREL) display a solar hydrogen conversion efficiency of 12.4%. The theoretical limit of solar-to-hydrogen generation has been estimated to be ~ 30% [22]. Currently, a system is developed to convert CO_2 and water into alcohols, such as methanol or ethanol, which are already being used as chemical fuels, a material system that uses solar energy to convert CO_2 and water into chemical fuels with reasonable efficiency and could have substantial implications for the energy industry. The ability to harness solar power in an effective and economic manner has improved dramatically in the last decade. While solar photovoltaics, solar thermal, and biofuels are already being used on industrial scales, the next generation of disruptive solar technologies may be based on artificial photosynthesis technology [23].

The conversion efficiency (?) of a solar cell is the most commonly used parameter to define a cell's performance and compare it to that of another cell. Efficiency is defined as the ratio of energy output from the solar cell to input energy from the sun. In addition to reflecting the performance of the solar cell itself, the efficiency depends on the spectrum and intensity of the incident sunlight and the temperature of the solar cell [24]. Therefore, the conditions under which efficiency is measured must be carefully controlled to compare the performance of one device to another. The efficiency of a solar cell is determined by the fraction of incident power that is converted into electricity and is defined as Eq. (15.1):

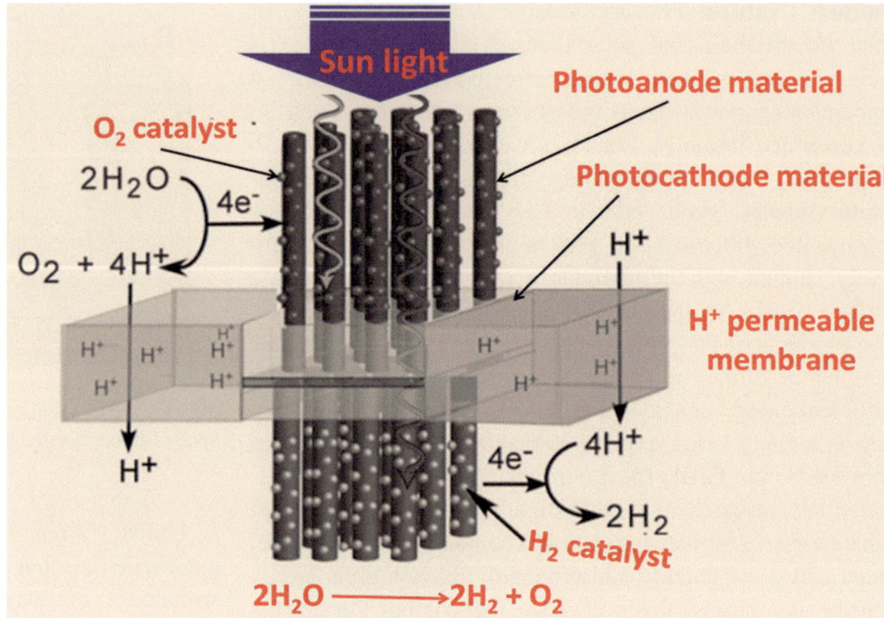

Fig. 15.3 Schematic of artificial photosynthesis for water splitting

$$\eta = \frac{P_{\mathrm{max}}}{P_{\mathrm{in}}} = \left[\frac{V_{\mathrm{OC}} \times J_{\mathrm{SC}} \times \mathrm{FF}}{P_{\mathrm{in}}} a\right] \times 100, \tag{15.1}$$

where V_{OC} is the open-circuit voltage, J_{SC} is the short-circuit current, and FF is the fill factor. The open-circuit voltage (V_{OC}) is the maximum voltage available from a solar cell, and this occurs at zero current. The open-circuit voltage corresponds to the amount of forward bias on the solar cell due to the bias of the solar cell junction with the light-generated current. The short-circuit current (J_{SC}) is the current through the solar cell when the voltage across the solar cell is zero (i.e., when the solar cell is short-circuited). The short-circuit current is equal to the absolute number of photons converted to hole-electron pairs due to the generation and collection of light-generated carriers. For an ideal solar cell with, at most, moderate resistive loss mechanisms, the short-circuit current and the light-generated current are identical. Therefore, the short-circuit current is the largest current that may be drawn from the solar cell [25]. The short-circuit current and the open-circuit voltage are the maximum current and voltage, respectively, from a solar cell. At both of these operating points, however, the power from the solar cell is zero. FF is a parameter that, in conjunction with V_{OC} and J_{SC}, determines the maximum power of a solar cell. FF is defined as the ratio of the maximum power from the solar cell to the product of V_{OC} and J_{SC}. A commonly used expression for FF can be determined empirically as Eq. (15.2).

$$\mathrm{FF} = \frac{P_{\mathrm{max}}}{V_{\mathrm{OC}} \times J_{\mathrm{SC}}} = \frac{V_{\mathrm{max}} \times J_{\mathrm{max}}}{V_{\mathrm{OC}} \times J_{\mathrm{SC}}}. \tag{15.2}$$

In general, solar cells can be sorted according to the semiconductor materials used for the active layer. Generally, the solar cells can be categorized as silicon-based solar cells, dye-sensitized solar cells (DSSCs), organic solar cells, including organic unit molecules and polymers, quantum-dot (QD) solar cells, and perovskite solar cells. In a specific cell, each of the factors,V_{OC}, J_{SC}, and FF, is affected by the material species, cell structure, fabrication process, and other technologies. To increase the cell efficiency by improving V_{OC}, J_{SC}, and FF, therefore, many approaches are used, such as the development of novel materials, light-harvesting device structures, and processing and other techniques [26]. The silicon-based solar cells have high ⍰ above 20%, but they have limitations of brittleness and difficult handling. Otherwise, other solar cells based on the coating process can be fabricated onto various electrodes, made of metal oxides (e.g., indium tin oxide [ITO], fluorine-doped tin oxide [FTO]), metals (Au, Ag, etc.), carbons (e.g., CNT and graphene), and coated substrates (i.e., glass, polymer films, metal wire, and textile). To fabricate "All-in-one energy-harvesting and storage devices" through hybridization, the understanding and application of devices in terms of materials, electrodes, and processes utilized in fabrication are essential tasks because the physical properties of flexibility, bendability, wiring, stacking, and so on are important for hybridization between devices [27].

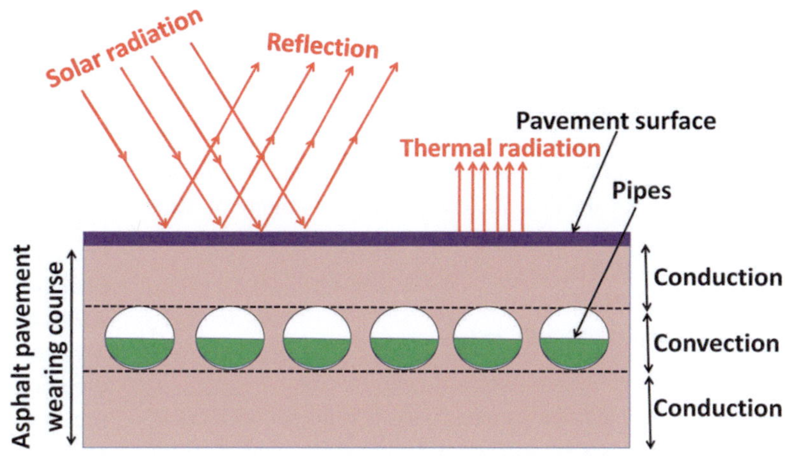

Fig. 15.4 Heat transfer mechanism in asphalt solar collector

15.2.1.1 Basic principles of Solar Collector System

A solar collector system consists of a network of pipes under road pavement with circulating fluid inside. As the pavement absorbs radiation from the sun and atmosphere, the pavement temperature increases, and heat is transferred to the fluid inside the piping system due to temperature gradients. There are three basic heat-balance processes involved in pavement solar collector system: conduction, convection, and radiation, as shown in Fig. 15.4. Conduction happens between pavement and pipe walls [28]. Energy convection occurs when there are temperature differences between the ambient air, pavement, pipe walls, and the fluid circulating in the pipes. Radiation via electromagnetic waves can occur without any material medium, including solar radiation transfer to the pavement and thermal radiation between the ambient atmosphere and the pavement. The heat captured by the piping system can be used in thermoelectric generators to generate electric energy or stored in energy reservoirs. During winter, the stored heat can be used to melt snow on roads, produce electricity, and warm nearby buildings. Another benefit associated with pavement solar collector system is its ability to mitigate urban heat island effects in metropolitan areas by reducing pavement temperature. The cooling effect also helps to retard pavement deterioration and maintain pavement performance under high-temperature weather conditions [29].

15.2.2 Thermal Energy Harvest

Thermal energy can be obtained from heat present in the ambience or from heat generated during some process (persons and animals, machines, or other natural sources). Either thermoelectric or pyroelectric effects can be used to harvest energy. Thermoelectric effects such as Peltier effect, Seeback effect and Thomson effects can generate power as long as a heat source is present [30]. Extraction of energy from a thermal source requires a thermal gradient and conversion efficiency. This mainly depends on the temperature difference between the heat source and the environment (i.e., the cold and the hot side). A greater temperature difference leads to a better output. Thermal gradients arise as a result of many processes, including the burning of fuel and geothermal processes. Thermoelectric energy harvesters extract electrical energy from temperature gradients. The energy extraction is enabled by the Seebeck effect in which the temperature difference across a conductor or semiconductor results in a net flow of electrons from the high-temperature side to the low-temperature side so that an electrical current (and voltage) is generated [31]. The thermal energy-harvesting technologies are currently based upon the thermoelectric, pyroelectric, thermomagnetic, and thermoelastic effects [32].

Thermoelectric effect: There are two types of energy harvesters using thermal energy. One is based on the Seebeck effect. The Seebeck effect is defined as the generation of an electric field when there is a temperature gradient at two ends of a thermoelectric generator device. The temperature gradient of the conductor and the electric current generation are reversible. The Seebeck effect is also known as the thermoelectric (TE) effect. The thermoelectric effect, known since the nineteenth century, provides an interesting perspective for the conversion of heat to electrical energy. Given a thermal gradient, a thermoelectric generator (TEG) is able to convert heat into electrical energy even with small temperature differences. TEG is simple, compact, robust, and very reliable because they contain no moving mechanical parts [33]. For all these reasons, TEG is attractive for a large variety of applications, in particular in the fields of green and renewable energy harvesting. The Seebeck effect can be expressed as Eq. (15.3):

$$V = \alpha \Delta T, \tag{15.3}$$

where V is the thermoelectric voltage, ΔT is the temperature gradient, and α is the Seebeck coefficient. Based on the thermoelectric effect, the efficiency of thermoelectric devices is determined by the thermoelectric material's figure of merit, ZT, which is a function of several transport coefficients as follows, Eq. (15.4).

$$\mathrm{ZT} = \frac{\sigma S^2 T}{K_e + K_l}, \tag{15.4}$$

where σ is the electrical conductivity, S is the seebeck coefficient, T is the mean operating temperature, and K is the thermal conductivity. The subscripts e and l on K signify the electronic and latic contributions, respectively. A high figure of merit corresponds to the high efficiency of the TEG. Therefore, there is serious interest in improving the figure of merit of thermoelectric materials for many industrial and energy applications. The most common material for thermoelectricity has been Bi_2Te_3 since 1954 because its high electrical conductivity after doping and low thermal conductivity give a ZT of about 0.7–0.8 at room temperature. Furthermore, various thermoelectric materials, such as $BiSbTe_3$, PbTe, $CoSb_3$, SiGe, and Mg_2Si, have been reported for TEG [34].

A TEG consists of a thermocouple, comprising a p-type and n-type semiconductor connected electrically in series and thermally in parallel as shown in Fig. 15.5. The TE module usually consists of two parallel n-type and p-type semiconductors with heat source and heat sink on each side. The thermogenerator (based on the Seebeck effect) produces an electrical current proportional to the temperature difference between the hot and the cold junctions [35]. An electrical load is connected in series with the thermogenerator creating an electric circuit. The Seebeck coefficient is positive for p-type materials and negative for n-type materials. The heat that enters or leaves a junction of a thermoelectric device has two reasons: (1) the presence of a temperature gradient at the junction and (2) the absorption or liberation of energy due to the Peltier effect. Carnot efficiency puts an upper limit on the heat energy that can be recovered. In the case of a temperature difference between the human body and the environment, for example, room temperature (20 °C), Starner estimates that the Carnot efficiency with this temperature condition is 5.5% [36]. In a warmer environment, the Carnot efficiency drops while in a colder environment the Carnot efficiency rises. The recoverable energy yields 3.7–6.4 W of power. However, evaporative heat loss accounts for 25% of the total heat dissipation, and therefore the maximum power available drops to 2.8–4.8 W [37]. The previous calculations are made assuming that all the heat radiated by the human body can be recovered and transformed into electrical power so that the obtained power is overestimated. Another problem is the location of the device dedicated to the capture of the heat of the human body.

Thermoelectric effect helps in a variety of applications in space vehicle industries and marine industries. In aircraft and helicopters, a considerable amount of heat is released into the atmosphere from turbine engines. To obtain a significant reduction of the gas pollutant into the environment, it is necessary a remarkable reduction of electricity consumption and utilization of the available energy in these types of vehicles. Implicitly, their operating costs are reduced [38]. To power these space vehicles, TEG systems are used (e.g., on fixed-wing aircraft). The backup TEG is a type of static thermoelectric energy-harvesting system with a significant temperature difference across the TEG around 100 °C. TEG for energy harvesting uses the available temperature gradient and collects sufficient energy to power up an energy wireless sensor node (WSN) to be autonomous. This WSN is used for health monitoring systems (HMS) in an aircraft structure. The main components of a WSN are the energy source and the wireless sensor unit [39]. A TEG energy harvesting captures enough energy for a wireless sensor. One side of the TEG is fixed directly to the fuselage and the other side is attached to a phase-change material (PCM) heat

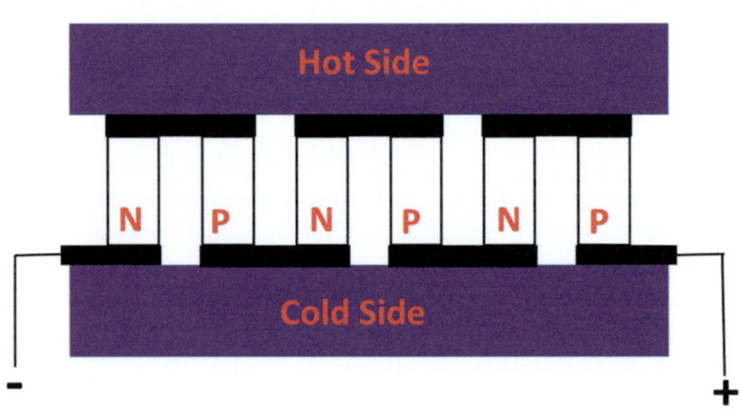

Fig. 15.5 A typical TEG

storage unit to obtain a temperature difference during take-off and landing. PCM is considered an essential element for the heat storage unit because it can maximize the ΔT of the TEG system to solve the low TEG conversion efficiency. In this case, electrical energy is generated. Water is an adequate PCM for heat storage. The temperature difference across the TEG is obtained from the slow-changing temperature of the heat storage unit and the rapidly changing temperature of the aircraft fuselage. A lot of energy is produced during the PC, through latent heat. An application of Bi_2Te_3 modules on turbine nozzles has been addressed in [40]. Although the electric power that can be harvested may be significant, the weight of the cold exchanger is still excessive for the specific application. Future applications in aircraft may be envisioned in locations where there are hot and cold heat flows, especially with the use of light thermoelectric materials. However, one of the main issues remains the weight of the heat exchangers. In the marine industry, there is a lack of clear and stringent international rules at the global level. Marine transport has a significant influence on climate change because there is large amount of greenhouse gas emissions at a particular waiting place. The naval transport generates a wide amount of waste heat, used to provide thermal energy on board and seldom electrical energy [41].

Radioisotope Thermoelectric Generators: The radioisotope thermoelectric generator (RTG) is a solid and highly reliable source of electrical energy to power space vehicles being capable of operating in vacuum and resisting high vibrations. RTGs are used to power space vehicles for distant NASA space expeditions (e.g., several years or several decades) where sunlight is not enough to supply solar panels. The natural radioactive decay of plutonium-238 releases huge amounts of heat, which is suitable for utilization in RTGs to convert it into electricity. The thermoelectric materials used in the thermocouples of the RTGs are adequate for high temperatures considering that the heat source temperature is about 1000 °C. These semiconductor materials can be silicon germanium (Si Ge); lead-tin telluride (PbSnTe), tellurides of antimony, germanium, and silver (TAGS); and lead telluride (PbTe) [42].

Pyroelectric Effect: The other energy-harvesting technology using thermal energy is pyroelectric, based on the change in spontaneous polarization in certain anisotropic solids due to temperature fluctuation. Usually, harvesting thermal energy mainly relies on the Seebeck effect. When the temperature varies in a time-dependent way, however, without a spatial gradient, the Seebeck effect is not able to harvest thermal energy. In this case, the pyroelectric effect could be used to harvest waste thermal energy. The working mechanism of a pyroelectric nanogenerator (PNG) will be explained for two different cases: the primary pyroelectric effect and the secondary pyroelectric effect. The primary pyroelectric effect is related to the charge generation because of the change in polarization with temperature when the dimensions of the pyroelectric material are fixed. The secondary pyroelectric effect is an additional contribution of piezoelectrically induced charge by thermal expansion of a pyroelectric material with temperature change [43]. The total pyroelectric effect is the sum of the primary and secondary pyroelectric effects. The pyroelectric coefficient (e) can be explained by Eq. (15.5).

$$e = \frac{d\rho}{dT}, \tag{15.5}$$

where ρ is the spontaneous polarization and T is the temperature.

The electric current generated by the pyroelectric effect is expressed as Eq. (15.6).

$$I = \frac{\mathrm{dQ}}{\mathrm{dT}} = \mu e A \frac{\mathrm{dT}}{\mathrm{dt}}, \tag{15.6}$$

where Q is the induced charge, μ is the absorption coefficient of radiation, A is the surface area, and $\frac{\mathrm{dT}}{\mathrm{dt}}$ is the rate of temperature change. Therefore, when pyroelectric materials are heated or cooled ($\frac{\mathrm{dT}}{\mathrm{dt}} > 0$, or $\frac{\mathrm{dT}}{\mathrm{dt}} < 0$), the overall polarization in the dipole moment is decreased or increased, which causes current to flow in the circuit. Usually, ferroelectric materials such as lead zirconate titanate (PZT), $BaTiO_3$, P (VDF-TrFE), and $KNbO_3$, and some piezoelectric materials that have spontaneous polarization, such as ZnO and CdS, are used to fabricate pyroelectric generators [44]. By harvesting the waste thermal energy, pyroelectric nanogenerators have potential applications such as environmental monitoring, temperature imaging, medical diagnostics, and personal electronics.

Geothermal Energy Harvesting: Geothermal energy is the heat energy coming from deep inside the earth, that is, the energy naturally stored in the planet. Geothermal heat pumps and underground thermal energy storage play important roles in the application of geothermal energy. Heat pumps are heat transfer devices that can enhance the heat output of the fluid when receiving the relatively low-temperature geothermal heat input. Figure 15.6 illustrates the main concept of geothermal energy under pavement surface using embedded pipes [45]. The basic principle of the most common heat pumps is vapor compression with the use of a compressor; the temperature of the gas increases when it is compressed without the loss of heat. For underground thermal energy storage, the key is to minimize energy loss that is affected by storage time, temperature, volume, and thermal properties of the storage medium.

For example, geothermal water or steam has been used as a heat source for snow melting in the United States since 1948. The first system was constructed in Klamath Falls,

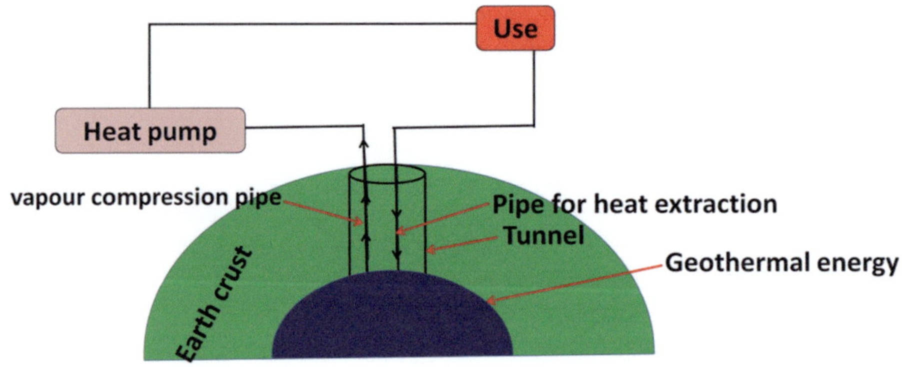

Fig. 15.6 Working principle of geothermal energy harvesting

Oregon, which consisted of an iron piping system with 50% ethylene glycol-eater solution circulating inside. However, after near 50 years of operation, external corrosion and serious leaking led to system failure. These days, metal pipes of the snow-melting system are replaced with plastic pipes to avoid corrosion [46]. Gradually, the geothermal water and gradual advancement of circulating materials are dragging interest on this energy-harvesting technique. A study on bridge deicing system has been studied with different heat sources including solar energy and geothermal heat. The combination of pavement solar collectors with geothermal heat storage is an efficient energy-harvesting technology. In Japan, the snow-melting system used underground geothermal as heat sources in winter and had a heat storage reservoir for solar heat collected in the summer. Recently, researchers found that the use of geothermal heat pumps combined with permeable pavement had the benefit of improving stormwater quality [47].

15.2.3 Mechanical Energy/Vibrational Energy Harvest

Mechanical sources provide a promising alternative to harvest energy where vibration source is the best. Vibrations in some situations can be very large, like in the case of the vibrations of civil structures, such as tall buildings, rail roads, ocean waves, and even human motions. Sources for conversion of vibration energy into electrical energy include strain on a piezoelectric material, electrostatic, or magnetic field [48, 49]. Figure 15.7a shows the PVDF harvester that captures energy from the arterial wall deformation regarding methods to harvest energy from arterial wall deformations; harvesters on blood vessels are prohibited to hamper blood flow and arterial movement. Therefore, soft piezoelectric materials such as PVDF films are preferred. Figure 15.7b shows various types of vibration-assisted piezoelectric application.

Piezoelectric Effect: Mechanical energy such as stress/strain can be converted into electrical energy by the piezoelectric effect (Fig. 15.8). This stress/strain can come from many different sources that exist everywhere, such as vibrations, body motion, acoustic noise, and airflow. When mechanical stress (or strain) is applied to a piezoelectric material, the material is deformed causes a potential difference, and there is a movement of electrical charges [50]. The polarization charge density due to the electrical moment is proportional to the applied mechanical stress, which is given by Eq. (15.7).

$$\rho = \mathrm{d}X, \tag{15.7}$$

where ρ is the polarization charge density, d is the piezoelectric coefficient, and X is the applied stress. Then, the charge density results in electric field and potential as follows in Eq. (15.8).

$$\nabla E = \frac{\rho}{\varepsilon}, \tag{15.8}$$

where ∇E is the divergence of the electric field, ρ is the charge density, and ε is the permittivity. Therefore, by designing a device that comprises a piezoelectric material, electrodes, and an external circuit, the electric potential due to the piezoelectric effect can generate electric current and can be used as a source of electrical energy. The so-called piezoelectric nanogenerator (PENG) was first invented in 2006 by Prof. Wang in the Georgia Institute of Technology by sweeping an atomic force microscope (AFM) tip across a vertically grown ZnO nanowire. Various semiconducting piezoelectric nanostructures in materials, such as ZnO, GaN, CdS, and InN, and insulating piezoelectric materials, such as $PbZrTiO_3$, $BaTiO_3$, $KNbO_3$, $NaNbO_3$, and poly(vinylidene-fluoride)-cotrifluoroethylene [P(VDF-TrFE)], have been intensively applied to fabricate PENGs [51]. These fabricated PENGs are subjected to various forms of external mechanical energy such as vibrations, sound, raindrops, pressing, bending, stretching, heartbeats, muscle movements, inhalation, and wind to generate electrical energy.

One simple example of piezoelectric energy harvest is by footwear. Energy harvester mounted on footwear that generated electrical energy from pressure on the shoe sole using multilayer laminates of PVDF or PZT unimorphor rotary electromagnetic generator [52]. Strain or deformation

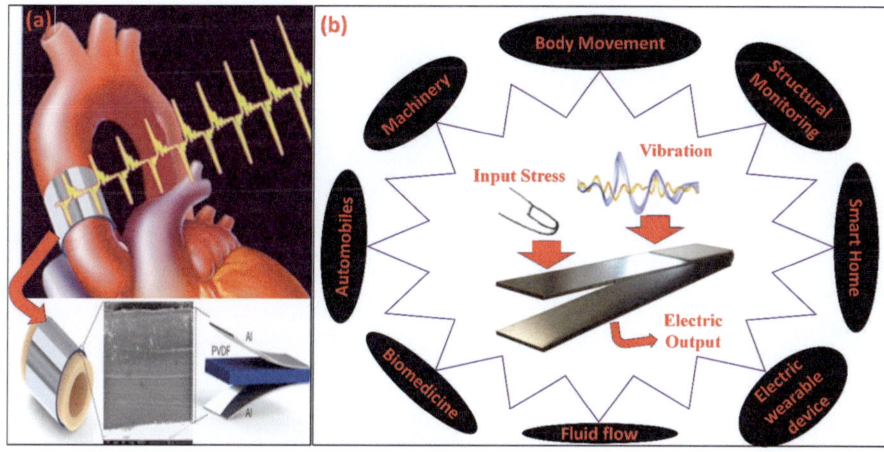

Fig. 15.7 (**a**) A PVDF Harvester that captures energy from the arterial wall, and (**b**) Various piezoelectric applications

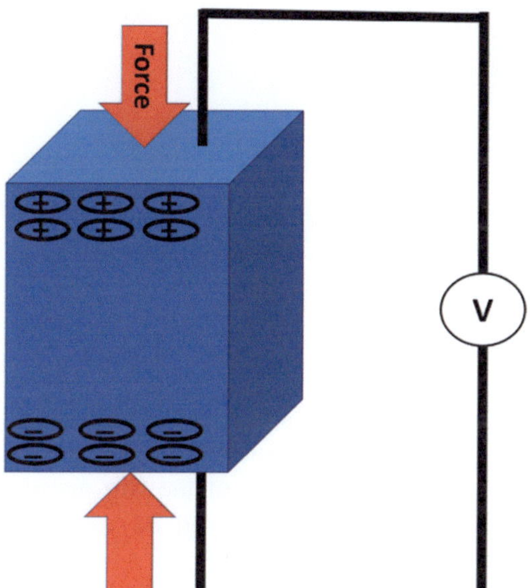

Fig. 15.8 Schematics of energy harvesters based on the piezoelectric effect

of a piezoelectric material causes charge separation across the device, producing an electric field and consequently a voltage drop proportional to the stress applied. A mini-scale electromagnetic energy harvester prototype that consists of a coil & a silicon wafer cantilever beam shown in Fig. 15.9a, with four pole magnets as its proof mass, was developed by Beeby et al. [53]. The harvester was able to produce considerably high power over its size. The mechanical and electrical equivalent of the energy harvester design is shown in Fig. 15.9b and c respectively. The high-temperature energy harvester, incorporating silicon carbide electronics and a PZT energy harvester, can operate at 300 °C. The electricity produced is thereafter formatted by a static converter before supplying a storage system or the load. By employing the capacitive impedance, mechanical vibration of varying amplitude can be harvested into energy. For optimal power flow purposes, an energy-harvesting circuit is proposed which consists of an AC–DC rectifier with an output capacitor, an electrochemical battery, and a switch-mode DC–DC converter that controls the energy flow into the battery [54].

Another general example is the piezoelectric crystal dance floor that harnesses the energy created by dancer's steps as shown in Fig. 15.10. The floor is based on the piezoelectric effect. Dance Floor modules flex slightly when stepped on which creates a movement that can be transformed into electric power by a small internal generator [55]. Each module by the size of 75 × 75 × 20 cm can produce up to 35 W of sustained output, between 5 and 20 W/person. The floor can trigger or power external systems, such as music, video screens, charging mobile phones, and charging a Christmas tree. Some nightclub uses the same principle for its dance floor, which the owners hope will one day generate 60% of the club's electricity [56].

Piezoelectric materials are also used in tire pressure monitoring system (TPMS). The automotive industry has a great interest in small-scale sensors for control and safety applications. Currently, each vehicle is equipped with about 60 sensors, and an increasing number of sensors are being integrated into vehicles along with the rapid development of self-driving technology and electric cars. One main barrier to the trend in this context is the cabling and connectors of these sensor modules. The total cable length in a modern car is more than 4 km (Auzanneau 227), which adds additional weight, occupies much space, and lowers the vehicle's reliability as a whole. These troubles can be avoided by updating sensors to be wireless [57]. However, wireless sensor modules cannot operate automatically for a long time due to the limited capacity of embedded batteries. To address the power source issue, researchers have started to exploit on-site wasted energy in vehicles. By harvesting the tremendous vibration and deformation energy in running vehicles, energy

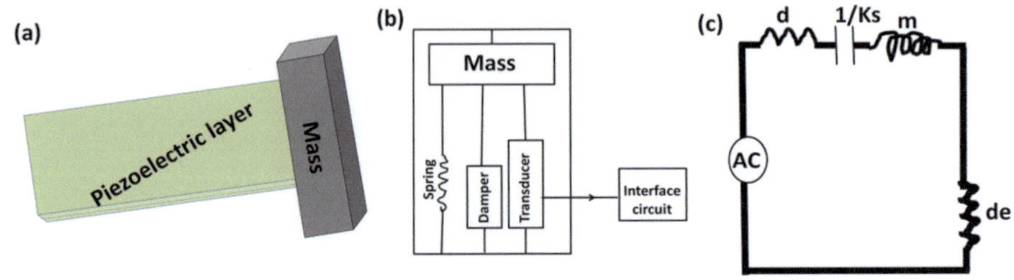

Fig. 15.9 (**a**) PZT-coated cantilever beam, (**b**) Mechanical equivalent model of piezoelectric energy harvester, and (**c**) Electrical equivalent model of a piezoelectric energy harvester (Here, Ks = Stiffness of the Spring, m = Mass, z(t) = Mass Position, d = Damping)

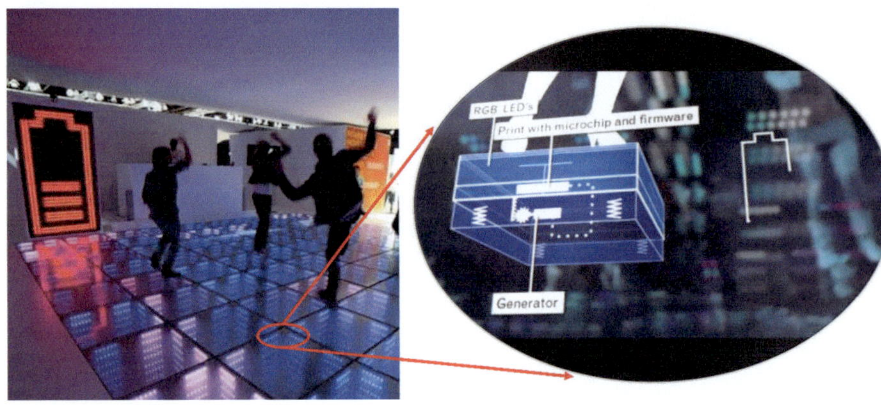

Fig. 15.10 Mechanism of dance floors with piezoelectric crystals installed

harvesters can effectively extend the battery life, with great potential of achieving lifelong autonomous operations of these small-size sensor modules. A variety of energy-harvesting methods have been proposed for different applications in the automotive industry. The tire is a component in a vehicle that directly contacts the ground. They carry out multiple key tasks including supporting the vehicle load, absorbing shocks from the road, and providing grip during steering and braking; thus, tires play a determinant role in driving safety and comfort. The TPMS was initially proposed to enhance safety but also contributes to higher fuel efficiency and longer tire life [58]. TPMS is a very promising application for energy harvesters and may become one of the first commercially available energy-harvesting products (Fig. 15.11) [59].

Triboelectric Effect: Newly invented triboelectric nanogenerators (TENG), based on the well-known contact electrification effect, provide a new approach to generating electricity from mechanical energy to operate small electronic devices (Fig. 15.12). In 2012, TENGs are invented and explained based on electrostatic and contact electrification physics. In the contact electrification effect, a material surface becomes electrically charged after it comes into contact with a different material through friction, owing to charge transfer between the two materials [60]. These transferred charges remain for a long time on their respective surfaces. An electrostatically charged material causes a potential, and it drives induced electrons to flow between the electrodes by periodic contact and separation of the two materials. The generated electric potential (V) can be calculated using Eq. (15.9).

$$V = -\frac{\rho d}{\epsilon_0}, \tag{15.9}$$

where ρ is the triboelectric charge dendsit y, ϵ_0 is the vacuum permittivity, and d is the interlayer distance in a given state. The current (I) generated across an external load can be defined as follows, Eq. (15.10).

$$I = C\frac{\partial V}{\partial t} + V\frac{\partial C}{\partial t}, \tag{15.10}$$

where C denotes the capacitance of the system and V is the voltage across the two electrodes. The first term is the change in potential between the top and the bottom electrodes due to the triboelectric charges. The second term is the variation in the capacitance of the system when the distance between two electrodes is changed due to mechanical deformation. Numerous advantages of TENG, including superior power output performance, many material options, easy tailoring of device structures, cost-effectiveness, the facile fabrication of large areas for applications, and stability and robustness, as

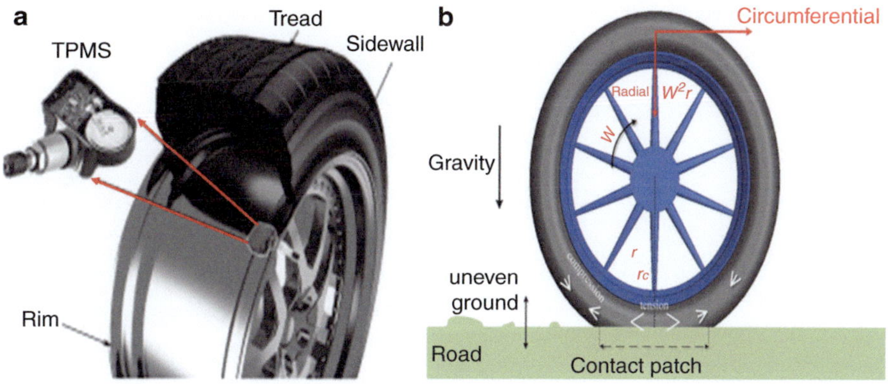

Fig. 15.11 (**a**) TPMS module and its installation on a wheel. (**b**) The load conditions of a wheel rotating on a road. Direct TPMS uses a sensor mounted on the wheel to measure air pressure in each tire. When air pressure drops 25% below the manufacturer's recommended level, the sensor transmits that information to the car's computer system and triggers your dashboard indicator light

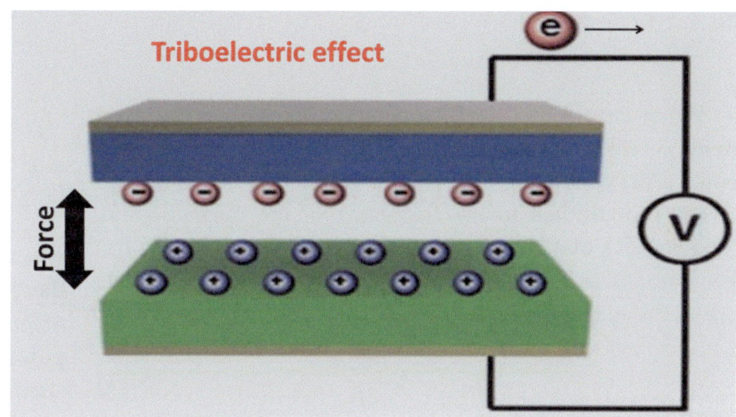

Fig. 15.12 The triboelectric effect

well as environmental friendliness, will bring more opportunities to our daily lives in the near future. Depending on the configuration of the electrodes and the different ways in which the triboelectric layers can be arranged, four operation modes of TENG have been developed: the vertical contact mode, the lateral sliding mode, the single electrode mode, and the free-standing mode [61, 62]. Their applications as various self-powered nanosystems, such as acceleration sensors, motion vector sensors, biomedical monitoring systems, electrochromic devices, sound recording systems, pressure sensors, angle measurement sensors, active tactile sensor systems, tactile imaging devices, electroluminescent systems, and mercury-ion detection systems, have been recently demonstrated. The triboelectric output can be further enhanced through control of electron affinity as well as the work function, chemical structure, pressure, and surface roughness of the materials [63].

15.2.4 Electromagnetic Energy Harvesting

Electromagnetic (EM) waves are created as a result of vibrations between an electric field and a magnetic field. In other words, EM waves are composed of oscillating magnetic. EM waves travel with a constant velocity of 3.00 × 108 m/s in a vacuum. They are deflected neither by the electric field nor by the magnetic field. In many applications, an RF energy harvester can supply power when other energy sources such as light, vibrations, and thermal gradients are not available. Electromagnetic energy conversion is a conventional technology being used in dynamos for macroscale electricity generation [64]. Conventional wind and water turbines were invented based on this effect. Modern electromagnetic energy harvesters with microscale electrical output are designed to harvest other forms of kinetic energy, for example, vibrations. One of the popular configurations of an electromagnetic energy harvester is based on a cantilever structure [65].

Electromagnetic energy can harvest through kinetic energy. Electromagnetic harvesters produce energy by means of electromotive force that a varying magnetic flux induces through a conductive coil according to Faraday's law, Fig. 15.13. The magnetic flux (B) source is obtained with a permanent magnet. The motion of a seismic mass attached to either a coil or magnet produces the variation in magnetic flux necessary to induce a current in the coil [66].

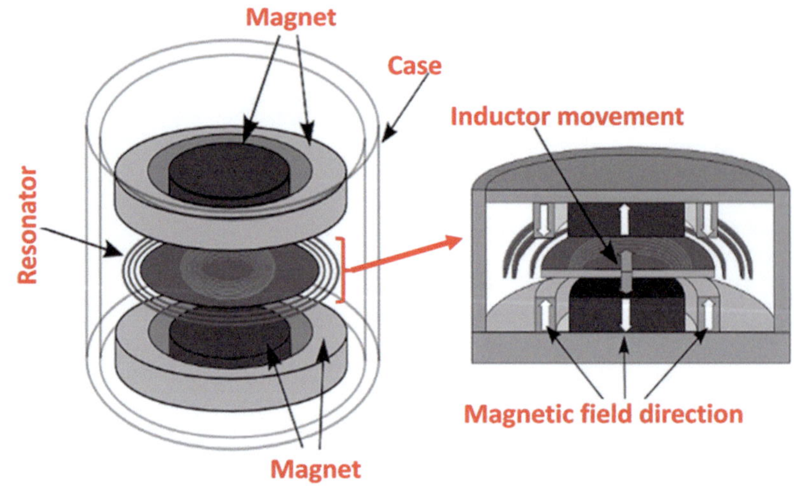

Fig. 15.13 Magnetic kinetic harvester

When an electric conductor moves through a magnetic field, an electromotive force (*Emf*) is induced between the ends of the conductor. The voltage induced in the conductor (V) is proportional to the frequency of the magnetic flux linkage ($\varnothing$) of the circuit, as shown in Eq. (15.11). The generator is a multi-turn coil (N), and permanent magnets create the magnetic field.

$$V = -N\frac{\mathrm{d}\varnothing}{\mathrm{d}t} \tag{15.11}$$

There are two possible cases: one is linear vibration and another one is time-varying magnetic field (B). In the linear vibration case, there is a relative motion between the coil and the magnet, and the voltage induced in the coil can be expressed as the product of a flux linkage gradient and the speed of movement as in Eq. (15.12). In the time-varying magnetic field (B) case, the flux density is uniform over the area (A) of the coil, and then the induced voltage depends on the angle (α) between the coil area and the direction of the flux density, as in Eq. (15.13).

$$V = -N\frac{\mathrm{d}\varnothing}{\mathrm{d}x}\frac{\mathrm{d}x}{\mathrm{d}t} \tag{15.12}$$

$$V = -\mathrm{NA}\frac{dB}{dt}\sin\alpha \tag{15.13}$$

Power is extracted from the generator by connecting the coil to a load resistance, RL. The induced current in the coil generates a magnetic field, which is opposed to the original magnetic field generated by the permanent magnets, according to the Faraday-Lenz law of electromagnetic induction [67]. This electromagnetic induction results in an electromotive force, F_{em}, that opposes the generator motion, which transfers mechanical energy into electrical energy. Thus, F_{em} is proportional to the current and the speed and it is defined in the Eq. (15.14).

$$F_{\mathrm{em}} = D_{\mathrm{em}}\frac{\mathrm{d}x}{\mathrm{d}t}, D_{\mathrm{em}} = \frac{1}{R_L + R_c + j\omega L_c}\left(\frac{d\varnothing}{dx}\right)^2, \tag{15.14}$$

where D_{em} is the electromagnetic damping, R_L is the load, R_c the coil resistance, $L_c$the coil inductance, and $\mathrm{d}\Phi/\mathrm{d}x$ is the magnetic flux. Therefore, to obtain the maximum electrical power output, the generator design must maximize D_{em} and speed. Increasing D_{em} implies maximizing the flux linkage gradient and minimizing the coil impedance. The flux linkage gradient is a function of the strength of the magnets, their relative position with the coil and the direction of movement, and the area and number of turns for the coil. The type of magnetic material determines the magnetic field strength. Properties of some common magnetic materials are given in Table 15.1. Usually, permanent magnets are made of ferromagnetic materials. Table 15.2 illustrates conductor material characteristics that are used for electromagnetic generators [68].

Figure 15.14 represents the equivalent circuit model for vibration-driven harvester using electromagnetic damping. The components on the primary side model the mechanical parts, where the current source represents the energy flux; the capacitor the mass; the inductor the spring; and the resistance the parasitic damping. The electronic components on the secondary side model are the self-inductance of the coil in the electromagnetic device.

These offer a well-established technique of electrical power generation and the effect has been used for many years in a variety of electrical generators.

Electromagnetic-radio Wave Energy Harvesting: Radiofrequency (RF) is any of the electromagnetic wave frequencies that lie in the range extending from around 3 kHz to 300 GHz, which include those frequencies used for communications or radar signals (Fig. 15.15). RF usually

Table 15.1 Properties of some common magnetic materials

Material	(BH)max (kJ/m^3)	Flux density (mT)	Max. work temperature (°C)	Curie temp. (°C)
Ceramic	26	100	250	460
AlNiCo	42	130	550	860
SmCo	208	350	300	750
NdFeB(N38H)	306	450	120	320

Table 15.2 Conductor materials resistivity and conductivity properties

Material	ρ(Ω.m) at 20 °C	σ (S/m) at 20 °C
Silver	1.59×10^{-8}	6.30×10^{7}
Copper	1.68×10^{-8}	5.96×10^{7}
Annealed copper	1.72×10^{-8}	5.80×10^{7}
Gold	2.44×10^{-8}	4.10×10^{7}
Aluminium	2.65×10^{-8}	3.77×10^{7}
Calcium	3.36×10^{-8}	2.98×10^{7}
Tungsten	5.60×10^{-8}	1.79×10^{7}
Zinc	5.90×10^{-8}	1.69×10^{7}
Nickel	6.99×10^{-8}	1.43×10^{7}
Lithium	9.28×10^{-8}	1.08×10^{7}
Iron	9.71×10^{-8}	1.00×10^{7}

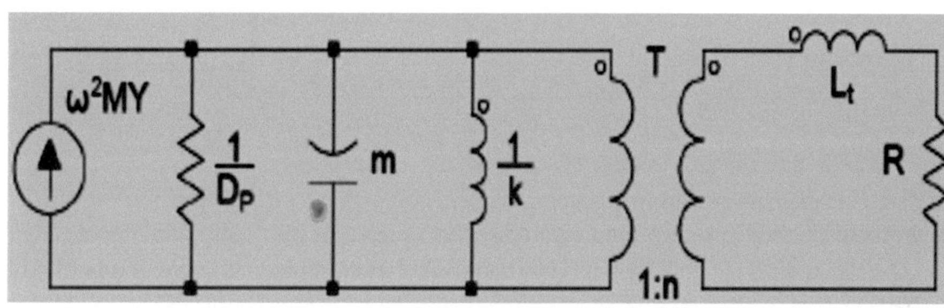

Fig. 15.14 Equivalent circuit of an electromagnetic force harvester

refers to electrical rather than mechanical oscillations [68]. RF radiation is employed to power ID cards by directing high-power electromagnetic energy to the devices from a nearby source. In addition to energy, it is possible to send information as well. However, the term energy harvesting implies that it is the same device that gets its energy from the environment. In cities and very populated areas, there are a large number of potential RF sources: broadcast radio and tv, mobile telephony, wireless networks, etc. The problem is collecting all these disparate sources and converting them into useful energy. The conversion is based on a rectifying antenna (*rectenna*), constructed with a Schottky diode located between the antenna dipoles. The energy levels actually present are so low that no present electronic device can use them. However, future technologies may allow the fabrication of lower power devices that would "recycle" RF energy generated for other purposes by different elements [69]. In the presence of electronics or radiated power sources, such as TV/radio transmitters and mobile base stations, there are typically electromagnetic fields. Another group of energy harvesters can generate electrical power by capturing energy from the electromagnetic field in the free space. The high-frequency electromagnetic energy can be converted into electrical current by using a dipole antenna to capture the electromagnetic waves and a diode connected across the dipole to rectify the AC current into DC current. Although electromagnetic energy sources usually have low-power densities (0.01–0.1 W/cm^2) when compared to other sources, the power in free space can be useful for remotely distributed sensors [70]. However, to capture enough electromagnetic energy from waves in free space, independent antennas and matching circuits need to be specifically designed, which is challenging for electronics requiring small device volumes. Electrization of the dielectric can be produced by four effects: (1) migration of internal charges to the surfaces of the dielectric, (2) migration of charges to the various layers of the dielectric, (3) migration of charges within the molecules of the dielectric, and (4) orientation of molecular diploes within the dielectric [71].

Radiofrequency energy harvesting (RFEH) is suitable for long-range wireless power transfer, that is, centimeter range

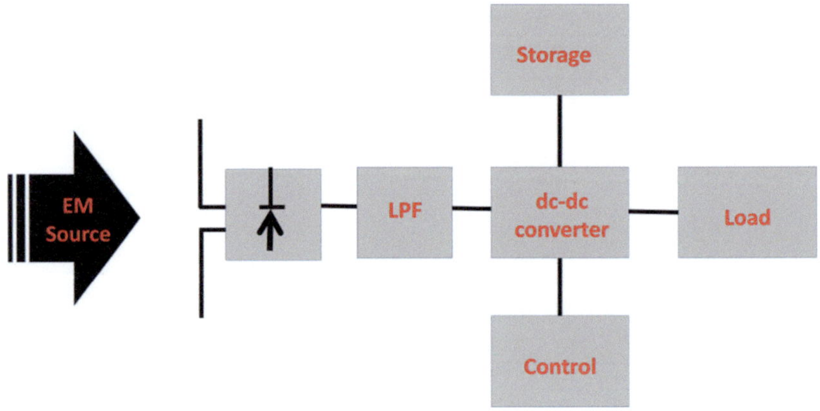

Fig. 15.15 EM energy harvest schematics

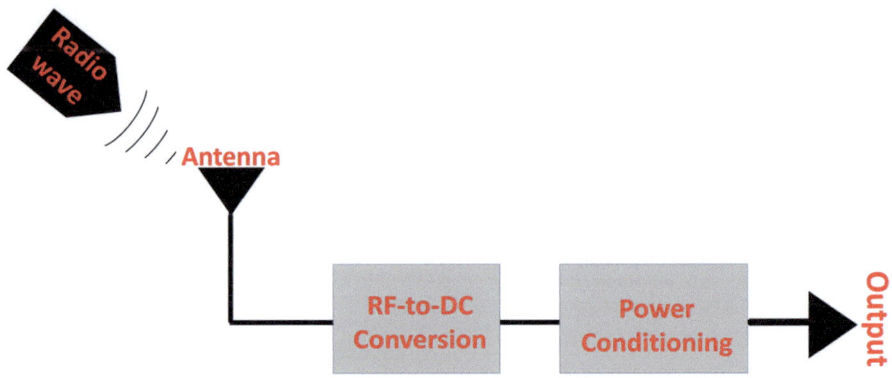

Fig. 15.16 Radiofrequency energy harvest schematics

for high-frequency on-chip antennas to several meter range for off-chip antennas (Fig. 15.16). This makes RFEH suitable for battery-less sensors in a WSN remotely powered by a hub (i.e., RF source). RFEH suits many applications, such as a smart house, smart grid, Internet of things (IoT), and wireless body area networks (WBAN) [72]. Especially, in the last few years, the WBAN application is gaining importance due to the growing importance of health care in society, as health needs to be continuously monitored to identify chronic diseases or prevent illness. Examples of WBANs are a sensor array for monitoring ExG signals and a disposable battery-less band-aid sensor. In WBAN applications, sensors may require power about micro-watts, depending on how they operate. For example, a temperature sensor is not required to update its momentary value very often as the temperature is a slowly varying quantity in most applications. On the other hand, the peak power consumption of a duty-cycled sensor might be significantly larger than the harvested (average) power. In such a case, the energy provided by the RFEH can be stored in a capacitor or battery that periodically supplies energy to the sensor. In this example, the RFEH is periodically connected to a capacitive load; thus, efficient energy transfer is required during capacitor charging to minimize losses and charging time [73]. Obviously, the energy supplied by the RFEH over time should be greater than the energy consumed by the sensor. An antenna-rectifier co-design example is given to realize a high-performance RF energy harvester. The design is verified by measurements in an anechoic chamber and with energy harvesting from an ambient RF source [74].

The obvious appeal of harvesting ambient RF energy is that it is essentially “free” energy. The number of radio transmitters, especially for mobile base stations and handsets, continues to increase. ABI Research and iSupply estimate the number of mobile phone subscriptions has recently surpassed 5 billion, and the ITU estimates there are over 1 billion subscriptions for mobile broadband. Mobile phones represent a large source of transmitters from which to harvest RF energy and will potentially enable users to provide power-on-demand for a variety of close-range sensing applications. Also, consider the number of WiFi routers and wireless end devices such as laptops [75]. In some urban environments, it is possible to literally detect hundreds of WiFi access points from a single location. At short range, such as within the same room, it is possible to harvest a tiny amount of energy from a typical WiFi router transmitting at a power level of 50–100 mW. For longer range operation, larger antennas with higher gain are needed for practical harvesting of RF energy from mobile base stations and broadcast radio towers [76]. In 2005, Powercast demonstrated ambient RF energy harvesting at 1.5 miles (~2.4 km) from a small, 5 kW AM radio station. RF energy can be broadcasted in unlicensed bands such as 868 MHz, 915 MHz, 2.4 GHz, and 5.8 GHz when more power or more predictable energy is needed than what is available from ambient sources. At 915 MHz, government regulations limit the output power of radios using unlicensed frequency bands to 4 W effective isotropic radiated power (EIRP), as in the case of radiofrequency-identification (RFID) interrogators [77]. As a comparison, earlier generations of mobile phones based on analog technology had maximum transmission power of 3.6W, and Powercast's TX91501 transmitter that sends power and data is 3 W.

RF energy-harvesting devices, such as Powercast's Powerharvester® receivers, convert RF energy into DC power. These components are easily added to circuit board designs and work with standard or custom 50-ohm antennas. With the current RF sensitivity of the P2110 Power harvester receiver at -11 dBm, powering devices or charging batteries at distances of 40–45 ft from a 3 W transmitter is easily achieved and can be verified with Powercast's development kits. Improving the RF sensitivity allows for RF-to-DC power conversion at greater distances from an RF energy source [78]. However, as the range increases, the available power and rate of charge decrease. An important performance aspect of an RF energy harvester is the ability to maintain RF-to-DC conversion efficiency over a wide range of operating conditions, including variations in input power and output load resistance. For example, Powercast's RF energy-harvesting components do not require additional energy-consuming circuitry for maximum power point tracking (MPPT) as is required with other energy-harvesting technologies. Powercast's components maintain high RF-to-DC conversion efficiency over a wide operating range that enables scalability across applications and devices. RF energy-harvesting circuits that can accommodate multi-band or wideband frequency ranges, and automatic frequency tuning, will further increase the power output, potentially expand mobility options, and simplify installation [79].

RF energy can be used to charge or operate a wide range of low-power devices. At close range to a low-power transmitter, this energy can be used to trickle charge a number of devices, including GPS or RLTS tracking tags, wearable medical sensors, and consumer electronics such as e-book readers and headsets. At longer range, the power can be used for battery-based or battery-free remote sensors for HVAC control and building automation, structural monitoring, and industrial control [80]. In some applications simply augmenting the battery life or offsetting the sleep current of a microcontroller is enough to justify adding RF-based **wireless power** and energy harvesting technology.

15.2.5 Electrostrictive Energy Harvesting

Electrostrictive energy harvesters are made of electrostrictive polymers (e.g., elastomers) to which are applied DC bias electric fields to induce statics or polarizations within materials. The electrostrictive energy harvesters work as either electrostatic or pseudo-piezoelectric energy harvesters. Compared to other kinetic energy harvesters, electrostrictive energy harvesters have advantages in terms of stretchability and flexibility because of their use of polymers. Their disadvantage is the necessity for an external voltage source, similar to that of electrostatic energy harvesters [81]. The important factors affecting their energy harvesting capability and applicability are the dielectric properties (e.g., permittivity, the higher the better) of the electrostrictive polymers and the operating voltage (external bias DC voltage, the lower the better), respectively. Developed composites made of polymer matrix and dielectric or other fillers are typical ways to pursue. Yin et al. reported a plasticizer-modified electrostrictive polymer. An energy conversion efficiency of 34% and an output power density of 4.31 mW/cm^3 could be achieved with a DC bias electric field of 300 kV/cm [82]. Tugui et al. developed a highly stretchable free-standing electrode, PDMS-carbon black. By integrating such electrodes with a commercial silicone elastomer (Elastosil), an energy density of 1.1 mJ/cm^3 was achieved under 200% strain [83]. Electrostatic harvesters are made of a variable capacitor whose plates are electrically isolated from each other by air, vacuum, or an insulator**,** Fig. 15.17. The external mechanical vibrations cause the gap between the plates to vary, changing the capacitance. To harvest energy, the plates must be charged. In these conditions, mechanical vibrations oppose the electrostatic forces present in the device. Therefore, if a voltage V biases the capacitor and if load circuitry is linear, the motion of the movable electrode produces electrical power. The fundamental equations that model this operation are Eqs. (15.15) and (15.16).

$$C = \frac{Q}{V} = \epsilon \frac{A}{d} \quad \text{i.e.} \quad V = \frac{Qd}{\varepsilon_0 A}, \tag{15.15}$$

$$E = 0.5QV = 0.5CV^2 = 0.5\frac{Q^2}{C}, \tag{15.16}$$

where C is the capacitance in Farads, V is the voltage in volts, Q is the charge in coulombs, A is the area of the plates in m^2, d is the gap between plates in m, ε is the permittivity of the material between the plates, ε_0 is the permittivity of free space, and E is the stored energy in Joules. Electrostatic generators can be either voltage or charge-constrained [84]. Voltage-constrained devices have a constant voltage applied to the plates; therefore, the charge stored on the plates varies with changes in capacitance. This usually involves an operating cycle that starts with the capacitor at its maximum

Fig. 15.17 Electrostatic kinetic harvester

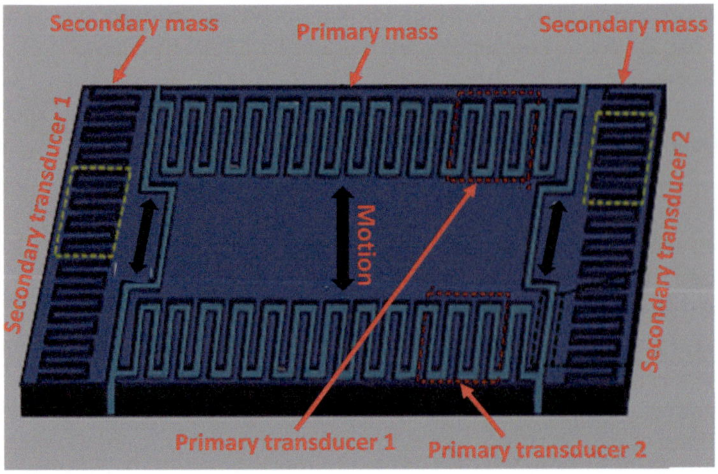

capacitance value (C_{max}). Then, the capacitor is charged up to a specified voltage (V_{max}) from a reservoir while the capacitance remains constant. The voltage is held constant while the plates move apart until the capacitance reaches its minimum value (C_{min}). The excess charge flows back to the reservoir as the plates move apart gaining energy. This energy is determined by Eq. (15.17).

$$E = \frac{1}{2}(C_{max} - C_{min})V^2{}_{max} \tag{15.17}$$

Alternatively, a fixed charge can be obtained using electrets materials, such as Teflon or Praylene. In either case, the mechanical work against the electrostatic forces is converted into electrical energy. The net energy gained is determined by Eq. (15.18).

$$E = \frac{1}{2}(C_{max} - C_{min})V_{max}V_{start} \tag{15.18}$$

In both equations, V_{max}must be compatible with the electronics and the fabrication technology. The previous two approaches have different strengths and weaknesses. The electrostatic generator can be categorized into one of three types: (1) Out of plane, gap varying, voltage constrained; (2) In a plane, overlap varying, charge constrained; and (3) In a plane, the gap is varying.

The electrostatic energy harvesters have many advantages over other methods of vibrating energy harvesting, such as high-quality factor Q, low noise, wide tuning range, and contained size. However, electrostatic harvesters produce less energy than other kinetic harvesters, and their application range is more limited due to their operating characteristics. Figure 15.18 represents the equivalent circuit model for vibration-driven harvester using electrostatic damping. The circuit on the primary side of the transformer models the mechanical behavior of the harvester [85]. The voltage source represents the vibration source; the capacitor represents the mass; the inductor represents the spring; and the resistor represents the parasitic damping. The electrical elements of the generator are on the secondary side, where the capacitor models the terminal capacitance of the piezoelectric material or the moving capacitor.

Other materials such as lead magnesium niobate, lead magnesium niobate-lead titanate, and lead lanthanum zirconate titanate are electrostrictive materials, and these can be used in electrostrictive energy-harvesting process.

15.2.6 Magnetostrictive Energy Harvesters

The magnetostrictive effect is the phenomenon where ferromagnetic materials change their shape during magnetization, that is, with the variation in orientation or intensity of an external magnetic field. The change in length or shape of a ferromagnetic substance under the application of an external magnetic field is known as the Joule effect [86]. Its reverse effect is also possible which is popularly known as the Villari effect. This effect tells that whenever mechanical stress such as vibration is imposed to a magnetostrictive material, it changes the magnetic domain orientation in that material which in turn changes the magnetic flux. According to Faraday's law of electromagnetic induction, if there is a change in magnetic flux with respect to time, then there will be an induced emf. By extracting this induced voltage, we can generate electricity too. So the source of that electricity is ambient vibration. For a magnetostrictive energy harvester, the coupling coefficient for electrical output from mechanical input is related to the rate of flux change within the magnetostrictive core and the transducer electrical impedance characteristics. The basis of energy conversion from the rate of flux change to electrical output is Faraday–Lenz's law. If it is possible to wound copper coil with few turns across the magnetostrictive rod, then the change in flux through the area surrounded by the turns of wire will induce a voltage in the

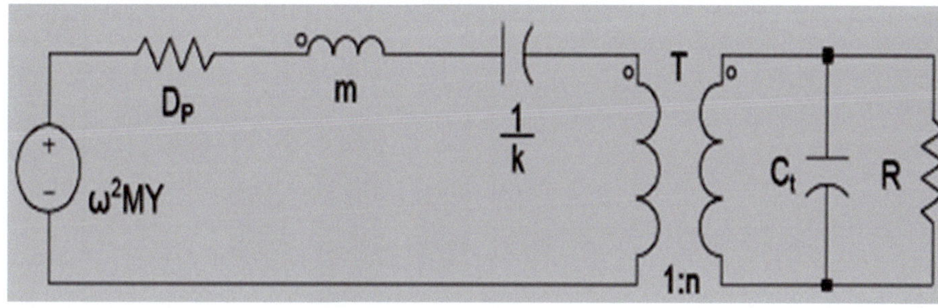

Fig. 15.18 Equivalent circuit of an electrostatic force harvester

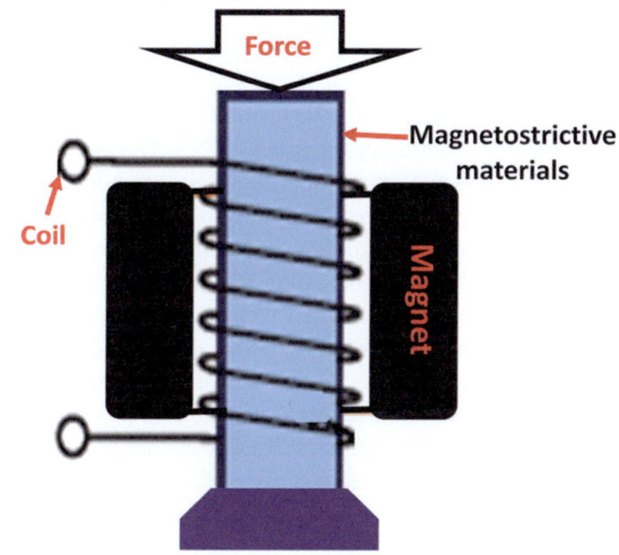

Fig. 15.19 Configuration of axial-type magnetostrictive energy harvester

wire which will oppose the change in flux. Therefore, a stress-induced change in the magnetostrictive rod flux density will produce a voltage proportional to the magnitude change in flux in the coil which should be placed concentrically around the magnetostrictive rod. For a given change in flux, a coil with a greater number of turns will produce greater voltage output than a coil with fewer turns (Fig. 15.19). However, as the number of turns of the coil increases, so does the resistance of the coil [87]. Hence, for the design of an optimized coil, there will be a trade-off in performance associated with the wire gauge and the number of turns to maximize power. The scaling effects of magnetostrictive materials also occur and the voltage induced across the Galfenol rod can be described as:

$$V = 2\pi fNBS, \tag{15.19}$$

where N is number of turns of the coil, B is Flux density variation, S is cross-sectional area of the Galfenol rod, f is first resonant frequency.

The resistance of the coil,

$$R = \rho\frac{L}{\pi r^2}, \tag{15.20}$$

where ρ is resistivity, L is coil wire length, and r is coil wire radius.

The power is measured as,

$$P = \frac{V^2}{R}. \tag{15.21}$$

Now if the dimension is multiplied by k time,

$$V_k = NB\left(K^2S\right)2\pi\left(\frac{f}{K}\right) = KV \tag{15.22}$$

$$R_K = \rho\frac{KL}{\pi(Kr)^2} = \frac{R}{K} \tag{15.23}$$

Therefore, power becomes,

$$P_K = \frac{{V_K}^2}{R_K} = K^3P \tag{15.24}$$

So, if the dimension is multiplied by a factor K, then the power becomes a cube of that factor. So power generation in a magnetostrictive energy harvester is directly proportional to the volume of the energy harvester. Using this concept, huge power can be harvested in the future from ambient vibrations.

Magnetostrictive materials have specific properties that show a coupling relationship between strain and stress mechanical quantities and magnetic and induction field strength. Magnetostrictive materials have a constitutive relationship that directly couples mechanical and/or thermal variables to magnetic variables, and they are used to build actuators or sensors. Magnetostrictive harvesters are divided into two main categories: direct force or force-driven, and inertial or velocity-driven [88, 89], as shown in Fig. 15.20. The figure includes two conceptual implementations of the mechanical part. Figure 15.20a shows where the active material is used between the source of the vibrations and a reference frame. The magnetostrictive rod is bound to a rigid frame and undergoes a time-variable, uniform vertical force

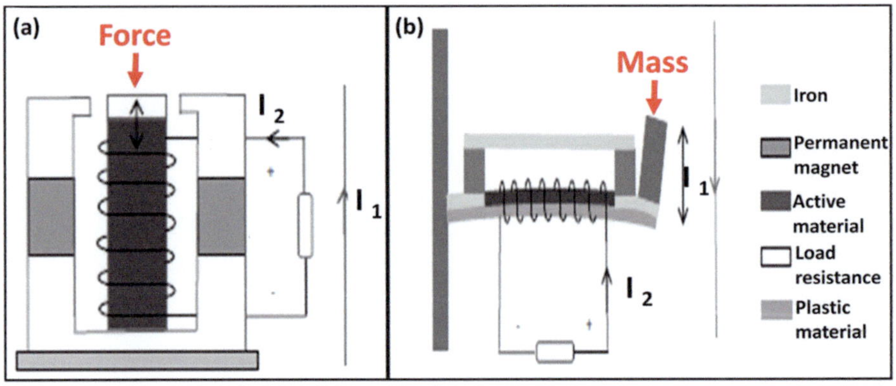

Fig. 15.20 Two types of magnetostrictive energy harvester (**a**) force-driven, and (**b**) velocity-driven

(z axis). A z-axis-directed compressive stress then appears, and the material generates a time-variable magnetization. Figure 15.20b is suitable when a vibrating frame is available. Here, one end of a magnetostrictive cantilever beam is rigidly connected to the vibrating frame; the other end is attached to a heavier mass. Because of the induced oscillations over the mass, the material undergoes longitudinal stress that leads to time-variable magnetization. Both methods share some common needs: a coil wrapped around the magnetostrictive material and a magnetic circuit to convey and close the magnetic flux lines. In brief, vibration energy harvesting is considered one of the most promising real solutions to provide electrical energy from any low-power electronic devices [90].

Magnetostrictive materials can be combined with piezoelectric materials and permanent magnets and used for vibration energy harvesting. The general process is that the vibration of permanent magnets causes shape changes in the magnetostrictive materials and then applying the resulting stress or strain to the piezoelectric components. As this is a two-stage energy conversion—kinetic to magnetic to electric energy—extra energy loss during the process may decrease the output and efficiency [91]. On the other hand, it enables two energy sources—magnetic field and kinetic energy to be harvested

15.2.7 Chemical Energy

Chemical energy is the energy stored in chemical substances, which form its energy inside molecules and atoms. Essentially, every compound contains some chemical energy, which can be released when its chemical bonds are broken. Examples of substances containing chemical energy include wood, food, fossil fuels, and batteries. It is one of the most convenient forms we have for storing energy. Chemical energy comes in different forms and may be released during a chemical reaction, usually in the form of heat. In chemical reactions, bonds of chemical substances are broken down, resulting in the formation of new bonds. Light energy and heat energy are released in the process [92]. The energy can then be used for different purposes. There are six primary types of chemical reactions: (1) synthesis (when two simple elements combine to form a more complex substance in a chemical reaction), (2) combustion (chemical energy in the form of heat is produced when oxygen mixes with other materials to form carbon dioxide and water), (3) single displacement (a chemical reaction in which a substance gives out some of its atoms to another), (4) double displacement (a chemical reaction in which a substance exchanges some of its atoms with atoms of another substance), (5) decomposition (when a complex substance is broken down into simpler substances in a chemical reaction), and (6) digestion (the process of breaking down food to release chemical energy is known as digestion) [93, 94]. Chemical energy is stored in certain inorganic compounds. For instance, you can use the chemical energy of the phosphorous compound on the head of a matchstick to produce light and heat energy.

Various examples of chemical energy harvesting have been discussed below:

From Fossil Fuel: Fossil fuels such as gas and methane are some of the most important forms of chemical energy in the world's economy. All it needs to do is to provide a source of ignition to the fuels. This will instantly transform the liquid fuels from their chemical state, generating massive amount of energy in the process. There are numerous ways to harness that energy especially for the purposes of transportation. For instance, when you step on your car's accelerator, the gas is transformed into mechanical energy. The mechanical energy subsequently sets your car in motion, which then generates kinetic energy [95].

From Food: Food is a form of stored chemical energy. It is actually the energy the body uses to move around and perform various functions. The sun provides solar energy for plants to grow. The energy is then converted into chemical energy in the plant tissues. When cooking food, some of the energy is released from the food's chemical bonds due to heat energy being applied. Once one eats the food, the digestive process further converts the chemical energy into a form that the body can utilize [96].

Explosives: Explosives also store chemical energy. Their molecules are composed of atoms that can reorganize themselves into other molecules, which have much less energy. When that happens, the excess is released as heat and light. The existing bonds would break with just a small stimulus. The atoms will then rearrange themselves into molecules with much less energy. The light and heat released, along with the rapid transformation of a substance into gases, generates a ferocious explosion [97].

Batteries: While atoms of some elements can easily donate electrons, others like to receive electrons. This is the concept under which batteries function. Namely, two different substances can be arranged in such a way that electrons flow from one substance to the other when they are connected in a circuit, producing an electric current. A vast range of different materials can be used to transform chemical energy into electricity in this way. This is why we have various types of batteries that can be used to power different types of gadgets and electronics, including phones, computers, drones, cameras, and so on. [98].

Photosynthesis: Plants harvest solar energy from the sun and convert it into chemical energy through a process known as photosynthesis. This process involves chemical reactions where the solar energy is tapped by the plant molecule before being transformed into chemical energy. Eventually, the chemical energy is consumed in the form of glucose [99].

Respiration: When consuming plant material, the glucose molecules are broken down to produce water and carbon dioxide. Carbon dioxide and water, together, have much less energy compared to sugar. Hence, the excess energy is released. The released energy is stored in a molecule known as adenosine triphosphate (ATP). This is done by adding a phosphate group to another molecule known as adenosine diphosphate (ADP). The energy can be released again when needed [100].

Cold Packs: Cold packs used in sports are another great example of chemical energy. The inner pouch is filled with water. When it is broken, it reacts with the ammonium nitrate granules, forming new bonds in the process. It also absorbs energy from the surroundings during the process. Chemical energy is stored in the new bonds, resulting in the decrease of the cold pack's temperature [101].

Coal: Coal is one of the best examples of chemical energy, especially when it is used to generate electricity. Coal stored in a reservoir is fed into a boiler, where it is burned at exceedingly high temperatures. This results in the release of chemical energy in the coal in the form of thermal energy. The thermal energy is then used to boil water in tanks to produce steam. Subsequently, the steam is directed through tubes attached to spinning shafts. The shafts are connected to a generator, which produces electricity from the process [102].

Wood: Wood is a readily available source of chemical energy and has been used since ancient times to generate heat and energy. When you burn wood, the chemical bonds in its structure are broken down. As a result, both light energy and heat energy are produced. In the process, the wood is changed into ashes, a chemical material with completely different properties [103].

15.2.8 Wind Energy Harvest

Wind energy is another method of harvesting energy. This wind flow, or motion energy, when harvested using huge wind turbines, can be used to generate electricity on a large scale. The use of wind power to pump water and grind grain has been around for centuries. In a basic windmill, the force of the wind pushes against the front side of the sails of the windmill causing them to turn. This rotation is mainly caused by drag. The sails are attached to a horizontal windshaft so when the sails turn, the windshaft turns [104]. This rotary motion is then converted into mechanical power and can turn the wheel of a grist mill or pump water. When people talk about harvesting wind energy today, however, they tend to be talking about converting the wind's kinetic energy into electricity (Fig. 15.21). This is a similar process to the ancient windmill, except the mechanical power in the rotor shaft is used to rotate an electric generator, producing electricity [105].

The specialty windmills that are used to convert wind energy into electricity are called wind turbines. There are two different types of wind turbines, horizontal-axis and vertical-axis (Fig. 15.22). In vertical-axis wind turbines, the main rotor shaft (similar to the windshaft of older windmills) is set vertically or perpendicular to the ground. In horizontal-axis wind turbines, the main rotor shaft is set horizontally and the blades are perpendicular to the ground. Most commercial wind turbines that are connected to the electrical grid are horizontal-axis wind turbines [106]. Modern horizontal-axis wind turbines are tall towers that typically have two or three blades. At the top of the tower is a weather vane which is connected to a computer. This keeps the turbine facing the wind. Like with the windmill, the wind moves the blades of the wind turbines, although the blades of the turbine work more like the wing of an airplane than windmill sails [107]. When the wind blows, a pocket of low-pressure air is formed on the back side of the blade, accompanying an area of high-pressure air on the front. This creates a pressure

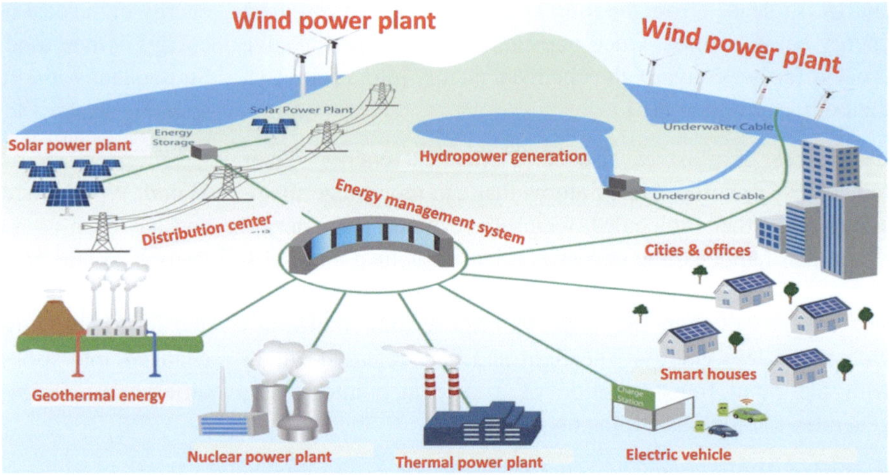

Fig. 15.21 Wind energy harvest and utilization

Fig. 15.22 (**a**) Vertical wind mill, and (**b**) Horizontal wind mill

differential and an aerodynamic force known as lift. Lift causes the blades to spin which, in turn, spins the main rotor shaft that is connected to a generator. The electric generator converts the mechanical energy of the rotor shaft into electricity [108].

One of the challenges of wind power is the intermittent nature of the wind. Even in the optimum locations for wind harvesting, there is no way to guarantee the wind will be providing enough energy to meet the demand for electricity at any given time as in the case of Fig. 15.23. In this application, if no train is coming, then the natural wind is able to rotate the S-rotor and if the train is coming then piston wind is able to rotate the H-rotor. One-way bearing-assisted S-rotor/H-rotor generates energy and stores it in the supercapacitor. Now this energy is used for self-power applications, such as railway, maintain track sensor, emergency repair, and so on. Currently, wind energy is a supplemental source of energy for electrical power grids. To ensure consistent and reliable electricity, wind energy needs to be able to be stored. The wind itself cannot be stored, but there are few ways to store wind energy [109]. Many storage solutions for wind energy have a high initial cost. At the moment, it is far less expensive to keep wind energy as one piece of a varied and flexible energy grid than it is to store wind energy. According to the American Wind Energy Association, wind turbines currently produce enough electricity to power over 15 million homes in America. While impressive, the wind could power much more. In fact, the U.S. wind energy potential exceeds the nation's electricity needs by more than 10 times. With the right energy storage technology, these needs could be met from only wind energy [110].

The rotating speed of wind turbines depends on the wind velocity. In the tunnel, the wind velocity is unstable. As the wind velocity changes, the rotating speed of wind turbines also changes, and therewith the rotating speed of the generator. In the wind energy harvesting arrangement, a three-phase

Fig. 15.23 Wind energy harvesting arrangement at rail runway

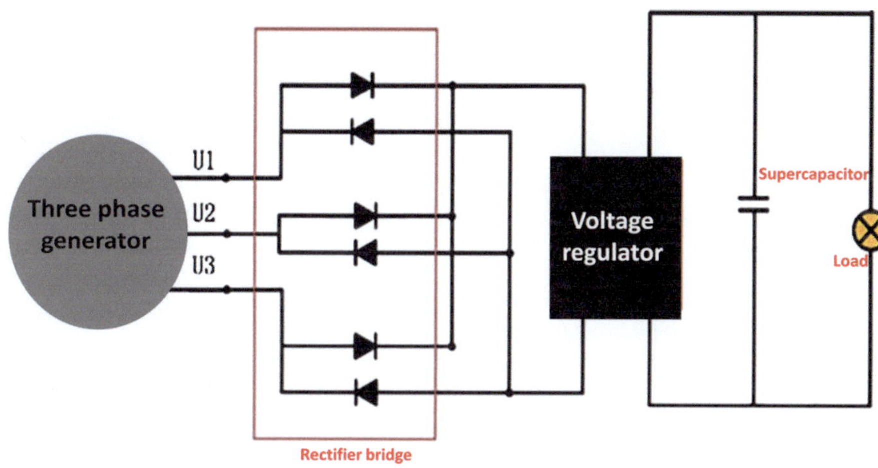

Fig. 15.24 Rectifying circuit for the wind energy harvesting arrangement

AC PMSM is used to generate a three-phase AC. As a result, the output current becomes unstable [111]. Therefore, it is important to rectify the output current and regulate the output voltage. A three-phase bridge rectifier circuit is designed to rectify the current as shown in Fig. 15.24. Electrochemical energy storage systems play an important role in many applications since they are the enabling technology for securing energy supplies. However, batteries are not suitable for the proposed system because of their capacity and the unstable and instantaneous output of the generator. Compared to batteries, supercapacitors usually manage high-power rates, which provide hundreds of times higher power in the same volume. Given the long cycle life, high-power density, simple principles, and quick charging, a supercapacitor is used in the proposed system to store the energy. The stored energy is supplied to railway maintenance, sensors, and emergency repair [112].

The wind flow harvester has two parts, one mechanical and one electrical. This system converts wind energy into mechanical energy, then into electromagnetic, and finally into electrical energy. The electromagnetic wind generators are reliable and have small mechanical damping and magnets suitable to operate at low wind velocities (Fig. 15.25). When the air flow passes through the system structure, the air flow force pushes the blades and because of that they rotate around a pivoting axis. The blades have magnets attached (rotor) and their movement generates a variable magnetic flux. Consequently, the magnetic field created is harvested as a current induced in the coils of the generator (stator) [113].

As the miniature horizontal axis wind turbine (MHAWT) spins with the wind, so the rotor does and captures and transforms the kinetic energy of the incoming wind into mechanical energy. By the aerodynamic equation of Ibrahim [114], the available kinetic power from the air flow is given in Eq. (15.25).

$$P_{\text{wind}} = \frac{\rho A v^3}{2}, \tag{15.25}$$

where ρ the air density, A is the area swept by the rotor of wind turbine, and v the wind speed. However, the conversion of wind power into rotational power in the rotor of the turbine is a complex aerodynamic phenomenon. Ideally, the power obtained from the ambient wind (P_{aero}) is expressed as in Eq. (15.26).

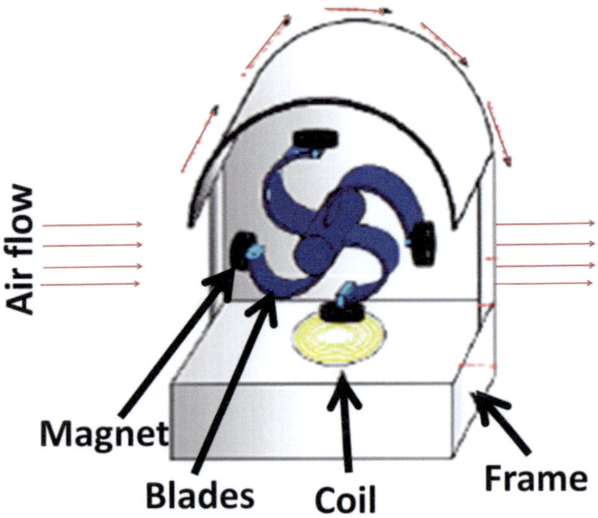

Fig. 15.25 Microwind harvester

$$P_{\text{aero}} = \frac{C_p(\lambda,\theta)\rho A v^3}{2}, \tag{15.26}$$

where $C_p$is the aerodynamic efficiency of the rotor or the power coefficient, which has a nonlinear dependence with the pitch angle θ of the turbine blades and the tip speed ratio λ. This final parameter is expressed in Eq. (15.27).

$$\lambda = \frac{\omega r}{v}, \tag{15.27}$$

where ω is the angular velocity and r the radius of the rotor. In addition, to calculate the power coefficient $C_p(\lambda, \theta)$ for small wind turbines, approximation equations are proposed by the Eqs. (15.28) and (15.29).

$$C_p(\lambda,\theta) = C_1\left[\left(\frac{C_2}{\lambda_i}\right) - C_3\theta - C_4\theta^{C_5} - C_6\right]e^{\left(\frac{C_7}{\lambda_i}\right)} \tag{15.28}$$

$$\lambda_i = \frac{1}{\left[\frac{1}{(\lambda+C_8\theta)}\right] - \left[\frac{C_9}{(\theta^3+1)}\right]}, \tag{15.29}$$

where, $C_1 to$ C_9 are power coefficients. In addition, the aerodynamic profiles of the turbine blades have a significant influence on the efficiency of the spinning. The blades determine directly the system torque force, which influences the output power level. The number of blades also conditions the performance of the energy harvester. System maximum efficiency and output power also depend on the impedance matching between the load, the torque force, and the wind flow.

15.2.9 Tide Energy

Tidal power or tidal energy is the form of hydropower that converts the energy obtained from tides into useful forms of power, mainly electricity. *Tidal energy* is an "alternative energy" that can also be classed as a "renewable energy source," as the Earth uses the gravitational forces of both the moon and the sun every day to move vast quantities of water around the oceans and seas producing tides. As the earth, the moon, and the sun rotate around each other in space, the gravitational movement of the moon and the sun with respect to the earth causes millions of gallons of water to flow around the earth's oceans creating periodic shifts in these moving bodies of water. These vertical shifts of water are called "tides" [115]. Tides are more predictable than the wind and the sun. Among sources of renewable energy, tidal energy has traditionally suffered from relatively high cost and limited availability of sites with sufficiently high tidal ranges or flow velocities, thus constricting its total availability. However, many recent technological developments and improvements, both in design (e.g., dynamic tidal power and tidal lagoons) and turbine technology (e.g. new axial turbines and cross flow turbines), indicate that the total availability of tidal power may be much higher than previously assumed and that economic and environmental costs may be brought down to competitive levels. Historically, tide mills have been used both in Europe and in the Atlantic coast of North America. The incoming water was contained in large storage ponds, and as the tide went out, it turned waterwheels that used the mechanical power it produced to mill grain [116]. The earliest occurrences date from the middle ages or even from Roman times. The process of using falling water and spinning turbines to create electricity was introduced in the United States and Europe in the nineteenth century. The world's first large-scale tidal power plant was the Rance Tidal Power Station in France, which became operational in 1966. It was the largest tidal power station in terms of output until Sihwa Lake Tidal Power Station opened in South Korea in August 2011. The Sihwa station uses sea wall defense barriers complete with 10 turbines generating 254 MW [117].

Tidal energy is environment-friendly energy and does not produce greenhouse gases. As 71% of Earth's surface is covered by water, there is scope to generate this energy on large scale. We can predict the rise and fall of tides as they follow cyclic fashion. The efficiency of tidal power is far greater when compared to coal, solar, or wind energy. Its efficiency is around 80%. Although the cost of construction of tidal power is high, the maintenance costs are relatively low. Tidal energy does not require any kind of fuel to run. The life of tidal energy power plant is very long. The energy density of tidal energy is relatively higher than other renewable energy sources. The intensity of sea waves is

Fig. 15.26 Tidal stream turbine (**a**) Side view (before installation), and (**b**) Front view inside the sea

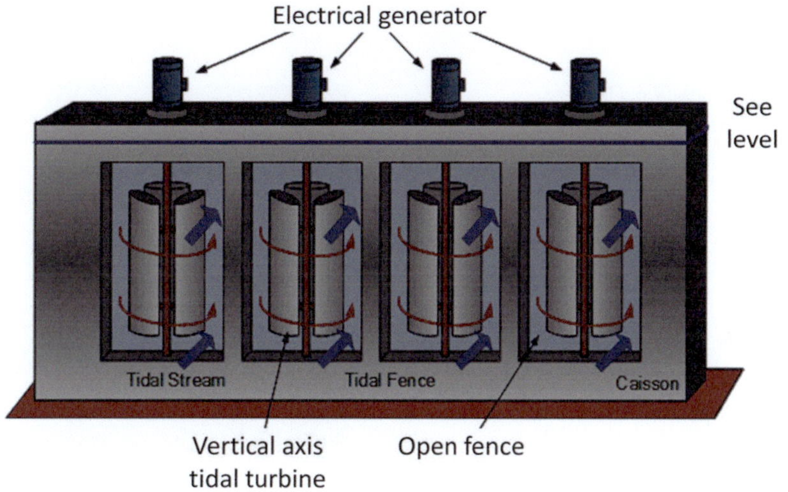

Fig. 15.27 Vertical-axis tidal turbine energy harvesting

unpredictable, and there can be damage to power generation units [118].

15.2.9.1 Types of Tide Energy to Harvest

Described subsequently are several popular methods and technologies that are leading the way in capturing this renewable ocean energy.

15.2.9.1.1 Tidal Stream Turbines

A *tidal stream generation* system (Fig. 15.26) uses turbine generators beneath the surface of the water. Major tidal flows and ocean currents, like the Gulf Stream, can be exploited to extract its tidal energy using underwater rotors and turbines. Single rotors are set with fixed or adjustable-pitch blades that can align with the tidal currents as they fluctuate. One of the world's largest arrays, built by SIMEC Atlantis Energy, is feeding Scotland's power grid with steady, reliable power. With blades up to 65 feet in diameter, its four 1.5-MW turbines generated more than 7 GWh of renewable energy in 2019 [119].

Tidal stream generation is very similar in principle to wind power generation except that time water currents flow across turbines rotor blades which rotates the turbine, much like how wind currents turn the blades for wind power turbines. In fact, tidal stream generation areas on the sea bed can look just like underwater wind farms. Unlike off-shore wind power that can suffer from storms or heavy sea damage, tidal stream turbines operate just below the sea surface or are fixed to the sea bed. Tidal streams are formed by the horizontal fast-flowing volumes of water caused by the ebb and flow of the tide as the profile of the sea bed causes the water to speed up as it approaches the shoreline. As water is much denser than air and has a much slower flow rate, tidal stream turbines have much smaller diameters and higher tip speed rates compared to an equivalent wind turbine. Tidal stream turbines generate tidal power on both the ebb and the flow of the tide. One of the disadvantages of tidal stream generation is that as the turbines are submerged under the surface of the water they can create hazards to navigation and shipping [120]. Other forms of tidal energy include tidal fences which use individual vertical-axis turbines (Fig. 15.27) that are mounted within a fence structure, known as the caisson, which completely blocks a channel and forces water through them [121].

15.2.9.1.2 Archimedes Screws

These devices are shaped like a corkscrew and have helical surfaces that surround a central cylindrical shaft. Water from

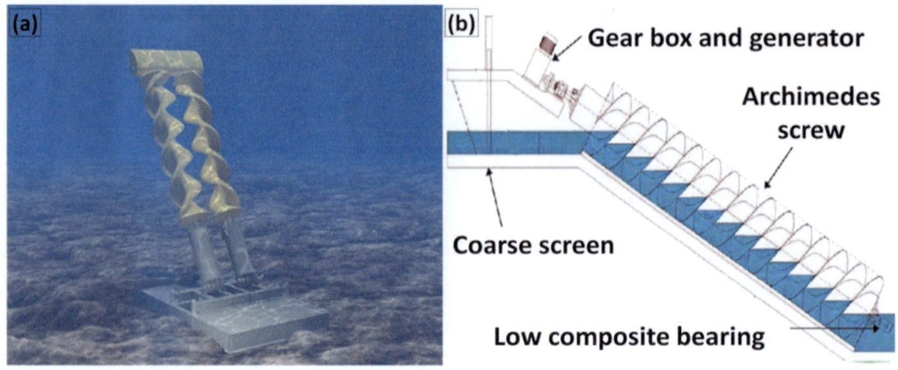

Fig. 15.28 Two types of Archimedes screws rotor (**a**) system inside water level and (**b**) a part of water enters inside the tube by screw

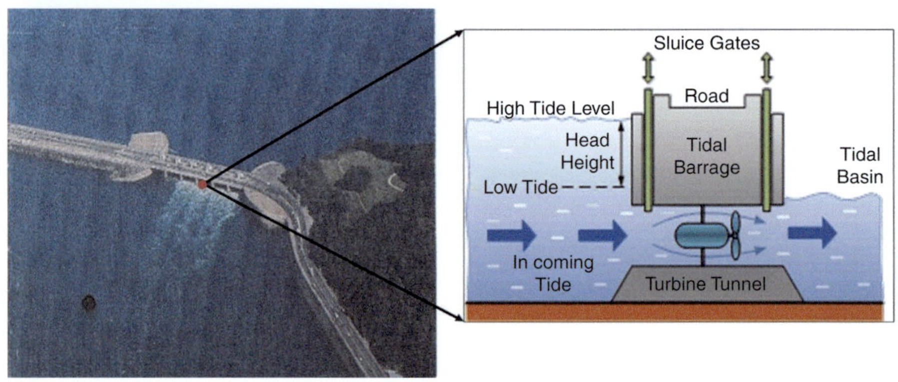

Fig. 15.29 Tidal barrage generation

the tidal stream is drawn up through the spiral, turning the turbines. Archimedes screws can be deployed as fully submerged tidal stream devices or can be enclosed in pipe systems. Compared to larger tidal stream turbines, Archimedes screws harm fewer fish and other forms of marine life [122]. Figure 15.28 shows the two types of Archimedes screws rotor: one can be installed inside water level, and the other one is used to allow a part of water to enter inside the tube by screw structure.

15.2.9.1.3 Tidal Dams/Barrages

Tidal barrages are long concrete structures, usually built across river estuaries, that have tunnels containing turbines. A *tidal dam/barrage* is a type of tidal power generation that involves the construction of a fairly low dam wall, known as a "barrage" and hence its name, across the entrance of a tidal inlet or basin creating a tidal reservoir (Fig. 15.29). This dam has a number of underwater tunnels cut into its width allowing sea water to flow through them in a controllable way using "sluice gates." Fixed within the tunnels are huge water turbine generators that spin as the water rushes past them generating tidal electricity [123]. Tidal barrages generate electricity using the difference in the vertical height between the incoming high tides and the outgoing low tides. As the tide ebbs and flows, sea water is allowed to flow in or out of the reservoir through a one-way underwater tunnel system. This flow of tidal water back and forth causes the water turbine generators located within the tunnels to rotate producing tidal energy with special generators used to produce electricity on both the incoming and the outgoing tides. The one disadvantage of tidal barrage generation is that it can only generate electricity when the tide is actually flowing either "in" or "out" as during high- and low-tide times, the tidal water is stationary [124]. However, because tides are totally predictable, other power stations can compensate for this stationary period when there is no tidal energy being produced. Another disadvantage of a tidal barrage system is the environmental and ecological effects that a long concrete dam may have on the estuaries they span [125].

15.2.9.1.4 Floating Structures

For these systems, turbines are mounted to the underside of a floating platform or barge (Fig. 15.30). They still tap the tidal currents but for greatly reduced installation and maintenance costs. Orbital Marine Power, a Scottish company, has developed a floating device that generated about 3 GWh of power. It is a 73-m-long floating superstructure that supports two

Fig. 15.30 Orbital O_2 is a 73-m-long floating device, supporting two 1-MW turbines at either side

1-MW turbines on each side that can generate twice as much energy and be distributed in arrays on the ocean's surface [126].

15.2.9.1.5 Tidal Kites

Tidal kites are tethered to the seabed and carry a turbine below the wing (Fig. 15.31). When the kite "flies" as shown in the figure in the tidal stream, it maximizes the volume of water passing through the turbine. Minesoto, a marine technology company, has created "Deep Green," a tidal kite that can create several hundred times more electricity than a stationary turbine on the sea floor. Almost all large tidal energy systems are anchored to the sea floor and require tidal currents of 2.5 m/s or faster to produce cost-effective electricity. Deep Green can also produce electricity from slower currents in the 1.2–2.4 m/s range, making it more versatile [127].

15.2.9.1.6 Wave Riding Arms

Another alternative way of harnessing tidal power is by using an "oscillating tidal turbine." This is basically a fixed wing called a hydroplane positioned on the sea bed. The hydroplane uses the energy of the tidal stream flowing past it to oscillate its giant wing, similar to a whale's flipper, up and down with the movement of the tidal currents. This motion is then used to generate electricity. The angle of the hydroplane to the flow of the tide can be varied to increase efficiency. These machines or devices use the oscillating motion of tidal wave patterns which usually drives some sort of piston up and down within the device. Well, waves in the ocean are constant, and they are going to be there no matter the type of weather conditions or what time it is. With this constant oscillation pattern 24/7, we can assume that this piston device is always moving and is creating some sort of work to generate clean energy, work in the form of suction and release of air to rotate turbine or by compression of piezoelectric materials [128]. Figure 15.32 shows the wave riding arms energy-harvesting system.

15.2.9.1.7 Artificially Intelligent Turbines

Nova Innovation, another Scottish company, is leading the development of a major European project called Element. The designers will integrate artificial intelligence equipment that will optimize the efficiency of the turbines and adjust them to changing conditions in real time. The artificial intelligence system will utilize data derived from wind energy that is captured at the surface and transmit to the turbine system to maximize turbine performance and efficiency, reducing lifetime costs for this tidal energy by nearly 20% [129].

15.2.9.2 Tidal Energy Generation

Since the position of the earth and the moon with respect to the sun changes throughout the year, we can utilize the potential energy of the water contained in the daily movement of the rising and falling sea levels to generate electricity. The generation of electricity from tides is similar in many ways to hydroelectric generation. The difference is that the water flows in and out of the turbines in both directions instead of in just one forward direction. The cost of reversible electrical generators is more expensive than single-direction generators. Tidal energy, just like hydroenergy, transforms water in motion into clean energy. The motion of the tidal water, driven by the pull of gravity, contains large amounts of kinetic energy in the form of strong tidal currents called tidal streams. The daily ebbing and flowing, back and forth of the oceans tides along a coastline and into and out of small inlets, bays, or coastal basins, is little different from the water flowing down a river or stream [130].

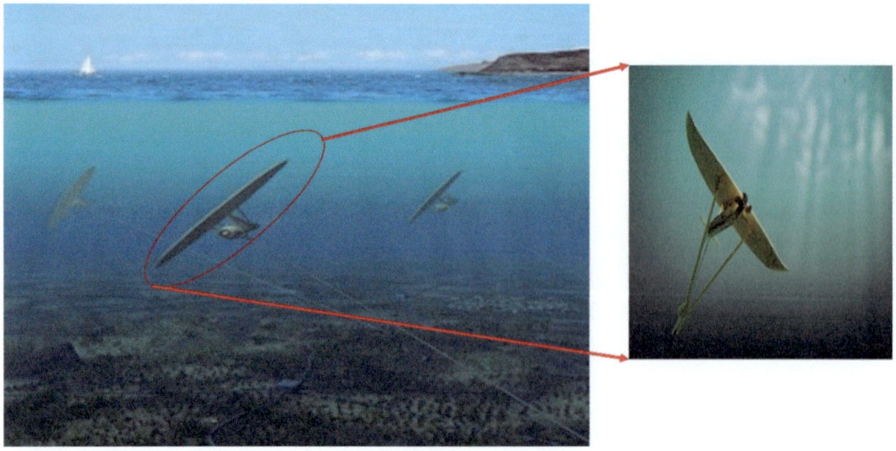

Fig. 15.31 Tidal kites energy harvesting

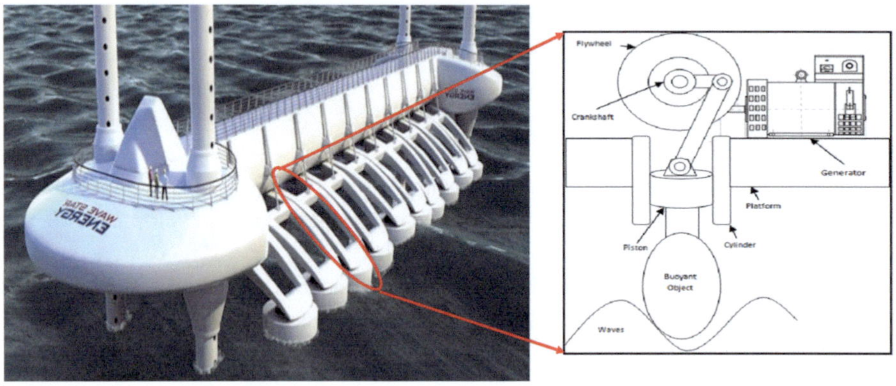

Fig. 15.32 Wave riding arms energy harvesting

15.2.9.3 Advantages and Disadvantages of Tidal Energy

Although tidal energy is still in early development and not yet cost-competitive with more mature renewable energy technologies such as wind and solar, the ever-increasing scope of new projects and new technologies will soon make tidal energy a major player in some areas around the world. Tidal energy is a renewable energy resource and produces no waste by-products. Tidal energy has the potential to produce a great deal of free and green energy. Tidal energy is not expensive to operate and maintain compared to other forms of renewable energies: low visual impact as the tidal turbines is mainly if not totally submerged beneath the water [131]; low noise pollution as any sound generated is transmitted through the water; and high predictability as high and low tides can be predicted years in advance, unlike wind. Tidal barrages protect against flooding and land damage. Tidal energy is not always a constant energy source as it depends on the strength and flow of the tides which themselves are affected by the gravitational effects of the moon and the sun. Tidal energy requires a suitable site, where the tides and tidal streams are consistently strong; must be able to withstand forces of nature resulting in high capital, construction, and maintenance costs; and high-power distribution costs to send the generated power from the submerged devices to the land using long underwater cables. Intermittent power generation only generates power 10 h a day and during the ebb and flow of the tides. Changes to the estuary ecosystem and an increase in coastal erosion where the tides are concentrated. Build-up of silt, sediments, and pollutants within the tidal barrage from rivers and streams flowing into the basin as it is unable to flow out into the sea. Danger to fish and other sea life, as they get stuck in the barrage or sucked through the tidal turbine blades [132].

All above-discussed process has been compared in Table 15.3 with their core materials, mode of operation, power density, and efficiency of use.

15.3 Energy Storage

Energy storage is the capture of energy produced at one time for use at a later time. In general, energy can be stored in a capacitor, supercapacitor, or battery. Some technologies provide short-term energy storage, while others can endure for much longer. Bulk energy storage is currently dominated by

Table 15.3 A comparative analysis between different types of energy-harvesting processes

Energy source	Energy-generating core material and mode of operation	Utilization	Power density	Efficiency	Advantages	Disadvantages
Solar energy	Photosensitive materials (photovoltaic effect)	Indoor	100 μW/cm^2	10–25%	• Is available every day of the year, even cloudy days produce some power • Low maintenance costs	• Solar Energy Storage is expensive • Uses a lot of space
		Outdoor	100 mW/cm^2			
Thermal energy	Thermoelectric materials (Seebeck effect), Pyroelectric materials (Olsen cycle)	Human	60 μW/cm^2	0.1–3%	• The fuel used (coal) is cheap • Thermal power plant can be set up anywhere near fuel and water supply • Require less space as compared to the hydroelectric power station	• Thermal power plants use fossil fuels to generate electricity, and these fossil fuels causes pollution • A large water source is required to convert water into steam • The maintenance cost is high
		Industrial	10 mW/cm^2			
Mechanical vibration	Piezoelectric materials (Piezoelectric effect)	Hz-Human	4 μW/cm^2	25–50%	• No external voltage source • High voltage of 1–10 V • Compact configuration • Compatible with MEMS • High coupling with single crystal	• Depolarization • Brittleness in bulk piezolayer • Charge leakage • High output impedance
		KHz-machine	800 μW/cm^2			
Electromagnetic	Rf generating materials (Ampere, Maxwell, and Faraday's law)	GSM 900 MHz	0.1 μW/cm^2	50%	• No need of smart materials • No external voltage source	• Bulky size of magnet and pick-up coil • Difficult to integrate with MEMS • Maximum voltage of 0.1V
		WiFi 2.4 GHz	0.001 μW/cm^2			
Chemical energy	Cation & anion transferring materials	Industrial	Depends on materials use	>90%	• Most forms of chemical energy are released through combustion • It is reasonably efficient • It is reasonably efficient	• It isn't a sustainable form of energy • It can be expensive • It can produce harmful waste
Wind & water	Any materials (Faraday's law)	Small scale	1.16 mW/cm^3	• 0.6–17.6%	• These are clean fuel source • These are sustainable	• Wind resource development might not be the most profitable use of the land • Turbines might cause noise and aesthetic pollution • Tidal power set up disrupt the sensitive marine life

hydroelectric dams, both conventional and pumped. Capacitors are used when the application needs to provide huge energy spikes. Batteries leak less energy and are therefore used when the device needs to provide a steady flow of energy. Compared to batteries, supercapacitors have virtually unlimited charge–discharge cycles and can therefore operate forever enabling a maintenance-free operation in IoT and wireless sensor devices. It is important to note that ambient energy is often random and intermittent. In most cases, it is, therefore, necessary to include a way of storing electrical energy after it is generated, which must also manage the various energy levels required by each electronic component to function properly [133]. The component that allows electrical energy to be temporarily stored is called an energy storage unit. Electricity is a secondary form of energy in the sense that once it has been created, it is more difficult to store than other forms of energy. The generated electricity is immediately consumed, lost, or stored in a capacitor. To utilize energy resources as fully as possible, and minimize losses, it is essential to store the energy harvested from the environment. Storing energy is a fundamentally challenging technical problem, as it often involves complicated processes that also require energy [134]. Regardless of the technology employed, energy storage units are essentially characterized by three values: (1) weight (or volume) energy density (measured in watt-hours per kilogram or watt-hours per liter)—this is the quantity of energy that can be stored per unit weight (or volume) of the accumulator; (2) weight power density (measured in watts per kilogram)—this is the power (electrical energy produced per unit time) yield per unit weight of the accumulator; and (3) the deep recharge cycle number (simply measured as a number of cycles, characterizing the lifetime of the storage unit)—this is the number of times that the unit can output an amount of energy higher than 80% of its nominal energy capacity. The operating cycle of an energy storage unit is comprised of two processes: loading and unloading. In each cycle, energy is converted twice [135, 136]. The yield of the storage unit is therefore particularly important and strongly depends on the type of storage and physical hardware. With the discovery of new materials and advancements in manufacturing technology, new energy storage processes are constantly being developed.

15.3.1 Types of Energy Storage

According to the application, three types of energy storage systems can be seen: thermal system, mechanical system (flywheel and hydropower), and battery. Thermal systems use heating and cooling methods to store and release energy. For example, molten salt stores solar-generated heat for use when there is no sunlight. Ice storage in buildings reduces the need to run compressors while still providing air conditioning over a period of several hours. Other systems use chilled water and dispatchable hot water heaters. In all cases, excess energy charges the storage system (heat the molten salts, freeze the water, etc.) and is later released as needed [137]. Flywheels store energy in a rapidly spinning mechanical rotor and are capable of absorbing and releasing high power for typically 15 min or less, although longer duration systems are being developed. These systems can balance fluctuations in electricity supply and demand where they respond to a control signal adjusted every few seconds. They also recapture braking energy from electric trains in some installations or provide short-term power until a backup generation comes online during a grid outage, such as in a critical manufacturing process where the product would be lost by a momentary electric interruption. Pumped hydroelectric facilities are the most common form of energy storage on the grid and account for over 95% of the storage in use today. During off-peak hours, turbines pump water to an elevated reservoir using excess electricity. When electricity demand is high, the reservoir opens to allow the retained water to flow through turbines and produce electricity. Siting these systems can be difficult because of the terrain needed (an upper and lower pool of water) and the large footprint [138]. Compressed air, superconducting magnets, underground pumped storage, and hydrogen storage are all forms of emerging energy storage that are in different stages of development. There are various forms of batteries, including lithium-ion, flow, lead-acid, sodium, and others, designed to meet specific power and duration requirements. Initially used for consumer products, lithium-ion batteries now have a range of applications including smaller residential systems and larger systems that can store multiple megawatt hours (MWh) and can support the entire electric grid. These systems typically house a large number of batteries together on a rack, combined with monitoring and management units. These systems have a small footprint for the amount of energy they store. For example, a system the size of a small refrigerator could power an average home for several days [139].

The integration of energy harvesting and energy storage in one device not only enables the conversion of ambient energy into electricity but also provides a sustainable power source for various electronic devices and systems. There are various energy harvesters integrated with energy storage devices such as capacitors or batteries to demonstrate practical applications such as light-emitting diodes, liquid crystal displays, electroluminescence, micro-heaters, mobile phones, sensors, pacemakers, deep brain stimulation, and so on. Solar cells and thermoelectric generators can be connected directly with energy storage devices without any electrical circuit because of their DC behavior [140]. Nevertheless, piezoelectric nanogenerator, triboelectric nanogenerators, and pyroelectric nanogenerator should be connected to energy storage devices with rectification diodes due to their AC behavior. To effectively charge an energy storage device, the charging voltage and current should be constant with an appropriate value, whereas energy harvesters generate irregular and unstable electricity. Therefore, an optimized circuit including a capacitor filter, AC-DC, and DC-DC converter is needed to economically charge the energy storage device using energy harvesters [141].

15.3.2 Batteries

Batteries are the main energy storage devices used in modern society to store the energy harvested from various sources such as solar, wind, thermal, strain, and inertia and to use the harvested energy efficiently when needed. Typically, a battery consists of five major components: an anode, cathode, the current collectors these may sit on, electrolyte, and separator, as shown in Fig. 15.33. The purpose of the anode is to hold the active ions in a high-energy state [142]. The higher the energy state, the higher the eventual voltage of the cell. In

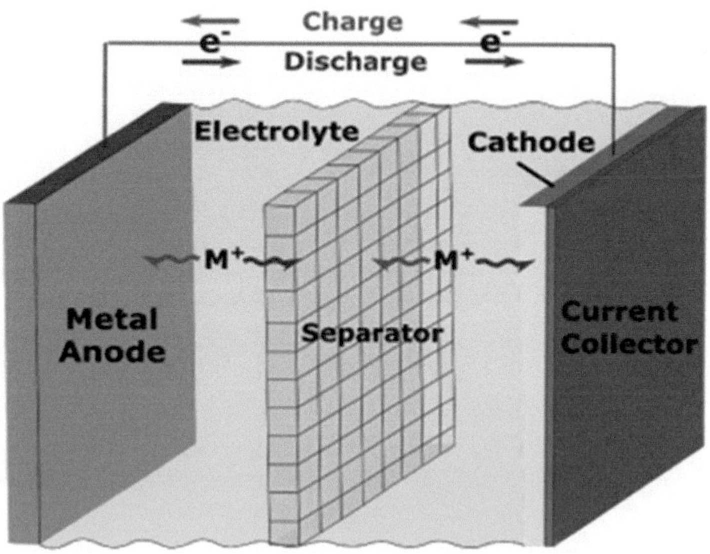

Fig. 15.33 Charging and discharging processes of typical cells are indicated in green and in red, respectively. During the discharge, high-potential metal atoms oxidize, and the resultant ions move toward and interact with the reducing cathode

principle, pure metal is the best anode material due to the metal being the highest possible energy state to hold the metal in as well as holding the largest amount of metal atoms in the smallest amount of space. However, dendrite formation often hampers the use of pure metal. Hence, other solutions, which can prevent this dendrite formation, are often used instead, such as inserting the ions into the interlayer spacings in graphite [143]. Conversely, when the metal is oxidized and the resultant ion moves over to the cathode during discharge, the ion-inserted cathode needs to be in a much lower energy state than when it was at the anode. It is this difference in energy that gives the thermodynamic push for electrons to move through a circuit and do work. The larger the difference, the higher the voltage, and the battery will store more energy. The electrolyte-soaked separator must help facilitate this ionic movement while being electrically insulating to prevent short-circuiting. They should also be spectators on the electrochemical stage. If they do participate, they will be rapidly decomposed and/or coat either electrode in a solid electrolyte interphase (SEI), which can impede ionic movement, thus degrading the cell. The electrolyte can be engineered with additives to form stable ionically conductive SEI. An ion battery operates by transferring charge using freely moving ions shuttling back and forth between the two electrodes. This charge transfer occurs via different mechanisms depending on how the active ion interacts with an electrode material. The amount of charge that a battery can store depends on the potential difference in the electrochemical reactions of the active ion that occur at the two electrodes and the number of electrons involved in each of these reactions. One can either increase the potential difference between the two electrode processes or use ions involving multiple electrons to increase the amount of charge stored or the capacity of the system. Both these factors are governed by the mechanism with which the charge is transferred. In all of the applications, the energy density of the battery, which is the amount of energy stored per unit volume (Wh/L) or per unit weight (W h/kg), is a critical parameter [144]. The amount of energy stored depends on the capacity (amount of charge stored) per unit volume (A.h/L) or per unit weight (A.h/kg) and the voltage (V) each cell can deliver. In addition, rechargeability, charge–discharge cycle life, and the rate at which the cell can be charged and discharged are also important parameters. Moreover, cost and environmental impact considerations need to be taken into account. All of these parameters and properties are related to the battery chemical composition and materials involved.

Currently, the most used anode materials are graphite, lithium metal, sodium metal, magnesium metal, aluminum metal, and alloying electrodes; cathode materials are sulfur, air cathodes, metal chalcogenides, titanium oxides, vanadium oxides, and transition metal sulfides; and electrolyte materials are organic electrolytes, aqueous electrolytes, water in salt electrolytes, ionic liquid electrolytes, solid inorganic electrolytes, and solid organic electrolytes. Battery engineers have two broad strategies to achieve low-cost cells: materials and morphology. Low-cost, abundant materials that can be economically engineered into the appropriate form are required for low-cost cells. Therefore, any fabrication process that is itself inherently expensive, despite using abundant materials, must be excluded when engineering low-cost cells. Figure 15.34 shows the materials that are available in the periodic table for the production of an efficient cell as well as a suitable low-cost cell [145].

Various types of batteries are discussed in the following sections:

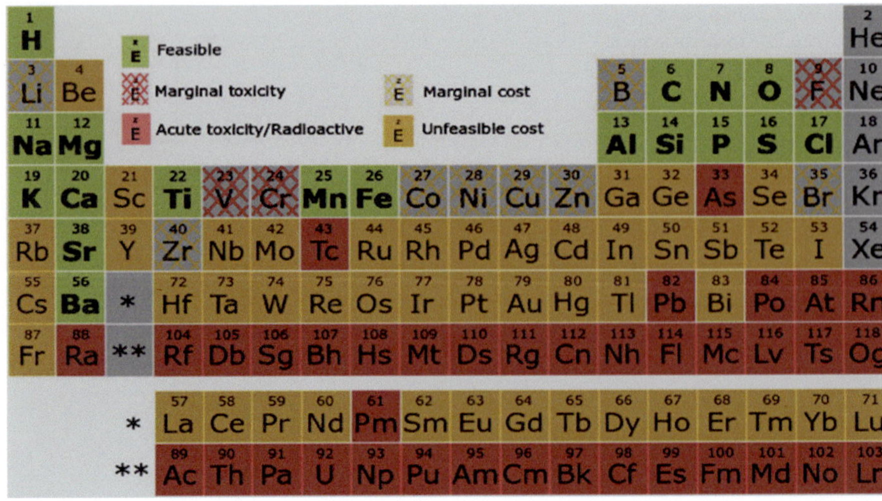

Fig. 15.34 The periodic table for cheap, safe batteries

15.3.2.1 Lithium-Ion Batteries

Lithium-ion batteries involve a reversible insertion/extraction of lithium ions into/from a host matrix during the discharge–charge process as shown in Fig. 15.35. The host matrix is called a lithium insertion compound and serves as the electrode material in the cell. The present generation of Li-ion cells mostly uses graphite and layered $LiCoO_2$ as the Li insertion compounds, which serve, respectively, as the anode and the cathode materials. A Li-containing salt such as $LiPF_6$ dissolved in a mixture of aprotic solvents like ethylene carbonate (EC) and diethyl carbonate (DEC) is used as the electrolyte [146]. During the charging process, the Li-ions are extracted from the layered $LiCoO_2$ cathode, flow through the electrolyte, and get inserted into the layers of the graphite anode, while the electrons flow through the external circuit from the $LiCoO_2$ cathode to the graphite anode to maintain charge balance. Thus, the charging process is accompanied by an oxidation reaction (Co^{3+} to Co^{4+}) at the cathode and a reduction reaction at the anode.

During discharge, exactly the reverse reactions occur at the anode and cathode with the flow of Li-ions (through the electrolyte) and electrons from the anode to the cathode. The free energy change involved in the relevant chemical reaction (Eq. 15.30) is taken out as electrical energy during the discharge process:

$$LiCoO_2 \rightarrow Li_{1-x}CoO_2 + Li_xC_6 \tag{15.30}$$

The open-circuit voltage V_{OC} of such a Li-ion cell has been given in Eq. (15.31).

$$V_{OC} = \frac{\mu_{Li(c)} - \mu_{Li(a)}}{F}, \tag{15.31}$$

where the lithium chemical potential between the cathode $\mu_{Li(c)}$ and the anode $\mu_{Li(a)}$, F, is the Faraday constant. Figure 15.36 shows a schematic energy diagram of a cell at the open circuit. The cell voltage $V_{OC}$is determined by the energies involved in both the electron transfer and the Li^+ transfer. While the energy involved in electron transfer is related to the work functions of the cathode and the anode, the energy involved in Li^+ transfer is determined by the crystal structure and the coordination geometry of the site into/from which Li^+ ions are inserted/extracted. Although the cell voltage is a function of the separation between the redox energies of the cathode (Ec) and the anode (Ea), thermodynamic stability considerations require the redox energies Ec and Ea to lie within the bandgap Eg of the electrolyte, as shown in Fig. 15.37, so that no unwanted reduction or oxidation of the electrolyte occurs during the charge–discharge process [147]. Thus, the electrochemical stability requirement imposes a limitation on the cell voltage as follows:

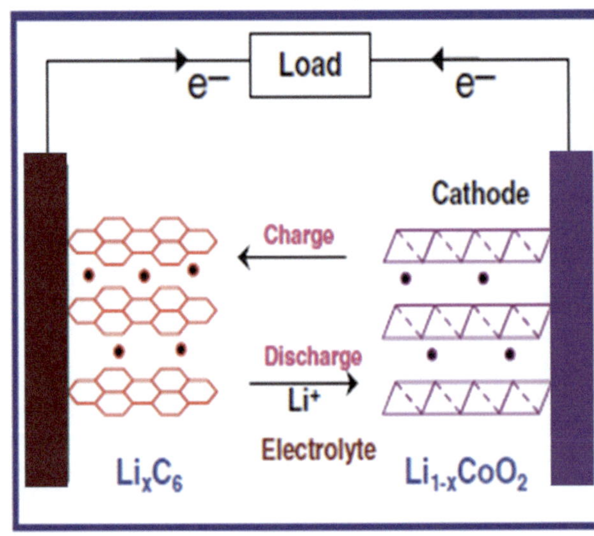

Fig. 15.35 The charge/discharge process involved in a Li-ion cell consisting of graphite as an anode and layered $LiCoO_2$ as a cathode

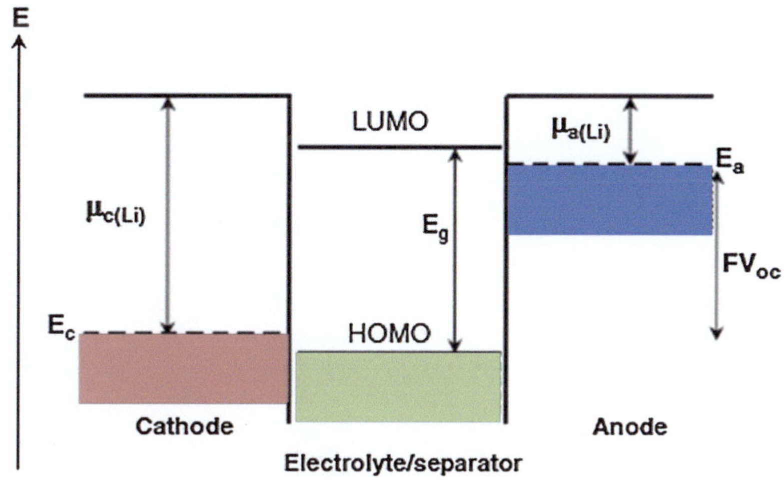

Fig. 15.36 Schematic energy diagram of a Li cell at open circuit. HOMO (highest occupied molecular orbital) and LUMO (lowest unoccupied molecular orbital)

$$FV_{OC} = \mu_{Li(c)} - \mu_{Li(a)} < E_g \tag{15.32}$$

Several criteria should be satisfied for a lithium insertion compound to be successful as a cathode or an anode material in a rechargeable lithium cell. Some of the most important criteria are listed below.

1. The cathode should have a low-lithium chemical potential ($\mu_{Li(c)}$) and the anode should have a high-lithium chemical potential ($\mu_{Li(a)}$) to maximize the cell voltage (V). This implies that the transition metal ion (Mn^+) in the lithium insertion compound $Li_xM_yX_z$ should have a high oxidation state to serve as a cathode and a low oxidation state to serve as an anode [148].
2. The lithium insertion compound $Li_xM_yX_z$ should allow an insertion/extraction of a large amount of lithium per unit weight or per unit volume to maximize the cell capacity (A.h/L or A.h/kg). This depends on the number of lithium sites available in the lithium insertion/extraction host for reversible lithium insertion/extraction and the accessibility of multiple valences for M in the lithium insertion/extraction host. A combination of high capacity and cell voltage can maximize the energy density (W.h/L or W.h/kg), which is given by the product of the cell capacity and cell voltage [149].
3. The lithium insertion compound $Li_xM_yX_z$ should support a reversible insertion/extraction of lithium with no or minimal changes in the host structure over the entire range of lithium insertion/extraction to provide good cycle life for the cell. This implies that the insertion compound $Li_xM_yX_z$ should have good structural stability without breaking any M-X bonds [150].
4. The lithium insertion compound should support both high-electronic conductivity (σe) and high Li-ion conductivity (σ_{Li}) to facilitate fast charge/discharge (rate capability) and offer high-power capability, that is, it should support mixed ionic-electronic conduction. This depends on the crystal structure, arrangement of the MX_n polyhedra, geometry, and interconnection of the lithium sites, nature, and electronic configuration of the Mn^+ ion, and the relative positions of the Mn^+ and Xn^- energies. The insertion compound should be chemically and thermally stable without undergoing any reaction with the electrolyte over the entire range of lithium insertion/extraction [151].
5. The redox energies of the cathode and the anode in the entire range of lithium insertion/extraction process should lie within the bandgap of the electrolyte to prevent any unwanted oxidation or reduction of the electrolyte [152].
6. The lithium insertion compound should be inexpensive, environmentally benign, and lightweight from a commercial point of view. This implies that the Mn^+ ion should preferably be from the 3d transition series [153].

In addition to the criteria outlined earlier for the cathode and the anode materials, the electrolyte should also satisfy several criteria. The electrolyte should have high lithium-ion conductivity but should be an electronic insulator to avoid internal short-circuiting. A high-ionic conductivity in the electrolyte is essential to minimize IR drop or ohmic polarization and realize a fast charge–discharge process (high rate or power capability). With a given electrolyte, the IR drop due to electrolyte resistance can be reduced and the rate capability can be improved by having a higher electrode interfacial area and thin separators. The electrolyte should also have good chemical and thermal stabilities without undergoing any direct reaction with the electrodes. It should act only as a medium to transport efficiently the Li^+ ions between the two electrodes (anode and cathode) [154]. Additionally, the engineering involved in cell design and fabrication plays a critical role in the overall cell performance, including electrochemical utilization, energy density, power

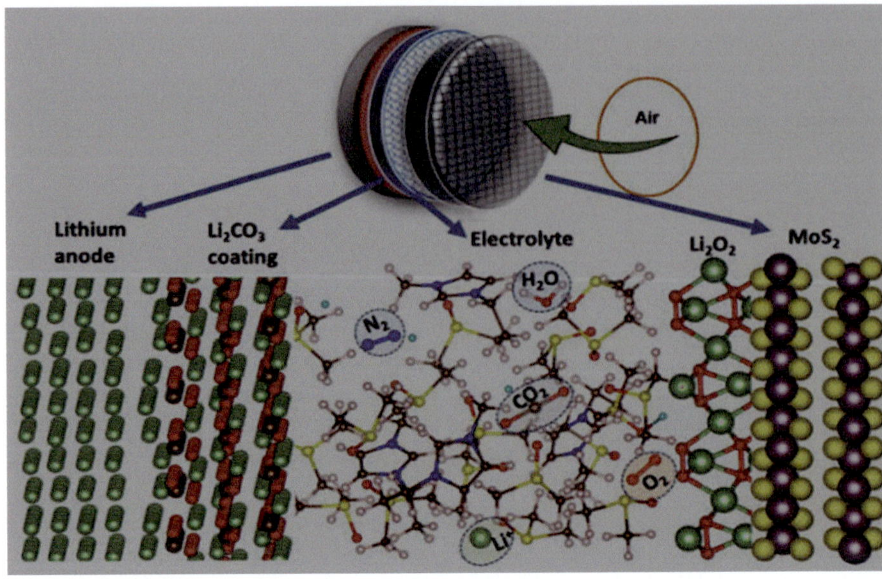

Fig. 15.37 Lithium-air battery operation

density, and cycle life. For example, although ideally high-electronic conductivity and Li-ion diffusion rate in the electrodes are preferred to minimize cell polarization, the electronic conductivity of the electrodes can be improved by adding electronically conducting additives such as carbon. However, the amount of additives should be minimized to avoid any undue sacrifice in cell capacity and energy density. Finally, cell safety, environmental factors, and raw materials and processing/manufacturing costs are also important considerations in materials selection, cell design, and cell fabrication [155].

15.3.2.2 Lithium-Air Batteries

Lithium-air batteries are believed to have the capacity to hold up to five times more energy than the same lithium-ion batteries powering today's phones, laptops, and electric vehicles. The lithium-air battery (Li-air) is a metal–air electrochemical battery that uses oxidation of lithium at the anode and reduction of oxygen at the cathode to induce a current flow (Fig. 15.37). Pairing lithium and ambient oxygen can theoretically lead to electrochemical cells with the highest possible specific energy. Indeed, the theoretical specific energy of a nonaqueous Li-air battery, in the charged state with Li_2O_2 product and excluding the oxygen mass, is ~40.1 MJ/kg. This is comparable to the theoretical specific energy of gasoline, ~46.8 MJ/kg. In practice, Li-air batteries with a specific energy of ~6.12 MJ/kg at the cell level have been demonstrated [156]. A 60 kg Li-air battery has capacity to run a 2000 kg EV for ~500 km (310 miles) on one charge. However, the practical power and life cycle of Li-air batteries need significant improvements before they can find a market niche. Significant electrolyte advances are needed to develop a commercial implementation. Four approaches are active: aprotic, aqueous, solid-state, and mixed aqueous-aprotic. Metal-air batteries, specifically zinc-air, have received attention due to potentially high energy densities. A major market driver for batteries is the automotive sector. The energy density of gasoline is approximately 13 kWh/kg, which corresponds to 1.7 kWh/kg of energy provided to the wheels after losses. Theoretically, lithium-air can achieve 12 kWh/kg (43.2 MJ/kg) excluding the oxygen mass [157].

Design and operation: In general, lithium ions move between the anode and the cathode across the electrolyte. Figure 15.38 shows the schematic of lithium-air battery charge and discharge cycles. Under discharge, electrons follow the external circuit to do electric work, and the lithium ions migrate to the cathode. During charge, the lithium metal plates onto the anode, freeing O_2 at the cathode. Both non-aqueous (with Li_2O_2 or LiO_2 as the discharge products) and aqueous (LiOH as the discharge product) Li-O_2 batteries have been considered. The aqueous battery requires a protective layer on the negative electrode to keep the Li metal from reacting with water [158]. Figure 15.39 shows the schematic of the artificial versus spontaneous electrolyte interface.

Anode: Lithium metal is the typical anode choice. At the anode, electrochemical potential forces the lithium metal to release electrons via oxidation (without involving the cathodic oxygen). The half-reaction is:

$$Li \rightleftharpoons Li^+ + e^- \tag{15.33}$$

Lithium has a high specific capacity (3840 mAh/g) compared to other metal-air battery materials (820 mAh/g for Zn, 2965 mAh/g for Al). Several issues affect such cells. The main challenge in anode development is preventing the anode from reacting with the electrolyte. Alternatives include new

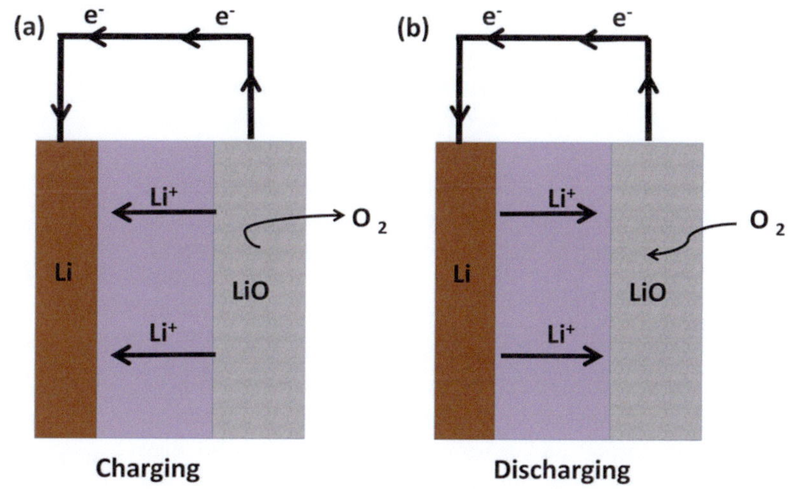

Fig. 15.38 Schematic of lithium-air battery charge and discharge cycles

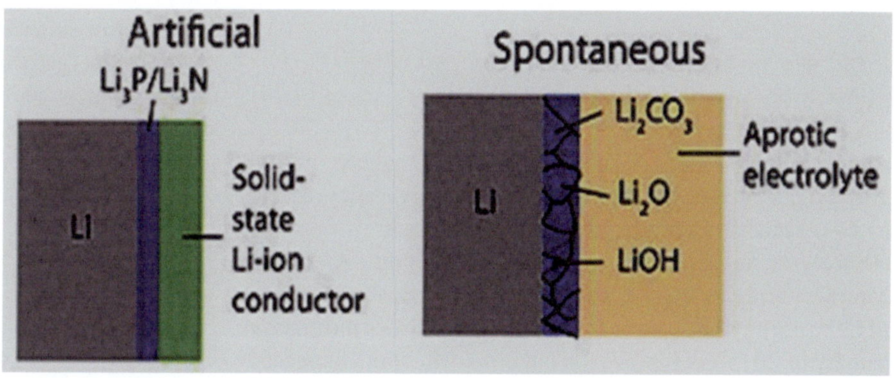

Fig. 15.39 Schematics of artificial vs. spontaneous electrolyte interface

electrolyte materials or redesigning the interface between electrolyte and anode. Lithium anodes risk dendritic lithium deposits, decreasing energy capacity or triggering a short circuit. The effects of pore size and pore size distribution remain poorly understood. Upon charging/discharging in aprotic cells, layers of lithium salts precipitate onto the anode, eventually covering it and creating a barrier between the lithium and electrolyte [159]. This barrier initially prevents corrosion but eventually inhibits the reaction kinetics between the anode and the electrolyte. This chemical change in the solid–electrolyte interface (SEI) results in varying chemical compositions across the surface, causing the current to vary accordingly. The uneven current distribution furthers branching dendrite growth and typically leads to a short circuit between the anode and the cathode. In aqueous cells, problems at the SEI stem from the high reactivity of lithium metal with water. Several approaches attempt to overcome these problems: formation of a Li-ion protective layer using di- and tri-block copolymer electrolytes. According to Seeo, Inc., such electrolytes (e.g., polystyrene with the high Li-ion conductivity of a soft polymer segment, such as a poly(ethylene oxide (PEO) and Li-salt mixture)) combine the mechanical stability of a hard polymer segment with the high ionic conductivity of the soft polymer-lithium-salt mixture. The hardness inhibits dendrite shorts via mechanical blocking. Li-ion conducting glass or glass-ceramic materials are readily reduced by lithium metal, and therefore, a thin film of a stable lithium conducting material, such as Li 3P or Li_3N, can be inserted between the ceramic and the metal. This ceramic-based SEI inhibits the formation of dendrites and protects the lithium metal from atmospheric contamination [160].

Cathode: At the cathode during charge, oxygen donates electrons to the lithium via reduction. Mesoporous carbon has been used as a cathode substrate with metal catalysts that enhance reduction kinetics and increase the cathode's specific capacity. Manganese, cobalt, ruthenium, platinum, silver, or a mixture of cobalt and manganese are potential metal catalysts. Under some circumstances, Mn-catalyzed cathodes performed best, with a specific capacity of 3137 mA·H/g carbon, and co-catalyzed cathodes performed second best, with a specific capacity of 2414 mA·H/g carbon. Based on the first pore-scale modeling of Li-air batteries, the microstructure of the cathode significantly affects battery capacity in both non-pore-blocking and pore-blocking regimes [161]. Most Li-air battery limits are at the cathode, which is also the source of its potential advantages. Atmospheric oxygen must be present at the cathode, but contaminants such as water vapor can damage it. Incomplete discharge due to blockage of the porous carbon cathode with discharge products such as lithium peroxide (in aprotic designs) is the most serious. Catalysts have shown promise in creating preferential nucleation of Li_2O_2 over Li_2O, which is irreversible with respect to lithium. Li-air performance is limited by the efficiency of the reaction at the cathode because most of the voltage drop occurs there. Multiple chemistries have been assessed, distinguished by their electrolyte. This discussion focuses on aprotic and aqueous electrolytes as solid-state electrochemistry is poorly understood [162]. In a cell with an aprotic electrolyte, lithium oxides are produced through reduction at the cathode:

$$Li^+ + e^- + O_2{}^+* \rightarrow LiO_2 * Li^+ + e^- + LiO_2* \rightarrow Li_2O_2 *, \qquad (15.34)$$

where "*" denotes a surface site on Li_2O_2 where growth proceeds, which is essentially a neutral Li vacancy in the Li_2O_2 surface. Lithium oxides are insoluble in aprotic electrolytes, which lead to cathode clogging. A MnO_2 nanowire array cathode augmented by a genetically modified M13 bacteriophage virus offers two to three times the energy density of 2015-era lithium-ion batteries. The virus increased the size of the nanowire array, which is about 80 nm across [163]. The resulting wires had a spiked surface. Spikes create more surface area to host reaction sites. The viral process creates a cross-linked 3D structure, rather than isolated wires, stabilizing the electrode. The viral process is water-based and takes place at room temperature.

Electrolyte: Efforts in Li-air batteries have focused on four electrolytes: aqueous acidic, aqueous alkaline, non-aqueous protic, and aprotic. In a cell with an aqueous electrolyte the reduction at the cathode can also produce lithium hydroxide:

Aqueous electrolyte: An aqueous Li-air battery consists of a lithium metal anode, an aqueous electrolyte, and a porous carbon cathode. The aqueous electrolyte combines lithium salts dissolved in water. It avoids the issue of cathode clogging because the reaction products are water-soluble. Figure 15.40 shows the schematic of an aqueous type Li-air battery design. The aqueous design has a higher practical discharge potential than its aprotic counterpart. However, Li metal reacts violently with water and thus the aqueous design requires a solid electrolyte interface between the lithium and the electrolyte. Commonly, a lithium-conducting ceramic or glass is used, but conductivity is generally low (on the order of 10^{-3} S/cm at ambient temperatures) [164].

15.3.2.2.1 Acidic Electrolyte

$$2Li + ½\,O_2 + 2H^+ \rightarrow 2Li^+ + H_2O \qquad (15.35)$$

A conjugate base is involved in the reaction. The theoretical maximal Li-air cell-specific energy and energy density are 1400 W h/kg and 1680 W h/l, respectively.

15.3.2.2.2 Alkaline Aqueous Electrolyte

$$2Li + ½\,O_2 + H_2O \rightarrow 2LiOH \qquad (15.36)$$

Water molecules are involved in the redox reactions at the air cathode. The theoretical maximal Li-air cell-specific energy and energy density are 1300 W h/kg and 1520 W h/l, respectively. New cathode materials must account for the accommodation of substantial amounts of LiO_2, Li_2O_2, and/or LiOH without causing the cathode pores to block and employ suitable catalysts to make the electrochemical reactions energetically practical. Dual pore system materials offer the most promising energy capacity. The first pore system serves as an oxidation product store and the second pore system serves as oxygen transport [165].

15.3.2.2.3 Aprotic Electrolyte

Non-aqueous Li-air batteries usually use mixed ethylene carbonate+propylene carbonate solvents with $LiPF_6$ or Li bis-sulfonimide salts like conventional Li-ion batteries, however, with a gelled rather than the liquid electrolyte. The voltage difference upon constant current charge and discharge is usually between 1.3 and 1.8 V (with an OCP of 4.2 V) even at such ridiculously low currents such as 0.01–0.5 mA/cm^2 and 50–500 mA/g of C on the positive electrode. However, the carbonate solvents evaporate and get oxidized due to a high overvoltage upon charge. Other solvents, such as end-capped glymes, DMSO, dimethylacetamide, and ionic liquids, have been considered. The carbon cathode gets oxidized above +3.5 V and Li during charge, forming Li_2CO_3, which leads to an irreversible capacity loss [166].

Most efforts involved aprotic materials, which consist of a lithium metal anode, a liquid organic electrolyte, and a porous carbon cathode. The electrolyte can be made of any organic liquid able to solvate lithium salts, such as $LiPF_6$, $LiAsF_6$, $LiN(SO_2CF_3)_2$, and $LiSO_3CF_3$), but typically consisted of carbonates, ethers, and esters. The carbon cathode is usually made of high-surface-area carbon material with a nanostructured metal oxide catalyst (commonly MnO_2 or Mn_3O_4). A major advantage is the spontaneous formation of a barrier between anode and electrolyte (analogous to the barrier formed between the electrolyte and carbon-lithium anodes in conventional Li-ion batteries) that protects the

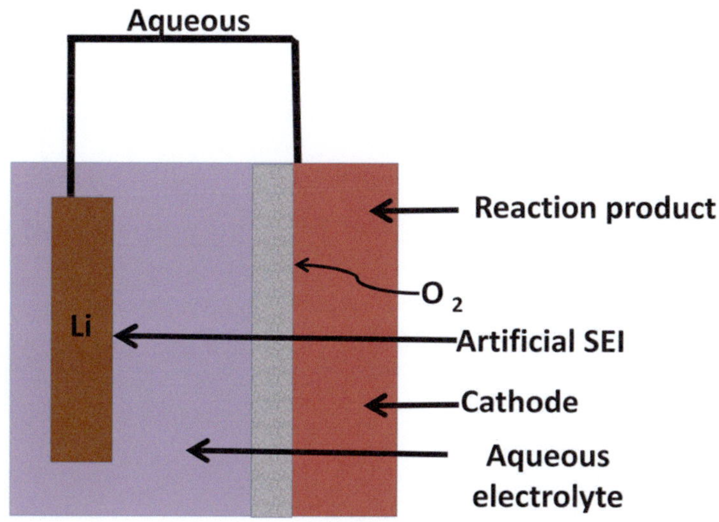

Fig. 15.40 Schematic of aqueous type Li-air battery design

lithium metal from further reaction with the electrolyte. Although rechargeable, the Li_2O_2 produced at the cathode is generally insoluble in the organic electrolyte, leading to buildup along the cathode–electrolyte interface. This makes cathodes in aprotic batteries prone to clogging and volume expansion that progressively reduces conductivity and degrades battery performance. Another issue is that organic electrolytes are flammable and can ignite if the cell is damaged [167].

Although most studies agree that Li_2O_2 is the final discharge product of nonaqueous Li-O2 batteries, considerable evidence that its formation does not proceed as a direct 2-electron electroreduction to peroxide O_2^{-2} (which is the common pathway for O_2 reduction in water on carbon) but rather via a one-electron reduction to superoxide O^{-2}, followed by its disproportionation:

$$2LiO_2 \rightleftharpoons Li_2O_2 + O_2 \tag{15.37}$$

Superoxide (O^{-2}) has traditionally been considered a dangerous intermediate in aprotic oxygen batteries due to its high nucleophilicity, basicity, and redox potential. However, reports suggest that LiO_2 is both an intermediate during the discharge to peroxide (Li_2O_2) and can be used as the final discharge product, potentially with an improved cycle life albeit with lower specific energy (a little heavier battery weight). Indeed, it was shown that under certain conditions, the superoxide can be stable on the scale of 20–70 h at room temperature. An irreversible capacity loss upon disproportionation of LiO_2 in the charged battery was not addressed. Pt/C seems to be the best electrocatalyst for O_2 evolution and Au/C for O_2 reduction when Li_2O_2 is the product. Nevertheless, "the performance of rechargeable lithium-air batteries with nonaqueous electrolytes is limited by the reactions on the oxygen electrode, especially by O_2 evolution. Conventional porous carbon air electrodes are unable to provide mAh/g and mAh/cm^2 capacities and discharge rates at the magnitudes required for really high energy density batteries for EV applications." The capacity (in mAh/cm^2) and the cycle life of nonaqueous Li-O_2 batteries are limited by the deposition of insoluble and poorly electronically conducting LiO_x phases upon discharge. (Li_3O_4 is predicted to have better Li^+ conductivity than the LiO_2 and Li_2O_2 phases) [168]. This makes the practical specific energy of Li-O_2 batteries significantly smaller than the reagent-level calculation predicts. It seems that these parameters have reached their limits, and further improvement is expected only from alternative methods (Fig. 15.41).

Mixed Aqueous-aprotic: The aqueous-aprotic or mixed Li-air battery design attempts to unite the advantages of the aprotic and aqueous battery designs (Fig. 15.42). The common feature of hybrid designs is a two-part (one part aqueous and one part aprotic) electrolyte connected by a lithium-conducting membrane. The anode abuts the aprotic side while the cathode is in contact with the aqueous side. A lithium-conducting ceramic is typically employed as the membrane joining the two electrolytes. The use of a solid electrolyte is one such alternative approach that allows for a combination of a lithium metal anode with an aqueous cathode. An example of ceramic solid electrolytes is $Li_{1-x}(Al/Sc/Y)_x(Ti/Ge)_{2-x}(PO_4)_3$ [169]. In 2015, researchers announced a design that used highly porous graphene for the anode, an electrolyte of lithium bis(trifluoromethyl) sulfonylimide/dimethoxyethane with added water and lithium iodide for use as a "mediator." The electrolyte produces lithium hydroxide (LiOH) at the cathode instead of lithium peroxide (Li_2O_2) [170]. The result offered an energy efficiency of 93% (voltage gap of 0.2) and cycled more than 2,000 times with little

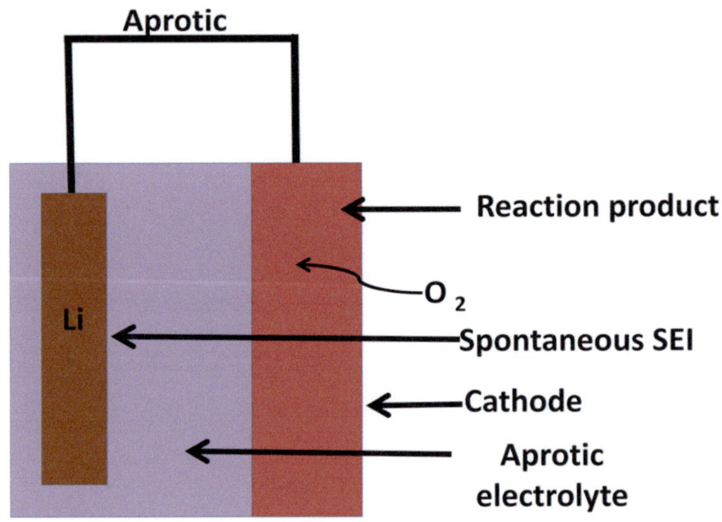

Fig. 15.41 Schematic of aprotic type Li-air battery design

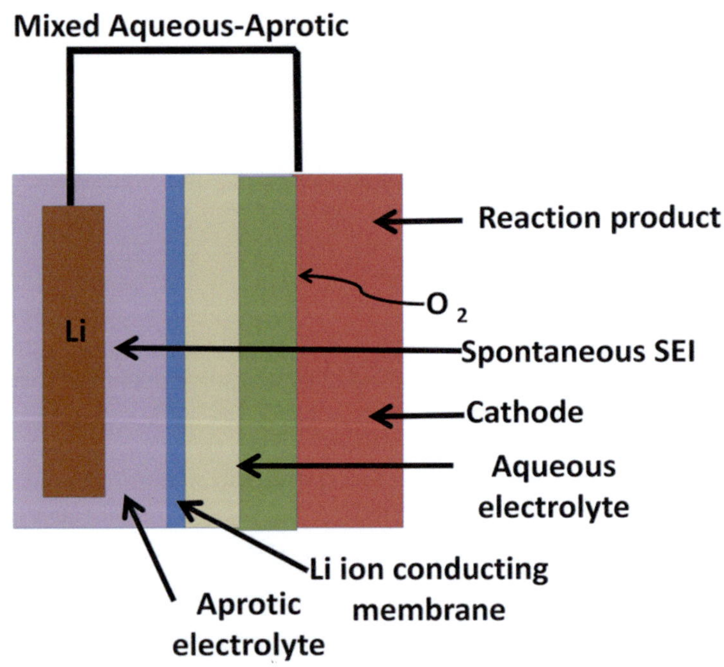

Fig. 15.42 Schematic of a mixed aqueous-aprotic type Li-air battery design

impact on output. However, the design required pure oxygen rather than ambient air.

Solid State: A solid-state battery design is attractive for its safety, eliminating the chance of ignition from rupture. Current solid-state Li-air batteries use a lithium anode, a ceramic, glass, or glass-ceramic electrolyte, and a porous carbon cathode. The anode and cathode are typically separated from the electrolyte by polymer–ceramic composites that enhance charge transfer at the anode and electrochemically couple the cathode to the electrolyte. The polymer–ceramic composites reduce overall impedance. The main drawback of the solid-state battery design is the low conductivity of most glass-ceramic electrolytes. The ionic conductivity of current lithium fast-ion conductors is lower than liquid electrolyte alternatives [171]. Figure 15.43 shows the schematic of solid-state type Li-air battery design.

Among the various rechargeable battery chemistries known to date, lithium-ion batteries offer the highest energy density when compared to the other rechargeable battery systems, such as lead-acid, nickel-cadmium, and nickel-metal hydride batteries. The higher volumetric and gravimetric energy densities of the lithium-ion cells are due to the higher cell voltages ($\sim$4 V) achievable by the use of non-aqueous electrolytes in contrast to $<$2 V achievable with most of the aqueous electrolyte-based cells [172].

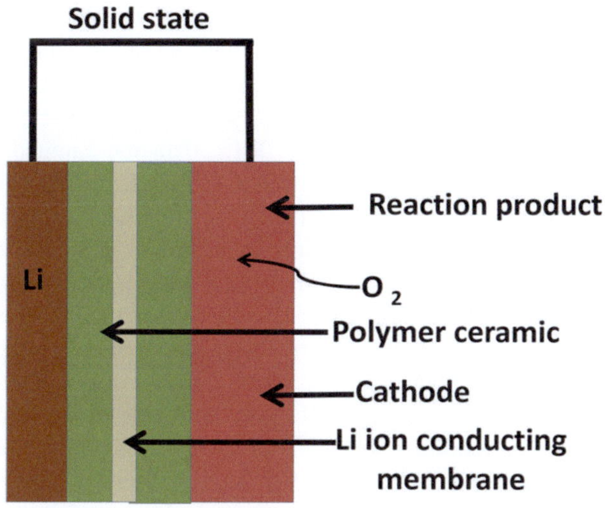

Fig. 15.43 Schematic of solid-state type Li-air battery design

15.3.2.3 Lithium-Polymer Battery

The lithium-polymer differentiates itself from conventional battery systems in the type of electrolyte used. The original design, dating back to the 1970s, uses a dry solid polymer electrolyte. This electrolyte resembles a plastic-like film that does not conduct electricity but allows ions exchange (electrically charged atoms or groups of atoms). The polymer electrolyte replaces the traditional porous separator, which is soaked with electrolyte. The dry polymer design offers simplifications with respect to fabrication, ruggedness, safety, and thin-profile geometry [173]. With a cell thickness measuring as little as 1 mm (0.039 in.), equipment designers are left to their own imagination in terms of form, shape, and size. Unfortunately, the dry lithium-polymer suffers from poor conductivity. The internal resistance is too high and cannot deliver the current bursts needed to power modern communication devices and spin up the hard drives of mobile computing equipment. Heating the cell to 60 °C and higher increases the conductivity, a requirement that is unsuitable for portable applications. To compromise, some gelled electrolyte has been added [174]. The commercial cells use a separator/electrolyte membrane prepared from the same traditional porous polyethylene or polypropylene separator filled with a polymer, which gels upon filling with the liquid electrolyte. Thus, the commercial lithium-ion polymer cells are very similar in chemistry and materials to their liquid electrolyte counterparts. Lithium-ion-polymer has not caught on as quickly as some analysts had expected. Its superiority to other systems and low manufacturing costs has not been realized. No improvements in capacity gains are achieved, and in fact, the capacity is slightly less than that of the standard lithium-ion battery. Lithium-ion-polymer finds its market niche in wafer-thin geometries, such as batteries for credit cards and other such applications [175].

Advantages are very low profile (batteries resembling the profile of a credit card are feasible), flexible form factor (manufacturers are not bound by standard cell formats. With high volume, any reasonable size can be produced economically), lightweight (gelled electrolytes enable simplified packaging by eliminating the metal shell), and improved safety (more resistant to overcharge; less chance for electrolyte leakage). However, some limitations are lower energy density and decreased cycle count compared to lithium-ion, expensive to manufacture, no standard sizes (most cells are produced for high volume consumer markets), and higher cost-to-energy ratio than lithium-ion [176]. There are restrictions on lithium content for air travel. Air travelers ask the question, "How much lithium in a battery am I allowed to bring onboard?" We differentiate between two battery types: lithium metal and lithium-ion. Most lithium metal batteries are nonrechargeable and are used in film cameras. Lithium-ion packs are rechargeable and power laptops, cellular phones, and camcorders. Both battery types, including spare packs, are allowed as carry-on but cannot exceed the following lithium content: 2 g for lithium metal or lithium alloy batteries whereas 8 g for lithium-ion batteries.

15.3.2.4 Sodium-Ion Batteries

The **sodium-ion battery** (**NIB**) is a type of rechargeable battery analogous to the lithium-ion battery but using sodium ions (Na^+) as the charge carriers. Its working principle and cell construction are identical with that of the commercially widespread lithium-ion battery with the only difference being that the lithium compounds are swapped with sodium compounds: in essence, it consists of a cathode based on sodium-containing material, an anode (not necessarily a sodium-based material), and a liquid electrolyte containing dissociated sodium salts in polar protic or aprotic solvents [177]. During charging, Na^+ is extracted from the cathode and inserted into the anode while the electrons travel through the external circuit; during discharge, the reverse process occurs where the Na^+ is extracted from the anode and reinserted in the cathode, with the electrons traveling through the external circuit doing useful work. Ideally, the anode and cathode materials should be able to withstand repeated cycles of sodium storage without degradation. Development of the sodium-ion battery took place side-by-side with that of the lithium-ion battery in the 1970s and early 1980s; however, its development was superseded by that of the lithium-ion battery in the 1990s and 2000s. From 2011, research interest in sodium-ion batteries has been revived [178].

Anodes: The dominant anode used in commercial lithium-ion batteries, graphite, cannot be used in sodium-ion batteries, as it cannot store the larger sodium ion in appreciable quantities. Instead, a disordered carbon material consisting of a nongraphitizable, noncrystalline, and amorphous carbon structure (called 'hard carbon') is the current preferred sodium-ion anode of choice. In hard carbon's sodium storage, the anode was shown to deliver 300 mAh/g with a sloping potential profile above 0.15 V vs Na/Na^+, roughly accounting for half of the capacity and a flat potential profile (a potential plateau) below 0.15 V vs Na/Na^+ [179]. Such a storage performance is similar to that seen for lithium storage in graphite anode for lithium-ion batteries, where capacities of 300–360 mAh/g are typical. The first sodium-ion cell using hard carbon was hence demonstrated in 2003 which showed a high 3.7 V average voltage during discharge. There are now several companies offering hard carbon commercially for sodium-ion applications. While hard carbon is clearly the most preferred anode due to its excellent combination of high capacity, lower working potentials, and good cycling stability, there have been a few other notable developments in lower performing anodes. Incidentally, it was discovered that graphite could store sodium through solvent cointercalation in ether-based electrolytes in 2015: Low capacities around 100 mAh/g were obtained with the working potentials being relatively high between 0 and 1.2 V versus Na/Na^+. Some sodium titanate phases such as $Na_2Ti_3O_7$, or $NaTiO_2$, can deliver capacities around 90–180 mAh/g at low working potentials (<1 V vs Na/Na^+), although cycling stability is currently limited to a few hundred cycles [180]. There have been numerous reports of anode materials storing sodium via an alloy reaction mechanism and/or conversion reaction mechanism. However, the severe stress–strain experienced on the material in the course of repeated storage cycles severely limits their cycling stability, especially in large-format cells, and is a major technical challenge that needs to be overcome by a cost-effective approach.

Cathodes: Significant progress has been achieved in devising high-energy density sodium-ion cathodes since 2011. Similar to all lithium-ion cathodes, sodium-ion cathodes also store sodium via an intercalation reaction mechanism. Owing to their high tap density, high operating potentials, and high capacities, cathodes based on sodium transition metal oxides have received the greatest attention. Furthermore, from a desire to keep costs low, significant research has been geared toward avoiding or reducing costly elements, such as Co, Cr, Ni, or V in the oxides [181]. A P_2-type $Na_{2/3}Fe_{1/2}Mn_{1/2}O_2$ oxide from earth-abundant Fe and Mn resources was demonstrated to reversibly store 190 mAh/g at an average discharge voltage of 2.75 V vs Na/Na^+ utilizing the $Fe^{3+/4+}$ redox couple in 2012. Such energy density was on par or better than commercial lithium-ion cathodes such as $LiFePO_4$ or $LiMn_2O_4$. However, its sodium-deficient nature meant sacrifices in energy density in practical full cells. To overcome sodium deficiency inherent in P2 oxides, significant efforts were expended in developing Na richer oxides. A mixed P3/P2/O3-type $Na_{0.76}Mn_{0.5}Ni_{0.3}Fe_{0.1}Mg_{0.1}O_2$ was demonstrated to deliver 140 mAh/g at an average discharge voltage of 3.2 V vs Na/Na^+ in 2015. Faradion Limited, a sodium-ion company based in the United Kingdom, has patented the highest energy density oxide-based cathodes currently known for sodium-ion applications. In particular, the O3-type NaNi1/4Na1/6Mn2/12Ti4/12Sn1/12O2 oxide can deliver 160 mAh/g at an average voltage of 3.22 V vs Na/Na^+, while a series of doped Ni-based oxides of the stoichiometry $NaNi_{(1-x-y-z)}Mn_xMg_yTi_zO_2$ can deliver 157 mAh/g in a sodium-ion "full cell" with the anode being hard carbon (contrast with the "half-cell" terminology used when the anode is sodium metal) at average discharge voltage of 3.2 V utilizing the $Ni^{2+/4+}$ redox couple. Such performance in full cell configuration is better or on par with commercial lithium-ion systems [182].

Apart from oxide cathodes, there has been tremendous research interest in developing cathodes based on polyanions. While these cathodes would be expected to have lower tap density than oxide-based cathodes (which would negatively impact energy density of the resulting sodium-ion battery), on account of the bulky anion, for many of such cathodes, the stronger covalent bonding of the polyanion translates to a more robust cathode that positively impacts cycle life and safety. Among such polyanion-based cathodes, sodium vanadium phosphate and fluorophosphate have demonstrated excellent cycling stability and, in the case of the latter, an acceptably high capacity (≈120 mAh/g) at high average discharge voltages (≈3.6 V vs Na/Na^+). There have also been several promising reports on the use of various Prussian Blue Analogues (PBAs) as sodium-ion cathodes, with the patented rhombohedral $Na_2MnFe(CN)_6$ particularly attractive displaying 150–160 mAh/g in capacity and a 3.4 V average discharge voltage [183]. Novasis Energies Inc. is currently working to commercialize sodium-ion batteries based on this material and hard carbon anode.

Electrolytes: Sodium-ion batteries can use aqueous as well as nonaqueous electrolytes. Aqueous electrolytes, owing to the limited electrochemical stability window of water, result in sodium-ion batteries of lower voltages and, hence, limited energy densities. To extend the voltage range of sodium-ion batteries, the same nonaqueous carbonate ester polar aprotic solvents used in lithium-ion electrolytes, such as ethylene carbonate, dimethyl carbonate, diethyl carbonate, propylene carbonate, and so on can be used. The current most widely used nonaqueous electrolyte utilizes sodium hexafluorophate as the salt dissolved in a mixture of the aforementioned solvents. Additionally, electrolyte additives can be used which can beneficially affect a host of performance metrics of the battery [184].

Table 15.4 Comparison of Na-ion battery with the Li-ion battery [185–188]

Items	Na-ion battery	Li-ion battery
Energy density	Moderate/high	High
Materials	Earth abundant	Scarce
Cycling stability	High (negligible self-discharge)	High (negligible self-discharge)
Efficiency	High (>90%)	High (>90%)
Temperature range	−40–60 °C	−25–40 °C
Transportation	Less mature technology; easy transportation	Transportation restrictions at discharged state
Cost	Low	High

Comparison between the two commercial batteries such as Na-ion battery and Li-ion battery has been given in Table 15.4.

15.3.2.5 Magnesium Batteries

Magnesium is under research as a possible replacement or improvement on lithium-ion battery in certain applications. In comparison to lithium as an anode material magnesium has a (theoretical) energy density per unit mass under half that of lithium (18.8 MJ/kg vs. 42.3 MJ/kg), but a volumetric energy density around 50% higher (32.731 GJ/m^3 vs. 22.569 GJ/m^3). Magnesium batteries are batteries that utilize magnesium cations as the active charge transporting agent in solution and as the elemental anode of an electrochemical cell. Both nonrechargeable primary cell and rechargeable secondary cell chemistries have been investigated. Magnesium primary cell batteries have been commercialized and have found use as reserve and general use batteries. Magnesium secondary cell batteries are an active topic of research, specifically as a possible replacement or improvement over lithium-ion–based battery chemistries in certain applications. A significant advantage of magnesium cells is their use of a solid magnesium anode, allowing a higher energy density cell design than that made with lithium, which in many instances requires an intercalated lithium anode [189]. Primary magnesium cells have been developed since the early twentieth century. A number of chemistries for reserve battery types have been researched, with cathode materials including silver chloride, copper(I) chloride, palladium(II) chloride, copper (I) iodide, copper(I) thiocyanate, manganese dioxide, and air (oxygen) [190]. In comparison to metallic lithium anodes, magnesium anodes do not exhibit dendrite formation, which may allow magnesium metal to be used without an intercalation compound at the anode; the ability to use a magnesium anode without an intercalation layer raises the theoretical maximum relative volumetric energy density to around five times that of a lithium-ion cell. Additionally, magnesium-based batteries may have a cost advantage over lithium due to the abundance of magnesium on earth and the relative scarcity of lithium deposits [191, 192]. The potential use of a Mg-based battery had been recognized as early as the 1990s based on a V_2O_5, TiS_2, or Ti_2S_4 cathode materials and magnesium metal anodes. However, observation of instabilities in the discharge state and uncertainties on the role of water in the electrolyte limited progress was reported [193].

Anodes and electrolytes: A key drawback to using a metallic magnesium anode is the tendency to form a passivating (non conducting) layer when recharging, blocking further charging (in contrast to lithium's behavior); The passivating layers were thought to originate from the decomposition of the electrolyte during magnesium ion reduction. Common counter ions such as perchlorate and tetrafluoroborate were found to contribute to passivation as were some common polar aprotic solvents such as carbonates and nitriles [194]. Early attempts to develop magnesium batteries explored the use of "magnesium insertion electrodes," based on the reversible insertion of magnesium metal into metal alloy anode (such as Bismuth/Antinomy or Tin). These are shown to prevent anode surface passivation but suffered from anode destruction due to volumetric changes on insertion as well as slow kinetics of insertion. Examples of insertion anode types researched include Sn and Mg_2Sn. The compound $Mg(BPh_2Bu_2)_2$ was used in the first demonstrated rechargeable magnesium battery, and its usefulness was limited by electrochemical oxidation (i.e. a low anodic limit of the voltage window). Other electrolytes researched include borohydrides, phenolates, alkoxides, amido-based complexes (e.g., based on hexamethyldisilazane), carborane salts, fluorinated alkoxyborates, an $Mg(BH_4)(NH_2)$ solid-state electrolyte, and gel polymers containing $Mg(AlCl_2EtBu)_2$ in tetraglyme/PVDF [195].

The current wave of interest in magnesium-metal batteries started in 2000 when an Israeli group reported reversible magnesium plating from mixed solutions of magnesium chloride and aluminum chloride in ethers, such as THF. The primary advantage of this electrolyte is a significantly larger positive limit of the voltage window (and, thus, a higher battery voltage) than the previously reported Mg plating electrolytes. Since then, several other Mg salts, less corrosive than chloride, have been reported [196]. One drawback compared to lithium is magnesium's higher charge (+2) in solution, which tends to result in increased viscosity and reduced mobility in the electrolyte. In solution, a number of species may exist depending on counter ions/complexing agents, these often include singly charged species (e.g. $MgCl^+$ in the presence of chloride), although dimers are often formed (e.g. Mg_2Cl^{3+}). The movement of the magnesium ion into cathode host lattices is also (as of 2014) problematically slow [197]. In 2018, a chloride-free electrolyte together with a quinone-based polymer cathode demonstrated promising performance, with up to 243 Wh (870 kJ) per kg energy density, up to 3.4 kW/kg power density, and up to 87% retention at 2500 cycles. The absence of chloride in the electrolyte was claimed to improve ion kinetics, thus reducing the amount of electrolyte used and increasing performance density figures [198].

A promising approach could be the combination of an Mg anode with a sulfur/carbon cathode. Therefore, a non-nucleophilic electrolyte is necessary which does not convert the sulfur into sulfide just by its reducing properties. Such electrolytes have been developed based on chlorine-containing and chlorine-free complex salts. The electrolyte is Mg, salt-containing Mg cation and two boron-hexafluoroisoproplylate groups as anions. This system is easy to synthesize, it shows an ionic conductivity similar to that of Li ion cells, its electrochemical stability window is up to 4.5 V, it is stable in air, and versatile toward different solvents [199].

Cathode Materials: For cathode materials, a number of different compounds have been researched for the suitability, including those used in magnesium primary batteries. New cathode materials investigated or proposed include zirconium disulfide, cobalt(II,III) oxide, tungsten diselenide, vanadium pentoxide, and vanadate-based cathodes. Cobalt-based spinels showed inferior kinetics to insertion compared to their behavior with lithium. In 2000, the chevrel phase form of Mo6S8 was shown to have good suitability as a cathode, enduring 2000 cycles at 100% discharge with a 15% loss; drawbacks were poor low-temperature performance (reduced Mg mobility, compensated by substituting Selenium) as well as a low voltage, c. 1.2 V, and low energy density (110 mAh/g). A molybdenum disulfide cathode showed improved voltage and energy density, 1.8 V and 170 mAh/g. Transition metal sulfides is considered promising candidates for magnesium ion battery cathodes. A hybrid magnesium cell using a mixed magnesium/sodium electrolyte with sodium insertion into a nanocrystallineiron(II) disulfide cathode was reported in 2015 [200].

Manganese dioxide–based cathodes have shown good properties but deteriorated on cycling. Modified manganese-based spinels ("post spinels") are an active topic of research (2014) for magnesium-ion insertion cathodes. In 2014, a rechargeable magnesium battery was reported utilizing an ion-exchange, olivine type $MgFeSiO_4$ cathode with a bis(trifluoromethylsulfonyl)imide/triglyme electrolyte, the cell showed a capacity of 300 mAh/g with a voltage of 2.4 V. $MgMnSiO_4$ has also been investigated as a potential Mg^{2+} insertion cathode.

Cathodic materials other than non-inorganic metal oxide/sulfide types have also been investigated. In 2015, a cathode based on a polymer incorporating anthraquinone was reported; and other organic, and organo-polymer cathode materials capable of undergoing redox reactions have also been investigated, such as poly-2,2′-dithiodianiline. Quinone-based cathodes also formed the cathode, a high-energy density magnesium battery reported by researchers in 2019.

15.3.2.6 Zinc-Ion Batteries

A zinc-ion battery (ZIB) uses Zn^{2+} as charge carriers. ZIB utilizes Zn as the anode, Zn-intercalating materials as the cathode, and a Zn-containing electrolyte. The increasing demands for environmentally friendly grid-scale electric energy storage devices with high energy density and low cost have stimulated the rapid development of various energy storage systems, due to the environmental pollution and energy crisis caused by traditional energy storage technologies [201]. As one of the new and most promising alternative energy storage technologies, zinc-ion rechargeable batteries have recently received much attention owing to their high abundance of zinc in natural resources, intrinsic safety, and cost-effectiveness when compared to the popular, but unsafe and expensive, lithium-ion batteries. In particular, the use of mild aqueous electrolytes in ZIB demonstrates high potential for portable electronic applications and large-scale energy storage systems. Moreover, the development of superior electrolyte operating at either high temperature or sub-zero conditions is crucial for practical applications of ZIBs in harsh environments, such as aerospace, airplanes, or submarines [202].

The utilization of manganese oxides in rechargeable ZIB is attracting research interest owing to the low cost, moderate discharge potentials, and acceptable rate/cycle performances, together with a high theoretical capacity of about 308 mAh/g ($Zn_{0.5}MnO_2$). The unique two-electron redox electrolysis reaction of Mn^{4+}/Mn^{2+} is used in a high-voltage electrolytic Zn-MnO_2 battery, with a theoretical voltage of about 2 V and an energy density of approximately 700 Wh/kg. In general, three charge storage mechanisms have been reported thus far: (1) reversible Zn ion insertion/extraction in/from bulk materials; (2) reversible proton reaction accompanied by deposition of zinc hydroxide sulfate species; and (3) the successive uptake/removal of both H^+ and Zn^{2+} at different charge/discharge stages. However, these mechanisms make use of only one effective electron redox process of Mn^{4+}/

Mn^{3+} in the applied cathode, limiting both their capacities and output voltages. Therefore, new Zn-Mn electrochemistry is desired. Essentially, the multivalency of Mn (+2, +3, and +4) holds remarkable potential for two-electron Mn^{4+}/Mn^{2+} reactions with a theoretical capacity of about 616 mAh/g and high voltage. New Zn-Mn cell designs are expected to realize these promises in providing a high working window of about 2 V and, consequently, high energy/power in aqueous Zn-based batteries [203]. The as-designed electrolytic Zn-MnO_2 battery enables a high discharge plateau at about 1.95 V, an excellent rate capability of 60 mA/cm^2 in 100 s, long-term cycling stability over 1800 cycles, and a record energy density of approximately 409 Wh/kg. The cost of the electrolytic Zn-MnO_2 battery was estimated at <US$10/kWh. This is significantly less than that for current Li-ion batteries of US$300/kWh, ZIBs (US$65/kWh), Ni-Fe batteries (US$72/kWh), and lead-acid batteries (US$48/kWh) [204].

15.3.2.7 Zinc-Air Batteries

Zinc-air batteries have some properties of fuel cells as well as batteries: the zinc is the fuel, the reaction rate can be controlled by varying the air flow, and oxidized zinc/electrolyte paste can be replaced with fresh paste [205]. Zinc-air batteries (non-rechargeable) and zinc-air fuel cells (mechanically rechargeable) are metal-air batteries powered by oxidizing zinc with oxygen from the air. These batteries have high energy densities and are relatively inexpensive to produce. Sizes range from very small button cells for hearing aids, larger batteries used in film cameras that previously used mercury batteries, to very large batteries used for electric vehicle propulsion and grid-scale energy storage. During discharge, a mass of zinc particles forms a porous anode, which is saturated with an electrolyte. Oxygen from the air reacts at the cathode and forms hydroxyl ions that migrate into the zinc paste and form zincate ($Zn(OH)^{2-}{}_4$), releasing electrons to travel to the cathode. The zincate decays into zinc oxide and water returns to the electrolyte. The water and hydroxyl from the anode are recycled at the cathode, so the water is not consumed. The reactions produce a theoretical 1.65 V, but this is reduced to 1.35–1.4 V in available cells [206]. The chemical equations for the zinc-air cell are:

At anode:

$$Zn + 4OH^- \rightarrow Zn(OH)_4^{2-} + 2e^- \; (E_0 = -1.25\ V) \quad (15.38)$$

At fluid:

$$Zn(OH)_4^{2-} \rightarrow ZnO + H_2O + 2OH^- \quad (15.39)$$

At cathode:

$$1/2\, O_2 + H_2O + 2e^- \rightarrow 2OH^- \; (E_0 = 0.34\ V\ pH = 11) \quad (15.40)$$

Overall:

$$2Zn + O_2 \rightarrow 2ZnO \; (E_0 = 1.59\ V) \quad (15.41)$$

Zinc-air batteries cannot be used in a sealed battery holder since some air must come in; the oxygen in 1 liter of air is required for every ampere-hour of capacity used. Zinc-air batteries have a higher energy density and specific energy (and weight) ratio than other types of batteries because atmospheric air is one of the battery reactants. The air is not packaged with the battery so that a cell can use more zinc in the anode than a cell that must also contain, for example, manganese dioxide. This increases capacity for a given weight or volume. As a specific example, a zinc-air battery of 11.6 mm diameter and height 5.4 mm from one manufacturer has a capacity of 620 mAh and weight 1.9 g; various silver oxide (SR44) and alkaline cells of the same size supply 150–200 mAh and weigh 2.3–2.4 g [207].

Zinc-air cells have a long shelf-life if sealed to keep air out; even miniature button cells can be stored for up to 3 years at room temperature with little capacity loss if their seal is not removed. Industrial cells stored in a dry state have an indefinite storage life. The operating life of a zinc-air cell is a critical function of its interaction with its environment. The electrolyte loses water more rapidly in conditions of high temperature and low humidity [208]. Low humidity increases water loss; if enough water is lost the cell fails. Because the cathode does not change properties during discharge, terminal voltage is quite stable until the cell approaches exhaustion. Power capacity is a function of several variables: cathode area, air availability, porosity, and the catalytic value of the cathode surface. Oxygen entry into the cell must be balanced against electrolyte water loss; cathode membranes are coated with (hydrophobic) Teflon material to limit water loss. Button cells have a limited current drain; for example, an IEC PR44 cell has a capacity of 600 milliamp-hours (mAh) but a maximum current of only 22 milliamps (mA). Pulse load currents can be much higher since some oxygen remains in the cell between pulses. Low temperature reduces primary cell capacity, but the effect is small for low drains. A cell may deliver 80% of its capacity if discharged over 300 h at 0 °C (32 °F) but only 20% of capacity if discharged at a 50-h rate at that temperature. The lower temperature also reduces cell voltage [209].

15.3.2.8 K-Ion Batteries

A potassium-ion battery (KIB) is a type of battery and analog to lithium-ion batteries, using potassium ions for charge transfer [210]. Same as the case of lithium-ion batteries, graphite could also accommodate the intercalation of potassium within electrochemical process, while with different kinetics, graphite anodes suffer from low capacity retention during cycling within potassium-ion batteries. Thus, the approach of structural engineering of graphite anode is needed to achieve stable performance. Other types of carbonaceous materials besides graphite have been used as anode material for potassium-ion battery, such as expanded graphite, carbon nanotubes, carbon nanofibers, and also nitrogen or phosphorus-doped carbon materials. Conversion anodes that can form a compound with potassium ion with boosted storage capacity and reversibility have also been studied to fit for potassium-ion battery. To buffer the volume change of conversion anode, a carbon material matrix is always applied such as MoS_2@rGO, Sb_2S_3-SNG, SnS_2-rGO, and so on. Classic alloying anodes such as Si, Sb, and Sn that can form an alloy with lithium-ion during the cycling process are also applicable for the potassium-ion battery. Among them, Sb is the most promising candidate due to its low cost and theoretical capacity of up to 660 mAh/g. Other organic compounds are also being developed to achieve strong mechanical strength as well as maintaining decent performance [211]. A series of potassium transition metal oxide such as $K_{0.3}MnO_2$, $K_{0.55}CoO_2$have been demonstrated as cathode material with a layered structure. Polyanionic compounds with inductive defects could provide the highest working voltage among other types of cathode for potassium-ion batteries. During the electrochemical cycling process, its crystal structure will be distorted to create more induced defects upon the insertion of potassium ions [212]. Due to the chemical activity higher than lithium, an electrolyte for potassium ion battery requires more delicate engineering to address safety concerns. Commercial ethylene carbonate (EC) and diethyl carbonate (DEC) or other traditional ether/ester liquid electrolytes showed poor cycling performance and fast capacity degradation due to the Lewis acidity of potassium, also the highly flammable feature of it has prevented further application. Ionic liquid electrolyte offers a new way to expand the electrochemical window of potassium ion battery with much negative redox voltage, and it is especially stable with a graphite anode. Recently, solid polymer electrolytes for all-solid-state potassium-ion batteries have attracted much attention due to its flexibility and enhanced safety [213].

The K-ion has certain advantages over similar Li-ion batteries: the cell design is simple and both the material and the fabrication procedures are cheaper. The key advantage is the abundance and low cost of potassium in comparison with lithium, which makes potassium batteries a promising candidate for large-scale batteries such as household energy storage and electric vehicles. Another advantage of K-ion battery over Li-ion battery is the possibility for charging faster. This means that the next generation of portable devices based on K-ion batteries could be charged within a few minutes. The chemical diffusion coefficient of K^+ in the cell is higher than that of Li^+ in lithium batteries due to a smaller Stokes radius of solvated K^+. Since the electrochemical potential of K^+ is identical to that of Li^+, the cell potential is similar to that of lithium-ion. Potassium batteries can accept a wide range of cathode materials that can offer rechargeability lower cost. One noticeable advantage is the availability of potassium graphite, which is used as an anode material in some lithium-ion batteries. Its stable structure guarantees a reversible intercalation/de-intercalation of potassium ions under charge/discharge [214]. The price of potassium metal is relatively high compared with sodium; however, the price of potassium salt, that is, the raw materials for electrode fabrication (K_2CO_3), is similar to that of Na_2CO_3, which is much cheaper compared to Li_2CO_3. In addition, aluminum foil can be used as a current collector in PIBs instead of the copper foil found in LIBs, which will not only notably reduce the price of the PIB but also reduce the weight of the current collector and address over-discharge problems. Although potassium has the largest atomic radius (1.38 Å) compared to lithium (0.68 Å) and sodium (0.97 Å), K^+ has the smallest Stokes' radius (3.6 Å) compared to Li^+ (4.8 Å) and Na^+ (4.6 Å) in propylene carbonate (PC) solvents, indicating that it has the highest ion mobility and ion conductivity. Based on the advantages mentioned earlier, replacing Li^+ with K^+ would enable us to enhance the rate capability and realize high mass loading electrodes without sacrificing specific capacity [215].

15.3.2.9 Aluminum-Ion Batteries

Aluminum-ion batteries are a class of rechargeable batteries in which aluminum ions provide energy by flowing from the negative electrode of the battery, the anode, to the positive electrode, the cathode. In Al-ion batteries, two electrodes are connected by an electrolyte, an ionically (but not electrically) conductive material acting as a medium for the flow of charge carriers. When recharging, aluminum ions return to the negative electrode and can exchange three electrons per ion. This means that the insertion of one Al^{3+} is equivalent to three Li^+ ions in conventional intercalation cathodes. Thus, since the ionic radii of Al^{3+} (0.54 A) and Li^+ (0.76 A) are similar, significantly higher models of electrons and Al^{3+} ions can be accepted by the cathodes without much pulverization. The trivalent charge carrier, Al^{3+} is both the advantage and disadvantage of this battery. While transferring three units of charge by one ion significantly increases the energy storage capacity, the electrostatic intercalation of the host materials with a trivalent cation is too strong for well-defined electrochemical behavior. Rechargeable Al-based batteries offer the possibilities of low cost and low flammability, together with three-electron-redox properties leading to high capacity. The inertness of Al and the ease of handling in an ambient environment is expected to offer significant safety improvements for this kind of battery. In addition, Al possesses a higher volumetric capacity than Li, K, Mg, Na, Ca, and Zn owing to its high density (2.7 g/cm^3 at 25 °C) and ability to exchange three electrons. This again means that the energy stored in aluminum batteries on a per-volume basis is higher than that in other metal-based batteries. Hence, Al-batteries are expected to be smaller in size. Al-ion batteries also have a higher number of charge–discharge cycles. Thus, Al-ion batteries have the potential to replace Li-ion batteries [216].

Unlike lithium-ion batteries, where the mobile ion is Li^+, aluminum forms a complex with chloride in most electrolytes and generates an anionic mobile charge carrier, usually $AlCl^{-4}$ or Al_2Cl^{-7}. The amount of energy or power that a battery can release is dependent on factors, including the battery cell's voltage, capacity, and chemical composition. A battery can maximize its energy output levels by increasing the chemical potential difference between the two electrodes by reducing the mass of reactants and by preventing the electrolyte from being modified by the chemical reactions. The electrochemical associated with the Al batteries are:

Anode reaction:

$$Al + 7AlCl_4^- \leftrightharpoons 4Al_2Cl^-{}_7 + 3e^- \tag{15.42}$$

Catrhode reaction:

$$2MnO_2 + Li^+ + e^- \leftrightharpoons 4Al_2Cl^-{}_7 + 3LiMn_2O_4 \tag{15.43}$$

Combining the two half reactions yields the following reaction

$$Al + 7AlCl_4^- + 6MnO_2 + 3Li^+ \leftrightharpoons LiMn_2O_4 \tag{15.44}$$

The theoretical voltage for aluminum-ion batteries is lower than lithium-ion batteries, 2.65 V and 4 V, respectively, and the theoretical energy density potential for aluminum-ion batteries is 1060 Wh/kg in comparison to lithium-ion batteries is 406 Wh/kg limit. Today's lithium-ion batteries have high power density (fast discharge) and high energy density. It can also develop dendrites, similar to splinters, that can short-circuit a battery and lead to a fire. Aluminum also transfers energy more efficiently. Inside a battery, atoms of the element-lithium or aluminum give up some of their electrons, which flow through external wires to power a device. Because of their atomic structure, lithium ions can only provide one electron at a time; aluminum can give three at a time. Aluminum is also more abundant than lithium, lowering material costs. Using a selection algorithm for the evaluation of suitable materials, the concept of a rechargeable, high-valent all-solid-state aluminum-ion battery appears promising, in which metallic aluminum is used as the negative electrode. On the one hand, this offers the advantage of a volumetric capacity four times higher (theoretically) compared to lithium analog. On the other hand, aluminum is the most abundant metal in the earth's crust. There is a mature industry and recycling infrastructure, making aluminum very cost-efficient. This would make the aluminum-ion battery an important contribution to the energy transition process, which has already started globally [217, 218].

15.3.2.10 Nickel-Bismuth Batteries

The exploration of a stable and high-rate anode is of pivotal importance for achieving advanced aqueous rechargeable batteries. Owing to the beneficial properties of high conductivity, suitable negative working voltage, and three-electron redox, bismuth (Bi) is considered a promising anode material, but it suffers from poor stability. Here, we successfully endow Bi nanoflakes (NFs) with prominent cycling performance by a one-step surface oxidation approach to remarkably boost its reversibility. As a result, the partially oxidized Bi NFs (BiOx) show an admirable capacity (0.38 mAh/cm^2 at 2 mA/cm^2), good rate capability, and superior long-term stability (almost no capacity decay after 20,000 cycles). Furthermore, a durable aqueous Ni//Bi battery is constructed based on the optimized BiOx anode, which exhibits excellent durability with 96% capacity retention after 5000 cycles. This study could open a new avenue for the rational design of efficient anodes for eco-friendly and reliable aqueous rechargeable batteries [219].

15.3.2.11 Organic Batteries

An organic radical battery (ORB) is a type of battery first developed in 2005. As of 2011, this type of battery was generally not available for the consumer, although their development at that time was considered to be approaching practical use. ORB is potentially more environmentally friendly than conventional metal-based batteries because they use organic radical polymers (flexible plastics) to provide electrical power instead of metals. ORB is considered to be a high-power alternative to the Li-ion battery. Functional prototypes of the battery have been researched and developed by different research groups and corporations including the Japanese corporation NEC. The organic radical polymers used in ORBs are examples of stable radicals, which are stabilized by steric and/or resonance effects. For example, the nitroxide radical in (2,2,6,6-

tetramethylpiperidin-1-yl)oxyl (TEMPO), the most common subunit used in ORB, is a stable oxygen-centered molecular radical. Here, the radical is stabilized by the delocalization of electrons from the nitrogen onto the oxygen. TEMPO radicals can be attached to polymer backbones to form poly (2,2,6,6-tetramethyl-piperidenyloxyl-4-yl methacrylate) (PTMA). PTMA-based ORBs have a charge density slightly higher than that of conventional Li-ion batteries, which should theoretically make it possible for an ORB to provide more charge than a Li-ion battery of similar size and weight. As of 2007, ORB research was being directed mostly toward Hybrid ORB/Li-ion batteries because organic radical polymers with appropriate electrical properties for the anode are difficult to synthesize [220].

Organic radical batteries are much more environmentally friendly than Li-ion batteries because ORBs do not contain any metals that pose the problem of proper disposal. ORBs are nontoxic and nonflammable and do not require additional care when handling. Burning nitroxide radical polymers yields carbon dioxide, water, and nitrogen oxide without ash or odor. While being environmentally friendly, they have properties that are otherwise comparable to Li-ion batteries: ORBs have a theoretical capacity of 147 mAh/g, which is slightly higher than that of Li-ion batteries with 140 mAh/g. ORBs also show comparable charge times and retain charge-discharge capacity well, matching lithium-ion batteries at 75% of their initial charge after 500 cycles. Additionally, radical concentrations in ORBs are stable enough at ambient conditions to remain unchanged for over a year. ORBs are also more flexible than Li-ion batteries, which would make them more adaptable to different design constraints, such as curved devices [221, 222]. A major difficulty in the development of ORBs is the difficulty of synthesizing an appropriate negative electrode. This disadvantage arises because the redox reaction of the negative electrode is not fully reversible. Hybrid ORB/Li-ion batteries, in which the negative electrode is replaced by the one found in a Li-ion battery, have been proposed as a compromise to overcome this difficulty. Polymerization reactions of the stable radical-containing monomer have also proved to be an area of difficulty in development. The stable organic radicals that are crucial to the functioning of the battery are sometimes consumed in side reactions of various polymerization reactions. A research group has, however, successfully synthesized a cross-linked organic radical polymer while only losing 0.4% of the organic radicals in the synthesis of the polymer [223, 224]. Theoretically, ORBs could replace Li-ion batteries as more environmentally friendly batteries of similar or higher charge capacity and similar or shorter charge time. This would make ORBs well-suited for handheld electronic devices [225].

15.4 Summary

This is the high time to harvest the energy from nature without hampering environmental activities, which is described in this chapter. Solar energy, wind energy, tide energy, piezoelectric energy, magnetostrictive energy, and chemical energy have been discussed successfully with their materials and mechanisms. In between the energy conversion toward human help, to maintain the time of use, energy storage materials plays a big role. Currently, all are focusing on the use of the different types of batteries that is discussed in this chapter.

References

1. Huidong, L., Tian, A.U., Deng, C., Daniel, Z.: Energy harvesting from low frequency applications using piezoelectric materials. Appl. Phys. Rev. **1**, 041301 (2014). https://doi.org/10.1063/1.4900845
2. Zhou, G., Huang, L., Li, W., Zhu, Z.: Harvesting ambient environmental energy for wireless sensor networks: a survey. J. Sens. **2014**, 20 (2014). https://doi.org/10.1155/2014/815467. 815467
3. See: https://en.wikipedia.org/wiki/Energy_harvesting. 01.04.2020
4. Prauzek, M., Konecny, J., Borova, M., Janosova, K., Hlavica, J., Musilek, P.: Energy harvesting sources, storage devices and system topologies for environmental wireless sensor networks: a review. Sensors (Basel). **18**(8), 2018 (2446). https://doi.org/10.3390/s18082446
5. Tang, X., Wang, X., Cattley, R., Gu, F., Ball, A.D.: Energy harvesting technologies for achieving self-powered wireless sensor networks in machine condition monitoring: a review. Sensors (Basel). **18**(12), 4113 (2018). https://doi.org/10.3390/s18124113
6. Enescu, D.: Thermoelectric energy harvesting: basic principles and applications, green energy advances. IntehOpen, London (2019). https://doi.org/10.5772/intechopen.83495
7. Jaziri, N., Boughamoura, A., Müller, J., Mezghani, B., Tounsi, F., Ismail, M.: A comprehensive review of thermoelectric generators: technologies and common applications. Energy Rep. (2019). https://doi.org/10.1016/j.egyr.2019.12.011. ISSN 2352–4847
8. Mizuguchi, M., Nakatsuji, S.: Energy-harvesting materials based on the anomalous Nernst effect. Sci. Technol. Adv. Mater. **20**(1), 262–275 (2019). https://doi.org/10.1080/14686996.2019.1585143
9. Scott, W., Ronald, D.: Thermoelectric energy harvesting from diurnal heat flow in the upper soil layer. Energy Convers. Manag. **64**, 397–402 (2012). https://doi.org/10.1016/j.enconman.2012.06.015
10. Mizuguchi, M., Nakatsuji, S.: Energy-harvesting materials based on the anomalous Nernst effect. Sci. Technol. Adv. Mater. **20**(1), 262–275 (2019). https://doi.org/10.1080/14686996.2019.1585143
11. Nechibvute, A., Chawanda, A., Luhanga, P.: Piezoelectric energy harvesting devices: an alternative energy source for wireless sensors. Smart Mater. Res. **2012**, 13 (2012). https://doi.org/10.1155/2012/853481. 853481
12. See: https://ren21.net/Portals/0/documents/Resources/GSR/2013/GSR2013_lowres.pdf. 05.04.2020
13. See: https://en.wikipedia.org/wiki/Solar_cell. 05.04.2020
14. See: https://www.science.org.au/curious/technology-future/solar-pv. 05.04.2020

15. See: https://www.seia.org/initiatives/concentrating-solar-power. 05.04.2020
16. Nathan, G.J., Jafarian, M., Dally, B.B., Saw, W.L., Ashman, P.J., Hu, E., Steinfeld, A.: Solar thermal hybrids for combustion power plant: A growing opportunity. Prog. Energy Combust. Sci. **64**, 4–28 (2018). https://doi.org/10.1016/j.pecs.2017.08.002. ISSN 0360–1285
17. Khan, M.I., Shin, J.H., Kim, J.D.: The promising future of microalgae: current status, challenges, and optimization of a sustainable and renewable industry for biofuels, feed, and other products. Microb Cell Fact. **17**(1), 36 (2018). https://doi.org/10.1186/s12934-018-0879-x
18. See: https://en.wikipedia.org/wiki/Photosynthetic_efficiency. 05.04.2020
19. Zhang, B., Sun, L.: Artificial photosynthesis: opportunities and challenges of molecular catalysts. Chem. Soc. Rev. **48**, 2216–2264 (2019). https://doi.org/10.1039/C8CS00897C
20. National Center for Biotechnology Information: PubChem Compound Summary for CID 9989226, Aluminum oxide (2020) Retrieved August 29, 2020 from https://pubchem.ncbi.nlm.nih.gov/compound/Aluminum-oxide
21. American Chemical Society National Historic Chemical Landmarks. Hall Process: Production and Commercialization of Aluminum. http://www.acs.org/content/acs/en/education/whatischemistry/landmarks/aluminumprocess.html, 1997
22. James, Y., Myles, S., Henning, D., Ryan, F., John, T., Todd, D.: Direct solar-to-hydrogen conversion via inverted metamorphic multi-junction semiconductor architectures. Nature Energy. **2**, 17028 (2017). https://doi.org/10.1038/nenergy.2017.28
23. Olle, I., Villy, S.: Solar energy for electricity and fuels. Ambio. **45**, 15–23 (2015). https://doi.org/10.1007/s13280-015-0729-6
24. See: https://www.pveducation.org/pvcdrom/solar-cell-operation/solar-cell-efficiency. 05.04.2020
25. Lee, J.-H., Kim, J., Kim, T.Y., Md, S.A.H., Kim, S.-W., Kim, J.H.: All-in-one energy harvesting and storage devices. J. Mater. Chem. A. **4**, 7983–7999 (2016). https://doi.org/10.1039/C6TA01229A
26. Arvind, S., Milan, V., Johannes, M., Meillaud, F., Joelle, G., Diego, F., Corinne, D., Niquille, X., Sylvie, F., Evelyne, V., Vanessa, T.-D., Bailat, J.: Basic efficiency limits, recent experimental results and novel light-trapping schemes in a-Si:H, ??c-Si:H and 'micromorph tandem' solar cells. J. Non-Crystalline Solids. **338–340**, 639–645 (2004). https://doi.org/10.1016/j.jnoncrysol.2004.03.074
27. Xiong, P., Weiguo, H., Zhong, W.: Toward wearable self-charging power systems: the integration of energy-harvesting and storage devices. Small. **14**, 1702817 (2017). https://doi.org/10.1002/smll.201702817
28. Bobes-Jesus, V., Pascual-Muñoz, P., Castro-Fresno, D., Rodriguez-Hernandez, J.: Asphalt solar collectors: a literature review. Appl. Energy. **102**, 962–970 (2013). https://doi.org/10.1016/j.apenergy.2012.08.050
29. Chiarelli, A., Al-Mohammedawi, A., Dawson, A., García, A.: Construction and configuration of convection-powered asphalt solar collectors for the reduction of urban temperatures. Int. J. Thermal Sci. **112** (2016). https://doi.org/10.1016/j.ijthermalsci.2016.10.012
30. Yildiz, F., Coogler, K.: Low power energy harvesting with a thermoelectric generator through an air conditioning condenser. ASEE Annual Conference and Exposition, Conference Proceedings. (2014) 34.
31. See: https://www.britannica.com/science/thermoelectricity. 08.04.2020
32. Kishore, R., Priya, S.J.: A review on low-grade thermal energy harvesting: materials, methods and devices. Materials. **11**, 1433 (2018). https://doi.org/10.3390/ma11081433
33. Freer, R., Powell, A.V.: Realising the potential of thermoelectric technology: a Roadmap. J. Mater. Chem. C. **8**, 441–463 (2020). https://doi.org/10.1039/C9TC05710B
34. Jiang, B., Liu, X., Wang, Q., Cui, J., Jia, B., Zhu, Y., Feng, J., Qiu, Y., Gu, M., Ge, Z., He, J.: Realizing high-efficiency power generation in low-cost PbS-based thermoelectric materials. Energy Environ. Sci. **13**, 579–591 (2020). https://doi.org/10.1039/C9EE03410B
35. See: https://thermoelectricsolutions.com/how-thermoelectric-generators-work/. 08.04.2020
36. See: https://www.mdpi.com/1424-8220/18/6/1927/pdf. 08.04.2020
37. Prakash, S., Agarwal, A.: Human power harvesting. Int. J. Eng. Res. Technol. **3**(02) (2014)
38. See: https://www.energy.gov/eere/electricvehicles/reducing-pollution-electric-vehicles. 08.04.2020
39. Noel, A., Abdaoui, A., Badawy, A., El-Fouly, T., Ahmed, M., Shehata, M.: Structural health monitoring using wireless sensor networks: a comprehensive survey. IEEE Commun. Surveys Tutorials, 1 (2017). https://doi.org/10.1109/COMST.2017.2691551
40. Damilola, S., Otterpohl, T., Kluge, M., Schmid, U., Becker, T.: Aircraft-specific thermoelectric generator module. J. Electron. Mater. **39**, 2092–2095 (2010). https://doi.org/10.1007/s11664-009-0997-7
41. Mondejar, M.E., Andreasen, J.G., Pierobon, L., Larsen, U., Thern, M., Haglind, F.: A review of the use of organic Rankine cycle power systems for maritime applications. Renew. Sustain. Energy Rev. **91**, 126–151 (2018). https://doi.org/10.1016/j.rser.2018.03.074. ISSN 1364–0321
42. He, W., Zhang, G., Zhang, X., Ji, J., Li, G., Zhao, X.: Recent development and application of thermoelectric generator and cooler. Appl. Energy. **143**, 1–25 (2015). https://doi.org/10.1016/j.apenergy.2014.12.075. ISSN 0306–2619
43. Morozovska, A.N., Eliseev, E.A., Svechnikov, G.S., Kalinin, S.V. L.: Pyroelectric response of ferroelectric nanowires: size effect and electric energy harvesting. AIP Publishing, Melville, NY (2010). https://doi.org/10.1063/1.3474964
44. Yang, Y., Wang, S., Zhang, Y., Wang, Z.: Pyroelectric nanogenerators for driving wireless sensors. Nano Lett. **12** (2012). https://doi.org/10.1021/nl303755m
45. See: https://www.electricaleasy.com/2015/12/geothermal-energy-and-geothermal-power-plant.html. 10.04.2020
46. Wasim, M., Shoaib, S., Inamuddin, I., Asiri, A.M.: Factors influencing corrosion of metal pipes in soils. Environ. Chem. Lett., 1–19 (2018). https://doi.org/10.1007/s10311-018-0731-x
47. Tota-Maharaj, K., Paul, P.: Sustainable approaches for stormwater quality improvements with experimental geothermal paving systems. Sustainability. **7**, 1388–1410 (2015). https://doi.org/10.3390/su7021388
48. Kalyani, V., Piaus, A., Vyas, P.: Harvesting electrical energy via vibration energy and its applications. J. Manag. Eng. Inf. Technol. **2**, 2394–8124 (2015)
49. Zuo, L., Tang, X.: Large-Scale vibration energy harvesting. J. Intell. Mater. Syst. Struct. **24**, 1405–1430 (2013). https://doi.org/10.1177/1045389X13486707
50. See: https://en.wikipedia.org/wiki/Piezoelectricity. 10.04.2020
51. Wang, Z., Hu, J., Suryavanshi, A., Yum, K., Yu, M.-F.: Voltage generation from individual BaTiO3 nanowires under periodic tensile mechanical load. Nano Lett. **7**, 2966–2969 (2007). https://doi.org/10.1021/nl070814e
52. Kymissis, J., Kendall, C., Paradiso, J., Gershenfeld, N.: Parasitic power harvesting in shoes. IEEE Int. Symp. Wrbl Co. **24**, 132–139 (1998). https://doi.org/10.1109/ISWC.1998.729539
53. Zhu, D., Beeby, S.: Kinetic energy harvesting. In: Energy harvesting systems: principles, modeling and applications.

Springer, New York, NY (2011). https://doi.org/10.1007/978-1-4419-7566-9_1

54. Sodano, H., Park, G., Inman, D.: Estimation of electric charge output for piezoelectric energy harvesting. Strain. **40**, 49–58 (2004). https://doi.org/10.1111/j.1475-1305.2004.00120.x
55. Aswal, P., Singh, S., Thakur, A.: Generation of electricity by piezoelectric crystal in dance floor. Springer, New York, NY (2017). https://doi.org/10.1007/978-981-10-1708-7_52
56. Kalyani, V., Piaus, A., Vyas, P.: Harvesting electrical energy via vibration energy and its applications. J. Manag. Eng. Inf. Technol. **2**, 2394–8124 (2015)
57. See: https://semiengineering.com/shedding-pounds-in-automotive-electronics/. 12.04.2020
58. See: https://en.wikipedia.org/wiki/Tire-pressure_monitoring_system. 12.04.2020
59. Yang, Z., Zhou, S., Zu, J., Daniel, I.: High-performance piezoelectric energy harvesters and their applications. JOULE. **2**(4), 642–697 (2018). https://doi.org/10.1016/j.joule.2018.03.011. ISSN 2542–4351
60. Wang, Z., Wang, A.: On the origin of contact-electrification. Mater. Today. **30** (2019). https://doi.org/10.1016/j.mattod.2019.05.016
61. Khan, U., Hinchet, R., Ryu, H., Kim, S.-W.: Research update: Nanogenerators for self-powered autonomous wireless sensors. APL Mater. **5** (2017). https://doi.org/10.1063/1.4979954
62. Qiang, Z., Shi, B., Wang, Z.: Recent progress on piezoelectric and triboelectric energy harvesters in biomedical systems. Adv. Sci. **4**, 1700029 (2017). https://doi.org/10.1002/advs.201700029
63. Wang, Z.L., Chen, J., Lin, L.: Progress in triboelectric nanogenerators as a new energy technology and self-powered sensors. Energy Environ. Sci. **8**, 2250–2282 (2015). https://doi.org/10.1039/C5EE01532D
64. Bai, Y., Jantunen, H., Juuti, J.: Adv. Mater. **30**, 1707271 (2018). https://doi.org/10.1002/adma.201707271
65. Zhao, L., Yang, Y.: Toward small-scale wind energy harvesting: design, enhancement, performance comparison, and applicability. Shock Vib. **2017**, 31 (2017). https://doi.org/10.1155/2017/3585972. 3585972
66. See: https://www.electronics-tutorials.ws/electromagnetism/electromagnetic-induction.html. 14.04.2020
67. See: https://en.wikipedia.org/wiki/Lenz%27s_law#:~:text=Lenz's%20law%2C%20named%20after%20the,the%20initial%20changing%20magnetic%20field. 14.04.2020
68. See: https://www.electrical4u.com/materials-used-for-transmission-line-conductor/. 14.04.2020
69. Tran, L., Cha, H., Park, W.: RF power harvesting: a review on designing methodologies and applications. Micro Nano Syst. Lett. **5**, 14 (2017). https://doi.org/10.1186/s40486-017-0051-0
70. https://www.nrcan.gc.ca/maps-tools-publications/satellite-imagery-air-photos/remote-sensing-tutorials/introduction/electromagnetic-spectrum/14623. 15.04.2020
71. See: https://en.wikipedia.org/wiki/Thermoelectric_generator. 15.04.2020
72. Bakkali, A., Pelegri-Sebastia, J., Sogorb, T., Llario, V., Bou-Escriva, A.: A dual-band antenna for RF energy harvesting systems in wireless sensor networks. J. Sens. **2016**, 8 (2016). https://doi.org/10.1155/2016/5725836. 5725836
73. Lee, H.M., Ghovanloo, M.: A power-efficient wireless capacitor charging system through an inductive link. IEEE Trans. Circuits Syst. II Expr. Briefs. **60**(10), 707–711 (2013). https://doi.org/10.1109/TCSII.2013.2278104
74. Serdijn, W., Mansano, A., Stoopman, M.: Introduction to RF energy harvesting. Elsevier, Amsterdam (2014). https://doi.org/10.1016/B978-0-12-418662-0.00019-2
75. See: https://www.mouser.in/applications/rf_energy_harvesting/. 15.04.2020
76. See: https://www.mwrf.com/technologies/components/article/21841498/receivers-caps-harvest-rf-energy. 15.04.2020
77. Cui, L., Zhang, Z., Gao, N., Meng, Z., Li, Z.: Radio frequency identification and sensing techniques and their applications-a review of the state-of-the-art. Sensors (Basel). **19**(18), 4012 (2019). https://doi.org/10.3390/s19184012
78. Divakaran, S.K., Krishna, D.D., Nasimuddin: RF energy harvesting systems: an overview and design issues. Int. J. RF Microw. Comput. Aided Eng. **29**, e21633 (2019). https://doi.org/10.1002/mmce.21633
79. See: https://www.eeworldonline.com/rf-energy-harvesting-perpetually-powers-wireless-sensors/. 15.04.2020
80. Sung, G.M., Chung, C.K., Lai, Y.J., Syu, J.Y.: Small-area radiofrequency-energy-harvesting integrated circuits for powering wireless sensor networks. Sensors (Basel). **19**(8), 1754 (2019). https://doi.org/10.3390/s19081754
81. Jean-Mistral, C., Basrour, S., Chaillout, J.-J.: Comparison of electroactive polymers for energy scavenging applications. Smart Materials and Structures. **19**, 085012 (2010). https://doi.org/10.1088/0964-1726/19/8/085012
82. Yin, X., Lallart, M., Cottinet, P.-J., Guyomar, D., Capsal, J.-F.: Mechanical energy harvesting via a plasticizer-modified electrostrictive polymer. Appl. Phys. Lett. **108**, 042901 (2016). https://doi.org/10.1063/1.4939859
83. Tugui, C., Ursu, C., Sacarescu, L., Asandulesa, M., Stoian, G., Gabriel, A., Cazacu, M.: Stretchable energy harvesting devices: attempts to high-performance electrodes. ACS Sustain. Chem. Eng. **5**, 7851–7858 (2017). https://doi.org/10.1021/acssuschemeng.7b01354
84. Koul, S., Ahmed, S., Kakkar, V.: A comparative analysis of different vibration based energy harvesting techniques for implantables. Int. Conf. Comput. Commun. Autom. (2015). https://doi.org/10.1109/CCAA.2015.7148517
85. Pozo, B., Garate, J.I., Araujo, J.Á., Ferreiro, S.: Energy harvesting technologies and equivalent electronic structural models-review. Electronics. **8**(5), 486 (2019). https://doi.org/10.3390/electronics8050486
86. Ekreem, N., Olabi, A.G., Prescott, T., Rafferty, A., Hashmi, M.S.J.: An overview of magnetostriction, its use and methods to measure these properties. J. Mater. Process. Technol. **191**, 96–101 (2007). https://doi.org/10.1016/j.jmatprotec.2007.03.064
87. See: https://courses.lumenlearning.com/boundless-physics/chapter/magnetic-flux-induction-and-faradays-law/. 15.04.2020
88. Clemente, C. S., Mahgoub, A., Davino, D., &Visone, C. (2017). Multiphysics circuit of a magnetostrictiveenergy harvesting device. J. Intell. Mater. Syst. Struct., 28(17), 2317–2330. https://doi.org/10.1177/1045389X16685444
89. Apicella, V., Clemente, C.S., Davino, D., Leone, D., Visone, C.: Review of modeling and control of magnetostrictive actuators. Actuators. **8**(2), 45 (2019). https://doi.org/10.3390/act8020045
90. Dong, L., Closson, A.B., Jin, C., Trase, I., Chen, Z., Zhang, J.X.J.: Vibration-energy-harvesting system: transduction mechanisms, frequency tuning techniques, and biomechanical applications. Adv. Mater. Technol. **4**, 1900177 (2019). https://doi.org/10.1002/admt.201900177
91. Wang, L., Yuan, F.-G.: Vibration energy harvesting by magnetostrictive material. Smart Mater. Struct. **17**, 45009–45014 (2008). https://doi.org/10.1088/0964-1726/17/4/045009
92. See: https://courses.lumenlearning.com/introchem/chapter/energy-changes-in-chemical-reactions/. 16.04.2020
93. See: https://www.britannica.com/science/chemical-reaction/Energy-considerations. 16.04.2020
94. See: https://chem.libretexts.org/Courses/Valley_City_State_University/Chem_121/Chapter_5%3A_Introduction_to_Redox_Chemistry/5.3%3A_Types_of_Chemical_Reactions. 16.04.2020

95. See: https://www.eia.gov/energyexplained/electricity/how-electricity-is-generated.php. 16.04.2020
96. See: https://www.ncbi.nlm.nih.gov/books/NBK26882/. 16.04.2020
97. See: https://en.wikipedia.org/wiki/Explosive. 16.04.2020
98. See: https://en.wikipedia.org/wiki/Electric_battery. 16.04.2020
99. See: https://en.wikipedia.org/wiki/Photosynthesis, 16.04.2020
100. See: https://en.wikipedia.org/wiki/Adenosine_diphosphate. 16.04.2020
101. See: https://www.thoughtco.com/cold-packs-and-endothermic-reactions-3976046. 16.04.2020
102. See: https://www.solarschools.net/knowledge-bank/energy/types/chemical. 16.04.2020
103. See: https://en.wikipedia.org/wiki/Fuel. 16.04.2020
104. See: https://www.renewableresourcescoalition.org/windmills/. 16.04.2020
105. See: https://ctmmagnetics.com/wind-power-basics-how-to-harvest-and-store-wind-energy/. 16.04.2020
106. See: https://www.energy.gov/articles/how-wind-turbine-works. 16.04.2020
107. See: https://en.wikipedia.org/wiki/Wind_turbine_design. 16.04.2020
108. See: https://www.renewableenergyworld.com/types-of-renewable-energy/wind-power-tech/. 16.04.2020
109. See: https://www.energy.gov/eere/wind/advantages-and-challenges-wind-energy. 16.04.2020
110. See: https://www.ucsusa.org/resources/how-wind-energy-works. 16.04.2020
111. Pan, H., Li, B., Zhang, T., Laghari, A.A., Zhang, Z., Qian, B.: A portable renewable wind energy harvesting system integrated S-rotor and H-rotor for self-powered applications in high-speed railway tunnels. Energy Convers. Manag. **196**, 56–68 (2019)
112. Panda, P.K., Grigoriev, A., Mishra, Y.K., Ahuja, R.: Progress in supercapacitors: roles of two dimensional nanotubular materials. Nanoscale Adv. **2**, 70–108 (2020). https://doi.org/10.1039/C9NA00307J
113. Tao, J.X., Nguyen, V., Carpinteri, A., Wang, Q.: Energy harvesting from wind by a piezoelectric harvester. Eng. Struct. **133**, 74–80 (2017). https://doi.org/10.1016/j.engstruct.2016.12.021
114. Xu, F., Yuan, F.-G., Hu, J., Qiu, Y.: Design of a miniature wind turbine for powering wireless sensors. Proc. SPIE Int. Soc. Opt. Eng. **7647** (2010). https://doi.org/10.1117/12.847429
115. See: https://www.lpi.usra.edu/education/explore/marvelMoon/background/moon-influence/. 18.04.2020
116. See: https://en.wikipedia.org/wiki/Tidal_power. 18.04.2020
117. See: https://en.wikipedia.org/wiki/Sihwa_Lake_Tidal_Power_Station. 18.04.2020
118. See: https://www.power-technology.com/features/tidal-energy-advantages-and-disadvantages/. 18.04.2020
119. See: https://www.cnbc.com/2020/01/27/tidal-project-generates-electricity-to-power-nearly-4000-homes.html. 18.04.2020
120. See: https://www.alternative-energy-tutorials.com/tidal-energy/tidal-energy.html. 18.04.2020
121. See: http://www.energybc.ca/tidal.html. 18.04.2020
122. Waters, S., Aggidis, G.A.: Over 2000 years in review: revival of the archimedes screw from pump to turbine. Renew. Sustain. Energy Rev. **51** (2015). https://doi.org/10.1016/j.rser.2015.06.028
123. Ilzarbe, J., Amaral Teixeira, J.: Recent patents on tidal power extraction devices. Recent Patents Eng. **3**, 178–193 (2009). https://doi.org/10.2174/187221209789117780
124. See: http://tidalpower.co.uk/disadvantages-of-tidal-power. 18.04.2020
125. See: https://www.eesi.org/topics/water-hydropower-wave-power/description. 18.04.2020
126. See: https://www.cnbc.com/2019/08/13/scottish-firm-bags-contract-for-the-most-powerful-tidal-turbine.html. 18.04.2020
127. See: https://spectrum.ieee.org/energywise/green-tech/geothermal-and-tidal/underwater-kite-harvests-energy-from-slow-currents. 18.04.2020
128. See: https://sigearth.com/the-future-of-renewable-energy-tidal-wave-power/. 19.04.2020
129. Watson, S., Moro, A., Reis, V., Baniotopoulos, C., Barth, S., Bartoli, G., Bauer, F., Boelman, E., Bosse, D., Cherubini, A., Croce, A., Fagiano, L., Fontana, M., Gambier, A., Gkoumas, K., Golightly, C., Latour, M.I., Jamieson, P., Kaldellis, J., Macdonald, A., Murphy, J., Muskulus, M., Petrini, F., Pigolotti, L., Rasmussen, F., Schild, P., Schmehl, R., Stavridou, N., Tande, J., Taylor, N., Telsnig, T., Wiser, R.: Future emerging technologies in the wind power sector: A European perspective. Renewable and Sustainable Energy Reviews. **13**, 109270 (2019). https://doi.org/10.1016/j.rser.2019.109270. ISSN 1364–0321
130. See: https://www.alternative-energy-tutorials.com/tidal-energy/tidal-energy.html#:~:text=Tidal%20Energy%20Generation&text=Tidal%20energy%2C%20just%20like%20hydro,motion%20into%20a%20clean%20energy.&text=The%20daily%20ebbing%20and%20flowing,down%20a%20river%20or%20stream. 19.04.2020
131. See: https://www.world-nuclear.org/information-library/energy-and-the-environment/renewable-energy-and-electricity.aspx. 19.04.2020
132. See: https://gov.wales/sites/default/files/publications/2019-07/marine-renewable-energy-strategic-framework-risk-to-fish.pdf. 19.04.2020
133. Gür, T.M.: Review of electrical energy storage technologies, materials and systems: challenges and prospects for large-scale grid storage. Energy Environ. Sci. **11**, 2696–2767 (2018). https://doi.org/10.1039/C8EE01419A
134. See: https://en.wikipedia.org/wiki/Energy_storage. 19.04.2020
135. Khan, N., Dilshad, S., Khalid, R., Kalair, A.R., Abas, N.: Review of energy storage and transportation of energy. Energy Storage., e49 (2019). https://doi.org/10.1002/est2.49
136. Kumar, A., Rajak, D.K., Kumar, R.: Optimization of packed bed solar air heaters: a thermo-hydraulic approach. Energy Storage. (2020). https://doi.org/10.1002/est2.119
137. See: https://www.nyserda.ny.gov/All-Programs/Programs/Energy-Storage/Energy-Storage-for-Your-Business/Types-of-Energy-Storage. 19.04.2020
138. See: https://en.wikipedia.org/wiki/Pumped-storage_hydroelectricity. 19.04.2020
139. See: https://www.solarpowerworldonline.com/2018/11/common-battery-types-used-in-solarstorage/. 19.04.2020
140. Dennison, B., Gayathri, R., Anand, V., Raju, L., Rajamanickam, K., Ramakrishna, S.: Thermoelectric power generation using solar energy, 19.04.2020 (2017)
141. Zi, Y., Wang, J., Wang, S., Li, S., Guo, H., Wang, Z.: Effective energy storage from a triboelectric nanogenerator. Nat. Commun. **7**, 10987 (2016). https://doi.org/10.1038/ncomms10987
142. See: https://link.springer.com/content/pdf/10.1007%2F978-0-387-76464-1_14.pdf. 19.04.2020
143. Borah, R., Hughson, F.R., Johnston, J., Nann, T.: On battery materials and methods. Mater. Today Adv. **6**, 100046 (2020). https://doi.org/10.1016/j.mtadv.2019.100046. ISSN 2590–0498
144. See: https://onlinelibrary.wiley.com/doi/10.1002/aenm.201803170. 20.04.2020
145. Borah, R., Hughson, F., Johnston, J., Nann, T.: On battery materials and methods. Mater. Today Adv. **6** (2020). https://doi.org/10.1016/j.mtadv.2019.100046
146. Ponnuchamy, V., Mossa, S., Skarmoutsos, I.: Solvent and salt effect on lithium ion solvation and contact ion pair formation in organic carbonates: a quantum chemical perspective. J. Phys. Chem. C. **122**(45), 25930–25939 (2018). https://doi.org/10.1021/acs.jpcc.8b09892

147. Hausbrand, R., Cherkashinin, G., Ehrenberg, H., Groting, M., Albe, K., Hess, C., Jaegermann, W.: Fundamental degradation mechanisms of layered oxide Li-ion battery cathode materials: methodology, insights and novel approaches. Mater. Sci. Eng. B. **192**, 3–25 (2015). https://doi.org/10.1016/j.mseb.2014.11.014. ISSN 0921–5107
148. Feixiang, W., Maier, J., Yu, Y.: Guidelines and trends for next-generation rechargeable lithium and lithium-ion batteries. Chem. Soc. Rev. **49**, 1569–1614 (2020). https://doi.org/10.1039/C7CS00863E
149. Wu, F., Yushin, G.: Conversion cathodes for rechargeable lithium and lithium-ion batteries. Energy Environ. Sci. **10**, 435–459 (2017). https://doi.org/10.1039/C6EE02326F
150. See: https://shodhganga.inflibnet.ac.in/bitstream/10603/102722/8/08_chapter%201.pdf. 20.04.2020
151. Tang, Y., Zhang, Y., Li, W., Ma, B., Chen, X.: Rational material design for ultrafast rechargeable lithium-ion batteries. Chem. Soc. Rev. **44**, 5926–5940 (2015). https://doi.org/10.1039/C4CS00442F
152. Tang, Y., Zhang, Y., Li, W., Ma, B., Chen, X.: Rational materials design for ultrafast rechargeable lithium-ion batteries. Chem. Soc. Rev. https://doi.org/10.1039/x0xx00000x
153. Lee, S., Song, M., Kim, S., Mathew, V., Sambandam, B., Hwang, J.-Y., Kim, J.: High lithium storage properties in a manganese sulfide anode via an intercalation-cum-conversion reaction. J. Mater. Chem. A. (2020). https://doi.org/10.1039/D0TA05758D. Advance Article
154. An, S.J., Li, J., Daniel, C., Mohanty, D., Nagpure, S., Wood, D.L.: The state of understanding of the lithium-ion-battery graphite solid electrolyte interphase (SEI) and its relationship to formation cycling. Carbon. **105**, 52–76 (2016). https://doi.org/10.1016/j.carbon.2016.04.008. ISSN 0008–6223
155. See: https://www.tms.org/pubs/journals/JOM/0809/daniel-0809.html. 24.04.2020
156. See: https://www.evrus.net/Li-air.html. 24.04.2020
157. Caramia, V., Bozzini, B.: Materials science aspects of zinc-air batteries: a review. Mater. Renew. Sustain. Energy. **3** (2014). https://doi.org/10.1007/s40243-014-0028-3
158. Capsoni, D., Bini, M., Ferrari, S., Quartarone, E.: Recent advances in the development of Li–air batteries. J. Power Sources. **220**, 253–263 (2012). https://doi.org/10.1016/j.jpowsour.2012.07.123
159. Li, L., Li, S., Lu, Y.: Suppression of dendritic lithium growth in lithium metal-based batteries. Chem. Commun. **54**, 6648–6661 (2018). https://doi.org/10.1039/C8CC02280A
160. Li, Y., Zhou, W., Chen, X., Lu, X., Cui, Z., Xin, S., Xue, L., Jia, Q., Goodenough, J.: Mastering the interface for advanced all-solid-state lithium rechargeable batteries. Proc. Natl. Acad. Sci. U. S. Am. **113** (2016). https://doi.org/10.1073/pnas.1615912113
161. Bi, X., Amine, K., Lu, J.: The importance of anode protection towards lithium oxygen batteries. J. Mater. Chem. A. **8**, 3563–3573 (2020). https://doi.org/10.1039/C9TA12414D
162. Moon, J.S., Kim, W.G., Kim, C., et al.: M13 bacteriophage-based self-assembly structures and their functional capabilities. Mini. Rev. Org. Chem. **12**(3), 271–281 (2015). https://doi.org/10.2174/1570193X1203150429105418
163. Mohan, K., Weiss, G.A.: Chemically modifying viruses for diverse applications. ACS Chem. Biol. **11**(5), 1167–1179 (2016). https://doi.org/10.1021/acschembio.6b00060
164. Zhang, T., Imanishi, N., Takeda, Y., Yamamoto, O.: Aqueous lithium/air rechargeable batteries. Chem. Lett. **40**, 668–673 (2011). https://doi.org/10.1246/cl.2011.668
165. Jung, J., Cho, S.-H., Nam, J., Kim, I.-D.: Current and future cathode materials for non-aqueous Li-air (O2) battery technology – a focused review. Energy Storage Mater. **24** (2019). https://doi.org/10.1016/j.ensm.2019.07.006
166. Wang, F., Chen, H., Wu, Q., et al.: Study on the Mixed Electrolyte of N,N-dimethylacetamide/sulfolane and its application in aprotic lithium-air batteries. ACS Omega. **2**(1), 236–242 (2017). https://doi.org/10.1021/acsomega.6b00254
167. Mccloskey, B., Burke, C., Nichols, J., Renfrew, S.: ChemInform abstract: mechanistic insights for the development of Li-O2 battery materials: addressing Li2O2 conductivity limitations and electrolyte and cathode Instabilities. Chem. Commun. **51** (2015). https://doi.org/10.1039/C5CC04620C
168. See: https://en.wikipedia.org/wiki/Lithium%E2%80%93air_battery. 25.04.2020
169. Samson, A.J., Hofstetter, K., Bag, S., Thangadurai, V.: A bird's-eye view of Li-stuffed garnet-type Li7La3Zr2O12 ceramic electrolytes for advanced all-solid-state Li batteries. Energy Environ. Sci. **12**, 2957–2975 (2019). https://doi.org/10.1039/C9EE01548E
170. Guo, H., Luo, W., Chen, J., Chou, S.-L., Liu, H.-K., Wang, J.: Review of electrolytes in nonaqueous lithium-oxygen batteries. Adv. Sustain. Syst. **2**, 1700183 (2018). https://doi.org/10.1002/adsu.201700183
171. Wang, Y., Song, S., Xu, C., Hu, N., Molenda, J., Lu, L.: Development of solid-state electrolytes for sodium-ion battery: a short review. Nano Mater. Sci. **1**(2), 91–100 (2019). https://doi.org/10.1016/j.nanoms.2019.02.007. ISSN 2589–9651
172. Manthiram, A.: An outlook on lithium ion battery technology. ACS Cent Sci. **3**(10), 1063–1069 (2017). https://doi.org/10.1021/acscentsci.7b00288
173. See: https://en.globtek.com/lithium-polymer-battery-packs/. 25.04.2020
174. See: http://www.energy-without-carbon.org/LithiumPolymerBattery. 25.04.2020
175. See: https://batteryuniversity.com/learn/archive/is_lithium_ion_the_ideal_battery. 25.04.2020
176. See: https://en.wikipedia.org/wiki/Lithium-ion_battery. 25.04.2020
177. Roberts, S., Kendrick, E.: The re-emergence of sodium ion batteries: testing, processing, and manufacturability. Nanotechnol. Sci. Appl. **23-33**(2018), 11 (2018). https://doi.org/10.2147/NSA.S146365
178. Hwang, J.-Y., Myung, S.-T., Sun, Y.-K.: Sodium-ion batteries: present and future. Chem. Soc. Rev. **46**, 3529–3614 (2017). https://doi.org/10.1039/C6CS00776G
179. Ghimbeu, C., Górka, J., Simone, V., Simonin, L., Martinet, S., Vix-Guterl, C.: Insights on the Na+ ion storage mechanism in hard carbon: Discrimination between the porosity, surface functional groups and defects. Nano Energy. **44**, 327–335 (2018). https://doi.org/10.1016/j.nanoen.2017.12.013
180. Mukherjee, S., Bin Mujib, S., Soares, D., Singh, G.: Electrode materials for high-performance sodium-ion batteries. Materials (Basel). **12**(12), 1952 (2019). https://doi.org/10.3390/ma12121952
181. Clément, R.J., Lun, Z., Ceder, G.: Cation-disordered rocksalt transition metal oxides and oxyfluorides for high energy lithium-ion cathodes. Energy Environ. Sci. **13**, 345–373 (2020). https://doi.org/10.1039/C9EE02803J
182. Zhang, Y., Zhang, R., Huang, Y.: Air-Stable Na x TMO2 cathodes for sodium storage. Front Chem. **7**, 335 (2019). https://doi.org/10.3389/fchem.2019.00335
183. Gutierrez, A., Benedek, N., Manthiram, A.: Crystal-chemical guide for understanding redox energy variations of M2+/3+ couples in polyanion cathodes for lithium-ion batteries. Chem. Mater. **25**, 4010–4016 (2013). https://doi.org/10.1021/cm401949n
184. Haregewoin, A.M., Wotango, A.S., Hwang, B.-J.: Electrolyte additives for lithium ion battery electrodes: progress and perspectives, Energy Environ. Sci. **9**, 1955–1988 (2016). https://doi.org/10.1039/C6EE00123H
185. See: https://www.chemistryworld.com/features/a-battery-technology-worth-its-salt/3010966.article. 26.04.2020

186. Tomaszewska, A., Chu, Z., Feng, X., O'Kane, S., Liu, X., Chen, J., Ji, C., Endler, E., Li, R., Liu, L., Li, Y., Zheng, S., Vetterlein, S., Gao, M., Du, J., Parkes, M., Ouyang, M., Marinescu, M., Gregory, O., Wu, B.: Lithium-ion battery fast charging: a review. eTransportation. **1**, 100011 (2019). https://doi.org/10.1016/j.etran.2019.100011. ISSN 2590–1168
187. Liu, T., Zhang, Y., Chen, C., Lin, Z., Zhang, S., Lu, J.: Sustainability-inspired cell design for a fully recyclable sodium ion battery. Nat. Commun. **10**(1), 1965 (2019). https://doi.org/10.1038/s41467-019-09933-0
188. See: https://en.wikipedia.org/wiki/Sodium-ion_battery. 26.04.2020
189. See: https://batteryindustry.tech/dictionary/magnesium-battery/. 26.04.2020
190. Mallela, V.S., Ilankumaran, V., Rao, N.S.: Trends in cardiac pacemaker batteries. Indian Pacing Electrophysiol. J. **4**(4), 201–212 (2004)
191. See: sciencedaily.com/releases/2018/12/181221123724.htm,. 26.04.2020
192. See: https://wikivisually.com/wiki/Magnesium_battery. 26.04.2020
193. Wang, P., Buchmeiser, M.R.: Rechargeable magnesium–sulfur battery technology: state of the art and key challenges. Adv. Funct. Mater. **29**, 1905248 (2019). https://doi.org/10.1002/adfm.201905248
194. Arya, A., Sharma, A.L.: Polymer electrolytes for lithium ion batteries: a critical study. Ionics. **23**, 497–540 (2017). https://doi.org/10.1007/s11581-016-1908-6
195. Asano, T., Sakai, A., Ouchi, S., Sakaida, M., Miyazaki, A., Hasegawa, S.: Adv. Mater. **30**, 1803075 (2018). https://doi.org/10.1002/adma.201803075
196. Ran, A., Michael, S., Hirsch, B., Goffer, Y., Aurbach, D.: Anode-electrolyte interfaces in secondary magnesium batteries. Joule. **3** (1); 27–52 (2019). https://doi.org/10.1016/j.joule.2018.10.028. ISSN 2542–4351
197. Cho, J.-H., Jin Kim, S., Jinwoo, O., Ha, J.H., Kim, K.-B., Lee, K.-Y., Lee, J.K.: Strategic design of highly concentrated electrolyte solutions for Mg2+/Li+ dual-salt hybrid batteries. J. Phys. Chem.y C. **122**(49), 27866–27874 (2018). https://doi.org/10.1021/acs.jpcc.8b09080
198. Bitenc, J., Pavčnik, T., Košir, U., Pirnat, K.: Quinone based materials as renewable high energy density cathode materials for rechargeable magnesium batteries. Materials (Basel). **13**(3), 506 (2020). https://doi.org/10.3390/ma13030506
199. See: https://thereaderwiki.com/en/Magnesium_battery. 27.04.2020
200. Mohtadi, R., Mizuno, F.: Magnesium batteries: current state of the art, issues and future perspectives. Beilstein J. Nanotechnol. **5**, 1291–1311 (2014). https://doi.org/10.3762/bjnano.5.143
201. Guozhao, F., Zhou, J., Pan, A., Liang, S.: Recent advances in aqueous zinc-ion batteries. ACS Energy Lett. **3** (2018). https://doi.org/10.1021/acsenergylett.8b01426
202. Xu, W., Wang, Y.: Recent progress on zinc-ion rechargeable batteries. Nano-Micro Lett. **11**, 90 (2019). https://doi.org/10.1007/s40820-019-0322-9
203. Li, G., Huang, Z.-X., Chen, J., Yao, F., Liu, J., Li, O.L., Sun, S., Shi, Z.: Rechargeable Zn-ion batteries with high power and energy density: a two-electron reaction pathway in birnessite MnO2 cathode materials. J. Mater. Chem. A. **8** (2020). https://doi.org/10.1039/C9TA11985J
204. Chao, D., Zhou, W., Ye, C., Zhang, Q., Chen, Y., Gu, L., Davey, K., Qiao, S.: An electrolytic Zn-MnO_2 battery demonstrated for high-voltage and scalable energy storage. Angew. Chemie. **131** (2019). https://doi.org/10.1002/ange.201904174
205. See: https://www.ctc-n.org/products/zinc-air-battery. 27.04.2020
206. Dongmo, S., Stock, D., Alexander Kreissl, J.J., et al.: Implications of testing a zinc-oxygen battery with zinc foil anode revealed by operando gas analysis. ACS Omega. **5**(1), 626–633 (2019). https://doi.org/10.1021/acsomega.9b03224
207. Zhang, J., Zhou, Q., Tang, Y., Zhang, L., Li, Y.: Zinc–air batteries: are they ready for prime time? Chem. Sci. **10**, 8924–8929 (2019). https://doi.org/10.1039/C9SC04221K
208. See: https://steelguru.com/metal/new-rechargeable-zinc-air-batteries-could-threaten-li-ion-based-battery/488429. 27.04.2020
209. See: https://batteryuniversity.com/learn/article/discharging_at_high_and_low_temperatures. 27.04.2020
210. See: https://batteryindustry.tech/dictionary/potassium-ion-battery-or-k-ion-battery/. 28.04.2020
211. Sultana, I., Rahman, M.M., Chen, Y., Glushenkov, A.: M. Adv Funct Mater. **28**, 1703857 (2018). https://doi.org/10.1002/adfm.201703857
212. See: https://en.wikipedia.org/wiki/Potassium-ion_battery. 28.04.2020
213. Zhang, W., Liu, Y., Guo, Z.: Approaching high-performance potassium-ion batteries via advanced design strategies and engineering. Sci. Adv. **5**(5), eaav7412 (2019). https://doi.org/10.1126/sciadv.aav7412
214. Eftekhari, A. Potassium secondary cell based on Prussian blue cathode J. Power Sources. 2004;126:221–228.10.1016/j.jpowsour.2003.08.007
215. Pham, T., Kweon, K., Samanta, A., Lordi, V., Pask, J.: Solvation and dynamics of sodium and potassium in ethylene carbonate from Ab initio molecular dynamics simulations. J. Phys. Chem. C. **121** (2017). https://doi.org/10.1021/acs.jpcc.7b06457
216. Das, S.K., Mahapatra, S., Lahan, H.: Aluminium-ion batteries: developments and challenges. J. Mater. Chem. A. **5**, 6347–6367 (2017). https://doi.org/10.1039/C7TA00228A
217. Leisegang, T., Meutzner, F., Zschornak, M., et al.: The aluminum-ion battery: a sustainable and seminal concept? Front Chem. **7**, 268 (2019). https://doi.org/10.3389/fchem.2019.00268
218. See: https://ui.adsabs.harvard.edu/abs/2019FrCh....7..268L/abstract. 28.04.2020
219. Zeng, Y., Wang, M., He, W., Fang, P.-P., Wu, M., Tong, Y., Chen, M., Lu, X.: Engineering high reversibility and fast kinetics of Bi nanoflakes by surface modulation for ultrastable nickel-bismuth batteries. Chem. Sci. **10** (2019). https://doi.org/10.1039/C8SC04967J
220. See: https://en.wikipedia.org/wiki/Organic_radical_battery. 28.04.2020
221. Li, N., Chen, Z., Ren, W., Li, F., Cheng, H.M.: Flexible graphene-based lithium ion batteries with ultrafast charge and discharge rates. Proc. Natl. Acad. Sci. U. S. A. **109**(43), 17360–17365 (2012). https://doi.org/10.1073/pnas.1210072109
222. See: https://batteryuniversity.com/learn/article/how_to_prolong_lithium_based_batteries. 29.04.2020
223. Zhang, K., Monteiro, M.J., Jia, Z.: Stable organic radical polymers: synthesis and applications. Polym. Chem. **7**, 5589–5614 (2016). https://doi.org/10.1039/C6PY00996D
224. Zhang, Y., Park, A., Cintora, A., et al.: Impact of the synthesis method on the solid-state charge transport of radical polymers. J. Mater. Chem. C Mater. **6**(1), 111–118 (2018). https://doi.org/10.1039/C7TC04645F
225. Janoschka, T., Hager, M.D., Schubert, U.S.: Powering up the future: radical polymers for battery applications. Adv. Mater. **24**, 6397–6409 (2012). https://doi.org/10.1002/adma.201203119

16 Advanced Semiconductor/Conductor Materials

Abstract

The scope of application of electrical and semiconductor materials is very vast. These materials find utilities are not only electrical machines, equipments, devices, etc., but are also used as components, circuits, and other auxiliaries related to electronics, computers, and instrumentation fields. Their importance is also realized in cable networking, wireless networking, satellites, optical devices, etc. They find very useful applications even in medical, mechanical, nuclear, biotechnological fields. This chapter is specially designed for advanced semiconductor/conductor materials such as supercapacitors, superconducting materials, advanced semiconductor materials, high-mobility organic transistors. Here, the mechanism for each material with their clear vision in an application has been discussed.

Keywords

Supercapacitor · Batteries · Fuel cells and supercapacitors · Electrodes materials · Electrolytes · Separators · Collectors and housing · Superconducting materials · Advanced semiconductor materials · High-mobility organic transistors

16.1 Supercapacitor

A capacitor is a device that stores electrical energy in an electric field. It is a passive electronic component with two terminals. The effect of a capacitor is known as capacitance. While some capacitance exists between any two electrical conductors in proximity in a circuit, a capacitor is a component designed to add capacitance to a circuit. The capacitor was originally known as a condenser or condensator [1]. This name and its cognates are still widely used in many languages, but rarely in English, one notable exception being condenser microphones, also called capacitor microphones. A supercapacitor (SC), also known as ultracapacitor, is a high-capacity capacitor with a capacitance value much higher than other capacitors, but with lower voltage limits, that bridges the gap between electrolytic capacitors and rechargeable batteries [2]. It typically stores 10–100 times more energy per unit volume or mass than electrolytic capacitors, can accept and deliver charge much faster than batteries, and tolerates many more charge and discharge cycles than rechargeable batteries [3].

Supercapacitors are used in applications requiring many rapid charge/discharge cycles, rather than long-term compact energy storage in automobiles, buses, trains, cranes, and elevators, where they are used for regenerative braking, short-term energy storage, or burst-mode power delivery. Smaller units are used as power backup for static random-access memory (SRAM) [4]. Unlike ordinary capacitors, supercapacitors do not use the conventional solid dielectric, but rather, they use electrostatic double-layer capacitance (EDLC) and electrochemical pseudocapacitance (EP), both of which contribute to the total capacitance of the capacitor [5, 6]. EDLC uses carbon electrodes or derivatives with much higher electrostatic double-layer capacitance than electrochemical pseudocapacitance, achieving separation of charge in a Helmholtz double layer at the interface between the surface of a conductive electrode and an electrolyte. The separation of charge is of the order of a few angstroms (0.3–0.8 nm), much smaller than in a conventional capacitor [7]. EP uses metal oxide or conducting polymer electrodes with a high amount of electrochemical pseudocapacitance additional to the double-layer capacitance. Pseudocapacitance is achieved by Faradaic electron charge transfer with redox reactions, intercalation, or electrosorption [8]. Hybrid capacitors, such as the lithium-ion capacitor, use electrodes with a difference in characteristics: one exhibiting mostly electrostatic capacitance and the other mostly electrochemical capacitance. The electrolyte forms an ionic conductive connection between the two electrodes which distinguishes them from conventional electrolytic

A. Behera, *Advanced Materials*, https://doi.org/10.1007/978-3-030-80359-9_16

capacitors where a dielectric layer always exists [9]. Supercapacitors are polarized by design with asymmetric electrodes, or, for symmetric electrodes, by a potential applied during manufacture.

16.2 History of Supercapacitor

In the early 1950s, General Electric engineers began experimenting with porous carbon electrodes in the design of capacitors, from the design of fuel cells and rechargeable batteries. Activated charcoal is an electrical conductor that is an extremely porous "spongy" form of carbon with a high specific surface area [10]. In 1954, H. Becker developed a "Low voltage electrolytic capacitor with porous carbon electrodes". He reported that the energy was stored as a charge in the carbon pores as in the pores of the etched foils of electrolytic capacitors. Because the double layer mechanism was not known by him at the time, he wrote in the patent: "It is not known exactly what is taking place in the component if it is used for energy storage, but it leads to an extremely high capacity" [11]. General Electric did not immediately pursue this work. In 1966 researchers at Standard Oil of Ohio (SOHIO) developed another version of the component as "electrical energy storage apparatus", while working on experimental fuel cell designs. The nature of electrochemical energy storage was not described in this patent [12]. Even in 1970, the electrochemical capacitor patented by Donald L. Boos was registered as an electrolytic capacitor with activated carbon electrodes. Early electrochemical capacitors used two aluminum foils covered with activated carbon—the electrodes—that were soaked in an electrolyte and separated by a thin porous insulator. This design gave a capacitor with a capacitance on the order of one farad, significantly higher than electrolytic capacitors of the same dimensions. This basic mechanical design remains the basis of most electrochemical capacitors. SOHIO did not commercialize their invention, licensing the technology to NEC, who finally marketed the results as "supercapacitors" in 1971, to provide backup power for computer memory [13].

Between 1975 and 1980 Brian Evans Conway conducted extensive fundamental and development work on ruthenium oxide electrochemical capacitors. In 1991 he described the difference between "supercapacitor" and "battery" behavior in electrochemical energy storage [14]. In 1999 he defined the term "supercapacitor" to make reference to the increase in observed capacitance by surface redox reactions with faradaic charge transfer between electrodes and ions. His "supercapacitor" stored electrical charge partially in the Helmholtz double layer and partially as a result of faradaic reactions with "pseudocapacitance" charge transfer of electrons and protons between electrode and electrolyte [15]. The working mechanisms of pseudocapacitors are redox reactions, intercalation, and electrosorption (adsorption onto a surface). With his research, Conway greatly expanded the knowledge of electrochemical capacitors. The market expanded slowly. That changed around 1978 as Panasonic marketed its Goldcaps brand. This product became a successful energy source for memory backup applications. Competition started only years later. In 1987 ELNA "Dynacap's" entered the market [16]. First-generation EDLCs had relatively high internal resistance that limited the discharge current. They were used for low-current applications such as powering SRAM chips or for data backup. At the end of the 1980s, improved electrode materials increased capacitance values. At the same time, the development of electrolytes with better conductivity lowered the equivalent series resistance (ESR) increasing charge/discharge currents. The first supercapacitor with low internal resistance was developed in 1982 for military applications through the Pinnacle Research Institute (PRI) and was marketed under the brand name "PRI Ultracapacitor" [17]. In 1992, Maxwell Laboratories (later Maxwell Technologies) took over this development. Maxwell adopted the term Ultracapacitor from PRI and called them "Boost Caps" to underline their use for power applications [18].

Since capacitors' energy content increases with the square of the voltage, researchers were looking for a way to increase the electrolyte's breakdown voltage. In 1994 using the anode of a 200 V high voltage tantalum electrolytic capacitor, David A. Evans developed an "Electrolytic-Hybrid Electrochemical Capacitor" [19]. These capacitors combine features of electrolytic and electrochemical capacitors. They combine the high dielectric strength of an anode from an electrolytic capacitor with the high capacitance of a pseudocapacitive metal oxide (ruthenium (IV) oxide) cathode from an electrochemical capacitor, yielding a hybrid electrochemical capacitor. Evans' capacitors, coined Capattery, had an energy content about a factor of 5 higher than a comparable tantalum electrolytic capacitor of the same size. Their high costs limited them to specific military applications. Recent developments include lithium-ion capacitors. These hybrid capacitors were pioneered by FDK in 2007 [20]. They combine an electrostatic carbon electrode with a pre-doped lithium-ion electrochemical electrode. This combination increases the capacitance value. Additionally, the pre-doping process lowers the anode potential and results in a high cell output voltage, further increasing specific energy. Research departments active in many companies and universities are working to improve characteristics such as specific energy, specific power, and cycle stability and to reduce production costs.

16.3 Batteries, Fuel Cells, and Supercapacitors

Energy consumption/production that relies on the combustion of fossil fuels is forecast to have a severe future impact on world economics and ecology. Electrochemical energy production is under serious consideration as an alternative energy/power source, as long as this energy consumption is designed to be more sustainable and more environmentally friendly. Systems for electrochemical energy storage and conversion include batteries, fuel cells, and electrochemical capacitors (ECs). Although the energy storage and conversion mechanisms are different, there are "electrochemical similarities" between these three systems. Common features are that the energy-providing processes take place at the phase boundary of the electrode/electrolyte interface and that electron and ion transport are separated [21]. Figures 16.1a–c) shows the basic operation mechanisms of the three systems. Note that batteries, fuel cells, and supercapacitors all consist of two electrodes in contact with an electrolyte solution. In batteries and fuel cells, electrical energy is generated by the conversion of chemical energy via redox reactions at the anode and cathode. As reactions at the anode usually take place at lower electrode potentials than at the cathode, the terms negative and positive electrode are used. The more negative electrode is designated the anode, whereas the cathode is the more positive one. The difference between batteries and fuel cells is related to the locations of energy storage and conversion. Batteries are closed systems, with the anode and cathode being the charge-transfer medium and taking an active role in the redox reaction as "active masses." In other words, energy storage and conversion occur in the same compartment [22].

Fuel cells are open systems where the anode and cathode are just charge-transfer media and the active masses undergoing the redox reaction are delivered from outside the cell, either from the environment, for example, oxygen from air, or from a tank, for example, fuels such as hydrogen and hydrocarbons. Energy storage (in the tank) and energy conversion (in the fuel cell) are thus locally separated. In electrochemical capacitors (or supercapacitors), energy may not be delivered via redox reactions and, thus the use of the terms anode and cathode may not be appropriate but are in common usage. By orientation of electrolyte ions at the electrolyte/electrolyte interface, so-called electrical double layers (EDLs) are formed and released, which results in a parallel movement of electrons in the external wire, that is, in the energy-delivering process [23].

In comparison to supercapacitors and fuel cells, batteries have found by far the most application markets and have an established market position. Whereas supercapacitors have found niche markets as memory protection in several

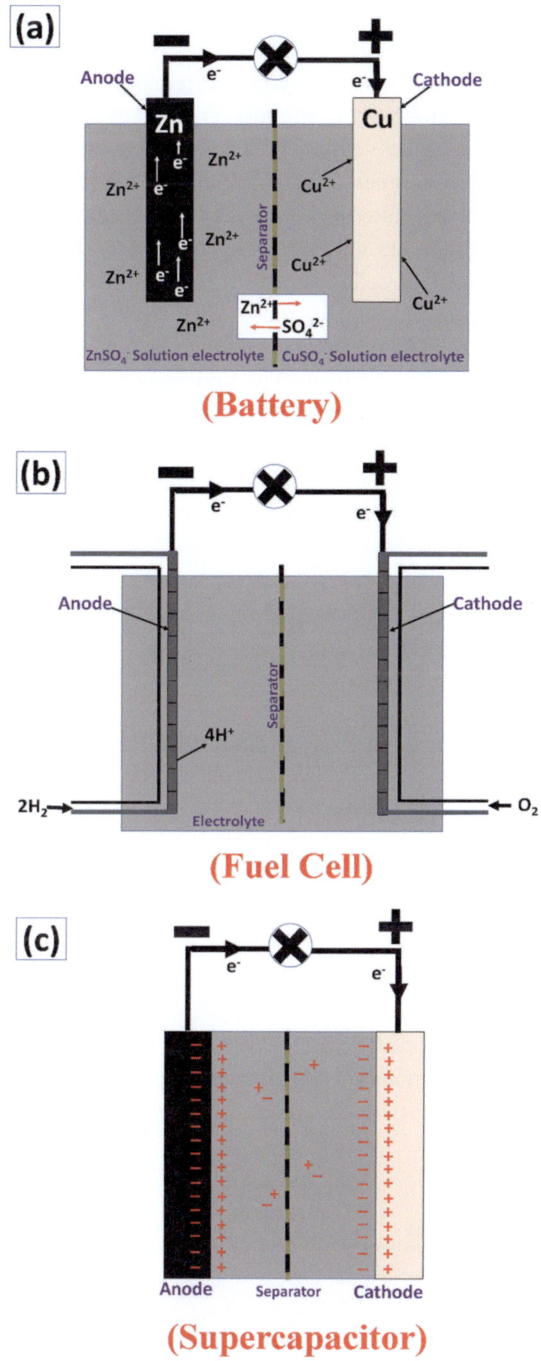

Fig. 16.1 Working of (**a**) batteries, (**b**) fuel cells, and (**c**) supercapacitors

electronic devices, fuel cells are basically still in the development stage and are searching to find the best materials that allow their penetration into the market. Fuel cells established their usefulness in space applications with the advent of the Gemini and Apollo space programs. The most promising future markets for fuel cells and supercapacitors are in the same application sector as batteries. In other words, supercapacitor and fuel cell development aim to compete with, or even to replace, batteries in several application areas [24]. Thus, fuel cells, which originally were intended to replace combustion engines and combustion power sources due to possible higher energy conversion efficiencies and lower environmental impacts, are now under development to replace batteries to power cellular telephones and notebook computers and for stationary energy storage. The motivation for fuel cells to enter the battery market is simple. Fuel cells cannot compete today with combustion engines and gas/steam turbines because of much higher costs, inferior power and energy performance, and insufficient durability and lifetime. With operation times of typically <3000 h and, at least to an order of magnitude, similar costs, batteries are less strong competitors for fuel cells. Fuel cells can be considered high-energy systems, whereas supercapacitors are considered to be high-power systems. Batteries have intermediate power and energy characteristics. There are some overlap in energy and power of supercapacitors, or fuel cells, with batteries [25].

16.4 Work and Processing of Ultracapacitor

16.4.1 Basic Design

Typical construction of a supercapacitor (electrochemical capacitor) consists of two electrodes that separated by an ion-permeable membrane (separator), and an electrolyte ionically connecting both electrodes (Fig. 16.2). When the electrodes are polarized by an applied voltage, ions in the electrolyte form electric double layers of opposite polarity to the electrode's polarity. For example, positively polarized electrodes will have a layer of negative ions at the electrode/electrolyte interface along with a charge-balancing layer of positive ions adsorbing onto the negative layer. The opposite is true for the negatively polarized electrode. Additionally, depending on electrode material and surface shape, some ions may permeate the double layer becoming specifically adsorbed ions and contribute with pseudocapacitance to the total capacitance of the supercapacitor.

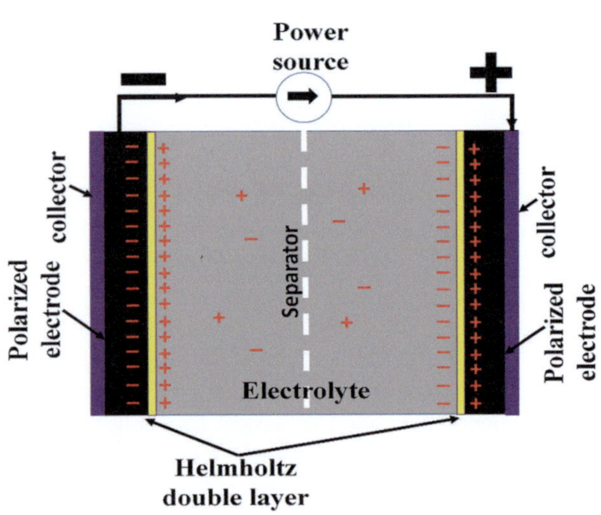

Fig. 16.2 Schematics of a supercapacitor showing various parts

The two electrodes form a series circuit of two individual capacitors C_1 and C_2. The total capacitance C_{total} is given in Eq. (16.1).

$$C_{total} = \frac{C_1 . C_2}{C_1 + C_2} \tag{16.1}$$

Supercapacitors may have either symmetric or asymmetric electrodes. Symmetry implies that both electrodes have the same capacitance value, yielding a total capacitance of half the value of every single electrode (if $C_1 = C_2$, then $C_{total} = ½\, C_1$). For asymmetric capacitors, the total capacitance can be taken as that of the electrode with the smaller capacitance (if $C_1 >> C_2$, then $C_{total} \approx C_2$).

16.4.2 Storage Principles

Electrochemical capacitors use the double-layer effect to store electric energy; however, this double layer has no conventional solid dielectric to separate the charges. There are two storage principles in the electric double layer of the electrodes that contribute to the total capacitance of an electrochemical capacitor [26]: (1) Double-layer capacitance (electrostatic storage of the electrical energy achieved by separation of charge in a Helmholtz double layer) and (2) Pseudocapacitance, electrochemical storage of the electrical energy achieved by faradaic redox reactions with charge transfer. Both capacitances are only separable by measurement techniques. The amount of charge stored per unit voltage in an electrochemical capacitor is primarily a function of the electrode size, although the amount of capacitance of each storage principle can vary extremely. Practically, these storage principles yield a capacitor with a capacitance value in the order of 1–100 farad.

Supercapacitors do not support alternating current (AC) applications. Supercapacitors have advantages in applications where a large amount of power is needed for a relatively short time, where a very high number of charge/discharge cycles or a longer life time is required. Typical applications range from milliamp currents or milliwatts of power for upto a few minutes to several amps current or several hundred kilowatts power for much shorter periods. The time t, a supercapacitor can deliver a constant current I can be calculated as [27]:

$$t = \frac{C \cdot \left(U_{\text{charge}} - U_{\text{min}}\right)}{I} \tag{16.2}$$

As the capacitor voltage decreases from U_{charge} down to U_{min}. If the application needs a constant power (P) for a certain time (t) this can be calculated as:

$$t = \frac{1}{2P} \times C \times \left({U^2}_{\text{charge}} - {U^2}_{\text{min}}\right) \tag{16.3}$$

Wherein also the capacitor voltage decreases from U_{charge} down to U_{min}.

16.4.3 Potential Distribution

Conventional capacitors (also known as electrostatic capacitors), such as ceramic capacitors and film capacitors, consist of two electrodes separated by a dielectric material (Fig. 16.3a). When charged, the energy is stored in a static electric field that permeates the dielectric between the electrodes. The total energy increases with the amount of stored charge, which in turn correlates linearly with the potential (voltage) between the plates. The maximum potential difference between the plates (the maximal voltage) is limited by the dielectric's breakdown field strength. The same static storage also applies for electrolytic capacitors in which most of the potential decreases over the anode's thin oxide layer. The somewhat resistive liquid electrolyte (cathode) accounts for a small decrease of potential for "wet" electrolytic capacitors, while electrolytic capacitors with solid conductive polymer electrolyte this voltage drop is negligible.

In contrast, supercapacitors consist of two electrodes separated by an ion-permeable membrane (separator) and electrically connected via an electrolyte (Fig. 16.3b). When both electrodes have approximately the same resistance (internal resistance), the potential of the capacitor decreases symmetrically over both double layers, whereby a voltage drop across the equivalent series resistance (ESR) of the electrolyte is achieved. For asymmetrical supercapacitors like hybrid capacitors, the voltage drop between the electrodes could be asymmetrical. The maximum potential across the capacitor (the maximal voltage) is limited by the electrolyte decomposition voltage [28]. Both electrostatic and electrochemical energy storage in supercapacitors is linear with respect to the stored charge, just as in conventional capacitors (Fig. 16.3c). The voltage between the capacitor terminals is linear with respect to the amount of stored energy. Such linear voltage gradient differs from rechargeable electrochemical batteries, in which the voltage between the terminals remains independent of the amount of stored energy, providing a relatively constant voltage [29]. The voltage behavior of supercapacitors and batteries during the charging/discharging has been given in Fig. 16.4.

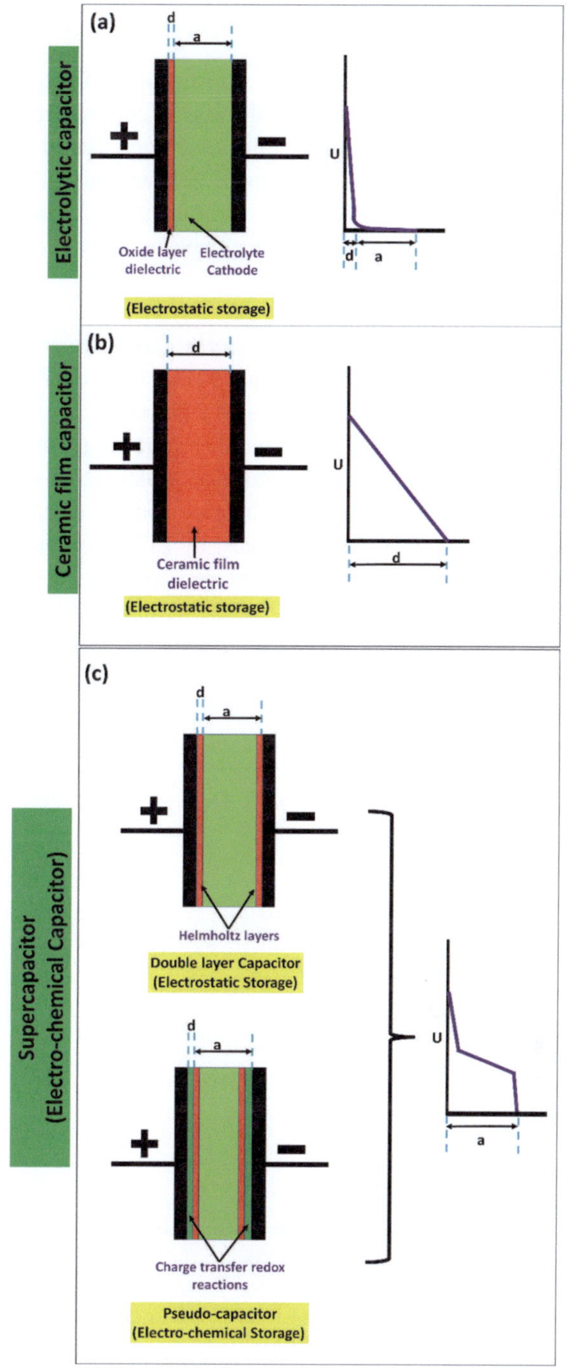

Fig. 16.3 Charge storage principles of different capacitor types and their internal potential distribution: (**a**) Electrolytic capacitor, (**b**) Ceramic film capacitor, (**c**) Supercapacitor

16.4.4 Types of Supercapacitor

The principle of operation of the supercapacitor is based on energy storage and distribution of the ions coming from the electrolyte to the surface area of the electrodes. Based on the energy storage mechanism supercapacitors are classified into three classes: Electrochemical double-layer capacitors, pseudocapacitors, and hybrid supercapacitors as shown in Fig. 16.5 below.

16.4.4.1 Electrostatic Double-layer Capacitance

Electrostatic double-layer capacitance (EDLC) is constructed using two carbon-based materials as electrodes, an electrolyte, and a separator. EDLCs can either store charge electrostatically or via non-faradic process, which involves no transfer of charge between the electrode and the electrolyte. The principle of energy storage used by EDLCs is the electrochemical double layer. When voltage is applied, there is an accumulation of charge on electrode surfaces, due to the difference in potential there is an attraction of opposite charges, these results to ions in electrolyte diffusing over the separator and onto pores of the opposite charged electrode. To avoid recombination of ions at electrodes a double layer of charge is formed. The double layer, combined with the increase in specific surface area and distances between electrodes decreased, allows EDLC to attain higher energy density [30]. At each of the two electrode surfaces originates an area in which the liquid electrolyte contacts the conductive metallic surface of the electrode. This interface forms a common boundary among two different phases of matter, such as an insoluble solid electrode surface and an adjacent liquid electrolyte. In this interface occurs a very special phenomenon of the double layer effect (Fig. 16.6).

These double layers consist of two layers of charges: one electronic layer is in the surface lattice structure of the electrode, and the other with opposite polarity, emerges from dissolved and solvated ions in the electrolyte. The two layers are separated by a monolayer of solvent molecules, *e.g.*, for water as solvent by water molecules, called the inner Helmholtz plane (IHP). Solvent molecules adhere by physical adsorption on the surface of the electrode and separate the oppositely polarized ions from each other, and can be idealized as a molecular dielectric. In the process, there is

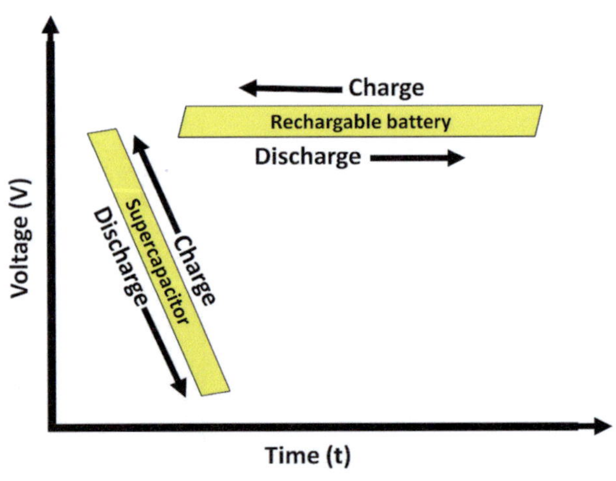

Fig. 16.4 The voltage behavior of supercapacitors and batteries during the charging/discharging differs clearly

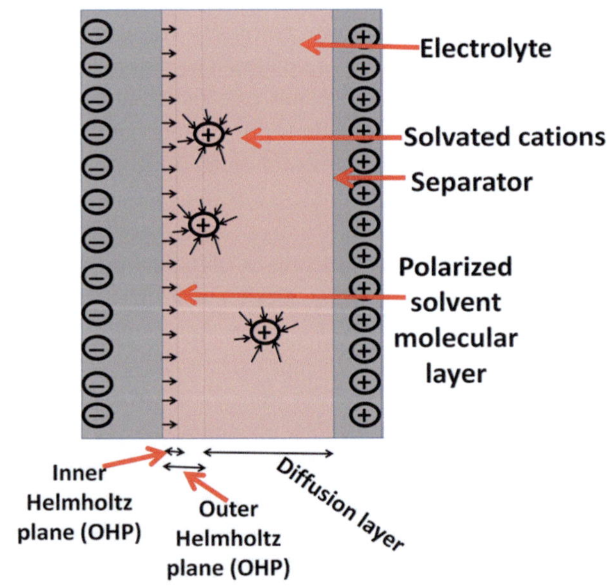

Fig. 16.6 Electrostatic double-layer capacitance

Fig. 16.5 Taxonomy of supercapacitors

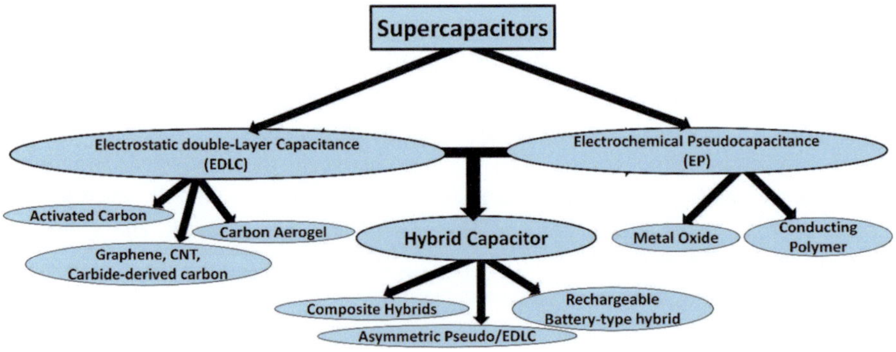

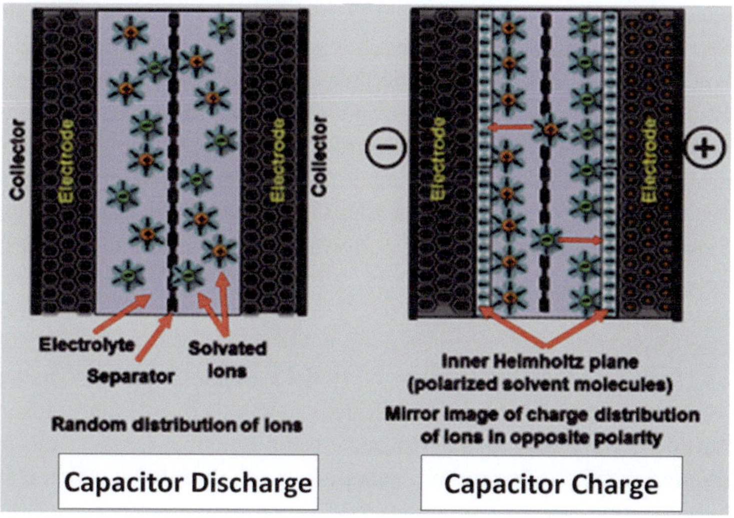

Fig. 16.7 Charge-discharge operation of capacitor

no transfer of charge between electrode and electrolyte, so the forces that cause the adhesion are not chemical bonds, but physical forces, e.g., electrostatic forces. The adsorbed molecules are polarized, but, due to the lack of transfer of charge between electrolyte and electrode, suffered no chemical changes [31]. The amount of charge in the electrode is matched by the magnitude of countercharges in the outer Helmholtz plane (OHP). This double-layer phenomena stores electrical charge is same as in a conventional capacitor. The double-layer charge forms a static electric field in the molecular layer of the solvent molecules in the IHP that corresponds to the strength of the applied voltage [32] (Fig. 16.7).

Applying a voltage to the capacitor at both electrodes a Helmholtz double layer will be formed separating the ions in the electrolyte in a mirror charge distribution of opposite polarity. The double layer serves approximately as the dielectric layer in a conventional capacitor, though with the thickness of a single molecule [33]. Thus, the standard formula for conventional plate capacitors can be used to calculate their capacitance is given in Eq. (16.4):

$$C = \frac{\epsilon_0 \times \varepsilon_r \times A}{d} \tag{16.4}$$

where A is the surface area of the electrode; ϵ_0 is the permittivity of free space; ε_r is the relative permittivity of the dielectric material; and d is the distance between two oppositely biased electrodes. According to the fundamental relationship given in Eq. (16.4), the capacitance of a standard capacitor can be increased by an increase in the dielectric constant of the material and surface area and the decrease of interplanar thickness. However, such an increase can be achieved by further modifying the material system and capacitor design [52]. For instance, one may change the particle size to the sub-nanometer scale where the quantum confinement limits are almost reached. This leads to the material having extraordinary electrochemical performance. Alternatively, metal ion doping i.e., Fe, Mn, Cr, and Co may increase the electrical conductivity of the electrode material which consequently increases the capacitance as well which applies to the design of capacitors. For instance, if a capacitor has symmetric electrodes or its working principle is based on faradaic reactions this supercapacitor can eventually have enhanced electrochemical performance [34]. The main drawback of carbon electrodes of double-layer supercapacitors is small values of quantum capacitance which act in series with capacitance of ionic space charge. Therefore, further increase of density of capacitance in supercapacitors can be connected with increasing quantum capacitance of carbon electrode nanostructures.

The amount of charge stored per unit voltage in an electrochemical capacitor is primarily a function of the electrode size. The electrostatic storage of energy in the double layers is linear with respect to the stored charge and corresponds to the concentration of the adsorbed ions. Also, while the charge in conventional capacitors is transferred via electrons, capacitance in double-layer capacitors is related to the limited moving speed of ions in the electrolyte and the resistive porous structure of the electrodes. Since no chemical changes take place within the electrode or electrolyte, charging and discharging electric double layers in principle is unlimited. Real supercapacitor's lifetimes are only limited by electrolyte evaporation effects [35]. Additionally, due to the EDLCs storage mechanism, this allows for very fast energy uptake, delivery, and better power performance. Due to the non-faradic process, that is no chemical reaction. It eliminates swelling observed in the active material which batteries demonstrate during charging and discharging. A few differences between EDLCs and batteries can be noticed as (1) EDLCs can withstand millions of cycles, unlike batteries that can withstand few thousand at best. (2) Charge storage mechanism does not involve solvent of the electrolyte; in Li-ion batteries, it contributes to solid electrolyte interphase when high-potential cathodes are used or graphite anodes. However, due to the electrostatic surface charging mechanism, EDLCs devices experience a limited energy density, which is why today's EDLCs research is mainly focused on increasing energy performance and improving temperature range where batteries cannot operate. The performance of EDLC can be adjusted depending on the type of electrolyte used [36].

There are three main types of EDLCs in terms of the carbon content which leads to different functions or roles in the device. One may distinguish such useful functions by the properties of carbonaceous materials, i.e., morphology, hybridization, and structural defects:

1. Carbon aerogels (nanopores), carbon foams (micropores) and, carbide-derived carbon (CDC) (controllable pore size).
2. Carbon nanotubes (CNTs) and graphene, and
3. Activated carbon.

16.4.4.2 Electrochemical Pseudocapacitance

Compared to EDLCs, that store charge electrostatically, pseudocapacitors store charge via a faradic process which involves the transfer of charge between electrode and electrolyte (Fig. 16.8). A simplified view of a double layer with specifically adsorbed ions, which have transferred their charge to the electrode to explain the faradaic charge transfer of the pseudocapacitance. Applying a voltage at the electrochemical capacitor terminals moves electrolyte ions to the opposite polarized electrode and forms a double layer in which a single layer of solvent molecules acts as a separator.

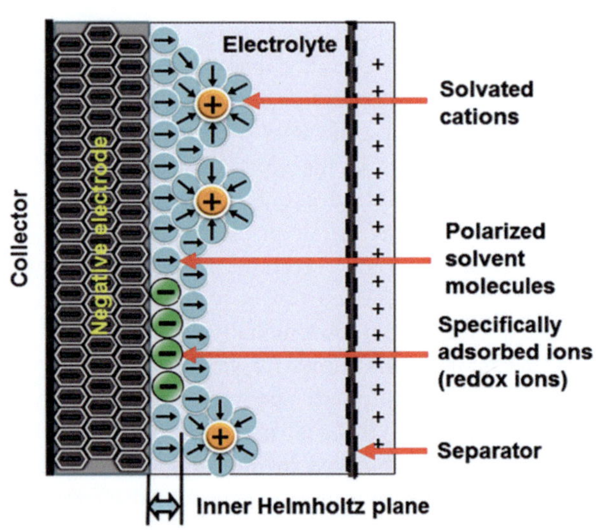

Fig. 16.8 Pseudocapacitors

Pseudocapacitance can originate when specifically adsorbed ions out of the electrolyte pervade the double layer. This pseudocapacitance stores electrical energy by means of reversible faradaic redox reactions on the surface of suitable electrodes in an electrochemical capacitor with an electric double layer [37].

Pseudocapacitance is accompanied with an electron charge transfer between electrolyte and electrode coming from a de-solvated and adsorbed ion whereby only one electron per charge unit is participating. This faradaic charge transfer originates by a very fast sequence of reversible redox, intercalation, or electrosorption processes. The adsorbed ion has no chemical reaction with the atoms of the electrode (no chemical bonds arise) since only a charge transfer takes place.

A cyclic voltammogram shows the fundamental differences between static capacitance and pseudocapacitance (Fig. 16.9). The electrons involved in the faradaic processes are transferred to or from valence electron states (orbitals) of the redox electrode reagent. They enter the negative electrode and flow through the external circuit to the positive electrode where a second double layer with an equal number of anions has formed. The electrons reaching the positive electrode are not transferred to the anions forming the double layer, instead they remain in the strongly ionized and "electron hungry" transition-metal ions of the electrode's surface. As such, the storage capacity of faradaic pseudocapacitance is limited by the finite quantity of reagent in the available surface. A faradaic pseudocapacitance only occurs together with a static double-layer capacitance, and its magnitude may exceed the value of double-layer capacitance for the same surface area by factor 100, depending on the nature and the structure of the electrode, because all the pseudocapacitance reactions take place only with de-solvated ions, which are much smaller than solvated ion with their solvating shell. The amount of pseudocapacitance has a linear function within narrow limits determined by the potential-dependent degree of surface coverage of the adsorbed anions. The ability of electrodes to accomplish pseudocapacitance effects by redox reactions, intercalation, or electrosorption strongly depends on the chemical affinity of electrode materials to the ions adsorbed on the electrode surface as well as on the structure and dimension of the electrode pores. Materials exhibiting redox behavior for use as electrodes in pseudocapacitors are transition-metal oxides like RuO_2, IrO_2, or MnO_2 inserted by doping in the conductive electrode material such as active carbon, as well as conducting polymers such as polyaniline or derivatives of polythiophene covering the electrode material [38].

16.4.4.3 Hybrid Capacitors

As we have seen EDLC offers good cyclic stability, good power performance while in the case of pseudocapacitance

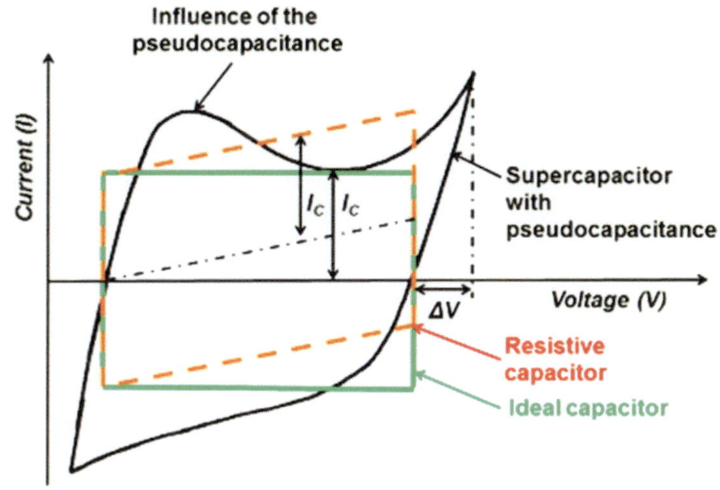

Fig. 16.9 Voltammogram shows the fundamental differences between static capacitance (rectangular) and pseudocapacitance (curved).

that offers greater specific capacitance. Hybridizing these two can offer a combination of both (EDLC, PC), that is by combining the energy source of battery-like electrode, with a power source of capacitor-like electrode in the same cell. With a correct electrode combination, it is possible to increase the cell voltage, which in turn leads to an improvement in energy and power densities. Several combinations have been tested in the past with both positive and negative electrodes in aqueous and inorganic electrolytes. Generally, the faradic electrode results in an increase of energy density at the cost of cyclic stability, which is the main drawback of hybrid devices compared to EDLCs, it is imperative to avoid turning a good supercapacitor into an ordinary battery. This type of supercapacitor consists of polarizable electrodes (carbon) and nonpolarizable electrodes (metal or conducting polymer) to store charges. It uses both faradaic and non-faradaic processes i.e. making use of these properties to obtain high-energy storage through both the battery type and the capacitor type electrode resulting in better cycling stability and lower costs than EDLC. Currently, researchers have focused on the three different types of hybrid supercapacitors, which can be distinguished by their electrode configurations: Composite, Asymmetric, and Battery type [39].

Because double-layer capacitance and pseudocapacitance both contribute inseparably to the total capacitance value of an electrochemical capacitor, a correct description of these capacitors only can be given under the generic term. The concepts of supercapattery and supercabattery have been recently proposed to better represent those hybrid devices that behave more like the supercapacitor and the rechargeable battery, respectively.

Composite Hybrid Supercapacitor (CHS): The purpose of composite hybrid supercapacitors is to have synergistic outcomes of specific capacitance, cycling stability, and high conductivity. As seen in EDLC above, carbon-based supercapacitors have a great surface area, no Joule heating, low resistance, and great mechanical strength. However, carbon itself shows a poor energy density compared to commercially used lead-acid batteries and lithium-ion batteries, whereas, metal oxides, which are being investigated, have poor conductivity, with the exception of RuO_2, and experience Joule heating, with a poor surface area and poor structural stability when under strain. But they are competent in storing charge, and thus energy. Composite hybrid supercapacitors combine the properties of both carbon and metal oxides, merging to provide the synergistic characteristics that are sought after, which include specific capacitance, cycling stability, and high conductivity [40]. Together carbon will provide a channel for charge transport and the metal oxide will store charge via redox reactions contributing to high specific capacitance and high-energy density. While the conductivity of the composite is highly tunable, it depends on the structure of the carbon whether microporous, or, mesoporous and or, macroporous, which means that the pore diameter is an essential factor to consider as it determines whether the ions will or will not be adsorbed to the surface of the electrode, defining its EDLC characteristic for charging/discharging. If not, it will show low conductivity. There are limitations of the composites as well for example when vanadium oxide was layered on carbon nanofibers, the efficiency started to decrease when the layer became thicker (18 nm). This has arisen due to an imbalance between the redox site and the conductivity of the composite itself. Also, another challenge arises when the successive ion diffusion declines, due to protruding nanowhiskers that grow on the carbon nanofibres, even though the surface area of the metal oxide increases. This indicates how the limitations and benefits of the composites are highly dependent on the constituents, their combination, and the electrolyte [41].

Asymmetric Hybrid Supercapacitors: Amongst other supercapacitors, this type is distinctive because of its two dissimilar electrodes. They are designed to work simultaneously to address the power and energy density requirement, as one works as a capacitive electrode and the other as a faradaic electrode. Mostly carbon-derived materials serve as the negative electrode while a metal or metal oxide electrode serves as the anode. Metal electrodes are said to have high intrinsic volumetric capacity leading to an increase in energy densities. These types of capacitors have the potential to display higher energy density and cycling stability than symmetric supercapacitors. This is the case seen for carbon and MnO_2 on a nickel foam-based electrode. Self-discharge of a capacitor is a major issue in all capacitors. One way to tackle it is to incorporate in an asymmetric capacitor the simple rocking chair mechanism. Here is where the maximum potential is ensured at zero current. It is observed overtime with nearly all electrolytes a depletion of ions and electrodes causes a decrease in the conductivity along with an increase in the internal resistance, and the newly designed electrolytes could feasibly evade this pressing concern. Another challenge is to increase the working voltage of supercapacitors. If the negative electrode is activated carbon or a carbon-derived sample effectively p-doped, it has the potential to improve the voltage range along with the rocking-chair mechanism. The carbon microporous layout is also a critical feature to be perfected [42]. Once amended it would allow convenient ion transport to increase the capacitance of the device.

Rechargeable Battery-type Hybrid Supercapacitors: The prospects of this category of supercapacitors lies in the struggle to rise through the midway diagonal of the Ragone plot which holds promising traits of higher specific capacitance, energy density, and power density that still need to be excelled towards, compared to those of present supercapacitors. Some considered factors discussed here include surface modification, synthesis of a perfect nanocomposite material, and microstructure optimization. As the creation of electroactive nanoparticles leads to faster reactions with the electrolyte, it should effectively lead to faster reactions, by undergoing redox with electroactive nanoparticles, as mentioned before in this study. However, a challenging matter arises because it is also possible, that it will face spurious reactions with the electrolyte as well. In building nanocomposite materials with some metal oxides multiple challenges arise [43]. One example of such a challenge lies with $LiMnPO_4$, which has a higher potential than its Fe counterpart, then again is difficult to coat with a carbon layer like $LiFePO_4$. But a remarkable strategic approach to overcome it was by creating a multilayered structure of carbon layered over Fe over Mn. Electroactive species are suitable for their valuable property in providing faster electrode reactions. One enhancing technique lies in developing granules in disproportioned fractions on the electrode surface. And so "fractal granularity," at the electrolyte–electrode interface, causes a higher surface area exposure leading to greater chances of raising the total energy that the capacitor provides. Another major improvement lies in applying the double layer concept to this type of supercapacitor. It creates a Helmholtz double layer where the charge is stored at the interface between the carbon electrode and the electrolyte. It occurs due to like charges being repelled from each other at the interface and attraction of counterions, causing a physical charge storage mechanism to occur simply from the upsurge of polar ions. This had an effect of changing the energy density by several orders of magnitude [44].

16.4.5 Electrodes Materials

The properties of supercapacitors come from the interaction of their internal materials. Especially, the combination of electrode material and type of electrolyte determine the functionality and thermal and electrical characteristics of the capacitors. Supercapacitor electrodes are generally thin coatings applied and electrically connected to a conductive, metallic current collector. Electrodes must have good conductivity, high-temperature stability, long-term chemical stability (inertness), high corrosion resistance, and high surface areas per unit volume and mass. Other requirements include environmental friendliness and low cost [45]. The amount of double layer, as well as pseudocapacitance stored per unit voltage in a supercapacitor, is predominantly a function of the electrode surface area. Therefore, supercapacitor electrodes are typically made of porous, spongy material with an extraordinarily high specific surface area, such as activated carbon. Additionally, the ability of the electrode material to perform faradaic charge transfers enhances the total capacitance. Generally, smaller is the electrode's pore; greater is the capacitance and specific energy. However, smaller pores increase equivalent series resistance (ESR) and decrease specific power. Applications with high peak currents require larger pores and low internal losses, while applications requiring high specific energy need small pores [46].

16.4.5.1 Electrodes for EDLC

The most commonly used electrode material for supercapacitors is carbon in various manifestations such as activated carbon (AC), carbon fiber cloth (AFC), carbide-derived carbon (CDC), carbon aerogel, graphite (graphene), graphane, and carbon nanotubes (CNTs). Carbon-based electrodes exhibit predominantly static double-layer capacitance, even though a small amount of pseudocapacitance may also be present depending on the pore size distribution. Pore sizes in carbons typically range from micropores (less than 2 nm) to mesopores (2–50 nm), but only micropores (<2 nm) contribute to pseudocapacitance. As pore size approaches the solvation shell size, solvent molecules are excluded and only unsolvated ions fill the pores (even for large ions), increasing ionic packing density and storage capability by faradaic H_2 intercalation. Activated carbon was the first material chosen for EDLC electrodes [47]. Even though its electrical conductivity is approximately 0.003% of that of metals (1250–2000 S/m), it is sufficient for supercapacitors. Activated carbon is an extremely porous form of carbon with a high specific surface area, a common approximation is that 1 g (0.035 oz) has a surface area of roughly 1000–3000 m^2 (11,000–32,000 $ft.^2$), about the size of 4–12 tennis courts. The bulk form used in electrodes is low density with many pores, giving high double-layer capacitance. Solid activated carbon, also termed consolidated amorphous carbon (CAC) is the most used electrode material for supercapacitors and may be cheaper than other carbon derivatives [48]. It is produced from activated carbon powder pressed into the desired shape, forming a block with a wide distribution of pore sizes. An electrode with a surface area of about 1000 m^2/g results in a typical double-layer capacitance of about 10 $\mu F/cm^2$ and a specific capacitance of 100 F/g. As of 2010, virtually all commercial supercapacitors use powdered activated carbon made from coconut shells. Coconut shells produce activated carbon with more micropores than does charcoal made from wood [49]. Activated carbon fibers (ACF) are produced from activated carbon and have a typical diameter of 10 μm. They can have micropores with a very narrow pore-size distribution that can be readily controlled. The surface area of ACF woven into a textile is about 2500 m^2/g. Advantages of ACF electrodes include low electrical resistance along the fiber axis and good contact with the collector. As for activated carbon, ACF electrodes exhibit predominantly double-layer capacitance with a small amount of pseudocapacitance due to their micropores. Carbon aerogel is a highly porous, synthetic, ultralight material derived from an organic gel in which the liquid component of the gel has been replaced with a gas. Aerogel electrodes are made via pyrolysis of resorcinol–formaldehyde aerogels and are more conductive than most activated carbons. They enable thin and mechanically stable electrodes with a thickness in the range of several hundred micrometers (μm) and with uniform pore size. Aerogel electrodes also provide mechanical and vibration stability for supercapacitors used in high-vibration environments. Researchers have created a carbon aerogel electrode with gravimetric densities of about 400–1200 m^2/g and volumetric capacitance of 104 F/cm^3, yielding specific energy of 325 kJ/kg (90 Wh/kg) and specific power of 20 W/g. Standard aerogel electrodes exhibit predominantly double-layer capacitance [50]. Aerogel electrodes that incorporate composite material can add a high amount of pseudocapacitance. Carbide-derived carbon (CDC), also known as tunable nanoporous carbon, is a family of carbon materials derived from carbide precursors, such as binary SiC and TiC, that are transformed into pure carbon via physical, e.g., thermal decomposition or chemical, e.g., halogenation) processes. Carbide-derived carbons can exhibit high surface area and tunable pore diameters (from micropores to mesopores) to maximize ion confinement, increasing pseudocapacitance by faradaic H_2 adsorption treatment. CDC electrodes with tailored pore design offer as much as 75% greater specific energy than conventional activated carbons. As of 2015, a CDC supercapacitor offered specific energy of 10.1 Wh/kg, 3500 F capacitance, and over one million charge-discharge cycles [51].

Graphene is an atomic-scale honeycomb lattice made of carbon atoms. Graphene is a one-atom-thick sheet of graphite, with atoms arranged in a regular hexagonal pattern, also called "nanocomposite paper." Graphene has a theoretical specific surface area of 2630 m^2/g which can theoretically lead to a capacitance of 550 F/g. In addition, an advantage of graphene over activated carbon is its higher electrical conductivity. As of 2012, a new development used graphene sheets directly as electrodes without collectors for portable applications. In one embodiment, a graphene-based supercapacitor uses curved graphene sheets that do not stack face-to-face, forming mesopores that are accessible to and wettable by ionic electrolytes at voltages up to 4 V. A specific energy of 85.6 Wh/kg (308 kJ/kg) is obtained at room temperature equaling that of a conventional nickel-metal hydride battery, but with 100–1000 times greater specific power. The two-dimensional structure of graphene improves charging and discharging. Charge carriers in vertically oriented sheets can quickly migrate into or out of the deeper structures of the electrode, thus increasing currents. Such capacitors may be suitable for 100/120 Hz filter applications, which are unreachable for supercapacitors using other carbon materials [52].

Carbon Nanotubes (CNTs), also called buckytubes, are carbon molecules with a cylindrical nanostructure. They have a hollow structure with walls formed by one-atom-thick sheets of graphite. These sheets are rolled at specific and discrete ("chiral") angles, and the combination of chiral angle and radius controls the properties such as electrical conductivity, electrolyte wettability, and ion access. Nanotubes are categorized as single-walled nanotubes (SW-CNTs) or multi-walled nanotubes (MW-CNTs). The latter has one or more outer tubes successively enveloping an SW-CNT, much like the Russian matryoshka dolls. SW-CNTs have diameters ranging between 1 and 3 nm. MW-CNTs have thicker coaxial walls, separated by spacing (0.34 nm) that is close to graphene's interlayer distance. Nanotubes can grow vertically on the collector substrate, such as a silicon wafer. Typical lengths are 20–100 μm. Carbon nanotubes can greatly improve capacitor performance, due to the highly wettable surface area and high conductivity. The ion-size effect and the electrode–electrolyte wettability are the dominant factors affecting the electrochemical behavior of flexible SW-CNT supercapacitors in different 1 molar aqueous electrolytes with different anions and cations. CNTs can store about the same charge as activated carbon per unit surface area, but nanotubes' surface is arranged in a regular pattern, providing greater wettability. SW-CNTs have a high theoretical specific surface area of 1315 m^2/g, while that for MW-CNTs is lower and is determined by the diameter of the tubes and degree of nesting, compared with a surface area of about 3000 m^2/g of activated carbons. Nevertheless, CNTs have higher capacitance than activated carbon electrodes, e.g., 102 F/g for MWNTs and 180 F/g for SW-CNTs. MW-CNTs have mesopores that allow for easy access of ions at the electrode–electrolyte interface. As the pore size approaches the size of the ion solvation shell, the solvent molecules are partially stripped, resulting in a larger ionic packing density and increased faradaic storage capability. However, the considerable volume change during repeated intercalation and depletion decreases their mechanical stability [53].

16.4.5.2 Electrodes for Pseudocapacitors

MnO_2 and RuO_2 are typical materials used as electrodes for pseudocapacitors since they have the electrochemical signature of a capacitive electrode (linear dependence on current versus voltage curve) as well as exhibiting faradaic behavior. Additionally, the charge storage originates from electron-transfer mechanisms rather than the accumulation of ions in the electrochemical double layer. Pseudocapacitors were created through faradaic redox reactions that occur within the active electrode materials. More research was focused on transition-metal oxides such as MnO_2 since transition-metal oxides have a lower cost compared to noble metal oxides such as RuO_2. Moreover, the charge storage mechanisms of transition-metal oxides are based predominantly on pseudocapacitance [54]. Two mechanisms of MnO_2 charge storage behavior were introduced.

The first mechanism implies the intercalation of protons (H^+) or alkali metal cations (C^+) in the bulk of the material upon reduction followed by deintercalation upon oxidation.

$$MnO_2 + H^+(C^+) + e^- \rightleftharpoons MnOOH(C) \tag{16.5}$$

The second mechanism is based on the surface adsorption of electrolyte cations on MnO_2.

$$(MnO_2)_{surface} + C^+ + e^- \rightleftharpoons (MnO_2{}^- \, C^+)_{surface} \tag{16.6}$$

Not every material that exhibits faradaic behavior can be used as an electrode for pseudocapacitors, such as $Ni(OH)_2$ since it is a battery-type electrode (nonlinear dependence on current vs. voltage curve).

Metal Oxides: Brian Evans Conway's research described the electrodes of transition metal oxides that exhibited high amounts of pseudocapacitance. Metal oxides have high specific capacitance and conductivity, making them suitable for electrode fabrication focused on high energy and high-power supercapacitors. There are several different metal oxide materials used for electrode fabrication such as RuO_2, IrO_2, MnO_2, NiO, Co_2O_3, SnO_2, V_2O_5, or MoO_x [55]. Ruthenium oxide (RuO_2) is one of the most explored electrode materials has the highest specific capacitance among pseudocapacitive materials, about 1000 F/g. Besides, it has a wide potential window, highly reversible redox reactions, high-proton conductivity, good thermal stability, long cycle life, metallic-type conductivity, and high-rate capability. It also has three oxidation states accessible within 1.2 V. The most studied ones are RuO_2 and MnO_2. Oxides of transition metals or sulfides such as TiS_2 alone or in combination generate strong faradaic electron-transferring reactions combined with low resistance. RuO_2 in combination with H_2SO_4 electrolyte provides a specific capacitance of 720 F/g and high specific energy of 26.7 Wh/kg (96.12 kJ/kg). Charge/discharge takes place over a window of about 1.2 V per electrode. This pseudocapacitance of about 720 F/g is roughly 100 times higher than for double-layer capacitance using activated carbon electrodes. These transition metal electrodes offer excellent reversibility, with several hundred-thousand cycles. However, ruthenium is expensive and the 2.4 V voltage window for this capacitor limits their applications to military and space applications. The highest capacitance value of 1715 F/g was obtained for ruthenium oxide-based supercapacitor with electrodeposited ruthenium oxide onto porous single-wall carbon nanotube film electrode. RuO_2 supercapacitor anchored on a graphene foam electrode delivered specific capacitance of 502.78 F/g and areal capacitance of 1.11 F/cm^2 leading to a specific energy of 39.28 Wh/kg and specific power of 128.01 kW/kg over 8000 cycles with constant performance [56]. Manganese oxides appear to be an alternative to RuO_2 due to their relatively low cost, low toxicity, and environmental safety, and theoretical high capacitances going up to1100–1300 F/g. The main pseudocapacitive energy storage mechanism in this material is attributed to reversible redox transitions involving the exchange of protons and/or cations with the electrolyte and transitions between different oxidation states, Mn(III)/Mn (II), Mn (IV)/Mn(III), and Mn(VI)/Mn(IV) [57].

Conductive Polymers: Another approach uses electron-conducting polymers as pseudocapacitive material. Although mechanically weak, conductive polymers have high conductivity, resulting in a low ESR and relatively high capacitance. Such conducting polymers include polyaniline, polythiophene, polypyrrole, and polyacetylene. Such electrodes also employ electrochemical doping or de-doping of the polymers with anions and cations. Electrodes made from or coated with conductive polymers have costs comparable to carbon electrodes. Conducting polymer electrodes generally suffer from limited cycling stability. However, polyacene electrodes provide up to 10,000 cycles, much better than batteries. Aromatic polyimides (PI) are interesting as the matrix for conducting composites because of their thermal stability, good mechanical properties, and environmental stability. Composites made of intrinsically conducting polymers (ICPs) and PI matrix have improved mechanical properties, chemical stability, and electrical properties under varying temperatures. For example, polypyrrole/polyimides (PPy/PI) composites show enhanced temperature and environmental stability. The excellent miscibility between PPy carbonyls and PI's NH is caused by hydrogen bonding. This composite also has an excellent electroactivity. A thin (less than 100 μm) high surface area conducting polymer can be grown on a current collector in order to make an electrode. During this electrochemical formation process, the electrode can be p-doped or n-doped. On charging or discharging the electrode the dopant ions move in or out of the polymer electrode forming an electric double layer. These materials' charging mechanism is claimed to be pseudocapacitive rather than double layer charging, reaching very high capacitances (400–500 F/g of active material if the surface area is large enough). PPy is electroactive at positive electrode potentials. PIs are electroactive at negative potentials, and under certain conditions also at positive potentials. This makes it possible to use PPy/PI composites for supercapacitors [58].

16.4.5.3 Electrodes for Hybrid Capacitors

All commercial hybrid supercapacitors are asymmetric. They combine an electrode with a high amount of pseudocapacitance with an electrode with a high amount of double-layer capacitance. In such systems, the faradaic pseudocapacitance electrode with its higher capacitance provides high specific energy while the non-faradaic EDLC electrode enables high specific power. An advantage of the hybrid-type supercapacitors compared with symmetrical EDLCs is their higher specific capacitance value as well as their higher rated voltage and correspondingly their higher specific energy. Composite electrodes for hybrid-type supercapacitors are constructed from carbon-based material with incorporated or deposited pseudocapacitive active materials like metal oxides and conducting polymers. As of 2013 most research for supercapacitors explores composite electrodes. CNTs give a backbone for a homogeneous distribution of metal oxide or electrically conducting polymers (ECPs), producing good pseudocapacitance and good double-layer capacitance [59]. These electrodes achieve higher capacitances than either pure carbon or pure metal oxide or polymer-based electrodes. This is attributed to the accessibility of the nanotubes' tangled mat structure, which allows a uniform coating of pseudocapacitive materials and three-dimensional charge distribution. The process to anchor pseudocapacitive materials usually uses a hydrothermal process. However, precipitation of MnO_2 on an SWNT film makes an organic-electrolyte-based supercapacitor. Another way to enhance CNT electrodes is by doping with a pseudocapacitive dopant as in lithium-ion capacitors. In this case, the relatively small Li atoms intercalate between the layers of carbon. The anode is made of Li-doped carbon, which enables lower negative potential with a cathode made of activated carbon. This results in a larger voltage of 3.8–4 V that prevents electrolyte oxidation. As of 2007, they had achieved capacitance of 550 F/g. and reach a specific energy upto 14 Wh/kg (50.4 kJ/kg). Battery-type electrodes and Asymmetric electrodes (pseudo/EDLC) are used in the hybride system. Rechargeable battery electrodes influenced the development of electrodes for new hybrid-type supercapacitor electrodes as for Li-ion capacitors [60]. Together with a carbon EDLC electrode in an asymmetric construction offers this configuration higher specific energy than typical supercapacitors with higher specific power, longer cycle life, and faster charging and recharging times than batteries. Recently some asymmetric hybrid supercapacitors were developed in which the positive electrode was based on a real pseudocapacitive metal oxide electrode (not a composite electrode), and the negative electrode on an EDLC activated carbon electrode. An advantage of this type of supercapacitors is their higher voltage and correspondingly their higher specific energy (up to 10–20 Wh/kg (36–72 kJ/kg)). As far as known no commercial offered supercapacitors with such kind of asymmetric electrodes are on the market.

16.4.6 Electrolytes

Electrolytes consist of a solvent and dissolved chemicals that dissociate into positive cations and negative anions, making the electrolyte electrically conductive. The more ions the electrolyte contains, the better its conductivity. In supercapacitors, electrolytes are the electrically conductive connection between the two electrodes. Additionally, in supercapacitors, the electrolyte provides the molecules for the separating monolayer in the Helmholtz double layer and delivers the ions for pseudocapacitance. The electrolyte determines the capacitor's characteristics: its operating voltage, temperature range, ESR, and capacitance. With the same activated carbon electrode, an aqueous electrolyte achieves capacitance values of 160 F/g, while an organic electrolyte achieves only 100 F/g. The electrolyte must be chemically inert and not chemically attack the other materials in the capacitor to ensure long-time stable behavior of the capacitor's electrical parameters. The electrolyte's viscosity must be low enough to wet the porous, sponge-like structure of the electrodes. An ideal electrolyte does not exist, forcing a compromise between performance and other requirements.

Aqueous: Water is a relatively good solvent for inorganic chemicals. Treated with acids such as sulfuric acid (H_2SO_4), alkalis such as potassium hydroxide (KOH), or salts such as quaternary phosphonium salts, sodium perchlorate ($NaClO_4$), lithium perchlorate ($LiClO_4$), or lithium hexafluoride arsenate ($LiAsF_6$), water offers relatively high conductivity values of about 100–1000 mS/cm. Aqueous electrolytes have a dissociation voltage of 1.15 V per electrode (2.3 V capacitor voltage) and a relatively low operating temperature range. They are used in supercapacitors with low specific energy and high specific power [61].

Organic: Electrolytes with organic solvents such as acetonitrile, propylene carbonate, tetrahydrofuran, diethyl carbonate, γ-butyrolactone and solutions with quaternary ammonium salts or alkyl ammonium salts such as tetraethylammonium tetrafluoroborate($N(Et)_4BF_4$) or triethyl(metyl)-tetrafluoroborate ($NMe(Et)_3BF_4$) are used. They are more expensive than aqueous electrolytes, but they have a higher dissociation voltage of typically 1.35 V per electrode (2.7 V capacitor voltage) and a higher temperature range. The lower electrical conductivity of organic solvents (10–60 mS/cm) leads to a lower specific power, but since the specific energy increases with the square of the voltage, higher specific energy [62].

Ionic Liquids: Ionic electrolytes consists of liquid salts that can be stable in a wider electrochemical window, enabling capacitor voltages above 3.5 V. Ionic electrolytes typically have an ionic conductivity of a few mS/cm, lower than aqueous or organic electrolytes. Low-temperature ionic liquids (ILs), which are the type of ILs of interest to supercapacitors, are pure organic salts containing no solvents with melting points below 100 °C. If the liquid state is maintained at ambient temperature, they are termed room temperature ionic liquids (RTILs). RTILs are of interest to supercapacitors because they are nonvolatile, poorly combustible, and heat-resistant, with these properties being very peculiar and unachievable with conventional solvents [63]. In RTILs, at least one ion usually has a delocalized charge (very often aromatic structures) and one component is organic, which prevents the formation of a stable crystal lattice. Properties such as melting point, viscosity, and, conductivity are controlled by both the substituents on the organic ion and by the counterion. Many ionic liquids can be and have been developed with the extensive variation of physicochemical properties. For this reason, ionic liquids have been termed “designer solvents”.

Ionic liquid gel polymer electrolytes (ILGPEs) are also developed by incorporating ionic liquids into a polymer matrix. These are mechanically strong, electrochemically and thermally stable, and highly conductive. Ionic liquids are resistant to the reduction and the oxidation in a wide voltage potential window, which depends on the counterion, providing a cell voltage of around 4.5 V, with some of them being able to reach 6 V. Since ILs are solvent-free, there is no salvation shell, so the ion size is better known. The most important properties of ionic liquids from the point of view of electrochemistry are, conductivity, viscosity, and the potential range of electrochemical stability. The main drawback of IL electrolytes is their low electrical conductivity, which is typically less than 10 mS/cm, and is significantly lower compared to aqueous electrolytes. In order to overcome the low conductivity, a dilut ion with organic solvents is sometimes applied [64].

16.4.7 Separators

Separators have to physically separate the two electrodes to prevent a short circuit by direct contact. It can be very thin (a few hundredths of a millimeter) and must be very porous to the conducting ions to minimize ESR. Furthermore, separators must be chemically inert to protect the electrolyte's stability and conductivity. Inexpensive components use open capacitor papers. More sophisticated designs use nonwoven porous polymeric films like polyacrylonitrile or Kapton, woven glass fibers, or porous woven ceramic fibers [65].

16.4.8 Collectors and Housing

Current collectors connect the electrodes to the capacitor's terminals. The collector is either sprayed onto the electrode or is a metal foil. They must be able to distribute peak currents of upto 100 A. If the housing is made out of a metal (typically Al) the collectors should be made from the same material to avoid forming a corrosive galvanic cell.

16.4.9 Synthesis Approach for Electrode Materials

The method of synthesis of electrode materials plays an important role in controlling the structures and properties of the materials. Some synthesis methods are described here briefly:

16.4.9.1 Solgel Method

Solgel is a facile method to prepare materials with greater purity and homogeneity. The solgel method is so named, as in it microparticles in the solution (sol) agglomerate and link together in regulated conditions to form an integrated network (gel). Two basic variations of the solgel method are the colloidal method and the polymeric or the alkoxide method, which are different from each other on the type of precursors used. In both methods, the precursor is mixed in a liquid (usually water is used for the colloidal method and alcohol for the polymeric method) and is then activated with the addition of an acid or a base [66]. Then, as obtained activated precursor reacts forming a network, which develops with temperature and time maximally up to the container size. This process provides the advantage of preparing materials of different morphologies. The electrode material prepared by this process possesses high SSA with better electrochemical behavior which can also be controlled by temperature, change of surfactants, solvents, and reaction time. Activated carbon fiber material (ACFM)-Ni$(OH)_2$ composite, $NiCo_2O_4$ films, NiO/$LaNiO_3$ electrode fabricated by spin coating on Pt/Ti/SiO_2/Si (100) substrate, can be formulated by solgel which exhibits the Cs of 370–380 F/g, Cs of 2157 F/g, and 2030 F/g, respectively. This superior electrochemical response can be related to high porosity, well-connected network structures with reduced mass transfer resistance between electrolyte and ion which facilitates the electron hopping in nanoparticles [67].

16.4.9.2 Electro-polymerization/Electrodeposition

This is a common synthesis technique which provides precise regulation over the thickness of films and on the rate of polymerization. By suitable choice of deposition solution, nanostructured films with different mass loading and morphologies can be prepared by this method. This technique involves simple processing conditions and not much toxic chemicals are used in it. It is generally used for preparing CPs such as PANI, PEDOT, PPy, etc. MnO_2-PEDOT: PSS composite can be prepared by a co-electrodeposition strategy which exhibits an areal Cs of 1670 mF/cm^2 at 0.5 mA/cm^2 and excellent mechanical robustness. Also, an ultra-thin (<200 μm) asymmetric supercapacitor (ASC) is fabricated with high Ed, Pd, and rate capability. Nanosized MnO_2 electrodes on Au nanowire stems are grown electrochemically that exhibit high Cs (1130 F/g at 2 mV/s), high Ed (15 Wh/kg at 50 A/g), high Pd (20 kW/kg at 50 A/g), and long-term stability (90% of Cs left after 5000 cycles). ZnO@Ni_3S_2 core-shell nanorods are formed by the electrodeposition method that exhibits a Cs of 1529 F/g at 2 A/g and retains 42% of initial Cs after 2000 cycles. Stretchable CNT-PPy films can be prepared by electrochemical deposition [68].

16.4.9.3 In Situ Polymerization

In this process, monomers are dispersed into an aqueous solution using the sonication process. Then an oxidizing agent is mixed to initialize the polymerization in the aqueous solution and the sample is obtained by filtering the solution. Earlier this method yielded only irregular aggregates with a little portion of nanofibres, but with slight modification, nanoparticles, nanorods, and nanofibres were reported with better solution processability and better physical and chemical properties. A simple strategy for the growth of PEDOT structures on carbon fiber cloth (CFC) by in situ polymerization is reported. When a supercapacitor device is fabricated with these nanostructures, it exhibits a Cs of 203 F/g at 5 mV/s, an Ed of 4.4 Wh/kg, and Pd of 40.25 kW/kg in 1 M H_2SO_4 electrolyte. Also, it possesses 86% Cs retention after 12,000 cycles. Wang et al. have deposited PANI nanowires within the multi-walled carbon nanotubes (MWCNTs) by in situ electro-polymerization [69]. The aligned MW-CNTs provide support to the organic polymers along with providing a pathway for the transfer of charge. Also, confined MW-CNT channels limit the structural changes in PANI chains while charging–discharging and enhance the lifetime of the structure. The films made with CPs encapsulated in MW-CNTs showed a Cs of 296 F/g at 1.6 A/g. As prepared PEDOT: MWCNT composite reveals interconnected network due to the Π–Π interaction of PEDOT with non-covalent functionalized MWCNT and exhibits a Cs of 199 F/g at 0.5 A/g [70].

16.4.9.4 Direct Coating

This technique is employed for the fabrication of those supercapacitor electrodes in which active material, in the form of slurry, is applied directly on the substrate. Often, additives such as carbon black, polyvinylidene fluoride, acetylene black, polytetrafluoroethylene are introduced as binders to provide maximum adhesion along with retaining electrical conductivity. The working electrode is fabricated with 90 wt% electrode materials (NiO) and 10 wt% PVA in millipore water as a solvent and the slurry obtained is pasted on the Pt disc. Du et al. synthesized supercapacitor electrode by coating the slurry formed by adding active material with acetylene black and PTFE onto Ni foam [71].

16.4.9.5 Chemical Vapour Deposition (CVD)

CVD technique is generally used where the porosity is very important. This process is performed under vapor phase, where the initial material is prepared in vapor form, flowed, and subjected to a high temperature (800–1000 °C). The as-prepared structures have even morphology. Among various synthesis methods of graphene, for instance, mechanical cleavage of graphite, chemical exfoliation of graphite (in organic solvents), manufacturing of multilayered graphene by arc discharge, reduction of graphene oxide (GO) synthesized from the oxidation of graphite, graphene synthesized by CVD provides better results owing to their large crystal domains, monolayered structure and fewer defects in the sheets, which are helpful for enhancing carrier mobility. Lobiak et al. prepared hybrid carbon materials consisting of MW-CNTs and graphitic layers, produced by CVD, over MgO assisted metal catalyst, such materials provide fast charge transport in the cell [72].

16.4.9.6 Vacuum Filtration Technique

This quick and proficient technique uses the simple concept of vacuum filtration to prepare nanocomposites from physical combination of different materials. Generally, a mixture of materials is prepared followed by simple vacuum filtration and drying the filtrate. In this method, the composition can be simply altered by varying the concentration or the weight percentage of each constituent in the mixture. Graphene suspension, developed by vacuum filtration deposition by Zhang et al. for fabricating graphene-based Ni foam electrodes, shows a higher Ed and Pd along with good cycling performance. Xu et al. have synthesized a nanocomposite of graphene/AC/PPy by vacuum filtration method. As prepared electrode exhibits the Cs of 178 F/g at 0.5 mA/cm^2 and retains 64.4% of Cs after 5000 charge/discharge cycles. Y. Gao has used this technique to prepare graphene/polymer electrode on Ni foam in which the vacuum pressure and its duration controls the distribution of graphene [73].

16.4.9.7 Hydrothermal/Solvothermal Method

The hydrothermal process can be ascribed as environment-friendly superheated aqueous solution dispensation. In addition, this provides controlled diffusivity within a closed system. The process has superiority over other techniques as it is ideal for preparing designer particulates (particles with high purity, crystallinity, quality, and controlled chemical and physical characteristics). Also, this is a low-temperature sintering process with a small energy requirement which is simple to implement and scaleup. However, this process has lesser control over nanoparticle aggregation. The solvent properties (e.g., dielectric constant, solubility) change radically in the supercritical phase. Thus, the supercritical phase gives a favorable condition for particle formation owing to increased reaction rate and great supersaturation. If some other solvent is used instead of water, then the method is called solvothermal synthesis. A lot of supercapacitor electrodes have been fabricated using this process such as rod-like hollow $CoWO_4$/Co1-xS, Cobalt disulfide-reduced graphene oxide (CoS_2-rGO), hexagonal $NiCo_2O_4$ nanoparticles, etc. [74].

16.4.9.8 Coprecipitation Method

This is a facile method for the large-scale production of powder samples. For precipitation to take place, the concentration of one solute should be more than the solubility limit and the temperature should be high enough for fast separation into precipitates. Here, it is difficult to regulate the morphology of prepared samples due to the fast rate of precipitation. Various supercapacitor structures have been reported using this method such as $CoFe_2O_4$-magnetic nanoparticles with different precursors, $Ni_3(PO4)_2$@GO composite which exhibits a Cs of 1329.59 F/g at a 0.5 A/g and 88% of the Cs retention after 1000 cycles [75].

16.4.9.9 Dealloying Method

Dealloying method, also known as selective dissolution, is an easy, flexible and economical technique to produce nanoporous metallic materials (NPMs) with structures like core–shell, hollow core–shell, and porous nanoparticles. In this method, the more active material is removed from a solution of binary metallic solid by electrolytic dissolution thus producing an interconnected porous structure. Such structures possess higher surface area, good mechanical and compression strength along size-scale dependent elastic modulus. Much attention has been given to NPMs prepared by this method since the important work of Erlebacher et al. and has become a very important method to produce NPMs in the last decade. Fixed voltage dealloying of AgAu alloy particles in the size range of 2–6 nm and 20–55 nm, demonstrated that only the core–shell structures (2–6 nm in diameter) evolved above the potential corresponding to Ag^+/Ag equilibrium. CuS nanowire on nanoplate network with improved electrochemical performance has been prepared using an improved dealloying method at two contrasting reaction temperatures. Cu_2O has been synthesized by oxidation assisted dealloying method. Free-dealloying method has been used for the synthesis of Cu-based metallic glasses in HF and HCl solutions. With this technique, extensive elements can be fabricated with tunable pore sizes along with full recovery of the evaporated component. Flexible electrodes of Co_3O_4 flakes and γ-Fe_2O_3 nanoparticles have been prepared by oxidation-assisted dealloying method [76].

16.4.9.10 Other Synthesis Methods

Several other synthesis methods have been reported for supercapacitor electrodes. The microwave-assisted method has been used for rapid synthesis of tin selenide. Nitrogen functionalized carbon nanofibres (N-CNFs) are prepared by carbonizing PPy-coated nanofiber (NF), which in turn are obtained by "electrospinning" and deacetylation of electrospun cellulose acetate NFs and PPy polymerization. An additive free, cost-effective and scalable "successive ionic layer adsorption and reaction (SILAR) method" has been quoted to prepare Ni–Co binary hydroxide on rGO. The pulsed layer deposition method is used to fabricate NiO on graphene foam. Free-standing 3D porous rGO and PANI hybrid foam has been fabricated by "dipping and dry method." Hierarchical porous carbon microtubes (HPCNTs) have been synthesized by carbonization along with KOH activation [77].

16.4.10 Selection of Supercapacitor

To replace the batteries in the system, choosing the correct values and ratings of supercapacitors is very important. Also, it is important to relate capacitance and energy in terms of Watts per hour. Battery charging and discharging time are calculated based on the ampere-hour rating, an Eq. (16.7) relating ampere-hour and capacitance is given below.

$$\mathrm{Ah} = \left[\frac{V_{\min} - V_{\max}}{2}\right] \times \left[\frac{F}{3600}\right] \tag{16.7}$$

where Ah is Ampere hour, F is Farad, $V_{\min}$ and $V_{\max}$ are terminating voltage levels. Battery equivalent capacitance ratings are mentioned in Table 16.1. Stacked and rolled type supercapacitor has different effective characteristics, choosing between them is also important [78]. An ideal supercapacitor must have a high energy density, high power density, high pulse current, high capacitance, and low resistance.

16.4.11 Comparative analysis of Supercapacitor and Other Storage Devices

Compared to the battery or electrolytic capacitors, SC has higher energy density and higher power density together with smaller volume and weight. SC has a long life cycle compared to batteries, up to 500,000 times. It can be said that battery and SC are complementary because batteries are limited in the power levels they can support but have high energy to weight ratio, whereas SC can support various power levels but has lower energy to weight ratio. Modern applications are high power rated, this has led to the manufacturing of batteries with high power which in turn requires to sacrifice energy density and life cycle [79]. Similarly with the capacitors, they now suffer from low energy density and higher self-discharge. Compared to individual SC and batteries, hybrid energy storage systems can achieve better energy and power performances. There have been several hybrid models proposed which show superiority over battery-only systems. Battery-supercapacitor, fuel cell-supercapacitor hybrid models are some examples [80] (Table 16.2).

Table 16.1 Battery equivalent capacitance rating

Sl. No.	Battery rating (mAh)	Equivalent Capacitance (Farad)
1	1000	782
2	1250	978
3	1500	1173
4	1800	1408
5	2100	1643

16.4.12 Applications

In applications with fluctuating loads, such as laptop computers, PDAs, GPS, portable media players, handheld devices, and photovoltaic systems, supercapacitors can stabilize the power supply. Supercapacitors deliver power for photographic flashes in digital cameras and for LED flashlights that can be charged in much shorter periods of time, e.g., 90 s. Some portable speakers are powered by supercapacitors. A cordless electric screwdriver with supercapacitors for energy storage has about half the run time of a comparable battery model, but can be fully charged in 90 s. It retains 85% of its charge after 3 months left idle [81]. Supercapacitor used as a grid power buffer**.** Numerous nonlinear loads, such as EV chargers, HEVs, air-conditioning systems, and advanced power conversion systems cause current fluctuations and harmonics. These current differences create unwanted voltage fluctuations and therefore power oscillations on the grid. Power oscillations not only reduce the efficiency of the grid, but can cause voltage drops in the common coupling bus, and considerable frequency fluctuations throughout the entire system. To overcome this problem, supercapacitors can be implemented as an interface between the load and the grid to act as a buffer between the grid and the high pulse power drawn from the charging station [82].

Supercapacitors provide backup or emergency shutdown power to low-power equipment such as RAM, SRAM, micro-controllers, and PC Cards. They are the sole power source for low-energy applications such as automated meter reading (AMR) equipment or event notification in industrial electronics. Supercapacitors buffer power to and from rechargeable batteries, mitigating the effects of short power interruptions and high current peaks. Batteries kick in only during extended interruptions, e.g., if the mains power or a fuel cell fails, which lengthens battery life. Uninterruptible power supplies (UPS) may be powered by supercapacitors, which can replace much larger banks of electrolytic

Table 16.2 Comparison between supercapacitor, electrolytic capacitor, and battery performance

Storage device characteristics	Capacitor	Supercapacitor	Battery (Li-ion batteries)
Charging time	$10^{-3} < t < 10^{6}$ S	1–30 s	1 <t <5 h
Discharging time	$10^{-3} < t < 10^{6}$ S	1–30 s	t > 0.3 h
Energy density	<0.1 Wh/kg	1–10 Wh/kg	10–100 Wh/kg
Life cycle number	10^{6}	10^{6}	1000
Power density	>1,000,000 W/kg	10,000 W/kg	<1000 W/kg
Charge/discharge efficiency	>0.95	0.85–0.98	0.7–0.85

capacitors. This combination reduces the cost per cycle, saves on replacement and maintenance costs, enables the battery to be downsized, and extends battery life. Supercapacitors provide backup power for actuators in wind turbine pitch systems, so that blade pitch can be adjusted even if the main supply fails [83]. Supercapacitors can stabilize voltage fluctuations for power lines by acting as dampeners. Wind and photovoltaic systems exhibit fluctuating supply evoked by gusting or clouds that supercapacitors can buffer within milliseconds. Also, similar to electrolytic capacitors, supercapacitors are also placed along the power lines to consume reactive power and improve the AC power factor in a lagging power flow circuit. This would allow for a better used real power to produce power and make the grid overall more efficient [152]. Microgrids are usually powered by clean and renewable energy. Most of this energy generation, however, is not constant throughout the day and does not usually match demand. Supercapacitors can be used for microgrid storage to instantaneously inject power when the demand is high and the production dips momentarily, and to store energy in the reverse conditions. They are useful in this scenario because microgrids are increasingly producing power in DC, and capacitors can be utilized in both DC and AC applications. Supercapacitors work best in conjunction with chemical batteries. They provide an immediate voltage buffer to compensate for quick changing power loads due to their high charge and discharge rate through an active control system. Once the voltage is buffered, it is put through an inverter to supply AC power to the grid. It is important to note that supercapacitors cannot provide frequency correction in this form directly in the AC grid [84].

Supercapacitors are suitable temporary energy storage devices for energy harvesting systems. In energy harvesting systems, the energy is collected from ambient or renewable sources, e.g., mechanical movement, light or electromagnetic fields, and converted to electrical energy in an energy storage device. For example, it was demonstrated that energy collected from RF (radio frequency) fields (using an RF antenna as an appropriate rectifier circuit) can be stored in a printed supercapacitor. The harvested energy was then used to power an application-specific integrated circuit (ASIC) circuit for over 10 h [85]. The UltraBattery is a hybrid rechargeable lead-acid battery and a supercapacitor. Its cell construction contains a standard lead-acid battery positive electrode, standard sulfuric acid electrolyte, and a specially prepared negative carbon-based electrode that stores electrical energy with double-layer capacitance. The presence of the supercapacitor electrode alters the chemistry of the battery and affords it significant protection from sulfation in a high rate partial state of charge use, which is the typical failure mode of valve-regulated lead-acid cells used this way. The resulting cell performs with characteristics beyond either a lead-acid cell or a supercapacitor, with charge and discharge rates, cycle life, efficiency, and performance all enhanced [86].

Sado City, in Japan's Niigata Prefecture, has street lights that combine a stand-alone power source with solar cells and LEDs. Supercapacitors store the solar energy and supply 2 LED lamps, providing 15 W power consumption overnight. The supercapacitors can last more than 10 years and offer stable performance under various weather conditions, including temperatures from +40 to below −20 °C [87]. Toyota's Yaris Hybrid-R concept car uses a supercapacitor to provide bursts of power. PSA Peugeot Citroën has started using supercapacitors as part of its stop-start fuel-saving system, which permits faster initial acceleration. Mazda's i-ELOOP system stores energy in a supercapacitor during deceleration and uses it to power onboard electrical systems while the engine is stopped by the stop-start system. Maxwell Technologies, an American supercapacitor-maker, claimed that more than 20,000 hybrid buses use the devices to increase acceleration. Guangzhou, In 2014 China began using trams powered with supercapacitors that are recharged in 30 seconds by a device positioned between the rails, storing power to run the tram for up to 4 km more than enough to reach the next stop, where the cycle can be repeated. In 2005, aerospace systems and controls company Diehl Luftfahrt Elektronik GmbH chose supercapacitors to power emergency actuators for doors and evacuation slides used in airliners, including the Airbus 380. Supercapacitors' low internal resistance supports applications that require short-term high currents. Among the earliest uses were motor startup (cold engine starts, particularly with diesel) for large engines in tanks and submarines. Supercapacitors buffer the battery, handling short current peaks, reducing cycling, and extending battery life. Further military applications that require high specific power are phased array radar antennae, laser power supplies, military radio communications, avionics displays and instrumentation, backup power for airbag deployment, and GPS-guided missiles and projectiles [88].

Supercapacitor/battery combinations in electric vehicles (EV) and hybrid electric vehicles (HEV) are well investigated. A 20–60% fuel reduction has been claimed by recovering brake energy in EVs or HEVs. The ability of supercapacitors to charge much faster than batteries, their stable electrical properties, broader temperature range, and longer lifetime are suitable. Supercapacitors can be used to supplement batteries in starter systems in diesel railroad locomotives with diesel-electric transmission. The capacitors capture the braking energy of a full stop and deliver the peak current for starting the diesel engine and acceleration of the train and ensure the stabilization of line voltage. Depending on the driving mode upto 30% energy saving is possible by recovery of braking energy. Low maintenance and environmentally friendly materials encouraged the choice of supercapacitors. Mobile hybrid Diesel-electric rubber tyred gantry cranes move and stack containers within a terminal. Lifting the boxes requires large amounts of energy. Some of the energy could be recaptured while lowering the load, resulting in improved efficiency. A triple hybrid forklift truck uses fuel cells and batteries as primary energy storage and supercapacitors to buffer power peaks by storing braking energy. They provide the forklift with peak power over 30 kW. The triple-hybrid system offers over 50% energy savings compared with Diesel or fuel-cell systems. Supercapacitor-powered terminal tractors transport containers to warehouses. They provide an economical, quiet, and pollution-free alternative to Diesel terminal tractors. Supercapacitors make it possible not only to reduce energy but to replace overhead lines in historical city areas, so preserving the city's architectural heritage. This approach may allow many new light rail city lines to replace overhead wires that are too expensive to fully route [89–91].

The first hybrid bus with supercapacitors in Europe came in 2001 in Nuremberg, Germany. It was MAN's so-called Ultracapbus, and was tested in real operation in 2001/2002. The vehicle was equipped with a diesel-electric drive in combination with supercapacitors. The system was supplied with 8 Ultracap modules of 80 V, each containing 36 components. The system worked with 640 V and could be charged/discharged at 400 A. Its energy content was 0.4 kWh with a weight of 400 kg. The supercapacitors recaptured braking energy and delivered starting energy. Fuel consumption was reduced by 10–15% compared to conventional diesel vehicles. Other advantages included reduction of CO_2 emissions, quiet and emissions-free engine starts, lower vibration, and reduced maintenance costs [92]. The FIA, a governing body for motor racing events, proposed in the *Power-Train Regulation Framework for Formula 1* version 1.3 of May 23, 2007, that a new set of power train regulations be issued that includes a hybrid drive of up to 200 kW input and output power using "superbatteries" made with batteries and supercapacitors connected in parallel (KERS). About 20% tank-to-wheel efficiency could be reached using the KERS system. The Toyota TS030 Hybrid LMP1 car, a racing car developed under Le Man's Prototype rules, uses a hybrid drivetrain with supercapacitors. In the 2012 24 Hours of Le Mans race a TS030 qualified with the fastest lap only 1.055 s slower (3:24.842 versus 3:23.787) than the fastest car, an Audi R18 e-tron quattro with flywheel energy storage. The supercapacitor and flywheel components, whose rapid charge–discharge capabilities help in both braking and acceleration, made the Audi and Toyota hybrids the fastest cars in the race. In the 2012 Le Man's race the two competing TS030s, one of which was in the lead for part of the race, both retired for reasons unrelated to the supercapacitors. The TS030 won three of the eight races in the 2012 FIA World Endurance Championship season. In 2014 the Toyota TS040 Hybrid used a supercapacitor to add 480 horsepower from two electric motors [93–95].

16.4.13 Advantages and Limitations of Supercapacitor

Batteries are dangerous when mistreated, overheating may cause batteries to explode. SC does not overheat because of low internal resistance. The lifecycle of batteries is low, comparatively, SC has a lifetime of virtual infinity. This makes these devices useful where it is subjected to frequent charging and discharging cycles. Shorting terminals of fully charged SC will cause it to discharge quickly, which may result in electrical arching, which might damage the device. Some of the features of supercapacitors are low ESR, low leakage current, higher life cycle, a wide range of operating temperature, higher useable capacity.

16.5 Superconducting Materials

Superconductivity is a phenomenon displayed by some materials when they are cooled below a certain temperature, known as the superconducting critical temperature, T_c. Below T_c, superconducting materials exhibit two characteristic properties: (1) zero electrical resistance, (2) perfect diamagnetism (the Meissner effect) [96, 97]. Zero electrical resistance means that no energy is lost as heat as the material conducts electricity. The second of these properties, perfect diamagnetism, means that the superconducting material will exclude a magnetic field and this is known as the Meissner effect (after its discoverer). The Meissner effect is used to levitate a superconductor above a magnet (Fig. 16.10).

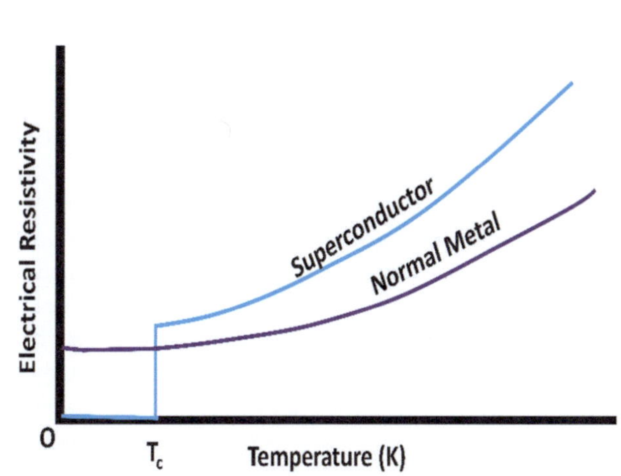

Fig. 16.10 The electrical resistance comparative diagram between the superconductor and normal metal

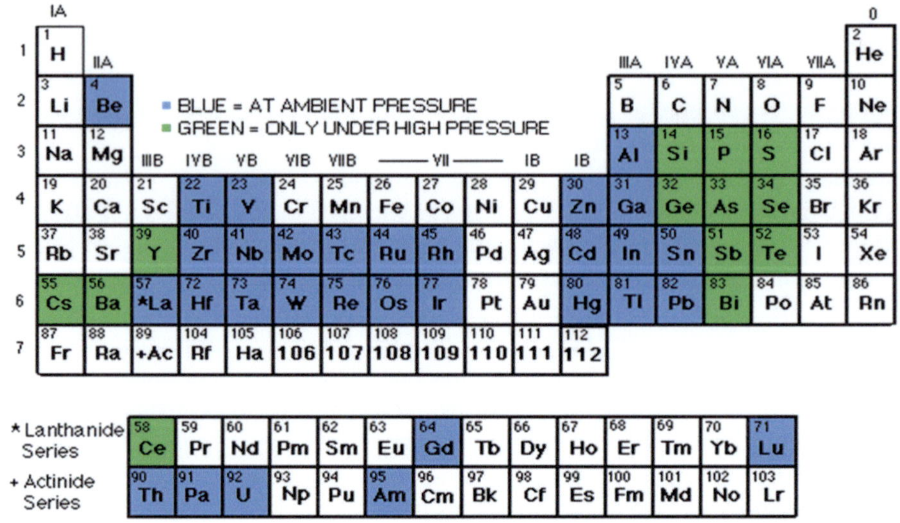

Fig. 16.11 Elements for superconductive materials

16.5.1 History of Superconductor

Superconductivity was first discovered in 1911 when mercury was cooled to approximately 4 K by Dutch physicist Heike Kamerlingh Onnes, which earned him the 1913 Nobel Prize in physics. In the years since this field has greatly expanded and many other forms of superconductors have been discovered, including Type 2 superconductors in the 1930s [98]. The basic theory of superconductivity, BCS Theory, earned the scientists—John Bardeen, Leon Cooper, and John Schrieffer—the 1972 Nobel Prize in physics. A portion of the 1973 Nobel Prize in physics went to Brian Josephson, also for work with superconductivity [99]. Like ferromagnetism and atomic spectral lines, superconductivity is a quantum mechanical mystery. It is characterized by the Meissner effect, the complete ejection of magnetic field lines from the interior of the superconductor during its transitions into the superconducting state [100]. The occurrence of the Meissner effect indicates that superconductivity cannot be understood simply as the idealization of perfect conductivity in classical physics. In January 1986, Karl Muller and Johannes Bednorz made a discovery that revolutionized how scientists thought of superconductors. Prior to this point, the understanding was that superconductivity manifested only when cooled to near absolute zero, but using an oxide of barium, lanthanum, and copper, they found that it became a superconductor at approximately 40 K. This initiated a race to discover materials that functioned as superconductors at much higher temperatures. In the decades since, the highest temperatures that had been reached were about 133 K (though you could get up to 164 K if applied a high pressure). In August 2015, a paper published in the journal Nature reported the discovery of superconductivity at a temperature of 203 K when under high pressure [101].

16.5.2 Classification of Superconducting Materials

Superconducting materials can be categorized into two types: Type I Superconductors and Type II Superconductors, as given in Fig. 16.11. Both types exhibit perfect electrical conductivity and can be restored to "normal" conductors in the presence of a sufficiently strong magnetic field. The threshold temperature below which a material transition into a superconductor state is designated as T_c, which stands for critical temperature. Not all materials turn into superconductors, and the materials that do each have their own value of *Tc*.

16.5.2.1 Type I Superconductors

This superconductor totally excludes all applied magnetic fields. Most elemental superconductors are Type I. *Type I superconductors* act as conductors at room temperature, but when cooled below T_c, the molecular motion within the material reduces enough that the flow of current can move unimpeded [102]. At present, type I superconductors have Tc s between 0.000325 °K and 7.196 K at standard pressure, as given in Table 16.3. Some type I superconductors require incredible amounts of pressure in order to reach the superconductive state. One such material is sulfur which requires a pressure of 9.3 million atmospheres (9.4×10^{11} N/m^2) and a temperature of 17 °K to reach superconductivity. The Type 1 category of superconductors is mainly comprised of metals and metalloids that show some conductivity at room temperature. They require incredible cold to slow down molecular vibrations sufficiently to facilitate unimpeded electron flow in accordance with what is known as BCS theory. BCS theory suggests that electrons team up in "Cooper pairs" in order to help each other overcome molecular obstacles much like race cars on a track drafting each other in order to go faster.

Table 16.3 Type-I Superconducting elements with their superconducting critical temperature

Elements	Tc (K)	Elements	Tc (K)
Pb	7.196	Zn	0.85
La	4.88 K	Os	0.66
Ta	4.47	Zr	0.61
Hg	4.15	Am	0.60
Sn	3.72	Cd	0.517
In	3.41	Ru	0.49
Pd	3.3	Ti	0.40
Cr	3	U	0.20
Tl	2.38	Hf	0.128
Re	1.697	Ir	0.1125
Pa	1.40	Be	0.023
Th	1.38	W	0.0154
Al	1.175	Pt	0.0019
Ga	1.083	Li	0.0004
Mo	0.915	Rh	0.000325

Scientists call this process phonon-mediated coupling because of the sound packets generated by the flexing of the crystal lattice [103].

Type 1 superconductors—characterized as the "soft" superconductors—were discovered first and require the coldest temperatures to become superconductive. They exhibit a very sharp transition to a superconducting state and "perfect" diamagnetism, the ability to repel a magnetic field completely. Table 16.3 is a list of known Type 1 superconductors along with the critical transition temperature (Tc) below which each superconducts. The 3rd column gives the lattice structure of the solid that produced the noted Tc. Surprisingly, copper, silver, and gold, three of the best metallic conductors, do not rank among the superconductive elements [104].

16.5.2.2 Type II Superconductors

This superconductor totally excludes low applied magnetic fields, but only partially excludes high applied magnetic fields; their diamagnetism is not perfect but mixed in the presence of high fields. Niobium is an example of an elemental Type II superconductor. Type 2 superconductors are not particularly good conductors at room temperature, the transition to a superconductor state is more gradual than Type 1 superconductors. The mechanism and physical basis for this change in state are not, at present, fully understood. Type 2 superconductors are typically metallic compounds and alloys [105]. They reach a superconductive state at much higher temperatures when compared to type I superconductors. The cause of this dramatic increase in temperature is not fully understood. The highest T_c reached at standard pressure, to date, is 135 °K or −138 °C by a compound ($HgBa_2Ca_2Cu_3O_8$) that falls into a group of superconductors known as cuprate perovskites. This group of superconductors generally has a ratio of two copper atoms to three oxygen atoms and is considered to be a ceramic. Type II superconductors can also be penetrated by a magnetic field whereas type I cannot [106].

Except for the elements vanadium, technetium, and niobium, the Type 2 category of superconductors is comprised of metallic compounds and alloys. The recently discovered superconducting "perovskites" (metal-oxide ceramics that normally have a ratio of two metal atoms to every three oxygen atoms) belong to this Type 2 group. They achieve higher T_c's than Type 1 superconductors by a mechanism that is still not completely understood. Conventional wisdom holds that it relates to the planar layering within the crystalline structure. Although, other recent research suggests the holes of hypocharged oxygen in the charge reservoirs are responsible. (Holes are positively charged vacancies within the lattice.) The superconducting cuprates (copper oxides) have achieved astonishingly high Tc's when you consider that by 1985 known Tc's had only reached 23 K. To date, the highest Tc attained at ambient pressure for a material that will form stoichiometrically (by direct mixing) has been 147 K [107]. And the highest Tc overall is 216 °C for a material which does not form stoichiometrically (Table 16.4). It is almost certain that other, more synergistic compounds still await discovery among the high-temperature superconductors. The first superconducting Type 2 compound, an alloy of Pb and Bi, was fabricated in 1930 by W. de Haas and J. Voogd. But, was not recognized as such until later, after the Meissner effect had been discovered. This new category of superconductors was identified by L.V. Shubnikov at the Kharkov Institute of Science and Technology in Ukraine in 1936 when he found two distinct critical magnetic fields (known as H_{c1} and H_{c2}) in $PbTl_2$. The first of the oxide superconductors was created in 1973 by

Table 16.4 Type-II Superconducting compound with their superconducting critical temperature

Compound	Tc	Compound	Tc
$Sn_{12}SbTe_{11}Ba_2V_2Mg_{24}O_{50+}$	+216 °C	$ZnMgO_2$	115 K
$Sn_{11}SbTe_{10}Ba_2V_2Mg_{22}O_{46+}$	+209 °C	$CuMgO_2$	114 K
$Sn_{11}SbTe_{10}Ba_2VMg_{23}O_{46+}$	+202 °C	$SrCaMg_2O_x$	111 K
$Sn_{10}SbTe_9Ba_2MnCu_{21}O_{42+}$	+187 °C	$SrCaCu_2O_4$	110 K
$Sn_9SbTe_8Ba_2MnCu_{19}O_{38+}$	+178 °C	$YSrCa_2Cu_4O_8$	101 K
$Sn_8SbTe_7Ba_2MnCu_{17}O_{34+}$	+167 °C	$(Ba,Sr)CuO_2$	90 K
$Sn_7Te_7Ba_2MnCu_{15}O_{30+}$	+158 °C	$BaSr_2CaCu_4O_8$	90 K
$Sn_7SbTe_6Ba_2MnCu_{15}O_{30+}$	+155 °C	$GaBa_2O_x$	90 K
$Sn_{10}SbTe_4Ba_2MnCu_{16}O_{32+}$	+141 °C	BMg_2O_x	85 K
$Sn_9SbTe_4Ba_2MnCu_{15}O_{30+}$	+136 °C	$ScBa_2O_x$	84 K
$Sn_8SbTe_4Ba_2MnCu_{14}O_{28+}$	+129 °C	YBa_2O_x	83 K
$Sn_9SbTe_3Ba_2MnCu_{14}O_{28+}$	+121 °C	$TlMg_2O_x$	81 K
$Sn_9Te_3Ba_2MnCu_{13}O_{26+}$	+119 °C	$TiBa_2O_x$	77 K
$Sn_6Sb_6Ba_2MnCu_{13}O_{26+}$	+110 °C	$(La,Sr)CuO_2$	42 K
$Sn_5Te_5Ba_2VMg_{11}O_{22+}$	+109 °C	Pb_3MgO_5	307 K
$Sn_5Sb_5Ba_2MnCu_{11}O_{22+}$	+95 °C	$Pb_3Sr_4Ca_3Cu_6O_x$	106 K
$Sn_4Te_4Ba_2MnMg_9O_{18+}$	+87 °C	$Pb_3Sr_4Ca_2Cu_5O_{15}$	101 K
$Tl_7Sn_2Ba_2MnCu_{10}O_{20+}$	+77 °C	$(Pb_{1.5}Sn_{1.5})Sr_4Ca_2Cu_5O_{15}$	~95 K
$Tl_7Sn_2Ba_2TiCu_{10}O_{20+}$	+65 °C	$Pb_2Sr_2(Ca, Y)Cu_3O_8$	70 K
$Tl_6Sn_2Ba_2TiCu_9O_{18+}$	+56 °C	$AuBa_2Ca_3Cu_4O_{11}$	99 K
$Tl_7Sn_2Ba_2SiCu_{10}O_{20+}$	+53 °C	$AuBa_2(Y, Ca)Cu_2O_7$	82 K
$Tl_6Ba_4SiCu_9O_{18+}$	+48 °C	$AuBa_2Ca_2Cu_3O_9$	30 K
$Tl_5Ba_4SiCu_8O_{16+}$	+44 °C	$CdNbBa_9Cu_{10}O_{20+}$	337 K
$(Tl_5Sn_2)Ba_2SiCu_8O_{16+}$	+42 °C	$TiBa_9Cu_{10}O_{20+}$	327 K
$(Tl_5Pb_2)Ba_2SiCu_8O_{16+}$	+38 °C	$VBa_9Cu_{10}O_{20+}$	314 K
$(Tl_5Pb_2)Ba_2Si_{2.5}Cu_{8.5}O_{17+}$	+35 °C	$ZrBa_9Cu_{10}O_{20+}$	313 K
$(Tl_5Pb_2)Ba_2Mg_{2.5}Cu_{8.5}O_{17+}$	+30 °C	$NbBa_9Cu_{10}O_{20+}$	307 K
$(Tl_5Pb_2)Ba_2Mg_2Cu_9O_{18+}$	+28 °C	$TeBa_{10}Cu_{11}O_{22+}$	255 K
$(Tl_5Pb_2)Ba_2MgCu_{10}O_{20+}$	+18 °C	$TaBa_9Cu_{10}O_{20+}$	177 K
$(Tl_4Pb)Ba_2MgCu_8O_{13+}$	+3 °C	$Y_2Ba_{10}Cu_{12}O_{25+}$	106 K
$(Tl_4Ba)Ba_2MgCu_8O_{13+}$	265 K	$TiBa_7Cu_8O_x$	104 K
$(Tl_4Ba)Ba_2Mg_2Cu_7O_{13+}$	258 K	$TaBa_7Cu_8O_x$	104 K
$(Tl_4Ba)Ba_2Ca_2Cu_7O_{13+}$	254 K	$TeBa_7Cu_8O_x$	99 K
$(Tl_4Ba)Ba_4Ca_2Cu_{10}O_y$	242 K	$YBa_3Cu_4O_x$	97 K
$Tl_5Ba_4Ca_2Cu_{10}O_y$	233 K	$YCaBa_3Cu_5O_{11+}$	96 K
$(Sn_5In)Ba_4Ca_2Cu_{11}O_y$	218 K	$(Y_{0.5}Lu_{0.5})Ba_2Cu_3O_7$	96 K
$(Sn_5In)Ba_4Ca_2Cu_{10}O_y$	212 K	$(Y_{0.5}Tm_{0.5})Ba_2Cu_3O_7$	96 K
$Sn_6Ba_4Ca_2Cu_{10}O_y$	200 K	$Y_3Ba_5Cu_8O_{18+}$	95 K
$(Sn_{1.0}Pb_{0.5}In_{0.5})Ba_4Tm_6Cu_8O_{22+}$	195 K	$Y_2Ba_5Cu_8O_{17+}$	94 K
$(Sn_{1.0}Pb_{0.5}In_{0.5})Ba_4Tm_5Cu_7O_{20+}$	185 K	$TeCaBa_4Cu_6O_{16+}$	92 K
$(Sn_{1.0}Pb_{0.5}In_{0.5})Ba_4Tm_4Cu_6O_{18+}$	163 K	$Y_3CaBa_4Cu_8O_{18+}$	90 K
$Sn_3Ba_4Ca_2Cu_7O_y$	160 K	$(Y_{0.5}Gd_{0.5})Ba_2Cu_3O_7$	89 K
$(Hg_{0.8}Tl_{0.2})Ba_2Ca_2Cu_3O_{8.33}^{*}$	138 K	$Y_2CaBa_4Cu_7O_{16}$	62 K
$HgBa_2Ca_2Cu_3O_8$	133–135 K	$GaSr_2(Ca_{0.5}Tm_{0.5})Cu_2O_7$	99 K
$HgBa_2Ca_3Cu_4O_{10+}$	125–126 K	$Ga_2Sr_4Y_2CaCu_5O_x$	85 K
$HgBa_2(Ca_{1-x}Sr_x)Cu_2O_{6+}$	123–125 K	$Ga_2Sr_4Tm_2CaCu_5O_x$	81 K
$HgBa_2CuO_4+$	94–98 K	$La_2Ba_2CaCu_5O_{9+}$	79 K
$Tl_2Ba_2TeCu_3O_8$	147 K	$(Sr,Ca)_5Cu_4O_{10}$	70 K
$Tl_2Ba_2YCu_2O_6$	139 K	$GaSr_2(Ca, Y)Cu_2O_7$	70 K
$Tl_2Ba_2Ca_2Cu_3O_{10}$	127–128 K	$(In_{0.3}Pb_{0.7})Sr_2(Ca_{0.8}Y_{0.2})Cu_2O_x$	60 K
$(Tl_{1.6}Hg_{0.4})Ba_2Ca_2Cu_3O_{10+}$	126 K	$(La,Sr,Ca)_3Cu_2O_6$	58 K
$TlBa_2Ca_2Cu_3O_{9+}$	123 K	$La_2CaCu_2O_{6+}$	45 K
$(TlSn)Ba_4TmCaCu_4O_{14+}$	121 K	$(Eu,Ce)_2(Ba,Eu)_2Cu_3O_{10+}$	43 K

(continued)

Table 16.4 (continued)

Compound	Tc	Compound	Tc
$(Tl_{0.5}Pb_{0.5})Sr_2Ca_2Cu_3O_9$	118–120 K	$(La_{1.85}Sr_{0.15})CuO_4$	40 K
$Tl_2Ba_2CaCu_2O_6$	118 K	SrNdCuO	40 K
$TlBa_2Ca_3Cu_4O_{11}$	112 K	$(La,Ba)_2CuO_4$	35–38 K
$(Tl_{0.5}Pb_{0.5}Sn)Ba_4Tm_3Cu_5O_{16+}$	105 K	$(Nd,Sr,Ce)_2CuO_4$	35 K
$TlBa_2CaCu_2O_{7+}$	103 K	$Pb_2(Sr,La)_2Cu_2O_6$	32 K
$Tl_2Ba_2CuO_6$	95 K	$(La_{1.85}Ba_{0.15})CuO_4$	30 K
$TlSnBa_4Y_2Cu_4O_x$	86 K	GdFeAsO1-x	53.5 K
$Sn_4Ba_4(Tm_2Ca)Cu_7O_x$	~127 K	$(Ca,Sr,Ba)Fe_2As_2$	38 K
$Sn_4Ba_4TmCaCu_6O_{16+}$	~115 K	LiFeAs	18 K
$SnInBa_4Tm_3Cu_5O_x$	~113 K	B-doped Q-carbon	57 K
$Sn_3Ba_4Tm_3Cu_6O_x$	109 K	MgB_2	39 K
$SnBa_4Y_2Cu_5O_x$	107 K	$Ba_{0.6}K_{0.4}BiO_3$	30 K
$Sn_4Ba_4Tm_2YCu_7O_x$	~104 K	Nb_3Ge	23.2 K
$Sn_4Ba_4TmCaCu_4O_x$	~100 K	Nb_3Si	19 K
$Sn_4Ba_4Tm_3Cu_7O_x$	~98 K	Nb_3Sn	18.1 K
$Sn_2Ba_2(Y_{0.5}Tm_{0.5})Cu_3O_{8+}$	~96 K	Nb_3Al	18 K
$Sn_3Ba_4Y_2Cu_5O_x$	~91 K	V_3Si	17.1 K
$SnInBa_4Tm_4Cu_6O_x$	87 K	Ta_3Pb	17 K
$Sn_2Ba_2(Sr_{0.5}Y_{0.5})Cu_3O_8$	86 K	V_3Ga	16.8 K
$Sn_4Ba_4Y_3Cu_7O_x$	~80 K	Nb_3Ga	14.5 K
$Bi_2Sr_2TeCu_3O_8$	139 K	V_3In	13.9 K
$Bi_{1.6}Pb_{0.6}Sr_2Ca_2Sb_{0.1}Cu_3O_x$	115 K	$PuCoGa_5$	18.5 K
$Bi_2Sr_2Ca_2Cu_3O_{10}$	110 K	NbN	16.1 K
$Bi_2Sr_2CaCu_2O_9$	110 K	$Nb_{0.6}Ti_{0.4}$	9.8 K
$Bi_2Sr_2(Ca_{0.8}Y_{0.2})Cu_2O_8$	95–96 K	$MgCNi_3$	7–8 K
$Bi_2Sr_2CaCu_2O_8$	91–92 K	C	15 K
$BiSnBa_4TmCaCu_4O_{14}$	83 K	Nb	9.25 K
Cd_5MgO_6	310 K	Tc	7.80 K
$Ba_{12}ZnO_{13}$	306 K	V	5.40 K
$Cd_3CaCu_4O_8$	187 K	$RuSr_2(Gd,Eu,Sm)Cu_2O_8$	Tc ~58 K
$Cd_2CaCu_3O_6$	153 K	$EuFe_2(As_{0.79}P_{0.21})2$	Tc 24 K
Zn_3MgO_4	152 K	$ErNi_2B_2C$	Tc 10.5 K
Cu_3MgO_4	147 K	$YbPd_2Sn$	Tc ~2.5 K
Sr_3CaO_4	139 K	UGe_2	Tc ~1 K
Zn_2MgO_3	132 K	$URhGe_2$	Tc ~1 K
Cu_2MgO_3	130 K	$AuIn_3$	Tc 50 μK
Sr_2CaO_3	129 K	$Sr_{0.08}WO_3$	2–4 K
$CdCaMg_2O_x$	124 K	$Tl_{0.30}WO_3$	2.0–2.14 K
$CdCaCu_2O_4$	123 K	$Rb_{0.27–0.29}WO_3$	1.98 K
$SrCaO_2$	117 K	$SrTiO_3$	0.35 K

DuPont researcher Art Sleight when $Ba(Pb, Bi)O_3$ was found to have a Tc of 13 K. The superconducting oxocuprates followed in 1986 [108].

Type 2 superconductors also known as the "hard" superconductors differ from Type 1 in that their transition from a normal to a superconducting state is gradual across a region of "mixed state" behavior. Since a Type 2 will allow some penetration by an external magnetic field into its surface, this creates some rather novel mesoscopic phenomena like superconducting "stripes" and "flux-lattice vortices."

16.5.3 Applications of Superconductors

The supercurrents that flow through the superconductors generate an intense magnetic field, through electromagnetic induction, that can be used to accelerate and direct the team as desired. In addition, superconductors exhibit the Meissner effect in which they cancel all magnetic flux inside the material, becoming perfectly diamagnetic. In this case, the magnetic field lines actually travel around the cooled superconductor. It is this property of superconductors which is

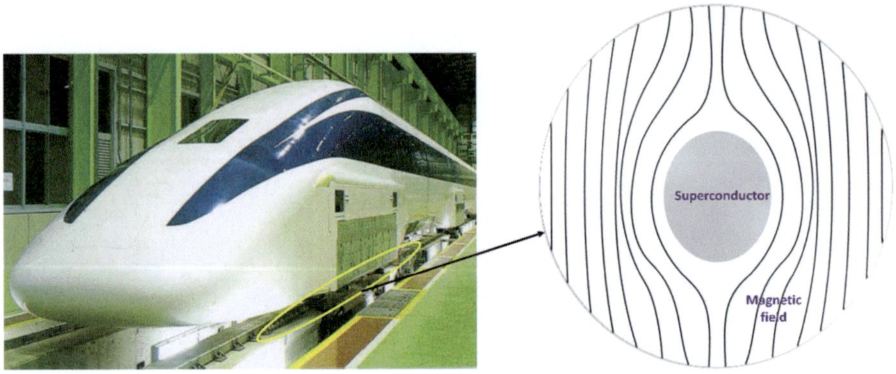

Fig. 16.12 Quantum locking in quantum levitation

frequently used in magnetic levitation experiments, such as the quantum locking seen in quantum levitation. Superconductors play a role in modern advancements in magnetic levitation trains, which provide a powerful possibility for high-speed public transport that is based on electricity (which can be generated using renewable energy) in contrast to nonrenewable current options like airplanes, cars, and coal-powered trains. A landmark for the commercial use of MAGLEV technology occurred in 1990 when it gained the status of a nationally funded project in Japan. The Minister of Transport authorized construction of the Yamanashi Maglev Test Line which opened on April 3, 1997 [109]. In April 2015, the MLX01 test vehicle (Fig. 16.12) attained an incredible speed of 374 mph (603 kph). Although the technology has now been proven, the wider use of MAGLEV vehicles has been constrained by political and environmental concerns (strong magnetic fields can create a biohazard). The world's first MAGLEV train to be adopted into commercial service, a shuttle in Birmingham, England, shut down in 1997 after operating for 11 years. A Sino-German maglev is currently operating over a 30 km course at Pudong International Airport in Shanghai, China [110].

An area where superconductors can perform a life-saving function is in the field of biomagnetism. Doctors need a noninvasive means of determining what's going on inside the human body. By impinging a strong superconductor-derived magnetic field into the body, hydrogen atoms that exist in the body's water and fat molecules are forced to accept energy from the magnetic field. They then release this energy at a frequency that can be detected and displayed graphically by a computer. Magnetic Resonance Imaging (MRI) was actually discovered in the mid of 1940s. But, the first MRI exam on a human being was not performed until July 3, 1977. And, it took almost 5 h to produce one image, today's faster computers process the data in much less time [111]. The Korean Superconductivity Group within KRISS has carried biomagnetic technology a step further with the development of a double-relaxation oscillation SQUID (Superconducting Quantum Interference Device) for use in Magnetoencephalography. SQUIDs are capable of sensing a change in a magnetic field over a billion times weaker than the force that moves the needle on a compass (compass: 5e-5T, SQUID: e-14T.). With this technology, the body can be probed to certain depths without the need for the strong magnetic fields associated with MRIs [112].

Electric generators made with superconducting wire are far more efficient than conventional generators wound with copper wire. In fact, their efficiency is above 99% and their size is about half that of conventional generators. These facts make them very lucrative ventures for power utilities. General Electric has estimated the potential worldwide market for superconducting generators in the next decade at around \$20–30 billion dollars. Late in 2002, GE Power Systems received \$12.3 million in funding from the U.S. Department of Energy to move high-temperature superconducting generator technology toward full commercialization [113]. Other commercial power projects in the works that employ superconductor technology include energy storage to enhance power stability. American Superconductor Corp. received an order from Alliant Energy in late March 2000 to install a Distributed Superconducting Magnetic Energy Storage System (D-SMES) in Wisconsin. Just one of these 6 D-SMES units has a power reserve of over 3 million watts, which can be retrieved whenever there is a need to stabilize line voltage during a disturbance in the power grid. AMSC has also installed more than 22 of its D-VAR systems to provide instantaneous reactive power support [114]. Recently, power utilities have also begun to use superconductor-based transformers and "fault limiters." The Swiss-Swedish company ABB was the first to connect a superconducting transformer to a utility power network in March of 1997. ABB also recently announced the development of a 6.4 MVA (mega-volt-ampere) fault current limiter—the most powerful in the world. This new generation of HTS superconducting fault limiters is being called upon due to their ability to respond in just thousandths of a second to limit tens of thousands of amperes of current. Advanced Ceramics Limited is another of several companies that make BSCCO type fault limiters. Intermagnetics General recently completed

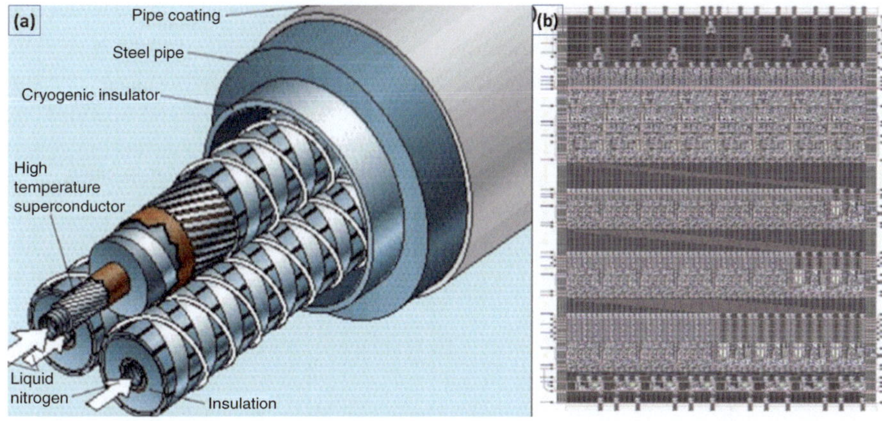

Fig. 16.13 (**a**) BSCCO cable-in-conduit cooled with liquid nitrogen, (**b**) Hypres Superconducting Microchip, Incorporating 6000 Josephson Junctions.

tests on its largest (15 kV class) power-utility-size fault limiter at a Southern California Edison (SCE) substation near Norwalk, California. And, both the US and Japan have plans to replace underground copper power cables with superconducting BSCCO cable-in-conduit cooled with liquid nitrogen (Fig. 16.13a). By doing this, more current can be routed through existing cable tunnels. In one instance 250 pounds of superconducting wire replaced 18,000 pounds of vintage copper wire, making it over 7000% more space efficient [115].

An idealized application for superconductors is to employ them in the transmission of commercial power to cities. However, due to the high cost and impracticality of cooling miles of superconducting wire to cryogenic temperatures, this has only happened with short "test runs". In May of 2001 some 150,000 residents of Copenhagen, Denmark, began receiving their electricity through HTS (high-temperature superconducting) material [116]. That cable was only 30 meters long but proved adequate for testing purposes. In the summer of 2001, Pirelli completed the installation of three 400-foot HTS cables for Detroit Edison at the Frisbie Substation capable of delivering 100 million watts of power. This marked the first time commercial power has been delivered to customers of a US power utility through the superconducting wire. Intermagnetics General has announced that its IGC-SuperPower subsidiary has joined with BOC and Sumitomo Electric in a $26 million project to install an underground, HTS power cable in Albany, New York, in Niagara Mohawk Power Corporation's power grid. Sumitomo Electric's DI-BSCCO cable was employed in the first in-grid power cable demonstration project sponsored by the U.S. Department of Energy and New York Energy Research & Development Authority. After connecting to the grid successfully in July 2006, the DI-BSCCO cable has been supplying power to approximately 70,000 households without any problems. Currently, the longest run of superconductive power cable was made in the AmpaCity project near Essen, Germany, in May 2014. That cable was a kilometer in length.

The National Science Foundation, along with NASA and DARPA, and various universities, is currently researching "petaflop" computers. A petaflop is a thousand-trillion floating-point operation per second. Currently, the fastest in the world is the Summit (OLCF-4) Supercomputer, capable of 200 petaflops per second. It has been conjectured that devices on the order of 50 nm in size along with unconventional switching mechanisms, such as the Josephson junctions associated with superconductors, will be necessary to achieve the next level of processing speeds. These Josephson junctions are incorporated into field-effect transistors which then become part of the logic circuits within the processors. Recently it was demonstrated at the Weizmann Institute in Israel that the tiny magnetic fields that penetrate Type 2 superconductors can be used for storing and retrieving digital information. It is, however, not a foregone conclusion that computers of the future will be built around superconducting devices. Competing technologies, such as quantum (DELTT) transistors, high-density molecule-scale processors, and DNA-based processing also have the potential to achieve petaflop benchmarks [117].

In the electronics industry, ultrahigh-performance filters are now being built. Since superconducting wire has near-zero resistance, even at high frequencies, many more filter stages can be employed to achieve the desired frequency response. This translates into an ability to pass desired frequencies and block undesirable frequencies in high-congestion rf (radio frequency) applications such as cellular telephone systems. ISCO International and Superconductor Technologies are companies currently offering such filters.

Superconductors have also found widespread applications in the military. HTSC SQUIDS are being used by the U.S. NAVY to detect mines and submarines. And, significantly smaller motors are being built for NAVY ships using superconducting wire and "tape." In mid-July, 2001, American Superconductor unveiled a 5000-horsepower motor made with superconducting wire (below). An even larger 36.5 MW HTS ship propulsion motor was delivered to the U.S. Navy in late 2006. The newest application for HTS wire is in the degaussing of naval vessels. American Superconductor has announced the development of a superconducting degaussing cable. Degaussing of a ship's hull eliminates residual magnetic fields which might otherwise give away a ship's presence. In addition to reduced power requirements, HTS degaussing cable offers reduced size and weight. Superconductors used in power generation are shown to operate with efficiencies that exceed 99%. Two systems were analyzed in detail and compared to existing conventional technologies to stress this point—a hybrid wing-body aircraft and a single-aisle turboelectric with fuselage boundary layer ingestion. The hybrid wing model was designed to be all electric with two superconducting turbogenerators at 22.4 MW power with 15 superconducting turbofans mounted in the nacelle for increased boundary layer ingestion; it showed a reduction of fuel burn by 70–72% with compromising payload, range, or cruising speed. The single-aisle turboelectric design calls for a 2.6 MW motor fed from two 1.45 MW hydro-carbon-burning generators [118]. The military is also looking at using superconductive tape as a means of reducing the length of very low-frequency antennas employed on submarines. Normally, the lower the frequency, the longer an antenna must be. However, inserting a coil of wire ahead of the antenna will make it function as if it were much longer. Unfortunately, this loading coil also increases system losses by adding the resistance in the coil's wire. Using superconductive materials can significantly reduce losses in this coil. The Electronic Materials and Devices Research Group at the University of Birmingham (UK) is credited with creating the first superconducting microwave antenna. Applications engineers suggest that superconducting carbon nanotubes might be an ideal nanoantenna for high-gigahertz and terahertz frequencies, once a method of achieving zero "on tube" contact resistance is perfected. The most ignominious military use of superconductors may come with the deployment of "E-bombs." These are devices that make use of strong, superconductor-derived magnetic fields to create a fast, high-intensity electromagnetic pulse (EMP) to disable an enemy's electronic equipment. Such a device saw its first use in wartime in March 2003 when U.S. Forces attacked an Iraqi broadcast facility [119].

Among emerging technologies are a stabilizing momentum wheel (gyroscope) for earth-orbiting satellites that employs the "flux-pinning" properties of imperfect superconductors to reduce friction to near zero. Superconducting X-ray detectors and ultrafast, superconducting light detectors are being developed due to their inherent ability to detect extremely weak amounts of energy. Already Scientists at the European Space Agency (ESA) have developed what's being called the S-Cam, an optical camera of phenomenal sensitivity (see above photo). And, superconductors may even play a role in Internet communications soon. In late February 2000, Irvine Sensors Corporation received a $1 million contract to research and develop a superconducting digital router for high-speed data communications upto 160 GHz. Since Internet traffic is increasing exponentially, superconductor technology may be called upon to meet this super need. Another impetus to the wider use of superconductors is political in nature. The reduction of greenhouse gas (GHG) emissions has become a topical issue due to the Kyoto Protocol which requires the European Union (EU) to reduce its emissions by 8%. Physicists in Finland have calculated that the EU could reduce carbon dioxide emissions by upto 53 million tons if high-temperature superconductors were used in power plants [120].

16.6 Advanced Semiconductor Materials

Besides above, electronic and semiconductor materials are also used in the fields of aeronautics, marine, defense, chemical, metallurgical, and automobile engineering applications. These are the solids having an energy gap (E_g) lying in between the conductor and insulator. Their conductivity is more than that of the dielectrics but less than that of conductors. They are basically electronic materials. Semiconductors are available in the forms: (1) elements such as silicon (Si) and germanium (Ge); (2) compounds such as GaAs, InP, AlSb, CdTe, ZnSe, etc.; and (3) Alloys such as GaAsx P1-x, HgCdxTe1-x, etc.

16.6.1 Classification of Semiconductor Materials

Semiconductors are also classified as Intrinsic and Extrinsic semiconductors based on energy Gap (E_g or Band) (Fig. 16.14). The elemental forms of pure Si and pure Ge are intrinsic. In intrinsic form, they are not useful. They are, therefore, doped by dopants to make extrinsic semiconductors. Extrinsic forms are directly useful and are widely employed in manufacturing of the solid-state devices. They belong to the category of alloys and compounds. The electronics industry requires purity better than $1{:}10^9$ in pure Si and Ge. Extrinsic semiconductors can be classified into n-type and p-type on the basis of their dopant. They may be in a compound form such as GaAs, CdTe, etc, or in alloy form such as $HgCd_xTe_{1-x}$. A number of semiconducting compounds are available in oxide, halide, and sulfide forms also [121]. The n- and p-types are used to make n–p junction for diode, n–p–n and p–n–p for the transistor, etc. Semiconductors are generally hard and brittle, and possess a negative coefficient of temperature resistance. In contrast with conductors whose conductivity increases with purification, the conductivity of semiconductors decreases with

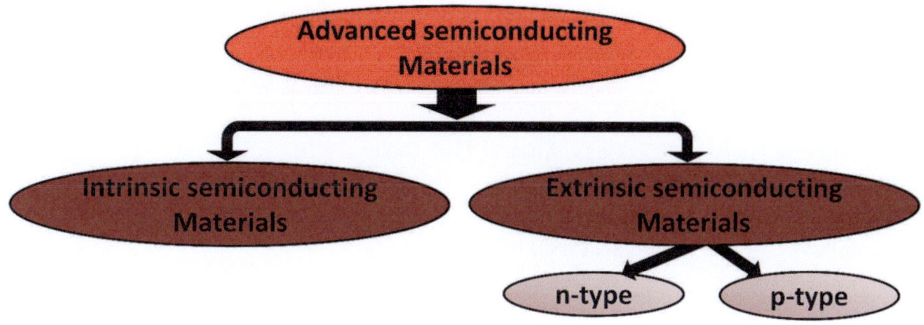

Fig. 16.14 Classification of semiconductors

Table 16.5 Classification of semiconductor materials from periodic table

Type of semiconductor		Example
Elemental form	Column II	Zn, Cd
	Column III	B, Al, Ga, In
	Column IV	C, Si, Ge
	Column V	N, P, As, Sb
	Column VI	S, Se, Te
Binary compound	II–VI compound	ZnS, ZnSe, SnTe, CdS, CdSe, CdTe
	III–V compound	AIP, AlAs, AlSb, GaN, GaP, GaAs, GaSb, InP, InAs, InSb
	IV–IV compound	SiC, SiGe
Ternary Compound		GaAsP, HgCdTe, AlGaAs
Quaternary compound		InGaAsP
Ternary Alloys		$Al_{0.3}Ga_{0.7}As$, $In_{0.11}Ga_{0.8}As$, $GaAs_{0.88}Sb_{0.12}$
Quaternary alloys		$In_xGa_{1-x}As_yP_{1-y}$

purification. Semiconductors are widely used as rectifiers, amplifiers, photocells, etc. and their properties are of greater importance in telecommunication, power electronics, computer hardware, etc.

Again, a detailed classification of them is given below in Table 16.5. In elemental form they belong to various columns of the periodic table; in the compound form, they are binary, ternary, and quaternary; and in alloy form, they may be homogeneous or heterogeneous types.

16.6.2 Semiconducting Devices

Semiconductors are solid materials of crystalline nature. They possess several properties that are influenced by different factors such as temperature, voltage, impurity, electric field, light illumination, optical behavior, etc. Depending upon the influencing factor, the semiconductors find use in the following important devices.

- **Thermistors:** These utilize the temperature dependency effect of semiconductors.
- **Varistors:** These utilize the voltage dependency effect of semiconductors.
- **Rectifiers:** These utilize the impurity dependency effect of semiconductors.
- **Strain gauges:** These utilize the change in resistance effect of semiconductors.
- **Zener diodes:** These utilize the electric field effect of semiconductors.
- **Transistors:** These utilize the amplification effects of semiconductors.
- **Photoconductive cells:** These utilize the light illumination effect of semiconductors.
- **Photovoltaic cells:** These utilize the optical characteristics of semiconductors.
- **Hall effect generators:** These utilize the carrier drift effect in semiconductors.

Various semiconducting materials mentioned above possess varying properties, and hence are suitable for application in vivid devices. Main among these materials and their suitability for different devices are given in Table 16.6.

16.6.3 Alloy of II–VI Semiconductors with Magnetic Materials

Diluted magnetic semiconductors (DMS) are compounds of alloy semiconductors containing a large fraction of magnetic ions (Mn^{+2}, Cr^{+2}, Fe^{+2}, Co^{+2}) and are studied mainly on II–VI-based materials such as CdTe and ZnSe etc. This is

Table 16.6 Various semiconducting elements in industrial application

Devices	Suitable semiconductors
Solar Cell/batteries	Photovoltaic action materials such as Se
Diodes	Si, Ge, GaAs
Semiconductor lasers	GaAs, AlGaAs, GaP, GaSb
CPU, microprocessor chips	AlGaAs
Transistors	Si, Ge
Photocells	Se, CdS, PbS
Rectifiers	Si, Se, CuO
IC chips	Si, GaAs, Si
Sensor elements for guided missiles	HgCdTe
Light detectors	InSb, CdSe, PbTe, HgCdTe
High-frequency devices	Ge
Fluorescent screens	Zn, ZnS, Cd, Be
Infrared detector	Si, Ge
Nuclear radiation detectors	Si, Ge
Gunn diode in microwave	GaAs, InP
LED	For visible green light: GaP, CdS; For visible red light: GaAsP, CdSe; For visible yellow light: GaP; For visible blue light: SiC; For invisible infrared light: GAaS, InSb; For ultraviolet region of light: ZnS
Photon detector	InP, InAs, InSb
Photoconductors	CdS (in green light), CdSe (in red light), CdTi (in infrared region)
Cinematography	Photocell effect based materials such as CdS, $PbSO_4$
Xerox-type photocopier	Se
Hyper high-speed computer chips	AlGaAs
Avalanche photodiode	InAlAs, InGaAs, GaAsSb
Automatic door opener	Photoconductivity based materials such as CdSe, CdS, CdTe
Stroboscope disk	Optoelectronic polymer
Red phosphor for TV tube	Y
Diamond transistors	P-doped diamond film n-type semiconductor

because such +2 magnetic ions are easily incorporated into the host II–VI crystals by replacing group II cations. In such II–VI-based DMS such as (CdMn)Se, magneto-optic properties were extensively studied, and optical isolators were recently fabricated. Although this phenomenon makes these DMS relatively easy to prepare in bulk form as well as thin epitaxial layers, II–VI-based DMS is difficult to dope to create p- and n-type, which makes the material less attractive for applications. The magnetic interaction in II–VI DMSs is dominated by the antiferromagnetic exchange among the Mn spins, which results in the paramagnetic, antiferromagnetic, or spins glass behavior of the material. Recently, the II–VI compound semiconductors ZnO, GaN have attracted revival attention since it was found that high-quality epitaxial thin-film display excitonic ultraviolet laser action at room temperature. In addition, the energy gap of this compound can be extended up to *4 eV by synthesizing alloy compounds of $Mg_xZn_{1-x}O$. Heavy electron doping (>1021/cm^3) was readily achieved in contrast to the other II–VI compound semiconductors [122].

16.6.4 Alloys of III–V Semiconductors with Ferromagnetic Properties

An approach compatible with the semiconductors used in present-day electronics is to make nonmagnetic III–V semiconductors magnetic, and even ferromagnetic, by introducing a high concentration of magnetic ions. The III–V semiconductors such as GaAs are already in use in a wide variety of electronic equipment in the form of electronic and optoelectronic devices, including cellular phones (microwave transistors), compact disks (semiconductor lasers), and in many other applications. The major obstacle in making III–V semiconductors magnetic has been the low solubility of magnetic elements (such as Mn) in the compounds. Because the magnetic effects are roughly proportional to the concentration of the magnetic ions, one would not expect a major change in properties with limited solubility of magnetic impurities, of the order of 1018/cm^3 or less. Using molecular beam epitaxy (MBE), a thin-film growth technique in the vacuum that allows one to work far from equilibrium, made a breakthrough. When a high concentration of magnetic elements is introduced in excess of the solubility limit, the formation of the second phase occurs if conditions are near equilibrium [123].

16.6.5 Polymer Semiconductor Crystals

From a fundamental viewpoint, polymer semiconductor single crystals are critical for understanding the physics of polymer crystallization. They are also important tools for elucidating macromolecular interactions and solid-state packing in benchmark materials, i.e., poly(3-hexylthiophene) (P3HT), and in newly synthesized polymer semiconductors. Polymer single crystals are also important for studying intrinsic charge transport and determining the performance limitation of the material in question. These highly organized solids will 1 day helps to answer the agelong question regarding the mechanism of charge transport in conjugated polymers. From a technical point of view, polymer crystals may play an important role in consumer electronic applications such as inflexible displays based on high mobility transistors, solar cells, gadgets, and toys, or possibly in technical applications not yet realized [124]. Organic single crystals grown from small-molecule semiconductors have attracted much attention in recent years because of their utilization in both fundamental and applied science. One of the long-standing challenges in the field of polymer semiconductors is to figure out how long interpenetrating and entangled polymer chains self-assemble into single crystals from the solution phase or melt. The ability to produce these crystalline solids has fascinated scientists from a broad range of backgrounds including physicists, chemists, and engineers. Scientists are still on the hunt for determining the mechanism of crystallization in these information-rich materials [125].

16.6.6 Oxide Semiconductor

The first evidence of the semiconductor properties in a metal oxide was done on copper oxide, more precisely Cu_2O, cuprous oxide, in 1917 by Kennard et al. Solid-state devices based on Cu_2O semiconductors are known for more than 90 years even before the era of germanium and silicon devices. Rectifier diodes based on this semiconductor were used industrially as early as 1926 and most of the theory of semiconductors was developed using the data on Cu_2O devices. Oxides of copper are known to show p-type conductivity and are attracting renewed interest as promising semiconductor materials for a wide range of optoelectronic devices. There are two common forms of copper oxide: cuprous oxide or cuprite (Cu_2O) and cupric oxide or tenorite (CuO). Both the CuO (monoclinic) and Cu_2O (cubic) are p-type semiconductors with a bandgap of 1.9–2.1 and 2.1–2.6 eV, respectively, and in some experimental conditions, Cu_2O shows mobilities exceeding 100 $cm^2/V/s$. Besides that, Cu_2O was regarded as one of the most promising materials for application in solar cells due to its high absorption coefficient in the visible region, non-toxicity, abundant availability, and low-cost production. The potential for solar cell application has been recognized since 1920, nevertheless, at that time and until the beginning of space explorations, the energy production from the sun by photovoltaic effect was just a curiosity. The p-type character of Cu_2O is attributed to the presence of negatively charged copper vacancies (V_{Cu}), which introduce an acceptor level at about 0.3 eV above the valence band [126]

Tin oxides are known to be wide bandgap oxide semiconductors and are present in two well-known forms: tin monoxide (SnO) and tin dioxide (SnO_2). SnO_2 (and impurity-doped SnO_2) is a typical functional material with multiple applications including transparent conducting oxides, low emission windows coatings, and solid-state gas-sensing material. In contrast, the physical properties of SnO have not been well explored. SnO_2 has a tetragonal structure and the unit cell contains two tin and four oxygen atoms. Each tin atom is at the center of six oxygen atoms placed approximately at the corners of a regular octahedron, while each oxygen atom is surrounded by three tin atoms at the corners of an equilateral triangle. SnO has a specific electronic structure associated with the presence of divalent tin, Sn(II), in a layered crystal structure. SnO_2 is an intrinsic n-type semiconductor and the electrical conduction results from the existence of defects, which may act as donors or acceptors. These defects are generally due to oxygen vacancies or interstitial tin atoms and are responsible for making electrons available at the conduction band. In the past decades, due to its technological applications SnO has been used in a variety of applications like anode materials for lithium rechargeable batteries, coatings, catalysts for several acids, and precursors for the production of SnO_2 [127].

Transparent electronics is today one of the most advanced topics for a wide range of device applications. The key components are wide bandgap semiconductors, where oxides of different origins play an important role, not only as passive components but also as active components, similar to what is observed in conventional semiconductors like silicon. Transparent electronics has gained special attention during the last few years and is today established as one of the most promising technologies for leading the next generation of flat panel display due to its excellent electronic performance. Recent progress in n- and p-type oxide-based thin-film transistors (TFT) and n-type TFTs processed by physical vapor deposition methods is discussed: The p-type oxide TFTs, mainly centered on two of the most promising semiconductor candidates: copper oxide and tin oxide, and complementary metal-oxide semiconductor devices based on n- and p-type oxide TFT [128]. Oxide semiconductors, especially the amorphous ones, are a promising class of TFT materials that have made an impressive progress particularly in display applications challenging silicon not only in conventional applications but opening doors to completely new and disruptive areas like paper electronics. This new class of semiconductor materials, amorphous oxides semiconductors (AOS) exhibits a stimulating combination of high optical transparency, high electron mobility, and amorphous microstructure. Besides that, AOS do not have grain boundaries, thereby obviating the primary limitation of mobility in polycrystalline semiconductors, which is a huge advantage for process integration. Other advantages include low-temperature deposition routes and ultra-smooth surfaces for suppressing interface traps and scattering centers. A significant worldwide interest appeared, especially for "active matrix for organic light-emitting diodes (AMOLED)" technology. Samsung has released at the end of the year 2010 what they boast proudly as one of the world's finest and largest 3D Ultra Definition (UD) TV with 70 in. (175 cm) diagonal and a sports resolution of 240 Hz that will meet the demands of 3D capability. Samsung has claimed that the making of this device is the first of the kind on oxide semiconductor TFT technology that supports high pixel density, with a resolution of 3840 $\times$ 2160, which is equivalent to 8 MP. Although the first transparent display has been done using the conventional α-Si: H technology in 2005. The enormous success of n-type oxide semiconductors and their application to TFTs has motivated the interest in p-type oxide-based semiconductors also to be applied to TFT. Organic materials have been studied and the significant improvements in the semiconducting properties associated with the discovery of electroluminescence in organic diode structures make these materials as excellent candidates for low-cost electronic and optoelectronic structures. The n-type organic semiconductors have much lower mobilities (10^{-2}–10^{-1} $cm^2/V/s$) limiting their field of applications, such as in CMOS [129].

16.6.7 Semiconductor Materials for Magnetoelectronics at Room Temperature

Most of the semiconductor materials are diamagnetic by nature and therefore cannot take an active part in the operation of the magnetoelectronic devices. Semiconductors and magnetic materials are two very important materials in electronic industries. On combining both the properties and form of both these, new materials are formed in which the performance of devices can be improved. Those materials are known as "diluted magnetic semiconductors" (DMS). Thus the concept of "magnetoelectronics" has come up i.e. the electronic chips consisting of DMS materials. DMS are expected to play an important role in interdisciplinary materials science and future electronics because the charge and spin degrees of freedom accommodated into a single material exhibit interesting magnetic, magneto-optical, magneto-electronic, and other properties. In DMS a fraction of host atoms of a semiconductor is substituted by transition metal atoms such as V, Cr, Mn, Fe, Co, and Ni. In the class of II–VI-based diluted magnetic semiconductors, much focus has been paid to $Mn_xCa_{1-x}S$ as the material CaS can accommodate a large percentage of Mn atoms and possesses a bandgap suitable for optical applications. Most of these DMS exhibit very high electron and hole mobility and thus useful for high-speed electronic devices. The recent DMS materials reported are (CdMn)Te, (GaMn)As, (GaMn)Sb, ZnMn(or Co)O, TiMn(or Co)O, etc. They have been produced as thin films by MBE and other methods [130]. It is expected that magnetoelectronic chips will be used in quantum computers. An inherent advantage of magnetoelectronics over electronics is the fact that magnets tend to stay magnetized for long. Hence this arises interest in industries to replace the semiconductor-based components of the computer with magnetic ones, starting from RAM. The new magnetic RAM will retain data even when the computer is turned off. And most important advantage will be eliminating the time-consuming process of "booting up" information from hard drive to a processor like a TV set, all the information would be there. One challenge in realizing magnetic RAM involves addressing individual memory elements, flipping their spins up or down to yield the zero and ones of binary computer logic. DMS are semiconductors in which a fraction of the component ions are replaced by those of transition metals or rare earth. Most importantly, the state of magnetization changes the electronic properties and vice versa through the spin-exchange interaction between local magnetic moments and carriers. In order to be practical, magnetoelectronics will need to use semiconductors that maintain their magnetic properties at room temperature. This is a challenge because most magnetic semiconductors lose their magnetic properties at temperatures well below the room temperature, and would require expensive and impractical refrigeration in order to work in actual computers [131].

16.6.8 Spintronics and Spintronic Semiconductor Materials

Spintronics (i.e. spin electronics) is one of the fast-emerging fields related to spin-dependent phenomena which are manipulated to achieve a desired electronic outcome, such as quantum computing. The spintronics devices store information into spins of up and down orientations, which are then attached to the mobile electrons to carry them along a wire, to be read at a terminal. They are/maybe bipolar devices such as spin diodes, spin transistors, spin-polarized solar cells, magnetic diodes, etc. These devices are attractive for the purpose of magnetic sensor and memory storage applications. Other aspects of spintronics are in the use of NSOM (Near Field Scanning Optical Microscopy) to detect electrons in semiconductor quantum dots, in quantum computing, and in SET (Single Electron Transistor) as a single electron spin detector [132]. Modern researches on spintronics concentrate on the development of advanced materials. Magnetic metal multilayer, $Ga_{1-x}Mn_xAs$ type ferromagnetic semiconductor, GaN, etc, are prominent among them. However, the ferromagnetic semiconductors are the most favorable material system, which combines the characteristics of ferromagnetism and semiconductor [133]. Some major fields of current spintronic research are the following: (1) development of spin-based devices such as p–n junctions and amplifiers, (2) spin relaxation behavior in metals and semiconductors, (3) spin-polarized transportation through the semiconductor/semiconductor interfaces, (4) spin-based quantum computation, (5) quantum computer hardware.

Spin orientation of conduction electrons survives for nanoseconds (10^{-9} s), whereas the time elapsed in the decay of electron momentum is tens of femtoseconds (10^{-15} s). Consequently, the spintronic devices are especially useful for quantum computing. Here, the electron spin represents a bit of information, which is known as "qubit." The understanding of spin relaxation and efforts needed to enhance the spin memory of conduction electrons in metals and semiconductors are essential for the study of spintronic devices. On application of magnetic field in normal metals, there is a change in their resistance. This change is normally of the order of 1% for the magnetic field of 1 tesla (T). In some ferromagnetic metals, the direction of magnetization can be reversed when a magnetic field of the order of 0.0001 T is applied. This result in a phenomenon called magnetoresistance (MR), which is utilized in the commercial production of small magnetic "read heads." They can sense very small magnetic fields in written information on hard discs. This causes a considerable decrease in the space required for storing the bits of information. Thus the storage capacity of hard discs increases. The utility of semiconductor spintronics devices depends on the availability of suitable materials which retain ferromagnetic properties above room temperature. Some recent and important materials in this list are: half-metallic materials such as Fe_3O_4, CrO_2, etc.; colossal magneto-resistive materials such as Sr-doped $LaMnO_3$, double perovskites; Heusler alloys having a composition of X_2YZ, where X and Y are transition elements and Z is an element of group III, IV or V; half-Heuslar alloys having a composition of XYZ; Ferromagnetic semiconductors [134, 135].

16.6.9 Application of Advanced Semiconducting Materials

Pure silicon is the basis for most integrated circuits. It provides the base (or substrate) for the entire chip and is chemically doped to provide the "n" and "p" regions that make up the integrated circuit components. The silicon of about 1 ppb impurity is used. Typical n-type dopants include P and Ar, whereas p-type dopant includes B and Ga. Aluminum is commonly used as a connector between the various IC components. The thin wire leads from the integrated circuit chip to its mounting package maybe Al or Au. The mounting package itself may be made from ceramic or plastic materials [136, 137]. Basically, the success of Si is the success of MOSFET, which with scaling and extreme integration has driven the industry.

One application of semiconductor is LED Stumps in Cricket. Powered by hidden low-voltage batteries, once the wicket is hit, the bails instantaneously bring flash red LED lights and send a radio signal to the stumps (Fig. 16.15) which also light up. The particular semiconductors used for LED manufacture are GaAs, GaP, or GaAsP. The impurities commonly added are Zn or nitrogen; but Si, Ge, and Te are also used. Au and Ag compounds most commonly used for contact purpose, because they form a chemical bond with the Ga at the surface of the wafer [138].

In the disk drive industry, both revolutionary and evolutionary changes are ongoing for recording head scaling to provide density and low cost. The scaling trend in this industry is leading to a continued increase in the electrostatic discharge (ESD) sensitivity of the recording elements. With the introduction of the giant magneto-resistor (GMR)

Fig. 16.15 Semiconductor stump detecting the point of impact

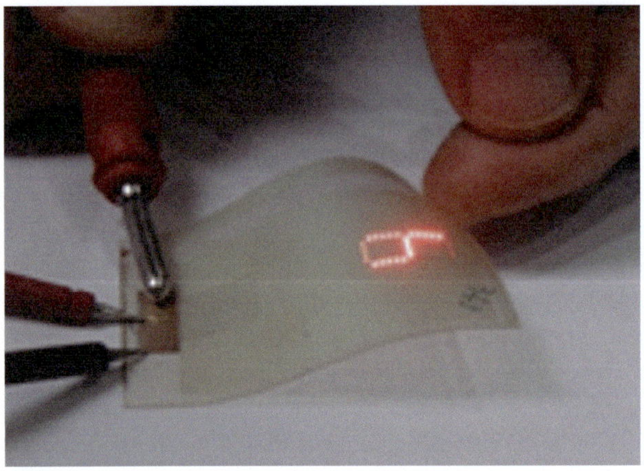

Fig. 16.16 Transparent passive matrix chip LED display using indium zinc oxide electrode

recording head, the ESD sensitivity continued to increase. The tunneling magneto-resistor (TMR) will be used in future disk drives as a read element [139]. Semiconductors are used in displays with oxide-based backplanes. When transparency of the overall display is not persuaded, oxide TFTs are of great interest. Their high mobility and large area uniformity can yield high-end displays, such as the 70″ ultra-definition (UD) LCD 3DTV with a high scanning frequency (240 Hz) demonstrated by Samsung in 2010. In most of the reported displays employing oxide TFTs, the channel layer is produced by sputtering. But despite the early stage of research on solution-processed oxide TFTs, LCD, and OLED displays were already fabricated using this lower cost processing technology [140, 141]. One example of advanced semiconductor application is in transparent passive matrix chip LED display as shown in Fig. 16.16.

Semiconducting materials such as ZnO, Cu_2O, Iron oxides, Sulfides, Ternary oxides, Semiconductor chalcogenide, Bismuth oxyhalides are used as photocatalysis in various green earth and renewable energy projects to prevent environmental pollution. Detection of pollutant, toxic, refining, combustible, and process gases is important for system and process control, safety monitoring, and environmental protection. Various methods can be used to accomplish gas sensing including gas chromatography, Fourier-transform infrared spectroscopy, chemiluminescence detectors, mass spectrometry, semiconductor gas sensors, and others. Gas sensors based on solid-state semiconductor materials offer considerable advantages in comparison to other gas-sensing methods. Semiconductor sensors are inexpensive to produce, easy to miniaturize, rugged, reliable, and can be designed to operate over a range of conditions including high temperatures. Semiconductor sensors can be produced in arrays to allow sensing of multiple species simultaneously and with advances insensitivity; detectivity limits are approaching ppm levels for some species. Tin dioxide semiconductor gas sensors were first proposed and patented in 1962. Tin dioxide (SnO_2) is the most important material for use in gas sensing applications. It is the dominant choice for solid-state gas detectors in domestic, commercial, and industrial applications [142].

In recent years, semiconductor nanostructures have become the choice for investigations of electrical conduction on short-length scales. This development was made possible by the availability of semiconducting materials of unprecedented purity and crystalline perfection. Such materials can be structured to contain a thin layer of highly mobile electrons. Motion perpendicular to the layer is quantized so that the electrons are constrained to move in a plane. As a model system, this two-dimensional electron gas (2DEG) combines a number of desirable properties, not shared by thin-metal films. It has a low electron density, which may be readily varied by means of an electric field (because of the large screening length). The low density implies a large Fermi wavelength (typically 40 nm), comparable to the dimensions of the smallest structures (nanostructures) that can be fabricated. The electron's mean free path can be quite large (exceeding 10 μm). Finally, the reduced dimensionality of the motion and the circular Fermi surface form simplifying factors. Quantum transport is conveniently studied in a 2DEG because of the combination of a large Fermi wavelength and a large mean free path. The quantum mechanical phase coherence characteristic of a microscopic object can be maintained at low temperatures (below 1 K) over distances of several microns, which one would otherwise have classified as macroscopic. The physics of these systems have been referred to as mesoscopic, a word borrowed from statistical mechanics. Semiconductor nanostructures are unique in offering the possibility of studying quantum transport in an artificial potential landscape. This is the regime of ballistic transport, in which scattering with impurities can be neglected. The transport properties can then be tailored by varying the geometry of the conductor, in much the same way as one would tailor the transmission properties of a waveguide. The physics of this transport regime could be called as electron optics in the solid state. The formal relation between conduction and transmission, known as the Landauer formula, has demonstrated its real power in this context. For example, the quantization of the conductance of a quantum point contact (a short and narrow constriction in the 2DEG) can be understood using the Landauer formula as resulting from the discreteness of the number of propagating modes in a waveguide [143].

Nanostructure-based solar energy is attracting significant attention as a possible candidate for achieving drastic improvement in photovoltaic energy conversion efficiency. Although such solar energy is expected to be more expensive, there is a growing need for efficient and lightweight solar cells in aerospace and related industries. It is required to rule the energy sector when the breakeven of high performance is achieved and its cost becomes comparable with other energy sources. Various approaches have been intended to enhance the efficiency of solar cells. Applications of nanotechnology help us to solar devices more economically. Nano photovoltaic cells are used to improve the efficiency to create effective systems for conversion cost, efficient solar energy storage systems, or solar energy on a large scale. A solar cell performs two major functions: photogeneration of charge carriers in a light-absorbing material and separation of the charge carriers to a conductive contact that will transmit the electricity. Solar cells are electronic devices used for the direct conversion of solar energy to electricity, using the photovoltaic (PV) effect. Fundamental properties of nanostructured materials are currently extensively studied because of their potential application in numerous fields which include electronic devices, optoelectronics, optics, tribology, biotechnology, human medicine, and others [144].

Two nanotechnologies-semiconductor and metallic nanostructures are the most advanced in this field and have been extensively investigated for clinical use. Due to advanced properties, the incorporation of nanostructures in detection schemes or the use the nanostructures as contrast agents could improve the specificity and accuracy of current diagnostics. Furthermore, nanostructures can improve the effectiveness of drugs in the treatment of a variety of diseases when they are used as a delivery agent because of the increase in therapeutic payload. Semiconductor nanocrystals synthesized via organometallic methods contain the hydrophobic ligand tri-n-octylphosphine oxide (TOPO). These nanocrystals can subsequently be made water-soluble by covering the TOPO with the amphiphilic polymer polyacrylic acid grafted with octylamine. The hydrophilic carboxylic acid groups of the polymer coat render the nanocrystal stable in an aqueous environment. Furthermore, these groups can be used for conjugation to polymers and/or biological molecules such as antibodies and oligonucleotides via the formation of peptide bonds [145].

16.7 High-mobility Organic Transistors

Especially for the field-effect transistor (FET), a lot of efforts have been done to develop new organic materials to improve device performance with high charge carrier mobility and good air stability. Organic field-effect transistors (OFET) have been considered as a key component of organic integrated circuits for application in flexible smart cards, low-cost radio frequency identification (RFID) tags, sensor devices, organic active matrix displays, etc. OFET is of interest since the solution processability of the semiconductor material raises the possibility of lower device fabrication costs. Substantial progress on molecular design, processing conditions and device architecture have helped to achieve high mobility (μ) [146]. One processing strategy for improving operational properties and lowering the cost of organic electronic devices has been to blend organic semiconductors with commodity insulating polymers. In addition to reducing materials costs, these commonly available polymers have the potential to enhance environmental stability and improve mechanical properties when blended with the semiconductor component. As an example, mobilities of 0.4 $cm^2/V/s$ were obtained for 5 wt.% poly[2,5-bis(3-tetradecylthiophen-2-yl) thieno[3,2-b]thiophene] blended with polystyrene (PS) and in the presence of TCNQ. Higher μ may still be possible. Donor-acceptor polymers, while well-studied and among the best performers in OPV and OFET applications, have not been fully explored within the context of polymer-insulator blends. By incorporating a high-performing donor-acceptor polymer into a majority-insulator blend, we expect to see high μ maintained at low polymer concentrations provided that the materials phase separate appropriately [147].

The mobilities of organic semiconductors have achieved significant progress in OFETs from the initially reported 10^{-5} $cm^2/V/s$ for polythiophene in 1986 to 10 $cm^2/V/s$ for present diketopyrrolopyrrole (DPP)-based polymers. The high mobility of organic semiconductors over conventional amorphous silicon indicates the large potential application of organic electronic devices. The remarkable progress of organic semiconductors provides a road for the organic electronic industry. Generally, for high-performance organic semiconductors, some critical factors, such as molecular structure, molecular packing, electronic structure, energy alignment, and purity, play important roles. Among them, tuning the molecular packing is especially important for high-performance semiconductors since the charge carrier transport is along with the molecular orbitals. Hence, the overlap degree of neighboring molecular orbitals significantly determines the charge carrier mobility. Molecular packing with strong intermolecular interactions is favorable for efficient charge transport and high field-effect mobility. The electronic structure and energy levels are crucial for the materials and device stability. In order to obtain high-performance and stable organic semiconductors, structural modification with electron donors and acceptors are necessary [148].

16.7.1 P-type Semiconductors

In the last two decades, p-type semiconductor materials have achieved much progress because of their simple design and synthetic approach. P-type organic semiconductors mainly contain acene, heteroacene, thiophenes, as well as their correlated oligomers and polymers, and two-dimensional

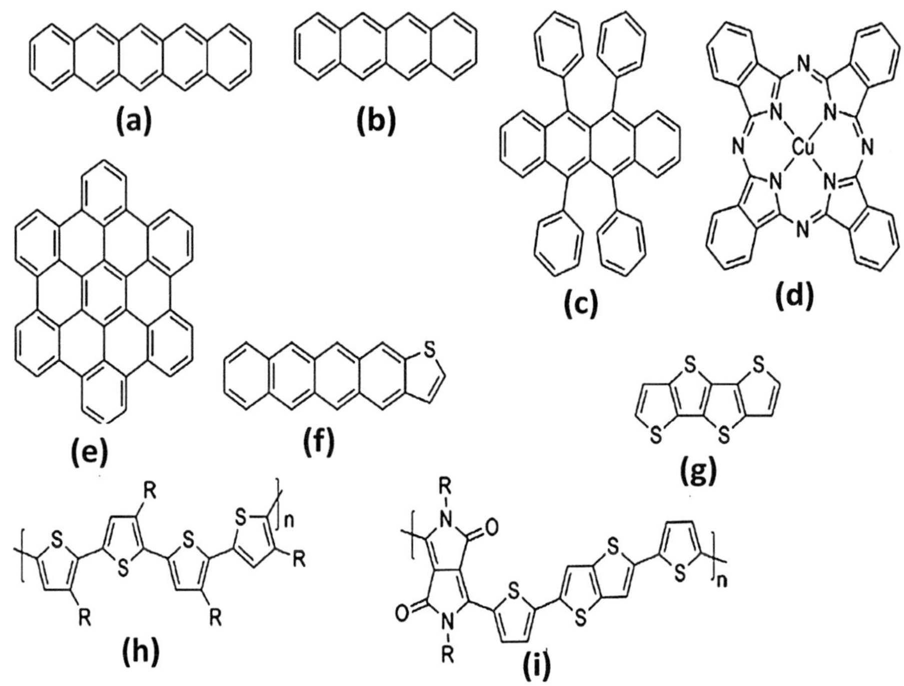

Fig. 16.17 Chemical structures of some p-type semiconducting materials: (**a**) pentacene, (**b**) tetracene, (**c**) rubrene, (**d**) phthalocyanine, (**e**) Hexa-peri-benzocoronenes, (**f**) tetraceno-thiophene, (**g**) tetrathienoacene, and (**h**) poly(3-hexylthiophene), and (**i**) PDDPT-TT

(2D) disk-like molecules [149]. Among them, the polycyclic aromatic hydrocarbons are most representative of the class of compounds due to their unique features. Some representative p-type semiconductors are given in Fig. 16.17.

Pentacene (Fig. 16.17a), as the benchmark of organic semiconductors, was first reported in 1970s, but the numerous OFET applications were only conducted recently. With strong intermolecular interactions and herringbone packing motif, pentacene exhibits efficient charge transport. Hence, a polycrystalline thin film of pentacene (Fig. 16.17a) and tetracene (Fig. 16.17b) showed surprisingly high mobility approaching 0.1 cm^2/V/s and 3.0 cm^2/V/s, respectively. The substituted tetracene derivative rubrene (Fig. 16.17c) showed the highest charge carriers mobility with 20 cm^2/V/s for a single crystal device in the FET configuration. This implies that the conjugated acene is a good building block for the p-type semiconductors. Later on, phthalocyanines (Fig. 16.17d) and more core-extended Hexa-peri-benzocoronenes (HBC) (Fig. 16.17e) containing 2D aromatic core showed typically discotic columnar liquid crystalline phases. As a result, the HBC showed enhanced mobility along the column due to the solid-state organization. Moreover, HBC-based OFETs by zone casting method exhibited high mobility up to 0.01 cm^2/V/s. The chemistry based on acene has paved the way for designing efficient p-type semiconducting materials. The sulfur-containing heteroacenes and their derivatives constitute another large group of p-type aromatic hydrocarbons. The thienoacenes and their derivatives were also synthesized and investigated as semiconductors for p-type materials. The asymmetric oligoacene, such as the tetraceno-thiophene (Fig. 16.17f), was also synthesized and showed similar mobility (0.3 cm^2/V/s) compared to their centrosymmetric counterparts processed in the same conditions. The tetrathienoacene (Fig. 16.17g) with aryl groups had a higher mobility up to 0.14 cm^2/V/s by vapor deposition. The sulfur–sulfur interaction in the packing motif was believed to enhance the charge carrier transport. The introduction of sulfur and other heteroatoms induced different energy alignments and crystal packing, which promotes the development of p-type materials. Among the p-type polymers, poly (3-hexylthiophene) (P3HT) (Fig. 16.17h) has shown high mobility due to its good crystalline properties and well-ordered lamella structure which facilitates efficient charge transport. And it has been widely used as electron donor in organic solar cells. DPP-based polymers, such as PDDPT-TT (Fig. 16.17i), have been shown as high-performance semiconductor materials with hole mobility over 10 cm^2/V/s [150, 151].

16.7.2 n-type Semiconductors

Although the p-type semiconductor materials have achieved much progress, the development of n-type organic semiconductors still lags behind that of p-type organic semiconductors due to low device performance, ambient instability, and complex synthesis. Owing to their important roles in organic electronics, such as p–n junctions, bipolar transistors, and complementary circuits, it is desirable to develop stable n-type semiconductor materials with high charge carrier mobility for organic field-effect transistors

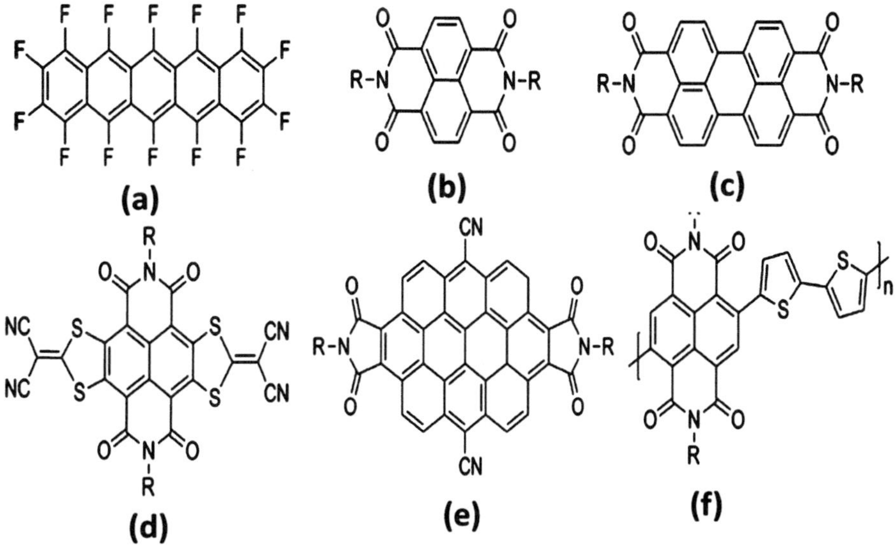

Fig. 16.18 Chemical structures of some n-type semiconducting materials: (**a**) perfluoropentacene, (**b**) Naphthalene diimide, (**c**) aromatic diimide, (**d**) ovalene diimide, (**e**) 2-(1,3-dithiol-2-ylidene)malononitrile, and (**f**) P(NDI2OD-T2)

[152]. Most of the n-type materials are still air unstable in ambient conditions due to their high lowest unoccupied molecular orbital (LUMO) energy level. The air unstable problem is due to the redox reaction with oxygen and water. The LUMO energy level should be lower than −3.97 eV in order to be stable toward water and oxygen. n-type organic semiconductors mainly contain halogen or cyano-substituted n-type semiconductors that could be converted from p-type materials, perylene derivatives, naphthalene derivatives, fullerene-based materials, and so on (Fig. 16.18) [153].

The important n-type semiconductor material is perfluoropentacene. This molecule adopted similar crystal packing to pentacene, and transistors fabricated from vacuum-deposited films showed high mobility up to 0.11 $cm^2/V/s$ and an on/off ratio of 10^5. It was thought that attaching fluorine atoms could lower the LUMO energy level of this compound. However, the LUMO energy level is not low enough to make the OFET device stable in the ambient condition. Similarly, the 2, 5, 8, 11, 14, 17-hexafluoro-hexa-perihexabenzocoronene was synthesized from hexakis (4-fluorophenyl)-benzene. This fluorinated compound was also suitable for the fabrication of n-channel transistors due to the decreased LUMO energy level, showing mobility of 1.6×10^{-2} $cm^2/V/s$ and an on/off ratio of 10^4. Based on these results, they showed that halogen substitution is a proper way to obtain n-type semiconductors [154]. Naphthalene diimide and perylene diimide derivatives are two of the most studied n-type materials used in OFETs. Simple naphthalene and perylene diimides can be prepared from bis anhydrides and primary amines. Generally, the aromatic diimide in transistors shows an n-type character due to imide functionalization. Then, cyano or halogen was introduced to improve air stability. Naphthalene diimide substituted with electron-withdrawing CN groups at the core position. This molecule showed mobility as high as 0.11 $cm^2/V/s$ as well as good ambient stability compared to unsubstituted compounds [155]. The good air stability was also observed, which indicates that cyano substituent is another efficient way to lower the LUMO energy level and achieve stable n-type materials. Later on, the core-expand NDI bearing two 2-(1,3-dithiol-2-ylidene) malononitrile moieties at the core need to be mentioned due to its good solution processability and good air stability. Based on these results, it could be concluded that there is an efficient way to achieve stable n-type materials by combining imide functionalization and cyano or halogen substitutions. Electron-deficient aromatic diimides, such as ovalene diimide (ODI-CN), have attracted increasing attention as promising n-type semiconductors for OFETs. The materials of this class showed not only a highly planar conjugated backbone but also easily tunable electronic properties through core and imide-nitrogen substituents with electron-withdrawing groups and alkyl chains, respectively. The most promising result for n-type polymers is a naphthalene-based polymer (P(NDI2OD-T2)), which exhibited an unprecedented high performance, with mobility of >0.1 $cm^2/V/s$ (up to 0.85 $cm^2/V/s$) and on/off ratio of 10^6 and excellent air stability in ambient conditions [156].

16.8 Summary

In near future, the exponential growth of electronic industries is achieved with the help of advanced semiconductor/conductor materials. This chapter described the emerging advanced electronic materials such as supercapacitors, superconducting materials, advanced semiconductor materials, high-mobility organic transistors, etc. with their various classification, mechanisms, and applications.

References

1. Hassan, E.S.F., Elsherbiny, S.M.: Effect of adding a passive condenser on solar still performance. Energy Convers. Manag. **34**(1), 63–72 (1993). https://doi.org/10.1016/0196-8904(93)90008-X
2. Wang, Y., Shi, Z., Huang, Y., Ma, Y., Wang, C., Chen, M., Chen, Y.: Supercapacitor devices based on graphene materials. J. Phys. Chem. C. **113**(30), 13103–13107 (2009). https://doi.org/10.1021/jp902214f
3. Winter, M., Brodd, R.J.: What are batteries, fuel cells, and supercapacitors. Chem. Rev. **104**(10), 4245–4270 (2004). https://doi.org/10.1021/cr020730k
4. See: https://www.einnews.com/pr_news/522630430/global-supercapacitor-market-size-study-by-type-electrode-material-application-and-regional-forecasts-2020-2027
5. Chen, J., Han, Y., Kong, X., Deng, X., Park, H.J., Guo, Y., Jin, S., Qi, Z., Lee, Z., Qiao, Z., Ruoff, R.S., Ji, H.: The origin of improved electrical double-layer capacitance by inclusion of topological defects and dopants in graphene for supercapacitors. Angew. Chem. Int. Ed. **55**(44), 13822–13827 (2016). https://doi.org/10.1002/anie.201605926
6. Jeong, H.-K., Jin, M., Ra, E.J., Sheem, K.Y., Han, G.H., Arepalli, S., Lee, Y.H.: Enhanced electric double layer capacitance of graphite oxide intercalated by poly(sodium 4-styrensulfonate) with high cycle stability. ACS Nano. **4**(2), 1162–1166 (2010). https://doi.org/10.1021/nn901790f
7. Pandolfo, A.G., Hollenkamp, A.F.: Carbon properties and their role in supercapacitors. J. Power Sources. **157**(1), 11–27 (2006). https://doi.org/10.1016/j.jpowsour.2006.02.065
8. Lang, X., Hirata, A., Fujita, T., et al.: Nanoporous metal/oxide hybrid electrodes for electrochemical supercapacitors. Nat. Nanotech. **6**, 232–236 (2011). https://doi.org/10.1038/nnano.2011.13
9. Zhang, S., Li, C., Zhang, X., Sun, X., Wang, K., Ma, Y.: High performance lithium-ion hybrid capacitors employing Fe3O4-graphene composite anode and activated carbon cathode. ACS Appl. Mater. Interfaces. **9**(20), 17136–17144 (2017). https://doi.org/10.1021/acsami.7b03452
10. See: https://www.skeletontech.com/skeleton-blog/whats-the-difference-between-an-ultracapacitor-and-a-supercapacitor. 06.05.2020.
11. Becker, H.I.: Low voltage electrolytic capacitor, 1954, patent no: US2800616A
12. Yassine, M., Fabris, D.: Performance of commercially available supercapacitors. Energies. **10**, 1340 (2017)
13. Choudhary, R.B., Ansari, S., Purty, B.: Robust electrochemical performance of polypyrrole (PPy) and polyindole (PIn) based hybrid electrode materials for supercapacitor application: a review. J. Energy Storage. **29**, 101302 (2020). https://doi.org/10.1016/j.est.2020.101302
14. Hossain, A., Bandyopadhyay, P., Guin, P.S., Roy, S.: Recent developed different structural nanomaterials and their performance for supercapacitor application. Appl. Mater. Today. **9**, 300–313 (2017). https://doi.org/10.1016/j.apmt.2017.08.010
15. Zhu, M., Meng, W., Huang, Y., Huang, Y., Zhi, C.: Proton-insertion-enhanced pseudocapacitance based on the assembly structure of tungsten oxide. ACS Appl. Mater. Interfaces. **6**(21), 18901–18910 (2014). https://doi.org/10.1021/am504756u
16. See: http://mkollaras.com/uploads/1/3/0/5/130539202/papaloborafisojezafu.pdf
17. Burke, A.: Ultracapacitors: why, how, and where is the technology. J. Power Sources. **91**(1), 37–50 (2000). https://doi.org/10.1016/S0378-7753(00)00485-7
18. See: https://www.latimes.com/archives/la-xpm-1997-05-06-fi-55809-story.html. 06.05.2020
19. Rantho, M.N., Madito, M.J., Oyedotun, K.O., Tarimo, D.J., Manyala, N.: Hybrid electrochemical supercapacitor based on birnessite-type MnO2/carbon composite as the positive electrode and carbonized iron-polyaniline/nickel graphene foam as a negative electrode. AIP Adv. **10**(6) (2020). https://doi.org/10.1063/5.0011862
20. See: https://www.banaao.co.in/super-capacitors/. 06.05.2020.
21. Eftekhari, A., Fang, B.: Electrochemical hydrogen storage: opportunities for fuel storage, batteries, fuel cells, and supercapacitors. Int. J. Hydrog. Energy. **42**(40), 25143–25165 (2017). https://doi.org/10.1016/j.ijhydene.2017.08.103
22. Misra, A.: Energy storage for electrified aircraft: the need for better batteries, fuel cells, and supercapacitors. IEEE Electrif. Mag. **6**(3), 54–61 (2018). https://doi.org/10.1109/MELE.2018.2849922
23. Caizán-Juanarena, L., Borsje, C., Sleutels, T., et al.: Combination of bioelectrochemical systems and electrochemical capacitors: principles, analysis and opportunities. Biotechnol. Adv. **39**, 107456 (2020). https://doi.org/10.1016/j.biotechadv.2019.107456
24. Yue, M., Jemei, S., Gouriveau, R., Zerhouni, N.: Review on health-conscious energy management strategies for fuel cell hybrid electric vehicles: degradation models and strategies. Int. J. Hydrog. Energy. **44**(13), 6844–6861 (2019). https://doi.org/10.1016/j.ijhydene.2019.01.190
25. Burke, A.F.: Batteries and ultracapacitors for electric, hybrid, and fuel cell vehicles. Proc. IEEE. **95**(4), 806–820 (2007). https://doi.org/10.1109/JPROC.2007.892490
26. Conway, B.E., Pell, W.G.: Double-layer and pseudocapacitance types of electrochemical capacitors and their applications to the development of hybrid devices. J. Solid State Electrochem. **7**, 637–644 (2003). https://doi.org/10.1007/s10008-003-0395-7
27. Shin, D., Kim, Y., Wang, Y., Chang, N., Pedram, M.: Constant-current regulator-based battery-supercapacitor hybrid architecture for high-rate pulsed load applications. J. Power Sources. **205**, 516–524 (2012). https://doi.org/10.1016/j.jpowsour.2011.12.043
28. Yoon, S., Lee, C.W., Oh, S.M.: Characterization of equivalent series resistance of electric double-layer capacitor electrodes using transient analysis. J. Power Sources. **195**(13), 4391–4399 (2010). https://doi.org/10.1016/j.jpowsour.2010.01.086
29. Conway, B.E.: Transition from "Supercapacitor" to "Battery" behavior in electrochemical energy storage. J. Electrochem. Soc. **138**. https://doi.org/10.1149/1.2085829
30. Jet-Sing, M.L., Briggs, M.E., Hu, C.-C., Cooper, A.I.: Controlling electric double-layer capacitance and pseudocapacitance in heteroatom-doped carbons derived from hypercrosslinked microporous polymers. Nano Energy. **46**, 277–289 (2018). https://doi.org/10.1016/j.nanoen.2018.01.042
31. Deyang, Q., Shi, H.: Studies of activated carbons used in double-layer capacitors. J. Power Sources. **74**(1), 99–107 (1998). https://doi.org/10.1016/S0378-7753(98)00038-X
32. Libich, J., Máca, J., Vondrák, J., Čech, O., Sedlaříková, M.: Supercapacitors: properties and applications. J. Energy Storage. **17**, 224–227 (2018). https://doi.org/10.1016/j.est.2018.03.012
33. Béguin, F., Presser, V., Balducci, A., Frackowiak, E.: Carbons and electrolytes for advanced supercapacitors. Adv. Mater. **26**(14), 2219–2251 (2014). https://doi.org/10.1002/adma.201304137
34. Le Yu, B.G., Xiao, W., Wen (David) Lou, X.: Formation of yolk-shelled Ni-Co mixed oxide nanoprisms with enhanced electrochemical performance for hybrid supercapacitors and lithium ion batteries. Adv. Energy Mater. **5**(21), 500981 (2015). https://doi.org/10.1002/aenm.201500981
35. Peng, X., Liu, H., Yin, Q., et al.: A zwitterionic gel electrolyte for efficient solid-state supercapacitors. Nat. Commun. **7**, 11782 (2016). https://doi.org/10.1038/ncomms11782
36. Anneser, K., Reichstein, J., Braxmeier, S., Reichenauer, G.: Carbon xerogel based electric double layer capacitors with polymer gel

electrolytes-Improving the performance by adjusting the type of electrolyte and its processing. Electrochim. Acta. **278**, 196–203 (2018). https://doi.org/10.1016/j.electacta.2018.05.046
37. Li, H., Yu, M., Wang, F., et al.: Amorphous nickel hydroxide nanospheres with ultrahigh capacitance and energy density as electrochemical pseudocapacitor materials. Nat. Commun. **4**, 1894 (2013). https://doi.org/10.1038/ncomms2932
38. Farsi, H., Gobal, F., Raissi, H., et al.: On the pseudocapacitive behavior of nanostructured molybdenum oxide. J. Solid State Electrochem. **14**, 643–650 (2010). https://doi.org/10.1007/s10008-009-0830-5
39. Dong, L., Yang, W., Yang, W., Li, Y., Wu, W., Wang, G.: Multivalent metal ion hybrid capacitors: a review with a focus on zinc-ion hybrid capacitors. J. Mater. Chem. A. **7**, 13810–13832 (2019). https://doi.org/10.1039/C9TA02678A
40. Wang, R., Lang, J., Zhang, P., Lin, Z., Yan, X.: Fast and large lithium storage in 3D porous VN nanowires-graphene composite as a superior anode toward high-performance hybrid supercapacitors. Adv. Funct. Mater. **25**(5), 2270–2278 (2015). https://doi.org/10.1002/adfm.201404472
41. Liu, W., Li, J., Feng, K., Sy, A., Liu, Y., Lim, L., Lui, G., Tjandra, R., Rasenthiram, L., Chiu, G., Yu, A.: Advanced Li-Ion hybrid supercapacitors based on 3D graphene–foam composites. ACS Appl. Mater. Interfaces. **8**(39), 25941–25953 (2016). https://doi.org/10.1021/acsami.6b07365
42. Karthikeyan, K., Aravindan, V., Lee, S.B., Jang, I.C., Lim, H.H., Park, G.J., Yoshio, M., Lee, Y.S.: Electrochemical performance of carbon-coated lithium manganese silicate for asymmetric hybrid supercapacitors. J. Power Sources. **195**(11), 3761–3764 (2010). https://doi.org/10.1016/j.jpowsour.2009.11.138
43. Ramulu, B., Nagaraju, G., Chandra Sekhar, S., Hussain, S.K., Narsimulu, D., Jae Su, Y.: Synergistic effects of cobalt Molybdate@Phosphate core–shell architectures with ultrahigh capacity for rechargeable hybrid supercapacitors. ACS Appl. Mater. Interfaces. **11**(44), 41245–41257 (2019). https://doi.org/10.1021/acsami.9b11707
44. Morita, M., Kaigaishi, T., Yoshimoto, N., Egashira, M., Aida, T.: Effects of the electrolyte composition on the electric double-layer capacitance at carbon electrodes. Electrochem. Solid State Lett. **9** (8). https://doi.org/10.1149/1.2208013
45. Wu, S., Chen, R., Zhang, S., et al.: A chemically inert bismuth interlayer enhances long-term stability of inverted perovskite solar cells. Nat. Commun. **10**, 1161 (2019). https://doi.org/10.1038/s41467-019-09167-0
46. Turgut, M.G.: Review of electrical energy storage technologies, materials and systems: challenges and prospects for large-scale grid storage. Energy Environ. Sci. **11**, 2696–2767 (2018). https://doi.org/10.1039/C8EE01419A
47. Hurilechaoketu, Wang, J., Cui, C., Qian, W.: Highly electroconductive mesoporous activated carbon fibers and their performance in the ionic liquid-based electrical double-layer capacitors. Carbon. **154**, 1–6 (2019). https://doi.org/10.1016/j.carbon.2019.07.093
48. Singh, D., Gupta, M., Singh, R.C., Pandey, S.P., Karn, R.K., Singh, P.K.: Polyvinylpyrrolidone (PVP) with ammonium iodide (NH4i) and 1-hexyl-3-methylimidazolium iodide ionic liquid doped solid polymer electrolyte for efficient supercapacitors. Macrolol. Symp. **388**(1), 1900036 (2019). https://doi.org/10.1002/masy.201900036. Special Issue: National Conference on Exotic Materials and Devices - NCEMD 2019
49. Presser, V., Heon, M., Gogotsi, Y.: Carbide-derived carbons – from porous networks to nanotubes and graphene. Adv. Funct. Mater. **21**(5), 810–833 (2011). https://doi.org/10.1002/adfm.201002094
50. Hanzawa, Y., Kaneko, K., Pekala, R.W., Dresselhaus, M.S.: Activated Carbon Aerogels. Langmuir. **12**(26), 6167–6169 (1996). https://doi.org/10.1021/la960481t
51. Gleb, N.Y., Hoffman, E.N., Nikitin, A., Ye, H., Barsoum, M.W., Gogotsi, Y.: Synthesis of nanoporous carbide-derived carbon by chlorination of titanium silicon carbide. Carbon. **43**(10), 2075–2082 (2005). https://doi.org/10.1016/j.carbon.2005.03.014
52. Xu, L., Jia, M., Li, Y., et al.: High-performance MnO2-deposited graphene/activated carbon film electrodes for flexible solid-state supercapacitor. Sci. Rep. **7**, 12857 (2017). https://doi.org/10.1038/s41598-017-11267-0
53. Pyrzyńska, K., Bystrzejewski, M.: Comparative study of heavy metal ions sorption onto activated carbon, carbon nanotubes, and carbon-encapsulated magnetic nanoparticles. Colloids Surf. A Physicochem. Eng. Asp. **362**(1–3), 102–109 (2010). https://doi.org/10.1016/j.colsurfa.2010.03.047
54. Brezesinski, K., Wang, J., Haetge, J., Reitz, C., Steinmueller, S.O., Tolbert, S.H., Smarsly, B.M., Dunn, B., Brezesinski, T.: pseudocapacitive contributions to charge storage in highly ordered mesoporous group V transition metal oxides with iso-oriented layered nanocrystalline domains. J. Am. Chem. Soc. **132**(20), 6982–6990 (2010). https://doi.org/10.1021/ja9106385
55. Liu, T., Pell, W.G., Conway, B.E.: Self-discharge and potential recovery phenomena at thermally and electrochemically prepared RuO2 supercapacitor electrodes. Electrochim. Acta. **42**(23–24), 3541–3552 (1997). https://doi.org/10.1016/S0013-4686(97)81190-5
56. Fic, K., Frackowiaka, E., Béguin, F.: Unusual energy enhancement in carbon-based electrochemical capacitors. J. Mater. Chem. **22**, 24213–24223 (2012). https://doi.org/10.1039/C2JM35711A
57. Zhong Shuai, W., Wang, D.W., Ren, W., Zhao, J., Zhou, G., Li, F., Cheng, H.M.: Anchoring hydrous RuO2 on graphene sheets for high-performance electrochemical capacitors. Adv. Funct. Mater. **20**(20), 3595–3602 (2010). https://doi.org/10.1002/adfm.201001054
58. Levin, K.L., Pshchelko, N.S.: Electrochemical properties of a polypyrrole-polyimide composite. Polym. Sci. Ser. A. **53**, 510–520 (2011). https://doi.org/10.1134/S0965545X11060083
59. Md Moniruzzaman, S.K., Yue, C.Y., Ghosh, K., Jena, R.K.: Review on advances in porous nanostructured nickel oxides and their composite electrodes for high-performance supercapacitors. J. Power Sources. **308**, 121–140 (2016). https://doi.org/10.1016/j.jpowsour.2016.01.056
60. Jayaseelan, S.S., Radhakrishnan, S., Saravanakumar, B., Seo, M.-K., Khil, M.-S., Kim, H.-Y., Kim, B.-S.: Mesoporous 3D NiCo2O4/MWCNT nanocomposite aerogels prepared by a supercritical CO2 drying method for high performance hybrid supercapacitor electrodes. Colloids Surf. A Physicochem. Eng. Asp. **538**, 451–459 (2018). https://doi.org/10.1016/j.colsurfa.2017.11.037
61. Khomenko, V., Raymundo-Piñero, E., Frackowiak, E., et al.: High-voltage asymmetric supercapacitors operating in aqueous electrolyte. Appl. Phys. A. **82**, 567–573 (2006). https://doi.org/10.1007/s00339-005-3397-8
62. Azaïs, P., Duclaux, L., Florian, P., Massiot, D., Lillo-Rodenas, M.-A., Linares-Solano, A., Peres, J.-P., Jehoulet, C., Béguin, F.: Causes of supercapacitors ageing in organic electrolyte. J. Power Sources. **171**(2), 1046–1053 (2007). https://doi.org/10.1016/j.jpowsour.2007.07.001
63. Balducci, A., Bardi, U., Caporali, S., Mastragostino, M., Soavi, F.: Ionic liquids for hybrid supercapacitors. Electrochem. Commun. **6** (6), 566–570 (2004). https://doi.org/10.1016/j.elecom.2004.04.005
64. Mallakpour, S., Rafiee, Z.: Ionic liquids as environmentally friendly solvents in macromolecules chemistry and technology, Part II. J. Polym. Environ. **19**, 485 (2011). https://doi.org/10.1007/s10924-011-0291-7

65. Liu, M., Turcheniuk, K., Fu, W., Yang, Y., Liu, M., Yushin, G.: Scalable, safe, high-rate supercapacitor separators based on the Al2O3 nanowire polyvinyl butyral nonwoven membranes. Nano Energy. **104627**, 71 (2020)
66. Ye Qin, W., Chen, X.Y., Ji, P.T., Zhou, Q.Q.: Sol–gel approach for controllable synthesis and electrochemical properties of NiCo2O4 crystals as electrode materials for application in supercapacitors. Electrochim. Acta. **56**(22), 7517–7522 (2011). https://doi.org/10.1016/j.electacta.2011.06.101
67. William, R.V., Marikani, A., Madhavan, D.: Dielectric behavior and magnetical response for porous BFO thin films with various thicknesses over Pt/Ti/SiO2/Si substrate. Ceramics Int. **42**(6), 6807–6816 (2016). https://doi.org/10.1016/j.ceramint.2016.01.058
68. Chee, W.K., Lim, H.N., Zainal, Z., Huang, N.M., Harrison, I., Andou, Y.: Flexible graphene-based supercapacitors: a review. J. Phys. Chem. C. **120**(8), 4153–4172 (2016). https://doi.org/10.1021/acs.jpcc.5b10187
69. Zhang, Y., Zhen, Z., Zhang, Z., Lao, J., Wei, J., Wang, K., Kang, F., Zhu, H.: In-situ synthesis of carbon nanotube/graphene composite sponge and its application as compressible supercapacitor electrode. Electrochim. Acta. **157**, 134–141 (2015). https://doi.org/10.1016/j.electacta.2015.01.084
70. Bai, X., Hu, X., Zhou, S., Yan, J., Sun, C., Chen, P., Li, L.: In situ polymerization and characterization of grafted poly (3,4-thylenedioxythiophene)/multiwalled carbon nanotubes composite with high electrochemical performances. Electrochim. Acta. **87**, 394–400 (2013). https://doi.org/10.1016/j.electacta.2012.09.079
71. Liu, Z., Wu, Z.-S., Yang, S., Dong, R., Feng, X., Müllen, K.: Ultraflexible in-plane micro-supercapacitors by direct printing of solution-processable electrochemically exfoliated graphene. Adv. Mater. **28**(11), 2217–2222 (2016). https://doi.org/10.1002/adma.201505304
72. Atchudan, R., Edison, T.N.J.I., Perumal, S., RanjithKumar, D., Lee, Y.R.: Direct growth of iron oxide nanoparticles filled multiwalled carbon nanotube via chemical vapour deposition method as high-performance supercapacitors. Int. J. Hydrog. Energy. **44**(4), 2349–2360 (2019). https://doi.org/10.1016/j.ijhydene.2018.08.183
73. Zhang, Q., Levi, M.D., Chai, Y., Zhang, X., Xiao, D., Dou, Q., Ma, P., Ji, H., Yan, X.: Vacuum filtration-and-transfer technique helps electrochemical quartz crystal microbalance to reveal accurate charge storage in supercapacitors. Small Methods. **3**(11), 1900246 (2019). https://doi.org/10.1002/smtd.201900246
74. Yang, J., Duan, X., Qina, Q., Zheng, W.: Solvothermal synthesis of hierarchical flower-like β-NiS with excellent electrochemical performance for supercapacitors. J. Mater. Chem. A. **1**, 7880–7884 (2013). https://doi.org/10.1039/C3TA11167A
75. Omar, F.S., Numan, A., Duraisamy, N., Bashir, S., Ramesha, K., Ramesh, S.: Ultrahigh capacitance of amorphous nickel phosphate for asymmetric supercapacitor applications. RSC Adv. **6**, 76298–76306 (2016). https://doi.org/10.1039/C6RA15111F
76. Wang, R., Qi, J.Q., Sui, Y.W., Chang, Y., He, Y.Z., Wei, F.X., Meng, Q.K., Sun, Z., Zhao, Y.L.: Fabrication of nanosheets Co3O4 by oxidation-assisted dealloying method for high capacity supercapacitors. Mater. Lett. **184**, 181–184 (2016). https://doi.org/10.1016/j.matlet.2016.08.049
77. Cai, J., Niu, H., Li, Z., Yong, D., Cizek, P., Xie, Z., Xiong, H., Lin, T.: High-performance supercapacitor electrode materials from cellulose-derived carbon nanofibers. ACS Appl. Mater. Interfaces. **7**(27), 14946–14953 (2015). https://doi.org/10.1021/acsami.5b03757
78. Muzaffar, A., Ahamed, M.B., Deshmukh, K., Thirumalai, J.: A review on recent advances in hybrid supercapacitors: design, fabrication and applications. Renew. Sustain. Energy Rev. **101**, 123–145 (2019). https://doi.org/10.1016/j.rser.2018.10.026
79. Yan, J., Wang, Q., Wei, T., Fan, Z.: Recent advances in design and fabrication of electrochemical supercapacitors with high energy densities. Adv. Energy Mater. **4**(4), 1300816 (2014). https://doi.org/10.1002/aenm.201300816
80. Şahin, M.E., Blaabjerg, F.: A hybrid PV-battery/supercapacitor system and a basic active power control proposal in MATLAB/Simulink. Electronics. **9**(1), 129 (2020). https://doi.org/10.3390/electronics9010129
81. Kim, B.K., Sy, S., Yu, A., Zhang, J.: Electrochemical supercapacitors for energy storage and conversion. In: Handbook clean energy syst. Wiley, Hoboken, NJ. https://doi.org/10.1002/9781118991978.hces112
82. Bentley, P., Stone, D.A., Schofield, N.: The parallel combination of a VRLA cell and supercapacitor for use as a hybrid vehicle peak power buffer. J. Power Sources. **147**(1–2), 288–294 (2005). https://doi.org/10.1016/j.jpowsour.2005.01.016
83. Stepanov, A., Galkin, I., Bisenieks, L.: Implementation of supercapacitors in uninterruptible power supplies. In: 2007 European Conference on Power Electronics and Applications, Aalborg, pp. 1–7 (2007). https://doi.org/10.1109/EPE.2007.4417559
84. Sathishkumar, R., Kollimalla, S.K., Mishra, M.K.: Dynamic energy management of micro grids using battery super capacitor combined storage. In: 2012 Annual IEEE India Conference (INDICON), Kochi, pp. 1078–1083 (2012). https://doi.org/10.1109/INDCON.2012.6420777
85. Lehtimäki, S., Li, M., Salomaa, J., Pörhönen, J., Kalanti, A., Tuukkanen, S., Heljo, P., Halonen, K., Lupo, D.: Performance of printable supercapacitors in an RF energy harvesting circuit. Int. J. Electr. Power Energy Syst. **58**, 42–46 (2014). https://doi.org/10.1016/j.ijepes.2014.01.004
86. Fairweather, A.J., Stone, D.A., Foster, M.P.: Evaluation of UltraBattery™ performance in comparison with a battery-supercapacitor parallel network. J. Power Sources. **226**, 191–201 (2013). https://doi.org/10.1016/j.jpowsour.2012.10.095
87. Varma, S.J., Kumar, K.S., Seal, S., Rajaraman, S., Thomas, J.: Fiber-type solar cells, nanogenerators, batteries, and supercapacitors for wearable applications. Adv. Sci. **5**(1–2), 1800340 (2018). https://doi.org/10.1002/advs.201800340
88. Gee, A.M., Robinson, F.V.P., Dunn, R.W.: Analysis of Battery Lifetime Extension in a Small-Scale Wind-Energy System Using Supercapacitors. In: IEEE Transactions on Energy Conversion, vol. 28, no. 1, pp. 24–33 (2013). https://doi.org/10.1109/TEC.2012.2228195.
89. Wieschemann, A., Eichner, H.C.: Battery Electric Tractor Trailer for ISO Container Transport (2012), patent no: DE102012108768A1
90. Steiner, M., Scholten, J.: Energy storage on board of railway vehicles. In: 2005 European Conference on Power Electronics and Applications, Dresden, pp. 10 (2005). https://doi.org/10.1109/EPE.2005.219410
91. Schaltz, E., Khaligh, A., Rasmussen, P.O.: Influence of Battery/Ultracapacitor Energy-Storage Sizing on Battery Lifetime in a Fuel Cell Hybrid Electric Vehicle. In: IEEE Transactions on Vehicular Technology, vol. 58, no. 8, pp. 388–3891 (Oct. 2009). https://doi.org/10.1109/TVT.2009.2027909.
92. https://www.alstom.com/press-releases-news/2019/9/alstom-presents-srs-ground-based-recharging-system-electric-buses. 08.05.2020
93. See: https://wikivisually.com/wiki/Capa_vehicle. 10.05.2020
94. Sharma, R., Walia, S., Raghav, S:. Electric vehicles (EV) and hybrid electric vehicles (HEV) in regenerative antibraking system: a literature review. https://doi.org/10.37628/jpnc.v4i1.606
95. Tie, S.F., Tan, C.W.: A review of energy sources and energy management system in electric vehicles. Renew. Sustain. Energy Rev. **20**, 82–102 (2013). https://doi.org/10.1016/j.rser.2012.11.077

96. Larbalestier, D.: Superconducting materials: a review of recent advances and current problems in practical materials. In: IEEE Transactions on Magnetics, vol. 17, no. 5, pp. 1668–1686 (September 1981). https://doi.org/10.1109/TMAG.1981.1061365.
97. Maletta, H., Malozemoff, A.P., Cronemeyer, D.C., Tsuei, C.C., Greene, R.L., Bednorz, J.G., Müller, K.A.: Diamagnetic shielding and Meissner effect in the high Tc superconductor Sr0.2La1.8CuO4. Solid State Commun. **88**(11–12), 837–840 (1993). https://doi.org/10.1016/0038-1098(93)90252-I
98. Faisal, W.M.: High Tc superconducting fabrication of loop antenna. Alexandria Eng. J. **51**(3), 171–183 (2012). https://doi.org/10.1016/j.aej.2012.02.004
99. See: https://www.nobelprize.org/prizes/physics/1973/summary/. 10.05.2020
100. Hirsch, J.E.: The origin of the Meissner effect in new and old superconductors. Phys. Scripta. **85**(3). https://doi.org/10.1088/0031-8949/85/03/035704
101. Smith, F.W., Baratoff, A., Cardona, M.: Superheating, supercooling, surface superconductivity and Ginzburg-Landau parameters of pure type-I superconductors and their alloys. Phys. Kondens Mater. **12**, 145–192 (1970). https://doi.org/10.1007/BF02422376
102. Jones, A., Lam, S.K.H., Jia, D., Rubanov, S., Pan, A.V.: Guided vortex motion control in superconducting thin films by sawtooth ion surface modification. ACS Appl. Mater. Interfaces. **12**(23), 26170–26176 (2020). https://doi.org/10.1021/acsami.0c04658
103. See: https://electronicsistechnology.blogspot.com/2010/12/type-1-superconductors-and-periodic.html. 12.05.2020
104. Wesche, R.: High-temperature superconductors. In: Kasap, S., Capper, P. (eds.) Springer handbook of electronic and photonic materials. Springer handbooks. Springer, Cham (2017). https://doi.org/10.1007/978-3-319-48933-9_50
105. Brandt, E.H., Indenbom, M.: Type-II-superconductor strip with current in a perpendicular magnetic field. Phys. Rev. B. **48**, 12893 (1993). https://doi.org/10.1103/PhysRevB.48.12893
106. Alden, T.H., Livingston, J.D.: Ferromagnetic particles in a type-II superconductor. J. Appl. Phys. **37**(9). https://doi.org/10.1063/1.1708900
107. See: http://repository.sustech.edu/handle/123456789/19437. 12.05.2020
108. See: http://electronicsistechnology.blogspot.com/2010/12/. 12.05.2020.
109. Yan, L.: Development and Application of the Maglev Transportation System. In: IEEE Transactions on Applied Superconductivity, vol. 18, no. 2, pp. 92–99 (June 2008). https://doi.org/10.1109/TASC.2008.922239.
110. Janic, M.: Multicriteria evaluation of high-speed rail, transrapid maglev and air passenger transport in Europe. Transp. Plan. Technol. **26**(6) (2003). https://doi.org/10.1080/0308106032000167373
111. Joyce, K.A.: From numbers to pictures: The development of magnetic resonance imaging and the visual turn in medicine. Sci. Cult. **15**(1) (2006). https://doi.org/10.1080/09505430600639322
112. See: http://www.superconductors.org/Uses.htm. 18.05.2020
113. See: http://ffden-2.phys.uaf.edu/webproj/212_spring_2014/Dallon_Knight/web/uses%20of%20superconductors.html. 18.05.2020
114. See: https://ir.amsc.com/static-files/d199e518-fd51-45df-aeb0-e0a6f9b83e7a. 18.05.2020
115. See: https://new.abb.com/news/detail/13724/abb-first-to-connect-superconducting-transformer-to-a-utility-power-network. 18.05.2020
116. See: https://helenthehare.org.uk/2015/08/21/13112/. 18.05.2020
117. Md. Atikur, R., Md. Zahidur, R.: A review on high-Tc superconductors and their principle applications. J. Adv. Phys. **4** (2), 87–100(14) (2015). https://doi.org/10.1166/jap.2015.1175
118. Susner, M.A., Haugan, T.J.: A review of the state-of-the-art superconductor technology for high power applications, 2018 AIAA/IEEE electric aircraft technologies symposium, pp. 1–17. EATS, Cincinnati, OH (2018)
119. Moralee, D.: Defence electronics. EMP: the forgotten threat? In: Electronics and Power, vol. 27, no. 7.8, pp. 524–525 (1981). https://doi.org/10.1049/ep.1981.0260.
120. Hartikainen, T., Lehtonen, J., Mikkonen, R.: Role of HTS devices in greenhouse gas emission reduction. Supercond. Sci. Technol. **16** (8). https://doi.org/10.1088/0953-2048/16/8/324
121. Mazzeo, M.P., Restuccia, L.: Thermodynamics of n-type extrinsic semiconductors. Energy. **36**(7), 4577–4584 (2011). https://doi.org/10.1016/j.energy.2011.02.055
122. Giebultowicz, T.M., Rhyne, J.J.: Inelastic neutron scattering studies of II-VI diluted magnetic semiconductors (invited). J. Appl. Phys. **67**, 5096 (1990). https://doi.org/10.1063/1.344683
123. Ohno, H.: Properties of ferromagnetic III–V semiconductors. J. Magn. Magn. Mater. **200**(1–3), 110–129 (1999). https://doi.org/10.1016/S0304-8853(99)00444-8
124. Lim, J.A., Liu, F., Ferdous, S., Muthukumar, M., Briseno, A.L.: Polymer semiconductor crystals. Mater. Today. **13**(5), 14–24 (2010). https://doi.org/10.1016/S1369-7021(10)70080-8
125. Wang, C., Dong, H., Jiang, L., Wenping, H.: Organic semiconductor crystals. Chem. Soc. Rev. **47**, 422–500 (2018). https://doi.org/10.1039/C7CS00490G
126. Minami, T., Miyata, T., Nishi, Y.: Cu2O-based heterojunction solar cells with an Al-doped ZnO/oxide semiconductor/thermally oxidized Cu2O sheet structure. Solar Energy. **105**, 206–217 (2014). https://doi.org/10.1016/j.solener.2014.03.036
127. Lee, C.-H., Nam, B.-A., Choi, W.-K., Lee, J.-K., Choi, D.-J., Oh, Y.-J.: Mn:SnO2 ceramics as p-type oxide semiconductor. Mater. Lett. **65**(4), 722–725 (2011). https://doi.org/10.1016/j.matlet.2010.11.021
128. Wang, Z., Nayak, P.K., Caraveo-Frescas, J.A., Alshareef, H.N.: Recent developments in p-type oxide semiconductor materials and devices. Adv. Mater. **28**(20), 3831–3892 (2016). https://doi.org/10.1002/adma.201503080. Special Issue: Metal Oxide Heterointerfaces in Hybrid Electronic Platforms
129. Wangying, X., Li, H., Xu, J.-B., Wang, L.: Recent advances of solution-processed metal oxide thin-film transistors. ACS Appl. Mater. Interfaces. **10**(31), 25878–25901 (2018). https://doi.org/10.1021/acsami.7b16010
130. Kamilla, S.K., Basu, S.: New semiconductor materials for magnetoelectronics at room temperature. Bull. Mater. Sci. **25**, 541–543 (2002). https://doi.org/10.1007/BF02710546
131. Furdyna, J.K.: Diluted magnetic semiconductors. J. Appl. Phys. **64** (4) (1988). https://doi.org/10.1063/1.341700
132. Ivanov, V.A., Aminov, T.G., Novotortsev, V.M.: Spintronics and spintronics materials. Russ. Chem. Bull. **53**, 2357–2405 (2004). https://doi.org/10.1007/s11172-005-0135-5
133. Flatte, M.E.: Spintronics. In: IEEE Transactions on Electron Devices, vol. 54, no. 5, pp. 907–920 (May 2007). https://doi.org/10.1109/TED.2007.894376.
134. Gupta, K.M., Gupta, N.: Semiconductor materials: their properties, applications, and recent advances. In: Advanced semiconducting materials and devices. Engineering materials. Springer, Cham (2016). https://doi.org/10.1007/978-3-319-19758-6_1
135. Furdyna, J.K., semiconductors, D.m.: An interface of semiconductor physics and magnetism. J. Appl. Phys. **53**, 7637 (1982). https://doi.org/10.1063/1.330137
136. Lodha, A., Singh, R.: Prospects of manufacturing organic semiconductor-based integrated circuits. In: IEEE Transactions on Semiconductor Manufacturing, vol. 14, no. 3, pp. 281–296 (2001). https://doi.org/10.1109/66.939830.
137. Wang, F., Wang, Z., Jiang, C., Yin, L., Cheng, R., Zhan, X., Xu, K., Wang, F., Zhang, Y., He, J.: Progress on electronic and

optoelectronic devices of 2D layered semiconducting materials. Small. **13**(35), 1604298 (2017). https://doi.org/10.1002/smll.201604298. Special Issue: Beyond Graphene-Booming Development of Novel 2D Nanomaterials

138. See: https://www.redbull.com/in-en/how-the-zing-cricket-wicket-was-made. 22.05.2020
139. Vinal, A.: Considerations for applying solid state sensors to high density magnetic disc recording. In: IEEE Transactions on Magnetics, vol. 20, no. 5, pp. 681–686 (September 1984). https://doi.org/10.1109/TMAG.1984.1063228
140. Zhu, H., Eun-Sol, S.A., Dongseob, L., Yong, J., Noh, X.Y.-Y.: Printable semiconductors for backplane TFTs of flexible OLED displays. Adv. Funct. Mater. **30**(20), 1904588 (2020). https://doi.org/10.1002/adfm.201904588. Special Issue: Emerging Thin-Film Transistor Technologies and Applications
141. Lee, S., Chen, Y., Kim, J., Kim, H.M., Jang, J.: Transparent AMOLED display driven by split oxide TFT backplane. J. Soc. Inf. Disp. **26**(3), 164–168 (2018). https://doi.org/10.1002/jsid.646
142. Chakraborty, S., Mandal, I., Ray, I., Majumdar, S., Sen, A., Maiti, H.S.: Improvement of recovery time of nanostructured tin dioxide-based thick film gas sensors through surface modification. Sens. Actuators B Chem. **127**(2), 554–558 (2007). https://doi.org/10.1016/j.snb.2007.05.005
143. Beenakker, C.W.J., van Houten, H.: Quantum transport in semiconductor nanostructures. In: Ehrenreich, H., Turnbull, D. (eds.) Solid state physics, vol. 44, pp. 1–228. Academic Press, Cambridge (1991). https://doi.org/10.1016/S0081-1947(08)60091-0. ISSN 0081-1947, ISBN 9780126077445
144. Xianluo, H., Li, G., Jimmy, C.Y.: Design, fabrication, and modification of nanostructured semiconductor materials for environmental and energy applications. Langmuir. **26**(5), 3031–3039 (2010). https://doi.org/10.1021/la902142b
145. Salata, O.: Applications of nanoparticles in biology and medicine. J. Nanobiotechnol. **2**, 3 (2004). https://doi.org/10.1186/1477-3155-2-3
146. Jurchescu, O.D., Popinciuc, M., van Wees, B.J., Palstra, T.T.M.: Interface-controlled, high-mobility organic transistors. Adv. Mater. **19**(5), 688–692 (2007). https://doi.org/10.1002/adma.200600929
147. Ford, M.J., Wang, M., Patel, S.N., Phan, H., Segalman, R.A., Nguyen, T.-Q., Bazan, G.C.: High mobility organic field-effect transistors from majority insulator blends. Chem. Mater. **28**(5), 1256–1260 (2016). https://doi.org/10.1021/acs.chemmater.5b04774
148. Ruiz†, C., García-Frutos, E.M., Hennrich, G., Gómez-Lor, B.: Organic semiconductors toward electronic devices: high mobility and easy processability. J. Phys. Chem. Lett. **3**(11), 1428–1436 (2012). https://doi.org/10.1021/jz300251u
149. Paterson, A.F., Singh, S., Fallon, K.J., Hodsden, T., Han, Y., Schroeder, B.C., Bronstein, H., Heeney, M., McCulloch, I., Anthopoulos, T.D.: Recent progress in high-mobility organic transistors: a reality check. Adv. Mater. **30**(36), 1801079 (2018). https://doi.org/10.1002/adma.201801079. This article also appears in: Advanced Materials Hall of Fame
150. Syed Abthagir, P., Ha, Y.-G., You, E.-A., Jeong, S.-H., Seo, H.-S., Choi, J.-H.: Studies of tetracene- and pentacene-based organic thin-film transistors fabricated by the neutral cluster beam deposition method. J. Phys. Chem. B. **109**(50), 23918–23924 (2005). https://doi.org/10.1021/jp054894r
151. Gao, X., Zhao, Z.: High mobility organic semiconductors for field-effect transistors. Sci. China Chem. **58**, 947–968 (2015). https://doi.org/10.1007/s11426-015-5399-5
152. Kanimozhi, C., Yaacobi-Gross, N., Chou, K.W., Amassian§, A., Anthopoulos, T.D., Patil, S.: Diketopyrrolopyrrole-diketopyrrolopyrrole-based conjugated copolymer for high-mobility organic field-effect transistors. J. Am. Chem. Soc. **134**(40), 16532–16535 (2012). https://doi.org/10.1021/ja308211n
153. Lei, T., Dou, J.-H., Cao, X.-Y., Wang, J.-Y., Pei, J.: Electron-deficient poly(p-phenylene vinylene) provides electron mobility over 1 cm2 V–1 s–1 under ambient conditions. J. Am. Chem. Soc. **135**(33), 12168–12171 (2013). https://doi.org/10.1021/ja403624a
154. Yang, J., Zhao, Z., Geng, H., Cheng, C., Chen, J., Sun, Y., Shi, L., Yi, Y., Shuai, Z., Guo, Y., Wang, S., Liu, Y.: Isoindigo based polymers with small effective masses for high-mobility ambipolar field effect transistors. Adv. Mater. **29**(36), 1702115 (2017). https://doi.org/10.1002/adma.201702115
155. Fallon, K.J., Wijeyasinghe, N., Manley, E.F., Dimitrov, S.D., Yousaf, S.A., Ashraf, R.S., Duffy, W., Guilbert, A.A.Y., Freeman, D.M.E., Al-Hashimi, M., Nelson, J., Durrant, J.R., Chen, L.X., McCulloch, I., Marks, T.J., Clarke, T.M., Anthopoulos, T.D., Bronstein, H.: Indolo-naphthyridine-6,13-dione thiophene building block for conjugated polymer electronics: molecular origin of ultrahigh n-type mobility. Chem. Mater. **28**(22), 8366–8378 (2016). https://doi.org/10.1021/acs.chemmater.6b03671
156. Gao, X., Zhao, Z.: High mobility organic semiconductors for field-effect transistors. Sci. China Chem. **58**(6), 947–968 (2015). https://doi.org/10.1007/s11426-015-5399-5

Ultrafine-Grained Materials 17

Abstract

This chapter discusses about emerging advanced materials: ultrafine grain materials. The concept of this material has been discussed with respect to the grain size and grain boundary behavior toward the low- and high-temperature processing. Different superplastic deformation processes such as equal-channel angular pressing, high-pressure torsion, accumulative roll bonding, friction stir processing, multidirectional forging, cyclic extrusion and compression, repetitive corrugation and straightening, twist extrusion along with the machining process have been discussed. Currently, revealed properties and the developing product have been discussed.

Keywords

Ultrafine-grained materials · Grain size · Grain boundaries · Second phases · Internal stress

17.1 What Is Ultrafine-Grained Materials

Ultrafine-grained (UFG) metallic materials are at the cutting edge of modern materials science as they exhibit outstanding properties which make them very interesting for prospective structural or functional engineering applications. Grain size is a key factor which affecting nearly all aspects of the physical, mechanical, and chemical behavior of polycrystalline metals to the surrounding media [1]. Hence, modification of grain size can able to design materials with desired properties. Physical, mechanical, and chemical properties can benefit greatly from the reduction of grain size. Ultrafine-grained (UFG)/nanocrystalline metals exhibit extraordinary properties [2]. Grain refinement of metallic materials to submicron or nanoscale by severe plastic deformation (SPD) has been well-recognized by the community of materials scientists (Fig. 17.1). The advent of SPD has broken the wall of the minimum grain size achievable by the conventional thermomechanical processing and realized the fabrication of materials with the so-called ultrafine grain structure in a bulk form for load-carrying structural components [3]. Among the different strengthening mechanisms used for advanced steels, grain refinement is one of the few methods to improve both strength and toughness simultaneously [4]. Therefore, ultrafine-grained steels with relatively plain chemical compositions, strengthened primarily by grain refinement, have great potential for replacing some conventional low alloyed high-strength steels. This applies for both (mainly) single, and massively alloyed advanced high-strength multiphase steels [5]. The main benefits behind such an approach are to avoid additional alloying elements; to avoid additional heat treatments like soft annealing, quenching and tempering; and to improve weldability owing to lower required carbon contents and other alloying elements when compared with other high-strength steels. A further high-potential domain for such ultrafine-grained steel is the possibility for high strain rate superplasticity at medium and elevated temperatures [6].

However, it has been considered impossible to fabricate bulk UFG/nanocrystalline metals by casting, a technology that has been used for over 6000 years, partly due to its slow cooling (e.g., less than 100 K/s) [7]. Conventional microstructure refinement methods, such as fast cooling (thousands to millions of kelvin per second) and inoculation, have reached certain fundamental or technical limits. Fast cooling substantially restricts the size and complexity of the as-solidified materials. The minimum grain size achievable by inoculation in casting falls in the range of tens of micrometers [8]. Grain refinement in metals during solidification is of great interest due to the enhanced mechanical properties, more homogeneous microstructure, and improved processability of refined microstructures. Nanocrystalline (NC) materials are polycrystals with the mean grain size below 10 nm. Submicrocrystalline polycrystals with a mean grain size of about 100 nm are often called ultrafine-grained

A. Behera, *Advanced Materials*, https://doi.org/10.1007/978-3-030-80359-9_17

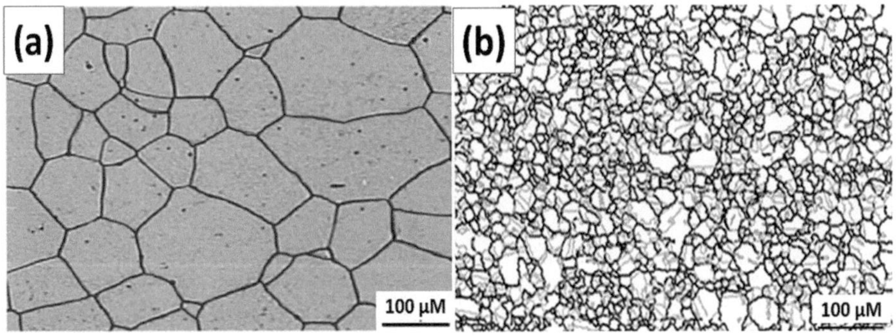

Fig. 17.1 (**a**) Coarse grained and (**b**) ultrafine grain structure of stainless steel

(UFG) materials. In recent years, the NC, as well as the UFG materials, has attracted much interest of researchers in the field of materials science [9].

17.2 Historical Background to UFG Metals

The procedure of achieving superior properties in metals through the application of severe plastic deformation may be traced back more than 2,000 years to the Han dynasty (200 BC) and the three states dynasty (280 AD) of ancient China. Specifically, there is evidence that a "folding and forging" technique was introduced for the fabrication of high-quality steel swords as early as about 500 BC wherein the steel was consecutively forged, folded, and then forged again in order to produce high strength. The continuous repetition of this process led ultimately to the development of Bai-Lian steel around 220–280 AD, where this term denotes the use of a multiple repetitive processing operations through 100 separate consecutive operations [10]. Archeological artifacts available from this period, such as steel swords and knives, have inscriptions that are now believed to record the precise numbers of repetitive operations conducted during processing. For example, a 50-Lian steel sword is believed to have undergone 50 separate smeltings or repetitive forgings and foldings during the processing operation. Wootz steel is an ultra-high carbon steel that was developed in India in the period from approximately 300 BC to 300 AD [11]. This ancient steel may be regarded as an advanced material because of both the high impact hardness and the superplastic properties at elevated temperatures. The export of wootz steel to the Middle East led ultimately to the development of the world-famous Damascus steel which was manufactured primarily in the vicinity of Damascus, Syria, until approximately the middle of the eighteenth century when the fabrication technique was essentially lost. The various fabrication techniques employed in the ancient world provide important information on the historical development of SPD processing but they necessarily lack the scientific approach that is needed to understand and fully optimize the various techniques [12].

The first detailed scientific studies were undertaken by Professor P. W. Bridgman at Harvard University in the United States immediately after the Second World War and the results of these experiments were subsequently summarized in a classic text describing the large plastic flow. This early work is very relevant to the procedures currently under investigation in many laboratories around the world. For example, Bridgman introduced a methodology that would now be regarded as analogous to the well-established testing by high-pressure torsion (HPT). Thereafter, various processing techniques were introduced and evaluated including, most importantly, the procedure of equal-channel angular pressing (ECAP) that was first developed by Segal and coworkers at an institute in Minsk in the former Soviet Union (now the capital of Belarus). As is well known, processing by ECAP represents the most widely used and important procedure for the fabrication of bulk UFG solids at the present time [13]. The historical research and development on various UFG materials in the scientific community has been shown in Table 17.1.

17.3 Concept on Ultrafine-Grained Materials

It is well known that the room temperature mechanical properties and high-temperature superplastic behavior of polycrystalline metals depend on microstructural characteristics and especially, on the grain size. Superplastic elongations are achieved through the relative displacements of adjacent grains within the polycrystalline matrix [14]. Although the dominant flow process in superplasticity is grain boundary sliding that is accommodated by intragranular slip within the grains in order to prevent the development of internal cavities. A decrease in the grain size leads to enhanced strength of the metal at room temperature, according to the well-known Hall–Petch relationship [15]. A reduction in the grain size also provides superplastic forming capability at elevated temperatures. The grain size is one of the most important characteristics of a polycrystalline metal microstructure. The common grain size of the metals in the industry is >10 μm and are called coarse-grained materials

Table 17.1 The gradual development of various ultrafine materials

Year	Patent title
1994	Sputtering target with ultrafine oriented grains and method of making same
1994	On the structure and strength of ultrafine-grained copper produced by severe plastic deformation
1999	A two-step SPD processing of the ultrafine-grained titanium
2000	Method for producing ultrafine-grained materials using repetitive corrugation and straightening
2000	Ultrafine-grained titanium for medical implants
2001	Method for the preparation of ultrafine-grained titanium and titanium alloy articles and articles prepared by said method
2002	Method of preparing ultrafine grain metallic articles and metallic articles prepared thereby
2003	Medium and high-carbon steel shaft parts of ultrafine crystal grain and its production technology
2004	Continuous processing of ultrafine-grained Al by ECAP-conform
2005	Method and apparatus for an equal channel angular pressing (ECAP) consolidation process for cryomilled nanocrystalline metal powders
2005	Method for preparing pre-coated, ultrafine, submicron grain high-temperature aluminum and aluminum-alloy components and components prepared thereby
2006	Producing bulk ultrafine-grained materials by severe plastic deformation
2007	Achieving ultrafine grain size in Mg–Al–Zn alloy by friction stir processing
2007	Continuous frictional angular extrusion and its application in the production of ultrafine-grained sheet metals
2008	Wire hydrostatic extrusion device and method for extruding superfine grain wire using the device
2009	Mechanical properties and microstructures of ultrafine-grained pure aluminum by asymmetric rolling
2009	Pin tool for friction stir welding and ultrafine grain preparation method thereof
2010	Large plasticizing deformation method for producing ultrafine crystal material
2013	The device of the rough standby block body ultrafine grain metal material of reciprocating extrusion pier
2013	Method for preparing large-size ultrafine-grained material
2013	The preparation method of ultrafine grain rare earth magnesium alloy
2014	High-strength high-toughness ultrafine-grained high-entropy alloy and preparation method thereof
2014	Mold for preparing ultrafine-grained bulk material
2006	Producing bulk ultrafine-grained materials by severe plastic deformation
2016	Method for obtaining ultrafine-grained titanium concentrate
2017	A kind of crimp processing mold preparing ultrafine-grained beta-titanium alloy and technique
2017	The blanking type variable cross section of ultrafine-grained steel bar back and forth squeezes and turns round upsetting the manufacturing process
2018	Equidistant spiral rolling method for large-size titanium alloy ultrafine-grained bar
2018	It is a kind of prepare bulk ultrafine-grained materials squeeze upsetting mold and method repeatedly
2018	A kind of cycle extrusion mold preparing bulk ultrafine-grained materials and method
2019	Composite large plastic deformation method for preparing ultrafine-grained aluminum and aluminum alloy
2019	High-performance ultrafine-grained hot-rolled TRIP steel material and preparation method thereof
2020	Equidistant spiral rolling method for large-size titanium alloy ultrafine-grained bar

[16]. Fine-grained metals that are processed using industrial thermomechanical processing have a grain size of ~1–10 μm. It is hard to achieve a metal with grain size <1 μm (called ultrafine grained) or <100 nm (called nanograined) when using conventional thermomechanical and conventional metal-forming processes. The main reasons may be the limited amount of plastic strain that results from limitations in the reduced cross section, limited values of hydrostatic compressive stresses, and the lack of high-angle grain boundaries. In general, the term *ultrafine grain* is used here in the context of average grain sizes between 1 μm and 2 μm in diameter; *submicron* refers to grain sizes between 100 nm (0.1 μm) and 1000 nm (1 μm); while *nano-structured* refers to grain sizes below 100 nm (<0.1 μm) [17].

Grain size represents a critical parameter in all polycrystalline metal solids because it plays a major role in determining both the strength and ductility of the metals. In broad terms, the strength at room temperature is increased and the forming properties at elevated temperatures are enhanced when the grain size is reduced. Accordingly, much attention is devoted in industrial practice to the use of thermomechanical treatments, which are designed to introduce limited grain refinement to reduce the grain size of metals to values below ~2–5 μm [18]. Because of this limitation, considerable attention has been directed toward devising alternative procedures, such as the process using severe plastic deformation (SPD) techniques, which produce very high strains without the introduction of any significant changes in the overall dimensions of the workpieces. In practice, SPD processing provides the potential for exceptional grain refinement leading to ultrafine-grained (UFG) metals with grain sizes within the nanometer range [19].

At low temperatures, the grain size controls the strength such that the strength increases as the grain size is reduced, whereas at high temperatures the superplastic forming capability increases with decreasing grain size, because small grain sizes are a prerequisite for superplastic flow. These characteristics mean that a reduction in grain size should introduce two desirable properties: higher strength and increased formability [20]. Thus, in the low-temperature regime, the yield stress (σ_y) as well as the hardness (H) depends on the mean grain size (d), can be described by the Hall–Petch relationship:

$$\sigma_y = \sigma_0 + \frac{k_y}{\sqrt{d}} \tag{17.1}$$

or

$$H = H_0 + \frac{k}{\sqrt{d}} \tag{17.2}$$

where σ_0 is the lattice friction stress. k_y, H_0 and k are materials constant. The mechanical behavior of materials with smaller grains is still being discussed in terms of various models. Some studies have revealed materials softening with decreasing grain size (negative slop of Hall–Petch-like relationship), while the usual relation (positive slop) has been found in others (Fig. 17.2).

Conversely, in the high-temperature regime, the creep rate under steady-state conditions ε is expressed by Eq. (17.3), which is developed in 1994, which incorporated both grain boundary sliding in materials with larger grain sizes and superplastic flow in materials with very small grain sizes,

$$\varepsilon = \frac{ADGb}{kT}\left(\frac{b}{d}\right)^P\left(\frac{\sigma}{G}\right)^n \tag{17.3}$$

where D is the appropriate diffusion constant= $D_0 \exp\left(\frac{-Q}{RT}\right)$, D_0 is the frequency factor, Q is the activation energy, for the flow process, R is the gas constant, and T is the absolute temperature, G is the shear modulus, b is the burger vector, k is Boltzmann's constant, σ is the flow stress, n and p are the exponents of the stress and the inverse grain size, respectively, and A is a dimensionless constant.

It follows from inspection of Eqs. (17.1) and (17.2) that a small grain size is advantageous because it leads to a significantly higher strength. Additionally, Eq. (17.3) shows that a small grain size leads to faster strain rates, and this provides the possibility of achieving a superplastic forming capability at rapid rates that may be readily employed in industrial forming operations. Thus, grain refinement is an important processing tool for achieving optimum properties in metallic materials.

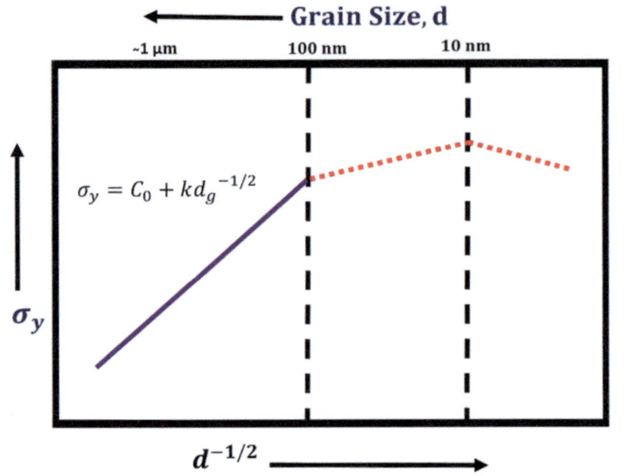

Fig. 17.2 The variation of Hall–Petch relationship with a varying range of grain size

17.4 Methods for Producing UFG Materials

Severe plastic deformation (SPD) methods are advanced metal-forming processes that apply high levels of hydrostatic stress without changing the material cross section and are capable of producing ultrafine-grained metals with high-angle grain boundaries. A large number of SPD processes have been proposed during the last 20 years for producing ultrafine-grained and nanograined metals. At the present time, typical SPD methods are equal-channel angular pressing (ECAP), high-pressure torsion (HPT), high-pressure sliding (HPS), multiaxial forging, and accumulative roll bonding (ARB) [18, 21–24]. All of these SPD methods can be repeated several times in order to introduce a large plastic deformation into bulk materials. The main difference between various SPD methods is their effectivity in microstructure refinement. The differences between them are mainly relevant to the deformation behavior, the shape of the workpiece, the strain imposed per pass, and the processing load required. From the workpiece shape point of view, SPD methods may be classified as suitable for bulk, for sheet, or for tubular components. The ECAP, cyclic extrusion–compression, and HPT are the oldest and main SPD processes suitable for bulk metals. There are many additional methods for dealing with bulk materials (twist extrusion, torsional–ECAP, multiaxial forging, cyclic expansion–extrusion, cyclic close-die forging, repetitive forging using inclined punch, repetitive upsetting or repetitive extrusion and upsetting, accumulative back extrusion, expansion ECAP, constrained groove pressing, equal-channel forward extrusion, parallel-ECAP, pure shear extrusion, vortex extrusion, friction stir processing, elliptical cross section spiral equal-channel extrusion, hydrostatic ECAP, half channel angular extrusion, C-shape equal-channel reciprocating extrusion, accumulative press bonding and the tandem process of simple shear extrusion and twist extrusion). The SPD processes listed above are suitable for the processing of ultrafine-grained and nanograined metals at laboratory scales. Scaling up these SPD methods remains a challenge for the scientific community. During the last decade, industrial demand for large, high-strength ultrafine-grained samples has pushed researchers to develop methods suitable for processing large samples using continuous processes like incremental high-pressure torsion, continuous high-pressure torsion, ECAP Conform, incremental ECAP, and equal channel.

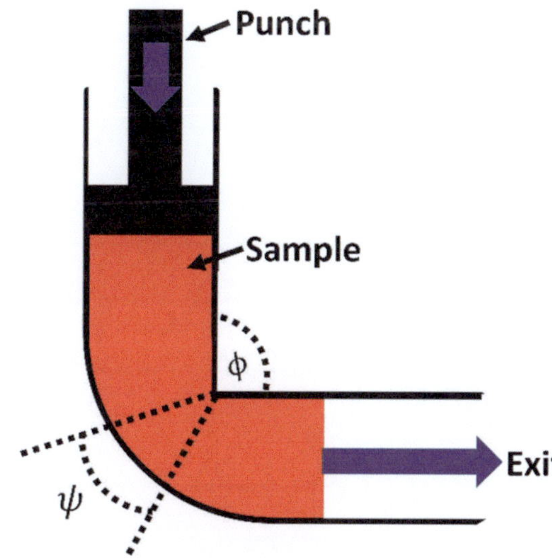

Fig. 17.3 Schematic of the ECAP process

17.4.1 Equal-Channel Angular Pressing

Equal-channel angular pressing (ECAP) imposes large plastic strains on massive billets via a pure shear strain state. The approach was developed by Segal et al. in the early 1980s. Its goal was to introduce intense plastic strain into materials without changing the cross-sectional area of the deformed billets. Owing to this characteristic, repeated deformation is possible. At the beginning of the 1990s, this method was further developed and applied as an SPD method for the processing of microstructures with submicron grain sizes. The ECAP method was mainly applied for nonferrous alloys (e.g., Al and Mg alloys) and some low carbon steels. The finest ferrite grain size obtained by use of this method is reportedly about 0.2 μm. The general principles of ECAP are shown schematically in Fig. 17.3. Processing by ECAP uses a die containing a channel that is bent through a sharp angle near the center of the die. The sample is machined to fit within the channel, and it is then pressed through the die using a plunger and an applied pressure P. Ultimately, the sample emerges from the die. In practice, the channel is defined by two angles: the channel angle (**U**) represents the angle between the two parts of the channel (equal to 90^0), and the curvature angle (**W**) represents the angle at the outer arc of curvature where the two parts of the channel intersect. The cross-sectional dimensions of the sample are not changed during processing. This means that the sample may be pressed repetitively through the die in order to attain a very high strain. Furthermore, it is possible to initiate different slip systems by rotating the sample between each pass [25].

In ECAP, the processing routes of billets were shown in the Fig. 17.4:

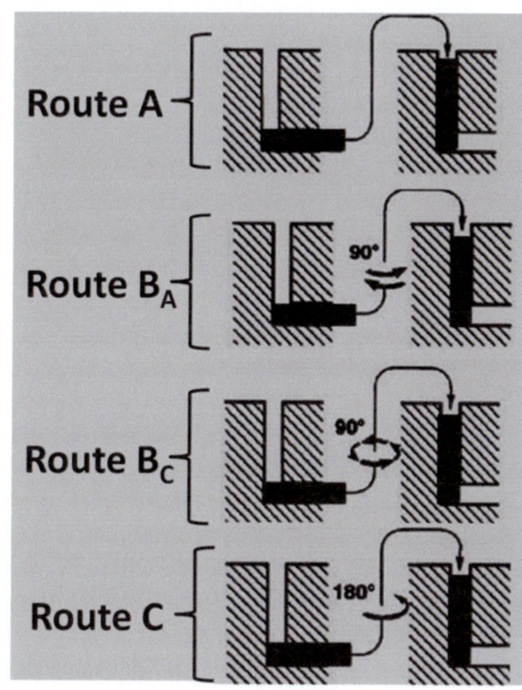

Fig. 17.4 Schematic of the ECAP process

Route A: Here, the sample is pressed repetitively without any rotation

Route B_A: The sample is rotated by 90° in alternate directions between consecutive passes

Route B_C: The sample is rotated in the same sense by 90° between each pass

Route C: The sample is rotated by 180° between passes.

The given routes are distinguished in their shear directions at repeat passes of a billet through intersecting channels. Due

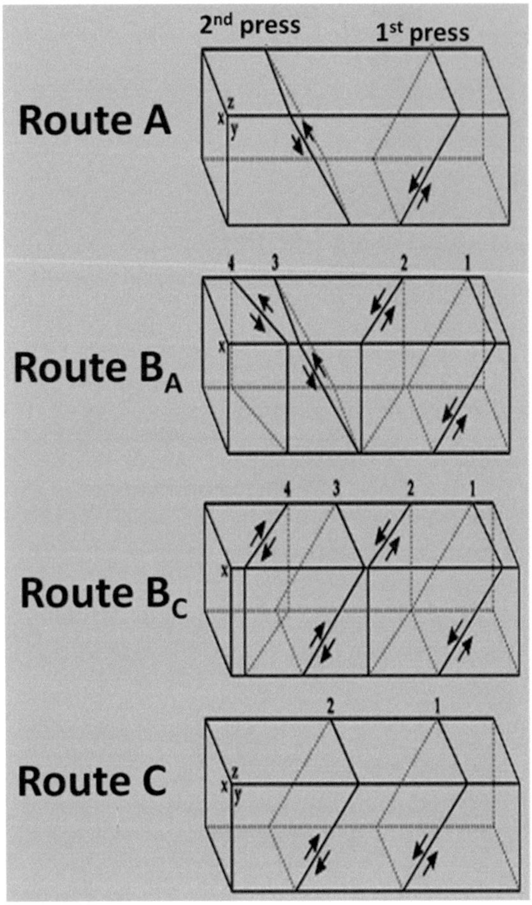

Fig. 17.5 The slip systems viewed on the *X*, *Y*, and *Z* planes for consecutive passes using processing routes A, B_A, B_C, and C

to that, during ECAP a change in a spherical cell within a billet body occurs. The different slip systems associated with these various processing routes are depicted schematically in Fig. 17.5 where *X*, *Y*, and *Z* planes correspond to the three orthogonal planes, and slip is shown for different passes in each processing route. Thus, the planes labeled 1 through 4 correspond to the first 4 passes of ECAP.

In route C, the shearing continues on the same plane in each consecutive passage through the die but the direction of shear is reversed on each pass. Thus, route C is termed a redundant strain process and the strain is restored after every even number of passes. It is apparent that route B_C is also a redundant strain process because the slip in the first pass is canceled by the slip in the third pass and the slip in the second pass is canceled by the slip in the fourth pass. By contrast, routes A and B_A are not redundant strain processes and there are two separate shearing planes intersecting at an angle of 90° in route A and four distinct shearing planes intersecting at angles in route B_A. In routes A and B_A, there is a cumulative buildup of additional strain on each separate pass through the die [26].

Experiments on FCC metals have shown that route B_C is the optimum processing route for producing an array of equiaxed ultrafine grains separated by boundaries with high angles of misorientation. Furthermore, experiments on Cu where samples were processed for up to 25 passes using an ECAP die showed that samples processed by route B_C exhibited the highest yield stress, the maximum ultimate tensile stress, and the highest ductilities after pressing through 10 passes or more. The reason for an optimization using route B_C is that this processing route contains the largest angular ranges for the slip occurring on each of the three orthogonal planes within the ECAP billet. The strain imposed in each pass of ECAP is dependent primarily upon the angle U and, to a lesser extent, on the angle W. It can be shown from first principles that the shear strain ε_N is given by a relationship in Eq. (17.4).

$$\varepsilon_N = \frac{N}{\sqrt{3}}\left[2\cot\left(\frac{\phi}{2}+\frac{\psi}{2}\right)+\psi\,\mathrm{cosec}\left(\frac{\phi}{2}+\frac{\psi}{2}\right)\right] \quad (17.4)$$

where *N* is the number of passes through the die. In conventional ECAP, it is generally assumed that the billet fills the corner of the die at the intersection of the two parts of the channel, and this produces a uniform microstructure throughout the billet. Nevertheless, experiments have demonstrated that a strain inhomogeneity generally forms near the lower surface of the ECAP billet owing to the development of a corner gap or dead zone at the outer corner when the billet passes through the die. The formation of a dead zone has been widely reported in ECAP both through direct experimental observations and through the predictions of finite element modeling [27].

17.4.2 High-Pressure Torsion

Processing by high-pressure torsion (HPT) is generally conducted using samples in the form of thin disks, although some recent experiments have described the use of small cylindrical specimens and samples in the form of rings. The principle of processing by HPT is illustrated schematically in Fig. 17.6. HPT refers to the processing of metals in which samples are subjected to compressive force and concurrent torsional straining under a high hydrostatic pressure (>2 GPa). Samples located between two anvils are imposed on a compressive applied pressure and plastic torsional straining is achieved through rotation of the lower anvil. Surface friction forces make samples shear deformation under a quasi-hydrostatic pressure. In practice, there are three different types of HPT, which depend upon the geometry of the anvils and the degree of restriction imposed on any lateral flow during the processing operation. In unconstrained

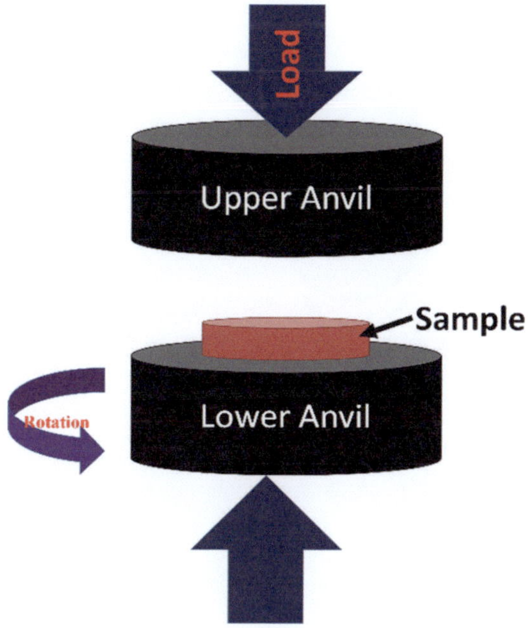

Fig. 17.6 Schematics of HPT process

HPT, the anvils are flat so that the material flows outwards in an unconstrained manner during processing. In constrained HPT, the disk is placed within a cavity in the lower anvil, a plunger from the upper anvil enters the cavity, and there is no lateral flow during processing. In practice, however, most of the HPT processing is now conducted under quasi-constrained conditions where the disk is contained within depressions on the inner surfaces of the upper and lower anvils. The disk thickness is slightly larger than the combined depths of the two depressions, and some limited outward flow occurs during torsional straining. The differences between these various types of HPT are important because careful experiments on pure Zr have shown that the occurrence of the allotropic phase transformation is influenced by the geometry of the die used for the HPT processing [28].

The equivalent von Mises strain $\boldsymbol{\varepsilon}_{eq}$ imposed on the disk in HPT is given by the Eq. (17.5).

$$\varepsilon_{\mathrm{eq}} = \frac{2\pi N r}{h\sqrt{3}} \tag{17.5}$$

where N is the number of turns of torsional straining, r is the radial distance measured from the center of the disk, and h is the initial height (or thickness) of the sample. From Eq. (17.5), the strain varies across the disk and there is a maximum value at the outer edge and a minimum value of zero strain at the center of the disk where $r = 0$. This relationship suggests, therefore, that HPT processing will produce materials containing very significant inhomogeneities.

There are two important problems associated with HPT processing that requires careful attention. First, there is a possibility that the disk may slip during processing so that there is a difference between the measured torsional rotation of the disk and the overall rotation imposed by the HPT facility. This can be checked very easily by placing parallel marker lines on the upper and lower surfaces of a disk, torsionally straining the disk through a fraction of a rotation, such as one-half of a rotation, and then measuring the angle between the two lines. Experiments of this type show that the extent of slippage depends upon the material used for the disk, with little or no slippage for Al but some significant slippage for harder materials such as Fe. These results also show that the extent of slippage increases when using faster rotational speeds or lower values for the imposed pressure. Second, there is the problem that the two large anvils of the HPT facility may be slightly misaligned prior to the torsional straining. It was shown that shear vortices and double swirls are visible on the disk surfaces when the anvils are initially misaligned but no shear vortices are visible when the anvils are in perfect alignment. There are now several reports describing the occurrence of swirls and shear vortices on the surfaces of deformed disks. Thus, the conclusion from these experiments is that care must be exercised in all experiments to ensure that the anvils are in good alignment.

The torsion straining achieves a substantial degree of substructure refinement and controls the evolution of large crystallographic misorientations among adjacent grains. The HPT technique also has the advantage of being able to refine the grain size during powder consolidation, making it possible to produce bulk nanomaterials from micrometer-sized metallic powders [29].

17.4.3 Accumulative Roll Bonding

Accumulative roll bonding (ARB) utilizes a conventional rolling facility. ARB essentially involves repeated application of conventional rolling. This process involves simultaneous bonding and deformation. In the ARB method, the rolled material is cut, stacked to the initial thickness, and rolled again. Owing to this approach, multiple repetitions are possible to achieve huge strains. A natural limit of this approach lies in the increase in strength and the gradually reduced surface quality of the roll-bonded sheets. Figure 17.7 schematically represents the ARB process. Two sheets surface treated in advance are stacked together, and then the stacked sheets are rolled to one-half thickness of a pre-rolled condition by a traditional roll bonding process. The rolled sheet is cut into two halves which are degreased and wire-brushed before being stacked again. Thus, a series of surface treatment, stacking, rolling, and cutting operations are repeated to achieve a large plastic strain in the sheet. The whole rolling process should be at an elevated temperature where there is no recrystallization to ensure the accumulative strain. The CP–Ti was processed by ARB process up to eight cycles (equivalent strain of 6.4) at ambient temperature. In ARB processed samples, two kinds of

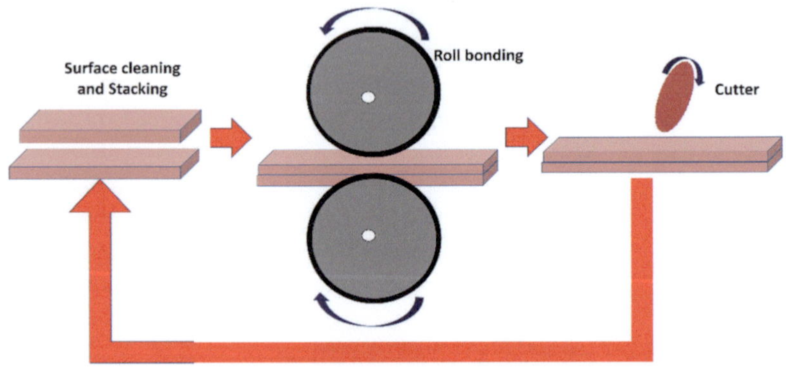

Fig. 17.7 Schematics of ARB process

the UFG microstructures were observed. One was the lamellar boundary structure elongated along the rolling direction. The lamellar boundary interval decreased sharply with increasing the accumulative plastic strain; the other was equiaxed grains. Both the grain size reached approximately 80 nm after five cycles. Furthermore, the volume of equiaxed grains went an upward trend with the increase of strain and arrived at 90% after eight cycles. When ARB was applied in hot rolled CP–Ti with 7 cycles at 450 MPa, a predominantly equiaxed ultrafine grain structure with an average grain size of 100 nm was formed. The tensile strength doubled of UFG alloy as the initial one, from 450 to 900 MPa. Furthermore, jump in tensile strength and grain refinement was obtained after the first cycle, indicating that the most effective ARB processing was attained at low cycles [30].

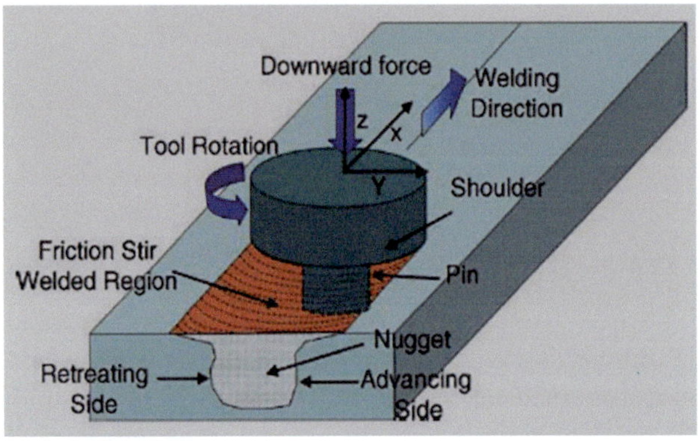

Fig. 17.8 Schematics of FSP

17.4.4 Friction Stir Processing (FSP)

The Friction stir processing (FSP) based on the principle of friction stir welding, is an effective solid-state processing method providing localized modification of the surface layers. Schematics of FSP is given in Fig. 17.8. A rotating tool with a pin and shoulder is inserted in a single piece of material. The tool heats the samples and localized heating softens the material around the pin and the combination of tool rotation and translation gives rise to the movement of material from the front to the back of the pin. The heats rise from friction between tool and workpieces and plastic deformation of the samples. The advantage of the technique is that we can optionally control the depth of the processed zone by adjustment of the length of the pin tool. Until now, literature about the FSP-processed biomedical alloys have been only confined to the TiNi and Ti-6Al-4V alloys [31].

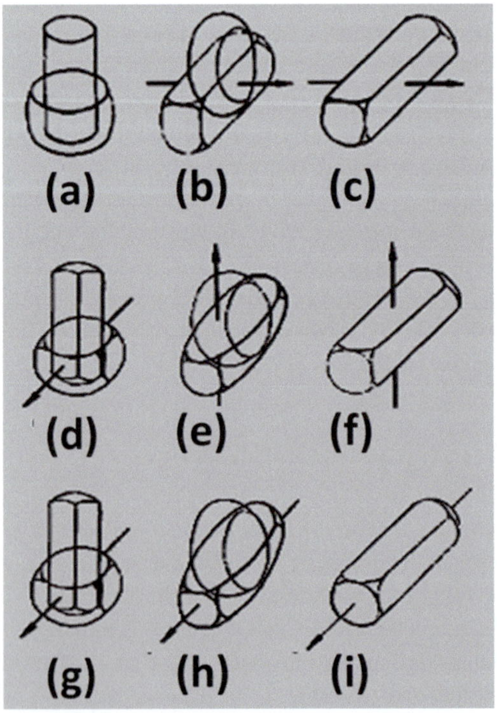

Fig. 17.9 The principle of MDF: (**a**–**c**) show setting and pulling along the first axis, (**d**–**f**) show setting and pulling along the second axis, and (**g**–**i**) show setting and pulling along the third axis

17.4.5 Multi-Directional Forging

Multidirectional forging (MDF) was applied for the first time in the first half of the 1990s for the formation of UFG structures in bulk billets. The process of MDF is usually associated with dynamic recrystallization in single-phase metals/alloys. MDF process has been given in Fig. 17.9. It assumes multiple repeats of free-forging operations including setting and pulling with changes of the axes of the applied load. The homogeneity of the strain produced by MDF is lower than in ECAP and HPT. However, the method can be used to obtain a nanostructured state in rather brittle materials because processing starts at elevated temperatures, and the specific loads on tooling are relatively low. The choice of the appropriate temperature–strain rate regimes of deformation leads to the desired grain refinement. The operation is usually realized over the temperature interval of 0.1–0.5 T_m. It is useful for producing large-sized billets with nanocrystalline structures [32].

17.4.6 Cyclic Extrusion and Compression

Cyclic extrusion and compression (CEC) are performed by pushing a sample from one cylindrical chamber of diameter do to another with equal dimensions through a die with diameter d_m which is markedly smaller than d_o; the principle is illustrated in Fig. 17.10. Thus, the processing induces extrusion and the chambers provide compression so that, during one cycle, the material is pushed to first experience compression, then extrusion, and finally compression again [33]. The true strain produced in one cycle can be calculated by Eq. (17.6).

$$\Delta\varepsilon = 4\ln\left(\frac{d_m}{d_0}\right) \tag{17.6}$$

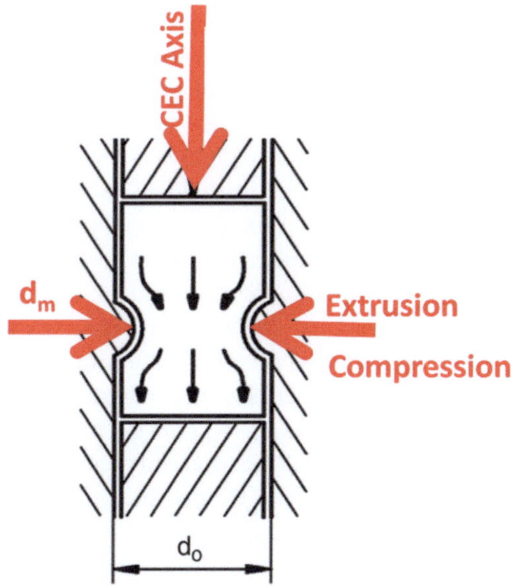

Fig. 17.10 The principle of CEC

In the second cycle, the extrusion direction is reversed, leading to the same sequence of deformation modes. The processes can be repeated N times by pushing the sample back and forth to give an accumulated true strain of ($N\Delta\varepsilon$). With a diameter ratio of typically $d_m/d_o = 0.9$, the strain imposed on the material in one cycle is $\Delta\varepsilon = 0.4$. Accumulated true strains of up to 90 have been reported with sample dimensions of about 25 mm in length and 10 mm in diameter. The deformation speed is as low as ~0.2 mm/s in order to limit heating of the specimen to <5 K. Although the strains reached with this method are much higher than those with any unidirectional SPD technique, the microstructure and/or mechanical properties are similar because of the extra annihilation of dislocations due to the cyclic character of the straining [34].

17.4.7 Repetitive Corrugation and Straightening

The repetitive corrugation and straightening (RCS) process has been shown in Fig. 17.11. Here, a repetitive two-step process, the workpiece is initially deformed to a corrugated shape and then straightened between two flat platens using a processing cycle that may be repeated many times. The RCS facility subjects the workpiece to both bending and shear, which promotes grain refinement. Processing by RCS was used to produce nanostructures in a copper sample with an average initial grain size of 760 μm. A similar procedure was used later for grain refinement of aluminum [35]. An advantage of RCS is that it can be adapted easily to current industrial rolling facilities. It is not difficult to machine a series of corrugating teeth into the rollers of a conventional rolling mill, thus enabling the RCS process, and this has the potential

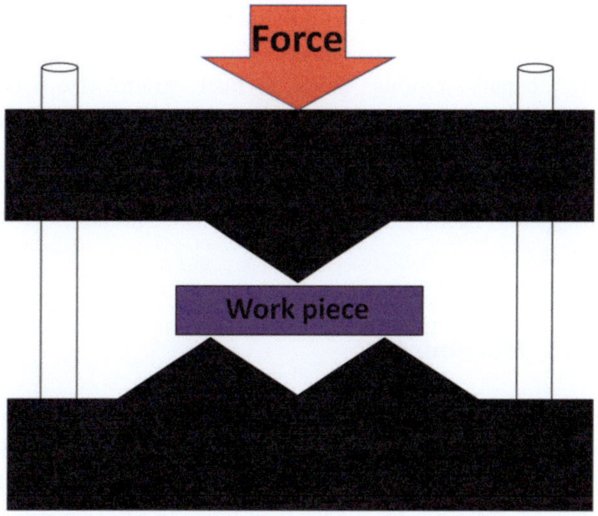

Fig. 17.11 Schematics of RCS process

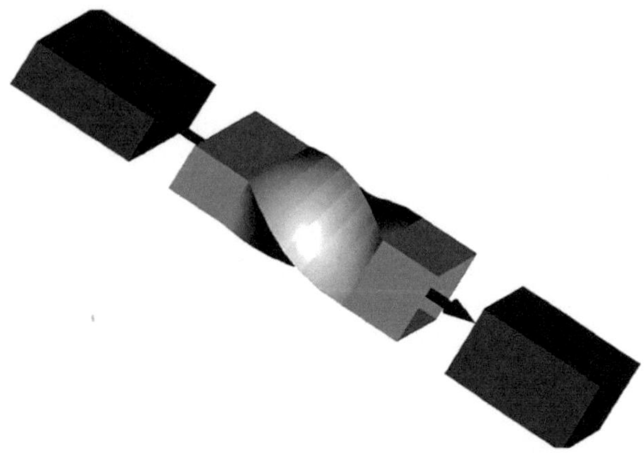

Fig. 17.12 The principle of TE

of producing nanostructured materials in a continuous and economical way. The RCS technique is currently in the early stages and further research is needed to develop the process to a mature SPD technique for producing nanostructured materials. One critical issue is the need to design equipment and processing schedules for improving microstructural homogeneity [36].

17.4.8 Twist Extrusion

The twist extrusion (TE) is illustrated in Fig. 17.12. During TE, a workpiece is pushed through an extrusion die whose cross section maintains its shape and size while it is twisted through a designated angle around its longitudinal axis. As a result, the workpiece regains its shape and size after each TE pass and thus it is possible to repetitively process a sample for excellent grain refinement. A variety of cross-sectional shapes, but not circular geometries, are possible with this technique. In practice, and by analogy to HPT, the plastic strain is not uniform across the cross section but the plastic strain increases with the distance from the axis so that the more distant regions have a finer grain size. This microstructural heterogeneity leads to inhomogeneous mechanical properties with the cross-sectional center having the lowest strength. It is anticipated that the microstructural homogeneity) may improve with increasing numbers of TE passes [37].

17.4.9 Machining

Machining is also a typical deformation process involving large strain. During machining, the total strain can increase up to 13 when a larger negative rake angle is adopted. The values of strain are far higher than those in SPD operations such as equal channel angular extrusion (ECAE) [38]. The chips produced in the machining of a material were entirely or primarily made up of ultrafine grained/nanocrystalline, and they could be more valuable than the material from which the part is being machined. Machining is suitable for producing nanocrystalline chips of strong alloys and metals such as hardened steel and superalloys. The machining operation also provides the possibility of reusing the enormous quantities of chips produced in manufacturing operations in high-value products. Furthermore, no negative impact was found on the machined articles, nanocrystalline chips can be produced as a useful byproduct. Microstructure of the machined carbon steel chip is characterized as ultrafine-grained material with equiaxed submicron grains of 100–300 nm in size. Finite element model is established to study large plastic deformation in the workpiece subjected to the orthogonal machining process. The huge plastic strain imposed in the deformation zone contributes to the grain refinement and the formation of the ultrafine-grained chip. It is feasible to take machining as a method to preparing ultrafine-grained materials and a type of experiment method to study SPD [39].

17.5 Role of Grain Size

Ultrafine grain size metals can exhibit a combination of high strength along with good ductility. Valiev et al. have demonstrated this paradox for Cu and Ti after severe plastic deformation by either ECAP or HPT, respectively. The grain sizes found were 100–200 nm in diameter. More recently, Wang et al. have reported on the preparation and tensile testing of ultrafine-grained Cu [40]. They rolled the Cu to 93% at liquid nitrogen temperatures and then annealed it at low temperatures up to 200 °C. The original heavily cold worked Cu had a high-dislocation density along with some resolvable grains less than 200 nm in size. Annealing resulted in the development of well-defined grains with high-angle boundaries. The annealing treatment (3 min at 200 °C) that optimized strength and ductility produced a mixture of ultrafine grains (80–200 nm) along with about 25% volume fraction of coarser grains (1–3 μm). The coarser grains were the result of secondary recrystallization. The excellent combination of strength and ductility is the result of: (1) multiaxial stress states in the confined grains, (2) twinning in the larger grains, and (3) preferential accommodation of strain in the larger grains [41]. The grain size distribution-spanning the deformation regimes allows for significant strain hardening which prevents localized deformation and premature fracture. By varying the microstructure of ultrafine grain size Zn by changing the milling times at either liquid nitrogen or room temperature, a very dramatic modulated cyclic variation of hardness was observed as a function of milling time at liquid nitrogen temperature. Large variations in the dislocation densities and grain-size distributions were observed as a function of milling time. These observations suggest that dynamic recrystallization takes place in the larger grains (d > 50 nm) when the dislocation density due to strain-hardening reaches a critical level. The major valleys in hardness as a function of milling time (at 2 and 4 h) occur when an anomalous decrease in the volume fraction of larger grains (d > 50 nm) is observed in the otherwise monotonic decrease with milling time. This was consistent with the TEM observations of dynamic recrystallization [42].

From the thermodynamic principle of electrochemistry, UFG materials have lower dissolution potential and therefore higher tendency to dissolution in aggressive media than coarse-grained counterparts because the former has a high density of grain boundaries and higher internal energy. However, the kinetics of corrosion is related with the mode of corrosion such as general or uniform corrosion, intergranular and pitting corrosion, and detailed micromechanism of electrochemical reaction is related with microstructure including inhomogeneity of impurity, solute elements, and precipitation as well as grain size [43]. Let us consider a simple pure metal composed of grain boundaries and grain interior with different dissolution potentials, and it is under an active state in an acidic media. When the acidic media is aggressive, and the oxidizing power, i.e., the half-cell electrode potential of oxidizing agent, is higher than the dissolution potential of both grain boundaries and grain interior, then dissolution proceeds in a uniform manner with little localization [44]. However, oxidizing power is weaker, and its equilibrium electrode potential is ideally lower than dissolution potential of grain interior, and higher than that of grain boundaries, then dissolution proceeds in a local manner, by forming a local cell between grain boundaries and grain interior. In the former case, corrosion rate may increase with decreasing grain size and thus increasing grain boundary area because of high-energy state of the atoms at the defective structures. On the other hand, in the latter case, corrosion rate may start to decrease with decreasing grain size when grain size becomes smaller than a certain critical size [45]. Under the critical grain size, the cathode area cannot counter the increasing grain boundary area as the anode site, resulting in the suppression of the cathode reaction. This grain size dependency becomes pronounced when corrosion kinetics is controlled by cathodic reactions such as diffusion of dissolved oxygen. This may be the reason why the corrosion of UFG pure copper in Livingston's dislocation etchant is slower than that of coarse grain copper [46].

17.6 Role of Grain Boundaries

The role of grain boundary can be understood by the bimodal grain size distribution, with micrometer-sized grains embedded in a sea of nanocrystalline and ultrafine grains, is conducive for good ductility. It was claimed that the nanoscale fraction of the grain population was responsible for high strength, while the coarse grains contributed to ductility by promoting strain hardening necessary to stabilize the tensile deformation against strain the localizations [47]. The interaction between dislocations and grain boundaries determines a large number of important aspects of the mechanical performance of materials, including the strength and ductility. The role of the dynamic interplay between dislocation and grain boundary processes increases progressively with grain refinement, and finally, it dominates overall responses as grain sizes approach the nanoscale [48]. Numerous results show the tremendous influence that the grain boundary state, distribution of misorientations, and grain sizes has on the mechanical response of a material after SPD and subsequent heat treatment. Indeed, the grain boundaries present effective obstacles to dislocation motion, since dislocations coming upon a boundary generally do not have the proper Burgers vector and slip plane to glide into the next crystal. Most commonly, the elastic interaction between lattice dislocations and grain boundaries is repulsive, and consequently, the dislocations tend to pile up or accumulate at the boundary. Alternatively, dislocations may be transmitted directly across the boundary if the slip planes on both sides intersect along a line that lies in the boundary plane or dislocation may be absorbed by the boundary without emission of a dislocation in the adjacent grain. Providing the case that the misorientation angle between two neighboring domains is small (low-angle grain boundary, LAGB), the misorientation is achieved by lattice distortions induced by dislocation arrays [49]. The mobility of dislocations constituting LAGB is restricted. Under the influence of an external force or thermal fluctuations, they can travel short distances, elastically interact with each other and lattice dislocations, and annihilate or unpin from the boundary. Therefore, these boundaries are thermally and mechanically unstable and they take an active part in the recovery process. If the dynamic recovery rate is controlled by dislocation annihilation in the LAGB and if this rate is high as is observed in UFG metals after modest strain imposed during processing, the hardening rate reduces sharply. Therefore, the necking critical condition is fulfilled at relatively small strains. In contrast, the high-angle grain boundaries, which cannot be associated with dislocations, are rather stable [50]. The ductility increasing with increasing large strains in the course of SPD is associated with an increasing fraction of HAGBs. Zhao et al. have convincingly demonstrated that the presence of general HAGBs combined with a low dislocation density is more effective in enhancing the uniform elongation of UFG metals than LAGBs and a high intragranular dislocation density. The capacity of grain boundaries to absorb lattice dislocations and, thus, remove them from the hardening process or transmit the slip without dislocation storage is strongly dependent on their atomic structure [51].

Grain boundaries in UFG materials processed by SPD are often regarded as "non-equilibrium" ones. Such boundaries are characterized by excess grain boundary energy, the presence of long-range elastic stresses, and enhanced free volume, which is believed to lead to much greater diffusivities than those of equilibrium grain boundaries in well-annealed coarse-grained materials. The importance of a nonequilibrium state of HAGB for UFG materials with unusual properties has long been recognized. Nonetheless, there is still no evidence about the significance of nonequilibrium grain boundaries for enhanced ductility. One can speculate, however, that the ultrafast diffusivity of nonequilibrium grain boundaries and a strong driving force to recovery promotes their interaction with lattice dislocations. This, in turn, increases the rate of recovery reducing, thereby, the uniform elongation as is observed regularly in experiments. Among the HAGB, the so-called coincidence site lattice (CSL) or special boundaries, possess the lowest energy and extraordinary stability. Particularly, twin boundaries including coherent twins have been proven effective in attempts to optimize strength and ductility [52].

Lu et al. have identified three essential structural characteristics for grain boundaries engineering: (1) coherency with surrounding matrix, (2) thermal and mechanical stability, and (3) smallest feature size finer than 100 nm [53]. Such boundaries can be intentionally generated in materials through a variety of processing routes to tune the desired properties. Electrodeposition, plastic deformation at cryogenic or low homologous temperature, dynamic deformation, and conventional annealing are among those most often used for this purpose. One can conclude overall that in parallel with grain refinement, tailoring the grain boundaries, their distribution, structure, and properties, constitutes a nowadays paradigm shift in the design of multifunctional UFG and bulk nanomaterials [54, 55].

Mostly high-angle grain boundaries leading to grain refinement can be formed after optimization of SPD processing routes and these grain boundaries possess specific nonequilibrium structures. Nonequilibrium grain boundaries exist in UFG materials and these specific grain boundaries possess an increased free energy density, increased width, high density of dislocations (full or partial) associated with the near-boundary region, and correspondingly large residual microstrain. The structural width of nonequilibrium grain boundaries is significantly smaller than 10 nm, if the rotational component of the strain gradient across the interface is used as a measure of the width. It reaches a value of 1.5–2.0 nm being at least twice as large as the width of relaxed HAGB in annealed coarse-grained materials. Figure 17.13 shows optical micrographs illustrating the grain and grain boundaries microstructural of the Al-4%Cu alloy with ECAP deformation [56].

17.7 Diffusion along Grain Boundaries

The diffusion is a highly sensitive probe for the investigation of structural modifications on the atomic scale since the thermally activated diffusivity depends exponentially on the corresponding activation barriers which are determined by the interatomic potentials and the atomic environment. The measurement of the atomic mobility at low temperatures, when diffusion within undisturbed regions of the crystal lattice is frozen, can be used for analyzing the structural modifications of short-circuit diffusion paths, such as grain boundaries. The preceding analysis demonstrated that SPD processing modifies the structure and thermodynamics of interfaces introducing various defects (e.g. abundant GB vacancies and dislocations at and near interfaces) and sometimes induces segregation [57]. First direct measurements of grain boundary diffusion in severely deformed materials yielded ambiguous results—both similar and enhanced rates of atomic transport were deduced with respect to the grain boundary diffusivities in reference to coarse-grained materials. Systematic measurements by the radiotracer technique discovered a hierarchic nature of internal interfaces, which are developing as a result of strong dislocation activity during SPD processing and, presumably, of localization of plastic flow. The existence of a hierarchy of interfaces in plastically deformed metals has been pointed out by Hansen and coworkers by introducing the so-called extended or geometrically necessary boundaries (GNBs) and incidental dislocation boundaries (IDBs) [58]. The diffusion studies indicate another type of hierarchy which corresponds to other kinds of involved interfaces, since the diffusivity of dislocation or low-angle dislocation grain boundaries is definitely lower than that of general high-angle interfaces. Generalizing these findings, the following hierarchy of interfaces in SPD materials can be proposed [59, 60]: (1) nonequilibrium interfaces (probably of different types and representing a certain spectrum of diffusivities and structures), (2) general high-angle grain boundaries (with diffusivities and, probably, structure being similar to those of relaxed high-angle grain boundaries), (3) highly defected (nonequilibrium) twin boundaries with diffusivities similar to those of the previous level. Note that diffusion along relaxed twin boundaries is hardly measurable, (4) low-angle boundaries, dislocation walls, single dislocations. SPD processing and grain refinement have to include all these levels of the hierarchy depend on processing routes and regimes (temperature, strain rate, applied pressure, etc.). It is important that these levels correspond to different scales with the mesh size ranging from several micrometers (the nonequilibrium boundaries) down to a hundred (dislocation walls) or even tens of nanometers (nano-twins). The appearance of general HAGB in SPD materials depends obviously

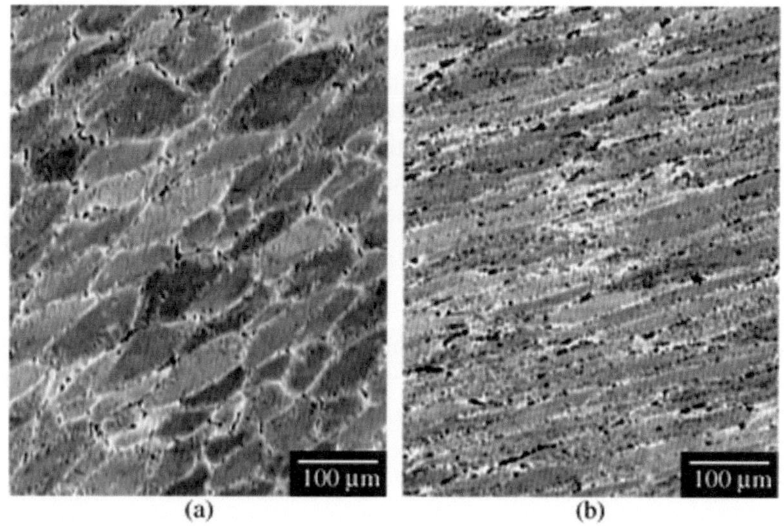

Fig. 17.13 Optical micrographs illustrating the microstructural evolution of the Al-4% Cu alloy with ECAP deformation (**a**) before and (**b**) after processing

on processing temperature because this fact might be a clear indication of dynamic recovery processes during SPD. The phenomena that contribute to the nonequilibrium state of grain boundaries in SPD processed materials are: (1) abundant vacancies and vacancy-like defects in interfaces produced by severe deformation; (2) redistribution of the related excess free volume, release of local strains/stresses; (3) chemical effects (ordering) may be important in alloys and compounds affecting the atomic redistribution and retarding, e.g. the stress/strain relaxation; and (4) segregation can be especially important in alloys involving even 2D compound formation along interfaces.

17.8 Influence of Second Phases

Most engineering structural materials are multiphase with, often nanoscale, precipitates or dispersoids act as the source of strengthening. Classic examples include age-hardened Al-based alloys and some steels. In terms of ductility enhancement in nanostructured metals, there has been little systematic work on the influence of second-phase particles in bulk nanostructured matrices [61]. However, the force instability limitation to ductility is common to metallic glasses and at least to some nanostructured metals. There has been significant success in the addition of second-phase particles to bulk metallic glasses which can interact with shear bands and enhance ductility. The strengthening effects of second-phase particles are well understood in conventional grain size metals with regard to the size, distribution, and mechanical properties of the second phase [62]. Hard dispersion particles increase the yield stress due to the Orowan mechanism and also increase subsequent strain hardening rates due to dislocation storage processes required for compatible plastic strain. While strain hardening effects may be absent for soft, shearable particles, they have been shown to be effective in modifying shear band propagation in bulk metallic glasses, thereby increasing ductility [63].

17.9 Effect of Internal Stress

Another important factor which may impact corrosion resistance and may occur commonly in SPD is residual stress or internal stress. The tensile and compressive stress fields are distributed between the surface and interior of components [64]. The residual stress arises from inhomogeneous thermal treatment and plastic strain that occurs in a common form of thermomechanical processing. On the other hand, the notion of internal stress is rather new and considered to arise from deformation-induced microstructures, and its tensile and compressive stress fields are distributed in grain-size scale. The high internal stress has often attributed to nonequilibrium grain boundaries [65]. Grain boundaries generated by plastic deformation are at a high-energy state with excessive grain boundaries dislocations, or alternatively called extrinsic grain boundary dislocations that extend long-range stress field into the grain interior. The high energy nonequilibrium grain boundaries cause fast diffusion, and resultant room temperature grain boundary sliding. Under the high internal stress, atoms at the surface are at a high free energy state with a low electron work function of the surface [66]. This facilitates either anodic dissolution or the formation of the passive film. The high internal stress can be relieved by annealing for the short time and at low temperature with little grain size change, possibly accompanying the transition from the nonequilibrium to an equilibrium state by dissipating extrinsic grain boundary dislocations. Miyamoto et al. examined the corrosion behavior of UFG copper by ECAP in a modified Livingston's dislocation etchant, and it is surprising that corrosion current increase by the post-ECAP annealing of the temperature of 200 °C for 90 s with little grain size change [67]. The annealing relieves the internal stress and decreases the free energy of grain interiors, which increases the energy difference between the two, resulting in an intergranular mode of corrosion. It is unlikely that macroscopic residual stress caused by plastic strain can be relieved by this short-time annealing. However, internal stress plays dual-edge roles, i.e., acts as an enhancer by facilitating the passive film formation, or a degrader by introducing crack or defective film. As the internal stress has dual-edge roles, so does the annealing intended for stress relief. Indeed, Kim et al. reported improvement of corrosion resistance in pure titanium and AZ61 by annealing [68].

17.10 Effect on Mechanical Behavior

The elastic properties of Cu (99.98% purity) have been studied using an ultrasound method. The samples were subjected to severe deformation up to $\varepsilon = 4$ and subsequently annealed for 1 hour at various temperatures within the interval 100–350 °C. The grain growth took place from 0.2 to 4 μm after annealing at higher than 175 °C. The measurements of the rate of longitudinal and transverse waves allowed us to estimate the values of Young's modulus E and the shear modulus G and their dependence on the annealing temperature. For both curves within the region 150–175 °C, an abrupt change in the E and G values occurs: for large grain sizes, the elastic properties depend weakly on these sizes, whereas for small grain sizes, as in typical nanocrystals, the elastic properties are considerably lower than in coarse-grained samples. The structural investigations showed that the observed variations of the elastic parameters were caused by the change in the GB structure state in the region of very small grain sizes [69]. Actually, in the Cu samples subjected to severe deformation and subsequently annealed at temperatures up to 150 °C, the mean grain size was about 0.2 μm and nonequilibrium GBs prevailed. At the same time, in the samples annealed at 175 °C and higher, the GBs were mainly at equilibrium. A comparative mechanical strength behavior of ultrafine grain materials with fine grain and coarse grain materials has been given in Fig. 17.14.

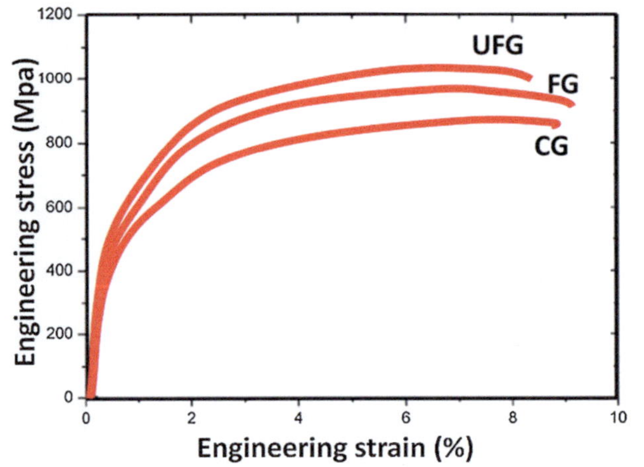

Fig. 17.14 Mechanical strength comparison of UFG, fine grain, and coarse grain materials

The cyclic deformation and fatigue behavior of UFG materials prepared by ECAP has only been studied more systematically quite recently, with special reference to UFG copper as a so-called model material, compared to the early research. In this context, the fatigue resistance of technical UFG materials, compared to that of their conventional grain-sized (CG) counterparts, is of considerable interest [70]. It has been recognized that the behavior of materials near the crack tip will ultimately form the basis for all fracture theories. In line with this particular basic thought, a (high strain) LCF life prediction model for ultrafine-grained metals was proposed. The microstructure of a UFG metal was treated as a two-phase "composite" in which all the grain interiors are considered as a "soft" matrix and all the grain boundaries are viewed as the "hard" reinforcement. This so-called composite model of the microstructure of the UFG metals is illustrated. The dislocation strengthening of the grain interiors is considered as the major strengthening mechanism in the case of UFG metals. The proposed model is based upon the assumption that there is a fatigue-damaged zone (FDZ) ahead of the crack tip within which the actual degradation of the UFG metal takes place. In high-strain LCF conditions, the fatigue-damaged zone is described as the region in which the local cyclic stress level approaches the ultimate tensile strength of the UFG metal, with the plastic strain localization caused by a dislocation sliding-off process within it [71].

Wear resistance is an important property for UFG materials in order to evaluate their potential for use as structural components. The wear of sliding surfaces can occur by one or more wear mechanisms, including adhesion, abrasion, fatigue wear, corrosive wear, and fretting. For the metallic materials, the wear volume under abrasive and some adhesive wear models are generally assumed to be inversely proportional to the hardness of the materials according to the traditional Archard relationship which is given by Eq. (17.7) [72].

$$V = K\frac{\mathrm{LN}}{H} \tag{17.7}$$

where V is the wear loss of the volume, N is the applied force, L is the sliding distance, K is the wear coefficient, and H is the hardness on the wear surface of the material. Because the UFG materials processed by SPD techniques normally have much higher hardness values than the conventional coarse-grained materials, it is critical to have superior wear resistance for UFG materials. There are a number of investigations reporting an improved wear resistance in

UFG materials produced by ECAP and HPT. For example, the dry sliding wear tests of an aluminum alloy processed by the ECAP method showed that the wear mass loss decreased significantly with increasing of the number of ECAP passes. An investigation of friction and wear behavior revealed that grain size was the important factor determining the transition from elastohydrodynamic lubrication to the boundary lubrication regions. An investigation of the aluminum bronze alloy processed by ECAP demonstrated that the coefficient of friction decreased with increasing number of ECAP passes and accordingly the wear resistance was improved significantly after ECAP processing. Similarly, a characterization of the dry sliding wear behaviors of Cu-0.1 wt% Zr alloy and AZ31 alloy processed by ECAP was investigated, and the wear volume loss of the samples processed by ECAP becomes much lower than the annealed alloy due to the higher microhardness introduced by ECAP processing. Processing by ECAP can produce bulk materials with significantly enhanced mechanical properties due to the grain refinement, and therefore the wear loss of the ECAP-processed alloy is much smaller than for the annealed alloy. Compared with the coarse-grained pure Ti, the wear resistance of pure Ti processed by HPT was improved significantly both in dry and wet sliding tests. On the other hand, there are also some other contradictory results on wear property in UFG materials. For example, the wear resistance of some UFG materials processed by ECAP was lower than for the as-received coarse-grained materials [73]. For example, the dry sliding wear tests of an Al-1050 alloy were conducted with the as-received condition and UFG materials with grain size of ~1.3 μm after ECAP processing through eight passes. The UFG samples have a similar coefficient of friction (COF) and a higher wear loss than the as-received sample although the microhardness value is improved significantly after ECAP processing. An investigation of UFG AISI 1024 steel processed by a warm multiaxial forging technique showed that there is no obvious improvement on wear resistance property though the strength property can be enhanced significantly due to the effects of higher density of grain boundaries and submicrometer sized cementite particles. There is a surprising result that there is no corresponding improvement in the wear resistance in pure titanium processed due to the occurrence of oxidative wear with an abrasive effect. As a consequence of these varying reports, it is readily apparent that further investigations should be further conducted in order to evaluate the wear behavior of UFG materials processed by SPD techniques [74].

17.11 Corrosion Behavior

The effect of grain size reduction on corrosion resistance is mostly positive in stainless steel and aluminum alloys, whereas the effect is marginal in copper and titanium alloys. However, there are several contradictory reports involving the same materials and environment [75]. The limited available literature on the effect of grain size on stress corrosion cracking (SCC) reports an increasing resistance to SCC with decreasing grain size. Edmund investigated the effect of grain size on the SCC of α-brass in an ammonia environment using a constant-load test and reported increasing fracture time with decreasing grain size [76]. The time to fracture by SCC increased with decreasing grain size down to 1μm but then decreased with further decreases in grain size into the submicron scale. In other words, there was a critical grain size above which the susceptibility began to increase, and this grain size matched the grain size that divides the two Hall–Petch relationships. Stress corrosion cracks propagated intergranular regardless of grain size. SPD-induced grain boundaries have a high sensitivity to chemical reactions and intergranular SCC. SCC sensitivity of UFG materials by SPD has far less been studied as compared to corrosion property. One reason for this limited SCC studies of UFG materials is that structural materials which exhibit SCC are generally high alloys having high strength such as austenitic stainless steel and Cu–Zn alloys, and hard to be processed by SPD [77].

17.12 Applications

Application and commercialization of UFG materials are on the way. Fine wires have a wide array of applications in fine mesh, fine brush, semiconductor, medical stent, jewelry, automotive, etc. Here, the fine wires are manufactured from several different materials, such as tungsten, copper, stainless steel, magnesium, and gold by a multi-pass wire drawing process [78–80]. The research on this class of materials is being carried out for application where load-carrying capacity is dominant. The first commercial application of bulk UFG metals was in sputtering targets for physical vapor deposition. Honeywell Electronic Materials, a division of Honeywell International Inc., offers UFG Al and Cu sputtering targets up to 300 mm in diameter which are produced from plates by ECAP [79]. The main advantages of UFG sputtering targets, compared to their coarse-grained (CG) counterparts, are: (1) the life span increased by 30% due to stronger material which allows the use of monolithic targets, and (2) a more uniform deposited coating which results from reduced arcing (Fig. 17.15a). Bolt and micro bolt are manufactured with titanium alloys processed by ECAP are used in the automobile and aircraft industries (Fig. 17.15b) [80].

Figure 17.16a is showing the ultrafine-grained stainless steel strip commercially produced by Tokushu Kinjoku excel co. ltd., Japan and Fig. 17.16b, c showing the ultrafine nickel mesh and ultrafine grain size cemented carbide H5H6 grinding rod, respectively, produced by China [81, 82].

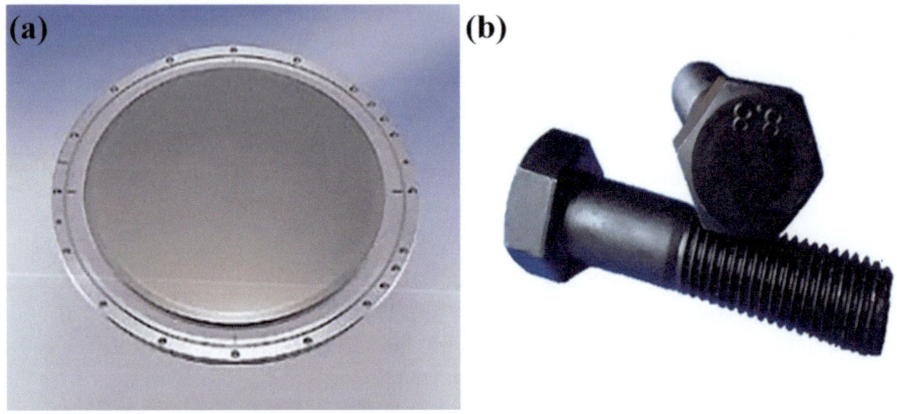

Fig. 17.15 (**a**) Sputtering target, (**b**) Microbolt

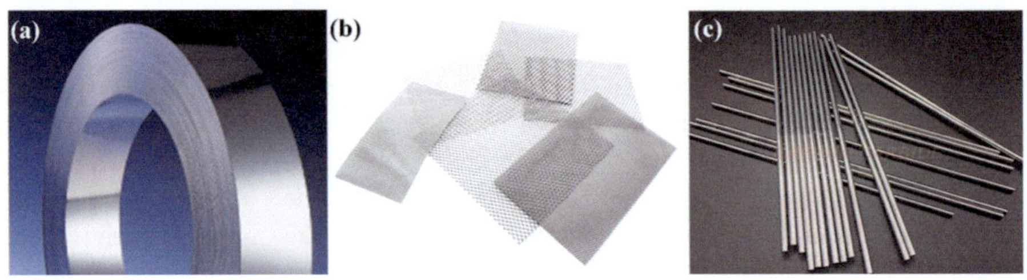

Fig. 17.16 (**a**) UFG stainless steel strips, (**b**) ultrafine nickel mesh, (**c**) ultrafine grain size cemented carbide

The UFG pure titanium processed by ECAP-Conform from the Ufa State Aviation Technical University under the management of professor Valiev has been used as a trademark application to manufacture dental implants in the company "Timplant" (Ostrava, Czech Republic) since 2006 [83]. The UFG Ti with an ultimate strength of 1350 MPa enabled the design of a thin dental implant with diameter of 2.0 mm, which serves as fully functional pillar, and it can be inserted into very thin bones. Another advantage of smaller dental implants is less damage induced into the jawbone during surgical intervention. Another dental implant product with UFG Ti produced by SPD was manufactured and sold by basic implant systems under the trademark Biotanium in the USA beginning in 2011 [84]. Thus, the small-diameter dental implants made from UFG Ti are possible to replace standard ones made from Ti-4Al-6V alloy, since the UFG pure Ti is characterized not only by the improved mechanical strength and fatigue life but also by better biocompatibility compared to the conventional Ti-4Al-6V alloy [85].

Producers of sports devices/equipment can also benefit from the UFG metals, particularly where high strength and low weight are required. The UFG materials could find applications in high-performance golf, bicycles, tennis, hockey, mountain equipment, etc. [86–88]. One of the important examples is nano-dynamic high-performance golf balls (Fig. 17.17), which have a hollow nanostructured titanium core. The core material is manufactured using the UFG chip from Purdue University. The Institute for Metals Superplasticity Problems (Russia) has developed a technology for the fabrication of golf club components from UFG Ti-6Al-4 V alloy with a grain size of 200 nm. The method for producing the goffer-type face using UFG or nanostructured metals and inserts provided processing faces characterized by enhanced strength and high-impact efficiency. This technology allowed a reduction in the weight of a golf club along with the increase of ball's flight distance due to the increased restitution factor. These application results demonstrate wide commercial potentialities for applying UFG materials processed by SPD.

The Potential application of UFG materials is in micro-forming technology. Micro-forming is defined as the production of parts or structures having at least two dimensions in the submillimeter range, which becomes an attractive option in the manufacturing of UFG products because of its advantages for mass production with controlled forming quality, high production rate, and low cost. Nevertheless, although the knowledge of tool design and fabrication techniques are now well developed for the conventional macroforming, there is an evidence that the occurrence of size effects may lead to a breakdown in these basic plastic deformation theory when the specimen dimensions are scaled down to the micro/mesoscale [89]. In practice, if there are only a few grains in the micro-parts, the response to the applied forces will show significant variations, and the reproducibility of the mechanical properties will become a serious problem in any micro-forming processes. Hopefully, there is a way to solve grain size effects in micro/meso-forming by applying UFG materials with submicrometer or even nanoscale grain sizes produced by SPD techniques because ultrafine grains can improve the micro-formability, surface roughness, and good mechanical properties of the MEMS components [90]. However, microdeformation behavior

Fig. 17.17 Components of golf club made from nanostructured Ti-6Al-4V alloy

Fig. 17.18 UFG material-assisted lighter armor for military vehicles

changes from dislocation dominated in large grains to grain boundary dominated in small-grain regimes when the grain size decreases to the submicron range. For example, the deformation behavior in UFG pure aluminum processed by ECAP and post-annealed specimens at room temperature was investigated, and the results show that different work hardening behaviors were observed during the macro-compression test when the grain size increased from 0.35 to 45 μm. The strain rate also has an obvious effect on the micro-compression behavior of UFG pure aluminum, and the results demonstrate that a lower strain rate causes activation of micro-shear banding, and the deformation mechanism may be related to grain boundary sliding in UFG pure aluminum [91].

The defense industry could benefit from two large-scale applications of UFG metals, which are armor plates and armor penetrators [92]. Lighter armor for military vehicles (Fig. 17.18) is crucial for the reduction of fuel consumption, higher speed, better maneuverability, longer operation range, and air-transport of vehicles to remote locations. At the same time, the ballistic performance must not be reduced. This can be achieved by the nanostructuring of aluminum or titanium alloys traditionally used for light armored vehicles [93]. A good example is a UFG Al 5083 plate, which was obtained by cryogenic ball milling, consolidation by HIP, forging or extrusion, and finally rolling. With the yield strength of 600–700 MPa and elongation of 11%, the material exhibited a 33% improvement in the ballistic performance or a similar mass reduction compared to the standard plate. Improvements in ballistic performance are also reported for the electrodeposited nanocrystalline Ni–Fe alloys produced by Integran Technologies [94]. Armor structures are usually fabricated by welding of plates. However, traditional welding based on melting is destructive to the UFG material. An alternative technique is a solid-state process of friction stir welding, which has the ability to refine grain structure. This results in the weld hardness being only marginally reduced compared to the initial hardness of a UFG material.

17.13 Summary

Material development is basically dealing with the requirement, the adaptation of process with selecting proper elements in the periodic table. Ultrafine materials now following the same stage in this era. This chapter described the concept of the name "ultrafine" which is mainly post-processing dependent. Grain and grain boundary plays the superior role to enhance the mechanical strength of the materials. Various influencing parameters such as diffusion along grain boundary, second phase formation, residual stress, internal stress, etc. have been discussed. In the last section of this chapter is trying to show the current market scenario of UFG materials with various applications.

References

1. Valiev, R.Z., Estrin, Y., Horita, Z.: Producing bulk ultrafine-grained materials by severe plastic deformation. JOM. **58**, 33–39 (2006). https://doi.org/10.1007/s11837-006-0213-7
2. Huang, Y., Langdon, T.G.: Advances in ultrafine-grained materials. Mater. Today. **16**(3), 85–93 (2013). https://doi.org/10.1016/j.mattod.2013.03.004
3. Tsuji, N., Saito, Y., Lee, S.H., Minamino, Y.: ARB (Accumulative Roll-Bonding) and other new techniques to produce bulk ultrafine grained materials. Adv. Eng. Mater. **5**(5), 338–344 (2003). https://doi.org/10.1002/adem.200310077
4. Terence, G.L.: Twenty-five years of ultrafine-grained materials: Achieving exceptional properties through grain refinement. Acta Mater. **61**(19), 7035–7059 (2013). https://doi.org/10.1016/j.actamat.2013.08.018
5. Kawasaki, M., Langdon, T.G.: Review: achieving superplastic properties in ultrafine-grained materials at high temperatures. J. Mater. Sci. **51**, 19–32 (2016). https://doi.org/10.1007/s10853-015-9176-9
6. Nguyen, N.T., Asghari-Rad, P., Sathiyamoorthi, P., et al.: Ultrahigh high-strain-rate superplasticity in a nanostructured high-entropy alloy. Nat. Commun. **11**, 2736 (2020). https://doi.org/10.1038/s41467-020-16601-1
7. Cao, V.C., GongchengYao, L.J., Sokoluk, M., Wang, X., Ciston, J., Javadi, A., Guan, Z., De Rosa, I., Xie, W., Lavernia, E.J., Schoenung, J.M., Li, X.: Bulk ultrafine grained/nanocrystalline metals via slow cooling. Sci. Adv. **5**(8), eaaw2398 (2019). https://doi.org/10.1126/sciadv.aaw2398
8. Czerwinski, F.: Thermomechanical processing of metal feedstock for semisolid forming: a review. Metall. Mater. Trans. B Process Metall. Mater. Process. Sci. **49**, 3220–3257 (2018). https://doi.org/10.1007/s11663-018-1387-4
9. Mughrabi, H., Höppel, H.W.: Cyclic deformation and fatigue properties of very fine-grained metals and alloys. Int. J. Fatigue. **32**(9), 1413–1427 (2010). https://doi.org/10.1016/j.ijfatigue.2009.10.007
10. Terence, G.L.: Ultrafine-grained materials: a personal perspective. Int. J. Mater. Res. **98**(4), 251–254 (2007). https://doi.org/10.3139/146.101473
11. See: http://damascus.free.fr/f_damas/f_quest/f_wsteel/indiaw.htm. 18.05.2020
12. Bruder, E.: Formability of ultrafine grained metals produced by severe plastic deformation–an overview. Adv. Eng. Mater. **21**(1), 1800316 (2019). https://doi.org/10.1002/adem.201800316
13. Kawasaki, M.: Processing of ultrafine-grained materials through the application of severe plastic deformation. Metall. Mater. Trans. A. **42**, 3035–3045 (2011). https://doi.org/10.1007/s11661-010-0501-2
14. Kawasaki, M., Langdon, T.G.: Principles of superplasticity in ultrafine-grained materials. J. Mater. Sci. **42**, 1782–1796 (2007). https://doi.org/10.1007/s10853-006-0954-2
15. Luo, P., McDonald, D.T., Xu, W., Palanisamy, S., Dargusch, M.S., Xia, K.: A modified Hall–Petch relationship in ultrafine-grained titanium recycled from chips by equal channel angular pressing. Scr. Mater. **66**(10), 785–788 (2012). https://doi.org/10.1016/j.scriptamat.2012.02.008
16. Valiev, R.Z., Korznikov, A.V., Mulyukov, R.R.: Structure and properties of ultrafine-grained materials produced by severe plastic deformation. Mater. Sci. Eng. A. **168**(2), 141–148 (1993). https://doi.org/10.1016/0921-5093(93)90717-S
17. Valiev, R.Z., Kozlov, E.V., Ivanov, Y.F., Lian, J., Nazarov, A.A., Baudelet, B.: Deformation behaviour of ultra-fine-grained copper. Acta. Metall. Mater. **42**(7), 2467–2475 (1994). https://doi.org/10.1016/0956-7151(94)90326-3
18. Hayes, R.W., Tellkamp, V., Lav, E.J., Hayes, R., Tellkamp, V., Lavernia, E.: Creep behavior of a cryomilled ultrafine-grained Al–4% Mg alloy. J. Mater. Res. **15**(10), 2215–2222 (2000). https://doi.org/10.1557/JMR.2000.0318
19. Kawasaki, M., Ahn, B., Kumar, P., Jang, J.-i., Langdon, T.G.: Nano- and micro-mechanical properties of ultrafine-grained materials processed by severe plastic deformation techniques. Adv. Eng. Mater. **19**(1), 1600578 (2017). https://doi.org/10.1002/adem.201600578
20. Saray, O., Purcek, G., Karaman, I., Maier, H.J.: Improvement of formability of ultrafine-grained materials by post-SPD annealing. Mater. Sci. Eng. A. **619**(014), 119–128 (2014). https://doi.org/10.1016/j.msea.2014.09.016
21. Iwahashi, Y., Furukawa, M., Horita, Z., et al.: Microstructural characteristics of ultrafine-grained aluminum produced using equal-channel angular pressing. Metall. Mater. Trans. A. **29**, 2245–2252 (1998). https://doi.org/10.1007/s11661-998-0102-5
22. Matsunoshita, H., Edalati, K., Furui, M., Horita, Z.: Ultrafine-grained magnesium–lithium alloy processed by high-pressure torsion: low-temperature superplasticity and potential for hydroforming. Mater. Sci. Eng. A. **640**, 443–448 (2015). https://doi.org/10.1016/j.msea.2015.05.103
23. Zherebtsov, S., Kudryavtsev, E., Kostjuchenko, S., Malysheva, S., Salishchev, G.: Strength and ductility-related properties of ultrafine grained two-phase titanium alloy produced by warm multiaxial forging. Mater. Sci. Eng. A. **536**, 190–196 (2012). https://doi.org/10.1016/j.msea.2011.12.102
24. Park, K.-T., Kwon, H.-J., Kim, W.-J., Kim, Y.-S.: Microstructural characteristics and thermal stability of ultrafine grained 6061 Al alloy fabricated by accumulative roll bonding process. Mater. Sci. Eng. A. **316**(1–2), 145–152 (2001). https://doi.org/10.1016/S0921-5093(01)01261-8
25. Segal, V.: Review: modes and processes of severe plastic deformation (SPD). Materials. **11**, 1175 (2018). https://doi.org/10.3390/ma11071175
26. Melicher, R.: Numerical simulation of plastic deformation of aluminiumworkpiece induced by ECAP technology. Appl. Comput. Mech. **3**, 319–330 (2009)
27. Kim, H.S.: Finite element analysis of equal channel angular pressing using a round corner die. Mater. Sci. Eng. A. **315**(1–2), 122–128 (2001). https://doi.org/10.1016/S0921-5093(01)01188-1
28. Čížek, J., Procházka, I., Smola, B., Stulíková, I., Kužel, R., Matěj, Z., Cherkaska, V., Islamgaliev, R.K., Kulyasova, O.: Microstructure and thermal stability of ultra fine grained mg-based alloys prepared by high-pressure torsion. Mater. Sci. Eng. A. **462**(1–2), 121–126 (2007). https://doi.org/10.1016/j.msea.2006.01.177
29. Ning, J.-l., Courtois-Manara, E., Kurmanaeva, L., Ganeev, A.V., Valiev, R.Z., Kübel, C., Ivanisenko, Y.: Tensile properties and work hardening behaviors of ultrafine grained carbon steel and pure iron processed by warm high pressure torsion. Mater. Sci. Eng. A. **581**, 8–15 (2013). https://doi.org/10.1016/j.msea.2013.05.008
30. Milner, J.L., Abu-Farha, F., Bunget, C., Kurfess, T., Hammond, V. H.: Grain refinement and mechanical properties of CP-Ti processed by warm accumulative roll bonding. Mater. Sci. Eng. A. **561**, 109–117 (2013). https://doi.org/10.1016/j.msea.2012.10.081
31. Hofmann, D.C., Vecchio, K.S.: Submerged friction stir processing (SFSP): an improved method for creating ultra-fine-grained bulk materials. Mater. Sci. Eng. A. **402**(1–2), 234–241 (2005). https://doi.org/10.1016/j.msea.2005.04.032

32. Takayama, A., Yang, X., Miura, H., Sakai, T.: Continuous static recrystallization in ultrafine-grained copper processed by multi-directional forging. Mater. Sci. Eng. A. **478**(1–2), 221–228 (2008). https://doi.org/10.1016/j.msea.2007.05.115
33. Swaminathan, S., Shankar, M.R., Lee, S., Hwang, J., King, A.H., Kezar, R.F., Rao, B.C., Brown, T.L., Chandrasekar, S., Compton, W.D., Trumble, K.P.: Large strain deformation and ultra-fine grained materials by machining. Mater. Sci. Eng. A. **410–411**, 358–363 (2005). https://doi.org/10.1016/j.msea.2005.08.139. ISSN 0921-5093
34. Ensafi, M., Faraji, G., Abdolvand, H.: Cyclic extrusion compression angular pressing (CECAP) as a novel severe plastic deformation method for producing bulk ultrafine grained metals. Mater. Lett. **197**, 12–16 (2017). https://doi.org/10.1016/j.matlet.2017.03.142
35. Richert, M.W.: Features of cyclic extrusion compression: method, structure & materials properties Solid state phenomena, vol. 114, pp. 19–28. Trans Tech Publications, Ltd., Freienbach (2006). https://doi.org/10.4028/www.scientific.net/ssp.114.19
36. Huang, J., Zhu, Y.T., Alexander, D.J., Liao, X., Lowe, T.C., Asaro, R.J.: Development of repetitive corrugation and straightening. Mater. Sci. Eng. A. **371**(1–2), 35–39 (2004). https://doi.org/10.1016/S0921-5093(03)00114-X
37. Beygelzimer, Y., Varyukhin, V., Synkov, S., Orlov, D.: Useful properties of twist extrusion. Mater. Sci. Eng. A. **503**(1–2), 14–17 (2009). https://doi.org/10.1016/j.msea.2007.12.055
38. Deng, W.J., Xia, W., Li, C., Tang, Y.: Ultrafine grained material produced by machining. Mater. Manuf. Processes. **25**(6) (2010). https://doi.org/10.1080/10426910902748024
39. Morehead, M., Yong Huang, K., Hartwig, T.: Machinability of ultrafine-grained copper using tungsten carbide and polycrystalline diamond tools. Int. J. Mach. Tools Manuf. **47**(2), 286–293 (2007). https://doi.org/10.1016/j.ijmachtools.2006.03.014
40. Fan, G.J., Choo, H., Liaw, P.K., Lavernia, E.J.: Plastic deformation and fracture of ultrafine-grained Al–Mg alloys with a bimodal grain size distribution. Acta Mater. **54**(7), 1759–1766 (2006). https://doi.org/10.1016/j.actamat.2005.11.044
41. Koch, C.C.: Optimization of strength and ductility in nanocrystalline and ultrafine grained metals. Scr. Mater. **49**(7), 657–662 (2003). https://doi.org/10.1016/S1359-6462(03)00394-4
42. Delincé, M., Bréchet, Y., Embury, J.D., Geers, M.G.D., Jacques, P. J., Pardoen, T.: Structure–property optimization of ultrafine-grained dual-phase steels using a microstructure-based strain hardening model. Acta Mater. **55**(7), 2337–2350 (2007). https://doi.org/10.1016/j.actamat.2006.11.029
43. Song, D., Ma, A.B., Jiang, J.H., Lin, P.H., Yang, D.H., Fan, J.F.: Corrosion behaviour of bulk ultra-fine grained AZ91D magnesium alloy fabricated by equal-channel angular pressing. Corros. Sci. **53** (1), 362–373 (2011). https://doi.org/10.1016/j.corsci.2010.09.044
44. Miyamoto, H., Yuasa, M., Rifai, M., Fujiwara, H.: Corrosion behavior of severely deformed pure and single-phase materials. Mater. Trans. **60**(7), 1243–1255 (2019). https://doi.org/10.2320/matertrans.MF201935
45. Kim, H.S., Yoo, S.J., Ahn, J.W., Kim, D.H., Kim, W.J.: Ultrafine grained titanium sheets with high strength and high corrosion resistance. Mater. Sci. Eng. A. **258**(29–30), 8479–8485 (2011). https://doi.org/10.1016/j.msea.2011.07.074
46. Orłowska, M., Ura-Bińczyk, E., Olejnik, L., Lewandowska, M.: The effect of grain size and grain boundary misorientation on the corrosion resistance of commercially pure aluminium. Corros. Sci. **148**, 57–70 (2019). https://doi.org/10.1016/j.corsci.2018.11.035
47. Shimokawa, T., Tanaka, M., Kinoshita, K., Higashida, K.: Roles of grain boundaries in improving fracture toughness of ultrafine-grained metals. Phys. Rev. B. **83**, 214113 (2011). https://doi.org/10.1103/PhysRevB.83.214113
48. Li, Y.J., Zeng, X.H., Blum, W.: Transition from strengthening to softening by grain boundaries in ultrafine-grained Cu. Acta Mater. **52**(17), 5009–5018 (2004). https://doi.org/10.1016/j.actamat.2004.07.003
49. Sauvage, X., Wilde, G., Divinski, S.V., Horita, Z., Valiev, R.Z.: Grain boundaries in ultrafine grained materials processed by severe plastic deformation and related phenomena. Mater. Sci. Eng. A. **540**, 1–12 (2012). https://doi.org/10.1016/j.msea.2012.01.080
50. Vinogradov, A.: Mechanical properties of ultrafine-grained metals: new challenges and perspectives. Adv. Eng. Mater. **17**(12), 1710–1722 (2015). https://doi.org/10.1002/adem.201500177. Special Issue: Bulk Nanostructured Materials, December
51. Zhao, Y.H., Bingert, J.F., Zhu, Y.T., Liao, X.Z., Valiev, R.Z., Horita, Z., Langdon, T.G., Zhou, Y.Z., Lavernia, E.J.: Tougher ultrafine grain Cu via high-angle grain boundaries and low dislocation density. Appl. Phys. Lett. **92**(8) (2008). https://doi.org/10.1063/1.2870014
52. Kobayashi, S., Tsurekawa, S., Watanabe, T., Palumbo, G.: Grain boundary engineering for control of sulfur segregation-induced embrittlement in ultrafine-grained nickel. Scr. Mater. **62**(5), 294–297 (2010). https://doi.org/10.1016/j.scriptamat.2009.11.022
53. Lu, K., Lu, L., Suresh, S.: Strengthening materials by engineering coherent internal boundaries at the nanoscale. Science. **324**(5925), 349–352 (2009). https://doi.org/10.1126/science.1159610
54. Canadinc, D., Biyikli, E., Niendorf, T., Maier, H.J.: Experimental and numerical investigation of the role of grain boundary misorientation angle on the dislocation–grain boundary interactions. Adv. Eng. Mater. **13**(4), 281–287 (2011). https://doi.org/10.1002/adem.201000229
55. Azushima, A., Kopp, R., Korhonen, A., Yang, D.Y., Micari, F., Lahoti, G.D., Groche, P., Yanagimoto, J., Tsuji, N., Rosochowski, A., Yanagida, A.: Severe plastic deformation (SPD) processes for metals. CIRP Ann. **57**(2), 716–735 (2008). https://doi.org/10.1016/j.cirp.2008.09.005
56. El Aal, M.I.A., El Mahallawy, N., Shehata, F.A., El Hameed, M.A., Yoon, E.Y., Kim, H.S.: Wear properties of ECAP-processed ultrafine grained Al–Cu alloys. Mater. Sci. Eng. A. **527**(16–17), 3726–3732 (2010). https://doi.org/10.1016/j.msea.2010.03.057
57. Amouyal, Y., Divinski, S.V., Estrin, Y., Rabkin, E.: Short-circuit diffusion in an ultrafine-grained copper–zirconium alloy produced by equal channel angular pressing. Acta Mater. **55**(17), 5968–5979 (2007). https://doi.org/10.1016/j.actamat.2007.07.026
58. Liu, Q., Hansen, N.: Geometrically necessary boundaries and incidental dislocation boundaries formed during cold deformation. United States: N. p., 1995. Web. https://doi.org/10.1016/0956-716X(94)00019-E
59. Zhang, X., Wang, H., Scattergood, R.O., Narayan, J., Koch, C.C., Sergueeva, A.V., Mukherjee, A.K.: Studies of deformation mechanisms in ultra-fine-grained and nanostructured Zn. Acta Mater. **50**(19), 4823–4830 (2002). https://doi.org/10.1016/S1359-6454(02)00349-X
60. Mishin, O.V., Gertsman, V.Y., Valiev, R.Z., Gottstein, G.: Grain boundary distribution and texture in ultrafine-grained copper produced by severe plastic deformation. Scr. Mater. **35**(7), 873–878 (1996). https://doi.org/10.1016/1359-6462(96)00222-9
61. Calcagnotto, M., Adachi, Y., Ponge, D., Raabe, D.: Deformation and fracture mechanisms in fine- and ultrafine-grained ferrite/martensite dual-phase steels and the effect of aging. Acta Mater. **59**(2), 658–670 (2011). https://doi.org/10.1016/j.actamat.2010.10.002
62. Alizadeh, M., Paydar, M.H., Terada, D., Tsuji, N.: Effect of SiC particles on the microstructure evolution and mechanical properties of aluminum during ARB process. Mater. Sci. Eng. A. **540**, 13–23 (2012). https://doi.org/10.1016/j.msea.2011.12.026
63. Kou, H., Lu, J., Li, Y.: High-strength and high-ductility nanostructured and amorphous metallic materials. Adv. Mater. **26** (31), 5518–5524 (2014). https://doi.org/10.1002/adma.201401595. Special Issue: Materials Science and Engineering Research in Hong Kong

64. Nazarov, A.A., Romanov, A.E., Valiev, R.Z.: On the nature of high internal stresses in ultrafine grained materials. Nanostruct. Mater. **4** (1), 93–101 (1994). https://doi.org/10.1016/0965-9773(94)90131-7
65. Liu, H., Pantleon, W., Mishnaevsky, L.: Non-equilibrium grain boundaries in titanium nanostructured by severe plastic deformation: computational study of sources of material strengthening. Comput. Mater. Sci. **83**, 318–330 (2014). https://doi.org/10.1016/j.commatsci.2013.11.009
66. Kim, H.-K., Choi, M.-I., Chung, C.-S.: Dong Hyuk shin, fatigue properties of ultrafine grained low carbon steel produced by equal channel angular pressing. Mater. Sci. Eng. A. **340**(1–2), 243–250 (2003). https://doi.org/10.1016/S0921-5093(02)00178-8
67. Miyamoto, H., Harada, K., Harada, K., Mimaki, T., Vinogradov, A., Hashimoto, S.: Corrosion of ultra-fine grained copper fabricated by equal-channel angular pressing. Corros. Sci. **50**(5), 1215–1220 (2008). https://doi.org/10.1016/j.corsci.2008.01.024
68. Kim, H.S., Kim, W.J.: Annealing effects on the corrosion resistance of ultrafine-grained pure titanium. Corros. Sci. **89**, 331–337 (2014). https://doi.org/10.1016/j.corsci.2014.08.017
69. Farrokh, B., Khan, A.S.: Grain size, strain rate, and temperature dependence of flow stress in ultra-fine grained and nanocrystalline Cu and Al: synthesis, experiment, and constitutive modeling. Int. J. Plast. **25**(5), 715–732 (2009). https://doi.org/10.1016/j.ijplas.2008.08.001
70. Höppel, H.W., Kautz, M., Xu, C., Murashkin, M., Langdon, T.G., Valiev, R.Z., Mughrabi, H.: An overview: fatigue behaviour of ultrafine-grained metals and alloys. Int. J. Fatigue. **28**(9), 1001–1010 (2006). https://doi.org/10.1016/j.ijfatigue.2005.08.014
71. Ding, H.-Z., Mughrabi, H., Höppel, H.W.: A low-cycle fatigue life prediction model of ultrafine-grained metals. Fatique Fract. Eng. Mater. Struct. **25**(10), 975–984 (2002). https://doi.org/10.1046/j.1460-2695.2002.00564.x
72. Edalati, K., Ashida, M., Horita, Z., Matsui, T., Kato, H.: Wear resistance and tribological features of pure aluminum and Al–Al2O3 composites consolidated by high-pressure torsion. Wear. **310**(1–2), 83–89 (2014). https://doi.org/10.1016/j.wear.2013.12.022
73. Gao, N., Wang, C.T., Wood, R.J.K., et al.: Tribological properties of ultrafine-grained materials processed by severe plastic deformation. J. Mater. Sci. **47**, 4779–4797 (2012). https://doi.org/10.1007/s10853-011-6231-z
74. Padap, A.K., Chaudhari, G.P., Nath, S.K.: Mechanical and dry sliding wear behavior of ultrafine-grained AISI 1024 steel processed using multiaxial forging. J. Mater. Sci. **45**, 4837–4845 (2010). https://doi.org/10.1007/s10853-010-4430-7
75. Gurao, N.P., Manivasagam, G., Govindaraj, P., et al.: Effect of texture and grain size on bio-corrosion response of ultrafine-grained titanium. Metall. Mater. Trans. A. **44**, 5602–5610 (2013). https://doi.org/10.1007/s11661-013-1910-9
76. Asabe, T., Rifai, M., Yuasa, M., Miyamoto, H.: Effect of grain size on the stress corrosion cracking of ultrafine grained Cu-10 wt% Zn alloy in ammonia. Int. J. Corros. **2017**, 2893276 (2017). https://doi.org/10.1155/2017/2893276
77. Huang, C.X., Hu, W.P., Wang, Q.Y., Wang, C., Yang, G., Zhu, Y.T.: An ideal ultrafine-grained structure for high strength and high ductility. Mater. Res. Lett. **3**(2), 88–94 (2014). https://doi.org/10.1080/21663831.2014.968680
78. Guo, Y.Z., Li, Y.L., Pan, Z., Zhou, F.H., Wei, Q.: A numerical study of microstructure effect on adiabatic shear instability: application to nanostructured/ultrafine grained materials. Mech. Mater. **42**(11), 1020–1029 (2010). https://doi.org/10.1016/j.mechmat.2010.09.002
79. Muley, S.V., Vidvans, A.N., Chaudhari, G.P., Udainiya, S.: An assessment of ultra fine grained 316L stainless steel for implant applications. Acta Mater. **30**, 408–419 (2016). https://doi.org/10.1016/j.actbio.2015.10.043
80. Li, B., Joshi, S., Azevedo, K., Ma, E., Ramesh, K.T., Figueiredo, R.B., Langdon, T.G.: Dynamic testing at high strain rates of an ultrafine-grained magnesium alloy processed by ECAP. Mater. Sci. Eng. A. **517**(1–2), 24–29 (2009). https://doi.org/10.1016/j.msea.2009.03.032
81. Suzuki, Y., Yang, M., Murakawa, M.: Optimum clearance in the microblanking of thin foil of austenitic stainless steel JIS SUS304 studied from shear cut surface and punch load. Materials. **13**, 678 (2020)
82. Wang, Z., Jia, J., Wang, B., Wang, Y.: Two-step spark plasma sintering process of ultrafine grained WC-12Co-0.2VC cemented carbide. Materials (Basel). **12**(15), 2443 (2019). https://doi.org/10.3390/ma12152443
83. See: https://www.comtesfht.cz/media/document/sbornik_full_texty_final.pdf. 20.05.2020
84. Fernandesa, D.J., Eliasa, C.N., Valiev, R.Z.: Properties and performance of ultrafine grained titanium for biomedical applications. Mater. Res. **18**(6), 1163–1175 (2015). https://doi.org/10.1590/1516-1439.005615
85. Semenova, I.P., Modina, J., Polyakov, A.V., Klevtsov, G.V., Klevtsova, N.A., Pigaleva, I.N., Valiev, R.Z.: Charpy absorbed energy of ultrafine-grained Ti-6Al-4V alloy at cryogenic and elevated temperatures. Mater. Sci. Eng. A. **743**, 581–589 (2019). https://doi.org/10.1016/j.msea.2018.10.076
86. Bagherpour, E., Pardis, N., Reihanian, M., Ebrahimi, R.: An overview on severe plastic deformation: research status, techniques classification, microstructure evolution, and applications. Int. J. Adv. Manuf. Technol. **100**, 1647–1694 (2019). https://doi.org/10.1007/s00170-018-2652-z
87. Valiev, R.Z., Estrin, Y., Horita, Z., et al.: Producing bulk ultrafine-grained materials by severe plastic deformation: ten years later. JOM. **68**, 1216–1226 (2016). https://doi.org/10.1007/s11837-016-1820-6
88. Sanusi, O., Makinde, D., Oliver, J.: Equal channel angular pressing technique for the formation of ultra-fine grained structures. S. Afr. J. Sci. **108**(9–10), 1–7 (2012) [cited 2020-08-31]. Available from: <http://www.scielo.org.za/scielo.php?script=sci_arttext&pid=S0038-23532012000500019&lng=en&nrm=iso>. ISSN 1996-7489
89. Xiao, G., Xia, Q., Cheng, X., et al.: New forming method of manufacturing cylindrical parts with nano/ultrafine grained structures by power spinning based on small plastic strains. Sci. China Technol. Sci. **59**, 1656–1665 (2016). https://doi.org/10.1007/s11431-016-0206-6
90. Qiao, X.G., Gao, N., Moktadir, Z., Kraft, M., Starink, M.J.: Fabrication of MEMS components using ultrafine-grained aluminium alloys. J. Micromech. Microeng. **20**(4). https://doi.org/10.1088/0960-1317/20/4/045029
91. Zhao, G., Xu, S., Luan, Y., Guan, Y., Lun, N., Ren, X.: Grain refinement mechanism analysis and experimental investigation of equal channel angular pressing for producing pure aluminum ultra-fine grained materials. Mater. Sci. Eng. A. **437**(2), 281–292 (2006). https://doi.org/10.1016/j.msea.2006.07.138
92. Lin, T., Yang, Q., Tan, C., et al.: Processing and ballistic performance of lightweight armors based on ultra-fine-grain aluminum composites. J. Mater. Sci. **43**, 7344–7348 (2008). https://doi.org/10.1007/s10853-008-2977-3
93. Piers Newbery, A., Nutt, S.R., Lavernia, E.J.: Multi-scale Al 5083 for military vehicles with improved performance. JOM. **58**, 56–61 (2006). https://doi.org/10.1007/s11837-006-0216-4
94. Erb, U., Aust, K.T., Palumbo, G.: 6 - electrodeposited nanocrystalline metals, alloys, and composites. In: Koch, C.C. (ed.) Nanostructured materials, 2nd edn, pp. 235–292. William Andrew Publishing, Norwich, NY (2007). https://doi.org/10.1016/B978-081551534-0.50008-7

18 Alloys Based on Intermetallic Compounds

Abstract

In this era, intermetallic technology going to take a broad advantage with its presence in high-temperature processing materials. This chapter gives a brief idea of the intermetallic compound. This chapter shows how their presence can improve the materials' properties. Various structures of the intermetallic compound and their classification have been discussed. Applications in aerovehicle industries, automobile industries, and electrical industries have been stated here.

Keywords

Intermetallic alloy · Hume-Rothery phases · Frank–Kasper phases · A15 phase · Laves phases · Sigma Phase · mu phase · Kurnakov phases · Zintl phases · Nowotny phases · B2 phase · L_{12} phase

18.1 What Is an Intermetallic Alloy?

Intermetallics are formed in many alloy systems when the concentration exceeds the solubility limit. These materials exhibit properties totally different from the common metals. Intermetallic forms a separate group of metallic materials with outstanding properties used in various fields of technology, especially for high-temperature applications [1]. An intermetallic is also known as intermetallic compound (IMC), intermetallic alloy, ordered intermetallic alloy, and a long-range-ordered alloy [2]. It can be defined as a solid phase compound constituted by two or more metallic elements, with optionally one or more nonmetallic elements exhibiting defined stoichiometry and ordered crystal structure. It does not show continuously variable proportions as in solid solutions [3, 4]. In accordance with their definition, IMCs are metallic compounds whose crystallographic structure is different from basic metals and their stacking is long distance. These stacked compounds exist usually in a narrow range of concentrations around the stoichiometric ratio. Although the term "intermetallic compounds", as it applies to solid phases, has been in use for many years, its introduction was regretted, for example by Hume-Rothery in 1955 [5]. Figure 18.1 shows a typical understanding of IMC formation with the formation of two types of solid solution. The former, typically substitutional solid-solution alloys, consist of metals of similar atomic size and electronic character with a crystal structure identical to that of the parent metal with random atomic arrangements (Fig. 18.1a). If the atoms of one element are sufficiently small to fit within the lattice void of the counterpart element, interstitial solid-solution alloys can be formed (Fig. 18.1b). Figure 18.1c shows significantly different characters and comprise distinct crystal structures with highly ordered atomic arrangements [6].

Metals are usually crystalline solids. In most cases, they have a relatively simple crystal structure distinguished by a close packing of atoms and a high degree of symmetry. Typically, the atoms of metals contain less than half the full complement of electrons in their outermost shell. Because of this characteristic, metals tend not to form compounds with each other [7]. They do, however, combine more readily with nonmetals (e.g., oxygen and sulfur), which generally have more than half the maximum number of valence electrons. Metals differ widely in their chemical reactivity. The most reactive include lithium, potassium, and radium, whereas those of low reactivity are gold, silver, palladium, and platinum [8]. The high electrical and thermal conductivities of the simple metals (i.e., the non-transition metals of the periodic table) are best explained by reference to the free-electron theory. According to this concept, the individual atoms in such metals have lost their valence electrons to the entire solid, and these free electrons that give rise to conductivity move as a group throughout the solid [9]. In the case of the more complex metals (i.e., the transition elements), conductivities are better explained by the band theory, which takes into account not only the presence of free electrons but also their interaction with so-called *d* electrons

A. Behera, *Advanced Materials*, https://doi.org/10.1007/978-3-030-80359-9_18

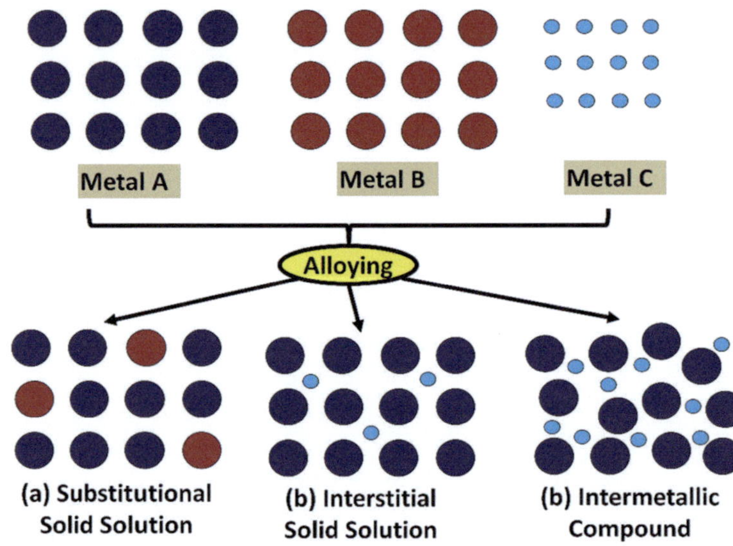

Fig. 18.1 Structure of bimetallic alloys: (**a**) Substitutional solid solution, (**b**) Interstitial solid solution, and (**c**) Intermetallic compound. The word "alloying" suggests the formation of alloy phase (**a**, **b**, or **c**) or simply, the mixing of two metal elements at an atomic level

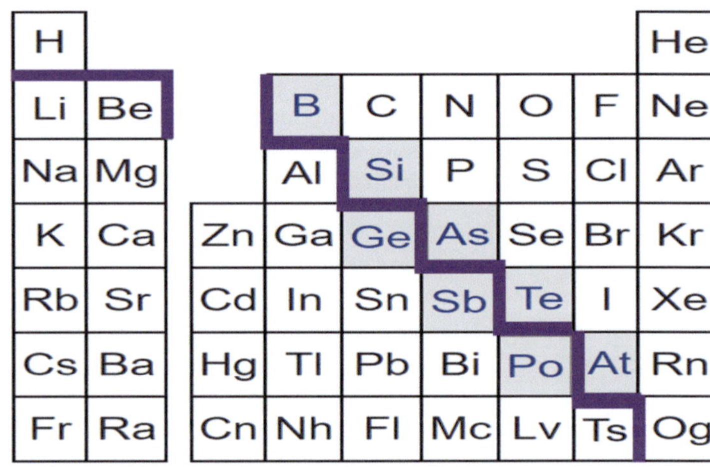

Fig. 18.2 Showing the Zintl line metal–nonmetal dividing line (arbitrary): between Li and H, Be and B, Al, and Si, Ge and As, Sb and Te, Po and At, Ts and Og

[10]. Intermetallic compounds, that is, compounds comprising two or more elements located left and around the Zintl line in Fig. 18.2, realize crystal structures that are completely or at least partly ordered and different from those of the constituent elements [11].

Intermetallics typically includes post-transition metals (Al, Ga, In, Th, Sn, and Pb) and metalloids (Si, Ge, Ar, Sb, and Te) to form electron (or Hume-Rothery) compounds, size packing phases (Laves phases, Frank–Kasper phases, and Nowotny phases), and Zintl phases [12–15]. Homogeneous and heterogeneous solid solutions of metals, and interstitial compounds (such as carbides and nitrides), are excluded under this category. However, interstitial intermetallic compounds are included, as are alloys of intermetallic compounds with a metal [16, 17].

Many intermetallic compounds display an attractive combination of physical and mechanical properties, including high melting point, low density, and good oxidation or corrosion resistance, creep resistance, and so on [18]. Many compounds have been found to have unique physical properties such as superconductivity, hydrogen storage capacity, and high saturation magnetization. The main groups within which they can be classified are nickel aluminides, iron aluminides, titanium aluminides, and others such as silicides, and refractory metal aluminides [19]. Aluminides are especially interesting because, for most of them, their ordering temperature is close to the melting temperature. Within this group, many exist over a range of compositions, but the degree of order decreases as the deviation from stoichiometry increases; iron, nickel, and titanium-based intermetallics provide a good mechanical strength/mass ratio and oxidation resistance to high temperatures [20]. As it appears, they cannot compete with superalloys due to their relatively low creep resistance, apart from their poor ductility. However, some researchers intend to solve this problem by precipitation and solid solution hardening or even by dispersing particles, which affect creep properties by performing as dislocation obstacles [21]. In general, aluminides of transition metals possess a sufficient concentration of aluminum to form an adherent alumina layer when exposed to air or oxygen atmospheres. This interesting property leads to the fact that they exhibit outstanding oxidation and corrosion resistance in aggressive chemical environments and even at temperatures as high as 1000 °C or more [22].

18.2 Structure of IMC

Intermetallic compounds are generally brittle and have high melting points. They often offer a compromise between ceramic and metallic properties when hardness and/or resistance to high temperatures is important enough to sacrifice some toughness and ease of processing. They can also display desirable magnetic, superconducting, and chemical properties, due to their strong internal order and mixed (metallic and covalent/ionic) bonding, respectively [23]. Bonds between different atoms are stronger than between atoms of the same kind, IMCs, therefore, form special crystallographic structures in which atoms of the same kind are organized in some preferential positions surrounded by atoms of another element. Crystallographic structure is characterized by the strength and character of the bond, which of course depends on electron configuration [24]. When two metals form a continuous range of solid solutions, one or more intermediate phases may be formed. These phases, differing in crystal structure from the component metal are referred to as intermetallic compounds. They are not true compounds but have variable compositions. Each compound may be assigned a formula. The two metals usually combine in stoichiometric proportions but do not follow any valence rules, for example, Cu_5Zn, $Cu_{31}Sn_8$, Fe_3C, Pt_5Zn_{21}. Intermetallic compounds differ structurally from the component metals and are stable over a limited compositional range [25]. These compounds' component metals are divided into three classes: B-class, M-class, and T-class as in Fig. 18.3. Thus the binary system may be MM, MB, MT, TT, TB. The intermetallic compounds of the type TB, according to the Hume-Rothery have formulae and structures that are determined by the ratio of the number of valence electrons to the total number of atoms of both kinds are often referred to as the "electron compounds" [26].

Large number of intermetallic compounds has been found to exist. Some of the defined IMC has been given below:

18.2.1 Hume-Rothery Phases

Hume-Rothery rules, named after William Hume-Rothery, are a set of basic rules that describe the conditions under which an element could dissolve in metal, forming a solid solution (Fig. 18.4) [27]. The Hume-Rothery phase includes the most well-known electron compounds: cubic B2 structure (type β-brass, FeAl, NiAl, CoAl), complex A13 (Zn_3Co, Cu_5Si), HTU A3 (type Mg) with electron concentration 3/2 (Cu_3Ga, Ag_3Al), or 7/4 ($CuZn_3$, Ag_5Al_3), complex cubic structure D82 (type γ brass) with electron concentration 21/13 (Cu_5Zn_8, Fe_5Zn_2) [28, 29]. The bonds in these intermetallics are not purely metallic, for example in NiAl, the covalent bond with slight metallic character has been observed without ionic part. Hume-Rothery rules describe only simple cases of bonds [30].

18.2.2 Frank–Kasper Phases

Topologically close pack (TCP) phases, also known as Frank–Kasper (FK) phases, are one of the largest groups of intermetallic compounds, known for their complex crystallographic structure and physical properties [31]. Owing to their combination of the periodic structure, some TCP phases belong to the class of quasicrystals. Based on the tetrahedral units, FK crystallographic structures are classified into low and high polyhedral groups denoted by their coordination numbers (CN) referring to the number of atoms centering

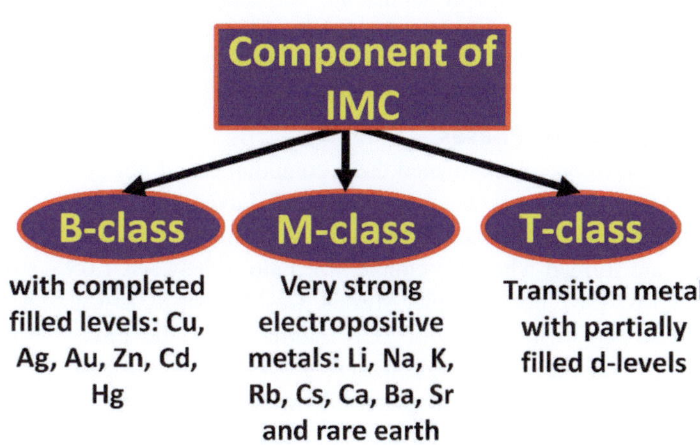

Fig. 18.3 Subcategory of binary IMC constitute elements

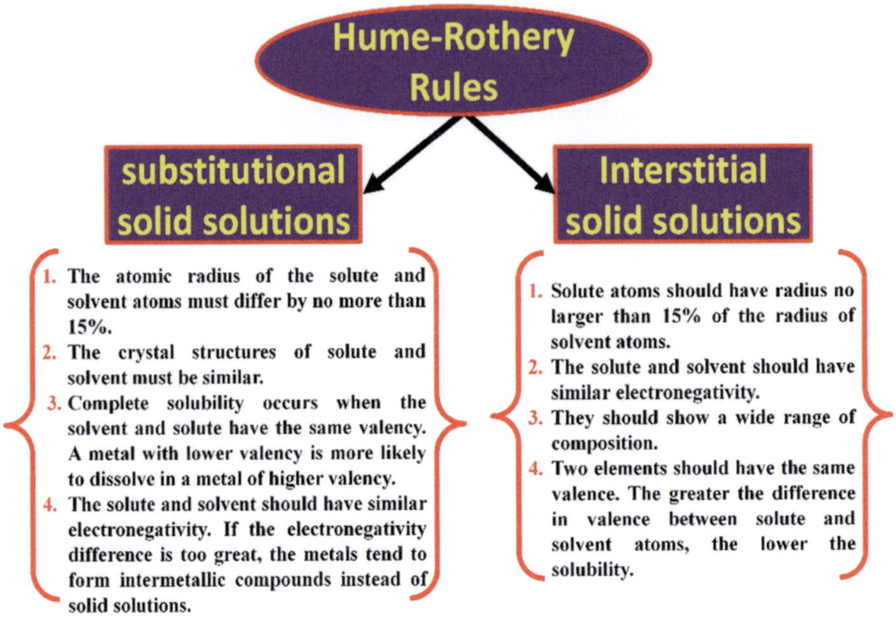

Fig. 18.4 Hume-Rothery rules for two types of solid solution: substitutional and interstitial solid solution

the polyhedron. Some atoms have an icosahedral structure with low coordination, labeled CN12 [32]. Some others have higher coordination numbers of 14, 15, and 16, labeled CN14, CN15, and CN16, respectively. These atoms with higher coordination numbers form uninterrupted networks connected along with the directions where the fivefold icosahedral symmetry is replaced by sixfold local symmetry. The most common members of the FK-phases family are the A15 phase, Laves phases, σ phase, and μ phase [33–35].

18.2.2.1 A15 Phase

The A15 phases are series of intermetallic compounds with the chemical formula A_3B (where A is a transition metal and B can be any element). The first intermetallic compound discovered with typical A_3B composition was Cr_3Si. A typical structure of the A15 phase is given in Fig. 18.5. A15 phases are intermetallic alloys with an average coordination number of 13.5 and eight A_3B stoichiometry atoms per unit cell where two B atoms are surrounded by CN12 polyhedral (icosahedra), and six A atoms are surrounded by CN14 polyhedral. Other examples of A_3B intermetallic compounds are Nb_3Ge, V_3Si, Nb_3Ge, Nb_3Ti, Nb_3Sn [36, 37].

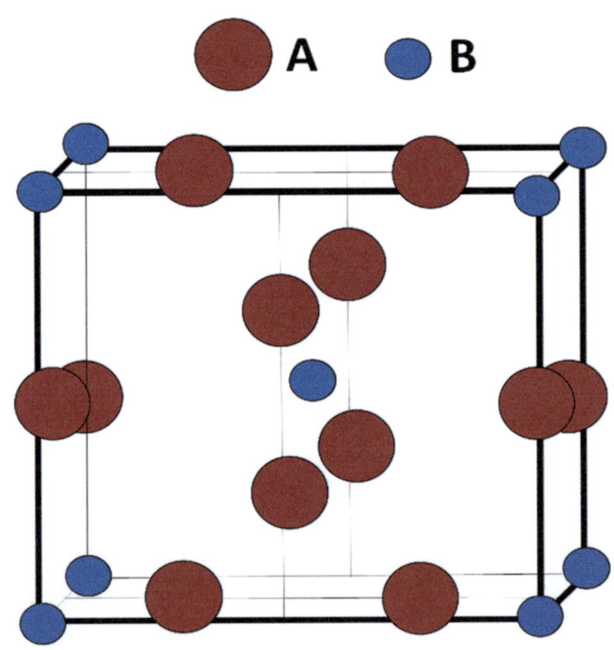

Fig. 18.5 Structure of A15 phases

18.2.2.2 Laves Phases

Laves have been investigated these compounds and referred to AB_2 type phases as "Laves phases." Laves phases are intermetallic compounds that have a stoichiometry of AB_2 and are formed when the atomic size ratio is between 1.05 and 1.67 [38]. For example, $MgZn_2$, Cu_2Mg, Na_2K, Li_2Ca, and so on. In the group of TCP intermetallics, the Laves phases constitute the single largest group. There are three types of Lave phases: hexagonal C14 (hP12; $MgZn_2$ prototype), cubic C15 (cF24; Cu_2Mg prototype), or dihexagonal C36 (hP24; $MgNi_2$ prototype) structures. The C14, C15, and C36 crystal structures differ only by the particular stacking of the same two-layered structural units, which allows structure transformations between these structures and twinning by "synchroshear" [39]. The stability of the three crystal structures is controlled by both the atomic size ratio of the A atoms (blue color) and B atoms (purple color) and by the valence electron concentration of the Laves phase. In these compounds, the A atoms take up ordered positions as in diamond, hexagonal diamond, or related structure while the B atoms take up tetrahedral positions around A atoms

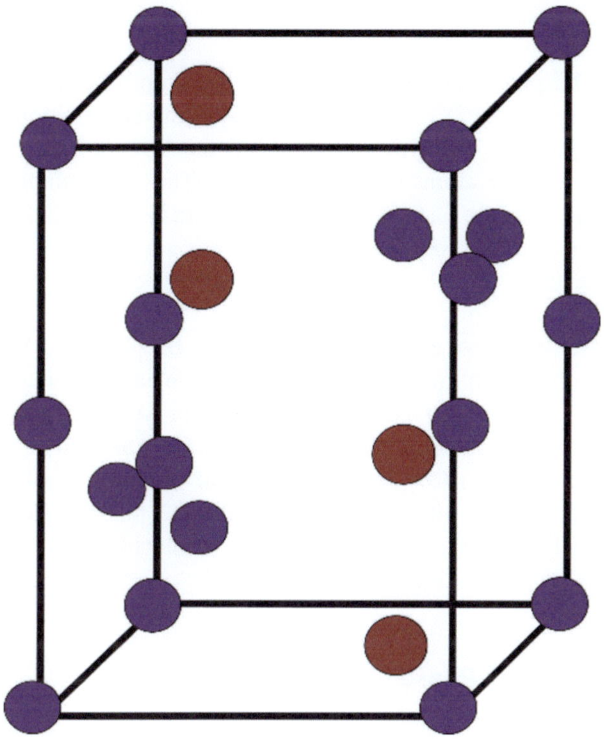

Fig. 18.6 A typical structure of the Lave phase

(Fig. 18.6). In case the atomic size ratio of A and B atoms is around 1.225, they form topologically tetrahedral close-packed structures with an overall packing density of 0.71 [40]. Some Laves phases have been regarded as promising for both functional and structural applications, such as superconducting C15 (Hf, Zr)V_2, magnetic C15 TFe_2 (where T = Ti, Zr, Hf, Nb, and Mo), and materials for hydrogen storage C15 ZrV_2, $ZrCr_2$, and C14 $ZrMn_2$ materials [41].

18.2.2.3 Sigma Phase

The sigma (σ) phase is an intermetallic compound known as the one without definite stoichiometric composition and formed at the electron/atom ratio range of 6.2–7. It has a primitive tetragonal unit cell with 30 atoms. The σ phase is usually observed in Cr-containing steels and has a typical composition of equiatomic FeCr with a tetragonal structure. The σ phase has also been observed with equiatomic CoCr or FeMo in binary CoCr and FeMo alloys. Figure 18.7 shows a typical structure of the sigma phase [42].

18.2.2.4 Mu Phase

The mu (μ) phase has an ideal A_6B_7 stoichiometry, with its prototype W_6Fe_7, containing a rhombohedral cell with 13 atoms. While many other Frank–Kasper alloy types have been identified. The alloy $Nb_{10}Ni_9Al_3$ is the prototype for the M phase. It has an orthorhombic space group with 52 atoms per unit cell [43]. The alloy $Cr_9Mo_{21}Ni_{20}$ is the prototype for

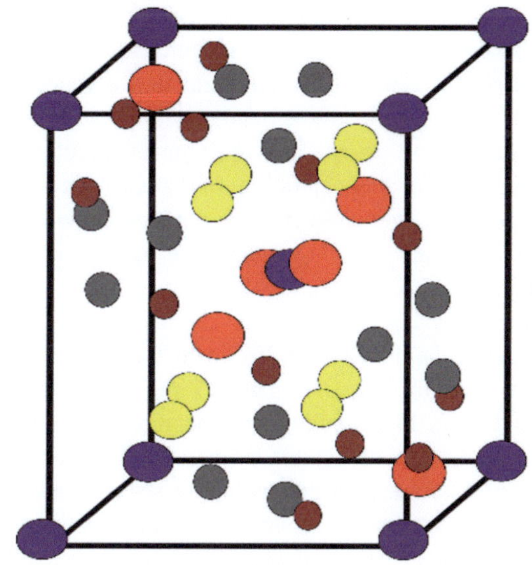

Fig. 18.7 A typical Sigma phase

the P-phase. It has a primitive orthorhombic cell with 56 atoms. The alloy $Co_5Cr_2Mo_3$ is the prototype for the R-phase which belongs to the rhombohedral space group with 53 atoms per cell [44].

18.2.3 Kurnakov Phases

These are formed during phase transitions from a solid solution of the constitution element at high temperature (but under melting point), their formation decreases crystal symmetry. In the simplest cases, the phase change is caused by the reaction of lattice stacking of the solid solution, where a superlattice is formed: Fe_3Al (D03), Ni_3Fe (L12), CuAu (L10) Cu_3Au (L12) [45]. The arrangement of atoms during superlattice formation is a result of a stronger bond between atoms of different elements in comparison with atoms of the same elements (in account of interaction bonding energy, arrangement energy). To predict the stability of these phases, it is not sufficient to know this interaction energy or concentration of valence electrons depending on the constituent elements. Prediction of bond type and crystallographic structure has to be based on calculations of quantum mechanics (ab-ignition calculations) for specific phases while taking into account the electron configuration of the constitute elements [46].

18.2.4 Zintl Phases

These are formed between metals on the left and right side of the periodic table and are characterized by the total occupation of electron orbitals (electron octet), they are therefore studied as valence compounds with ionic bonds, typical for salts—NaTl (cubic B_{32}), $MgSi_2$ (cubic C_1) [47]. However, all kinds of bond types have been observed: ionic, metallic, and covalent or mixed, which is related to specific electron distribution; Zintl phases can be considered electron compounds where the energy of the electronic band structure makes up a big part of the overall energy and the type of structure is related to the specific concentration of valence electrons [48]. Zintl phase is the product of a reaction between a group 1 (alkali metal) or group 2 (alkaline earth) and any post-transition metal or metalloid (i.e., from group 13, 14, 15, or 16). It is named after the German chemist Eduard Zintl who investigated them in the 1930s, with the term "Zintl Phases" first used by Laves in 1941. Zintl phases are a subgroup of brittle, high-melting intermetallic compounds which are diamagnetic or exhibit temperature-independent paramagnetism, and are poor conductors or semiconductors [49]. Zintl noted that there was an atomic volume contraction when these compounds were formed and realized this could indicate cation formation. He suggested that the structures of Zintl phases were ionic, where there was complete electron transfer from the more electropositive metal. The structure of the anion (nowadays called the Zintl ion) should then be considered on the basis of the resulting electronic state [50].

18.2.5 Nowotny Phases

Nowotny chimney ladder phase (NCL phase) is a particular intermetallic crystal structure found with certain binary compounds. NLC phases are generally tetragonal and are composed of two separate sublattices [51]. The first is a tetragonal array of transition metal atoms, generally from group 4 through group 9 of the periodic table. Contained within this array of transition metal atoms is a second network of main group atoms, typically from group 13 (boron group) or group 14 (carbon group). The transition metal atoms form a chimney with a helical zigzag chain [52]. The main-group elements form a ladder spiraling inside the transition metal helix. The phase is named after one of the early investigators H. Nowotny. Examples are $RuGa_2$, Mn_4Si_7, Ru_2Ge_3, Ir_3Ga_5, Ir_4Ge_5, $V_{17}Ge_{31}$, $Cr_{11}Ge_{19}$, $Mn_{11}Si_{19}$, $Mn_{15}Si_{26}$, Mo_9Ge_{16}, $Mo_{13}Ge_{23}$, $Rh_{10}Ga_{17}$, and $Rh_{17}Ge_{22}$ [53]. In RuGa2 the ruthenium atoms in the chimney are separated by 329 pm. The gallium atoms spiral around the Ru chimney with a Ga–Ga intrahelix distance of 257 pm. The view perpendicular to the chimney axis is that of a hexagonal lattice with gallium atoms occupying the vertices and ruthenium atoms occupying the center. Each gallium atom bonds to 5 other gallium atoms forming a distorted trigonal bipyramid. The gallium atoms carry a positive charge and the ruthenium atoms have a formal charge of −2 (filled 4d shell). In Ru_2Sn_3 the ruthenium atoms spiral around the tin inner helix. In two dimensions, the Ru atoms form a tetragonal lattice with the tin atoms appearing as triangular units in the Ru channels [54].

18.2.6 B2 Phase

The B2 phase is an ordered structure based on BCC (Pearson symbol of cP2), wherein the body-centered position is occupied by one type of atom and the body corners are occupied by atoms of another kind. The most common compounds are CsCl, CuZn, and NiAl [55]. The B2 phase has been observed either as a major or as a minor phase in a large number of HEAs. In some cases, it has been observed to precipitate during heat treatment of these alloys from a BCC phase. In almost all the cases where the B2 phase has been observed, the alloys contained 3d transition elements such as Ti, Cr, Mn, Fe, Co, Ni, and Cu along with Al. Among these 3d transition elements, Al has a strong affinity to form the B2 phase with Fe, Co, and Ni. A careful observation of the constituents of alloys that showed the B2 phase indicates that all have either one of these three elements (Fe, Co, and Ni) along with Al [56].

18.2.7 L12 Phase

The L12 phase is an ordered structure based on FCC (Pearson symbol of cP4), wherein the face-centered positions are occupied by one type of atom and the body corners are occupied by another kind of atom. The most common compounds that have this structure are $AuCu_3$ and Ni_3Al [57]. All these alloys containing both Al and Ni possess the FCC phase and also show the L12 phase in the matrix. In HEAs, the L12 phase is the multicomponent version (including Ni, Co, Fe, Ti, etc.) of γ' (Ni_3Al) that is usually observed in Ni-base superalloys [58].

18.3 Structure Defects of IMC

18.3.1 Point Defects

In intermetallic phases, point defects are numerous, for example, vacancies, interstitial, impurities, anti-positions, and their combinations (Fig. 18.8), their migration characteristics are not always easy to determine [59].

18.3.2 Structure of Antiphase Boundaries and Domains

Antiphase boundaries and antiphase domains (APB and APD for short) are characteristic microstructure phenomena in compound arrangements, they cannot be observed within unarranged alloys. In arranged alloys, the motion of dislocations can contain APB (Fig. 18.9). APBs are an effective barrier for dislocation movements [60]. If APBs close off a certain area, APD is formed. APD morphology influences

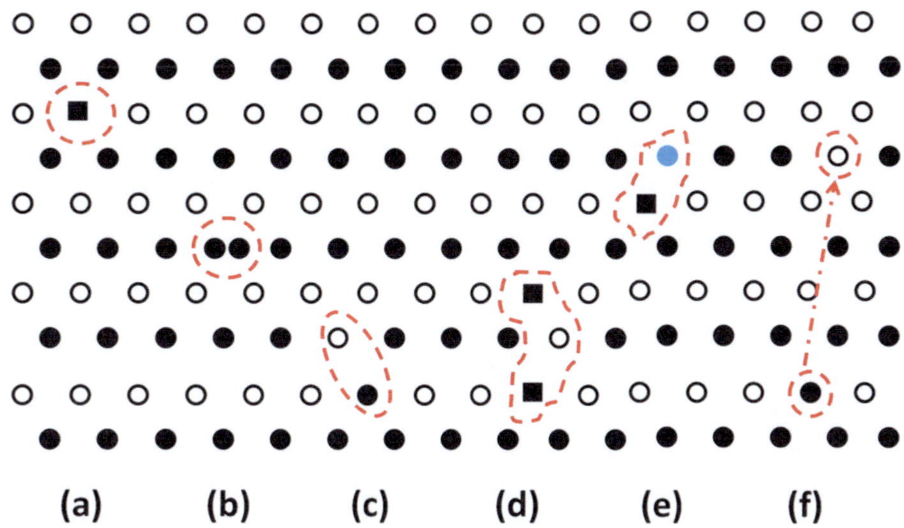

Fig. 18.8 Typical point defects in binary IMCs: (**a**) Vacancies, (**b**) Dual interstitials, (**c**) Bond of two anti-positions, (**d**) Bond between three defects (two vacancies and one anti-position), (**e**) Bond between vacancies and impurities, and (**f**) Free bond between wrong atoms

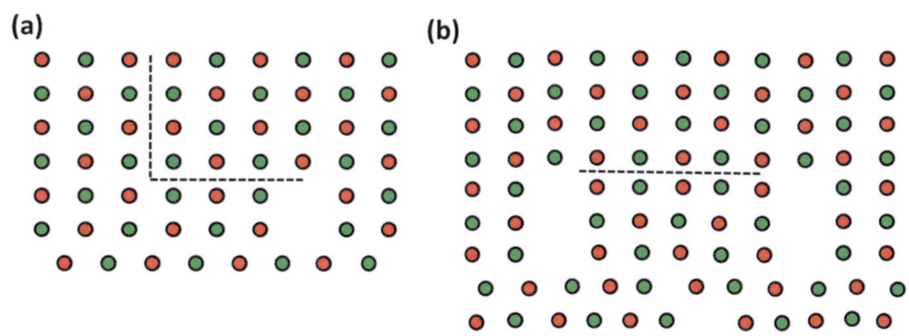

Fig. 18.9 Dislocations in an arranged structure: (**a**) Formation of APB after the passing of one dislocation and (**b**) formation of layer defect between two super-partial dislocations

the mechanical properties. The formation of APD is to a certain point controllable by thermal processing and alloying. Electromagnetic properties are also significantly influenced by APB and APD through the effect on the band structure. With IMCs, it can encounter various situations in its structure, that are (1) formation of APB and APD during the solidification of the alloy, (2) the transition from disarranged into an arranged state (during cooling over some transition temperature), and (3) dissociation of dislocations and formation of APBs in the shape of narrow bands connecting two or more super-partial dislocations [61, 62].

18.3.3 Superlattice Dislocations

Anomalous growth of yield strength with temperature, which was mentioned in the case of Ni-based superalloys is related to the occurrence of phase Ni_3Al making up more than 40 vol.% (Chap. 7), and this is related to the dislocation structure and yield systems activated at certain temperatures. In an arranged structure, superlattice dislocations (superdislocations) can be observed; movement of these dislocations is easier if their dissociation into super-partials occurs and stacking faults are created [63]: superlattice intrinsic stacking fault (SISF), superlattice extrinsic stacking fault (SESF), complex stacking fault (CSF), and antiphase boundaries (APB) (Fig. 18.10).

1. Dislocation of superlattice dislocations $\bar{1}01$ on plane $\{111\}$ into four Shockley's partial dislocations and formation of APB and CSF [64]:

$$a[\bar{1}01] \rightarrow \frac{a}{6}[\bar{1}\bar{1}2] + \frac{a}{6}[\bar{2}11] + \frac{a}{6}[\bar{2}11] + \frac{a}{6}[\bar{1}\bar{1}2] \quad (18.1)$$

2. Same as (1), with other partial dislocations identical but with opposite sign $\frac{1}{6} < 112 >$, which change APB into SISF:

$$a[\bar{1}01] \rightarrow \frac{a}{6}[\bar{1}\bar{1}2] + \frac{a}{6}[\bar{2}11] + \frac{a}{6}[\bar{1}2\bar{1}] + \frac{a}{6}[\bar{1}21] + \frac{a}{6}[\bar{1}\bar{1}2] + \frac{a}{6}[\bar{2}11] \quad (18.2)$$

3. Dissociation of superlattice dislocations $[\bar{1}01]$ on plane $\{111\}$ into super-partial dislocations $\frac{1}{2}[\bar{1}01]$ and formation of APB;

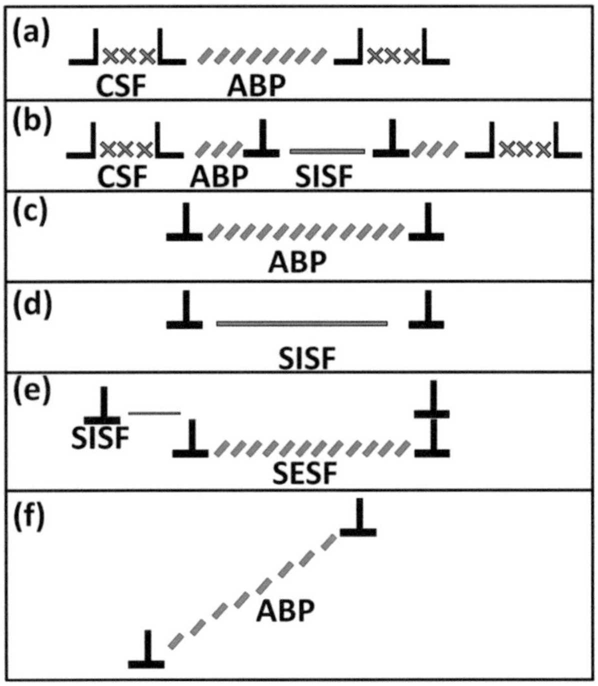

Fig. 18.10 A schematic of dissociation of superdislocations and formation of stacking faults

$$a\left[\bar{1}01\right] \rightarrow \frac{a}{2}\left[\bar{1}01\right] + \frac{a}{2}\left[\bar{1}01\right] \tag{18.3}$$

$$a\left[\bar{1}01\right] \rightarrow \frac{a}{3}\left[\bar{2}11\right] + \frac{a}{3}\left[\bar{1}\bar{1}2\right] \tag{18.4}$$

4. Dislocation $\left[\bar{1}01\right]$ dissociated on plane $\{111\}$ into two superpartial dislocations $\frac{1}{3} < 112 >$ and formation of SISF.
5. Dislocation $\left[\bar{2}11\right]$ dissociated into 4 superpartial dislocations $\frac{1}{3} < 112 >$, while forming SISF a SESF on two adjacent planes.

$$a\left[\bar{2}11\right] \rightarrow \frac{a}{3}\left[\bar{1}2\bar{1}\right] + \frac{a}{3}\left[\bar{1}\bar{1}2\right] + \frac{a}{3}\left[\bar{2}11\right] + \frac{a}{3}\left[\bar{2}11\right] \tag{18.5}$$

6. Dislocation of superlattice $\left[\bar{1}01\right]$ on plane $\{010\}$ into super-partial dislocations $\frac{1}{2}\left[\bar{1}01\right]$

$$a\left[\bar{1}01\right] \rightarrow \frac{a}{2}\left[\bar{1}01\right] + \frac{a}{2}\left[\bar{1}01\right] \tag{18.6}$$

18.4 Structure of Grain Boundaries and Brittleness of IMC

The structure of grain boundaries plays a very important role in many properties of an IMC-based polycrystal material. The purity of initial and final material is determined by chemical composition on the grain boundaries as well as close to them. On the grain boundaries, segregation of foreign elements and inclusions occurs (such as boron, sulfur, oxygen, or hydrogen) [65]. Normal bonds between various atoms do not have to be preserved on the grain boundaries, they may be deformed or bonds between similar atoms may be formed. The long-range order (LRO) differs from the order on the grain boundaries. Because of this, the grain boundaries represent the weakest link of an alloy and may significantly influence mechanical properties, processes of recrystallization, electrical conductivity, or corrosion resistance [66]. The brittleness of intermetallic polycrystal compounds at room temperature is a critical problem, which limits their processing and use; a number of intermetallics are usually easily damaged at normal temperature by a light hit. However, the brittleness in IMC is still much lower than in ceramics because the bonds between atoms in IMC are (at least partly) metallic, whereas in ceramics they are primarily covalent or ionic. Due to the existence of strong atomic bonds and therefore arranged states and complex crystallographic structures, plastic deformation in IMC is much more difficult than in metals or regular alloys. In general, the three main reasons for damaging the material in decreasing order of importance: high deformation stress relative to fracture stress, inadequate active yield systems, and weakening of grain boundaries [67].

Insufficient number and small mobility of dislocations and an insufficient number of yield systems is a parameter that is dependent not only on crystallographic symmetry but also on other specific parameters of concrete phases, therefore in choosing the phase and chemical changes by alloying, the aim is to achieve low energy and high mobility and at least five independent deformation modes (yield systems, twinning systems), which are in accordance with von Mises' criterion necessary for overall balanced plastic deformation. In some phases of IMC, it has been observed that their single crystals are formable only in some orientations [68]. Weakening of grain boundaries and the presence of microstructural nonhomogeneities lead to local concentration deformations and tension, which can be another cause of brittleness of IMC. As an example, it will state Ni_3Al, which has single crystals formable in all orientations at normal temperature, but in a polycrystal, Ni_3Al fractures form even at negligible plastic deformation [69]. A theory has been agreed on, that low ductility (almost zero) is the result of a premature breach of grain boundaries; their strength is lower than the strength in the crystal volume, which makes them break sooner than the interior of the crystal begins to plastically deform. Intermetallic compounds are also sensitive to the segregation of impurities on the grain boundaries. Impurities such as sulfur and oxygen have higher electronegativity and weaken the metal–metal bonds on the grain boundary, which lowers the cohesive strength and supports intercrystal damage. Sulfur causes brittleness at normal temperatures, oxygen at higher temperatures. A very specific factor that influences grain boundaries is hydrogen. It has been proved however that polycrystal stoichiometric and unalloyed alloy Ni_3Al exhibits intercrystal damage even if it is prepared from metals of high purity. The phenomenon of "weakening" of grain boundaries is therefore related to the damage of lattice structure in the areas of grain boundaries and grain boundary energies in low-angle and high-angle boundaries [70].

18.5 Optical Properties of Intermetallic Compound

Optical measurements of LiAl in the NaTl structure suggest that this compound is an excellent free-electron metal with only a small interband transition around 0.55 eV, whereas resistivity measurements present conflicting data, with values around 10 μΩ cm compared to silver and gold with values of 1.77 μΩcm and 2.66 μΩ cm, respectively [71]. Some of the alkali noble intermetallics had an optical gap to plasma frequency ratios greater than 1, indicating that it was very unlikely that interband transitions would disrupt the optical response of these materials. Unfortunately, calculations of the DC resistivity indicated that the Drude phenomenological scattering rate was too high for these compounds to compete with silver and gold. Rivory et al. also investigated the effect of short-range ordering on the stoichiometric alloy $AuCu_3$ [72]. With increasing order, a new peak appears at approximately 3.6 eV. Slightly different results were reported by Scott et al., where electropolishing of samples post-annealing was shown to cause the peak to appear at 3.28 eV but Skriver and Lengkeek noted that electropolishing preferentially etched grains in their polycrystalline sample and reported the peak at 3.6 eV. They also remove the intraband contribution from their experimental data using a Drude fit and note two additional peaks: one at 0.8 eV and another at 1.2 eV. $CoPt_3$ and $MnPt_3$ crystallize in the same structure as $AuCu_3$ but partially occupied d-states result in many low energy transition mechanisms, resulting in $Q_{LSP} < 1$ between 1.5 and 5.0 eV. The noble-group III alloys crystallize in the CaF_2 structure. The noble atoms occupy sites on an FCC cell and the group III atoms form a simple cubic structure in the center of the FCC cell. The discovery of the purple colored, gold aluminum alloy $AuAl_2$ is often attributed to Roberts-Austen [73]. It has since received a great deal of attention, not only due to its color and applications in jewelry, but also for its possible applications as an energy-efficient window coating. Cortie et al. measure and calculate the reflectance spectra of $AuAl_2$ using density functional theory. They show an experimentally determined reflectance minima at 2.5 eV, which has been shown to persist for Al:Au ratios of between 3.2:1 and 1:1. Minor discrepancies appear between the position of the measured and calculated reflectance minima due to self-interaction errors. They also measure the reflectance of $PtAl_2$ films, which show a reflectance maximum at around 1.9 eV of 55% that steadily decreases into the infrared. Vishnubhatla et al. studied the optical properties of $AuAl_2$, $AuGa_2$, and $AuIn_2$ [74]. They note that interband transitions appear at 2.2 eV in $AuAl_2$ and are responsible for the reflectance minimum at 2.5 eV. Hsu et al. note that this transition is not due to Au 5d bands as they lie too far below the Fermi energy. The onset of interband transitions decreases as the atomic number of the group 13 compound increases. $AuGa_2$ has much broader experimental interband transitions in the region around 2 eV than $AuAl_2$. Calculations show that this broadening is caused by an additional transition at approximately 1.6 eV. Of all the alloys studied here, CsCl structured binary intermetallics show the most promise, with large optical gaps and in some cases, bare plasma frequencies comparable to silver and gold [75].

NiAl, CoAl, and FeAl all have major interband contributions to the imaginary part of the permittivity, with the minimum value of ε'' being 10 for CoAl and FeAl at 2.75 eV and 3.25 eV, respectively. In $LaSn_3$ the onset of interband transitions occurs at approximately 0.5 eV with the main peak at approximately 1.5 eV. All the transitions between 1 and 5 eV can be explained due to the mixing of 4f character into states near the Fermi energy, with the dominant transition mechanism arising from La 5d to hybrid f-d-p states. In $CeSn_3$ the situation is even worse due to the increasing 4f character of conduction states, which causes partial occupation of f-d-p states allowing for additional transition mechanisms. $ThPd_3$ and UPd_3 both exhibit the $TiNi_3$ structure, UPd_3 having partially occupied 5f states and $ThPd_3$ having completely unoccupied 5f states; the onset of interband transitions in both materials is very low. The interband component of the optical spectra for all these materials is too great to allow for reasonable plasmonic activity [76]. MgAuSn exhibits the cubic AlLiSi structure and is colored purple due mainly to very strong interband transitions around 3 eV. The transitions are likely due to the parallel band effect, which gives aluminum its strong interband component at 1.5 eV. Because the transition in MgAuSn becomes steep at higher energy, and the material has a lower effective Drude plasma frequency, Kramers–Kronig integration forces the real part of the permittivity to be positive over the region 2.2–3.1 eV. The optical properties of the magnetic Heusler alloy Cu_2MnAl have been calculated using a tight-binding plane wave. Damping to plasma frequency ratios of between 8.85 and 10.7 have been measured in samples of Ni_2MnGa depending on the annealing temperature, however, both Ni_2MnGa and Ni_2MnIn have large low energy interband transitions. Fe_2TiAl has a plasma frequency of 0.22 eV, lower than the value of 1.32 eV for Fe_2VGa [77].

The Li_2AgIn alloy crystallizes in the NaTl-type structure (Zintl phase), with 16 atoms per unit cell such that a group of 8 BCC cells make up a cube. The corner atoms of each BCC cell are composed of alternating lithium and indium atoms and the centers of alternating BCC cells are lithium and silver atoms. For this compound, at exactly 2 eV the imaginary part of the permittivity is zero. This would have phenomenal optical properties, most notably, an infinitely sharp resonance [78].

18.6 Processing of IMC

To prepare an alloy based on an intermetallic compound, the following preparation methods are available: net shape casting, ingot casting, vacuum melting and controlled solidification, directional solidification, single crystals growth, rapid solidification, thermal spraying, CVD, PVD, electrodeposition, epitax growth, interdiffusion, mechanical alloying, powder metallurgy methods (reaction sintering, HIP), and so on [79, 80], and are largely associated with heat treatment processes. The principal difficulty is associated with the higher processing temperature required to deform and consolidate or to melt and infiltrate, materials intended for high-temperature structural use [81].

18.7 Most Used Intermetallic Compounds

Most used IMC in various purposes has been discussed subsequently:

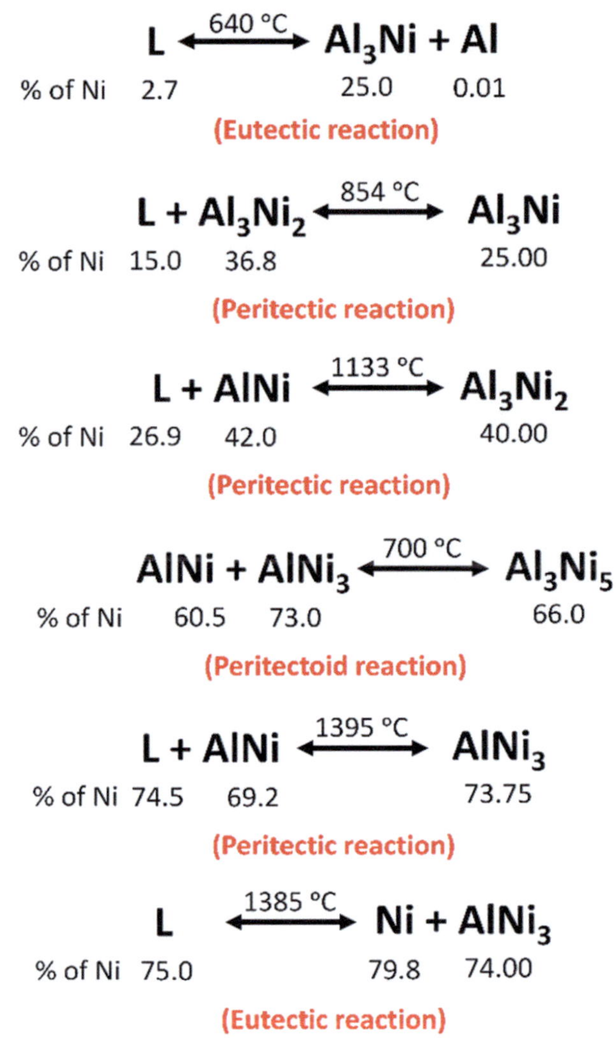

Fig. 18.11 Different reaction associated with NiAl intermetallics

18.7.1 NiAl Intermetallics

Ni_3Al has an L12 crystal structure, a derivative of the FCC crystal structure. This unit cell contains 4 atoms. The Ni atoms occupy face-centered positions and the Al occupies the corners of the unit cell [82]. On the contrary, NiAl presents the same crystal structure as FeAl, B2. In general, the properties of Ni-based intermetallic alloys are resistance to oxidation and carburization atmosphere; fatigue resistance superior to that of Ni-based superalloys; excellent wear resistance at high temperature; and good tensile and compressive yield strength at 650–1100 °C [83]. Figure 18.11 shows relevant information about the Ni–Al system and its characteristic reactions. In NiAl, Ni_3Al, Ni_2Al_3, $NiAl_3$ the weight percentage of Al are 31.49, 13.28, 40.81, 57.96 respectively with the heat of formation 28.3 ± 1.2 kcal/mol, −66.6 ± 1.2 kcal/mol, −67.5 ± 4.0 kcal/mol, −36.0 ± 2.0 kcal/mol respectively. Ni_3Al shows an increase of its elastic moduli, when temperatures reach a maximum between 600 and 800 °C, depending on small additions of alloying elements [84]. Similar to FeAl, it also forms an alumina layer that is responsible for its corrosion resistance. Like other intermetallics, Ni_3Al exhibits hydrogen embrittlement and grain boundary fracture, especially in its polycrystalline form. This mode of fracture has been attributed to the interaction of the material with water vapor, especially at elevated temperatures, where Al_2O_3 film appears to be insufficient to prevent rapid intergranular crack propagation. The addition of Cr and the production of an elongated grain structure could ameliorate the problem at intermediate temperatures by reducing the oxygen penetration [85]. At room temperature, the source of the intergranular fracture is weak grain boundaries, rather than environmental effects. Boron has been demonstrated to improve ductility by changing the fracture from intergranular to transgranular. Its low creep resistance appears to be another obstacle for its use, however, instead of superalloys. Hf and Zr are proven to be effective. Furthermore, regarding grain size sensitivity, coarse grain materials show better creep properties. Looking at NiAl properties, it appears to be more feasible for high-temperature applications than Ni_3Al. Nevertheless, poor ductility will be always a matter of concern at room temperature [86]. Here, the major cause would be the lack of sufficient slip systems. Surprisingly, a decrease in grain size does not result in better ductility behavior of NiAl. Boron added to NiAl serves as a potential solid-solution strengthener rather than as an effective suppressor of intergranular fracture. To obtain NiAl, the powders mixture reacts gradually and exothermically over a long period of continuous milling or in other way, the intermetallic formation occurs by opening the milling vessel after a short period of milling [87]. It is also known, among other applications, the suitability of Ni–Al system for bond coating, especially the composition Ni-5wt %Al. Ni–Al powders have been sprayed by Air Plasma Spray in air, Wire Arc spray, HVOF, and cold spray. The product produces coatings with very low porosity and oxide content [88].

18.7.2 FeAl Intermetallics

FeAl intermetallics have been intensively studied to use in high temperature or stainless steels and superalloys in some applications. They possess enough high concentrations of aluminum to form a continuous and adherent alumina layer on the surface when exposed to air or oxygen atmospheres [89]. In FeAl, Fe_3Al, $FeAl_2$, Fe_2Al_5 the weight percentage of Al are 32.57, 13.87, 49.1, 54.70, respectively with the heat of formation −12.0 kcal/mol, −16.0 kcal/mol, −18.9 kcal/mol, −34.3 kcal/mol, respectively. The most frequently studied compositions are 20 at.% and 40 at.% Al. These are related to the Fe_3Al and FeAl chemical formulae [90]. Actually, as can be extracted from the phase diagram, FeAl (single phase) exists over the composition 36–50 at.% Al. With regards to Fe_3Al, the transition temperature between the two observed ordered structures (D03 and B2) decreases and the B2 ordering temperature increases as the aluminum content is enhanced. FeAl presents a B2 (cP2) crystal lattice. The unit cell contains 8 aluminum atoms, 1 in each corner, which is shared with the other 8 unit cells surrounding the 1 aluminum atom, which is typical for the BCC structure. As the cP2 structure can be seen as 2 interpenetrating primitive cubic cells, the iron or nickel atoms are supposed to occupy the corner of the second sublattice [91]. Fe_3Al has the ordered cubic D03 (cF16) crystal structure. This unit cell contains 8 bcc type subcells and it may be thought of as being composed of 4 interpenetrating FCC lattices. In each subcell, the Fe atoms occupy corners, and thus, each of them is shared with 8 neighboring subcells, resulting 1/8 of the iron atom per subcell. Furthermore, 4 iron atoms occupy the centers of 4 subcells, so there are 12 iron atoms in the D03 unit cell. Including the 4 aluminum atoms, which also occupy the centers of 4 subcells, the total number of atoms per unit cell comes to 16. Fe_3Al transforms to a B2 (cP2) structure at temperatures above 540 °C. In general, the characteristics of Fe-based intermetallic alloys are better resistance to sulfidation than any other Fe- or Ni-based alloys; high electrical resistivity, which increases with temperature; good corrosion resistance; low material cost; and poor creep, and high-temperature strength above 500 °C [92].

Figure 18.12 shows relevant FeAl characteristic reactions. Regarding the mechanical properties, as well as for NiAl, the yield strength and fracture mode depend on the aluminum content; above or below 40 at.% Al content, this intermetallic shows intergranular or transgranular fracture, respectively. Regardless, the main reasons behind the brittle fracture in Fe_3Al are the environmental effects, while in FeAl, it must be also considered grain boundary weakness and vacancy hardening. However, contrary to NiAl behavior, FeAl strength and hardness increase near to stoichiometric composition [93]. While coarse-grained FeAl-based materials (below 40 at.% Al) exhibit a small increase of yield strength with temperatures reaching a maximum of 550–650 °C, an atomic aluminum concentration above 40% does not show such behavior. It is well known that the reactions to form iron aluminides from pure Fe and Al powders are exothermic, which is a great concern due to the uncontrollable nature of the process. The process enables to obtain FeAl hardened with Fe_2Al_5 and $FeAl_2$ inclusions, avoiding the tendency toward hydrogen embrittlement when being sprayed onto a substrate. When dealing with bulk intermetallics, it has been observed that when FeAl contains solute additions (Cr, Mo), second phase, or precipitates (Nb, Zr, C, B) the strength can be improved over 1000 MPa. The drawback of poor creep resistance, a characteristic of iron aluminides, is caused by the ease of movement of dislocations in the open BCC lattice. This can also be improved by stable particles such as carbides, dispersion boride particles, or oxides formed in mechanically alloyed materials [94].

FeAl alloys were for many years considered to be intrinsically brittle at room temperature because of their brittle transgranular fracture mode and the difficulty of cross-slip in ordered lattices. Liu et al. (1989) reported that FeAl alloys containing less than 40% aluminum were intrinsically quite ductile and that the low tensile ductility and brittle cleavage fracture were caused mainly by environmental embrittlement involving moisture in the air. Substantial improvements are possible in the tensile ductility of FeAl alloys. FeAl (35.8% aluminum) is brittle (2.2% ductility) owing to environmental embrittlement. The refinement of grain size by adding 0.05% zirconium and 0.24% boron increased the room-temperature ductility of FeAl (35.8% aluminum) from 2.2% to 10.7% [95]. The addition of boron enhanced grain-boundary cohesion and suppressed intergranular fracture of the alloy. Usually, preoxidation leads to a decrease in the ductility of metals

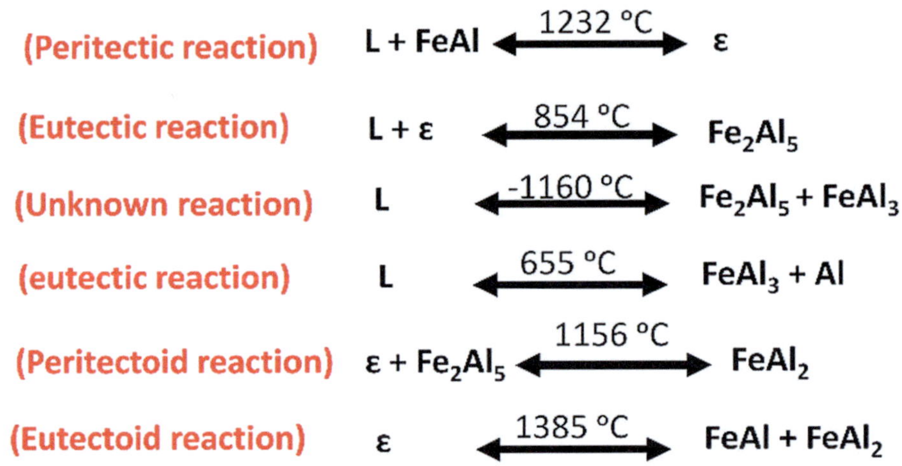

Fig. 18.12 Different reaction associated with Fe–Al intermetallics

and alloys. However, the formation of protective oxide scales by preoxidation at 700 °C reduced moisture absorption and further improved the tensile ductility of the FeAl alloy from 10.7% to 14.2%, that is, a 33% increase in room-temperature ductility. A series of FeAl alloys based on Fe ± (36 ± 40) Al ± 0.2Mo doped with 0.01 ± 0.15% carbon and boron, with improved mechanical properties at room and elevated temperatures, have been developed for structural use in oxidizing and sulfidizing environments [96].

18.7.3 TiAl Intermetallics

Titanium aluminide, TiAl, commonly gamma titanium, is an intermetallic compound. It is lightweight and resistant to oxidation and heat, however, it suffers from low ductility. The density of γ-TiAl is about 4.0 g/cm^3 [97]. The development of TiAl based alloys began in 1970. The alloys have been used in various applications only since about 2000. Titanium aluminide has three major intermetallic compounds [98]: γ-TiAl, α_2-Ti_3Al, and $TiAl_3$. Among the three, gamma TiAl has received the most interest and applications. Compared with Ti_3Al, the Al-rich so-called Υ titanium aluminide TiAl is regarded as more promising because of its lower density, higher strength at higher temperatures, and higher oxidation resistance. Thus the worldwide efforts to develop a lightweight alloy for application as high-temperature structural alloys are concentrated presently on TiAl. The α_2 alloys and super-α_2 alloys have been developed by alloying Ti_3Al with large amounts of Nb which show sufficient ductility at room temperature and aim at applications in the aircraft industry [99]. γ-TiAl has excellent mechanical properties and oxidation and corrosion resistance at elevated temperatures (over 600 °C), which makes it a possible replacement for traditional Ni-based superalloy components in aircraft turbine engines. Some γ-TiAl alloys retain strength and oxidation resistance to 1000 °C, which is 400 °C higher than the operating temperature limit of conventional Ti-alloys [100].

18.7.4 CrAl Intermetallics

Raynor examined solutions of Cr in Al and concluded that the transition elements were not contributing to the sea of delocalized electrons but instead were withdrawing electrons to fill their d-shells [101]. He proposed that 4.66 and 3.66 electrons per atom were absorbed by Cr and Mn, respectively; these values represent the number of vacancies in the atomic orbitals according to the Pauling theory of metals. These large negative valencies were not corroborated by electron counting based on measured X-ray diffraction intensities. It was clear, however, that some absorption of the conduction electrons was taking place in contrast with the original view of Hume-Rothery that the transition metals in the metallic state are zero-valent; that is, they neither consume nor contribute to the supply of conduction electrons [102]. The paper by Koster et al. provides information, based largely on magnetic evidence, on the number of 3d electrons on Cr in the several CrAl compounds. Six chromium–aluminum compounds are known to exist—Cr_2AI, Cr_4A1_9, Cr_6Al_6, CrAl Cr_2Al_{II}, and $CrAl_4$ [103]. In all cases, except Cr_4A1_9, the susceptibility was found either to increase with rising temperature, as is the case with elemental Cr or to be sensibly independent of temperature [104, 105].

18.7.5 NiTi Intermetallics

NiTi, one of the equilibrium intermetallic phases in the NiTi system, is interesting because it has two different crystal structures, depending on the temperature due to a martensitic transformation [106]. This transformation leads to a shape memory effect (detail given in Chap. 1). The high-temperature NiTi "austenitic" phase has a B2 (CsCl) structure, while the low-temperature "martensite" has a more complex monoclinic B-19′-type structure. Other intermetallics of the NiTi systems are $NiTi_2$, Ni_3Ti, $NiTi_3$, Ni_3Ti_4, Ni_2Ti, Ni_4Ti_3 [107]. In general, the properties of Ti-based intermetallic alloys are excellent high-temperature

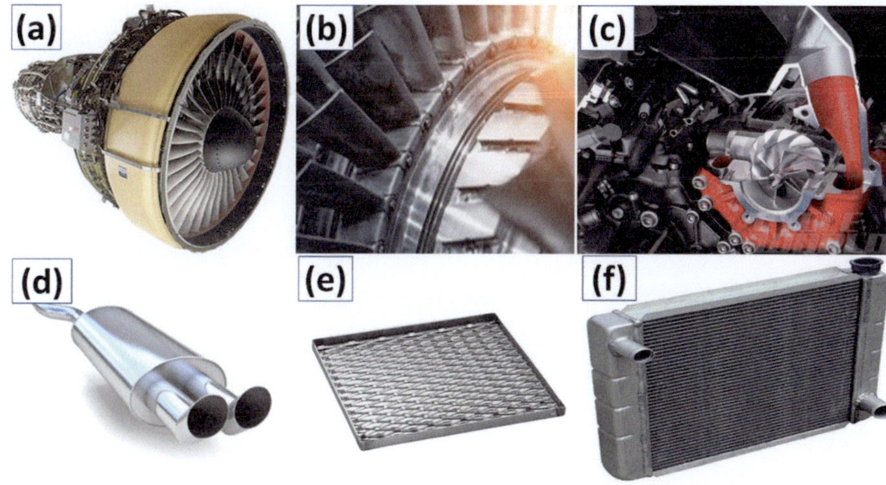

Fig. 18.13 (**a**) TiAl IMC low-pressure turbine blade (CF6-80C2, GE90), (**b**) TiAl IMC blade dampers (high-pressure blade turbine), (**c**) TiAl turbocharger wheel for automotive engine, and (**d**) FeAl IMC automotive exhaust systems), (**e**) Fe–Al heat-treating trays, and (**f**) Radiant tubes

mechanical properties; significantly lower density in comparison to superalloys; good creep resistance up to 1000 °C; and oxidation resistance up to 900 °C [108].

18.7.6 Compounds Containing Lanthanide Metals and Yttrium

Another area of very considerable activity during the year 1963 was that dealing with compounds involving the lanthanide metals and the chemically similar metal yttrium. This has been the most active area in the entire field of intermetallic compounds for the past decade, due undoubtedly to the fact that the lanthanide metals first became readily available commercially about 10 or 15 years ago [109]. Of particular interest in the work dealing with these compounds have been the studies of the magnetic properties of the lanthanides combined with Mn, Fe, Co, and Ni. In these, we have two magnetically active species and it has been of interest to see how the two moments couple [110]. Three types of compounds seem to merit attention—those possessing the cubic Laves phase structures, $MgCu_2$ (C1S), the CaCu (D2d) structure, and the CsCI (B2) structure. The magnetic behavior of 32 compounds represented by the formula $LnNi_2$, $LnFe_2$, and $LnCo_2$, and the implications in respect to the electronic makeup of these materials have been presented by Wallace and Skrabek [111]. In the Ni- and Co-series Ln was cerium, praseodymium, neodymium, samarium, gadolinium, terbium, dysprosium, holmium, erbium, thulium, lutetium and yttrium. The Fe-series was somewhat abbreviated in that the cerium, praseodymium, neodymium, and dysprosium compounds were not included. The Y-, Ce-, and Lu–Ni compounds were observed to be weakly paramagnetic. The $GdNi_2$ can be explained in terms of the moment of its gadolinium content, suggesting that Ni is also nonmagnetic in $GdNi_2$ [112, 113].

18.8 Application Fields of IMC Alloys

IMC have potentially applied in various high strength structures as well as in high-temperature application. In structural applications generally based on Aluminides (Ni_3Al, NiAl, Ti_3Al, TiAl, FeAl, Fe_3Al, Co_3Al), Silicides ($MoSi_2$, V_3Si, Fe_3Si, Ni_3Si), and the composites with matrix based on intermetallic alloys [114]. TiAl based IMC used in aircraft engines such as low-pressure turbine blade (CF6-80C2, GE90), carbon seal support (F414), transition duct beam (GE90), blade dampers (high-pressure blade turbine), compressor blade (Allison 14th stage), diffusion casting (advanced engine) (Fig. 18.13) [115–117]. These IMCs are also used in high-speed civil transport (divergent flap, nozzle sidewall face sheet, hot ducts and chute door, etc.), automotive engine (turbocharger wheel, exhaust valve), biomedical equipment (knee and hip implant) [118]. Possible industrial applications of Fe–Al system are heat-treating trays, immersion heaters, porous filters, automotive piston valves, and automotive exhaust systems, centrifugally cast tubes, radiant tubes for heat exchangers, and catalytic conversion vessels. General Electric uses γ-TiAl for the low-pressure turbine blades on its GEnx engine, which powers the Boeing 787 and Boeing 747-8 aircraft. This was the first large-scale use of this material on a commercial jet engine when it entered service in 2011 [119]. The TiAl LPT blades are cast by Precision Castparts Corp. and Avios.p.a. Machining of Stage 6 and Stage 7 LPT blades is performed by Moeller Manufacturing. An alternate pathway for the production of the γ-TiAl blades for the GEnx and GE9x engines using additive manufacturing is being explored.

Possible industrial applications of NiAl system are in transfer rolls, heat-treating trays, centrifugally cast tubes, rails for walking beam furnaces, die blocks, nuts and bolts, corrosion resistance tool bits, single-crystal turbine blades,

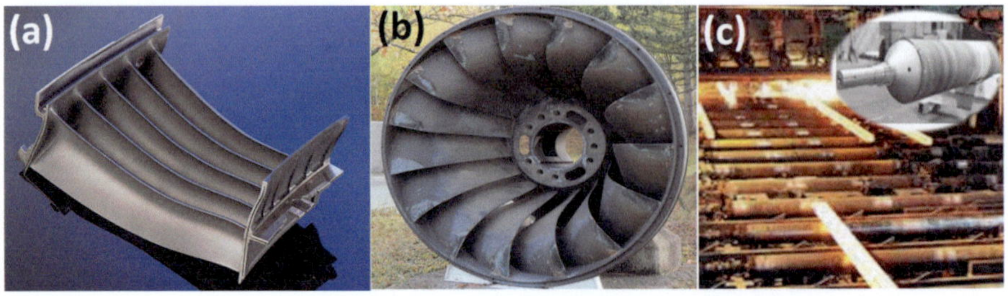

Fig. 18.14 (**a**) Turbine vanes made of IC 6 Ni_3Al-based alloy in an advanced air engine, (**b**) Water Turbine Rotors and Water Pumps, and (**c**) Transfer rolls in furnaces in steel Industry

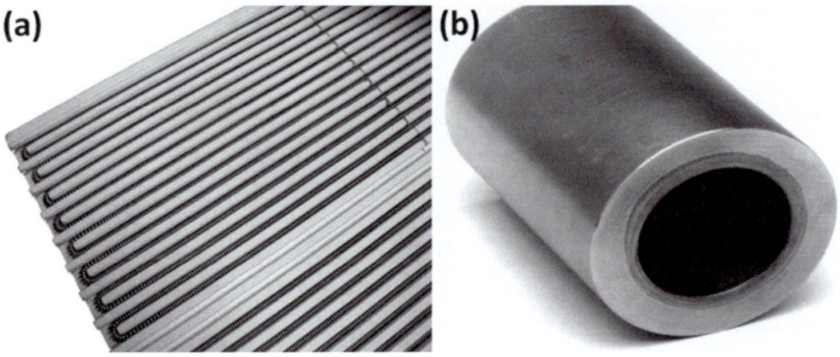

Fig. 18.15 (**a**) Chromium-free resistive heating elements of FeAl, (**b**) Extruded tube of Cr-free FeAl

aircraft fasteners, automotive turbochargers, pistons and valves, expansions joints to be used in corrosive environments and permanent molds [120]. The use of Ni_3Al as elements of aircraft engines is a classic example, commonly used to demonstrate the potential of future applications. The directionally solidified Ni_3Al base alloy with commercial name IC6 has been developed for advanced jet-engine turbine blades and vanes operating at the temperature range of 1050–1100 °C (Fig. 18.6). The Ni_3Al based alloys are still regarded as candidates for advanced high-temperature structural materials in aerospace applications, for example, turbine engine components. An introduction of this Ni_3Al intermetallic alloy can increase the maximum operation temperature on rotor blades and nozzle guide vane in turbine engines by 50–100 °C, leading to approximately 10% mass reduction and an improvement of heat resistance [121]. Consequently, it is believed that service life may increase two to three times. The Ni_3Al intermetallics are potential candidates as materials for turbochargers rotors in diesel-engine trucks. IC-221M alloy may substitute popular IN-713C Ni-superalloy that exhibits a worse fatigue strength, a higher density, and is more expensive (Fig. 18.14a). Intermetallic alloys have a much better cavitation and erosion resistance than conventional materials [122]. It is expected that IC-50 alloy may successfully replace actually applied materials. A water turbine rotor made of a Ni_3Al based alloy (Ni–Al 10.9-Zr 0.22-Cr 6.9-Mo 1.22-Fe 12-B 0.03 (wt%)) exhibits a definitely longer lifetime than its counterpart made of stainless steel. Due to high corrosion resistance, these materials can be used also for working elements in a seawater environment (Fig. 18.14b). Figure 18.14c showing the transfer rolls used in furnaces in the steel industry. The piston rings and valves of internal combustion engines of a car are completely made of a Ni_3Al based alloy or the Ni_3Al based composites strengthened by ceramic particles, for example, Al_2O_3, Cr_3C_2, Cr_2O_3, and SiC. Ni_3Al alloys can be used not only as automotive body material and also as elements with superior strength or absorbing energy. However, due to its cost, Ni_3Al intermetallic alloys may only be applied to higher end models, for example, Audi, Mercedes, and BMW. There are applications of Ni_3Al alloys in rolls in a continuous casting process. Ni_3Al is also used in radiant burner tubes, link belts for heat treating furnace, high-temperature reaction vessel, tube hanger, pump impeller for slurries, forging die, and so on [123]. The Cr-free resistive heating elements of FeAl and extruded tube of chromium-free FeAl have been given in Fig. 18.15.

A number of intermetallic compounds have become important for their hydrogen sorption properties. MgNi and LaNi IMC have been investigated on the better hydrogen sorption properties (Fig. 18.16a) [124]. Mechanical alloying has been used in the past to synthesize certain high magnetic field A15 superconductors such as NbA1 and NbSn (Fig. 18.16b). Possible industrial applications of NiTi IMCs are turbocharger rotor, piston head for diesel engines, jet engine compressor rotor, high-pressure turbine stator, turbine blades, and vane support rings. Special NiTi applications could be endovascular stents, vena cava filters, and dental files; as actuators, temperature controllers, safety gears in household appliances, and radiators; and in railroad wheel, or as contact tires on railroad wheels (Fig. 18.16c). The overview of alloys based on IMC compounds is only brief, it should give an idea of the number

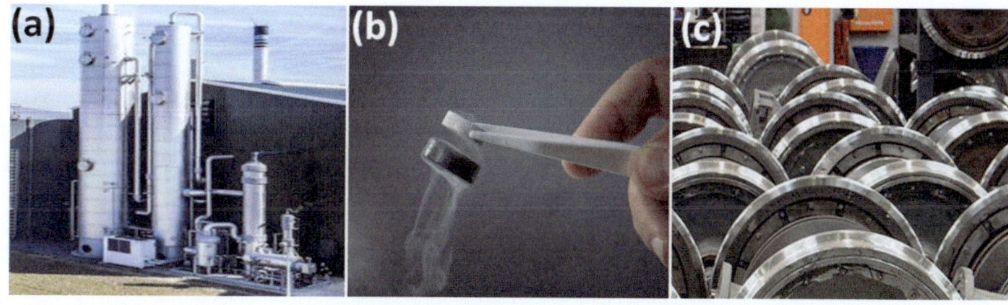

Fig. 18.16 (**a**) Hydrogen sorption MgNi and LaNi IMC, (**b**) Superconductors such as NbA1 and NbSn, (**c**) NiTi contact tires on railroad wheels

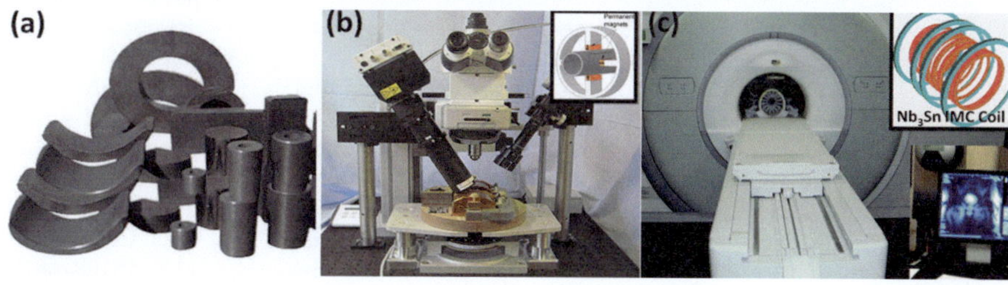

Fig. 18.17 (**a**) CoZrB are used in hard magnet, (**b**) GaMnN based IMCs used in Magneto-optical applications, (**c**) Nb_3Sn high-end MRI scanners and NMR spectrometers

and variability of IMCs in structural, magnetic, semiconductor, and other alloys [125].

FeNi (Isoperm), Fe_3Al (Alfenol), $Fe_3(Al,Si)$ (Sendust), FeCo (Permendur), $Co_{20}Al_3B_6$ (structure D84) etc. are used in soft magnet whereas CuNiFe (Cunife), CuNiCo (Cunico), (Fe, Al, Ni+Co, Cu, Ti, Nb) (Alnico), FeCoVSi (Vicalloy), AgMnAl (Silmanal), FePt, MnAl, MnAlGe, MnBi, CoZrB are used in hard magnet (Fig. 18.17a) [126]. IMC based on CdTe, GaAs used in semiconductors for photovoltaic applications. GaAs, HgCdTe, $LiTaO_3$ based IMC used in optical applications (Fig. 18.17b) [126]. Fe–Ni, Fe–Co, GaMnN based IMCs used in magneto-optical applications. PbTe, SiGe, BiTe based IMCs used in thermoelectric and electric applications. Mg_2Ni, Ti_2Ni, $LaNi_{5,}$ and so on. Most of the above-discussed IMCs are illegible for high-temperature coatings for gas turbines by plasma spraying, CVD methods, and other PVD methods. Nb_3Ge is a superconductor with an A15 structure. Many of these compounds have superconductivity at around 20 K (−253 °C), which is comparatively high, and remain superconductive in magnetic fields of tens of teslas (hundreds of kilogauss). This kind of superconductivity (Type-II superconductivity) is an important area of study as it has several practical applications [127]. V_3Si shows superconductivity at around 17 K. Niobium–germanium held the record for the highest temperature of 23.2 K from 1973 until the discovery of the cuprate superconductors in 1986 [128]. Though some A15 phase materials can withstand higher magnetic field intensity and have higher critical temperatures than the NbZr and NbTi alloys, NbTi is still used for most applications due to easier manufacturing. Nb_3Sn is used for some high field applications, for example, high-end MRI scanners and NMR spectrometers (Fig. 18.17c) [129].

18.9 Summary

This chapter successfully extracts the advantages of intermetallics, the new class of materials science that was previously considered as an unwanted phase in a high-end structure. This chapter revealed different structures of intermetallic compounds and their advantages in various applications. IMC limitations are low ductility, brittle fracture at room temperature, and processing problems that have been discussed here. Still, much research is required to extract and improve the property of IMC.

References

1. See: https://en.wikipedia.org/wiki/Intermetallic. 01.06.2020
2. Cinca, N., Lima, C.R.C., Guilemany, J.M.: An overview of intermetallics research and application: status of thermal spray coatings. J. Mater. Res. Technol. **2**(1), 75–86 (2013)
3. See: https://www.britannica.com/science/intermetallic-compound. 01.06.2020
4. Sayyadi, R., Naffakh-Moosavy, H.: The role of intermetallic compounds in controlling the microstructural, physical and mechanical properties of Cu-[Sn-Ag-Cu-Bi]-Cu solder joints. Sci. Rep. **9**, 8389 (2019). https://doi.org/10.1038/s41598-019-44758-3
5. See: http://snst-hu.lzu.edu.cn/zhangyi/ndata/Intermetallic.html. 01.06.2020
6. Goldschmidt, H.J.: Metal physics of interstitial alloys. In: Interstitial alloys. Springer, Boston, MA (1967). https://doi.org/10.1007/978-1-4899-5880-8_2
7. See: https://www.britannica.com/science/metal-chemistry. 01.06.2020
8. See: https://chem.libretexts.org/Bookshelves/Inorganic_Chemistry/Modules_and_Websites_(Inorganic_Chemistry)/Descriptive_Chemistry/Periodic_Trends_of_Elemental_Properties/Periodic_Properties_of_the_Elements. 01.06.2020

9. See: https://www.encyclopedia.com/science-and-technology/technology/technology-terms-and-concepts/electrical-conductivity. 01.06.2020
10. See: https://chem.libretexts.org/Bookshelves/General_Chemistry/Map%3A_General_Chemistry_(Petrucci_et_al.)/11%3A_Chemical_Bonding_II%3A_Additional_Aspects/11.7%3A_Bonding_in_Metals. 01.06.2020
11. Bende, D., Wagnerand, F.R., Grin, Y.: 8 – *N* rule and chemical bonding in main-group MgAgAs-type compounds. Inorg. Chem. **54**(8), 3970–3978 (2015). https://doi.org/10.1021/acs.inorgchem.5b00135
12. See: https://www.wikizero.com/en/Intermetallics. 01.06.2020
13. See: https://en.wikipedia.org/wiki/Post-transition_metal. 01.06.2020
14. See: https://www.chem.fsu.edu/~latturner/Researchpg.htm. 01.06.2020
15. Vernon, R.E.: Organising the metals and nonmetals. Found. Chem. **22**, 217–233 (2020). https://doi.org/10.1007/s10698-020-09356-6
16. Ovchinnikov, A., Smetana, V., Mudring, A.-V.: Metallic alloys at the edge of complexity: structural aspects, chemical bonding and physical properties. J. Phys.: Condens. Matter. **32**, 26 (2020) Article ID 243002
17. Armbrüster, M., Schlögl, R., Grin, Y.: Intermetallic compounds in heterogeneous catalysis-a quickly developing field. Sci. Technol. Adv. Mater. **15**(3), 034803 (2014)
18. Stoloff, N.S., Liu, C.T., Deevi, S.: Emerging applications of Intermetallics. Intermetallics. **8**, 1313–1320 (2000). https://doi.org/10.1016/S0966-9795(00)00077-7
19. Sauthoff, G.: Intermetallics. In: Ullmann's Encyclopedia of Industrial Chemistry, (Ed.) (2006). https://doi.org/10.1002/14356007.e14_e01.pub2
20. Cinca, N., Guilemany, J.M.: Thermal spraying of transition metal aluminides: an overview. Intermetallics. **24** (2012). https://doi.org/10.1016/j.intermet.2012.01.020
21. Zhang, W., Sundar, R., Seetharama, D.: Improvement of the creep resistance of FeAl-based alloys. Intermetallics. **12**, 893–897 (2004). https://doi.org/10.1016/j.intermet.2004.02.020
22. Sadeghi, E., Markocsan, N., Joshi, S.: Advances in corrosion-resistant thermal spray coatings for renewable energy power plants: part II-effect of environment and outlook. J. Therm. Spray Tech. **28**, 1789–1850 (2019). https://doi.org/10.1007/s11666-019-00939-0
23. Robert, L.F.: Intermetallic compounds for high-temperature structural use unique iridium and ruthenium compounds. Platin. Met. Rev. **36**(3), 138 (1992)
24. Antoncik, E.: On the lattice location of implanted impurities in silicon. Nucl. Instrum. Methods Phys. Res., Sect. B. **14**(2), 193–203., ISSN 0168-583X (1986). https://doi.org/10.1016/0168-583X(86)90044-3
25. Ferro, R., Saccone, A.: Structural characteristics of intermetallic phases. In: Pergamon materials series, Pergamon, vol. 13, pp. 81–218. Elsevier, Amsterdam (2008). https://doi.org/10.1016/S1470-1804(08)80005-3. ISSN 1470-1804, ISBN 9780080440996
26. Du, J., Dong, C., Melnik, R., Kawazoe, Y., Wen, B.: Hidden electronic rule in the cluster-plus-glue-atom model. Sci. Rep. **6**, 33672 (2016). https://doi.org/10.1038/srep33672
27. Michalak, M., Bytomski, G.: Solid solutions-a hume-rothery condition in the context of the set-theoretic notion of binary relation (2017) https://doi.org/10.5593/sgem2017/61
28. Kocjan, A., Kelhar, L., Gradišek, A., Likozar, B., Žagar, K., Ghanbaja, J., Kobe, S., Dubois, J.-M.: Solid solubility in Cu5Gd1−xCax system: structure, stability, and hydrogenation. Adv. Mater. Sci. Eng. **2017**, 9 (2017). https://doi.org/10.1155/2017/9203623. Article ID 9203623
29. Luo, L., Liu, Y., Duan, M.: Phase formation of Mg-Zn-Gd alloys on the Mg-rich corner. Materials (Basel). **11**(8), 1351 (2018). https://doi.org/10.3390/ma11081351
30. Schultz, P.A., Davenport, J.W.: Calculations of systematics in B2 structure 3d transition metal aluminides. J. Alloys Compd. **197**(2), 229–242 (1993). https://doi.org/10.1016/0925-8388(93)90045-O. ISSN 0925-8388
31. See: https://en.wikipedia.org/wiki/Frank%E2%80%93Kasper_phases. 02.06.2020
32. Ovchinnikov, A., Smetana, V., Mudring, A.-V.: Metallic alloys at the edge of complexity: structural aspects, chemical bonding and physical properties. J. Phys. Condens. Matter, 32 (2020). https://doi.org/10.1088/1361-648X/ab6b87
33. Prasad, B.V., Schmid, M.F.: Principles of virus structural organization. Adv. Exp. Med. Biol. **726**, 17–47 (2012). https://doi.org/10.1007/978-1-4614-0980-9_3
34. Rayment, I., Baker, T.S., Caspar, D.L.: A description of the techniques and application of molecular replacement used to determine the structure of Polyoma virus capsid at 22.5 Å resolution. Acta Crystallogr. B. **39**(4), 505–516 (1983). https://doi.org/10.1107/S0108768183002785
35. See: https://wikivisually.com/wiki/Frank%E2%80%93Kasper_phases. 02.06.2020
36. Huang, B., Song, C., Liu, Y., Gui, Y.: Microstructure characterization and Wear-resistant properties evaluation of an intermetallic composite in Ni–Mo–Si system. Materials (Basel). **10**(2), 130 (2017). https://doi.org/10.3390/ma10020130
37. Demchyna, R., Leoni, S., Rosner, H., Schwarz, U.: High-pressure crystal chemistry of binary intermetallic compounds. Z. Kristallogr. **221**, 420–434 (2006). https://doi.org/10.1524/zkri.2006.221.5–7.420
38. Johnston, R.L., Hoffmann, R.: Structure-bonding relationships in the laves phases. Z. Anorg. Allg. Chem. **616**, 105–120 (1992). https://doi.org/10.1002/zaac.19926161017
39. Stein, F., Palm, M., Sauthoff, G.: Structure and stability of laves phases. Part I. critical assessment of factors controlling laves phase stability. Intermetallics. **12**, 713–720 (2004). https://doi.org/10.1016/j.intermet.2004.02.010
40. See: https://kundoc.com/pdf-intermetallics-interstitial-compounds-and-metallic-glasses-in-high-entropy-alloy.html. 02.06.2020
41. Sahlberg, M., Karlsson, D., Zlotea, C., Jansson, U.: Superior hydrogen storage in high entropy alloys. Sci. Rep. **6**, 36770 (2016). https://doi.org/10.1038/srep36770
42. Hsieh, C.-C., Wu, W.: Overview of intermetallic sigma (?) phase precipitation in stainless steels. Int. Sch. Res. Notices. **2012**, 16 (2012). https://doi.org/10.5402/2012/732471. Article ID 732471
43. Ungar, G., Zeng, X.: Frank–Kasper, quasicrystalline and related phases in liquid crystals. Soft Matter. **1**, 95–106 (2005). https://doi.org/10.1039/B502443A
44. Bendersky, L.A., Roytburd, A., Boettinger, W.J.: Transformation of BCC and B2 high temperature phases to HCP and orthorhombic structures in the Ti-Al-Nb system. Part I: microstructural predictions based on a subgroup relation between phases. J. Res. Natl. Inst. Stand. Technol. **98**(5), 561–583 (1993). https://doi.org/10.6028/jres.098.038
45. Ali, Roushown & Yashima, Masatomo. Space group and crystal structure of the perovskite CaTiO3 from 296 to 1720 K. J. Solid State Chem. 2005;1782867–2872. https://doi.org/10.1016/j.jssc.2005.06.027.
46. Zhou, D.W., Liu, J.S., Peng, P., Chen, L., Hu, Y.J.: A first-principles study on the structural stability of Al2Ca Al4Ca and Mg2Ca phases. Mater. Lett. **62**, 206–210 (2008). https://doi.org/10.1016/j.matlet.2007.04.110
47. See: https://chem.libretexts.org/Bookshelves/Organic_Chemistry/Map%3A_Organic_Chemistry_(Vollhardt_and_Schore)/01._

Structure_and_Bonding_in_Organic_Molecules/1.3%3A_Ionic__and__Covalent_Bonds_-_The__Octet__Rule. 02.06.2020
48. Savin, A., Nesper, R., Wengert, S., Fässler, T.F.: ELF: the electron localization function. Angew. Chem. Int. Ed. Engl. **36**, 1808–1832 (1997). https://doi.org/10.1002/anie.199718081
49. See: https://en.wikipedia.org/wiki/Zintl_phase. 04.06.2020
50. Wang, F., Miller, G.J.: Revisiting the Zintl–Klemm concept: alkali metal trielides. Inorg. Chem. **50**(16), 7625–7636 (2011). https://doi.org/10.1021/ic200643f
51. Likhanov, M.S., Sytov, N.V., Zheng, W., Dikarev, E.V., Shevelkov, A.V.: Nowotny chimney ladder phases with group 5 metals: crystal and electronic structure and relations to the $CrSi_2$ structure type. Crystals. **10**(8), 670 (2020). https://doi.org/10.3390/cryst10080670
52. John, S.T.: A chemical perspective on high pressure crystal structures and properties. Natl. Sci. Rev. **7**(1), 149–169 (2020). https://doi.org/10.1093/nsr/nwz144
53. Higgins, J.M., Schmitt, A.L., Guzei, I.A., Jin, S.: Higher manganese silicide nanowires of Nowotny chimney ladder phase. J. Am. Chem. Soc. **130**(47), 16086–16094 (2008). https://doi.org/10.1021/ja8065122
54. Yannello, V., Fredrickson, D.: Generality of the 18-n rule: intermetallic structural chemistry explained through isolobal analogies to transition metal complexes. Inorg. Chem. **54** (2015). https://doi.org/10.1021/acs.inorgchem.5b02016
55. See: https://homepage.univie.ac.at/michael.leitner/lattice/struk/b2.html. 04.06.2020
56. Christofidou, K.A., Pickering, E.J., Orsatti, P., Mignanelli, P.M., Slater, T.J.A., Stone, H.J., Jones, N.G.: On the influence of Mn on the phase stability of the CrMnxFeCoNi high entropy alloys. Intermetallics. **92**, 84–92 (2018). https://doi.org/10.1016/j.intermet.2017.09.011
57. See: https://homepage.univie.ac.at/michael.leitner/lattice/struk/l1_2.html. 04.06.2020
58. Amouyal, Y., Mao, Z., Seidman, D.: Effects of tantalum on the partitioning of tungsten between the γ- and γ'-phases in nickel-based superalloys: linking experimental and computational approaches. Acta Mater. **58**, 5898–5911 (2010). https://doi.org/10.1016/j.actamat.2010.07.004
59. Amouyal, Y., Mao, Z., Seidman, D.: Effects of tantalum on the partitioning of tungsten between the γ- and γ'-phases in nickel-based superalloys: linking experimental and computational approaches. Acta Mater. **58**, 5898–5911 (2010). https://doi.org/10.1016/j.actamat.2010.07.004
60. Koizumi, Y., Ogata, S., Minamino, Y., Tsuji, N.: Energies of conservative and non-conservative antiphase boundaries in Ti 3 Al: A first principles study. Philos. Mag. **86**, 1243–1259 (2006). https://doi.org/10.1080/14786430500380126
61. Sun, J., Wang, W., Yue, Q.: Review on microwave-matter interaction fundamentals and efficient microwave-associated heating strategies. Materials. **9**(4), 231 (2016). https://doi.org/10.3390/ma9040231
62. Zhao, P.Y., Wang, H.Y., Wang, G.S.: Enhanced electromagnetic absorption properties of commercial Ni/MWCNTs composites by adjusting dielectric properties. Front Chem. **8**, 97 (2020). https://doi.org/10.3389/fchem.2020.00097
63. Nembach, E.: The high temperature peak of the yield strength of γ'-strengthened superalloys. Materials science and engineering A: structural materials properties microstructure and processing. Mater. Sci. Eng. A Struct. Mater. **429**, 277–286 (2006). https://doi.org/10.1016/j.msea.2006.05.032
64. Kroupa, F., Paidar, V.: Structure of DO_3 superlattice dislocations in Fe-Si alloys. Phys. Stat. Sol. (a). **33**, 555–561 (1976). https://doi.org/10.1002/pssa.2210330214
65. Shery, L.W., Kapoor, M., Underwood, O.D., Martens, R.L., Thompson, G.B., Jeffrey, L.E.: Influence of grain boundary character and annealing time on segregation in commercially pure nickel. J. Mater. **2016**, 15 (2016). https://doi.org/10.1155/2016/4597271
66. Vanswygenhoven, H., Derlet, P.: Grain-boundary sliding in nanocrystalline FCC metals. Phys. Rev. B. **64** (2001). https://doi.org/10.1103/PhysRevB.64.224105
67. Zhang, P., Li, S.X., Zhang, Z.: General relationship between strength and hardness. Mater. Sci. Eng. A. **529**, 62–73 (2011). https://doi.org/10.1016/j.msea.2011.08.061
68. Freudenberger, J., et al.: Materials science and engineering. In: Grote, K.H., Antonsson, E. (eds.) Springer handbook of mechanical engineering. Springer Handbooks. Springer, Berlin (2009). https://doi.org/10.1007/978-3-540-30738-9_3
69. Ovid'ko, I.A., Valiev, R.Z., Zhu, Y.T.: Review on superior strength and enhanced ductility of metallic nanomaterials. Prog. Mater. Sci. **94**, 462–540 (2018). https://doi.org/10.1016/j.pmatsci.2018.02.002. ISSN 0079-6425
70. Barrett, C., Imandoust, A., Oppedal, A., Inal, K., Tschopp, M., Kadiri, H.: Effect of grain boundaries on texture formation during dynamic recrystallization of magnesium alloys. Acta Mater. **128** (2017). https://doi.org/10.1016/j.actamat.2017.01.063
71. Derkachova, A., Kolwas, K., Demchenko, I.: Dielectric function for gold in plasmonics applications: size dependence of Plasmon resonance frequencies and damping rates for nanospheres. Plasmonics. **11**, 941–951 (2016). https://doi.org/10.1007/s11468-015-0128-7
72. Blaber, M., Arnold, M., Ford, M.: A review of the optical properties of alloys and intermetallics for plasmonics. J. Phys. Condens. Matter. **22**, 143201 (2010). https://doi.org/10.1088/0953-8984/22/14/143201
73. Mukherjee, D., Das, S.: SiO2–Al2O3–CaO glass-ceramics: Effects of CaF2 on crystallization, microstructure and properties. Ceramics Int. **39**, 571–578 (2013). https://doi.org/10.1016/j.ceramint.2012.06.066
74. Zhang, Y., Alarco, J.A., Best, A.S., Snook, G.A., Talbot, P.C., Nerkar, J.Y.: Re-evaluation of experimental measurements for the validation of electronic band structure calculations for LiFePO4 and FePO4. RSC Adv. **9**, 1134–1146 (2019). https://doi.org/10.1039/C8RA09154D
75. Maurya, D.K., Saini, S.M.: Structural, electronic and optical properties of YNi4Si-type RNi4Si compounds (R = La and Gd): a new orthorhombic derivative of CaCu5 structure. J. Mater. Sci. **51**, 868–875 (2016). https://doi.org/10.1007/s10853-015-9411-4
76. See: https://www.phys.ufl.edu/~tanner/Wooten-OpticalPropertiesOfSolids-2up.pdf. 06.06.2020
77. See: https://www.semanticscholar.org/paper/Electronic-and-Magnetic-Properties-of-Half-Metallic-Sharma-Kaphle/566b55bbd51be4b0c81cbe120e1ff17f1c279aad. 06.06.2020
78. Jahanshahi, P., Ghomeishi, M., Adikan, F.R.M.: Study on dielectric function models for surface plasmon resonance structure. Sci. World J. **2014**, 6 (2014). https://doi.org/10.1155/2014/503749. Article ID 503749
79. Moore, B., Asadi, E., Lewis, G.: Deposition methods for microstructured and nanostructured coatings on metallic bone implants: a review. Adv. Mater. Sci. Eng. **2017**, 9 (2017). https://doi.org/10.1155/2017/5812907. Article ID 5812907
80. Moore, B., Asadi, E., Lewis, G.: Deposition methods for microstructured and nanostructured coatings on metallic bone implants: a review. Adv. Mater. Sci. Eng. **2017**, 9 (2017). https://doi.org/10.1155/2017/5812907. Article ID 5812907
81. Amico, DG.: Si-SiC based materials obtained by infiltration of silicon: study and applications (2015) https://doi.org/10.13140/RG.2.1.1617.1920
82. See: http://web.boun.edu.tr/jeremy.mason/teaching/ME212/chapter_03_04_sol.pdf. 09.06.2020

83. Deevi, S.C., Sikka, V.K., Liu, C.T.: Processing, properties, and applications of nickel and iron aluminides. Prog. Mater. Sci. **42** (1–4), 177–192 (1997). https://doi.org/10.1016/S0079-6425(97)00014-5. ISSN 0079-6425
84. Omori, T., Oikawa, K., Sato, J., Ohnuma, I., Kattner, U., Kainuma, R., Ishida, K.: Partition behavior of alloying elements and phase transformation temperatures in Co–Al–W-base quaternary systems. Intermetallics. **32**, 274–283 (2013). https://doi.org/10.1016/j.intermet.2012.07.033
85. Zheng, L., Schmitz, G., Meng, Y., Chellali, M., Schlesiger, R.: Mechanism of intermediate temperature embrittlement of Ni and Ni-based superalloys. Crit. Rev. Solid State Mater. Sci. **37**, 181–214 (2012). https://doi.org/10.1080/10408436.2011.613492
86. Krupp, U., Kane, W., Pfaendtner, J.A., Liu, X., Lair, C., McMahon Jr., C.J.: Oxygen-induced intergranular fracture of the nickel-base alloy IN718 during mechanical loading at high temperatures. Mater. Res. **7**(1) (2004). https://doi.org/10.1590/S1516-14392004000100006
87. Moshksar, M.M., Mirzaee, M.: Formation of NiAl intermetallic by gradual and explosive exothermic reaction mechanism during ball milling. Intermetallics. **12**, 1361–1366 (2004). https://doi.org/10.1016/j.intermet.2004.03.018
88. Tejero-Martin, D., Rezvani Rad, M., McDonald, A., et al.: Beyond traditional coatings: a review on thermal-sprayed functional and smart coatings. J. Therm. Spray Tech. **28**, 598–644 (2019). https://doi.org/10.1007/s11666-019-00857-1
89. Babu, N., Balasubramaniam, R., Ghosh, A.: High-temperature oxidation of Fe3Al-based iron aluminides in oxygen. Corros. Sci. **43**, 2239–2254 (2001). https://doi.org/10.1016/S0010-938X(01)00035-X
90. Enayati, M.H., Salehi, M.: Formation mechanism of Fe3Al and FeAl intermetallic compounds during mechanical alloying. J. Mater. SCI. **40**, 3933–3938 (2005). https://doi.org/10.1007/s10853-005-0718-4
91. See: https://www.nde-ed.org/EducationResources/CommunityCollege/Materials/Structure/metallic_structures.htm. 09.06.2020
92. Li, Z., Gao, W.: High temperature corrosion of intermetallics. Intermetall. Res. Prog., 1–64 (2008)
93. Nagpal, P., Baker, I., Liu, F., Munroe, P.: Room temperature strength and fracture of FeAl And NiAl. MRS Proc., 213 (2011). https://doi.org/10.1557/PROC-213-533
94. Morris, D., Muñoz-Morris, M., Chao, J.: Development of high strength, high ductility and high creep resistant iron aluminide. Intermetallics. **12**, 821–826 (2004). https://doi.org/10.1016/j.intermet.2004.02.032
95. Jiao, Z.B., Luan, J.H., Liu, C.T.: Strategies for improving ductility of ordered intermetallics. Prog. Nat. Sci. Mater. Int. **26**(1), 1–12 (2016). https://doi.org/10.1016/j.pnsc.2016.01.014
96. See: https://www.imetllc.com/training-article/annealing-increase-metal-ductility/. 02.06.2020
97. Wu, X.: Review of alloy and process development of TiAl alloys. Intermetallics. **14**, 1114–1122 (2006). https://doi.org/10.1016/j.intermet.2005.10.019
98. See: https://www.tms.org/Superalloys/10.7449/1992/Superalloys_1992_381_389.pdf. 02.06.2020
99. Imayev, R., Imayev, V., Oehring, M., Appel, F.: Alloy design concepts for refined gamma titanium aluminide based alloys. Intermetallics. **15**, 451–460 (2007). https://doi.org/10.1016/j.intermet.2006.05.003
100. Raji, S.A., Popoola, A.P.I., Pityana, S.L., Popoola, O.M.: Characteristic effects of alloying elements on β solidifying titanium aluminides: a review. Heliyon. **6**(7), e04463 (2020). https://doi.org/10.1016/j.heliyon.2020.e04463
101. Ghosh, G., Korniyenko, K., Velikanova, T., Sidorko, V.: Aluminium - chromium - iron. Landolt Börnstein. (2008). https://doi.org/10.1007/978-3-540-69761-9_5
102. Liu, Z., Chen, W.-F., Zhang, X., Zhang, J., Koshy, P., Sorrell, C. C.: Structural and microstructural effects of Mo3+/Mo5+ codoping on properties and photocatalytic performance of nanostructured TiO2 thin films. J. Phys. Chem. C. **123**(18), 11781–11790 (2019). https://doi.org/10.1021/acs.jpcc.9b02667
103. Myrtille, O.J.Y.H., Harada, Y., Miyawaki, J., Wang, J., Meijerink, A., de Groot, F.M.F., van Schooneveld, M.M.: J. Phys. Chem. A. **122**(18), 4399–4413 (2018). https://doi.org/10.1021/acs.jpca.8b00984
104. Jürgen Buschow, K.H.: Permanent magnet materials. In: Cahn, R. W., Haasen, P., Kramer, E.J. (eds.) Materials science and technology (2006). https://doi.org/10.1002/9783527603978.mst0045
105. See: https://www.chemguide.co.uk/atoms/properties/3d4sproblem.html. 02.06.2020
106. Frenzel, J., George, E., Dlouhy, A., Somsen, C., Wagner, M., Eggeler, G.: Influence of Ni on martensitic phase transformations in NiTi shape memory alloys. Acta Mater. **58**, 3444–3458 (2010). https://doi.org/10.1016/j.actamat.2010.02.019
107. Khanlari, K., Ramezani, M., Kelly, P.: 60NiTi: a review of recent research findings, potential for structural and mechanical applications, and areas of continued investigations. Trans. Indian Inst. Metals. **71**, 781–799 (2018). https://doi.org/10.1007/s12666-017-1224-5
108. Małecka, J.: Transformation and precipitation processes in a metal substrate of oxidized TiAl-based alloys. Oxid. Met. **91**, 365–380 (2019). https://doi.org/10.1007/s11085-019-09886-1
109. Riva, S., Yusenko, K., Lavery, N., Jarvis, D., Brown, S.G.R.: The scandium effect in multicomponent alloys. Int. Mater. Rev. **61**, 1–26 (2016). https://doi.org/10.1080/09506608.2015.1137692
110. Liu, F., Velkos, G., Krylov, D.S., et al.: Air-stable redox-active nanomagnets with lanthanide spins radical-bridged by a metal-metal bond. Nat Commun. **10**(1), 571 (2019). https://doi.org/10.1038/s41467-019-08513-6
111. Duc, N.: Formation of 3d-Moments and Spin Fluctuations in the Rare Earth - Transition Metal Intermetallics (2014)
112. See: https://en.wikipedia.org/wiki/Actinide. 10.06.2020
113. Mizumaki, M., Yano, K., Umehara, I., Ishikawa, F., Sato, K., Koizumi, A., Sakai, N., Muro, T.: Verification of Ni magnetic moment in GdNi_ {2} laves phase by magnetic circular dichroism measurement. Phys. Rev. B. **67** (2003). https://doi.org/10.1103/PhysRevB.67.132404
114. Knaislová, A., Novák, P., Kopeček, J., Průša, F.: Properties Comparison of Ti-Al-Si Alloys Produced by Various Metallurgy Methods. Materials (Basel). **12**(19), 3084 (2019). https://doi.org/10.3390/ma12193084
115. See: http://www.freepatentsonline.com/6732502.html. 10.06.2020
116. See: https://www.flightglobal.com/power-house/67846.article. 10.06.2020
117. See: https://www.geaviation.com/commercial/engines/cf6-engine. 10.06.2020
118. Kamarudin, M., Anasyida, A., Mohd Sharif, N.: Effect of aluminium and silicon to IMC formation in low Ag-SAC solder. Mater. Sci. Forum. **819**, 63–67 (2015). https://doi.org/10.4028/www.scientific.net/MSF.819.63
119. Bewlay, B., Weimer, M., Kelly, T., Suzuki, A., Subramanian, P.: the science, technology, and implementation of TiAl alloys in commercial aircraft engines. MRS Proc. **1526** (2013). https://doi.org/10.1557/opl.2013.44
120. Scheppe, F., Sahm, P.R., Hermann, W., Paul, U., Preuhs, J.: Nickel aluminides: a step toward industrial application. Mater. Sci. Eng. A. **329–331**, 596–601 (2002). https://doi.org/10.1016/S0921-5093(01)01587-8

121. Kablov, E., Lomberg, B., Buntushkin, V., Golubovskii, E., Muboyadzhyan, S.: Intermetallic Ni3Al-base alloy: a promising material for turbine blades. Met. Sci. Heat Treat. **44**, 284–287 (2002). https://doi.org/10.1023/A:1021251703416
122. Jóźwik, P., Polkowski, W., Bojar, Z.: Applications of Ni3Al based intermetallic alloys-current stage and potential perceptivities. Materials. **8**, 2537–2568 (2015). https://doi.org/10.3390/ma8052537
123. Czeppe, T., Wierzbinski, S.: Structure and mechanical properties of NiAl and Ni3Al-based alloys. Int. J. Mech. Sci. **42**(8), 1499–1518 (2000). https://doi.org/10.1016/S0020-7403(99)00087-9
124. Dantzer, P.: Properties of intermetallic compounds suitable for hydrogen storage applications. Mater. Sci Eng. A. **34**, 313–320 (2002). https://doi.org/10.1016/S0921-5093(01)01590-8
125. Laurila, T., Vuorinen, V., Paulasto-Krö, M.: Impurity and alloying effects on interfacial reaction layers in Pb-free soldering. Mater. Sci. Eng. R Rep. **68** (2010). https://doi.org/10.1016/j.mser.2009.12.001
126. See: https://www.sigmaaldrich.com/technical-documents/articles/material-matters/soft-magnetic-nanocrystalline-alloys.html. 15.06.2020
127. Robertm A.H., Thomas L.F., Donald H.L.: Magnetic susceptibility of superconductors and other spin systems. https://doi.org/10.1007/978-1-4899-2379-0.
128. See: https://en.wikipedia.org/wiki/A15_phases. 15.06.2020
129. See: https://www.kobelco.co.jp/english/ktr/pdf/ktr_34/072-077. 15.06.2020

Metal-Organic Frameworks

19

Abstract

This chapter describes the novel step in organic/inorganic combined science that can solve many problems. A clear picture of the combination of metal-organic framework and their behavior has been discussed. All the synthesis procedures to fabricate these materials such as hydrother mal or solvothermal techniques, microwave-assisted synthesis, sonochemical synthesis, mechanochemical synthesis, electrochemical synthesis, surfactant-assisted synthesis, and microfluidic MOF synthesis have been discussed. Various types of working mechanism involved with the specific type of MOF materials have been discussed. The two broad principles, that is, adsorptive separation and membrane separation, have been described with their potential application.

Keywords

Metal-Organic Frameworks · Adsorptive separation · Membrane separation

19.1 What Is Metal-Organic Framework

Metal-organic framework (MOFs) is crystalline and porous in nature which is composed of both organic and inorganic components having a rigid periodic networked structure. MOFs are simply defined as a crystalline porous hybrid of organic–inorganic materials with an open-pore framework that consists of a regular array of positively charged metal ions strongly bonded with rigid or semi-rigid polytypic organic ligand molecules [1, 2]. Nodes formed by metal ions can bind the arms of linkers together to produce a repeating, cage-like structure. Single ions or clusters of ions of inorganic nodes for MOF can form one-, two-, or three-dimensional structures. Coordination compound extends by repeating coordination entities in two or three dimensions; a coordination polymer with a coordination compound extends coordination network by repeating coordination entities, in one dimension form, but in the form of cross-links between two or more individual chains, loops, or spiro-links [3, 4]. MOFs can be defined as porous coordination polymers, metal-organic polymers, hybrid organic–inorganic materials, and organic zeolite analogs. MOFs is an extraordinarily large internal surface area having a hollow in structure. Researchers are able to synthesize MOFs that have a surface area about more than 7,000 square meters per gram [5]. MOFs (Fig. 19.1) suggest the unique structural variety in contrast to the other porous materials uniformity in its porous structures, flexibility in network topology, tunable porosity, atomic-level structural uniformity, low density, higher loading capacities, geometry, dimension, and chemical functionality.

Various structures of MOFs can be synthesized depending on the kinds of metal ions and organic ligands. This is an exciting architecture in nanotechnology for MOFs. During removal of the guest molecules, the pores are stable due to some solvent, and the pores could be refilled with other compounds. Due to this property, MOFs may intend for the storage of gases such as hydrogen and carbon dioxide. Sometimes, they are used in gas purification, as conducting solids, in gas separation, as supercapacitors, in heterogeneous catalysis, sensing, and application in drug delivery. MOFs are ideal materials for capturing, storing, and/or delivering applications. In comparison to other nanocarriers, MOFs are aspiring to extensive applications because of high pore volume with a regular porosity whether the tuneable organic groups allow the easy modulation of the framework as well as of the pore size.

19.2 History and Background of MOF

Before the recognition of the MOF, coordination chemistry has a long history of investigating compounds that were labeled as coordination polymers since they were formed by

A. Behera, *Advanced Materials*, https://doi.org/10.1007/978-3-030-80359-9_19

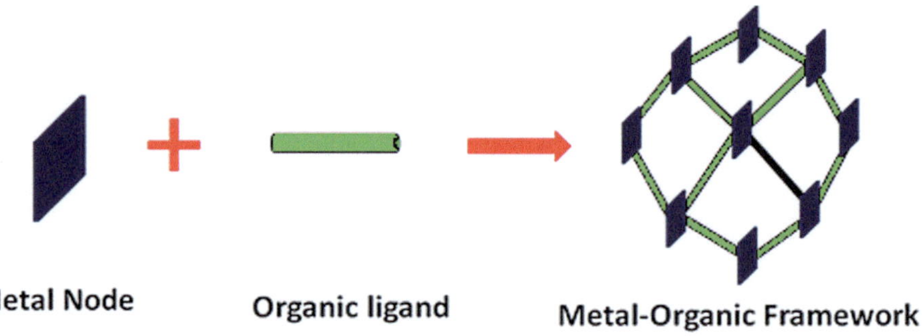

Fig. 19.1 Schematic representation of the metal-organic framework

the connection of metal ions as connectors and organic ligands as linkers. This area was first reviewed in 1960, that is, on separating water from saline solutions [6]. The interest in porous coordination polymers surfaced much later, although some complexes such as Prussian blue compounds and Hofmann clathrates had been known to show reversible sorption properties [7]. Thus, the interest in porous coordination polymers and MOFs started only around the early stage of 1980 [8]. MOF was pioneered around 1995 by Prof. Omar Yaghi at UC Berkeley [9]. After that, the MOF field has become a rapidly growing research field. In 1997, a 3D MOF was reported by Kitagawa et al. that exhibited gas sorption properties at room temperature. In 1999, the syntheses of MOF-518 and HKUST-119 were reported that in the following years and up to now are among the most studied MOFs. Starting in 2002, Ferey et al. reported on nonflexible as well as flexible porous MOFs, that is, MIL-4720 and MIL-5321/MIL-88, respectively [10]. The concept of isoreticular chemistry was made popular in 2002 for a series of Zn dicarboxylates but has also been extended to other materials. Especially, the mixed-linker compounds [M_2(dicarboxylate)$_2$(diamine)] (M = Zn, Cu) have shown to be a versatile class, and numerous studies have been carried out by varying the two organic components. The first structural model for the mixed-linker compounds was published in 2001. In 2002, the family of MOFs was extended to imidazolate-based compounds that are nowadays known as zeolitic imidazole frameworks (ZIFs) [11]. There is a large number of exciting compounds that are currently investigated due to their high stability, for example, Ni-CPO-27, UiO-66, Cr-MIL-100, Cr-MIL-101, MIL-125, CAU-1, and ZIF-8. The gradual development of various MOF applications has been mentioned in Table 19.1.

19.3 Structure of MOF

MOFs are composed of two major components: a metal ion or cluster of metal ions (also called secondary building units, SBUs) and organic linkers that are mostly carboxylic acid or nitrogen-containing ligands. For this reason, the materials are also known as hybrid organic–inorganic materials. The organic units are typically mono-, di-, tri-, or tetravalent ligands. Based on the dimension for the inorganic and organic ligands, the classification of the MOF has been given in Table 19.2. The choice of metal and linker dictates the structure and hence properties of the MOF. For example, the metal's coordination preference influences the size and shape of pores by dictating how many ligands can bind to the metal and in which orientation [12]. For example, Fig. 19.2 represents the SBU (metal node) and organic linker used for the synthesis of two well-known MOFs, namely, MOF-5 and HKUST-1. The combination of these structures results in a huge number of possible applications. When compared to other porous materials, such as zeolites and carbons, MOFs have the ability to tune their shape, size, and functionality by suitably selecting the metal ion and organic linkers for specific applications. MOFs have pore sizes ranging from microporous to macroporous and can accommodate diverse species, such as single metal atoms, nanoparticles, metal complexes, organic dyes, polyoxometalates, polymers, and small enzymes [13].

19.4 Synthesis of MOF

MOFs have become an interesting group of materials for researchers because their flexibility increases the scope of their application. MOFs can be prepared on large scale through low-priced or simple synthesis methods. Different methods are being used for the synthesis of MOFs, such as hydrothermal or solvothermal techniques, microwave synthesis, sonication synthesis, mechanochemical synthesis, and electrochemical synthesis (Fig. 19.3) [14–18].

19.4.1 Solvothermal or Hydrothermal Techniques

Solvothermal synthesis refers to reactions taking place under organic solvent or organic-water solvent mixture at temperatures higher than 100 °C and at a pressure above the ambient pressure. Here crystals are slowly grown from a hot solution. The reaction is carried out by heating the starting reaction mixture in a tightly capped vessel either Teflon-lined stainless steel autoclave or glass in an oven to

Table 19.1 Gradual development of MOF application in various industries

Year of concept developed	MOF materials development as per the patent
1960	High-flow porous membranes for separating water from saline solutions
1970	Molecular sieve adsorbent with alumina sol binder
1977	Multicomponent membranes for gas separations
1983	Anisotropic microporous supports impregnated with polymeric ion-exchange materials
1984	Oxygen permeable membrane
1985	Separation of fluids by means of mixed matrix membranes
1989	Hydrogen recovery by adsorbent membranes
1990	Novel multicomponent fluid separation membranes
1991	Process for removing mercury from organic media which also contain arsenic
1992	Extraction of organic compounds of aqueous solutions.
1995	Crystalline metal-organic microporous materials
1996	[Co (en) 3] CoSb4S8: a novel non-centrosymmetric lamellar heterometallic sulfide with large-framework holes
1997	Chiral 3D-framework material: d-Co (en) 3 [H3Ga2P4O16]
1998	Cyano-bridged Re6Q8 (Q = S, Se) cluster-metal framework solids
1999	$[R(L)_n]_m$Mn Metallo-organic polymers for gas separation and purification
2000	Carbon molecular sieves
2001	Mixed matrix membranes incorporating chabazite type molecular sieves
2002	Layered silicate material and applications of layered materials with porous layers
2003	Cross-linked polybenzimidazole membrane for gas separation
2004	Mixed matrix membranes using electrostatically stabilized suspensions
2005	Gas adsorbents based on microporous coordination polymers (metal bispyrimidinolate type of sodalite)
2006	Ligand-directed strategy for zeolite-type metal-organic frameworks: Zinc (II) imidazolates with unusual zeolitic topologies
2007	UV cross-linked polymer functionalized molecular sieve/polymer mixed matrix membranes
2008	Suspensions of silicate shell microcapsules
2009	Polybenzoxazole polymer-based mixed matrix membranes
2010	Halogen bonding in metal-organic-supramolecular networks
2011	High proton conduction in a chiral ferromagnetic metal-organic quartz-like framework
2012	Solvent-induced controllable synthesis, single-crystal to single-crystal transformation, and encapsulation of Alq3 for modulated luminescence in (4, 8)-connected metal-organic frameworks
2012	H_2 storage in isostructural UiO-67 and UiO-66 MOFs
2013	Silver nanoparticle/chitosan oligosaccharide/poly (vinyl alcohol) nanofibers as wound dressings
2014	Inherent anhydrous and water-assisted high proton conduction in a 3D metal-organic framework
2014	Porous metal-organic frameworks for gas storage and separation: What, how, and why?
2014	Encapsulation of strongly fluorescent carbon quantum dots in metal-organic frameworks for enhancing chemical sensing
2014	Light-emitting compound of metal-organic coordination polymers, and method of preparing the same
2015	Fragrance-containing cyclodextrin-based metal-organic frameworks
2016	Bioinspired design of ultrathin 2D bimetallic metal-organic framework nanosheets used as biomimetic enzymes
2017	Enhanced gas separation performance using mixed matrix membranes containing zeolite T and 6FDA-durene polyimide
2017	Porous scaffolds for electrochemically controlled reversible capture and release of alkenes
2018	Transition metal-organic framework having antibacterial properties
2018	Metal-organic frameworks for electrochemical detection of analytes
2019	$\{Zn_6\}$ cluster-based metal-organic framework with enhanced room-temperature phosphorescence and optoelectronic performances

Table 19.2 Classification of hybrid materials based on dimensionality

		Dimensionality of Organic			
		0	1	2	3
Dimensionality of inorganic	0	Molecular complexes	Chain coordination polymers	Layered coordination polymer	3D coordination polymer
	1	Hybrid inorganic chain	Mixed inorganic-organic layer	Mixed inorganic-organic 3D framework	–
	2	Hybrid inorganic layers	Mixed inorganic-organic 3D framework	–	–
	3	3D inorganic hybrids	–	–	–

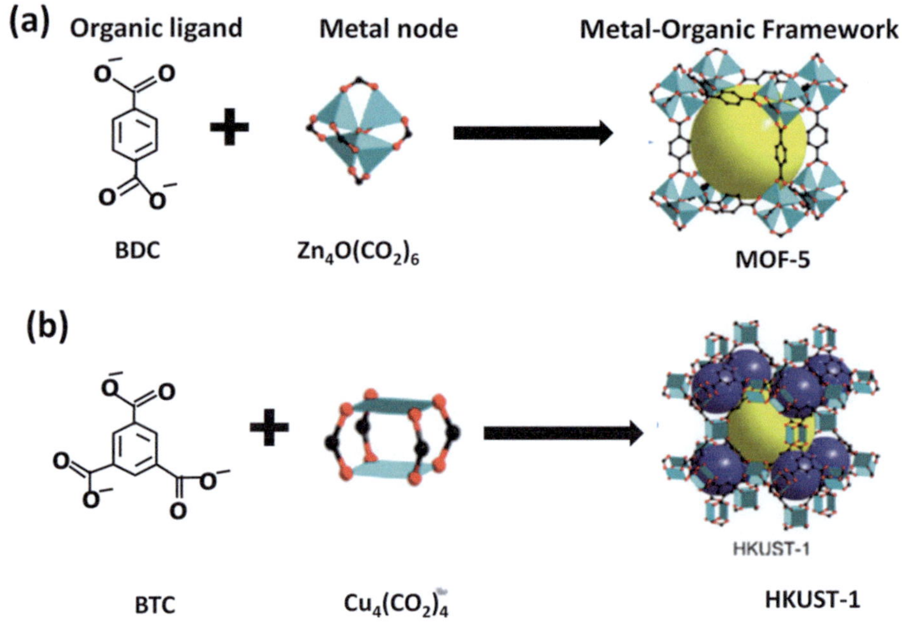

Fig. 19.2 Metal node (SBU) and organic linkers used for the synthesis of (**a**) MOF-5 and (**b**) HKUST-1 (Yellow and blue spheres represent the free spaces in the framework)

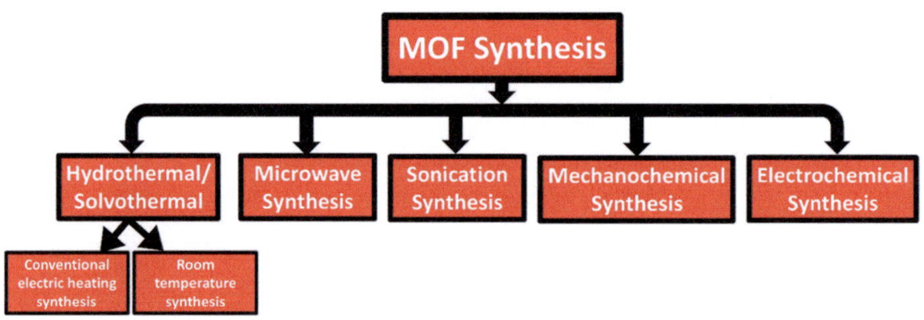

Fig. 19.3 Various synthesis methods of MOF

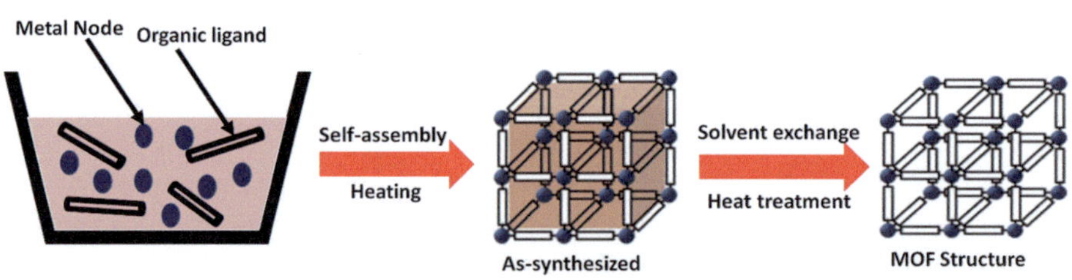

Fig. 19.4 Ionothermal synthesis of MOF structures

generate autogenous pressure. As the temperature increases beyond the atmospheric pressure, the autogenous pressure inside the closed vessel increases and often increases up to 15 bar. This increases the solubility and interaction of reactants in the solvent during the event of MOF synthesis. Solvents that are commonly used in solvothermal synthesis are alcohol, dialkyl formamide, and pyridine. In some cases, the addition of amine, such as triethylamine, in the synthesis favors the reaction without altering the final framework structures. Reaction time for this type of synthesis generally varies from a few hours to few days depending on the precursors used. Hydrothermal synthesis is a heterogeneous reaction that involves water as the solvent at a temperature below or above 100 °C and at a pressure above 1 bar. The water solvent could accelerate the reaction and alter physicochemical properties of reactants and products [19].

However, the solvothermal and hydrothermal methods have some limitations that are detrimental to the crystallization process. The former method involves the use of nonpolar organic solvent, making it difficult to dissolve inorganic precursors during the synthesis process. The latter method, on the other hand, involves the use of water which is good for dissolution of inorganic precursors but forms hydrogen bonds that hinder the interactions of framework species with templates and eventually prevents the nucleation process. These limitations can be overcome by employing "ionothermal synthesis" using conventional heating (Fig. 19.4). Ionothermal synthesis is an improved version of solvothermal synthesis that utilizes ionic liquid as both solvent and structure-directing agent (template) for the dissolution of starting precursors. Ionic liquids are salts that melt below the temperature used in the synthesis (150–220 °C). Moreover, the negligible vapor pressure of ionic liquids allows the synthesis to be carried out under ambient pressure which eliminates the safety concerns associated with the high autogenous pressure generated [20].

In 2002, Yan and coworkers revealed the rapid synthesis of MOF-5 materials via direct mixing of starting materials at room temperature. Room temperature synthesis is an attractive method to prepare MOF by mixing metal salts and organic linkers. This method offers a simple, rapid, and high-yield route for the preparation of MOF, as it does not require any energy source to trigger nucleation and growth. Indeed, this method has solved some problems associated with conventional electric heating, such as slow reaction time and the requirement of thermally insensitive starting materials for the preparation of MOFs. ZIF-8, for instance, is one of the most studied MOFs that can be synthesized at room temperature [21].

19.4.1.1 Microwave-Assisted Synthesis

In recent years, microwave-assisted synthesis has received increasing attention as an alternative method for the synthesis of MOFs. For almost half a century, microwave irradiation has been used for cooking food and now is widely used in organic synthesis, inorganic synthesis, analytical chemistry, and polymer curing. Microwaves are a form of electromagnetic radiation with frequency ranging from 0.3 to 300 GHz and wavelength between 1 mm and 1 m that lies between radio wave and infrared in the electromagnetic radiation region. Two important fundamental mechanisms that are responsible for transferring energy from microwaves to the molecules being heated are dipole polarization and ionic conduction. Figure 19.5 shows the microwave-assisted synthesis process. Microwave heating is generally based on the interaction of an electric component of microwaves with mobile electric charges, such as polar solvent molecules or conducting ions in a solvent or in a solid. The former mechanism is applied on polar molecules such as water molecules that will align themselves with the rapidly changing electric field of microwave, and energy is lost as heat due to rotation, friction, and collision of molecules [22]. For the latter mechanism, the conducting ions present in the solution will be moving in constantly changing directions through the solution, and heat is generated as a result of molecular friction and dielectric loss. In addition, the heating effect on semiconducting and conducting materials arises when the electric current is produced from ions or electrons within them due to the electrical resistance of the materials. Microwave-assisted synthesis is usually performed at a temperature above 100 °C, and the reaction time ranges from 1 min to 1 h. This method offers several advantages such as short crystallization time, phase selectivity, narrow particle size distribution, and ease of morphological control. Another important feature of microwave synthesis is that the microwaves provide efficient internal heating in the reaction vessel (often Teflon, borosilicate glass, or quartz), as it interacts directly with the molecules in the reaction mixture without heating the entire reactor. In contrast, conventional electric heating is a time-consuming and inefficient method for transferring energy into the system, as it heats the reactor walls followed by a reaction mixture through conduction or convection [23].

19.4.1.2 Sonochemical Synthesis

Sonochemical synthesis has been known for many years for its applications in organic synthesis and synthesis of nanostructured materials. Figure 19.6 shows the sonochemical synthesis of MOF structures. Sonochemical reactions can be performed using equipments such as ultrasonic cleaning baths and direct immersion ultrasonic horns. This approach is generally associated with the interaction of high-intensity ultrasound (20 kHz to 5 MHz) with the reaction mixture. The chemical effects of ultrasound mainly arise from acoustic cavitation, that is, the formation, growth, and implosive collapse of bubbles in liquids. The liquid will experience a dynamic tensile stress and the density will change accordingly with the alternating expansive and compressive acoustic waves when ultrasound interacts with a liquid. The generated bubbles will oscillate with the applied sound field. Subsequently, they will grow to size of usually tens of micrometers and at the same time effectively accumulate ultrasonic energy. The bubbles will then overgrow and collapse to release the energy accumulated in the bubble within a very short time. Formation and collapse of bubbles formed in the solution after sonication, termed acoustic cavitation, produces very high local temperatures (around 500 °C) and ultrasonic frequencies (>20 kHz) and results in extremely fast heating and cooling rates producing fine crystallites [24].

Applications of ultrasonic cleaning baths in sonochemical synthesis are limited owing to its insufficient intensity for most applications. Yet, cleaning bath is utilized in applications such as speeding up the dissolution of solids, activating highly reactive metals, forming emulsions, and exfoliating layered materials. Laboratory ultrasonic horn is alternative equipment for sonochemical reactions that can deliver ultrasonication intensity of 100 times greater than that of the ultrasonic cleaning bath. Moreover, an ultrasonic horn is directly immersed in the sample container, whereas the reaction vessel is immersed in the ultrasonic cleaning bath [25].

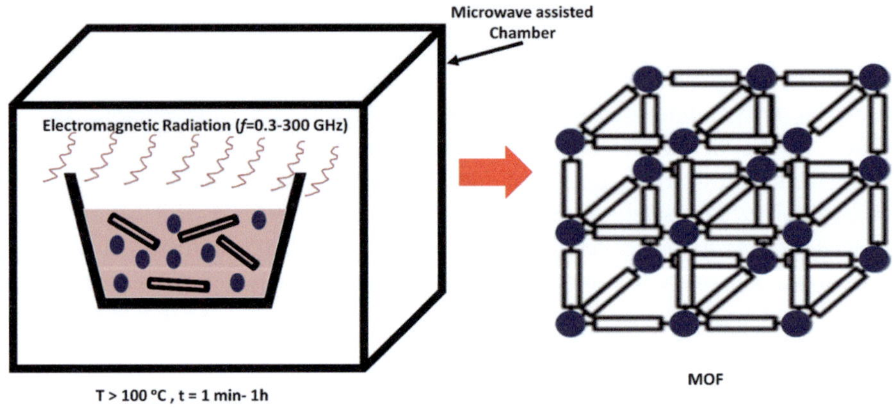

Fig. 19.5 Microwave-assisted synthesis process

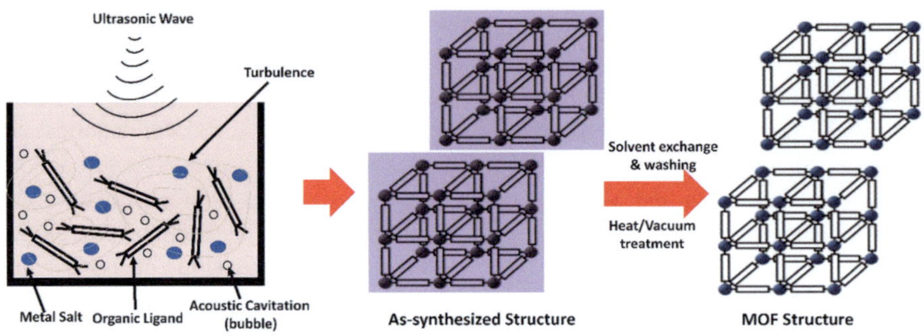

Fig. 19.6 Sonochemical synthesis of MOF structures

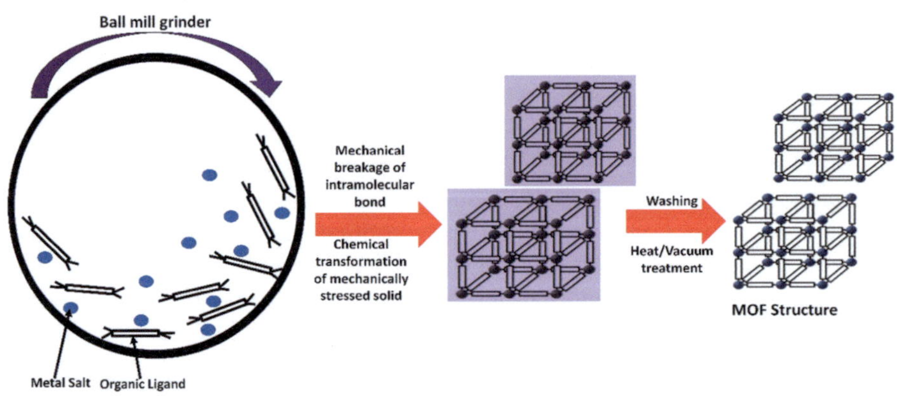

Fig. 19.7 Mechanochemical synthesis of MOF structures

19.4.1.3 Mechanochemical Synthesis

Compared tp conventional electric heating, mechanochemical synthesis offers advantages such as simplicity, low cost, reduction in reaction time, less or no solvent used, and high yield. Mechanochemical reactions involve the direct absorption of mechanical energy. Mechanical breakage of intramolecular bonds followed by a chemical transformation takes place in mechanochemical synthesis. Figure 19.7 shows the mechanochemical synthesis of MOF structures. This synthesis usually involves only grinding, either manually using a mortar and pestle or mechanically using a mixer or planetary mill such as ball mill. Ball milling is usually used when higher energy is needed and when the grinding time requires hours to days. The mechanical energy generated from grinding leads to a series of transformations on the crystalline solids such as reduction in particle size, production of new interfaces and crystal defects, heating, amorphization, and phase transformations in polymorphic materials [26]. MOF [Cu(isonicotinic acid)$_2$] was obtained by grinding copper acetate and isonicotinic acid together in a solvent-free condition. Liquid-assisted grinding (LAG), which is first mentioned in 2006 by Jones and coworkers, is a grinding approach involving the use of a small amount of solvent. Quantitative yields of small MOF particles can be obtained in short reaction times, normally in the range of 10–60 min. In many occasions, metal oxides were found to be preferred over metal salts as a starting material, which results in water as the only side product. The critical contribution of moisture in the mechanochemical synthesis of pillared-type MOFs was recently reported by Kitagawa group [27]. The addition of small amounts of solvents in liquid-assisted grinding (LAG) can lead to the acceleration of mechanochemical reactions due to an increase in mobility of the reactants on the molecular level. The liquid can also work as a structure-directing agent. More recently, the extension of the method to ion- and liquid-assisted grinding (ILAG) was reported to be highly efficient for the selective construction of pillared-layered MOFs. HKUST-1 was successfully produced by mechanochemical synthesis method without using a solvent. While a mechanochemical reaction between H3BTC and copper acetate produces HKUST-1, reaction using copper formate resulted in a previously unknown phase, potentially due to templating effects of the different acid by-products formed. Recently, a mechanochemical approach was also applied for ZIF synthesis using combinations of ZnO and imidazole (HIm), 2-methylimidazole (HMeIm), and 2-ethylimidazole (HEtIm) as the starting material within 30–60 min reaction time [28].

19.4.1.4 Electrochemical Synthesis

The electrochemical synthesis of MOFs uses metal ions continuously supplied through anodic dissolution as a metal source instead of metal salts, which react with the dissolved linker molecules and a conducting salt in the reaction medium. Figure 19.8 shows the electrochemical synthesis of MOF structures. The metal deposition on the cathode is avoided by employing protic solvents, but in the process, H_2 is generated. The electrochemical route is also possible to run a continuous process to obtain higher solid content compared to normal batch reactions [29]. In 2005, BASF researchers have reported the first electrochemical synthesis of MOFs in a patent. Their work focuses on the synthesis of Zn-, Cu-, Co-, and Mg-based MOFs through anodic oxidation of their corresponding metals. The main benefit of this approach is that MOFs can be produced continuously without the use of any metal salts, making it desirable for industrial applications. Recently, HKUST-1, ZIF-8, Al-MIL-100, Al-MIL-53, and Al-MIL-53-NH2 were synthesized via anodic dissolution in an electrochemical cell. It was claimed that electrochemical MOF synthesis has several advantages: faster synthesis at lower temperatures than conventional synthesis, metal salts are not needed and therefore separation of anions such as NO^{3-} or Cl^- from the synthesis solution is not needed prior to solvent recycle, and virtual total utilization of the linker can be achieved in combination with high Faraday efficiencies [30].

19.4.1.5 Surfactant-Assisted Synthesis

Surfactant-assisted synthesis is a soft templating approach traditionally used in the synthesis of mesoporous silicas and metal oxides, but now, this approach has been widely adopted in MOF synthesis. This approach generally involves the use of surfactants as a template to direct the formation of MOFs. Proper selection of surfactants is critical in the synthesis of MOF, as the structure and nature of surfactant will greatly affect the MOF materials obtained at the end of the reaction in terms of their structure, pore size, and surface area [31]. Surfactants generally can be classified into cationic, anionic, and nonionic surfactants. During the reaction, surfactant molecules aggregate to form micelles before forming rod micelles. These micelles will act as a template in the synthesis. After that, the removal of the template is needed to yield the final product. Iron-based MOF, Fe-MIL-88B-NH_2 [$Fe_3O(H_2NBDC)_3$] nanocrystals can be prepared using the nonionic surfactant triblockcopolymer Pluronic F127 as the stabilizing agent and acetic acid as the deprotonating agent. F127/Fe^{3+}molar ratio and CH_3COOH/Fe^{3+} molar ratio were varied to investigate the role of Pluronic F127 andCH_3COOH in controlling the size of Fe-MIL-88B-NH_2 nanocrystals. In the absence of F127, microsized crystals were obtained instead of nanocrystals. It is important to point out that the Pluronic F127 not only coordinates with the Fe^{3+} ions but also plays a crucial role in the formation of nanocrystals by stabilizing the MOF nuclei in the early stage of the synthesis process. Nevertheless, the size of the nanocrystals is not dependent on the amount of Pluronic F127 used. The addition of a higher concentration of acetic acid will lead to the formation of larger nanocrystals as a result of a lower degree of deprotonation of the carboxylate linkers which suppress the rate of nucleation and crystal growth. Very recently, surfactant-assisted synthesis of ZIF-8 nanocrystals with a particle size of ~100 nm and a high specific surface area of 1360m^2/g has been found [32].

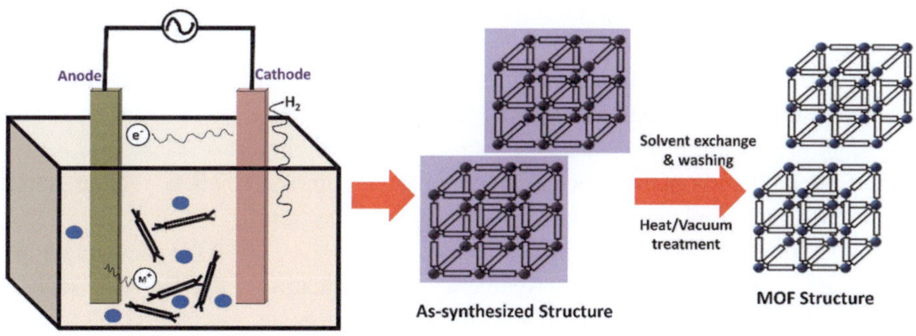

Fig. 19.8 Electrochemical synthesis of MOF structures

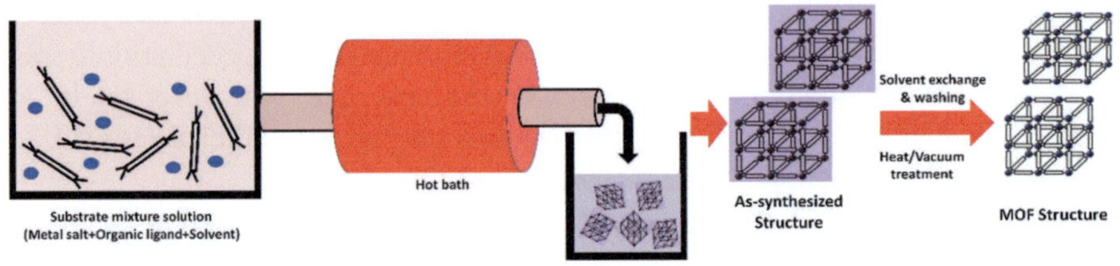

Fig. 19.9 Microfluidics synthesis of MOF structures

19.4.1.6 Microfluidic MOF Synthesis

The development of continuous, faster, and viable processes for MOF synthesis would be desirable to meet commercial and industrial requirements, and microfluidic synthesis of MOFs can be considered for this purpose (Fig. 19.9). Recently, preparation of HKUST-1 was investigated using a microfluidic device [33]. Initially, two kinds of substrate mixtures were prepared; the aqueous phase was prepared by adding $Cu(OAc)_2{\cdot}H_2O$ and polyvinylacohol in water (CuO particles formed were removed by centrifuging after 30 min stirring) at room temperature, whereas the organic ligand solution was prepared by dissolving H3BTC in 1-octanol and heated to 60 °C. Both immiscible liquids are supplied by syringe pumps to a T-junction, where the formation of aqueous solution droplets in the continuous organic phase takes place. The HKUST-1 capsule shell is formed at the liquid–liquid interface while these droplets travel through hydrophobic polytetrafluoroethylene tubing, before collection in ethanol. The remaining 1-octanol was removed by repeatedly exchanging the supernatant with fresh ethanol. The resulting HKUST-1 capsule walls were approximately 4 mm thick and had a macroporous structure outside, but the inner layer of the shell still consisted of densely packed crystallites [34].

19.5 Post-Synthetic Modification

Although the three-dimensional structure and internal environment of the pores can be in theory controlled through proper selection of nodes and organic linking groups, the direct synthesis of such materials with the desired functionalities can be difficult due to the high sensitivity of MOF systems. Thermal and chemical sensitivity, as well as high reactivity of reaction materials, can make forming desired products challenging to achieve. The exchange of guest molecules and counterions and the removal of solvents allow for some additional functionality but are still limited to the integral parts of the framework. The post-synthetic exchange of organic linkers and metal ions is an expanding area of the field and opens up possibilities for more complex structures, increased functionality, and greater system control [35].

Ligand Exchange: Post-synthetic modification techniques can be used to exchange an existing organic linking group in a prefabricated MOF with a new linker by ligand exchange or partial ligand exchange. This exchange allows for the pores and, in some cases, the overall framework of MOFs to be tailored for specific purposes. Some of these uses include fine-tuning the material for selective adsorption, gas storage, and catalysis. To perform ligand exchange, prefabricated MOF crystals are washed with solvent and then soaked in a solution of the new linker. The exchange often requires heat and occurs on a time scale of a few days. Post-synthetic ligand exchange also enables the incorporation of functional groups into MOFs that otherwise would not survive MOF synthesis, due to temperature, pH, or other reaction conditions, or hinder the synthesis itself by competition with donor groups on the loaning ligand [36].

Metal Exchange: Post-synthetic modification techniques can also be used to exchange an existing metal ion in a prefabricated MOF with a new metal ion by metal ion exchange. The complete metal metathesis from an integral part of the framework has been achieved without altering the framework or pore structure of the MOF. Similar to post-synthetic ligand exchange, post-synthetic metal exchange is performed by washing prefabricated MOF crystals with solvent and then soaking the crystal in a solution of the new metal. Post-synthetic metal exchange allows for a simple route to the formation of MOFs with the same framework yet different metal ions [37].

Stratified Synthesis: In addition to modifying the functionality of the ligands and metals themselves, post-synthetic modification can be used to expand upon the structure of the MOF. Using post-synthetic modification, MOFs can be converted from a highly ordered crystalline material toward a heterogeneous porous material. Using post-synthetic techniques, it is possible for the controlled installation of domains within a MOF crystal that exhibits unique structural and functional characteristics. Core-shell MOFs and other layered MOFs have been prepared where layers have unique functionalization but in most cases are crystallographically compatible from layer to layer [38].

19.6 Separation With MOF Materials

As a new class of porous materials, the investigation of MOFs for selective adsorption and separation greatly extended the scope of MOF applications in various industries and daily life. Separation processes play significant roles and are used for three primary functions: concentration, fractionation, and purification. These processes usually include distillation, crystallization, extraction, absorption, adsorption, and membrane separations, of which distillation accounts for 90–95% of all processes in the chemical industry. Distillation is however not always feasible because of some inherent limitations, such as the decompositions of some materials at high temperatures that are often required for the system. Adsorption- and membrane-based separations are alternatives in which a supporting medium, usually a porous material, is required for the separations. A lot of porous materials, including zeolites, carbon and metal-oxide molecular sieves, aluminophosphates, activated carbon, activated alumina, carbon nanotubes, silica gel, pillared clays, inorganic and polymeric resins, porous organic materials, and porous metal-organic composites, have been explored as adsorbents, and some of them have been implemented as membrane (filler) materials for various separations. Clearly, MOFs are feasible candidates as medium materials in adsorption and membrane-based separations.

19.6.1 Adsorptive Separation

Adsorptive separations refer to the process by which a mixture is separated based on differences in adsorption/desorption behavior of distinct components in the mixture and have a wide range of applications in industries. Several adsorptive separation mechanisms have been proposed, and they are highly dependent on the properties of adsorbates and adsorbents used as well as the associated interactions. Generally, steric, kinetic, or equilibrium effects or combinations thereof are present in guest adsorption by MOF [39]. Similar to those in zeolites, ion-exchange and reactive adsorption are also possible in some MOF. Separation in adsorptive processes is derived from the differences in adsorption behaviors of components of the mixture, which are directly related to the properties of the adsorbent in terms of adsorption equilibria and kinetics. The adsorption capacity and selectivity of an adsorbent are the principal properties relevant to adsorptive separation. The former depends on the nature of the adsorbate, the nature of the pores in the adsorbent, and the working conditions (such as temperature and pressures in gas separations, and solvent system in liquid separations). The latter is a more complicated, process-dependent property, although it is still related to the operational environment and the properties of the adsorbent and the adsorbate [40].

The separation of mixtures is frequently based on adsorption, a phenomenon that occurs whenever a bulk phase is exposed to an interface. Far from the interface, the thermodynamic state of the bulk is specified by a temperature, pressure, and a composition y_j, while the adsorbed phase has a specific density (usually per unit area or per unit mass of the adsorbent) and a composition x_i. Separation processes exploit the preferential concentration of the desired component at the interface. To quantify this effect, one defines the selectivity (alternatively known as the separation factor or partition coefficient) of component i with respect to component j [41]:

$$S_i = \frac{x_i}{y_j} \tag{19.1}$$

In addition to acceptable mechanical properties, a promising adsorbent should possess not only good adsorption capacity and selectivity but also favorable desorption kinetics. Regeneration of the adsorbent is another important issue in practical separation processes and is directly related to the cost of the processes. In gas-phase adsorption, the adsorbed species are most often removed by changing the temperature and the pressure of the system, resulting in temperature swing adsorption (TSA) and pressure swing adsorption (PSA) processes [42]. For liquid systems, a desorbent is required that preferentially displaces adsorbed species from the adsorbent and can then be easily separated from the adsorbate by other methods. Experimentally, the separation performance of an adsorbent can be evaluated by several techniques, including adsorption isotherms, pulse testing, breakthrough measurement, and chromatography separation. These methods also allow scientists and engineers to optimize operating parameters for practical mixture separations.

19.6.2 Membrane Separation

Membrane separations have broad coverage in science and engineering. In a general sense, a membrane is a barrier that selectively allows certain molecules to pass across it. Membrane-based separations have the advantages of low cost, high energy efficiency, ease of processing, excellent reliability, and usually smaller footprints when compared to other technologies, such as distillation and adsorption. When limiting our focus to the subject of MOFs acting as membrane materials for separations (primarily for gas separations), the separation of a mixture is mainly dominated by one (or more) of five separation mechanisms [43–47]: (1) molecular sieving, (2) Knudson diffusion, (3) surface diffusion, (4) solution diffusion, and (5) capillary condensation. These different mechanisms were developed based on different types of membranes and objects being separated.

Two parameters, permeance (or flux) (P) and selectivity (α), are usually used to evaluate the performance of a membrane in gas or liquid separations. The first one estimates the transport rate of species through a membrane, while the latter evaluates the ability of the membrane to separate components of a mixture. Experimental evaluations usually involve comparing permeances of pure fluids and mixtures and there from deducing the corresponding ideal selectivity and mixture separation factors. The ideal selectivity is defined as the ratio of permeability coefficients of pure fluid, $\alpha_{A/B} = P_A/P_B$, and can be calculated from pure-fluid measurements. However, for a mixture, the presence of one fluid can influence the transport of the others, therefore, deviating from the ideal selectivity. For a binary mixture system, the separation factor is defined as

$$\alpha^*{}_{A/B} = \frac{y_A/y_B}{x_A/x_B} \tag{19.2}$$

where y_A and y_B are the mole fractions of the components in the permeation, and x_A and x_B are their corresponding mole fractions in the feed phase. Similar to that of adsorptive separation, the efficiency of membrane separations largely depends on the membrane material that is used. In principle, all materials that form sufficiently thin films can be used to create membranes. Polymeric membranes, using organic polymers as membrane materials, are widely studied and being used in industry for gas separations. When compared to other membranes, they are in an advanced stage of development, much cheaper, easily fabricated into commercially useful shapes, stable at high pressures, and easily mass-produced. A lot of polymers have been used as membrane materials; among them, cellulose acetate (CA), polysulfone (PSf), and polyimide (PI) are the most widely used for industrial-scale applications. Inorganic materials that have been explored for membrane technologies include zeolites, alumina, carbon, ceramic, and others. Their membranes often show better separation performance due to their well-defined, rigid pores. Inorganic members also usually have higher chemical and thermal stability than polymeric membranes. However, high cost, poor mechanical properties (leading to defects), difficult synthesis using templates, and processing difficulties have limited their industrial applications. An alternative approach is to embed these inorganic porous materials, as fillers, into polymeric membranes to obtain a hybrid composite membrane, also call mixed-matrix membranes (MMMs). These hybrid membranes are expected to combine the advantages of inorganic porous materials, such as uniform and fixed pore shape and size, with those of organic polymers. However, achieving highly compatible connections between the two phases is a challenging issue. Clearly, ultrathin separation membranes fabricated by materials with uniform pores are much more desired than other membranes for specific separations because the pore size distribution plays a key role in their separation performance. Despite a long development, however, applications of zeolite membranes are hampered by various aspects, such as their synthesis using a template and limited choice of structure types. MOFs, just like zeolites, satisfy the requirement of well-defined pore size and shape and are therefore promising candidates for advanced membranes. Two types of membranes as supporting materials have been studied for separations using MOF: (1) pure MOF membranes (or thin film) and (2) mixed-matrix membranes with MOFs acting as fillers in organic polymers.

19.7 MOFs for Gas-Phase Adsorptive Separations

Separation of gas mixtures based on the differences in adsorption capacity is being widely used in various living and production-related activities, for example, gas drying, air separation, synthesis gas production, carbon dioxide capture, pollution control, and so on. Adsorptive separation is the more important in gas-phase separations, especially for light gases [48].

19.7.1 Selective Adsorptions and Separations of Gas

Efficient separations of light gases (CO_2, O_2, H_2, N_2, and CH_4) are becoming increasingly important from energetic, biological, and environmental standpoints. Selective adsorption and separation of these gases by MOF technology have been discussed subsequently.

19.7.1.1 Carbon Dioxide (CO_2)

CO_2 is one of the predominant greenhouse gas that is responsible for global emission. Reducing the CO_2 emission in the atmosphere has become primary environmental issue in the world. One option for reducing these harmful CO_2 emissions has been termed carbon capture and storage (CCS) [49], which entails the reduction of emissions primarily from large single point sources, such as power plants. In addition, the separation of CO_2 from CH_4 is also an important process in natural gas upgrading. MOFs have already been demonstrated to be promising materials in the separation of CO_2 from other gases. In fact, selective CO_2 capture from gas mixtures (especially from CO_2/CH_4 andCO_2/N_2) has attracted most attention [50].

The smaller kinetic diameter of CO_2 (3.3 Å) when compared to most other gases (except H_2, He, Ne, H_2O, and NH_3) has established the method of controlling the pore size of a MOF as a highly efficient way of achieving high CO_2 selectivity. This concept of designing a MOF with a pore size between the kinetic diameters of CO_2 and other gases allows only CO_2 to pass. In some cases, the actual pore size in a MOF is larger than the sizes of all tested gases; however, only

CO_2 adsorption is observed, which is usually attributed to a kinetic sieving effect. For example, Mn-$(HCO_2)_2$ has a porous framework structure containing cages with a diameter of about 5.5 Å connected to each other via small windows of about 4.5 Å. This MOF showed almost no N_2 and Ar sorption at 78 K but significant CO_2 uptake at 195 K [51]. The observed selective adsorption can primarily be attributed to the kinetic sieving effect, where the small windows limit the diffusion of larger N_2 and Ar molecules into pores resulting in no adsorption over the diffusion time of the measurement, while smaller CO_2 molecules are allowed to enter into the pore under the given conditions. This kinetic molecular sieving effect has been widely used in gas separations by using traditional porous materials, such as zeolites in both adsorption- and membrane-based separations. Furthermore, even if all of the tested gases can be adsorbed by a MOF, this diffusion limitation may have a significant effect on the uptake amount of different gases under given conditions [52].

19.7.1.2 Oxygen (O_2)

The separation of O_2 and N_2 from air is an important industrial process, presently carried out by using mainly cryogenic distillation to obtain large volumes of pure products. Other technologies include air separation using polymeric membranes or porous zeolites at ambient temperatures or using specialized ceramic membranes, suitable for small volumes of specific applications, at high temperatures. Cryogenic distillation of air is obviously highly energy consuming, thus leading to a long-term effort to develop other technologies that might enable this process to be carried out with lower energy costs. Adsorptive and membrane-based separation has already been shown to be energy saving when compared to distillation given that proper porous materials are available. However, obtaining a porous material capable of efficiently separating O_2 from air is not easy because of the similarity in its shape to N_2 and size to Ar. MOFs, with highly tailorable pore size, shape, and surface properties, are promising candidates for this task. With the large variety of MOFs already available, one can expect these novel materials to be capable of increasing selectivity for O_2, therefore improving energy efficiency and reducing the costs involved in air separation. Some MOFs have demonstrated selective adsorption of O_2 over N_2 and other gases. For example, at low temperature, selective adsorption of O_2 over CO and N_2 has been observed in Mg_3(2,6-ndc)$_3$ (2,6-ndc = 2,6-naphthalenedicarboxylate), $Zn_4O(H_2O)_3$(9,10-adc)$_3$ (PCN-13,9,10-adc = 9,10-anthracenedicarboxylate), and Ln_4(H2O)(tatb)8/3$(SO_4)_2$ (PCN-17, Ln = Yb, Y, Er, or Dy; tatb = 4,40,400-s-triazine-2,4,6-triyltribenzoate). Cu(bdt)(bdt = 1,4-benzeneditetrazolate) revealed adsorption selectivity of O_2 over N_2 and H_2. MOFs showing selective adsorption of O_2 over N_2 at low temperature include CUK-1,25[Ni(cyclam)]$_2$-(mtb) (cyclam = 1,4,8,11-tetraazacyclotetradecane; mtb = methanetetrabenzoate), Zn(dtp)(H_2dtp = 2,3-di-1H-tetrazol-5-ylpyrazine), Zn_3(OH)(dcbdc)2.5(DMF)4-x (dcbdc = 2,5-dichloro-1,4-benzenedicarboxylate; DMF = *N*,*N*-dimethylformamide), Mn-$(HCO_2)_2$, SNU-25, SNU-150, SNU-9, Zn(TCNQ-TCNQ)(bipy) (TCNQ = 7,7,8,8-tetracyano-p-quinodimethane), and Cu(bdtri)(DEF) (H2bdtri = 1,4-benzenedi(1H-1,2,3-triazole); DEF = diethylformamide). These MOFs are thus potentially useful for air separation, although further work is required such that these observed selectivities are conducted only at low temperatures. Selectivities could be achieved at higher pressures. Selective adsorption of O_2 over N_2 at ambient temperature has also been observed in mesoporous MOF Zn_4O(bdc)(btb)4/3 (UMCM-1, H3btb = 1,3,5-tris(4-carboxyphenyl)benzene)), which gave an O_2/N_2 selectivity of about 1.64 at 298 K and 0.96 bar. However, as the pressure was increased to 24.2 bar, the selectivity decreased to 1.1, contrary to the proposed trend in MOF-177. Also, selective adsorption is proposed to be a result of preferential adsorption due to a higher magnetic susceptibility of O_2 when compared to N_2; on the other hand, a dynamic effect is also possible. In contrast, HKUST-1, again, with a large pore size when compared to O_2 and N_2 but containing unsaturated metal sites, displayed adsorption preference for N_2 over O_2 at ambient temperature. The unsaturated metal sites are proposed to be responsible for the stronger interaction with the higher quadrupole moment of N_2 than that of O_2; the mechanism is similar to that observed in zeolites for selective adsorption of N_2 over O_2. These results revealed that adsorption selectivities of N_2 and O_2 in MOFs can be manipulated by the presence or absence of open metal sites. Similarly, the selective adsorption of O_2 over N_2 at room temperature was also observed in $Co_4(OH)_2$(dcdd)$_3$(H_2dcdd = 1,12-dihydroxy-carbonyl-1,12-dicarba-closo-dodecaborane) and Co(dcdd), both of which contain unsaturated metals sites in their structures. The evaluated O_2/N_2 selectivity in the two MOFs is about 6.5 and 3.5 at low pressure, respectively. Furthermore, the highly selective adsorption of O_2 over N_2 was observed in rigid $Cr_3(btc)_2$ at 298 K due to the direct bonding of O_2 with the metal sites in the MOF. Based on the charge transfer interaction between O_2 and the framework, high selective adsorption of O_2 over N_2, Ar, CO, CO_2, and C_2H_2 was also observed in a flexible MOF, Zn(TCNQ-TCNQ)(bipy) at low temperature. The selective adsorption is thus a result of combining the structural flexibility and electron-donating function of the soft framework [53, 54].

19.7.1.3 Hydrogen (H_2)

H_2 is considered to be one of the best alternative fuels to fossil resources because of its natural abundance, exceptional energy density, and nonpolluting nature when used in fuel cells. MOFs have attracted considerable attention in the hydrogen storage arena in recent years because of their high adsorption capacities. Because H_2 has the smallest size when compared to other gases from which it is usually separated, including CO_2, CO, Ar, N_2, and O_2, molecular sieving effects could be the primary mechanism responsible for the observed selective adsorption of H_2 over the others in MOFs with small pores. The hydrogen used in fuel cells is now primarily produced from steam reforming of natural gas. The resulting synthetic gas contains CH_4 and CO_2, which must be removed before H_2 can effectively be used. In principle, any porous material with a large enough pore size allowing H_2 and CO_2 to diffuse could be used to selectively adsorb CO_2 over H_2 at temperatures above 195 K because CO_2 is highly polarizable and has a higher quadruple moment as compared to H_2, which results in stronger adsorption interactions between CO_2 and the pore surface of the framework when compared to H_2. For gate-type flexible MOFs, this strong interaction can also be exploited to open the gate and allow more CO_2 to be adsorbed, whereas H_2 remains excluded. Long et al. tested several representative MOFs including MOF-177, $Be_{12}(OH)_{12}$-$(btb)_4$, Co (bdp) (bdp = 1,4-benzenedipyrazolate), $H_3[(Cu4Cl)_3$-$(BTTri)_8]$, and Mg-MOF-74 for their applications in the separation of CO_2 and H_2 via pressure swing adsorption. It is important that these measurements were performed under conditions (at 313 K and pressures up to 40 bar) close to practical application for H_2 purification and precombustion CO_2 capture [55, 56].

19.7.1.4 Gaseous Olefin and Paraffin

The separation of olefin and paraffin, which presently relies on energy-intensive cryogenic distillation-based technologies, is one of the most important separation processes in the petrochemical industry. Adsorptive separation is considered a more energy-efficient alternative; however, it still remains a challenge because of the similar molecular properties of olefin and paraffin pairs having the same number of carbon atoms. To date, only a very limited number of zeolites have shown the potential for kinetically separating olefins and paraffins. Other conventional adsorbents such as activated carbon, porous alumina, and silica have not shown a good selectivity. The development of new adsorbents with desired separation performance has thus become a key for the efficient separation of these chemicals. MOF has revealed potential in these separations [57]. Steam cracking of ethane is one of the routes for ethylene production, in which separation of ethylene and ethane is required. Now, HKUST-1 MOF can be useful in the separation of ethylene and ethane mixtures. Sorption isotherms for ethylene and ethane on HKUST-1 measured at 295 K demonstrated the preferential adsorption of ethylene over ethane, particularly in the low-pressure range. This adsorption selectivity was proposed to be due to interactions between π-electrons of the double bond in ethylene molecules and partial positive charges of coordinatively unsaturated Cu (II) sites in the framework. Besides the electrostatic interactions, their calculations also showed that stronger hydrogen-bonding interactions of ethylene with framework oxygen atoms play an important role in the preferential adsorption. Ethane can be selectively adsorbed over ethylene on ZIF-7(Zn(bim)2, bim = benzimidazolate). This is a microporous solid displaying the selective adsorption of paraffins over olefins. The efficient separation of ethane and ethylene by ZIF-7 was further confirmed by breakthrough experiments of the gas mixture. It was proposed that the separation performance of ZIF-7 can be attributed to a gate opening effect in which specific opening pressures control the uptake and release of different gas molecules. The opening pressure is usually determined by the interaction between guest molecules and the pore surface. In this case, the adsorption process was believed to be dominated by the interaction between the adsorbate gases and the benzene rings in the narrow ZIF-7 windows, which are also capable of selectively discriminating between the two molecules based on their different shapes. Based on the same mechanism, this work also showed that ZIF-7 is capable of separating propane and propylene, which is performed by the cryogenic distillation in the industry. Paraffin and acetylene are widely used as chemical feed stocks in the synthesis of many chemical products, in most of which the high purity of starting materials is a prerequisite. Their similar physical properties render removing acetylene from ethylene a difficult task. The adsorptive separation of ethylene and acetylene can be done in two flexible isostructural MOFs, Zn_3-$(cdc)_3$[Cu(SalPycy)] and Zn3-$(bdc)_3$[Cu(SalPycy)], with rationally tuned micropores in a 3D structure. At 195 K, $Zn_3(bdc)_3$[Cu(SalPycy) adsorbed both gases to give similar isotherms but with a higher acetylene uptake over ethylene (to give an equilibrium selectivity of 1.6) [58].

19.7.1.5 Harmful and Unsafe Gases

Most of the harmful or unsafe gases (such as H_2S, SO_2, Cl_2, CNCl, NH_3, NO_x, CO, C_2H_2, ethylene oxide, tetrahydrothiophene, and octane vapor) are released as waste gases from industrial processes into the environment. Effective capture of these harmful gases is extremely important to the protection of the environment. For the adsorption and separation of these gases, some MOF are successfully tested such as MOF-5, $Zn_4O(2$-NH_2-$bdc)_3$(IRMOF-3307), Zn-MOF-74, MOF-177, HKUST-1, and Zn_4O-$(dabb)_3$ (IRMOF-62, dabb = diacetylene-1,4-bis(4-benzoate)). Kinetic breakthrough measurements revealed that pore functionality, in particular by active adsorption sites, such as functional groups (NH_2- in IRMOF-3) and open metal sites (in MOF-74 and HKUST-1), played a key role in determining the dynamic adsorption performance of these MOFs. For example, HKUST-1 revealed high efficacy equal to or greater than BPL carbon against all gases tested except Cl_2. It was also found that CO cannot be captured effectively with any of the MOFs that were tested. Furthermore, M-MOF-74 (M = Zn, Co, Ni, or Mg) analogs, all with advantageous open metal sites, were tested to remove toxic gases including NH_3, CNCl, SO_2, and octane vapor (a physically adsorbed compound) from air in both dry and humid conditions [59]. Based on experimental breakthrough data, it was found that all of these MOFs are capable of removing toxic gases in dry environments but failed to do so in humid conditions, where the competitive adsorption of water nearly eliminated the adsorption capacity of all gases except NH_3. NH_3 was adsorbed under all conditions and in fact surpassed the capacities and retention of traditional adsorbent materials [60]. The removal of sulfur odorant components from natural gas can be done using HKUST-1. A fixed bed breakthrough adsorption experiment of a gas stream of methane odorized with tetrahydrothiophene (THT) at room temperature showed clearly that the sulfur odorant can completely be removed from natural gas by HKUST-1. It was also demonstrated that other electron-rich molecules, including amines, ammonia, water, alcohols, and oxygenates, can be successfully removed from natural gas by adsorption on HKUST-1. The adsorption of these polar molecules induced a color change of the HKUST-1 sample from a deep blue into a light green, which allows for visual detection of contaminants. Furthermore, after removal of the adsorbates by treatment under vacuum or heating, the original color reappeared, indicating practical regeneration of the adsorbent. A further example of harmful gas separation was demonstrated in flexible Zn (bchp) (H2bcph = 2,20-bis(4-carboxyphenyl)hexafluoropropane), in which H_2S and SO_2 were selectively adsorbed over N_2, CO_2, and CH_4 at room temperature. The adsorption of H_2S in MIL-53(Al, Cr, Fe), MIL-47, MIL-100(Cr), and MIL-101 also revealed that all of these MOFs are stable toward this corrosive gas and have high loading capacities and are easily regenerable, thus making them potential candidates in the purification of natural gas.

CO is a notoriously toxic gas because of its strong binding with metal sites of hemachrome. Bearing this in mind, MOFs with similar CUMs in their framework could selectively adsorb CO over other gases, which have comparatively weak interactions with open metal sites, such as N_2, H_2, and CH_4. HKUST-1 is quite selective for CO over H_2 and N_2 from their mixtures at 298 K. It was demonstrated that electrostatic interactions between the CO dipole and the partial charges on the open Cu(II) sites dominate the adsorption performances. NO is a biological signaling molecule and is also interesting in scientific research because of its environmental applications in gas separation and its use in NO_x traps for "lean-burn" engines. Two MOFs have shown ultrahigh selectivity for NO over other slight gases. A flexible Cu2(OH)(sip) (Cu-SIP-3, sip = 5-sulfoisophthalate) MOF found that Cu-SIP-3 did not adsorb N_2, H_2, CO_2, CO, N_2O, and CH_4 up to 10 bar, even at low temperature. The same is true for NO at very low pressures (<275 mbar); however, above a gate-opening pressure of 275 mbar, this material began to adsorb NO up to 0.88 NO molecules per formula unit at 1 bar. On reduction of the NO pressure, Cu-SIP-3 retained the light gas and showed a large desorption hysteresis, which was attributed to the relatively strong coordination of the gas to CUMs in the structure. It is more interesting that the NO-loaded MOF underwent controllable NO release upon the addition of water as the hydrated phase of the material was formed. Based on a similar mechanism, the selective adsorption of NO over Ar, N_2, CO, CO_2, and C_2H_2 was observed in another flexible MOF, Zn (TCNQ_TCNQ)(bipy). Clearly, the pore surface of this MOF, decorated by noncoordinated O atoms of the carboxylate ligands, dominated the C_2H_2 arrangement, where each O atom was hydrogen-bonded to one H atom of each C_2H_2 molecule. As a result of this strong interaction, highly selective adsorption of C_2H_2 over CO_2 as well as a probably safe storage of C_2H_2 by this MOF was realized. In addition, this group also reported C_2H_2 adsorption and storage in six other MOFs with a common molecular formula of M2(L)2(dabco) (M = Cu or Zn; L = bdc, 1,4-naphthalenedicarboxylate (1,4-ndc), and 9,10-adc) [61].

19.7.1.6 Nobel Gases and Others

Nobel gases, including He, Ne, Ar, Kr, Xe, and Rn, have a lot of important industrial applications. For example, Kr is applied as filler in the lamp industry, and Xe is used as a narcotic medicinal gas. These gases possess very similar properties and occur naturally as mixtures with other gases; thus, the separation of them is important but challenging [62]. In industry, Ne, Ar, Kr, and Xe are separated from air and He from natural gas by using cryogenic distillation. The separation of noble gases by MOFs as carriers is promising but largely unexplored. A simple process of pressure-swing adsorption is feasible in the separation of noble gases by using MOF HKUST-1 as an adsorbent. HKUST-1 can selectively separate Xe and Kr by continuous adsorption. It was found that Xe was adsorbed preferentially, while Kr was adsorbed to a much lesser extent. The calculated capacity of HKUST-1 for Xe is more than 60 wt %, almost twice as much as that on a high surface-area activated carbon [63]. HKUST-1 is also useful for the adsorption of Ar. The selective adsorption of N_2O over N_2 and O_2 and C_2H_4 over CO can be achieved in HKUST-1 at 295 K. The CPL-11 can selectively adsorb Xe and CCl_4 over O_2, N_2, and Ar. MAMS-1-4 are capable of selectively adsorbing CH_4 over C_2H_4 at 143 K, C_2H_4 over C_3H_6 at 195 K, and C_3H_6 over is o-C_4H_{10} at 231 K [64].

19.7.2 Selective Adsorptions and Separations of Chemical in Vapor Phase

Apart from light gases, vapors of various liquid compounds were also tested for their selective adsorption and separations in MOFs. Experimental approaches including vapor adsorption at room temperature or high temperature, breakthrough experiments of binary mixtures, gas chromatography (GC), and others have been discussed below.

19.7.2.1 Small Solvent Molecules

Separation and purification of organic solvents, particularly the removal of water, are of importance in a wide range of applications. A lot of MOFs have been tested for the adsorption of these small molecules in the vapor phase, and some of them revealed significant adsorption selectivities and separation potential. There are essentially two mechanisms, (1) size/shape exclusion and (2) different interactions between sorbates and pore surfaces of MOFs, that can be used to understand the observed preferential adsorption of these solvent vapors in MOFs. In most cases, both mechanisms operate together and complement each other. In these cases, the adsorption of guest molecules can lead to structural transformations and even changes in the composition of these soft materials, such as opening from no or small pores to large pores, to accommodate guest molecules [65]. For the size/shape-based selective adsorption in a rigid MOF, a typical example is the preferential adsorption of benzene over cyclohexane observed in $NH_4[Cu_3(OH)(4\text{-cpz})_3]$. The NH^{4+} cations reside in large cages and are exchangeable with other cations. It was found that when this MOF was exposed to benzene/cyclohexane (1:1) mixtures benzene was significantly enriched to give a ratio of 5:1 for benzene–cyclohexane in the adsorbed phase. Even more interesting is that, after exchanging NH^{4+} with larger Et_3NH^+ or $Li(H_2O)^{4+}$ cations, the resulting materials displayed a further enhanced preferential uptake for benzene over cyclohexane, with enriched ratios of 8:1 and 12:1, respectively. This increase can be attributed to the greater bulk of the cations and correspondingly resulting smaller pores in the MOF. Thus, this MOF is a viable candidate for the separation of benzene and cyclohexane, which cannot be separated by a distillation process due to their close boiling points of 80.1 and 80.7 °C [66].

Although some flexible MOF undergo structural changes during guest adsorption to produce open or closed pores, both forms have definite pore metrics and are therefore still available for size and/or shape-selective adsorption. For example, the selective adsorption of H_2O and MeOH over EtOH, THF, and Me_2CO vapors was observed in flexible Cd(pzdc)(bpee) at room temperature. Similarly, based on vapor phase measurements, the size/shape-dependent selective adsorptions of some small-molecule solvents have also been observed in rigid $Zn(SiF6)(pyz)_2$, Ag(tmpes)3BF4(tmpes = tetrakis [(4-methylthiophenyl)ethynyl]phenylsilane), Zn4(OH)2 $(1,2,4\text{-btc})_2$, Cd(abppt)2 3 (ClO4)2 (abppt = 4-amino-3,5-bis (4-pyridyl-3-phenyl)-1,2,4-triazole), and flexible MIL-96 and Cu(inaip)(inaip = 5-(isonicotinamido)isophthalate). Zn(tbip) (tbip = 5-tert-butyl isophthalate), on the other hand, is a rigid 3D framework containing close-packed 1D hydrophobic channels, 4.5 Å in size. It was demonstrated that these hydrophobic channels only allow MeOH and DME molecules to be adsorbed while excluding H_2O. MOFs with hydrophilic pore surfaces, usually decorated by specific functional groups or adsorption sites, prefer to selectively adsorb small, hydrophilic molecules. Active adsorption sites can form favorable interactions, such as coordination, hydrogen bonding, or π-π stacking, with guest molecules. These interactions induced a structural transformation of the flexible framework from an amorphous to crystalline or crystalline to crystalline phase. A typical example is Ln(tci) (Ln = Ce or Pr; H3tci = tris (2-carboxyethyl)isocyanurate), which has a nonporous 3D structure. Upon exposure to water vapor, these MOFs adsorbed water molecules to form a 2D layer structure with some of the adsorbed molecules coordinated to metal atoms, while MeOH vapor was not adsorbed. MOFs showing selective vapor adsorption combined with a structural transformation induced by H-bonding interactions between adsorbate molecules [67].

19.7.2.2 C8 Alkylaromatic Isomers

C8 alkylaromatic isomeric compounds consisting of o-xylene (oX), m-xylene (mX), p-xylene (pX), and ethylbenzene (EB) are important industrial feedstock chemicals: pX is used to produce polyethylene terephtalate for the polyester industry; oX is used to obtain phthalicanhydride (a plasticizer); mX is used to produce isophthalic acid, used in PET resin blends; and EB is used to produce styrene. Unfortunately, in industry, these isomers are always produced as mixtures containing all four components, which must be separated before further use. The separation of these isomers remains a challenge, particularly the separation of pX from mX and EB by, for example, a distillation process, because of their close boiling points. Alternately, fractional crystallization and/or adsorption have been used in industry. Commercially, these four C8 aromatics are currently separated mainly via fractional crystallization; cation exchanged zeolites are the only adsorbents currently being used in the bulk separation of these isomers. MOFs are promising adsorbents for the separation of C8 alkyl aromatic isomers, although only a limited number have thus far been tested. Experimental exploration of separating C8 alkylaromatic components in the vapor phase with MOFs. The selectivities evaluated from the mixed vapor adsorptions follow a trend of oX>pX > mX > EB, which is in agreement with the observations from single-component isotherms. The high selectivities for pX over EB (1.83) and pX over mX (2.07) were achieved in the binary separation at 70 °C. It was also found that the adsorption selectivity increased with an increasing degree of pore filling in each case, indicating that selectivity is pressure-dependent. Furthermore, in terms of mX and pX, the selectivity decreased at higher temperature.

The efficient separation of these components at a high pore filling in the vapor phase is a result of differences in packing modes of these molecules in the pores of MIL-47 [68].

19.7.2.3 Aliphatic Isomers

Similar to the separation of alkyl aromatic isomers discussed earlier, the separation of aliphatic isomers is also crucial in industry. As an example, the separation of hexane isomers to boost octane ratings in gasoline has been a very important process in the petroleum industry, which is currently accomplished by cryogenic distillation. Alternative processes, such as adsorption, are expected to lower the energy consumption involved in the distillation. Separation of aliphatic isomers can be done by using Zn(bdc)(bipy)0.5 (MOF-508). MOF-508 has a double-interpenetrated 3D framework with changeable pores based on the movement of single networks with respect to each other, producing open and dense forms. The framework transformation is reversible and dependent on the uptake and removal of guest molecules [69].

19.7.2.4 Others

Apart from the above-discussed vapor-phase selective adsorption and separation of arbitrarily classified and relatively widely explored chemicals on MOFs, there were several reports that dealt with other vapor-phase selective adsorptions. MIL-47 was found to be capable of selectively adsorbing thiophene from CH_4, a process related to natural-gas cleanup. It was demonstrated early that upon guest adsorption and desorption, the channels of MIL-47 can open and close reversibly, which was induced by the interactions between the guest molecules and the host framework. This selective removal of sulfur-containing molecules from CH_4 can thus be attributed to the noncovalent oriented weak interactions in the packing of thiophene molecules within the channels of the MOF, which are strong enough to open the channels to allow these molecules to be adsorbed [70].

The adsorptive separation of hydrocarbons can be done in the vapor phase with Cu(hfipbb)(H2hfipbb)0.5 (H2hfipbb = 4,40-(hexafluoroisopropylidene)-bis(benzoic acid)). This MOF has a 3D framework structure with hydrophobic channels consisting of large elliptic chambers connected by small necks [71]. It was revealed that this MOF adsorbed methanol, propane, and n-butane vapors rapidly but did not soak up any n-pentane, 2-methylpropane, 3-methylbutane, n-hexane, and 3-methylpentane at room temperature. This observed selectivity for shorter molecules over longer ones was attributed to the limited chamber length of 7.3 Å in the MOF. This chamber space is slightly longer than the length of n-C4 ($\sim$6.9 Å) but shorter than the length of n-C5 (8.1 Å). The small neck in the structure was otherwise responsible for the observed adsorption selectivity of normal over branched molecules. The diameter of the neck is approximately 3.2 Å, which is too small to allow branched alkanes with diameters of around 3.9 Å to pass. The adsorptive separation of a solid mixture of naphthalene and anthracene can be done (through sublimation) by Ni2(μ2-OH2)(1,3-bdc)2(tpcb) (tpcb = tetrakis(4-pyridyl)cyclobutane). This MOF has a diamond-typetopological framework with 1D channels composed of larger chambers linked by small kite-shaped windows. It was found that at room temperature this MOF can selectively adsorb sublimed naphthalene, which was accompanied by a single-crystal-to-single-crystal transformation, but completely excluded anthracene. Because of the same kinetic diameter of the two molecules, the observed selective adsorption of naphthalene over anthracene can be attributed to suitable guest–host interactions that are indeed shape-dependent on the guest molecules. This has been confirmed by the following single-crystal structural refinement of the guest-loaded MOF. It is also interesting that the adsorbed naphthalene molecules can be easily exchanged with EtOH; after removal of EtOH, the MOF is thus regenerated and thereby reusable [49].

19.8 MOFs for Liquid-Phase Adsorptive Separations

In general, the separation of liquids, such as various solvents, is usually accomplished by distillation whereas separation of chemical isomers with same boiling points is achieved by MOF technology. Adsorptive separation or membrane-based separations are the basis for this mixture separation [72]. Liquid-phase adsorptive separation process includes two main events: adsorption and desorption, with little difference in implementation from gas-phase separation. Adsorption of adsorbate onto the adsorbent is dictated by the characteristics of the adsorbate–adsorbent interaction. The difference arises during desorption in that a desorbent, which should be a suitable liquid, capable of displacing the adsorbate from the adsorbent, is required. After desorption, the desorbent will be separated from the extracted product usually by fractionation or evaporation, and then recycled back into the system. A lot of chemicals have been produced by liquid-phase separation or purification upon selective adsorption on porous materials, such as zeolites, activated carbon, and metal oxides. When compared to the gas phase, less attention has been paid to the use MOF for liquid-phase separations [51].

19.8.1 Selective Adsorptions and Separations of Chemically Different Species

Liquid-phase selective adsorption, inclusion, and separations of chemically different compounds, mainly organic compounds including small solvents and large organic molecules, have been tested on some MOFs. Based on the properties of the sorbates and the adsorption performance observed, four groups of chemical species were observed: organic molecules with different functional groups; organic molecules with different shape and size; organosulfur compounds; and ionic species.

19.8.1.1 Organic Molecules with Different Properties/Functional Group

One of the advantages of using MOFs as porous materials for adsorption-related applications is the ease of modification of their pore surfaces. This leads to the different adsorption preferences for different guest molecules, especially those with special chemical functional groups. These groups can introduce different/preferred interactions with the host frameworks, leading to the so-called selective adsorption. Among various host–guest interactions, π-π stacking, H-bonding, and the coordination with metal sites play important roles in the preferred adsorption and selective recognition. In addition, the hydrophilic/hydrophobic or polar/apolar properties of guest molecules are also at play when contacting the pore surface of a MOF [73]. It should be pointed out that in some cases, the strong interactions between guest molecules and the framework can induce structural change or framework transformation, which consequently affects the selectivity for these guests. These properties based on flexibility of the structure are unique to MOFs and inaccessible to other solid sorbent materials, such as zeolites. The MOFs with this property indeed possess great potential in liquid-phase separations. The first exploration of MOFs for selective binding of guest molecules in the liquid phase was performed on Co(Hbtc)(py)2, which has a sheet structure constructed by Hbtc2-ligands linking Co(II) atoms. These sheets stack to give alternating Co(II)-carboxylate and py(coordinated to Co(II)) layers in its 3D structure. It is important that the py ligands hold these layers together by π–π stacking to create a rigid 3D structure with rectangular channels in which guest molecules reside. After the removal of guest molecules, the channel structure remains unaltered, which allows the guest to be adsorbed by these pores. Adsorption experiments (typically by suspending the guest-free MOF sample in a mixture including several solvents for a given time and then filtrating the solvent and drying the MOF) demonstrated the selective adsorption of aromatic molecules, including benzene, nitrobenzene, cyanobenzene, and chlorobenzene over nonaromatic components including acetonitrile, nitromethane, and dichloroethane from their binary mixture. The selective adsorption of C6-C8 aromatics (including benzene, toluene, ethylbenzene, chlorobenzene, and three xyleneisomers) in the liquid phase can be done on a MIL-53 analog, Mn2(bdc)2(bpno) (MIL-53(MnII), bpno = 4,40-bipyridine-N,N0-dioxide). Single-component adsorptions of benzene, toluene, xylenes, ethylbenzene, and chlorobenzene by MIL-53(MnII) showed that only C6–C7 molecules could be intercalated into this MOF as confirmed by single-crystal structure determination. In the case of a two-component mixture of benzene and toluene, only benzene was selectively adsorbed. These observed selective adsorptions were mainly attributed to the different degrees of π–π interactions including those of guest–guest and guest–linker and other noncovalent interactions. A higher adsorption selectivity of chlorobenzene over benzene was also observed in this MOF, probably due to C-H-Cl hydrogen bonding between each chlorobenzene and the ligands. In addition, the packing efficiency, where increasing the number of electron-donating substituent groups can lead to an increase of intermolecular repulsions and thus lower packing, may also be a contributing factor in filling guest molecules into the channels of the MIL-53(MnII) [74].

19.8.1.2 Organic Molecules With Different Shape and Size

Shape- and size-based selective adsorption is another popular phenomenon and has been the foundation of an adsorptive separation of various molecular sieves and thin-film membrane separations. MOFs with easily controllable and adjustable structural metrics have great potential for shape- and size-based separations in the liquid phase. MOF plays a great role in selective adsorptions or separations of liquid chemicals based on different shapes or sizes of components. Besides the preferential adsorption due to the different functional groups of guest molecules observed in PIZA-1, this MOF also exhibited size and shape selectivity toward organic small molecules. By increasing the size of a series of aromatic amines, adsorption results in a decreased uptake of pyridine>aniline>2,4,6-trimethylpyridine. The selectivity was also detected by comparing the adsorption capacity of cyclohexylamine (8.9 guest molecules per unit cell of host) to that of dicyclohexylamine (2.3 guests per unit cell). The steric influence on the shape selectivity of this MOF was also observed in the picoline series (4-picoline > 3-picoline> 2-picoline) and a series of butyl-substituted amines (n-butylamine>di-n-butylamine>di-iso-, di-s-, and di-t-butylamines). The proposed reason that the bulky organic substituents encroach upon the hydrophilic group resulting in declining adsorption was further supported by the adsorption of simple alcohol molecules, which showed decreasing uptakes in the order of methanol > ethanol > propanol, butanol > hexanol. Direct adsorption comparisons between linear and branched alcohols (e.g., 1-propanol vs 3-propanol, and 1-butanol vs t-butanol) again proved the correlation between the increased steric hindrance and the decreased adsorption uptake [75]. Similarly, selective inclusion of alcohol molecules with different sizes and shapes has also been observed in $Zn_2(btc)(NO_3)$. It was found that this MOF can adsorb small nonhindered alcohols, including methanol,

ethanol, 1-propanol, isopropyl alcohol, 1-butanol, and tert-butyl alcohol, but rejected sterically hindered t-butylphenol. Finally, the overall order of alcohol inclusion selectivity is C1, C2 > C3, C4 > C5, C7, being directly dependent on their shape and size. A similar situation has also been found in a flexible MOF, Cd(4-btapa)$_2$3(NO$_3$)$_2$, which exhibited guest inclusion coupled with structural transformation for short-chain alcohols, including methanol, ethanol, n-propanol, and n-butanol, but not for long-chain alcohols, such as n-pentanol and n-hexanol. The size-selective inclusion of alcohol molecules has also been observed in Cu(in)$_2$. When soaking this MOF in an ethanol/n-propanol mixture, only ethanol molecules were adsorbed into its channels. In addition, the selective adsorption of water over methanol, due to their different sizes, from a 1:1 liquid mixture was observed in Cu(R-gla-Me)(bipy)0.5 (R-gla-Me = R-2-methylglutarate). This MOF has a 3D framework structure with narrow pores of about 2.8–3.6 Å, which can block the entrance of methanol into its channels, thus making it a potential drying agent [76].

19.8.1.3 Organosulfur Compound

Sulfur and organosulfur compounds are widely known contaminants in petroleum refining and in fuels (including gasoline, kerosene, diesel, and fuel oil). Desulfurization is understandably a subject of renewed interest because of environmental protection issues, the development of fuel cells that rely on reforming of hydrocarbons to hydrogen, and concerns of catalyst poisoning in petroleum refining. Typical organosulfur compounds include mercaptans (RSH), organic sulfides (R-S-R), organic disulfides (R-SS-R), carbon disulfides (S-C-S), thiophene, and substituted thiophenes (benzothiophenes, alkylthiophenes, alylbenzothiophenes, and alkyldibenzothiophenes) [77]. The removal of organosulfur compounds from hydrocarbon streams by adsorption has already been implemented in the refining industry by using other porous adsorbents, such as activated carbons and zeolites, but only a very limited number of MOFs have been explored to date. The liquid-phase adsorption of organosulfur compounds and desulfurization in five selected MOFs including Cu3(bpt)2(UMCM-150, bpt = biphenyl-3,40,5-tricarboxylate) MOF-505, HKUST-1, MOF-5, and MOF-177 [144]. These MOFs have different pore sizes, shapes, and metal clusters, thus offering adsorption behaviors for the liquid-phase adsorption of benzothiophene (BT), dibenzothiophene (DBT), and 4,6-dimethyldibenzothiophene (DMDBT), which are typical fuel contaminants. It was found that all of these MOFs exhibited large uptake capacities for the three compounds at high concentrations, even if saturation was not reached in each case. Pore size was also found to be a factor in deciding the adsorption capacity for given compounds [78, 79].

19.8.1.4 Cations and Anions

Selective exchange and sensing of cations or anions is another important application of porous materials. Some MOFs have been demonstrated to be responsive to different ions, showing selective exchange or recognition from solution. A typical example of selective cation exchange in MOFs is NaLa(H4pmtp) (H8pmtp = 1,4-phenylenbis (methylidyne)tetrakis(phosphonic acid)). This MOF features a flexible anionic framework with a remarkable charge and size selectivity for cations. In an aqueous solution, the Na^+ ions in the channels of the structure can be exchanged with other monovalent ions including Li^+, K^+, and Rb^+ having ionic radii ranging from 0.76 Å (Li^+) to 1.52 Å (Rb^+), while divalent ions (Mg^{2+}, Ca^{2+}, Sr^{2+}, Ba^{2+}, Ni^{2+}, Cu^{2+}, Zn^{2+}, and Mn^{2+}) in the same size range and larger Cs^+ ions were rejected. This charge-dependent selectivity was attributed to the site-specific role of the guest cation, which may affect the equilibrium between the expanded and the contracted forms of the flexible framework. The monovalent cations were located at specific sites in the framework, where they can satisfy their coordination requirements, whereas divalent ions could occupy only one-half of these sites. The size selectivity observed is most likely related to the pore size of the MOF, which is, even in its expanded form, not big enough to accommodate Cs^+ or its hydrate [80].

The recognition of anions by using Tb(btc) (MOF-76414) in methanol solution was explored. MOF-76 has a 3D open framework structure with 1D channels, in which the terminally coordinating solvent molecules partially occupy the pores. After immersing the activated MOF-76 in methanol solutions containing varying amounts of sodium salts with different anions (F^-, Cl^-, Br^-, CO_3^{2-}, and SO_4^{2-}), different quantities of these salts were adsorbed into the pores of the MOF. The adsorption of anions led to an enhanced luminescence, different for each anion, of the MOF in the solid-state. Fluoride ion showed the highest enhancement in the luminescent intensity, underlining the potential of MOF-76 for anion sensing. This luminescence enhancement was proposed to be a result of differences in hydrogen-bonding interactions between the anions and terminal OH moieties in the framework of the MOF [81]. Similarly, luminescent MOF, Tb$_2$(mucicate)$_3$, also showed ability in the selective adsorption of different anionic sodium salts from water solution. This MOF has a 2D layer structure connected by hydrogen bonds to form a 3D framework with square channels, in which –OH groups of the mucicate ligands decorated the pore surfaces. With an experimental method similar to that used in MOF-76, different luminescent enhancements of the solid samples were observed upon the adsorption of different anionic salts (sodium salt of I^-, Br^-, Cl^-, F^-, CN^-, CO_3^{2-}, NO_3^-, NO^{2-}, SO_4^{2}-, and PO_4^{3-}) from aqueous solutions. Among them, CO_3^{2-} led to the largest enhancement, and

NO_3^-, although similar in size to NO^{2-}, induced very different intensities, showing the excellent sensing performance of this MOF. These anionic responses were attributed to H-bonding interactions between anions and –OH groups in mucicate ligands [82].

19.8.2 Selective Adsorptions and Separations of Structural Isomer

A lot of important chemicals or chemical raw materials coexist with their isomers in natural sources or the early stages of refined products, such as petroleum and coal. The difficulties arising in the separation of these isomers mixtures are due to similarities in boiling and melting points and propensity to cocrystallize. To separate individual isomeric compounds is one of the most intense and challenging areas of industrial chemical research. An alternative to distillation and recrystallization, adsorption by porous materials, provides an efficient method for the separation of isomers based on their sizes, shapes, chiralities, as well as differences in affinities with pore surfaces of adsorbents. The separations of some structural isomers have been achieved in the industry by selective adsorption or membrane penetration relying on MOFs.

19.8.2.1 Aromatic Compound

Several MOFs have been tested for the separation of aromatic hydrocarbon compounds in the liquid phase, mainly focused on the C8 alkylaromatic isomeric compounds including the three xylene isomers (oX, mX, and pX) and ethylbenzene (EB). In a mixture of these compounds, only oX (bp = 144 °C) can easily be separated from the other isomers by distillation because of the similar boiling points of the remaining compounds (pX, 138 °C; mX, 138–139 °C; and EB, 136 °C). The liquid-phase adsorption and separation of C8 alkyl aromatic compounds use HKUST-1, MIL-53(Al)ht., and MIL-47. Competitive adsorption of a mixture of each of the two C8 isomers in hexane showed that HKUST-1 has low selectivities for the isomeric pairs except mX over oX, but MIL-53(Al)ht and MIL-47 have much higher selectivities for all C8 compound pairs, particularly, the prominent preference for pX over EB. Moreover, MIL-47 also preferred pX over mX, while MIL-53(Al)ht did not discriminate two isomers very effectively. Breakthrough experiments gave average selectivities of 2.5 for the separation of pX and mX and 7.6 for pX and EB. The regeneration of the MIL-47 column can be easily conducted by using hexane as the desorbent [83]. With the departure of uncoordinated terephthalic acid during calcination, the uptakes of both xylenes initially increased and then decreased. The selectivity of pX over mX decreased sharply and then flattened out, indicating that the presence of some terephthalic acid in the pore of MIL-47enhanced the selectivity between the two isomers. The higher selectivity with the presence of terephthalic acid in the pores was explained as: (1) the partly evacuated framework may be more flexible, therefore allowing an efficient parking of pX, and (2) some specific interactions between xylene molecules and terephthalic acid guests in the pores may lead to the improved selectivity. After removal of all uncoordinated terephthalic acid, the completely activated MIL-47 sample was further tested for the selective adsorption and separation of xylene and other disubstituted aromatic isomers, including ethyltoluene, dichlorobenzene, toluidine, and cresol. Pulse chromatography experiments revealed that xylene, dichlorobenzene, and cresol isomers have the same elution order of their respective three isomers: them-isomer eluting first and the p-isomer last. This adsorption preference for the p-isomer over the m-isomer was also observed for the ethyltoluene and toluidine isomers. Selectivities were also confirmed to be phase concentration dependent as supported by the observation that selectivities increased with increased bulk phase concentration for the xylene and dichlorobenzene isomers. Different from those in xylene isomers, molecular packing seems not to be the key factor in determining the p/m selectivity for ethyl toluene, toluidine, and cresol isomers. In the case of toluidines and cresols, the formation of H-bonds between guest and framework was believed to be the dominant factor. For ethyltoluenes, the large size of the molecule indeed did not allow for efficient packing in the pores. The breakthrough experiment also supported the validity of the p- and m-dichlorobenzene separation, giving a calculated average selectivity of 5.0. These results showed that the activated MIL-47 is capable of selectively adsorbing the p-isomer from p-m mixtures of these disubstituted aromatics, although different selectivity mechanisms appear to be at work. The selective adsorption of three dichlorobenzene isomers on HKUST-1 was also reported. Competitive adsorption tests in batches for each of the two isomers gave adsorption selectivities of 1.4, 6.2, and 9.0 form- over p-, m- over o-, and p- over o-isomer, respectively. The different affinities were attributed to either the differences in polarity or the steric packing effects of each isomer or both [84].

19.8.2.2 Aliphatic Compound

A large number of aliphatic compounds are also very important constituents of raw chemicals in the petroleum and chemical industries. The separation of their isomers is a major component in several industrial processes, such as the separation of linear alkanes from branched isomers in petroleum refining. Some related separations through adsorption have been achieved by using zeolites as adsorbents. The liquid-phase separation of C5-diolefins including isoprene, *cis*-piperylene, and *trans*-piperylene on MIL-96 was recently reported. This MOF has a strong adsorption preference for *trans*-piperylene over isoprene and *cis*-piperylene. The uptake and degree of pore filling reached the highest values for *trans*piperylene. It has been evaluated that each cage in the MIL-96 structure accommodated two *trans*-piperylene molecules at maximal uptake, but each cage was loaded with only 0.5 isoprene molecules [85]. The large uptake of *trans*-piperylene was attributed to the efficient packing of the guest molecules in the pores, which was supported by the observed similarities in adsorption enthalpies at a low degree of pore filling and variations in Henry equilibrium constants at low-coverage (with an order of isoprene≈*cis*-piperylene > *trans*-piperylene). Competitive batch experiments of all three isomers also showed preferential adsorption for *trans*-piperylene, with an uptake quantity similar to that in single-component adsorption measurements. The uptake of *cis*-piperylene was found to be identical to that for isoprene in competitive conditions. In breakthrough experiments, the elution order corresponded to the results of the competitive experiments, with *trans*-piperylene being retained a much longer time than the other two isomers. Moreover, the regeneration of the MOF can be easily achieved by flushing the column with pure heptane. These results indicated that MIL-96 is capable of separating *trans*-piperylene and *cis*-piperylene or isoprene isomersin the liquid phase [86].

19.8.3 Selective Adsorptions and Separations of Stereoisomer

Selective adsorption and separations of stereoisomers are isomeric molecules that have the same bond connection and sequence of constituent atoms but differ only in the 3D orientations of the atoms in space. Stereoisomers include enantiomers where different isomers are nonsuperimposable mirror images of each other and diastereomers (including *cis-trans* isomers and conformers). When compared to structural isomers, stereoisomers have much closer physical properties, such as nearly identical size, boiling, and melting points. Here MOFs are used for selective adsorption and separations of stereoisomers, with a central focus on enantioselective separations, that is, enantio-separations (or chiral separations) using homochiral MOFs.

19.8.3.1 Enantiomers (Enantio-Separation)

Enantiomers usually coexist as racemic mixtures in an achiral environment, thus requiring a chiral reagent for their separation. Doing this, usually, the racemates to be separated are put in a chiral environment where a chiral element (termed chiral selector) is capable of interacting enantio selectively [87]. Several techniques including different chromatographic techniques and electromigration have been developed and used in the separation of enantiomers. Homochiral materials have already been widely used as chiral selectors in a lot of separation processes; however, porous solids with highly uniform pores, such as zeolites and crystalline inorganic oxides, are not successful because the preparation of these materials in an enantiopure form is very difficult. This situation has prompted the exploration of other homochiral porous solids that is homochiral MOF for enantio-selective separations [88]. MOFs are typically synthesized under mild conditions, which allows for the facile construction of homochiral frameworks through the judicious choice of chiral building blocks or by using chiral induction. Clearly, a chiral pore with proper size and shape will give excellent enantio-selectivity. To realize this target is much easier in MOFs than in zeolites because of the modular building approach of MOFs. Despite the construction of a large number of homochiral MOFs, only very limited members have thus far been explored in enantio-selective adsorption and separations. The first example of enantio-selective inclusion of chiral molecules into the well-defined pores of a homochiral MOF, on Zn3 (μ3-O)(L-H)6(POST-1, L = (4S,5S)- or (4R,5R)-2,2-dimethyl-5-[(4-pyridinylamino)carbonyl]-1,3-dioxolane-4-carboxylic acid). This MOF has a 2D layer structure consisting of edge-sharing hexagons with trinuclear SBUs at each corner. The 2D layers stack to form a 3D framework with triangular, homochiral channels (that are 13.4 Å per side in length) in the stacking direction [89].

19.8.3.2 *Cis-Trans* Isomer

The separation of *cis-*/*trans*-isomers is another challenging issue, and only very limited progress has been made both in traditional porous sorbents, such as zeolites, and MOFs. MIL-96 was shown to be capable of separating *cis*-piperylene and *trans*-piperylene in the liquid phase. Single-component adsorption showed that *trans*-piperylene uptake is much higher than that of *cis*-piperylene from their heptane solutions, giving an evaluated pore occupation ratio of 2:0.6 for the two isomers. Competitive batch experiments also confirmed the preferential uptake of *trans*-piperylene from the mixture solution. Furthermore, it was found that the calculated separation factors increase with increasing concentration of the isomer mixture in solution, consistent with the assumption that packing effects determine the selective adsorption. In addition, the regeneration of the MOF adsorbent can be easily achieved by flushing the column with pure heptane. By contrast, there is a remarkable adsorption preference of HKUST-1toward *cis*-olefins over *trans*-olefins was observed. Several olefins with different chain lengths, including 2-butene, 2-pentene, 2-hexene, 2-heptene, 2-octene, 4-octene, 4-nonene, 5-decene, and methyl-9-octadecenoate, were tested for the adsorption from their binary equimolar mixtures of *cis*- and *trans*-isomers in hexane. The results gave separation factors of *cis*- over *trans*-isomers of 1.9, 4.9, 1.2, 3.4, 6.6, 2.6, 2.1,4.3, and 2.4, respectively. The presence of CUMs led to this MOF's ability to concentrate olefins through π-complexation in its pores. It was suggested that after adsorption, a double bond in the *cis* configuration would be more easily accommodated on the Cu(II) sites for steric reasons. As a representative example, the competitive uptake of *cis*- and *trans*-2-pentene was further investigated as a function of equilibrium bulk-phase concentration. The results indicated that with increased concentration the separation factor decreases, again a cooperative result from the steric hindrance and the strong adsorption through π-complexation of double bonds with the open Cu(II) sites in the framework. In addition, the selective capture of *cis*-crotononitrile from a mixture with its *trans*-isomer by flexible Mn(pmai)(H_2O) (pmai = 5-(pyridin-4-ylmethylamino) isophthalate) has been reported. This MOF has a 3D porous structure with water molecules coordinated to metal sites of its pore surface. After immersing crystals of the MOF in a 2:3 mixture of *cis*-crotononitrile and *trans*-crotononitrile for several days, it was found that only the *cis*-isomer was selectively captured. After adsorption, it was found that the coordinated water molecules and free solvent molecules in the as-synthesized crystals were replaced by *cis*-crotononitrile, accompanied by a change of the structural parameters of the MOF. Besides the coordination of *cis*-crotononitrile to metal sites, several other weak interactions between the adsorbed molecules and host framework were also observed. However, even if the crystals are immersed in pure *trans*-crotononitrile for several days, no uptake of the guest molecules was observed, which may be a result of shape mismatches between the *trans*-isomer and the pore or steric hindrance [90].

19.9 MOFs Membrane-Based Separations

As with adsorptive separation, distillation, and crystallization, using membranes for separation has its inherent advantages, including high energy efficiency, low cost, ease of processing, and excellent reliability. MOFs, like zeolites, are seen as feasible materials for membrane-based separation due to their well-defined, highly regular pore structures. The implementation of zeolite membranes in a broad range of applications is facing serious setbacks and difficulties because of a number of drawbacks not only in the materials themselves (such as the limited range in pore sizes accessible to zeolites) but also in the fabrication of the membranes [91]. MOFs cover a much wider range of pore sizes, shapes, and surface properties than zeolites and are usually synthesized under mild conditions. Some MOFs have been tested for their applications in membrane-based separations not only as thin films but also as porous adducts in mixed-matrix membranes [92].

19.9.1 Separations with MoF Thin Film

Crystalline thin films, due to their high permeability and selectivity, have attracted tremendous interest in membrane-based separations. MOF thin films are very promising for various separations in both the gas and the liquid phases. Separations of some gases by MOF thin film are given below.

19.9.1.1 H2 Separation

For the separation of H_2 from other gases, a MOF with small micropores is considered to be an ideal membrane material. However, some MOF thin films with larger pore openings have also presented excellent separation performance. Despite the large pore size of this MOF, the membrane presented an excellent H_2 separation ability from its binary mixture with CO_2, N_2, and CH_4 at room temperature. Further experiments also showed that the separation performances are temperature-dependent: the H_2 permeation flux increased and the separation selectivities decreased when the temperature increased from 273 to 343 K. The H_2/N_2 separation factor reached a maximum at 298 K whereas that of H_2/CO_2 continued to grow until 313 K [93]. HKUST-1-based membranes are used for H_2 separation from CO_2, N_2, O_2, and CH_4. The separation performance of the membranes at different temperatures was evaluated by single gas permeation measurements. The results revealed ideal selectivities of about 3.7, 2.4, and 3.5 at room temperature for H_2 over N_2, CH_4, and CO_2, respectively. This deviation may be due to the effects of the porous supports or nonselective intercrystalline diffusion through grain boundaries. As the temperature was increased, the selectivity of H_2 increased initially and then reached a plateau with maximum ideal selectivities of H_2 over N_2, CH_4, and CO_2 of about 7.5, 5.7, and 5.1, respectively. Regarding permeability, it was found that as the temperature was increased, the permeance values of all gases generally decreased. Furthermore, the permeance value of CO_2 became larger than those of CH_4 and N_2 with increased temperature, indicating the effect of the affinity between the quadrapolar CO_2 and the framework, in which the accessible Cu(II) sites became vacated when the coordinated solvent molecules were removed at high temperature. Another popular MOF, MOF-5, has also been fabricated into thin films for H_2 separation [94].

19.9.1.2 CO_2 Separation

The separation of CO_2 from other gases is another part of carbon capture and natural gas purification. Generally, CO_2 separation using MOF-based membranes in the spectra of carbon capture (mainly CO_2/N_2 separation) and natural gas purification (mainly CO_2/CH_4 separation) can be done. CO_2 separation from N_2 by a MOF thin film occurs due to high permselectivity of CO_2 over N_2 in humid conditions in a Zn (mimc)2 (SIM-1, Hmimc = 4-methyl-5-imidazolecarboxaldehyde) membrane fabricated on an asymmetric α-alumina tube. It was found that the ideal selectivity of CO_2/N_2, calculated from single gas permeances at 303 K, was 1.1. For a ternary mixture, $CO_2/N_2/H_2O$(10/87/3 vol%), this membrane presented a CO_2/N_2 separation factor of 4.5 at 324 K and 4 bar. It was thought that surface transport took place in the membrane, which allowed the separation of the two gases by preferential adsorption; that is, the most-adsorbed component reduced the diffusion of the other. CO_2/CH_4 separation using ZIF-8 membranes was prepared by in situ crystallization on tubular porous α-alumina supports. It was also found that with increasing thickness of the membranes the CO_2 permeance decreased, while the CO_2/CH_4 selectivity and separation index decreased at first but then increased, probably due to cracks in the membranes. The high separation indices were attributed to the small pores of ZIF-8, which favor the diffusion of CO_2 over CH_4. For example, Cu_2(bza)4(pyz)single-crystal membranes also exhibited high selectivity toward CO_2 over CO and CH_4, with a calculated selectivity factor of 10 and 25, respectively [95].

19.9.1.3 Other Gas and Vapor Separation

The adsorptions and separations of water and organic solvent vapors can be performed by nano ZIF-8 thin films with a tunable thickness. These thin films were prepared by a precisely controlled chemical solution deposition technique. Adsorption measurements of the membranes showed that only organic molecules such as alcohols and THF were adsorbed, but water was not due to the hydrophobic properties of the ZIF-8 framework. Furthermore, the high stability of these membranes was confirmed by running cycles of isopropanol adsorption, which showed no decrease in guest uptakes after several cycles. These ZIF-8 membranes are thus potentially applicable in the vapor-phase separation of organic solvents and water [96].

19.9.2 Separation with Mixed-Matrix MOF Membrane

An alternative route to introduce MOFs into membrane-based applications is to incorporate MOFs into polymers to obtain mixed-matrix membranes (MMMs), also called hybrid membranes, which are conceptually comprised of porous material particles (additives or fillers) dispersed in a polymer matrix. By using zeolites to fabricate MMMs, the smooth integration (without breaks) between zeolite particles and organic matrices is very difficult to control. As alternative materials to zeolites and other porous solids that could be used in MMMs, MOFs possess two distinct advantages: (1) MOFs with countless different structures and compositions can be synthesized, and (2) the organic linkers provide a useful platform for chemical modifications of the surface that can improve their adhesion to polymer matrices, thus making MOFs promising in MMM applications.

19.9.2.1 Gas Separation

The incorporation of MOFs into a polymer matrix to fabricate MMMs for gas separations has been suitably adopted. The tested MMM made from the incorporation of Cu(bpdc)-ted into poly(3-acetoxyethylthiophene) (PAET) showed an improvement in CH_4 permeability and selectivity when compared to a pure polymer membrane. The fabrication and gas separation of a $Cu_2(PF_6)(NO_3)(bipy)_4$ 3 2PF6 (Cubipy)-MMM, in which the MOF was dispersed into amorphous glassy polysulfone (PSf), has been reported. It was found that the loading amount of the MOF has a significant influence on the uniformity of the resulting MMM membrane. At lower than 5% loading, the MOF was well-dispersed in the polymer matrix and formed membranes with high uniformity. Gas permeation experiments revealed that the MMMs have lower permeabilities than a pure PSf membrane, which was attributed to an increase in the diffusion path length and a decrease in the effective cross-sectional area available for gas transport. However, as expected, the evaluated ideal selectivities of He, H_2, N_2, and O_2 over CH_4 were higher than those from the pure PSf membrane. A significant increase in the H_2/CH_4 and N_2/CH_4 selectivities relative to those of the larger gases was also observed, reportedly a result of a molecular sieving effect contributed by the small pores of the MOF filler. At a 5 wt% loading, the observed ideal selectivity of the MMM for He/CH_4, H_2/CH_4, O_2/CH_4, and N_2/CH_4 was about 230, 200, 20, and 10, respectively [97].

Gas separations of several MMMs with two different MOFs, HKUST-1 and Mn(HCO2)2, in polydimethylsiloxane (PDMS) and PSf, were also observed. The dependence of gas permeability and diffusion on the presence of the MOFs in the polymer matrices and their loading amounts was observed in these MMMs. Single gas permeation test showed that the permeabilities of all gases (H_2, CO_2, O_2, N_2, and CH_4) increased with MOF-5 loading [98].

19.9.2.2 Liquid Separation

Apart from exploring gas separations by MOF-based MMMs, liquid separations have also been attempted. Alcohol/water separation was tested by a pervaporation technique in a MOF-based MMM fabricated by dispersing microcrystals of $Cu_2(bza)_4(pyz)$ (bza = benzoate) into PDMS. The 3 wt% loaded membranes showed enhanced separation selectivities for MeOH and EtOH from a water solution containing 5 wt% alcohol (selectivity factors increased from 2.0 and 2.3 to 6.5 and 6.2, respectively) when compared to pure PDMS. Furthermore, the flux values of the MMM for MeOH and EtOH were slightly higher than those of the pure polymer membrane, indicating that the adsorbed alcohol molecules in the membrane diffused through the MOF crystals without being blocked. Other examples of liquid-phase separations are MMMs with HKUST-1, MIL-47, MIL-53(Al), or ZIF-8 as fillers in a PDMS matrix. Additionally, N-methyl-N-(trimethylsilyl)-trifluoroacetamide (MSTFA) was also used to modify these MOF crystals to improve the compatibility between the fillers and the polymer matrix. Overall, the MMMs with unmodified MOF fillers showed increased permeance for RB when compared to a pure PDMS membrane. It was also found that increasing MOF loading from 5% to 25% resulted in a slight enhancement in permeances in each case, which was attributed to the presence of a larger number of nonselective voids in the high-loaded membranes [59].

19.10 Potential Application of MOF

Uniform structures, adjustable porosity, and a wide variety of chemical functionalities offer solutions to various industries and many applications.

19.10.1 As a Catalyst

The high surface area and atomic metal sites feature of MOFs make them a suitable candidate for electrocatalysts, especially energy-related ones. Developing an effective catalyst that can be easily created, is not expensive, has low emission, and is greatly active, is a considerable goal in electro-catalytic research. Until now, MOFs have been used extensively as electrocatalysts for water splitting (H_2 and O_2 evolution reaction), carbon dioxide reduction, and oxygen reduction reaction. Currently, there are two routes: one is using MOFs as precursors to prepare electrocatalysts with carbon support [192]. Another one is using MOFs directly as electrocatalysts. However, some results have shown that some MOFs are not stable under an electrochemical environment. Creating effective, constant, and inexpensive catalysts for oxygen evolution reaction (OER) is extremely preferred in metal-air batteries and water splitting. Metal NPs-embedded MOF nanosheets are effective bifunctional electrochemical catalysts for energy uses. Improvement in the durability and activity can be attained for both oxygen evolution reaction and oxygen reduction reaction applying the reciprocal modification impacts on the metal nodes in MOFs and electron structures of Pt NPs [99]. For the Biocatalyst of S-adenosylmethionine, a magnetic responsive Ni-based metal, MOF, is used. Convenient preparation of the Ni-based metal organic framework nanorods (Fe_3O_4/Ni-BTC) with high magnetic responsiveness is carried out using a one-pot hydrothermal process. With a simple mixing step and under high-temperature condition (70 °C), about 95% activity recovery was obtained by S-adenosylmethionine synthetase from cell lysate in the biosynthesis of S-adenosylmethionine. For selective methane oxidation to methanol, MOF catalysts that were stimulated by Particulate methane mono-oxygenase were designed and synthesized. MOF-808 was applied by judicious selection of a framework with appropriate topology and chemical functionality to post-synthetically install ligands bearing imidazole units for subsequent metalation with Cu(I) in the presence of dioxygen. Under isothermal conditions at 150 °C, high selectivity for methane oxidation to methanol is exhibited by the catalysts. Bis(m-oxo) dicopper species are proposed as an active site of the catalysts. Under the solvothermal condition, a porous Cu (II)-organic framework synthesized according to a Y-shaped tricarboxylic ligand 30-nitro-[1,10-biphenyl]-3,40,5-tricarboxylic acid (H_3nbpt). According to the analysis of crystal structure, there was a $[Cu_7(OH)_4]_{10}$ secondary building unit in the compound, which was linked by the nbpt3-ligands into a 3D framework with 1D nanosized channels running along the b axis [100].

A normal method to enhance the MOFs' stability and catalytic activity is incorporating the catalytically active but unstable nanoparticles within porous MOFs. One of the most important factors in proton exchange membrane fuel cells is oxygen reduction reaction (ORR). Because of the need for high loading of Pt catalysts to accelerate the reaction due to the inactive kinetics of ORR, there is an unwanted escalation in the cost and hindering in the large-scale commercial use of fuel cells. To manufacture hierarchical porous structured metal nanoparticles/carbon composites through pyrolysis in an inert atmosphere, MOFs have currently proposed a novel technique to play the role of templates and precursors. The compositions of MOFs framework and pore structure can highly influence the performance of catalysts, such as shape-selective and bifunctional catalysis. It is needed to determine the dynamics of MOFs, their catalytic sites, and the intrinsic kinetics of catalytic reactions to advance the guidelines of optimum synthesis of the catalysts [101].

19.10.2 For Pollution Control

One of the phenomena majorly involved in global climate change is greenhouse gas emission. Because of their small, tunable pore sizes and high void fractions, MOFs are a promising potential material for use as an adsorbent to capture CO_2. MOFs could provide a more efficient alternative to traditional amine solvent-based methods in CO_2 capture from coal-fired power plants. MOFs could be employed in each of the main three carbon capture configurations for coal-fired power plants: pre-combustion, post-combustion, and oxy-combustion. However, since the post-combustion configuration is the only one that can be retrofitted to existing plants, it garners the most interest and research. In post-combustion carbon capture, the flue gas from the power plant would be fed through a MOF in a packed-bed reactor setup. Flue gas is generally 40–60 °C with a partial pressure of CO_2 at 0.13–0.16 bar. CO_2 can bind to the MOF surface through either physisorption, which is caused by Van der Waals interactions, or chemisorption, which is caused by covalent bond formation. Once the MOF is saturated with CO_2, the CO_2 would be removed from the MOF through either a temperature swing or a pressure swing. This process is known as regeneration. In a temperature swing regeneration, the MOF would be heated until CO_2 desorbs. To achieve working capacities comparable to the amine process, the MOF must be heated to around 200 °C. In a pressure swing, the pressure would be decreased until CO_2 desorbs [102]. Despite the great use of zeolites and activated carbons, these traditional solid adsorbents have low CO_2 uptake capacity and selective separation. MOFs are perfect materials for the capture and separation of CO_2 due to their high surface area, tunable pore sizes, and designable and controllable framework structures and topologies. The porous MOFs act as solid adsorbents for the capture and separation of CO_2 under different conditions. One particular MOF material exhibits an unprecedented cooperative mechanism for carbon dioxide capture and release with only small shifts in temperature. This structure of the MOF, with CO_2 adsorbed, closely resembles the RuBis CO enzyme found in plants, which captures CO_2 from the atmosphere for conversion into nutrients. The discovery paves the way for designing more efficient materials that dramatically reduce the overall energy cost of carbon capture. Such materials could be used for carbon capture from fossil fuel-based power plants as well as from the atmosphere, mitigating the greenhouse effect [103]. MFM-300(Al) MOF not only effectively filters harmful NO_2 gas but it also has outstanding capabilities for ammonia storage. Nanostructured Fe-Co–based MOF-74 is an adsorbent for the extraction of arsenic in water [104]. MOF is used as a desalination/ion separation system in mining industries. MOF membranes can mimic substantial ion selectivity. This offers the potential for use in desalination and water treatment. The mining industry uses membrane-based processes to reduce water pollution and to recover metals. MOFs could be used to extract metals such as lithium from seawater and waste streams. MOF membranes such as ZIF-8 and UiO-66 membranes with uniform subnanometer pores consisting of angstrom-scale windows and nanometer-scale cavities displayed ultrafast selective transport of alkali metal ions. The windows acted as ion selectivity filters for alkali metal ions, while the cavities functioned as pores for transport. The ZIF-8 and UiO-66 membranes showed a LiCl/RbCl selectivity of ~4.6 and ~ 1.8, respectively, much higher than the 0.6 to 0.8 selectivity in traditional membranes [105].

19.10.3 MOF Sensors

Because of its fast, inexpensive, and sensitive ability to detect, significant attention has been paid to electrochemical sensors. However, there are some disadvantages in the majority of electrocatalysts, including lack of structural design, low density of active sites, and low surface area. Therefore, it has become a major issue to create advanced materials so that the electrochemical performance could be improved. One promising way to make small, inexpensive, and energy-efficient gas sensors involve porous materials such as MOF. MOFs' high surface area is also a beneficial aspect for high-performance gas sensors. One example is a thin-film tailor-made MOF, coated onto an electrode, that forms an electronic sensor, which could detect traces of sulfur dioxide gas [62]. A potential application for MOFs is biological imaging and sensing via photoluminescence. A large subset of luminescent MOFs uses lanthanides in the metal clusters. Lanthanide photoluminescence has many unique properties that make them ideal for imaging applications, such as characteristically sharp and generally non-overlapping emission bands in the visible and near-infrared (NIR) regions of the spectrum, resistance to photobleaching or "blinking," and long luminescence lifetimes. However, lanthanide emissions are difficult to sensitize directly because they must undergo La Porte forbidden f-f transitions. Indirect sensitization of lanthanide emission can be accomplished by employing the "antenna effect," where the organic linkers act as antennae and absorb the excitation energy, transfer the energy to the excited state of the lanthanide, and yield lanthanide luminescence upon relaxation. A prime example of the antenna effect is demonstrated by MOF-76, which combines trivalent lanthanide ions and 1,3,5-benzenetricarboxylate (BTC) linkers to form infinite rod SBUs coordinated into a three-dimensional lattice. As demonstrated by multiple research groups, the BTC linker can effectively sensitize the lanthanide emission, resulting in a MOF with variable emission wavelengths depending on the lanthanide identity. Additionally, it has shown that Eu^{3+}- and Tb^{3+}- MOF-76 can be used for selective detection of acetophenone from other volatile monoaromatic hydrocarbons. Upon acetophenone uptake, the MOF shows a very sharp decrease, or quenching, in the luminescence intensity. For use in biological imaging, however, two main obstacles must be overcome: (1) MOFs must be synthesized on the nanoscale so as not to affect the target's normal interactions or behavior and (2) the absorbance and emission wavelengths must occur in regions with minimal overlap from sample autofluorescence, other absorbing species, and maximum tissue penetration [106].

To enhance sensing features of humidity sensor, MIL-101-NH_2-SO_3H was synthesized as an inorganic nanofiller. Through the estimation of the complex impedance spectra of sensors at various relative humidity, humidity-sensitive features of sensors were assessed. Small hysteresis (<3% RH, almost eighth of SPEEK sensor) and rapid response (absorption: 9 s) were exhibited as the best sensing features by the organic–inorganic humidity sensor based on the composite material (SPEEK/MNS-30%). Through hydrothermal, MOF-derived, and solvothermal synthesis, three types of TiO_2 were created with nanosphere, hollow ball, and nanoflower, respectively. Spraying three types of TiO_2 samples on QCM chips led to the creation of humidity sensors. Superparamagnetic Fe_3O_4 core encapsulated into a MOF shell, known as Fe_3O_4@MIL-100 (Fe) used as an electrochemical sensor for chlorogenic acid [107]. MOF can be used in implantable Nutrient Sensors. By integrating MOFs with flexible electronics, the electrochemical detection of nutrients without using enzymes becomes possible. MOF sensors can be used to detect a trace of ascorbic acid, L-Tryptophan, glycine, and glucose, all of which are nutrients closely involved in the metabolism and circulation processes. These sensor can be implanted and, as MOFs are very stable, the new technique could potentially be used to conduct long-term monitoring of biomolecules at different locations simultaneously. They can be used as implants to monitor biomolecules at different locations of various organs. When integrated with more stimulation and measurement functions, this type of device can be used to control animal behaviors, reveal the underlying mechanism of biological processes, monitor health conditions, and treat diseases [108].

19.10.4 Energy Storage Materials

The development toward improvement and refining the ability of electrochemical energy storage systems can be continued by the reduction of nonrenewable resources and the application of intermittent energy from green sources. As the human population continues to rise, the two ways to deal with the increase in energy demand are using accessible renewable energy sources with proper procedures and storing them appropriately for future requirements. Electrically conductive thin films of MOFs ($Cu_3(BTC)_2$ (also known as HKUST-1; BTC, benzene-1,3,5-tricarboxylic acid) infiltrated with the molecule 7,7,8,8-tetracyanoquinododimethane) that could be used in applications including photovoltaics, sensors, and electronic materials and a path toward creating semiconductors. The team demonstrated tunable, air-stable electrical conductivity with values as high as 7 siemens per meter, comparable to bronze. Ni_3(2,3,6,7,10,11-hexaiminotriphenylene)$_2$ was shown to be a metal-organic graphene analog that has a natural band gap, making it a semiconductor, and can self-assemble. Graphene must be doped to give it the properties of a semiconductor. $Ni_3(HITP)_2$ pellets had a conductivity of 2 S/cm, a record for a metal-organic compound [109]. Application of Ag_2O on MOF-derived N-doped carbon nanosheet arrays (NC) with a poly (3,4-ethylenedioxythiophenepoly (styrenesulfonate)) (PEDOT:PSS) buffer layer as cathode led to the successful creation of quasi solid-state, fiber-shaped Zn-Ag_2O battery [110]. Microporous structure and multicomponents were produced by oxygen, phosphorus, carbon, nickel, and nitrogen in the MOF for producing supercapacitors with high performance. Showing the moderate electrochemical capacitance of 979.8 F/g at a current density of 1 A/g, supercapacitor electrode materials are the same pristine sample that can be exploited directly. Sporadic hierarchical Ni/P/N/C composites, signified as Ni/P/N/C-500, Ni/P/N/C-600, Ni/P/N/C-700, Ni/P/N/C-800, were resulting from the parent MOF by the meek treatment of one-step pyrolysis in the nitrogen atmosphere at various annealing temperatures (500 °C, 600 °C, 700 °C, and 800 °C). At a current density of 1 A/g, the maximum precise capacitance of Ni/P/N/C-500 electrode could extent to 2887.87 F/g, which is higher than other hierarchical composites and can establish a novel standard in the related area. The hybrid constituents that were applied as the potential electrode-active compounds in supercapacitors were produced by the exclusive benefit of exploiting the predesigned MOFs as a template. In addition, an effective path was provided to produce the higher performance energy storage devices. In research, Li et al. placed the conductive two-dimensional metal-organic framework (Ni-CAT) nanocones on the flower petals surface of layered double hydroxide (NiCo-LDH)-based nanoflowers. After the synergistic between Ni-CAT and NiCo-LDH, specific capacitance of 882 F/g was exhibited by the composite structure at a current density of 1 A/g. Used as a positive electrode, an assembled asymmetric supercapacitor using NiCo-LDH@Ni-CAT shows a high energy density of 23.5 Wh/kg at a power density of 394.6 W/kg and long-term cycling stability up to 82% after 10,000 cycles [111].

MOF can be used as a Fuel cell. Created by iron (II) cations with four axially coordinated water molecules and chains of alternating bistriazolate-p-benzoquinone anions, a proton-conducting iron (II)-MOF [$Fe(C_6N_6O_2)$(-H_2O)4].5H_2O of an unusual structure was exhibited. To create a 3D grid-type network with channel pores occupied with water molecules, these chains accumulate between the aromatic units by p-p stacking. The highest proton conductivity of 3.3×10^{-3} S/cm at 94% relative humidity and 22 °C was shown by the iron(II)-MOF, which can be used in proton-exchange membrane fuel cells as membrane materials. Introducing an oxalate-bridged and Zn(II)-based 3D MOF, $\{[(Me_2NH_2)3(SO_4)]2[Zn_2(ox)_3]\}_n$, into a Nafion membrane can lead to the enhancement of proton conductivity under ambient conditions. In this regard, a high proton conductivity of 13×10^{-2} S/cm was observed at 100% RH with the 1 wt.% MOF/Nafion composite membrane. The potential of the composite is elevated by the facile fabrication of the composite and the inexpensive starting materials for the synthesis of MOF so that it could be applied as proton exchange membranes for fuel cell uses. Some MOFs also exhibit spontaneous electric polarization, which occurs due to the ordering of electric dipoles (polar linkers or guest molecules) below a certain phase transition temperature. If this long-range dipolar order can be controlled by the external electric field, a MOF is called ferroelectric. Some ferroelectric MOFs also exhibit magnetic ordering making them single structural phase multiferroics. This material property is highly interesting for the construction of memory devices with high information density. The coupling mechanism of $[(CH_3)_2NH_2][Ni(HCOO)_3]$ molecular multiferroic is spontaneous elastic strain mediated indirect coupling [112].

19.10.5 Biomedical Application

MOFs have shown potential for medical applications due to their nature of low toxicity and biocompatibility. Many groups have synthesized various low-toxicity MOFs and have studied their uses in loading and releasing various therapeutic drugs for potential medical applications. A variety of methods exist for inducing drug release, such as pH-response, magnetic-response, ion-response, temperature-response, and pressure response. MOF is applied as delivery vehicles for therapeutic agents and bioactive gases [113]. A biomedical CD-MOF-1 is synthesized from cheap edible natural products. CD-MOF-1 consists of repeating base units of 6 γ-cyclodextrin rings bound together by potassium ions. γ-cyclodextrin (γ-CD) is asymmetrical cyclic oligosaccharide that is mass-produced enzymatically from starch and consists of eight asymmetric α-1,4-linked D-glucopyranosyl residues. The molecular structure of these glucose derivatives, which approximates a truncated cone, bucket, or torus, generates a hydrophilic exterior surface, and a non-polar interior cavity. Cyclodextrins can interact with appropriately sized drug molecules to yield an inclusion complex. Two different methods of loading have been investigated on CD-MOF-1 with ibuprofen; crystallization using the potassium salt of ibuprofen as the alkali cation source for the production of MOF, and absorption and deprotonation of the free-acid of ibuprofen into the MOF. From there, the group performed in vitro and in vivo studies to determine the applicability of CD-MOF-1 as a viable delivery method for ibuprofen and other NSAIDs. In vitro studies showed no toxicity or effect on cell viability up to 100 μM. In vivo studies in mice showed the same rapid uptake of ibuprofen as the ibuprofen potassium salt control sample with a peak plasma concentration observed within 20 min, and the cocrystal has the added benefit of double the half-life in blood plasma samples. The increase in half-life is due to CD-MOF-1 increasing the solubility of ibuprofen compared to the pure salt form [114]. MOF ZIF-8 (zeolitic imidazolate framework-8) has been used in antitumor research to control the release of an autophagy inhibitor, 3-methyladenine (3-MA), and prevent it from dissipating in a large quantity before reaching the target [115]. Graphene oxide (GO) with carboxymethylcellulose (CMC) biopolymer and Zinc-based metal-organic framework (MOF-5) are also used for the delivery, solubility, and meticulous release of the drug into a precise place. The doxorubicin's effective loading on and leasing from the CMC/MOF-5/GO nanocomposite was demonstrated by drug loading and releasing tests. Exhibition of apoptosis to K562 cells and distinguished cytotoxicity by DOX@CMC/MOF-5/GO was carried out by the MTT assay. Water-soluble and positively charge characteristics of CMC/MOF-5/GO are responsible for these features produced by the alternation in GO with CMC and MOF-5. Results introduced the DOX-loaded CMC/MOF-5/GO as a great choice for targeted delivery and managed release of drug for cancer treatment [116].

MOF vaccines are based on a biocompatible polymer framework that "freezes" proteins inside vaccines. The proteins then dissolve when injected into human skin. This innovation could help health care providers transport and administer vaccines in remote areas with unreliable power. MOF vaccines are crystals that contain an antigen like the protein on the surface of influenza, except that since they are frozen inside a crystalline lattice, they can not denature or change shape. Structural advantages of MOFs allow them to perform better at room temperature than artificial encasings like silica. Specifically, MOFs' porous structure allows them to function as a semipermeable barrier to transport biological matter like proteins or antigens in vaccines [117]. MOFs are also applied as biomedical microrobots. By applying concepts developed in micro- and nanorobotics, researchers have demonstrated the controlled motion and delivery of cargo payloads embedded in MOFs. These helical MOF-based micromachines, termed MOFBOTs, are propelled by artificial bacterial flagella, can swim, and follow complex trajectories in three dimensions under the control of weak rotational magnetic fields [118].

19.10.6 Other Applications

The development of new pathways for efficient nuclear waste administration is essential in wake of increased public concern about radioactive contamination, due to nuclear plant operation and nuclear weapon decommission. The synthesis of novel materials capable of selective actinide sequestration and separation is one of the current challenges acknowledged in the nuclear waste sector. MOFs are a promising class of materials to address this challenge due to their porosity, modularity, crystallinity, and tenability. It can be used every building block of MOF structure for actinide incorporation. First, synthesize the MOF starting from actinide salts. In this case, actinides go to the metal node. In addition, in terms of metal nodes, either metal nodes extension or cation exchange and can also use organic linkers and functionalize it with a group capable of actinide uptake. And at last, it can use the porosity of MOFs to incorporate contained guest molecules and trap them in a structure by the installation of additional or capping linkers. In another case, at nuclear power plants and legacy waste sites, a particularly difficult-to-capture hazard is radioactive organic iodides. These compounds are made of hydrocarbons and iodine. By chemically modifying MOFs with binding sites that have reactive nitrogen that can bind to organic iodides, scientists have built MOF traps that exhibit a high methyl iodide capacity over three times higher than the currently used industrial adsorbent under identical conditions [119].

MOF can be used in space cooling applications. A prototype has been developed that captures water vapor from the air and then releases it with the application of a smaller amount of heat compared to existing commercially available technologies. Such MOFs could also be used to increase energy efficiency in room temperature space cooling applications. Space cooling was responsible for approximately 3% of the 2016 total world primary energy use, and demand in developing countries is increasing at ever greater rates. Therefore, air conditioning efficiency is a very desirable area for reducing future increases in energy consumption and CO_2 production from producing that energy. When cooling outdoor air, a cooling unit must deal with both the sensible heat and latent heat of the outdoor air. Typical vapor-compression-air-conditioning (VCAC) units manage the latent heat of the water vapor in air through cooling fins held below the dew point temperature of the moist air at the intake [120]. These fins condense the water, dehydrating the air and thus reducing the heat content of the air substantially. Unfortunately, the energy usage of the cooler is highly dependent on the temperature of the cooling coil and would be improved greatly if the temperature of this coil could be raised above the dew point. This makes it desirable to handle dehumidification through means other than condensation. One such means is by adsorbing the water from the air into a desiccant coated onto the heat exchangers, using the waste heat exhausted from the unit to desorb the water from the sorbent and thus regenerate the desiccant for repeated usage. This is accomplished by having two condenser/evaporator units through which the flow of refrigerant can be reversed once the desiccant on the condenser is saturated, thus making the condenser the evaporator and vice-versa. The conclusion is that a desiccant that can absorb a large amount of water and then easily release that water would be ideal for this application. There are a variety of MOF chemistry options that can help tune the optimal relative humidity for adsorption/desorption, and the sharpness of the water uptake [121]. MOFs can be efficiently used in the refrigeration system.

19.11 Summary

Metal-organic framework is a superior part in reticular chemistry and proved its high demands in material science. It can find out various ways to flexible materials world. This chapter explained a clear direction on synthesis, mechanism of MOFs with their potential, and possible application. MOF as catalyst, pollution control system, energy storage material, biomedical equipment, and gas storage materials proved its potential application. More research is necessary on this emerging MOF field to reveal various properties of MOF for new applications.

References

1. Lee, Y., Kim, J., Ahn, W.: Synthesis of metal-organic frameworks: a mini review. Korean J. Chem. Eng. **30**, 1667–1680 (2013). https://doi.org/10.1007/s11814-013-0140-6
2. Getman, R.B., Bae, Y., Wilmer, C.E., Snurr, R.Q.: Review and analysis of molecular simulations of methane, hydrogen, and acetylene storage in metal-organic frameworks. Chem. Rev. **112**(2), 703–723 (2012). https://doi.org/10.1021/cr200217c
3. Langmi, H.W., Ren, J., North, B., Mathe, M., Bessarabov, D.: Hydrogen storage in metal-organic frameworks: a review. Electrochim. Acta. **128**, 368–392 (2014). https://doi.org/10.1016/j.electacta.2013.10.190
4. Seth, S., Matzger, A.J.: Metal-organic frameworks: examples, counterexamples, and an actionable definition. Cryst. Growth Des. **17**(8), 4043–4048 (2017). https://doi.org/10.1021/acs.cgd.7b00808
5. Zhua, Q.-L., Xu, Q.: Metal-organic framework composites. Chem. Soc. Rev. **43**, 5468–5512 (2014). https://doi.org/10.1039/C3CS60472A
6. Dash, D.C., Mahapatra, A., Mohapatra, R.K., Ghosh, S., Naik, P.: Synthesis and characterization of Dioxouranium(VI), thorium(IV), Oxozirconium(IV) and Oxovanadium(IV) complexes with 1,11-dihydroxy-1,4,5,7,8,11-hexaaza-2,3,9,10-tetramethyl-1,3,8,10-decatetraene-6-thione and their derivatives with chloroaceticacid. Indian J. Chem. **47A**, 1009–1013 (2008)
7. Dash, D.C., Mahapatra, A., Naik, P., Naik, S.K., Mohapatra, R.K., Ghosh, S.: Synthesis and characterization of UO22+, ZrO2+ and Th4+ complexes with 3-(p-substituted aryl)-2-thiohydantoin. J. Indian Chem. Soc. **85**, 595–599 (2008)
8. Dash, D.C., Mohapatra, R.K., Ghosh, S., Naik, P.: Synthesis and characterization of UO2(VI), ZrO(IV) and VO(IV) complexes with a new Schiff base macro cyclic tetra dentate ligand and their derivatives with chloroaceticacid. J. Indian Chem. Soc. **86**, 121–126 (2009)

9. Jiang, J., Yaghi, O.M.: Brønsted acidity in metal-organic frameworks. Chem. Rev. **115**(14), 6966–6997 (2015). https://doi.org/10.1021/acs.chemrev.5b00221
10. Haldar, D., Duarah, P., Purkait, M.K.: MOFs for the treatment of arsenic, fluoride and iron contaminated drinking water: a review. Chemosphere. **251**, 126388 (2020). https://doi.org/10.1016/j.chemosphere.2020.126388
11. Dewi, W.L., Ruiz-Salvador, A.R., Gómez, A., Rodriguez-Albelo, L.M., Coudert, F.-X., Slater, B., Cheethamd, A.K., Mellot-Draznieks, C.: Zeolitic imidazole frameworks: structural and energetics trends compared with their zeolite analogues. Cryst. Eng. Comm. **11**, 2272–2276 (2009). https://doi.org/10.1039/B912997A
12. Stock, N., Biswas, S.: Synthesis of metal-organic frameworks (MOFs): routes to various MOF topologies, morphologies, and composites. Chem. Rev. **112**(2), 933–969 (2012). https://doi.org/10.1021/cr200304e
13. Liu, P., Gao, S., Huang, W., Ren, J., Yu, D., He, W.: Hybrid zeolite imidazolate framework derived N-implanted carbon polyhedrons with tunable heterogeneous interfaces for strong wideband microwave attenuation. Carbob. **159**, 83–93 (2020). https://doi.org/10.1016/j.carbon.2019.12.021
14. Choi, E.-Y., Park, K., Yang, C.-M., Kim, H., Son, J.-H., Lee, S.W., Lee, Y.H., Min, D., Kwon, Y.-U.: Benzene-templated hydrothermal synthesis of metal–organic frameworks with selective sorption properties. Chemistry. **10**(21), 5535–5540 (2012). https://doi.org/10.1002/chem.200400178
15. Lu, C.-M., Liu, J., Xiao, K., Harris, A.T.: Microwave enhanced synthesis of MOF-5 and its CO2 capture ability at moderate temperatures across multiple capture and release cycles. Chem. Eng. J. **156**(2), 465–470 (2010). https://doi.org/10.1016/j.cej.2009.10.067
16. Son, W.-J., Kim, J., Kim, J., Ahn, W.-S.: Sonochemical synthesis of MOF-5. Chem. Commun., 6336–6338 (2008). https://doi.org/10.1039/B814740J
17. Lv, D., Chen, Y., Li, Y., Shi, R., Wu, H., Sun, X., Xiao, J., Xi, H., Xia, Q., Li, Z.: Efficient mechanochemical synthesis of MOF-5 for linear alkanes adsorption. J. Chem. Eng. Data. **62**(7), 2030–2036 (2017). https://doi.org/10.1021/acs.jced.7b00049
18. Van Assche, T.R.C., Desmet, G., Ameloot, R., De Vos, D.E., Terryn, H., Denayer, J.F.M.: Electrochemical synthesis of thin HKUST-1 layers on copper mesh. Microporous Mesoporous Mater. **158**, 209–213 (2012). https://doi.org/10.1016/j.micromeso.2012.03.029
19. McKinstry, C., Cathcart, R.J., Cussen, E.J., Fletcher, A.J., Patwardhan, S.V., Sefcik, J.: Scalable continuous solvothermal synthesis of metal organic framework (MOF-5) crystals. Chem. Eng. J. **285**, 718–725 (2016). https://doi.org/10.1016/j.cej.2015.10.023
20. Russell, E.M.: Ionothermal synthesis-ionic liquids as functional solvents in the preparation of crystalline materials. Chem. Commun., 2990–2998 (2009). https://doi.org/10.1039/B902611H
21. Huang, L., Wang, H., Chen, J., Wang, Z., Sun, J., Zhao, D., Yan, Y.: Synthesis, morphology control, and properties of porous metal–organic coordination polymers. Microporous Mesoporous Mater. **58**(2), 105–114 (2003). https://doi.org/10.1016/S1387-1811(02)00609-1
22. Klinowski, J., Filipe, A., Paz, A., Silvab, P., Rocha, J.: Microwave-assisted synthesis of metal-organic frameworks. Dalton Trans. **40**, 321–330 (2011). https://doi.org/10.1039/C0DT00708K
23. Zheng, N., Masel, R.I.: Rapid production of metal-organic frameworks via microwave-assisted solvothermal synthesis. J. Am. Chem. Soc. **128**(38), 12394–12395 (2006). https://doi.org/10.1021/ja0635231
24. Dastbaz, A., Karimi-Sabet, J., Moosavian, M.A.: Sonochemical synthesis of novel decorated graphene nanosheets with amine functional Cu-terephthalate MOF for hydrogen adsorption: Effect of ultrasound and graphene content. Int. J. Hydrog. Energy. **44**(48), 26444–26458 (2019). https://doi.org/10.1016/j.ijhydene.2019.08.116
25. Kamali, M., Davarazar, M., Aminabhavi, T.M.: Single precursor sonochemical synthesis of mesoporous hexagonal-shape zero-valent copper for effective nitrate reduction. Chem. Eng. J. **384**, 123359 (2020). https://doi.org/10.1016/j.cej.2019.123359
26. Klimakow, M., Klobes, P., Thünemann, A.F., Rademann, K., Emmerling, F.: Mechanochemical synthesis of metal−organic frameworks: a fast and facile approach toward quantitative yields and high specific surface areas. Chem. Mater. **22**(18), 5216–5221 (2010). https://doi.org/10.1021/cm1012119
27. Sakamoto, H., Matsuda, R., Kitagawa, S.: Systematic mechanochemical preparation of a series of coordination pillared layer frameworks. Dalton Trans. **41**, 3956–3961 (2012). https://doi.org/10.1039/C2DT12012G
28. Yang, H., Orefuwa, S., Goudy, A.: Study of mechanochemical synthesis in the formation of the metal-organic framework Cu3 (BTC)2 for hydrogen storage. Microporous Mesoporous Mater. **143**(1), 37–45 (2011). https://doi.org/10.1016/j.micromeso.2011.02.003
29. Ameloot, R., Stappers, L., Fransaer, J., Alaerts, L., Sels, B.F., De Vos, D.E.: Patterned growth of metal-organic framework coatings by electrochemical synthesis. Chem. Mater. **21**(13), 2580–2582 (2009). https://doi.org/10.1021/cm900069f
30. Hui-Min, Y., Xian, L., Xiu-Li, S., Tai-Lai, Y., Zhen-Hai, L., Cai-Mei, F.: In situ electrochemical synthesis of MOF-5 and its application in improving photocatalytic activity of BiOBr. Trans. Nonferrous Metals Soc. China. **25**(12), 3987–3994 (2015). https://doi.org/10.1016/S1003-6326(15)64047-X
31. Imaging, M., Taylor, K.M.L., Jin, A., Lin, W.: Surfactant-assisted synthesis of nanoscale gadolinium metal–organic frameworks for potential. Angew. Chem. **47**(40), 7722–7725 (2008). https://doi.org/10.1002/anie.200802911
32. Vadivel, M., Ramesh Babu, R., Ramamurthi, K., Arivanandhan, M.: CTAB cationic surfactant assisted synthesis of CoFe2O4 magnetic nanoparticles. Ceram. Int. **42**(16), 19320–19328 (2016). https://doi.org/10.1016/j.ceramint.2016.09.101
33. Faustini, M., Kim, J., Jeong, G.-Y., Kim, J.Y., Moon, H.R., Ahn, W.-S., Kim, D.-P.: Microfluidic approach toward continuous and ultrafast synthesis of metal-organic framework crystals and hetero structures in confined microdroplets. J. Am. Chem. Soc. **135**(39), 14619–14626 (2013). https://doi.org/10.1021/ja4039642
34. Tannert, N., Gökpinar, S., Hastürk, E., Nießinga, S., Janiak, C.: Microwave-assisted dry-gel conversion-a new sustainable route for the rapid synthesis of metal–organic frameworks with solvent re-use. Dalton Trans. **47**, 9850–9860 (2018). https://doi.org/10.1039/C8DT02029A
35. José, L.S., Royuela, S., Ramos, M.M.: Post-synthetic modification of covalent organic frameworks. Chem. Soc. Rev. **48**, 3903–3945 (2019). https://doi.org/10.1039/C8CS00978C
36. Hong, D.H., Suh, M.P.: Enhancing CO2 separation ability of a metal-organic framework by post-synthetic ligand exchange with flexible aliphatic carboxylates. Chemistry. **20**(2), 426–434 (2014). https://doi.org/10.1002/chem.201303801
37. Zheng, Y., Wan, S., Yang, J., Kurmoo, M., Zeng, M.-H.: Recent advances in post-synthetic modification of metal-organic frameworks: New types and tandem reactions. Coord. Chem. Rev. **378**, 500–515 (2019). https://doi.org/10.1016/j.ccr.2017.11.015
38. Xiong, Z., Lic, S., Xia, Y.: Highly stable water-soluble magnetic nanoparticles synthesized through combined co-precipitation, surface-modification, and decomposition of a hybrid hydrogel. New J. Chem. **40**, 9951–9957 (2016). https://doi.org/10.1039/C6NJ02051H

39. Xue, D.-X., Belmabkhout, Y., Shekhah, O., Jiang, H., Adil, K., Cairns, A.J., Eddaoudi, M.: Tunable rare earth fcu-MOF platform: access to adsorption kinetics driven gas/vapor separations via pore size contraction. J. Am. Chem. Soc. **137**(15), 5034–5040 (2015). https://doi.org/10.1021/ja5131403
40. Li, J.-R., Sculley, J., Zhou, H.-C.: Metal-organic frameworks for separations. Chem. Rev. **112**(2), 869–932 (2012). https://doi.org/10.1021/cr200190s
41. Finsy, V., Verelst, H., Alaerts, L., De Vos, D., Jacobs, P.A., Baron, G.V., Denayer, J.F.M.: Pore-filling-dependent selectivity effects in the vapor-phase separation of xylene isomers on the metal-organic framework MIL-47. J. Am. Chem. Soc. **130**(22), 7110–7118 (2008). https://doi.org/10.1021/ja800686c
42. Bahamon, D., Vega, L.F.: Systematic evaluation of materials for post-combustion CO2 capture in a Temperature Swing Adsorption process. Chem. Eng. J. **284**, 438–447 (2015). https://doi.org/10.1016/j.cej.2015.08.098
43. Li, H., Song, Z., Zhang, X., Huang, Y., Li, S., Mao, Y., Ploehn, H. J., Yu, B., Yu, M.: Ultrathin, molecular-sieving graphene oxide membranes for selective hydrogen separation. Science. **342**(6154), 95–98. https://doi.org/10.1126/science.1236686
44. Gilron, J., Soffer, A.: Knudsen diffusion in microporous carbon membranes with molecular sieving character. J. Membr. Sci. **209**(2), 339–352 (2002). https://doi.org/10.1016/S0376-7388(02)00074-1
45. Rao, M.B., Sircar, S.: Nanoporous carbon membranes for separation of gas mixtures by selective surface flow. J. Membr. Sci. **85**(3), 253–264 (1993). https://doi.org/10.1016/0376-7388(93)85279-6
46. Koros, W.J., Fleming, G.K., Jordan, S.M., Kim, T.H., Hoehn, H. H.: Polymeric membrane materials for solution-diffusion based permeation separations. Prog. Polym. Sci. **13**(4), 339–401 (1988). https://doi.org/10.1016/0079-6700(88)90002-0
47. Uhlhorn, R.J.R., Keizer, K., Burggraaf, A.J.: Gas transport and separation with ceramic membranes. Part I. Multilayer diffusion and capillary condensation. J. Membr. Sci. **66**(2–3), 259–269 (1992). https://doi.org/10.1016/0376-7388(92)87016-Q
48. Sircar, S.: Basic research needs for design of adsorptive gas separation processes. Ind. Eng. Chem. Res. **45**(16), 5435–5448 (2006). https://doi.org/10.1021/ie051056a
49. Van de Voorde, B., Bueken, B., Denayer, J., De Vos, D.: Adsorptive separation on metal-organic frameworks in the liquid phase (review article). Chem. Soc. Rev. **43**, 5766–5788 (2014). https://doi.org/10.1039/C4CS00006D
50. Li, Y.-Z., Wang, G.-D., Yang, H.-Y., Hou, L., Wang, Y.-Y., Zhu, Z.: Novel cage-like MOF for gas separation, CO2 conversion and selective adsorption of an organic dye. Inorg. Chem. Front. **7**, 746–755 (2020). https://doi.org/10.1039/C9QI01262A
51. Alaerts, L., Maes, M., van der Veen, M.A., Jacobs, P.A., De Vos, D.E.: Metal-organic frameworks as high-potential adsorbents for liquid-phase separations of olefins, alkylnaphthalenes and dichlorobenzenes. Phys. Chem. Chem. Phys. **11**, 2903–2911 (2009). https://doi.org/10.1039/B823233D
52. Belmabkhout, Y., Bhatt, P.M., Adil, K., et al.: Natural gas upgrading using a fluorinated MOF with tuned H2S and CO2 adsorption selectivity. Nat. Energy. **3**, 1059–1066 (2018). https://doi.org/10.1038/s41560-018-0267-0
53. Marie, V.P., Sava Gallis, D.F., Greathouse, J.A., Nenoff, T.M.: Effect of metal in M3(btc)2 and M2(dobdc) MOFs for O2/N2 separations: a combined density functional theory and experimental study. Phys. Chem. C. **119**(12), 6556–6567 (2015). https://doi.org/10.1021/jp511789g
54. Kumar, R., Ahmadi, M.H., Rajak, D.K., Nazari, M.A.: A study on CO2 Absorption Using Hybrid Solvents in Packed Columns. Int. J. Low-Carbon Technol. (2019). https://doi.org/10.1093/ijlct/ctz051
55. Li, Y.-S., Liang, F.-Y., Bux, H., Feldhoff, A., Yang, W.-S., Caro, J.: Molecular sieve membrane: supported metal–organic framework with high hydrogen selectivity. Angew. Chem. Int. Ed. **122**(3), 558–561 (2010). https://doi.org/10.1002/ange.200905645
56. Eric, D.B., Queen, W.L., Hudson, M.R., Mason, J.A., Xiao, D.J., Murray, L.J., Flacau, R., Brown, C.M., Long, J.R.: Hydrogen storage and selective, reversible O2 adsorption in a metal-organic framework with open chromium(II) sites. Angew. Chem. Int. Ed. **128**(30), 8747–8751 (2016). https://doi.org/10.1002/ange.201602950
57. Kim, H., Park, J., Jung, Y.: The binding nature of light hydrocarbons on Fe/MOF-74 for gas separation. Phys. Chem. Chem. Phys. **15**, 19644–19650 (2013). https://doi.org/10.1039/C3CP52980K
58. Böhme, U., Barth, B., Paula, C., Kuhnt, A., Schwieger, W., Mundstock, A., Caro, J., Hartmann, M.: Ethene/ethane and propene/propane separation via the olefin and paraffin selective metal-organic framework adsorbents CPO-27 and ZIF-8. Langmuir. **29**(27), 8592–8600 (2013). https://doi.org/10.1021/la401471g
59. Ismail, A.F., Goh, P.S., Sanip, S.M., Aziz, M.: Transport and separation properties of carbon nanotube-mixed matrix membrane. Sep. Purif. Technol. **70**(1), 12–26 (2009). https://doi.org/10.1016/j.seppur.2009.09.002
60. Khan, N.A., Hasan, Z., Jhung, S.H.: Adsorptive removal of hazardous materials using metal-organic frameworks (MOFs): a review. J. Hazard. Mater. **244–245**, 444–456 (2013). https://doi.org/10.1016/j.jhazmat.2012.11.011
61. Woellner, M., Hausdorf, S., Klein, N., Mueller, P., Smith, M.W., Kaskel, S.: Adsorption and detection of hazardous trace gases by metal-organic frameworks. Adv. Mater. **30**(37), 1704679 (2018). https://doi.org/10.1002/adma.201704679. Special Issue: Metal-Organic Frameworks
62. Dolgopolova, E.A., Rice, A.M., Martina, C.R., Shustova, N.B.: Photochemistry and photophysics of MOFs: steps towards MOF-based sensing enhancements. Chem. Soc. Rev. **47**, 4710–4728 (2018). https://doi.org/10.1039/C7CS00861A
63. Wang, H., Yao, K., Zhang, Z., Jagiello, J., Gong, Q., Yu, H., Li, J.: The first example of commensurate adsorption of atomic gas in a MOF and effective separation of xenon from other noble gases. Chem. Sci. **5**, 620–624 (2014). https://doi.org/10.1039/C3SC52348A
64. Xiong, S., Liu, Q., Wang, Q., Li, W., Tang, Y., Wang, X., Hu, S., Chen, B.: A flexible zinc tetrazolate framework exhibiting breathing behaviour on xenon adsorption and selective adsorption of xenon over other noble gases. J. Mater. Chem. A. **3**, 10747–10752 (2015). https://doi.org/10.1039/C5TA00460H
65. Madhab, C.D., Guo, Q., He, Y., Kim, J., Zhao, C.-G., Hong, K., Xiang, S., Zhang, Z., Thomas, K.M., Krishna, R., Chen, B.: Interplay of metalloligand and organic ligand to tune micropores within isostructural mixed-metal organic frameworks (M'MOFs) for their highly selective separation of chiral and achiral small molecules. J. Am. Chem. Soc. **134**(20), 8703–8710 (2012). https://doi.org/10.1021/ja302380x
66. Li, J.-R., Kupplera, R.J., Zhou, H.-C.: Selective gas adsorption and separation in metal–organic frameworks. Chem. Soc. Rev. **38**, 1477–1504 (2009). https://doi.org/10.1039/B802426J
67. Khatua, S., Goswami, S., Biswas, S., Tomar, K., Jena, H.S., Konar, S.: Stable multiresponsive luminescent MOF for colorimetric detection of small molecules in selective and reversible manner. Chem. Mater. **27**(15), 5349–5360 (2015). https://doi.org/10.1021/acs.chemmater.5b01773
68. Miguel, I.G., Kapelewski, M.T., Bloch, E.D., Milner, P.J., Reed, D.A., Hudson, M.R., Mason, J.A., Barin, G., Brown, C.M., Jeffrey, R.: Long, separation of xylene isomers through multiple metal site

interactions in metal–organic frameworks. J. Am. Chem. Soc. **140** (9), 3412–3422 (2018). https://doi.org/10.1021/jacs.7b13825
69. Yang, X., Li, C., Qi, M., Liangti, Q.: Graphene-ZIF8 composite material as stationary phase for high-resolution gas chromatographic separations of aliphatic and aromatic isomers. J. Chromatogr. A. **1460**, 173–180 (2016). https://doi.org/10.1016/j.chroma.2016.07.029
70. Permatasari S.,Artik, S., Angkawijaya, E., Bundjaja, V., Soetaredjo, F.E., Ismadji, S.: Chapter 12: Metal-organic frameworks and their hybrid composites for adsorption of volatile organic compounds, In: Inamuddin, Boddula, R., Mohd Imran A., Asiri, A.M. (eds.) Applications of metal-organic frameworks and their derived materials, John Wiley & Sons, Hoboken (2020) https://doi.org/10.1002/9781119651079.ch12
71. Yang, W., Lin, X., Blake, A.J., Wilson, C., Hubberstey, P., Champness, N.R., Schröder, M.: self-assembly of metal-organic coordination polymers constructed from a bent dicarboxylate ligand: diversity of coordination modes, structures, and gas adsorption. Inorg. Chem. **48**(23), 11067–11078 (2009). https://doi.org/10.1021/ic901429u
72. Zhang, Q., Cui, Y., Qian, G.: Goal-directed design of metal–organic frameworks for liquid-phase adsorption and separation. Coord. Chem. Rev. **378**, 310–332 (2019). https://doi.org/10.1016/j.ccr.2017.10.028
73. Weber, E., Czugler, M.: Functional group assisted clathrate formation — scissor-like and roof-shaped host molecules. In: Weber, E. (ed.) Molecular inclusion and molecular recognition-clathrates II. Topics in current chemistry, vol. 149. Springer, Berlin (1988). https://doi.org/10.1007/3-540-19338-3_2
74. Colpani, G.L., Dal'Toe, A.T.O., Zanetti, M., Zeferino, R.C.F., Silva, L.L., de Mello, J.M.M., Fiori, M.A.: Photocatalytic adsorbents nanoparticles. IntechOpen, London (2018). https://doi.org/10.5772/intechopen.79954
75. Zhang, X., Zhang, X., Zou, K., Lee, C.-S., Lee, S.-T., Nanoribbons, S.-C.: Nanotubes, and nanowires from intramolecular charge-transfer organic molecules. J. Am. Chem. Soc. **129**(12), 3527–3532 (2007). https://doi.org/10.1021/ja0642109
76. Chen, B., Ji, Y., Xue, M., Fronczek, F.R., Hurtado, E.J., Mondal, J. U., Liang, C., Dai, S.: Metal-organic framework with rationally tuned micropores for selective adsorption of water over methanol. Inorg. Chem. **47**(13), 5543–5545 (2008). https://doi.org/10.1021/ic8004008
77. Blanco-Brieva, G., Campos-Martin, J.M., Al-Zahrani, S.M., Fierro, J.L.G.: Effectiveness of metal–organic frameworks for removal of refractory organo-sulfur compound present in liquid fuels. Fuel. **90**(1), 190–197 (2011). https://doi.org/10.1016/j.fuel.2010.08.008
78. Katie, A.C., Wong-Foy, A.G., Matzger, A.J.: Liquid phase adsorption by microporous coordination polymers: removal of organosulfur compounds. J. Am. Chem. Soc. **130**(22), 6938–6939 (2008). https://doi.org/10.1021/ja802121u
79. Böhle, T., Mertens, A.F.: [Cu2(bdc)2(dabco)] coated GC capillary column for the separation of light hydrocarbons and the determination thermodynamic and kinetic data thereof. Microporous Mesoporous Mater. **183**, 162–167 (2014). https://doi.org/10.1016/j.micromeso.2013.09.001
80. Chen, S., Shi, Z., Qin, L., Jia, H., Zheng, H.: Two new luminescent Cd(II)-metal-organic frameworks as bifunctional chemosensors for detection of cations Fe3+, anions CrO42-, and Cr2O72- in aqueous solution. Cryst. Growth Des. **17**(1), 67–72 (2017). https://doi.org/10.1021/acs.cgd.6b01197
81. Sheng, D., Lin, Z., Xu, C., Xiao, C., Wang, Y., Wang, Y., Chen, L., Diwu, J., Chen, J., Chai, Z., Albrecht-Schmitt, T.E., Wang, S.: Efficient and selective uptake of TcO4- by a cationic metal-organic framework material with open Ag+ sites. Environ. Sci. Technol. **51** (6), 3471–3479 (2017). https://doi.org/10.1021/acs.est.7b00339
82. Wong, K.-L., Law, G.-L., Yang, Y.-Y., Wong, W.-T.: A highly porous luminescent terbium–organic framework for reversible anion sensing. Adv. Mater. **18**(8), 1051–1054 (2006). https://doi.org/10.1002/adma.200502138
83. Huang, W., Jiang, J., Wu, D., Xu, J., Xue, B., Kirillov, A.M.: A Highly stable nanotubular MOF rotator for selective adsorption of benzene and separation of xylene isomers. Inorg. Chem. **54**(22), 10524–10526 (2015). https://doi.org/10.1021/acs.inorgchem.5b01581
84. Peralta, D., Barthelet, K., Pérez-Pellitero, J., Chizallet, C., Chaplais, G., Simon-Masseron, A., Pirngruber, G.D.: Adsorption and separation of xylene isomers: CPO-27-Ni vs HKUST-1 vs NaY. J. Phys. Chem. C. **116**(41), 21844–21855 (2012). https://doi.org/10.1021/jp306828x
85. Lauren, K.M., Mensforth, E.J., Babarao, R., Konstas, K., Telfer, S. G., Doherty, C.M., Tsanaktsidis, J., Batten, S.R., Hill, M.R.: CUB-5: a contoured aliphatic pore environment in a cubic framework with potential for benzene separation applications. J. Am. Chem. Soc. **141**(9), 3828–3832 (2019). https://doi.org/10.1021/jacs.8b13639
86. Tan, H., Chen, Q., Chen, T., Liu, H.: Selective adsorption and separation of xylene isomers and benzene/cyclohexane with microporous organic polymers POP-1. ACS Appl. Mater. Interfaces. **10** (38), 32717–32725 (2018). https://doi.org/10.1021/acsami.8b11657
87. Zhang, Y., Hidajat, K., Ray, A.K.: Determination of competitive adsorption isotherm parameters of pindolol enantiomers on α1-acid glycoprotein chiral stationary phase. J. Chromatogr. A. **1131**(1–2), 176–184 (2006). https://doi.org/10.1016/j.chroma.2006.07.052
88. Li, H., Jiang, X., Xu, W., Chen, Y., Yu, W., Xu, J.: Numerical determination of non-Langmuirian adsorption isotherms of ibuprofen enantiomers on Chiralcel OD column using ultraviolet–circular dichroism dual detector. J. Chromatogr. A. **1435**, 92–99 (2016). https://doi.org/10.1016/j.chroma.2016.01.048
89. Cui, Y.L.W.X.Y.: Engineering homochiral metal-organic frameworks for heterogeneous asymmetric catalysis and enantioselective separation. Adv. Mater. **22**(37), 4112–4135 (2010). https://doi.org/10.1002/adma.201000197
90. Paschke, R.F., Tolberg, W., Wheeler, D.H.: Cis, trans isomerism of the eleostearate isomers. J. Am. Oil Chem. Soc. **30**, 97–99 (1953). https://doi.org/10.1007/BF02638657
91. Denny, M., Moreton, J., Benz, L., et al.: Metal-organic frameworks for membrane-based separations. Nat. Rev. Mater. **1**, 16078 (2016). https://doi.org/10.1038/natrevmats.2016.78
92. Smith, S., Ladewig, B., Hill, A., et al.: Post-synthetic Ti exchanged UiO-66 metal-organic frameworks that deliver exceptional gas permeability in mixed matrix membranes. Sci. Rep. **5**, 7823 (2015). https://doi.org/10.1038/srep07823
93. Deepu, J.B., He, G., Villalobos, L.F., Agrawal, K.V.: Crystal engineering of metal-organic framework thin films for gas separations. ACS Sustain. Chem. Eng. **7**(1), 49–69 (2019). https://doi.org/10.1021/acssuschemeng.8b05409
94. Galizia, M., Chi, W.S., Smith, Z.P., Merkel, T.C., Baker, R.W., Freeman, B.D.: 50th anniversary perspective: polymers and mixed matrix membranes for gas and vapor separation: a review and prospective opportunities. Macromolecules. **50**(20), 7809–7843 (2017). https://doi.org/10.1021/acs.macromol.7b01718
95. Daglar, H., Keskin, S.: Computational screening of metal-organic frameworks for membrane-based CO2/N2/H2O separations: best materials for flue gas separation. J. Phys. Chem. C. **122**(30), 17347–17357 (2018). https://doi.org/10.1021/acs.jpcc.8b05416
96. Keskin, S., Sholl, D.S.: Assessment of a Metal–Organic Framework Membrane for Gas Separations Using Atomically Detailed Calculations: CO2, CH4, N2, H2 Mixtures in MOF-5. Ind. Eng. Chem. Res. **48**(2), 914–922 (2009). https://doi.org/10.1021/ie8010885

97. Goh, P.S., Ismail, A.F., Sanip, S.M., Ng, B.C., Aziz, M.: Recent advances of inorganic fillers in mixed matrix membrane for gas separation. Sep. Purif. Technol. **81**(3), 243–264 (2011). https://doi.org/10.1016/j.seppur.2011.07.042
98. Aroon, M.A., Ismail, A.F., Matsuura, T., Montazer-Rahmati, M. M.: Performance studies of mixed matrix membranes for gas separation: a review. Sep. Purif. Technol. **75**(3), 229–242 (2010). https://doi.org/10.1016/j.seppur.2010.08.023
99. Raoof, J.-B., Hosseini, S.R., Ojani, R., Mandegarzad, S.: MOF-derived Cu/nanoporous carbon composite and its application for electro-catalysis of hydrogen evolution reaction. Energy. **90** (Part 1), 1075–1081 (2015). https://doi.org/10.1016/j.energy.2015.08.013
100. He, J., Sun, S., Zhou, Z., Yuan, Q., Liu, Y., Liang, H.: Thermostable enzyme-immobilized magnetic responsive Ni-based metal-organic framework nanorods as recyclable biocatalysts for efficient biosynthesis of S-adenosylmethionine. Dalton Trans. **48**, 2077–2085 (2019). https://doi.org/10.1039/C8DT04857F
101. Bibi, R., Huang, H., Kalulu, M., Shen, Q., Wei, L., Oderinde, O., Li, N., Zhou, J.: Synthesis of amino-functionalized Ti-MOF derived yolk–shell and hollow heterostructures for enhanced photocatalytic hydrogen production under visible light. ACS Sustain. Chem. Eng. **7**(5), 4868–4877 (2019). https://doi.org/10.1021/acssuschemeng.8b05352
102. Fernández Bertos, M., Simons, S.J.R., Hills, C.D., Carey, P.J.: A review of accelerated carbonation technology in the treatment of cement-based materials and sequestration of CO2. J. Hazard. Mater. **112**(3), 193–205 (2004). https://doi.org/10.1016/j.jhazmat.2004.04.019
103. de Richter, R., Caillol, S.: Fighting global warming: The potential of photocatalysis against CO2, CH4, N2O, CFCs, tropospheric O3, BC and other major contributors to climate change. J. Photochem, Photobiol. C Photochem. Rev. **12**(1), 1–19 (2011). https://doi.org/10.1016/j.jphotochemrev.2011.05.002
104. Chen, C., Feng, X., Zhu, Q., Dong, R., Yang, R., Cheng, Y., He, C.: Microwave-assisted rapid synthesis of well-shaped MOF-74 (Ni) for CO2 efficient capture. Inorg. Chem. **58**(4), 2717–2728 (2019). https://doi.org/10.1021/acs.inorgchem.8b03271
105. Srimuk, P., Su, X., Yoon, J.: Charge-transfer materials for electrochemical water desalination, ion separation and the recovery of elements. Nat Rev Mater. **5**, 517–538 (2020). https://doi.org/10.1038/s41578-020-0193-1
106. Mikhail, Y.B., Achilefu, S.: Fluorescence lifetime measurements and biological imaging. Chem. Rev. **110**(5), 2641–2684 (2010). https://doi.org/10.1021/cr900343z
107. Li, Y., Xie, M., Zhang, X., Liu, Q., Lin, D., Xu, C., Xie, F., Sun, X.: Co-MOF nanosheet array: a high-performance electrochemical sensor for non-enzymatic glucose detection. Sens. Actuators B Chem. **278**, 126–132 (2019). https://doi.org/10.1016/j.snb.2018.09.076
108. Ling, W., Liew, G., Li, Y., Hao, Y., Pan, H., Wang, H., Ning, B., Xu, H., Huang, X.: Materials and techniques for implantable nutrient sensing using flexible sensors integrated with metal-organic frameworks. Adv. Mater. **30**(23), 1800917 (2018). https://doi.org/10.1002/adma.201800917
109. Biemmi, E., Scherb, C., Bein, T.: Oriented growth of the metal organic framework Cu3(BTC)2(H2O)3·xH2O tunable with functionalized self-assembled monolayers. J. Am. Chem. Soc. **129**(26), 8054–8055 (2007). https://doi.org/10.1021/ja0701208
110. Kumar, R., Shin, J., Yin, L., You, J.-M., Meng, Y.S., Wang, J.: All-printed, stretchable Zn-Ag2O rechargeable battery via hyperelastic binder for self-powering wearable electronics. Adv. Eng. Mater. **7**(8), 1602096. https://doi.org/10.1002/aenm.201602096
111. Sheberla, D., Bachman, J., Elias, J., et al.: Conductive MOF electrodes for stable supercapacitors with high areal capacitance. Nat. Mater. **16**, 220–224 (2017). https://doi.org/10.1038/nmat4766
112. Jin, M., Lu, S.-Y., Zhong, X., Liu, H., Liu, H., Gan, M., Ma, L.: Spindle-like MOF derived TiO2@NC-NCNTs composite with modulating defect site and graphitization nanoconfined Pt NPs as superior bifunctional fuel cell electrocatalysts. ACS Sustain. Chem. Eng. **8**(4), 1933–1942 (2020). https://doi.org/10.1021/acssuschemeng.9b06329
113. Stuart, R.M., Heurtaux, D., Baati, T., Horcajada, P., Grenèche, J.-M., Serre, C.: Biodegradable therapeutic MOFs for the delivery of bioactive molecules. Chem. Commun. **46**, 4526–4528 (2010). https://doi.org/10.1039/C001181A
114. Kumar, S., Jain, S., Nehra, M., Dilbaghi, N., Marrazza, G., Kim, K.-H.: Green synthesis of metal-organic frameworks: a state-of-the-art review of potential environmental and medical applications. Coord. Chem. Rev. **420**, 213407 (202). https://doi.org/10.1016/j.ccr.2020.213407
115. Li, T.-T., Cen, X., Ren, H.-T., Wu, L., Peng, H.-K., Wang, W., Gao, B., Lou, C.-W., Lin, J.-H.: Zeolitic imidazolate framework-8/polypropylene-polycarbonate barklike meltblown fibrous membranes by a facile in situ growth method for efficient PM2.5 capture. ACS Appl. Mater. Interfaces. **12**(7), 8730–8739 (2020). https://doi.org/10.1021/acsami.9b21340
116. Wang, F., He, H., Barnes, T.J., Barnett, C., Prestidge, C.A.: Oxidized mesoporous silicon microparticles for improved oral delivery of poorly soluble drugs. Mol. Pharm. **7**(1), 227–236 (2010). https://doi.org/10.1021/mp900221e
117. Mahapatro, A., Singh, D.K.: Biodegradable nanoparticles are excellent vehicle for site directed in-vivo delivery of drugs and vaccines. J. Nanobiotechnol. **9**, 55 (2011). https://doi.org/10.1186/1477-3155-9-55
118. Wang, X., Chen, X.-Z., Alcantara, C.C.J., Sevim, S., Hoop, M., Terzopoulou, A., de Marco, C., Hu, C., de Mello, A.J., Falcaro, P., Furukawa, S., Nelson, B.J., Puigmartí-Luis, J., Pané, S.: MOFBOTS: Metal-Organic-Framework-Based Biomedical Microrobots. Adv. Mater. **31**(27), 1901592 (2019). https://doi.org/10.1002/adma.201901592
119. Ekaterina, A.D., Ricea, A.M., Shustova, N.B.: Actinide-based MOFs: a middle ground in solution and solid-state structural motifs. Chem. Commun. **54**, 6472–6483 (2018). https://doi.org/10.1039/C7CC09780H
120. Peter, G.Y., Dakkama, H., Mahmoud, S.M., AL-Dadah, R.K.: Experimental investigation of adsorption water desalination/cooling system using CPO-27Ni MOF. Desalination. **404**, 192–199 (2017). https://doi.org/10.1016/j.desal.2016.11.008
121. Stefan, K.H., Habib, H.A., Janiak, C.: MOFs as adsorbents for low temperature heating and cooling applications. J. Am. Chem. Soc. **131**(8), 2776–2777 (2009). https://doi.org/10.1021/ja808444z

20 Additive Manufacturing Materials

Abstract

Currently, additive manufacturing heavily impacted on the production industries and tends to change the global economy quickly, contributing greatly to their overall success. Additive manufacturing is a production process by which a 3D CAD model is converted into a physical object by joining material layer by layer. In this chapter, the main concept of the process and how it can be advantageous than the conventional subtractive processes have been described. Various types of additive manufacturing processes, such as material extrusion, VAT photopolymerization, material jetting, powder bed fusion, directed energy deposition, binder jetting, sheet lamination, have been discussed. The most used metal, ceramics and polymer, composites, intermetallic compound, high entropy alloys, and bulk metallic glass has been discussed. The concept behind 3D, 4D, and 5D printing has been given here. At last, various types of most used industrial practices have been explained here.

Keywords

Additive manufacturing · Layer-by-layer · Material extrusion · VAT photopolymerization · Material jetting · Powder bed fusion · Directed energy deposition · Binder jetting · Sheet lamination · 3D printing · 4D printing · 5D printing

20.1 Introduction

Additive manufacturing (AM) is an emerging technology that has the potential to revolutionize the entire manufacturing ecosystem. AM is also known by different synonymy, that are, additive processes, additive techniques, additive layer manufacturing, layered manufacturing, layer-by-layer manufacturing, rapid prototyping, rapid manufacturing, and freeform fabrication [1, 3]. In 2010, the American Society for Testing and Materials (ASTM) group "ASTM F42-Additive Manufacturing," formulated a set of standards that classify the range of additive manufacturing processes into seven categories (Standard Terminology for Additive Manufacturing Technologies, 2012) [4]. According to current practice, various kinds of additive manufacturing processes have been shown in Fig. 20.1. Metal additive manufacturing is "the process of joining materials to make objects from 3D computer-aided design (CAD) model data, usually layer upon layer, as opposed to subtractive manufacturing technologies" [5]. A variety of materials such as metal, ceramic, polymer, intermetallic can be used in AM process. All seven individual AM processes, cover the use of various materials. Materials are often produced in powder form or in wire feedstock or laminar form. It is essentially feasible to print any structure (paper, chocolate, tissue, etc.) in this layer-by-layer method, and the final quality will be largely determined by the material.

The first form of creating layer-by-layer a three-dimensional object using CAD was rapid prototyping, developed in the 1980s for creating models and prototype parts. This technology was created to help the realization of what engineers have in mind. Rapid prototyping is one of the earlier additive manufacturing (AM) processes that quickly create a scale model of a part or finished product, using CAD software. The steps involved in product development using rapid prototyping are shown in Fig. 20.2. In addition, it is important to notice that rapid manufacturing became possible by other technologies, which are computer-aided design (CAD), computer-aided manufacturing (CAM), and computer numerical control (CNC). These three technologies combined together made possible the printing of three-dimensional objects [6].

The processes above can also change the microstructure of a material due to high temperatures and pressures, therefore material characteristics may not always be completely similar post-manufacture, when compared to other manufacturing processes. Although the term "3D Printing" is a synonym for all additive manufacturing processes, there are actually lots of individual processes which vary in their method of

A. Behera, *Advanced Materials*, https://doi.org/10.1007/978-3-030-80359-9_20

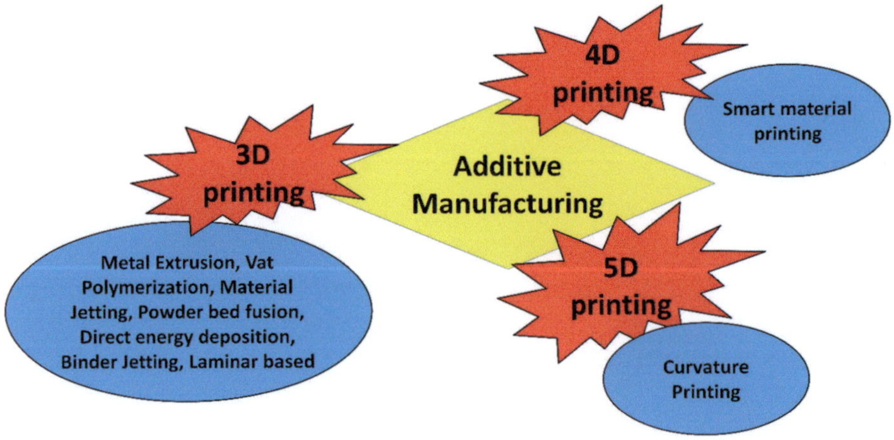

Fig. 20.1 Categories of AM processes

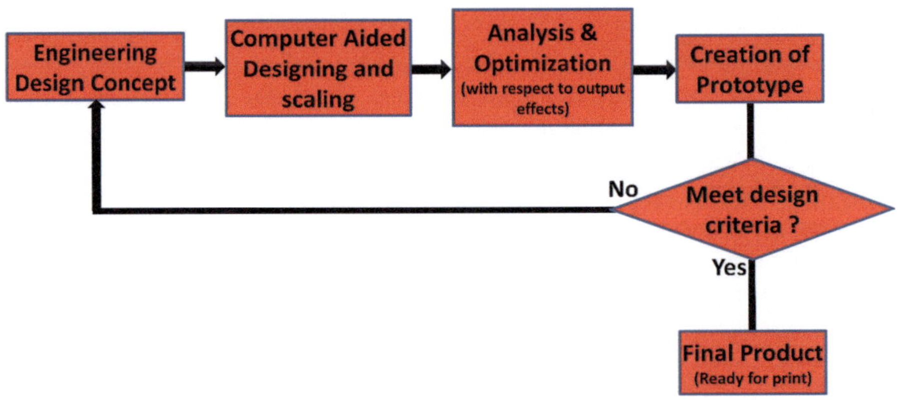

Fig. 20.2 Process involves from initial step (prototype designing) to final step (final product) in additive manufacturing

layer manufacturing. Individual processes will differ depending on the material and machine technology used. 3D printing and rapid prototyping are often confused as the same thing, however, they are somewhat different. 3D printing is a method of additive manufacturing, whereas rapid prototyping is an application of this technology. 3D printing is a newer, more cost-effective method of additive manufacturing. In the 1990s, 3D-printing techniques were considered suitable only for the production of functional or aesthetic prototypes and more appropriate term for it was rapid prototyping [7]. As of 2019, the precision, repeatability, and material range have increased to the point that some 3D-printing processes are considered viable as an industrial-production technology, whereby the term additive manufacturing can be used synonymously with "3D printing." Hence, the 3D-printing process is to build a three-dimensional object from a computer-aided design (CAD) model, usually by successively adding material layer-by-layer, which is why it is also called additive manufacturing.

20.2 Additive Manufacturing Market

Currently, the demands of additive manufactured products are rises with a higher rate. The additive manufacturing material market is expected to reach $1082.0 million by 2020, at a CAGR of 20.9% from 2014 to 2020 (Fig. 20.3) [8]. On the basis of application, technology, and material market, the current key players are in Europe, United States, and Asia-Pacific. The major application market for additive manufacturing can be attributed to the aerospace industry, automotive industry, consumer products, healthcare products, defense, and others (arts, architecture, and forensics) [9]. Major players in the global additive manufacturing market include 3D Systems Inc. (U.S.), Stratasys Ltd. (U.S.), Ex One (U.S.), Arcam AB (Sweden), Envision TEC (Germany), EOS (Germany), Materialise NV (Belgium), and MCor Technologies Ltd. (Ireland), among others. The stereolithography technology dominates other technologies with the highest share in the market.

20.3 Additive Manufacturing Advantages Over Conventional Manufacturing

The development of innovative, advanced additive manufacturing technologies has progressed quickly yielding broader and high-value applications. This accelerating trend has been due to the benefits of additive manufacturing compared to conventional manufacturing processes. Some of these benefits are described below [10, 11]:

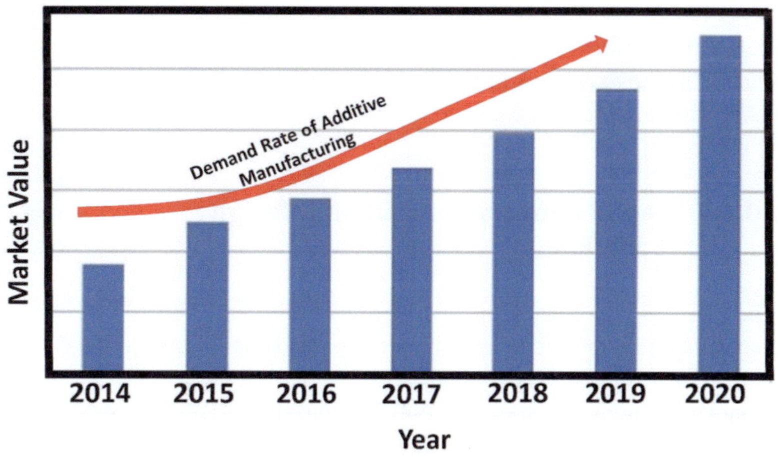

Fig. 20.3 Global Additive Manufacturing Market, 2014–2020 ($ Million)

Wide Range of Material Choices: Wide range of materials (polymers, metals, ceramics, etc.) has been processed in the same procedure. No additional external parameters and equipment are required. Only energy input is the focus of the point.

Less Dedicated Tooling: No expensive tooling requirements. Unlike many widely used manufacturing techniques such as injection molding, no tooling is required, which can be a barrier to production due to the high cost.

Elimination of Design Constraints: Flexibility in design. It is possible to quickly modify the design during production. Additive Manufacturing allows for highly complex structures which can still be extremely light and stable. It provides a high degree of design freedom, the optimization, and integration of functional features, the manufacture of small batch sizes at reasonable unit costs, and a high degree of product customization even in serial production.

Reduced Development Costs and Time to Market: Small footprint for manufacturing and continually shrinking equipment costs. Compared to conventional techniques with more geometric limitations, additive manufacturing can produce models quickly, in hours, not weeks. Fewer resources for machines and little skilled labor are required when compared to conventional model-making craftsmanship.

Anywhere Manufacture: Parts can be sent digitally and printed in homes or locations near to consumers, reducing the requirement and dependence on transport.

Less Waste: AM process is considered as a green manufacturing and clean process. In advanced machine, automatic clean up and sieving has been done and stored in the hopper for next printing. Hence it produces zero waste.

Customization of Parts: Particularly within the medical sector, where parts can be fully customized to the patient and their individual requirements, where other manufacturing facilities are not available such as on ships or in space. Well suited to the manufacture of high-value replacement and repair parts.

20.4 Steps Involved in AM Processes

As previously stated the numerous AM systems share some similarities but have a number of distinctions. The first AM system to be commercialized was stereolithography, whereby a concentrated beam of the ultraviolet lamp is used to solidify a liquid photopolymer by tracing at two-dimensional (2D) layers in the form of a contour and then an infill. Once the beam has completed a single layer, the build platform will then move downward in the z-axis, a new layer of photopolymerized is a tribute and the process is repeated until the final layer is completed. Laser sintering and laser melting processes work in a similar manner, whereby polymer or metallic powders are selectively melted in 2D layers, through high-power lasers until a solid part is complete. Another popular process, particularly with the makers, is the FDM process. In this method, materials, usually polymer filaments are extruded through a heated nozzle to print 2D layers successively, one on top of another, until the part is complete. Whether through melting of metallic powders or through the extrusion of polymer filaments, all AM processes share the additive principle of building components [12, 13]. It is also easy when one is using 3D printers to give life to the imagination. Knowing all 3D printing basics is essential if wishing to start 3D printing at home or for the business. The major steps in this generic process of CAD to part are discussed below:

20.4.1 Step 1: Conceptualization and CAD Designing a 3D Model

Need a blueprint to start 3D printing. This works as an input for the 3D printer just like the document file works for an inkjet printer. In 3D printing, you will only require CAD software which is also known as Computer-Aided Software. This software is a 3D modeling software usually required to create blueprints of the building, and every architect uses it to create floor plans and other specifications. For example, Fig. 20.4 shows a blueprint of a design with the help of CAD software that can create a 3D model.

20.4.2 Step 2: Conversion of Digital Design of STL File

The STL file was created in 1987 by 3D Systems Inc. When they first developed the stereolithography, and the STL file stands for this term. It is also called as Standard Tessellation Language. There are other types of files, but the STL file is the standard for every additive manufacturing process. STL file is the digital footprint of the model one wants to build using 3D printers. The STL file creation process mainly converts the continuous geometry in the CAD file into a header, small triangles, or coordinates triplet list of x, y, and z coordinates and the normal vector to the triangles. This process is inaccurate and the smaller the triangles the closer to reality. The interior and exterior surfaces are identified using the right-hand rule and vertices cannot share a point with a line. Hence, creating an STL file is a very crucial part of the entire process. So, how can one create an STL file? There are two ways to get an STL file as discussed below. Figure 20.5 shows the data flow in the STL file creation.

1. **Create It from Scratch:** If one is interested to design his own file, he can choose from a plethora of 3D modeling tools available. Such as Sketch Up, Mesh Lab, Free CAD, and many others. Many of these tools are free and some are designed especially for those who have no experience with 3D modeling. Using the easy-to-use tools with the drag and drop feature, one can create STL files. There are many tips available online. It may be a tough task for beginners but gets easier as one starts getting to know these tools better. With few trials and errors, one can certainly learn to create STL files on their own, shown in Fig. 20.5.
2. **Download It:** If one does not want to get into the technicalities of 3D modeling, he can directly download it from online repositories for example Cults 3D. You can find many free STL files and designs that can be printed right away ready for various materials printing. Once you have the STL file, you can build the model, layer-by-layer, using your 3D printer.

20.4.3 Step 3: Slicing Using a 3D Printer Slicer Software and Manipulation of STL File

Additional edges are added when the figure is sliced. The slicing process also introduces inaccuracy to the file because

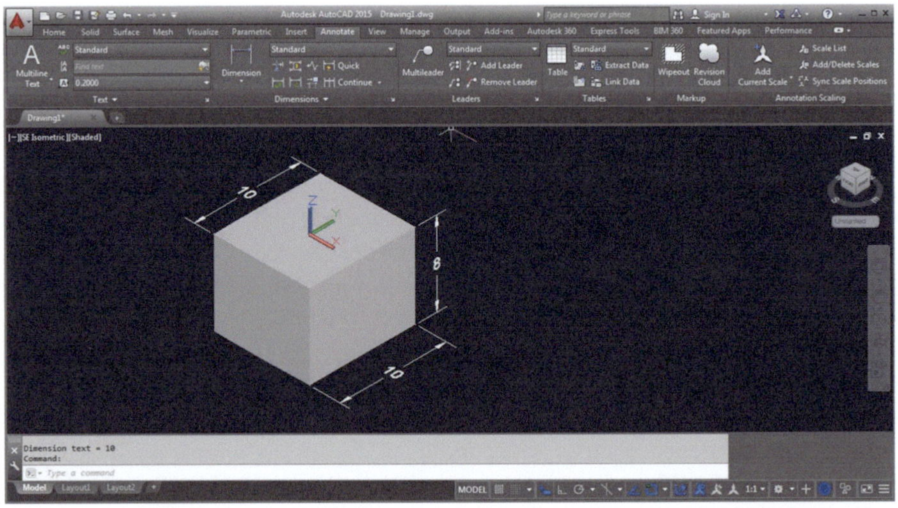

Fig. 20.4 CAD blueprint scale-up

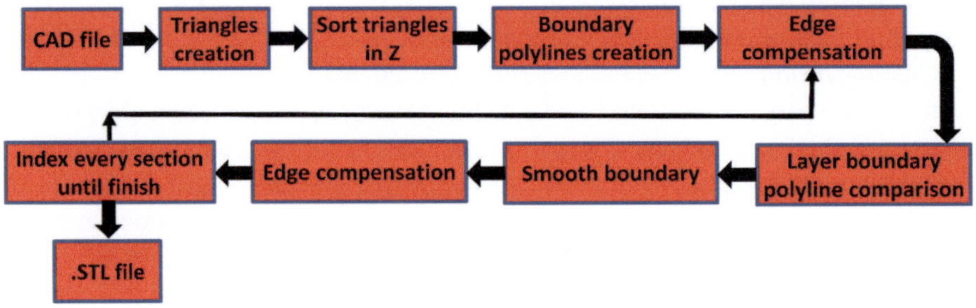

Fig. 20.5 Data flow in STL file creation

here the algorithm replaces the continuous contour with discrete stair steps. After completing the design, prepare the design for slicing by exporting it as an STL file format. For doing so, it will require slicer software. You can change the settings of the printer using this software. For example, which nozzle to use or which material to be used and much more. The slicer will then create instructions for the 3D printer. It will break the 3D designs into horizontal layers and hence is named slicer. By using slicer software, there is a control of different printing processes. The job of the slicer is very crucial. It plans ahead of the best way that the printer should be moving to create fine objects. It plans about the movement while creating the first slice, then the second and next till the object is completely printed. To start using a slicer, one must drag and drop the STL file or open it to use the 3D design and make changes to it. Once your design opens in the slicer, you can check its rotation, change the size and other inputs, select nozzle, and other details of how the object would be filled from inside. All these settings help one in creating the exact 3D design. Once it is done, simply save the file in an SD card for a 3D printer to print the design.

Slicer: There are times when there is no slicer software found with the 3D printer as it is not included by the manufacturer. Or sometimes, may not find satisfactory results with the included slicer software with the printer. In that case, one can use free software available online. One such software is Cura slicing software which can be used for free.

To reduce inaccuracy, the technique for a feature that has a small radius in relation to the dimension of the part is to create STL files separately and to combine them later. The dimension in z-direction should be designed to have a multiple of the layer thickness value. Figure 20.6 is shown the position of the STL file creation in the data flow of a rapid prototyping process. Other types of files are stereolithography contour (SLC) and SLI from 3D Systems, CLI from EOS, Hewlett-Packard graphics language (HPGL) from Hewlett-Packard, and F&S from Fockele and Schwarze and initial graphics exchange specifications (IGES).

20.4.4 Step 4: Machine Parametric Setup

Each machine has its own requirements for how to prepare for a new print job. This includes refilling the materials to be print, binders, and other consumables the printer will use. It also covers adding a tray to serve as a foundation or adding the material to build temporary water-soluble supports. Here, the parametric setup is totally depends on the materials to be used. On the basis of materials, the energy of the source, scanning speed, chord height, and angle is fixed.

20.4.5 Step 5: Build

This is the final step of the printing process and the simplest among all as well. Connect the printer to your computer. Once done, click print and sit back. Each layer is usually about 0.1 mm thick, though it can be much thinner or thicker (Fig. 20.6). Depending on the object's size, the machine, and the materials used, this process could take hours or even days to complete. Be sure to check on the machine periodically to make sure there are no errors. In this step, the printer will start printing the design to create a physical object. On occasions, one may have to load the G code file to the designated memory to let your 3D printer work without any halts. As the command reaches the 3D printer to print, it starts working to create the object layer-by-layer until the complete object is created.

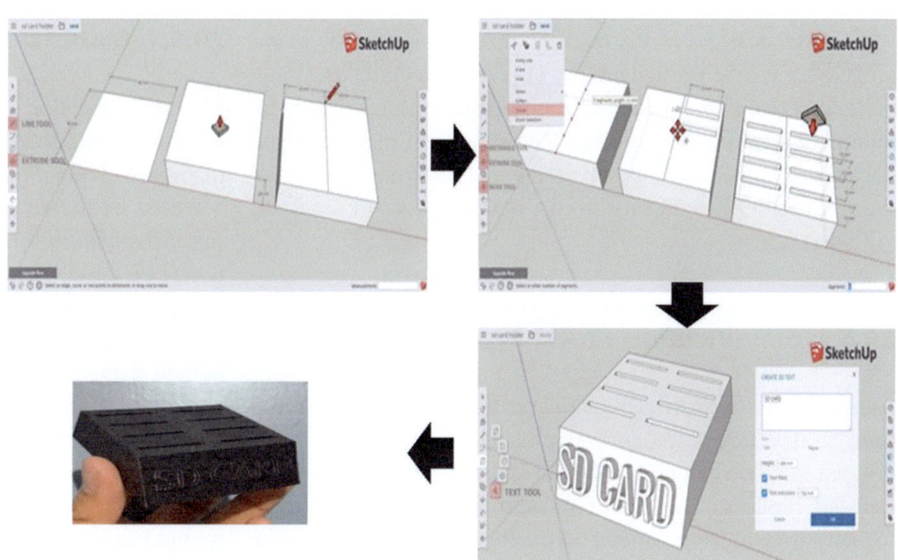

Fig. 20.6 Layer-by-layer direction

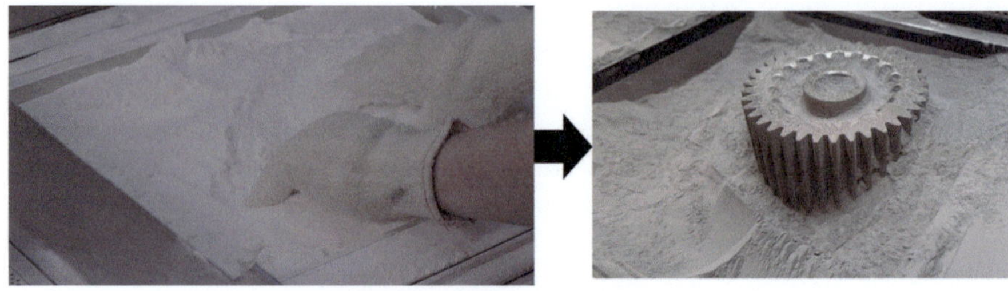

Fig. 20.7 3D printed product removal

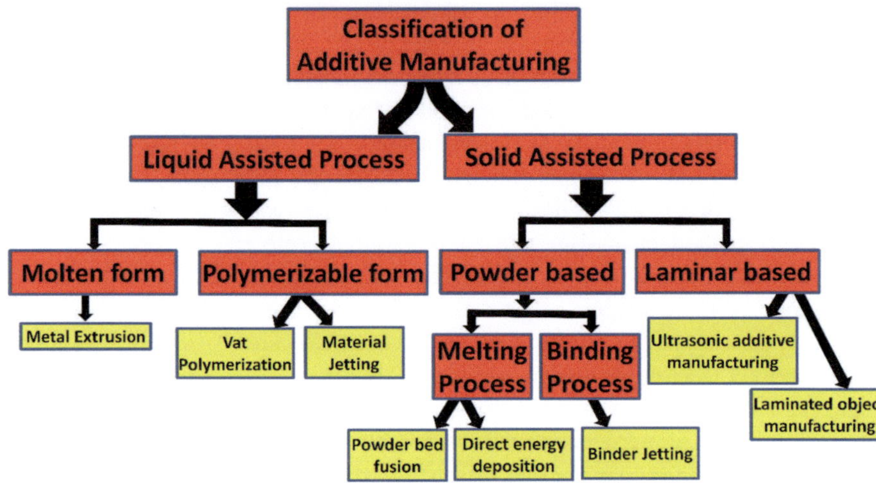

Fig. 20.8 Classifications of AM processes

20.4.6 Step 6: Removal of Product

Remove the printed object or multiple objects from the machine. Be sure to take any safety precautions to avoid injuries, such as wearing gloves to protect the associated one from hot surfaces or toxic chemicals. Figure 20.7 shows the 3D printed product removal from the printing tray.

20.4.7 Step 7: Post-Processing

Many 3D printers will require some amount of post-processing for the printed object. This could include brushing off any remaining powder or bathing the printed object to remove water-soluble supports. The new print may be weak during this step since some materials require time to cure, so caution might be necessary to ensure that it doesn't break or fall apart.

20.5 Classification of AM Processes

Various types of additive manufacturing processes have been used to develop different structural and functional part are divided into two categories: liquid-assisted process and solid-assisted process. The liquid-assisted processes are of two forms: molten form (metal extrusion) and polymerizable form (vat polymerization and materials jetting). The solid-assisted processes can be divided into powder base and laminar base. The powder base is subdivided into powder bed melting process (powder bed fusion, direct energy deposition) and powder bed binding process (binder jetting). The laminar base processes are ultrasonic additive manufacturing and laminated object manufacturing. The categorizations of AM processes are given in Fig. 20.8 and a comparative statement has been given in Table 20.1.

20.5.1 Material Extrusion

Material extrusion is becoming one of the most prominent additive manufacturing processes. In this process, the part is made by depositing an extruded material layer by layer. Material extrusion additive manufacturing is also known as fused deposition modeling (FDM) or fused filament fabrication (FFF). FDM is a common material extrusion process and is trademarked by the company Stratasys. In this technique, the material is drawn through a nozzle, where it is heated and is then deposited layer by layer. The nozzle can move horizontally and a platform moves vertically down after each new layer is deposited. It is a commonly used technique used on many inexpensive, domestic, and hobby 3D printers. The process has many factors that influence the final model quality but has great potential and viability when these factors are

Table 20.1 Comparative table of additive manufacturing processes [14–60]

Type of AM process	Other name/Related Techniques	Advantages	Disadvantages
Metal extrusion	FDM, FFF, MJS, PEM, PED, 3DFD, SEF, and LDM	• Least expensive.	• Only use low-temperature materials. • Occurrence of shrinkage. • Relatively low dimensional accuracy and mechanical strength.
Vat polymerization	SLA, DLP, CLIP, and 3SP	• High dimensional accuracy. • Good surface finish. • Ability to produce the complex part.	• Only use of UV-activated polymer. • Time-consuming materials changing, material contamination and waste in the process,
Material jetting	DOD, and MJM	• Good surface finish, • High resolution, • Fast process, a wide range of materials, materials mixing on droplet scale.	• Limited to jettable materials, clogging problem, low viscosity prevents buildup in 3D.
Powder bed fusion	DMLS, EBM, SHS, SLM, and SLS	• Wide range of materials, great material properties, high material strength. • Zero waste.	• Thermal stress, degradation. • Require inert gas. • High energy cost.
Direct energy deposition	LENS, LC, DMD, LMD, and EBF3	• High control of microstructure in build part. • Wide range of materials, great material properties.	• Low-dimensional accuracy. • Complex and expensive.
Binder jetting	3DP and DOP	• Low-temperature process • Wide range of materials • Materials mixing on droplet scale	• Limited to jettable photopolymers, • Clogging problem,
Sheet lamination	UAM, LOM, and SDL	• Easy materials handling. • Can be operated at an open atmosphere. • Low cost.	• Complex geometry is difficult to produce. • Less accuracy than all other AM. • Great amount of scrap.

controlled successfully. Whilst FDM is similar to all other 3D printing processes, it varies in the fact that material is added through a nozzle under constant pressure and in a continuous stream. This pressure must be kept steady and at a constant speed to enable accurate product. Material layers can be bonded by temperature control or through the use of chemical agents. Layers are fused together upon deposition as the material is in a melted state [61]. Material is often added to the machine in spool form as shown in Fig. 20.9.

This process includes low melting temperature materials, which can produce models with good structural properties, close to a final production model. In low-volume cases, this can be a more economical method than using injection molding. Materials extrusion is one of the simplest and least expensive additive manufacturing processes. In fact a toy 3D printer including software that will be on the market in the fall of 2019 for a price of $300. It is a widespread and inexpensive process. However, the process requires many factors to control in order to achieve a high-quality finish. As with most heat-related post-processing processes, shrinkage is likely to occur and must be taken into account if a high tolerance is required. The size of the nozzle radius limits and reduces the final quality. FDM printers use two kinds of materials during construction: (1) the modeling material, which constitutes the finished object, and (2) the support material, which acts as a scaffolding to support the object as it is being printed. Support materials are usually water-soluble wax or brittle thermoplastics, like polyphenylsulfone (PPSF) [62]. Once an object comes off the FDM printer, its support materials are removed either by soaking the object in a water and detergent solution or, in the case of thermoplastic supports, snapping the support material off by hand. Objects may also be sanded, milled, painted, or plated to improve their function and appearance. Other extrusion-based techniques are multiphase jet solidification (MJS), precise extrusion manufacturing (PEM), precision extrusion

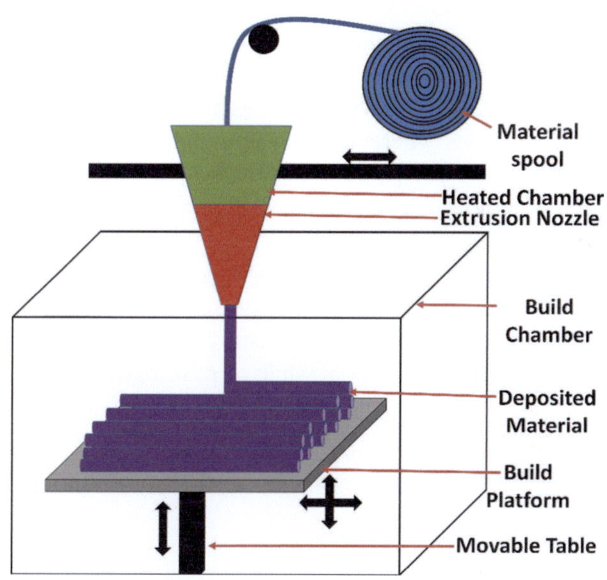

Fig. 20.9 Material extrusion process

deposition (PED), 3D fiber deposition (3DFD), solvent-based extrusion free-forming (SEF), and low-temperature deposition manufacturing (LDM). ABS is the most widely used polymer, but other polymers have also be used such as polycarbonate, high-density polyethylene, polycarbonate/abs blends, polyphenylsulfone, high-impact polystyrene, and biopolymers such as polylactic acid, etc. Multiple materials can be used for both the product build and support. Because thermoplastics are environmentally stable, part accuracy (or tolerance) doesn't change with ambient conditions or time. This enables FDM parts to be among the most dimensionally accurate.

20.5.2 VAT Photopolymerization

Vat polymerization uses a vat of liquid photopolymer resin, out of which the model is constructed layer by layer. A UV light cures the resin layer by layer. The platform continues to move downwards and additional layers are built on top of the previous (Fig. 20.10) [63]. As the process uses liquid to form objects, there is no structural support from the material during the build phase, unlike powder-based methods, where support is given from the unbound material. In this case, support structures will often need to be added. Resins are cured using a process of photo-polymerization or UV light, where the light is directed across the surface of the resin with the use of motor-controlled mirrors. The light-activated polymer quickly solidifies wherever the beam strikes the surface of the liquid. After completion, the vat is drained of resin and the object removed. The self-adhesive property of the photopolymer causes the layers to bond to one another, and eventually, a complete three-dimensional object is fully deposited and hardened. Designs are then immersed in a chemical bath in order to remove any excess resin and post-cured in an ultraviolet oven. It is also possible to print objects "bottom up" by using a vat with a somewhat flexible, transparent bottom, and focusing the UV upward through the bottom of the vat.

This process produces a high level of accuracy and good finish and also a relatively quick process. Photopolymers used in 3D imaging processes must be designed to have low volume shrinkage on polymerization in order to avoid distortion of the solid object. Common monomers utilized include multifunctional acrylates and methacrylates combined with a non-polymeric component in order to reduce volume shrinkage. A competing composite mixture of epoxide resins with cationic photoinitiators is becoming increasingly used since their volume shrinkage upon ring-opening polymerization is significantly below those of acrylates and methacrylates. Free-radical and cationic polymerizations composed of both epoxide and acrylate monomers have also been employed, gaining a high rate of polymerization from the acrylic monomer and better mechanical properties from the epoxy matrix. Advantages are the high resolution and accuracy, ability to produce complex parts, smooth surface, accommodates large build areas. The Vat polymerization process uses Plastics and Polymers (UV-curable photopolymer resin). Although photopolymerization can be used to produce virtually any design, it is often costly. The cost of resin and stereolithography machines was once very high. Recently, interest in 3D printable products has inspired the design of several models of 3D printers which feature drastically reduced prices (less than $10,000 for an industrial-sized printer. Several companies are now producing photopolymerizable resins at prices as low as $80 per liter [64]. Major applications are prototyping, consumer toys, electronics, guides, and fixtures.

Another form of photopolymerization technology is stereolithography (SL), digital light processing (DLP), continuous liquid interphase production (CLIP), scan, spin, and selectively photocure (3SP). The CLIP was introduced that

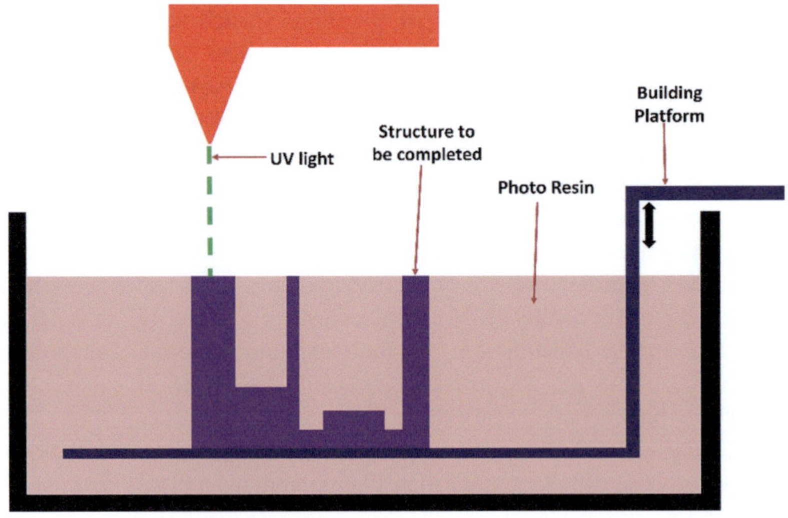

Fig. 20.10 Vat photopolymerization

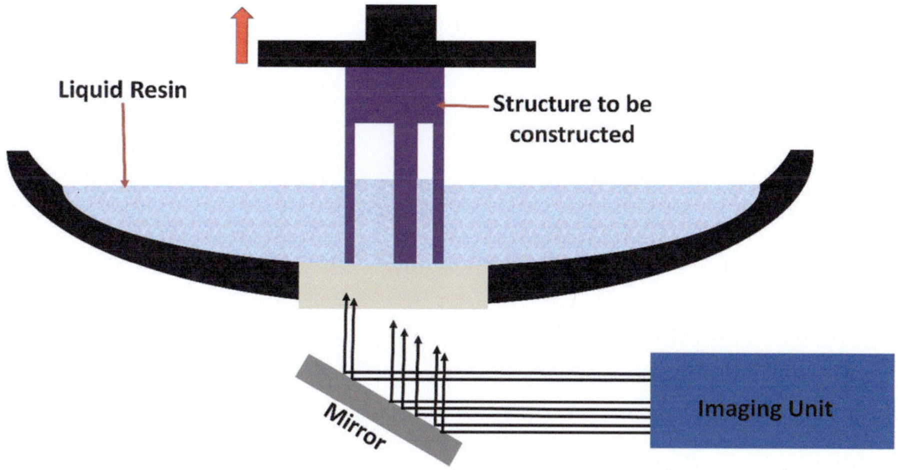

Fig. 20.11 Continuous liquid interface production process

claims speed 25–100 times faster than traditional 3D printing (Fig. 20.11) [65]. The CLIP process works by carefully balancing the interaction of UV light (which initiates photopolymerization) and oxygen (which inhibits the reaction). Part production is achieved with an oxygen-permeable window below the UV image projection plane. This creates a "dead-zone" where photopolymerization is inhibited between the window and the elevating polymerizing part. In this way parts that usually take hours to manufacture can be made in minutes [66].

20.5.3 Material Jetting

This process was the origin of the term "3D printing" (3DP). Material jetting creates objects in a similar method to a two-dimensional inkjet printer. Material is jetted onto a build platform using either a continuous or drop on demand (DOD) approach. DOD is used to dispense material onto the required surface. The print head is positioned above the build platform. The material is deposited from a nozzle which moves horizontally across the build platform. Material is jetted onto the build surface or platform, where it solidifies and the model is built layer by layer (Fig. 20.12). Machines vary in complexity and in their methods of controlling the deposition of material. The material layers are then cured or hardened using ultraviolet (UV) light. Droplets of material are deposited from the print head onto the surface where required, using either the thermal or piezoelectric method. Polymers and waxes are suitable and commonly used materials, due to their viscous nature and ability to form drops. This is a continuous system which allows for a high level of droplet control and positioning. Droplets that are not used are recycled back into the printing system [67].

The nature of using droplets limits the number of materials available to use. Polymers and waxes are often used and are suitable due to their viscous nature and ability to form drops. Viscosity is the main determinant in the process; there is a need to refill the reservoir quickly and this, in turn, affects print speed. Unlike a continuous stream of material, droplets are dispensed only when needed, released by a pressure change in the nozzle from thermal or piezoelectric actuators. Thermal actuators deposit droplets at a very fast rate and use a thin-film resistor to form the droplet. The piezoelectric method is often considered better as it allows a wider range of materials to be used. Machines vary in complexity and in their methods of controlling the deposition of the material. multi-jet modeling (MJM) is in the category of material jetting process [68]. Advantages of this technique are good surface finish, high resolution, allows full-color parts, enables multiple materials. As material must be deposited in drops, the number of materials available to use is limited (low melting temperature materials). The material jetting process uses plastics and polymers (Polypropylene, HDPE, PS, PMMA, PC, ABS, HIPS, and EDP). The major applications are high-resolution prototypes, circuit boards, and other electronics, consumer products, tooling.

20.5.4 Powder Bed Fusion

The powder bed fusion process (PBF) technique uses high-temperature materials such as metal, not plastic. Various thermal energy sources can be used, and as a result, there are several variations of this process: direct metal laser sintering (DMLS), electron beam melting (EBM), selective heat sintering (SHS), selective laser melting (SLM), selective laser sintering (SLS). PBF methods use either a laser or electron beam to melt and fuse the material powder together (Fig. 20.13). Electron beam melting (EBM), methods require a vacuum but can be used with metals and alloys in the creation of functional parts. All PBF processes involve the spreading of the powder material over previous layers. The layer, typically 0.1 mm thick of material is spread over the

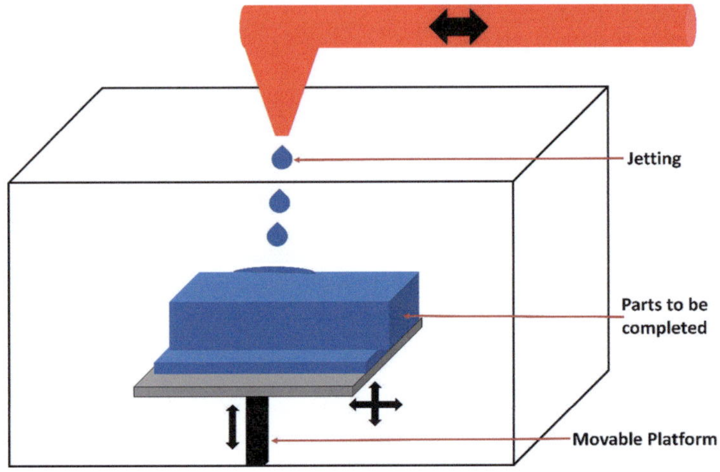

Fig. 20.12 Material jetting

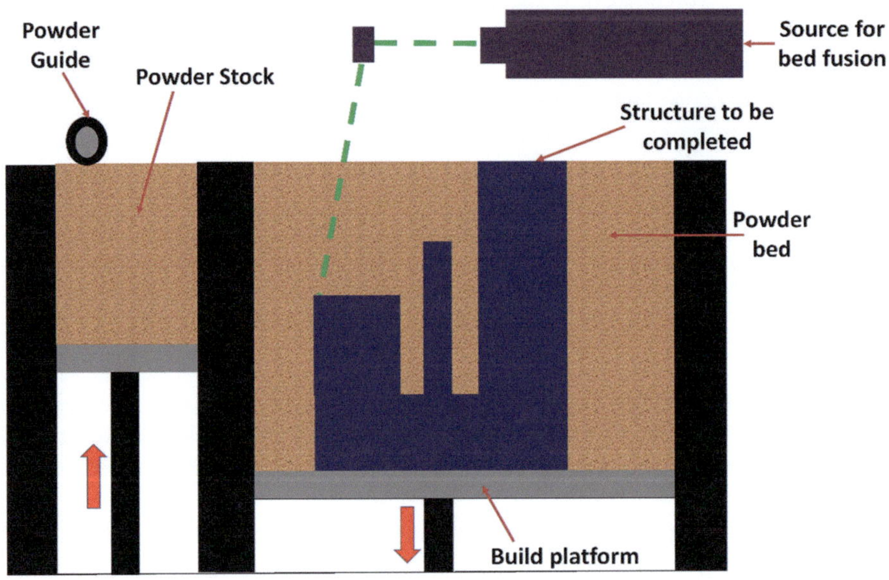

Fig. 20.13 Powder bed fusion additive manufacturing process

build platform. There are different guide mechanisms to enable this, including a roller or a blade. A hopper or a reservoir below of aside the bed provides fresh material supply. A source fuses the first layer or first cross section of the model. A new layer of powder is spread across the previous layer using a guide. Further layers or cross sections are fused and added. The process repeats until the entire model is created. The platform lowers the model accordingly. The loose, unfused powder remains in position but is removed during post-processing [69].

The SLS machines are made up of three components: a heat source to fuse the material, a method to control this heat source and a mechanism to add new layers of material over the previous. The SLS process benefits from requiring no additional support structure, as the powder material provides adequate model support throughout the build process [70]. The build platform is within a temperature-controlled chamber, where the temperature is usually a few degrees below that of the material melting point, reducing the dependency of the laser to fuse layers together. The chamber is often filled with nitrogen to avoid oxidation. Models require a cool-down period to ensure a high tolerance and quality of fusion. Some machines monitor the temperature layer by layer and adapt the power and wattage of the laser, respectively, to improve quality. In comparison with SLS, SLM is often faster, but requires the use of inert gas, has higher energy costs, and typically has a poor energy efficiency of 10–20%. The process uses either a roller or a blade to spread new layers of powder over previous layers [71]. SHS uses a heated thermal print head to fuse powder material together. As before, layers are added with a roller in between the fusion of layers. The process is used in creating concept prototypes and less so structural components. The use of a thermal print head and not a laser benefits the process by reducing significantly the heat and power levels required. Thermoplastics powders are used and as before act as support material

[14]. DMLS uses the same process as SLS, but with the use of metals and not plastic powders. The process sinters the powder layer by layer and a range of engineering metals are available [15]. EBM Layers are fused using an electron beam to melt metal powders. Electromagnetic coils are used to control the beam and a vacuum pressure (typically 1×10^{-5} MPa). EBM provides models with very good strength properties due to an even temperature distribution during fusion. The high quality and finish that the process allows for makes it suited to the manufacture of high-standard parts used in aeroplanes and medical applications. The process offers a number of benefits over traditional methods of implant creation, including hip stem prosthesis. Compared to CNC machining, using EBM with titanium and a layer thickness of 0.1 mm, can achieve better results, in a faster time and can reduce the cost by up to 35% [16].

Post-processing requirements include removing excess powder and further cleaning and CNC work. One advantage and common aim of post-processing is to increase the density and therefore the structural strength of a part. Liquid phase sintering is a method of melting the metal powder or powder combination in order to achieve homogenization and a more continuous microstructure throughout the material, however, shrinking during the process must be accounted for. Hot isotactic pressing is another method to increase density; a vacuum-sealed chamber is used to exert high pressures and temperatures of the material. Although this is an effective technique to improve strength, the trade-off is a longer and more expensive build time. The powder bed fusion process uses any powder-based materials. The common materials used in SHS are Nylon; in DMLS, SLS, SLM are Stainless Steel, Ti, Al, CoCr, Steel powder; and in EBM are Ti, CoCr, Stainless steel, Al and Co. Polymer powders also can be used in PBF processes such as polyamide, glass-filled polyamide, Polyetheretherketone, and Polystyrene. PBF process form low waste, complex structures are possible, a wide range of materials, no support required, high heat and chemical resistant materials, whereas relatively slow speed, have size limitations, high power usage, the finish is dependent on powder grain size and high equipment cost [17]. Major applications of PBF products are in aerospace, automotive, medical products, tooling, and dental implants.

20.5.5 Directed Energy Deposition

Directed energy deposition uses a laser beam to melt and fuse particles of the powder material delivered from the material deposition head. Directed energy deposition is similar to powder bed fusion except the material is first injected into an energy field and the common substrates are metal, metal

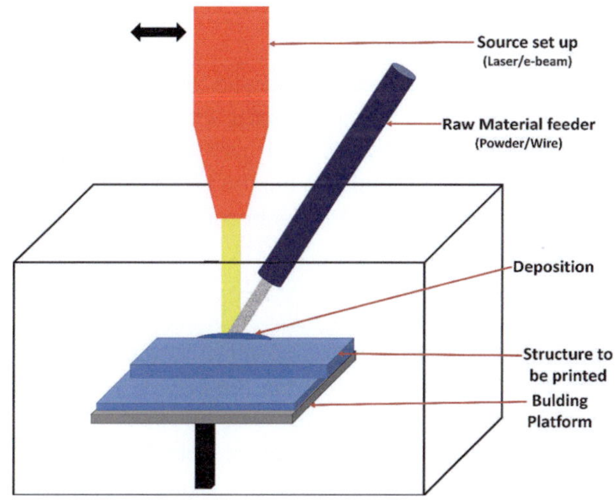

Fig. 20.14 Directed-energy deposition system

wire, glass, and ceramics [18]. Figure 20.14 shows the directed energy deposition system. There are various types of this technology such as laser engineering net shape (LENS), laser cladding (LC), and direct metal deposition (DMD), laser metal deposition (LMD), electron beam free-form fabrication (EBF3) [19–23]. Multiple-material directed energy deposition employs a nozzle to feed multiple powders and these are melted on a substrate by a laser beam to form fully dense objects. The key feature of this process is the powder feeding mechanism that can change or mix materials when fabricating multi-material structures. A wide range of different metals and alloys such as tool steel, stainless, Ti, Ni-base superalloys, and Co–Cr–Mo alloy has been deposited using this method. Independently controllable multiple powder feeders in the LENS process enable variations of composition and porosity simultaneously in one operation and enable the manufacture of novel structures. Functionally graded structures with a hard and wear-resistant CoCrMo alloy coating on a porous Ti6Al4V alloy with a metallurgically sound interface have been produced using LENS. The graded structures exhibited good bonding between the individual layers without any gross porosity, cracks, or lack of fusion defects. For electronic components and conductive lines, the laser micro-cladding method is used to fabricate electronic pastes on insulated boards that will be useful in the electronic manufacturing industry and other fields such as MEMS. The main advantage of this process is its ability to produce a highly controllable microstructure inbuilt parts, because it can exhibit deep structured-phase transformations to fabricate a fully dense part. The primary restriction of this process is poor resolution and surface roughness. Part geometries of these processes are limited because they cannot build free-hanging features or internal overhang features which require rigid support since there are no support materials in these processes. Additionally, the build times of these processes can be very long.

Directed energy deposition processes generally do not use polymeric materials but employ metal wire or powder. High energy heating sources such as a laser are directed at the material to melt it and build up the product. Directed energy deposition is considered to be a more complex and expensive additive manufacturing process. The main advantages of directed energy deposition are, (1) it can operate in an open atmosphere, (2) multiple materials can be used, (3) large parts are possible, (4) high single point deposition rates, and (5) not limited by direction or axis. Some limitations of the DED process are, (1) expensive equipment, (2) lower resolutions and reduced ability to manufacture complex parts, and (3) final machining is often required. The major applications are to repair or build up of high-volume parts.

20.5.6 Binder Jetting

The binder jetting process uses two materials; a powder-based material and a binder (Fig. 20.15). The binder is usually in liquid form and the build material in powder form. A print head moves horizontally along the x- and y-axes of the machine and deposits alternating layers of the build material and the binding material [24]. After each consecutive layer, the object being printed is lowered on its build platform. The print head deposits the binder adhesive on top of the powder where required. The process is repeated until the entire object has been made. The unbound powder remains in position surrounding the object. Due to the method of binding, the material characteristics are not always suitable for structural parts and despite the relative speed of printing, additional post-processing can add significant time to the overall process. As with other powder-based manufacturing methods, the object being printed is self-supported within the powder bed and is removed from the unbound powder once completed. Other names of this process are Drop on Powder (DOP), Powder Bed Printing.

The binder jetting process allows for color printing and uses metal, polymers, and ceramic materials. The process is generally faster than others and can be further quickened by increasing the number of print head holes that deposit material. The two-material approach allows for a large number of different binder-powder combinations, and various mechanical properties of the final model to be achieved by changing the ratio and individual properties of the two materials. The process is therefore well suited for when the internal material structure needs to be of a specific quality [25]. After construction of the final structure, the overall process time is extended as it requires the binder to set and the part is often allowed to cool in the machine to fully solidify to achieve a high-quality finish. This post-processing is often required to make the part stronger and give the binder material better mechanical and structural properties. Additionally, the parts produced via binder jetting are supported by the loose powder, thus eliminating the need for a build plate. Spreading speeds for binder jetting outperform other processes. Binder jetting has the ability to print large parts and is often more cost-effective than other additive manufacturing methods. All types of materials can be used with the binder jetting process. The two-material method allows for a large number of different binder–powder combinations and various mechanical properties. This process produces a rough surface. Other additive techniques utilize a heat source which can create residual stresses in the parts. These stresses must be relieved

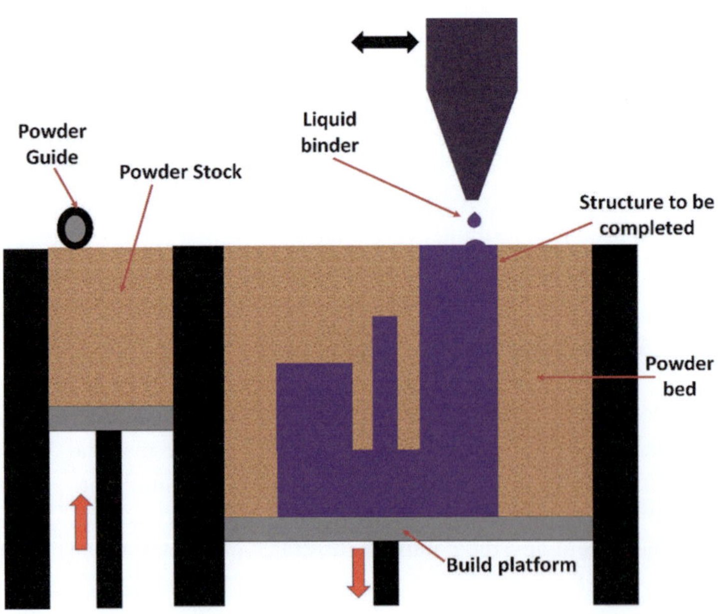

Fig. 20.15 Binder jetting process

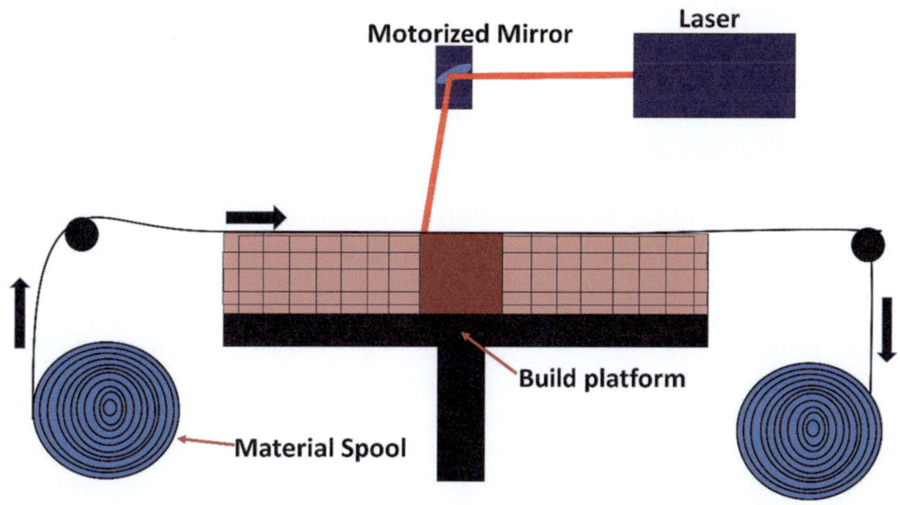

Fig. 20.16 Sheet lamination process

in a secondary post-processing operation. Major applications of this process are prototyping, tooling.

20.5.7 Sheet Lamination

In sheet lamination processes the material is positioned in place on the cutting bed. The material is bonded in a place, over the previous layer, using the adhesive (Fig. 20.16). The required shape is then cut from the layer, by laser or knife [26]. The next layer is added. Sheet lamination processes include ultrasonic additive manufacturing (UAM), laminated object manufacturing (LOM), and selective deposition lamination (SDL) [27–29]. The ultrasonic additive manufacturing process uses sheets or ribbons of metal, which are bound together using ultrasonic welding. The process does require additional CNC machining and removal of the unbound metal, often during the welding process. Laminated object manufacturing (LOM) uses a similar layer-by-layer approach but uses paper as material and adhesive instead of welding. LOM is one of the first additive manufacturing techniques created and uses a variety of sheet material, namely paper. The LOM process uses a cross-hatching method during the printing process to allow for easy removal post build. Laminated objects are often used for aesthetic and visual models and are not suitable for structural use. UAM uses metals and includes Al, Co, Ti, and stainless steel. The process is low temperature and allows for internal geometries to be created. The process can bond different materials and requires relatively little energy, as the metal is not melted. The process does require additional CNC machining of the unbound metal. Unlike LOM, the metal cannot be easily removed by hand and unwanted material must be removed by machining. Materials are bonded and helped by plastic deformation of the metals. Plastic deformation allows more contact between surfaces and backs up existing bonds.

Post-processing requires the extraction of the part from the surrounding sheet material. With LOM, cross hatching is used to make this process easier, but as the paper is used, the process does not require any special tools and is time-efficient. Whilst the structural quality of parts is limited, adding adhesive, paint, and sanding can improve the appearance, as well as further machining. The most commonly used material is A4 paper. Polymers are often used but paper or metal foils are also typically processed and find application in cases where heat-sensitive materials cannot be used and low costs must be realized. Almost any polymer can be used as long as it is available in thin sheet form and can be bonded by either adhesives or heat. Benefits of this process include speed, low cost, ease of material handling, but the strength and integrity of models are reliant on the adhesive used. Cutting can be very fast due to the cutting route only being that of the shape outline, not the entire cross-sectional area. Finishes can vary depending on paper or plastic material but may require post-processing to achieve the desired effect. Fusion processes require more research to further advance the process into a more mainstream positioning [30]. The main advantages of sheet lamination are lower materials cost, the process does not require a closed environment, high volumetric build rates, allows for the combination of materials and embedding components. The primary disadvantages are that complex geometries are difficult to produce, and they can be less accurate than all other AM processes. Major applications are large parts, tooling (Table 20.1).

20.6 Materials for AM Processes

Various types of materials such as metal, ceramic, polymer, intermetallic are used in AM processes are discussed below:

20.6.1 Metal

Metals are very important material when it comes to 3D printing. Because of their great mechanical properties, metals can be used to create huge parts for multiple applications for different niches. Either talked about tooling implants or engine parts, metals can be utilized everywhere. The best part is that 3D printing can help realize designs of metals with complex structures. Metallic AM has three different categories of material processing depend on AM process used, build volume, and energy source used. Those categories are powder bed systems, powder feed systems, and wire feed systems. Other classifications of AM metallic processes depend on the way of bonding between metal particles; indirect or direct method [31]. The examples of AM processes used in the indirect technique are SLS, 3DP, FDM, SLA, and LOM, while the examples used for the direct techniques are SLM, LMD, and EBM. Additionally, the additive manufacturing process itself introduces new options for materials that were impossible using more conventional methods. For example, some metal 3D printing methods enable layering of different metals such as Al, Ta, and Ni together in a single part. On the other hand, the 3D printing process also introduces new potential issues and sources of error, including porosity, residual stress, and warpage. As a general rule, if a metal welds or casts well, then it should also be amenable to additive manufacturing. Current market demand structural and functional metal are Ti, Ni, Al, Co, Cu, precious metals (Au, Ag, Pt), Ta, W, stainless steel, tool steel, and their alloys. The most used additive manufactured metals are discussed below:

Ti and Their Alloys: Titanium is one of the most popular materials for 3D printing in production, particularly for aerospace and medical applications. It is used in the DMLS process [32]. It can be alloyed with Fe, Al, V, Mo among other elements, to produce strong, lightweight alloys for aerospace (jet engines, missiles, and spacecraft), military, industrial processes (chemicals and petrochemicals, desalination plants, pulp, and paper), automotive, agriculture (farming), medical prostheses, orthopedic implants, sporting goods, jewelry, mobile phones, and other applications. It combines with Ni to produce smart materials which have a higher demand in this advanced age. However, these advantages are offset by titanium's relatively high cost. For this reason, the potential for waste reduction makes additive manufacturing an attractive option for titanium parts. Powdered Ti is pyrophoric and reacts explosively with water at temperatures in excess of 700 °C. For this reason, Ti powder has to be 3D printed in a vacuum or Ar gas chamber. It is also possible to 3D print titanium using EBM with a wire feedstock, which eliminates the risk of explosive reaction. A typical $\alpha + \beta$ titanium alloy Ti-6.5Al-3.5Mo-1.5Zr-0.3Si was fabricated by laser melting deposition process. The as-deposited sample exhibits a strong tensile anisotropy due to the columnar grains morphology and the strong texture of β <001> parallel to the deposition direction. Despite the anisotropy, the $\alpha + \beta$ annealing can increase the mechanical properties in both strength and ductility to levels that are much higher than the specification limits of the alloy. The two most common Ti alloys used in additive manufacturing are Ti-6Al-4 V and equiatomic NiTi [33].

Ni Alloys: Ni-based materials are the primary requirement for the production of additive manufactured superalloys using typically SLM, EBM. Superalloys are generally characterized by having a good combination of high tensile, creep, and rupture strength with excellent tensile, fatigue corrosion, and thermal fatigue resistance. AM-Inconel provides outstanding resistance to thermal fatigue as well as exceptional rupture strength at high temperatures. This alloy is widely used in the chemical process industry, sea water, and power plant scrubber, which requires high pitting and crevice corrosion resistance and high-temperature aerospace, chemical process, and power industry applications. Inconel 718 suits for a variety of high-end applications in aircraft turbine engines and land-based turbines (blades, rings, casings, fasteners, and instrumentation parts). Hastelloy X is also resistant to stress-corrosion cracking. It is applied in industrial furnaces, for petrochemical applications, and in the chemical process industry (catalyst support grids, muffles, retorts, furnace baffles, and tubing for pyrolysis). Hastelloy applications can be found in aerospace technology, gas turbine parts, and combustion zone components such as transition ducts, combustor cans, spray bars, flame holders, afterburners, tailpipes, and cabin heaters [34].

Al and Their Alloys: Powders of Al alloys such as $AlSi_{12}$ and AlSi10Mg are relatively suitable for laser melting due to the small difference between their liquidus (melting) and solidus temperature (solidification point) compared to high-strength wrought aluminum alloys. An Al alloy is a chemical composition where other elements are added in order to enhance its properties, primarily to increase its strength. These other elements include Fe, Si, Co, Mg, Mn, and Zn at levels that combined may make up as much as 15% of the alloy by weight. $AlSi_{12}$ is a metal powder for additive manufacturing of lightweight parts with good thermal properties. Typical applications are thin-walled parts such as in heat exchangers, or other industrial-grade prototypes, production parts, or spare parts for automotive, aerospace, and aviation. The silicon/magnesium combination results in a significant increase in strength and hardness. This Al alloy is used for parts with thin walls, complex geometries and is ideal for applications in need of good thermal properties and low weight. $AlSi_{10}Mg$ can also be heat treated to improve the mechanical properties. ALM parts are fully dense, with similar properties to cast or wrought parts. 3D print aluminum is a lightweight and versatile metal that can be used in aerospace components as well as parts for auto racing. Though not as strong as steel, aluminum is much lighter and more resistant to corrosion. It is also more expensive, though not as expensive as titanium. The biggest advantage aluminum offers for 3D printing is the ability to produce parts with fine detail and thin walls—as thin as 50 μm. Aluminum parts made with additive manufacturing tend to have a textured, matte surface, as opposed to the milled surface that is typical of machined aluminum parts. The most common aluminum alloy used for 3D printing is $AlSi_{10}Mg$ [35].

Co Alloys: Cobalt–chromium alloy classes of superalloys present high strength, superior corrosion resistance, nonmagnetic behavior, and good biocompatibility. Co28Cr6Mo is an alloy commonly used for surgical implants as artificial joints including knee and hip joints due to high wear-resistance, biocompatibility, and Ni-free (<0.1% Ni content) composition. After adequate heat treatment (hot isostatic pressing +homogenization) Co–Cr–Mo alloys built by EBM show superior properties compared to wrought or cast alloys. It is also used for engine components, wind turbines, and many other industrial components as well as in the fashion industry to make jewelery [36].

Cu Alloys: Processing pure copper by laser-aided additive manufacturing is challenging by nature. The high thermal conductivity rapidly evacuates heat away from the melt pool while the high optical reflectivity of the powder diverts large amounts of power to the surroundings. As a consequence, higher laser power is required and alloying copper with other metals is a way to mitigate these difficulties. $CuSn_{10}$ is popularly known as bronze, this alloy has excellent thermal and electrical conduction properties. It is a fine choice for heat management applications where the excellent heat conductivity of the copper can be combined with the AM design freedom to give rise to a complex internal structure and cooling channels. This enables to cool more effectively tool inserts in molds, for example, or more generally, hot spots such as in semiconductor devices. $CuSn_{10}$ is the preferred choice for AM parts in this copper alloy can be used in micro-heat exchangers, with thin walls and complex shapes [37].

Au, Ag, and Their Alloys: These materials are generally used in the jewelry sector. These metals use the DMLS or SLM process for printing. These filaments are sturdy materials and are processed in powder form. Printing with gold and silver is expensive. It takes a lot of effort and time to get it right. Both gold and silver are difficult to work with lasers because of their high reflectivity and high thermal conductivity. Since extremely high temperature is needed to print these materials, a regular FDM 3D printer is not suitable to use [38].

Mg Alloys: As one of the most promising lightweight structural materials, Mg alloys have been widely used in the aeronautic industry. Additive manufacturing of Mg alloys is of growing interest in the community due to enabling design capabilities not achievable with traditional manufacturing and its potential for the development of biodegradable implants. Additive manufacturing of magnesium has been demonstrated using powder bed fusion, wire arc AM, paste extrusion deposition, friction stir AM, and jetting technologies. These processes have different process mechanics and forms of raw materials. Each process yields AM components having different structural properties. By manufacturing components in this way, AM can be used to develop highly complex geometries that are either difficult or impossible to make using conventional machining processes [39].

Stainless Steel, Tool Steels, and Maraging Steels: Compared to Al, Ti, and most of the other metals on our list, stainless steel is by far the most affordable option. It can be used to 3D print waterproof parts with high strength and density for use in extreme environments, such as jet engines and rockets. There have also been several studies indicating the viability of using 316 L stainless steel to produce nuclear pressure vessels made via additive manufacturing. Stainless steel is printed by fusion or laser sintering. There are two possible technologies that can be used for this material. It can be DMLS or SLM technologies. Since stainless steel is all about strength and detail, it is perfect to use for miniatures, bolts, and key chains. Building time for 3D printing using these metals is much longer. Printing with stainless steel is expensive and printing size is limited. The most common stainless steels used in additive manufacturing are 17-4PH, 15-5-PH, ASM 316L, and 304L [40]. Tool steel refers to a variety of carbon and alloy steels particularly well-suited to be transformed into tools. Their suitability comes from their excellent hardness, resistance to abrasion and deformation, and their ability to hold a cutting edge at elevated temperatures. H13 is a hot work tool steel often used for dies, able to withstand the process conditions for an undetermined period of time. "Martensitic-aging" steels are known for combining high strength, toughness, and dimensional stability during aging. They differ from other steels as they are hardened by a metallurgical reaction, which does not involve carbon but the precipitation of intermetallic compounds, rich in Ni, Co, and Mo. Due to the high hardenability and wear resistance, Maraging 300 is suitable for many tooling applications (injection molding, die casting of light metal alloys, punching, and extrusion) but also for various high performance industrial and engineering parts (aerospace, high strength airframe parts, and motor racing applications) [41].

20.6.2 Ceramic

Ceramics is one of the newest materials that are used in 3D printing. It is more durable than metal and plastic since it can withstand extreme heat and pressure without even breaking or warping it. Moreover, these types of materials are not prone to corrosion like other metals or wear away like plastics do. This material is generally used in binder jetting technology, stereolithography, and digital light processing (DLP). Additive manufactured ceramics have high-precision components with a smooth and glossy surface, higher resistance to acid and heat. Since ceramic is fragile, it has limitations in printing objects with enclosed and interlocking parts. It is not ideal for the piece assembly process so that taking advantages of additive manufacturing [42].

20.6.3 Polymer

The history of polymer additive manufacturing can be said to have begun with the patenting of stereolithography in 1984 by Charles Hull. Hull found that he could create a solid 3D structure by curing consecutive layers of photopolymer one atop another [43]. Other processes for producing 3D structures would be developed over the coming decades. While the end result of each process was the same, a 3D structure, the fabrication method varied. As mentioned, SL cures photopolymer liquid in layers to produce a part. Binder jetting and material jetting use inkjet print heads to produce parts. Extrusion-based methods melt a thermoplastic feedstock and extrude the molten material as thin layers to build apart. Powder bed fusion uses a laser or electron beam to melt or sinter plastic powder. Each method has strengths and weaknesses and is often selected with consideration of its idiosyncrasies. Polymer AM proved a method for quickly producing prototype parts. This use became so prevalent that polymer AM is alternatively referred to as rapid prototyping [44]. The list of 3D-printable polymer materials is even longer than the one for metals, but some of the most popular polymers used are acetal, acrylic, acrylonitrile butadiene styrene (ABS), acrylonitrile styrene acrylate (ASA), high impact polystyrene (HIPS), nylon, polycarbonate (PC), polyether ether ketone (PEEK), polyethylene terephthalate (PET), polyethylene trimethylene terephthalate (PETT), polyethylene yerephthalate glycol-modified (PETG), polylactic acid (PLA), polypropylene (PP), polyvinyl alcohol (PVA), thermoplastic elastomer (TPE).

Some major used polymer has been discussed below:

Acrylonitrile Butadiene Styrene (ABS): ABS is by far the most popular material in production applications of 3D printing. Although PLA is more popular overall, ABS is almost always a better choice for manufacturing, due to its strength, durability, and low cost. ABS needs to be heated to a relatively high range of 230–250 °C to be printable on a 3D printer, and as such, also generally requires a heated print bed to ensure proper cooling and prevent warping. ABS parts can be additively manufactured using FDM, SLA, or poly jetting. The major drawback to ABS is the fact that it is toxic and emits fumes when it reaches its melting point. 3D-printed ABS parts are most often found as casings in end-use products or in rapid tooling applications. ABS is highly available and has a wide variety of colors. This material has a longer lifespan compared to Nylon [45].

Nylon: Nylon—generically known as polyamide—is a synthetic thermoplastic linear polymer that offers more strength than ABS at an increased cost. It is also flexible and demonstrates excellent material memory. Layer adhesion for 3D-printed parts made with nylon is also above average. Nylon's moisture sensitivity requires that it be additively manufactured either in a vacuum or at high temperatures, and it should also be stored in air-tight containers. Some nylon parts can be prone to shrinkage, making them less accurate than ABS. The most popular types of Nylon for additive manufacturing are Taulman 618, Taulman 645, and Bridge Nylon. It is a well-known 3D printing filament because of its flexibility, durability, low friction, and corrosion resistance. Nylon is also a popular material used in the manufacturing of clothes and accessories. Nylon is suitable to use when creating complex and delicate geometries. It is primarily used as filaments in FDM or Fused Filament Fabrication 3D printers. This material is inexpensive and recognized as one of the toughest plastic material. This material can shrink during cooling, thus, prints may be less precise and printer suitability also varies [46].

Polylactic Acid (PLA): Polylactic Acid is made from renewable resources such as sugarcane or corn starch. It is also called "green plastic." It is mostly used material since it is safe to use and easy to print with. It is also used in FDM desktop printing. PLA is easy to print since it has low warping and can be printed on a cold surface. It can print with sharper corners and features compared to ABS material. This material is available in different colors. PLA materials are not very sturdy and can deform when exposed to extreme heat [47].

PET/PETG: Like nylon, PET or Polyethylene terephthalate is also one of the frequently used plastic. This material is used in thermoforming processes. It can also be combined with other materials like glass fiber to create engineering resins. PETG is a modified version of PET where G stands for "glycol-modified." As a result, a filament that is less brittle, clearer, and easier to use than PET is formed. This filament is applicable in FDM or FFF technologies. This material is durable and is impact-resistant and recyclable. It can also be sterilized. It has an excellent layer adhesion. It has a combined functionality of ABS (temperature resistant, stronger) and PLA (easy to print). The material can be weakened by UV light, prone to scratching [48].

High-Impact Polystyrene (HIPS): HIPS are plastic filaments that are used for support structures in FDM printers. It is comparable to ABS when it comes to ease of use. The only difference is its ability to dissolve. HIPS is completely soluble in a liquid hydrocarbon called limonene. It has good machinability. It can also be used to make complex structures. It is very smooth and lightweight. It is water resistant and impact resistant and is inexpensive. It produces strong fumes. Thus, it is recommended to be used in a ventilated area during additive manufacturing. Without constant heat flow, this material can clog up the nozzle and delivery tubes of the printer [49].

Polycarbonate: Polycarbonate plastics (PC) also known under the trade name Lexan, are both light and dense, with excellent tensile strength. Its transparency enables PC to be used in a variety of applications, including sunglasses. Carbon-reinforced PC can be used to create intake manifolds and other parts subject to high temperatures. PC is soluble in dichloromethane and melts at temperatures of 260–300 °C, which is quite high for 3D printing. Though it's naturally transparent, PC can be colored if necessary. Like ABS, it requires a heated print bed to ensure adhesion and reduce the chance of warping [50].

Resin: Resin is one of the most used materials in 3D printing. It is mainly used in technologies such as SLA, DLP, Multijet, or CLIP technologies. There are various types of resins that can be used in 3D printing such as castable resins, tough resins, flexible resins, etc. It has low shrinkage. Resin materials have high chemical resistance. It is expensive. This type of filament also expires and needs to be stored securely due to its high photo-reactivity. When exposed to heat, it can cause premature polymerization [51].

20.6.4 Composite

Composite materials offer unusual combinations of properties and AM processes have many advantages for fabrication complex shapes and sustaining or improving product characteristics as well. Investigation of AM technology effects on composite materials is still in the area of research, but number of researchers have tried to study those effects through different categories of composite materials. Different types of composites such as metal matrix, ceramic matrix, polymer matrix, structural composites, and nano-composite are discussed here.

Metal Matrix Composites and Metallic Alloys: Metal matrix composites (MMC) are composites in which the matrix phase is a ductile metal. The metal matrix provides ductility and thermal stability for the composite at elevated temperatures, while the fiber may increase the strength, the stiffness, enhance the resistance to creep or abrasion, and improve the thermal conductivity. Aluminum and its alloys, copper, titanium, and magnesium are the most common metals used in MMCs [52]. Metal matrix composites can be divided into three different categories: related to the type of reinforcement, fiber, or particulate. The most common fabrication processes for MMC are powder metallurgy, spray deposition, and squeeze casting, whereas, in additive manufacturing, ultrasonic additive manufacturing process is extensively studied. For example, Al 3003-H18 is used as a matrix phase and with prestrained NiTi embedded inside. The constitutive models of the NiTi element and Al matrix have investigated average interface shear strength of 7.28 MPa and an effective coefficient of thermal expansion of zero at 135 °C. Furthermore, interface failure temperatures can be increased as the embedded fiber length is increased. Ultrasonic additive manufacturing (rapid prototyping process) based on ultrasonic metal welding has been used to develop active aluminum matrix composites. This composites material is consists of aluminum matrices and embedded shape memory NiTi, magnetostrictive Galfenol, and electro-active PVDF phases. One of the most important advantages of this process is working at temperatures as low as 25 °C during fabrication rather than other metal-matrix fabrication processes which require temperatures of 500 °C. Ti, Ni, and Fe-based alloys powder are considered as the most common mature phase of additive manufacturing practical applications. A complete melting mechanism using LM or LMD is the most basic processing technique for those types of alloys because of easy process controllability. The main effective type of Ti-based alloys is Ti-6Al-4 V because it has a lot of applications in aerospace and medical fields [53].

Ceramic Matrix Composites and Ceramic Components: It is obtained that fabrication of ceramic matrix composite is difficult by using conventional techniques. On the other hand, AM technology has the ability to deal with those types of materials without the need for molds or part-specific tools [72]. The most common processes used for producing ceramic components, in general, are SLM, SL, and DIP. Selective laser gelation (SLG) has been used to fabricate one of the ceramic matrix composites which consist of stainless steel powder and a silica sol at a proportion of 65–35 wt. %. The gelled silica matrix with embedded metal particles was used to form a 3D composite part and distributed over the silica gelled layer using Nd: YAG laser technique. The advantage of this processing approach is an optimal saving of laser-forming energy (low the better) and fabrication speed (higher the better) [54].

AM technology was used to fabricate and develop fully dense ceramic freeform components through high-strength oxide ceramics (ZrO_2–Al_2O_3 ceramic) with improved mechanical properties. Complete melting of ZrO_2–Al_2O_3 by using SLM can be obtained experimentally; 100% density and 500 MPa of flexural strength were observed without any sintering processes or any post-processing crack-free specimens. A number of ceramic suspensions such as; 3Y-TZP, Al_2O_3, and ZTA (oxide) or Si_3N_4, and $MoSi_2$ (non-oxide) have been studied using DIP to investigate processing possibilities through microstructures, laminates, three-dimensional specimens, and dispersion ceramics. This process was modified by using aqueous ceramic suspensions of high solids content instead of ink. It is found that wall thicknesses of about 200 μm can be achieved using the proposed DIP. Good bonding between layers of the same material was observed in the processing of ZrO_2 and for different materials as well (ZTA/3Y-TZP). Structural and functional parts by layer-wise buildup via DIP were produced using aqueous inks of Si_3N_4 and $MoSi_2$ with high solids content and excellent mechanical characteristics were noticed. It is concluded that optimization of the process technology and material parameters are recommended to create fast, reliable, and flexible production of complex-shaped non-oxide ceramic parts. Another technique of AM technology used for dense and strong ceramic components is investigated. This technique is called lithography-based ceramic manufacturing (LCM) and it is considered as a dynamic mask exposure process based on the selective curing of photosensitive slurry. LCM has been applied for the fabrication of strong, dense, and accurate alumina ceramics. No geometrical limitation has been observed and it is investigated that over 99.3% of a theoretical alumina density was achieved. Furthermore, a bending strength of 427 MPa was obtained and very smooth surfaces were created. In terms of mechanical properties, additive manufacturing offers equivalent to ceramic materials structured by conventional processes [55].

Polymer Matrix Composites: AM technology has an impact in processing effective and potential polymer composites such as; carbon fiber–polymer composites as it has the ability to handle complex shapes with great design flexibility. Short fiber (0.2–0.4 mm) reinforced ABS composites were fabricated using 3D printing. In comparison with traditional compression molded composites, 3D printed samples achieved higher percentages around 115%, 700%, 91.5% for the tensile strength, tensile modulus, and yielding, respectively. The use of carbon fiber-reinforced feedstock with optimized orientation, good dispersion capabilities, and improving interfacial adhesion between fibers and matrix via surface modification has an impact advantage in the industry of load-bearing composite parts [56]. There is a list of barriers that limit the widespread adoption of AM technology of polymer composites such as: (1) extremely low production rate; (2) the small physical size of the parts; (3) the mechanical properties limitations. On the other hand and especially for carbon fibers, more developed ways are obtained to solve these barriers. In terms of improving specific strength, there is use of carbon–fiber-reinforced polymers. Decreasing the distortion and warping of the material during deposition by using carbon fiber additions which provide a high deposition rate manufacturing. The integration between carbon fiber and AM technologies has an important advantage of creating complex components that would not be possible with any conventional technology alone. On the other hand, quasicrystalline polymers have a high commercialization impact on the development of polymer composites through AM technology [57].

Nano-composites: It is investigated that more concern is required to fabricate composite materials in one process cycle which provides sufficient material characteristics without any creating, reconditioning, and treatment of the bulk material. One possibility of the additive technologies to achieve the previous requirement is using nano-size structures of constructional materials [73]. It is obtained that 3D composite structures based on nanomaterials have a number of potential applications in temperature drop operations and aggressive chemical or biological environments applications. On the other hand, SLS is considered as a very efficient process to create metal composites functional parts with less labor effort, shorter time, and the ability for complex or internal cavities geometry. Another example of nanocomposite AM fabrication is the fabrication of nanocrystalline titanium carbide (TiC) reinforced with Inconel 718 matrix bulk form using the SLM process [58, 59]. Another nano additive manufacturing (ANM) technologies are dip-pen lithography (DPN), electro-hydrodynamic jet printing (EHD), optical tweezers, and electrokinetic nanomanipulation, and direct laser writing (DLM). The most significant nano additive manufacturing process variables are minimum feature size (resolution), deposition speed, and material selection [60].

20.6.5 Intermetallic Compound

Intermetallic compounds are used in various additive manufacturing processes such as SEBM, SLM, and DMD to produce the near net shape or net-shape structure. For example, TiAl and NbSi-based alloy [74]. TiAl-based alloys are highly attractive for the aeronautical industry as new high temperature lightweight structural materials due to their low density (3.7–3.9 g/cm^3), high elastic moduli, excellent tensile strength, and creep resistance at high temperature. Therefore, TiAl alloys have shown a potential for application with a service temperature of 760–850 °C. However, the room temperature tensile elongation is very poor and only about 1%, which makes the TiAl-base alloy difficult to be processed into components especially the complex-shaped ones. Compared to the traditional manufacturing methods such as cast, extrusion, and forging, the AM technologies are the emerging near-net technologies and will be applied to produce the near-net-shape TiAl-based alloys components. Nb–Si-based alloys prepared by AM technique have been carried out and extremely finer microstructure, higher hardness, higher room temperature toughness and better oxidation resistance at high temperature could be obtained by AM. However, due to the intrinsic brittleness of Nb–Si-based alloys, and the rapid melting and solidification process during AM, microcracks tend to appear in the preparation of larger scale samples. Further research work is highly needed for composition optimization by adjusting the level of Si and adding the alloying elements of Ti, Hf, etc., as well as processing optimization such as preheating during AM. Nb-22Si-26Ti-6Cr-3Hf-2Al (at.%), Nb-17Si-23Ti (at.%), Nb-26Ti-22Si-3Hf-2Al (at.%), Nb-18Si-24Ti-2Cr-2Al-2Hf (at.%) alloy has been successfully fabricated by SLM using pre-alloyed powder [75].

20.6.6 High Entropy Alloys and Bulk Metallic Glass

High entropy alloy (HEA) has superior properties such as high strength, excellent high-temperature strength, good corrosion resistance, and good wear resistance. However, it is difficult to overcome the inherent complexity and high levels of control required to produce homogeneous bulk alloys industrially using a conventional casting method [76]. AM which facilitates a high level of local process control, generates rapid solidification cooling rates, and enables the production of complex geometries, may be a suitable way of utilizing HEAs as engineering materials. The study on AM of HEA just began several years ago. Cube-shaped SEBM specimens (20 mm × 20 mm × 16 mm) have been manufactured from gas-atomized AlCoCrFeNi powders. The SEBM specimens exhibit superior ductility and lower yield strength than the as-cast specimens. However, the SEBM specimens exhibited the FCC phase precipitation at grain boundaries of the B2/BCC grains whereas the cast specimens showed almost exclusively B2/BCC phases and nearly no FCC phase. The average grain sizes of the B2/BCC phases of the cast and SEBM specimens are approximately 300 μm and 10 μm, respectively. The grain refinement by SEBM is assumed to be related to the rapid solidification cooling rates during the SEBM process. The fraction of the FCC phase in the bottom part (29.7%) is significantly higher than that of the top part (7.1%). The bottom part of the SEBM specimens was maintained at 950 °C for about 13 h during fabrication which promoted the formation of the precipitation of the FCC phase at grain boundaries. The hardness of the FCC phase is much lower than that of the B2/BCC mixture. The finer grains and larger amount of FCC phases of the SEBM specimens contribute to better ductility and lower yield strength compared to the cast specimens. HEA coatings on the aluminum substrate have been done for the improvement in corrosion resistance by laser additive manufacturing, using Al, Fe, Co, Cr, and Ni elemental powders. Further, Al*x*CoCrFeNi ($x = 0.3$, 0.6, and 0.85) HEA claddings produced by coaxial direct laser deposition (DLD) on a 253MA austenitic steel substrate using a mixture of blended elemental powders. The variation of Al content (i.e., $x = 0.3$, 0.6 and 0.85) resulted in different phase constitutions in the HEA claddings, principally FCC phase for $x = 0.3$, mixed FCC + BCC phases for $x = 0.6$ and BCC phase for $x = 0.85$. The cladding cross section was composed of coarse columnar grains, mostly enclosed by high angle grain boundaries (HAGBs), and the mean grain size was ~40.9 ± 3.7, 38.2 ± 3.1 (FCC grain size) and 32.5 ± 1.5 μm for the Al0.3, Al0.6 and Al0.85 HEA claddings, respectively. Co1.5CrFeNi1.5Ti0.5Mo0.1 HEA fabricated and studied the effect of solution treatment on the tensile properties and corrosion resistance. The SEBM specimens exhibited superior tensile properties to those of the corresponding casting specimen. The solution-treated SEBM specimens exhibited both high strength and high pitting potential, which in combination were superior to the conventional alloys used in severe corrosion environments. The strength of the SEBM specimens air-cooled by solution treatment was higher than that of the SEBM specimens water-quenched by solution treatment. On the other hand, the elongation was lower than that of the SEBM specimens water-quenched by solution treatment [77].

20.7 Processability in AM

A wide range of materials and alloys are considered as "processable" for AM. The term "processability," however, contains several parameters, which affect the additive build processes, and the final product properties, such as shape and size of particles, materials feed rate, deposition speed, nozzle angle, nozzle distance, chamber pressure, and bed temperature. As a powder-bed AM process, the shape and size of particles are two crucial factors for the printing process and performance of the resulting 3DP part. Sometimes, finer powders are required to achieve finer microscale features, better surface finishes, thinner layers and to promote printability, especially for powder bed printing. However, fine powders may exhibit a significant aggregation which is principally determined by van der Waals forces, it was found that the smaller sized particles of ceramic powders, with dimensions smaller than 5 μm, are prone to agglomerate, thus forming defects and craters on the surface of the powder during layer preparation. However, large particles (>90 μm) reduce the maximum layer packing density available in SLM. Furthermore, it has been realized that the particle size distribution of the powders has an important effect on their flowability and thermal behavior as well as the mechanical properties of the manufactured product. Therefore the forming quality in AM can be determined by particle size, size distribution, surface roughness, and shape. After the optimization of the building block filling materials, the choice of energy input from the source is the most important parameter to get a better result. The energy of the source depends on the size of the spot, scanning speed, angle of the impingement from the source, etc. [78].

20.8 4D Printing

4D printing is a targeted evolution of the 3D printed structure and is based on 3D printing technology, but requires additional stimulus and stimulus-responsive materials to change shape over time. The fourth dimension is described here as the transformation over time, emphasizing that printed structures are no longer simply static, dead objects; rather, they are programmably active and can transform independently [79]. 4D printing pioneered by Skylar Tibbits at the Self-Assembly Lab of MIT in collaboration with Stratasys, Inc., uses time as the fourth dimension for 3D fabricated "smart" structures. Based on certain interaction mechanisms between the stimulus and smart materials, as well as the appropriate design of multi-material structures from mathematical modeling, 4D printed structures evolve as a function of time and exhibit intelligent behavior. Hence, 4D printing can be defined as the additive manufacturing of objects able to self-transform, as a function of time, when are exposed to a predetermined stimulus such as temperature, pressure, ultraviolet light, magnetic field, and P^H. The product is capable of achieving self-assembly, multi-functionality, and self-repair. It is time-dependent, printer-independent, and predictable. Research into 4D printing has attracted unprecedented interest since 2013 when the idea was first introduced. This technique offers a streamlined path from idea to reality with performance-driven functionality built directly into the materials. Adaptive and dynamic responses for structures and products are now plausible without adding time, cost, or extra components to make systems "smarter" [80].

This stimuli-responsive behavior, programmable by means of CAD and the appropriate choice of printing materials, significantly expands the range of conventional smart materials, shape memory alloys, and multifunctional materials systems. 4D printing has been used to enable active shape in which a flat sheet automatically folds into complex shapes. The original flat plate shape can afterward be thermally recovered. Furthermore, 4D printing of glassy shape memory polymer fibers in an elastomeric matrix affords contoured shapes with spatially varying curvature, folded shapes, as well as twisted, bent, and coiled strips. 4D printed multi-material grippers that can grab and release objects based on temperature have been shown in Fig. 20.17 [81].

They utilized a single strand to make a rigid wireframe 3D cube. At each of the joints, two rigid discs were printed that acted as angle limiters, which when folded and touching one another forced the strand to stop at 90° angles. This cube is the first generation of a fractal Hilbert curve, where a single line is drawn through all eight points of the cube without overlapping or intersecting. Again, they also generated surface transformations self-folded Hilbert cube. In this case, a two-dimensional flat plane was printed, with both rigid and active materials. This flat plane represents the six unfolded surfaces of a cube. At each of the joints, a long strip of active and rigid materials was printed that describes a 90° angle limiter that stops the surface from folding when it reaches the final-state condition. When submerged in water, the surface folds into a closed-surface cube with filleted edges. A wide range of other 1D, 2D, and 3D transformations are also possible including self-folding smart structures, self-healing structures where holes close after encountering water, and other global geometric reconfigurations. In a recent advance, ultrafast 4D printing (<30 s) of multidimensional responsive acrylic polymers such as hydrogels and shape memory polymers have been achieved. MIT and Stratasys have developed a variety of physical prototypes, each demonstrating transformation from one-dimensional and two-dimensional flexible shapes into rigid structures.

The main governing factors to be counted during additive manufacturing of 4D materials are the design to be print, shape-shifting behavior (swelling ratio in polymer and coefficient of expansion/contraction in metal), shape triggering limit, interaction mechanism, and mathematical modeling (Fig. 20.18). In some cases, the desired shape of a 4D printed structure is not directly achieved by simply exposing the smart materials to the stimulus. The stimulus needs to be

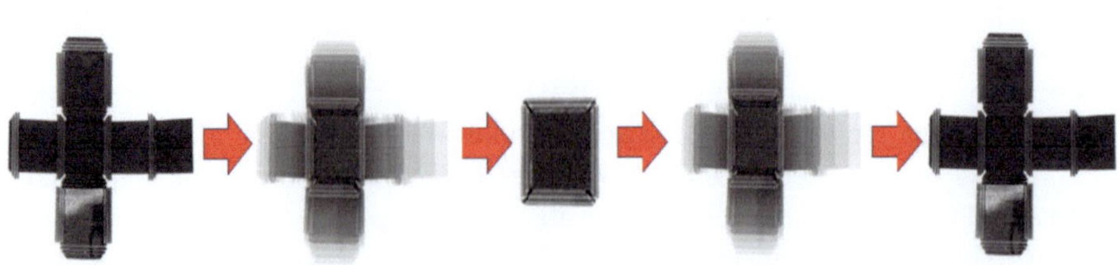

Fig. 20.17 4D printed structure actuation

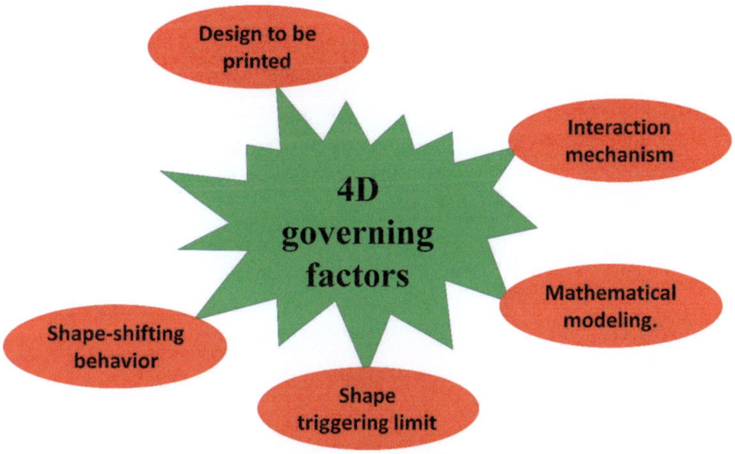

Fig. 20.18 Governing factors for 4D manufactured materials

applied in a certain sequence for an appropriate amount of time. Modeling is necessary for 4D printing in order to design the material distribution and structure needed to achieve the desired change in shape, property, or functionality. Theoretical and numerical models need to be developed to establish the connections between four core elements: material structure, desired final shape, material properties, and stimulus properties [82]. Chapter 1 has been given on all shape responsive materials, which are the target of future additive manufacturing products.

20.9 5D Printing

5D printing is a new branch of additive manufacturing. About 2 years ago, the first talks around 5D printing started to circulate in American Universities but one company had actually implemented this new technology partially, namely Mitsubishi Electric Research Labs (MERL) [83]. A 5D printer allows for objects to not be printed from one point upwards but from five axes. Hence, where the number five in 5D comes from. In this technology, the print head and the printable object have five degrees of freedom. The print head moves around from 5 different angles and instead of the flat layer, it produces curved layers while printing. These movements allow for the printer head to come in from many different angles, otherwise not achieved with 3D printing [84]. These new angles result in the printing head being able to follow the path of the object's shape and outline. By not having to follow a straight path on a static plateau—and using the shape of the object instead, the printed parts can be created with curved layers instead of flat layers. These curved layers allow for stronger parts that have a complex design to be printed [85]. Instead of three axes used in 3D printing, 5D printing technologies use a five-axis printing technique which produces objects in multiple dimensions. In this five-axis printing, the print bed can move back and forth on two axes besides of X, Y, and Z axis of the 3D printing technologies. Thus this technology is highly capable of producing stronger products in comparison to parts made through 3D printing. A

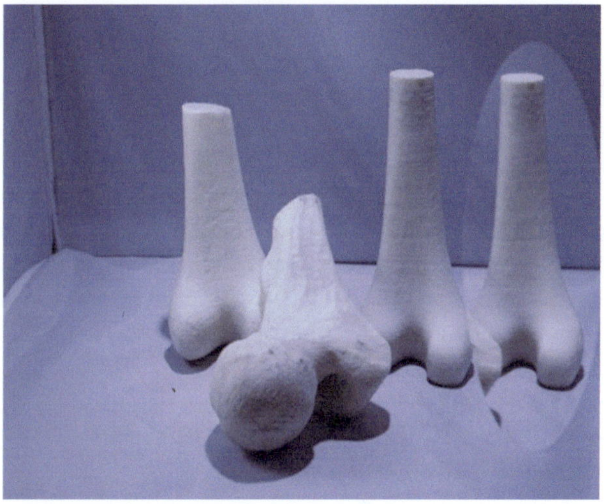

Fig. 20.19 5D printed bone implant

5D printed model provides the potential to fabricate artificial bone for surgery. Because human bones are not flat and having a curved surface, so there is a requirement to manufacture artificial bones with 5D printing to provide excellent strength to these bone implants. This technology has great potential to fulfill this primary requirement (Fig. 20.19). 3D printing is not very good in the manufacturing of complex curved Orthopaedic implants as it uses flat layers. Results of the test conducted show that 5D printed objects are three to five times stronger than a 3D printed object. There is the force of material during curved layers that make printable objects stronger. Therefore, there is a lesser requirement of raw material as compared to 3D printing for making implants of the same strength.

20.10 Differences Between 3D, 4D, 5D Printing, and Other

The main difference between 3D printing and 5D printing is that 5D printing creates a stronger part with a curved layer whereas 3D printing creates a part with a flat surface. Rest both the processes use the same set of technologies such as

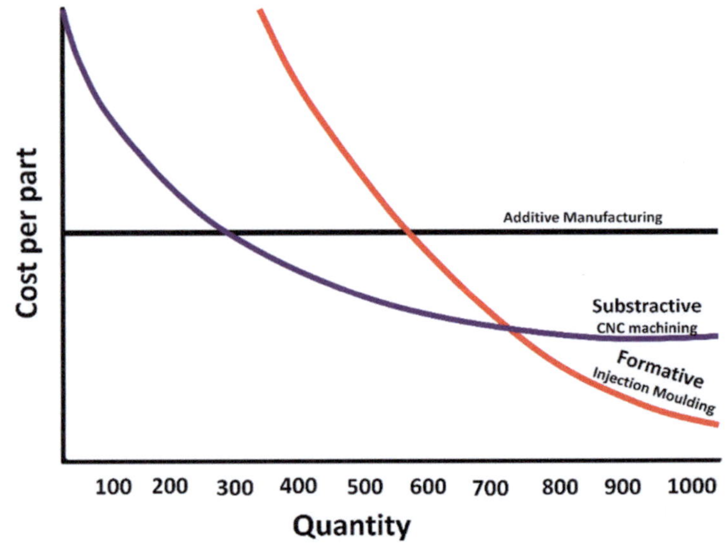

Fig. 20.20 Comparative cost per quantity of different manufacturing processes

the input of 3D CAD file, 3D scanner, and 3D printing material. However, 4D printing is different from these two technologies. It uses different kind of material which is programmable and can change its shape and function concerning time and temperature. These materials are called smart materials. Currently, research is going on to develop smart materials by 5D technology and can alternatively called 6D printing (5D printing + time of sensing).

20.11 Advantages and Limitations of Additive Manufacturing

Additive manufacturing techniques and materials tend to be more expensive than traditional counterparts for large production runs, and thus they are most competitive for applications where flexibility and fast product development cycles are needed. Examples markets include those for customized parts and small production runs down to one (i.e., prototypes or fully custom products). Presently, low build speeds and technical limitations tend to limit the size of building the materials (small parts, such as one cubic foot or below, are needed). High materials costs, as shown in Fig. 20.20, additive manufacturing competes well in high-value markets. The creation of some complex shapes and geometric features is difficult, if not impossible, to achieve using traditional methods, additive manufacturing is more competitive where part complexity is desirable because any part that can be modeled digitally can be built with little or no additional cost related to complexity. It is also possible to use AM to consolidate several parts without the need for assembly. Any time a part, or a batch of parts, can be produced without tooling, substantial savings are possible. This usually occurs in situations where the production volume, part size, and part complexity combine to give AM an advantage over tool production on a cost-per-part basis [86]. However, significant improvements in speed, size, and cost, without the need for an environmental chamber, are likely to open up new applications in other.

20.12 Applications of AM

Additive manufacturing is a fast-moving industry that is currently generating significant attention in the market. With the introduction of AM machines selling for under $2000, it is becoming increasingly possible for individuals or groups of makers to purchase and operate additive manufacturing machines. Here various parts used in different major industries have been described below.

20.12.1 Aero Industries

Within the aerospace industry, AM can help significantly reduce the high buy-to-fly ratios of the cast, forged, and machined components. In these cases, the causes of higher costs are time, highly skilled labor (e.g., mold making), and high levels of scrapped material. AM can reduce and sometimes eliminate the need for tooling, thus helping to accelerate the development cycle for new parts. Managing spare parts for military weapon systems and space missions is a complicated, time-consuming, and expensive task involving large inventories [87]. Many military systems, including aircraft, are increasingly being used beyond their designed life expectancy, resulting in parts that are in danger of failure. Given that many of these parts are out of production, remaking them using traditional methods of manufacturing can often take multiple years, not including additional time for qualification and delivery. These problems not only

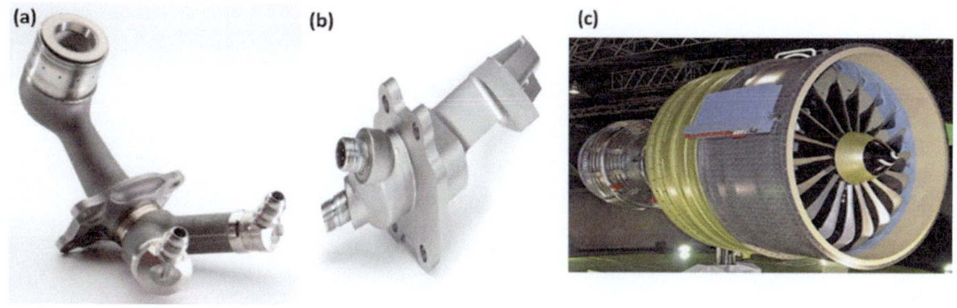

Fig. 20.21 3D-printed (**a**) Fuel nozzle, (**b**) Sensor housing, and (**c**) LEAP A1 engine

Fig. 20.22 Engine ducting systems

require billions of dollars to support. These requirements are fulfilled by 3D printing in the aerospace industry [88].

The 3D printing of both jet engine prototypes and end-use parts is already having a significant impact on development and production. The GE90 series of high-bypass turbofan aircraft engines first entered service in 1995 [89]. Now, the next-generation GE9X from GE Aviation will use 19 3D-printed fuel nozzles to help power the next generation of wide-body aircraft like the Boeing 777X (Fig. 20.21a) [90]. In April 2015, GE Aviation received its first FAA clearance to use a 3D-printed part in a commercial jet engine. The sensor housing for a compressor inlet temperature sensor is produced by a selective laser sintering system (Fig. 20.21b). The GE9X is the largest and most powerful jet engine ever manufactured—the front fan alone measures 11 ft. in diameter. While it is approximately 10% more fuel-efficient than its predecessor, it still generates 100,000 pounds of thrust at takeoff. In April 2016, Airbus took delivery of the first two LEAP-1A engines for its next-generation A320 passenger jet (Fig. 20.21c). GE Aviation and France's Safran are equal partners in the engine's manufacturer, CFM International [91]. GE contracted with Boeing to use the part in more than 400 GE90-94B engines to be used on Boeing 777 jets. Jet engine combustors are notoriously difficult to manufacture using traditional means [92]. Using an AM process with cobalt–chrome alloy powder as the build material, high-quality components are fabricated that protect the highly sensitive electronics in the temperature sensor from surging airflows and significant icing inside the engine. AM processes are also being used to fabricate other aircraft parts, including those used in ducting systems (Fig. 20.22). Additive manufacturing also produces sophisticated tools for the aerospace industry [93]. In November 2017, the Tubesat-POD (TuPOD) satellite completed its mission. The tube-shaped deployment satellite was the first 3D-printed satellite launched from the ISS. AM experiments on the International Space Station (ISS) test the feasibility of 3D printing in space. In 2014, astronauts aboard the International Space Station printed their first plastic part on a 3D printer. Objects printed aboard the ISS were returned to earth to determine if microgravity impacted mechanical properties or other essential characteristics in any material way. Print-on-demand technology could be pivotal in carrying out plans to travel to Mars [94].

Some major parts of space vehicles are described below:

Plane Seat: A lighter plane seat has been 3D printed by Andreas Bastian, an engineer at Autodesk which weighs 40% less (766 gm) than a conventional plane seat (Fig. 20.23a). He created the ceramic mold after creating the plastic mold using 3D printing to obtain the final piece [95].

Drones: Stratasys collaborated with Aurora Flight Sciences to create an advanced unmanned series of aerial vehicles with jet propulsion in the year 2015 which can fly faster than 150 miles/h and is called the UAV. More than 75% of the vehicle's parts have been 3D printed manufactured through fused deposition modeling technique (Fig. 20.23b). US Army

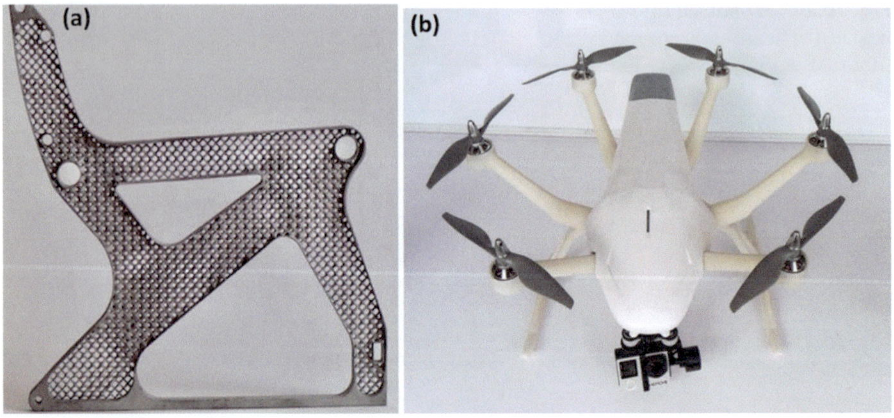

Fig. 20.23 (**a**) Aeroplane seat and (**b**) Drone by AM

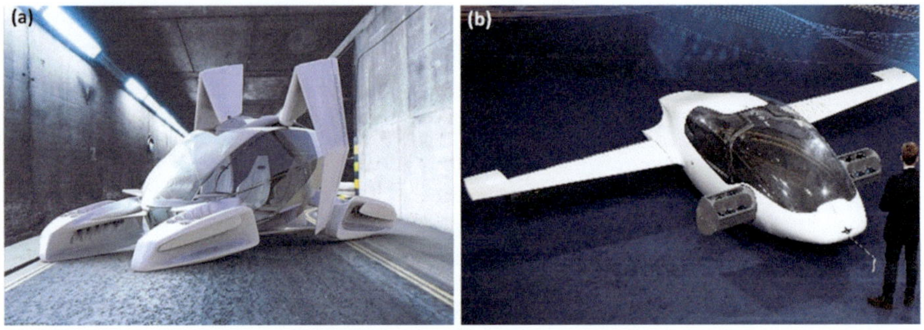

Fig. 20.24 (**a**) Closed form, and (**b**) Open form

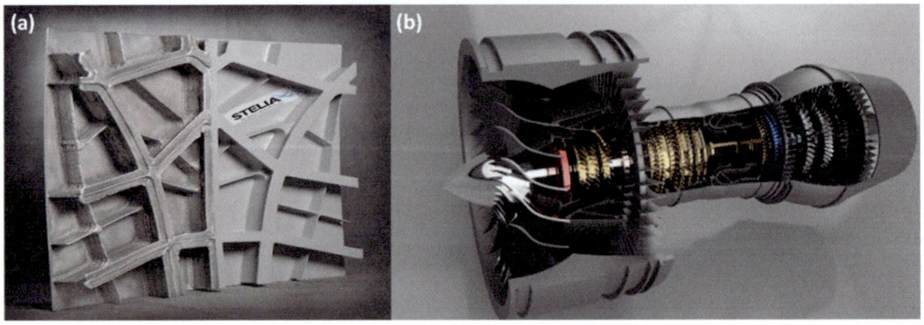

Fig. 20.25 (**a**) Fuselage Panel of STELIA, (**b**) Pratt & Whitney turbofan Engines

in collaboration with researchers at MIT designed "Perdix" drones and tested them successfully. The US Army have previously created concrete barracks using 3D printing. 103 AM drones perform collectively as one brain and don't act individually in their work position. In order to avoid crashing, they use sensors to maintain a safe flight distance. They have the ability to jam enemy radars [96].

Hoversurf's Flying Car: Hoversurf is known for creating unique hovercrafts, Scorpion-3 being their best development so far which can fly one person. Hoversurf has announced to launch their new car Formula by next year whose parts are all 3d printed. It can attain a speed up to 300 km/h and will carry 5 passengers. The best feature of Formula is it can be parked in a normal parking space as its wings are folded up. It is an electrically driven car [97]. Both closed form and open form is shown in Fig. 20.24.

Safran Helicopter Engines: Safran Helicopters recently launched a new range of helicopter engines. The Anteo-1K engines have 3D printed parts, including parts inside the combustion chamber. Additive manufacturing has enabled Safran to reduce production costs without compromising engine performance. These 3D printed engines created are almost 30% more powerful than those previously manufactured. This increased performance helps helicopters in departments such as search and rescue missions [98].

Fuselage Panel of STELIA: STELIA aerospace has recently created their first 3D printed reinforced fuselage panel (Fig. 20.25a). They carried out the project using Wire and Arc Additive Manufacturing (WAAM) technology. Additive manufacturing makes it very easy and flexible to design the stiffeners of the fuselage panels, offering more design flexibilities [99].

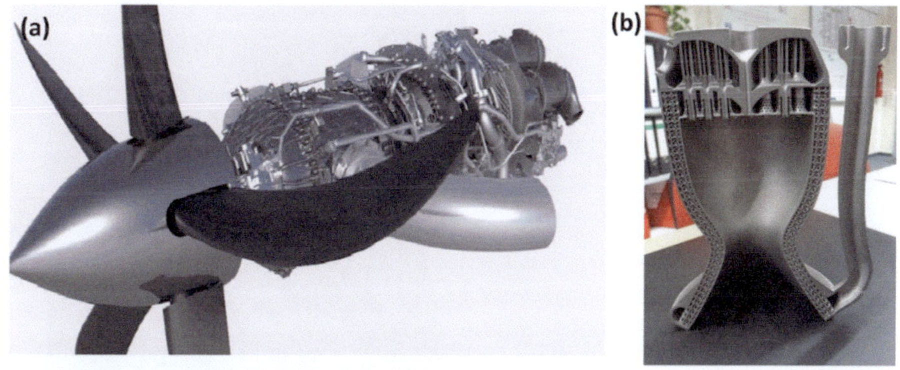

Fig. 20.26 3D-printed (**a**) rocket engine and (**b**) propulsion system of Falcon heavy rockets

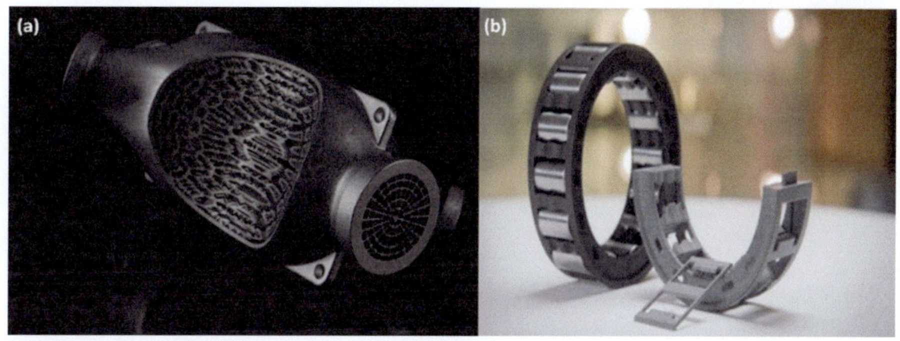

Fig. 20.27 (**a**) Heat Exchangers, (**b**) Production of bearings

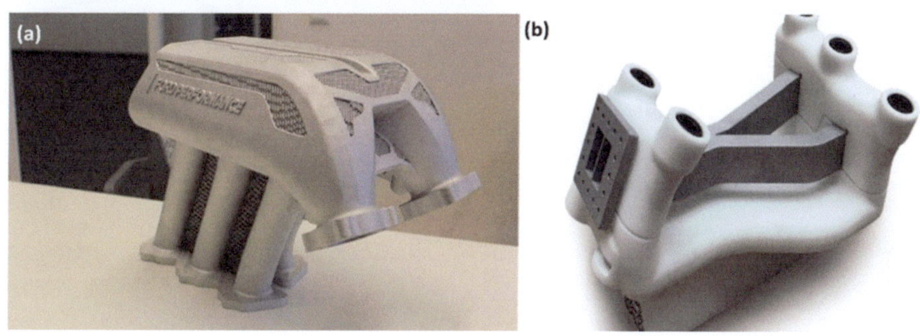

Fig. 20.28 (**a**) Intake manifold, (**b**) Jigs & fixtures

Pratt & Whitney Turbofan Engines: Almost 12 parts of a Pratt & Whitney engine have been created using AM, engines that now equip Bombardier aircraft and carriers (Fig. 20.25b). These are mainly fasteners and injection nozzles 3D printed from Ti and Ni. Pratt & Whitney has saved almost 15 months over the entire design process and the final weight of the part has come out 50% less than the conventional. The engine manufacturer has used electron beam melting and direct metal laser sintering technologies [100].

Raptor System: The Raptor system is SpaceX's most powerful engine and uses many 3D-printed parts, including the turbopump. One of the Raptor Engines tested by SpaceX last year was made up of over 40% 3D printed parts by mass. This propulsion system is used in the upper stage of the Falcon 9 and Falcon Heavy rockets, and will likely be used in SpaceX's future missions to Mars [101]. Figure 20.26 shows the 3D printed rocket engine engines.

20.12.2 Automobile

Additive manufacturing in the automotive industry is a competitive advantage and many of the major automakers are keeping their key advances under wraps in order to maintain a competitive edge. Some key examples of parts they are adopting AM are listed below.

Heat Exchangers: Heat exchangers consist of a series of tubes, which contain fluid that must be either heated or cooled. A second fluid runs over the tubes that are being heated or cooled so that it can either provide the heat or absorb the heat required. A set of tubes is called the tube bundle and can be made up of several types of tubes: plain, longitudinally finned, etc. The entire heat exchanger setup can be fabricated by the help of additive manufacturing (Fig. 20.27a) [102].

Bearings: A bearing is a machine element that constrains relative motion to only the desired motion, and reduces

Fig. 20.29 3D print tire molds

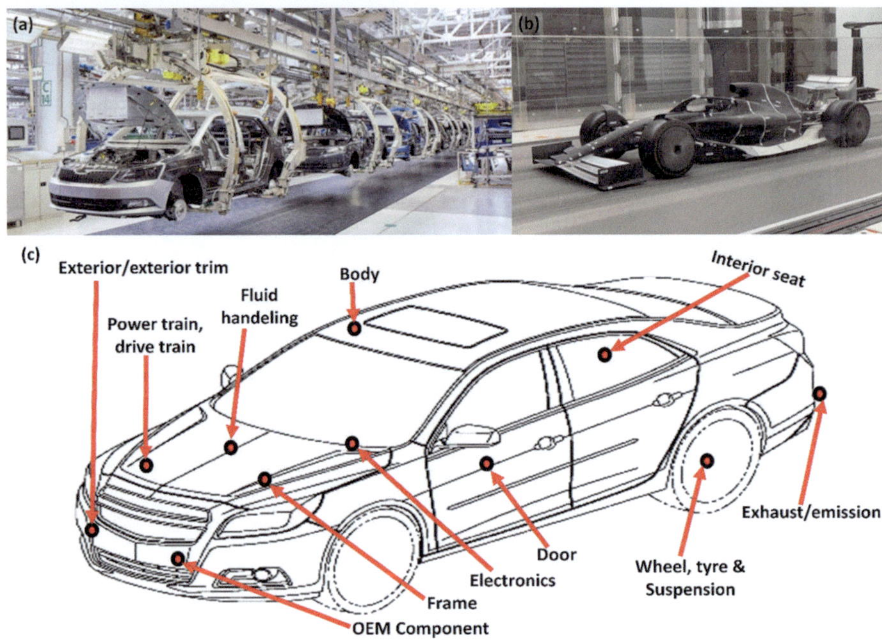

Fig. 20.30 (**a**) Automobile body frame, (**b**) Total part of FIA, and (**c**) AM parts used in the automobile parts

friction between moving parts. The design of the bearing may, for example, provide for free linear movement of the moving part or for free rotation around a fixed axis; or, it may prevent a motion by controlling the vectors of normal forces that bear on the moving parts. Most bearings facilitate the desired motion by minimizing friction [103]. Figure 20.27 is showing the 3D printed bearings.

Intake Manifold: The intake manifold supplies the air–fuel mixture to the cylinders. 3D printed intake manifold has given in Fig. 20.28a. Ford motors used AM to manufacture the manifold which was used in motorsport competition. Despite not being a production part this proves that the technology is viable for engine parts that are typically exposed to high stress and temperature [72].

Jigs and Fixtures: This is an often-overlooked facet of AM in the automotive industry as the parts made specifically for the production environment are not reported as they provide a competitive advantage in terms of improving production efficiency. However, jigs and tooling are often manufactured in the house with AM machines as this is a low-risk way of trying out new tooling and techniques without investing in long lead times and overly complex tooling (Fig. 20.28b) [104].

Michelin Tire Molds: Michelin has leveraged additive manufacturing in the automotive industry to streamline the manufacture of their tire molds. The technology has enabled the design and manufacture of tires with intricate tread patterns that change their behavior as the tire wears in order to maintain optimal performance [105] (Fig. 20.29).

Formula 1 and FIA use additive manufacturing to build 50% scaled models to test their 2021 Car Regulations [106]. In this earlier twentieth century BMW, Ford, Volvo, GM, Porsche already using AM technique to manufacture automobile parts (Fig. 20.30a, b). Figure 20.30a shows the automobile body frame and the total part of FIA. The application of every part given in Fig. 20.30c in the automobile has explained in Table 20.2.

Table 20.2 AM processes used for different part in automobile industries

AM parts	AM technology	Materials
Powertrain, drivetrain (engine component)	SLM, EBM	Al-, Ti- alloys
Frame, body, door (body panels)	SLM	Al alloys
Fluid handling (pump, valve)	SLM, EBM	Al alloys
Wheels, tires, and suspension (hubcaps, tires, suspension springs)	SLS, inkjet, SLM	Polymer, Al alloys
Interior and seating (dashboards, seat frame)	SLS, Stereolithography	Polymer
Exterior/exterior trim (bumpers, wind breaker)	SLM	Polymer
Electronics (embedded components such as sensor, single part control panel)	SLM	Polymer
OEM components (body in white)	SLM, EBM	Al alloys, steel
Exhaust/emissions (cooling vent)	SLM	Al alloys
Manufacturing processes (prototyping, customized tooling, investment casting)	FDM, inkjet, SLS, SLM	Polymer, wax, hot worksheet

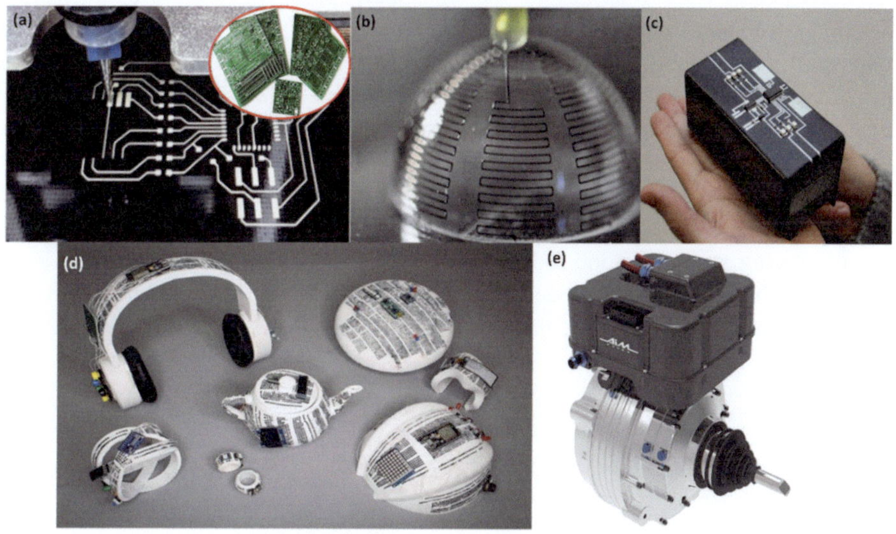

Fig. 20.31 (**a**) Printed circuit, (**b**) Complex-design based printing, (**c**) Electronics gadgets, (**d**) Control unit embedded battery, and (**e**) 3D-printed electric motor

20.12.3 Electrical Industries

The production of conformal electronics with additive manufacturing shows potential. The types of components that might be printed to conform to the shape of a product include energy storage devices (batteries), electronic sensors (e.g., RFIDs, strain gauges, and thermocouples), and electronic controls. The materials for these electronics would be deposited within the body of the housing, enclosure, or another section of a part as it is being manufactured. While no commercially available additive manufacturing process currently prints conformal electronics, the concept was demonstrated in 2005 in a joint project between Sandia National Laboratory and the University of Texas at El Paso (UTEP). Cornell University has created parts that embed electronics, such as conductors and LEDs, using its Fab@Home system. The system was also used to produce a zinc battery and a polymer actuator. The battery powered the actuator, causing it to move. In January 2011, UTEP opened the Structural and Printed Electronics Center, a facility that will conduct research to combine additive manufacturing and printed electronics technologies [107]. Printed electronics are being used in many commercial applications today with the main applications being the printing of PV solar bus bars, glucose test strips, force sensors, touch screen electrodes, membrane circuits, and heating elements. Of the different printing processes available, screen printing dominates the commercial printed electronics-based products today and it continues to improve in terms of printed line width size due to demand from the PV and touch screen industries to have thinner printed lines. In touch screens, thinner edge electrodes translate to thinner edge bezels around displays, something desired by consumers [108]. Figure 20.31 has shown the 3D printed circuit, complex-designed printing, electronics gadgets, control unit embedded battery, and electric motor.

20.12.4 Biomedical Industries

Additive manufacturing printing technologies have vast applications in the medical and health care industries. They are transforming the practice of medicine through the possibilities of making rapid prototypes and very high-quality bone transplants and models of damaged bone of the patients for analysis. Additive manufacturing printing methods permit to scan and build a physical model of defective bones from patients and give doctors a better idea of what to expect and plan better the procedure, this will save cost and time and help achieve a better result [109]. Bone transplants now can be done by printing them and additive manufacturing methods make it possible to have a transplant that is practically identical to the original. Because of the limitless form or shape of what could be built, doctors have the option to create a porous-controlled material that will permit osteoconductivity or to create a precise metal transplant identical to the original depending on the bone to be replaced [110]. Characteristics of the transplants such as density, pore shape and size, and pore interconnectivity are important parameters that will manipulate tissue in the growth and mechanical properties of the implant bone. The mechanical strength of these implants is three to five times higher than others produced by conventional processes and the possibility of inflammation caused by micro debris that breaks during the procedure is reduced. Additive manufacturing is a very good tool for dentists because they can easily build a plaster model of a patient's mouth or replace the teeth, which have a unique form with the process like stereolithography, selective laser sintering, and electron beam melting [73]. According to PC Magazine, an 83-year-old Belgian woman became the first-ever person to receive a transplant jaw bone tailor-made for her face using a 3D printer and the surgery time and recovery were a lot less than other patients that received the same procedure. The shapes of bones differ too much between each person and additive manufacturing printing produces transplants that fit better, and are easier to insert and secure, reducing the time for the procedure and produce a better cosmetic result [111].

Stereolithography is being used to manufacture prosthetic sockets. By using this technology to ensure that the form of the socket adapts better to the patient while being more cost-effective than hand or machined methods. Not only hard parts like bones can be produced, but also it is possible to print cells in a 3D array that with the possibility of printing complex shapes and arrays human tissue can be printed. This technology will help patients that have lost tissue in accidents or from other reasons to recover faster and with better cosmetic results [112]. In addition, 3D cell printing technologies offer the possibility of printing artificial blood vessels that can be used in coronary bypass surgery or any other blood vessel procedure or diseases, like cardiovascular defects and medical therapy. Research in this area, also called bioprinting

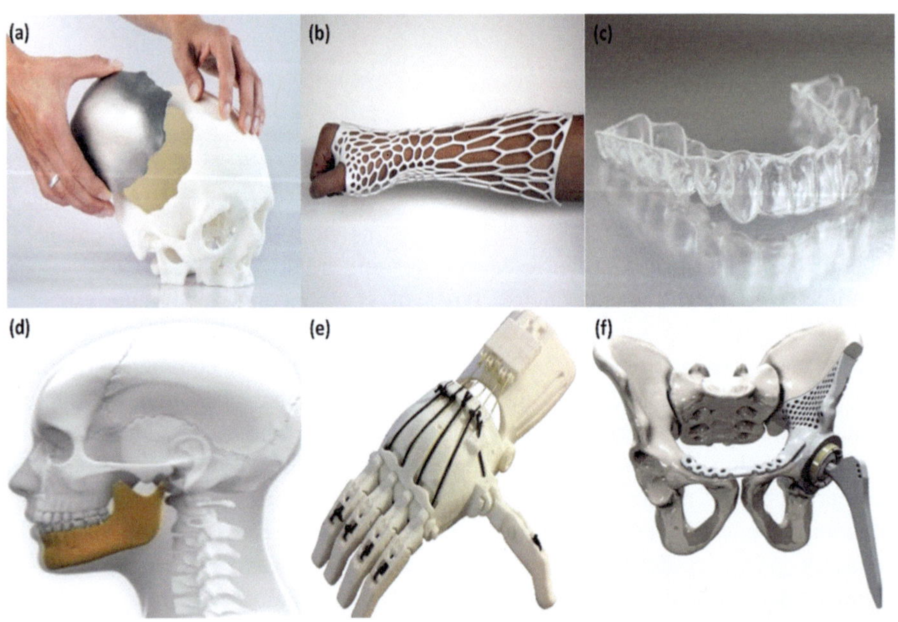

Fig. 20.32 3D-printed (**a**) skull bones, (**b**) hand supporter, (**c**) dental cover, (**d**) lower jaw, (**e**) artificial hand, (**f**) heap implant respectively

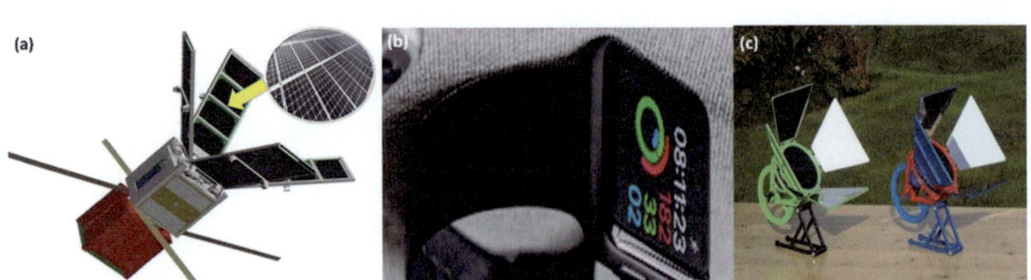

Fig. 20.33 (**a**) Solar cell in satellite, (**b**) digital watch, and (**c**) portable solar cell

organs, will eventually lead to printed organs. Cell printing is not limited to print human tissue; it is also used in the field of molecular electronics. The precision of high-resolution processes like nanolithography and photolithography permits the creation of biochips and biosensors [113]. Figure 20.32a–f is showing cortex 3D-printed skull bones, hand supporter, dental cover, lower jaw, an artificial hand, heap biometal, respectively.

20.12.5 Energy Harvesting Industries

AM processes use to improving the manufacturing of fuel cells. Figure 20.33 shows various 3D-assisted parts in the solar cell in satellite, digital watch, and portable solar cell. Additive manufacturing technologies can be used in processes that require a very precise thin film of a certain material. In the manufacture of polymer electrolyte membrane fuel cells (PEMFCs), it is necessary to precisely deposit a very thin layer of platinum, needed for the oxidation and reduction reactions, with high utilization efficiency of the platinum [114]. Energy storage is an integral part of mobile electronics and there is a continued demand for ever smaller yet more powerful batteries. Over the years, tremendous efforts have been put into exploring new electrode materials, electrolytes, cell structures, and novel fabrication approaches with the goal of improving the electrochemical performance of batteries, reducing manufacturing cost, and expanding their application. 3D-printing technologies can fabricate an entire device, including the battery and structural and electronic components, in almost any shape [2]. Figure 20.34 is showing various 3D-printed battery systems in microbattery, lithium-ion mobile batteries, piezoelectric materials for smart energy harvesting, and 3D-printed tree charge smartphones.

20.12.6 Other Industries

The first attempts to use AM technologies for the processing of food were reported in 2001 in a patent claiming the additive fabrication of a 3D-designed cake. No physical prototype was reported, and it was not until a few years,

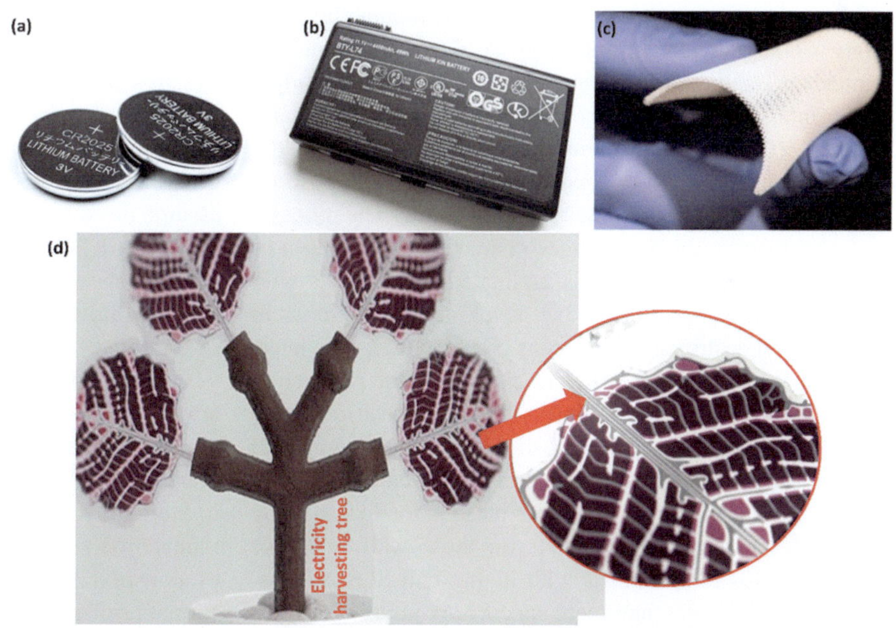

Fig. 20.34 (**a**) Micro battery, (**b**) lithium-ion mobile batteries, (**c**) piezoelectric materials for smart energy harvesting, and (**d**) 3D-printed tree charges smartphones with solar power

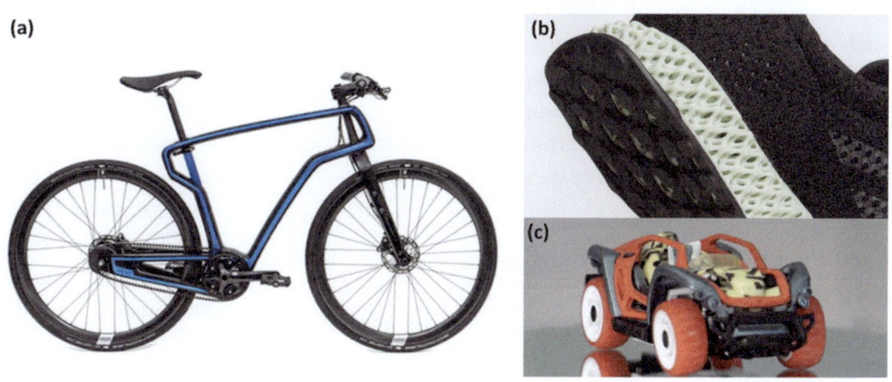

Fig. 20.35 (**a**) 3D printed composite bicycle at 3D Printing USA, (**b**) 3D Print footwear, (**c**) 3D printed toy cars

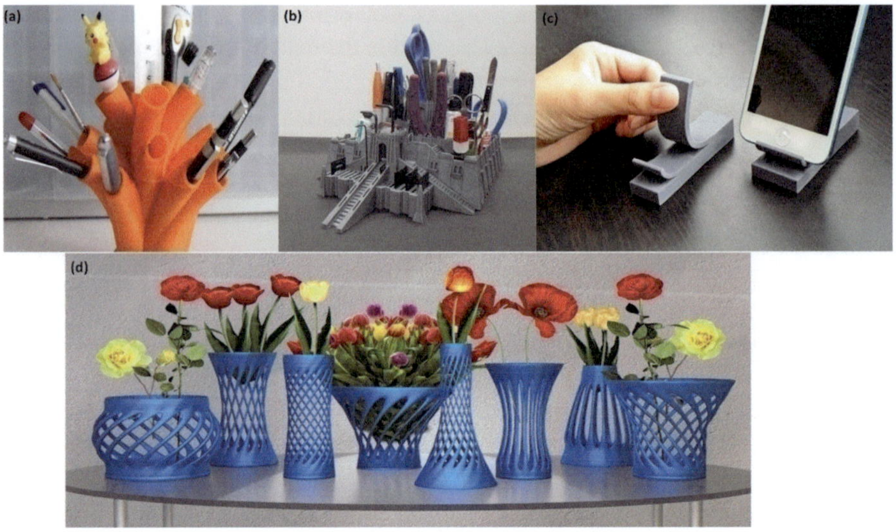

Fig. 20.36 3D printed household items (**a**, **b**) Designed pen stand with a showpiece, (**c**) Mobile stand, and (**d**) Complex designed flower vase

with the introduction of affordable open-source 3D printers, that further developments in AM of food came [115]. One of the first experimental implementations was based on selective sintering of sugar reported by the Candy Fab project in 2008. The choice of sugar as a building material was not based on edibility alone, but rather on the low price, availability, and safety of sugar. The introduction of the Fab@home 3D printer, one of the most influential open-source, low-budget DIY printers and based on the deposition of paste-like material, led to further and more purposeful activities in the direction of printing food materials such as cake frosting, processed cheese, peanut butter, and chocolate. The process' low investment cost, open-source character, and freedom of using nonproprietary materials facilitated experiments with novel formulations similar to the development of recipes in conventional gastronomy. In recent years, an increased interest in the application of AM technologies for the production of food for industrial and domestic use has been observed [44]. Figure 20.35 is showing the 3D printed composite bicycle, footwear, and various children's toy cars.

PLA and ABS are the principal plastics for FDM, while polyamides or metals are options for those who outsource their print jobs to specialists with higher end SLS printers. Large format printers, and more specifically industrial robot-arm extruders, have been used to print custom-designed furniture from recycled ABS [116]. The advantage with such extruders is that no filament is required and the plastic can be directly dropped into a hopper and passed through a large (4–10 mm diameter) extruder. This level of feature resolution is acceptable for furniture because it also means that a single customized piece can be printed in a few hours instead of a few days.

Clothing design is an exceptional example of functional art and one where 3D printing is being adopted with fervor. In 2013, Michael Schmidt and Francis Bitonti used CAD to contour a dress for the model Dita Von Teese. The dress was printed in 17 separate sections from polyamide using SLS and afterward painted and adorned with crystals. While many other haute dressmaking fashion designers have embraced 3D printing, the technology has been slow to transfer to a wider audience. The prime hindrances are print time and cost as well as the mechanical rigidity of plastics for AM (ABS, PLA, and polyamide). The company Nervous Systems addresses all of these points by using computer simulations to fold a 3D dress image into a smaller volume for printing. The dress (termed kinematic) consists of thousands of interlocking polygons, which after printing assume the shape of a full-size dress but also allow flexibility [44]. Figure 20.36 is showing the 3D printed household Items such as showpiece pen stand, mobile stand, and complex designed flower vase.

Architects have been using CAD on a daily basis for many years, and this has eased the adoption of 3D printing. The advantage of 3D printing is that it is faster than making models by hand and allows clients to better visualize the architect's design [117]. This latter point is also important to the builder because 3D printed models can improve communication and thus reduce construction delays and the risk of potential structural failures. The development of larger format printers, additive manufacturing is being used increasingly not just for design but for the actual construction of buildings and houses [118]. AM construction techniques can be divided into two general forms: extrusion methods such as contour crafting and concrete printing, and powder binder methods. Contour crafting is based on a large extruder head (15 mm nozzle) with trowels supported by a gantry or crane, which disperses concrete in a manner similar to FDM. Concrete printing is similar but lacks trowels and differs in the handling of overhangs. D-shape is the most common powder binder AM method for construction and relies on a chlorinated liquid to selectively bind a dry concrete powder [119].

20.13 Summary

This chapter discussed on origin, opportunities, challenges, and future look of additive manufacturing processes. The seven primary types of additive manufacturing processes and their difficulties are well discussed along with their most used industrial materials. The advantages and disadvantages of additive manufacturing given here will give a new direction for future researchers to come over in new possibilities in the production process. Most AM adopted major industrial sectors, such as aerospace, automobile, electrical, biomedical, energy harvesting industries, have been discussed here.

References

1. Petrovic, V., Gonzalez, J.V.H., Ferrando, O.J., Gordillo, J.D., Puchades, J.R.B., Griñan, L.P.: Additive layered manufacturing: sectors of industrial application shown through case studies. Int. J. Prod. Res. **49**(4), 1061–1079 (2011). https://doi.org/10.1080/00207540903479786
2. Behera, A.: 3D Print Battery. In: Nanobatteries and nanogenarators. Elsevier Publisher, Amsterdam (2020) (In Press). ISBN: ors/song/978-0-12-821548-7
3. Behera, A.: Processes and application in additive manufacturing: practices in aerospace, automobile, medical and electronic industries. In: Additive Manufacturing applications for metals and composites, p. 23. IGI Publisher, Hershey, PA (2020). https://doi.org/10.4018/978-1-7998-4054-1.ch002
4. Kumbhar, N.N., Mulay, A.V.: Post processing methods used to improve surface finish of products which are manufactured by additive Manufacturing technologies: A review. J. Inst. Eng. India Ser. C. **99**, 481–487 (2018). https://doi.org/10.1007/s40032-016-0340-z
5. Huang, Y., Leu, M.C., Mazumder, J., Donmez, A.: Additive manufacturing: current state, future potential, gaps and needs, and recommendations. ASME. J. Manuf. Sci. Eng. **137**(1), 014001 (2015). https://doi.org/10.1115/1.4028725
6. Nannan, G.U.O., Ming, C.: LEU, additive manufacturing: technology, applications and research needs. Front. Mech. Eng. **8**(3), 215–243 (2013). https://doi.org/10.1007/s11465-013-0248-8
7. Lee, J.-Y., An, J., Chua, C.K.: Fundamentals and applications of 3D printing for novel materials. Appl. Mater. Today. **7**, 120–133 (2017). https://doi.org/10.1016/j.apmt.2017.02.004
8. See: https://www.marketsandmarkets.com/Market-Reports/additive-manufacturing-material-market-167268760.html#:~:text=The%20global%20additive%20manufacturing%20market,rapid%20prototyping%20and%20rapid%20manufacturing.&text=The%20additive%20manufacturing%20application%20in,CAGR%20in%20the%20forecast%20period. 01.02.2018
9. See: https://www.smartechanalysis.com/news/2019-additive-manufacturing-market-growth/. 01.02.2018
10. Huang, S.H., Liu, P., Mokasdar, A., et al.: Additive manufacturing and its societal impact: a literature review. Int. J. Adv. Manuf. Technol. **67**, 1191–1203 (2013). https://doi.org/10.1007/s00170-012-4558-5
11. Pereira, T., Kennedy, J.V., Potgieter, J.: A comparison of traditional manufacturing vs additive manufacturing, the best method for the job. Proc. Manuf. **30**, 11–18 (2019). https://doi.org/10.1016/j.promfg.2019.02.003
12. See: https://engineeringproductdesign.com/additive-manufacturing-process-steps/. 04.03.2018
13. Deepak Kumar, S., Ghose, J., Jha, S.K., Behera, A., Mandal, A.: Optimization and simulation of additive manufacturing processes: challenges and opportunities, book title. In: Additive manufacturing applications for metals and composites, p. 23. IGI Publisher, Hershey, PA (2020). https://doi.org/10.4018/978-1-7998-4054-1.ch010
14. See: https://www.lboro.ac.uk/research/amrg/about/the7categoriesofadditivemanufacturing/powderbedfusion/. 06.03.2018
15. Lin, W.S., Starr, T.L., Harris, B.T., Zandinejad, A., Morton, D.: Additive manufacturing technology (direct metal laser sintering) as a novel approach to fabricate functionally graded titanium implants: preliminary investigation of fabrication parameters. Int. J. Oral Maxillofac. Implants. **28**(6), 1490–1495 (2013). https://doi.org/10.11607/jomi.3164
16. Murr, L.E.: Metallurgy of additive manufacturing: examples from electron beam melting. Addit. Manuf. **5**, 40–53 (2015). https://doi.org/10.1016/j.addma.2014.12.002
17. Haleem, A., Javaid, M.: Polyether ether ketone (PEEK) and its manufacturing of customised 3D printed dentistry parts using additive manufacturing. Clin. Epidemiol. Glob. Health. **7**(4), 654–660 (2019). https://doi.org/10.1016/j.cegh.2019.03.001
18. Saboori, A., Gallo, D., Biamino, S., Fino, P., Lombardi, M.: An overview of additive Manufacturing of titanium components by directed energy deposition: microstructure and mechanical properties. Appl. Sci. **7**, 883 (2017)
19. Gasser, A., Backes, G., Kelbassa, I., Weisheit, A., Wissenbach, K.: Laser metal deposition (LMD) and selective laser melting (SLM) in turbo-engine applications. Laser Addit. Manuf. **7**(2), 58–63 (2010). https://doi.org/10.1002/latj.201090029
20. Bax, B., Rajput, R., Kellet, R., Reisacher, M.: Systematic evaluation of process parameter maps for laser cladding and directed energy deposition. Addit. Manuf. **21**, 487–494 (2018). https://doi.org/10.1016/j.addma.2018.04.002
21. Heigel, J.C., Michaleris, P., Reutzel, E.W.: Thermo-mechanical model development and validation of directed energy deposition additive manufacturing of Ti-6Al-4V. Addit. Manuf. **5**, 9–19 (2015). https://doi.org/10.1016/j.addma.2014.10.003
22. Wang, Z., Palmer, T.A., Beese, A.M.: Effect of processing parameters on microstructure and tensile properties of austenitic stainless steel 304L made by directed energy deposition additive manufacturing. Acta Mater. **110**, 226–235 (2016). https://doi.org/10.1016/j.actamat.2016.03.019
23. See: https://ntrs.nasa.gov/citations/20140005339. 12.05.2018
24. Gokuldoss, P.K., Kolla, S., Eckert, J.: Additive manufacturing processes: selective laser melting, electron beam melting and binder jetting-selection guidelines. Materials. **10**, 672 (2017)
25. Gaytan, S.M., Cadena, M.A., Karim, H., Delfin, D., Lin, Y., Espalin, D., MacDonald, E., Wicker, R.B.: Fabrication of barium titanate by binder jetting additive manufacturing technology. Ceramics Int. **41**(5, Part A), 6610–6619 (2015). https://doi.org/10.1016/j.ceramint.2015.01.108
26. Prahar, M.B., Kabir, A.M., Peralta, M., Bruck, H.A., Gupta, S.K.: A robotic cell for performing sheet lamination-based additive manufacturing. Addit. Manuf. **27**, 278–289 (2019). https://doi.org/10.1016/j.addma.2019.02.002
27. Dehoff, R.R., Babu, S.S.: Characterization of interfacial microstructures in 3003 aluminum alloy blocks fabricated by ultrasonic additive manufacturing. Acta Mater. **58**(13), 4305–4315 (2010). https://doi.org/10.1016/j.actamat.2010.03.006
28. Luong, D.X., Subramanian, A.K., Lopez Silva, G.A., Yoon, J., Cofer, S., Yang, K., Owuor, P.S., Wang, T., Wang, Z., Lou, J., Ajayan, P.M., Tour, J.M.: Laminated object manufacturing of

3D-printed laser-induced graphene foams. Adv. Mater. **30**(28), 1707416 (2018). https://doi.org/10.1002/adma.201707416
29. See: https://www.sculpteo.com/en/glossary/selective-deposition-lamination-definition/. 15.04.2018
30. Jiang, J., Xu, X., Stringer, J.: Support structures for additive manufacturing: a review. J. Manuf. Mater. Process. **2**, 64 (2018)
31. Saalfrank, P.: Quantum dynamics of bond breaking in a dissipative environment: Indirect and direct photodesorption of neutrals from metals. J. Chem. Phys. **105**(6) (1996). https://doi.org/10.1063/1.472112
32. Manfredi, D., Calignano, F., Krishnan, M., Canali, R., Ambrosio, E.P., Atzeni, E.: From powders to dense metal parts: characterization of a commercial AlSiMg alloy processed through direct metal Laser sintering. Materials. **6**, 856–869 (2013). https://doi.org/10.3390/ma6030856
33. Zhu, Y., Tian, X., Li, J., Wang, H.: The anisotropy of laser melting deposition additive manufacturing Ti-6.5Al-3.5Mo-1.5Zr-0.3Si titanium alloy. Mater. Des. **67**, 538–542 (2015). https://doi.org/10.1016/j.matdes.2014.11.001
34. Shao, S., Khonsari, M.M., Guo, S., Meng, W.J., Li, N.: Overview: additive manufacturing enabled accelerated design of Ni-based alloys for improved fatigue life. Addit. Manuf. **29**, 100779 (2019). https://doi.org/10.1016/j.addma.2019.100779
35. Nesma, T.A., Simonelli, M., Parry, L., Ashcroft, I., Tuck, C., Hague, R.: 3D printing of Aluminium alloys: additive manufacturing of aluminium alloys using selective laser melting. Prog. Mater. Sci. **106**, 100578 (2019). https://doi.org/10.1016/j.pmatsci.2019.100578
36. Chaudhary, V., Yadav, N.M.S.K.K., Mantri, S.A., Dasari, S., Jagetia, A., Ramanujan, R.V., Banerjee, R.: Additive manufacturing of functionally graded Co-Fe and Ni-Fe magnetic materials. J. Alloys Compd. **823**, 153817 (2020). https://doi.org/10.1016/j.jallcom.2020.153817
37. Ahuja, B., Karg, M., Nagulin, K.Y., Schmidt, M.: Fabrication and characterization of high strength Al-Cu alloys processed using laser beam melting in metal powder bed. Phys. Proc. **56**, 135–146 (2014). https://doi.org/10.1016/j.phpro.2014.08.156
38. Gorsse, S., Hutchinson, C., Gouné, M., Banerjee, R.: Additive manufacturing of metals: a brief review of the characteristic microstructures and properties of steels, Ti-6Al-4V and high-entropy alloys. Sci. Technol. Adv. Mater. **18**(1), 584–610 (2017). https://doi.org/10.1080/14686996.2017.1361305
39. Salehi, M., Maleksaeedi, S., Sapari, M.A.B., Nai, M.L.S., Meenashisundaram, G.K., Gupta, M.: Additive manufacturing of magnesium–zinc–zirconium (ZK) alloys via capillary-mediated binderless three-dimensional printing. Mater. Des. **169**, 107683 (2019). https://doi.org/10.1016/j.matdes.2019.107683
40. Wu, A.S., Brown, D.W., Kumar, M., et al.: An experimental investigation into additive Manufacturing-induced residual stresses in 316L stainless steel. Metall. Mater. Trans. A. **45**, 6260–6270 (2014). https://doi.org/10.1007/s11661-014-2549-x
41. Wei, D., Bai, Q., Zhang, B.: A novel method for additive/subtractive hybrid Manufacturing of metallic parts. Proc. Manuf. **5**, 1018–1030 (2016). https://doi.org/10.1016/j.promfg.2016.08.067
42. Travitzky, N., Bonet, A., Dermeik, B., Fey, T., Filbert-Demut, I., Schlier, L., Schlordt, T., Greil, P.: Additive manufacturing of ceramic-based, materials. Adv. Eng. Mater. **16**(6), 729–754 (2014). https://doi.org/10.1002/adem.201400097. Special Issue: Advanced Ceramics and Coating Processing
43. Ligon, S.C., Liska, R., Stampfl, J., Gurr, M., Mulhaupt, R.: Polymers for 3D printing and customized additive manufacturing. Chem. Rev. **117**(15), 10212–10290 (2017). https://doi.org/10.1021/acs.chemrev.7b00074
44. Ligon, S.C., Liska, R., Stampfl, J., Gurr, M., Mülhaupt, R.: Polymers for 3D printing and customized additive manufacturing. Chem. Rev. **117**(15), 10212–10290 (2017). https://doi.org/10.1021/acs.chemrev.7b00074
45. Mohammed, M.I., Wilson, D., Gomez-Kervin, E., Tang, B., Wang, J.: Investigation of closed-loop manufacturing with acrylonitrile butadiene styrene over multiple generations using additive manufacturing. ACS Sustain. Chem. Eng. **7**(16), 13955–13969 (2019). https://doi.org/10.1021/acssuschemeng.9b02368
46. Miguel, M., Leite, M., Ribeiro, A.M.R., Deus, A.M., Reis, L., Vaz, M.F.: Failure of polymer coated nylon parts produced by additive manufacturing. Eng. Fail. Anal. **101**, 485–492 (2019). https://doi.org/10.1016/j.engfailanal.2019.04.005
47. Cicala, G., Giordano, D., Tosto, C., Filippone, G., Recca, A., Blanco, I.: Polylactide (PLA) filaments a biobased solution for additive manufacturing: correlating rheology and thermomechanical properties with printing quality. Materials. **11**, 1191 (2018)
48. Nicole E.Z.: Recycled polymer feedstocks for material extrusion additive manufacturing. In: Polymer-based additive manufacturing: recent developments, Chapter 3. ACS Symposium Series, vol. 1315, pp. 37–51 (2019). https://doi.org/10.1021/bk-2019-1315.ch003. ISBN13: 9780841234260eISBN: 9780841234253
49. Lopez, D.M.B., Ahmad, R.: Tensile mechanical behaviour of multi-polymer sandwich structures via fused deposition modelling. Polymers. **12**, 651 (2020). https://doi.org/10.3390/polym12030651
50. Reich, M.J., Woern, A.L., Tanikella, N.G., Pearce, J.M.: Mechanical properties and applications of recycled polycarbonate particle material extrusion-based additive manufacturing. Materials. **12** (10), 1642 (2019). https://doi.org/10.3390/ma12101642
51. Alexander, W.B., Honnig, A.E., Breyta, C.M., Dunn, I.C., La Scala, J.J., Stanzione, J.F.: Vanillin-based resin for additive manufacturing. ACS Sustain. Chem. Eng. **8**(14), 5626–5635 (2020). https://doi.org/10.1021/acssuschemeng.0c00159
52. Hu, Y., Cong, W.: A review on laser deposition-additive manufacturing of ceramics and ceramic reinforced metal matrix composites. Ceramics International. **44**(17), 20599–20612 (2018). https://doi.org/10.1016/j.ceramint.2018.08.083
53. Bermingham, M.J., McDonald, S.D., Dargusch, M.S.: Effect of trace lanthanum hexaboride and boron additions on microstructure, tensile properties and anisotropy of Ti-6Al-4V produced by additive manufacturing. Mater. Sci. Eng. A. **719**, 1–11 (2018). https://doi.org/10.1016/j.msea.2018.02.012
54. Lu, Z., Cao, J., ORCID Icon, Song, Z., Li, D., Lu, B.: Research progress of ceramic matrix composite parts based on additive manufacturing technology. Virtual Phys. Prototyp. **14**(4), 333–348 (2019). https://doi.org/10.1080/17452759.2019.1607759
55. Scheithauer, U., Schwarzer, E., Moritz, T., et al.: Additive manufacturing of ceramic heat exchanger: opportunities and limits of the lithography-based ceramic manufacturing (LCM). J. Mater. Eng. Perform. **27**, 14–20 (2018). https://doi.org/10.1007/s11665-017-2843-z
56. Yasa, E., Ersoy, K.: Additive manufacturing of polymer matrix composites. IntechOpen, London (2018). https://doi.org/10.5772/intechopen.75628
57. Hussien, A.H.: Design for additive manufacturing of composite materials and potential alloys: a review. Manuf. Rev. **3**, 17 (2016). https://doi.org/10.1051/mfreview/2016010. Special Issue - Additive Manufacturing Materials & Devices
58. Wu, H., Fahy, W.P., Kim, S., Kim, H., Zhao, N., Pilato, L., Kafi, A., Bateman, S., Koo, J.H.: Recent developments in polymers/polymer nanocomposites for additive manufacturing. Prog. Mater. Sci. **111**, 100638 (2020). https://doi.org/10.1016/j.pmatsci.2020.100638
59. Dongdong, G., Wang, H., Zhang, A.G.: Selective laser melting additive manufacturing of Ti-based nanocomposites: the role of

nanopowder. Metall. Mater. Trans. A. **45a** (2014). https://doi.org/10.1007/s11661-013-1968-4
60. Changhai, R., Luo, J., Xie, S., Sun, Y.: A review of non-contact micro- and nano-printing technologies. J. Micromech. Microeng. **24**(5) (2014). https://doi.org/10.1088/0960-1317/24/5/053001
61. Park, S.-I., Rosen, D.W., Choi, S.-k., Duty, C.E.: Effective mechanical properties of lattice material fabricated by material extrusion additive manufacturing. Addit. Manuf. **1-4**, 12–23 (2014). https://doi.org/10.1016/j.addma.2014.07.002
62. Singh, S., Ramakrishna, S., Singh, R.: Material issues in additive manufacturing: a review. J. Manuf. Processes. **25**, 185–200 (2017). https://doi.org/10.1016/j.jmapro.2016.11.006
63. Carve, M., Wlodkowic, D.: 3D-printed chips: compatibility of additive Manufacturing Photopolymeric substrata with biological applications. Micromachines. **9**(2), 91 (2018). https://doi.org/10.3390/mi9020091
64. Hofstättera, T., Spangenberga, J., Pedersena, D.B., Toselloa, G., Hansen, H.N.: Flow characteristics of a thermoset fiber composite photopolymer resin in a vat polymerization additive manufacturing process. AIP Conf. Proc. **2065**(1) (2019). https://doi.org/10.1063/1.5088257
65. de Beer, M.P., van der Laan, H.L., Cole, M.A., Whelan, R.J., Burns, M.A., Scott, T.F.: Rapid, continuous additive manufacturing by volumetric polymerization inhibition patterning. Sci. Adv. **5**(1), 8723 (2019). https://doi.org/10.1126/sciadv.aau8723
66. Li, W., Mille, L.S., Robledo, J.A., Uribe, T., Huerta, V., Zhang, Y. S.: Recent advances in formulating and processing biomaterial inks for vat polymerization-based 3D printing. Adv. Healthcare Mater. **9**(15), 2000156 (2020). https://doi.org/10.1002/adhm.202000156. Special Issue: Biomaterial Inks, 2020
67. Yap, Y.L., Wang, C., Sing, S.L., Dikshit, V., Yeong, W.Y., Wei, J.: Material jetting additive manufacturing: An experimental study using designed metrological benchmarks. Precis. Eng. **50**, 275–285 (2017). https://doi.org/10.1016/j.precisioneng.2017.05.015
68. Dilag, J., Chen, T., Li, S., Bateman, S.A.: Design and direct additive manufacturing of three-dimensional surface micro-structures using material jetting technologies. Addit. Manuf. **27**, 167–174 (2019). https://doi.org/10.1016/j.addma.2019.01.009
69. Gong, H., Rafi, K., Hengfeng, G., Starr, T., Stucker, B.: Analysis of defect generation in Ti–6Al–4V parts made using powder bed fusion additive manufacturing processes. Addit. Manuf. **1–4**, 87–98 (2014). https://doi.org/10.1016/j.addma.2014.08.002
70. Shirazi, S.F.S., Gharehkhani, S., Mehrali, M., Yarmand, H., Metselaar, H.S.C., Kadri, N.A.: A Review on Powder-Based Additive Manufacturing for Tissue Engineering: Selective Laser Sintering and Inkjet 3D Printing. Sci. Technol. Adv. Mater. (2015). https://doi.org/10.1088/1468-6996/16/3/033502
71. Gokuldoss, P.K., Kolla, S., Eckert, J.: Additive manufacturing processes: selective laser melting, electron beam melting and binder jetting-selection guidelines. Materials (Basel). **10**(6), 672 (2017). https://doi.org/10.3390/ma10060672
72. See: https://www.additiveindustries.com/markets/automotive/manifolds. 10.05.2018
73. Javaid, M., Haleem, A.: Additive manufacturing applications in medical cases: a literature based review. Alexandria J. Med. **54**(4), 411–422 (2018). https://doi.org/10.1016/j.ajme.2017.09.003
74. Subramanian, P.R., Mendiratta, M.G., Dimiduk, D.M.: The development of Nb-based advanced intermetallic alloys for structural applications. JOM. **48**, 33–38 (1996). https://doi.org/10.1007/BF03221360
75. Li, N., Huang, S., Zhang, G., Qin, R., Liu, W., Xiong, H., Shi, G., Blackburn, J.: Progress in additive manufacturing on new materials: a review. J. Mater. Sci. Technol. **35**(2), 242–269 (2019). https://doi.org/10.1016/j.jmst.2018.09.002
76. Li, X.: Additive manufacturing of advanced multi-component alloys: bulk metallic glasses and high entropy alloys. Adv. Eng. Mater. **20**(5), 1700874 (2018). https://doi.org/10.1002/adem.201700874
77. Mohanty, A., Sampreeth, J.K., Bembalge, O., Hascoet, J.Y., Marya, S., Immanuel, R.J., Panigrahi, S.K.: High temperature oxidation study of direct laser deposited AlXCoCrFeNi (X=0.3,0.7) high entropy alloys. Surface Coat. Technol. **380**, 125028 (2019). https://doi.org/10.1016/j.surfcoat.2019.125028
78. Murr, L.E., Martinez, E., Amato, K.N., Gaytan, S.M., Hernandez, J., Ramirez, D.A., Shindo, P.W., Medina, F., Wicker, R.B.: Fabrication of metal and alloy components by additive manufacturing: examples of 3D materials science. J. Mater. Res. Technol. **1**(1), 42–54 (2012). https://doi.org/10.1016/S2238-7854(12)70009-1
79. Tibbits, S.: 4D printing: multi-material shape change. Archit. Des. **84**(1), 116–121 (2014). https://doi.org/10.1002/ad.1710. Special Issue: High Definition: Zero Tolerance in Design and Production
80. Mitchell, A., Lafont, U., Hołyńska, M., Semprimoschnig, C.: Additive manufacturing: a review of 4D printing and future applications. Addit. Manuf. **24**, 606–626 (2018). https://doi.org/10.1016/j.addma.2018.10.038
81. Boydston, A.J., Cao, B., Nelson, A., Ono, R.J., Saha, A., Schwartz, J.J., Thrasher, C.J.: Additive manufacturing with stimuli-responsive materials. J. Mater. Chem. A. **6**, 20621–20645 (2018). https://doi.org/10.1039/C8TA07716A
82. Zhang, Z., Demir, K.G., Gu, G.X.: Developments in 4D-printing: a review on current smart materials, technologies, and applications. International Journal of Smart and Nano Materials. **10**(3), 205–244 (2019). https://doi.org/10.1080/19475411.2019.1591541
83. See: https://www.sculpteo.com/blog/2018/05/07/5d-printing-a-new-branch-of-additive-manufacturing/. 28.04.2018
84. See: https://www.aniwaa.com/blog/2-5d-3d-4d-5d-printing-five-dimensions-3d-printing-explained/. 28.04.2018
85. See: https://3dprintingindustry.com/news/5d-printing-83337/. 28.04.2018
86. See:https://www.businesswire.com/news/home/20200106005416/en/Global-Additive-Manufacturing-Material-Market-Analysis-Trends. 28.04.2018
87. Liu, R., Wang, Z., Sparks, T., Liou, F., Newkirk, J.: 13 - Aerospace Applications of Laser Additive Manufacturing. In: Brandt, M. (ed.) Woodhead Publishing series in electronic and optical materials, laser additive manufacturing, pp. 351–371. Woodhead Publishing, Sawston (2017). https://doi.org/10.1016/B978-0-08-100433-3.00013-0. ISBN 9780081004333
88. Uriondo, A., Esperon-Miguez, M., Perinpanayagam, S.: The present and future of additive manufacturing in the aerospace sector: a review of important aspects. Proc. Inst. Mech. Eng. G J. Aerospace Eng. **229**(11), 2132–2147 (2015). https://doi.org/10.1177/0954410014568797
89. https://www.ge.com/additive/additive-manufacturing/industries/aviation-aerospace
90. Mohd Yusuf, S., Cutler, S., Gao, N.: Review: the impact of metal additive manufacturing on the aerospace industry. Metals. **9**, 1286 (2019)
91. See: https://m.blog.naver.com/siriusb/221009042383. 02.05.2018
92. Herderick, E.D.: Additive manufacturing in the minerals, metals, and Materials community: past, present, and exciting future. JOM. **68**, 721–723 (2016). https://doi.org/10.1007/s11837-015-1799-4
93. See: https://www.longdom.org/open-access/additive-manufacturing-for-the-aircraft-industry-a-review-18967.html. 02.05.2018
94. See: https://www.nasa.gov/mission_pages/station/research/news/3d-printing-in-space-long-duration-spaceflight-applications/. 02.05.2018
95. See: https://www.autodesk.com/campaigns/additive/airplane-seat. 02.05.2018

96. See: https://www.avascent.com/wp-content/uploads/2017/07/White_Paper_3D_printing_07192017.pdf. 02.05.2018
97. See: https://www.industr.com/en/additive-manufacturing-growing-its-grasp-in-the-aerospace-sector-2378021. 02.05.2018
98. See: https://www.safran-helicopter-engines.com/media/add-demonstrateur-avec-30-de-pieces-en-fabrication-additive-20190618. 02.05.2018
99. See: https://www.tctmagazine.com/additive-manufacturing-3d-printing-news/stelia-aerospace-wire-arc-additive-manufacturing/. 02.05.2018
100. See: https://additivemanufacturingtoday.com/pratt-whitney-pw1500g-engines-to-feature-additively-manufactured-parts. 10.05.2018
101. See: https://www.engineering.com/AdvancedManufacturing/ArticleID/14218/Additive-Manufacturing-in-the-Aerospace-Industry.aspx. 10.05.2018
102. See: https://www.fastradius.com/resources/improving-heat-exchanger-models/#:~:text=With%20additive%20manufacturing%2C%20you%20can,using%20smaller%20volumes%20of%20fluid. 10.05.2018
103. See: https://auto.economictimes.indiatimes.com/news/auto-technology/schaeffler-partners-dmg-mori-for-additive-manufacturing-of-rolling-bearing/55700981. 10.05.2018
104. See: https://www.intamsys.com/jigs-fixtures-automotive-industry/. 10.05.2018
105. See: https://www.metal-am.com/articles/metal-3d-printing-gains-ground-in-the-tyre-industry/. 10.05.2018
106. See: https://manufactur3dmag.com/formula-1-fia-use-additive-manufacturing-to-build-50-scaled-models-to-test-their-2021-car-regulations/. 10.05.2018
107. Behera, A., Rajak, D.K., Hussain, P.B.: 3D Printing and Nanosensors. In: Nanosensors for smart manufacturing. Elsevier Publisher, Amsterdam (2020) (In Press). ISBN: 978-0-12-823358-0
108. Lamichhane, T.N., Sethuraman, L., Dalagan, A., Wang, H., Keller, J., Paranthaman, M.P.: Additive manufacturing of soft magnets for electrical machines-a review. Mater. Today Phys. **15**, 100255 (2020). https://doi.org/10.1016/j.mtphys.2020.100255
109. Harun, W.S.W., Manam, N.S., Kamariah, M.S.I.N., Sharif, S., Zulkifly, A.H., Ahmad, I., Miura, H.: A review of powdered additive manufacturing techniques for Ti-6al-4v biomedical applications. Powder Technol. **331**, 74–97 (2018). https://doi.org/10.1016/j.powtec.2018.03.010
110. Mahmoud, D., Elbestawi, M.A.: Lattice structures and functionally graded Materials applications in additive Manufacturing of orthopedic implants: a review. J. Manuf. Mater. Process. **1**, 13 (2017)
111. See: https://www.todaysmedicaldevelopments.com/article/additive-manufacturing-reality-biomedical-industry/. 12.05.2018
112. Wong, K.V., Hernandez, A.: A review of additive manufacturing. Int. Sch. Res. Notices, 208760 (2012). https://doi.org/10.5402/2012/208760
113. See: https://www.carecloud.com/continuum/3d-printing-future-of-medicine/. 12.05.2018
114. See: https://www.e3s-conferences.org/articles/e3sconf/pdf/2016/05/e3sconf_seed2016_00127.pdf. 12.05.2018
115. See: https://matmatch.com/blog/how-3d-printed-food-takes-us-one-step-closer-to-the-future/. 12.05.2018
116. See: https://hal.archives-ouvertes.fr/hal-02560191/file/Plastic_recycling_in_additive_manufactur.pdf. 12.05.2018
117. See: https://space10.com/project/digital-in-architecture/. 2.07.2018
118. See: https://www.constrofacilitator.com/3d-printing-in-construction-advantages-and-innovation/. 2.07.2018
119. Paolini, A., Kollmannsberger, S., Rank, E.: Additive manufacturing in construction: a review on processes, applications, and digital planning methods. Addit. Manuf. **30**, 100894 (2019). https://doi.org/10.1016/j.addma.2019.100894

Index

A. Behera, *Advanced Materials*, https://doi.org/10.1007/978-3-030-80359-9

E

H

N

MIX
Papier aus verantwortungsvollen Quellen
Paper from responsible sources
FSC® C105338

If you have any concerns about our products,
you can contact us on
ProductSafety@springernature.com

In case Publisher is established outside the EU,
the EU authorized representative is:
Springer Nature Customer Service Center GmbH
Europaplatz 3, 69115 Heidelberg, Germany

Printed by Libri Plureos GmbH
in Hamburg, Germany